S	Svedberg unit
SAM	S-adenosylmethionine
Ser	serine
SH	sulfhydryl
SNARE	synaptosome-associated protein receptor
snRNA	small nuclear RNA
snRNP	small nuclear ribonucleoprotein
SSB	single-strand binding protein
T (or Thy)	thymine
TCA	tricarboxylic acid cycle
TF	transcription factor
TG	triacylglycerol
THF	tetrahydrofolate
Thr	threonine
TPP	thiamin pyrophosphate
Trp	tryptophan
TTP	thymidine trisphosphate
Tyr	tyrosine
U (or Ura)	uracil
UDP	uridine diphosphate
UDP-galactose	uridine diphosphate galactose
UDP-glucose	uridine diphosphate glucose
UMP	uridine monophosphate
UTP	uridine trisphosphate
Val	valine
VLDL	very low density lipoprotein
YAC	yeast artificial chromosome

Textbook of BIOCHEMISTRY With Clinical Correlations

TO ALL THE RESEARCH SCIENTISTS
whose contributions to our knowledge
are the source of the information presented here,

and

TO MARJORIE
for her enduring encouragement, support, and love.

Textbook of BIOCHEMISTRY

With Clinical Correlations

Sixth Edition

EDITED BY
THOMAS M. DEVLIN
PROFESSOR EMERITUS
DEPARTMENT OF BIOCHEMISTRY
AND MOLECULAR BIOLOGY
COLLEGE OF MEDICINE
DREXEL UNIVERSITY

A JOHN WILEY & SONS, INC., PUBLICATION

Cover Illustration: The front and back covers contain two views of the complete structure of the yeast RNA polymerase II elongation complex. The views are a ribbon model of the twelve polymerase subunits and nucleic acids. The polymerase subunits Rpb1-Rpb12 are colored silver, gold, red, light red, magenta, cyan, blue, green, orange, dark blue, yellow, and light green, for Rpb1, Rpb2, Rpb3, Rpb4, Rpb5, Rpb6, Rpb7, Rpb8, Rpb9, Rpb10, Rpb11, and Rpb12, respectively. Template DNA, nontemplate DNA, and product RNA are in dark blue, light blue and red, respectively. See page 184 for discussion of RNA polymerase II. Figures were adapted from Kettenberger, H., Armache, K.-J., and Cramer, P. Molecular Cell 16: 955, 2004, and kindly supplied by Dr. H. Kettenberger, Dr. K. Armache, and Dr. Patrick Cramer, Managing Director, Gene Center, Ludwig-Maximilians-University of Munich, Feodor-Lynen-Str. 25, 81377, Munich, Germany.

Published simultaneously in Canada

Library of Congress Cataloging-in-Publication Data:

Textbook of biochemistry : with clinical correlations / edited by Thomas M. Devlin. - - 6 th ed.
p. ; cm.
Includes bibliographical references and index.
ISBN-13 978-0-471-67808-3 (cloth)
ISBN-10 0-471-67808-2 (cloth)
1. Biochemistry. 2. Clinical biochemistry.
[DNLM: 1. Biochemistry. QU 4 T355 2006] I. Devlin, Thomas. M.
QP514.2.T4 2006
612'.015—dc22 2005008549

Printed in the United States of America

10 9 8 7 6 5 4 3 2 1

CONTENTS IN BRIEF

CONTENTS

PART III FUNCTIONS OF PROTEINS

28 | PRINCIPLES OF NUTRITION II: MICRONUTRIENTS, 1091

Stephen G. Chaney

PREFACE

The objectives of the sixth edition of the *Textbook of Biochemistry with Clinical Correlations* remain unchanged from those for the previous editions. They are: (1) to present a clear and precise discussion of the biochemistry of eukaryotic cells, particularly those of mammalian tissues; (2) to relate biochemical events at the cellular level to physiological processes in the whole animal; and (3) to cite examples of abnormal biochemical processes in human disease. The text continues to have an emphasis on the biochemistry of mammalian cells because of the importance of biochemistry in understanding human diseases. Information from biochemical investigations of prokaryotes and other eukaryotes, however, is presented when these studies are the primary source of knowledge about a topic. The unraveling of many biological and physiological processes at the molecular level is due in part to the multiplicity of new innovative techniques and research approaches from different disciplines. The differences and research approaches between biochemistry, molecular biology, cell biology, cell physiology, and molecular pharmacology are becoming indistinct. Thus as biochemistry has spread through other disciplines, so have these disciplines permeated biochemistry. The continuing expansion of information in the biological sciences and the integration of disciplines have had a significant impact on the content of biochemistry courses, requiring inclusion of many topics heretofore presented in a more general descriptive manner. It also has had an impact on the content of textbooks. In the preparation of this revision, every chapter was updated with inclusion of new information and deletion of some material. Totally new topics are included in some chapters, such as sections on the basal lamina protein complex, and molecular motors, and two new chapters, **Fundamentals of Signal Transduction**, and **Cell Cycle, Programmed Cell Death and Cancer**, have been added.

The scope and depth of presentation in this book should fulfill the requirements of most upper-level undergraduate, graduate-level, and especially professional school courses in biochemistry. Topics for inclusion were selected to cover the essential areas of both biochemistry and physiological chemistry. The textbook is organized and written such that any sequence of topics considered most appropriate by an instructor can be presented. The content of the sixth edition is divided into five major parts, in which related topics are grouped together. **Part I, Structure of Macromolecules**, contains an introductory chapter on cell structure, followed by chapters on nucleic acid and protein structure. **Part II, Transmission of Information**, describes the synthesis of the major cellular macromolecules, that is, DNA, RNA, and protein. A chapter on biotechnology is included because information from this area has had such a significant impact on the development of our current knowledge. Part II concludes with a chapter on the Regulation of Gene Expression in which mechanisms of both prokaryotes and eukaryotes are presented. **Part III, Functions of Proteins**, opens with a presentation of the structure-function relationship of four major families of proteins. This is followed by a discussion of enzymes, including a separate chapter on the cytochromes P450, then a chapter on membrane structure and transmembrane transport mechanisms. Part III concludes with a chapter covering the fundamentals of cellular signal transduction mechanisms. **Part IV, Metabolic Pathways and Their Control**, begins with a chapter on bioenergetics and oxidative metabolism, then separate chapters describing the synthesis and degradation of carbohydrates, lipids, amino acids, purine and pyrimidine nucleotides, and heme. Each chapter highlights the mechanism of control of the individual metabolic pathways. A chapter on the integration of these metabolic pathways in humans completes this part. **Part V, Physiological Processes**, covers those areas unique to mammalian cells and tissues beginning with a chapter on hormones that emphasizes their biochemical functions as messengers, and a chapter on molecular cell biology containing discussions of four major physiological signal transducing systems: the nervous system, the eye, muscle contraction and molecular motors, and blood coagulation. The textbook concludes with presentations of the biochemistry of digestion and absorption of basic nutritional constituents, and principles of human nutrition.

In each chapter, the relevancy of the topic being discussed to that of human life processes is presented in separate **Clinical Correlations**, which describe the aberrant biochemistry of disease states. A number of new correlations have been included because the genetic and biochemical bases of an ever increasing number of diseases have been described. There has been no attempt, however, to review all of the major diseases; rather the purpose of the Clinical Correlations is to present examples of disease processes where the ramifications of deviant biochemical processes are well established. References are included in the Clinical Correlations to facilitate exploration of the topic in more detail. In some instances, the same clinical condition is presented in different chapters, but each from a different perspective. All pertinent biochemical information is presented in the main text, and an understanding of the material does not require a reading of the Clinical Correlations. In some cases, clinical conditions are discussed as part of the primary text because of the significance of the medical condition to an understanding of the biochemical process.

Every chapter contains a **Bibliography** that serves as an entry point to the research literature; references are generally to review articles and seminal publications. A set of **Questions and Answers** concludes each chapter. They include multiple choice questions, some with clinical vignettes as they are the type used in national medical examinations, and problem solving questions. Brief annotated answers are given.

Illustrations were updated and new figures added including a number of protein structures. The adage "A picture is worth a thousand words" is appropriate and the reader is encouraged to study the illustrations because they are meant to clarify confusing aspects of a topic.

The Appendix, **Review of Organic Chemistry**, is designed as a ready reference for the nomenclature and structures of organic molecules encountered in biochemistry; it is not intended as a comprehensive review of organic chemistry. The material is presented in the Appendix rather than at the beginning of chapters dealing with the different biologically important molecules. The reader should become familiar with the content of the Appendix and then use it as a ready reference when reading related sections in the main text. A **Glossary** has been compiled for the sixth edition; the ever expanding language of the biochemical sciences indicated a need for a ready reference to the most common words. Pertinent **Biochemical Abbreviations** and **Normal Clinical Laboratory Values for Blood and Urine** are presented on the inside of the front and back covers, respectively.

We still believe that a **multi-contributor textbook** is the best approach to achieve an accurate and up-to-date presentation of biochemistry. Each contributor is involved actively in teaching biochemistry in a medical and/or graduate school, and has an active research interest in the field in which he or she has written. Thus, each has the perspective of the classroom instructor, with the experience to select the topics and determine the emphasis required for students in a biochemistry course. Every contributor, however, brings to the book an individual approach, leading to some differences in presentation. Every chapter, however, was edited to have a consistent writing style and to eliminate unnecessary repetitions and redundancies. A few topics are presented in two different places in the book in order to make the individual discussions complete and self contained. This repetition should facilitate the learning process.

The contributors prepared their chapters for a **teaching book**, selecting the important and relevant information in their subject matter. The textbook is not intended as a compendium of biochemical facts or a review of the current literature, but each chapter, however, contains sufficient detail on the subject to make it useful as a resource. Contributors were requested not to reference individual researchers and not to dwell on the historical aspects of their topic; our apologies to the many scientists who deserve recognition for their outstanding research contributions.

One person must accept the responsibility for the final product in any project. The decisions concerning the selection of topics and format, reviewing the drafts, and responsibility for the final checking of the book were entirely mine. I welcome comments, criticisms, and suggestions from students, faculty, and professionals. It is our hope that this work will be of value to those embarking on the exciting experience of learning biochemistry for the first time as well as those returning to a topic in which the information is expanding so rapidly.

THOMAS M. DEVLIN

Berwyn, Pennsylvania
September, 2005

ACKNOWLEDGMENTS

The efforts and encouragement of many people have made possible the publication of this sixth edition of the Textbook of Biochemistry. As in the past, I am personally indebted to each of the contributors for accepting the challenge of preparing the chapters, sharing their ideas to improve the book, accepting so readily suggestions to modify their contributions, and cooperating throughout the period of preparation. To each I extend my deepest appreciation for a job well done. The individual contributors have received the support of colleagues and students in the preparation of their chapters. Thus, to everyone who gave unselfishly of their time and shared in the objective and critical evaluation of the text, we extend a sincere thank you. In addition, every contributor has been influenced by former teachers and colleagues, various reference resources, and, of course, the scientific literature. We are very indebted to these many sources of inspiration.

I am deeply grateful to Dr. Francis Vella, former Professor of Biochemistry, University of Saskatchewan, Canada, who gave me invaluable assistance in editing the text, and who made significant suggestions for clarifying and improving the presentation. Dr. Vella is a distinguished biochemist who has made a major personal effort to improve the teaching of biochemistry throughout the world. I extend to him my appreciation and thanks for his participation and friendship. Our gratitude is extended to Dr. Patrick Cramer, Managing Director, Gene Center, Ludwig-Maximilians-University of Munich, for supplying us with the figures of the structure of RNA polymerase II elongation complex presented on the cover.

I extend my sincerest appreciation and thanks to the members of the staff of the STM Division of John Wiley & Sons who participated in the preparation of this edition. Again, it has been a pleasure to work with an extremely intelligent, professional, and encouraging group of individuals. My deepest gratitude to Dr. Darla P. Henderson, Senior Editor, Chemistry and Biochemistry Books, who conscientiously guided the planning of this edition and who made many valuable suggestions. She has been a constant support. Many thanks to Christine J. Moore, Editorial Assistant, who handled promptly and efficiently the administrative details and my special requests. My appreciation to Janet D. Bailey, Vice President, Scientific, Technical, and Medical Books, for her unqualified support of the project. I am in debt to Lisa M. Van Horn, Senior Production Editor, who patiently and meticulously oversaw the production. Lisa kept me well informed, managed the many details, acted promptly to my suggestions and concerns, and kept us on schedule. It has been a pleasure to work with an efficient, knowledgeable, and conscientious professional, as well as a very pleasant individual; to her I extend my heartfelt thanks. The design of the book is the work of Lee Goldstein, Designer, to whom I extend a very special thanks. My appreciation to Robert Golden, Copy-editor, and Coughlin Indexing Services; both did an excellent job.

A textbook is only as good as its illustrations and for ours we extend our deepest appreciation and thanks to Dean Gonzalez, Illustration Manager. He oversaw the preparation and revision of the artwork, many times making corrections himself to expedite production. He patiently worked with the contributors on the many revisions of the graphics. Also, many thanks to J.C. Morgan and the staff at Precision Graphics who prepared the new illustrations.

We are indebted to Kimi Sugeno, Senior Manager, Online Book Production, for the development of the sixth edition for WileyPLUS, the online version. No book is successful without the activities of a Marketing Department. Special recognition and my thanks to Elizabeth Seth, Marketing Manager, and my appreciation to Kim McDonnell, Program Marketing Manager, Fred Filler, Associate Marketing Director, and Ellen Nichols, Director of Marketing for their ideas and effort.

Finally, a very special note of gratitude to my wife, Marjorie, who had the foresight many years ago to encourage me to undertake the preparation of a Textbook, who supported me during the days of intensive work, and who created an environment in which I could devote the many hours required for the preparation of this textbook. To Marjorie, my deepest and sincerest thank you.

T. M. D.

CONTRIBUTORS

CAROL N. ANGSTADT, PH.D.
Professor Emerita
School of Nursing and Health Professions
Drexel University
490 S. Old Middletown Road
Media, PA 19063
Email: angstadtc@drexel.edu

WILLIAM AWAD, JR., M.D., PH.D.
Professor
Departments of Medicine and of Biochemistry and Molecular Biology
University of Miami School of Medicine
PO Box 016960
Miami, FL 33101
Email: w.awad.jr@miami.edu

DIANA S. BEATTIE, PH.D.
Professor and Chair
Department of Biochemistry and Molecular Pharmacology
West Virginia University School of Medicine
PO Box 9142
Morgantown, WV 26506
Email: dbeattie@hsc.wva.edu

STEPHEN G. CHANEY, PH.D.
Professor
Departments of Biochemistry and Biophysics and of Nutrition, CB# 7260
School of Medicine
University of North Carolina at Chapel Hill
Mary Ellen Jones Building
Chapel Hill, NC 27599
Email: stephen_chaney@med.unc.edu

MARGUERITE W. COOMES, PH.D.
Associate Professor
Department of Biochemistry and Molecular Biology
College of Medicine
Howard University
520 W Street, N.W.
Washington, DC 20059
Email: mcoomes@fac.howard.edu

JOSEPH G. CORY, PH.D.
Professor and Chair
Department of Biochemistry
Brody School of Medicine
East Carolina University
Greenville, NC 27858
Email: coryjo@mail.ecu.edu

DAVID W. CRABB, M.D.
John B. Hickam Professor and Chair
Department of Medicine
Indiana University School of Medicine
545 Barnhill Drive
Indianapolis, IN 46202
Email: dcrabb@iupui.edu

THOMAS M. DEVLIN, PH.D.
Professor Emeritus and Former Chair
Department of Biochemistry and Molecular Biology
College of Medicine
Drexel University
159 Greenville Court
Berwyn, PA 19312
Email: tdevlin@drexel.edu

JOHN E. DONELSON, PH.D.
Professor and Head
Department of Biochemistry
Carver College of Medicine
University of Iowa
Iowa City, IA 52242
Email: john-donelson@uiowa.edu

GEORGE R. DUBYAK, PH.D.
Professor
Department of Physiology and Biophysics
Case School of Medicine
Case Western Reserve University
2109 Adelbert Road
Cleveland, OH 44106
Email: george.dubyak@case.edu

Howard J. Edenberg, Ph.D.
Chancellor's Professor
Departments of Biochemistry and Molecular Biology and of Medical and Molecular Genetics
Indiana University School of Medicine
635 Barnhill Drive
Indianapolis, IN 46202
Email: edenberg@iupui.edu

Robert H. Glew, Ph.D.
Professor
Department of Biochemistry and Molecular Biology
School of Medicine
University of New Mexico
915 Camino de Salud NE
Albuquerque, NM 87131
Email: rglew@salud.unm.edu

Dohn G. Glitz, Ph.D.
Professor Emeritus
Department of Biochemistry
UCLA School of Medicine
11260 Barnett Valley Road
Sebastopol, CA 95472
Email: dglitz@mednet.ucla.edu

Richard W. Hanson, Ph.D.
Leonard & Jean Skeggs Professor
Department of Biochemistry
Case School of Medicine
Case Western Reserve University
Cleveland, OH 44106
E-mail: rwh@cwru.edu

Robert A. Harris, Ph.D.
Distinguished Professor
Showalter Professor of Biochemistry and Former Chair
Department of Biochemistry and Molecular Biology
Indiana University School of Medicine
1345 W. 16th Street
Indianapolis, IN 46202
Email: raharris@iupui.edu

Ulrich Hopfer, M.D., Ph.D.
Professor
Departments of Physiology and Biophysics, and of Medicine
Case School of Medicine
Case Western Reserve University
10900 Euclid Ave.
Cleveland, OH 44106
Email: ulrich.hopfer@case.edu

Michael N. Liebman, Ph.D.
Executive Director
Windber Research Institute
620 Seventh Street
Windber, PA 15963
Email: m.liebman@wriwindber.org

Gerald Litwack, Ph.D.
Professor Emeritus and Former Chair
Department of Biochemistry & Molecular Pharmacology
Jefferson Medical College
Thomas Jefferson University
Visiting Scholar
Department of Biological Chemistry
David Geffen School of Medicine, UCLA
4610 Ledge Avenue
Toluca Lake, CA 91602
Email: gerry.litwack@mail.tju.edu

Bettie Sue Siler Masters, Ph.D., D.Sc., M.D. (Hon.)
Robert A. Welch Professor of Chemistry
Department of Biochemistry
University of Texas Health Science Center at San Antonio
7703 Floyd Curl Drive
San Antonio, TX 78229
Email: masters@uthscsa.edu

J. Denis McGarry, Ph.D. (deceased)
Professor
Departments of Internal Medicine and of Biochemistry
University of Texas Southwestern Medical Center at Dallas
5323 Harry Hines Blvd
Dallas, TX 75235-9135

Linda J. Roman, Ph.D.
Assistant Professor
Department of Biochemistry
University of Texas Health Science Center at San Antonio
703 Floyd Curl Drive
San Antonio, TX 78229
Email: roman@uthscsa.edu

Francis J. Schmidt, Ph.D.
Professor
Department of Biochemistry
Univ. of Missouri-Columbia
M743 Medical Sciences
Columbia, MO 65212
Email: schmidtf@missouri.edu

Thomas J. Schmidt, Ph.D.
Professor
Department of Physiology and Biophysics
Carver College of Medicine
University of Iowa
Iowa City, IA 52242
Email: thomas-schmidt@uiowa.edu

Richard M. Schultz, Ph.D.
Professor and Former Chair, Department of Biochemistry
Division of Molecular and Cellular Biochemistry
Department of Cell Biology, Neurobiology, and Anatomy
Stritch School of Medicine
Loyola University of Chicago
2160 South First Avenue
Maywood, IL 60153
Email: rschult@lumc.edu

Nancy B. Schwartz, Ph.D.
Professor
Departments of Pediatrics and of
Biochemistry and Molecular Biology
University of Chicago,
5841 S. Maryland Ave.
Chicago, IL 60637
Email: n-schwartz@uchicago.edu

David R. Setzer, PH. D.
Professor
Division of Biological Sciences
University of Missouri
410 Tucker Hall
Columbia, MO 65211
E-mail: setzerd@missouri.edu

Thomas E. Smith, Ph.D.
Professor and Former Chair
Department of Biochemistry and Molecular Biology
College of Medicine
Howard University
520 W Street, N.W.
Washington, DC 20059
Email: tsmith@fac.howard.edu

Martin D. Snider, Ph.D.
Associate Professor
Department of Biochemistry
Case School of Medicine
Case Western Reserve University
10900 Euclid Ave
Cleveland, OH 44106
E-mail: mds@cwru.edu

Gerald Soslau, Ph.D.
Professor
Department of Biochemistry
and Molecular Biology
College of Medicine
Drexel University,
245 North 15th Street
Philadelphia, PA 19102
Email: Gerald.Soslau@drexel.edu

Francis Vella, M.D., Ph.D.
Professor-Retired
Department of Biochemistry
University of Saskatchewan
18 Leyden Crescent
Saskatoon, Saskatchewan
SK S7J 2S4, Canada
E-mail: f.vella@sasktel.net

Daniel L. Weeks, Ph.D.
Professor
Department of Biochemistry
Carver College of Medicine
University of Iowa
Iowa City, IA 52242
Email: daniel_weeks@uiowa.edu

Henry Weiner, Ph.D.
Professor
Department of Biochemistry
Purdue University
175 S. University Street
West Lafayette IN 47907
Email: hweiner@purdue.edu

Stephen A. Woski, Ph.D.
Associate Professor
Department of Chemistry
University of Alabama
Box 870336
Tuscaloosa, AL 35487
Email: swoski@bama.ua.edu

PART I STRUCTURE OF MACROMOLECULES

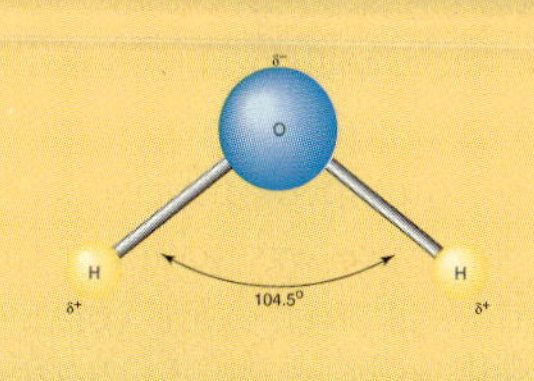

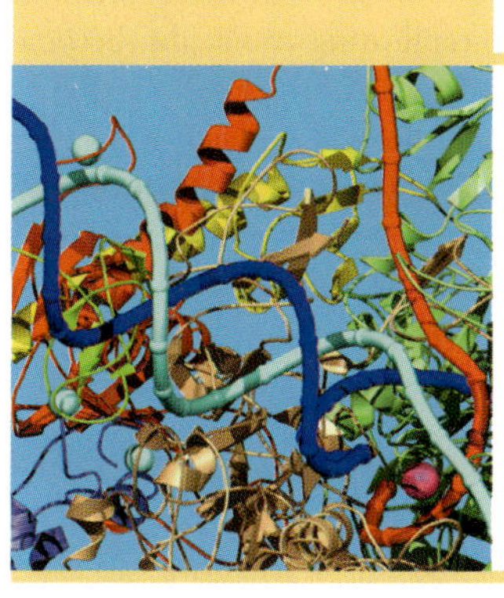

1 EUKARYOTIC CELL STRUCTURE

Thomas M. Devlin

Textbook of Biochemistry With Clinical Correlations, Sixth Edition, Edited by Thomas M. Devlin

1.1 OVERVIEW: CELLS AND CELLULAR COMPARTMENTS

Over three and half billion years ago, under conditions not entirely clear and in a time span difficult to comprehend, the elements carbon, hydrogen, oxygen, nitrogen, sulfur, and phosphorus formed simple chemical compounds. These combined, dispersed, and recombined to form a variety of larger molecules until a structure was achieved that was capable of replicating itself. With continued formation of ever more complex molecules, the environment around some of these self-replicating molecules became enclosed by a membrane. This development gave these primordial structures the ability to control their own environment to some extent. A form of life had evolved, and a unit of three-dimensional space—a **cell**—had been established. These life forms were the ancestors of all life on earth. With the passing of time, a diversity of cells evolved, and their structure and chemistry became more complex. They could extract nutrients from the environment, convert these nutrients to sources of energy or to complex molecules, control chemical processes that they catalyzed, and carry out cellular replication. The earliest cells, now classified as archaea, meaning ancient, lacked a defined nucleus and were the precursors of the extensive class of organisms called bacteria. At some point in time, perhaps 3 billion years ago, an extraordinary development occurred that led to a totally different life form containing a defined nucleus. The vast diversity of life observed today, from the simplest bacteria to complex multicellular organisms such as plants and animals, including humans, are the products of these evolutionary changes. The basic unit of life in all living things is still, however, the cell, regardless of the complexity of the organism.

All cells have a limiting outer membrane, **plasma membrane**, composed of amphipathic phospholipids and protein that delineates the space occupied and separates a variable and potentially hostile environment outside from a relatively constant milieu within. Protein receptors, channels, and transporting systems link the interior of the cell to the outside by controlling the movement of substances in and out of the cell.

Living cells are classified as either prokaryotes, which do not have a nucleus or internal membrane structures, or eukaryotes, which have a defined nucleus and intracellular organelles surrounded by membranes. **Prokaryotes**, which include bacteria and archaea, are usually unicellular (Figure 1.1*a*) but in some cases form colonies or filaments. They have a variety of shapes and sizes and can live under a variety of conditions, some very extreme. The plasma membrane is often invaginated. **Deoxyribonucleic acid**, DNA, of prokaryotes is single-stranded and often segregated into a discrete mass, the **nucleoid** region that is not surrounded by a membrane or envelope. Even without defined membrane compartments, the intracellular milieu of prokaryotes is organized into functional compartments. **Eukaryotes**, which include yeasts, fungi, plants, and animals, have a volume 1000 to 10,000 times larger than that of prokaryotes. They have a well-defined membrane surrounding a central nucleus containing the bulk of the cell's DNA, along with a variety of intracellular structures and organelles (Figure 1.1*b*). Intracellular membrane systems establish distinct cellular compartments, as described in Section 1.3, permitting a unique degree of subcellular organization. By compartmentalization, different chemical reactions that require different environments can occur simultaneously. In addition, many reactions occur in or on specific membranes that create additional environments for diverse cellular functions.

Besides structural distinctions between prokaryotic and eukaryotic cells (Figures 1.1*a* and 1.1*b*) there are significant differences in chemical composition and biochemical activities. As an example, eukaryotes but not prokaryotes contain histones, a highly conserved class of proteins in all eukaryotes that complex with DNA (see p. 54). There are also differences in enzyme content and in the ribonucleic acid–protein complexes, called ribosomes, involved in biosynthesis of proteins. The many similarities, however, are equally striking. Emphasis throughout this book is on the biochemistry of eukaryotes, particularly mammals, but much of our knowledge of the biochemistry of living cells has come from studies of prokaryotic and nonmammalian eukaryotic cells. The basic chemical components and fundamental chemical reactions of all living cells are very

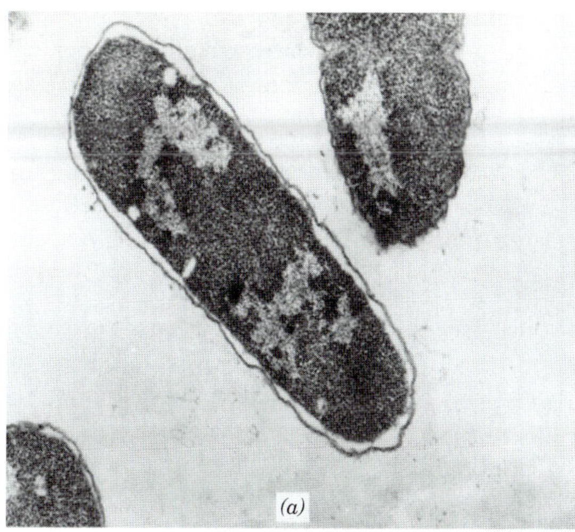

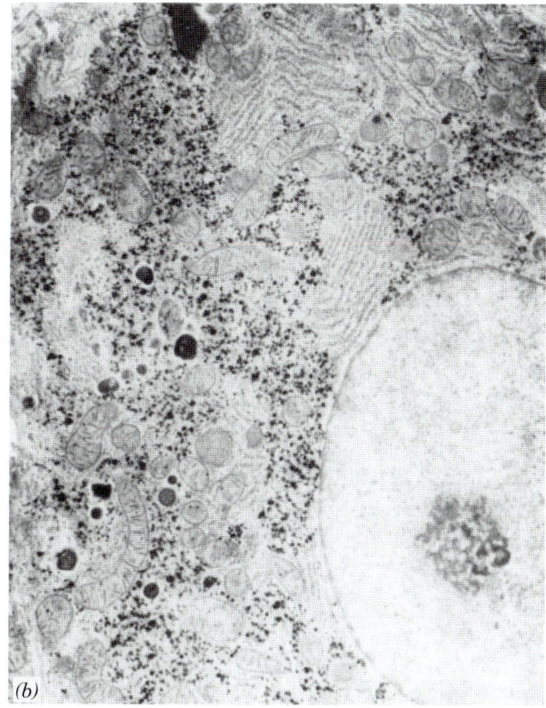

FIGURE 1.1

Cellular organization of prokaryotic and eukaryotic cells. (*a*) Electron micrograph of *Escherichia coli*, a representative prokaryote; approximate magnification ×30,000. There is little apparent intracellular organization and no membrane-enclosed organelles. Chromatin is condensed in a nucleoid but not surrounded by a membrane. Prokaryotic cells are much smaller than eukaryotic cells. (*b*) Electron micrograph of a thin section of a liver cell (rat hepatocyte), a representative eukaryotic cell; approximate magnification ×7500. Note the distinct nuclear membrane, different membrane-bound organelles or vesicles, and extensive membrane systems. Photograph (a) generously supplied by Dr. M. E. Bayer, Fox Chase Cancer Institute, Philadelphia, PA; photograph (b) reprinted with permission of Dr. K. R. Porter, from Porter, K. R. and Bonneville, M. A. In: *Fine Structure of Cells and Tissues*. Philadelphia: Lea & Febiger, 1972.

similar. The universality of many biochemical phenomena permits many extrapolations from prokaryotes to humans.

The intracellular environment, particularly the presence of water, places constraints on many cellular activities. Thus, it is appropriate to review some of the chemical and physical characteristics of this environment. The activities and roles of subcellular mammalian compartments are also described.

1.2 WATER, pH, AND SOLUTES: THE AQUEOUS ENVIRONMENT OF CELLS

All living cells contain essentially the same inorganic ions, small organic molecules, and types of macromolecules. The general classes of cellular components are presented in Table 1.1.

Intracellular concentrations of inorganic ions in different mammalian cells are essentially similar but very different from those of the extracellular milieu (see p. 12).

TABLE 1.1 Chemical Components of Biological Cells

Component	Range of Molecular Weights
H_2O	18
Inorganic ions	23–100
Na^+, K^+, Cl^-, $SO_4{}^{2-}$, $HPO_4{}^{2-}$, $HCO_3{}^-$ Ca^{2+}, Mg^{2+}, etc.	
Small organic molecules	100–1200
Carbohydrates, amino acids, lipids, nucleotides, peptides	
Macromolecules	6000–1,000,000
Proteins, polysaccharides, nucleic acids	

Microenvironments, created within eukaryotic cells by organelles as well as around macromolecules and membranes, leads to variations in concentration of components throughout a cell. **Water** is the one common component of these microenvironments, and substances required for the cell's existence are dissolved or suspended in water. Life on earth exists because of the unique physicochemical properties of water.

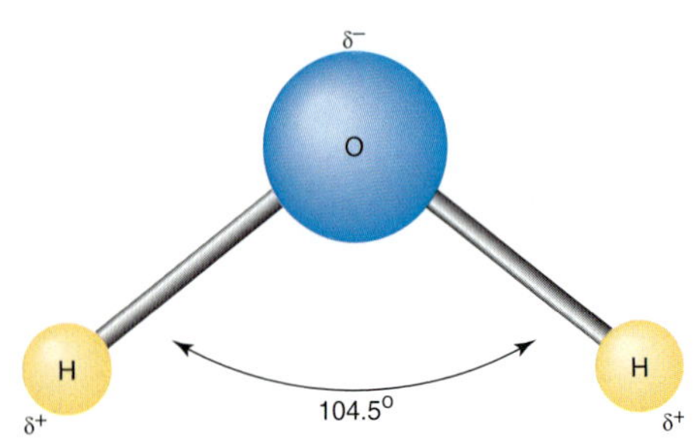

FIGURE 1.2

Structure of a water molecule. The H—O—H bond angle is 104.5°. Both hydrogen atoms carry a partial positive and the oxygen a partial negative charge, creating a dipole.

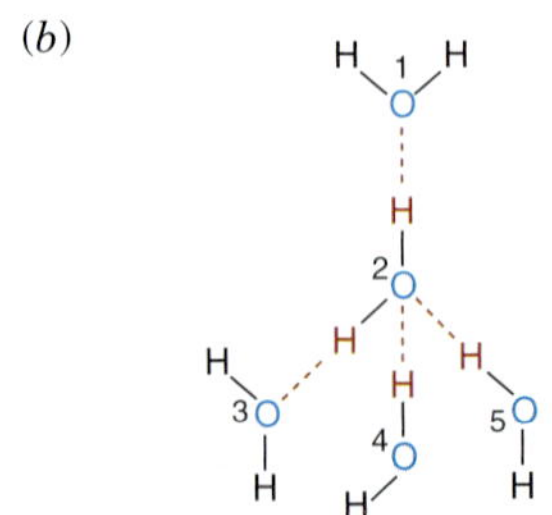

FIGURE 1.3

Hydrogen bonding in water. (*a*) Hydrogen bonding, indicated by dashed lines, between two water molecules. (*b*) Tetrahedral hydrogen bonding of five water molecules. Water molecules 1, 2, and 3 are in the plane of the page, 4 is below, and 5 is above.

Hydrogen Bonds Form between Water Molecules

Two hydrogen atoms share their electrons with an unshared pair of electrons of an oxygen atom to form a water molecule. Water is a polar molecule because the oxygen nucleus has a stronger attraction for shared electrons than hydrogen, and positively charged hydrogen nuclei are left with an unequal share of electrons. This creates a partial positive charge on both hydrogens and a partial negative charge on oxygen. The bond angle between hydrogens and oxygen is 104.5°, making the molecule electrically asymmetric and producing an electric dipole (Figure 1.2).

Water molecules interact with each other because positively charged hydrogen atoms on one molecule are attracted to a negatively charged oxygen atom on another, with formation of a weak bond between two molecules (Figure 1.3*a*). This bond, indicated by a dashed line, is a **hydrogen bond**. Recent studies, however, suggest that the bond between two water molecules is partially covalent. A detailed discussion of noncovalent interactions, including electrostatic, van der Waals, and hydrophobic, between molecules is presented on page 110. Five molecules of water can form a tetrahedral structure (Figure 1.3*b*), with each oxygen sharing its electrons with four hydrogen atoms and each hydrogen sharing its electrons with another oxygen. A tetrahedral lattice structure is responsible for the crystalline structure of ice. In the transition from ice to liquid water, only some hydrogen bonds are broken. Hydrogen bonds are relatively weak compared to covalent bonds, but their large number is the reason for the stability of liquid water. Water has a rapidly changing structure as hydrogen bonds break and new bonds form; the half-life of hydrogen bonds in water is less than 1×10^{-10} s. Even at 100°C, liquid water contains a significant number of hydrogen bonds, which accounts for its high heat of vaporization. In the transformation from liquid to vapor state, hydrogen bonds are disrupted. Many models for the structure of liquid water have been proposed, but none adequately explains all its properties.

The structure of pure water is altered when its atoms hydrogen bond to other chemical structures. The water environment in cells is not uniform. Interaction of water molecules with the surface of lipid leads to substantial orientation of water along the surface, and those near the surface of membranes are more ordered because of the amphiphilic nature of phospholipid-containing membranes (see p. 447). These microenvironments alter the activities of substances present in this milieu. A similar change can occur when water is present within protein or nucleic acid molecules, stabilizing these macromolecules.

Hydrogen bonding also occurs between other molecules and within a molecule wherever electronegative oxygen or nitrogen atoms come in close proximity with hydrogen covalently bonded to another electronegative atom. Representative hydrogen bonds are presented in Figure 1.4. Intramolecular hydrogen bonding occurs extensively in large macromolecules such as proteins and nucleic acids and is partially responsible for their structural stability.

Water Has Unique Solvent Properties

The polar nature and ability to form hydrogen bonds are the basis for the unique **solvent properties** of water. Polar molecules are readily dispersed in water. **Salts**, in which a crystal lattice is held together by attraction of positively and negatively charged atoms or groups, dissolve in water because electrostatic forces in the crystal can be overcome by attraction of charged components to the dipole of water. NaCl dissolves in water because the electrostatic attraction of individual Na^+ and Cl^- ions is overcome by (a) interaction of Na^+ with the negative charge on oxygen atoms of water and (b) interaction of Cl^-

with the positive charge on hydrogen atoms. Thus, a shell of water surrounds individual ions. The number of weak charge-charge interactions between water and Na^+ and Cl^- ions is sufficient to maintain separation of the charged ions.

Many organic molecules contain nonionic but weakly polar groups and are soluble in water because of attraction of these groups to molecules of water. Sugars and alcohols are readily soluble for this reason. **Amphipathic** molecules, compounds that contain both polar and nonpolar groups, disperse in water if attraction of the polar group for water can overcome hydrophobic interactions of nonpolar portions of the molecules. Very hydrophobic molecules, such as lipids containing long hydrocarbon chains, however, do not readily disperse as single molecules in water but interact with one another to exclude the polar water molecules.

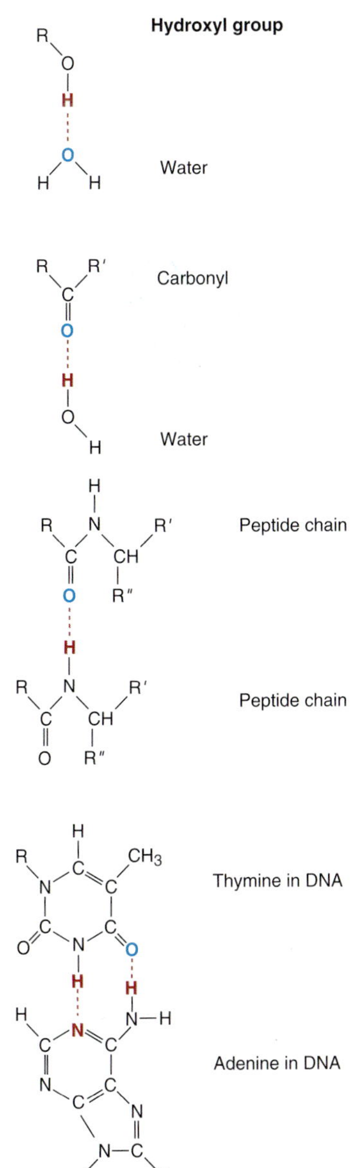

FIGURE 1.4

Representative hydrogen bonds of importance in biological systems.

Some Molecules Dissociate to Form Cations and Anions

Substances that dissociate in water into a **cation** (positively charged ion) and an **anion** (negatively charged ion) are **electrolytes** because these ions facilitate conductance of an electrical current. Sugars or alcohols are **nonelectrolytes** because they dissolve readily in water but do not carry a charge or dissociate into charged species.

At low concentrations of salts of alkali metals (e.g., Li, Na, and K) dissociate completely when dissolved in water but not necessarily at high concentrations. It is customary, however, to consider such compounds and salts of organic acids (e.g., sodium lactate) to be dissociated totally in biological systems because their concentrations are low. The dissociated anions of an organic acid (e.g., lactate ions) react to a limited extent with a proton from water to form undissociated acid (Figure 1.5). If a solution contains several different salts (e.g., NaCl, K_2SO_4, and Na lactate), these molecules do not exist as such in solution, only the dissociated ions (e.g., Na^+, K^+, Cl^-, SO_4^{2-}, and lactate$^-$).

Many acids, however, when dissolved in water, do not dissociate totally but establish an equilibrium between undissociated and dissociated components. Thus lactic acid, an important metabolic intermediate, partially dissociates into a lactate anion and a proton (H^+) as follows:

$$CH_3-CHOH-COOH \rightleftharpoons CH_3-CHOH-COO^- + H^+$$

A dynamic equilibrium is established in which the products of the reaction reform the undissociated reactant while other molecules dissociate. The degree of dissociation of such an electrolyte depends on the affinity of the anion for a H^+. There will be more dissociation if the weak dipole forces of water that interact with the anion and cation are stronger than the electrostatic forces between anion and H^+. On a molar basis such compounds, termed **weak electrolytes**, have a lower capacity to carry an electrical charge in comparison to those that dissociate totally, termed **strong electrolytes**.

In partial dissociation of a weak electrolyte, represented by HA, the concentration of various species can be determined from the equilibrium equation

$$K'_{eq} = \frac{[H^+][A^-]}{[HA]} \tag{1.1}$$

where K'_{eq} is a physical constant, A^- represents the dissociated anion, and square brackets indicate concentration of each component in units such as moles per liter (mol L^{-1} or M) or millimoles per liter (millimol L^{-1} or mM). The **activity** of each species rather than concentration should be employed in the equilibrium equation, but since most compounds of interest in biological systems are present in low concentration, the value for activity approaches that of concentration. The equilibrium constant, however, is indicated as $\boldsymbol{K'_{eq}}$ to indicate that it is an apparent constant based on concentrations. The dissociation of an acid increases with increasing temperatures. From the dissociation equation, it is apparent that K'_{eq} will be a small number if the degree of dissociation of a substance is small (large denominator in Eq. 1.1) but large if the degree of dissociation is large (small denominator). A K'_{eq} cannot be determined for strong electrolytes because at equilibrium there is no remaining undissociated solute.

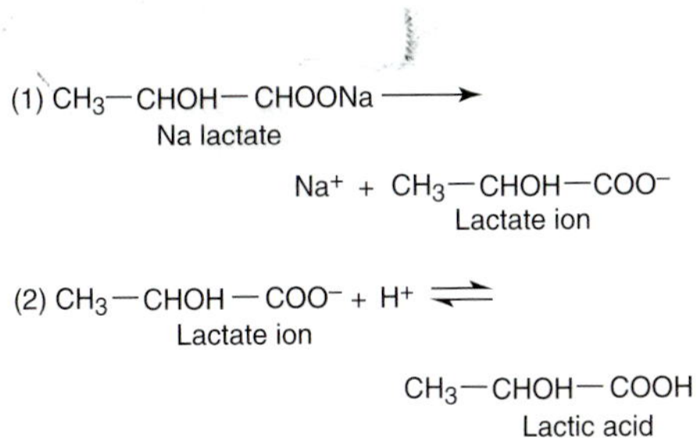

(1) $CH_3-CHOH-CHOONa \longrightarrow Na^+ + CH_3-CHOH-COO^-$
Na lactate → Lactate ion

(2) $CH_3-CHOH-COO^- + H^+ \rightleftharpoons CH_3-CHOH-COOH$
Lactate ion ⇌ Lactic acid

FIGURE 1.5

Reactions that occur when sodium lactate is dissolved in water.

Water Is a Weak Electrolyte

Water dissociates as follows:

$$\text{HOH} \rightleftharpoons \text{H}^+ + \text{OH}^-$$

Protons that dissociate interact with oxygens of other water molecules to form clusters of water molecules, $H^+(H_2O)_n$, where n has been determined to be from 6 to 27. This hydration of H^+ is often presented as H_3O^+, the **hydronium ion**. It is a generally accepted practice, and one that will be employed in this book, to present the proton as H^+ rather than H_3O^+, while recognizing that $H^+(H_2O)_n$ is the actual chemical species. At 25°C the value of K'_{eq} for dissociation of water is about 1.8×10^{-16}:

$$K'_{eq} = 1.8 \times 10^{-16} = \frac{[[H^+][OH^-]}{[H_2O]} \tag{1.2}$$

With such a small K'_{eq}, an extremely small number of water molecules actually dissociate relative to the number of undissociated molecules. Thus the concentration of water, which is 55.5 M, is essentially unchanged by the dissociation and thus is a constant. Equation 1.2 can be rewritten as follows:

$$K'_{eq} \times [H_2O] = [H^+][OH] \tag{1.3}$$

$K'_{eq} \times [55.5]$ is a constant and is termed the **ion product of water**. Its value at 25°C is 1×10^{-14}. In pure water the concentration of H^+ equals OH^-, and by substituting $[H^+]$ for $[OH^-]$ in Eq. 1.3, $[H^+]$ is 1×10^{-7} M. Similarly, $[OH^-]$ is also 1×10^{-7} M. The equilibrium reaction of H_2O, H^+, and OH^- always exists in solutions regardless of the presence of dissolved substances. If dissolved material alters the H^+ or OH^- concentration, as occurs on addition of an acid or base, a concomitant change in the other ion must occur in order to satisfy the equilibrium relationship of water. Using the equation for ion product, $[H^+]$ or $[OH^-]$ can be calculated if the concentration of one ion is known. The importance of hydrogen ions in biological systems will become apparent in discussions of enzyme activity (see p. 393) and metabolism.

For convenience, $[H^+]$ is usually expressed in terms of pH, defined as

$$\mathbf{pH} = \mathbf{log}\frac{\mathbf{1}}{[\mathbf{H}^+]} \tag{1.4}$$

In pure water, $[H^+]$ and $[OH^-]$ are both 1×10^{-7} M, and pH = 7.0. $[OH^-]$ can be expressed as the pOH. For the equation describing dissociation of water, $1 \times 10^{-14} = [H^+][OH^-]$; taking negative logarithms of both sides, the equation becomes 14 = pH + pOH. Table 1.2 presents the relationship between pH and $[H^+]$.

pH values of different biological fluids are presented in Table 1.3. In blood plasma, $[H^+]$ is 0.00000004 M or a pH of 7.4. Other cations are between 0.001 and 0.10 M, well over 10,000 times higher than $[H^+]$. An increase in hydrogen ion to 0.0000001 M (pH 7.0) or a decrease to 0.00000002 M (pH 7.8) of blood leads to serious medical consequences and are life-threatening.

TABLE 1.2 Relationships Between $[H^+]$, pH, $[OH^-]$, and pOH

$[H^+]$ (M)	pH	$[OH^-]$ (M)	pOH
1.0	0	1×10^{-14}	14
0.1 (1×10^{-1})	1	1×10^{-13}	13
1×10^{-2}	2	1×10^{-12}	12
1×10^{-3}	3	1×10^{-11}	11
1×10^{-4}	4	1×10^{-10}	10
1×10^{-5}	5	1×10^{-9}	9
1×10^{-6}	6	1×10^{-8}	8
1×10^{-7}	7	1×10^{-7}	7
1×10^{-8}	8	1×10^{-6}	6
1×10^{-9}	9	1×10^{-5}	5
1×10^{-10}	10	1×10^{-4}	4
1×10^{-11}	11	1×10^{-3}	3
1×10^{-12}	12	1×10^{-2}	2
1×10^{-13}	13	0.1 (1×10^{-1})	1
1×10^{-14}	14	1.0	0

TABLE 1.3 pH of Some Biological Fluids

Fluid	pH
Blood plasma	7.4
Interstitial fluid	7.4
Intracellular fluid	
Cytosol (liver)	6.9
Lysosomal matrix	Below 5.0
Gastric juice	1.5–3.0
Pancreatic juice	7.8–8.0
Human milk	7.4
Saliva	6.4–7.0
Urine	5.0–8.0

Many Biologically Important Molecules Are Weak Acids or Bases

The definitions of an acid and a base proposed by Lowry and Brønsted are most convenient in considering biological systems. An **acid** is a **proton donor**, and a **base** is a **proton acceptor**. Hydrochloric acid (HCl) and sulfuric acid (H_2SO_4) are strong acids because they dissociate totally releasing H^+. OH^- ion is a base. Addition of either an acid or base to water will lead to establishment of a new equilibrium of $OH^- + H^+ \rightleftharpoons H_2O$. When a strong acid and OH^- are combined, H^+ from the acid and OH^- interact nearly totally and neutralize each other nearly totally.

Anions produced when strong acids dissociate, such as Cl^- from HCl, are not bases because they do not associate with protons in solution. Organic acids are usually **weak acids** because they dissociate partially, establishing an equilibrium between HA (proton donor), an anion of the acid (A^-), and a proton as follows:

$$\text{HA} \rightleftharpoons \text{A}^- + \text{H}^+$$

TABLE 1.4 Some Conjugate Acid–Base Pairs of Importance in Biological Systems

Proton Donor (Acid)		*Proton Acceptor (Base)*
$CH_3—CHOH—COOH$ (lactic acid)	$\rightleftharpoons$	$H^+ + CH_3—CHOH—COO^-$ (lactate)
$CH_3—CO—COOH$ (pyruvic acid)	$\rightleftharpoons$	$H^+ + CH_3—CO—COO^-$ (pyruvate)
$HOOC—CH_2—CH_2—COOH$ (succinic acid)	$\rightleftharpoons$	$2H^+ + {}^-OOC—CH_2—CH_2—COO^-$ (Succinate)
${}^+H_3NCH_2—COOH$ (glycine)	$\rightleftharpoons$	$H^+ + {}^+H_3N—CH_2—COO^-$ (glycinate)
H_3PO_4	$\rightleftharpoons$	$H^+ + H_2PO_4^-$
$H_2PO_4^-$	$\rightleftharpoons$	$H^+ + HPO_4^{2-}$
HPO_4^{2-}	$\rightleftharpoons$	$H^+ + PO_4^{3-}$
Glucose 6-PO_3H^-	$\rightleftharpoons$	H^+ + glucose 6-PO_3^{2-}
H_2CO_3	$\rightleftharpoons$	$H^+ + HCO_3^-$
NH_4^+	$\rightleftharpoons$	$H^+ + NH_3$
H_2O	$\rightleftharpoons$	$H^+ + OH^-$

The anion formed in this dissociation is a base because it can accept a proton to reform the acid. A weak acid and its base (anion) formed on dissociation are referred to as a **conjugate pair**; examples of some biologically important conjugate pairs are presented in Table 1.4. Ammonium ion (NH_4^+) is an acid because it dissociates to yield H^+ and uncharged ammonia (NH_3), a conjugate base. Phosphoric acid (H_3PO_4) is an acid and PO_4^{3-} is a base, but $H_2PO_4^-$ and HPO_4^{2-} are either a base or acid depending on whether the phosphate group is accepting or donating a proton.

The tendency of a **conjugate acid** to release H^+ can be assessed from the K'_{eq} (Eq. 1.1). The smaller the value of K'_{eq}, the less the tendency to give up a proton and the weaker the acid; the larger the value of K'_{eq}, the greater the tendency to dissociate and the stronger the acid. Water is a very weak acid with a K'_{eq} of 1.8×10^{-16} at 25°C.

A convenient method of stating the K'_{eq} is in the form of pK', as

$$\mathbf{p}K' = \log \frac{1}{K'_{eq}} \qquad (1.5)$$

Note the similarity of this definition with that of pH; as with pH and $[H^+]$, the relationship between pK' and K'_{eq} is an inverse one: The smaller the K'_{eq}, the larger the pK'. Representative values of K'_{eq}, and pK' for conjugate acids of importance in biological systems are presented in Table 1.5.

Carbonic Acid

Carbonic acid (H_2CO_3) is a weak acid important in medicine. Carbon dioxide, when dissolved in water, is involved in the following equilibrium reactions:

$$CO_2 + H_2O \xrightleftharpoons{K'_2} H_2CO_3 \xrightleftharpoons{K'_1} H^+ + HCO_3^-$$

Carbonic acid has a pK'_1 of 3.77, which is comparable to organic acids such as lactic acid. The equilibrium equation for this reaction is

$$K'_1 = \frac{[H^+][HCO_3^-]}{[H_2CO_3]} \qquad (1.6)$$

H_2CO_3 is, however, in equilibrium with dissolved CO_2, and the equilibrium equation for this reaction is

$$K'_2 = \frac{[H_2CO_3]}{[CO_2][H_2O]} \qquad (1.7)$$

TABLE 1.5 Apparent Dissociation Constant and pK′ of Some Compounds of Importance in Biochemistry

Compound	*Structures*	K'_{eq}(*M*)	*pK′*
Acetic acid	CH_3—COOH	1.74×10^{-5}	4.76
Alanine	CH_3—CH(NH_3^+)—COOH	4.57×10^{-3} 2.04×10^{-10}	2.34 (COOH) 9.69 (NH_3^+)
Citric acid	HOOC—CH_2—COH(COOH)—CH_2—COOH	8.12×10^{-4} 1.77×10^{-5} 3.89×10^{-6}	3.09 3.74 5.41
Glutamic acid	HOOC—CH_2—CH_2—CH(NH_3^+)—COOH	6.45×10^{-3} 5.62×10^{-5} 2.14×10^{-10}	2.19 (COOH) 4.25 (COOH) 9.67 (NH_3^+)
Glycine	CH_2(NH_3^+)—COOH	4.57×10^{-3} 2.51×10^{-10}	2.34 (COOH) 9.60 (NH_3^+)
Lactic acid	CH_3—CHOH—COOH	1.38×10^{-4}	3.86
Pyruvic acid	CH_3—CO—COOH	3.16×10^{-3}	2.50
Succinic acid	HOOC—CH_2—CH_2—COOH	6.46×10^{-5} 3.31×10^{-6}	4.19 5.48
Glucose 6-PO_3H^-	$C_{12}H_{11}O_5$ PO_3H^-	7.76×10^{-7}	6.11
	H_3PO_4	1×10^{-2}	2.0
	$H_2PO_4^-$	2.0×10^{-7}	6.7
	HPO_4^{2-}	3.4×10^{-13}	12.5
	H_2CO_3	1.70×10^{-4}	3.77
	NH_4^+	5.62×10^{-10}	9.25
	H_2O	1.8×10^{-16}	15.74

Solving Eq. 1.7 for H_2CO_3 and substituting for the H_2CO_3 in Eq. 1.6, the two equilibrium reactions can be combined into one equation:

$$K'_1 = \frac{[H^+][HCO_3^-]}{K'_2[CO_2][H_2O]} \tag{1.8}$$

Rearranging to combine constants, including the concentration of H_2O, simplifies the equation and yields a new combined constant, K'_3, as follows:

$$K'_1K'_2[H_2O] = K'_3 = \frac{[H^+][HCO_3^-]}{[CO_2]} \tag{1.9}$$

It is common practice in medicine to refer to dissolved CO_2 as the conjugate acid; it is the acid anhydride of H_2CO_3. The term K'_3 has a value of 7.95×10^{-7} and $pK'_3 = 6.1$. If the aqueous system is in contact with an air phase, dissolved CO_2 will also be in equilibrium with CO_2 in the air phase. Decrease or increase of one component—that is, CO_2 (air), CO_2 (dissolved), H_2CO_3, H^+, or HCO_3^-—causes a change in every component. The CO_2–HCO_3^- system is extremely important for maintaining pH homeostasis in animals. CO_2 is constantly produced in catabolic reactions in tissues and is removed by the lungs in expired air.

Henderson–Hasselbalch Equation Defines the Relationship between pH and Concentrations of Conjugate Acid and Base

A change in concentration of any component in an equilibrium reaction necessitates a concomitant change in every component. For example, an increase in $[H^+]$ decreases the concentration of conjugate base (e.g., lactate ion) with an equivalent increase in the conjugate acid (e.g., lactic acid). This relationship is conveniently expressed by rearranging the equilibrium equation and solving for H^+, as shown for the following dissociation:

$$\text{Conjugate acid} \rightleftharpoons \text{conjugate base} + H^+$$

$$K'_{eq} = \frac{[H^+][\text{conjugate base}]}{[\text{conjugate acid}]} \quad (1.10)$$

Rearranging Eq. 1.10 by dividing through by both $[H^+]$ and K'_{eq} leads to

$$\frac{1}{[H^+]} = \frac{1}{|K'_{eq}|} \cdot \frac{[\text{conjugate base}]}{[\text{conjugate acid}]} \quad (1.11)$$

Taking the logarithm of both sides gives

$$\log \frac{1}{[H^+]} = \log \frac{1}{K'_{eq}} + \log \frac{[\text{conjugate base}]}{[\text{conjugate acid}]} \quad (1.12)$$

Since $\text{pH} = \log 1/[H^+]$ and $\text{p}K' = \log 1/K'_{eq}$, Eq. 1.12 becomes

$$\mathbf{pH} = \mathbf{p}\boldsymbol{K'} + \mathbf{log}\frac{\mathbf{[conjugate\ base]}}{\mathbf{[conjugate\ acid]}} \quad (1.13)$$

Equation 1.13, developed by **Henderson and Hasselbalch**, is a convenient way of viewing the relationship between pH of a solution and relative amounts of conjugate base and acid present. Analysis of Eq. 1.13 demonstrates that when the ratio of [base]/[acid] is 1:1, pH equals the pK' of the acid because $\log 1 = 0$. If pH is one unit less than pK', the [base]/[acid] ratio is 1:10, and if pH is one unit above pK', the [base]/[acid] ratio is 10:1. Figure 1.6 is a plot of ratios of conjugate base to conjugate acid versus pH of several weak acids; note that ratios are presented on a logarithmic scale.

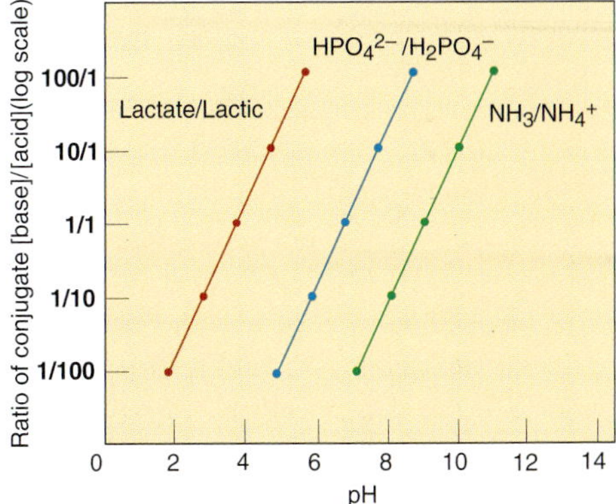

FIGURE 1.6

Ratio of conjugate [base]/[acid] as a function of pH. When the ratio of [base]/[acid] is 1, pH equals pK' of weak acid.

Buffering Is Important to Control pH

When NaOH is added to a solution of a weak acid such as lactic acid, the ratio of [conjugate base]/[conjugate acid] changes. NaOH dissociates totally and the OH^- formed is neutralized by H^+ from dissociated acid to form H_2O. The decrease in $[H^+]$ causes further dissociation of weak acid to comply with requirements of its equilibrium reaction. The amount of weak acid dissociated will be so nearly equal to the amount of OH^- added that it is considered equal. Thus, the decrease in amount of conjugate acid is equal to the amount of conjugate base that is formed. The events are represented in titration curves of two weak acids presented in Figure 1.7. When 0.5 equiv of OH^- is added, 50% of the weak acid is dissociated and the [acid]/[base] ratio is 1.0; pH at this point is equal to pK' of the acid. Shapes of individual titration curves are similar but displaced due to differences in pK' values. There is a steep rise in pH when only 0.1 equiv of OH^- is added, but between 0.1 and 0.9 equiv of added OH^- the pH change is only ~2. Thus, a large amount of OH^- is added with a relatively small change in pH. This is called **buffering** and is defined as the ability of a solution to resist a change in pH when an acid or base is added. If weak acid is not present, a small amount of OH^- would lead to a high pH because there would be no source of H^+ to neutralize the OH^-.

The best buffering range for a conjugate pair is in the pH range near the pK' of the weak acid. Starting from a pH one unit below to a pH one unit above pK', about 82% of a weak acid in solution will dissociate, and therefore an amount of base equivalent to about 82% of original acid can be neutralized with a change in pH of 2. Maximum buffering ranges for conjugate pairs are considered to be between 1 pH unit below and 1 pH unit above pK'. The conjugate pair lactate ion/lactic acid with a p$K' = 3.86$ is an effective buffer between pH 3 and pH 5, with no significant buffering capacity at pH = 7.0. The $HPO_4^{2-}/H_2PO_4^-$ pair with p$K' = 6.7$ is an effective buffer at pH = 7.0. Thus at the pH of the cell's cytosol (~ 7.0), the lactate–lactic acid pair is not an effective buffer but $HPO_4^{2-}/H_2PO_4^-$ is.

Buffering capacity also depends on concentrations of conjugate acid and base. The higher the concentration of conjugate base, the more added H^+ with which it can react.

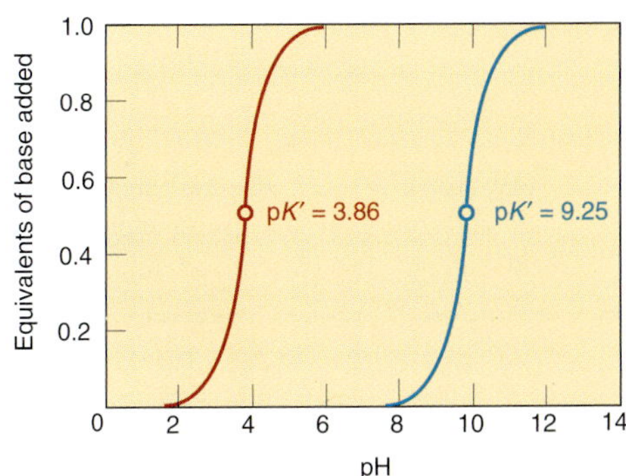

FIGURE 1.7

Acid–base titration curves for lactic acid (pK' 3.86) and NH_4^+ (pK' 9.25). At pH equal to respective pK' values, there will be an equal amount of acid and base for each conjugate pair.

CLINICAL CORRELATION 1.1

Blood Bicarbonate Concentration in Metabolic Acidosis

In a normal adult, buffers maintain blood pH at about 7.40; if pH should drop below 7.35, the condition is referred to as an acidosis. A blood pH of near 7.0 could lead to serious consequences and possibly death. In acidosis, particularly that caused by a metabolic abnormality, it is important to monitor the acid–base parameters of a patient's blood. Values of interest to a clinician include the blood pH, HCO_3^- and CO_2 concentrations. Normal values for these are pH = 7.40, $[HCO_3^-]$ = 24.0 mM, and $[CO_2]$ = 1.20 mM.

Blood values of a patient with a metabolic acidosis were pH = 7.03, and $[CO_2]$ = 1.10 mM. What is the patient's blood $[HCO_3^-]$ and how much of the normal $[HCO_3^-]$ has been used in buffering the acid causing the condition?

1. The Henderson–Hasselbalch equation is

$$\mathrm{pH} = \mathrm{p}K' + \log([HCO_3^-]/[CO_2])$$

pK' for $[HCO_3^-]/[CO_2]$ is 6.10.

2. Substitute given values in the equation.

$$7.03 = 6.10 + \log([HCO_3^-]/1.10\ \mathrm{mM})$$

or

$$7.03 - 6.10 = 0.93$$
$$= \log([HCO_3^-]/1.10\ \mathrm{mM})$$

Antilog of 0.93 = 8.5; thus,

$$8.5 = [HCO_3^-]/1.10\ \mathrm{mM}$$

or

$$[HCO_3^-] = 9.4\ \mathrm{mM}$$

3. Thus, there has been a decrease of 14.6 mmol of HCO_3^- per liter of blood. If much more HCO_3^- is lost, this important buffer would be unavailable to buffer any more acid in the blood and the pH would drop rapidly, very likely leading to death of the patient.

The more conjugate acid, the more added OH^- can be neutralized by dissociation of the acid. A case in point is blood plasma at pH 7.4. The pK' for $HPO_4^{2-}/H_2PO_4^-$ of 6.7 would suggest that this conjugate pair would be an effective buffer; the concentration of the phosphate pair, however, is low compared to that of the HCO_3^-/CO_2 system with a pK' of 6.1, which is present at a 20-fold higher concentration and accounts for most of the buffering capacity. Both pK' and concentration of a conjugate pair must be taken into account when considering buffering capacity. Most organic acids are relatively unimportant as buffers in cellular fluids because their pK' values are more than several pH units lower than the pH of the cell, and their concentrations are too low in comparison to buffer systems such as $HPO_4^{2-}/H_2PO_4^-$ and HCO_3^-/CO_2.

Figure 1.8 presents some typical problems using the Henderson–Hasselbalch equation. Control of pH and buffering in humans cannot be overemphasized; Clinical Correlation 1.1 is a representative problem in clinical practice.

1. Calculate the ratio of HPO_4^{2-} / $H_2PO_4^-$ (pK=6.7) at pH 5.7, 6.7, and 8.7.

Solution: $$\mathrm{pH} = \mathrm{p}K' + \log\frac{[HPO_4^{2-}]}{[H_2PO_4^-]}$$

5.7 = 6.7 + log of ratio; rearranging

5.7– 6.7 = –1 = log of ratio

The antilog of –1 = 0.1 or 1/10. Thus, $HPO_4^{2-}/H_2PO_4^-$ = 1/10 at pH 5.7. Using the same procedure, the ratio at pH 6.7 = 1/1 and at pH 8.7 = 100/1.

2. If the pH of blood is 7.1 and the HCO_3^- concentration is 8 mM, what is the concentration of CO_2 in blood (pK' for HCO_3^-/CO_2 = 6.1)?

Solution: $$\mathrm{pH} = \mathrm{p}K + \log\frac{[HCO_3^-]}{[CO_2]}$$

7.1 = 6.1 + log (8 mM / $[CO_2]$); rearranging

7.1 – 6.1 = 1 = log (8 mM / $[CO_2]$).

The antilog of 1 = 10. Thus, 10 = 8 mM / $[CO_2]$, or $[CO_2]$ = 8 mM/10 = 0.8 mM.

3. At a normal blood pH of 7.4, the sum of $[HCO_3^-]$ + $[CO_2]$ = 25.2 mM. What is the concentration of HCO_3^- and CO_2 (pK' for HCO_3^-/CO_2 = 6.1)?

Solution: $$\mathrm{pH} = \mathrm{p}K + \log\frac{[HCO_3^-]}{[CO_2]}$$

7.4 = 6.1 + log ($[HCO_3^-]$ / $[CO_2]$); rearranging

7.4 – 6.1 = 1.3 = log ($[HCO_3^-]$ / $[CO_2]$).

The antilog of 1.3 is 20. Thus $[HCO_3^-]$ / $[CO_2]$ = 20. Given $[HCO_3^-]$ + $[CO_2]$ = 25.2, solve these two equations for $[CO_2]$ by rearranging the first equation:

$[HCO_3^-]$ = 20 $[CO_2]$.

Substituting in the second equation,

20 $[CO_2]$ + $[CO_2]$ = 25.2

or

CO_2 = 1.2 mM

Then substituting for CO_2, 1.2 + $[HCO_3^-]$ = 25.2, and solving, $[HCO_3^-]$ = 24 mM.

FIGURE 1.8

Typical problems of pH and buffering.

1.3 COMPOSITION OF EUKARYOTIC CELLS: FUNCTIONAL ROLES OF SUBCELLULAR ORGANELLES AND MEMBRANE SYSTEMS

Eukaryotic cells contain well-defined **cellular organelles** such as a nucleus, mitochondria, lysosomes, and peroxisomes, each delineated by a membrane (Figure 1.9). Membranes form a tubular network throughout the cell, the endoplasmic reticulum and Golgi complex, enclosing an interconnecting space or cisternae, respectively. The lipid–protein nature of **cellular membranes** (see p. 461) prevents rapid movement of many molecules, including water, from one compartment to another. Specific mechanisms for translocation of small or large, charged or uncharged molecules allow the various membranes to modulate concentrations of substances in their compartments. The cytosol and fluid compartment of organelles have a distinguishing composition of

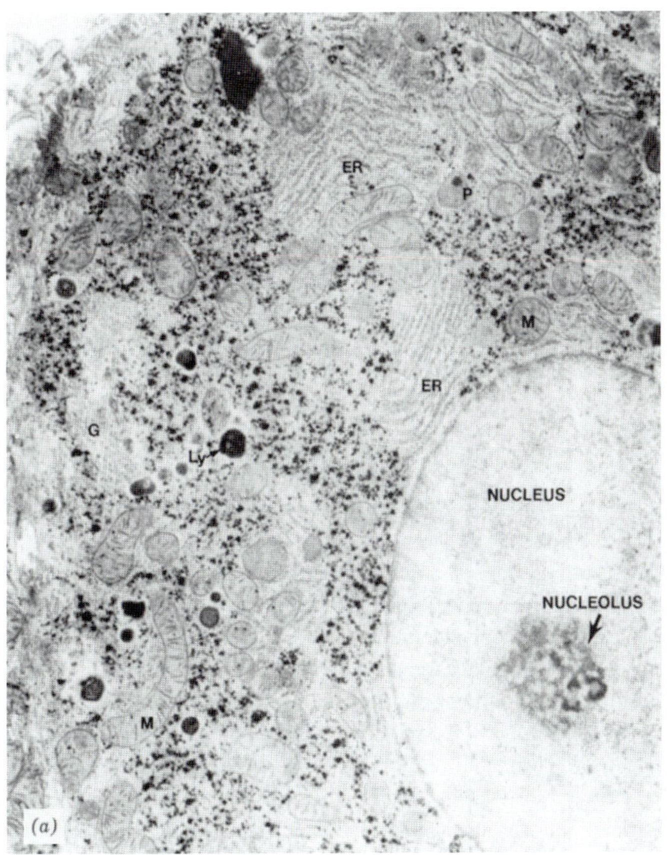

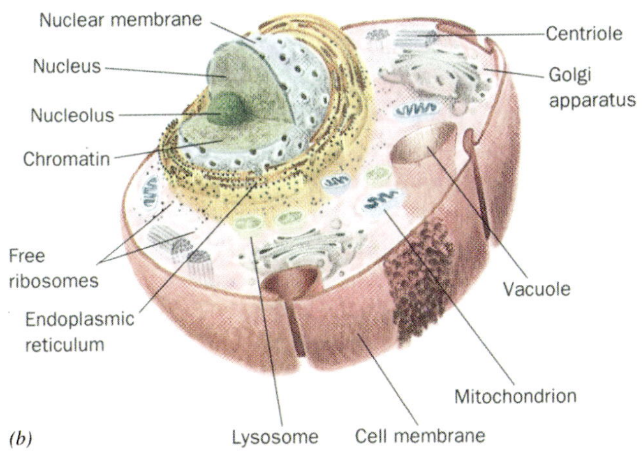

FIGURE 1.9

(a) Electron micrograph of a rat liver cell labeled to indicate the major structural components of eukaryotic cells and (b) a schematic drawing of an animal cell. Note the number and variety of subcellular organelles and the network of interconnecting membranes enclosing channels or cisternae. All eukaryotic cells are not as complex in their appearance, but most contain the major structures shown. ER, endoplasmic reticulum; G, Golgi zone; Ly, lysosome; P, peroxisome; M, mitochondrion.
Photograph (**a**) reprinted with permission of Dr. K. R. Porter from Porter, K. R. and Bonneville, M. A. In: *Fine Structure of Cells and Tissues*. Philadelphia: Lea & Febiger, 1972. Schematic (**b**) reprinted with permission from Voet, D. and Voet, J. G. *Biochemistry*, 2nd ed. New York: Wiley, 1995. © (1995) John Wiley & Sons, Inc.

inorganic ions, organic molecules, proteins, and nucleic acids. Partitioning of activities and components in membrane-enclosed spaces has several advantages for the economy of the cell, including (a) sequestering of substrates, cofactors, and enzymes for greater metabolic efficiency and (b) adjustment of pH and ionic composition for maximum activity of biological processes.

The activities and composition of cellular structures and organelles are determined on intact cells by histochemical, immunological, and fluorescent staining methods. Continuous observations in real time of cellular events in intact viable cells are possible. As an example, changes of pH and ionic calcium concentration can be studied in the cytosol by use of ion-specific indicators. Individual organelles, membranes, and components of the cytosol can be isolated and analyzed after disruption of the plasma membrane. Techniques for disrupting membranes include use of detergents, osmotic shock, or homogenization of tissues where shearing forces break down the plasma membrane. In appropriate isolation media, cell organelles and membrane systems can be separated by centrifugation because of differences in size and density. These techniques permit isolation of cellular fractions from most mammalian tissues. In addition, components of organelles, such as mitochondria and peroxisomes, can be isolated after disruption of the organelle membrane. By the use of these various techniques, the activities and functions of the various cellular compartments have been studied.

Most isolated structures and cellular fractions appear to retain the chemical and biochemical characteristics of the structure *in situ*. But biological membrane systems are subject to damage even under very mild conditions, and alterations that occur during isolation lead to changes in composition. Damage to a membrane alters its permeability properties and allows transfer of substances that would normally be excluded by the membrane barrier. In addition, many proteins are loosely associated with membranes and dissociate easily (see p. 457).

Not unexpectedly, there are differences in structure, composition, and activities of cells from different tissues due to their diverse functions. The major biochemical activities, however, are fairly constant from tissue to tissue. Differences between cell types are usually in their distinctive specialized activities.

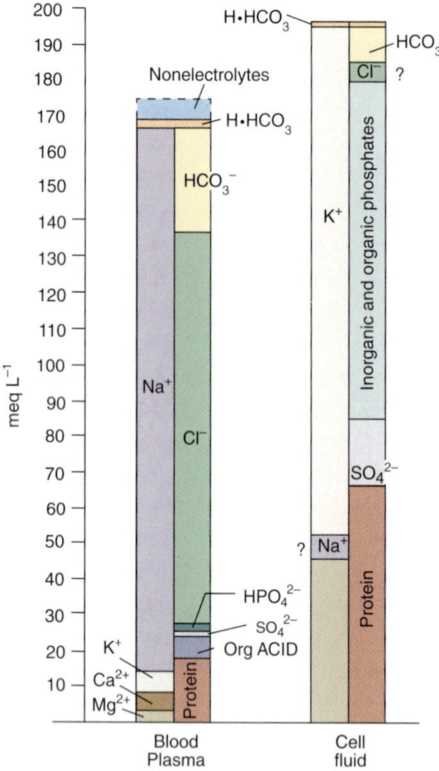

FIGURE 1.10

Major chemical constituents of blood plasma and cell fluid. Height of left half of each column indicates total concentration of cations; that of right half, concentration of anions. Both are expressed in milliequivalents per liter (meq L^{-1}) of fluid. Note that chloride and sodium values in cell fluid are questioned. It is probable that, at least in muscle, the cytosol contains some sodium but no chloride.

Adapted from Gregersen, M. I. In: P. Bard (Ed.), *Medical Physiology*, 11th ed. St. Louis: Mosby, 1961, p. 307.

General Chemical Composition of Cells

Each cellular compartment has an aqueous fluid or **matrix** that contains various ions, small molecules, and a variety of proteins, and in some cases nucleic acids. Cellular localization of most enzymes and major metabolic pathways, as well as that of nucleic acids, has been determined, but the exact ionic composition of the matrix of specific organelles is still uncertain. Each presumably has a distinctive ionic composition and pH. An approximation of the ionic composition of intracellular fluid, considered to represent primarily the cytosol, compared to blood plasma is presented in Figure 1.10. The volume and degree of hydration of cells are dynamic; thus, these concentrations are only estimates of the actual concentrations. **Na^+** is the major extracellular cation, with a concentration of ~140 meq L^{-1} (mM); very little Na^+ is present in intracellular fluid. **K^+** is the major intracellular cation. **Mg^{2+}** is present in both extra- and intracellular compartments at concentrations much lower than that of Na^+ and K^+. The major extracellular anions are **Cl^-** and **HCO_3^-**, with smaller amounts of phosphate and sulfate. Most proteins have a negative charge at pH 7.4 (see p. 86), being anions at the pH of tissue fluids. Major intracellular anions are inorganic phosphate, organic phosphates, and proteins. Other inorganic and organic anions and cations are present in concentrations well below the meq L^{-1} level. Except for very small differences created by membranes and leading to development of membrane potentials, *total cation equals total anion concentration in the different fluids.*

Intracellular concentrations of most small molecular weight organic molecules, such as sugars, organic acids, amino acids, and phosphorylated intermediates, are in the range of 0.01–1.0 meq L^{-1} but some are significantly lower. Coenzymes, organic molecules required for activity of some enzymes, are in the same range of concentration. The overall cellular concentrations of substrates for enzymes are relatively low in concentration in comparison to inorganic ions; localization in specific organelles or cellular microenvironments, however, can increase their concentrations significantly.

It is not very meaningful to determine the molar concentration of individual proteins in cells. In many cases, they are localized in specific structures or in combination with other proteins to create functional units. It is in a restricted compartment that individual proteins carry out their role, whether structural, catalytic, or regulatory. It is of interest that blood plasma contains thousands of distinct proteins, ranging in concentration from about 10^{-3} M (albumin) to 10^{-13} M (parathyroid hormone), with some proteins with even lower concentrations.

Functional Role of Subcellular Organelles and Membrane Systems

Some metabolic pathways are located in a single cellular compartment. As examples, the tricarboxylic acid cycle (see p. 543) is entirely in the mitochondria, glycolysis in the cytosol (see p. 583), and DNA replication in the nucleus and mitochondria. Some pathways are divided between two locations, with the pathway intermediates diffusing or being translocated from one compartment to another. The enzymes for heme synthesis (see p. 835) are split between mitochondria and cytosol. Organelles have very specific functions, and their specific enzymatic activities serve as markers for the organelle during isolation.

The following describes briefly some major roles of eukaryotic cell structures, and it is intended to indicate the complexity and organization of cells. A summary of functions and division of labor within eukaryotic cells is presented in Table 1.6, and the structures are identified in Figure 1.9.

Plasma Membrane Is Limiting Boundary of a Cell

The outer surface of **plasma membranes** is in contact with a variable external environment, and the inner surface is in contact with a relatively constant environment provided by the cell's cytoplasm. The two sides of mammalian plasma membranes have different chemical compositions and functions. The outer surface makes adhesive interactions with the extracellular matrix and other cells through **integrins**, which are integral transmembrane proteins. Integrins bind to the cytoskeleton inside the cells and participate in signal transduction across the membrane. Through cytoskeletal elements the plasma membrane is involved in determining cell shape and movement. The lipid nature of membranes excludes many substances, including proteins and other macromolecules, but specific transport mechanisms or pores permit transmembrane movement of selective ions and organic molecules. Through the plasma membrane, cells communicate with other cells; the membrane contains many specific protein receptors for binding extracellular signals, such as hormones and neurotransmitters. Plasma membranes of eukaryotes have different proteins in distinct **membrane domains**, and their protein composition varies between cell types. Details of membrane structure and biochemistry are presented in Chapter 12.

Nucleus Is Site of DNA and RNA Synthesis

Early microscopists described the interior of eukaryotic cells as containing a **nucleus**, a large membrane-bound compartment, and a **cytoplasm**. The nucleus is surrounded by two membranes, termed the **nuclear envelope**, with the outer membrane being continuous with membranes of the endoplasmic reticulum. The nuclear envelope contains numerous pores—called nuclear pore complexes, about 90 Å in diameter—that permit controlled movement of particles and large molecules between the nuclear matrix and the cytoplasm. Transport of macromolecules requires expenditure of energy, and it is facilitated by a diverse series of cargo proteins and soluble transport factors, which cycle between the two compartments. The nucleus is compartmentalized, but there is controlled flow of material between compartments. A major subcompartment, seen clearly in electron micrographs, is the **nucleolus**. **Deoxyribonucleic acid**, **DNA**, the repository of genetic information, is located in the nucleus as a DNA–protein complex,

TABLE 1.6 Summary of Eukaryotic Cell Compartments and Their Major Functions

Compartment	*Major Functions*
Plasma membrane	Transport of ions and small molecules Recognition Receptors for small and large molecules Cell morphology and movement
Nucleus	DNA synthesis and repair RNA synthesis
Nucleolus	RNA processing and ribosome synthesis
Endoplasmic reticulum	Membrane synthesis Synthesis of proteins and lipids for some organelles and for export Lipid and steroid synthesis Detoxification reactions
Golgi apparatus	Modification and sorting of proteins for incorporation into membranes and organelles, and for export Export of proteins
Mitochondria	Production of ATP Cellular respiration Oxidation of pyruvate, amino acids, and fatty acids Urea and heme synthesis
Lysosomes	Cellular digestion: hydrolysis of proteins, carbohydrates, lipids, and nucleic acids
Peroxisomes	Lipid oxidation Oxidative reactions involving O_2 Utilization of H_2O_2
Microtubules, intermediate filaments, and microfilaments	Cell cytoskeleton Cell morphology Cell motility Intracellular movements
Cytosol	Metabolism of carbohydrates, amino acids, and nucleotides; synthesis of fatty acids Protein synthesis

chromatin, that is organized into chromosomes. The nucleus contains the proteins and enzymes involved in replication of DNA before mitosis and in repair of DNA that has been damaged (see Chapter 4). Transcription of genetic information in DNA into a form that can be translated into cell proteins (see Chapter 5) involves synthesis and processing into a variety of forms of ribonucleic acid, RNA, in the nucleus. The processing of ribosomal RNA and formation of small and large ribosomal subunits occurs in the nucleolus.

Endoplasmic Reticulum Has a Role in Protein Synthesis and Many Synthetic Pathways

The cytoplasm of eukaryotic cells contains an extensive network of interconnecting membranes that extend from the nuclear envelope to the plasma membrane. This extensive structure, termed **endoplasmic reticulum** (**ER**), consists of membranes with a smooth (**smooth ER** or **SER**) appearance in some parts and rough (**rough ER** or **RER**) in other places. The ER encloses the ER lumen, where newly synthesized proteins are modified for export or incorporation into organelles and membranes. The rough appearance is due to the presence of **ribonucleoprotein particles**, called **ribosomes**, attached on its cytosolic side. During cell fractionation the endoplasmic reticulum network is disrupted and the membrane reseals to form small vesicles or **microsomes**, which can be isolated by differential centrifugation. Microsomes, as such, do not occur in cells.

A major function of ribosomes on rough endoplasmic reticulum is biosynthesis of proteins for incorporation into membranes and cellular organelles, as well as for export to the outside of the cell. Smooth endoplasmic reticulum is involved in lipid synthesis and contains enzymes termed **cytochromes P450** (see p. 414) that catalyze hydroxylation of a variety of endogenous and exogenous compounds. These enzymes are important in biosynthesis of steroid hormones and removal of toxic substances. Endoplasmic reticulum and Golgi apparatus are involved in formation of other cellular organelles such as lysosomes and peroxisomes.

Golgi Complex Is Involved in Secretion of Proteins

Golgi complex is a network of flattened smooth membrane sacs (**cisternae**) and vesicles responsible for secretion from the cell of proteins such as hormones, blood plasma proteins, and digestive enzymes. It works in consort with endoplasmic reticulum where proteins for certain destinations are synthesized. Enzymes in Golgi membranes catalyze transfer of carbohydrate units to proteins to form glycoproteins, a process that is important in determining the proteins eventual destination. Golgi is the major site of new membrane synthesis and participates in formation of lysosomes and peroxisomes. Membrane vesicles shuttle proteins between the cisternae; they are pinched off one cisterna and fuse with another with the aid of a family of proteins, termed SNARE proteins. Vesicles originating from Golgi complex transport proteins to the plasma membrane for secretion.

Mitochondria Supply Most of the Cellular Need for ATP

In electron micrographs of cells, **mitochondria** appear as spheres, rods, or filamentous bodies that are usually about 0.5–1 μm in diameter and up to 7 μm in length. The internal matrix, the **mitosol**, is surrounded by two membranes, distinctly different in appearance and biochemical function. The inner membrane convolutes into the matrix to form **cristae** and contains numerous small spheres attached by stalks extending into the mitosol; these structures contain the enzymes for the synthesis of over 90% of the adenosine triphosphate (ATP) required by the cell. The structure of cristae varies from tubular to lamellar depending on the tissue and functional state of mitochondria. Like other cellular components, turnover of mitochondria occurs and biogenesis is a continuing process. Exercise promotes an increase in the number of mitochondria in skeletal muscle. The outer and inner membranes are very different in their function and enzyme activities. Components of the electron transport system and oxidative phosphorylation are part of the inner membrane (see p. 550). Metabolic pathways for oxidation of pyruvate produced by glycolysis, fatty acids, and amino acids, and some reactions in biosynthesis of urea and heme are located in the mitosol. The outer membrane contains **porins** (see p. 464) that permit access to larger molecules, but the inner membrane is highly selective and contains a variety of transmembrane systems for translocation of various metabolites. Mitochondria have a key role in aging, and cytochrome c (a component of the mitochondrial electron transport system) is an initiator of programmed cell death, termed apoptosis (see p. 1021).

Mitochondria have a role in their own replication. They contain several copies of a circular DNA (**mtDNA**), with genetic information for 13 mitochondrial proteins and some RNAs. The DNA is similar in size (16,569 base pairs) to that of bacterial chromosomes. The presence of an independent genome and the similarity to bacteria have led to the widely accepted hypothesis that these organelles were bacteria that were taken up by a more advanced cell and, instead of being destroyed, formed a mutual dependency. It is estimated that this event might have occurred some 3 billion years ago. The inheritance of mitochondria is by maternal transmission, and it has been possible to study the global movement of humans by evaluating variations in mtDNA. Mitochondria also have the requisite RNAs (see p. 68) and enzymes to catalyze the synthesis of some proteins. Most mitochondrial proteins are, however, derived from genes present in nuclear DNA and synthesized on free ribosomes in the cytosol, then imported into the organelle. There are several hundred genetic diseases of mitochondrial activities; some result from mutations of nuclear DNA coding for mitochondrial proteins, whereas others result from mutation of mitochondrial DNA (see Clin. Corr. 1.2).

CLINICAL CORRELATION 1.2
Mitochondrial Diseases

The first disease (Luft's disease) specifically involving mitochondrial energy transduction was reported in 1962. A 30-year-old patient had general weakness, excessive perspiration, a high caloric intake without increase in body weight, and an excessively elevated basal metabolic rate (a measure of oxygen utilization). She had a defect in the mechanism that controls mitochondrial oxygen utilization (see Chapter 14). Since then, several hundred genetic abnormalities have been identified that lead to alterations in mitochondrial enzymes, ribonucleic acids, electron transport components, and membrane transport systems. Mutations of mtDNA as well as of nuclear DNA lead to mitochondrial genetic diseases. The first disease to be identified as due to a mutation of mtDNA was Leber's Hereditary Optic Neuropathy, which leads to sudden blindness in early adulthood. Many mitochondrial diseases involve skeletal muscle and central nervous systems. Mitochondrial DNA damage may occur due to free radicals (superoxides) formed in the mitochondria. Abnormalities in mitochondria have been implicated in the pathophysiology of schizophrenia, bipolar disorder, and age-related degenerative diseases, such as Parkinson's disease, Alzheimer's disease, and cardiomyopathies. Recently it has been suggested that a single mutation for a mitochondrial t-RNA leads to a constellation of symptoms including hypertension, high blood cholesterol, and decreased levels of plasma Mg^{2+}. See the following Clinical Correlations for details of mitochondrial diseases: 6.5, p. 221; 14.4, p. 574; 14.5, p. 575; 14.6, p. 575; and 14.7, p. 577.

Source: Luft, R. The development of mitochondrial medicine. *Proc. Natl. Acad. Sci. USA* 91:8731,1994; Chalmers, R. M. and Schapira, A. H. V. Clinical, biochemical and molecular genetic features of Leber's hereditary optic neuropathy. *Biochim. Biophys. Acta* 1410:147, 1999; Wallace, D. C. Mitochondrial DNA in aging and disease. *Sci. Am.* 280:40, 1997; and Wallace, D. C. Mitochondrial diseases in man and mouse. *Science* 283:1482, 1999.

TABLE 1.7 Representative Lysosomal Enzymes and Their Substrates

Type of Substrate and Enzyme	*Specific Substrate*
POLYSACCHARIDE-HYDROLYZING ENZYMES	
α-Glucosidase	Glycogen
α-Flucosidase	Membrane fucose
β-Galactosidase	Galactosides
α-Mannosidase	Mannosides
β-Glucuronidase	Glucuronides
Hyaluronidase	Hyaluronic acid and chondroitin sulfates
Arylsulfatase	Organic sulfates
Lysozyme	Bacterial cell walls
PROTEIN-HYDROLYZING ENZYMES	
Cathepsins	Proteins
Collagenase	Collagen
Elastase	Elastin
Peptidases	Peptides
NUCLEIC ACID-HYDROLYZING ENZYMES	
Ribonuclease	RNA
Deoxyribonuclease	DNA
LIPID-HYDROLYZING ENZYMES	
Lipases	Triacylglycerol and cholesterol esters
Esterase	Fatty acid esters
Phospholipase	Phospholipids
PHOSPHATASES	
Phosphatase	Phosphomonoesters
Phosphodiesterase	Phosphodiesters
SULFATASES	
	Heparan sulfate
	Dermatan sulfate

Lysosomes Are Required for Intracellular Digestion

Lysosomes are responsible for **intracellular digestion** of extracellular and intracellular substances. With a single limiting membrane, they maintain an acidic matrix pH of about 5. Encapsulated in these organelles is a class of glycoprotein enzymes—hydrolases—that catalyze hydrolytic cleavage of carbon–oxygen, carbon–nitrogen, carbon–sulfur, and oxygen–phosphorus bonds in proteins, lipids, carbohydrates, and nucleic acids. A partial list of lysosomal enzymes is presented in Table 1.7. Lysosomal hydrolases are most active at acidic pHs and split complex molecules into simple low-molecular-weight compounds that can be reutilized. The relationship between pH and enzyme activity is discussed on p. 375.

The enzyme content of lysosomes varies in different tissues and depends on specific tissue functions. The lysosomal membrane contains specific protein mediators and receptors as well as transporters for translocation of substances across it. Isolated lysosomes do not catalyze hydrolysis of added substrates until the lysosomal membrane is disrupted. Disruption of the lysosomal membrane within cells leads to cellular digestion. Various pathological conditions have been attributed to release of lysosomal enzymes, including arthritis, allergic responses, several muscle diseases, and drug-induced tissue destruction (see Clin. Corr. 1.3).

Lysosomes are involved in digestion of intra- and extracellular substances that must be removed. Through **endocytosis**, material from the exterior is taken into cells and encapsulated in membrane-bound vesicles (Figure 1.11). The process occurs by several different mechanisms; one involves several cytosolic proteins, including clathrin (see p. 917), to form coated pits, which bud off segments of membrane. The plasma membrane invaginates around foreign particles, such as microorganisms, by **phagocytosis** or extracellular fluid that contains suspended material by **pinocytosis**. Vesicles containing external material fuse with lysosomes to form **secondary lysosomes** or **digestive vacuoles**, which contain material to be digested and lysosomal hydrolases to carry out the digestion. They are identified microscopically by their size and often by the presence of partially digested structures. **Primary lysosomes** are those in which the enzymes are not involved in the digestive process.

Cellular proteins, nucleic acids, and lipid as well as organelles, such as mitochondria, are in a dynamic state of synthesis and degradation. Lysosomes are responsible for hydrolysis of these cellular components, a process called **autophagy**. Substances destined to be degraded are identified and taken up by lysosomes or are first encapsulated within membrane vesicles that fuse with a lysosome. An outline of the process is presented in Figure 1.11.

Products of lysosomal digestion are released from lysosomes and are reutilized by the cell. Indigestible material accumulates in vesicles called residual bodies whose contents

CLINICAL CORRELATION 1.3

Lysosomal Enzymes and Gout

Catabolism of purines, nitrogen-containing heterocyclic compounds found in nucleic acids, leads to formation of uric acid, which is excreted in urine (see p. 799). Gout is an abnormality in which uric acid is produced in excess, leading to an increase of uric acid in blood and deposition of urate crystals in joints. Clinical manifestations include inflammation, pain, swelling, and increased warmth of some joints, particularly the big toe. Uric acid is rather insoluble, and some of the clinical symptoms of gout can be attributed to damage done by urate crystals. Crystals are phagocytized by cells in the joint, and they accumulate in intracellular digestive vacuoles that contain lysosomal enzymes. Crystals cause physical damage to the vacuoles, releasing lysosomal hydrolases into the cytosol. Even though the pH optimum of lysosomal enzymes is lower than the pH of the cytosol, they have some hydrolytic activity at the higher pH, which causes digestion of cellular components, release of substances from the cell, and cellular autolysis.

Source: Weissmann, G. Crystals, lysosomes and gout. *Adv. Intern. Med.* 19:239, 1974; and Burt, H. M., Kalkman, P. H., and Mauldin, D. Membranolytic effects of crystalline monosodium urate monohydrate. *J. Rheumatol.* 10:440, 1983.

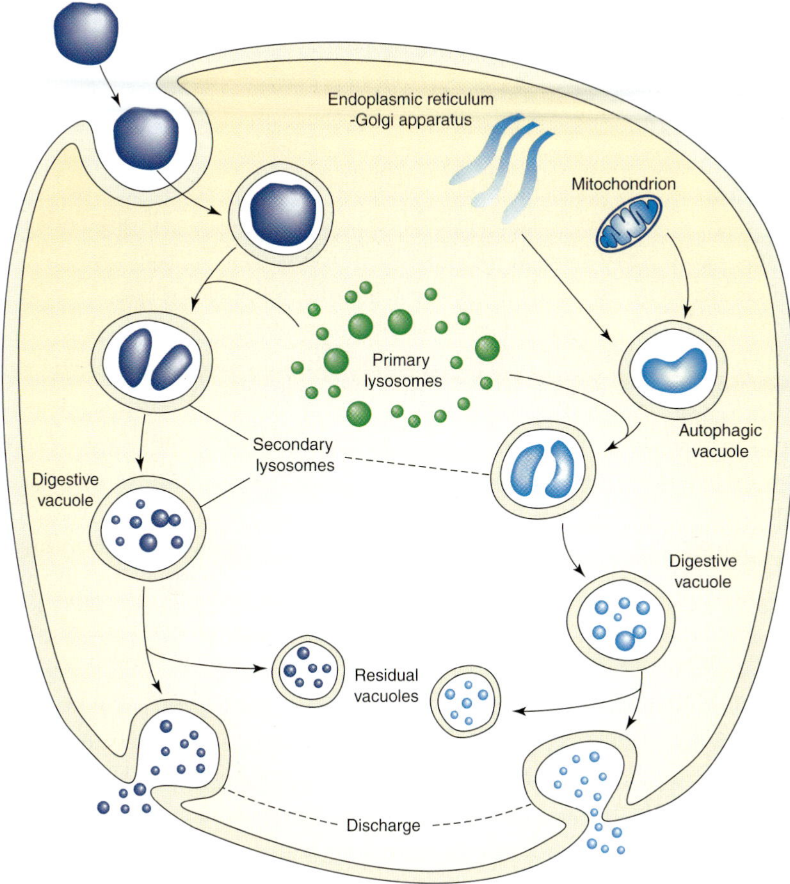

FIGURE 1.11

Diagrammatic representation of the role of lysosomes in intracellular digestion of substances internalized by phagocytosis (heterophagy) and of cellular components (autophagy). In both processes, substances to be digested are enclosed in a membrane vesicle, followed by fusion with a primary lysosome to form a secondary lysosome where digestion occurs.

are normally removed from the cell by exocytosis. Some residual bodies contain high concentrations of a pigmented substance that is chemically heterogeneous and contains polyunsaturated fatty acids and proteins, termed **lipofuscin**, or "**age pigment**," or "**wear and tear pigment**," that accumulate in cells. Lipofuscin has been observed particularly in postmitotic neurons and muscle cells, and it has been implicated in the aging process.

Some lysosomal enzymes are normally secreted from the cell for digestion of extracellular material in connective tissue and prostate gland. In a number of genetic diseases, classified as **lysosomal storage diseases** (see p. 726), individual lysosomal enzymes are missing, leading to accumulation of the substrate of the missing enzyme. The lysosomes become enlarged with undigested material, thereby interfering with normal cell processes (see Clin. Corr. 1.4). One such case is **I cell disease**, in which the cellular mechanism for directing lysosomal enzymes into lysosomes during their synthesis is defective and the hydrolytic enzymes are exported out of the cell and damage the extracellular matrix (see Clin. Corr. 6.8, p. 230).

Peroxisomes Have an Important Role in Lipid Metabolism

Most mammalian, protozoan, and plant cells have organelles, designated **microbodies** or **peroxisomes**, the latter because of their ability to produce or utilize **hydrogen peroxide** (H_2O_2). They are small (0.3–1.5 μm in diameter) and are spherical or oval, with a fine network of tubules in their matrix. They vary in their function and number between cells. Over 50 enzymes, catalyzing oxidative and biosynthetic reactions, have been identified in peroxisomes from different tissues.

Peroxisomes are essential for oxidation of very long-chain fatty acids (see p. 690) and synthesis of glycerolipids, glycerol ether lipids (plasmalogens), and isoprenoids (see p. 705). In mice, peroxisomes have been demonstrated to have a significant physiological role in lipid synthesis and degradation. They also contain enzymes that oxidize D-amino acids, uric acid, and various 2-hydroxy acids using molecular O_2 with formation of

CLINICAL CORRELATION 1.4
Lysosomal Acid Lipase Deficiency

Human lysosomal acid lipase (hLAL) hydrolyzes triacylglycerol to free fatty acids and glycerol and hydrolyzes cholesteryl esters to cholesterol and fatty acids. It is a critical enzyme in cholesterol metabolism, serving to make cholesterol available for the needs of cells. Cholesteryl ester storage disease (CESD) and Wolman's disease are two distinct phenotypic forms of a genetic deficiency of hLAL; both are rare autosomal recessive diseases. CESD is usually diagnosed in adulthood and is evidenced by hypercholesterolemia, hepatomegaly, and early onset of severe atherosclerosis. Affected individuals have hLAL activity but at a very low level (<5% of normal). Apparently, this level is sufficient to hydrolyze triacylglycerol but not cholesteryl esters. Patients are homozygous for a mutation at a splice junction on both alleles (see p. 1194), leading to a catalytically defective and unstable enzyme.

Wolman's disease is manifested in infants and is usually fatal by age one. There is no detectable activity of hLAL. Both triacylglycerol and cholesteryl esters accumulate in tissues. Patients are homozygous for a mutation that leads to an absence of active enzyme.

Source: Hegele, R. A., Little, J. A., Vezina, C., et al. Hepatic lipase deficiency. Clinical, biochemical and molecular genetic characteristics. *Atheroscler. Thromb.* 13:720, 1993; Lohse, P., Maas, S., Lohse, P., Elleder, M., Kirk, J. M., Besley, G. T., and Seidel, D. Compound heterozygosity for a Wolman mutation is frequent among patients with cholesteryl ester storage disease. *J. Lipid Res.* 41:23, 2000; and Anderson, R. A. Bryson, G. M., and Parks, J. S. Lysosomal acid lipase mutations that determine phenotype in Wolman and cholesteryl ester storage disease. *Mol. Genet. Metab.* 68:333, 1999.

$$(1)\ 2H_2O_2 \longrightarrow 2H_2O + O_2$$

$$(2)\ RH_2 + H_2O_2 \longrightarrow R + 2H_2O$$

FIGURE 1.12
Reactions catalyzed by catalase.

H_2O_2. **Catalase** catalyzes the conversion of H_2O_2 to water and oxygen, and it catalyzes oxidation of various compounds by H_2O_2 (Figure 1.12). By having both peroxide-producing and peroxide-utilizing enzymes in one compartment, cells protect themselves from the toxicity of H_2O_2.

Proteins responsible for the biogenesis of peroxisomes have been identified. Several classes of **xenobiotics** (i.e., foreign substances), including aspirin, lead to proliferation of peroxisomes in liver. Over 25 diseases involve peroxisomes; they are grouped together as disorders of peroxisome biogenesis (see Clin. Corr. 1.5).

Cytoskeleton Organizes the Intracellular Contents

Eukaryotic cells contain microtubules, intermediate filaments, and actin filaments (microfilaments) as parts of a cytoskeletal network (see p. 981). The **cytoskeleton** has a role in maintenance of cellular morphology, intracellular transport, cell motility, and cell division. **Microtubules** are multimers of **tubulin**, a protein that rapidly assembles into tubular structures and disassembles depending on the needs of cells. Actin and myosin form very important cellular filaments in striated muscle where they are responsible for muscular contraction (see p. 980). Three separate superfamilies of mechanochemical proteins—myosin, **dynein**, and **kinesin**—convert chemical energy into mechanical energy for movement of cellular components (see p. 990). These **molecular motors** are associated with the cytoskeleton. Dynein is involved in ciliary and flagellar movement, whereas kinesin is a driving force for the movement of vesicles and organelles along microtubules especially in neuronal axons. These motor proteins have a significant role in the pathogenesis of a variety of diseases.

Cytosol Contains Soluble Cellular Components

The least complex in structure, but not in chemistry, is the organelle-free cell sap, or **cytosol**. It is here that many reactions and pathways of metabolism (as an example, glycolysis, glycogenesis, glycogenolysis, and fatty acid synthesis) occur. Although there is no apparent structure to the cytosol, the high protein content precludes it from being a truly homogeneous mixture of soluble components, and it is believed that there are functional compartments throughout the cytosol. Many reactions are localized in selected areas where substrate availability is more favorable. The actual physicochemical state of the cytosol is poorly understood. A major role of cytosol is to support synthesis of proteins on both free ribosomes often in a polysome form and those bound to endoplasmic reticulum by supplying required intermediates.

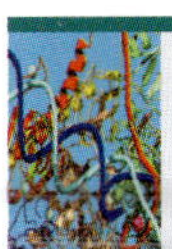

CLINICAL CORRELATION 1.5
Peroxisome Biogenesis Disorders (PBD)

Peroxisomes are responsible for a number of important metabolic reactions, including synthesis of glycerol ethers, shortening very long-chain fatty acids so that mitochondria can completely oxidize them, and oxidation of the side chain of cholesterol needed for bile acid synthesis. Peroxisome biogenesis disorders (PBDs) comprise more than 25 genetically and phenotypically related disorders that involve enzymatic activities of peroxisomes. They are rare autosomal recessive diseases characterized by decreased levels of glycerol–ether lipids (plasmalogens), increased levels of very long-chain fatty acids (C_{24} and C_{26}), and cholestanoic acid derivatives (precursors of bile acids). The disease can affect the liver, kidney, brain, and skeletal system. The most severe is Zellweger's syndrome, which is due to absence of functional peroxisomes; death frequently occurs by age 6 months. In this condition, the genetic defect is in the mechanism for importing enzymes into the matrix of peroxisomes. Some PBD conditions are caused by donor splice or missense mutations (see p. 160); in some, there is the absence of a single metabolic enzyme or defect in a membrane transport component. In some instances the disease can be diagnosed prenatally by assay of peroxisomal enzymes or fatty acids in cells of amniotic fluid.

Source: Wanders, R. J., Schutgens, R. B., and Barth, P. G. Peroxisomal disorders: A review. *J. Neuropathol. Exp. Neurol.* 54:726, 1995; FitzPatrick, D. R. Zellweger syndrome and associated phenotypes. *J. Med. Genet.* 33:863, 1996; and Warren, D. S., Wolfe, B. D., and Gould, S. J. Phenotype–genotype relationships in PEX10-deficient peroxisome biogenesis disorder patients. *Hum. Mutat.* 15:509, 2000.

Some enzymes occur as soluble proteins when the cytosol is isolated, but in the intact cell many of them may be loosely attached to membrane structures or to cytoskeletal components.

1.4 INTEGRATION AND CONTROL OF CELLULAR FUNCTIONS

A eukaryotic cell is a complex structure that maintains an intracellular environment that permits many complex reactions and functions to occur as efficiently as possible. Cells of multicellular organisms also participate in maintaining the well-being of the whole organism by exerting influences on each other to maintain balance between tissue and cellular activities. Intracellular processes and metabolic pathways are tightly controlled and integrated in order to achieve this balance. Very few functions operate totally independently; changes in one function can exert an influence, positive or negative, on other functions. As will be described throughout this textbook, control of function is mediated at many levels, from the expression of a gene, to altering the concentration of an enzyme or a protein effector, to changes in substrate or coenzyme levels to adjust the rate of a specific enzyme reaction. The integration of many cellular processes is controlled by proteins that serve as activators or inhibitors that maintain cellular homeostasis. Many cellular processes are programmed to occur under specific conditions; for example, cell division in normal cells occurs only when the processes required for cell division are activated (see p. 1015). Then and only then is there an orderly and integrated series of reactions culminating in the division of a cell into two daughter cells. A fascinating process is **apoptosis**, programmed cell death, also referred to as cell suicide (see p. 1020). This highly regulated process occurs in cells of all mammalian tissues, but individual steps in the process vary from tissue to tissue. Many diseases are due to a failure in specific control mechanisms. As one continues to comprehend the complexity of biological cells, one becomes amazed that there are not many more errors occurring and not many more individuals with abnormal conditions. Thus as we proceed to study the separate chemical components and activities of cells in subsequent chapters, it is important to keep in mind the concurrent and surrounding activities, constraints, and influence of the environment. Only by bringing together all the parts and activities of a cell—that is, reassembling the puzzle—will we appreciate the wonder of living cells.

BIBLIOGRAPHY

Water and Electrolytes

Eienberg, D. and Kauzmann, W. *The Structures and Properties of Water*. Fairlawn, NJ: Oxford University Press, 1969.

Gestein, M. and Levitt, M. Simulating water and the molecules of life. *Sci. Am.* 279:100, 1998.

Klein, M. L. Water on the move. *Science* 291:2106, 2001.

Miyazaki, M., Fujii, A, Ebata, T., and Mikami, N. Infrared spectroscopic evidence for protonated water clusters forming nanoscale cages. *Science* 304:1134, 2004.

Morris, J. G. *A Biologist's Physical Chemistry*. Reading, MA: Addison-Wesley, 1968.

Westof, E. *Water and Biological Macromolecules*. Boca Raton, FL: CRC Press, 1993.

Wiggins, P. M. Role of water in some biological processes. *Microbiol. Rev.* 54:432, 1990.

Cell Structure

Alberts, B., Johnson, A., Lewis, J., Raff, M., Roberts, K., and Walter, P. *Molecular Biology of the Cell*, 4th ed. New York: Garland Publishing, 2002.

Becker, W. M. and Deamer, D. W. *The World of the Living Cell*, 2nd ed. Redwood City, CA: Benjamin, 1991.

DeDuve, C. *Guided Tour of the Living Cell*, Vols. 1 and 2. New York: Scientific American Books, 1984.

Dingle, J. T., Dean, R. T., and Sly, W. S. (Eds.). *Lysosomes in Biology and Pathology*. Amsterdam: Elsevier (a serial publication covering all aspects of lysosomes).

Holtzman, E. and Novikoff, A. B. *Cells and Organelles*, 3rd ed. New York: Holt, Rinehart & Winston, 1984.

Hoppert, M. and Mayer, F. Principles of macromolecular organization and cell function in bacteria and archaea. *Cell Biochem. Biophys.* 31:247, 1999.

Misteli, T. Protein dynamics: Implications for nuclear architecture and gene expression. *Science* 291:843, 2001.

Porter, K. R. and Bonneville, M. A. *Fine Structure of Cells and Tissues*. Philadelphia: Lea & Febiger, 1972.

Vale, R. D. Intracellular transport using microtubule-based motors. *Annu. Rev. Cell. Biol.* 3:347, 1987.

Cell Organelles

Baumann, O. and Walz, B. Endoplasmic reticulum of animal cells and its organization into structural and functional domains. *Int. Rev. Cytol.* 205:149, 2001.

Benedetti, A. *Endoplasmic Reticulum: A Metabolic Compartment*. New York: Ios Press, 2005.

Brocard, J. B., Rintoul, G. L., and Reynolds, I. J. New perspectives on mitochondrial morphology in cell function. *Biol. Cell* 95:239, 2003.

Cuervo, A. M. Autophagy: In sickness and in health. *Trends Cell Biol.* 14:70, 2004.

De Duve, C. The birth of complex cells. *Sci. Am.* April:50, 1996.

Eckert, J. H. and Erdmann, R. Peroxisome biogenesis. *Rev. Physiol. Biochem. Pharm.* 147:75, 2003.

Enns, G. M. The contribution of mitochondria to common disorders. *Mol. Genet. Metab.* 80:11, 2003.

Fujiki, Y. Peroxisome biogenesis and peroxisome biogenesis disorders. *Fed. Eur. Biochem. Lett.* 476:42, 2000.

Hirokawa, N. and Takemura, R. Biochemical and molecular characterization of diseases linked to motor proteins. *Trends Biochem. Sci.* 28:558, 2003.

Holtzman, E. *Lysosomes*. New York: Plenum Press, 1989.

Latruffe, N. and Bugaut, M. *Peroxisomes*. New York: Springer-Verlag, 1994.

Lemasters, J. J. and Nieminen, A. *Mitochondria in Pathogenesis*. New York: Kluwer, 2001.

Luzio, J. P., Poupon, V., Lindsay, M. R., Mullock, B. , M., Piper, R. C. and Pryor, P. R. Membrane dynamics and the biogenesis of lysosomes. *Mol. Membr. Biol.* 20:141, 2003.

Moser, H. W. Molecular genetics of peroxisomal disorders. *Frontiers Biosci.* 5:D298, 2000.

Osteryoung, K. W. and Nunnari, J. The division of endosymbiotic organelles. *Science* 302:1698, 2003.

Pavelka, M. *Functional Morphology of the Golgi Apparatus*. New York: Springer-Verlag, 1987.

Preston. T. M., King, C. A., and Hyams, J. S. *The Cytoskeleton and Cell Motility*. New York: Chapman and Hall, 1990.

Rothman, J. E. and Orci, L. Budding vesicles in living cells. *Sci. Am.* 274:70, 1996.

Scheffler, I. E. *Mitochondria*. New York: Wiley, 1999.

Strauss, P. R. and Wilson, S. H. *The Eukaryotic Nucleus: Molecular Biochemistry and Macromolecular Assemblies*. Caldwell, NJ: Telford Press, 1990.

Van der Klei, I. J. and Veenhuis, M. Peroxisomes: Flexible and dynamic organelles. *Curr. Opin. Cell Biol.* 14:500, 2002.

Web Sites

Endoplasmic reticulum: http://cellbio.utmb.edu/cellbio/rerlinks.html

Golgi: http://cellbiol.utmb.edu/cellbio/golgi.htm

Lysosomes and Peroxisomes: HTTP://www.ultranet.com/jkmball/biology/BiologyPages/L/lysosomes.html

Mitochondria: http://mit.edu:8001/esgbio/cb/org/organelles.html

QUESTIONS | CAROL N. ANGSTADT

Multiple Choice Questions

1. Prokaryotic cells, but not eukaryotic cells, have:
   A. endoplasmic reticulum.
   B. histones.
   C. nucleoid.
   D. a nucleus.
   E. a plasma membrane.

2. Factors responsible for a water molecule being a dipole include:
   A. the similarity in electron affinity of hydrogen and oxygen.
   B. the tetrahedral structure of liquid water.
   C. the magnitude of the H—O—H bond angle.
   D. the ability of water to hydrogen bond to various chemical structures.
   E. the difference in bond strength between hydrogen bonds and covalent bonds.

3. Hydrogen bonds can be expected to form only between electronegative atoms such as oxygen or nitrogen and a hydrogen atom bonded to:

A. carbon.
B. an electronegative atom.
C. hydrogen.
D. iodine.
E. sulfur.

4. Which of the following is both a Brønsted acid and a Brønsted base in water?
A. $H_2PO_4^-$
B. H_2CO_3
C. NH_3
D. NH_4^+
E. Cl^-

5. Biological membranes are associated with all of the following *except*:
A. prevention of free diffusion of ionic solutes.
B. release of proteins when damaged.
C. specific systems for the transport of uncharged molecules.
D. sites for biochemical reactions.
E. free movement of proteins and nucleic acids cross the membrane.

6. Analysis of the composition of the major fluid compartments of the body shows that:
A. the major blood plasma cation is K^+.
B. the major cell fluid cation is Na^+.
C. one of the major intracellular anions is Cl^-.
D. one of the major intracellular anions is phosphate.
E. plasma and cell fluid are all very similar in ionic composition.

Questions 7 and 8: A patient with Luft's disease presented with general weakness, excessive perspiration, a high caloric intake without increase in body weight, and an excessively elevated basal metabolic rate. Luft's disease was the first disease involving a defect in mitochondria to be described. It is a defect in the mechanism that controls oxygen utilization in mitochondria.

7. Mitochondria are associated with all of the following *except*:
A. ATP synthesis.
B. DNA synthesis.
C. protein synthesis.
D. hydrolysis of various macromolecules at low pH.
E. apoptosis.

8. Which of the following is/are characteristics of mitochondria?
A. The inner membrane forms cristae and contains small spheres attached by stalks on the inner surface.
B. The outer membrane is highly selective in its permeability.
C. Components of the electron transport system are located in the mitosol.
D. Mitochondrial DNA is similar to nuclear DNA in size and shape.
E. All of the above are correct.

Questions 9 and 10: Gout is a condition in which excessive production of uric acid leads to deposition of urate crystals in joints. Clinical manifestations include inflammation, pain, and swelling of joints, especially the joint of the big toe. Crystals are phagocytosed by cells in the joint and accumulate in digestive vacuoles that contain lysosomal enzymes. Crystals cause physical damage to the vacuoles, releasing lysosomal enzymes into the cytosol.

9. Lysosomal enzymes:
A. are hydrolases.
B. usually operate at acidic pH.
C. are normally isolated from their substrates by the lysosomal membrane.
D. can lead to cellular digestion if the lysosomal membrane is disrupted.
E. all of the above are correct.

10. Individual lysosomal enzymes are missing in a number of genetic diseases referred to as *lysosomal storage diseases*. In these diseases:
A. defect is an inability to direct enzymes to the lysosome after synthesis.
B. lysosomes of affected cells become enlarged with undigested materials.
C. lipofuscin, or "wear and tear pigment," accumulates in cells.
D. any material taken into the lysosome will accumulate.
E. residual bodies contain the products of digestion.

Questions 11 and 12: Zellweger's syndrome is one of a class called *peroxisome biogenesis disorders* (PBDs). PBDs are characterized by abnormalities of the liver, kidney, brain, and skeletal system. Zellweger's syndrome is particularly severe, and death usually occurs by age 6 months. There is an absence of functional peroxisomes.

11. Peroxisomes have a role in all of the following *except*:
A. oxidation of very long-chain fatty acids.
B. synthesis of glycerolipids.
C. hydrolysis of cholesteryl esters.
D. oxidation of D-amino acids.
E. oxidation of uric acid.

12. One of the roles of peroxisomes is to render H_2O_2 nontoxic by:
A. oxidizing amino acids with O_2.
B. the action of the enzyme catalase.
C. transporting H_2O_2 from blood into peroxisomes.
D. the action of cathepsins.
E. shortening of a fatty acid chain.

Problems

13. If a weak acid is 91% neutralized at pH 5.7, what is the pK' of the acid?

14. Metabolic alkalosis is a condition in which the blood pH is higher than its normal value of 7.40 and can be caused, among other reasons, by an increases in $[HCO_3^-]$. If a patient has the following blood values, pH = 7.45, $[CO_2] = 1.25$ mM, what is the $[HCO_3^-]$? The pK' for the bicarbonate/carbonic acid system is 6.1. How does the calculated value of bicarbonate compare with the normal value?

ANSWERS

1. **C** Prokaryotic DNA is organized into a structure which also contains RNA and protein, called nucleoid. A, B, and D are found in eukaryotic cells, and E is an element of both prokaryotic and eukaryotic cells.

2. **C** Water is a polar molecule because the bonding electrons are attracted more strongly to oxygen than to hydrogen. The bond angle gives rise to asymmetry of the charge distribution; if water were linear, it would not be a dipole. A: Hydrogen and oxygen have

very different electron affinity. B and D are consequences of water's structure, not factors responsible for it.

3. **B** Only hydrogen atoms bonded to one of the electronegative elements (O, N, F) can form hydrogen bonds. A hydrogen participating in hydrogen bonding must have an electronegative element on both sides of it.
4. **A** $H_2PO_4^-$ can donate a proton to become HPO_4^{2-}. It can also accept a proton to become H_3PO_4. B and D are Brønsted acids; C is a Brønsted base. The Cl^- ion in water is neither.
5. **E** These molecules are too large to cross the membrane freely unless the membrane is damaged (B). A: Ionic solutes do not cross the lipid membrane readily. C: Most substances require transport across the membrane. D: Different membranes have different reactions.
6. **D** Phosphate and protein are the major intracellular anions. A, B, E: Plasma and cell fluid are strikingly different. The Na^+ ion is the major cation of plasma. C: Most chloride is extracellular.
7. **D** This is a lysosomal function. A: This is a major role of mitochondria. B and C: Mitochondria synthesize some of their own proteins. E: Cytochrome c, part of the mitochondrial electron transport system, is an initiator of programmed cell death.
8. **A** This is very different from the outer membrane. B: The inner membrane is highly selective, the outer membrane relatively permeable. C: The electron transport system is in the inner membrane. D: Mitochondrial DNA (mtDNA) is small and circular.
9. **E** A: There are hydrolases for all classes of macromolecules. B, D: Lysosomal pH is usually about 5 but the enzymes retain some activity at higher pH. C: Substrates and enzymes are brought together by phagocytosis or pinocytosis.
10. **B** Engorgement interferes with normal cell processes. A: This is I cell disease and affects all of the lysosomal enzymes exported out of the cell. C: This is a normal process. Lipofuscin is a chemically heterogeneous pigmented substance and not caused by a missing lysosomal enzyme. D: Each Lysosomal Storage disease affects a single enzyme so only substrates of that enzyme would be affected. E: Residual bodies contain indigestible material.
11. **C** This occurs in lysosomes.
12. **B** H_2O_2 is converted to O_2 and H_2O. A, E: These produce H_2O_2. C: H_2O_2 is produced and utilized within peroxisomes. D: These are enzymes of lysosomes.
13. If a weak acid is 91% neutralized, 91 parts are present as conjugate base and 9 parts remain as the undissociated acid. Thus the conjugate base/acid ratio is 10:1. Substituting this ratio into the Henderson–Hasselbalch equation gives $5.7 = pK' + \log(10/1)$. Solving the equation for pK' gives an answer of 4.7. The acid could be β-hydroxybutyric acid, an important physiological acid, which has this pK'.
14. One need simply substitute the given values into the Henderson–Hasselbalch equation.

$$7.45 = 6.1 + \log(x/1.25)$$

$$1.35 = \log(x/1.25)$$

Antilog of 1.35 = 22.39. Therefore, $[HCO_3^-] = 22.39 \times 1.25 = 27.98$ mM

Normal $[HCO_3^-] = 24.0$ mM so there is an increase of 3.98 mmol (per liter) of HCO_3^-.

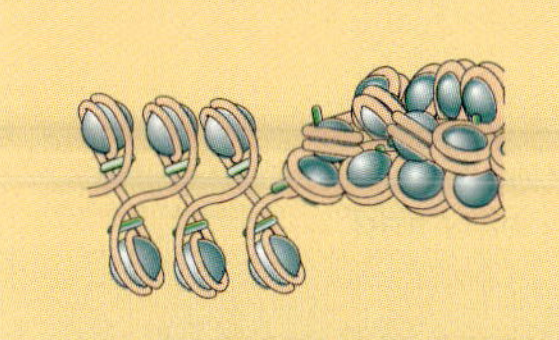

2

DNA AND RNA: COMPOSITION AND STRUCTURE

Stephen A. Woski and Francis J. Schmidt

Textbook of Biochemistry With Clinical Correlations, Sixth Edition, Edited by Thomas M. Devlin

2.1 OVERVIEW

One of the hallmarks of living organisms is their ability to reproduce. The information that makes each individual life form unique must be preserved and passed on to progeny. All life on Earth uses **nucleic acids** for storage of genetic information. With the exception of some viruses, the biomolecule utilized for information storage is **deoxyribonucleic acid** (**DNA**). This molecule is remarkably well suited for its task because of its chemical stability and its ability to encode vast amounts of information using a simple four-letter code.

Central Dogma of Molecular Biology

Because of its role as the repository of genetic information, DNA can arguably be considered the most important biomolecule in living systems. However, DNA is only a part of the core architecture of life. **Ribonucleic acid** (**RNA**) and protein play equally important roles. The interrelationship of these three classes of molecules constitutes the "**Central Dogma of Molecular Biology**" (Figure 2.1), which holds that DNA stores information that determines the sequence of RNA, which in turn determines the structure of protein. Much of the structure and biochemistry of cells is due to the properties of their constituent proteins. These properties are determined by portions of the sequence of DNA. Information, however, cannot flow directly from DNA to protein, but depends on ribonucleic acid (RNA) to transport the information. Genetic information is transmitted from DNA to RNA by **transcription**. The sequence of RNA is then **translated** into a protein sequence at the ribosome. DNA also plays an essential role in heredity by serving as the template for its own **replication**. Several discoveries have begun to blur the distinct roles of each of these biomolecules in the central dogma. For example, RNA can act as a catalyst or **ribozymes** in biochemical reactions and as a template for DNA synthesis (**reverse transcription**). In addition, RNA constitutes the active component of ribosomes, the protein synthesis machinery of the cell. DNA and RNA interact closely with proteins in many of their functions, and it has been suggested that proteins can store inheritable information. Nevertheless, the central role that nucleic acids play in life is undeniable.

DNA can Transform Cells

These universally accepted principles were rejected outright not long ago. In fact, prior to the 1950s, the general view was that nucleic acids were substances of limited cellular

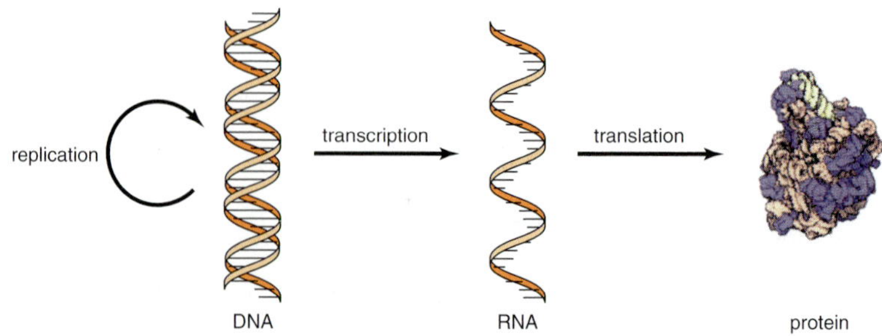

FIGURE 2.1
Central dogma of molecular biology.

CLINICAL CORRELATION **2.1**

DNA Vaccines

Traditional procedures of vaccination have used purified components of an infectious organism, viruses, or dead or attenuated intact cells to elicit production of specific antibodies that can provide individuals with active immunity. Such vaccines have successfully provided protection against diseases such as polio, smallpox, whooping cough, typhoid fever, and diphtheria. However, the development of immunizations against infectious pathogens such as HIV and malaria has proven to be difficult. The enormous worldwide impact of these pathogens, the potential bioterror uses of others, and the worsening problem of antibiotic resistance have prompted the efforts to produce new vaccines.

A recent approach to immunization has used DNA containing a sequence of the pathogen's genome. This DNA is typically a bacterial plasmid engineered to include the sequence of an antigenic protein from the pathogen. This DNA can enter a number of cell types, and it can be expressed there using cellular transcription and translation machinery. In this respect, DNA vaccines act much like viruses. However, these DNAs contain only a very limited amount of genetic information and cannot become infectious. The mechanisms of uptake and induction of the immune response are not yet clear. However, promising results have been observed against viruses, bacteria, and parasitic microorganisms. This approach holds great promise for development of effective vaccines against intractable diseases including HIV/AIDS, tuberculosis, and malaria.

DNA vaccines may also be useful for vaccination against cancers. While the antigens presented by tumor cells are only weakly immunogenic, model studies using plasmid DNA have shown promising results. Mice vaccinated by oral delivery of plasmids grown within attenuated *Salmonella* bacteria were able to slow or completely stop the growth of a lethal dose of carcinoma cells. Death of the bacteria presumably releases large numbers of the plasmids that are taken up by antigen-presenting cells of immune system. Extension of this work to humans is currently under investigation.

Source: Lewis, P. J. and Babiuk, L. A. DNA vaccines: a review. *Adv. Virus Res.* 54:129, 1999; Kim, J. J., Choo, A. Y., and Weiner, D. B. DNA vaccines. In: N. S. Templeton (Ed.), *Gene and Cell Therapy*, 2nd ed. New York: Dekker, 2004, pp. 681–706; and Reisfeld, R. A., Niethammer, A. G., Luo, Y., and Xiang, R. DNA vaccines suppress tumor growth and metastases by the induction of anti-angiogenesis. *Immunol. Rev.* 199:181, 2004.

importance. The first evidence that DNA is the genetic material was found during the 1920s, but definitive demonstration of DNA's role was not accomplished until 1944. The key experiments involved two strains of pneumococcus, a bacterium that causes a form of pneumonia. When cultured, one strain formed smooth colonies and the other formed rough colonies; these were labeled the S- and R-forms, respectively. The S-form is virulent while the R-form is nonvirulent, and the two forms are genetically distinct and cannot interconvert spontaneously. Treatment of R-form bacteria with pure DNA extracted from the S-form resulted in its **transformation** into the S-form. The transformation was inherited permanently by subsequent generations. This demonstrated that DNA was the transforming agent and the material responsible for transmitting genetic information from one generation to the next. Clinical Correlation 2.1 describes vaccines based on transformation of mammalian cells with DNA.

Information Capacity of DNA is Enormous

A striking characteristic of DNA is its ability to encode an enormous quantity of biological information. For example, a human cell contains information for the synthesis of about 20,000 to 25,000 proteins. This information is stored in the cell nucleus, a package roughly 0.00001 m in diameter. Despite this compactness, information in DNA is readily accessed and duplicated on command. The ability to store large amounts of information on molecules and to access it readily is still far beyond modern information technology. The capacity of nucleic acids to maintain and transmit the archived information efficiently arises directly from their chemical structure.

2.2 STRUCTURAL COMPONENTS OF NUCLEIC ACIDS: NUCLEOBASES, NUCLEOSIDES, AND NUCLEOTIDES

Nucleic acids are linear polymers of **nucleotide** units. Each nucleotide consists of a phosphate ester, a pentose sugar, and a heterocyclic nucleobase. In RNA, the sugar is **D-ribose**; in DNA, it is the closely related sugar, **2-deoxy-D-ribose**. In either case, the base is attached to the 1-position of the sugar through a β-N-glycosidic bond. Two classes of major nucleobases are found in nucleic acids: **purines** and **pyrimidines** (Figure 2.2). The major purines are **guanine** and **adenine** which occur in both DNA and RNA and are attached to the sugar at N9. The three major pyrimidine nucleobases are **cytosine, uracil**, and **thymine**. Cytosine is present in both DNA and RNA. However, uracil is generally found only in RNA, and thymine is generally found only in DNA. Each pyrimidine is linked to the sugar through the N1 position. A nucleobase glycosylated with either pentose sugar is a **nucleoside**. Nucleosides that contain ribose are **ribonucleosides** (Figure 2.3), whereas those with deoxyribose are **deoxyribonucleosides** (Figure 2.4). Four ribonucleosides are commonly found in RNA—**adenosine (A), guanosine (G), cytidine (C)**, and **uridine (U)**—and four deoxyribonucleosides in DNA—**deoxyadenosine (dA), deoxyguanosine (dG), deoxycytidine (dC)**, and **deoxythymidine (dT)**. In addition, more than 80 minor nucleosides have been found in naturally occurring nucleic acids.

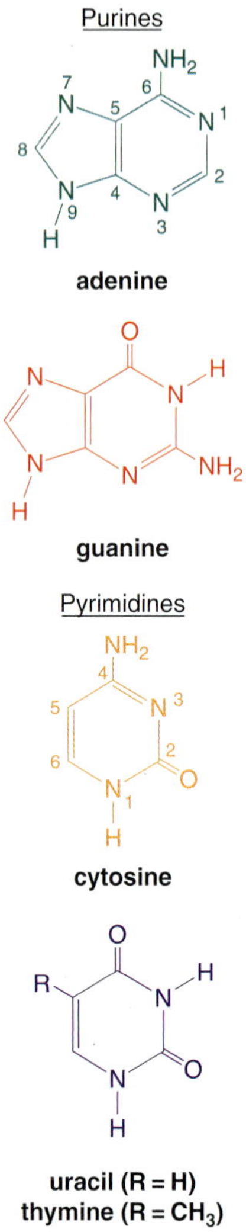

Nucleotides are phosphate esters of nucleosides (Figure 2.5). Any of the hydroxyl groups on their sugars can be phosphorylated, but the bases are not. Nucleotides that contain a phosphate monoester are nucleoside monophosphates. For example, the 5′-nucleotide of deoxycytidine is deoxycytidine 5′-monophosphate (5′-dCMP). More than one phosphate can be linked by an anhydride linkage, resulting in the corresponding di-, tri-, and tetraphosphate esters. The 5′-triphosphate of guanosine is guanosine 5′-triphosphate (GTP). Phosphate diesters are also possible, including the important "second messengers" adenosine-3′,5′-cyclic monophosphate (cAMP) and guanosine-3′,5′-cyclic monophosphate (cGMP).

Physical Properties of Nucleosides and Nucleotides

To varying degrees, nucleobases, nucleosides, and nucleotides are soluble in water over a wide range of pHs. Because of the hydrophilic nature of the sugars, nucleosides are more soluble than the corresponding nucleobases. However, at biologically relevant pHs (pH 6–8), nucleosides are neutral species. Only at very low pHs or very high pHs do nucleosides become charged. A further increase in aqueous solubility is seen upon phosphorylation of the nucleosides. The pK_a of phosphate mono- and diesters is ~1; thus, phosphates carry a negative charge at physiological pHs. The second pK_a of a phosphate monoester is ~6.5; thus, an equilibrium exists between the mono-anion and the dianion at neutral pHs. Nucleotides with di- and triphosphates carry multiple

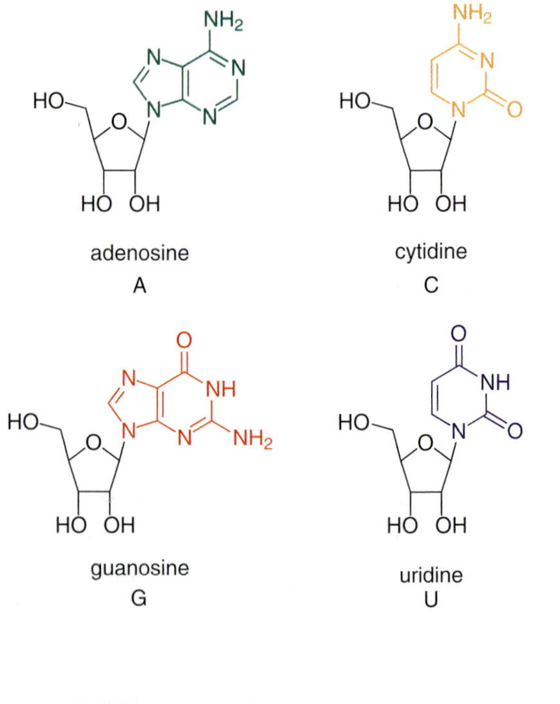

Purines

adenine

guanine

Pyrimidines

cytosine

uracil (R = H)
thymine (R = CH_3)

FIGURE 2.2

Major purine and pyrimidine bases found in DNA and RNA.

adenosine
A

cytidine
C

guanosine
G

uridine
U

FIGURE 2.3

Structures of ribonucleosides. Shown are one-letter abbreviations for each compound. These abbreviations are also used for the corresponding bases and nucleotides and, in some instances, for the deoxyribonucleosides.

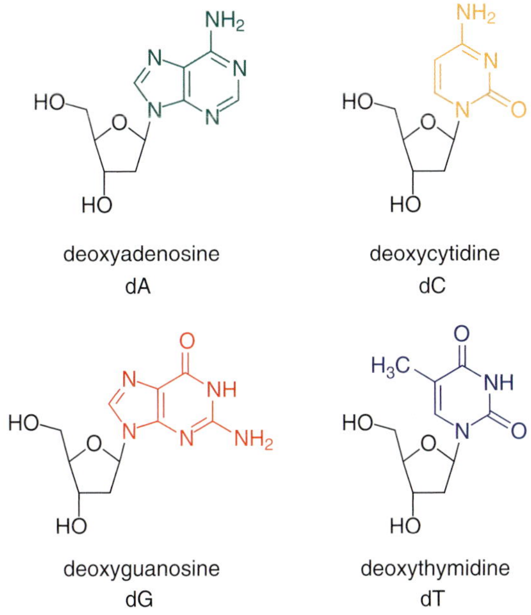

FIGURE 2.4

Structures of deoxyribonucleosides. Presence of 2-deoxyribose is abbreviated by a "d" preceding the one letter notation. Note that thymidine (T) is often used interchangeably with deoxythymidine (dT).

negative charges. The presence of charged phosphate groups on nucleotides provides sites for electrostatic interactions with positively charged sites on proteins and metal ions.

Nucleosides and nucleotides are stable over a wide range of pHs. Under strongly basic conditions, hydrolysis of phosphate esters slowly occurs. *N*-Glycosidic bonds of nucleosides and nucleotides are stable under these conditions, but under acidic conditions they are considerably more labile. At elevated temperatures, protonation of purine bases (G and A) results in rapid scission of the bond between the sugar and the base. Pyrimidine (C, T, U) nucleosides and nucleotides are much more resistant to acid treatment. Conditions that break the glycosidic bond (e.g., 60% perchloric acid at 100°C) also lead to destruction of the sugar.

Molecules that contain purine or pyrimidine bases strongly absorb UV light. Purines and purine-containing nucleosides and nucleotides have higher molar extinction coefficients (absorb light more strongly) than pyrimidine derivatives. The UV wavelength where maximum absorption occurs varies but is usually near 260 nm. Because proteins and other cellular components typically do not absorb strongly in this region, a strong absorbance at 260 nm is suggestive of the presence of nucleobases. Because of the high extinction coefficients of nucleobases and their high concentration in nucleic acids, the absorbance at 260 nm can be used to quantitate accurately the amount of nucleic acid (DNA or RNA) present in a sample. Further analysis and separation of nucleobases, nucleosides, nucleotides, and nucleic acids can be accomplished by use of a variety of techniques, including high-performance liquid chromatography (HPLC), thin-layer chromatography (TLC), paper chromatography, and electrophoresis.

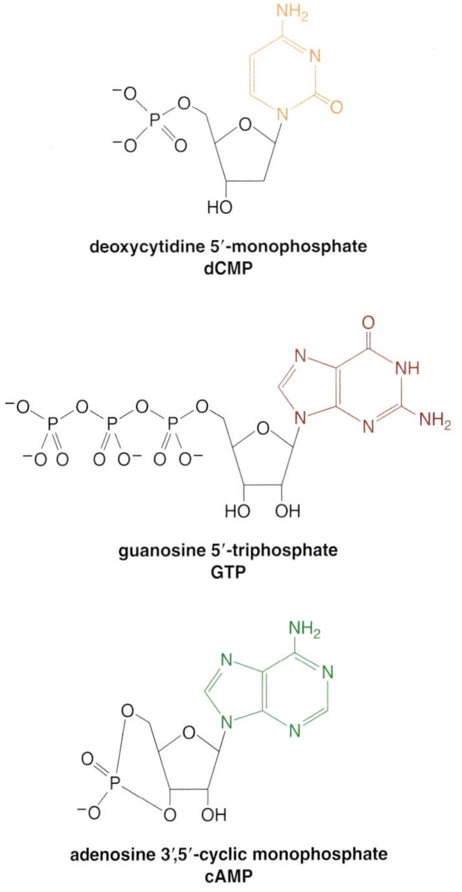

FIGURE 2.5

Structures of some representative nucleotides.

Structural Properties of Nucleosides and Nucleotides

A striking feature of nucleosides and nucleotides is the considerable number of possible conformations that they can adopt. Unlike six-membered rings that have relatively high barriers to conformational change, five-membered rings are highly flexible. The five ring atoms in a pentose sugar are not coplanar due to a number of unfavorable interactions. To alleviate these, one or two of the atoms of the ring twist out of plane. In cyclopentane rings, there are several "envelope" and "half-chair" conformations that rapidly interconvert. Because of the asymmetric substitution pattern of the pentose ring in nucleosides, two conformations are preferred (Figure 2.6). These modes of **sugar puckering** are defined by the displacement of the 2′- and 3′-carbons "above" the plane of the C1′–O4′–C4′ atoms. By convention, "above" is the direction in

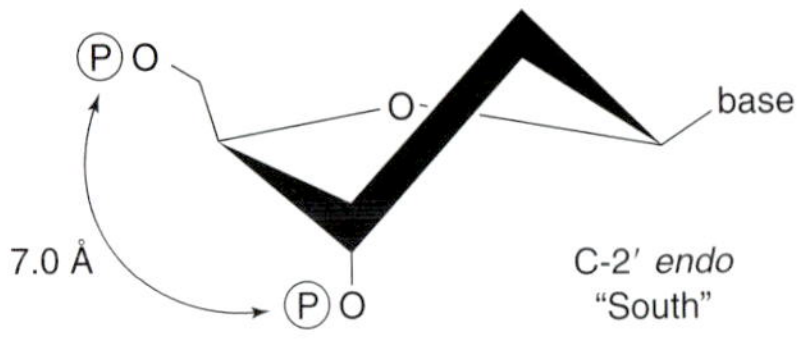

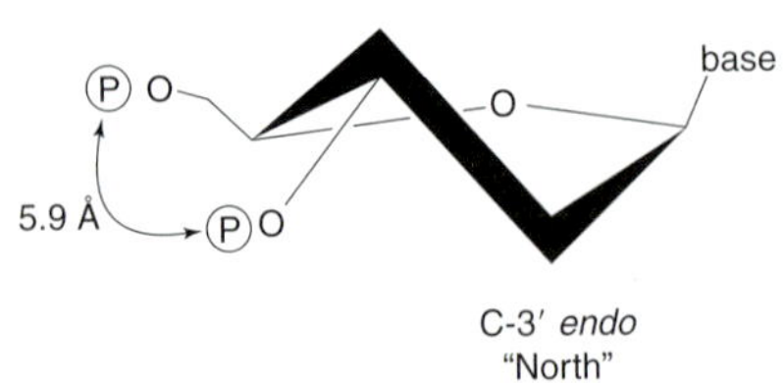

FIGURE 2.6

Preferred conformations of pentose sugars. The two conformations produce variations in relative orientation of the base (with respect to the sugar) and in the distance between 3′- and 5′-phosphate groups (P). Ultimately, these differences effect overall conformation of the double-helical complex.

which the base and C5′ project from the ring, and is the endo face of the pentose. If C2′ is displaced above the pentose ring, the conformation is C2′-endo. When viewed edge-on, the pentose carbon atoms in this arrangement resemble the letter S; thus, this is sometimes called the South-conformation. In the second common pucker, C3′ is displaced toward the endo-face and is called C3′-*endo*. The pentose carbons trace the letter N, producing the North-conformation. Notably, the groups attached to the sugar have very different orientations in each of these conformations. For example, 5′- and 3′-phosphate groups are much farther apart in the C2′-*endo* pucker than in the C3′-*endo* pucker. The orientation of the glycosidic bond also changes significantly in the two conformations.

C2′-endo and C3′-*endo* conformations are in rapid equilibrium. An electronegative substituent at the 2′-position of the pentose favors the C3′-endo conformation. Thus, ribonucleosides in RNA prefer this sugar pucker. Additional factors such as hydrogen-bonding between the 2′-OH group and the O4′ atom of the neighboring residue shift the equilibrium toward the C3′-*endo* conformation. However, the 2′-deoxynucleosides of DNA contain a hydrogen in place of the 2′-OH group, and the C2′-*endo* conformation is preferred.

The bases in nucleosides are planar. While free rotation around the glycosidic bond is possible, two orientations of the base with respect to the sugar predominate (Figure 2.7). In purines, the ***anti* conformation** places H8 over the sugar, whereas the ***syn* conformation** positions this atom away from the sugar and the bulk of the bicyclic purine over the sugar. In pyrimidines, the H6 atom is above the pentose ring in the anti conformation, and the larger O2 atom is above the ring in the *syn* glycosidic conformation. Pyrimidines, therefore, show a large preference for the less sterically hindered *anti* conformation. Purines rapidly interconvert between the two conformations but favor the *anti* orientation. However, guanine 5′-nucleotides are exceptions. In these cases, favorable interactions between the 2-NH_2 group and the 5′-phosphate group stabilize the *syn* conformation. 2′-Deoxyguanosine 5′-monophosphate (dGMP), for example, prefers the *syn* glycosidic conformation. This preference has also been observed in double-stranded DNA with sequences of alternating Gs and Cs. The *syn* conformation of the G residues in such DNA results in formation of an unusual left-handed helix (see p. 40).

FIGURE 2.7

Glycosidic conformations of purines and pyrimidines. In pyrimidines, steric clashes between the sugar and the O2 of the base strongly disfavors the *syn* conformation. In purines, the *anti* and *syn* conformations readily interconvert, with *anti* being more stable in most cases. The *syn* conformation is stabilized in guanosine 5′-phosphates because of favorable interactions between the 2-NH_2 group and the phosphate oxygens.

2.3 STRUCTURE OF DNA

Polynucleotide Structure

Nucleic acids are strands of nucleotides linked by **phosphodiester** bonds (Figure 2.8). The length of these strands varies considerably, ranging from two residues to hundreds of millions of residues. Typically, strands of nucleic acids containing ≤ 50 nucleotides are **oligonucleotides**, whereas those that are longer are **polynucleotides.** The phosphodiester bond links the 5′-hydroxyl group of one residue to the 3′-hydroxyl group of the next. Linkages between two 5′-OHs or two 3′-OHs are not seen in naturally occurring DNA. The **directionality** of this bond means that linear oligo- and polynucleotides have one end that terminates in a 5′-OH and another that terminates in a 3′-OH. These ends are the **5′-terminus** and **3′-terminus**, respectively. In many polynucleotides, one or both termini are chemically modified with moieties such as phosphate groups or amino acids. Circular polynucleotides have no free termini, and they are formed by joining the 5′-terminus of a linear polynucleotide with its own 3′-terminus through a phosphodiester bond.

In both linear and circular polynucleotides, the sugar–phosphodiester backbone is highly uniform, and chemical diversity arises primarily from the bases. The **sequence** of bases provides each molecule with a unique chemical identity (Figure 2.9). This arrangement is analogous to that in polypeptides and proteins (see p. 90) where variations in chemical structure primarily arise from the identity of the amino acid side chains.

Polynucleotide Conformations

The bases are largely responsible for the conformations of polynucleotides. The edges of the bases contain nitrogen and oxygen atoms (—NH_2, =N-, and =O groups) that can interact with other polar groups or surrounding water molecules. The faces of the rings, however, cannot participate in such interactions and tend to avoid contact with water. Instead they interact with one another, producing a stacked conformation. **Base stacking** reduces the hydrophobic surface area that must be solvated by polar water molecules. Release of these water molecules into the bulk solvent is entropically

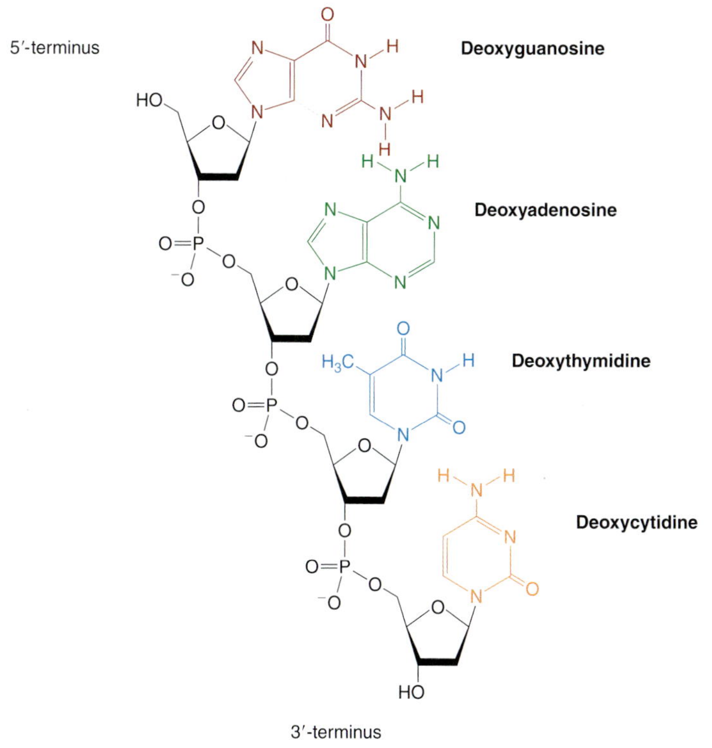

5′-terminus

Deoxyguanosine

Deoxyadenosine

Deoxythymidine

Deoxycytidine

3′-terminus

FIGURE 2.8

Structure of a DNA polynucleotide. Shown is a tetranucleotide. Generally, nucleic acids less than 50 nucleotides long are referred to as oligonucleotides. Longer nucleic acids are called polynucleotides.

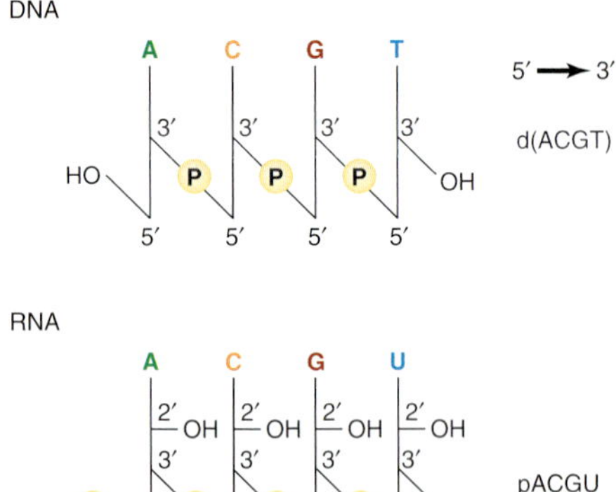

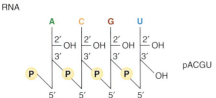

FIGURE 2.9

Shorthand notations for structure of oligonucleotides. The convention used in writing the structure of an oligo- or polynucleotide is a perpendicular bar represents the sugar moiety, with the 5′-OH position of the sugar located at the bottom of the bar and the 3′-OH (and 2′-OH, if present) at a midway position. Bars joining 3′- and 5′- positions represent the 3′, 5′-phosphodiester bond, and the P on the left or right side of the perpendicular bar represents a 5′-phosphate or 3′-phosphate ester, respectively. The base is represented by its initial placed at the top of the bar. An alternative shorthand form is to use the one-letter initials for the bases written in the 5′ → 3′ direction from left to right. Internal phosphodiester groups are assumed, and terminal phosphates are denoted with a "p". Oligonucleotide sequences containing deoxyribose sugars are preceded by a "d".

favorable. The stacked arrangement of bases is further stabilized by favorable electronic interactions (van der Waals forces). Like other aromatic systems, the bases possess highly delocalized π-orbitals above and below the planes of their rings. This electron density can be polarized by nearby dipoles. The resulting induced dipole can then favorably interact with the polarizing group. These weak interactions are further enhanced by attractive London dispersion forces (induced dipole-induced dipole). Because the strength of these electronic interactions is strongly dependent upon distance, no empty space remains between the stacked bases.

Polynucleotides adopt conformations that maximize favorable stacking interactions between neighboring bases. The constraints imposed by the structure of the sugar–phosphodiester backbone favor helical structures (Figure 2.10). The overall stability of these helical structures is dependent upon factors that include sequence-dependent base stacking, pH, salt concentration, and temperature. For example, some synthetic polynucleotides, including poly(C) and poly(A), form right-handed helices in which the bases are highly stacked. Many other polynucleotides remain in a largely disordered **random coil** conformation. Large polynucleotides may have both helical and disordered regions, depending upon the local sequence. The characteristics of the solution also determine whether polynucleotides adopt stacked conformations. For example, high concentrations of cations, especially polyvalent ions such as Mg^{2+}, stabilize the helical conformation by shielding the charges of the phosphodiester groups in the backbone. Without this shielding, the negative charges on the phosphates would destabilize the helix by electrostatic repulsion.

Stability of the Polynucleotide Backbone

Polynucleotides are relatively stable in aqueous solutions near neutral pH. It has been estimated that the half-life for hydrolysis of phosphodiester linkages in DNA is about 200 million years. This high stability makes DNA suitable for the long-term storage of genetic information. In contrast, RNA is much more prone to hydrolysis. The presence of the 2′-OH group provides an internal nucleophile for transesterification of the 3′, 5′-phosphodiester linkage (Figure 2.11). The result is scission of the polynucleotide backbone, leaving a 2′, 3′-cyclic phosphate on one fragment and a free 5′-OH group on the other. The greater hydrolytic lability of RNA makes it less suitable as a genetic material than DNA.

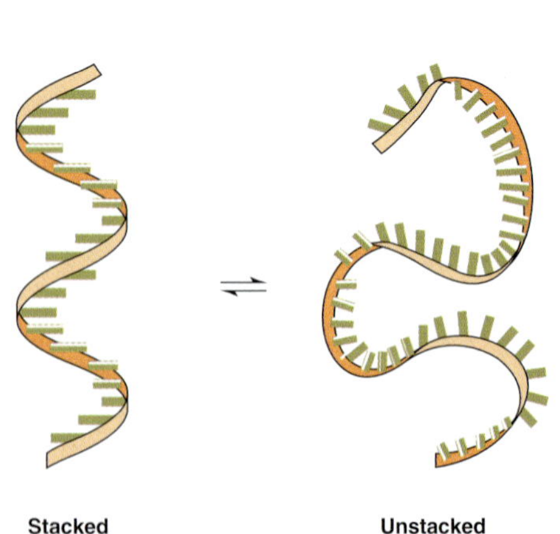

FIGURE 2.10

Stacked and random coil conformations of a single-stranded polynucleotide. The helical band represents the sugar–phosphate backbone of the polynucleotide. Bases are shown in a side view as lines in contact with their neighbors. Unstacking of bases leads to a more flexible structure with bases oriented randomly.

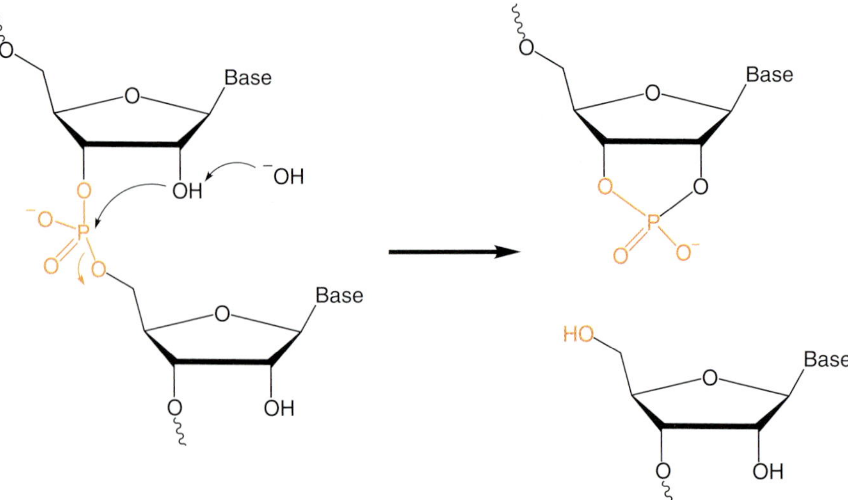

FIGURE 2.11

Hydrolysis of RNA is accelerated by participation of the 2′-OH group. Intramolecular nucleophilic attack by the 2′-OH group of ribose greatly increases the rate of hydrolytic cleavage of the phosphodiester backbone, especially under basic conditions. The resulting 2′, 3′-cyclic phosphodiester can subsequently react with water or hydroxide to a mixture of 2′- and 3′-phosphate monoesters.

While the DNA backbone is relatively stable to hydrolysis, numerous enzymes catalyze phosphodiester scission. These nucleases are characterized by the types of polynucleotides they hydrolyze and the specific bonds that are broken (Figure 2.12). **Exonucleases** cleave the last nucleotide residue at either the 5′- or 3′- terminus of a polynucleotide. Stepwise removal of individual nucleotides from one end of a polynucleotide can result in its complete degradation. Nucleases can sever one of the two nonequivalent bonds in the phosphodiester linkage, leaving either a free 5′-OH group and a 3′-phosphate or a 5′-phosphate and a 3′-OH. For example, treatment of an oligodeoxyribonucleotide with snake venom phosphodiesterase yields deoxyribonucleoside 5′-phosphates. In contrast, treatment with a spleen phosphodiesterase produces deoxyribonucleoside 3′-phosphates. **Endonucleases** cleave phosphodiester bonds located in the interior of polynucleotides. They do not require a free terminus; therefore, they can also cleave circular polynucleotides. Endonucleases such as DNase I and DNase II hydrolyze DNA with little sequence selectivity. The **restriction endonucleases** recognize and cleave very specific sequences. Restriction endonucleases have been particularly useful in the development of methodologies for sequencing of DNA polynucleotides and provide the basis for recombinant DNA techniques. Nucleases also exhibit specificity with respect to the overall structure of polynucleotides. For instance, some nucleases act on both single- or double-stranded polynucleotides, whereas others discriminate between these two structures. Some act on both DNA or RNA, whereas others are active toward only one type of polynucleotide.

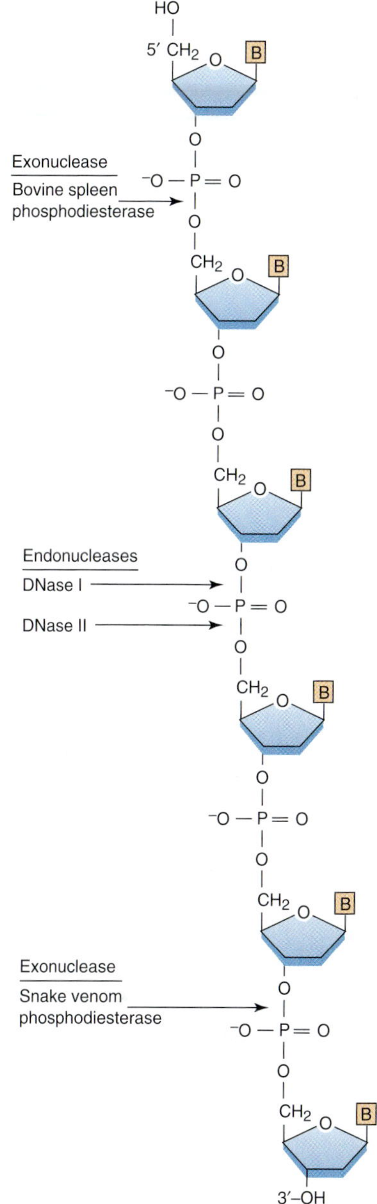

FIGURE 2.12

Specificity of nucleases. Exonucleases remove nucleotide residues from either end of a polynucleotide, depending on their specificity. Endonucleases hydrolyze interior phosphodiester bonds. Both endo- and exonucleases typically show specificity regarding which oxygen–phosphorous bond is hydrolyzed.

Double-Helical DNA

With the recognition in the early and mid-twentieth century that DNA served as the carrier of genetic information, work to establish the structural basis for information storage intensified. However, several misconceptions slowed progress. The most serious of these was the erroneous experimental observation that DNA contained equal amounts of each of the four nucleosides (deoxyadenosine, deoxyguanosine, deoxycytidine, and deoxythymidine). This led to the proposal that DNA consisted of a cyclic tetranucleotide containing one of each of the nucleosides (Figure 2.13). A glaring problem with this structure was that DNA was found to have molecular masses thousands of times larger than that of a tetranucleotide. It was not obvious how the monotonous structure of repeating tetranucleotides could achieve the complexity needed to convey an enormous number of hereditary traits.

Synthetic and X-ray diffraction experiments led to the acceptance of a linear polynucleotide structure for DNA. However, the three-dimensional structure of DNA remained a mystery. Three key pieces of information were necessary for deduction of the structure of DNA. The first was the determination that DNA did not contain equal amounts of the four nucleosides, but contained variable amounts in different organisms. However, it was found that the abundance of deoxyadenosine always equaled that of deoxythymidine and that the abundance of deoxyguanosine always equaled that of deoxycytidine. This led to consideration of structures with the nucleosides specifically paired together (dA with dT and dG with dC). The second was that X-ray diffraction data suggested that DNA contained **double-helical** structures, and symmetry suggested that the two polynucleotide strands were oriented **antiparallel** to each other. The third was the suggestion that the nucleobases were in the keto and amino **tautomeric** forms rather than enol and imino forms. Tautomers are isomers of a molecule which differ only in the position of a hydrogen atom. Each of the four nucleobases has two or more possible tautomeric forms that are in equilibrium (Figure 2.14). The incorrect assignment of the predominant tautomeric structures meant that researchers were attempting to form base pairs using incorrect hydrogen bonding patterns. With the correct tautomers, a model for DNA rapidly fell into place.

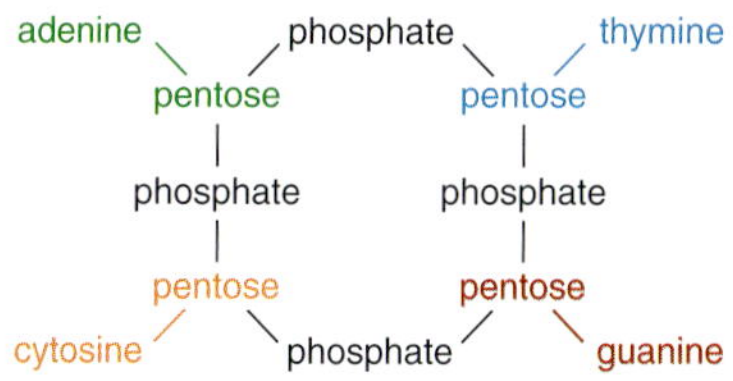

The structure that Watson and Crick proposed in 1953 for double-helical DNA was attractive because of its simplicity and symmetry (Figure 2.15). Moreover, it explained all of available structural data for DNA and immediately led to hypotheses regarding a mechanism for storage of genetic information. The Watson–Crick double helix can

adenine — phosphate — thymine
pentose — pentose
phosphate — phosphate
pentose — pentose
cytosine — phosphate — guanine

FIGURE 2.13

Tetranucleotide structure for DNA proposed in the 1930s.

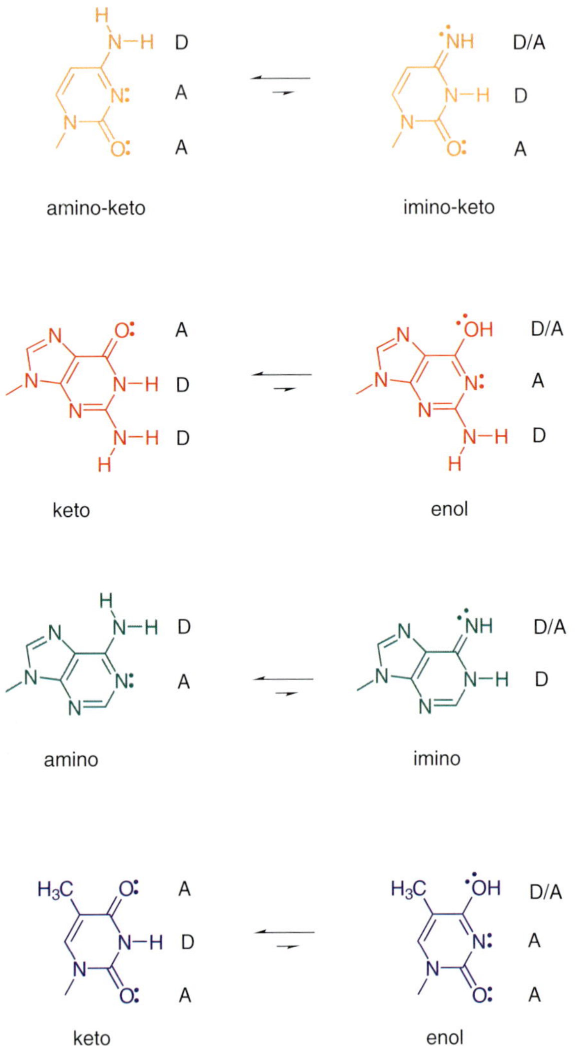

FIGURE 2.14

Tautomeric forms of bases. The pattern of hydrogen bonding donating (D) and accepting (A) groups changes, depending on the tautomer of the base that is present. Shown are predominant tautomeric forms (left) and one of the alternate forms (right) for each base. The hydrogen bonding pattern for each tautomer is also shown; ambiguity at the —OH and ═NH positions of the minor tautomers arises from rotation or isomerism of these groups. The equilibrium ratio of the predominant form to all others is greater than 99:1.

be visualized as the interwinding of two right-handed helical polynucleotide strands around a common axis. The strands achieve contact through hydrogen bonds formed at the hydrophilic edges of the bases. The N—H groups of the nucleobases are good hydrogen-bond donors, and the electron pairs on the sp^2-hybridized oxygens of the C═O groups and nitrogens of the ═N— groups are good hydrogen-bond acceptors. Pairing occurs when an acceptor and a donor are in a position to form a hydrogen bond. These bonds extend between purine residues in one strand and pyrimidine residues in the other, and the matching of hydrogen bond donors and acceptors results in two types of **base pairs**: adenine–thymine and guanine–cytosine (Figure 2.16). A direct consequence of these hydrogen bonding specificities is that **double-stranded DNA (dsDNA)** must contain ratios of nucleosides that agree with experimental observations (dA = dT and dG = dC). Finally, the geometries of the dA/dT and dG/dC base pairs result in similar C1′—C1′ distances (~10.6 Å) and glycosidic bond orientations. This **structural isomorphism** means that any of the four possible base pairs (dA-dT, dT-dA, dG-dC, and dC-dG) can be placed into the double helix without significantly changing the structure of the backbone.

The relationship between bases on opposing strands within the double helix is described as **complementarity**. Bases are complementary because every nucleobase of one strand is matched by shape and hydrogen bonding to a complementary base on the other strand (Figure 2.17). For instance, for each adenine projecting toward the axis of the double helix, a thymine must be projected from the opposite strand to hydrogen bond and exactly fill the space between the strands. Neither cytosine nor guanine fits

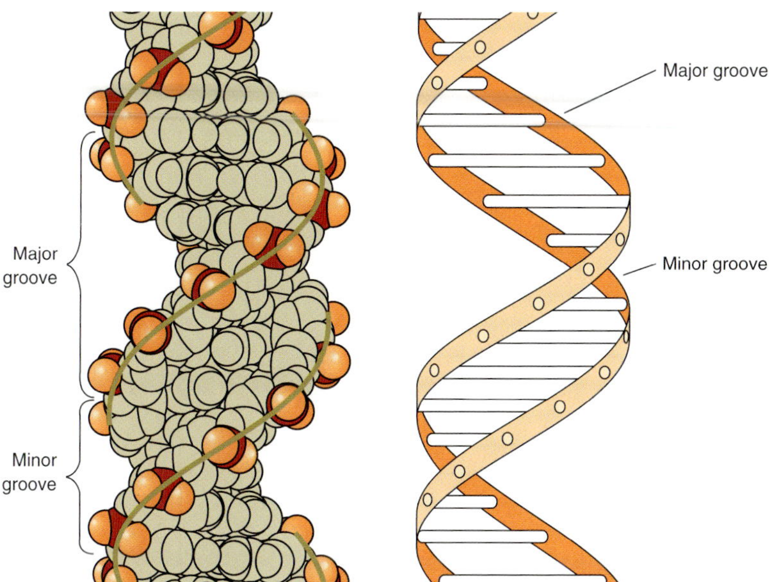

FIGURE 2.15

The Watson–Crick model of DNA. On the left is a space-filling model of DNA; on the right is an idealized ribbon model. Bases are stacked in the interior of the helix, whereas hydrophilic sugar-phosphodiester backbone is located on the exterior. Redrawn from Rich, A. J. *Biomol. Struct. Dyn.* 1:1, 1983.

precisely in the available space across from adenine in a manner that allows formation of hydrogen bonds across strands. These hydrogen bonding specificities ensure that the entire base sequence of one strand is complementary to that of the other strand.

The exterior of the double helix consists of the sugar–phosphate backbones of its component strands. Polynucleotides are asymmetric structures because of their intrinsic polarity. The two strands are aligned in opposite directions; if two adjacent nucleobases in the one strand—for example, thymine and cytosine—are connected in the $5' \rightarrow 3'$ direction, their complementary nucleobases on the other strand, adenine and guanine, will be linked in the $3' \rightarrow 5'$ direction. This antiparallel alignment produces a stable association between strands to the exclusion of the alternate parallel arrangement.

Interwinding of the two antiparallel strands produces a structure that has two distinct helical grooves between the sugar–phosphate backbones (Figure 2.15). The **major groove** is much wider than the **minor groove**. This disparity arises from the geometry of the base pairs: The glycosidic bonds between the bases and the backbone pentose are not arranged directly opposite to each other but are displaced toward the minor groove. Significantly, the nucleotide sequence of DNA can be discerned without dissociating the double helix by looking inside these grooves. Each base always displays the same atoms into each of the grooves on the double helix. These atoms then constitute an important means of sequence-specific recognition of DNA by proteins and small molecules. For example, N7 of purines is always displayed in the major groove, and it can serve as a hydrogen bond acceptor in interactions with donor groups on proteins (Figure 2.16). Similarly, the exocyclic 2-NH_2 group of guanine always projects into the minor groove and can form a steric blockade to the binding of small molecules.

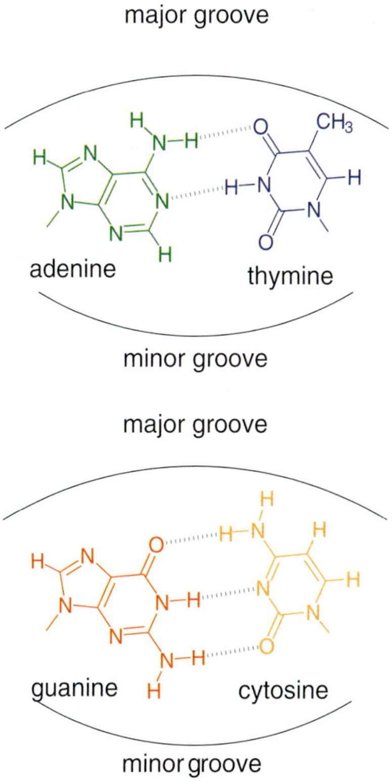

Factors that Stabilize Double-Helical DNA

Stacking interactions that stabilize the helical structures of single-stranded polynucleotides are also instrumental in stabilizing the double helix. The separation between the hydrophobic core of the stacked nucleobases and the hydrophilic exterior of the charged sugar–phosphate groups is even more pronounced in the double helix than in single-stranded helices. The stacking tendency of single-stranded polynucleotides can be viewed as resulting from a tendency of the bases to reduce their contact with water. The double-stranded helix is a more favorable arrangement, essentially removing the nucleobases from the aqueous environment while permitting the hydrophilic phosphate backbone to be highly solvated by water.

Stacking interactions, a combination of hydrophobic forces and van der Waals interactions, are estimated to generate 4–15 kcal mol^{-1} of stabilization energy for each

major groove

adenine thymine

minor groove

major groove

guanine cytosine

minor groove

FIGURE 2.16

Watson–Crick base pairs. Selective base pairs are formed between adenine and thymine and between guanine and cytosine. Note the formation of two hydrogen bonds in an A-T base pair and three in a G-C base pair.

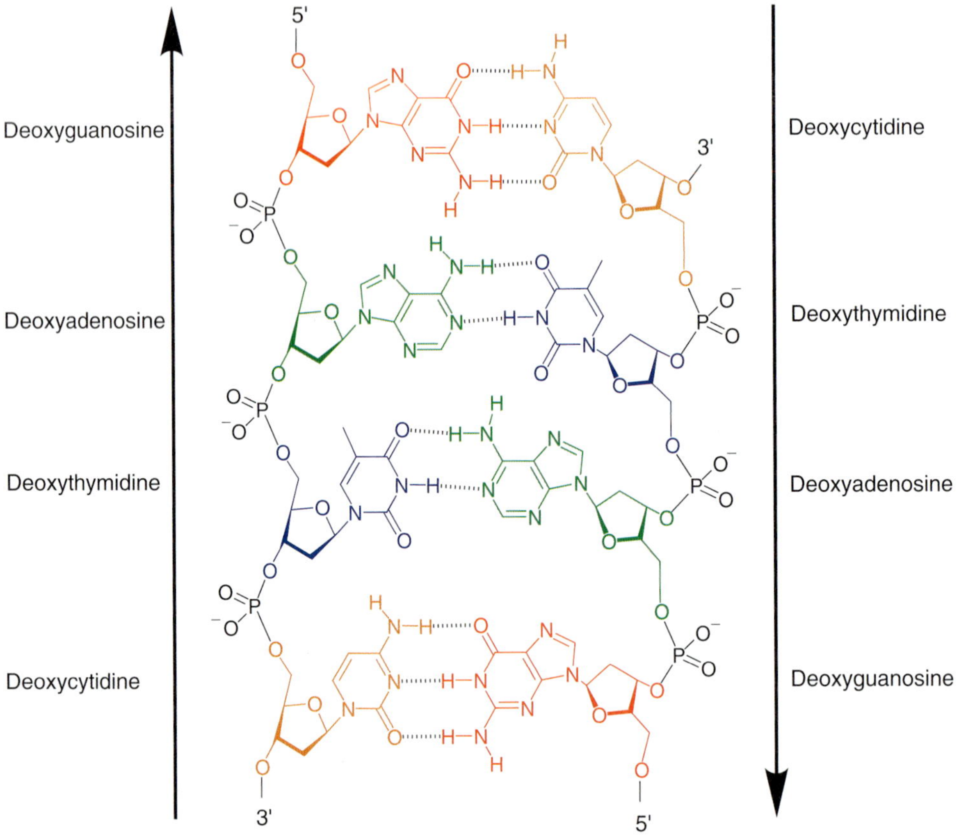

FIGURE 2.17

Formation of hydrogen bonds between complementary bases in double-stranded DNA. Interaction between polynucleotide strands is highly selective. Complementarity depends not only on the geometric factors that allow the proper fitting between the complementary bases of the two strands, but also on the formation of specific hydrogen bonds between complementary bases. Note the antiparallel orientation of the strands of a double-stranded DNA. Geometry of the helices does not prevent a parallel alignment, but such an arrangement is not found in DNA.

adjacent pair of stacked nucleobases (Table 2.1). Additional stabilization of the double helix results from extensive networks of cooperative hydrogen bonds. Generally, these bonds are weak (3–7 kcal mol^{-1}), and they are even weaker in DNA (2–3 kcal mol^{-1}) because of geometric constraints within the double helix. It might appear that hydrogen bonds between nucleobases could cumulatively provide substantial stabilization for the double helix. However, the hydrogen bonds in a double helix merely replace energetically similar ones between the nucleobases and water in single-stranded polynucleotides. Therefore, hydrogen bonding is not the "glue" that maintains double-helical structures. Yet, in contrast to stacking forces, hydrogen bonds are highly directional and help discriminate between correct and incorrect base pairs. Because of their directionality, hydrogen bonds tend to orient the nucleobases to favor stacking. Therefore, the contribution of hydrogen bonds is indirectly vital for the stability of the double helix.

The relative importance of stacking interactions versus hydrogen bonding in stabilizing the double helix was not always appreciated. Indeed, hydrogen bonds have been considered to be the "glue" that holds the two strands together. However, experiments with reagents that reduce the stability of the double helix (**denaturants**) illustrate the greater relative importance of stacking interactions (Table 2.2). These results show that the destabilizing effect of a reagent is not related to its ability to break hydrogen bonds but is determined by the solubility of the free bases in the reagent. As the reagent becomes a better solvent for the bases, the driving force for stacking diminishes, and the double helix is destabilized.

Electrostatic forces also have an effect on stability and conformation of the double helix. Phosphodiester groups are ionized at physiological pH; thus, the exterior of the double helix carries two negative charges per base pair. The interstrand electrostatic repulsion between negatively charged phosphates is destabilizing and tends to separate the complementary strands. In distilled water, DNA strands separate at room temperature. However, cations such as Mg^{2+}, spermine^{4+} (a tetramine), and the basic side chains of proteins can shield the phosphate groups and decrease repulsive forces.

TABLE 2.1 Base-Pair Stacking Energies

Dinucleotide Base Pairs	*Stacking Energies (kcal mol^{-1} per Stacked Pair)*[a]
(GC) · (GC)	−14.59
(AC) · (GT)	−10.51
(TC) · (GA)	−9.81
(CG) · (CG)	−9.69
(GG) · (CC)	−8.26
(AT) · (AT)	−6.57
(TG) · (CA)	−6.57
(AG) · (CT)	−6.78
(AA) · (TT)	−5.37
(TA) · (TA)	−3.82

[a] Data from Ornstein, R. L., Reim, R., Breen, D. L., and McElroy, R. D. *Biopolymers* 17:2341, 1978.

TABLE 2.2 Effects of Various Reagents on the Stability of the Double Helix[a]

Reagent	*Adenine Solubility* $\times 10^{-3}$ *(in 1 M Reagent)*	*Molarity Producing 50% Denaturation*
Ethylurea	22.5	0.60
Propionamide	22.5	0.62
Ethanol	17.7	1.2
Urea	17.7	1.0
Methanol	15.9	3.5
Formamide	15.4	1.9

Source: Data from Levine, L., Gordon, J., and Jencks, W. P. *Biochemistry* 2:168, 1963.

[a] The destabilizing effect of the reagents listed on the double helix is independent of the ability of these reagents to break hydrogen bonds. Rather, the destabilizing effect is determined by the solubility of adenine. Similar results would be expected if the solubilities of the other bases were examined.

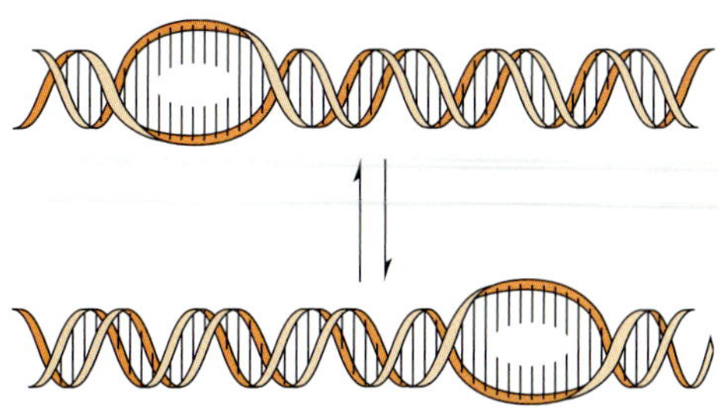

FIGURE 2.18

Migration of bubbles through double-helical DNA. DNA contains short open-stranded sections of that can "move" along the helix.

Denaturation and Renaturation

The double helix is disrupted during almost every important biological process in which DNA participates, including DNA replication, transcription, repair, and recombination. The forces that hold the two strands together are adequate for providing stability but weak enough to allow facile strand separation. The double helix is stabilized relative to the single strands by about 1 kcal mol^{-1} per base pair so that a relatively minor perturbation can produce a local disruption in a short section of the double helix. These base pairs can close up again, releasing free energy that can then cause the adjacent base pairs to unwind. In this manner, minor disruptions of the double helix can migrate along its length (Figure 2.18). Thus, at any particular moment, the large majority of bases remain hydrogen bonded, but all bases can transiently pass through the single-stranded state. The ability of the DNA double helix to "breathe" is an essential prerequisite for its biological functions.

Separation of DNA strands can be studied by increasing the temperature of a solution. At relatively low temperatures a few base pairs are disrupted, creating one or more open-stranded bubbles (Figure 2.19). These bubbles form initially in sections that contain relatively higher proportions of adenine–thymine pairs because of the lower stacking energies of such pairs. As the temperature is raised, the size of the bubbles increases, and the thermal motion of the polynucleotides eventually overcomes the forces that stabilize the double helix. At even higher temperatures, the strands separate and acquire a random-coil conformation. The process is most appropriately described as a helix-to-coil transition, but is commonly called **denaturation** or **melting**.

Denaturation is accompanied by several physical changes in the DNA-containing solution, including an increase in buoyant density, a reduction in viscosity, a change in ability to rotate polarized light, and changes in absorbance of UV light. The last are frequently used to follow the process of denaturation experimentally (Figure 2.20). Because of the strong absorbance by the purine and pyrimidine bases, DNA absorbs strongly in the UV region with a maximum near 260 nm. However, the absorbance by individual bases is reduced by electronic interactions between stacked bases. The total absorbance may be reduced by as much as 40% compared to an unstacked state. This reduction in the extinction coefficient of the aggregate bases is termed **hypochromicity**. Stacking interactions decrease gradually as the ordered structure of the double helix is disrupted by increasing temperatures. Therefore, a completely disordered polynucleotide approaches an absorbance comparable to the sum of the absorbances of its purine and pyrimidine constituents.

Measurement of the absorbance of a DNA complex at 260 nm while slowly increasing the temperature provides a means to observe denaturation. In a thermal denaturation experiment monitored using absorption spectroscopy, the polynucleotide

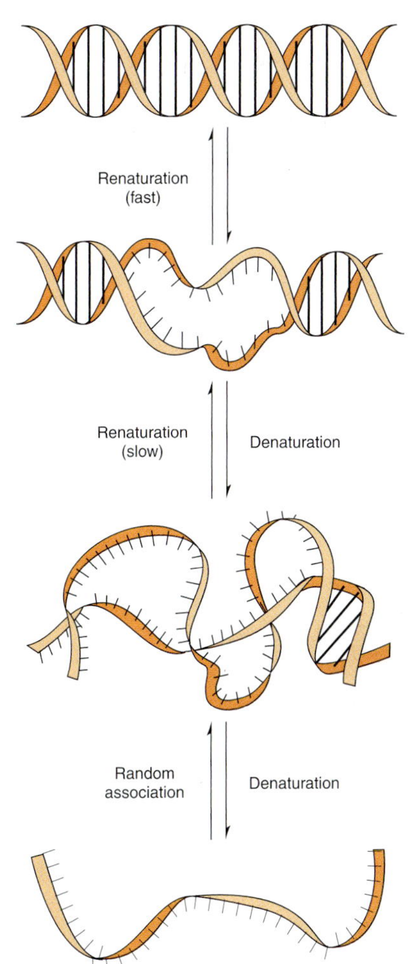

FIGURE 2.19

Denaturation of DNA. At high temperatures the double-stranded structure of DNA is completely disrupted, with eventual separation of strands and formation of single-stranded open coils. Denaturation also occurs at extreme pH ranges or at extreme ionic strengths.

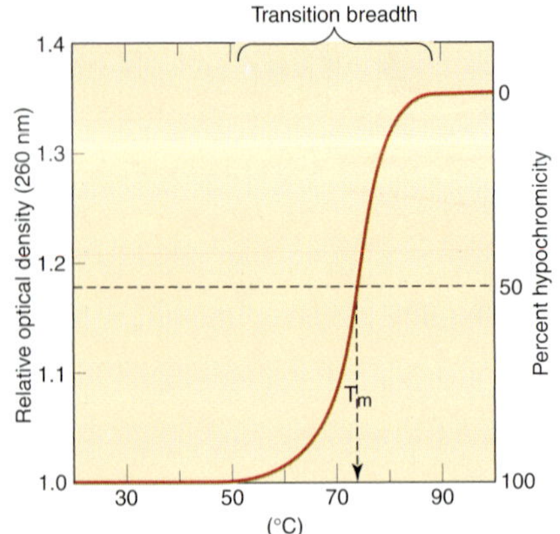

FIGURE 2.20

Temperature-optical density profile for DNA. When DNA is heated, the absorbance at 260 nm increases with rising temperature. A graph in which absorbance versus temperature is plotted is called a melting curve. Relative optical density is the ratio of the absorbance at any temperature to that at low temperatures. The temperature at which one-half of the maximum optical density is reached is the midpoint temperature (T_m).

absorbance typically changes very slowly at first, then rises rapidly to a maximum value (Figure 2.20). Before the rise, the DNA is double-stranded. In the rising section of the curve, an increasing number of base pairs are interrupted. Complete strand separation occurs at temperatures corresponding to the upper plateau of the curve. The midpoint temperature of this transition, the $\boldsymbol{T_m}$, is characteristic of the base content of DNA under standard conditions of concentration and ionic strength. The higher the guanine–cytosine content, the higher the transition temperature between double-stranded helix and single strands. This difference in T_m values is attributed to increased stability of guanine–cytosine pairs, which arises from more favorable stacking interactions.

DNA becomes denatured at pH >11.3 as the N–H groups on the bases become deprotonated, preventing them from participating in hydrogen bonding. Alkaline denaturation is often used to prevent damage to the DNA that can occur at a high temperature or low pH. Denaturation can also be induced at low ionic strengths, because of enhanced interstrand repulsion between negatively charged phosphates, and by various denaturing agents (compounds that can effectively hydrogen bond to the bases while disrupting hydrophobic stacking interactions). A complete denaturation curve similar to that shown in Figure 2.20 is observed at a relatively low constant temperature by variation of the concentration of an added denaturant such as urea.

Complementary DNA strands, separated by denaturation, can reform a double helix if appropriately treated. This is called **renaturation** or **reannealing**. If denaturation is not complete and a few nucleobases remain hydrogen bonded between the two strands, the helix-to-coil transition is rapidly reversible. Annealing is possible even after complementary strands have been completely separated. Under these conditions the renaturation process depends upon the DNA strands meeting in a manner that can lead to reformation of the original structure. Not surprisingly, this is a slow, concentration-dependent process. As renaturation begins, some of the hydrogen bonds formed are extended between short tracts of polynucleotides that may have been distant in the original native structure. These randomly base-paired structures are short-lived because the bases that surround the short complementary segments cannot pair and, thus, cannot form a stable fully hydrogen-bonded structure. Once the correct nucleobases begin to pair by chance, the double helix is rapidly reformed over the entire DNA molecule.

Sudden onset of denaturation/renaturation reveals the "all-or-none" nature of helix-to-coil/coil-to-helix transitions. The renaturation process requires formation of a short double-helical region to initiate formation of the double helix (Figure 2.21). This begins with formation of a single base pair that is rather unstable. However, formation of a second neighboring base pair is enhanced because the process is now less entropically unfavorable. As new base pairs begin to form a stacked structure, formation of subsequent base pairs is further facilitated. Unpaired nucleotides of each strand begin to stack on the growing helix, positioned optimally for base pairing. After formation of a double helix with four to five base pairs, a stable double helix forms, and the remainder of the complex will rapidly and spontaneously "zip up." After formation of the initial base pair, stacking and base pairing are not independent events but are influenced by the neighboring pairs. Such a process is called **cooperative**. The presence of a short double helix serves as a nucleation site for annealing by facilitating formation of subsequent base pairs. Denaturation is the same process in reverse: A bubble serves as an initiation site for unstacking of the nucleobases and rapid unzipping of the helix.

Hybridization

A technique based upon the association of complementary polynucleotide strands, **hybridization**, has been developed for the detection and quantitation of specific sequences of target nucleic acid. These are important basic tools in contemporary molecular biology and are being used for determining (1) whether a certain sequence occurs in the DNA of a particular organism, (2) a genetic or evolutionary relatedness between different organisms, (3) the number of genes transcribed in a particular mRNA, and (4) the location of any given DNA sequence. They are based on annealing a

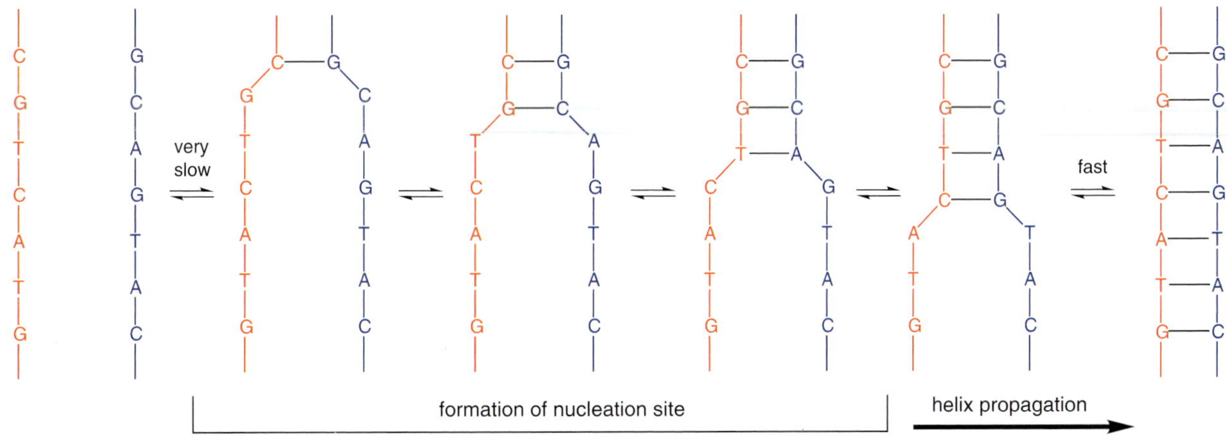

FIGURE 2.21

Cooperativity of renaturation/denaturation of DNA double helices. During renaturation, formation of the first base pair is very slow. Annealing of neighboring pairs is facilitated, especially after formation of a 3- to 5-bp nucleation site. Denaturation follows a similar course, but order of steps is reversed. Redrawn from W. Saenger, *Principles of Nucleic Acid Structure*. New York: Springer-Verlag, 1984, p. 141.

complementary polynucleotide, called a **probe**, which is appropriately tagged for easy detection of the hybrid double helix (see p. 258).

In a typical experiment, DNA or RNA to be tested for hybridization is denatured and immobilized by binding to a suitable insoluble matrix. Labeled DNA probes are then allowed to hybridize to complementary sequences bound to the matrix (Figure 2.22). Probes are short single-stranded RNA or DNA oligonucleotides that are complementary to specific sequences of interest. Under the proper conditions, probes interact only with the segment of interest, indicating whether it is present in a particular sample of DNA. Probe molecules are generally >20 nucleotides long. Appropriate labels include radioactive elements, fluorescent chromophores, and biotin. Because the double-helical complex containing the hybridized probe is usually bound to an insoluble matrix, unhybridized probes can be washed away. Detection of bound labels allows direct quantitation of the sequence of interest. Determination of the maximum amount of DNA that can be hybridized can establish the degree of homology between DNA of different species since the base sequences in each organism are unique. The observed homologies serve as indices of evolutionary relatedness and have been particularly useful for defining phylogenesis in prokaryotes. Hybridization studies between DNA and RNA have provided very useful information about the biological role of DNA, particularly the mechanism of transcription. Arrays of probes are useful for definitive and rapid diagnosis of genetic disorders, infectious disease, and cancer as described in Clinical Correlation 2.2.

Conformations of Double-Helical DNA

The early X-ray diffraction studies showed that there was more than one conformation of DNA (Figure 2.23). One of these, A-DNA, was found under conditions of low humidity and high salt concentration. Adding organic solvents such as ethanol reduced the "humidity" of these aqueous solutions. A second distinct form, B-DNA, appeared under conditions of high humidity and low salt concentration and was the basis of the Watson–Crick structure. Eleven distinct conformations of double-helical DNA have since been described. They vary in orientation of the bases relative to the helix and to each other and in other geometric parameters of the double helix. One form, Z-DNA, incorporates a left-handed helix rather than the usual right-handed variety.

The structural polymorphism of double-helical DNA depends on the base composition and on physical conditions. The local structure of DNA is sufficiently flexible to allow for changes in conformation that maximize stacking while minimizing unfavorable steric interactions. The stacking preferences of bases can favor one conformation

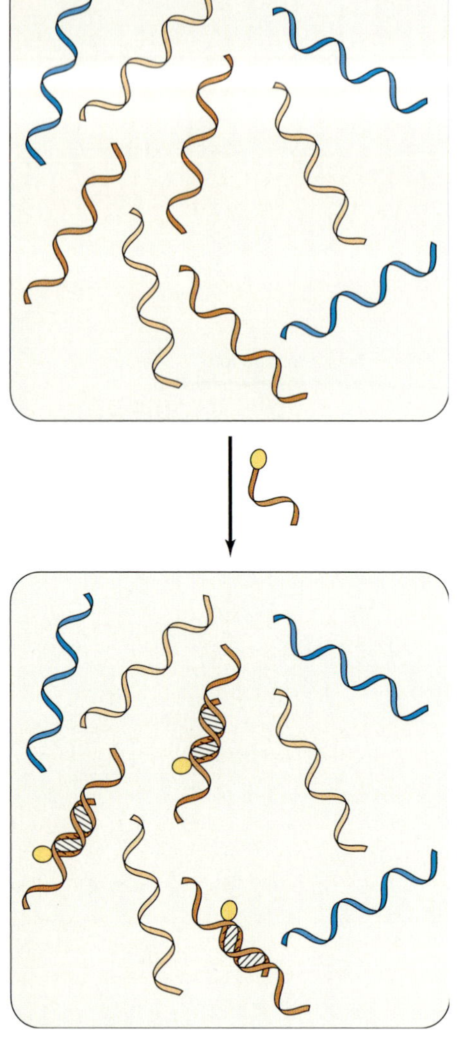

FIGURE 2.22

General representation of hybridization experiments. A mixture of denatured DNAs is treated with a DNA probe bearing a label. The probe can hybridize with those DNAs with complementary sequences. Detection of the double-helical complexes allows for detection and quantitation of DNA that contains the sequence of interest. Specific applications often feature steps to separate and immobilize the different DNAs in the mixture to be probed.

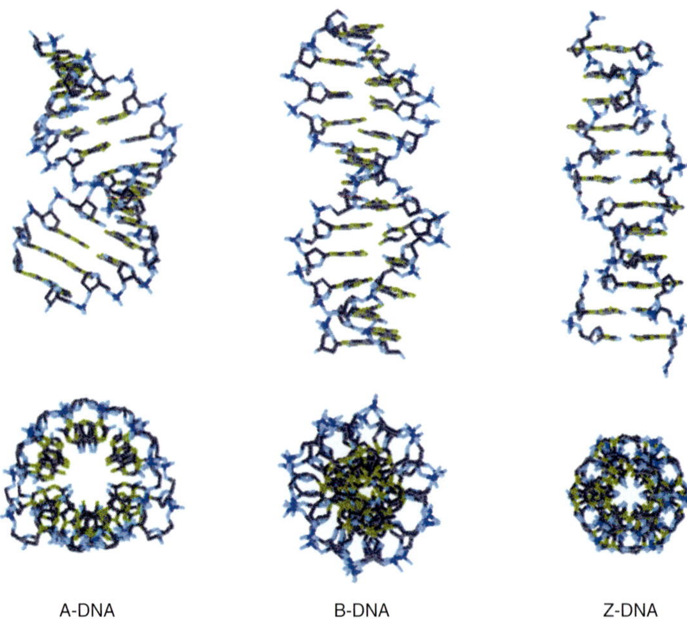

FIGURE 2.23

The varied geometries of double-helical DNA. Depending on conditions and base sequence, the double helix can acquire various distinct geometries. There are three main families of DNA conformations: A, B, and Z. The right-handed forms, B-DNA and A-DNA, differ in sugar pucker, which leads to differing helical structures. The A-form is underwound compared to the B-form, and the resulting helix is shorter and fatter. The Z-DNA structure is a left-handed helix with a zigzagging backbone. The sugar puckers and glycosidic conformations alternate from one residue to the next, producing a local reversal in chain direction.

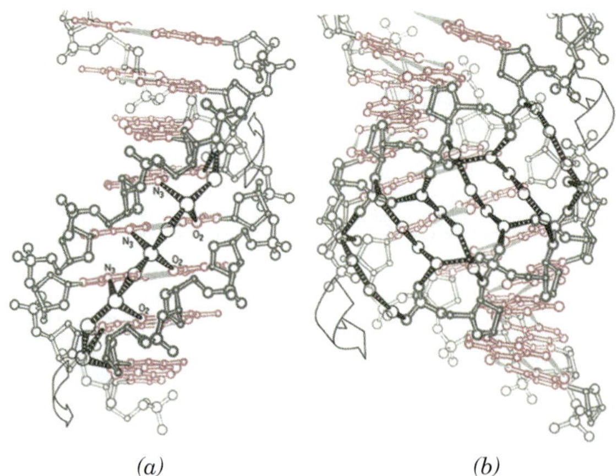

FIGURE 2.24

Hydration of the grooves of DNA. (*a*) An organized spine of hydration fills the minor groove of B-form DNA. (*b*) The phosphates of lining the major groove are spanned by a network of waters in A-form DNA. Reproduced with permission from Saenger, W. *Principles of Nucleic Acid Structure*. New York: Springer-Verlag, 1984, p. 379.

over others. For example, consecutive guanines on one strand favor A-DNA-like conformations. The solution conditions also play a key role in determining the favored conformation. Water molecules interact differently with double helices in different conformations. For example, the phosphate groups in B-DNA are more accessible to water molecules than in A-DNA. Also, polar groups on the bases are better hydrated in a B-DNA conformation. In fact, in AT-rich sequences, an ordered array of water molecules occupies the narrow minor groove of B-DNA (Figure 2.24*a*). With a decrease in humidity, the available water molecules solvate the highly polar phosphate groups in preference to the bases. The major groove narrows, allowing water molecules to bridge the phosphates (Figure 2.24*b*), thereby stabilizing the A-DNA conformation.

CLINICAL CORRELATION 2.2
Diagnostic Use of DNA Arrays in Medicine and Genetics

With completion of the Human Genome Project, a wealth of genetic information is rapidly becoming available. Application of this knowledge to medicine requires development of new techniques to monitor gene expression and to analyze rapidly genes for mutations.

Oligonucleotide arrays consist of a number of gene-specific oligonucleotide probes immobilized at specific sites on a solid matrix (chip). They can contain thousands of unique probe molecules, each within a spatially fixed address. Gene chips can then be treated with labeled target nucleic acids (DNA or RNA) derived from cells of an organism. Hybridization of the targets with complementary probe sequences allows for immobilization of the label at specific sites on the chip. The presence of specific sequences can be determined, and the amount of labeled target hybridized to a site can be quantitated.

Such techniques may lead to diagnostics for rapid screening of genomic DNA for disease-associated mutations. For example, high-density DNA arrays with thousands of oligonucleotide probes has been used to detect mutations leading to ataxia telangiectasia, a recessive disease characterized by neurological disorder, recurrent respiratory infection, and dilated blood vessels in the skin and eyes. Similar studies have examined mutations in the hereditary breast and ovarian cancer gene BRCA. Comparable assays can be used to identify accurately pathogens present in a clinical sample. The ability to quantitate the amounts of mRNAs in various cells allows for the profiling of gene expression. This may lead to techniques to evaluate illnesses such as cancer and select individualized treatments.

High density DNA arrays for mutation analysis

Arrays for the ataxia telangiectasia (A-T) gene are interrogated separately with a reference (unmutated) sample and a test sample.

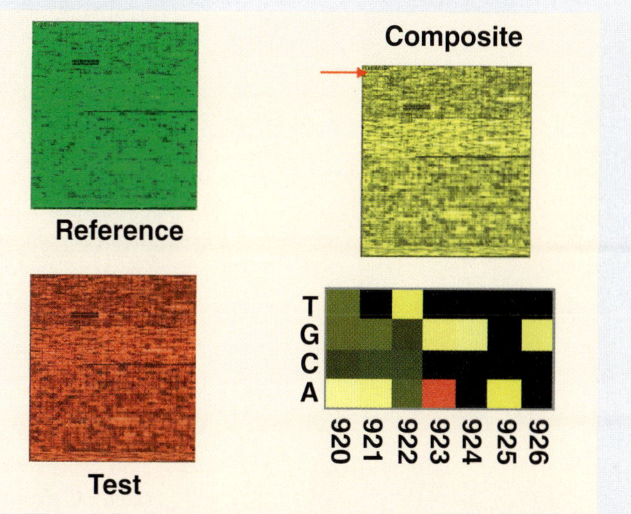

The composite image is made by combination of the images from the two experiments. At sites where the test and reference DNAs have identical sequences, the composite image is yellow (red^{+} green). In this sample, the person has a mutation at position 923, which appears red. This mutation, the replacement of a G with an A, causes premature termination of protein synthesis. Because this sample also has a yellow spot at position 923, the mutation has occurred on only one copy of the gene (e.g., it is heterozygous). In recessive disorders such as A-T, people with heterozygous mutations are carriers, but will not develop the disease themselves.

Source: Freeman, W. M., Robertson, D. J., and Vrana, K. E. Fundamentals of DNA hybridization arrays for gene expression analysis. *BioTechniques* 29:1042, 2000; Stover, A. G., Jeffery, E., Xu, J., Persing, D. H. Hybridization array technology. In: D. H. Persing (Ed.), *Molecular Microbiology*. Herndon, VA: ASM, 2004, pp. 619–639; and Hacia, J. G., Brody, L. C., Chee, M. S., Fodor, S. P., Collins, F. S. Detection of heterozygous mutations in BRCA1 using high density oligonucleotide arrays and two-colour fluorescence analysis. *Nat. Genet.* 14:441, 1996.

The different conformations of DNA can be grouped into three families: A-DNA, B-DNA, and Z-DNA. The parameters for these conformations, listed in Table 2.3, have been determined by X-ray diffraction methods. It must be emphasized that the average overall structure of DNA in living organisms is believed to be B-DNA-like. Notably, the B-conformation, unlike the A- and Z-forms, is highly flexible. In native B-DNA, considerable local variation in conformation of individual nucleotides may occur. Such variations may be important in regulation of gene expression, since they can influence the extent of DNA binding with various types of regulatory proteins.

DNA conformations in the B-family feature base pairs that are nearly perpendicular to the helical axis, which passes through the base pairs. The major and minor grooves are roughly the same depth, and the minor groove is relatively narrow. The helix is long and thin, with approximately 10 base pairs per helical turn. The rise per residue is 3.4 Å, the approximate thickness of the bases. In contrast, the A-DNA structure is

TABLE 2.3 Structural Features of A-, B-, and Z-DNA

Features	A-DNA	B-DNA	Z-DNA
Helix rotation	Right-handed	Right-handed	Left-handed
Base pairs per turn (crystal)	10.7	9.7	12
Base pairs per turn (fiber)	11	10	—
Base pairs per turn (solution)	—	10.5	—
Pitch per turn of helix	24.6 Å	33.2 Å	45.6 Å
Proportions	Short and broad	Longer and thinner	Elongated and thin
Helix packing diameter	25.5 Å	23.7 Å	18.4 Å
Rise per base pair (crystal)	2.3 Å	3.3 Å	3.7
Rise per base pair (fiber)	2.6	3.4 Å	—
Helix axis	Major groove	Through base pairs	Minor groove
Sugar ring conformation (crystal)	C3′ *endo*	Variable	Alternating
Sugar ring conformation (fiber)	C3′ *endo*	C2′ *endo*	—
Glycosyl bond conformation	*anti*	*anti*	*anti* at C, *syn* at G

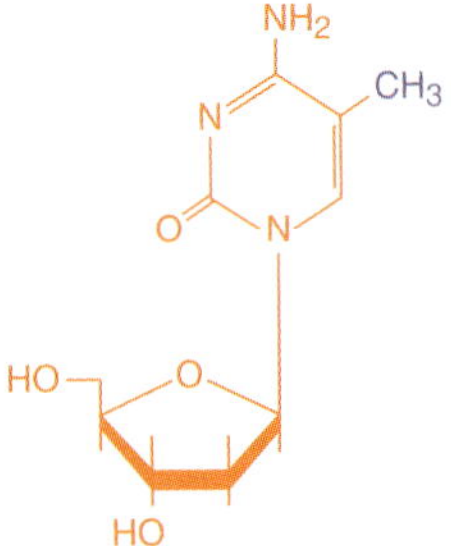

FIGURE 2.25
Structure of 5-methylcytidine.

shorter and thicker. There are about 11 base pairs per helical turn, with a vertical rise of 2.56 Å per residue. In order to accommodate the thickness of the bases, the base pair is tilted 20° from the plane perpendicular to the helical axis. The helical axis is displaced to the major groove side of the base pairs. This results in a very deep major groove and a shallow minor groove, and it forms a hole ~3 Å in diameter that runs through the center of the helix. Low humidity favors the A-DNA structure that exposes more hydrophobic surface to the solvent than B-DNA.

Z-DNA is a radically different left-handed, double-helical conformation for double-stranded DNA. It is generally observed in sequences of alternating purines and pyrimidines, particularly d(GC)$_n$. The designation "Z" was chosen because the phosphodiester backbone assumes a zigzag arrangement compared to the smooth conformation that characterizes A-DNA and B-DNA. The Z-DNA structure is longer and much thinner than that of B-DNA and completes one turn in 12 base pairs. The minor groove is very deep and contains the helical axis. The base pairs are displaced so far into the major groove that a distinct channel no longer exists. These changes place the stacked nucleobases on the outer part of Z-DNA rather than in their conventional positions in the interior of the double helix.

Some evidence exists that suggests that Z-DNA influences gene expression and regulation. Apparently small stretches of DNA that contain alternating purines and pyrimidines are more commonly found at the 5′-ends of genes, regions that regulate transcriptional activities. Also, methylation of either guanine residues in C8 and N7 positions or cytosine residues in C5 position (Figure 2.25) stabilizes the Z-form. Sequences that are not strictly alternating purines and pyrimidines may also acquire the Z-conformation because of methylation. The suggestion that Z-DNA may have a role in gene regulation is supported by modifications in methylation patterns that accompany the process of gene expression. However, Z-DNA has not yet been detected in DNA *in vivo*.

Inverted Repeat
5′ GGAATCGATCTTAAGATCGATTCC 3′
3′ CCTTAGCTAGAATTCTAGCTAAGG 5′

Mirror Repeat
5′ GGAATCGATCTTTTCTAGCTAAGG 3′
3′ CCTTAGCTAGAAAAGATCGATTCC 5′

Direct Repeat
5′ GGAATCGATCTTGGAATCGATCTT 3′
3′ CCTTAGCTAGAACCTTAGCTAGAA 5′

FIGURE 2.26
Symmetry elements in DNA sequences. Three types of symmetry elements for double-stranded DNA sequences are shown. Arrows illustrate the special relationship of these elements in each one of these sequences. In inverted repeats, or palindromes, each DNA strand is self-complementary within the inverted region that contains the symmetry elements. A mirror repeat is characterized by the presence of identical base pairs equidistant from a center of symmetry within the DNA segment. Direct repeats are regions of DNA in which a particular sequence is repeated. The repeats need not be adjacent to one another.

Noncanonical DNA Structures

A-, B-, and Z-DNA are associated mainly with variation in the conformation of the nucleotide constituents of DNA. It is now recognized that even canonical B-DNA is not a straight, monotonous, and uniform structure. Instead, DNA forms unusual structures such as cruciforms or triple-stranded arrangements and bends as it interacts with certain proteins. Such variations in DNA conformation appear to be an important recurring theme in the process of molecular recognition of DNA by proteins and enzymes. Variations in DNA structure or conformation are favored by specific DNA sequence motifs such as inverted repeats, palindromes, mirror repeats, direct repeats (Figure 2.26), as well as homopurine–homopyrimidine sequences, phased A tracts, and G-rich regions.

AT-rich sequences, which are prone to easy strand separation, exist near the origins of DNA replication. The human genome is rich in homopurine–homopyrimidine sequences and alternating purine–pyrimidine tracts.

Bent DNA

DNA sequences with runs of 4 to 6 adenines separated by 10-base-pair (bp) spacers produce bent conformations. Structural studies have indicated that minor grooves of these sequences are compressed. However, it is not clear whether bending arises from this feature or from the boundary between an unusual conformation and normal B-DNA. DNA bending appears to be a fundamental element in the interaction between DNA sequences and proteins that catalyze central processes, such as replication, transcription, and site-specific recombination. Bending induced by interactions of DNA with enzymes and other proteins, such as histones, does not require the nucleotide sequence conditions that are needed for bending of protein-free DNA. Bending also occurs because of photochemical damage or mispairing of bases and serves as a recognition signal for initiation of DNA repair. Antitumor antibiotics that produce bent structures are discussed in Clinical Correlation 2.3.

Cruciform DNA

Inverted repeats are quite widespread within the human genome and often occur near putative control regions of genes or at origins of DNA replication. It has been speculated that inverted repeats may function as molecular switches for replication and transcription. Disruption of hydrogen bonds between the complementary strands and formation of intrastrand hydrogen bonds within the region of an inverted repeat produce a cruciform structure (Figure 2.27). The loops generated by cruciform formation require the unpairing of 3–4 bases at the end of the "hairpin." Depending upon the sequence, these structures may be only slightly destabilizing because residues in the loop can

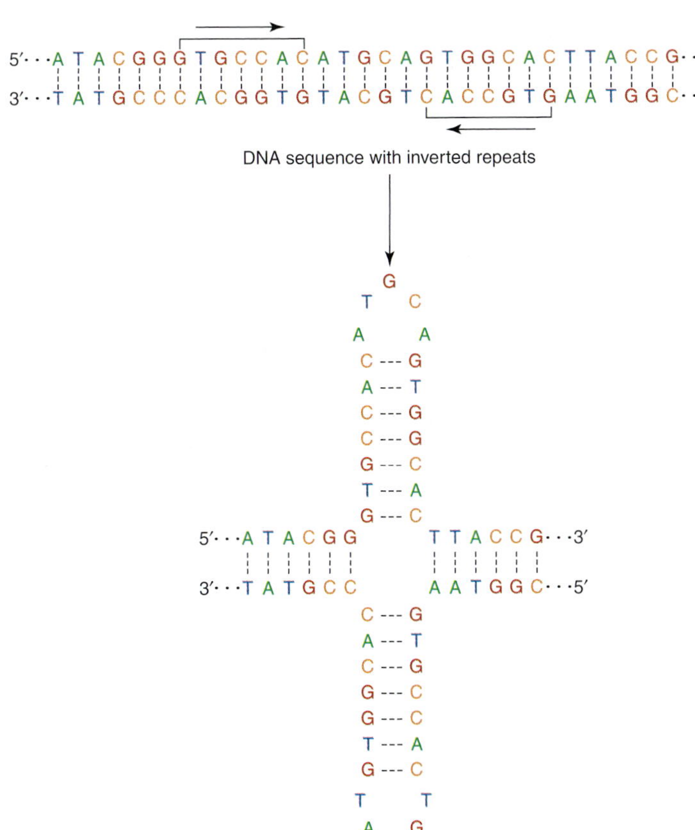

FIGURE 2.27

Formation of cruciform structures in DNA. The existence of inverted repeats in double-stranded DNA is a necessary but not a sufficient condition for the formation of cruciform structures. In relaxed DNA, cruciforms are not likely to form because the linear DNA accommodates more hydrogen-bonded stacked base pairs than the cruciform structure, making the formation of the latter thermodynamically unfavored. Unwinding is followed by intrastrand hydrogen bond formation between the two symmetrical parts of the repeat to produce the cruciform structure. Formation of cruciform structures is not favored at DNA regions that consist of mirror repeats because such cruciforms would be constructed from parallel rather than antiparallel DNA strands. Instead, certain mirror repeats tend to form triple helices.

CLINICAL CORRELATION 2.3
Antitumor Antibiotics that Change the Shape of DNA

The local three-dimensional structure of DNA is important in interactions with proteins involved in repair, transcription, recombination, and chromatin condensation. It has been proposed that antibiotics can induce formation of DNA structures that can recruit these proteins with cytotoxic results. The best-studied example is the antitumor drug cisplatin, a tetra-coordinate platinum complex (*cis*-$Pt(NH_2)_2Cl_2$). Cisplatin is used alone or in combination with other antitumor agents to treat a variety of tumors including testicular, ovarian, bone, and lung cancers. It forms inter- and intrastrand cross-links in double-stranded DNA with the latter adduct comprising 90% of DNA lesions. These bonds arise from displacement of chloride ligands on platinum by N7 atoms of two neighboring guanines. Structural studies on intrastrand cross-linked DNA adduct show that the double helix is strongly bent toward the major groove.

Bent structures of the DNA–cisplatin adduct are specifically recognized by several DNA-binding proteins such as nucleotide excision repair (NER) proteins and nonhistone DNA-binding proteins such as HMG-1. The cytotoxicity of cisplatin adducts is a complicated process mediated by specific interactions with these proteins. Cellular processes such as transcription and apoptosis are also affected by formation of cisplatin–DNA adducts. The lesions themselves and the adduct–protein complexes are likely to interfere with transcription. NER proteins are recruited to repair the lesion, but excision repair is prone to produce DNA strand breaks. Accumulation of these breaks will ultimately induce apoptosis as the DNA becomes too damaged to function. Similar mechanisms have been proposed to account for cytotoxicity of other DNA-binding drugs such as ditercalinium, a bifunctional molecule that forms noncovalent adducts with DNA that are also highly bent. Cytotoxicity is thought to arise from induction of abortive repair pathways that lead to DNA strand breaks. Interactions of the cisplatin–DNA adduct with HMG proteins may also contribute to its cytotoxicity. Binding of HMG proteins may incorrectly signal that the damaged region of DNA is transcriptionally active and prevent condensation into folded chromatin structures. These complexes also perpetuate the lesion by shielding the DNA–cisplatin adduct from repair.

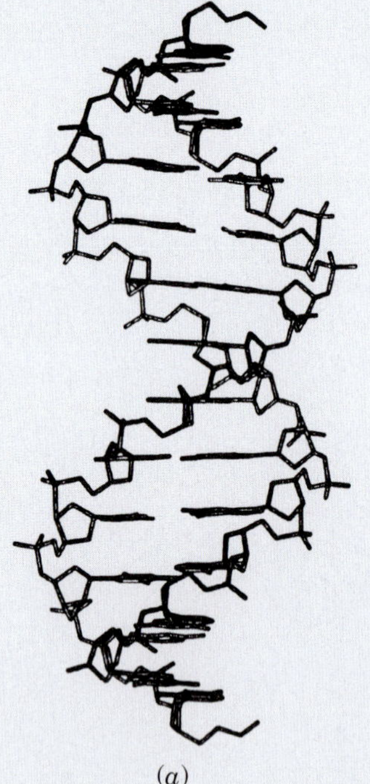

(*a*)

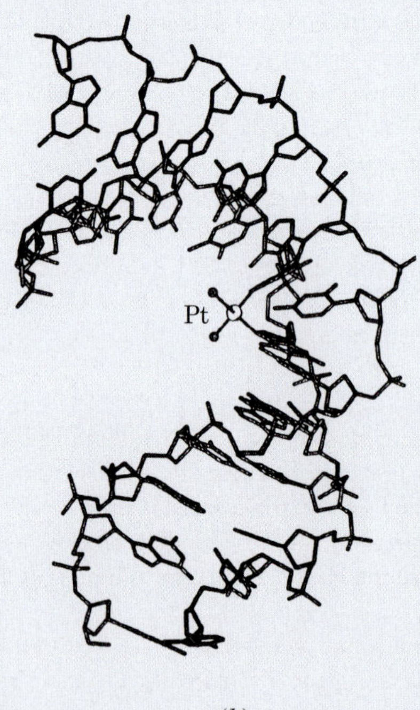

(*b*)

Source: Zamble, D. B. and Lippert, S. J. The response of cellular proteins to cisplatin-damaged DNA. In: B. Lippert (Ed.), *Cisplatin: Chemistry and Biochemistry of a Leading Anticancer Drug*. New York: Wiley-VCH, 1999, pp. 73–134; and Lambert, B., Segal-Bendirdjian, E., Esnault, C., Le Pecq, J.-B., Roques, B. P., Jones, B., and Yeung, A. T. Recognition by the DNA repair system of DNA structural alterations induced by reversible drug–DNA interactions. *Anti-Cancer Drug Des.* 5:43, 1990.

remain stacked at the end of the helix. Overall, a biological function for cruciforms has not been established.

Triple-Stranded DNA

Some polynucleotides such as poly(dA) and poly(dT) combine to form triple-stranded complexes rather than the expected double helices. The stoichiometries of these complexes require that one strand contain a homopurine sequence while the other two have homopyrimidine sequences. Even when participating in Watson–Crick base pairing, purines possess two potential hydrogen bonding sites in the major groove: N7 and O6 for guanine, N7 and 6-NH_2 for adenine. Nucleobases appropriately positioned in the major groove can form specific **Hoogsteen** base triplets (Figure 2.28). For example, a thymine can selectively form two Hoogsteen hydrogen bonds to the adenine of an A-T pair. Likewise, a protonated cytosine can form two Hoogsteen H bonds with the guanine of a G-C pair, resulting in a base triplet isomorphous to the T-A-T. The pK_a of cytidine is approximately 4.5, and triple helices containing C-G-C triplets show a strong dependence on pH. However, the templating effect of the triplet raises the apparent pK_a of the cytosine, making it possible to form triple helices even in solutions that are only mildly acidic (pH 6). The unique Hoogsteen hydrogen bonding patterns of guanine and adenine provide for specificity similar to the Watson–Crick pairs. Changing the orientation of the third-strand nucleobases in the major groove allows for the formation of **reversed-Hoogsteen** triplets (Figure 2.28). Selective triplets can be formed between an adenine and the A-T pair as well as between a guanine and a G-C pair. A reverse-Hoogsteen triplet can also be formed between thymine and A-T, but the resulting triplet is not isomorphous to the pu·pu·py triplets. However, the backbone is able to accommodate the distortions that result from incorporation of these triplets.

As in double-helical complexes, base stacking plays the key role in stabilizing the triplex structure. However, bringing three negatively charged backbone strands together increases electrostatic repulsion. Thus, the triple-helical complex is less stable than the associated Watson–Crick double helix. The presence of Mg^{2+} or other multivalent cations stabilizes the triple helix by shielding the phosphate charges.

Intramolecular triple helices can be formed by disruption of double-helical DNA with polypurine sequences in mirror repeats. A mirror repeat is a region such as AGGGGA that has the same base sequence when read in either direction from a central point. Refolding generates a triple-stranded region and a single-stranded loop in a structure called **H-DNA** (Figure 2.29). Even though the formation of H-DNA is thermodynamically unfavorable because of a reduction of stacking interactions, intramolecular triple helices have been detected in cellular DNA when under superhelical stress. DNA supercoiling provides the energy to drive the unwinding of DNA that is necessary for the formation of the triple helix. Triple-strand formation produces a relaxation of negative supercoils. Also, the binding of proteins to the single-stranded DNA may further stabilize the H-DNA structure and prevent degradation of the loop by nucleases.

Many sequences in eukaryotic genomes have the potential to form triple-stranded DNA structures. Such regions occur with much higher frequency than expected from probability considerations alone. Polypurine tracts over 25 nucleotides long constitute as much as 0.5% of some eukaryotic genomes. These potential triple-helical regions are especially common near sequences involved in gene regulation. Because of this, it has been proposed that H-DNA may play a role in control of transcription (synthesis of RNA from DNA). A hereditary disorder where intramolecular triplex formation within DNA segments having Hoogsteen base-pairing ability may play a role is discussed in Clinical Correlation 2.4. The abilities of triple helical-DNA to interfere with transcription have also led to efforts to use intermolecular triple helices to control artificially RNA synthesis and subsequent protein synthesis. Other potential biological tasks have been proposed for triple-helical DNA, including possible roles in initiation and termination of replication and recombination. However, there is no definitive evidence to prove a biological role for triple-helical DNA.

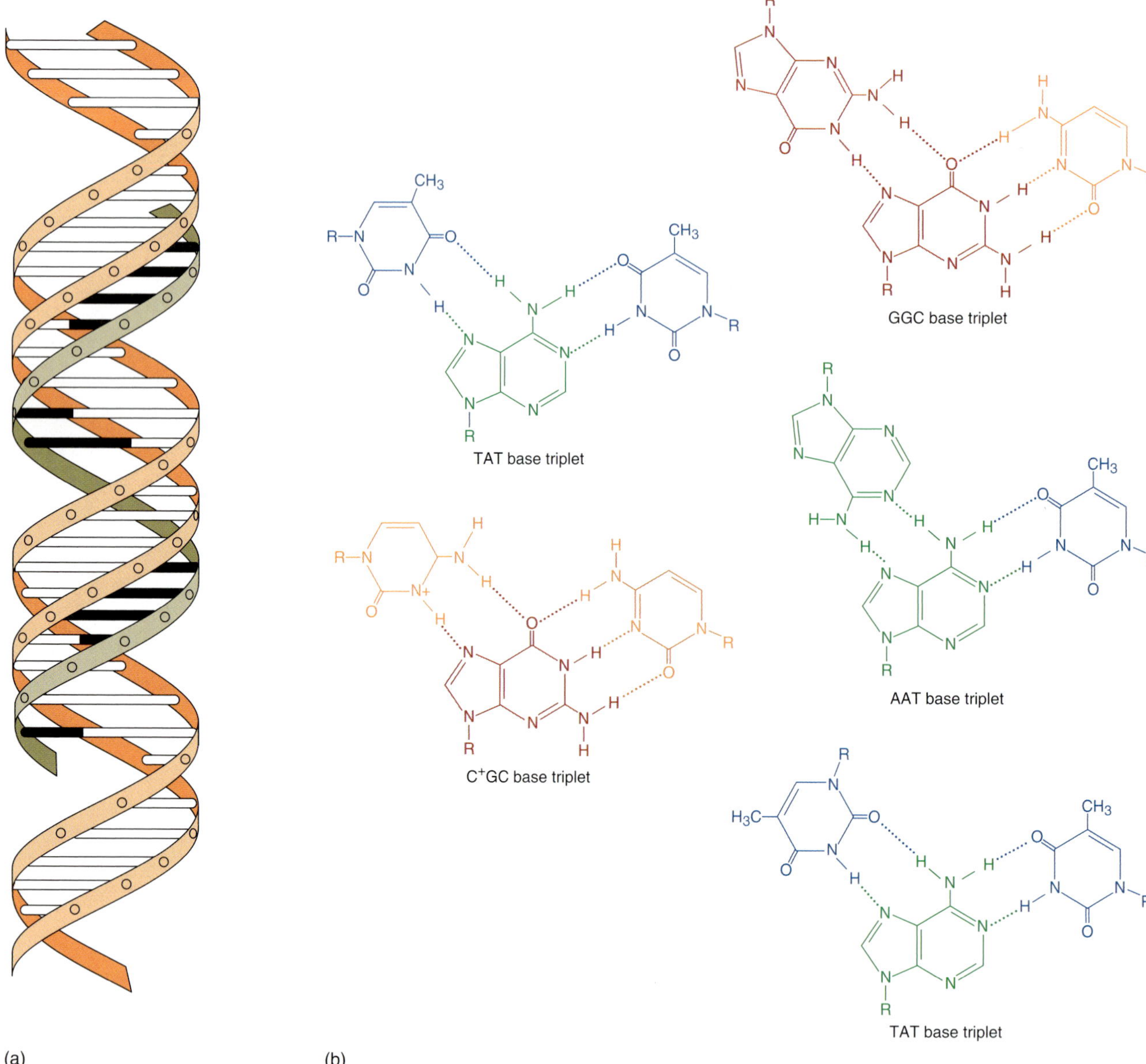

FIGURE 2.28

Triple helices. (*a*) Hoogsteen triple helix. Hoogsteen hydrogen bonds between a homopurine strand of a double helix and a parallel homopyrimidine strand in the major groove. The resulting isomorphous base triplets, TAT and C^+GC, provide for sequence selective binding. (*b*) Reverse Hoogsteen triple helix. Triple helices can be formed by antiparallel binding of an oligonucleotide in the major groove to the homopurine strand of a Watson–Crick double helix. Reverse Hoogsteen hydrogen bonding produces three possible triplets: GGC, AAT, and TAT. The last triplet is not isomorphous with the others because the sugars (represented by R) are positioned differently with respect to the Watson–Crick base pair.

Four-Stranded DNA

Guanine nucleotides and highly G-rich polynucleotides form novel tetrameric structures called **G-quartets** that contain a planar array of guanines connected by Hoogsteen hydrogen bonds (Figure 2.30). Polynucleotides can interact to form **tetraplexes** where G-quartets stack upon each other to form a multilayered structure (Figure 2.31). These structures are stabilized by Na^+ and K^+, which interact with guanine oxygens (O6) in the center of the quartet plane or between two adjacent planes. These cations stabilize the complex both electrostatically by balancing out the negative charges on the four polynucleotide strands and entropically by releasing a large number of water molecules into the bulk solution.

CLINICAL CORRELATION 2.4
Hereditary Persistence of Fetal Hemoglobin

Hereditary persistence of fetal hemoglobin (HPFH) is a group of conditions in which fetal hemoglobin synthesis is not terminated at birth but continues into adulthood. The homozygous form of the condition is extremely uncommon, being characterized by changes in red blood cells similar to those found in the genetic blood disorder β-thalassemia. HPFH, in either the homozygous or the heterozygous state, is associated with mild clinical or hematologic abnormalities. Mild musculoskeletal pains may occur infrequently, but HPFH patients are frequently asymptomatic.

The condition results from failure to stop transcription of human ${}^{G}\gamma$- and ${}^{A}\gamma$-globin genes, leading to elevated levels of fetal hemoglobin. The formation of an intramolecular DNA triple-helical structure located about 200 bp upstream from the initiation site for transcription of the γ-globin genes (between positions −194 and −215) acts as a brake for their expression. Hemoglobin genes of patients contain mutations in one or more positions in this region, decreasing the stability of the triple helix and reducing its ability to inhibit the protein synthesis.

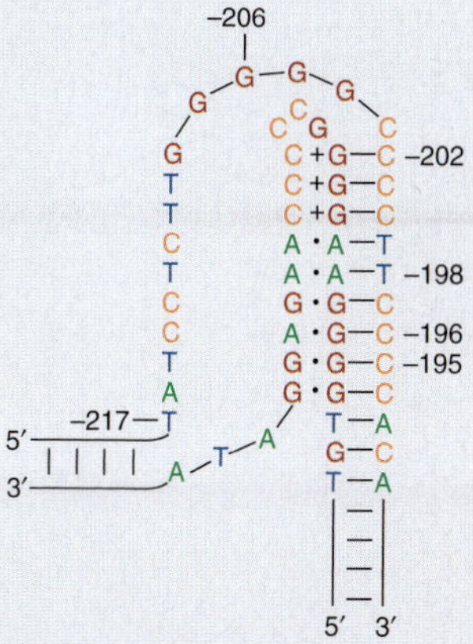

Source: Ulrich, M. J., Gray, W. J., and Ley, T. J. An intramolecular DNA triplex is disrupted by point mutations associated with hereditary persistence of fetal hemoglobin. *J. Biol. Chem.* 267:18649, 1992; and Bacolla, A., Ulrich, M. J., Larson, J. E., Ley, T. J., and Wells, R. D. An intramolecular triplex in the human gamma-globin 5′-flanking region is altered by point mutations associated with hereditary persistence of fetal hemoglobin. *J. Biol. Chem.* 270:24556, 1995.

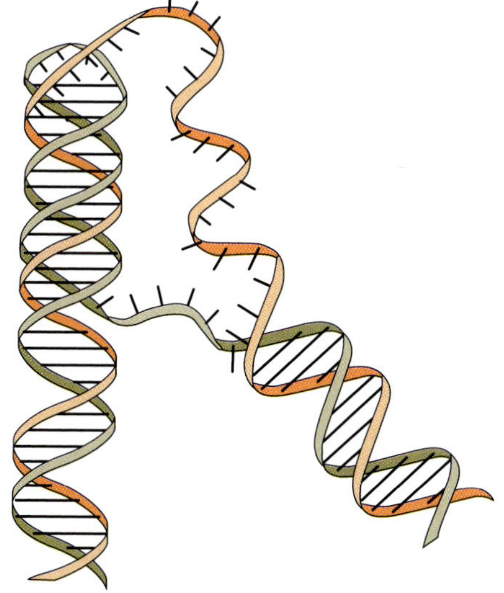

FIGURE 2.29

Intramolecular triple helices: H-DNA. Polypurine–polypyrimidine regions of DNA with a mirror repeat symmetry can form an intramolecular triple helix in which the third strand lays in the major groove, whereas its complementary strand acquires a single-stranded conformation.
Redrawn based on figure in Sinden, R. R. *DNA Structure and Function*. New York: Academic Press, 1994.

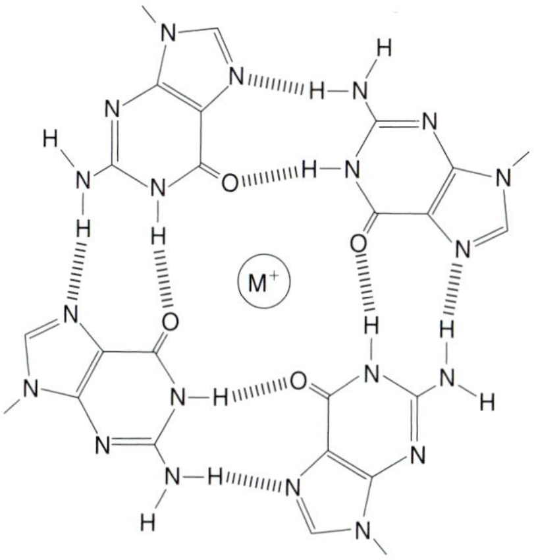

FIGURE 2.30

The structure of a G-quartet. The four coplanar guanines form a tetrameric structure by formation of Hoogsteen hydrogen bonds. The cavity in the center of the quartet can accommodate a sodium or potassium ion with coordination by the four O-6 oxygens.

While G-tetraplexes have been observed by X-ray diffraction and NMR spectroscopy, their existence *in vivo* has not been proven. However, the ends of eukaryotic chromosomes (**telomeres**) contain repetitive G-rich sequences. Human telomeres contain 800–2400 copies of the hexameric repeat sequence d(TTAGGG)$_n$ and terminate in a single-stranded overhang that is roughly 150 nucleotides long. *In vitro*, oligonucleotides

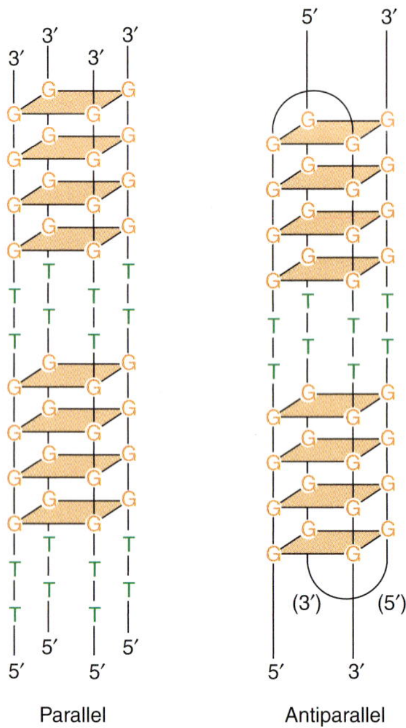

FIGURE 2.31

G-Quadruplex DNA. Four-stranded structures can arise from stacking of G-quartets. Quadruplex structures may be parallel or antiparallel (one possible arrangement shown). The latter may be formed by the G-rich sequences of telomeric DNA.

Redrawn based on figure in Sinden, R. R. *DNA Structure and Function*. New York: Academic Press, 1994.

with this sequence can form tetraplex structures. The role tetraplexes play in telomere functions is unknown. However, telomeres are attracting attention as targets for new anticancer chemotherapies (Clin. Corr. 2.5). In addition, G-tetraplexes have been implicated in recombination of immunoglobulin genes and in dimerization of double-stranded genomic RNA of the human immunodeficiency virus (HIV).

Another four-stranded DNA structure involves cytosine-rich sequences. X-ray diffraction and NMR experiments have shown that C-rich oligodeoxyribonucleotides can form **i-DNA** (Figure 2.32), which is composed of two duplexes that mutually intercalate to form a four-stranded structure. Each duplex is formed by C-C base pairs. These stack on C-C pairs from a second duplex, forming an interdigitated four-stranded complex. Protonation of one of the cytosine bases at N3 allows for the formation of three hydrogen bonds in the C-C base pair, and it partly counteracts the interstrand phosphate–phosphate repulsions. Although C-rich telomeric sequences form i-DNA complexes, their lower stability at pH 7.4 casts doubt upon their physiological relevance.

Slipped DNA

DNA regions with direct repeat symmetry can form slipped, mispaired DNA (SMP-DNA). Their formation involves unwinding of the double helix, realignment, and subsequent pairing of one copy of the direct repeat with an adjacent copy on the other strand. This generates two single-stranded loops. Two isomeric structures of an SMP-DNA are possible (Figure 2.33). Although SMP-DNA has not yet been identified *in vivo*, genetic evidence suggests that it is involved in spontaneous frameshift mutagenesis that results in base addition or deletion within runs of single bases (Figure 2.34). Deletions and duplications of DNA segments that are longer than a single base can occur during DNA replication between direct repeats, causing slipped-looped structures. Duplication of certain simple triplet repeats that are the basis of several human genetic diseases (Clin. Corr. 2.6) may also occur by this mechanism.

2.4 HIGHER-ORDER DNA STRUCTURE

With the exception of RNA-containing viruses, all life on Earth uses DNA to store genetic information. The length of DNA varies from species to species and ranges from

CLINICAL CORRELATION 2.5

Telomerase as a Target for Anticancer Agents

Telomeres, the ends of linear eukaryotic chromosomes, are critical for maintaining the stability of the genome. They are progressively shortened during each cycle of cell division; upon reaching a critical length, they trigger apoptosis. Telomerase acts to maintain or lengthen the telomeres but is not active in normal somatic cells. Telomerase activity in most tumor cell lines may be responsible for their immortalization; when increased, it correlates to poorer clinical prognosis.

Two approaches are being examined for selective inhibition of telomerase. The first involves targeting of the RNA-containing portion of the enzyme. This RNA serves as the template for extension of the telomeric repeat sequence. Nucleic acids with chemically modified sugar–phosphate backbones bind to telomerase RNA in immortal human cells, inhibit activity, and ultimately cause cell death. Modified nucleic acids were used to resist nuclease degradation and provide high affinity for forming double-helical complexes with RNA. A second approach involves drugs that bind to G-quadruplex DNA. Large aromatic molecules such as porphyrins and anthraquinones selectively bind and stabilize G-quadruplex DNA structures. Although *in vivo* existence of G-quadruplexes has not been demonstrated, some of these compounds have been shown to inhibit telomerase activity.

Source: Herbert, B.-S., Pitts, A. E., Baker, S. I., Hamilton, S. E., Wright, W. E., Shay, J. W., and Corey, D. R. Inhibition of human telomerase in immortal human cells leads to progressive telomere shortening and cell death. *Proc. Natl. Acad. Sci. USA* 96:14276, 1999; and Cuesta, J., Read, M. A., and Neidle, S. The design of G-quadruplex ligands as telomerase inhibitors. *Mini-Rev. Med. Chem.* 3:11, 2003.

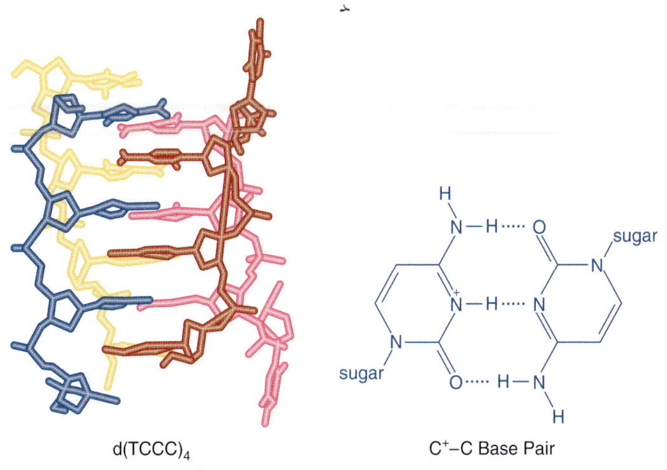

FIGURE 2.32

i-DNA. i-DNA consists of two interdigitated pairs of duplexes containing several C^+-C base pairs. The favorable protonation of one of the cytosines in each pair explains the greater stability of this structure at acidic pHs.
Reproduced with permission from Feigon, J. *Curr. Biol.* 3:611, 1993.

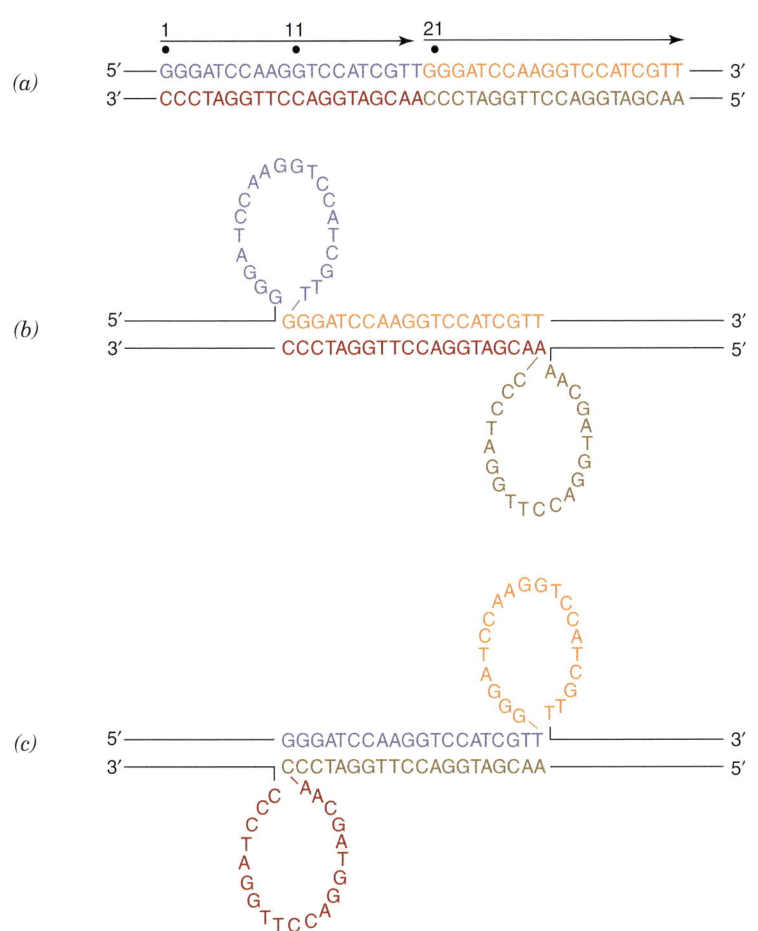

FIGURE 2.33

Slipped, mispaired DNA. Presence of two adjacent tandem repeats (a) can give rise to two isomers of slipped, mispaired DNA. In (b), the second copy of the direct repeat in top strand pairs with first copy of repeat on bottom strand. In (c), pairing of first copy of direct repeat in top strand with second copy of direct repeat in bottom strand. Two single-stranded loops are generated in each isomer.

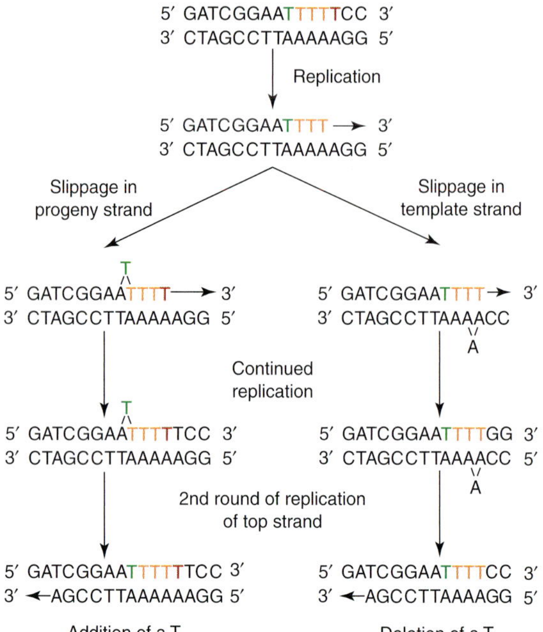

FIGURE 2.34

Frameshift mutagenesis by DNA slippage. DNA replication within a run of a single base can produce a single base frameshift. In this example, a run of five As is replicated and, depending on whether a slippage occurs in progeny strand or template strand, a T may be added or deleted from the DNA.

a few thousand base pairs for small viruses, to millions of base pairs in bacteria, and to billions of base pairs in plants and animals. In all organisms, the **contour length** (length of the DNA assuming a B-form double helix) of genomic DNA is usually much larger than the size of the cell that contains it. For example, a medium-sized virus, such as λ-phage, contains 4.8×10^4 bp of DNA, which is 16.5 μm long. However, the viral particle is only 0.19 μm long. The common bacterium *Escherichia coli* is approximately 2 μm long, has a single chromosome with 4.6×10^6 bp, and has a contour length of 1.5 mm. A single diploid human cell contains 6×10^9 bp of chromosomal DNA packaged in 46 chromosomes. The contour length of this DNA approaches 2 m, which is packed within a nucleus about 10 μm in diameter. It is clear that the DNA of all organisms must be exquisitely packaged.

Genomic DNA may be Linear or Circular

With the exception of some small bacteriophages (such as φX174) that can acquire a single-stranded form, most DNAs exist as double-helical complexes. Depending upon the source, the complexes can be linear or circular. For example, DNAs of several small viruses are linear double-stranded helices. Some of these DNAs contain naturally occurring interior single-stranded breaks. The double-helical structure is maintained because the breaks in one strand are generally in different locations from those in the complementary strand. DNAs in most higher organisms are also linear. Each of the 46 chromosomes in a diploid human cell is a linear double-helical DNA complexed with proteins.

Circular DNA results from the formation of phosphodiester bonds between the 3′- and 5′-termini of linear polynucleotides. The circular nature of the single-stranded phage φX174 DNA was suspected from studies that showed that no ends were available for reaction with exonucleases. In addition, endonuclease cleavage at a single site yielded only one polynucleotide. These results were later confirmed by direct observation with electron microscopy.

Some organisms have DNAs that are either linear or circular at different points in their life cycles. Before entering *E. coli*, the λ phage DNA is linear with single-stranded overhangs on the 5′-termini. These overhangs are approximately 20 nucleotides long and have complementary sequences. Upon infection, circularization occurs by hybridization of the ends to each other. Formation of phosphodiester bonds between the 3′- and 5′-ends of each strand by **DNA ligase** produces a covalently closed circle (Figure 2.35). The

CLINICAL CORRELATION 2.6
Expansion of DNA Triple Repeats and Human Disease

The presence of reiterated 3-bp DNA sequences occurs in a number of human genetic diseases including fragile X syndrome, myotonic dystrophy, X-linked spinal and bulbar muscular atrophy (Kennedy's syndrome), Friedreich's ataxia, and Huntington's disease. These diseases are associated with expansion of nucleotide triplet repeats that appear within or near specific genes. Kennedy's disease was found to contain a CAG repeat in the first exon of the androgen receptor gene. Triplet repeats can also be found within untranslated regions of the gene: In Friedreich's ataxia a repeated GAA is found within an intron, and in myotonic dystrophy a CTG repeat is found in the 3′-untranslated region of its gene. Fragile X syndrome, a leading cause of mental retardation, is characterized by expansion of a GCC triplet on the 5′-side of the FMR-1 gene. This repeat expands from 30 copies to thousands of copies. The disease develops when normal expression of FMR-1 gene is turned off by methylation of CpG sites. In all cases, expansion of the triplet interferes with normal functioning of the related protein. Often, there is a loss of protein function, but occasionally a gain of a deleterious function occurs. These diseases are characterized by an increase in severity of the disease with each successive generation, which is known as anticipation.

Triplet expansion may result from slipped mispairing during DNA synthesis. Because of massive amplification that characterizes these diseases, repeated or multiple slippage would have to be involved to explain the high degree of expansion. One possible mechanism involves slippage of nascent DNA during lagging strand synthesis. This process may be aided by formation of a stable hairpin structure by the slipped loop. Repetition of this process leads to accumulation of large numbers of triplet repeats.

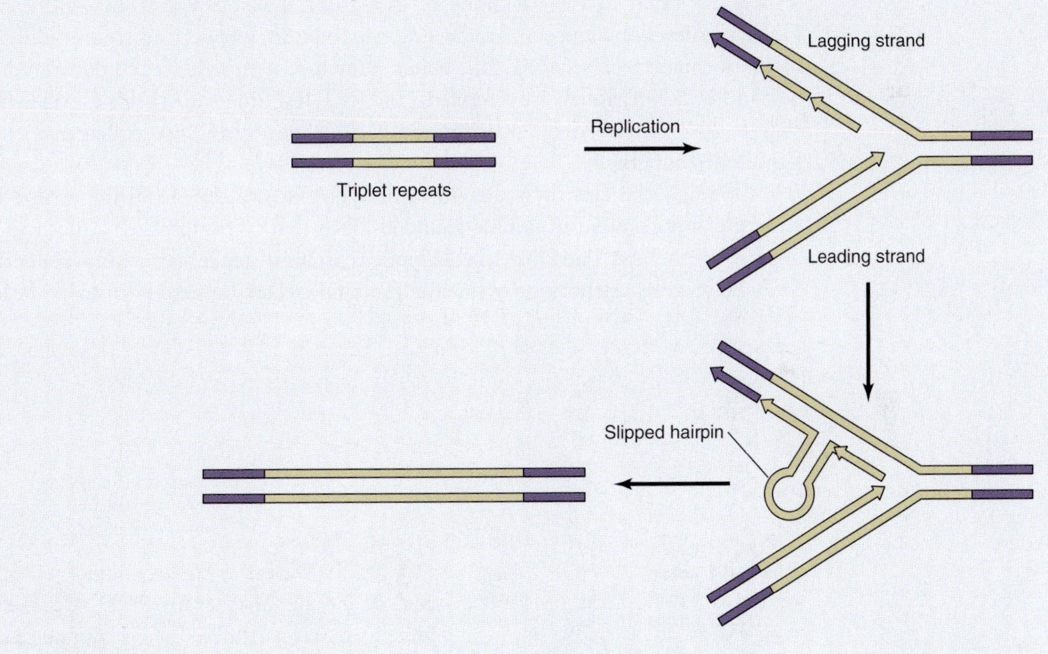

Source: Timchenko, L. T., and Caskey, C. T. Triplet repeat disorders: Discussion of molecular mechanisms. *Cell. Mol. Life Sci.* 55:1432, 1999; Lieberman, A. P. and Fischbeck, K. H. Triplet repeat expansion in neuromuscular disease. *Muscle Nerve* 23:843, 2000; Patel, P. I. and Isaya, G. Friedreich ataxia: From GAA triplet-repeat expansion to frataxin deficiency. *Am. J. Hum. Genet.* 69:15, 2001.

strands of a circular DNA cannot be irreversibly separated by denaturation because they exist as intertwined closed circles. The absence of 3′- or 5′-termini endows the circular DNA with complete resistance toward exonucleases, improving the longevity of DNA.

Most DNA in bacteria exists solely as closed circles. This includes the bacterial chromosomes and the smaller extrachromosomal **plasmid** DNAs. The latter are a few thousand base pairs long and encode accessory genes such as those for antibiotic resistance. Plasmids are maintained and replicated separately from chromosomal DNA and may number in the hundreds within a bacterium. Circular DNA also exists in higher organisms. Yeasts can also carry circular plasmids. Mitochondria and chloroplasts in higher eukaryotic cells contain circular DNAs $200-1500 \times 10^3$ bp long which encode

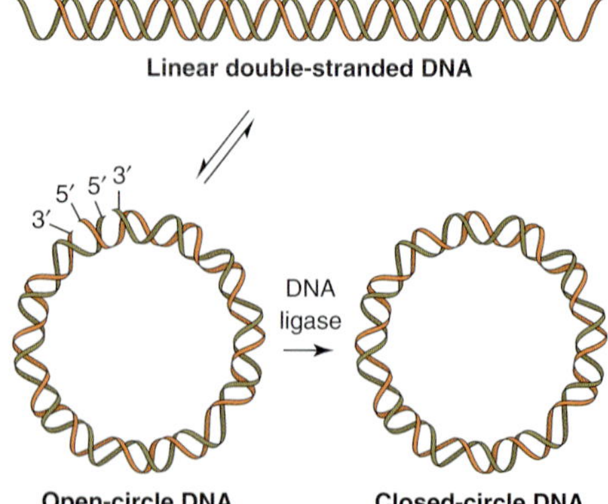

FIGURE 2.35

Circularization of λ DNA. DNA of bacteriophage λ exists in linear and circular forms, which are interconvertible. Circularization of λ DNA is possible because the 5′-overhangs of the linear form are complementary sequences.

unique proteins used by the organelles. The presence of an independent genome and a marked similarity to cyanobacteria have led to the hypothesis that these organelles arose billions of years ago from a symbiotic relationship between protoeukaryotic cells and bacteria.

DNA is Superhelical

Circular double-stranded DNA formed by ligating the free termini of a linear DNA is **relaxed**; that is, it has the thermodynamically favored structure of B-DNA. This relaxed DNA has greatly reduced activity in replication, translation, and recombination. The biologically active form of DNA is **superhelical**, a topologically strained isomer created by underwinding or overwinding the double helix. Before circularization of a linear B-form DNA, the double helix contains about 10.5 base pairs per complete turn. If the DNA is untwisted before sealing the circle, the resulting structure will be strained (Figure 2.36). The untwisting reduces the total number of helical turns present in the circular structure. The underwinding of the helix can be accommodated by disrupting the base pairing over a small region to produce a pair of single-stranded loops in a relaxed circular structure. Loss of base stacking in this structure is energetically unfavorable. However, if all base stacking is maintained, the underwinding generates a torsional strain in the backbone of the double helix. Twisting the entire circular DNA in direction opposite to the original rotation relieves the strain. This results in the formation of a coiled coil, better known as a superhelix. Underwound DNA is said to be negatively supercoiled, and the resulting superhelix is right-handed (Figure 2.37). Conversely, overwound DNA is positively supercoiled and forms a left-handed superhelix.

Although the closed-circular form of DNA is ideal for acquiring a superhelical structure, any segment of double-stranded DNA that is immobilized at both ends can also be superhelical. The DNA of eukaryotic cells, for instance, can acquire a superhelical form because its anchoring by nuclear proteins creates numerous closed **topological**

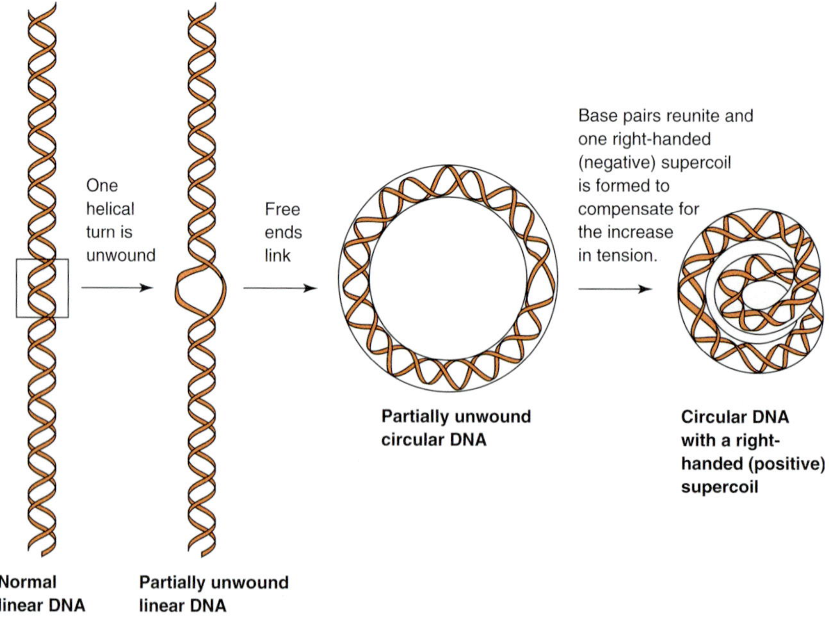

FIGURE 2.36

Negative DNA supercoiling. Right-handed supercoils (negatively supercoiled DNA) are formed if relaxed DNA is partially unwound. Unwinding may lead to a disruption of hydrogen bonds or may produce negative supercoils. The negative supercoils are formed to compensate for the increase in tension that is generated when disrupted base pairs are reformed.
Redrawn from Darnell, J., Lodish, H., and Baltimore, D. *Molecular Cell Biology*. New York: W. H. Freeman, 1986.

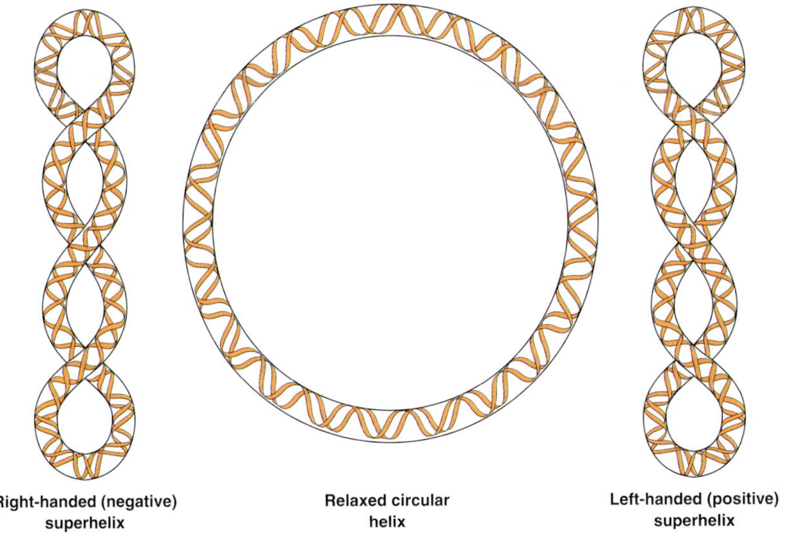

FIGURE 2.37

Relaxed and supercoiled DNA. Relaxed DNA can be converted to either right- or left-handed superhelical DNA. Right-handed DNA (negatively supercoiled DNA) is the form normally present in cells. Left-handed DNA may be transiently generated as DNA is subjected to enzyme-catalyzed transformations (replication, recombination, etc.) and is present in certain bacterial species. The distinctly different patterns of folding for right- and left-handed DNA superhelices are apparent in this representation.

domains. A topological domain is defined as a DNA segment contained in a manner that restrains rotation of the double helix (Figure 2.38). Overall, whether DNA is circular or linear, existence of negative superhelicity appears to be an important feature. Supercoiling promotes packaging of DNA within the confines of the cell by facilitating formation of compact structures. Superhelices also possess a tendency to generate regions with disrupted hydrogen bonding (bubbles), which may be instrumental in facilitating the process of localized DNA strand separation during DNA repair, synthesis, and recombination.

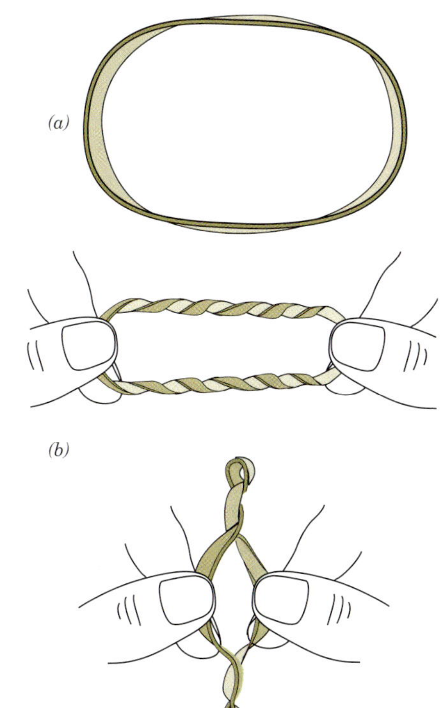

FIGURE 2.38

Superhelical model for DNA. A rubber band can demonstrate the topological properties of double-stranded circular DNA. The relaxed form of the band, shown in (*a*), can be twisted to generate two distinct topological domains, separated by the pair of "thumb-forefinger anchors," as shown in (*b*). Left-handed (counterclockwise) turns are introduced into the upper section of the band, with compensating right-handed (clockwise) turns formed in the bottom section. When the "anchors" are brought into close proximity with each other as shown in (*c*), the upper section that contained the left-handed turns forms a right-handed superhelix. The bottom section produces a left-handed superhelix. Superhelicity is not the property of a DNA molecule as a whole but rather a property of specific DNA domains.
Redrawn from Sinden, R. R. and Wells, R. D. *Curr. Opin. Biotech.* 3:612, 1992.

Topoisomerases

Topoisomerases regulate the formation of superhelices. These enzymes catalyze the concerted breakage and rejoining of DNA strands, producing a DNA that is more or less superhelical than the original (Table 2.4). The precise regulation of the cellular level of DNA superhelicity is important to facilitate protein interactions with DNA. The cellular ATP-to-ADP ratio may play a role in this process, because it influences the activity of some topoisomerases. Compounds that inhibit topoisomerases and gyrases are effective antibacterial and antitumor agents (Clin. Corr. 2.7).

Topoisomerases I make a transient single-strand break in a negatively supercoiled DNA double helix; passage of the unbroken strand through the gap eliminates one supercoil from DNA (Figure 2.39). During the reaction, the enzyme remains bound to DNA by a covalent bond between a tyrosyl residue and a phosphoryl group at the incision site. This conserves the energy of the interrupted phosphodiester bond for the subsequent repair of the nick. The two subclasses of topoisomerase I differ in the formation of 5′-phosphotyrosine (class IA) or 3′-phosphotyrosine (class IB) bonds.

Topoisomerases II are dimeric proteins that bind to a double-helical DNA and cleave both strands (Figure 2.40). Passage of another double-helical DNA segment through the break removes or adds two supercoils. All eukaryotic and many prokaryotic topoisomerases II only relax supercoiled DNA. ATP hydrolysis is required for turnover of the enzyme, but not actually for the relaxation reaction. The **gyrases**, a subset of type II topoisomerases, are the only enzymes that add negative supercoils into DNA. They occur only in bacteria; eukaryotes use the wrapping of DNA around chromosomal proteins for introduction of negative supercoils. The gyrase reaction requires the hydrolysis of ATP as an energy source and can add negative supercoils at a rate of about 100 per minute. Gyrases bind to DNA in a conformation that introduces a positive and negative superhelical loop, leaving the topological state of the DNA unchanged. Double-strand breakage and passage of DNA through the gap introduces two negative supercoils.

Topoisomerases also catalyze other topological isomerizations. Bacterial type III topoisomerases have type I topoisomerase properties; that is, they relax supercoils

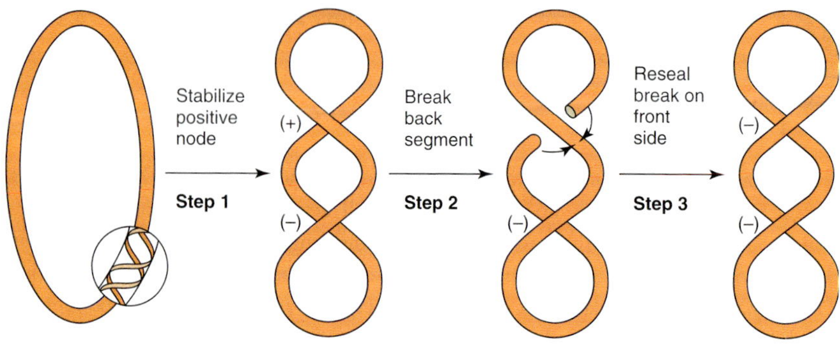

FIGURE 2.40

Mechanism of action of topoisomerases II. Topoisomerases II can relax or, in case of gyrases, introduce negative superhelices into DNA. The mechanism of action of gyrase is illustrated using the conversion of a relaxed DNA molecule to a molecule that contains first two supercoils, one positive and one negative (step 1). Passage of a DNA segment through the positive supercoil shown on the right most part of the figure (step 3) produces a molecule that contains two negative supercoils. This reaction requires the hydrolysis of ATP as an energy source.
Redrawn with permission from Brown, P. O. and Cozzarelli, N. R. *Science* 206:1081, 1979. Copyright (1979) AAAS.

of a single supercoiled DNA molecule organized into about 40 loops, each approximately 10^5 bp in size, that merge into a scaffold rich in protein and RNA. A role for RNA in chromosome packing is unique to prokaryotes. As a result of formation of a nucleoid with a diameter of 2 μm, the *E. coli* genome can easily be fit into the cell. Bacterial chromosomes are dynamic structures reflecting the need for rapid DNA synthesis, cell division, and transcription.

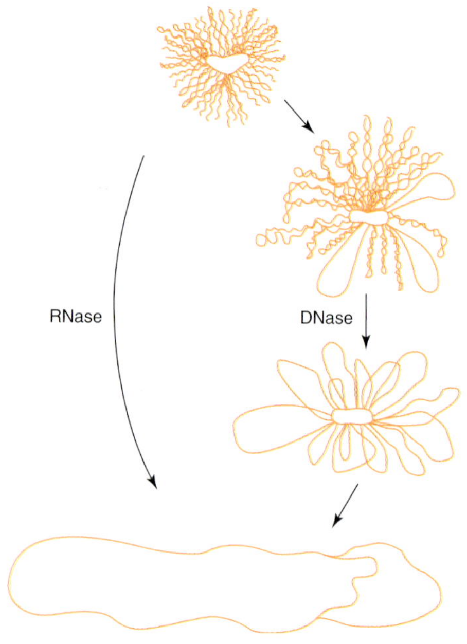

FIGURE 2.41

Bacterial chromosomes are packaged in nucleoids. The circular chromosome of a bacterium is compacted into about 40–50 loops of supercoiled DNA organized by a central RNA–protein scaffold. DNase relaxes the structure progressively by opening individual loops, one at a time. RNase completely unfolds the chromosome in a single step by disrupting the nucleoid core.
Redrawn from Worcel, A. and Burgi, E. *J. Mol. Biol.* 71:127, 1972.

Organization of Eukaryotic Chromatin

The enormous length of the genome of most eukaryotes necessitates the division of genetic information into several independent chromosomes. Human cells contain 23 pairs of chromosomes with an average length of 1.3×10^8 bp or approximately 43 mm. Each human chromosome consists of a DNA molecule varying in size from 263×10^6 bp for chromosome 1 to less than 50×10^6 bp for the Y chromosome. For this amount of DNA to fit within a cell nucleus with a diameter of approximately 10 μm requires a condensation ratio over five orders of magnitude (Figure 2.42).

DNA in eukaryotic cells is associated with numerous proteins to form **chromatin.** In nondividing (interphase) cells, chromatin is amorphous and dispersed throughout the nucleus. Just prior to cell division (metaphase), it becomes organized into highly compacted structures called **chromosomes.** Each chromosome is characterized by a **centromere,** which is the site for attachment to proteins that link the chromosome to the mitotic spindle. **Telomeres** define the termini of linear chromosomes. Chromosomes also contain sequences required for initiation of DNA replication (**origin of replication**).

Histones are the most numerous proteins in chromatin. There are five classes of these proteins; **histones H1, H2A, H2B, H3, and H4.** Because of their unusually high content of the basic amino acids lysine and arginine, histones are highly polycationic and interact with the polyanionic phosphate backbone of DNA to produce uncharged nucleoproteins. The amino acid sequences of the histones are very highly conserved between species. Histones H4 from peas and cows differ by only two amino acids, although these species diverged more than a billion years ago. Histones H3 are also very highly conserved, while the nonbasic regions of histones H2A and H2B are less highly conserved. H1 is larger, more basic, and by far the most tissue-specific and species-specific histone. In some cell types of vertebrates, a sixth histone, H5, functions as a replacement for H1. A heterogeneous group of proteins with high species and organ specificity is also present in chromatin. These nonhistone proteins consist of

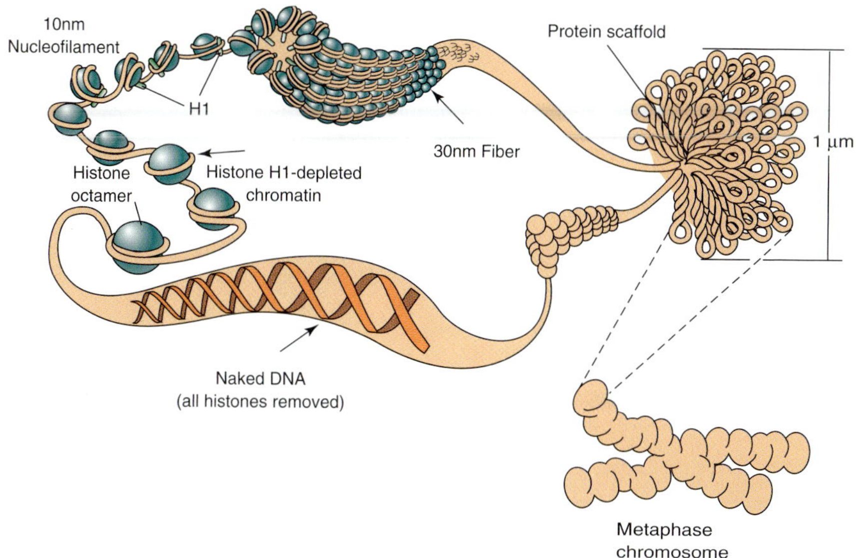

FIGURE 2.42

Organization of DNA into chromosomes. A drawing showing the stepwise condensation of DNA into chromatin. The DNA initially wraps around the histones of the nucleosome core. Condensation with histone H1 produces the 100-nm nucleofilament, which is subsequently packaged into a twisted, looped structure attached to a protein scaffold within the chromosome.

several hundred members, most of which are present in trace amounts. Many of them are associated with functions of chromosomes such as replication, gene expression, and chromosome organization.

Nucleosomes and Polynucleosomes

Histones interact with DNA to form a periodic "beads-on-a-string" structure, called a polynucleosome, in which the elementary unit is a **nucleosome** (Figure 2.43). Each nucleosome is disk-shaped, about 11 nm in diameter and 6 nm in height, and consists of a DNA segment and a histone cluster composed of two molecules each of H2A, H2B, H3, and H4 histones. Each octameric cluster consists of the tetramer $(H3)_2-(H4)_2$ with two H2A–H2B dimers. Each histone is characterized by a central nonpolar domain, which forms a globular structure and is responsible for histone–histone interactions. The neighboring N-terminal regions contain most of the positively charged amino acid residues that provide favorable electrostatic interactions with DNA. These tail sequences appear to extend radially out from the histone core and may be involved in interactions between nucleosomes. The DNA is wrapped around the octamer with the $(H3)_2-(H4)_2$ core interacting with the central 70–80 bp of the surrounding DNA (Figure 2.44). Approximately 146 bp of DNA wraps around the histone octamer. The histones are in contact with the minor groove of DNA and leave the major groove available for interaction with proteins that regulate gene expression and other DNA functions. The structure of the nucleosome core explains why eukaryotic cells lack gyrases that can

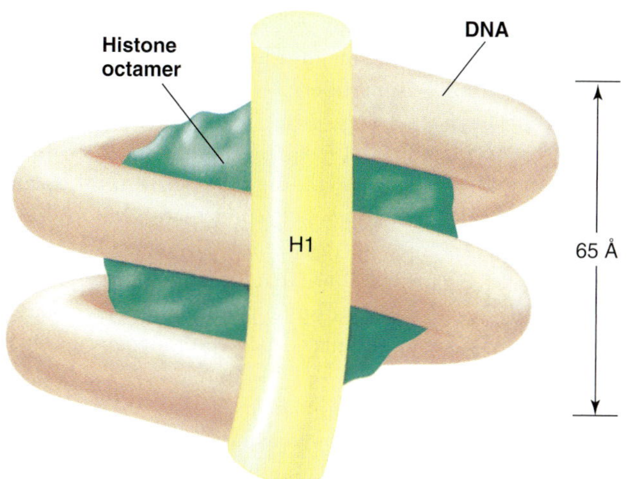

FIGURE 2.43

Models of the nucleosome complex. The nucleosome consists of approximately 146 bp of DNA corresponding to $1\frac{3}{4}$ superhelical turns wound around a histone octamer. The chromatosome consists of about 166 bp of DNA (two superhelical turns). The H1 histone is associated with the linker DNA. Reprinted with permission from Voet, D. and Voet, J. G. Biochemistry, 3rd edition, New York, Wiley, 2004. © (2004) Donald and Judith Voet.

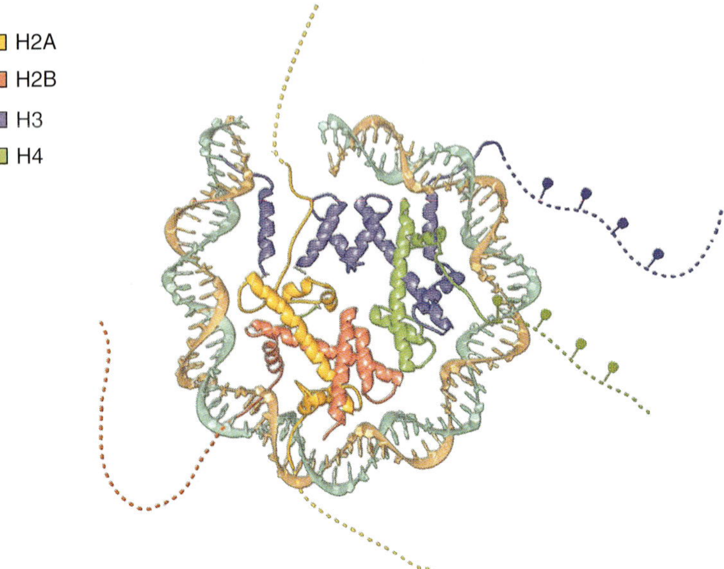

FIGURE 2.44

X-ray structure of the nucleosome core. The disk of the nucleosome core has been split in half to show one turn of DNA wrapped around four histone molecules. The dashed lines represent unstructured parts of the histone tails.
Reproduced with permission from Kornberg, R. D. and Lorch, Y. *Cell* 98:285, 1999.

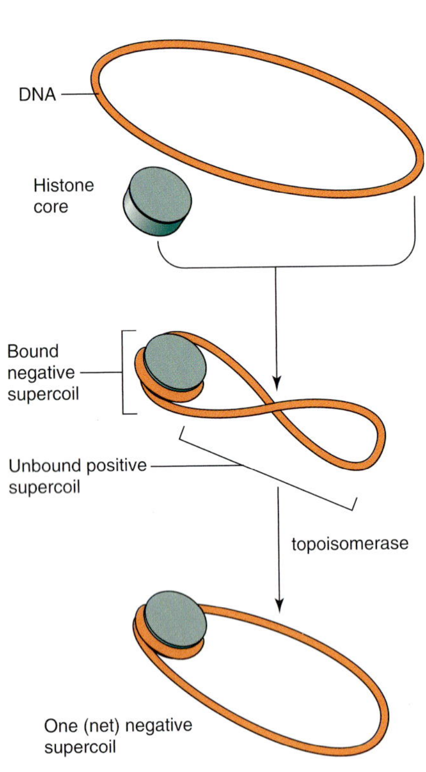

FIGURE 2.45

Generation of negative supercoiling in eukaryotic DNA. The binding of a histone octamer to a relaxed, closed-domain DNA forces the DNA to wrap around the octamer, generating a negative supercoil. In the absence of any strand breaks, the domain remains intact and a compensating positive supercoil must be generated elsewhere within the domain. The action of a eukaryotic topoisomerase subsequently relaxes the positive supercoil, leaving the closed domain with one net negative supercoil.

underwind relaxed DNA. Negative superhelicity is introduced because of DNA forming a coil around the histone core (Figure 2.45). Such wrapping removes approximately one helical turn from DNA, generating a negative supercoil within the region wrapped around the histone core and a compensating positive supercoil elsewhere in the molecule. Subsequent relaxation of the positive supercoil by eukaryotic topoisomerases leaves one net negative supercoil within the nucleosomal region. Between nucleosomes is 20–90 bp of **linker DNA**. It is associated with histone H1 that locks the coiled DNA in place; the resulting complex is a **chromatosome** (Figure 2.43). The periodicity of nucleosome distribution along the polynucleosome structure has been determined by controlled digestion by a nuclease that preferentially attacks linker DNA. The distribution of nucleosomes is not random with respect to the DNA base sequence. DNA does not bend uniformly but rather bends gently and then more sharply around the histone octamers. This suggests that DNA binding is sequence dependent and that positioning of nucleosomes may be influenced by the DNA sequence. In fact, nucleosomes tend to form preferentially in certain DNA regions. DNA that contains long A tracts or G-C repeats does not usually form nucleosomes. In contrast, certain bent DNA regions—for instance, periodically phased A tracts—associate strongly with histones. Histone octamers can migrate along the DNA sequence. This mobility allows access to the DNA by polymerases and other proteins necessary for transcription and replication.

Polynucleosome Packing into Higher Structures

Wrapping of DNA around histones to form nucleosomes results in a tenfold reduction in the apparent length of DNA and the formation of a **10-nm nucleofilament**. Condensation of 10-nm nucleofilaments into a **solenoid** arrangement involving six to seven chromatosomes per turn forms **30-nm fibers** (Figure 2.46). Histone H1 molecules bind to one another cooperatively, bringing the neighboring nucleosomes closer together in 30-nm fibers. At physiological salt concentrations, the 30-nm fibers form spontaneously, but at low ionic strength they dissociate into 10-nm nucleofilaments. Formation of polynucleosomes and their condensation into 30-nm fibers provides for a DNA compaction ratio of up to two orders of magnitude. The 30-nm fibers form only over selected regions of DNA that are characterized by the absence of binding with other sequence-specific nonhistone proteins. The presence of these and their effects on formation of 30-nm fibers may depend on the transcriptional status of the regions of DNA involved.

Models of the higher levels of packing of 30-nm fibers are based on indirect evidence obtained from the lampbrush chromosomes of vertebrate oocytes and polytene chromosomes of fruit fly giant secretory cells. These chromosomes are exceptional in

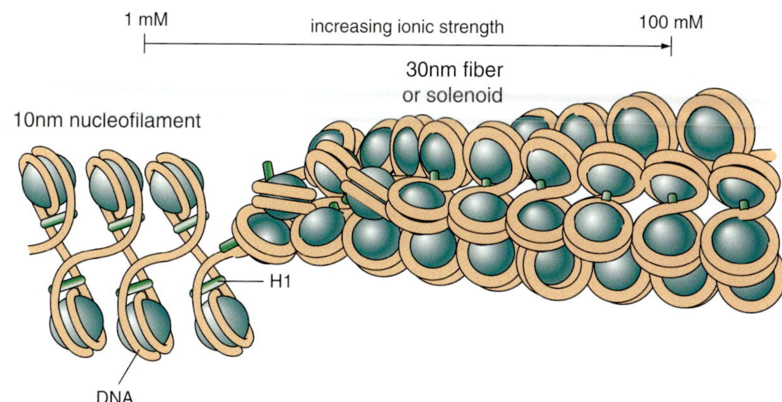

FIGURE 2.46

Nucleofilament structure. Histone H1 attached to the linker regions between nucleosomes results in condensation into 10-nm fibers. At higher ionic strengths the nucleofilament forms a very compact helical structure or a helical solenoid. The H1 histones interact strongly with one another in this structure.
Adapted from Kornberg, R. D. and Klug, A. *The Nucleosome.* San Diego, CA: Academic Press, 1989.

that they maintain precisely defined higher-order structures in interphase—that is, when cells are in a resting (nondividing) state. By extrapolation, the structural features of interphase lampbrush chromosomes have led to the proposal that chromosomes in general are organized as a series of looped, condensed domains of 30-nm fibers of variable size for different organisms. These loops may contain 5000 to 120,000 bp with an average of about 20,000. Thus, the haploid human genome would contain about 60,000 loops. A loop may contain a few linked genes. The loops are bound to a protein scaffold consisting of H1 histone and several nonhistone proteins, including two major scaffold proteins Sc1 (a topoisomerase II) and Sc2. They are fixed at their bases and can therefore accumulate supercoils. Specific AT-rich regions known as SARs (scaffold attachment regions) are preferentially associated with the scaffold. SARs also contain topoisomerase II binding sites. The presence of the topoisomerases suggests that changes in supercoiling within these domains are biologically important. Formation of looped domains may account for an additional 200-fold condensation in the length of DNA and an overall packing ratio of more than four orders of magnitude. Each loop can be coiled and then supercoiled into 0.4 μm of a 30-nm fiber. Since a chromosome is about 1 μm in diameter, packing of the 20-nm fiber into a chromatid would require just one more order of folding. The next level of chromosomal organization involves the packing of loops. This may be achieved by arranging the loops of 30-nm fiber into tightly stacked helical coils. **Chromatids**, the two linked arms of replicated DNA in the highly condensed metaphase chromosomes, may consist of helically packed loops of 30-nm fibers.

Packing changes at various stages of the cell cycle appear to be partially controlled by covalent modification of core histones. Histones H3 and H4 undergo cell-cycle-dependent reversible acetylation on the ϵ-amino group of lysine by a histone acetylase and a histone deacylase. Acetylation is linked with transcriptional activity. It changes the charges of the histones and may affect the negative superhelical tension within domains and the binding of certain transcription factors. Methylation of basic amino acid side chains (lysine and arginine) and phosphorylation of serine and threonine are also involved in regulating the activity of nucleosomal DNA. For example, acetylation and phosphorylation can drastically change the charge on the core histones, causing them to bind less tightly to DNA. These modifications may also modulate interactions between nucleosomes by the histone tails and linker histones. The resultant unraveling of 30-nm fibers and decondensation of chromatin produces the loosely packed **euchromatin**. The change from compact to decondensed chromatin is also promoted by HMG proteins (high-mobility-group proteins), which interact preferentially with transcriptionally active euchromatin, that is, the 10-nm fiber. Control of eukaryotic transcription and replication involves both histone and nonhistone protein. While dissociation of histones from chromosomal DNA may be a prerequisite for transcription, nonhistone proteins provide more finely tuned transcription controls. These proteins control gene expression during differentiation and development and may serve as sites for binding of hormones and other regulatory molecules.

Although active genes must be accessible to regulatory proteins, permanently repressed genes must remain inaccessible. Throughout the cell cycle, these genes, as well as other untranscribed regions such as centromeres, telomeres, and "junk" DNA, remain in a highly condensed form called **heterochromatin**. This inactive form of chromatin is characterized by increased methylation and by decreased acetylation of the histones. Some modifications play specific roles in shutting down transcriptional activity. For example, methylation of a tail lysine in histone H3 may provide a binding site for HP1, a protein that blocks access of transcription factors to the DNA. Maintaining inactive DNA in a highly condensed state may also determine accessibility of DNA to DNA-damaging agents.

2.5 DNA SEQUENCE AND FUNCTION

The size and average base composition of DNA vary widely between species. What makes the DNA of a species unique is the nucleotide sequence. Until recently, direct determination of nucleotide sequences in genomic DNA was an intimidating undertaking. The technology developed in connection with the Human Genome Project has accelerated the rate at which DNA sequences are determined. Many complete sequences for a number of species have already become available for analysis.

Restriction Endonucleases and Palindromes

One key event that enabled development of methods to sequence genomic DNA was the discovery of bacterial **restriction endonucleases**. These cleave double-stranded DNA at a specific sequence by cutting each strand (Figure 2.47). Bacteria developed restriction enzymes as a defense against infection by phages. Cleavage exposes the viral DNA to eventual degradation by nonspecific bacterial exonucleases. Bacterial DNA can be protected from cleavage by sequence-specific methylation. The recognition sites for restriction methylases correspond to those of the endonucleases. Methylation of specific bases within the recognition site prevents cleavage by the cognate nuclease. The most common sites for base methylation are the 5-position of cytosine and the 6-NH_2 group of adenine. Notably, base methylation is also critical in gene regulation in higher organisms.

Many hundreds of restriction endonucleases have been purified. With few exceptions, they recognize sequences four to six nucleotides long. Rare cutters, endonucleases with unusually long recognition sites, are valuable because of the relative infrequency of cleavage of very large DNAs such as eukaryotic chromosomes. Not I, for example, has an eight-nucleotide recognition sequence. The recognized sequences are often symmetrical inverted repeats or palindromes: The order of the bases is the same when the two complementary strands of the palindrome are read $5' \rightarrow 3'$. For EcoR1 from *E. coli*, the sequence is 5′-GAATTC-3′. Restriction endonucleases types I and III cut in

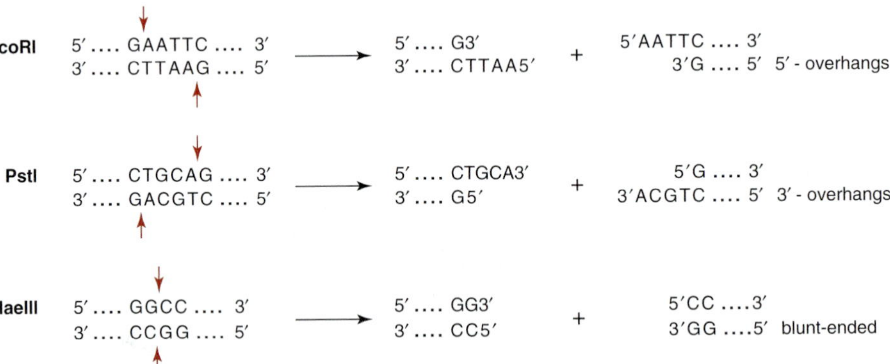

FIGURE 2.47

Types of products generated by type II restriction endonucleases. Enzymes exemplified by EcoRI and PstI nick on both sides of the center of symmetry of the palindrome, generating single-stranded overhangs. Many commonly used enzymes generate 5′-overhangs, although some produce ends with 3′-overhangs as shown for PstI. Other restriction nucleases cut across the center of symmetry of the recognition sequence, to produce flush or blunt ends, as exemplified by HaeIII.

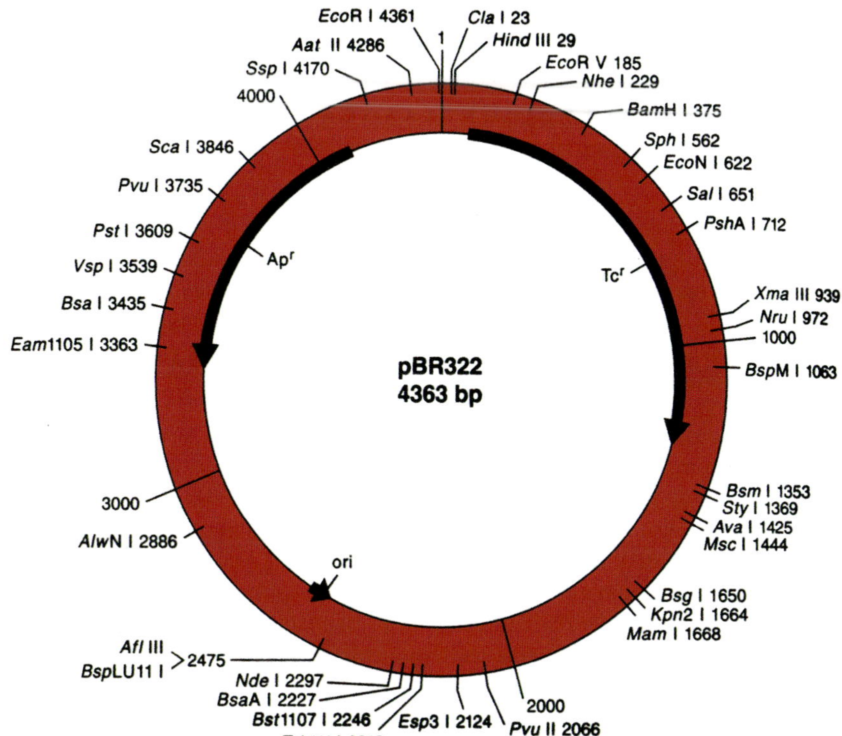

FIGURE 2.48

Restriction map for plasmid DNA. Bacterial plasmid pBR322 is a 4363-bp circular DNA containing an origin of replication (ori) and genes for resistance to the antibiotics ampicillin (Ap^r) and tetracycline (Tc^r). The locations of unique DNA sequences that are recognized by a number of restriction endonuclease are marked. Other endonucleases may cleave this plasmid many times or not at all. For example, the recognition sequence for Bgl I appears in three locations on this plasmid, whereas Bgl II does not cleave this plasmid at all.

the vicinity of the recognition site, while type II specifically cleaves DNA within the recognition sequence.

Restriction endonucleases recognize specific sequences that have relatively low frequencies and fragment DNA very selectively. For example, a typical bacterial DNA with 3×10^6 bp will be cleaved into a few hundred fragments. Plasmid DNA may have few or no cutting sites at all for a particular restriction endonuclease (Figure 2.48). Thus, a particular restriction enzyme generates a unique family of fragments, or restriction digest, for any given DNA molecule. Availability of restriction enzymes and development of gel electrophoresis techniques for separating DNA fragments have made determination of sequences a simple matter (see p. 249).

Most Prokaryotic DNA Codes for Specific Proteins

In prokaryotes a large percentage of the DNA codes for specific proteins. The entire *E. coli* genome consists of about 4.6×10^6 bp of DNA and contains ~4200 genes. Not all of the genes code for proteins. For example, 80 genes code for tRNA molecules, and some may not encode functional molecules at all. *E. coli* DNA is densely packed with sequence information, and there is little repetition of information within it. As much as 1% is composed of multiple copies of short repetitive sequences known as repeated extragenic palindromic elements (REP elements). These are present at sites of DNA interaction with functional proteins—for example, in the region of initiation of DNA replication (referred to as OriC). At OriC, REP elements have a consensus sequence of 34 nucleotides and bind topoisomerase II. REP elements with the sequence GCTGGTGG (Chi sites) bind the enzyme RecBCD, which initiates DNA recombination. Chi sites are regularly spaced at intervals separated by about 4000 bp. Genetic information is even more densely organized in smaller organisms, such as bacteriophages, where the primary sequence of DNA reveals that structural genes—nucleotide sequences coding for protein—do not always have exclusive physical locations. Rather, they frequently overlap with one another, as illustrated by the partial sequence of phage **φX174** (Figure 2.49). This overlap makes the efficient and economic utilization of the limited DNA present and may control the order of gene expression.

Approximately 1–15% of eukaryotic genomic DNA consists of sequences typically shorter than 20 nucleotides reiterated thousands or millions of times. Most highly reiterated sequences have a characteristic base composition, and they can be isolated by shearing the DNA into segments of a few hundred nucleotides long and separating the fragments by density gradient centrifugation. These fragments are termed **satellite DNA** because they appear as satellites of the band of bulk DNA after centrifugation. Other highly reiterated sequences, which cannot be isolated by centrifugation, can be identified by their property of rapid reannealing. These highly reiterated DNAs are also called simple-sequence DNA. Simple sequences are typically present in the DNA of most, if not all, eukaryotes. In some species, one major sequence is present, while in others, several simple sequences are repeated up to one million times. Simple sequence DNA can often be isolated as satellite DNA. That found in the centromeres of higher eukaryotes consists of thousands of tandem copies of one or a few short sequences. Satellite sequences are only 5–10 bp long and are a constituent of telomeres where they have a well-defined role in DNA replication. Some longer, simple-sequence DNA has been identified. For instance, in the genome of the African green monkey a 172-bp segment which contains some sequence repetitions is highly reiterated.

Inverted repeats are a structural motif of DNA. Short inverted repeats, consisting of up to six nucleotides long (e.g., the palindromic sequence GAATTC), occur by chance about once for every 3000 nucleotides. Such short repeats cannot form a stable cruciform structure as is formed by longer palindromic sequences. Inverted repeat.sequences that are long enough to form stable cruciforms are not likely to occur by chance, and they should be classified as a separate class of eukaryotic sequence. In human DNA about two million inverted repeats are present, with an average length of about 200 bp; however, inverted sequences longer than 1000 bp have been detected. Most inverted repeat sequences are repeated 1000 or more times per cell.

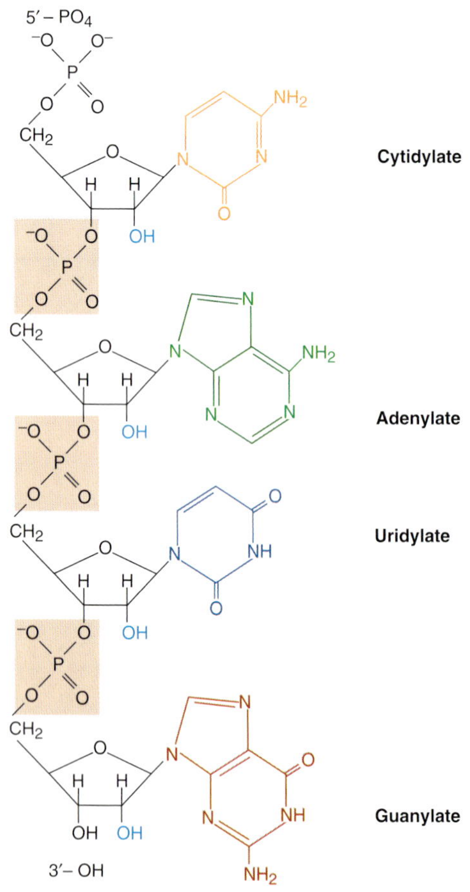

FIGURE 2.51

Structure of the 3′,5′-phosphodiester bonds between ribonucleotides forming a single strand of RNA. Phosphate joins the 3′-OH group of one ribose with the 5′-OH group of the next ribose. This linkage produces a polyribonucleotide having a sugar–phosphate "backbone." Purine and pyrimidine bases extend away from the axis of the backbone and may pair with complementary bases to form double-helical base-paired regions.

2.6 RNA STRUCTURE

RNA is a Polymer of Ribonucleoside 5′-Monophosphates

RNA is a linear polymer of ribonucleoside monophosphates. The purine bases in RNA are adenine and guanine; the pyrimidines are cytosine and uracil. Except for uracil, which replaces thymine, these are the same bases found in DNA. A, C, G, and U nucleotides are incorporated into RNA during transcription. Many RNAs also contain modified nucleotides that are produced by processing. Modified nucleotides are especially characteristic of stable RNA species (i.e., tRNA and rRNA); however, some methylated nucleotides are also present in eukaryotic mRNA. For the most part, modified nucleotides in RNA have "fine tuning" rather than indispensable roles in the cell.

The 3′,5′-phosphodiester bonds of RNA form a backbone from which the bases extend (Figure 2.51). Eukaryotic RNAs vary from approximately 20 nucleotides long to more than 200,000 nucleotides long. Each RNA is complementary to the base sequences of specific portions of only one strand of DNA. Thus, unlike the base composition of DNA, molar ratios of (A + U) and (G + C) in RNA are not equal. Cellular RNA is linear and single-stranded, but double-stranded RNA is present in some viral genomes.

Chemically, RNA is similar to DNA. Both contain negatively charged phosphodiester bonds, and the bases are very similar chemically. The chemical differences between DNA and RNA are largely due to two factors. First, RNA contains ribose rather than 2′-deoxyribose as the nucleotide sugar component, and, second, RNAs are generally single-stranded rather than double-stranded.

The 2′-hydroxyl group makes the phosphodiester bonds of an RNA molecule more susceptible to chemical hydrolysis, especially in alkaline solution, than those of DNA The chemical instability of RNA is reflected in its metabolic instability. Some RNAs, such as bacterial mRNA, are synthesized, used, and degraded within minutes. Others, such as human rRNA, are more stable metabolically, with a lifetime measured in days. Nevertheless, even the most stable RNAs are much less stable than DNA.

Secondary Structure of RNA Involves Intramolecular Base-Pairing

Since RNA molecules are single-stranded, they do not usually form extensive double helices. Rather, the secondary structure of an RNA molecule results from relatively short regions of intramolecular base-pairing. Even nonpaired sequences of single-stranded RNAs may contain considerable helical structure (Figure 2.52). Helices within RNA are usually the A-type with 11 nucleotides per turn in the double helix.

Double-helical stem-loop regions in RNA often form "hairpins." There are considerable variations in the fine structural details of "hairpin" structures, including the length of base-paired regions and the size and number of unpaired loops (Figure 2.53). Transfer RNAs are excellent examples of base stacking and hydrogen bonding in a single-chain RNA molecule (Figure 2.54). About 60% of bases are paired in four double-helical stems and two loops interact with each other, resulting in an L-shaped molecule. The anticodon region in tRNA is an unpaired, base-stacked loop of seven nucleotides. The partial helix caused by base stacking in this loop binds, by base-pairing, to a complementary codon in mRNA so that translation can occur.

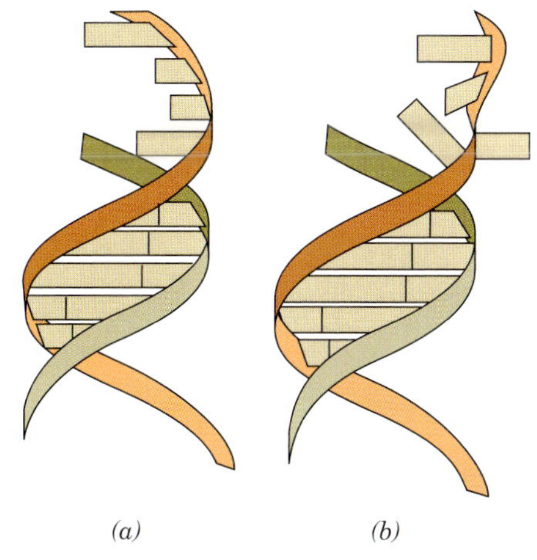

FIGURE 2.52

Helical structure of RNA. Models indicating a helical structure due to (*a*) base stacking in the CCA terminus of tRNA and (*b*) lack of an ordered helix when no stacking occurs in this non-base-paired region.
Redrawn from Sprinzl, M. and Cramer, F. *Prog. Nucl. Res. Mol. Biol.* 22:9, 1979.

RNA Molecules have Tertiary Structures

Structures of functional RNA molecules are more complex than is suggested by the base-stacked and hydrogen-bonded helices mentioned above. Particularly, the bases in unpaired regions can form hydrogen bonds or otherwise interact with each other and with helical portions of the molecule. *In vivo* RNAs are dynamic molecules that can change their conformations during synthesis, processing, and functioning. Proteins associated with RNA molecules often lend stability to the RNA structure. In fact, cellular RNA functions as RNA–protein complexes, rather than as free RNA molecules.

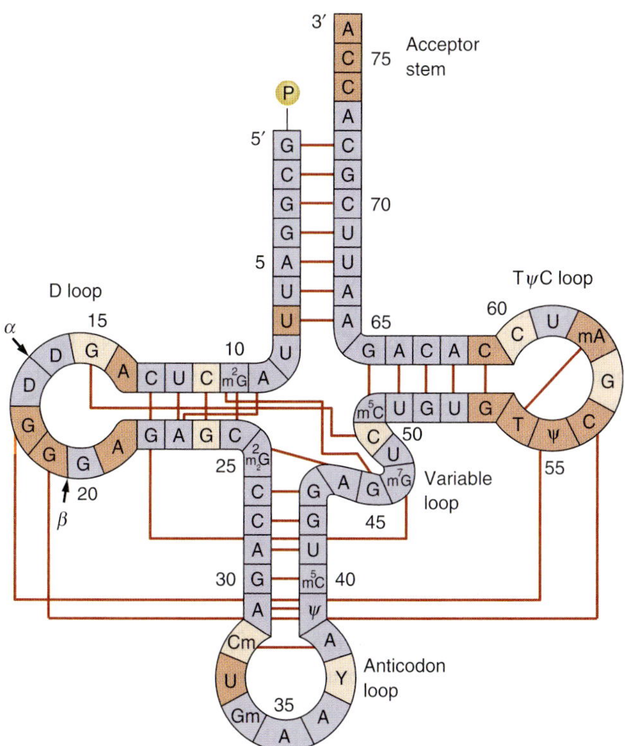

FIGURE 2.53

Cloverleaf structure of tRNA. Cloverleaf diagram of the two-dimensional structure and nucleotide sequence of yeast $tRNA^{Phe}$. Red lines connecting circled nucleotides indicate hydrogen-bonded bases. Rose squares indicate constant nucleotides; tan squares indicate a constant purine or pyrimidine. Insertion of nucleotides in the D loop occurs at positions α and β for different tRNAs.
Redrawn with permission from Quigley, G. J. and Rich, A. *Science* 194:797, 1976. Copyright (1976) AAAS.

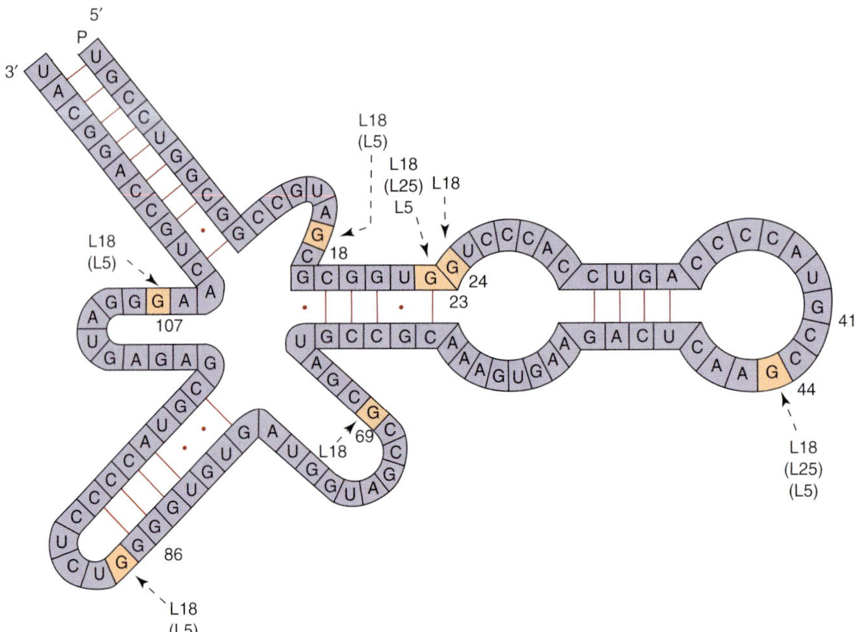

FIGURE 2.56

Secondary, base-paired structure proposed for 5S rRNA. Arrows indicate regions protected by proteins in the large ribosomal subunit.
Combined information from Fox, G. E. and Woese, C. R. *Nature (London)* 256:505, 1975, and Gray, P. N.et al. J. Mol. 73:133, 1973.

of tRNAs. tRNAs that accept the same amino acid are called isoacceptors. A tRNA that accepts phenylalanine would be written as $tRNA^{Phe}$, whereas one accepting tyrosine would be written $tRNA^{Tyr}$.

tRNAs are 65–110 nucleotides long, corresponding to a molecular weight range of 22,000–37,000. The sequences of all tRNA molecules (thousands are known) can be arranged into a common cloverleaf secondary structure by complementary Watson–Crick base-pairing to form three stem and loop structures. The anticodon triplet sequence is at one “leaf” of the cloverleaf while the CCA acceptor stem is at the “stem.” This arrangement is preserved in the tertiary structure of $tRNA^{Phe}$ (see Figure 2.54), where the anticodon and acceptor stems are at each end of the L-shaped molecule. Additional non-Watson–Crick hydrogen bonds form the tertiary structure of the L-shaped molecule.

tRNAs contain modified nucleotides—for example, 7-methyguanine at position 46 in Figure 2.53. Modified nucleotides affect tRNA structure and stability but are not required for their basic functioning. For example, a modified base next to the anticodon makes codon recognition more efficient, but a tRNA without this modification can still be read correctly by the ribosome. Many structural features are common to all tRNA cloverleafs. Seven base pairs form the amino acid acceptor stem, which terminates with the nucleotide triplet CCA. This CCA triplet is not base-paired. The dihydrouracil or “D” stem has three or four base pairs, while the anticodon and T stems have five base pairs each. Both the anticodon loop and T-loop contain seven nucleotides. Differences in the number of nucleotides in different tRNAs are accounted for by the variable loop. Most tRNAs have small variable loops of 4–5 nucleotides, while others have larger loops of 13–21 nucleotides. The positions of some nucleotides are constant in all tRNAs (see the rose boxes in Figure 2.53).

Ribosomal RNA is Part of the Protein Synthesis Apparatus

Protein synthesis takes place on ribosomes. In eukaryotes, these complex assemblies are composed of four RNA molecules, representing about two-thirds of the particle mass, and 82 proteins. The smaller subunit, the 40S particle, contains one 18S RNA and 33 proteins. The larger subunit, the 60S particle, contains one copy each of three RNAs, 28S, 5.8S, and 5S rRNAs along with 49 proteins. The total assembly forms the 80S ribosome. Prokaryotic ribosomes are smaller; the 30S subunit contains a single 16S

CLINICAL CORRELATION **2.8**

Staphylococcal Resistance to Erythromycin

Bacteria exposed to antibiotics in clinical or agricultural settings often develop resistance to the drugs. This resistance can arise from a mutation in the target cell's DNA, which gives rise to resistant descendants. An alternative and clinically more serious mode of resistance arises when plasmids coding for antibiotic resistance proliferate through the bacterial population. These plasmids may carry multiple resistance determinants and render several antibiotics useless at the same time.

Erythromycin inhibits protein synthesis by binding to the large ribosomal subunit. *Staphylococcus aureus* can become resistant to erythromycin and similar antibiotics as a result of a plasmid-borne RNA methylase that converts a single adenosine in 23S rRNA to N^6-dimethyladenosine. Since the same ribosomal site binds lincomycin and clindamycin, the plasmid causes cross-resistance to these antibiotics as well. Synthesis of the methylase is induced by erythromycin.

A microorganism that produces an antibiotic must also be immune to it or else it would be killed by its own toxic product. The producer of erythromycin, *Streptomyces erythraeus*, itself possesses an rRNA methylase that acts at the same ribosomal site as the one from *S. aureus.* Which came first? It is likely that many of the resistance genes in target organisms evolved from those of producer organisms. In several cases, DNA sequences from resistance genes of the same specificity have been conserved in producer and target organisms. We may therefore look on plasmid-borne antibiotic resistance as a case of "natural genetic engineering" whereby DNA from one organism (e.g., *Streptomyces* producer) is appropriated and expressed in another (e.g., the *Staphylococcus* target).

Source: Cundliffe, E. How antibiotic-producing microorganisms avoid suicide. *Annu. Rev. Microbiol.* 43:207, 1989.

rRNA and 21 proteins while the larger 50S subunit contains one copy each of 5S and 23S rRNAs and 34 proteins (see p. 209 for a discussion of the structure of ribosomes).

rRNA accounts for 80% of cellular RNA and is metabolically stable. This stability is required for repeated functioning of the ribosome and is enhanced by association with ribosomal proteins. Eukaryotic 28S (4718 nucleotides), 18S (1874 nucleotides), and 5.8S (160 nucleotides) rRNAs are synthesized in the nucleolus. The 5S rRNA (120 nucleotides) is not transcribed in the nucleolus but rather from separate genes within the nucleoplasm (Figure 2.56). The three larger rRNAs are synthesized as part of one long polynucleotide chain which is then processed to a yield the individual molecules.

Nucleotide modifications in rRNA are primarily methylations to form 2′-*O*-methylribose. Methylation of rRNA has been directly related to bacterial antibiotic resistance in a pathogenic species (Clin. Corr. 2.8). A small number of N^6-dimethyladenines are present in 18S rRNA. The 28S rRNA has about 45 methyl groups, and the 18S rRNA has 30 methyl groups.

Ribosomal RNA is the catalytic component of the ribosome. The X-ray crystal structure of the large subunit of a bacterial ribosome species shows that the active site for peptide bond synthesis is composed of 23S RNA only: Any potential catalytic amino acid side chains are too far away to participate directly in the reaction. The chemical reactions necessary for peptide bond synthesis are catalyzed by the large subunit rRNA (see p. 69 for a discussion of catalytic RNAs).

Messenger RNAs Carry the Information for the Primary Structure of Proteins

mRNAs are the direct carriers of genetic information from genome to ribosomes. Each eukaryotic mRNA is **monocistronic**, that is contains information for only one polypeptide chain. In prokaryotes, mRNA species are often **polycistronic** encoding more than one protein. A cell's phenotype and functional state are related directly to its mRNA content.

In the cytoplasm, mRNAs have relatively short life spans. Some are synthesized and stored in an inactive or dormant state in the cytoplasm, ready for a quick response when protein synthesis is required. In various animals, immediately upon fertilization the egg undergoes rapid protein synthesis in the absence of transcription, indicating that preformed mRNA and ribosomes are present in its cytoplasm.

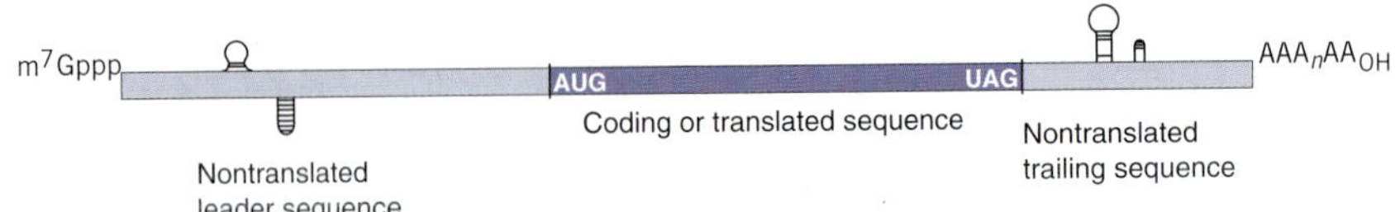

FIGURE 2.57

General structure for a eukaryotic mRNA. There is a "blocked" 5′-terminus, cap, followed by a nontranslated leader containing a promoter sequence. Coding region usually begins with the initiator codon AUG and continues to the translation termination sequence UAG, UAA, or UGA. This is followed by the nontranslated trailer and a poly(A) tail on the 3′-end.

Eukaryotic mRNAs have unique structural features (Figure 2.57). Since the information within mRNA lies in the linear sequence of the nucleotides, the integrity of this sequence is extremely important. Any loss or change of nucleotides could alter the structure of the protein being translated. Translation of mRNA on the ribosomes must also begin and end at specific sequences. In eukaryotes, a "cap" structure at the 5′-end of an mRNA specifies where translation should begin. The caps of eukaryotic mRNAs contain an inverted phosphodiester bond, a methylated base attached to the mRNA via a 5′-phosphate-5′-phosphate rather than the usual 3′, 5′-phosphodiester linkage. The cap is attached to the first transcribed nucleotide, usually a 2′*O*-methylated purine (Figure 2.58). The cap is followed by a nontranslated or "leader" sequence. Translation starts at the initiation codon, most often AUG, and continues through the coding sequence. At the end of the coding sequence a termination sequence signals termination of polypeptide formation and release from the ribosome. A second nontranslated or "trailer" sequence follows, terminated by a string of 20–200 adenine nucleotides, called a poly(A) tail, which makes up the 3′-terminus of the mRNA.

The 5′-cap has a positive effect on initiation of message translation. During the initiation of translation of a mRNA, the cap structure is recognized by a single ribosomal protein, an initiation factor (see p. 213). The poly(A) sequence affects the stability of the mRNA; degradation of an mRNA often begins with shortening of its poly(A) tail.

Mitochondria Contain Unique RNA Species

Mitochondria (mt) have their own protein synthesizing apparatus, including ribosomes, tRNAs, and mRNAs. Mitochondrial rRNAs, 12S and 16S, are transcribed from the mitochondrial DNA (**mtDNA**) as are 22 specific tRNAs and 13 mRNAs, most of which encode proteins of the electron transport chain and ATP synthase. Note that there are fewer mitochondrial tRNAs than prokaryotic or cytosolic tRNA species; there is generally only one mitochondrial tRNA species per amino acid. The mtRNAs account for 4% of the total cellular RNA. They are transcribed by a mitochondrial-specific RNA polymerase and are processed from a pair of RNA precursors that contain tRNA

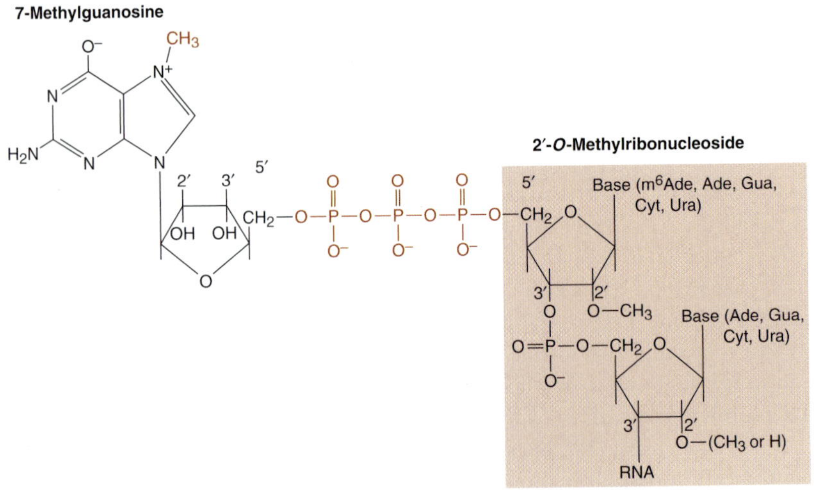

FIGURE 2.58

Diagram of "cap" structure or blocked 5′-terminus in mRNA. The 7-methylguanosine is inverted to form a 5′-phosphate to 5′-phosphate linkage with the first nucleotide of the mRNA. This nucleotide is often a methylated purine.

and mRNA sequences. Expression of the nuclear and mitochondrial genomes is tightly coordinated. Most of the amino acylating enzymes for the mitochondrial tRNAs and all of the mitochondrial ribosomal proteins are specified by nuclear genes, translated in the cytosol and transported into mitochondria. The modified bases in mitochondrial tRNA species are synthesized by enzymes encoded in nuclear DNA.

RNA in Ribonucleoprotein Particles

Other small, stable RNA species can be found in the nucleus, cytosol, and mitochondria. These small RNA species function as ribonucleoprotein particles (RNPs), with one or more protein subunits attached. Different RNP species have been implicated in RNA processing, RNA transport, and control of translation, as well as in the recognition of proteins due to be exported. The actual roles of these species, where known, are described more fully in the discussion of specific metabolic events.

Catalytic RNA: Ribozymes

RNA can be an enzyme. In several cases, the RNA component of a ribonucleoprotein particle is the catalytically active subunit of the enzyme. In other cases, catalytic reactions can be carried out *in vitro* by RNA in the absence of any protein. Enzymes whose RNA subunits carry out catalytic reactions are called ribozymes. There are five classes of ribozyme. Three of these RNA species carry out self-processing reactions while the others, ribonuclease P (RNase P) and rRNA, are true catalysts that act on separate substrates.

In the ciliated protozoan, *Tetrahymena thermophila*, an intron in the rRNA precursor is removed by a multistep reaction (Figure 2.59). A guanine nucleoside or nucleotide reacts with the intron–exon phosphodiester linkage to displace the donor exon from the intron. This reaction, a transesterification, is promoted by the folded intron itself. The free donor exon then similarly attacks the exon–intron phosphodiester bond at the acceptor end of the intron. Introns of this type (Group I introns) have been found in a variety of genes in fungal mitochondria, in bacteria, and in the bacteriophage T4. Although these introns are not true enzymes *in vivo* because they only work for one reaction cycle, they can be made to carry out catalytic reactions under specialized conditions.

Group II self-splicing introns are present in the mitochondrial RNA precursors of yeasts and other fungi. Even though these RNAs self-splice, there are many parallels between this reaction and the removal of introns from mRNA precursors during processing (see p. 192).

Group III of self-cleaving RNAs is found in the genomic RNAs of several viruses. These RNAs self-cleave during generation of single genomic RNA molecules from larger precursors. The three-dimensional structure of the hammerhead ribozyme, a member of this third class, has been determined (Figure 2.60). Although the catalytic cycle is not completely determined, the amino group of a cytosine base plays an essential role in the cleavage reaction. The phosphate of the cleaved bond is left at the 3′-hydroxyl position of the RNA product. A self-cleaving RNA is found in a small satellite virus, hepatitis delta virus, that is implicated in severe human infectious hepatitis. All of the above self-processing RNAs can be made to act as true catalysts (i.e., exhibiting multiple turnover) *in vitro* and *in vivo*.

The fourth type of ribozyme, Ribonuclease P, contains both a protein and an RNA component. It acts as a true enzyme in cells, cleaving tRNA precursors to generate the mature 5′-end of tRNA molecules. RNase P recognizes constant structures associated with tRNA precursors (e.g., the acceptor stem and CCA sequence) rather than using extensive base-pairing to bind the substrate RNA to the ribozyme. The product of cleavage contains a 5′-phosphate in contrast to the products of hammerhead and similar RNAs. In all of these events, the structure of the catalytic RNA is essential for catalysis.

Finally, the X-ray crystallographic structure of a bacterial ribosome reveals that it, too, is a ribozyme. The active site of the ribosome is on the 50S subunit. There are no protein functional groups close enough to catalyze peptide bond formation. Instead, residues within 23S RNA helps transfer a hydrogen ion during peptide bond synthesis.

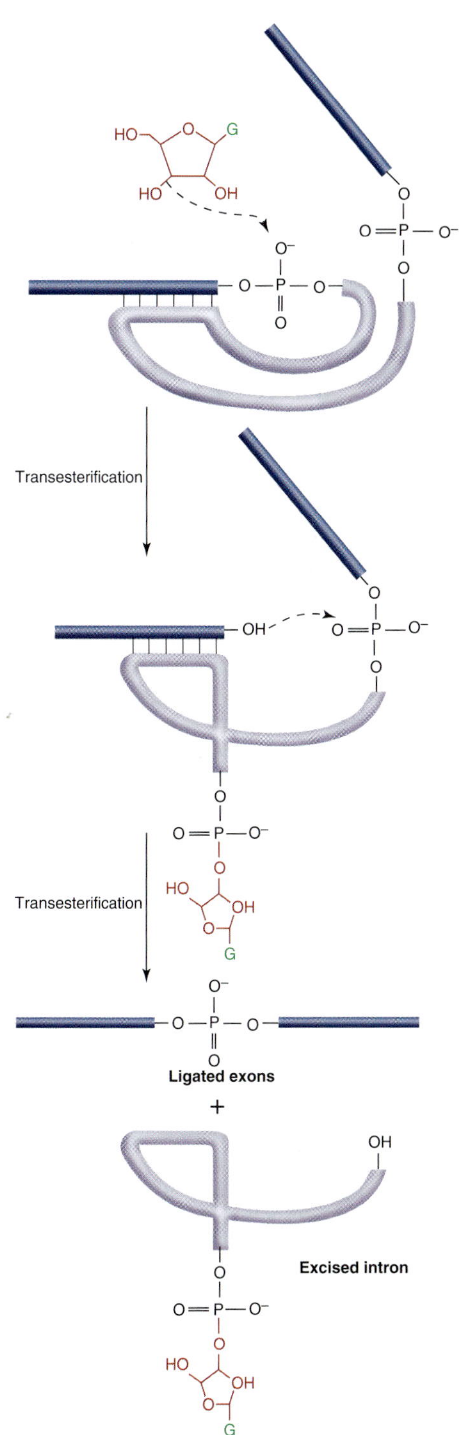

FIGURE 2.59

Mechanism of self-splicing of the rRNA precursor of *Tetrahymena*. Two exons of rRNA are denoted in dark blue. Catalytic functions reside in the intron (purple). This splicing function requires an added guanine nucleoside or nucleotide.

Reproduced from Cech, T. R. *JAMA* 260:308, 1988.

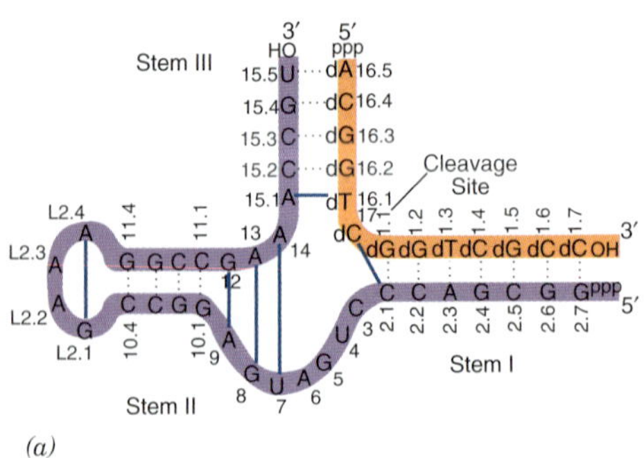

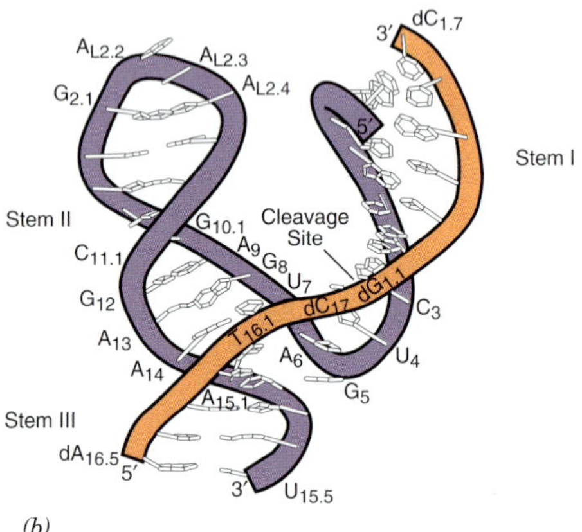

FIGURE 2.60

"Hammerhead" structure of viral RNA.
(*a*) "Hammerhead" structure of a self-cleaving viral RNA. This artificial molecule is formed by the base-pairing of two separate RNAs. Cleavage of the RNA sequence at the site indicated by the arrow in the top strand requires its base-pairing with the sequence at the bottom of the molecule. The boxed nucleotides are a consensus sequence found in self-cleaving RNA viral RNAs.
(*b*) Three-dimensional folding of the hammerhead catalytic RNA. Helices II and III stack to form an apparently continuous helix while non-Watson–Crick interactions position the noncomplementary bases in the hammerhead into a "uridine turn" structure identical to that found in tRNA.
Part (*a*) redrawn from Sampson, J. R., Sullivan, F. X., Behlen, L. S., DiRenzo, A. B., and Uhlenbeck, O. C. *Cold Spring Harbor Symp. Quant. Biol.* 52:267, 1987. Part (*b*) redrawn from Pley, H. W., Flaherty, K. M., and McKay, D. B. *Nature* 372:68, 1994.

Discovery of RNA catalysis has greatly altered our concepts of biochemical evolution and the range of allowable cellular chemistry. We now recognize that RNA can serve as both a catalyst and a carrier of genetic information. This has raised the possibility that the earliest living organisms were based entirely on RNA and that DNA and proteins evolved later. This model is sometimes called the "RNA world." Second, we know that many viruses, including human pathogens, use RNA genetic information; some of these RNAs have been shown to be catalytic. Thus, catalytic RNA presents opportunities for the discovery of RNA-based pharmaceuticals. Third, many of the information processing events in mRNA splicing require RNA components. These RNAs may also be acting as catalysts. Finally, ribozymes can be made to specifically recognize and cleave various cellular mRNAs. When introduced into cultured cells, they can prevent gene expression. This is only a research tool at present, but it has been a useful strategy for determining the function of genes whose metabolic activity can only be guessed at from genomic sequence information.

RNAs can Bind Other Molecules

Consideration of the RNA world has led to a new type of "combinatorial biochemistry" based on the large number of potential sequences (4^N) that could be made if A,C,G, or U were inserted randomly in each of N positions in a nucleic acid. A set of chemically synthesized, randomized, nucleic acid molecules 25 nucleotides long would contain $4^{25} = 10^{15}$ potential members. Individual molecules within this large collection of RNAs would be expected to fold into a similarly large collection of shapes. The large number of molecular shapes implies that some member of this collection should be capable of strong, specific binding to any ligand, much as Group I introns bind guanine nucleotides specifically. These binding RNAs are termed **aptamers.** Though a single molecule would be too rare to study within the original population, the aptamer RNA can be selected and preferentially replicated *in vitro*. For example, an RNA capable of distinguishing theophylline from caffeine was selected from a complex population (Figure 2.61). Theophylline is used in the treatment of chronic asthma, but the level must be carefully controlled to avoid side effects. Monitoring of theophylline concentrations by conventional antibody-based assay is difficult because caffeine and theophylline differ only by a single methyl group. Therefore, anti-theophylline antibodies show considerable cross-reaction with caffeine. RNA aptamers have been found that bind theophylline 10,000-fold more tightly than caffeine. Other extensions of the technology have used selection procedures to identify new, synthetic ribozymes and potential therapeutic RNAs.

RNAs Control Translation

Translation of an mRNA requires that it be bound to and move relative to the ribosome. Controlling the efficiency of these processes allows for translational control. One way to do this is through aptamer sites contained in mRNA. While most aptamers have been found by selection, a few bacterial mRNAs contain naturally occurring aptamer sites. These sites in the mRNA bind ligands such as amino acids or vitamins. Binding of the ligand prevents translation of the mRNA, thereby controlling gene expression. This is a form of regulation in **cis** because control is exerted on the same molecule that contains the aptamer site.

Double-stranded RNA is a negative regulator of eukaryotic gene expression in **trans**—that is, on another molecule. For example, double-stranded RNA is a necessary

Caffeine

Theophylline

FIGURE 2.61

Structures of theophylline and caffeine. Although these compounds differ only by a single methyl group, a specific synthetic RNA can bind to theophylline 10,000-fold more tightly than to caffeine.

intermediate in the replication of some RNA viruses. As a means of defense, the host processes the double-stranded RNA to produce short single-stranded RNAs about 23 nucleotides long. These RNAs, termed **microRNAs (miRNAs)**, base-pair to mRNA and either stop its translation or trigger its degradation (see p. 196).

BIBLIOGRAPHY

DNA

Bates, A. D. *DNA Topology*. Oxford, England: IRL Press, 1993.

Bell, J. I. The double helix in clinical practice. *Nature* 421:414, 2003.

Blackburn, G. M. and Gait, M. J. *Nucleic Acids in Chemistry and Biology*, 2nd ed. NY: Oxford University, 1996.

Cech, T. R. Life at the end of the chromosome: Telomeres and telomerase. *Angew. Chem. Int. Ed.* 39:34, 2000.

Champoux, J. J. DNA topoisomerases: Structure, function, and mechanism. *Annu. Rev. Biochem.* 70:369, 2001.

Crothers, D. M. and Shakked, Z. DNA bending by adenine–thymine tracts. In: S. Neidle (Ed.), *Oxford Handbook of Nucleic Acid Structure*. New York: Oxford University Press, 1999, p. 455.

De Bruijn, F. J., Lupski, J. R., and Weinstock, G. M. (Eds.). *Bacterial Genomes: Physical Structure and Analysis*. New York: Chapman & Hall, 1998.

Dickerson, R. E. DNA structure from A to Z. *Meth. Enzymol.* 211:67, 1992.

Felsenfeld, G. and Goudine, M. Controlling the double helix. *Nature* 421:448, 2003.

Hecht, S. M. (Ed.). *Bioorganic Chemistry: Nucleic Acids*. New York: Oxford University Press, 1996.

Hood, L. and Galas, D. The digital code of DNA. *Nature* 421:444, 2003.

Horn, P. J. and Peterson, C. L. Chromatin higher order folding: Wrapping up transcription. *Science* 297:1824, 2002.

Hud, N. V. and Plavec, J. A unified model for the origin of DNA sequence-directed curvature. *Biopolymers* 69:144, 2003.

Kornberg, R. D. and Lorch, Y. Twenty-five years of the nucleosome, fundamental particle of the eukaryote chromosome. *Cell* 98:285, 1999.

Leach, D. R. Long DNA palindromes, cruciform structures, genetic instability and secondary structure repair. *Bioessays* 16:893, 1994.

Mighell, A. J., Markham, A. F., and Robinson, P. A. Alu sequences. *FEBS Lett.* 417:1, 1997.

Rich, A. Speculation on the biological roles of left-handed Z-DNA. In: S. S. Wallace, B. van Houten, and Y. W. Kow (Eds.), *DNA Damage: Effects on DNA Structure and Protein Recognition*. New York: New York Academy of Sciences, 1994, p. 1.

Saenger, W. *Principles of Nucleic Acid Structure*. New York: Springer-Verlag, 1984.

Tenover, F. C. *DNA Probes for Infectious Diseases*. Boca Raton, FL: CRC, 1989.

Van Holde, K. and Zlatanova, J. Unusual DNA structures, chromatin and transcription. *Bioessays* 16:59, 1994.

Watson, J. D. and Crick, F. H. C. Molecular structure of nucleic acids: A structure for deoxyribose nucleic acid. *Nature* 171:737, 1953.

Wells, R. D. Unusual DNA structures. *J. Biol. Chem.* 263:1095, 1988.

Williamson, J. R. G-quartet structures in telomeric DNA. *Annu. Rev. Biophys. Biomol. Struct.* 23:703, 1994.

Yakubovskaya, E. A. and Gabibov, A. G. Topoisomerases: Mechanisms of DNA topological alterations. *Mol. Biol.* 33:318, 1999.

RNA

Cech, T. R. The ribosome is a ribozyme. *Science* 289:878, 2000.

Draper, D. E. A guide to ions and RNA structure. *RNA* 10:335, 2004.

Gesteland, R. F., Cech, T., and Atkins, J. F. (Eds.). *The RNA World*, 2nd ed. Cold Spring Harbor, NY: Cold Spring Harbor Laboratory Press, 1999.

Gold, L., Polisky, B., Uhlenbeck, O., and Yarus, M. Diversity of oligonucleotide functions. *Annu. Rev. Biochem.* 64:763, 1995.

Ibba, M., Becker, H. D., Stathopoulos, C., Tumbula, D. L., and Söll, D. The adaptor hypothesis revisited. *Trends Biochem. Sci.* 25:311, 2000.

Leontis, N. B. and Westhof, E. Analysis of RNA motifs. *Curr. Opin. in Struct. Biol.* 13:300, 2003.

McKnight, K. L. and Heinz, B. A. RNA as a target for developing antivirals. *Antiviral Chem. Chemotherapy.* 14:61, 2003.

Pace, N. R. and Brown, J. W. Evolutionary perspective on the structure and function of ribonuclease P, a ribozyme. *J. Bacteriol.* 177:1919, 1995.

Puerta-Fernandez, E., Romero-Lopez, C., Barroso-delJesus, A., and Berzal-Herranz, A. Ribozymes: Recent advances in the development of RNA tools. *FEMS Microbiol. Rev.* 27:75, 2003.

Westhof, E. and Fritsch, V. RNA folding: Beyond Watson–Crick pairs. *Structure* 8:R55, 2000.

QUESTIONS | CAROL N. ANGSTADT

Multiple Choice Questions

1. The best definition of an endonuclease is an enzyme that hydrolyzes:
   A. a nucleotide from only the 3′-end of an oligonucleotide.
   B. a nucleotide from either terminal of an oligonucleotide.
   C. a phosphodiester bond located in the interior of a polynucleotide.
   D. a bond only in a specific sequence of nucleotides.
   E. a bond that is distal to the base that occupies the 5′-position of the bond.

2. A palindrome is a sequence of nucleotides in DNA that:
   A. is highly reiterated.
   B. is part of the introns of eukaryotic genes.
   C. is a structural gene.
   D. has local symmetry and may serve as a recognition site for various proteins.
   E. has the information necessary to confer antibiotic resistance in bacteria.

3. The Z DNA helix:
   A. has fewer base pairs per turn than the B DNA.
   B. is favored by an alternating GC sequence.
   C. tends to be found at the 3′-end of genes.
   D. is inhibited by methylation of the bases.
   E. is a permanent conformation of DNA.

4. A nucleosome:
   A. is a regularly repeating structure of DNA and histone proteins.
   B. has a core of DNA with proteins wrapped around the outside.
   C. uses only one type of histone per nucleosome.
   D. is separated from a second nucleosome by nonhistone proteins.
   E. has histones in contact with the major groove of the DNA.

5. RNA:
   A. incorporates both modified and unmodified bases during transcription.
   B. does not exhibit any double-helical structure.
   C. structures exhibit base-stacking and hydrogen-bonded base-pairing.
   D. usually contains about 65–100 nucleotides.
   E. does not exhibit Watson–Crick base-pairing.

6. Ribozymes:
   A. are any ribonucleoprotein particles.
   B. are enzymes whose catalytic function resides in RNA subunits.
   C. carry out self-processing reactions but cannot be considered true catalysts.
   D. require a protein cofactor to form a peptide bond.
   E. function only in the processing of mRNA.

Questions 7 and 8: Nearly every process in which DNA participates requires that the DNA interact with proteins. Interaction with proteins depends on the local three-dimensional structure of DNA. One antitumor drug, cisplatin, is a tetracoordinate platinum complex. It forms intrastrand cross-links with DNA, causing the double helix to strongly bend toward the major groove. Cellular processes such as transcription and programmed cell death are affected.

7. Bent DNA:
   A. occurs only in the presence of external agents like the antitumor drugs.
   B. may be a fundamental element in the interaction between DNA sequences and proteins.
   C. occurs primarily in the presence of triple-stranded DNA.
   D. requires the presence of inverted repeats.
   E. occurs only in DNA that is in the Z-form.

8. Another unusual form of DNA is a triple-stranded complex. Triple-stranded DNA:
   A. generally occurs in DNA in regions that play no role in transcription.
   B. may involve Hoogsteen hydrogen bonding.
   C. is characterized by the presence of a string of alternating purine–pyrimidine bases.
   D. forms only intermolecularly.
   E. assumes a cruciform conformation.

Questions 9 and 10: Telomeres are guanine-rich repetitive sequences at the ends of linear eukaryotic chromosomes. They are progressively shortened during each cycle of cell division until a critical length is reached and apoptosis occurs. The ribonucleoprotein, telomerase, maintains the length of telomeres for a finite number of cell divisions and then it is turned off, but some tumor cells maintain telomerase activity. The higher the tumor telomerase activity, the poorer the clinical prognosis is. A current chemotherapeutic approach being studied is selective inhibition of telomerase. Certain drugs selectively bind to G-quadruplex DNA structure in telomerase and have been shown to inhibit telomerase activity. These drugs are now being studied for antitumor activity.

9. G-quadruplex DNA:
   A. is a stacked structure of guanine tetramers.
   B. is stabilized by metal cations.
   C. involves bases held together by Hoogsteen hydrogen bonds.
   D. can form either parallel or antiparallel arrangements.
   E. all of the above are correct.

10. Another approach to cancer treatment is drugs that inhibit topoisomerases. Topoisomerases:
    A. regulate the level of superhelicity of DNA in cells.
    B. always break only one strand of DNA.
    C. can create but not remove supercoils.
    D. must hydrolyze ATP for their action.
    E. of the subclass gyrases, introduce negative superhelices in eukaryotic DNA.

Questions 11 and 12: Eukaryotic DNA contains many sequences that can be repeated a few times, for certain coding genes, to millions of times for some relatively short sequences. Abnormal expansion of reiterated 3-bp DNA sequences can lead to a number of diseases. Fragile X syndrome, which causes mental retardation, is characterized by expansion of a GCC triplet near the FMR-1 gene, from a normal 30 copies to thousands of copies. Diseases of this type generally increase in severity with each successive generation. One possible mechanism of expansion involves slippage of DNA during synthesis of the lagging strand.

11. Slipped, mispaired DNA occurs when the DNA region has:
    A. direct repeats.
    B. homopurine–homopyrimidine sequences.
    C. inverted repeats.
    D. mirror repeats.
    E. palindromes.

12. Normally, certain kinds of reiterated sequences occur in a chromosome as an interspersion pattern that is:
    A. highly repetitive DNA sequences.
    B. the portion of DNA composed of single copy DNA.
    C. Alu sequences.
    D. alternating blocks of single copy DNA and moderately repetitive DNA.
    E. alternating blocks of short interspersed repeats and long interspersed repeats.

Problem

13. Which conformation of DNA—totally double helix, minimally unwound, or largely unwound—would have the highest relative absorbance at 260 nm? Would a molecule of DNA having a higher content of guanine and cytosine than of adenine and thymine have a higher or lower T_m than one with the reverse composition?

ANSWERS

1. **C** Both A and B describe exonucleases. D refers specifically to a restriction endonuclease and is not a definition of the general type. E: Both endo- and exonucleases show specificity toward the bond hydrolyzed.
2. **D** A palindrome reads the same forward and backward. Short palindromic segments of DNA are recognized by a variety of proteins. A: This is not likely since it would be incompatible with specific recognition. B: This is possible but has not been shown. C: Genes are thousands of base pairs in length, whereas palindromes are short segments. E: This would not be likely because palindromes are too short.
3. **B** The alternating purine–pyrimidine sequence is important. A: Z DNA is longer and thinner than the B-form because it has 12 bp per turn instead of 10. C: It is more likely to be found at the 5′-end, consistent with one of its proposed roles in transcriptional regulation. D: Methylation favors the Z-form in which the methyl is protected from water. E: B to Z transition is influenced by such things as methylation.
4. **A** The "beads-on-a-string" structure is called a polynucleosome. B: Histones form the core with DNA on the outside. C: All five types of histones are present—four in the core and one outside. D: The linker regions are DNA. E: Proteins that regulate gene expression and other activities bind to the major groove. Histones bind to the minor groove.
5. **C** Stacking stabilizes the single-strand helix. Folded portions of the structure have hydrogen-bonded base-pairing. A: Only the four bases A, G, U, and C are incorporated during transcription. B: Although single-stranded, RNA exhibits considerable secondary and tertiary structure. D: Only tRNA would be this small. E: This occurs in the intrachain helical regions.
6. **B** Ribozymes are a very specific type of particle. A: See previous comment. C: One of the four classes, RNase P, catalyzes a cleavage reaction. D: X-ray crystallography has shown that there is no amino acid chain sufficiently close to catalyze peptide bond formation. E: Ribozymes have been implicated in the processing of ribosomal and tRNAs.
7. **B** The bent DNA may be a recognition site for specific proteins to bind. A and D: Bent DNA occurs naturally either in runs of 4–6 adenosines separated by spacers or is induced by interactions of DNA with certain proteins. C: Bent DNA is double-stranded. E: Z DNA does not have a major groove, and the bending is toward the major groove.
8. **B** Hoogsteen bonding in TAT and GGC triplets are responsible for holding the third strand in the major groove. A: They are found frequently in regions involved in gene regulation. C: The required sequence is a homopurine string. D: They can also form intramolecularly by unfolding and refolding of the DNA. E: A cruciform is an alternate conformation of DNA but does not involve a third strand.
9. **E** A, C: The guanines are held together by Hoogsteen bonds and are called G-quartets. B: Sodium and potassium are most effective because they fit into the center of the quartet and bind to guanine oxygens. D: Parallel structures occur when four separate polynucleotides interact and antiparallel structures are favored when one polynucleotide provides two or more strands.
10. **A** Topoisomerase I relaxes DNA and topoisomerase II generates supercoils. B: Topoisomerase I breaks one strand but topoisomerase II nicks both strands. C: Topoisomerase I removes supercoils. D: Topoisomerase II uses ATP but topoisomerase I does not. E: Gyrases are found only in bacteria.
11. **A** When DNA unwinds and realigns, one copy of a direct repeat on one strand pairs with an adjacent copy on the other strand. B, C, D, E: These are other types of structures important in DNA.
12. **D** A, B: These are two of several kinds of DNA but do not constitute patterns. C, E: Alu is a type of short interspersed repeat. Short and long interspersed repeats are the two classes of moderately repetitive DNA.
13. Unstacked purine and pyrimidine bases have a higher relative optical density than stacked bases, so the most unwound form of DNA would have the highest value. T_m is the midpoint of the transition between double-stranded and separated DNA. Higher GC content means that the DNA is more stable, so its T_m would be higher.

3

PROTEINS I: COMPOSITION AND STRUCTURE

Richard M. Schultz and Michael N. Liebman

Textbook of Biochemistry With Clinical Correlations, Sixth Edition, Edited by Thomas M. Devlin

3.1 FUNCTIONAL ROLES OF PROTEINS IN HUMANS

Proteins perform a surprising variety of essential dynamic and structural functions in mammalian organisms. Dynamic functions include catalysis of chemical transformations, transport, metabolic control, and contraction. In their structural roles, proteins provide the matrix for bone and connective tissue, giving structure and form to the human organism.

Enzymes are dynamic proteins that catalyze chemical reactions, converting a substrate to a product at the enzyme's active site. Almost all of the thousands of chemical reactions in living organisms require a specific enzyme catalyst to ensure that these reactions occur at a rate compatible with life. The character of any cell is based on its particular chemistry, which is determined by its specific enzyme composition that establishes the cellular phenotype. Many genetic diseases result from altered levels of enzyme production or from specific alterations to their amino acid sequence.

Transport is another major dynamic function of proteins. Examples discussed in detail in this text are hemoglobin and myoglobin, which transport oxygen in blood and in muscle, respectively. Transferrin transports iron in blood. Other transport proteins bind and carry steroid hormones in blood from their site of synthesis to their site of action. Many drugs and toxic compounds are transported bound to proteins. Proteins also participate in contractile mechanisms. Myosin and actin function in muscle contraction and in changes in cell shape.

Proteins with a protective role include (a) immunoglobulins and interferon, which protect against bacterial or viral infection, and (b) fibrin, which stops the loss of blood on injury to the vascular system.

Many hormones are proteins or peptides. Insulin, thyrotropin, somatotropin (growth hormone), prolactin, luteinizing hormone, and follicle-stimulating hormone are proteins. Many polypeptide hormones have a low molecular weight (<5 kDa) and are referred to as peptides. In general, the term **protein** is used for molecules that contain over 50 amino acids, and **peptide** is used for those that contain less than 50 amino acids. Important peptide hormones include adrenocorticotropic hormone, antidiuretic hormone, glucagon, and calcitonin.

Proteins control and regulate gene transcription and translation. These include histones that are closely associated with DNA, repressor and enhancer transcription factors that control gene transcription, and components of the heteronuclear RNA particles and ribosomes.

Structural proteins function in "brick-and-mortar" roles. They include collagen and elastin, which form the matrix of bone and ligaments and provide structural strength and elasticity to organs and the vascular system. α-Keratin is present in hair and other epidermal tissue.

An understanding of normal functioning and pathology of the mammalian organism requires a clear understanding of the properties of proteins.

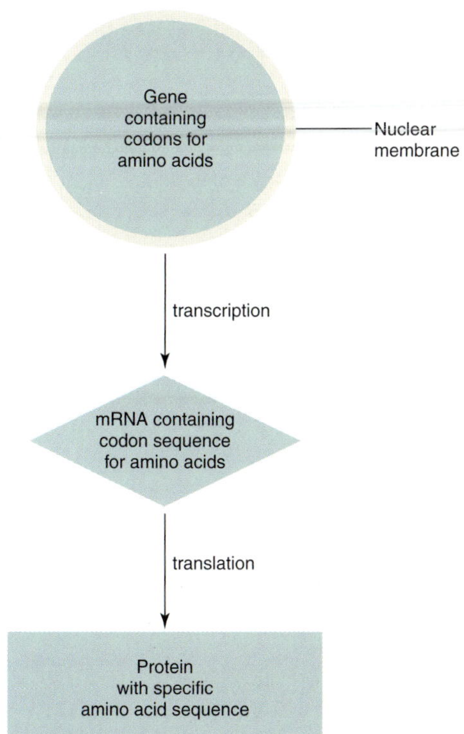

FIGURE 3.1
Genetic information is transcribed from a DNA sequence into mRNA and then translated to the amino acid sequence of a protein.

3.2 AMINO ACID COMPOSITION OF PROTEINS

All the different types of proteins are polymers of only 20 amino acids. These **common amino acids** are those for which at least one codon exists in the genetic code. Transcription and translation of the DNA code result in polymerization of amino acids into a specific linear sequence characteristic of a protein (Figure 3.1). Many proteins also contain **derived amino acids**, which are usually formed by enzymatic modification of a common amino acid after it has been incorporated into a protein. Examples of derived amino acids are cystine (see p. 79), desmosine, and isodesmosine found in elastin, hydroxyproline, and hydroxylysine of collagen, γ-carboxyglutamate in prothrombin, and phosphoserine, phosphothreonine, and phosphotyrosine.

Common Amino Acids

Common amino acids have the general structure depicted in Figure 3.2. They contain in common a central alpha (α)-carbon atom to which a carboxylic acid group, an amino group, and a hydrogen atom are covalently bonded. In addition, the α-carbon atom is bound to a specific chemical group, designated R and called the side chain, that uniquely defines each of the 20 common amino acids. Figure 3.2 depicts the ionized form of a common amino acid in solution at pH 7. Under physiological conditions, the α-amino group is protonated and in its ammonium ion form; the carboxylic acid group is in its unprotonated or carboxylate ion form.

$$\overset{+}{N}H_3 - \underset{\displaystyle R}{\overset{\displaystyle COO^-}{\overset{|}{\underset{|}{C_\alpha}}}} - H$$

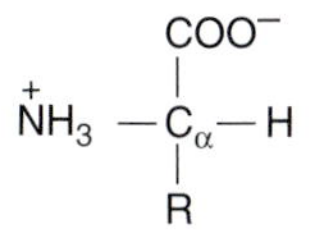

FIGURE 3.2
General structure of the common amino acids.

Side Chains Define The Chemical Nature and Structures of α-Amino Acids

Structures of the common amino acids are shown in Figure 3.3. Amino acids that contain alkyl group side chains include glycine, alanine, valine, leucine, and isoleucine. **Glycine** has the simplest structure, with R = H. **Alanine** contains a methyl (CH_3—) group. **Valine** has an isopropyl R group (Figure 3.4). The leucine and isoleucine R groups are isobutyl groups that are structural isomers of each other. In **leucine** the branching of the isobutyl side chain occurs on the gamma (γ)-carbon, and in **isoleucine** it is branched on the beta (β)-carbon.

The aromatic amino acids are phenylalanine, tyrosine, and tryptophan. Phenylalanine contains a benzene ring, **tyrosine** a phenol group, and **tryptophan** the heterocyclic structure, indole. In each case the aromatic moiety is attached to the α-carbon through a methylene (—CH_2—) carbon (Figure 3.3).

Sulfur-containing amino acids are cysteine and methionine. The **cysteine** side-chain group is a thiolmethyl ($HSCH_2$—). In **methionine** the side chain is a methyl ethyl thiol ether ($CH_3SCH_2CH_2$—).

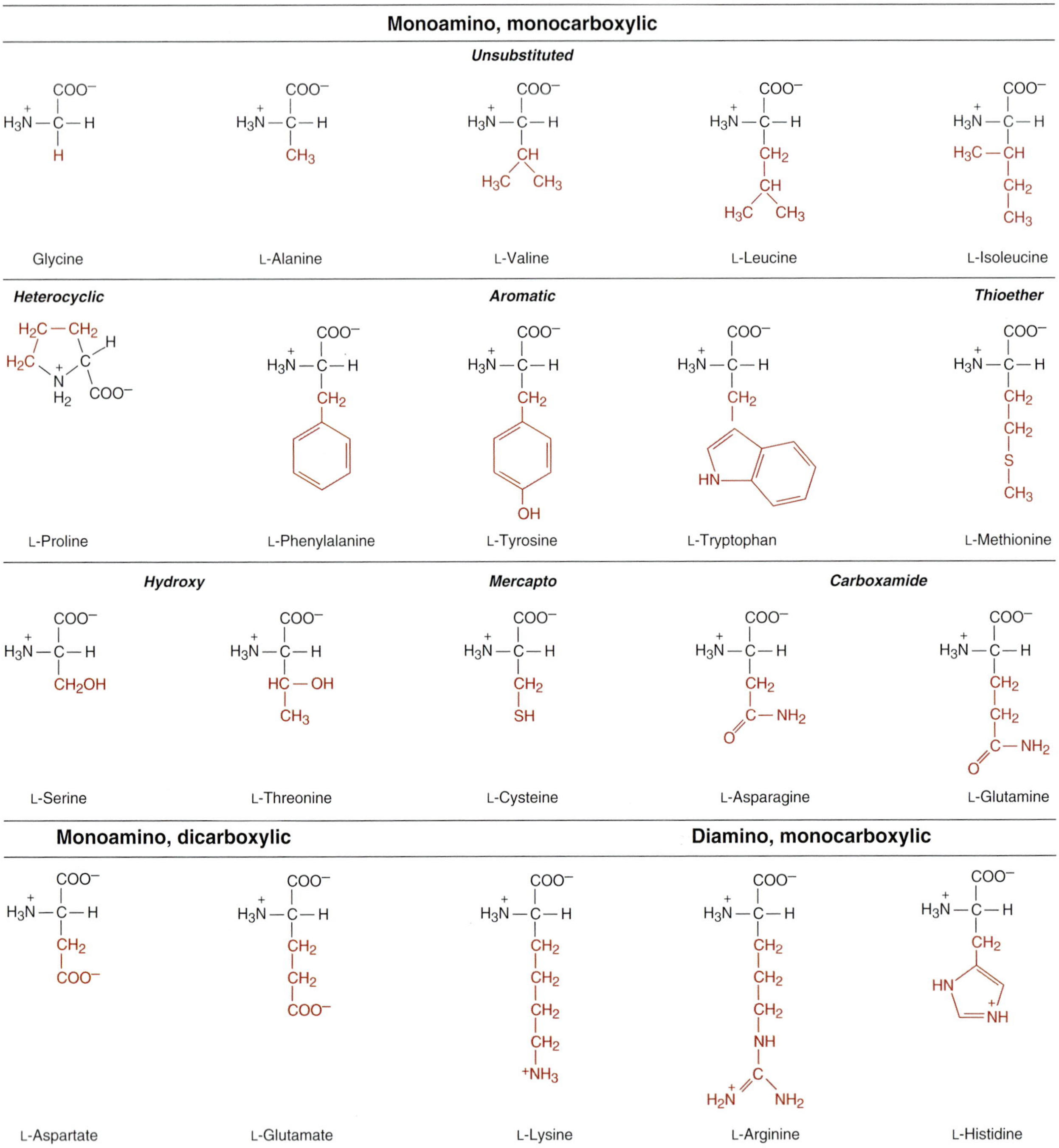

Monoamino, monocarboxylic

Unsubstituted

Glycine | L-Alanine | L-Valine | L-Leucine | L-Isoleucine

Heterocyclic

L-Proline

Aromatic

L-Phenylalanine | L-Tyrosine | L-Tryptophan

Thioether

L-Methionine

Hydroxy

L-Serine | L-Threonine

Mercapto

L-Cysteine

Carboxamide

L-Asparagine | L-Glutamine

Monoamino, dicarboxylic

L-Aspartate | L-Glutamate

Diamino, monocarboxylic

L-Lysine | L-Arginine | L-Histidine

FIGURE 3.3

Structures of the common amino acids. Charge forms are those present at pH 7.0.

The two hydroxy (alcohol)-containing amino acids are serine and threonine. In **serine** the side chain is a hydroxymethyl ($HOCH_2$—). In **threonine** an ethanol structure is connected to the α-carbon to produce a secondary alcohol (CH_3—CHOH—CH_α—).

Proline is unique in that it incorporates the α-amino group in its side chain and is more accurately classified as an α-imino acid, since its α-amine is a secondary amine with its α-nitrogen having two covalent bonds to carbon. Incorporation of the α-amino nitrogen into a five-membered ring constrains the rotational freedom around the —N_α—C_α— bond in proline and thereby limits proline participation to particular polypeptide chain conformations.

The amino acids discussed so far contain side chains that are uncharged at physiological pH. The **dicarboxylic-monoamino acids** contain a carboxylic group in their side chain. In **aspartate**, this group is separated by a methylene carbon (—CH_2—) from the α-carbon (Figure 3.5). In **glutamate** (Figure 3.5), the group is separated by two methylene (—CH_2—CH_2—) carbon atoms from the α-carbon (Figure 3.2). At physiological pH, these groups are unprotonated and negatively charged. The **dibasic-monocarboxylic** acids are lysine, arginine, and histidine (Figure 3.3). In these, the R group contains one or two nitrogen atoms that act as a base by binding a proton. In **lysine** the side chain is a *N*-butyl amine. In **arginine**, the side chain contains a guanidino group (Figure 3.6) separated from the α-carbon by three methylene carbon atoms. Both the guanidino group and the ϵ-amino group are protonated at physiological pH ($\sim$7) and positively charged. In **histidine** the side chain contains a five-membered heterocyclic structure, the imidazole group (Figure 3.6). The pK_a' of the imidazole group is approximately 6.0 in water; thus physiological solutions contain relatively high concentrations of both basic (imidazole) and acidic (imidazolium) forms of the histidine side chain.

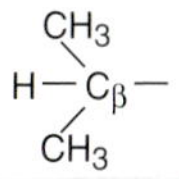

Isopropyl R group of valine

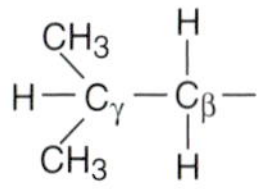

Isobutyl R group of leucine

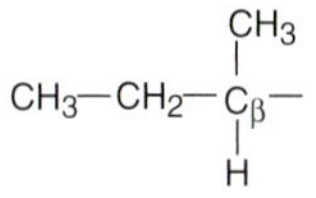

Isobutyl R group of isoleucine

FIGURE 3.4

Alkyl side chains of valine, leucine, and isoleucine.

Glutamine and **asparagine** are structural analogs of glutamic acid and aspartic acid with their carboxylic acid side chain groups amidated. Unique DNA codons exist for glutamine and asparagine separate from those for glutamic acid and aspartic acid. The amide side chains of glutamine and asparagine cannot be protonated and are uncharged at physiological pH.

To represent sequences of amino acids in proteins, three-letter and one-letter abbreviations for the common amino acids have been established (Table 3.1). These abbreviations are universally accepted and will be used throughout this book. The three-letter abbreviations of aspartic acid (Asp) and glutamic acid (Glu) should not be confused with those for asparagine (Asn) and glutamine (Gln). In determination of the amino acids of a protein by chemical procedures, one cannot easily differentiate between Asn and Asp, or between Gln and Glu, because the amide side-chain groups in Asn and Gln are often hydrolyzed in the procedure and generate Asp and Glu (see Section 3.9). The symbols Asx (for Asp or Asn) and Glx (for Glu or Gln) indicate this ambiguity in the analysis. A similar scheme is used with the one-letter abbreviations for Asp or Asn, and Glu or Gln.

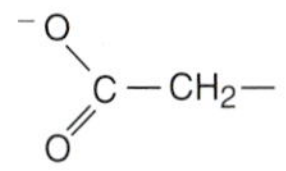

Aspartate R group

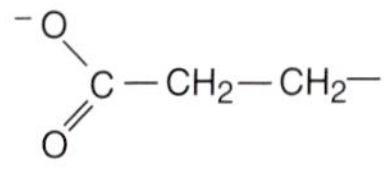

Glutamate R group

FIGURE 3.5

Side chains of aspartate and glutamate.

Cystine Is A Derived Amino Acid

A derived amino acid found in many proteins is cystine. It is formed by the oxidation of two cysteine thiol side chains to form a disulfide covalent bond (Figure 3.7). Within proteins, disulfide links of cystine formed from cysteines, separated from each other within a polypeptide chain (intrachain) or between two polypeptide chains (interchain), have an important role in stabilizing the folded conformation of proteins.

Amino Acids Have an Asymmetric Center

The common amino acids in Figure 3.2 have four substituents (R, H, COO^-, NH_3^+) covalently bonded to an α-carbon atom. A carbon atom with four different substituents in a **tetrahedral configuration** is asymmetric and exists in two enantiomeric forms. Thus, all amino acids exhibit optical isomerism except glycine, in which R = H and thus two of the four substituents on the α-carbon atom are hydrogen. The absolute configuration for an amino acid is depicted in Figure 3.8 using the **Fischer projection** to show the position in space of the tetrahedrally arranged α-carbon substituents. The α-COO^- group is directed up and behind the plane of the page, and the R group is directed down and behind the plane of the page. The α-H and α-NH_3^+ groups are directed toward the reader. An amino acid held in this way projects its α-NH_3^+ group to either the left or right of the α-carbon atom. By convention, if the α-NH_3^+ is projected to the left, the amino acid has an L absolute configuration. Its optical enantiomer, with α-NH_3^+ projected toward the right, has a D absolute configuration. Mammalian proteins contain amino acids of only L configuration. The L and D designations refer to the ability to rotate polarized light to the left (L, levo) or right (D, dextro) from its plane

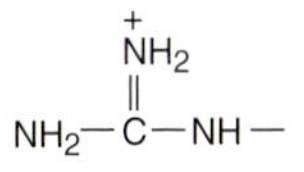

Guanidinium group (charged form) of arginine

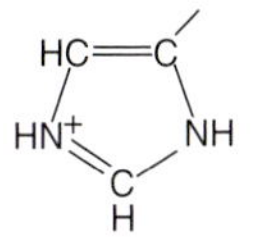

Imidazolium group of histidine

FIGURE 3.6

Guanidinium and imidazolium groups of arginine and histidine.

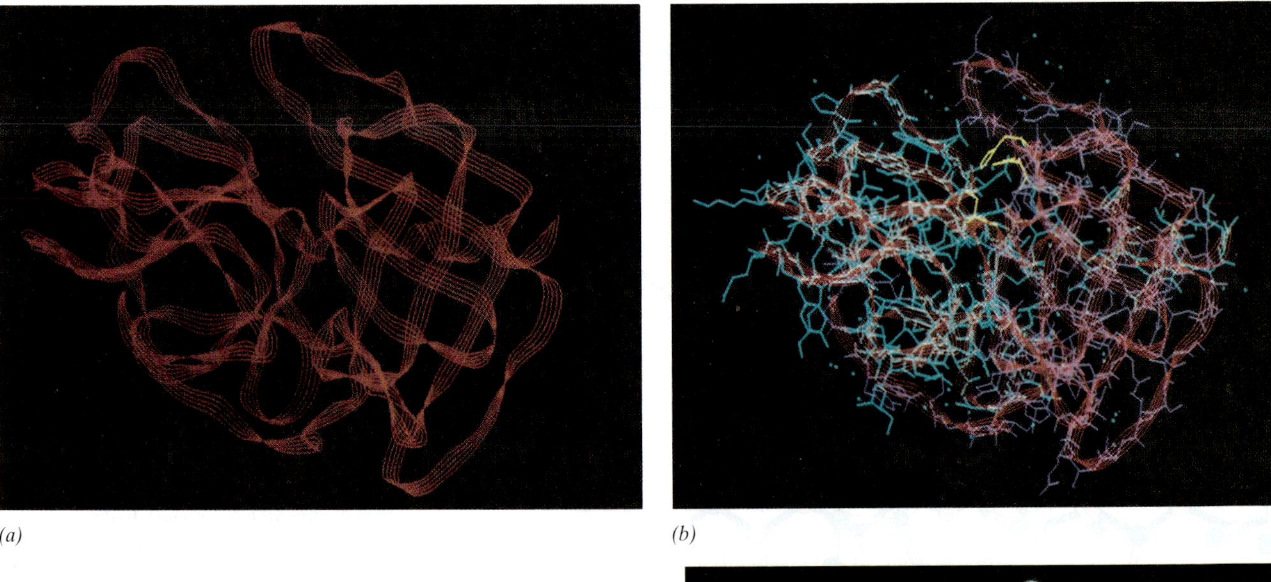

(a) (b)

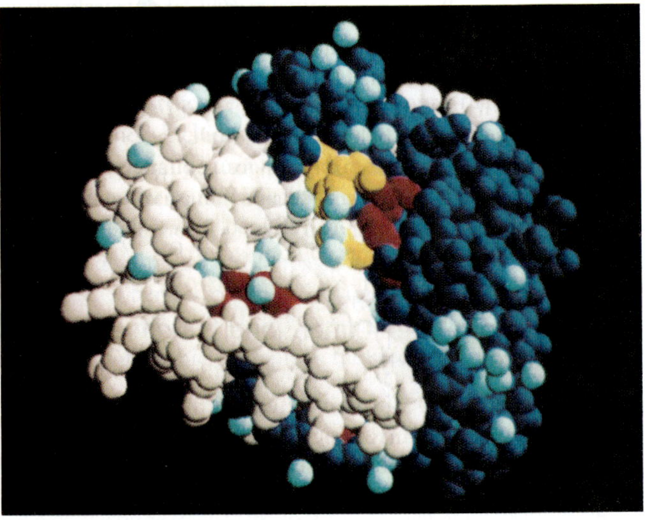

(c)

FIGURE 3.30

Tertiary structure of trypsin. (*a*) Ribbon structure outlines conformation of the polypeptide chain. (*b*) Structure shows side chains including active site residues (in yellow) with outline of polypeptide chain (ribbon) superimposed. (*c*) Space-filling structure in which each atom is depicted as the size of its van der Waals radius. Hydrogen atoms are not shown. Different domains are shown in dark blue and white. Active site residues are in yellow and intrachain disulfide bonds of cystine in red. Light blue spheres represent water molecules associated with the protein. This structure shows the density of packing within interior of the protein.

A long polypeptide often folds into multiple compact semi-independent regions or **domains**, each having a characteristic compact geometry with a hydrophobic core and polar surface. They typically contain 100–150 contiguous amino acids. Domains in a **multidomain protein** may be connected by a segment that lacks regular secondary structure. Alternatively, the dense spherical folded regions may be separated by a cleft or region less dense in tertiary structure (Figure 3.31). Trypsin contains two domains with a cleft in between that contains the substrate-binding catalytic site. An active site within an interdomain interface is characteristic of many enzymes. Different domains within a protein can move with respect to each other. Hexokinase (Figure 3.32), which catalyzes phosphorylation of glucose by adenosine triphosphate (ATP), has a glucose-binding site in a region between two domains. When glucose binds in the active site, the surrounding domains move to enclose the substrate to trap it for phosphorylation (Figure 3.32). In enzymes with more than one substrate or allosteric effector sites (see p. 401), the different sites may be located within different domains. In multifunctional proteins, each domain may perform a different task.

Quaternary Structure

Quaternary structure refers to the arrangement of polypeptide chains in a multichain protein. The subunits in a quaternary structure are associated noncovalently. α-Chymotrypsin contains three polypeptides covalently joined together by interchain

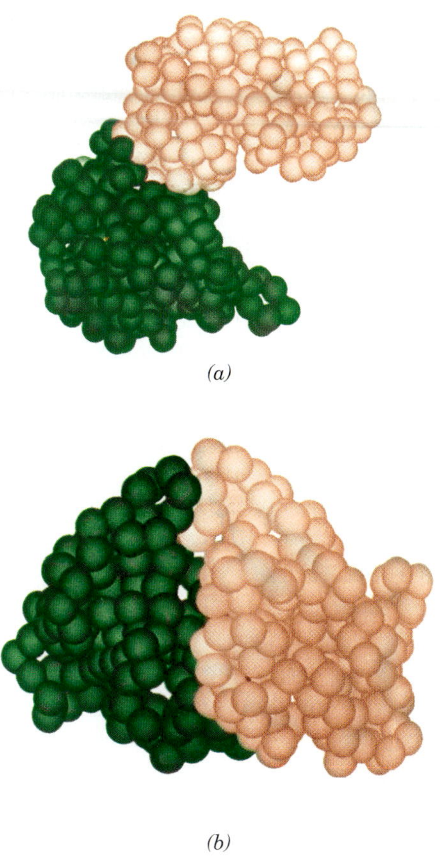

FIGURE 3.31

Globular domains within proteins. (*a*) Phosphoglycerate kinase has two domains with a relatively narrow neck in between. (*b*) Elastase has two tightly associated domains separated by a narrow cleft. Each sphere in the space-filling drawing represents the α-carbon position for an amino acid within the protein structure.

Reprinted with permission from Richardson, J. S. *Adv. Protein Chem.* 34:168, 1981.

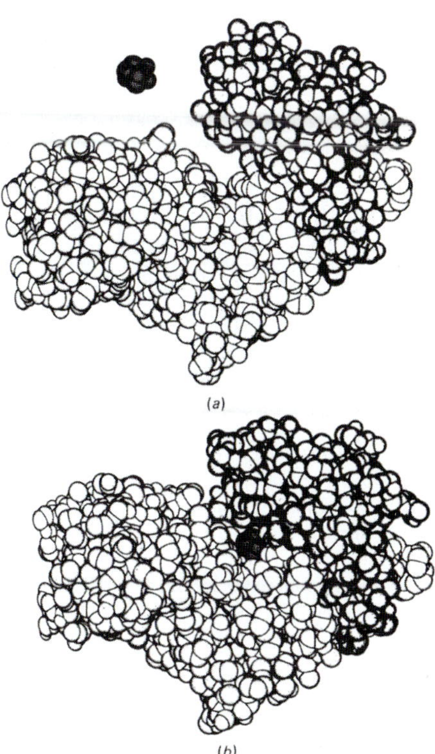

FIGURE 3.32

Drawings of (a) unliganded form of hexokinase and free glucose and (b) conformation of hexokinase with glucose bound. In this space-filling drawing, each circle represents van der Waals radius of an atom in the structure. Glucose is black, and each domain is differently shaded.

Reprinted with permission from Bennett, W. S. and Huber, R. *CRC Rev. Biochem.* 15:291, 1984.

disulfide bonds into a single covalent unit and therefore does not have a quaternary structure. Myoglobin consists of one polypeptide and has no quaternary structure. However, hemoglobin A contains four polypeptide subunits ($\alpha_2\beta_2$) held together noncovalently in a specific conformation as required for its function (see p. 338). Thus, hemoglobin has a quaternary structure. Bacterial aspartate carbamoyl transferase (see p. 752) has a quaternary structure comprised of 12 polypeptide subunits. The poliovirus coat protein contains 60 polypeptide subunits, and the tobacco mosaic virus protein contains 2120 noncovalently associated subunits.

Bioinformatics Relates Structure and Function of Protein Gene Products

Bioinformatics is a computationally based research area that focuses on integration and analysis of complex biological data with computer algorithms. A major emphasis of bioinformatics has been to identify patterns within nucleic acid or amino acid sequences that are signatures of structural features or motifs and signatures of the protein family or class to which the gene product belongs. These homology-searching algorithms are used to identify and classify gene products based on sequence similarity; and when the structure is known, homology searching may be based on structural similarity. The scoring for sequence similarity uses criteria from that of absolute identity of residues at equivalent positions to similarity based on polarity, hydrophobicity, and size. The algorithms allow for insertion or deletion of segments of polypeptide chain or structure to give the maximum possible overlap between two proteins.

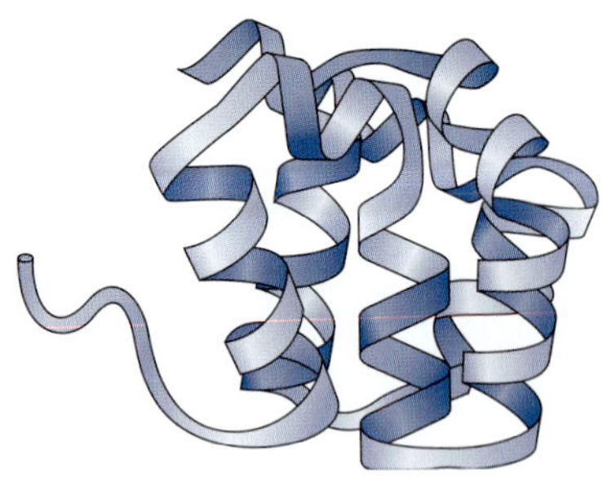

FIGURE 3.33

An example of an all α-domain globin fold in lysozyme. In this drawing and those that follow (Figures 3.34–3.36), only the outline of the polypeptide chain is shown. β-Strands are shown by arrows with direction of arrow showing N $\rightarrow$ C terminal direction; lightning bolts represent disulfide bonds, and circles represent metal ion cofactors (when present). All-α domain is the globin fold.
Redrawn with permission from Richardson, J. S. *Adv. Protein Chem.* 34:168, 1981.

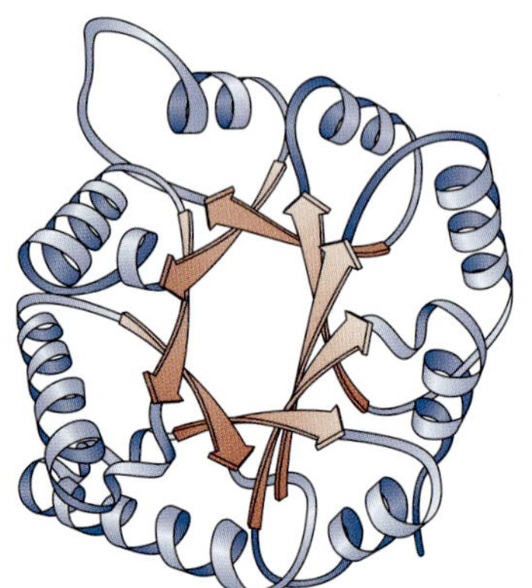

Triose Phosphate Isomerase

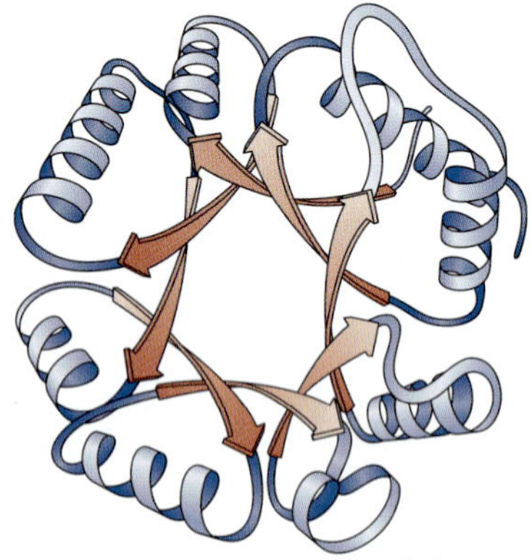

Pyruvate Kinase domain 1

FIGURE 3.34

Example of an α,β-domain fold in triose phosphate isomerase and in pyruvate kinase domain 1. See legend to Figure 3.33. In this commonly formed superfold the β strands form a β-barrel in the center of the domain while α-helix segments are on outside of the domain. β-Strands are in parallel directions. Regions of α-helix alternate with β-strands within the polypeptide chain.
Redrawn with permission from Richardson, J. S. *Adv. Protein Chem.* 34:168, 1981.

Protein domains are classified by class, fold, and family. The **class** is determined by the predominant type of secondary structure present in the protein. Some examples are mainly α-helix (all α), mainly β-strand (all β), and approximately equal amounts of α-helix and β-strand [alternating (α/β) and nonalternating (α + B)]. The **fold** is determined by the particular arrangement of secondary structure elements within the domain. The **family** is determined by the degree of sequence identity between proteins. Proteins that are members of the same family have a common evolutionary relationship; that is, they are derived from the same primordial gene. Proteins of the same family have the same fold and often have similar functions.

Of interest in clinical medicine is the finding of mutations in the amino acid sequence, which can give rise to significant alterations in function. At the molecular level, changes in even one amino acid can be significant. An amino acid mutation may perturb the native conformation, the conformational flexibility, the energetics, or motion of the molecule, and the selectivity of enzymes toward substrates and inhibitors. Examples of this are the hemoglobins for which there are extensive catalogs of single-site and many other types of mutations. Some hemoglobin mutations produce significant clinical symptoms, such as sickle cell anemia, where the substitution of a single amino acid enables the hemoglobin molecules to form aggregates, which precipitate within the red blood cell (see Clin. Corr. 3.3). Certain positions in the amino acid sequence are variant among populations. These sequence positions, when they involve single changes in the base codon for that amino acid, are termed **single nucleotide polymorphisms** (SNPs) and can lead to an understanding of the differences in response to a disease or therapeutic treatment among human populations.

Homologous Fold Structures are Often Formed from Nonhomologous Amino Acid Sequences

Although each native conformation is unique, comparison of tertiary structures of different proteins shows that similar arrangements of secondary structure motifs are often observed in the fold structures of domains. Folds of similar structure from proteins unrelated by function, sequence, or evolution are designated **superfolds**. They form because of the thermodynamic stability of their secondary structure arrangements or their kinetic accessibility.

A common all-α fold or domain is found in lysozyme and is designated the globin fold (Figure 3.33), as it was first reported in myoglobin and the subunits of hemoglobin (see p. 338). These all-α structures have seven or eight sections of α-helix joined by smaller segments that allow the helices to fold back upon themselves to form a characteristic globular shape. Because this fold is generated by proteins from different sequence homology families, the globin fold is a superfold. Another common superfold is the **α/β-domain structure** of triose phosphate isomerase (Figure 3.34) in which the strands (designated by arrows) form a central **ß-barrel** with each ß-strand in the interior interconnected by α-helical regions located on the outside of the fold. A similar fold forms a domain of pyruvate kinase (Figure 3.34), which has no sequence or functional homology. A different type of α,ß-superfold is present in the nonhomologous domain 1 of lactate dehydrogenase and domain 2 of phosphoglycerate kinase (Figure 3.35). In these, central sections participate in a **twisted ß-sheet**. Again, the ß-strand segments are joined by α-helical regions positioned on the outside to give the characteristic fold pattern. An **all-ß-domain** superfold is present in Cu, Zn superoxide dismutase, in which the antiparallel β-sheet forms a "Greek key" ß-barrel (Figure 3.36). A similar fold pattern occurs in each of the domains of the immunoglobulins. Concanavalin A (Figure 3.36) shows an all-ß-domain superfold in which the antiparallel ß-strands form a ß-barrel fold called a "jelly-roll." Proteins that are not water-soluble may contain different fold patterns (see Sections 3.6).

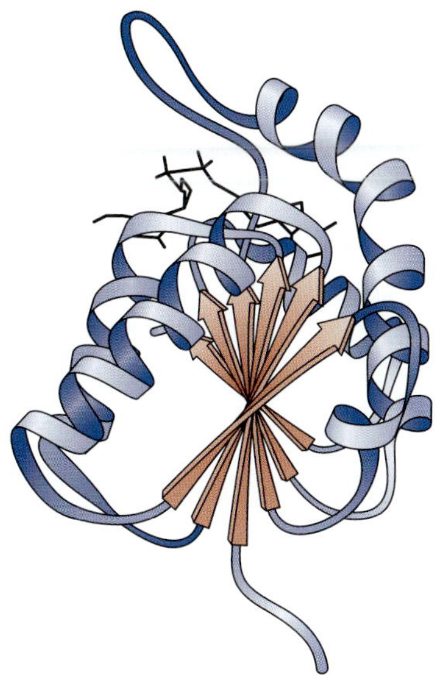

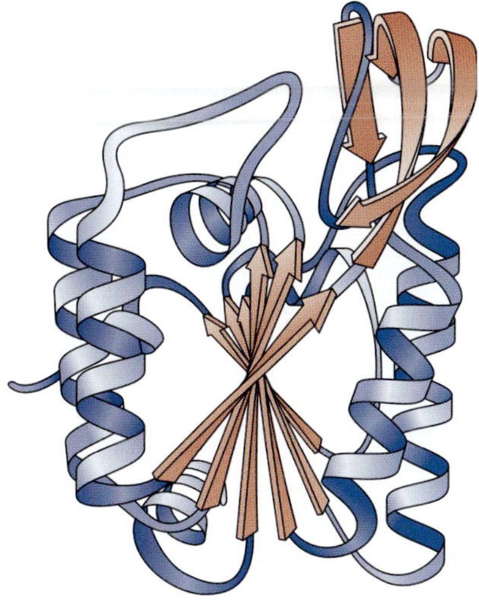

FIGURE 3.35

Example of an α/β-domain fold in which β strands form a classical twisted β-sheet of lactate dehydrogenase and phosphoglycerate kinase. See legend to Figure 3.33. As in previous α/β-domain fold, regions of α-helix alternate with regions of β-strand within polypeptide chain. β-Sheet structure is on the inside while the α-helical segments are on the outside. β-Strands are in parallel within the β-structure.
Redrawn with permission from Richardson, J. S. *Adv. Protein Chem.* 34:168, 1981.

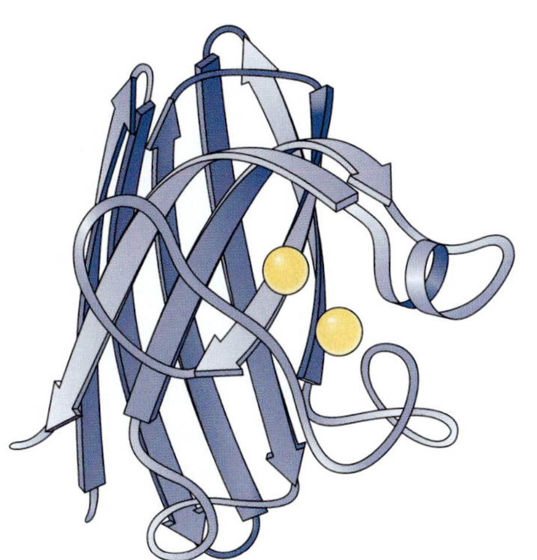

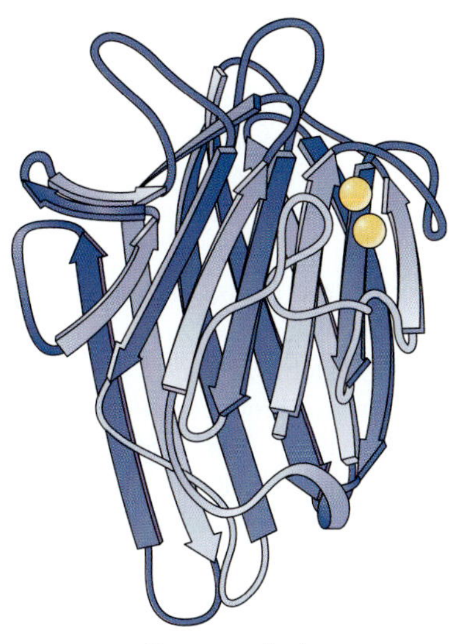

FIGURE 3.36

Examples of all-β-domain superfolds: the Greek key barrel and jelly roll folds shown in superoxide dismutase and concanavalin A (see legend to Figure 3.33). β-Strands are mostly antiparallel in all β-domain folds.
Redrawn with permission from Richardson, J. S. *Adv. Protein Chem.* 34:168, 1981.

3.6 OTHER TYPES OF PROTEINS

The characteristics of protein structure, already discussed, are based on observations on globular and water-soluble proteins. Proteins that do not conform to the globular soluble protein model are the fibrous proteins and membrane proteins. These are nonglobular and have a low water solubility. In addition, lipoproteins and glycoproteins contain lipid and carbohydrate nonprotein components and may or may not have globular structures.

Globular proteins have a spheroidal shape, vary in size, have relatively high water solubility, and function as catalysts, transporters, and regulators of metabolic pathways and gene expression.

Fibrous Proteins: Collagen, Elastin, Keratin, and Tropomyosin

Fibrous proteins characteristically contain larger amounts of regular secondary structure, a long cylindrical (rod-like) shape, low solubility in water, and a structural rather than a dynamic role. Examples of fibrous proteins with these characteristics are collagen, keratin, and tropomyosin.

CLINICAL CORRELATION 3.4

Diseases of Collagen Synthesis

Collagen is present in virtually all tissues and is the most abundant protein in the body. Certain organs depend heavily on it to function physiologically. Abnormal collagen synthesis or structure causes dysfunction of cardiovascular organs (aortic and arterial aneurysms and heart valve malfunction), bone (fragility and easy fracturing), skin (poor healing and unusual distensibility), joints (hypermobility and arthritis), and eyes (dislocation of the lens). Diseases caused by abnormal collagen synthesis include Ehlers–Danlos syndrome, osteogenesis imperfecta, and scurvy. These may result from abnormal collagen genes, abnormal posttranslational modification of collagen, or deficiency of cofactors needed by enzymes responsible for posttranslational modification of collagen.

Source: Aronson, D. Cross-linking of glycated collagen in the pathogenesis of arterial and myocardial stiffening of aging and diabetes. *J. Hypertens.* 21:3, 2003. Byers, P. H. Disorders of collagen biosynthesis and structure. In: C. R. Scriver, et al. (Eds.), *The Metabolic and Molecular Bases of Inherited Disease*, 8th ed. McGraw-Hill, 2001, Chapter 205. Hudson, B. G., et al., Alport's syndrome, Goodpasture's syndrome and type IV collagen, *N Engl J Med* 348:2543, 2003.

Collagen

Collagen is a family of proteins present in all tissues and organs and provides the framework that gives the tissues their form and strength. The percentage of collagen by weight for some representative human tissues and organs is liver 4%, lung 10%, aorta 12–24%, cartilage 50%, cornea 64%, whole cortical bone 23%, and skin 74% (see Clin. Corr. 3.4).

Amino Acid Composition of Collagen

The composition of type I skin collagen and of the globular proteins ribonuclease and hemoglobin are given in Table 3.10. Skin collagen is rich in glycine (33% of its amino acids), proline (13%), and the derived amino acids 4-hydroxyproline (9%) and 5-hydroxylysine (0.6%) (Figure 3.37). Hydroxyproline is unique to collagens being formed enzymatically from proline. Most of the hydroxyproline has the hydroxyl group in the 4-position (γ carbon), although a small amount of 3-hydroxyproline is also formed (Table 3.10). Collagens are glycoproteins with carbohydrate joined to 5-hydroxylysine, by an *O*-glycosidic bond through the δ-carbon hydroxyl group.

Amino Acid Sequence of Collagen

The collagen family is made up of polypeptides derived from 40 known collagen chain genes that produce about 20 types of collagen. Each mature collagen or tropocollagen molecule contains three polypeptide chains. Some types of collagen contain three identical polypeptide chains. In type I (Table 3.11), there are two $\alpha1$(I) chains and one $\alpha2$(I). Type V collagen contains $\alpha1$(V), $\alpha2$(V), and $\alpha3$(V). Collagens differ in amino acid sequence, but there are large regions of homologous sequence among all the different collagen types. In all the collagen types, there are regions with the tripeptides **Gly-Pro-Y** and **Gly-X-Hyp** (where X and Y are any amino acid) repeated in tandem several hundred times. In type I collagen polypeptides, the triplet sequences encompass over 600 of approximately 1000 amino acids per polypeptide. The collagens differ in their carbohydrate component. Some characteristics of collagen types I–VI are summarized in Table 3.11.

Structure of Collagen

Synthetic polyproline chains assume a regular secondary structure in aqueous solution, taking on the form of a tightly twisted extended helix with three residues per turn ($n = 3$). This helix with all trans-peptide bonds is the **polyproline type II** helix (see Figure 3.12 for differences between cis- and trans-peptide bonds). The polyproline helix closely resembles that found in collagen chains in regions that contain a proline or hydroxyproline at approximately every third position in the repeated tripeptide

TABLE 3.10 Comparison of Amino Acid Content of Human Skin Collagen (Type I) and Mature Elastin with That of Two Typical Globular Proteins[a]

Amino Acid	*Collagen (Human Skin)*	*Elastin (Mammalian)*	*Ribonuclease (Bovine)*	*Hemoglobin (Human)*
common amino acids		*percentoftotal*		
Ala	11	22	8	9
Arg	5	0.9	5	3
Asn			8	3
Asp	5	1	15	10
Cys	0	0	0	1
Glu	7	2	12	6
Gln			6	1
Gly	33	31	2	4
His	0.5	0.1	4	9
Ile	1	2	3	0
Leu	2	6	2	14
Lys	3	0.8	11	10
Met	0.6	0.2	4	1
Phe	1	3	4	7
Pro	13	11	4	5
Ser	4	1	11	4
Thr	2	1	9	5
Trp	2	1	9	2
Tyr	0.3	2	8	3
Val	2	12	8	10
derived amino acids				
Cystine	0	0	7	0
3-Hydroxyproline	0.1		0	0
4-Hydroxyproline	9	1	0	0
5-Hydroxylysine	0.6	0	0	0
Desmosine and isodesmosine	0	1	0	0

[a] Boxed numbers emphasize important differences in amino acid composition between the fibrous proteins (collagen and elastin) and typical globular proteins.

sequences, indicating that the thermodynamic forces leading to formation of the collagen helix are due to the properties of proline. In proline, the ϕ angle contributed to the polypeptide chain is part of the five-member cyclic side chain. This constrains the $C\alpha$—N bond to an angle compatible with the polyproline helix.

In the polyproline type II helix, the plane of each peptide bond is perpendicular to the axis of the helix, thus the peptide carbonyl group points toward neighboring chains and are correctly oriented to form strong interchain hydrogen bonds. This contrasts with the α-helix, in which the plane of the peptide bond is parallel to the axis where the NH and carbonyl oxygen of the peptides form only intrachain hydrogen bonds. The three chains of a collagen molecule, where each of the chains is in a polyproline type II helix conformation, wind around each other to form a superhelical structure (Figure 3.38). The three-chain collagen **superhelix** has a characteristic rise (d) and pitch (p) as does each single-chain helix. The collagen triple helix is stabilized by the interchain hydrogen bonds and because glycine occurs at every third position. With the polyproline type II helical conformation of each polypeptide chain with three residues per turn ($n = 3$), the glycines form an **apolar edge.** The glycine edges from the three chains form interactions between the chains that stabilize the three chain superhelix. Any side chain other than that of glycine would impede the adjacent chains from coming together into the superhelix structure (Figure 3.38).

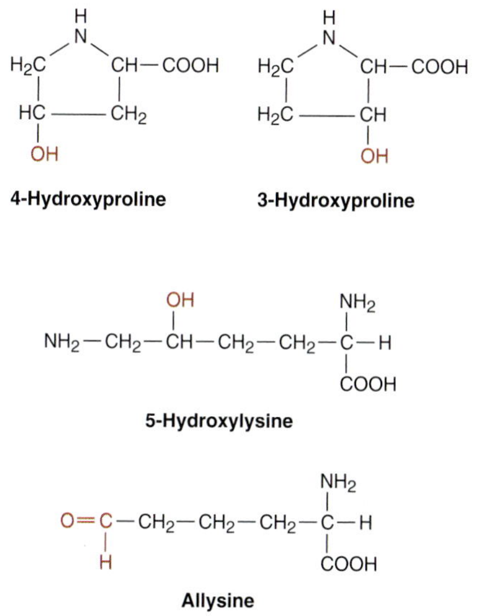

4-Hydroxyproline

3-Hydroxyproline

$NH_2-CH_2-CH(OH)-CH_2-CH_2-C(NH_2)(H)-COOH$

5-Hydroxylysine

$O{=}C(H)-CH_2-CH_2-CH_2-C(NH_2)(H)-COOH$

Allysine

FIGURE 3.37

Derived amino acids found in collagen. Carbohydrate is attached to 5-OH in 5-hydroxylysine by a type III glycosidic linkage (see Figure 3.45).

TABLE 3.11 Classification of Collagen Types

Type	Chain Designations	Tissue Found	Characteristics
I	$[\alpha 1(I)]_2\alpha 2(I)$	Bone, skin, tendon, scar tissue, heart valve, intestinal and uterine wall	Low carbohydrate; <10 hydroxylysines per chain
II	$[\alpha 1(II)]_3$	Cartilage, vitreous	10% carbohydrate; >20 hydroxylysines per chain
III	$[\alpha 1(III)]_3$	Blood vessels, newborn skin, scar tissue, intestinal and uterine wall	Low carbohydrate; high hydroxyproline and Gly; contains Cys
IV	$[\alpha 1(IV)]_3$ $[\alpha 2(IV)]_3$	Basement membrane, lens capsule	High 3-hydroxyproline; >40 hydroxylysines per chain; low Ala and Arg; contains Cys; high carbohydrate (15%)
V	$[\alpha 1(V)]_2\alpha 2(V)$ $[\alpha 1(V)]_3$ $\alpha 1(V)\alpha 2(V)\alpha 3(V)$	Cell surfaces or extracellular matrix associated with cell; widely distributed in low amounts	High carbohydrate, relatively high glycine, and hydroxylysine
VI	$\alpha 1(VI)\alpha 2(VI)$ $\alpha 3(VI)$	Aortic intima, placenta, kidney, and skin in low amounts	Relatively large globular domains in telopeptide region; high Cys and Tyr; molecular weight relatively low (~160 kDa); equimolar amounts of hydroxylysine and hydroxyproline

In type I collagen the triple helix extends for most of the sequence and only the carboxyl-terminus and amino-terminal segments (known as the **telopeptides)** are not in a triple-helical conformation. The type I collagen molecule is 3000 Å long and only 15 Å wide, a very long cylindrical structure. In other collagen types, superhelical regions may be periodically broken by globular regions.

Formation of Covalent Cross-Links in Collagen

An extracellular enzyme acts on procollagen molecules (see p. 234) to convert the ε-amino group of some lysine side chains to a δ-aldehyde (Figure 3.39). The derived amino acid is **allysine.** The newly formed aldehyde side chain spontaneously undergoes nucleophilic addition reactions with nonmodified lysine ε-amino groups and with the δ-carbon atoms of other allysine aldehydic groups. These covalent linkages can be between chains within the superhelical structure or between adjacent superhelical collagen molecules in a collagen fibril (see p. 234 for discussion of collagen biosynthesis).

Elastin is a Fibrous Protein with Allysine-Generated Cross-Links

Elastin gives tissues and organs the capacity to stretch without tearing. It is classed as a fibrous protein because of its structural function and relative insolubility in water; it is abundant in ligaments, lungs, walls of arteries, and skin. Elastin lacks a regular secondary structure. As in collagen, allysines form cross-links in **elastin.** An extracellular

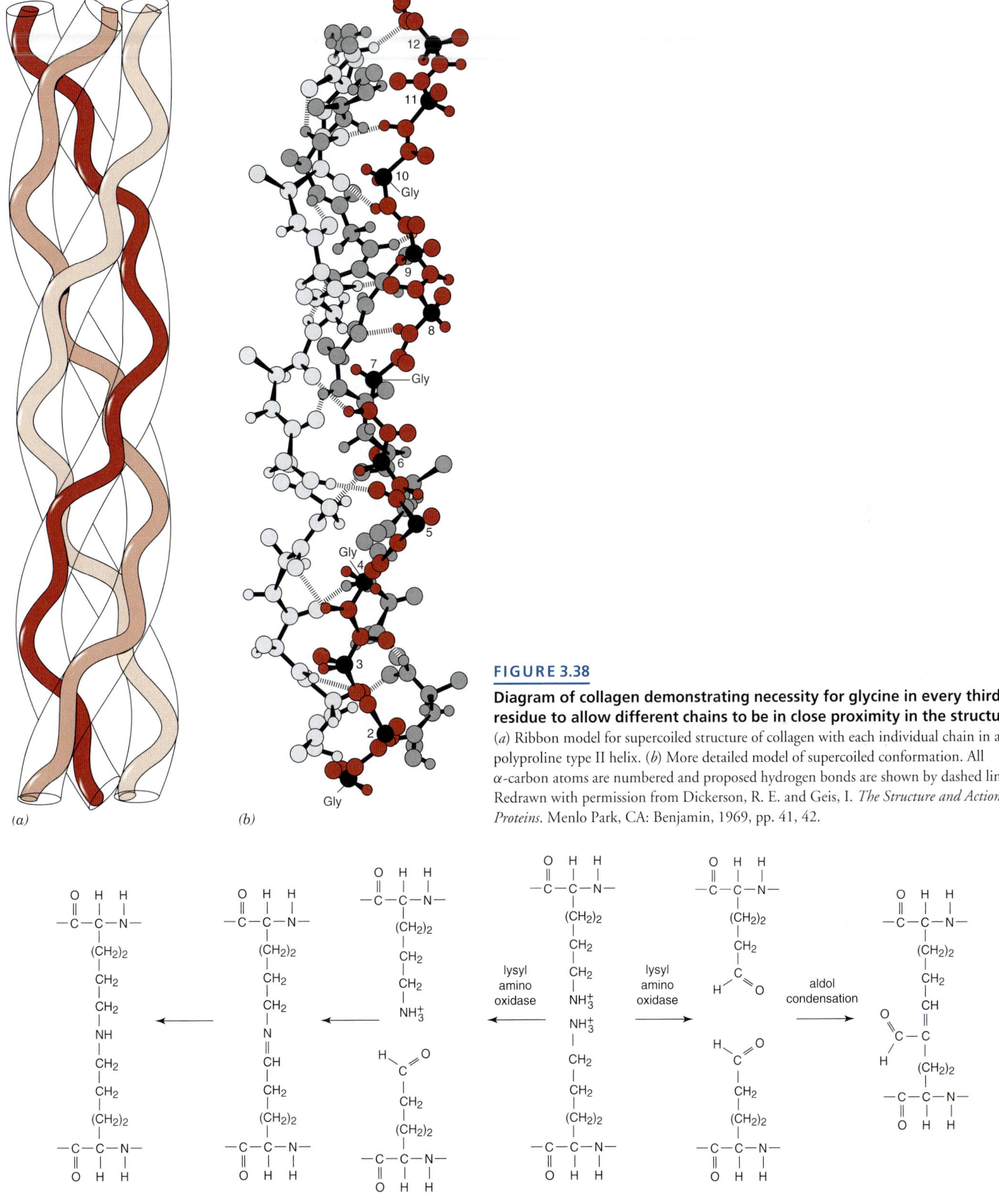

FIGURE 3.38

Diagram of collagen demonstrating necessity for glycine in every third residue to allow different chains to be in close proximity in the structure. (*a*) Ribbon model for supercoiled structure of collagen with each individual chain in a polyproline type II helix. (*b*) More detailed model of supercoiled conformation. All α-carbon atoms are numbered and proposed hydrogen bonds are shown by dashed lines. Redrawn with permission from Dickerson, R. E. and Geis, I. *The Structure and Actions of Proteins*. Menlo Park, CA: Benjamin, 1969, pp. 41, 42.

FIGURE 3.39

Covalent cross-links formed in collagen through allysine intermediates. Formation of allysines is catalyzed by lysyl amino oxidase.

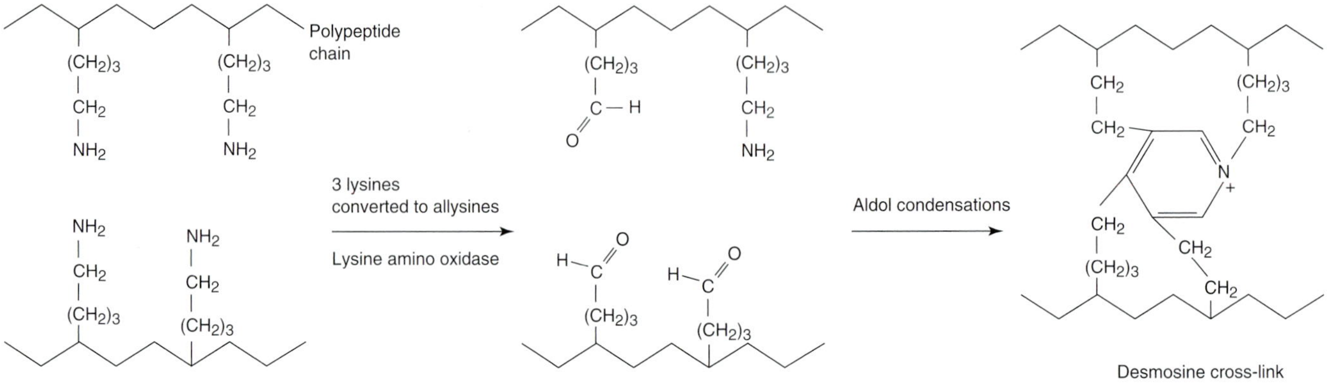

Polypeptide chain

3 lysines converted to allysines

Lysine amino oxidase

Aldol condensations

Desmosine cross-link

FIGURE 3.40

Formation of desmosine covalent cross-link in elastin from lysine and allysines. Polypeptide chain drawn schematically with intersections of lines representing placement of α carbons.

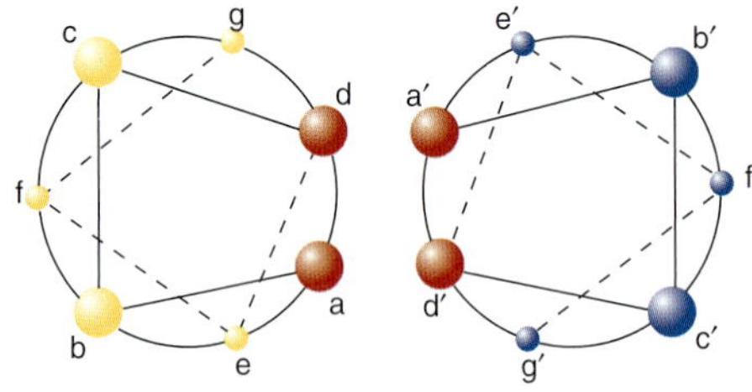

FIGURE 3.41

Interaction of apolar edges of two chains in α-helical conformation as in keratin and tropomyosin. Interaction of apolar d-a and a′-d residues of two α-helices aligned parallel in an NH_2 terminal (top) to COOH terminal direction is presented. Redrawn from McLachlan, A. D. and Stewart, M. J. *Mol. Biol.* 98:293, 1975.

lysine amino oxidase converts lysine side chains in the sequence -Lys-Ala-Ala-Lys- and -Lys-Ala-Ala-Ala-Lys- to allysines. Three allysines and an unmodified lysine from different regions in the polypeptide chains react to form the heterocyclic structure of **desmosine** or **isodesmosin**, which cross-link the polypeptide chains in elastin networks (Figure 3.40).

Keratin and Tropomyosin

Keratin and tropomyosin are fibrous proteins in which each polypeptide is α-helical. Keratin is found in the epidermal layer of skin, in nails, and in hair. Tropomyosin is a component of the thin filament in muscle tissue. The sequences in both proteins show tandem repetition of seven residue segments (heptad), in which the first and fourth amino acids have hydrophobic side chains and the fifth and seventh polar side chains. The reiteration of hydrophobic and polar side chains in heptad segments is symbolically represented by the formulation $(\text{a-b-c-d-e-f-g})_i$, where a and d are hydrophobic amino acids, and e and g are polar or ionized side chain groups. Since a seven-amino-acid segment represents two complete turns of an α-helix ($n = 3.6$), the apolar residues at a and d align to form an apolar edge along one side of the α-helix (Figure 3.41). This apolar edge interacts with polypeptide apolar edges of other α-keratin chains to form a **coiled-coil** superhelical structure containing two or three polypeptide chains. Each polypeptide helix also contains a polar edge, due to residues e and g, that interacts with water on the outside of the superhelix and stabilizes the superhelix.

Thus collagen, keratin, and tropomyosin are multistrand structures whose polypeptide chains have a highly regular secondary structure (polyproline type II helix or α-helix) and form rod-shaped multichain supercoiled conformations. In each case, the amino acid sequences of the chains generate edges on the cylindrical surfaces that stabilize hydrophobic interactions between the chains in the supercoiled conformations. In turn, these molecular structures are aligned into multimolecular fibrils stabilized, in some cases, by covalent cross-links.

Plasma Lipoproteins are Complexes of Lipids with Proteins

Plasma lipoproteins are complexes of proteins and lipids that form distinct aggregates with an approximate stoichiometry between protein and lipid components. Changes in their relative concentrations are predictive of atherosclerosis, a major human disease (see Clin. Corr. 3.5). They transport lipids from tissue to tissue and participate in lipid metabolism (see p. 711). Four classes exist in plasma of normal fasting humans (Table 3.12), and in the postabsorptive period a fifth class, chylomicrons, is also present. These are distinguished by their density, as determined by ultracentrifugation and by

CLINICAL CORRELATION 3.5
Hyperlipidemias

Hyperlipidemias are disorders of the rates of synthesis or clearance of lipoproteins from the bloodstream. Usually they are detected by measuring plasma triacylglycerol and cholesterol and are classified on the basis of which class of lipoproteins is elevated.

Type I hyperlipidemia is due to accumulation of chylomicrons. Two genetic forms are known: lipoprotein lipase deficiency and ApoCII deficiency. ApoCII is required by lipoprotein lipase for full activity. Patients with type I hyperlipidemia have exceedingly high plasma triacylglycerol concentrations (over 1000 mg dL^{-1}) and suffer from eruptive xanthomas (yellowish triacylglycerol deposits in the skin) and pancreatitis.

Type II hyperlipidemia is characterized by elevated LDL levels. Most cases are due to genetic defects in the synthesis, processing, or function of the LDL receptor. Heterozygotes have elevated LDL levels, hence the trait is dominantly expressed. Homozygous patients have very high LDL levels and may suffer myocardial infarctions before age 20.

Type III hyperlipidemia is due to abnormalities of ApoE, which interfere with the uptake of chylomicron and VLDL remnants. Hypothyroidism can produce a very similar hyperlipidemia. These patients have an increased risk of atherosclerosis.

Type IV hyperlipidemia is the commonest abnormality. The VLDL levels are increased, often due to obesity, alcohol abuse, or diabetes. Familial forms are also known.

Type V hyperlipidemia is, like type I, associated with high chylomicron triacylglycerol levels, pancreatitis, and eruptive xanthomas.

Hypercholesterolemia also occurs in certain types of liver disease in which biliary excretion of cholesterol is reduced. An abnormal lipoprotein called lipoprotein X accumulates. This disorder is not associated with increased cardiovascular disease from atherosclerosis.

Source: Havel, R. J. and Kane, J. P., Introduction: Structure and metabolism of plasma lipoproteins. In: C. R. Scriver, et al. (Eds.), *The Metabolic and Molecular Bases of Inherited Disease*, 8th ed. New York: McGraw-Hill, 2001, Chapter 114. Goldstein, J. L., Hobbs, H. H., and Brown, M. S. Familial hypercholesterolemia. In: C. R. Scriver, et al. (Eds.), *The Metabolic and Molecular Bases of Inherited Disease*, 8th ed. New York: McGraw-Hill, 2001, Chapter 120.

TABLE 3.12 Hydrated Density Classes of Plasma Lipoproteins

Lipoprotein Fraction	Density ($g\,mL^{-1}$)	Flotation Rate, S_f (Svedberg units)	Molecular Weight (daltons)	Particle Diameter (Å)
HDL	1.063–1.210		HDL_2, 4×10^5	70–130
			HDL_3, 2×10^5	50–100
LDL (or LDL_2)	1.019–1.063	0–12	2×10^6	200–280
IDL (or LDL_1)	1.006–1.019	12–20	4.5×10^6	250
VLDL	0.95–1.006	20–400	5×10^6–10^7	250–750
Chylomicrons	<0.95	>400	10^9–10^{10}	10^3–10^4

Source: Data from Soutar, A. K. and Myant, N. B. In: R. E. Offord (Ed.), *Chemistry of Macromolecules*, IIB. Baltimore, MD: University Park Press, 1979.

electrophoresis (Figure 3.42). Their protein components are termed **apolipoproteins**, and each class of lipoprotein having a characteristic apolipoprotein composition. The most prominent apolipoproteins (Table 3.13) are: apolipoprotein **ApoA-I** in high-density lipoproteins **(HDLs); ApoB** in low-density lipoproteins (**LDLs**), intermediate-density lipoproteins **(IDLs)**, and very-low-density lipoproteins **(VLDLs)**, and **ApoC** in IDLs and VLDLs. Each apolipoprotein class is genetically and structurally distinct (see Clin. Corr. 3.6). Apolipoproteins vary from 6 kDa (ApoC-I) to 550 kDa for ApoB-100. The latter is a long polypeptide (4536 amino acids) and occurs in truncated form (the N-terminal 2512 residues only) as apoB-48 in chylomicrons.

A model for a VLDL particle is shown in Figure 3.43. On the inside are neutral lipids such as cholesteryl esters and triacylglycerols. Surrounding this inner core of neutral lipids, is a shell ~20 Å thick, in which reside the proteins and the charged amphoteric lipids such as unesterified cholesterol and phosphatidylcholines (see p. 447). Amphoteric lipids and proteins in the outer shell place their hydrophobic apolar regions

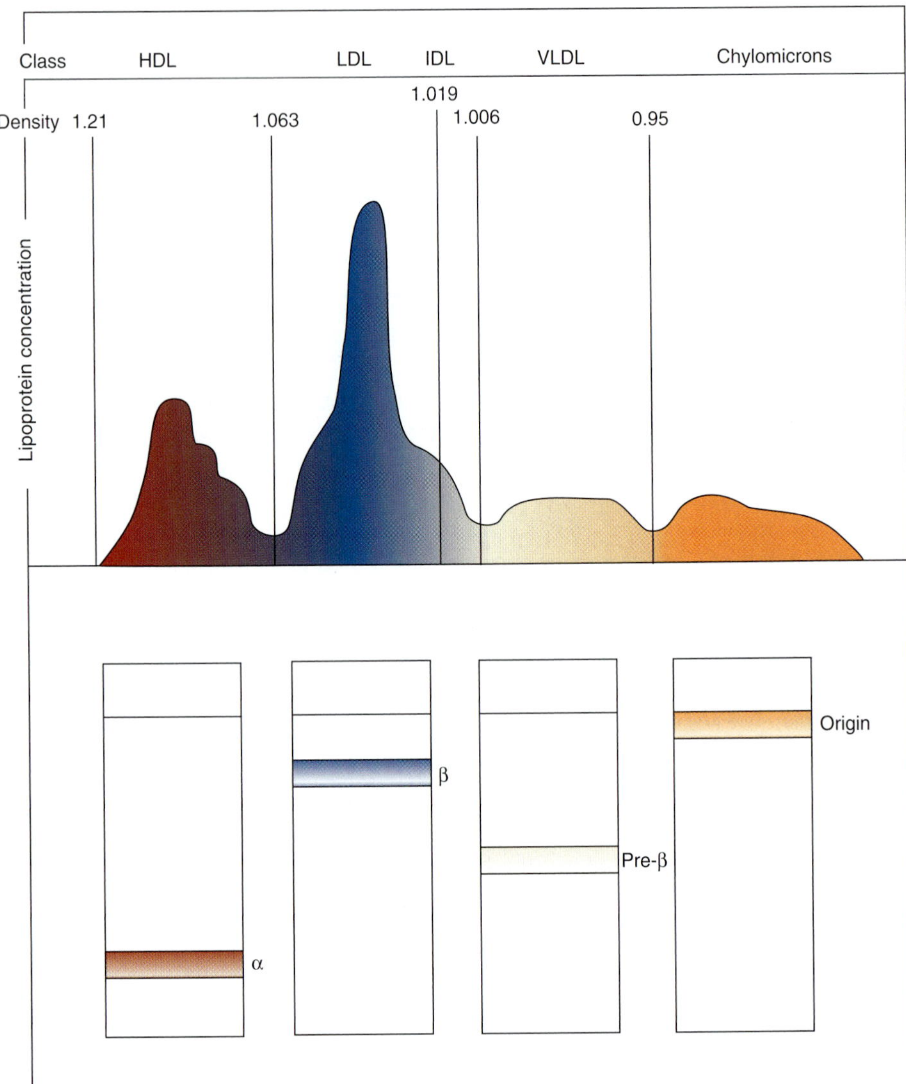

FIGURE 3.42

Correspondence of plasma lipoprotein density classes with electrophoretic mobility in a plasma electrophoresis. In upper diagram an ultracentrifugation Schlieren pattern is shown. At bottom, electrophoresis on a paper support shows mobilities of major plasma lipoprotein classes with respect to α- and β-globulin bands.

Reprinted with permission from Soutar, A. K. and Myant, N. B. In: R. E. Offord (Ed.), *Chemistry of Macromolecules*, IIB. Baltimore, MD: University Park Press, 1979.

TABLE 3.13 Apolipoproteins of Human Plasma Lipoproteins (Values in Percentage of Total Protein Present)[a]

Apolipoprotein	HDL_2	HDL_3	*LDL*	*IDL*	*VLDL*	*Chylomicrons*
ApoA-I	85	70–75	Trace	0	0–3	0–3
ApoA-II	5	20	Trace	0	0–0.5	0–1.5
ApoD	0	1–2			0	1
ApoB	0–2	0	95–100	50–60	40–50	20–22[b]
ApoC-I	1–2	1–2	0–5	<1	5	5–10
ApoC-II	1	1	0.5	2.5	10	15
ApoC-III	2–3	2–3	0–5	17	20–25	40
ApoE	Trace	0–5	0	15–20	5–10	5
ApoF	Trace	Trace				
ApoG	Trace	Trace				

Source: Data from Soutar, A. K, and Myant, N. B. In: R. E. Offord (Ed.), *Chemistry of Macromolecules*, IIB. Baltimore, MD: University Park Press, 1979; Kostner, G. M. *Adv. Lipid Res.* 20:1, 1983.

[a] Values show variability from different laboratories.

[b] Primarily ApoB-48.

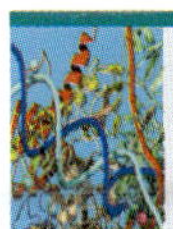

CLINICAL CORRELATION 3.6
Hypolipoproteinemias

Abetalipoproteinemia is a genetic disease that is characterized by absence of chylomicrons, VLDLs, and LDLs due to an inability to synthesize apolipoproteins apoB-100 and apoB-48. There is accumulation of lipid droplets in small intestinal cells, malabsorption of fat, acanthocytosis (spiny-shaped red cells), and neurological disease (retinitis pigmentosa, ataxia, and retardation).

Tangier's disease, an α-lipoprotein deficiency, is a rare autosomal recessive disease in which the HDL level is 1–5% of its normal value. Clinical features are due to the accumulation of cholesterol in the lymphoreticular system, which may lead to hepatomegaly and splenomegaly. The plasma cholesterol and phospholipids are greatly reduced.

Deficiency of the enzyme lecithin:cholesterol acyltransferase is a rare disease that results in the production of lipoprotein X (see Clin. Corr. 3.5). Also characteristic is the decrease in the α-lipoprotein and pre-β-lipoprotein bands, along with an increase in the β-lipoprotein because of presence of lipoprotein X on electrophoresis.

Source: Kane, J. P. and Havel, R. J. Disorders of the biogenesis and secretion of lipoproteins containing the β apolipoproteins. In: C. R. Scriver, A. L. Beaudet, W. S. Sly, and D. Valle (Eds.), *The Metabolic and Molecular Bases of Inherited Disease*, 8th ed., New York: McGraw-Hill, 2001, Chapter 115. Assmann, G., von Eckardstein, A., and Brewer, H. B. Jr. Familial high density lipoprotein deficiency: Tangier disease. In: C. R. Scriver, A. L. Beaudet, W. S. Sly, and D. Valle (Eds.), *The Metabolic and Molecular Bases of Inherited Disease*, 8th ed., New York: McGraw-Hill, 2001, Chapter 122.

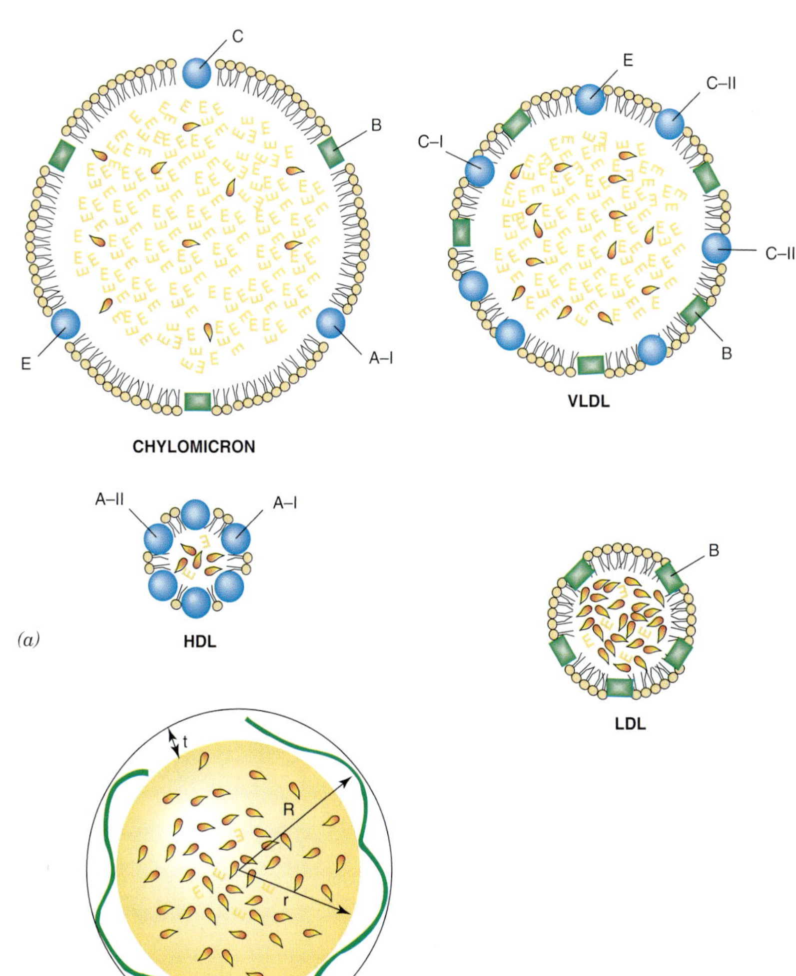

FIGURE 3.43

Generalized structure of plasma lipoproteins.
(*a*) Spherical particle model consisting of a core of triacylglycerols (yellow E's) and cholesterol esters (orange drops •-) with a shell ~20 Å thick of apolipoproteins (lettered), phospholipids, and unesterified cholesterol. Apolipoproteins are embedded with their hydrophobic edges oriented toward core and their hydrophilic edges toward the outside. (*b*) LDL particle showing ApoB-100 imbedded in outer shell of particle.
Part (*a*) from Segrest, J. P., et al. *Adv. Protein Chem.* 45:303, 1994. Part (*b*) from Schumaker, V. N., et al. *Protein Chem.* 45:205, 1994.

toward the inside of the particle and their charged groups toward the outside where they interact with each other and with water.

This lipoprotein particle model applies to all classes of plasma lipoproteins. As the diameter of a particle decreases, the outer shell makes up a greater percentage of the total particle. Thus, the smaller particles have a higher percentage of the surface protein and amphoteric lipid molecules than the larger particles. The HDL particles, which are much smaller than VLDL particles, thus consist of 45% protein and 55% lipid, while the VLDL particles consist of 10% protein with 90% lipid (Table 3.14).

The apolipoproteins have high α-helical content when in association with lipid. Their helical regions are amphipathic since every third or fourth amino acid is charged. These residues form a polar edge that associates with the polar heads of phospholipids and the aqueous environment. The opposite sides of the helices have hydrophobic side chains that are oriented toward the nonpolar core of the lipoprotein particle. The α-helical structure of part of ApoC-I is shown in Figure 3.44. ApoB contains α-helical and β-strand segments. A single long polypeptide of ApoB-100 circles the

TABLE 3.14 Chemical Composition of Plasma Lipoprotein Classes

			Percent Composition of Lipid Fraction			
Total Lipoprotein Class	***Total Protein (%)***	***Lipid (%)***	***Phospholipids***	***Esterified Cholesterol***	***Unesterified Cholesterol***	***Triacylglycerols***
HDL_2[a]	40–45	55	35	12	4	5
HDL_3[a]	50–55	50	20–25	12	3–4	3
LDL	20–25	75–80	15–20	35–40	7–10	7–10
IDL	15–20	80–85	22	22	8	30
VLDL	5–10	90–95	15–20	10–15	5–10	50–65
Chylomicrons	1.5–2.5	97–99	7–9	3–5	1–3	84–89

Source: Data from Soutar, A. K. and Myant, N. B. In: R. E. Offord (Ed.), *Chemistry of Macromolecules*, IIB. Baltimore, MD: University Park Press, 1979.
[a] Subclasses of HDL.

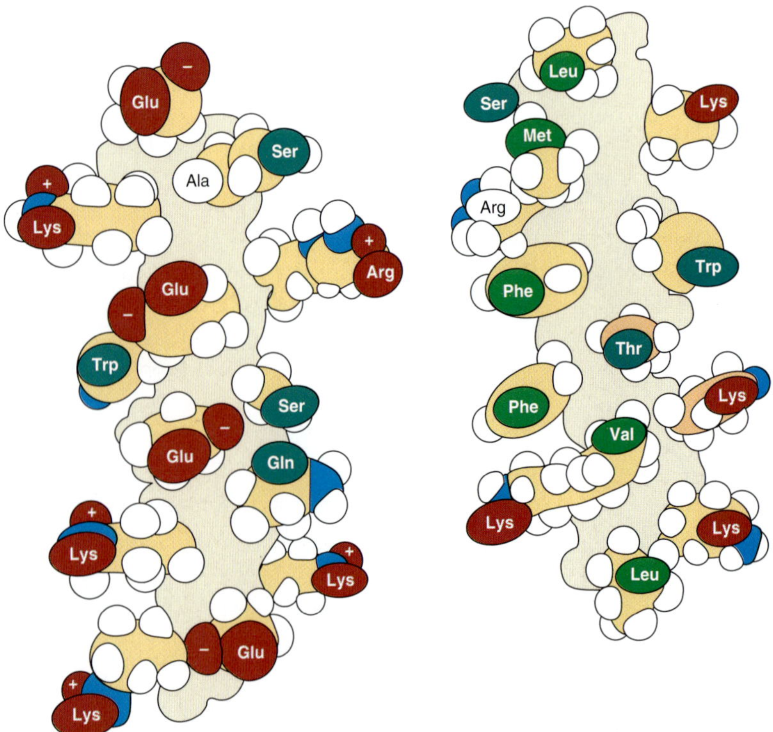

FIGURE 3.44
Side chains of a helical amphipathic segment of apolipoprotein C1 between residues 32 and 53. The polar face shows acid residues in the center and basic residues at the edge. On the other side of the helix, hydrophobic residues form a nonpolar longitudinal face. Redrawn with permission from Sparrow, J. T. and Gotto, A. M., Jr. *CRC Crit. Rev. Biochem.* 13:87, 1983. Copyright (1983) CRC Press, Inc. Boca Raton, FL.

circumference of the LDL particle like a belt weaving in and out of the phospholipid outer shell (Figure 3.43).

Glycoproteins Contain Covalently Bound Carbohydrate

Glycoproteins participate in many normal and disease-related functions of clinical relevance. The best-characterized glycoproteins are those of the outer surface of plasma membranes, the extracellular matrix, and blood plasma. Those on plasma membranes provide information for identification of cells by other cells and for regulation of cell growth by contact inhibition. They also provide the antigenic determinants of the various blood group systems (e.g., ABO and Rh) on erythrocytes, the histocompatibility and viability of tissue transplants, and receptors for hormones, neurotransmitters, and viruses. Tumorigenesis and malignant transformation are associated with changes in the glycoprotein composition of plasma membranes. In the extracellular matrix, many proteins (e.g., collagen, laminin) contain attached carbohydrate as do proteins of mucous secretions that lubricate and protect surfaces. The major plasma proteins (but not albumin) are glycoproteins. These include blood-clotting proteins, immunoglobulins, complement proteins, and follicle stimulating hormone, luteinizing hormone, and thyroid stimulating hormone.

The amount of carbohydrate in glycoproteins is variable. IgG contains 4% by weight, glycophorin of human erythrocyte membranes is 60%, and human gastric glycoprotein 82%. The carbohydrate can be distributed evenly along the polypeptide chain or concentrated in defined regions. In proteins of plasma membrane, carbohydrate is attached only on the extracellular portion. The attached carbohydrate usually contains less than 15 sugar residues and in some cases only one. Glycoproteins with the same function from different animal species often have homologous sequences but vary in their carbohydrate. Heterogeneity in carbohydrate content is common in the same protein even within a single organism. For example, pancreatic ribonuclease A and B forms have an identical amino acid sequence but differ in their carbohydrate composition.

The incorporation of carbohydrate occurs in a series of enzyme-catalyzed reactions as the polypeptide chain is transported through the endoplasmic reticulum and Golgi network (see p. 227).

Carbohydrate–Protein Covalent Linkages

The two most common carbohydrate linkages are (a) the ***N*-glycosidic linkage** (type I linkage) between the amide group of asparagine and a sugar and (b) the ***O*-glycosidic linkage** (type II linkage) between a hydroxyl group of serine or threonine and a sugar (Figure 3.45). In type I the bond is to asparagine within the sequence Asn-X-Thr(Ser). Mammalian glycoproteins also contain *O*-glycosidic bonds to 5-hydroxylysine (type III linkage). This occurs in collagens and in the serum complement protein C1q. Less common are attachments to the hydroxyl group of 4-hydroxyproline (type IV linkage), to a cysteine thiol (type V linkage), and to an α-amino group of a polypeptide chain (type VI linkage). High concentrations of type VI linkages are formed nonenzymatically with hemoglobin and blood glucose in uncontrolled diabetes. Assay of glycosylated hemoglobin is used to follow changes in blood glucose concentration (see Clin. Corr. 3.7).

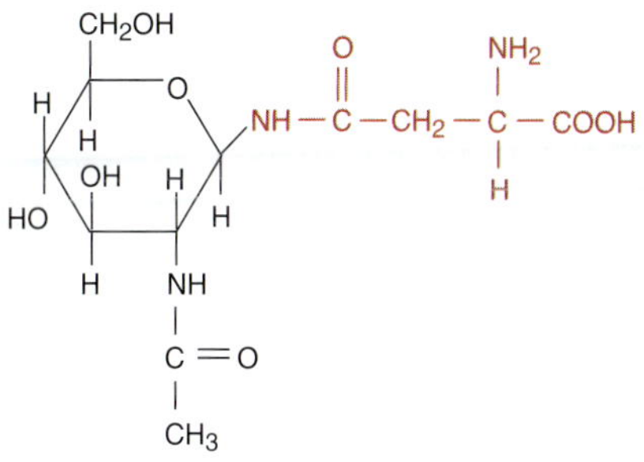

Type I *N*-Glycosyl linkage to asparagine

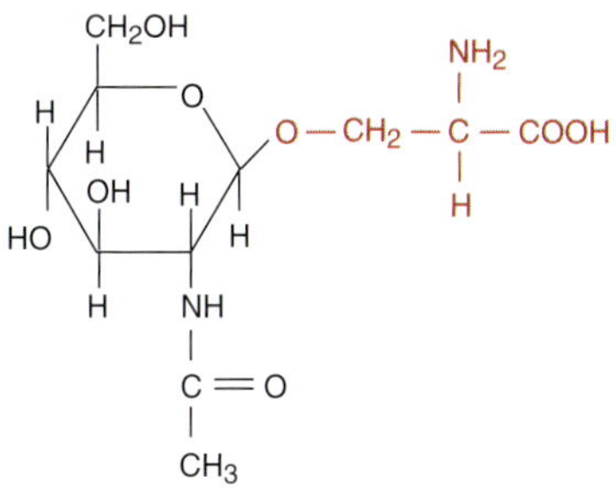

Type II *O*-Glycosyl linkage to serine

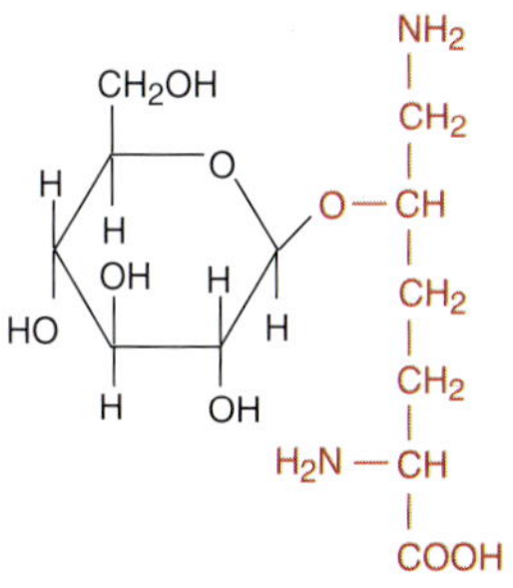

Type III *O*-Glycosyl linkage to 5-hydroxylysine

FIGURE 3.45

Examples of glycosidic linkages to amino acids in proteins. Type I is an *N*-glycosidic linkage through an amide nitrogen of Asn; type II is an *O*-glycosidic linkage through OH of Ser or Thr; and type III is an *O*-glycosidic linkage to 5-OH of 5-hydroxylysine. Amino acids in red.

3.7 FOLDING OF PROTEINS FROM RANDOMIZED TO UNIQUE STRUCTURES: PROTEIN STABILITY

The Protein Folding Problem: A Possible Pathway

The ability of a primary protein structure to fold spontaneously to its native secondary or tertiary conformation, without any information other than the amino acid sequence and the noncovalent forces that act on the sequence, has been demonstrated. Pancreatic

CLINICAL CORRELATION 3.7
Glycosylated Hemoglobin, HbA_{1c}

A glycosylated hemoglobin, designated HbA_{1c}, is formed nonenzymatically in red blood cells by combination of the NH_2 terminal amino groups of the hemoglobin β chains and glucose. The aldehyde form of glucose first forms a Schiff base with the NH_2 terminal amino group,

```
            OH
            |
  —N=C—C—
       |    |
       H   H
```

which then rearranges to a more stable amino ketone linkage,

```
        H   O
        |   ||
  —N—C—C—
    |   |
    H  H
```

by a spontaneous reaction known as the Amadori rearrangement. The concentration of HbA_{1c} depends on the concentration of glucose in the blood and the duration of hyperglycemia. In prolonged hyperglycemia, the concentration may rise to 12% or more of the total hemoglobin. Patients with diabetes mellitus have high concentrations of blood glucose and therefore high amounts of HbA_{1c}. The changes in the concentration of HbA_{1c} in diabetic patients can be used to follow the effectiveness of treatment for the diabetes.

Source: Bunn, H. F. Evaluation of glycosylated hemoglobin in diabetic patients. *Diabetes* 30:613, 1980. Brown, S. B. and Bowes, M. A. Glycosylated hemoglobins and their role in management of diabetes mellitus. *Biochem. Educ.* 13:2, 1985.

ribonuclease spontaneously refolds to its native conformation after being denatured, and its disulfide bonds are reduced without the hydrolysis of peptide bonds. Such observations led to the hypothesis that a polypeptide sequence promotes spontaneous folding to its unique active conformation under correct solvent conditions and in the presence of prosthetic groups that may be a part of its structure. As described below, chaperone proteins may facilitate the rate of protein folding. Quaternary structures also assemble spontaneously, after the tertiary structure of the subunits has formed.

It may appear surprising that a polypeptide chain folds into a single unique conformation given all the possible rotational conformations available around single bonds in its primary structure. For example, the α-chain of hemoglobin contains 141 amino acids, and there are at least four single bonds per amino acid residue around which free rotation can occur. If each bond about which free rotation occurs has two or more stable rotamer conformations accessible to it, then there are a minimum of 4^{141} or 5×10^{86} possible conformations for the α-chain polypeptide.

The conformation of a protein is the one of lowest Gibbs free energy accessible to its sequence within a physiological time frame. Thus folding is under thermodynamic and kinetic control. Although detailed knowledge of de novo folding of a polypeptide is at present an unattainable goal, the involvement of certain processes appears reasonable. There is evidence that in many cases folding is initiated by short-range noncovalent interactions between a side chain and its nearest neighbors that form secondary structures in small regions of the polypeptide. Particular side chains have a propensity to promote the formation of α-helices, ß-strands, and sharp turns or bends (ß-turns) in the polypeptide. Such regions of polypeptide, called **initiation sites**, spontaneously assume a secondary structure. The partially folded structures then interact with each other to form a **molten-globule** state. This is a condensed intermediate on the folding pathway that contains much of the secondary structure elements of the native conformation but many incorrect tertiary structure interactions. Regions of secondary structure in the molten-globule state are highly mobile relative to one another, and the molten globule is in rapid equilibrium with the fully unfolded state. The correct medium- and long-range interactions between different initiation sites are found by rearrangements within the molten globule; and the low free energy, native conformation of the polypeptide chain is formed. This is followed by formation of disulfide bonds (cystines). The rate-determining step for folding and unfolding of the native conformation lies between the molten globule and the native structure. For some proteins there may exist two or more thermodynamically stable folded conformations of low Gibbs free energy. Clinical Correlation 3.8 contains a discussion of proteins folding and prion infectious agents.

Chaperone Proteins May Assist the Protein Folding Process

Cells contain proteins that facilitate the folding process. These include *cis–trans*-prolyl isomerases, protein disulfide isomerases, and chaperone proteins. The ***cis–trans*-prolyl isomerases** interconvert cis- and trans-peptide bonds of proline residues. This allows the correct prolyl peptide bond conformation to form for each proline as required by the native conformation. **Protein disulfide isomerases** catalyze the breakage and formation of disulfide cystine linkages, so incorrect linkages are not stabilized and the correct arrangement of linkages for the native conformation is rapidly achieved.

Chaperone proteins were discovered as **heat shock proteins (hsps)**, a family of proteins whose synthesis is increased at elevated temperatures. The chaperones do not change the final outcome of the folding process but prevent protein aggregation prior to completion of folding and prevent formation of metastable dead-end or nonproductive intermediates. Chaperones of the hsp 70-kDa family bind to polypeptides as they are synthesized on the ribosomes, shielding the hydrophobic surfaces that would normally be exposed to solvent. This protects the protein from aggregation until the full chain is synthesized and folding can occur. Some proteins, however, cannot complete their folding process while in the presence of hsp70 chaperones and are delivered to the hsp60 family (GroEL in *Escherichia coli*) of chaperone proteins, also called **chaperonins**. The chaperonins are long cylindrical multisubunit structures that bind unfolded polypeptides

CLINICAL CORRELATION 3.8

Proteins as Infectious Agents: Prions and Human Transmissible Spongiform Encephalopathies (TSEs)

Proteins that act as infectious agents in the absence of DNA or RNA are known as prions. Creutzfeldt–Jakob disease (CJD) is the most common of the prion diseases. It is characterized by ataxia, dementia, and paralysis and is almost always fatal. Pathological examination of the brains shows amyloid plaques and spongiform (vacuolar) degeneration of the brain. The disease can be generated by an inherited mutation (familial disease), a sporadic mutation, or an infectious process. The infectious form was first observed among the Fore people of New Guinea, who exhibited a transmissible spongiform encephalopathy called Kuru, in which the prions were transmitted through ritual cannibalism. More recently, bovine spongiform encephalopathy or mad cow disease has been transmitted to humans through the ingestion of meat from cattle infected by bovine spongiform encephalopathy.

CJD is believed to be caused by a conformational change in the prion protein, PrP, from its normal soluble cellular conformation, PrP^c, to a toxic conformation, PrP^{Sc}. The PrP^{Sc} polymerizes into insoluble amyloid fibers, which cause the neurotoxic effects. The soluble PrP^c domain contains three α-helical and two small β-strand segments. The conversion to the PrP^{Sc} form is characterized by the conversion of two of the α-helical segments into β-strands. The PrP^{Sc} conformation polymerizes into amyloid fibers through β-strand interactions between monomers. Formation of the amyloid polymer is irreversible. An unknown factor, designated protein X (perhaps a chaperone protein), facilitates the conversion of the α-helical to β-strand conformation and/or promotes the polymerization of the β-strand conformation into the amyloid fiber.

Two mechanisms are suggested for amyloid plaque formation initiated by an infective prion. In the Nucleation-Polymerization Mechanism (A), PrP^c (green) is in rapid equilibrium with the PrP^{Sc} conformation (brown) (step 1), but the polymerization into an amyloid fiber is slow (step 2) in the absence of an initiator molecule. The introduction of a PrP^{Sc} infective prion (red) nucleates the polymerization process (step 3). The process is propagated by fragments from the newly generated polymer (step 5).

In the Template-Directed Mechanism (B), the prion, PrP^{Sc}, serves as a template to promote the change in conformation of endogenous PrP^c to PrP^{Sc} (steps 1 and 2). Once converted to the PrP^{Sc}, the PrP^{Sc} rapidly polymerize into insoluble fibrils. Familial forms of Creutzfeldt–Jakob disease are induced by mutant PrP proteins that have a higher tendency to spontaneously form the PrP^{Sc}.

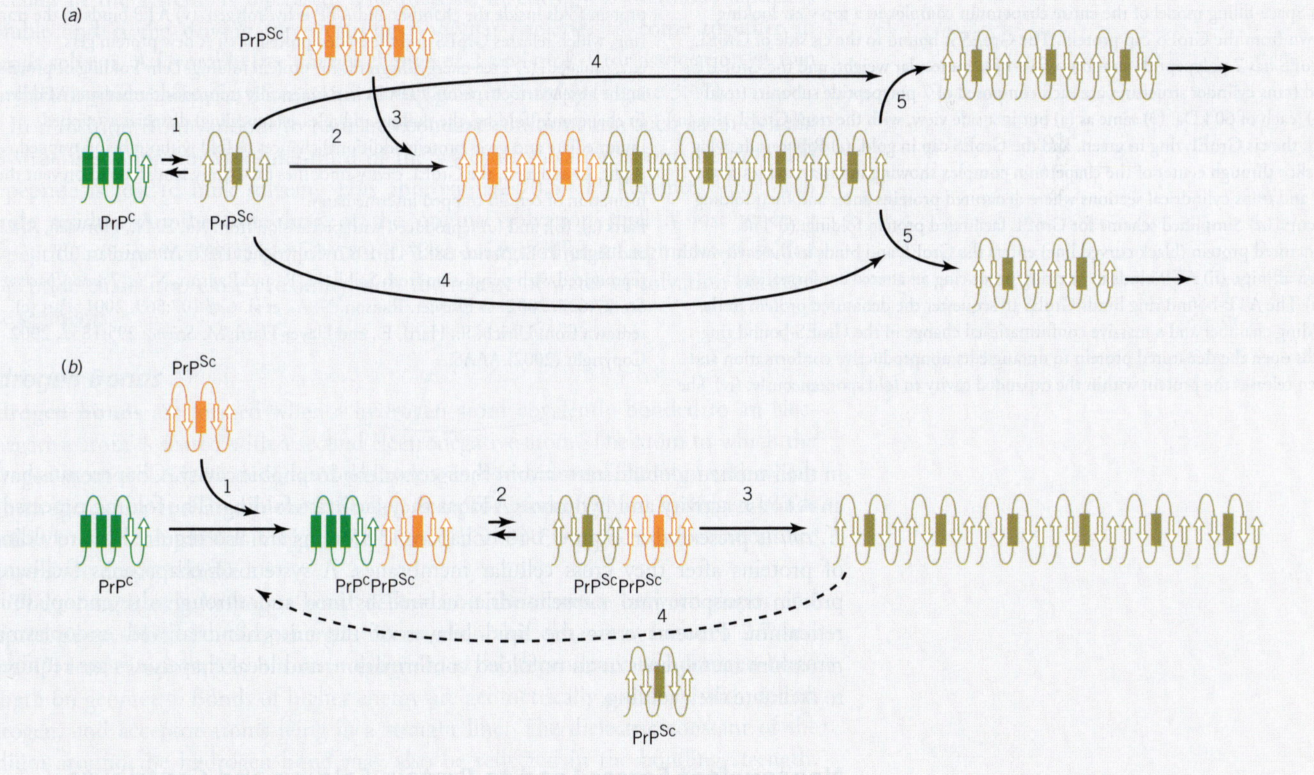

Source: Aguzzi, A. and Haass, C. Games played by rogue proteins in prion disorders and Alzheimer's disease. *Science* 302, 814, 2003. Aguzzi, A. and Polymenidou, M. Mammalian prion biology: One century of evolving concepts. *Cell* 116: 313, 2004. Horwich, A. L. and Weissman, J. S. Deadly conformations—protein misfolding in prion disease. *Cell* 89:499, 1997. Prusiner, S. B. Prion diseases and the BSE crises. *Science* 278:245, 1997.

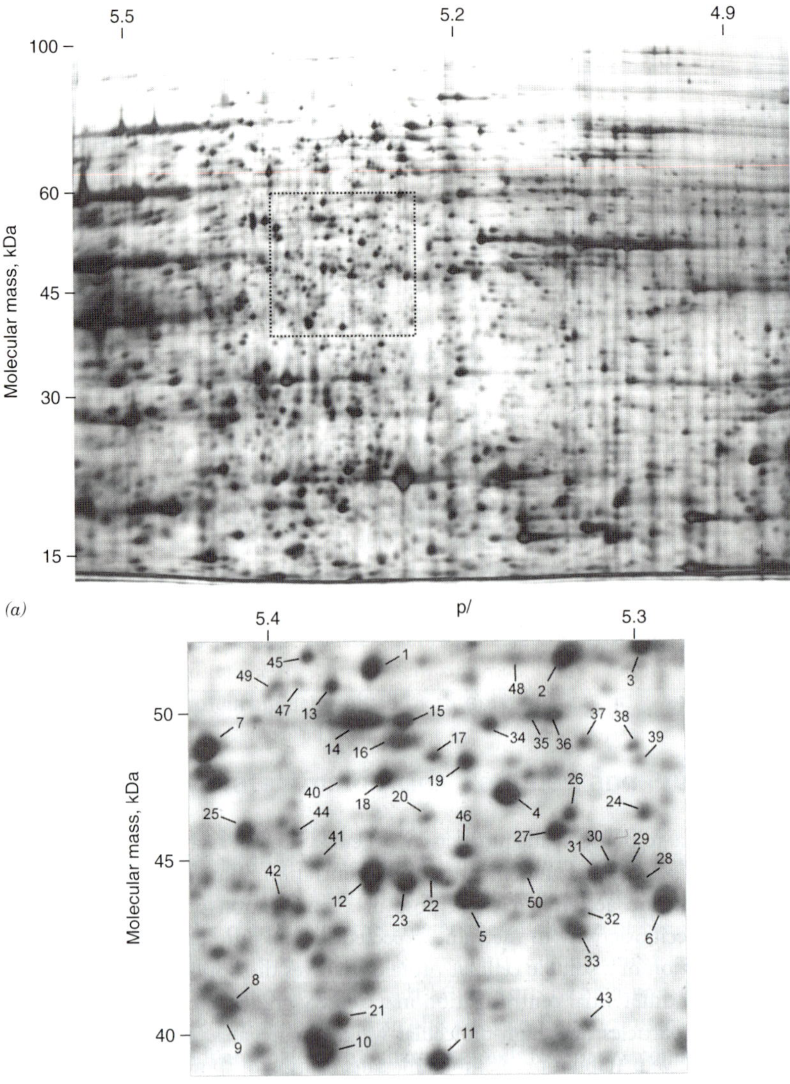

FIGURE 3.63

A two-dimensional (2-D) display of expressed proteins from cultured cells. (*a*) Soluble proteins from cellular extract (500 μg) loaded on the gel and separated by isoelectric focusing (pH 4.9–5.5) in the horizontal direction and by molecular mass in the vertical direction (electrophoresed in the presence of SDS detergent). More than 1500 proteins are observed in the gel by silver staining.
(*b*) A region from the gel expanded to show detail. Numbered proteins were analyzed by protease hydrolysis and mass spectrometry to determine their partial amino acid sequences, leading to their identification.
Reproduced with permission from Gygi, S. P., Corthals, G. L., Zhang, Y., Rochon, Y., and Aebersold, R. *Proc. Natl. Acad. Sci. USA* 97:9390, 2000.

molecular mass (see Polyacrylamide Gel Electrophoresis in the Presence of a Detergent). The resulting gel is stained for protein, and the intensity of each of the thousands of protein spots is measured in order to determine whether a particular protein is expressed and its concentration (Figure 3.63).

Determining the identity of each of the protein spots in a 2-D gel is not a trivial task. The 2-D gel pattern, if carried out under standard conditions, may be compared to patterns obtained by reference laboratories who have determined the identity of the majority of spots in the 2-D pattern from a particular cell type. These reference 2-D patterns are available over the Internet. More definitively, a spot may be extracted from the gel and then the protein partially hydrolyzed into smaller peptide fragments by proteolytic enzyme digestion (e.g., trypsin or chymotrypsin), and the peptide fragments are subjected to mass spectroscopy. The mass spectroscopy rapidly determines the amino acid sequence of many of the small fragments. This technique is called peptide **mass fingerprinting**. Utilizing these sequences to search protein sequence or gene sequence databases leads to the identification of the protein extracted from the 2-D gel. Robotic instruments now perform each of the steps in protein spot extraction and identification. In this way, the thousands of expressed proteins may be identified.

The technique currently fails to identify low abundance proteins in cells or tissues. In addition, certain types of proteins are difficult to analyze due to low solubility, low molecular charge, or very low molecular mass. For example, integral membrane proteins are highly hydrophobic and not soluble in the standard isoelectric focusing solvents.

Determination of Amino Acid Composition of a Protein

Determination of the amino acid composition is an essential component in study of a protein's structure and physiological properties. A protein is hydrolyzed to its constituent amino acids by heating it at 110°C in 6 N HCl for 18–36 h, in a sealed tube under vacuum to prevent degradation of oxidation-sensitive side chains by oxygen in air. Tryptophan is destroyed in this method, and alternative procedures are used for its analysis. Side-chain amides of asparagine and glutamine are hydrolyzed to aspartate and glutamate and free ammonia; they are included within the glutamic acid and aspartic acid content in the analysis.

Common procedures for amino acid identification use cation-exchange chromatography or reverse-phase HPLC to separate them and then react with ninhydrin, fluorescamine, dansyl chloride, or similar chromophoric or fluorophoric reagents for quantitation (Figure 3.62). With some types of derivatization, amino acids are identified at concentrations of 0.5×10^{-12} mol (pmol). Analysis of the amino acid composition of physiological fluids (i.e., blood and urine) is utilized in diagnosis of disease (see Clin. Corr. 3.9).

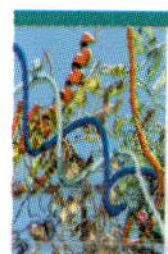

CLINICAL CORRELATION 3.9

Use of Amino Acid Analysis in Diagnosis of Disease

Elevated concentrations of amino acids are found in plasma or urine in a number of clinical disorders. An abnormally high concentration in urine is called an aminoaciduria. Phenylketonuria is a metabolic defect in which patients lack sufficient amounts of phenylalanine hydroxylase, which converts phenylalanine to tyrosine. As a result, large concentrations of phenylalanine, phenylpyruvate, and phenyllactate accumulate in the plasma and urine. Phenylketonuria occurs clinically in the first few weeks after birth, and if the infant is not placed on a special diet, severe mental retardation will occur (see Clin. Corr. 19.5). Cystinuria is a genetically transmitted defect in the membrane transport system for cystine and the basic amino acids (lysine, arginine, and the derived amino acid ornithine) in epithelial cells. Large amounts of these amino acids are excreted in urine. Other symptoms of this disease may arise from the formation of renal stones composed of cystine precipitated within the kidney (see Clin. Corr. 19.11). Hartnup disease is a genetically transmitted defect in epithelial cell transport of neutral-type amino acids (monoamino monocarboxylic acids), and high concentrations of these are found in the urine. The symptoms of the disease are primarily caused by a deficiency of tryptophan. These may include a pellagra-like rash (nicotinamide is partly derived from tryptophan precursors) and cerebellar ataxia (irregular and jerky muscular movements) due to the toxic effects of indole derived from the bacterial degradation of unabsorbed tryptophan present in large amounts in the gut (see Clin. Corr. 27.2). Fanconi's syndrome is a generalized aminoaciduria associated with hypophosphatemia and glucosuria. Abnormal reabsorption of amino acids, phosphate, and glucose by the tubular cells is the underlying defect.

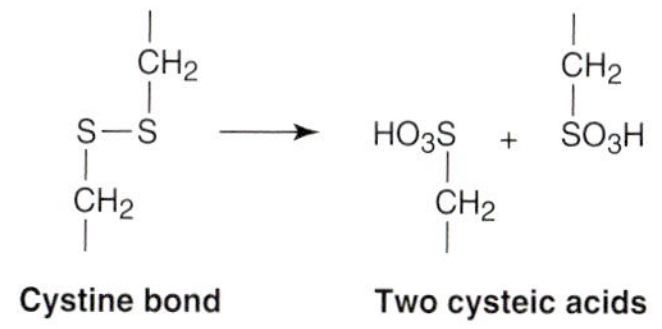

Determination of Amino Acid Sequence

The ability to clone genes has led to determination of the amino acid sequence of a protein from its DNA or mRNA sequence. This is a much faster method for obtaining an amino acid sequence. Sequencing of a protein, however, is required for the determination of modifications to the protein structure that occur after its biosynthesis, to identify a part of the protein sequence in order that its gene can be cloned, and to identify a protein as the product of a particular gene. Determination of primary structure of a protein requires a purified protein and determination of the number of chains in it. Individual chains are purified by the same techniques used in purification of the whole protein. If disulfide bonds join the chains, these bonds have to be broken (Figure 3.64).

Polypeptides are most commonly sequenced by the **Edman reaction** or by **mass spectroscopy**. In the Edman reaction, the polypeptide chain is reacted with phenylisothiocyanate, which reacts with the NH_2-terminal amino group. Acidic conditions catalyze intramolecular cyclization that cleaves the NH_2-terminal amino acid as a phenylthiohydantoin derivative (Figure 3.65). This amino acid derivative may be separated chromatographically and identified against standards. The remaining polypeptide is isolated, and the Edman reaction is repeated to identify the next NH_2-terminal amino acid. Theoretically, this can be repeated until the sequence of the entire polypeptide is determined but under favorable conditions can only be carried out for 30 or 40 amino acids into the polypeptide chain, when impurities generated from incomplete reactions in the reaction series make further **Edman cycles** unfeasible. Polypeptides longer than 30 or 40 amino acids are hydrolyzed into smaller fragments and sequenced in sections. For sequencing by mass spectroscopy, it is also necessary to break long polypeptide chains into smaller fragments.

Enzymatic and chemical methods are used to break polypeptide chains into smaller fragments (Figure 3.66). **Trypsin** preferentially cleaves the peptide bond on the COOH-terminal side of lysine and arginine within polypeptide chains. **Chymotrypsin** cleaves the peptide bond on the COOH-terminal side of large apolar side chains. Other enzymes cleave polypeptide chains on the COOH-terminal side of glutamic and aspartic acid. **Cyanogen bromide** cleaves peptide bonds on the COOH-terminal side of methionine residues (Figure 3.66). After **partial hydrolysis**, the segments are separated, and the sequence of each is determined by the Edman reaction or mass spectroscopy. To place the sequenced peptides correctly into the complete sequence, another sample of the original polypeptide is subjected to partial hydrolysis by a different hydrolytic reagent from that used initially, and the fragments are separated and sequenced. This generates overlapping sequences, which permit determination of the complete sequence (Figure 3.67).

FIGURE 3.64
Breaking of disulfide bonds by oxidation to cysteic acids.

Doolittle, R. F. The multiplicity of domains in proteins. *Annu. Rev. Biochem.* 64:287, 1995.

Fasman, G. D. Protein conformational prediction. *Trends Biochem. Sci.* 14:295, 1989.

Finkelstein, A. V., Gutun, A. M., and Badretdinov, A. Ya. Why are the same protein folds used to perform different functions? *FEBS Lett.* 325:23, 1993.

Jones, D. T. Learning to speak the language of proteins. *Science* 302:1347, 2003.

Orengo, C. A., Jones, D. T., and Thornton, J. M. Protein superfamilies and domain superfolds. *Nature* 372:631, 1994.

Richardson, J. S. The anatomy and taxonomy of protein structure. *Adv. Protein Chem.* 34:168, 1981.

Rose, G. D. and Wolfenden, R. Hydrogen bonding, hydrophobicity, packing, and protein folding. *Annu. Rev. Biophys. Biomol. Struct.* 22:381, 1993.

Srinivasan, R. and Rose, G. D. A physical basis for protein secondary structure. *Proc. Natl. Acad. Sci. U.S.A.* 96:14258, 1999.

Stevens, R. C. Long live structural biology. *Nature Structural Mol. Biol.* 11:293, 2004.

Protein Folding

Baldwin, R. L. and Rose, G. D. Is protein folding hierarchic? I. Local structure and peptide folding. *Trends Biochem. Sci.* 24:26, 1999.

Baldwin, R. L. and Rose, G. D. Is protein folding hierarchic? II. Folding intermediates and transition states. *Trends Biochem Sci.* 24:77, 1999.

Clark, P. L. Protein folding in the cell: Reshaping the folding funnel. *Trends Biochem. Sci.* 29:527, 2004.

Dinner, A. R., Sali, A., Smith, L. J., Dobson, C. M., and Karplus, M. Understanding protein folding via free-energy surfaces from theory and experiment. *Trends Biochem. Sci.* 25:331, 2000.

Dobson, C. M. Protein folding and misfolding. *Nature* 426:884, 2003.

Hartl, F. U. and Hayer-Hartl, M. Molecular chaperones in the cytosol: From nascent chain to folded protein. *Science* 295:1852, 2002.

Lins, L. and Brasseur, R. The hydrophobic effect in protein folding. *FASEB J.* 9:535, 1995.

Saibil, H. R. and Ranson, N. A. The chaperonin folding machine. *Trends Biochem. Sci.* 27:627, 2002.

Selkoe, D. J. Cell biology of protein misfolding: The examples of Alzheimer's and Parkinson's diseases. *Nature Cell Biology* 6:1054, 2004.

Shakhnovich, E., Abkevich, V., and Ptitsyn, O. Conserved residues and the mechanism of protein folding. *Nature* 379:96, 1996.

Proteomics and Protein Networks

Bray, D. Molecular networks: The top-down view. *Science* 301:1864, 2003.

Eisenberg, D., Marcotte, E. M., Xenarios, I., and Yeates, T. O. Protein function in the postgenomic era. *Nature* 405:823, 2000.

Gavin, A. C., Bosche, M., Krause, R., et al. Functional organization of the yeast proteome by systematic analysis of protein complexes. *Nature* 415:141, 2002.

Koonin, E. V., Wolf, Y. I., and Karev, G. P. The structure of the protein universe and genome evolution. *Nature* 420:218, 2002.

Pandey, A. and Mann, M. Proteomics to study genes and genomes. *Nature* 405:837, 2000.

Pawson, T. and Nash, P. Assembly of cell regulatory systems through protein interaction domains. *Science* 300:445, 2003.

Petricoin, E. F., III, Ardekani, A. M., Hitt, B. A, et al. Use of proteomic patterns in serum to identify ovarian cancer. *Lancet* 359:572, 2002.

Zhu, H., Bilgin, M., and Snyder, M. Proteomics. *Annu. Rev. Biochem.* 72:783, 2003.

Techniques for The Study of Proteins

Bax, A. and Grzesiek, S. Methodological advances in protein NMR. *Acc. Chem. Res.* 26:131, 1993.

Mann, M., Hendrickson, R. C., and Pandey, A. Analysis of proteins and proteomes by mass spectrometry. *Annu. Rev. Biochem.* 70:437, 2001.

Reif, O. W., Lausch, R., and Fritag, R. High-performance capillary electrophoresis of human serum and plasma proteins. *Adv. Chromatogr.* 34:1, 1994.

Rhodes, G. *Crystallography Made Crystal Clear*, 2nd ed. San Diego, CA: Academic Press, 2000.

Tugarinov, B., Hwang, P. M., and Kay, L. E. Nuclear magnetic resonance spectroscopy of high-molecular-weight proteins. *Annu. Rev. Biochem.* 73:107, 2004.

Dynamics in Folded Proteins

Daggett, V., and Levitt, M. Realistic simulations of native-protein dynamics in solution and beyond. *Annu. Rev. Biophys. Biomol. Struct.* 22:353, 1993.

Joseph, D., Petsko, G. A., and Karplus, M. Anatomy of a conformational change: hinged lid motion of the triosephosphate isomerase loop. *Science* 249:1425, 1990.

Glycoproteins

Drickamer, K. and Taylor, M. E. Evolving views of protein glycosylation. *Trends Biochem. Sci.* 23:321, 1998.

Grogan, M. J., Pratt, M. R., Marcaurelle, L. A., and Bertozzi, C. R. Homogeneous glycopeptides and glycoproteins for biological investigation. *Annu. Rev. Biochem.* 71:593, 2002.

Lis., H. and Sharon, N. Protein glycosylation. Structural and functional aspects. *Eur. J. Biochem.* 218:1, 1993.

Lipoproteins

Jonas, A. In: D. E. Vance and J. E. Vance (Eds.), Lipoprotein Structure in *Biochemistry of Lipids, Lipoproteins, and Membranes*, 4th ed. Amsterdam: Elsevier, 2002, p. 483.

Myers, G. L., Cooper, G. R., and Sampson, E. J. Traditional lipoprotein profile: Clinical utility, performance requirement, and standardization. *Atherosclerosis* 108:S157, 1994.

Segrest, J. P., Garber, D. W., Brouillette, C. G., Harvey, S. C., and Anantharamaiah, G. M. The amphipathic α helix: A multifunctional structural motif in plasma apolipoproteins. *Adv. Protein Chem.* 45:303, 1994.

Collagen

Brodsky, B., and Shah, N. K. The triple-helix motif in proteins. *FASEB J.* 9:1537, 1995.

Myllyharju, J. and Kivirikko, K. I. Collagens, modifying enzymes, and their mutations in humans, flies, and worms. *Trends Genetics* 20:33, 2004.

Prockop, D. J. and Kivirikko, K. I. Collagens: Molecular biology, diseases, and potentials for therapy. *Annu. Rev. Biochem.* 64:403, 1995.

QUESTIONS | CAROL N. ANGSTADT

Multiple Choice Questions

1. All of the following are correct about a peptide bond *except*:
   A. it exhibits partial double bond character.
   B. it is more stable in the *cis* configuration than in the *trans* configuration.
   C. it has restricted rotation around the carbonyl carbon to nitrogen bond.
   D. it is planar.
   E. in proline, the nitrogen is attached to the side chain.

2. In an α-helix:
   A. side-chain groups can align to give a polar face.
   B. each peptide bond forms two hydrogen bonds.
   C. there are 3.6 amino acids per turn.
   D. all of the above are correct.
   E. none of the above are correct.

3. Chaperone proteins:
   A. all require ATP to exert their effect.
   B. cleave incorrect disulfide bonds, allowing correct ones to subsequently form.
   C. guide the folding of polypeptide chains into patterns that would be thermodynamically unstable without the presence of chaperones.
   D. of the Hsp70 class are involved in transport of proteins across mitochondrial and endoplasmic reticulum membranes.
   E. act only on fully synthesized polypeptide chains.

4. Proteins may be separated according to size by:
   A. isoelectric focusing.
   B. polyacrylamide gel electrophoresis.
   C. ion exchange chromatography.
   D. molecular exclusion chromatography.
   E. reverse-phase HPLC.

5. All lipoprotein particles in the blood have the same general architecture which includes:
   A. a neutral core of triacylglycerols and cholesteryl esters.
   B. amphipathic lipids oriented with their polar head groups at the surface and their hydrophobic chains oriented toward the core.
   C. most surface apoproteins containing amphipathic helices.
   D. unesterified cholesterol associated with the outer shell.
   E. all of the above.

6. Glycoproteins:
   A. are found in cells but not in plasma.
   B. in a plasma membrane, typically have the carbohydrate portion on the cytosolic side.
   C. may have the carbohydrate portion covalently linked to the protein at an asparagine.
   D. that are carbohydrate to hydroxyl linked always have the linkage to hydroxylysine.
   E. of a given type always have identical carbohydrate chains.

Questions 7 and 8: Abnormalities in the synthesis or structure of collagen cause dysfunctions in cardiac organs, bone, skin, joints, and eyes. Problems may result from abnormal collagen genes, abnormal posttranslational modifications of collagen, or deficiency of cofactors needed by enzymes responsible for posttranslational modifications. Scurvy, a lack of vitamin C, is an example of the last type.

7. In collagen:
   A. intrachain hydrogen bonding stabilizes the native structure.
   B. three chains with polyproline type II helical conformation can wind about one another to form a superhelix because of the structure of glycine.
   C. the ϕ angles contributed by proline are free to rotate.
   D. regions of superhelicity comprise the entire structure except for the N- and C-termini.
   E. cross-links between triple helices form after lysine is converted to allysine.

8. The formation of covalent cross-links in collagen:
   A. occurs during synthesis of the peptide chain.
   B. uses hydroxyproline.
   C. involves glycine residues.
   D. requires conversion of some ε-amino groups of lysine to δ-aldehydes.

Questions 9 and 10: Hundreds of variants of human hemoglobin are known, with about 95% resulting from a single amino acid substitution. Many show no clinical effects, while others cause serious abnormalities. HbS causes sickle cell disease. In HbS, glutamate in the 6th position of the β-chain of HbA (normal adult hemoglobin) has been replaced by a valine. This replacement causes HbS, in its deoxy form, to polymerize. The result is a change in red cell structure to a sickle shape which leads to clogging of capillaries. Some other hemoglobin variants are Hiroshima (His 146 $\rightarrow$ Asp), Riverdale-Bronx (Gly 24 $\rightarrow$ Asp), and $M_{\text{Hyde Park}}$ (His 92 $\rightarrow$ Tyr).

9. Which of the following statements is correct?
   A. The substitution in HbS is conservative.
   B. All four hemoglobins have nonconservative substitutions.
   C. HbS has a nonconservative substitution but the other three are conservative.
   D. The substitution in $M_{\text{Hyde Park}}$ would not be likely to cause a change in structure.
   E. All amino acid substitutions in proteins cause changes in the protein's function.

10. HbS polymerizes because:
   A. the structure shifts to bury valine in the interior of the protein.
   B. the C-terminal end of the chain has been modified.
   C. a positive charge has been introduced on the surface to attract negative charges.
   D. the hydrophobic group on the surface is attracted to other hydrophobic groups.
   E. valine has a strong tendency to hydrogen bond.

Questions 11 and 12: Many pathological hyperlipidemias result from abnormalities in the rates of synthesis or clearance of lipoproteins in the blood. They are usually characterized by elevated levels of cholesterol and/or triacylglycerols in the blood. Type I has very high plasma triacylglycerol levels (>1000 g/Dl^{-1}) because of an accumulation of chylomicrons. Type II (familial hypercholesterolemia) has elevated cholesterol, specifically in the form of LDL. Another abnormality of lipoproteins is hypolipoproteinemia, in which lipoproteins are not formed because of the inability to make a particular apoprotein.

11. If the serum of a patient with Type I hyperlipidemia were centrifuged, the lipid band would be found:

A. at the top of the tube.
B. just below the top of the tube.
C. below the top but still in the upper half of the tube.
D. about the middle of the tube.
E. near the bottom of the tube.

12. Abetalipoproteinemia is a disease in which chylomicrons, VLDL, and LDL are absent from the blood because of an inability to synthesize:
A. ApoA-I.
B. ApoB-100.
C. ApoC-II.
D. ApoD.
E. ApoE.

Problems

13. In a plot of equivalents of OH^- versus pH, pH is ~2 when 0.5 equiv has been used, pH is ~6 for 1.5 equiv and pH is ~9.5 for 2.5 equiv. What amino acid has been titrated? In each case, the point indicated is midpoint of a steep part of the curve.

14. After purification, the Edman reaction was used to sequence a dodecapeptide, yielding the following data: The C-terminal amino acid is isoleucine; N-terminal amino acid is methionine; peptide fragments are Ala-Ala-Ile, Leu-Arg-Lys-Lys-Glu-Lys-Glu-Ala, Met-Gly-Leu, and Met-Phe-Pro-Met. What is the sequence of this peptide?

ANSWERS

1. **B** This would put both side chains on the same side of the bond, which is less favorable. C: This is a consequence of A. D: All atoms attached to the carbon and nitrogen of the bond are in a common plane. E: Proline has an imino group.
2. **D** A: They could also align to give a nonpolar face. B: One is to the fourth residue above the bond, and the other is to the fourth residue below. C: This is one of the characteristics of a right-handed α-helix.
3. **D** Proteins cross the membrane in an unfolded state and refold once they cross the membrane. A: The hsp60 family of chaperones is ATP-linked, but the hsp70 family is not. B: Disulfide isomerases catalyze this reaction. C: The final product is thermodynamically stable; chaperones prevent unfavorable intermediate interactions. E: Hsp70 chaperones react with nascent polypeptide chains as they are synthesized by the ribosome. The protein may then be delivered to a hsp60 chaperone for facilitation of final folding.
4. **D** Another method that separates on the basis of size is SDS PAGE. A–C: These separate molecules on the basis of charge. E: Reverse-phase HPLC effects separations based on polarity.
5. **E** All lipoproteins share these characteristics. C: The polar face interacts with water, and the hydrophobic face is oriented toward the core. D: Cholesterol's hydroxyl group is sufficiently polar that it orients toward the outer shell.
6. **C** In the N-linked type, asparagine is in the sequence Asn-X-Thr(Ser). A: Most plasma proteins, except albumin, are glycoproteins. B: Carbohydrate is on the outside of the cell for functions like cell–cell recognition. D: This link is in collagen only. Other O-linked carbohydrates have the linkage to serine or threonine. E: Carbohydrate structure is not determined by genes and is variable.
7. **B** Close contacts in the interior of the triple helix are possible only when the R group at that position is very small, that is, hydrogen. A: Hydrogen bonding in collagen is interchain. C: The ϕ angle is part of the proline ring and is not free to rotate. D: Although the statement is true of type I collagen, the superhelical regions in other collagen types may be broken by regions of globular domains. E: The conversion and cross-linking are extracellular.
8. **D** These then react which each other or unmodified ϵ-amino groups of lysine. A: It is posttranslational. B, C: This is not the role of hydroxyproline or glycine.
9. **B** A nonconservative substitution is one in which the polarity changes significantly. A: Negative charge to hydrophobic group is nonconservative. C: Hiroshima is positive to negative, Riverdale-Bronx is slightly hydrophobic to positive, and Hyde Park is positive to very hydrophobic. D: This mutation leads to loss of the ligand bond to the iron heme and loss of heme from the heme binding site in the beta chain. E: Change in function depends on whether the substitution changes the structure and whether it is in a critical position. These variants are pathological.
10. **D** These are called "sticky ends" since valine is strongly hydrophobic and attracts other hydrophobic groups. A: Glutamate's original position is on the exterior, and this is where the valine remains. B: Chains are numbered from the N-terminal, so position 6 is near the N-terminal. C: There is no positive charge. E: Valine cannot hydrogen bond.
11. **A** Chylomicrons have a density less than water so float on the surface. B: VLDL might be found here. C: This band would be LDL. D: Probably none of the lipoproteins remain at this point. E: HDL is found near the bottom of the tube because it is the most dense.
12. **B** ApoB-100 is the protein recognized by LDL receptors. Chylomicrons have apoB-48, which is formed from apoB-100. A: This is the major protein of HDL. C: ApoC-II is present in all three particles to activate lipoprotein lipase but accounts for only 2–15% of total protein. D: This is present only in chylomicrons and then only at 1%. E: This is found in all three particles but at 5–20% level.
13. This is a titration curve for an acid with three dissociable groups. The points indicated represent half-titration of each of these groups, and the pHs are the pK values for each group. With these pK values, the amino acid must be histidine.
14. You must have the correct N- and C-terminal amino acids and the correct total number (12) of amino acids in the peptide. Using these values and overlapping the fragments, you should get the sequence: Met-Phe-Pro-Met-Gly-Leu-Arg-Lys-Glu-Ala-Ala-Ile. Notice that the Lys-Glu fragment does not give any additional information for the sequencing.

PART II TRANSMISSION OF INFORMATION

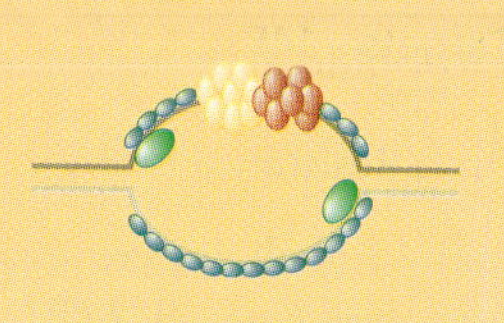

4

DNA REPLICATION, RECOMBINATION, AND REPAIR

Howard J. Edenberg

Textbook of Biochemistry With Clinical Correlations, Sixth Edition, Edited by Thomas M. Devlin

is added. It also leads to removal of some properly incorporated nucleotides, but that is a price that must be paid for the increased accuracy provided by proofreading. The ability to discriminate between properly and incorrectly incorporated nucleotides is compromised when the growing chain is very short, because the strength of base-pairing between a short chain and template is weak. This is the most likely explanation for the evolution of DNA polymerases unable to start chains *de novo*, and for the use of ribonucleotides to prime DNA synthesis.

Some mutated DNA polymerases are actually more accurate than the normal polymerase. This is paradoxical if one assumes that there is an evolutionary advantage to ever-increasing accuracy. However, there are reasons to expect that beyond a certain level, higher accuracy is not advantageous and may in fact be disadvantageous. First, extremely accurate proofreading is energetically costly. To ensure that virtually all misincorporated bases are removed, a polymerase would also have to remove a large number of correctly incorporated bases that have moved slightly with respect to the template due to normal thermal fluctuations. This would greatly slow replication and utilize much additional energy in the form of dNTPs. Second, the ability of DNA repair systems to deal with residual errors reduces the advantages of ever more accurate polymerases. It has also been argued that a low level of mutations provides the raw material for evolution, producing a population with some genetic variation that is better able to survive changing conditions.

Despite their very high accuracy, polymerases can incorporate nucleotide analogs. Nucleoside analogs are often used in chemotherapy to kill rapidly growing cancer cells or viruses responsible for serious illnesses. Nucleoside analogs that are phosphorylated to nucleotides can be incorporated into DNA, where they can inhibit further synthesis or lead to a high level of mutation. Differences between bacterial or viral polymerases and host polymerases in their ability to incorporate nucleotide analogs can provide a therapeutic window within which physicians can preferentially target infected cells.

Separating Parental Strands: The Replication Fork

Both prokaryotes and eukaryotes solve problems of replication in fundamentally similar ways. Separation of the two parental strands of DNA allows each to serve as a template for a new complementary strand. The separation of parental strands creates a structure called a **replication fork** (Figure 4.5). This separation requires considerable input of energy. Melting double-stranded DNA into two single strands normally occurs only at elevated temperatures, usually over 90°C. To separate the parental strands at physiological temperatures, cells use enzymes called **helicases**. Helicases bind to single-stranded DNA and rachet along it in a fixed direction, with each step requiring hydrolysis of ATP. This "pushes apart" the parental DNA to form a replication fork (Figure 4.5).

In the absence of additional proteins, parental strands would quickly reanneal behind the helicase, because the complementary strands are close and aligned in proper register. Reannealing is prevented by **single-stranded DNA-binding proteins (SSBs)**

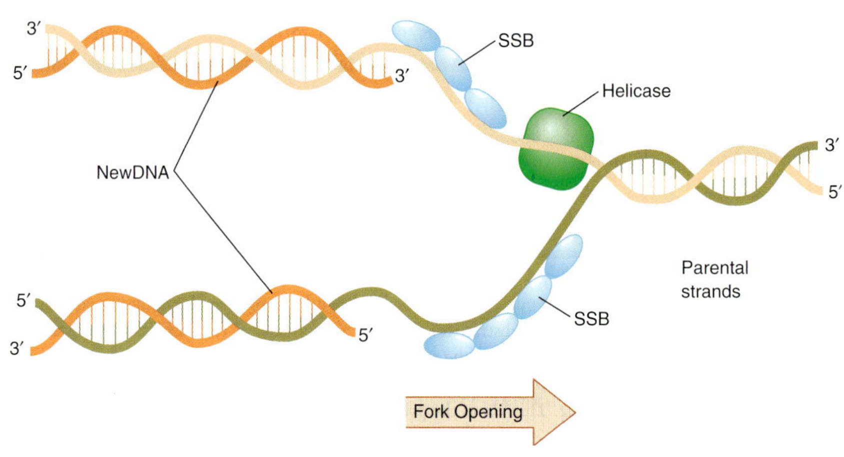

FIGURE 4.5

Separation of parental strands at a replication fork. DNA synthesis occurs at a structure called a replication fork, at which parental strands are separated. A helicase is needed to separate parental strands and allow the fork to progress. Single-stranded DNA-binding proteins (SSB) are needed to keep the strands apart. The parental strands twist around each other approximately every 10.5 bp, and they must untwist to separate; this creates topological problems. Note that one of the growing strands is oriented with its 3′-end toward the fork, while the other is oriented with its 5′-end toward the fork.

(Figure 4.5). SSBs keep the strands apart, reduce potential secondary structure (hairpins that might impede polymerization), and align the template strands for rapid DNA synthesis. SSBs are important not only in replication, but also in recombination and repair.

Solving The Polarity Problem: Semi-Discontinuous DNA Synthesis

Antiparallel strands at a replication fork present an immediate problem for the process of replication; one daughter strand is oriented with the 3′-end toward the fork, and the other strand has its 5′-end oriented toward the fork (Figure 4.5). The strand with its 3′-OH oriented toward the fork can be elongated simply by sequential addition of new nucleotides to this end (see Figure 4.3); this strand is called the **continuous, leading**, or **anteriograde** strand. The other strand presents the problem: No DNA polymerase catalyzes addition of nucleotides to the 5′-end of a growing chain. Yet when viewed on a larger scale, replication appears to proceed along both strands in the direction of fork movement. This problem is solved by synthesizing one strand as a series of short pieces, each made in the normal 5′ to 3′ direction, and later joining them (Figure 4.6). The strand made in pieces is called the **discontinuous, lagging**, or **retrograde** (backward going) strand; the small pieces from which the discontinuous strand is made are called **Okazaki fragments** in honor of Reiji Okazaki, who first demonstrated this process. Okazaki fragments in human cells average about 130–200 nucleotides in length. In *E. coli*, they are about 10 times larger. The overall process is called **semidiscontinuous** synthesis, because one strand is synthesized continuously and the other discontinuously.

Replication Fork Movement

Priming

DNA polymerases require primers. The **continuous** (leading) **strand** needs only a single priming event, which occurs during replication initiation (see below). But each Okazaki fragment of the discontinuous (retrograde, lagging) strand requires a separate primer. These primers are short stretches of RNA, synthesized by a special enzyme called a **primase**. In eukaryotic cells, the RNA primers are about 8–10 nucleotides long. They provide a free 3′-OH to which the first deoxyribonucleotide can be covalently added (Figure 4.7). Therefore, each Okazaki fragment carries a short stretch of RNA at its 5′-end, covalently attached to the DNA.

Strand Elongation

DNA chains grow by repeated addition of nucleotides to the 3′-OH ends of the chains by the mechanism shown in Figure 4.3. This reaction is catalyzed by a DNA polymerase.

Primer Removal

The RNA primers are removed from the 5′-ends of Okazaki fragments by enzymes with **RNase H** (**RNA hybridase**) activity. RNase H catalyzes hydrolysis of an RNA chain hydrogen-bonded to a DNA chain (i.e., an RNA–DNA hybrid). In eukaryotes, a special flap endonuclease (FEN1) also participates in this process; FEN1 cuts at a bend where the RNA portion has been peeled away from the template. All of the ribonucleotides are removed, leaving only DNA.

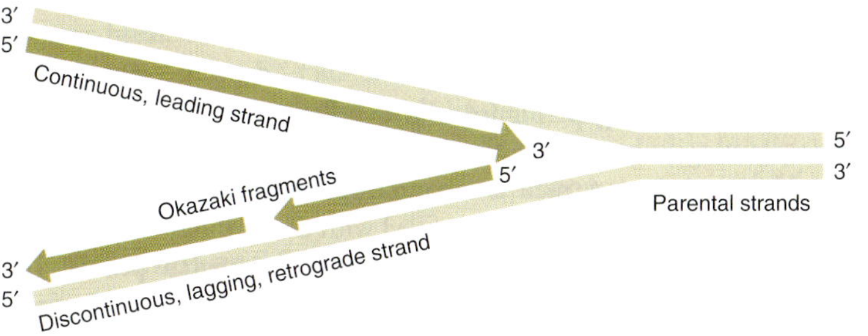

FIGURE 4.6

Solving the polarity problem: semidiscontinuous DNA synthesis. Both growing strands are elongated at their 3′-ends. The one with its 3′-end toward the fork can grow continuously; it is called the continuous or leading strand. Because no DNA polymerase can add nucleotides to the 5′-end of a growing chain, the strand with its 5′-end toward the fork is made as a series of short pieces, called Okazaki fragments, each synthesized in 5′ to 3′ direction and later joined. This is called the discontinuous, retrograde, or lagging strand.

encounters dNTPs at the 5′-end of the previously synthesized Okazaki fragment. The remaining nick is sealed by DNA ligase. The pol δ/PCNA complex of the replisome must then reattach to the next partly synthesized Okazaki fragment to repeat this process. As in *E. coli*, two molecules of the main replicative polymerase, pol δ in this case, are held together in a replisome or "replication factory" in which both strands are synthesized (Figure 4.11).

Eukaryotic DNA is packaged into **nucleosomes** that contain approximately 200 bp of DNA. Dissociation of nucleosomes is required for replication, and probably limits the rate of DNA synthesis. When a single nucleosome is dissociated, about 200 bp of parental DNA are available to be separated, and primer synthesis can occur somewhere in the exposed single-stranded DNA. This would explain the limited size of the Okazaki fragments in human cells.

Humans have both type I and type II topoisomerases. The human type I topoisomerase, called nicking–closing enzyme (Figure 4.9), can remove both positive and negative supercoils and functions during DNA replication and transcription. Type II topoisomerase is critical at the termination step and for segregation of chromosomes, which would otherwise be tangled together as the many replication bubbles are completed. Topo II is an abundant protein that also plays a role in attaching DNA to special sites in the nuclear matrix during interphase. Human type II topoisomerase is not a gyrase, in that it does not introduce negative supercoiling. Cancer chemotherapy often targets topoisomerases, using poisons that lead to double-strand breaks during replication. Rapidly replicating cells are more sensitive to these drugs than quiescent cells.

Initiation of Replication

In the previous discussion, we addressed progression of a replication fork. But, how does a replication fork get started? Replication begins from specific sites called **origins of replication** (**ori**). Known oris contain (a) multiple, short, repeated sequences that bind specific initiator proteins and (b) AT-rich regions at which the initial separation of parental strands occurs. The *E. coli* chromosome has a single origin of replication, oriC (origin of chromosomal replication), a region of approximately 245 bp. There are thousands of origins in eukaryotic cells (Figure 4.14). In yeast the oris are termed **autonomously replicating sequences** (**ARS**). In humans, specific sequences that serve as oris have not been identified.

In *E. coli*, oriC is bound by an **initiator protein** called **DnaA. DnaC** then associates and acts like a "matchmaker" to allow **DnaB**, the helicase, to bind and begin separating the parental strands to create a replication fork. At each ori, a replication bubble is

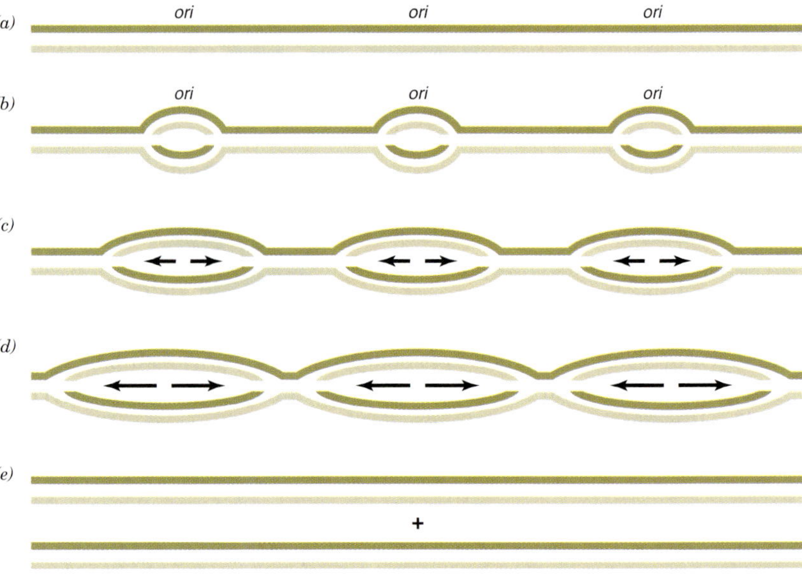

FIGURE 4.14

Tandem replicons and bidirectional replication in eukaryotes. (*a*) There are multiple origins of replication (ori) tandemly arrayed along each eukaryotic chromosome. In humans they are spaced at approximately 50,000- to 100,000-bp intervals. (*b*) Initiation occurs at each ori. Adjacent clusters of oris tend to function together. (*c*) Two diverging replication forks are established at each ori, so replication is bidirectional. The structures formed are called replication bubbles. (*d*) Replication bubbles enlarge as replication continues, until they are in close proximity. At that stage, termination of replication joins adjacent bubbles and unlinks parental DNA. Topoisomerase II is essential for termination of replication and segregation of the chromosomes. (*e*) The resulting duplicated chromosomes can then segregate into two daughter cells.

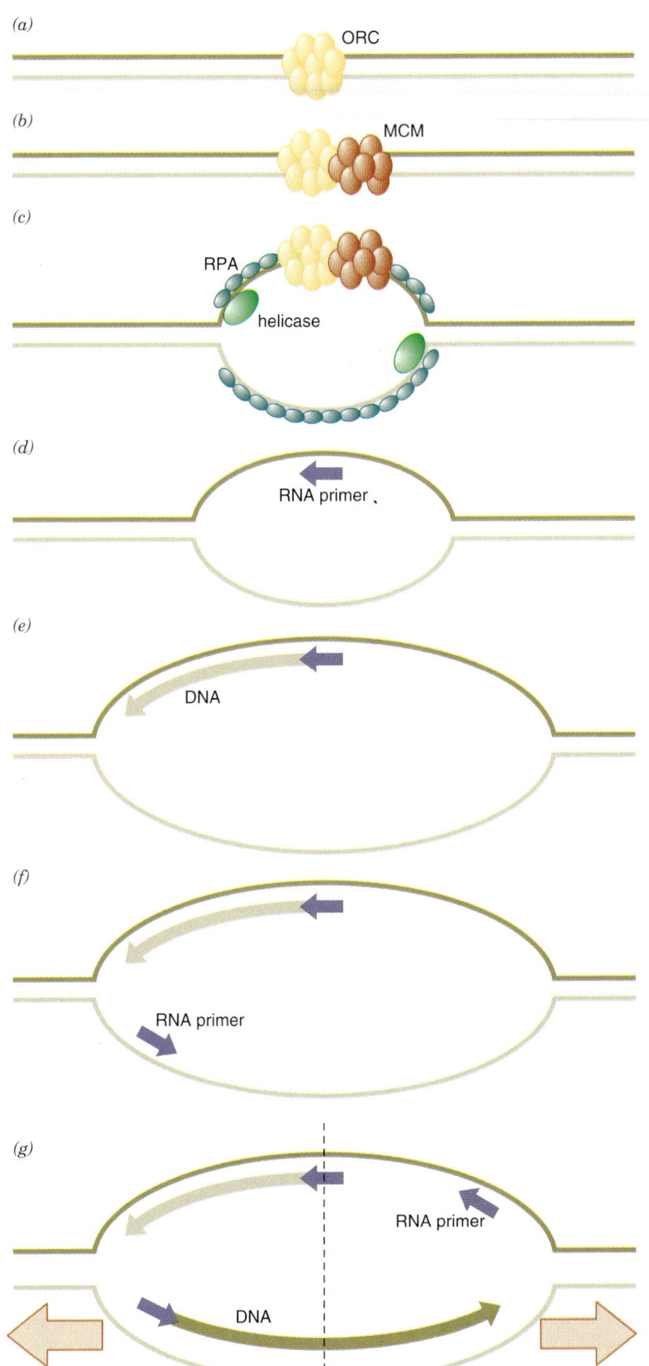

FIGURE 4.15

Initiation of replication at a eukaryotic replication origin. (*a*) The origin recognition complex (ORC) binds to an ori. (*b*) An MCM complex binds to this; cdc6p (cell division cycle 6 protein) is important in this assembly. (*c*) The initiator complex is activated and the helicase activity opens the parental strands to form a very small bubble. SSB binds to the exposed single strands, helicases are loaded onto the DNA, and the bubble is enlarged. (*d*) Polα/primase synthesizes the first RNA primer; after about 10 nt, the polα/primase switches to elongating the RNA primer with deoxyribonucleotides. [In this and subsequent panels, the focus is on processes occurring on DNA, and proteins are not shown.] (*e*) After about 15–30 deoxyribonucleotides are added, polα/primase leaves and the chain is elongated by DNA polymerase δ, which incorporates deoxyribonucleotides. To this stage, the process is like that occurring on the discontinuous (retrograde) strand (compare Figure 4.13). However, the elongating strand will not encounter a previous Okazaki fragment; it can continue elongating, becoming the continuous (leading) strand on the leftward-moving replication fork. (*f*) An RNA primer is synthesized on the discontinuous side of this replication fork, as in normal fork progression. (*g*) This primer is elongated as previously described. However, this elongating strand will not encounter a previous Okazaki fragment and can continue elongating, becoming the continuous (leading) strand on the rightward-moving replication fork. An RNA primer can be synthesized on this fork by the normal mechanism (compare Figure 4.13). The result is two replication forks diverging from the origin. The process is symmetrical around the axis indicated by the dashed line, with mirror-image forks diverging from ori.

formed and a pair of replication forks are established that move away from the ori, one in each direction. Thus, replication is **bidirectional**.

The **origin recognition complex (ORC) i**n eukaryotes assembles at multiple ori (Figure 4.15). Assembly of an ORC at an origin is necessary but not sufficient for initiation to occur. A second complex called **MCM** (**minichromosome maintenance proteins**), which has a weak helicase activity, must also bind, forming a pre-replicative complex; Cdc6p (the protein encoded by Cdc6) is required for this binding. Phosphorylation of Cdc6p by a **cyclin-dependent kinase** inactivates it and prevents reassembly of the complex; this is one of the ways in which reinitiation of replication is prevented. Activation of the ORC/MCM complex is regulated by **cyclins** and **cyclin-dependent protein kinases** (see p. 1015). Once the ORC/MCM complex bound at an ori is activated, it catalyzes the initial separation of parental strands to form a small **replication** bubble (Figure 4.15). **SSBs** (the human SSB is called RPA, replication protein A) bind and hold the separated strands apart. The **MCM** complex may then function as a helicase to allow replication fork unwinding.

The initiation of DNA synthesis uses most of the same mechanisms and enzymes as the movement of a replication fork (Figure 4.15). In eukaryotes, the polα/primase complex initiates synthesis of an RNA primer and then switches to elongating it with dNTPs. This short polynucleotide is elongated by a DNA polymerase δ in the same way as on the discontinuous (retrograde) strand at a replication fork. The replisome elongating this first strand can continue synthesis and forms the leading strand of the fork on one side of the replication bubble (Figure 4.15). On that half of the bubble, synthesis of the first Okazaki fragment on the retrograde strand is initiated by the same mechanism as is used in moving the replication fork (compare Figure 4.13). However, as this first Okazaki fragment is elongated, it will not run into a previous fragment so it continues to grow in the $5'$ to $3'$ direction; as it passes the origin, it becomes the leading strand of the replication fork moving in the opposite direction (Figure 4.15). Both leading strands continue to be elongated, and Okazaki fragments are synthesized on the opposite sides. The result is a pair of replication forks diverging from the origin.

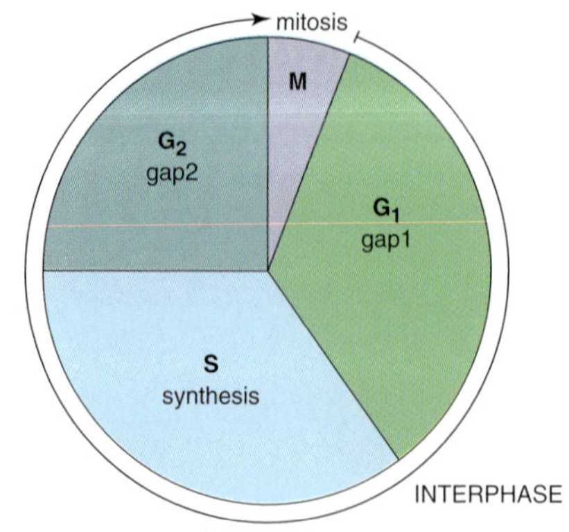

FIGURE 4.16

Cell cycle. In eukaryotic cells, DNA replication (S-phase) and mitosis (M-phase) are separated by two gaps, G_1 and G_2. Size of the segment represents the typical fraction of the cell division cycle time.

The Cell Cycle

In eukaryotes, there is a distinct and regulated pattern of activities that constitutes a **cell cycle** (Figure 4.16 and see p. 1014). The most striking morphological feature of the cell cycle is mitosis, the condensation, alignment, and separation of chromosomes followed by division of a cell into two cells. The phase of the cell cycle in which mitosis occurs is called **M-phase** (for mitotic phase). DNA replication takes place during the **S-phase** (Synthesis phase). **G_1-phase** (Gap 1) occurs between mitosis and S-phase, and **G_2-phase** (Gap 2) occurs between S-phase and mitosis. G_1, S, and G_2 together are called interphase, and they constitute most of the total time of a cell cycle. A higher-than-normal fraction of cells in mitosis is a histological indicator of cancer (Clin. Corr. 4.3).

For cell viability, it is important that DNA be completely replicated before mitosis. Cells have regulatory mechanisms that ensure coordination of the events of the cell cycle at several checkpoints. Cells are most sensitive to DNA damage during the process of replication, because replication forks can encounter the damage before it has been repaired (see Clin. Corr. 4.4). Cells have evolved sensors that recognize DNA damage and blockage of replication forks, and they cause slowing of S-phase progression and arrest of the cell cycle at either the G_1/S boundary or the G_2/M boundary. These mechanisms also induce synthesis of proteins that aid in DNA repair. In yeast, pol ε mutants are deficient in induction of damage response genes and checkpoint control. PCNA may play an important role in coordination of replication and repair functions. The interaction of PCNA with $p21^{cip1}$, an inhibitor of cyclin-dependent kinase, appears to inhibit replication but not repair; thus it may allow repair to proceed before replication continues.

In eukaryotes, replication is regulated by determining whether or not to start DNA replication. Once started, cells generally proceed through the entire replication process. Initiation of replication at oris must be coordinated to ensure that all regions of all chromosomes are replicated in each cell cycle and that no region is replicated more than once. This is, obviously, a difficult challenge in a cell with thousands of oris. Failure to replicate the entire genome would lead to chromosome instability and loss of genetic information. If a portion of the chromosomal DNA is not replicated, the two progeny chromosomes cannot separate properly, or the chromosome could be broken during the attempt. Broken chromosomes are unstable and trigger recombination, which cause further problems. Either the lack of key genes in a cell missing a chromosome due to mis-segregation or the excessive expression of other genes in a cell with an extra copy of a chromosome can be disastrous. Trisomy 21, the presence of one extra copy of the smallest human chromosome, causes Down's syndrome.

Different regions of the genome are replicated at different times during the S-phase of the cell cycle. Typically, regions that are actively transcribed are replicated early during

CLINICAL CORRELATION 4.3

Cancer and the Cell Cycle

Cancer is defined as the unregulated, excessive division of cells. It is recognized pathologically by a higher fraction of cells actively in the cell cycle than is expected for the normal tissue from which it arose. This includes a higher fraction of cells in mitosis, recognizable by microscopy, and a higher fraction of cells in S-phase.

Many kinds of drugs are used to disrupt replication. Antimetabolites interfere with synthesis of precursors of 5′-dNTPs. Methotrexate inhibits dihydrofolate reductase, needed to maintain reduced tetrahydrofolate required for nucleotide synthesis. 5-Fluorouracil inhibits thymidylate synthase; when metabolized into its triphosphate, it can be incorporated into DNA and lead to strand breaks. Vinca alkaloids inhibit microtubule assembly and thereby interfere with mitosis and chromosome segregation. Topoisomerase inhibitors slow or stop the process of replication by preventing untwisting of parental strands. Topoisomerase poisons inhibit resealing of the phosphodiester bond, leaving covalent protein–DNA junctions that are converted into strand breaks; this class of antineoplastic agents includes etoposide and doxorubicin.

CLINICAL CORRELATION **4.4**

Nucleoside Analogs and Drug Resistance in HIV Therapy

AIDS (acquired immune deficiency syndrome) is caused by infection with the retrovirus HIV (human immunodeficiency virus). A key step in the life cycle of this virus is synthesis of a DNA copy of the viral RNA genome, catalyzed by a reverse transcriptase. Reverse transcriptase is a major target of chemotherapy because it is not essential for normal cells.

AZT (3′-azido-2′,3′-dideoxythymidine; Zidovudine) was the first drug approved for treatment of HIV infection. It is a nucleoside analog with an azido group on the sugar. It can be phosphorylated into the triphosphate form, which competes with dTTP for incorporation into the reverse transcript. Once incorporated, it terminates the growing chain of the transcript because the azido group on the 3′ carbon of the sugar is not a substrate for nucleotide addition. AZT is much less efficient at competing with the more accurate cellular DNA polymerases, providing a therapeutic window in which the effect is primarily upon viral replication. Nevertheless, side effects include toxicity to bone marrow, which contains rapidly dividing cells, and myopathy that might be related to toxicity to mitochondria, which contain their own DNA polymerase, pol δ. Other nucleoside analogs have been used to treat HIV infection. DDI (2′, 3′ dideoxyinosine; didanosine) and dideoxycytidine (zalcitabine) also function as chain terminators after incorporation of their phosphorylated derivatives by the HIV reverse transcriptase.

Reverse transcriptases do not carry out proofreading, thus their error rate is much higher than that of cellular DNA polymerases. This high error rate complicates the treatment of AIDS, because the population of viruses carried by any one patient contains many mutants. Some of these mutants are likely to be resistant to any given therapeutic agent. Thus, many drugs initially reduce the viral load, but later become ineffective due to the selective growth of viruses in which the drug target is mutated to an insensitive form. Combination therapy, with multiple drugs that target different viral proteins, is an attempt to circumvent this problem.

Source: There is an interesting video on the HIV life cycle at http://www.hopkins-aids.edu/hiv_lifecycle/hivcycle_txt.html

S-phase. Conversely, heterochromatic regions (where the chromatin is condensed and not actively transcribed) are typically replicated late in S-phase.

Although the details of the process are not yet clear, some general principles can be stated. The current model is referred to as "**licensing**." A **licensing factor** must be bound to the origins to allow initiation of replication at that origin. This occurs before the start of S-phase, and it involves the MCM complex binding to the origin recognition complex, ORC, at the many oris. This binding is facilitated by "**matchmaker**" proteins, including Cdc6p (cell division cycle protein 6). Assembly of this pre-replication complex only occurs during G_1-phase, before DNA synthesis occurs. Once initiation of replication has occurred, the licensing factor is inactivated by dissociation of Cdc6 and the MCM complex. For that origin to be used again, a new licensing factor must bind, but the assembly of the complex cannot occur until the next G_1-phase. This prevents premature reinitiation during a single cell cycle. Cyclin-dependent protein kinases regulate these processes (see p. 1015).

Termination of Replication in Circular Genomes

Termination of replication of a circular genome generally occurs 180° away from the origin. Two converging replication forks meet, and the last portion of the genome is synthesized. Topological unlinking of the two new chromosomes must occur. This is a key function of type II topoisomerases. In some small viruses like SV40, termination occurs wherever replication forks meet; there is no special sequence involved. The *E. coli* genome contains special termination sequences that constrain termination to occur within a defined region, by preventing replication forks from proceeding past the region.

Termination of Replication in Linear Genomes: Telomeres

Human cells, and eukaryotic cells in general, have linear chromosomes. There are special difficulties in replicating the ends of linear chromosomes. What exactly is the problem in replicating the ends of linear chromosomes? Although the continuous strand can theoretically be synthesized to the very end of its template, the discontinuous strand

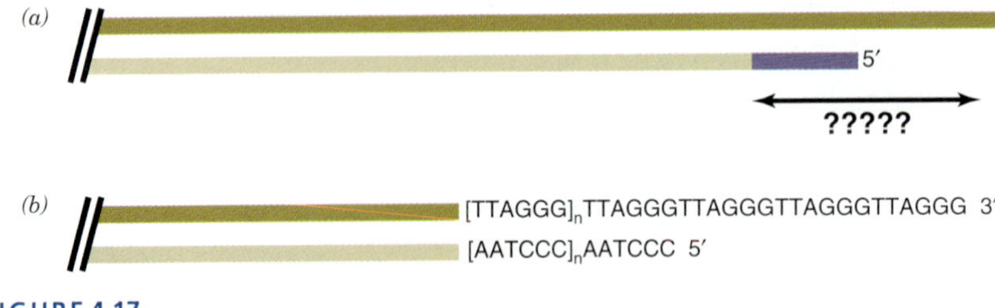

FIGURE 4.17

Human Telomeres. (*a*) The telomere replication problem. The 3′-end of one parental strand, the template for the discontinuous (retrograde) daughter strand, is shown in dark tan; the daughter strand is shown in light tan, and an RNA primer is in purple. The problem is that there is no place to synthesize a primer that would allow the daughter strand to be completed (region shown with ?????), so the daughter strand will be shorter than the parental strand. Removal of the last RNA primer makes it even shorter. (*b*) A telomere consists of many tandem repeats of a 6-nt sequence, TTAGGG in humans, with the G-rich strand extending beyond the C-rich strand by about 12–18 nt.

cannot. There is no place to synthesize a primer to which the nucleotides opposite the end of the template can be added (Figure 4.17*a*). Even if there were a primase that could start at the very end of the template, removal of the RNA would leave a short gap. While the failure to complete the lagging strand at the end of the chromosome would not be a problem in a single generation, over many cycles of replication chromosome ends would be shortened until essential genes were lost and the cell dies. Therefore, it is essential to prevent continued loss of DNA at the ends of chromosomes.

A second problem that eukaryotes face is that the ends of DNA molecules tend to trigger recombination (discussed below). To avoid both problems, the ends of eukaryotic linear chromosomes are special structures called **telomeres**, which contain many repeats of a six-nucleotide, G-rich repeated sequence. Human telomeres contain thousands of the repeat TTAGGG. The 3′-end of the chromosome extends about 18 nucleotides beyond the 5′-end (Figure 4.17*b*), leaving three repeats as an overhang. The overhanging 3′-end folds back on itself, forming non-Watson–Crick G-G hydrogen bonds, and binds proteins that define its length and protect the chromosome ends from undergoing recombination.

Telomerase

Telomeres are maintained by **telomerases**, enzymes that add new six-nucleotide repeats to the 3′-ends of the telomeres. Telomerases are ribonucleoprotein complexes containing a small RNA that serves as a template for addition of a new six-nucleotide repeat (Figure 4.18). A telomerase binds to the end of the 3′-strand, with part of the telomerase RNA hydrogen bonded to the last few nucleotides of the chromosome. A six-nucleotide repeat is synthesized, using the RNA as a template. Then the telomerase can dissociate and reassociate to add another hexamer.

Telomeres do not have to remain exactly the same length; some shortening is not a problem, because the repeats do not encode proteins. Telomeres undergo cycles of

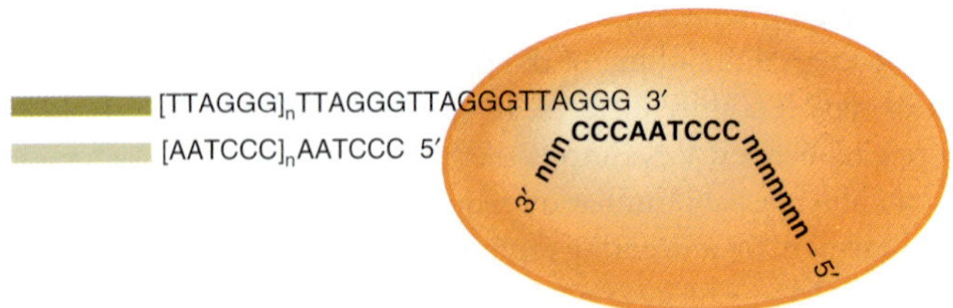

FIGURE 4.18

Telomerase. Telomerase is a ribonucleoprotein complex with a short RNA strand as an integral part; it catalyzes the addition of new 6-nt telomere repeats to the 3′-end of a DNA chain. Telomerase RNA partially base-pairs with the telomeric repeat and serves as the template for the reaction, while the protein component functions as a reverse transcriptase, synthesizing DNA using the RNA template. After a six-nucleotide repeat is added, the enzyme can dissociate and bind again and add additional 6-nt repeats.

shortening of the lagging strands due to the inability to complete synthesis (Figure 4.17*a*) and addition of new six-nucleotide repeats to the 3′-end by telomerase (Figure 4.18). Although the length of the telomere does not remain constant, progressive shortening is avoided by addition of repeats. Telomerases also reestablish the 3′-overhang characteristic of telomeres.

Cells that have differentiated and will divide only a limited number of times do not express telomerase. Thus, the telomeres shorten with each further division; this limits the number of times such cells can divide before the loss of telomeres triggers apoptosis—that is, programmed cell death (see p. 1020). Telomerase expression is generally reactivated in tumor cells, which allows them to continue division indefinitely without chromosome shortening. This makes telomerase an attractive target for cancer chemotherapy. It should be noted that inactivating telomerase in a tumor would not lead to a rapid halt in tumor growth; the effect would be delayed by many cell cycles, until the chromosome ends shorten significantly. Thus, telomerase inhibitors are likely to be useful only in conjunction with other therapies.

Replication of RNA Genomes

Some viruses have an RNA genome. Such genomes are replicated with much lower accuracy, and they can accumulate variations relatively rapidly. A particularly important example of this is the **human immunodeficiency virus (HIV)**, which causes **AIDS** (acquired immune deficiency syndrome). The HIV viral RNA is reverse-transcribed into DNA and then the DNA integrates into the chromosome. The HIV reverse transcriptase is a target for antiviral chemotherapy (Clin. Corr. 4.4).

Reverse transcription of the RNA genome of HIV is far less accurate than DNA synthesis, and the reduced accuracy leads to the rapid generation of a collection of variant viruses within an individual. A small fraction of these variants is likely to be resistant to any single drug being used to treat the infection. That fraction can continue to replicate in the presence of the therapeutic agent to become the dominant variants, which leads to loss of efficacy of the drug. Current combination therapies are designed to reduce the probability that a virus will be simultaneously resistant to all drugs in the combination. Current combination therapies target both the HIV reverse transcriptase and the HIV protease.

4.3 RECOMBINATION

Recombination is the exchange of genetic information. There are two basic types: **homologous recombination** and **nonhomologous recombination**. Homologous recombination (also called **general recombination**) occurs between identical or nearly identical sequences—for example, between paternal and maternal chromosomes of a pair. Chromosomes are not passed down intact from generation to generation (Figure 4.19); rather, each chromosome you inherit from your father contains portions from both of his parents, and likewise for chromosomes inherited from your mother. This is a normal part of the process of chromosome alignment and segregation necessary to ensure that each germ cell gets a single haploid set of chromosomes during meiosis.

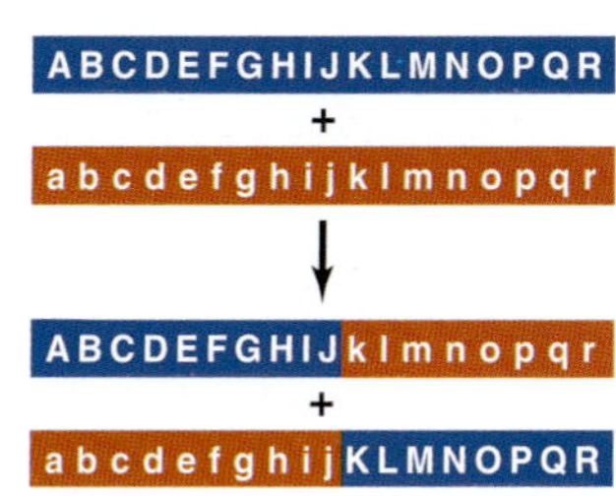

FIGURE 4.19

Homologous recombination. (*a*) The paternal chromosome is shown in blue with alleles shown in capital letters. The homologous maternal chromosome is shown in red with alleles shown in lowercase. (*b*) After homologous recombination between genes j and k, both chromosomes contain DNA from both parents. There has been an equal, reciprocal exchange between them.

Homologous recombination is reciprocal, transferring part of one chromosome to the other and vice versa. It shuffles the combination of genes before they are passed to the next generation, generating genetic diversity. The probability that a recombination event will occur between any two points on a chromosome is roughly proportional to the physical distance between them. This is the basis for genetic mapping. A 1% frequency of recombination between two genes or markers is defined as a genetic distance of one **centimorgan** (**cM**). In humans, 1 cM is approximately 1,000,000 bp along the chromosome.

Models for Homologous Recombination

Recombination is also important for DNA replication and repair (see p. 153). There are three major models for homologous recombination. They are related in many ways, and they have a common intermediate called a **Holliday junction**. The simplest is the Holliday model; understanding its key points will make it easy to see the distinctions of the others.

Holliday Model

The key features of the **Holliday model** are: (a) homology; (b) symmetry of both breaks and strand invasion; and (c) presence of a four-stranded "Holliday junction" as a key intermediate. In this model, recombination is initiated by a pair of single-strand breaks at homologous positions in the two aligned DNA duplexes (Figure 4.20). The strands of each duplex partly unwind, and each "invades" the opposite duplex. That is, a portion of one strand from each duplex base-pairs with the opposite duplex in a reciprocal, symmetrical fashion. These strands can be covalently joined to the opposite duplex, creating a joint molecule in which two strands cross between the DNA molecules.

Branch migration then occurs (Figure 4.20*c*); branch migration is the simultaneous unwinding and rewinding of the two duplexes such that the total number of hydrogen bonds remains constant but the position of the crossover moves. Because the number

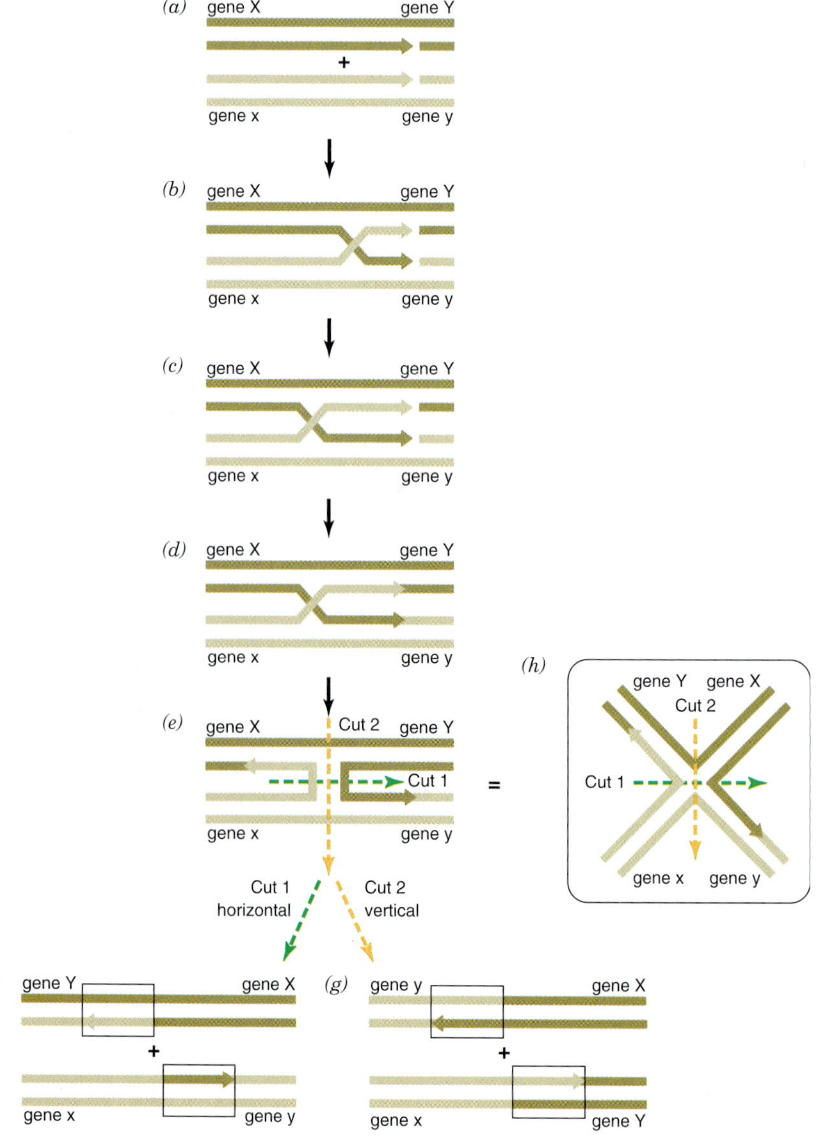

FIGURE 4.20

Holliday model of homologous recombination. (*a*) Synapsis (pairing): two homologous DNA molecules come together properly aligned. [X,x and Y,y represent different alleles of genes (or markers) X and Y, one on either side of the crossover.] Single-strand breaks are made at homologous positions in strands of the same polarity in each duplex. (*b*) Strands partly unwind and reciprocally "invade" (base pair with) the opposite molecule, forming a structure in which two intact strands are joined by two crossed strands. This is a symmetrical and reciprocal event. Regions in which one strand was from one duplex and the other from the homologous duplex are called heteroduplexes. (*c*) "Branch migration" occurs. (*d*) DNA ligase seals the nicks. Further branch migration can occur. All four strands are held together at one crossover point. (*e*) Rotation of this structure forms a "Holliday intermediate" shown in inset (*h*). (*f*, *g*) Resolution: Molecules are separated by a pair of symmetrical cuts in either of two directions (orange or green arrows). The ends are resealed by ligase. The direction of the cuts determines whether flanking regions are exchanged. (*f*) If crossed strands are cut (cut 1, horizontal), flanking genes or markers remain as they were (XY/xy). You would not detect recombination between these loci. (*g*) If uncrossed parental strands are cut (cut 2, vertical), flanking genes or markers are exchanged (Xy/xY). This would be detected as a recombination event. Note that there is a region of heteroduplex in either case (boxed); mismatch repair in this region can lead to gene conversion.

of hydrogen bonds remains constant and the thin, cylindrical DNA molecule is easy to rotate around its axis, branch migration requires little energy. Branch migration creates a region of **heteroduplex**, a region where one strand comes from one original duplex and the other comes from the other duplex. In humans, two individuals differ by approximately one base pair per thousand; therefore, a region of heteroduplex often contains at least one mismatched base pair.

The crossed strands are covalently joined to the other duplex (Figure 4.20*c*) The two DNA duplexes, joined by a single crossover point, can rotate as shown in Figure 4.20*e* and 4.20*h* to create a four-stranded Holliday junction. Its resolution into two duplexes can occur in either of two ways, depicted as vertical or horizontal cuts in Figure 4.20*e*. If the crossed strands are cut (cut 1), markers flanking the region of heteroduplex remain linked together in the same phase as originally (X with Y and x with y; Figure 4.20*f*); on a gross level, the chromosomes remain unchanged. If the intact strands are cut (cut 2), the flanking regions are exchanged between the two chromosomes, creating a new chromosome containing part from one parent and part from the other (X with y and X with y; Figure 4.20*g*). The heteroduplex may undergo mismatch repair (see below), converting one allele into another, a process called **gene conversion**.

Meselson and Radding Model

Once you understand the Holliday model, it is easy to understand the variant proposed by Meselson and Radding. It is initiated by a single nick and a single invading strand, rather than a pair of nicks and reciprocal invasion (Figure 4.21*a*). The nicked strand serves as a template–primer combination that a DNA polymerase elongates, and this elongation displaces part of the invaded duplex (Figure 4.21*b*). The displaced strand invades the opposite, aligned duplex, forming a structure called a D-loop (Figure 4.21*c*). The single-stranded portion of the D-loop is degraded, and repair synthesis and resealing result in crossing over of one strand. Branch migration occurs. The remaining nick is sealed, and the strands can swivel to form a Holliday junction identical to that described for the Holliday model above (Figure 4.21*e*), which is resolved in exactly the same way, resulting in an exchange of flanking markers (cut 2: X with y and X with y; Figure 4.21*g*) or the original configuration of markers (cut 1: X with Y and x with y; Figure 4.21*f*). The key difference in the **Meselson–Radding model** is the asymmetrical heteroduplex that results from the invasion by one strand from one of the two duplexes. Mismatch repair can lead to gene conversion, as in the Holliday model.

Double-Strand Break Model

The third model is the double-strand break model. In this, recombination is initiated by a double-strand break in one duplex (Figure 4.22*a*). The ends are resected (Figure 4.22*b*), resulting in the loss of genetic information from one of the two duplexes. There are alternative forms of this model. In one, the two single-stranded segments of the broken duplex both invade the intact duplex, forming a joint molecule that has two Holliday junctions. The broken ends can be repaired by a polymerase, using the intact strand as template, and a DNA ligase. Branch migration occurs as in the previous models. In an alternative model, the 3′-end of one of the two sides of the break invades the other strand and establishes a D-loop that allows copying of the intact duplex. This can result in (a) a small replication bubble with synthesis on both strands for a short distance and (b) resolution when the bubble encounters the other side of the break. This would account for extensive gene conversion.

Resolution of the two Holliday junctions is by a pair of cuts (Figure 4.22*f*). Because there are two possible directions in which to resolve each junction, there are four possible outcomes. Two leave flanking markers together in their original configuration (X with Y and x with y), and two lead to the exchange of flanking markers (X with y and X with y). Resolution by a type II topoisomerase can also occur. Because one duplex has lost DNA in the region of the break, closure of the gap required the genetic information for both resulting chromosomes to come from the intact duplex, therefore if there were any sequence differences between the chromosomes in the gap, gene conversion will have occurred.

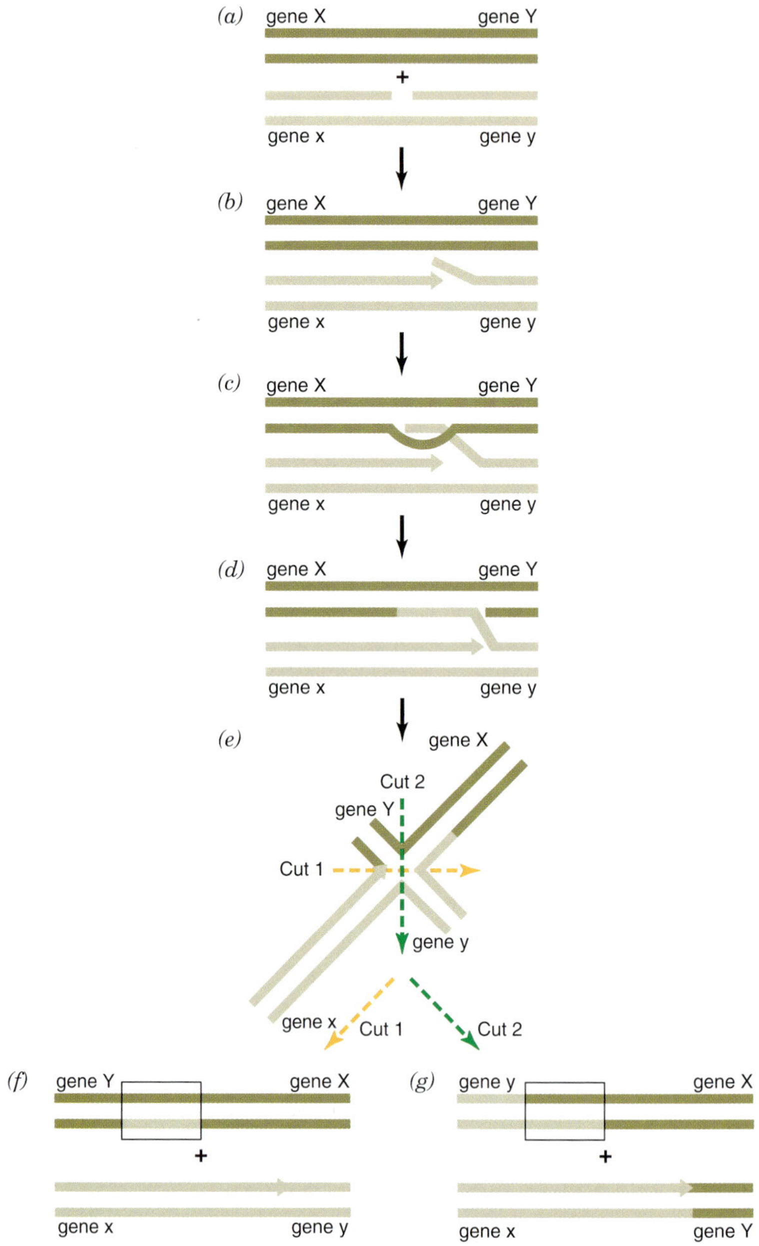

FIGURE 4.21

Meselson–Radding model of homologous recombination. This model is similar to the Holliday model, but initiation is by a single nick and a single invading strand. (*a*) Synapsis (pairing): Two homologous DNA molecules come together properly aligned. A single-strand break is made in one strand. (*b*) The 3′-OH end is elongated by a DNA polymerase, displacing the 5′-end. (*c*) The free 5′-end invades the homologous duplex, forming a "D-loop." (*d*) The D-loop is degraded, and repair synthesis and resealing by DNA ligase lead to a molecule in which one strand crosses between two duplexes. Branch migration occurs as one strand is elongated further. The elongating strand crosses over and is joined to the opposite duplex (not shown), leaving all four strands joined at a single crossover point, as in Figure 4.20. (*e*) Strands rotate to form a Holliday junction exactly as in Figure 4.20e and 4.20h. Resolution of the Holliday junction can occur in two ways, as shown in Figure 4.20, and the direction of the cut determines whether flanking regions are exchanged. Note that there is a region of heteroduplex in either case (boxed), but it is not reciprocal, it is only on one duplex; mismatch repair in this region can lead to gene conversion.

Key Proteins of Recombination in *E. Coli*

RecA

In *E. coli*, the key protein in the process of recombination is **RecA**, which binds in a cooperative manner to single-stranded DNA and forms a helical coil. This facilitates pairing of the single strand with a homologous duplex DNA to form a three-stranded D-loop structure. Strand switching can occur, leading to heteroduplex DNA. Eukaryotes have homologous proteins, the human version of which is called RAD51.

RecBCD, RuvAB, RuvC

RecBCD is an *E. coli* complex with multiple activities; it binds at a double-strand break and processes it into a substrate for recombination by chewing back the DNA and, upon encountering a special sequence called a χ site, leaving a protruding single strand with 3′-OH. RecA can bind to this single strand to initiate pairing with the homologous duplex. **RuvA** and **RuvB** form a complex with DNA helicase activity that catalyzes extensive branch migration, a key step in recombination. RuvC is an endonuclease that binds to this complex and catalyzes the resolution of Holliday junctions.

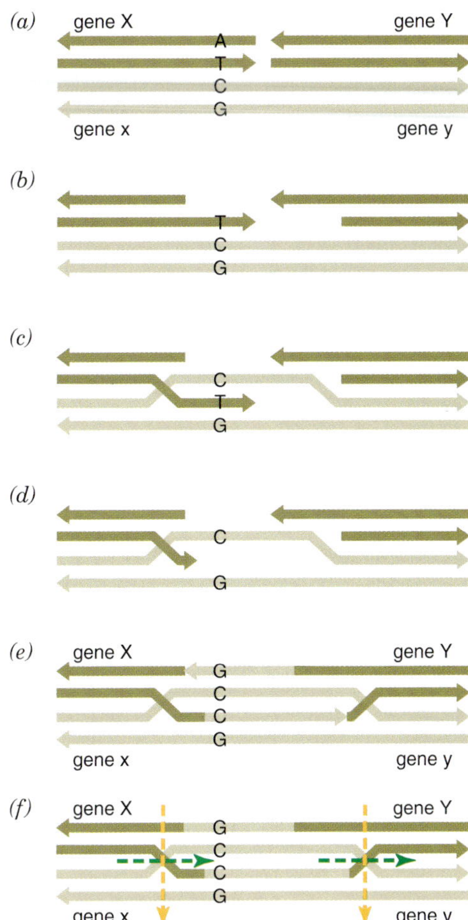

FIGURE 4.22

Double-strand break model of homologous recombination. This model is similar in many ways to the two models discussed previously, but initiation is by a double-strand break in one of the two DNAs. (*a*) Initiation is by a double-strand break in one DNA molecule. Two aligned duplexes are shown; flanking genes have different alleles (X,x;Y,y). Also shown is a position at which one duplex has an AT base pair and the other has a GC base pair. (*b*) DNA is resected from the break by an exonuclease that leaves 3′-OH overhangs. This resection may remove one of the nucleotides at the site that differs between duplexes. (*c*) One or both of these overhangs can invade the intact homologous duplex. (*d*) Where there is a mismatch, the free end can be resected further by enzymes of mismatch repair. (*e*) The free end can be extended by a DNA polymerase, using the undegraded strand as template and resulting in gene conversion. Remaining nicks are repaired, forming two Holliday junctions. (*f*) Both Holliday junctions are then resolved. Because there are two junctions and two ways to resolve each (as in Figure 4.20*b*), there are four possible results. Two of the four possible ways to resolve the Holliday junctions (vertical/vertical or horizontal/horizontal cuts) leave flanking markers as they were (XY/xy). The other two ways to resolve the Holliday junctions lead to exchange of flanking markers (Xy/xY). Resolution can also be carried out by topoisomerase.

Nonhomologous Recombination

Site-Specific Recombination

In site-specific recombination, enzymes catalyze the integration of a sequence into particular sites in the DNA. Integration of the bacteriophage λ into the *E. coli* chromosome is the best-studied example. The **λ-integrase** catalyzes specific nicking of λ DNA and of a special sequence in the *E. coli* chromosome, and it also catalyzes the resealing involved. This is a reversible process; the integrated λ-sequence can be excised from the chromosome. There are important examples of site-specific recombination that control processes such as (a) mating type conversion in yeast and (b) phase change (alterations in major surface protein) in trypanosomes. Phase change in trypanosomes helps the parasite to evade the immune response.

Transposition

Transposition is the movement of specific pieces of DNA in the genome. It resembles site-specific recombination in being catalyzed by special enzymes. There are two types of transposition events: (a) simple transposition in which a sequence is excised from one place in the genome and inserted elsewhere and (b) replicative transposition in which a copy of the sequence is inserted elsewhere in the genome without the original being lost. Transposases are the enzymes that catalyze transposition. They recognize and act at **insertion sequences**. Some pieces of DNA, called **transposons**, have both key features that enable them to move nearly anywhere into a target chromosome: They encode **transposases** flanked by insertion sequences. Transposons sometimes carry other genes with them when they transpose. Other pieces of DNA contain insertion sequences but do not encode their own transposase; these can move when a transposase encoded elsewhere is expressed.

Transposons can move to new places within a chromosome or integrate into other chromosomes. A significant fraction of the human genome has resulted from accumulation of transposons and insertion sequences. **LINE**s (**long interspersed elements**, of which the L1 element is most common) are transposons present in about 50,000 copies in the human genome. Alu sequences are **SINE**s (**short interspersed elements**), and they are present in about 500,000 copies per haploid human genome; they do not encode a reverse transcriptase. Some mutations that cause disease are due to insertion of a transposon into a gene, altering its coding or regulation.

Retrotransposons are sequences that were transcribed into RNA and then reverse-transcribed and integrated back into a random site in the genome. Retroviruses make DNA copies of their RNA genome (using **reverse transcriptase**, an enzyme that makes a DNA copy of an RNA) and insert them into the host chromosome by a transposition event catalyzed by a virus-encoded **integrase**.

CLINICAL CORRELATION 4.5

Gene Therapy

Gene therapy is the introduction of new or altered genes into cells to correct a genetic defect or treat a disease. As we learn more about specific diseases, the possibilities for genetic intervention increase. The easiest type of disease to cure by gene therapy is one in which the supply of a circulating protein can be provided from cells engineered to produce it, with no requirement that those cells be in a specific location. Slightly more difficult would be the regulated production of a protein—for example, insulin—that can be provided by cells located anywhere in the body. It will be much more difficult to replace a defective gene with a normal copy in particular cells in which it is expressed.

Several major limitations impede progress. One is the difficulty of introducing genes into the relevant cells; this is being addressed by more work on vectors for gene introduction and on stem cell biology. Another is the tendency of introduced genes to be incorporated randomly into the genome. DNA introduced into mammalian cells is most frequently integrated into the genome by nonhomologous end-joining. This random insertion can result in the creation of a new mutation if the transgene inserts into an inappropriate location. It can also create difficulties in sustaining gene expression over time when the gene is inserted into an inappropriate region of chromatin, affecting its regulation.

Nonhomologous End Joining

In **nonhomologous end joining (NHEJ**; sometimes called illegitimate recombination) no homology is needed, nor are there special DNA sequences. Rather, broken ends of a DNA duplex can recombine with another duplex in the absence of homology. This can lead to the repair of a double-strand break in DNA, and therefore be of considerable survival value. It can also lead to chromosomal translocations or the insertion of a DNA fragment anywhere into the genome. In mammalian cells, a heterodimer called **Ku** binds avidly to DNA ends and can move along the DNA. Ku associates with a large catalytic subunit to form **DNA-PK**, a DNA-dependent serine/threonine protein kinase. This is thought to play a key role in NHEJ by binding to and bringing together the two free ends of a double-strand break and protecting them from excessive nuclease action. It may unwind the ends until very short regions of "**microhomology**" (2–6 bp) can pair. The unpaired flaps could be removed by FEN1, the gaps filled by DNA polymerase and the nicks ligated by DNA ligase. Defects in Ku lead to X-ray sensitivity, due to reduced ability to repair the double-strand breaks caused by X-rays. Defects in Ku can also lead to severe combined immunodeficiency disease (SCID) because a second role of Ku is in V(D)J recombination, the site-specific cleavage and rejoining required to assemble functional immunoglobulin genes.

When genes are introduced into mammalian cells, they usually integrate randomly into the chromosome by NHEJ. This random integration of a fragment of DNA can disrupt genes and cause mutations or dysregulation. Therefore, it is possible that the attempt to correct one genetic problem by inserting a gene could cause another problem; this is a major limitation to gene therapy at present (see Clin. Corr. 4.5).

4.4 REPAIR

Although we have emphasized the accuracy of DNA replication, it is not 100%. Even at an error rate of 10^{-9}, a handful of errors is introduced during each round of replication. Also, DNA in cells is constantly being altered by cellular constituents, including active oxygen species that are byproducts of metabolism. Many environmental agents attack and modify DNA. Thus maintenance of the genetic information requires constant repair of DNA damage.

All free-living organisms have mechanisms to repair damage to DNA. In this section, we will discuss some of the most important lesions in DNA and the mechanisms by which they are repaired. Some lesions are directly repaired, but most are removed from DNA as part of the repair process. A key feature in most repair processes is the double-stranded nature of DNA, which allows restoration of the correct sequence on a damaged strand using the complementary genetic information on the other strand.

DNA Damage

In aqueous solution at 37°C, there is spontaneous **deamination** of C, A and G bases in DNA. C deaminates to form U (Figure 4.23*a*), A to hypoxanthine, and G to xanthine. Spontaneous **depurination** due to cleavage of the glycosyl bond connecting purines to the backbone (leaving the backbone of the DNA intact; Figure 4.23*b*) occurs at a substantial rate. It has been estimated that between 2000 and 10,000 purines are lost per mammalian cell in 24 h. These depurinated sites are called abasic (lacking a base) or **AP sites** (originally meaning apurinic, lacking a purine, but since generalized to lacking any base).

Bases are oxidized at a substantial rate by reactive oxygen species that are byproducts of oxidative metabolism; this creates altered bases—for example, 8-hydroxyguanine (8-oxo G; Figure 4.23*c*). Oxidative damage tends to increase with age. Products of lipid

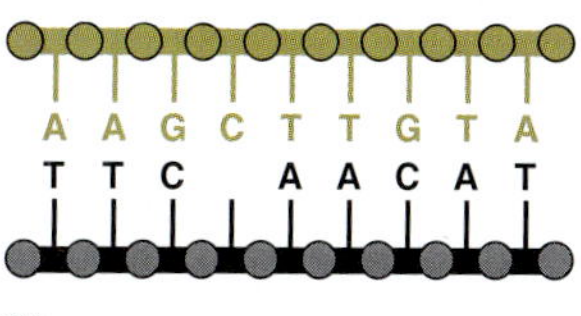

(*a*)

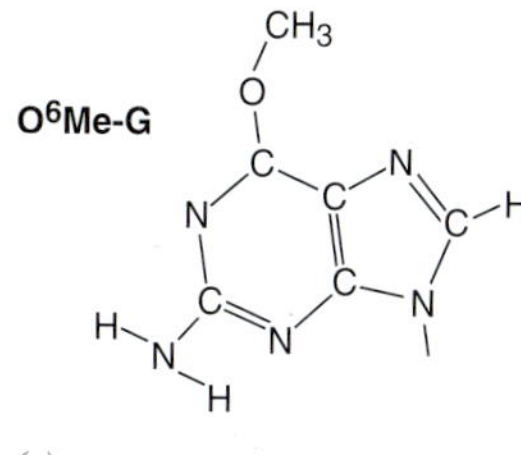

(*b*)

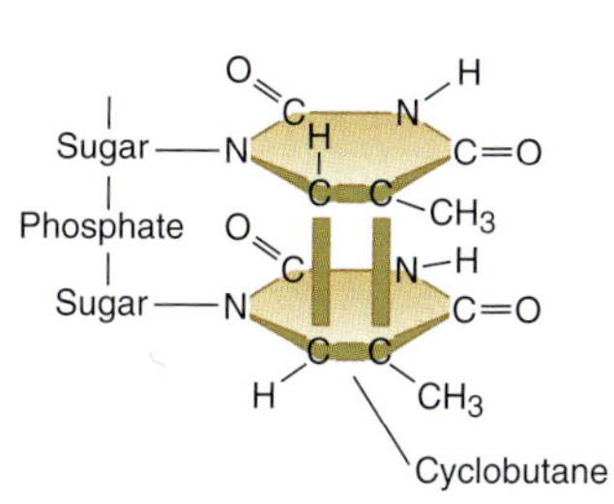

(*c*)

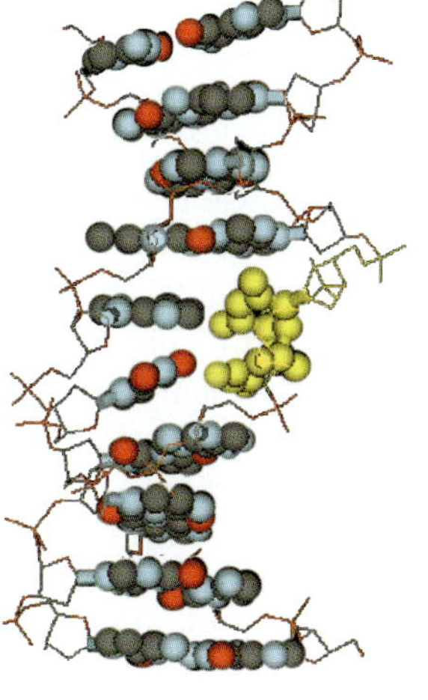

(*d*) (*e*)

FIGURE 4.23

DNA Damage. (*a*) Oxidative deamination converts C to U. (*b*) AP site. Depurination (or, less frequently, depyrimidination) is the cleavage of a glycosyl bond between the 1′-position of the sugar and the base. This leaves an abasic or AP site, but does not break sugar–phosphate backbone. (*c*) O^6-Methyl guanosine is a highly mutagenic lesion. (*d*) Ultraviolet light leads to cross-linking of adjacent pyrimidines along one strand of DNA. A *cis-syn*-cyclobutane thymine dimer is shown in schematic. (*e*) A cyclobutane dimer is shown within double-stranded DNA; backbone is shown as a stick figure to reveal distortion of the cross-linked bases (yellow) and complementary bases on the other strand.

CLINICAL CORRELATION 4.6
Chemotherapy, DNA Damage, and Repair

Many widely used chemotherapeutic drugs are alkylating agents, often ones with two functional groups that can create both intrastrand and inter-strand cross-links in DNA. Cyclophosphamide, busulfan, and nitrosoureas are alkylating agents. Alkylating agents are not only cytotoxic but also mutagenic, sometimes resulting in secondary cancers such as leukemias.

Cisplatin (*cis*-diamminedichloroplatinum),

$$\begin{array}{c} NH_3 \\ | \\ Cl-Pt-NH_3 \\ | \\ Cl \end{array}$$

used to treat metastatic testicular, ovarian, and a wide variety of other cancers, cross-links DNA. Among the serious side effects of cisplatin are bone marrow depletion, severe kidney impairment, and loss of hearing and balance. Disseminated testicular cancer used to be nearly always fatal (5% cure rate in 1973), until a combination chemotherapy, in which cisplatin was added to vinblastine and bleomycin, was developed by Drs. Larry Einhorn and John Donohue at the Indiana University School of Medicine; the new combination greatly increased survival. This regimen was later modified to a combination of cisplatin, etoposide, and bleomycin. Today, approximately 80% of patients survive. After treatment for advanced testicular cancer, world-leading cyclist Lance Armstrong went on to win the Tour de France a record seven times.

Source: Einhorn, L. H. Curing metastatic testicular cancer. *Proc. Natl. Acad. Sci. USA* 99:4592, 2002.

peroxidation can form adducts with bases, particularly G residues. Alkylating agents, including some used for chemotherapy, add methyl or other alkyl groups to the bases. *S*-adenosyl methionine, a carrier of methyl groups in normal metabolism, occasionally methylates a base in DNA.

Ultraviolet radiation (from sunlight or tanning lamps) covalently links adjacent pyrimidines along one strand of the DNA to form *cis-syn*-cyclobutane **pyrimidine dimers** (Figure 4.23*d*) and other photoproducts, including pyrimidine 6–4 pyrimidones. Ionizing radiation, including X-rays and radioactive decay, creates strand breaks and produces reactive oxygen species that damage DNA.

Carcinogens and mutagens attack DNA. Some are direct acting and others are **pro-carcinogens**. Pro-carcinogens in their native form do not damage DNA, but they can be activated by metabolic processes (e.g., oxidized by cytochrome P450s) into carcinogens that damage DNA. This process is called **metabolic activation. Benzo(*a*)pyrene**, a constituent of coal tar, is an extremely potent pro-carcinogen that requires oxidation to the epoxide to attack DNA. Many chemotherapeutic agents attack DNA; some cause bases to become alkylated, others, including cisplatin, cross-link the two strands (Clin. Corr. 4.6). The higher sensitivity of replicating cells to DNA damage means that rapidly dividing tumor cells are more sensitive to these damaging agents than are normal cells, but their damage to normal cells limits their dosing and use.

Mutations

Mutations are inheritable changes in the DNA sequence. They can result from replication errors, from damage to the DNA, or from errors during repair of damage. Mutations that are changes of a single base pair are called **point mutations**. Point mutations can be categorized by the nature of the bases altered. **Transitions** are point mutations in which one purine is substituted for another (e.g., A for G or G for A) or one pyrimidine is substituted for another (e.g., T for C or C for T). Deamination of C, if unrepaired, would lead to a transition. The frequency of transitions is increased by base analogs, including 2-amino purine (Clin. Corr. 4.7). **Transversions** are point mutations in which a purine is substituted for a pyrimidine, or vice versa (e.g., A for C or C for A).

Point mutations can also be characterized by their effect upon a coding sequence. **Missense mutations** are point mutations that change a single base pair in a codon such that the codon now encodes a different amino acid (Figure 4.24*a*). **Nonsense mutations** are point mutations that change a single base pair in a codon to a stop codon that terminates translation (Figure 4.24*b*). Nonsense mutations usually have more severe effects than missense mutations, because they lead to synthesis of truncated (and generally unstable) polypeptides. Silent or **synonymous mutations** do not alter the amino acid encoded; these include many changes in the third nucleotide of a codon.

Insertions or deletions of one or more base pairs (if the number of base pairs is not a multiple of 3) lead to **frameshifts** that disrupt the coding of a protein (Figure 4.24*c,d*). Translation of an mRNA does not have punctuation; rather, once the initiation codon is determined, successive triplets are read as codons. Therefore, addition (or deletion) of a multiple of three base pairs in a coding region would add (or subtract) amino acids to a protein, but addition of other numbers of base pairs shifts the reading frame from that point onward. Frameshifts change the amino acids encoded beyond the point of the insertion or deletion. Frameshifts usually lead to premature termination (or more rarely elongation) of the encoded polypeptide chain, when stop codons are generated or removed by the frameshift. Some chemicals, including **acridines** and **proflavin**, intercalate into the DNA; that is, they insert between adjacent base pairs. This usually leads to insertions or deletions of a single base pair. Mutations may also result from large-scale changes including the insertion of transposons. Although rare in any one generation, mutations have accumulated in populations over millions of years, such that two people differ by about 1 bp per 1000 along their genomes. Many of these genetic differences are without effect, but others affect our physiology, susceptibility to disease, and response to treatment (Clin. Corr. 4.8).

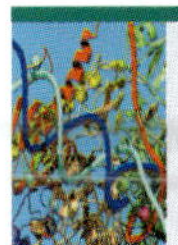

CLINICAL CORRELATION 4.7
Nucleoside Analogs as Drugs: Thiopurines

6-Mercapto purine (6-MP) is an orally administered purine analog that is useful for chemotherapy of acute leukemias and for immunosuppression after organ transplantation. It acts through several mechanisms, including inhibiting purine biosynthesis and toxicity after incorporation into DNA. It is metabolized into the 6-MP riboside 5′-phosphate, which has a short half-life because it is degraded by xanthine oxidase. The rate of degradation is greatly reduced in patients being treated with allopurinol (an inhibitor of xanthine oxidase) for gout-related hyperuricemia, so the dose must be drastically reduced in such patients.

Another enzyme that metabolizes 6-MP (and the related antimetabolite 6-thioguanine) is thiopurine methyl transferase (TPMT). Some patients (about 10% of the population) are heterozygous for a polymorphism that inactivates the enzyme, and therefore have approximately 50% activity, and 1/300 people have no TPMT activity and are at extremely high risk for severe immunosuppression and death if treated with 6-MP. Conversely, people who metabolize the drugs more rapidly may not get a sufficient therapeutic dose. This pharmacogenetic difference, therefore, has serious implications for treatment with thiopurines.

Source: Sanderson, J., Ansari, A., Marinaki, T., and Duley, J. Thiopurine methyltransferase: Should it be measured before commencing thiopurine drug therapy? *Ann. Clin. Biochem.* 41:294, 2004.

CLINICAL CORRELATION 4.8
Individualized Medicine

Individuals differ in the sequence of DNA in their genome in subtle ways, as a result of the very gradual accumulation of mutations in the population over many generations. About once per 1000 bp, there is a difference between individuals in a single base pair; this is called a single base pair polymorphism, or SNP. Most polymorphisms do not affect cell functions, but others contribute to the many differences that we see. Some affect susceptibility to specific diseases, some affect metabolism of drugs and other compounds (pharmacokinetics), and still others affect response to these drugs (pharmacodynamics).

As part of the human genome project, there is a major effort to find and catalog polymorphisms. As more is learned about the effects of these sequence differences on disease and treatment, medicine will become more tailored to the individual. Technology already exists to determine rapidly which polymorphisms are present at specific sites in the genome of an individual. When these technologies are made cheaper and more routine, determining the alleles present in key genes will become a routine part of diagnosis and planning of treatment. For example, there are significant differences between individuals in the rate at which they metabolize drugs (compare Clin. Corr. 4.7). A standard dose based upon body weight may be too low to be effective in those who metabolize the drug rapidly, but toxic to those who metabolize it very slowly. Determination of the relevant polymorphisms and routine testing could lead to more effective therapies with fewer side effects.

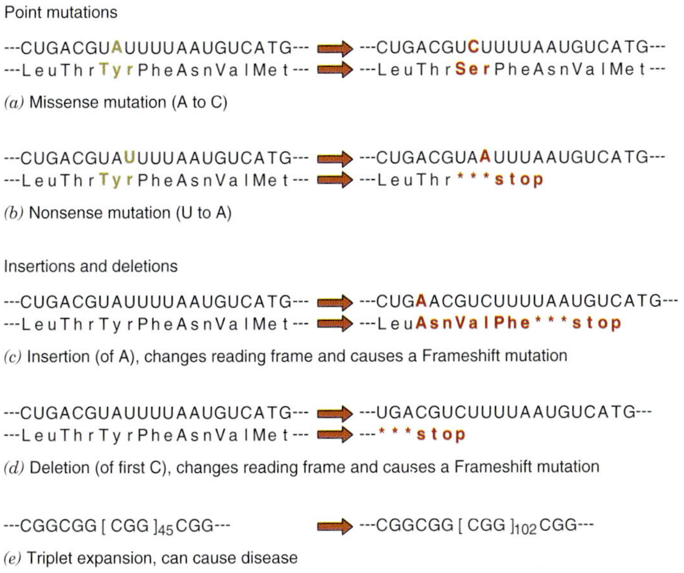

Point mutations

---CUGACGUAUUUUAAUGUCATG--- → ---CUGACGUCUUUUAAUGUCATG---
---LeuThrTyrPheAsnValMet--- → ---LeuThrSerPheAsnValMet---

(a) Missense mutation (A to C)

---CUGACGUAUUUUAAUGUCATG--- → ---CUGACGUAAUUUAAUGUCATG---
---LeuThrTyrPheAsnValMet--- → ---LeuThr***stop

(b) Nonsense mutation (U to A)

Insertions and deletions

---CUGACGUAUUUUAAUGUCATG--- → ---CUGAACGUCUUUUAAUGUCATG---
---LeuThrTyrPheAsnValMet--- → ---LeuAsnValPhe***stop

(c) Insertion (of A), changes reading frame and causes a Frameshift mutation

---CUGACGUAUUUUAAUGUCATG--- → ---UGACGUCUUUUAAUGUCATG---
---LeuThrTyrPheAsnValMet--- → ---***stop

(d) Deletion (of first C), changes reading frame and causes a Frameshift mutation

---CGGCGG [CGG $]_{45}$CGG--- → ---CGGCGG [CGG $]_{102}$CGG---

(e) Triplet expansion, can cause disease

FIGURE 4.24

Mutations. Point mutations. *(a)* A missense mutation changes a single amino acid in the encoded polypeptide. *(b)* A nonsense mutation changes a codon for an amino acid into a stop codon, terminating synthesis of the encoded polypeptide. Insertion/deletion mutations. *(c)* Insertion of a single nucleotide changes the reading frame of all codons beyond the point of insertion; this usually leads to formation of a new stop codon that terminates synthesis. *(d)* Deletion of a single nucleotide changes the reading frame of all codons beyond the point of insertion; this usually leads to formation of a new stop codon that terminates synthesis. *(e)* Triplet expansion is a great increase in the number of triplet repeats. Triplet expansion causes many diseases, including Huntington's disease and Fragile X disease, by adding long stretches of a single amino acid to the encoded polypeptide or by disrupting regulation of a gene.

The genome contains many stretches of repeating 2, 3, 4, and 5 nucleotide sequences. These simple repeats have a higher than average mutation rate because they can occasionally misalign during replication and recombination, leading to the insertion or deletion of a number of repeats. Thus over many generations, different lengths of these sequences have accumulated in the human population; these are called simple sequence length polymorphisms (also called microsatellites or simple sequence repeat polymorphisms). There is a special kind of mutation that results from a great increase in the number of repeating triplets (e.g., CGGCGG[CGG]$_n$CGG), called **Triplet expansion** (Figure 4.24*e*). When the length of a triplet repeat gets much longer than normal, it can cause disease by encoding a much longer than normal stretch of the same amino acid residue in the encoded polypeptide, as in **Huntington's disease**. Triplet expansion can also affect regulation of a gene, as in **Fragile X syndrome**, in which the greatly lengthened triplet sequence in the promoter alters gene expression. For these triplet expansion diseases, there is an unusual phenomenon: The disease often gets more severe in succeeding generations, as a result of even further increases in triplet length, because the longer the repeating sequence is, the more likely there can be misalignment during replication or unequal crossing over to lead to dramatically longer repeats.

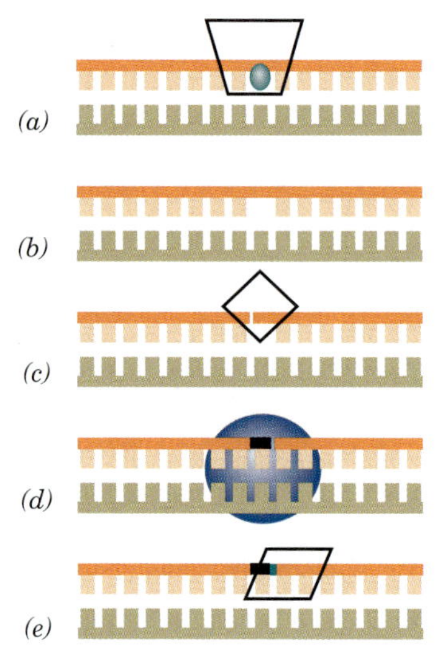

FIGURE 4.25

Base excision repair. (*a*) The damaged site (blue dot) is recognized by a DNA glycosylase. (*b*) The base is removed by cleavage of the glycosyl bond that connects it to the deoxyribose sugar; the sugar–phosphate backbone is not broken. This leaves an AP site. (*c*) The sugar–phosphate backbone is cleaved 5′ to the abasic site by an AP endonuclease. There is also a cleavage at the 3′-side of this site to remove the sugar residue; this can be done by some AP endonucleases or an AP lyase activity. (*d*) The single nucleotide gap is filled by a DNA polymerase. (*e*) The remaining nick is sealed by a DNA ligase.

Excision Repair

Excision repair is a general mechanism that can repair many different kinds of damage. The key, defining characteristic of excision repair is removal of the damaged nucleotide(s), leaving a gap in the DNA, followed by resynthesis using the genetic information on the opposite strand, and then ligation to restore continuity of the DNA. There are two major modes of excision repair: **base excision repair** (BER) and **nucleotide excision repair** (NER) (Table 4.3).

Base Excision Repair

Base excision repair is an essential process that repairs many different types of damaged bases, including methylated bases, deaminated bases (e.g., U resulting from deamination of C), oxidized bases, and abasic (AP) sites. The initial step is removal of a single damaged base from the DNA backbone by a DNA **glycosylase**, an enzyme that cuts the *N*-glycosyl bond between the sugar and the base (Figure 4.25). This step does not break the sugar–phosphate backbone of the DNA; rather it leaves an abasic deoxyribose in the backbone, an AP site that must be removed. Two different activities are required to remove this sugar: (a) an **AP endonuclease** that cleaves the phosphodiester bond at the 5′-side but leaves the sugar still attached to the next nucleotide and (b) an **AP lyase** that cuts 3′ to the AP site to remove the sugar. The resulting single nucleotide gap has a free 3′-hydroxyl. The gap is filled by a DNA polymerase and ligated by DNA ligase (Figure 4.25).

There are at least eight different glycosylases in humans, each of which recognizes certain types of damaged bases. One of these, **uracil DNA glycosylase (UNG)**, is

TABLE 4.3 Excision Repair

Basic steps

- Recognize damage
- Remove damage by excising part of one strand
- Resynthesize to fill gap; uses genetic information from other strand
- Ligate to restore continuity of DNA backbone

Base excision repair	Nucleotide excision repair
• Glycosylase removes base, leaves backbone intact • AP endonuclease cuts backbone, AP lyase removes sugar	• Double excision removes damage as part of an oligonucleotide (12–13 nt in *E. coli*, 27–29 nt in humans)
• DNA polymerase fills gap • DNA ligase seals nick	• DNA polymerase fills gap • DNA ligase seals nick

specialized to remove U from DNA. U is not a normal constituent of DNA but is formed when C is deaminated, and it can occasionally be incorporated into DNA during replication. The frequent deamination of C to U noted above (Figure 4.23*a*) would lead to a high rate of mutation; therefore, it is not surprising that a special repair mechanism evolved to remove U. But many C residues in mammalian DNA, in the sequence CG, are methylated at the 5-position; this is important in the regulation of gene expression. Deamination of a 5-methyl C leaves a T, rather than a U, in the DNA; these T residues are not recognized by uracil DNA glycosylase, but there are other glycosylases that recognize the resulting T:G mismatch and preferentially remove the T, leaving an AP site (Figure 4.23*b*). Sometimes, however, the mismatch repair system removes the G instead of the T, creating a mutation. Over millions of years, this has led to a much lower than expected frequency of CG dinucleotides in mammalian DNA.

The most important **AP endonuclease** in humans is APE1. After cleaving the backbone at the 5′-side of an AP site to leave a free 3′-OH, APE1 recruits DNA polymerase β (pol β) to the site. Pol β has **AP lyase** activity that removes the abasic sugar phosphate remaining at the site of the strand break, leaving a single-nucleotide gap with a free 3′-OH. Pol β also has DNA polymerase activity that can fill the gap. For many types of damage, the gap is filled with only one or two nucleotides; this is called **short patch repair**. In cases where AP lyase cannot readily remove the sugar phosphate, strand displacement can create a flap of several nucleotides and the endonuclease FEN1 (which also serves to help remove the RNA from Okazaki fragments as noted above) can cleave it. The somewhat larger gap is filled by a polymerase such as pol β, or possibly pol δ (or pol ε) stimulated by PCNA. In either case, the remaining nick is sealed by a DNA ligase. Protein–protein interactions coordinate this process, keeping the nick and the intermediates protected until the next enzyme has been recruited to the site.

Nucleotide Excision Repair

Nucleotide excision repair acts upon a wide variety of damage, typically involving large adducts or distortion of the double-helical structure of DNA. The best studied are pyrimidine dimers, the major products of ultraviolet damage. Chemical adducts with carcinogens such as **benzo(*a*)pyrene** and **aflatoxin**, and with chemotherapeutic agents such as **cisplatin**, are removed by this pathway, as are some mismatched bases and small loops in DNA. The damage is removed by an enzyme complex that cuts several nucleotides away on both sides of the damaged base(s), so the damage is released as part of an oligonucleotide (Figure 4.26).

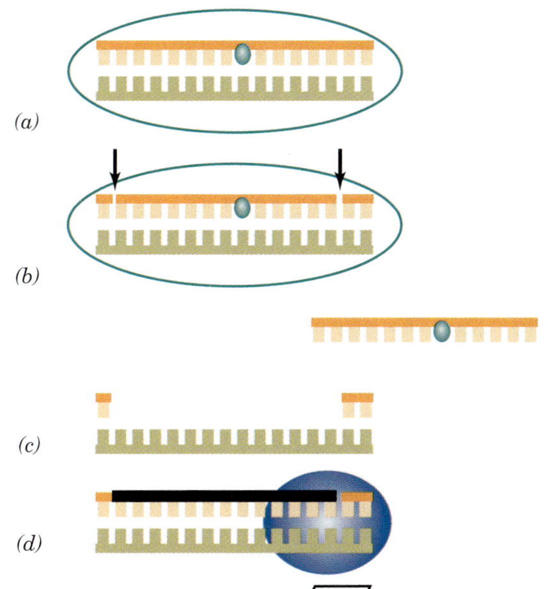

FIGURE 4.26

Nucleotide excision repair. (*a*) The damaged base (blue dot) is recognized by a DNA repair complex. (*b*) The segment around the damaged base is excised by an enzyme complex that makes two nicks in the damaged strand, one on either side of the damage. (*c*) An oligonucleotide (27–29 nt in humans, 12–13 nt in *E. coli*) is released. (*d*) Resynthesis: A DNA polymerase fills the gap, using the opposite strand as template. (*e*) DNA ligase seals the remaining nick.

CLINICAL CORRELATION 4.9
Xeroderma Pigmentosum

Xeroderma pigmentosum (XP) was the first disease demonstrated to be caused by defective DNA repair. Patients are photosensitive and highly susceptible to skin cancers in sun-exposed areas of the body; rates of skin cancer are 2000–5000-fold higher than average. Many patients also have neurological problems.

XP is a rare, autosomal recessive disease that can be caused by defects in any of eight different genes, reflecting the complexity of DNA repair of pyrimidine dimers, the most common damage introduced by exposure to ultraviolet light. Mutations in seven of the XP genes lead to defects in the initial incision step in nucleotide-excision repair of pyrimidine dimers. This explains the extreme photosensitivity that is characteristic of this disease. An additional class, called XP-V (for variant) is not detectably deficient in excision repair, but has a defect in translesion synthesis after ultraviolet damage. Cells from XP patients may also be hypersensitive to carcinogens in cigarette smoke.

XPA is critical for the recognition of pyrimidine dimers and interacts with other repair proteins. XPB and XPD encode proteins that function as subunits of TFII-H, a general transcription factor that is also required for transcription-coupled DNA repair. TFII-H can function as a helicase, opening a region around the dimer. ERCC1-XPF is a complex that binds to TFII-H and incises at the 5′-side of the dimer. XPG incises at the 3′-side.

Source: A good reference for genetic diseases is the Online Mendelian Inheritance in Man, OMIM, at http://www.ncbi.nlm.nih.gov/entrez/query.fcgi?db=OMIM

In *E. coli*, a damage excision complex containing UvrA, UvrB, and UvrC recognizes the damage and makes two endonucleolytic cuts, one 3 to 5 nucleotides 3′ to the lesion and the other 8 nucleotides 5′ to the lesion. This excises the lesion as part of an oligonucleotide 12–13 nucleotides long. Damage recognition by UvrA appears to start the process, and opening of the strands by UvrB (a helicase) requires energy in the form of ATP hydrolysis. UvrC binds and catalyzes the dual cleavage, with one cut on either side of the damage. A DNA polymerase fills in the gap, followed by ligation of the nick.

In humans, several large enzyme complexes are involved in damage recognition, opening of DNA, and the two cleavages. The rare, autosomal recessive disease **xeroderma pigmentosum** (XP), which renders patients photosensitive and susceptible to skin cancers, results from mutation in any of seven genes involved in excision repair or in one gene involved in translesion DNA synthesis (Clin. Corr. 4.9). XPC is part of a complex that binds tightly to lesions that distort DNA structure. XPA is a key factor involved in damage recognition and assembly of the excision complex; mutations in the XPA gene render individuals essentially unable to remove UV-induced pyrimidine dimers. XPA recruits a large complex called **TFII-H**, a general transcription factor involved in initiation of transcription. TFII-H contains XPB and XPD subunits, both of which are ATPases with limited helicase activity. The DNA is partially unwound by the helicase activity of TFII-H, creating a bubble of about 25 bp. Then an XPF/ERCC1 (excision repair cross-complementing 1) heterodimer makes an endonucleolytic cut on the damaged strand approximately 22–24 nucleotides 5′ to the lesion, and XPG makes an endonucleolytic cleavage approximately 5 nucleotides 3′ to the lesion. These two cuts liberate a lesion-containing oligonucleotide approximately 27 to 29 nucleotides long. The single-stranded DNA-binding protein RPA is required for excision repair. The gap in DNA is bound by RPA and filled by pol δ (or pol ϵ) stimulated by PCNA. The remaining nick is sealed by a DNA ligase, probably LIG1. These latter steps resemble completion of an Okazaki fragment.

Transcription-Coupled Repair

Many lesions block transcription in the same manner that they block replication. Thus cells face an immediate problem even if they are not synthesizing DNA: They must be able to make key proteins in order to survive. **Transcription-coupled repair** directs the immediate repair response to the template strand of transcribed regions, so that lesions there are repaired faster than lesions elsewhere and transcription can resume; the survival value of this is obvious. Transcription-coupled repair is a form of excision repair triggered by an RNA polymerase complex halted at a lesion. The large size of the transcription complex could block access to the lesion, preventing its repair. In eukaryotic cells, a complex containing CSA and CSB (mutations in which cause Cockayne's syndrome, another DNA repair deficiency disease) recognizes the stalled RNA polymerase, causing it to back up away from the lesion, and recruits repair proteins. Then, many kinds of damage that block transcription are repaired by the nucleotide excision repair pathway involving XPA, TFII-H, and other repair factors. Recent evidence suggests that transcription-coupled repair also acts on oxidative lesions, which are repaired by base excision repair after the transcription complex is backed away from the lesion. Once repair is complete, transcription can resume.

Mismatch Repair

Mismatch repair is a specialized form of nucleotide excision repair that removes replication errors. Mismatches are not like DNA damage: There is no damaged or modified base present, just the wrong one of the four bases. Thus, recognition of mismatches relies upon the distortion of the double-helical structure. A major difference between repair of DNA damage and repair of mismatches is in the choice of which base to excise. Enzymes can recognize damaged bases specifically and remove them, either individually in base excision repair or as part of an oligonucleotide in nucleotide excision repair. But when a mismatch is recognized, both bases are normal. Which one should be excised? In newly replicated DNA, removing the newly synthesized base would preserve the genetic information, whereas excising the base on the parental strand would permanently alter the DNA, producing a mutation. Randomly choosing one would lead to mutations half of the time, which is unacceptably high. Therefore, the challenge is to recognize the newly synthesized strand and then remove the mismatched base on that strand.

The mismatch repair system is best understood in *E. coli*. DNA in most organisms (but not all) is methylated at specific positions. In *E. coli*, the most frequent methylation is of A in a GATC sequence (Figure 4.27*a*), a palindromic sequence (meaning that the opposite strand has the same sequence when read from 5′ to 3′); both strands are methylated. This specific methylation does not affect base pairing. The A residues are incorporated into the DNA in unmodified form, and later they are methylated by a **maintenance methylase** that recognizes **hemimethylated** sequences in DNA (sequences methylated on only one strand, the parental strand; Figure 4.27*b*) and methylates the appropriate base. During the very brief period in which the daughter strand is not yet methylated, the *E. coli* mismatch repair system can recognize the newly synthesized strand in the region of a mismatch and remove the mismatched nucleotide from the new strand. Mismatches are recognized by a MutS homodimer, which creates a loop that contains the mismatch (Figure 4.28). A MutL homodimer binds and coordinates the subsequent cleavage and excision. MutH nicks the unmethylated, newly synthesized strand on either side of the mismatch. A helicase (UvrD) unwinds the strand from the site of the nick through the site of the mismatch, and the nicked single strand is degraded by either a 5′ to 3′ exonuclease or a 3′ to 5′ exonuclease, depending on the free end. The long (hundreds of nucleotides) gap that is created is filled by the action of DNA polymerase, adding nucleotides to the 3′-end of the nicked strand. When the gap has been filled, the remaining nick is sealed by DNA ligase.

Humans have a similar mismatch repair system with homologous proteins, although the system is more complex and the mechanism by which it recognizes the new strand is not clear. Human DNA is methylated on the 5-position of C in CG sequences (Figure 4.27*c*), but it is not clear whether the hemimethylated CG sequences, transient single-strand breaks from unligated Okazaki fragments, or binding to PCNA play roles in the recognition of the newly synthesized strand. There are several MutS homologs (MSH proteins) and MutL homologs (MLH proteins and PMS proteins), which work as heterodimers. MutSα, a heterodimer of MSH2 and MSH6, recognizes base–base mismatches and short insertion/deletions; MutSβ, a heterodimer of MSH2 and MSH3, recognizes short insertions/deletions. MutLα, dimers containing MLH1 and PMS2, are the main complexes required for the incision step. The enzymes responsible for excision of the nicked strand to a point beyond the mismatch have not been identified. As in *E. coli*, the excised region is hundreds of nucleotides long. PCNA, the sliding clamp, is required for mismatch repair. In addition to its role as a processivity factor for the DNA polymerase, it appears to function during the excision step. The resynthesis of the large gap requires DNA pol δ (or pol ϵ) and PCNA. The nick that remains is sealed by DNA ligase.

Mismatch repair can lead to gene conversion during recombination by removing one strand of a heteroduplex and resynthesizing it using the other strand as a template; it can also act to reduce the frequency of recombination. Mismatch repair proteins are also involved in transcription-coupled repair by the NER pathway, although the mechanism is unknown.

Defects in mismatch repair cause **hereditary nonpolyposis colon cancer** (HNPCC) (see Clin. Corr. 4.10). Between 60% and 70% of the cases of HNPCC are due to mutations in MLH1 or MSH2, with other cases due to mutations in other mismatch repair genes. Some other sporadic cancers (particularly some colorectal, gastric, and endometrial carcinomas) also appear to be related to suppression or functional inactivation of the mismatch repair system.

Mismatch repair plays a paradoxical role in response of cells to cytotoxic agents such as methylating agents (*N*-nitrosourea) and cisplatin, which are used in cancer chemotherapy. The recognition of alkylation damage by mismatch repair proteins can trigger cell death by apoptosis; cells that are deficient in mismatch repair may escape this process, and thereby be resistant to chemotherapeutic agents. Mismatch repair-deficient cells have higher rates of mutation and can thus evolve to tumors that are more aggressive.

```
       m
       |
5′---GATC---3′
3′---CTAG---5′
        |
        m
```

(*a*) Methylated DNA in *E. coli*

```
       m
       |
5′---GATC---3′
3′---CTAG---5′
```

(*b*) Hemi-methylated DNA in *E. coli*

```
      m
      |
5′---CG---5′
3′---GC---3′
       |
       m
```

(*c*) Methylated DNA in human cells

FIGURE 4.27

Methylated DNA. (*a*) DNA methylation in *E. coli* is on the A within a GATC sequence. (*b*) Hemimethylated *E. coli* DNA. Parental strand methylated, daughter strand not yet methylated. (*c*) Methylation in human cells is on the C within a CG sequence. Note that in both cases the sequence is a palindrome that reads the same (5′ to 3′) along both strands, and the residues in both strands are eventually methylated.

CLINICAL CORRELATION 4.10
Mismatch Repair and Cancer

Defects in mismatch repair cause hereditary nonpolyposis colon cancer (HNPCC), and they are important in several other cancers. This was initially suggested by the finding that some colon tumors showed frequent mutations in micro-satellites (simple sequence length polymorphisms, short repeating sequences, particularly mono- and dinucleotides), a phenotype called microsatellite instability.

HNPCC is an autosomal dominant disease. A gene contributing to the risk for HNPCC was mapped to a region that contains a mismatch repair gene. This led to studies that demonstrated the primary defect to be a mutation in one of the mismatch repair genes. Later, defects in other mismatch repair genes were also determined to cause HNPCC; the most common are MLH1 and MSH2.

Why does the inheritance of a single defective allele in a mismatch repair gene lead to HNPCC? Cells lining the intestinal tract actively divide throughout a person's life. There is a high probability that at some time in one's life, a somatic mutation will occur in one allele of a mismatch repair gene in at least one colon cell. Cells in which one of the alleles at any of these loci is inactive can still carry out mismatch repair, so this does not generally lead to a tumor. However, in an individual who inherits one inactive allele, a somatic mutation that inactivates the single active copy leads to defective mismatch repair. Mutation rates can be elevated 100- to 1000-fold in cells with inactivated mismatch repair. The result is the accumulation of mutations that eventually lead to tumor formation through mutations in growth-regulating genes.

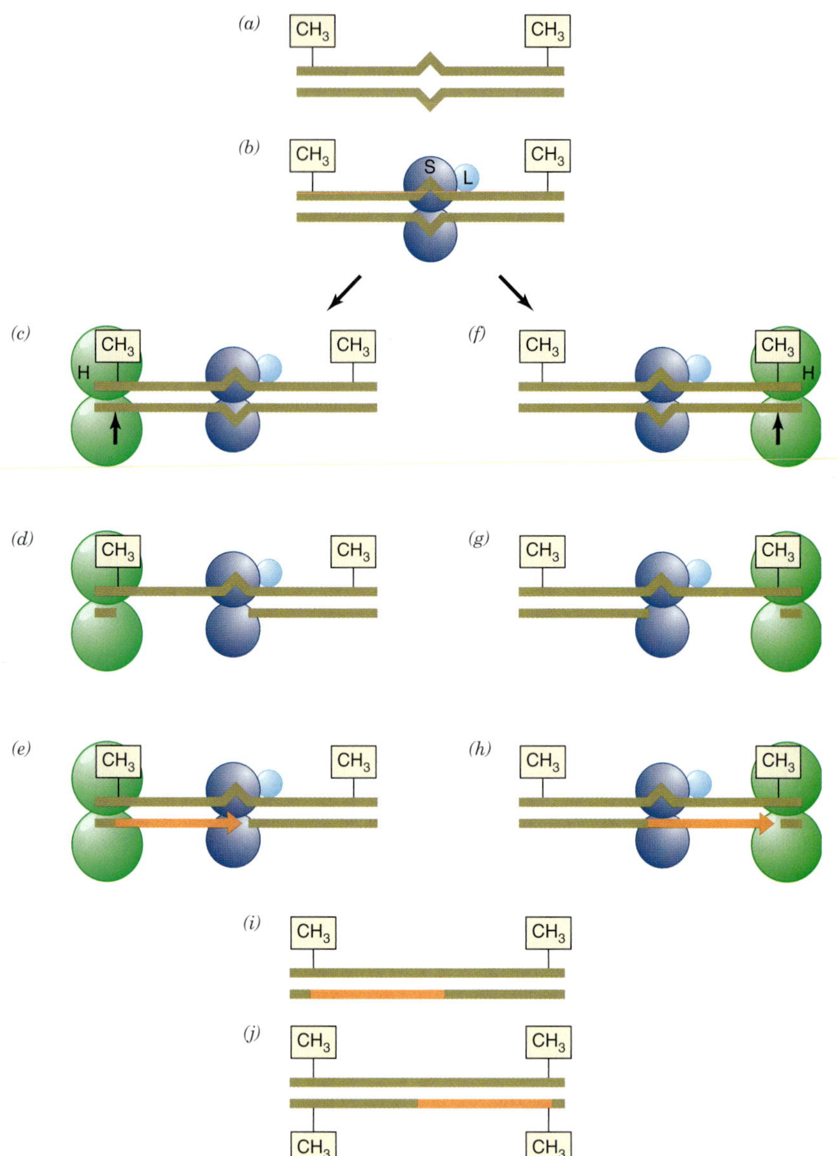

FIGURE 4.28

Mismatch repair in *E. coli*. (*a*) A mismatch in newly replicated DNA. The parental strand is methylated, but for the first few minutes after it is synthesized, the new strand is not yet methylated; such a site, with only one of the strands methylated, is called hemimethylated. (*b*) MutS and MutL bind to the mismatch. ATP is hydrolyzed as they extrude a short loop of DNA. (*c*, *f*) MutH binds at a hemimethylated site (which could be to either the left or right side of the mismatch). (*d*, *g*) The unmethylated strand is nicked and then excised, extending back past site of mismatch. This can be in either direction (*d* and *g*). (*e*, *h*) The resulting long gap is filled by a DNA polymerase, adding nucleotides to 3′-end of the strand (red). (*i*) Remaining nick is sealed by a DNA ligase (strand from (*e*) is shown). (*j*) Hemimethylated sites are methylated (strand from (*h*) is shown).

Direct Demethylation

In addition to the normally methylated bases in DNA, some bases become alkylated inappropriately, by carcinogens or by interaction with the normal methyl carriers in cells (e.g., *S*-adenosyl methionine). Alkylation on positions that affect base pairing is highly mutagenic. For example, O^6-methyl-guanine base-pairs with T rather than with C (Figure 4.29*a*), so its replication would lead to a mutation. Cells have glycosylases that recognize inappropriately methylated bases and trigger base excision repair, but they also have special proteins that recognize O^6-methyl guanine in DNA and directly remove the methyl group, leaving the DNA intact. **O^6-methyl guanine methyl transferase** (MGMT) recognizes O^6-methyl guanine and transfers the methyl group from the

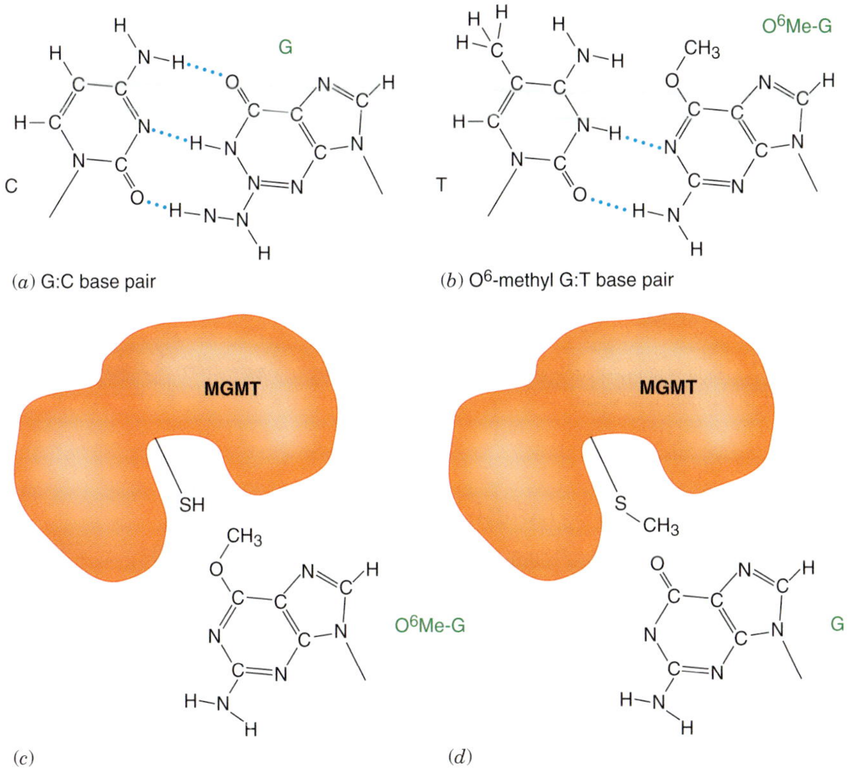

(*a*) G:C base pair

(*b*) O^6-methyl G:T base pair

(*c*)

(*d*)

FIGURE 4.29

Direct demethylation. (*a*) Normal G:C base pair. (*b*) O^6-methyl G:T base pair. This base pairing would allow incorporation of a T opposite O^6-methyl G during DNA replication, thereby leading to a mutation in next generation. (*c*) MGMT (methyl guanine methyl transferase) can bind to O^6-methyl G. (*d*) MGMT transfers the methyl group from the O^6 of guanine to a sulfhydryl group on the MGMT protein itself, leaving an intact G in DNA. The methylated MGMT protein is inactive, becomes ubiquitinated, and is then degraded.

guanine to a cysteine on the protein itself (Figure 4.29*b*). A single molecule of MGMT can remove only one methyl group, because methylation of the cysteine inactivates the protein and targets it for degradation through the ubiquitin proteolytic pathway. The cost to the cell of removing a single methyl group through this pathway is high, because cells must replicate and transcribe the gene for this enzyme, then process and translate the mRNA, and all of these steps require considerable energy and material. This reemphasizes how important it is to maintain the integrity of the genome, even at great metabolic cost.

Photoreactivation

Photoreactivation is a specific mechanism for repair of cyclobutane pyrimidine dimers, the major lesions produced by ultraviolet irradiation. It is a direct reversal of the damage. A specialized enzyme called **photolyase** binds to the damage (Figure 4.30). Upon absorbing light, photolyase catalyzes the reversal of the bonds between the adjacent pyrimidines, leaving the DNA exactly as it was before the dimer formed. This is a direct reversal of the damage. Photoreactivation, while of great importance in most organisms, is not significant in mammals.

Lesions Can Block Replication

The accuracy and proofreading abilities of DNA polymerases generally prevent them from inserting nucleotides opposite damaged bases; they require very accurate base pairing before they insert a nucleotide and move on. Thus when a replication complex reaches a lesion in the template strand, it halts. There are three general ways for replication to proceed: The complex can either (1) disassemble or back-up to allow repair of the lesion and then resume synthesis, (2) directly bypass the lesion, or (3) skip past the lesion and resume synthesis beyond it (leaving a gap in the daughter strand). All three mechanisms are used, and all present some advantages and some problems.

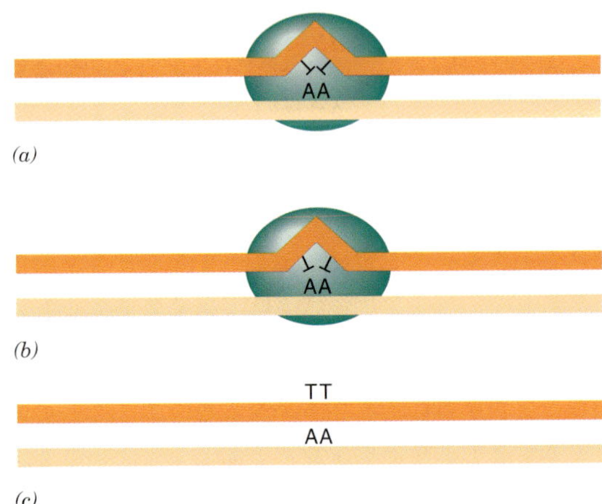

FIGURE 4.30

Photoreactivation. (*a*) Photolyase binds to a cyclobutane pyrimidine dimer; light is not needed for this step. (*b*) The complex absorbs light, which results in cleavage of the bonds linking adjacent pyrimidines. (*c*) The enzyme then dissociates from the DNA.

Bypass (Translesion) Synthesis

A very accurate, proofreading polymerase cannot insert nucleotides opposite damaged bases, and therefore would arrest synthesis (or leave a gap, as described below). A process called **bypass synthesis** or **translesion synthesis** can, however, allow replication to proceed past the lesion (Figure 4.31). There is a family of "error-prone" or "relaxed stringency" DNA polymerases that lack proofreading ability (Table 4.2). Due to their relaxed stringency, these polymerases can insert nucleotides opposite the damaged bases and thereby allow replication to continue. Their reduced accuracy increases the probability that the wrong nucleotide will be incorporated opposite the lesion, as well as in any other DNA made by these polymerases. The harmful consequences of this are evidently less than those of a long (or permanent) block to replication.

In *E. coli*, DNA polymerases IV (DinB) and V ($UmuD'_2C$ complex) are induced as part of the SOS response described below. This ensures that the error-prone bypass (translesion) synthesis mechanism functions only when there is sufficient damage to the DNA to trigger the SOS response, limiting the risk of mutations to situations in which the cell's viability is seriously threatened. Both pol IV and pol V are much less accurate than the normal replicative DNA polymerases III and I. They have low processivity, generally synthesizing fewer than 10 nucleotides before dissociating from the DNA, which serves to limit the extent of mutations introduced.

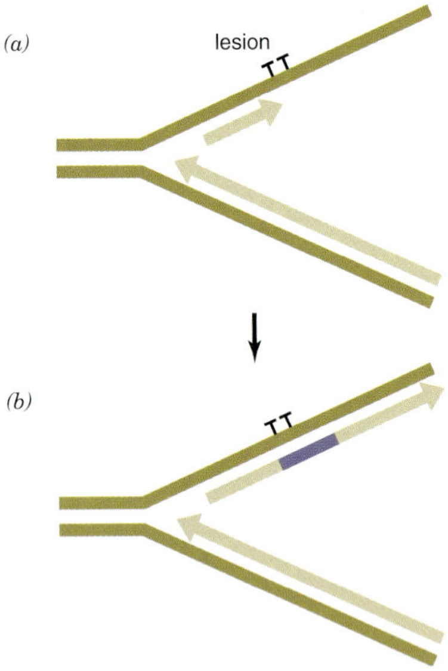

Eukaryotes also have error-prone polymerases that can carry out bypass (translesion) synthesis (Table 4.2). DNA polymerase-η predominantly incorporates A opposite the thymine dimers introduced by UV irradiation; because A is the appropriate nucleotide to incorporate opposite the T of the dimer, the probability of mutagenesis at those lesions is reduced (but not eliminated). Polymerase-η lacks intrinsic proofreading capability and incorporates one mismatch per 20–400 nucleotides synthesized, suggesting that its activity must be tightly controlled. DNA polymerase ι incorporates nucleotides opposite other highly distorting lesions or lesions such as abasic sites that cannot normally serve as templates. These polymerases appear to function with DNA polymerase ζ, which doesn't itself add nucleotides opposite lesions but can extend the structures formed by the other bypass polymerases until the normal replicative polymerase can take over. These bypass polymerases are distributive enzymes, so the region in which mismatches are likely to be introduced is typically small, limiting the introduction of mutations.

FIGURE 4.31

Bypass (translesion) synthesis. (*a*) Damage to the template generally halts replicative DNA polymerases, since they cannot add a nucleotide opposite a noncoding or strand-distorting lesion. (*b*) Specialized DNA polymerases with relaxed specificity (see Table 4.2) can insert nucleotides opposite lesions in the template, thereby allowing bypass of damage and synthesis of non-gapped daughter strands. The relaxed specificity and lack of proofreading in these polymerases means that errors (potential mutations) are often introduced. These polymerases readily dissociate from the DNA, so the length of DNA they synthesize is usually short.

Daughter Strand Gap Repair

If DNA synthesis is blocked by a lesion but restarts beyond the lesion, a gap is left in the DNA (Figure 4.32). This situation is easiest to picture on the discontinuous strand where it can occur simply by dissociation of the polymerase from the blocked site and initiation of the next Okazaki fragment by the normal replication mechanism. Gaps in DNA are potentially lethal. A lesion opposite a gap is not a substrate for excision repair, because there is no intact strand opposite the lesion to provide coding information necessary for repair. **Daughter-strand gap repair** (previously called post-replication recombination) does not remove the lesion that caused the gap, but does repair the gap. This protects the cell from the potential damage that gaps can cause, and it also provides a substrate for excision repair to act on the lesion itself.

Daughter-strand gap repair is by recombination. Single-stranded DNA, such as that opposite a gap, is recombinogenic; that is, it stimulates recombination. Recombination between the two newly formed duplexes allows transfer of a piece of the normal parental strand into the gap (Figure 4.32*a*; the parental strand on one side of the fork has the same polarity as the daughter strand on the other side). This leaves a gap in the parental strand that donated the patch (Figure 4.32*b*); but because that strand has a normal strand of DNA opposite, that gap can readily be filled by DNA polymerase and sealed by a DNA ligase (Figure 4.32*c*). The result is that the gap is repaired, although the lesion remains. As noted above, in this configuration the lesion is a substrate for excision repair.

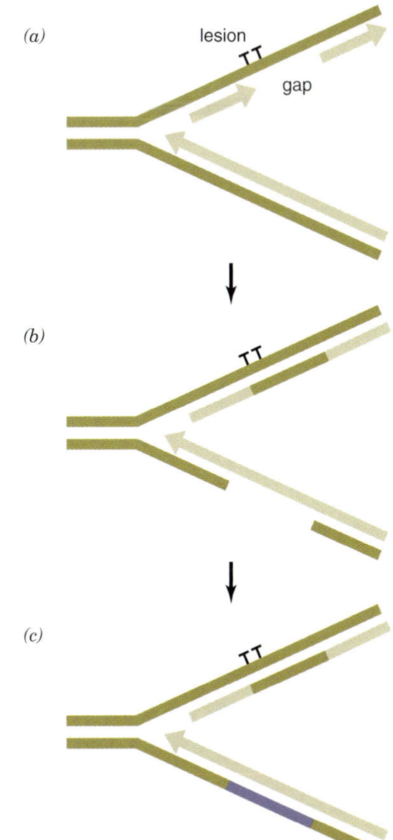

FIGURE 4.32

Daughter-strand-gap repair. (*a*) Gaps are left in newly replicated DNA where replication forks are halted by lesions. Because the lesion does not have intact DNA opposite, it is not a substrate for excision repair. (*b*) Recombination allows the isopolar parental strand (dark tan) to fill the gap in the daughter strand (light tan). This leaves a gap in that parental strand. (*c*) The gap in the parental strand can be filled (purple) by a DNA polymerase, because there is an intact template opposite it. The result is repair of gap in daughter strand, but the lesion remains. Note that the lesion can now be repaired by excision repair, because it is now opposite an intact strand.

Rewinding and Repair of Replication Forks

It is easy for a replication complex that has stopped synthesis of the discontinuous strand at a lesion to resume synthesis to form the next Okazaki fragment, using the normal mechanisms of DNA replication (see above). Reinitiating the continuous strand beyond a block is more difficult, because initiations are normally tightly controlled at replication origins. It appears that in this case, replication forks can regress by annealing of the daughter strands, which allows repair of the lesion and restarting of replication. In *E. coli*, RecA binds to the single-stranded regions at blocked replication forks and maintains the integrity of the DNA. The replication fork can regress by branch migration while it is protected by the RecA, allowing NER or BER repair mechanisms to remove the lesion. Then, a reversal of the branch migration or degradation of the displaced strand could allow synthesis to resume. Details are still unclear, including whether the replication apparatus could remain with the displaced strand or whether the replication complex would need to be reassembled in an origin-independent manner.

Double-Strand Break Repair

Double-strand breaks (DSBs) in DNA are potentially lethal events. They stimulate genetic recombination that can lead to chromosomal translocations, and unrepaired DSBs lead to broken chromosomes and cell death. Double-strand breaks are caused by many agents, particularly ionizing radiation, reactive oxygens species, and chemotherapeutic agents that generate oxidative free radicals (e.g., bleomycin) or are topoisomerase poisons. When a replication fork encounters a single-strand break, that break is converted into a double-strand break. Cells have mechanisms, including checkpoint controls, that slow or arrest replication in the presence of damage to allow time for repair and to minimize the likelihood that a single-strand break would be encountered. Nevertheless, it has been estimated that about 10 double-strand breaks are formed during a single cycle of replication in mammalian cells.

The major pathway for cell survival involves double-strand-break repair by homologous recombination (described above; Figure 4.24) or by nonhomologous end-joining. In yeast, homologous recombination predominates; in mammalian cells, nonhomologous end-joining predominates.

Regulation of DNA Repair: The SOS Regulon

In *E. coli*, damage to DNA triggers the **SOS response**, a coordinated change in gene expression that aids in recovery from damage. The SOS response is a coordinated induction of many genes whose transcription is at least partly regulated by a common repressor, LexA. A group of operons that are controlled by a common repressor is called a **regulon**. Among the genes in the SOS regulon are uvrA, uvrB, uvrC, recA, and lexA itself. There is a low, constitutive level of many of the genes in the SOS regulon, because

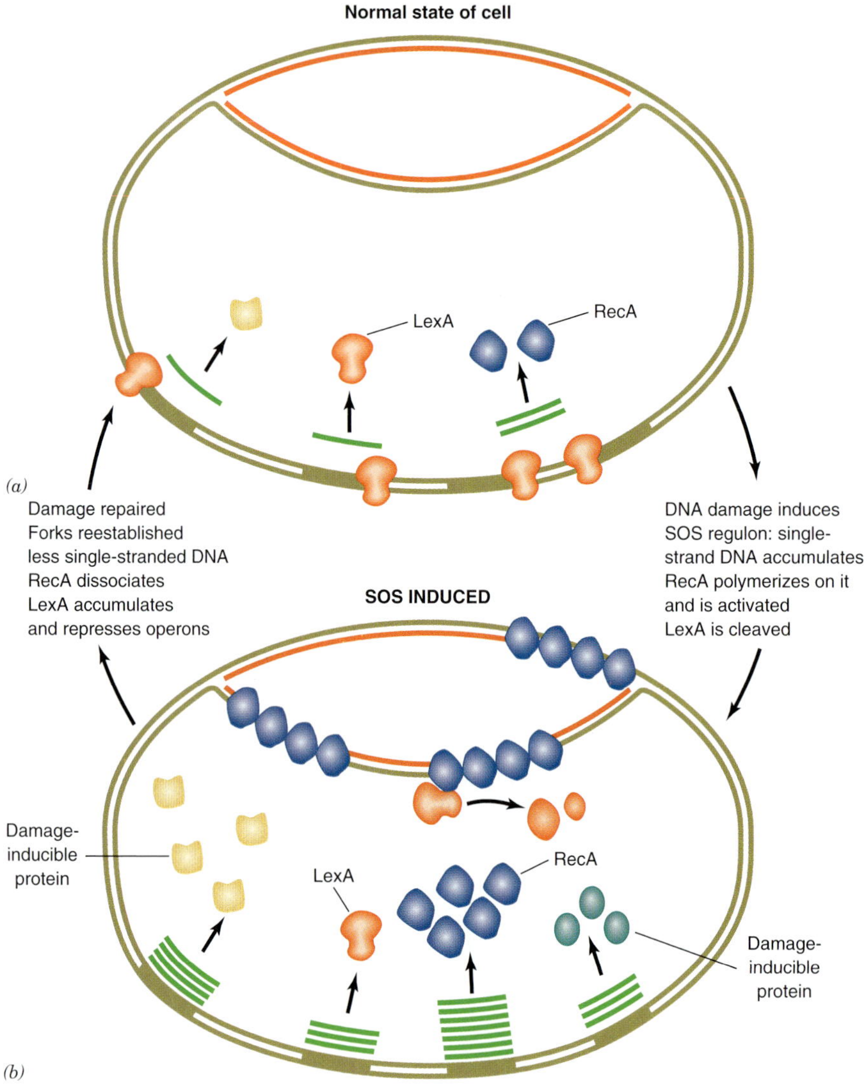

FIGURE 4.33

Regulation of DNA repair and recovery in *E. coli*: The SOS response. (*a*) In undamaged cells, a group of operons that constitute the SOS regulon is regulated by a common repressor, LexA. This regulon includes the lexA gene itself (autoregulation) and the recA gene, along with many genes encoding enzymes that act to repair damage. There is a low, constitutive transcription of many of these genes, including lexA and recA, even when LexA protein is present. (*b*) Damage blocks replication forks, thereby leaving single-stranded DNA. RecA binds to single-stranded DNA and is "activated." In that form, it aids in cleavage of LexA. The cleaved fragments of LexA cannot bind to operators, so the entire set of operons is induced. Much more RecA is made, along with other proteins including LexA. As long as the damage remains and RecA remains activated, LexA continues to be cleaved so the operons remain on. When damage is repaired, RecA is "deactivated" and no longer aids the cleavage of LexA, so active LexA builds up and shuts off the genes in this regulon.

repression by LexA is not complete and some of these genes have alternative promoters not repressed by LexA.

The **SOS response** is triggered by long stretches of single-stranded DNA such as those in gaps left by blocked replication (Figure 4.33). When such a stretch is formed, the low, constitutive level of RecA protein can cooperatively polymerize along single-stranded DNA to form a nucleoprotein filament. In this state, it is "activated" and it binds LexA, increasing the rate at which LexA is cleaved into inactive fragments. This relieves the repression and allows rapid synthesis of the genes normally repressed by LexA. RecA participates in daughter-strand gap repair and protection of the replication fork, as noted above. The other damage-inducible genes (din genes) produce proteins that aid in DNA repair and cell recovery and inhibit initiation of new replication forks and cell division until repair is completed (Figure 4.33).

Autoregulation of lexA is based upon a simple feedback loop that effectively allows a rapid but transient response to DNA damage. When LexA protein is cleaved, among the many genes of the SOS regulon that are induced is lexA itself, so the amount of LexA produced increases significantly. However, while DNA damage is still present and RecA protein is still activated, this LexA protein is cleaved and it cannot repress the many genes of the SOS regulon. As the damage is repaired, the amount of activated RecA drops and therefore cleavage of LexA decreases. The increasing levels of LexA bind to the operators of the din genes and gradually shut down the SOS response.

BIBLIOGRAPHY

Baynton, K. and Fuchs, R. P. P. Lesions in DNA: Hurdles for polymerases. *Trends Biochem. Sci.* 25:75, 2000.

Berger, J. M., Gamblin, S. J., Harrison, S. C., and Wang, J. C. Structure and mechanism of DNA topoisomerase II. *Nature* 379:225, 1996.

Blow, J. J. and Tada, S. A new check on issuing the license. *Nature* 404:560, 2000.

Cooke, M. S., Evans, M. D., Dizdaroglu, M., and Lunec, J. Oxidative DNA damage: Mechanisms, mutation, and disease. *FASEB J.* 17:1995, 2003.

Cox, M. M., Goodman, M. F., Kreuzer, K. N., Sherratt, D. J., Sandler, S. J., and Marians, K. J. The importance of repairing stalled replication forks. *Nature* 404:37, 2000.

Evans, A. R., Limp-Foster, M., and Kelley, M. R. Going Ape over Ref-1. *Mutat. Res.* 461:83, 2000.

Flores-Rozas, H. and Kolodner, R. D. Links between replication, recombination, and genome instability in eukaryotes. *Trends Biochem. Sci.* 25:196, 2000.

Friedberg, E. C. DNA damage and repair. *Nature* 421:436, 2003.

Friedberg, E. C. How nucleotide excision repair protects against cancer. *Nature Rev. Cancer* 1:22, 2001.

Gerson, S. L. MGMT: Its role in cancer aetiology and cancer therapeutics. *Nature Rev. Cancer* 4:296, 2004.

Haber, J. E. DNA recombination: the replication connection. *Trends Biochem. Sci.* 25:271, 2000.

Hanawalt, P. C. Transcription-coupled repair and human disease. *Science* 266:1957, 1994.

Hanawalt, P. C. The bases for Cockayne syndrome. *Nature* 405:415, 2000.

Hubscher, U., Maga, G., and Spadari, S. Eukaryotic DNA polymerases. *Annu. Rev. Biochem.* 71:133, 2002.

Human genome project. http://www.nhgri.nih.gov/HGP/

Human genome sequencing. http://www.ncbi.nlm.nih.gov/genome/seq/

Johnson, R. E., Washington, M. T., Haracska, L., Prakash, S., and Prakash, L. Eukaryotic polymerases ι and ζ act sequentially to bypass DNA lesions. *Nature* 406:1015, 2000.

Kowalczykowski, S. C. Initiation of genetic recombination and recombination-dependent replication. *Trends Biochem. Sci.* 25:156, 2000.

Lieber, M. R., Ma, Y., Pannicke, U., and Schwarz, K. Mechanism and regulation of human non-homologous DNA end joining. *Nature Rev. Mol. Cell Biol.* 4:712, 2003.

Lindahl, T. and Wood, R. D. Quality control by DNA repair. *Science* 286:1897, 1999.

Modrich P. Mismatch repair, genetic stability, and cancer. *Science* 266:1959, 1994.

Naktinis, V., Turner, J., and O'Donnell, M. A molecular switch in a replication machine defined by an internal competition for protein rings. *Cell* 84:137, 1996.

Online Mendelian Inheritance in Man (OMIM) http://www.ncbi.nlm.nih.gov/entrez/query.fcgi?db=OMIM

Rattray, A. J. and Strathern, J. N. Error-prone DNA polymerases: When making a mistake is the only way to get ahead. *Annu. Rev. Genet.* 37:31, 2003.

Sancar, A. Excision repair in mammalian cells. *J. Biol.Chem.* 270:15915, 1995.

Stillman, B. Cell cycle control of DNA replication. *Science* 274:1659, 1996.

Stahl, F. Meiotic recombination in yeast: Coronation of the double-strand-break repair model. *Cell* 87:965, 1996.

Ulaner, G. A. Telomere maintenance in clinical medicine. *Am. J. Med.* 117:262, 2004.

Von Hippel, P. H. and Jing, D. H. Bit players in the trombone orchestra. *Science* 287:2435, 2000.

Waga, S. and Stillman, B. Anatomy of a DNA replication fork revealed by reconstitution of SV40 replication *in vitro*. *Nature* 369:207, 1994.

Watson, J. D. and Crick, F. H. C. Genetical implications of the structure of deoxyribonucleic acid. *Nature* 171:964, 1953.

Wood, R. D., Mitchell, M., Sgouros, J., and Lindahl, T. Human DNA repair genes. *Science* 291:1284, 2001.

Yu, Z., Chen, J., Ford, B. N., Brackley, M. E., and Glickman, B. W. Human DNA repair systems: An overview. *Environ. Mol. Mutagenesis* 33:3, 1999.

Zakian, V. A. Telomeres: Beginning to understand the end. *Science* 270:1601, 1995.

QUESTIONS

Carol N. Angstadt

Multiple Choice Questions

1. Replication:
   A. is semiconservative.
   B. requires only proteins with DNA polymerase activity.
   C. uses 5′ to 3′ polymerase activity to synthesize one strand and 3′ to 5′ polymerase activity to synthesize the complementary strand.
   D. requires a primer in eukaryotes but not in prokaryotes.
   E. must begin with an excision step.

2. In eukaryotic DNA replication:
   A. only one replisome forms because there is a single origin of replication.
   B. the Okazaki fragments are 1000 to 2000 nucleotides in length.
   C. helicase dissociates from DNA as soon as the initiation bubble forms.
   D. FEN1 (flap endonuclease 1) is involved in removing the primer.
   E. the process occurs throughout the cell cycle.

3. All of the following statements about telomerase are correct *except*:
   A. the RNA component acts as a template for the synthesis of a segment of DNA.
   B. it adds telomeres to the 5′-ends of the DNA strands.
   C. it provides a mechanism for replicating the ends of linear chromosomes.
   D. it recognizes a G-rich single strand of DNA.
   E. it is a reverse transcriptase.

4. A transition mutation:
   A. occurs when a purine is substituted for a pyrimidine, or vice-versa.
   B. results from the insertion of one or two bases into the DNA chain.
   C. decreases in frequency in the presence of base analogs.
   D. results from substitution of one purine for another or of one pyrimidine for another.
   E. always is a missense mutation.

5. Homologous recombination:
   A. occurs only between two segments from the same DNA molecule.
   B. requires that a specific DNA sequence be present.
   C. requires one of the duplexes undergoing recombination be nicked in both strands.
   D. may result in strand exchange by branch migration.
   E. is catalyzed by transposases.

6. All of the following are true about transpositions *except*:
   A. transposons move from one location to a different one within a chromosome.
   B. both the donor and target sites must be homologous.
   C. transposons have insertion sequences that are recognized by transposases.
   D. the transposon may either be excised and moved, or be replicated with the replicated piece moving.
   E. they may either activate or inactivate a gene.

Questions 7 and 8: Retroviruses, like HIV which causes AIDS, have their genetic information in the form of RNA. Reverse transcriptase synthesizes a DNA copy of the viral genome. One drug used in treating AIDS is AZT, an analog of deoxythymidine, which has an azido group at the 3′ position of the sugar. It can be phosphorylated and competes with dTTP for incorporation into the reverse transcript. Once incorporated, its presence terminates chain elongation.

7. The growing chain is terminated because:
   A. the analog can not hydrogen bond to RNA.
   B. the presence of the AZT analog inhibits the proofreading ability of reverse transcriptase.
   C. AZT does not have a free 3′-OH.
   D. the analog causes distortion of the growing chain inhibiting reverse transcriptase.
   E. dTTP can no longer be added to the growing chain.

8. There is a window in which the effect is primarily on viral replication since AZT is much less effective at competing with dTTP for incorporation by cellular DNA polymerases because of the proofreading ability of DNA polymerases. Proofreading activity to maintain the fidelity of DNA synthesis:
   A. occurs after the synthesis has been completed.
   B. is a function of 3′ to 5′ exonuclease activity intrinsic to or associated with DNA polymerases.
   C. requires the presence of an enzyme separate from the DNA polymerases.
   D. occurs in prokaryotes but not eukaryotes.
   E. is independent of the polymerase activity in prokaryotes.

Questions 9 and 10: Patients with the rare genetic disease xeroderma pigmentosum (XP) are very sensitive to light and are highly susceptible to skin cancers. Study of such patients has enhanced our knowledge of DNA repair because XP is caused by defective DNA repair–nucleotide excision repair. (A variant, XP-V, is deficient in post-replication repair.)

9. All of the following are true about nucleotide excision repair *except*:
   A. removal of the damaged bases occurs on only one strand of the DNA.
   B. it removes thymine dimers generated by UV light.
   C. it involves the activity of an excision nuclease, which is an endonuclease.
   D. it requires a polymerase and a ligase.
   E. only the damaged nucleotides are removed.

10. Another type of DNA repair is base excision repair. Base excision repair:
    A. is used only for bases that have been deaminated.
    B. uses enzymes called DNA glycosylases to generate an abasic sugar site.
    C. removes about 10 to 15 nucleotides.
    D. does not require an endonuclease.
    E. recognizes a bulky lesion.

Questions 11 and 12: Interfering with topoisomerases is one way of inhibiting DNA replication. Certain antibiotics target DNA gyrase (type II topoisomerase) of *E. coli* inhibiting catalytic activity. Topoisomerase poisons prevent resealing of the phosphodiester bond, leaving covalent protein-DNA junctions. These compounds are used in treating infections and as chemotherapeutic agents.

11. DNA gyrase:
    A. ATP subunits hydrolyze ATP to form new phosphodiester bonds.
    B. removes negative supercoils.
    C. swivelase subunits create and reseal transient nicks on both strands.
    D. increases the linking number.
    E. occurs in eukaryotes as well as prokaryotes.

12. All of the following are correct about double strand breaks in DNA *except* they:
    A. can lead to loss of genetic information.
    B. are always involved in homologous recombination.
    C. are involved in nonhomologous recombination.
    D. are associated with a heterodimer (Ku) in mammals.
    E. can lead to mutations or improper regulation of gene expression.

Problems

13. Mismatch repair removes replication errors by excising incorrect bases. There is no DNA damage or modified bases present. How does the cell distinguish the newly synthesized strand and preserve the correct parental DNA strand?

14. In the coding strand of DNA for the alpha gene of normal hemoglobin (HbA), the three bases that correspond to codon 142 of the mRNA are TAA and the alpha chain has 141 amino acids. In the coding strand of the gene for the alpha chain of Hemoglobin Constant Spring, the three bases are CAA and the chain contains 172 amino acids. Explain the mutation that has occurred.

ANSWERS

1. **A** New DNA has one parent and one new strand. B and D: Replication requires proteins like primase, ligases, helicases, and others. C: Replication involves Okazaki fragments because synthesis occurs only in the 5′ to 3′ direction. E: Excision is the recognition step for DNA repair.

2. **D** It removes the last ribonucleotide. A: There are multiple initiation sites. B: This is the size of prokaryotic Okazaki fragments; eukaryotic fragments are about 200 nucleotides. C: Helicase activity is also necessary for the continuation of synthesis. E: Replication is confined to the S phase.

3. **B** Telomeres are at the 3′ end of each strand so that the 5′ ends can be replicated. A and C: Telomerase both positions itself at the 3′ end of the DNA and provides the template for extending that end. D: This is a characteristic of the 3′ end. E: It is using an RNA template to synthesize DNA.
4. **D** This is the definition. B: This is frame-shift; transitions are point mutations. C: Frequency of mutation increases. E: It could be a missense mutation if the change coded for a different amino acid or a nonsense mutation if the code was changed to a stop signal.
5. **D** This is just one of the events in this complex process. A and B: It may occur between two distinct DNA molecules if the two sequences are homologous. C: Nicks can be on a single strand. E: These are the enzymes of transpositional site-specific recombination.
6. **B** Only the donor site requires a specific nucleotide sequence; homology is not required. A: This is the definition. C: This is one of the key features. D: Both types of event are catalyzed by transposases. E: Insertion into the middle of a gene would inactivate it; insertion of a promoter next to a gene may activate it.
7. **C** The chemistry of nucleotide formation requires a free 3′-OH at the end of the chain for attachment of the next nucleotide. A: The presence of the azido group does not affect hydrogen bonding. B: Reverse transcriptase does not have proofreading properties. D: This does not happen. E: The AZT analog competes with dTTP, it does not eliminate it.
8. **B** This activity removes a newly added base if there is a mismatch with the template. A: This is called repair. C: Most polymerases are multifunctional and have proofreading ability. D: Not all eukaryotic polymerases have 3′ to 5′ exonuclease activity but some do. E: The polymerase seems to detect the mismatch and repair it.
9. **E** Cuts are made several nucleotides on either side of damaged bases. A: The uncut strand serves as template for repair. B: Thymine dimers are only one cause of bulky lesions. C and D: The excision nuclease is a complex of proteins needed to unwind the DNA and remove the lesion. A polymerase and ligase fill in the gap.
10. **B** These catalyze the first step of the process. A: Methylated and other chemically modified bases can also be removed. C and E: These are characteristics of a different repair system. D: The abasic sugar phosphate must be removed.
11. **C** This enables unwinding of the DNA. A: ATP hydrolysis triggers conformational changes. B: It removes positive supercoils and introduces negative ones. D: Linking number is reduced. E: DNA gyrase is specifically prokaryotic; eukaryotes have different topoisomerases.
12. **B** This can also occur with single strand nicks. A: This can be repaired but covalent protein-DNA junctions caused by topoisomerase poisons can be lethal. C: Broken ends of a DNA duplex can recombine with another duplex. D: This plays a role in non-homologous end joining. E: Random integration of a fragment of DNA by non-homologous joining can disrupt genes or disrupt promoters.
13. In *E. coli*. DNA is methylated on the A of a GATC sequence. Since this is a palindrome, both strands are methylated. Methylation occurs after the unmethylated bases are incorporated into DNA. During synthesis, the DNA will be hemimethylated for a short period. The mismatch repair system recognizes the hemimethylated state and can remove the mismatch on the unmethylated (new) strand. Eukaryotes have a similar repair system although the details of recognizing the new strand are not yet known.
14. The coding strand of the gene has the same sequence as the mRNA (except U replaces T in the RNA). In HbA, the codon at position 142 of mRNA is a stop codon so the last amino acid added is 141. In Hb Constant Spring, a point mutation has mutated the DNA so that the mRNA codon at 142 now codes for an amino acid instead of stop. Translation continues until a stop codon appears at position 173 (so 172 amino acids).

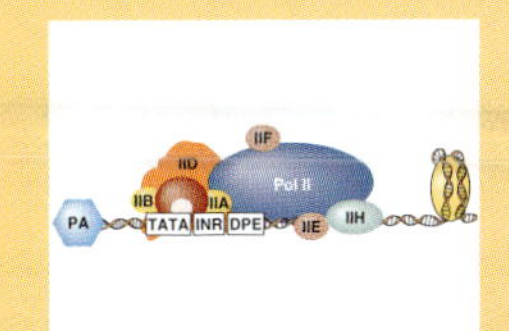

5

RNA: TRANSCRIPTION AND RNA PROCESSING

Francis J. Schmidt and David R. Setzer

Textbook of Biochemistry With Clinical Correlations, Sixth Edition, Edited by Thomas M. Devlin

5.1 OVERVIEW

Synthesis of an RNA molecule involves copying one strand of a template sequence using Watson–Crick base pairing between nucleotides of the template (usually DNA) and the nucleotides that are being incorporated into the **transcript**. The initiation of transcription by RNA polymerase is perhaps the most important event in the control of gene expression. Transcription initiation requires specialized DNA sequences, called **promoters**, which signal where RNA synthesis should begin. The recognition of promoter sequences involves molecular contacts between DNA and protein **factors**. Factors bind both to RNA polymerase enzyme and to DNA nucleotides through hydrogen bonds and other contacts. During elongation, **RNA processing** reactions remove, add, or modify nucleotides in the primary transcript. These processing reactions can occur cotranscriptionally. In other words, they occur on parts of the transcript while downstream sequences are still being transcribed.

5.2 MECHANISMS OF TRANSCRIPTION

Initial Process of RNA Synthesis is Transcription

Transcription is the process by which RNA chains are made from DNA templates. Transcription reactions take the following form:

$$\text{DNA template} + n(\text{NTP}) \rightarrow \text{pppN(pN)}_{n-1} + (n-1)\text{PP}_i + \text{DNA template}$$

Enzymes that catalyze this reaction are the **RNA polymerases**; it is important to recognize that, like DNA polymerases, they are absolutely template-dependent. Unlike DNA polymerases, however, RNA polymerases do not require a primer molecule. The energetics favoring the RNA polymerase reaction are twofold: First, the $5'\alpha$-nucleotide phosphate of the ribonucleoside triphosphate is converted from a phosphate anhydride to a phosphodiester bond with a change in free energy ($\Delta G'$) of approximately 3 kcal (12.5 kJ) mol^{-1} under standard conditions. Second, the released pyrophosphate, PP_i, is cleaved to two phosphates by pyrophosphatase so that its concentration is low and phosphodiester bond formation is more favored relative to standard conditions (see p. 532).

Since a DNA template is required for RNA synthesis, eukaryotic transcription takes place in the nucleus or in the matrix of mitochondria and chloroplasts. Structural changes occur in DNA during its transcription. In polytene chromosomes of *Drosophila*, transcriptionally active genes are visualized in the light microscope as puffs distinct from the condensed, inactive chromatin. In fact, active genes exist in modified chromatin structures that are less compacted than are inactive genes, as revealed by the accessibility of these genes to chemical reagents or enzymes. In both prokaryotes and eukaryotes, the DNA double helix is transiently opened (unwound) as the transcription complex proceeds along the DNA.

These openings and unwindings are necessary because DNA is a double helix and the two strands of DNA must be separated from one another to permit Watson–Crick base pairing between the template strand of DNA and nucleotides in the newly synthesized RNA. Local opening and unwinding of the DNA allows transcription to proceed without wrapping the RNA product around the DNA template.

The process of transcription is generally divided into three parts: Initiation refers to the recognition of a specific DNA sequence by RNA polymerase and the beginning of the bond formation process. Elongation is the actual synthesis of the RNA chain, which is followed by chain termination and release.

DNA Sequence Information Signals RNA Synthesis

Much of the human genome is never transcribed into RNA. Even in bacteria, where almost all the DNA specifies gene products, not all genes are expressed at any one time.

RNA polymerase molecules must, therefore, distinguish between genes and non-genes in the DNA, as well as between genes that need to be expressed and those that do not.

Initiation of transcription starts with recognition of a promoter sequence in the DNA template. Typically, **promoters** are located a short distance upstream of the nucleotide where transcription starts. Different promoters often contain similar sequences, permitting identification of a "**consensus**" promoter sequence (Figure 5.1), from which actual promoters vary to differing extent. Although prokaryotic and eukaryotic promoters differ from one another, conserved sequences are found within both categories of promoters. Some DNA sequences stimulate transcription but are located further away from the initiation site, and some of these can stimulate transcription independently of distance, location, or orientation relative to the transcriptional start site. Such regulatory sequences are called enhancers. Enhancers stimulate the synthesis of some prokaryotic RNAs and of most, if not all, eukaryotic mRNAs. Enhancers work by binding specific protein factors, called activators. When an activator binds to an enhancer, a structural change (often a looping or bending) in the DNA template allows interaction of the activator with other factors or with RNA polymerase. This interaction facilitates transcription by "recruiting" RNA polymerase and other factors to form an initiation complex.

RNA Polymerase Catalyzes the Transcription Process

RNA polymerases synthesize RNA in the $5' \rightarrow 3'$ direction using a DNA template, being similar to the template-dependent DNA polymerases (see p. 136). Unlike DNA polymerases, however, RNA polymerases initiate polymerization at promoter sequences without the need for a DNA or RNA primer.

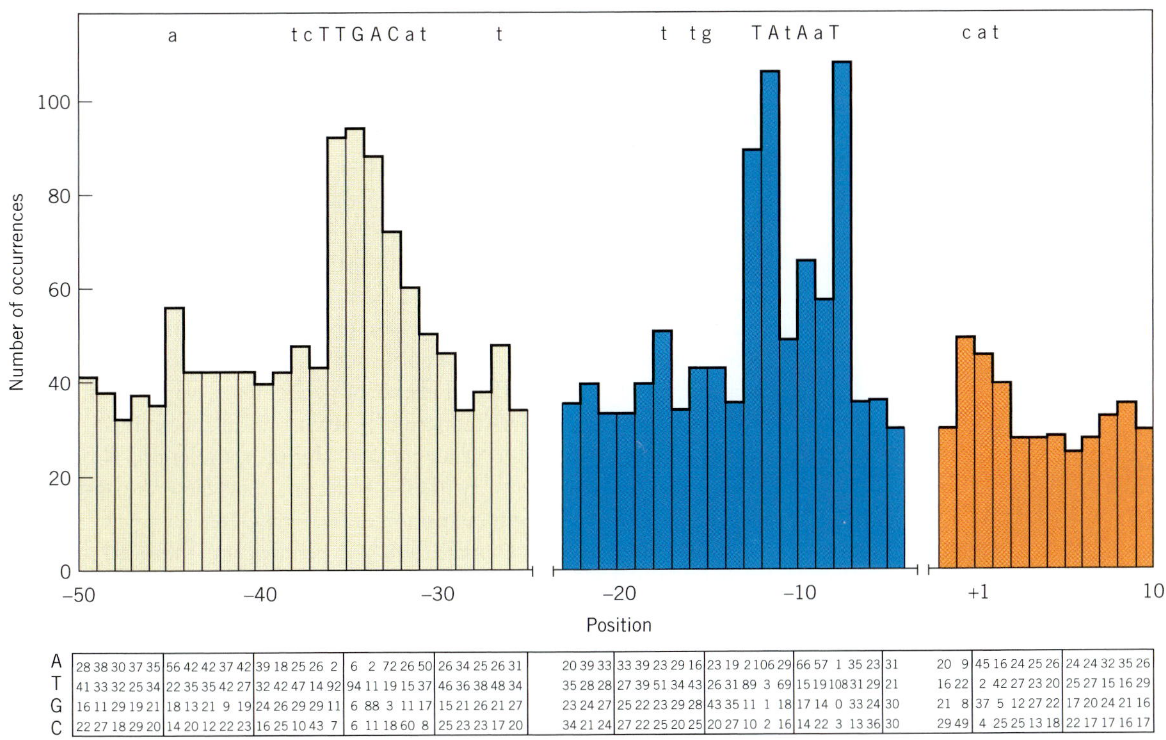

FIGURE 5.1

A consensus sequence for prokaryotic promoters. The sequences of a large number of promoters from *E. coli* were compiled, and the number of times each base occurs at individual positions in the promoter was tabulated. The histogram displays the number of occurrences of the most frequent base at each position. This analysis permits the definition of a consensus sequence for *E. coli* promoters recognized by the form of RNA polymerase holoenzyme containing σ^{70}. This consensus sequence is shown at the top, with the most highly conserved bases in capital letters and the more weakly conserved bases in lowercase.
Redrawn from Hawley, D. K. and McClure, W. R. *Nucleic Acids Res.* 11:2237, 1983.

TABLE 5.1 Subunit Composition of Some RNA Polymerases[a]

Yeast Pol I	*Yeast Pol II*	*Yeast Pol III*	*E. coli RNA Polymerase Core*
A190	B220 (Rpb1)	C160	β'
A135	B150 (Rpb2)	C128	β
AC40	B45 (Rpb3)	AC40	α
AC19	B12.5 (Rpb11)	AC19	ω
ABC23	ABC23 (Rpb6)	ABC23	
ABC27	ABC27 (Rpb5)	ABC27	
ABC14.5	ABC14.5 (Rpb8)	ABC14.5	
ABC10α	ABC10α ((Rpb12)	ABC10α	
ABC10β	ABC10β (Rpb10)	ABC10β	
A12.2	B12.6 (Rpb9)	C11	
A14	B32 (Rpb4)	C17	
A43	B16 (Rpb7)	C25	
A49			
A34.5			
		C82	
		C53	
		C37	
		C34	
		C31	

[a] Homologous subunits are shown in the same row. Note that five subunits are shared between RNA polymerases I, II, and III from yeast, and that two additional subunits are shared between RNA polymerases I and III. In addition, RNA polymerase I contains two, and RNA polymerase III contains five, subunits that have no obvious homologues in the other polymerases. Four subunits of each of the eukaryotic polymerases are homologous to the four different subunits of *E. coli* core RNA polymerase. The letters A, B, and C in the subunit names refer to the occurrence of that subunit in RNA polymerase I, II, or III, respectively, and the number in the subunit name refers to the subunit's molecular weight in kilodaltons. For RNA polymerase II, the gene encoding each of the subunits is given in parentheses.

Prokaryotic and eukaryotic RNA polymerases are large multisubunit enzymes whose mechanisms are only partially understood. RNA polymerase from *Escherichia coli* consists of five subunits with an aggregate molecular weight of over 500,000 (Table 5.1). Two α subunits, one β subunit and one β' subunit constitute the **core enzyme**, which is capable of faithful transcription but not of specific (i.e., promoter-initiated) RNA synthesis. Addition of a fifth protein subunit, designated σ, forms the **holoenzyme** that is capable of specific RNA synthesis *in vitro* and *in vivo*. That σ is involved in the specific recognition of promoters has been borne out by a variety of biochemical studies. Specific σ factors can recognize different classes of genes. Thus, a specific σ factor recognizes promoters for genes that are induced as a result of heat shock. In sporulating bacteria, specific σ factors recognize genes induced during sporulation. Some bacteriophages (viruses that infect bacteria) synthesize σ factors that allow the appropriation of the cell's RNA polymerase for transcription of the viral DNA.

The common prokaryotic RNA polymerases are inhibited by the antibiotic rifampicin (used in treating tuberculosis), which binds to the β subunit (Clin. Corr. 5.1). Eukaryotic nuclear RNA polymerases are distinguished because they are inhibited differentially by α-**amanitin**, which is synthesized by the poisonous mushroom *Amanita phalloides*. Three nuclear RNA polymerase classes can be distinguished since very low concentrations of α-amanitin inhibit the synthesis of mRNA and some small nuclear RNAs (snRNAs); higher concentrations inhibit the synthesis of tRNA, 5S rRNA, and other snRNAs. rRNA synthesis (other than 5S rRNA synthesis) is not inhibited much even by very high concentrations of the toxin. The purified enzymes are differentially inhibited by α-amanitin, so it is possible to conclude that mRNA synthesis is the function of RNA polymerase II, the most sensitive of the purified RNA polymerase forms. Synthesis of tRNA, 5S rRNA, U6 snRNA, and other small cellular and viral RNAs is carried out by RNA polymerase III. rRNA genes are transcribed by RNA polymerase I, which is concentrated in the nucleolus. (The numbers refer to the order of elution of the enzymes from a chromatography column.) Each enzyme has a highly complex structure (Table 5.1).

An RNA polymerase in mitochondria is responsible for synthesis of mitochondrial mRNA, tRNA, and rRNA species. This enzyme, like bacterial RNA polymerase, is inhibited by rifampicin (Clin. Corr. 5.1).

Steps of Transcription in Prokaryotes

Chromosomal DNA is usually transcribed in only one direction. This is illustrated as follows:

DNA: $5' \rightarrow 3'$

DNA: $3' \rightarrow 5'$

RNA: $5' \rightarrow 3' \cdots$

The DNA strand that serves as the template for RNA synthesis is complementary to the RNA transcript. Conventionally, the template strand is usually the "bottom" strand of a double-stranded DNA as written. The other strand, the non-template or "top" strand, has the same direction as the transcript when read in the $5'$ to $3'$ direction; this strand is sometimes called the coding strand. When only a single DNA sequence is given in this book, the coding strand is represented. Its sequence can be converted to the RNA transcript of a gene by simply substituting U (uracil) for T (thymine) bases.

Promoter Recognition

Prokaryotic transcription begins with binding of RNA polymerase to a gene's promoter (Figure 5.2). RNA polymerase holoenzyme binds to one face of the DNA extending ~45 bp upstream and 10 bp downstream from the RNA initiation site. Two short sequences in this region are highly conserved (Figure 5.1). One sequence that is located about 10 bp upstream from the transcription start is the consensus sequence (sometimes called a "−10" or "Pribnow" box):

$$T^{*}A^{*}TAAT^{*}$$

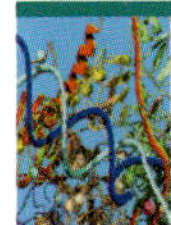

CLINICAL CORRELATION **5.1**

Antibiotics and Toxins that Target RNA Polymerase

RNA polymerase is an essential enzyme for life since transcription is the first step of gene expression. Lack of RNA polymerase activity means no other proteins are made. Two natural products illustrate this principle; in both cases, inhibition of RNA polymerase leads to death of the organism.

The "death cap" or "destroying angel" mushroom, *Amanita phalloides*, is highly poisonous and still causes several deaths each year despite widespread warnings to amateur mushroom hunters (it is reputed to taste delicious, incidentally). The most lethal toxin, α-amanitin, inhibits eukaryotic RNA polymerase II, thereby inhibiting mRNA synthesis. The poisoning starts with relatively mild, gastrointestinal symptoms, followed about 48 h later by massive liver failure as essential mRNAs and their proteins are degraded but not replaced by newly synthesized molecules. The only therapy is supportive, including liver transplantation.

More benign (at least from the point of view of our own species) is the action of the antibiotic rifampicin to inhibit the RNA polymerases of a variety of bacteria, most notably in the treatment of tuberculosis. *Mycobacterium tuberculosis*, the causative agent, is insensitive to many commonly used antibiotics but it is sensitive to rifampicin, the product of a soil streptomycetes. Since mammalian RNA polymerase differs from the prokaryotic variety, inhibition of the latter enzyme is possible without great toxicity to the host. This implies a good therapeutic index for the drug—that is, the ability to treat a disease without causing undue harm to the patient. Together with improved public health measures, antibiotic therapy with rifampicin and isoniazid (an antimetabolite) has greatly reduced the morbidity due to tuberculosis in industrialized countries. Unfortunately the disease is still endemic in impoverished populations in the United States and in other countries. In increasing numbers, immunocompromised individuals (especially AIDS patients), have active tuberculosis.

Source: Mitchel, D. H. *Amanita* mushroom poisoning. *Annu. Rev. Med.* 31:51, 1980. A. G. Gilman, T. W. Rall, A. S. Nies, and P. Taylor (Eds.). *The Pharmacological Basis of Therapeutics*, 8th ed. New York: Pergamon Press, 1990, p. 129 and K. M. DeCock, B. Soro, I. M. Colibaly, and S. B. Lucas. Tuberculosis and HIV infection in sub-Saharan Africa. *JAMA* 268:1581, 1992.

The positions marked with an asterisk are the most conserved; indeed, the last T residue is always found in *E. coli* promoters.

A second consensus sequence is located upstream from the **Pribnow** or "−10" **box**. This "−35 sequence"

T*T*G*ACA

is centered about 35 bp upstream from the transcription start; the nucleotides with asterisks are most conserved. The spacing between the "−35" and "−10" sequences' is crucial, with 17 bp being highly conserved. The TTGACA and TATAAT sequences are asymmetrical; that is, they do not have the same sequence if the complementary sequence is read. Thus, the promoter sequence itself determines that transcription will proceed in only one direction.

What difference do these consensus sequences make to a gene? Measurements of RNA polymerase binding affinity and initiation efficiency on various promoter sequences show that the most active promoters fit the consensus sequences most closely. The degree of similarity of a particular promoter sequence to the consensus correlates well with the measured "strength" of a promoter—that is, its ability to initiate transcription with purified RNA polymerase.

Bases flanking the "−35" and "−10" sequences, bases near the transcription start, and bases located near the "−16" position are weakly conserved. In some of these weakly conserved regions, RNA polymerase may require that a particular nucleotide not be present or that local variations in DNA helical structure be present. Promoters for *E. coli* heat shock genes have different consensus sequences at the "−35" and "−10" regions. This is consistent with their being recognized by a different σ factor.

An RNA transcript usually starts with a purine ribonucleoside triphosphate (i.e., pppG. . . or pppA. . .), but pyrimidine starts are also known (Figure 5.1). The position of transcription initiation differs slightly among various promoters, but usually is 5–8 bp downstream from the invariant T of the Pribnow box.

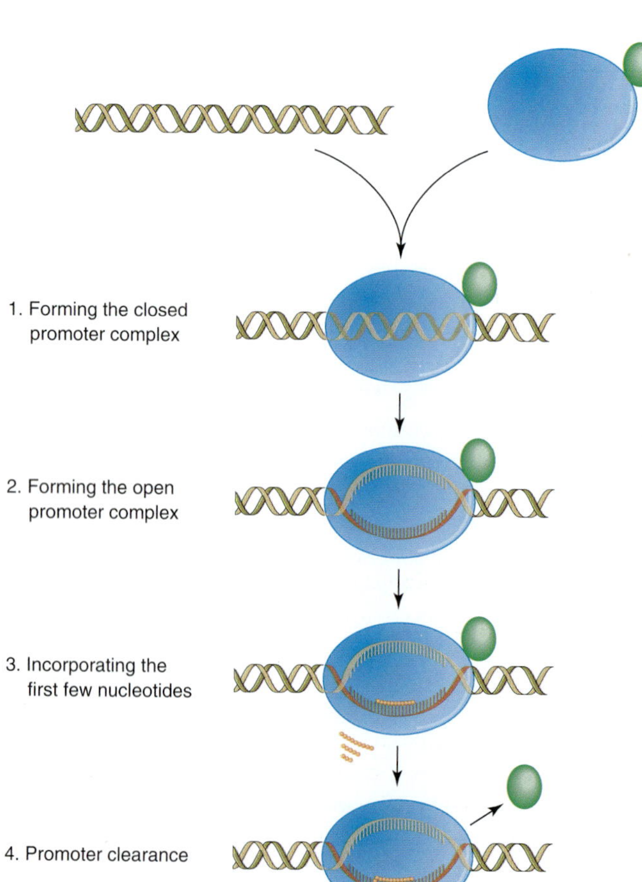

FIGURE 5.2

Early events in prokaryotic transcription. Promoter-containing DNA is shown in double-helical form, and *E. coli* RNA polymerase holoenzyme is shown as two ovals, the smaller of which represents sigma factor and the larger of which represents the core. The nascent RNA is shown within the transcription bubble annealed to the DNA template strand, and abortive initiation products are shown as products released from the initiation complex. Although this figure depicts release of the sigma subunit as RNA polymerase enters productive elongation mode, the fate of sigma factor at this stage of the transcription process remains somewhat controversial.

Start of Synthesis

Two kinetically distinct steps are required for RNA polymerase to initiate synthesis of an RNA transcript (Figure 5.2). In the first step, described above, RNA polymerase holoenzyme binds relatively weakly to the promoter DNA to form a "closed complex." In the second step, the holoenzyme forms a more tightly bound "open complex," characterized by a local opening of about 10 bp of the DNA double helix. Since the consensus Pribnow box is A-T-rich, it can facilitate this local unwinding: Its base pairs are more easily disrupted during opening. As discussed in Chapter 2, opening 10 bp of DNA is topologically equivalent to the relaxation of a single negative supercoil. As might be predicted from this observation, the activity of some promoters depends on the superhelical state of the DNA template. Some promoters are more active on highly supercoiled DNA, while others are more active when the superhelical density of the template is lower. The unwound DNA binds the initiating nucleoside triphosphate, and RNA polymerase then catalyzes formation of the first phosphodiester bond. The enzyme translocates to the next position (this is the rifampicin-inhibited step) and continues synthesis. After several nucleotides have been added to the growing RNA chain, the enzyme enters an elongation mode characterized by a very stable association with the DNA template and does not disengage until specific sequence signals for transcription termination are encountered. Release of the sigma subunit may be associated with entry into this highly processive elongation state. Once the RNA polymerase has cleared the promoter region, other RNA polymerase molecules can bind and initiate at the promoter so that the gene can be transcribed by many polymerases at the same time (Figure 5.3).

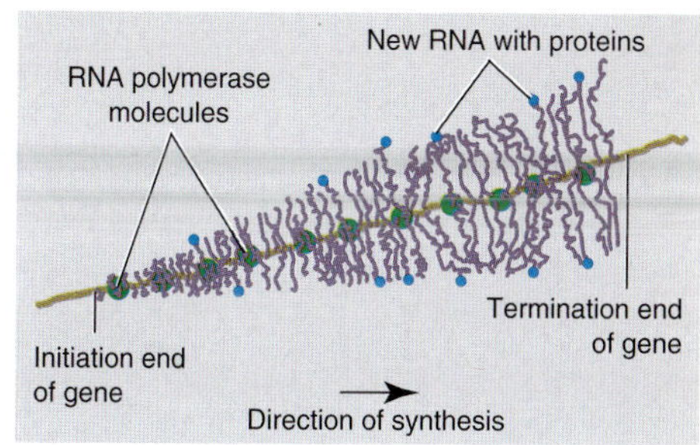

FIGURE 5.3

Simultaneous transcription of a gene by many RNA polymerases, depicting increasing length of nascent RNA molecules.
Courtesy of Dr. O. L. Miller, University of Virginia. Reproduced with permission from Miller, O. L. and Beatty, B. R. *J. Cell Physiol.* 74:225, 1969.

Elongation

RNA polymerase continues the nucleotide-binding-bond formation-translocation cycle at an average rate of about 40 nucleotides per second. However, many examples are known where RNA polymerase pauses or slows down at particular sequences. As discussed below, these pauses can be linked to transcription termination.

As RNA polymerase continues along the double helix, it continues to separate the two strands of the DNA template. This process allows the template strand of the DNA to base-pair with the growing RNA chain. Thus, a single mechanism of information transfer (**Watson–Crick base pairing**) serves several processes: DNA replication, DNA repair, and transcription of genetic information into RNA. Base pairing is essential for translation as well (see p. 204). The process of unwinding and restoring the DNA double helix leads to changes in superhelical density in the DNA, and the superhelical state of the DNA is controlled by the activity of DNA topoisomerases I and II.

Changes in the transcription complex during the elongation phase can affect subsequent termination events. These changes depend on the binding of other cellular proteins (NusA protein, for example) to core RNA polymerase. Failure to bind these protein factors can result in an increased frequency of termination and, consequently, a reduced level of gene expression.

Termination

The RNA polymerase complex also recognizes the ends of genes (Figure 5.4). Transcription termination can occur in either of two modes, depending on whether or not it is dependent on the protein **factor ρ**. Terminators are thus classified as ρ-independent or ρ-dependent. **Rho-independent terminators** are well-characterized (Figure 5.4). A consensus-type sequence is involved: a G-C-rich palindrome (inverted repeat) which precedes a sequence of 6–7 U residues in the RNA chain. As a result, the RNA chain forms a stem-and-loop structure just upstream of the oligoU residues. This secondary structure of the stem and loop is crucial for termination, as demonstrated by experiments showing that base change mutations that disrupt pairing in the RNA also reduce termination. The stem and loop left after termination stabilizes prokaryotic mRNA against nucleolytic degradation.

Rho-dependent terminators are less well-defined. Rho factor is a hexameric protein that has an essential RNA-dependent ATPase activity. The sequences of ρ-dependent termination sites feature C-rich sequences located some distance upstream

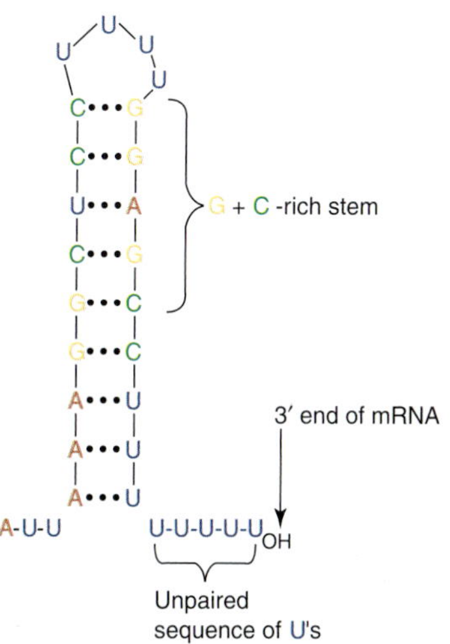

FIGURE 5.4

Stem-loop structure of RNA transcript that determines rho-independent transcriptional termination. Note the two components of the structure: the G + C-rich stem and loop, followed by a sequence of U residues.

of the termination site, which is often a rather broad region within which termination occurs, rather than a specific site. The ρ factor is thought to translocate along the nascent RNA chain in an ATP-dependent fashion until it catches up with the elongating RNA polymerase. When that occurs, rho in some way destabilizes the elongation complex, leading to release of both the template DNA and the completed RNA chain from each other and from the polymerase.

Prokaryotic ribosomes usually attach to nascent mRNA while it is being transcribed. This coupling between transcription and translation is important in gene control by attenuation (see p. 301).

5.3 TRANSCRIPTION IN EUKARYOTES

Initiation of eukaryotic transcription differs substantially from its prokaryotic counterpart. While the definition of a promoter is the same—DNA sequence information that specifies the start of transcription—the molecular events required for transcription initiation are more complex. First, chromatin containing the promoter sequence must be made accessible to the transcription machinery. Second, transcription factors distinct from RNA polymerase must bind to DNA sequences in the promoter region for a gene to be active. Third, enhancers and other cis-acting transcriptional control elements bind other protein factors (activators) to stimulate transcription. Eukaryotic transcription factors bind to DNA and recruit RNA polymerase to the promoter. This contrasts with the action of bacterial σ factors, which do not bind DNA without first binding to the RNA polymerase core enzyme. Eukaryotic RNA polymerase consists of three distinct enzyme forms, with each specific form capable of synthesizing different classes of cellular RNA. By contrast, all prokaryotic genes are transcribed by a single form of core RNA polymerase, although different σ factors may be involved in initiation of different genes.

Nature of Active Chromatin

The structural organization of eukaryotic chromosomes is discussed in Chapter 2 and 8. Although **chromatin** is organized into nucleosomes whether or not it is capable of being transcribed, an active gene has a generally "looser" conformation than transcriptionally inactive chromatin. This difference is most striking in the promoter sequences, parts of which are not organized into nucleosomes at all (Figure 5.5). The lack of nucleosomes is manifested experimentally by the enhanced sensitivity of promoter sequences to external reagents that cleave DNA, such as the enzyme DNase I. This enhanced accessibility of promoter sequences (termed DNase I hypersensitivity) ensures that transcription factors can bind to appropriate regulatory sequences. Although the transcribed sequences of a gene may be organized into nucleosomes, the nucleosomes are less tightly bound than nucleosomes are in an inactive gene. Various covalent modifications of the histones are associated with changes in the transcriptional state of genes organized into nucleosomes. One such modification that is usually associated with transcriptional activation is the

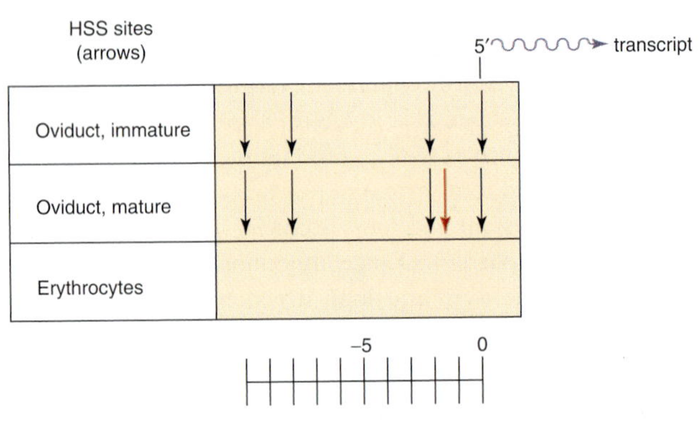

FIGURE 5.5

DNase-hypersensitive sites (HSS) upstream of the promoter for the chick lysozyme gene, a typical eukaryotic transcriptional unit. Hypersensitive sites—that is, sequences around the lysozyme gene that are particularly susceptible to nuclease digestion even when packaged into chromatin—are indicated by arrows. These sites are likely to be free of nucleosomes. Note that some hypersensitive sites are found in the lysozyme promoter in both mature and immature oviduct, even though lysozyme synthesis does not occur in immature oviduct. Induction of lysozyme synthesis in mature oviduct is accompanied by the appearance of a new hypersensitive site. In contrast, no hypersensitive sites are present in nucleated erythrocytes that never synthesize lysozyme. Adapted from Elgin, S. C. R. *J. Biol. Chem.* 263:1925, 1988.

acetylation of histones by **histone acetyltransferases**, a reaction that transfers acetyl groups from acetyl-CoA to histones, especially to the N-terminal regions of histones H4 and H3. Other histone modifications that influence gene activity include methylation, phosphorylation, and ubiquitination. Combinations of these modifications occurring at different specific positions in histones may constitute a "**histone code**" that couples histone modification, chromatin compaction, DNA modification, and gene activity (Clin. Corr. 5.2). An overall theme is that partially unfolded chromatin is necessary but not sufficient for transcription.

Transcription Activation Operates by Recruitment of RNA Polymerase

Eukaryotic protein factors, regardless of the sequence to which they bind, operate in a fundamentally different way than *E. coli* σ factor. Rather than first forming part of a protein complex and then seeking out the relevant DNA sequence, the factors bind to a specific site (sequence) on DNA and then bind to RNA polymerase (with or without involvement of intermediary factors). This mechanism is termed "recruitment." Recruitment is a minor means of gene activation in prokaryotes and the major mechanism in eukaryotes.

Enhancers

Enhancers increase expression of a gene many-fold. Enhancer-binding transcription factors are called **activators**. Activator proteins have at least two domains, one of which binds to the enhancer sequence, while the other binds to other protein factors or to RNA polymerase. The most accepted model for these effects is that chromatin forms a "loop" that allows the enhancer and the promoter to be close together in space, even though they are separated by a relatively long sequence of DNA.

CLINICAL CORRELATION **5.2**

Fragile X Syndrome: An RNA-Chromatin Disease?

Fragile X syndrome is the single most common form of inherited mental retardation, affecting 1/1250 males and 1/2000 females. A variety of anatomical and neurological symptoms result from inactivation of the FMR1 gene, located on the X chromosome. The genetics of the syndrome are complex due to the molecular mechanism of the fragile X mutation.

The fragile X condition results from expansion of a trinucleotide repeat sequence, CGG, found at the 5′ untranslated region of the FRM1 gene. Normally, this repeat is present in 30 copies, although normal individuals can have up to 200 copies of the repeat. In individuals with fragile X syndrome, the FMR1 gene contains many more copies, from 200 to thousands, of the CGG repeat. The complex genetics of the disease result from the potential of the CGG repeat sequence to expand from generation to generation.

The presence of an abnormally high number of CGG repeats induces extensive DNA methylation of the entire promoter region of FMR1. Methylated DNA is transcriptionally inactive, so FMR1 mRNA is not synthesized. Absence of FMR1 protein leads to the pathology of the disease. FMR1 protein normally is located in the cytoplasm in all tissues of the early fetus and, later, especially in fetal brain and neural tissue. The FMR protein affects translation of a number of mRNAs and one hypothesis is that the protein aids in the translation and localization of specific mRNAs during development. A recently discovered possibility is that the FMR1 protein is involved in RNA interference and its loss can cause widespread abnormal gene regulation (see Section 5.7). It is likely that several signaling pathways are linked in the pathology of this very complex disease.

Source: Warren, S. L. and Nelson, D. L. Advances in molecular analysis of fragile x syndrome. *JAMA* 271:536, 1994. Caskey, C. T. Triple repeat mutations in human disease. *Science* 256:784, 1992. Jin, P., Zarnescu, D. C., Ceman, S., Nakamoto, M., Mowrey, J., Jongens, T. A., Nelson, D. L., Moses, K., and Warren, S. T. Biochemical and genetic interaction between the fragile X mental retardation protein and the micro RNA pathway. *Nature Neurosci.* 7:113, 2004. Miyashiro, K. and Eberwine, J. Fragile X syndrome: (What's) lost in translation? *Proc. Natl. Acad. Sci. USA* 101:17329, 2004. Fragile Site Mental Retardation 1 Gene; FMR1 in Online Mendelian Inheritance in Man, http://www.ncbi.nlm.nih.gov/entrez/dispomim.cgi?id=309550

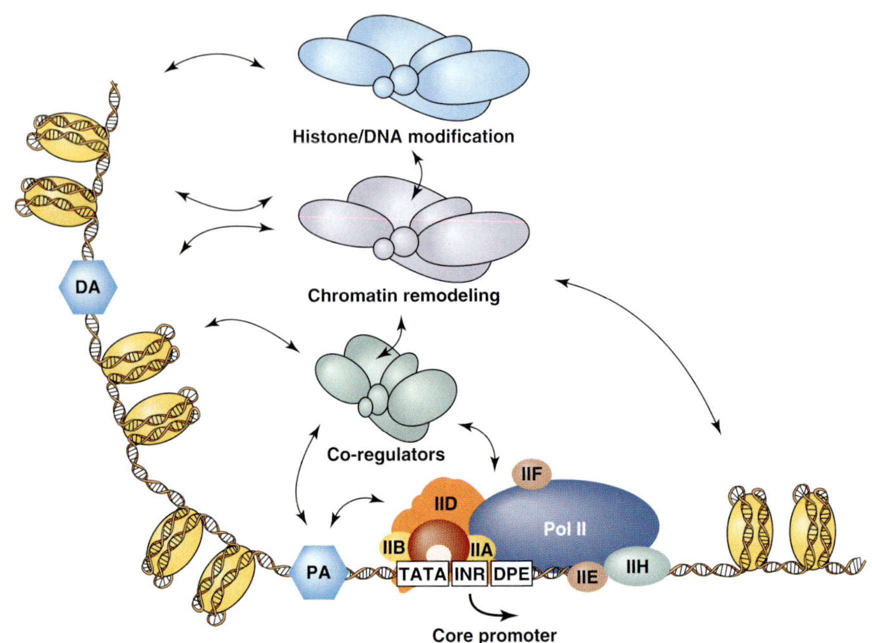

FIGURE 5.6

Interaction of transcription factors and chromatin modifying factors with eukaryotic promoters. Cis-acting transcriptional control elements located near the transcription start site may include a TATA box, an initiator element (INR), or a downstream promoter element (DPE) and are shown here interacting with the core factors required for specific transcription by RNA polymerase II. Promoters typically contain only a subset of these elements, and not all three. The core factors include TFIIA, TFIIB, TFIID, TFIIE, TFIIF, and TFIIH, in addition to RNA polymerase II. Transcriptional activator proteins can bind to promoter-proximal (PA) or promoter-distal (DA) sequences and activate transcription through interactions with the core factors, co-regulators, or chromatin modifying complexes, as shown. Binding of either the core factors or transcriptional activators may require or be accompanied by modifications of chromatin structure, including loss of nucleosomes.
Reproduced with permission from Hochheimer, A. and Tjian, R. *Genes Dev.* 17:1309, 2003.

Transcription by RNA Polymerase II

RNA polymerase II is responsible for synthesis of mRNA in the nucleus. Several common themes have emerged from research on a large number of genes (Figure 5.6). (1) The DNA sequences that control transcription are complex; a single gene may be controlled by dozens of DNA sequence elements in addition to the promoter, which includes the site at which a multiprotein complex containing RNA polymerase assembles. This pre-initiation complex is made up of a number of basal-level, or general, transcription factors, whose function is to place RNA polymerase II at the correct site for initiating transcription, to assist in separation of DNA strands at the start site, and to control the transition of polymerase from initiation to elongation mode. Without additional transcription control proteins interacting with enhancers and other cis-acting sequence elements, this process is generally quite inefficient, however. Controlling sequence elements function in combination to give a finely tuned pattern of control. (2) The effect of controlling sequences on transcription is mediated by binding of proteins to each sequence element. These transcription factors recognize the nucleotide sequence of the appropriate controlling sequence element. (3) Bound transcription factors interact with each other and ultimately act to recruit RNA polymerase, often acting through factors in the pre-initiation complex or other intermediates. The DNA binding and activation activities of the factors reside in separate domains of the proteins. (4) RNA polymerase II is modified during the transcription reaction. The modified polymerase recruits other nuclear components, including RNA processing enzymes, during the elongation phase of transcription.

Promoters for mRNA Synthesis

In contrast to prokaryotic RNA polymerase that recognizes only a single promoter sequence, transcription factors acting in conjunction with RNA polymerase II can recognize several classes of consensus sequences upstream from the mRNA start site. The first and most prominent of these, sometimes called the **TATA box**, has the sequence

TATA(A/T)(A/T) A

where the nucleotides in parentheses can be either an A or a T.

The TATA box is centered about 25 bp upstream from the transcription unit. Experiments in which it was deleted suggest that it is required for efficient transcription of many genes, but may only specify the correct transcriptional start site in others. Some promoters lack it entirely.

Other promoter-proximal control elements are often found in genes transcribed by RNA polymerase II, but are not universal. One such sequence is the **CAAT box**, with the sequence

GG(T/C) CAATCT

Other sequences, illustrated in Figure 5.6, may also promote transcription.

CAAT and TATA boxes, as well as other sequences shown in Figure 5.6, do not contact RNA polymerase II directly. Rather, they require the binding of specific transcription factors to function. Note how protein factors bind not only to their recognition sequences but also to each other and to RNA polymerase, itself a very large and complex enzyme. Mutated forms of some of these transcription factors are products of oncogenes (Clin. Corr. 5.3).

Transcription by RNA Polymerase I

Ribosomal RNAs are synthesized in a specific nuclear body, the nucleolus, and the rRNA genes are located in a specific chromosomal region termed the "nucleolar organizer."

CLINICAL CORRELATION 5.3
Involvement of Transcriptional Factors in Carcinogenesis

Conversion of a normally well-regulated cell into a cancerous one requires a number of independent steps whose end result is a transformed cell capable of uncontrolled growth and metastasis. Insights into this process have come from recombinant DNA studies of the genes, termed oncogenes, whose mutated or overexpressed products contribute to carcinogenesis. Oncogenes were first identified as products of DNA or RNA tumor viruses, but they are also present in normal cells. The normal, nonmutated cellular analogs of oncogenes are termed protooncogenes. Their products are components of the many pathways that regulate growth and differentiation of a normal cell; mutation into an oncogenic form involves a change that makes the regulatory product less responsive to normal control.

Some protooncogene products are involved in transduction of hormonal signals or recognition of cellular growth factors, and they act cytoplasmically. Other protooncogenes have a nuclear site of action; their gene products are often associated with the transcriptional apparatus, and they are synthesized in response to growth stimuli. It is easy to visualize how overproduction or permanent activation of such a positive transcription factor could aid the transformation of a cell to malignancy: Genes normally transcribed at a low or controlled level would be overexpressed by such a deranged control mechanism.

A more subtle genetic effect predisposing to cancer is exemplified by the human tumor suppressor protein p53. This protein is a the product of a dominant oncogene. A single copy of the mutant gene causes Li–Fraumeni syndrome, an inherited condition predisposing to carcinomas of the breast and adrenal cortex, sarcomas, leukemia, and brain tumors.

Somatic mutations in p53 can be identified in about half of all human cancers. Mutations represent a loss of function, affecting either the stability or DNA-binding ability of p53. Thus, wild-type p53 functions as a tumor suppressor. The wild-type protein helps to control the checkpoint between the G1 and S phases of the cell cycle, activates DNA repair, and, in other circumstances, leads to programmed cell death (apoptosis). Thus, the biochemical actions of p53 serve to keep cell growth regulated, maintain the information content of the genome, and, finally, eliminate damaged cells. All of these functions would counteract neoplastic transformation of a cell.

These varied roles are a function of the action of p53s as a transcription factor, inhibiting some genes and activating others. For example, p53 inhibits transcription of genes with TATA sequences, perhaps by binding to the complex formed between transcription factors and the TATA sequence. Alternatively, p53 is a site-specific DNA-binding protein and promotes transcription of some other genes—for example, those for DNA repair.

The three-dimensional structure of p53 has been determined. Mutations found in p53 from tumors affect the DNA-binding domain of the protein. For example, nearly 20 per cent of all mutated residues involve mutations at two positions in p53. The crystal structure of the protein–DNA complex shows that these two amino acids, both arginines, form hydrogen bonds with DNA. Arginine 248 forms hydrogen bonds in the minor groove of the DNA helix with a thymine oxygen and with a ring nitrogen of adenine. Mutation disrupts this H-bonded network and therefore the ability of p53 to regulate transcription.

Source: Weinberg, R. A. Oncogenes, antioncogenes, and the molecular basis of multistep carcinogenesis. *Cancer Res.* 49:3713, 1989. Cho, Y., Gorina, S., Jeffrey, P. D., and Pavletich, N. P. Crystal structure of a p53 tumor suppressor–DNA complex: Understanding tumorigenic mutations. *Science* 265:346, 1994. Friend, S. p43: A glimpse at the puppet behind the shadow play. *Science* 265:334, 1994. Harris, C. C., and Hollstein, M. Clinical implications of the p53 tumor-suppressor gene. *N. Engl. J. Med.* 329:1318, 1993. Tumor Protein p53; TP53 in Online Mendelian Inheritance in Man, http://www.ncbi.nlm.nih.gov/entrez/dispomim.cgi?id=191170

Each transcriptional unit contains sequences for 28S, 5.8S, and 18S rRNAs, in that order. Several hundred copies of each transcriptional unit occur tandemly (one after another) in the chromosome. Transcriptional units are separated by spacer sequences. Spacer sequences include sequences that specify binding of RNA polymerase I and of Class I transcription factors, which promote RNA polymerase I activity. Figure 5.7 is a diagram of this arrangement. Each repeat unit is transcribed as a unit, yielding a primary transcript that contains one copy each of the 28S, 5.8S, and 18S sequences, ensuring synthesis of equimolar amounts of these three RNAs. The primary transcript is then processed by ribonucleases and modifying enzymes to the three mature rRNA species. Termination of transcription occurs within the nontranscribed spacer region before RNA polymerase I reaches the promoter of the next repeat unit and occurs by a mechanism different from those discussed previously for *E. coli* RNA polymerase.

The promoter recognized by RNA polymerase I is located within the nontranscribed spacer, from about positions −40 to +10 and from −150 to −110. A transcription factor binds to the promoter and thereby directs recognition of the promoter sequence by RNA polymerase I. In addition, an enhancer element is located about 250 bp upstream from the promoter in human ribosomal DNA. The size of nontranscribed spacer varies considerably among organisms, as does the position of the enhancer element.

Transcription of rRNA can be very rapid; this reflects the fact that synthesis of ribosomes is rate-limiting for cell growth and the demand for rRNA can therefore be very high. In other situations, when growth is not so rapid, only some of the rDNA repeats are transcriptionally active.

Transcription by RNA Polymerase III

Many of the themes elaborated above for transcription of Class I and Class II promoters hold for the synthesis of 5S RNA, tRNA, and other small cellular and viral RNAs by **RNA polymerase III**. Transcription factors bind to DNA and direct the action of RNA polymerase. One unusual feature of RNA polymerase III action is the location of transcription factor-binding sequences; these can be located within the DNA sequence encoding the RNA and are often referred to as the "internal control region." In some cases, the DNA in the region immediately 5′ to the transcribed region of the gene can be substituted by other sequences with relatively minor effects on transcription. In other cases, however, sequences upstream of the transcription start site can play a major role in establishing efficient transcription. The differences appear to result from the relative importance of various DNA–protein contacts in stabilizing the entire transcription complex made up of multiple proteins in addition to RNA polymerase III. The relationship of these various proteins to the underlying DNA sequence is shown for a 5S rRNA gene and for a tRNA gene in Figure 5.8.

The Common Enzymatic Basis for RNA Polymerase Action

Our understanding of the molecular basis of RNA polymerase action has grown recently, as the three-dimensional structures of a σ factor, bacterial core RNA polymerase, the RNA polymerase holoenzyme, a eukaryotic RNA polymerase II, and even an RNA polymerase II engaged in transcription elongation have been determined by X-ray crystallography. Despite all the differences in subunit composition, size, and mechanism, RNA polymerases seem to interact with their DNA templates, nucleotide substrates, and nascent RNA chains in a similar fashion.

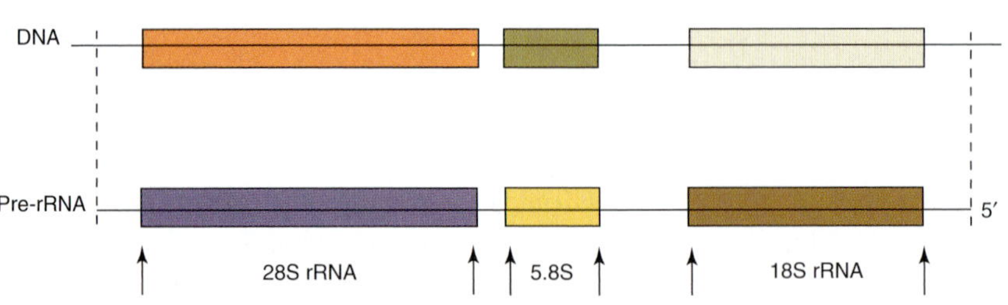

FIGURE 5.7
Structure of a rRNA transcription unit. Ribosomal RNA genes are arranged with many copies one after another. Each copy is transcribed separately, and each transcript is processed into three separate RNA species. Promoter and enhancer sequences are located in the nontranscribed regions of the tandemly repeated sequences.

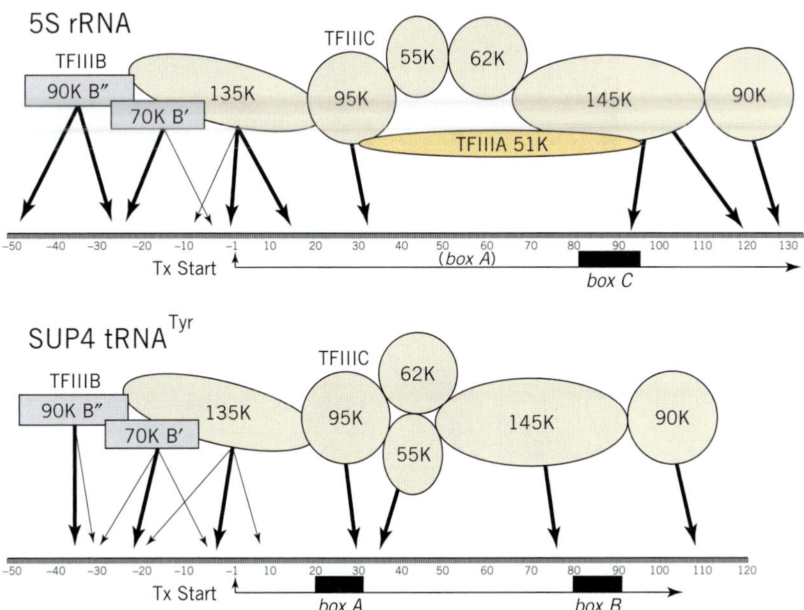

FIGURE 5.8

The sequences of the non-template strands of yeast genes encoding 5S rRNA and a tRNA are shown, along with schematic representations of the various subunits of TFIIIA, TFIIIB, and TFIIIC. The approximate locations of these subunits relative to the DNA sequences of the two genes are indicated by arrows. Note that TFIIIA is required to form a transcription complex on 5S rRNA genes, but not on tRNA genes. TFIIIA is responsible for much of the sequence-specific recognition of the internal control region of the 5S rRNA gene, whereas TFIIIC performs this function for a tRNA gene. Nonetheless, both TFIIIC and TFIIIB are required for transcription of both 5S rRNA and tRNA genes.
Redrawn from Braun, B. R., Bartholomew, B., Kassavetis, G. A., and Geiduschek E. P. Topography of transcription factor complexes on the *Saccharomyces cerevisiae* 5 S RNA gene. *J. Mol. Biol.* 228:1063, 1992.

Remarkably, the shapes of the prokaryotic and eukaryotic enzymes are strikingly similar, particularly within the enzymes' catalytic core. Both bacterial core RNA polymerase and RNA polymerase II from yeast are shaped rather like the claw of a crab (Figure 5.9). DNA comes in at one end of the molecule, and it moves through a deep cleft in the polymerase before making a 90-degree bend and exiting from the polymerase. The catalytic site—that is, the site at which a new nucleotide is added to the 3′-end of the growing RNA chain—is located at the end of the channel. The two strands of the double-helical DNA molecule are separated by the enzyme to form a "**transcription bubble**," and a transient 9-bp RNA–DNA hybrid forms within the bubble. Nucleotides come in and RNA exits through two other, separate channels in the molecule. Despite the differences in sequence and subunit composition, the prokaryotic and eukaryotic RNA polymerases have retained similar structures to accomplish the same goals leading to synthesis of a new RNA chain. These include: (1) threading the DNA template through a channel in the enzyme; (2) separating the DNA strands to form a bubble within which an RNA–DNA hybrid exists with the 3′-end of the RNA chain positioned in the active site; (3) bringing nucleotides into the active site through a pore in the enzyme; (4) using a metal ion to assist in catalyzing the formation of a new phosphate ester bond between the α phosphate of the incoming nucleoside triphosphate and the 3′-end of the nascent RNA chain; (5) extruding the RNA product through another pore; and (6) directing reannealing of the two DNA strands as they exit from the enzyme.

Recall that the σ subunit of the prokaryotic holoenzyme is responsible for binding to the −10 and −35 boxes of promoters. Structural studies have shown that sigma is an oblong molecule, composed of a bundle of α-helical residues, packed into an open "V" shape. One arm of the "V" contains critical residues for promoter recognition and core polymerase binding. One side of this arm contains an α-helix that binds to the "−10" sequence, and the other face binds to core polymerase through hydrophobic interactions.

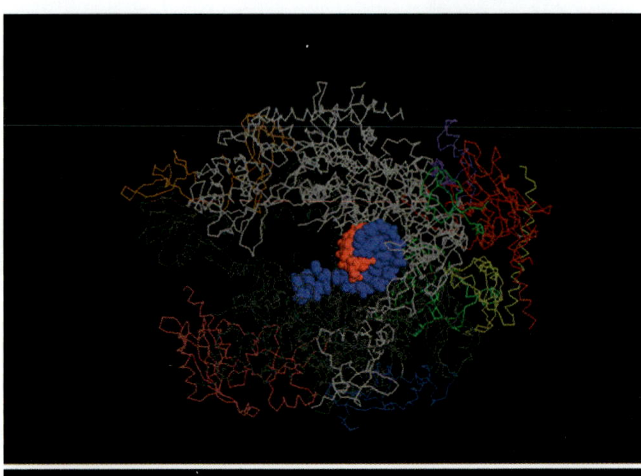

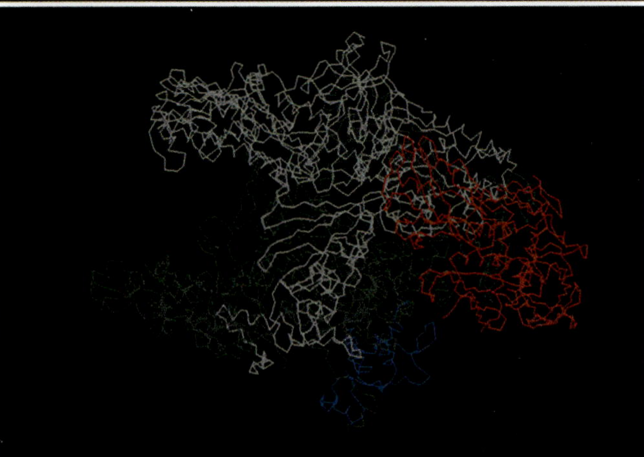

FIGURE 5.9

(a) The three-dimensional structure of yeast RNA polymerase II engaged in transcription elongation. (b) The structure of the RNA polymerase core from *Thermus aquaticus*. The various subunits are shown in different colors, with the largest subunit shown in gray and the second largest subunit in white. The DNA template strand is blue and the nascent RNA is red, both shown in a space-filling representation. This structure is missing two subunits of the RNA polymerase (products of the Rpb4 and Rpb7 genes), and the non-template strand of DNA was not resolved in the structural determination. The β' subunit is shown in gray, the β subunit in white, the two α subunits in red, and the ω subunit in blue. Note the similarities in the overall shapes of the two enzymes.
Prepared using RasWin and coordinates available in public databases. Data from Gnatt, A. I., Cramer, P., Fu, J., Bushnell, D. A., and Kornberg, R. D. Structural basis of transcription: An polymerase II elongation complex at 3.3 resolution. *Science* 292:876, 2001. Minakhin, L., Bhagat, S., Brunning, A., Campbell, E. A., et al. Bacterial RNA polymerase subunit omega and eukaryotic RNA polymerase subunit RPB6 are sequence, structural, and functional homologs and promote RNA polymerase assembly. *Proc. Natl. Acad. Sci. USA* 98:892, 2001.

5.4 RNA PROCESSING

RNA copies of DNA sequences must be modified to mature, functional, molecules in prokaryotes and eukaryotes. The reactions of RNA processing can include removal of extra nucleotides, base modification, addition of nucleotides, and separation of different RNA sequences by the action of specific nucleases. Processing reactions can occur either co-transcriptionally (while the RNA is still being transcribed) or posttranscriptionally (after the transcript is released by RNA polymerase). Finally, in eukaryotes, RNAs are exported from the nucleus.

Transfer RNA is Modified by Cleavage, Addition, and Base Modification

Cleavage

The primary transcript of a tRNA gene contains extra nucleotide sequences both 5′ and 3′ to the tRNA sequence. These primary transcripts may also contain **introns** in the anticodon region. Posttranscriptional processing reactions occur in a closely defined but not necessarily rigid temporal order. First, the primary transcript is trimmed in a relatively nonspecific manner to yield a precursor molecule with shorter 5′- and 3′-extensions. Then **ribonuclease P**, a ribozyme (see p. 69), removes the 5′-extension by endonucleolytic cleavage. The 3′-end is trimmed exonucleolytically, followed by synthesis of the CCA terminus. Synthesis of the modified nucleotides occurs in any order relative to the nucleolytic trimming. Intron removal is dictated by the secondary structure of the precursor (see Figure 5.10) and is carried out by a soluble, two-component enzyme system; one enzyme removes the intron and the other reseals the nucleotide chain.

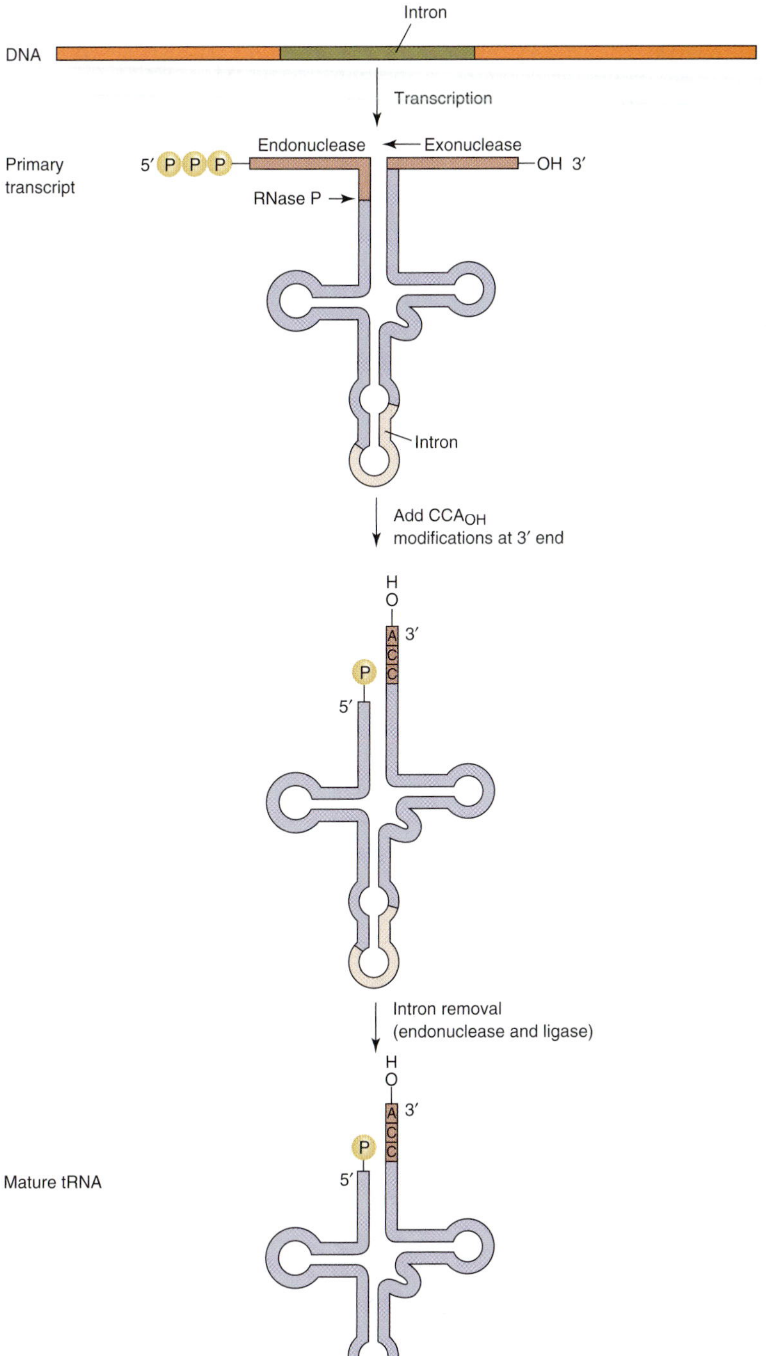

FIGURE 5.10

Scheme for processing a eukaryotic tRNA. Primary transcript is cleaved by RNase P and a 3′-exonuclease, and the terminal CCA is synthesized by tRNA nucleotidyltransferase before the intron is removed, if necessary.

3′-End Addition

Each functional tRNA has the sequence CCA at its 3′-terminus. This sequence is essential for tRNA to accept amino acids. In most instances, it is added sequentially by the enzyme tRNA nucleotidyltransferase. Nucleotidyltransferase uses ATP and CTP as substrates and always incorporates them into tRNA at a ratio of 2C/1A. The CCA ends are found on both cytosolic and mitochondrial tRNAs.

Modified Nucleosides

Transfer RNA nucleotides are the most highly modified of all nucleic acids. More than 60 different modifications to the bases and ribose, requiring well over 100 different enzymatic reactions, have been found in tRNA. Many are simple, one-step methylations, but others involve multistep synthesis. Formation of some modified bases actually requires severing of the β-glycosidic bond between ribose and the base. Modifying enzymes produce the same specific modification in more than one species of tRNA; however, the modifying enzymes are location-specific. Most modifications are completed before the tRNA precursors have been cleaved to the size of mature tRNA.

Ribosomal RNA Processing Releases Several RNAs from a Longer Precursor

The primary product of rRNA gene transcription is a long RNA, termed 45S RNA, which contains the sequences of 28S, 5.8S, and 18S rRNAs. Processing of 45S RNA occurs in the nucleolus, and it is carried out by large multisubunit ribonucleoprotein assemblies. Processing of the rRNAs follows a sequential order (Figure 5.11).

Processing of pre-rRNA in prokaryotes also involves cleavage of high-molecular-weight precursors to smaller molecules. At an early stage of processing, some bases in 35S rRNA are modified by methylation on the ring nitrogens of the bases and by the formation of pseudouridine. These reactions are specified by small nucleolar ribonuclear protein particles (snoRNPs) in the nucleolus. Each snoRNP contains a **guide RNA** that base-pairs to the rRNA transcript at the site of modification, specifying where a methylation or formation of pseudouridine should occur. A similar mechanism operates for modification of some other small RNAs.

Messenger RNA Processing Ensures the Correct Coding Sequence

Most eukaryotic mRNAs have distinctive structural features added in the nucleus by enzyme systems other than RNA polymerase. These include the 3′-terminal poly(A) tail, methylated internal nucleotides, and the 5′-terminal cap. Mature mRNAs are shorter than their primary transcripts, which can contain additional terminal and

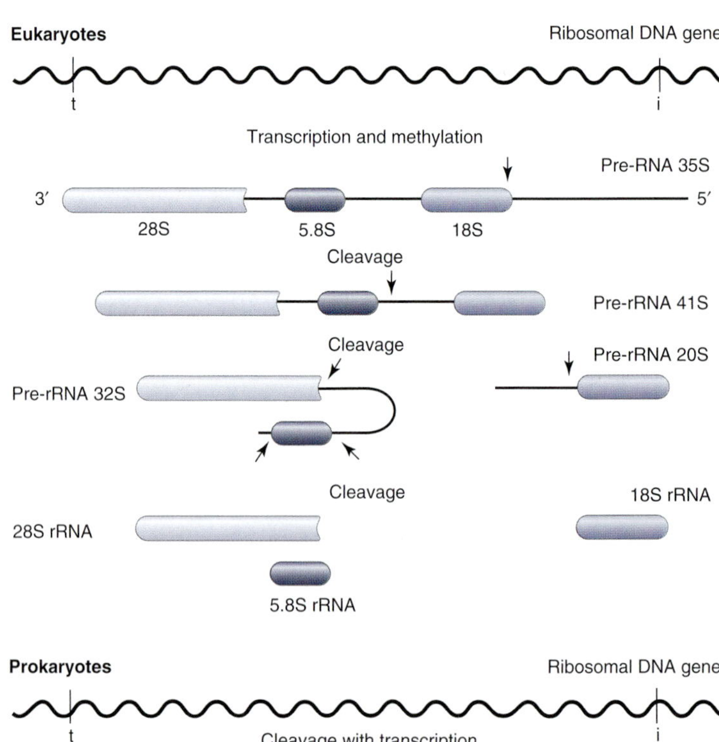

FIGURE 5.11

Schemes for transcription and processing of rRNAs.
Redrawn from Perry, R. *Annu. Rev. Biochem.* 45:611, 1976. Copyright (1976) Annual Reviews; www.annualreviews.org.

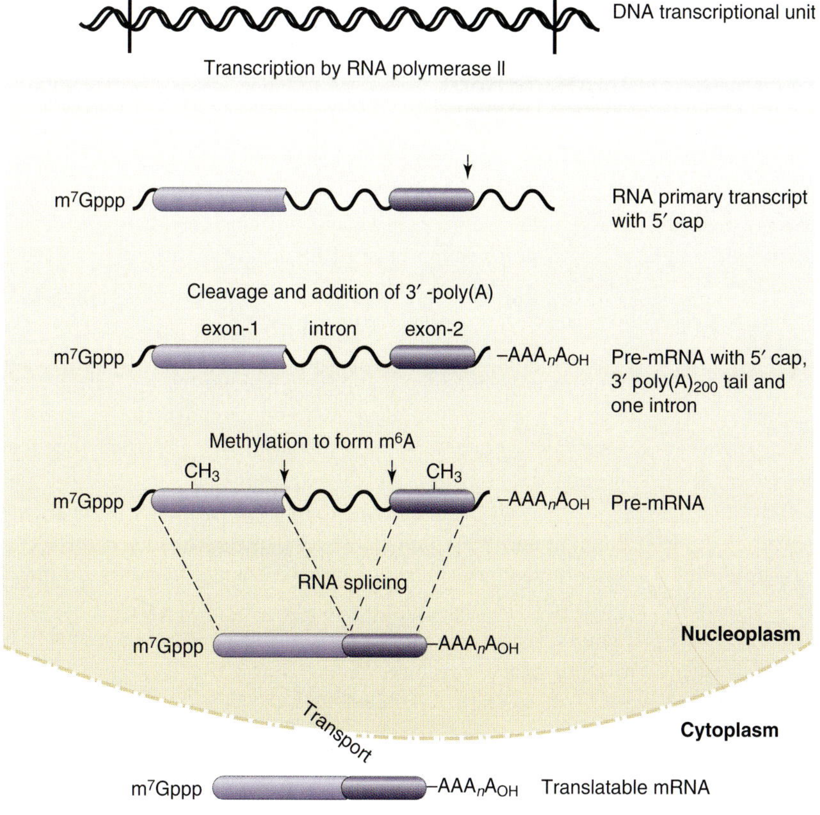

FIGURE 5.12

Scheme for processing mRNA. Points for initiation and termination of transcription are indicated on the DNA. Arrows indicate cleavage points. Many proteins associated with the RNA and tertiary conformations are not shown.

internal sequences. Noncoding sequences present within pre-mRNA, but not present in mature mRNAs, are called intervening sequences or introns. The retained sequences are called exons. The general pattern for mRNA processing is depicted in Figure 5.12. Incompletely processed mRNA make up a large part of the **heterogeneous nuclear RNA** (HnRNAs).

Processing of eukaryotic pre-mRNA involves a number of molecular reactions, all of which must be carried out with exact fidelity. This is most clear in removal of introns from an mRNA transcript. An extra nucleotide in the coding sequence of mature mRNA would cause the reading frame of that message to be shifted, and the resulting protein will almost certainly be nonfunctional. Indeed, mutations that interfere with intron removal are a major cause of human genetic diseases—for example, β-thalassemia (Clin. Corr. 5.4). The task for cells becomes even more daunting since some important human genes consist of over 90% intron sequences. The complex reactions to remove introns are accomplished by multicomponent ribonucleoprotein enzyme systems in the nucleus; after these reactions are completed, the mRNA is exported to the cytoplasm where it interacts with ribosomes to initiate translation. Most processing is cotranscriptional.

RNA Polymerase II Recruits Processing Enzymes during Transcription in Eukaryotes

The largest subunit of RNA polymerase II contains a **C-terminal domain** (**CTD**) that functions to couple transcription and processing. When RNA polymerase II is in an initiation mode, the CTD is not modified. When the transcript is about 25 nucleotides long, the CTD is extensively phosphorylated. Phosphorylated CTD successively recruits the capping enzyme, splicing, and polyadenylation complexes.

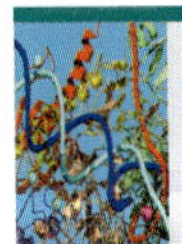

CLINICAL CORRELATION 5.4

Thalassemia Due to Defects in Messenger RNA Synthesis

Thalassemias are genetic defects in the coordinated synthesis of α- and β-globin peptide chains; a deficiency of β-globin chains is termed β-thalassemia, while a deficiency of α globin chains is termed α-thalassemia. A patient suffering from β-thalassemia present with anemia at about 6 months of age as HbF synthesis ceases and HbA synthesis becomes predominant. The severity of symptoms leads to the classification of the disease into either thalassemia major, where a severe deficiency of globin synthesis occurs, or thalassemia minor, which represents a less severe imbalance. Occasionally an intermediate form is seen. Therapy for thalassemia major involves frequent blood transfusions, leading to a risk of complications from iron overload. Unless chelation therapy is successful, the deposition of iron in peripheral tissues, termed hemosiderosis, can lead to death before adulthood. Carriers of the disease are typically asymptomatic, or nearly so. Ethnographically, the disease is common in persons of Mediterranean, Arabian, and East Asian descent. As in sickle-cell anemia (HbS) and glucose 6-phosphate dehydrogenase deficiency, the abnormality of the carriers' erythrocytes affords some protection from malaria. Maps of the regions where one or another of these diseases is frequent in the native population superimpose over the areas of the world where malaria is endemic.

The most common form of α-thalassemia is due to a genetic deletion, which presumably results from unequal crossing over between adjacent, duplicated α-globin alleles. In contrast, β-thalassemia can result from a wide variety of mutations. Known events include mutations leading to frameshifts in the β-globin coding sequence, as well as mutations leading to premature termination of peptide synthesis. Many β-thalassemias result from mutations affecting the biosynthesis of β-globin mRNA. Genetic defects are known that affect the promoter of the gene, leading to inefficient transcription. Other mutations result in aberrant processing of the nascent transcript, either during splicing out of its two introns from the pre-mRNA or during polyadenylation of the mRNA precursor. Examples where the molecular defect illustrates a general principle of mRNA synthesis are discussed in the text.

Source: Orkin, S. H. Disorders of hemoglobin synthesis: The thalassemias. G. Stamatoyannopoulis, A. W. Nienhuis, P. Leder, and P. W. Majerus (Eds.). *The Molecular Basis of Blood Diseases.* Philadelphia: Saunders, 1987. Weatherall, D. J., Clegg, J. B., Higgs, D. R., and Wood, W. G. The hemoglobinopathies. In C. R. Scriver, A. L. Beaudet, W. S. Sly, and D. Valle (Eds.), *The Metabolic Basis of Inherited Disease*, 7th ed. New York: McGraw-Hill, 1995. Hemoglobin-Beta locus HBB, in Online Mendelian Inheritance in Man website. http://www.ncbi.nlm.nih.gov/entrez/dispomim.cgi?id=141900

Capping

As the transcription complex moves along the DNA, the **capping** enzyme complex modifies the 5′-end of the nascent mRNA. Capping involves the synthesis of a 5′–5′ triphosphate bond with the addition of a G residue. This structure is further modified by methylation.

Removal of Introns from mRNA Precursors

As pre-mRNA is extruded from the RNA polymerase complex, it is rapidly bound by small nuclear ribonucleoproteins, **snRNPs** ("**snurps**"), which carry out the dual steps of RNA splicing: (1) breakage of the intron at the 5′ donor site and (2) joining the upstream and downstream exon sequences together. All introns begin with a GU sequence and end with AG; these are termed the donor and acceptor intron–exon junctions, respectively. Not all GU or AG sequences are spliced out of RNA, however. How does the cell know which GU sequences are in introns (and therefore must be removed) and which are destined to remain in mature mRNA? This discrimination is accomplished by formation of base pairs between U1 RNA and the sequence of the mRNA precursor surrounding the donor GU sequence (Figure 5.13) (see Clin. Corr. 5.5). Another snRNP, containing U2 RNA, recognizes important sequences at the 3′-acceptor end of the intron. Still other snRNP species, among them U5 and U6, then bind to the RNA precursor, forming a large complex termed a spliceosome (by analogy with the large ribonucleoprotein assembly involved in protein synthesis, the ribosome). The spliceosome uses ATP to carry out accurate removal of the intron. First, the phosphodiester bond between the exon and the donor GU sequence is broken, leaving a free 3′-OH group at the end of the first exon and a 5′-phosphate on the donor G of the intron. This pG is then used to form an unusual linkage with the 2′-OH group of an adenosine within the intron to form a branched or lariat RNA structure,

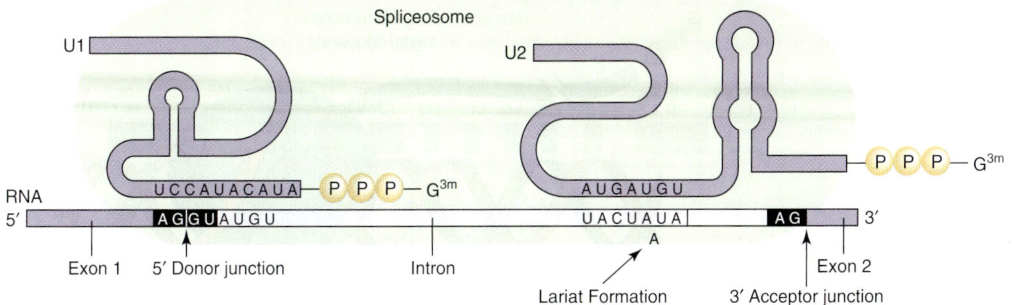

FIGURE 5.13

Mechanism of splice junction recognition. Recognition of the 5′ splice junction involves base pairing between the intron–exon junction and the U1 RNA snRNP. This base pairing targets the intron for removal. Adapted from Sharp, P. A. *JAMA* 260:3035, 1988.

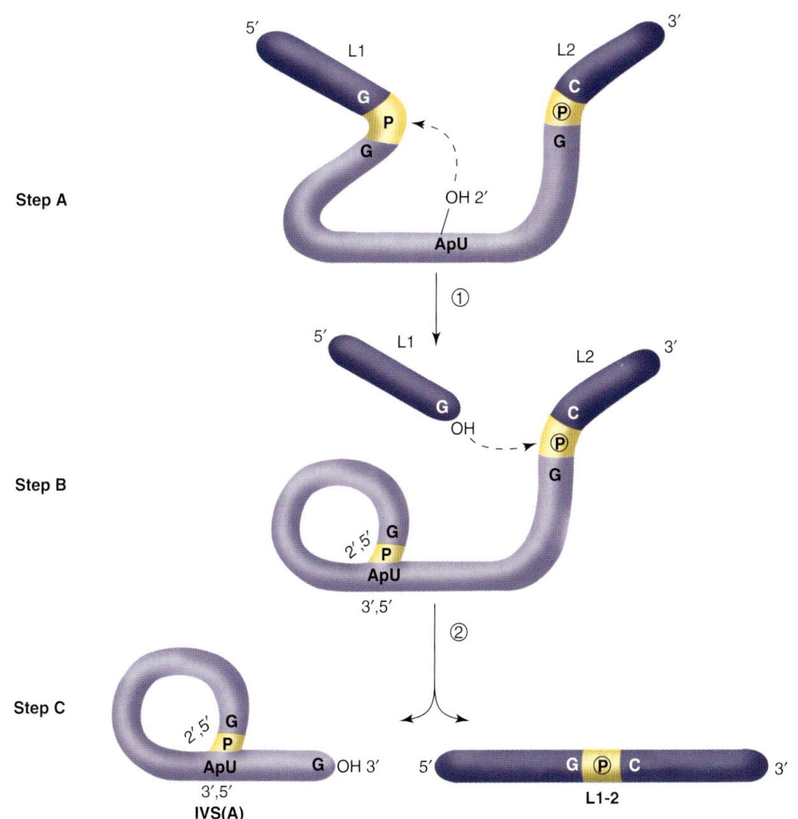

FIGURE 5.14

Proposed scheme for mRNA splicing to include the lariat structure. A messenger RNA is depicted with two exons (in dark blue) and an intervening intron (in light blue). A 2′-OH group of the intron sequence reacts with 5′-phosphate of the intron's 5′-terminal nucleotide producing a 2′–5′ linkage and the lariat structure. Simultaneously, the exon 1–intron phosphodiester bond is broken, leaving a 3′-OH terminus on this exon free to react with 5′-phosphate of the exon 2, displacing the intron and creating the spliced mRNA. The released intron lariat is subsequently digested by cellular nucleases.

as shown in Figure 5.14. After the lariat is formed, the second step of splicing occurs. The phosphodiester bond immediately following the AG is cleaved and the two exon sequences are ligated together.

Polyadenylation

RNA polymerase continues transcribing the gene until a polyadenylation signal sequence is reached (Figure 5.15). This sequence, which has the consensus AAUAAA, appears in the mature mRNA but does not form part of its coding region. Rather, it signals cleavage of the nascent mRNA precursor about 20 or so nucleotides downstream. The poly(A) sequence is then added by polymerase to the free 3′-end produced by

CLINICAL CORRELATION 5.5

Autoimmunity in Connective Tissue Disease

Humoral antibodies in sera of patients with various connective tissue diseases recognize a variety of ribonucleoprotein complexes. Patients with systemic lupus erythematosus exhibit a serum antibody activity designated Sm, and those with mixed connective tissue disease exhibit an antibody designated RNP. Each antibody recognizes a distinct site on the same RNA–protein complex, U1 RNP, that is involved in mRNA processing in mammalian cells. The U1–RNP complex contains U1 RNA, a 165-nucleotide sequence highly conserved among eukaryotes, which at its 5′-terminus includes a sequence complementary to intron-exon splice junctions. Addition of this antibody to *in vitro* splicing assays inhibits splicing, presumably by removal of the U1 RNP from the reaction. Sera from patients with other connective tissue diseases recognize different nuclear antigens, nucleolar proteins, and or chromosomal centromere. Sera of patients with myositis have been shown to recognize cytoplasmic antigens such as aminoacyl-tRNA synthetases. Although humoral antibodies have been reported to enter cells via Fc receptors, there is no evidence that this is part of the mechanism of autoimmune disease.

efficient on regions of highly ordered RNA structure. Thus, tRNAs are preferentially cleaved in unpaired regions of the sequence. On the other hand, many RNases involved in RNA processing require the substrate to have a defined three-dimensional structure for recognition.

Messenger RNAs are initially degraded in the cytoplasm. The rates vary for different mRNA species, raising the possibility of control by differential degradation. The pathway for mRNA degradation is usually triggered by shortening of the 3′-polyA, which then triggers removal of the 5′-cap. The decapped mRNA is then degraded from both ends by exonucleases. The exonucleases that remove sequences from the 3′-end are found in a large complex called the exosome, consisting of a number of 3′-exonucleases and an RNA helicase (unwinding enzyme). Endonucleases can also trigger mRNA degradation.

Two examples of the role of RNA stability in gene control illustrate how the stability of mRNA influences gene expression. Tubulin is the major component of the microtubules of the cytoskeleton of many cell types. When there is an excess of tubulin in the cell, the monomeric protein binds to and promotes degradation of tubulin mRNA, thereby reducing tubulin synthesis. A second example is provided by herpes simplex viruses (HSV), the agent causing cold sores and some genital infections. An early event in the establishment of HSV infection is the ability of the virus to destabilize all cellular mRNA molecules, thereby reducing the competition for free ribosomes. Thus, viral proteins are more efficiently translated.

BIBLIOGRAPHY

Allmang, C., Kufel, J., Chanfreau, G., Mitchell, P., Petfalski, E., and Tollervey, D. Functions of the exosome in rRNA, snoRNA and snRNA synthesis. *EMBO J.* 18:5399, 1999.

Bentley, D. Coupling RNA polymerase II transcription with pre-mRNA processing. *Curr. Opin. Cell Biol.* 11:347, 1999.

Berk, A. J. Activation of RNA polymerase II transcription. *Curr. Opin. Cell Biol.* 11:330, 1999.

Blanc, V. and Davidson, N. O. C-to-U RNA editing: Mechanisms leading to genetic diversity *J. Biol. Chem.* 278:1395, 2002.

Carmo-Fonseca, M., Mendes-Soares, L., and Campos, I. To be or not to be in the nucleolus. *Nature Cell Biol.* 2:E107, 2000.

Caskey, C. T. Triple repeat mutations in human disease. *Science* 256:784, 1992.

Citterio, E., Vermeulen, W., and Hoeijmakers, J. H. J. Transcriptional healing. *Cell* 101:447, 2000.

Conaway, J. W. and Conaway, R. C. Transcription elongation and human disease. *Annu. Rev. Biochem.* 68:301, 1999.

Cramer, P. Multisubunit RNA polymerases. *Curr. Opin. Struct. Biol.* 12:89, 2002.

Daneholt, B. A look at messenger RNP moving through the nuclear pore. *Cell* 88:585, 1997.

Decatur, W. A. and Fournier, M. J. RNA-guided nucleotide modification of ribosomal and Other RNAs. *J. Biol. Chem.* 278:695, 2002.

Friend, S. A glimpse at the puppet behind the shadow play. *Science* 265:334, 1994.

Gesteland, R. F., Cech, T., and Atkins, J. F. (Eds.). *The RNA World.*, 2nd ed. Cold Spring Harbor, NY: Cold Spring Harbor Laboratory Press, 1998.

Hawley, D. K. and McClure, W. R. Compilation and analysis of Escherichia coli promoter DNA sequences. *Nucleic Acids Res.* 11:2237, 1983.

Hochheimer, A. and Tjian, R. Diversified transcription initiation complexes expand promoter selectivity and tissue-specific gene expression. *Genes Dev.* 17:1309, 2003.

Holbrook J. A., Neu-Yilik, G., Hentze, M. W., and Kulozik, A. E. Nonsense-mediated decay approaches the clinic. *Nat. Genet.* 36:801, 2004.

Izaurralde, E., Kann, M., Panté, Sodeik, B., and Hohn, T. Viruses, microorganisms and scientists meet at the nuclear pore. *EMBO J.* 18:289, 1999.

Koleske, A. J. and Young, R. A. The RNA polymerase II holoenzyme and its implications for gene regulation. *Trends Biochem. Sci.* 20:113, 1995.

Larson, D. E., Zahradka, P., and Sells, B. H. Control points in eucaryotic ribosome biogenesis. *Biochem. Cell. Biol.* 69:5, 1991.

Lewin, B. *Genes VII*. New York: Oxford University Press, 2000.

Maas, S., Rich, A., and Nishikura, K. A-to-I RNA editing: Recent news and residual mysteries. *J. Biol. Chem.* 278:1391, 2003.

Meister, G. and Tuschl, T. Mechanisms of gene silencing by double-stranded RNA. *Nature* 431:343, 2004.

Mooney, R. A. and Landick, R. RNA polymerase unveiled. *Cell* 98:687, 1999.

Parker, R. and Song, H. The enzymes and control of eukaryotic mRNA turnover. *Nature Struct. Mol. Biol.* 11:121. 2004.

Paule, M. R. and White, R. J. Survey and summary: Transcription by RNA polymerases I and III. *Nucleic Acids Res.* 28:1283, 2000.

Proudfoot, N. Connecting transcription to messenger RNA processing. *Trends Biochem. Sci.* 25:290, 2000.

Ptashne, M. and Gann, A. Transcriptional activation by recruitment. *Nature* 386:569, 1997.

Samuel, C. E. RNA editing minireview series. *J. Biol. Chem.* 278:1389, 2002.

Smith, C. W. J. and Valcárcel, J. Alternative pre-mRNA splicing: The logic of combinatorial control. *Trends Biochem. Sci.* 25:381, 2000.

van Hoof, A. and Parker, R. The exosome: A proteosome for RNA? *Cell* 99:347, 1999.

Weatherall, D. J., Clegg, J. B., Higgs, D. R., and Wood, W. G. The hemoglobinopathies. In C. R. Scriver et al. (Eds.), *The Metabolic and Molecular Bases of Inherited Disease*, 8th ed. New York: McGraw-Hill, 2001. p. 4571.

Weinstein, L. B. and Steitz, J. A. Guided tours: From precursor snoRNA to functional snRNP. *Curr. Opin. Cell Biol.* 11:378, 1999.

QUESTIONS | CAROL N. ANGSTADT

Multiple Choice Questions

1. In eukaryotic transcription:
   A. RNA polymerase does not require a template.
   B. all RNA is synthesized in the nucleolus.
   C. consensus sequences are the only known promoter elements.
   D. phosphodiester bond formation is favored because there is pyrophosphate hydrolysis.
   E. RNA polymerase requires a primer.

2. The sigma (σ) subunit of prokaryotic RNA polymerase:
   A. is part of the core enzyme.
   B. binds the antibiotic rifampicin.
   C. is inhibited by α-amanitin.
   D. must be present for transcription to occur.
   E. specifically recognizes promoter sites.

3. Termination of a prokaryotic transcript:
   A. is a random process.
   B. requires the presence of the rho subunit of the holoenzyme.
   C. does not require rho factor if the end of the gene contains a G-C-rich palindrome.
   D. is most efficient if there is an A-T-rich segment at the end of the gene.
   E. requires an ATPase in addition to rho factor.

4. Eukaryotic transcription:
   A. is independent of the presence of upstream consensus sequences.
   B. may involve a promoter located within the region transcribed rather than upstream.
   C. requires a separate promoter region for each of the three ribosomal RNAs transcribed.
   D. requires that the entire gene be in the nucleosome form of chromatin.
   E. is affected by enhancer sequences only if they are adjacent to the promoter.

5. All of the following are correct about a primary transcript in eukaryotes *except* that it:
   A. is usually longer than the functional RNA.
   B. may contain nucleotide sequences that are not present in functional RNA.
   C. will contain no modified bases.
   D. could contain information for more than one RNA molecule.
   E. contains a TATA box.

6. In the cellular turnover of RNA:
   A. repair is more active than degradation.
   B. regions of extensive base pairing are more susceptible to cleavage.
   C. endonucleases may cleave the molecule starting at either the 5′- or 3′-end.
   D. the products are nucleotides with a phosphate at either the 3′- or 5′-OH group.
   E. all species except rRNA are cleaved.

Questions 7 and 8: Fragile X syndrome is a common form of inherited mental retardation. The mutation in the disease allows the increase of a CGG repeat in a particular gene from a normal of about 30 repeats to 200–1000 repeats. This repeat is normally found in the 5′-untranslated region of a gene for the protein FMR1. FMR1 might be involved in the translation of brain-specific mRNAs during brain development. The consequence of the very large number of CGG repeats in the DNA is extensive methylation of the entire promoter region of the FMR1 gene.

7. Methylation of bases in DNA usually:
   A. facilitates the binding of transcription factors to the DNA.
   B. makes a difference in activity only if it occurs in an enhancer region.
   C. prevents chromatin from unwinding.
   D. inactivates DNA for transcription.
   E. results in increased production of the product of whatever gene is methylated.

8. Transcription of eucaryotic genes requires the presence of a promoter and usually the presence of enhancers. An enhancer:
   A. is a consensus sequence in DNA located where RNA polymerase first binds.
   B. may be located in various places in different genes.
   C. may be on either strand of DNA in the region of the gene.
   D. functions by binding RNA polymerase.
   E. stimulates transcription in all prokaryotes and eukaryotes.

Questions 9 and 10: The synthesis of normal adult hemoglobin (HbA) requires the coordinated synthesis of α-globin and β-globin. β-Thalassemia is a genetic disease leading to a deficiency of β-globin chains and an inability of the blood to deliver oxygen properly. β-Thalassemia can result from a wide variety of mutations.

9. All of the following could lead to lack of or nonfunctional β-globin *except*:
   A. a frameshift mutation leading to premature termination of protein synthesis.
   B. mutation in the promoter region of the β-globin gene.
   C. mutation toward the 3′-end of the β-globin gene that codes for the polyadenylation site.
   D. mutation in the middle of an intron that is not at an A base.
   E. all of the above could lead to this result.

10. One mutation leading to β-thalassemia occurs at a splice junction. Which of the following statements about removing introns is/are correct?
    A. Small nuclear ribonucleoproteins (snRNP) are necessary for removing introns.
    B. The consensus sequences at the 5′- and 3′-ends of introns are identical.
    C. Removal of an intron does not require metabolic energy.
    D. The exon at one end of an intron must always be joined to the exon at its other end.
    E. The nucleoside at the end of the intron first released forms a bond with a 3′-OH group on one of the nucleotides within the intron.

Questions 11 and 12: Protooncogenes produce products that have specific roles in regulating growth and differentiation of normal cells. Mutations can turn these genes into oncogenes whose products are less responsive to normal control. Unmutated protein p53, a tumor suppressor, is a transcription factor, inhibiting some genes and activating others. P53 inhibits genes with TATA sequences and activates genes for DNA repair.

11. The TATA sequence:
    A. occurs about 25 bp downstream from the start of transcription.
    B. binds directly to RNA polymerase.
    C. binds transcription factors which bind RNA polymerase.
    D. binds p53.
    E. is an enhancer sequence.

12. Transcription-coupled DNA repair:
    A. occurs because a DNA lesion causes the RNA polymerase to stall during transcription.
    B. occurs when a DNA lesion forms in a region of DNA which is highly compacted.
    C. leads to a normal transcript because transcription continues after the repair is made.
    D. occurs because both the template and coding strands of the DNA have lesions.
    E. all of the above are correct.

Problems

13. What events are involved in the processing of transfer RNA?

14. (a) How could you experimentally determine whether a purified preparation of an RNA polymerase is from a prokaryotic or eukaryotic source? (b) A purified preparation of RNA polymerase is sensitive to inhibition by α-amanitin at a concentration of 10^{-8} M. The synthesis of what type or types of RNA is inhibited?

ANSWERS

1. **D** This is an important mechanism for driving reactions. A, B: Transcription is directed by DNA, generating rRNA precursors in the nucleolus and mRNA and tRNA precursors in nucleoplasm. C: Eukaryotic transcription may have internal promoter regions as well as enhancers. E: This is a difference from DNA polymerase.
2. **E** A, D, E: Sigma factor is required for correct initiation and dissociates from the core enzyme after the first bonds have been formed. Core enzyme can transcribe but cannot correctly initiate transcription. B, C: Rifampicin binds to the β subunit, and α-amanitin is an inhibitor of eukaryotic polymerases.
3. **C** Rho-independent termination involves secondary structure, which is stabilized by high G-C content. A, B, E: There is a rho-dependent as well as a rho-independent process. Rho is a separate protein from RNA polymerase and appears to possess ATPase activity. D: G-C-rich region is required.
4. **B** RNA polymerase III uses an internal promoter. A: RNA polymerase II activity involves the TATA and CAAT boxes. C: RNA polymerase I produces one transcript, which is later processed to yield three rRNAs. D: Parts of the promoter are not in a nucleosome. E: Enhancers may be thousands of bp away.
5. **E** The TATA box is part of the promoter, which is not transcribed. A–D: Modification of bases, cleavage, and splicing are all important events in posttranscriptional processing to form functional molecules.
6. **D** The location of the -OH depends on which side of the phosphodiester bond is split. A: There are no RNA repair processes. B: Most degradative enzymes are less efficient on an ordered structure. C: An endonuclease cleaves an interior phosphodiester bond. E: All RNA turns over.
7. **D** Methylation–demethylation of DNA is one form of control, with the methylated DNA being inactive for transcription. A, E: These would indicate that methylation activates DNA. B: Altering the enhancer region could make a difference, but alteration of the promoter region certainly would. C: Unfolding of chromatin in this region is necessary for transcription, but methylation of DNA is not directly involved in the process.
8. **B** B, C: Enhancer sequences seem to work whether they are at the beginning or end of the gene, but they must be on the same DNA molecule as the transcribed gene. A: RNA polymerase first binds at the promoter. D: They seem to function by binding proteins which themselves bind RNA polymerase.
9. **D** Mutation at either terminus of the intron would lead to a splicing error but a mutation in the middle of the intron should not. The exception would be if the specific mutation site were at the adenosine that forms the branch point. A–C: These are all correct.
10. **A** snRNPs are responsible both for cleaving the intron at the 5′-end and joining the exons together. B: The sequences at the two ends of an intron are different: GU at the 5′-end and AG at the 3′-end. C: The spliceosome requires ATP to remove the intron. D: Nonconsecutive exons can be joined together, leading to multiple proteins from one gene. E: The lariat forms at a 2′-OH.
11. **C** B, C: RNA polymerase does not bind directly to TATA, but to transcription factors that bind TATA. A, E: TATA is part of the promoter, so it is upstream from the transcription start. D: p53 probably binds to the complex formed between transcription factors and TATA, interfering with transcription.
12. **A** This signals the excision repair system to repair the lesion. B: Repair is targeted to actively transcribed genes where chromatin is more loosely organized. C: For repair to occur, RNA polymerase must dissociate and the transcript is sacrificed. D: A typical lesion is thymine dimers in the template strand. The coding strand is normal and acts as the template for repair.
13. The primary transcript is longer than the functional molecule, and extra bases are cleaved from both ends (and sometimes elsewhere). Extensive modification such as methylation occurs on specific bases. The sequence CCA is added at the 3′-end by nucleotidyl transferases. Some transcripts have an intron in the anticoding region, which must be removed.
14. (a) Test the preparation for inhibition by α-amanitin at a high concentration where all forms of eukaryotic RNA polymerase are inhibited but prokaryotic enzyme is not. If the preparation is not inhibited by the amanitin, confirm that it is inhibited by rifampicin. It will not be possible to distinguish eukaryotic mitochondrial RNA polymerase from prokaryotic enzyme because they have the same sensitivity. (b) RNA polymerase II is sensitive to α-amanitin at a very low concentration, so mRNA synthesis and some snRNA would be inhibited.

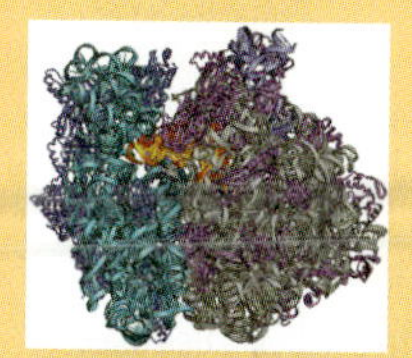

6

PROTEIN SYNTHESIS: TRANSLATION AND POSTTRANSLATIONAL MODIFICATIONS

Dohn Glitz

Textbook of Biochemistry With Clinical Correlations, Sixth Edition, Edited by Thomas M. Devlin

6.1 OVERVIEW

Genetic information is stored and transmitted in the four-letter alphabet and language of DNA and ultimately expressed in the 20-letter alphabet and language of proteins. Protein biosynthesis is called translation because it involves the biochemical translation of information between these languages. Translation requires many protein "factors" and an energy source. Other needs center on functional RNA species: an informational messenger RNA molecule that is exported from the nucleus, many "bilingual" transfer RNA species that read the message, RNA-rich ribosomes that act as catalytic and organizational centers for protein synthesis, and small RNA species that help regulate the process. Polypeptides are formed by stepwise addition of amino acids in the specific order determined by information carried in the nucleotide sequence of the mRNA. Correct folding into a correct tertiary structure may be assisted by a protein chaperone. A protein is often matured or processed by modifications that may target it to a specific intracellular location or to secretion from the cell, or might modulate its stability, activity, or function. These complex processes are carried out with considerable speed and precision, and they are regulated both globally and for specific proteins. Finally, if a protein is mis-translated, becomes nonfunctional, or is no longer needed, it is degraded, and its amino acids are catabolized or recycled into new proteins.

Cells vary in their need and ability to synthesize proteins. Terminally differentiated red blood cells lack the apparatus for translation and have a life span of only 120 days. Other cells need to maintain concentrations of enzymes and other proteins and, to remain viable, carry out this protein synthesis. Growing and dividing cells must synthesize much larger amounts of protein, and some cells also synthesize proteins for export. For example, liver cells synthesize many enzymes needed in multiple metabolic pathways plus proteins for export including serum albumin, the major protein of blood plasma. Liver cells are protein factories that are particularly rich in the machinery for translation.

6.2 COMPONENTS OF THE TRANSLATIONAL APPARATUS

Messenger RNA Transmits Information Encoded in DNA

Genetic information is inherited and stored in the nucleotide sequences of DNA. Selective expression of this information requires its transcription into mRNA that carries a specific and precise message from the nuclear "data bank" to the cytoplasmic site of protein synthesis. In eukaryotes, mRNAs are usually synthesized as significantly larger precursor molecules that are processed prior to their export from the nucleus (see p 190). Eukaryotic mRNA is almost always **monocistronic**, i.e. it encodes a single polypeptide. The 5′-end is capped with **7-methylguanosine** linked through a 5′-triphosphate bridge to the 5′-end of the mRNA (see p. 68). A 5′-nontranslated region of up to a few hundred nucleotides in length separates the cap from the translation **initiation signal**. Usually, but not always, this is the first AUG sequence encountered as the message is read 5′ → 3′. Uninterrupted 3-nucleotide-long **codon** sequences that specify a unique polypeptide sequence follow the initiation codon until a translation termination codon is reached. This is followed by a 3′-untranslated sequence (or UTR), usually about 100 nucleotides long, before the mRNA terminates in a 100- to 200-nucleotide-long polyadenylate tail.

Prokaryotic mRNA differs significantly. The 5′-end of the mRNA is not capped, but it retains a terminal triphosphate from its synthesis by RNA polymerase. Most mRNAs are **polycistronic**; they encode several polypeptides and include more than one initiation AUG codon. A ribosome-positioning sequence is located about 10 nucleotides upstream of the AUG initiation codon. A UTR follows the termination signal but there is no polyadenylate tail.

Transfer RNA Is a Bilingual Translator Molecule

All tRNA molecules have several common structural characteristics including the 3′-terminal —CCA sequence to which an amino acid is bound, a conserved "cloverleaf" secondary structure, and an L-shaped tertiary structure (see p. 63). However, each of the many tRNA species has a unique nucleotide sequence that allows great specificity in interactions with mRNA and with the aminoacyl tRNA synthetase that couples one specific amino acid to it.

The Genetic Code Uses a Four-Letter Alphabet of Nucleotides

Genetic information exists as linear sequences of nucleotides in DNA, analogous to the linear sequence of letters in these words. The DNA language uses a **four-letter alphabet** comprised of two purines, A and G (adenine and guanine), and two pyrimidines, C and T (cytosine and thymine). The information in mRNA is transcribed into a similar four-letter alphabet, but U (uracil) replaces T. The language of RNA is thus a dialect of the genetic language of DNA. Genetic information is **expressed** mostly in the form of proteins that derive their properties from their sequence of amino acids. Thus, in protein biosynthesis the four-letter language of nucleic acids is **translated** into the 20-letter language of proteins. Implicit in the analogy to language is the directionality of these sequences. By convention, nucleic acid sequences are written in a 5′ → 3′ direction and protein sequences from the amino terminus to the carboxy terminus. These directions correspond in their reading and biosynthetic senses.

Codons in mRNA Are Three-Letter Words

A 1:1 correspondence of bases to amino acids would only permit mRNA to encode four different amino acids, while a 2:1 correspondence would encode $4^2 = 16$ amino acids. Neither one is sufficient since 20 different amino acids occur in most proteins. The three-letter genetic code has $4^3 = 64$ permutations or words; this is also sufficient to encode start and stop signals, equivalent to punctuation. The triplets are called **codons** and are customarily shown in the form of Table 6.1. Two amino acids are designated by single codons: methionine by AUG and tryptophan by UGG. The rest are designated by two, three, four, or six codons. Multiple codons for a single amino acid represent **degeneracy** in the code. The genetic code is nearly universal, since the same code words are used in all organisms. An exception to universality occurs in mitochondria, in which a few codons have a different meaning than in the cytosol of the same organism (Table 6.2).

TABLE 6.1 The Genetic Code[a]

5′ Base	U	C	A	G	3′ Base
U	UUU Phe	UCU Ser	UAU Tyr	UGU Cys	U
U	UUC Phe	UCC Ser	UAC Tyr	UGC Cys	C
U	UUA Leu	UCA Ser	UAA Stop	UGA Stop	A
U	UUG Leu	UCG Ser	UAG Stop	UGG Trp	G
C	CUU Leu	CCU Pro	CAU His	CGU Arg	U
C	CUC Leu	CCC Pro	CAC His	CGC Arg	C
C	CUA Leu	CCA Pro	CAA Gln	CGA Arg	A
C	CUG Leu	CCG Pro	CAG Gln	CGG Arg	G
A	AUU Ile	ACU Thr	AAU Asn	AGU Ser	U
A	AUC Ile	ACC Thr	AAC Asn	AGC Ser	C
A	AUA Ile	ACA Thr	AAA Lys	AGA Arg	A
A	AUG Met	ACG Thr	AAG Lys	AGG Arg	G
G	GUU Val	GCU Ala	GAU Asp	GGU Gly	U
G	GUC Val	GCC Ala	GAC Asp	GGC Gly	C
G	GUA Val	GCA Ala	GAA Glu	GGA Gly	A
G	GUG Val	GCG Ala	GAG Glu	GGG Gly	G

[a] The genetic code comprises 64 codons, which are permutations of four bases taken in threes. Note the importance of sequence: three bases, each used once per triplet codon, give six permutations: ACG, AGC, GAC, GCA, CAG, and CGA, for threonine, serine, aspartate, alanine, glutamine, and arginine, respectively.

TABLE 6.2 Nonuniversal Codon Usage in Mammalian Mitochondria

Codon	Usual Code	Mitochondrial Code
UGA	Termination	Tryptophan
AUA	Isoleucine	Methionine
AGA	Arginine	Termination
AGG	Arginine	Termination

Punctuation

Four codons act partially or fully as punctuation. The **start signal, AUG**, also specifies methionine. An AUG at an appropriate site in mRNA signifies methionine as the initial, amino-terminal residue. AUG codons later in the mRNA specify methionine residues within the protein. Three codons, UAG, UAA, and UGA, are **stop signals** that specify no amino acid and are **termination codons** or, less appropriately, **nonsense codons**.

Codon–Anticodon Interactions Permit Reading of mRNA

Translation of the codons of mRNA involves their direct interaction with complementary **anticodon sequences** in tRNA. Each tRNA species carries a unique amino acid, and each has a specific anticodon triplet. Like base pairing in DNA, codon–anticodon base pairing is **antiparallel**, as shown in Figure 6.1. Codons are read in a **sequential, nonoverlapping reading frame**. Anticodon and amino acid-acceptor sites are at opposite ends of the L-shaped tRNA molecule (see p. 64). The tRNA both conceptually and physically bridges the gap between the nucleotide sequence of the ribosome-bound mRNA and the site of protein assembly on the ribosome.

Since 61 codons designate 20 amino acids, it might seem necessary to have exactly 61 different tRNA species. This is not the case. Many amino acids can be carried by more than one tRNA species, and degenerate codons can be read by more than one tRNA (but always one carrying the correct amino acid). Variation from standard base pairing is common in codon–anticodon interactions. Much of the complexity is explained by the **wobble hypothesis**, which permits less stringent base pairing between the first position of the anticodon and the **degenerate** (third) position of the codon. A frequent anticodon nucleotide is inosinic acid (I), the nucleotide of hypoxanthine, which base-pairs with U, C, or A. Wobble base-pairing rules are shown in Table 6.3. Modified nucleotides at or beside the first nucleotide of the anticodon in many tRNA species also modulate codon–anticodon interactions.

If the wobble rules are followed, the 61 nonpunctuation codons could be read by 31 tRNA molecules, but most cells have 50 or more tRNA species. Some codons are read more efficiently by one anticodon than another. Not all codons are used equally, some being used very rarely. Examination of many mRNA sequences has allowed construction

TABLE 6.3 Wobble Base-Pairing Rules

3′ Codon Base	5′ Anticodon Bases Possible
A	U or I
C	G or I
G	C or U
U	A or G or I

of "codon usage" tables that show that different organisms preferentially use different codons to generate similar polypeptide sequences.

"Breaking" the Genetic Code

The genetic code (Table 6.1) was determined before methods were developed to sequence natural mRNA. Code-breaking experiments used simple synthetic mRNAs or trinucleotide codons to provide insight into how proteins are encoded.

Polynucleotide phosphorylase catalyzes the template-independent reaction:

$$x\text{NDP} \leftrightarrow \text{polynucleotide } (\text{pN})_x + x\text{P}_i$$

where NDP is any nucleoside 5′-diphosphate or a mixture of two or more. If the nucleoside diphosphate is UDP, a polymer of U, poly(U), is formed. *In vitro* protein synthesis with poly(U) as mRNA generates polyphenylalanine. Similarly, poly(A) encodes polylysine and poly(C) polyproline. An mRNA with a random sequence of only U and C produces polypeptides that contain not only proline and phenylalanine as predicted, but also serine (from UCU and UCC) and leucine (from CUU and CUC). Degeneracy in the code and the complexity of products made experiments with random sequence mRNAs difficult to interpret, so mRNAs of defined sequence were transcribed from simple repeating synthetic DNAs. Thus poly(AU), transcribed from a repeating poly(dAT), produces only a repeating copolymer of Ile-Tyr-Ile-Tyr, read from successive triplets AUA UAU AUA UAU, and so on. Poly(CUG) contains possible codons CUG for Leu, UGC for Cys, and GCU for Ala, each repeating itself once the **reading frame** has been set. Selection of the initiation codon is random in these *in vitro* experiments, so three different homopolypeptides are produced: polyleucine, polycysteine, and polyalanine. A perfect poly(CUCG) produces a polypeptide with the sequence (-Leu-Ala-Arg-Ser-) whatever the initiation point. These relationships, summarized in Table 6.4, show codons to be triplets read in exact sequence, without overlap or omission. Other experiments used synthetic trinucleotide codons as minimal messages. No proteins were made, but ribosome binding of only one amino acid (as a tRNA conjugate) was stimulated by a given codon. The meaning of each possible codon was later verified by determination of mRNA sequences.

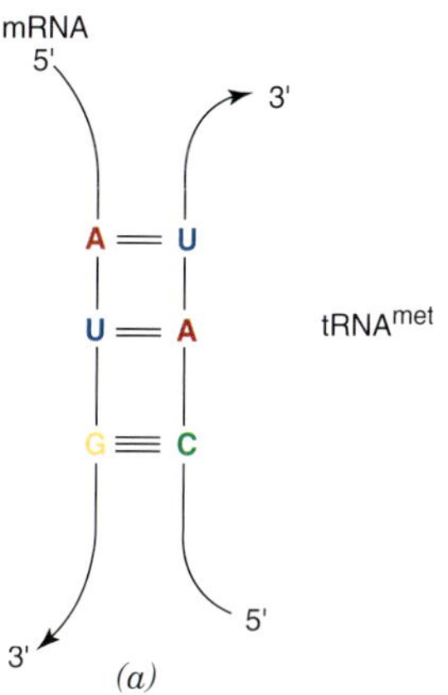

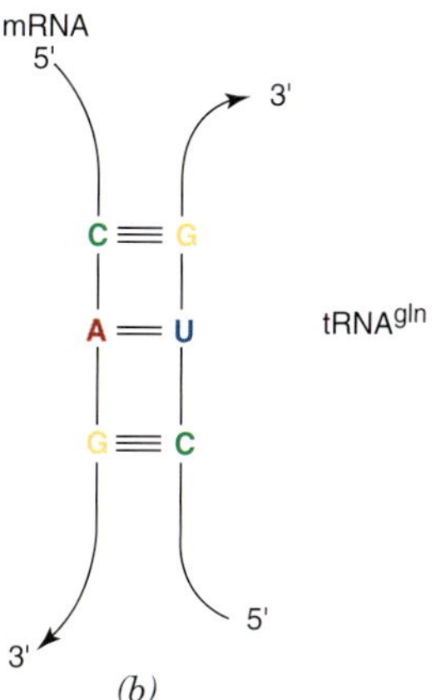

FIGURE 6.1

Codon–anticodon interactions. Shown are interactions between the AUG (methionine) codon and its CAU anticodon (a) and the CAG (glutamine) codon and a CUG anticodon (b). Note that base pairing of mRNA with tRNA is antiparallel.

Mutations

Understanding the genetic code and how it is read provides a basis for understanding the nature of mutations. A mutation is simply a heritable change in a gene. Point mutations involve a change in a single base pair in the DNA, and thus a single base in the corresponding mRNA. If this change occurs in the third position of a degenerate codon, there may be no change in the amino acid specified (e.g., UCC to UCA still codes for serine). Such **silent mutations** are detected by gene sequence determination. They are commonly seen in comparison of genes for similar proteins—for example, hemoglobins from different species. **Missense mutations** arise from a base change that causes incorporation of a different amino acid in a protein (see Clin. Corr. 6.1). Point mutations can also form or destroy a termination codon and thus change the length

TABLE 6.4 Polypeptide Products of Synthetic mRNAs[a]

mRNA	*Codon Sequence*	*Products*
—(AU)$_n$—	—AUA UAU AUA UAU—	—(Ile-Tyr)$_{n/3}$—
—(CUG)$_n$—	—CUG CUG CUG CUG—	—(Leu)$_n$—
	—UGC UGC UGC UGC—	—(Cys)$_n$—
	—GCU GCU GCU GCU—	—(Ala)$_n$—
—(CUCG)$_n$—	CUC GCU CGC UCG	—(Leu-Ala-Arg-Ser)$_{n/3}$—

[a] The horizontal brackets accent the reading frame.

CLINICAL CORRELATION 6.4

Programmed Frameshifting in Biosynthesis of HIV Proteins

Maintaining the reading frame during translation is central to the accuracy and fidelity of translation. However, many retroviruses, including HIV, the human immunodeficiency virus that causes AIDS, take advantage of mRNA slippage and a change in reading frame to generate different proteins from the same message. A single mRNA, designated gag-pol, encodes two polyproteins that overlap by about 200 nucleotides and are in different reading frames. The gag polyprotein is translated from the initiation codon to an in-frame termination codon near the gag-pol junction; gag polyprotein is then cleaved to generate several structural proteins of the virus. However, about 5% of the time a one-nucleotide frameshift occurs within the overlapping segment of mRNA and the termination codon is bypassed because it is no longer in the reading frame. A gag-pol fusion polyprotein is produced; proteolytic cleavage of the pol polyprotein produces the viral reverse transcriptase and other proteins needed in virus reproduction.

Source: Jacks, T., Power, M. D., Masiarz, F. R., Luciw, P. A., Barr, P. J., and Varmus, H. E. Characterization of ribosomal frameshifting in HIV-1 gag-pol expression. *Nature* 331:280, 1988.

$$\text{H—N(H)—CH(R)—C(=O)—OH} + \text{ATP} + \text{E} \rightleftharpoons [\text{H—N(H)—CH(R)—C(=O)} \sim \text{AMP} \cdot \text{E}] + \text{PP}_i \quad (1)$$

$$[\text{H—N(H)—CH(R)—C(=O)} \sim \text{AMP} \cdot \text{E}] + \text{tRNA} \rightleftharpoons \text{H—N(H)—CH(R)—C(=O)—tRNA} + \text{AMP} + \text{E} \quad (2)$$

Sum:

$$\text{H—N(H)—CH(R)—C(=O)—OH} + \text{ATP} + \text{tRNA} \rightleftharpoons \text{H—N(H)—CH(R)—C(=O)—tRNA} + \text{AMP} + \text{PP}_i$$

Brackets surrounding the aminoacyl-AMP–enzyme complex indicate that it is a transient, enzyme-bound intermediate. The aminoacyladenylate, a mixed acid anhydride with carboxyl and phosphoryl components, is a "high-energy" intermediate. The aminoacyl ester linkage in tRNA is lower in energy than in aminoacyl-adenylate, but still higher than in the carboxyl group of the free amino acid. Pyrophosphatases cleave the pyrophosphate that is produced, and the equilibrium is strongly shifted toward formation of aminoacyl-tRNA. From the viewpoint of precision in translation, the amino acid, which had only its side chain (R-group) to distinguish it, becomes linked to a large, complex, and easily recognized carrier.

Specificity and Fidelity of Aminoacylation Reactions

Almost all cells contain 20 different aminoacyl-tRNA synthetases, each specific for one amino acid, and at most a small family of carrier tRNAs for that amino acid. Since codon–anticodon interactions define the amino acid to be incorporated, an incorrect amino acid carried by the tRNA will cause an error in the protein. Correct selection of both tRNA and amino acid by the synthetase is therefore central to fidelity in protein synthesis. Aminoacyl-tRNA synthetases share a common mechanism, and many are normally associated within the cell. Nevertheless, they are a diverse group of proteins that may contain one, two, or four identical subunits or pairs of dissimilar subunits. Separate structural domains can be involved in aminoacyl-adenylate formation, tRNA recognition, and, if it occurs, subunit interactions. Each enzyme is capable of almost error-free formation of correct aminoacyl-tRNA combinations. While some amino acids are easily recognized by their bulk (e.g., tryptophan), or lack of bulk (glycine), or by

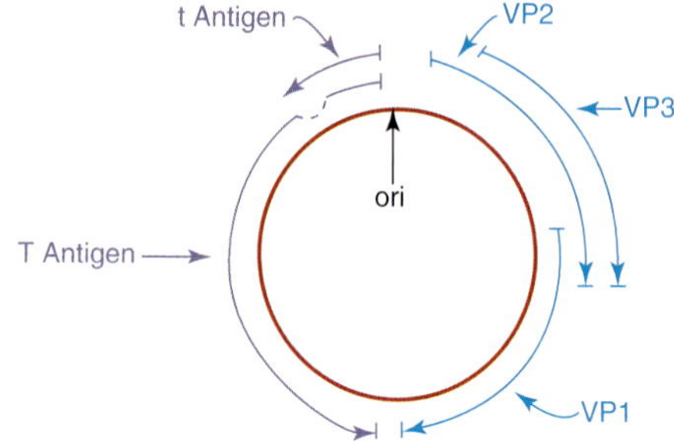

FIGURE 6.2

Genome of simian virus 40 (SV40). DNA of SV40, shown in red, is a double-stranded circle of slightly more than 5000 bp that encodes all information needed for viral survival and replication within a host cell. It demonstrates extremely efficient use of the protein-coding potential of a genome. Proteins VP1, VP2, and VP3 are structural proteins of the virus; VP2 and VP3 are translated from different initiation points to the same carboxyl terminus. VP1 is translated in a different reading frame so that its N-terminal section overlaps the VP2 and VP3 genes but its amino acid sequence in the overlapping segment is different from that of VP2 and VP3. Two proteins, the large T and small t tumor antigens which promote transformation of infected cells, are encoded in a common mRNA precursor. Both proteins have identical N-terminal sequences; the C-terminal segment of small t protein is encoded by a segment of mRNA that is spliced out of the large T message, and the carboxyl-terminal sequence of large T is encoded in RNA that follows termination of small t. The origin of DNA replication (ori) is outside of coding regions.

positive or negative charges on the side chains (e.g., lysine and glutamate), others are much more difficult to discriminate. For example, recognition of valine rather than threonine or isoleucine by the valyl-tRNA synthetase is difficult since the side chains differ by either an added hydroxyl or single methylene group.

The amino acid recognition and activation sites of each enzyme have great specificity, but sometimes misrecognition does occur. An additional proofreading step increases discrimination. This most often occurs through hydrolysis of the aminoacyladenylate intermediate, with the release of amino acid and AMP. Valyl-tRNA synthetase efficiently hydrolyzes threonyl-adenylate, and it hydrolyzes isoleucyl-adenylate in the presence of bound (but not aminoacylated) $tRNA^{val}$. Alternatively, misacylated tRNA is recognized and deacylated. Valyl-, phenylalanyl-, and isoleucyl-tRNA synthetases deacylate tRNAs that have been mischarged with threonine, tyrosine, and valine, respectively. Proofreading is performed by many but not all aminoacyl-tRNA synthetases. The net result is an average level of misacylation of one in 10^4 to 10^5.

Each synthetase must also correctly recognize one to several tRNA species that can carry the same amino acid, while rejecting incorrect tRNA species. This may be simpler than selection of a single amino acid. However, recall the conformational similarity and common sequence elements of all tRNAs (p. 64). Different synthetases recognize different elements of tRNA structure. Usually, multiple structural elements contribute to recognition, but many of these elements are not absolute determinants.

One logical element of tRNA recognition is the anticodon, specific to one amino acid. In the case of $tRNA^{met}$, changing the anticodon also alters recognition by the synthetase. In other instances, this is only partly true, and sometimes the anticodon is not a recognition determinant. This is shown by **suppressor mutations** that "suppress" the expression of classes of nonsense mutations. A point mutation in a glutamine (CAG) codon produces a termination (UAG) codon, which causes premature termination of the encoded protein. A second suppressor mutation in the anticodon of a $tRNA^{tyr}$, in which the normal GUA anticodon is changed to CUA, allows "read through" of the termination codon. The initial mutation is suppressed and a nearly normal protein is made, with the affected glutamine replaced by tyrosine. Aminoacylation of the mutant $tRNA^{tyr}$ with tyrosine shows that in this case the anticodon does not determine synthetase specificity. Other tRNA-identification features may include elements of the acceptor stem and parts of the variable loop or the D-stem/loop. In *E. coli* $tRNA^{ala}$, the primary recognition characteristic is a G3-U70 base pair in the acceptor stem; even if no other changes in the $tRNA^{ala}$ occur, any variation at this position destroys its acceptor ability with alanine-$tRNA^{ala}$ synthetase. The X-ray structure of glutaminyl synthetase-tRNA complex shown in Figure 6.3 shows binding at the concave tRNA surface, which is typical and compatible with the biochemical observations.

Ribosomes Are Workbenches for Protein Biosynthesis

Ribosomes are complex **ribonucleoprotein** particles made up of two dissimilar subunits, each of which contains RNA and many proteins. With one exception, each protein is present in a single copy per ribosome, as is each RNA species. Ribosome architecture and function have been conserved in evolution, and similarities between ribosomes and subunits from different sources are more important than the differences. The composition of major ribosome types is shown in Table 6.7.

Ribosomes and their subunits can be visualized by electron microscopy of stained or frozen preparations. Locations of most ribosomal proteins, some elements of the RNA, and functional sites on each subunit were shown by electron microscopy of subunits in complexes with antibodies against a single ribosomal component. Chemical cross-linking identified near neighbors within the structure, and neutron diffraction measurements quantitated the distance between pairs of proteins. Sequence comparisons, and chemical and enzymatic probes give information about RNA conformation. Because prokaryotic ribosomes can self-assemble, native structures can be reconstituted from mixtures of purified individual proteins and RNAs. Reconstitution of subunits from mixtures in

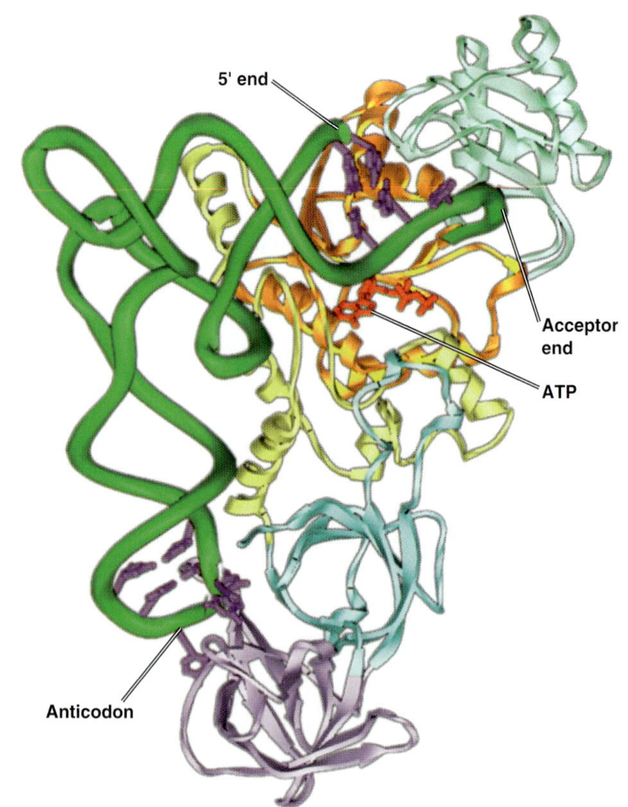

FIGURE 6.3

Interaction of a tRNA with its aminoacyl-tRNA synthetase. The sugar–phosphate backbone of *E. coli* glutaminyl-tRNA is shown in green, and the peptide backbone of the glutaminyl-tRNA synthetase is shown in multiple colors. Note the interactions of the synthetase with partially unwound acceptor stem and anticodon loop of the tRNA; also note the placement of ATP, shown in red, within a few angstroms of the 3′-end of tRNA. Space-filling models of the enzyme and tRNA show both molecules as solid objects with several sites of direct contact.
Adapted from Perona, J., Rould, M., and Steitz, T. *Biochemistry* 32:8758, 1993.

TABLE 6.7 Ribosome Classification and Composition

		Subunits	
Ribosome Source	***Monomer Size***	***Small***	***Large***
Eukaryotes			
Cytosol	80S	40S: 34 proteins 18S RNA	60S: 50 proteins 28S, 5.8S, 5S RNAs
Mitochondria			
Animals	55S–60S	30S–35S: 12S RNA	40–45S: 16S RNA
		70–100 proteins	
Higher plants	77S–80S	40S: 19S RNA	60S: 25S, 5S RNAs
		70–75 proteins	
Chloroplasts	70S	30S: 20–24 proteins 16S RNA	50S: 34–38 proteins 23S, 5S, 4.5S RNAs
Prokaryotes			
Escherichia coli	70S	30S: 21 proteins 16S RNA	50S: 34 proteins 23S, 5S RNAs

which a single component is omitted or modified can show if that protein is important for assembly of the subunit or for some specific function. Total reconstitution of eukaryotic subunits has not succeeded, but conclusions about how prokaryotic ribosomes function apply equally to eukaryotic systems.

Correlations of structural data with functional measurements in protein synthesis have allowed development of ribosome models, features of which are described in Figure 6.4. Some types of ribosomes and their subunits can also be crystallized, and

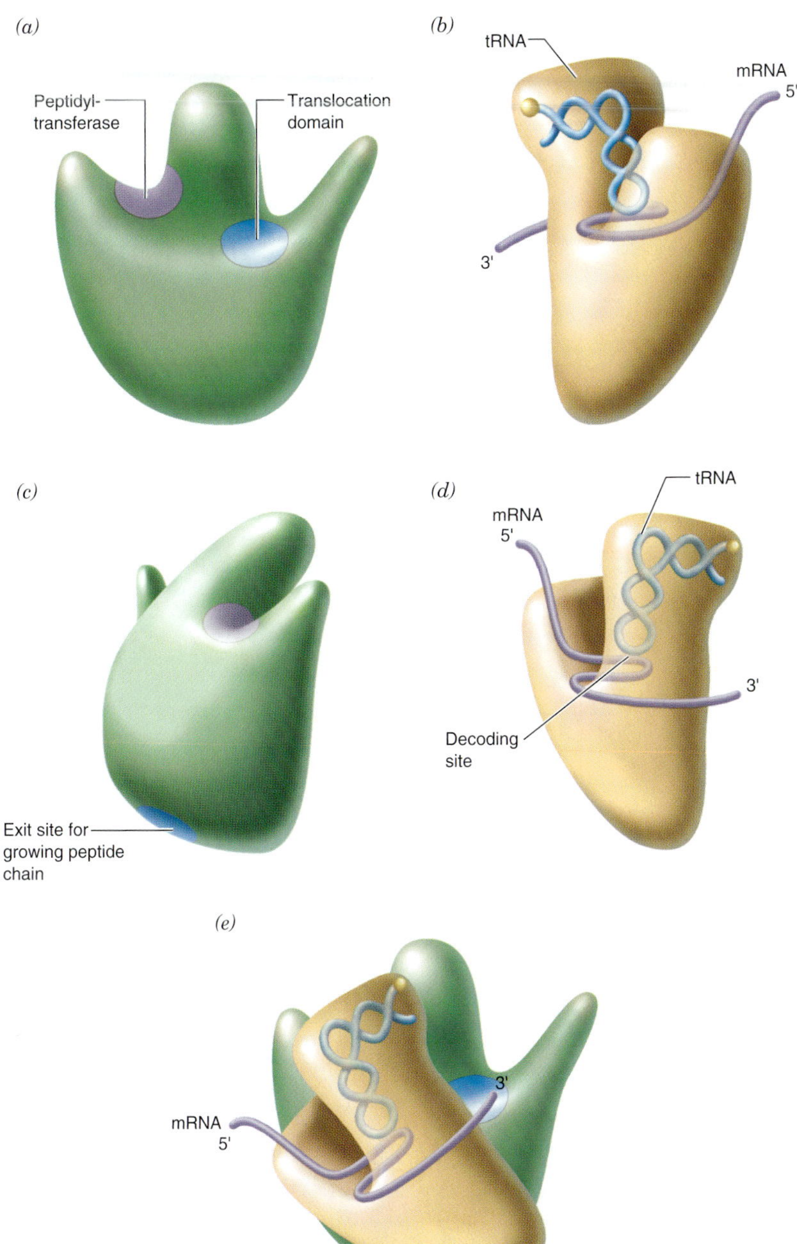

FIGURE 6.4

Model of ribosome structure and functional sites. Top row shows the face of each subunit that interacts in the functional ribosome. In (*a*) the large subunit sites of peptide bond formation (peptidyl transferase) and of elongation factor binding (translocation domain) are seen on opposite sides of the central protuberance. The stalk projects to the right. In (*b*) the small subunit platform protrudes toward the reader. mRNA and tRNAs interact in the decoding site, in a cleft between the platform and the subunit body. In (*c*) the large subunit is rotated 90°, the stalk projects into the page, and the exit site for the nascent peptide is seen near the base of the subunit. This site contacts membrane in the bound ribosomes of rough endoplasmic reticulum. The site of peptide bond formation is distant from the exit site; the growing peptide passes through a groove or tunnel in the ribosome to reach the exit site. In (*d*) the small subunit has been rotated 90°; the platform projects toward the face of the large subunit. In (*e*) the subunits have been brought together to show their orientation in the ribosome. Aminoacyl-tRNA bound by the small subunit is oriented with the acceptor end near peptidyl transferase while the translocation domain is near the decoding site.

X-ray structural determination confirms the validity of the models. Figure 6.5 shows the structure of a prokaryotic ribosome at 5.5-Å resolution, which is sufficient to distinguish RNA elements within each subunit, proteins at their surfaces, and tRNA molecules between the subunits. Higher-resolution structures of isolated subunits give further insight into both structure and function.

Ribosomes are organized in two more ways. Several ribosomes often translate a single mRNA molecule simultaneously. Purified mRNA-linked **polysomes** can be visualized by electron microscopy (Figure 6.6). In eukaryotic cells some ribosomes occur free in the cytosol, but many are bound to membranes of the rough endoplasmic reticulum. In general, **free ribosomes** synthesize proteins that remain within the cytosol or become targeted to the nucleus, mitochondria, or some other organelle. **Membrane-bound ribosomes** synthesize proteins that will be secreted from the cell or sequestered and function in the endoplasmic reticulum, Golgi complex, or lysosomes. In cell homogenates, membrane fragments with bound ribosomes constitute the **microsome** fraction; detergents that disrupt membranes release these ribosomes.

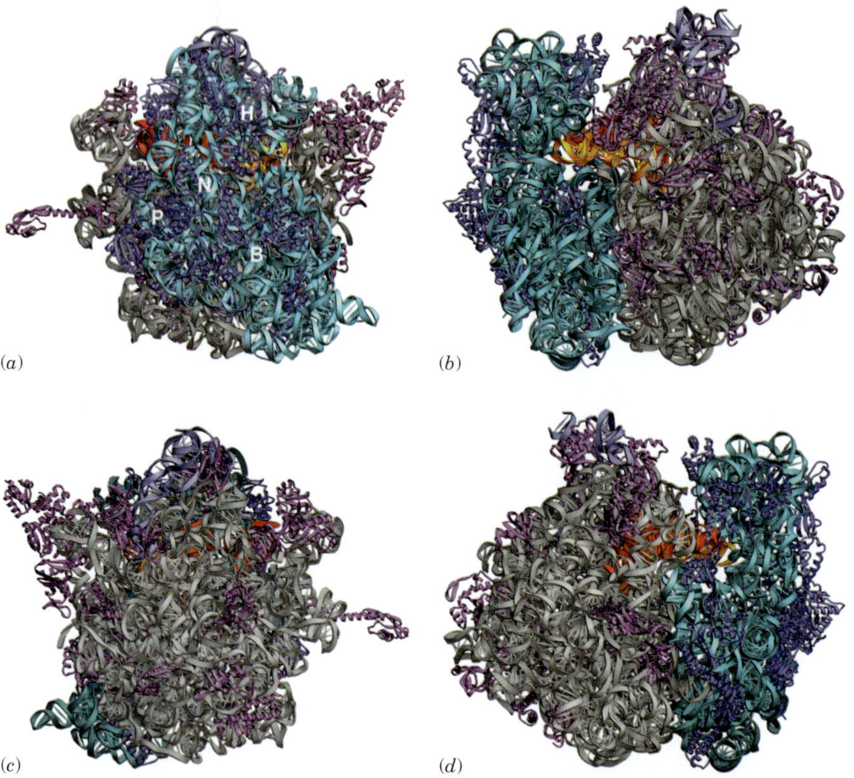

FIGURE 6.5

Crystallographic structure of a 70 S ribosome. The structure of the *Thermus thermophilus* ribosome is shown at 5.5-Å resolution. Successive views are rotated 90° around the vertical axis. (*a*) The small subunit lies atop the large subunit, as in the model of Figure 6.4*e*. Small subunit features include the platform (P), head (H), body (B), and a connecting neck (N). The 16 S RNA is colored cyan, and small subunit proteins are dark blue. (*b*) The large subunit is on the right; 23 S RNA is gray, 5 S RNA is light blue, and large subunit proteins are magenta. A tRNA molecule (gold) spans the subunits. (*c*). The large subunit lies on top with its stalk protruding to the left. (*d*) The large subunit is on the left and tRNA elements are visible in the subunit interface.
Reprinted with permission from Yusupov, M. M., Yusupova, G. Z., Baucom, A., Lieberman, K., Earnest, T. N., Cate, J. H. D., and Noller, H. F. *Science* 292:883, 2001. Copyright (2001) AAAS. Figure generously supplied by Drs. A. Baucom and H. Noller.

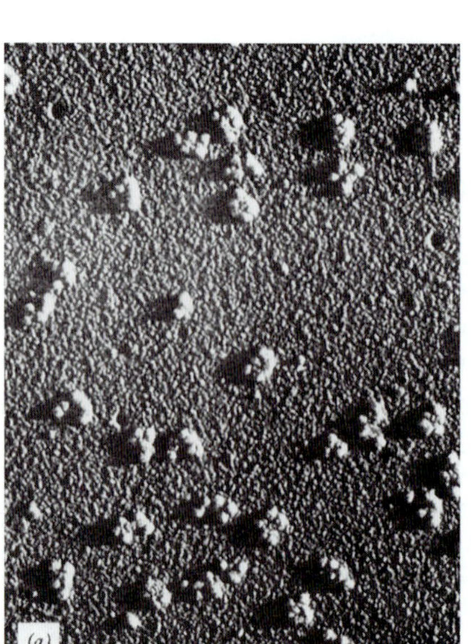

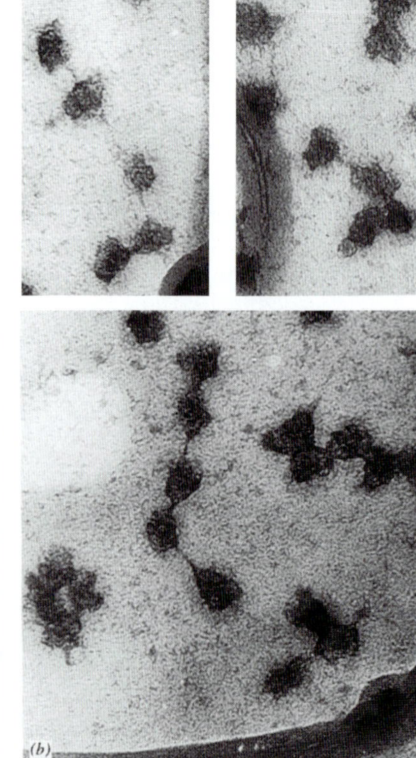

FIGURE 6.6

Electron micrographs of polysomes. (a) Reticulocyte polyribosomes shadowed with platinum are seen in clusters of three to six ribosomes, a number consistent with the size of mRNA for a globin chain. (b) Uranyl acetate staining and visualization at a higher magnification shows polysomes in which parts of the mRNA are visible.
Courtesy of Dr. Alex Rich, MIT.

6.3 PROTEIN BIOSYNTHESIS

Translation Is Directional and Collinear with mRNA

Messenger RNA sequences are written (and transcribed) $5' \rightarrow 3'$, and during translation they are read in the same direction. Amino acid sequences are both written and synthesized from the amino-terminal residue to the carboxy terminus. A ribosome remains bound to an mRNA molecule and moves along the length of the mRNA until a stop codon is reached. Comparison of mRNA sequences with sequences of the proteins they encode shows a perfect, collinear, nonoverlapping and gap-free correspondence of the mRNA coding sequence and that of the polypeptide synthesized. (For a rare exception, see Clin. Corr. 6.4.) In fact, it is common to deduce the sequence of a protein solely from the nucleotide sequence of its mRNA or the DNA of its gene. However, the deduced sequence may differ from the genuine protein because of posttranslational modifications.

A story can be analyzed in terms of its beginning, its development or middle section, and its ending. Protein biosynthesis will be described in a similar framework: initiation of the process, elongation during which the great bulk of the protein is formed, and termination of synthesis and release of the completed polypeptide. We will then examine posttranslational modifications that a protein may undergo.

Initiation of Protein Synthesis Is a Complex Process

Initiation requires bringing together a small (40 S) ribosomal subunit, an mRNA, and a tRNA complex of the amino-terminal amino acid, all in a proper orientation. This is followed by association of the large (60 S) subunit to form a completed initiation complex on an 80 S ribosome. This process requires a group of transiently bound proteins, known as **initiation factors**, that act only in initiation. The specific function of some eukaryotic initiation factors remain unclear, but prokaryotic protein synthesis provides a simpler model for comparison. The initiation of translation is shown in Figure 6.7. As a first step**, eukaryotic initiation factor 2a** (**eIF-2a**) binds to GTP and the initiator tRNA, Met-$tRNA_i^{met}$, to form a ternary complex. No other aminoacyl-tRNA can replace the initiation-specific Met-$tRNA_i^{met}$ in this step. Prokaryotes also utilize a specific initiator tRNA whose methionine is modified by formylation of its amino group. Only fMet-$tRNA_i^{met}$ is recognized by prokaryotic IF-2.

The second step requires 40 S ribosomal subunits associated with a very complex protein, **eIF-3**. Mammalian eIF-3 contains eight different polypeptides and has a mass of 600–650 kDa; it binds to the 40 S subunit surface that will contact the 60 S subunit and physically blocks subunit association. Hence eIF-3 is a ribosome **anti-association factor**—as is eIF-6, which binds to 60 S subunits. A complex that includes eIF-2a-GTP, Met-$tRNA_i^{met}$, eIF-3-40 S, and additional protein factors now forms. In the third step the **pre-initiation complex** is formed; mRNA, **eIF-4f**, also called the **cap binding complex, eIF-4a**, an RNA helicase that unwinds secondary structure in the untranslated leader sequence of the mRNA, **PAB**, a polyA-binding protein that loops the 3′-end of the mRNA near to the 5′-cap, and several other proteins are needed. The mRNA is then scanned $5' \rightarrow 3'$ until the first AUG triplet is found. Rarely, the first AUG is not used and a later AUG, characterized by its secondary structure within the mRNA, is selected for initiation. GTP is hydrolyzed by eIF-2a with the help of eIF-5, and eIF-2a-GDP, and other factors are released. The eIF-2a-GDP interacts with the guanine nucleotide exchange factor eIF-2b and GTP to regenerate eIF-2a-GTP for another round of initiation.

The final step requires binding of this complex by a 60 S subunit and an additional factor, **eIF-5b–GTP**. GTP is hydrolyzed, and eIF-5b–GDP and other factors are released. The complete **initiation complex** is an 80 S ribosome with the mRNA and initiator tRNA correctly positioned to begin translation.

Prokaryotes use fewer initiation factors to form a similar initiation complex. Their 30 S subunits, complexed with a simpler IF-3, can first bind either mRNA or a ternary complex of IF-2, fMet-$tRNA_i^{met}$, and GTP. Orientation of the mRNA relies in part

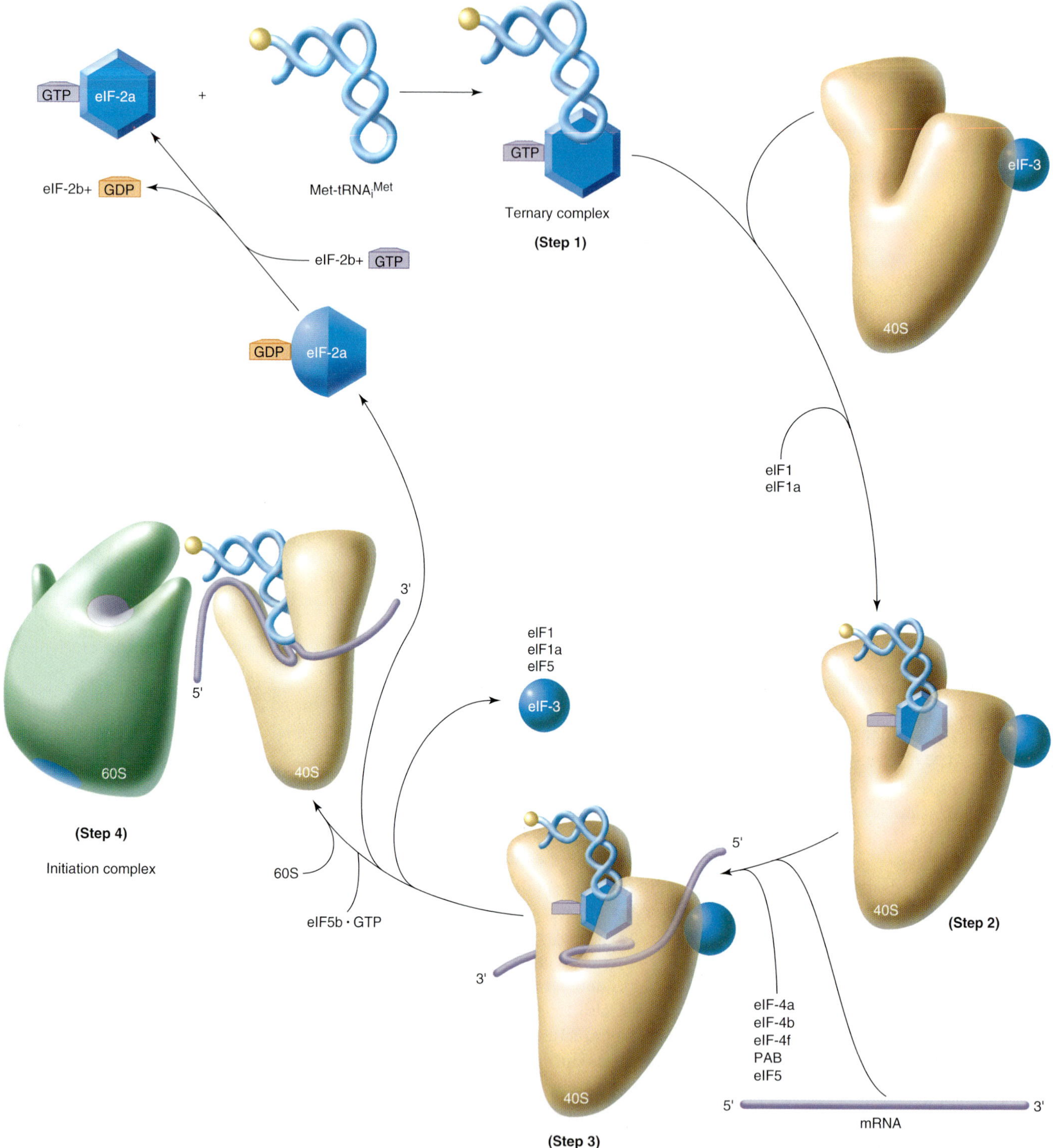

FIGURE 6.7

Initiation of translation in eukaryotes. Details in text. A ternary complex that includes the initiator tRNA (step 1) combines with a small ribosomal subunit (step 2). Interaction with mRNA forms a pre-initiation complex (step 3). Large subunit binding completes formation of the initiation complex (step 4). The different shape of eIF2a in complexes with GTP and GDP indicates that conformational changes occur upon hydrolysis of triphosphate. After elongation has begun, additional small subunits will complex with the same mRNA to form polysomes.

on base-pairing between a pyrimidine-rich sequence of eight nucleotides in 16 S rRNA and a purine-rich "Shine–Dalgarno" sequence about 10 nucleotides upstream of the initiator AUG codon. Complementarity between rRNA and mRNA may include several mismatches, but greater complementarity usually leads to initiation that is more efficient. Initiation factor IF1 also acts in formation of the preinitiation complex. Finally, 50 S subunit is bound, GTP is hydrolyzed, and the initiation factors are released. It is interesting that prokaryotes use RNA–RNA base-pairing to position mRNA, while eukaryotes use many protein factors to attain the same result.

Elongation Is the Stepwise Formation of Peptide Bonds

Protein synthesis occurs by stepwise elongation to form a polypeptide chain. At each step, ribosomal **peptidyl transferase** transfers the growing peptide (or in the first step the initiating methionine residue) from its carrier tRNA to the α-amino group of the amino acid residue of the aminoacyl-tRNA specified by the following codon.

$$\cdots\mathrm{NH{-}CH(R_n){-}C({=}O){-}tRNA_n} + \mathrm{H_2N{-}CH(R_{n+1}){-}C({=}O){-}tRNA_{n+1}} \longleftrightarrow \cdots\mathrm{NH{-}CH(R_n){-}C({=}O){-}NH{-}CH(R_{n+1}){-}C({=}O){-}tRNA_{n+1}}$$

peptidyl-tRNA (***n*** residue length) + aminoacyl-tRNA ⟷ peptidyl-tRNA (***n* + 1** residue length)

Efficiency and fidelity are enhanced by non-ribosomal protein elongation factors that utilize the energy released by GTP hydrolysis to ensure placement of the proper aminoacyl-tRNA species and to move the mRNA and associated tRNAs through the ribosome. Elongation is illustrated in Figure 6.8.

During translation, up to three tRNA molecules may be bound at specific sites that span both ribosomal subunits. The initiating methionyl-tRNA is placed in position so that its methionyl residue can be transferred (or donated) to the free α-amino group of the incoming aminoacyl-tRNA; it thus occupies the **donor site**, also called the **peptidyl site** or **P site** of the ribosome. The aminoacyl-tRNA specified by the next codon of the mRNA is bound at the **acceptor site**, also called the **aminoacyl site** or **A site** of the ribosome. Selection of the correct aminoacyl-tRNA is enhanced by **elongation factor EF-1**. A component of EF-1, EF-1α, first forms a ternary complex with aminoacyl-tRNA and GTP. The EF-1α–aminoacyl-tRNA–GTP complex binds to the ribosome and if codon–anticodon interactions are correct, the aminoacyl-tRNA is placed at the A site, GTP is hydrolyzed, and the EF-1α–GDP complex dissociates. The initiating methionyl-tRNA and the incoming aminoacyl-tRNA are now juxtaposed on the ribosome. Their anticodons are paired with successive codons of the mRNA in the P and A sites of the small subunit, and their amino acids are beside one another at the peptidyl transferase site of the large subunit. Peptidyl transferase catalyzes attack of the α-amino group of aminoacyl-tRNA on the carbonyl carbon of methionyl-tRNA. The result is transfer of methionine to the amino group of the aminoacyl-tRNA, which then occupies a "hybrid" position on the ribosome. The anticodon remains in the 40 S A site, while the acceptor end and the attached peptide are in the 60 S P site. The anticodon of the deacylated tRNA remains in the 40 S P site, and its acceptor end is located in the 60 S **exit** or **E site**.

The mRNA and the dipeptidyl-tRNA at the 40 S A site must be repositioned to permit another round of elongation to begin. This is done by **elongation factor 2 (EF-2)**, also called **translocase**. EF-2 moves the mRNA and dipeptidyl-tRNA, in codon–anticodon register, from the 40 S A site to the P site. In the process, GTP is hydrolyzed, providing energy for the movement, and the A site is vacated. As the dipeptidyl-tRNA moves to the P site, the deacylated donor (methionine) tRNA also moves to the E site on the 60 S subunit. The ribosome can now enter a new round. The next aminoacyl-tRNA specified by the mRNA is delivered by EF-1α–GTP to the A site, and the deacylated tRNA in the E site is released. Peptide transfer again

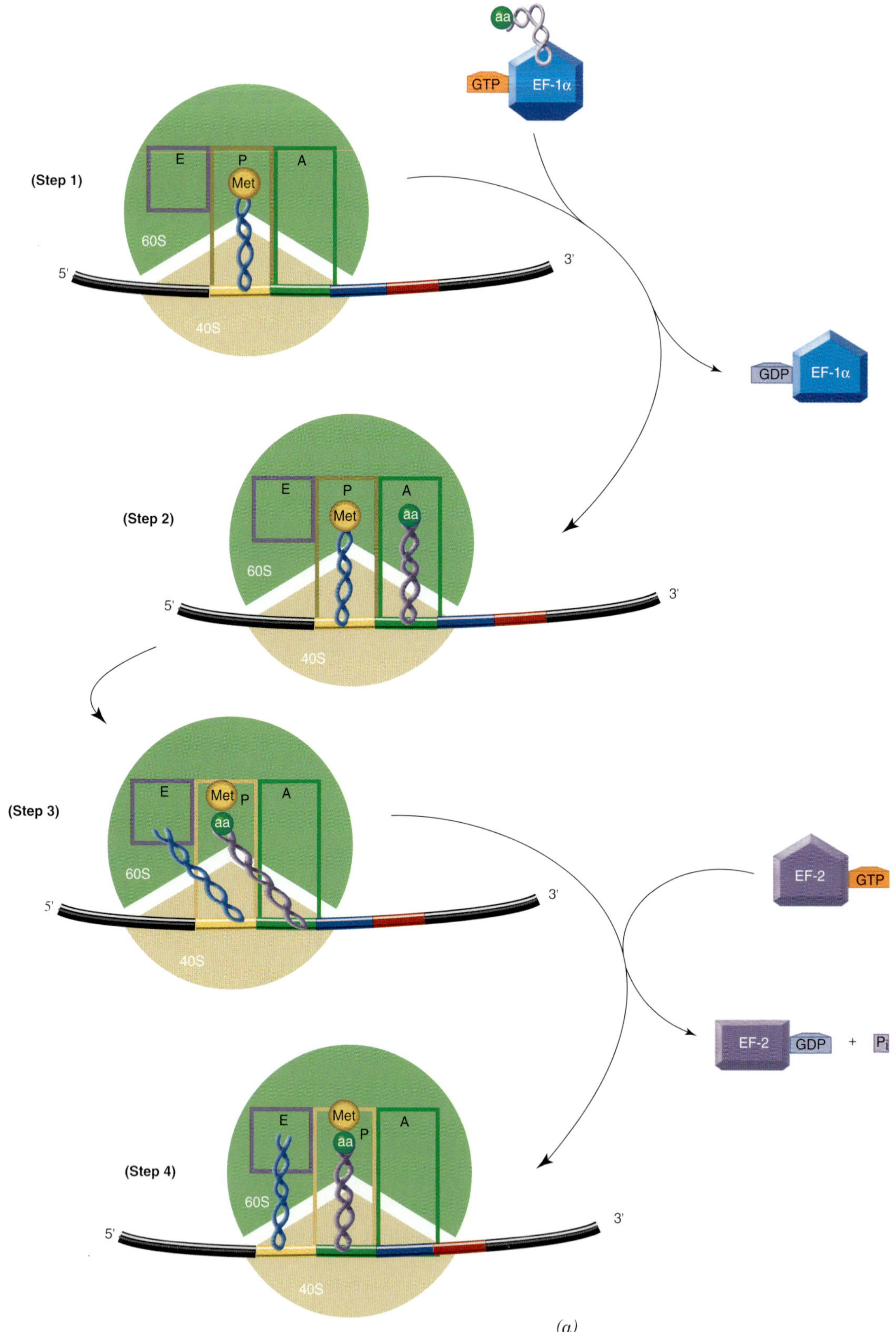

FIGURE 6.8

Elongation steps in eukaryotic protein synthesis.

(*a*) The first round of elongation is shown. Step 1: Initiation complex with methionyl $tRNA_i^{met}$ in 80 S P site. Step 2: EF1α has placed an aminoacyl-tRNA in the A site. GTP hydrolysis results in a conformational change in EF1α. Step 3: The first peptide bond has been formed, new peptidyl tRNA occupies a hybrid (A/P) site on the ribosome, and the acceptor stem of deacylated $tRNA_i^{met}$ is in the large subunit E site. Step 4: The mRNA–peptidyl tRNA complex has been translocated to the P site while deacylated initiator tRNA is moved to E site. (*b*) Further rounds of elongation. Step 5: Binding of aminoacyl-tRNA in A site causes release of deacylated tRNA from the E site. Step 6: Formation of peptide bond results in the new peptidyl RNA occupying a hybrid A/P site on ribosome. Step 7: Translocation moves mRNA and new peptidyl tRNA in register into P site. Additional amino acids are added by successive repetitions of the cycle. For further details see text.

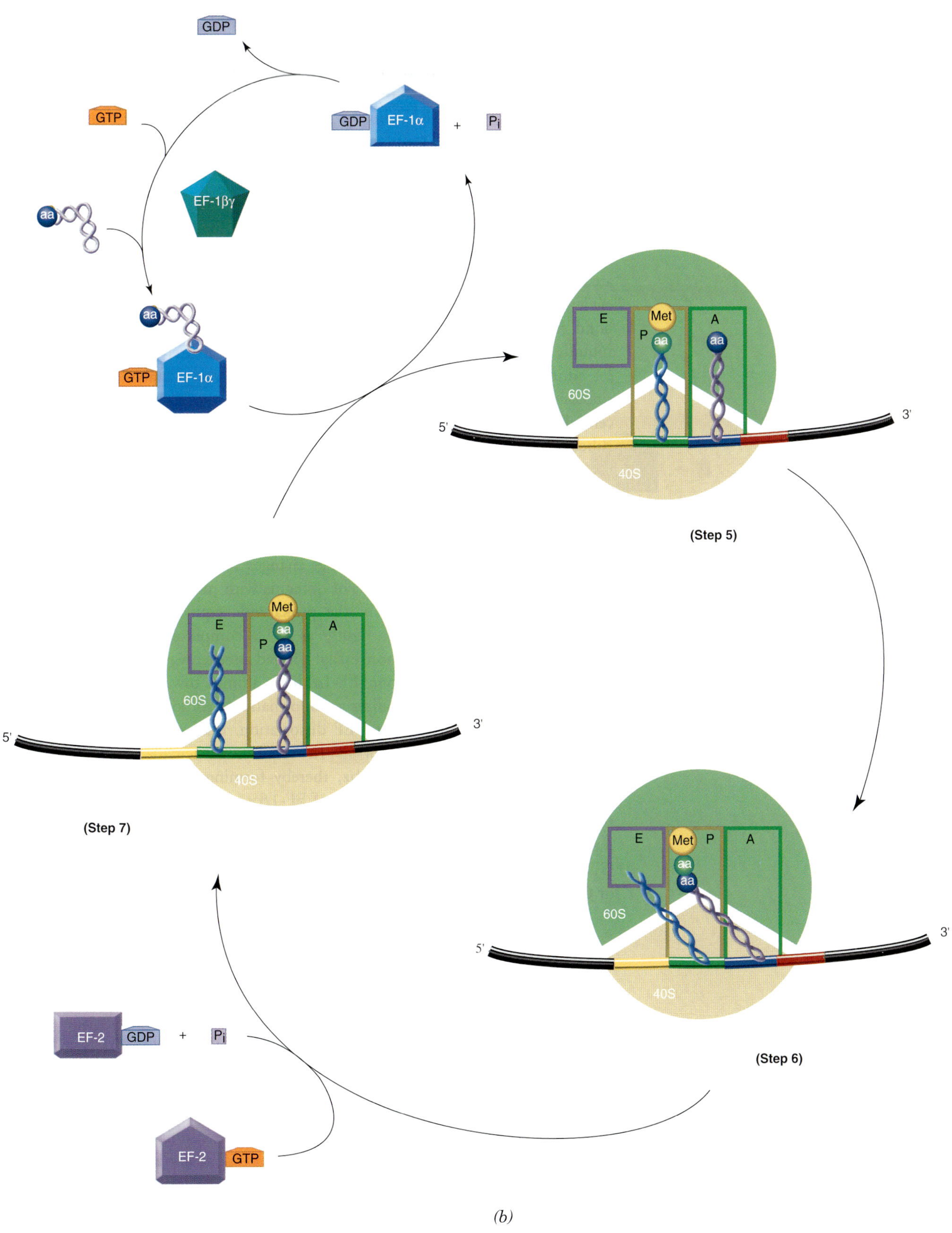

(b)

FIGURE 6.8
(continued)

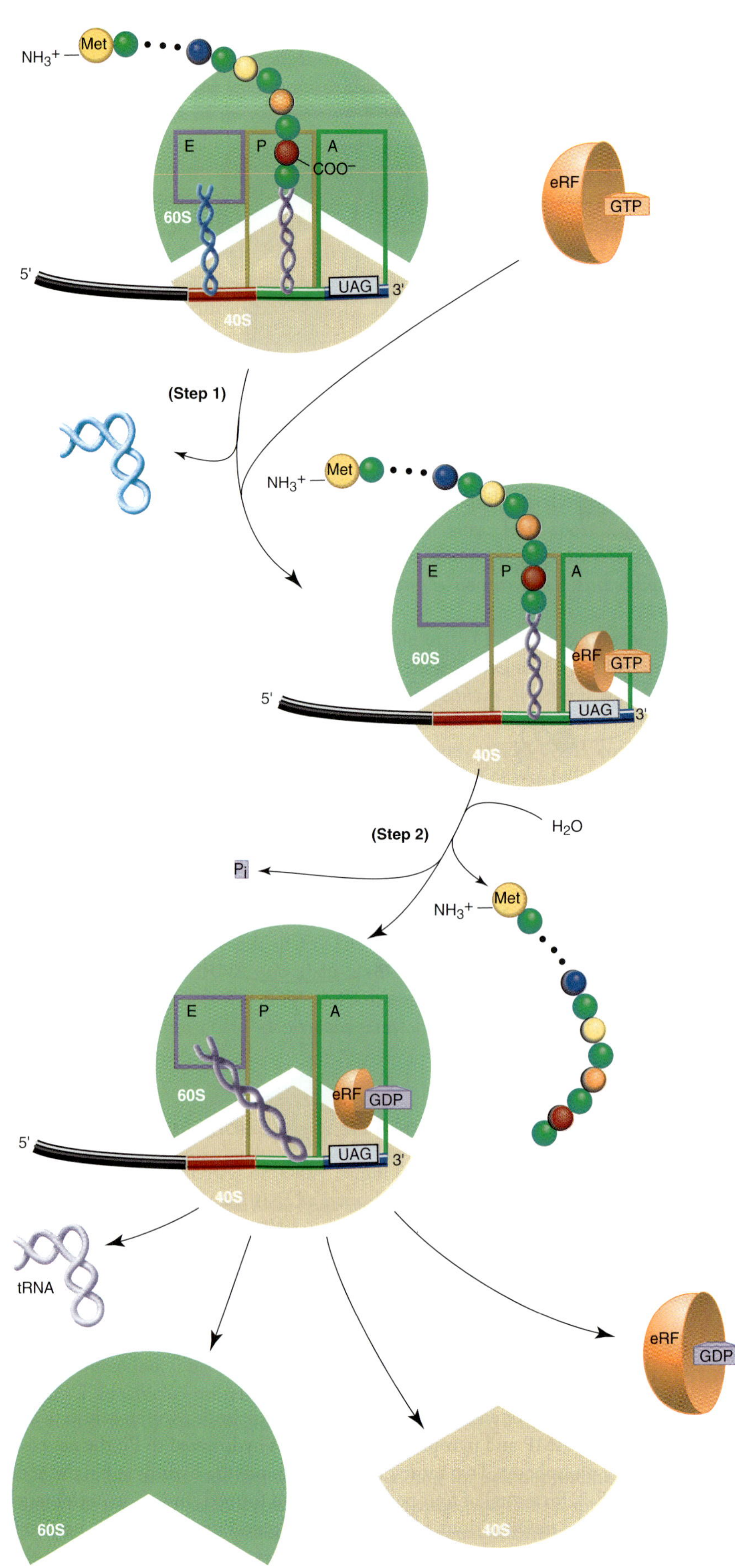

FIGURE 6.10

Termination of protein biosynthesis. When a termination codon in mRNA occupies the A site, binding of a release factor-GTP complex occurs (Step 1) and deacylated tRNA is released. Step 2: peptidyl transferase acts as a hydrolase; hydrolysis of the ester bond linking protein to tRNA releases the protein. The acceptor end of deacylated tRNA is probably displaced, GTP is hydrolyzed, and release factor-GDP dissociates. Dissociated components can now enter additional rounds of protein synthesis.

Protein Synthesis in Mitochondria Differs Slightly

Many characteristics of the mitochondrion suggest that it is a descendant of an aerobic prokaryote that was engulfed and formed a symbiotic relationship within a eukaryotic cell. Some independence and prokaryotic character are retained. Human mitochondria have a circular DNA genome of 16,569 bp that encodes 13 proteins, 22 tRNA species, and mitochondrion-specific 12 S and 16 S rRNA species. Their apparatus for protein synthesis includes RNA polymerase, aminoacyl-tRNA synthetases, tRNAs, and ribosomes that are unique to the mitochondrion, but some details of translation differ from translation in the cytosol. Mitochondrial ribosomes are smaller and the rRNAs are shorter than those of either the eukaryotic cytosol or of prokaryotes (Table 6.7). An initiator Met-tRNA$_i^{met}$ is modified by a **transformylase** that uses N^{10}-formyl H_4-folate to produce **fMet-tRNA$_i^{met}$**. The genetic code is also slightly different (Table 6.2). Most mitochondrial proteins are encoded in nuclear DNA, synthesized in the cytosol, and imported into the organelle. Cells must coordinate protein synthesis within mitochondria with the cytosolic synthesis of proteins destined for import into mitochondria. The importance of mitochondrial protein synthesis is indicated in Clinical Correlation 6.5.

Some Antibiotics and Toxins Inhibit Protein Biosynthesis

Protein biosynthesis is central to the life and reproduction of cells. Because an organism can gain a biological advantage by interfering with the ability of its competitors to synthesize proteins, many antibiotics and toxins function in this way. Some are selective for prokaryotic protein synthesis and thus are extremely useful in clinical practice. Examples showing several mechanisms of antibiotic action are listed in Table 6.8.

Streptomycin binds the small subunit of prokaryotic ribosomes, interferes with the initiation of protein synthesis, and causes misreading of mRNA (see Clin. Corr. 6.5). Streptomycin alters interactions of tRNA with the ribosomal subunit and mRNA, probably by affecting subunit conformation. Mutations in a ribosomal protein or the rRNA of the small subunit can confer streptomycin resistance or dependence. Other aminoglycoside antibiotics, such as the **neomycins** and **gentamycins**, also cause mistranslation; they interact with the small ribosomal subunit, but at different sites than streptomycin. **Kasugamycin** inhibits the initiation of translation; sensitivity to kasugamycin depends on base methylation that normally occurs on two adjacent adenines of small subunit rRNA. **Tetracyclines** bind to ribosomes and interfere in aminoacyl-tRNA binding. **Puromycin** (Figure 6.11) resembles an aminoacyl-tRNA; it binds at the large subunit A site and acts as an acceptor of the nascent peptide in the peptidyl transferase reaction. Peptidyl-puromycin is not translocated and it cannot serve as a peptide donor. Translation is prematurely terminated, and peptidyl-puromycin is released. **Chloramphenicol** inhibits peptidyl transferase by binding at the transferase center; no transfer occurs, and peptidyl-tRNA remains associated with the ribosome. **Erythromycin**, a macrolide antibiotic, interferes with translocation on prokaryotic ribosomes.

Eukaryotic translocation is inhibited by **diphtheria toxin**, a protein produced by *Corynebacterium diphtheriae*; the toxin binds at the cell membrane and a subunit

CLINICAL CORRELATION 6.5

Mutation in Mitochondrial Ribosomal RNA Results in Antibiotic-Induced Deafness

Instances of permanent deafness have been linked to use of normally safe and effective amounts of aminoglycoside antibiotics such as streptomycin, paromomycin, and gentamycin. The unusual sensitivity to aminoglycosides is maternally transmitted, suggesting a mitochondrial locus (sperm do not contribute mitochondria to the zygote). Aminoglycosides target bacterial ribosomes, so the mitochondrial ribosome is a logical place to look for a mutation site. A single A → G point mutation in mitochondrial DNA at nucleotide 1555 of the 12 S rRNA gene has been identified in families with this aminoglycoside susceptibility. The mutation occurs in a highly conserved region of the rRNA sequence that is involved in aminoglycoside binding; some mutations in the same region confer resistance to the antibiotics, and the RNA region is part of the ribosomal decoding site. It is likely that the mutation increases affinity for aminoglycosides and their ability to inhibit protein synthesis. Reduced mitochondrial protein synthesis decreases levels of enzymes of the oxidative phosphorylation system, so affected cells become starved of ATP.

Source: Fischel-Ghodsian, N., Prezant, T., Bu, X., and Öztas, S. Mitochondrial ribosomal RNA gene mutation in a patient with sporadic aminoglycoside ototoxicity. *Am. J. Otolaryngol.* 14:399, 1993; Guan, M.-X., Fischel-Ghodsian, N., and Attardi, G. A biochemical basis for the inherited susceptibility to aminoglycoside ototoxicity. *Hum. Mol. Genet.* 9:1787, 2000.

TABLE 6.8 Some Inhibitors of Protein Biosynthesis

Inhibitor	*Processes Affected*	*Site of Action*
Streptomycin	Initiation, elongation	Prokaryotes: 30S subunit
Neomycins	Translation	Prokaryotes: multiple sites
Tetracyclines	Aminoacyl-tRNA binding	30S or 40S subunits
Puromycin	Peptide transfer	70S or 80S ribosomes
Erythromycin	Translocation	Prokaryotes: 50S subunit
Fusidic acid	Translocation	Prokaryotes: EF-G
Cycloheximide	Elongation	Eukaryotes: 80S ribosomes
Ricin	Multiple	Eukaryotes: 60S subunit

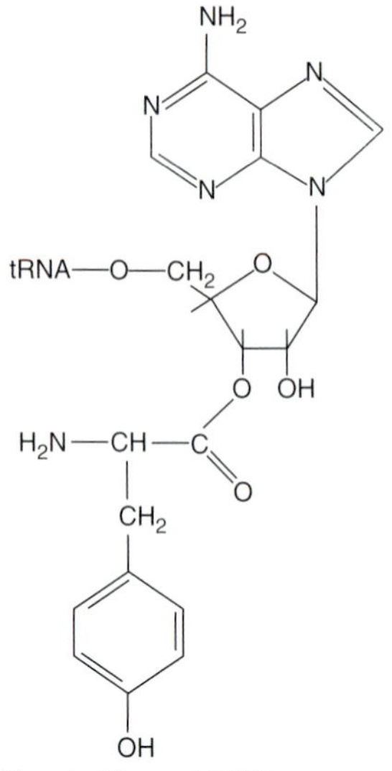

3' end of tyrosyl-tRNA

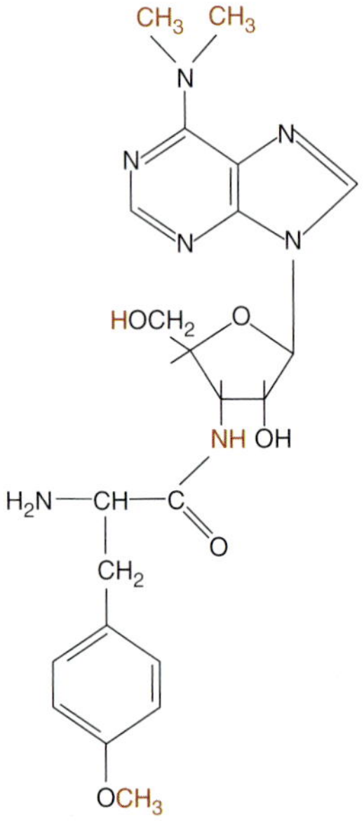

Puromycin

FIGURE 6.11

Puromycin (bottom) terminates translation by binding at the A site and acting as an analog of aminoacyl-tRNA [here tyrosyl-tRNA (top)]. The α-amino group is an acceptor in the peptidyltransferase reaction, and peptidyl-puromycin dissociates from the ribosome.

enters the cytoplasm and catalyzes the ADP-ribosylation and inactivation of EF-2, as represented in the reaction

$$\underset{\text{(active)}}{\text{EF-2} + \text{NAD}} \rightleftharpoons \underset{\text{(inactive)}}{\text{ADP-ribosyl EF-2}} + \text{nicotinamide} + \text{H}^+$$

ADP-ribose is attached to EF-2 at a **diphthamide** residue, which is a posttranslationally modified form of histidine (see Figure 6.20, p. 235).

Ricin (from castor beans) and related toxins are *N*-glycosidases that cleave a single adenine from the large subunit rRNA backbone. A fungal toxin, α-**sarcin**, cleaves large subunit rRNA at a single site; in both cases the ribosome is inactivated. Some *E. coli* strains make extracellular toxins that affect other bacteria. One of these, **colicin E3**, is a ribonuclease that cleaves 16 S RNA near the mRNA-binding sequence and decoding site; it inactivates the small subunit and halts protein synthesis in competitors of the colicin-producing cell.

6.4 PROTEIN MATURATION: FOLDING, MODIFICATION, SECRETION, AND TARGETING

Some proteins emerge from the ribosome ready to function, while others undergo a variety of **posttranslational modifications**. The result may be conversion to a functional form, direction to a specific subcellular compartment, secretion from the cell, or an alteration in activity or stability. Information that determines the posttranslational fate of a protein resides in its structure; the sequence and conformation of the protein determine whether it will be a substrate for a modifying enzyme and/or identify it for direction to a subcellular or extracellular location.

Chaperones Aid in Protein Folding

Many proteins can spontaneously generate their native conformation. For example, fully denatured pancreatic ribonuclease can, under appropriate conditions, refold and generate correct disulfide bridges and full activity. Many newly synthesized proteins fold correctly as they emerge from the ribosome. However, other proteins require the assistance of **chaperones** (see p. 110), which reversibly bind hydrophobic regions of nascent proteins and proteins in an intermediate stage of folding. Chaperones can stabilize intermediates, maintain proteins in an unfolded state to allow passage through membranes, help unfold misfolded segments, prevent formation of incorrectly folded intermediates, and prevent aggregation or other inappropriate interactions with other proteins. These actions can help a protein attain its functional, usually most compact conformation. Failure to fold correctly usually leads to rapid protein degradation (see Clin. Corr. 6.6). Accumulation of misfolded proteins can result in protein aggregation and serious disease (see Clin. Corr. 6.7).

Proteins for Export Follow the Secretory Pathway

Proteins destined for export are synthesized on membrane-bound ribosomes of the **rough endoplasmic reticulum** (**ER**) (Figure 6.12). Such proteins have a hydrophobic **signal peptide**, usually at or near the amino terminus. There is no unique sequence; signal peptide characteristics include a positively charged N-terminal region and a core of 8–12 hydrophobic amino acids in an α-helix followed by a more polar C-terminal segment that eventually serves as a cleavage site for excision of the signal peptide. Initiation and elongation begin on free cytosolic ribosomes, and as the signal peptide of 15–30 amino acids emerges from the ribosome, it binds a **signal recognition particle** (**SRP**) (see Figure 6.13). SRP is composed of six different proteins and a small (7 S) RNA molecule. The signal peptide binds in a hydrophobic pocket of the SRP with the positively charged N-terminal segment in contact with SRP RNA. Translation halts and the complex

CLINICAL CORRELATION **6.6**

Deletion of a Codon, Incorrect Posttranslational Modification, and Premature Protein Degradation: Cystic Fibrosis

Cystic fibrosis is the most common autosomal recessive disease in Caucasians, with a frequency of about 1 per 2000. The CF gene is 230 kb long and includes 27 exons that encode a protein of 1480 amino acids. The protein known as the cystic fibrosis transmembrane conductance regulator or CFTR (see p. 479) is a member of a family of ATP-dependent transport proteins. It contains two membrane-spanning domains, each having six transmembrane regions, two ATP-binding domains, and one regulatory domain that includes several phosphorylation sites. CFTR functions as a cyclic AMP-regulated chloride channel. CF epithelia are characterized by defective electrolyte transport. The organs most strongly affected include the lungs, pancreas, and liver, and the most life-threatening effects involve thick mucous secretions that lead to chronic obstructive lung disease and persistent infections of lungs.

In about 70% of patients, there is a deletion of the three nucleotides that encode phenylalanine 508, normally located in ATP-binding domain 1 on the cytoplasmic side of the plasma membrane. As with several other CF mutations, the Phe 508-deletion protein does not fold correctly in the ER, and it is not properly glycosylated or transported to the cell surface. Instead, it is returned to the cytoplasm to be degraded within proteasomes. One therapeutic approach, not yet applied to patients, uses drugs that foster chaperone interactions with mutant CFTR and aid in its folding and transport to the membrane.

Source: Ward, C., Omura, S., and Kopito, R. Degradation of CFTR by the ubiquitin–proteasome pathway. *Cell* 83:121, 1995. Plemper, R. K. and Wolf, D. H. Retrograde protein translocation: Eradication of secretory proteins in health and disease. *Trends Biochem. Sci.* 24:266, 1999. Egan, M. E., Pearson, M., Weiner, S. A., Rajendran, V., Rubin, D., Glockner-Pagel, J., et al. Curcumin, a major constituent of turmeric, corrects cystic fibrosis defects. *Science* 304:600, 2004.

moves to the ER where the SRP recognizes and binds to an **SRP receptor** or **docking protein**, on the cytosolic surface of the ER membrane. The ribosome is transferred to a **translocon**, a ribosome receptor that crosses the membrane and serves as a passageway through the membrane. The SRP is released, relieving the translational block. The translocon passageway opens to allow transit of the nascent protein, the hydrophobic signal sequence is inserted, and translation and extrusion into or through the translocon are now coupled. Even very hydrophilic or charged segments are directed through the membrane into the ER lumen. The signal peptide is excised by **signal peptidase**, an integral membrane protein at the luminal surface of the ER. The protein can fold into a three-dimensional conformation, either co- or posttranslationally. Disulfide bonds can form, and components of multisubunit proteins may assemble. Other steps may include proteolytic processing and glycosylation that occur within the ER lumen and during transit of the protein through the Golgi apparatus and into secretory vesicles.

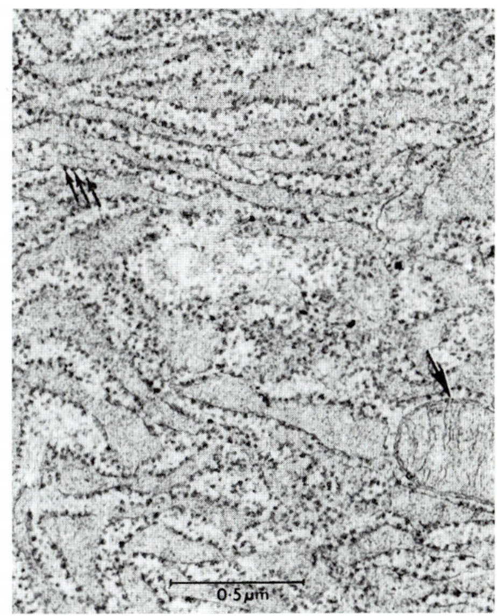

FIGURE 6.12

Rough endoplasmic reticulum. Three parallel arrows indicate three ribosomes among the many attached to the membranes. Single arrow indicates a mitochondrion for comparison.
Courtesy of Dr. U. Jarlfors, University of Miami.

Glycosylation of Proteins Occurs in the Endoplasmic Reticulum and Golgi Apparatus

Glycosylation of proteins to form glycoproteins (see p. 345) is important for many reasons. Glycosylation alters the properties of proteins, changing their stability, solubility, and physical bulk. In addition, the carbohydrate moieties act as recognition signals that can direct protein targeting and influence cell–cell interactions and development of the organism. Glycosylation sites are determined by the type of amino acid, its neighboring sequence in the protein, and the availability of enzymes and substrates for the reactions. Up to 100 different **glycosyltransferases**, classes of which are summarized in Table 6.9, carry out a similar basic reaction in which a sugar is transferred from an activated donor substrate to an appropriate acceptor, usually another sugar residue that is part of an oligosaccharide under construction. The enzymes show specificity for the **monosaccharide** that is transferred, the structure and sequence of the **acceptor**, and the site and configuration of the **anomeric linkage** formed.

CLINICAL CORRELATION 6.7

Protein Misfolding and Aggregation: Creutzfeldt–Jacob Disease, Mad Cow Disease, Alzheimer's Disease, and Huntington's Disease

Misfolded proteins are usually degraded, as occurs with mutant CFTR protein in cystic fibrosis (see Clin. Corr. 6.6). Several neurological diseases illustrate the result of cellular accumulation of aggregates of misfolded proteins or their partial-degradation products.

Prions (proteinaceous infectious agents) lack nucleic acid but have some characteristics of viral or microbial pathogens. The first prion-caused disease identified was Scrapie, a disorder of sheep. Human disorders include Kuru, Creutzfeldt–Jacob disease (CJD), and Variant CJD, arising from bovine spongiform encephalopathy (BSE), popularly known as "mad cow disease" in cattle. These diseases vary in the course and rate of development and can be inherited or spontaneous or can result from infection, but all cause neurodegeneration and eventual death. Autopsy shows spongiform degeneration of brain and the intracellular accumulation of rod-shaped deposits of amyloid plaque.

Disease results from a conformational transition in a normal 254 residue prion protein, designated PrP^{C}. The protein is found mostly on the outer surface of neurons, but its function is still unclear. Structural studies show it to be rich in α-helix and devoid of β-sheet; it is soluble in mild detergent solutions and has little tendency to form large aggregates. The infectious form of prion protein, designated PrP^{Sc}, is rich in β-sheet and is much less soluble. A protease-resistant fragment is particularly likely to aggregate and form plaque, and several point mutations in PrP^{C} are associated with formation of infectious PrP^{Sc}. Infection occurs by consumption of neural tissue that contains PrP^{Sc}. Ritual consumption of human brain by the Fore in New Guinea spread kuru, and inclusion of infected offal in animal feed helped spread Scrapie and BSE. Humans have acquired variant CJD by eating BSE-infected beef. PrP^{Sc} acts by altering the conformation of normal cellular PrP^{C}; PrP^{Sc} appears to function with or as a chaperone, fostering adoption of the PrP^{Sc} conformation in existing or newly synthesized PrP^{C}. The protease-resistant form and products of its partial proteolysis accumulate and form the aggregates characteristic of the disease (but not necessarily the cause of its symptoms).

Alzheimer's disease is characterized by progressive memory loss and cognition, and it is most common in elderly people. Mitochondrial disfunction in neural cells is prevalent. Alzheimer's disease is not caused by a prion or other external infectious agent, but it too is characterized by accumulation of intraneuronal and extracellular bundles and filaments that form plaques. The major component of the plaque is β-amyloid, a 39- to 43-residue peptide derived from a larger amyloid precursor protein. Other components are highly ubiquitinylated, but protease-resistant in compacted and tangled fibers.

Huntington's disease is a neurodegenerative disorder that is also associated with accumulation of protein aggregates. In this case, a variant of a cellular protein called huntingtin is involved; a CAG codon within the gene is expanded, resulting in polyglutamine repeats of up to 180 residues within huntingtin. Protein or peptides derived from the protein accumulate in inclusions in the nucleus, and transcription is affected. Several other neurodegenerative disorders are also linked to polyglutamine tract expansion and protein deposition.

Source: Baldwin, M. A., Cohen, F. A., and Prusiner, S. P. Prion protein isoforms, a convergence of biological and structural investigations. *J. Biol. Chem.* 270:19197, 1995. Weissmann, C. Molecular genetics of transmissible spongiform encephalopathies. *J. Biol. Chem.* 274:3, 1999. Aguzzi, A. and Haass, C. Games played by rogue proteins in prion disorders and Alzheimer disease. *Science* 302:814, 2003. Taylor, J. P., Hardy, J., and Fishbeck, K. H. Toxic proteins in neurodegenerative disease. *Science* 296:1991, 2002. Hardy, J. and Selkoe, D. J. The amyloid hypothesis of Alzheimer's disease: Progress and problems on the road to therapeutics. *Science* 297:353, 2002. Dunah, A. W., Jeong, H., Griffin, A., Kim, Y.-M., Standaert, D. G., Hersch, S. M., et al. Sp1 and TAFII130 transcriptional activity disrupted in early Huntington's disease. *Science* 296:2238, 2002.

In **N-linked glycosylation** an oligosaccharide is linked through the amide nitrogen of asparagine. (The antibiotic tunicamycin prevents *N*-glycosylation.) Formation of N-linked oligosaccharides begins in the ER lumen and continues in the Golgi apparatus. A specific sequence, **Asn-X-Thr** (or Ser) in which X can be any amino acid except proline or aspartic acid, is required. Not all Asn-X-Thr/Ser sequences are glycosylated because some may be unavailable to the glycosyltransferase.

N-linked glycosylation is initiated with a lipid-linked intermediate (Figure 6.14). **Dolichol phosphate** at the cytoplasmic surface of the ER membrane serves as glycosyl acceptor for *N*-acetylglucosamine. Stepwise glycosylation and formation of a branched $(Man)_5(GlcNAc)_2$-pyrophosphoryl-dolichol on the cytosolic side of the membrane follows. This intermediate is then **reoriented** to the lumenal surface of the ER membrane, and four additional mannose and three glucose residues are added. The complete oligosaccharide is then transferred from its dolichol carrier to an asparagine residue of the polypeptide as it emerges into the ER lumen. Thus, *N*-glycosylation is

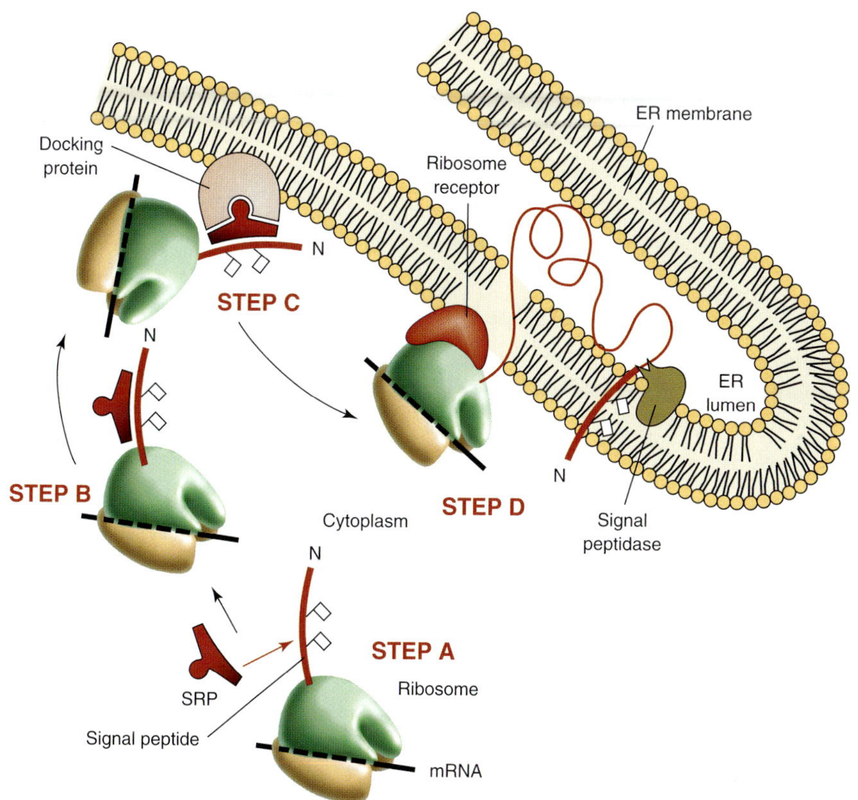

FIGURE 6.13

Secretory pathway: signal peptide recognition. Step A: A hydrophobic signal peptide emerges from a free ribosome in the cytosol. Step B: Signal recognition particle (SRP) binds signal peptide and elongation is temporarily halted. Step C: The ribosome moves to ER membrane where docking protein binds SRP. Step D: Ribosome is transferred to a translocon, elongation is resumed, and newly synthesized protein is extruded through the membrane into the ER lumen.

TABLE 6.9 Glycosyltransferases in Eukaryotic Cells

Sugar Transferred	*Abbreviation*	*Donor*	*Glycosyltransferase*
Mannose	Man	GDP-Man Dolichol-Man	Mannosyltransferase
Galactose	Gal	UDP-Gal	Galactosyltransferase
Glucose	Glc	UDP-Glc Dolichol-Glc	Glucosyltransferase
Fucose	Fuc	GDP-Fuc	Fucosyltransferase
N-Acetylgalactosamine	GalNAc	UDP-GalNac	*N*-acetylgalactosaminyltransferase
N-Acetylglucosamine	GlcNAc	UDP-GlcNAc	*N*-acetylglucosaminyltransferase
N-Acetylneuraminic acid (or sialic acid)	NANA or NeuNAc SA	CMP-NANA CMP-SA	*N*-Acetylneuraminyltransferase (sialyltransferase)

cotranslational; it occurs as the protein is being synthesized and it can affect protein folding.

Glycosidases now remove some sugar residues from the newly transferred oligosaccharide. Glucose residues, which were required for transfer of the oligosaccharide from the dolichol carrier, are removed. The folded glycoprotein is then transported to the Golgi apparatus where further trimming may occur or additional sugars can be added. The resulting N-linked oligosaccharides are diverse, but two classes are distinguishable. Each has a common core region ($GlcNAc_2Man_3$) linked to asparagine and originating from the dolichol-linked intermediate. The **high-mannose type** includes mannose residues in a variety of linkages and shows less processing from the dolichol-linked intermediate. The **complex type** is more highly processed and diverse, with a larger variety of sugars and linkages (see Figure 6.15 for examples).

In **O-linked glycosylation**, oligosaccharides are bound through hydroxyl groups of serine or threonine in proteins (see p. 345). O-linked glycosylation occurs only after the protein has reached the Golgi apparatus; hence *O*-glycosylation is **posttranslational**

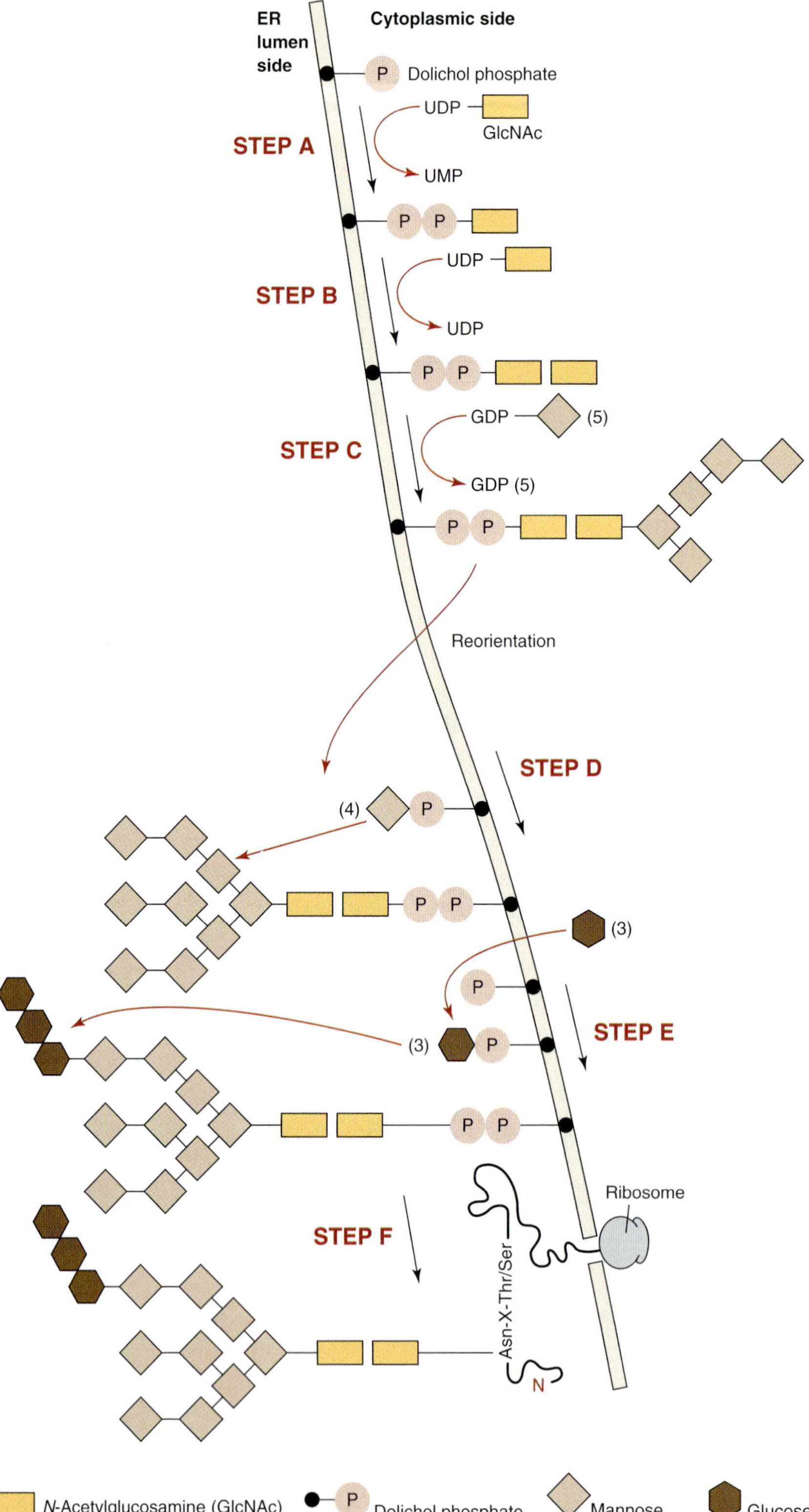

FIGURE 6.14

Biosynthesis of N-linked oligosaccharides at the surface of the endoplasmic reticulum. Step A: Synthesis begins on the cytoplasmic face of the ER membrane with transfer of *N*-acetylglucosamine phosphate to a dolichol acceptor. Step B: Formation of the first sugar-to-sugar glycosidic bond occurs upon transfer of a residue of *N*-acetylglucosamine. Step C: Five residues of mannose (from GDP mannose) are added sequentially and the lipid-linked oligosaccharide is reoriented to the lumenal side of the membrane. Step D: Additional mannose and (step E) glucose residues are transferred from dolichol-linked intermediates. Dolichol-sugars are generated from cytosolic nucleoside diphosphate sugars. Step F: The completed oligosaccharide is transferred to a nascent polypeptide at the membrane surface; signal peptide may have already been cleaved at this point.

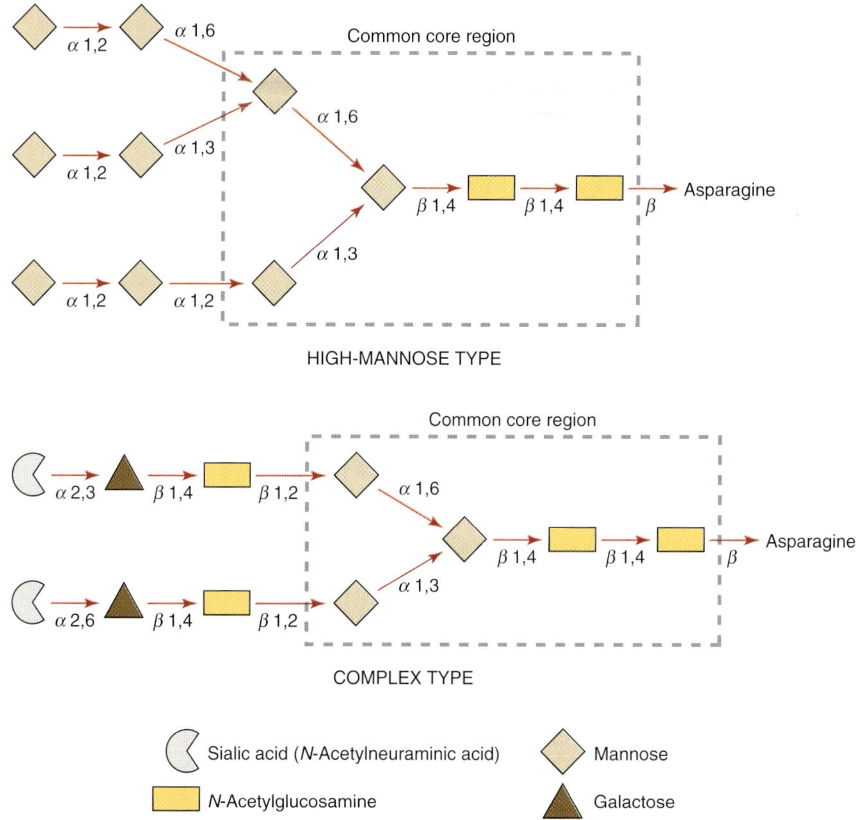

FIGURE 6.15

Structure of N-linked oligosaccharides. Basic structures of both types of N-linked oligosaccharides are shown. In each case the structure is derived from the initial dolichol-linked oligosaccharide through action of glycosidases and glycosyltransferases. Note the variety of glycosidic linkages involved in these structures.

and occurs only on folded proteins. There is no specific amino acid sequence in which the serine or threonine must occur, but only residues on the protein surface serve as acceptors for the GalNAc-transferase that attaches *N*-acetyl galactosamine. Stepwise addition of sugars to the GalNAc acceptor follows. The oligosaccharides synthesized depend on the types and amounts of glycosyltransferases in a given cell. If an acceptor is a substrate for more than one transferase, the amounts of each transferase influence the competition between them. Oligosaccharide growth proceeds until structures are formed that are not acceptors for any of the glycosyltransferases present. This can result in different oligosaccharide structures on otherwise identical polypeptides, so **heterogeneity** in glycoproteins is common (see Figure 6.16).

6.5 MEMBRANE AND ORGANELLE TARGETING

Protein transport from the ER to and through the Golgi apparatus and beyond uses carrier vesicles. Only proteins correctly folded are recognized as **cargo** for transport, and chaperones in the ER assist folding and foster correct disulfide formation. Misfolded or damaged proteins are exported to the cytosol for degradation. Sorting of proteins for their ultimate destinations occurs in conjunction with their glycosylation and proteolytic trimming as they pass through the cis, medial, and trans elements of the Golgi apparatus. Families of vesicle and receptor proteins provide specificity for membrane targeting and fusion.

Sorting of Proteins in the Secretory Pathway

Targeting of specific glycoproteins to **lysosomes** is well understood. In the cis Golgi some aspect of tertiary structure causes lysosomal proteins to be recognized by a glycosyltransferase that attaches *N*-acetylglucosamine phosphate (GlcNAc-P) to high-mannose-type oligosaccharides. A glycosidase then removes the GlcNAc, forming an

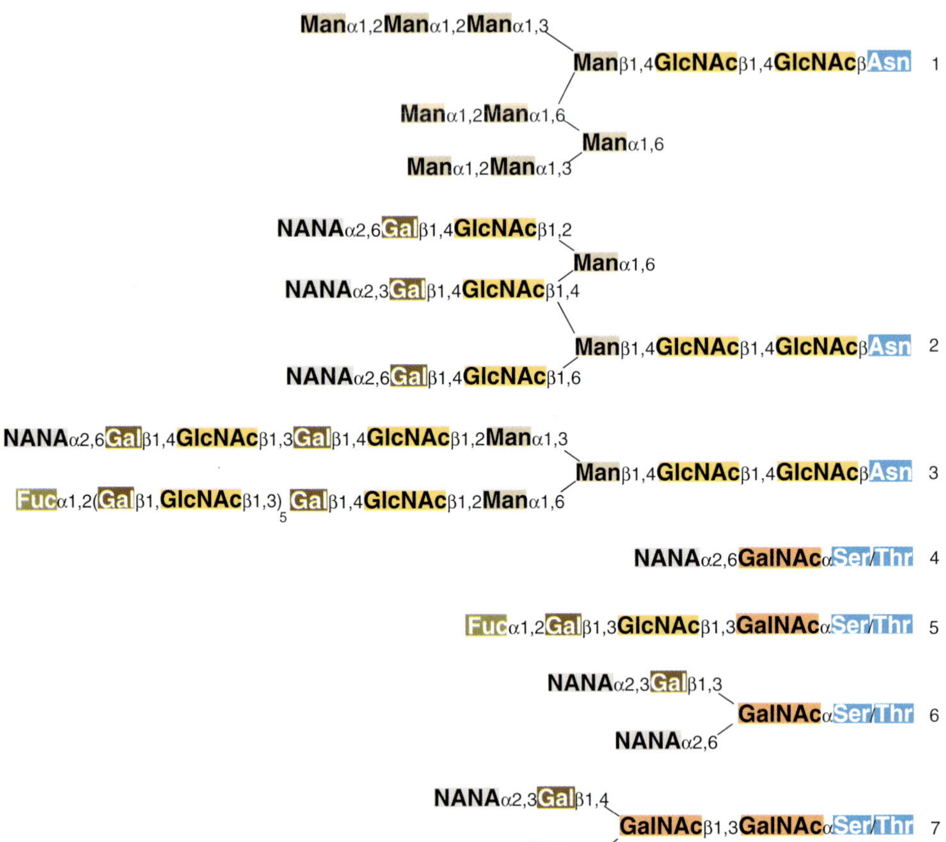

FIGURE 6.16

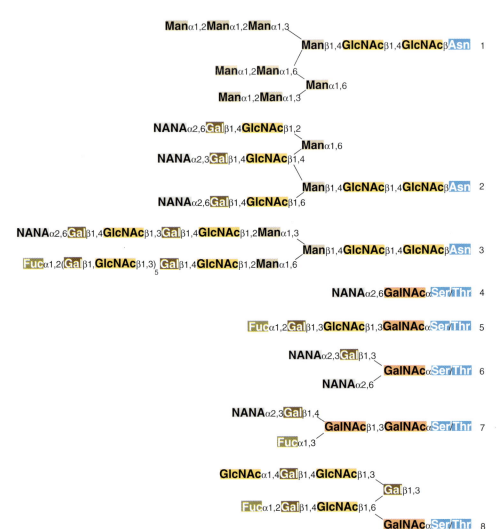

Examples of oligosaccharide structure. Structures 1–3 are typical N-linked oligosaccharides of high-mannose (1) and complex types (2, 3); note the common core structure from the protein asparagine residue through the first branch point. Structures 4–8 are common O-linked oligosaccharides that may be simple or highly complex. Note that although the core structure (GalNAc-Ser/Thr) is unlike that of N-linked oligosaccharides, the termini can be quite similar (e.g., structures 2 and 6, 3, and 7). Abbreviations: Man = mannose; Gal = galactose; Fuc = fucose; GlcNAc = *N*-acetylglucosamine; GalNAc = *N*-acetylgalactosamine; NANA = *N*-acetylneuraminic acid (sialic acid). Adapted from Paulson, J. *Trends Biochem. Sci.* 14:272, 1989

oligosaccharide that contains **mannose 6-phosphate** (Figure 6.17) that is responsible for compartmentation and vesicular transport of these proteins to lysosomes. Other oligosaccharide chains on the proteins may be further processed to form complex-type structures, but the mannose 6-phosphate determines their lysosomal destination. Patients with **I-cell disease** lack the GlcNAc-P glycosyltransferase, cannot mark lysosomal enzymes for their destination, and secrete them from the cell (see Clin. Corr. 6.8).

Other sorting signals, some not yet deciphered, determine how proteins are directed to Golgi subcompartments, various storage and secretory granules, and specific elements of the plasma membrane. For example, soluble proteins are retained in the ER lumen in response to a C-terminal **KDEL** (Lys-Asp-Glu-Leu) sequence. A different sequence in an exposed C-terminus signals retention in the ER membrane. Some transmembrane domains result in retention in the Golgi. Polypeptide-specific glycosylation and sulfation of some glycoprotein hormones in the anterior pituitary mediates their sorting into storage granules. Polysialic acid modification of a neural cell adhesion protein is specific to the protein and regulated in development.

The secretory pathway directs proteins to lysosomes, to the plasma membrane, or for secretion from the cell. Proteins of the ER and Golgi apparatus are targeted through partial use of the pathway. For example, localization of proteins on either side of or spanning the ER membrane can utilize the signal recognition particle in slightly different

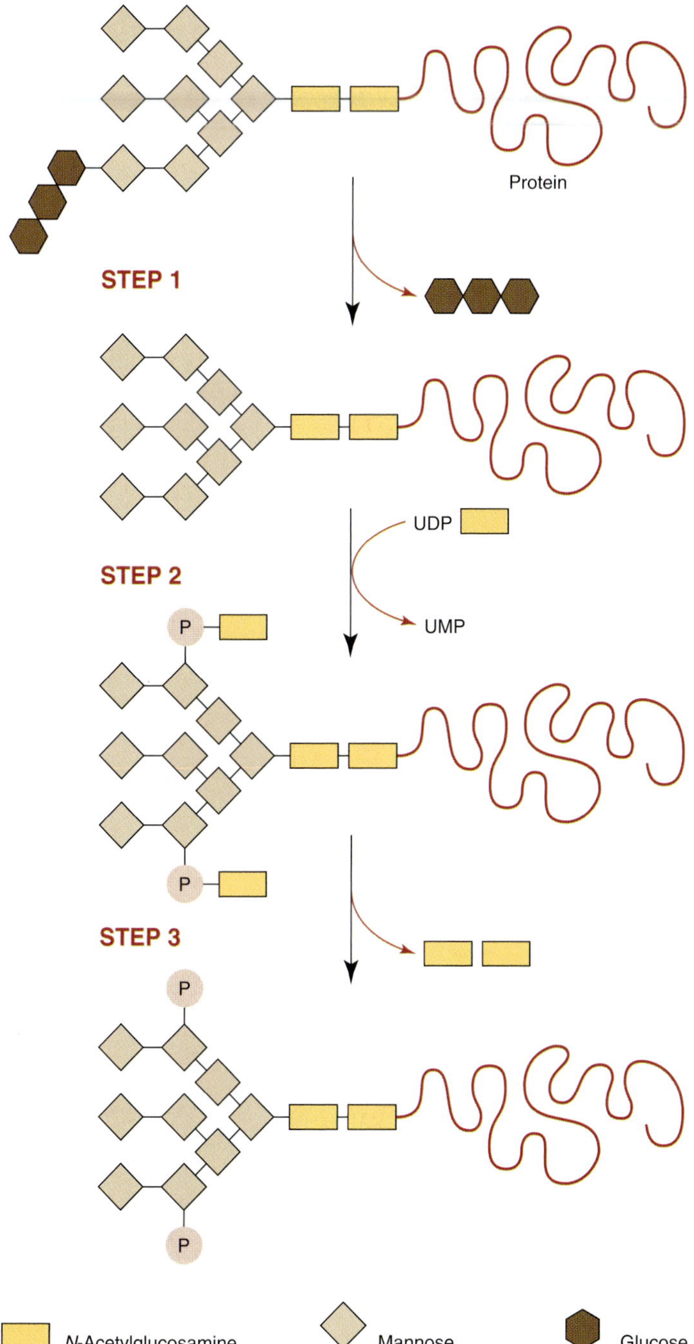

FIGURE 6.17

Targeting of enzymes to lysosomes. Complete folded N-linked glycoprotein is released from the ER membrane and, prior to transport to the Golgi apparatus, glycosidases remove glucose residues (Step 1). A mannose residue may also be removed. Step 2: In the Golgi apparatus a glycosyl transferase links one or sometimes two *N*-acetylglucosamine phosphate residues to the oligosaccharide. Step 3: A glycosidase removes *N*-acetylglucosamine, leaving one or two mannose 6-phosphate residues on the oligosaccharide. The protein is then recognized by a mannose 6-phosphate receptor and directed to vesicles that are targeted to lysosomes.
Adapted from Kornfeld, R. and Kornfeld, S. *Annu. Rev. Biochem.* 54:631, 1985.

ways (Figure 6.18). Sometimes a signal sequence is not near the amino terminus of the protein but somewhere downstream where it is bound by SRP. The amino end is not inserted into the ER membrane but remains on the cytoplasmic surface, and the signal sequence and C-terminal segment are passed into and through the translocon. Hydrophobic **anchoring sequences** in a protein can embed in the membrane yet allow much of the sequence to remain at the cytosolic surface or to be retained on the luminal surface of the ER membrane. Multiple anchoring sequences in a single polypeptide can cause it to span the membrane several times, and thus it will be largely buried in the membrane. Such hydrophobic sequences are separated by polar loops whose orientation is determined by positively charged flanking residues that predominate on the cytoplasmic side of the membrane.

CLINICAL CORRELATION 6.8
Diseases of Lysosome Function

I-cell disease (mucolipidosis II) and pseudo-Hurler polydystrophy (mucolipidosis III) are related diseases that arise from defects in lysosomal enzyme targeting because of deficiency of the enzyme that transfers *N*-acetylglucosamine phosphate to the high-mannose-type oligosaccharides of proteins destined for the lysosome. Fibroblasts from affected individuals show dense inclusion bodies (hence I-cells) and lack multiple lysosomal enzymes that are secreted into the medium. Abnormally high levels of lysosomal enzymes are present in plasma and other body fluids of patients. The disease is characterized by severe psychomotor retardation, many skeletal abnormalities, coarse facial features, and restricted joint movement. Symptoms are usually present at birth and progress until death, usually by age 8. In Pseudo-Hurler polydystrophy, onset is usually delayed until the age of 2–4 years, the disease progresses more slowly, and patients survive into adulthood. Prenatal diagnosis of both diseases is possible, but there is as yet no definitive treatment.

Mucopolysaccharidosis type I is caused by defects in α-L-iduronidase, a lysosomal enzyme that acts in the degradation of glycosaminoglycans. More than 70 mutations have been identified, and symptoms vary from severe mental retardation and early childhood death (Hurler syndrome) to milder effects including hearing loss and corneal clouding, but normal intelligence and lifespan. Trials of enzyme replacement therapy using recombinant normal human enzyme show promise.

Source: Kornfeld, S. Trafficking of lysosomal enzymes in normal and disease states. *J. Clin. Invest.* 77:1, 1986. Shields, D. and Arvan, P. Disease models provide insights into post-Golgi protein trafficking, localization and processing. *Curr. Opin. Cell Biol.* 11:489, 1999. Yogalingam, G., Guo, X.-L., Muller, V. J., Brooks, D. A., Clements, P. R., Kakkis, E. D., and Hopwood, J. J. Identification and molecular characterization of α-L-iduronidase mutations present in mucopolysaccharidosis type I patients undergoing enzyme replacement therapy. *Hum. Mutat.* 24:199, 2004. Wraith, J. E., Clarke, L. A., Beck, M., Kolodny, E. H., Pastores, G. M., Muenzer, J., et al. Enzyme replacement therapy for mucopolysaccharidosis I: A randomized, double-blinded, placebo-controlled, multinational study of recombinant human α-L-iduronidase (Laronidase). *J. Pediatr.* 144:581, 2004.

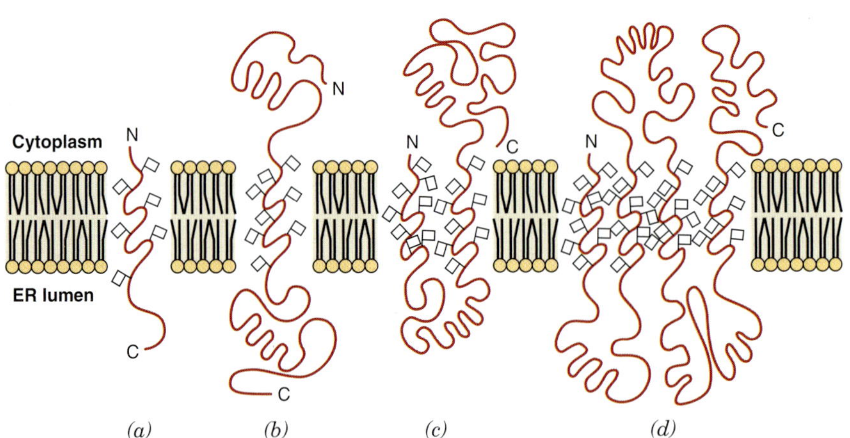

FIGURE 6.18

Topology of proteins at membranes of endoplasmic reticulum. Proteins are shown in several orientations with respect to membrane. In (*a*) the protein is anchored to luminal surface of membrane by an uncleaved signal peptide. In (*b*) the signal sequence is not at the N-terminus; a domain of the protein was synthesized before emergence of signal peptide. Insertion of internal signal sequence, followed by completion of translation, resulted in a protein with a cytoplasmic N-terminal domain, a membrane-spanning central segment, and a C-terminal domain in the ER lumen. (*c*) A protein with the opposite orientation: An N-terminal signal sequence, which might also have been cleaved by signal peptidase, resulted in extrusion of a segment of protein into the ER lumen. A hydrophobic anchoring sequence remained membrane associated and prevented further passage of protein through the membrane, thus formatting a C-terminal cytoplasmic domain. In (d) several internal signal and anchoring sequences allow various segments of the protein to be oriented on each side of membrane.

Import of Proteins by Mitochondria Requires Specific Signals

Mitochondria provide a complex targeting problem. Most mitochondrial proteins are synthesized in the cytosol on free ribosomes as larger **preproteins**. N-terminal **presequences** mark the protein for the mitochondrial matrix; the targeting signal is not a specific sequence, but rather a positively charged amphiphilic α-helix that is recognized by a mitochondrial receptor. The preproteins are transported (unfolded) with the aid of chaperones to the mitochondrion and are translocated across both membranes and into the mitochondrial matrix in an energy-dependent reaction. Passage occurs at adhesion sites where inner and outer membranes are close together, and several intermembrane proteins are involved (see Clin. Corr. 6.5). Proteases remove the matrix targeting signal, but may leave other sequences that further sort the protein within the mitochondrion to the inner or outer membrane or the intermembrane space.

For example, in response to a hydrophobic sequence a clipped precursor of cytochrome b_2 is moved back across the inner membrane where further proteolysis frees it in the intermembrane space. In contrast, cytochrome c apoprotein (without heme) binds at the outer membrane and is passed into the intermembrane space. There it acquires its heme and undergoes a conformational change that prevents return to the cytosol. Localization in the outer membrane can use the matrix targeting mechanism to translocate part of the protein, but a large apolar sequence blocks full transfer and leaves a membrane-bound protein with a C-terminal domain on the outer mitochondrial membrane surface.

Targeting to Other Organelles Requires Specific Signals

The nucleus must import proteins that are synthesized in the cytosol. Nuclear proteins are targeted by localization signals that include clusters of basic amino acids. They interact with carrier proteins (e.g., importins) that transport them through cylindrical **nuclear pore complexes** that span the nuclear membrane. Release from the carrier and export of the carrier to the cytosol require GTP hydrolysis. Peroxisomes also import proteins that are synthesized in the cytosol and transported with the help of chaperones. One targeting signal is a carboxy-terminal tripeptide, **Ser-Lys-Leu** (SKL). An N-terminal nonapeptide targeting signal also functions, and others may exist.

Some proteins reside in more than one subcellular compartment. A suboptimal localization signal can lead to inefficient targeting and a dual location, as in the partial secretion of an inhibitor of the plasminogen activator. A protein can contain two targeting signals, resulting in dual localization. Gene duplication and divergence can result in different targeting signals in closely related mature polypeptides. Alternative transcription initiation sites or alternative splicing of pre-mRNA can generate different messages from a single gene. Alternatively spliced mRNAs for calcium/calmodulin-dependent protein kinase differ with respect to the presence of an internal segment that encodes a nuclear localization signal. Without this segment, the protein remains in the cytosol.

Finally, the first initiation codon in an mRNA is sometimes bypassed. **Alternative translation initiation** leads to two forms of rat liver fumarase; one includes a mitochondrial targeting sequence, while the other does not and remains in the cytosol.

6.6 FURTHER POSTTRANSLATIONAL MODIFICATIONS

Additional maturation events can modify proteins to generate their final, functional forms. Some are very common, while others are highly restricted. Many reversible modifications regulate protein activity.

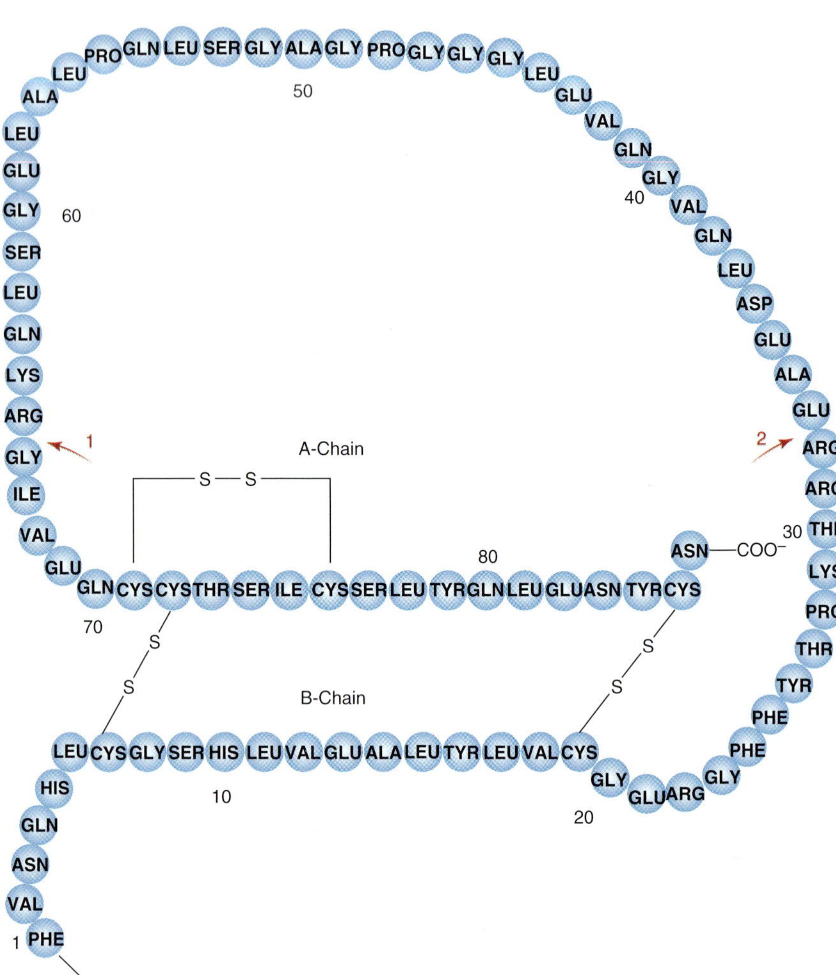

FIGURE 6.19

Maturation of human proinsulin. After cleavage at the two sites indicated by arrows, arginine residues 31, 32, and 65 and lysine residue 64 are removed to produce insulin and C-peptide.
Redrawn from Bell, G. I., Swain, W. F., Pictet, R., Cordell, B., Goodman, H. M., and Rutter, W. *J. Nature* 282:525, 1979.

Partial Proteolysis Releases Insulin and Activates Zymogens

Partial proteolysis is a common maturation step. Sequences can be removed from either end or from within the protein. Proteolysis in the ER and Golgi apparatus helps to mature the protein hormone insulin (Figure 6.19). Preproinsulin encoded by mRNA is inserted into the ER lumen. Signal peptidase cleaves the signal peptide to generate proinsulin, which folds to form the correct disulfide linkages. Proinsulin is transported to the Golgi apparatus, where it is packaged into secretory granules. An internal **connecting peptide** (**C peptide**) is removed by proteolysis, and mature insulin is secreted. In familial hyperproinsulinemia processing is incomplete (see Clin. Corr. 6.9). This pathway for insulin biosynthesis ensures production of equal amounts of A and B chains. Moreover, proinsulin folds into a conformation in which the cysteine residues are positioned for correct disulfide bond formation. Reduced and denatured proinsulin can refold correctly, while renaturation of reduced and denatured insulin is inefficient and leads to incorrect disulfide linkages.

Cleavage of precursor proteins is a common means of enzyme activation. Digestive proteases are classic examples (see p. 1053). Inactive **zymogen** precursors are packaged in storage granules and activated by proteolysis after secretion. Thus, **trypsinogen** is cleaved to give **trypsin** plus an amino-terminal hexapeptide, and **chymotrypsinogen** is cleaved to form **chymotrypsin** and two peptides (see p. 1053).

Amino Acids Can Be Modified After Incorporation into Proteins

Although 20 amino acids are encoded genetically, **posttranslational modification** leads to formation of many different amino acid derivatives in proteins. Modification may be permanent or easily reversible. While the number of modified amino acids in a protein may be small, they often have a major role in its function. Examples are listed in Table 6.10.

Protein amino-termini are frequently modified. Protein synthesis is initiated using methionine, but in many cases proteolysis removes one to several amino-terminal residues. The new amino-terminus is then often altered (e.g., by acetylation). The α subunits of signal transducing G-proteins (see p. 509) are derivatized with myristic or palmitic acid. In some proteins, amino-terminal glutamines spontaneously cyclize to form a pyroglutamyl residue. The amino-terminus can also be lengthened by addition of an amino acid (see Section 6.8 on protein degradation).

Interchain and intrachain disulfide bond formation is posttranslational and is catalyzed by **disulfide isomerase**. Disulfide formation can prevent unfolding of proteins and their passage across membranes, so it also becomes a means of localization. As in the case of insulin, disulfide bonds can covalently link separate polypeptides and be necessary for function. Cysteine modification also occurs; *S*-palmitoylation aids membrane interactions and trafficking of some proteins. The γ subunits of the heterotricyclic G-proteins (see p. 511) are modified by thioester linkage of an isoprenoid to a cysteine at or near the carboxy-terminus. Multiple sulfatase deficiency arises from reduced ability to carry out a posttranslational modification of cysteine (Clin. Corr. 6.10).

Acetylation and **methylation** of lysine ϵ-amino groups occurs in histones and modulates their interactions with DNA. A fraction of the H2A histone is also modified

TABLE 6.10 Modified Amino Acids in Proteins[a]

Amino Acid	*Modifications Found*
Amino terminus	Formylation, acetylation, aminoacylation, myristoylation, glycosylation
Carboxyl terminus	Methylation, glycosyl-phosphatidylinositol anchor formation, ADP-ribosylation
Arginine	*N*-Methylation, ADP-ribosylation
Asparagine	*N*-Glycosylation, *N*-methylation, deamidation
Aspartic acid	Methylation, phosphorylation, hydroxylation
Cysteine	Cystine formation, palmitoylation, linkage to heme, *S*-glycosylation, prenylation, ADP-ribosylation
Glutamic acid	Methylation, γ-carboxylation, ADP-ribosylation
Glutamine	Deamidation, cross-linking, pyroglutamate formation
Histidine	Methylation, phosphorylation, diphthamide formation, ADP-ribosylation
Lysine	*N*-acetylation, *N*-methylation, oxidation, hydroxylation, cross-linking, ubiquitination, allysine formation, biotinylation
Methionine	Sulfoxide formation
Phenylalanine	β-Hydroxylation and glycosylation
Proline	Hydroxylation, glycosylation
Serine	Phosphorylation, glycosylation, acetylation, α-formylglycine formation
Threonine	Phosphorylation, glycosylation, methylation
Tryptophan	β-Hydroxylation, dione formation
Tyrosine	Phosphorylation, iodination, adenylation, sulfonylation, hydroxylation

Source: Adapted from Krishna, R. G. and Wold, F. Post-translational modification of proteins. In: A. Meister (Ed.), *Advances in Enzymology*, Vol. 67. New York: Wiley-Interscience, 1993, pp. 265–298.

[a] The listing is not comprehensive and some of the modifications are very rare. Note that no derivatives of alanine, glycine, isoleucine, and valine have been identified in proteins.

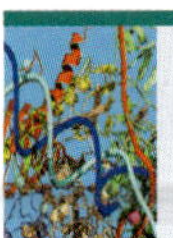

CLINICAL CORRELATION 6.9
Familial Hyperproinsulinemia

Familial hyperproinsulinemia, an autosomal dominant condition, results in approximately equal amounts of insulin and an abnormally processed proinsulin being released into the circulation. Although affected individuals have high concentrations of proinsulin in their blood, they are apparently normal in terms of glucose metabolism, being neither diabetic nor hypoglycemic. The defect was originally thought to result from a deficiency of one of three proteases that process proinsulin: (a) endopeptidases that cleave the Arg31-Arg32 and Lys64-Arg65 peptide bonds and (b) a carboxypeptidase. In several families the defect is the substitution of Arg65 by His or Leu, which prevents cleavage between the C-peptide and the A chain of insulin resulting in secretion of a partially processed proinsulin. In one family a point mutation (His10 → Asp10) causes the hyperproinsulinemia, but how this mutation interferes with processing is not known.

Source: Zhou, A., Webb, G., Zhou, X., and Steiner, D. F. Proteolytic processing in the secretory pathway. *J. Biol. Chem.* 274:20745, 1999.

CLINICAL CORRELATION **6.10**

Absence of Posttranslational Modification: Multiple Sulfatase Deficiency

Many biological molecules are sulfated (e.g., glycosaminoglycans, steroids, and glycolipids). Ineffective sulfation of the glycosaminoglycans chondroitin sulfate and keratan sulfate (see p. 652) of cartilage results in major skeletal deformities. Degradation of sulfated molecules depends on the activity of several related sulfatases, most of which are lysosomal.

Multiple sulfatase deficiency is a rare lysosomal storage disorder. Affected individuals develop slowly and from their second year of life lose the ability to stand, sit, or speak; physical deformities and neurological deficiencies develop, and death before age 10 is usual. There is a severe lack of all the sulfatases. In contrast, deficiencies in individual sulfatases are also known, and several distinct diseases are linked to single enzyme defects.

Multiple sulfatase deficiency arises from a defect in a posttranslational modification that is common to all sulfatase enzymes and is necessary for their enzymatic activity. In each case a cysteine residue of the enzyme is normally converted to C_{α}-formylglycine. Fibroblasts from individuals with multiple sulfatase deficiency catalyze this modification with significantly lowered efficiency, and the unmodified sulfatases are catalytically inactive.

Source: Schmidt, B., Selmer, T., Ingendoh, A., and von Figura, K. A novel amino acid modification in sulfatases that is deficient in multiple sulfatase deficiency. *Cell* 82:271, 1995. Dierks, T., Schmidt, B., Borissenko, L. V., Peng, J., Preusser, A., Mariapaan, M., and von Figura, K. Multiple sulfatase deficiency is caused by mutations in the gene encoding the human C_{α}-formylglycine generating enzyme. *Cell* 113:435, 2003.

through isopeptide linkage of a lysine ϵ-amino group to the C-terminal glycine of ubiquitin, a small protein with many regulatory functions. Biotin is also linked to a few carboxylases through amide linkages to lysine (see p. 1106).

Serine and threonine hydroxyl groups are major sites of glycosylation and of reversible **phosphorylation** by protein kinases and protein phosphatases. A classic example is phosphorylation of a serine residue of glycogen phosphorylase by phosphorylase kinase (see p. 619). Tyrosine residues are phosphorylated by highly specific tyrosine kinases. Kinase activity is a property of many growth factor receptors in which growth factor binding stimulates cell division. Oncogenes, responsible in part for cell transformation and proliferation of tumor cells, often have tyrosine kinase activity and show strong homology with normal growth factor receptors. Dozens of other examples exist; together the protein kinases and protein phosphatases control the activity of many proteins that are central to normal and abnormal cellular development.

ADP-ribosylation of EF2 occurs on **diphthamide,** a modified histidine residue (Figure 6.20) of the protein. ADP-ribosylation of the diphthamide by **diphtheria toxin** inhibits EF2 activity. Physiological ADP-ribosylation, not mediated by bacterial toxins, usually involves modification of arginine and cysteine residues.

Formation of **γ-carboxyglutamate** from glutamic acid residues occurs in several blood-clotting proteins including prothrombin and factors VII, IX, and X. The γ-carboxyglutamate chelates calcium ion, which is required for blood clotting (see p. 1005). This modification requires vitamin K. It can be blocked by coumarin derivatives (see p. 1006), which antagonize vitamin K. Coumarin derivatives are used in therapy to increase clotting times.

Collagen Biosynthesis Requires Many Posttranslational Modifications

Collagen, the most abundant family of related proteins in the human, is a fibrous protein that provides the structural framework for tissues and organs. It undergoes a wide variety of posttranslational modifications that directly affect its structure and function, and defects in its modification result in serious diseases.

Different types of collagen, designated Types I, II, III, IV, and so on (see p. 100 for details of structure), are encoded on several chromosomes and expressed in different tissues. Their amino acid sequences differ, but a repeating Gly-X-Y sequence of about 1000 residues predominates. Every third residue is glycine, about one-third of the X positions are occupied by proline, and a similar number of Y positions are 4-hydroxyproline, a posttranslationally modified form of proline. Proline and hydroxyproline residues impart considerable rigidity to the structure that forms a polyproline type II helix (Figure 6.21). Each collagen polypeptide is designated an α-chain; a collagen molecule has three α-chains intertwined in a collagen triple helix in which glycine residues occupy the center of the structure (see p. 99).

Procollagen Formation in the Endoplasmic Reticulum and Golgi Apparatus

Collagen α-chain synthesis starts in the cytosol, where the amino-terminal signal sequences bind signal recognition particles. Precursor forms—designated, for example, prepro α 1(I)—are extruded into the ER lumen and the signal peptides are cleaved. Hydroxylation of proline and lysine residues occurs cotranslationally, before assembly of a triple helix. Prolyl 4-hydroxylase requires an -X-Pro-Gly- sequence (hence, 4-hydroxyproline is found only at Y positions in the -Gly-X-Y- sequence). Prolyl 3-hydroxylase modifies a smaller number of proline residues, and lysyl hydroxylase modifies some of the Y-position lysine residues. These hydroxylases require Fe^{2+} and ascorbic acid (vitamin C); the extent of modification depends on the specific α-chain type. Proline hydroxylation stabilizes collagen, and lysine hydroxylation provides sites for interchain cross-linking and for glycosylation by specific glycosyltransferases of the ER.

ADP-ribose

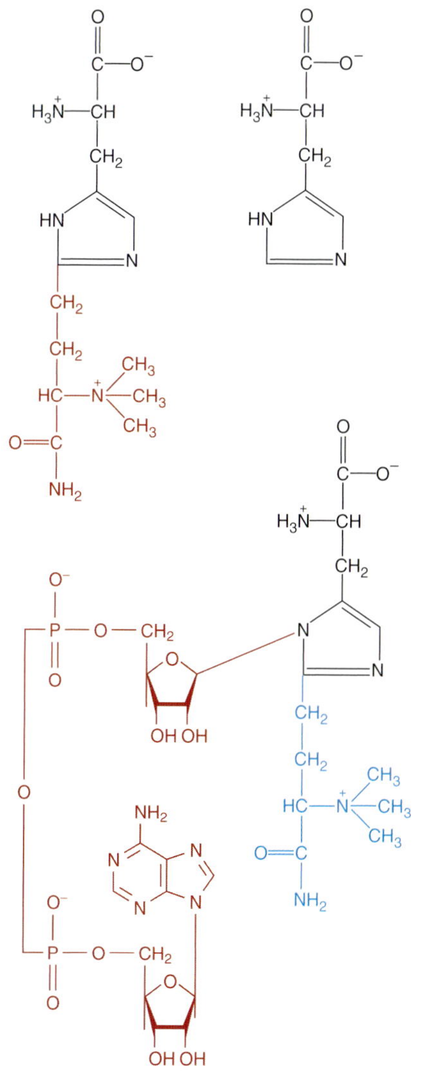

FIGURE 6.20

Diphthamide (top left) is a posttranslational modification of a specific residue of histidine (top right) in EF2. (bottom) ADP-ribosyl-diphthamide.

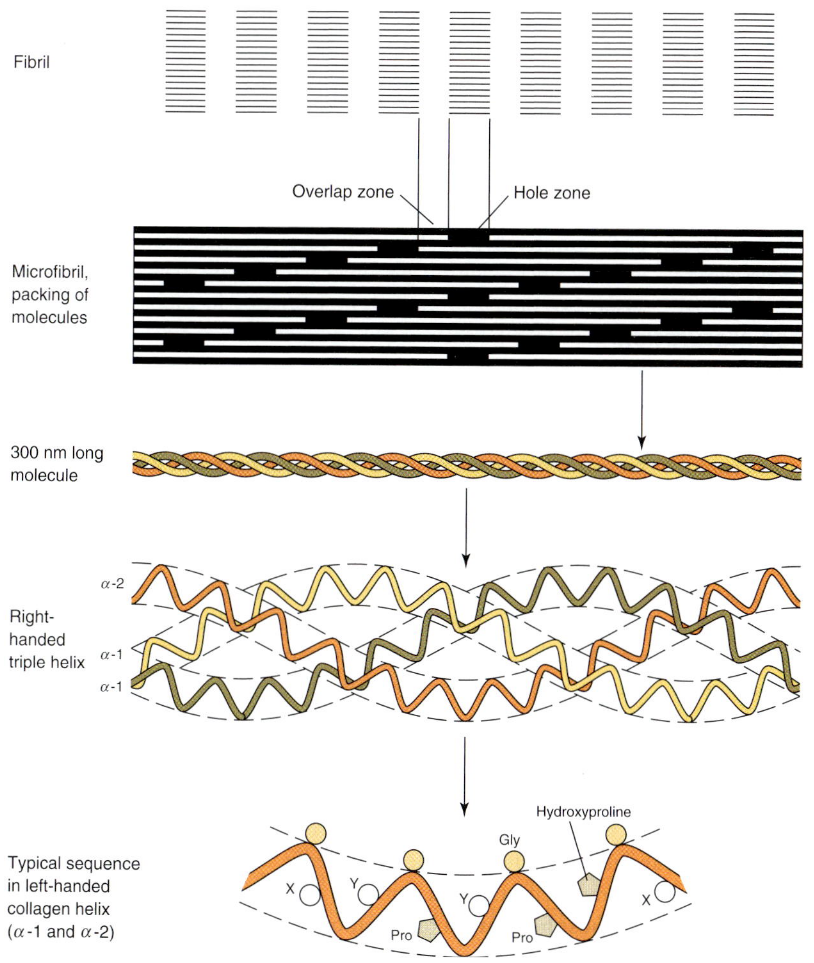

FIGURE 6.21

Collagen structure, illustrating (bottom to top) the regularity of primary sequence in a left-handed polyproline type II helix; the right-handed triple helix; the 300-nm molecule; and the organization of molecules in a typical fibril, within which collagen molecules are cross-linked.

Li, M., Brooks, C. L., Wu-Baer, F., Chen, D., Baer, R., and Gu, W. Mono- versus polyubiquitinylation: Differential control of p53 fate by Mdm2. *Science* 302:1972, 2003.

Löwe, J., Stock, D., Jap, B., Zwickl, P., Baumeister, W., and Huber, R. Crystal structure of the 20 S proteasome from the archaeon *T. acidophilum* at 3.4 Å resolution. *Science* 268:33, 1995.

Miller, M. H., Finger, A., Schweiger, M., and Wolf, D. H. ER degradation of a misfolded luminal protein by the cytosolic ubiquitin–proteasome pathway. *Science* 273:1725, 1996.

Miller, S. L. H., Malotky, E., and O'Bryan, J. P. Analysis of the role of ubiquitin-interacting motifs in ubiquitin binding and ubiquitinylation. *J. Biol. Chem.* 279:33 528, 2004.

Perrin, B. J. and Huttenlocher, A. Calpain. *Int. J. Biochem. Cell Biol.* 34:722, 2002.

Rock, K., Gramm, C., Rothstein, L., Clark, K., Stein, R., Dick, L., et al. Inhibitors of the proteasome block the degradation of most cell proteins and the generation of peptides presented on MHC class II molecules. *Cell* 78:761, 1994.

Rogers, S., Wells, R., and Rechsteiner, M. Amino acid sequences common to rapidly degraded proteins: The PEST hypothesis. *Science* 234:364, 1986.

Wolf, B. B. and Green, D. R. Suicidal tendencies: Apoptotic cell death by caspase family proteinases. *J. Biol. Chem.* 274:20 049, 1999.

QUESTIONS | CAROL N. ANGSTADT

Multiple Choice Questions

1. Degeneracy of the genetic code denotes the existence of:
   A. multiple codons for a single amino acid.
   B. codons consisting of only two bases.
   C. base triplets that do not code for any amino acid.
   D. different systems in which a given triplet codes for different amino acids.
   E. codons that include one or more of the "unusual" bases.

2. In the formation of an aminoacyl-tRNA:
   A. ADP and P_i are products of the reaction.
   B. aminoacyl adenylate appears in solution as a free intermediate.
   C. aminoacyl-tRNA synthetase is believed to recognize and hydrolyze incorrect aminoacyl-tRNA's it may have produced.
   D. separate aminoacyl-tRNA synthetases exist for every amino acid in the functional protein.
   E. there is a separate aminoacyl-tRNA synthetase for every tRNA species.

3. During initiation of protein synthesis:
   A. methionyl-tRNA appears at the A site of the 80 S initiation complex.
   B. eIF-3 and the 40 S ribosomal subunit participate in forming a preinitiation complex.
   C. eIF-2 is phosphorylated by GTP.
   D. the same methionyl-tRNA is used as is used during elongation.
   E. a complex of mRNA, 60 S ribosomal subunit, and certain initiation factors is formed.

4. During the elongation stage of eukaryotic protein synthesis:
   A. the incoming aminoacyl-tRNA binds to the P site.
   B. a new peptide bond synthesized by peptidyl transferase requires GTP hydrolysis.
   C. the peptidyl-tRNA is translocated to a different site on the ribosome.
   D. streptomycin can cause premature release of the incomplete peptide.
   E. peptide bond formation occurs by the attack of the carboxyl group of the incoming amino acyl-tRNA on the amino group of the growing peptide chain.

5. Formation of mature insulin includes all of the following *except*:
   A. removal of a signal peptide.
   B. folding into a three-dimensional structure.
   C. disulfide bond formation.
   D. removal of a peptide from an internal region.
   E. γ-carboxylation of glutamate residues.

6. Chaperones:
   A. are always required to direct the folding of proteins.
   B. when bound to protein increase the rate of protein degradation.
   C. usually bind to strongly hydrophilic regions of unfolded proteins.
   D. sometimes maintain proteins in an unfolded state to allow passage through membranes.
   E. foster aggregation of proteins into plaques.

Questions 7 and 8: Cystic fibrosis is a frequent genetic disease of Caucasians. The CF gene codes for a protein called the cystic fibrosis transmembrane conductance regulator (CFTR) which functions as a cAMP-regulated chloride channel. The protein has two membrane-spanning domains, two domains that interact with ATP and one regulatory domain. The most common defect is deletion of three consecutive bases in the gene for one of the ATP binding domains. The result is a protein that does not fold correctly in the endoplasmic reticulum, is not properly glycosylated, and is not transported to the cell surface. Rather, it is degraded in the cytosol within proteasomes.

7. The particular mutation described above most likely leads to a protein that:
   A. must be longer than it should be.
   B. must be several amino acids shorter than it should be.
   C. would always have a different amino acid sequence from the point of mutation to the end of the protein compared to the unmutated situation.
   D. has a substitution of one amino acid for another compared to the protein from the unmutated gene.
   E. has the deletion of one amino acid compared to the protein from the unmutated gene.

8. Targeting a protein to be degraded within proteasomes usually requires ubiquitin. In the function of ubiquitin all of the following are true *except*:
   A. ATP is required for activation of ubiquitin.
   B. a peptide bond forms between the carboxyl terminal of ubiquitin and an ϵ-amino group of a lysine.
   C. linkage of a protein to ubiquitin does not always mark it for degradation.

D. the N-terminal amino acid is one determinant of selection for degradation.
E. ATP is required by the enzyme that transfers the ubiquitin to the protein to be degraded.

Questions 9 and 10: Collagen is unusual in its amino acid composition and requires a wide variety of posttranslational modifications to convert it to a functional molecule. Because of the complexity of collagen synthesis, there are many diseases, resulting in structural weaknesses in connective tissue, caused by defects in the process. Scurvy leads to a less stable collagen lacking sufficient hydroxyproline.

9. 4-Hydroxylation of specific prolyl residues during collagen synthesis requires all of the following *except*:
A. Fe^{2+}.
B. a specific amino acid sequence at the site of hydroxylation.
C. ascorbic acid.
D. co-hydroxylation of lysine.
E. individual α chains, not yet assembled into a triple helix.

10. Much of procollagen formation occurs in the endoplasmic reticulum and Golgi apparatus which requires signal peptide. All of the following statements about targeting a protein for the ER are true *except*:
A. signal peptide usually has a positively charged N-terminus and a stretch of hydrophobic amino acids.
B. signal peptide emerging from a free ribosome, binds signal recognition particle (SRP).
C. signal peptide is usually cleaved from the protein before the protein is inserted into the ER membrane.
D. docking protein is actually an SRP receptor and serves to bind the SRP to the ER.
E. SRP and docking protein do not enter the ER lumen but are recycled.

Questions 11 and 12: Since protein synthesis is necessary for cells to survive and reproduce, any interference with this will have profound effects. Many antibiotics and toxins function by interfering with protein synthesis. Antibiotics selective for prokaryotic protein synthesis are useful clinically. Agents affecting eukaryotic synthesis are not.

11. Streptomycin binds the small subunit of prokaryotic ribosomes and:
A. causes premature release of the incomplete peptide.
B. prevents binding of the 40 S and 60 S subunits.
C. interferes with initiation of protein synthesis.
D. inhibits peptidyl transferase activity.
E. acts as an *N*-glycosidase.

12. Diphtheria toxin:
A. acts catalytically.
B. releases incomplete polypeptide chains from the ribosome.
C. activates translocase.
D. prevents release factor from recognizing termination signals.
E. attacks the RNA of the large subunit.

Problems

13. I- cell (inclusion bodies) disease results from a defect in the enzyme that transfers *N*-acetylglucosamine phosphate to proteins containing high mannose type oligosaccharides. What enzymes would be deficient in I-cell patients? How does the disease name correspond to the problem in these patients?

14. What is the significance of the following structures? A positively charged amphiphilic α-helix, a cluster of lysine and arginine residues, and a carboxy-terminal Ser-Lys-Leu (SKL) sequence.

ANSWERS

1. **A** This is the definition of degeneracy. B and E are not known to occur, although sometimes tRNA reads only the first two bases of a triplet (wobble), and sometimes unusual bases occur in anticodons. C denotes the stop (nonsense) codons. D is a deviation from universality of the code, as found in mitochondria.
2. **C** Bonds between a tRNA and an incorrect smaller amino acid may form but are rapidly hydrolyzed. A and B: ATP and the amino acid react to form an enzyme-bound aminoacyl adenylate; PP_i is released into the medium. D: Some amino acids, such as hydroxyproline and hydroxylysine, arise by co- or posttranslational modification. E: An aminoacyl-tRNA synthetase may recognize any of several tRNA's specific for a given amino acid.
3. **B** This then binds the mRNA. A: Methionyl-$tRNA_i^{met}$ appears at the P site. C: Phosphorylation of eIF-2 inhibits initiation. D: methionyl-$tRNA_c^{met}$ is used internally. E: mRNA associates first with the 40 S subunit.
4. **C** This is necessary to free the A site for the next incoming tRNA. A: Incoming amino acyl tRNA binds to the A site. B: Peptide bond formation requires no energy source other than the aminoacyl-tRNA. D: Streptomycin inhibits formation of the prokaryotic 70 S initiation complex and causes misreading of the genetic code. E: The electron pair of the amino group carries out a nucleophilic attack on the carbonyl carbon.
5. **E** γ-Carboxylation is of special importance in several blood-clotting proteins but not in insulin formation. A: Preproinsulin is inserted into the ER. B: All proteins, except fibrous ones, have to fold into a three-dimensional structure. C: Proinsulin folds and forms disulfide bonds before the chain is cleaved. D: This is called the C-peptide.
6. **D** This is only one of many functions chaperones serve. A: Many proteins spontaneously fold correctly. B: Misfolded proteins are recognized and rapidly degraded. C: Chaperones bind to hydrophobic regions of unfolded proteins. E: Misfolded proteins are more likely to do this.
7. **E** Deletion of an entire codon means that particular amino acid doesn't appear. A: The only way this would happen is if the codon was a stop codon. B: The new protein is one amino acid shorter but not several. C: This is an example of a frameshift mutation that arises from a one- or two-nucleotide deletion. The only way this could happen with three nucleotides would be for the three to span two codons which is not the usual process. D: This arises from a point mutation.
8. **E** ATP is required in the ubiquitin activation and the protease steps but not here. B, D: These are both correct. C: Linkage to histones does not result in their degradation.
9. **D** Lysine is hydroxylated but by a different enzyme. A and C: Prolyl hydroxylase requires both Fe^{+2} and ascorbic acid. B: the sequence is -X-Pro-Gly-. E: Hydroxylation is a cotranslational event.
10. **C** Signal peptidase is located on the luminal surface of the ER. A: These are common features along with a polar segment that signal peptidase recognizes. B, D, E: These are all essential features of the process.

11. **C** By altering interactions of tRNA, mRNA and ribosomal subunit it interferes with initiation and also causes misreading. A: Puromycin, resembling aminoacyl-tRNA, does this. B: These are eukaryotic subunits. D: Chloramphenicol does this. E: This is the action of ricin and related toxins.
12. **A** This toxin catalyzes the formation of an ADP ribosyl derivative of translocase (EF-2), which irreversibly inactivates the translocase.
13. Multiple enzymes are severely deficient in the lysosomes and are found in abnormally high levels in sera and other body fluids. Mannose 6-phosphate is the signal to target enzymes to lysosomes. The first step is addition of *N*-acetylglucosamine phosphate to high-mannose oligosaccharides. Subsequent cleavage produces the mannose 6-phosphate signal. Lack of lysosomal enzymes means that undegraded products accumulate in lysosomes, thus forming inclusion bodies.
14. These structures are all targeting signals to send proteins to particular subcellular particles. The three indicated are targets for mitochondria, nucleus, and peroxisomes, respectively.

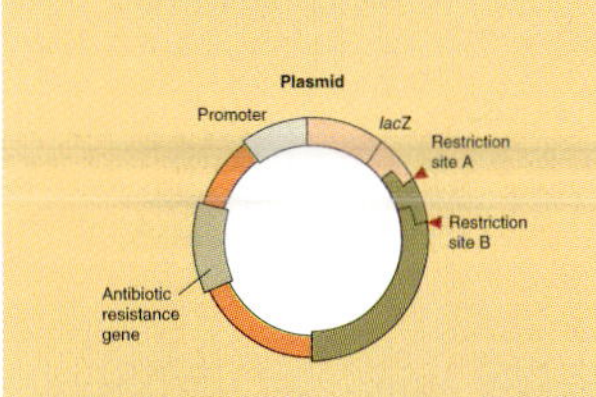

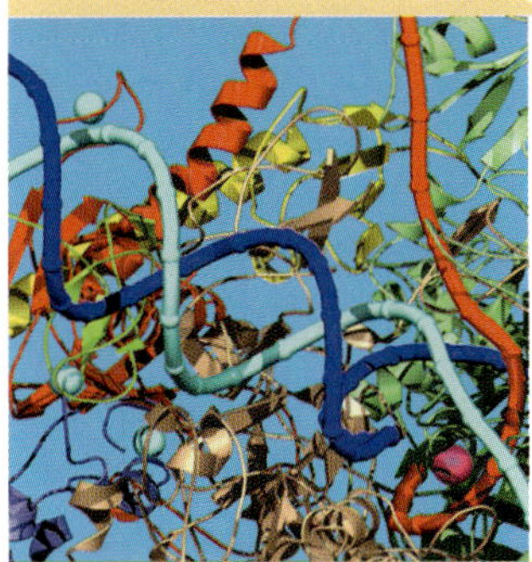

7 RECOMBINANT DNA AND BIOTECHNOLOGY

Gerald Soslau

Textbook of Biochemistry With Clinical Correlations, Sixth Edition, Edited by Thomas M. Devlin

7.1 OVERVIEW

By 1970, the stage was set for modern molecular biology based on the studies of numerous scientists in the previous 30 years during which ignorance of what biochemical entity orchestrated the replication of life forms with such fidelity gave way to a state where sequencing and manipulation of the expression of genes would become commonplace. The relentless march toward a full understanding of gene regulation under normal and pathological conditions has moved with increasing rapidity since then. Deoxyribonucleic acid, composed of only four different nucleotides covalently linked by a sugar–phosphate backbone, is deceptively complex because of the nonrandom sequence of its bases, multiple conformations that exist in equilibrium in the biological environment, and specific proteins that recognize and bind to selected regions. From established knowledge of the cellular processes and their macromolecular components, it was clear that gene expression was highly regulated. Enzymes were involved in DNA replication, RNA transcription had been purified, and their functions were defined. The genetic code had been broken. Genetic maps of prokaryotic chromosomes had been established from gene linkage studies of thousands of mutants. Finally, RNA species could be purified, enzymatically hydrolyzed into discrete pieces, and laboriously sequenced. However, further understanding of gene regulation would require techniques to cut selectively DNA into homogeneous pieces. Since small, highly purified viral DNA genomes were too complex to decipher, the thought of tackling the human genome with more than 3×10^9 base pairs was all the more onerous.

Identification, purification, and characterization of restriction endonucleases permitted the development of recombinant DNA methodologies. Development of DNA sequencing revealed the secrets within the organization of the diverse biological genomes. Genes could finally be sequenced, as could the flanking regions that regulate their expression. This led to defined consensus sequences such as those of promoters, enhancers, and many binding sites for regulatory proteins (see p. 308). While some DNA regulatory sites lie just upstream of the transcription initiation site, other regulatory regions are hundreds to thousands of bases removed and still others are downstream.

This chapter presents many of the sophisticated techniques that allow for the dissection of complex genomes into defined fragments with the complete analysis of the nucleotide sequence and function of these DNA fragments. The modification and manipulation of genes—that is, genetic engineering—facilitates the introduction and expression of genes in both prokaryotic and eukaryotic cells. Many aspects of genetic engineering have been greatly simplified by the employment of a method that rapidly amplifies selected regions of DNA, the polymerase chain reaction (PCR). Proteins for experimental and clinical uses are readily produced by these procedures; in the not too distant future, these methods should allow for numerous treatment modalities for genetic diseases by gene replacement therapy. The significance to our society of the advancements in the understanding of genetic macromolecules and their manipulation cannot be overstated.

7.2 THE POLYMERASE CHAIN REACTION

Until recently, analysis of the sequence and function of a selected region of DNA requires relatively large amounts of the purified DNA segment. The rapid production of large quantities of a specific DNA sequence took a leap forward with the development of the polymerase chain reaction (PCR). This requires two nucleotide oligomers that hybridize to the complementary DNA strands in a region of interest. The oligomers serve as primers for a DNA polymerase that extends each strand. Repeated cycling of the PCR yields large amounts of each DNA molecule of interest in a matter of hours as opposed to days and weeks required for cloning techniques.

The amplification of a specific DNA sequence by PCR can be accomplished with purified DNA or with a complex mixture of DNA. The principles of the reaction are shown in Figure 7.1. The nucleotide sequence of the DNA to be amplified must be known or it must be cloned in a vector (see p. 266) for which the sequence of the flanking DNA has been established. The product of PCR is a double-stranded DNA (dsDNA) molecule, and the reaction is completed in each cycle when all of the template molecules have been copied. In order to initiate a new round of replication, the sample is heated to melt the dsDNA and, in the presence of excess oligonucleotide primers, cooled to permit hybridization of the single-stranded template with free oligomers. DNA replication will start in the presence of DNA polymerase and all four dNTPs. Heating to about 95°C as required for melting DNA inactivates most DNA polymerases, but a heat-stable polymerase, termed Taq DNA polymerase, isolated from *Thermus aquaticus*, is now employed and obviates the need for fresh polymerase being added after each cycle. This has permitted the automation of PCR with each DNA molecule capable of being amplified one million-fold.

When the DNA to be amplified is present in very low concentrations relative to the total DNA in the sample, the DNA region of interest along with other spurious sequences can be amplified. In this situation, the specificity of the amplification can be enhanced by **nested PCR**. After conducting the first PCR with one set of primers for 10–20 cycles, a small aliquot is removed for a second PCR. However, this is conducted with primers that are complementary to the template DNA just downstream of the first primers, or "nested" between the original set of primers. This process amplifies the DNA region of interest twice with a greatly enhanced specificity. PCR has many applications in gene diagnosis, forensic investigations for which only a drop of dried blood or a single hair is available, and evolutionary studies on preserved biological material (see Clin. Corr. 7.1).

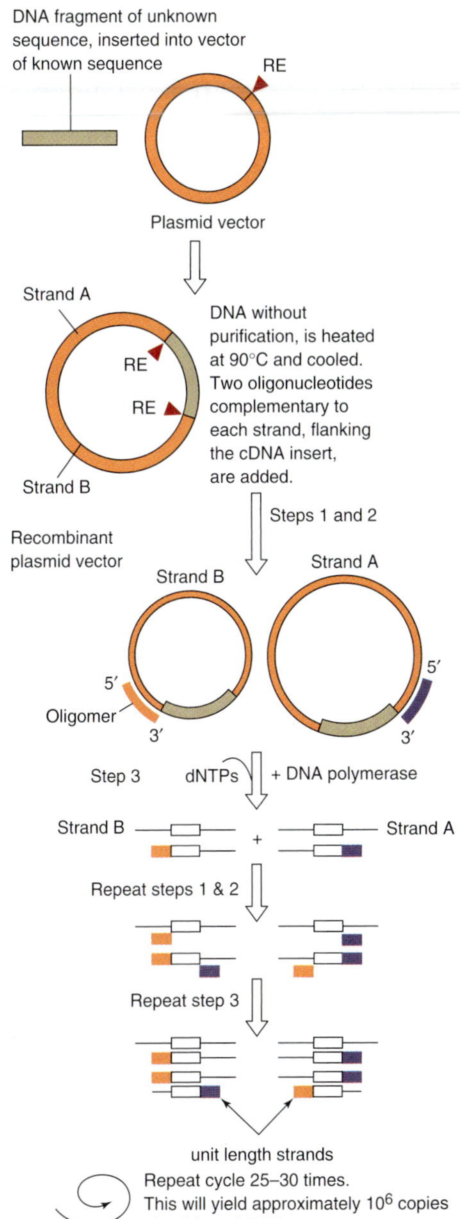

FIGURE 7.1

Polymerase chain reaction (PCR). A DNA fragment of unknown sequence is inserted into a vector of known sequence by normal recombinant methodology. The recombinant DNA of interest does not need to be purified. The DNA is heated to 90°C to dissociate the double strands and cooled in the presence of excess amounts of two different complementary oligomers that hybridize to the known vector DNA sequences flanking the foreign DNA insert. Only recombinant single-stranded DNA species can serve as templates for DNA replication yielding double-stranded DNA fragments of foreign DNA bounded by the oligomer DNA sequences. The heating-replication cycle is repeated many times to rapidly produce large amounts of the original foreign DNA. The DNA fragment of interest can be purified from the PCR mixture by cleaving it with the original restriction endonuclease (RE), electrophoresing the DNA mixture through an agarose gel, and eluting the band of interest from the gel.

CLINICAL CORRELATION **7.1**

Polymerase Chain Reaction

PCR in Screening for Human Immunodeficiency Virus

The use of PCR to amplify minute quantities of DNA has revolutionized the detection and analysis of DNA species. With PCR it is possible to synthesize sufficient DNA for analysis. Conventional detection and identification of the human immunodeficiency virus (HIV) e.g. by Southern blot–DNA hybridization and antigen analysis, is labor-intensive, and expensive and has low sensitivity. An infected individual, with no sign of AIDS (acquired immune deficiency syndrome), may test false negative for HIV by these procedures. Early detection of HIV infections is crucial to initiate treatment and/or monitor the progression of the disease. In addition, a sensitive method is required to ensure that blood contributed by donors does not contain HIV. PCR amplification of potential HIV DNA sequences within DNA isolated from an individual's white blood cells permits detection of infection prior to appearance of antibodies, the so-called seronegative state. Current methods are too costly to apply this testing to large-scale screening of donor blood samples. PCR can also be used to detect and characterize DNA sequences of any other human infectious pathogen.

Nested PCR to Detect Microchimerism

Donor leukocytes transferred to patients during blood transfusion have been shown to survive in the recipient's peripheral blood. The significance, if any, of this microchimeric population of leukocytes remains to be resolved. One of the best ways to detect donor-derived cells in the recipient's blood is to use PCR detection of polymorphisms in the HLA-DR region of the major histocompatibility complex (MHC). A nested PCR assay was at least 100-fold more sensitive than a standard PCR assay. However, because of the increased sensitivity, nonspecific products may appear due to mispriming events that are generally associated with pseudogenes. As such, it is essential to establish a baseline pattern with pre-transfusion blood samples. Once potential false positives are excluded with these baseline patterns, the detection of donor leukocytes is greatly enhanced.

Source: Kwok, S. and Sninsky, J. J. Application of PCR to the detection of human infectious diseases. In H. A. Erlich (Ed.), *PCR Technology*. New York: Stockton Press, 1989, p. 235.

Carter, A. S., Cerundolo, L., Bunce, M., Koo, D. D. H., Welsh, K. I., Morris, P. J., and Fuggle, S. V. Nested polymerase chain reaction with sequence-specific primers typing for HLA-A, -B, and -C alleles: Detection of microchimerism in DR-matched individuals. *Blood* 94:1471, 1999.

7.3 RESTRICTION ENDONUCLEASE AND RESTRICTION MAPS

Restriction Endonucleases Selectively Hydrolyze DNA

Restriction endonucleases are capable of selectively dissecting DNA molecules of many sizes and origins into smaller fragments. They confer some protection on bacteria against invading viruses (bacteriophage). Bacterial DNA sequences normally recognized by a restriction endonuclease are protected from cleavage in host cells by methylation of bases within the palindrome. The unmethylated viral DNA is recognized as foreign and is hydrolyzed. Numerous Type II restriction endonucleases are now commercially available (see p. 58 for discussion of restriction endonuclease activities).

Restriction endonucleases permit construction of a restriction map, in which the site of cleavage within the DNA is identified. Purified DNA species that contain susceptible sequences are subjected to restriction endonuclease cleavage. By regulating the time of exposure of the purified DNA for cleavage, a population of DNA fragments of different sizes is generated. Separation of these fragments by agarose gel electrophoresis allows for the construction of a restriction map, an example is presented in Figure 7.2. The sequential use of different restriction endonucleases has permitted a detailed restriction map of numerous circular DNA species including bacterial plasmids, viruses, and mitochondrial DNA. The method is equally amenable to linear DNA fragments that have been purified to homogeneity.

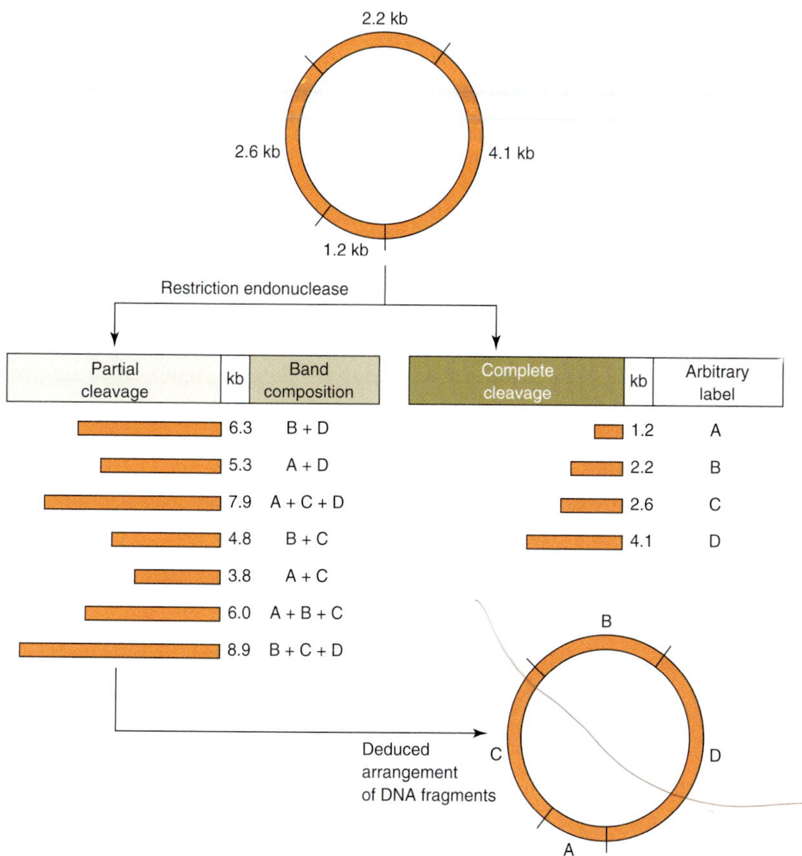

FIGURE 7.2

Restriction endonuclease mapping of DNA. Purified DNA is subjected to restriction endonuclease digestion for varying times which generates partially to fully cleaved DNA fragments. The fragments are separated by agarose gel electrophoresis and stained with ethidium bromide. The bands are visualized with a UV light source and photographed. The size of the fragments is determined by the relative migration through the gel as compared to coelectrophoresed DNA standards. The relative arrangement of each fragment within the DNA molecule can be deduced from the size of the incompletely hydrolyzed fragments.

Restriction Maps Permit Routine Preparation of Defined Segments of DNA

Restriction maps may yield little information as to the genes or regulatory elements within the various DNA fragments. They have been used to demonstrate sequence diversity of organelle DNA, such as mitochondrial DNA, within species (see Clin. Corr. 7.2). They can also be used to detect deletion mutations in which a DNA fragment from the parental strain migrates as a smaller fragment in the mutated strain. Most importantly, the restriction endonucleases cut DNA into defined homogeneous fragments that can be readily purified. These maps are crucial for cloning and for sequencing of genes and their flanking DNA regions.

7.4 DNA SEQUENCING

In the late 1970s two different sequencing techniques were developed, one by A. Maxam and W. Gilbert, the chemical-cleavage approach, and the other by F. Sanger, the enzymatic approach. Both procedures employ labeling of a terminal nucleotide, followed by the separation and detection of generated oligonucleotides. The Sanger method has become the method of choice due to the relative ease of the procedure and its ability to sequence longer stretches of DNA (400 bases) relative to the Maxam and Gilbert method (250 bases).

CLINICAL CORRELATION 7.2
Restriction Mapping and Evolution

In the past, evolutionary studies of species have depended solely on anatomical changes observed in fossil records and on carbon dating. These studies are being supported by the molecular analysis of the sequence and size of selected genes or whole DNA molecules. Alterations in a selected DNA molecule from different species can be rapidly assessed by restriction mapping, which requires a pure preparation of DNA. Mammalian mitochondria contain a closed circular DNA molecule of 16,569 bps that can rapidly be purified from cells. The mitochondrial DNA (mtDNA) can be employed directly for the study of evolutionary changes in DNA without the need of cloning a specific gene.

Mitochondrial DNA has been purified from the Guinea baboon, rhesus macaque, guenon, and human and has been cleaved with 11 different restriction endonucleases. Restriction maps were aligned relative to the direction and the nucleotide site where DNA replication is initiated. A comparison of restriction endonuclease sites permitted the calculation of the degree of divergence in nucleotide sequence between species. The rate of base substitution (calculated from the degree of divergence versus the time of divergence) has been about 10-fold greater than in the nuclear genome. This high rate of mutation of the readily purified mtDNA molecule makes this an excellent model to study evolutionary relationships between species.

Analysis of mtDNA can be used to study the migration patterns of people who settled in diverse geographic regions but possess the same restriction enzyme generated patterns. Native Americans have been grouped into four major haplogroups based upon their mtDNA restriction patterns. These groups appear to have migrated from Asia and Siberia some 18,000 years ago, just before the major ice age period.

Source: Brown, W. M., George, M., Jr., and Wilson, A. C. Rapid evolution of animal mitochondrial DNA. *Proc. Natl. Acad. Sci. USA*, 76:1967, 1979. Schurr, J. G. Mitochondrial DNA and the people of the new world. *Am. Sci.* 88:246, 2000.

BASE

Deoxynucleoside triphosphate

BASE

Dideoxynucleoside triphosphate

FIGURE 7.3

Structure of deoxynucleoside triphosphate and dideoxynucleoside triphosphate. The 3′-OH group is lacking on the ribose component of the dideoxynucleoside triphosphate (ddNTP). This molecule can be incorporated into a growing DNA molecule through a phosphodiester bond with its 5′-phosphates. Once incorporated the ddNTP blocks further synthesis of the DNA molecule since it lacks the 3′-OH acceptor group for an incoming nucleotide.

Interrupted Enzymatic-Cleavage Method: Sanger Procedure

The **Sanger procedure** is based on the random termination of a DNA chain during enzymatic synthesis. The technique depends on the dideoxynucleotide analog of each of the four normal nucleotides (Figure 7.3) being incorporated into a growing DNA chain by DNA polymerase and blocking further elongation. The ribose of the **dideoxynucleoside triphosphate (ddNTP)** lacks the OH group at both the 2′ and 3′ positions, whereas dNTP lacks only the OH group at the 2′ position. Thus, the ddNTP incorporated into the growing chain cannot form a phosphodiester bond with another dNTP because the 3′ position of the ribose does not contain a OH group. The growing DNA molecule can be terminated at random points by including in the reaction system the normal dNTP and the ddNTP (e.g., dATP and ddATP) at concentrations such that the two nucleotides compete for incorporation. Identification of DNA fragments requires labeling of the 5′-end of the DNA molecules or the incorporation of labeled nucleotides during synthesis. The technique, outlined in Figure 7.4, is best conducted with pure single-stranded DNA; however, denatured double-stranded DNA can also be used. The DNA to be sequenced is frequently isolated from a recombinant single-stranded bacteriophage (see p. 266) in which a region that flanks the DNA of interest contains a sequence that is complementary to a universal primer. The primer can be labeled with either ^{32}P or ^{35}S nucleotide. Primer extension is accomplished with one of several DNA polymerases; one with great versatility is a genetically engineered form of the bacteriophage T 7 DNA polymerase. The reaction mixture, composed of the target DNA, labeled primer, and all four dNTPs, is divided into four tubes, each containing a different ddNTP. ddNTPs are randomly incorporated during the enzymatic synthesis of DNA and cause chain termination. Since the ddNTP is present in the reaction tube at a low level, relative to the corresponding dNTP, termination of DNA synthesis occurs randomly at all possible complementary sites of the DNA template. This yields DNA molecules of varying sizes, labeled at the 5′-end, that can be separated by polyacrylamide gel electrophoresis. The labeled species are detected by X-ray autoradiography, and the sequence is read.

Initially, this method required a single-stranded DNA template, production of a specific complementary oligonucleotide primer, and a relatively pure preparation of the Klenow fragment of *E. coli* DNA polymerase I. These difficulties have been overcome, and modifications have simplified the procedure.

The PCR and the Sanger methods can be combined for **direct sequencing** of small DNA regions of interest. The double-stranded PCR product is employed directly as template. Conditions are set so that one strand of melted DNA (template) anneals with

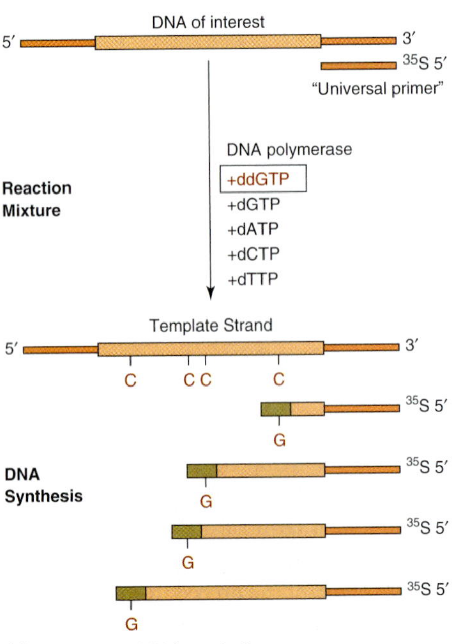

(*a*) Recombinant M13 bacteriophage

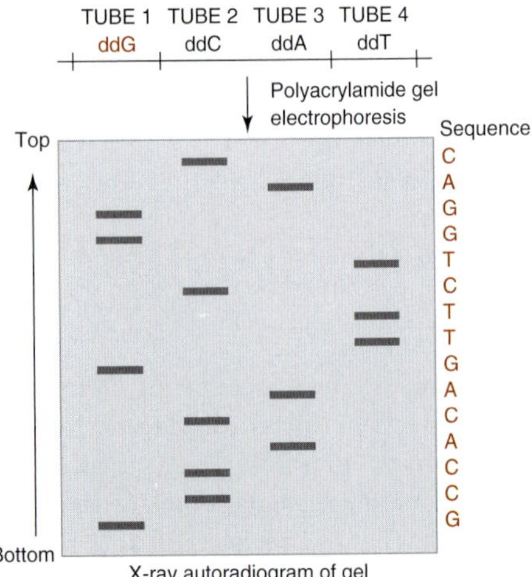

(*b*) Polyacrylamide gel electrophoresis of reaction mixture

FIGURE 7.4

The Sanger dideoxynucleoside triphosphate method to sequence DNA. The DNA region of interest is inserted into bacteriophage DNA molecule. Replicating bacteriophage produce a single-stranded recombinant DNA molecule that is readily purified. The known sequence of the bacteriophage DNA downstream of the DNA insert serves as a hybridization site for an end-labeled oligomer with a complementary sequence, a universal primer. Extension of this primer is catalyzed with a DNA polymerase and all four dNTPs plus one ddNTP—for example, ddGTP. Synthesis stops whenever a ddNTP is incorporated into the growing molecule. Note that the ddNTP competes for incorporation with the dNTP. This generates end-labeled DNA fragments of all possible lengths that are separated by electrophoresis. The DNA sequence can then be determined from the electrophoretic patterns.

the primer in preference to reannealing with the complementary strand. Sequencing then follows the standard dideoxy chain termination reaction (typically with Sequenase replacing of the Klenow polymerase) with synthesis of random-length chains occurring as extensions of the PCR primer. This method has been successfully employed for diagnosis of genetic disorders (see Clin. Corr. 7.3).

7.5 RECOMBINANT DNA AND CLONING

DNA from Different Sources Can Be Ligated to Form a New DNA Species: Recombinant DNA

The ability to cleave a population of DNA molecules selectively with restriction endonucleases led to the technique for joining two different DNA molecules termed **recombinant DNA**. This procedure, combined with techniques for replication, separation, and identification, permits the production of large quantities of purified DNA fragments. The combined techniques, referred to as recombinant DNA technology, allow removal of a piece of DNA out of a larger complex molecule, such as the genome of a virus or human, and amplification of the DNA fragment. Recombinant DNAs have been prepared that combine DNA fragments from bacteria with fragments from humans, viruses with viruses, and so on. The joining of two different pieces of DNA is achieved by a restriction endonuclease and a DNA ligase. Many restriction endonucleases, varying in their nucleotide sequence specificity, can be used (Section 7.3). Some hydrolyze the two strands of DNA in a staggered fashion to produce "sticky or cohesive" ends (Figure 7.5), while others cut both strands symmetrically to produce a blunt end. A specific restriction enzyme cuts DNA at exactly the same sequence site regardless of source of DNA (bacteria, plant, mammal, etc.). A DNA molecule may have none or numerous recognition sites for a particular restriction endonuclease. The staggered cut results in DNA fragments with single-stranded ends. When different DNA fragments generated by the same restriction endonuclease are mixed, their single-stranded ends can hybridize or can anneal together. DNA ligase joins the two fragments to produce a recombinant DNA molecule.

The DNA fragments that contain blunt ends can also be ligated, but with much less efficiency. The efficiency can be increased by adding enzymatically poly(dA) tails

CLINICAL CORRELATION 7.3

Direct Sequencing of DNA for the Diagnosis of Genetic Disorders

The X-linked recessive hemorrhagic disorder, hemophilia B, is caused by coagulation factor IX deficiency. The factor IX gene contains 8 exons spanning 34 kb that encode a glycoprotein secreted by the liver. Over 300 mutations of the gene have been discovered of which about 85% are single base substitutions and the rest are complete or partial gene deletions. Several methods to identify carriers of a defective gene copy and for prenatal diagnoses have been costly, time-consuming, and all too often inaccurate. Direct sequencing of PCR-amplified genomic DNA circumvents these shortcomings. Between 0.1 and 1 μg of genomic DNA can readily be isolated from patient blood samples and each factor IX exon can be PCR amplified with appropriate primers. Amplified DNA can be used for direct sequencing to determine if a mutation in the gene exists that would be diagnostic of one of the forms of hemophilia B. For example, a patient with a moderate hemophilia B (London 6) had an A to G transition at position 10442 that led to a substitution of Asp 64 by Gly.

Source: Green, P. M., Bentley, D. R., Mibashan, R. S., Nilsson, I. M., and Gianelli, F. Molecular pathology of hemophilia B. *EMBO J.* 8:1067, 1989.

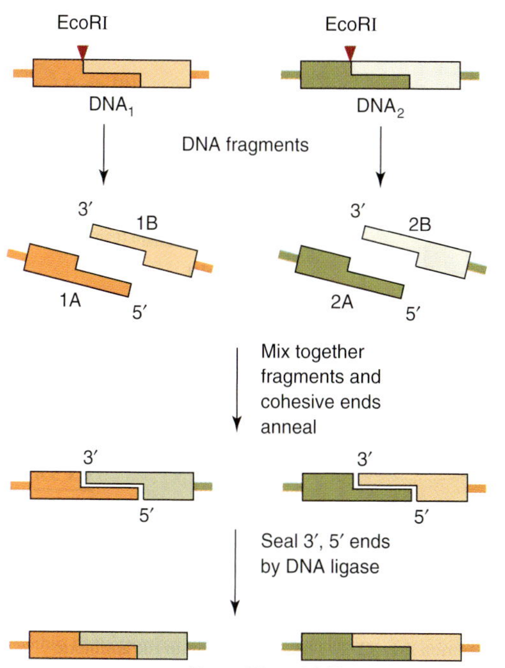

FIGURE 7.5

The formation of recombinant DNA from restriction endonuclease-generated fragments that contain cohesive ends. Many restriction endonucleases hydrolyze DNA in a staggered fashion, yielding fragments with single-stranded regions at their 5′- and 3′-ends. DNA fragments generated from different molecules with the same restriction endonuclease have complementary single-stranded ends that can be annealed and covalently linked together with a DNA ligase. All different combinations are possible in a mixture. When two DNA fragments of different origin combine, a recombinant DNA molecule results.

of different bacterial colonies. The plasmid construct pBR322 and its modifications carry two genes that confer antibiotic resistance and are sensitive to several restriction endonucleases. When a fragment of foreign DNA is inserted into a restriction site within one of them, the gene becomes nonfunctional and the bacteria carrying this recombinant plasmid are sensitive to the antibiotic (Figure 7.9). The second antibiotic resistance gene within the plasmid, however, remains intact and the bacteria will be resistant to this antibiotic. This **insertional inactivation** of plasmid gene products permits selection of bacteria that carry recombinant plasmids.

pBR322 contains genes that confer resistance to **ampicillin** (*amp*^{r}) and **tetracycline** (*tet*^{r}). A gene library of cellular DNA fragments inserted within the *tet*^{r} gene can be selected and screened in two stages (Figure 7.9). First, the bacteria are grown in an ampicillin containing growth medium. Bacteria that are not transformed because they lack a normal or recombinant plasmid will not grow in the presence of the antibiotic, and this population of bacteria is eliminated. This, however, does not indicate which of the remaining viable bacteria carry a recombinant plasmid vector rather than a plasmid with no DNA insert. The second step is to identify bacteria that carry recombinant vectors with nonfunctional *tet*^{r} genes, which are, therefore, sensitive to tetracycline. The bacteria resistant to ampicillin are plated and grown on agar plates containing ampicillin (Figure 7.9). Replica plates are made by touching the colonies on the original agar plate with a filter and then touching additional sterile plates with the filter. All the plates will contain portions of each original colony at identifiable positions on the plates. The replica plate can contain tetracycline, which will kill bacteria harboring a disrupted *tet*^{r} gene. Comparison of replica plates with and without tetracycline indicates which colonies on the original ampicillin plate contain recombinant plasmids.

Either DNA or RNA probes (see pp. 258 and 284) can be used to identify the DNA of interest. Ampicillin-resistant bacterial colonies on agar can be replica plated onto a nitrocellulose filter and adhering cells lysed with NaOH (Figure 7.9). DNA within the lysed bacteria is also denatured by the NaOH and becomes firmly bound to the filter. A labeled DNA or RNA probe that is complementary to the DNA of interest will hybridize with the nitrocellulose-bound DNA. The filter is then tested by X-ray autoradiography, which will detect any colony that carries the cloned DNA of interest. These spots indicate the colonies on the original agar plate that can be grown in a large-scale culture for further manipulation.

Cloned and amplified DNA fragments usually do not contain a complete gene and are not expressed. The DNA inserts, however, can readily be purified for sequencing or used as probes to detect genes within a mixture of genomic DNA, to assay transcription levels of mRNA, and to detect disease-producing mutations.

α-Complementation for Selecting Bacteria Carrying Recombinant Plasmids

Vectors have been constructed (the pUC series) such that selected bacteria transformed with these vectors carrying foreign DNA inserts, can be identified visually (Figure 7.10). The pUC plasmids contain the regulatory sequences and part of the 5′-coding sequence (N-terminal 146 amino acids) for the β-galactosidase gene (lac Z gene) of the lac operon (see p. 293). The translated N-terminal fragment of β-galactosidase is an inactive polypeptide. Mutant *E. coli*, which code for the missing inactive carboxy-terminal portion of β-galactosidase, can be transformed using the pUC plasmids. The translation of the host cell and plasmid portions of the β-galactosidase in response to an inducer, **isopropyl thio-β-d-galactoside**, complements each other to yield an active enzyme. The process is called **α-complementation**. When these transformed bacteria are grown in the presence of a chromogenic substrate (5-bromo-4-chloro-3-indolyl-β-d-galactoside [X-gal]) for β-galactosidase, they form blue colonies. If, however, a foreign DNA fragment has been inserted into the sequence for the N-terminal portion of β-galactosidase, the active enzyme cannot be formed. Bacteria transformed with these recombinant plasmids and grown on X-gal yield white colonies and can be selected visually from nontransformed blue colonies.

1 Insertional Inactivation

amp^r
pBR322
tet^r
Genomic DNA
Restriction endonuclease
Restriction endonuclease
amp^r
n DNA fragments
amp^r
tet^r
amp^r
tet^r
tet^r
Reannealed plasmid without a foreign DNA fragment
Recombinant plasmid with DNA insert inactive *tet*^r gene
Recombinant DNA linear arrays

2 Transformation of Interest

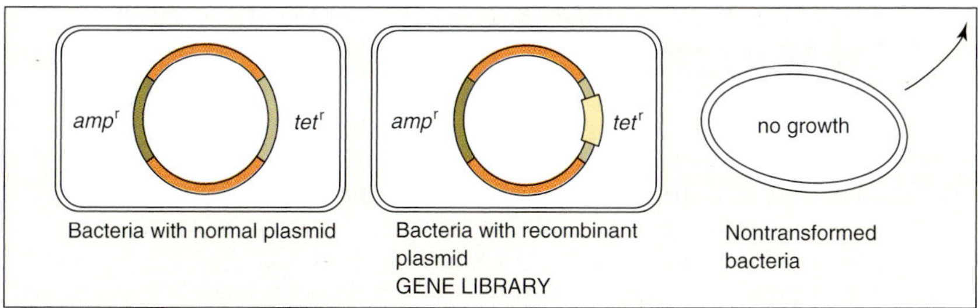

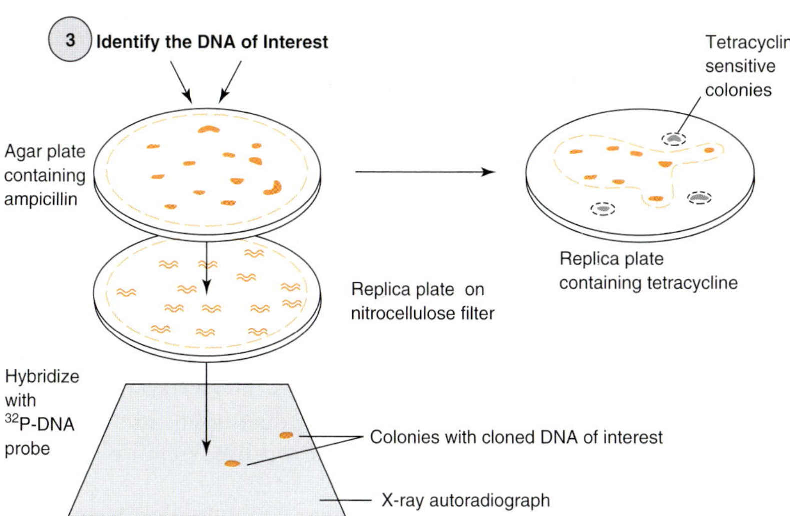

FIGURE 7.9

Insertional inactivation of recombinant plasmids and detection of transformed bacteria carrying a cloned DNA of interest. When the insertion of a foreign DNA fragment into a vector disrupts a functional gene sequence, the resulting recombinant DNA does not express the gene. The gene that codes for resistance to tetracycline (*Tet*^r) is destroyed by DNA insertion while the ampicillin resistance gene (*Amp*^r) remains functional. Destruction of one resistance gene and the retention of a second antibiotic resistance gene permits the detection of bacterial colonies carrying the foreign DNA of interest within the replicating recombinant vector.

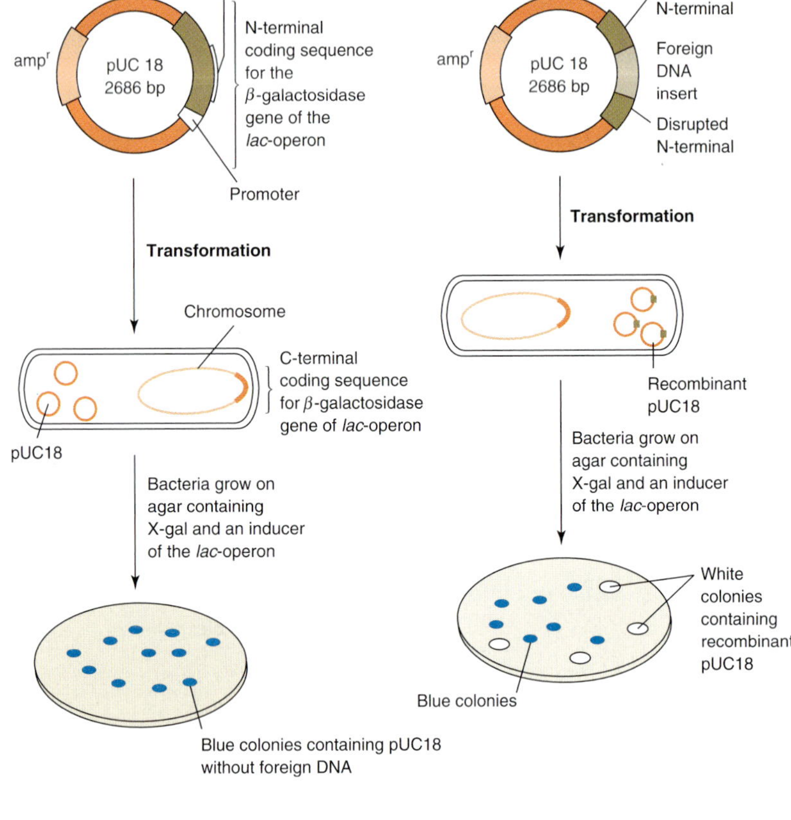

FIGURE 7.10

α-Complementation for detection of transformed bacteria. A constructed vector (pUC 18) expresses the N-terminal coding sequence for β-galactosidase of the lac operon. Bacterial mutants coding for the C-terminal portion of β-galactosidase are transformed with pUC 18. These transformed bacteria grown in the presence of a special substrate for the intact enzyme (X-gal) result in blue colonies because they contain the enzyme to react with substrate. The functional N-terminal and C-terminal coding sequences complement each other to yield a functional enzyme. If, however, a foreign DNA fragment insert disrupts the pUC 18 N-terminal coding sequence for β-galactosidase, bacteria transformed with this recombinant molecule will not produce a functional enzyme. Bacterial colonies carrying these recombinant vectors can then be visually detected as white colonies.

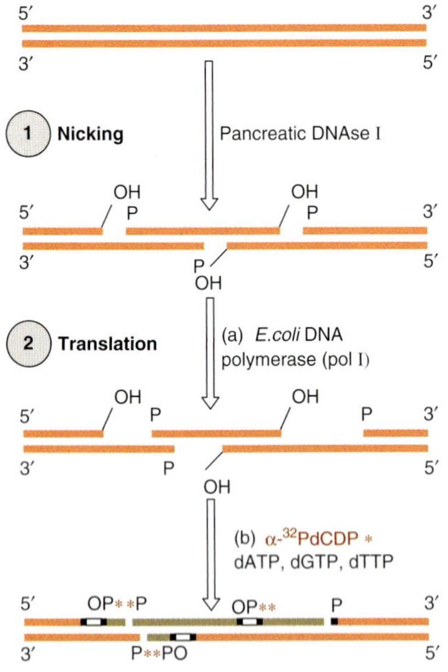

FIGURE 7.11

Nick translation to label DNA probes. Purified DNA molecules can be radioactively labeled and used to detect, by hybridization, the presence of complementary RNA or DNA in experimental samples. (1) Nicking step: introduces random single-stranded breaks in DNA. (2) Translation step: *E. coli* DNA polymerase (pol I) has both (a) 5′–3′ exonuclease activity that removes nucleotides from the 5′-end of the nick; and (b) polymerase activity that simultaneously fills in the single-stranded gap with radioactively labeled nucleotides using the 3′-end as a primer.

7.7 DETECTION AND IDENTIFICATION OF NUCLEIC ACIDS AND DNA-BINDING PROTEINS

Nucleic Acids as Probes for Specific DNA or RNA Sequences

DNA and RNA probes are used for selection of bacteria that harbor recombinant DNA of interest, for analysis of mRNA expressed in a cell, or for identification of DNA sequences within a genome. They contain sequences complementary to the target nucleic acid and hybridize with the nucleic acid of interest. The degree of complementarity determines the tightness of binding of the probe. The probe does not need to contain the entire complementary sequence of the DNA. The probe can be labeled, usually with ^{32}P or with nonradioactive labels that depend on enzyme substrates coupled to nucleotides, which when incorporated into the nucleic acid can be detected by an enzyme-catalyzed reaction.

Labeled probes can be produced by **nick translation** of double-stranded DNA. Nick translation (Figure 7.11) involves the random cleavage of a phosphodiester bond in the backbone of a DNA strand by DNase I; the breaks are called nicks. *E. coli* DNA polymerase I, with its $5' \rightarrow 3'$ exonucleolytic activity and its DNA polymerase activity, creates single-strand gaps by hydrolyzing nucleotides from the 5′-side of the nick and then filling in the gaps with its polymerase activity. The reaction is usually carried out in the presence of an α-^{32}P-labeled dNTP and the other three unlabeled dNTPs. The DNA employed in this method is usually purified and is derived from cloned DNA, viral DNA, or cDNA.

Random primer labeling of DNA has distinct advantages over the nick translation method. The random primer method typically requires only 25 ng of DNA as opposed to 1–2 μ g of DNA for nick translation and results in labeled probes with a specific activity ($>10^9$ cmp ug^{-1}) approximately 10 times higher. This method generally produces longer labeled DNA probes. The double-stranded probe is melted and hybridized with a mixture of random hexanucleotides containing all possible sequences (ACTCGG, ACTCGA, ACTCGC, etc.). Hybridized hexanucleotides serve as primers for DNA synthesis with a DNA polymerase, such as the Klenow enzyme, in the presence of one or more radioactively labeled dNTPs.

Labeled RNA probes have advantages over DNA probes. For one, relatively large amounts of RNA can be transcribed from a template, which may be available in very limited quantities. A double-stranded DNA (dsDNA) probe must be denatured prior to hybridization with the target DNA and rehybridization with itself competes for hybridization with the DNA of interest. No similar competition occurs with single-stranded RNA probes that hybridize with complementary DNA or RNA molecules. Synthesis of an RNA probe requires DNA as a template. To be transcribed, the template must be covalently linked to an upstream promoter that can be recognized by a DNA-dependent RNA polymerase. Vectors have been constructed that are well-suited for this technique.

A labeled DNA or RNA probe can be hybridized to nitrocellulose-bound nucleic acids and identified by the detection of the labeled probe. Nucleic acids of interest can be transferred to nitrocellulose from bacterial colonies grown on agar or from agarose gels on which they have been electrophoretically separated by size.

Southern Blot Technique for Identifying DNA Fragments

The transfer of DNA species separated by agarose gel electrophoresis to a filter for analysis was developed in the 1970s by E. M. Southern and is an indispensable tool. The method is called the **Southern blot technique** (Figure 7.12). A mixture of restriction endonuclease-generated fragments can be separated according to size by electrophoresis through an agarose gel. The DNA is denatured by soaking the gel in alkali. The gel is then placed on absorbent paper, and a nitrocellulose filter is placed directly on top of the gel. Several layers of absorbent paper are placed on top of the nitrocellulose filter. The absorbent paper under the gel is kept wet with a concentrated salt solution that is pulled up through the gel, the nitrocellulose, and into the absorbent paper layers above by capillary action. The DNA is eluted from the gel by the upward movement of the high salt solution onto the nitrocellulose filter directly above, where it becomes bound. The position of the DNA bound to the nitrocellulose filter is the same as that present in the agarose gel. In its single-stranded membrane-bound form, the DNA can be analyzed with labeled probes.

The Southern blot technique is invaluable for detection and determination of the number of copies of particular sequences in complex genomic DNA, for confirmation of DNA cloning results, and for demonstration of polymorphic DNA in the human genome that correspond to pathological states. An example of the use of Southern blots is shown in Figure 7.12. Whole human genomic DNA, isolated from three individuals, was digested separately with a restriction endonuclease generating thousands of fragments. These were distributed throughout the agarose gel according to size in an electric field. The DNA was transferred (blotted) to a nitrocellulose filter and hybridized with a ^{32}P-labeled DNA or RNA probe for a gene of interest. The probe detected two bands in all three individuals, indicating that the gene of interest was cleaved at one site within its sequence. Individuals A and B presented a normal pattern while patient C had one normal band and one lower-molecular-weight band. This is an example of detecting altered restriction endonuclease generated DNA fragments from different individuals within a single species, **restriction fragment length polymorphism** (**RFLP**). The detection of a reduced-molecular-weight fragment in patient C implies that deletion of a segment of the gene occurred and may be associated with a pathological state. The gene from this patient can be cloned, sequenced, and fully analyzed to characterize the altered nature of the DNA (see Clin. Corr. 7.5).

Other techniques that employ the principles of Southern blot are the transfer of RNA (Northern blots), as described below, and of proteins (Western blots) to nitrocellulose filters or nylon membranes.

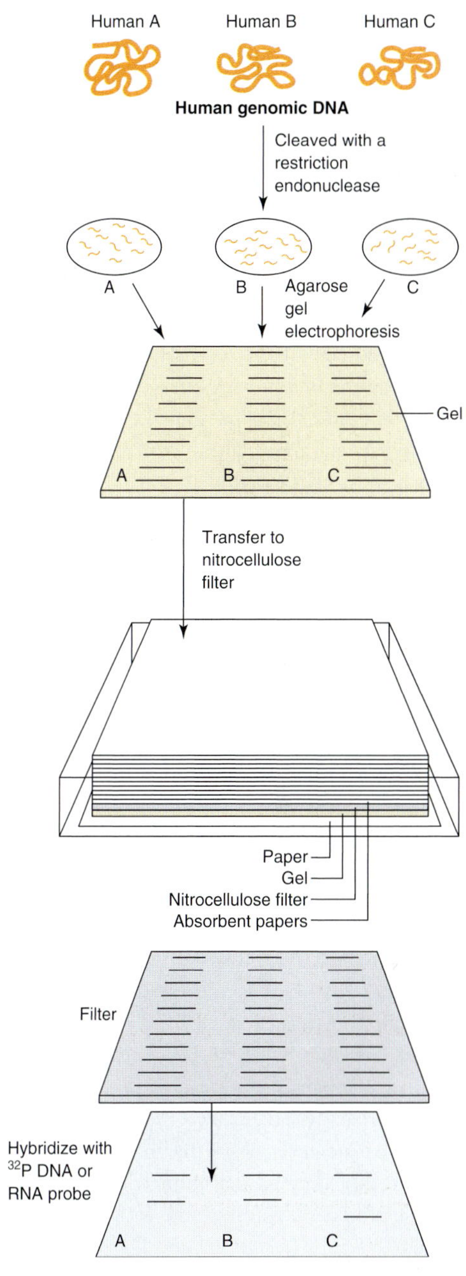

FIGURE 7.12

Southern blot to transfer DNA from agarose gels to nitrocellulose. Transfer of DNA to nitrocellulose, as single-stranded molecules, allows for the detection of specific DNA sequences within a complex mixture of DNA. Hybridization with nick-translated labeled probes can demonstrate if a DNA sequence of interest is present in the same or different regions of the genome.

CLINICAL CORRELATION 7.5

Restriction Fragment Length Polymorphisms Determine the Clonal Origin of Tumors

It is generally assumed that most tumors are monoclonal in origin, that is, a rare event alters a single somatic cell genome such that the cells grow abnormally into a tumor mass with all-daughter cells carrying the identically altered genome. Proof that a tumor is monoclonal versus polyclonal can help to distinguish hyperplasia (increased production and growth of normal cells) from neoplasia (growth of new or tumor cells). The detection of restriction fragment length polymorphisms (RFLPs) in Southern blotted DNA samples allows one to define the clonal origin of human tumors. If tumor cells were collectively derived from different parental cells, they should contain a mixture of DNA markers characteristic of each cell of origin. However, an identical DNA marker in all tumor cells would indicate a monoclonal origin. The analysis is limited to females to take advantage of the fact that each cell carries only one active X chromosome of either paternal or maternal origin, with the second X chromosome being inactivated. Activation occurs randomly during embryogenesis and is faithfully maintained in all daughter cells, with one-half the cells carrying an activated maternal X chromosome and the other one-half an activated paternal X chromosome.

Analysis of the clonal nature of a human tumor depends on the fact that activation of an X chromosome involves changes in the methylation of selected cytosine (C) residues within the DNA molecule. Several restriction endonucleases, such as Hha I, which cleaves DNA at GCGC sites, do not cleave DNA at their recognition sequences if a C is methylated within this site. Therefore, the methylated state (activated versus inactivated) of the X chromosome can be probed with restriction endonucleases. Also, the paternal X chromosome can be distinguished from the maternal X chromosome in a significant number of individuals by differences in the electrophoretic migration of restriction endonuclease generated fragments derived from selected regions of the chromosome. These fragments are identified on a Southern blot by hybridization with a DNA probe that is complementary to this region of the X chromosome. An X-linked gene that is amenable to these studies is the hypoxanthine guanine phosphoribosyltransferase (HGPRTase) gene. The HGPRTase gene consistently has two Bam HI restriction endonuclease sites (B_1 and B_3 in accompanying figure), but in some individuals a third site (B_2) is also present (see figure).

The presence of site B_2 in only one parental X chromosome HGPRTase permits detection of RFLPs. Therefore, a female cell may carry one X chromosome with the HGPRTase gene possessing 2Bam HI sites (results in one DNA fragment of 24 kb) or 3Bam HI sites (results in a single detectable DNA fragment of 12 kb). This figure depicts the expected results for the analysis of tumor cell DNA to determine its monoclonal or polyclonal origin. As expected, three human tumors examined by this method were shown to be of monoclonal origin.

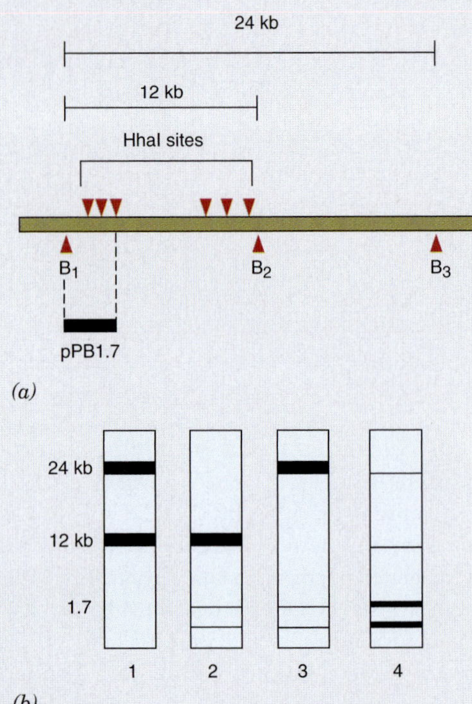

Analysis of genomic DNA to determine the clonal origin of tumors. (*a*) The X chromosome-linked hypoxanthine guanine phosphoribosyltransferase (HGPRTase) gene contains two invariant Bam HI restriction endonuclease sites (B1 and B3), while in some individuals a third site, B2, is also present. The gene also contains several Hha I sites; however, all of these sites, except H1, are usually methylated in the active X chromosome. Therefore, only the H1 site would be available for cleavage by Hha I in the active X chromosome. A cloned, labeled probe, pPB1.7, is used to determine which form of the HGPRTase gene is present in a tumor and if it is present on an active X chromosome. (*b*) Restriction endonuclease patterns predicted for monoclonal versus polyclonal tumors are as follows: (1) Cleaved with Bam HI alone. Shown here are a 24-kb fragment derived from a gene containing only B1 and B3 sites and a 12-kb fragment derived from a gene containing extra B2 site. Pattern is characteristic for heterozygous individual. (2) Cleaved with Bam HI plus Hha I. Monoclonal tumor with the 12 kb derived from an active X chromosome (methylated). (3) Cleaved with Bam HI plus Hha I; monoclonal tumor with the 24-kb fragment derived from an active X chromosome (methylated). (4) Cleaved with Bam HI plus Hha I. Polyclonal tumor. All tumors studied displayed patterns as in Lane 2 or Lane 3.

Source: Vogelstein, B., Fearon, E. R., Hamilton, S. R., and Feinberg, A. B. Use of restriction fragment length polymorphism to determine the clonal origin of tumors. *Science* 227:642, 1985.

Single-Strand Conformation Polymorphism

Southern blot analysis and detection of base changes in DNA from different individuals by RFLP analysis is dependent on alteration of a restriction endonuclease site. A base substitution, deletion, or insertion rarely occurs within a restriction endonuclease site. However, these modifications can still be detected by **single-strand conformation polymorphism (SSCP)**. This technique takes advantage of the fact that single-stranded DNA, smaller than 400 bases long, subjected to electrophoresis through a polyacrylamide gel migrates with a mobility partially dependent on its conformation. A single base alteration usually modifies the DNA conformation sufficiently to be detected as a mobility shift on electrophoresis through a non-denaturing polyacrylamide gel.

The analysis of a small region of genomic DNA or cDNA for SSCP can be accomplished by PCR- amplification of the region of interest. Sense and antisense oligonucleotide primers are synthesized that flank the region of interest, and this DNA is amplified by PCR in the presence of radiolabeled nucleotide(s). The purified radiolabeled double-stranded PCR product is then heat-denatured in 80% formamide and immediately loaded onto a non-denaturing polyacrylamide gel. The mobilities of control products are compared to samples from experimental studies or patients. Detection of mutations in samples from patients can identify genetic lesions. These procedures were successfully applied to the analysis of genes associated with the long-QT syndrome that has been implicated in the sudden infant death syndrome (SIDS) (see Clin. Corr. 7.6). The method depends on prior knowledge of the sequence of the gene/gene fragment of interest while analysis by RFLP requires only restriction map analysis of DNA.

Detection of mRNA

Analysis of the amount of mRNA species in tissue or total RNA preparations is often critical for our understanding gene regulation of cell growth and tissue differentiation. One can assess timing, level, and site within a tissue of gene expression through the analysis of specific mRNA species. Several techniques are available to assay the amount of a specific mRNA species in an RNA preparation: (1) **Northern blot analysis** requires the electrophoretic separation of RNA species, by size, on an agarose gel followed

CLINICAL CORRELATION **7.6**

Single-Strand Conformational Polymorphism for Detection of Spontaneous Mutations that May Lead to SIDS

The sudden infant death syndrome (SIDS) is a major cause of death during the first year of life in the United States. Prospective study of more then 34,000 newborns who were monitored by electrocardiography indicated a strong correlation between increased risk of SIDS and prolonged QT interval in their heart EKG. Based on this study, it was decided to look for a mutation in one or more of the genes known to be related to the Long QT syndrome in a 44-day-old infant that presented cyanotic, apneic, and pulseless to a hospital emergency room. The child's arrhythmia with a prolonged QT interval was stabilized with multiple electrical DC shocks followed by drug treatment. Genomic DNA was prepared from peripheral blood lymphocytes from the infant and his parents. One gene associated with the Long QT syndrome contained the substitution of AAC for TCC at position 2971 to 2972 (protein associated with the sodium channel) by single-strand conformational polymorphism (SSCP) analysis in the child's but not the parents' DNA sample. The mutation replaced a serine residue with an asparagine in a highly conserved region of the protein that is presumed to participate in the function of the sodium channel. The mutation was not detected in 200 control subjects. The conclusion was that the child had a spontaneous mutation in a gene that was associated with a prolonged QT interval and that this contributed to a SIDS-like event. After treatment, the child was symptom free at nearly five years of age. This study points to the potential value of neonatal electrocardiographic screening to reduce infant mortality due to SIDS.

Source: Schwartz, P. J., Priori, S. G., Dumaine, R., Napolitano, C., Antzelevitch, C., Stramba-Badiale, M., Richard, T. A., Berji, M. R., and Bloise, R. A molecular link between the sudden infant death syndrome and the Long-QT syndrome. *N. Engl. J. Med.* 343:262, 2000.

by the transfer and cross-linking to a membrane as in the Southern blot technique. The fixed RNA species are then hybridized with a labeled probe specific for the mRNA species of interest. If the probe is radiolabeled, hybrids can be detected by autoradiography. Nonradioactive probes are also available. Northern blot analysis allows ready determination of the size of the mRNA and identification of potential alternatively spliced transcripts and/or the presence of multigene family transcripts. (2) **RT-PCR** (see Section 7.8), unlike the Northern blot, allows one to measure the amount of an mRNA species present in an RNA sample. Measurement is based on comparison of the relative amount of the PCR product of an internal control transcript to the amount of a specific mRNA in the same sample. Typical internal control transcripts include GAPDH (glyceraldehyde 3-phosphate dehydrogenase) and β-actin. (3) The **nuclease protection assay** (NPA) does not give any information about the size of the mRNA of interest, but is ideally suited for the simultaneous analysis of multiple mRNA species. Radiolabeled or nonisotopic probe(s) are hybridized with the RNA sample in solution. Hybridization in solution is more efficient than hybridization to membrane-fixed RNA, as used in Northern blots. Any remaining unhybridized probe and RNA are then hydrolyzed with nucleases. The mixture of hybridized species is then separated through a low percentage (6%) acrylamide gel. The size of each hybridized species is determined by the size of the probe employed. NPA is commonly used to detect mRNA transcripts in different tissues (Figure 7.13). (4) ***In situ* hybridization** is the only method described in this section to define which cells in a sample of formalin-fixed tissue expresses a specific gene. In this method, RNA is not isolated nor subjected to electrophoretic separations. Rather tissues of interest are fixed, thin-sectioned, and mounted onto microscope slides. Proteins are digested with protease K to increase accessibility of the labeled probe to hybridize with the cellular mRNA of interest. This method can support findings of the nuclease protection assays and also define which cell types within the tissues are expressing the mRNA of interest.

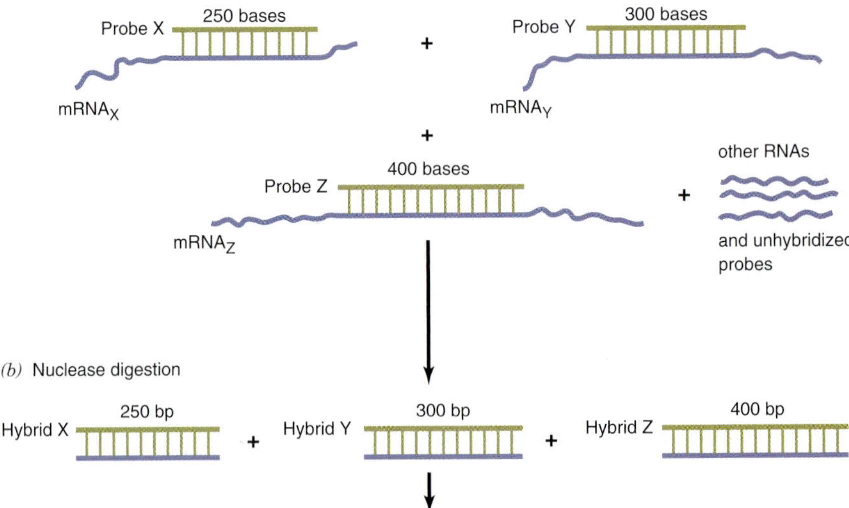

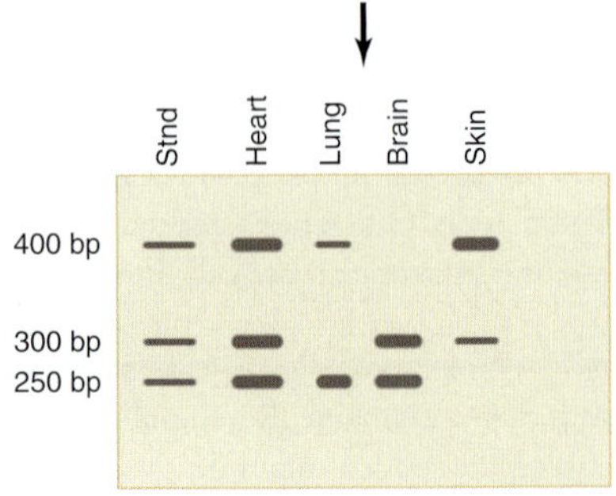

FIGURE 7.13

Nuclease protection assay. Total cellular mRNA can be isolated from different tissues. Single-stranded DNA probes that are complementary to known sequences of different gene transcripts (mRNAx, mRNAy, mRNAz) are hybridized with the RNA mixture. Nuclease digestion with a ribonuclease will hydrolyze single-stranded RNA regions of mRNA not hybridized with the DNA probe and all nonhybridized RNA species. Only the DNA–RNA hybrids protected against nuclease will remain for analysis by acrylamide gel electrophoresis. Differential expression of genes in specific tissues is then readily observed.

Detection of Sequence-Specific DNA-Binding Proteins

Regulatory proteins bind to specific DNA sequences that flank genes that up- or down-regulate gene expression. The **electrophoretic mobility shift assay** (EMSA) or gel retardation method has been employed extensively to analyze sequence-specific DNA-binding proteins along with the DNA sequence required for binding. Proteins with potential DNA-binding characteristics are prepared from whole-cell extracts, nuclear extracts, purified protein preparations, or recombinant proteins from genetically engineered expression systems. The DNA probe is radiolabeled. The DNA employed may be DNA fragments, double-stranded synthetic oligonucleotides, or cloned DNA containing a known protein-binding site(s) or potential protein-binding site(s). The DNA must be double-stranded since single-stranded DNA would bind nonspecific single-strand binding proteins (with cell or nuclear extracts) that would interfere with the interpretation of the results. The purified labeled DNA probe and protein sample are preincubated to form a stable protein–DNA complex prior to the electrophoretic analysis of the sample.

Figure 7.14 schematically depicts the expected results where the radiolabeled DNA complexes with a protein, resulting in its retarded movement through the gel relative to the unreacted DNA. If an antibody to the DNA-binding protein of interest is added to the preincubation tube, one of two possible reactions can occur. If the antibody binds to an epitope that does not impede protein binding to the DNA, the complex size is increased, resulting in a **supershift** (greater retardation) in the gel migration. Conversely, if the antibody blocks the protein's DNA-binding site, a protein–antibody complex will form and the labeled DNA probe will be unmodified in its migration through the gel. In either case the result with the antibody would confirm the identity of the DNA-binding protein.

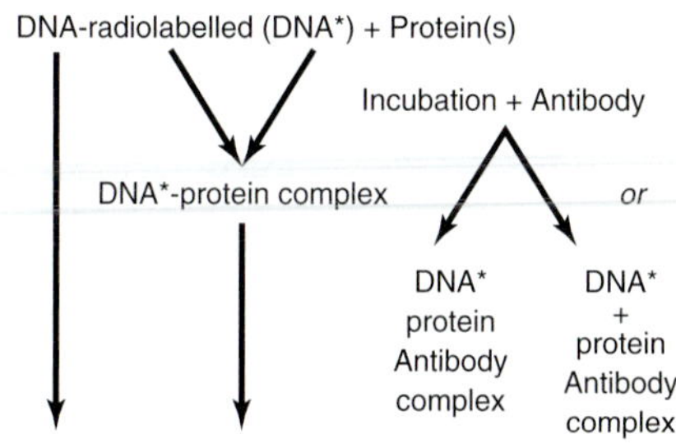

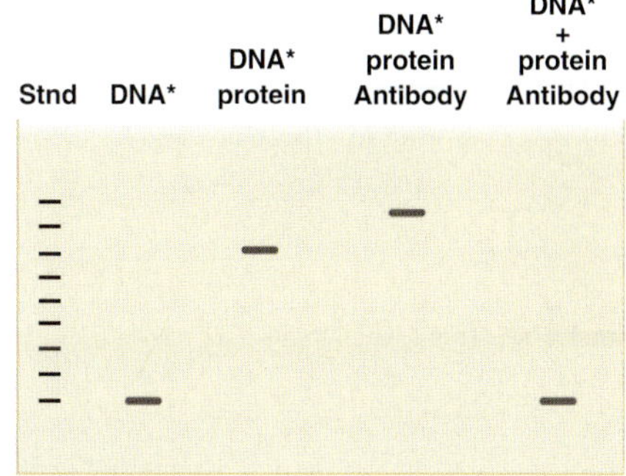

FIGURE 7.14

Electrophoretic mobility shift assay. Purified DNA migrates through a gel when subjected to an electric field based upon its charge and molecular mass. If a protein(s) is added to this purified DNA that complexes with the DNA, the total apparent molecular mass of the DNA is increased. This would slow its migration through the gel—a mobility shift. The addition of an antibody that reacts with the protein complexed with the DNA, but does not interfere with the DNA–protein interaction, will further increase the size of the DNA–protein complex and slow its migration. If the antibody reacts with the DNA-binding region of the protein, a DNA–protein complex will not form. In both cases the antibody helps to identify the DNA-binding protein.

7.8 COMPLEMENTARY DNA AND COMPLEMENTARY DNA LIBRARIES

Insertion of specific functional eukaryotic genes into vectors that can be expressed in a prokaryotic cell could produce large amounts of "genetically engineered" proteins with significant medical, agricultural, and experimental potential. Hormones and enzymes, including insulin, erythropoietin, thrombopoietin, interleukins, interferons, and tissue plasminogen activator, are currently produced by these methods. It is not possible to clone functional genes from genomic DNA, except in rare instances. One reason is that most genes within the mammalian genome yield transcripts that contain introns that must be spliced out of the primary mRNA transcript. Prokaryotic systems cannot splice out the introns to yield functional mRNA transcripts. This problem can be circumvented by synthesizing **complementary DNA (cDNA)** from functional eukaryotic mRNA.

mRNA as Template for DNA Synthesis Using Reverse Transcriptase

Messenger RNA can be reverse-transcribed to cDNA and the cDNA inserted into a vector for amplification, identification, and expression. Mammalian cells normally contain 10,000–30,000 different species of mRNA molecules at any time during the cell cycle. In some cases, however, a specific mRNA species may approach 90% of the total mRNA, such as mRNA for globin in reticulocytes. Many mRNAs are normally present at only a few (1–14) copies per cell. A cDNA library can be constructed from the total cellular mRNA; but if only a few copies per cell of mRNA of interest are present, the cDNA may be very difficult to identify. Methods that enrich the population of mRNAs or their corresponding cDNAs permit reduction in number of different cDNA species within a cDNA library and greatly enhances the probability of identifying the clone of interest.

Desired mRNA can be Enriched by Separation Techniques

Messenger RNA can be separated by size by gel electrophoresis or centrifugation. Isolation of mRNA in a specific molecular size range enriches severalfold an mRNA of interest. Knowledge of the molecular weight of the protein encoded by the gene of interest gives a clue to the approximate size of the mRNA transcript or its cDNA; variability in the predicted size, however, arises from differences in the length of the untranslated regions of the mRNAs.

Enrichment of a specific mRNA molecule can also be accomplished by immunological procedures, but it requires the availability of antibodies against the encoded protein by the gene of interest. Antibodies added to an *in vitro* protein synthesis mixture react with the growing polypeptide chain associated with polysomes and precipitate it. The mRNA can then be purified from the immunoprecipitated polysomal fraction.

Complementary DNA Synthesis

An mRNA mixture is used as a template to synthesize complementary strands of DNA using RNA-dependent DNA polymerase, reverse transcriptase (Figure 7.15). A primer is required; advantage is taken of the poly(A) tail at the 3′-terminus of eukaryotic mRNA. An oligo(dT) with 12–18 bases is employed as the primer that hybridizes with the poly(A) sequence. After cDNA synthesis, the hybrid is denatured or the

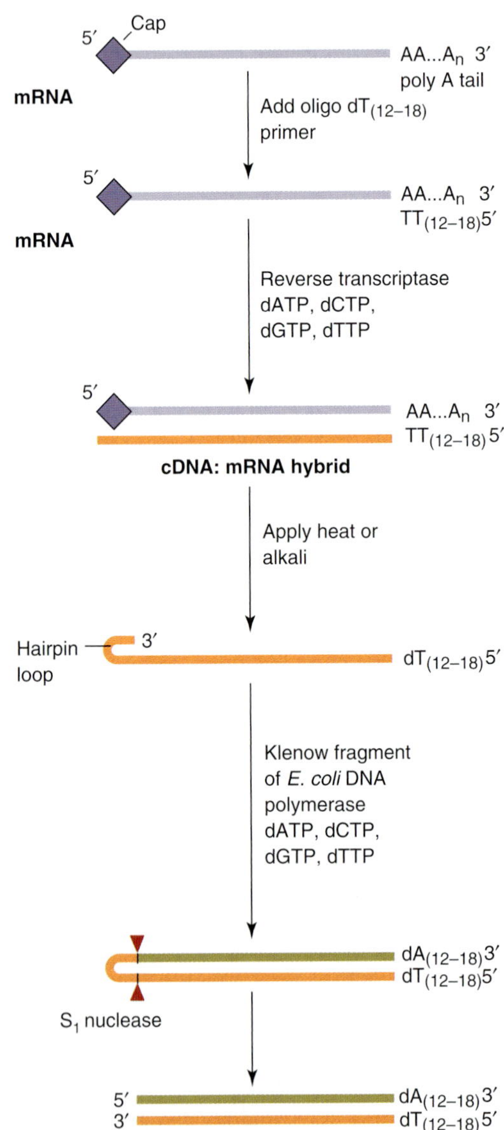

FIGURE 7.15

Synthesis of cDNA from mRNA. The 3′ poly(A) tail of mRNA is hybridized with an oligomer of dT (oligo(dT)12–18) that serves as a primer for reverse transcriptase which catalyzes the synthesis of the complementary DNA (cDNA) strand in the presence of all four deoxynucleotide triphosphates (dNTPs). The resulting cDNA:mRNA hybrid is separated into single-stranded cDNA by melting with heat or hydrolyzing the mRNA with alkali. The 3′-end of the cDNA molecule forms a hairpin loop that serves as a primer for the synthesis of the second DNA strand catalyzed by the Klenow fragment of *E. coli* DNA polymerase. The single-stranded unpaired DNA loop is hydrolyzed by S_1 nuclease to yield a double-stranded DNA molecule.

mRNA hydrolyzed in alkali to obtain the single-stranded cDNA. The 3′-termini of single-stranded cDNAs form a hairpin loop that serves as a primer for the synthesis of the second strand of the cDNA by the Klenow fragment or a reverse transcriptase. The resulting double-stranded cDNA contains a single-stranded loop that is recognized and digested by S1 nuclease. The ends of the cDNA must be modified prior to cloning in a vector. One method involves incubating blunt-ended cDNA molecules with linker molecules and a bacteriophage T4 DNA ligase that catalyzes the ligation of blunt-ended molecules (Figure 7.16). The synthetic linker molecules contain restriction endonuclease sites that can now be cleaved with the appropriate enzyme for insertion of the cDNA into a vector cleaved with the same endonuclease.

Bacteriophage DNA (see p. 266) is the most convenient and efficient vector to create cDNA libraries because they can readily be amplified and stored indefinitely. Two bacteriophage vectors, **λgt10** and **λgt11**, and their newer constructs, have been employed to produce cDNA libraries. The cDNA libraries in λgt10 can be screened only with labeled nucleic acid probes when the sequence of the DNA is unknown, whereas those in λgt11, an expression vector, can also be screened with antibody for the production of the protein or antigen of interest. If the sequence of the desired cDNA is known, then PCR can be used to screen the recombinant bacteriophage.

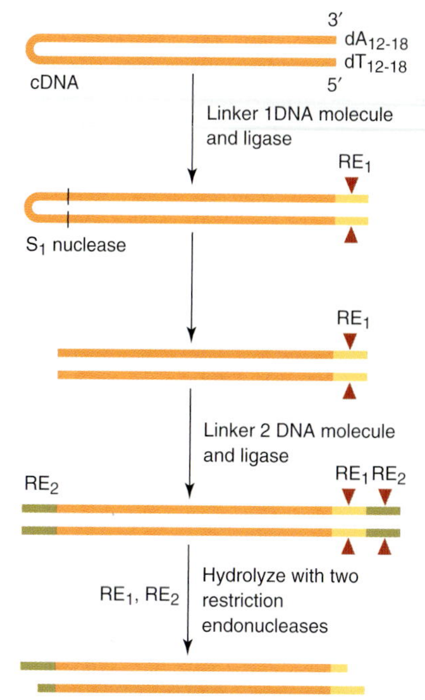

FIGURE 7.16

Modification of cDNA for cloning. The procedure begins with double-stranded DNA that contains a hairpin loop.

A linker DNA that contains a restriction endonuclease site (RE_1) is added to the free end of the cDNA by blunt end ligation. The single-stranded hairpin loop is next hydrolyzed with S_1 nuclease. A second linker with a different restriction endonuclease site within (RE_2) is blunt-end ligated to the newly created free cDNA. The second linker will probably bind to both ends but will not interfere with the first restriction endonuclease site. The modified DNA is hydrolyzed with the two restriction endonucleases and can be inserted into a plasmid or bacteriophage DNA by directional cloning.

Total Cellular RNA as Template for DNA Synthesis using RT–PCR

Alternative methods to construct cDNA libraries employ a **reverse transcriptase–PCR (RT–PCR)** technique and obviate the need to purify mRNA. One such strategy is depicted in Figure 7.17 and begins with the reverse transcriptase production of a DNA–mRNA hybrid. The method uses terminal transferase to add a dG homopolymer tail to the 3′-end, followed by hydrolysis of the mRNA. PCR primers are synthesized to hybridize with the dG and dA tails and terminate with two different restriction

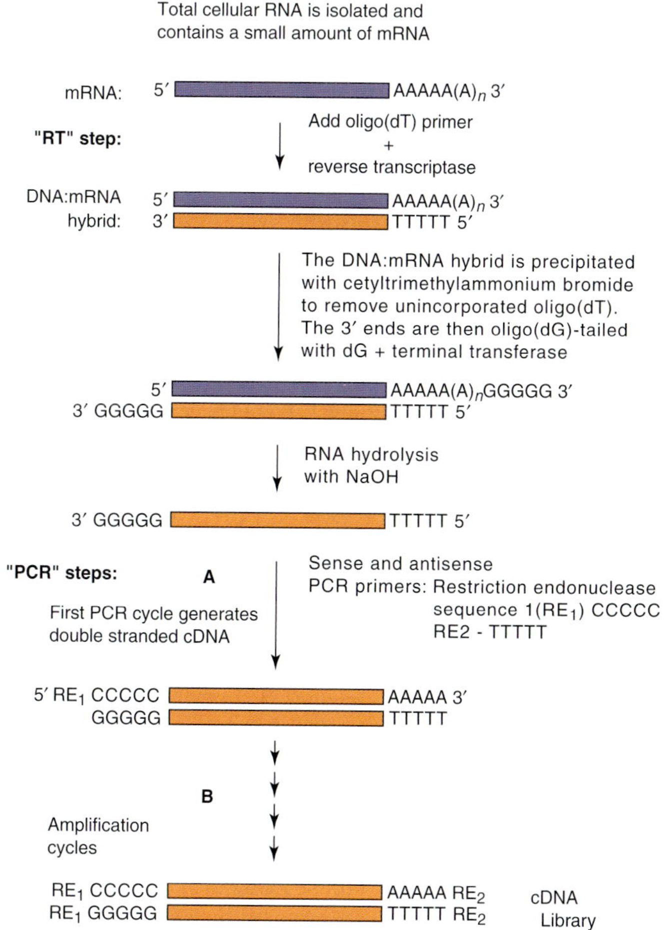

FIGURE 7.17

Generation of cDNA by reverse transcriptase–PCR (RT–PCR). Total cellular RNA or mRNA can be used to generate cDNA by RT–PCR. The mRNA with an oligo(rA) tail is reverse-transcribed with an oligo(dT) primer. An oligo(dG) tail is added to the 3′-ends of the RNA and DNA strands, and the RNA strand is subsequently hydrolyzed with NaOH. Sense and antisense primers, modified with restriction site sequences, are then employed to amplify the cDNA by the PCR. The products can be hydrolyzed with the specific restriction endonucleases (RE_1 and RE_2) for cloning and subsequent studies.

endonuclease sequences. The resulting PCR-amplified cDNA can then be hydrolyzed with two different restriction endonucleases for directional cloning (see p. 253) into an appropriate vector.

7.9 BACTERIOPHAGE, COSMID, AND YEAST CLONING VECTORS

Detection of noncoding sequences (i.e., introns) in most eukaryotic genes and distant regulatory regions that flank the genes necessitated cloning strategies to package larger DNA fragments than can be cloned in plasmids. Plasmids can accommodate foreign DNA inserts 5–10 kb (kilobases) long. Recombinant DNA fragments larger than this are randomly deleted during replication of the plasmid within the bacterium. Thus, alternate vectors have been developed.

Bacteriophage as Cloning Vectors

Bacteriophage λ (**λ phage**), a virus that infects and replicates in bacteria, is an ideal vector for DNA inserts about 15 kb long. The λ phage selectively infects bacteria and replicates by a lytic or nonlytic (lysogenic) pathway. The λ phage contains a self-complementary 12-base single-stranded tail (cohesive termini) at both ends of its 50-kb double-stranded DNA. On infection of the bacteria the cohesive termini (cos sites) of a single λ phage DNA self-anneal, and the ends are covalently linked by the host cell DNA ligase. The circular DNA serves as a template for transcription and replication. Lambda phage, with restriction endonuclease generated fragments representing a cell's whole genomic DNA inserted into it, are used to infect bacteria. Recombinant bacteriophage, released from the lysed cells, are collected and constitute a genomic library in λ phage. The phage library can be screened more rapidly than a plasmid library because of the increased size of the DNA inserts.

Numerous λ-phage vectors have been constructed for different cloning strategies. Only a generic λ-phage vector will be described. A 15- to 25-kb segment of the phage DNA can be replaced without impairing its replication in *E. coli* (Figure 7.18). The length of phage DNA to be packaged into the virus particle must be about 50 kb long. The linear λ-phage DNA can be digested with a specific restriction endonuclease that generates small terminal fragments with their cos sites (arms). These are then separated from larger intervening fragments. Cellular genomic DNA is partially digested by appropriate restriction enzymes to permit annealing and ligation with the phage arms. Genomic DNA is now enzymatically hydrolyzed completely in order to randomly generate fragments that can be properly packaged into phage particles. The DNA fragments smaller or larger than 15–25 kb can hybridize with the cos arms but are excluded from being packaged into infectious bacteriophage particles. All the information required for phage infection and replication in bacteria is carried within the cos arms. The recombinant phage DNA is mixed with λ-phage proteins *in vitro*, which assemble into infectious virions. These are then propagated in an appropriate *E. coli* strain to yield a λ-phage library. Many *E. coli* strains have been genetically altered to sustain replication of specific recombinant virions.

Screening Bacteriophage Libraries

A bacteriophage library can be screened by plating the virus on a continuous layer of bacteria (a bacterial lawn) grown on agar plates (Figure 7.19). Individual phage will infect, replicate, and lyse one cell. The progeny virions will then infect and subsequently lyse bacteria immediately adjacent to the site of the first infected cell, creating a clear region or plaque in the opaque bacterial lawn. Phage, within each plaque, can be picked up on a nitrocellulose filter (as for replica plating) and the DNA fixed to the filter with NaOH. The location of cloned DNA fragments of interest are located by hybridizing the filter-bound DNA with a labeled DNA or RNA probe followed by autoradiography. The bacteriophage in the plaque corresponding to the labeled filter-bound hybrid are

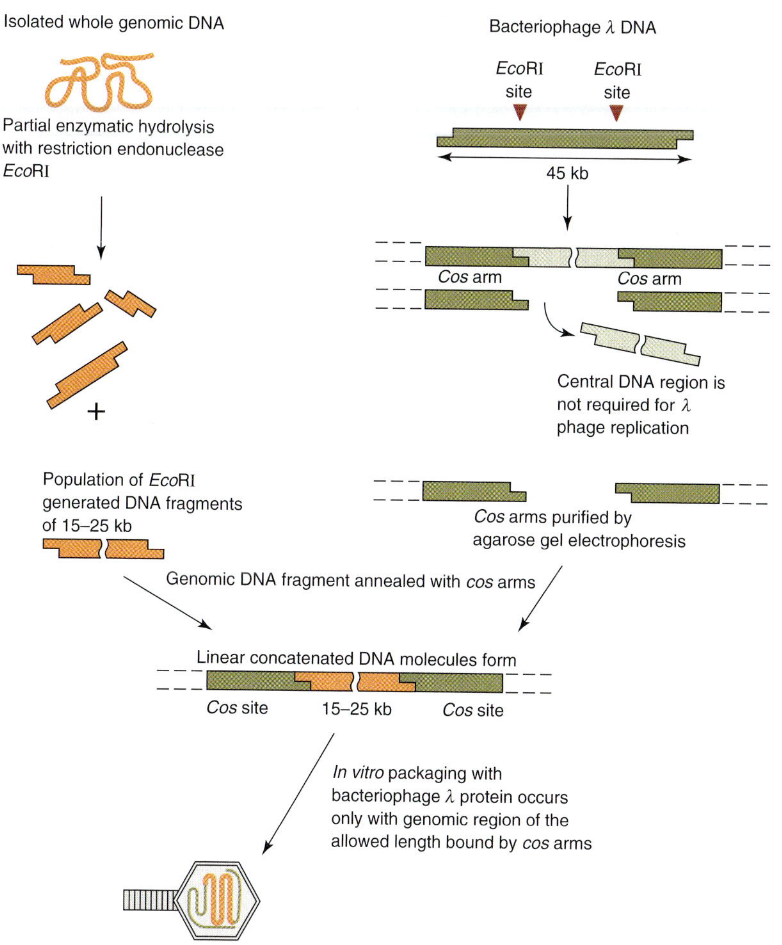

FIGURE 7.18

Cloning genomic DNA in bacteriophage λ. Whole genomic DNA is incompletely digested with a restriction endonuclease (e.g., EcoR1). This results in DNA of random-size fragments with single-stranded sticky ends. DNA fragments called Cos arms are generated with the same restriction endonuclease from bacteriophage DNA. The purified Cos arms carry sequence signals required for packaging DNA into a bacteriophage virion. The genomic fragments are mixed with the Cos arms, annealed and ligated, forming linear concatenated DNA arrays. The *in vitro* packaging with bacteriophage λ proteins occurs only with genomic DNA fragments of allowed lengths (15–25 kb) bounded by Cos arms.

picked up and amplified in bacteria for further analysis. PCR can also be employed if the full or partial sequence is known of the desired DNA. Here one takes portions of several plaques and if the DNA of interest is present in one or more, a region can be amplified with an appropriate primer pair by PCR. The PCR product is then detected by gel electrophoresis. Complementary DNA libraries in bacteriophage are also constructed that contain the phage cos arms. If the cDNA is recombined with phage DNA that permits expression of the gene, such as λgt 11, plaques can be screened immunologically with antibodies specific for the antigen of interest.

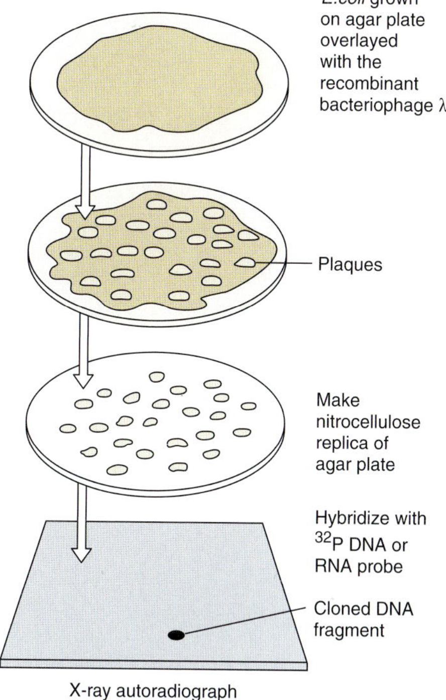

FIGURE 7.19

Screening genomic libraries in bacteriophage λ. Competent *E. coli* are grown to confluence on an agar plate and then overlaid with the recombinant bacteriophage. Plaques develop where bacteria are infected and subsequently lysed by the phage λ. Replicas of the plate can be made by touching the plate with a nitrocellulose filter. The DNA is denatured and fixed to the nitrocellulose with NaOH. The fixed DNA is hybridized with a ^{32}P-labeled probe and exposed to X-ray film. The autoradiograph identifies the plaque(s) with recombinant DNA of interest.

Cloning DNA Fragments into Cosmid and Artificial Chromosome Vectors

Even though λ phage is the most commonly used vector to construct genomic DNA libraries, the length of many genes exceeds the maximum size of DNA that can be inserted between the phage arms. A **cosmid vector** can accommodate foreign DNA inserts of about 45 kb. **Bacterial artificial chromosomes (BACs) and yeast artificial chromosomes (YACs)** have been developed to clone DNA fragments 100–200 and 200–500 kb long, respectively. While cosmid and YAC vectors are difficult to work with, their libraries permit the cloning of large genes with their flanking regulatory sequences, as well as families of genes or contiguous genes.

Cosmid vectors are a cross between plasmid and bacteriophage vectors. Cosmids contain an antibiotic resistance gene for selection of recombinant DNA molecules, an origin of replication for propagation in bacteria, and a cos site for packaging of recombinant molecules in bacteriophage particles. The bacteriophage with recombinant cosmid DNA can infect E. coli and inject its DNA into the cell. Cosmid vectors contain only approximately 5 kb of the 50-kb bacteriophage DNA and, therefore, cannot direct its replication and assembly of new infectious phage particles. Instead, the recombinant cosmid DNA circularizes and replicates as a large plasmid. Bacterial colonies with recombinants of interest can be selected and amplified by methods similar to those described for plasmids.

Standard cloning procedures and some novel methods are employed to construct YACs. Very large foreign DNA fragments are joined to yeast DNA sequences, one that functions as telomeres (distal extremities of chromosome arms) and another that functions as a centromere and as an origin of replication. The recombinant YAC DNA is introduced into the yeast by transformation. The YAC constructs are designed so that yeast transformed with recombinant chromosomes grow as visually distinguishable colonies. This facilitates the selection and analysis of cloned DNA fragments.

7.10 ANALYSIS OF LONG STRETCHES OF DNA

Subcloning Permits Definition of Large Segments of DNA

Complete analysis of functional elements in a cloned DNA fragment requires sequencing of the entire molecule. Current techniques can sequence 200–400 bases, yet cloned DNA inserts are frequently much larger. Restriction maps of the initial DNA clone are essential for cleaving the DNA into smaller pieces to be **subcloned** for further analysis. The sequence of each subcloned DNA fragment can be determined. Overlapping regions of the subcloned DNA properly align and confirm the entire sequence of the original DNA clone.

Sequencing can often be accomplished without subcloning. Antisense primers are used that are complementary to the initially sequenced 3′-ends of the cloned DNA. This process is repeated until all of the cloned DNA has been sequenced. This method obviates the need to prepare subclones but requires synthesis/purchase of numerous primers. On the other hand, the subcloned DNA is always inserted back into the same region of the plasmid. Therefore, one set of primers complementary to the plasmid DNA sequences flanking the inserted DNA can be used for all of the sequencing reactions with subcloned DNA.

Chromosome Walking Defines Gene Arrangement in Long Stretches of DNA

Knowledge of how genes and their regulatory elements are arranged in a chromosome should lead to an understanding of how sets of genes may be coordinately regulated. It is difficult to clone DNA fragments large enough to identify contiguous genes. The combination of several techniques allows for analysis of stretches of DNA 50–100 kb long. The method, **chromosome walking**, is possible because λ-phage or cosmid libraries contain partially cleaved genomic DNA cut at specific restriction endonuclease sites. The cloned fragments will contain overlapping sequences with other cloned fragments. Overlapping regions are identified by restriction mapping, subcloning, screening λ-phage or cosmid libraries, and sequencing procedures.

The procedure of chromosome walking is shown in Figure 7.20. A λ-phage library is screened with a DNA or RNA probe for a sequence of interest. The selected DNA is cloned and restriction-mapped, and a small segment is subcloned in a plasmid, amplified, purified, and labeled by nick translation. This labeled probe is then used to rescreen the λ-phage library for complementary sequences, and these are then cloned. The newly identified overlapping cloned DNA is then treated in the same fashion as

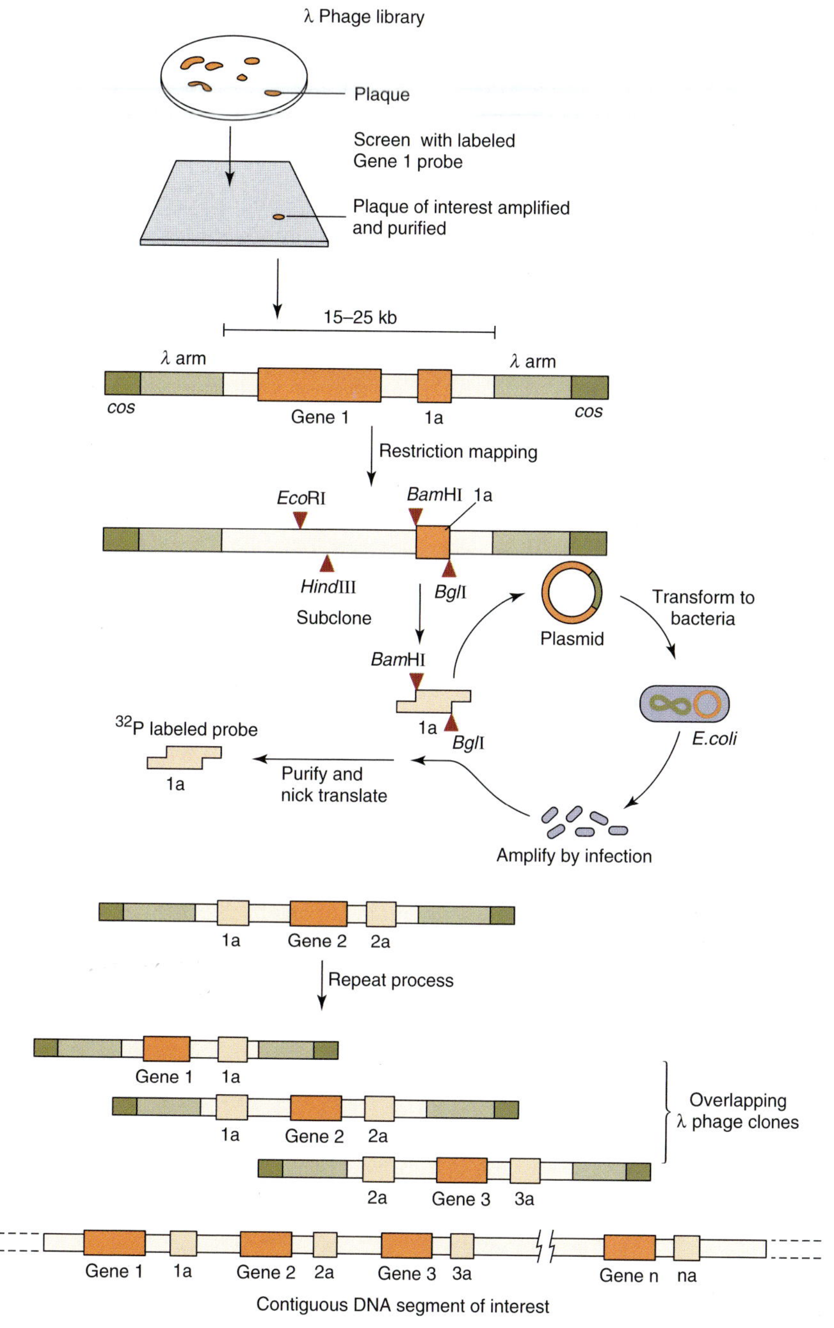

FIGURE 7.20

Chromosome walking to analyze contiguous DNA segments in a genome. Initially, a DNA fragment is labeled by nick translation to screen a library for recombinant λ phage carrying a gene of interest. The amplified DNA is mapped with several restriction endonucleases to select a new region (1a) within the original cloned DNA that can be recloned (subcloned). The subcloned DNA (1a) is used to identify other DNA fragments within the original library that would overlap the initially amplified DNA region. The process can be repeated many times to identify contiguous DNA regions upstream and downstream of the initial DNA (gene 1) of interest.

the initial DNA clone to search for other overlapping sequences. In higher eukaryotic genomes caution must be taken so that subcloned DNA does not contain one of the large numbers of repeating DNA sequences. If a subcloned DNA probe contains a repeat sequence, it hybridizes to numerous bacteriophage plaques and prevents identification of a specific overlapping clone.

7.11 EXPRESSION VECTORS AND FUSION PROTEINS

The recombinant DNA methodology described so far has dealt primarily with screening, amplification, and purification of cloned DNA species. An important goal of recombinant DNA studies is to have a foreign gene expressed in bacteria with the product in a biologically active form. Sequencing of many bacterial genes and their flanking regions has identified the spatial arrangement of regulatory sequences required for expression of genes. A promoter and other regulatory elements upstream of the gene are required to transcribe a gene (see p. 294). The mRNA transcript of a recombinant eukaryotic gene, however, is not translated in a bacterial system because it lacks the bacterial recognition sequence, the Shine–Dalgarno sequence, required to orient it properly with a functional bacterial ribosome. **Expression vectors** facilitate the functional transcription of DNA inserts. A foreign gene can be inserted into such a vector downstream of a regulated promoter but within a bacterial gene, commonly the lacZ gene. The mRNA transcript of the recombinant DNA contains the lacZ Shine–Dalgarno sequence, codons for a portion of the 3′-end of the lacZ protein, followed by the codons of the complete foreign gene of interest. The protein product is a **fusion protein** that contains a few N-terminal amino acids of the lacZ protein and the complete sequence of the foreign protein product.

Foreign Genes Expressed in Bacteria Allow Synthesis of Their Encoded Proteins

Many plasmid and bacteriophage vectors permit the expression of eukaryotic genes in bacterial cells. Rapidly replicating bacteria become a biological factory to produce large amounts of specific proteins, which have research, clinical, and commercial value. As an example, recombinant technologies have produced human protein hormones for treatment of patients with aberrant or missing hormone production. Figure 7.21 depicts a generalized plasmid vector for the expression of a mammalian gene. Recall that the inserted gene must be in the form of cDNA from its corresponding mRNA since the bacterial system cannot remove the introns in the pre-mRNA transcript. The DNA must be inserted in register with the codons of the 3′-terminal region of the bacterial protein when creating a fusion protein. That is, insertion must occur after a triplet codon of the bacterial protein and at the beginning of one for the eukaryotic protein to ensure proper translation. Finally, to yield a functional transcript, the foreign gene must

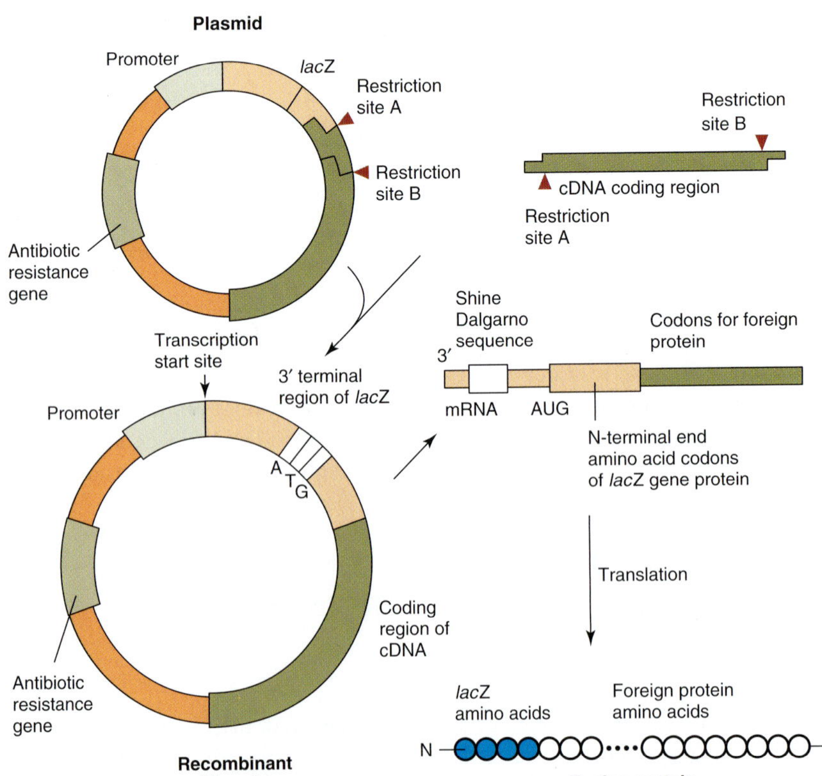

FIGURE 7.21

Construction of a bacterial expression vector. A cDNA for a protein of interest is inserted downstream of bacterial regulatory sequences (promoter (P)) for the lacZ gene, the coding sequence for the mRNA Shine–Dalgarno sequence, the AUG codon, and a few codons for the N-terminal amino acids of the LacZ protein. The mRNA produced from this expression vector directs synthesis in the bacterium of a fusion protein with a few of its N-terminal residues identical to those of the LacZ protein.

be inserted in the proper orientation relative to the promoter. This can be achieved by directional cloning.

Eukaryotic proteins synthesized within bacteria are often unstable and are degraded by intracellular proteases. Fusion protein products, however, are usually stable. The amino acids encoded by the prokaryotic genome may be cleaved from the purified fusion protein by enzymatic or chemical procedures. A strategy to circumvent the intracellular instability of some proteins is to produce a foreign protein that is secreted. This requires cloning the foreign gene in a vector so that the fusion protein synthesized contains a signal peptide that can be recognized by the bacterial signal peptidase that properly processes the protein for secretion.

7.12 EXPRESSION VECTORS IN EUKARYOTIC CELLS

Genetic diseases in mammals result from missing or defective intracellular proteins. To use recombinant techniques to treat these diseases, vectors have to be incorporated into mammalian cells. Vectors have to be selective for the affected tissue or cells containing the aberrant protein. Numerous vectors permit the expression of foreign DNA genes in mammalian cells grown in tissue culture. They have been used extensively for elucidation of the posttranslational processing and synthesis of proteins in cultured eukaryotic cells. The targeting of foreign genes to specific tissues or at specific developmental stages has met with very limited success.

Expression vectors have been developed that allow replication, transcription, and translation of foreign genes in cultured eukaryotic cells. They include both RNA and DNA viral vectors. Shuttle vectors contain both bacterial and eukaryotic replication signals, thus permitting replication of the vector in both bacteria and mammalian cells. Shuttle vectors allow a gene to be cloned and purified in large quantities from a bacterial system and then transferred for expression in mammalian cells. Some expression vectors become integrated into the host cell genome and are stably expressed in the daughter cells. Other expression vectors remain as extra-chromosomal or episomal DNA and permit only transient expression of their foreign gene prior to cell death.

Introduction of foreign DNA, such as viral expression vectors, into the cultured eukaryotic cells is called **transfection**, a process that is analogous to transformation of DNA into bacterial cells. Two common methods involve the formation of a complex of DNA with calcium phosphate or diethylaminoethyl (DEAE)-dextran, which is then endocytosed and transferred to the nucleus, where it is replicated and expressed. Both methods are employed to establish transiently expressed vectors, while the calcium phosphate procedure is also used for permanent expression of foreign genes. Typically, 10–20% of the cells in culture can be transfected by these procedures.

DNA Elements Required for Expression of Vectors in Mammalian Cells

Expression of recombinant genes in mammalian cells requires the presence of appropriate controlling elements within the vector. To be expressed the cloned gene is inserted in the vector in the proper orientation relative to control elements which include a promoter, polyadenylation signals and an enhancer. Expression may be improved by the inclusion of an intron. Some or all of these DNA elements may be present in the recombinant gene if whole genomic DNA is used for cloning. A particular cloned fragment generated by restriction endonuclease cleavage, however, may not contain these elements. A cDNA would not contain these required DNA elements. Therefore, it is necessary that the expression vector to be used in mammalian cells be constructed such that it contains all of the required controlling elements.

The controlling elements can be inserted into the vector by recombinant technologies. Enhancer and promoter elements should be recognized by a broad spectrum of cells in culture for the greatest applicability of the vector. Controlling elements derived from viruses with a broad host range—for example, **papovavirus, simian virus 40(SV40), Rous sarcoma virus**, or the **human cytomegalovirus**—are used. The vector must contain a viral origin of replication (ori) sequence. Specific protein factors, encoded by genes inserted in the vector or previously introduced into the host genome, recognize and interact with the ori sequence to initiate DNA replication.

Selection of Transfected Eukaryotic Cells by Utilizing Mutant Cells and Specific Nutrients

It is important to grow the transfected cells selectively since they often represent only 10–20% of the cell population. A gene can be incorporated into a cell that confers resistance to a drug or confers selective growth capability. This is achieved by **cotransfection** in which two different vectors are efficiently taken up by cells, one that carries the selectable marker and the other that carries the gene of interest. In most cases, more than 90% of transfected cells carry both vectors. Two commonly used selectable markers are the thymidine kinase (tk) and the dihydrofolate reductase (dhfr) genes. Thymidine kinase is expressed in most mammalian cells and where it participates in the salvage pathway of thymidine. Several mutant cell lines are available that lack a functional thymidine kinase gene (tk−) and do not survive in growth medium that contains hypoxanthine, aminopterin, and thymidine. Only those tk− mutant cells cotransfected with a vector carrying a tk gene, usually from herpes simplex virus, will grow in the medium. In most instances, these cells will have been cotransfected with the gene of interest.

Dihydrofolate reductase maintains cellular concentrations of tetrahydrofolate for nucleotide biosynthesis (see p. 1107). Cells lacking this enzyme survive only in media that contains thymidine, glycine, and purines. Mutant cells (dhfr−), transfected with the dhfr gene, can, therefore, be selectively grown in a medium that lacks these supplements. Expression of foreign genes in mutant cells, cotransfected with selectable markers, is limited to cell types with the required gene defect that can be isolated.

Normal cells transfected with a vector that carries the dhfr gene are resistant to methotrexate, an inhibitor of dihydrofolate reductase, and can be selected for by growth in methotrexate. Similarly, normal cells transfected with a bacterial gene coding for aminoglycoside 3′-phosphotransferase (APH) are resistant to aminoglycoside antibiotics such as neomycin and kanamycin, which inhibits protein synthesis in both prokaryotes and eukaryotes. Vectors that carry an APH gene can be used as a selectable marker in both bacterial and mammalian cells.

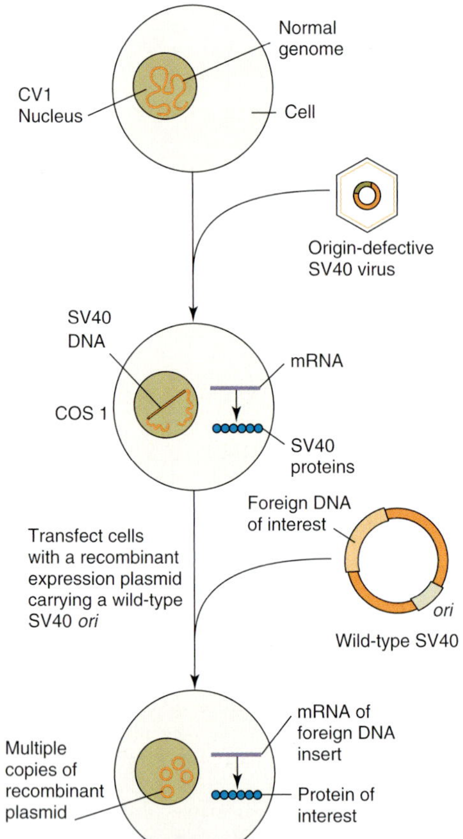

FIGURE 7.22

Expression of foreign genes in eukaryotic COS cells. CV1, an established tissue culture cell line of simian origin, can be infected and supports the lytic replication of the simian DNA virus, SV40. Cells are infected with an origin (ori)-defective mutant of SV40 whose DNA permanently integrates into the host CV-1 cell genome. The defective viral DNA continuously codes for proteins that can associate with a normal SV40 ori to regulate replication. Due to its defective ori, the integrated viral DNA will not produce viruses. The SV40 proteins synthesized in the permanently altered CV-1 cell line, COS-1, can, however, induce the replication of recombinant plasmids carrying a wild-type SV40 ori to a high copy number (as high as 105 molecules per cell). The foreign protein synthesized in the transfected cells may be detected immunologically or enzymatically.

Foreign Genes Expressed in Virus-Transformed Eukaryotic Cells

Figure 7.22 depicts the transient expression of a transfected gene in COS cells, a commonly used system to express foreign eukaryotic genes. The COS cells are permanently cultured simian cells, transformed with an origin-defective SV40 genome. The defective viral genome has become integrated into the host cell genome and constantly expresses viral proteins. Infectious viruses that are normally lytic to infected cells are not produced because the viral origin of replication is defective. The SV40 proteins expressed by the transformed COS cell interact with a normal SV40 ori carried in a transfected vector and therefore, promote the repeated replication of the vector that may reach a copy number in excess of 10^5 molecules/cell. Transfected COS cells die after 3–4 days probably from toxic overload of the episomal vector DNA.

7.13 SITE-DIRECTED MUTAGENESIS

By mutating selected regions or single nucleotides within cloned DNA, it is possible to define the role of DNA sequences in gene regulation and amino acid sequences in protein function.

Site-directed mutagenesis is the controlled alteration of selected regions of a DNA molecule. It may involve insertion or deletion of selected DNA sequences or replacement of a specific nucleotide with a different base. A variety of chemical methods mutates DNA usually at random sites within the molecule.

Role of DNA Flanking Regions Evaluated by Deletion and Insertion Mutations

Site-directed mutagenesis can be carried out in various regions of a DNA sequence including the gene itself or the flanking regions. Figure 7.23 depicts a simple deletion mutation strategy where the sequence of interest is selectively cleaved with restriction endonuclease, the specific sequences removed, and the altered recombinant vector recircularized with DNA ligase. The role of the deleted sequence can be determined by comparing the level of expression (translation) of the gene product, measured immunologically or enzymatically, to the unaltered recombinant expression vector. A similar technique is used to insert new sequences at the site of cleavage. Deletion of a DNA sequence within the flanking region of a cloned gene can indicate its regulatory role in gene expression. The spatial arrangement of regulatory elements relative to one another, to the gene, and to its promoter may also be important in the regulation of gene expression (see Chapter 8).

Analysis of a potential regulatory sequence involves inserting the sequence upstream of a reporter gene in an expression vector. A reporter gene, usually of prokaryotic origin, encodes a protein that can be distinguished from proteins normally present in the nontransfected cell and which can be conveniently and rapidly assayed. A commonly used reporter gene is the chloramphenicol acetyltransferase (CAT) gene of bacteria. This enzyme acetylates and inactivates chloramphenicol, an inhibitor of protein synthesis in prokaryotic cells. The effect of a regulatory element on expression of the CAT gene then can be determined. The regulatory element can be mutated prior to insertion into the vector carrying the reporter gene to determine its spatial and sequence requirements.

Regulatory elements may lack restriction endonuclease sites. In such a case, deletion mutations can be made by linearizing cloned DNA by restriction endonuclease cleavage downstream of the regulatory sequence. The DNA can then be systematically truncated with an exonuclease that removes nucleotides from the free end of both strands of the linearized DNA. Longer digestion generates smaller DNA fragments. Figure 7.24 demonstrates how larger deletion mutations (yielding smaller fragments) can be tested for functional activity. Hydrolysis of the DNA occurs at both ends of the linearized vector and destroys the original restriction endonuclease site (RE2). A unique restriction endonuclease site is reestablished to recircularize the truncated DNA molecule for further manipulations to evaluate the function of the deleted sequence. This is accomplished by ligating the blunt ends with a linker DNA, a synthetic oligonucleotide containing one or more restriction endonuclease sites. The ligated linkers are cut with the appropriate enzyme permitting recircularization and ligation of the DNA.

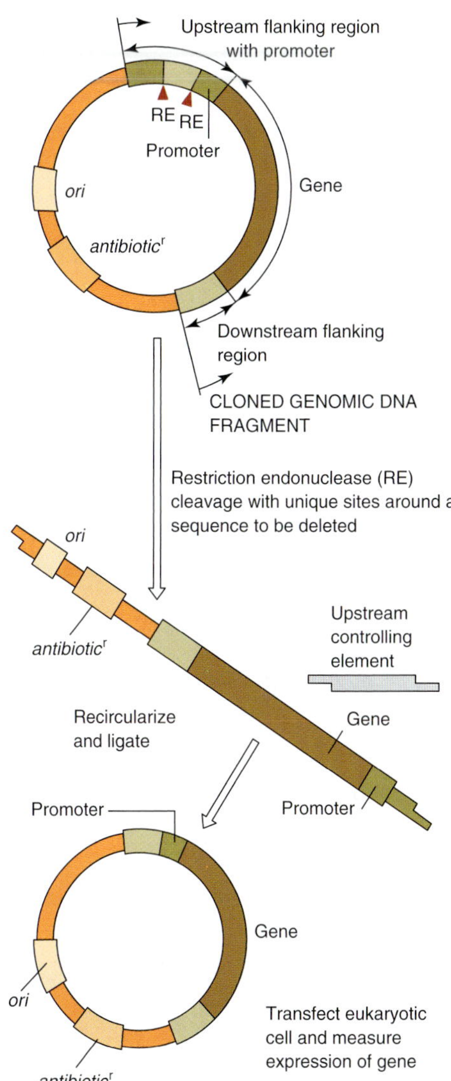

FIGURE 7.23

Use of expression vectors to study DNA regulatory sequences. The gene of interest with upstream and/or downstream DNA flanking regions is inserted and cloned in an expression vector, and the baseline gene expression is determined in an appropriate cell. Defined regions of potential regulatory sequences can be removed by restriction endonuclease cleavage, and the truncated recombinant DNA vector can be recircularized, ligated, and transfected into an appropriate host cell. The level of gene expression in the absence of the potential regulator is determined and compared to controls to ascertain the regulatory role of the deleted flanking DNA sequence.

Site-Directed Mutagenesis of a Single Nucleotide

The procedures discussed previously can elucidate the functional roles of DNA sequences. Frequently, however, one wants to evaluate the role of one nucleotide at a selected site within the DNA so as to evaluate the role of specific amino acids in a protein (see Clin. Corr. 7.7). This method also permits creation or destruction of a restriction endonuclease site at specific locations. Site-directed mutagenesis of a specific nucleotide is a multistep process that begins with cloning the normal gene in a bacteriophage (Figure 7.25). The M13 series of recombinant bacteriophage vectors are commonly employed. This filamentous bacteriophage specifically infects male *E. coli* that express sex pili (F factor) coded on a plasmid. The M13 bacteriophage contains DNA in a single-stranded or replicative form, which is replicated to double-stranded DNA within an infected cell. The double-stranded form is isolated from infected cells and used for cloning the gene to be mutated. The plaques of interest can be visually identified by α-complementation (see p. 256).

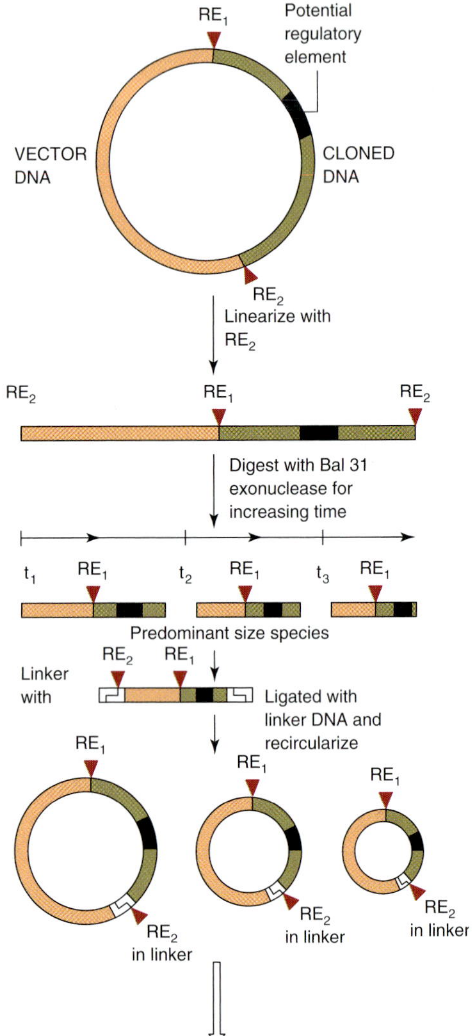

FIGURE 7.24

Enzymatic modification of potential DNA regulatory sequences. A purified recombinant DNA molecule with a suspected gene regulatory element within flanking DNA regions is cleaved with a restriction endonuclease (RE_2). The linearized recombinant DNA is digested for varying periods with the exonuclease, Bal31, reducing the size of the DNA flanking the potential regulatory element. The resulting recombinant DNA molecules of varying reduced sizes have small DNA oligomers (linkers) containing a restriction endonuclease sequence for RE_2 ligated to their ends. The linker-modified DNA is hydrolyzed with RE_2 to create complementary single-stranded sticky ends that permit recircularization. The potential regulatory element, bounded by various reduced-sized flanking DNA sequences, can be amplified, purified, sequenced, and inserted upstream of a competent gene in an expression vector. Modification of the expression of the gene in an appropriate transfected cell can then be monitored to evaluate the role of the potential regulatory element placed at varying distances from the gene.

The recombinant M13 is used to infect susceptible *E. coli*. The progeny bacteriophage released into the growth medium contain single-stranded DNA. An oligonucleotide (18–30 nucleotides long) is synthesized that is complementary to a region of interest except for the nucleotide to be mutated. This oligomer, with one mismatched base, hybridizes to the single-stranded gene and serves as a primer. Primer extension by bacteriophage T4 DNA polymerase results in double-stranded DNA that can be transferred into susceptible *E. coli* where the mutated DNA strand serves as a template for new (+) strands that carry the mutated nucleotide.

The bacteriophage plaques, which contain mutated DNA, are screened by hybridization with a labeled probe of the original oligonucleotide carrying the altered nucleotide. By adjusting the wash temperature of the hybridized probe, only the perfectly matched

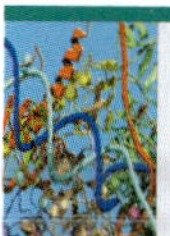

CLINICAL CORRELATION 7.7
Site-Directed Mutagenesis of HSV IgD

The structural and functional roles of a carbohydrate moiety covalently linked to a protein can be studied by site-directed mutagenesis. The gene for a glycoprotein in which an asparagine is normally glycosylated (N-linked) must first be cloned. The herpes simplex virus, type I (HSV I) glycoprotein D (gD) may contain up to three N-linked carbohydrate groups. This viral envelope-bound protein appears to play a central role in virus adsorption and penetration. Carbohydrate groups may play a role in these processes.

The gene modified by site-directed mutagenesis at each, or all, of the three potential glycosylation sites were cloned within an expression vector and were transfected into eukaryotic cells (COS-1), where the gD protein was transiently expressed. The mutated protein, lacking one or all of the carbohydrate groups, was tested with several monoclonal anti-gD antibodies to determine if immunological epitopes (specific sites on a protein recognized by an antibody) were altered. Mutations at two of the glycosylation sites altered the native conformation of the protein so that it was less reactive with the antibodies. Alteration at a third site had no apparent effect, while loss of the carbohydrate at all three sites did not prevent normal processing of the protein.

Source: Sodora, D. L., Cohen, G. H., and Eisenberg, R. J. Influence of asparagine-linked oligosaccharides on antigenicity, processing, and cell surface expression of herpes simplex virus type I glycoprotein D. *J. Virol.* 63:5184, 1989.

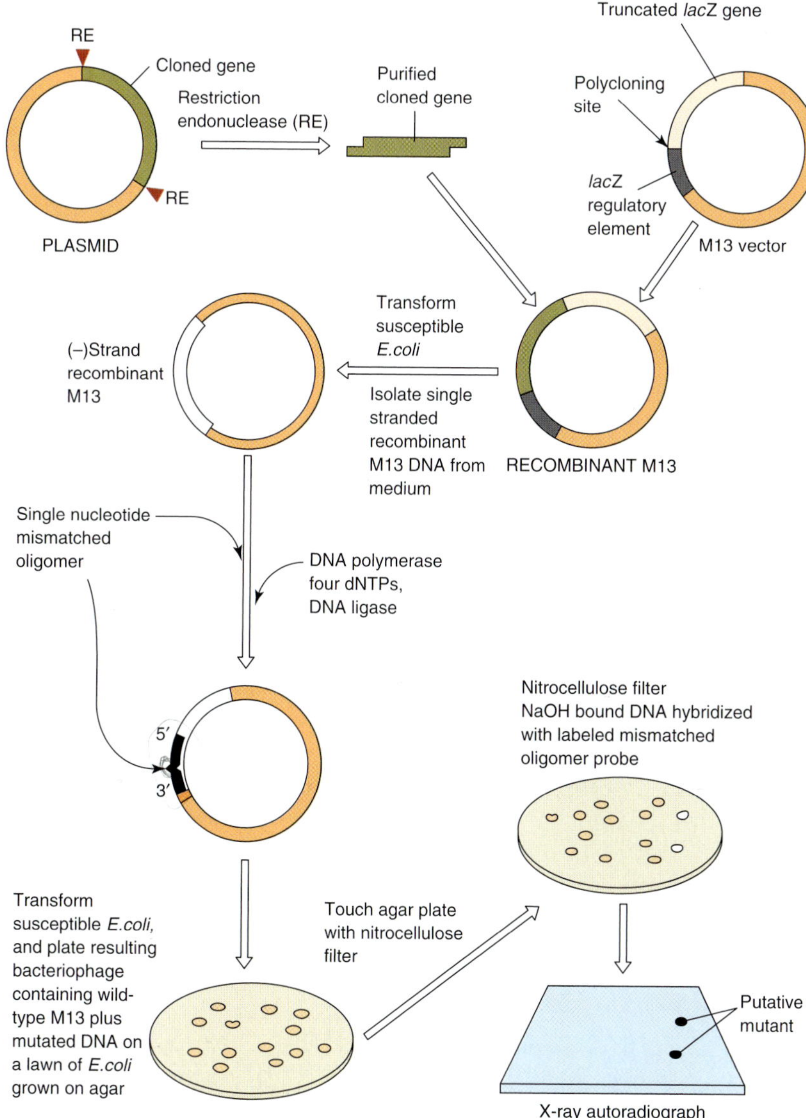

FIGURE 7.25

Site-directed mutagenesis of a single nucleotide and detection of the mutated DNA. The figure is a simplified overview of the method. This process involves the insertion of an amplified pure DNA fragment into a modified bacteriophage vector, M13. Susceptible *E. coli*, transformed with the recombinant M13 DNA, synthesize the (+) strand DNA packaged within the bacteriophage proteins. The bacteriophages are isolated from the growth medium, and the single-stranded recombinant M13 DNA is purified. The recombinant M13 DNA serves as a template for DNA replication by DNA polymerase, deoxynucleoside triphosphates (dNTPs), DNA ligase, and a special primer. The DNA primer (mismatched oligomer) is synthesized to be exactly complementary to a region of the DNA (gene) of interest except for the one base intended to be altered (mutated). The newly synthesized M13 DNA, therefore, contains a specifically mutated base which when reintroduced into susceptible *E. coli* will be faithfully replicated. The transformed *E. coli* are grown on agar plates with replicas of the resulting colonies picked up on a nitrocellulose filter. DNA associated with each colony is denatured and fixed to the filter with NaOH and the filter bound DNA is hybridized with a ^{32}P-labeled mismatched DNA oligomer probe. The putative mutants are then identified by exposing the filter to X-ray film.

hybrid remains complexed with the mutated DNA while the oligomer with mismatched nucleotide will dissociate from the wild type DNA. The M13 that carries the mutated gene is then replicated in bacteria, the DNA purified, and the mutated region of the gene sequenced to confirm the identity of the mutation. A modified method to selectively replicate the mutated strand has been developed to improve the efficiency of site-directed mutagenesis. The M13 bacteriophage, replicated in a mutant *E. coli*, incorporates some uracil residues in place of thymine because of a metabolic defect in synthesis of dTTP from dUTP and lack of an enzyme that normally removes uracil from DNA. The purified single-stranded M13 uracil-containing DNA is hybridized with a complementary oligomer containing a mismatched base at the nucleotide to be mutated. The oligomer serves as the primer for DNA replication *in vitro* with the template (+) strand containing uracils and the new (−) strand containing thymines. When the double-stranded M13 DNA is transformed into a wild-type *E. coli* the uracil-containing strand is destroyed and the mutated (−) strand serves as the template for the progeny bacteriophage, most of which will carry the mutation of interest.

The PCR can also be employed for site-directed mutagenesis. Strategies have been developed to incorporate a mismatched base into one of the oligonucleotides that primes the PCR. Some of these procedures employ M13 bacteriophage and follow the principles described in Figure 7.26. A variation of these PCR methods, **inverse PCR**

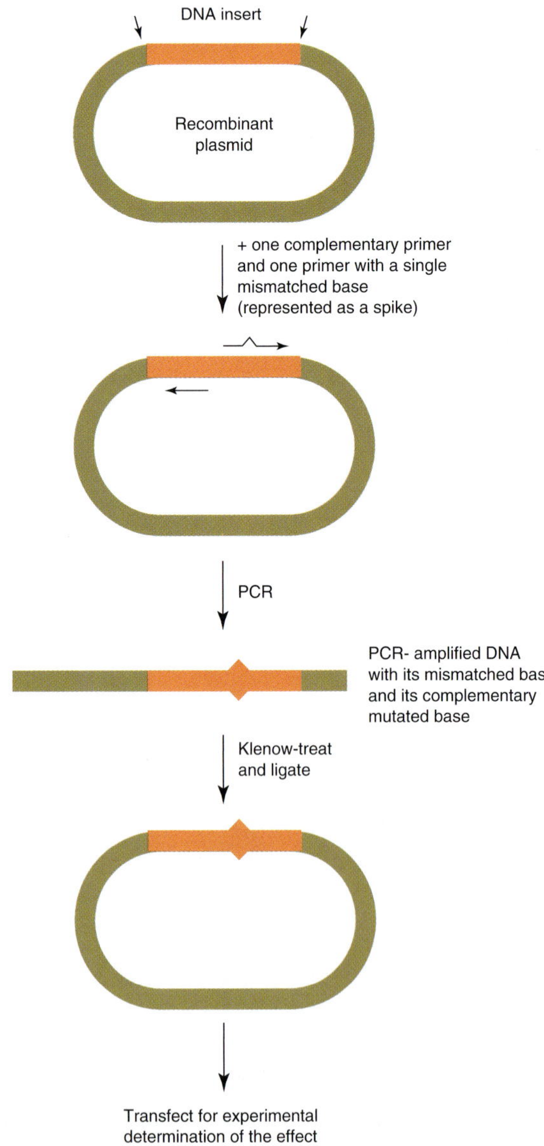

FIGURE 7.26

Inverse PCR mutagenesis. A single base can be mutated in recombinant DNA plasmids by inverse PCR. Two primers are synthesized with their antiparallel 5′-ends complementary to adjacent bases on the two strands of DNA. One of the two primers carries a specific mismatched base that is faithfully copied during the PCR amplification steps, yielding ultimately a recombinant plasmid with a single mutated base.

mutagenesis, has been applied to small recombinant plasmids (4–5 kb) (Figure 7.25). The method is very rapid, with 50–100% of the generated colonies containing the mutant sequence. The two primers are synthesized so that they anneal back-to-back with one primer carrying the mismatched base.

7.14 APPLICATIONS OF RECOMBINANT DNA TECHNOLOGIES

Recombinant DNA methods are applicable to numerous biological disciplines including agriculture, studies of evolution, forensic biology, and clinical medicine. Genetic engineering can introduce new or altered proteins into crops (e.g., corn), so that they contain amino acids essential to humans but often lacking in plant proteins. Toxins lethal to specific insects but harmless to humans can be introduced into crops to protect plants thus avoiding the use of environmentally destructive pesticides. The DNA from cells in the amniotic fluid of a pregnant woman can be analyzed for genetic defects in the fetus. Minuscule quantities of DNA from biological samples preserved in ancient tar pits or frozen tundra can be amplified and sequenced for evolutionary studies. DNA from a single hair, a drop of blood, or sperm from a rape victim can be isolated, amplified, and mapped for forensic purposes. Improvement in current technologies should permit selective introduction of genes into defective cells and of nucleic acids into specific cells to inhibit selectively the expression of detrimental genes.

Antisense Nucleic Acids in Research and Therapy

Antisense nucleic acids (RNA or DNA), can be used to study the intracellular expression and function of specific proteins. Natural and synthetic antisense nucleic acids that are complementary to mRNA inactivate it and block translation. The introduction of antisense nucleic acids into cells has opened new avenues to explore how selectively repressed proteins function within that cell. This method also holds great promise in control of viral infections. Antisense technology and site-directed mutagenesis are part of reverse genetics (from gene to phenotype) which selectively modifies a gene to evaluate its function, while classical genetics depends on the isolation and analysis of cells carrying random mutations that can be identified. A second use of the term **reverse genetics** refers to the mapping and ultimate cloning of a human gene associated with a disease for which the molecular agents are still unknown.

Antisense RNA can be introduced into cells by common cloning techniques. Figure 7.27 demonstrates a gene cloned in an expression vector downstream of a promoter but in the opposite direction to normal. This places the complementary or antisense strand of the DNA under the control of the promoter with expression or transcription yielding antisense RNA. Transfection of cells with the antisense expression vector introduces antisense RNA that hybridizes with normal mRNA. The mRNA–antisense RNA complex is not translated due to a number of reasons such as its inability to bind to ribosomes, blockage of normal processing, and rapid enzymatic degradation.

Synthetic antisense DNA oligonucleotides have been produced that inhibit viral infection including that by the human immunodeficiency virus (HIV). It is conceivable that antisense HIV nucleic acids will be introduced into bone marrow cells removed from AIDS patients. These "protected" cells will then be reintroduced into the patient and replace those cells normally destroyed by the virus. Progress is being made with antisense nucleic acids that regulate the expression of oncogenes, genes involved in cancer formation. Harnessing antisense technologies holds great promise for the treatment of human diseases. (see Clin. Corr. 7.8).

Molecular Techniques Applied to the Whole Animal

Many *in vitro* and intracellular systems have been described here that facilitate the purification, sequencing, and modification of genes or their cDNAs and DNA flanking

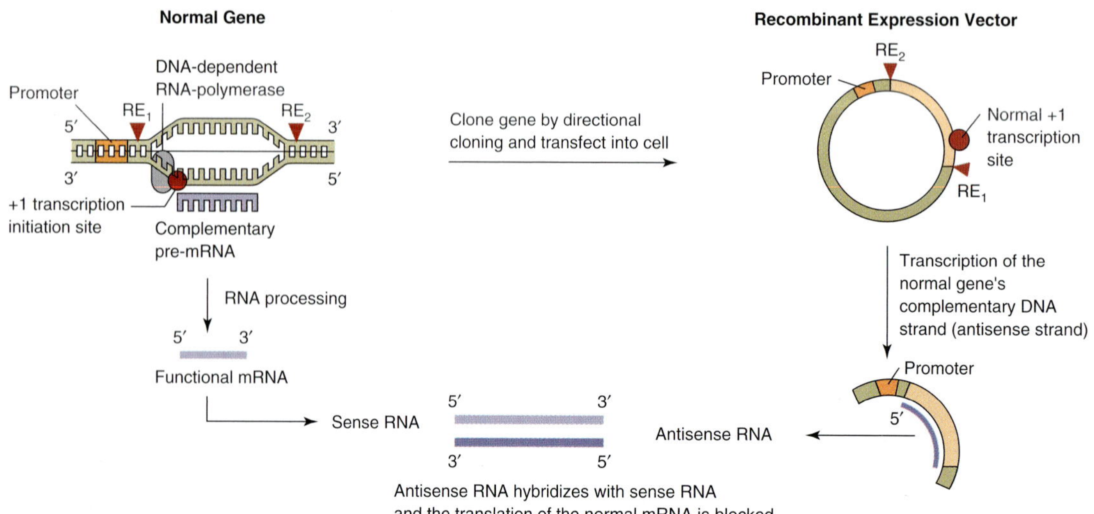

FIGURE 7.27

Production of antisense RNA. A gene, or a portion of it, is inserted into a vector downstream of a promoter by directional cloning and in the reverse orientation to that normally found. Transfection of this recombinant DNA into the parental cell that carries the normal gene results in the transcription of RNA (antisense RNA) from the cloned reversed-polarity DNA along with a normal cellular mRNA (sense RNA) transcript. The two antiparallel complementary RNAs hybridize within the cell resulting in blocked expression (translation) of the normal mRNA transcript.

CLINICAL CORRELATION 7.8

RNA-Mediated Inhibition of HIV

RNA molecules with catalytic activity, **ribozymes**, were targeted to specific HIV-1 RNA coding sequence for tat and rev, two essential genes for HIV-1 expression. The ribozymes were expressed from retroviral vectors in transduced peripheral $CD34^+$ hematopoietic progenitor cells. HIV-1-infected individuals were infused with these genetically modified CD 34^+ cells, which are resistant to HIV-1 expression. These cells persisted for up to one year post bone marrow transplant and then disappeared, indicating that more efficient manipulations of these stem cells are required. Current work toward this goal is the coexpression of the anti-HIV ribozymes with siRNA constructs. Small interfering RNAs (siRNA) are small double-stranded RNA molecules (21–23 nucleotides) that have antisense sequences to a targeted mRNA (such as tat and rev HIV-1 RNA) and form an RNA-induced silencing complex that leads to the destruction of the targeted RNA. These siRNA constructs have been very effective in regulating HIV-1 expression in cultured cells. The ultimate goal for an RNA-dependent anti-HIV gene therapy seems attainable in the near future.

Source: Michienzi, A., Castanotto, D., Lee, N., Li, S., Zaia, J. A., and Rossi, J. J. RNA-mediated inhibition of HIV in a gene therapy setting. In Y. S. Cho-Chung, A. M. Gewirtz, and C. A. Stein (Eds.). *Therapeutic Oligonucleotides. Ann. NY Acad. Sci.* 1002:63, 2003.

regions of genes with potential regulatory functions. Introduction of foreign genes into a whole animal, deletion of a gene from the whole animal genome, and cloning of a whole animal has been achieved. While clear bioethical issues and hurdles must still be addressed when using some of these methodologies, the methods themselves hold great promise for future gene therapy.

Gene Therapy: Normal Genes can be Introduced into Cells with a Defective Gene

Individuals who possess a defective gene resulting in a debilitating or fatal condition can theoretically be treated by supplying their cells with a normal gene. **Gene therapy**, the transfer of a normal gene to humans, has been accomplished using retroviral vectors (see Clin. Corr. 7.9). The success of gene transfer depends, in part, on the integration of the gene into the host genome which is directed by the retroviral integration

CLINICAL CORRELATION 7.9

Gene Therapy: Normal Genes Can Be Introduced into Cells with Defective Genes

More than 4000 genetic diseases are known, many of which are debilitating or fatal. Most are currently incurable. With the advent of new technologies in molecular biology, the clinical application of gene transfer and gene therapy is becoming a reality. Diseases that result from a deficiency in adenosine deaminase (ADA) or a mutation in the gene that encodes a subunit of several cytokine receptors, γc, are but two of many genetic diseases that may be readily cured by gene therapy.

ADA is important in purine salvage, catalyzing the conversion of adenosine to inosine or deoxyadenosine to deoxyinosine. It contains 363 amino acids and has highest activity in thymus and other lymphoid tissues. Over 30 mutations in the ADA gene are associated with severe combined immune deficiency disease (SCID), an autosomal recessive disorder. These immune-compromised children usually die in the first few years of life from overwhelming infections. The first authorized gene therapy in humans began on September 14, 1990 with the treatment of a 4-year-old girl with ADA deficiency. The patient's peripheral blood T cells were cultured with appropriate growth factors. The ADA gene was introduced within these cells by retroviral mediated gene transfer. A modified retrovirus was constructed to contain the human ADA gene such that it would be expressed in human cells without virus replication. The modified T cells carrying a normal ADA gene were then reintroduced to the patient by autologous transfusion. Levels of ADA as low as 10% of normal are sufficient to normalize the patient. The patient, now 10 years later at age 13, is alive and well and has maintained circulating gene-corrected T cells at a level of 20–25% of total T cells.

The X-linked inherited disorder, SCID-X1, results in the blocked differentiation of T and natural killer (NK) lymphocytes. SCID-X1 results from a mutation in the cytokine receptor subunit, γc, common to interleukin-2, -4, -7, -9, and -15 receptors. The methodology of treatment for an 8- and an 11-month-old patient was similar to that employed for the ADA-deficient SCID patients, which tends to be a less severe disease than the SCID-X1. The SCID-X1 studies transduced CD34+ stem cells with a retroviral vector carrying the normal γc cDNA. This resulted in a much higher level of gene transduction than observed in earlier studies. Both infants had their immune systems normalized sufficiently after 3 months to leave the protective isolation of the hospital. A moratorium on gene therapy for SCID-X1 was recently triggered by the induction of a leukemia-like syndrome in 2 of 14 European children treated for this disease. The T-cell leukemia was caused by retroviral insertion and up-regulation of the LMO2 gene that encodes a transcription factor required for hematopoiesis. It is anticipated that future modifications in the vector used for gene therapy should prevent the apparent insertional preference of the retrovirus near active genes.

Source: Blaese, R. M. Progress toward gene therapy. *Clin. Immun. Immunopathol.* 61:574, 1991. Mitani, K., Wakamiya, M., and Caskey, C. T. Long-term expression of retroviral- transduced adenosine deaminase in human primitive hematopoietic progenitors. *Human Gene Therapy* 4:9, 1993. Anderson, W. F. The best of times, the worst of times. *Science* 288:627, 2000. Cavazzana-Calvo, M., Hacein-Bey, S., de Saint Basile, G., Gross, F., Yvon, E., Nusbaum, P., et al. Gene therapy of human severe combined immunodeficiency (SCID)-X1 disease. *Science* 288:669, 2000. Hacein-Bey-Abina, S., VonKalle, C., Schmidt, M., McCormack, N. Wulffraat, P., et al. LMO2-associated clonal T-cell proliferation in two patients after gene therapy for SCID-X1. *Science* 302:415, 2003.

system. Integration, however, is normally a random event that could have deleterious consequences. Exciting studies indicate that the viral integration machinery can be selectively tethered to specific target sequences within the host DNA by protein–protein interactions to obviate these potential problems.

Transgenic Animals

In order to investigate the role of a selected gene product in growth and development of a whole animal, the gene must be introduced into the fertilized egg. Foreign genes can be inserted into the genome of a fertilized egg. Animals that develop from such a fertilized egg carry the inserted gene in every cell and are referred to as **transgenic animals**.

Figure 7.28 outlines a popular method to create transgenic animals. The gene of interest is usually a cloned recombinant DNA molecule with its own promoter or is cloned with a different promoter that can be selectively regulated. Multiple copies of the gene are microinjected into the pronucleus of the fertilized egg. The foreign DNA inserts randomly within the chromosomal DNA. If an insert disrupts, a critical cellular gene the embryo will die. Usually, nonlethal mutagenic events result from the insertion of foreign DNA into the chromosome.

Transgenic animals are currently being used to study DNA regulatory elements, expression of proteins during differentiation, tissue specificity, and the role of oncogene

CLINICAL CORRELATION 7.10
Transgenic Animal Models

Transgenic animal models hold promise to correct genetic diseases early in fetal development. They are used to study the regulation of expression and the function of specific gene products in an animal and may lead to creation of new breeds of commercially valuable animals. Transgenic mice have been produced from fertilized mouse eggs with rat growth hormone (GH) genes microinjected into their male pronuclei (see p. 279). Approximately 600 copies of the gene, fused to the mouse metallothionein-I (MT-I) promoter region, were introduced into each egg, which was then inserted into the reproductive tract of a foster mother mouse. The resulting transgenic mouse carried the rat GH gene within its genome. The diet of the animals was supplemented with $ZnSO_4$ at 33 days post parturition to activate the mouse MT-I promoter and initiate transcription of the rat GH gene. Continuous overexpression of rat GH in some transgenic animals produced mice nearly twice the size of littermates that did not carry the rat GH gene. A transgenic mouse transmitted the rat GH gene to one-half of its offspring, indicating that it had stably integrated into the germ cell genome and that new breeds of animals can be created.

Source: Palmiter, R. D., et al. Dramatic growth of mice that develop from eggs microinjected with metallothionein-growth hormone fusion genes. *Nature (London)* 300:611, 1982.

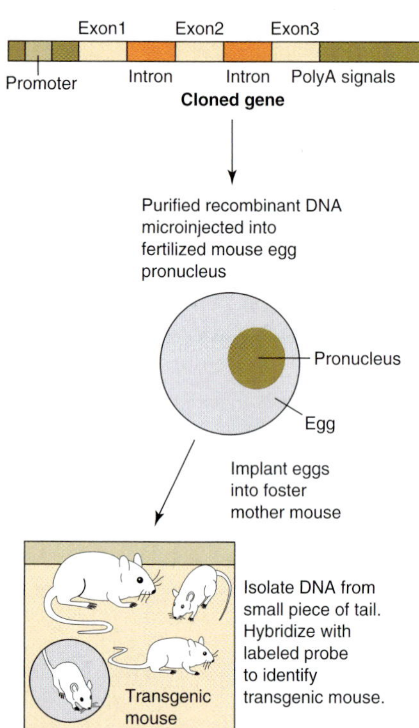

FIGURE 7.28

Production of transgenic animals. Cloned, amplified, and purified functional genes are microinjected into several fertilized mouse egg pronuclei *in vitro*. The eggs are implanted into a foster mother. DNA is isolated from a small piece of each offspring pup's tail and hybridized with a labeled probe to identify animals carrying the foreign gene (transgenic mouse). The transgenic mice can be mated to establish a new strain of mice. Cell lines can also be established from tissues of transgenic mice to study gene regulation and the structure/function of the foreign gene product.

products on growth, differentiation and induction of tumorigenesis. Such technologies should permit replacing defective genes in the developing embryo (see Clin. Corr. 7.10).

Knockout Mice

The creation of an animal with a selected gene "destroyed" in every cell allows researches to define the biological role of the gene if its loss is nonlethal. The basic principles behind creating a **null or knockout mouse** involves: inactivating a recombinant purified gene; introducing this altered gene into an embryonic stem (ES) cell such that it replaces the normal gene; injecting the modified ES cell into a developing blastocyst; implanting the blastocyst into a foster mother; and selecting offspring lacking a normal gene. A common approach to inactivate a selected gene is to insert an antibiotic resistance gene, known as *neo*, into the cloned gene of interest. This both inactivates the gene of interest and allows identification of ES cells that have been successfully transfected. The normal gene in a very low percentage of ES cells may be replaced by the *neo*-disrupted recombinant gene by **homologous recombination**. The nucleotide sequences of the altered/disrupted gene align with the homologous sequence of the normal gene and swap places, or may insert within the normal gene. In either case the net result is to destroy/knockout the normal ES cellular gene. Selection of altered ES cells, microinjection into a mouse blastocyst, birth of chimeric offspring, and breeding of offspring homozygous for the altered gene (Figure 7.29) is an extremely labor-intensive process. The process has been applied in numerous experiments to evaluate the role of selected genes (see Clin. Corr. 7.11).

Dolly, a Lamb Cloned from an Adult Cell

The cloning of animals raises numerous ethical questions such as, Is it man's purview to select the genetic composition of animals and to propagate these animals? Clearly, there are many experimental advantages to developing animals with identical genetic information. However, the application of whole animal cloning techniques to humans raises ethical issues that most consider unacceptable.

Cloning viable offspring from adult mammalian cells is very labor intensive. The method that produced "Dolly," the cloned lamb, began with the culturing of mammary gland cells from a six-year-old Finn Dorset ewe. Cells were arrested in G_0 by being maintained in culture medium with a low serum content. The nuclei from quiescent diploid donor cells were then transferred to anucleated oocytes derived from a Scottish Blackface ewe. The artificially fertilized eggs developed to the morula or blastocyst stage in culture and then transferred to a recipient ewe who carried the fetus to term. Only a small percentage of implanted embryos survive to term. However, the methodology to clone offspring from adult somatic cells has been successfully extended to cattle, mice, and monkeys.

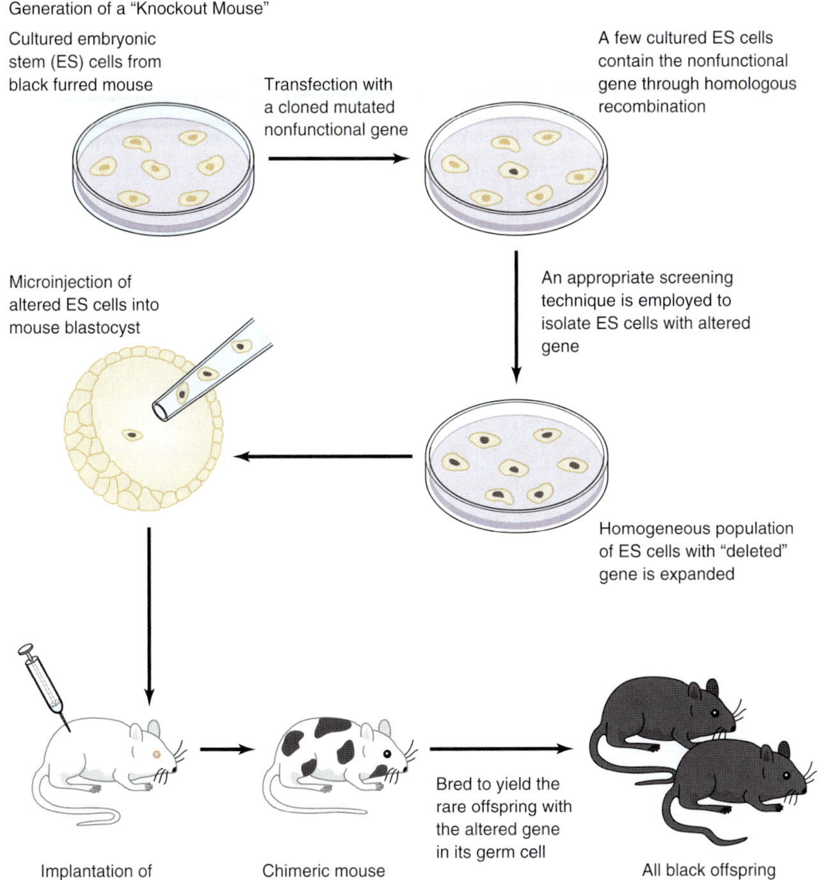
Generation of a "Knockout Mouse"
Cultured embryonic stem (ES) cells from black furred mouse
Transfection with a cloned mutated nonfunctional gene
A few cultured ES cells contain the nonfunctional gene through homologous recombination
An appropriate screening technique is employed to isolate ES cells with altered gene
Microinjection of altered ES cells into mouse blastocyst
Homogeneous population of ES cells with "deleted" gene is expanded
Bred to yield the rare offspring with the altered gene in its germ cell
Implantation of blastocyst into albino foster mother
Chimeric mouse visually observed with black fur
All black offspring with altered gene

FIGURE 7.29

Generation of a knockout mouse. Cultured embryonic cells can be manipulated through recombinant technology to carry a defective or "deleted" gene. The altered cells can be introduced into a blastocyst that is then implanted into a foster mother. Offspring can then be selected that have the nonfunctional "deleted" gene in all of their cells—the knockout animal.

CLINICAL CORRELATION 7.11

Knockout Mice to Define a Role for the $P2Y_1$ Purinoceptor

Platelets play a central role in hemostasis and thrombosis. An important physiologic agonist that induces platelet aggregation is ADP. Studies in the 1970s established the dogma that ADP, bound to a G-protein-coupled receptor, induced aggregation while extracellular ATP, acting at the same receptor site, competitively inhibits aggregation. Sometimes rules, based on limited evidence, become firmly entrenched and take years to be reworked. In the 1990s it became clear that platelets possess at least three different purinoceptors with different binding specificities for ATP versus ADP. An understanding of how these potentially interactive receptors function is very important clinically. ATP and ADP are released into the cardiovascular system/blood under a variety of normal and pathological conditions. The balance between the agonist action of ADP and the antagonist action of ATP may play a significant role in the regulation of platelet responses. Defining the mechanisms of action may lead to the production of new drugs to combat thrombosis. The recent development of a knockout (null) mouse model allowed for the assessment of the role of the purine receptor, $P2Y_1$.

While the P2Y1 receptor is found in many tissues other than platelets, the knockout mouse had no apparent abnormalities. However, normal levels of ADP failed to induce their platelets to aggregate, while high concentrations of ADP did. Aggregated platelets did not show the usual change in shape. The $P2Y_1$-null mice were resistant to thromboembolism, and thus this receptor may be a good target for antithrombotic drugs.

Source: Soslau, G., and Youngprapakorn, D. A possible dual physiological role of extracellular ATP in the modulation of platelet aggregation. *Biochim. Biophys. Acta* 1355:131, 1997. Leon, C., Hechler, G., Freund, M., Eckly, A., Vial C., et al. Defective platelet aggregation and increased resistance to thrombosis in purinergic $P2Y_1$ receptor-null mice. *J. Clin. Invest.* 104:1731, 1999.

Dolly was euthanized in 2004 due to a progressive lung disease associated with what appeared to be a premature aging process. It is not clear if these defects were a result of the cloning process and/or environmental factors. However, cloning techniques continue to be explored with the intent that it will have great therapeutic value. Recently a female mouse was created from two female eggs by bypassing a process called **imprinting**. Imprinting dictates that some active genes must be contributed by the mother and others from the father. Normally mammalian embryos engineered with two sets of female or male chromosomes will not develop. The female mouse created from two mothers lived to adulthood and bore her own litter of pups.

With the success of creating Dolly and the isolation of human embryonic stem cells (hESC), scientists began their quest to combine the two techniques to create embryonic stem cells genetically tailored for a particular patient; this is known as **therapeutic cloning**. A major step in this direction has been accomplished by the Herculean effort that created a single hESC line derived from a cloned blastula. The cloned blastula was created by inserting the nucleus of a human cumulus cell (cells surrounding eggs in an ovary) into an anucleated human egg and subjecting it to the appropriate chemical environment to induce replication. The hESC line grows normally in culture and should be able to differentiate into all cell types of the body. These cells could be induced to form normal cells of any tissues to replace defective tissues in patients.

Recombinant DNA in Agriculture Has Commercial Impact

Perhaps the greatest gain to humanity would be the practical use of recombinant technology to improve agricultural crops. Genes must be identified and isolated that code for properties such as higher crop yield, rapid plant growth, resistance to adverse conditions such as arid or cold periods, and plant size. New genes, not common to plants, may be engineered into plants so as to confer resistance to insects, fungi, or bacteria. Finally, genes for structural proteins can be modified to contain essential amino acids not normally present, without modifying the function of the protein. Production of plants with new genetic properties requires introduction of genes into plant cells that can differentiate into whole plants.

New genetic information carried in crown gall plasmids can be introduced into plants infected with soil bacteria known as agrobacteria. Agrobacteria naturally contain a crown gall or Ti (tumor-inducing) plasmid whose genes integrate into an infected cell's chromosome. The plasmid genes direct the host plant cell to produce new amino acid species called opines (such as the amino acid derivatives of arginine and pyruvate, octopine, or arginine and alpha-keto glutarate, nopaline), which are required for bacterial growth. A crown gall, or tumor mass of undifferentiated plant cells, develops at the site of bacterial infection. New genes engineered into the Ti plasmid can be introduced into plant cells on infection with the Agrobacteria. The cells can then be grown in culture and induced to redifferentiate into whole plants. Every cell would contain the new genetic information and would represent a transgenic plant.

Some limitations in producing plants with improved genetic properties must be overcome before significant advances in our world food supply can be realized. Clearly, genes for desired characteristics must be identified and isolated. Also, important crops such as corn and wheat cannot be transformed by Ti plasmids, therefore, other vectors must be identified. Significant success has been achieved in designing crop plants resistant to insects and viruses. Recently genetic engineering inserted a foreign gene into pea plants to produce a protein that inhibits the feeding of weevil larvae on the pea seeds. This will permit storage of peas and other legume seeds without the need of protective chemical fumigants (currently Brazilian farmers lose 20–40% of their stored beans to pests).

7.15 GENOMICS, PROTEOMICS, AND MICROARRAY ANALYSIS

Genomics is the study of the molecular characteristics of the whole genome. This includes the sequence of the genome, identification of genes and their regulatory

TABLE 7.1 A Partial Listing of Prokaryotic and Eukaryotic Species Whose Genomes Have Been Sequenced

	Species	Genome Size
Prokaryotes (over 100 microbes):	*Hemophilus influenzae*	1.83 Mb
	Chlamydia pneumoniae	1.23 Mb
	Neisseria meningitides	2.27 Mb
	Vibrio cholerae	4.0 Mb
		Number of Chromosomes
Simple Eukaryotes:	*Saccharomyces cerevisiae* (yeast)	16
	Plasmodium falciparum (malarial protozoan)	14
	Anopheles gambiae (mosquito)	3
	Drosophila melanogaster (fruit fly)	5
Plants:	*Avena sativa* (oat)	7
	Hordeum vulgare (barley)	7
	Oryza sativa (rice)	12
	Zea mays (corn)	10
Vertebrates:	*Danio rerio* (zebra fish)	25
	Gallus gallus (chicken)	39
	Mus musulus (mouse)	21
	Rattus norvegicus (rat)	22
	Homo sapiens (human)	24

sequences, and the pattern of gene expression. Dozens of prokaryotic and eukaryotic genomes, as well as over 1000 viruses, have been sequenced (Table 7.1). The sequencing of numerous genomes of marine microbes has recently added approximately one million new genes to the 180,000 documented genes. It is estimated that there are more than 10 billion genes in the Earth's repertoire. A singe gene may give rise to multiple protein species due to splice choice of RNA transcript and variable posttranslational processing of the protein product. The complement of proteins within a cell is referred to as its **proteome**. Proteomics is the study of proteins expressed in tissues under various conditions, protein interactions, and protein modifications. Clinical Correlation 7.12 describes how proteomics has been used to diagnose early stages of cancer by identifying tumor-associated biomarkers.

CLINICAL CORRELATION 7.12
Microarray Analysis of Breast Cancer

The treatment modality of human breast cancer patients depends, in large part, on the estrogen receptor (ER) status of the tumor cells. Estrogen-receptor-positive tumors generally respond well to adjuvant hormone/drug therapy, while ER negative tumors usually do not. Patients with ER-negative tumors have variable responses to chemotherapy, and there are no known indicators to predict post-operative prognoses. The amplified cellular RNA derived from 10 ER-negative patients who died within 5 years of breast cancer and 10 ER negative patients who were alive, disease-free, was analyzed on a cDNA microarray consisting of 25,344 genes. The analysis detected 71 genes with higher expression in the group that died than in those who lived, and 15 genes were expressed at higher levels in the survivors than in those who died. A scoring system was developed based on gene markers to classify ER-negative breast cancer patients into a poor prognosis versus a good prognosis group. The scoring system appears to have a high degree of accuracy and may help direct treatment protocols and define new gene targets for drug treatment.

Source: Nagahata, T., Onda M, Emi, M., Nagai, H., Tsumagari, K. et al. Expression profiling to predict postoperative prognosis for estrogen receptor-negative breast cancer by analysis of 25,344 genes on a cDNA microarray. *Cancer Sci.* 95:218, 2004.

Microarray Analysis

Analysis of genomes and proteomes has been greatly facilitated by the development of microarray technologies developed first for analyzing nucleic acids. Nucleic acid hybridization and antigen–antibody complex formation have been employed extensively in techniques described in preceding sections for the analysis of one or few macromolecular species. However, techniques with a very high throughput are required to analyze simultaneously the hundreds to thousands of genes expressed in a cell. Microarray techniques depend upon these same macromolecular interactions described previously such as in Southern, Northern, and Western blots. The fixation of DNA probes to solid impermeable surfaces reduces the solution volumes and times required for hybridization. DNA microarrays are produced by computer-controlled additions of microdrops of cloned or PCR synthesized DNA sequences to a glass slide with a poly-L-lysine substrate to which these DNA probes are covalently cross-linked. DNA or RNA to be analyzed is isotopically or fluorescently labeled and then hybridized with the fixed probes on a glass slide (Figure 7.30). Positive reactions are detected and quantified by high-resolution imaging and computer analysis. Many DNA microarrays are available that focus on selected genes involved in discreet processes or pathways such as signal transduction and apoptosis. A specific example of altered gene expression in malaria-infected cells is depicted in Figure 7.31. DNA oligonucleotides (usually 20 or more bases) can also be synthesized on micromatrices fixed to a glass slide. The size of the micromatrix can vary from 1 × 10 × 5 μm to 100 × 100 × 20 μm. Thousands of

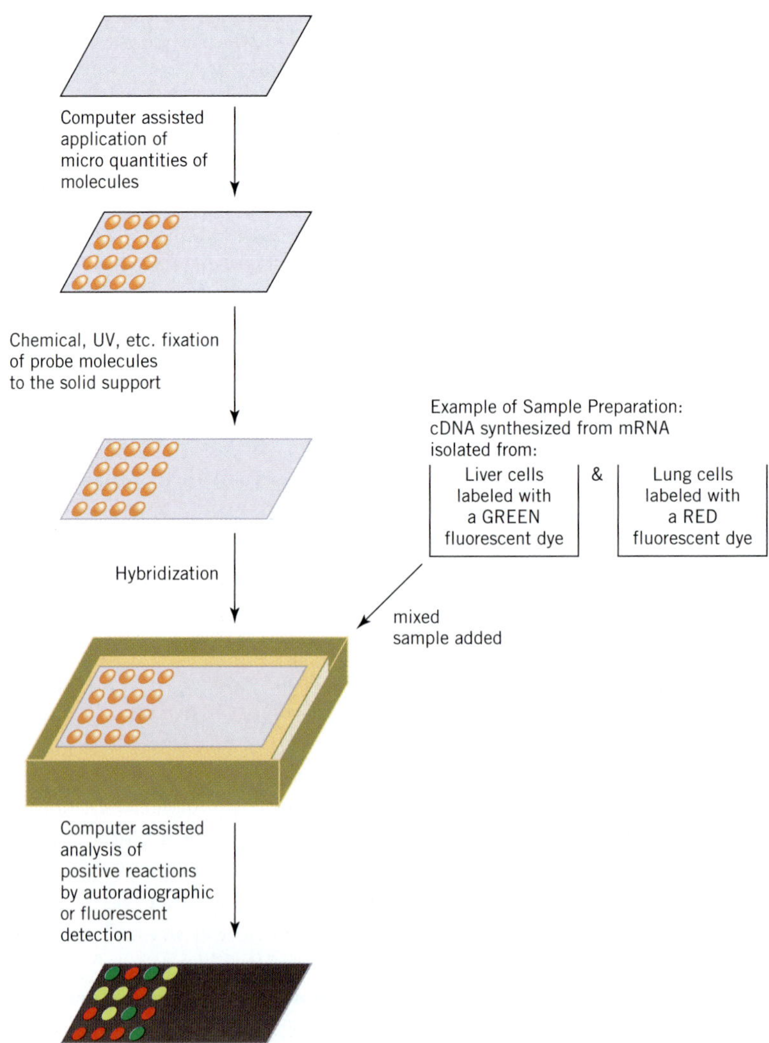

FIGURE 7.30

DNA microarray analysis. The surface of an impermeable support, typically glass or polypropylene, is prepared to apply multiple samples of DNA or oligonucleotides. Slide with oligonucleotides or nucleic acids representing hundreds to thousands of genes is incubated in a chamber under appropriate conditions for hybridization followed by appropriate washes to remove nonhybridized cDNA. Green fluorescent spots would indicate genes expressed in liver cells, red fluorescent spots would indicate genes expressed in lung cells, and yellow fluorescent spots would indicate genes expressed in both tissues.

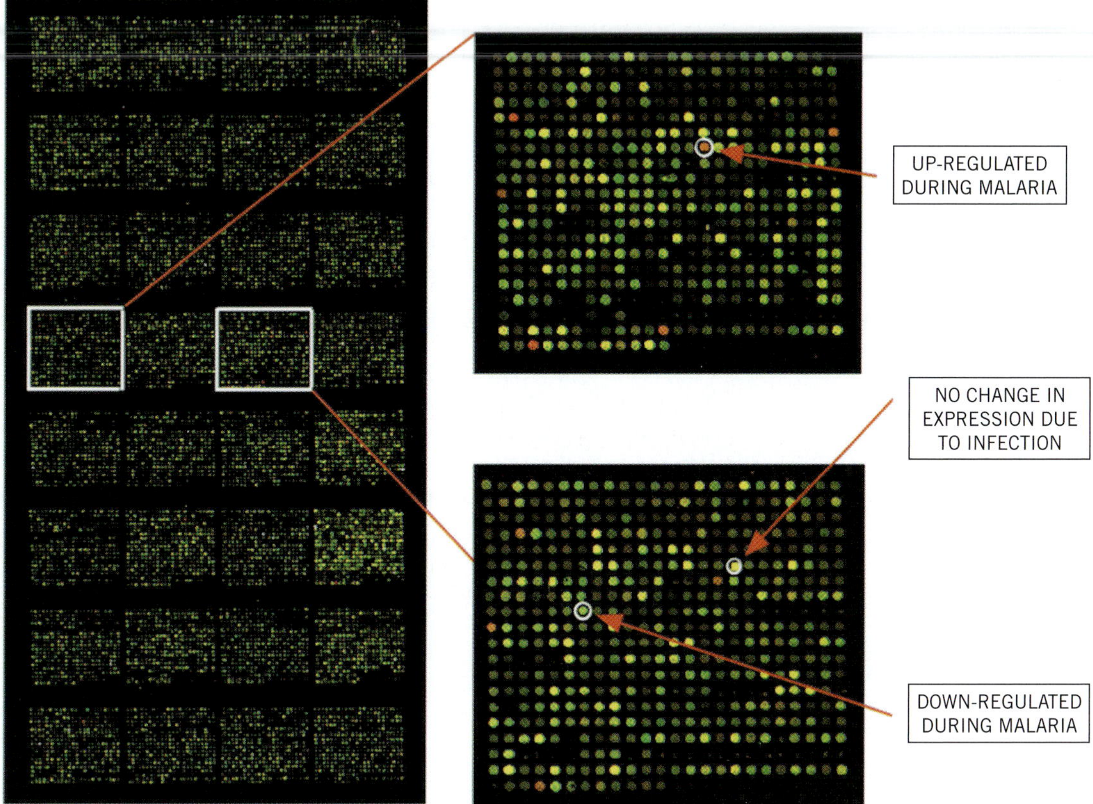

FIGURE 7.31

Microarray analysis of gene expression in mice infected with a rodent malaria parasite. DNA, RNA, or oligonucleotides, representing part or all of the cellular genome or proteins—or antibodies, representing part of the cellular proteome—are fixed on an impermeable support. The fixed probes are reacted with appropriate labeled cellular components that will form a stable complex through hybridization or protein-protein interactions. Identification of nucleic acids/proteins that associate with the fixed samples in the microarrays allows for the analysis of cellular gene expression in cells maintained under a variety of conditions.

Mouse DNA microarrays were produced for the mouse genome, which contains 13,443 synthetic oligonucleotides representing well-characterized mouse genes. Each oligonucleotide is 70 bases in length derived from the 3′-end of each gene, optimized to minimize cross-hybridization to all known mouse genes. The array includes a comprehensive set of immune-system-related genes encoding a lymphocyte subset and activation markers, cytokines, chemokines, their receptors, adhesion molecules, co-stimulatory molecules, immunoregulatory factors, transcription factors, and signal transduction proteins and others.

Samples: Splenocyte RNA from uninfected mice was used to synthesize cDNA fluorescently labeled with Cy3 (green). Splenocyte RNA from mice on day 8 of infection with a rodent malaria parasite was used to synthesize cDNA fluorescently labeled with Cy5 (red). The mixed fluorescently labeled cDNA probes are hybridized with the target oligonucleotide array. Genes that are up-regulated in mice infected with malaria are red, and those that are down-regulated in these animals are green. Genes that are unaffected by the malarial parasite are yellow.

Data are unpublished work from the laboratory of Dr. James Burns, Department of Microbiology, Drexel University College of Medicine.

different oligonucleotides, each complementary to a specific gene, can be immobilized on what is termed a **DNA chip**. A single-chip microarray for the human genome, which will be commercially available, contains 1 million oligonucleotide probes. This DNA chip will detect about 50,000 RNA transcripts from approximately 30,000 human genes. These biological microchips are compared to electronic microchips since both are capable of performing multiple reactions in parallel in a high-throughput fashion.

Microarrays allow for the analysis of expression of the whole genome or genes in a selected pathway. Genes expressed in a specific tissue, or under a selected metabolic or diseased state could give clues as to which genes are active in the specific condition.

Unfortunately, a single microarray plate gives a static snapshot of genomic genes expressed under a particular condition. Regulatory genes involved in the final altered expression cannot be readily deduced from this single picture since many genes may be turned on or off by down stream pathways during the transition to the final cellular state. Co-regulatory genes involved in the switch between different metabolic or disease states could be determined by microarray analysis at multiple time points between the initial and final stage of altered gene expression. Computer analysis of microarrays generated over the time course can cluster genes that appear to be coordinately expressed. **Cluster analysis** allows identification of potential genes that co-regulate the alteration of the cell from one state to another.

Protein microarrays are produced with antibodies or nonantibody proteins fixed to slides or filters that can capture target proteins in a sample in a fashion similar to those used for DNA microarrays. Other protein probes are being developed for protein microarrays such as short stretches of synthesized DNA or RNA (**aptamers**) that selectively bind specific proteins. Protein arrays are generally produced with recombinant proteins that may be generated from tissue specific cDNA expression libraries. These microarrays can then be readily analyzed with specific antibodies that are fluorescently labeled. This approach has been successfully employed to detect the antibody repertoire in human plasmas associated with some autoimmune diseases.

Microarray techniques are critical for the analysis of genomic expression and the proteomic state of cells in normal cells as compared to the patterns in cells that are altered due to changes such as a modified metabolic environment, differentiation, response to drugs, response to growth factors, or a diseased state.

Human Genome

By 1996 the genome of yeast, a eukaryote, consisting of approximately 14×10^6 bps distributed among 16 chromosomes was sequenced. The genome of the nematode, *C. elegans*, sequenced by 1998, contains approximately 18,000 genes as compared to the 6000 genes of yeast. In 2000, a **shotgun sequencing** approach was employed to sequence essentially the whole 180×10^6 bps genome of the fruit fly *Drosophila melanogaster*. This method involves the breakage of the whole genome into small pieces, sequencing the pieces by a rapid automated procedure, and assembling the sequenced fragments into the proper order assisted by overlapping sequences and a supercomputer. The availability of previously cloned sequences facilitated the sequencing. The shotgun sequencing approach was employed to speed up the sequencing of the human genome. Several human chromosomes had already been mapped and thousands of cDNA clones had been sequenced or were ready to be sequenced, which would ultimately provide landmarks of the huge human genetic map. In 2000, J. C. Venter and F. S. Collins jointly announced that their groups had sequenced 97% of the human genome.

The determination of a genomic sequence is only a part of the genome puzzle. The start-and-stop site of each gene must be determined along with what the gene codes for—a process referred to as **annotation**. Eukaryotic genomes usually contain permanently condensed or heterochromatic regions around the centromeres. These regions contain repetitive DNA that cannot be cloned and sequenced by any of the current sequencing methods. It may be decades or more before the complete determination of all the genes and regulatory elements are resolved.

The completion of the human genome project generated enormous amounts of data. The achievement of the project is only the beginning of our understanding of the molecular underpinnings of how a single cell functions, let alone an organism. It is now considered that the human genome encodes about 30,000 genes, far fewer genes than anticipated, and only about 20% more than a plant genome and about 40% more than *C. elegans* (roundworm). However, the RNA transcripts of the "static" genome are relatively "dynamic" as are their translated protein products. Splice choice of the primary RNA transcript results in different protein products from the same gene. Protein products may also undergo numerous chemical modifications and association with other peptide subunits or proteins. It is estimated that the 30,000 genes ultimately give rise to 200,000 to 2 million protein products whose expression is dependent, in part, upon tissue specificity, stage of development, and metabolic conditions. The task before us now is to be able to analyze transcripts and their final translation products at the structure/function level.

The delineation of all of the human genes and their regulatory sequences should greatly enhance our understanding of many genetic diseases. Technological advances to manipulate genes and gene expression should open new avenues to regulate and/or cure many diseases. Genetic diseases should eventually be curable by gene replacement therapy; limited clinical trials in gene therapy have been initiated. Considering the enormous advances made in molecular biology in just the past three decades, it is reasonable to believe that the "when"' will not be that far off. The old challenge confronting scientists was how to sequence the human genome; the new challenge is how to manipulate effectively that knowledge to benefit humankind.

BIBLIOGRAPHY

Adams, D. M., Celniker, S. E., Holt, R. A., plus 192 collaborators. The genome sequence of *Drosophila melanogaster*. *Science* 287:2185, 2000.

Brown, W. M., George, M., Jr., and Wilson, A. C. Rapid evolution of animal mitochondrial DNA. *Proc. Natl. Acad. Sci. USA* 76:1967, 1979.

Bushman, F. Targeting retroviral integration. *Science* 267:1443, 1995.

Capecchi, M. R. Altering the genome by homologous recombination. *Science* 244:1288, 1989.

C. elegans Sequencing Consortium. Genome sequence of the nematode *C. elegans*: A platform for investigating biology. *Science* 282:2012, 1998.

Chilton, M. D. A vector for introducing new genes into plants. *Sci. Am.* 51:36 June 1983.

Davis, L. G., Kuehl, W. M., and Battey, J. F. *In vitro* amplification of DNA using the polymerase chain reaction (PCR). *Basic Methods in Molecular Biology*, 2nd ed. Appleton and Lange, Norwalk, CT: 1994, p. 111.

Feinberg, A. and Vogelstein, B. Addendum: A technique for radiolabeling DNA restriction endonuclease fragments to high specific activity. *Anal. Biochem.* 137:266, 1984.

Fields, S. Proteomics in genomeland. *Science* 291:1221, 2001.

Goffeau, A., et al. Life with 6,000 genes. *Science* 274:546, 1996.

International Human Genome Sequencing Consortium. Initial sequencing of the human genome. *Nature* 409:860,2001.

Jaenisch, R. Transgenic animals. *Science* 240:1468, 1988.

Kerr, L. D. Electrophoretic mobility shift assay. *Methods Enzymol.* 254:619, 1995.

Kreeger, K. Y. Influential consortium's cDNA clones praised as genome research time-saver. *Scientist* 9:1, 1995.

Kunkel, T. A. Rapid and efficient site-specific mutagenesis without phenotypic selection. *Proc. Natl. Acad. Sci. USA* 82:488, 1985.

McPherson, M. J., Quirke, P., and Taylor, G. R., eds. *PCR. A Practical Approach*, Vol. 1. Oxford, England: Oxford University Press, 1994.

Marshall, E. Gene therapy's growing pains. *Science* 269:1050, 1995.

Maxam, A. M., and Gilbert, W. A new method of sequencing DNA. *Proc. Natl. Acad. Sci. USA* 74:560, 1977.

Mulligan, R. C. The basic science of gene therapy. *Science* 260:926, 1993.

Palmiter, R. D., et al. Dramatic growth of mice that develop from eggs microinjected with metallothionein-growth hormone fusion genes. *Nature* 300:611, 1982.

Reddy K. S. and Perrotta P. L. Proteomics in transfusion medicine. *Transfusion* 44:601, 2004.

Rigby, P. W. J., Dieckmann, M., Rhodes, C., and Berg, P. Labelled deoxyribonucleic acid to high specific activity *in vitro* by nick translation with DNA polymerase I. *J. Mol. Biol.* 113:237, 1977.

Sambrook, J., Fritsch, E. F., and Maniatis, T. *Molecular Cloning. A Laboratory Manual*, 2nd ed. New York: Cold Spring Harbor Laboratory Press, 1989.

Sanger, F., Nicklen, S., and Coulson, A. R. DNA sequencing with chain-terminating inhibitors. *Proc. Natl. Acad. Sci. USA* 74:5463, 1977.

Simon R., Mirlacher, M., and Sauter, G. Tissue microarrays. *BioTechinques* 36:98, 2004.

Southern, E. M. Detection of specific sequences among DNA fragments separated by gel electrophoresis. *J. Mol. Biol.* 98:503, 1975.

Southern, E. M. Microarrays. In: J. B. Rampal, Ed., *DNA Arrays, Methods and Protocols, Methods in Molecular Biology*, Vol. 170, Totawa, NJ: Humana Press, 2001, p. 1.

Venter, J. C., Adams, M. D., Myers, E. W., Li, P. W., et al. The sequence of the human genome. *Science* 291:1304, 2001.

Venter, J. C., Remington, K, Heidelberg, J. F., Halpern, A. L., et al. Environmental genome shotgun sequencing of the Sargasso Sea. *Science* 304:66, 2004.

Vogelstein, B., Fearon, E. R., Hamilton, S. R., and Feinberg, A. P. Use of restriction fragment length polymorphism to determine the clonal origin of human tumors. *Science* 227:642, 1985.

Watson, J. D., Tooze, J., and Kurtz, D. T. *Recombinant DNA: A Short Course*. Scientific American Books. San Francisco: Freeman, 1983.

Weintraub, H. M. Antisense RNA and DNA. *Sci. Am.* 262:40, 1990.

Wilmut, I., Schnieke, A. E, McWhir, J. Kind, A. J., and Campbell, K. H. S. Viable offspring derived from fetal and adult mammalian cells. *Nature* 385:810, 1997.

Zhang, Y., and Yunis, J. J. Improved blood RNA extraction microtechnique for RT-PCR. *BioTechniques* 18:788, 1998.

QUESTIONS | CAROL N. ANGSTADT

Multiple Choice Questions

1. Development of recombinant DNA methodologies is based upon discovery of:
   A. the polymerase chain reaction (PCR).
   B. restriction endonucleases.
   C. plasmids.
   D. complementary DNA (cDNA).
   E. yeast artificial chromosomes (YACs).

2. Construction of a restriction map of DNA requires all of the following *except*:
   A. partial hydrolysis of DNA.
   B. complete hydrolysis of DNA.
   C. electrophoretic separation of fragments on a gel.
   D. staining of an electrophoretic gel to locate DNA.
   E. cyclic heating and cooling of the reaction mixture.

3. Preparation of recombinant DNA requires:
   A. restriction endonucleases that cut in a staggered fashion.
   B. restriction endonucleases that cleave to yield blunt-ended fragments.
   C. poly(dT).
   D. DNA ligase.
   E. cDNA.

4. In the selection of bacterial colonies that carry cloned DNA in plasmids, such as pBR322, that contain two antibiotic resistance genes:
   A. one antibiotic resistance gene is nonfunctional in the desired bacterial colonies.
   B. untransformed bacteria are antibiotic resistant.
   C. both antibiotic resistance genes are functional in the desired bacterial colonies.
   D. radiolabeled DNA or RNA probes play a role.
   E. none of the above.

5. A technique for defining gene arrangement in very long stretches of DNA (50–100 kb) is:
   A. RFLP.
   B. chromosome walking.
   C. nick translation.
   D. Southern blotting.
   E. SSCP.

6. Antisense nucleic acids:
   A. complementary to mRNA would enhance translation.
   B. could result if a gene is inserted downstream of a promoter but in opposite direction to normal.
   C. can have no clinical uses.
   D. react only with DNA.
   E. are necessary for recombinant DNA technology.

Questions 7 and 8: In this country, a major cause of death of babies during the first year is sudden infant death syndrome (SIDS). One study showed a strong correlation for an increased risk of SIDS with a prolonged QT interval in their electrocardiograms. In one child, a gene associated with the Long QT syndrome had a substitution of AAC for TCC. This gene codes for a protein associated with the sodium channel. The mutation was detected by single-strand conformation polymorphism (SSCP).

7. Which of the following statements about SSCP is/are correct?
   A. The electrophoretic mobility on polyacrylamide gel of small, single-stranded DNA during the SSCP technique depends partly on the secondary conformation.
   B. There must be a restriction endonuclease site in the region studied for SSCP to work.
   C. Radiolabeling is not used in this technique.
   D. It is not necessary to know the sequence of the DNA to be studied.
   E. All of the above are correct.

8. In order to get enough DNA to analyze, DNA is amplified by a polymerase chain reaction (PCR). The essential property of the DNA polymerase employed is that it:
   A. does not require a primer.
   B. is unusually active.
   C. is thermostable.
   D. replicates double-stranded DNA.
   E. can replicate both eucaryotic and procaryotic DNA.

Questions 9 and 10: The purpose of gene therapy is to introduce a normal gene into cells containing a defective gene. The first authorized human gene therapy was given to a 4-year-old girl with adenosine deaminase deficiency (ADA). ADA children have a severe combined immune deficiency (SCID) and usually die within the first few years from overwhelming infections. A modified retrovirus was constructed to contain the human ADA gene which could be expressed in human cells without replication of the virus. The child's isolated T cells were "infected" with the retrovirus to transfer the normal gene to them. The modified T cells were reintroduced into the patient. The girl is now 13 years old and doing well.

9. Expression of recombinant genes in mammalian cells:
   A. will not occur if the gene contains an intron.
   B. occurs most efficiently if cDNA is used in the vector.
   C. does not require directional insertion of the gene into the vector.
   D. requires that the vector have enhancer and promoter elements engineered into the vector.
   E. does not require that the vector have an origin of replication (ori).

10. Another clinical use for recombinant DNA technology is to have rapidly replicating bacteria produce large amounts of specific proteins (e.g., hormones). Expression of a eucaryotic gene in prokaryotes involves:
    A. a Shine–Delgarno (SD) sequence in mRNA.
    B. absence of introns.
    C. regulatory elements upstream of the gene.
    D. a fusion protein.
    E. all of the above.

Questions 11 and 12: Whole genomic DNA was isolated from three individuals, digested separately with a restriction endonuclease to fragments and the fragments separated on agarose gel in an electric field. The gene of interest was isolated and analyzed using Southern blot technique. Each individual sample showed two bands. The bands were identical for two of the individuals. For the third, one band was identical to one band of the other two, but the second band was of lower molecular weight than the others. This is an example of restriction fragment length polymorphism (RFLP).

11. A reasonable explanation for this RFLP is that the gene in the third individual:
    A. lacked the recognition sequence for the restriction endonuclease.
    B. had a mutation at the cleavage site.
    C. had an additional recognition site for the endonuclease.
    D. had a deletion of a segment of the gene.
    E. underwent a random cleavage.

12. The Southern blot technique:
    A. transfers DNA fragments from agarose gel to a nitrocellulose filter.
    B. requires that the DNA fragments remain double-stranded.
    C. requires that the DNA is radiolabeled prior to addition to the agarose gel.

D. alters the position of the DNA fragments during the process.
E. amplifies the amount of DNA material.

Problems

13. The X-ray autoradiogram of one strand of a fragment of DNA sequenced by the Sanger method was obtained. The 5′-end of the nucleotide was labeled with ^{32}P. Numbering from bottom to top of the autoradiogram, the lane from the ddG tube showed bands at positions 4 and 11; ddC tube had bands at 1, 2, 7, 8, 9, and 13; ddA at 5, 10, and 15; ddT at 3, 6, 12, and 14. What is the sequence of the fragment? Construct the autoradiogram pattern of the complementary sequence.

14. What would be the best vector to use to carry a segment of DNA that is 250 kb in size? Would any other type of vector work? Why?

ANSWERS

1. **B** The ability to cleave DNA predictably at specific sites is essential to recombinant DNA technology. A, C, D, and E. These techniques or structures are involved in recombinant DNA technology but came later.
2. **E** Cyclic heating and cooling is part of the PCR process, not of restriction mapping. A, B: Restriction mapping involves all degrees of hydrolysis. Partial hydrolysis gives fragments of varying sizes, and complete hydrolysis gives the smallest possible fragments. C, D: Fragments are electrophoretically separated by size on agarose gel, which is stained to reveal the DNA.
3. **D** DNA ligase covalently connects fragments held together by interaction of cohesive ends. A: This is the most desirable type of restriction endonuclease to use, but it is not essential. B: Restriction nucleases that make blunt cuts can also be used if necessary. C: This is used in conjunction with poly dA if restriction endonucleases that make blunt cuts are employed, but it is not essential to all of recombinant DNA preparation. E: cDNA is only one type that can be used.
4. **A** The foreign DNA is inserted into one antibiotic resistance gene, thus destroying it. B: Resistance is due to the plasmids. C: See the comment for A above. D: Radiolabeling detects the DNA of interest, not the colonies that contain cloned DNA.
5. **B** A: Restriction fragment length polymorphism (RFLP) is a characteristic of DNA, not a technique. C: Nick translation is used to label DNA during chromosome walking. D: Southern blotting is a method for analyzing DNA. E: Single-strand conformation polymorphism (SSCP) is a method for detecting base changes in DNA that do not alter restriction endonuclease sites.
6. **B** This would put the antisense strand of DNA under control of the promoter with subsequent transcription of antisense mRNA. A: Antisense nucleic acid binds mRNA blocking translation. C: If not now, in the future. Antisense DNA nucleotides have been produced that inhibit viral infection. D: React with both DNA and RNA.
7. **A** A single base substitution usually modifies the conformation enough to shift the mobility as detected by SSCP. B: SSCP is the method of choice if there is no restriction endonuclease site. C: DNA is amplified by PCR in the presence of radiolabeled nucleotides. There has to be a detection method for the bands. D: SSCP requires prior knowledge of the sequence.
8. **C** PCR requires cycling between low temperatures, where hybridization of template DNA and oligomeric primers occurs, and high temperatures, where DNA melts. The Taq DNA polymerase, isolated from a thermophilic organism discovered in a hot spring, is stable at high temperatures, and it makes the cycling possible with no addition of fresh polymerase after each cycle.
9. **D** Eukaryotic systems require controlling elements, which are not necessary in bacterial systems. A: Expression may be improved if an intron is present. B: cDNA does not possess the required controlling elements. C: The gene must be inserted in the proper orientation relative to the control elements. E: The vector must replicate, so it needs the sequence to promote replication.
10. **E** A: The SD sequence is necessary for the bacterial ribosome to recognize the mRNA. B: Bacteria do not have the intracellular machinery to remove introns from mRNA. C: Appropriate regulatory elements are necessary to allow the DNA to be transcribed. D: A fusion protein may be a product of the reaction.
11. **D** This would account for one band being of lower molecular weight. A, B, E: The cleavage site has a specific recognition sequence. C: This would yield more than two bands.
12. **A** This occurs by elution from the gel by a high salt solution. B: DNA is hydrolyzed by alkali and single-strand DNA binds to the filter. C: The probe to identify the bands on the filter is radiolabeled. D: The positions on the filter are identical to the gel. E: Doesn't happen.
13. The sequence is C C T G A T C C C A G T C T A reading 5′ to 3′. The autoradiograph of the complementary strand should have bands for ddG at 1, 2, 7, 8, 9, and 13; ddC at 4 and 11; ddA at 3, 6, 12, and 14; ddT at 5, 10, 15.
14. YACs (yeast artificial chromosomes) would be the best vector since they can accept DNA between 200 and 500 kb. Cosmids accept the next largest piece, but it is only 45 kb. Plasmid and bacteriophages accept much smaller pieces.

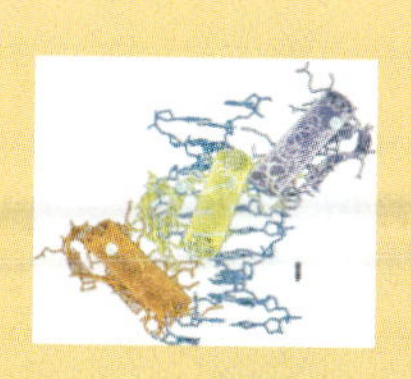
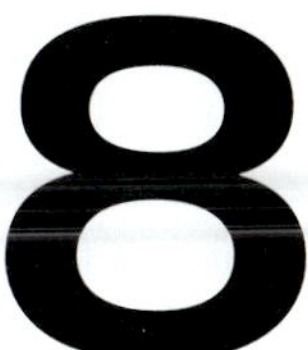

8

REGULATION OF GENE EXPRESSION

Daniel L. Weeks and John E. Donelson

8.1 OVERVIEW

To survive, a living cell must respond to changes in its environment. One way that cells adjust to these changes is to alter expression of specific genes, which affects the number of the corresponding protein molecules in the cell. This chapter focuses on some of the molecular mechanisms that determine when a given gene is expressed and to what extent. Understanding how expression of genes is regulated is one of the most active areas of biochemical research today.

Escherichia coli (E. coli) contain genes for about 4300 different proteins, but it does not need to synthesize all of them at the same time. The classic illustration of bacterial gene regulation is that of the gene for **β-galactosidase**, which converts the disaccharide lactose to glucose and galactose. When *E. coli* grows in a medium containing glucose as the carbon source, only about five molecules of the enzyme are present in a cell. When lactose is the sole carbon source, however, there are 5000 or more molecules. Thus, when bacteria need to metabolize lactose, they increase synthesis of β-galactosidase. If lactose is removed from the medium, synthesis of this enzyme stops as rapidly as it began.

Eukaryotic cells have more extensive mechanisms of gene regulation than prokaryotic cells. Differentiated cells of higher organisms have a highly complicated structure and often a specialized biological function that is determined by expression of their genes. For example, **preproinsulin** is synthesized in β cells of the pancreas but not in kidney cells even though the nuclei of all cells of the body contain the gene for preproinsulin. During development of the organism, the presence of proteins in specific cell types is tightly controlled with respect to timing and order of developmental events.

Much more is understood about gene regulation in prokaryotes than in eukaryotes. However, studies on prokaryotes often provide exciting new ideas that can be tested in eukaryotic cells and are important for developing rational therapies for diseases caused by bacteria.

Several well-studied examples of gene regulation in bacteria will be discussed, followed by examples of the many protein transcription factors regulating expression of genes in the human genome.

8.2 UNIT OF TRANSCRIPTION IN BACTERIA: THE OPERON

The *E. coli* chromosome is a circular double-stranded DNA molecule of 4.6 million base pairs whose entire sequence has been determined. Most of its approximately 4300 genes are not distributed randomly; instead, genes encoding enzymes of a specific metabolic pathway are clustered in one region of the DNA. Similarly, genes for associated structural proteins, such as the 70 or so proteins that comprise the **ribosome**, are frequently adjacent to one another. A set of clustered genes is usually coordinately regulated; that is, the genes are transcribed together into a "polycistronic" mRNA species that contains the coding sequences for the various proteins. The term **operon** describes the complete regulatory unit of a set of clustered genes. It includes the adjacent **structural genes** for the related enzymes or associated proteins, a **regulatory gene** or genes encoding regulator protein(s), and **control elements** or sites on the DNA near the structural genes at which regulator proteins act. Figure 8.1 shows a partial genetic map of the *E. coli* genome that gives locations of structural genes of some *E. coli* operons.

When transcription of the structural genes of an operon increases in response to a specific substrate in the medium, the effect is known as **induction**. An example is the increase in transcription of the β-galactosidase gene when lactose is the sole carbon source. Bacteria also respond to nutritional changes by quickly turning off synthesis of enzymes no longer needed. For example, *E. coli* synthesizes tryptophan as the end product of a specific biosynthetic pathway. However, if tryptophan is present in the medium, synthesis of enzymes for this pathway stops. The process is called **repression** and it prevents bacteria from using their energy to make unnecessary proteins.

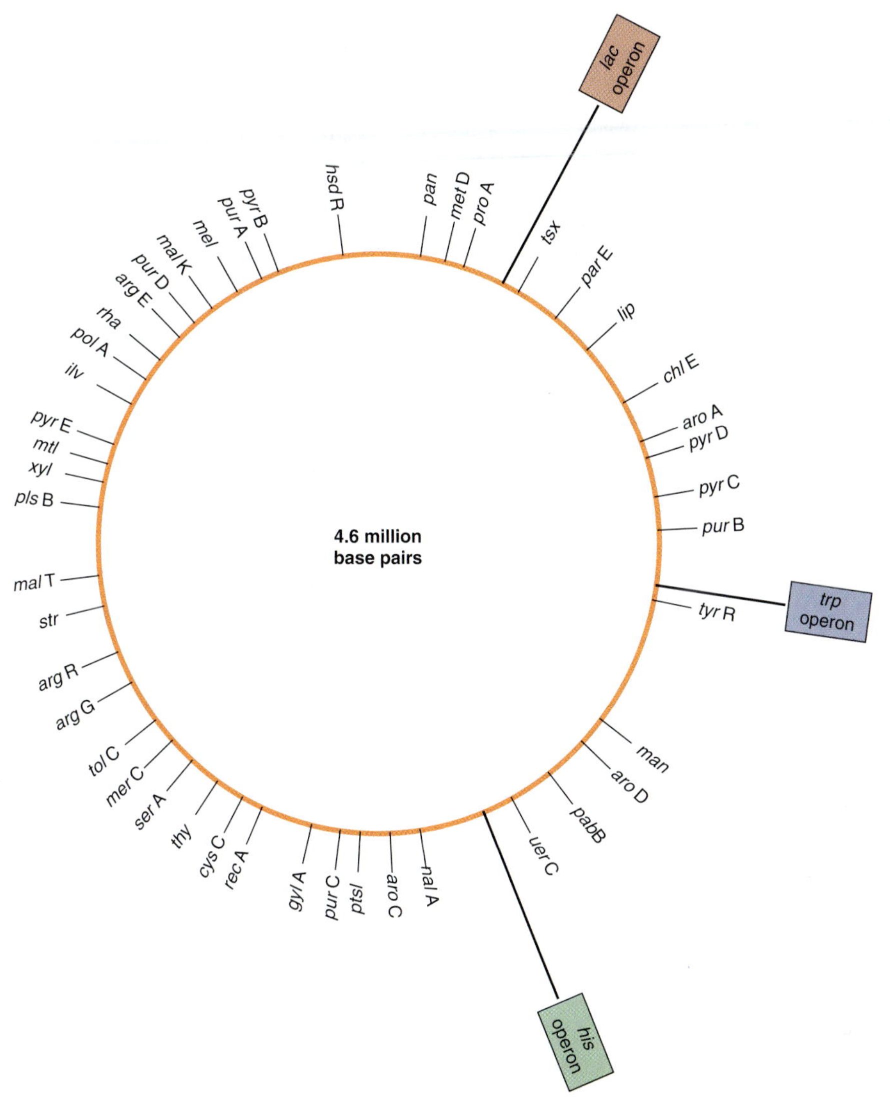

FIGURE 8.1

Partial genetic map of *E. coli*. Locations of a few of the 4300 genes in the circular *E. coli* genome are shown. Three operons discussed in this chapter are indicated. Reproduced with permission from Stent, G. S., and Calendar, R. *Molecular Genetics, An Introductory Narrative*. San Francisco: Freeman, 1978, p. 289. Modified from Bachmann, B. J., Low, K. B., and Taylor, A. L. *Bacteriol. Rev.* 40:116, 1976.

Induction and repression are manifestations of the same phenomenon. The signal for each is the small molecule that is a substrate for the metabolic pathway or a product of the pathway, respectively. These small molecules are called **inducers** when they stimulate induction and **corepressors** when they cause repression.

8.3 LACTOSE OPERON OF *E. COLI*

The lactose operon contains three adjacent structural genes (Figure 8.2). *Lac*Z codes for β-galactosidase, which is composed of four identical subunits of 1021 amino acids each. *Lac*Y codes for a permease, a 275-amino-acid protein of the cell membrane that participates in the transport of sugars including lactose, across the membrane. *Lac*A codes for β-galactoside transacetylase, a 275-amino-acid enzyme that transfers an acetyl group from acetyl CoA to β-galactosides. Of the three proteins, only β-galactosidase participates in a known metabolic pathway. The permease helps lactose move across the cell membrane. The transacetylase may be associated with detoxification and excretion reactions of nonmetabolized analogs of β-galactosides.

Mutations in *lac*Z or *lac*Y that inactivate β-galactosidase or permease prevent cells from cleaving lactose or acquiring it from the medium, respectively. Mutations in *lac*A that inactivate transacetylase do not seem to have an effect on cell growth and division.

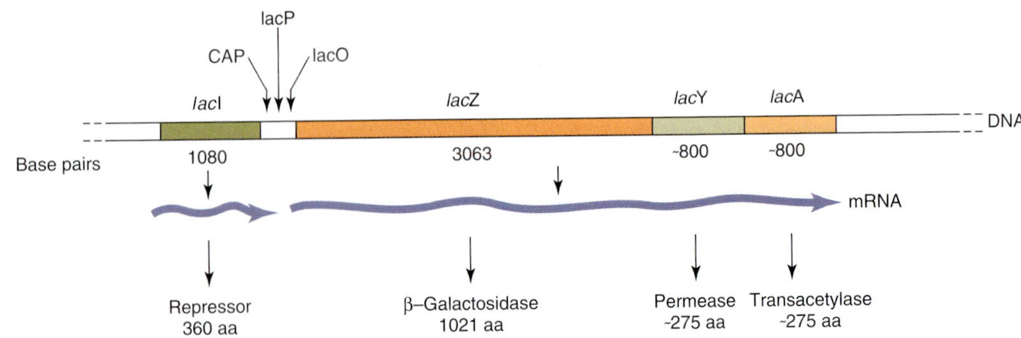

FIGURE 8.2

Lactose operon of *E. coli*. Lactose operon is composed of *lac*I gene, which codes for a repressor, control elements of CAP, *lac*P, and *lac*O, and three structural genes, *lac*Z, lacY, and *lac*A, which code for β-galactosidase, a permease, and a transacetylase, respectively. The *lac*I gene is transcribed from its own promoter. The three structural genes are transcribed from the promoter, *lac*P, to form a polycistronic mRNA from which the three proteins are translated.

A single mRNA species containing the coding sequences of all three structural genes is transcribed from a **promoter**, called *lac*P, located just upstream from *lac*Z. Induction of these genes occurs during initiation of their transcription. Without the inducer, transcription occurs at a very low level. In the presence of inducer, transcription begins at *lac*P and proceeds through all three genes to a transcription terminator located slightly beyond *lac*A. Thus, the genes are **coordinately expressed**; all are transcribed in unison or none is transcribed.

The presence of three coding sequences on the same mRNA molecule suggests that the relative amount of the three proteins is always the same under varying conditions of induction. The inducer is a molecular switch that influences synthesis of one mRNA species for all three genes. However, the amount of each of the three proteins is often different because of differences in rates of translation or of degradation of the proteins themselves.

The mRNA induced by lactose has a half-life of about 3 min. Therefore, expression of the operon can be altered very quickly. Transcription ceases as soon as inducer is no longer present, existing mRNA molecules disappear within minutes, and cells stop making the proteins.

Repressor of Lactose Operon is a Diffusible Protein

The regulatory gene of the lactose operon, *lac*I, codes for the lactose **repressor** whose only function is to control initiation of transcription of the *lac* structural genes. *Lac*I is located just upstream of the controlling elements for the *lac*ZYA cluster. However, a regulatory gene need not be close to the gene cluster it regulates. Transcription of *lac*I is not regulated; this gene is transcribed from its own promoter at a low rate that is relatively independent of the cell's status.

Lactose repressor is synthesized as a monomer of 360 amino acids, which forms an active homotetramer. Usually there are about 10 tetramers per cell. It has strong affinity for a DNA sequence called the **operator**, or *lac*O, that lies between *lac*P and *lac*Z. *Lac*O partly overlaps *lac*P, so bound repressor prevents RNA polymerase from binding to *lac*P and initiating transcription.

Besides recognizing and binding to *lac*O, the repressor has strong affinity for inducer molecules of the lactose operon. Each monomer binds to an inducer molecule, which causes a **conformational change** that greatly lowers the repressor's affinity for *lac*O (Figure 8.3). Thus, when inducer is present, the repressor no longer binds to *lac*O and RNA polymerase begins transcription from the promoter.

Study of the lactose operon was greatly facilitated by the discovery that some small molecules, such as isopropylthiogalactoside (**IPTG**), fortuitously serve as inducers but are not metabolized by β-galactosidase. These **gratuitous inducers** bind like inducers to the repressor molecule.

The repressor protein acts in trans; that is, it can diffuse to its site of action. Some *lac*I mutations change or delete amino acids of the repressor that are part of the binding site for the inducer. These changes do not affect the affinity of repressor for

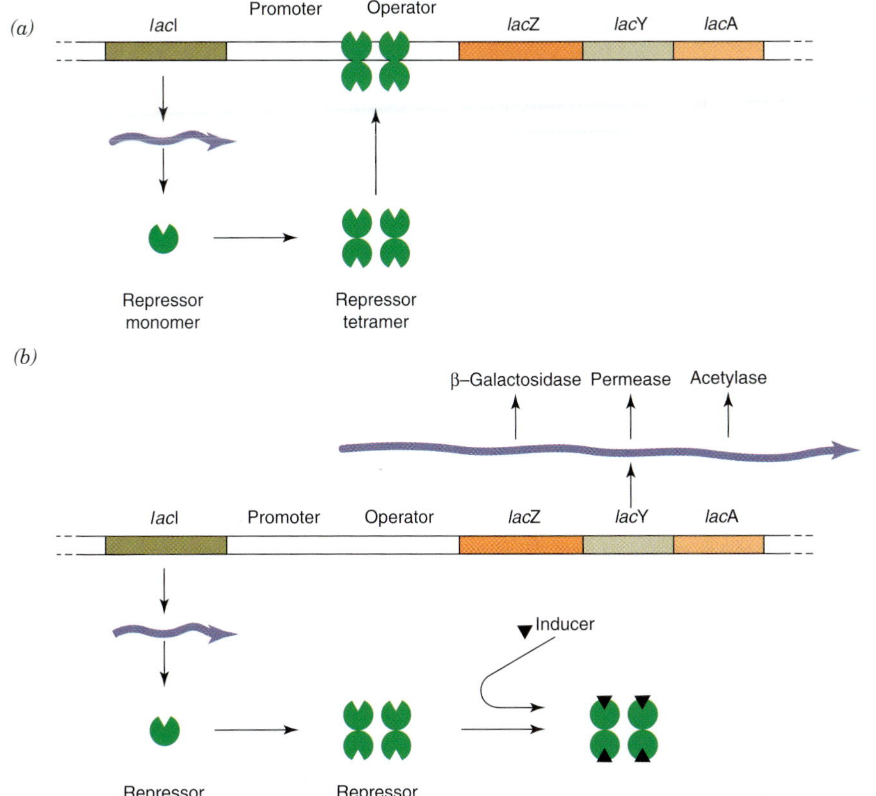

FIGURE 8.3

Control of lactose operon. (*a*) Repressor tetramer binds to operator and prevents transcription of structural genes. (*b*) Inducer binds to repressor tetramer, which prevents repressor from binding to operator. Transcription of the three structural genes can occur from the promoter.

the operator, so that the repressor is always bound to the operator, even in the presence of inducer, and the *lac*ZYA genes are never transcribed above a very low basal level. Other *lac*I mutations change the amino acids in the operator binding site and lessen the affinity of repressor for operator. Thus, the repressor does not bind the operator, and the *lac*ZYA genes are continuously transcribed. Such mutations are **repressor-constitutive mutations** because the *lac* genes are permanently expressed. A few rare *lac*I mutants increase the affinity of repressor for operator. In these cases, inducer molecules can bind to repressor, but they are less effective in releasing repressor from the operator.

Repressor-constitutive mutants illustrate the features of a negative control system. An active repressor, in the absence of an inducer, shuts off expression of the *lac* structural genes. An inactive repressor results in the constitutive, unregulated, expression of these genes. Using recombinant DNA techniques described in Chapter 7, a plasmid containing wild-type *lac*I (but not the rest of the lactose operon) can be introduced into constitutive *lac*I mutant cells. Such cells synthesize both active and inactive repressor molecules. Under these conditions, normal wild-type regulation of the lactose operon occurs. Thus, in genetic terms, wild-type induction is dominant over mutant constitutivity.

Operator Sequence of Lactose Operon is Contiguous with a Promoter and Three Structural Genes

The control elements upstream of the structural genes of the lactose operon are the operator and the promoter. The operator was identified, like *lac*I, by mutations that affect transcription of the *lac*ZYA region. Some of these mutations result in the constitutive synthesis of *lac* mRNA; that is, they are **operator-constitutive mutations**. In these cases the operator DNA sequence has one or more base pair changes so the repressor no longer binds as tightly to the sequence. Thus, the repressor is less effective in preventing RNA polymerase from initiating transcription.

In contrast to mutations in *lac*I that encodes the diffusible repressor, mutations in the operator do not change a diffusible protein. They exert their influence on transcription of only the three *lac* genes immediately downstream of the operator on the same DNA molecule. If a second lactose operon is introduced into a bacterium on a recombinant plasmid, the operator of one operon has no influence action on the other operon. Thus, an operon with a wild-type operator will be repressed under the usual conditions, whereas in the same engineered bacterium a second operon with an operator-constitutive mutation will be continuously transcribed.

Operator mutations are frequently referred to as **cis-dominant** to emphasize that they affect only adjacent genes on the same DNA molecule. Cis-dominant mutations occur in DNA sequences that are recognized by proteins rather than in DNA sequences that code for the diffusible proteins. **Trans-dominant** mutations occur in genes that specify diffusible products. Therefore, cis-dominant mutations also occur in promoter and transcription termination sequences, whereas trans-dominant mutations also occur in genes for subunit proteins of RNA polymerase, ribosomal proteins, and so on.

Figure 8.4 shows the sequence of the *lac* operator and promoter. The operator sequence has an axis of **dyad symmetry**. The sequence of the upper strand on the left side of the operator is nearly identical to that of the lower strand on the right side; only three differences occur. This symmetry in the DNA recognition sequence reflects symmetry in the tetrameric repressor and facilitates tight binding of the repressor subunits to the operator. Dyad symmetry in the double-stranded DNA sequence is a common feature of many protein binding or recognition sites, including most recognition sites for restriction enzymes.

The *lac* operator of 30 bp is an extremely small fraction of the total *E. coli* genome and occupies an even smaller fraction of the volume of the cell. However, the approximately 10 tetrameric repressors are also confined to a small fraction of the cell volume. Since the repressor gene is very close to the *lac* operator, the repressor does not have far to diffuse if its translation begins before its mRNA is fully synthesized. More importantly, the repressor has a low general affinity for all DNA sequences. When the inducer binds to the repressor, its affinity for the operator is reduced about 1000-fold, but its low affinity for random DNA sequences is unaltered. Therefore, all of the lactose repressors in the cell are in loose association with DNA. When binding of inducer releases a repressor molecule from the operator, it quickly binds a nearby DNA region. Therefore, induction redistributes the repressor on the DNA rather than generating freely diffusing repressor molecules.

How does lactose enter a cell in the first place if the *lac*Y gene encoding the permease is repressed, yet permease is required for lactose transport across the cell membrane? Even in the repressed state, there is a very low basal level of transcription of the *lac* operon that provides five or six molecules of permease per cell. Perhaps this is just enough to get a few molecules of lactose inside the cell and begin the process.

Another curious observation is that lactose is not the natural inducer of the lactose operon. When the repressor is isolated from fully induced cells, the small molecule bound to each repressor monomer is **allolactose**, not lactose. Allolactose, like lactose, is composed of galactose and glucose, but the linkage between the two sugars is different.

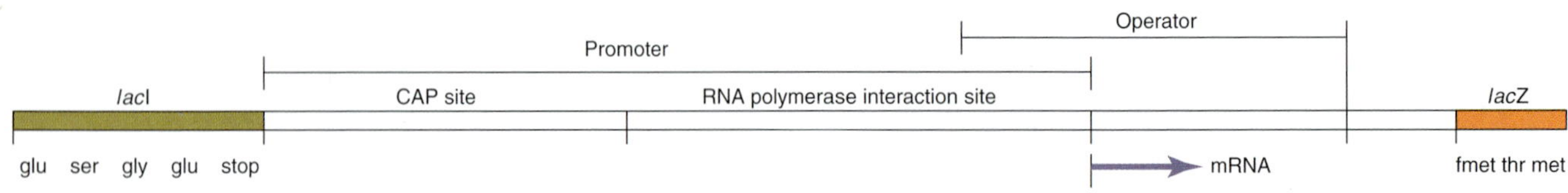

FIGURE 8.4

Nucleotide sequence of control elements of lactose operon. The end of *lac*I gene (coding for the lactose repressor) and beginning of *lacZ* gene (coding for β-galactosidase) are also shown. Lines above and below the sequence indicate symmetrical sequences within the CAP site and operator.

A side reaction of β-galactosidase (which normally breaks down lactose to galactose and glucose) converts these two products to allolactose.

Therefore, a few molecules of lactose are taken up and converted by β-galactosidase to allolactose, which binds to the repressor and induces the operon. Further confirmation that lactose is not the real inducer comes from experiments indicating that binding of lactose to purified repressor actually increases the repressor's affinity for the operator. Therefore, in the induced state a small amount of allolactose must be present to overcome this "anti-inducer" effect of lactose.

RNA Polymerase and a Regulator Protein Recognize Promoter Sequence of Lactose Operon

Immediately upstream of the *lac* operator is the promoter. This sequence contains the binding sites for RNA polymerase and **catabolite activator protein (CAP**; also called **cAMP receptor protein** or **CRP**) (Figure 8.4). The site to which RNA polymerase binds has been identified by genetic and biochemical approaches. Point mutations and deletions (or insertions) in this region dramatically affect RNA polymerase binding. The end points of the sequence to which RNA polymerase binds were identified by DNase protection experiments. Purified RNA polymerase was bound to the *lac* promoter region cloned in a bacteriophage DNA or a plasmid, and this protein–DNA complex was digested with DNase I. The DNA segment protected from degradation by DNase was recovered and its sequence determined. The ends of this protected segment varied slightly with different DNA molecules but corresponded closely to the boundaries of the RNA polymerase binding site shown in Figure 8.4.

The RNA polymerase binding site does not have symmetrical elements like the operator sequence. This is not surprising since RNA polymerase must associate with DNA in an asymmetrical fashion for transcription to be initiated in only one direction. However, the part of the promoter sequence recognized by CAP does contain some symmetry.

Catabolite Activator Protein Binds Lactose Promoter

E. coli prefers glucose over other sugars as a carbon source. For example, if both glucose and lactose are in the medium, the bacteria selectively metabolize glucose. Figure 8.5 shows that β-galactosidase does not appear until the glucose is depleted. This indicates that glucose interferes with induction of the lactose operon, a process called **catabolite repression** since it occurs during the catabolism of glucose. An identical effect is seen with other inducible operons, including the arabinose and galactose operons. This is probably a general coordinating system for turning off synthesis of unwanted enzymes when glucose is present.

Catabolite repression occurs because glucose inhibits adenylate cyclase, which synthesizes **cyclic AMP (cAMP)**, leading to a lower intracellular concentration of cAMP. cAMP forms a complex with **catabolite activator protein (CAP)**, which then binds the CAP regulatory site at promoters of lactose (and other) operons (Figure 8.6). The CAP–cAMP complex exerts a positive control on transcription by causing the DNA helix to bend, or kink, about 90 degrees at the site of the binding. This DNA bend and the interaction between CAP–cAMP and RNA polymerase activate transcription initiation. If the CAP site is not occupied, RNA polymerase has difficulty binding to the promoter and transcription initiation is much less efficient.

The lactose operon demonstrates how bacteria can coordinate a general response to a metabolic condition (the need to use a sugar as an energy or carbon source) and a specific response to that condition (the need to utilize lactose as the sugar).

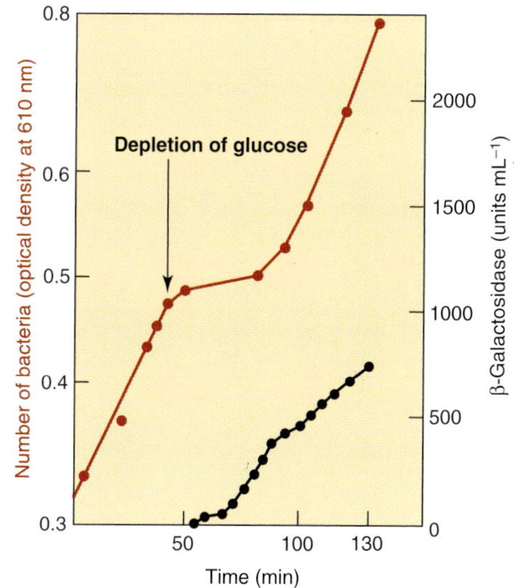

FIGURE 8.5

Lack of synthesis of β-galactosidase in *E. coli* when glucose is present. Bacteria are growing in a medium containing initially 0.4 mg glucose mL^{-1} and 2 mg lactose mL^{-1}. Left-hand ordinate indicates optical density of growing culture, an indicator of the number of bacterial cells. Right-hand ordinate indicates units of β-galactosidase per milliliter. Note that appearance of β-galactosidase is delayed until the glucose is depleted. Redrawn from Epstein, W., Naono, S., and Gros, F. *Biochem. Biophys. Res. Commun.* 24:588, 1966.

8.4 TRYPTOPHAN OPERON OF *E. COLI*

Bacteria must have the proper amount and relative balance of the 20 amino acids that make up proteins. **Tryptophan**, for example, is required for the synthesis of all proteins

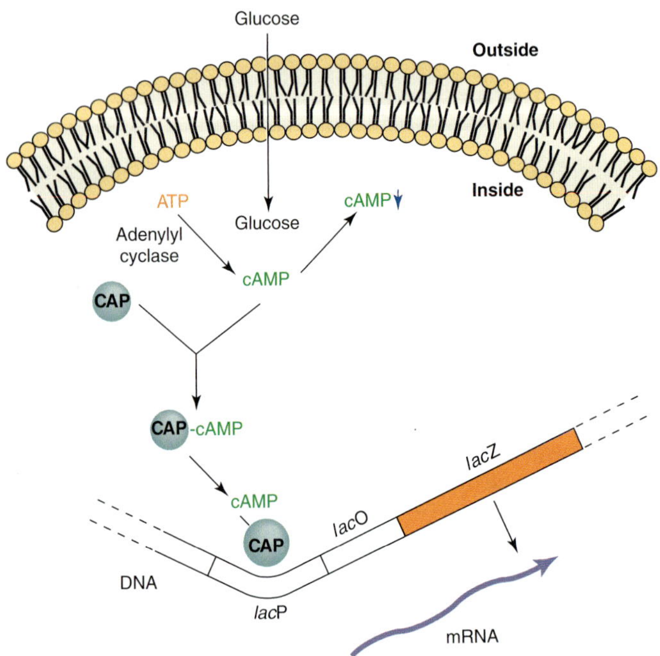

FIGURE 8.6

Control of *lac*P by cAMP. A CAP–cAMP complex binds to CAP site and enhances transcription at *lac*P. Catabolite repression occurs when glucose lowers intracellular concentration of cAMP. This reduces the amount of the CAP–cAMP complex and decreases transcription from *lac*P and from promoters of several other operons.

that contain it. Therefore, if tryptophan is not present in sufficient quantity in the medium, the bacterial cell has to make it. In contrast, lactose is not absolutely required for the cell's growth; many other sugars can substitute for it. As a result, synthesis of biosynthetic enzymes for tryptophan is regulated differently than synthesis of proteins encoded by the lactose operon.

Tryptophan Operon is Controlled by a Repressor Protein

In *E. coli*, tryptophan is synthesized from chorismic acid in a five-step pathway catalyzed by three enzymes (Figure 8.7). The **tryptophan operon** contains five structural genes that code for these three enzymes (two of which contain two different subunits). Upstream from this gene cluster is a promoter where transcription begins and an operator to which binds a repressor protein encoded by a separate *trp*R gene. Transcription of the lactose operon is generally "turned off," or repressed, unless induced by a small molecule inducer. The tryptophan operon, on the other hand, is always active, or derepressed, unless it is repressed by a small molecule **corepressor**, which for the tryptophan operon is tryptophan itself.

Tryptophan biosynthesis is regulated by both the synthesis and the activity of enzymes that catalyze the pathway. For example, anthranilate synthetase, which catalyzes the first step, contains two subunits encoded by *trp*E and *trp*D, and its enzyme activity is regulated by **feedback inhibition**. This is a common short-term means of regulating the first committed step in a metabolic pathway. Tryptophan can bind to an allosteric site on anthranilate synthetase and prevent its enzymatic activity. Thus, as the concentration of tryptophan builds up, it inhibits anthranilate synthetase. Tryptophan also is a corepressor that shuts down transcription of the tryptophan operon. Feedback inhibition is a short-term control with an immediate effect on the pathway, whereas repression takes longer but has the more permanent effect of repressing transcription.

The tryptophan repressor is a homodimer whose subunits have 108 amino acids each. Under normal conditions, about 20 molecules of the repressor dimer are present. The repressor must be complexed with two molecules of tryptophan to bind to the operator. Recall that the lactose repressor binds its operator only in the absence of its inducer. The tryptophan repressor also regulates transcription of *trp*R, its own gene. As tryptophan accumulates in cells, the repressor–tryptophan complex binds to a region upstream of *trp*R, turning off its transcription and maintaining the equilibrium of 20 repressors per cell. The *trp* operator occurs entirely within the *trp* promoter rather

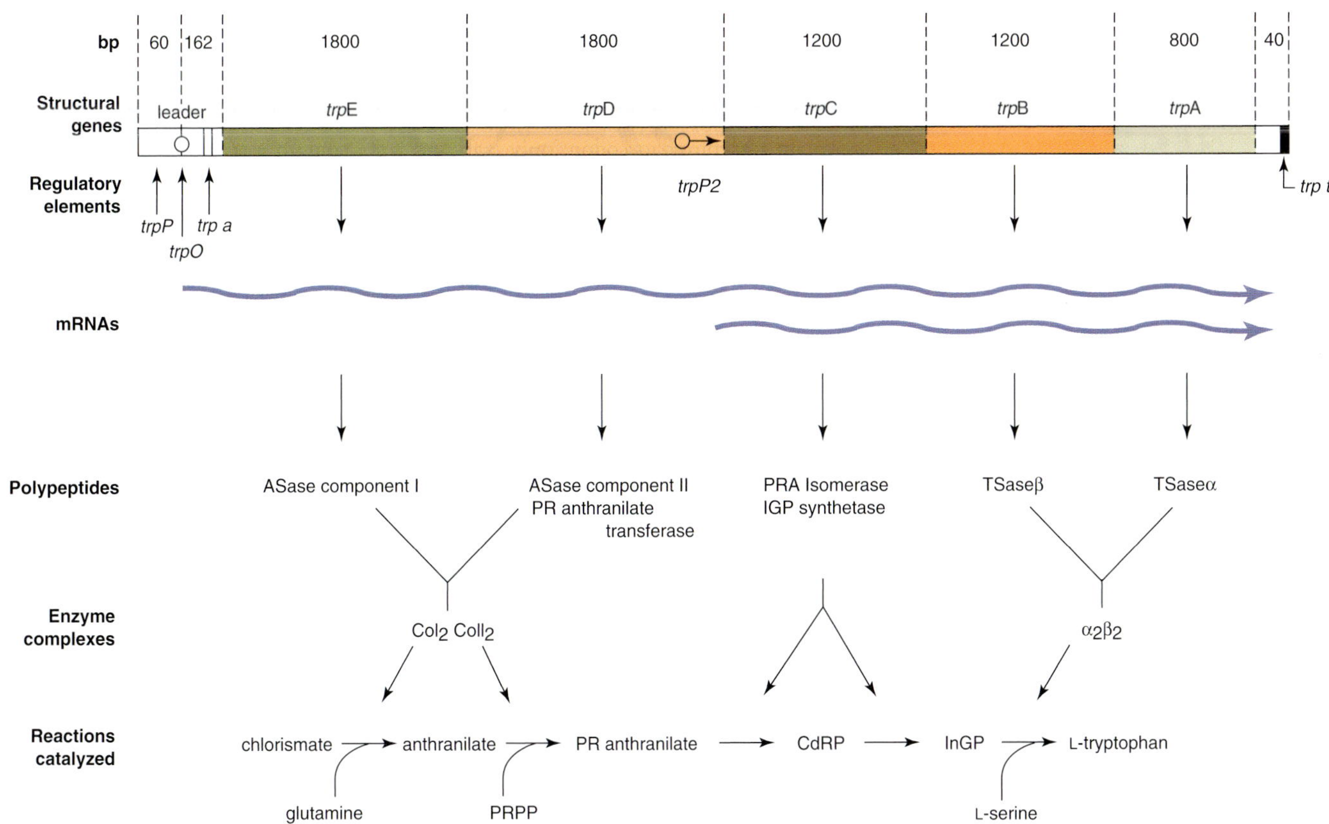

FIGURE 8.7

Genes of tryptophan operon of *E. coli*. Regulatory elements are the primary promoter (*trp*P), operator (*trp*O), attenuator (*trp*a), secondary internal promoter (*trp*P2), and terminator (*trp*t). Direction of mRNA synthesis is indicated by wavy arrows representing mRNAs. CoI_2 and $CoII_2$ signify components I and II, respectively, of the anthranilate synthetase (ASase) complex; PR-anthranilate is *N*-5′-phosphoribosyl-anthranilate; CdRP is 1-(*o*-carboxy-phenylamino)-1-deoxyribulose-5-phosphate; InGP is indole-3-glycerol phosphate; PRPP is 5-phosphoribosyl-1-pyrophosphate; and TSase is tryptophan synthetase.

Redrawn from Platt, T. The tryptophan operon. In: J. H. Miller and W. Resnikoff (Eds.), *The Operon*. Cold Spring Harbor, NY: Cold Spring Harbor Laboratory Press, 1978, p. 263.

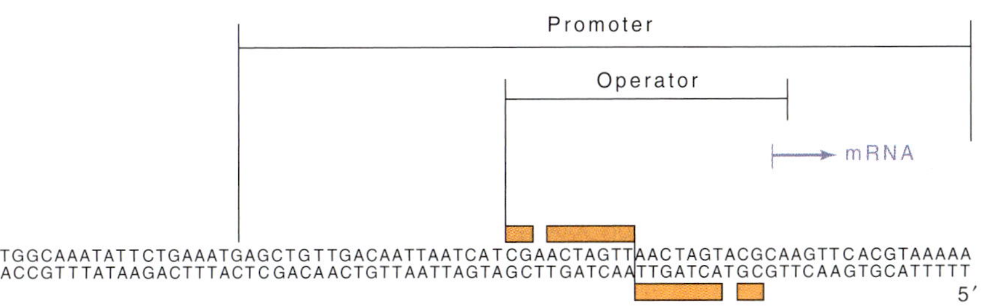

FIGURE 8.8

Nucleotide sequence of control elements of tryptophan operon. Boxes above and below sequence indicate symmetrical sequences within operator.

than adjacent to it (Figure 8.8). The operator is a region of dyad symmetry, and the mechanism for preventing transcription is the same as in the lactose operon; that is, binding of the repressor-corepressor complex to the operator prevents binding of RNA polymerase.

Repression decreases the rate of transcription initiation at the *trp* promoter by about 70-fold. (The basal level of lactose operon gene products is about 1000-fold lower than the induced level.) However, the tryptophan operon contains additional regulatory elements. A secondary promoter designated *trp*P2 is located within the coding sequence of *trp*D (see Figure 8.7) and is not regulated by the tryptophan repressor. Transcription from it occurs constitutively at a relatively low rate to generate an mRNA that contains

Operon	Leader peptide sequence	Regulatory amino acids
his	Met-Thr-Arg-Val-Gln-Phe-Lys-His-His-His-His-His-His-His-Pro-Asp	His
pheA	Met-Lys-His-Ile-Pro-Phe-Phe-Phe-Ala-Phe-Phe-Phe-Thr-Phe-Pro	Phe
thr	Met-Lys-Arg-Ile-Ser-Thr-Thr-Ile-Thr-Thr-Thr-Ile-Thr-Ile-Thr-Thr-Gly-Asn-Gly-Ala-Gly	Thr Ile
leu	Met-Ser-His-Ile-Val-Arg-Phe-Thr-Gly-Leu-Leu-Leu-Leu-Asn-Ala-Phe-Ile-Val-Arg-Gly-Arg-Pro-Val-Gly-Gly-Ile-Gln-His	Leu
ilv	Met-Thr-Ala-Leu-Leu-Arg-Val-Ile-Ser-Leu-Val-Val-Ile-Ser-Val-Val-Val-Ile-Ile-Ile-Pro-Pro-Cys-Gly-Ala-Ala-Leu-Gly-Arg-Gly-Lys-Ala	Leu, Val, Ile

FIGURE 8.11

Leader peptide sequences specified by biosynthetic operons of *E. coli*. All of the leader peptide sequences contain multiple copies of amino acid(s) synthesized by enzymes encoded by that operon.

(i.e., stalled at) the attenuator site, transcription of the downstream structural genes is enhanced. If the ribosome is not bound, transcription of these genes is greatly reduced.

Transcription of some operons, shown in Figure 8.11, can be attenuated by more than one amino acid. For example, the thr operon is attenuated by threonine or isoleucine, while the ilv operon is attenuated by leucine, valine, or isoleucine. This effect can be explained in each case by stalling of the ribosome at the corresponding codon, which interferes with formation of a termination hairpin. It is possible that in longer leader sequences, stalling at more than one codon is necessary to achieve maximal transcription through the attenuation region.

8.5 OTHER BACTERIAL OPERONS

Synthesis of Ribosomal Proteins is Regulated in a Coordinated Manner

Many bacterial operons possess the same general regulatory mechanisms as the *lac, trp*, and *his* operons. However, each operon has its own distinctive characteristics. One example is the structural genes for the 70 or more proteins that comprise a ribosome (Figure 8.12). Each ribosome contains one copy of each **ribosomal protein** (except for protein L7–L12, which is probably present in four copies). Therefore, all 70 proteins are required in equimolar amounts, and it makes sense that their synthesis is regulated in a coordinated fashion. Six different operons, containing about one-half of the ribosomal protein genes, occur in two major gene clusters. One cluster contains four adjacent operons (*Spc*, *S10*, *str* and *a*), and the other contains two operons (*L11* and *rif*) located elsewhere in the *E. coli* chromosome. There is no obvious pattern in the distribution of genes among these operons. Some operons code for proteins of one ribosomal subunit; others code for proteins of both subunits. These operons also contain genes for other

Operon	Regulator protein	Proteins specified by the operon
Spc	S8	L14-L24-L5-S14-S8-L6-L18-S5-L15-L30
S10	L4	S10-L3-L2-L4-L23-S19-L22-S3-S17-L16-L29
str	S7	S12-S7-EF.G-EF.Tu
α	S4	S13-S11-S4-α-L17
L11	L1	L11-L1
rif	L10	L10-L7-β-β′

FIGURE 8.12

Operons containing genes for *E. coli* ribosomal proteins. Genes for protein components of the small (S) and large (L) ribosomal subunits of *E. coli* are clustered on several operons. Some of these operons also contain genes for RNA polymerase subunits α, β, and β' and protein synthesis factors EF.G and EF.Tu. At least one protein product of each operon usually regulates expression of that operon (see text).

(related) proteins. For example, the *str* operon contains genes for soluble translation elongation factors, EF-Tu and EF-G, and genes for some proteins in the 30S ribosomal subunit. The *a* operon contains genes for proteins of both ribosomal subunits and a gene for the α subunit of RNA polymerase. The rif operon has genes for the β and β' subunits of RNA polymerase and for ribosomal proteins.

A common feature of the six operons for ribosomal proteins is that their expression is regulated by one of their own structural gene products; that is, they are **self-regulated**. In some cases, regulation occurs at the level of translation, not transcription as discussed for the *lac* and *trp* operons. After the polycistronic mRNA is made, the "regulatory" ribosomal protein binds to this mRNA and determines which regions, if any, are translated. In general, the ribosomal protein that regulates expression of its own operon associates with ribosomal RNA (rRNA) in the ribosome. This protein has a high affinity for rRNA and a lower affinity for one or more regions of its own mRNA. Therefore, competition occurs between rRNA and the operon's mRNA for binding with the protein. As the protein accumulates to a higher level than free rRNA, it binds to its own mRNA and prevents protein synthesis at one or more of the coding sequences on this mRNA (Figure 8.13). As more ribosomes are formed, the excess of this protein is used up and translation of its mRNA can begin again.

Stringent Response Controls Synthesis of rRNAs and tRNAs

Bacteria respond in several ways to **extreme general stress**. One such situation is when there are insufficient amino acids to maintain protein synthesis. Under this condition the cell invokes the **stringent response** that reduces synthesis of rRNAs and tRNAs about 20-fold. Synthesis of mRNAs also decreases about threefold.

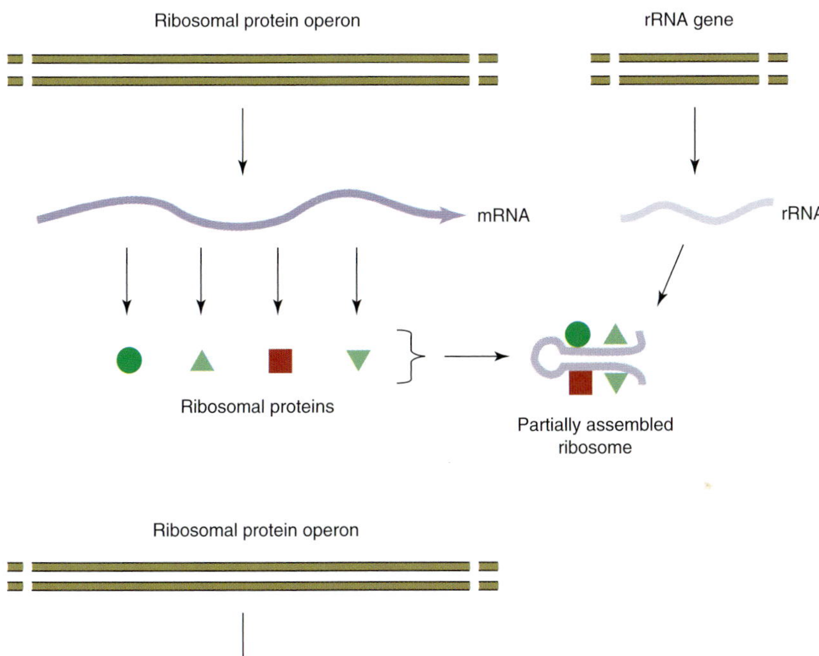

FIGURE 8.13

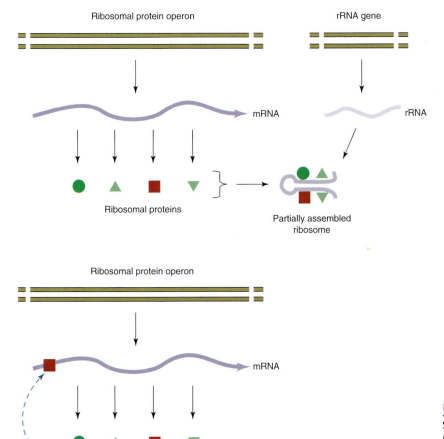

Self-regulation of ribosomal protein synthesis. If free rRNA is not available for assembly of new ribosomal subunits, individual ribosomal proteins bind to polycistronic mRNA from their own operon, blocking further translation.

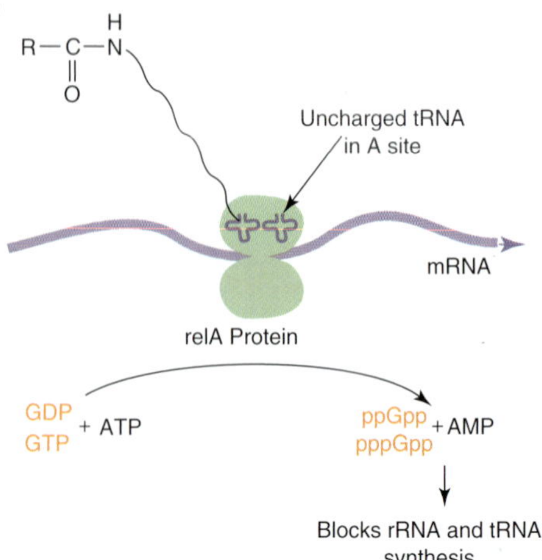

FIGURE 8.14

Stringent control of protein synthesis in *E. coli*. During extreme amino acid starvation, an uncharged tRNA in the A site of the ribosome activates relA protein to synthesize ppGpp and pppGpp, which are involved in decreasing transcription of the genes coding for rRNAs and tRNAs.

The stringent response is triggered by the presence of an uncharged tRNA in the A site of the ribosome, which occurs when the concentration of the corresponding charged tRNA is very low. The first result is that further peptide elongation by the ribosome stops. This stoppage causes a protein called the **stringent factor**, the product of the relA gene, to synthesize **guanosine tetraphosphate** (ppGpp) and **guanosine pentaphosphate** (pppGpp), from ATP and GTP or GDP as shown in Figure 8.14. Stringent factor is loosely associated with some ribosomes of the cell. Perhaps a conformational change in the ribosome is induced by the presence of an uncharged tRNA in the A site, which, in turn, activates the associated stringent factor. The exact functions of ppGpp and pppGpp are not known. However, they inhibit transcription initiation of rRNA and tRNA genes and they affect transcription of some operons more than others.

8.6 BACTERIAL TRANSPOSONS

Transposons are Mobile Segments of DNA

The vast majority of bacterial genes have fixed locations in the chromosome. In fact, the genetic maps of *E. coli* and *Salmonella typhimurium* are very similar, indicating the lack of much evolutionary movement of most genes within bacterial chromosomes. There is a small class of bacterial genes, however, in which newly duplicated gene copies "jump" from one genomic site to another with a frequency of about 10^{-7} per generation, the same rate as spontaneous point mutations. These mobile segments of DNA are called **transposable elements** or **transposons** (Figure 8.15). Some genes within bacterial transposons control the presence and transposition of the transposon itself, whereas others, usually antibiotic-resistance genes, provide the bacterium with a selective advantage against other bacteria.

Transposons were first detected as rare insertions of foreign DNA into structural genes of bacterial operons. Usually, these insertions interfere with expression of the affected structural gene and all downstream genes of the operon. This is not surprising since insertions can destroy the translation reading frame, introduce transcription termination signals, affect the mRNA stability, and so on. Many transposons and the sites into which they insert have been isolated using recombinant DNA techniques and have been extensively characterized.

Some transposons consist of a few thousand base pairs and contain two or three genes; others are much longer and contain many genes. Sometimes, small transposons can occur within a large transposon. All active transposons contain at least one gene that codes for a **transposase**, which is required for the transposition or "jumping" event. Often they contain genes that code for resistance to antibiotics or heavy metals. Most transpositions involve generation of an additional copy of the transposon and insertion of this copy into another location. The original copy is unaffected by insertion of its duplicate into a new target site. Transposons contain short inverted **terminal repeat sequences** that are essential for the insertion mechanism, and they are often used to define the two boundaries of a transposon. Most target sites are fairly random in sequence, although some transposons have a propensity for insertion at specific "hot spots." The duplicated transposon can be located in a different DNA molecule than its donor. Frequently, transposons are found on plasmids that pass from one bacterial strain to another and are the source of a suddenly acquired resistance to one or more antibiotics by a bacterium (Clin. Corr. 8.1).

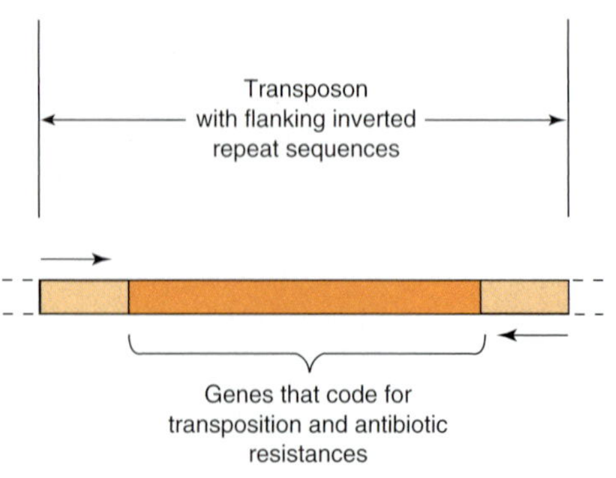

FIGURE 8.15

General structure of transposons. Transposons are relatively rare mobile segments of DNA that contain genes coding for their own rearrangement and (often) genes that specify resistance to various antibiotics.

Transposon *TN3* Contains Three Structural Genes

An example of a transposon is **transposon *Tn3***, which contains 4957 bp including inverted repeats of 38 bp at each end (Figure 8.16). Three genes are present in *Tn3*. One codes for β-lactamase, which hydrolyzes ampicillin and renders the cell resistant to this antibiotic. The others, *tnp*A and *tnp*R, code for a transposase and a repressor protein, respectively. The transposase contains 1021 amino acids and binds to single-stranded

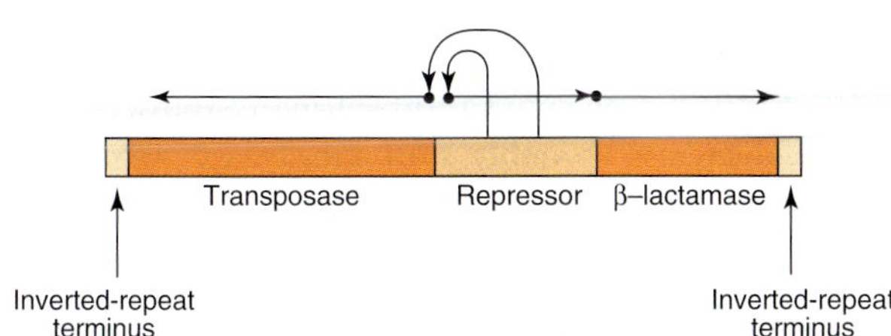

FIGURE 8.16

Functional components of the transposon *Tn3*. Genetic and DNA sequence analyses show there are four distinct regions: the inverted repeat termini; a gene for the enzyme β-lactamase, which confers resistance to ampicillin and related antibiotics; a gene encoding an enzyme required for transposition (transposase); and a gene for a repressor protein that controls transcription of the genes for the transposase and the repressor itself. Horizontal arrows indicate directions of transcription.
Redrawn from Cohen, S. N. and Shapiro, J. A. *Sci. Am.* 242:40, 1980. W. H. Freeman and Company, Copyright © 1980.

DNA. It recognizes the repetitive ends of the transposon and participates in cleavage of the target site into which the new transposon copy inserts. The repressor protein of 185 amino acids controls transcription of both the transposase gene and its own gene. *tnp*A and *tnp*R are transcribed divergently from a 163-bp control region located between them that binds the repressor. The repressor also participates in the insertion of the new transposon, but does not affect transcription of the ampicillin-resistance gene.

Mutations in *tnp*A that inactivate the transposase decrease the frequency of *Tn*3 transposition. Those in *tnp*R that inactivate the repressor increase the frequency of transposition. These mutations derepress *tnp*A, resulting in more molecules of the transposase; this enhances the formation of more duplicated transposon copies. They also derepress *tnp*R but since the repressor is inactive, it has no effect.

Transposons located on bacterial plasmids are of increasing importance in clinical uses of antibiotics. Bacterial plasmids that have not been altered for experimental use usually contain genes that facilitate their transfer from one bacterium to another. As these plasmids transfer among different infecting bacterial strains, their transposons containing **antibiotic-resistance genes** are moved into new bacterial strains. Once inside a new bacterium, the transposon can be duplicated onto the chromosome and become permanently established in that cell's lineage. The result is that more and more pathogenic bacterial strains have become resistant to an increasing number of antibiotics.

8.7 GENE EXPRESSION IN EUKARYOTES

Gene transcription in eukaryotic organisms is also regulated to provide the appropriate response to biological needs. In addition to modulating gene expression in response to nutritional or environmental conditions, multicellular organisms regulate the expression of specialized genes that drive cellular differentiation and typify specific cell types. Some genes (so-called housekeeping genes) are expressed in most cells, other genes are activated upon demand, and still other genes are rendered permanently inactive in all but a few cell types. In eukaryotic cells, the nuclear membrane serves as a barrier that selectively allows some proteins access to DNA while keeping others in the cytosol. In bacteria, one RNA polymerase is responsible for transcription of all RNAs (tRNA, rRNA, and mRNA). In eukaryotic organisms, three different RNA polymerases are used (see p. 184). RNA polymerase I transcribes the rRNA genes, RNA polymerase II transcribes the protein-encoding genes whose transcripts become the mRNAs, and RNA polymerase III transcribes the genes for tRNAs and for most other small RNAs. Although some principles for eukaryotic gene activation and control are applicable for all three RNA polymerases, in this section the focus will be on transcription by RNA polymerase II. RNA polymerase II is comprised of at least 10 different subunits, ranging in size from 10 to 220 kDa. Some of the subunits are also part of the RNA polymerase I and III complexes, while others are unique to RNA polymerase II. The largest subunit of RNA polymerase II has, depending on the species, as many as 52 repeats of the amino acid motif PTSPSYS in its C-terminal domain (CTD). A distinguishing feature of these repeats is that threonine (T), serine (S), and tyrosine (Y) can be phosphorylated.

To better understand the regulation of transcription in eukaryotes, it is useful to recall the organization of DNA into **chromatin** and the role of DNA modification, specifically **methylation** of cytosine, on gene activation. In addition, we'll consider how

CLINICAL CORRELATION 8.1
Transmissible Multiple Drug Resistance

An alarming trend is that pathogenic bacteria are becoming increasingly resistant to a large number of antibiotics. Many cases have been documented in which a bacterial strain in a patient being treated with one antibiotic suddenly became resistant to that antibiotic and, simultaneously, to several other antibiotics even though the bacterial strain had never been previously exposed to these other antibiotics. This occurs when a bacterium suddenly acquires from another bacterial strain a plasmid that contains several different transposons, each containing one or more antibiotic-resistance genes. Examples include: the genes encoding β-lactamase, which inactivates penicillins and cephalosporins; chloramphenicol acetyltransferase, which inactivates chloramphenicol; and phosphotransferases that modify aminoglycosides such as neomycin and gentamycin.

Source: Neu, H. C. The crisis in antibiotic resistance. *Science* 257:1064, 1992.

RNA polymerase II is positioned at the appropriate spot in the promoter of a gene to transcribe that gene by the formation of a preinitiation complex that involves assembly of **general transcription factors (TFs)** with RNA polymerase II. Next, we'll look at how specific gene activity can be regulated through the use of enhancers, **transcription factor binding sites**, and **RNA polymerase assembly sites**. Finally, we will discuss the activation of transcription by specific transcription factors, some of their general characteristics, and how they are regulated.

Eukaryotic DNA is Bound by Histones to Form Chromatin

Segments of eukaryotic DNA are wrapped around an octamer of histone proteins that contain two molecules each of histone H2A, H2B, H3, and H4 to form **nucleosomes** (see p. 55). Histones associate with DNA through electrostatic interactions. Histones have a large number of positively charged lysine residues that interact with the negatively charged phosphodiester backbone of DNA. In most cells, histone H1 or H5 binds when the octamer–DNA association is established. About 200 bp of DNA make up a single nucleosomal unit, with about 130–160 bp in direct contact with the octamer core. Some of the remaining DNA binds histones H1 and H5, and the rest is the **linker** DNA between nucleosomes. Although the histone octamer–DNA interaction is not sequence-specific, some sequence-dependent patterns of association have been noted. When, for instance, the double-stranded DNA helix bends because of the presence of an AT-rich region, the minor groove generally faces the nucleosome core particle. In GC-rich regions, the minor groove faces away from the histone octamer core. Despite these tendencies, there is no clear way to predict the sequences within a DNA duplex that will comprise the major and minor grooves on the outside of the nucleosome complex. This positioning is important because assembly of the transcription initiator complex and binding of other proteins that influence gene expression need access to the grooves in the DNA duplex. Sequence-specific interactions occur via hydrogen bonds formed with the edges of the base pairs in the major or minor groove. When DNA is wrapped around a histone octamer, the specific DNA sequences that bind transcription factors may be occluded. The complete removal of DNA from the nucleosome is not required to allow access to specific sequences. By rolling the DNA helix relative to the histone octamer, different major and minor groove contacts can be made available on the outer surface of the nucleosome (Figure 8.17).

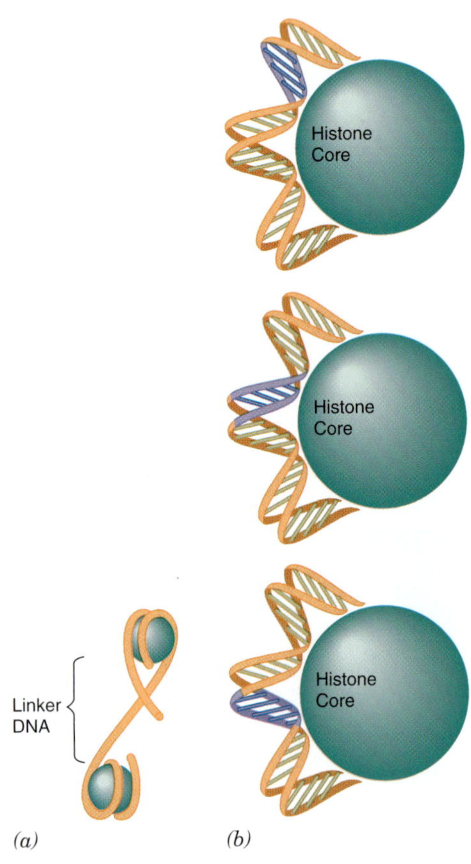

FIGURE 8.17

Access to specific sequences in major and minor groove of the DNA depends upon nucleosome positioning. (*a*) DNA wound around the core histone octamer (two molecules each of histone H2A, H2B, H3, and H4) is called the nucleosome core particle. Under most conditions, histone H1 or H5 (not shown) associates with DNA where the winding around the histone octamer begins. DNA between nucleosome core particles is called linker DNA. When exposed to agents like micrococcal nuclease, linker regions are more accessible than DNA wound around the histone core particle. (*b*) Within the nucleosome, the side of the helix facing away from the histone core is accessible, but the side facing in is not. If we rotate the helix with respect to the histone core of a nucleosome and follow the position of the same five base pairs, the access to the major and minor groove to other interactions is changed dependent on whether the base pair is on the inside or the outside of the wrapped nucleosome core particle.
Redrawn from Wolffe, A. P. *Chromatin: Structure and Function*, 3rd ed. New York: Academic Press, 1998.

Genes that are not transcribed within a particular cell form highly condensed heterochromatin. In contrast, transcriptionally active regions of DNA have a less condensed, more open structure. Part of this difference is due to posttranslational modification of histones. The acetylation of the ϵ-amino group of the lysine residues near the N-termini of histones reduces net positive charge and thus weakens the electrostatic attraction between the histones and the DNA. In general, acetylation of histones leads to activation of gene expression, while deacetylation reverses the effect. Acetylation of histones, and the subsequent repositioning of the nucleosome by the destabilization of histone DNA interactions, is important in providing sequence-specific access to DNA (Figure 8.18).

The relative position of a nucleosome can also be influenced by the SWI/SNF complex, first identified and best characterized in the yeast *S. cerevisiae* and named after mating-type **swi**tching and **s**ucrose **n**on-**f**ermenting strains. This complex contains about 10 proteins and interacts with the C-terminal domain of the large subunit of RNA polymerase II disrupting nucleosomal arrays in an ATP-dependent manner. The result is to open regions of DNA for interaction with transcription factors, and thus facilitate gene activation. There are fewer SWI/SNF complexes in the cell than genes being transcribed, suggesting that the SWI/SNF complex acts in a catalytic manner (see Clin. Corr. 8.2).

Methylation of DNA Correlates with Gene Inactivation

Methylation of human DNA centers on the formation of 5-methylcytosine (Figure 8.19). The reaction is sequence specific with 5′CpG3′ (Convention: p represents bridging phosphate and fixes C at 5′ end.) as a substrate. The result is that both

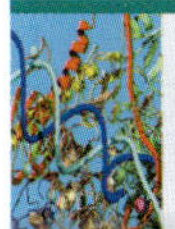

CLINICAL CORRELATION 8.2
Rubinstein–Taybi Syndrome

Regulation of histone acetylation influences activation and inactivation of gene expression. Two major enzymes responsible for acetylation of histones in mammalian cells are the acetyltransferases p300 and CBP. p300 was named for its molecular mass and CBP stands for CREB binding protein, with CREB being a transcription factor that is a cAMP regulatory-element binding protein. These acetyltransferases allow activation of appropriate genes in association with CREB and other transcription factors, leading to the opening of chromatin structure through weakening of histone–DNA interactions.

Rubinstein–Taybi Syndrome, characterized by mental retardation and other developmental abnormalities, is caused by point mutations, small deletions, and rearrangements in the CBP gene. More severe developmental defects correspond to substantially more mutations. The complete loss of the CBP gene product is probably lethal.

Source: F. Petrij, Giles, R. H., Dauwerse, H. G., Saris, J. J., et al. Rubinstein–Taybi syndrome caused by mutations in the transcriptional co-activator CBP. *Nature* 376:348, 1995. Wolffe, A. P. The cancer chromatin connection. *Sci. Med.* 6:28, 1999.

strands of the duplex are methylated. About 70% of CpG sequences in human DNA are methylated. Methylation is implicated in the **imprinting** of genomic DNA. In imprinting, the methylation patterns of DNA inherited from the sperm or egg correlate with choice of allelic expression. Methylation patterns are conserved after replication by a hemimethylase (which methylates only one of the two strands containing the CpG). More recently it has been suggested that methylation of DNA that results in gene silencing can be induced by the action of siRNA (short interfering RNA) that are complementary to the sequence that is methylated.

A terminology often encountered in the DNA methylation literature is "CpG islands" or "CpG-rich regions." The sequence CpG is under represented in the genome. This is likely due to the accumulation, over many generations, of C to T transitions caused by the deamination of 5-methylcytosine. Maintenance of CpGs are presumably due to selective pressure to maintain a regulatory region. In fact, CpG islands are most commonly found in gene promoter regions. Methylation of DNA often correlates with lack of transcriptional activity. This is thought to be mediated by proteins that recognize and bind methylated DNA, which prevents binding of transcription factors. Methylation of DNA correlates with deacetylation of histones, providing two different means of repression of transcription at a specific location (Figure 8.20).

Hypermethylation of DNA is a common feature in cells of cancerous tissue. There is increasing evidence that methylation occurs in genes encoding proteins that would direct abnormally dividing cells into programmed cell death (apoptosis). Conversely, in mammalian totipotent and pluripotent cells, such as a fertilized egg and cells of the very early embryo, DNA undergoes fairly global demethylation.

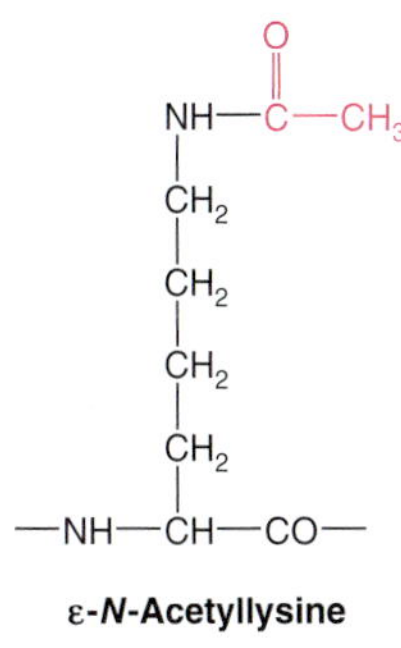

(*a*)

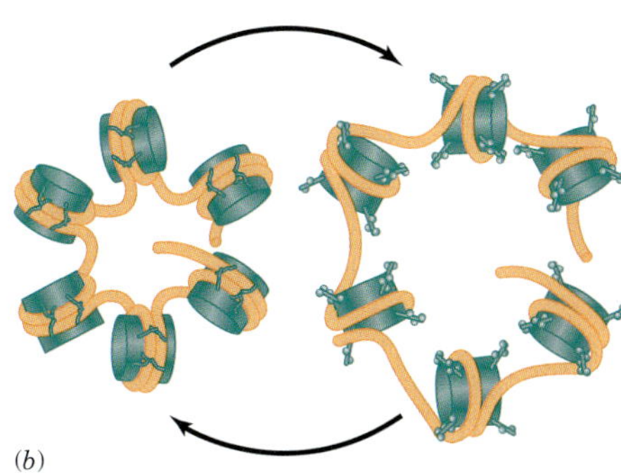

(*b*)

FIGURE 8.18

Strength of histone DNA association is modified by acetylation of lysine residues in the N-terminus of histone proteins. Histones that make up the core octamer contain lysine residues that, by virtue of their positively charged side group, can promote interaction with the negatively charged phosphodiester linkages of DNA. Modification of the lysines by acetylation replaces the positive charge with a neutral acetyl group, and it weakens the electrostatic interaction between the octamer core and the DNA. This process is reversible, and the acetylation and deacetylation of histones provides a way to loosen or tighten chromatin structure.
Redrawn from Wolffe, A. P. The cancer-chromatin connection. *Sci. Med.* 6:28, 1999.

8.8 PREINITIATION COMPLEX IN EUKARYOTES: TRANSCRIPTION FACTORS, RNA POLYMERASE II, AND DNA

Unlike bacterial RNA polymerase, eukaryotic RNA polymerase II does not undergo sequence-specific binding to eukaryotic promoters. Rather, an initiation complex is formed through the initial contact of the promoter with the general transcription factor TFIID. TFIID is one of at least six general transcription factors (TFIIA, TFIIB, TFIID, TFIIE, TFIIF, and TFIIH) required for basal transcription by RNA polymerase II. The nomenclature TFII reflects that they are transcription factors involved in the assembly of the RNA polymerase II preinitiation complex (Table 8.1).

TFIID is a multi-subunit complex that contains the TBP (TATA binding protein) and a number of different TAFs (TBP-associated factors). TFIID binds in the minor groove of DNA at a consensus sequence called the TATA box, located about 27 bp upstream of the transcription start site. The TATA box must be accessible to TFIID, so it cannot be on the inner face of a nucleosome core particle if the preinitiation complex is to form. Interaction of TBP with the DNA causes a large distortion in the DNA duplex (Figure 8.21). TBP was once thought to be essential for transcription using any RNA polymerase, but the recent identification of additional TLFs (TBP-like factors) suggests that some genes use an alternate protein. After binding to the TATA box, TBP directs assembly of the preinitiation complex by ordered addition of several general transcription factors and RNA polymerase II (Figure 8.22). Recall that the large subunit of RNA polymerase II has a C-terminal domain (CTD) that contains many repeats of the amino acid sequence PTSPSYS. The entry of RNA polymerase II into the preinitiation complex occurs when its CTD is not phosphorylated. However, movement of RNA polymerase II away from the assembly site is correlated with phosphorylation of the CTD. Thus, the assembly of the preinitiation complex does not ensure transcription; further signaling, such as phosphorylation, must occur for transcription to begin.

Formation of the preinitiation complex allows the ATP-dependent helicase activity associated with TFIIH to open the two strands of the DNA, and it provides a template for the subsequent elongation phase of RNA synthesis. TFIID remains behind, still bound to the TATA box, to promote the assembly of additional preinitiation complexes. TFIIF moves with RNA polymerase II during the elongation phase, but the other general transcription factors dissociate from the elongation complex.

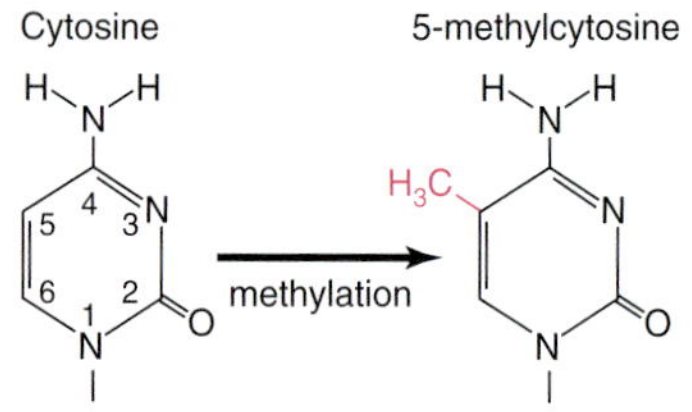

FIGURE 8.19

The most common methylated base in humans is 5-methylcytosine.

Eukaryotic Promoters and Other Sequences that Influence Transcription

Promoters of eukaryotic genes transcribed by RNA polymerase II are operationally defined as those sequences that influence the initiation of gene transcription. The hallmark of eukaryotic promoters is their utilization of multiple transcription factor

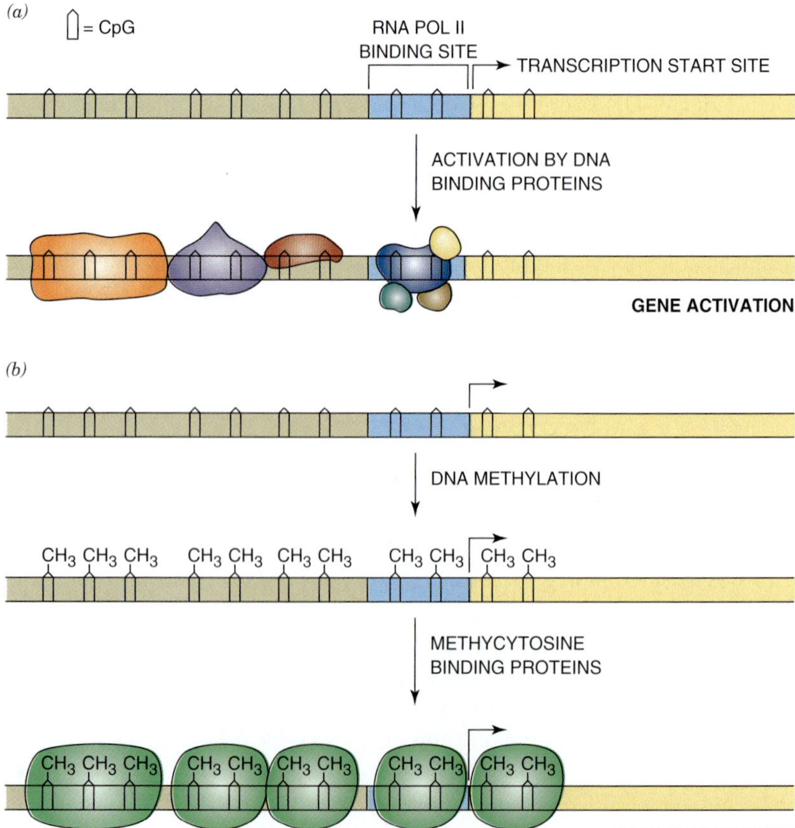

FIGURE 8.20

Methylation of DNA leads to altered gene activity. For maximal transcription of most genes, transcription factors must recognize and bind to specific sequences of DNA in the promoter region. Their interaction with the DNA and the general transcription factors in the RNA polymerase II initiation complex leads to the expression of a gene. Methylation of DNA, specifically the formation of 5-methyl cytosine, provides a new target for protein DNA interaction. The association of 5-meC DNA-binding proteins with methylated DNA may block the ability of other transcription factors to bind to DNA. This inhibition is usually not through the sequence-specific competition for DNA binding, but rather by steric hindrance. Redrawn after Alberts, B., Bray, D., Lewis, J., Raff, M., Roberts, K., and Watson, J. *Molecular Biology of the Cell.* New York: Garland Publishing, 1994.

binding sites to regulate gene activity. In general, these binding sites are relatively close to the TATA box that marks the site of assembly of the preinitiation complex. Other consensus sequences often found in eukaryotic promoters include the CAAT and GC box (Figure 8.23). The exact position of the CAAT box and the position, orientation, and number of GC boxes varies among promoters. The CAAT box is a binding site for several different transcription factors, including NF1. The presence of a CAAT box usually indicates a strong promoter. The GC box provides a binding site for the transcription factor SP1 and is characteristic of many housekeeping genes.

Binding of transcription factors to other sequences in promoters may be influenced by a signaling molecule, such as hormones associated with steroid response transcription factors, or by modification such as phosphorylation of amino acids. DNA sequences that bind regulated transcription factors are often referred to as **response elements.** Transcription factors may be unique to specific tissues or appear at developmentally specific times. The relative effectiveness of regulatory sequences in promoters depends upon their orientation or directionality, and it may be diminished by altering their distance from the TATA box.

Many cellular and viral genomes also have **enhancer elements** that bind proteins that activate transcription. Enhancers differ from promoter sequences in two important respects. First, enhancers can be located many thousands of base pairs away from the preinitiation complex assembly site and can be either upstream or downstream of the transcription start site. Second, the enhancer element can act in either orientation. Transcription factors that bind to response elements close to the transcription initiation site may act directly with the initiation complex or through the use of proteins that make a bridge to the initiator complex (see Figure 8.30). It is quite common for proteins influencing transcription to serve this bridging function. However, the linear distance between an enhancer and the preinitiation complex may be very large.

Transcription factors bound to enhancers are brought together with the preinitiation complex by looping of the DNA. This helps to explain how the enhancer can act at a distance and why the orientation of the enhancer sequence does not matter. Once the protein binds to the enhancer, the proper alignment of the protein with the preinitiation complex is adjusted as the DNA loops back.

TABLE 8.1 General Transcription Factors Found in Eukaryotes

Factor	*Number of Subunits*	*Mass (kDa)*
TFIID		
TBP	1	38
TAFs	12	15–250
TFIIA	3	12, 19, 35
TFIIB	1	15
TFIIE	2	34, 57
TFIIF	2	30, 74
TFIIH	9	35–89

Source: Roeder, R. G. *Trends Biochem. Sci.* 21:329, 1996.

Modular Design of Eukaryotic Transcription Factors

Eukaryotic transcription factors have multiple domains that carry out specific interactions. The DNA recognition domains contribute site-specific binding, and the activation domains contact general transcription factors, RNA polymerase II, or other regulators of transcription. Many transcription factors also have dimerization domains that promote the formation of homodimers or heterodimers, protein interaction domains that allow association with proteins like histone acetylase, or domains that bind to co-activators.

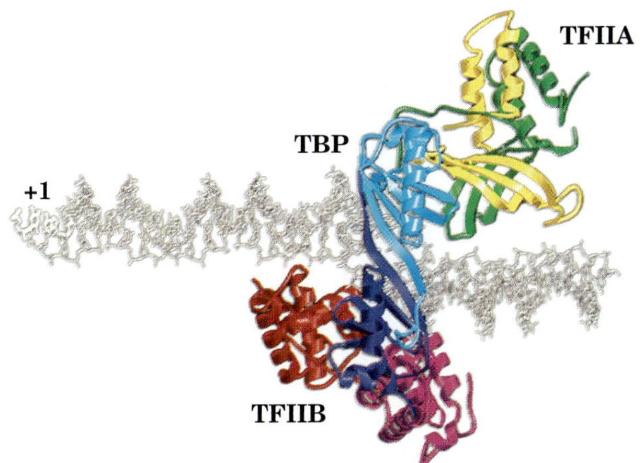

FIGURE 8.21

The TATA-binding protein (TBP) has been co-crystallized with DNA. The first step in forming the transcription complex that allows RNA polymerase II-mediated gene transcription is the association of the general transcription factor TFIID with DNA. In most cases, this is mediated through one of the proteins of the TFIID protein complex, TBP, which binds to DNA through contacts with the sequence TATAA. The co-crystallization of TBP with DNA allowed a closer look at the interaction and indicated that binding of TBP introduced a significant bend in the DNA. Included in the figure are two other general transcription factors, TFIIA and TFIIB, neither of which are involved in the initial contact with DNA. TFIIA binds to TBP and stabilizes the TBP:DNA interaction. TFIIB binds to TBP, leads to the recruitment of RNA polymerase II to the initiation complex, and is involved in identification of the transcription start site.
Figure reproduced with permission from Voet, D., Voet, J., and Pratt, C. W. *Fundamentals of Biochemistry*. New York: Wiley, 1999. © (1999) John Wiley & Sons, Inc.

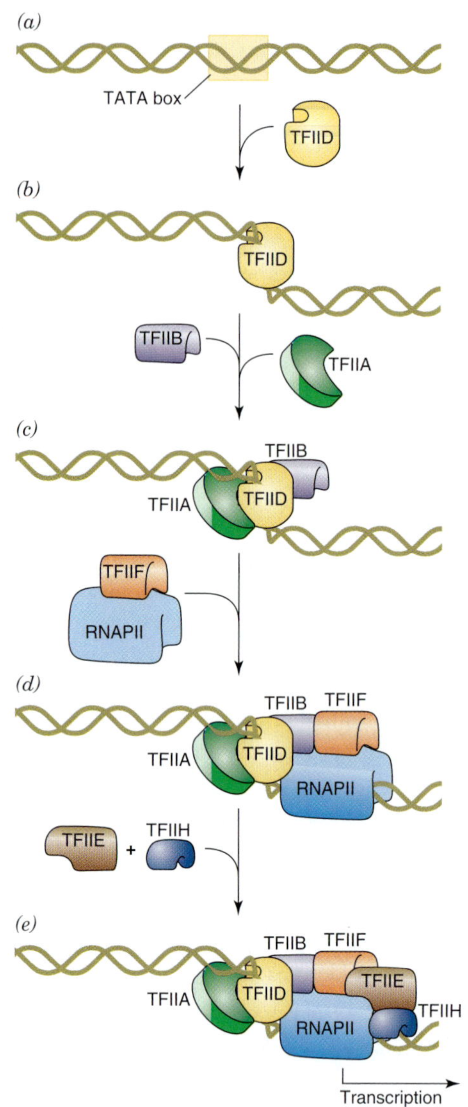

FIGURE 8.22

Formation of the initiation complex for genes transcribed by RNA polymerase II is ordered. The current models of initiation complex assembly include the following steps. (*a*) The promoter region of a gene with a TATA box before TFIID binding. (*b*) TFIID, which includes TBP binds at the TATA box and bends the DNA. (*c*) The DNA:TFIID complex serves as a coordination site for binding other general transcription factors. The TFIID complex associates with TFIIA and TFIIB. TFIIA serves to stabilize the TFIID DNA interaction, and TFIIB provides an appropriate interaction site for the binding of RNA polymerase II. (*d*) TFIIF and RNA polymerase II join the complex. At this stage the carboxy-terminal domain (CTD) of RNA polymerase II is unphosphorylated. TFIIF is thought to destabilize nonspecific RNA polymerase II DNA interactions, thus targeting the RNA polymerase II to the growing initiation complex. (*e*) Two additional general transcription factors join the complex. TFIIE recruits TFIIH to the complex. TFIIH has helicase, ATPase, and kinase activity. TFIIH is thought to mediate transient opening of the DNA duplex and also the phosphorylation of the CTD of RNA polymerase II, allowing the polymerase to leave the initiation complex and start transcription.
Redrawn from Voet, D., Voet, J., and Pratt, C. W. *Fundamentals of Biochemistry*. New York: Wiley, 1999.

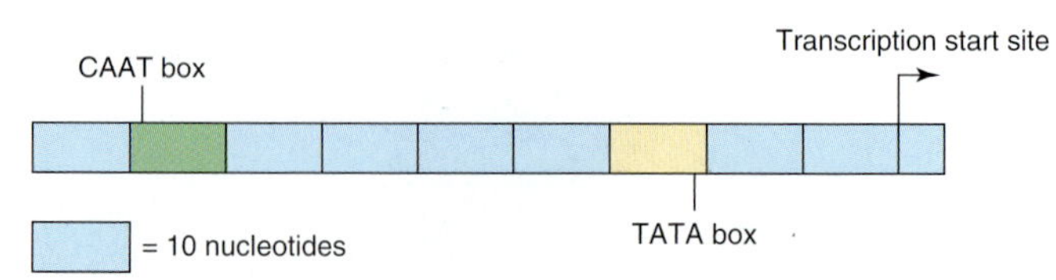

FIGURE 8.23

DNA sequences in genes transcribed in eukaryotes typically contain a TATA box, and they may contain a CAAT box, GC boxes, and multiple transcription factor binding sites.

Co-activators include steroid hormones and cAMP that change the ability of the transcription factor to bind DNA or serve as an activator. The domains involved in such activities can be grouped into a few characteristic motifs.

Common Motifs in Proteins that Bind DNA and Regulate Transcription

Several amino acid motifs are commonly found in transcription factors (trans-acting regulators). They include helix–turn–helix (HTH), zinc finger, helix–loop–helix (HLH), and basic region–leucine zipper (bZIP) proteins. These motifs are present in about 80% of known sequence-specific binding proteins.

The lactose repressor of *E. coli* (see Section 8.3) and a group of developmentally important transcription factors called homeodomain proteins are among the proteins with the helix–turn–helix (HTH) motif. The homeodomain is the DNA recognition portion of these transcription factors. Homeodomain proteins were first identified in the fruit fly when mutations in genes encoding them led to homeosis, or the replacement of one part of the body for another. For instance, one mutation in a homeodomain protein can cause legs to form where antennae should be. Homeodomain proteins are key regulators in mammalian development as well. The HTH motif is about 20 amino acids and forms a relatively small part of a much larger protein. Within the motif the first seven amino acids form one helix followed by a four-amino-acid turn and then a nine-amino-acid helix. The second helix is the "recognition helix" that binds in a sequence-specific manner in the major groove, while hydrophobic interactions between the first and second helix stabilize the structure. The rest of the protein may have allosteric sites that bind to regulatory molecules and regions that allow other protein–protein interactions (Figure 8.24).

TFIIIA, a general transcription factor that binds to promoters of RNA polymerase III-transcribed genes, SP1 (which gives 10- to 20-fold stimulation of all genes with GC boxes), Gal4 in yeast, and the steroid hormone receptor superfamily are all examples of zinc finger proteins. Different subclasses of zinc finger proteins are defined by the specific amino acids that coordinate Zn binding. For instance, in TFIIIA two Cys and two His coordinate Zn binding and are in the C_2H_2 class, whereas the steroid hormone receptor transcription factors use four Cys for each Zn and are of the C_4 class. The zinc finger motif binds in the major groove of the DNA in a sequence-specific manner, mediated by an α-helix formed on one side of the finger region. As was seen with HLH proteins, the α-helix fits in the major groove and amino acid side chains form hydrogen bonds with the bases for sequence-specific binding.

The two classes of zinc finger proteins have binding sites characteristic of the way each class positions itself on DNA. For TFIIIA, sequential zinc fingers follow the major groove, each forming hydrogen bonds with specific bases. Zinc finger proteins of the C_4 class, such as the steroid hormone receptor transcription factors, use one zinc finger to bind to DNA and use a second zinc finger to stabilize DNA binding by the first. The steroid hormone receptor transcription factors associate with DNA as dimers. Their binding sites consist of two palindromic "half sites" spaced to accommodate the two DNA binding fingers of the dimer (Figure 8.25).

Basic region–leucine zipper (bZIP) proteins include fos, jun, and CREB. bZIP proteins contain leucine residues at every seventh position in an α-helix. These leucines form hydrophobic interactions with a second protein that also has this pattern of leucine repeats to form either homo- or a heterodimers (Figure 8.26).

bZip proteins have a DNA-binding domain comprised of a basic region, defined by the presence of arginine and lysine, located seven amino acids before the first leucine and α-helix. The basic amino acids stabilize the DNA–protein association through electrostatic interactions with the negatively charged DNA backbone in addition to forming hydrogen bonds in the major groove. Homodimer-binding sites have dyad symmetry, whereas this symmetry is not found in heterodimer-binding sites.

The helix–loop–helix class of transcription factors includes myoD, myc, and max. Two amphipathic α-helical segments separated by an intervening loop characterize

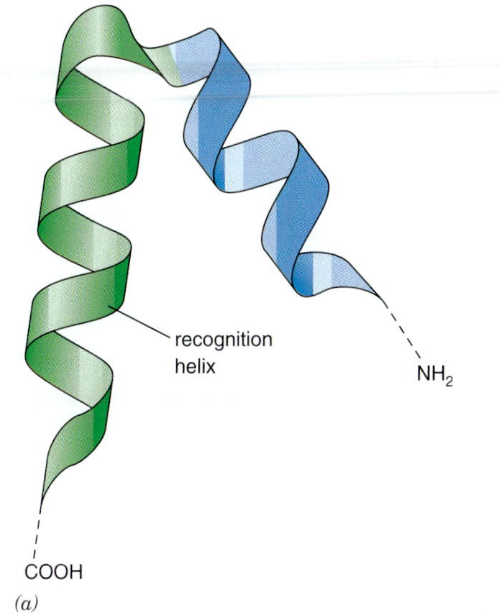

(*a*)

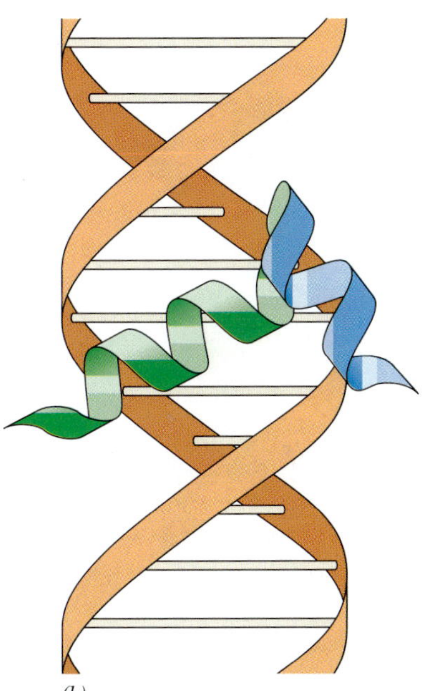
(*b*)

FIGURE 8.24

Helix-turn-helix proteins use one helix to bind in the major groove while the other supports that binding through hydrophobic interaction. (*a*) The HTH domain of a protein typically contains about 20 amino acids that form two α-helices, joined by a nonhelical turn. (*b*) The dimensions of the α-helix of the protein allow it to fit into the major groove of the DNA helix. The helix that interacts directly with the DNA (the recognition helix) includes amino acids with side chains capable of forming hydrogen bonds with specific bases that are exposed in the major groove. The second helix does not directly interact with DNA, but stabilizes the binding of the recognition helix through hydrophobic interactions. Thus, both helices include amino acids (like valine or leucine) that allow these hydrophobic interactions to occur.

Redrawn from Alberts, B., Bray, D., Lewis, J., Raff, M., Roberts, K., and Watson, J. *Molecular Biology of the Cell.* New York: Garland, 1994.

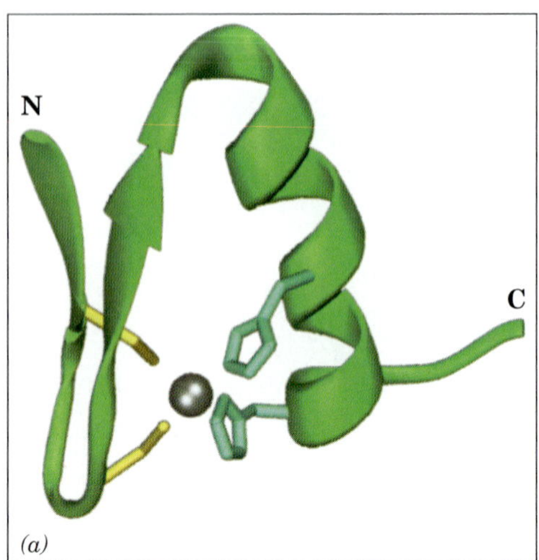

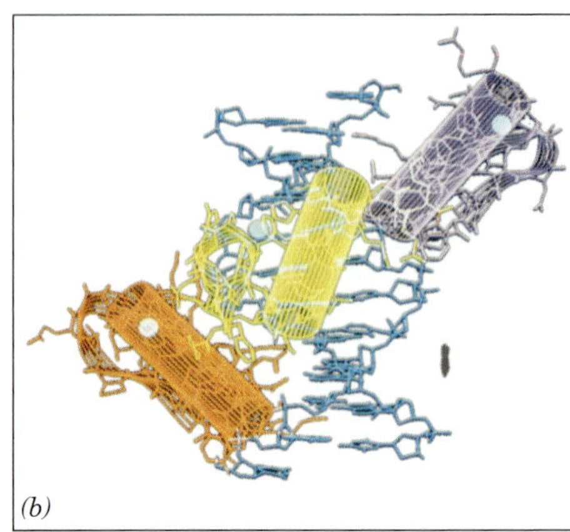

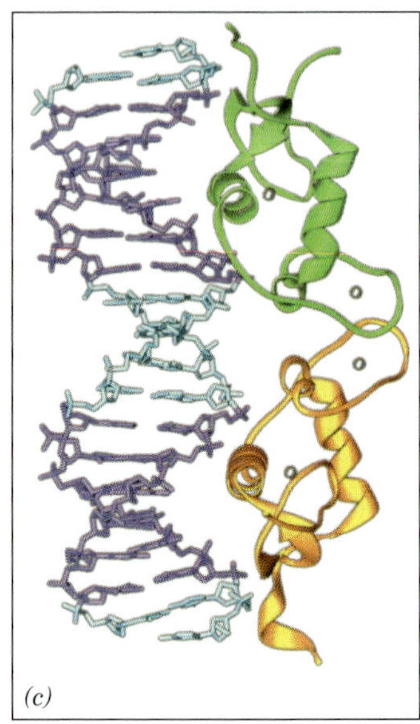

FIGURE 8.25

Two different Zn finger motifs are found in transcription factors. Although both Zn finger proteins use the coordinate binding of a Zn molecule to assume their final structure, there are recognizable differences in the two major motifs that have been identified. Both classes take advantage of the formation of α-helical structure to form domains that bind in the major groove of DNA. (*a*) The C_2H_2 class includes proteins that may have many Zn finger domains. Each α-helix from a Zn finger has the potential to bind in a sequence-specific manner to sites along the major groove. The result may be a procession of protein–DNA interactions, each dependent on the particular array of amino acid side chains found in each α-helical domain of the Zn finger protein. (*b*) C_x class of Zn finger proteins commonly has two Zn finger domains. The α-helix from one Zn fingers binds the DNA in the major groove, while the α-helix in the other Zn finger supports that interaction by hydrophobic interactions with the domain binding to the DNA. The C_x class of Zn finger proteins normally binds DNA by forming dimers. Shown is the interaction of the estrogen receptor dimer, with each monomer contacting the DNA. Zn molecules are indicated by spheres. Reproduced from Voet, D. and Voet, J. G. *Biochemistry*, 2nd ed. New York: Wiley, 1995. Reprinted with permission of John Wiley & Sons, Inc. Part (c) courtesy of C. Pabo, MIT.

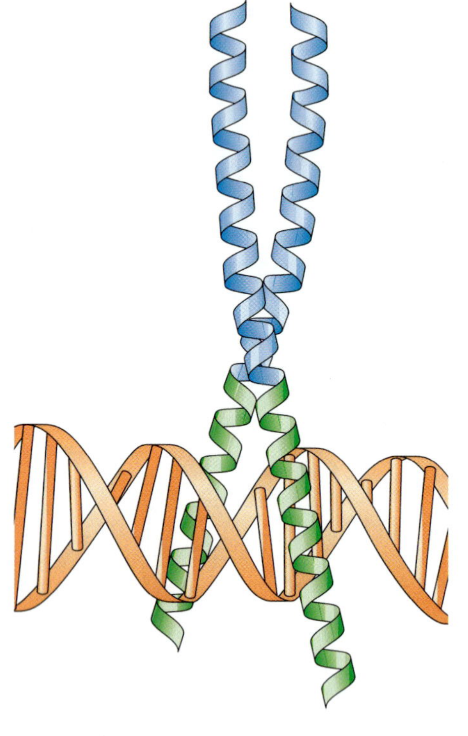

FIGURE 8.26

Leucine zipper proteins bind to DNA as dimers. Leucine zipper proteins form dimers by virtue of a leucine every seventh residue along an α-helix. These leucines form a hydrophobic face that interacts with the hydrophobic face of a similar α-helix. The protein–protein interaction domains are in blue. The α-helical regions may continue beyond the protein–protein interaction domain, allowing each monomer to bind to the major groove of DNA (green). For homodimers, the DNA-binding site is characterized by two recognizable and symmetric half sites. Modified from Alberts, B., Bray, D., Lewis, J., Raff, M., Roberts, K., and Watson, J. *Molecular Biology of the Cell.* New York: Garland, 1994.

helix–loop–helix proteins. The helices are not responsible for DNA binding, as in the zinc finger proteins, but for dimerization with another protein. As was described for the bZip proteins, the dimers formed can be homo- or heterodimers. The DNA-binding domain is an extension of one of the α-helices that form the four-helix bundle generated by dimerization (Figure 8.27).

8.9 REGULATION OF EUKARYOTIC GENE EXPRESSION

As indicated above, a relatively small region of a regulatory protein may be dedicated to sequence-specific binding of DNA, while other domains are involved in protein–protein or ligand interactions. Several characteristic activation domains have been identified, including acidic domains (high concentration of amino acids with acidic side chains), glutamine-rich domains, and proline-rich domains. Experimentally, overproduction of any of these domains by recombinant techniques, even without their corresponding DNA-binding domain, can lead to inappropriate activation of transcription from a variety of genes. They appear to activate transcription by increasing the rate of assembly of the preinitiation complex. Some interact directly with TFIID to enhance binding to the TATA box, while others interact with TFIIB or TAFs that are part of the TFIID complex. When multiple transcription factors bind to a promoter, they can have a combinatorial effect on the binding and assembly of the preinitiation complex. Activation domains of many transcription factors target the same proteins in the

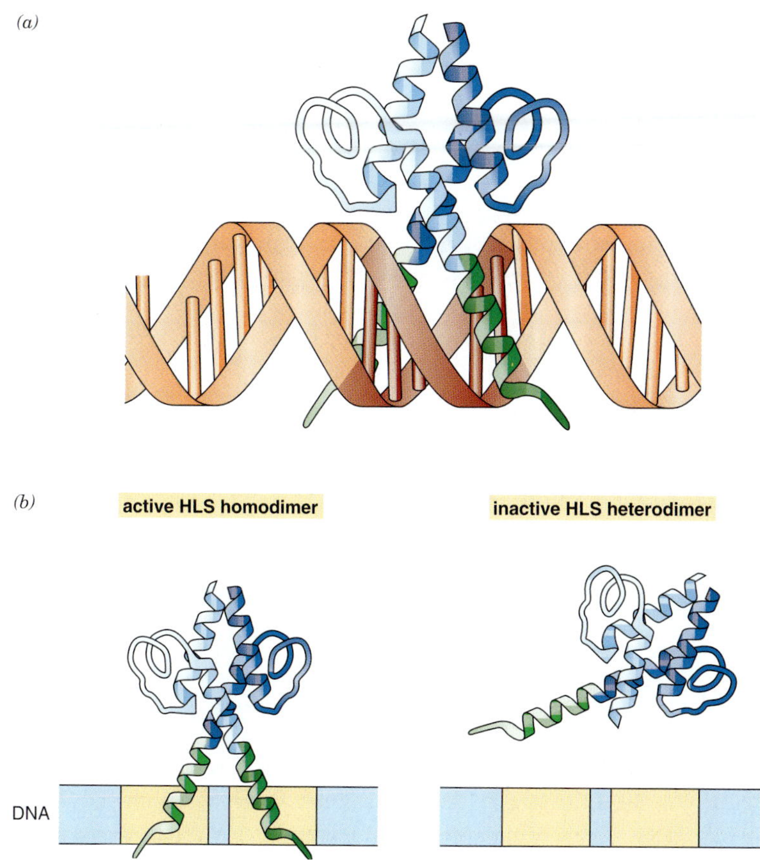

FIGURE 8.27

Transcription factor dimer formation is mediated through helix loop helix interactions. The helix–loop–helix motif brings together two monomers to form a dimer that binds to DNA. (*a*) Each monomer has two helices joined by a loop. One helix is used for protein–protein interaction, while the other is used to bind the major groove of DNA. Thus, the dimer consists of a four-helix bundle. If the dimer is formed by two identical monomers, then the DNA-binding sites are expected to be very similar or identical; however, if the monomers are different proteins (forming heterodimers), then the DNA-binding sites may be unrelated. (*b*) When transcription factors bind as dimers, the presence of a truncated monomer can prevent DNA binding even in the presence of full-length monomers. For example, if the protein dimerization helix is made without the DNA-binding domain, the dimerization with a full-length monomer produces a product unable to bind effectively to DNA.
Modified from Alberts, B., Bray, D., Lewis, J., Raff, M., Roberts, K., and Watson, J. *Molecular Biology of the Cell.* New York: Garland, 1994.

preinitiation complex. Thus, transcription regulation is linked to the placement of DNA-binding sites and not to protein–protein interactions unique to a specific gene.

Transcription factors can also regulate gene expression by recruitment of other proteins to the promoter area. These proteins may not bind DNA but form a bridge between the DNA-bound transcription factor and the initiation complex. Alternatively, the interacting protein may bring chromatin modification enzymes, such as histone acetylase (e.g., the CBP/p300 complex), to specific genes. Acetylation of histones loosens chromatin structure. Some regulated transcription factors that recruit the CBP/p300 complex, and the molecules or events to which they respond, are: CREB and cAMP, SREBP and cholesterol, NFkB and cytokines, and p53 and growth arrest.

Transcription factors that reduce gene expression may preemptively bind to the DNA, either to the same sequence as a positive factor or more globally, such as the negative effect of methylated DNA-binding proteins. Alternatively, negative factors may inhibit binding or assembly of the preinitiation complex or bind to a positive transcription factor to prevent its binding to DNA or speed its degradation.

Regulating the Regulators

Transcription factors are regulated in a variety of ways. Probably the simplest examples are those that are only synthesized in specialized cells, or at specific times in development. Some are regulated by cofactor binding, which may either inhibit or stimulate their binding to DNA. For transcription factors that form dimers, the formation of nonproductive homo- or heterodimer complexes can alter DNA binding or the protein–protein interactions that allow gene activation. Posttranslational modification, such as protein processing or phosphorylation, may affect transcription factor binding to DNA or other proteins and may also affect their ability to move from the cytoplasm into the nucleus (Clin. Corr. 8.3 and 8.4).

Since enhancers can act over long distances, and in either orientation, there needs to be a way to ensure that they are acting on the correct gene. This is accomplished

CLINICAL CORRELATION 8.3
Tamoxifen and Targeting of Estrogen Receptor

Tamoxifen, a drug used to treat breast cancer, is a competitive inhibitor of the estrogen receptor (ER). Breast cancer cells respond to normal estrogen by increasing their proliferation rate. Estrogen activates the estrogen receptor, whereas tamoxifen prevents normal activation and reduces transcription from genes regulated by the estrogen receptor, and hence reduces growth of breast cancer cells. The National Cancer Institute in 2004 stated, "The benefits of tamoxifen as a treatment for breast cancer are firmly established and far outweigh the potential risks." However, tamoxifen treatment can increase the risk of uterine cancer. The apparent cause lies in the presence of two estrogen receptor subtypes (α and β). Both subtypes exist in breast tissue, but estrogen receptor-α predominates in uterine tissue. In uterine tissue, tamoxifen apparently activates receptor-α, rather than inhibiting it. The result is stimulation of transcription in conjunction with transcription factors fos and jun.

Source: Paech, K., Webb, P., Kuiper, G., Nilsson, S., Gustafsson, J.-A., Kushner, P. J., and Scanlan, T. S. Differential ligand activation of estrogen receptors ERα and ERβ at AP1 sites. *Science* 277:1508, 1997.

CLINICAL CORRELATION 8.4
Transcription Factors and Cardiovascular Disease

Among genes recently identified as causing human cardiovascular disease are two that encode transcription factors. One, Nkx2-5, is a homeodomain-containing protein. Homeodomain proteins often are involved in regulating cell fate in embryonic cells, and Nkx2-5 regulates genes involved in heart formation. Mutations in one allele of Nkx2-5 are linked to atrial septal defects and conduction defects. The null mutation in the mouse is embryonic lethal. Mutations in the other gene, Tbx, are responsible for Holt–Oram syndrome. Holt–Oram syndrome often includes atrial septal defects (ASD) ventricular septal defects (VSD) conduction defects and defects in hand and arm development.

Source: Barinaga, M. Tracking down mutations that can stop the heart. *Science* 128:32, 1998. Schott, J. J., Benson, D. W., Basson, C. T., Pease, W., et al. Congenital heart disease caused by mutations in the transcription factor NKX2-5. *Science* 281:108, 1998. Li Q. Y., Newbury-Ecob, R. A., Terrett J. A., et al. Holt–Oram syndrome is caused by mutations in TBX5, a member of the Brachyury (T) gene family. *Nature (Genetics)* 15:21, 1997.

by the use of insulator sequences that set the boundaries of influence of an enhancer. Insulators are DNA sequences that, when bound to the Zn finger protein CCTF (CCCTC-binding factor) and positioned between an enhancer and the TATA box, prevent enhancer-mediated gene activation. An additional activity of insulators is to prevent the spread of heterochromatin that would lead to gene silencing,

Activation of Transcription of the LDL Receptor Gene Illustrates Many Features Found in Eukaryotic Gene Regulation

The description of the regulation of expression of specific eukaryotic genes can rapidly turn into an alphabet soup of transcription factors and regulators of transcription factors that make sense to researchers studying that gene, but are confusing to most others. However, many features of transcription factors described above are illustrated by the transcriptional control of the gene for the low-density lipoprotein (LDL) receptor. This gene is transcribed in response to the lack of cellular cholesterol. Increased transcription leads to an increased amount of the LDL receptor and enhanced uptake of LDLs and their cholesterol in the blood (see p. 711).

The promoter of the LDL receptor gene has a TATA box and binding sites for a variety of transcription factors (Figure 8.28). The TATA box provides the site for TFIID binding and the formation of the preinitiation complex with RNA

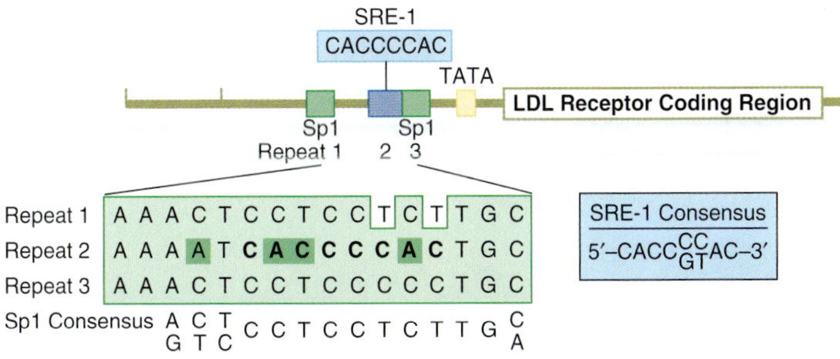

FIGURE 8.28

A schematic of the LDL receptor gene promoter. The LDL receptor gene promoter, like most eukaryotic genes, has several different transcription-factor-binding sites. Shown here are the principle sites involved in regulating the LDL receptor gene in response to cholesterol levels. They include the TATA box, just upstream from the transcription initiation site, several GC boxes (Sp1 binding sites), and the response element (SRE) where SREBP binds.
Modified from Goldstein, J. L. and Brown, M. S. Regulation of the mevalonate pathway. *Nature* 343:425, 1990.

polymerase II. There are three consensus binding sites for the zinc-finger-containing transcription factor Sp1 (recall that Sp1 binds to GC boxes in promoters). Sp1 has a glutamine-rich activation domain, which is thought to help recruit TFIID to the TATA box. This recruitment needs an additional factor called the **c**ofactor **r**equired for **Sp**1 activation (CRSP). However, the binding of Sp1 is not sufficient to activate transcription of the gene. Activation requires the participation of a second transcription factor called SREBP-1a (**s**terol **r**esponsive **e**lement-**b**inding **p**rotein 1a). SREBP-1a is a helix–loop–helix–leucine zipper protein that binds to the sterol response element between the Sp1-binding sites in the LDL receptor gene.

The capacity to respond to the concentration of cholesterol is controlled by the release of SREBP from a membrane-bound state that prevents it from being transported into the nucleus. SREBP-1a contains two domains that loop through the membrane of the endoplasmic reticulum (ER), leaving the domains capable of serving as a transcription factor exposed to the cytoplasm but tethered to the ER. In order for SRCBP-1a to move to the nucleus, it must be cleaved by two proteolytic steps, the first of which is carried out by a protease found in an active form in a post-ER compartment called the cis-Golgi. Transport of SRCBP-1a to the cis-Golgi requires **S**REBP **c**leavage **a**ctivating **p**rotein (**SCAP**), which also is partially embedded in the ER membrane. SCAP has a cholesterol-sensing region and when membrane cholesterol levels are low, the SCAP-SRCBP-1a complex is transported to the cis-Golgi, where

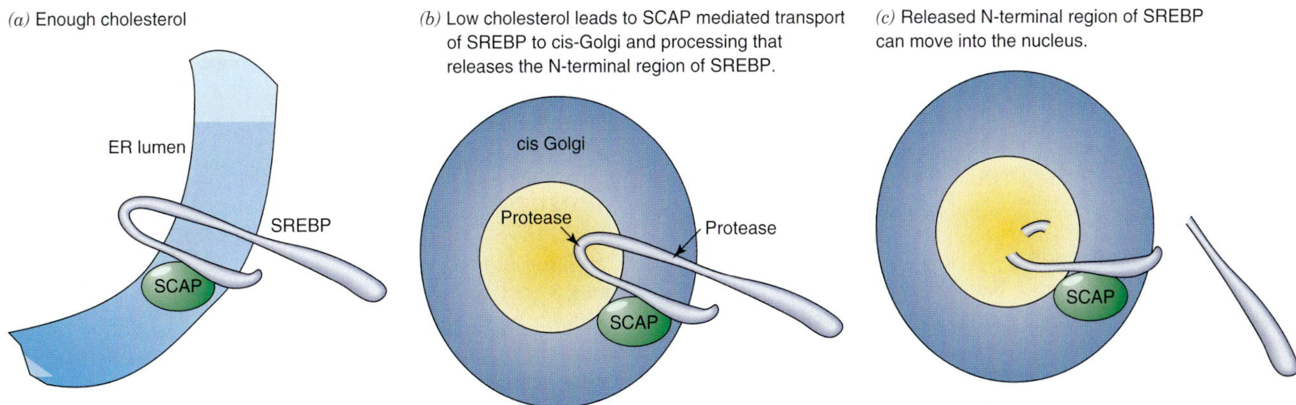

FIGURE 8.29

SREBP is released from a membrane-bound precursor by protease action. (*a*) The sterol responsive element-binding protein (SREBP) is synthesized as a precursor protein that must be proteolytically processed before it acts as a transcription factor. The precursor is tethered to the membrane of the ER by two membrane-spanning regions. It is in close association with SREBP cleavage activating protein (SCAP), but remains unprocessed when cholesterol levels in the ER membrane are normal. (*b*) SCAP senses low levels of cholesterol and moves with SREBP to the cis-Golgi where two different proteases cleave SREBP, allowing the cytoplasmic domain to move to the nucleus. In this way the SREBP does not activate the LDL receptor gene unless cholesterol levels are low.
Modified from Brown, M. S., Ye, J., Rawson, R. B., and Goldstein, J. L. Regulated Intermembrane proteolysis: A control mechanism conserved from bacteria to humans. *Cell* 100:391, 2000.

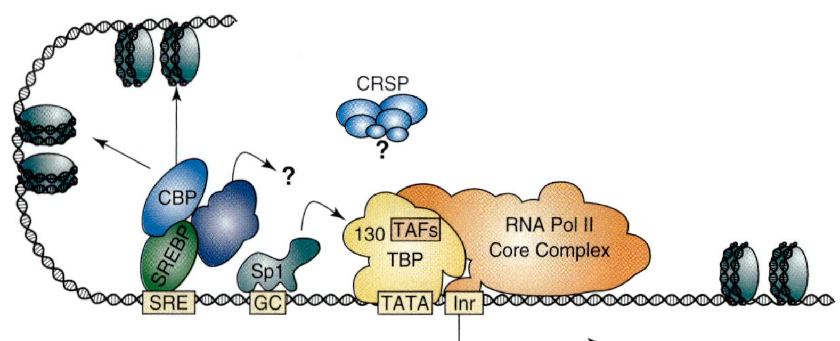

FIGURE 8.30

LDL receptor gene is activated through coordinate effect of several transcription factors. Activation of the LDL receptor gene requires several factors. Once SREBP, Sp1, and the cofactor required for Sp1 activation (CRSP) bind, the histone acetylase CBP is recruited to the promoter region. This combination of factors provides the positive signal needed to enhance binding of the initiation complex to the TATA box and, through recruitment of CBP and its histone acetyl transferase activity, also affects nearby chromatin structure.
Modified from Naar, A. M., Ryu, S., and Tijian, R. Cofactor requirements for transcriptional activation by Sp1. *Cold Spring Harbor Symp. Quant. Biol.* LXIII:189, 1998.

proteolysis of SRCBP-1a occurs, releasing the cytoplasmic domain, the transcription factor, for entry into the nucleus (Figure 8.29). Once there, it binds the steroid response element (SRE) on the LDL promoter and recruits a histone acetyl transferase called CBP and other proteins to the promoter area. Then, acting with SP1 and CRSP, the LDL receptor gene is transcribed (Figure 8.30). This illustrates the composite nature of eukaryotic gene regulation, in which activation of transcription factors, coordination of multiple transcription factors, and recruitment of chromatin remodeling enzymes work in concert to regulate gene expression.

BIBLIOGRAPHY

Prokaryotic Gene Expression

Blattner, F. R., Plunkett, G., Bloch, C. A., Perna, N. T., et al. The complete genome sequence of *Escherichia coli* K-12. *Science* 277:1453, 1997.

Cohen, S. N. and Shapiro, J. A. Transposable genetic elements. *Sci. Am.* 242:40, 1980.

Miller, J. H. The lac gene: Its role in lac operon control and its use as a genetic system. In: J. H. Miller and W. S. Resnikoff (Eds.), *The Operon.* Cold Spring Harbor, NY: Cold Spring Harbor Laboratory Press, 1978, p. 31.

Platt, T. Regulation of gene expression in the tryptophan operon of *Escherichia coli*. In: J. H. Miller and W. S. Resnikoff (Eds.), *The Operon.* Cold Spring Harbor, NY: Cold Spring Harbor Laboratory Press, 1978, p. 263.

Schultz, S. C., Shields, G. C., and Steitz, T. A. Crystal structure of a CAP–DNA complex: The DNA is bent by 90 degrees. *Science* 253:1001, 1991.

Eukaryotic Gene Expression

Alberts, B., Johnson, A., Lewis, J., Raff, M., Roberts, K., and Walter, P. *Molecular Biology of the Cell*. New York: Garland, 2002.

Brown, M. S., Ye, J., Rawson, R. B., and Goldstein, J. L. Regulated intramembrane proteolysis: A control mechanism conserved from bacteria to humans. *Cell* 100:391, 2000. Brown, C. E., Lechner, T., Howe, L., and Workman, J. L. The many HATs of transcription coactivators. *Trends Biochem. Sci.* 25:15, 2000.

Burley, S. K. and Roeder, R. Biochemistry and structural biology of transcription factor IID (TFIID). *Annu. Rev. Biochem.* 65:769, 1996.

Dantonel, J. -C., Wurtz, J. -M., Poch, J. M., Moras, D., and Tora, L. The TBP-like factor: An alternative transcription factor in Metazoa. *Trends Biochem. Sci.* 24:335, 1999.

Goldstein, J. L. and Brown, M. S.. Regulation of the mevalonate pathway. *Nature* 343:425, 1990.

Kawasaki, H. and Taira, K. Induction of DNA methylation and gene silencing by short interfering RNAs in human cells. *Nature* 431:211, 2004.

Kornberg, R. D. Mechanism and regulation of yeast RNA polymerase II transcription. *Cold Spring Harbor Symp. Quant. Biol.* LXIII:229, 1998.

Latchman, D. S. (1998). *Eukaryotic Transcription Factors*, 3rd ed. New York: Academic Press.

Lemon, B., and Tjian, R. Orchestrated response: a symphony of transcription factors for gene control. *Genes Dev.* 14:2551, 2000.

Lewin, B. *Genes VII*. New York: Oxford University Press, 2000.

Naar, A. M., Ryu, S., and Tijian, R. Cofactor requirements for transcriptional activation by Sp1. *Cold Spring Harbor Symp. Quant. Biol.* LXIII:189, 1998.

Ng, H. and Bird, A. Histone deacetylases: silencers for hire. *Trends Biochem. Sci.*:121. 2000.

Pruss, D., Hayes, J. J., and Wolffe, A. Nucleosomal anatomy—Where are the histones? *BioEssays* 17:161, 1995.

Stewart, S. and Crabtree, G. Regulating the regulators. *Nature* 408:46, 2000.

West, A. G., Gaszner, M., and Felsenfeld, G. Insulators: Many functions, many mechanisms. *Genes Dev.* 16:271, 2002.

Wolffe, A. P. *Chromatin: Structure and Function*, 3rd ed. New York: Academic Press, 1998.

Wolffe, A. P. The cancer-chromatin connection. *Sci. Med.* 6:28, 1999.

Woychik, N., and Young, R. RNA polymerase II: Subunit structure and function. *Trends Biochem. Sci.* 15:347, 1990.

QUESTIONS | CAROL N. ANGSTADT

Multiple Choice Questions

1. Full expression of the *lac* operon requires:
   A. lactose and cAMP.
   B. allolactose and cAMP.
   C. lactose alone.
   D. allolactose alone.
   E. absence or inactivation of the *lac* corepressor.

2. In an operon:
   A. each gene of the operon is regulated independently.
   B. control may be exerted *via* induction or *via* repression.
   C. operator and promoter may be *trans* to the genes they regulate.
   D. the structural genes are either not expressed at all or they are fully expressed.
   E. control of gene expression consists exclusively of induction and repression.

3. The *E. coli lac*ZYA region will be up-regulated if:
   A. there is a defect in binding of the inducer to the product of the *lac*I gene.
   B. glucose and lactose are both present, but the cell cannot bind the CAP protein.
   C. glucose and lactose are both readily available in the growth medium.
   D. the operator has mutated so it can no longer bind repressor.
   E. the *lac* corepressor is not present.

4. All of the following describe an operon *except*:
   A. control mechanism for eukaryotic genes.
   B. includes structural genes.
   C. expected to code for polycistronic mRNA.
   D. contains control sequences such as an operator.
   E. can have multiple promoters.

5. In bacteria, amino acid starvation is associated with the production of guanosine tetraphosphate and guanosine pentaphosphate. This situation is referred to as:
   A. attenuation.
   B. corepression.
   C. repression.
   D. self-regulation.
   E. stringent response.

6. Tryptophan as a corepressor for the *trp* operon binds to the:
   A. operator.
   B. promoter.
   C. repressor protein.
   D. RNA polymerase.
   E. upstream region from the promoter.

7. In eukaryotic transcription by RNA polymerase II, formation of a preinitiation complex:
   A. begins with the binding of a protein (TBP) to the TATA box of the promoter.
   B. involves the ordered addition of several transcription factors and the RNA polymerase.
   C. allows an ATP-dependent opening of the two strands of DNA.
   D. requires that the C-terminal domain of RNA polymerase II not be phosphorylated.
   E. all of the above are correct.

8. Enhancers:
   A. are sequences in the promoter that bind to hormone-transcription factor complexes.
   B. are more effective the closer they are to the TATA box.
   C. may be thousands of base pairs away from preinitiation complex assembly.
   D. must be upstream of the site of the preinitiation complex assembly.
   E. bind transcription factors that act directly with the initiation complex.

Questions 9 and 10: The problem of pathogenic bacteria becoming resistant to a large number of antibiotics is a serious public health concern. A bacterial strain in a patient being treated with one antibiotic may suddenly become resistant not only to that antibiotic but to others as well even though it has not been exposed to the other antibiotics. This occurs when the bacteria acquire a plasmid from another strain that contains several different transposons.

9. All of the following phrases describe transposons *except*:
   A. a means for the permanent incorporation of antibiotic resistance into the bacterial chromosome.
   B. contain short inverted terminal repeat sequences.
   C. code for an enzyme that synthesizes guanosine tetraphosphate and guanosine pentaphosphate, which inhibit further transposition.
   D. include at least one gene that codes for a transposase.
   E. contain varying numbers of genes.

10. In the operation of transposons:
    A. typically the transposon moves from its original site and relocates to a different site.
    B. a duplicated transposon must be inserted into the same DNA molecule as the original.
    C. all transposons are approximately the same size.
    D. the insertion sites must be in a consensus sequence.
    E. the transposase may recognize the repetitive ends of the transposon and participate in the cleavage of the recipient site.

Questions 11 and 12: Genes present in a region of DNA which is in the highly condensed heterochromatin form cannot be transcribed. To be transcribed, this region of DNA must change to a more open structure which may occur by acetylation of histones by an acetyltransferase like CBP, CREB binding protein. CREB is a transcription factor. Rubinstein–Taybi syndrome is caused by mutations of the CBP gene. Rubinstein–Taybi patients are mentally retarded and have other developmental abnormalities, the severity of which depends on the extent of mutation.

11. In chromatin:
    A. a nucleosome consists of four molecules of histones surrounding a DNA core.
    B. DNA positioned so that the major and minor grooves are on the outside of the nucleosome is more accessible for transcription than if the grooves face the interior.
    C. DNA must be completely removed from the nucleosome structure for transcription.
    D. linker DNA is the only DNA capable of binding transcription factors.
    E. the histone octamer consists of eight different kinds of histone proteins.

12. Acetylation of histones can lead to a more open DNA structure by:
    A. weakening the electrostatic attraction between histones and DNA.
    B. causing histones to interact with the C-terminal domain (CTD) of RNA polymerase.
    C. causing electrostatic repulsion between histones and DNA.
    D. facilitating methylation of DNA.
    E. attracting transcription factors to DNA.

Problems

13. What will be the status of transcription of the *lac* operon in (a) the presence of glucose and (b) the absence of glucose, in each case with lactose present, if there is a mutation that produces an inactive adenyate cyclase?

14. In an operon for synthesis of an amino acid which is controlled wholly or in part by attenuation, why does the presence of the amino acid prevent transcription of the whole operon while the absence of the amino acid permits it?

ANSWERS

1. **B** The true inducer is allolactose, which is usually formed from lactose by the action of β-galactosidase. A: See above. C, D: In addition to the sugar binding to the repressor, cAMP must bind to the CAP protein for positive control of transcription. E: The *lac* operon does not involve corepression.
2. **B** Induction and repression are among the mechanisms used to control operons. A: The structural genes are under coordinate control. C: The operator and promoter are elements of the same strand of DNA as the operon they control, not diffusible. D: Typically, regulation of operators is somewhat leaky. E: Another mechanism for regulation of an operon is attenuation.
3. **D** If the operator is unable to bind repressor, the rate of transcription is greater than the basal level. A: The product of the *lac*I gene is the repressor protein. When this protein binds an inducer, it no longer binds to the operator and transcription increases. Failure to bind an inducer prevents this sequence. B, C: In the presence of glucose, catabolite repression occurs. Glucose lowers the intracellular level of cAMP, so there is no CAP–cAMP complex to activate transcription. E: The *lac* operon does not involve corepression.
4. **A** Operons are procaryotic mechanisms. B–D: An operon is the complete regulatory unit of a set of clustered genes, including the structural genes, regulatory genes, and control elements, such as the operator. **E**: An operon may have more than one promoter, as does the tryptophan operon of *E. coli*.
5. **E** The mechanism is unclear, but transcription initiation of rRNA is inhibited.
6. **C** The repressor must be complexed with two molecules of tryptophan to bind to the operator and inhibit transcription.
7. **E** A: TBP is part of the general transcription factor TFIID. B: Many proteins are involved. C: The helicase activity is associated with one of the transcription factors (TFIIH). D: The CTD must be in the unphosphorylated form for RNA polymerase to enter the preinitiation complex.
8. **C** Transcription factors bound to enhancers are brought together with the preinitiation complex by looping of the DNA to bring distant portions together. A, B, E: These are characteristics of response elements such as the steroid response element. D: Enhancers can act in either orientation, again because the DNA loops back on itself.
9. **C** These guanosine phosphates are synthesized by the product of the relA gene; they inhibit initiation of transcription of the rRNA and tRNA genes, shutting off protein synthesis in general. This is the stringent response.
10. **E** This has been demonstrated for transposon Tn3. A: Most transpositions involve generation of an additional copy of the transposon which is then inserted somewhere else in DNA. B: The copy could be inserted into a different DNA molecule. C: Length varies considerably, depending on the number of genes incorporated into the transposon. D: Most target sites seem to be fairly random in sequence, although some transposons tend to insert at specific "hot spots."
11. **B** Assembly of the transcription initiation complex and other proteins involved in gene expression need access to the grooves of DNA. A, E: The core of the nucleosome is an octamer of two molecules each of four different histones. C: A 5-base-pair shift in the association of DNA with the histone core is sufficient to change the orientation of the grooves and allow access. D: It seems to be the DNA associated with the nucleosome that is involved in binding transcription factors and other proteins.
12. **A** Acetylation of an ϵ-amino group of lysines near the N-terminus of histones changes a positive charge to a neutral species so there is less attraction to the negative phosphates of DNA. B: Histones do not interact directly with CTD of RNA polymerase II. C: The change is from positive to neutral, not negative, so there is no electrostatic repulsion. D: Methylation of DNA does affect transcription but is a separate phenomenon from acetylation of histones. E: The nucleosome undergoes repositioning.
13. Normally, glucose lowers the intracellular level of cAMP. There is then little or no CAP–cAMP complex to activate transcription, so transcription of the operon is low. The same situation would exist in this mutation. Normally, in the absence of glucose, cAMP is high, positive control is exerted and the *lac* operon is transcribed. In this mutation, cAMP cannot form since the enzyme for its formation is defective. There is no positive control, and the *lac* operon will not be transcribed effectively.
14. Such operons will code for a leader peptide that has one or more codons for the amino acid in question. Once the RNA for the leader peptide has been synthesized, it can form different secondary structures depending on whether or not the ribosome is stalled at this region. If the ribosome does not stall because there is enough amino acid (and thus charged tRNA), the secondary structure is a termination signal and synthesis stops. If the ribosome stalls because of insufficient charged tRNA, the secondary structure that forms is not recognized as a termination signal and transcription through the operon continues.

PART III FUNCTIONS OF PROTEINS

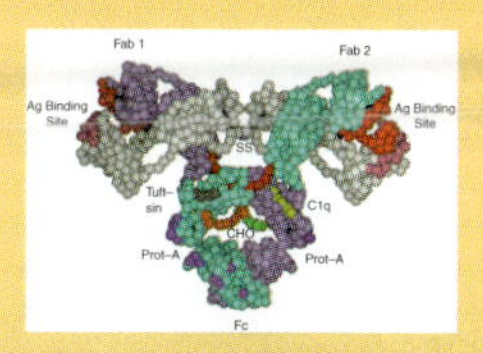

PROTEINS II: STRUCTURE–FUNCTION RELATIONSHIPS IN PROTEIN FAMILIES

Richard M. Schultz

Textbook of Biochemistry With Clinical Correlations, Sixth Edition, Edited by Thomas M. Devlin

9.1 OVERVIEW

The fundamentals of protein architecture were presented in Chapter 3. In this chapter there is a discussion of the specific relationship between structure and function in three **protein families**: immunoglobulins, serine proteases, and hemoglobins. We also describe the organization of proteins forming the basement membrane of the extracellular matrix that surrounds cells and tissues. We pursue this study through examination of amino acid sequence, structural organization, and biological function.

Immunoglobulins provide examples of multidomain architecture that supports recognition and binding to foreign molecules and leads to their sequestration. Diversity among family members is the source of specific recognition of molecular and individual binding capabilities.

Serine proteases provide examples of enzymes that appear to have diverged to perform unique physiological functions, frequently within highly organized enzyme cascades. Similarities in their catalytic mechanism and three-dimensional structure are a common link.

Hemoglobins offer examples of highly fine-tuned proteins that simultaneously perform multiple highly regulated physiological functions. They can accommodate small substitutions or mutations, many of which have clinical implications, and still retain their physiological functions. They reveal the diversity of amino acid sequence substitutions that can be tolerated and allow the protein to function effectively.

The **basement membrane** is an extracellular protein formed by cell-secreted proteins that bind with each other to form a molecular network in extracellular space that provides the boundaries and regulatory environment of the cells and tissues. It provides an example of a complex network composed of multiple different protein subunits that spontaneously associate into a specific molecular arrangement. Such networks provide essential intracellular as well as extracellular functions.

9.2 ANTIBODY MOLECULES: IMMUNOGLOBULIN SUPERFAMILY OF PROTEINS

Antibody molecules are produced in response to invasion by foreign compounds that can be proteins, carbohydrate polymers, and nucleic acid. They noncovalently associate with foreign substances, initiating a process by which the foreign substance is eliminated from the organism. Molecules that induce antibody production are **antigens** and may contain multiple antigenic determinants—that is, small regions of the antigen molecule that elicit the production of a specific antibody to which the antigen binds. In proteins, an antigenic determinant may comprise only six or seven amino acids.

A **hapten** is a small molecule that cannot alone elicit production of specific antibodies but, when covalently attached to a larger molecule, acts as an antigenic determinant and induces antibody synthesis. Whereas haptens need attachment to a larger molecule to elicit antibody synthesis, in their free state they bind strongly to antibody.

Each human can potentially produce about 1×10^8 different antibody structures. All antibodies, however, have a similar overall structure. This has been determined from studies of their amino acid sequence and X-ray diffraction of the antibody molecule alone or in complex with antigen. Structural studies require pure homogeneous preparations of proteins. Antibodies are extremely difficult to purify from plasma because of the wide diversity of antibody molecules present. Homogeneous antibodies can be obtained, however, by the monoclonal hybridoma technique in which mouse myeloma cells are fused with mouse-antibody-producing B lymphocytes to construct immortalized hybridoma cells that express a single antibody.

Antibody Molecules Contain Four Polypeptide Chains

Antibody molecules are glycoproteins composed of two **light chains (L)** of identical sequence combined with two identical **heavy chains (H)** to form the structure $(LH)_2$. In the most common immunoglobulin class, IgG, the H chains contain approximately 440 amino acids (50 kDa). The L chains contain about 220 amino acids (25 kDa). The four chains are covalently interconnected by disulfide bonds (Figures 9.1 and 9.2). Each H chain is associated with an L chain such that the NH_2-terminal ends of both chains are near each other. Since the L chain is half the size of the H chain, only the NH_2-terminal half of the H chain is associated with the L chain.

In the other classes of immunoglobulins (Table 9.1) the H chains are slightly longer than in the IgG class. A variable amount of carbohydrate (2–12%, depending on class) is attached to the H chain.

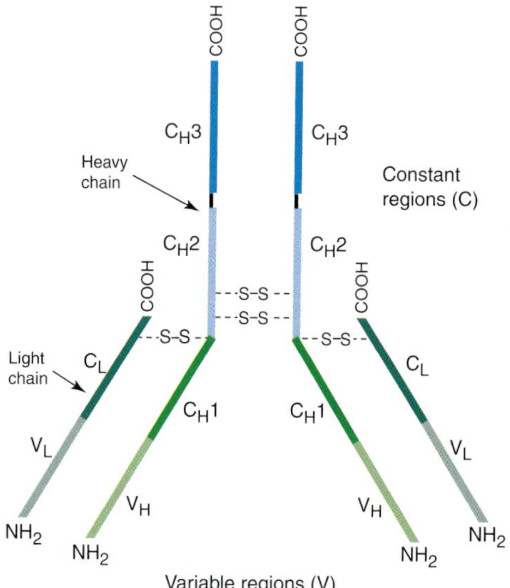

FIGURE 9.1

Linear representation of IgG antibody molecule. Two heavy (H) chains and two light (L) chains are co-oriented parallel to each other. Interchain disulfide bonds link H chains to each other, and they link L chains to H chains. Ig fold repeats in the constant (C) region of the H chain are C_H1, C_H2, and C_H3. The constant region of L chain is designated C_L, and variable (V) regions are V_H and V_L of H and L chains, respectively.
Based on figure by Burton, D. R. In: F. Calabi and M. S. Neuberger (Eds.), *Molecular Genetics of Immunoglobulin*. Amsterdam: Elsevier, 1987, p. 1.

Constant and Variable Sequence Regions of Primary Structure

Comparison of amino acid sequences of antibody molecules elicited by different antigens shows regions of sequence homology and others of sequence variability. In particular, sequences of the NH_2-terminal half of L chains and the NH_2-terminal quarter of H chains are highly variable. These are the **variable (V) regions** designated V_H and V_L domains of H and L chains, respectively. Within these V domains, certain regions are "hypervariable." Three **hypervariable regions** of between 5 and 7 residues in the V_L domain and three or four hypervariable regions of between 6 and 17 residues in the V_H domain are common. They are the **complementarity-determining regions (CDRs)** that form the antigen-binding site complementary to the topology of the antigen.

TABLE 9.1 Immunoglobulin Classes

Classes of Immunoglobulin	*Approximate Molecular Mass*	*H-Chain Isotype*	*Carbohydrate by Weight (%)*	*Concentration in Serum (mg 100 mL^{-1})*
IgG	150,000	γ, 53,000	2–3	600–1800
IgA	170,000–720,000[a]	α, 64,000	7–12	90–420
IgD	160,000	δ, 58,000		0.3–40
IgE	190,000	ϵ, 75,000	10–12	0.01–0.10
IgM	950,000[a]	μ, 70,000	10–12	50–190

[a] Forms polymer structures of basic structural unit.

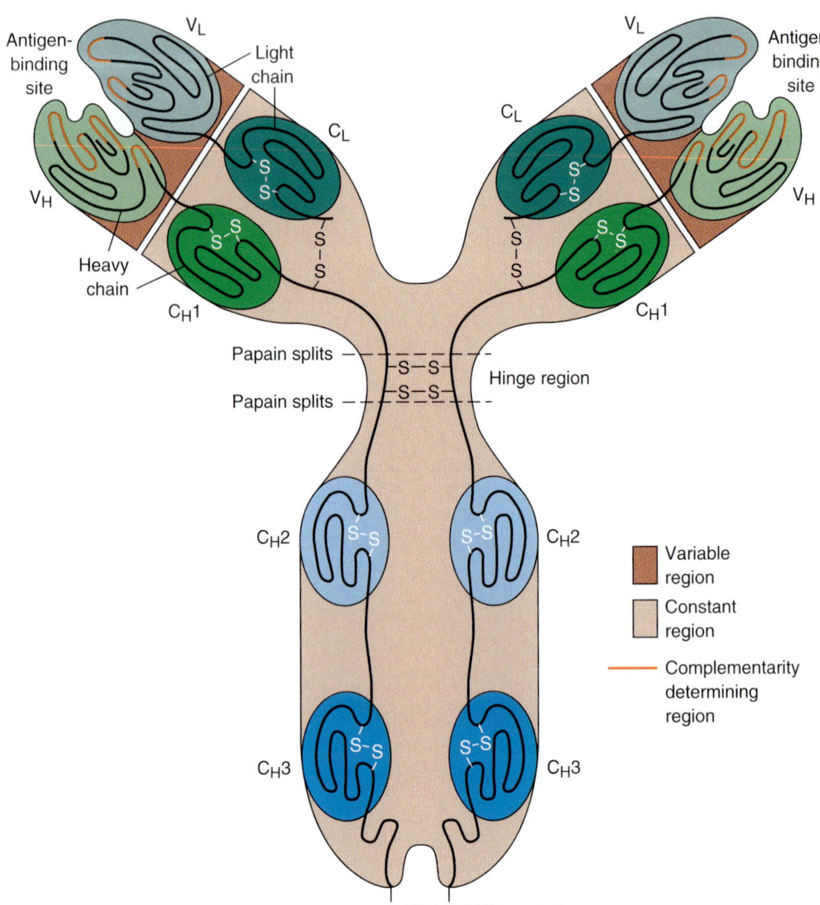

FIGURE 9.2

Diagrammatic structure of IgG. Light chains (L) are divided into Ig folds V_L (variable amino acid sequence) and C_L (constant amino acid sequence). Heavy chains (H) are divided into V_H and C_H1, C_H2, and C_H3. Antigen-binding sites are V_H–V_L. "Hinge" polypeptides interconnect domains. Positions of inter- and intrachain cystine bonds are shown.

From Cantor, C. R. and Schimmel, P. R. *Biophysical Chemistry*, Part I, San Francisco: Freeman, 1980. Reprinted with permission of Mr. Irving Geis, New York.

In contrast, the COOH-terminal three-quarters of H chains and the COOH-terminal half of L chains are homologous in sequence with other H or L chains of the same class. These **constant (C) regions** are designated C_H and C_L in the H and L chains, respectively.

The C_H regions determine the antibody class, provide for binding of complement proteins (see Clin. Corr. 9.1), and contain the site necessary for antibodies to cross the placental membrane. The V regions determine the antigen specificity of the antibody.

FIGURE 9.3

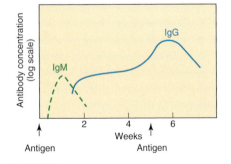

Time course of specific antibody IgM and IgG response to added antigen.

Based on a figure in Stryer, L. *Biochemistry*. San Francisco: Freeman, 1988, p. 890.

Immunoglobulins in a Class Contain Common Homologous Sequence Regions

Differences in sequence of the C_H regions between immunoglobulin classes determine the characteristics of each class. In some classes, the C_H sequence promotes polymerization of molecules of the basic structure $(LH)_2$. Thus antibodies of the **IgA** class are often covalently linked dimeric structures $[(LH)_2]_2$. Similarly, IgM molecules are pentamers $[(LH)_2]_5$. The different H chains, designated γ, α, δ, ϵ, and μ, occur in IgG, IgA, IgM, IgD, and IgE classes, respectively (Table 9.1; see Clin. Corr. 9.2). Two types of L chain sequences are synthesized, designated lambda (λ) and kappa (κ), either of which are found combined with each of the five classes of H chains.

IgG is the major immunoglobulin in plasma. Biosynthesis of a specific IgG in significant concentrations takes about 10 days after exposure to a new antigen (see Clin. Corr. 9.3). Antibodies of the IgM class are synthesized at a faster rate and are the first line of defense until large quantities of IgG are produced (Figure 9.3; see Clin. Corr. 9.3).

CLINICAL CORRELATION 9.1
The Complement Proteins

At least 11 distinct complement proteins exist in plasma (see p. 992) for their designations). They are activated by IgG or IgM antibody binding to antigens on the outer surface of invading bacterial cells, protozoa, or tumor cells. After this binding event, the complement proteins are sequentially activated and associate with the cell membrane to lyse the membrane and kill the target cell.

Many complement proteins are inactive precursors of proteolytic enzymes. Upon activation, they activate a succeeding protein of the pathway by cleavage of a specific peptide bond, leading to a cascade effect. Activation by hydrolysis of a specific peptide bond is an important method for activation of extracellular enzymes. For example, the enzymes that catalyze blood clot formation induce fibrinolysis of blood clots, and digest dietary proteins in the gut are all activated by specific proteases (see pp. 994 and 1041).

Upon binding to a cellular antigen the complement-binding site in the antibody's F_c region becomes exposed and causes binding of the C1 complement proteins, a complex composed of the proteins: Clq, Clr, and Cls. Proteins Clr and Cls undergo a conformational change and become active enzymes on the cell surface. The activated C1 complex (Cla) hydrolyzes a peptide bond in C2 and C4, which then also associate on the cell surface. The now active C2—C4 complex hydrolyzes a peptide bond in C3. Activated C3 protein binds to the cell surface and the activated C2—C4—C3 complex activates C5. Activated C5 associates with C6, C7, C8, and six molecules of C9, which bind to the cell surface and initiate membrane lysis.

This constitutes a cascade in which amplification of the trigger event occurs. In summary, activated C1 can activate many molecules of C4—C2—C3, and each activated C4—C2—C3 complex can, in turn, activate many molecules of C5 to C9. The reactions of the classical complement pathway are summarized below, where "a" and "b" designate the proteolytically modified proteins and a line above a protein indicates an enzyme activity.

$$\text{IgG or IgM} \xrightarrow{\text{Clq, Clr, Cls}} \overline{\text{Cla}} \xrightarrow{\text{C2, C4}}$$

$$\text{C4b} \cdot \overline{\text{C2a}} \xrightarrow{\text{C3}}$$

$$\text{C4b} \cdot \overline{\text{C2a}} \cdot \text{C3b} \xrightarrow{\text{C5, C6, C7, C8, C9}} \text{lysis}$$

An "alternative pathway" for C3 complement activation is initiated by aggregates of IgA or by bacterial polysaccharide in the absence of immunoglobulin binding to cell membrane antigens. This pathway involves the proteins properdin, C3 proactivator convertase, and C3 proactivator.

A major role of the complement systems is to generate opsonins, proteins that stimulate phagocytosis by neutrophils and macrophages. The major opsonin is C3b; macrophages have specific receptors for this protein. Patients with inherited deficiency of C3 suffer from repeated bacterial infections.

Source: Colten, H. R., and Rosen, F. S. Complement deficiencies. *Annu. Rev. Immunol.* 10:809, 1992. Morgan, B. P. Physiology and pathophysiology of complement: Progress and trends. *Crit. Rev. Clin. Lab. Sci.* 32:265, 1995.

CLINICAL CORRELATION 9.2
Functions of Different Antibody Classes

The IgA class of immunoglobulins is found primarily in mucosal secretions (bronchial, nasal, and intestinal mucous secretions, tears, milk, and colostrum). They are the initial defense against invading viral and bacterial pathogens prior to their entry into plasma or other internal space.

The IgM class is found primarily in plasma. They are the first antibodies elicited in significant quantity on exposure to a foreign antigen. They promote phagocytosis of microorganisms by macrophage and polymorphonuclear leukocytes and are also potent activators of complement (see Clin. Corr. 9.1). They occur in many external secretions but at levels lower than those of IgA.

The IgG class occurs in high concentration in plasma. Their synthesis in response to foreign antigens takes a longer period of time than that of IgM. At maximum concentration they are present in significantly higher concentration than IgM. Like IgM, IgG antibodies promote phagocytosis in plasma and activate complement.

The normal biological functions of the IgD and IgE classes are not known;, however, the IgE play an important role in allergic responses such as anaphylactic shock, hay fever, and asthma.

Immunoglobulin deficiency usually causes increased susceptibility to infection. X-linked agammaglobulinemia and common variable immunodeficiency are examples. The commonest disorder is selective IgA deficiency, which results in recurrent infection of sinuses and the respiratory tract.

Source: Rosen, F. S., Cooper, M. D., and Wedgewood, R. J. P. The primary immunodeficiencies. *N. Engl. J. Med.* 311:235 (Part 1); 300 (Part II), 1984.

CLINICAL CORRELATION 9.3
Immunization

An immunizing vaccine consists of killed bacterial cells, inactivated viruses, killed parasites, nonvirulent forms of live bacteria, denatured bacterial toxins, or recombinant protein. The introduction of a vaccine into a human leads to protection against a virulent form of the microorganisms or toxic agents that contain the same antigen. Antigens in nonvirulent material cause differentiation of lymphoid cells so that they produce antibody toward the foreign antigen and differentiation of some lymphoid cells into memory cells. Memory cells do not secrete antibody but place antibodies to the antigen onto their outer surface, where they act as future sensors for the antigen. These memory cells are like longstanding radar for the potentially virulent antigen. On reintroduction of the antigen later, binding of the antigen to the cell surface antibody in the memory cells stimulates the memory cell to divide into antibody-producing cells and new memory cells. Once introduction of an antigen occurs, the memory cells reduce the time required for antibody production and increase the concentration of antigen-specific antibody produced. This is the basis for the protection provided by immunization.

Vaccines recently introduced for adults include pneumococcal vaccine (to prevent pneumonia due to *Diplococcus pneumoniae)*, hepatitis B vaccine, and influenza vaccine. The composition of the latter changes each year to account for antigenic variation in the influenza virus.

Source: Flexner, C. New approaches to vaccination. *Adv. Pharmacol.* 21:51, 1990. Sparling, P. F., Elkins, C., Wyrick, P. B., and Cohen, M. S. Vaccines for bacterial sexually transmitted infections: a realistic goal? *Proc. Natl. Acad. Sci. USA* 91:2456, 1994.

Repeating Sequences Generate Homologous Three-Dimensional Folds within An Antibody

Within each chain of an antibody molecule is a repeating pattern of amino acid sequences. For the IgG class, the repetitive pattern contains approximately 110 amino acids within both L and H chains. This homology is far from exact, but clearly a number of amino acids match identically on alignment of these segments. Other residues are matched in the sequence by having similar nonpolar or polar side chains. In the H chains, the repetition occurs in the V_H region and C_H regions (designated C_H1, C_H2, and C_H3) (see Figures 9.1 and 9.2). The L chain contains one V_L and one C_L region. Each repeat contains an intrachain disulfide bond (Figure 9.2).

The repeat segments form structural folds of similar tertiary structure as shown by X-ray diffraction studies. Each 110-amino-acid segment has a similar arrangement of antiparallel β-strands known as an **immunoglobulin fold** (Figure 9.4). This motif consists of 7 to 9 polypeptide strands in two antiparallel β-sheets aligned face-to-face. The strong interactions between two immunoglobulin folds form globular domains (Figure 9.5). The fold interactions are between V_L and V_H and between C_L and C_H1 in the H and L chains, respectively. In the C-terminal half of the H chains, the two chains associate to generate domains C_H2—C_H2 and C_H3—C_H3 (Figure 9.2). A "hinge" region interconnects the two C_H—C_L1 domains with the C_H2—C_H2 domain of the H chains (Figure 9.2). Thus the antibody structure exhibits six domains, each domain formed from the interaction of two immunoglobulin folds (Figures 9.2 and 9.6). The NH_2-terminal V_L—V_H domains contain a shallow crevice in the center of a hydrophobic core that binds the antigen. **Hypervariable sequences** in the V-domain crevices form loops that come close together and are the complementarity binding site (complementarity determining regions, CDRs) for the antigen (see Figures 9.6 and 9.7). The sequences of the hypervariable loops provide a unique three-dimensional conformation for each antibody that makes it specific for its antigenic specificity. Small changes in conformation of the CDRs occur on binding of antigen to V_L—V_H domains, indicating that the antigen binding induces an optimal fit within the variable CDR site. Antigen binding may also induce conformational changes between the V_L—V_H domains and the other domains to activate effector sites, such as the binding of complement to a site within the C_H2—C_H2 domain (see C1q site in Figure 9.6). The strength of binding between antibody and antigen is due to noncovalent forces (see p. 112). Complementarity of the structures of the antigenic determinant and antigen-binding

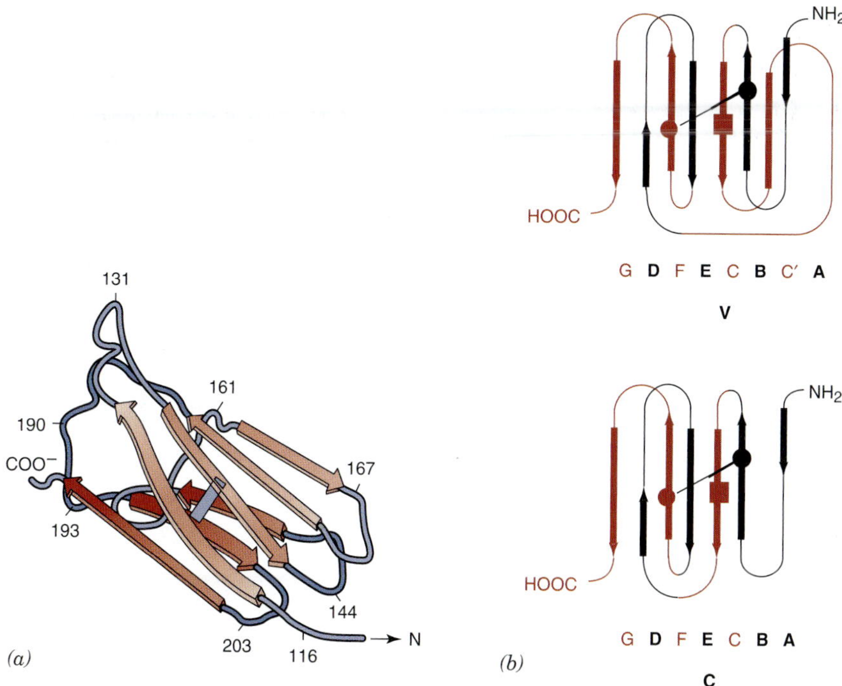

FIGURE 9.4

Immunoglobulin fold. (a) Schematic diagram of folding of a C_L domain, showing β-pleated sheet structure. Arrows show strands of β-sheet, and bar (blue) shows position of cystine bond. Pink arrows are for β-strands in plane above, and dark red arrows are β-strands in plane below. (b) Diagrammatic outline of arrangement (topology) of β-strands in immunoglobulin fold motif. Examples are for IgG variable (upper diagram) and constant (lower diagram) regions. Thick arrows indicate β-strands and thin lines loops that connect the β-strands. Circles indicate cysteines that form intradomain disulfide bond. Squares show positions of tryptophan residues that are an invariant component of the core of the immunoglobulin fold. Boldface black letters indicate strands that form one plane of the sheet, while other letters form the parallel plane behind the first plane.

Part (a) reproduced with permission from Edmundson, A. B., Ely, K. R., Abola, E. E., Schiffer, M., and Pavagiotopoulos, N. *Biochemistry* 14:3953, 1975. Copyright (1975) American Chemical Society. Part (b) Redrawn from Calabi, F. In: F. Calabi and M. S. Neuberger (Eds.), *Molecular Genetics of Immunoglobulin*. Amsterdam: Elsevier, 1987, p. 203.

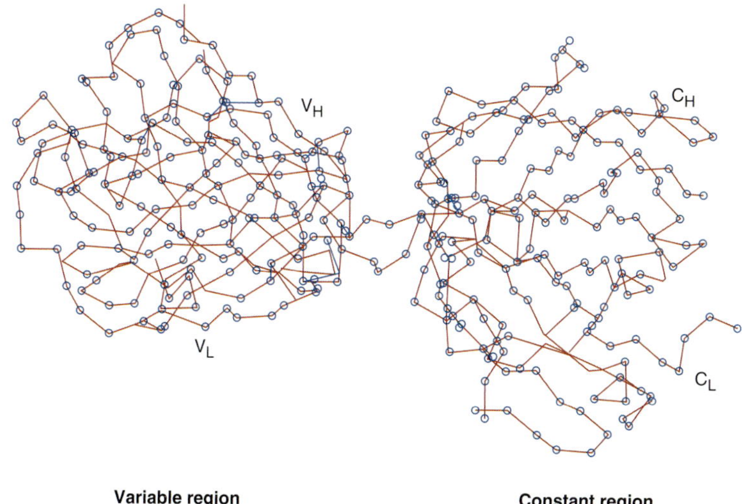

FIGURE 9.5

α-Carbon (○) structure of F_{ab} fragment of IgG KOL showing V_L–V_H and C_L–C_H1 domains interconnected by the hinge regions.

Redrawn with permission from Huber, R., Deisenhofer, J., Coleman, P. M., Matsushima, M., and Palm, W. In: *The Immune System*, F. Melchers and K. Rajwsky (Eds.) 27th Mosbach Colloquium. Berlin: Springer-Verlag, 1976, p. 26.

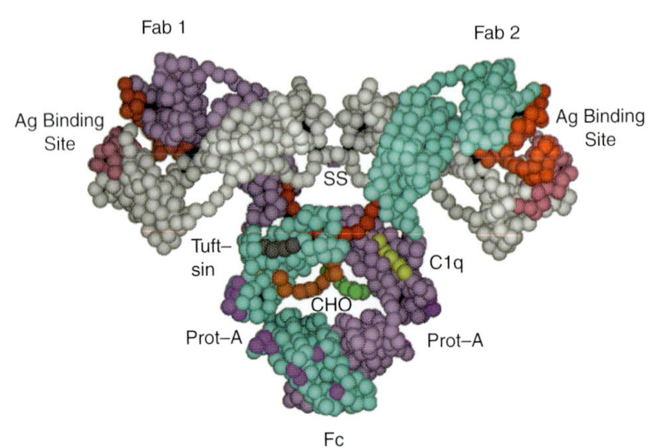

FIGURE 9.6

Model of an IgG antibody molecule. Only the α carbons of the structure appear. The two L chains are represented by light gray spheres and H chains by lavender and blue spheres. Carbohydrates attached to the two C_H2 domains are colored light green and orange. The CDR regions of the V_H-V_L domains are dark red in the H chains and pink in L chains. The interchain disulfide bond between the L and H chains is a magenta ball-and-stick representation (partially hidden). The heptapeptide hinge between C_H1 and C_H2 domains, connecting the Fab and Fc units, are dark red. The center of the C1q binding site in the C_H2 domains is yellow, the protein A docking sites at the junction of C_H2 and C_H3 are magenta, and the tuftsin binding site in C_H2 is gray. Tuftsin is a natural tetrapeptide that induces phagocytosis by macrophages and may be transported bound to an immunoglobulin. Protein A is a bacterial protein with a high affinity for immunoglobulins.

Photograph generously supplied by Dr. Allen B. Edmundson, from Guddat, L. W., Shan, L., Fan, Z-C., et al. *FASEB J.* 9:101, 1995.

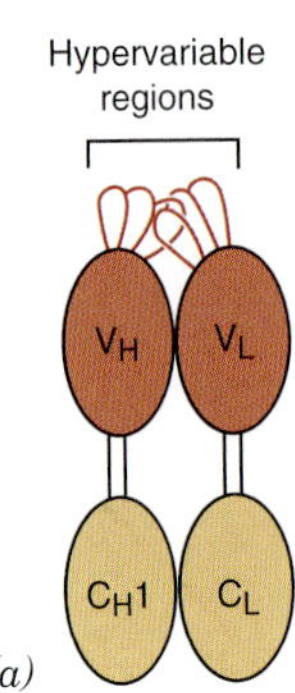

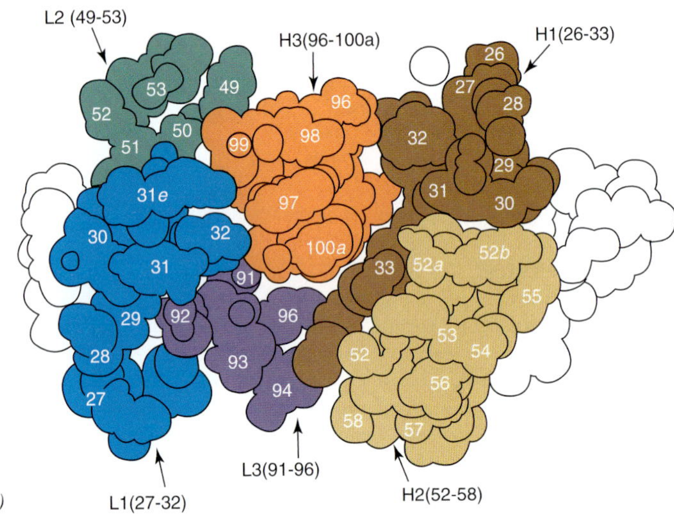

FIGURE 9.7

Hypervariable loops in immunoglobin. (*a*) Schematic diagram showing hypervariable loops (CDRs) in V_L-V_H domain that form the antigen-binding site. (*b*) A cut through an antigen-binding site showing contributions of different CDRs using CPK space-filling models of the atoms. Numbers refer to amino acid residue numbers in the sequence of the hypervariable loop regions of the light (L) and heavy (H) chains.

Part (*a*) redrawn from Branden, C. and Tooze, J. *Introduction to Protein Structure*. New York: Garland, 1991, p. 187. Part (*b*) redrawn from Branden, C., and Tooze, J. *Introduction to Protein Structure*. New York: Garland, 1991, p. 189. Attributed to Chothia, C. and Lesk, A. *J. Mol. Biol.* 196:914, 1987.

site results in extremely high equilibrium affinity constants, between 10^5 and 10^{10} M^{-1} (strength of 7–14 kcal mol^{-1}) for this noncovalent association.

There Are Two Antigen-Binding Sites Per Antibody Molecule

The NH_2-terminal V domains of the L and H chains (V_L—V_H) comprise an antigen-binding site; thus there are two antigen-binding sites per antibody molecule. The existence of an antigen-binding site in each LH pair is demonstrated by treating antibody molecules with the proteolytic enzyme papain, which hydrolyzes a peptide bond in the hinge region of each H chain (see Figures 9.2 and 9.8) to release three fragments. Two are identical, each consisting of the NH_2-terminal half of the H chain (V_H—C_H1) associated with the full L chain (Figure 9.8). Each of these fragments binds antigen with a similar affinity to the intact molecule and is designated an $\mathbf{F_{ab}}$ (antigen binding) **fragment.** The other fragment is the COOH-terminal half of the H chains (C_H2—C_H3) joined together by a disulfide bond. This is the $\mathbf{F_c}$ (crystallizable) **fragment**, which does not bind the antigen. The L chain can be dissociated from its H chain partner within the F_{ab} fragment by reduction of disulfide bonds, which eliminates antigen binding. Accordingly, each antigen-binding site must be formed from components of both the L-chain (V_L) and the H-chain (V_H) domains acting together.

The major features of antibody structure and antibody–antigen interactions are as follows: (1) The polypeptide chains fold into several immunoglobulins folds. (2) Two immunoglobulin folds on separate chains associate to form the six domains of the basic immunoglobulin structure. The V_L and V_H associate to form the two NH_2-terminal

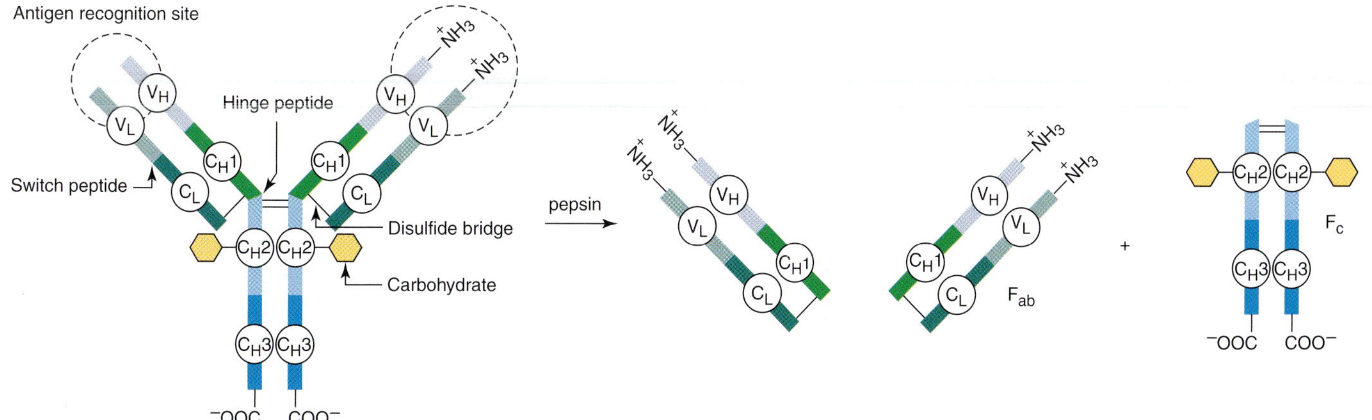

FIGURE 9.8
Hydrolysis of IgG into two F_{ab} and one F_c fragments by papain, a proteolytic enzyme.

domains that bind to antigen. (3) The antigen-binding site is generated by hypervariable loops (CDRs), which form a continuous surface with a topology complementary to the antigenic determinant. (4) The strong interactions between antigen and antibody CDRs are noncovalent and include van der Waals, hydrogen bonding, hydrophobic interactions, and, to a smaller extent, ionic salt bridges. (5) Small conformational changes occur in the V_L—V_H domain upon binding of antigen, indicating an "induced fit." (6) The binding of antigen to the V_L—V_H domains induces conformational changes in distant domains of the antibody. These allosteric changes alter the binding affinity of effector sites in the constant domains such as that for binding of complement protein C1q to the C_H2—C_H2 domain (see Clin. Corr. 9.1).

Genetics of The Immunoglobulins

The V and C regions of the L and H chains are specified by distinct genes. There are four unique genes that code for the C domains of the H chain in the IgG antibody class. Each gene codes for a complete constant region. They are known as gamma (γ) genes—that is, γ_1, γ_2, γ_3, and γ_4—that give rise to **IgG isotypes** IgG_1, IgG_2, IgG_3, and IgG_4. Figure 9.9 presents the sequences of three γ-gene proteins. There is a 95% homology in amino acid sequence among the genes.

It is likely that a primordial gene coded for a single immunoglobulin fold sequence of approximately 110 amino acids, and that **gene duplication** events resulted in the three repeating units within the same IgG H γ gene. Mutations modified the individual sequences so that an exact sequence correspondence no longer exists. However, each immunoglobulin fold has a similar length and folding pattern stabilized by a cystine linkage. Later in evolution, further gene duplications led to the multiple genes (γ_1, γ_2, γ_3, and γ_4) that code for the constant regions of the IgG classes.

Immunoglobulin Fold Is Found in a Large Family of Proteins with Different Functional Roles

The immunoglobulin fold is present in many nonimmunological proteins, which exhibit widely different functions. Based on their structural homology, they are grouped into a **protein superfamily** (Figure 9.10). For example, the Class I major histocompatibility complex proteins are in this superfamily; they have immunoglobulin fold motifs that consist of two stacked antiparallel β-sheets enclosing an internal space filled mainly by hydrophobic amino acids. Two cysteines in the structure form a disulfide bond linking the facing β-sheets. Transcription factors Nf-κB and p53 also contain an immunoglobulin fold motif. It can be speculated that gene duplication during evolution led to distribution of the structural motif in the functionally diverse proteins.

```
Constant Region C1:
Cγ1: AlaSerThrLysGlyProSerValPheProLeuAlaProSerSerLysSerThrSerGlyGlyThrAlaAlaLeuGly
Cγ2                                         C     R            E  S
Cγ4

Cγ1  CysLeuValLysAspTyrPheProGluProValThrValSerTrpAsnSerGlyAlaLeuThrSerGlyValHisThr
Cγ2
Cγ4

Cγ1  PheProAlaValLeuGlnSerSerGlyLeuTyrSerLeuSerSerValValThrValProSerSerSerLeuGly
Cγ2                                                                         N  F
Cγ4

Cγ1  ThrGlnThrTyrIleCysAsnValAsnHisLysProSerAsnThrLysValAspLysLysVal
Cγ2              T           D                                 T
Cγ4     K                                                      R


Hinge Region H:
Cγ1  GluProLysSerCysAspLysThrHisThrCysProProCysPro
Cγ2     R     C     V  E  C  P  P        -  -  -
Cγ4     S     Y  G  P  P        S        -  -  -


Constant Region C2:
Cγ1  AlaPro   GluLeuLeuGlyGlyProSerValPheLeuPheProProLysProLysAspThrLeuMetIleSerArg
Cγ2           -  P  V  A
Cγ4        E  F     G

Cγ1  ThrProGluValThrCysValValValAspValSerHisGluAspProGluValLysPheAsnTrpTyrValAspGly
Cγ2                                                        Q
Cγ4                                       Q

Cγ1  ValGluValHisAsnAlaLysThrLysProArgGluGluGlnTyrAsnSerThrThrArgValValSerValLeuThr
Cγ2                                            F           F
Cγ4                                                        Y

Cγ1  ValLeuHisGlnAspTrpLeuAsnGlyLysGluTyrLysCysLysValSerAsnLysAlaLeuProAlaProIleGlu
Cγ2     V                                                        G
Cγ4                                                                       S  S

Cγ1  LysThrIleSerLysAlaLys
Cγ2                 T
Cγ4


Constant Region C3:
Cγ1  GlyGlnProArgGluProGlnValTyrThrLeuProProSerArgAspGluLeuThrLysAsnGlnValSerLeuThr
Cγ2                                                E     M
Cγ4                                             Q

Cγ1  CysLeuValLysGlyPheTyrProSerAspIleAlaValGluTrpGluSerAsnGlyGlnProGluAsnAsnTyrLys
Cγ2
Cγ4

Cγ1  ThrThrProProValLeuAspSerAspGlySerPhePheLeuTyrSerLysLeuThrValAspLysSerArgTrpGln
Cγ2              M
Cγ4                                                  R

Cγ1  GlnGlyAsnValPheSerCysSerValMetHisGluAlaLeuHisAsnHisTyrThrGlnLysSerLeuSerLeuSer
Cγ2
Cγ4  E

Cγ1  ProGlyLysStop
Cγ2
Cγ4  L
```

FIGURE 9.9

Amino acid sequence of heavy chain constant regions of IgG heavy chain γ_1, γ_2, and γ_4 genes. Sequences of C_H1, hinge region H, C_H2, and C_H3 regions are presented. Sequence for γ_1 is complete and differences in γ_2 and γ_4 from γ_1 sequence are shown using single-letter amino acid abbreviations. Dashed line (—) indicates absence of an amino acid in position correlated with γ_1, in order to better align sequences to show maximum homology.
Sequence of γ_1 chain from Ellison, J. W., Berson, B. J., and Hood, L. E. *Nucleic Acid Res.* 10:4071, 1982. Sequences of the γ_2 and γ_4 genes from Ellison, J. and Hood, L. *Proc. Natl. Acad. Sci. USA* 79:1984, 1982.

9.3 PROTEINS WITH A COMMON CATALYTIC MECHANISM: SERINE PROTEASES

Serine proteases are a family of enzymes that use a uniquely activated serine residue in their substrate-binding site to catalytically hydrolyze peptide bonds. This serine can be characterized by the irreversible reaction of its side-chain hydroxyl group with diisopropylfluorophosphate (DFP) (Figure 9.11). Of all the serines in the protein, DFP reacts only with the catalytically active serine to form a phosphate ester.

Proteolytic Enzymes Are Classified by Their Catalytic Mechanism

Proteolytic enzymes are classified according to their catalytic mechanism. Besides serine proteases, other classes utilize cysteine (**cysteine proteases**), aspartate (**aspartate proteases**), or metal ions (**metalloproteases**) to perform their catalytic function. Those that hydrolyze peptide bonds in the interior of a polypeptide are **endopeptidases**, and those that cleave the peptide bond of either the COOH- or NH_2-terminal amino acid are **exopeptidases**.

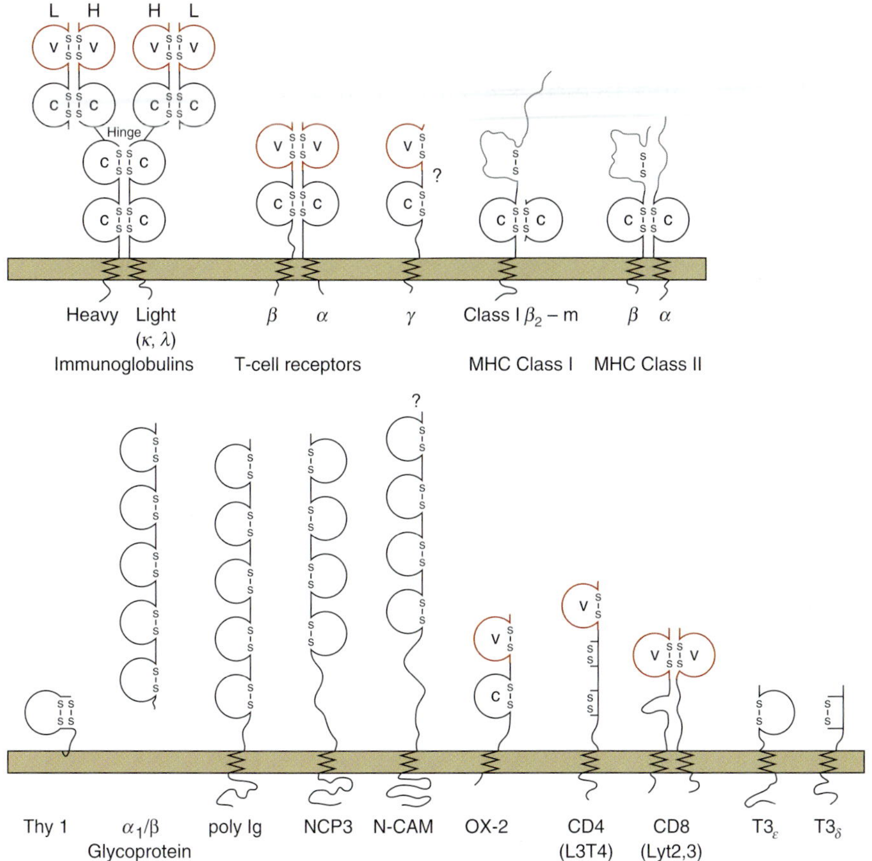

FIGURE 9.10

Diagrammatic representation of immunoglobulin fold structure found in different proteins of immunoglobulin gene superfamily. Proteins presented include heavy and light chains of immunoglobulins, T-cell receptors, major histocompatibility complex (MHC) Class I and Class II proteins, T-cell accessory proteins involved in Class I (CD8) and Class II (CD4) MHC recognition and possible ion channel formation, a receptor responsible for transporting certain classes of immunoglobulin across mucosal membranes (poly-Ig), β_2-microglobulin, which associates with class I molecules, a human plasma protein with unknown function (α_1/β-glycoprotein), two molecules of unknown function with a tissue distribution that includes lymphocytes and neurons (Thy-1, OX-2), and two brain-specific molecules, neuronal cell-adhesion molecule (N-CAM) and neurocytoplasmic protein 3 (NCP3).
Redrawn from Hunkapiller, T. and Hood, L. *Nature* 323:15, 1986.

Serine proteases often activate other serine proteases from their inactive precursor form, termed a **zymogen**, by cleavage of a specific peptide bond. This mechanism of zymogen activation allows serine proteases to participate in carefully controlled physiological processes such as blood coagulation (see Clin. Corr. 9.4), fibrinolysis, complement activation (see Clin. Corr. 9.1), fertilization, and hormone production (Table 9.2). The serine protease activations are examples of **limited proteolysis** because only one or two specific peptide bonds of the hundreds in the protein substrate are hydrolyzed. With denaturation, however, these same proteases hydrolyze multiple peptide bonds that lead to digestion of peptides, proteins, and even self-digestion (autolysis). Several diseases, such as emphysema, arthritis, thrombosis, cancer metastasis (see Clin. Corr. 9.5), and some forms of hemophilia, are thought to result from the lack of regulation of a specific serine protease.

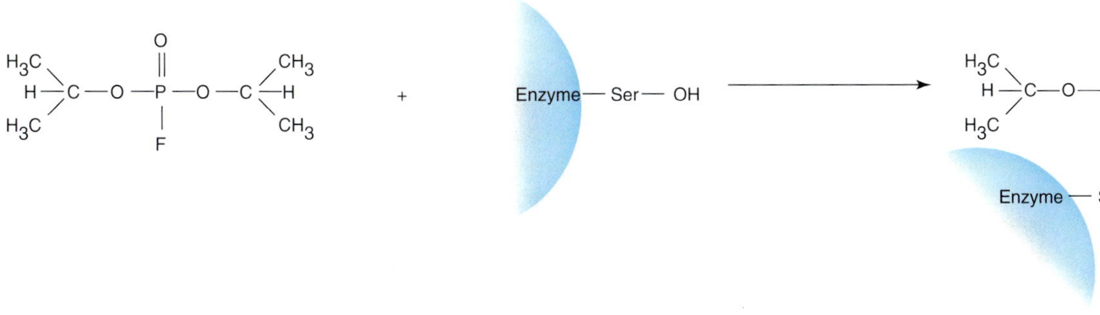

$$(H_3C)_2CH{-}O{-}P(=O)(F){-}O{-}CH(CH_3)_2 + \text{Enzyme}{-}\text{Ser}{-}\text{OH} \longrightarrow (H_3C)_2CH{-}O{-}P(=O)(O{-}\text{Ser}{-}\text{Enzyme}){-}O{-}CH(CH_3)_2$$

FIGURE 9.11

Reaction of diisopropylfluorophosphate (DFP) with the active-site serine in a serine protease

CLINICAL CORRELATION 9.4

Fibrin Formation in a Myocardial Infarct and Use of Recombinant Tissue Plasminogen Activator (rt-PA)

Coagulation is an enzyme cascade process in which inactive precursor serine proteases (zymogens) are activated by other serine proteases in a stepwise manner (the coagulation pathway is described in Chapter 24). These activation events generate active proteases with a dramatic amplification of the initial signal of the pathway. The end product of the coagulation pathway is a cross-linked fibrin clot. The zymogen components include factor II (prothrombin), factor VII (proconvertin), factor IX (Christmas factor), factor X (Stuart factor), factor XI (plasma thromboplastin antecedent), and factor XII (Hageman factor). The Roman numeral designation indicates the order of their discovery and not their order of action within the pathway. Upon activation, the enzymes are denoted with the suffix "a." Thus, prothrombin and thrombin are denoted as factor II and factor IIa, respectively.

The main function of coagulation is to maintain the integrity of the closed circulatory system after blood vessel injury. The process, however, can become dangerously activated in a myocardial infarction and can decrease blood flow to heart muscle. About 1.5 million individuals suffer heart attacks each year, which result in 600,000 deaths.

A fibrinolysis pathway degrades fibrin clots and also utilizes serine proteases activated from zymogens. The end reaction is formation of plasmin, a serine protease, which acts directly on fibrin to degrade the fibrin clot. Tissue plasminogen activator (t-PA) is a plasminogen activator. Recombinant t-PA (rt-PA) is produced by gene cloning technology (see Chapter 18). Clinical studies show that the administration of rt-PA shortly after a myocardial infarct significantly enhances recovery. Plasminogen activators such as urokinase and streptokinase are also effective.

Source: The GUSTO investigators. An international randomized trial comparing four thrombolytic strategies for acute myocardial infarction. *N. Engl. J. Med.* 329:673, 1993. International Study Group. In hospital mortality and clinical course of 20,891 patients with suspected acute myocardial infarction randomized between alteplase and streptokinase with or without heparin. *Lancet* 336:71, 1990. Gillis, J. C., Wagstaff, A. J., and Goa, K. L. Alteplase. A reappraisal of its pharmacological properties and therapeutic use in acute myocardial infarction. *Drugs* 50: 102, 1995.

CLINICAL CORRELATION 9.5

Involvement of Serine Proteases in Tumor Cell Metastasis

The serine protease urokinase is believed to be required for metastasis of cancer cells. Metastasis is the process by which a cancer cell leaves a primary tumor and migrates through the blood or lymph systems to another tissue or organ, where a secondary tumor grows. Increased synthesis of urokinase has been correlated with an increased ability to metastasize in many cancers.

Urokinase activates plasminogen to form plasmin. Plasminogen is ubiquitously located in the extracellular space and its activation to plasmin can cause degradation of proteins in the extracellular matrix through which the metastasizing tumor cells migrate. Plasmin can also convert procollagenase to collagenase and can promote degradation of collagen in the basement membrane surrounding the capillaries and lymph system. This promotion of proteolytic activity by urokinase secreted by tumor cells allows them to invade the target tissue and form secondary tumor sites.

Source: Dano, K., Andreasen, P. A., Grondahl-Hansen, J., Kristensen, P., Nielsen, L. S., and Skriver, L. Plasminogen activators, tissue degradation and cancer. *Adv Cancer Res.* 44:139, 1985. Yu, H. and Schultz, R. M. Relationship between secreted urokinase plasminogen activator activity and metastatic potential in murine B16 cells transfected with human urokinase sense and antisense genes. *Cancer Res.* 50:7623, 1990. Fazioli, F. and Blasi, F. Urokinase-type plasminogen activator and its receptor: new targets for anti-metastatic therapy? *Trends Pharmacol. Sci.* 15:25, 1994.

Serine Proteases Exhibit Remarkable Specificity in Peptide Bond Hydrolysis

Many serine proteases prefer to hydrolyze peptide bonds next to a particular type of amino acid. Thus trypsin preferentially cleaves on the carboxyl side of the basic amino acids arginine and lysine, and chymotrypsin cleaves on the carboxyl side of large hydrophobic amino acid residues such as tryptophan, phenylalanine, tyrosine, and leucine. Elastase cleaves on the carboxyl side of small hydrophobic residues such as

TABLE 9.2 Some Serine Proteases and Their Biochemical and Physiological Roles

Protease	Action	Possible Disease Due to Deficiency or Malfunction
Plasma kallikrein Factor XIIa Factor XIa Factor IXa Factor VIIa Factor Xa Factor IIa (thrombin) Activated protein C	Coagulation (see Clin. Corr. 9.4)	Cerebral infarction (stroke), coronary infarction, thrombosis, bleeding disorders
Factor Clr Factor Cls Factor D Factor B C3 convertase	Complement (see Clin. Corr. 9.1)	Inflammation, rheumatoid arthritis, autoimmune disease
Trypsin Chymotrypsin Elastase (pancreatic) Enteropeptidase	Digestion	Pancreatitis
Urokinase plasminogen activator Tissue plasminogen activator Plasmin	Fibrinolysis, cell migration, embryogenesis, menstruation	Clotting disorders, tumor metastasis (see Clin. Corr. 9.5)
Tissue kallikreins	Hormone activation	
Acrosin	Fertilization	Infertility
α-Subunit of nerve growth factor γ-Subunit of nerve growth factor	Growth factor activation	
Granulocyte elastase Cathepsin G Mast cell chymases Mast cell tryptases	Extracellular protein and peptide degradation, mast cell function	Inflammation, allergic response

alanine. A serine protease is trypsin-like, chymotrypsin-like, or elastase-like if it has a specificity similar to the named enzymes. The specificity for a certain type of amino acid only indicates an enzyme's relative preference. Trypsin can also cleave peptide bonds following hydrophobic amino acids, but at a much slower rate than for the basic amino acids. Thus specificity for hydrolysis of a peptide bond may not be absolute, but may be more accurately described as a range of most likely targets. Each of the identical amino acid hydrolysis sites within a protein substrate is not equally susceptible. Trypsin hydrolyzes each of the arginine peptide bonds in a particular protein at a different catalytic rate, and some may require a conformational change to make them accessible.

The specificity of serine proteases for particular peptide bonds has been determined with synthetic substrates with fewer than 10 amino acids (Table 9.3). Because these are significantly smaller than the natural ones, they interact only with the catalytic site (primary binding site S_1; see below) and are said to be **active-site directed.** These studies with small substrates and inhibitors indicate that the site of hydrolysis is flanked by approximately four residues in both directions that bind to the enzyme and impact on the reactivity of the bond hydrolyzed. The two residues in the substrate that contribute the hydrolyzable bond are designated $\mathbf{P_1{-}P'_1}$. Thus in trypsin substrates the P_1 residue likely will be lysine or arginine, and in chymotrypsin substrates the P_1 residue will be a

TABLE 9.3 Reactivity of α-Chymotrypsin and Elastase Toward Substrates of Various Structures

Structure	Variation of Side Chain Group in S_1 Site (Chymotrypsin)	Relative Reactivity[a]
Glycyl	H—	1
Leucyl	$(H_3C)_2CH-CH_2-$	1.6×10^4
Methionyl	$CH_3-S-CH_2-CH_2-$	2.4×10^4
Phenylalanyl	$-CH_2-$	4.3×10^6
Hexahydrophenylalanyl	S $-CH_2-$	8.2×10^6
Tyrosyl	HO— $-CH_2-$	3.7×10^7
Tryptophanyl	CH_2- N H	4.3×10^7
Variation in chain length (elastase hydrolysis of Ala N-terminal amide)[b]		
	Ac-Ala-NH_2	1
	Ac-Pro-Ala-NH_2	1.4×10^1
	Ac-Ala-Pro-Ala-NH_2	4.2×10^3
	Ac-Pro-Ala-Pro-Ala-NH_2	4.4×10^5
	Ac-Ala-Pro-Ala-Pro-Ala-NH_2	2.7×10^5

[a] Calculated from values of k_{cat}/K_m found for *N*-acetyl amino acid methyl esters in chymotrypsin substrates.

[b] Calculated from values of k_{cat}/K_m in Thompson, R. C. and Blout, E. R. *Biochemistry* 12:57, 1973.

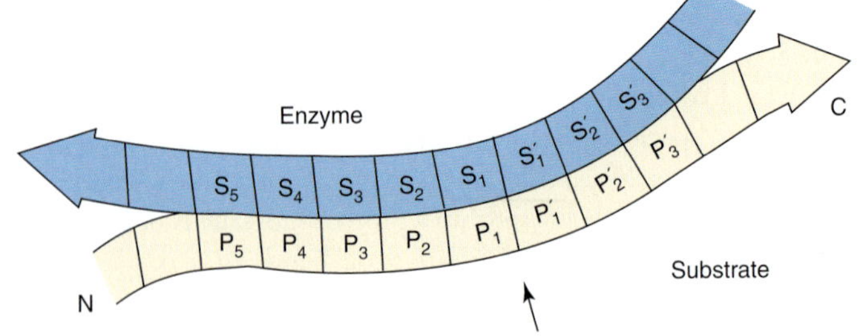

FIGURE 9.12

Schematic diagram of binding of a polypeptide substrate to binding site in a proteolytic enzyme. $P_5-P'_3$ are amino acid residues in the substrate that bind to subsites $S_5-S'_3$ in the enzyme with peptide hydrolysis occurring between $P_1-P'_1$ (arrow). NH_2-terminal direction of substrate polypeptide chain is indicated by N, and COOH-terminal direction by C.
Redrawn from Polgar, L. In: A. Neuberger and K. Brocklehurst (Eds.), *Hydrolytic Enzymes*. Amsterdam: Elsevier, 1987, p. 174.

hydrophobic amino acid. The other interacting amino acid residues of the protein substrate are labeled P_4-P_2 and $P'_2-P'_4$ on either end of the scissile bond, respectively. The complementary regions in the enzyme that bind these residues in the substrate are designated $\mathbf{S_4-S'_4}$ (Figure 9.12). It is the **secondary interactions** with the substrate outside $S_1-S'_1$ that ultimately determines a protease's specificity toward a particular protein substrate. Thus, the coagulation serine protease factor Xa only cleaves at a particular arginine in prothrombin to produce thrombin. It is the secondary interaction that allows factor Xa to recognize the particular arginine it cleaves in prothrombin. The interactions of P_4–P'_4 with the subsites S_4–S'_4 are noncovalent. The extended binding site in the substrate interacts with the binding site in the enzyme to produce a β-sheet structure between the enzyme and the substrate, which places the scissile peptide bond of the substrate in the S_1–S'_1 position (Figure 9.13).

Serine Proteases Are Synthesized As Zymogens

Serine proteases are synthesized as **zymogens** that require limited proteolysis to produce the active enzyme. The coagulation zymogens are synthesized in liver cells and are secreted into the blood for subsequent activation following vascular injury by other serine proteases. Zymogens are usually designated by adding the suffix -ogen to the enzyme name; as in trypsin*ogen* and chymotrypsin*ogen*. In some cases the zymogen is referred to as a **proenzyme**; for example, the zymogen of thrombin is prothrombin.

Several plasma serine proteases have zymogen forms that contain **multiple non-similar domains**. Protein C, involved in a fibrinolysis pathway in blood, has four distinct domains (Figure 9.14). The NH_2-terminal domain contains the derived amino acid, **γ–carboxyglutamic acid** (Figure 9.15), which is formed by carboxylation of glutamic acid residues in a vitamin K-dependent reaction (see pp. 1006 and 1099). The γ-carboxyglutamates chelate Ca^{2+} and form part of a site that binds membranes. The COOH-terminal segment contains the catalytic domains. Activation requires peptide cleavage outside the catalytic domains (Figure 9.14) and is controlled by the binding of the γ-carboxyglutamate-containing NH_2-terminal end through calcium ions to a membrane site.

Specific Protein Inhibitors of Serine Proteases

Evolution of the serine protease family for participation in physiological processes promoted a parallel evolution of inhibitors. Specific proteins inhibit the activity of serine proteases when their physiological role has ended (Table 9.4). Thus, coagulation is limited to the site of vascular injury and complementation activation leads to lysis only of cells exhibiting foreign antigens. Inability to control these proteases, such as by a deficiency of a specific inhibitor, can lead to undesirable consequences, such as thrombus formation in myocardial infarction and stroke or uncontrolled reactions of complement in autoimmune disease. These natural inhibitors of serine proteases are termed **serpins** for *ser*ine *p*rotease *in*hibitors. They occur in animals that produce the proteases, but surprisingly they are also found in plants that lack proteases.

Serine Proteases Have Similar Structure–Function Relationships

The relationships between structure and physiological function of the serine proteases can be summarized as follows: (1) Only one serine residue is catalytically active and

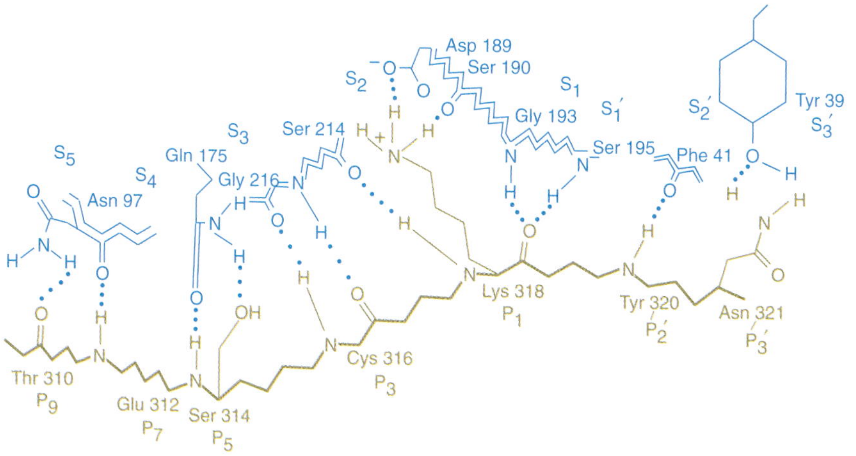

FIGURE 9.13

Schematic drawing of binding of pancreatic trypsin inhibitor to trypsin. Binding-site region of trypsin (blue) shown in complex with region of pancreatic trypsin inhibitor (gold). The inhibitor mimics the binding of a protein substrate. Dotted lines are hydrogen bonds. The inhibitor has an extended conformation so that amino acids P_9, P_7, P_5, P_3, P_1–P'_3 interact with binding subsites S_5–S'_3 in trypsin. Potentially hydrolyzable bond in inhibitor is between P_1–P'_1.
The diagram is based on X-ray diffraction of the complex of pancreatic trypsin inhibitor with trypsinogen. The binding region of trypsinogen in the complex assumes a conformation like that of active trypsin bound to a protein substrate.
Redrawn from Bolognesi, M., Gatti, B., Menegatti, E., Guarneri, M., Papamokos, E., and Huber, R. *J. Mol. Biol.* 162:839, 1983.

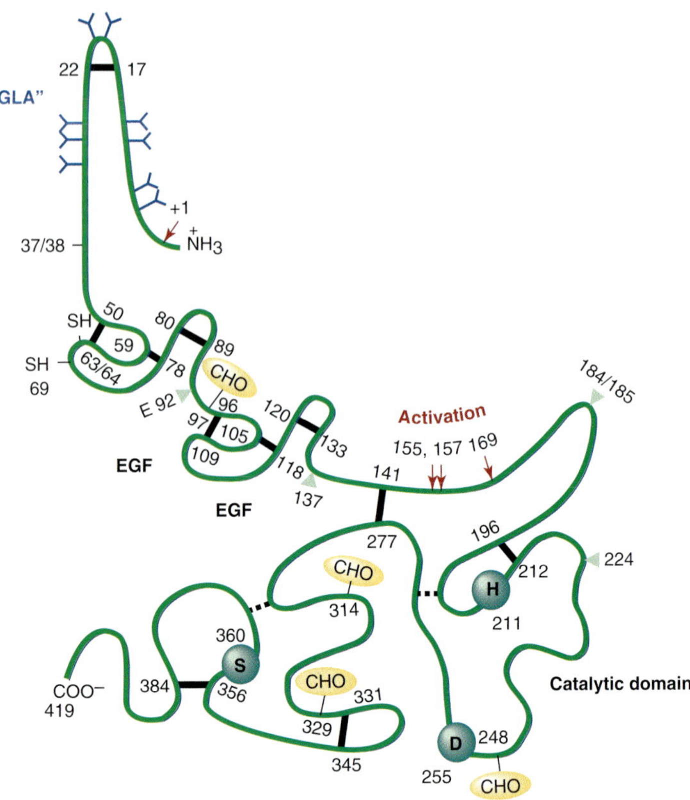

FIGURE 9.14

Schematic of domain structure for protein C showing multidomain structure. "GLA" refers to the γ-carboxy-glutamic residues (indicated by tree structures) in the NH_2-terminal domain, disulfide bridges are indicated by thick bars, EGF indicates positions of epidermal growth factor-like domains, and CHO indicates positions where sugar residues are joined to the polypeptide chain. Proteolytic cleavage sites leading to catalytic activation are shown by arrows. Amino acid sequence is numbered from NH_2-terminal end, and catalytic sites of serine, histidine, and aspartate are shown in the catalytic domains by circled one-letter abbreviations S, H, and D, respectively.
Redrawn from Long, G. L. *J. Cell. Biochem.* 33:185, 1987.

participates in peptide bond cleavage. Bovine trypsin contains 34 serine residues with only one catalytically active or able to react with the inhibitor DFP (see Figure 9.11). (2) X-ray diffraction and **sequence homology** studies demonstrate that a histidine and an aspartate are always associated with the activated serine in the catalytic site. Based on their positions in chymotrypsinogen, these three invariant residues are named Ser 195, His 57, and Asp 102 and called the "catalytic triad." This numbering, based on their sequence number in chymotrypsinogen, is used to identify these residues irrespective of their exact position in any serine protease. (3) Eukaryotic serine proteases exhibit high

TABLE 9.4 Some Human Proteins that Inhibit Serine Proteases

Inhibitor	*Action*
α_1-Proteinase inhibitor	Inhibits tissue proteases including neutrophil elastase; deficiency leads to pulmonary emphysema
α_1-Antichymotrypsin	Inhibits proteases of chymotrypsin-like specificity from neutrophils, basophils, and mast cells including cathepsin G and chymase
Inter-α-trypsin inhibitor	Inhibits broad range of serine protease activities in plasma
α_2-Antiplasmin	Inhibits plasmin
Antithrombin III	Inhibits thrombin and other coagulation proteases
C_1 Inhibitor	Inhibits complement reaction
α_2-Macroglobulin	General protease inhibitor
Protease nexin I	Inhibits thrombin, urokinase, and plasmin
Protease nexin II	Inhibits growth factor-associated serine proteases, identical to NH_2-terminal domain of amyloid protein secreted in Alzheimer's disease
Plasminogen activator inhibitor I	Inhibits plasminogen activators
Plasminogen activator inhibitor II	Inhibits urokinase plasminogen activator

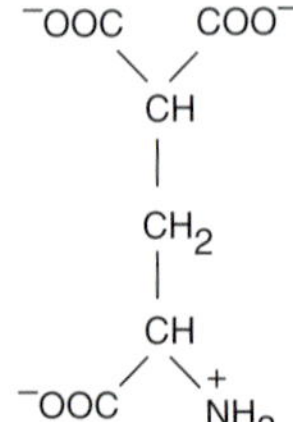

FIGURE 9.15

Structure of the derived amino acid γ-carboxyglutamic acid (abbreviation Gla), found in NH_2-terminal domain of many coagulation proteins.

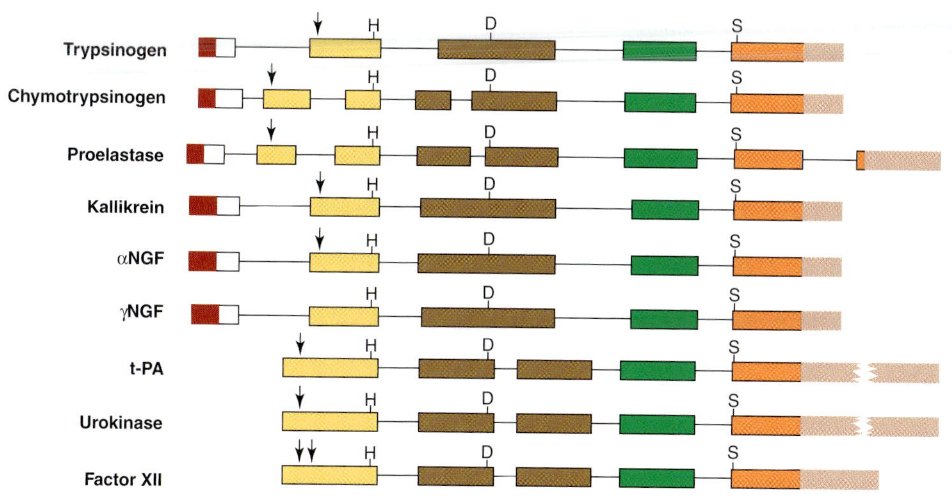

FIGURE 9.16

Organization of exons and introns in genes for serine proteases. t-PA is a tissue plasminogen activator, and NGF is a nerve growth factor that contains a structure homologous to a serine protease. Exons are shown by boxes and introns by connecting lines. Position of the codons for active-site serine, histidine, and aspartate are denoted by S, H, and D, respectively. Red boxes, on left, show regions that code for NH_2-terminal signal peptide that is cleaved before secretion. Light-colored boxes, on right, represent part of gene sequence transcribed into messenger RNA (mRNA), but not translated into protein. Arrows show codons for residues at which proteolytic activation of zymogen occurs. Redrawn and modified from Irwin, D. M., Roberts, K. A., and MacGillivray, R. T. *J. Mol. Biol.* 200:31, 1988.

sequence and structural similarity. (4) Genes that code for serine proteases are organized similarly (Figure 9.16). Their **exon–intron patterns** show that each of the catalytically essential residues (Ser 195, His 57, and Asp 102) are on a different exon. The essential histidine and serine are all almost adjacent to their exon boundary. The cross-species homology in gene structure supports the concept that the serine proteases evolved from a common primordial gene. (5) The catalytic portion of serine proteases comprises two domains, of approximately equal size. The catalytic site is within their interface or the crevice between the two domains. (6) Serine proteases that interact with membranes typically have an additional domain to provide for this specific function. (7) Natural protein substrates and inhibitors of serine proteases bind through an extended specificity site. (8) Specificity for natural protein inhibitors is marked by extremely tight binding. The binding constant for trypsin to pancreatic trypsin inhibitor is about 10^{13} M^{-1}, reflecting a binding free energy of approximately 18 kcal mol^{-1}. (9) Natural protein inhibitors are usually poor substrates with strong inhibition requiring hydrolysis of a peptide bond in the inhibitor by the protease. (10) Serine proteases in eukaryotes are synthesized as zymogens to permit their transport in an inactive state to their sites of action. (11) Zymogen activation frequently involves hydrolysis by another serine protease. (12) Several serine proteases undergo **autolysis** or self-hydrolysis. Sometimes this leads to specific peptide bond cleavage and further activation of the catalytic activity. At other times, autolysis leads to inactivation of the protease.

Sequence Homology in Serine Proteases

Much early knowledge of the serine protease family came from trypsin and chymotrypsin purified from bovine materials. This has yielded a useful but nonintuitive nomenclature, which uses a sequence alignment against the amino acid sequence of chymotrypsin, to name and number residues of other serine proteases. As mentioned previously, the catalytically essential residues are Ser-195, His-57, and Asp-102. Insertions and deletions of the amino acids in another serine protease are compared to the numbering of residues in chymotrypsin. Alignment is made by algorithms that maximize sequence homology, with exact alignment of the essential serine, histidine, and aspartate residues. These three residues are invariant in all serine proteases and the sequences surrounding them are invariant among the serine proteases of the chymotrypsin family (Table 9.5).

Ser-195 in chymotrypsin reacts with the inhibitor diisopropylfluorophosphate (DFP), with a 1:1 enzyme:DFP stoichiometry (p. 328 and Figure 9.11). The three-dimensional structure of chymotrypsin reveals that the Ser-195 is situated within an internal pocket, with access to the solvent interface. His-57 and Asp-102 are oriented so that they participate with the Ser-195 in the catalytic mechanism (see Chapter 10).

TABLE 9.5 Invariant Sequences Around Catalytically Essential Serine (S) and Histidine (H)

Enzyme	Sequence (Residues Identical to Chymotrypsin Are in Bold)																								
	Residues Around Catalytically Essential Histidine																								
Chymotrypsin A	F	**H**	**F**	**C**	**G**	**G**	**S**	**L**	**I**	**N**	**E**	**N**	**W**	**V**	**V**	**T**	**A**	**A**	**H***	**C**	G	V	T	T	S	D
Trypsin	Y	**H**	**F**	**C**	**G**	**G**	**S**	**L**	**I**	**N**	S	Q	**W**	**V**	**V**	S	**A**	**A**	**H**	**C**	Y	K	S	G	I	Q
Pancreatic elastase	A	**H**	T	**C**	**G**	**G**	T	**L**	**I**	R	Q	**N**	**W**	**V**	M	**T**	**A**	**A**	**H**	**C**	V	D	R	E	L	T
Thrombin	E	L	L	**C**	**G**	A	**S**	**L**	**I**	S	D	R	**W**	**V**	L	**T**	**A**	**A**	**H**	**C**	L	L	Y	P	P	W
Factor X	E	G	**F**	**C**	**G**	**G**	T	I	L	**N**	**E**	F	Y	**V**	L	**T**	**A**	**A**	**H**	**C**	L	H	Q	A	K	R
Plasmin	M	**H**	**F**	**C**	**G**	**G**	T	**L**	**I**	S	P	E	**W**	**V**	L	**T**	**A**	**A**	**H**	**C**	L	E	K	S	P	R
Plasma kallikrein	S	F	Q	**C**	**G**	**G**	V	**L**	V	**N**	P	K	**W**	**V**	L	**T**	**A**	**A**	**H**	**C**	K	N	D	N	Y	E
Streptomyces trypsin	—	—	—	**C**	**G**	**G**	A	**L**	Y	A	Q	D	I	**V**	L	**T**	**A**	**A**	**H**	**C**	V	S	G	S	G	N
Subtilisin	V	G	G	A	S	F	V	A	G	E	A	Y	N	T	D	G	N	G	**H**	G	T	H	V	A	G	T
	Residues Around Catalytically Essential Serine																								
Chymotrypsin A	**C**	**A**	**G**	—	—	—	A	S	**G**	**V**	—	—	**S**	**S**	**C**	**M**	**G**	**D**	**S***	**G**	**G**	**P**	**L**	**V**		
Trypsin	**C**	**A**	**G**	Y	—	—	L	E	**G**	G	K	—	D	**S**	**C**	Q	**G**	**D**	**S**	**G**	**G**	**P**	V	**V**		
Pancreatic elastase	**C**	**A**	**G**	—	—	—	G	N	**G**	**V**	R	—	**S**	G	**C**	Q	**G**	**D**	**S**	**G**	**G**	**P**	**L**	**H**		
Thrombin	**C**	**A**	**G**	Y	K	P	G	E	**G**	K	R	G	D	A	**C**	E	**G**	**D**	**S**	**G**	**G**	**P**	F	**V**		
Factor X	**C**	**A**	**G**	Y	—	—	D	T	**Q**	P	E	—	D	A	**C**	Q	**G**	**D**	**S**	**G**	**G**	**P**	H	**V**		
Plasmin	**C**	**A**	**G**	H	—	—	L	A	**G**	G	T	—	D	**S**	**C**	Q	**G**	**D**	**S**	**G**	**G**	**P**	**L**	**V**		
Plasma kallikrein	**C**	**A**	**G**	Y	—	—	L	P	**G**	G	K	—	D	T	**C**	**M**	**G**	**D**	**S**	**G**	**G**	**P**	**L**	I		
Streptomyces trypsin	**C**	**A**	**G**	Y	—	P	D	T	**G**	G	V	—	D	T	**C**	Q	**G**	**D**	**S**	**G**	**G**	**P**	M	F		
Subtilisin	A	G	V	Y	S	T	Y	P	T	N	T	Y	A	T	L	N	**G**	T	**S**	M	A	S	P	H		

Source: Barrett, A. J. In: A. J. Barrett and G. Salvesen (Eds.), *Proteinase Inhibitors*. Amsterdam: Elsevier, 1986, p. 7.

Members of the chymotrypsin family also occur in prokaryotes. Thus, bacterial serine proteases from *Streptomyces griseus* and *Myxobacteria* 450 are homologous with chymotrypsin. However, a separate class of serine proteases first discovered in bacteria has no structural homology to the mammalian chymotrypsin family. The serine protease subtilisin, isolated from *Bacillus subtilis*, hydrolyzes peptide bonds and contains an activated serine with a histidine and aspartate in its active site, but its active site arises from structural regions of the protein that have no sequence or structural homology with the chymotrypsin family. This is an example of **convergent evolution** of an enzyme catalytic mechanism. Apparently a completely different gene evolved the same catalytic mechanism for peptide hydrolysis.

Tertiary Structures of Serine Protease Family Are Similar

Structure determinations by X-ray crystallography have been carried out on many members of this protein family (Table 9.6), including active enzyme forms, zymogens, the same enzyme in multiple species, enzyme–inhibitor complexes, and a particular enzyme at different temperatures and in different solvents. A high-resolution analysis of trypsin has yielded a structure at better than 1.7-Å resolution, which can resolve atoms at a separation of 1.3 Å such as the C=O of the carbonyl group (1.2 Å). This resolution, however, is not uniform over the entire structure. Different regions vary in their tendency to be localized, and for some atoms their exact position cannot be as precisely defined as for others. The structural disorder is especially apparent in surface residues not in contact with neighboring molecules.

Trypsin is globular and contains two distinct domains of approximately equal size (Figure 9.17), which do not penetrate one another. There is little α-helix, except in the COOH-terminal region. The structure is predominantly β-structure, with each of the domains in a "deformed" β-barrel. Loop regions protrude from the barrel ends, being almost symmetrically presented by each of the two folded domains. These loops

TABLE 9.6 Serine Protease Structures Determined by X-Ray Crystallography

Enzyme	*Species Source*	*Inhibitors Present*	*Resolution (Å)*
Chymotrypsin[a]	Bovine	Yes[b]	1.67[c]
Chymotrypsinogen	Bovine	No	2.5
Elastase	Porcine	Yes	2.5
Kallikrein	Porcine	Yes	2.05
Proteinase A	*S. griseus*	No	1.5
Proteinase B	*S. griseus*	Yes	1.8
Proteinase II	Rat	No	1.9
Trypsin[a]	Bovine	Yes[b]	1.4[c]
Trypsinogen[a]	Bovine	Yes[b]	1.65[c]

[a] Structure of this enzyme molecule independently determined by two or more investigators.

[b] Structure obtained with no inhibitor present (native structure) and with inhibitors. Inhibitors used include low molecular weight inhibitors (i.e., benzamidine, DFP, and tosyl) and protein inhibitors (i.e., bovine pancreatic trypsin inhibitor).

[c] Highest resolution for this molecule of the multiple determinations.

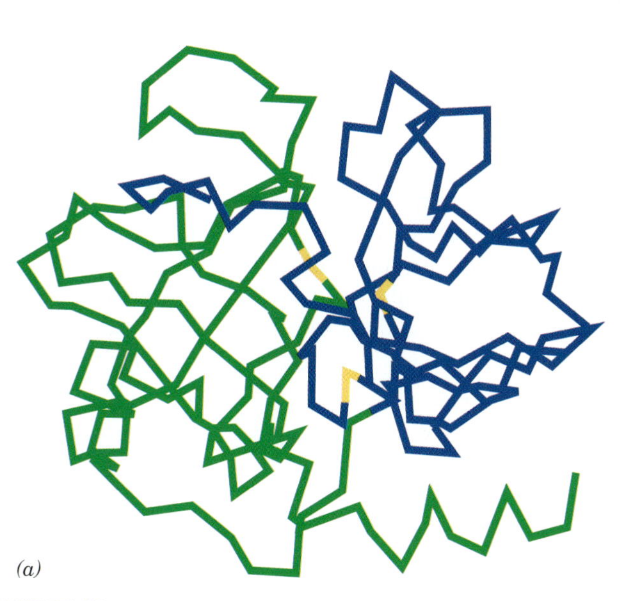

(a)

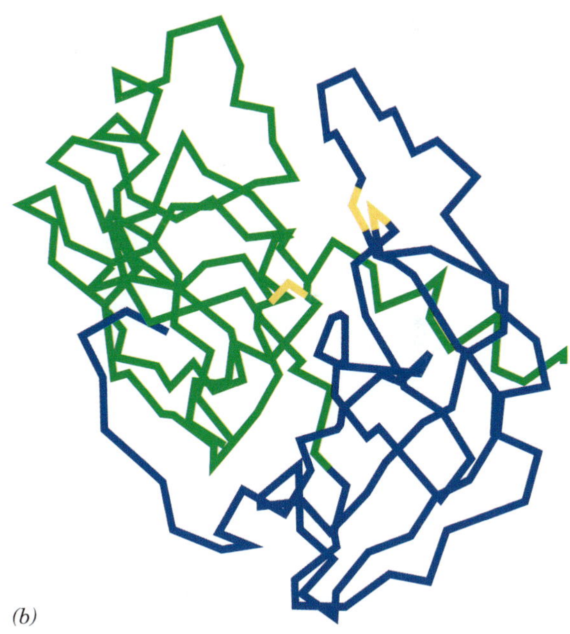

(b)

FIGURE 9.17

Two views of structure of trypsin showing fold structure of its catalytic domains. Active-site serine, histidine, and aspartate are indicated in yellow.

combine to form a surface region that extends outward, above the catalytic site, and are functionally similar to the CDRs of immunoglobulins.

Alignment of three-dimensional structures can be performed on serine proteases. Table 9.7 shows for sets of two proteases the number of amino acids that are superimposable in the three-dimensional structure and the number of amino acids that are chemically identical in comparison of the sequences. Structurally superimposable amino acids, even if they differ chemically, cannot be distinguished from one another at the resolution of the X-ray diffraction. The proteins are more homologous in their structures than in sequence. The regions of greatest difference are in the CDR-like loop regions. Altering these loops changes the **macromolecular binding specificity** of the protease. The structure of the loop in factor Xa, for example, allows it to bind specifically to

TABLE 9.7 Structural Superposition of Selected Trypsin Family Serine Proteases and the Resultant Amino Acid Sequence Comparison

Comparison	Number of Amino Acids in Sequence: Protease 1	Number of Amino Acids in Sequence: Protease 2	Number of Structurally Equivalent Residues	Number of Chemically Identical Residues
Trypsin–elastase	223	240	188	81
Trypsin–chymotrypsin	223	241	185	93
Trypsin–mast cell protease	223	224	188	69
Trypsin–prekallikrein	223	232	194	84
Trypsin–*S. griseus* protease	223	180	121	25

prothrombin. Serpins interact with different proteases through their affinity for the loop structures. Bacterial proteases related to the eukaryotic serine protease family contain the same two domains but lack most of the loops. This agrees with their lack of a requirement for complex interactions and the observation that bacterial proteases are not synthesized as zymogens.

Thus the serine protease family constitutes a structurally related series of proteins that use a catalytically active serine. During evolution, the basic two-domain structure and the catalytically essential residues have been maintained, but the regions of the secondary interactions (loop regions) have changed to provide for different specificities toward substrates, activators, and inhibitors, related to their diverse physiological roles.

9.4 HEMOGLOBIN AND MYOGLOBIN

Hemoglobins are globular proteins, present in high concentrations in red blood cells, that bind oxygen in the lungs and transport it to cells. They also transport CO_2 and H^+ from the tissues to the lungs, and they carry and release nitric oxide (NO) in the blood vessels of the tissues. NO is a potent vasodilator and inhibitor of platelet aggregation. The molecular structure and chemistry of the hemoglobins beautifully fit these complex and interrelated physiological roles.

Human Hemoglobin Occurs in Several Forms

A hemoglobin molecule consists of four polypeptide chains, two each of two different sequences. Each chain contains a heme prosthetic group that binds oxygen. In the binding of O_2 in hemoglobin, the four sites of O_2-binding act in a cooperative manner. The major form of human adult hemoglobin, **HbA$_1$**, consists of two α and two β chains ($\alpha_2\beta_2$). The α-polypeptide and the β-polypeptide contain 141 and 146 amino acids, respectively. The fetal form (**HbF**) (Figure 9.18) contains the

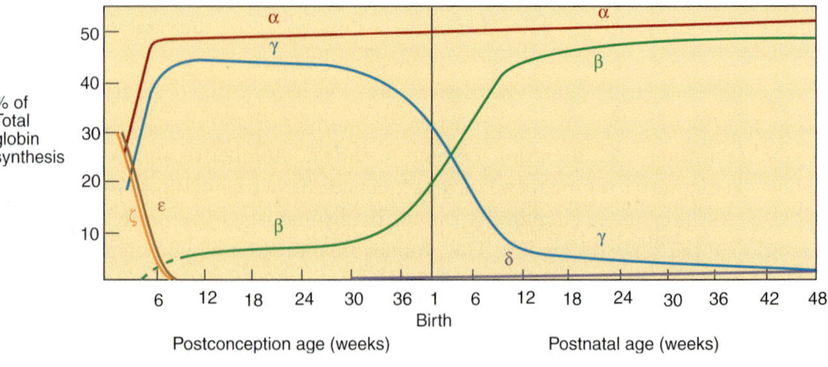

FIGURE 9.18

Changes in globin chain production during development.

Redrawn from Nienhuis, A. W. and Maniatis, T. In: G. Stamatoyannopoulos, A. W. Nienhuis, P. Leder, and P. W. Majerus (Eds.), *The Molecular Basis of Blood Diseases*. Philadelphia: Saunders, 1987, p. 68, where the following reference is acknowledged: Weatherall, D. J. and Clegg, J. B., *The Thalassemia Syndromes*, 3rd ed., Oxford: Blackwell Scientific Publications, 1981.

TABLE 9.8 Chains of Human Hemoglobin

Developmental Stage	*Symbol*	*Chain Composition*
Adult	HbA_1	$\alpha_2\beta_2$
Adult	HbA_2	$\alpha_2\delta_2$
Fetus	HbF	$\alpha_2\gamma_2$
Embryo	Hb Gower-1	$\zeta_2\epsilon_2$
Embryo	Hb Portland	$\zeta_2\gamma_2$

CLINICAL CORRELATION 9.6
Hemoglobinopathies

There are over 800 different mutant human hemoglobins. Mutations cause instability in hemoglobin structure, increased or decreased oxygen affinity, or an increase in the rate of oxidation of the heme ferrous iron (Fe^{2+}) to the ferric state (Fe^{3+}). A nonfunctional ferric hemoglobin is designated methemoglobin and is symbolized HbM.

Unstable hemoglobins arise by the substitution of proline for an amino acid within an α-helical region of the globin fold. Prolines do not participate in α-helical structure and the breaking of an α-helical segment causes an unstable hemoglobin (examples are $Hb_{Saki}{}^{\beta Leu14 \rightarrow Pro}$ and $Hb_{Genova}{}^{\beta 28Leu \rightarrow Pro}$). Other unstable hemoglobins arise from substitution by an amino acid that is either too large or small to make appropriate contacts, or by placement of a charged or polar group on the inside of a domain. Unstable hemoglobins easily denature and precipitate to form Heinz bodies which damage the erythrocyte membrane. Patients with unstable hemoglobins may develop anemia, reticulocytosis, splenomegaly, and urobilinuria.

Some mutant hemoglobins have an increased oxygen affinity (lower P_{50}). An interesting example is $Hb_{Cowtown}{}^{\beta His146 \rightarrow Leu}$, in which the histidine that dissociates 50% of the Bohr effect protons is lost. This mutation impedes the regulation of oxygen dissociation by hydrogen ion concentration and destabilizes the T conformation relative to the R conformation, causing an increase in oxygen affinity and decreased release of O_2 to the tissues. Hemoglobin mutations that impair DPG binding also increase oxygen affinity. Hemoglobinopathies of increased oxygen affinity are often characterized by hemolytic anemia and Heinz body formation.

Mutant hemoglobins that form methemoglobin include $HbM_{Iwate}{}^{\alpha 87His \rightarrow Tyr}$ and $HbM_{Hyde\ Park}{}^{\beta 92His \rightarrow Tyr}$ where the proximal histidines F8 in the α and β chains, respectively, are mutated. In $HbM_{\beta oston}{}^{\alpha 58His \rightarrow Tyr}$ and $HbM_{Saskatoon}{}^{\beta 63His \rightarrow Tyr}$ the distal histidines E7 are mutated. Mutations of amino acids that pack against the heme or form the oxygen binding site often lead to methemoglobinemia. Patients with high concentrations of methemoglobin show cyanosis (bluish color of the skin).

Two of the most prevalent mutations occur at the same amino acid position, the β6Glu. When this glutamate is replaced by valine, the result is $HbS^{\beta 6Glu \rightarrow Val}$; whereas replacement by lysine yields $HbC^{\beta 6Glu \rightarrow Lys}$. Homozygotes with HbS express sickle cell anemia in which the hemoglobin molecules precipitate as tactoids or long arrays, which produce the sickle shape of the erythrocyte (see Clin. Corr. 3.3). HbC forms a different aggregate structure consisting of blunt-ended crystalloids. This reduces the survival time of the red blood cells but produces less hemolysis than HbS. This form of hemoglobinopathy exhibits more limited pathological effects. As both HbS and HbC are commonly found among certain black African populations, it is not unusual to find individuals heterozygous for both mutant genes among this population. Individuals with HbSC will have an intermediate anemia between that observed of homozygotes for HbS and HbC.

Source: Dickerson, R. E. and Geis, I. *Hemoglobin: Structure, Function, Evolution, and Pathology*. Menlo Park, CA: Benjamin/Cummings, 1983. Arcasoy, M. O. and Gallagher, P. G. Molecular diagnosis of hemoglobinopathies and other red blood cell disorders. *Semin. Hematol.* 36:328, 1999.

same α chains as HbA_1, and γ chains that differ in sequence from the β chain (Table 9.8). Two other hemoglobin forms appear in the first months after conception (embryonic) in which the α chains are substituted by ζ chains of different sequence and the ε chains serve as the β chains. A minor adult hemoglobin, $\mathbf{HbA_2}$, comprises about 2% of normal adult hemoglobin and contains two α chains and two chains designated δ (Table 9.8). A discussion of abnormal hemoglobins is presented in Clinical Correlation 9.6.

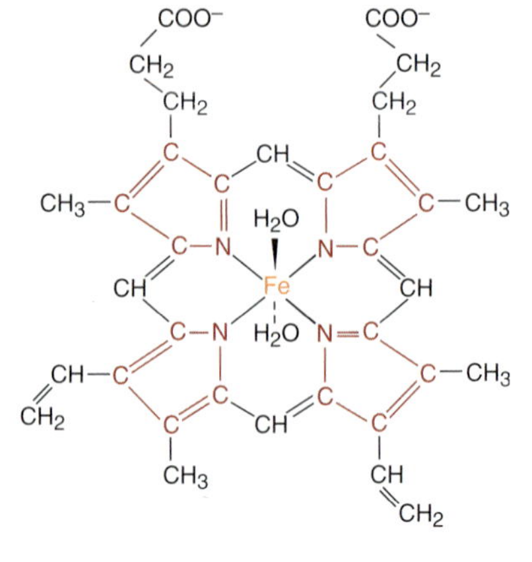

Heme

FIGURE 9.19

Structure of heme.

Myoglobin: A Single Polypeptide with One O_2-Binding Site

Myoglobin (Mb) is an O_2 carrying protein that binds and releases O_2 with changes in the O_2 concentration in the sarcoplasm of skeletal muscle cells. In contrast to hemoglobin, which has four chains and four O_2-binding sites, myoglobin contains only a single polypeptide chain and one O_2-binding site. Myoglobin is a model for what occurs when a single protomer molecule acts alone without the interactions exhibited among the four O_2-binding sites in the more complex tetramer molecule of hemoglobin.

Heme Prosthetic Group Is Site of O_2 Binding

The four globin subunits in hemoglobin and the one of myoglobin each contain a heme prosthetic group. A **prosthetic group** is a nonpolypeptide moiety that forms a functional part of a protein. Without its prosthetic group, a protein is designated an **apoprotein**. With its prosthetic group, it is a **holoprotein**.

Heme is protoporphyrin IX (see p. 833) with an iron atom in its center (Figure 9.19). The iron is in the ferrous (2^+ charge) state in functional hemoglobin and myoglobin, where it can form five or six covalent bonds, depending on whether or not O_2 is bound to it. Four bonds are to the pyrrole nitrogen atoms of the porphyrin. As the pyrrole rings and connecting carbons are part of the same aromatic system, these atoms lie in a common plane. The porphyrin bonded iron will tend to lie in this same porphyrin plane. The fifth and the potentially sixth bonds to iron are directed along an axis perpendicular to the plane of the porphyrin ring (Figure 9.20). The fifth bond is to a nitrogen of a histidine imidazole. This is designated the **proximal histidine** in hemoglobin and myoglobin structures (Figures 9.20 and 9.21). O_2 forms the sixth bond; the O_2 is located between the ferrous atom and a second histidine imidazole, designated the **distal histidine**. In deoxyhemoglobin, the sixth position is unoccupied.

The heme is positioned within a hydrophobic pocket of each globin subunit, with approximately 80 interactions provided by about 18 residues mostly between apolar side chains and the apolar regions of the porphyrin. As discussed in Chapter 3, the driving force for these interactions is the expulsion of water of solvation on association of the hydrophobic heme with the apolar side chains in the heme pocket. Additional interactions in myoglobin are made between the negatively charged propionate groups of the heme and positively charged arginine and histidine side chains. However, in hemoglobin chains a difference in the amino acid sequence in this region of the heme binding site leads to stabilization of the porphyrin propionates by interaction with an

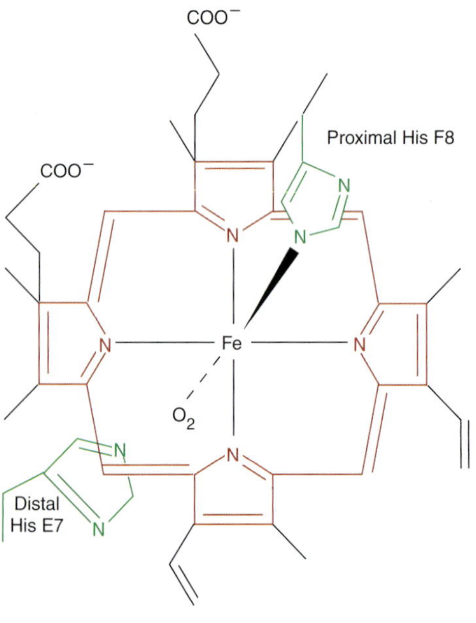

FIGURE 9.20

Ligand bonds to ferrous atom in oxyhemoglobin.

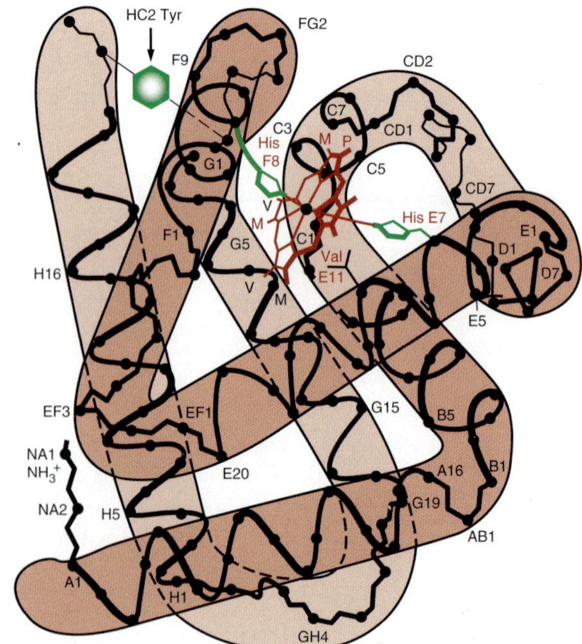

FIGURE 9.21

Secondary and tertiary structure characteristics of globins of hemoglobin.
Proximal His F8, distal His E7, and Val E11 side chains are shown. Other amino acids of polypeptide chain are represented by α-carbon positions only; the letters M, V, and P refer to the methyl, vinyl, and propionate side chains of the heme.
Reprinted with permission from Perutz, M. *Br. Med. Bull.* 32:195, 1976.

uncharged histidine imidazole and with water molecules of solvent toward the outer surface of the molecule.

X-Ray Crystallography Has Defined the Structure of Hemoglobin and Myoglobin

The structures of deoxy and oxy forms of hemoglobin and myoglobin have been resolved by X-ray crystallography. In fact, sperm whale myoglobin was the first globular protein whose structure was determined by this technique, followed by horse hemoglobin. These structures show that each globin consists of multiple α-helical regions connected by turns that allow folding into a spheroidal shape characteristic of the globin fold (Figure 9.21; also see Figure 3.33). The mechanism for the cooperative association of O_2 is based on the X-ray structures of oxyhemoglobin, deoxyhemoglobin, and a variety of hemoglobin derivatives.

Primary, Secondary, and Tertiary Structures of Myoglobin and Hemoglobin Chains

The amino acid sequences of the globin of myoglobin of many animal species have been determined. All myoglobins contain 153 amino acids, of which 83 are invariant. Only 15 invariant residues are identical to the invariant residues of the mammalian hemoglobins. However, the changes are mostly conservative and preserve the physical properties of the residues (Table 9.9). Since myoglobin is a monomer, many of its surface positions interact with water and prevent another molecule of myoglobin from associating. In contrast, surface residues of the individual subunits in hemoglobin provide hydrogen bonds and nonpolar contacts with other subunits in the quaternary structure. The proximal and distal histidines are preserved in all the globin chains, as are the hydrophobic heme pockets that form essential nonpolar contacts with the heme.

While there is surprising variability in sequences, the secondary and tertiary structures are almost identical (Figure 9.22). Any significant differences in the physiological properties between α, β, γ, and δ chains of hemoglobins and the chain of myoglobin are due to rather small specific changes in their structures. The similarity in tertiary structure shows that the same fold structure of a polypeptide can be arrived at by different sequences.

Approximately 70% of the residues participate in the α-helices of the globin fold, generating seven α-helices in the α chain and eight in the β chain. These latter eight helical regions are commonly labeled A-H, starting from the NH_2-terminal end. The interhelical regions are designated as AB, BC, CD, . . ., GH, respectively. The nonhelical region at the NH_2-terminal end is designated NA; and that at the COOH-terminal end is HC (Figure 9.21). Amino acid residues are sequentially numbered from 1 in each of the lettered helices and interhelical regions. This numbering system allows discussion of particular regions and residues that have similar functional and structural roles in hemoglobin and myoglobin.

A Simple Equilibrium Defines O_2 Binding to Myoglobin

The association of oxygen to myoglobin is characterized by a simple equilibrium constant (Eqs. 9.1 and 9.2). In Eq. 9.2, $[MbO_2]$ is the solution concentration of oxymyoglobin, [Mb] is that of deoxymyoglobin, and $[O_2]$ is that of oxygen, in units of moles per liter. The equilibrium constant, K_{eq}, also has units of moles per liter. As for any true equilibrium constant, the value of K_{eq} is dependent on pH, ionic strength, and temperature.

$$\text{Mb} + \text{O}_2 \overset{K_{eq}}{\rightleftharpoons} \text{MbO}_2 \tag{9.1}$$

$$K_{eq} = \frac{[\text{Mb}][\text{O}_2]}{[\text{MbO}_2]} \tag{9.2}$$

Since oxygen is a gas, it is more convenient to express its concentration as the pressure of oxygen in torr (1 torr equals the pressure of 1 mm Hg at 0°C and standard gravity).

TABLE 9.9 Amino Acid Sequences of Hemoglobin Chains and Myoglobin[a]

	NA 1	2	3	A 1	2	3	4	5	6	7	8	9	10	11	12	13	14	15	A 16	AB 1	B 1	2	3	4	5	6
MYOGLOBIN	Val	...	Leu	Ser	Glu	Gly	Glu	Trp	Gln	Leu	Val	Leu	His	Val	Trp	Ala	Lys	Val	Glu	Ala	Asp	Val	Ala	Gly	His	Gly
HEMOGLOBIN Horse α	Val	...	Leu	Ser	Ala	Ala	Asp	Lys	Thr	Asn	Val	Lys	Ala	Ala	Trp	Ser	Lys	Val	Gly	Gly	His	Ala	Gly	Glu	Tyr	Gly
β	Val	Gln	Leu	Ser	Gly	Glu	Glu	Lys	Ala	Ala	Val	Leu	Ala	Leu	Trp	Asp	Lys	Val	Asn	...	...	Glu	Glu	Glu	Val	Gly
Human α	Val	...	Leu	Ser	Pro	Ala	Asp	Lys	Thr	Asn	Val	Lys	Ala	Ala	Trp	Gly	Lys	Val	Gly	Ala	His	Ala	Gly	Glu	Tyr	Gly
β	Val	His	Leu	Thr	Pro	Glu	Glu	Lys	Ser	Ala	Val	Thr	Ala	Leu	Trp	Gly	Lys	Val	Asn	...	...	Val	Asp	Glu	Val	Gly
γ	Gly	His	Phe	Thr	Glu	Glu	Asp	Lys	Ala	Thr	Ilu	Thr	Ser	Leu	Trp	Gly	Lys	Val	Asn	...	...	Val	Glu	Asp	Ala	Gly
δ	Val	His	Leu	Thr	Pro	Glu	Glu	Lys	Thr	Ala	Val	Asn	Ala	Leu	Trp	Gly	Lys	Val	Asn	...	...	Val	Asp	Ala	Val	Gly
	7	8	9	10	11	12	13	14	15	16	C 1	2	3	4	5	6	7	CD 1	2	3	4	5	6	7	8	D 1
MYOGLOBIN	Gln	Asp	Ilu	Leu	Ilu	Arg	Leu	Phe	Lys	Ser	His	Pro	Glu	Thr	Leu	Glu	Lys	Phe	Asp	Arg	Phe	Lys	His	Leu	Lys	Thr
HEMOGLOBIN Horse α	Ala	Glu	Ala	Leu	Glu	Arg	Met	Phe	Leu	Gly	Phe	Pro	Thr	Thr	Lys	Thr	Tyr	Phe	Pro	His	Phe	...	Asp	Leu	Ser	His
β	Gly	Glu	Ala	Leu	Gly	Arg	Leu	Leu	Val	Val	Tyr	Pro	Trp	Thr	Gln	Arg	Phe	Phe	Asp	Ser	Phe	Gly	Asp	Leu	Ser	Gly
Human α	Ala	Glu	Ala	Leu	Glu	Arg	Met	Phe	Leu	Ser	Phe	Pro	Thr	Thr	Lys	Thr	Tyr	Phe	Pro	His	Phe	...	Asp	Leu	Ser	His
β	Gly	Glu	Ala	Leu	Gly	Arg	Leu	Leu	Val	Val	Tyr	Pro	Trp	Thr	Gln	Arg	Phe	Phe	Glu	Ser	Phe	Gly	Asp	Leu	Ser	Thr
γ	Gly	Glu	Thr	Leu	Gly	Arg	Leu	Leu	Val	Val	Tyr	Pro	Trp	Thr	Gln	Arg	Phe	Phe	Asp	Ser	Phe	Gly	Asn	Leu	Ser	Ser
δ	Gly	Glu	Ala	Leu	Gly	Arg	Leu	Leu	Val	Val	Tyr	Pro	Trp	Thr	Gln	Arg	Phe	Phe	Glu	Ser	Phe	Gly	Asp	Leu	Ser	Ser
	2	3	4	5	6	7	E 1	2	3	4	5	6	7	8	9	10	11	12	13	14	E 15	16	17	18	19	20
MYOGLOBIN	Glu	Ala	Glu	Met	Lys	Ala	Ser	Glu	Asp	Leu	Lys	Lys	His	Gly	Val	Thr	Val	Leu	Thr	Ala	Leu	Gly	Ala	Ilu	Leu	Lys
HEMOGLOBIN Horse α	...	...	...	...	...	Gly	Ser	Ala	Gln	Val	Lys	Ala	His	Gly	Lys	Lys	Val	Ala	Asp	Gly	Leu	Thr	Leu	Ala	Val	Gly
β	Pro	Asp	Ala	Val	Met	Gly	Asn	Pro	Lys	Val	Lys	Ala	His	Gly	Lys	Lys	Val	Leu	His	Ser	Phe	Gly	Glu	Gly	Val	His
Human α	...	...	...	...	...	Gly	Ser	Ala	Gln	Val	Lys	Gly	His	Gly	Lys	Lys	Val	Ala	Asp	Ala	Leu	Thr	Asn	Ala	Val	Ala
β	Pro	Asp	Ala	Val	Met	Gly	Asn	Pro	Lys	Val	Lys	Ala	His	Gly	Lys	Lys	Val	Leu	Gly	Ala	Phe	Ser	Asp	Gly	Leu	Ala
γ	Ala	Der	Ala	Ilu	Met	Gly	Asn	Pro	Lys	Val	Lys	Ala	His	Gly	Lys	Lys	Val	Leu	Thr	Ser	Leu	Gly	Asp	Ala	Ilu	Lys
δ	Pro	Asp	Ala	Val	Met	Gly	Asn	Pro	Lys	Val	Lys	Ala	His	Gly	Lys	Lys	Val	Leu	Gly	Ala	Phe	Ser	Asp	Gly	Leu	Ala
	EF 1	2	3	4	5	6	7	8	F 1	2	3	4	F 5	6	7	8	9	FG 1	2	3	4	5	G 1	2	3	4
MYOGLOBIN	Lys	Lys	Gly	His	His	Glu	Ala	Glu	Leu	Lys	Pro	Leu	Ala	Gln	Ser	His	Ala	Thr	Lys	His	Lys	Ilu	Pro	Ilu	Lys	Tyr
HEMOGLOBIN Horse α	His	Leu	Asp	Asp	Leu	Pro	Gly	Ala	Leu	Ser	Asp	Leu	Ser	Asn	Leu	His	Ala	His	Lys	Leu	Arg	Val	Asp	Pro	Val	Asn
β	His	Leu	Asp	Asn	Leu	Lys	Gly	Thr	Phe	Ala	Ala	Leu	Ser	Glu	Leu	His	Cys	Asp	Lys	Leu	His	Val	Asp	Pro	Glu	Asn
Human α	His	Val	Asp	Asp	Met	Pro	Asn	Ala	Leu	Ser	Ala	Leu	Ser	Asp	Leu	His	Ala	His	Lys	Leu	Arg	Val	Asp	Pro	Val	Asn
β	His	Leu	Asp	Asn	Leu	Lys	Gly	Thr	Phe	Ala	Thr	Leu	Ser	Glu	Leu	His	Cys	Asp	Lys	Leu	His	Val	Asp	Pro	Glu	Asn
γ	His	Leu	Asp	Asp	Leu	Lys	Gly	Thr	Phe	Ala	Gln	Leu	Ser	Glu	Leu	His	Cys	Asp	Lys	Leu	His	Val	Asp	Pro	Glu	Asn
δ	His	Leu	Asn	Asp	Leu	Lys	Gly	Thr	Phe	Ser	Gln	Leu	Ser	Glu	Leu	His	Cys	Asp	Lys	Leu	His	Val	Asp	Pro	Glu	Asn
	5	6	7	8	G 9	10	11	12	13	14	15	16	17	18	19	GH 1	2	3	4	5	6	H 1	2	H 3	4	5
MYOGLOBIN	Leu	Glu	Phe	Ilu	Ser	Glu	Ala	Ilu	Ilu	His	Val	Leu	His	Ser	Arg	His	Pro	Gly	Asn	Phe	Gly	Ala	Asp	Ala	Gln	Gly
HEMOGLOBIN Horse α	Phe	Lys	Leu	Leu	Ser	His	Cys	Leu	Leu	Ser	Thr	Leu	Ala	Val	His	Leu	Pro	Asn	Asp	Phe	Thr	Pro	Ala	Val	His	Ala
β	Phe	Arg	Leu	Leu	Gly	Asn	Val	Leu	Ala	Leu	Val	Val	Ala	Arg	His	Phe	Gly	Lys	Asp	Phe	Thr	Pro	Glu	Leu	Gln	Ala
Human α	Phe	Lys	Leu	Leu	Ser	His	Cys	Leu	Leu	Val	Thr	Leu	Ala	Ala	His	Leu	Pro	Ala	Glu	Phe	Thr	Pro	Ala	Val	His	Ala
β	Phe	Srg	Leu	Leu	Gly	Asn	Val	Leu	Val	Cys	Val	Leu	Ala	His	His	Phe	Gly	Lys	Glu	Phe	Thr	Pro	Pro	Val	Gln	Ala
γ	Phe	Lys	Leu	Leu	Gly	Asn	Val	Leu	Val	Thr	Val	Leu	Ala	Ilu	His	Phe	Gly	Lys	Glu	Phe	Thr	Pro	Glu	Val	Gln	Ala
δ	Phe	Arg	Leu	Leu	Gly	Asn	Val	Leu	Val	Cys	Val	Leu	Ala	Arg	Asn	Phe	Gly	Lys	Glu	Phe	Thr	Pro	Gln	Met	Gln	Ala
	6	7	8	9	10	11	12	13	14	15	16	17	18	19	20	H 21	22	23	2	HC 1	2	3	4	5		
MYOGLOBIN	Ala	Met	Asn	Lys	Ala	Leu	Glu	Leu	Phe	Arg	Lys	Asp	Ilu	Ala	Ala	Lys	Tyr	Lys	Glu	Leu	Gly	Tyr	Gln	Gly		
HEMOGLOBIN Horse α	Ser	Leu	Asp	Lys	Phe	Leu	Ser	Ser	Val	Ser	Thr	Val	Leu	Thr	Ser	Lys	Tyr	Arg								
β	Ser	Tyr	Gln	Lys	Val	Val	Ala	Gly	Val	Ala	Asn	Ala	Leu	Ala	His	Lys	Tyr	His								
Human α	Ser	Leu	Asp	Lys	Phe	Leu	Ala	Ser	Val	Ser	Thr	Val	Leu	Thr	Ser	Lys	Tyr	Arg								
β	Ala	Tyr	Gln	Lys	Val	Val	Ala	Gly	Val	Ala	Asn	Ala	Leu	Ala	His	Lys	Tyr	His								
γ	Ser	Trp	Gln	Lys	Met	Val	Thr	Gly	Val	Ala	Ser	Ala	Leu	Ser	Ser	Arg	Tyr	His								
δ	Ala	Tyr	Gln	Lys	Val	Val	Ala	Gly	Val	Ala	Asn	Ala	Leu	Ala	His	Lys	Tyr	His								

Source: Based on diagram in Dickerson, R. E. and Geis, I. *The Structure and Function of Proteins*. New York: Harper & Row, 1969, p. 52.

[a] Myoglobin from sperm whale. Residues that are identical are enclosed in box. A, B, C, ... designate different helices of tertiary structure (see text).

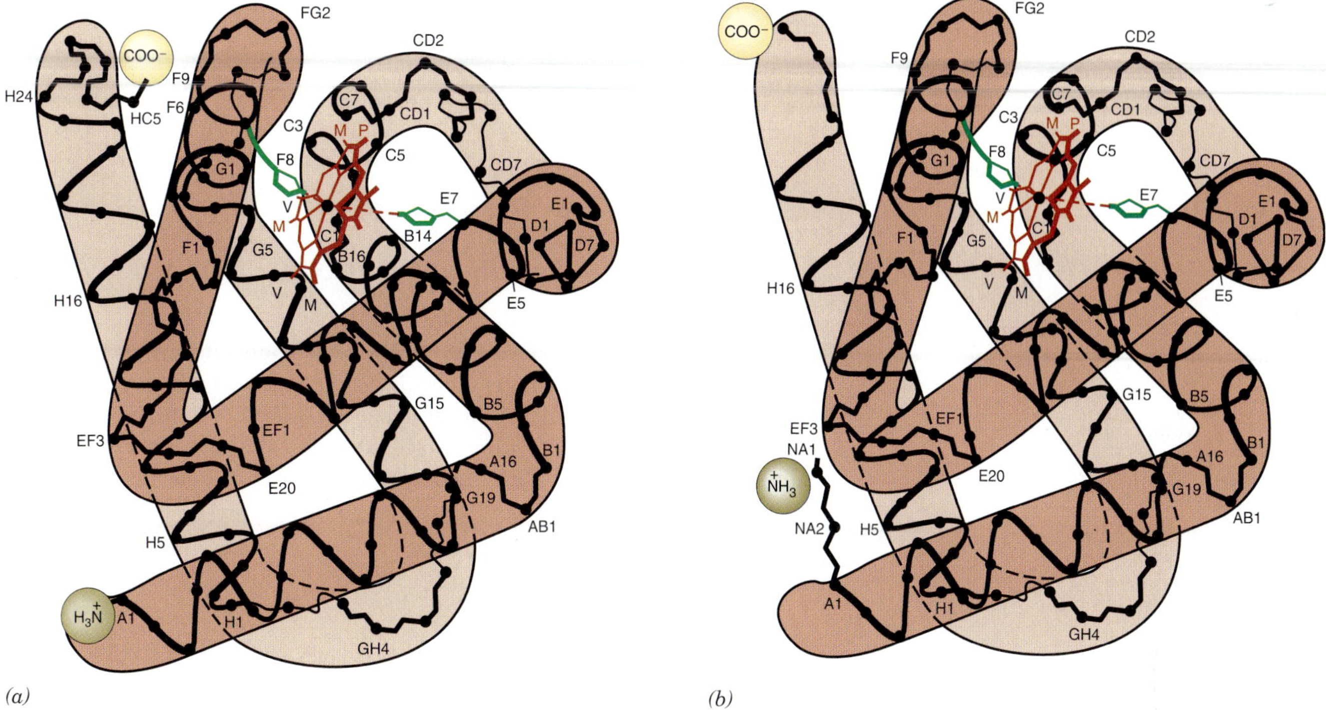

FIGURE 9.22

Comparison of conformation of (a) myoglobin and (b) β chain of HbA_1. Overall structures are very similar, except at NH_2-terminal and COOH-terminal ends.
Reprinted with permission from Fersht, A. *Enzyme Structure and Mechanism*. San Francisco: Freeman, 1977, pp. 12, 13.

In Eq. 9.3 this conversion of units has been made as shown by the change in symbols: $\boldsymbol{P_{50}}$, the equilibrium constant, and pO_2, the concentration of oxygen, being expressed in torr.

$$P_{50} = \frac{[\mathrm{Mb}] \cdot pO_2}{[\mathrm{MbO_2}]} \tag{9.3}$$

In an oxygen-binding or saturation curve the fraction of oxygen-binding sites that contain oxygen (Y, Eq. 9.4) is plotted on the ordinate versus pO_2 (oxygen concentration) on the abscissa. The Y value is defined for myoglobin by Eq. 9.5. Substitution into Eq. 9.5 of the value of $[MbO_2]$ obtained from Eq. 9.3, and then dividing through by [Mb], results in Eq. 9.6, which shows the dependence of Y on the value of the equilibrium constant, P_{50}, and on pO_2. It is seen from Eqs. 9.3 and 9.6 that the value of P_{50} is equal to pO_2, when $Y = 0.5$ (50% of the available sites occupied)—hence the designation of the equilibrium constant as P_{50}.

$$Y = \frac{\text{number of binding sites occupied}}{\text{total number of binding sites in solution}} \tag{9.4}$$

$$Y = \frac{[\mathrm{MbO_2}]}{[\mathrm{Mb}] + [\mathrm{MbO_2}]} \tag{9.5}$$

$$Y = \frac{pO_2}{P_{50} + pO_2} \tag{9.6}$$

A plot of Eq. 9.6 for Y versus pO_2 generates an oxygen–saturation curve for myoglobin and has the form of a rectangular hyperbola (Figure 9.23).

A simple algebraic manipulation of Eq. 9.6 leads to Eq. 9.7. Taking the logarithm of both sides of Eq. 9.7 results in Eq. 9.8, the **Hill equation**. A plot of $\log([Y/(1-Y)]$ versus $\log pO_2$, according to Eq. 9.8, yields a straight line with a slope equal to 1 for

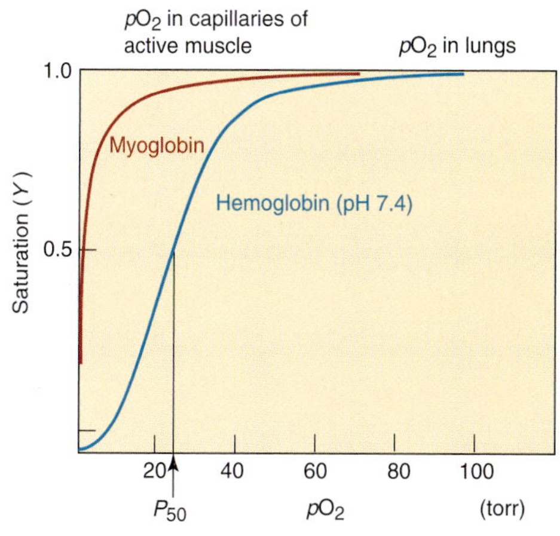

FIGURE 9.23

Oxygen-binding curves for myoglobin and hemoglobin.

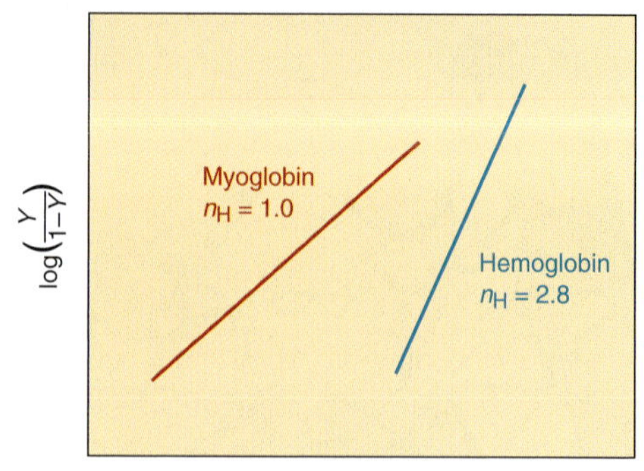

FIGURE 9.24

Hill plots for myoglobin and hemoglobin A_1.

myoglobin (Figure 9.24). This is the Hill plot, and the slope (n_H) is the **Hill coefficient** (see Eq. 9.9).

$$\frac{Y}{1-Y} = \frac{pO_2}{P_{50}} \tag{9.7}$$

$$\log \frac{Y}{1-Y} = \log pO_2 - \log P_{50} \tag{9.8}$$

Binding of O_2 to Hemoglobin Involves Cooperativity between Subunits

Binding of four O_2 to hemoglobin manifests as **positive cooperativity**, since binding of the first O_2 facilitates binding of O_2 to the other subunits. Conversely, dissociation of the first O_2 from $Hb(O_2)_4$ will make dissociation of O_2 from the other subunits easier.

Because of this cooperativity, the oxygen saturation curve for hemoglobin differs from that for myoglobin. A plot of Y versus pO_2 for hemoglobin is sigmoidal, indicating cooperativity in oxygen association (Figure 9.23). A plot of the Hill equation (Eq. 9.9) gives a slope (n_H) equal to 2.8 (Figure 9.24).

$$\log \frac{Y}{1-Y} = n_H \log pO_2 - \text{constant} \tag{9.9}$$

The meaning of the Hill coefficient to cooperative O_2 association can be evaluated quantitatively as presented in Table 9.10. A **cooperativity index**, R_x, is calculated, which shows the fold-change of pO_2 required to change Y from a value of $Y = 0.1$ (10% of sites filled) to a value of $Y = 0.9$ (90% of sites filled) for Hill coefficient values found experimentally. For myoglobin $n_H = 1$ and an 81-fold change in oxygen concentration is required to change from $Y = 0.1$ to $Y = 0.9$. For hemoglobin, $n_H = 2.8$ and only a 4.8-fold change in oxygen concentration is required for the same change in Y.

TABLE 9.10 Relationship Between Hill Coefficient (n_H) and Cooperativity Index (R_x)

n_H	R_x	*Observation*
0.5	6560	Negative substrate cooperativity
0.6	1520	
0.7	533	
0.8	243	
0.9	132	
1.0	81.0	Noncooperativity
1.5	18.7	Positive substrate cooperativity
2.0	9.0	
2.8	4.8	
3.5	3.5	
6.0	2.1	
10.0	1.6	
20.0	1.3	

Source: Based on Table 7.1 in Cornish-Bowden, A. *Principles of Enzyme Kinetics*. London: Butterworths Scientific Publishers, 1976.

Molecular Mechanism of Cooperativity in O_2 Binding

X-ray diffraction data on deoxyhemoglobin show that the ferrous atoms actually sit out of the plane of their porphyrins by about 0.4–0.6 Å. The electronic configuration of the five-coordinated ferrous atom in deoxyhemoglobin has a slightly larger radius than the distance from the center of the porphyrin to each of the pyrrole nitrogen atoms. Accordingly, the iron can be placed in the center of the porphyrin only with some distortion of the porphyrin conformation. Probably more important is that if the iron atom sits in the plane of the porphyrin, the proximal His F8 imidazole will interact unfavorably with the porphyrin. This unfavorable steric interaction is due, in part, to conformational constraints on the His F8 and the porphyrin in the deoxyhemoglobin conformation that forces the approach of the His F8 toward the porphyrin to a particular path (Figure 9.25). These constraints become less significant in oxyhemoglobin.

With the iron atom out of the plane of the porphyrin, the conformation is unstrained and energetically favored for the five-coordinate ferrous atom. When O_2 forms the sixth bond of the iron, however, this conformation becomes strained. A more energetically favorable conformation for the O_2-bonded iron is that in which the iron atom is within the plane of the porphyrin.

The binding energy of O_2 overcomes the repulsive interaction between His F8 and porphyrin, and the ferrous atom moves into the plane of the porphyrin. This is the most thermodynamically stable position for the oxy-heme, the six-bonded iron atom having one axial ligand on either side of the plane of the porphyrin, and the steric repulsion of one axial ligand is balanced by the repulsion of the second axial ligand on the opposite side when the ferrous atom is in the center. Also, the radius of the iron atom with six bonds is reduced so that it can just fit into the center of the porphyrin without distortion of the porphyrin conformation.

Since steric repulsion between porphyrin and His F8 in the deoxy conformation must be overcome on O_2 association, binding of the first O_2 is characterized by a relatively low affinity constant. However, on binding of O_2 to the first heme, the change of the iron into the center of the porphyrin triggers a conformational change in other

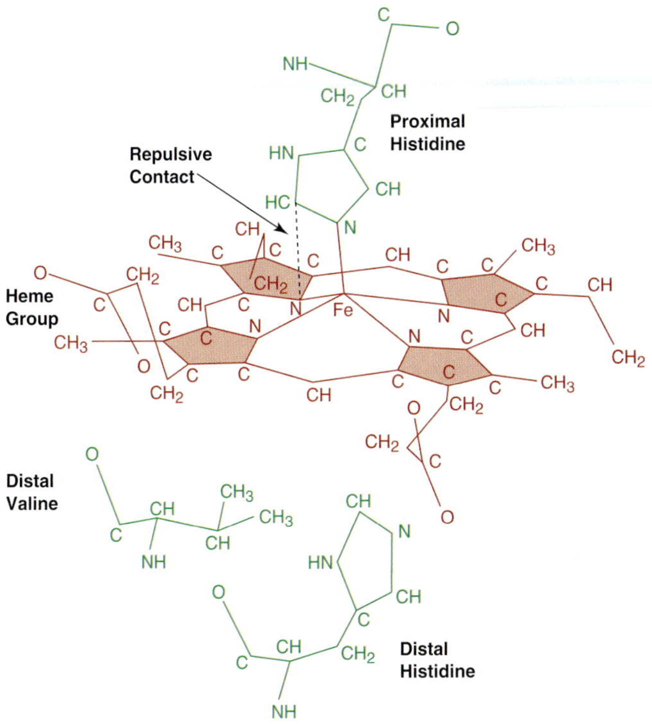

FIGURE 9.25

Steric hindrance between proximal histidine and porphyrin in deoxyhemoglobin.

Redrawn from Perutz, M. *Sci. Am.* 239:92, 1978.

subunits of the hemoglobin molecule. This change results in a greater affinity of O_2 to other empty heme sites.

The conformation change involves the following sequence of events: (i) The binding of O_2 pulls the Fe^{2+} into the porphyrin plane. (ii) The connected His F8 moves toward the porphyrin. (iii) The binding of O_2 changes the position of the F-helix, of which the His F8 is a part. (iv) The position of the FG corner of the subunit changes, destabilizing the FG noncovalent interaction with the C-helix of the adjacent subunit at an $\alpha_1\beta_2$ or $\alpha_2\beta_1$ subunit interface (Figures 9.26 and 9.27). (v) The adjacent globin structure changes to a conformation that allows O_2 to bind to its heme with higher affinity.

The FG to C intersubunit contacts act as a "switch" between two different arrangements between the FG corner of one subunit and the C-helix of the adjacent subunit. The new conformation allows the His F8 residues to approach their porphyrins on O_2 association with less steric repulsion (Figure 9.27). Thus, O_2 binds to the empty hemes in the modified conformation with higher affinity. In addition, ionic interactions that stabilize the deoxy-conformation (Figure 9.28) are broken by the conformation change induced by the binding of O_2 to one of the hemes, allowing the R conformation in other subunits to more easily form.

The deoxy conformation of hemoglobin is the "tense" or **T conformational state.** The oxy form with higher O_2 affinity is the "relaxed" or **R conformational state.** The allosteric mechanism describes how initial binding of the O_2 to one of the heme subunits of the tetrameric molecule pushes the molecular conformation from the T to R conformational state. The affinity constant for O_2 is greater in the R state by a factor of 150–300, depending on the solution conditions.

Hemoglobin Facilitates Transport of CO_2 and NO

The survival of cells relies on the ability of hemoglobin in red blood cells to deliver O_2 for cellular metabolism and to facilitate the transport of CO_2 away from the cells to the lung. Hemoglobin also binds the vasodilator, NO, and delivers it to the blood vessel walls in the tissues. The T to R conformational equilibrium of the hemoglobin molecule controls the delivery of O_2, CO_2, and NO to their appropriate sites. The $T \leftrightarrows R$ equilibrium is regulated by the hydrogen ion concentration, which links the delivery of O_2 and NO to the tissues with the transport of CO_2 away from the tissues.

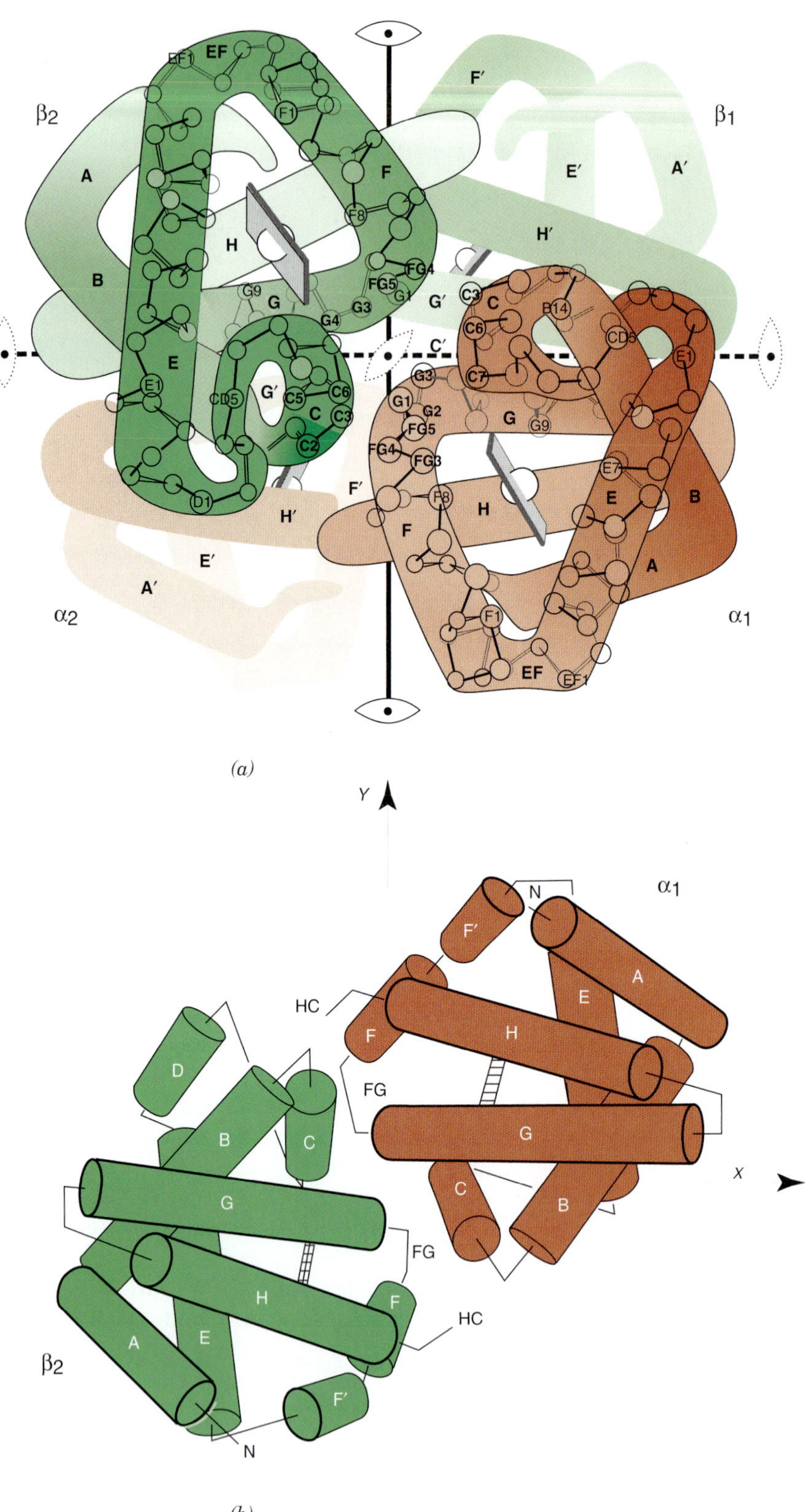

FIGURE 9.26

Quaternary structure of hemoglobin showing FG corner C-helix interactions across α_1–β_2 interface. (*a*) $\alpha_1\beta_2$ interface contacts between FG corners and C helix are shown. (*b*) Cylinder representation of α_1 and β_2 subunits in hemoglobin showing α_1 and β_2 interface contacts between FG corner and C-helix, viewed from opposite side of x–y plane from (a).

Part (*a*) reprinted with permission from Dickerson, R. E., and Geis, I. The Structure and Action of Proteins., Menlo Park, CA: Benjamin, 1969, p. 56. Part (*b*) reprinted with permission from Baldwin, J., and Chothia, C. *J. Mol. Biol.* 129:175, 1979.

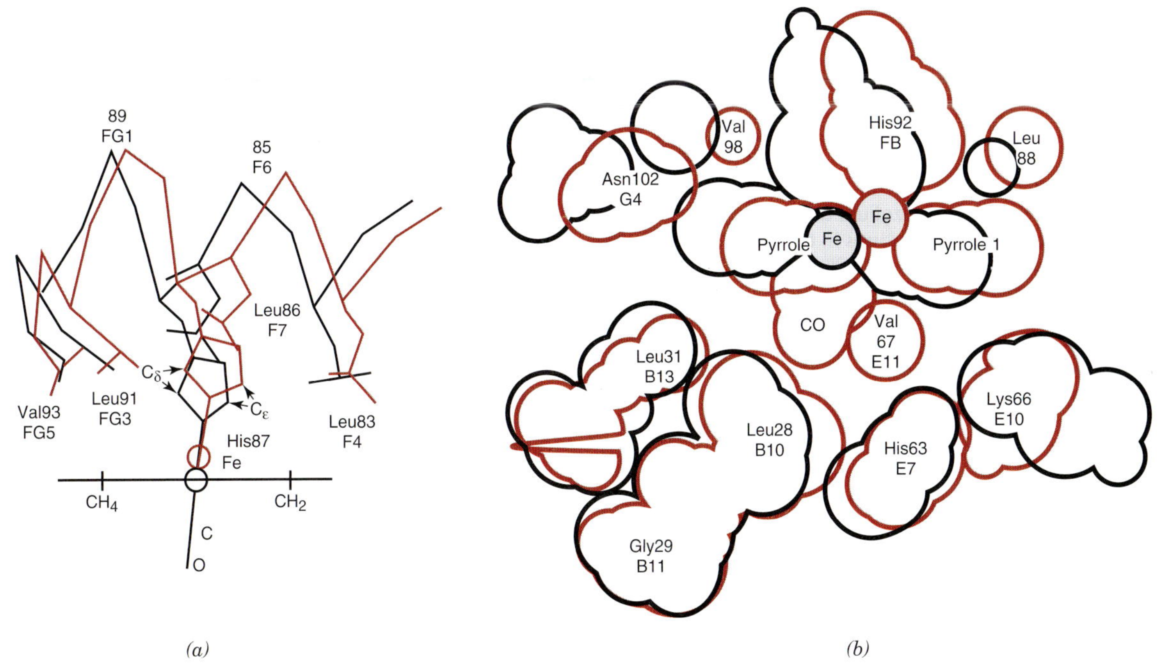

(a)

(b)

FIGURE 9.27

Stick and space-filling diagrams drawn by computer graphics showing movements of residues in heme environment on transition from deoxyhemoglobin into oxyhemoglobin. (*a*) Black line outlines position of polypeptide chain and His F8 in carbon monoxide hemoglobin, a model for oxyhemoglobin. Red line outlines the positions of the same regions for deoxyhemoglobin. Position of iron atom shown by circle. Movements are for an α subunit. (*b*) Similar movements amino acid residue position in a β subunit using space-filling diagram shown. Residue labels centered in density for the deoxy conformation.

Redrawn with permission from Baldwin, J., and Chothia, C. *J. Mol. Biol.* 129:175, 1979.

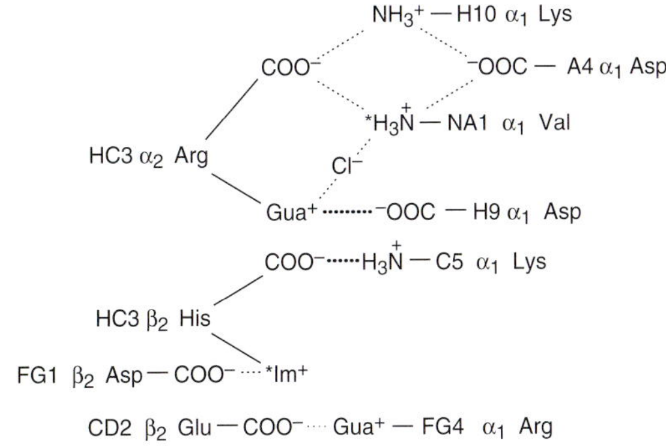

FIGURE 9.28

Salt bridges between subunits in deoxyhemoglobin that are broken in oxyhemoglobin. Im^+ is imidazolium; Gua^+ is guanidinium; starred residues account for approximately 60% of alkaline Bohr effect.

Redrawn from Perutz, M. *Br. Med. Bull.* 32:195, 1976.

Decrease in pK_A of Acid Groups with Change from T to R Conformation Releases Protons

Equation 9.10 shows the release of protons as the T conformation is converted to the R conformation. This release of protons at blood pH is the alkaline **Bohr effect** and is of physiological importance.

$$\underset{\text{T}}{\text{Hb}} + 4\text{O}_2 \rightleftharpoons \underset{\text{R}}{\text{Hb}(\text{O}_2)_4} + n\text{H}^+ \qquad (9.10)$$

The value of n in Eq. 9.10 (equivalents of protons released when one molecule of deoxy-Hb is converted to oxy-Hb) varies between 1.2 and 2.7, depending on solution

(a) Red Blood Cell in Capillaries of Tissues

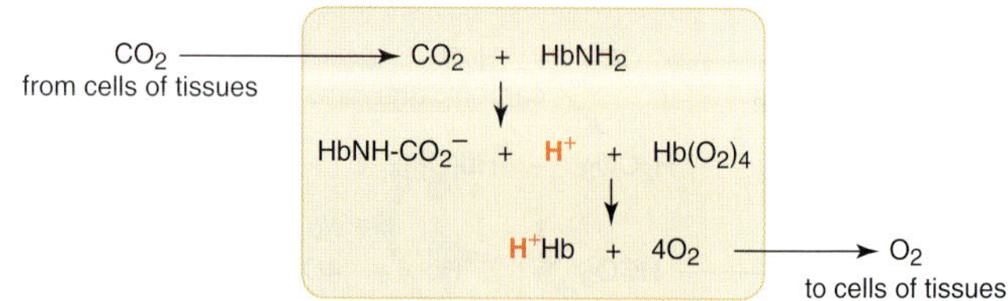

(b) Red Blood Cell in Capillaries of Lung

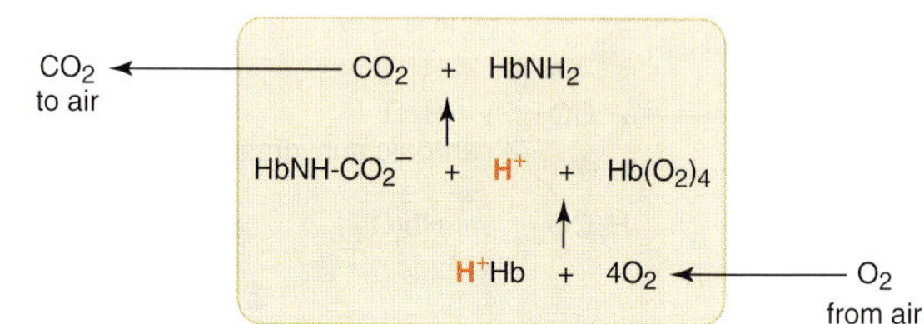

FIGURE 9.32

Transport of CO_2 as carbamino-hemoglobin. *(a)* Reactions in the red blood cell at the tissues. CO_2 diffuses into red blood cell and reacts with the NH_2-terminal amino group of hemoglobin chains to form carbamino-hemoglobin. The reaction releases a proton (red H^+) that promotes the dissociation of O_2 from hemoglobin. The O_2 diffuses out of the red blood cell to the tissues. *(b)* Reactions in the red blood cell at the level of the lung. In the lung, the high O_2 pressure forces the reactions in the opposite direction leading to the expiration of CO_2 from the lung. Reactions are the reverse of those in the capillaries.

FIGURE 9.33

Structure of 2,3-bisphosphoglycerate (BPG). Molecule has a charge of −5 at pH 7.4.

2,3-Bisphosphoglycerate (BPG) in Red Blood Cells Modulates Oxygen Release from Hemoglobin

An important modulator of the hemoglobin equilibrium is **2,3-bisphosphoglycerate (BPG)** or **2,3-diphosphoglycerate (DPG)** (Figure 9.33). BPG is formed in a minor pathway for glucose metabolism and is present in small amounts in all cells. However, in the red blood cell this pathway is highly active, and BPG concentrations are approximately equimolar to that of hemoglobin. BPG binds to the deoxy (T) but not the oxy (R) conformation. Binding of BPG to Hb stabilizes the T conformation, and increases its concentration relative to the R conformation. BPG dissociates as deoxy-Hb is converted to oxy-Hb (Eq. 9.14).

$$H^+BPG \cdot Hb + 4O_2 \rightleftharpoons Hb(O_2)_4 + BPG + nH^+ \qquad (9.14)$$

Increased concentrations of BPG force the equilibrium (Eq. 9.14) to the left, and they correspondingly shift the saturation plot to the right with an increase in P_{50} (Figure 9.34). In contrast, lowered concentrations of BPG force this equilibrium to the right, and correspondingly shift the saturation plot to the left with a lower P_{50}.

BPG binds within a pocket formed at the $\beta_1-\beta_2$-subunit interface. This site contains eight positively charged groups (the His-143(β), Lys-82(β), His-2(β), and NH_2-terminal ammonium residues from both β chains) (Figure 9.35). The BPG has a charge of minus 5 (Figure 9.33) and is strongly attracted to the positive charges of the $\beta-\beta$-interface binding site. In the R conformation, the size of the binding pocket is decreased, making the BPG unable to fit easily into the binding pocket.

Conditions that cause hypoxia (deficiency of oxygen) such as anemia, smoking, and high altitude increase BPG levels in the red blood cells. In turn, conditions leading to hyperoxia result in lower levels of BPG. Changes in red blood cell levels of BPG are slow and occur over hours and days to compensate for chronic changes in pO_2 levels.

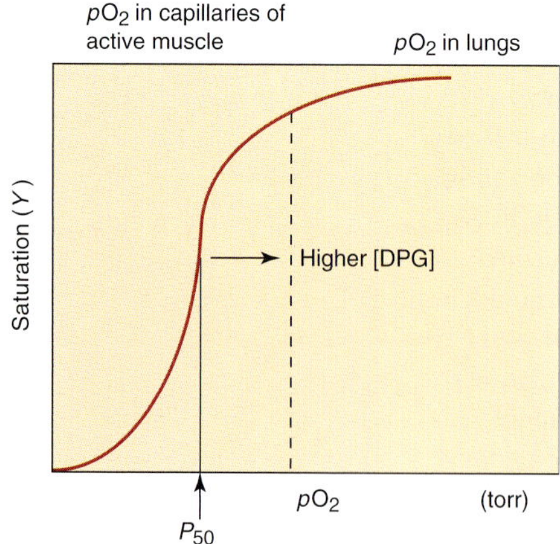

FIGURE 9.34

Change in oxygen–hemoglobin saturation curve to higher P_{50} value with increase in BPG concentration.

Hemoglobin Delivers Nitric Oxide (NO) to the Capillary Wall of Tissues Where It Promotes O_2 Delivery

Hemoglobin reversibly binds nitric oxide (NO) a potent vasodilator, with a very short lifetime in blood. By binding NO, hemoglobin sequesters it from rapid destruction. Hb releases NO by transferring it to small sulfhydryl (X-SH) molecules as hemoglobin changes from oxy (R) to the deoxy (T) conformation. A common X-SH molecule is glutathione (Figure 9.36). The released NO in the form of X-S-NO stabilizes NO against rapid degradation and allows efficient delivery of a bioactive NO equivalent (X-S-NO) to the NO receptors in cells of the blood vessel wall, promoting relaxation of

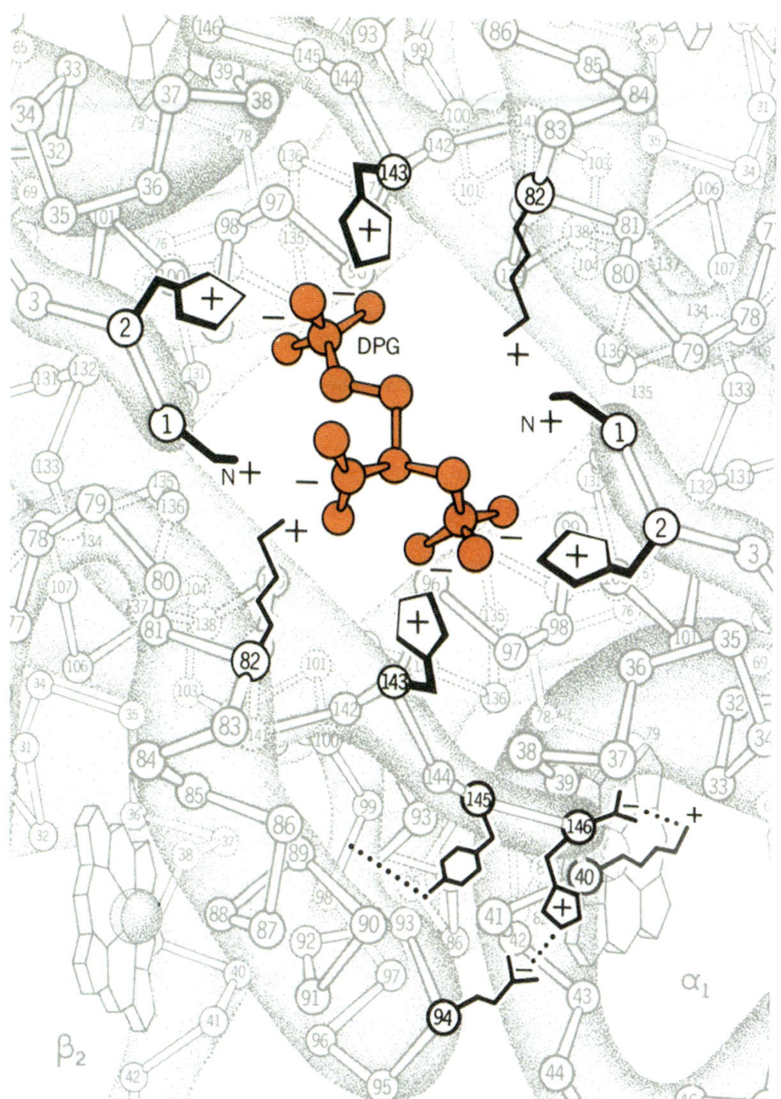

FIGURE 9.35

2,3-Bisphosphoglycerate binding site at the β–β interface of deoxy-hemoglobin. Shown are the positively charged side chains of two βVal-1 amino-terminus ammonium, βHis-2 imidazolium, βLys-82 ϵ-ammonium, and β143-His imidazolium. The negatively charged 2,3-bisphosphoglycerate binds in the middle of the ring of positively charged groups. Reprinted with permission from Dickerson, R. E. and Geis, I. *Hemoglobin: Structure, Function, Evolution, and Pathology*. Menlo Park, CA: Benjamin-Cummings, 1983. Illustration by Irving Geis, Geis Archives Trust. Rights owned by Howard Hughes Medical Institute.

the vascular wall. This facilitates the transfer of gases between the blood and the cells of the tissues. The molar concentration of NO in blood is 1/70th the concentration of Hb. However, this low concentration is physiological significant because of the potent biological activity of X-S-NO.

NO is first captured by the Fe^{2+} at the heme and then transferred to the sulfhydryl of the β-chain cysteine-93 (βCys^{93}) residues. The heme iron preferentially binds NO when in the T conformation. The NO is transferred to βCys^{93} when hemoglobin is in the R conformation. Then when the R changes again to the T conformation to deliver O_2 to the tissues, the NO is transferred to small S-XH molecules such as glutathione (Figures 9.36 and 9.37). The net effect is the conversion of unstable free NO to stable X-S-NO.

Figure 9.38 shows the region of hemoglobin that contains βCys^{93}-S-NO. In the T conformation the S-NO is on the outside and accessible to glutathione molecules. This allows *trans*-nitrosylation to occur. In contrast, in the R conformation the βCys^{93}-SH is pointed toward the heme iron and inaccessible to the outside solvent and the distance is optimal for transfer of NO from the heme to the βCys^{93}-SH. Thus the conformational changes in the hemoglobin molecule, induced by changes in oxygen pressure between the lungs and tissues, regulate NO uptake and release from the βCys^{93}. The net effect is delivery of bioactive NO to the tissue capillaries under low oxygen tension.

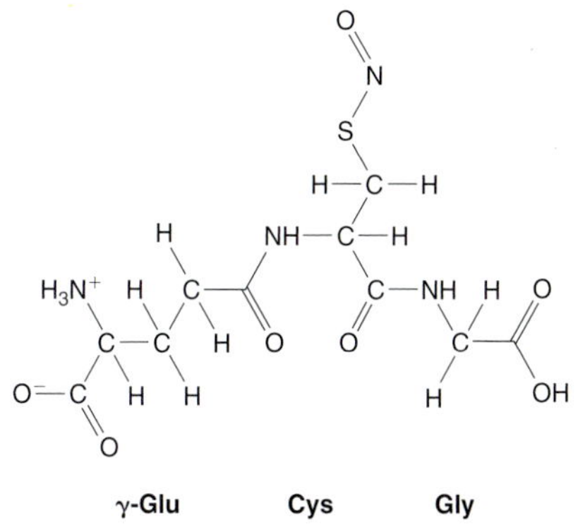

γ-Glu Cys Gly

FIGURE 9.36

The X-S-NO transporter glutathione (γ-glutamyl-cysteinyl-glycine) transports NO bound to the sulfhydryl side chain of its cysteine.

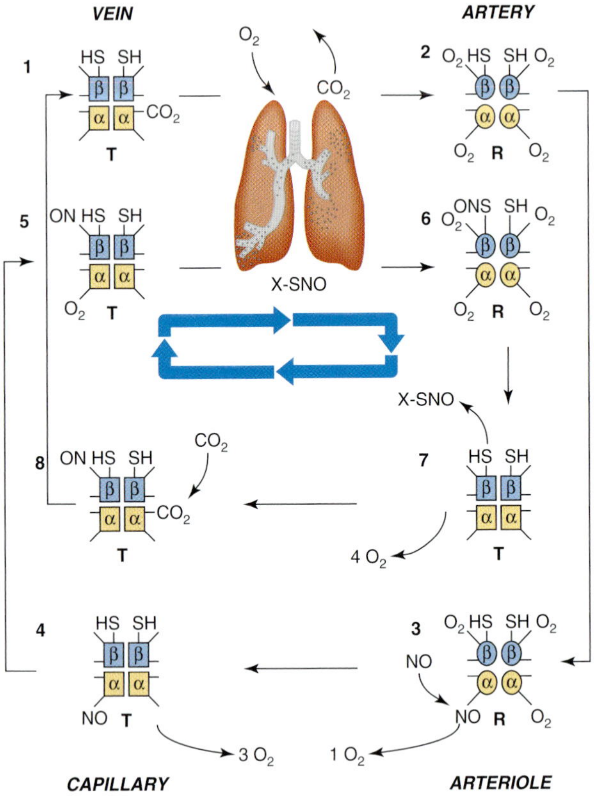

FIGURE 9.37

Binding and release of NO by hemoglobin during the respiratory cycle. The model shows the binding and dissociation of NO, O_2, and CO_2 as a hemoglobin molecule makes two complete cycles through the circulation. The first cycle involves intermediates 1–4 and the second cycle intermediates 5–8. The T and R conformations are shown and the SH groups are from the side chain of βCys^{93}. The NO is either directly bound to a heme iron or to the βCys^{93} SH. The key steps in the NO transport are (i) its initial binding to a heme in intermediate 3 and transfer from a β-subunit heme to βCys^{93} in intermediate 6 (R conformation) and (ii) its transfer to a small molecular thiol X-SH in intermediate 7 (T conformation) when hemoglobin is converted from R to T. The hemoglobin molecule depicted may be only 1 in 1000 hemoglobin molecules that are cycling, due to the low relative molar concentration of NO in blood.
Redrawn from Gross, S. S. and Lane, P. *Proc. Natl. Acad. Sci. USA* 96:9967, 1999.

9.5 THE BASAL LAMINA PROTEIN COMPLEX

A **basal lamina** is a highly structured complex of extracellular matrix proteins first observed microscopically as an amorphous densely packed region of about 50- to 100-nm thickness surrounding tissues or cells (Figure 9.39). The term **basement membrane** is used to describe both the basal lamina and connecting fibrillar collagens attached to its outside. The basement membrane gives support to tissues and regulates access of cells to the interstitial stroma. It also participates in determining the properties of the cells that are attached to it, including the critical processes of cell division, death (apoptosis), differentiation, and migration. All cells produce constituents of basement membrane, and each basement membrane has characteristics of the cell type from which it is synthesized. A basement membrane underlies sheets of epithelial and endothelial cells, and it surrounds other cell types (Figure 9.39). A basement membrane separates two sheets of cells in the kidney glomerulus, where it acts as a selective filter.

The basal lamina is formed by the noncovalent associations between specific sites located in binding domains of the associated proteins. Many basement membrane proteins also bind to cells through cellular binding domains in the proteins and cell receptors in the outer cell membrane. The structure of many of the proteins is composed of module units with sequence and fold homology to common superfolds, such as the immunoglobulin (Ig) and the epidermal growth factor (EGF) folds. These folds are found repetitively and are the building blocks of extracellular matrix proteins. While the EGF modules within these proteins do not appear to have a growth factor function, growth factor and cytokine proteins are found within the basal lamina, particularly in association with the carbohydrate parts of the proteoglycan components. During turnover of the basement membrane induced by proteases and heparanases, these growth factors and cytokines are freed to act on nearby cells. In addition, many basal lamina proteins hide cryptic activities that are activated when cleaved out of the full sequence by action of proteases (see endostatin, pp. 1028). The pro-protease plasminogen is ubiquitously present within extracellular matrix and is activated by cellular secretions of plasminogen activators (see p. 1004).

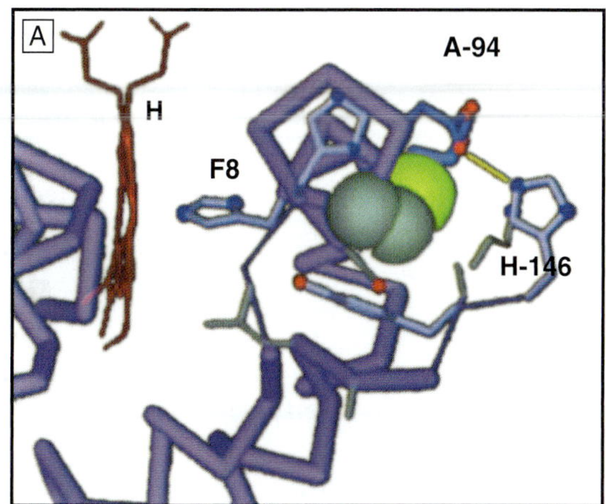

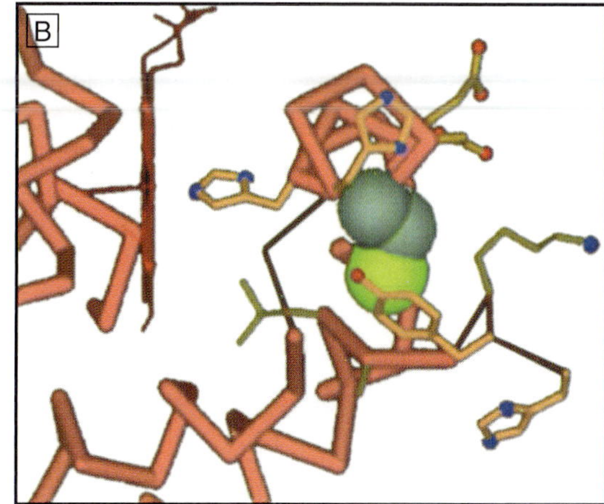

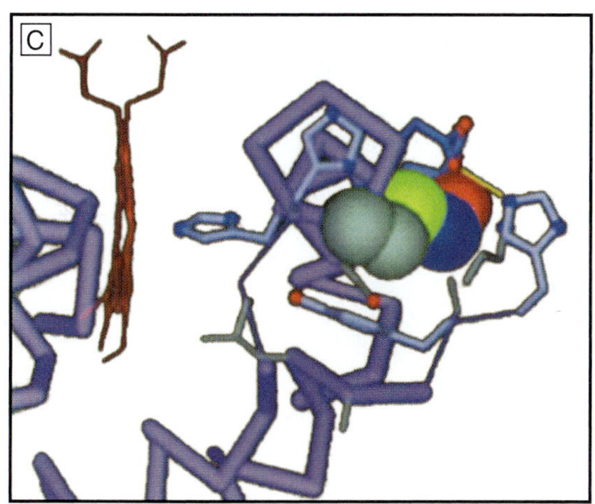

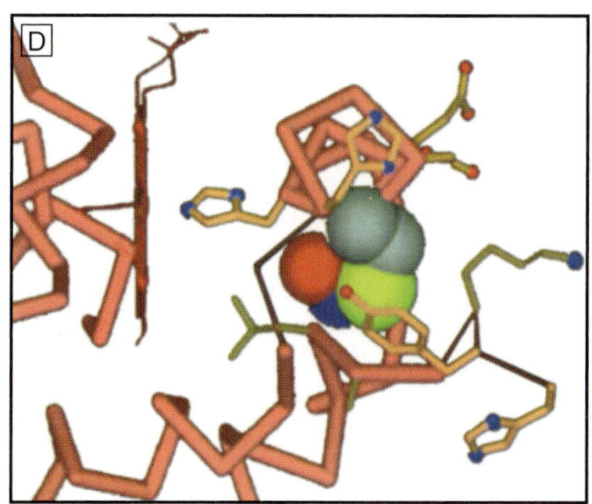

FIGURE 9.38

Structure of βCys^{93} and βCys^{93}-S-NO in the T and R conformations. In all structures the heme (H, red) edge is pointed towards reader and is shown to be bonded to proximal histidine imidazole (F8). The two carbons (blue-gray) and sulfur (yellow-green) of the Cys-93 side chain are shown with space-filling models. In panels C and D, NO is bonded to the sulfur of βCys^{93} with the N atom colored blue and the O atom colored red. (*A*) The deoxy(T) conformation with the βCys^{93} side chain (—CH_2—SH) is on the surface of the molecule away from the heme. The cysteine side chain is prevented from entering the heme-binding site by the βHis^{146}-imidazolium to βAsp^{94}-carboxylate hydrogen bonded ion pair (H-bond between groups shown in yellow) on the upper right. (*B*) The oxy(R) conformation with the βCys^{93} side chain pointed toward the heme and away from the solvent on the outside of the molecule. The βHis^{146} to βAsp^{94} salt bridge is broken in the R conformation, which allows the folding of the βCys-SH toward the heme pocket. (*C*) Model of βCys^{93}-SNO in the deoxy(T) conformation. SNO is positioned on the outside accessible to react with X-SH small molecules in the solvent. As in (A), the cysteine side chain is prevented from entering heme site by the βHis^{146}-βAsp^{94} ion pair. (*D*) Model of βCys^{93}-SNO in oxy(R) conformation. SNO is buried near the heme and away from the outside solvent. This conformation facilitates the transfer of NO from the heme iron to the cysteine SH and prevents the reaction of the βCys^{93}-SNO with X-SH molecules in the solvent.

Reprinted with permission from Stamler, J. S., Jia, L., Eu, J. P., McMahon, T. J., Demchenko, I. T., Bonaventura, J., Gernert, K., and Piantadosi, C. A. *Science* 276: 2034, 1997. Copyright (1997) AAAS.

Protein Composition of the Basal Lamina

The basal lamina is composed of type IV collagen, laminin, nidogen (also referred to as entactin), and perlecan, the heparin sulfate proteoglycan. In addition, minor amounts of perhaps 50 other proteins may be present including osteopontin (also referred to as BM-40 or SPARC), fibulin, type XV collagen, type XVIII collagen, and the proteoglycan agrin. The diversity and tissue specificity of a basement membrane is determined by the isoforms of type IV collagen and laminin and the types of minor proteins present. **Type IV** collagen isoforms are produced by seven different type IV collagen genes. These isoforms share domain structure homology, but differ by 30–50% in their amino acid sequences. There are at least 12 different **laminin** isoforms. The isoforms of type IV collagen and laminin expressed are characteristic of the cell type and tissue that synthesizes the associated basement membrane.

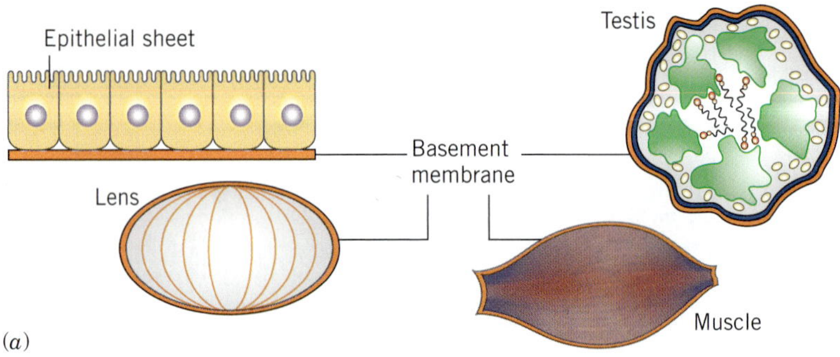

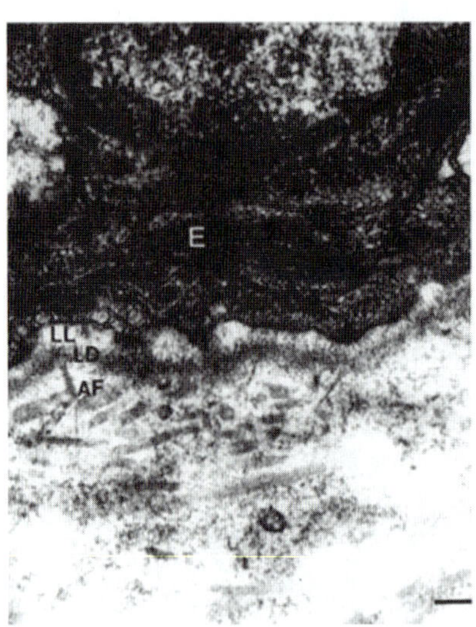

FIGURE 9.39

Basement membranes. (*a*) Diagram of the basement membranes surrounding various tissues and cell types. (*b*) Electron micrograph showing ultrastructure of extracellular matrix with basement membrane adjacent to epithelial cell (E). The lamina lucida (LL) and lamina densa (LD) or basal lamina of the basement membrane is shown. Below lamina densa is stroma of extracellular matrix. Bar represents 100 nm.
(*a*) Part reprinted with permission from Kalluri, *Nature Rev. Cancer* 3:422, 2003. Copyright (2003) Nature. Part (*b*) reprinted with permission from Bosman, F. T. and Stamenkovic, I. *J. Pathol.* 200:423, 2003. Copyright (2003) Pathological Society of Great Britain and Ireland. Reproduced with permission. Permission is granted by John Wiley & Sons, Ltd. on behalf of the PathSoc.

Molecular Structure of Basal Lamina Is Formed from Networks of Laminin and of Type IV Collagen

The structure of the basal lamina is produced by the connecting of planar networks of laminin and of type IV collagen. Nidogen/entactin and perlecan proteoglycan molecules link these networks. The laminin polymer structure initiates formation of the basement membrane, facilitated by its attachment to cellular receptors (Figure 9.40).

Laminin Network

Laminin is composed of three-polypeptide chains (α, β, γ) cross-linked by disulfide bonds. Each chain contains about 1500 amino acids and the molecular weight of the laminin-1 isoform is approximately 800 kD. The overall structure appears as an asymmetric cross (Figure 9.41). The cruciform structure contains three short arms, each formed by a different chain, and a long arm composed of all three chains. There are five distinct genes that code for an α chain and three each for the β and γ chains, which combine to form at least 12 distinct $\alpha\beta\gamma$ isoforms. Different domains in the structure bind to perlecan and nidogen, and at least two domains contain binding sites for cell surface receptors. Near the cell surface, the laminin molecules self-associate to form a sheet-like network structure through interactions between binding sites in the domains of the short arms of the cruciform structure (Figure 9.42).

Type IV Collagen Network

Collagen molecules contain three polypeptide chains that initially associate through noncovalent interactions (see p. 112). Fibrillar forming collagens, such as type I collagen, contain long stretches of the repeating sequences $(\text{Gly-Pro-X})_n/(\text{Gly-Y-HyPro})_n$ containing a proline or hydroxyproline (HyPro) and a glycine approximately every third residue (see p. 100). The sequence contains the derived amino acid hydroxyproline, formed by proline hydroxylase in the endoplasmic reticulum on

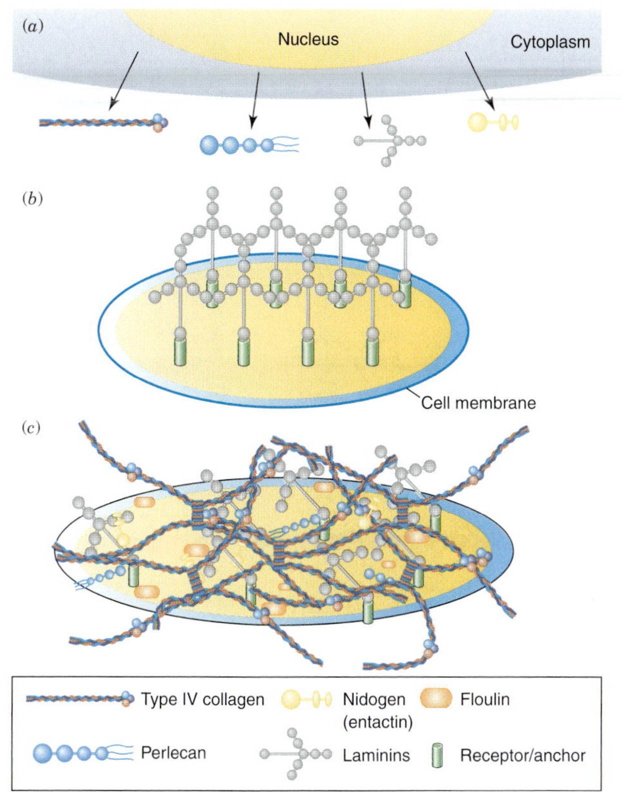

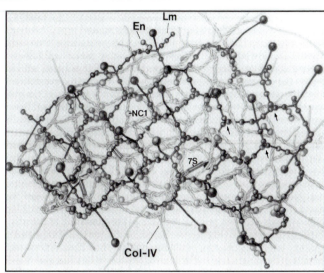

FIGURE 9.40

Molecular structure of basal lamina.
The laminin and Type IV collagen molecules self-associate to form sheet-like network structures, which are linked by the proteins nidogen/entactin and by the heparan sulfate proteoglycan perlecan. (*a*) Synthesis of basement membrane proteins by cell. (*b*) Assembly of laminin network initiated by binding of laminin to cell surface receptor. (*c*) Assembly of Type IV collagen network and association of proteins linking the laminin and Type IV collagen scaffolds. (*d*) Type IV collagen network joined through 7S and NC1 domain interactions and with entactin/nidogen (En) bridging the collagen and laminin (Lm) networks.
Parts (*a*)–(*c*) reprinted with permission from Kalluri, R. *Nature Rev. Cancer* 3:422, 2003. Copyright (2003) Nature. Part (*d*) reprinted with permission from Yurchenco, P. D. Assembly of basement membrane networks. In: P. D. Yurchenco, D. E. Birk, and R. P. Mecham (Eds.), *Extracellular Matrix Assembly and Structure*. New York: Academic Press, 1994, p. 351. Copyright (1994) Elsevier.

FIGURE 9.41

Structure of laminin (isoform 1). (*a*) Diagrammatic structure of laminin-1. The α, β, and γ laminin chains each form a short arm of the cruciform structure. The long arm is formed by a coiled-coil structure of α-helical regions of all three chains with the α-laminin chain extending out to form the COOH-terminal globular G domain. The short arms are composed of globular domains, which are separated by epidermal growth factor-like (EGF-like) repeats. (*b*) Contrast-reversed glycerol rotary shadowed replica of laminin molecule. α-Chain NH_2-terminal and α chain globular (G) domain indicated.
Part (*a*) redrawn based on figure from Yurchenco, P. D. Assembly of basement membrane networks. In: P. D. Yurchenco, D. E. Birk, and R. P. Mecham (Eds.), *Extracellular Matrix Assembly and Structure*. New York: Academic Press, 1994, p. 351. Part (*b*) reprinted with permission from Yurchenco, P. D. Assembly of basement membrane networks. In: P. D. Yurchenco, D. E. Birk, and R. P. Mecham (Eds.), *Extracellular Matrix Assembly and Structure*. New York: Academic Press, 1994, p. 351. Copyright © (1994) Elsevier.

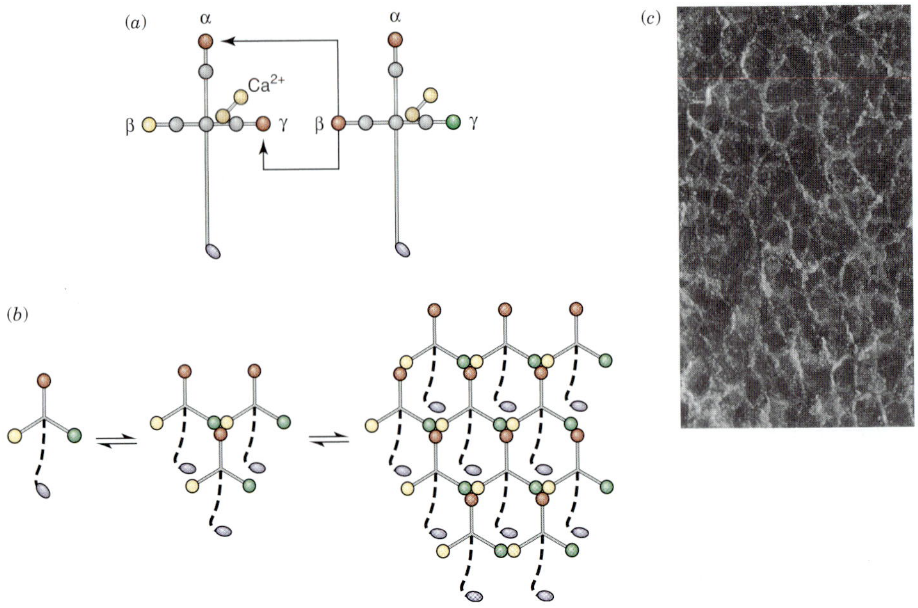

FIGURE 9.42

Formation of laminin network structure. (*a*, *b*) Binding interactions between short arms of different laminin molecules lead to sheetlike network structure. Entactin is shown bound to a short arm of laminin, but is not required for laminin network polymerization. Long arm is believed to be free to participate in other interactions. (*c*) Laminin network visualized by high angle platinum replicas.
Figures from Yurchenco, P. D. Assembly of basement membrane networks. In: P. D. Yurchenco, D. E. Birk, and R. P. Mecham (Eds.), *Extracellular Matrix Assembly and Structure*. New York: Academic Press, 1994, p. 351. Parts (*a*) and (*b*) redrawn and Part (*c*) reprinted with permission.

prolines within the collagen chain. The high content of proline and hydroxyproline generates a helical conformation for this sequence known as the polyproline type II helix (see p. 100). This helix is characterized by three residues per helix turn ($n = 3$), and with a Gly every third residue the helix forms a longitudinal glycine edge along one side that promotes self-association of three polypeptides, each in a polyproline helical conformation, into a triple helical or superhelical structure. This three-chain molecule comprises the collagen protomer unit.

Type IV collagen does not form fibrillar structures as type I, but a sheet-like network. Type IV collagen polypeptides contain regions of $(\text{Gly-X-Pro})_n/(\text{Gly-HyPro-Y})_n$ sequence that form rod-shaped triple-helical regions, but different from type I these regions are interspersed by linker and globular domains that break up the superhelical regions (Figure 9.43). The globular domain at the C-terminal end of the protomer is called the NC1 (*n*on-*c*ollagenous 1) domain. This C-terminal domain is preceded by a triple-helical (TH) central region, and a hinge connecting to a smaller triple-helical (7S) region at the N-terminal end.

In assembling the type IV collagen network, each protomer binds to another through a NC1 to NC1 head-to-head interaction (Figure 9.44). Then the dimers interact through their 7S regions to form tetramers, which further self-aggregate to form the network. The 7S interactions are the nodes of the collagen network structure (Figure 9.44).

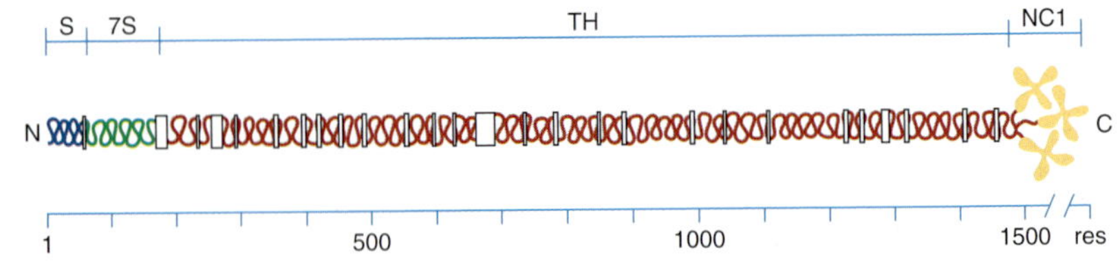

FIGURE 9.43

Diagram of the Type IV collagen protomer composed of three polypeptide chains. Top line shows the location in the primary structure of the NC1 region, the central triple helical (TH) region with a superhelical structure, the N-terminal 7S region, and the signal peptide(S). Black bars, lines, and boxes show locations of the multiple interruptions of the Gly-X-Y sequence in the chains of type IV collagen.
Redrawn from Yurchenco, P. D. Assembly of basement membrane networks. In P. D. Yurchenco, D. E. Birk, and R. P. Mecham (Eds.), *Extracellular Matrix Assembly and Structure*. New York: Academic Press, 1994, p. 351.

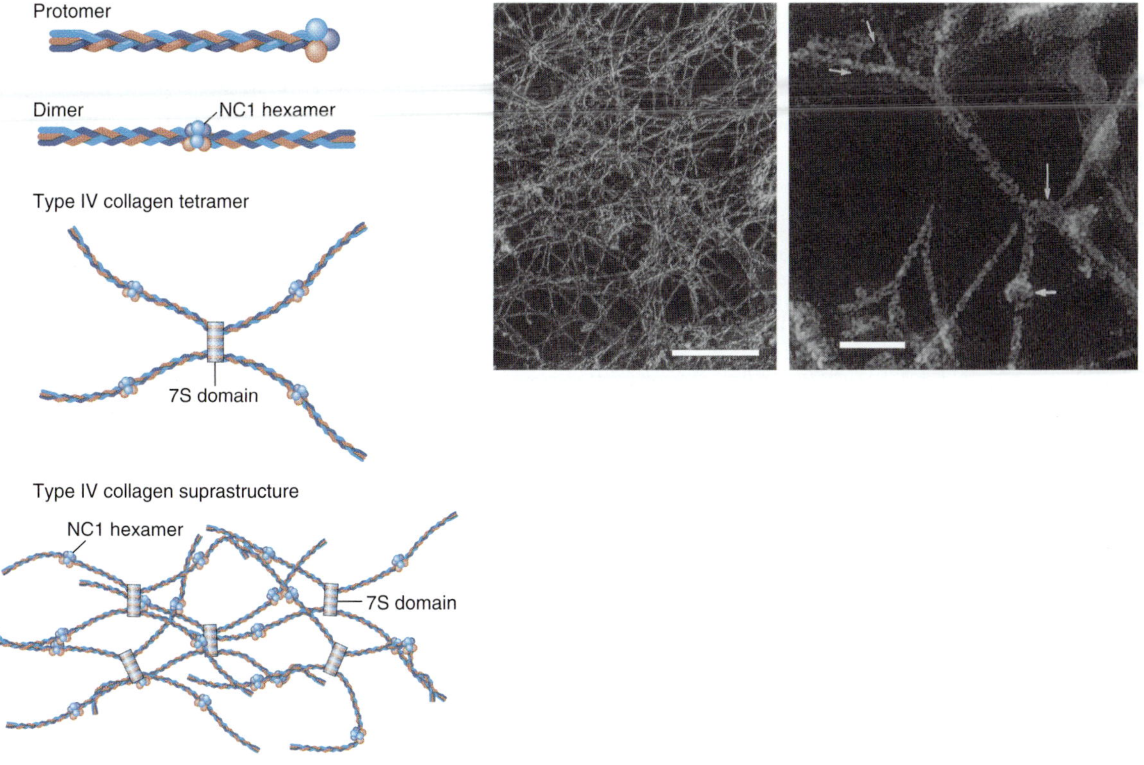

FIGURE 9.44

Formation of type IV collagen network structure. (*A*) Diagrammatic depiction of network formation. (*a*) Protomer type IV collagen molecule is formed from three polypeptide chains (red, purple, blue) in the endoplasmic reticulum and Golgi bodies of cell. Collagen chains are aligned with their NH_2-terminal 7 S regions on one end and COOH-terminal NC1 regions at other end. (*b*) Two protomers are joined by their NC1 domains to form a dimer of protomers (each protomer has three chains so that interaction is between NC1 domains of six chains). (*c*). Two protomer dimers bind through 7 S regions. (*d*). Polymerization of tetramers of protomers forms type IV collagen network. (*B*) Type IV collagen networks viewed by a high-angle replicate of amniotic basement membrane observed *in situ*.

Part (*A*) redrawn from Kalluri, R. *Nature Rev. Cancer* 3:422, 2003. Part (*B*) reprinted with permission from Yurchenco, P. D. and Ruben, J. *Cell Biol.* 105:2559, 1987. Modified from Yurchenco, P. D. Assembly of basement membrane networks. In: P. D. Yurchenco, D. E. Birk, and R. P. Mecham (Eds.), *Extracellular Matrix Assembly and Structure*. New York: Academic Press, 1994, p. 351. Copyright (1994) Elsevier.

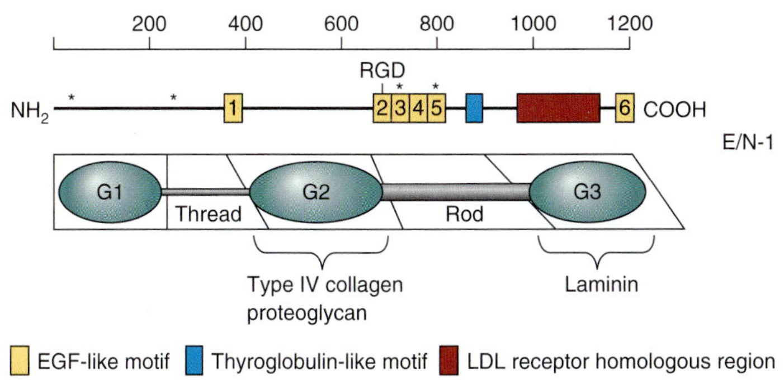

FIGURE 9.45

Structure of nidogen. Schematic structure of the entactin/nidogen-1 molecule. The potential calcium binding sites are marked by stars. Binding sites for type IV collagen and proteoglycan are in G2 domain, and the binding site for laminin is in the G3 domain.

Redrawn from Erickson A. C. and Couchman, J. R. *J. Histochem Cytochem.* 48:1291, 2000.

Nidogen/Entactin Interconnects Laminin and Type IV Collagen Networks

The two isoforms, **nidogen-1** and **nidogen-2**, have a 46% homology. Both forms are composed of three globular domains (G1, G2, and G3) separated by a link region between G1 and G2 and a longer rod region between G2 and G3 (Figure 9.45). Domain G3 is a ***β* propeller fold** composed of antiparallel β-strands that appear as blades of an

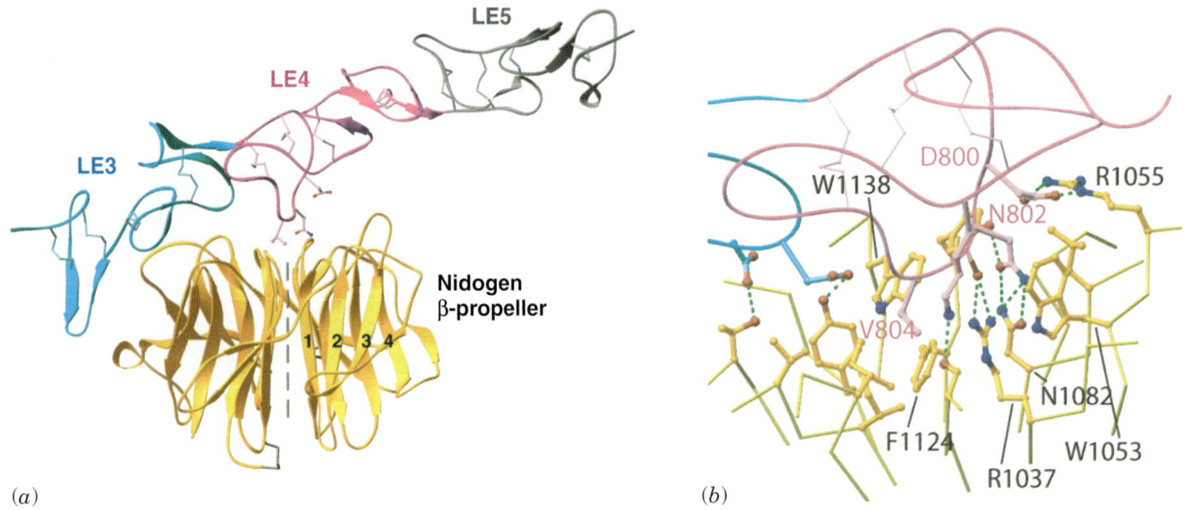

FIGURE 9.46

Molecular details of molecular interactions between nidogen and laminin. (*a*) Ribbon diagram of the nidogen β-propeller complex of domain G3 with laminin modules LE3-5. β-Strands are numbered on the propeller β-sheet. (*b*) View of the nidogen interaction with laminin domain LE4 and the adjacent portion of laminin domain LE3. Portions of the nidogen backbone as C_α-trace and side chains forming the interaction are shown in gold. Green dashed lines show hydrogen bonds.
Reprinted with permission from Takagi, I., Yang, Y., Lu, J., Wang, H., and Springer, T. A. *Nature* 424:969, 2003. Copyright (2003) Nature.

airplane propeller. It is similar to a fold of the LDL receptor protein. Nidogens bind to laminin by interaction of the β-propeller interface with laminin domains III and IV (Figure 9.46). Other sites in nidogen form complexes with type IV collagen and perlecan (Figure 9.47).

Heparan Sulfate Proteoglycan Perlecan Interconnects Laminin and Type IV Collagen Networks

Perlecan (MW 600 kDa) is a major proteoglycan of the *basal lamina*. Proteoglycans have a polypeptide core attached from serine side chains through a tetrasaccharide to a glycosaminoglycan (GAG) chain, composed of a disaccharide of an amino sugar (*N*-acetylglucosamine or *N*-acetylgalactosamine) and a uronic acid (glucuronic or iduronic). In perlecan the GAG is the heavily sulfated heparin sulfate (Figure 9.48). Each perlecan contains 2–15 heparin sulfate **(GAG)** chains (Figure 9.49). The sulfated GAG chains are highly negatively charged and bind cations and water to form gels. This provides a swelling pressure that enables the basement membrane to withstand compressive forces. Perlecan interacts with the other three major components of basement membrane through either its core protein to bind type IV collagen or through its heparan sulfate chains to bind laminin. Other sites in the heparin sulfate bind to cellular receptors and still others to growth factors such as TGF-β.

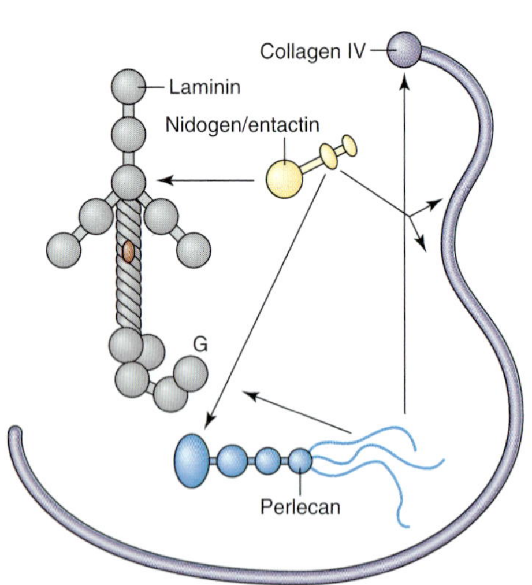

FIGURE 9.47

Interactions of nidogen and the proteoglycan perlecan. Arrows show domain–domain binding interactions between the major proteins in basement membrane suprastructure.
Redrawn from Kalluri R. *Nature Rev. Cancer* 3:422, 2003.

Protein Complex Interactions Form Communication Networks

Structured molecular complexes of multiple protein subunits are common in biology. The interior of living cells may be described as an aqueous slurry of protein molecular complexes. Many proteins are present in more than one complex. Such a sharing of a protein connects the two complexes and generates a **communication network** (Figure 9.50).

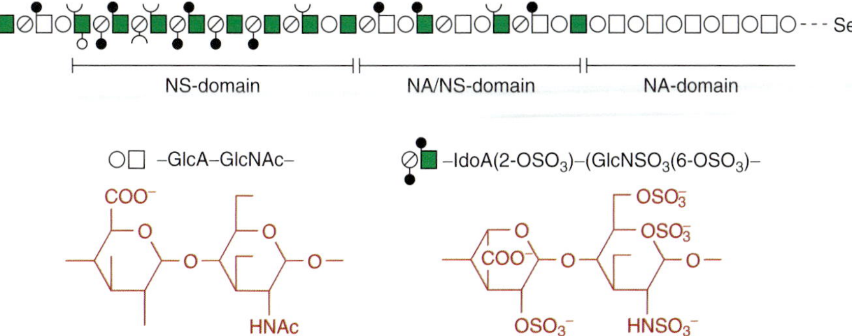

FIGURE 9.48

Structure of a heparan sulfate chain. The GAG disaccharide polymers contain alternating hexuronic acid (D-glucuronic acid (GlcA) or L-iduronic acid (IdoA)) and D-glucosamine (GlcN) units. The GlcN is *N*-acetylated in regions of the chain (NA region). In further processed regions of the heparan polymer the *N*-acetyl groups have been partially replaced by *N*-sulfate (NA/NS regions) or the *N*-acetyl groups have been completely replaced by *N*-sulfate (NS regions). Open circle is 3-*O*-sulfate.
The polymers are further modified by C5 epimerization of GlcA to IdoA residues, and, finally incorporation of *O*-sulfate groups at various positions. The heparan chain is joined to the protein core by forming a bond to a serine side chain.
Redrawn from Lindahl, U., et al. *J. Biol. Chem.* 273:24979, 1998.

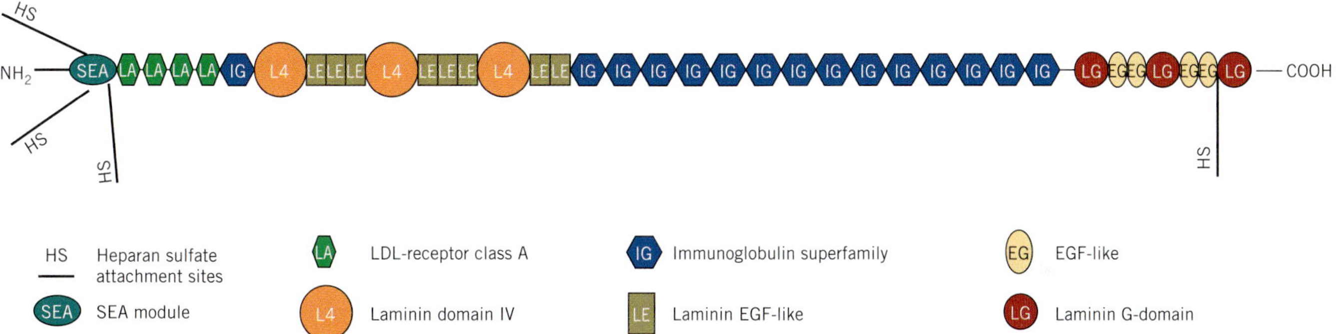

FIGURE 9.49

Module organization of domains in the protein core of perlecan. Domain organization of mouse perlecan. The attachment of the heparan sulfate (HS) chains is indicated. The domains present are SEA (homologous to the domain found in sea urchin sperm protein, enterokinase, and agrin); LA (homologous to LDL receptor type A domain); L4 (homologous to laminin domain IV), IG (immunoglobulin-like domain); LE (homologous to the laminin type epidermal growth factor-like domain); EG (epidermal growth factor-like); and LG (homologous to laminin G-like domain).
Redrawn from Hopf, M., Göhring, W., Kohfeldt, E., Yamada, Y., and Timpl, R. *Eur. J. Biochem.* 259:917–925 (1999) and Kvansakul, M., Hopf, M., Ries, A., Timpl, R., and Hohenester, E. *EMBO J.* 20:5342, 2001.

The basement membrane is an extracellular protein complex that is connected to the intracellular cytoskeletal protein complex, as both complexes are connected to the same receptor proteins present in the focal contact regions of cellular membranes. These protein–protein binding interactions connect the extracellular matrix to the cellular cytoskeleton and are critical for the regulation of intracellular processes, such as cell mobility and cell death. In most cancer cells, this binding of the membrane receptor protein complex to both the extracellular *basal lamina* and the cytoskeletal protein complexes is broken.

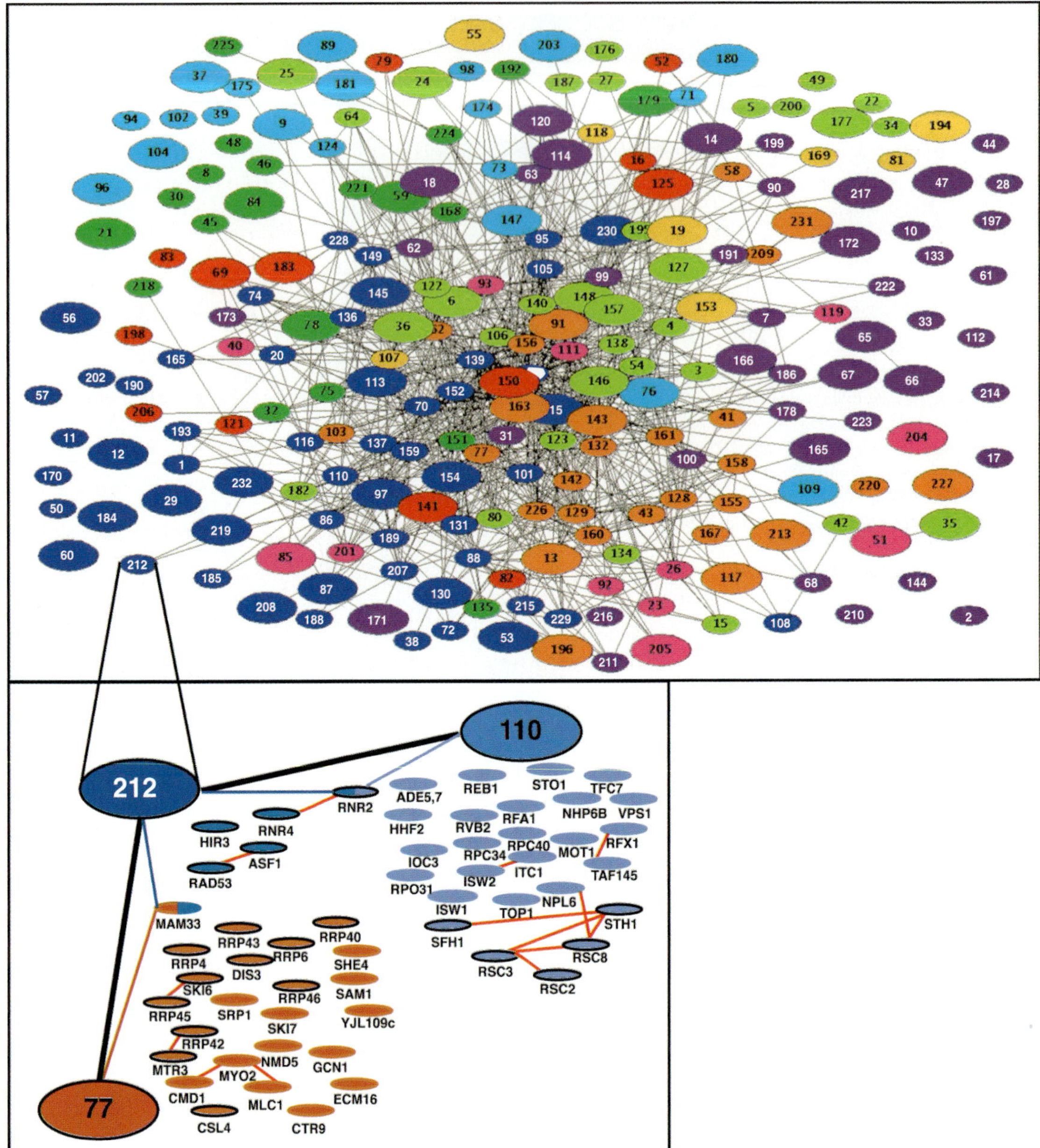

FIGURE 9.50

Interconnected Protein Complexes Form Cellular Network.
Diagram shows network of connected protein complexes found in a partial analysis of the *E. coli* yeast proteome (expressed proteins). The number of different proteins in a complex varies from 2 to 83 (average number is 12) per complex. Lines link complexes in which the same protein was found in both complexes. The more connected complexes are more centrally located, and complexes with at least 50% of their protein orthologous to human proteins are shown as double-sized. Complexes are color-coded by their function: red, cell cycle; dark green, signaling; dark blue, transcription, DNA maintenance, chromatin structure; pink, protein and RNA transport; orange, RNA metabolism; light green, protein synthesis and turnover; light blue, membrane biogenesis and traffic. Lower panel shows example of the linking of a complex with two others by shared proteins. Red lines show interactions previously listed in the Yeast Protein Database (YPD). Reproduced from Gavin, A.-C., Bösche, M., Krause, R., Grandi, P., et al. *Nature* 415:141, 2002. Copyright (2002) Nature. Figure generously supplied by Dr. A.-C. Gavin.

BIBLIOGRAPHY

Immunoglobulins

Alzari, P. M., Lascombe, M. B., and Poljak, R. J. Structure of antibodies. *Annu. Rev. Immunol.* 6:555, 1988.

Barclay, A. N. Ig-like domains: Evolution from simple interaction molecules to sophisticated antigen recognition. *Proc. Natl. Acad. Sci. USA* 96:14,672, 1999.

Chothia, C., Lesk, A. M., Tramontano, A., Levitt, M., Smith-Gill, S. J., Air, G., Sheriff, S., Padlan, E. A., Davies, D., Tulip, W. R., Colman, P. M., Spinelli, S., Alzari, P. M., and Poljak, R. J. Conformations of immunoglobulin hypervariable regions. *Nature* 342:877, 1989.

Davies, D. R., Padlan, E. A., and Sheriff, S. Antibody–antigen complexes. *Acc. Chem. Res.* 26:421, 1993.

Guddat, L. W., Shan, L., Fan, Z. -C., Andersen, K. N., Rosauer, R., Linthicum, D. S., and Edmundson, A. B. Intramolecular signaling upon complexation. *FASEB J.* 9:101, 1995.

Hunkapiller, T. and Hood, L. Diversity of the immunoglobulin gene superfamily. *Adv. Immunol.* 44:1, 1989.

Padlan, E. A. Anatomy of the antibody molecule. *Mol. Immunol.* 31:169, 1994.

Rini, J. M., Schultze-Gahmen, U., Wilson, I. A. Structural evidence for induced fit as a mechanism for antibody–antigen recognition. *Science* 255:959, 1992.

Serine Proteases

Birk, Y. Proteinase inhibitors. In: A. Neuberger and K. Brocklehurst (Eds.), *Hydrolytic Enzymes*. Amsterdam: Elsevier, 1987, p. 257.

Liebman, M. N. Structural organization in the serine proteases. *Enzyme* 36:115, 1986.

Neurath, H. Proteolytic processing and physiological regulation. *Trends Biochem. Sci.* 14:268, 1989.

Perona, J. J. and Craik, C. S., Structural basis of substrate specificity in the serine proteases. *Protein Sci.* 4:337, 1995.

Polgar, L. Structure and function of serine proteases. In: A. Neuberger and K. Brocklehurst (Eds.), *Hydrolytic Enzymes*, series in *New Comprehensive Biochemistry*, Vol. 16. Amsterdam: Elsevier, 1987, p. 159.

Hemoglobin

Website of interest: The Globin Gene Server database of sequence alignments and experimental results for the β-like globin gene cluster of mammals. http://globin.cse.psu.edu/

Baldwin, J. and Chothia, C. Haemoglobin: the structural changes related to ligand binding and its allosteric mechanism. *J. Mol. Biol.* 129:175, 1979.

Benz, E. J., Jr. Genotypes and phenotypes-another lesson from the hemoglobinopathies. *N. Engl. J. Med.* 351:1490, 2004.

Dickerson, R. E. and Geis, I. *Hemoglobin: Structure, Function, Evolution and Pathology*. Menlo Park, CA: Benjamin-Cummings, 1983.

Gross, S. S. and Lane, P. Physiological reactions of nitric oxide and hemoglobin: A radical rethink. *Proc. Natl. Acad. Sci. USA* 96:9967, 1999.

Hsia, C. C. W. Mechanisms of disease: Respiratory function of hemoglobin. *N. Engl. J. Med.* 338:239, 1998.

Perutz, M. Hemoglobin structure and respiratory transport. *Sci. Am.* 239:92, 1978.

Perutz, M. F., Fermi, G., and Shih, T. -B. Structure of deoxy cowtown [His HC3(146)beta to Leu]: origin of the alkaline Bohr effect. *Proc. Natl. Acad. Sci. USA* 81:4781, 1984.

Perutz, M. F., Wilkinson, A. J., Paoli, M., and Dodson, G. G. The stereochemical mechanism of the cooperative effects in hemoglobin revisted. *Annu. Rev. Biophys. Biomol. Struct.* 27:1, 1998.

Takashi Yonetani, T., Park, S., Tsuneshige, A., Imai, K., and Kanaori, K. *Global allostery model of hemoglobin; modulation of O_2 affinity, cooperativity, and Bohr effect by heterotropic allosteric effectors. J. Biol. Chem.* 277:34,508, 2002.

Veeramachaneni, N. K., Harken, A. H., and Cairns, C. B. Clinical implications of hemoglobin as a nitric oxide carrier. *Arch. Surg.* 134:434, 1999.

Basal Lamina and Basement Membrane

Annes, J. P., Munger, J. S., and Rifkin, D. B. Making sense of latent TGFβ activation. *J. Cell Science* 116:217, 2003.

Brooke, B. S., Karnik, S. K., and Li, D. Y. Extracellular matrix in vascular morphogenesis and disease: Structure versus signal. *Trends Cell Biol.* 13:51, 2003.

Dityatev, A. and Schachner, M. Extracellular matrix molecules and synaptic plasticity. *Nature Rev. Neurosci.* 4:456, 2003.

Erickson, A. C., and Couchman, J. R. Still more complexity in mammalian basement membranes. *J. Histochem. Cytochem.* 48:1291, 2000.

Kalluri, R. Basement membranes: Structure, assembly and role in tumour angiogenesis. *Nature Cancer Rev.* 3:422, 2003.

Yurchenco, P. D. Assembly of basement membrane networks. In: P. D. Yurchenco, D. E. Birk, and R. P. Mecham (Eds.), *Extracellular Matrix Assembly and Structure*. New York: Academic Press, 1994. p. 351.

QUESTIONS | CAROL N. ANGSTADT

Multiple Choice Questions

1. Haptens:
   A. can function as antigens.
   B. strongly bind to antibodies specific for them.
   C. may be macromolecules.
   D. never act as antigenic determinants.
   E. can directly elicit the production of specific antibodies.

2. In the three-dimensional structure of immunoglobulins:
   A. β-Sheets align edge to edge.
   B. in each chain (H and L) the C and V regions fold onto one another, forming CV associations.
   C. C_L–V_L associations form the complementary sites for binding antigens.
   D. free —SH groups are preserved to function in forming tight covalent bonds to antigens.
   E. hinge domains connect globular domains.

3. The active sites of all serine proteases contain which of the following amino acid residues?
   A. asparagine
   B. γ-carboxyglutamate
   C. histidine
   D. lysine or arginine
   E. threonine

4. Hemoglobin and myoglobin both have all of the following characteristics *except*:
   A. subunits that provide hydrogen bonds to and nonpolar interaction with other subunits.
   B. highly α-helical.
   C. bind one molecule of heme per globin chain.
   D. bind heme in a hydrophobic pocket.
   E. can bind one O_2 per heme.

5. Isohydric transport of carbon dioxide from tissues to lung refers to:
   A. free CO_2 dissolved in plasma.

B. HCO_3^- in plasma.
C. carbamino compounds.
D. the Bohr effect.
E. the presence of carbonic anhydrase in plasma.

6. Basal lamina structure is produced by connecting planar networks of laminins and type IV collagen. Which of the following statements about these is/are correct?
   A. Both laminin and type IV collagen are composed of three polypeptide chains.
   B. Laminin has a cruciform structure.
   C. Laminin and type IV collagen are interconnected by a heparin sulfate proteoglycan.
   D. Type IV collagen contains repeating sequences of $(\text{Gly-HyPro-Y})_n$.
   E. All of the above are correct.

Questions 7 and 8: Immunoglobulins are a family of proteins that are produced in response to invasion by foreign substances, initiating a process by which the foreign substance can be eliminated from the organism. Binding of antibodies to antigens on the cell membrane of the invading organism like bacteria, activates the complement cascade. Activation of the complement system leads to the production of proteins that stimulate phagocytosis by neutrophils and macrophages. An individual with either an immunoglobulin deficiency or a deficiency of one of the complement proteins is susceptible to recurrent infections.

7. In immunoglobulins all of the following are true *except*:
   A. there are four polypeptide chains.
   B. there are two copies of each type of chain.
   C. all chains are linked by disulfide bonds.
   D. carbohydrate is covalently bound to the protein.
   E. immunoglobulin class is determined by the C_L regions.

8. IgG:
   A. is found primarily in mucosal secretions.
   B. is the first antibody elicited when a foreign antigen is introduced into the host's plasma.
   C. has the highest molecular weight of all the immunoglobulins.
   D. contains carbohydrate covalently attached to the H chain.
   E. plays an important role in allergic responses.

Questions 9 and 10: When a blood vessel is injured, the coagulation process is initiated to prevent loss of blood and maintain the integrity of the circulatory system. Coagulation is a process in which zymogens are converted to active serine proteases in a stepwise, cascade process with the final result the production of a cross-linked fibrin clot. In a myocardial infarction, the process can be activated to such an extent that blood flow to the heart muscle is decreased. The fibrinolysis pathway to dissolve fibrin clots also involves activating zymogens to active serine proteases, the final step being the activation of plasminogen to plasmin which acts directly on the fibrin clot. One of the current treatments for myocardial infarctions is rapid administration of t-PA (tissue plasminogen activator). Actually recombinant t-PA, produced by gene cloning technology, is used.

9. Serine proteases:
   A. hydrolyze peptide bonds involving the carboxyl groups of serine residues.
   B. are characterized by having several active sites per molecule, each containing a serine residue.
   C. are inactivated by reacting with one molecule of diisopropylfluorophosphate per molecule of protein.
   D. are exopeptidases.
   E. recognize only the amino acids that contribute to the bond to be broken.

10. All of the following are characteristic of serine proteases as a class *except*:
   A. only one serine residue is catalytically active.
   B. natural protein substrates and inhibitors bind very tightly to the protease.
   C. the genes that code for them are organized in a similar fashion.
   D. catalytic units exhibit two structural domains of dramatically different size.
   E. conversion of zymogen to active enzyme usually involves one or more hydrolytic reactions.

Questions 11 and 12: There are over 800 mutant human hemoglobins. Mutations may cause instability leading to rapid degradation, alterations in oxygen affinity, or more rapid oxidation of heme Fe^{2+} to Fe^{3+}. All of these alter hemoglobin's ability to carry out its physiological functions. In the mutation $Hb_{Cowtown}$, histidine 146 that contributes to 50% of the Bohr effect is lost. Regulation of oxygen dissociation by protons is impeded, the T confirmation is destabilized relative to the R conformation and oxygen affinity is increased.

11. All of the following are believed to contribute to the stability of the deoxy or T conformation of hemoglobin *except*:
   A. a larger ionic radius of six-coordinated ferrous ion as compared to five-coordinated ion.
   B. unstrained steric interaction of His F8 with the porphyrin ring when iron is above the plane.
   C. interactions between the FG corner of one subunit and the C-helix of the adjacent subunit.
   D. ionic interactions.

12. When hemoglobin is converted from the deoxy (T) form to oxyhemoglobin (R):
   A. it becomes more acidic and releases protons.
   B. carbamino formation is promoted.
   C. binding of 2,3-bisphosphoglycerate (BPG) is favored.
   D. bound NO is transferred to glutathione.
   E. all of the above are correct.

Problems

13. One of the adaptations to high altitude is an increase in the concentration of BPG. What effect does this have on a saturation versus pO_2 curve? Why does increasing [BPG] increase the delivery of O_2 to tissues?

14. Both myoglobin and hemoglobin consist of globin (protein) bound to a heme prosthetic group. One heme group binds one O_2. Why is the oxygen saturation curve (saturation versus pO_2) of myoglobin a rectangular hyperbola while that of hemoglobin is sigmoidal?

ANSWERS

1. **B** Haptens are small molecules and cannot alone elicit antibody production; thus they are not antigens. They can act as antigenic determinants if covalently bound to a larger molecule, and free haptens may bind strongly to the antibodies thereby produced.
2. **E** See Figure 9.2. A: The β-sheets align face-to-face. B: The V and C regions are adjacent to each other. C: The complementarity regions are the variable regions of both the heavy and light chains (V(H)–V(L). D: Antigen binding is noncovalent.
3. **C** Histidine participates in the catalytic mechanism. A: Aspartate, not asparagine, is involved. B: γ-Carboxyglutamate is essential to some of the serine proteases, but it is not at the active site. D: These are the substrate specificities of the trypsin-like proteases. E: This is not involved.
4. **A** Hemoglobin has four chains and four oxygen binding sites, whereas myoglobin has one chain and one oxygen binding site. Each oxygen binding site is a heme.
5. **B** Protons released as H_2CO_3 dissociates to HCO_3^- and H^+ bind to oxyhemoglobin, facilitating release of O_2 to tissues. A: Very little CO_2 is in this form. C: This is another way to carry CO_2. D: Bohr effect releases H^+ as deoxy Hb picks up O_2. E: Carbonic anhydrase is in red cells.
6. **E** All of these contribute to the basil lamina which with connecting fibrillar collagen attached to the outside is called the basement membrane.
7. **E** The C_H regions determine class. A: There are two copies of each of two types of polypeptide chain.
8. **D** All immunoglobulins are glycoproteins. A: IgA is associated with mucosal secretions. B: IgM arises first; IgG takes longer to appear but eventually is present in higher concentration. C: IgM has the highest molecular weight. E: IgE plays an important role in allergic responses.
9. **C** This is the distinguishing characteristic of the serine proteases, and of the serine hydrolases in general. A: They have various specificities. B: There is only one active site per molecule. D: They are all endopeptidases. E: An "extended active site" containing the hydrolyzable bond and about four amino acids on either side is responsible for specificity.
10. **D** The domains are of about equal size.
11. **A** Six-coordinated ferrous ion has a smaller ionic radius than the five-coordinated species and just fits into the center of the porphyrin ring without distortion.
12. **A** Positively charged histidine is no longer stabilized by close proximity to an aspartate carboxylate group. This is the Bohr effect. B: Increased H^+ favors dissociation of carbamino groups which is what happens in the lung. C: BPG does not bind effectively to oxyhemoglobin because the binding pocket is too small. D: In oxyhemoglobin, NO is bound to cysteine and it gets transferred to glutathione when O_2 is released (R form reverts to T form).
13. Binding of BPG to hemoglobin stabilizes the T conformation so for any given amount of O_2, less O_2 will be bound to hemoglobin. This shifts the saturation curve to the right and increases P_{50}. The equilibrium shown illustrates that increasing BPG causes the release of O_2 which must happen in order for O_2 to enter tissues.

$$H^+BPG{\cdot}Hb + 4O_2 \rightleftharpoons Hb(O_2)_4 + BPG + nH^+$$

(Eq. 9.14 from text)

14. Myoglobin is a monomer with one globin and one heme binding one O_2. O_2 binding is a simple equilibrium with a Hill coefficient of 1. Hemoglobin is a tetramer with four O_2 bound to the four hemes. Hemoglobin without oxygen is in the T conformation in which binding of O_2 is difficult. Thus, the curve starts with a slow increase as pO_2 increases. Binding of the first O_2 shifts the conformation of the whole molecule to favor the R form which binds O_2 more readily. Thus, the slope rises steeply giving it a sigmoidal shape. This is an example of cooperativity. The Hill coefficient for hemoglobin is 2.8.

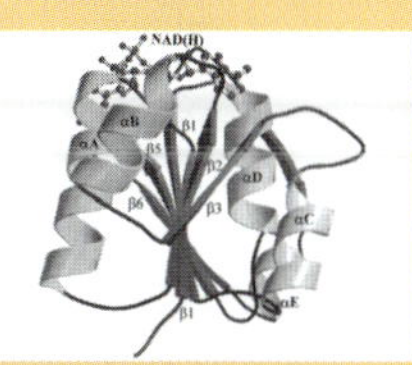

10

ENZYMES: CLASSIFICATION, KINETICS, AND CONTROL

Henry Weiner

Textbook of Biochemistry With Clinical Correlations, Sixth Edition, Edited by Thomas M. Devlin

10.1 OVERVIEW

Enzymes are specialized proteins that catalyze biological reactions. Virtually every reaction that occurs within a cell requires the action of an enzyme because most reactions will not occur under the physical conditions (pH, temperature, and ionic milieu) of the cell. A catalyst increases the rate of a chemical reaction and can be an inorganic or organic compound. Enzymes are efficient catalysts: They not only increase the rate of conversion of substrate to product, but also recognize a specific structure in the presence of similar structures to produce a unique product. There are six major classes of enzymes, each class catalyzing a different chemical reaction. An enzyme speeds up the rate of product formation by lowering the **energy of activation.** This term is related to how much of a barrier there is for the substrate to become the product. Changing the energy of activation is accomplished when the enzyme interacts with the substrate, stabilizing the substrate in a form that will allow formation of product. This is referred to as the **transition state**. The more stable the transition state, the less energy it will take for conversion of substrate to product, and the faster will be the reaction. The transition state of many enzymes has been described and is useful in design of drugs. Each substrate and product binds to the enzyme but with different affinities.

Derivatives of vitamins, called coenzymes, are important in many enzyme catalyzed reactions. Some must be present for the reaction to occur but are unchanged at the end of the reaction. In other cases, the coenzyme is chemically altered during the reaction. Coenzymes may be covalently bound to the enzyme or free to associate and dissociate from the protein. The enzyme without its cofactor is called the **apoenzyme** and **haloenzyme** when the cofactor is bound. Metal ions also serve as cofactors for many enzymes.

An enzyme is much larger than the substrate, and the portion of the enzyme where the substrate binds is referred to as the **active site or substrate-binding site**. Some enzymes are promiscuous in that they can bind many different substrates while others are very specific and will bind only one unique structure. The velocity that an enzyme converts substrate to product can be described by the **Michaelis–Menten equation**, in which one term, K_m, is related to how well the substrate binds to the enzyme and another term, k_{cat}, is related to how fast the enzyme converts the substrate to product.

In addition to the substrate and cofactor binding sites on the enzyme, some enzymes possess additional sites, referred to as **allosteric sites,** that regulate the catalytic activity. The binding ligands are called **allosteric modifiers or effectors**. Often their binding causes the enzyme to change its three-dimensional structure to become either more or less efficient as a catalyst. The catalytic property of some enzymes is modified by covalent modification, such as phosphorylation, of the protein. Thus, the cell alters the activities of some enzymes to coincide with the metabolic needs of the cell. Some reactions are catalyzed by different enzymes, which are products of different genes. These proteins share very similar amino acid sequences and are called **isozymes**.

Not all biological catalysts, however, are enzymes. Catalytic RNAs (see p. 69) have been identified that participate in processing introns (see page 192) and tRNA, while others perform a self-processing step by hydrolyzing a phosphodiester bond within its own nucleotide chain. Artificial enzymes, termed **abzymes**, have been synthesized by making antibodies against chemicals that are transition state analogs. Abzymes have been designed to catalyze over 100 chemical reactions.

The following sections will present the types of enzymes, some general mechanistic concepts to illustrate how catalysis takes place, and how the catalytic properties of enzymes can be regulated. Chemical and kinetic properties, as well as the way inhibitors affect enzymes, will be described.

10.2 CLASSIFICATIONS OF ENZYMES

With the advent of genomic sequences, many enzymes have been identified. By mid-2004, information is available for over 83,000 different enzymes from 9800 different organisms. All enzymes are classified as belonging to one of six classes, defined by the chemical reaction they catalyze. A simple numbering system, consisting of four numbers for each enzyme, has been developed by the International Union of Biochemistry and Molecular Biology (IUBMB) to characterize each enzyme. The first number defines the type of reaction that is catalyzed, followed by numbers to define details of the reaction. It is interesting that the entire chemistry of life can be reduced to six different types of chemical reactions. The classes of enzymes are listed in Table 10.1.

Arbitrary names are frequently given to enzymes; for example, aldolase was the name given and still used for the enzyme that catalyzes an aldol-condensation reaction between glyceraldehyde-3-phosphate and dihydroxyacetone phosphate (see p. 587), and lysozyme was the term used for an enzyme that breaks down bacterial cell walls.

Systematic names include the substrate and the type of reaction. Alcohol dehydrogenase, an enzyme that oxidizes an alcohol to an aldehyde, has the IUBMB number 1.1.1.1. This indicates that the enzyme is involved in an oxidation–reduction reaction (first number), which removes hydrogen as a hydride ion with NAD^+ as the electron

TABLE 10.1 Summary of the Enzyme Classes and Major Subclasses

1. Oxidoreductases	2. Transferases
Dehydrogenases	Transaldolase
Oxidases	and transketolase
Reductases	Acyl, methyl,
Peroxidases	glucosyl, and
Catalase	phosphoryltransferases
Oxygenases	Kinases
Hydroxylases	Phosphomutases
3. Hydrolases	4. Lyases
Esterases	Decarboxylases
Glycosidases	Aldolases
Peptidases	Hydratases
Phosphatases	Dehydratases
Thiolases	Synthases
Phospholipases	Lyases
Amidases	6. Ligases
Deaminases	Synthetases
Ribonucleases	Carboxylases
5. Isomerases	
Racemases	
Epimerases	
Isomerases	
Mutases (not all)	

acceptor (second number), with alcohol being the substrate (third number). The last number is reserved to differentiate each enzyme that catalyzes the same overall reaction but with different substrates. Lactate dehydrogenase (1.1.1.27) catalyzes an identical reaction as alcohol dehydrogenase, but the substrate is lactate. Enzymes are often named for the direction of the reaction of importance in the cell. There has been an effort to drop the trivial in favor of the systematic name of many enzymes, but the literature continues to contain references to both names for many enzymes.

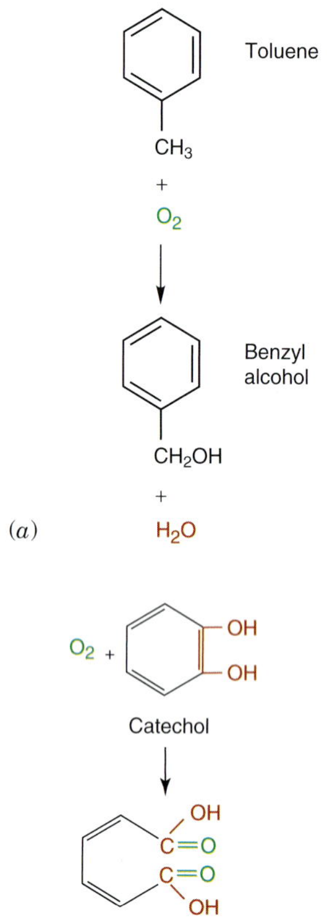

FIGURE 10.1

Enzyme-catalyzed oxidation reactions. (*a*) Hydroxylation of toluene. (*b*) Oxygenation of catechol by an oxygenase.

Class 1: Oxidoreductases

Class 1 enzymes catalyze oxidation–reduction reactions and are referred to as oxidoreductases. Oxidation means the loss of electrons while reduction the addition of electrons. Electrons are removed from one substrate (donor or reductant) that is oxidized, and they are added to a second substrate (acceptor or oxidant) that is reduced. Many different electron acceptors are used in biological systems. Dehydrogenases are typically named in the oxidizing direction. Examples are lactate dehydrogenase rather than pyruvate reductase and glyceraldehyde-3-phoshpate dehydrogenase rather than 1,3-bisphosphoglyerate reductase.

Oxidations and reduction reactions include the following pairs of donors–acceptors: saturated–unsaturated carbon-carbon bonds, alcohols–aldehydes, aldehydes–acids, and amines–imines. Molecular oxygen (O_2) as an electron acceptor is involved in a variety of irreversible oxidation–reduction reactions. Monooxygenases catalyze formation of a hydroxy group and dioxygenases incorporate both atoms of O_2 into a substrate. Cytochromes P450 (see p. 414) are an important group of enzymes that use oxygen in the metabolism of xenobiotics such as drugs and toxins; these enzymes convert a saturated compound into an alcohol, a hydroxylation reaction, using one of the oxygen atoms in the O_2 molecule (Figure 10.1).

Class 2: Transferases

Members of this large family transfer a chemical group from one molecule to another, thus they have two substrates and produce two products. Hexokinase transfers a phosphate group from ATP to an acceptor such as glucose (Figure 10.2). An enzyme that transfers phosphate to another molecule using a nucleotide triphosphate as the donor is referred to as a **kinase**; the acceptor can be a small molecule or a protein. Another example of a transferase is acyltransferase that catalyzes the transfer of an acetyl group from acetyl CoA to an acceptor such as lysine or serine (Figure 10.3).

An important amine transfer reaction occurs between an amino acid and a carbonyl-containing compound. These enzymes, **aminotransferases**, transfer the amino group and convert the amino acid to a keto acid; the accepting substrate, a keto acid, is transformed in to an amino acid (Figure 10.4). Aminotransferases require pyridoxal phosphate, derived from vitamin B_6, as a coenzyme (see p. 747).

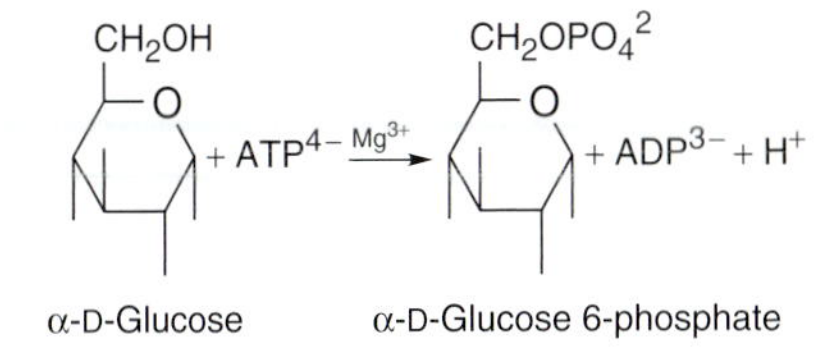

α-D-Glucose + ATP^{4-} $\xrightarrow{Mg^{2+}}$ α-D-Glucose 6-phosphate + ADP^{3-} + H^+

FIGURE 10.2

Phosphorylation reaction. Phosphorylation of glucose by ATP catalyzed by hexokinase.

Class 3: Hydrolases

Hydrolases catalyze hydrolysis reactions, which is the addition of water to a chemical bond (Figure 10.5). This is essentially a transfer of an —OH from water to the substrate, but the enzymes are not classified as transferases. Hydrolysis is essentially an irreversible reaction. The substrate is typically an ester or an amide—for example, an ester between an alcohol and a carboxylic acid. Such a reaction is the hydrolysis of a cholesterol ester to produce cholesterol and a fatty acid. The acid need not be a carboxylic acid; sulfuric acid, phosphoric acid, or any acid can form an ester bond. RNase and DNases are enzymes that hydrolyze the phospho–ester bond (see p. 31).

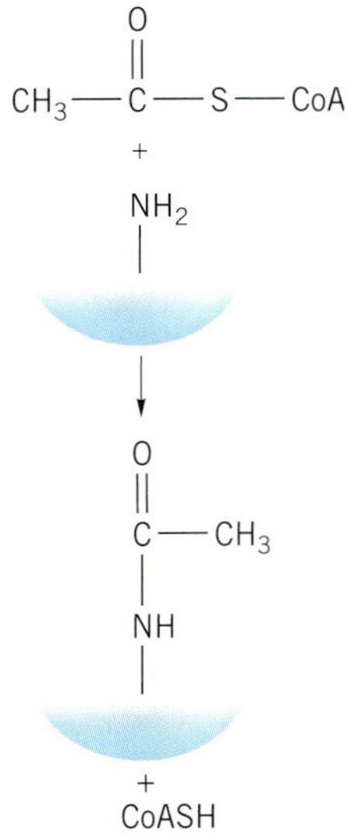

FIGURE 10.3

Acetylation of amino acid residues in a protein.

Class 4: Lyases

Lyase implies a breaking apart. These enzymes usually catalyze a carbon–carbon bond cleavage. Other enzymes in this class can form or break a carbon–nitrogen bond or release a CO_2 from a β-keto acid. Some of the reactions are reversible so a carbon–carbon bond can form, as illustrated by the reaction catalyzed by aldolase (Figure 10.6). Some lyases require pyridoxal phosphate as the cofactor (see p. 1104). Important neurochemical reactions including the formation of dopamine (Figure 10.7) and serotonin are examples of this reaction. If a nucleotide triphosphate is not involved in forming the new bond, the enzyme is called a **synthase**; but if the energy released when ATP is hydrolyzed is required, the enzyme is referred to as a **synthetase**.

Class 5: Isomerases

The enzymes in this class are involved with moving a group or a double bond within the same molecule. These include exchanging the position of a hydroxyl and a carbonyl, such as found when glucose-6-phosphate is converted to fructose-6-phosphate (see p. 588), and moving of a double bond from one position to the adjacent one, as found in fatty acid metabolism. The enzyme is called a mutase when a phosphate is moved from one carbon to another within the same molecule such as occurs with phosphoglycerate mutase that converts 2-phosphoglycerate to 3-phosphoglycerate (Figure 10.8). Isomerases and epimerases may change the stereochemistry at a carbon atom. The conversion of D-lactate to L-lactate is an example of an isomerase and D-xylulose 5-phosphate to D-ribulose 5-phosphate an epimerase (Figure 10.9). The

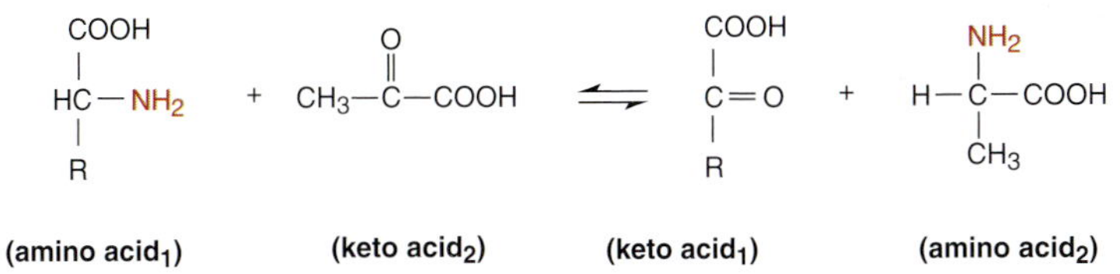

(amino $acid_1$) + (keto $acid_2$) ⇌ (keto $acid_1$) + (amino $acid_2$)

FIGURE 10.4

Example of a reaction catalyzed by an aminotransferase. Aminotransferase reactions require pyridoxal phosphate (PLP).

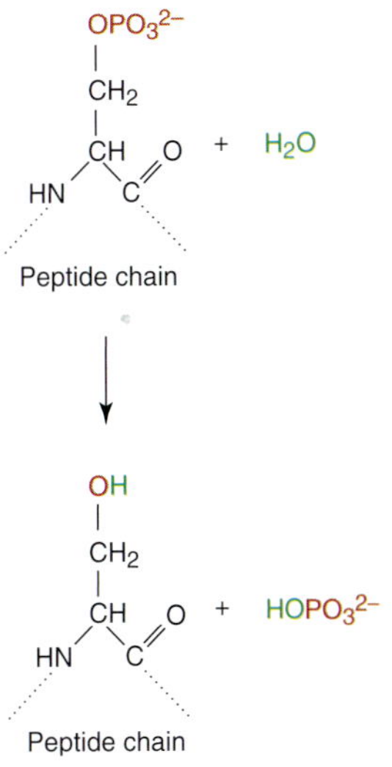

FIGURE 10.5

Hydrolysis of a phosphorylated protein by a protein phosphatase.

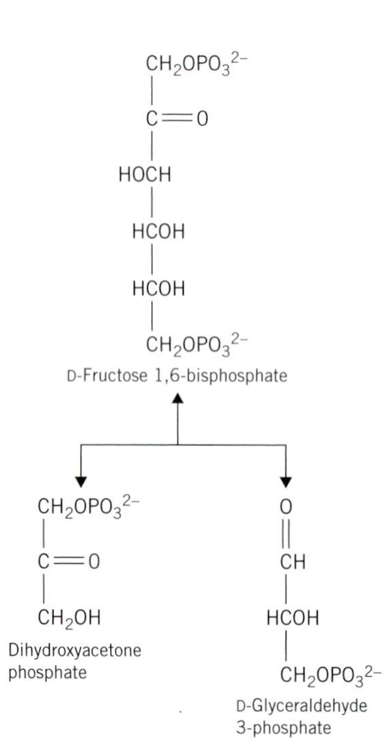

FIGURE 10.6

Lyase reaction catalyzed by aldolase.

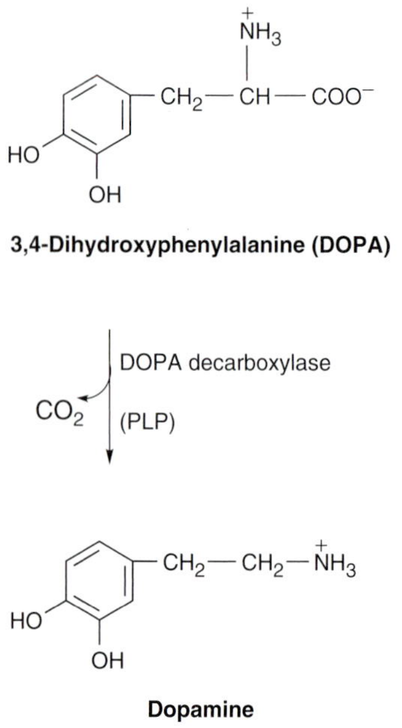

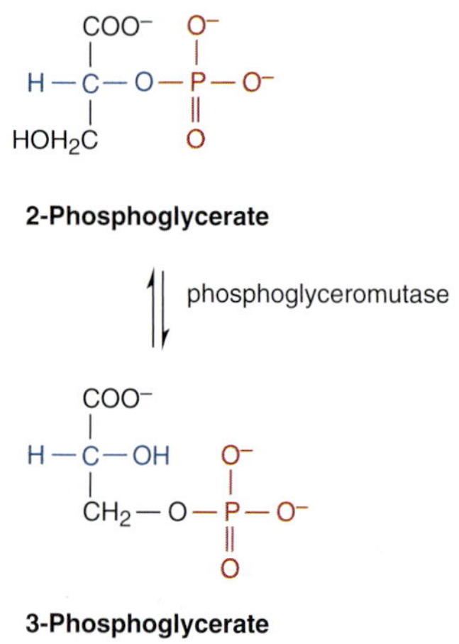

FIGURE 10.8

Interconversion of 2- and 3-phosphoglycerates.

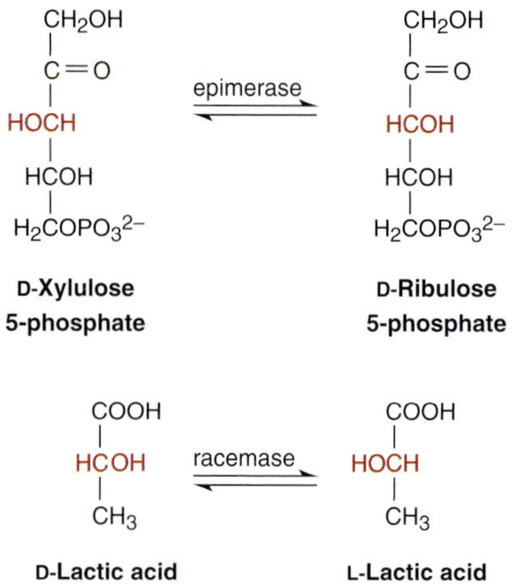

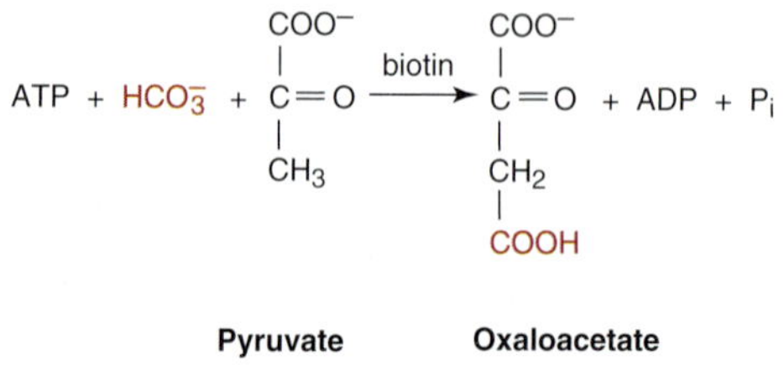

FIGURE 10.10

Pyruvate carboxylase reaction.

FIGURE 10.7

Dopamine synthesis involves a lyase reaction. Synthesis of dopamine requires pyridoxal phosphate (PLP).

FIGURE 10.9

Examples of reactions catalyzed by an epimerase and a racemase.

UDP-glucose 4-epimerase-catalyzed reaction requires the involvement of NAD, so it is assumed that an oxidation–reduction reaction occurs during the reaction but there is no net oxidation–reduction of the coenzyme.

Class 6: Ligases

Ligases join carbon atoms together, but unlike lyases, Class 4, they require energy for the reaction to occur. They are referred to as **synthetases**. Typically, the energy comes from the involvement of ATP. For example, to add CO_2 to pyruvate, CO_2 is incorporated into the coenzyme biotin, which requires hydrolysis of ATP (Figure 10.10). The CO_2 is transferred to pyruvate by the same enzyme. The addition of an amino acid to tRNA is another example of an enzyme in this class (see p. 207). There must be some interaction between the components of the reaction to utilize the energy released from the hydrolysis of ATP for synthesis of the new bond.

10.3 GENERAL CONCEPTS OF ENZYME MECHANISMS

Thermodynamic Considerations

Energy is required to form the starting material of a reaction and to form the product of a reaction. This energy is referred to as the **heats of formation**. There is a **thermodynamic relationship** between the difference in energy needed to form the starting material (reactants) and the energy required to form the products. The energy difference is $\boldsymbol{\Delta G_0}$ and the units are kilocalories/mole (kcal/mol). When the equilibrium of the reaction is established, the concentration of each is related as indicated in Eq. 10.1:

$$\Delta G_0 = -2.3\,RT \times \log\,[\text{Products}]/[\text{Reactants}] \qquad (10.1)$$

where R is the universal gas constant (1.38 cal/mol-°K) and T is the temperature in degrees K. When added to a reaction a catalyst increases the rate of reaching equilibrium between the starting material and the product. The ratio of [Products]/[Reactants] at equilibrium is the equilibrium constant, $\boldsymbol{K_{eq}}$. This implies that the difference in the free energy (heats of formation) of the products and reactants will determine the amount of each that is present when equilibrium is reached. It does not dictate the rate at which equilibrium is reached. A catalyst increases the rate of the reaction to reach equilibrium but does not change the equilibrium. K_{eq} is governed by the heats of formation, which is a constant for each reaction.

Many reactions are considered essentially irreversible; even in the presence of an enzyme, it is not possible for products of irreversible reactions to be reconverted to starting material. In this case, essentially 100% of the starting material will become product. This statement is not completely valid because even with essentially irreversible reactions the ratio of starting material to products might be 1:200,000 or even 1:1,000,000. The reaction reaches equilibrium, but the amount of starting material present at equilibrium is beyond the level of detection, so the reaction is considered essentially irreversible.

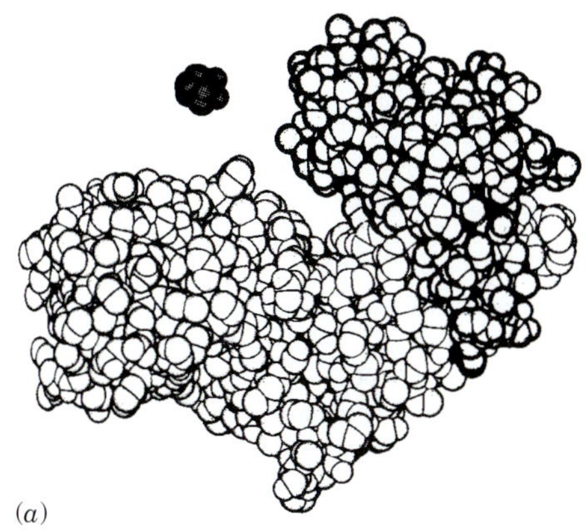

(*a*)

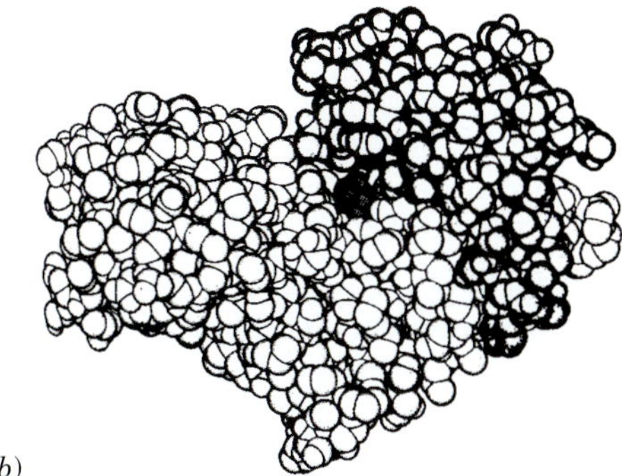

(*b*)

FIGURE 10.11

Glucose induced conformational change of hexokinase. Drawings of (*a*) unliganded form of hexokinase and free glucose and (*b*) conformation of hexokinase with glucose bound. In this space-filling, drawing each circle represents van der Waals radius of an atom in the structure. Glucose is black, and each domain is differently shaded.

Reprinted with permission from Bennett, W. S. and Huber, R. *CRC Rev. Biochem.* 15:291, 1984. Copyright (1984) CRC Press, Inc.

Binding of Substrate by an Enzyme

For an enzyme to function as a catalyst it must first bind the substrate(s), then lower the energy of activation so that the reaction will proceed at a faster rate than it would in the absence of the catalyst. Two prevalent theories exist to explain how substrate binds to the enzyme. The classical one is the **lock-and key-theory**, which implies that the binding site for substrate on the enzyme is a rigid entity and only a compound with a particular shape will fit, analogous to how a lock (enzyme) allows only one key (substrate) to make the proper contact. An alternative theory, developed in the 1960s, is the **induced fit theory**. It assumes that the enzyme is flexible and after substrates bind to the enzyme the conformation of the protein changes so that a stable binary complex forms. Both of the theories were presented prior to knowing the actual structure of any protein. Now that three-dimensional structures of many enzymes are known, it is found that substrate binding causes a small movement of the peptide bonds supporting the induced-fit hypothesis (Figure 10.11).

The altered structure of the enzyme, albeit similar in most cases to the enzyme structure found in the absence of substrate, allows for complementary interactions between substrate and enzyme to take place. That is, a hydrophilic portion of the substrate will lie near amino acid residues that allow a hydrogen bond to be shared and a charged portion of the substrate will interact such that either its charge is neutralized by an opposite charge or it is stabilized by forming a hydrogen bond. Hydrophobic portions of the substrate bind such that they are in a hydrophobic portion of the protein, often referred to as a **hydrophobic pocket**.

Transition State

The theory to describe how a chemical reaction occurs is based upon the transition state model. The basic concept is that the starting material (substrate) has certain structural and chemical properties, and a product has others. The transition state is a structure where the bond(s) undergoing a transformation are such that they are not like either the starting material or the product. This can be illustrated in the reaction of ammonia adding to *trans*-cinnamic acid to produce phenylalanine (Figure 10.12) where the double bond in the starting material is planar. The product has the amine attached so the carbon that was part of the double bond (planar) is now tetrahedral in structure. The transition state is where the bond is starting to form and the geometry is changing. It takes energy to reach the transition state, and the velocity of a reaction is inversely proportional to the energy needed to reach the transition state. Presented in Figure 10.13 is a reaction energy plot used to depict the energy change necessary to convert a substrate to the transition state. At the transition state, an entity exists that has never been observed or isolated. The transition state entity can proceed to produce product or it can return to the starting material. A catalyst simply lowers the energy of the transition state (ΔE^*), thus speeding up the rate of its formation from the starting material.

Enzymes stabilize the transition state by lowering the energy of activation, hence speeding up the reaction. This is accomplished by having residues in the enzyme interact with the molecule that is in the transition state.

The hydrolysis of a peptide bond to produce a carboxylic acid and a free amino group illustrates how it might be possible for the enzyme to stabilize the transition state (Figure 10.14). As water adds to the peptide bond, the carbonyl group becomes tetrahedral and hence has to change its bond angle and shape. The transition state is not the tetrahedral intermediate but is the conformation that would lead to it should

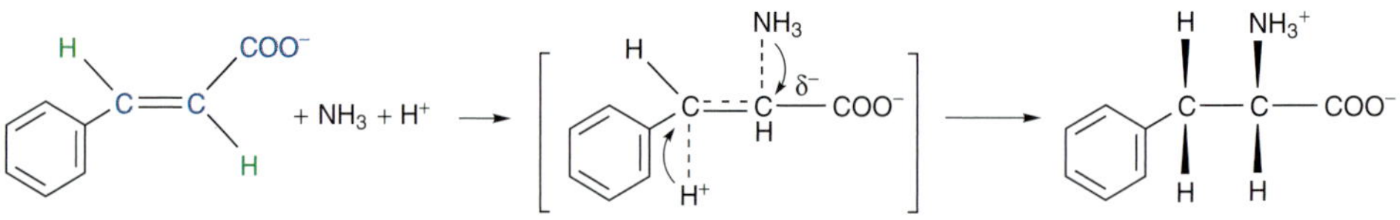

FIGURE 10.12

Transition state. The transition state structure shows electrons in the double bond moving to accommodate the incoming proton, which makes the other carbon positive so the pair of electrons on the nitrogen can attack.

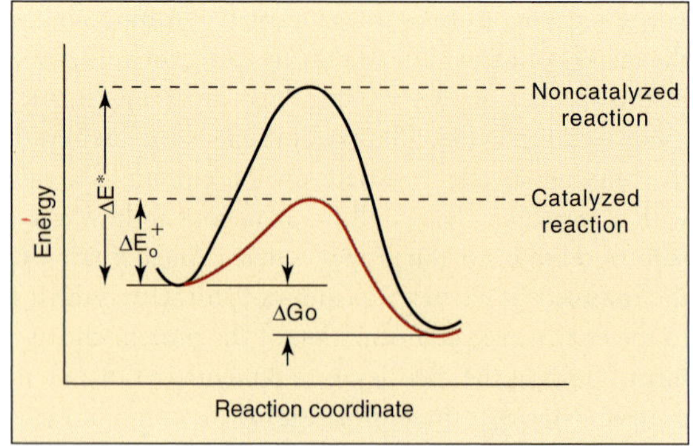

FIGURE 10.13

Energy diagrams for catalyzed versus noncatalyzed reactions. Diagrammatic representation of the energy differences between the reactant and product. The ground state is indicated by ΔG^0 and the transition state by ΔE^*. The velocity of the reaction is inversely proportional to ΔE^* while K_{eq} is related to the difference in ΔG^0. ΔE^* is the energy required to raise the ground state to the hypothetical transition state.

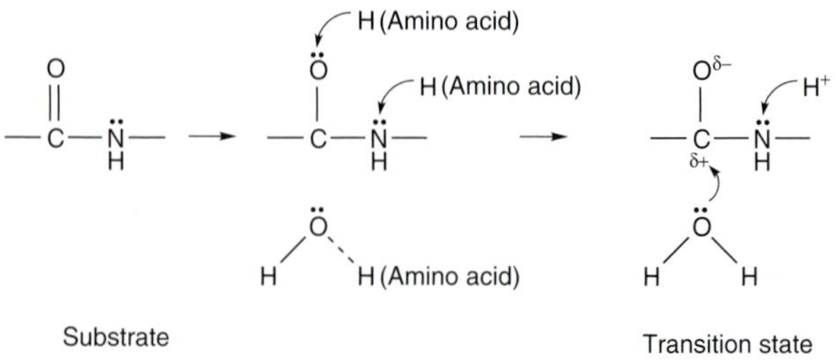

FIGURE 10.14

Transition state for the hydrolysis of a peptide bond. The carbonyl group in the planar peptide bond becomes tetrahedral as water (or OH^-) attacks and the nitrogen is protonated, followed by hydrolysis of the peptide bond. Amino acids in the active site furnish a H^+ to stabilize the developing negative charge on the carbonyl oxygen atom and protonate the nitrogen so it will leave as an amine. A different amino acid in the protein removes a proton from water producing the more reactive OH^-.

the covalent bond actually form. During the reaction, not only does the carbon–oxygen bond of the carbonyl group on the substrate change from being planar to being tetrahedral, but a negative charge develops on the oxygen that needs to be stabilized.

Tight Binding in the Transition State

An enzyme binds the transition state tightly but not the substrate or final product. If it binds substrate tightly then it takes more energy to reach the transition state. Similarly, the enzyme cannot bind the product too tightly for if it did, it would become too difficult for the product to dissociate from the enzyme. The enzyme needs to release product so it can again interact with another molecule of substrate. These concepts are illustrated in Figure 10.15.

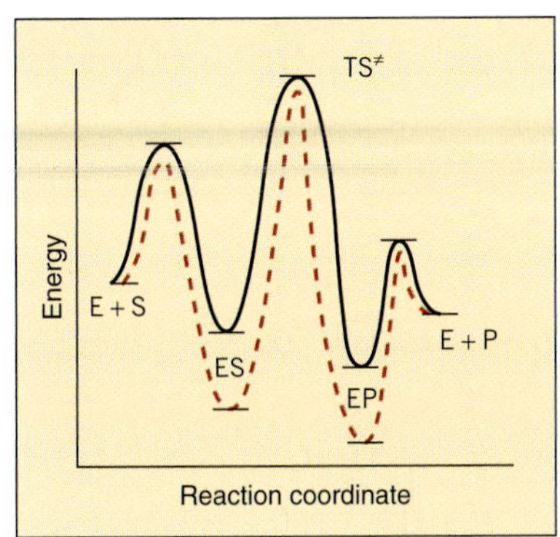

FIGURE 10.15

Energy diagrams for transition state of tightly bound substrate. Red line indicates tight binding of substrate in comparison to a less tightly bound substrate. More energy is required if the substrate is tightly bound, hence a slower reaction to reach the transition state. A similar condition occurs if the product is bound too tightly; more energy would be required to reach the transition state for releasing product. There is a transition state for each step in the reaction.

Ionic Reactions Need Not Involve Ions

The majority of reactions found in nature can be explained using ionic models. That is, most cellular reactions require positive and negative charges within the molecules to allow the components to interact. The enzyme can add or remove protons from the substrate changing its charge, or a neutral group of the enzyme becomes ionic. The removal of a proton makes the group more nucleophilic—that is, a group (nucleophile) that will attack a positive center. For example, OH^- is more nucleophilic than is water, and ionized cysteine is more nucleophilic than is the un-ionized sulfur atom. Similarly, enzymes can protonate a group to make it positive so it will be more susceptible to nucleophilic attack. In the hydrolysis of a peptide bond the more nucleophilic —OH^- group attacks the bond rather than water. In addition, the attack will be on a polarized carbonyl group, where the carbon atom would have a partial positive charge, rather than on an unpolarized carbonyl group (Figure 10.14). The addition and removal of protons is referred to as **general acid** and **general base** catalysis, respectively. The advantage of enzymes using general acids and bases is that at a neutral pH they carry out acid/base chemistry even though the concentration of OH^- and H^+ is very low.

Partially Charged Bonds

A common way for forming a carbon–carbon bond is to have a carbanion attack a carbonyl group, as illustrated with the formation of fructose 1,6-bisphosphate from two three-carbon compounds (Figure 10.16). The presence of the carbonyl group in dihydroxyacetone phosphate allows the hydrogen atoms on the carbon adjacent to be more acidic. The enzyme removes a proton, leaving a negative charge on the carbon. The nature of the carbonyl group is such that the electrons reside closer to the oxygen so that the carbon carries a partial positive charge. This stabilizes the carbanion since the electrons are delocalized and not just associated with one carbon. The carbanion

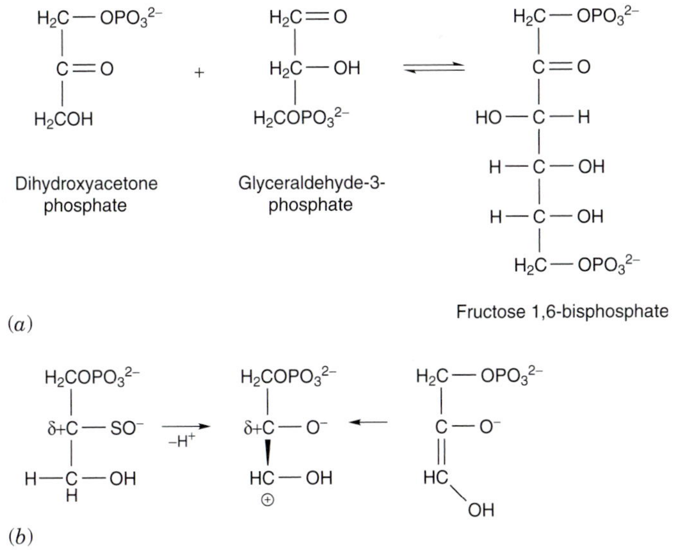

(a)

Dihydroxyacetone phosphate + Glyceraldehyde-3-phosphate ⇌ Fructose 1,6-bisphosphate

(b)

FIGURE 10.16

A carbonyl group stabilizes a negative charge on the adjacent carbon. (*a*) Reaction of dihydroxyacetone phosphate and glyceraldehyde 3-phosphate to form fructose 1,6-bisphosphate. (*b*) Dihydroxyacetone phosphate transition states. The partial positive charge on the carbon can be attacked by a carbanion to form a new carbon–carbon bond. The presence of the carbonyl group will allow a proton to be removed from the adjacent carbon atom, allowing the carbanion to be produced.

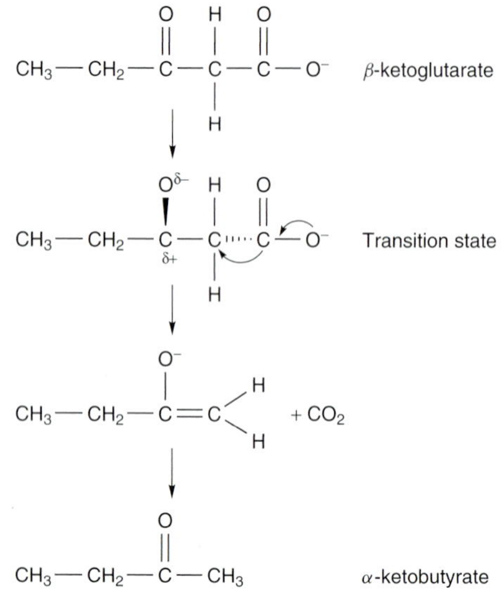

FIGURE 10.17

Decarboxylation of β-ketoglutarate. The decarboxylation of β-ketoglutarate occurs because the partial positive charge on the carbon of the carbonyl group stabilizes the developing negative charge that forms as the bond to the carboxyl group starts to break. A carbonyl adjacent to the leaving CO_2 cannot stabilize the charge.

then can attack the carbonyl group to form the new bond. The enzyme both holds the two substrates near to each other, increasing the likelihood of favorable collision, and is involved in polarizing the carbonyl bond and removing a proton.

Importance of the Carbonyl Group

Many metabolic transformations are based upon the special nature of the carbonyl group and its ability to stabilize a negative charge on the carbon adjacent to it. For example, β-keto acids such as β-ketoglutarate (Figure 10.17) can be decarboxylated, but it is energetically more difficult for an α-keto acid such as pyruvate to be decarboxylated. As the $—CH_2—COO^-$ bond breaks, the electrons in the bond are left behind on the β-carbon. The developing negative charge on what was the β-carbon can be stabilized by the partial positive charge on the carbonyl carbon. The enzyme stabilizes the transition state by having a proton interact with the carbonyl oxygen that is partially negative due to the resonance of the carbonyl carbon and the developing negative charge on the β-carbon.

In contrast, pyruvate cannot undergo a simple decarboxylation reaction, because there would be no way to stabilize the negative charge that would result from the pair of electrons being left as the CO_2 group departs. For decarboxylation of α-keto acids, coenzymes are required (see p. 539).

Oxidations

Even oxidation reactions have an ionic component. During the oxidation of an alcohol to an aldehyde or a carbon–carbon saturated bond to an unsaturated bond the electrons are moved not as free electrons but as a hydride ion—that is, a proton with two electrons carrying a negative charge (Figure 10.18). Dehydrogenases stabilize the transition state for the developing charges and shapes.

Addition and Removal of Protons

Most enzymes efficiently add and remove protons, but often in a unique way. Proteases can polarize the carbonyl group of the peptide bond so that the bond is more easily hydrolyzed, catalyzing general base catalysis, or acid catalysis, or metal ion catalysis (see p. 383).

Phosphorylation of proteins, especially on the hydroxyl group of a serine, threonine, or tyrosine residues by ATP or GTP, proceeds by having the —OH group attack phosphate (Figure 10.19). It is necessary to make the —OH more nucleophilic by having it lose a proton while the phosphate–oxygen bond is being polarized so the phosphorous atom is partially positive and is more susceptible to nucleophilic attack. Nearly every enzyme (kinases) catalyzing this type of reaction uses Mg^{2+}. Divalent ions can bind to two or three of the phosphates. The presence of the positive ion causes the electrons in the P=O bond to reside closer to the oxygen atom, thus leaving the phosphate ion with a slightly positive charge.

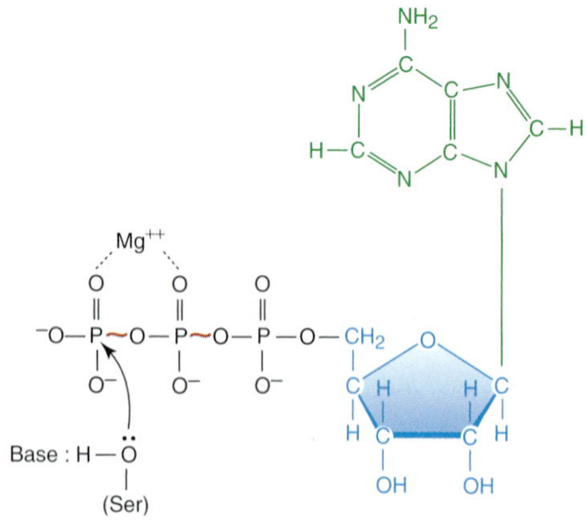

FIGURE 10.19

Phosphorylation is an attack of a nucleophile on a nucleoside triphosphate (NTP). ATP or another NTP is attacked by the pair of electrons on the oxygen of an alcohol such as serine, threonine, or tyrosine. An Mg^{2+} ion when bound by the NTP serves to pull electrons away from the phosphorous group, making it more susceptible to attack by the nucleophile.

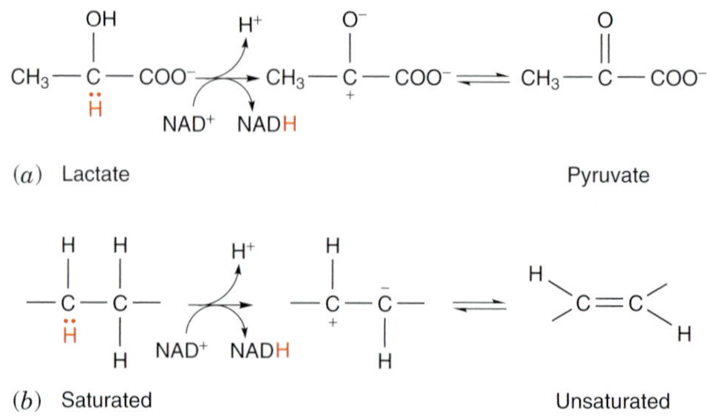

FIGURE 10.18

Oxidation of lactate and a saturated carbon bond. (*a*) Lactate to pyruvate. (*b*) Saturated bond to an unsaturated bond. These reactions undergo an ionic type reaction even though a hydride ion (H^-) is involved.

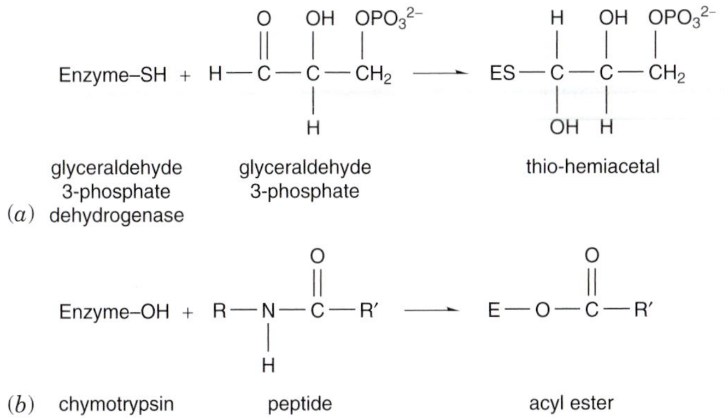

(a) Enzyme–SH + H–C(=O)–CH(OH)–$CH_2OPO_3^{2-}$ ⟶ ES–CH(OH)–CH(OH)–$CH_2OPO_3^{2-}$

glyceraldehyde 3-phosphate dehydrogenase; glyceraldehyde 3-phosphate; thio-hemiacetal

(b) Enzyme–OH + R–NH–C(=O)–R′ ⟶ E–O–C(=O)–R′

chymotrypsin; peptide; acyl ester

FIGURE 10.20

Some substrates form a covalent bond with an enzyme. (*a*) Glyceraldehyde 3-phosphate dehydrogenase and (*b*) chymotrypsin function by first forming a covalent intermediate with substrate.

Covalent Attachment of Substrate to Enzyme

Enzymes not only function in performing general acid and base catalysis, but in some cases the substrate forms a covalent bond with the enzyme. Both chymotrypsin and glyceraldehyde 3-phosphate dehydrogenase are examples of such enzymes (Figure 10.20). In each case, a nucleophilic amino acid attacks a carbonyl group. Even some kinases function with covalent catalysis; ATP initially binds to the enzyme and the phosphate is transferred to a nucleophilic residue, forming a phosphorylated enzyme. The chemical reaction is essentially identical to the one where the sugar attacks ATP, only in this case a hydroxyl group on a serine or threonine is activated and does the initial attacking. Substrate then enters, and it is activated by the general base on the enzyme; the base attacks the phosphate group forming the final product.

pH Affects the Reaction by Affecting General Acids and Bases

Since virtually all enzymes involve general acid/base catalysis, it is expected that the reaction rates would be affected by pH. As an example, an enzyme requiring a histidine to donate a proton in the reaction will not proceed at pH above 8, where the imidazole group of histidine is essentially unprotonated because its pK_a is near 7 (Figure 10.21). All enzymes have a pH versus rate profile (see p. 393), which in some cases gives an insight as to the groups involved in the active site. Since the overall structure of the protein is important in the catalytic activity, if a change in pH alters the overall protein conformation, then the activity will also be changed.

10.4 ACTIVE SITE OF AN ENZYME

The active site of an enzyme is small compared to the overall size of the protein molecule. Most of the amino acid residues are not in contact with the substrate (Figure 10.22). The distances and angles between the catalytic residues of the enzyme and the substrate must be exact to permit catalysis to occur.

Most of the amino acids in the protein serve as a large scaffold to allow for the proper alignment of the functional groups of the substrates. A mutation of a residue a distant from the active site can cause the enzyme to have altered activity. This is illustrated in Figure 10.23 for a point mutation that was introduced into human liver aldehyde dehydrogenase.

The active site is specific for the substrate. As an example, subtle geometric constraints prevent an enzyme from hydrolyzing a β-glycosidic linkage between two glucose molecules while allowing the enzyme to hydrolysis the α-linkage (Figure 10.24). Enzymes also differentiate between optical isomers; though both the D and L forms of many compounds exist, mammalian cells utilize only one stereoisomer. D-Lactate is hardly recognized by mammalian lactate dehydrogenase, which is specific for the L-isomer (Figure 10.25). The binding can also react with pro-chiral centers to produce

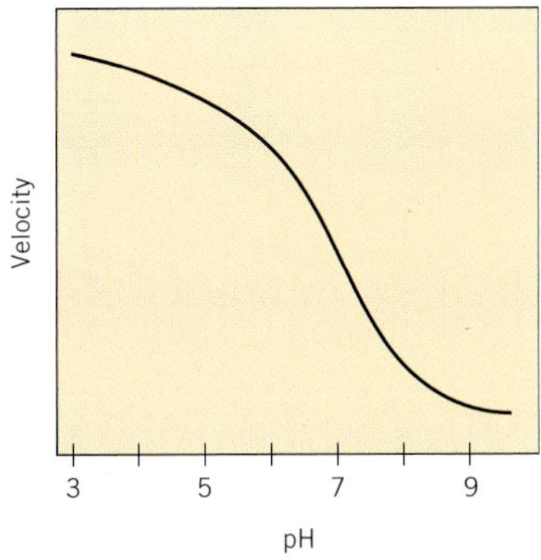

FIGURE 10.21

Velocity of an enzyme-catalyzed reaction can be affected by pH if a group dissociates a H^+. If an imidazole of histidine needs to be protonated for catalytic activity of an enzyme, the velocity of the reaction will decrease as the pH increases above the pK_a (assume 7) of the residue. If the group needed to be unprotonated for the enzyme to function, then the curve would have the reverse shape, with maximum activity above pH 8.

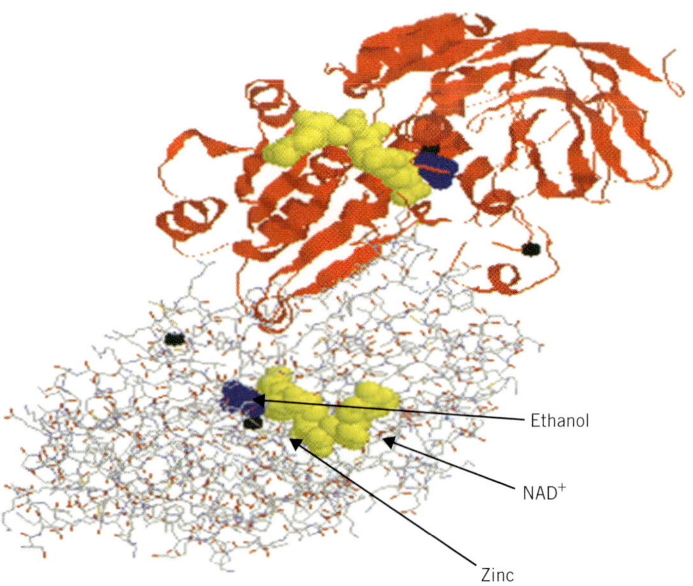

FIGURE 10.22

Enzymes are much larger than their substrates. Alcohol dehydrogenase with ethanol bound to a Zn^{2+} and an NAD^+ at the active site. Most of the amino acids of the protein are not in contact with the substrate. One subunit of the dimer is shown as ribbons and the other as atoms.

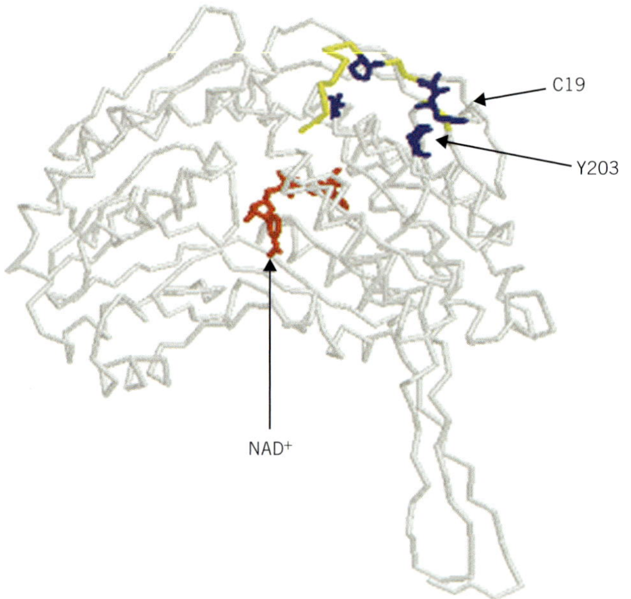

FIGURE 10.23

Amino acids at a distance from the active site are critical. An amino acid residue on the surface of an enzyme might be necessary to keep the enzyme in its proper shape. Residue 19, a cysteine, of a subunit of aldehyde dehydrogenase (500 amino acids) was changed to tyrosine. The resulting mutant was insoluble. Residue 19 normally bonds to residue 203, which is apparently necessary to maintain the proper three-dimensional structure.

an optically active compound. A chiral carbon has four different groups attached and is optically active; a **pro-chiral center** is a carbon that is not optically active but becomes optically active upon a stereo-specific addition of a group. The non-optically active compound binds to the enzyme in just one orientation forcing the chemical reaction to produce an optically pure product. Glycerol, a non-optically active compound, is phosphorylated by glycerol kinase and only the L-product is produced. The enzyme binds the compound such that just one of the primary alcohols is near ATP. This

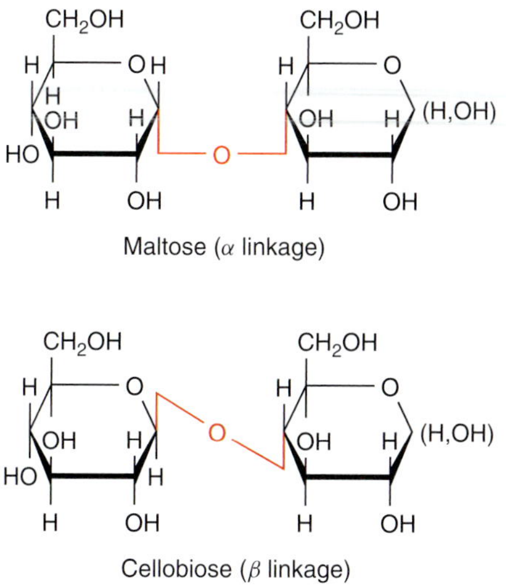

Maltose (α linkage)

Cellobiose (β linkage)

FIGURE 10.24

Enzymes can differentiate between α- and β-linkages. An enzyme that hydrolyzes maltose does not hydrolyze cellobiose.

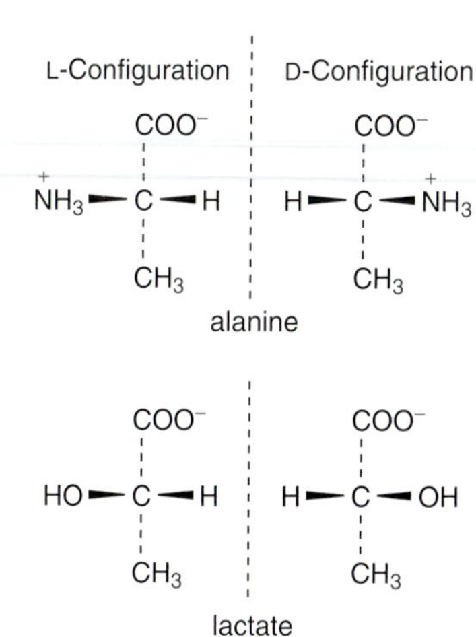

FIGURE 10.25

Enzymes can differentiate between optical isomers.

occurs because the surface of the enzyme is asymmetric since the amino acids are all L configuration and the helices are all right-handed. Though perhaps an oversimplification, it is often stated that if there were a three-point attachment of the substrate to the enzyme, then an optically pure product would be formed from a pro-chiral center (Figure 10.26).

Stereochemistry Helps Explain the Mechanism of Enzyme-Catalyzed Reactions

An enzyme-catalyzed carbonium ion reaction will produce a product with retention of stereochemical configuration. The carbonium ion would not be free in solution but would be associated with the enzyme, so the incoming nucleophilic attack could come only from the same direction as the leaving group. Retention of configuration can also be achieved by an enzyme that functions with covalent catalysis. Reactions catalyzed by a ping-pong mechanism (see p. 394) involve two inversion steps that produce a product with retention of configuration. As an example, the phosphorolysis of a disaccharide starts with the sugar in the β conformation and ends with a phosphorylated product that is also β. The attachment of the substrate to the enzyme causes inversion to an α configuration, but phosphate attacks to give another inversion, resulting in retention of configuration. If the reaction occurred by a direct attack of the phosphate on the substrate, then the product would have been inverted. Inversion of configuration at an optically active carbon usually occurs when there is a ternary complex and the nucleophile attacks the substrate (see Clin. Corr. 10.7, p. 397).

Influence of Groups on the Substrate Distal to the Bond Being Modified

The active site is involved with one place on the substrate, but frequently the adjoining portions of the substrate are necessary for the reaction to occur. For example, the mitochondrial processing protein peptidase requires at least 12 residues before the site of hydrolysis in order to function. Lysozyme, an enzyme that hydrolyzes bacterial cell walls, requires sugar residues located on both sides of the bond to be split (Figure 10.27). The binding of the adjacent sugar moieties helps create a change in the conformation of the sugar at the bond being broken. Thus, the active site is more than just the amino acid residues involved in the bond breakage or formation; it is a complex region on the enzyme that can adjust its structure to accommodate the proper substrate and stabilize the transition state.

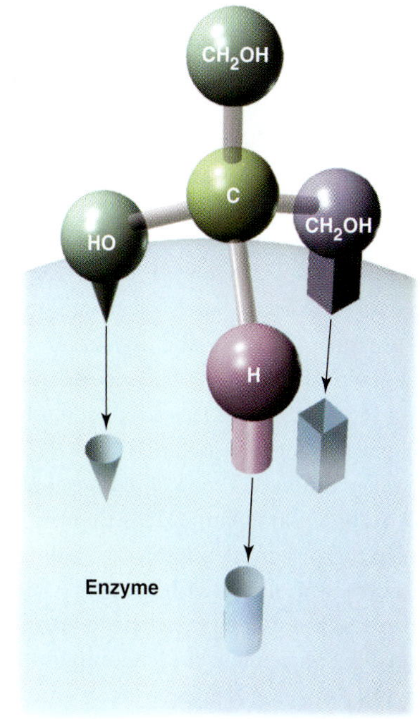

FIGURE 10.26

Three-point attachment of a symmetrical substrate to an asymmetric substrate-binding site. By virtue of dissimilar binding sites for the -H and -OH group of glycerol, glycerol kinase binds only the α'-hydroxymethyl group to the active site. One stereoisomer results from the kinase reaction, L-glycerol 3-phosphate. Active site is box on the right.

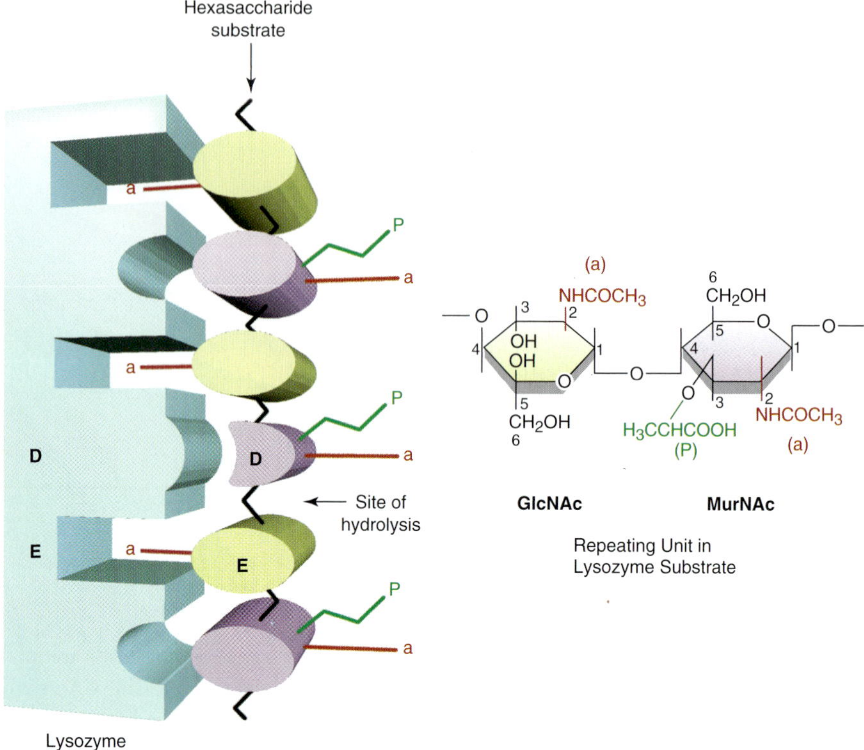

FIGURE 10.27

Hexasaccharide binding at active site of lysozyme. In the model substrate pictured, the ovals represent individual pyranose rings of the repeating units of the lysozyme substrate shown to the right. Ring D is strained by the enzyme to the half-chair conformation, and hydrolysis occurs between the D and E rings. Six subsites on the enzyme bind the substrate. Alternate sites are specific for acetamido groups (a) but are unable to accept the lactyl (P) side chains, which occur on the *N*-acetylmuramic acid residues. Thus the substrate can bind to the enzyme in only one orientation.

Redrawn based on model proposed by Imoto, T., et al. In: P. Boyer (Ed.), *The Enzymes*, Vol. 7, 3rd ed. New York: Academic Press, 1972, p. 713.

CLINICAL CORRELATION 10.1 Mutation of a Coenzyme-Binding Site Results in Clinical Disease

Cystathioninuria is a genetic disease in which γ-cystathionase is either deficient or inactive. Cystathionase catalyzes the reaction

$$\text{Cystathionine} \rightarrow \text{cysteine} + \alpha\text{-ketobutyrate}$$

Deficiency of the enzyme leads to accumulation of cystathionine in the plasma. Since cystathionase is a pyridoxal phosphate-dependent enzyme, vitamin B6 was administered to patients whose fibroblasts contained material that cross-reacted with antibody against cystathionase. Many responded to B_6 therapy with a fall in plasma levels of cystathionine. These patients produce the apoenzyme that reacted with the antibody. In one patient the enzyme activity was undetectable in fibroblast homogenates but increased to 31% of normal with the addition of 1 mM of pyridoxal phosphate to the assay mixture. It is thought that the K_m for pyridoxal phosphate binding to the enzyme was increased because of a mutation in the binding site. Activity is partially restored by increasing the concentration of coenzyme. Apparently these patients require a higher steady-state concentration of coenzyme to maintain γ-cystathionase activity.

Source: Pascal, T. A., Gaull, G. E., Beratis, N. G., Gillam, B. M., Tallan, H. H., and Hirschhorn, K. Vitamin B6-responsive and unresponsive cystathionuria: Two variant molecular forms. *Science* 190:1209, 1975.

10.5 COENZYMES, CO-SUBSTRATES, AND COFACTORS

Many enzymes require the participation of a coenzyme, co-substrate, or cofactor in the catalytic reaction. **Coenzymes** are small organic molecules, often derivatives of vitamins (see p. 1102). They may or may not be modified (e.g., oxidized or reduced) in the reaction. Those that are altered are also termed co-substrates. For some reactions the energy from hydrolysis of ATP is required without incorporation of phosphate in the product. ATP in these reactions is a **co-substrate**. Metal ions are often required for enzyme reactions and referred to as a **cofactor**.

Coenzymes

Table 10.2 lists the coenzymes and the vitamins from which they are derived. Coenzymes participate in enzyme-catalyzed reactions, but they are not the primary compounds being modified. Some coenzymes participate in the action of many different enzymes, while others participate only in a limited number of reactions. They can have affinities for the enzyme similar to the substrate, can be tightly bound, or can be covalently attached. Some are modified during a reaction but are in their original state at the end of the reaction (cyclic change), while others remain modified at the end. If they are modified at the end (e.g., oxidized or reduced), they must participate in another reaction to be returned to their original state. Coenzymes are present in cells in a reasonably constant concentration and play a dynamic role in metabolism. Clinical Correlation 10.1 points

TABLE 10.2 Coenzymes

Coenzyme	*Vitamin*	*Reaction Mediated*
Biotin	Biotin	Carboxylation
Cobalamin (B_{12})	Cobalamin (B_{12})	Alkylation
Coenzyme A	Pantothenate	Acyl transfer
Flavin coenzymes	Riboflavin (B_2)	Oxidation–reduction
Lipoic acid		Acyl transfer
Nicotinamide coenzymes	Nicotinamide	Oxidation–reduction
Pyridoxal phosphate	Pyridoxine (B_6)	Amino group transfer
Tetrahydrofolate	Folic acid	One-carbon group transfer
Thiamine pyrophosphate	Thiamine (B_1)	Carbonyl transfer

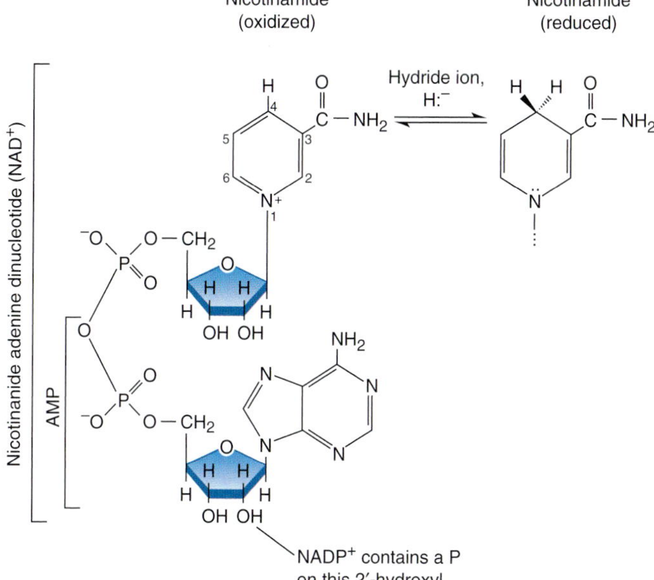

FIGURE 10.28

Nicotinamide adenine dinucleotide (NAD) and nicotinamide adenine dinucleotide phosphate (NADP).

out the importance of the coenzyme-binding site and how alterations in this site cause metabolic dysfunction.

NAD and NADP are Coenzyme Forms of Niacin

Nicotinamide adenine dinucleotide (NAD) and nicotinamide adenine dinucleotide phosphate (NADP) (Figure 10.28) are derived from niacin (see p. 1103) and are involved in oxidation–reduction reactions catalyzed by dehydrogenases. The difference between NAD and NADP is the presence of a phosphate group on the 2′ carbon of the adenine ribose. The notation NAD and NADP are used generically to indicate the coenzymes whether oxidized or reduced, whereas NAD^+ and $NADP^+$ specify the oxidized forms while NADH and NADPH specify the reduced forms. A generalized dehydrogenase reaction involving NAD is presented in Figure 10.29. Though most dehydrogenases are specific for one or the other coenzyme, some can use either form. Most dehydrogenases are readily reversible.

The functional portion of the coenzyme is the nicotinamide ring that has a nitrogen atom at position 1 and carries a positive charge due to the attachment to a ribose ring. The positive charge makes it possible for the nicotinamide to accept a hydride ion, transferring two electrons (Figure 10.30). A ternary complex between enzyme, coenzyme, and substrate forms, and a hydride ion is transferred from substrate to $NAD(P)^+$ and a proton released to the medium. The reactions in which they participate can be considered as ionic, though the ion they transfer is a hydride ion that carries a negative charge. Hydride ions do not exist in the free state, so the ion is directly

COO⁻
HOCH
CH₂
COO⁻
L-Malate
NAD⁺
NADH + H⁺
COO⁻
C=O
CH₂
COO⁻
Oxaloacetate

FIGURE 10.29

Reaction catalyzed by malate dehydrogenase. The enzyme forms a ternary complex by binding both NAD^+ and malate before the hydride ion is passed to NAD^+.

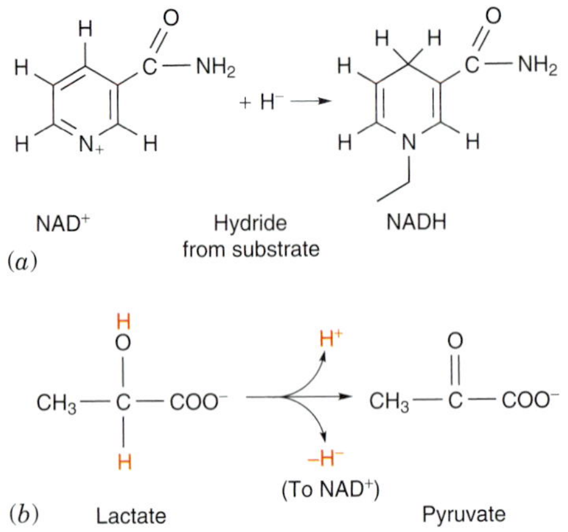

NAD$^+$ + H$^-$ (Hydride from substrate) → NADH

(*a*)

Lactate → Pyruvate (+ H$^+$, −H$^-$ (To NAD$^+$))

(*b*)

FIGURE 10.30

NAD is involved in oxidation reduction reactions. (*a*) Nicotinamide group in NAD$^+$ accepts a hydride ion (H^-) while oxidizing substrate. (*b*) Lactate donates the hydride ion when oxidized to pyruvate.

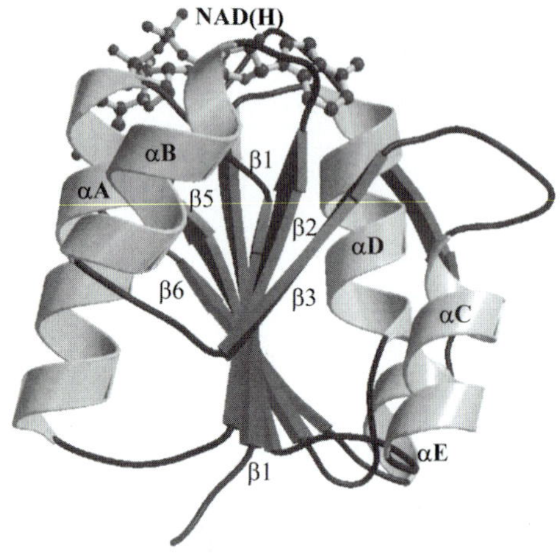

FIGURE 10.31

The "Rossmann Fold" in liver alcohol dehydrogenase. NAD(H) binds to the "Rossmann fold" in most dehydrogenases.

transferred between substrate and coenzyme and does not dissociate. After accepting or donating the hydride ion, the coenzyme dissociates from the enzyme and enters the cellular pools of NAD and NADP, being available for other reactions. The ratios of NAD$^+$/NADH and NADP$^+$/NADPH are independent of a specific dehydrogenase but are measures of the oxidation–reduction state of the cytosol or cellular compartment (see p. 531). There is typically a high ratio of NAD$^+$/NADH and a low ratio of NADP$^+$/NADPH. NAD-dependent dehydrogenases are more often involved in catabolism, while NADP-dependent ones are usually involved in anabolism.

Nearly all NAD(P)-dependent dehydrogenases possess a similar structural motif involved in the coenzyme-binding domain, referred to as the **Rossmann fold;** this motif for alcohol dehydrogenase is presented in Figure 10.31.

FMN and FAD are Coenzyme Forms of Riboflavin

The two coenzyme forms of riboflavin are **FMN** (**flavin mononucleotide**) and **FAD** (**flavin adenine dinucleotide**). The vitamin riboflavin consists of the heterocyclic ring, isoalloxazine (flavin) connected through N10 to the alcohol ribitol (Figure 10.32). FMN has a phosphate esterified to the 5′-OH group of ribitol. FAD is structurally analogous to FMN in having adenosine linked by a pyrophosphate linkage to riboflavin (Figure 10.33). Both FAD and FMN function in oxidoreduction reactions by accepting and donating $2e^-$ in the isoalloxazine ring in the form of a hydride. In some cases, these coenzymes are $1e^-$ acceptors, which lead to flavin semiquinone formation (a free radical). Flavin coenzymes are not found free in solution but are bound either very tightly or covalently to an enzyme. They would be considered a **prosthetic group** in the latter case (see p. 340). This requires that an enzyme involved in an oxidation reaction generating reduced bound flavin must also carry out the reduction of another substrate so that oxidized flavin will be regenerated. The oxidation of saturated carbon–carbon bonds such as in oxidation of succinate to fumarate (see p. 546) (Figure 10.34) or a saturated fatty acid to an unsaturated one are representative flavin-dependent dehydrogenase reactions. In the case of monoamine oxidase, a flavin-containing enzyme, a hydride ion is removed from the amine, but the terminal acceptor of electrons is molecular oxygen.

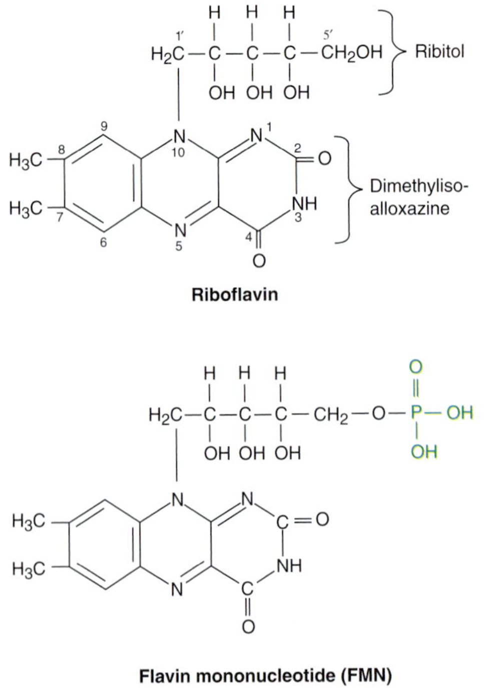

Ribitol

Dimethylisoalloxazine

Riboflavin

Flavin mononucleotide (FMN)

FIGURE 10.32

Riboflavin and flavin mononucleotide.

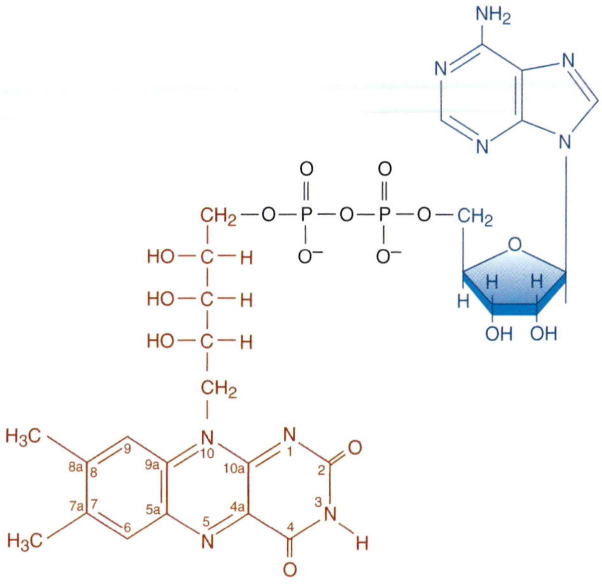

Flavin adenine dinucleotide (FAD)

FIGURE 10.33
Flavin adenine dinucleotide (FAD).

Succinic acid + FAD ⇌ Fumaric acid + $FADH_2$

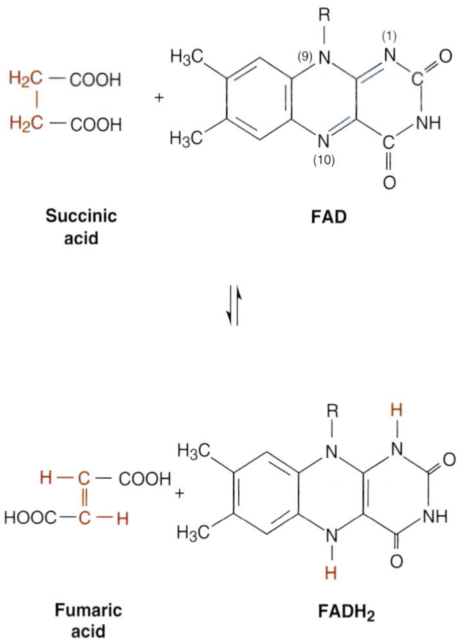

FIGURE 10.34
FAD is the coenzyme in the succinic dehydrogenase reaction.

Pyridoxal Phosphate is the Coenzyme Form of Pyridoxal

Pyridoxal phosphate (Figure 10. 35) is primarily involved in reactions involving amino acids. A major reaction utilizing pyridoxal phosphate is transamination, a reaction (Figure 10.4) that introduces an amino group into a compound. The acceptor is a carbonyl group such as an α- keto acid, and the donor is an amino acid such as glutamate or aspartate (see Figure 19.8, p. 747). The amino acid bound to pyridoxal phosphate transfers the amino group to it, making an enzyme-bound pyridoxamine. After the α-keto acid leaves, the enzyme and the carbonyl containing acceptor binds to the enzyme, followed by the transfer of the amine from pyridoxamine to the new α-keto acid. Pyridoxal is thus regenerated. This is an example of the modification of a coenzyme during the reaction but returning to its original state at the end.

Resonance in the pyridine ring of the coenzyme stabilizes a negative charge at what was the α-carbon. The basis of the reaction is that the aldehyde in the pyridoxal moiety binds with the amino acid, making what is called a Schiff base. It is on this covalent intermediate that the enzyme catalyzes the removal of the group. A pair of electrons is left on the carbon, and their charge is stabilized by resonance with the pyridoxal ring. The formation of dopamine and serotonin, important neurotransmitters, involves a pyridoxal phosphate enzyme (see p. 775).

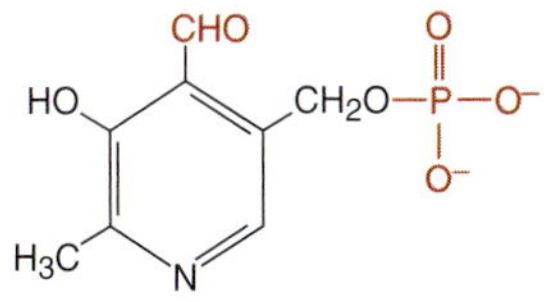

Pyridoxal phosphate

FIGURE 10.35
Pyridoxal phosphate. Pyridoxal phosphate binds to the amino group of an amino acid and assists in the removal of the R-group, the H, or the CO_2.

Adenosine Triphosphate may be a Second Substrate or a Modulator of Activity

Adenosine triphosphate (ATP) (Figure 10.36) functions as co-substrate—that is, a second substrate—but it can also serve as a cofactor in modulation of the activity of specific enzymes. This compound is central in biochemistry and is synthesized *de novo* in all mammalian cells. The nitrogenous heterocyclic ring is adenine. The biochemically functional end is the reactive triphosphate. The terminal phosphate–oxygen bond has a high free energy of hydrolysis, which means that the phosphate can be transferred from ATP to other acceptor groups. For example, a co-substrate ATP is utilized by the kinases for the transfer of the terminal phosphate to various acceptors. A typical example is the reaction catalyzed by hexokinase (see Figure 10.2, p. 369).

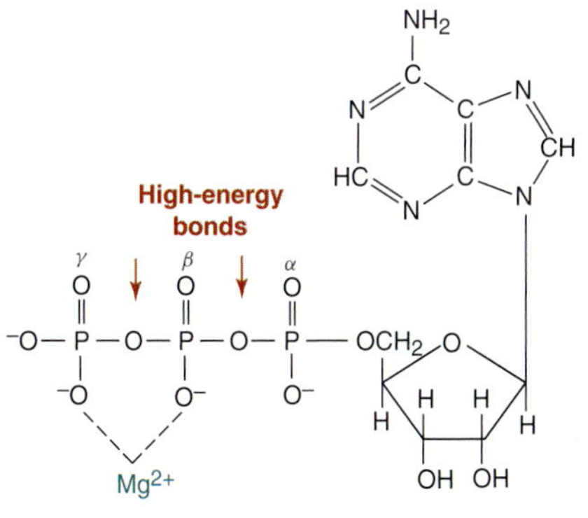

High-energy bonds

D-Ribose

Adenosine 5′-triphosphate

FIGURE 10.36
Structure of Mg^{2+}-ATP.

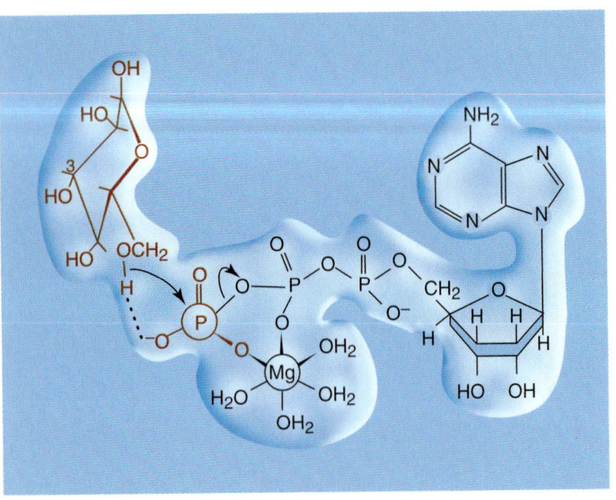

FIGURE 10.37

Role of Mg^{2+} as a substrate-bridged complex in the active site of the kinases. In hexokinase the terminal phosphate of ATP is transferred to glucose, yielding glucose 6-phosphate. Mg^{2+} coordinates with the ATP to form the true substrate and may labilize the terminal P–O bond of ATP to facilitate transfer of the phosphate to glucose. There are specific binding sites (light blue) on the enzyme (darker blue) for glucose (upper left, in red) as well as for the adenine and ribose moieties of ATP (black).

ATP also serves as a modulator of the activity of some enzymes. These enzymes have binding sites for ATP, occupancy of which changes the affinity or reactivity of the enzyme toward its substrates. In these cases, ATP acts as an **allosteric effector** (see p. 401).

Metal Ion Cofactors

Many enzymes require the presence of a metal ion for their activity. The metal ion can serve a structural role, or function as a Lewis acid (a positive ion that can bind to unpaired electrons), or as an electron acceptor/donor in oxidation–reduction reactions.

ATP-dependent reactions require a divalent ion, typically Mg^{2+}. The ion interacts with phosphates on ATP to form a complex; thus the actual substrate is the ATP-Mg^{2+} complex (Figure 10.36). Mg^{2+} does not interact directly with the enzyme; rather its role is to polarize the P=O bond, making the phosphate more susceptible to nucleophilic attack. Ternary complexes of this conformation are known as "**substrate-bridged**" complexes (Enz-S-M), and nearly all kinases are substrate-bridged complexes. A scheme for the binding of Mg^{2+}-ATP and glucose in the active site of hexokinase is presented in Figure 10.37.

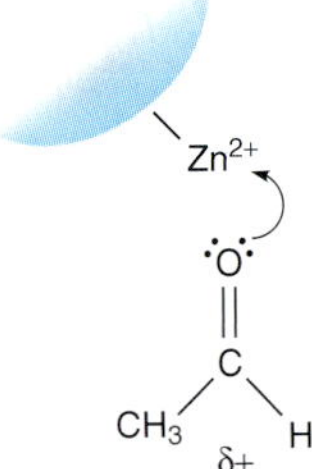

FIGURE 10.38

Binding of a carbonyl to Zn^{2+}. Zn^{2+} binding of the carbonyl of an aldehyde results in making the carbon atom partially positive and more receptive to accepting a hydride ion from NADH.

Metals that chelate the substrate or ATP to the enzyme form a "**metal-bridged**" ternary complex (Enz-M-S). Mg^{+} chelates ATP to muscle pyruvate kinase and phosphoenolpyruvate carboxykinase. Other **metalloenzymes** contain a tightly bound transition metal, such as Zn^{2+} or Fe^{2+}, that facilitates binding of substrate by forming a metal-bridged complex. Metals function either by binding substrate and promoting electrophilic (nucleophilic) catalysis at the site of bond cleavage or by stabilizing intermediates in the reaction pathway. In alcohol dehydrogenase, an enzyme-bound Zn^{2+} interacts with the oxygen in the substrate, changing the properties of the carbon to which the oxygen is attached (Figure 10.38). Zn^{2+} functions in **carboxypeptidase** and **thermolysin,** proteases with identical active sites, to generate a hydroxyl group from water and to stabilize the transition state resulting from attack of the hydroxyl on the peptide bond. Figure 10.39 depicts the generation of the active-site hydroxyl by Zn^{2+}. Glu 270 functions as a base necessary to remove the proton from water. Stabilization of the tetrahedral transition state by Zn^{2+} is shown in Figure 10.40. The Zn^{2+} provides a counterion to stabilize the negative oxygen on the tetrahedral carbon.

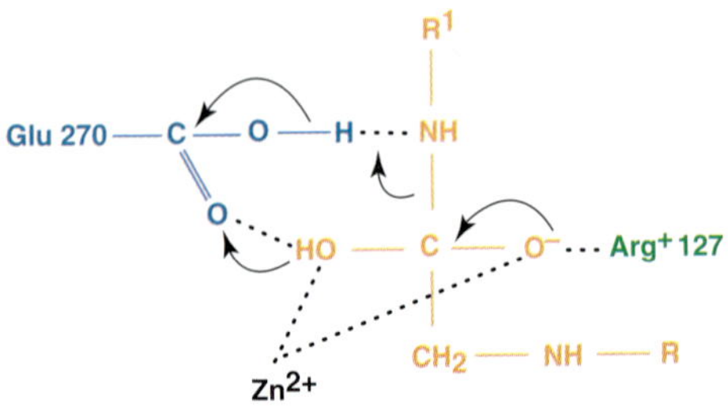

FIGURE 10.39

Stabilization of the transition state of the tetrahedral intermediate by Zn^{2+}. Positive charge on the Zn^{2+} stabilizes the negative charge that develops on the oxygens of the tetrahedral carbon in the transition state. The tetrahedral intermediate then collapses as indicated by the arrows, resulting in breakage of the peptide bond.

In addition to the role of binding enzyme and substrate, metals also bind directly to the enzyme to stabilize it in the active conformation or perhaps to induce the formation of an active site. Mn^{2+} has such a role as do weakly bound alkali metals (Na^{+} or K^{+}). In pyruvate kinase, K^{+} induces an initial conformational change, which is necessary, but not sufficient, for ternary complex formation. Upon substrate binding, K^{+} induces a second conformational change to the catalytically active ternary complex as indicated

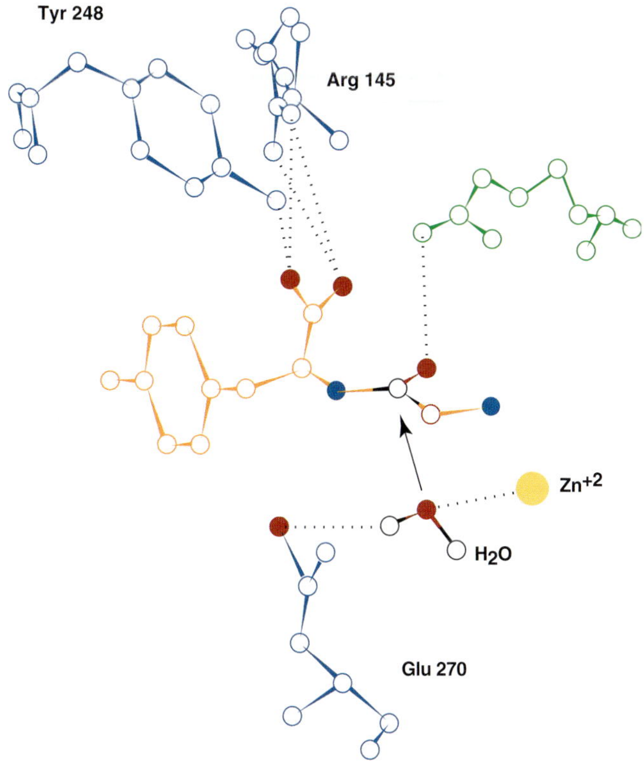

FIGURE 10.40

Zn^{2+} in the mechanism of reaction of carboxypeptidase A. Enzyme-bound Zn^{2+} generates a hydroxyl nucleophile from bound water, which attacks the carbonyl of the peptide bond as indicated by the arrow. Glu 270 assists by pulling the proton from the zinc-bound water.
Redrawn from Lipscomb, W. N. *Robert A. Welch Found. Conf. Chem. Res.* 15:140, 1971.

in Figure 10.41. Mammalian alcohol dehydrogenase is unique in that it possesses two molecules of Zn^{2+} per subunit; one is involved in catalysis while the other has a structural role.

Role of Metals in Oxidation and Reduction

Metal ions such as copper and iron, with less than a full complement of electrons in their outer shell, can accept or donate electrons in oxidation–reduction reactions, but monovalent ions (Na^+ and K^+) or divalent ions (Zn^{2+}, Mg^{2+}, or Ca^{2+}) with their outer shell of electrons complete do not.

Iron–sulfur proteins, often referred to as nonheme iron proteins, are a unique class of metalloenzymes in which the active center consists of one or more clusters of sulfur-bridged iron chelates (see p. 826). In some cases the sulfur atoms comes only from cysteine and for others it comes from both cysteine and free ionic sulfur. These nonheme iron enzymes have reasonably low reducing potentials (E_0') and function in electron-transfer reactions (see p. 555). **Cytochromes** are heme iron proteins where the iron undergoes reversible one electron transfers. In some cases, heme is bound to the apoprotein by coordination of amino acid side chains to the iron of heme. Thus, the metal serves not only a structural role but also participates in the chemical event (see p. 553). Copper and iron also have a role in activation of molecular oxygen. Copper is an active participant in several oxidases and hydroxylases. For example, **dopamine β-hydroxylase** catalyzes the introduction of one oxygen atom from O_2 into dopamine to form norepinephrine (Figure 10.42). The active enzyme contains one atom of cuprous ion that reacts with oxygen to form an activated oxygen–copper complex. The copper–hydroperoxide complex is thought to be converted to a copper (II)–O^- species that serves as the "active oxygen" in the hydroxylation of DOPA. Other species of "active oxygen" are generated in metalloenzymes and used for hydroxylation.

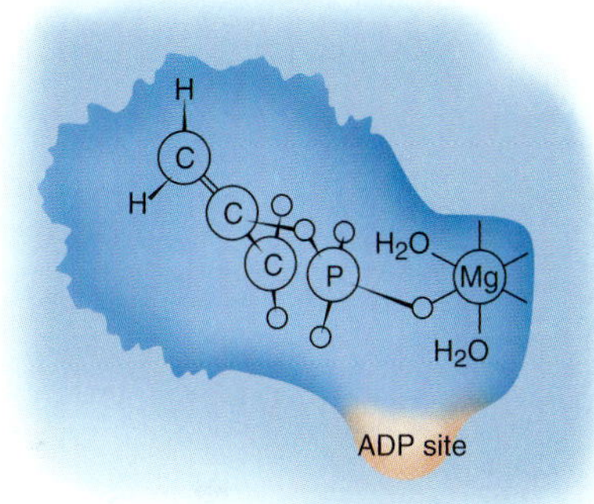

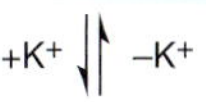

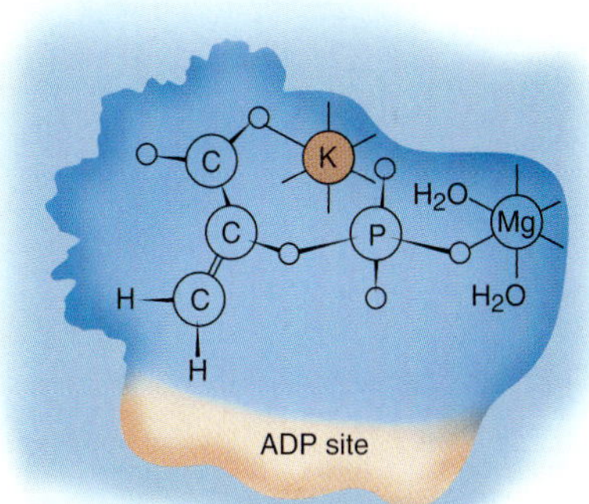

FIGURE 10.41

Model of the role of K^+ in the active site of pyruvate kinase. Pyruvate kinase catalyzes the reaction: phosphoenolpyruvate + ADP → ATP + pyruvate. Initial binding of K^+ induces conformational changes in the kinase, which result in increased affinity for phosphoenolpyruvate. In addition, K^+ orients the phosphoenolpyruvate in the correct position for transfer of its phosphate to ADP (not shown), the second substrate. Mg^{2+} coordinates the substrate to the enzyme active site.
Modified with permission from Mildvan, A. S. *Annu. Rev. Biochem.* 43:365, 1971. Copyright (1971) Annual Reviews, Inc.

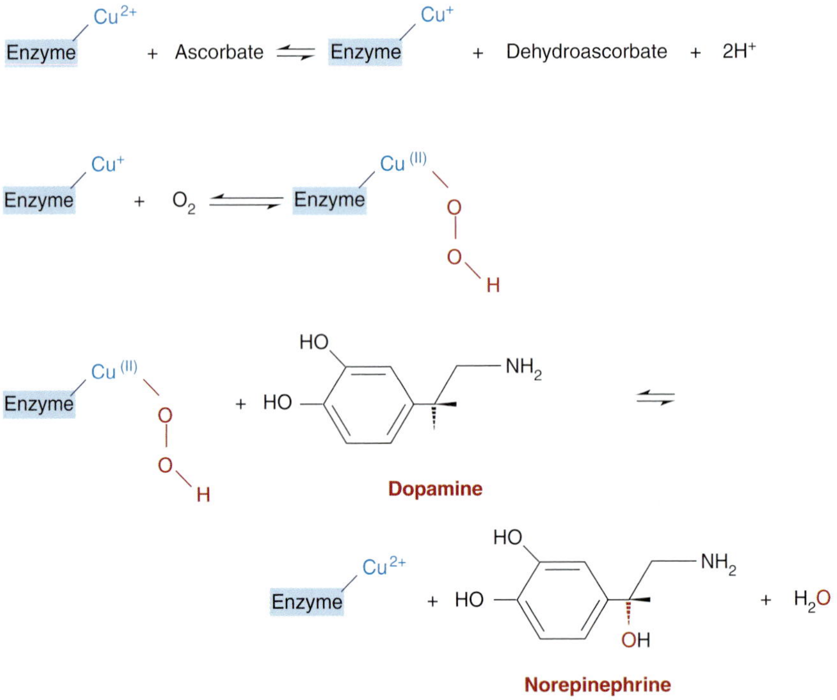

FIGURE 10.42

Role of copper in activation of molecular oxygen by dopamine hydroxylase. The normal cupric form of the enzyme is not reactive with oxygen but, on reduction by the co-substrate, ascorbate, generates a reactive enzyme-copper bound oxygen radical that then reacts with dopamine to form norepinephrine and an inactive cupric enzyme.

10.6 KINETICS OF CHEMICAL REACTIONS

Rate of Product Formation

The rates of all chemical reactions can be expressed in mathematical terms. In developing the rate equations, either the formation of product or the disappearance of initial reactants can be determined; in most cases, product formation is measured. From the mathematical relationships between the rate(s) of product(s) formation (termed velocity) as a function of substrate(s) concentration(s), some aspects of the mechanism of enzyme catalysis can be determined. An equation to describe the rate of product formation in any reaction contains a rate constant (k) and the concentration of all the substrates that change during the course of the reaction. In the reaction $A + B \rightarrow P$, the change in concentration of product as a function of time is dP/dt, where P is concentration of product at any time t. The basic equation relating velocity to reactants is

$$d[P]/dt = k[A][B] \tag{10.2}$$

where k is the rate constant and is a constant for a particular reaction. [A] and [B] are the concentrations of the substrates at any time. The rate constant is related to the energy of activation to reach the transition state and is changed by the addition of a catalyst. Rate constants can be altered only by changing the reaction conditions—that is, addition of catalyst, or change of pH or ion composition.

First- and Second-Order Reactions

A reaction with only one substrate ($A \rightarrow P$) is a first-order reaction with

$$d[P]/dt = k[A] \tag{10.3}$$

If there are two substrates, the reaction is a second-order reaction (Eq. 10.2). The units of k are different in the equations for a first- and second-order reaction. Since $d[P]/dt$ has the units moles/minute (or any other unit of concentration or time), k in Eq. 10.2 would have the units of moles^{-1} × min^{-1} while in Eq. 10.3 k would be in min^{-1}. If it takes two molecules of compound A to form product P, then the equation would be second order because the velocity of making P is really related to a reaction of [A] + [A]. The equation would be $d[P]/dt = k$ [A][A] or k[A]^2, which is identical to a second-order reaction even though only one compound is involved. For $d[P]/dt = k[A][B]$ the reaction is first order with respect to both A and B, but the overall reaction is second order.

The concentration terms in the equation are only for those substrates whose concentrations change during the course of the reaction. In reactions involving water, the concentration of water is not altered (or not measurably). Thus if B is water in the reaction $A + B \rightarrow C$, the value of [B] is a constant and the reaction is first order (Eq. 10.3).

Rate of Substrate Disappearance

For the simple reaction $A \rightarrow P$, the velocity of the reaction can also be given by disappearance of substrate as a function of time, $-d[A]/dt$. The minus sign is necessary since the concentration of A is decreasing with time. But $d[P]/dt = -d[A]/dt$; thus the velocity equation for this reaction would look identical to the one written in Eq. 10.3:

$$d[P]/dt = -d[A]/dt = k[A] \quad (10.4)$$

After rearranging, the equation becomes

$$d[A]/dt = -k[A] \quad (10.5)$$

Since the integral of dA/A is ln A; integrations leads to

$$\ln[A] = 2.3 \log [A] = -kt + \text{constant} \quad (10.6)$$

Equation 10.6 is the mathematical relationship for the loss of substrate as a function of time. At zero time, [A] is the initial concentration, $[A_0]$, so the equation can be rewritten as

$$\ln[A/A_0] = -kt \quad (10.7)$$

A plot of the ln[A] versus time is a straight line with a slope of $-k$ (Figure 10.43).

For comparison purposes, the term half-life is often used. The half-life ($t_{1/2}$ or t_{50}%) is the time it would take for 50% of the substrate to be converted to product. Since the value $[A/A_0]$ would be 0.5 and the $\ln(0.5) = -0.69$, this leads to

$$-0.69 = -kt_{50}\% \quad (10.8)$$

or half-life is equal to $0.69/k$. Thus, the larger the rate constant, the faster the reaction will deplete half the substrate.

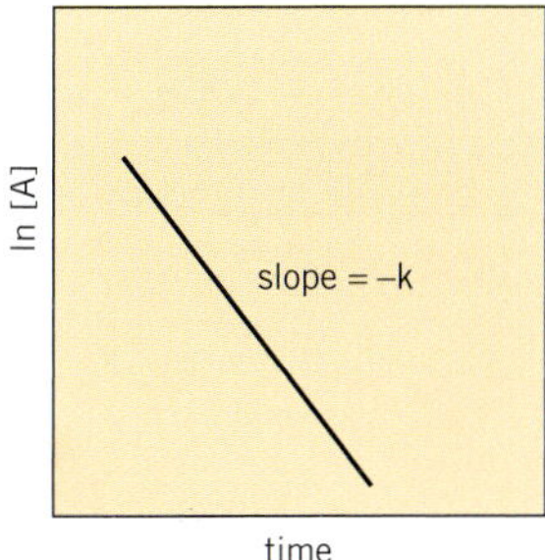

FIGURE 10.43

A plot of the ln[A] versus time for a first-order reaction.

Reversible Reactions

Many reactions are reversible—that is, $A \underset{k_2}{\overset{k_1}{\rightleftharpoons}} P$. One can write a velocity equation for the formation of product P or of starting material A.

$$d[P]/dt = k_1[A] - k_2[P] \quad (10.9)$$

$$d[A]/dt = k_2[P] - k_1[A] \quad (10.10)$$

At equilibrium, where there is no net change in the concentrations of components, the rate of formation of one component is equal to the rate of formation of the other. Figure 10.44 is a plot of substrate disappearance and product formation. The velocity equation for product formation reflects that the material is being produced ($k_1[A]$) but

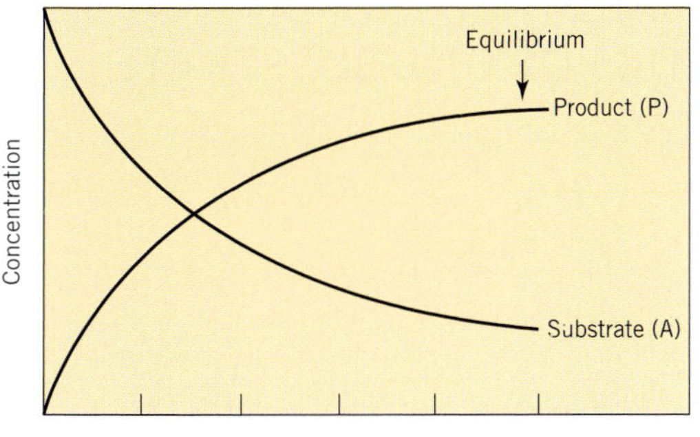

FIGURE 10.44

Plot of substrate disappearance and product formation. For a reversible reaction, no net change in substrate or product concentration occurs at equilibrium.

is also being lost as it goes back to starting material ($-k_2$[P]). The negative sign shows that the material is being removed. A similar equation is written for the change in starting material. Since at equilibrium

$$d[\text{P}]/dt = d[\text{A}]/dt \tag{10.11}$$

then

$$k_1[\text{A}] = k_2[\text{P}] \quad \text{so} \quad k_1/k_2 = [\text{P}]/[\text{A}] \tag{10.12}$$

The ratio ([P]/[A]) is the equilibrium constant K_{eq} and is equal to k_1/k_2. Even though the rate constant for a reaction is independent of the equilibrium constant, the ratio of the constants is constrained by the equilibrium constant.

Complex Reactions

Nucleophilic displacement is a common organic reaction that also occurs in biochemical reactions. A general form of the reaction is R-X + Y → Y-R + X (where R represents part of an organic compound and X or Y represents a component such as an alcohol or amine group). The reaction can occur in two different ways, each leading to a different rate equation.

The first possibility is that the compound R-X dissociates, forming $R^+ + X^-$. Next Y can attack R^+ to form the final product, R-Y. The second possibility is that Y will attack R-X directly displacing X to form the same final product. The kinetic equations for the two reactions are different.

The velocity for the first step in the reaction of R-X going to $R^+ + X^-$ is

$$d[\text{R-X}]/dt = -k[\text{R-X}] \tag{10.13}$$

For the second half of the reaction, $R^+ + Y$, the velocity is given by k′ [R^+][Y]. If the first step is very slow compared to the second step, then the overall velocity will be governed by only the first step. In kinetics, only the slow steps are measured. To illustrate, assume that R-X goes to $R^+ + X^-$ at a rate of 10 μM per minute and the attack of Y on R^+ proceeds at 1000 μM per minute. Thus, as a molecule of R^+ is formed, it very rapidly becomes the final product. Increasing the fast step will not alter the rate of product formation because it cannot be formed faster than the velocity of the slowest step. This means that the rate of formation of product will be described by the rate equation $d[\text{R-X}]/dt = -k_1[\text{R-X}]$. This reaction is independent of the concentration of component Y. That is, the reaction is a first-order reaction. It can be said that the reaction is **zero order** with respect to [Y], meaning that the reaction is independent of the concentration of Y. (Recall that $[\text{Y}]^0 = 1$.) Increasing the concentration of Y will have no effect on the velocity of the reaction. In the reaction of Y attacking R-X the velocity term contains both the concentrations of both substrates. That is, $d[\text{R-Y}]/dt = k_2[\text{R-X}][\text{Y}]$, a traditional second-order reaction. The major difference between the two reactions pathways is that in one the slowest step involves the transition state for R-X going to $R^+ + X^-$. In the other example, the transition step involves Y attacking R-X (Figure 10.45). To reiterate, kinetics tells something about the slower steps that occur during a reaction. The slowest step that occurs during an enzyme catalyzed reaction is called the **rate-limiting step**.

10.7 ENZYME KINETICS OF ONE-SUBSTRATE REACTIONS

An enzyme-catalyzed reaction is like a second order-reaction. Hence, as the concentration of substrate is increased, the rate of reaction increases as shown in Figure 10.46. The velocity slows down during the course of the reaction because the concentration of substrate is becoming depleted or product is starting to accumulate. Assuming the reaction is reversible, it will reach equilibrium. By calculating the initial velocity of the

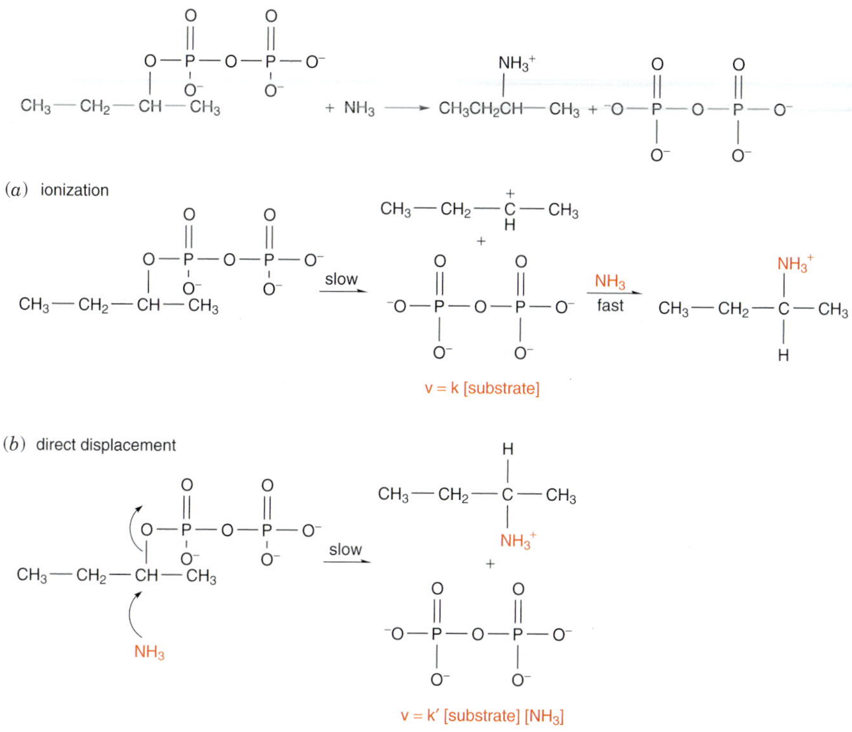

FIGURE 10.45

Nucleophilic displacement reactions. (*a*) Ionization mechanism. (*b*) Direct displacement mechanism. Note that in the ionization mechanism, the velocity of the reaction is a function of only the one substrate concentration, whereas in the direct displacement the velocity depends on both substrate concentrations.

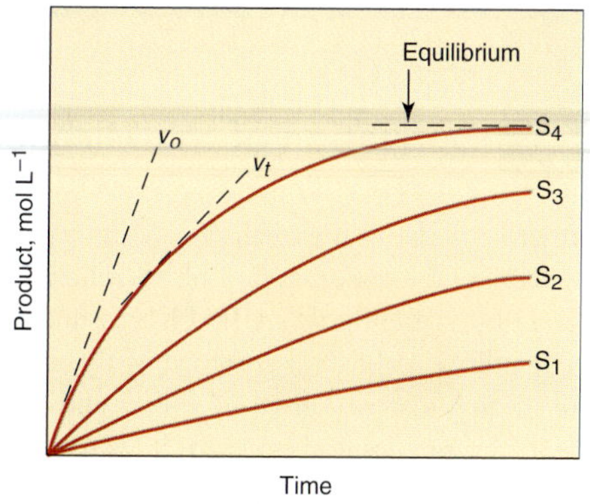

FIGURE 10.46

Progress curves for an enzyme-catalyzed reaction. The initial velocity (v_0) of the reaction is determined from the slope of the progress curve at the beginning of the reaction. The initial velocity increases with increasing substrate concentration (S1–S4) but reaches a limiting value characteristic of each enzyme. The velocity at any time, t, is denoted as v_t.

reaction (v_o) at different substrate concentrations, one can obtain the curve presented in Figure 10.47. The velocity becomes zero order with respect to substrate at high concentration of substrate. That is, the velocity no longer increases with an increase in [S]. The reason is that the enzyme and substrate form a complex and the rate of product formation is proportional to the concentration of the complex. All of the enzyme will be in the complex at high substrate concentrations. The complex can return to the starting material, by reversibly dissociating, or proceed to form products. The overall reaction is written as

$$E + S \underset{k_2}{\overset{k_1}{\rightleftharpoons}} ES \xrightarrow{k_3} E + P \tag{10.14}$$

Rate constants (k) will be numbered such that the ones leading to products will be odd numbers (k_1, k_3, and k_5), while the reversible steps will be even ones (k_2, k_4, and k_6); some are first-order and others second-order constants. *The units of first- and second-order constants are different and cannot be added to each other but can be multiplied or divided.*

The rate of formation of product is the term leading to product times the rate constant.

$$dP/dt = k_3[ES] \tag{10.15}$$

To determine the rate of product formation, it is necessary to find the value of ES as a function of the concentration of enzyme and substrate. Equation 10.15 can be solved for the rate of the reaction (see Box 10.1 for derivation) and the final form is

$$\boldsymbol{v = \frac{k[E][S]}{K + [S]}} \tag{10.16}$$

where k is a rate constant and K is a constant which includes the individual rate constants relating to the substrate's interaction with the enzyme.

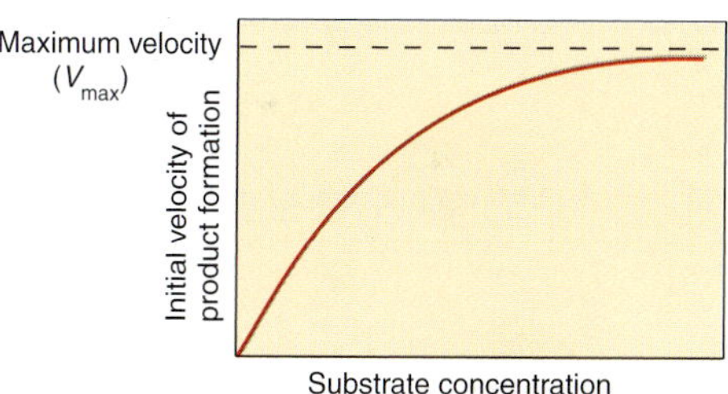

FIGURE 10.47

Plot of velocity versus substrate concentration for an enzyme-catalyzed reaction. Initial velocities are plotted against the substrate concentration at which they were determined. The curve is a rectangular hyperbola, which asymptotically approaches the maximum velocity possible with a given amount of enzyme.

BOX 10.1 Derivation of the Michaelis–Menten Equation

In order to derive the Michaelis–Menten equation, it is necessary to determine the concentration of ES. The rate of formation of ES is given by $k_1[E_f][S]$, where $[E_f]$ is the concentration of free enzyme. The utilization of ES is governed by two reactions; (1) ES returning to starting material, given by $k_2[ES]$ and (2) ES going to products, $k_3[ES]$. Combining these terms leads to an equation describing the change in concentration of ES.

$$d[ES]/dt = k_1[E_f][S] - k_2[ES] - k_3[ES]$$

Assuming that the concentration of ES is constant except for the first few milliseconds after contact of substrate and enzyme, then $d[ES]/dt = 0$. This is referred to as the "steady-state assumption," meaning that the concentration of ES during the initial time period of the enzyme catalyzed reaction (period for measuring the initial rate). The validity of the steady-state condition can be proven mathematically. Substituting $[E_f]$ with $[E_t] - [ES]$, where E_t is the total enzyme concentration and rearranging, leads to

$$k_1([E_t] - [ES])\,[S] = (k_2 + k_3)\,[ES]$$

Solving for [ES], one obtains

$$[ES] = \frac{k_1[E_t][S]}{((k_2 + k_3) + k_1[S])}$$

Since the rate (v) of the reaction is $k_3[ES]$, multiplying the above equation by k_3 gives

$$v = \frac{k_3 k_1[E_t][S]}{((k_2 + k_3) + k_1[S])}$$

To convert this equation into the Michaelis–Menten equation, the numerator and denominator are divided by k_1.

$$v = \frac{k_3[E_t][S]}{\frac{(k_2+k_3)}{k_1} + [S]} \quad \text{or} \quad v = \frac{V_m[S]}{K_m + [S]}$$

k_3E_t is the maximum velocity of the reaction (V_m), with all the enzyme in the ES form, and therefore k_3 is k_{cat}. The constants in the denominator are defined as K_m. It is apparent that K_m is not a simple dissociation constant. If k_3 were very small compared to k_2 then the resulting ratio would be k_2/k_1, which is K_d, the dissociation constant.

Michaelis–Menten Equation

The simplest enzyme reaction where one substrate goes to one product is

$$E + S \underset{k_2}{\overset{k_1}{\rightleftharpoons}} ES \underset{k_4}{\overset{k_3}{\rightleftharpoons}} E + P \qquad (10.17)$$

If the reaction in Eq. 10.17 is essentially irreversible ($k_4 = 0$), the equation for the rate of the reaction is

$$v = \frac{k_3[E][S]}{\frac{(k_3 + k_2)}{k_1} + [S]} \qquad (10.18)$$

Combining constants in the denominator into the term K_m and if the reaction ES → E + P in Eq. 10.16 is the rate-limiting step, then k_3 defines the catalytic activity of the enzyme (k_{cat}), and the equation becomes

$$v = \frac{k_{cat}[E_t][S]}{K_m + [S]} \qquad (10.19)$$

Note the similarity of Eqs. 10.16 and 10.19. Since k_{cat} (the rate constant) and $[E_t]$ (total enzyme concentration) are constants, $k_{cat}[E_t] = V_{max}$. The equation can be written as

$$v = \frac{V_m[S]}{K_m + [S]} \qquad (10.20)$$

and is referred to as the **Michaelis–Menten equation.** For reactions that are more complicated, k_{cat} is composed of a collection of rate constants (see p. 386). V_m in Eq. 10.19 is the **maximum velocity** of the reaction at a defined enzyme concentration and has the units of moles of product per unit time. Thus V_m (i.e., k_3 $[E_t]$ or k_{cat} $[E_t]$) is a constant for a fixed concentration of enzyme. If the enzyme concentration is increased or decreased, V_{max} will increase or decrease, respectively (Figure 10.48). K_m is composed of individual rate constants; however, from Eq. 10.19, it can be derived that K_m *is the concentration of substrate where the velocity of the reaction (v) is one-half of the maximum velocity* (V_m), that is, $v/V_m = 0.5$ (Figure 10.49). When v is equal to $\frac{1}{2}V_m$, one-half of

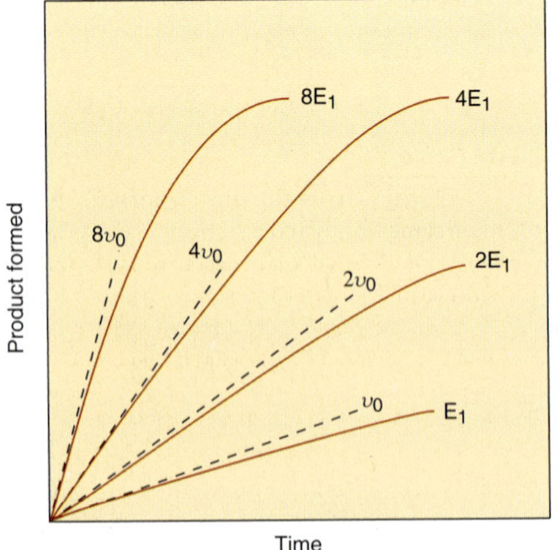

FIGURE 10.48

Progress curves at variable concentrations of enzyme and saturating concentrations of substrate. The initial velocity (v_0) doubles as the enzyme concentration doubles. Since the substrate concentrations are the same, the final equilibrium concentrations of product will be identical in each case; however, equilibrium will be reached at a slower rate in those assays containing lower amounts of enzyme.

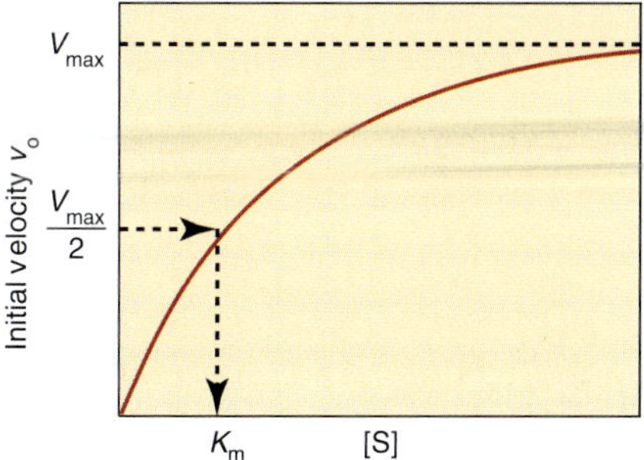

FIGURE 10.49

Graphic estimation of K_m for the v_0 versus [S] plot. K_m is the substrate concentration at which the enzyme has half-maximal activity.

the total enzyme is in a complex with substrate and one-half is free. In contrast, when $[S] \gg K_m$ essentially 100% of the enzyme is in the ES complex. Under this condition the reaction reaches V_m since the velocity would be $v = V_m$. It is usually considered that V_m is approached when [S] is greater than 10 times $[K_m]$. The biphasic nature of enzyme mechanisms—that is, binding followed by modification of substrate—is reinforced in Clinical Correlation 10.2.

If the reaction is more complex such that

$$\mathrm{E} + \mathrm{S} \underset{k_2}{\overset{k_1}{\rightleftharpoons}} \mathrm{ES} \xrightarrow{k_3} \mathrm{ES'} \xrightarrow{k_5} \mathrm{E} + \mathrm{P} \tag{10.21}$$

where ES′ represents a modified but independent form of the complex, then the velocity equation becomes

$$v = \frac{k_3 k_5/(k_3 + k_5)\ [\mathrm{E}_t]\ [\mathrm{S}]}{k_5(k_2 + k_3)/k_1(k_3 + k_5) + [\mathrm{S}]} \tag{10.22}$$

Inspection of Eqs. of 10.20 and 10.21 reveals that the generalized form of the equation is

$$v = \frac{\mathrm{Constant}[\mathrm{E}_t][\mathrm{S}]}{\mathrm{Constant} + [\mathrm{S}]} \tag{10.23}$$

This is the same general equation as in Eq. 10.19. The numerator rate constants are grouped to become the k_{cat} term while the constants in the denominator are defined as K_m. Thus, both k_{cat} and K_m are typically composed of a complex set of individual rate constants. The specific constants that make up the terms are related to the actual kinetic mechanism and magnitude of the individual rate constants. As an example, if the value of k_3 is much larger than k_5 in Eq. 10.22, k_{cat} would reduce to k_5 because the sum of k_3 and k_5 would be essentially the value of k_3. If, though, the two rate constants had similar values, then k_{cat} would be composed of many terms as would K_m whose units are concentration.

Concentration of Free Enzyme

The derivation of the Michaelis–Menten equation requires a careful account of all forms of the enzyme. It is assumed that the concentration of substrate is greater than that of the enzyme so during the initial course of the reaction the value of [S] does not change. Obviously, various values of [S] can be used, but for each it will remain constant during the reaction. For the simple reaction $\mathrm{E} + \mathrm{S} \rightleftharpoons \mathrm{ES} \rightarrow \mathrm{E} + \mathrm{P}$, the concentration of enzyme is always the sum of the free enzyme and the substrate bound enzyme; that is,

$$[\mathrm{E}_t] = [\mathrm{E}_f] + [\mathrm{ES}] \tag{10.24}$$

where $[\mathrm{E}_t]$ is the total concentration of enzyme and $[\mathrm{E}_f]$ is the concentration of free enzyme. It is E_f that is necessary in a reaction to interact with substrate $(\mathrm{E}_f + \mathrm{S})$. The concentration of free enzyme is not directly measured but rather the total enzyme concentration is known. This means

$$[\mathrm{E}_t] - [\mathrm{ES}] = [\mathrm{E}_f] \tag{10.25}$$

CLINICAL CORRELATION 10.2

A Case of Gout Demonstrates Two Phases in the Mechanism of Enzyme Action

The two phases of the Michaelis–Menten model of enzyme action, binding followed by modification of substrate, are illustrated by studies on a family with gout. The patient excreted three times the normal amount of uric acid per day and had markedly increased levels of 5′-phosphoribosyl-α-pyrophosphate (PRPP) in his red blood cells. PRPP is an intermediate in the biosynthesis of AMP and GMP, which are converted to ATP and GTP. Uric acid arises directly from degradation of AMP and GMP. Assays *in vitro* revealed that the patient's red cell PRPP synthetase activity was increased threefold. The pH optimum and the K_m of the enzyme for ATP and ribose 5-phosphate were normal, but V_{max} was increased threefold! This increase was not due to an increase in the amount of enzyme; immunologic testing with an antibody specific to the enzyme revealed similar quantities of the enzyme protein as in normal red cells. This finding demonstrates that the binding of substrate (as reflected by K_m) and the subsequent chemical event in catalysis, which is reflected in V_{max}, are separate phases of the overall catalytic process. This situation holds only for those enzyme mechanisms in which $k_3 \gg k_2$.

Source: Becker, M. A., Kostel, P. J., Meyer, L. J., and Seegmiller, J. E. Human phosphoribosylpyrophosphate synthetase: Increased enzyme specific activity in a family with gout and excessive purine synthesis. *Proc. Natl. Acad. Sci. USA* 70:2749, 1973.

CLINICAL CORRELATION **10.3**

Physiological Effect of Changes in Enzyme K_m Values

The unusual sensitivity of Asians to alcoholic beverages has a biochemical basis. In some Japanese and Chinese, much less ethanol is required to produce vasodilation that results in facial flushing and rapid heart rate than is required to achieve the same effect in Europeans. The physiological effects are due to acetaldehyde generated by liver alcohol dehydrogenase. Acetaldehyde is normally removed by aldehyde dehydrogenase (ALDH), which converts acetaldehyde to acetate. One amino acid at position 487 in the 500-amino-acid subunit of the tetrameric enzyme is changed in some of the affected individuals. The active enzyme has a glutamate, while the inactive variant has a lysine. The Asian variant was found to have a very low activity, and the K_m for NAD^+ increased from 30 μM to 7000 μM. Even though the enzyme was active, it would have very little activity in the liver because the K_m was so high and k_{cat} was low.

In addition, affected individuals were heterozygotic, having genes for both the active glutamate enzyme and the essentially inactive lysine enzyme. It was expected that their enzyme would be 50% active, yet it had very low activity. ALDH forms tetramers with the two monomers (E_4, E_3K, E_2K_2, EK_3, and K_4), where E_4 and K_4 are homotetrameric forms of the glutamate- and lysine-containing subunits, respectively, and the others were are the heterotetramers. E_3K had 50% of the total activity, not 75%, while EK_3 had essentially no activity, not the 25% one could expect from the one active subunit. The K-subunit was dominant in that it could inactivate the subunit to which it was paired as the residue at position 487 interacted with an arginine at position 475. When a glutamate was at position 487, a stable salt bond formed, but when a lysine was at position 487, it caused the arginine to move. The movement disrupted the NAD binding pocket even though residue 487 was not in contact with the Rossmann fold. This example illustrates the fact that a point mutation on the enzyme can affect the active site even though the residue is not in contact with the region. It further shows how one subunit can be dominant over another.

Source: Zhou J. and Weiner, H. Basis for half-of-the-site reactivity and the dominance of the K487 oriental subunit over the E487 subunit in heterotetrameric human liver mitochondrial aldehyde dehydrogenase. *Biochemistry* 39:12019, 2000.

If there were other forms of the enzyme such as a form with an inhibitor bound (see p. 395), then Eq. 10.25 would have to be expanded to include each form of enzyme.

Meaning of K_m

K_m consists of a number of rate constants and is not a derived dissociation constant (K_d). Its value, however, is used often to indicate how well a substrate interacts with an enzyme. A K_m of 10^{-7} M indicates that the substrate has a greater affinity for the enzyme than if the K_m is 10^{-5} M. The smaller the value of K_m, the tighter is the interaction between substrate and enzyme. Even though K_m is not a dissociation constant, the smaller value means it takes less substrate to bind 50% of the enzyme (Clin. Corr. 10.3).

An example of the importance of K_m is the physiological utilization of glucose. Glucose can be phosphorylated by two different kinases to form glucose 6-phosphate. Liver contains both hexokinase and glucokinase that catalyze the identical reaction of glucose + ATP → glucose 6-phosphate + ADP. For hexokinase the K_m for glucose is 0.1 mM, while for glucokinase it is 5 mM. When the concentration of blood sugar is low, as occurs in the fasted state, hexokinase is used to phosphorylate glucose; but when blood glucose increases after feeding, the high K_m enzyme also functions (see p. 597).

Turnover Number (k_{cat})

In Eq. 10.19, k_{cat} in the numerator has units of $time^{-1}$. This constant is also referred to as the **turnover number** of the enzyme—that is, the number of molecules of substrate converted to product per unit time per molecule of enzyme. The larger the value of k_{cat} for an enzyme, the "faster" the reaction will be. In general, this term represents the slowest step(s) in the reaction. Like K_m, k_{cat} is composed of a number of individual rate constants.

Significance of k_{cat} in the Michaelis–Menten Equation

When Substrate Concentration is much Larger than K_m

Equation 10.19 is in the form of a general hyperbolic equation. When the value of [S] is much larger than the value of K_m, the value of the denominator approaches the value of [S], and the equation can be approximated by

$$v = \frac{k_{cat}[E][S]}{[S]} = k_{cat}[E] \tag{10.26}$$

that is, the velocity becomes independent of the concentration of S; the reaction becomes zero order with respect to S. This is found in the portion of curve where the velocity essentially levels off and is not increased as the concentration of [S] increases (slope = 0)(Figure 10.47). The reaction is proceeding at its maximum velocity under these conditions because essentially 100% of the enzyme is in the ES complex. The instant a molecule of product is made, it leaves the enzyme and another molecule of S binds enzyme, always keeping the enzyme saturated with S. Under these conditions the rate is governed strictly by the terms that govern the reaction of ES going to product (ES $\rightarrow$ E + P).

When Substrate Concentration is much Smaller than K_m

When [S] is very small compared to K_m, Eq. 10.18 can be approximated as

$$v = \frac{k_{cat}[Et][S]}{K_m} \quad \text{or} \quad v = \frac{V_{max}[S]}{K_m} \tag{10.27}$$

A plot of v versus [S] would be linear, a condition that exists only when [S] $\ll K_m$ value. Since for any chemical reaction the velocity of product formation is a rate constant times the concentration of the reactants the term k_{cat}/K_m can be considered as a second-order rate constant since there are two concentration terms in the rate equation ([E_t] and [S]). The fact that this ratio is equivalent to a second-order rate constant puts certain constraints upon the ratio. For example, a rate constant cannot be faster than the rate two molecules diffuse in solution. Even if there were no energy of activation required, the two molecules have to collide with each other in order for a reaction to occur. The rate constant for a second-order diffusion reaction is around 10^9 mole^{-1} s^{-1}. Thus, if k_{cat} is very large, K_m cannot be very small or it would exceed the diffusion rate. As an example, catalase, an enzyme that destroys hydrogen peroxide, has a k_{cat} of 4×10^7 s^{-1}. The K_m for hydrogen peroxide, however, is 1 molar, a very high value. The k_{cat}/K_m ratio is just less than a factor of 25 (10^9 versus 4×10^7) of the theoretical maximum velocity that a second-order reaction can attain. Catalase is one of the fastest known enzymes when the term for k_{cat}/K_m is measured.

k_{cat} can be only a very low number (such as 1 to 10 s^{-1}) if the K_m for substrate is very low (as an example, 1×10^9 M). This slow rate means that one molecule of enzyme would produce just 1–10 molecules of product per second. The reaction is so slow because the substrate binds very tightly to the enzyme. A great deal of energy is required to convert the bound substrate to the transition state. An extremely small K_m (high affinity) for substrate requires the chemical reaction to be slow, whereas a large K_m (low affinity) means that k_{cat} can be high. The values of k_{cat}/K_m for enzymes range from 10 to 10^7 s^{-1}.

Reversible Reactions

Another limitation to the ratio of k_{cat}/K_m occurs when the reaction is reversible.

$$E + S \underset{k_r}{\overset{k_f}{\rightleftharpoons}} E + P \tag{10.28}$$

k_f and k_r are the forward and reverse rate constants (the terms ES and EP are ignored because they are the same for both directions). It can be derived that k_f is $(k_{cat}/K_m)_f$ and k_r is $(k_{cat}/K_m)_r$. For a reversible reaction the equilibrium constant K_{eq} = ([P]/[S]) is k_f/k_r or

$$K_{eq} = \frac{(k_{cat}/K_m)_f}{(k_{cat}/K_m)_r} \tag{10.29}$$

The values for k_{cat} and K_m terms can not change randomly since their ratio has to be equal to K_{eq}. This implies that if a mutant of an enzyme has a greater value of k_{cat} than the original enzyme, some other constant must change to satisfy the equilibrium situation. These equilibrium-type relationships are called **Haldane** relationships.

TABLE 10.3 K_m and K_{cat} Relationships[a]

K_m (M)	k_{cat} (s^{-1})	Rate (s^{-1})
10^{-6}	1	1
10^{-5}	10	9
10^{-4}	10^2	90
10^{-3}	10^3	500
10^{-2}	10^4	909
10^{-1}	10^5	990
1	10^6	999

[a] Reaction conditions: $K_{cat}/K_m = 10^6 M^{-1}s^{-1}$ and $[S] = 10^3 M$.

Low K_m Versus High k_{cat}

Since the ratio of k_{cat}/K_m has an upper limit, the question can be raised as to whether or not it would be preferable for an enzyme to have a very low K_m or a very high k_{cat}. Frequently it is considered that a low K_m is most desirable and compounds with the lowest K_m are reported as the best substrates. Actually, the compound with the highest k_{cat}/K_m is the best substrate. At a constant k_{cat}/K_m, it is better to have a high k_{cat} rather than a low K_m. An example is presented in Table 10.3. At a fixed concentration of substrate, 1 mM in the example, the overall velocity is always faster when K_m is high because this forces k_{cat} to be larger to satisfy the k_{cat}/K_m ratio. Thus, *when evaluating different substrates for an enzyme the substrates with the highest k_{cat}/K_m value should be considered the better substrate for the enzyme not the one with the lowest K_m.* In a cell it is not just the k_{cat} and K_m terms that are important, but one must take into account the total concentration of enzyme. The velocity term is first order with respect to enzyme, so a few picomoles of an enzyme with a k_{cat}/K_m of $10^7\ s^{-1}\ mol^{-1}$ in the cell would be less efficient at converting a substrate to product than would a nanomole of an enzyme with a k_{cat}/K_m of $10^5\ s^{-1}\ mol^{-1}$.

Calculating the Constants

The K_m can be approximated from plots of v versus [S]. Graphical presentations allow for the calculation of the kinetic constants. The most common graphical presentation is based on the **Lineweaver–Burk equation,** which is the reciprocal of the Michaelis–Menten equation. The reciprocal of Eq. 10.20 is

$$\frac{1}{v} = \frac{1}{V_m} + \frac{K_m}{V_m} \times \frac{1}{[S]} \tag{10.30}$$

A plot of $1/v$ versus $1/[S]$ is a straight line (Figure 10.50); the intercept on the y-axis is $1/V_m$, the intercept on the x-axis is $-1/K_m$, and the slope of the line is K_m/V_m. Thus from the graph of $1/v$ at different concentrations of $1/[S]$, two key parameters of the enzyme catalyzed reaction can be determined. Computer programs can calculate the k_{cat} and K_m from the velocity and substrate data.

Substrate and Product Bind to the Same Site

The presence of product will act as an inhibitor since its presence can prevent substrate from binding. It will be a competitive inhibitor since both S and P bind to E_f. The equation relating the initial velocity of the reaction in the presence of product is

$$v = \frac{V_m[S]}{K_m(1 + [P]/K_{mP}) + [S]} \tag{10.31}$$

where K_{mp} is the K_m for product, assuming the reaction could be reversible. If it is a nonreversible reaction, then the K_{mp} term would be K_{dp}, the dissociation constant for product from the enzyme. When running an assay, *in vitro* product is initially not present but starts to accumulate with time. A graph of product formation as a function of time will initially be linear, but the slope will decrease with time as product accumulates and acts as an inhibitor (Figure 10.46, p. 387).

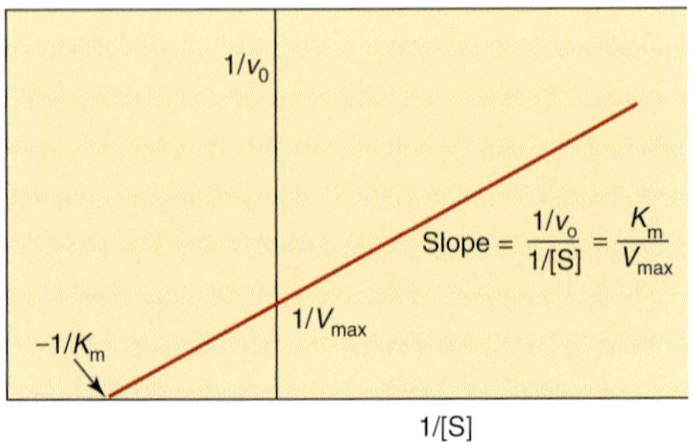

Effect of Assay Conditions

When running a reaction under V_m conditions essentially 100% of the enzyme is in the ES complex. In contrast, when a reaction is studied where $[S] \ll K_m$, the effect being measured would reflect what is occurring with both free and substrate bound enzyme. Thus to study the effect of an environmental change such as pH, the assay is usually conducted under V_{max} conditions. The effect of experimental conditions under V_{max} conditions gives information about the k_{cat} term for the total enzyme concentration. To evaluate the effect of pH on the free enzyme, then k_{cat}/K_m as function of pH would be determined.

FIGURE 10.50

Determination of K_m and V_{max} from the Lineweaver–Burk double-reciprocal plot. Plots of the reciprocal of the initial velocity versus the reciprocal of the substrate concentration used to determine the initial velocity yield a line whose x-intercept is $-1/K_m$.

External conditions, such as pH, temperature, and salt concentration, affect enzyme activity. These effects are probably not important *in vivo* under normal conditions but are very important in setting up enzyme assays *in vitro* to measure enzyme activity in samples of a patient's plasma or tissue.

Temperature

Plots of velocity versus temperature for most enzymes reveal a bell-shaped curve with an optimum between 40°C and 45°C for mammalian enzymes (Figure 10.51). Above this temperature, heat denaturation of the enzyme occurs. Between 0°C and 40°C, many enzymes show a twofold increase in activity for every 10°C rise. Mutation of an enzyme to a thermolabile form can have serious consequences (Clin. Corr. 10.4).

pH

Nearly all enzymes show a bell-shaped pH-velocity profile, but the maximum (**pH optimum**) varies greatly with different enzymes. Alkaline and acid phosphatases with very different pH optima are both found in humans (Figure 10.52). The bell-shaped curve and its position on the *x*-axis are dependent on the particular ionized state of the substrate that will be optimally bound to the enzyme. This in turn is related to the ionization of specific amino acid residues that constitute the substrate-binding site. In addition, amino acid residues involved in catalyzing the reaction must be in the

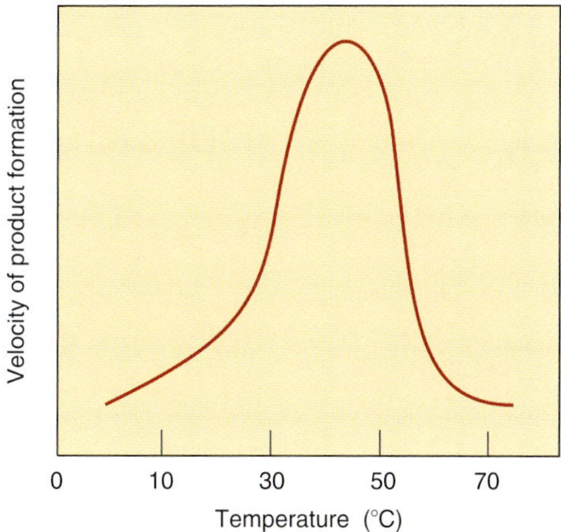

FIGURE 10.51

Temperature dependence of a typical mammalian enzyme. To the left of the optimum, the rate is low because the environmental temperature is too low to provide enough kinetic energy to overcome the energy of activation. To the right of the optimum, the enzyme is inactivated by heat denaturation.

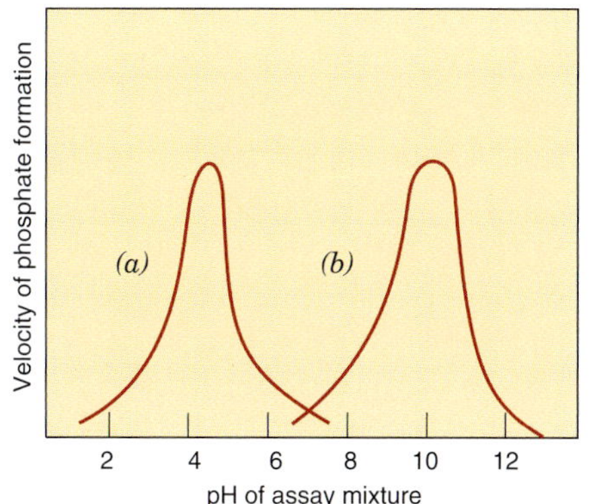

FIGURE 10.52

The pH dependence of (*a*) acid and (*b*) alkaline phosphatase reactions. In each case the optimum represents the ideal ionic state for binding of enzyme and substrate and the correct ionic state for the amino acid side chains involved in the catalytic event.

CLINICAL CORRELATION 10.4

Thermal Lability of Glucose 6-Phosphate Dehydrogenase Results in Hemolytic Anemia

In red cells, glucose 6-phosphate dehydrogenase (G6PD) is an important enzyme in the red cell for the maintenance of the membrane integrity. A deficiency or inactivity of this enzyme leads to a hemolytic anemia. In other cases, a variant enzyme is present that normally has sufficient activity to maintain the membrane but fails under conditions of oxidative stress. A mutation of this enzyme leads to a protein with normal kinetic constants but a decreased thermal stability. This condition is especially critical to the red cell, since it is devoid of protein-synthesizing capacity and cannot renew enzymes as they denature. The end result is a greatly decreased lifetime for those red cells that have an unstable G6PD. These red cells are also susceptible to drug-induced hemolysis.

Source: Luzzato, L., and Mehta, A., Glucose-6-phosphate dehydrogenase deficiency. In: C. R. Scriver, A. L. Beaudet, W. S. Sly, and D. Valle (Eds.), *The Metabolic and Molecular Bases of Inherited Disease*, 7th ed. New York: McGraw-Hill, 1995, p. 3369.

CLINICAL CORRELATION 10.5

Alcohol Dehydrogenase Isoenzymes with Different pH Optima

In addition to the change in aldehyde dehydrogenase isoenzyme composition in some Asians different alcohol dehydrogenase isoenzymes are also observed. Alcohol dehydrogenase (ADH) is encoded by three genes, which produce three different polypeptides: α, β, and γ. Three alleles are found for the β-gene that differ in a single nucleotide base, which causes substitutions for arginine. The substitutions are shown below:

	Residue 47	Residue 369
β_1	Arg	Arg
β_2	His	Arg
β_3	Arg	Cys

The liver β_3 form has ADH activity with a pH optimum near 7, compared with 10 for β_1 and 8.5 for β_2. The rate-determining step in alcohol dehydrogenase is the release of NADH. NADH is held on the enzyme by ionic bonds between the phosphates of the coenzyme and the arginines at positions 47 and 369. In the β_1 isozyme this ionic interaction is not broken until the pH is quite alkaline and the guanidinium group of arginine starts to dissociate H^+. Substitution of amino acids with lower pK values, as in β_2 and β_3, weakens the interaction and lowers the pH optimum. Since the release of NADH is facilitated, the V_{max} values for β_2 and β_3 are also higher than for β_1.

Source: Burnell, J. C., Carr, L. G., Dwulet, F. E., Edenberg, H. J., Li, T-K., and Bosron, W. F. The human β_3 alcohol dehydrogenase subunit differs from β_1 by a cys- for arg-369 substitution which decreases NAD(H) binding. *Biochem. Biophys. Res. Commun.* 146:1227, 1987.

correct charge state to be functional. Clinical Correlation 10.5 points out the effect of a mutation leading to a change in the pH optimum of a physiologically important enzyme.

10.8 KINETICS OF TWO-SUBSTRATE REACTIONS

Sequential Mechanism

There are two mechanisms of interactions for a two-substrate reaction as illustrated in Figure 10.53. In the **sequential mechanism,** both substrates bind to the enzyme in a sequential manner to form a ternary complex. With substrates A and B, the ternary complex would be E-A-B (A, B, C . . . are substrates and P, Q, R . . . are products). If substrate A binds prior to B, then this is called an **ordered-sequential reaction**, while if either A or B can bind first, the reaction is a called a **random-sequential reaction**. The equation for the sequential reaction contains K_m terms from both substrates A and B (K_a and K_b) along with a term K_{ia}, the dissociation constant for the first substrate. To find the various constants a double reciprocal plot of $1/v$ versus $1/[A]$ is plotted for the reaction done with different fixed concentrations of substrate B (Figure 10.54). The resulting plot shows a series of lines that intersect with each other. The individual constants can be found from the slopes and intercepts.

Ping-Pong Mechanism

The second mechanism for a two-substrate reaction is called a **ping-pong mechanism**. Substrate A binds to the enzyme and is converted to product P, leaving a modified enzyme. The modified enzyme binds substrate B, which is then converted to product Q. Enzymes that exhibit ping-pong kinetics are modified by the substrate. Kinases that follow a ping-pong mechanism do so by having ATP or another nucleotide triphosphate first bind to and then phosphorylate the enzyme. ADP dissociates and the second substrate binds to the phosphorylated enzyme followed by phosphate transfer to produce the phosphorylated acceptor. Enzymes that use covalent catalysis or have coenzymes that do not dissociate employ ping-pong reactions (see Clin. Corr. 10.2).

The equation relating the velocity to the concentrations of A and B in the ping-pong mechanism is similar to the equation for the sequential reaction, but, most significantly, it does not have the K_{ia} term as found in the sequential reaction. Double reciprocal plots for this reaction result in a series of parallel lines (Figure 10.55). Again, the individual K_m and V_m values can be obtained from the slopes and intercepts.

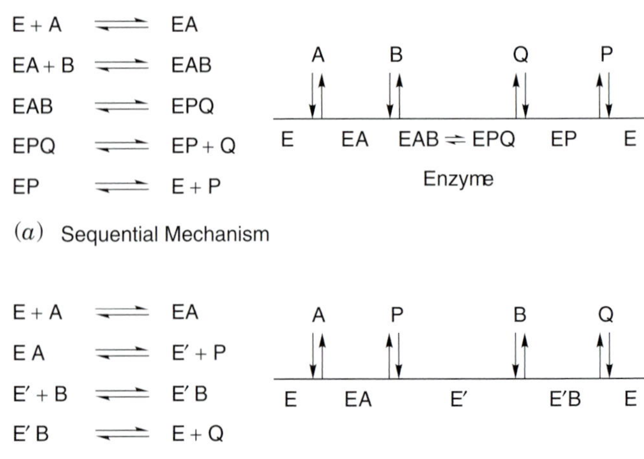

FIGURE 10.53

Mechanisms of interaction for two substrate reactions. (a) Sequential mechanism. (b) Ping-Pong mechanism.

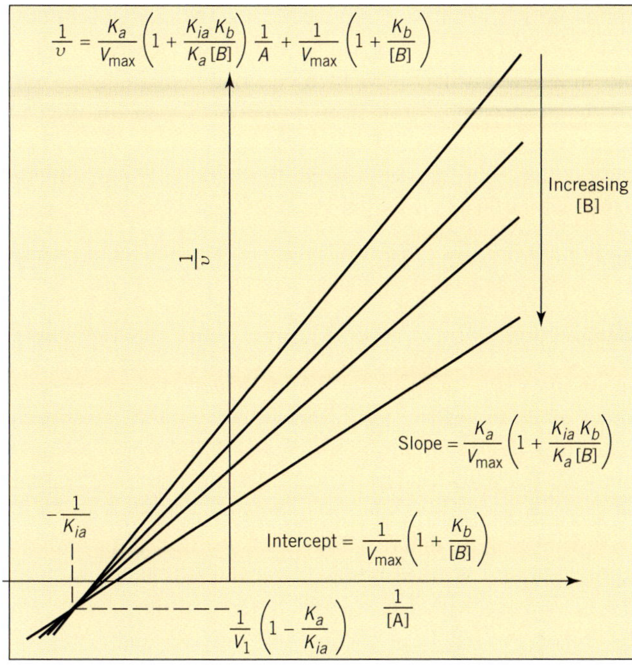

FIGURE 10.54

Double reciprocal plot of initial velocity for a sequential reaction. Substrate B is fixed at specified concentrations and substrate A varied. The rate equation is

$$\frac{1}{v} = \frac{K_a}{V_{max}}\left(1 + \frac{K_{ia}K_b}{K_aB}\right)\frac{1}{A} + \frac{1}{V_{max}}\left(1 + \frac{K_b}{B}\right)$$

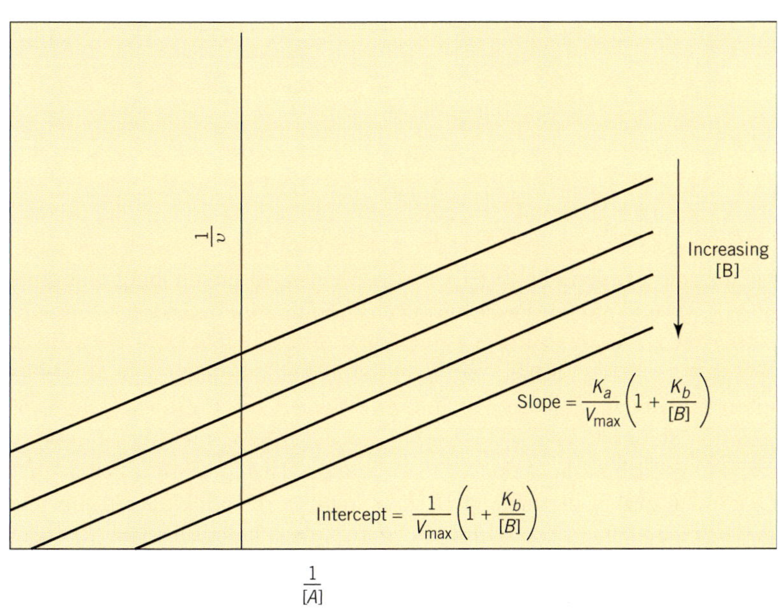

FIGURE 10.55

Double reciprocal plot of initial velocity for a ping-pong reaction. Substrate B is fixed at specified concentrations and substrate A varied. The rate equation is

$$\frac{1}{v} = \frac{K_a}{V_{max}}\left(1 + \frac{K_b}{B}\right) + \frac{1}{V_{max}}\left(1 + \frac{K_b}{B}\right)$$

K_{mapp}

Often the K_m for just one of the substrates is desired. Performing the reaction in the presence of very high concentrations of substrate A and varying the concentration of B permits determination of a K_{mapp} for substrate B. This is possible because when $[A] \gg K_a$ the velocity equation reduces to a simple Michaelis–Menten equation.

10.9 INHIBITORS

Much of our basic knowledge of metabolic pathways was determined by using inhibitors of specific enzymes. Pharmaceutical compounds are now designed as inhibitors of specific enzymes and many naturally occurring compounds are inhibitors (Clin. Corr. 10.6). Inhibitory compounds can bind to the free enzyme or to the ES complex and affect

CLINICAL CORRELATION 10.6
Inhibitors of Xanthine Oxidase Isolated from Plants

Gout is an inflammatory disease that results from the overproduction or undersecretion of uric acid, a compound derived from purines (see p. 799). The terminal step in purine metabolism is catalyzed by xanthine oxidase; a common treatment is the use of allopurinol, a structural analogue of xanthine and an inhibitor of the enzyme. Many older societies used medicinal plants and have maintained records of which are effective. Records of Vietnamese traditional medicine indicate that 96 plants, including *Chrysanthemum sinense*, have been used to treat gout. Extracts from these plants were examined for their ability to inhibit mammalian xanthine oxidase, and a few were found that were nearly as effective as was allopurinol. Their structures are quite different from allopurinol and xanthine. The results indicate that study of compounds from traditional medicine may lead to new pharmaceuticals.

Source: Nguyen, M. T. T., Awale, S., Tezuka, Y., Tran, Q. L. Watanabe, H. and Kadota, S. Xanthine oxidase inhibitory activity of Vietnamese medicinal plants. *Biol. Pharm. Bull.* 27:1414, 2004.

the velocity of the reaction. With an inhibitor, EI and/or ESI (where I is inhibitor) can be present. An equation relating the total concentration of enzyme to all its possible complexes results in a new Michaelis–Menten equation containing the term $(1 + [I]/K_i)$, where K_i is the dissociation constant of EI or ESI. k_{cat} in the presence of the inhibitor is an apparent k_{cat} (k_{catapp}) and K_m will be an apparent K_m (K_{mapp}). The inhibitor concentration does not change during the course of the reaction, therefore [I] is a constant. Several classes of inhibitors are defined based on these reaction kinetics.

Competitive Inhibition

Competitive inhibition occurs when an inhibitor competes with the substrate for binding to the free enzyme. Some **competitive inhibitors** are structurally similar to the substrate or product, while others are structurally quite different. For example, malonate is a competitive inhibitor of succinic dehydrogenase (Figure 10.56 and p. 545), binding presumably to the binding site; however, pyrazole is a competitive inhibitor of alcohol dehydrogenase, but its structure is quite different from the substrate. If an inhibitor binds to the active site and prevents S from binding, the reaction would be as in Figure 10.57. The inhibitor competes with substrate for binding to the enzyme. The amount of enzyme in the ES complex will be a function of [S], [I], K_m and K_i (k_5/k_6). Since $E_t = E_f + ES + EI$, Eq. 10.32 is obtained.

$$v = \frac{k_3 k_1 [E_t][S]}{((k_2 + k_3)(1 + [I]/K_i) + k_1[S])} \tag{10.32}$$

Dividing all terms by k_1 yields Eq. 10.33 since $(k_2 + k_3)/k_1$ is K_m in the absence of I.

$$v = \frac{k_3[E][S]}{K_m(1 + [I/K_i]) + [S]} \tag{10.33}$$

Since the [I] is constant, the term $(1 + [I]/K_i)$ is a constant. K_m will be an K_{mapp} because it is the true K_m times $(1 + [I]/K_i)$. The presence of the inhibitor will cause the reaction to become slower at lower substrate concentrations because the constants in the denominator are larger than if no inhibitor is present. Only when [S] is much greater than K_{mapp} would V_{max} be obtained as the denominator would be just [S]. Under this condition, virtually no inhibitor will bind to enzyme (Clin. Corr. 10.7). Both the substrate and the inhibitor compete for binding; thus increasing the concentration of the substrate will overcome the action of the inhibitor. Therapeutically, competitive inhibitors are useful only if the value of $[I]/K_i$ is high, so the inhibitor can function in the presence of even high concentrations of substrate. Low values for K_i (tight binding) are desired so large amounts of inhibitor need not be required to inhibit the enzyme.

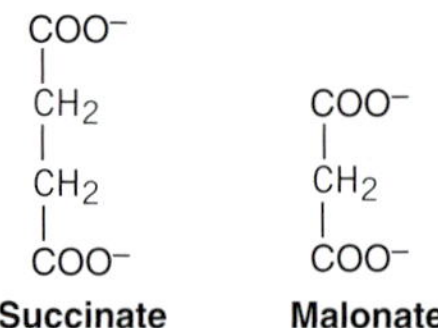

FIGURE 10.56

Substrate and inhibitor of succinate dehydrogenase.

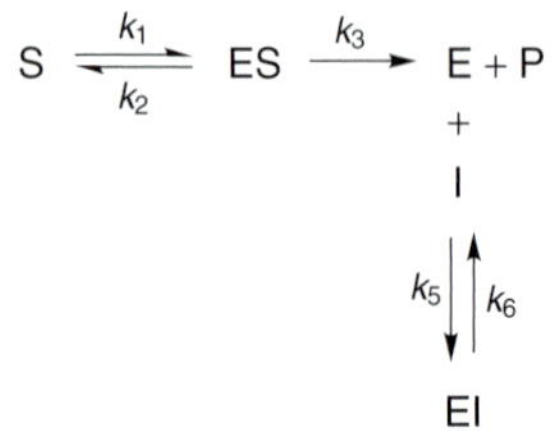

$$S \underset{k_2}{\overset{k_1}{\rightleftharpoons}} ES \xrightarrow{k_3} E + P$$
$$+ \; I \; \underset{k_6}{\overset{k_5}{\rightleftharpoons}} \; EI$$

FIGURE 10.57

Reaction of a competitive inhibitor reacting with free enzyme.

Uncompetitive Inhibition

An inhibitor can bind to the ES complex rather than the free enzyme, as indicated Figure 10.58. The Michaelis–Menten equation will be

$$v = \frac{\dfrac{V_{max}[S]}{(1 + [I]/K_i)}}{\dfrac{K_m + [S]}{(1 + [I]/K_i)}} \tag{10.34}$$

k_{cat} is now decreased, so the V_{max} will be decreased compared to the value in the absence of inhibitor and K_{mapp} term is also changed. This situation is called **uncompetitive inhibition**. In contrast to competitive inhibition, increasing the concentration of substrate will not overcome the effect of the inhibitor because it is binding to ES complex and not free enzyme.

CLINICAL CORRELATION 10.7
Design of a Selective Inhibitor

Prostaglandins are a very important class of paracrine hormones derived from long-chain unsaturated fatty acids (see p. 730). One of the early steps in their synthesis involves an enzyme, cyclooxygenase (COX). Among the many physiological effects of the prostaglandins are the pain and inflammation associated with arthritis. Traditionally, arthritis has been treated with aspirin and other nonsteroidal anti-inflammatory drugs that have been shown to work by inhibiting the cyclooxygenases (see p. 733), thus decreasing the levels of "bad" prostaglandins causing the symptoms. A serious problem with these drugs is the development of gastrointestinal bleeding and perforation caused by a concomitant decrease in the "good" prostaglandins which are protective of the gastrointestinal mucosa.

There are two different cyclooxygenases, COX-1 and COX-2 (see p. 732), which have different tissue distributions. The COX-2 enzyme is inducible by mediators from the arthritic condition, whereas COX-1 constitutively makes the mucosa-sparing prostaglandins. The only significant difference in the active sites of the two enzymes is that COX-2 has a valine rather than isoleucine at residue 523. The smaller valine allows for preferential binding of custom-designed inhibitors, resulting in selective inhibition of COX-2. Such a custom-designed drug, VIOXX™ (in green), is shown modeled into the COX-1 active site.

Key amino acid side chains forming the substrate-binding pocket are shown in red; the heme, which participates in prostaglandin synthesis, is shown in blue. It has been shown that aspirin acetylates Serine 530 in both COX-1 and -2, whereas VIOXX™ selectively binds to and inhibits COX-2 thus, preserving the production of "good" prostaglandins by COX-1. The phenyl group of aspirin occupies the identical location in the active site as the phenyl group of VIOXX. Initially, VIOXX-like drugs appear to be competitive inhibitors; but with COX-2, but not with COX-1, they show a time-dependent irreversible inhibition that does not involve a covalent linkage. VIOXX came on the market in 2000 and made a tremendous difference in the quality of life for arthritic patients. Unfortunately, significant side effects involving the heart have developed with some patients, and the drug was withdrawn from the market.

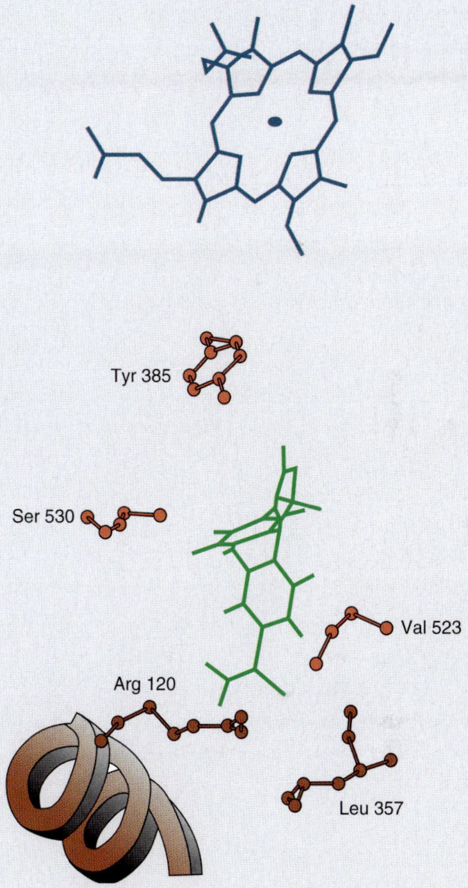

Source: Gierse, J. K., McDonald, J. J., Hauser, S. D., Rangwala, S. H., Koboldt, C. M., and Seibert K. A single amino acid difference between cyclooxygenase-1 (COX-1) and -2 (COX-2) reverses the selectivity of COX-2 specific inhibitors. *J. Biol. Chem.* 271:15810, 1996.

Noncompetitive Inhibition

A more complicated situation occurs if a reversible inhibitor binds to both the free enzyme and the ES complex. A variety of situations can occur depending upon how the presence of I affects the binding of substrate and the ability of the enzyme to catalyze the reaction Figure 10.59. The Michaelis-Menten equation is very complex as both ES and ES-I can yield products. If I binds with the same affinity to both E and ES and k_3' is zero, then the following equation can be derived.

$$v = \frac{\dfrac{V_{\max}[\mathrm{S}]}{(1 + [\mathrm{I}]/K_i)}}{K_m + [\mathrm{S}]} \tag{10.35}$$

This mode of binding leads to an inhibition of activity in the presence of even high concentrations of [S]. If, k_3' is not zero, then a different situation arises since the ES-I complex will produce product.

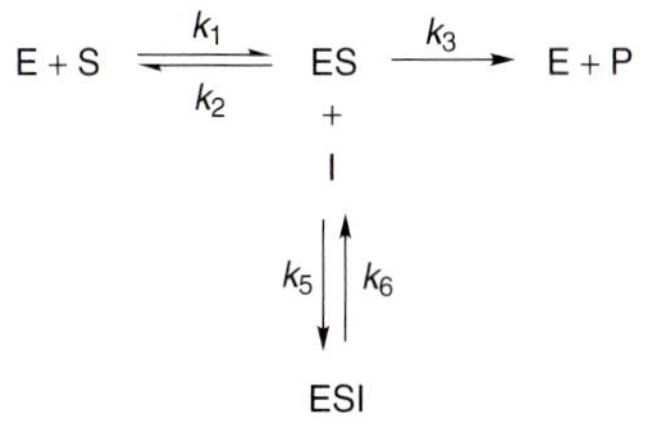

FIGURE 10.58
Reaction of an uncompetitive inhibitor reacting with the enzyme–substrate complex.

$$
\begin{array}{ccccc}
E + S & \rightleftharpoons & E\text{-}S & \xrightarrow{k_3} & E + P \\
+I \updownarrow & & +I \updownarrow & & \\
E\text{-}I & \rightleftharpoons & E\text{-}S\text{-}I & \xrightarrow[k'_3]{} & E + P
\end{array}
$$

FIGURE 10.59

Reaction of a noncompetitive inhibitor reacting with an enzyme.

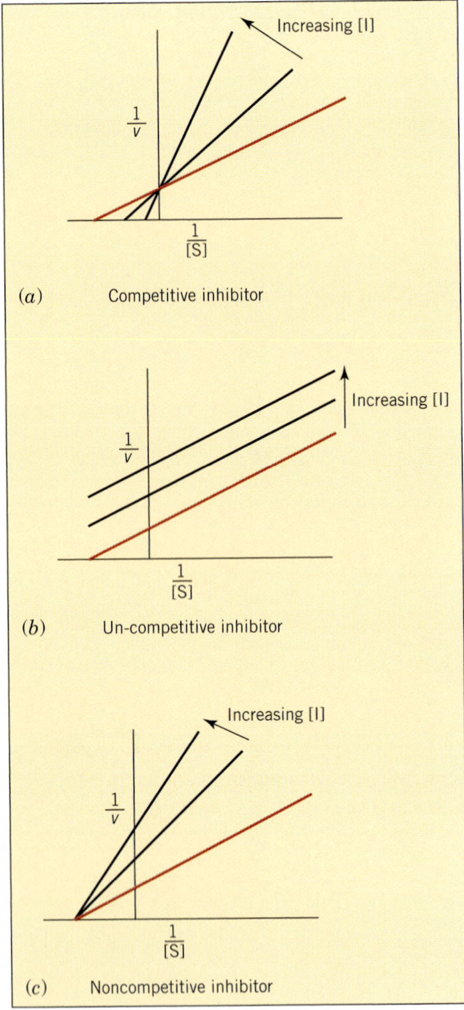

FIGURE 10.60

Double-reciprocal plots for competitive, uncompetitive, and reversible noncompetitive inhibition. A competitive inhibitor binds at the substrate-binding site and effectively increases the K_m for the substrate. An uncompetitive inhibitor causes an equivalent shift in both V_{max} and K_m, resulting in a line parallel to that given by the uninhibited enzyme. A reversible noncompetitive inhibitor binds with both E and ES, presumably at a site other than the substrate-binding site; therefore the effective K_m does not change, but the apparent V_{max} decreases.

Lineweaver–Burk Plots in the Presence of Inhibitors

Competitive and Noncompetitive Inhibitors

The Michaelis–Menten equation for an enzyme reaction performed in the presence of the three different types of reversible inhibitors leads to different double reciprocal plots (Figure 10.60). The reason is that the term $(1 + [I]/K_i)$ is found associated with the K_m or k_{cat} term, depending upon what type of inhibitor is being used. In the case of a competitive inhibitor (Figure 10.60a), V_{max} is not changed so the velocity lines in the presence and absence of inhibitor intersect on the y-axis. With uncompetitive inhibition as the inhibitor concentration is changed, a series of parallel lines forms since the slope of the line is V/K and both V and K terms contain the $(1 + [I[/K_i)$ term (Figure 10.60b). With noncompetitive inhibition the inhibitor binds to both E and ES and the Lineweaver–Burk plot is a family of lines that intersect on the y-axis.

The double-reciprocal plots can be used to determine the mode of inhibitor binding. From the intercepts on the x- and y-axis, V_{mapp} and K_{mapp} can be calculated. If the K_m for a substrate is known, then the K_i for the inhibitors can be calculated.

Noncompetitive inhibitors effectively reduce the concentration of enzyme, and thus V_{max} is decreased (Figure 10.60c). Even if substrate were saturating ($[S] \gg K_m$), the observed V_{max} will be lower than it would be in the absence of the inhibitor. Compounds that are these types of inhibitors are more useful as drugs since they can inhibit the enzyme independent of the concentration of substrate.

Inhibitors of Two-Substrate Reactions

A competitive inhibitor of a two-substrate reaction usually inhibits only one of the two substrates. If the compound binds to only free enzyme, it will be competitive against substrate A. If the inhibitor binds to the E-A complex, then it will be competitive against substrate B. The K_i value for inhibitors of two substrate reactions can be determined by varying the concentration of one substrate in the presences of an excess of the other substrate. The reactions for noncompetitive inhibitors would be the same as with a single substrate

Other Inhibitors

Transition State Inhibitors

As described previously, enzymes lower the energy of activation by stabilizing the transition state not the binding of substrate (see p. 372). Thus, if a compound resembled the transition state, it may bind more tightly to the enzyme than would a compound that resembles the substrate. These inhibitors have a K_i value much lower than the K_m for substrate. No one has ever seen a transition state, so proposed structures are based on chemical intuition. An example of a transition state analog is an inhibitor of proline racemase (Figure 10.61). The tetrahedral substrate will become planar during the conversion of D-proline to L-proline, and pyrrole 2-carboxylate structurally resembles the proposed transition state.

Suicide Inhibitors

Enzymes perform complex reactions where many transient intermediates occur as substrate is transformed to product. Compounds have been synthesized where the enzyme performs an initial step in the reaction but then converts the substrate into a form that remains covalently attached to the enzyme. The activity of the enzyme will be destroyed—hence the name suicide inhibitor.

Irreversible Inhibitors

Compounds that chemically modify and inactivate an enzyme are classified as **irreversible inhibitors.** These react with amino acids residues (see Figures 10.62 and 10.63). Aspirin and Antabuse, a drug given to deter people from abusive alcohol consumption, are examples of irreversible enzyme inhibitors used clinically (Clin. Corrs. 10.8 and 10.9).

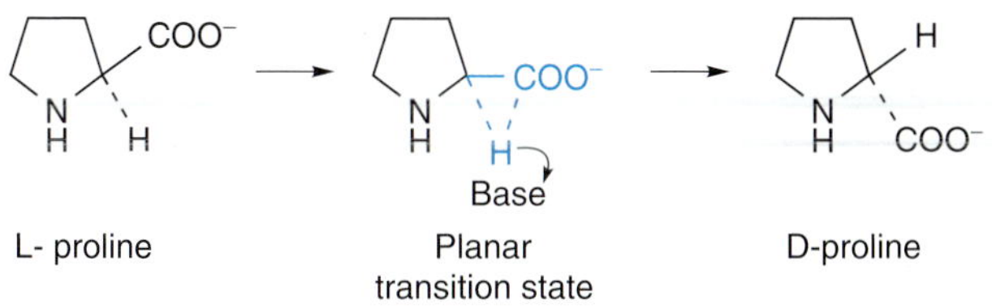

L- proline → Planar transition state → D-proline

Base

(*a*) Proline racemase reaction

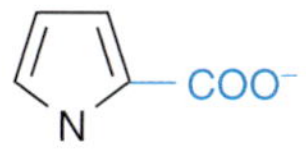

(*b*) Pyrrole-2-carboxylate

FIGURE 10.61

Pyrrole 2-carboxylate mimics the planar transition state and is an excellent inhibitor of proline racemase.

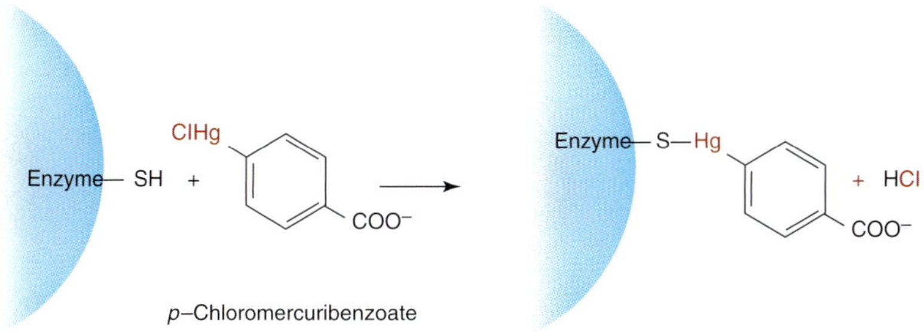

Enzyme— SH + ClHg–C$_6$H$_4$–COO$^-$ → Enzyme—S—Hg–C$_6$H$_4$–COO$^-$ + HCl

p–Chloromercuribenzoate

FIGURE 10.62

Enzyme inhibition by covalent modification of an active site cysteine.

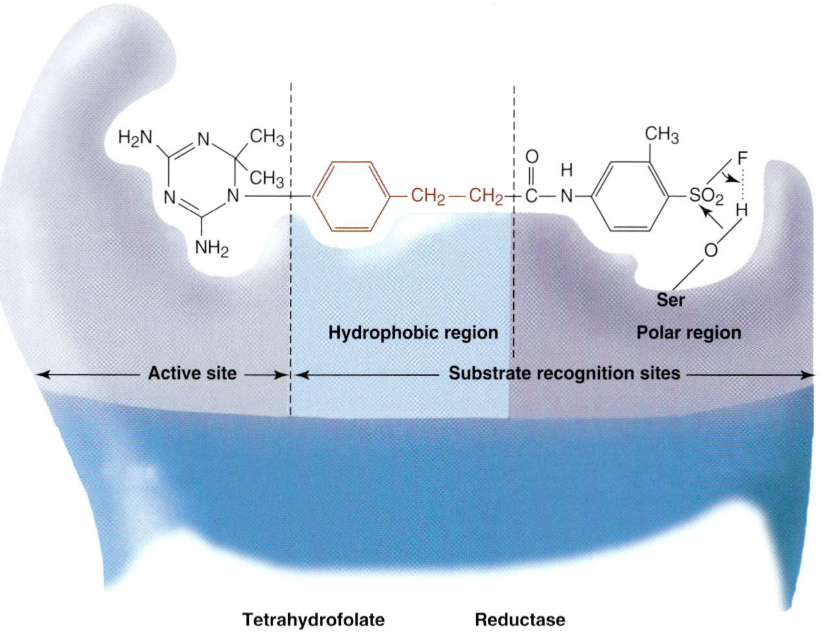

FIGURE 10.63

Site-directed inactivation of tetrahydrofolate reductase. The irreversible inhibitor, a substituted dihydrotriazine, structurally resembles dihydrofolate and binds specifically to the dihydrofolate site on dihydrofolate reductase. The triazine portion of the inhibitor resembles the pterin moiety and therefore binds to the active site. The ethylbenzene group (in red) binds to the hydrophobic site normally occupied by the *p*-aminobenzoyl group. The reactive end of the inhibitor contains a reactive sulfonyl fluoride that forms a covalent linkage with a serine hydroxyl on the enzyme surface. Thus, this inhibitor irreversibly inhibits the enzyme by blocking access of dihydrofolate to the active site. This is not a suicide inhibitor but is instead an example of an inhibitor that binds to the active site and has a reactive group.

CLINICAL CORRELATION 10.8
A Case of Poisoning

Emergency room personnel encounter many instances of pesticide poisoning and must be equipped to recognize and treat these cases. Many of the common insecticides are organophosphate compounds that irreversibly inhibit the action of acetylcholine esterase, AChE, in the postsynaptic fibers of the cholinergic neurons (p. 957) by forming stable phosphate esters with a specific serine in the active site of the esterase. Inhibition of AChE prevents the hydrolysis of acetylcholine in the synapse, resulting in constant stimulation of the end organs of these neurons. The most prominent effects of pesticide poisoning in humans are paralysis of the respiratory muscles and pulmonary edema. If given early enough, a drug like Pralidoxime can displace the alkyl phosphate from the pesticide bound to the active site serine and regenerate an active AChE:

In the disease myasthenia gravis, poisoning the AChE is good medical practice. In this condition there is a functional decrease in the amount of postsynaptic acetylcholine receptor. One way to overcome this deficit is to increase levels of acetylcholine by treating the patient with an AChE inhibitor.

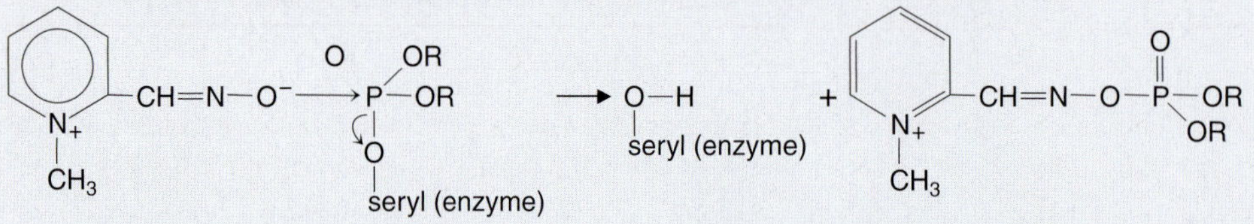

Pralidoxime + Inhibited AChE (seryl (enzyme)–O–P(=O)(OR)(OR)) → Active AChE (seryl (enzyme)–O–H) + Inactive inhibitor (pyridinium–CH=N–O–P(=O)(OR)(OR), N^+–CH_3)

Pralidoxime | Inhibited AChE | Active AChE | Inactive inhibitor

Source: R. Main in E. Hodgson and F. E. Guthrie, ed., *Introduction to Biochemical Toxicology*. New York: Elsevier, 1980, pp. 193–223.

CLINICAL CORRELATION 10.9
Mushrooms and Alcohol Metabolism

Coprine, a toxin produced by the Inky Cap mushroom, forms a covalent bond with aldehyde dehydrogenase, an enzyme involved in the metabolism of alcohol. Coprine is not usually very toxic except in individuals who consume an alcoholic beverage while eating the mushrooms. The person gets a toxic reaction to the accumulated acetaldehyde formed in the oxidation of ethanol. Disulfuram, a drug used to deter abusive consumption of alcohol, is sold under the name Antabuse, which is also a covalent inhibitor of aldehyde dehydrogenase. The drug makes the patient ill if they consume alcohol. Disulfuram was originally used as an antioxidant in the rubber industry, where it was observed that workers became ill when having a drink after work. The inhibitor does not dissociate from the enzyme, and the only way to restore the activity is by biosynthesis in the tissue of new enzyme.

Source: Wiseman, J. S. and Abeles, R. H. Mechanism of inhibition of aldehyde dehydrogenase by cyclopropanone hydrate and the mushroom toxin coprine. *Biochemistry* 18:427, 1979.

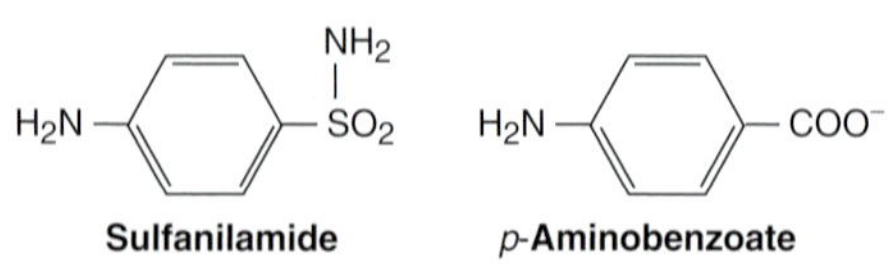

Enzyme Inhibitors as Drugs

As indicated in the discussion above, many drugs have their effect by inhibiting specific enzymes. New pharmacologically active compounds are screened for their effect on a variety of enzymes. Drugs designed to inhibit enzymes unique to a microorganism will produce fewer side effects in patients. A classic example is sulfa drugs because bacteria convert *p*-aminobenzoic acid to folic acid whereas humans do not have the enzyme for this reaction (Figure 10.64). Subtle differences that exist between isozymes of mammalian enzymes are being exploited in order to produce specific inhibitors. An example is **sildenafil** (trade name **Viagra**), a drug for erectile dysfunction, which inhibits a specific isozyme of phosphodiesterase (see p. 436). However, an agent may be found to be specific for an enzyme *in vitro*, but it may have effects on other untested enzymes.

H_2N–C_6H_4–SO_2–NH_2 **Sulfanilamide**

H_2N–C_6H_4–COO^- ***p*-Aminobenzoate**

FIGURE 10.64
Structure of *p*-aminobenzoate and sulfanilamide, a competitive inhibitor of a bacterial enzyme involved in the synthesis of folic acid.

10.10 REGULATION OF ENZYME ACTIVITY

Covalent Modification

Cells have the ability to regulate the activity of key enzymes by covalent modification or reversible binding of ligands. Many examples of covalent modification, such as phosphorylation, will be described in other chapters. There are many protein kinases that catalyze phosphorylation of serine, threonine, or tyrosine residues on specific proteins and enzymes (see p. 496) and protein phosphorylases that remove the phosphate by hydrolysis. Depending on the protein, phosphorylation can increase or decrease the enzymatic activity. Groups other than phosphate can be added to an enzyme to alter its activity, including sulfate and acetate. Covalent modification of proteins is a rapid and effective way of controlling a protein or enzyme activity.

Allosteric Control of Enzyme Activity

Although the substrate binding and active site of an enzyme are well-defined structures, the activity of many enzymes can be modulated by ligands acting in ways other than as competitive or noncompetitive inhibitors. Ligands can be activators, inhibitors, or even the substrates of enzymes. Those ligands that change enzymatic activity, but are unchanged because of enzyme action, are referred to as effectors, modifiers, or modulators. Most of the enzymes subject to modulation by ligands are rate-determining enzymes in metabolic pathways.

Enzymes that respond to modulators have additional site(s) known as allosteric site(s). **Allosteric** is derived from the Greek root allo, meaning "the other." An allosteric site is a unique region of the enzyme quite different from the substrate-binding site. The existence of allosteric sites is illustrated in Clin. Corr. 10.10. The ligands that bind at the allosteric site are called allosteric effectors or modulators. Binding of an allosteric effector causes a conformational change of the enzyme so that the affinity for the substrate or other ligands also changes. Positive (+) allosteric effectors increase the enzyme affinity for substrate or other ligand. The reverse is true for negative (−) allosteric effectors. The allosteric site at which the positive effector binds is referred to as an activator site; the negative effector binds at an inhibitory site.

Allosteric enzymes are divided into two classes based on the effect of the allosteric effector on the K_m and V_{max}. In the **K class** the effector alters the K_m but not V_{max}, whereas in the **V class** the effector alters V_{max} but not K_m. K class enzymes give double-reciprocal plots like those of competitive inhibitors, and V class enzymes give double-reciprocal plots like those of noncompetitive inhibitors. The terms competitive and noncompetitive are inappropriate for allosteric enzyme systems because the mechanism of the effect of an allosteric inhibitor on a V or K enzyme is different from that of a simple competitive or noncompetitive inhibitor. For example, in the K class, the negative effector binding at an allosteric site affects the affinity of the substrate-binding site for the substrate, whereas in simple competitive inhibition the inhibitor competes directly with substrate for the site. In V class enzymes, positive and negative allosteric modifiers increase or decrease the rate of breakdown of the ES complex to products. There are a few enzymes in which both K_m and V_{max} are affected.

In theory, a monomeric enzyme can undergo an allosteric transition in response to a modulating ligand. In practice, only two monomeric allosteric enzymes are known, ribonucleoside diphosphate reductase and pyruvate-UDP-*N*-acetylglucosamine transferase. Most allosteric enzymes are **oligomeric**; that is, they consist of several subunits. As a consequence of the oligomeric nature of allosteric enzymes, binding of ligand to one protomer can affect the binding of ligands on the other protomers in the oligomer. Such ligand effects are referred to as **homotropic interactions**. Transmission of the homotropic effects between protomers is one aspect of cooperativity, considered later. Substrate influencing substrate, activator influencing activator, or inhibitor influencing inhibitor binding are homotropic interactions. Homotropic interactions are almost always positive.

CLINICAL CORRELATION 10.10

A Case of Gout Demonstrates the Difference Between an Allosteric and Substrate-Binding Site

The realization that allosteric inhibitory sites are separate from allosteric activator sites as well as from the substrate-binding and the catalytic sites is illustrated by a study of a gouty patient whose red blood cell PRPP level was increased. It was found that the patient's PRPP synthetase had normal K_m and V_{max} values, along with sensitivity to activation by phosphate. The increased PRPP levels and hyperuricemia arose because the end products of the pathway (ATP, GTP) were not able to inhibit the synthetase through the allosteric inhibitory site (I). It was suggested that a mutation in the inhibitory site or in the coupling mechanism between the inhibitory and catalytic site led to failure of the feedback control mechanism.

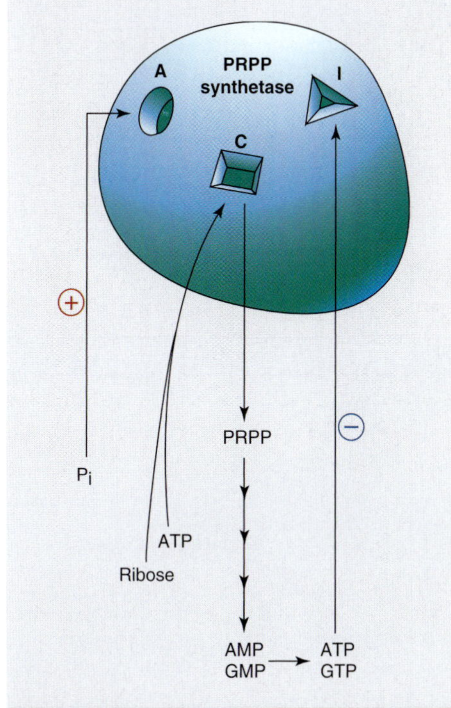

Source: Sperling, O., Persky-Brosh, S., Boen, P., and DeVries, A. Human erythrocyte phosphoribosyl pyrophosphate synthetase mutationally altered in regulatory properties. *Biochem. Med.* 7:389, 1973.

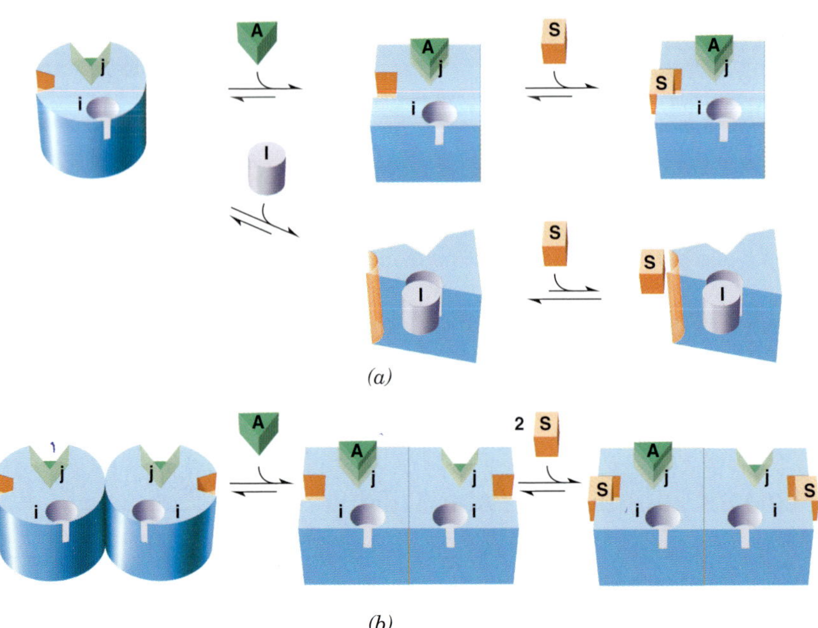

FIGURE 10.65

Models of allosteric enzyme systems. (*a*) Model of a monomeric enzyme. Binding of a positive allosteric effector, A (green), to the activator site, j, induces a new conformation to the enzyme, one that has a greater affinity for the substrate. Binding of a negative allosteric effector (purple) to the inhibitor site, i, results in an enzyme conformation having a decreased affinity for substrate (orange). (*b*) A model of a polymeric allosteric enzyme. Binding of the positive allosteric effector, A, at the j site causes an allosteric change in the conformation of the protomer to which the effector binds. This change in the conformation is transmitted to the second protomer through cooperative protomer–protomer interactions. The affinity for the substrate is increased in both protomers. A negative effector decreases the affinity for substrate of both protomers.

A **heterotropic interaction** is the effect of one ligand on the binding of a different ligand. For example, the effect of a negative effector on the binding of substrate or on binding of an allosteric activator are heterotropic interactions. Heterotropic interactions can be positive or negative and can occur in monomeric allosteric enzymes. Heterotropic and homotropic effects in oligomeric enzymes are mediated by cooperativity between subunits.

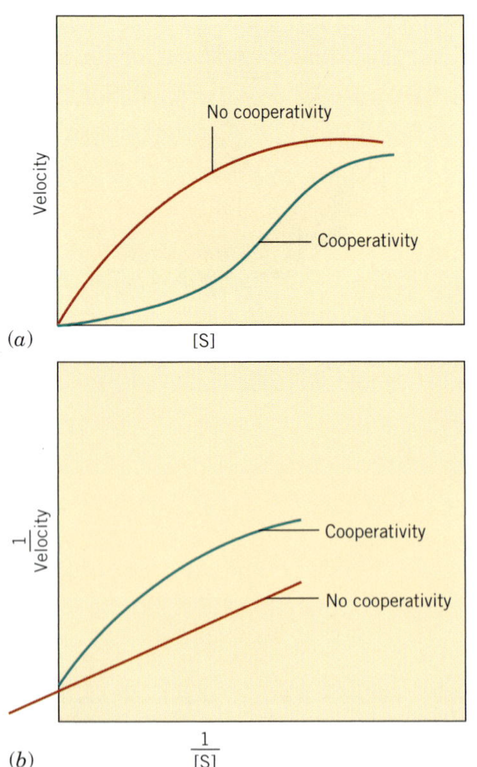

FIGURE 10.66

Plots for enzymes that exhibit cooperativity are atypical. (*a*) Velocity versus substrate plot and (*b*) double-reciprocal plot. Cooperative curves are for positive cooperativity.

Multi-subunit Enzymes: Cooperativity

Many enzymes are multimers consisting of either the same or different monomers. Subunits of some multimeric enzymes act independently of each other, with events taking place in one subunit not influencing the activity of the other subunits. k_{cat} or V_{max} values can be divided by four to obtain the actual turnover number per subunit. In other multi-subunit enzymes, an event in one subunit affects the ability of the adjacent subunits to function. In these situations, the K_m for substrate is usually changed but in some cases k_{cat} is altered (Clin. Corr. 10.5, p. 394). One subunit in a multimer can affect the properties of an adjacent subunit by changing the conformation of the subunits after substrate binding (Figure 10.65a). The altered monomer could bind substrate more easily and hence be more saturated. If the binding becomes easier (decrease K_m), then the enzyme is said to exhibit **positive cooperativity**, analogous to hemoglobin binding oxygen (p. 344). When the structural change makes binding of substrate to the second subunit more difficult (increase K_m), then the enzyme is said to exhibit **negative cooperativity**.

Test for Cooperativity

The test for an enzyme exhibiting cooperativity between subunits is a plot of velocity versus [S]. There will be a normal hyperbolic curve that satisfies the Michaelis–Menten equation for a noninteracting (noncooperative) enzyme. When cooperativity exists, then the curve will not fit the simple equation because the K_m will change as the concentration of substrate changes. The curve will be sigmoidal rather than hyperbolic (Figure 10.66a), and double-reciprocal plots will not be linear (Figure 10.66b).

It is very difficult to calculate precise K_m values for an enzyme exhibiting either positive or negative cooperativity. The value, however, is estimated from the concentration of substrate that allows v to be 50% of V_{max} and is referred to as K_{mapp}.

The Hill equation, used to explain the cooperative binding of oxygen to hemoglobin (see p. 344), is used to estimate the degree of cooperativity between interacting subunits

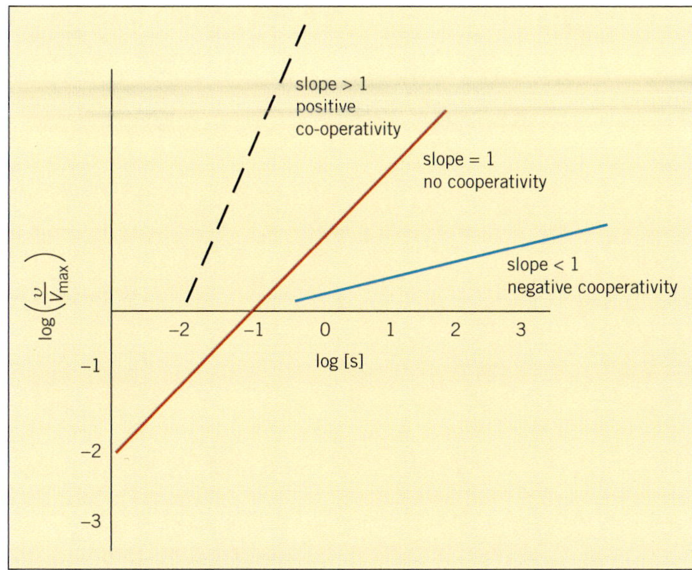

FIGURE 10.67
Hill plot for an allosteric enzyme.

of an enzyme, and the Hill coefficient is an estimate of the degree of cooperativity that exists in the enzyme. Values of less than one demonstrate negative cooperativity and greater than one positive cooperativity. Representative examples of positive and negative cooperativity are illustrated in Figure 10.67.

Models to Explain Cooperativity

Two models were independently developed to explain protein–protein cooperativity. Neither has been proven unequivocally to be correct. Both use the premise that there are interactions between the subunits, and the conformation of the subunit binding a ligand is different from that of the subunit in the absence of ligand. The major difference between the models relates to whether or not the two conformations are preexisting.

The sequential model presented by Koshland, Némethy, and Filmer (KNF) proposes that as ligand binds to one subunit, there is a change in the conformation of the subunit, which then induces a conformational change in a contiguous subunit. The effect of ligand binding is sequentially transmitted through the interface between subunits producing increased or decreased affinity for the ligand by contiguous protomers (Figure 10.68a). In this model numerous hybrid states occur, giving rise to cooperativity and sigmoid plots of velocity versus [S]. Both positive and negative cooperativity can be accommodated by the model. In addition, positive allosteric modulators can induce a conformation in the subunit that has an increased affinity for the substrate; a negative modulator induces a different conformation in the subunit, resulting in decreased affinity for substrate. Both effects are cooperativity transmitted to adjacent subunits. A similar argument can be made for enzymes whose k_{cat} is changed.

The concerted model proposed by Monod, Wyman, and Changeux proposes that the two conformations are in equilibrium in the absence of ligand and, as ligand binds, equilibrium is driven toward the ligand bound conformation. The two states in the concerted model are called the T (tense or taut) and the R (relaxed) (Figure 10.68b). Activators and substrates bind to the R state and shift the preexisting equilibrium toward the R state. With more of the protein in the R state, it is easier for the next molecule of substrate or activator to bind. Conversely, inhibitors favor the T state. In the presence of an inhibitor, it is more difficult for the substrate to bind to the T conformation. Although this model accounts for the kinetic behavior of many enzymes, it cannot easily account for negative cooperativity.

Regulatory Subunits Modulate the Activity of Catalytic Subunits

In the foregoing an allosteric site was considered to reside on the same subunit as the catalytic site, and all subunits were identical. In several very important enzymes a

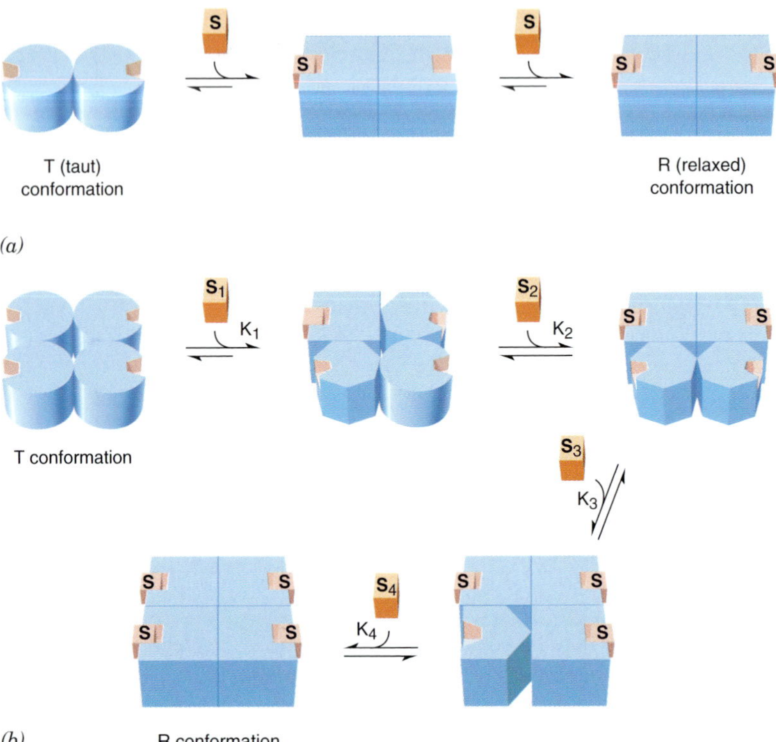

FIGURE 10.68

Models of cooperativity. (*a*) The concerted model. The enzyme exists in only two states, the T (tense or taut) and R (relaxed) conformations. Substrates and activators have a greater affinity for the R state and inhibitors for the T state. Ligands shift the equilibrium between the T and R states. (*b*) The sequential induced-fit model. Binding of a ligand to any one subunit induces a conformational change in that subunit. This conformational change is transmitted partially to adjoining subunits through subunit–subunit interaction. Thus the effect of the first ligand bound is transmitted cooperatively and sequentially to the other subunits (protomers) in the oligomer, resulting in a sequential increase or decrease in ligand affinity of the other protomers. The cooperativity may be either positive or negative, depending on the ligand.

distinct regulatory protein subunit exists. These regulatory subunits have no catalytic function, but their binding with the catalytic subunit modulates the activity of the catalytic subunit through an induced conformational change. Protein kinase A (PKA) is regulated by this mechanism (Figure 10.69). Each regulatory subunit (R) has a segment of its primary sequence that is a pseudo-substrate for the catalytic subunit (C). In the absence of cAMP (see p. 516), the R subunit binds to the C subunit at its active site through the pseudo-substrate sequence, and this causes the inhibition of the protein kinase activity. When cellular cAMP levels rise, cAMP binds to a site on the R subunits, causing a conformational change. This removes the pseudo-substrate sequence from the active site of the C subunit. The C subunits are now able to accept other protein substrates containing the pseudo-substrate sequence.

Calmodulin (see p. 478), a Ca^{2+}-binding protein, is a regulatory subunit for enzymes using Ca^{2+} as a modulator of their activity. Binding of Ca^{2+} to calmodulin causes a conformational change, allowing it to bind to the Ca^{2+} dependent enzyme, which induces a conformational change in the enzyme and restores enzymatic activity.

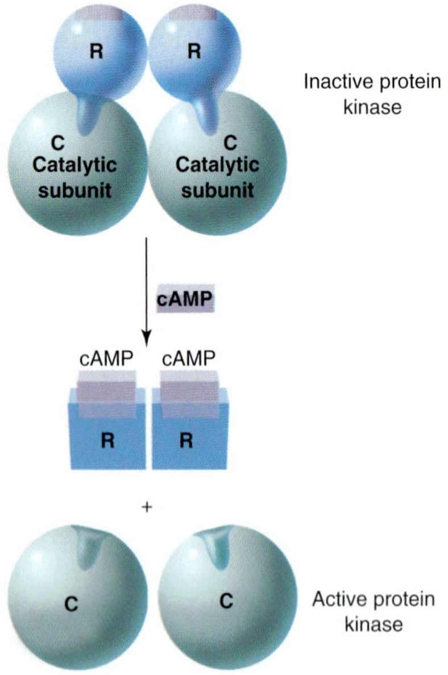

FIGURE 10.69

Model of allosteric enzyme with separate catalytic (C) and regulatory (R) subunits. The regulatory subunit of protein kinase A contains a pseudo-substrate region in its primary sequence that binds to the substrate site of the catalytic subunit. In the presence of cAMP the conformation of the R subunit changes so that the pseudo-substrate region can no longer bind, resulting in release of active C subunits.

10.11 REGULATION OF METABOLIC PATHWAYS

The physiological integration of many enzymes into a metabolic pathway, as well as the interrelationship of the products of one pathway with the activity of other pathways, is controlled. It is not necessary to regulate the activity of every enzyme in a pathway. Rather modulation of the activity of one or more key enzymes of the pathway is sufficient. One enzyme-catalyzed reaction is usually slower than the others—that is, the **rate-limiting step**—so that the rate of flux of substrates in the sequence is dependent on one enzyme in the entire pathway. Altering its kinetic properties has a large effect of the rate of formation of the final product. Another important place to regulate is at the enzyme catalyzing the **commitment step** of the pathway, the first irreversible reaction that is unique to the pathway. The rate-limiting enzyme is not necessarily the one catalyzing the committed step. In addition, one substrate might serve as an intermediate for two or more metabolic transformations each leading to different products. Specific examples of these regulatory enzymes will be described in later chapters. The activity of the enzyme associated with the committed step or with the rate-limiting step can be regulated in a number of ways.

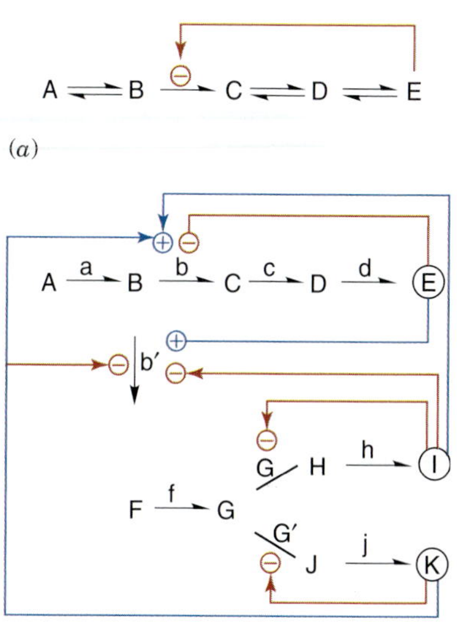

FIGURE 10.70

Control of metabolic pathways. (*a*) Feedback inhibition. Product E could inhibit the enzyme that catalyzes reaction of B to C. This may be an irreversible or rate-limiting reaction. (*b*) Regulation of a hypothetical complex pathway. Product E could inhibit enzyme b while activating enzyme b′. Similarly, product I and K could inhibit their synthesis by inhibiting enzymes b′, g, and g′ while activating enzyme b.

Cellular Enzyme Concentration

Many key enzymes have relatively short half-lives—for example, less than 1 h. Cells regulate enzyme concentration by changing the rate of *de novo* synthesis at either the transcriptional or translational level. For example, glucose represses the *de novo* synthesis of phosphoenolpyruvate carboxykinase, a rate-limiting enzyme in the conversion of pyruvate to glucose. If sufficient glucose is available in the blood, there is a low level of the carboxykinase in the liver. Low blood levels of glucose lead to an increase in the enzyme and increase in glucose synthesis (see p. 616).

Activators, Inhibitors, and Covalent Modification

Short-term regulation of an enzyme occurs through modulation of the activity by activators, inhibitors, and covalent modification. When the cellular concentration of deoxyribonucleotides increases such that the cell has sufficient amounts for synthesis of DNA, the key enzyme of the synthetic pathway is inhibited by the end products of the pathway, resulting in a slow-down of the synthetic pathway. This is referred to as **feedback inhibition** (Figure 10.70a). In addition to feedback within a pathway, feedback inhibition of other pathways can also occur. This is referred to as **cross-regulation**. Here a product of one pathway serves as an inhibitor or activator of an enzyme occurring early in another pathway (Figure 10.70b). A good example is the cross-regulation of the production of the four deoxyribonucleotides for DNA synthesis (see p. 808).

Compartmentation

Finally, the activity of a pathway can be regulated by physically partitioning reactions in specific cellular compartments separated by impermeable membranes (see p. 461). The transport of substrates, products, and cofactors across the membrane controls their concentration in the compartment. This regulates their availability and can limit enzyme activity. This is referred to as **compartmentation.** For example, acetyl CoA is formed in the mitochondria and must be transported out to the cytosol for use in fatty acid synthesis. Acetyl CoA carboxylase, the first step in fatty acid synthesis, requires acetyl CoA. The substrate cannot diffuse into the cytosol, so a complex mechanism transports it across the membrane (see p. 672). Anabolic and catabolic pathways are often segregated in different organelles in order to better control each pathway. There would be no point to having oxidation of fatty acids occurring at the same time and in the same compartment as biosynthesis of fatty acids. By maintaining fatty acid biosynthesis in the cytoplasm and oxidation in the mitochondria, control can be exerted by regulating transport of common intermediates across the mitochondrial membrane.

10.12 CLINICAL APPLICATIONS OF ENZYMES

Enzymes found in plasma are of two types: (1) those that are normally present and have a functional role in plasma and (2) those released from tissues. The latter can be present normally at very low levels but have no functional role in the plasma; however, they are most important for diagnostic purposes. Normally, the plasma content of intracellular enzymes is low or absent; however, when cells are damaged, the contents can appear in blood. A disease process may cause changes in cell membrane permeability or increased cell death (see p. 1020), resulting in release of intracellular enzymes. Lower-molecular-weight components can also diffuse out of cells when there is an abnormal increase in their cellular concentration. The greater the degree of tissue damage, the greater the amount of released cellular components. Intracellular enzymes are cleared from the plasma at varying rates, depending on the stability of the enzyme and its uptake by the reticuloendothelial system. Analyzing blood for cellular components represents a convenient way for clinicians to assess damage occurring elsewhere in the body. In the diagnosis of specific organ involvement in a disease process, it would be ideal if an enzyme unique to each organ could be easily identified. Unfortunately, identical enzymes are found in most tissues, though some organ-specific enzymes or isozymes exist. For example, an isozyme of alcohol dehydrogenase is liver-specific, and an acid phosphatase form found primarily in prostate is useful for specific identification of diseases in these organs. For some enzymes the tissue-specific isozymes possess different kinetic properties, so determination of K_m and k_{cat} values can be used to determine from what tissue they arose.

Evaluation of the timing of appearance and disappearance of particular enzymes in plasma permits a diagnosis of specific organ involvement. Figure 10.71 illustrates the time-dependent release of the plasma activities of enzymes from the myocardium following a heart attack. Such profiles allow one to establish when the attack occurred and whether treatment is effective. Clinical Correlation 10.11 demonstrates how diagnosis of a specific enzyme defect led to a rational clinical treatment that restored the patient to health.

Clinical Enzyme Assays

Studies of the kinetics of appearance and disappearance of plasma enzymes require a highly reproducible enzyme assay. A good assay is based on temperature and pH control, as well as controlling the concentration of substrates, co-substrates, and cofactors. To accomplish the latter, the K_m must be known for the enzyme being evaluated. Typically, assays are performed under V_{max} conditions, though they need not be. Clinical laboratory assay conditions are optimized for the properties of the normal enzyme. The assay may not correctly measure levels of a variant enzyme because pH optima and/or the K_m for substrate and cofactors may be different from the normal in a variant (Clin. Corr. 10.12).

It is necessary to measure the concentration of a product formed or substrate remaining as a function of time. In some reactions, the products or substrates absorb light. For these it is relatively straightforward to follow the change in light absorption as a function of time. Many synthetic substrates have been designed so that they or their products are chromophoric compounds (absorb light or emit fluorescent light). Enzymes that employ the coenzymes NAD(P) and FAD are easily measured because their reduced forms have a readily measured absorption in the visible range of light. Many enzymes that do not directly reduce NAD^+ or FAD generate products that can be acted upon by a NAD(P) or FAD-linked dehydrogenase. Thus by coupling two enzyme reactions, the activity of enzyme of interest is determined.

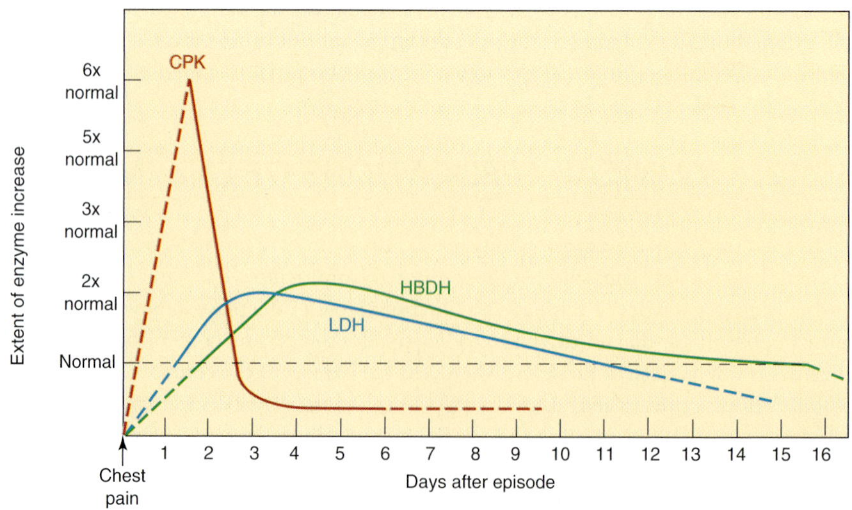

FIGURE 10.71

Kinetics of release of cardiac enzymes into serum following a myocardial infarction. CPK, creatine kinase; LDH, lactic dehydrogenase; HBDH, α-hydroxybutyric dehydrogenase. Such kinetic profiles allow one to determine where the patient is with respect to the infarct and recovery. Note: CPK rises sharply but briefly; HBDH rises slowly but persists.

Reprinted with permission from Coodley, E. L. *Diagnostic Enzymes*. Philadelphia: Lea & Febiger, 1970, p. 61.

CLINICAL CORRELATION 10.11
Identification and Treatment of an Enzyme Deficiency

Enzyme deficiencies usually lead to increased accumulation of specific intermediary metabolites in plasma and hence in urine. Recognition of the intermediates that accumulate in biological fluids is useful in pinpointing possible enzyme defects. After the enzyme deficiency is established, metabolites that normally occur in the pathway but are distal to the block may be supplied exogenously in order to overcome the metabolic effects of the enzyme deficiency. In hereditary orotic aciduria there is a double enzyme deficiency in the pyrimidine biosynthetic pathway, leading to accumulation of orotic acid. Both orotate phosphoribosyltransferase and orotidine 5′-phosphate decarboxylase are deficient, causing decreased *in vivo* levels of CTP and TTP. The two activities are deficient because they reside in separate domains of a bifunctional polypeptide of 480 amino acids. dCTP and dTTP, which arise from CTP and TTP, are required for cell division. In these enzyme deficiency diseases, the patients are pale and weak and fail to thrive. Administration of the missing pyrimidines as uridine or cytidine promotes growth and general well-being and also decreases orotic acid excretion. The latter occurs because the TTP and CTP formed from the supplied uridine and cytidine repress carbamoyl-phosphate synthetase, the committed step, by feedback inhibition, resulting in a decrease in orotate production.

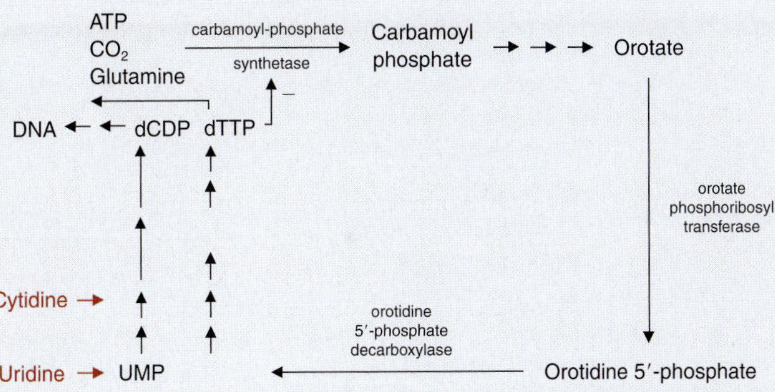

Source: Webster, D. R., Becroft, D. M. O., and Suttie, D. P. Hereditary orotic aciduria and other diseases of pyrimidine metabolism. In: C. R. Scriver, A. L. Beaudet, W. S. Sly, and D. Valle (Eds.), *The Metabolic and Molecular Bases of Inherited Disease*, 7th ed. New York: McGraw-Hill, 1995, p. 1799.

CLINICAL CORRELATION 10.12
Ambiguity in the Assay of Mutated Enzymes

Structural gene mutations leading to production of enzymes with increases or decreases in K_m are frequently observed. A case in point is a patient with hyperuricemia and gout, whose red blood cell hypoxanthine-guanine-phosphoribosyltransferase (HGPRT) showed little activity in assays *in vitro*. This enzyme is involved in the salvage of purine bases and catalyzes the reaction

$$\text{Hypoxanthine} + \text{PRPP} \rightarrow \text{Inosine monophosphate} + \text{PPi}$$

where PRPP is phosphoribosyl pyrophosphate.

The absence of HGPRT activity results in a severe neurological disorder known as Lesch–Nyhan syndrome (see p. 798), yet this patient did not have the clinical signs of this disorder. Immunological testing with a specific antibody to the enzyme revealed as much cross-reacting material in the patient's red blood cells as in normal controls. The enzyme was therefore being synthesized but was inactive in the assay *in vitro*. Increasing the substrate concentration in the assay restored full activity in the patient's red cell hemolysate. This anomaly is explained as a mutation in the substrate-binding site of HGPRT, leading to an increased K_m. Neither the concentration of substrate in the assay nor that in the red blood cells was high enough to bind to the enzyme. This case reinforces the point that an accurate enzyme determination is dependent on zero-order kinetics—that is, the enzyme being saturated with substrate.

Source: Sorenson, L. and Benke, P. J. Biochemical evidence for a distinct type of primary gout. *Nature* 213:1122, 1967.

Enzymes are used in clinical analyzers for screening purposes. As an example, assays for cholesterol and triacylglycerols can be completed in a few minutes using small amounts of plasma. Cholesterol oxidase is employed for cholesterol determination and lipase for triacylglycerols. The enzymes are immobilized in a bilayer along with the necessary cofactors and indicator reagents, which become chromophores when reduced. In the case of cholesterol oxidase, hydrogen peroxide is a product of the oxidase reaction, and the peroxide formed oxidizes a colorless dye to a colored product mediated by a another enzyme, a peroxidase.

Enzyme-Linked Immunoassays Employ Enzymes as Indicators

Modern clinical chemistry has benefited from the merging of enzyme chemistry and immunology. Antibodies specific to a protein antigen are coupled to an indicator enzyme such as horseradish peroxidase to generate a very specific and sensitive assay. After binding of the peroxidase-coupled antibody to the antigen, the peroxidase is used to generate a colored product that is measurable and whose concentration is related to the amount of antigen in a sample. Because of the catalytic nature of the enzyme the system greatly amplifies the signal. This assay has been given the acronym **ELISA** for **e**nzyme-**l**inked **i**mmunoad**s**orbent **a**ssay. With the advent of monoclonal antibody production, it is possible be generate an antibody that is specific to one isozyme or to one member of a family of very similar enzymes.

Horseradish peroxidase is often used as an indicator in ELISA such as testing for the presence of human immunodeficiency virus (HIV) coat protein antigens. The procedure is schematically diagramed in Figure 10.72. This assay amplifies the signal because of the catalytic nature of the reporter group, the enzyme peroxidase. Such amplified enzyme assays allow the measurement of remarkably small amounts of antigens.

Measurement of Isozymes is used Diagnostically

Isozymes can be separated from each other by electrophoresis. In some cases, different tissues express different isozymes, and in other cases the same isozymes occur in all tissues. If the enzyme is a multimer, there could be many combined forms. Isozymes that have wide clinical application are lactate dehydrogenase, creatine phosphokinase, and alkaline phosphatase. **Creatine phosphokinase** (CPK) (see p. 985) occurs as a dimer with two types of subunits, M (muscle type) and B (brain type). In brain both subunits are electrophoretically the same and are designated B. In skeletal muscle the subunits are both of the M type. Isozymes containing both M- and B-type subunits (MB) are found

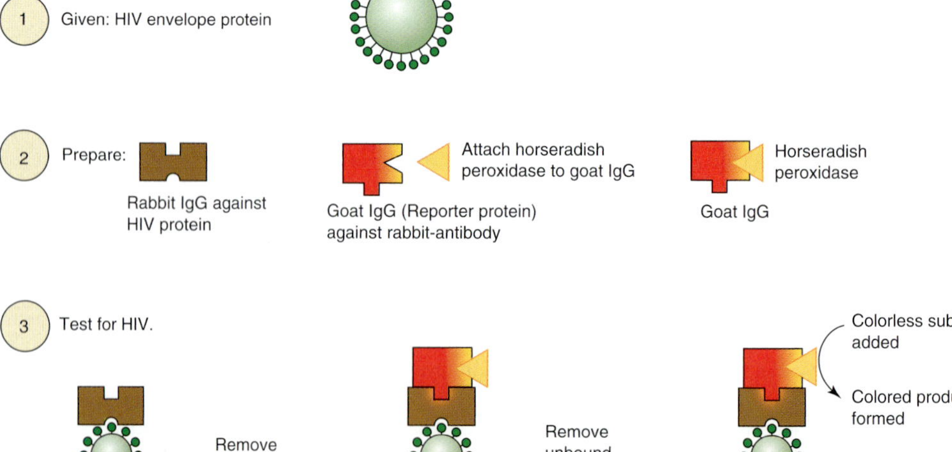

FIGURE 10.72
Schematic of ELISA (enzyme-linked immunoadsorbent assay) for detecting the human immunodeficiency virus (HIV) envelope proteins.

only in the myocardium. Other tissues contain variable amounts of the MM and BB isozymes. The isozymes are numbered beginning with the species migrating the fastest to the anode on electrophoresis—that is, CPK_1 (BB), CPK_2 (MB), and CPK_3 (MM).

Two genes code for tissue-specific isozymes of lactic dehydrogenase (LDH). The enzyme is a tetramer, thus five forms of the enzyme exist since apparently the subunits are free to dissociate and reform. One isozyme is found almost exclusively in heart muscle (abbreviated H), and the homotetramer (H_4) is LDH_1. The various isoforms of LDH are presented in Table 10.4. Liver and skeletal muscle have the M-subunit, and the homotetramer (M_4) is LDH_5. Normally there is little LDH_1 in blood, but after a heart attack the level increases. The H- and M-subunits have different isoelectric points, so each tetramer has a different charge and all five can be separated by gel electrophoresis and identified by measurement of their activity directly on the gel.

To illustrate how isozyme analysis could be used clinically, the isozyme pattern for lactate dehydrogenase is presented in Figure 10.73 as a function of time after a myocardial infarction. Not only does the total activity increase, but also there is a change in the isozyme pattern. More LDH_1 is found. In this example, LDH_5 is increased, indicative of liver damage to the patient. Thus secondary complications of heart failure can be monitored. After two weeks the isozyme pattern returns to that of a healthy patient.

TABLE 10.4 Lactate Dehydrogenase Isozymes

Type	*Composition*	*Location*
LDH_1	HHHH	Myocardium and RBC
LDH_2	HHHM	Myocardium and RBC
LDH_3	HHMM	Brain and kidney
LDH_4	HMMM	
LDH_5	MMMM	Liver and skeletal muscle

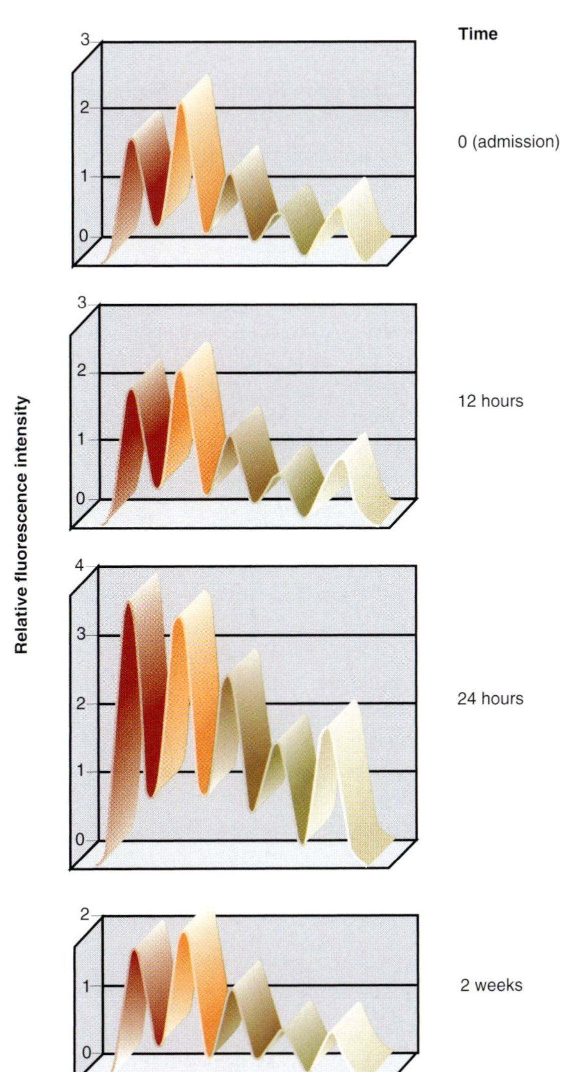

FIGURE 10.73

Tracings of densitometer scans of LDH isozymes at time intervals following a myocardial infarction. Total LDH increases and LDH1 becomes greater than LDH2 between 12 and 24 h. Increase in LDH5 is diagnostic of a secondary congestive liver involvement. Note the Y axis scales are not identical. After electrophoresis on agarose gels, the LDH activity is assayed by measuring the fluorescence of the NADH formed in LDH-catalyzed reaction. Courtesy of Dr. A. T. Gajda, Clinical Laboratories, University of Arkansas for Medical Science.

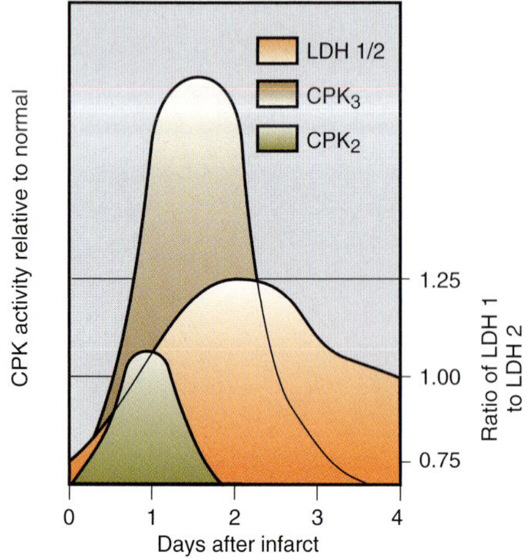

FIGURE 10.74

Characteristic changes in serum CPK and LDH isozymes following a myocardial infarction. CPK2 (MB) isozyme increases to a maximum within 1 day of the infarction. CPK3 lags behind CPK2 by about 1 day. Total LDH level increases more slowly. The increase of LDH1 and LDH2 within 12–24 h coupled with an increase in CPK2 is diagnostic of myocardial infarction.

Kinetic analyses of both creatine kinase and lactate dehydrogenase are useful for diagnostic purposes. CPK and LDH isozymes are plotted in Figure 10.74 as a function of time after damage to the heart. The tissue damage releases CPK_2 into the blood within the first 6–18 h after an infarct, but LDH release lags behind the appearance of CPK_2 by 1–2 days. Normally, the activity of the LDH_2 isozyme is higher than that of LDH_1; however, in the case of infarction the activity of LDH_1 becomes greater than LDH_2, at about the time that CPK_2 levels are back to baseline (48–60 h). The LDH isozyme "switch" coupled with increased CPK_2 is diagnostic of myocardial infarct (MI) in virtually 100% of the cases (see Clin. Corr. 10.9).

Some Enzymes are used as Therapeutic Agents

In a few cases, enzymes have been administered as therapeutic agents. **Streptokinase,** an enzyme mixture prepared from a streptococcus, is useful in clearing blood clots that occur in myocardial infarcts and in the lower extremities. It activates the fibrinolytic pre-enzyme **plasminogen** (see p. 1004) that is normally present in plasma. The activated enzyme is plasmin, a serine protease that cleaves the insoluble fibrin in blood clots into several soluble components. Another serine protease, human tissue plasminogen activator, t-PA, is being commercially produced by bioengineered *E. coli* for use in dissolving blood clots in patients suffering myocardial infarction (see p. 1004). t-PA also functions by activating plasminogen. Pronase, another bacterial enzyme, is used in debridement of wounds.

Most enzymes have a short half-life in blood; consequently, unreasonably large amounts of enzyme are required to maintain therapeutic levels. Work is in progress to enhance enzyme stability by coupling enzymes to solid matrices and implanting these materials in areas that are well-perfused.

BIBLIOGRAPHY

Blackburn, G. M., Kang, A. S., Kingsbury, G. A., and Burton, D. R. Review of abzymes. *Biochem. J.* 262:381, 1989.

Bugg, T. *An Introduction to Enzyme and Coenzyme Chemistry*. London: Blackwell Press, 1997.

Cornish-Bowden, A. *Fundamentals of Enzyme Kinetics*. London: Portland Press, 1995.

Eliot, A. C. and Kirsch, J. F. Pyridoxal phosphate enzymes: Mechanistic, structural, and evolutionary considerations. *Annu. Rev. Biochem.* 73:383, 2004.

Fersht, A. *Enzyme Structure and Mechanism in Protein Science: A Guide to Enzyme Catalysis and Protein Folding*. New York: Freeman, 1999.

Holden, H. M., Rayment, I., and Thoden, J. B. Structure and function of enzymes of the Leloir pathway for galactose metabolism. *J. Biol. Chem.* 278:43885, 2003.

Keffer, J. H. Myocardial markers of injury. *Am. J. Clin. Pathol.* 105:305, 1996.

Knowles, J. R. and Alberty, W. J. Evolution of enzyme function and the development of catalytic efficiency. *Biochemistry* 15:5631, 1976.

Koshland Jr., D. E. A new model for protein stereospecificity. *Nature* 403:614, 2000.

Koshland, Jr., D. E. and Hamadani, K. Proteomics and models for enzyme cooperativity. *J. Biol. Chem.* 277:46841, 2002.

Kraut, D. A., Carroll, K. S., and Herschlag, D. Challenges in enzyme mechanism and energetics. *Annu. Rev. Biochem.* 72:517, 2003.

Kraut, J. How do enzymes work? *Science* 242:533, 1988.

Kyte, J. *Mechanism in Protein Chemistry*. New York: Garland, 1995.

Lerner, R. A., Benkovic, S. J., and Schultz. P. G. At the crossroads of chemistry and immunology: Catalytic antibodies. *Science* 252:659, 1991.

Lilley, D. M. C. The origins of RNA catalysis in ribozymes. *Trends Biochem. Sci.* 28:495, 2003.

Nevinsky, G. A. and Buneva, V. N. Catalytic antibodies in healthy humans and patients with autoimmune and viral diseases. *J. Cell Mol. Med.* 7:265, 2003.

Plowman, K. M. *Enzyme Kinetics*. New York: McGraw-Hill, 1972.

Silverman, R. B. *The Organic Chemistry of Enzyme-Catalyzed Reactions*. San Diego: Academic Press, 2000.

Wackett, L. P. Evolution of enzymes for the metabolism of new chemical inputs into the environment. *J. Biol. Chem.* 279:41259, 2004.

QUESTIONS

Carol N. Angstadt

1. In all enzymes the active site:
   A. contains the substrate-binding site.
   B. is contiguous with the substrate binding site in the primary sequence.
   C. lies in a region of the primary sequence distant from the substrate-binding site.
   D. contains a metal ion as a prosthetic group.
   E. contains the amino acid side chains involved in catalyzing the reaction.

2. Although enzymic catalysis is reversible, a given reaction may appear irreversible:
   A. if the products are thermodynamically far more stable than the reactants.
   B. under equilibrium conditions.
   C. if a product accumulates.
   D. at high enzyme concentrations.
   E. at high temperatures.

3. Metal cations may do all of the following *except*:
   A. donate electron pairs to functional groups found in the primary structure of the enzyme protein.
   B. serve as Lewis acids in enzymes.
   C. participate in oxidation–reduction processes.
   D. stabilize the active conformation of an enzyme.
   E. form chelates with the substrate, with the chelate being the true substrate.

4. Enzymes may be specific with respect to all of the following *except*:
   A. chemical identity of the substrate.
   B. the atomic mass of the elements in the reactive group (e.g., ^{12}C but not ^{14}C).
   C. optical activity of product formed from a symmetrical substrate.
   D. type of reaction catalyzed.
   E. which of a pair of optical isomers will react.

5. In the reaction sequence below, the best point for controlling production of Compound 6 is reaction:

$$\begin{array}{c} \text{Cpd 3} \\ \updownarrow \text{B} \\ \text{Cpd 1} \xrightarrow{A} \text{Cpd 2} \overset{C}{\rightleftharpoons} \text{Cpd 4} \xrightarrow{D} \text{Cpd 5} \overset{E}{\rightleftharpoons} \text{Cpd 6} \end{array}$$

   A. A
   B. B
   C. C
   D. D
   E. E

6. If the plasma activity of an intracellular enzyme is abnormally high all of the following may be a valid explanation *except*:
   A. the rate of removal of the enzyme from plasma may be depressed.
   B. tissue damage may have occurred.
   C. the enzyme may have been activated.
   D. determination of the isozyme distribution may yield useful information.
   E. the rate of synthesis of the enzyme may have increased.

Questions 7 and 8: A man of Japanese ancestry found himself to be experiencing severe flushing and a very rapid heart rate after consuming one alcoholic beverage. His companion, a Caucasian male, did not have the same symptoms even though he had finished his second drink. These physiological effects are related to the presence of acetaldehyde (CH_3CHO) generated from the alcohol. Acetaldehyde is normally removed by the reaction of mitochondrial aldehyde dehydrogenase which catalyzes the reaction:

$$CH_3CHO + NAD^+ \rightleftharpoons CH_3COO^- + NADH + H^+$$

7. Acetaldehyde dehydrogenase is:
   A. an oxidoreductase.
   B. a transferase.
   C. a hydrolase.
   D. a lyase.
   E. a ligase.

8. The explanation for the difference in physiological effects is that the Japanese man was missing the normal mitochondrial aldehyde dehydrogenase and had only a cytosolic isozyme. The cytosolic isozyme:
   A. does not react with CH_3CHO.
   B. activates the enzyme that produces CH_3CHO.
   C. differs from the mitochondrial enzyme in that it has a higher K_m for CH_3CHO.
   D. would be expected to have a greater affinity for the substrate than the mitochondrial enzyme.
   E. produces a low steady-state level of acetaldehyde following alcohol consumption.

Question 9 and 10: A research technician who is working with organophosphate compounds is required to have a weekly blood test for acetylcholine esterase activity. Typically, esterase activity remains relatively constant for some time and then abruptly drops to zero. If this happens, the technician must immediately stop working with the organophosphate compounds. The organophosphate compounds form stable esters with a critical serine hydroxyl group in the esterase.

9. In the esterase, serine transfers a proton to a histidine residue. Which of the following is correct?
   A. Serine is acting as a general acid.
   B. Histidine is acting as a general acid.
   C. Serine and histidine form a covalent intermediate.
   D. The enzyme would be relatively insensitive to pH changes.
   E. Serine is acting as a transition stabilization catalyst.

10. Organophosphate compounds inactive the esterase by:
   A. competitive inhibition.
   B. uncompetitive inhibition.
   C. noncompetitive inhibition.
   D. suicide inhibition.
   E. irreversible inhibition.

Questions 11 and 12: Gout is a disease in which uric acid is high in blood and urine. One patient excreted three times normal uric acid and had very high blood levels of PRPP, an intermediate in biosynthesis of AMP and GMP, which are precursors of ATP and GTP. Degradation of these products produces uric acid. The patient's PRPP synthetase had normal K_m and V_{max} values but was insensitive to regulation by the end products of the pathway (ATP, GTP). These are negative allosteric modifiers of PRPP synthetase.

11. All of the following statements about allosteric effectors are correct *except*:
   A. they may increase the enzyme's affinity for its substrate.
   B. they may decrease the enzyme's affinity for its substrate.
   C. they bind at the substrate-binding site.
   D. they cause a conformational change in the enzyme.
   E. they can change either the K_m or the V_{max} of the reaction.

12. Most allosteric enzymes:
   A. are monomers.
   B. have more than one subunit.
   C. exhibit only homotropic interactions.
   D. exhibit only heterotropic interactions.
   E. bind the allosteric effector with no effect on binding other ligands.

Problems

13. An experiment measuring velocity versus substrate concentration was run, first in the absence of substance A, and then in the presence of substance A. The following data were obtained:

[S], μM	Velocity in absence of A, μmol/min	Velocity in presence of A, μmol/min
2.5	0.32	0.20
3.3	0.40	0.26
5.0	0.52	0.36
10.0	0.69	0.56

Is substance A an activator or an inhibitor? If it is an inhibitor, what kind of inhibitor is it?

14. For the experiment in question 13, calculate the K_m and V_{max} both in the absence and in the presence of substance A. Are these results consistent with your answer for question 13?

ANSWERS

1. **E** The active site contains all of the machinery, including the amino acid side chains, involved in catalyzing the reaction. A–D are all possible, but none is necessarily true.
2. **A** Stable products do not react in the reverse direction at an appreciable rate. B: At equilibrium the forward and reverse reactions proceed at identical rates. C: Product accumulation would tend to reverse the reaction. D: Enzymes merely catalyze reactions, and they do not affect the equilibrium of the reaction. E: Temperature affects the rates of reactions, and may also affect the position of the equilibrium, but does not interconvert reversible and irreversible reactions.
3. **A** Metal cations are electron-deficient and may accept electron pairs, serving as Lewis acids, but they do not donate electrons to other functional groups. C: On the contrary, they sometimes accept electron pairs from groups in amino acid side chains. D: In doing so, they may become chelated, which may stabilize the appropriate structure. E: Sometimes they are chelated by the substrate, with the chelate being the true substrate.
4. **B** Enzymes do not distinguish among different nuclides of an element, although the rate of reaction of a heavier nuclide might be less than that of a lighter one. A, D: Enzymes are specific for the substrate and the type of reaction. C, E: The asymmetry of the binding site generally permits only one of a pair of optical isomers to react, and only one optical isomer is generated when a symmetric substrate yields an asymmetric product.
5. **D** Reaction D is irreversible; if it were not controlled, Cpd 5 might build up to toxic levels. A: Control of reaction A would control production of both Cpds 3 and 6. B: Reaction B is not on the direct route. C, E: Reactions C and E are freely reversible, so do not need to be controlled.
6. **E** Since appearance of intracellular enzymes in plasma arises from leakage, typically from damaged or destroyed cells, changes in their rates of synthesis within the cell would not be expected to affect plasma concentration. A: This would lead to elevated levels. B: Intracellular enzymes may appear in abnormal amounts when tissues are damaged. C: Laboratory assays depend on all factors except amount of enzyme being constant. D: Different tissues have characteristic distributions of isozymes.
7. **A** Aldehyde to acid conversion is an oxidation. This is of the subclass dehydrogenases as indicated by the presence of the NAD^+ .
8. **C** The lower affinity (higher K_m) makes it more difficult for the enzyme to remove CH_3CHO. A: Isozymes, by definition, catalyze the same reaction. B: It is highly unlikely that one enzyme would activate another enzyme. D: With a greater affinity (lower K_m) the reaction would be occurring at a rapid rate, thus removing CH_3CHO. E: These effects are due to a high steady-state concentration.
9. **A** Acids donate protons. General acids are weakly ionized at physiological pH. The environment of the serine in the protein facilitates its ability to transfer a proton and thus act as a nucleophile. B: Bases accept protons. C: No covalent bonds are formed. D: The ability of histidine to accept a proton is very pH-dependent. E: This is not likely.
10. **E** The phosphate ester does not dissociate and the enzyme is not restored. A–C: These types of inhibitors bind to the enzyme and/or enzyme substrate complex reversibly. D: A suicide inhibitor is formed from the substrate and remains covalently attached to the enzyme. Organophosphates are not substrates for the esterase.
11. **C** "Allo" means other. A and B: Allosteric effectors may be either positive or negative. D: This alters the affinity for the substrate. E: K class alters K_m and V class alters V_{max}.
12. **B** Most are oligomers. A: Only two monomeric allosteric enzymes are known. C and D: Both homotropic and heterotropic interactions are common. E: The principle of allosterism is that binding one ligand affects binding of other ligands.
13. Take the reciprocals of both [S] and v and construct a Lineweaver–Burke plot. You should find that the two curves cross the y-axis at the same point but the curve in the presence of A crosses the x-axis closer to the origin. This pattern indicates that A is a competitive inhibitor.
14.

	$-1/K_m$	K_m	$1/V_{max}$	V_{max}
Absence of A:	−0.14	7.1	0.8	1.25
Presence of A:	−0.08	12.5	0.8	1.25

With a competitive inhibitor, the V_{max} remains constant (be sure you understand why) but the apparent K_m is larger. It takes more substrate to reach a given velocity because the substrate has to compete with the inhibitor.

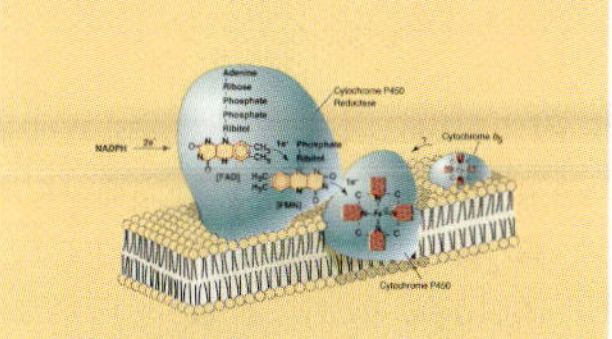

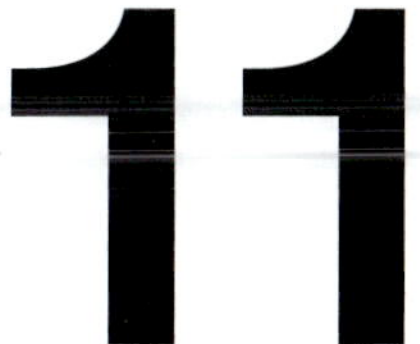

11

THE CYTOCHROMES P450 AND NITRIC OXIDE SYNTHASES

Linda J. Roman and Bettie Sue Siler Masters

Textbook of Biochemistry With Clinical Correlations, Sixth Edition, Edited by Thomas M. Devlin

11.1 OVERVIEW

Cytochromes P450 constitute a unique family of heme proteins, present in bacteria, fungi, insects, plants, fish, mammals and primates, that catalyze the monooxygenation (i.e., insertion of one atom of molecular oxygen) of a large variety of structurally diverse compounds. Substrates for this enzyme system include both endogenously synthesized compounds, such as cholesterol, steroid hormones, and fatty acids, and exogenous compounds, such as drugs, food additives, components of cigarette smoke, pesticides, and chemicals, that are ingested, inhaled, or absorbed through the skin. The cytochromes P450 are involved in: (a) production of steroid hormones; (b) metabolism of fatty acids, prostaglandins, leukotrienes, and retinoids; (c) inactivation or activation of therapeutic agents; (d) conversion of chemicals to highly reactive molecules that may produce unwanted cellular damage; and (e) enzyme inhibition or induction resulting in drug–drug interactions and adverse effects.

The nitric oxide synthases are bi-domain enzymes, consisting of a heme-containing or oxygenase domain and a flavin-containing or reductase domain. These enzymes catalyze the formation of nitric oxide, a highly reactive gaseous free radical molecule that functions in neurotransmission, in hemodynamic regulation, or in the immune response, depending upon the tissue in which it is produced. Like the cytochromes P450, nitric oxide synthases are cysteine thiolate-liganded, heme-containing enzymes that catalyze the monooxygenation of their substrates using similar catalytic mechanisms. The following will discuss both the cytochromes P450 and the nitric oxide synthases in detail, addressing both the similarities and differences between the two systems.

11.2 CYTOCHROMES P450: PROPERTIES AND FUNCTION

Cytochromes P450 are integral membrane proteins containing a single iron **protoporphyrin IX** (heme) prosthetic group (see p. 833). The heme iron of cytochromes P450 is capable of forming six bonds, four to the four pyrrole nitrogen atoms of the porphyrin ring and two to axial ligands. One of the axial ligands (so-called proximal) is a sulfhydryl group from a cysteine residue located toward the carboxyl end of the cytochrome P450 molecule (Figure 11.1), and the other (distal) ligand can be open (producing a penta-coordinated, high-spin heme) or occupied by oxygen, CO, NO, water, or other ligand (producing a hexa-coordinated, low-spin heme). The spin state of the heme is a description of the occupancy of the *d*-electronic shells within the iron atom determined by the ligand state; these can be distinguished spectroscopically.

All cytochromes P450 are triangle-shaped molecules, closely resembling that shown in Figure 11.1. Half of the enzyme consists of α-helical structure, and the other half is β-sheet and other structures; a long I-helix runs throughout the enzyme behind the heme.

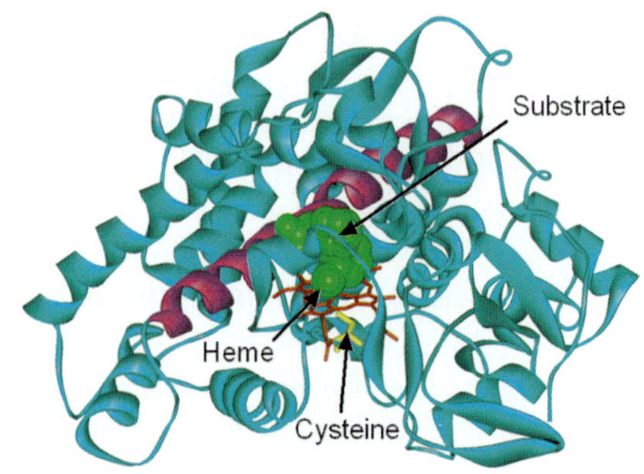

FIGURE 11.1

Model of the active site of mammalian CYP2C5 showing the protoporphyrin IX prosthetic group (red) with the cysteine thiolate ligand (yellow) attached to the heme iron. The substrate diclofenac, a nonsteroidal anti-inflammatory drug (green), is in the active site of the P450. The I-helix (purple) spans the molecule and is one of the most easily identifiable structural characteristics of cytochromes P450.

Generated from Protein Data Bank file 1NR6, deposited by M. R. Wester, E. F. Johnson, and C. D. Stout, using WebLab Viewer Lite (Molecular Simulations, Inc.)

This basic scaffolding is conserved among all known cytochromes P450, although relative placement may differ and some of the minor elements may be missing. The mammalian proteins have an N-terminal membrane anchor sequence and internal membrane association sequences that result in these proteins being deeply buried in the membrane.

Cytochromes P450 are so-called because of the unique absorbance spectrum that is produced when CO is bound to the reduced, ferrous, form of the heme. The spectrum exhibits a peak at approximately 450 nm (Figure 11.2)—hence the name P450 for a **p**igment with an absorbance at **450** nm. This spectrum is characteristic of a heme thiolate-liganded protein. Much as with the heme iron of hemoglobin, CO binds to the heme iron of cytochromes P450 with a much higher binding affinity than oxygen and thereby is a potent inhibitor of its function.

The general reaction catalyzed by a cytochrome P450 is

$$\mathrm{NADPH} + \mathrm{H}^+ + \mathrm{O}_2 + \mathrm{SH} \rightarrow \mathrm{NADP}^+ + \mathrm{H}_2\mathrm{O} + \mathrm{SOH}$$

where NADPH is a two-electron donor and the substrate (S) may be a steroid, fatty acid, drug, or other chemical that has an alkane, alkene, aromatic ring, or heterocyclic ring substituent that can serve as a site for oxygenation. The reaction is a monooxygenation and the enzyme is a monooxygenase because it incorporates only one of the two oxygen atoms into the substrate.

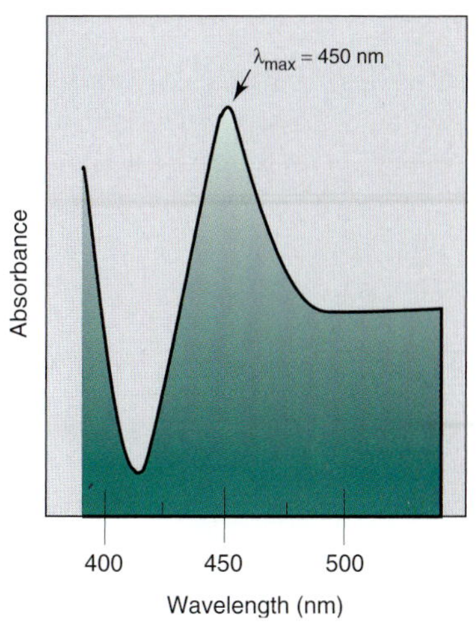

FIGURE 11.2

The absorbance spectrum of the carbon monoxide-bound cytochrome P450. The reduced form of cytochrome P450 binds carbon monoxide to produce a maximum absorbance at approximately 450 nm.

11.3 CYTOCHROME P450 REACTION CYCLE

The overall mechanism by which cytochromes P450 catalyze the **monooxygenation of** their substrates is shown in Figure 11.3. The reaction cycle is initiated by the binding of substrate to the native, low-spin, hexa-coordinate, ferric form of the heme, converting it to a high-spin, penta-coordinate ferric heme. The first electron can be transferred to the heme, reducing it to the high-spin, penta-coordinate ferrous form only after substrate has been bound. Electrons cannot be transferred to the heme in the absence of substrate because the reduction potential of the heme (−300 mV) in that state is more electronegative than that of the FMN moiety (−270 mV), making such a transfer thermodynamically unfavorable. Upon substrate binding, there is a conformational change of the protein structure surrounding the heme iron, and the reduction potential of the heme becomes more positive (−230 mV), thus allowing the P450 heme iron to be reduced (Figure 11.3). Oxygen can now bind to the ferrous heme, and the second

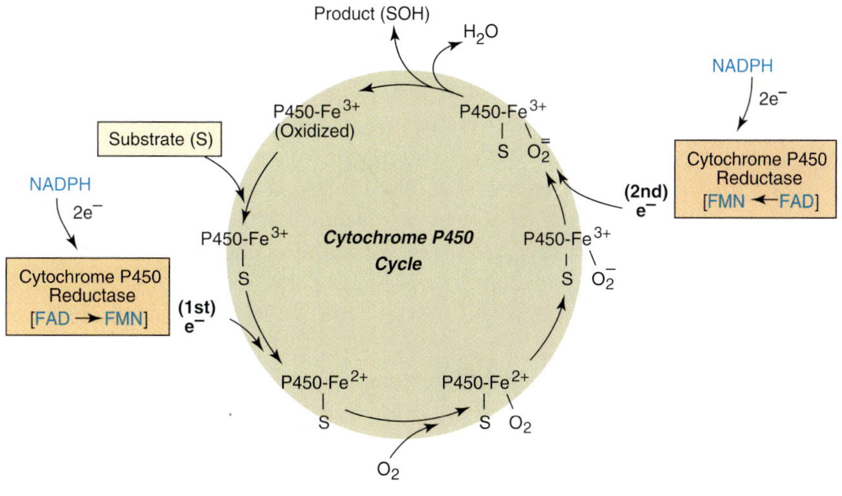

Redox Potentials (mV)
P450 from endoplasmic reticulum

NADPH	−324
FAD	−290
FMN	−270
P450 heme	−300
P450 heme + substrate	−230

FIGURE 11.3

Reaction cycle of cytochrome P450. Diagram shows the binding of substrate, transfer of the first and second electrons from NADPH-cytochrome P450 reductase, and binding of O_2. The reduction potentials of the various components are also shown.

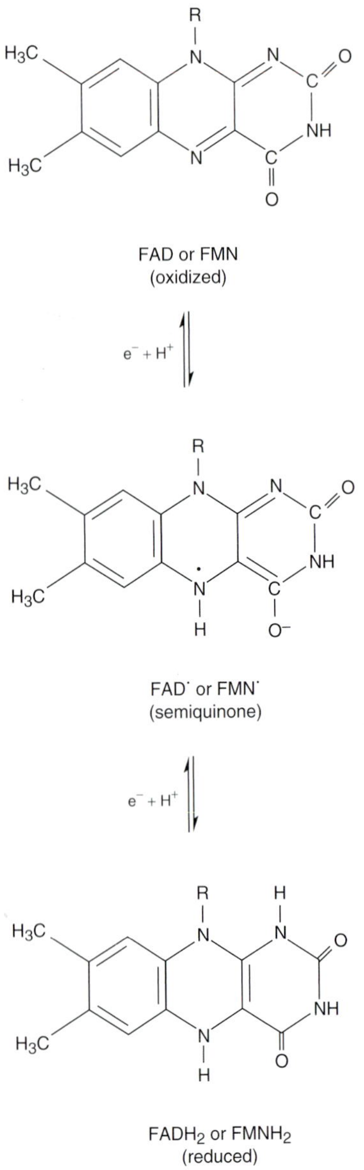

FIGURE 11.4

Isoalloxazine ring of FMN or FAD in its oxidized, semiquinone (1e⁻ reduced form), or fully reduced (2e⁻ reduced) state.

electron transfer is facilitated. Subsequently, molecular oxygen is activated and cleaved; one atom is inserted into the substrate (monooxygenation), and the other combines with two protons and two electrons to form H_2O.

11.4 CYTOCHROME P450 ELECTRON TRANSPORT SYSTEMS

The two electrons required for the overall monooxygenase reaction are provided by NADPH, but a basic mechanistic problem is created in that NADPH is a two-electron donor (see p. 379), but cytochrome P450, with its single heme iron, can accept only one electron at a time. This problem is solved by the presence of a **NADPH-dependent flavoprotein reductase**, which accepts two electrons from NADPH simultaneously but transfers the electrons individually, either to an intermediate iron–sulfur protein or directly to cytochrome P450. The active redox group of the flavoprotein is the isoalloxazine ring, which is uniquely suited to perform this chemical task since it can exist in fully oxidized or one- or two-electron reduced states (Figure 11.4). The transfer of electrons from NADPH to cytochromes P450 is accomplished by distinct electron transport systems present exclusively in either mitochondria or endoplasmic reticulum (microsomes).

NADPH-Cytochrome P450 Reductase Is the Flavoprotein Electron Donor in the Endoplasmic Reticulum

Most mammalian cytochromes P450 are found deeply embedded in the cytoplasmic side of the endoplasmic reticulum (microsome fraction) of hepatocytes, renal and adrenal cortical cells, ovarian and testicular cells, and cells of the respiratory tract. These microsomal cytochromes P450 use a single NADPH-cytochrome P450 reductase, a peripheral membrane protein of 76.6 kD containing both flavin adenine dinucleotide (FAD) and flavin mononucleotide (FMN) as prosthetic groups, for electron delivery. Until the recent characterization of nitric oxide synthases, the P450 reductase was the only mammalian flavoprotein known to contain both FAD and FMN. NADPH donates electrons to the FAD moiety of P450 reductase, and the FMN serves as the exit flavin, transferring two electrons, one at a time, to cytochrome P450. Because a single flavin molecule may exist as a one- or two-electron-reduced form and two flavin molecules are bound per reductase molecule, the enzyme may receive electrons from NADPH and store them between the two flavin molecules before transferring them to one of the numerous microsomal cytochromes P450 (Figure 11.5).

The binding of the P450 and reductase to one another is electrostatic rather than hydrophobic, and P450 levels exceed those of reductase in liver by a 10:1–20:1 ratio. If a particular P450 has a higher affinity for association with reductase, electrons will flow preferentially to those P450s. Consequently, P450s with lower affinities must have mechanisms that allow them to receive electrons. In some cases, the binding of substrate increases both the affinity and the rate of association of the P450/reductase complex; this mechanism makes sense because an "empty" P450 monopolizing the reductase would be nonproductive.

In some reactions, transfer of the second electron may not be directly from the P450 reductase, but may occur from cytochrome b_5, a small heme protein (15.2 kDa) that can be reduced by either the P450 reductase or another endoplasmic reticulum bound flavoprotein, NADH-cytochrome b_5 reductase. Some reactions catalyzed by specific cytochromes P450 require cytochrome b_5 for optimal enzymatic activity possibly because cytochrome b_5 (1) alters the affinity of the P450/reductase interaction, (2) alters the affinity of the P450 for a particular substrate, or (3) increases the rate of transfer of the second electron to the P450 heme.

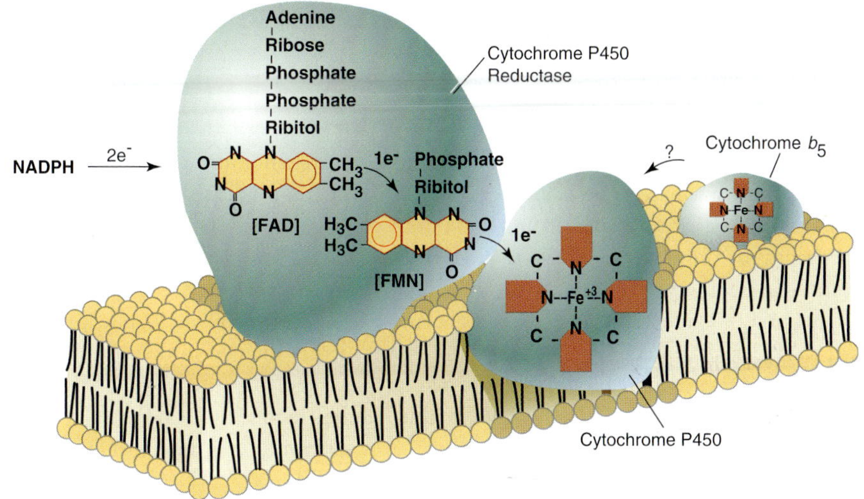

FIGURE 11.5

Components of the endoplasmic reticulum (microsomal) cytochrome P450 system. NADPH-cytochrome P450 reductase is bound by its hydrophobic tail to the membrane (a peripheral protein), whereas cytochrome P450 is deeply embedded in the membrane (an integral protein). Also shown is cytochrome b5, which may participate in selected cytochrome P450-mediated reactions.

NADPH-Adrenodoxin Reductase Is the Flavoprotein Electron Donor in Mitochondria

The mitochondrial cytochromes P450 are found in the inner membrane of mitochondria and are involved in steroid hydroxylation reactions. This electron transfer system requires two additional proteins, the NADPH-adrenodoxin reductase (51 kD), which contains a single FAD prosthetic group, and **adrenodoxin**, a 12.5-kD iron–sulfur protein, for product formation (Figure 11.6). Adrenodoxin reductase is only weakly associated with the inner mitochondrial membrane and cannot directly transfer either the first or second electron to the heme iron of cytochrome P450. A second protein, adrenodoxin, which is also weakly associated with the inner mitochondrial membrane, contains an iron–sulfur cluster that serves as a redox center for this molecule and functions as an electron shuttle between the FAD of adrenodoxin reductase and the heme iron of a mitochondrial cytochrome P450. One adrenodoxin molecule receives an electron from its mitochondrial flavoprotein reductase and interacts with a second adrenodoxin, which then transfers its electron to a specific cytochrome P450. Components of the mitochondrial cytochrome P450 system are synthesized in the cytosol as larger-molecular-weight precursors, transported into mitochondria, and processed by proteases into smaller-molecular-weight, mature proteins.

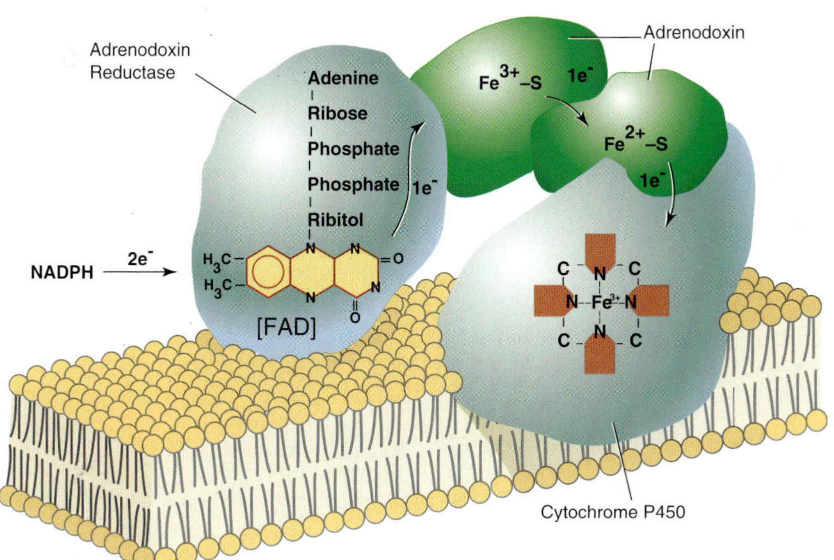

FIGURE 11.6

Components of mitochondrial cytochrome P450 system. Cytochrome P450 is an integral protein of the inner mitochondrial membrane. NADPH-adrenodoxin reductase and adrenodoxin are peripheral proteins, not embedded in the membrane.

11.5 CYTOCHROME P450: NOMENCLATURE AND ISOFORMS

Because of the large number of cytochromes P450 that have been identified (over 4500, as of January 2005), a system for classification of the enzymes into functional groups and nomenclature had to be developed. The system chosen is based on classification according to the relative identity of the amino acid sequences of the enzymes. The superfamily of cytochromes P450 is thus divided initially into families in which the amino acid sequence identity of the members is greater than 40%. The family is designated by the prefix "**CYP**", for **cy**tochrome **P**450, followed by an Arabic numeral (e.g., CYP1, CYP2, CYP3, etc.). The families are further divided into subfamilies in which the amino acid sequence identity of the members is greater than 55%. The subfamily is identified by an additional arabic, capital letter (e.g., CYP1A, CYP1B, CYP1C, etc.). The individual members of each subfamily are then numbered in the order in which they were identified (e.g., CYP1A1, CYP1A2, CYP1A3, etc.). Although cytochromes P450 are enzymes, the terms "isoenzyme" or "isozymes" are generally not used to describe these proteins; rather, the term "**form**" or "**isoform**" is used.

Tables 11.1 and 11.2 list the known human P450 isoforms. There are 57 isoforms split into 18 families and 41 subfamilies. The human genome also encodes for 58 CYP pseudogenes that do not form active protein. Table 11.1 lists the isoforms that utilize primarily exogenous compounds (i.e., drugs and xenobiotics); these isoforms each metabolize a wide variety of substrates. Table 11.2 lists the isoforms involved in metabolism of endogenous compounds; these isoforms generally recognize only one or two specific substrates and these are listed in the table.

TABLE 11.1 Human Cytochromes P450 Involved in Exogenous Metabolism

CYP Family	*Isoforms*	*Selected Substrate(s)*	*Selected Inhibitor(s)*	*Selected Inducer(s)*
1	1A1	Benzo(*a*)pyrene, diclofenac	Ketoconazole	Benzo(*a*)pyrene
	1A2	Benzo(*a*)pyrene, warfarin	Ciprofloxin	St. John's wort
	1B1	Benzo(*a*)pyrene, aflatoxin B1	Tamoxifen	NC[a]
2	2A6	Acetaminophen, nicotine	Cannabidol	Dexamethazone
	2A7	NC	NC	NC
	2A13	Hexamethylphosphoramide	NC	NC
	2B6	Diazepam, mephenytoin	Ketoconazole	Rifampicin
	2C8	Taxol, ibuprofen, verapamil	Quinine	Phenobarbital
	2C9	Amitriptyline, naproxen	Sulfaphenazole	Rifampicin
	2C18	Imipramine, methadone	NC	Rifampicin
	2C19	Diazepam, omeprazole	Isoniazid	Rifampicin
	2D6	Fluvastatin, codeine, risperidone	Quinidine	Dimethylsulfoxide
	2E1	Acetaminophen, halothane	Watercress	Isoniazid, ethanol
	2F1	Naphthalene, styrene	NC	NC
	2J2	Bufuralol	NC	NC
	2R1,2S1	NC	NC	NC
	2U1,2W1	NC	NC	NC
3	3A4	Erythromycin, nifedipine, codeine, warfarin, terfenadine	Troleandomycin, ketoconazole	Cortisol, rifampin, phenobarbital
	3A5	Verapamil, prevastatin	NC	Dexamethazone
	3A7	Retinoic acid, codeine, cortisol	DHEA	NC
	3A43	Testosterone	NC	NC

[a] NC, not well-characterized.

TABLE 11.2 Human Cytochromes P450 Involved in Endogenous Metabolism

CYP Family	Isoforms	Substrate(s)	Activity
4	4A11	Fatty acids	ω-Hydroxylase
	4B1	Arachidonic acid	12-Hydroxylase
	4F2	Arachidonic acid, leukotriene B_4	ω-Hydroxylase
	4F3	Leukotriene B_4	ω-Hydroxylase
	4F8	Arachidonic acid, prostaglandins	18-Hydroxylase
	4F12	Arachidonic acid	ω-Hydroxylase
	4A22,4F11,4F22,4V2,4X1,4Z1	Unknown	Unknown
5	5A1	Prostaglandin H_2	Thromboxin A_2 synthase
7	7A1	Cholesterol	7-α-Hydroxylase
	7B1	Pregnenolone, dehydroepiandrosterone (DHEA)	7-α-Hydroxylase
8	8A1	Prostaglandin H_2	Prostacyclin synthase
	8B1	Sterols	12-α-Hydroxylase
11	11A1	Cholesterol	Side-chain cleavage
	11B1	11-Deoxycortisol, 11-deoxycorticosterone	11-β-Hydroxylase
	11B2	Corticosterone	18-Hydroxylase
17	17A1	Pregnenolone, progesterone	17-α-Hydroxylase
		17-hydroxy pregnenolone, 17-hydroxy progesterone	17–20 Lyase
19	19A1	Androstenedione, testosterone	Aromatase
20	20A1	Unknown	Unknown
21	21A2	Progesterone, 17-hydroxy progesterone	21-Hydroxylase
24	24A1	25-Hydroxy vitamin D3	24-Hydroxylase
26	26A1	Retinoic acid	4-Hydroxylase
	26B1	Retinoic acid	Unknown
	26C1	Unknown	Unknown
27	27A1	Sterol	27-Hydroxylase
		Vitamin D3	25-Hydroxylase
	27B1	Vitamin D3	1-α-Hydroxylase
	27C1	Unknown	Unknown
39	39A1	24-Hydroxy cholesterol	7-Hydroxylase
46	46A1	Cholesterol	24-Hydroxylase
51	51A1	Lanosterol	14-α-Demethylase

11.6 CYTOCHROME P450: SUBSTRATES AND PHYSIOLOGICAL FUNCTIONS

Cytochromes P450 metabolize a variety of lipophilic compounds of endogenous or exogenous origin. As shown in Figure 11.7, cytochromes P450 may catalyze hydroxylation of the carbon atom of a methyl group, hydroxylation of the methylene carbon of an alkane, hydroxylation of an aromatic ring to form a phenol, or addition of an oxygen atom across a double bond to form an epoxide. They may also catalyze dealkylation reactions in which alkyl groups attached to oxygen, nitrogen, or sulfur atoms are removed. Oxidation of nitrogen, sulfur, and phosphorus atoms and dehalogenation reactions are also catalyzed by cytochrome P450 forms.

Both endogenous and exogenous substrates are metabolized by the cytochromes P450. Endogenous substrates include cholesterol, steroids, prostaglandins, and fatty acids. Exogenous substrates (xenobiotics, meaning "foreign to life") include many drugs, chemicals, environmental contaminants, and food additives. Cytochromes P450 oxidize these primarily lipophilic (hydrophobic) compounds to make them more water soluble (hydrophilic) for excretion.

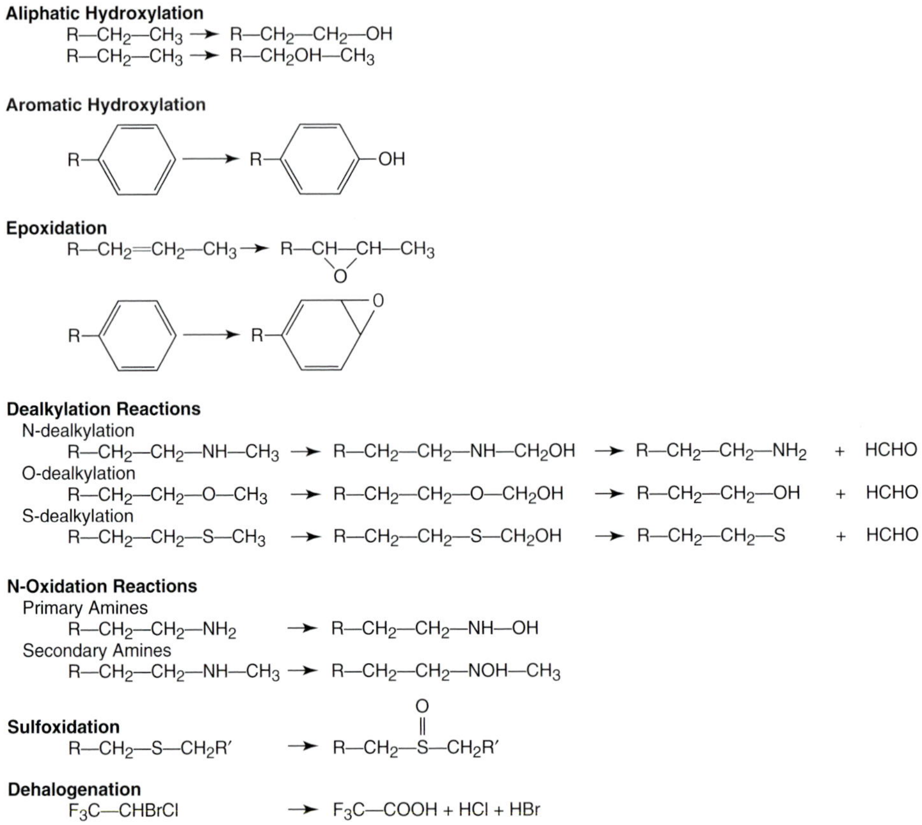

Aliphatic Hydroxylation

$R{-}CH_2{-}CH_3 \rightarrow R{-}CH_2{-}CH_2{-}OH$

$R{-}CH_2{-}CH_3 \rightarrow R{-}CH_2OH{-}CH_3$

Aromatic Hydroxylation

R— → R— —OH

Epoxidation

$R{-}CH_2{=}CH_2{-}CH_3 \rightarrow R{-}CH{-}CH{-}CH_3$ (O bridging the two CH)

R— → R— (O)

Dealkylation Reactions

N-dealkylation

$R{-}CH_2{-}CH_2{-}NH{-}CH_3 \rightarrow R{-}CH_2{-}CH_2{-}NH{-}CH_2OH \rightarrow R{-}CH_2{-}CH_2{-}NH_2 + HCHO$

O-dealkylation

$R{-}CH_2{-}CH_2{-}O{-}CH_3 \rightarrow R{-}CH_2{-}CH_2{-}O{-}CH_2OH \rightarrow R{-}CH_2{-}CH_2{-}OH + HCHO$

S-dealkylation

$R{-}CH_2{-}CH_2{-}S{-}CH_3 \rightarrow R{-}CH_2{-}CH_2{-}S{-}CH_2OH \rightarrow R{-}CH_2{-}CH_2{-}S + HCHO$

N-Oxidation Reactions

Primary Amines

$R{-}CH_2{-}CH_2{-}NH_2 \rightarrow R{-}CH_2{-}CH_2{-}NH{-}OH$

Secondary Amines

$R{-}CH_2{-}CH_2{-}NH{-}CH_3 \rightarrow R{-}CH_2{-}CH_2{-}NOH{-}CH_3$

Sulfoxidation

$R{-}CH_2{-}S{-}CH_2R' \rightarrow R{-}CH_2{-}S({=}O){-}CH_2R'$

Dehalogenation

$F_3C{-}CHBrCl \rightarrow F_3C{-}COOH + HCl + HBr$

FIGURE 11.7

Common reactions catalyzed by cytochromes P450.

11.7 CYTOCHROMES P450 PARTICIPATE IN THE SYNTHESIS OF STEROID HORMONES AND OXYGENATION OF ENDOGENOUS COMPOUNDS

The importance of cytochrome P450-catalyzed reactions with endogenous substrates is illustrated by the synthesis of steroid hormones from cholesterol in the adrenal cortex and sex organs. Both mitochondrial and microsomal (endoplasmic reticulum) cytochrome P450 systems are required to convert cholesterol to aldosterone and cortisol in adrenal cortex, testosterone in testes, and estradiol in ovaries.

Cytochromes P450 are responsible for several steps in the adrenal synthesis of aldosterone, the mineralocorticoid responsible for regulating salt and water balance, and cortisol, the glucocorticoid that governs protein, carbohydrate, and lipid metabolism. In addition, adrenal cytochromes P450 catalyze the synthesis of small quantities of the **androgen androstenedione**, a regulator of secondary sex characteristics and a precursor of both estrogens and testosterone. Figure 11.8 presents a summary of these pathways.

In adrenal mitochondria, a cytochrome P450 (CYP11A1) catalyzes the **side-chain cleavage reaction** converting cholesterol to pregnenolone, a committed step in steroid biosynthesis. This reaction involves sequential hydroxylation at carbon atoms 22 and 20 to produce 22-hydroxycholesterol and then 20,22-dihydroxycholesterol, respectively (Figure 11.9). CYP11A1 then cleaves the carbon–carbon bond between C20 and C22 to produce pregnenolone, a 21-carbon steroid. This reaction sequence requires three NADPH and three O_2 molecules, and it is catalyzed completely by CYP11A1.

Pregnenolone is produced in mitochondria, but is moved to the cytosol where it is oxidized by 3-β-hydroxysteroid dehydrogenase $\Delta^{4,5}$-isomerase to progesterone. Progesterone is hydroxylated at the C21 position of the steroid nucleus to deoxycorticosterone (DOC) by the microsomal P450 CYP21A2. DOC is subsequently hydroxylated at C11 by the mitochondrial P450 CYP11B1 to form corticosterone, which can then be hydroxylated by another mitochondrial cytochrome P450 enzyme, CYP11B2, at C18 to form the mineralocorticoid aldosterone (Figure 11.8; p. 832).

Cortisol is synthesized from either pregnenolone or progesterone beginning with the hydroxylation of C17 by microsomal P450 CYP17A1. Hydroxylation at C21 of 17α-hydroxyprogesterone by CYP21A2 produces 11-deoxycortisol, which is transported into the mitochondrion, where it is further hydroxylated by CYP11B1 at C11 to form cortisol (Figure 11.8). Congenital deficiencies resulting from mutations in the gene that codes for CYP21A2 are discussed in Clinical Correlation 11.1.

Synthesis of steroids containing 19 carbon atoms from 7α-hydroxypregnenolone or 17α-hydroxyprogesterone proceeds with the loss of the acetyl group at C17. This deacetylation reaction is catalyzed by CYP17A1, the same cytochrome P450 that hydroxylates steroids at C17. The factors that determine whether this cytochrome P450 performs only a single hydroxylation step to produce the 17-hydroxy product or proceeds further to cleave the C17–C20 bond have not been determined, although the presence of cytochrome b_5 in gonadal tissues appears to allow CYP17A1 to catalyze the cleavage reaction. The products formed after removal of the acetyl group are dehydroepiandrosterone (DHEA) from 17α-hydroxypregnenolone or androstenedione from 17-α-hydroxyprogesterone. DHEA may be dehydrogenated at the 3-OH group to androstenedione, a potent androgenic steroid that serves as the immediate precursor of testosterone.

The estrogens are synthesized from the androgens in a reaction called **aromatization** because an aromatic ring is produced in ring A of the steroid product. In this complex reaction, multiple hydroxylation reactions are catalyzed by a single cytochrome P450 enzyme, CYP19A1 or aromatase, to form the aromatic ring with removal of the

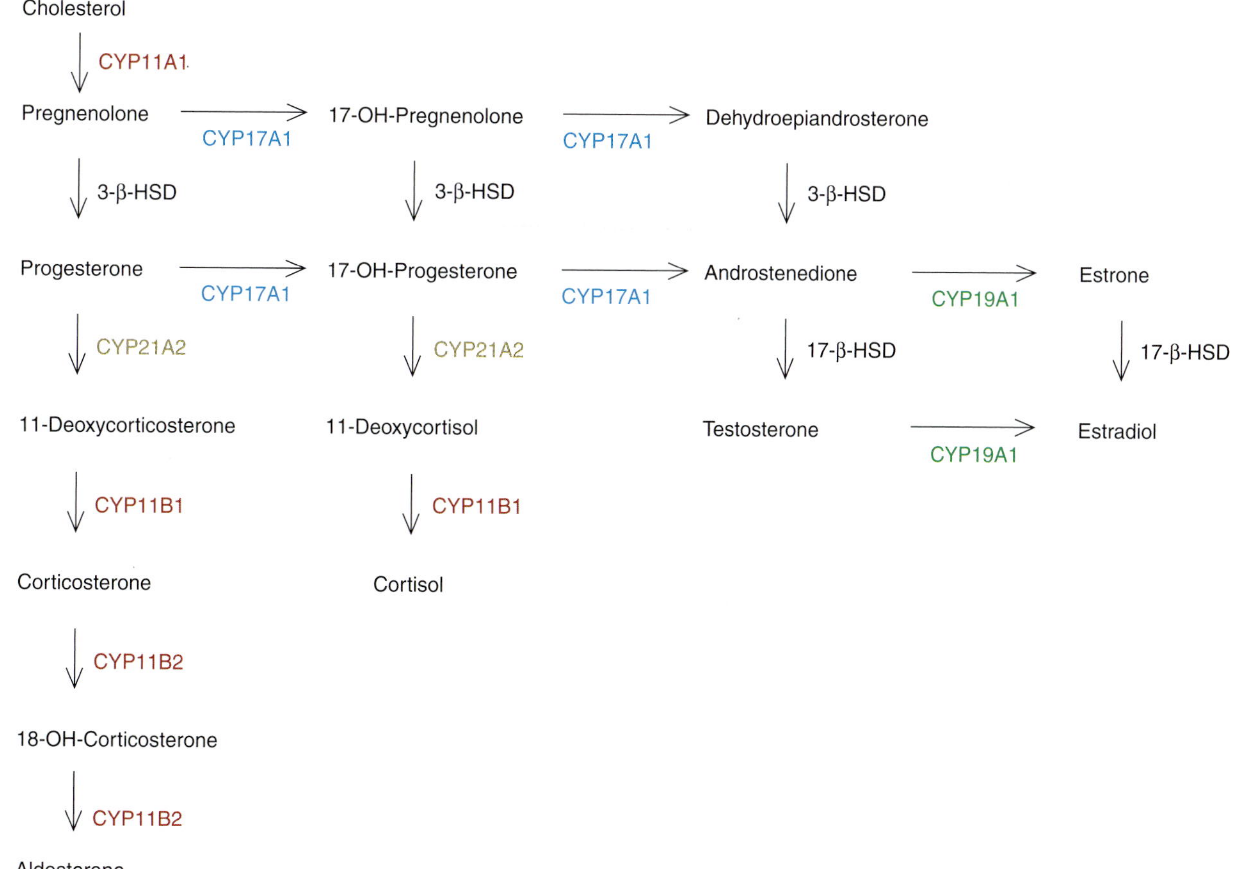

FIGURE 11.8

Steroid hormone synthesis in the adrenal gland. The cytochromes P450 involved are CYPs 11A1, 11B1, and 11B2 in the mitochondria (red) and CYPs 17A1, 19A1, and 21A2 in the endoplasmic reticulum (blue). The other enzymes shown, 3-β-HSD and 17-β-HSD, are hydroxysteroid dehydrogenases.

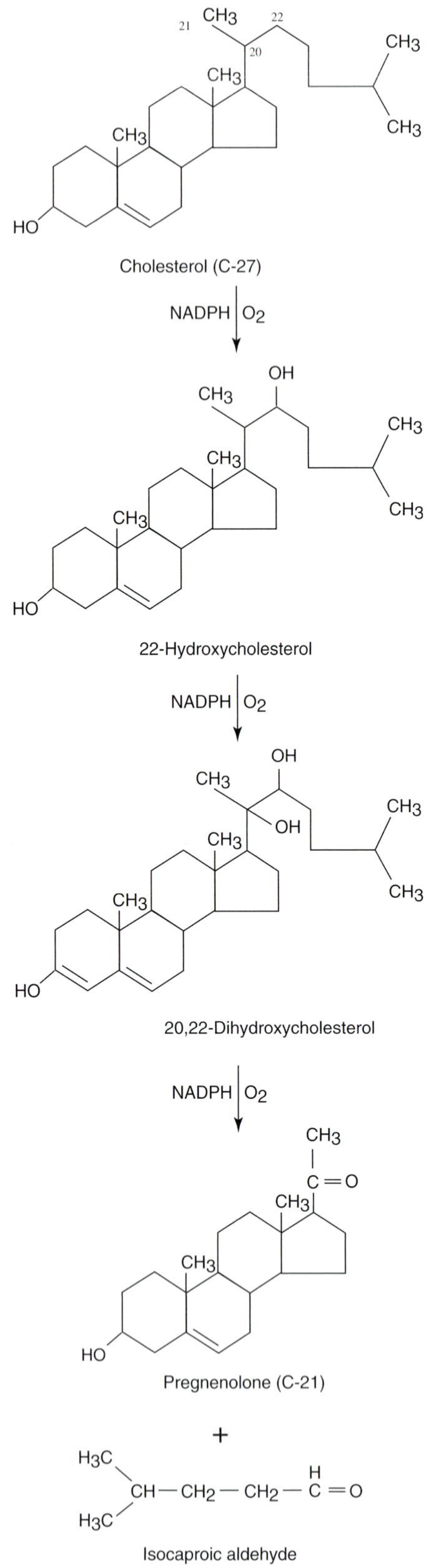

methyl group at C19. Figure 11.10 details the aromatization reaction. Two sequential hydroxylation reactions at the methyl C19 result in the introduction of an aldehyde group. The final step may involve a peroxidative attack at C19 with the loss of the methyl group and elimination of the hydrogen atom to produce the aromatic ring. All three hydroxylation reactions are catalyzed by CYP19A1. The complexity of steroid hormone production during pregnancy and the role of P450 forms are illustrated in Clinical Correlation 11.2.

Cytochromes P450 also metabolize other endogenous compounds that are involved in physiological processes and thus regulate their activity. For example, arachidonic acid is metabolized by cytochromes P450 to form 19- and 20-hydroxyeicosatetraenoic acids (HETEs), epoxyeicosatetraenoic acids (EETs), and dihydroxyeicosatetraenoic acids (DiHETEs). These metabolites function in the regulation of vascular, renal, pulmonary, and cardiac systems and in inflammatory and growth responses. 20-HETE, formed by the metabolism of arachidonic acid by CYP4As and CYP4Fs in humans, is a potent vasoconstrictor. The EETs, formed by CYP1A, CYP2B, CYP2C, and CYP2J in humans, are endothelium-derived vasodilators. The CYP4As also catalyze the hydroxylation of prostaglandins and fatty acids, thus activating or inactivating these important regulatory factors.

Cytochromes P450 produce vital metabolites for maintenance of homeostasis in other systems, as well. CYP7A1 catalyzes the first, and rate-limiting, step of bile acid synthesis in the liver (see pp. 717 and 1065). In liver, CYP27B1 catalyzes the 1-hydroxylation of 25-hydroxy-vitamin D3, formed in kidney, to produce the 1,25-dihydroxy metabolite, which is the active form of this hormone, important in developing and maintaining bone (see p. 1097), while CYP24A1 is involved in degradation or inactivation of vitamin D metabolites. In the immune system, leukotriene B4 is a polymorphonuclear leukocyte chemoattractant that is hydroxylated by CYP4F3 to the less active 20-hydroxy-leukotriene B4, vital in the inactivation of the immune response. Retinoic acid is metabolized by CYP26A1 to the 4-hydroxy derivative, which may have greater activity than retinoic acid in certain organs, where it plays a role in cell cycling and signaling. Table 11.2 lists the human cytochromes P450 that metabolize endogenous substrates and details their substrate specificities and functions.

FIGURE 11.9

Side-chain cleavage reaction of cholesterol. Three sequential reactions are catalyzed by CYP11A1 to produce pregnenolone and isocaproic aldehyde.

Cytochromes P450 Oxidize Exogenous Lipophilic Substrates

Exogenous substrates metabolized by cytochromes P450 include therapeutic drugs, chemicals used in the workplace, industrial byproducts that become environmental contaminants, and food additives. The human cytochromes P450 involved in xenobiotic metabolism are listed in Table 11.1. These cytochromes P450 exhibit broad substrate specificities, oxidizing a wide variety of xenobiotics, particularly lipophilic compounds. Addition of a hydroxyl group makes the compound more polar and, therefore, more soluble in the aqueous environment of the cell. Many exogenous compounds are highly lipophilic and will accumulate within cells over time, potentially interfering with cellular function, unless they are metabolized to more hydrophilic products.

It is estimated that CYP3A4 metabolizes approximately 50% of therapeutic drugs, CYP2D6 approximately 20%, CYP2C9 and 2C19 approximately 15%, and CYP1A2, CYP2A6, CYP2B6, and others metabolize the remaining 15% of the drugs. CYP3A4 is the major drug-metabolizing P450 in the human. It is present in the gastrointestinal tract as well as the liver and is responsible for the poor oral bioavailability of many drugs. Because cytochromes P450 have such broad substrate specificities, a compound may be metabolized by more than one cytochrome P450 isoform and/or at different sites, as shown in Figure 11.11 for the compound diazepam.

Drugs and xenobiotics are metabolized primarily via two pathways: Phase I (oxidative functionalization reactions) and Phase II (biosynthetic reactions). They may be metabolized by either or both pathways. In Phase I metabolism, a functional group, such as a hydroxyl group, is introduced into or exposed on the drug by cytochromes P450 (among other enzymes; Table 11.3). The enzymes in Phase II metabolism then connect this functional group to an endogenous species such as glucuronic acid, sulfate, glutathione, amino acids, or acetate. The net effect of either or both of these

CLINICAL CORRELATION 11.1
Congenital Adrenal Hyperplasia: Deficiency of CYP21A2

The adrenal cortex, a major site of steroid hormone production during fetal and adult life, is metabolically more active in fetal life and may produce 100–200 mg of steroids per day in comparison to 20–30 mg in the nonstressed adult adrenal gland. Enzyme deficiencies have been reported at all steps of cortisol production. Diseases associated with insufficient cortisol production are referred to as congenital adrenal hyperplasias (CAH). The most common enzyme deficiency in CAH is in CYP21A2, a 21-hydroxylase, resulting in the failure to metabolize 17α-hydroxyprogesterone to 11-deoxycortisol, which is converted by CYP11B1 to cortisol. Because there is not enough cortisol, secretion of ACTH, the pituitary hormone that regulates production of cortisol, increases. Prolonged periods of elevated ACTH levels cause adrenal hyperplasia and an increased production of the androgenic hormones, dehydroepiandrosterone (DHEA) and androstenedione.

Clinical problems arise because the additional production of androgenic steroids causes virilization in females, precocious sex organ development in prepubertal males, or diseases related to salt imbalance because of decreased levels of aldosterone. Clinical consequences of severe 21-hydroxylase deficiency may be recognizable at birth, particularly in females, because the excessive buildup of androgenic steroids may cause obvious irregular development of their genitalia. In male newborns, a deficiency in 21-hydroxylase activity may be overlooked, because male genitalia appear normal, but there will be precocious masculinization and physical development later on. In late onset CAH, clinical symptoms may vary considerably from early development of pubic hair, early fusion of epiphyseal growth plates causing premature cessation of growth, or male baldness patterns in females.

Source: Donohoue, P. A., Parker, K., and Migeon, C. J. Congenital adrenal hyperplasia. In: C. S. Scriver, et al. (Eds.), *The Metabolic and Molecular Bases of Inherited Disease*, Vol. II, 7th ed., New York: McGraw-Hill, 2001, p. 4077; and White, P. C. and Speiser, P. W. Congenital adrenal hyperplasia due to 21-hydroxylase deficiency. *Endocr. Rev.* 21:245, 2000.

CLINICAL CORRELATION 11.2
Steroid Hormone Production During Pregnancy

Cytochromes P450 play a major role in estrogen synthesis. During pregnancy, a unique interaction among cytochromes P450 in different organs is needed to synthesize the large quantities that are required. Steroid hormone production increases dramatically and, at term, reaches 15–20 mg of estradiol, 50–100 mg of estriol, and approximately 250 mg of progesterone per day, amounts that are a 1000 times greater than those produced by premenopausal, nonpregnant women.

The corpus luteum of the ovary is the major site for estrogen production in the first few weeks of pregnancy, but at approximately 4 weeks of gestation, the placenta begins synthesizing and secreting progesterone and estrogens. After 8 weeks of gestation, the placenta becomes the dominant source of progesterone. The human placenta lacks CYP17, which catalyzes both the 17α-hydroxylation reaction and the cleavage of the C17–C20 bond and thus cannot synthesize estrogens from cholesterol. The placenta, however, can catalyze the side-chain cleavage reaction to form pregnenolone from cholesterol and oxidizes pregnenolone to progesterone, which is secreted into the maternal circulation. How then does the placenta produce estrogens if it cannot synthesize DHEA or androstenedione from progesterone? The fetal adrenal gland, a highly active steroidogenic organ, catalyzes the synthesis of DHEA from cholesterol and releases it into the fetal circulation. A large proportion of fetal DHEA is metabolized by the fetal adrenal and liver to 16α-hydroxy-DHEA, which is converted by CYP19 in the placenta to the estrogen estriol. This is an elegant demonstration of the cooperativity of the cytochrome P450-mediated hydroxylating systems in the fetal and maternal organ systems, leading to the progressive formation of estrogens during gestational development.

Source: Cunningham, F. G., MacDonald, P. C. Gant, N. F., Leveno, K. J., Gilstrap, L. C., Hankins, G. D. V., and Clark, S. L. The placental hormones. In: *Williams Obstetrics*, 20th ed. East Norwalk, CT: Appleton and Lange, 1997, Chapter 6, p. 125.

modifications is to make the foreign compound more water-soluble and thus easier to excrete, generally through the kidneys, but also through the intestines via excretion into bile. Cytochromes P450 are involved in Phase I, but not Phase II, metabolism. General examples of xenobiotic-metabolizing enzymes and their reactions are shown in Table 11.3.

Metabolism of xenobiotics by cytochromes P450 can have one of three effects on the compound: **inactivation, activation**, or **formation of a toxic metabolite**. In inactivation, the active form of a drug is converted to an inactive form, thus decreasing bioavailability or, in the case of harmful xenobiotics, decreasing potential harmful effects (e.g., phenobarbital). Diazepam (brand name Valium, prescribed as a sedative), for example, is metabolized by CYPs 2C19 (*N*-dealkylation) and 3A4 (hydroxylation) to the biologically inactive oxazepam (Figure 11.11), which then undergoes Phase II conjugation with glucuronic acid before elimination.

Other drugs are biologically inactive until metabolized by a cytochrome P450 to a biologically active form. The drug terfenadine (brand name Seldane, a histamine H1 receptor antagonist prescribed for seasonal allergies) is an inactive form of a drug, or a prodrug, that is activated by sequential CYP3A4 hydroxylation to fexofenadine, the functional, desired form (Figure 11.12). Problems associated with interference of terfenadine activation are detailed in Clinical Correlation 11.3.

In addition to xenobiotic activation or inactivation, cytochrome P450 metabolism can inadvertently cause the formation of toxic metabolites. Benzo[*a*]pyrene is a weak carcinogen produced from the burning of coal, combustion of materials in tobacco products, food barbecued on charcoal briquettes, and industrial processing. It is converted by CYPs 1A1, 1A2, and 1B1 to benzo[*a*]pyrene-7,8-dihydrodiol-9,10-epoxide, a much more potent carcinogen (Figure 11.13). Benzo[*a*]pyrene is converted to the 7,8-epoxide form by CYP1A1. Subsequent hydrolysis by epoxide hydrolase forms the vicinal hydroxy derivative, benzo[*a*]pyrene-7,8-dihydrodiol, and a second

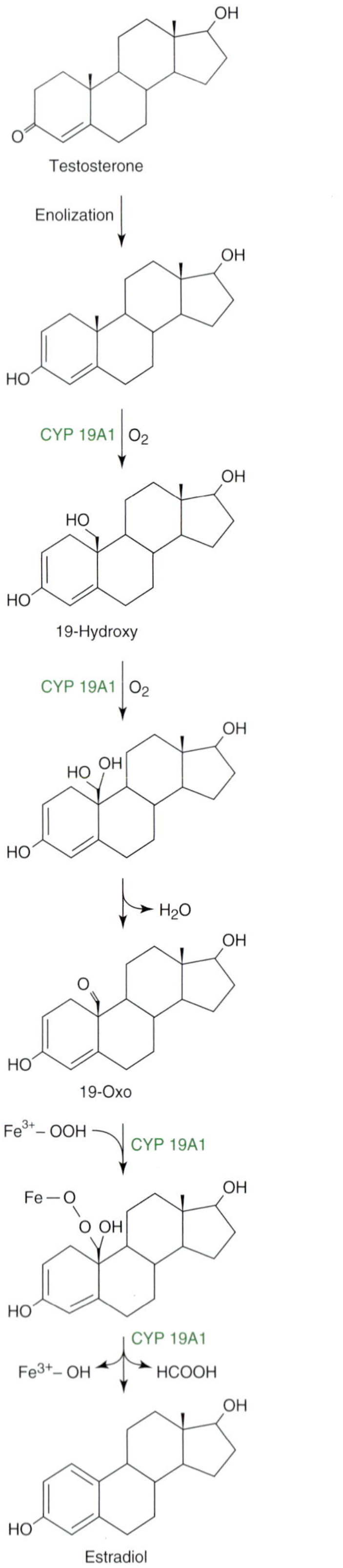

FIGURE 11.10

The sequence of reactions leading to the aromatization of androgens to estrogens. Adapted from Graham-Lorence, S., Amarneh, B., White R. E., Peterson, J. A., and Simpson E. R., A three dimensional model of aromatase cytochrome P450. *Protein Sci.* 4:1065, 1995.

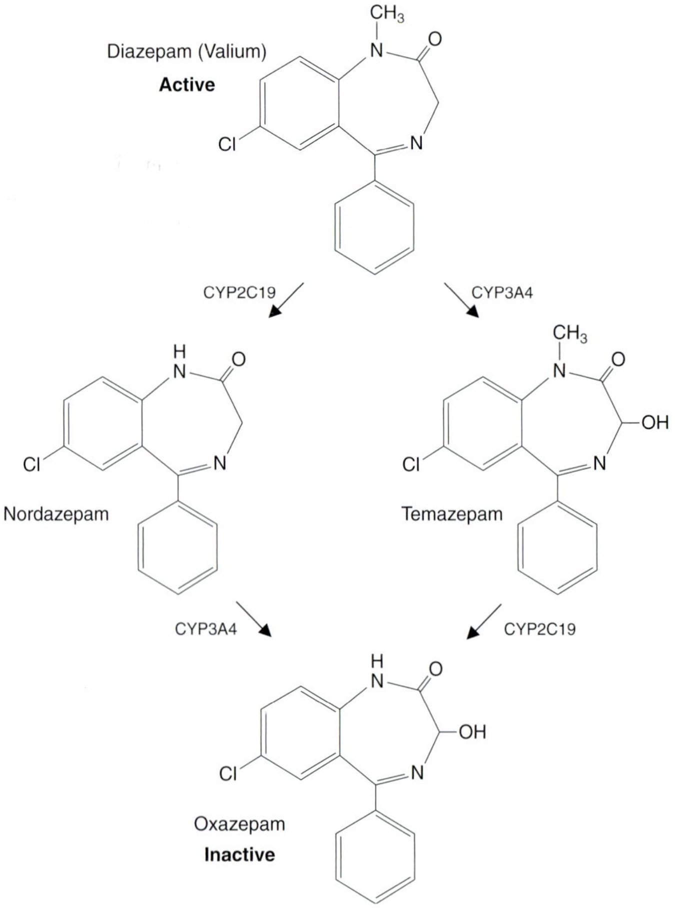

FIGURE 11.11

The metabolism of diazepam by CYPs 3A4 and 2C19. Diazepam may be metabolized by more than one P450. The metabolites that are formed from a compound will be affected by the amount of P450 forms that are expressed in tissues.

TABLE 11.3 Xenobiotic-Metabolizing Reactions and Enzymes

Type of Reaction	*General Enzyme Name*	*General Enzymatic Reaction*
Oxidation	Dehydrogenases	Oxidation of alcohol or aldehyde groups
	Flavin-containing monooxygenases (FMO)	Oxidation of primary, secondary, and tertiary amines, nitrones, thioethers, thiocarbamates and phosphines.
	Monoamine oxidases	Removal of an amine by oxidative deamination
	Cytochrome P450	See Tables 11.1 and 11.2 and Fig. 11.7 for reactions catalyzed by P450 enzymes
Reduction	Aldehyde and ketone reductases	Reduction of carbonyl group to a hydroxyl group
	Azoreductases	Reduction of azo group to form primary amines
	Quinone reductases	Reduction of quinones to hydroquinones
Hydrolysis	Epoxide hydrolases	Addition of water to arene oxide or epoxide groups to form vicinal hydroxy groups
	Carboxylesterases	Addition of water to an ester bond
	Amidases	Addition of water to an amide bond
Conjugation	*N*-acetyltransferases	Addition of CoA from acetyl-CoA to aromatic amines or compounds containing hydrazines
	UDP-glucuronosyl transferases	Addition of glucuronic acid from uridine-5′-diphosphate (UDP)–glucuronic acid to carboxyl, phenolic, aliphatic hydroxyl, aromatic or aliphatic amines, or sulfhydryl groups
	Sulfotransferases	Addition of sulfate group from phosphoadenosyl phosphosulfate to phenolic groups, aliphatic alcohols or aromatic amines
	Methyltransferases	Addition of methyl group from *S*-adenosyl methionine to compounds containing phenolic, catechol, aliphatic or aromatic amines, or sulfhydryl groups
	Glutathione *S*-transferase	Conjugation of compounds containing electrophilic heteroatoms, nitrenium ions, carbonium ions, free radicals, epoxide, or arene oxide groups

epoxidation reaction by CYP1A1 or 3A4 forms benzo[*a*]pyrene-7,8-dihydrodiol-9,10-epoxide, which can form guanine adducts on DNA, causing disruption of gene function leading to mutations. The interaction of benzo[*a*]pyrene-7,8-dihydrodiol-9,10-epoxide with the p53 gene is a critical step in the mechanism by which benzo[*a*]pyrene is carcinogenic in humans. Benzo[*a*]pyrene also binds to the aryl hydrocarbon receptor (AhR) and induces these P450 forms, thus increasing its own metabolism.

CLINICAL CORRELATION 11.3

Cytochrome P450 Inhibition: Drug–Drug Interactions and Adverse Effects

The roles that cytochromes P450 play in drug metabolism and the serious consequences of drug–drug interactions were clearly shown for two drugs, terfenadine (Seldane) and cisapride (Propulsid), when their metabolism was inhibited by other drugs. A small percentage of users experienced severe adverse effects that forced their manufacturers and the Food and Drug Administration (FDA) to issue warnings that the drugs should not be taken with other drugs that inhibit CYP3A4. The seriousness of these adverse effects forced the FDA to have both drugs removed from the market.

The FDA approved terfenadine, a second-generation H1 antihistamine, in 1985 to treat seasonal allergies. Terfenadine is rapidly metabolized in the liver by CYP3A4 to fexofenadine, as shown in Figure 11.12, resulting in low levels of the parent drug soon after ingestion, but the therapeutic effects of terfenadine are actually caused by fexofenadine. Because many other drugs are substrates or inhibitors of CYP3A4, the metabolism of terfenadine is potentially inhibitable. Individuals who took terfenadine with the macrolide antibiotic erythromycin or the anti-fungal agent, ketoconazole, both strong inhibitors of CYP3A4, had significantly elevated plasma levels of terfenadine. In a small percentage of these terfenadine users, serious cardiac problems developed because the parent drug caused alteration in cardiac potassium channels and increased the risk of a rare ventricular tachycardia called Torsades de Pointes. Some individuals died from cardiac problems that developed after taking terfenadine with erythromycin or ketoconazole. Since the therapeutic properties of terfenadine were associated with its nontoxic metabolite, fexofenadine, the manufacturer tested fexofenadine as a new medication and sought FDA approval for this metabolite, now marketed as Allegra.

Cisapride was approved in 1993 for patients who suffer from nighttime heartburn as a result of gastroesophageal reflux disease or GERD. The elimination of this drug from the body is dependent on CYP3A4 metabolism, and when administered alone or with other drugs that do not inhibit CYP3A4, the parent drug does not accumulate in the plasma. However, when taken with drugs that are substrates or inhibitors of CYP3A4, cisapride metabolism is reduced, and it accumulates with subsequent intake. In some individuals, increased cisapride levels cause cardiac arrhythmias, and, by the end of 1999, heart rhythm abnormalities were reported in 341 patients taking cisapride, resulting in 80 deaths. Accordingly, the manufacturer of cisapride stopped marketing this drug in the United States after 2000.

Source: Terfenadine: Proposal to withdraw approval of two new drug applications and one abbreviated new drug application. Fed. Reg. 62:1889, 1997. (This document may be accessed from the internet site for the Federal Register at http://www.access.gpo.gov/su_docs/aces/aces140.html); and Desta, Z., Soukhova, N., Mahal, S. K. and Flockhart, D. A. Interaction of cisapride with the human cytochrome P450 system: Metabolism and inhibition studies. *Drug Metab. Disp.* 28:789, 2000.

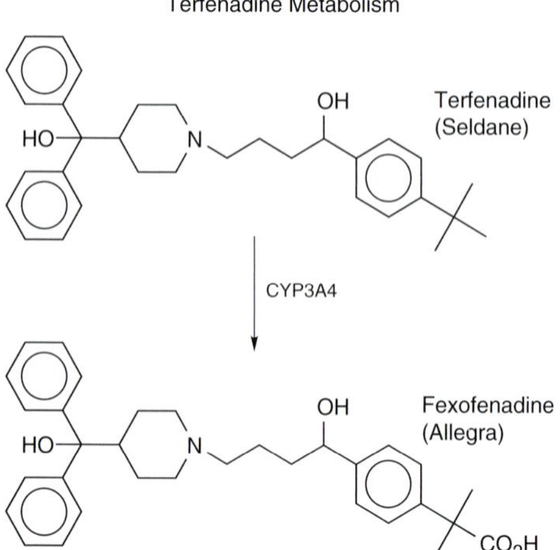

FIGURE 11.12

The metabolism of terfenadine to fexofenadine, the bioactive compound.

Acetaminophen, commonly used as an analgesic and antipyretic, is converted by CYP2E1 to *N*-acetyl-*p*-benzoquinone imine (NAPQI), a highly reactive compound leading to protein adducts, oxidative stress, and toxicity. Normally, acetaminophen is primarily metabolized by glucuronidation and sulfation pathways to polar, inactive conjugates that are readily excreted (Figure 11.14, upper and lower pathways, respectively). Acetaminophen is also metabolized by CYP2E1 to a highly reactive NAPQI (Figure 11.14, middle pathway). The amount of acetaminophen that is metabolized by CYP2E1 is normally low in comparison to glucuronidation and sulfation. The small amounts of NAPQI formed are rapidly conjugated by glutathione (see p. 782) to a nontoxic metabolite. However, if glutathione levels are depleted, NAPQI will escape glutathione conjugation and react with liver cell components to cause cell damage. Normally, the levels of CYP2E1 are low in comparison to other P450 forms and the production of NAPQI, when normal doses of acetaminophen are taken, is small. However, intake of large quantities of acetaminophen will cause an increase in the production of NAPQI that may cause severe liver damage. Another factor in acetaminophen-induced liver toxicity is that the consumption of alcoholic beverages induces CYP2E1, which will increase the production of NAPQI. However, alcohol is also a substrate for this cytochrome P450 and will inhibit the metabolism of other CYP2E1 substrates. The extent of liver damage thus depends on the timing and amount of acetaminophen taken. Acetaminophen overdose results in more calls to poison control centers in the United States than any other pharmacological substance. In addition, according to the American Liver Foundation, 35% of cases involving liver failure are caused by acetaminophen poisoning. A description of the effects of alcohol on acetaminophen metabolism is presented in Clinical Correlation 11.4.

As exemplified by **benzo[*a*]pyrene** and **acetaminophen**, the P450 system can produce toxic metabolites. Formation of toxic compounds by the cytochromes P450, however, does not mean that cell damage or a tumor will occur. Other cellular processes will determine the extent of cellular damage; for example, conjugation of the benzo[*a*]pyrene-7,8-dihydrodiol-9,10-epoxide by glutathione transferases plays a key protective role in inhibiting DNA adduct formation and reducing its carcinogenic activity. The actions of various detoxification enzyme systems, including conjugation reactions such as glucuronidation and sulfation, the status of the immune system, nutritional state, genetic predisposition, and exposure to various environmental chemicals all play key roles in determining the extent of P450-mediated cell damage and toxicity. Although it seems counterproductive for mammalian systems to possess enzymes that potentially create highly toxic compounds, the purpose of the cytochrome P450 system is to add or expose functional groups, making the molecule more polar and/or more susceptible to metabolism by other detoxification enzyme systems; the creation of toxic metabolites is an unexpected consequence of this process.

Although the vast majority of the substrates for these cytochromes P450 (Table 11.1) are exogenous compounds, some endogenous functions have also been postulated. CYP1A1 and CYP1A2 appear to be involved in heme catabolism and estradiol metabolism, respectively. There is also evidence that members of the CYP2A, CYP2B, CYP2C, and CYP3A subfamilies are involved in testosterone metabolism, and those of the CYP2D and CYP2E subfamilies are involved in catecholamine metabolism and gluconeogenesis, respectively.

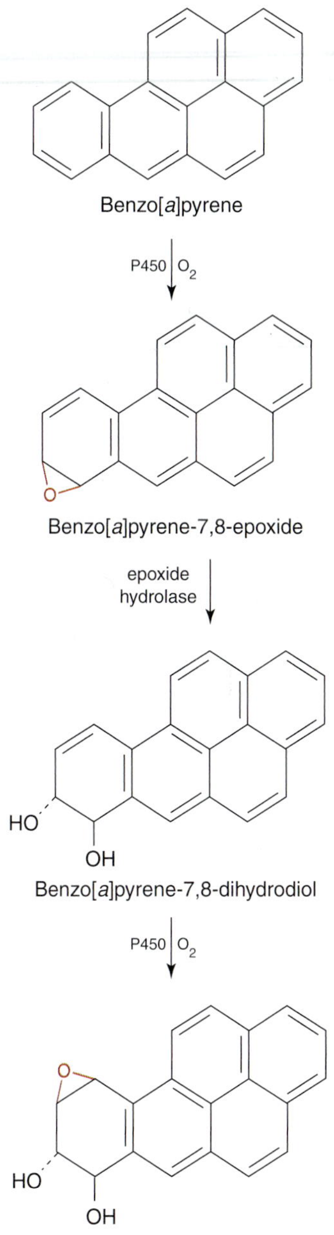

Benzo[*a*]pyrene

P450 | O_2

Benzo[*a*]pyrene-7,8-epoxide

epoxide hydrolase

HO OH

Benzo[*a*]pyrene-7,8-dihydrodiol

P450 | O_2

O HO OH

Benzo[*a*]pyrene-7,8-dihydrodiol-9,10-epoxide

FIGURE 11.13

The metabolism of benzopyrene by cytochrome P450 and epoxide hydrolase to form benzopyrene-7,8-dihydrodiol-9,10-epoxide.

HNCCH3 (O) — O—Glucuronide

HNCCH3 — OH — **Acetaminophen** — CYP2E1 → HNCCH3 — O — **NAPQI**

→ HNCCH3 — Protein — OH → Cell Death

GSH → HNCCH3 — GSH — OH

HNCCH3 — O—Sulfate

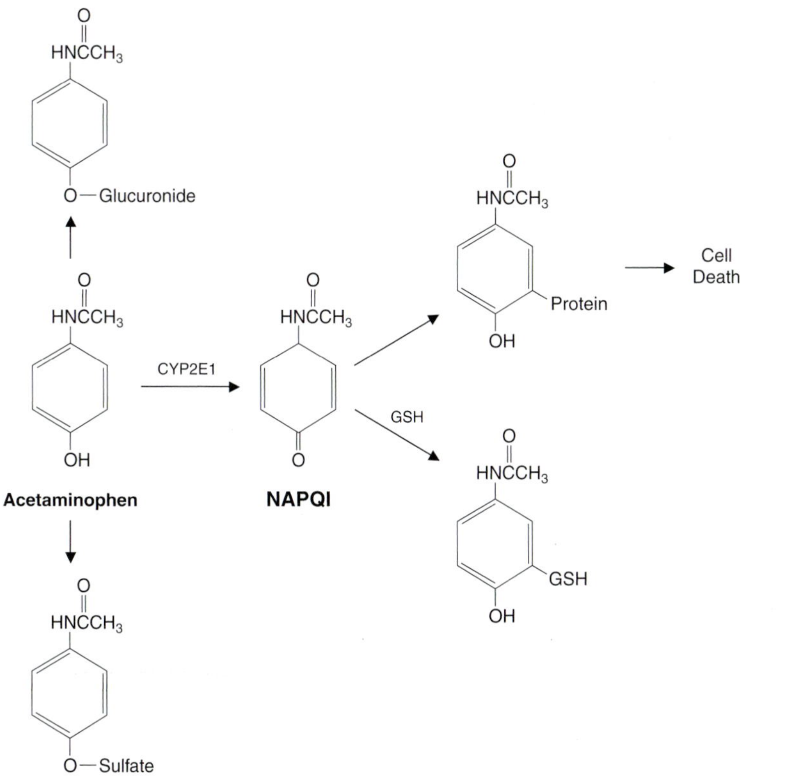

FIGURE 11.14

Acetaminophen may be metabolized by the Phase II enzymes, sulfotransferase and glucuronsyl transferase, and by CYP2E1. Metabolism by CYP2E1 will lead to the formation of *N*-acetyl-*p*-benzoquinoneimine (NAPQI). When NAPQI formation is low, NAPQI can be conjugated by glutathione to a nontoxic metabolite; however, if NAPQI is formed in excessive amounts, it may escape glutathione conjugation and react with liver cell components to cause cell damage.

CLINICAL CORRELATION 11.4
Role of CYP2E1 in Acetaminophen-Induced Liver Toxicity

Acetaminophen is one of the most commonly used analgesic (pain-relieving) and antipyretic (fever-reducing) over-the-counter medications. It is available alone or as a component of more than 100 other nonprescription medications. The pathway of acetaminophen metabolism is shown in Figure 11.14. The consumption of alcohol, an inducer and substrate of CYP2E1, has a profound effect on acetaminophen. Alcohol consumption equivalent to drinking a 750-mL bottle of wine, six 12-oz. cans of beer, or 9 oz. of 80-proof liquor over a 6- to 7-h period causes a 22% increase in CYP2E1-mediated metabolism of acetaminophen that manifests several hours post alcohol intake. Thus, the timing between the last consumption of alcohol and acetaminophen intake may be very critical in the development of acetaminophen-induced liver injury. In a person drinking alcohol, metabolism of acetaminophen taken concomitantly or shortly after cessation of drinking is delayed because ethanol, like acetaminophen, is a substrate for CYP2E1 and therefore competes for binding. In chronic alcohol consumption, CYP2E1 is induced and greater amounts of *N*-acetyl-*p*-benzoquinoneimine (NAPQI) are produced; however, alcohol may actually protect the liver from acetaminophen-induced injury, if taken at the same time or shortly before. If acetaminophen is taken several hours post alcohol intake, the alcohol has already been metabolized and increased levels of CYP2E1 are then available to metabolize acetaminophen to NAPQI, increasing the risk of liver damage.

The length of time and amount of alcohol consumed daily affect the amount of CYP2E1 expressed in hepatocytes. The amount of acetaminophen consumed is also a major determining factor for risk, as normally recommended doses do not usually generate sufficient NAPQI to produce liver cell damage. Higher doses of acetaminophen increase the risk of liver cell injury as more drug will be metabolized through the CYP2E1 pathway. Both the sulfation and glucuronidation pathways are essential in determining the amount of acetaminophen that will be metabolized by CYP2E1. In addition, the levels of intracellular glutathione are critical as glutathione forms conjugates with NAPQI to protect the liver from this reactive metabolite. In the presence of high levels of NAPQI, glutathione is rapidly depleted and can no longer protect against damage.

Source: Brunner, L. J., McGuinness, M. E., Meyer, M. M., and Munar, M. Y. Acute acetaminophen toxicity during chronic alcohol use. U.S. Pharmacist. Sept:HS11-HS19 1999. This article may be access from the website at http://www.uspharmacist.com/; and Thummel, K. E., Slattery, J. T., Ro, H., Chien, J. Y., Nelson, S. D., Lown, K. E., Watkins, P. B. Ethanol and production of the hepatotoxic metabolite of acetaminophen in healthy adults. *Clin. Pharmacol. Ther.* 67:591, 2000. Slattery, J. T., Nelson, S. D., and Thummel, K. E. The complex interaction between ethanol and acetaminophen. *Clin. Pharmacol. Ther.* 60:241, 1996.

11.8 CYTOCHROME P450 INDUCTION AND INHIBITION

Clearly, the levels and activity of the cytochromes P450 in any given person will greatly influence the effect of a particular drug or xenobiotic on the system. Thus, compounds that inhibit or induce P450 activity, as well as polymorphisms occurring in a P450 gene, can produce unexpected results.

Drug–Drug Interactions

Many drugs induce the formation of the cytochrome P450 that metabolizes them. Other drugs cause inhibition of P450 activity. Altering the metabolism of a particular drug from its predicted rate may cause unexpected and adverse effects and is of particular concern for individuals who take a combination of drugs. For drugs that are dependent on P450 metabolism for elimination from the body, inhibition will lead to accumulation of the parent drug to concentrations that may be toxic. Conversely, **induction** of the P450 isoform required to metabolize a particular drug may lead to over-metabolism and sub-effective concentrations.

The unintended effects that can occur when levels of cytochromes P450 are induced or inhibited by other drugs are called drug–drug interactions. If these inducing or inhibiting drugs are given with other drugs that are normally metabolized by P450 enzymes, the lifetime of these other drugs will be altered. If these drugs affect critical systems, the result can be fatal. For example, CYP2E1 is induced by alcohol or isoniazid. It also metabolizes anesthetics, such as halothane and enflurane, to compounds which damage liver proteins. The immune system then attacks these abnormal proteins, producing a form of hepatitis. People with higher than normal CYP2E1 activity produced by alcohol intake are at higher risk for this sort of hepatitis reaction to an anesthetic.

Drug–drug interactions also occur with CYP3A4, one of the most important drug metabolizing P450s and the P450 most abundantly present in liver and small intestine. CYP3A4 levels are induced by rifampicin, anticonvulsants such as carbamazepine and phenytoin, and glucocorticoids such as dexamethasone, and they are inhibited by antifungals such as ketoconazole, HIV protease inhibitors, calcium channel blockers, cyclosporine, antibiotics such as erythromycin, SSRI antidepressants, and even grapefruit juice. A very few of the many compounds metabolized by this enzyme are listed in Table 11.1.

It is readily apparent that mixing inducers and/or inhibitors of CYP3A4 with various drugs (which can themselves be inducers or inhibitors) may have unintended or even fatal effects. Taking rifampicin, a CYP3A4 inducer, with oral contraceptives, which are metabolized by CYP3A4, has resulted in lower efficacy of the contraceptives, leading to unintended pregnancies. Taking ketoconazole or even drinking grapefruit juice, both of which inhibit CYP3A4, along with the anticoagulant warfarin, could lead to excessive bleeding. Clinical consequences that relate to inhibition or induction of drug-metabolizing enzymes are presented in Clinical Correlations 11.3 and 11.5.

The mechanisms of induction of cytochromes P450 are at either the transcriptional or posttranscriptional (e.g., stabilization of mRNA or a decrease in protein degradation) level. It is not possible to predict the mode of induction based on the inducing compound. The complexity of the induction process is illustrated with CYP2E1 in the human versus rat liver. CYP2E1 protein is induced in both species by small organic molecules, such as ethanol, acetone, or pyrazole, or under fasting or diabetic conditions. In humans, the levels of mRNA for the protein are unaffected, and the induction is likely due to stabilization of human CYP2E1 from proteolytic degradation. However, in diabetic rats the sixfold induction of CYP2E1 protein is accompanied by a 10-fold increase in mRNA without involving an increase in gene transcription, suggesting stabilization of the mRNA.

Specific cytosolic receptor proteins have been indicated for some inducing agents. One of the most extensively studied is the **aryl hydrocarbon** (or **Ah**) **receptor**, which interacts with 2,3,7,8-tetrachlorodibenzo-*p*-dioxin (TCDD) and causes induction of CYP1A1, CYP1A2, and CYP1B1. Polycyclic aromatic hydrocarbons are ligands for the Ah receptor, and they produce a ligand–receptor complex that is translocated into the nucleus by the Ah receptor nuclear translocator (Arnt), where it binds to specific response elements in the upstream regulatory regions of cytochrome P450 genes. Figure 11.15 illustrates this process.

Another receptor system of P450 genes is the **pregnane X receptor** (PXR), which regulates the CYP3A gene. Pregnane refers to 21-carbon steroids such as the progesterone metabolite 5α-pregnane-3,20-dione, which is a ligand for PXR and a potent inducer of the human CYP3A4 gene. In addition to 21-carbon steroids, other structurally dissimilar compounds such as carbamazepine (a drug used for the treatment of convulsions), rifampin (an anti-tuberculosis drug), and St. John's wort (a nonprescription herbal agent used to treat mood disorders) are ligands for PXR. PXR binds the retinoid X receptor (RXR) to form a heterodimeric complex (PXR-RXR) that binds specific response elements in the 5′-flanking region of CYP3A genes. Recall that approximately 50% of prescription drugs are metabolized by this cytochrome P450 form.

Another inducer of cytochrome P450 genes is **phenobarbital**. This induction is mediated by the constitutive androstane receptor (CAR), which also forms a heterodimeric complex with RXR to interact with its specific response element in the 5′-flanking region of CYP2B genes. Two endogenous androstane steroids, androstanol and androstenol, are also ligands for CAR; when either ligand is bound to CAR, they prevent the interaction of the CAR-RXR receptor complex with the protein, SRC-1. SRC-1 is a coactivator of nuclear receptors and is required for transcriptional activation

CLINICAL CORRELATION **11.5**

Cytochrome P450 Induction: Drug–Drug Interactions and Adverse Effects

Induction of specific cytochromes P450 may decrease the therapeutic effects of drugs, because increases in liver P450 levels increase rates of metabolism and, therefore, inactivation and/or excretion of drugs. Induction of the P450 system may also have adverse effects due to the increased formation of toxic metabolites (also see Clin. Corr. 11.4).

Induction of CYP3A4 by, for example, the anti-tuberculosis drug rifampin, may greatly increase elimination of the oral contraceptive ethinyl estradiol, causing sub-effective plasma levels of the drug, which increases the risk of pregnancy. Treatment with the anticonvulsant drugs, phenytoin and carbamazepine, may also induce CYP3A4 and reduce the contraceptive potency of ethinyl estradiol. It is imperative that a woman taking ethinyl estradiol as an oral contraceptive simultaneously with a CYP3A4 inducer increase the dose of the contraceptive or use an alternative contraceptive method or drug.

The widely used herbal agent St. John's wort, which can be purchased without a prescription, also induces CYP3A4 activity, raising concerns that individuals who ingest the herb may not receive the therapeutic benefits of drugs that are CYP3A4 substrates. A reduction in plasma levels of oral contraceptives, HIV anti-protease drugs, the immunosuppressive drug cyclosporin, or certain statin drugs, used to lower cholesterol levels, can occur because each of these drugs is a substrate for CYP3A4. Because St. John's wort is considered a natural substance, rather than a drug, individuals may take this agent without notifying their physician or knowing its effects upon the efficacy of other drugs they are taking.

Source: Roby, C. A., Anderson, G. D., Kantor, E., Dryer, D. A., and Burstein, A. H. St. John's wort: Effect on CYP3A4 activity. *Clin. Pharmacol. Ther.* 67: 451, 2000; and Shader, R. I. and Oesterheld, J. R. Contraceptive effectiveness: Cytochromes and induction. *J. Clin. Psychopharmacol.* 20:119, 2000.

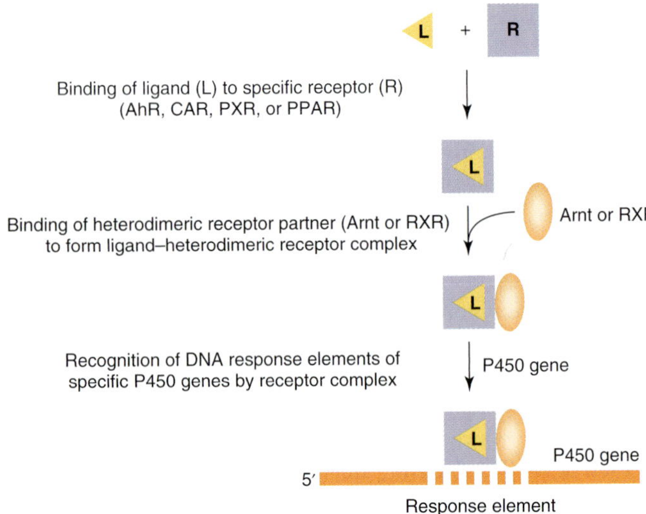

FIGURE 11.15

The interaction of a ligand with its receptor and receptor partner to form the heterodimeric receptor complex initiates induction of cytochrome P450 forms.

by CAR. Administration of phenobarbital or other phenobarbital-like inducers displaces the androstane ligands, permitting SRC-1 binding and transcription of the CYP2B gene.

Cytochrome P450 Genetic Polymorphisms

In addition to exposure to various inducing agents that may change the cytochrome P450 expression pattern in the liver and other organs, individuals may differ in their metabolism rates for particular drugs because of unique cytochrome P450 genes or alleles they possess—that is, polymorphisms. A **polymorphism** is a difference in DNA sequence found at 1% or higher in a population. These differences in DNA sequence can lead to differences in drug metabolism. Since these are genetic variations, many of them are present in specific ethnic groups, which may, therefore, experience more frequent adverse effects with certain drugs. It is estimated that 40% of the different human cytochrome P450 forms that metabolize xenobiotics are polymorphic.

Genetic polymorphisms in CYP2D6 have been studied extensively, and 65 different alleles have been identified. The population can be divided into "extensive metabolizers" or "poor metabolizers," depending on the levels of CYP2D6 expression. If there is no other way to clear a particular drug from the system, poor metabolizers may be at risk for adverse drug reactions. Alternatively, extensive metabolizers may find many drugs to be ineffective.

Many persons of Ethiopian and Saudi Arabian origin exhibit high expression of CYP2D6. This isoform metabolizes a variety of drugs, making them ineffective: Many antidepressants and neuroleptics are important examples. Conversely, prodrugs will be extensively activated: Codeine will be converted very quickly to high amounts of morphine.

In contrast, many individuals lack functional CYP2D6. These patients will be predisposed to drug toxicity caused by antidepressants or neuroleptics, but will find codeine to be ineffective due to lack of activation. Perhexiline (a calcium channel blocker) was withdrawn from the market due to the neuropathy it caused in CYP2D6-inactive patients. Even beta-blocker (e.g., propranolol) removal may be impaired in CYP2D6-deficient people. Clinical Correlation 11.6 discusses polymorphism effects.

Therapeutic Inhibition of Cytochrome P450

Because cytochromes P450 are present in almost all organisms and because they play such important roles in cellular homeostasis, selective inhibition of these enzymes is of clinical importance. Azole inhibitors (ketoconazole, itraconazole, and fluconazole),

CLINICAL CORRELATION 11.6
Genetic Polymorphisms of P450 Enzymes

Multiple alleles have been demonstrated for approximately 40% of human P450 genes. These genetic polymorphisms may produce defective P450 proteins. A description of the polymorphic CYP21 and its importance in the metabolism of an endogenous compound was given in Clinical Correlation 11.1. The absence of a cytochrome P450 activity may cause many problems due to the accumulation of therapeutic drugs that cannot be metabolized. The discovery of exaggerated hypotensive effects with the antihypertensive drug, debrisoquine, led to the characterization of individuals who inefficiently metabolized CYP2D6 substrates. Approximately 5–10% of the Caucasian, 8% of the African and African-American, and 1% of the Asian populations are deficient in catalytically active CYP2D6. In addition to debrisoquine, other drugs that are metabolized by CYP2D6 are sparteine, amitriptyline, dextromethorphan, and codeine. In the case of codeine, CYP2D6 in normal individuals catalyzes the *O*-demethylation of approximately 10% of the codeine to morphine. Individuals who lack functional CYP2D6 are unable to catalyze this reaction and so do not achieve the analgesic effects of codeine.

A genetic polymorphism associated with CYP2C19 has been demonstrated in individuals who are poor metabolizers of mephenytoin, a drug used in the treatment of epilepsy; CYP2C19 hydroxylates C4 of the S-enantiomer of mephenytoin to inactivate the physiological effect. Poor metabolizers of this drug suffer greater sedative effects at normal dosages. Approximately 14% to 22% of the Asian, 4% to 7% of the African and African-American, and 3% of the Caucasian populations lack the active form of CYP2C19.

CYP2C9 polymorphisms occur in less than 0.003% of the African or African-American, 0.08% of the Asian, and 0.36% of the Caucasian populations. The absence of a functional CYP2C9 may have serious consequences in the metabolism of S-warfarin, an orally administered drug used to inhibit blood coagulation in patients who have suffered a heart attack or a stroke to prevent the re-occurrence of life-threatening clots. Plasma levels of the drug must be maintained within a specific range because excessive amounts will cause uncontrolled bleeding and, possibly, death. Warfarin is metabolized by CYP2C9 for elimination. CYP2C9*3 is an allelic variant in which isoleucine is substituted for leucine at residue 359, resulting in substantial loss of enzymatic activity. Individuals deficient in CYP2C9 may require only 0.5–1 mg of warfarin per week in contrast to the 4–5 mg/day prescribed for an individual with normal CYP2C9. If a 5-mg dose of warfarin was taken by someone lacking functional CYP2C9, uncontrollable bleeding could result from a simple cut.

Source: Ingelman-Sundberg., M., Oscarson, M., and McLellan, R.A. Polymorphic human cytochrome P450 enzymes: An opportunity for individualized drug treatment. Trends Pharm. Sci. 20:342, 1999.

for example, are used to control yeast infections contracted by AIDS patients, cancer patients undergoing chemotherapy, and patients who are treated with immunosuppressive drugs. In yeast, CYP51 catalyzes the demethylation of sterol molecules in the synthesis of ergosterol, an important membrane lipid. In humans, a related P450 form, CYP51A1, catalyzes the demethylation of lanosterol, a reaction that is essential to cholesterol synthesis. Although the human and yeast CYP51 proteins are similar, sufficient differences exist such that azole antifungal agents are effective inhibitors of ergosterol formation. As discussed earlier, however, these agents may also induce or inhibit native human cytochromes P450, possibly leading to drug–drug interaction phenomena.

Certain tumors, such as breast tumors, are dependent on estrogen for their growth. Thus, the prevention of estrogen synthesis without inhibiting production of other steroid hormones might serve as a selective chemotherapeutic tool to reduce and eliminate tumors. CYP19A1, the cytochrome P450 that catalyzes estradiol formation from androstenedione, is an attractive target for this type of intervention. A particularly selective type of inhibitor is a mechanism-based, or suicide, inhibitor. These compounds bear strong resemblance to a substrate of an enzyme but form an irreversible inhibition product during catalytic turnover with the enzyme prosthetic group or protein. Such inhibitors would be highly specific for a particular cytochrome P450 form and would, therefore, be a good tactical approach to drug design.

It is readily apparent that without an understanding of drug metabolism and drug interactions mediated by the cytochrome P450 system, medication cannot be safely prescribed, particularly when multiple medications are required, as in older patients. It is critical to determine how therapeutic drugs are metabolized and whether cytochrome P450 forms can be inhibited by different compounds. The Food and Drug Administration's (FDA) drug approval process requires that extensive information be submitted on the metabolism of drugs before they are finally approved for general human use.

11.9 THE NITRIC OXIDE SYNTHASES: PROPERTIES AND FUNCTION

The **nitric oxide synthases** (NOSs) are heme- and flavin-containing enzymes that catalyze the synthesis of nitric oxide (NO) through two serial monooxygenase reactions analogous to those of the cytochrome P450 systems. The NOSs are functional dimers and consist of two major domains, a heme-containing oxygenase domain, which is structurally unrelated to cytochrome P450s, and a flavin-containing reductase domain, which structurally resembles the P450 reductase (Figure 11.16). Electrons are transferred from NADPH, through the flavins FAD and FMN, to the heme iron where molecular oxygen is bound and activated. The domains are connected by a calmodulin-binding site, occupation of which by calmodulin is required for electron transfer from the reductase to the oxygenase domain.

The heme moiety of the NOSs resembles that of cytochromes P450 in that it forms a thiolate ligand with a cysteine in the NOS protein, and it yields the same type of absorbance spectrum when CO is bound to the reduced, ferrous form of the heme—that is, an absorbance peak at about 450 nm. The NOS oxygenase domain is, however, structurally very different from the cytochromes P450 (compare the NOS oxygenase domain crystal structure in Figure 11.17 with that of the P450 in Figure 11.1). In addition to the heme moiety and substrate binding site, the oxygenase domains of NOSs also contain a binding site for tetrahydrobiopterin (H_4B), which apparently functions in electron transfer, and a tetra-coordinated zinc atom, which is important for structural integrity.

The general reaction catalyzed by the NOSs is sequential monooxygenation of the amino acid L-arginine to form NO and citrulline, as shown in Figure 11.18. The reaction cycle is initiated by the binding of calmodulin to "open the gate" between the

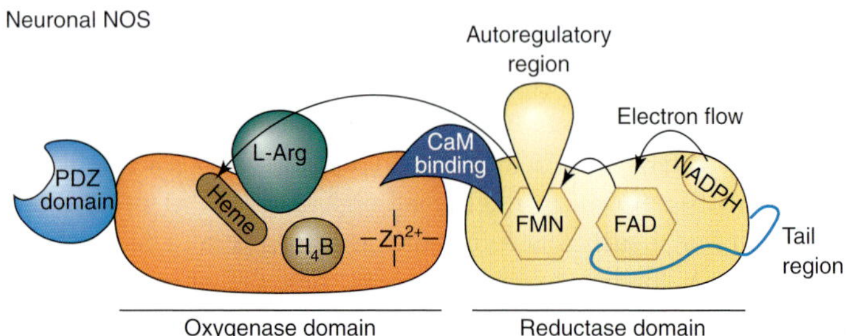

FIGURE 11.16

A modular structure of neuronal nitric oxide synthase showing approximate locations of prosthetic groups and cofactors.

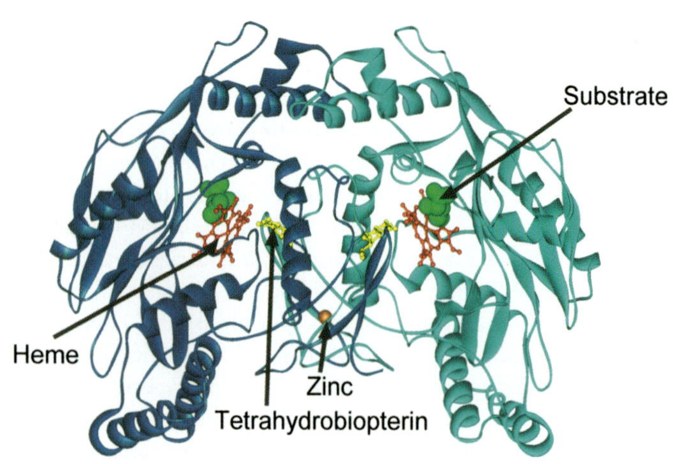

FIGURE 11.17

Model of the dimeric structure of bovine NOSIII showing the protoporphyrin IX (red) and tetrahydrobiopterin (yellow) prosthetic groups. The two monomers are depicted in light blue and dark blue, respectively. The substrate arginine (green) is in the active site of the NOSIII. A zinc atom is at the interface between the two monomers.

Generated from Protein Data Bank file 2NSE, deposited by C. S. Raman, H. Li, P. Martasek, V. Kral, B. S. Masters, and T. L. Poulos, using WebLab Viewer Lite (Molecular Simulations, Inc.)

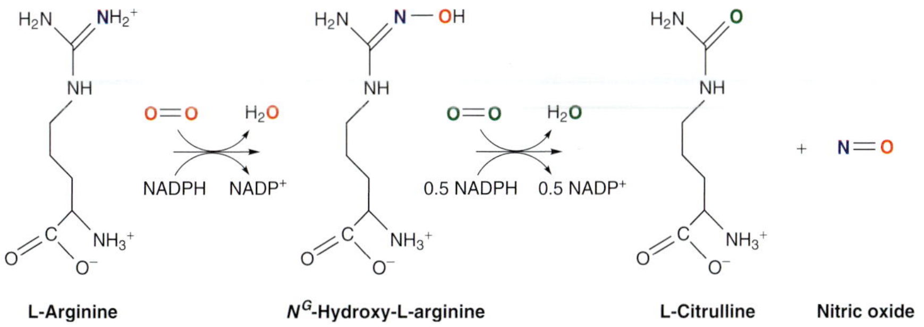

L-Arginine → ($O{=}O$, NADPH → H_2O, $NADP^+$) → N^G-Hydroxy-L-arginine → ($O{=}O$, 0.5 NADPH → H_2O, 0.5 $NADP^+$) → L-Citrulline + $N{=}O$ Nitric oxide

FIGURE 11.18

Production of nitric oxide. NOSs catalyze the formation of NO from L-arginine through two sequential monooxygenation steps that appear to be mechanistically similar to that of the microsomal cytochrome P450 system. The nitrogen atom (blue) of NO is derived from the guanidino group of the arginine side chain, and the oxygen atom (red) of NO is derived from molecular oxygen.

reductase and oxygenase domains. The electron donor is NADPH, which donates two electrons to FAD, which, in turn, reduces FMN. The FMN reduces the heme iron to its ferrous form, to which oxygen can bind. In the first step, arginine is oxidized to the stable intermediate *N*-hydroxy-L-arginine. In the second step, *N*-hydroxy-L-arginine is oxidized to NO and citrulline. Note that the source of NO is the guanidino group of the substrate, arginine. This entire process requires 2 oxygen molecules and 1.5 molecules of NADPH and, like that of the cytochromes P450, is inhibited by carbon monoxide.

The overall electron transfer chain and mechanism by which the NOSs form NO is similar to that of the cytochromes P450, but there are some significant differences in regulation of the electron transfer process. The most obvious is the requirement of the NOSs for **calmodulin** binding. Calmodulin is not bound to the enzyme at intracellular, basal calcium levels. It requires an influx of calcium to raise the calcium concentration, increasing the binding of calcium to calmodulin (see p. 478). The calcium–calmodulin complex binds to and activates NOSs. In the absence of calmodulin binding, heme reduction occurs only very slowly and does not support NO formation, and transfer between FAD and FMN is also slow. Calmodulin binding causes a conformational change in the enzyme, potentiating electron transfer between the flavins and allowing productive heme reduction. The reductase domains of the NOSs also contain several regulatory regions, shown in Figure 11.16, which are involved in transducing the signal of calmodulin binding through the enzyme.

Another difference between the NOSs and P450s is that substrate binding is not required by the NOSs to initiate the reaction cycle, and the binding of substrate does not alter the reduction potential of the heme. If substrate is not present, the electron transferred to the heme can activate molecular oxygen, which will be released as superoxide. If the second electron is transferred before the release occurs, the peroxy-form of oxygen is generated, which will dissociate as hydrogen peroxide. Thus, if substrate is not present, but the NOS is active, potentially destructive reactive oxygen species may be formed. The formation of these reduced oxygen species has been implicated in ischemia/reperfusion injury, where tissues are first depleted of oxygen and then reperfused with oxygen-rich blood. This type of injury occurs in cerebral thrombosis and in tissues removed for transplantation.

11.10 NITRIC OXIDE SYNTHASE ISOFORMS AND PHYSIOLOGICAL FUNCTIONS

There are three major NOS isoforms—neuronal (**NOSI** or **nNOS**), inducible (**NOSII** or **iNOS**), and endothelial (**NOSIII** or **eNOS**)—although variations of each of these types occur in many different organisms. Properties of these isoforms are summarized in Table 11.4.

Many of the physiological functions of NO are mediated through activation of soluble guanylate cyclase, a heterodimeric (α/β) heme-containing protein that converts GTP to cGMP. cGMP is a second messenger involved in many signal transduction cascades (see p. 518). The inactive form of soluble guanylate cyclase has a penta-coordinate heme, bound by a histidine on the β-subunit. NO binds the sixth ligand to

TABLE 11.4 Properties of the NOS Isoforms

Property	*NOSI*	*NOSII*	*NOSIII*
Molecular mass	160 kD	130 kD	135 kD
Expression	Constitutive	Inducible	Constitutive
Cell fraction	Cytoplasmic	Cytoplasmic	Membrane-bound
Dependence on calcium influx	Dependent	Independent	Dependent
Physiological action	Neurotransmission	Cytotoxicity	Vasodilation

form a hexa-coordinate heme and causes breakage of the heme-histidine bond, yielding the active, penta-coordinate nitrosyl complex of soluble guanylate cyclase (Figure 11.19). Activation of soluble guanylate cyclase causes up to a 400-fold increase in the rate of cGMP formation. The cGMP levels are regulated by a balance between guanylate cyclase activity and the phosphodiesterases (PDE), particularly PDE5, which is specific for cGMP, that hydrolyze cGMP to 5′-GMP, thus turning off the signal.

NOSI

NOSI is found primarily in skeletal muscle and in neurons of both the central and peripheral nervous systems. It is a soluble enzyme, although it is localized to the membrane via protein–protein interactions with a host of regulatory and localization proteins. It is constitutively expressed, meaning that it is not normally induced at the transcriptional level, and is activated by the influx of calcium. In the case of NOSI (and NOSIII, see below), calmodulin is not bound to the enzyme at intracellular, basal calcium levels. It requires an influx of calcium to raise the calcium concentration, allowing calmodulin to bind, thus activating NO formation. The amount of NO synthesized in the activated state is very low—that is, picomolar. The NO produced serves as a neurotransmitter in the central and peripheral nervous systems. In skeletal muscle, NO also serves as a mediator of contractile force.

In the central nervous system, NO is produced generally by NOSI in the postsynaptic neuron, but diffuses back across the synapse to the presynaptic neuron. NO synthesis is regulated by the influx of calcium through receptor-dependent channels—that is, following postsynaptic stimulation of *N*-methyl-D-aspartate (NMDA) receptors by the excitatory neurotransmitter glutamate. Guanylate cyclase is activated by NO, producing cGMP, which regulates synthesis of the neurotransmitters norepinephrine and glutamate, thus enhancing NO production (Figure 11.20). NO has been implicated in neural signaling, neurotoxicity, synaptic plasticity, learning and memory, and perception of pain.

In addition to its oxygenase and reductase domains, NOSI has an N-terminal 300-amino-acid PDZ domain, which mediates its interaction with other proteins. In the central nervous system, the PDZ domain of NOSI interacts with the postsynaptic density protein PSD-95. The NMDA receptor also interacts with PSD-95, thus bringing NOSI and the NMDA receptor into close proximity, such that NOSI is directly exposed to the calcium entering the ion channel of the activated NMDA receptor.

NO is very different from other known neurotransmitters (see p. 953). NO is synthesized on demand and cannot be stored in lipid vesicles, as it freely diffuses across membranes. The action of NO is not terminated by degradation or reuptake but is

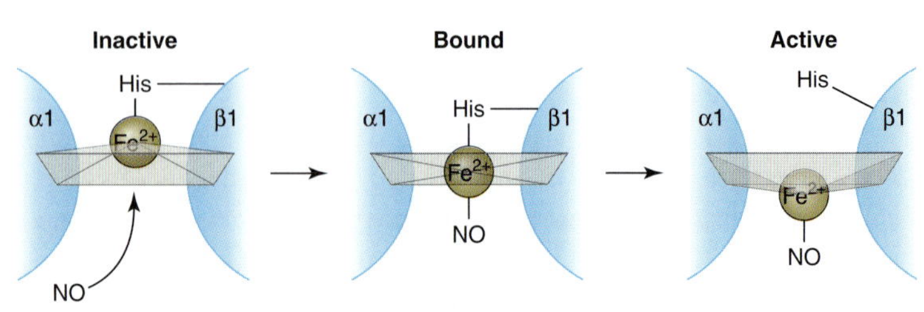

FIGURE 11.19

Activation of soluble guanylate cyclase by NO.
Redrawn based on figure in Bellamy, T. C. and Garthwaite, J. The receptor-like properties of nitric oxide-activated soluble guanylate cyclase in intact cells. *Mol. Cell. Biochem.* 230:165, 2002.

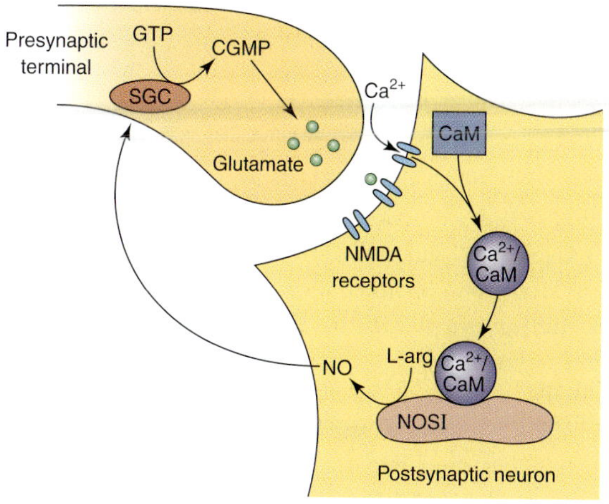

FIGURE 11.20

NO produced by NOSI in the central nervous system. In this instance, NO is produced by the postsynaptic cell and travels to the presynaptic cell where it activates sGC, creating cGMP that activates PKG. PKG phosphorylates proteins on the neurotransmitter vesicles, leading to more neurotransmitter release and potentiation of the signal.

removed by interaction with its target, which is an intracellular second messenger rather than a membrane-bound receptor.

In the peripheral nervous system, NO is produced by NOSI in myenteric neurons of the gastrointestinal, pulmonary, vascular, and urogenital systems. In the gastrointestinal system, NO is involved in gut motility and control of the pyloric sphincter. NO mediates the main bronchodilator pathway in the human pulmonary system and is involved in neural control of the vascular system, particularly important in cerebral blood flow. In the urogenital system, NO is involved in urethra and bladder control and in penile erection. During sexual excitement, NO is released from nerves adjacent to the blood vessels of the penis. Relaxation of these vessels causes blood to engorge the corpus cavernosum, producing erection. NOSI is activated in these systems by the influx of calcium through voltage-dependent calcium channels (see p. 468). It diffuses to a neighboring smooth muscle cell, causing activation of soluble guanylate cyclase. The smooth muscle relaxes via the mechanism described below for vascular smooth muscle. Clinical Correlation 11.7 describes the therapeutic manipulation of cGMP levels.

High levels of NOSI are expressed in the skeletal muscle, where it is involved in mediation of contractile force, innervation of developing muscle, and perhaps glucose uptake. Skeletal muscle relaxation by NO also occurs through the cGMP pathway via the mechanism described below for vascular smooth muscle. NOSI is targeted to the sarcolemma of the muscle due to its association with α-syntrophin, a member of the muscle dystrophin complex. NOSI is activated by influx of calcium through voltage-dependent calcium channels as well as from the sarcoplasmic reticulum.

NOSII

NOSII is found primarily in activated neutrophils and macrophages, astrocytes, and hepatocytes and is involved in the early immune response. It is a soluble enzyme and is induced at the transcriptional level by cytokines, such as interferon-γ, interleukin 1, and tumor necrosis factor-α, or endotoxins, such as lipopolysaccharides. Induction occurs over several hours via the inflammatory NFκB pathway. Calmodulin is bound to NOSII under basal physiological conditions (i.e., independent of calcium influx), so the enzyme is always activated once it is synthesized. The amount of NO synthesized by NOSII is in the nanomolar range, approximately 1000 times greater than that of either NOSI or NOSIII.

The NO produced by NOSII is a potent cytotoxin, and its principal role is to destroy pathogens engulfed by neutrophils and macrophages. NOSII controls intracellular pathogens such as parasites [i.e., plasmodia (malaria), schistosoma (schistosomiasis), leishmania (leishmaniasis), and toxoplasma (toxoplasmosis)] and microbials [i.e., bacteria and mycobacteria (tuberculosis and leprosy), fungi, and even tumor cells]. The cellular targets for the NO produced by NOSII are metal-containing heme proteins, such as cytochromes P450, and iron–sulfur proteins, such as aconitase and mitochondrial

CLINICAL CORRELATION 11.7
Mechanism of Action of Sildenafil

The levels of cyclic nucleotides (cGMP and cAMP) present in the cell are dependent not only on their formation by adenylate or guanylate cyclases, but also on their degradation by phosphodiesterases (PDEs), which hydrolyze them to 5′-nucleotides and so play a critical role in the modulation of these second messenger signaling pathways. The inhibition of PDE destruction of cyclic nucleotides would thus prolong the physiological response elicited by the cyclic nucleotides. Sildenafil (Viagra), a drug prescribed for erectile dysfunction in males, is a potent inhibitor of PDE5, which is specific for cGMP and is found primarily in the human corpus cavernosum and in vascular and gastrointestinal smooth muscle. In the penis, NO produces erection via a cGMP-dependent pathway, as discussed in the text. As the cGMP is degraded by PDE5, the vasodilatory signal attenuates and the erection subsides. Sildenafil inhibits PDE5, delaying cGMP degradation and thus prolonging erection.

Sildenafil is highly selective for PDE5 over PDE3, a cAMP-specific PDE involved in the regulation of cardiac contractility. This is important because inhibition of PDE3 increases the incidence of cardiac arrhythmias and decreases long-term mortality in heart failure patients. Interestingly, Sildenafil also inhibits, to a lesser extent, PDE6, which is localized to photoreceptors in the retina, and one of the side effects sometimes reported with Sildenafil use is a blue-green tint in the patient's vision.

Because PDE5 is present in vascular smooth muscle, Sildenafil causes a mild, clinically insignificant, reduction in blood pressure due to vasodilation, but, more importantly, it greatly exacerbates the effects of nitrates and external NO donors such as nitroglycerin, taken for angina, causing potentially fatal hypotension. These hypotensive effects occur regardless of the sequence and/or timing of the drugs. Sildenafil is therefore contraindicated in patients using nitrates in any form. In addition, Sildenafil is metabolized by both the CYP2C9 and 3A4 pathways. Inhibition of these pathways by other drugs may increase the plasma concentration of Sildenafil, intensifying side effects.

Source: Cheitlin, M. D., Hutter, A. M., Brindis, R. G., Ganz, P., Kaul, S., Russell, R. O., et al. Use of Sildenafil (Viagra) in patients with cardiovascular disease. *J. Am. Coll. Cardiol.* 33:273, 1999. Corbin, J. D., Francis, S. H., and Webb, D. J. Phosphodiesterase type 5 as a pharmacologic target in erectile dysfunction. *Urology* 60:4, 2002. Seftel, A. D. Phosphodiesterase type 5 inhibitor differentiation based on selectivity, pharmacokinetic, and efficacy profiles. *Clin. Cardiol.* 27: I14, 2004.

complexes I and II, all of which are inhibited. NO also reacts with thiol groups and tyrosines, causing nitrosation or oxidation of proteins, and with oxygen or superoxide, forming highly reactive nitrogen and oxygen intermediates (e.g., peroxynitrite). These reactive intermediates cause a variety of DNA alterations including strand cleavage and deamination. These modifications thus affect many crucial metabolic enzymes, leading to cytostatic or cytotoxic effects against these pathogens.

In addition to being produced during acute inflammation, NO is generated by NOSII during chronic inflammation. NOSII is present and active in a number of chronic inflammatory conditions including rheumatoid arthritis, Crohn's disease, and asthma, where it may potentially exacerbate some situations, although protective effects have also been reported.

NO is essential for the antimicrobial function of the macrophage, but overproduction of NO by NOSII has been implicated in septic/cytokine-induced circulatory shock. This occurs through the mechanism for smooth muscle relaxation via soluble guanylate cyclase discussed below. In this case, too much NO leads to massive systemic vasodilation and a precipitous and sometimes fatal drop in blood pressure and subsequent dysfunction of multiple organs. Hypotension in these patients is often refractory to conventional vasoconstrictor drugs, but therapeutic intervention by NOS inhibitors has shown some promise. Some clinical aspects of NO production are discussed in Clinical Correlation 11.8.

NOSIII

NOSIII is found primarily in vascular endothelial cells, which line all blood vessels, and cardiac myocytes. It is localized to the membrane due to cotranslational myristoylation and posttranslational palmitoylation of the N-terminus. Like NOSI, it is constitutively expressed and is activated by the influx of calcium. The amount of NO synthesized in the activated state is very low (i.e., picomolar), and the NO produced serves as a

CLINICAL CORRELATION 11.8
Clinical Aspects of Nitric Oxide Production

Although NO is essential in tumoricidal and bactericidal functions of macrophages, overproduction of NO (by NOSII) has been implicated in septic/cytokine-induced circulatory shock in humans through the activation of guanylate cyclase. This overproduction is responsible for profound hypotension in postoperative patients complicated by bacterial infections that produce endotoxins; this hypotension is often refractory to treatment with conventional vasoconstrictor drugs. Therapeutic intervention by NOS inhibitors is being investigated for treatment of septic shock, inflammatory diseases of the gastrointestinal tract (such as pancreatitis and ulcerative colitis), and arthritis.

NOSIII plays a critical role in maintaining a basic vasotonus in hemodynamic regulation such that a decrease in the production of NO could result in hypertension, thrombosis, or atherosclerosis. On the other hand, direct application of NO gas may be beneficial in the treatment of pulmonary hypertension. In selected patients—for example, hypoxic children and adults—inhaled NO improves arterial oxygenation and reduces pulmonary arterial hypertension. In newborn patients with hypoxic respiratory failure, it reduces the need for extracorporeal membrane oxygenation, which is both expensive and invasive. Inhaled NO has also been used to treat acute respiratory distress syndrome (ARDS), inducing pulmonary vasodilation, decreasing pulmonary arterial pressure (MPAP), and increasing cardiac index.

When the gene for NOSI has been knocked out in mice, the animals have distended stomachs due to constriction of the pyloric sphincter, which is relaxed via signaling by NOSI located in the neurons that innervate the myenteric plexus. This work has unexpectedly produced a model for infantile hypertrophic pyloric stenosis. These NOSI-deficient mice are also resistant to brain damage as a result of ischemic injury, suggesting involvement of NOSI in this injurious process. Thus, specific inhibition or induction of the isoforms of nitric oxide synthase is an ideal goal for therapeutic intervention.

Source: Hurford, W. E., Steudel, W., and Zapol, W. Clinical therapy with inhaled nitric oxide in respiratory diseases. In: Louis J. Ignarro (Ed.), *Nitric Oxide Biology and Pathobiology*. San Diego, CA: Academic Press, 2000, Chapter 56, p. 931; Kim, Y.-M., Tzeng, E., and Billiar, T. M. Role of NO and nitrogen intermediates in regulation of cell functions. In: Michael S. Goligorsky and Steven S. Gross (Eds.), *Nitric Oxide and the Kidney. Physiology and Pathophysiology*. New York: Chapman and Hall, 1997, Chapter 2, p. 22. Vallance, P. Nitric oxide: Therapeutic opportunities. *Fundam. Clin. Pharmacol.* 17:1, 2003.

vasodilator of vascular smooth muscle, both ligand-mediated and flow-dependent. The NO also serves anti-thrombotic and anti-inflammatory functions by inhibiting platelet and leukocyte adhesion and aggregation. In addition, NO also inhibits angiotensin II, a vasoconstrictor.

NO synthesis by NOSIII is activated in response to an increase in calcium following binding of ligands such as acetylcholine, bradykinin, histamine, or insulin or following shear stress, the mechanical force of blood flow on the luminal surface of the vascular endothelium. Increased blood flow velocity thus stimulates calcium release and increased NOSIII activity, leading to vasodilation.

NO diffuses out of the endothelial cell and into adjacent smooth muscle cells where it binds to and activates soluble guanylate cyclase, which makes cGMP (Figure 11.21). cGMP activates protein kinase G, a cGMP-dependent kinase that phosphorylates a variety of channels, including L-type calcium channels, and receptors, all leading to inhibition of calcium influx into the smooth muscle cell. Vasoconstriction in vascular smooth muscle requires the influx of calcium, which binds to calmodulin that, in turn, activates muscle light chain kinase (MLCK). MLCK then phosphorylates myosin light chain (MLC), leading to cross-bridge formation between the myosin heads and the actin filament, resulting in contraction (see p. 985). Inhibition of calcium influx leads to decreased calmodulin stimulation of MLCK and decreased phosphorylation of MLC, causing diminished development of smooth muscle tension and, thus, vasodilation. Increased cGMP also directly causes MLC dephosphorylation by activation of MLC phosphatase. As with NOSI, cGMP levels are regulated by a balance between guanylate cyclase activity and the phosphodiesterases, which hydrolyze cGMP to 5′-GMP and thus turn off protein kinase G activity and vasodilation. Clinical Correlation 11.7 discusses NO-mediated vasodilation by nitroglycerin, and Clinical Correlation 11.9 discusses the history of the use of nitroglycerin.

NOSIII has a unique N-terminal sequence that undergoes myristoylation at one site and palmitoylation at two other sites within this sequence, localizing the enzyme

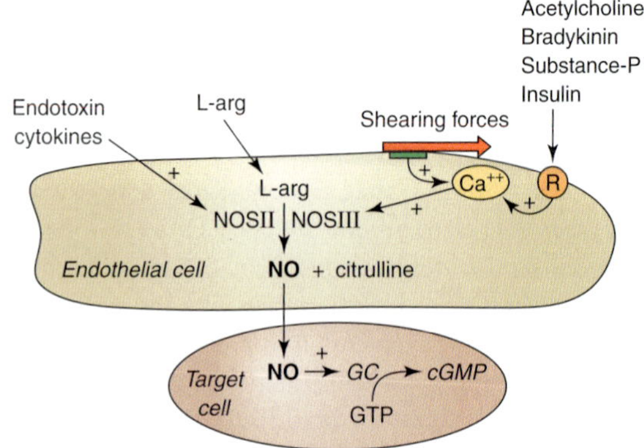

FIGURE 11.21

NO produced by NOSIII in the endothelial cell. Calcium concentration in the endothelial cell rises due to ligand binding or shear stress leading to activation of NO production by NOSIII. The NO diffuses across the membrane of the endothelial cell to a smooth muscle cell, where it activates sGC.
Redrawn based on figure by Klabunde, R. E. *Cardiovascular Physiology Concepts.* HTTP://cvphysiogy.com/blood%20Flow/BFO11.htm.

CLINICAL CORRELATION 11.9
History of Nitroglycerin

Nitroglycerin, more correctly called glyceryl trinitrate, was first synthesized by Ascanio Sobrero in 1846. Because nitroglycerin was extremely explosive and many fatal accidents occurred, Alfred Nobel developed a procedure, in 1864, to stabilize nitroglycerin by mixing it with diatomaceous earth. The fortune he made from dynamite is the bases of the Nobel Prizes, awarded yearly in Sweden for outstanding contributions to the fields of physics, chemistry, physiology or medicine, literature, economics, and world peace. Shortly thereafter, in 1867, Sir Thomas Lauder Brunton first reported the use of inhaled amyl nitrate in his patients with chest pains and, in 1879, William Murrell published an article describing the use of the longer-lasting and more convenient compound nitroglycerin to better alleviate the pain of angina pectoris. Nitroglycerin worked by dilating the arteries of the heart, thus allowing more oxygenation of the tissue. Ironically, Murrell died in 1912 from untreated heart failure, having never taken nitroglycerin, and Nobel refused to take it when prescribed by his doctor for heart disease.

It was not until 1987, over a century after Murrell's publication, that nitric oxide gas released from nitroglycerin metabolism was postulated to be the active agent in the dilation of the coronary blood vessels. Even today the actual mechanism of nitric oxide release from nitroglycerin is unclear.

Source: Marsh, N. and Marsh, A. A short history of nitroglycerine and nitric oxide in pharmacology and physiology. *Clin. Exp. Pharmacol. Physiol.* 27:313, 2000; and Napoli, C. and Ignarro, L. J. Nitric oxide-releasing drugs. *Annu. Rev. Pharmacol. Toxicol.* 43:97, 2003. Wells, W. From explosives to the gas that heals. Written for "Beyond Discovery: The Path from Research to Human Benefit" 2002. http://www.beyonddiscovery.org/content/view.article.asp?a=318.

to the membrane where it is targeted to plasmalemma caveolae. The caveolae are small invaginations of the plasma membrane, characterized by the presence of proteins (called caveolins) that serve to organize and attach signaling molecules such as receptors, G-proteins, and NOS to plasma membranes. At low cytoplasmic calcium concentrations, caveolin-1 binds to and inhibits NOSIII, maintaining it in an inactive state. Upon calcium influx, calmodulin competitively displaces the caveolin from NOSIII, resulting in activation of NOS synthesis.

Also localized to the caveolae are several other proteins important for NOSIII activity. The cationic amino acid transporter CAT-1 present in caveolae is involved in the uptake of L-arginine, thus ensuring a supply of substrate for NO synthesis. NOSIII interacts with porin, a voltage-dependent anion/cation channel, which localizes NOSIII near a source of calcium influx. The bradykinin B2 receptor is also present, and it binds to and inactivates NOSIII. Upon stimulation with bradykinin, this complex dissociates and NOSIII becomes activated.

Localization to the caveolae is an important regulator of NOSIII activity; if palmitoylation is inhibited, NOSIII is not found in the caveolae, and NO synthesis does not occur in those cells. Palmitoylation is a reversible process that is influenced by some agonists and is essential for membrane localization; cellular relocalization, resulting in deactivation, may occur upon depalmitoylation.

All of the NOSs produce NO, which serves such diverse functions as neurotransmission, vasodilation, muscle relaxation, and cytotoxicity. The consequences of NO can be beneficial, as in maintenance of vascular tone, acquisition of memory and learning, and protection against pathogen invasion, as well as detrimental, as with septic shock or chronic inflammatory and mutagenic responses. The net effect of the NO produced is dependent on where it is being produced, regulation of its production, how much is being produced, and which other reactive species are present.

BIBLIOGRAPHY

General References on Drug Metabolism and Cytochrome P450

Coon, M. J. Multiple oxidants and multiple mechanisms in cytochrome P450 catalysis. *Biochem. Biophys. Res. Commun.* 312:163, 2003.

Cupp, M. and Tracy, T. S. Cytochrome P450: New nomenclature and clinical implications. *Am. Family Phys.* 57:107, 1998. http://www.aafp.org/afp/980101ap/cupp.html

Estabrook, R. W., Cooper, D. Y., and Rosenthal, O. The light reversible carbon monoxide inhibition of the steroid C21-hydroxylase system of the adrenal cortex. *Biochem. Zeit.* 338:741, 1963.

Gibson, G. G. and Skett, P. *Introduction to Drug Metabolism*, 2nd ed. London: Academic Press, 1994.

Lewis, D. F. V. *Guide to Cytochromes P450 Structure and Function.* New York: Taylor and Francis, 2001.

Lewis, D. F. 57 varieties: The human cytochromes P450. *Pharmacogenomics* 5:305. 2004.

Nelson, D. R. Cytochrome P450 and individuality of species. *Arch. Biochem. Biophys.* 369:1, 1999.

Nelson, D. R. website: http://drnelson.utmem.edu/CytochromeP450.html

Substrate Specificity of Cytochromes P450: Physiological Functions

Cunningham, F. G., MacDonald, P. C., Gant, N. F., Leveno, K. J., Gilstrap, L. C., Hankins, G. D. V., and Clark, S. L. The placental hormones. In: *Williams Obstetrics*, 20th ed. Stamford, CT: Appleton and Lange, 1997, Chapter 6, p. 125.

Donohoue, P. A., Parker, K., and Migeon, C. J. Congenital adrenal hyperplasia. In: C. R. Scriver, et al. (Eds.), *The Metabolic and Molecular Bases of Inherited Disease*, Vol. II, 7th ed. New York: McGraw-Hill, 2001, Chapter 94, p. 4077.

Masters, B. S. S., Muerhoff, A. S., and Okita, R. T. Enzymology of extrahepatic cytochromes P450. In: F. P. Guengerich (Ed.), *Mammalian Cytochromes P450*. Boca Raton, FL: CRC Press, 1987, Chapter 3, p. 107.

New, M. I. Inborn errors of adrenal steroidogenesis. *Mol. Cell. Endocrinol.* 211:75, 2003.

Rendic, S. Summary of information on human CYP enzymes: Human P450 metabolism data. *Drug. Met. Rev.* 34:83, 2002.

Stratakis, C. A., and Bossis, I. Genetics of the adrenal gland. *Rev Endocr. Metab. Disord.* 5:53, 2004.

Cytochrome P450 Induction and Inhibition

Cascorbi, I. Pharmacogenetics of cytochrome P4502D6: Genetic background and clinical implication. *Eur. J. Clin. Invest.* 33:17, 2003.

Denison, M. S. and Whitlock, J. P. Xenobiotic-inducible transcription of cytochrome P450 genes. *J. Biol. Chem.* 270:18175, 1995.

Guengerich F. P. Cytochromes P450, drugs, and diseases. *Mol. Interv.* 3:194, 2003.

Handschin, C. and Meyer, U. A. Induction of drug metabolism: The role of nuclear receptors. *Pharmacol Rev.* 55:649. 2003.

Ingelman-Sundberg, M. Pharmacogenetics of cytochrome P450 and its applications in drug therapy: the past, present and future. *Trends Pharmacol. Sci.* 25:193, 2004.

Kalow, W. *Pharmacogenetics of Drug Metabolism.* New York: Pergamon Press, 1992.

Kalow, W. Pharmacogenetics in biological perspective. *Pharmacol. Rev.* 49:369, 1997.

Lin, J. H. and Lu, A. Y. H. Role of pharmacokinetics and metabolism in drug discovery and development. *Pharmacol. Rev.* 49:403, 1997.

Meyer, U. A. and Zanger, U. M. Molecular mechanisms of genetic polymorphisms of drug metabolism. *Annu. Rev. Pharmacol. Toxicol.* 37:269, 1997.

Schwarz, U. I. Clinical relevance of genetic polymorphisms in the human CYP2C9 gene. *Eur. J. Clin. Invest.* 33:23, 2003.

Shimada, T. and Fujii-Kuriyama, Y. Metabolic activation of polycyclic aromatic hydrocarbons to carcinogens by cytochromes P450 1A1 and 1B1. *Cancer Sci.* 95:1, 2004.

Thummel, K. E. and Wilkinson, G. R. *In vitro* and *in vivo* drug interactions involving human CYP3A. *Annu. Rev. Pharmacol. Toxicol.* 38:389, 1998.

Waxman, D. J. P450 gene induction by structurally diverse xenochemicals: Central role of nuclear receptors CAR, RXR, PPAR. *Arch Biochem. Biophys.* 369:11, 1999.

Zanger, U. M., Raimundo, S., and Eichelbaum, M. Cytochrome P450 2D6: Overview and update on pharmacology, genetics, biochemistry. *Naunyn Schmiedebergs Arch. Pharmacol.* 369:23, 2004.

Zhou, S., Chan, E., Pan, S. Q., Huang, M., and Lee, E. J. Pharmacokinetic interactions of drugs with St. John's wort. *J. Psychopharmacol.* 18(2):262, 2004.

Biochemistry and Physiology of Nitric Oxide Formation

Alderton, W. K., Cooper, C. E., and Knowles, R. G. Nitric oxide synthases: Structure, function and inhibition. *Biochem. J.* 357:593, 2001.

Bredt, D. S. Endogenous nitric oxide synthesis: biological functions and pathophysiology. *Free Radical Res.* 31:577, 1999.

Bredt, D. S. Nitric oxide signaling specificity—The heart of the problem. *J. Cell Sci.* 116:9, 2003.

Esplugues, J. V. NO as a signalling molecule in the nervous system. *Br. J. Pharmacol.* 135:1079, 2002.

Garthwaite, J. and Boulton, C. L. Nitric oxide signalling in the central nervous system. *Annu. Rev. Physiol.* 57:683, 1995.

Griffith, O. W. and Stuehr, D. J. Nitric oxide synthases—Properties and catalytic mechanism. *Annu. Rev. Physiol.* 57:707, 1995.

Huang, P. L. Endothelial nitric oxide synthase and endothelial dysfunction. *Curr. Hypertens. Rep.* 5:473, 2003.

Ignarro, L. J., Buga, G. M., Wood, K. S., Byrns, R. E., and Chaudhuri, G. Endothelium-derived relaxing factor produced and released from artery and vein is nitric oxide. *Proc. Natl. Acad. Sci. USA* 84:9265, 1987.

Khan, M. T. and Furchgott, R. F. Additional evidence that endothelium-derived relaxing factor is nitric oxide. In: M. J. Rand and C. Raper (Eds.), *Pharmacology*. Amsterdam: Elsevier, 1987, p. 341.

Kleinert, H., Schwarz, P. M., and Forstermann, U. Regulation of the expression of inducible nitric oxide synthase. *Biol. Chem.* 384:1343, 2003.

Liaudet, L., Soriano, F. G., Yaffe, M. B., and Fink, M. P. Biology of nitric oxide signaling. *Crit. Care Med.* 28:N37, 2000.

Masters, B. S. S. Nitric oxide synthases: Why so complex? *Annu. Rev. Nutr.* 14:131, 1994.

Palmer, R. M. J., Ferrige, A. G., and Moncada, S. Nitric oxide release accounts for the biological activity of endothelium-derived relaxing factor. *Nature* 327:524, 1987.

Raman, C. S., Martasek, P., and Masters, B. S. Structural themes determining function in nitric oxide synthases. The Porphyrin Handbook, Vol. 4. Academic Press: San Diego, 2000, p. 293.

Roman, L. J., Martasek, P., and Masters, B. S. Intrinsic and extrinsic modulation of nitric oxide synthase activity. *Chem. Rev.* 102:1179, 2002.

Vallance, P. Nitric oxide: Therapeutic opportunities. *Fund. Clin. Pharmacol.* 17:1, 2003.

Walford, G. and Loscalzo, J. Nitric oxide in vascular biology. *J. Thromb. Haemost.* 1:2112, 2003.

Website: http://metallo.scripps.edu/PROMISE/NOS.html

QUESTIONS | CAROL N. ANGSTADT

1. All of the following are correct about a molecule designated as a cytochrome P450 *except*:
   A. it contains a heme as a prosthetic group.
   B. it catalyzes the hydroxylation of a hydrophobic substrate.
   C. it may accept electrons from a substance such as NADPH.
   D. it undergoes a change in the heme iron upon binding a substrate.
   E. it comes from the same gene family as all other cytochromes P450.

2. Flavoproteins are usually intermediates in the transfer of electrons from NADPH to cytochrome P450 because:
   A. NADPH cannot enter the membrane.
   B. flavoproteins can accept two electrons from NADPH and donate them one at a time to cytochrome P450.
   C. they have a more negative reduction potential than NADPH so accept electrons more readily.
   D. as proteins, they can bind to cytochrome P450 while the nonprotein NADPH cannot.
   E. they contain iron–sulfur centers.

3. In the conversion of cholesterol to steroid hormones in the adrenal gland:
   A. all of the cytochrome P450 oxidations occur in the endoplasmic reticulum.
   B. all of the cytochrome P450 oxidations occur in the mitochondria.
   C. side-chain cleavage of cholesterol to pregnenolone is one of the cytochrome P450 systems that uses adrenodoxin reductase.
   D. cytochrome P450 is necessary for the formation of aldosterone and cortisol but not for the formation of the androgens and estrogens.
   E. aromatization of the first ring of the steroid does not use cytochrome P450 because it involves removal of a methyl group, not a hydroxylation.

4. Benzopyrene, a xenobiotic produced by combustion of a variety of substances:
   A. induces the synthesis of cytochrome P450.
   B. undergoes epoxidation by a cytochrome P450.
   C. is converted to a potent carcinogen in animals by cytochrome P450.
   D. would be rendered more water-soluble after the action of cytochrome P450.
   E. all of the above.

5. Genetic polymorphism in genes for cytochromes P450:
   A. could cause an individual to metabolize poorly certain drugs.
   B. could cause an individual to metabolize certain drugs more rapidly than normal.
   C. could cause a specific ethnic group to experience more effects with certain drugs.
   D. all of the above are correct.
   E. none of the above is correct.

6. In the presence of active nitric oxide synthase but absence of substrate:
   A. electrons can not be transferred to heme.
   B. calmodulin cannot bind Ca^{2+}.
   C. the oxygenase and reductase domains dissociate.
   D. cGMP production is increased.
   E. superoxide and/or hydrogen peroxide will be formed.

Questions 7 and 8: Because herbal remedies are not prescription drugs, many patients fail to inform their physicians that they are using such products. A patient was told by his physician that his blood cholesterol levels were not responding to the usually effective statin drug he was taking. The patient was also taking St. John's wort, an herbal agent, to improve his mood. St. John's wort induces a cytochrome P450 (CYP3A4) activity. Statins are one of the many drugs that are metabolized by CYP3A4.

7. The induction of cytochromes P450:
   A. occurs only by exogenous compounds.
   B. occurs only at the transcriptional level.
   C. necessarily results from increased transcription of the appropriate mRNA.
   D. necessitates the formation of an inducer-receptor protein complex.
   E. may occur by posttranscriptional processes.

8. Statins could be considered xenobiotics (exogenous substances metabolized by the body). Many xenobiotics are oxidized by cytochromes P450 in order to:
   A. make them carcinogenic.
   B. increase their solubility in an aqueous environment.
   C. enhance their deposition in adipose tissue.
   D. increase their pharmacological activity.
   E. all of the above.

Questions 9 and 10: Some patients show profound hypotension after abdominal surgery complicated by bacterial infections that produce endotoxins. Such hypotension is often refractory to treatment with conventional vasoconstrictor drugs. This hypotension may be caused by an overproduction of nitric oxide by the induced form of nitric oxide synthase (NOS). Administration of NOS inhibitors specific to this form might be an appropriate treatment for such patients.

9. Nitric oxide:
   A. is formed spontaneously by a reduction of NO_2.
   B. is synthesized only in macrophages.
   C. is synthesized from arginine.
   D. acts as a potent vasoconstrictor.
   E. has three isoforms.

10. Nitric oxide synthase:
   A. catalyzes a dioxygenase reaction.
   B. is similar to cytochromes P450 in binding zinc and tetrahydrobiopterin.
   C. accepts electrons from NADH.
   D. uses a flow of electrons from NADPH to FAD to FMN to heme-iron.
   E. is inhibited by Ca^{+2}.

Questions 11 and 12: A female infant was referred to a physician because of abnormalities in her genitalia. She was subsequently diagnosed with congenital adrenal hyperplasia (CAH), a deficiency in CYP21A2, a 21-hydroxylase of endoplasmic reticulum. The disease results in decreased cortisol, increased ACTH, increased androgenic hormones and problems related to salt imbalance. NADPH-cytochrome P450 reductase is the electron donor for the CYP21A2 reaction.

11. NADPH-cytochrome P450 reductase:
   A. uses both FAD and FMN as prosthetic groups.
   B. binds to cytochrome P450 by strong hydrophobic interactions.
   C. requires an iron–sulfur center for activity.
   D. always passes its electrons to cytochrome b_5.
   E. can use NADH as readily as NADPH.

12. Reactions after the 21-hydroxylation leading to cortisol and androgens occur in mitochondria which require NADPH-adrenodoxin reductase and adrenodoxin: NADPH-adrenodoxin reductase:
   A. contains both FAD and FMN.
   B. passes its electrons to a protein with iron-sulfur centers.
   C. is an integral protein of the membrane.
   D. reacts directly with cytochrome P450.
   E. reacts directly with cytochrome b_5.

Problems

13. Phenobarbital is a potent inducer of cytochrome P450. Warfarin, an anticoagulant, is a substrate for cytochrome P450 so the drug is metabolized more rapidly than normal. If phenobarbital is given to a patient, with no change in warfarin dosage, what would happen? What would happen if the warfarin dosage were adjusted for a proper response, and then phenobarbital withdrawn without adjusting the warfarin dosage?

14. Acetaminophen is primarily metabolized by sulfation and glucuronidation and excreted. It can also be metabolized by a cytochrome P450 (CYP2E1) to the highly reactive NAPQI, which can damage liver. Alcohol is an inducer and substrate for CYP2E1. If acetaminophen is taken several hours after drinking alcohol, liver toxicity is greatly enhanced. If acetaminophen and alcohol are consumed together, enhancement of toxicity is not observed. Why?

ANSWERS

1. **E** Several gene families are known. A: All cytochromes are heme proteins. B: The types of substrates are hydrophobic. It is classified as a monooxygenase. C: This is the usual electron donor. D: The change from hexa to penta coordinated gives the compound a more positive reduction potential.
2. **B** Heme can accept only one electron at a time while NADPH always donates two at a time. A: NADPH passes only electrons; it does not have to enter the membrane. C: If this were true, the flow of electrons would not occur in the way it does. D: Protein–protein binding is not known to play a role here. E: Iron–sulfur centers play a role in some, but not all, systems.
3. **C** This is a mitochondrial process. A, B: Hormone synthesis involves a series of reactions in mitochondria and endoplasmic reticulum. D, E: Removal of side chains frequently begins with oxidation reactions.
4. **E** A: Xenobiotics frequently induce synthesis of something that will enhance their own metabolism. B, C: Epoxidation is the first step in the conversion of this compound to one that is carcinogenic. D: Benzopyrene is highly hydrophobic; introducing oxygens increases water solubility.
5. **D** A: This occurs if the allelic variants code for inactive or less active proteins. B: If there is gene duplication, an individual may have larger amounts of active protein. C: There may be greater expression of specific allelic variants in specific ethnic populations.
6. **E** O_2 is activated leading to these toxic products. A: Superoxide results from the transfer. B, C: These do not happen. D: NO increases cGMP, and NO is not formed in absence of substrate.
7. **E** There may be a stabilization of mRNA or decrease in the degradation of the protein. A: Both endogenous and exogenous substances can induce cytochromes P450. B, C: Transcriptional modification is only one of the mechanisms of induction. D: This has been shown with induction by some compounds but not all.
8. **B** Xenobiotics oxidized by cytochrome P450 are usually highly lipophilic but must be excreted in the aqueous urine or bile. A: This may happen but is certainly not the purpose. C: They do that prior to oxidation. D: Oxidation tends to reduce pharmacological activity.
9. **C** The other product is citrulline. A: It is formed from arginine. B: One of the isoforms of NO synthase has been found in macrophages but neuronal and endothelial isoforms also exist. D: Nitric oxide is a vasodilator. E: Three isoforms of nitric oxide synthase have been identified but nitric oxide is a specific compound.
10. **D** This is one of two mammalian enzymes that use both FAD and FMN. A: The reaction is a monooxygenation. B: Cytochromes do not bind these. C: The donor is NADPH. E: The system requires Ca^{2+}-calmodulin, at least the neuronal and endothelial isoforms.
11. **A** This enzyme is one of two mammalian proteins known to do so. B: The binding is electrostatic. C: Some reductases do so, but not this one. D: Only certain reactions catalyzed by the enzyme do. E: There are NADH-dependent reductases but they are different enzymes.
12. **B** The iron–sulfur protein is adrenodoxin. A: It has only FAD. C: It is weakly associated with the membrane. D, E: Adrenodoxin acts as an electron shuttle between the reductase and heme of cytochrome P450.

13. If warfarin is metabolized and cleared more rapidly by cytochrome P450, its therapeutic efficiency is decreased. At the same dosage, it will be less effective as an anticoagulant. If the phenobarbital is withdrawn, the cytochrome P450 levels will decrease with time to the uninduced level. With no change in the warfarin level, eventually there will be an increased possibility of hemorrhaging.
14. The toxic effects of acetaminophen are due to the production of NAPQI by CYP2E1. In the presence of alcohol, CYP2E1 levels are increased. When acetaminophen and alcohol are consumed together, they compete for the enzyme and the metabolism of acetaminophen by this pathway is slowed. If alcohol is consumed alone, it induces the enzyme. When acetaminophen is later consumed, its metabolism by this pathway is more rapid (there is no competition with alcohol) and more NAPQI is formed. The level of acetaminophen consumed is also a factor. Small amounts may be able to be handled by the normal conjugation pathways in a nontoxic fashion.

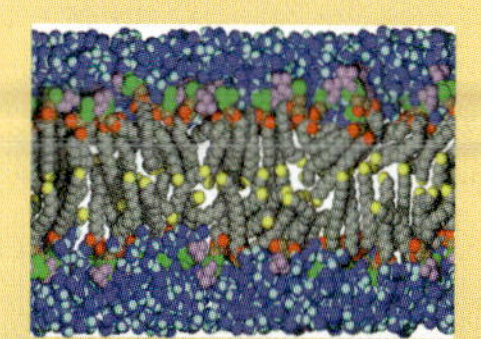

12

BIOLOGICAL MEMBRANES: STRUCTURE AND MEMBRANE TRANSPORT

Thomas M. Devlin

Textbook of Biochemistry With Clinical Correlations, Sixth Edition, Edited by Thomas M. Devlin

12.1 OVERVIEW

Membranes of eukaryotic or prokaryotic cells have the same classes of chemical components, are similar in structural organization, and have many properties in common. There are differences, however, in specific lipid, protein, and carbohydrate components. Membranes have a trilaminar appearance when viewed by electron microscopy (Figure 12.1), with two dark bands on each side of a light band. Most mammalian membranes have a width of 7–10 nm and a dense inner layer often thicker than an outer dense layer. Intracellular membranes are usually thinner than plasma membranes. There is a chemical asymmetry of membranes. The development of sophisticated techniques has revealed that at the molecular level, membrane surfaces are not smooth but dotted with protruding globular masses.

Membranes are dynamic structures that permit cells and subcellular structures to adjust their shape and to change position. The lipid and protein components are ideally suited for this dynamic role. Membranes are an organized sea of lipid in a fluid state, in which both proteins and lipids are able to diffuse and interact.

Cellular membranes control the composition of space that they enclose by excluding a variety of ions and molecules and by the presence of selective **transport systems** for transmembrane movement of substances. By controlling translocation of substrates, cofactors, and ions, the concentrations of such substances in cellular compartments are modulated, thereby exerting an influence on metabolic pathways. Hormones, growth regulators, and metabolic regulators bind to specific **protein receptors** on plasma membranes (see p. 494), and information of their presence is transmitted from the membrane receptors to the appropriate metabolic pathway by intracellular intermediates, termed "second messengers." Plasma membranes have a role in cell–cell recognition, in maintenance of cell shape, and in cell locomotion.

The discussion that follows is primarily of mammalian cells, but the chemistry and activities described are applicable to all biological membranes.

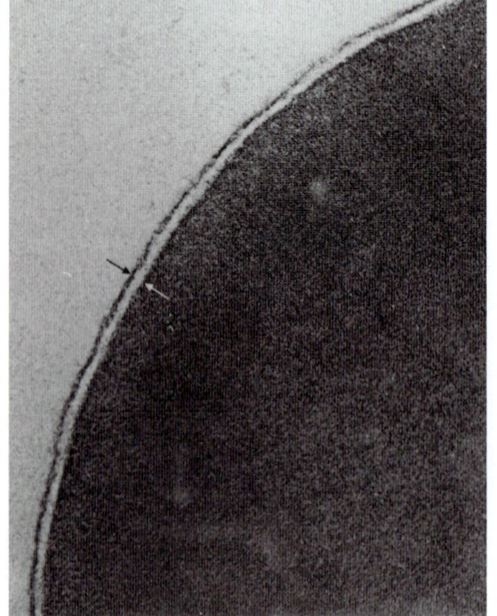

FIGURE 12.1
Electron micrograph of the erythrocyte plasma membrane showing the trilaminar appearance.
A clear space separates the two electron-dense lines. Electron microscopy has demonstrated that the inner dense line is frequently thicker than the outer line. Magnification about 150,000×.
Courtesy of J. D. Robertson, Duke University, Durham, North Carolina.

12.2 CHEMICAL COMPOSITION OF MEMBRANES

Lipids and proteins are the two major components of all membranes. Their concentrations vary greatly between membranes (Figure 12.2). Protein ranges from about 20% in the myelin sheath to over 70% in the inner membrane of mitochondria. Proteins are responsible for the function of membranes; thus, membranes from different cells have very specific complements of proteins. Integral membrane proteins represent about 25% of sequenced genes. Membranes contain a small amount of various polysaccharides as glycoprotein and glycolipid, but no free carbohydrate.

Lipids are Major Components of Membranes

The three major lipid components of eukaryotic membranes are glycerophospholipids, sphingolipids, and cholesterol. Glycerophospholipids and sphingomyelin, a sphingolipid that contains phosphate, are classified as phospholipids. Bacteria and blue-green algae contain glycerolipids in which a carbohydrate is attached directly to the glycerol. Some membranes contain small quantities of other lipids, such as triacylglycerol and diol derivatives (see p. 664 for structures), as well as lipid covalently linked to protein.

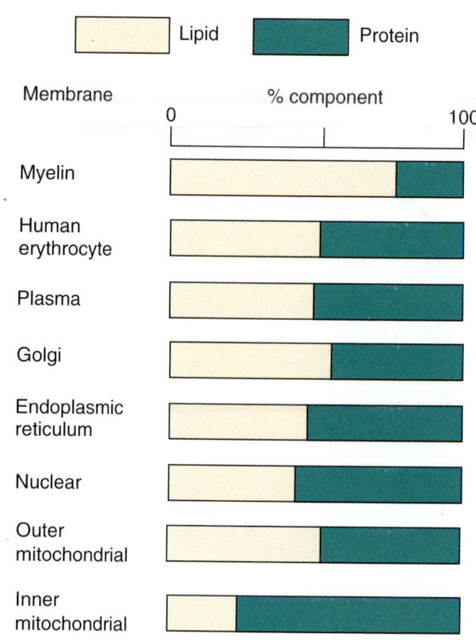

FIGURE 12.2

Percentage of lipid and protein in various cellular membranes. Values are for rat liver, except for the myelin and human erythrocyte plasma membrane. Values for liver from other species, including human, indicate a similar pattern.

Glycerophospholipids are the Most Abundant Lipids of Membranes

Glycerophospholipids (phosphoglycerides) have a glycerol molecule with a phosphate esterified at the α-carbon (Figure 12.3) and two long-chain fatty acids esterified to the remaining carbon atoms (Figure 12.4). Glycerol does not contain an asymmetric carbon, but the α-carbon atoms are not stereochemically identical. Esterification of a phosphate to an α carbon makes the molecule asymmetric. Naturally occurring glycerophospholipids are designated by the stereospecific numbering system (*sn*) (Figure 12.3) discussed on p. 696.

1,2-Diacylglycerol 3-phosphate, or **phosphatidic acid**, is the parent compound of the glycerophospholipids. **A phosphodiester bridge to glycerol attaches choline, ethanolamine, serine, glycerol, and inositol** (Figure 12.5). **Phosphatidylethanolamine** (**ethanolamine glycerophospholipids** or **cephalin**) and phosphatidylcholine (**choline glycerophospholipid** or **lecithin**) are the most common

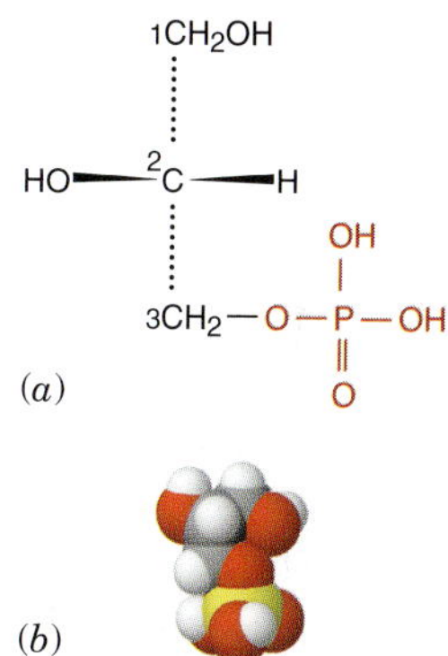

FIGURE 12.3

L-Glycerol 3-phosphate. (*a*) Stereochemical configuration of L-glycerol 3-phosphate (*sn*-glycerol 3-phosphate). The H and OH attached to C2 are above, and C1 and C3 are below the plane of the page. (*b*) Space-filling model of L-glycerol 3-phosphate. Courtesy of Dr. Daniel Predecki, Shippensburg University, Shippensburg, PA. Color scheme: H, white; C, gray; O, red; P, yellow.

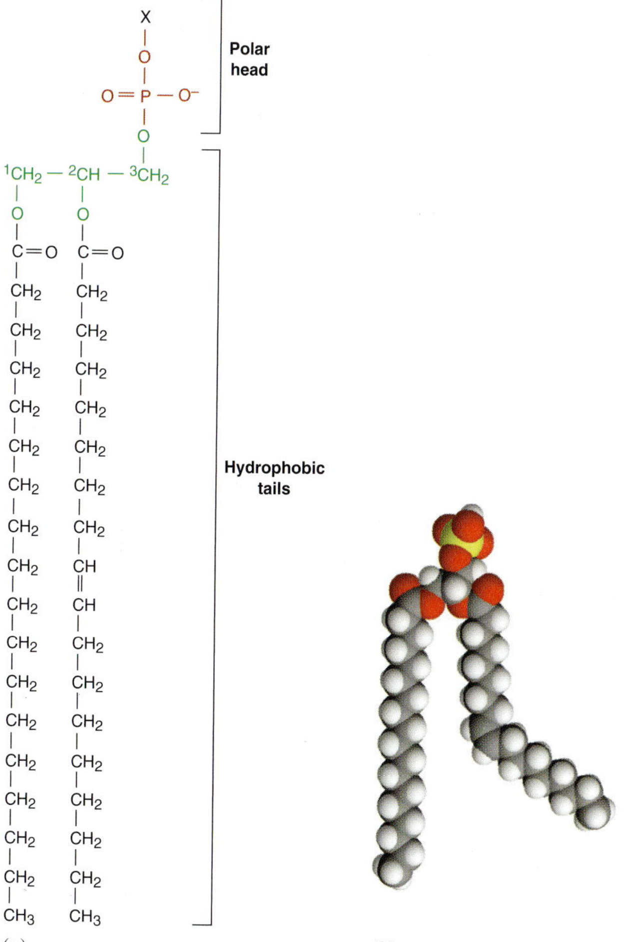

FIGURE 12.4

Structure of glycerophospholipid. (*a*) Long-chain fatty acids are esterified at C1 and C2 of the L-glycerol 3-phosphate. X can be a H (phosphatidic acid) or one of several alcohols presented in Figure 12.5. (*b*) Space filling model.
Courtesy of Dr. Daniel Predecki, Shippensburg University, Shippensburg, PA. See Figure 12.3 for color scheme.

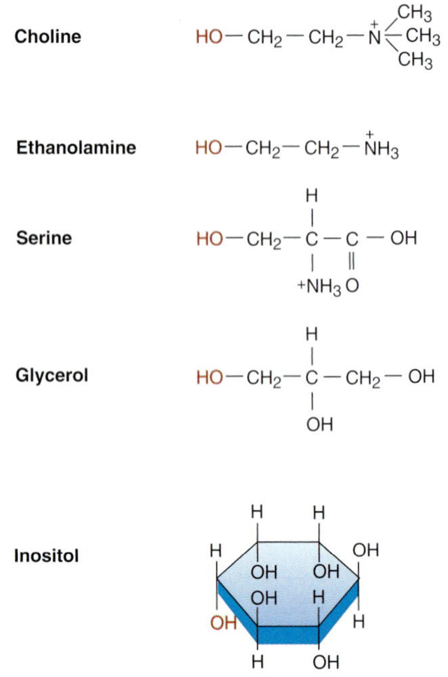

Choline $HO-CH_2-CH_2-\overset{+}{N}(CH_3)_3$

Ethanolamine $HO-CH_2-CH_2-\overset{+}{N}H_3$

Serine $HO-CH_2-CH(^{+}NH_3)-C(=O)-OH$

Glycerol $HO-CH_2-CH(OH)-CH_2-OH$

Inositol

FIGURE 12.5

Structures of major alcohols esterified to phosphatidic acid to form the glycerophospholipid.

Long aliphatic chains

Phosphatidylcholine

Long aliphatic chains

Phosphatidylethanolamine

(*a*) (*b*)

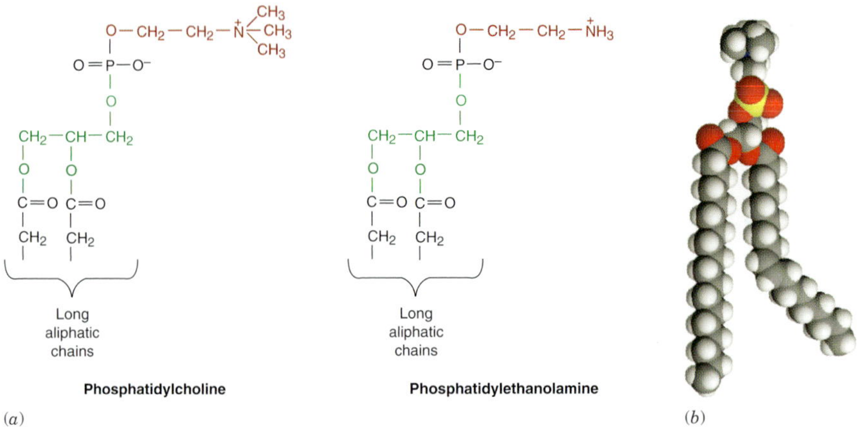

FIGURE 12.6

Structures of the two most common glycerophospholipids. (*a*) Phosphatidylcholine and phosphatidylethanolamine. (*b*) Space-filling model of phosphatidylcholine.
Courtesy of Dr. Daniel Predecki, Shippensburg University, Shippensburg, PA. See Figure 12.3 for color scheme.

glycerophospholipids in membranes (Figure 12.6). **Phosphatidylglycerol phosphoglyceride** (Figure 12.7) (**diphosphatidylglycerol** or **cardiolipin**) contains two phosphatidic acids linked by a glycerol and is found nearly exclusively in mitochondrial inner membranes and bacterial membranes.

Inositol, a hexahydroxy alcohol, is esterified to phosphate in **phosphatidylinositol** (Figure 12.8). The phosphatidylinositol 4-phosphate predominates in Golgi and the 4,5-bisphosphate derivative predominates in the plasma membrane; the latter is the source of inositol trisphosphate and diacylglycerol, "second messengers," involved in the action of some hormones (see p. 523). Phosphatidylinositol 3-phosphate and 3,5-bisphosphate are localized in endosomes and presumed to be involved in control of membrane traffic.

Glycerophospholipids contain two fatty acyl groups esterified to C1 and C2 of glycerol; some of the major fatty acids found in glycerophospholipids are presented in Table 12.1. A saturated fatty acid is usually found on C1 of the glycerol and an unsaturated fatty acid on C2. Designation of the different classes of glycerophospholipids does not specify which fatty acids they contain. Phosphatidylcholine usually contains palmitic or stearic acid in the C1 position and a C_{18} unsaturated fatty acid, oleic, linoleic, or linolenic, on the C2 carbon.

Phosphatidylethanolamine contains palmitic or oleic acid on *sn*-1 but a long-chain polyunsaturated fatty acid, such as arachidonic, on the C2 position. Each type of cell membrane has a distinctive composition of fatty acyl groups in its glycerophospholipids. Fatty acyl groups of the same tissue in different species are very similar. In addition, the acyl group composition can change under different nutritional and pathophysiological conditions.

A saturated fatty acid is a straight chain, as is a trans unsaturated fatty acid. Cis double bonds, which occur in most naturally occurring fatty acids, create a kink in their hydrocarbon chain (Figure 12.9). The presence of unsaturated fatty acids has a marked effect on the fluidity of the membrane (p. 458).

Glycerol ether phospholipids contain a long aliphatic chain in ether linkage to the glycerol at the C1 position (Figure 12.10). Ether phospholipids contain an alkyl group (alkyl acyl glycerophospholipid) or an α,β-unsaturated ether, termed a **plasmalogen**. Plasmalogens containing ethanolamine (**ethanolamine plasmalogen**) or choline (**choline plasmalogen**) esterified to the phosphate are abundant in nervous tissue and heart but not in liver. In human heart, more than 50% of the ethanolamine glycerophospholipids are plasmalogens. High levels of ether linked lipids in plasma membranes of very metastatic cancer cells have been reported, suggesting a role for the lipids in the invasive properties of these cells.

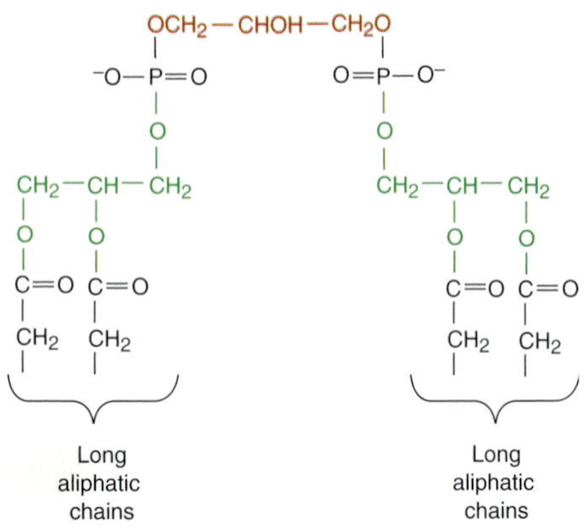

FIGURE 12.7

Phosphatidylglycerol phosphoglyceride (cardiolipin).

TABLE 12.1 Major Fatty Acids in Glycerophospholipids

Common Name	*Systematic Name*	*Structural Formula*
Myristic acid	*n*-Tetradecanoic	$CH_3{-}(CH_2)_{12}{-}COOH$
Palmitic acid	*n*-Hexadecanoic	$CH_3{-}(CH_2)_{14}{-}COOH$
Palmitoleic acid	*cis*-9-Hexadecenoic	$CH_3{-}(CH_2)_5{-}CH{=}CH{-}(CH_2)_7{-}COOH$
Stearic acid	*n*-Octadecanoic	$CH_3{-}(CH_2)_{16}{-}COOH$
Oleic acid	*cis*-9-Octadecenoic acid	$CH_3{-}(CH_2)_7{-}CH{=}CH{-}(CH_2)_7{-}COOH$
Linoleic acid	*cis,cis*-9,12-Octadecadienoic	$CH_3{-}(CH_2)_3{-}(CH_2{-}CH{=}CH)_2{-}(CH_2)_7{-}COOH$
Linolenic acid	*cis,cis,cis*-9,12,15-Octadecatrienoic	$CH_3{-}(CH_2{-}CH{=}CH)_3{-}(CH_2)_7{-}COOH$
Arachidonic acid	*cis,cis,cis,cis*-5,8,11,14-Icosatetraenoic	$CH_3{-}(CH_2)_3{-}(CH_2{-}CH{=}CH)_4{-}(CH_2)_3{-}COOH$

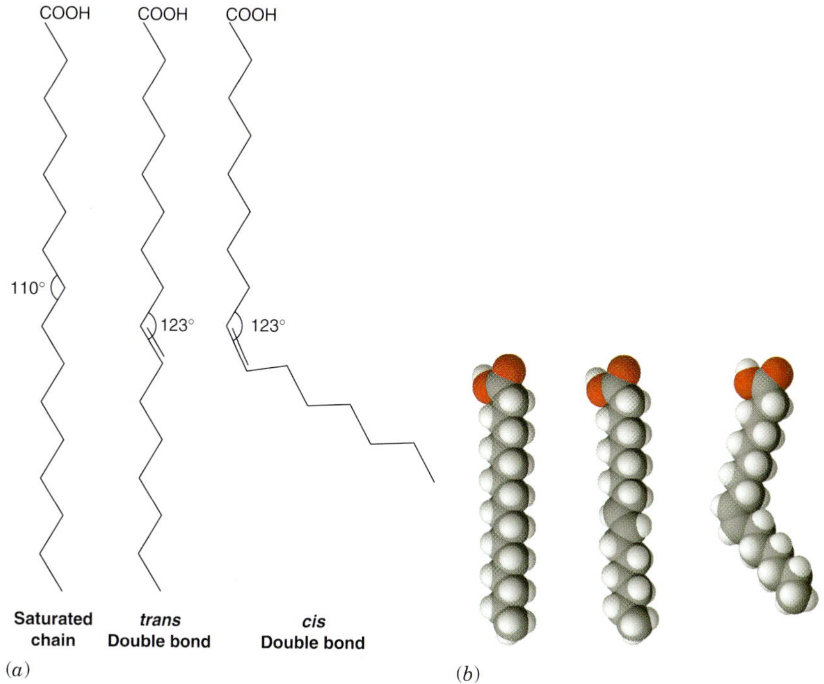

FIGURE 12.9

Conformation of fatty acyl groups in phospholipids. (*a*) Saturated (palmitic)and unsaturated fatty acids (palmitoleic) with trans double bonds are straight chains in their minimum energy conformation, whereas a chain (palmitoleic) with a cis double bond has a bend. The trans double bond is rare in naturally occurring fatty acids. (*b*) Space-filling models.

Courtesy of Dr. Daniel Predecki, Shippensburg University, Shippensburg, PA. See Figure 12.3 for color scheme.

May be present on C-4 or C-4 and C-5

Long aliphatic chains

FIGURE 12.8

Phosphatidylinositol. Phosphate groups are also found on C4, or C4 and C5, of the inositol. The additional phosphate groups increase the charge on the polar head of this glycerophospholipid.

Glycerophospholipids are Amphipathic

Glycerophospholipids contain both a polar and nonpolar end and therefore are **amphipathic**. The polar end (head group) consists of a charged phosphate (p$K \sim 2$), which is negatively charged at pH 7.0, and the charge of the substitutions on the phosphate (Table 12.2). Choline and ethanolamine glycerophospholipids are zwitterions at pH 7.0, because they carry a negative charge on phosphate and a positive charge on nitrogen. Phosphatidylserine has a positive charge on the α-amino group of serine and two negative charges, one on phosphate and one on the carboxyl group of serine, with a net charge of -1. In contrast, glycerophospholipids containing inositol and glycerol have one negative charge on phosphate. The 4-phosphoinositol and 4,5-bisphosphoinositol derivatives are very polar compounds with negative charges on phosphate. The nonpolar end of glycerophospholipids consists of the hydrophobic hydrocarbon chains of attached fatty acyl groups.

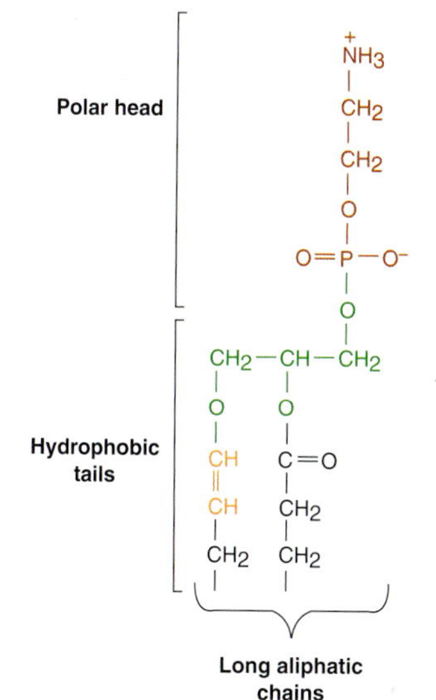

FIGURE 12.10

Ethanolamine plasmalogen. Note the ether linkage of the aliphatic chain on C1 of glycerol.

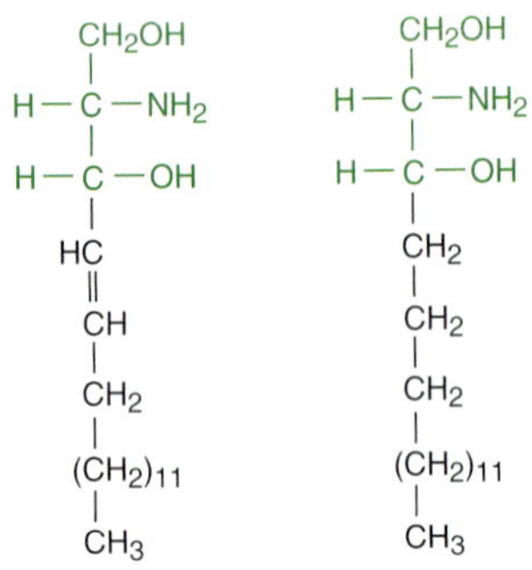

TABLE 12.2 Predominant Charge on Glycerophospholipids and Sphingomyelin at pH 7.0

Lipid	*Phosphate Group*	*Base*	*Net Charge*
Phosphatidylcholine	−1	+1	0
Phosphatidylethanolamine	−1	+1	0
Phosphatidylserine	−1	+1, −1	−1
Phosphatidylglycerol	−1	0	−1
Diphosphatidylglycerol (cardiolipin)	−2	0	−2
Phosphatidylinositol	−1	0	−1
Sphingomyelin	−1	+1	0

Sphingosine (D-4-sphingenine) **Dihydrosphingosine (D-sphingenine)**

FIGURE 12.11
Structures of sphingosine and dihydrosphingosine.

Sphingolipids are Present in Membranes

The amino alcohols **sphingosine** (D-4-sphingenine) and **dihydrosphingosine** (Figure 12.11) are the basis for the **sphingolipids. Ceramides** have a saturated or unsaturated long-chain fatty acyl group in amide linkage with the amino group of sphingosine (Figure 12.12). With its two nonpolar tails a ceramide is similar in structure to diacylglycerol. Various substitutions are found on the hydroxyl group at C1. The **sphingomyelins**, the most abundant sphingolipids in mammalian tissues, have phosphorylcholine esterified at C1 (Figure 12.13). Sphingomyelin resembles that of choline glycerophospholipid; both are classified as phospholipids and have many properties in common, including the fact that both are amphipathic. The fatty acid composition of sphingomyelin varies from tissue to tissue; myelin sphingomyelin contains predominantly longer chain fatty acids (C_{24}).

Glycosphingolipids contain a sugar linked by a β-glycosidic bond to C1 OH group of a ceramide; they lack phosphate and are uncharged. A subgroup is the **cerebrosides**, which contain either glucose (**glucocerebrosides**) or galactose (**galactocerebrosides**). **Phrenosine** (Figure 12.14) is a galactocerebroside that contains a C_{24} fatty acid in amide linkage. Galactocerebrosides predominate in brain and nervous tissue, whereas glucocerebrosides occur in small quantities in nonneural tissues. Galactocerebrosides that contain a sulfate group esterified on the C3 of the sugar are classified as **sulfatides** (Figure 12.15). Cerebrosides and sulfatides usually contain fatty acids with 22–26 carbons.

Neutral glycosphingolipids often have two (dihexosides), three (trihexosides), or four (tetrahexosides) sugar residues attached to the 1-OH group of ceramide. Diglucose and digalactose, ***N*-acetyldiglucosamine** and ***N*-acetyldigalactosamine**, respectively, are the usual sugar components.

Gangliosides, the most complex glycosphingolipids, contain an oligosaccharide group with one or more residues of ***N*-acetylneuraminic acid** (**sialic acid**) (see p. 724 for structure); they are amphipathic compounds with a negative charge at pH 7.0 and represent 5–8% of the total lipids in brain. Some 20 different types have been identified that differ in the number and relative position of the hexose and sialic acid residues. A detailed description of gangliosides is presented on p. 725. The carbohydrate moiety of gangliosides extends beyond the surface of the membrane, is involved in cell–cell recognition, and serves as a binding site for hormones and bacteria toxins such as cholera toxin (p. 514).

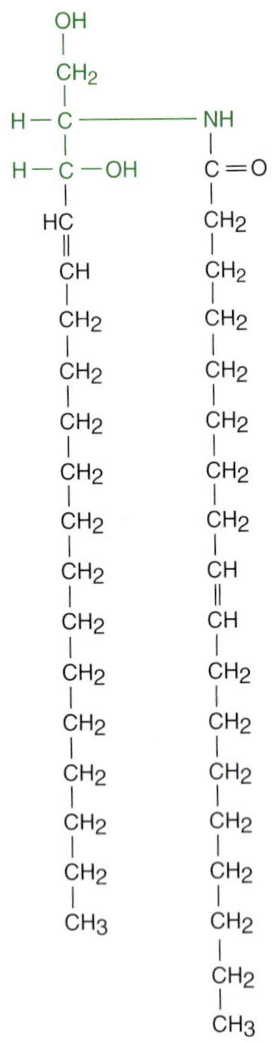

FIGURE 12.12
Structure of a ceramide.

Cholesterol is an Important Component of Plasma Membranes

Cholesterol is the third major lipid in membranes. With four fused rings and a C8 branched hydrocarbon chain attached to the D ring at position 17, cholesterol is a compact, rigid, hydrophobic molecule (Figure 12.16). It has a polar hydroxyl group at C3. Cholesterol alters the fluidity of membranes and participates in controlling the microstructure of the plasma membranes (p. 460).

Lipid Composition Varies Between Membranes

Tissue and various cell membranes have a distinctive lipid composition (Figure 12.17). The membranes of a specific tissue (e.g., liver) in different species contain very similar classes of lipids. The plasma membrane exhibits the greatest variation in percentage composition because the cholesterol content is affected by the nutritional state of the animal. Myelin membranes of neuronal axons are rich in sphingolipids and have a high proportion of glycosphingolipids. Intracellular membranes—for example, endoplasmic reticulum—contain primarily glycerophospholipids and little sphingolipids. The membrane lipid composition of mitochondria, nuclei, and rough endoplasmic reticulum is similar, with that of the Golgi complex being somewhere between that of other intracellular membranes and the plasma membrane. The amount of cardiolipin is high in the inner mitochondrial membrane and low in the outer membrane, with essentially none in other membranes. The choline containing lipids, phosphatidylcholine and sphingomyelin, are most common, followed by phosphatidylethanolamine. The constancy of composition of various membranes suggests a relationship between lipids and the specific functions of those membranes.

Membrane Proteins

Membrane proteins are classified based on the ease of removal of the protein from the membrane. Removal of **integral (intrinsic) membrane proteins** requires disruption of the membrane by detergents or organic solvents and, when isolated, usually contain tightly bound lipid. Integral proteins contain sequences rich in hydrophobic amino acids (see p. 88), which interact with the hydrophobic hydrocarbons of the lipids, thereby stabilizing the protein–lipid complex. They span the lipid bilayer and are in contact with the aqueous environment on both sides. Many integral membrane proteins are glycoproteins. **Peripheral (extrinsic) membrane proteins** are located on the membrane surface, and they can be removed with little or no disruption of membrane integrity. They are released by treatment with salt solutions of different ionic strength, extremes of pH, or cleavage of covalently bound lipid that serves to attach the protein to the membrane. The peripheral membrane proteins are subclassified on the basis of their mode of attachment to a membrane (see p. 457). Isolated peripheral proteins, many of which are enzymes, are typical water-soluble proteins.

Proteolipids are hydrophobic lipoproteins soluble in chloroform and methanol but insoluble in water. They are present in many membranes but particularly in myelin, where they represent about 50% of the protein component. **Lipophilin**, a major lipoprotein of brain myelin, contains over 65% hydrophobic amino acids and covalently bound fatty acids.

Membrane proteins have a variety of functions, including as mediators of transmembrane movement of charged and uncharged molecules, as receptors for the binding of hormones and growth factors, and as enzymes involved in transduction of signals (see p. 494). Some integral membrane proteins have a structural role to maintain the shape of the cell.

Membrane Carbohydrates are Part of Glycoproteins or Glycolipids

Carbohydrates are present in membranes as oligosaccharides covalently attached to proteins (**glycoproteins**) and to lipids (**glycolipids**). The sugars in the oligosaccharides include glucose, galactose, mannose, fucose, *N*-acetylgalactosamine, *N*-acetylglucosamine, and *N*-acetylneuraminic acid (sialic acid) (Figure 12.18 and Appendix for structures). Structures of glycoproteins and glycolipids are presented on pages 649 and 723, respectively. The carbohydrate is on the extracellular surface of plasma membrane and the luminal surface of endoplasmic reticulum. Roles for protein-bound carbohydrates of membranes include cell–cell recognition, adhesion, and receptor action. There is little or no free carbohydrate in membranes.

(*a*)

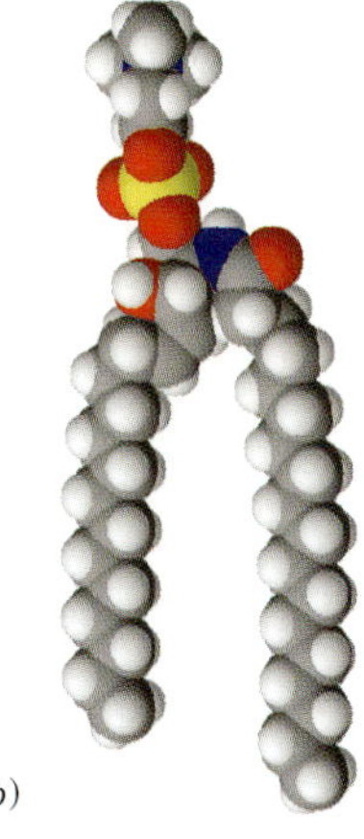

(*b*)

FIGURE 12.13

Choline containing sphingomyelin. (*a*) Structure of choline containing sphingomyelin. (*b*) Space-filling model.

Courtesy of Dr. Daniel Predecki, Shippensburg University, Shippensburg, PA. See Figure 12.3 for color scheme and N, blue.

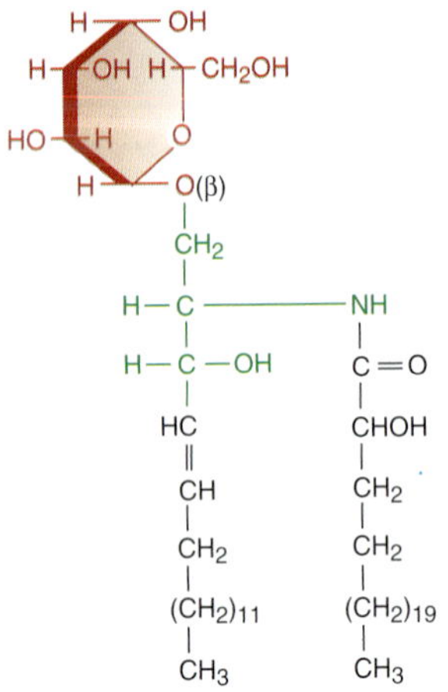

FIGURE 12.14

Structure of a galactocerebroside containing a C_{24} fatty acid.

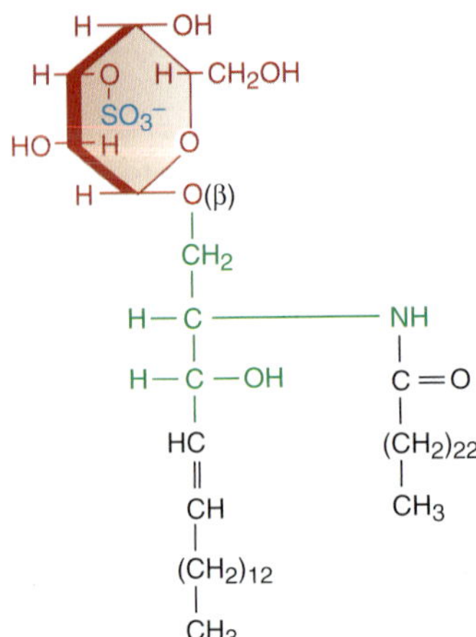

FIGURE 12.15

Structure of a sulfatide.

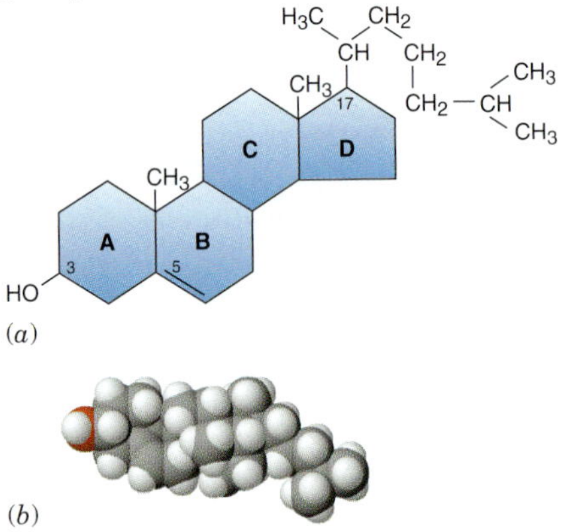

FIGURE 12.16

Cholesterol. (*a*) Structure of cholesterol. (*b*) Space-filling model.
Courtesy of Dr. Daniel Predecki, Shippensburg University, Shippensburg, PA. See Figure 12.3 for color scheme.

12.3 MICELLES, LIPID BILAYERS, AND LIPOSOMES

Lipids Form Vesicular Structures

The basic structural characteristic of membranes is due to the physicochemical properties of the glycerophospholipids and sphingolipids. These amphipathic compounds, with a hydrophilic head and a hydrophobic tail (Figure 12.19*a*), interact in aqueous systems *in vitro* to form spheres, termed **micelles** (Figure 12.19*b*). The charged polar head groups are on the outside of the sphere while the hydrophobic tails interact to exclude water. Micelles have only one polar surface, which is the side presented to the aqueous phase. Micelles can be prepared that contain a single lipid or a mixture of lipids. The concentration of lipid required for micelle formation is the **critical micelle concentration**. Formation of micelles also depends on the temperature and, if a mixture of lipids is present, on the ratio of concentrations of the different lipids (p. 1062).

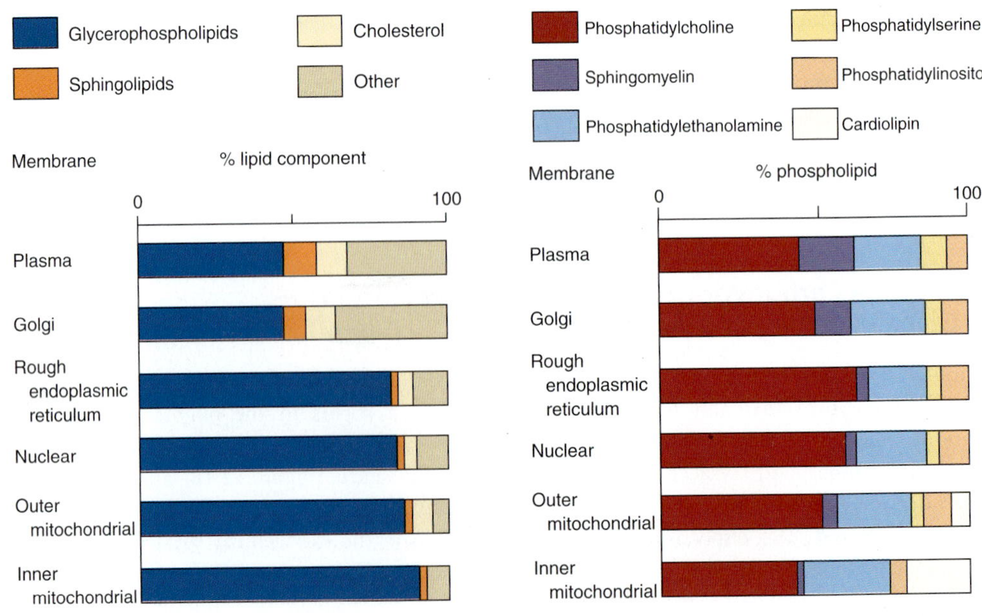

FIGURE 12.17

Lipid composition of cellular membranes isolated from rat liver. (*a*) Amount of major lipid components as percentage of total lipid. The area labeled "Other" includes mono-, di-, and triacylglycerol, fatty acids, and cholesterol esters. (*b*) Phospholipid composition as a percentage of total phospholipid.
Values from R. Harrison and G. G. Lunt, Biological Membranes. New York: Wiley, 1975.

The structure of a micelle is very stable because of hydrophobic interactions between hydrocarbon chains and attraction of the polar head groups to water. Micelles are important in intestinal digestion and absorption of lipids (p. 1061).

Synthetic Lipid Bilayers and Liposomes

Depending on experimental conditions, amphipathic lipids such as glycerophospholipids form a bilayer structure with two layers of lipid forming a structure in which there is minimal contact of hydrocarbon chains with water. The polar head groups are at the interface between the aqueous medium and the lipid, and the hydrophobic tails interact, creating an interior hydrophobic environment that excludes water (Figure 12.19*c*). This bilayer configuration is the basic lipid structure of all biological membranes. Lipid bilayers are extremely stable, being held together by hydrophobic forces of the hydrocarbon chains and ionic interactions of charged head groups with water. Lipid bilayers self-seal if disrupted.

A lipid bilayer under the proper conditions will close in on itself to form a spherical vesicle that separates the external environment from an internal aqueous compartment. Such vesicles, termed **liposomes** (Figure 12.19*d*), are prepared using purified lipids and lipids extracted from biological membranes. Depending on the procedure, unilamellar and multilamellar (vesicles within vesicles) vesicles of various sizes (20- to 1000-nm diameter) are prepared. The ability of amphipathic lipids to self-assemble into bilayers is an important property for formation of cell membranes.

General Properties of Lipid Bilayers

Individual phospholipid molecules exchange places with neighboring molecules in a bilayer, leading to rapid lateral diffusion in the plane of the membrane (Figure 12.20). Rotation occurs around the carbon–carbon bonds in fatty acyl chains; in fact, there is a greater degree of rotation nearer the methyl end, leading to greater motion at the center than the surface of the lipid bilayer. Individual lipid molecules do not migrate readily from one monolayer to the other—a transverse movement, termed **flip-flop**—because of thermodynamic constraints on movement of a charged head group through the lipophilic core. In addition, lipids do not readily escape from the bilayer. Lipid bilayers, therefore, have an inherent stability and **fluidity** in which individual molecules move rapidly in their own monolayer but do not readily exchange with an adjoining monolayer. Lipids are distributed randomly between monolayers in artificial membranes composed of different lipids.

Interaction of lipids in a bilayer is very different from that illustrated in Figure 12.19*c*. The interior of membrane bilayers is very fluid and in constant motion.

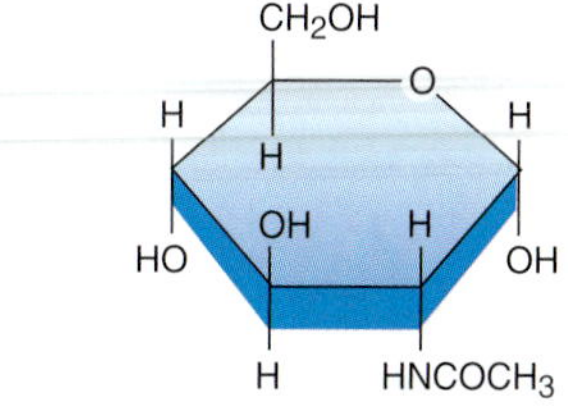

***N*-Acetyl-α-D-glucosamine**

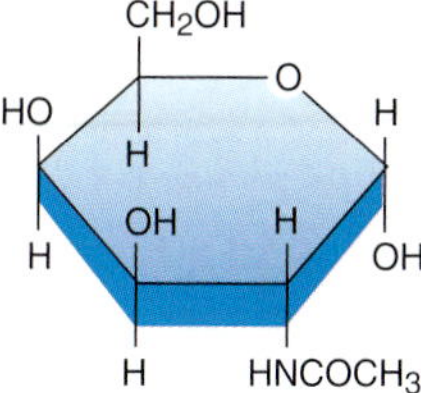

***N*-Acetyl-α-D-galactosamine**

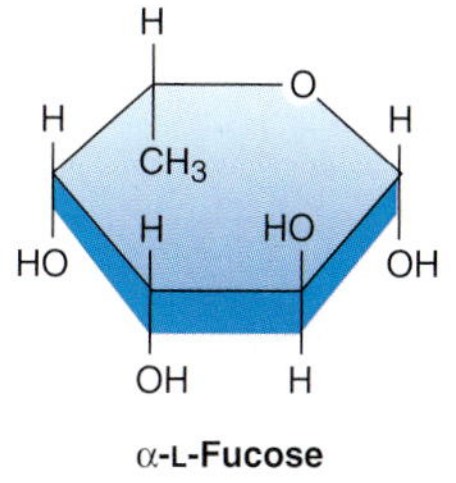

α-L-Fucose

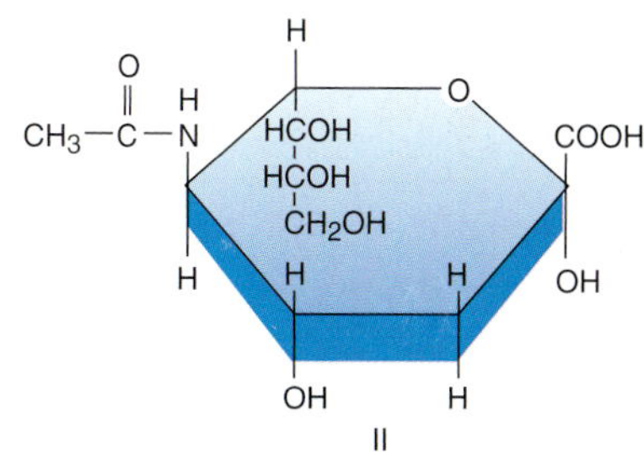

***N*-Acetyl-D-neuraminic acid**

FIGURE 12.18

Structures of some membrane carbohydrates.

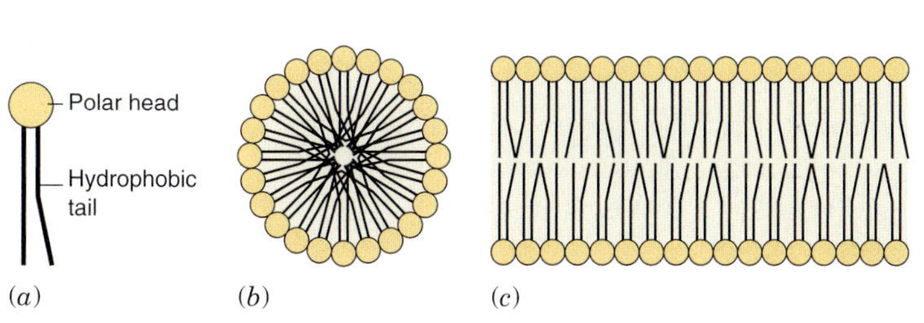

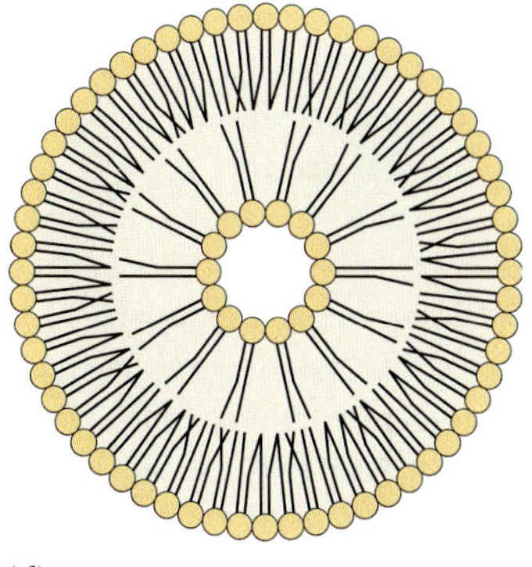

FIGURE 12.19

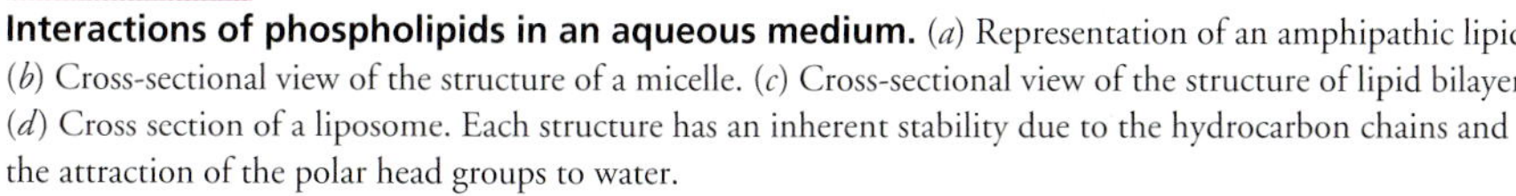
Interactions of phospholipids in an aqueous medium. (*a*) Representation of an amphipathic lipid. (*b*) Cross-sectional view of the structure of a micelle. (*c*) Cross-sectional view of the structure of lipid bilayer. (*d*) Cross section of a liposome. Each structure has an inherent stability due to the hydrocarbon chains and the attraction of the polar head groups to water.

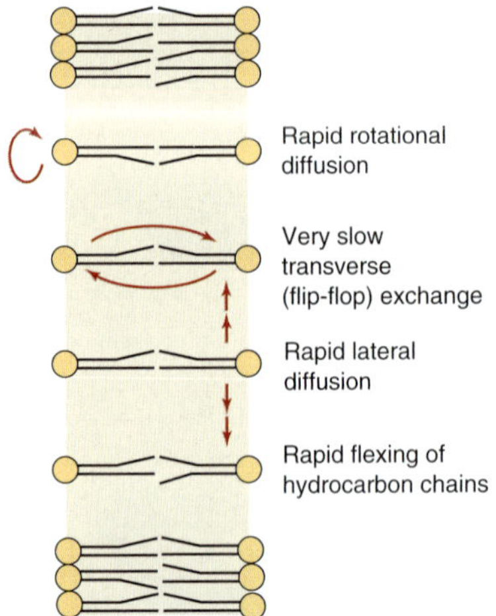

FIGURE 12.20
Mobility of lipid components in membranes.

Acyl chains of glycerophospholipids and sphingolipids have a random motion, intercalating with chains in the opposing monolayer. The viscosity increases significantly closer to lipid head groups. Cholesterol incorporated in the bilayer of liposomes reduces the fluidity near the surface of the membrane because cholesterol does not extend into the layer as far as the acyl chains of phospholipids. Figure 12.21 is a computer representation of the arrangement of phospholipids in a bilayer at a moment in time. Several layers of ordered water molecules that cover the surface of membranes influence the environment of the membrane.

The lipid bilayer is essentially impermeable to nonlipid (as an example, carbohydrates) and charged molecules but not to neutral hydrophobic molecules. Inorganic

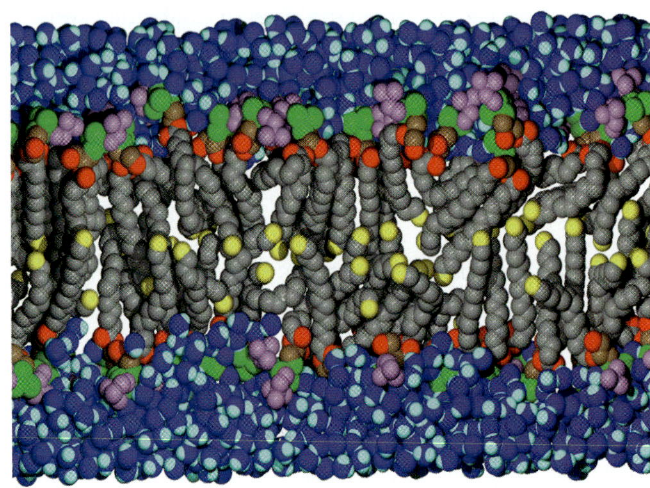

FIGURE 12.21
Model of a lipid bilayer stopped at a moment in time. An artificial bilayer consisting of dipalmitoyl phosphatidylcholine surrounded by water as modeled by computer. Atom colors are chain C gray (except terminal methyl C yellow and glycerol C brown), ester O red, P and O green, choline C and N pale violet, water O dark blue, and water H light blue. Lipid H has been omitted.
Courtesy of Richard Pastor and Richard Venable, FDA, Bethesda, Maryland.

CLINICAL CORRELATION 12.1
Liposomes as Carriers of Drugs and Enzymes

A major obstacle in the use of many drugs is lack of tissue specificity in the action of the drug. Administration of drugs orally or intravenously leads to a drug acting on many tissues and not exclusively on a target organ, resulting in toxic side effects. An example is the commonly observed suppression of bone marrow cells by anticancer drugs. Some drugs are metabolized rapidly, and their period of effectiveness is relatively short. Liposomes have been prepared encapsulated with drugs, enzymes, and DNA to be used as carriers for these substances to target organs. Liposomes prepared from purified phospholipids and cholesterol are nontoxic and biodegradable but are not specific for a target tissue. Attempts have been made to prepare liposomes for interaction at a specific target organ; these include incorporation into liposomes of compounds that change the surface charge or will interact with specific membrane receptors (as example, the folate receptor). Some of these approaches have enhanced drug incorporation and release. Antibiotic, antineoplastic, antimalarial, antiviral, antifungal, and anti-inflammatory agents are effective when administered in liposomes. Some drugs have a longer period of effectiveness when encapsulated in liposomes. It may be possible to prepare liposomes with a high degree of tissue specificity so that drugs and perhaps even enzyme replacement may be possible with this technique.

Source: Banerjee, R. Liposomes: Applications in medicine. *J. Biomater. Appl.* 16:3, 2001. Voinea, M. and Simionescu, M. Designing of intelligent liposomes for efficient delivery of drugs. *J. Cell. Mol. Med.* 6:465, 2002. Audony, S. A. L., de Leij, L. F. M. H., Hoekstra, D., and Molema, G. *In vivo* characteristics of cationic liposomes as delivery vectors for gene therapy. *Pharm. Res.* 19:1599, 2002. Smyth, T. N. Cationic liposomes as *in vivo* delivery vehicles. *Curr. Med. Chem.* 10:1279, 2003.

ions (Na^+, K^+, Cl^-, etc.), charged organic molecules, and macromolecules outside liposomes do not have access to the interior of liposomes. During liposome formation, nonpermeable solutes, however, can be trapped in the aqueous interior. Lipid-soluble compounds, such as triacylglycerol and undissociated organic acids, diffuse readily into the bilayer and are trapped in the hydrophobic environment. Both the external and internal environments of liposomes can be manipulated; and properties, including ability to exclude molecules, interaction with various substances, and stability under different conditions, can be evaluated. Liposomes are an important tool in studying aspects of natural membrane structure (e.g., membrane curvature) and biological processes such as membrane fusion and activities of membrane-bound proteins. Proteins involved in translocation of molecules across a membrane have been isolated and incorporated into the lipid bilayer of liposomes for assessment of their function. With some limited success, liposomes of varying lipid and protein composition have been tested in both animals and humans as a red cell substitute, to deliver encapsulated drugs, and as vectors for gene therapy (Clin. Corr. 12.1).

12.4 STRUCTURE OF BIOLOGICAL MEMBRANES

Fluid Mosaic Model of Biological Membranes

Knowledge of cellular membrane structure has developed based on evidence from a variety of physicochemical, microscopic, and biochemical investigations. All biological membranes are lipid bilayers with amphipathic lipids and cholesterol oriented so that their hydrophobic portions interact to minimize their contact with water or other polar groups. Their polar head groups are at the interface with the aqueous environment. This is the model proposed by J. D. Davson and J. Danielli in 1935 and refined by J. D. Robertson. In the early 1970s, S. J. Singer and G. L. Nicolson proposed the **fluid mosaic model** for membranes in which some proteins (integral) are actually immersed in the lipid bilayer while others (peripheral) are attached to the surface. It was suggested that some proteins span the lipid bilayer and are in contact with the aqueous environment on both sides. Both lipids and proteins were considered to diffuse laterally in the membrane. Figure 12.22 is a traditional representation of a biological membrane. Lipids in a membrane, however, can assume nonlamellar structural domains and acyl chains in one layer interdigitate with those of the opposite layer (Figure 12.21). The distribution of both lipids and proteins in cellular membranes is not homogeneous. The presence of ordered microdomains of lipids and proteins in cell membranes (see p. 460) transforms the classical fluid mosaic model of membranes into a more complex structure. The

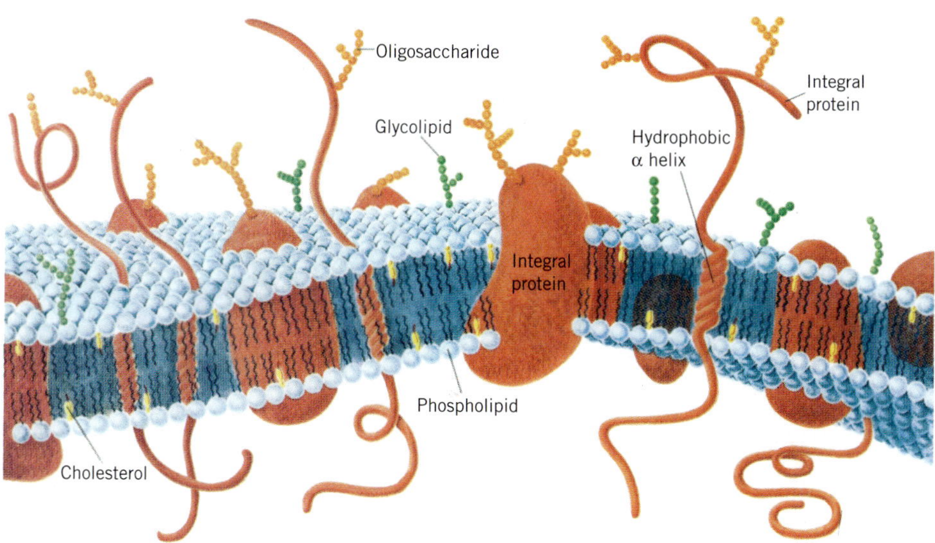

FIGURE 12.22

Fluid mosaic model of biological membranes. Figure reproduced with permission from D. Voet and J. Voet, *Biochemistry*, 2nd ed. New York: Wiley, 1995. Copyright (1995) John Wiley & Sons.

characteristics of the lipid bilayer, however, explain many of the properties of cellular membranes, including fluidity, flexibility that permits change of shape and form, ability to self-seal, and impermeability. The development of techniques for isolation of membrane proteins, for determination of their structure, and for identification of their specific functional domains has led to an understanding of the structural relationship between the hydrophobic lipid bilayer and membrane proteins.

Although membrane models suggest that some proteins are randomly distributed throughout or on the membrane, there is a high degree of functional organization with definite restrictions on the localization of some proteins. Proteins of the electron transport chain are organized into functional units in the inner membrane of mitochondria. Cells lining the lumen of kidney nephrons have specific plasma membrane proteins on the luminal surface, but these proteins do not occur on the contraluminal surface; enzymes restricted to a particular region of a membrane are there to meet a specific function of the cell. Thus, there is a high degree of molecular organization of biological membranes, which is not apparent from diagrammatic models.

Lipids are Asymmetrically Distributed in Membranes

There is an overall asymmetric distribution of lipid components across biological membranes. Each layer of the bilayer has a different composition of glycerophospholipids and sphingolipids. The asymmetric distribution of lipids in the human erythrocyte membrane is presented in Figure 12.23. Sphingomyelin is predominant in the outer layer, whereas phosphatidylethanolamine is predominant in the inner lipid layer. In contrast, some studies indicate that cholesterol is equally distributed in both layers. Phosphatidylinositols, not shown in Figure 12.23 because of their low amounts, are located primarily in the inner layer, but this varies depending on the membrane and tissue.

Uncatalyzed transverse movement from one side to the other (flip-flop) of glycerophospholipids and sphingolipids is slow but does occur. Asymmetry of lipids is maintained by **lipid transporters** that catalyze unidirectional movement of specific

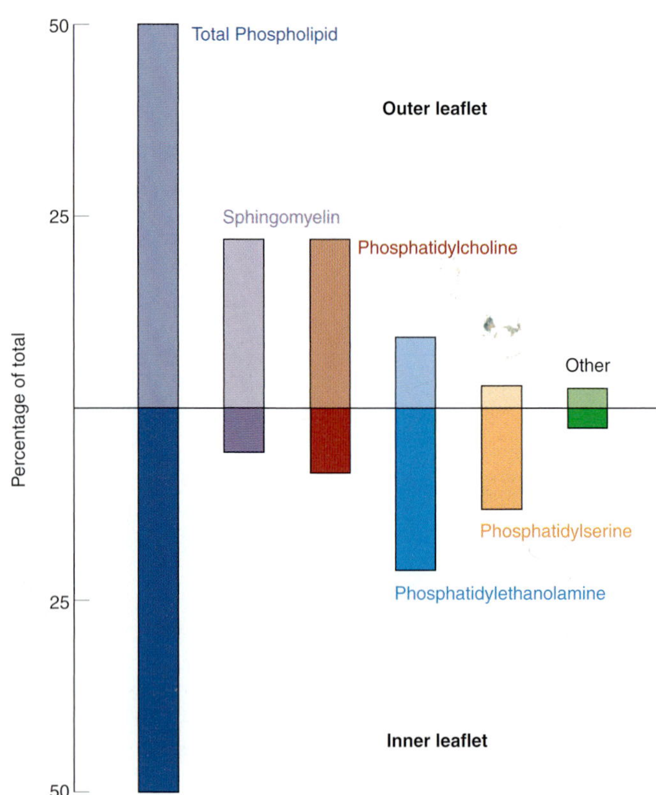

FIGURE 12.23

Distribution of phospholipids between inner and outer layers of the human erythrocyte membrane. Values are percentage of each phospholipid in the membrane. Other include phosphatidylinositol, phosphatidylinositol 4-phosphate, phosphatidylinositol 4,5-phosphate, and phosphatidic acid. Data from A. J. Verkeij, R. F. A. Zwaal, B. Roelofsen, P. Comfurius, D. Kastelijn, and L. L. M. Van Deenan. The asymmetric distribution of phospholipids in the human red cell membrane. *Biochim. Biophys. Acta* 323:178, 1973. Zachowski, A. Phospholipids in animal eukaryotic membranes: Transverse symmetry and movement. *Biochem. J.* 294:1, 1993.

lipids from one layer to the other, sometimes against their concentration gradient. An ATP-dependent lipid class of transporters, **aminophospholipid translocases** or **flippases**, specific for phosphatidylserine and phosphatidylethanolamine catalyze the transport of these aminoglycerolipids from the extracellular to the cytoplasmic layer. They are responsible for maintaining the low concentration of these phospholipids in the outer layer of the plasma membrane. An outward-directed ATP-dependent transporter, termed **floppase**, is nonspecific for phospholipids. **Multidrug resistance-associated protein** (**MRP1**) may function as a **floppase** in human red blood cells; they are responsible for drug resistance of some cancer cells (p. 479).

A third lipid transporter, **phospholipid scramblase 1 (PLSCR 1)**, facilitates bidirectional mixing of phospholipids between the two leaflets; it is nonspecific with respect to phospholipids and is stimulated by increases in intracellular Ca^{2+}. PLSCR 1 can be reversibly palmitoylated–depalmitoylated and contains a protein kinase C phosphorylation site for control of activity (see p. 523). The transporter has an important role in (a) the cell-mediated coagulation cascade (p. 993) and (b) recognition of dying cells (see p. 1020) for removal by the reticuloendothelial system. In both responses, the movement of phosphatidylserine from the inner to the outer leaflet, leading to exposure of phosphatidylserine on the extracellular surface, apparently initiates the response.

Integral Membrane Proteins are Immersed in Lipid Bilayer

Various modes of attachment of proteins to a biological membrane are illustrated in Figure 12.24. **Integral membrane proteins** (p. 449) are embedded and are asymmetric in the membrane. They have a defined, rather than a random, orientation. The orientation of proteins is determined by their primary structure; their bulkiness and thermodynamic restrictions prevents transverse (flip-flop) movement. As with nonmembrane proteins, they contain specific domains for ligand binding, for catalytic or transport activity, and for attachment of carbohydrate or lipid. For some proteins, the amino and carboxyl termini are both on one side of the membrane, whereas in other instances they are on opposite sides, frequently with the carboxyl terminus on the

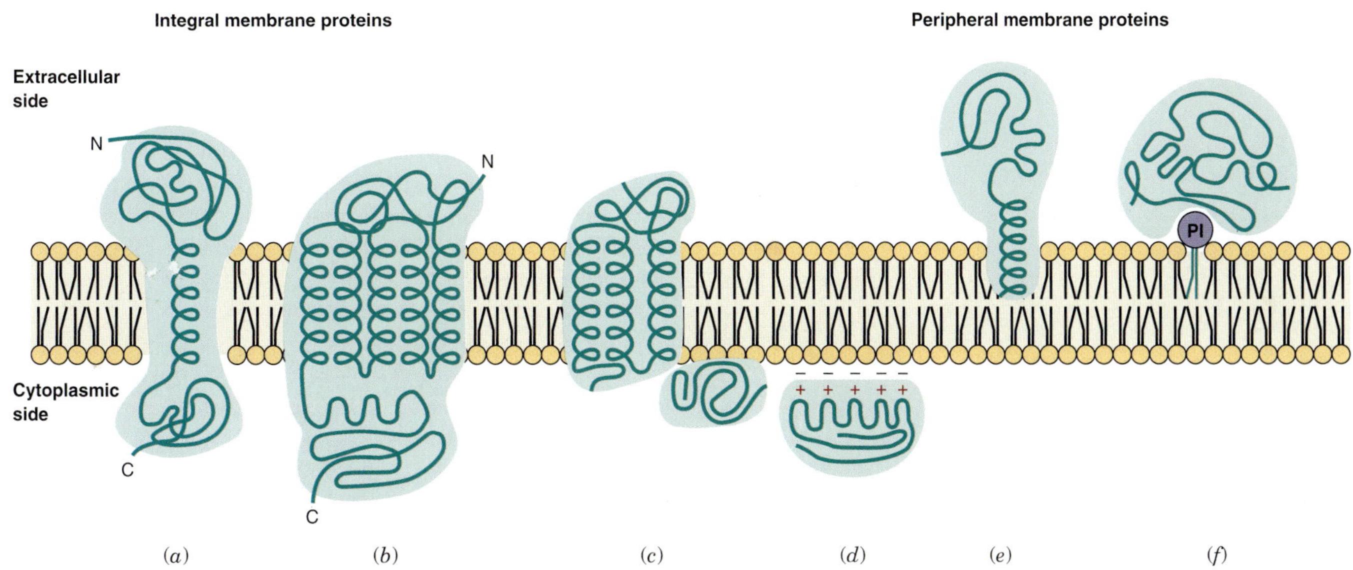

FIGURE 12.24

Interactions of membrane proteins with the lipid bilayer. Diagram illustrates the multiple types of binding of proteins in or to the lipid bilayer. (*a*) A single transmembrane segment; (*b*) multiple transmembrane segments; (*c*) bound to an integral protein; (*d*) bound electrostatically to the lipid bilayer; (*e*) attached by a short terminal hydrophobic sequence of amino acids; and (*f*) noncovalent binding to a phosphatidylinositol (PI) in the membrane.

cytoplasmic side. Some integral proteins form multiple subunit structures in order to carry out their function.

Measurements of the hydrophobicity of amino acid residues and partial proteolytic digestion of integral membrane proteins have revealed sequences of amino acids that are presumably embedded in the membrane. Some proteins contain a single transmembrane segment, as illustrated in Figure 12.24*a*, consisting of an α-helical structure primarily of hydrophobic amino acids (such as leucine, isoleucine, valine, and phenylalanine). An example is **glycophorin**, with 131 residues, present in the plasma membrane of human erythrocytes; the transmembrane sequence consists of amino acid residues 73–91. The amino-terminus is exterior to the cell and contains various oligosaccharides including the ABO and MN blood group determinants; glycophorin contains 60% by weight carbohydrate. The **integrins** also have a single transmembrane segment; this ubiquitous family of proteins links the cytoplasm to the extracellular matrix (see p. 358) and mediates transmembrane signaling in both directions. Other integral proteins have multiple transmembrane sequences that loop back and forth across the lipid bilayer (Figure 12.24*b*, *c*). Based on gene analysis, the amino acid sequence of a large number of integral membrane proteins have been determined even though they have not been isolated. It is possible to determine the number of putative transmembrane segments by plotting the degree of hydrophobicity of each amino acid residue in the protein (see p. 88). A hydropathy plot for aquaporin is presented in Figure 12.25; the plot indicates the presence of six putative transmembrane sequences. Proteins with 2–24 putative transmembrane sequences have been identified from such plots. The transmembrane domains of most integral membrane proteins are α-helical, but a β-barrel motif occurs in some. Multiple α-helical transmembrane segments of an integral protein are often organized to form a tubular structure, which can serve as a passage way for movement of molecules through the membrane (Figure 12.26). The **anion channel** of human erythrocytes, a glycoprotein with 926 amino acids, forms a tubular channel (see p. 461) and mediates the exchange of Cl^- and HCO_3^- across the membrane. It has two major domains: a hydrophilic amino-terminus domain on the cytoplasmic side of the membrane with binding sites for **ankyrin**, a protein that anchors the cytoskeleton and other cytoplasmic proteins, and a domain of 509 amino acids with 12 transmembrane sequences. With an even number of transmembrane segments, the carboxyl terminus is also on the cytoplasmic side.

The activities of many membrane enzymes depend on the surrounding lipid. D-β-Hydroxybutyrate dehydrogenase (see p. 687), an inner mitochondrial membrane enzyme, requires phosphatidylcholine for activity, and cardiolipin stabilizes respiratory chain complexes but is not necessary for activity (see p. 554). Maximum activity of diacylglycerol kinase occurs with phosphatidylcholine with C_{18} chains, and cholesterol has been implicated in the activity of various membrane ion pumps, including the Na^+/K^+-exchanging ATPase (see p. 476), Ca^{2+}-transporting ATPases (see p. 477), and acetylcholine receptors (see p. 470). Some modulating effects of lipids may reflect a change in ordering and fluidity of the membrane, but the lipid may also have a direct influence on the activity of the protein.

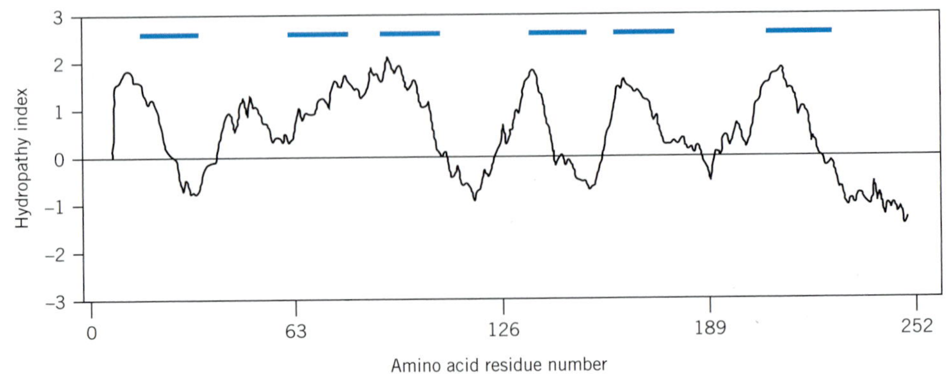

FIGURE 12.25

Hydropathy plot of aquaporin1 from human erythrocytes. Hydropathy index for an amino acid is a value indicating their hydrophobic and hydrophilic tendency. The more positive the value, the more hydrophobic the amino acid; a positive value is a good predictor of whether or not the amino acid will be in contact with an aqueous environment. Segments of the protein containing a high proportion of hydrophobic amino acids are considered to be putative transmembrane segments. Blue dashes indicate putative transmembrane segments.

Peripheral Membrane Proteins

Peripheral membrane proteins are on the surface of membranes and can be easily removed without disrupting the lipid bilayer. Their various methods of attachment are illustrated in Figure 12.24*c–f*. Some bind to integral membrane proteins, such as ankyrin binding to the anion channel in erythrocytes (Figure 12.24*c*). Negatively charged phospholipids of membranes interact with positively charged regions of proteins and produce electrostatic binding (Figure 12.24*d*); in some cases Ca^{2+} mediates the binding. Several peripheral proteins have short sequences of hydrophobic amino acids at one end that serve as a membrane anchor (Figure 12.24*e*); cytochrome b_5 is attached to the endoplasmic reticulum by such an anchor at the carboxyl terminus. Other peripheral membrane proteins have a specific conserved domain that binds noncovalently to the inositol 3-phosphate head group of phosphatidylinositol fixed in the membrane (Figure 12.24*f*). Some proteins have several different forms of attachment.

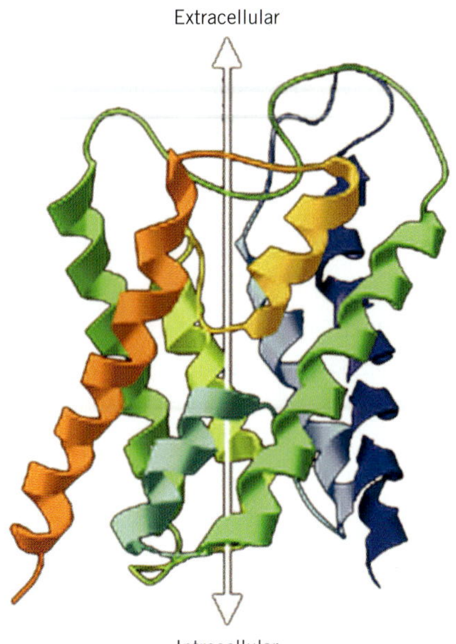

FIGURE 12.26

Multiple α-helical transmembrane segments. Ribbon model of human red cell AQP1 monomer. Multiple α-helical transmembrane segments form a tubular structure. The amino-terminus is blue and the carboxy-terminus is orange. Six membrane-spanning tilted helices surround two hemipores (loops with short helices—cyan and yellow) that meet in the center of the bilayer. The white arrow illustrates the aqueous channel through the protein.

Figure reproduced with permission from Kozono, D., Yasui, M., King, L. S., and Agre, P. Aquaporin water channels: atomic structure and molecular dynamics meet clinical medicine. *J. Clin. Invest.* 109:1395, 2002. Copyright (2002) *Amer. Soc. Clin. Invest.* via Copyright Clearance Center.

Lipid Anchored Membrane Proteins

A number of peripheral membrane proteins are tethered to the membrane by covalently linked lipid. The lipid is inserted into the lipid membrane, thus anchoring the protein to the membrane. The various types of **lipid anchors** are presented in Table 12.3 and Figure 12.27. One such anchor involves phosphatidylinositol (Figure 12.8, p. 447), attached to a **glycan** consisting of ethanolamine, phosphate, mannose, mannose, mannose, and glucosamine (see p. 652). The glycan is covalently bound to the carboxyl terminus of a protein by ethanolamine (Figure 12.27*a*) and the glucosamine is linked covalently to phosphatidylinositol. The fatty acyl groups of phosphatidylinositol are inserted into the lipid membrane, thus anchoring the protein. This form of attachment is referred to as a **glycosylphosphatidylinositol** (**GPI**) anchor. Additional carbohydrate can be attached to the last mannose. The fatty acyl groups of phosphatidylinositol appear to be specific for different proteins. GPI anchored proteins are attached on the external surface of plasma membranes and are found as part of sphingolipid- and cholesterol-rich microdomains (see p. 460). This form of anchoring is important because release and reattachment of the protein to the anchor can be controlled, thereby allowing regulation

TABLE 12.3 Types of Lipid Anchors

Type of Anchor	*Lipid Involved in Attachment*	*Attachment*	*Representative Proteins*
Phosphatidyl inositol (GPI)	Phosphatidylinositol	Glycan (ethanolamine, phosphate, mannose, mannose, mannose, and glucosamine)	Acetylcholine esterase Alkaline phosphodiesterase Carbonic anhydrase Cell–cell adhesion molecules Cell surface hydrolases Lipoprotein lipase Scrapie prion protein Surface antigens
Myristoyl	Myristic acid (C_{14})	Amide linkage at N-terminal glycine	β-Adrenergic receptor c-CAMP protein kinase Insulin receptor α-Subunit of G protein
Thioester and hydroxy ester	Myristic acid (C_{14}) Palmitic acid (C_{16}) Stearic acid (C_{18}) Oleic acid (C_{18})	—SH or —OH of Cysteine Serine Threonine	G-protein-coupled receptors Transferrin receptor
Thioether-linked	Isoprenoid lipid Farnesyl (C_{15}) Geranylgeranyl (C_{20}) Dolichol	Cysteine	Many GTP-binding proteins Nuclear lamins Protein kinases Protein phosphatases

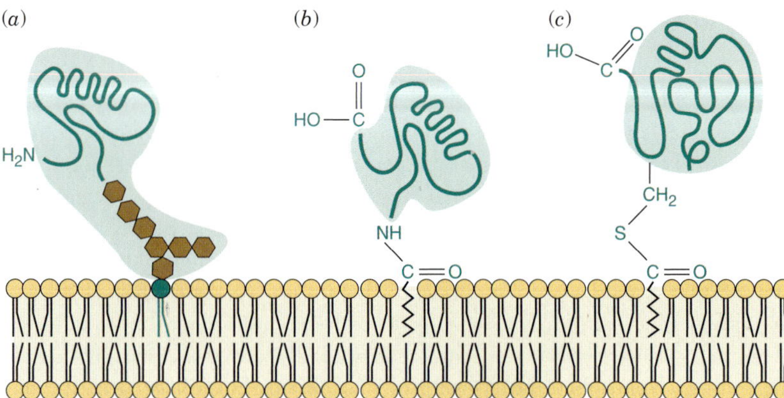

FIGURE 12.27

Types of lipid anchors for attachment of membrane proteins. (*a*) Glycosylphosphatidylinositol (GPI) anchor, (*b*) myristoyl anchor, and (*c*) thioester anchor.

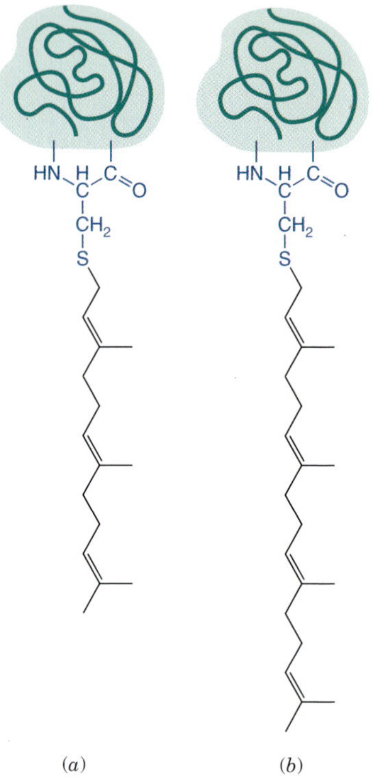

FIGURE 12.28

Isoprenoid thioether anchor for attachment of membrane proteins. (*a*) Farnesyl anchor and (*b*) geranylgeranyl anchor.

of the activity of the protein. A specific phosphatidylinositol-specific phospholipase C catalyzes the hydrolysis of the phosphate-inositol bond leading to release of the protein The GPI anchor also controls the localization of the protein on the membrane. Over 50 proteins including enzymes, antigens, and cell adhesion proteins are so attached (Table 12.3). The GPI anchor has been conserved through evolution, occurring in eukaryotes from protozoa to vertebrates.

The **thio-** and **hydroxy-ester linked acyl anchors** (Table 12.3) involve myristic acid (C_{14}), palmitic acid (C_{16}), stearic acid (C_{18}), or oleic acid (C_{18}) covalently linked to a cysteine, serine, or threonine residue in the protein. As with other lipid-attached proteins, the acyl groups serve to anchor the protein in the lipid bilayer (Figure 12.27*b*) and the specific fatty acid apparently directs the localization on the membrane. In another type of lipid attachment, myristic acid is attached by an amide linkage to an N-terminal glycine.

Another form of lipid attachment of a peripheral protein involves a thioether linkage between the protein and either **farnesyl** (C_{15}) or **geranylgeranyl** (C_{20}), isoprenoid lipids (Figure 12.28). These attachments are referred to as **CAAX prenylated proteins** because the isoprenoid is attached to a cysteine residue in the sequence CAAX (Cys-Aliphatic-Aliphatic-Any residue) close to the carboxyl terminus. The X apparently directs which isoprenoid is linked; glutamine, methionine, or serine is required for attachment of farnesyl, and leucine is required for geranylgeranyl. CAAX prenylated proteins are usually attached on the cytoplasmic surface.

Many lipid anchors are involved in intracellular signal transduction (see p. 494), where the rapid removal and reattachment of the lipid anchor can control activity of the protein.

Lipids and Proteins Diffuse in Membrane Leaflets

Interactions among different lipids and between lipids and proteins are very complex and dynamic. There is **fluidity** in the lipid portion of membranes, and individual lipids and proteins can move rapidly across the surface of membranes. The degree of fluidity is dependent on the temperature and membrane composition. At low temperatures, lipids are in a gel-crystalline state, being restricted in their mobility. As temperature is increased, there is a phase transition into a liquid-crystalline state, with an increase in fluidity (Figure 12.29). With liposomes prepared from a single phospholipid, the phase transition temperature, T_m, is rather precise; but with liposomes prepared from mixtures of lipids, T_m becomes less precise as individual clusters of lipids may be in either the gel-crystalline or the liquid-crystalline state. T_m is not precise for biological membranes because of their heterogeneous composition. Interactions between lipids and proteins lead to variations in the gel–liquid state throughout the membrane and to differences in fluidity in different areas.

Membranes with glycerophospholipids containing short-chain and unsaturated fatty acyl groups have greater fluidity. Cis-double bonds in unsaturated fatty acids of phospholipids lead to kinks in the hydrocarbon chain, which prevents tight packing

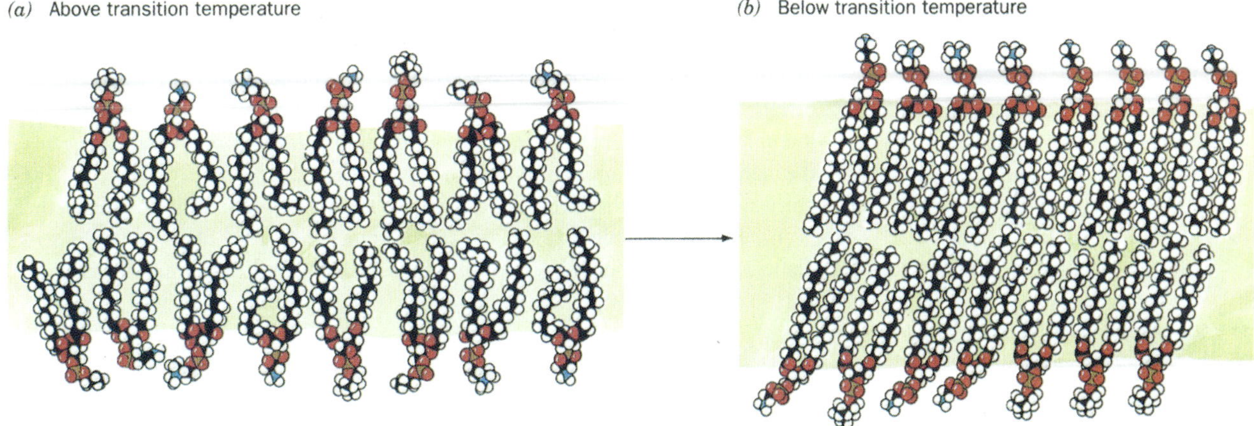

FIGURE 12.29

Structure of lipid bilayer above and below transition temperature. Figure reproduced with permission from Voet, D. and Voet, J. *Biochemistry*, 2nd ed. New York: Wiley, 1995. Copyright (1995) John Wiley & Sons.

CLINICAL CORRELATION **12.2**

Abnormalities of Cell Membrane Fluidity in Disease

Membrane fluidity can control the activity of membrane-bound enzymes, and it can also control functions such as (a) phagocytosis and (b) cell growth and death. A major factor in the fluidity of the plasma membrane in higher organisms and mammals is the presence of cholesterol. With increasing cholesterol content the lipid bilayers become less fluid on their outer surface but more fluid in the hydrophobic core. Erythrocyte membranes of individuals with spur cell anemia have increased cholesterol content and a spiny shape, and the cells are destroyed prematurely in the spleen. This condition occurs in severe liver disease such as alcoholic cirrhosis. Cholesterol content is increased 25–65%, and the fluidity of the membrane is decreased. Erythrocyte membranes require a high degree of fluidity to pass through the capillaries. Increased plasma membrane cholesterol in other cells leads to an increase in intracellular membrane cholesterol, which also affects their fluidity. The intoxicating effect of ethanol on the nervous system is probably due to modification of membrane fluidity, altering membrane receptors and ion channels. Individuals with abetalipoproteinemia have an increase in sphingomyelin content and a decrease in phosphatidylcholine in cellular membranes, with a decrease in membrane fluidity. Recently, changes in membrane fluidity have been suggested as a factor in lecithin:cholesterol acyltransferase deficiency, hypertension, and Alzheimer disease. As techniques for measurement and evaluation of cellular membrane fluidity improve, some of the pathological manifestations in disease will be explained based on changes in membrane structure and function.

Source: Cooper, R. A. Abnormalities of cell membrane fluidity in the pathogenesis of disease. *N. Engl. J. Med.* 297:371, 1977. Muller, W. E., Kirsch, C., and Eckert, G. P. Membrane-disordering effects of beta-amyloid peptides. *Biochem. Soc. Trans.* 29:617, 2001. Tsudo, K. and Nishio, I. Membrane fluidity and hypertension. *Am. J. Hypertens.* 16:259, 2003.

of the chains, and creates pockets in the hydrophobic regions. Cholesterol with its flat stiff ring structure reduces the coiling of the fatty acid chain and decreases fluidity. The steroid makes membranes more rigid toward the periphery because it does not reach into their central core. The clinical significance of high blood cholesterol on the fluidity of cell membranes is described in Clinical Correlation 12.2. Ca^{2+} ion decreases fluidity of membranes because of its interaction with negatively charged head groups of the phospholipids, reducing repulsion between polar groups and increasing packing of lipid molecules. This causes aggregation of lipids into clusters and reduces fluidity.

The asymmetric distribution of phospho- and sphingolipids leads to a difference in fluidity of the two leaflets. The hydrocarbon chains of the lipids are flexible, producing a greater degree of fluidity in the hydrophobic core than closer to the surface, where there are more constraints from stiffer portions of the hydrocarbon chains.

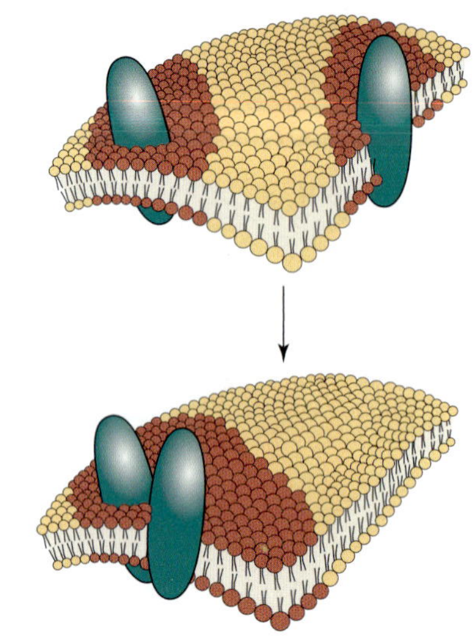

FIGURE 12.30

Schematic representation of lipid rafts of membrane. The diagram indicates the movement of lipid rafts with merging of rafts and association of two proteins.

The fluidity of cellular membranes responds to variations in diet and physiological state. The composition of fatty acyl chains of phospholipids and of cholesterol in membranes is altered by increased release of free radicals, changes in cellular Ca^{2+} homeostasis, and lipid peroxidation. Some pharmacological agents may have a direct effect on membrane fluidity. Anesthetics increase membrane fluidity *in vitro* and may act *in vivo* through their effect on membrane fluidity of specific cells. Structurally unrelated compounds induce anesthesia, but their common feature is lipid solubility. Finally, there have been reports of changes in fluidity of membranes in hypercholesteremia, hypertension, diabetes mellitus, obesity, alcoholism, schizophrenia, and Alzheimer disease.

Microdomains of Lipid–Protein Complexes are Present in Membranes

Microdomains composed of lipids and protein occur in both the outer and inner layer of the plasma membrane, and in cellular membranes destined to interact with the plasma membrane, such as the trans-side of the Golgi complex. Thus, biological membranes may actually be a fluid mosaic of microdomains, termed **lipid rafts**, rather than a fluid mosaic of individual lipids (Figure 12.30). Lipid rafts are small and variable in size, but may constitute a relatively large percentage of the total area of the plasma membrane. They apparently can diffuse freely and cluster to form larger, ordered subdomains in the membrane and in some cases are located in specific regions of the cell membrane. Lipid rafts are assembled in the Golgi complex in mammalian cells.

These microdomains are probably bilayer in structure and can have a different distribution of lipids than the total membrane. The outer leaflets of rafts are composed primarily of glycosphingolipids (see p. 448) and cholesterol, and the inner leaflet contains phospholipids and cholesterol. Glycosphingolipids usually contain longer-chain fatty acids, and their accumulation in the outer leaflet would lead to areas of increased thickness. Lipids in the rafts are more ordered and tightly packed than the surrounding bilayer. Electrostatic interactions of polar head groups, hydrophobic interactions of cholesterol with selected phospholipids or glycolipids, and protein–lipid interactions can constrain movement of lipids. The acyl chains of the phospholipids present in rafts are more highly saturated than the acyl chains in the non-raft areas, which allows close packing of the chains. The presence of cholesterol also leads to a less fluid ordered domain and it may have a major role in maintaining the integrity and function lipid rafts. Thus, the individual lipids in the rafts are constrained and do not diffuse readily throughout the entire membrane.

The partitioning of proteins in and out of rafts is tightly regulated. Some membrane proteins are mainly associated with rafts, some apparently move in and out of rafts, and others are present only in the liquid-disordered portions of the membrane. This partitioning serves to organize proteins in the membrane. Glycosylphosphatidylinositol (GPI)-anchored (see p. 457) proteins partition into lipid rafts and are markers for these subdomains, and CAAX prenylated anchored proteins (see p. 458) are in rafts on the inner leaflet. In some cells, the raft and the sequestered protein are attached to the cytoskeleton of the cell. Certain proteins involved in cell signaling are maintained in the inactive state in individual rafts; activation occurs when several rafts cluster together, forming a larger raft and bringing together the inactive proteins (Figure 12.30). Other proteins have a weak affinity for rafts in the unliganded state but, after binding to a ligand, undergo a conformational change or oligomerize increasing their affinity for a specific raft. Some peripheral membrane proteins are reversibly palmitoylated leading to association with a raft. They dissociate from the raft upon depalmitoylation.

Plasma membranes of cells from some tissues contain **Caveolae**, small flask-like invaginations whose lipid composition is similar to the lipid rafts. Like lipid rafts, they have a high content of cholesterol and glycosphingolipids. Caveolae, however, contain the cholesterol-binding proteins **caveolin**-1, -2, or -3, that may be responsible for stabilizing the structure.

Lipid rafts have important roles in many cellular processes, including membrane sorting and trafficking, cell polarization, and signal transduction processes. Regulation

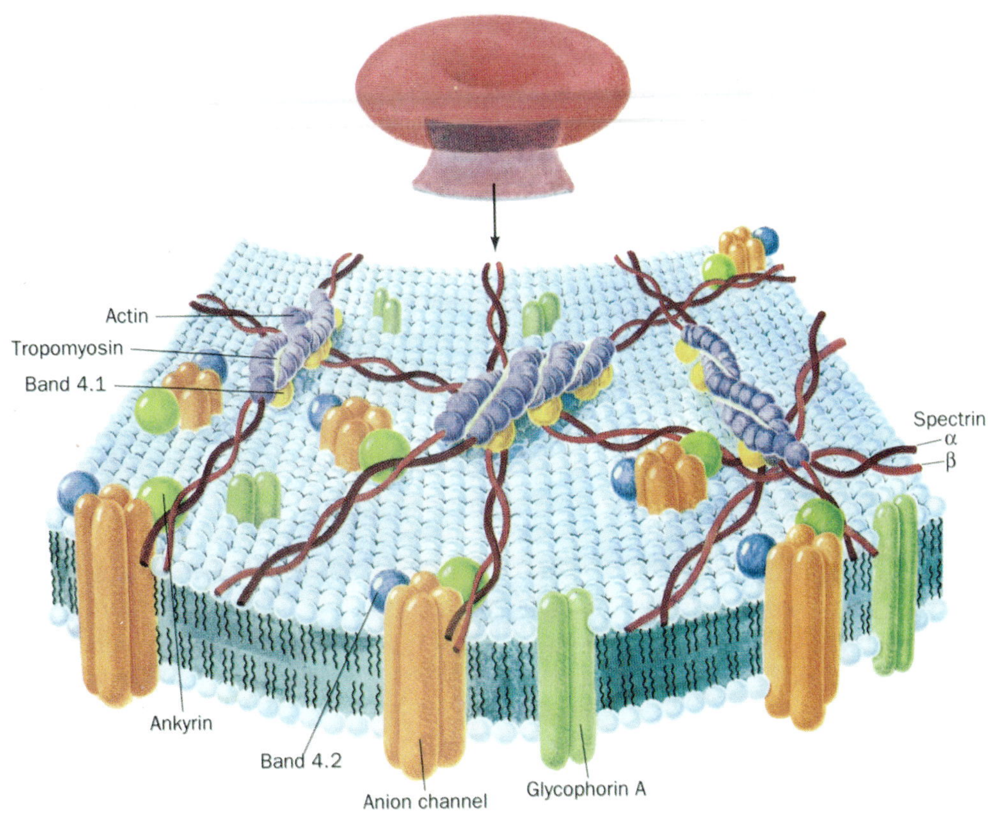

FIGURE 12.31

Schematic diagram of the erythrocyte membrane. Diagram indicates the relationship of four membrane-associated proteins with the lipid bilayer. Glycophorin is a glycoprotein that contains 131 amino acids. The anion channel contains over 900 amino acids and is involved in interacting with ankyrin and in the facilitated diffusion of Cl^- and HCO_3^-. Ankyrin and spectrin are part of the cytoskeleton and are peripheral membrane proteins. Ankyrin binds to the anion channel and spectrin is anchored to the membrane by ankyrin.
Figure reproduced with permission from Voet, D. and Voet, J. *Biochemistry*, 2nd ed. New York: Wiley, 1995. Copyright (1995) John Wiley & Sons.

of cell growth, survival, and apoptosis (see p. 1020) also involve membrane rafts and their clustering. Bacteria, prions, viruses, and parasites may utilize lipid rafts of other cells for their purposes.

Dynamic Nature of Membranes

The structure of the plasma membrane of the human erythrocyte has been investigated extensively because of the ease with which the membrane can be purified from other cellular components. Figure 12.31 is a representation of the interaction of some proteins in this membrane. Random protein movement is restricted. Cellular membranes are in a constantly changing state, with movement of proteins and lipids laterally, changes in association of lipid anchored proteins, and molecules moving into and out of the membrane. The membrane creates a number of microenvironments, from its hydrophobic core to the interface with the surrounding environments. Neither words nor illustrations can capture the time-dependent dynamic changes that occur in the structure of biological membranes.

12.5 MOVEMENT OF MOLECULES THROUGH MEMBRANES

The lipid nature of biological membranes severely restricts the types of molecules that can diffuse through the membrane. Inorganic ions and charged organic molecules do not diffuse at a significant rate because of their attraction to water molecules and exclusion of charged species by the hydrophobic environment, and macromolecules such as proteins and nucleic acids are precluded by their size and charge. The **rate of diffusion** of most nutrients and inorganic ions is too slow to accommodate cellular requirements for them. To overcome this, translocation across membranes of water, ions, nutrients, metabolic waste products, and macromolecules involve specialized channels and transporters. Integral membrane proteins mediate transmembrane transport processes. Several thousand individual channels and transporters have been identified, many by genomic technology. A large majority are present in prokaryotes, where there is a great diversity of transporters. Some of the many *families* of protein channels and transporters, and the variety of substances transported are listed in Table 12.4. There are multiple types of translocation systems for some substrates; in all cases, except for a bacterial process termed group translocation, the substance is unchanged during translocation. Transmembrane transport processes in eukaryotes sometimes function in conjunction with receptors or receptor domains on the extracellular surface and with energy-coupling and regulatory proteins domains on the cytoplasmic side of the protein. The discussion that follows emphasizes the basic mechanisms of transport systems particularly in eukaryotes. The physiological roles of specific translocation systems in cellular homeostasis are described in the appropriate sections of the book.

TABLE 12.4 Number of Protein Families of Channels and Transporters Based on Substrate

	Number of Protein Families		
Substrate	***Channels and Porins***	***Primary Transporters***	***Secondary Transporters***
Inorganic molecules	45	16	44
Sugars, organic acids, and other compounds	5	2	26
Amino acids	4	2	26
Bases, nucleosides and nucleotides	2	0	10
Vitamins and cofactors	2	1	11
Drugs and toxins	1	1	6
Macromolecules	14	8	4

Data from Saier, M. H., A functional-phylogenetic classification system for transmembrane solute transporters. *Micro. Mol. Biol. Rev.*, 64:354, 2000.

Some Molecules Diffuse through Lipid Bilayers

Diffusion of a solute through a membrane involves the solute: (1) leaving the aqueous environment on one side and entering the membrane; (2) traversing the membrane; and (3) leaving the membrane and entering a new environment on the opposite side (Figure 12.32). Each step involves equilibrium of solute between two states. Thermodynamic and kinetic constraints control the concentration equilibrium of a solute on two sides of a membrane and the rate at which it attains equilibrium. Diffusion of gases such as O_2, N_2, CO_2, NO, CO, and H_2S occurs rapidly through a lipid bilayer and the rate depends entirely on the concentration gradient. Water diffuses under osmotic forces, but to meet the needs for rapid equilibration of water across plasma membranes a family of protein transporters, termed aquaporins, facilitates transmembrane movement of water (see p. 465). For diffusion of a solute, the surrounding shell of water must be stripped away before it enters the lipid milieu and then is regained on leaving the membrane. Distribution of lipophilic substances between aqueous phase and lipid membrane depends on the degree of lipid solubility of the substance; very lipid-soluble materials will dissolve readily in the membrane. The rate of diffusion is directly proportional to its solubility and diffusion coefficient in lipids; the latter is a function of the size and shape of the substance.

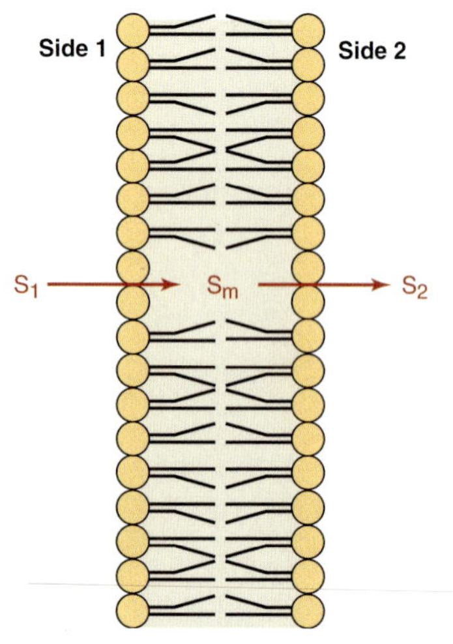

FIGURE 12.32

Diffusion of a solute molecule through a membrane. S_1 and S_2 are solute on each side of membrane, and S_m is solute in membrane.

Movement of solutes by diffusion is always from a higher to a lower concentration, and the rate is described by **Fick's first law of diffusion**:

$$J = -D\left(\frac{\delta c}{\delta x}\right) \quad (12.1)$$

where J is the net amount of substance moved per time, D is the diffusion coefficient, and $\delta c/\delta x$ is the chemical gradient of substance. As the concentration of solute on one side of the membrane is increased, there is an increasing *initial rate* of diffusion as illustrated in Figure 12.33. Net movement of molecules from one side to another will continue until the concentration in each is at chemical equilibrium. Continued exchange of solute molecules from one side to another occurs after equilibrium is attained, but there is no net accumulation on either side because this would recreate a concentration gradient.

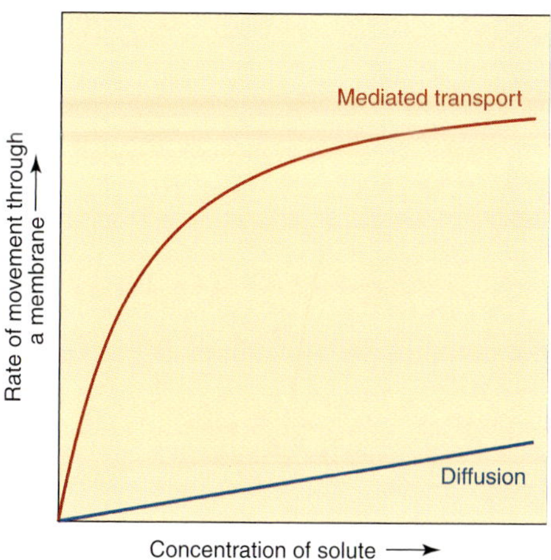

FIGURE 12.33
Kinetics of movement of a solute molecule through a membrane. Initial rate of diffusion is directly proportional to the concentration of solute. In mediated transport, rate will reach a V_{max} when carrier is saturated.

Classification and Nomenclature of Membrane Translocation Systems

Biological membranes have two major protein-mediated mechanisms for translocation of substances; they are **channels (pores)** and **transporters.** These mechanisms can be further classified based on other parameters including specific mechanisms, energetics, and substrate specificity. The International Union of Biochemistry and Molecular Biology (IUBMB) established a functional–phylogenetic system for classification of transport mechanisms. For the following discussion, however, the various systems are grouped together as presented in Table 12.5.

TABLE 12.5 Classification of Membrane Translocation Systems

Type	*Class*	*Example*
Channel	1. Voltage regulated	Na^+ channel
	2. Agonist regulated	Acetylcholine receptor
	3. cAMP regulated	Cl^- channel
	4. Other	Pressure sensitive
		Stretch sensitive
		Heat sensitive
Transporter	1. Facilitated diffusion	Glucose transporter
	2. Active mediated	
	a. Primary-redox coupled	Respiratory chain linked
	Primary-ATPases	Na^+, K^+-ATPase
	ATP binding cassette	Multidrug resistance protein transporter
	b. Secondary	Na^+-dependent glucose transport

A variety of names have been used in the scientific literature for the proteins involved in translocation, including transporter, translocase, translocator, permease, pump, transporter system, translocation mechanism, and mediated transport system. The IUBMB has recommended that the naming be limited to **channels, pores, transport system, transporter, porter, permease system, or permease.**

Channels and **pores** facilitate translocation of molecules or ions across cell membranes by creating a central aqueous channel in the protein that permits diffusion of substances in both directions. Substances move only in the direction of lower concentration. Channel proteins do not bind or sequester the molecules or ions in transit, and their specificity is based on the size and charge of the substance. Flow through channels can be inhibited and is regulated by various mechanisms that open or shut the passageway.

Transporters catalyze movement of a molecule or ion across the membrane by binding and physically moving it across. The activity can be evaluated in the same kinetic terms as an enzyme catalyzed reaction. Transporters have specificity for the substance to be transported, referred to as the substrate, have defined kinetics, catalyze a vectorial transport, and can be affected by both competitive and noncompetitive inhibitors. Some transporters move their substrate only down the substrates concentration gradient (referred to frequently as **passive transport, facilitated diffusion or protein-mediated**

CLINICAL CORRELATION 12.3

Cystic Fibrosis and the Cl^- Channel

Cystic fibrosis (CF), an autosomal recessive disease, is the commonest, serious, inherited disease of Caucasians, occurring with a frequency of 1 in 2000 live births. It is a multi-organ disease, with pulmonary obstruction as a principal manifestation; thick mucous secretions obstruct the small airways allowing recurrent bacterial infections. Exocrine pancreatic dysfunction occurs early and leads to steatorrhea (fatty stool); see p. 1060 for a discussion of the role of the pancreas in fat digestion and absorption. CF patients have reduced Cl^- permeability that impairs fluid and electrolyte secretion, leading to luminal dehydration. Diagnosis of CF is confirmed by a significant increase of Cl^- content of sweat of affected in comparison to normal individuals.

The gene responsible for CF was identified in 1989, and over 800 mutations leading to CF have been found. The most common mutation affects about 70% of the patients and is a deletion of a single phenylalanine at position 508 on the protein, but missense, nonsense, frameshift, and splice-junction mutations (see p. 160) have been reported. Many mutations lead to a change in protein folding. The CF gene product is the **c**ystic **f**ibrosis **t**ransmembrane conductance **r**egulator (CFTR). CFTR is a cAMP-dependent Cl^- channel, which may regulate other ion channels; it is expressed in epithelial tissues. Phosphorylation of a cytoplasmic regulatory domain by protein kinase A activates the channel. CFTR is a polypeptide of 1480 amino acids with structural homology to the superfamily of ATP-binding cassette (ABC) transporters. The gene has been cloned and a major effort is under way to treat the disease by gene therapy, using both viral and nonviral vectors including liposomes (see Clin. Corr. 12.1).

Source: Kunzelmann K. and Mall M. Pharmacotherapy of the ion transport defect in cystic fibrosis. *Clin. Exp. Pharm. Physiol.* 28:857, 2001. Zeitlin, P. L. Therapies directed at the basic defect in cystic fibrosis. *Clin. Chest Med.* 19:515, 1998; Frizzell R. A. Functions of the cystic fibrosis transmembrane conductance regulator protein. *Am. J. Respir. Crit. Care Med.* 151:54, 1995. Naren, A. P., Cormet-Boyaka, E., Fu, J., et al. CFTR chloride channel regulation by an interdomain interaction. *Science* 286:544, 1999.

AQP9); sites of expression of selected aquaporins in humans are presented in Table 12.6. Some tissues have several different isoforms with different functions; the kidney has at least seven isoforms located at distinct sites along the nephron and collecting duct. Clinical Correlation 12.4 describes the distribution of aquaporins in kidneys and examples of pathophysiological changes. Plant aquaporins contribute to water uptake by rootlets, transpiration, and seed desiccation.

Aquaporins are small, very hydrophobic, intrinsic membrane proteins with several highly conserved amino acid sequences. They assemble in membranes as homotetramers (Figure 12.35). Each monomer, consisting of six transmembrane α-helical domains, contains a distinct water pore. AQP1 of the human erythrocyte contains an asparagine–proline–alanine sequence, a highly conserved sequence motif within the aquaporins, in two separate loops. The loops fold into the lipid bilayer in an hourglass fashion, forming an aqueous channel. The selectivity is due to a constriction of the channel to about 2.8 Å, which limits the size of molecules that can pass through. The aqueous pathway of AQP1 (Figure 12.36) is lined with conserved hydrophobic residues

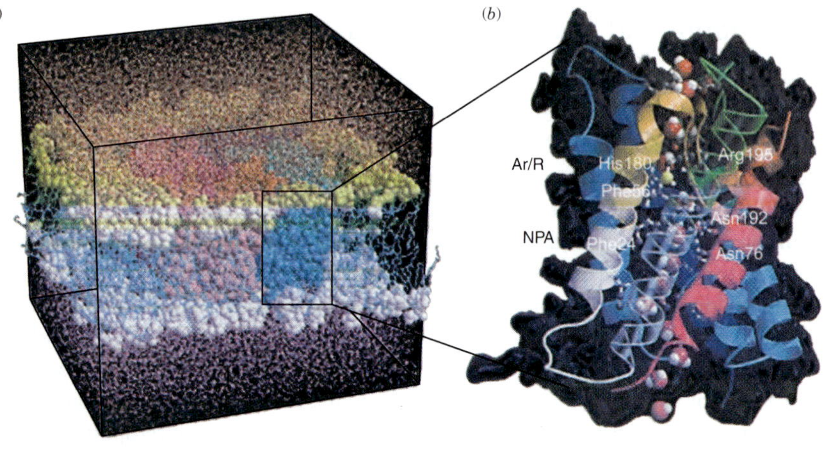

FIGURE 12.35

Simulation of water permeation through AQP1.
(*a*) The simulation of the AQP1 tetramer (orange/magenta/blue/cyan) embedded in a palmitoyl oleoyl phosphatidylethanolamine bilayer (yellow/green) surrounded by water (red/white) consisted of approximately 100,000 atoms. (*b*) Simulation of the pathway of one of the monomers of APQ1 for water permeation observed through the pore.
Figure reproduced with permission from Fujiyoshi, Y., Mitsuoka, k., de Groot, B. L., Philippsen, A., Grubmüller, H., Agre, P., and Engel, A. Structure and function of water channels. *Curr. Opin. Struct. Biol.* 12:509, 2002. © Elsevier.

TABLE 12.6 Site of Expression of Selected Aquaporins in Humans

Tissue	Location of Expression	Aquaporin	Function
Blood cells	Erythrocytes	AQP1	Osmotic protection
	Leukocytes	AQP9[a]	?
Brain	Choroid plexus	AQP1	Cerebral spinal fluid production
	Ependymal cells	AQP4	Cerebral spinal fluid balance
	Hypothalamus	AQP4	Osmolarity sensing (?)
Eye	Lens fiber cells	AQP0	Fluid balance in lens
	Ciliary epithelium	AQP1	Production of aqueous humor
Kidney	Proximal tubules	AQP1	Concentration of urine
	Collecting ducts	AQP2	Mediates ADH[b] activity
	Collecting ducts	AQP3[a]	Reabsorption of water into blood
	Collecting ducts	AQP4	Reabsorption of water
	Collecting ducts	AQP6	?
Lung	Alveolar epithelial cells	AQP1	Alveolar hydration
	Bronchial epithelium	AQP4	Bronchial fluid secretion
Mouth and throat	Trachea epithelial cells	AQP3[a]	Secretion of water into trachea
	Salivary glands	AQP5	Production of saliva

[a] An aquaglyceroporin.

[b] Antidiuretic hormone.

Source: King, L. S., Kazono, D., and Agre, P. From structure to disease: The evolving tale of aquaporin biology. *Nature Rev.: Mol. Cell Biol.* 5:687, 2004.

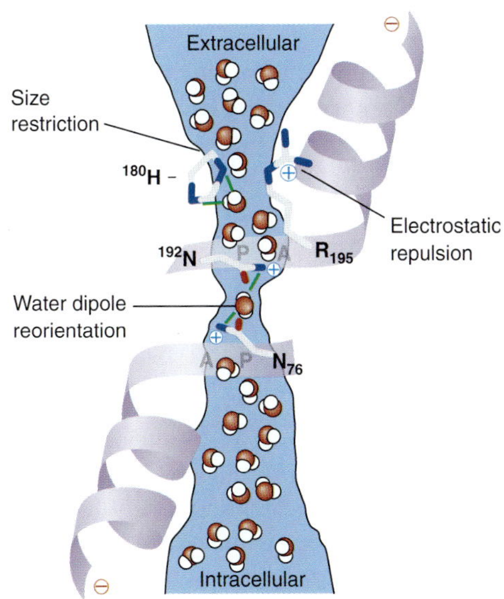

FIGURE 12.36

Water channel within an AQP1 subunit. The shape of the aqueous pore (blue) is derived from calculations based on the structure of bovine AQP1. Three features of the channel specify selectivity for water: (*a*) Size restriction. Eight angstroms above the midpoint of the channel, the pore narrows to a diameter of 2.8 Å (approximately the diameter of a water molecule). (*b*) Electrostatic repulsion. A conserved residue (R, Arg-195) at the narrowest constriction of the pore imposes a barrier to cations, including protonated water (H_3O^+). (*c*) Water dipole reorientation. Two partial helices meet at the midpoint of the channel, providing positively charged dipoles that reorient a water molecule as it traverses this point. Disrupting hydrogen bonding in the single-file chain water molecules prevents the formation of a proton conductance. Reproduced with permission from Kozono, D., Yasui, M., King, L. S. and Agre, P. Aquaporin water channels: Atomic structure and molecular dynamics meet clinical medicine. *J. Clin. Invest.* 109:1395, 2002. © Elsevier

CLINICAL CORRELATION 12.4
The Mammalian Kidney and Aquaporins

The primary role of the kidney is to regulate the pH of blood, eliminate toxic products of metabolism, and regulate water balance of the body. Each kidney contains about one million nephrons, the multicellular functional units of the tissue. Each nephron has several distinct regions (see diagram), each with very specific functions in the processing of the blood filtrate, which occurs in the glomerulus. A large quantity of blood is filtered (about 150 L/day), and the water must be reabsorbed in the nephron and collecting ducts except for about 1.5 L that is excreted as urine. Aquaporins are responsible for the recovery of water from the filtrate as it flows through the lumen of the nephron. As indicated in the diagram, at least seven different isoforms of aquaporins, located at different sites along the nephron and collecting ducts, are responsible for water reabsorption.

There are a number of renal disorders where the reabsorption of water is abnormal and where the expression and function of the aquaporins has been investigated. Several animal models of these conditions have been invaluable in determining the cause of some of these conditions. Low levels of AQP2 and polyuria (excessive excretion of urine) are found in acquired nephrogenic diabetes insipidus (NDI), acquired hypokalemia (low blood K^+), and hypercalcemia (increased blood Ca^{2+}). In many cases, NDI is caused by the inability of the kidney to respond to vasopressin (see p. 920); it is considered that this leads to a decreased expression and/or incorporation of AQP2 in the membrane. In other cases of NDI, there is a defect in the aquaporin gene; in some cases, this defect leads to an inability of the monomers to form normal tetramer structures. Levels of AQP1, AQP2, and AQP3 in animal models are reduced in tissue ischemia. In some conditions, such as congestive heart failure, liver cirrhosis, and pregnancy, there is an increase in the amount of AQP2 in the kidney, leading to an expansion of the extracellular fluid volume. Humans lacking AQP1 activity of the kidney apparently do not have any measurable problems under normal but may under stress conditions (dehydration). It is expected that additional clinical conditions will be attributed to changes in the other aquaporins.

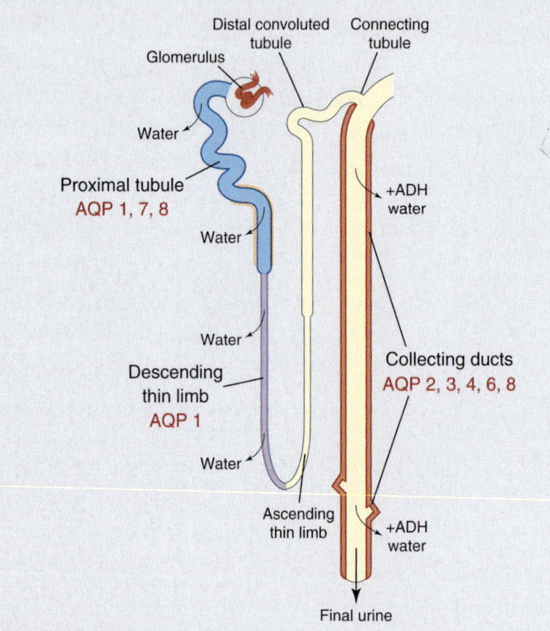

Localization of aquaporins in each segment of the nephron and collecting duct of the kidney.
Modified from figures in the citations below.

Source: King, L. S. and Yasui, M. Aquaporins and disease: Lessons from mice to humans. *Trends Endocrinol. Metab.* 13:355, 2002. Nielsen, S., Frokiaer, J., Marples, D., Kwon, T-H., Agre, P., and Knepper, M. A. Aquaporins in the kidney: From molecules to medicine. *Physiol. Rev.* 82: 205, 2002. King, L. S., Kozono, K., and Agre, P. From structure to disease: The evolving tale of Aquaporin biology. *Nature Rev. Mol. Cell Biol.* 5:687, 2004.

and a few hydrophilic residues that attract H_2O but not charged molecules such as H_3O^+. Water molecules apparently flow through the channel in single file.

In addition to water, AQP1 permits translocation of CO_2, AQP3 of glycerol, and AQP9 of glycerol, urea, polyols, purines, pyrimidines, nucleosides, and monocarboxylates. AQP1 is reversibly inhibited by $HgCl_2$. AQP2 in kidneys of humans is under hormonal control, and at least three others may be controlled by pH.

Na^+, Ca^{2+}, K^+, and Cl^- Voltage-Gated Ion Channels

The individual ion-specific ***v**oltage-gated **i**on **c**hannels* **(VIC)** for Na^+, Ca^{2+}, and K^+ are members of a large superfamily of ion channels found in bacteria, archaea, viruses, and eukaryotes. These channels are controlled by the transmembrane electrical potential, open and close remarkably fast (milliseconds), and in mammalian cells are responsible for the generation of conducted electrical signals in neurons and other excitable cells. Each ion channel consists of specific glycoprotein subunits as homo- or hetero-oligomeric structures (Table 12.7). The large α-subunits (about 260 kDa) contain the ion conducting pore. They have four homologous domains each with six transmembrane segments, connected by both intracellular and extracellular loops, for a total of 24 transmembrane segments (Figure 12.37). Accessory subunits (β, γ, or δ) do not share a common structure; some have several transmembrane segments and others are peripheral proteins located entirely intra- or extracellularly. They are usually involved in regulation of the channel; β-subunits may interact with cytoskeletal and extracellular matrix proteins. A model of the **Na^+ channel** is presented in Figure 12.38. One transmembrane segment has a positively charged residue at every third position and may serve as a voltage sensor; mechanical shift of this segment may lead to a conformational change in the protein, resulting in the opening of the channel. Two very different mechanisms have been suggested, but not proven, for how the channel and the voltage-sensor domains are coupled to initiate the opening of the channel. Voltage-gated ion channels can exist in one of three conformational and functional states: closed (resting), open (active), and inactive. In response to membrane depolarization, there is a transition from the closed to an open state. Both resting and inactive states are closed, and transition between them occurs on binding of ligands or by protein phosphorylation. In humans, there are six types of Ca^{2+} channels and at least 10 types of K^+ channels responding to different stimuli. Clinical Correlation 24.12, p. 986, describes changes in cation voltage-gated channels in myotonic muscle disorders.

TABLE 12.7 Subunit Composition of Voltage-Gated Cation Channels

Cation Channel	*Subunits*
Na^+	α, β_1, and β_2
K^+	4α, β
Ca^{2+}	α_1, α_2, δ, and β

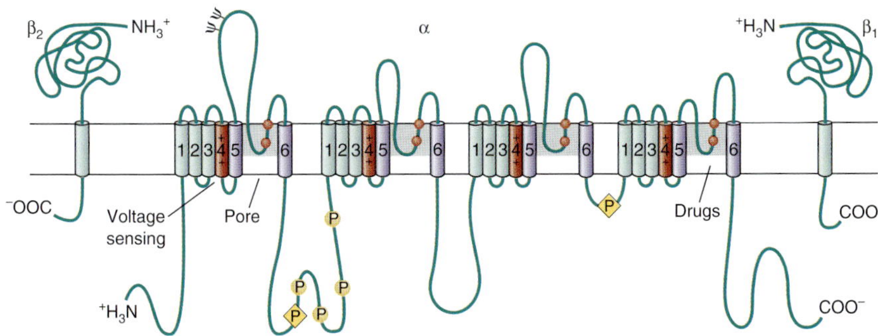

FIGURE 12.37

Subunit structure of the voltage-gated Na^+ channel. The primary structures of the subunits of the voltage-gated Na^+ channel are illustrated as a transmembrane folding diagram. Cylinders represent probable α-helical segments. The red cylinders represent the voltage sensor and the purple cylinders the segments in the pore (see Figure 12.38). Blue lines represent the polypeptide chains of each subunit, with length approximately proportional to the number of amino acid residues in the brain Na^+ channel subtypes. The extracellular domains of the β_1- and β_2-subunits are shown as globular structures. Protein phosphorylation sites are indicated with a P.
Redrawn from Catterall, W. A., Goldin, A. I., and Waxman, S. G. International Union of Pharmacology. XXXIX. Compendium of voltage-gated ion channels: Sodium channels. *Pharmacol. Rev.* 55:575, 2003.

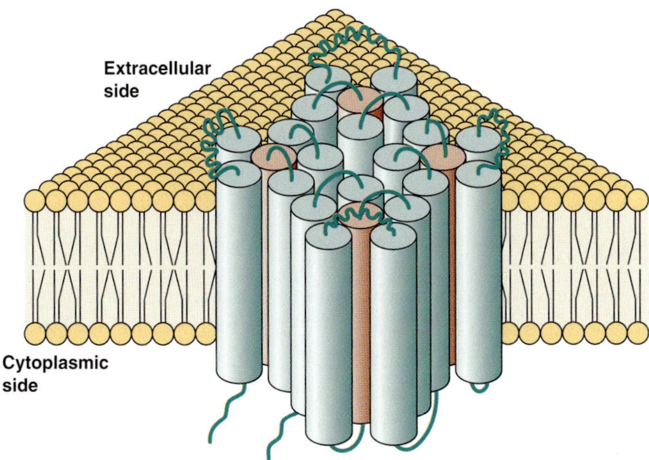

FIGURE 12.38

Model of the Na^+ channel. The single peptide consists of four repeating units with each unit folding into six transmembrane helices. The red transmembrane segments represent the voltage sensor.
Redrawn from Noda, M., et al., *Nature* 320:188, 1986.

(a)

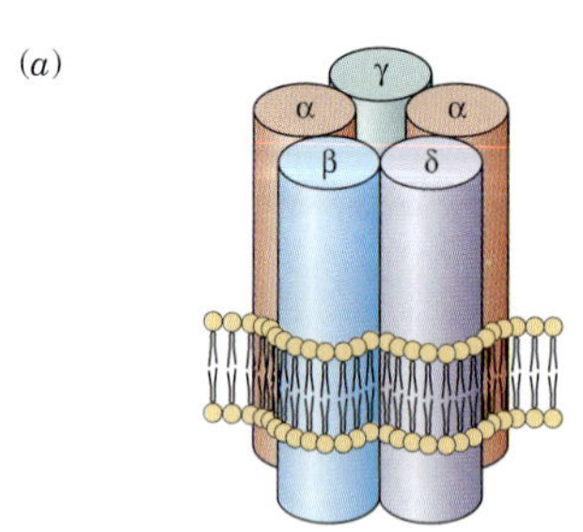

(b)

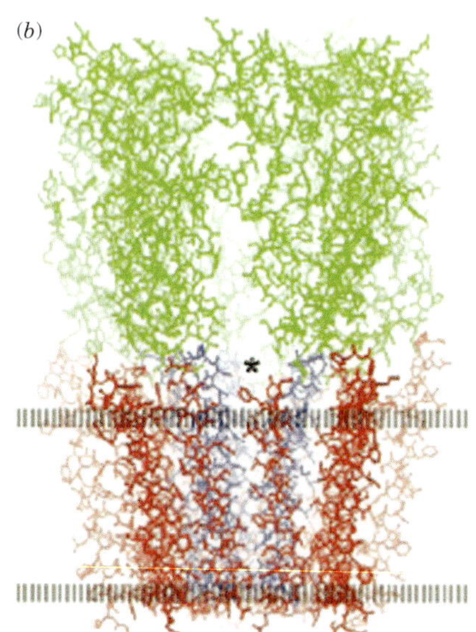

(c)

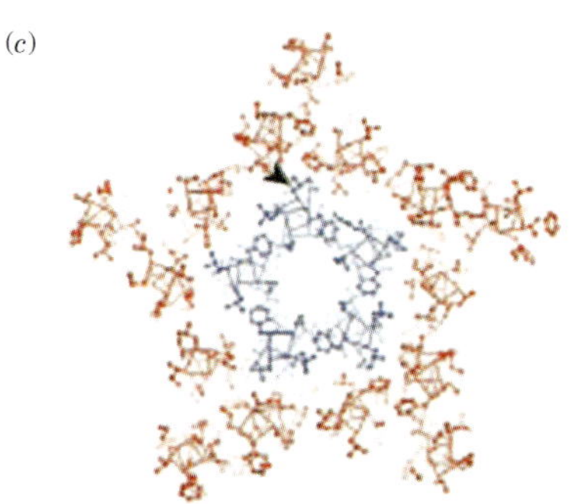

FIGURE 12.40

Model of acetylcholine (ACh) receptor. (*a*) The arrangement of the ACh receptor subunits and their common topology. (*b*) AcH receptor showing the α-helical pore structure (blue, pore-facing; red, lipid-facing helices) in relation to the membrane surfaces and the β-sheet structure (green) comprising the ligand-binding domain; the asterisk denotes open space at a subunit interface. (*c*) Cross-sectional view through the pentamer at the middle of the membrane, showing partitioning of the structure into pore- and lipid-facing parts, with intervening spaces.
Part (a) redrawn from Wilson, G. G. and Karlin, A. Acetylcholine receptor channel structure in the resting, open, and desensitized states probed with the substituted-cysteine accessibility method. *Proc. Nat. Acad. Sci. USA* 98:1241, 2001. Parts (b) and (c) reproduced with permission from Miyazawa, A., Fujiyoshi, Y., and Unwin, N. Structure and gating mechanism of the acetylcholine receptor pore. *Nature* 423:949, 2003. Copyright (2003) Nature.

$$CH_3-\overset{\overset{O}{\|}}{C}-O-CH_2-CH_2-\overset{+}{N}-(CH_3)_3$$

FIGURE 12.39

Structure of acetylcholine.

In contrast to the cation channels, not much is known about the structure of the **voltage-gated Cl^- channels** (see p. 953); they apparently have 12 transmembrane helical segments. Cl^- channels regulate cell volume, stabilize membrane potential, are involved in signal transduction, and regulate transepithelial transport. The cystic fibrosis transmembrane regulator is a Cl^- channel. The voltage-gated ion channels have a distinctive expression pattern in different tissues. In humans, there are a considerable number of inherited ion channel diseases, termed channelopathies, caused by mutations in the anion and cation channels (see Clinical Correlation 24.11, p. 986).

Nicotinic-Acetylcholine Channel (nAChR)

The **nicotinic-acetylcholine channel**, also referred to as the **acetylcholine receptor**, is an example of a ligand-regulated channel, in which binding of acetylcholine (Figure 12.39) opens the channel. The dual name is used to differentiate this receptor from other acetylcholine receptors, which function in a different manner (see p. 957). The channel is expressed at the nerve–muscle synapse (see p. 985) and is a member of a superfamily of transmitter-gated ion channels, which includes the serotonin, γ-aminobutyric-acid, and glycine receptors (see p. 959). Acetylcholine diffuses to the skeletal muscle membrane, where it interacts with the acetylcholine receptor, opening the channel, and allowing positively charged ions and small nonelectrolytes but not negatively charged ions to flow through the channel. These receptors are formed as pentameric complex of two α- and one each of β-, γ-, and δ-subunits (Figure 12.40). Each α-subunit is glycosylated and two other subunits contain covalently bound lipid. The subunits each have an extracellular N- and C-terminus, four membrane-spanning domains (M1, M2, M3, M4), and a long cytoplasmic loop between M3 and M4 that contains consensus sites for phosphorylation by protein kinases. Five juxtaposed M2 domains (one from each protein subunit) form the ion-conducting pore. The M2 domains of cation-selective receptor channels, such as the nicotinic acetylcholine receptor, are enriched in negatively charged amino acids at the intracellular "mouth" of the pore. This acts to attract cations like Na^+ and repel anions like Cl^-. Binding of the neurotransmitter to extracellular sites on the receptor complex causes very small and subtle rearrangements of how the M2 domains pack together. This near-instantaneous rearrangement removes the energetic barriers to ionic flow through the water-lined pore. Phosphorylation of the α-subunit is required for activity, and closure of the channel occurs within a millisecond due to (a) hydrolysis of acetylcholine to acetate and choline and (b) the release of acetate and choline from the protein.

The nicotinic-acetylcholine receptor is inhibited by several deadly neurotoxins including ***d*-tubocurarine**, the active ingredient of **curare**, and several toxins from snakes including **α-bungarotoxin, erabutoxin**, and **cobratoxin. Succinylcholine**, a muscle relaxant, opens the channel leading to depolarization of the membrane; succinyl choline is used in surgical procedures because its activity is reversible due to the rapid hydrolysis of the compound after cessation of administration.

Gap Junctions and Nuclear Pore Channels

Gap junctions and nuclear pore channels are relatively large multimeric protein channels. **Gap junctions**, found in the plasma membranes of mammalian cells, consist of transmembrane channels, **connexons**, which create a connecting channel between two cells. Connexons in the plasma membrane of one cell dock end-to-end with connexons in the membrane of a neighboring cell. Connexons consist of homo- or hetero-hexameric proteins of **connexins** (**gap junction proteins**). Over 15 connexin subunit isoforms

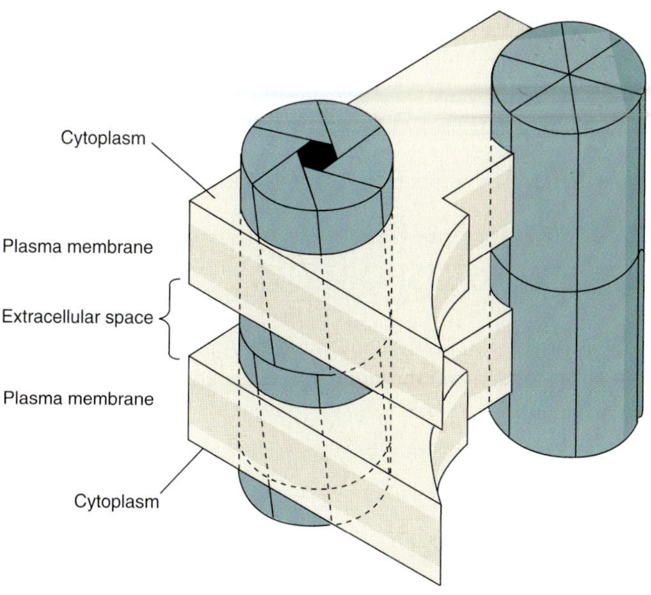

FIGURE 12.41
Model for a channel in the gap junction.

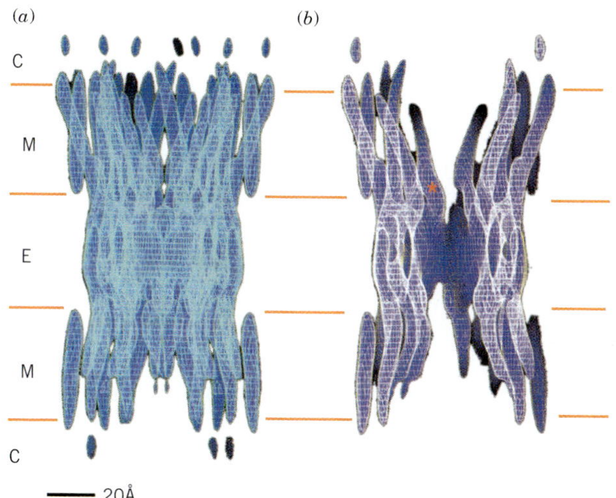

FIGURE 12.42
Molecular organization of recombinant cardiac gap junction. Model developed from electron crystallography. (*a*) A full side view is shown. (*b*) The density has been cropped to show the channel interior. M, membrane bilayer; E, the extracellular space; C, the cytoplasmic space.
Reprinted with permission from Unger, V. M., Kumar, N. M., Gilula, N. B., and Yeager, M. *Science* 283:1176, 1999. Copyright (1999) AAAS.

are known, and each has four putative α-helical transmembrane sequences. Twelve subunits, six in each cell, form a hexameric structure in each membrane (Figure 12.41). The molecular organization of a recombinant gap junction is presented in Figure 12.42. The diameter of the opening ranges from 1.2 to 2 nm. The aqueous channel mediates electrical coupling, as well as exchange of ions and metabolites (but not macromolecules), between cells. Gap junctions composed of different connexins may exhibit differing specificities for solutes. The channels are normally open. Closure occurs with increases in cytoplasmic Ca^{2+}, a change in metabolism, a drop in transmembrane potential, or acidification of the cytoplasm. Some gap junctions may be controlled by phosphorylation of the connexins. When the channel is open the subunits appear to be slightly tilted, but when closed they appear to be more perpendicular in the membrane, suggesting that subunits slide over each other.

The **nuclear pore complex** is present in the nuclear membranes of eukaryotic cells, spanning two membranes, creating an aqueous channel in the nuclear envelope. Pores are about 9–20 nm in diameter and 45 nm in length. The nuclear pore complex is composed of over 50 proteins, called nucleoporins. Large (proteins, RNA, and ribonucleoprotein particles) and small molecules are readily transported in both directions between the nucleoplasm and the cytoplasm. Details of the mechanism of macromolecule transport are not known. Proteins destined for the nucleus contain short amino acid sequences, termed **nuclear localization sequences**, which target them for nuclear import. Nuclear receptors, **karyopherins**, and cytosolic factors are involved in nuclear import and export of the proteins. Karyopherin β mediates passage of macromolecules by interacting with nucleoporins and a GTPase protein, **Ran**. ATP (or GTP) is required for nuclear export of karyopherin β but not for import. Transport through the nuclear pore complex occurs by facilitated diffusion of the soluble carrier proteins and carrier-macromolecule complexes.

12.7 MEMBRANE TRANSPORTERS

Membrane protein transporters facilitate the movement of a molecule or molecules through the lipid bilayer at a rate that is significantly greater than can be accounted for by simple diffusion. If S_1 is the solute on side 1 and S_2 on side 2, then the transporter promotes the establishment of equilibrium as follows:

$$[S_1] \rightleftharpoons [S_2]$$

Brackets represent the concentration of solute. If the transporter T is included, the reaction becomes

$$[S_1] + T \rightleftharpoons [S - T] \rightleftharpoons [S_2] + T$$

If no energy from another source is expended, the concentration on both sides of the membrane will be equal at equilibrium, but if there is an expenditure of energy, a concentration gradient can be established. Note the similarity of the transporter to that of an enzyme; in both cases, the protein increases the rate but does not determine the final equilibrium.

Table 12.8 lists major characteristics of membrane transporters. Protein-mediated transport demonstrates saturation kinetics (Figure 12.33; p. 463); as the concentration of the substance to be translocated increases, the initial rate of transport increases but the rate reaches a maximum when the substance saturates the protein transporter. Simple diffusion does not exhibit saturation kinetics. Constants such as $V_{\max}$ and K_m can be determined for transporters. Theoretically a transporter can facilitate movement of a solute in both directions across the membrane, but in most situations there is a unidirectional movement because of the large $\Delta G^{o'}$ for the reaction.

Most transporters have a high degree of structural and stereospecificity for substrate. In mediated transport of D-glucose in erythrocytes, the K_m for D-galactose is 10 times larger and for L-glucose 1000 times larger than for D-glucose, and the transporter has essentially no activity with D-fructose or disaccharides. Structural analogs of the substrate for a transpor inhibit competitively, and reagents that react with specific groups on proteins are noncompetitive inhibitors.

Four Steps in the Transport of Solute Molecules

There are four steps of mediated transport (Figure 12.43): (1) recognition by transporter of appropriate solute from amongst a variety of solutes in the aqueous environment, (2) translocation of solute across membrane, (3) release of solute by transporter, and (4) recovery of transporter to its original condition.

TABLE 12.8 Characteristics of Membrane Transporters

Passive Transport	*Active Transport*
1. Saturation kinetics	1. Saturation kinetics
2. Specificity for solute transported	2. Specificity for solute transported
3. Can be inhibited	3. Can be inhibited
4. Solute moves down concentration gradient	4. Solute can move against concentration gradient
5. No expenditure of energy	5. Requires input of energy

Recognition:	$S_1 + T_1 \rightleftharpoons S - T_1$
Transport:	$S - T_1 \rightleftharpoons S - T_2$
Release:	$S - T_2 \rightleftharpoons T_2 + S_2$
Recovery:	$T_2 \rightleftharpoons T_1$

FIGURE 12.43

Steps involved in mediated transport across a biological membrane. S_1 and S_2 are solutes on side 1 and 2 of the membrane, respectively; T_1 and T_2 are binding sites on the transporter on sides 1 and 2, respectively.

The first step, **recognition** of a specific substrate by the transporter, is the same as recognition of a substrate by an enzyme (p. 971). Specific binding sites on the protein recognize the correct structure of the solute. The second step, **translocation**, requires a conformational change in the transporter after binding of the substrate (Figure 12.44), which moves the substrate a short distance, perhaps only 2 or 3 Å to the environment of the opposite side of the membrane. It is not necessary for the transporter to move the molecule the entire distance across the membrane. Step 3, **release** of substrate, depends on the affinity of the transporter for the substrate. Without a change in affinity (K_{eq}) for substrate, the substrate will be released in the new environment if the concentration of substrate is lower in the new environment than in the initial environment because of a shift in the equilibrium between bound and free substrate. This occurs with passive transporters. For active transporters, which move a substrate against a concentration gradient, release of the substrate at the higher concentration requires a change to a lower affinity of the transporter for substrate. A conformational change of transporter can lead to a decrease in affinity. Finally in **recovery**, step 4, the transporter returns to its original conformation upon release of substrate.

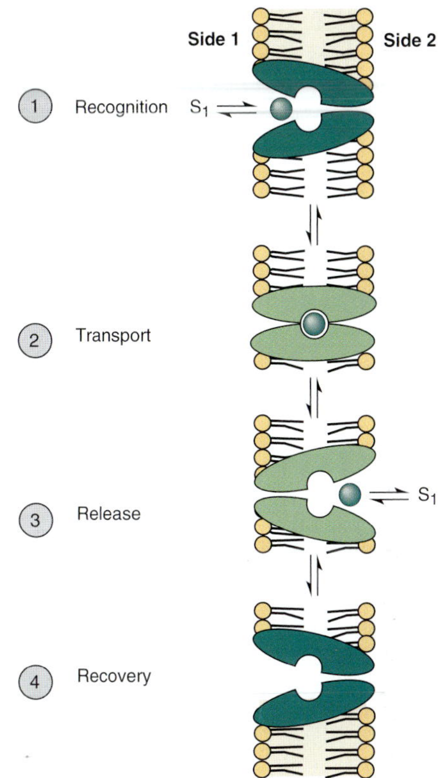

FIGURE 12.44

Model for a mediated transport system in a biological membrane. Model is based on the concept of specific sites for binding of substrate and a conformational change in the transporter to move bound solute a short distance but into the environment of the other side of the membrane. Once moved, solute is released from the transporter.

The discussion above centered on movement of a single substrate by a transporter (**uniporter**). There are systems that move two substrates simultaneously in one direction (**symporter**) and two substrates in opposite directions (**antiporter**) (Figure 12.45). When a charged substance, such as K^+, is translocated and no ion of the opposite charge is moved, a charge separation occurs across the membrane. This mechanism is termed **electrogenic transport** and leads to development of a membrane potential. If an oppositely charged ion is moved to balance the charge, the mechanism is called **neutral** or **electrically silent transport**.

Energetics of Membrane Transport Systems

Equation 12.2 gives the change in free energy when an uncharged molecule moves from concentration C1 to concentration C2 on the other side of a membrane.

$$\Delta G' = 2.3RT \log\left(\frac{C_2}{C_1}\right) \tag{12.2}$$

When $\boldsymbol{\Delta G'}$ is negative—that is, there is release of free energy—movement of solute occurs without the need for a driving force. When $\Delta G'$ is positive, as would be the case if C_2 is larger than C_1, then there needs to be input of energy to drive the transport. For a charged molecule (e.g., Na^+) both the electrical potential and concentrations of solute are involved in calculating the change in free energy as indicated in Eq. 12.3:

$$\Delta G' = 2.3RT \log\left(\frac{C_2}{C_1}\right) + Z\mathscr{F}\Psi \tag{12.3}$$

where $\Delta G'$ is electrochemical potential, Z is the charge of the transported species, $\mathscr{F}$ is Faraday constant (23.062 kcal V^{-1} mol^{-1}), and $\Delta\psi$ is the electrical potential difference in volts across the membrane. The magnitude and sign of both Z and $\Delta\psi$ influence $\Delta G'$.

Passive transport (facilitated diffusion or protein-mediated diffusion) occurs when $\Delta G'$ is negative and the movement of solute occurs spontaneously. When $\Delta G'$ is positive, input of energy from some source is required and the process is called

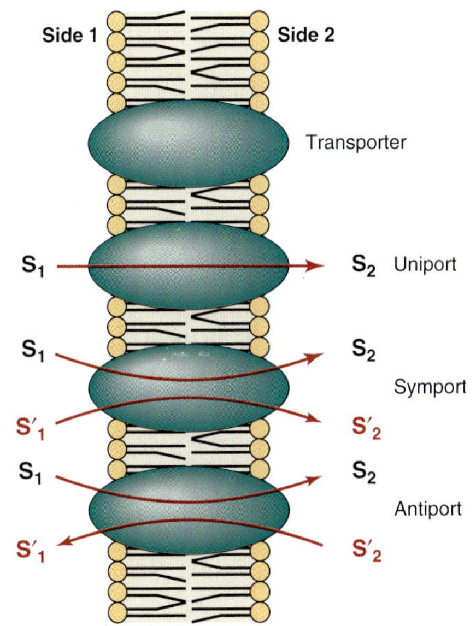

FIGURE 12.45

Uniport, symport, and antiport mechanisms for translocation of substances. S and S′ represent different molecules.

active transport (pump). Different chemical reactions supply the energy for active transport; prokaryotes have transporters driven by decarboxylation and methyl transfer reactions, and plants have light-absorption-driven transporters. Active transport in mammalian cells is driven by either hydrolysis of nucleoside triphosphate (e.g., ATP to ADP + phosphate) or utilization of an electrochemical gradient of Na^+ or H^+ across the membrane.

Some transporters maintain very large concentration gradients, such as the plasma membrane transporter that maintains the Na^+ and K^+ gradients (p. 12). A striking example is an active transporter in parietal cells of gastric glands, which is responsible for secretion of HCl into the lumen of the stomach (p. 1051). The pH of plasma is about 7.4 (4×10^{-8} M H^+), and the luminal pH of the stomach can reach a pH of 0.8 (0.15 M H^+). H^+ is transported against a concentration gradient of $1 \times 10^{6.6}$. Assuming there is no electrical component, the energy for H^+ secretion under these conditions can be calculated from Eq. 12.2 and is 9.1 kcal mol^{-1} of HCl.

12.8 PASSIVE TRANSPORT

Passive transport occurs without expenditure of metabolic energy (Table 12.8; p. 472). Hundreds of passive transporters have been reported, most in bacteria, with specificity for inorganic ions, sugars, amino acids, and tricarboxylic acid cycle and glycolytic intermediates. Many have 400–600 amino acid residues and have 2–24 putative α-helical transmembrane sequences. Passive-mediated transport demonstrates saturation kinetics, specificity for the class of substrate moving across the membrane, and specificity for inhibitors.

Glucose is Translocated by Passive Transport

A family of ***glu*cose *t*ransporters** (**GLUT**) facilitates transport of D-glucose across the plasma membranes of mammalian cells by a uniport mechanism. The genes for the GLUT family are expressed in a tissue specific manner. The glucose transporters, designated as GLUT1, GLUT2, and so on (Table 12.9), all have 12 transmembrane segments, with a significant amino acid similarity. Most are in the plasma membrane, where

TABLE 12.9 Major Isoforms of Glucose Transporter

Isoform	*Tissue*	*Cellular Localization*	*Comments*
GLUT1	Muscle Heart Blood-brain barrier Glia cells Placenta	Plasma membrane	Activity in muscle increased by insulin, hypoxia, and diet
GLUT2	Liver Pancreas Intestine Kidney	Plasma membrane	
GLUT3	Brain Kidney Placenta	Plasma membrane	
GLUT4	Muscle Adipose Heart Blastocysts	Plasma membrane	Activity in muscle and adipose tissue stimulated by insulin, and in muscle by hypoxia and diet
GLUT5	Muscle Spermatozoa	Sarcolemmal vesicles	Fructose as substrate

direction of movement of glucose is usually out to in. Depending on the concentration gradient, however, erythrocyte **GLUT1** facilitates transport in both directions demonstrating the reversibility of the system. The three high-affinity transporters (GLUT1, **GLUT3** and **GLUT4**) function at rates close to maximal velocity because their K_m values are below the normal blood glucose concentration. D-Galactose, D-mannose, D-arabinose, and several other D-sugars as well as glycerol are translocated by some isoforms. **GLUT5** of sarcolemmal membranes of skeletal muscle transports fructose preferentially. GLUT1 is important for supplying erythrocytes and brain cells with glucose. The rate of glucose uptake by the low-affinity GLUT2 increases in parallel with the rise in blood glucose concentration and has a major role in intestinal absorption of sugars (see p. 1059). In liver cells, **GLUT2** catalyzes both glucose influx and efflux, having responsibility for glucose export (p. 598); in pancreatic β-cells, GLUT2 is involved in sensing blood glucose levels. GLUT4 is an insulin-responsive transporter (p. 869) in adipose tissue, heart muscle, and skeletal muscle. Activity of some glucose transporters in muscle is stimulated by exercise and hypoxia apparently because there is a need for increased glucose utilization by the tissue. Several sugar analogs as well as phoretin and 2,4,6-trihydroxyacetophenone (Figure 12.46) are competitive inhibitors.

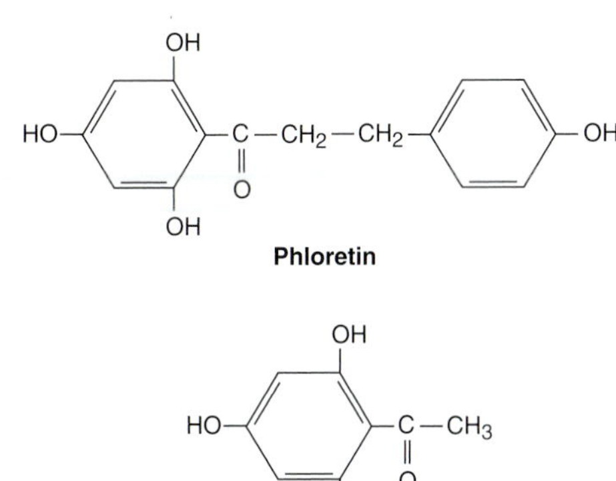

FIGURE 12.46
Inhibitors of passive transport of d-glucose in erythrocytes.

Cl^- and HCO_3^+ are Transported by an Antiport Mechanism

An anion transporter in erythrocytes and kidneys mediates the antiport movement of Cl^- and HCO_3^- (Figure 12.47). The transporter is referred to as the **Na^+ independent Cl^-–HCO_3^- exchanger** or **anion exchange protein**, and in erythrocytes as **band 3**, the latter because of its position in SDS polyacrylamide gel electrophoresis of erythrocyte membrane proteins. The direction of ion flow depends on the concentration gradients of the ions across the membrane. The transporter is important in adjusting the erythrocyte HCO_3^- concentration in arterial and venous blood. The kidney transporter is a truncated form of the erythrocyte protein and is responsible for base (HCO_3^-) efflux to balance ATP driven H^+ efflux.

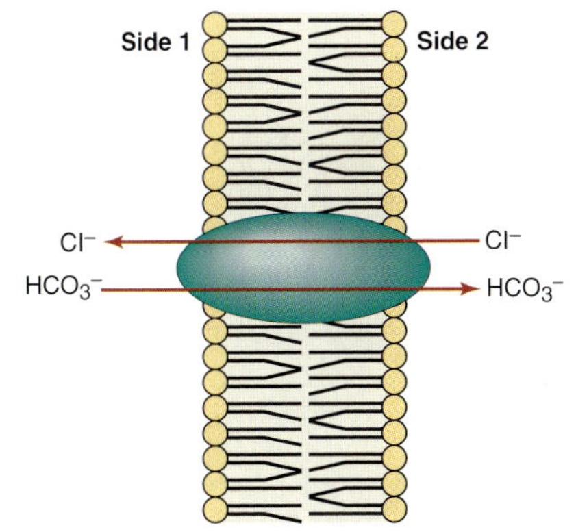

FIGURE 12.47
Passive anion antiport mechanism for movement of Cl^- and HCO_3^- across the erythrocyte plasma membrane.

Mitochondria Contain a Number of Membrane Transporters

The inner mitochondrial membrane contains transporters for exchange of anions between cytoplasm and mitochondrial matrix, including (1) an antiporter for exchange of ADP and ATP, (2) a symporter for transport of phosphate and H^+, (3) a dicarboxylate transporter, and (4) an antiporter for glutamate-aspartate (see p. 568 for a detailed discussion). These transporters are important in energy conservation and metabolism. In the absence of an input of energy, these transporters can mediate a passive exchange of metabolites down their concentration gradient to achieve a thermodynamic equilibrium. In an antiport mechanism, the concentration gradient of one solute can drive the movement of the other solute. In some cases, the transporter mediates the antiport movement of an equal number of electrical charges on the substrates with the mitochondrial membrane potential influencing the equilibrium.

12.9 ACTIVE TRANSPORT

Active transporters catalyze the translocation of a solute against its concentration gradient, leading to creation of a chemical or electrochemical gradient. Like passive transporters, they are characterized by saturation kinetics, specificity for one or a few related substrates, and an ability to be inhibited (Table 12.8; p. 472).

Primary active transporters, which are found in all living organisms, require an energy source to drive uptake or extrusion of solute. Mammalian cells require ATP or another nucleoside triphosphate as the energy source; the transport proteins may be transiently phosphorylated, but the substrate is not phosphorylated. Transporters utilizing ATP are also referred to as an **ATPase**, an enzyme activity, since ATP is

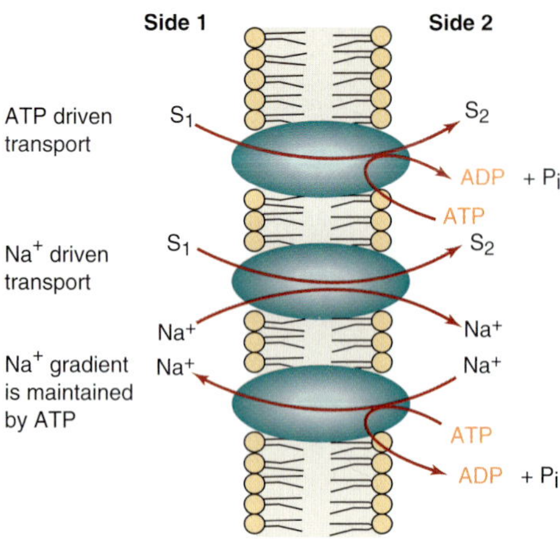

FIGURE 12.48

Involvement of metabolic energy (ATP) in active mediated transport systems. Chemical energy released on hydrolysis of ATP to ADP and inorganic phosphate is used to drive active transport of various substances, including Na^+. The transmembrane concentration gradient of Na^+ is also used for active transport of substances.

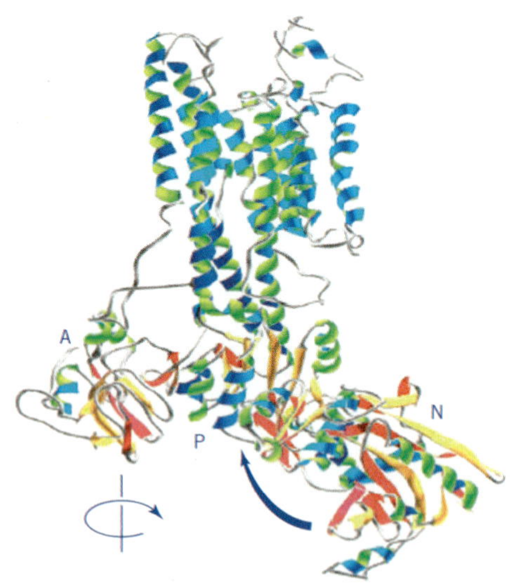

FIGURE 12.49

Structure of the α-subunit of Na^+/K^+-exchanging ATPase. Model of the sheep Na^+/K^+ -exchanging ATPase α-subunit. The predicted domains are the nucleotide-binding domain (*N*) and the phosphorylation domain (*P*). The α-helices are shown in shades of blue and green, β-sheets are shown in shades of red and yellow, and coils are shown in gray. Reproduced with permission from Kaplan, J. H. Biochemistry of Na, K-ATPase. *Annu. Rev. Biochem.* 71:511, 2002. Copyright (2002) Annual Reviews; www.annualreviews.org

hydrolyzed during translocation. In other organisms decarboxylation, methyl transfer, oxidoreduction, and light absorption can serve to drive transport.

Three families of primary active transporters catalyze the translocation of inorganic cations and are classified as P-, V-, or F-type transporters. **P-type transporters** are phosphorylated and dephosphorylated during transport of Na^+, K^+, Ca^{2+}, and so on; there are over 300 members of this family**. V-type transporters** (V for vacuole) are proton pumps, responsible for acidification of the interior of lysosomes, endosomes, Golgi vesicles, and secretory vesicles. **F-type transporters**, present in mitochondria, chloroplasts, and bacteria, translocate protons at the expense of ATP hydrolysis, but synthesize ATP when functioning in the reverse direction—that is, when protons are transported down their concentration gradient (see p. 565). The ATP-binding cassette (ABC) transporters are a separate and unique superfamily of ATP-dependent primary active transporters.

Secondary active transporters utilize a secondary source of energy such as a transmembrane electrochemical gradient of either Na^+ (**sodium motive force, SMF**) or H^+ (**proton motive force**, PMF) to drive translocation. The SMF (Figure 12.48) is generated at the expense of a primary energy source such as ATP hydrolysis and the PMF by oxidation–reduction reactions (see p. 564). Inhibition of ATP synthesis or a collapse of the PMF will dissipate the Na^+ and/or H^+ gradient, leading to a cessation of secondary active transport activity. Secondary active transporters translocate many different small organic molecules such as amino acids, sugars, and a variety of organic molecules.

Linked Translocation of Na^+ and K^+ is by Primary Active Transport

The external or plasma membrane of cells from all forms of life contain a P-type Na^+/K^+ antiporter, which utilizes the energy from hydrolysis of ATP to ADP + phosphate to drive translocation. The transporter is termed the **Na^+/K^+ -exchanging ATPase**, but it is frequently referred to as **Na^+/K^+ -ATPase** or **Na^+ pump**. This superfamily of multisubunit proteins found in bacteria, archaea, and eukaryotes have a large subunit that catalyzes both the ATPase activity and ion translocation. Translocators of the superfamily have been described in bacteria for uptake and/or efflux for K^+, Mg^{2+}, Ca^{2+}, Ag^+, Zn^{2+}, Co^{2+}, Pb^{2+}, Ni^{2+}, Cd^{2+}, and Cu^{2+}.

The plasma membranes of all mammalian cells catalyze the reaction

$$\text{ATP} \xrightarrow[\text{Mg}^{2+}]{\text{Na}^+ + \text{K}^+} \text{ADP} + \text{P}_\text{i} \tag{12.4}$$

with an absolute requirement for Na^+, for K^+ ions, and for Mg^{2+}, a cofactor for ATP-requiring reactions. The transporter is responsible for maintaining the high K^+ and low Na^+ concentrations in the cytoplasm of mammalian cells (p. 12). Excitable tissues, such as muscle and nerve, and cells actively involved in movement of Na^+ ion, such as those in the salivary gland and kidney cortex, have high activities of the Na^+/K^+-exchanging ATPase. This transporter uses about 60–70% of the ATP synthesized by nerve and muscle cells and uses about 35% of ATP generated in a resting individual.

The transporter consists of an α-subunit (110 kDa) and a β-subunit (55 kDa); there is some question whether the actual transporter may be a tetramer with 2 α- and 2 β-subunits. The α-subunit has 10 transmembrane α-helical segments (Figure 12.49), contains both the ATPase and translocation activity, and is phosphorylated and dephosphorylated on a specific aspartic acid residue to form a β-aspartyl phosphate during ion translocation. The cation- and ATP-binding sites are on different domains of the α-subunit. The β-subunit is heavily glycosylated, facilitates insertion of the α-subunit into the membrane after synthesis, and stabilizes the complex; it may also function actively in the transport activity. The transporter from the kidney contains a γ-subunit, which may influence kinetic parameters of the activity. Phosphorylation of the α-subunit requires Na^+ binding and Mg^{2+} but not K^+, whereas dephosphorylation requires K^+ binding but not Na^+ or Mg^{2+} (Figure 12.50). The protein complex has two distinguishable conformations, and is classified as an E_1–E_2-type transporter, to indicate the change in conformation.

Movement of Na^+ and K^+ is an electrogenic antiport process, with three Na^+ ions moving out and two K^+ ions into the cell for each ATP molecule hydrolyzed. This leads to an increase in external positive charge and is part of the mechanism for maintaining the transmembrane potential in cells. ATP hydrolysis by the transporter occurs only if Na^+ and K^+ are translocated. A model for movement of Na^+ and K^+ is presented in Figure 12.51. The protein goes through conformational changes during which the Na^+ and K^+ move short distances. This conformational transition decreases the affinity of the binding site for the cations, resulting in the release of the cation into a milieu where the concentration is higher than that from which it was transported.

Various isoforms of the transporter are regulated by different mechanisms, including substrate concentration, the cytoskeleton, circulating inhibitors, and a variety of hormones. The activity has a requirement for phospholipids; levels of docosahexaenoic acid (long-chain polyunsaturated fatty acid) in brain correlate with the ATPase activity. **Cardiotonic steroids**, such as **digitalis**, are inhibitors of the Na^+/K^+-exchanging ATPase and increase the force of contraction of heart muscle by altering the excitability of the tissue. **Ouabain** (Figure 12.52) is one of the most active inhibitors of Na^+/K^+-exchanging ATPase; it binds on the small subunit on the eternal surface of the membrane.

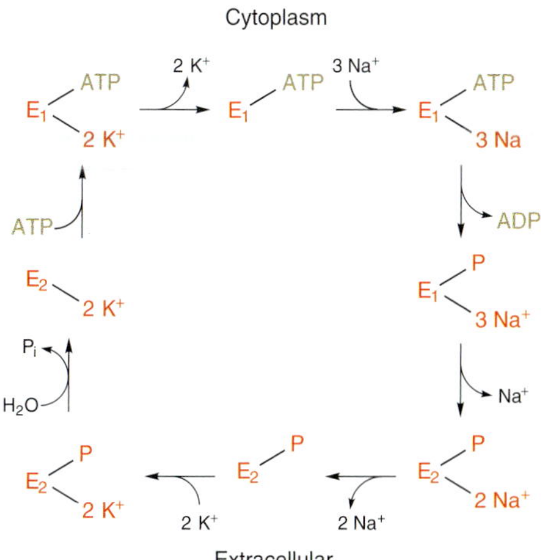

FIGURE 12.50

Proposed sequence of reactions and intermediates in hydrolysis of ATP by the Na^+/K^+-exchanging ATPase. E_1 and E_2 are different conformations of the enzyme. Phosphorylation of the enzyme requires Na^+ and Mg^{2+}, and dephosphorylation involves K^+.

Ca^{2+} Translocation is by Primary Active Transport

Calcium is an important intracellular messenger that regulates many cellular processes from carbohydrate metabolism to muscle contraction. Cytosolic Ca^{2+} is about 0.10 μM, over 10,000 times less than that of extracellular Ca^{2+}. Ca^{2+} in the cytosol increases rapidly by (1) release of Ca^{2+} from the endoplasmic or sarcoplasmic reticulum or (2) transient opening of Ca^{2+} channels in the plasma membrane, permitting flow of Ca^{2+} down the large concentration gradient (see p. 12). To reestablish low cytosolic levels, Ca^{2+} is actively transported by two different **Ca^{2+}-transporting ATPases (Ca^{2+}-ATPase)**, one transporting Ca^{2+} back into the lumen of the endoplasmic or sarcoplasmic reticulum, and the other out across the plasma membrane. The Ca^{2+}-ATPases are P-type transporters, phosphorylated by ATP on an aspartyl residue; they exist in two conformational states, thus are E1–E2-type transporters.

The Ca^{2+}-ATPase of muscle sarcoplasmic reticulum (SERCA) is involved in the contraction–relaxation cycles of muscle, representing 80% of the integral membrane protein of the sarcoplasmic reticulum and occupying one-third of the surface area (p. 985). It consists of a single polypeptide chain with 10 transmembrane helices (Figure. 12.53). Two Ca^{2+}s are translocated from the cytosol to the lumen against a large concentration gradient (10^4), in an antiport exchange for 2 H^+ per ATP hydrolyzed. In the E1 state (Figure 12.54) the binding sites have a high affinity for Ca^{2+} and are accessible from the cytosol, whereas in the E2 state the Ca^{2+}-binding sites have low affinity and are open to the lumen. The Ca^{2+}-free form shows large conformational differences from the Ca^{2+}-bound form (Figure 12.53). The two Ca^{2+} ions are bound side by side, surrounded by four transmembrane helices. It has been proposed that the sarcoplasmic reticulum Ca^{2+}-ATPase may also play a role in **nonshivering thermogenesis** by binding Ca^{2+}, followed by hydrolysis of ATP, but releasing the Ca^{2+} back into the cytosol without translocation of the cation. The energy

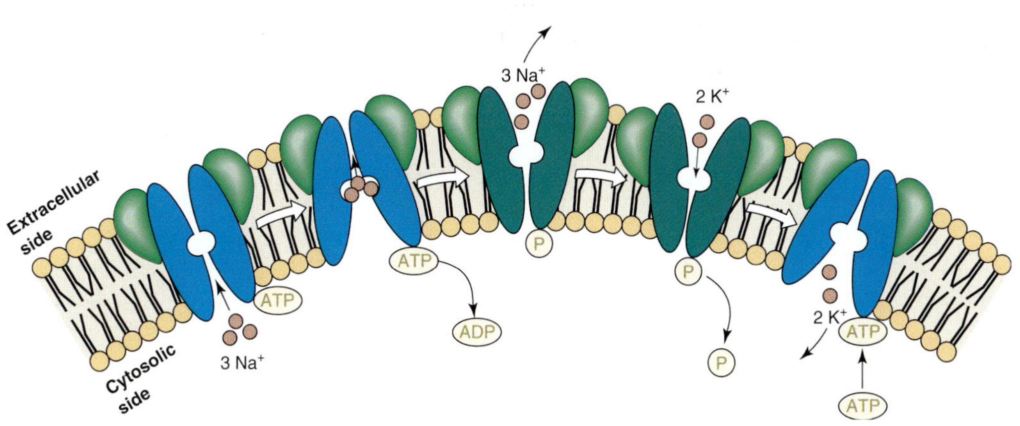

FIGURE 12.51

Model for translocation of Na^+ and K^+ across plasma membrane by the Na^+/K^+-exchanging ATPase. (1) Transporter in conformation 1 picks up Na^+. (2) Transporter in conformation 2 translocates and releases Na^+. (3) Transporter in conformation 2 picks up K^+. (4) Transporter in conformation 1 translocates and releases K^+. All steps not shown (see Figure 12.50).

of the ATP hydrolysis is released as heat. A small peptide, **sarcolipin**, increases heat production.

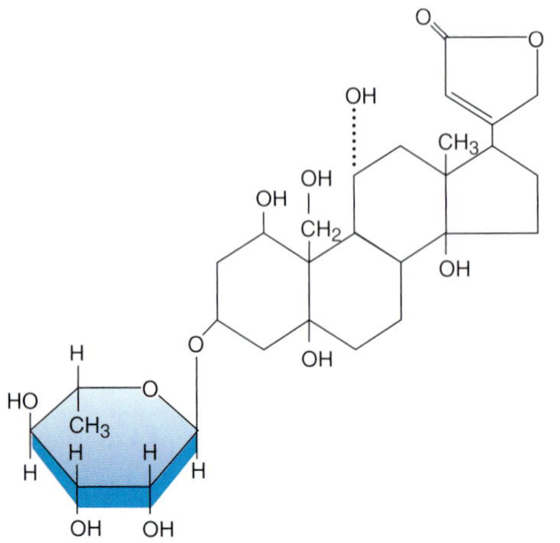

FIGURE 12.52

Structure of ouabain, a cardiotonic steroid, a potent inhibitor of the Na^+/K^+-exchanging ATPase.

Calmodulin Controls the Ca^{2+}-Transporting ATPase of Plasma Membranes

The Ca^{2+}-transporting ATPase of plasma membranes (PMCA) resembles that of the sarcoplasmic reticulum. Thirty isoforms of PMCA are known. In contrast to the SERCA, it transports only one Ca^{2+}/ATP hydrolyzed. In eukaryotic cells, the transporter is regulated by cytoplasmic Ca^{2+} levels through **calmodulin**, a Ca^{2+}-binding protein (Figure 12.55). As cytosolic Ca^{2+} levels increase, Ca^{2+} binds to calmodulin, which has a dissociation constant of ~1 μM for Ca^{2+}. The Ca^{2+}–calmodulin complex binds to the Ca^{2+} transporter increasing Ca^{2+} transport by lowering the K_m of the transporter for Ca^{2+} from about 20 to 0.5 μM. Increased activity reduces cytosolic Ca^{2+} to its normal resting level (~0.10 μM) at which concentration the Ca^{2+}–calmodulin complex dissociates and the rate of Ca^{2+} transport returns to its basal value. Calmodulin is one of several Ca^{2+}-binding proteins, including **parvalbumin** and **troponin C**, which have very similar structures. The Ca^{2+}–calmodulin complex also controls other cellular processes, which are affected by Ca^{2+}. Calmodulin (17 kDa) has the shape of a dumbbell with two globular ends connected by a seven-turn α-helix; it contains four Ca^{2+}-binding sites, two of high affinity on one end and two of low affinity on the other. Binding of Ca^{2+} to the lower-affinity binding sites causes a conformational change which reveals a hydrophobic region that interacts with a protein it controls. Each Ca^{2+}-binding site consists of a helix–loop–helix structural motif (Figure 12.55) with Ca^{2+} bound in the loop connecting the helices. A similar structure is found in other Ca^{2+}-binding proteins.

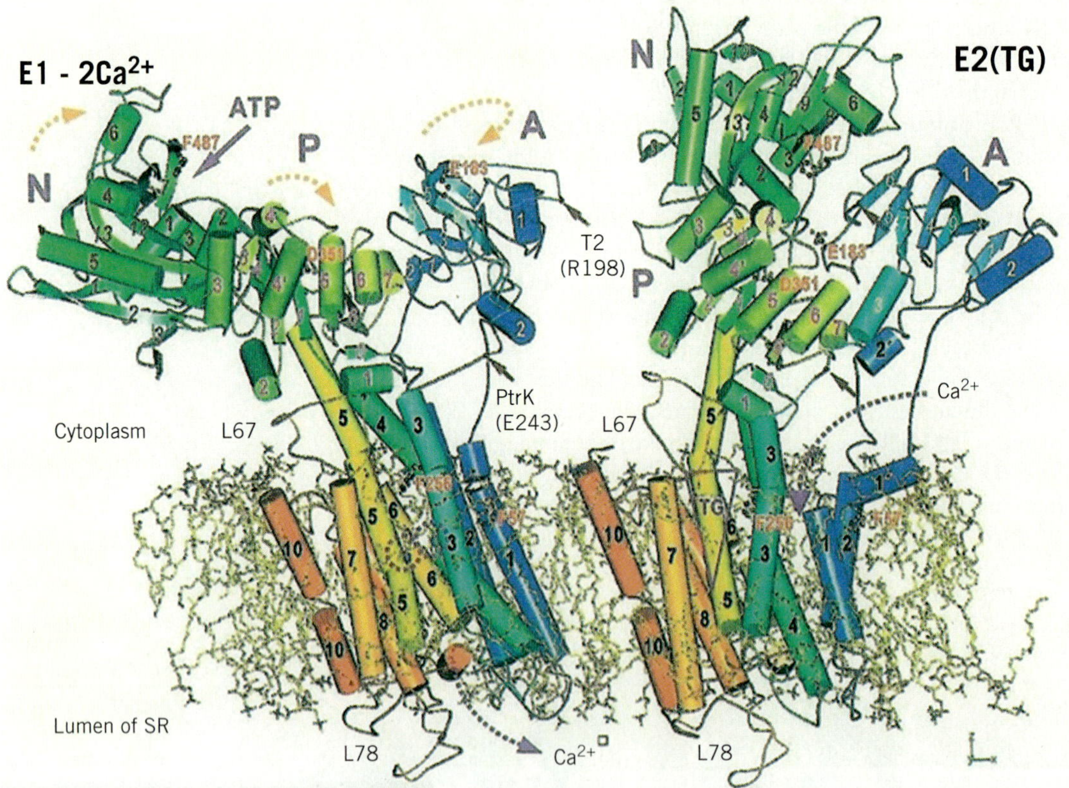

FIGURE 12.53

Structure of Ca^{2+}-transporting ATPase. Conformational changes of the Ca^{2+}-transporting ATPase of the sarcoplasmic reticulum. Points of entry and exit of Ca^{2+} are shown by the purple arrow. Figure on the left is with bound Ca^{2+}, and figure on the right is without Ca^{2+}.

Reproduced with permission from Toyoshima, C. and Inesi, G. Structural basis of ion pumping by Ca^{2+}-ATPase of the sarcoplasmic reticulum. *Annu. Rev. Biochem.* 73:269, 2004. Copyright (2004) Annual Reviews; www.annualreviews.org

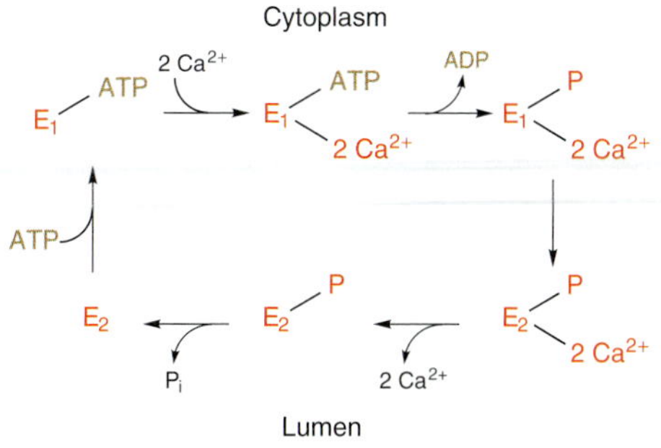

FIGURE 12.54

Proposed sequence of reactions and intermediates in translocating Ca^{2+} by Ca^{2+}-transporting ATPase. E_1 and E_2 are different conformations of enzyme.

The motif is called the EF hand, based on studies with parvalbumin where the Ca^{2+} binds between helices E and F of the protein.

The two families of Ca^{2+} transporters share some properties—for example, the high affinity for Ca^{2+}, the membrane topography, and the organization of the catalytic domain. They are different in structural and functional characteristics, however, principally in their regulation.

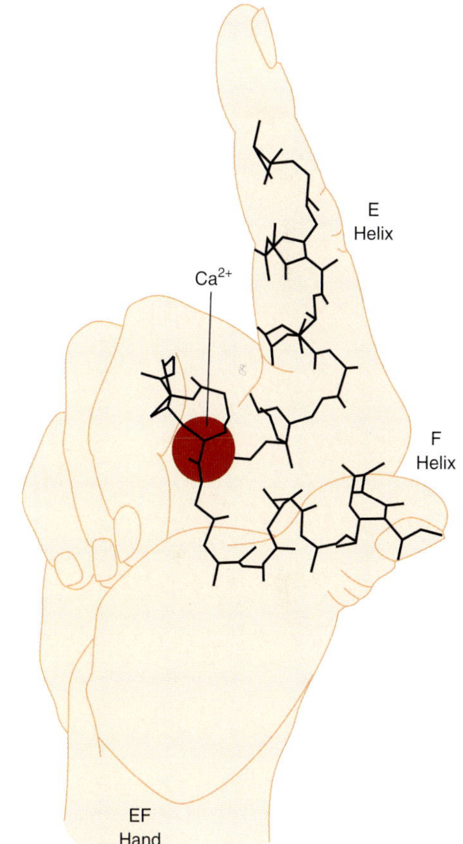

FIGURE 12.55

Binding site for Ca^{2+} in calmodulin. Calmodulin contains four Ca^{2+}-binding sites, each with a helix–loop–helix motif. Ca^{2+} ion is bound in the loop that connects two helices. This motif occurs in various Ca^{2+}-binding proteins and is referred to as the EF hand.

ATP-Binding Cassette (ABC) Transporters

Membranes from prokaryotes to eukaryotes have transporters that belong to the superfamily of **ATP-binding cassette (ABC) transporters** that catalyze an ATP-dependent vectorial movement of diverse substances. There are dozens of families within the superfamily and over 1000 genes, mostly from bacteria. Many ABC transporters have been cloned; the human genome contains at least 49 genes. ABC transporters are present in both the plasma and intracellular membranes of mammalian cells, catalyzing either influx or efflux of various lipids (phospholipids, long-chain fatty acyl CoA, bile salts, and cholesterol), peptides, and a variety of toxic organic molecules and chemotherapeutic agents. The ABC transporters in bacteria are very versatile, transporting ions, heavy metals, sugars, drugs, amino acids, peptides, and proteins. These transporters have a modular structure, with a six-transmembrane-segment domain and a cytosolic ATP-binding domain (or cassette). The ATP-binding domains in the superfamily share extensive sequence homology. Most subfamilies have a tandem repeat of two transmembrane domains and two ATP-binding domains (Figure 12.56). Individual ABC transporters function as either channels or transporters. ATP binding induces a conformational change in the transmembrane domains, but ATP hydrolysis does not lead to phosphorylation of the transporter. The mechanism of coupling the energy of ATP hydrolysis with movement of substrate is still not defined.

One subfamily of ABC transporters in humans is coded by the **multidrug resistance (Mdr) family of genes**. The gene products are glycoproteins, termed **P-glycoproteins**, some of which are responsible for extrusion from cells of a variety of xenobiotics. Overexpression of Mdr genes by tumor cells leads to increased resistance to a variety of drugs because chemotherapeutic agents are rapidly transported out of cells with the increased numbers of transporters. Therefore, the drug does not attain effective therapeutic concentrations. Other P-glycoprotein transporters are responsible for efflux of lipids such as phosphatidylcholine, cholesterol, and bile acids across the canalicular plasma membrane of hepatocytes (p. 1065). Other **multidrug resistance associated proteins (MRP)** are responsible for the efflux from cells of compounds conjugated to glutathione, glucuronate, and sulfate. One member has the important physiological role of transporting bilirubin glucuronides from the liver into the bile (see p. 842). Another ABC transporter is the **sulfonylurea receptor** linked to a K^+-channel, which functions in the regulation of glucose-induced insulin secretion.

The **cystic fibrosis transmembrane conductance regulator** (**CFTR**) is an ABC transporter but different from other ABC transporters in that it is both a conductance

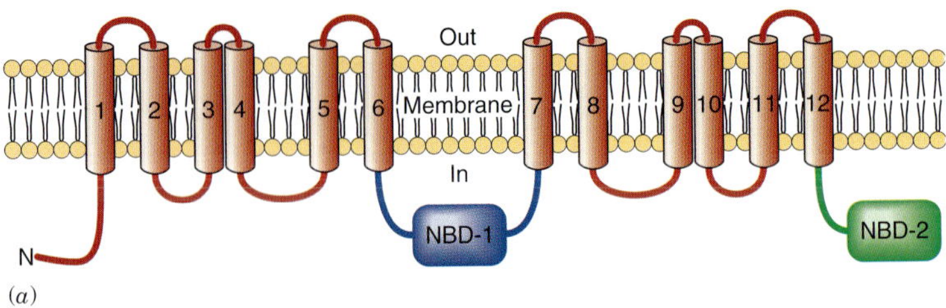

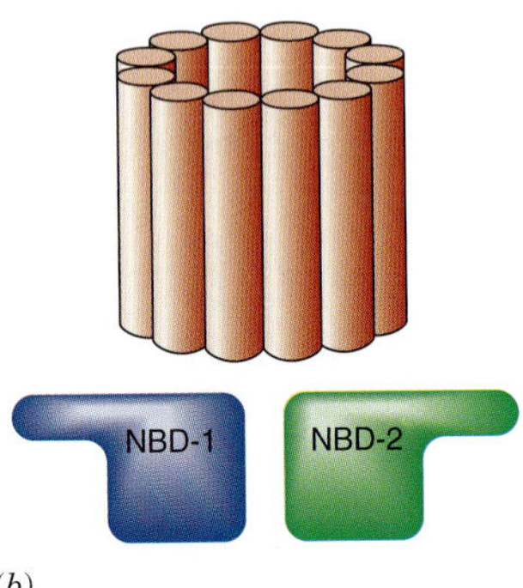

FIGURE 12.56

Model of a P-glycoprotein, an ATP-binding cassette (ABC) transporter. NBD-1 and NBD-2 are nucleotide-binding domains.

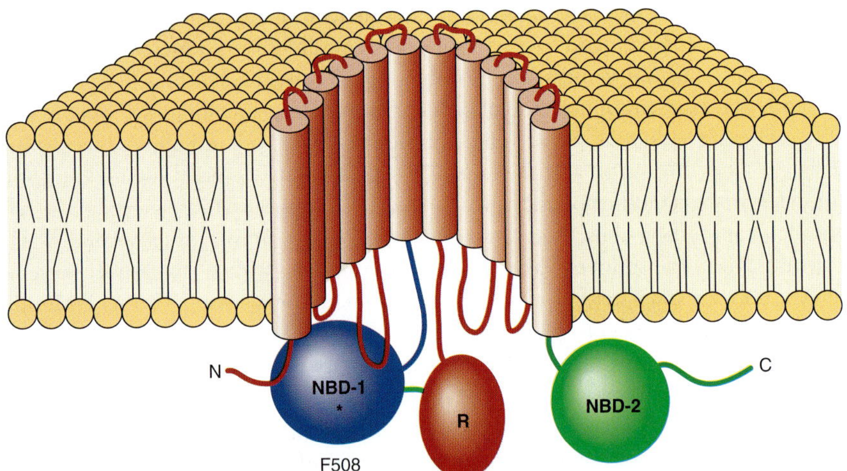

FIGURE 12.57

Schematic diagram of the cystic fibrosis transmembrane conductance regulator (CFTR), a member of the ABC transporter family. NBD-1 and NBD-2 are ATP-binding subunits and R is a regulatory subunit. F508 is the site of deletion of phenylalanine which occurs in about 70% of patients with cystic fibrosis.

Redrawn based on a figure in Ko, Y. H. and Pedersen, P. L. Frontiers in research on cystic fibrosis: Understanding its molecular and chemical basis and relationship to the pathogenesis of the disease. *Bioenerg. Biomembr.* **29**:417, 1997.

regulator and a Cl^- channel. The channel mediates transepithelial salt and liquid movement in the apical plasma membrane of epithelial cells. A dysfunction of CFTR causes the genetic disease cystic fibrosis. CFTR consists of five domains, two domains each of six transmembrane helices, two nucleotide-binding loops, and a regulatory globular domain (Figure 12.57). Phosphorylation of the regulatory domain regulates channel activity while ATP hydrolysis controls gating of the channel. A deletion of phenylalanine residue F508 in the ATP-binding loop NBD-1 (nucleotide-binding domain 1) is present in about 70% of the cases of cystic fibrosis. This deletion prevents the proper folding of NBD-1.

Other genetic diseases (see Clin. Corr. 12.5) attributed to defects in specific ABC transport proteins include adrenoleukodystrophy, Stargardt macular dystrophy, a hereditary degenerative disease of the macula lutea leading to rapid loss of visual acuity, and Dubin–Johnson syndrome.

Na^+-Dependent Secondary Active Transporters

Na^+-dependent secondary active transport is a widespread mechanism of solute transport across the outer membrane of all cells. There are over 400 expressed transporters in this family of proteins. The Na^+ electrochemical gradient across the plasma membrane (p. 12) is an energy source for symport movement of Na^+ with sugars, amino acids, ions, and other small molecules. The general mechanism of the ***sodium/glucose cotransporter* (SGLT)** is presented in Figure 12.58. The diagram represents transport of D-glucose driven by the movement of Na^+. In the process, two Na^+ move into the cell down the Na^+ electrochemical gradient by passive facilitated transport, and glucose is carried along against its concentration gradient. The Na^+ gradient dissipates in the process, but the Na^+/K^+-exchanging ATPase continuously reestablishes it (p. 476 and Figure 12.58). Thus, metabolic energy in the form of ATP is indirectly involved in cotransport because it is required to maintain the Na^+ gradient. A decrease in cellular ATP or inhibition of the Na^+/K^+-exchanging ATPase leads to a decrease in the Na^+ gradient, thus preventing glucose uptake by this mechanism.

Over 37 members of the SGLT gene family have been identified in animal cells, yeast, and bacteria; four are expressed in humans. Some isoforms are expressed in proximal renal tubules and small intestinal epithelium (see p. 1059). All SGLTs have a common core structure of 13 transmembrane helices, but different isoforms have additional transmembrane segments, including human SGLT1, which has 14. SGLTs can catalyze a uniport movement of Na^+ and water in the absence of glucose. A model for the translocation has been proposed in which Na^+ binds to the empty transporter, inducing a conformational alteration that increases the affinity of the transporter for glucose. The binding of glucose, forming a ternary complex, induces another structural change that exposes Na^+ and substrate to the other side of the membrane. Dissociation of Na^+ from the transporter because of the low intracellular Na^+ concentration leads to a return of the transporter to its original conformation, which has a decreased affinity of glucose. Thus glucose is released. The empty transporter reorientates in the membrane, allowing the cycle to begin again.

Amino acids are translocated by intestinal epithelial cells by **Na^+/amino acid cotransporters** by a symport mechanism; at least seven cotransporters have been identified for different classes of amino acids (p. 1054). The Na^+ gradient across the plasma membrane is also utilized to drive the transport of other ions, including a symport mechanism for uptake of Cl^- and an antiport mechanism for excretion of Ca^{2+}.

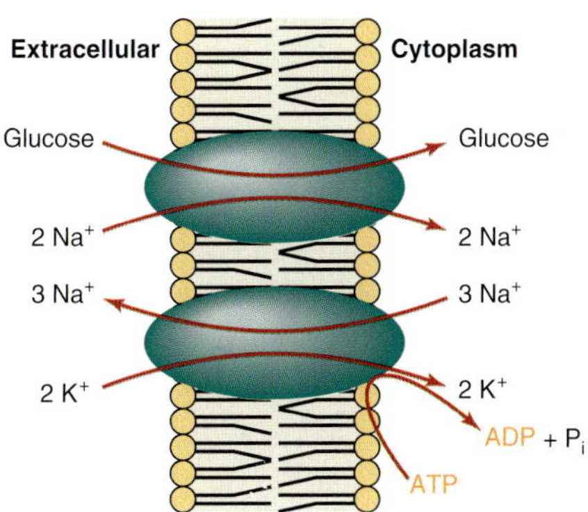

FIGURE 12.58

Na^+-dependent symport transport of glucose across the plasma membrane.

CLINICAL CORRELATION 12.5
Diseases Involving the Superfamily of ABC Transporters

ATP binding cassette (ABC) transporters are a superfamily of proteins with a number of subfamilies that all contain a conserved ATP-binding domain and diverse transmembrane domains. The members of the superfamily catalyze the ATP dependent transport of a wide variety of substances and are present in many different tissues. A number are found in mutant form in various diseases. The following lists selected ABC transporters, their substrates, and related diseases.

Transporter	Substrate	Disease
ABCA1	Phospholipid	Tangier disease
ABCG5	Plant sterols	Sitosterolemia
ABCR	Phosphatidyl-ethanolamine	Juvenile macular degeneration Retinitis pigmentosa
ABC7	Iron?	Sideroblastic anemia with ataxia
BSEP	Bile salts	Progressive familial intrahepatic cholestasis
MRP1	Xenobiotics	Multidrug resistance
MRP2	Bilirubin glucuronides	Dubin–Johnson syndrome
PGY3	Long-chain phosphatidyl-choline	Cholestasis of pregnancy
TAP1	Peptide transporter	Ankylosing spondylitis Insulin-dependent diabetes mellitus Celiac disease

Note that genetic diseases involving the ABC transporters involve a variety of pathophysiological conditions, involving a number of different tissues, and various biochemical systems.

Source: For a detailed list of the ABC transporters and their diseases, see: http://nutrigene.4t.com/humanabc.htm.

Borst, P. and Elferink, R. O. Mammalian ABC transporters in health and disease. *Annu. Rev. Biochem.* 71:537, 2002.

TABLE 12.10 Representative Transport Systems in Mammalian Cells[a]

Substance Transported	Mechanism of Transport	Tissues
Sugars		
Glucose	Passive	Widespread
	Active symport with Na^+	Small intestines and renal tubular cells
Fructose	Passive	Intestines and liver
Amino acids		
Amino acid-specific transporters	Active symport with Na^+	Intestines, kidney, and liver
Specific amino acids	Passive	Small intestine
Other organic molecules		
ATP-ADP	Antiport transport of nucleotides; can be active transport	Mitochondria
Ascorbic acid	Active symport with Na^+	Widespread
Biotin	Active symport with Na^+	Liver
Cholesterol	Active: ABC transporter	Widespread
Cholic acid deoxycholic acid, and taurocholic acid	Active symport with Na^+ and ABC transporter	Intestines and liver
Dicarboxylic acids	Active symport with Na^+	Kidney
Folate	Active	Widespread
Various drugs	Active: ABC transporter	Widespread
Lactate and monocarboxylic acids	Active symport with H^+	Widespread
Neurotransmitters (e.g., γ-aminobutyric acid, norepinephrine, glutamate, dopamine)	Active symport with Na^+	Brain
Organic anions (e.g., malate, α-ketoglutarate, glutamate)	Antiport with counterorganic anion	Mitochondria
Peptides (2–4 amino acids)	Active symport with H^+	Intestines
Urea	Passive	Erythrocytes and kidney
Inorganic ions		
H^+	Active	Mitochondria
H^+	Active; vacuolar ATPase	Widespread; lysosomes, endosomes, and Golgi complex
Na^+	Passive	Distal renal tubular cells
Na^+, H^+	Active antiport	Proximal renal tubular cells and small intestines
Na^+, K^+	Active: ATP driven	Plasma membrane of all cells
Na^+, $HPO_4{}^{2-}$	Active cotransport	Kidney
Ca^{2+}	Active: ATP driven	Plasma membrane and endoplasmic (sarcoplasmic) reticulum
Ca^{2+}, Na^+	Active antiport	Widespread
H^+, K^+	Active antiport	Parietal cells of gastric mucosa secreting H^+
Cl^-, $HCO_3{}^-$	Passive antiport	Erythrocytes and kidneys

[a] The transport systems are only indicative of the variety of transporters known. Most systems have been studied in only a few tissues and their localization may be more extensive than indicated. ABC, ATP-binding cassette.

Summary of Transport Systems

Mammalian cells have a wide variety of transport systems and in many instances different mechanisms to translocate the same substrate. Table 12.10 summarizes characteristics of some major transport systems. A membrane can have a variety of transporters and in some instances for the same substrate. The number of active transporters present in a membrane at a given time controls the rate of transport of a substance across a membrane. Many genetic diseases are attributable to defects in transport systems (Clin. Corr. 12.6).

CLINICAL CORRELATION 12.6

Diseases Due to Loss of Membrane Transport Systems

A number of pathological conditions are due to an alteration in a transport system for specific cellular components. Individuals with a decrease in glucose uptake from the intestinal track lack the specific sodium-coupled glucose–galactose transporter. Fructose malabsorption syndromes are caused by an alteration in the activity of the transport system for fructose. In Hartnup disease there is a decrease in the transport of neutral amino acids in the epithelial cells of the intestine and renal tubules.

In cystinuria, renal reabsorption of cystine and the basic amino acids lysine, arginine, and ornithine is abnormal, resulting in formation of renal cystine stones. In hypophosphatemic, vitamin D-resistant rickets, renal absorption of phosphate is abnormal. Recently there have been reports about genetic based diseases involving the substrate transporters for dopamine and creatine, and the mitochondrial aspartate glutamate transporter. The various ATP requiring active transporters have been implicated in a number of pathophysiological conditions.

Source: Evans, L., Grasset, E., Heyman, M., Dumontier, et. al. Congenital selective malabsorption of glucose and galactose. *J. Pediatr. Gastroenterol. Nutr.* 4:878, 1985. Mueckler, M. Facilitative glucose transporters. *Eur. J. Biochem.* 219:713, 1994. deGrauw, T. J., Cecil, K. M., Byars, A. W., Salomons., G. S., et al. The clinical syndrome of creatine transporter deficiency. *Mol. Cell. Biochem.* 244:45, 2003; and Pedersen, P. L. Transport ATPases in biological systems and relationship to human disease: a brief overview. *J. Bioenerg. Biomembr.* 34:327, 2002.

12.10 IONOPHORES

An interesting group of compounds synthesized and excreted by bacteria facilitates the translocation of inorganic ions across membranes of other cells. These molecules, referred to as **ionophores**, are members of the non-ribosomally synthesized channels (see p. 465) and are relatively low-molecular-weight compounds (up to several thousand daltons). There are two major subgroups: (1) mobile transporters that bind an ion and readily diffuse in a membrane and (2) channel formers. Some major ionophores are listed in Table 12.11.

Each mobile transporter has a definite ion specificity. **Valinomycin** (Figure 12.59) has an affinity for K^+ 1000 times greater than for Na^+, and **A23187** (Figure 12.60) has an affinity for Ca^{2+} 10 times greater than Mg^{2+}. Several of the mobile transporters have a cyclic structure, and the ion coordinates to oxygen atoms in the core of the structure; the periphery of the molecule consists of hydrophobic groups. When an ion is chelated by the ionophore, its water shell is stripped away and the ion is encompassed by the hydrophobic shell. The ionophore–ion complex freely diffuses across the membrane. Since interaction of ion and ionophore is an equilibrium reaction, a steady-state concentration of the ion develops on both sides of the membrane.

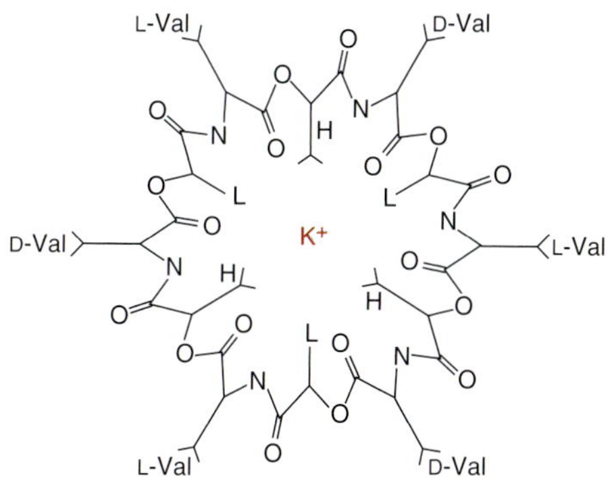

FIGURE 12.59

Structure of valinomycin-K^+ complex. Abbreviations: D-Val, D-valine; L-Val, L-valine; L, L-lactate; H, D-hydroxyisovalerate.

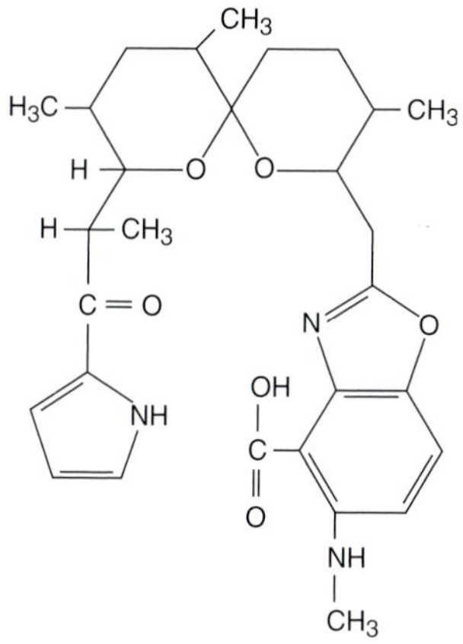

FIGURE 12.60

Structure of A23187, a Ca^{2+} ionophore.

TABLE 12.11 Major Ionophores

Compound	*Major Cations Transported*	*Action*
Valinomycin	K^+ or Rb^+	Uniport, electrogenic
Nonactin	$NH_4{}^+$, K^+	Uniport, electrogenic
A23187	Ca^{2+}/2 H^+	Antiport, electroneutral
Nigericin	K^+/H^+	Antiport, electroneutral
Monensin	Na^+/H^+	Antiport, electroneutral
Gramicidin	H^+, Na^+, K^+, Rb^+	Forms channels
Alamethicin	K^+, Rb^+	Forms channels

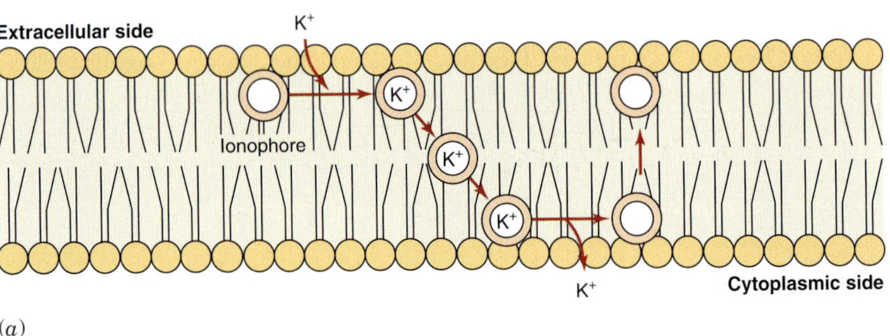

(*a*)

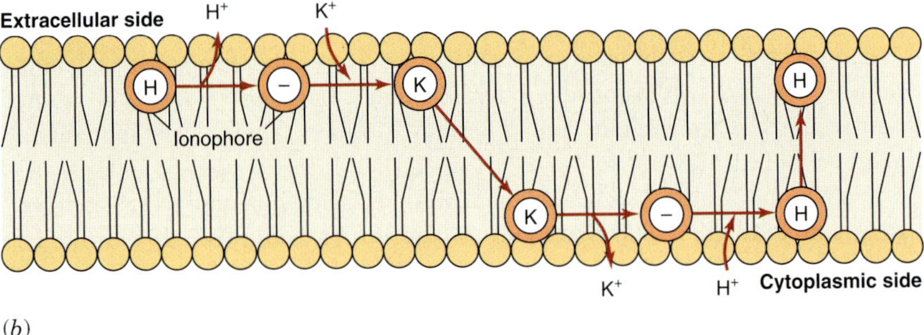

(*b*)

FIGURE 12.61

Proposed mechanism for ionophoretic activities of valinomycin and nigericin. (*a*) Transport by valinomycin. (*b*) Transport by nigericin. I represents ionophore. The valinomycin–K^+ complex is positively charged and translocation of K^+ is electrogenic, leading to creation of a charge separation across membrane. Nigericin translocates K^+ in exchange for an H^+ across membrane, and the mechanism is electrically neutral.
Diagram adapted from Pressman, B. C. *Annu. Rev. Biochem.* 45:501, 1976.

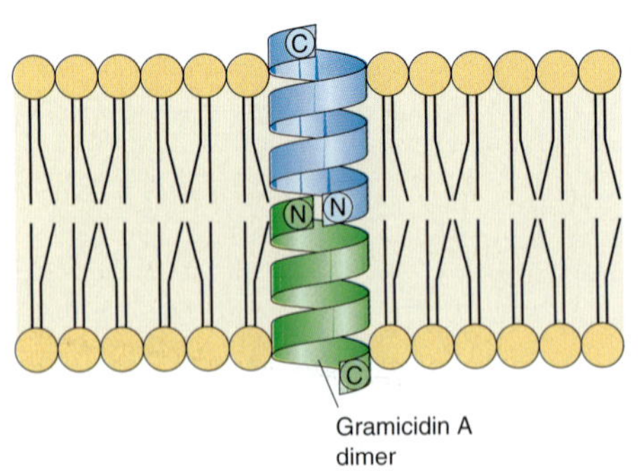

FIGURE 12.62

Action of gramicidin A. Two molecules of gramicidin A form a dimer by hydrogen binding at the amino ends of the peptides, creating a channel in a membrane.

Valinomycin transports K^+ by an electrogenic uniport mechanism that creates an electrochemical gradient across a membrane as it transports a positively charged K^+ (Figure 12.61*a*). **Nigericin** is an electrically neutral antiporter; its carboxyl group, when dissociated, binds a positive ion, such as K^+, leading to a neutral complex that moves across a membrane. It transports a proton back on diffusion through the membrane, leading to an exchange of K^+ for H^+ (Figure 12.61*b*).

Gramicidin A is a 15-residue peptide with alternating D- and L-amino acids. In membranes it forms a β-helix and can dimerize, forming a long-transmembrane segment (25 Å) and narrow-diameter (5 Å) channel (Figure 12.62). Polar residues line the channel and hydrophobic groups are on the periphery of the channel interacting with the lipid membrane. The structure permits the passage of water and univalent cations but not anions. Association and dissociation of the monomers control the rate of ion flux.

Ionophores have antibiotic activity because they disrupt the intracellular ionic balance. They are also valuable experimental tools in studies of ion translocation in biological membranes and for manipulation of the ionic compositions of cells.

BIBLIOGRAPHY

General

Dobler, M. *Ionophores and Their Structure*. New York: Wiley-Interscience, 1981.

Gil, T., Ipsen, J. H., Mouritsen, O. G., Sabra, M. C., Sperotto, M. M., and Zuckermann, M. J. The theoretical analysis of protein organization in lipid membranes. *Biochim. Biophys. Acta* 1376:245, 1998.

Jones, M. N. *Micelles, Monolayers, and Biomembranes*. New York: Wiley-Liss, 1995.

Maxfield, F. R. Plasma membrane microdomains. *Curr. Opin. Cell Biol.* 14:483, 2002.

Scott, H. L. Modeling the lipid components of membranes. *Curr. Opin. Struct. Biol.* 12:495, 2002.

Van Winkle, L. J. *Biomembrane Transport*. San Diego: Academic Press, 1999.

Membrane Structure

Daleke, L. D. and Lyles, J. V. Identification and purification of aminophospholipid flippase. *Biochim. Biophys. Acta* 1486:108, 2000.

Gregoriadis, G. Engineering liposomes for drug delivery: progress and problems. *Trends Biotechnol.* 13:527, 1995.

Ikezawa, H. Glycosylphosphatidylinositol (GPI)-anchored proteins. *Biol. Pharm. Bull.* 25:409, 2002.

McMurchie, E. J. Dietary lipids and the regulation of membrane fluidity and function. In: Aloia, R. C., Curtain, C. C., and Gordon, L. M. (Eds.), *Physiological Regulation of Membrane Fluidity*. New York: Liss, 1988, p. 189.

Pike, L. J. Lipid rafts: Heterogeneity on the high seas. *Biochem. J.* 378:281, 2004.

Roth, M. G. Phosphoinositides in constitutive membrane traffic. *Physiol. Rev.* 84:699, 2004.

Rothman, J. E. and Lenard, J. Membrane asymmetry. *Science* 195:743, 1977.

Simons, K. and Ehehalt, R. Cholesterol, lipid rafts, and disease. *J. Clin. Invest.* 110:597, 2002.

Transport Processes

Borgnia, M., Nielsen, S., Engel, A, and Agre, P. Cellular and molecular biology of the aquaporin water channels. *Annu. Rev. Biochem.* 68:425, 1999.

Borst, P. and Elferink, R. O. Mammalian ABC transporters in health and disease. *Annu. Rev. Biochem.* 71:537, 2002.

Carafoli, E. and Brini, M. Calcium pumps: Structural basis for and mechanism of calcium transmembrane transport. *Curr. Opinion Chem. Biol.* 4:152, 2000.

Chapman, D. E. TRP channels as cellular sensors. *Nature* 426:517, 2003.

Chin, D. and Means, A. R. Calmodulin: A prototypical calcium sensor. *Trends Cell Biol.* 10:322, 2000.

Davis, L. I. The nuclear pore complex. *Annu. Rev. Biochem.* 64:865, 1995.

Dutzler, R., Campbell, E. B., and MacKinnon. Gating the selectivity filter in ClC chloride channels. *Science* 300:108, 2003.

Facciotti, M. T., Rouhani-Manshadi, S., and Glaeser, R. M. Energy transduction in transmembrane ion pumps. *Trends Biochem. Sci.* 29:445, 2004.

Hatta, S., Sakamoto, J., and Horio, Y. Ion channels and diseases. *Med. Electron Micros.* 35:117, 2002.

Holland, I. B., Kuchler, K., Higgins, C, and Cole, S., eds. *ABC Proteins: From Bacteria to Man*. London: Academic, 2002.

Jung, H. The sodium/substrate symporter family: Structural and functional features. *FEBS Lett.* 529:73, 2002.

Kakuda, D. K. and MacLeod C. L. Na^+-independent transport (uniport) of amino acids and glucose in mammalian cells. *J. Exp. Biol.* 196:93, 1994.

Kaplan, J. H. Biochemistry of Na, K-ATPase. *Annu. Rev. Biochem.* 71:511, 2002.

Lee, A. G. A calcium pump made visible. *Curr. Opin. Struct. Biol.* 12:547, 2002.

Leo, T. W. and Clarke, D. M. Molecular dissection of the human multidrug resistance P-glycoprotein. *Biochem. Cell Biol.* 77:11, 1999.

Miyazawa, A., Fujiyoshi, Y., and Unwin, N. Structure and gating mechanism of the acetylcholine receptor pore. *Nature* 423:949, 2003.

Murata, K., Mitsuoka, K., Hirai, T., Walz, T., Agre, P., Heymann, J. B., Engel, A., and Fujiyoshi, Y. Structural determinants of water permeation through aquaporin-1. *Science* 407:599, 2000.

Nielsen, S., Frøkiær, J., Marples, D., Kwon, T., Agre, P., and Knepper, M. A. Aquaporins in the kidney: From molecules to medicine *Physiol. Rev.* 82:205, 2002.

Peracchia, C. (Ed.) *Gap Junctions: Molecular Basis of Cell Communication in Health and Disease*. San Diego: Academic Press, 2000.

Poole, R. C. and Halestrap, A. P. Transport of lactate and other monocarboxylates across mammalian plasma membranes. *Am. J. Physiol.* 264:C761, 1993.

Saier, M. H. A functional–phylogenetic classification system for transmembrane solute transporters. *Micro. Mol. Biol. Rev.* 64:354, 2000.

Scarborough, G. A. Structure and function of the P-type ATPases. *Curr. Opin. Cell Biol.* 11:517, 1999.

Toyoshima, C. and Inesi, G. Structural basis of ion pumping by Ca^{2+}-ATPase of the sarcoplasmic reticulum. *Annu. Rev. Biochem.* 73:269, 2004.

Wright, E. M. and Turk, E. The sodium/glucose cotransport family SLC5. *Eur. J. Physiol.* 447:510, 2004.

QUESTIONS

Carol N. Angstadt

1. All of the following are ways in which peripheral proteins bind to membranes *except*:
   A. binding to an integral protein.
   B. electrostatic binding between phospholipids and positive groups on the protein.
   C. by a short hydrophobic group at one end of the protein.
   D. attached by the charged carboxyl group at the carboxyl terminus of the protein.
   E. binding noncovalently to membrane phosphatidylinositol.

2. Characteristics of a mediated transport system include:
   A. nonspecific binding of solute to transporter.
   B. release of the transporter from the membrane following transport.
   C. a rate of transport directly proportional to the concentration of solute.
   D. release of the solute only if the concentration on the new side is lower than that on the original side.
   E. a mechanism for translocating the solute from one side of the membrane to the other.

3. The translocation of Ca^{2+} across a membrane:
   A. is a passive mediated transport.
   B. is an example of a symport system.
   C. involves the phosphorylation of a serine residue by ATP.
   D. may be regulated by the binding of a Ca^{2+}–calmodulin complex to the transporter.
   E. maintains [Ca^{2+}] very much higher in the cell than in extracellular fluid.

4. The glycerophospholipids and sphingolipids of membranes:
   A. all contain phosphorus.
   B. are all amphipathic.
   C. all have individual charges but are zwitterions (no net charge).
   D. are present in membranes in equal quantities.
   E. all contain one or more monosaccharides.

5. All of the following are correct about an ionophore *except* that it:
   A. requires the input of metabolic energy for mediated transport of an ion.
   B. may diffuse back and forth across a membrane.
   C. may form a channel across a membrane through which an ion may diffuse.
   D. may catalyze electrogenic mediated transport of an ion.
   E. will have specificity for the ion it moves.

6. Cells control the distribution of water across membranes by aquaporins and aquaglyceroporins because nonfacilitated diffusion is too slow. Aquaporins (AQP):
   A. permit translocation of water and small solutes.
   B. reduce the pH gradient because they transport H_3O^+.
   C. are peripheral proteins of the membrane.
   D. form channels through which water flows.
   E. have no specific controls to open the channel.

Questions 7 and 8: Two problems encountered with oral or intravenous administration of drugs are the lack of tissue specificity in the action of the drug and rapid metabolism, and therefore limited period of effectiveness, of some drugs. One attempt to circumvent these problems is the use of liposomes to encapsulate the drugs. Some drugs have a longer period of effectiveness when administered this way. Research is ongoing to prepare liposomes with a high degree of tissue specificity. Liposomes are also useful as a research tool to study the properties of biological membranes since they have a similar structure and properties. Much of our understanding of biological membranes has been obtained using liposomes.

7. Cell membranes typically:
   A. are about 90% phospholipid.
   B. have both integral and peripheral proteins.
   C. contain cholesteryl esters.
   D. contain free carbohydrate such as glucose.
   E. contain large amounts of triacylglycerols.

8. Which of the following statements concerning membranes is/are correct?
   A. Microdomains, called lipid rafts, are fixed in position in membranes.
   B. Lipid composition of the two layers of the membrane equilibrate.
   C. As demonstrated with liposomes, the membrane is most fluid at the surfaces.
   D. An increase in the cholesterol content of a membrane increases membrane fluidity.
   E. Lipid transporters catalyze unidirectional movement of specific lipids from one layer to the other.

Questions 9 and 10: Cystic fibrosis is a relatively common (1 in 2000 live births) genetic disease of Caucasians. Although it affects many organs, pulmonary obstruction is a major problem. CF patients have reduced Cl^- permeability and the disease can be diagnosed by elevated [Cl^-] in sweat. Genetic mutations lead to defects in cystic fibrosis transmembrane conductance regulator (CFTR) protein which is a cAMP-dependent Cl^- channel. CFTR has structural homology to the superfamily of ATP-binding cassette (ABC) transporters.

9. Membrane channels:
   A. have a large aqueous area in the protein structure so are not very selective.
   B. commonly contain amphipathic α-helices.
   C. are opened or closed only as a result of a change in the transmembrane potential.
   D. are the same as gap junctions.
   E. allow substrates to flow only from outside to inside of the cell.

10. ATP-binding cassette (ABC) transporters:
    A. all have both a membrane-spanning domain that recognizes the substrate and an ATP-binding domain.
    B. all effect translocation by forming channels.
    C. are found only in eukaryotes.
    D. all have two functions—forming a channel and conductance regulation.
    E. are all P-glycoproteins.

Questions 11 and 12: Alterations in membrane transport systems for specific components lead to a number of diseases. In Hartnup's disease there is a decrease in transport of neutral amino acids by intestine and renal tubules. Individuals with a decreased glucose uptake from the intestinal tract lack a specific glucose–galactose transporter. In these diseases the transport systems are Na^+/(amino acid) or (glucose) co-transporters.

11. This type of transport system:
    A. moves Na^+ and the amino acid or glucose in opposite directions across the membrane.
    B. uses the energy of the Na^+ gradient to concentrate the other substance against its gradient.
    C. results in the hydrolysis of ATP during the transport.
    D. is the same as coded for by the multidrug resistance (Mdr) family of genes.
    E. is the only type of system used to transport glucose across membranes.

12. A different type of transport system that maintains the Na^+ and K^+ gradients across the plasma membrane of cells:
    A. involves an enzyme that is an ATPase.
    B. is a symport system.
    C. moves Na^+ either into or out of the cell.
    D. is an electrically neutral system.
    E. in the membrane, hydrolyzes ATP independently of the movement of Na^+ and K^+.

Problems

13. Draw a curve illustrating the rate of movement of a solute through a membrane as a function of the concentration of the solute for a) O_2 and b) uptake of glucose into the erythrocyte.

14. What kind of transporter is the acetylcholine receptor of skeletal muscle membrane and how is it controlled?

ANSWERS

1. **D** A single electrostatic charge probably would not be sufficient. A: An example is ankyrin binding to the anion channel in erythrocytes. B: The negative charge of the phospholipids is involved. C: This would insert into the lipid bilayer. E: A specific domain binds to the inositol 3-phosphate head group.
2. **E** This is essential to having the solute transported across the membrane. A: Specificity of binding is an integral part of the process. B: Recovery of the transporter to its original condition is one of the characteristics of mediated transport. C: Only at low concentrations of solute; transporters show saturation kinetics. D: Active transport, movement against a gradient, is also mediated transport.
3. **D** This occurs with the eukaryotic plasma membrane. A, B: Ca^{2+} translocation is an active uniport. C: Like Na^+, K^+-ATPase, phosphorylation occurs on an aspartyl residue. E: Extracellular Ca^{2+} is about 10,000 times higher than intracellular.
4. **B** This property is essential to their function in membranes. A: Sphingolipids, except for sphingomyelins, do not. C: Cerebrosides are not charged; some glycerophospholipids and sphingomyelins are neutral; others have net negative charges. D: Glycerophospholipids are usually present in much larger quantities than sphingolipids. E: Only sphingolipids, except for sphingomyelins, contain carbohydrate.
5. **A** Ionophores transport by passive mediated mechanisms. B, C: These are the two major types of ionophores. D: Valinomycin transports K^+ by a uniport mechanism. There are also antiport systems that are electroneutral. E: Valinomycin has an affinity for K^+ 1000 times greater than for Na^+.
6. **D** They assemble as homotetramers with each monomer forming a specific pore. A: This is a description of aquaglyceroporins. B: They do not transport H_3O^+. C: They are integral proteins that span the membrane. E: Channels need controls.
7. **B** Some proteins are embedded in the membrane while others are at the surface. A: This is more than the total lipid. C: Cholesterol in membranes is unesterified. D: All carbohydrate in membranes is in the form of glycoproteins and glycolipids. E: This is a minor component.
8. **E** Asymmetry is maintained by ATP-dependent flippases and floppases. A: They diffuse freely and can cluster into larger domains. B: Moving a charged head group through the lipid core of the membrane is thermodynamically unfavorable. C: The area of greatest fluidity is the center of the membrane. D: Cholesterol is a rigid molecule which decreases fluidity.
9. **B** The helices typically form the channel with hydrophilic side chains lining the channel and hydrophobic ones facing the lipid core of the membrane. A: This describes a pore; channels are quite specific. C: Voltage-gated channels, like that for Na^+, are controlled this way but others, like the nicotinic acetylcholine channel, are chemically regulated. D: Clusters of membrane channels work together to form a gap junction. E: Substances may move in either direction as dictated by the concentration gradient.
10. **A** These are the common features. B: They operate by a variety of mechanisms, including transporters and receptors. C: Both prokaryotes and eukaryotes have them. D: CFTR is unusual in having both of these functions. E: Some, such as those coded for by the MDR gene family or the transporter for cholesterol out of cells are P-glycoproteins, but others are not.
11. **B** These are Na^+-dependent secondary active transporters. A: These are symports, moving both substances in the same direction. C: ATP is used to maintain the Na^+ gradient, not in the transport of the amino acid or glucose. D: Mdr are ABC transporters. E: Glucose can also be transported by a passive uniport system.
12. **A** The Na^+, K^+-transporter is the Na^+, K^+ -ATPase. It is an antiport, vectorial (Na^+ out), electrogenic ($3Na^+$, $2K^+$) system. E: ATP hydrolysis is not useless.
13. See Figure 12.33. (a) O_2 crosses membranes by simple diffusion, thus, rate increases with concentration as illustrated by the lower curve on the above figure. (b) Uptake of glucose by erythrocytes is a passive mediated transport so should show saturation kinetics as illustrated by the upper curve in the figure.
14. This is also called the nicotinic–acetylcholine channel and is an example of an agonist-regulated channel. Acetylcholine interacts with the acetylcholine receptor on the membrane, causing a conformational change in the protein which opens the channel and allows cations (but not anions) to flow through. Hydrolysis of acetylcholine with release of the products closes the channel.

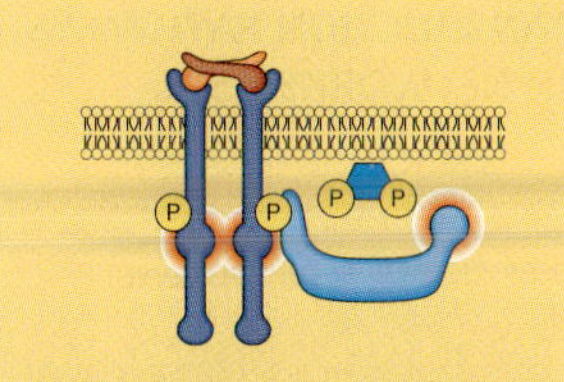

13

FUNDAMENTALS OF SIGNAL TRANSDUCTION

George R. Dubyak

Textbook of Biochemistry With Clinical Correlations, Sixth Edition, Edited by Thomas M. Devlin

13.1 OVERVIEW

Communication or signal transduction between cells and tissues consists of two phases. Intercellular signal transduction characterizes the passage of a signal from one cell to target cells via the extracellular environment. Intracellular signal transduction comprises the biochemical decoding of that signal upon receipt by the target cells, a process that involves stepwise regulation of intracellular signaling proteins that ultimately result in altered function of proteins involved in metabolism, gene regulation, membrane transport, and cell motility (Figure 13.1). The combination of intercellular and intracellular signaling allows cells in one organ or tissue to respond to changes in the function of cells in other organs or tissues to ensure homeostasis within the entire organism. The following will highlight the similarities and differences among endocrine, synaptic, paracrine, and autocrine pathways of intercellular signaling. The material will also examine the major elements used by cells for intracellular signaling including receptors, effectors, second messengers, G proteins, protein kinases, and mechanisms for signal termination.

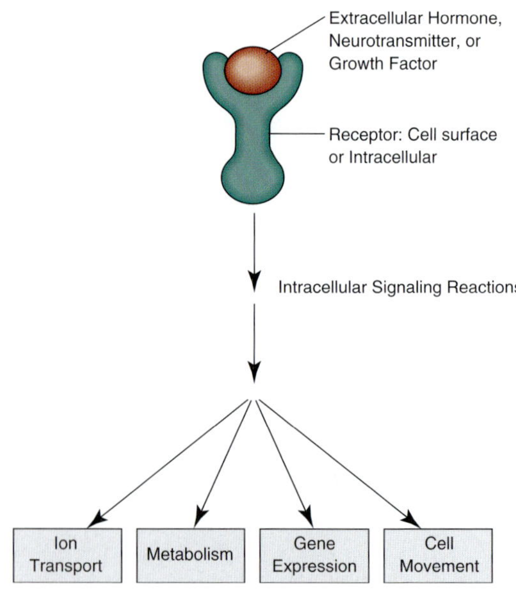

FIGURE 13.1
Basic elements of a signal transduction pathway at the cellular level.

13.2 INTERCELLULAR SIGNAL TRANSDUCTION

Two Fundamental Modes of Intercellular Signal Transduction

Intercellular signaling describes the mechanisms by which one cell (the "sender" cell) sends a message to change the function of another cell (the "target" or recipient cell). Two cells can communicate (i.e., transmit information to each other) in several fundamental ways (Figure 13.2). The first requires that they be in direct physical contact; this mode is termed **juxtacrine or contact-dependent**. The second mode of signaling involves release of **secreted molecules** from the sender cells and does not require direct contact between the sender cell and the target cell. Rather, the secreted molecule diffuses through extracellular space to the target cell where it interacts with receptor proteins, resulting in an activation of intracellular signaling and altered function. Thus, this latter mode can permit intercellular communication between cells separated by large distances—for example, cells in different tissues or organs or between cells separated by short distances. Signaling by secreted molecules includes four major modes termed **endocrine, paracrine, synaptic** for neuronal, and **autocrine**. These modes are distinguished by (1) the actual distance between the sender and target cell and (2) the biochemical nature of the secreted molecules.

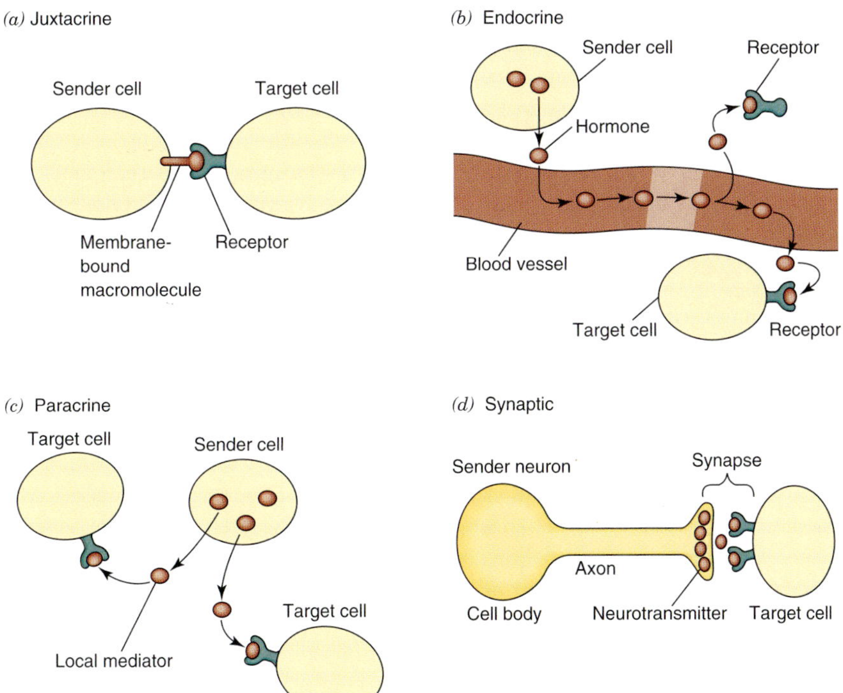

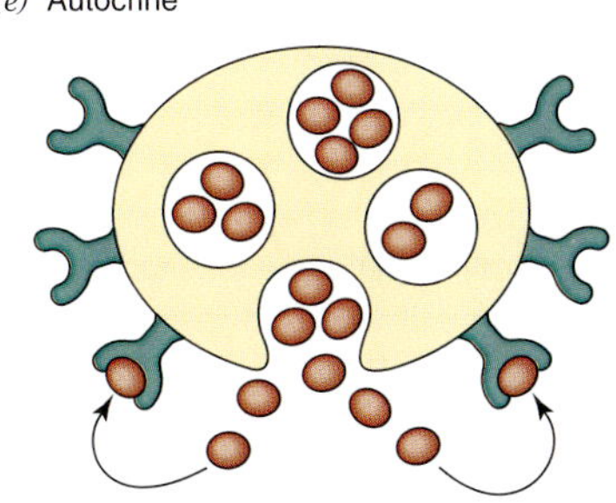

FIGURE 13.2

Four major modes of intercellular signal transduction.

Redrawn based on figures from Alberts, B., et al. *Essential Cell Biology*, 2nd ed. New York: Garland, 2004. Lodish, H. et al. *Molecular Cell Biology*, 4th ed. New York: W. H. Freeman, 2000.

Juxtacrine or Contact-Dependent Signaling

In **juxtacrine signaling**, cells communicate (1) through gap junction channels that allow direct transfer of small ions or metabolites between neighboring cells or (2) by the interaction of a protein expressed on the surface of sender cell with a receptor protein expressed on the surface of the target cell (Figure 13.2*a*). This interaction changes the conformation of the receptor protein and triggers a cascade of intracellular signaling events in the target cell that results in altered function of that cell. Important elements of juxtacrine signaling include a variety of proteins involved in cell–cell adhesion.

Endocrine Signaling

This mode involves signaling between cells separated by short or long distances. Specialized sender cells in glandular tissues synthesize, package, and secrete molecules called hormones (Figure 13.2*b*). By definition, hormones are molecules secreted into the blood for long distance transport to the target cells in various tissues. This is a broadcast type of remote signaling because blood carries the hormone to most cell types in most tissues. Although most cells in the body will be exposed to the hormone, only those cells that express specific receptors for that hormone will be target cells. Two important aspects of endocrine signaling should be noted. First, because a hormone is being secreted into the total blood volume, it will be diluted many-fold by the time it reaches target cells in peripheral tissues. The concentration of hormones in blood and interstitial spaces is always low (picomolar to nanomolar concentrations, i. e., $10^{-12}-10^{-9}$ M) even during maximal rates of secretion from the source gland. Because of the low concentration of hormones, the target cell receptors must have a very high affinity and selectivity for a particular hormone. A consequence of this high-affinity is that once the hormone is bound, it cannot easily dissociate from its receptor. A second important feature of endocrine signaling is that it takes a considerable amount of time to increase and/or to decrease the concentration of a hormone in blood because of the large volume or capacity of the blood. Following secretion, hormone concentrations in the blood and interstitial fluid can remain elevated for many minutes or even hours. This and the high affinity of hormone–receptor interactions have important consequences for the mechanisms by which target cells turn off their response to a hormone. It also ensures that intracellular signaling by the target cell will last for long times.

Paracrine Signaling

In this mode, a sender cell does not secrete signaling molecules into the blood for long-distance transport, but only secretes the molecules into its immediate or local environment (Figure 13.2*c*). This greatly restricts the distances over which the molecules can travel (by passive diffusion) and thus limits responses only to target cells in the immediate vicinity of the sender cell. Given their localized action, the secreted molecules used for paracrine signaling are generically termed "local mediators." Because the secreted local mediators are not massively diluted (as are secreted hormones), their concentration near target cells can be fairly high (nanomolar to micromolar range, i.e., $10^{-9}-10^{-6}$ M). Thus, target cell receptors for local mediators usually have lower affinity for these secreted molecules. Lower affinity means that the local mediator can rapidly dissociate from the receptor when the local, extracellular concentration of the mediator is reduced. Moreover, because the mediator is locally secreted into a small extracellular volume, its concentration at the target cell will be only transiently elevated due to rapid diffusion and dilution into the larger tissue space. Thus, paracrine signaling can be used for rapid and localized communication between cells.

Synaptic or Neuronal Signaling

An extreme type of paracrine signaling is used by nerve cells that send out specialized cellular extensions (axons) to the immediate vicinity of a single target cell that is usually another neuron or a muscle cell (Figure 13.2*d*). The synapses between a neuron and their target cells constitute a restricted extracellular space into which the neuron secretes signaling molecules called neurotransmitters. Given this special synaptic geometry, neurotransmitters only travel very short distances to the target cell. Because of this short intercellular distance and the restricted synaptic volume, the local concentration of neurotransmitter reaching the target cell can be very high (micromolar to millimolar range, i.e., 10^{-6}–10^{-3} M). Thus, receptors on the target cell can have relatively low affinity for the neurotransmitter. This low affinity ensures that neurotransmitters can rapidly dissociate from their receptors following decreases in their local concentration. This is essential for the very rapid (millisecond) termination of neurotransmission.

Autocrine Signaling

Another mode of paracrine signaling occurs when one cell type is both the sender cell and the target cell (Figure 13.2*e*). Autocrine signaling is often used by embryonic/neonatal organisms during tissue and organ development and by adult organisms as part of immune and inflammatory responses. The general features of autocrine signaling are the same as for nonsynaptic, paracrine signaling except that the sender cell provides both the secreted molecule (often a large polypeptide growth factor) and a receptor for that molecule.

Secreted Signaling Molecules

Many types of biological molecules, including proteins, glycoproteins, small peptides, amino acids, amines, lipids (fatty acids and steroids), nucleotides, nucleosides, and even gases (e.g., nitric oxide), are used as secreted signaling molecules by various cells. The major differences between the signaling molecules are size and relative water solubility. The hormones used for endocrine signaling are usually large proteins, glycoproteins, polypeptides or small peptides, and steroids. The neurotransmitters used for synaptic signaling are usually very small, simple molecules such as amino acids (e.g., glutamate, glycine), substituted amines (e.g., epinephrine, acetylcholine), nucleotides (e.g., ATP), or small peptides. The local mediators used for paracrine (nonsynaptic) and autocrine signaling are very diverse and include proteins, small peptides, simple organic molecules (e.g., histamine, ATP, and adenosine), and fatty acid derivatives (e.g., prostaglandins).

13.3 RECEPTORS FOR SECRETED MOLECULES

Receptors for secreted molecules can be divided broadly into two major classes: **intracellular receptors** and **cell-surface receptors** (Figure 13.3). Hydrophobic secreted molecules, such as steroid hormones and certain vitamins, readily permeate through the plasma membrane of target cells. Once within the cell, these hydrophobic molecules bind to **intracellular receptor** proteins that in bound form enter the nucleus and act as transcriptional regulators.

Intracellular receptors include those for steroid hormones, derivatives of vitamin D_3, retinoic acid, and thyroid hormone (Figure 13.3*a*). These receptors comprise a highly homologous group of structurally related proteins. Some receptors are cytosolic proteins that traffic to the nucleus only on binding their cognate hormone; others enter the nucleus even in the absence of bound hormone. In all cases, the hormone-bound receptors function as dimeric complexes (homomers or heteromers) that bind to specific DNA sequences within the upstream regulatory elements of various genes to enhance or repress transcription and thus alter expression of proteins encoded by these genes. Because of their effects on gene expression, activation of intracellular receptors produces long-lasting changes (hours to days) in the function of target cells. The structure and function of these receptors is described in more detail on p. 494.

Cell-surface receptors act as the recognition site for the vast majority of signaling molecules that are either too big (protein or polypeptide hormones) or too hydrophilic to rapidly cross the target cell plasma membrane. Such signaling molecules interact with the target cell by binding to cell-surface receptors that are integral membrane proteins (Figure 13.3*b*). Surface receptor proteins are coupled to a variety of intracellular biochemical reactions termed **signal transduction cascades or pathways.** The most

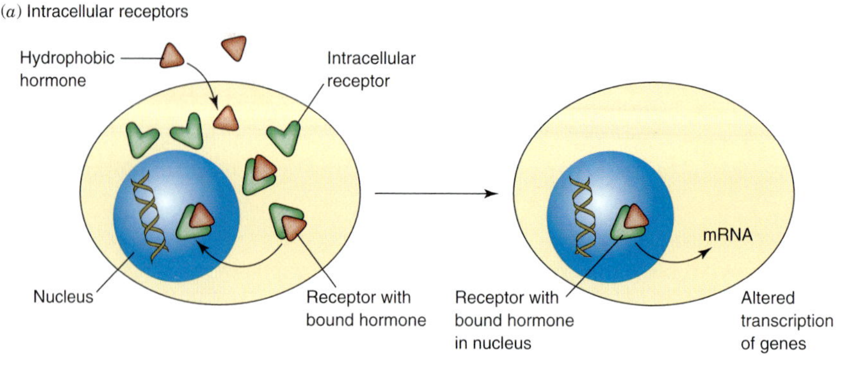

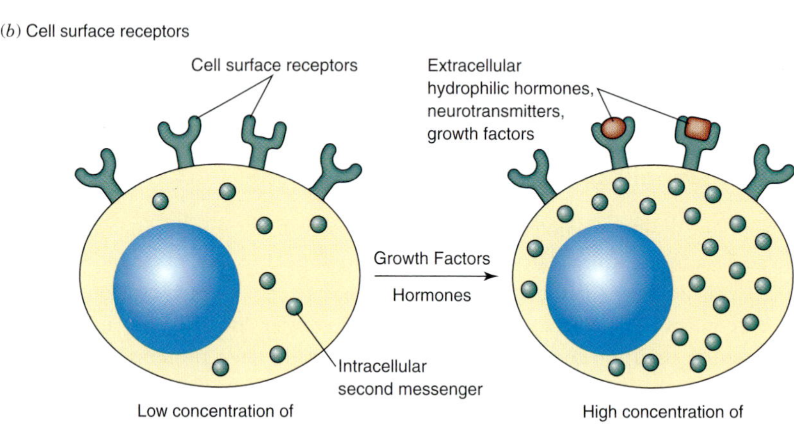

FIGURE 13.3

Basic properties of intracellular receptors versus cell surface receptors.

proximal or earliest signal transduction event triggered by binding of a signal molecule to such a receptor is a conformational change in the receptor that results in the following: (1) the generation of intracellular signaling molecules known as **second messengers** (the extracellular secreted molecule is the first messenger); or (2) a change in plasma membrane electrical potential; and (3) activation of enzymatic cascades involving protein kinases, protein phosphatases, or proteases. Four major superfamilies of structurally and functionally related cell-surface receptors mediate the vast majority of intercellular communication pathways (Figure 13.4); these include **ligand-gated ion channel receptors, enzyme-linked or catalytic receptors; cytokine family receptors; and G-protein-coupled receptors** or **GPCR.**

13.4 INTRACELLULAR SIGNAL TRANSDUCTION BY CELL-SURFACE RECEPTORS

Ligands, Receptors, and Receptor–Ligand Interactions

A **ligand** is any molecule that binds to a receptor protein. An **agonist** is a ligand that, on binding, activates signal transduction, while an **antagonist** is a ligand that prevents signal transduction when it binds to the receptor. A **physiological agonist** or **antagonist** is a naturally occurring molecule (like a hormone or neurotransmitter) that acts a receptor ligand. A **pharmacological agonist** or **antagonist** is a synthetic molecule that acts as a receptor ligand. Many therapeutic drugs are agonists or antagonists of receptors.

Certain physiological agonists can stimulate multiple types of receptors that are termed **receptor subtypes**. The key concept is that the same extracellular signaling molecule can bind to different receptor proteins that are the products of different genes. For example, acetylcholine can interact with a ligand-gated ion channel receptors (see p. 470) that cause contraction of skeletal muscle or with G-protein-coupled receptors (see

(*a*) Ligand-gated Ion Channel Receptor

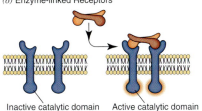

Plasma membrane
Ions
Neurotransmitter
Cytosol
Receptor

(*b*) Enzyme-linked Receptors

Inactive catalytic domain
Active catalytic domain

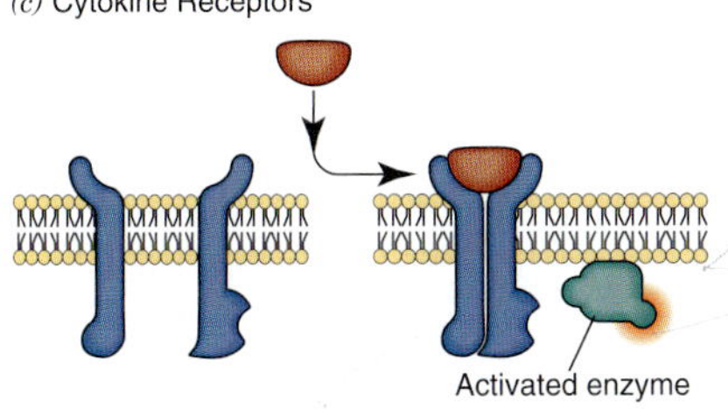

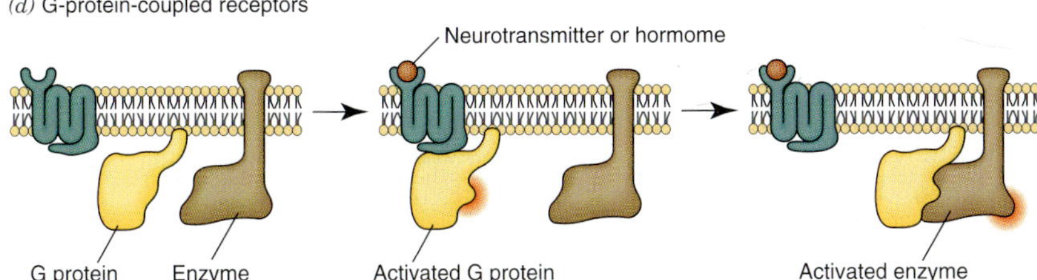

FIGURE 13.4

Major classes of cell surface receptors for secreted signaling molecules.

Redrawn based on figure from Alberts, B., et al. *Essential Cell Biology*, 2nd ed. New York: Garland, 2004.

p. 497) that cause relaxation of cardiac muscle (Figure 13.5). Some neurotransmitters, such as acetylcholine, serotonin, and dopamine, recognize a dozen or more receptor subtypes. Each receptor subtype can be differentially expressed in various tissues/cells or can be coexpressed in the same tissue/cell. Moreover, receptor subtypes often activate opposing signal transduction responses. This greatly complicates the design and testing of synthetic agonists and antagonists and is a major reason why most drugs aimed at receptors can have dangerous side effects when used improperly.

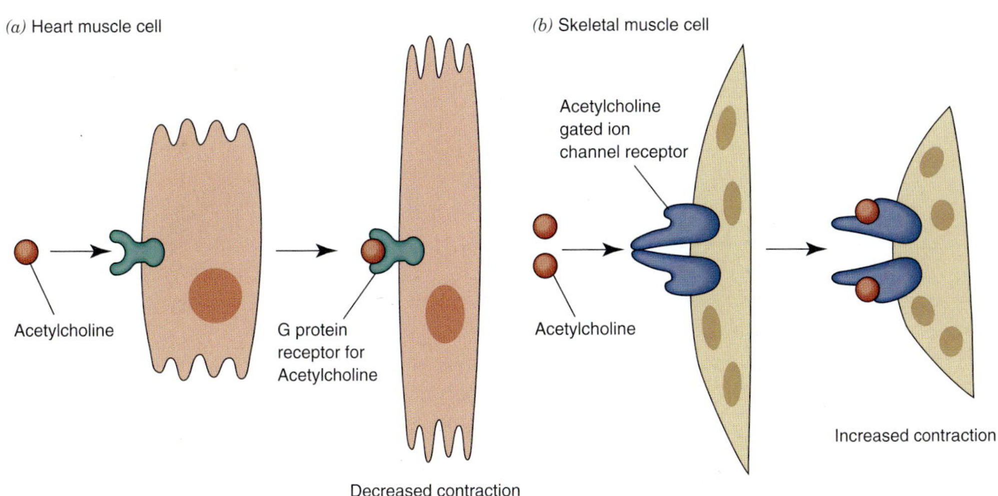

FIGURE 13.5

Basic characteristics of receptor subtypes as functionally distinct receptor proteins that bind a common extracellular signaling molecule.

Relationships between Receptors, Effectors, and Second Messengers

The first step in signaling by cell surface receptors is the binding of the extracellular agonist to the receptor protein that changes the intrinsic energy and structural conformation of the receptor protein. In this conformation, the receptor can interact with **effector proteins.** Effectors are signaling proteins (enzymes or ion channels) that are distal to, but activated by, the agonist-bound receptor (Figure 13.4). In some cases, the receptor protein itself can be the effector. Activation of an effector leads to generation of **second messengers** or to changes in membrane potential. Some receptors interact directly with effector proteins while others require **coupling or transducing proteins** to interact with the effector. GTP-binding regulatory proteins (G proteins) are a very important class of such coupling proteins.

Enzyme effectors can be an intrinsic component of a receptor protein (e.g., receptor tyrosine kinases or separate proteins (e.g., nonreceptor tyrosine kinases). They can be integral membrane proteins (e.g., adenylate cyclase) or cytosolic proteins (e.g., guanylate cyclase) and include nucleotide cyclases, phospholipases, protein kinases, and protein phosphatases. These will be discussed in later sections. Ion channel effectors are an intrinsic component of some receptors (ligand-gated ion channels) while others are indirectly regulated by receptors via by G proteins or second messengers.

Second messengers are small intracellular molecules that transmit and amplify the initial signal from agonist-activated receptors (Figure 13.6). They can be inorganic ions or organic products of enzyme-catalyzed reactions.

Ions: Ions passing through receptor-regulated channels can function as second messengers in two ways. If the ionic flux is sufficiently great, the cytosolic concentrations of certain regulatory ions can change appreciably; Ca^{2+} is the major intracellular regulatory ion. If the particular ionic flux is the predominant current across the plasma membrane, the membrane potential will be driven to the equilibrium potential for that ion. This will result in depolarization or hyperpolarization of the membrane.

Water-soluble second messengers: Certain effector enzymes catalyze the rapid production of water-soluble molecules, which act as second messengers—for example, cyclic AMP (cAMP), cyclic GMP (cGMP), and inositol trisphosphate (IP_3) (Figure 13.6).

Lipid second messengers: Certain phospholipase effector enzymes catalyze rapid production of hydrophobic metabolites that function as membrane-associated second messengers [e.g., diacylglycerol (DAG) or phosphatidylinositol 3,4,5-trisphosphate (PIP_3)] (Figure 13.6).

Protein Phosphorylation in Signal Transduction

Most (but not all) pathways of intracellular signal transduction involve changes in phosphorylation of certain proteins. Phosphorylation of proteins can change their

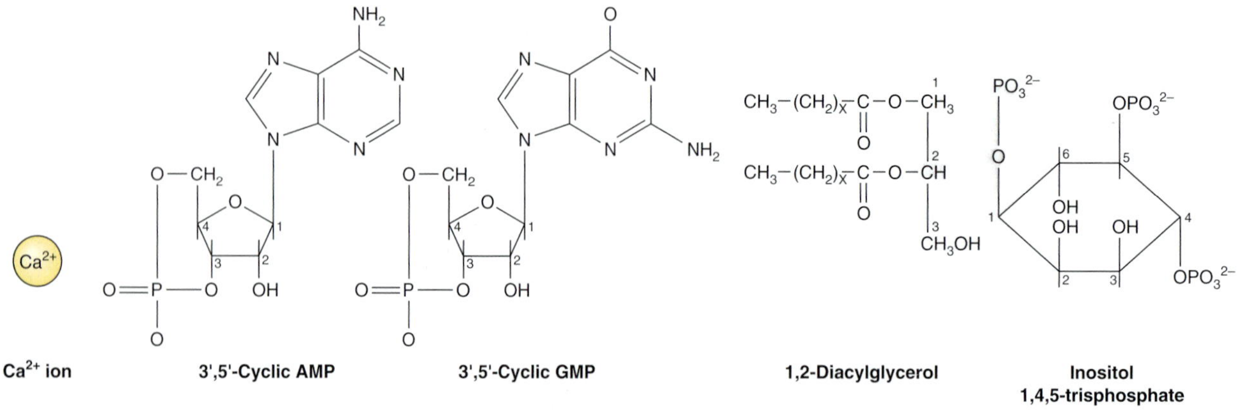

Ca^{2+} ion

3',5'-Cyclic AMP

3',5'-Cyclic GMP

1,2-Diacylglycerol

Inositol 1,4,5-trisphosphate

FIGURE 13.6

Structures of four second messenger molecules.

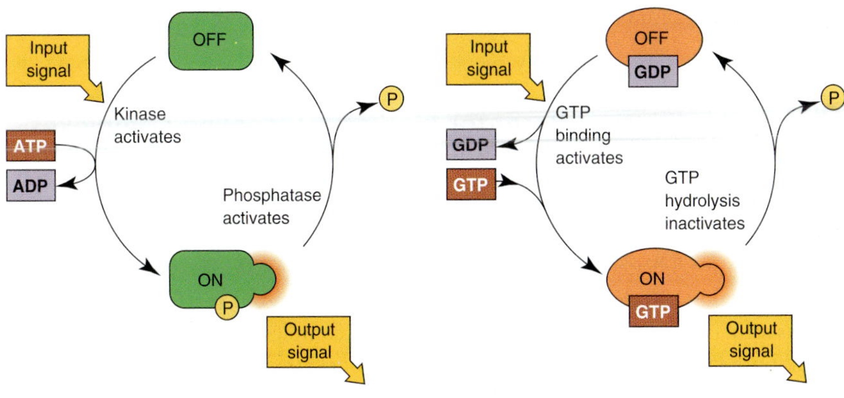

FIGURE 13.7

Protein phosphorylation versus guanine nucleotide binding as two major types of molecular switches for changing protein function during intracellular signaling.

secondary, tertiary, or quaternary structure, resulting in altered function. Protein phosphorylation involves the regulated activities of a variety of protein kinases and protein phosphatases (Figure 13.7*a*). Protein kinases are broadly classified based on their selectivity for particular amino acids. Most protein kinases activated by second messengers phosphorylate specific proteins on serine or threonine residues. Most growth factors and cytokines will additionally stimulate kinases that phosphorylate proteins on tyrosine residues. These protein tyrosine kinases play important roles in regulation of cell growth and differentiation. Protein substrates for either serine/threonine kinases or tyrosine kinases are very diverse and include metabolic enzymes, ion channels, transcriptional factors, or other protein kinases. Protein phosphatases also exhibit selectivity for serine/threonine phosphates or tyrosine phosphates. The ability of some protein kinases to phosphorylate and activate target proteins, which are themselves protein kinases, underscores the involvement of **protein kinase cascades** in the regulation of complex cellular responses. These protein kinase cascades are often critical elements in the **amplification** process that characterizes the intracellular signaling reactions initiated binding of ligands to cell-surface receptors.

Protein kinases can be regulated by receptors in several ways. Some tyrosine kinases are intrinsic components of receptor protein structure (e.g., receptor tyrosine kinases) while others are cytosolic proteins that transiently associate with activated receptors (e.g., nonreceptor tyrosine kinases binding to cytokine receptors). Most serine/threonine kinases are soluble proteins that are regulated by an increase in second-messenger concentration (e.g., cAMP-dependent protein kinase). Likewise, there are multiple types of phosphatases, which reverse the signals generated by receptor- or second-messenger activated kinases.

GTP-Binding Regulatory Proteins in Signal Transduction

The G proteins that are functionally coupled to the **"G protein-coupled receptors"** constitute a subgroup of a family of proteins, which can bind and hydrolyze GTP (Figure 13.7*b*). These **GTPases** have an important **"molecular switch"** role, being involved in critical biological functions such as initiation, elongation, and termination of protein synthesis, microtubule assembly, regulation of enzyme activity, and membrane trafficking. Many of these "switch" GTP-binding proteins exist as monomeric proteins, which distinguishes them from the trimeric, receptor-coupled G proteins. All GTP-binding proteins (monomeric and trimeric) molecules have different conformations and activities depending on whether they have guanosine-5′-diphosphate (GDP) or GTP bound to them. The GDP-bound forms interact with other molecules that catalyze the exchange of GDP for GTP. The GTP-bound forms interact with and change the activity of a third class of molecules, effector proteins. The GTPase activity converts the GTP-bound form to a GDP-bound form, which has decreased ability to interact with effector proteins and interacts with the nucleotide exchanger in order to resume its effector active conformation. A major function of many cell-surface receptors is to trigger, directly or indirectly, the GDP/GTP exchange reactions that result in G-protein activation.

Other Components of Receptor-Mediated Signaling Complexes and Cascades

In addition to effector proteins, second messengers, trimeric G proteins, and protein kinases, a wide variety of other intracellular proteins may be involved in the cell-surface receptor-mediated intracellular signaling pathways. By physically colocalizing multiple signaling proteins within a given signaling pathway, **scaffold and adaptor proteins** can increase the fidelity and speed of a complex signaling cascade. Other proteins termed **anchoring proteins** localize and concentrate soluble signaling proteins at particular subcellular locations. For example, a wide variety of AKAPs (A-kinase anchoring proteins) can localize cAMP-regulated protein kinases (protein kinase A or PKA) at the plasma membrane, in the cytoskeleton, or within the nucleus. The assembly of receptors, adaptor proteins, effector proteins, and other signaling components into signaling complexes often involves the specific interactions of particular domains within these proteins (Figure 13.8). **SH2 domains** (Src-Homology type 2) and **PTB domains** (phospho-tyrosine binding) specifically recognize and bind to specific phosphorylated tyrosine residues on a variety of signaling proteins. **SH3 domains** (Src-Homology type 3) bind to proline-rich domains that characterize certain signaling proteins. **PH domains** (Pleckstrin Homology) specifically recognize and bind to the certain types of inositol phospholipids that have important roles as either lipid second messengers or lipid substrates for phospholipase C.

Ligand–Receptor Interaction and Downstream Signaling Events

Binding of an **agonistic ligand** to a receptor is the first step in signal transduction. Interaction between receptors and agonists is characterized by binding **affinity** and **capacity**. Affinity describes the relationship between agonist concentration and formation of agonist–receptor complexes. Affinity is inversely related to the dissociation constant or K_D (measured in ligand molarity) that characterizes the agonist–receptor complex (Figure 13.9). The K_D is equal to the concentration of free agonist that permits

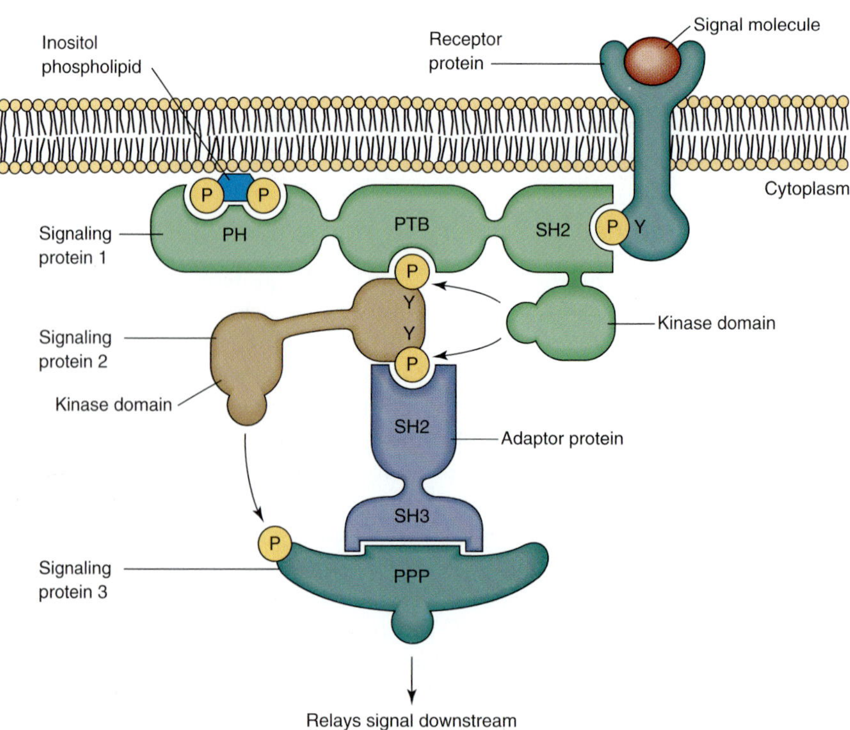

FIGURE 13.8

Roles of adapter proteins and protein–protein interaction domains in the assembly of intracellular signaling complexes.

Redrawn based on figure from Alberts, B., et al. *Molecular Biology of the Cell*, 4th ed. New York: Garland, 2002.

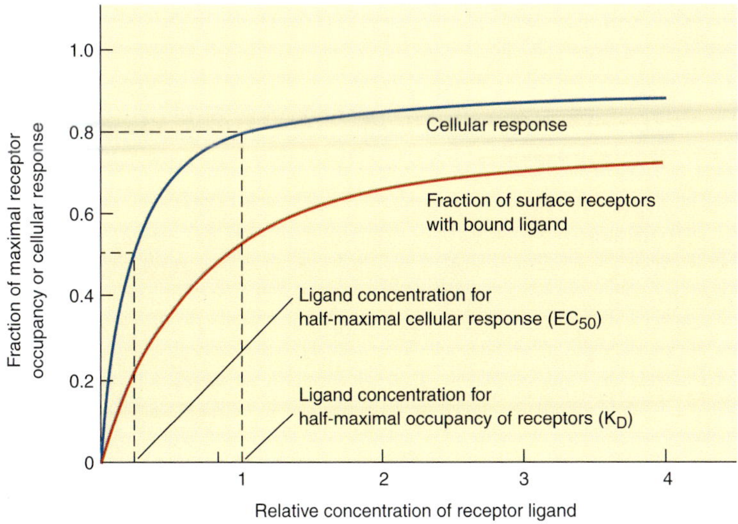

FIGURE 13.9

Relationship between K_D, as a measure of ligand–receptor binding affinity, and EC_{50}, as a measure of functional response to the formation of ligand–receptor complexes.

binding of 50% of the total number of binding sites on the receptor (half-maximal occupancy). **Potency** is another term used to measure of the relative affinity of an agonist for its receptor. An agonist with high potency can bind even when present at very low concentrations (high affinity, low K_D). The capacity of the interaction refers the absolute number of receptor sites. Because receptor proteins are usually expressed in relatively small amounts per cell, agonist–receptor binding curves show saturation. In addition to directly measuring the binding of an agonist to its receptor, one can indirectly measure the agonist–receptor interaction by assaying the various functional responses triggered by the agonist–bound receptor. In general, the magnitude of these responses is proportional of the amount of agonist–bound receptor complex. The dose–response curve measures the relationship between agonist concentration and biological response.

The concentration of agonist that cause half-maximal response is the half-maximal effective concentration or **EC_{50}**. The EC_{50} is directly proportional to the K_D of the agonist–receptor complex (Figure 13.9). However, EC_{50} is usually much lower than the K_D; that is, the agonist concentration required for half-maximal activation of the cell's biological response is lower than the agonist concentration required for half-maximal occupancy of the cell's total number of receptors. This means that a cell usually expresses more receptor than it needs for effective biological responses. These extra or "spare receptors" permit the cell to respond to agonist even after the inactivation or degradation of receptors activated by previous agonist occupation. As described below, receptor inactivation or degradation plays an important role in terminating signal transduction by receptors.

Termination of Signal Transduction by Cell-Surface Receptors

Cells also express multiple mechanisms that permit them to terminate or attenuate their responses to various hormones, neurotransmitters, and local mediators. These mechanisms are critical for allowing cells to return to their basal, pre-stimulus state and for preventing overactivation or inappropriate activation of the functions controlled by a particular receptor (Figure 13.10). These termination mechanisms include (1) reducing agonist availability in the extracellular vicinity of a target cell, (2) internalizing and degrading the agonist-bound receptor complex, and (3) rapidly modifying the receptor (e.g., by phosphorylation) so that it becomes inactive or desensitized.

The simplest mechanism for terminating signal transduction is to reduce the extracellular concentration of the secreted agonist. The effectiveness and rapidity of this termination mechanism is determined by (1) the stability of the secreted agonist, (2) the size or capacity of the extracellular compartment into which the agonist is secreted, and (3) the proximity of sender cells to target cells. It is difficult to terminate endocrine signaling by reducing agonist availability rapidly because hormones are secreted into a large

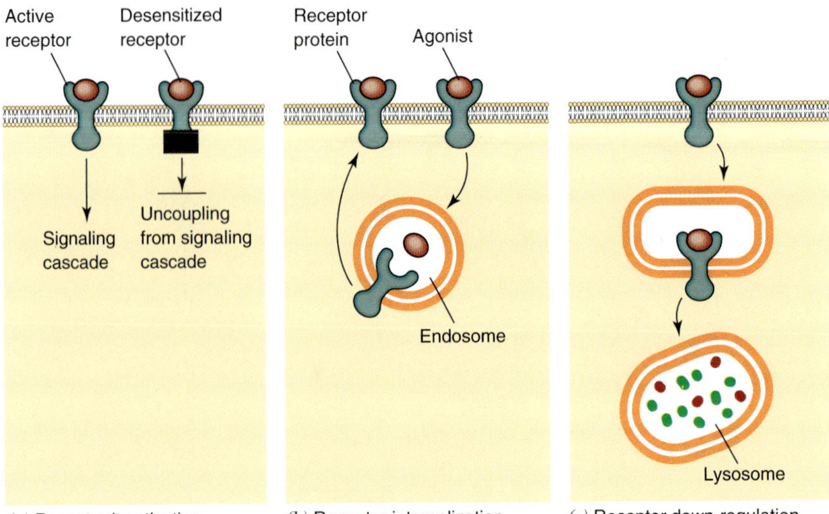

FIGURE 13.10

Major mechanisms for the termination of receptor-dependent signal transduction.

capacity extracellular environment. In contrast, rapid reduction in agonist availability plays a significant role in terminating synaptic signaling; neurotransmitters are locally secreted into a very low capacity space, the synaptic cleft, and they can rapidly diffuse away. Some cells can also actively reduce the concentration of nearby agonist by reaccumulating the released agonist (e.g., synaptic reuptake of neurotransmitters like serotonin and dopamine) or by metabolizing the extracellular agonist. Extracellular metabolism can be catalyzed by specific ecto-enzymes (e.g., the acetylcholinesterases) present on the surface of target cells. Certain neurotoxins act as inhibitors of neurotransmitter breakdown.

Cells can also terminate signal transduction by regulating the functional availability of receptors, a process known as **adaptation** or **desensitization**. If repeatedly exposed to a particular agonist, a cell adapts its sensitivity to that agonist. Over time, a cell will become less and less sensitive to a concentration of agonist that normally causes maximal activation. This desensitization results from several processes. First, many receptors are rapidly internalized when bound by their agonistic ligand. These receptor–ligand complexes are incorporated in endosomes, and subsequent acidification of endosomes causes dissociation of the ligand–receptor complexes. Some endosomes containing ligand-free receptor are routed back to the plasma membrane where the receptors can be reincorporated into the surface membrane. Other endosomes are routed to lysosomes where both the ligand and receptor protein are degraded. Thus, over time, there will be fewer receptors on the surface membrane.

Second, a much faster mechanism for desensitization involves **structural modifications** of the receptor that result in functional inactivation. These inactive receptors remain on the surface membrane and may even bind agonist. However, they cannot trigger the intracellular steps of signal transduction. In many cases, this functional inactivation of the receptors is due to phosphorylation of the receptor protein itself. Thus, while the agonist-occupied receptor stimulates a cascade of signal transduction events involving activation of protein kinases, some of these protein kinases phosphorylate the receptor itself, which then inactivates that receptor. This is a classic type of negative feedback regulation. Depending on the receptor, this functional inactivation can be reversed rapidly or slowly.

13.5 LIGAND-GATED ION CHANNEL RECEPTORS

Very rapid (milliseconds) signaling is required at nerve–nerve or nerve–muscle synapses. Thus, neurotransmitter receptors need to change cell responses with a minimum number of enzymatic steps because biochemical reactions are relatively slow (seconds to minutes).

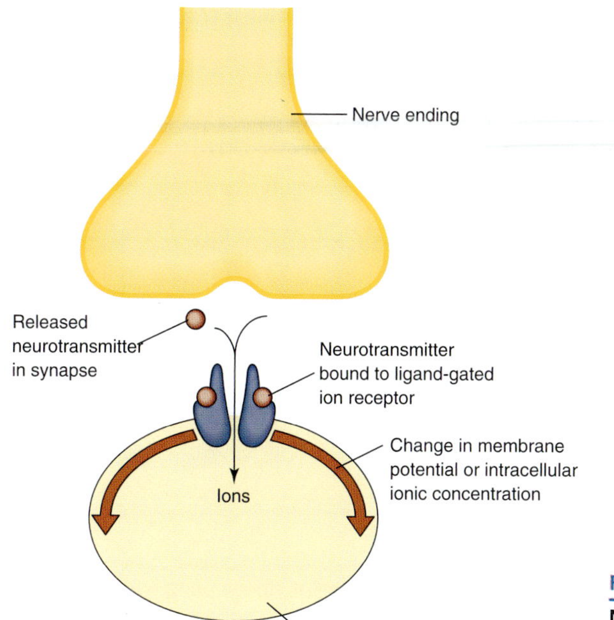

FIGURE 13.11

Neurotransmitter-gated ion channel receptors as major signal transduction elements at neuronal synapses.

Many (but not all) neurotransmitters bind to receptors that are ligand-gated ion channels (Figure 13.4*a*). The extracellular ligands for these receptors are invariably very small organic molecules like amino acids, substituted amines, or nucleotides. These are packaged in vesicles and released via rapid exocytosis at synapses. Because these ligands are so small, they rapidly diffuse (Figure 13.11). The ligands bind to their receptors with low affinity (micromolar–millimolar K_D). This permits rapid dissociation of the ligand–receptor complex upon reduction of the extracellular concentration of the neurotransmitter.

The binding of neurotransmitter to extracellular ligand-binding sites of such receptors triggers a near-instantaneous change in receptor conformation that permits ions to permeate through the receptor protein complex. The flux of these ions rapidly changes the cell's membrane potential. If a cation like Na^+ passes through the receptor channel, the cell will become depolarized; if an anion like Cl^- is the permeating species, the cell will become hyperpolarized (Figure 13.12). Cation-selective receptors are gated by the **excitatory neurotransmitters** (acetylcholine, glutamate, serotonin, and ATP) while anion-selective receptors are gated by the inhibitory neurotransmitters [GABÃ (γ-amino butyric acid) and glycine]. Activation of these neurotransmitter-gated ion channels changes the cell's membrane potential of the plasma membrane, resulting in secondary regulation (activation or inhibition) of many types of voltage-gated channels, including voltage-gated channels for Ca^{2+}. Changes in the activity of these channels will alter the cytosolic concentration of Ca^{2+}, a key second messenger. In turn, increased cytosolic Ca^{2+} drives acute responses in target cells, such as exocytotic release of neurotransmitter-containing vesicles from neurons or contraction of muscle cells, or long-term responses, such as activation of Ca^{2+}-sensitive gene expression. Most of the neurotransmitters that activate ligand-gated ion channel receptors can also interact with other receptor subtypes that are G-protein-coupled receptors. Thus, receptors for acetylcholine are either (a) nicotinic receptors that are ligand-gated channels or (b) muscarinic receptors that are G-protein-coupled.

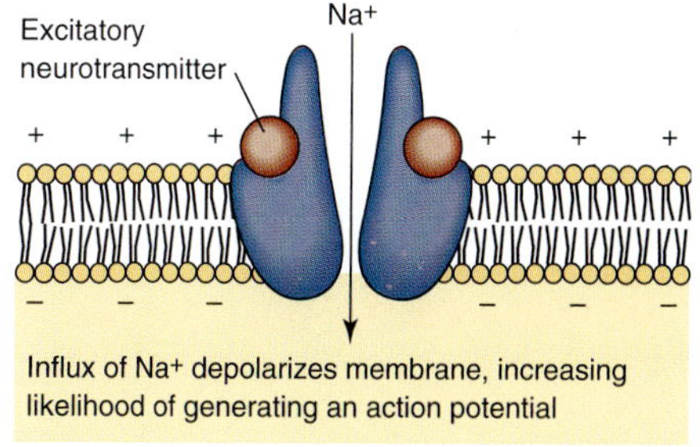

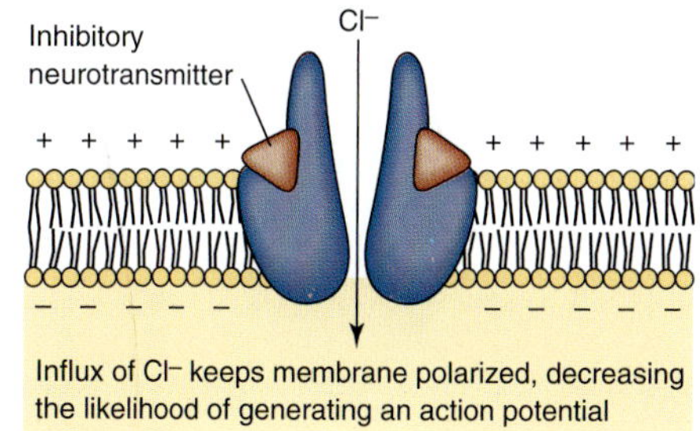

FIGURE 13.12

Excitatory versus inhibitory neurotransmitters as agonists for ligand-gated ion channel receptors.

Ion Channel Receptors

All **ion channel receptors** are oligomeric complexes of three to five protein subunits. Some are homomeric complexes of multiple copies of the same protein subunit (each subunit is the product of the same gene). Others are heteromeric complexes of different protein subunits (each subunit is the product of a different gene). Each individual

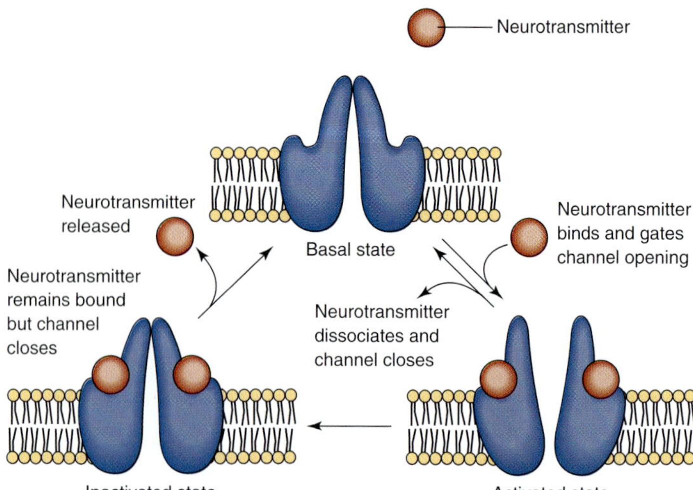

FIGURE 13.13

Conformational changes in ligand-gated ion channels during activation and inactivation of receptor function.
Redrawn based on figure from Alberts, B., et al. *Molecular Biology of the Cell*, 4th ed. New York: Garland, 2002.

subunit is an integral membrane protein with multiple transmembrane domains. **Nicotinic acetylcholine receptors** provide a good example of the important structural features that characterize ligand-gated ion channel receptors (see p. 470). These receptors are formed as a pentameric complex of subunits, each of which has an extracellular N- and C-terminus, and four membrane-spanning domains (M1, M2, M3, M4). The five juxtaposed M2 domains, one from each protein subunit, form the ion-conducting pore. Binding of the neurotransmitter to extracellular sites on the receptor complex causes very small and subtle rearrangements of how the M2 domains pack together. This near-instantaneous rearrangement removes the energetic barriers to ionic flow through the water-lined pore. In this state, the receptor is ligand-bound and activated. However, the activated state of ligand-bound receptor channels persists for only few seconds or minutes (it varies with different types of ligand-gated receptor channel) before a second rearrangement of the subunits occurs. This second rearrangement causes the channel to assume a nonconducting conformation even though the neurotransmitter remains bound (i.e., the receptor is ligand-bound but inactive). This rapid inactivation is a noncovalent form of receptor desensitization (Figure 13.13). Reactivation of the receptor channel occurs only after the neurotransmitter dissociates from the extracellular binding sites. Thus, these receptors cycle between three major functional states (no ligand/inactive > ligand-bound/active > ligand-bound/ inactive).

Termination of Signaling by Ion Channel Receptors

Several mechanisms exist for termination of signaling by ion channel receptors. First, the ligands for these receptors are secreted very transiently in very small, defined extracellular spaces (e.g., the synapse), permitting them to diffuse away from the receptor rapidly. Second, the extracellular concentration of the neurotransmitter is reduced by rapid breakdown by ecto-enzymes (like acetylcholinesterase) and rapid re-uptake into the secreting neuron. Finally, the formation of the ligand-bound but inactive state ensures very brief periods of signal transduction by these neurotransmitter receptors.

Although rapid gating of ion fluxes is the direct function of these receptors, these receptors also physically associate with a large number of other cellular proteins. Some of these associated proteins act as adapter proteins to colocalize efficiently these receptors with other downstream signaling proteins. They form "concentrated" **signaling complexes** at particular subcellular domains such as the plasma membrane at synaptic contacts. For example, the intracellular C-terminal tail of glutamate-gated ion channels associates with an **adapter protein** called PSD-95 at the postsynaptic density under the synaptic contact site. This interaction is facilitated by a protein interaction domain on PSD-95 called the PDZ domain. PSD-95 contains multiple PDZ domains (in addition to the specific PDZ site that binds to the glutamate-gated channels) that permit the colocalization of the receptor with various kinases, phosphatases, and cytoskeletal proteins.

Other families of intracellular proteins that associate with neurotransmitter-gated channels help to target and specifically cluster these receptors at the synaptic membrane. For example, nicotinic acetylcholine receptors are highly concentrated at the neuromuscular synapse via their physical interaction with the protein rapsyn.

Regulation of Ion Channels by Other Receptors

Receptors belonging to families other than the neurotransmitter-gated ion channels utilize ion channels for intracellular signal transduction. Certain types of channels are activated by direct protein–protein interaction with subunits from activated trimeric G proteins, which in turn are activated when upstream G-protein-coupled receptors are activated by agonistic ligands. Other types of ion channels are activated upon the binding of intracellular second-messenger ligands such as cAMP, cGMP, or Ca^{2+}. These second-messenger ligands accumulate upon activation of many types of cell-surface receptors including G-protein-coupled receptors, catalytic receptors for certain growth factors, and some cytokine receptors. Thus, these ligand-gated ion channels are indirectly regulated by cell surface receptors. Some very important second-messenger gated ion channels are Ca^{2+} channels localized in the endoplasmic reticulum rather than the plasma membrane. Finally, the activity of yet other ion channels can be increased (or decreased) upon phosphorylation by various second-messenger-regulated protein kinases. This is another example of indirect regulation of channel function by cell surface receptors.

13.6 ENZYME-LINKED RECEPTORS

Physiological Roles and Extracellular Ligands

The receptors for many polypeptide growth factors and hormones are transmembrane proteins that have intrinsic catalytic activity (Figure 13.4*b*). Enzyme-linked or catalytic receptors of this "superfamily" of related proteins are distinguished by several major structural features. Some are single subunit receptors but others (e.g., insulin receptor) exist as multimeric complexes. Their intracellular domains have catalytic capacities that may include protein kinase, protein phosphatase, protease, or nucleotide phosphodiesterase activities. These receptors primarily regulate long-term cell functions on a time-scale of minutes to hours by initiating intracellular signaling cascades that culminate in the activation or inhibition of gene expression. In turn, these changes in gene expression direct very fundamental pathways of integrated cellular response, such as **cell division, programmed cell death**, or **cell differentiation**. Their agonists are usually large, secreted proteins that function as paracrine/endocrine growth factors or differentiation factors. These secreted ligands generally bind to their cognate receptors with very high affinity (picomolar K_D). However, the agonistic ligands for some types of catalytic receptors are macromolecules expressed on the extracellular surfaces of adjacent cells (juxtacrine signaling). The most common catalytic receptors have tyrosine kinase activity. These **receptor tyrosine kinases (RTK)** include the receptors for epidermal growth factor (EGF), platelet-derived growth factor (PDGF), insulin, and many other polypeptide **growth factors**. A smaller number of catalytic receptors have **serine/threonine kinase** activity. Finally, certain hormones involved in blood pressure regulation bind to receptors with **guanylate cyclase** activity (see p. 518).

Receptor Tyrosine Kinases (RTKs)

Most RTKs are single subunit receptors but some (like the insulin receptor) exist as multimeric complexes (see Figure 23.27, p. 918). Each monomer has a single transmembrane (TM) spanning domain that comprises 25–28 amino acids. The NH_2-terminal ends of such receptors are extracellular and comprise a very large domain for

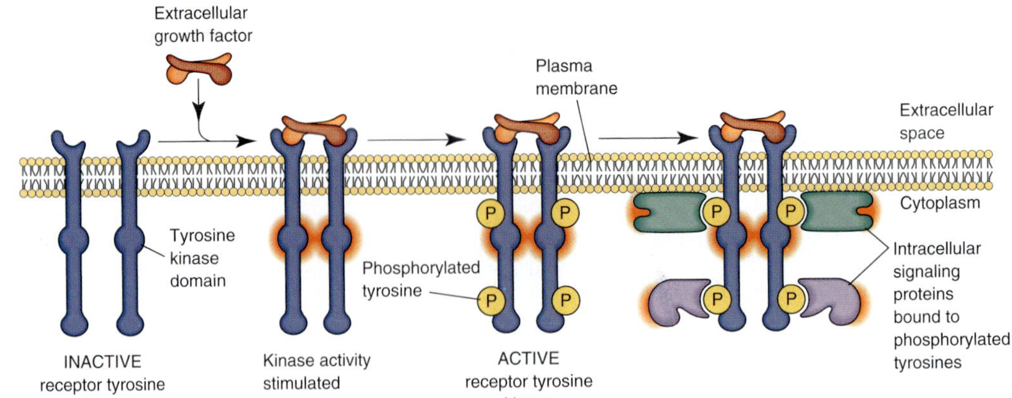

FIGURE 13.14

Conformational and functional changes in a receptor tyrosine kinase during activation by growth factor binding.

Redrawn based on figure from Alberts, B., et al. *Essential Cell Biology*, 2nd ed. New York: Garland, 2004.

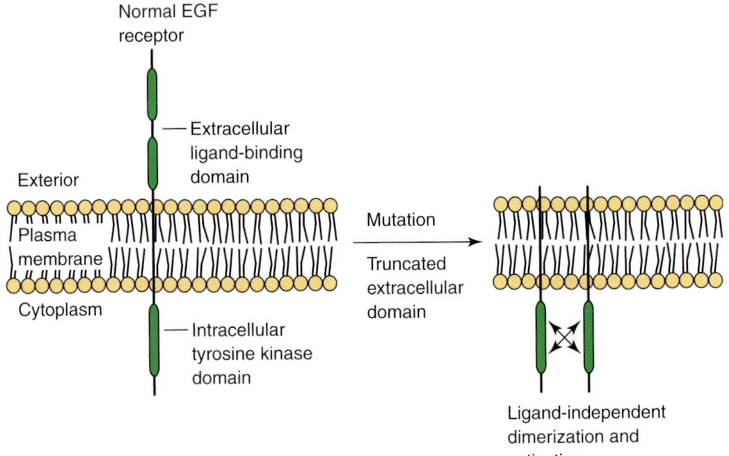

FIGURE 13.15

Mutated forms of receptor tyrosine kinases as the products of cancer-causing oncogenes.

binding the growth factor or hormone. The COOH-terminus are intracellular and comprise the domains responsible for the catalytic activities of the receptors.

When a growth factor binds to the extracellular domain of an RTK, it triggers dimerization with adjacent RTK subunits (Figure 13.14), which leads to rapid activation of the cytoplasmic kinase domain. The first protein substrate for this activated tyrosine kinase is the receptor itself. The intracellular part of the receptor becomes autophosphorylated on multiple tyrosine residues. Each of the phosphorylated tyrosine residues then acts as recognition or anchoring sites for other proteins that are substrates for the RTK. These protein substrates usually contain consensus sequence SH2-type protein–protein interaction domains, which recognize the phosphorylated tyrosines on the receptor. This permits tight binding to the receptor and subsequent phosphorylation of tyrosines on the substrate proteins. Thus, activation of RTK results in accumulation of many types of tyrosine-phosphorylated proteins. These "downstream" substrates act as effectors for the RTK and include signaling proteins such as other protein kinases, regulators of small GTPases, and enzymes that modify phospholipid synthesis and breakdown. When phosphorylated on tyrosine residues, these effector proteins become active links in signaling cascades that ultimately result in changes in the location or functional activity of transcription factors or other proteins involved in mitogenesis or differentiation.

Mutation of certain RTK genes can result in expression of receptors that assume activated conformations in the absence of growth factor stimulation (Figure 13.15). Mutated RTK genes can act as **oncogenes** (see p. 1024) or genes that contribute to the initiation or progression of cancer (see Clin. Corr. 13.1).

Receptor dimerization also triggers an endocytic internalization of the growth factor–RTK complex. Acidification of these endosomes leads to dissociation of the

CLINICAL CORRELATION 13.1

ErbB/HER-Family Receptor Tyrosine Kinases as Targets for Cancer Chemotherapy

The epidermal growth factor (EGF) receptor was the first with intrinsic tyrosine kinase activity to be identified and characterized in detail. This catalytic receptor is one member of a family of four related proteins, termed the ErbB/HER receptors because of their similarity to the v-erbB oncogene of avian erythroblastosis virus that induces erythroid leukemia in birds. This link between ErbB/HER proteins and cancer is also observed in humans. Overexpression of the human ErbB1 gene, which encodes the human EGF receptor (HER1), characterizes bladder, breast, kidney, prostrate, and non-small-cell lung cancers. A mutant ErbB1 gene produces a receptor lacking the extracellular EGF-binding domain, and it is highly expressed in the glioblastomas that account for 25% of the brain tumors in adult humans. Human ErbB3 and ErbB4, respectively, encode the HER3 and HER4 receptors that bind extracellular proteins belonging to the NRG (neuregulin/heregulin/neu) family of growth and differentiation factors. ErbB3 is often overexpressed in breast, colon, prostate, and stomach cancers, while ErbB4 overexpression has been noted in ovarian granulosa cell tumors. The remaining family member is the ErbB2 gene that encodes the HER2 protein that, surprisingly, lacks the ability to bind any known extracellular growth factor. Rather, HER2 receptors possess a basal conformation that permits them to form homodimers with other unliganded HER2 proteins or heterodimers with growth-factor-occupied HER1, HER3, or HER4 receptors. Because dimerization is the critical first step for activating the intrinsic tyrosine kinase activity of this protein, even modest overexpression of HER2 can alter normal cell growth regulation. Significantly, ErbB2 gene expression is amplified by up to two orders of magnitudes in 20–30% of human subjects with invasive breast cancer.

The aberrant expression of ErbB/HER protein in multiple human cancers has prompted the development of several drug therapies that target these receptors. One group of therapeutic agents includes monoclonal antibodies that bind to functionally significant extracellular domains of different HER subtypes. Trastuzumab (Herceptin® from Genentech) is an anti-HER2 antibody that has been used since 1998 for the treatment of those breast cancers characterized by overexpression of ErbB2/HER2. The mechanism underlying the anti-tumor actions of Trastuzumab/Herceptin® involves both the antibody-dependent recruitment of immune factors that kill the tumor cells and the attenuation of the HER2 ectodomain proteolysis (by native metalloproteases) which further potentiates the constitutive dimerization of these receptors. Cetuximab (IMC-C225 or Erbitux® from ImClone) is an antibody that interacts with the EGF-binding domain of the ErbB1/HER1 protein and thereby prevents ligand-induced activation of the receptor. This agent is being tested in trials with patients who have either squamous cell cancers of the head and neck or non-small-cell cancers of the lung. In addition to these antibody-based therapies, a variety of small molecule drugs has designed to target the intracellular tyrosine kinase domains of the ErbB/HER proteins. These reagents are 4-anilinoquinazoline-based compounds that act as competitive inhibitors of the ATP-binding sites of the kinases, particularly the ErbB1/HER1 subtype. At present, these drugs are mainly being tested in trials with patients suffering from non-small-lung cancers that have failed to respond to other chemotherapies.

Source: Roskowski, R., Jr. The ErbB/HER receptor protein-tyrosine kinases and cancer. *Biochem. Biophys. Res. Commun.* 319:1, 2004.

growth factor from the receptor; the internalized growth factor is degraded in the lysosomes. Although some of the internalized receptors are also degraded, most are rerouted back to the plasma membrane. This internalization of receptor–ligand complex, followed by dissociation of the ligand, is the major mechanism for terminating signal transduction by these receptors.

Ras GTPase and MAP Kinase

Ras is a small GTPase or monomeric GTP-binding regulatory protein that is a critical regulator in cell proliferation. About 30% of all human tumors involve cells expressing mutated **Ras oncogenes**. Among the proteins that directly bind to phosphotyrosine sites of activated RTK are adapter molecules that recruit and stimulate Ras-activating proteins (Figure 13.16). In this stimulated state, Ras-activating proteins act to enhance the exchange of GTP for GDP at the guanine nucleotide binding site of Ras itself; this results in accumulation of Ras molecules with active, GTP-bound conformations. In turn, these active Ras proteins transiently bind to and stimulate a family of serine/threonine protein kinases that trigger the **m**itogen-**a**ctivated **p**rotein kinase cascade or **MAP kinase cascade** (Figure 13.17). This amplification cascade involves the serial actions of three protein kinases; the initial Ras-activated kinase (or MAP kinase kinase kinase) activates an intermediary set of MAP kinase kinases, which activate the terminal **MAP kinase** effectors. When activated, the terminal MAP kinases

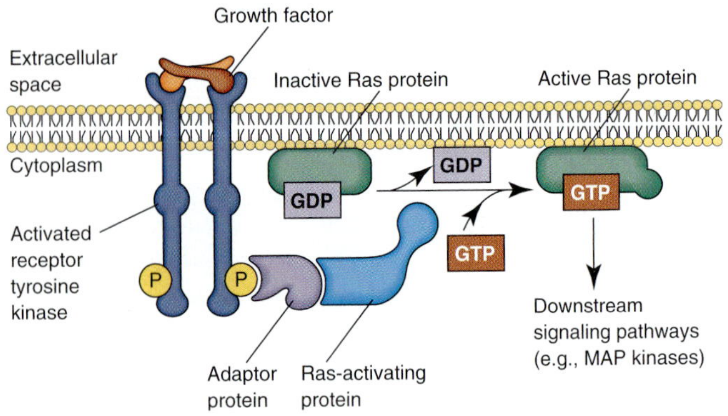

FIGURE 13.16

Role of the ras GTPase during intracellular signal transduction by an activated receptor tyrosine kinase.

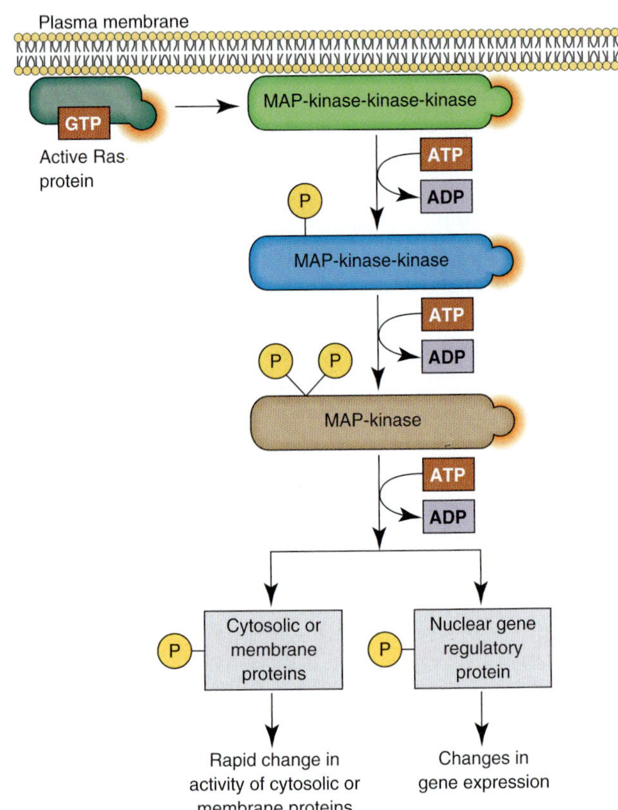

FIGURE 13.17

Role of the MAP kinase cascade during intracellular signal transduction by an activated receptor tyrosine kinase.

Redrawn based on figure from Alberts, B., et al. *Essential Cell Biology*, 2nd ed. New York: Garland, 2004.

phosphorylate multiple target proteins in both the cytosol and the nucleus, including transcription factors that regulate the expression of genes required for cell division, cell survival, or phenotypic differentiation.

Receptor Serine/Threonine Kinases

The growth regulatory factors **transforming growth factor-β (TGF-β)** and **bone morphogenetic proteins (BMP)** have receptors with intrinsic serine/threonine kinase activity. These growth factors have important signaling roles in tissue development at the fetal and neonatal stages and in the maintenance of a differentiated tissue phenotype in adult organisms. As for the receptor tyrosine kinases, mutations in TGF-β receptors and other serine/threonine kinase receptors contribute to the initiation or progression of certain cancers. Likewise, these receptors comprise plasma membrane proteins with an extracellular amino terminus, a single transmembrane-spanning region, and intracellular carboxy terminus that contains the catalytic domain. These receptors also form homodimeric complexes when their agonists bind to the tandem extracellular

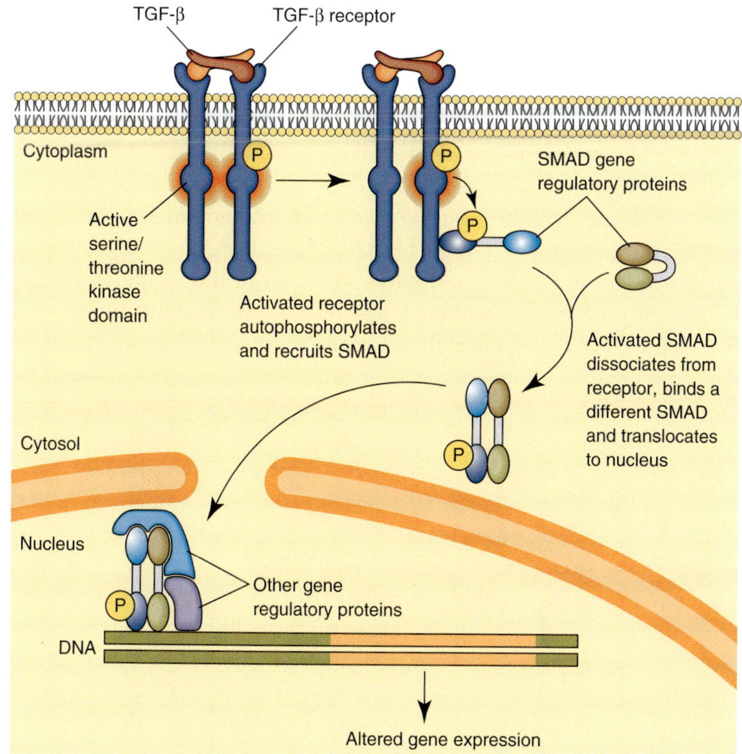

FIGURE 13.18

Intracellular signaling cascades triggered by a receptor serine/threonine kinase during activation by TGFβ (transforming growth factor β).

Redrawn based on figure from Alberts, B., et al. *Essential Cell Biology*, 2nd ed. New York: Garland, 2004.

domains (Figure 13.18). The resulting conformational change results in one catalytic subunit phosphorylating its partner subunit on serine/threonine residues. This initial phosphorylation allows the partner subunit to recruit and to phosphorylate cytoplasmic proteins termed **SMADs** that comprise a multimembered family of gene regulatory factors. In their dephosphorylated state, SMADs adopt a folded conformation, which prevents their interaction with other SMADs and also maintains their localization as cytosolic proteins. However, phosphorylation by activated receptor kinases allows certain SMAD subtypes to both unfold and to form dimeric complexes with other SMAD subtypes. This dimerization also exposes **nuclear localization sequences (NLS)** on the SMADs, resulting in translocation of the cytosolic complex into the nucleus wherein the SMAD dimers interact with other gene regulatory proteins to modulate the transcription of genes involved in organ development or tissue differentiation.

13.7 CYTOKINE RECEPTORS

Cytokines represent another diverse group of secreted polypeptides that act as autocrine/paracrine regulators of growth and differentiation. Many, but not all, cytokines control the growth and differentiation of **hematopoietic** (blood-forming) cells, including the various types of white blood cells or leukocytes. For this reason, some cytokines are known as **interleukins** because they regulate information transfer among different types of leukocytes during various stages of immune or inflammatory responses. Other cytokines are called **interferons** for their ability to interfere with the changes in function induced in cells and tissues infected with viral or bacterial **pathogens**. Like the growth factors that bind to catalytic receptors, cytokines bind to their cognate receptors with high affinity and trigger the rapid activation of protein kinase cascades and accumulation of phosphorylated signaling proteins that ultimately direct long-lasting changes in gene expression. However, the plasma membrane receptors for cytokines lack intrinsic tyrosine kinase or serine/threonine kinase activity (Figure 13.4*c*).

Cytokine Receptors: Structure and Function

Cytokine receptors are multimeric receptors and each subunit has a single transmembrane domain, a large extracellular amino-terminus domain for binding the cytokine, and an intracellular carboxy-terminus domain containing different types of protein–protein interaction motifs but lacking intrinsic enzyme activity. Functional cytokine receptors exist as a stable **hetero-oligomeric complex** of two or more different receptor subunits. High-affinity binding of the cytokine usually requires interaction with at least two different receptor subunits. This coordinated binding of extracellular cytokine ligands triggers conformational changes in the intracellular domains of the receptor subunits that facilitate their rapid association with other intracellular signal transducing enzymes. Cytokine-family receptors exhibit significantly greater diversity in the structures of their individual subunits, as well as in their assembled multimeric complexes. Thus, the different subfamilies of cytokine receptors can recruit a broad range of intracellular signaling proteins. Among the most important of these associated proteins are **nonreceptor tyrosine kinases**, such as the various src-family kinases and the **Janus** kinases **(JAKs)** (Figure 13.19). These latter tyrosine kinases are not integral membrane proteins, but exist as either soluble proteins or weakly membrane-associated proteins that become activated only when they associate with ligand-occupied cytokine receptors. They often phosphorylate the other subunits of the cytokine receptor on tyrosine residues. These phosphorylated tyrosines then act as recognition sites for the binding of other signaling proteins with SH2-interaction domains. Thus, several of the interleukin-type cytokine receptors recruit JAKs, which phosphorylate the receptors to generate binding sites for gene regulatory proteins known as **STAT**s (**signal transducers and activators of transcription**). Like SMAD proteins, unphosphorylated STAT proteins exist as cytosolic monomers. Subsequent tyrosine phosphorylation by activated JAKs induces dimerization, resulting in translocation of the complex to the nucleus and association with additional gene regulatory proteins.

Other types of cytokine receptors recruit adaptor proteins that initiate the assembly of protein kinase complexes that culminate in the cytosol-to-nucleus translocation of the **NFκB** (nuclear factor of activated B-cells) complex, which is an important transcription factor for regulation of expression of genes involved in multiple types of the immune and inflammatory responses. Many cytokine receptors also trigger the activation of the MAP kinase cascades.

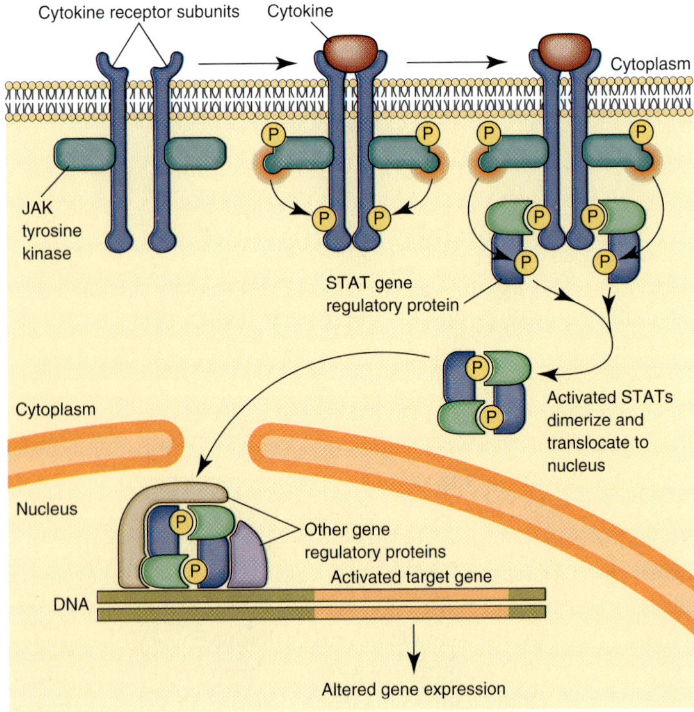

FIGURE 13.19

Intracellular signaling cascades triggered by prototypical noncatalytic cytokine receptor.

13.8 G-PROTEIN-COUPLED RECEPTORS

Physiological Roles and Extracellular Ligands

The superfamily family of **G-protein-coupled receptors** (GPCR) binds an extraordinarily diverse range of agonistic ligands including proteins, peptides, amino acid derivatives, catecholamines, lipids, nucleotides, and nucleosides (Figure 13.20). These ligands include hormones, neurotransmitters, and local mediators. The G-protein-coupled receptors play important roles in endocrine, synaptic, paracrine, or autocrine signaling in virtually all tissues and cell types. It is estimated that several percent of the genes in the human genome encode GPCR. Ligands bind to GPCRs with affinities in the nanomolar to micromolar range, which is intermediate between the high affinity of enzyme-linked receptors and the low affinity of the ligand-gated channel receptors. Many important sensory proteins are G-protein-coupled receptors including rhodopsin (see p. 968) and the many receptors that permit perception of taste and smell. However, the actual ligands for these taste and smell receptors are largely unknown. Many GPCR are involved in the acute regulation of critical physiological responses such as cardiac contractility, metabolism, and complex behavior. Thus, these receptors or their downstream signaling pathways are the targets for many of the drugs most widely used in the treatment of human diseases (see Clin. Corr. 13.2).

Structure of G-Protein-Coupled Receptors

When occupied by agonistic ligands, these receptors directly activate **heterotrimeric GTP-binding regulatory proteins (G proteins)** (Figure 13.20), which then directly interact with other signaling elements that are collectively called effector proteins; these effectors are usually enzymes or ion channels. These receptors do not directly regulate effector proteins. Thus, even the earliest signaling events requires the participation of at least five proteins: the receptor itself, the intermediary G protein (with its three distinct protein subunits), and the effector protein. The existence of multiple G-protein types that can differentially couple to various receptors and effectors provides for considerable molecular diversity in the types of signal transduction pathways controlled by these receptors.

Despite the extreme diversity of the ligands that can bind to **G-protein-coupled receptors (GPCR)**, the receptors per se show a high degree of structural but not sequence similarity (Figure 13.20). Each receptor is a single subunit membrane protein with seven α-helical transmembrane-spanning domains. (GPCR have also been termed **heptahelical** or **7-transmembrane domain receptors**.) Each is organized with an extracellular amino terminus, three extracellular loops, three intracellular loops, and an

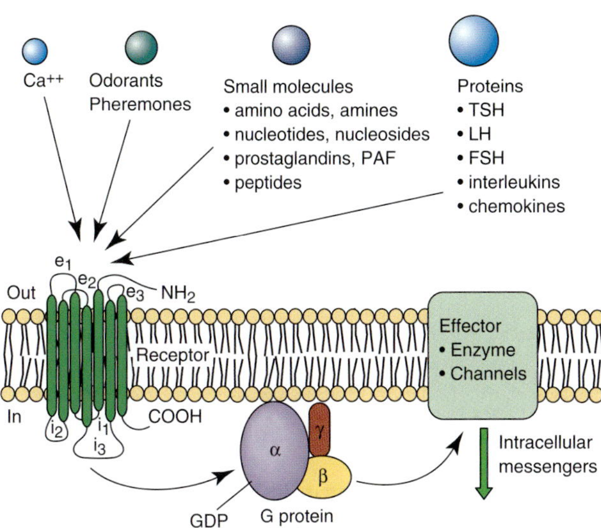

FIGURE 13.20

Major elements of signal transduction initiated by the seven-transmembrane-domain G-protein-coupled receptors.
Redrawn based on figure from Bockaert, J. and Pin, J-P. Molecular tinkering of G protein-coupled receptors: an evolutionary success. *EMBO J.* 18:1723, 1999.

CLINICAL CORRELATION 13.2
G-Protein-Coupled Chemokine Receptors as Targets for the Human Immunodeficiency Virus (HIV)

Among the 1000 or so human genes encoding different G-protein-coupled receptors are the subgroup of chemokine receptors that are highly expressed in leukocytes (white blood cells). Chemokines include approximately 50 polypeptide factors that are secreted by multiple cell types (e.g., epithelial cells and stromal cells) within various tissues. The different chemokines act as agonistic ligands for 19 types of chemokine receptors that are differentially expressed in the various classes of leukocytes (e.g., monocytes, T-lymphocytes, B-lymphocytes, natural killer cells) which execute critical immune and inflammatory responses within tissues invaded by foreign pathogens (protozoans, bacteria, or viruses). Chemokines function as chemotactic agents to recruit leukocytes to the infected tissue location, as acute regulators of G-protein-dependent signaling pathways involved in the rapid killing or sequestering of pathogens, and as inducers of gene expression that contribute to long-term adaptive immunity to pathogens. Not surprisingly, the disruption or disabling of chemokine receptor signaling can contribute to a wide range of infectious diseases. Of particular significance is the relatively recent (within the past decade) finding that human immunodeficiency virus (HIV) utilizes endogenous chemokine GPCR to infect and ultimately kill immune effector leukocytes.

HIV entry into leukocytes requires the direct interaction of the viral envelope glycoprotein (gp120) with extracellular surface proteins of the target host cell. Early studies identified the CD4 membrane protein, a critical immune receptor of helper T-lymphocytes as a major target for gp120 binding (CD4 is also expressed at much lower levels in other types of leukocytes). This explains why the numbers of helper T cells are drastically reduced in most HIV-infected patients who progress to develop the characteristic symptoms of AIDS (acquired immune deficiency syndrome). However, early studies also revealed that different strains of HIV preferentially infected T-lymphocytes (T-trophic virus) or monocyte-macrophages (M-trophic virus) while epidemiological analyses revealed a subpopulation of individuals who were highly resistant to infection by M-trophic strains of HIV. This resistance is due to mutations in the gene encoding CCR5, a chemokine receptor that is highly expressed in monocyte/macrophages; these mutations result in low or no expression of cell surface CCR5 protein. Thus, M-trophic strains of HIV require the presence of both the CCR5 and CD4, as co-receptors, to infect efficiently monocyte/macrophages. Other studies have demonstrated that CXCR4, a chemokine receptor highly expressed in T-lymphocytes, acts as a co-receptor with CD4 for binding the gp120 of T-trophic strains of HIV. Significantly, the ability of CCR5 and CXCR4 to bind gp120 and thereby facilitate entry of HIV into the host leukocyte does not require activation of the G_i-family proteins to which these receptors are normally coupled when occupied by their physiological chemokine ligands. Thus, treatment of leukocytes with pertussis toxin (which uncouples receptors from the downstream G_i proteins) does not reduce the extent of HIV internalization into leukocytes. Although G-protein activation by CCR5 and CXCR4 is not obligatory for HIV entry, gp120 binding to these receptors does induce functional activation of the downstream G-protein signaling cascades. Recent studies suggest that activation of these signaling pathways within the HIV-infected cells may facilitate the chemo-attraction of naive uninfected leukocytes and thus aid in propagation and dissemination of the virus within the host subject. The identification of CCR5 and CXCR4 as HIV co-receptors has also elicited intense interest in the development of antagonistic ligands that might selectively attenuate gp120 binding and thereby retard the colonization and spread of this devastating pathogen within infected subjects.

Source: Sodhi, A., Montaner, S., and Gutkind, J. S. Viral hijacking of G protein-coupled receptor signaling networks. *Nature Rev. Mol. Cell Biol.* 5:998, 2004.

intracellular carboxy terminus. The extracellular region provides sites for glycosylation, and it is quite divergent among different GPCR. The extracellular loops contain essential cysteine residues for formation of interloop disulfide bridges. The transmembrane (TM)-spanning domains are quite homologous, but not identical among different GPCR. These TM domains form a "pocket" that often contains the critical amino acid residues for ligand binding. The sequences and sizes of the intracellular loops are divergent with the third intracellular loop containing critical sequences for selective interaction with G proteins. The intracellular tail varies greatly in length and sequence among the different receptors; this domain contains sites involved in recognition/activation of G proteins. It also contains serine and threonine sites for phosphorylation by protein kinases. Phosphorylations of these residues play important roles in the termination of signal transduction by activated GPCR.

The tertiary structure of GPCR within biological membranes involves a clustering of the seven transmembrane helices. The exact clustering of these segments controls the conformations and apposition of the intracellular loops and the carboxy-terminus. Large ligands such as polypeptide hormones bind to the extracellular loops, while small ligands such as acetylcholine and epinephrine bind to the pocket formed by the seven membrane-spanning segments. In either case, binding causes a rearrangement of the membrane-spanning helices relative to each other and changes the conformations of the intracellular loops and carboxy terminus. These latter conformational changes expose recognition sites for the particular G proteins to which the receptor is functionally coupled. Thus, a GPCR occupied by an agonist can transiently form a complex with a G protein. An antagonist can bind to the receptor but this binding does not cause appropriate rearrangements of the membrane-spanning domains or exposure of the G protein recognition sites on the intracellular loops and tail.

The transient interaction between the agonist-occupied receptor and its G protein affects the function of both the receptor protein and the G protein. The functional changes in the G protein are described below. The major functional change in the receptor following its binding to a G protein is that its affinity for the agonistic ligand is significantly reduced, which usually leads to dissociation of the agonist. The agonist-free receptor assumes its inactive conformation that, in turn, eliminates its ability to interact with the G protein. The G protein dissociates from the receptor/G-protein complex returning the receptor to its ligand-free, inactive state.

Heterotrimeric G Proteins

G proteins that are functionally coupled to the seven-transmembrane domain receptors exist as heterotrimeric complexes consisting of **α-**, **β-**, and **γ-subunits** (Figure 13.21). The α-subunit, the largest subunit of each trimeric complex, is a relatively hydrophilic protein with molecular weights in the 39 to 46-kDa range. The α-subunit contains the guanine nucleotide-binding site and GTPase activity that is critical for proper regulation of the receptor/G-protein interaction. It also contains domains that interact with various effector proteins. Twenty distinct subtypes of α-subunits are grouped into several subfamilies based on both structural features and type of coupling to particular effector proteins and second messenger cascades. The **α_s**-subunits (stimulatory α-subunit) of the **Gs** family stimulate **adenylate cyclase**, which results in an increase in cellular **cAMP** (see Clin. Corr. 13.3). In contrast, the **α_i**-subunits (inhibitory α-subunits) of the **G_i** family inhibit adenylate cyclase, thereby lowering cellular levels of cAMP. The **α_o**- and

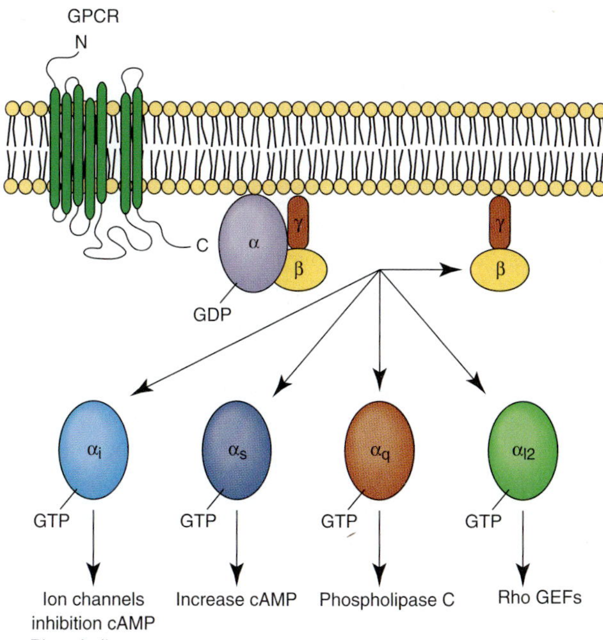

FIGURE 13.21

Major subclasses of heterotrimeric G proteins that are activated by G by the seven-transmembrane domain G-protein-coupled receptors.

$\boldsymbol{\alpha_z}$-subunits, which also belong to the Gi family, lack clearly defined effector protein targets. The G_i family also includes the $\boldsymbol{\alpha_t}$-subunits ("t" for transducin, the G proteins that transduce light perception by the rhodopsin-type GPCR; see p. 974), which stimulate **cGMP phosphodiesterase** resulting in rapid decreases in **cGMP** content. The $\mathbf{G_q}$ family includes various α-subunits (α_q, α_{11}, α_{15}, α_{16}) that are positively coupled to the **phospholipase C** effector enzyme that trigger increases in three second messengers: IP_3 (inositol trisphosphate), DAG (diacylglycerol), and Ca^{2+}. Finally, the α_{12}- and α_{13}-subunits of the $\mathbf{G_{12/13}}$ family stimulate effector proteins that function as **guanine nucleotide exchange factors** (**GEFs**) for a variety of small monomeric GTPases involved in cell migration, cell shape regulation, and cell division.

The **$\boldsymbol{\beta}$- and $\boldsymbol{\gamma}$-subunits** of G proteins are very hydrophobic proteins, which under non-denaturing conditions are isolated as a **dimeric complex**. The β-subunits (~35 kDa) and γ-subunits (10 kDa) of different G proteins are very similar. Purified $\beta\gamma$ complex isolated from one G protein can often be recombined with the unique, purified α-subunit of a second G protein to produce a heterotrimer, which is functionally indistinguishable from the native heterotrimer of the second G protein. The $\beta\gamma$ complex is necessary for the interaction of the α-subunit with receptors. In addition, the $\beta\gamma$

CLINICAL CORRELATION **13.3**

Gsα G-Protein Mutations in Pituitary Gland Tumors and Endocrine Diseases

The pituitary gland plays a central role in regulating the endocrine status of mammals by secreting multiple hormones that regulate the function of other organs and glands such as the liver, thyroid, adrenal, and reproductive glands. The release of hormones from the pituitary gland per se is regulated by neurotransmitters, neuropeptides, and other hormones released by the hypothalamic region of the brain. For example, the hypothalamus and pituitary cells known as somatotrophs act in concert to control the release of growth hormone (GH) which acts to regulate metabolism and cell growth in many peripheral tissues. A hypothalamus-derived factor called growth hormone releasing hormone (GHRH) is an agonist for a seven-transmembrane domain receptor expressed by pituitary somatotroph cells. The GHRH receptor is coupled to adenylate cyclase via the G_s heterotrimeric G protein. Increases in cAMP triggered by the GHRH receptor activate CREB-dependent transcription of the GH gene, resulting in increased synthesis and release of GH. Certain pituitary tumors can result in the hypersecretion of GH; this causes the endocrine syndrome known as acromegaly, which is characterized by abnormal growth patterns in many tissues. These tumors, which are generally nonmetastatic, are believed to arise from the replication of single, transformed pituitary cells that have acquired a growth or survival advantage due to spontaneous mutations in key regulatory genes.

Significantly, 30–40% of the GH-secreting pituitary tumors in humans express a mutated version of the GNAS1 gene that encodes α_s-subunit of G_s protein. This mutated genes, termed *gsp* oncogenes, result from single acid substitutions in either of two residues that play critical roles in the binding and hydrolysis of GTP. Both mutations induce expression of G_s-α protein with a greatly diminished GTPase activity and thereby cause the constitutive activation of G_s and its adenylate cyclase effector protein. In turn, the resulting increase in adenylate cyclase activity markedly increases cAMP generation, protein kinase A (PKA) activation, the PKA-dependent phosphorylation of CREB and other cAMP-sensitive gene regulatory proteins, and, finally, the expression of cAMP-inducible genes such as that encoding GH. Similar *gsp* mutations have been found in other pituitary tumor cells including corticotrophs that secrete ACTH (adrenocorticotropic hormone); the constitutive hypersecretion of ACTH from these transformed cells results in overproduction of glucocorticoid hormones by the adrenal gland.

It is interesting to note that the two amino acids altered in these naturally occurring Gsα mutants were identified by early *in vitro* biochemical studies as functionally significant residues for the regulation of GTPase activity. Arginine-201, which is replaced by cysteine in the most common *gsp* mutant, is the acceptor site for the covalent ADP-ribosylation induced by cholera toxin. This ADP-ribosylation of wild-type $G_{s\alpha}$ in the intestinal epithelial cells of cholera victims represses GTPase activity, leading to a greatly increased rate of cAMP generation and perturbation of cAMP-dependent salt and water transport. Glutamine-227, which is replaced by alanine or other amino acids in some *gsp* mutations, is a conserved glutamine located in the so-called switch-II domains of all G protein α subunits. X-ray crystallographic studies have shown that the switch-II domain undergoes relatively large conformational movements, depending on whether GTP or GDP is bound to the α-subunit. The engineered substitution of this glutamine residue within all known α-subunits results in decreased GTPase activity and it is widely used by molecular biologists to generate constitutively active versions of the various G proteins for transgenic animal studies.

Source: Lania, A., Mantovani, G., and Spada, A. Genetics of pituitary tumors: Focus on G protein mutations. *Exp. Biol. Med.* 228:1004, 2003.

complexes from some activated G proteins also directly interact with, and activate, a variety of downstream effector proteins. For example, G_i-derived $\beta\gamma$ dimers act to gate the opening of certain K^+ channels in the heart, thereby inducing cardiac myocyte hyperpolarization and a slowing of heart rate.

The G-Protein Cycle

In their basal state, G proteins exist as the $\alpha\beta\gamma$ heterotrimeric complex with GDP occupying the nucleotide binding site of the α-subunit (Figure 13.22). The rate of release of GDP in the presence of unoccupied receptors (basal conditions) is low (although not zero). The intrinsic GTPase activity present in the α-subunit catalyzes the hydrolysis of GTP into GDP and P_i. While the P_i is readily released, the GDP remains bound for a much longer time. This sequential release makes the GTPase reaction irreversible. The α-GDP displays high affinity for the $\beta\gamma$ dimer, and the heterotrimer is therefore regenerated. Consequently, in the unstimulated steady state, only a small fraction of the total number of G-protein molecules is in the effector active conformation (GTP-bound). Upon interaction of an $\alpha\beta\gamma$ heterotrimeric complex with agonist-occupied GPCR, the rate of release of GDP from the α-subunit is increased by over an order of magnitude. This allows GTP, which is much more abundant than GDP in the cytosol, to bind to the α-subunit, thereby triggering dissociation of α-GTP from the $\beta\gamma$ dimer. The released α-subunits, as well the released $\beta\gamma$ dimers, then interact with appropriate downstream effector enzymes to change second messenger levels. Activation of G proteins by agonist-occupied GPCR is catalytic as rather than stoichiometric. Thus, depending on the type of GPCR, one molecule of agonist-occupied receptor may serially interact with 10–100 molecules of trimeric G protein; this results in major signal amplification at the earliest stages of GPCR activation. Although the intrinsic GTPase

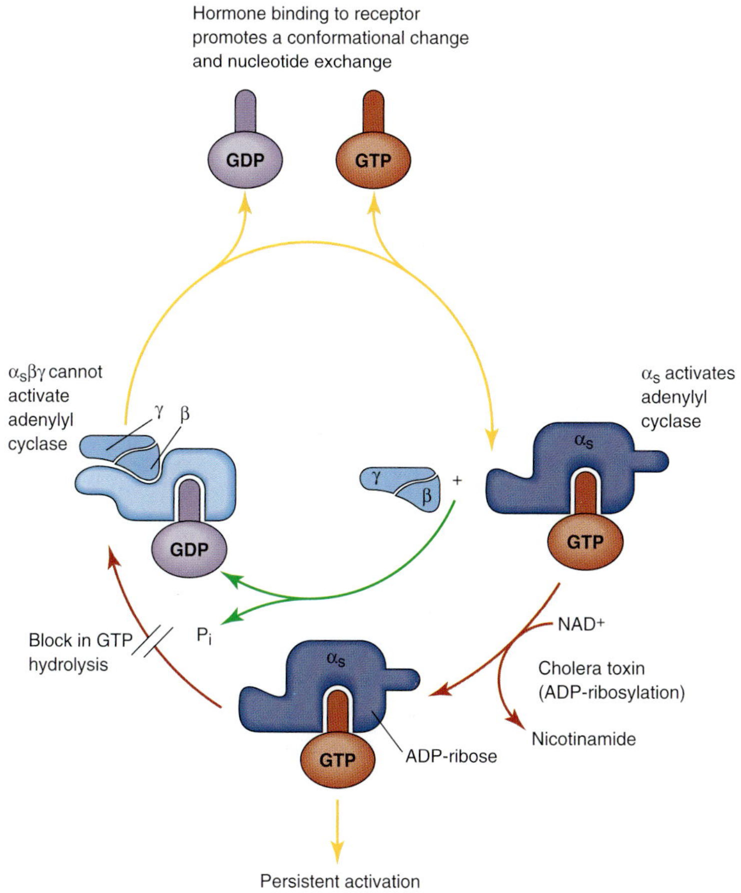

FIGURE 13.22

Activation and inactivation cycle of a heterotrimeric G protein.

Redrawn based on figure from Lodish, H. et al. *Molecular Cell Biology*, 4th ed. New York: W. H. Freeman, 2000.

activity of the released α-subunits ensures that the dissociation of $\alpha\beta\gamma$ heterotrimeric complex will be transient, the overall repetition rate of the cycle will be increased as long as there are agonist-bound GPCR in the plasma membrane of the cell.

Termination of Signaling by G-Protein-Coupled Receptors

Many of the small ligands for GPCR can be rapidly metabolized or inactivated by extracellular enzymes, and most receptors exhibit rapid **desensitization** or **adaptation**. Although some of this adaptation is due to receptor internalization and down-regulation, the primary mechanism for rapid GPCR desensitization involves phosphorylation of the cell-surface receptor that results in loss of function (Figure 13.23). The phosphorylated receptors cannot efficiently activate G proteins. Two families of protein kinases phosphorylate GPCR and induce desensitization. Second-messenger-dependent protein kinases, such as cAMP-dependent **protein kinase A (PKA)**, phosphorylate many types of GPCR. PKA can phosphorylate such receptors regardless of whether the receptor is in an agonist-occupied or agonist-free (empty) conformation. There are GPCR-specific protein kinases (**GRKs**) that are also known as **β ARKs** (βAR kinases) because the β-adrenergic receptor (βAR) was the first identified substrate (see Clin. Corr. 13.4). In contrast to PKA and other second messenger-regulated kinases, βARKs only phosphorylate GPCR when they are occupied by agonistic ligands; the agonist-free receptors are not substrates.

Effects of Bacterial Toxins on Heterotrimeric G Proteins

Some α-subunits of G proteins contain specific amino acids that can be ADP-ribosylated by **exotoxins** secreted from some bacterial pathogens, such as *Vibrio cholerae* or *Bordetella pertussis* (Figure 13.22). These toxins possess both **NAD glycohydrolase**, which catalyzes breakdown of NAD^+ into nicotinamide and ADP-ribose, and **ADP-ribosyl transferase** activity, which catalyzes the transfer of the ADP-ribose moiety to the side chains of specific residues of some G-protein α-subunits (each toxin displays a specific residue selection). These toxin-catalyzed ribosylation reactions interfere with the normal function of the G protein and thus result in abnormal signal transduction and regulation of cell function. For example, ADP-ribosylation of the Gs α-subunits by cholera toxin represses the intrinsic GTPase activity, thereby causing constitutive dissociation of the GTP-bound α subunits from the $\beta\gamma$ dimers even when GPCR is not occupied by agonist. This chronic elevation of GTP-bound α_s induces continuous activation of the adenylate cyclase and cAMP production. In the large intestine, this chronic elevation of cAMP results in a sustained PKA-mediated phosphorylation of chloride channels that normally regulate salt and water transport; the hyperactivation of these channels severely disrupts salt and water transport, resulting in life-threatening diarrhea.

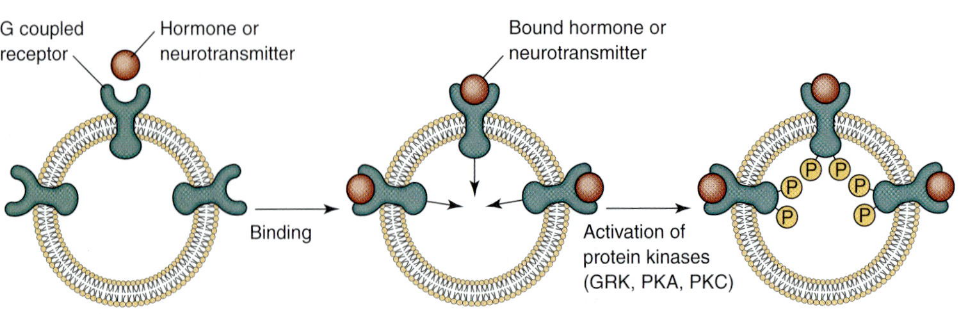

FIGURE 13.23
Receptor phosphorylation as a major mechanism for the desensitization/inactivation of G-protein-coupled receptors.

CLINICAL CORRELATION 13.4

Alterations in β-Adrenergic Receptor Signaling Proteins in Congestive Heart Failure

Congestive heart failure is a pathological condition in which the heart fails to pump blood in amounts sufficient for adequate perfusion of peripheral tissues. Heart failure is accompanied by adaptations of cardiac myocytes for the maintenance of cardiac output following myocardial injury or during development of excessive hemodynamic burdens, such as high blood pressure. These adaptations include myocardial hypertrophy, in which the amount of contractile tissue is augmented. They also include activation of sympathetic nerve output, especially the release of norepinephrine and epinephrine from cardiac adrenergic nerves, which augments myocardial contractility. Although these adaptive mechanisms may be adequate to maintain cardiac contractile performance at relatively normal levels during the initial development of heart failure, they ultimately become maladaptive and result in progressive decreases in adequate cardiac performance. This shift from adaptive to maladaptive response is particularly evident in changes with the β-adrenergic receptor signaling system.

In normal cardiac muscle cells, epinephrine activates the G-protein-coupled β1- and β2-adrenergic receptor subtypes. Both of these β-adrenergic receptors are coupled to Gs proteins, stimulation of adenylate cyclase, increased cAMP accumulation, and activation of protein kinase A (PKA). In cardiac muscle cells, PKA phosphorylates several target proteins including L-type, dihydropyridine-sensitive Ca^{2+} channels and phospholamban, an allosteric regulator of cardiac sarcoplasmic reticulum (SR) Ca^{2+}-translocating ATPase. Phosphorylation of the L-type C channels increases the amount of time these channels stay open in response to depolarization; this results in enhanced Ca^{2+} influx. In its unphosphorylated state, phospholamban represses the function of the SR Ca^{2+}-ATPase pump. When phospholamban is phosphorylated by PKA, this inhibitory effect is relieved resulting in an increased rate and extent of Ca^{2+} pumping by the Sr. The combined action of enhanced plasma membrane Ca^{2+} influx (through the L-type channels) and enhanced Ca^{2+} sequestration within the SR results in higher amounts of Ca^{2+}-induced Ca^{2+} release from the SR during subsequent cardiac action potentials. Thus, during adrenergic stimulation, more Ca^{2+} is released during each action potential resulting in a stronger contraction (positive inotropic effect). However, the released Ca^{2+} is also more rapidly sequestered by the SR resulting in a more rapid cycle of contraction/relaxation (positive chronotropic effect). Thus, at the whole-organ level, β-adrenergic stimulation of heart results in both stronger contractions and more blood pumped per heart contraction (stroke volume) and a more rapid contraction/relaxation cycle (heart rate). Cardiac output (stroke volume × heart rate) is increased, resulting in more blood delivery to the systemic circulation. However, this means that the heart will do more work (higher energy output) and will require a higher delivery of oxygen and metabolites (fatty acids and glucose) to generate that energy.

Because of reduced myocardial function due to chronic disease, heart failure triggers an elevated rate of secretion of epinephrine/norepinephrine to provide a compensatory hyperactivation of the inotropic signaling responses described above. While this initially results in increased activation of the β1 and β2 receptors, it also induces a chronic desensitization and down-regulation of the receptors. Moreover, these β-receptors increasingly activate "nonclassical" (i.e., other than Gs → adenylate cyclase → PKA) signaling cascades in addition to the signals for acute inotropic and chronotropic regulation. The β1 subtype appears to drive nonclassical cascades that are predominantly maladaptive such as decreased expression of survival genes and increased activation of pro-apoptotic signals programmed cell death). In contrast, the β2 receptor induces nonclassical signaling pathways that are protective or less maladaptive. Changes in the β-adrenergic signaling pathway that accompany the development of heart failure include a decrease in the ratio of β1-subtype receptors to β2-subtype receptors. Heart failure also induces an increased activity of the GRK2/βAR-kinase that phosphorylates the receptors and thereby uncouples them from activation of the classical G-protein-dependent signaling pathways.

Although it may seem paradoxical, current protocols for the treatment of heart failure are based on the use of β-adrenergic antagonists (β-blockers). At present, the most commonly used β-blockers for heart failure are antagonists, such as **metoprolol** or bisoprolol, which have high selectivity for the β1 receptor subtype. More recently, some large clinical studies have reported that another useful drug therapy for heart failure is **carvedilol**, which nonselectively blocks both β1- and β2-adrenergic receptors. Carvedilol additionally antagonizes α1-adrenergic receptors that are highly expressed in vascular smooth muscle cells; α1-adrenergic receptors couple through the Gq protein to phospholipase C, resulting in IP_3 accumulation and release of SR Ca^{2+} pools via the IP_3-gated Ca^{2+} release channels. The effectiveness of β-blocker therapy appears to involve repression of the nonclassical, maladaptive signaling pathways activated by β1 receptors.

Source: Lohse, M. J., Englehardt, S., and Eschenhagen, T. What is the role of β-adrenergic signaling in heart failure? *Circ. Res.* 93:896, 2003.

13.9 CYCLIC-AMP-BASED SIGNAL TRANSDUCTION

Regulation of Cyclic-AMP Synthesis and Degradation

Many metabolic and behavioral responses to different hormones and neurotransmitters are mediated by increases in intracellular cAMP. The use of cAMP as a second messenger is largely limited to G-protein-coupled receptors. Cells can actively regulate both the synthesis and degradation of this second-messenger. cAMP synthesis is catalyzed by **adenylate cyclase** effector enzymes that use ATP as a substrate to produce cAMP and pyrophosphate (PP_i) as products (Figure 13.24). The normal basal level of cAMP in cytosol is 10^{-7} M. During maximal hormonal stimulation, the cAMP concentration can increase 2- to 100-fold, depending on cell type. At least six genes encoding different subtypes of adenylate cyclase have been identified. All are integral membrane proteins with 12 membrane-spanning domains, one catalytic domain within the intracellular segment connecting transmembrane segments six and seven, and a second catalytic domain at the intracellular carboxy terminus. The various subtypes are expressed in tissue-specific patterns and can be differentially regulated by particular signaling proteins. All adenylate cyclase isoforms are positively regulated by the α_s-subunits of G_s proteins. However, the $\beta\gamma$-subunit complex from various G proteins (including G_s) can stimulate some adenylate cyclase while inhibiting other isoforms. The α_i-subunits from G_i proteins can directly inhibit subtypes of adenylate cyclase (Figure 13.25).

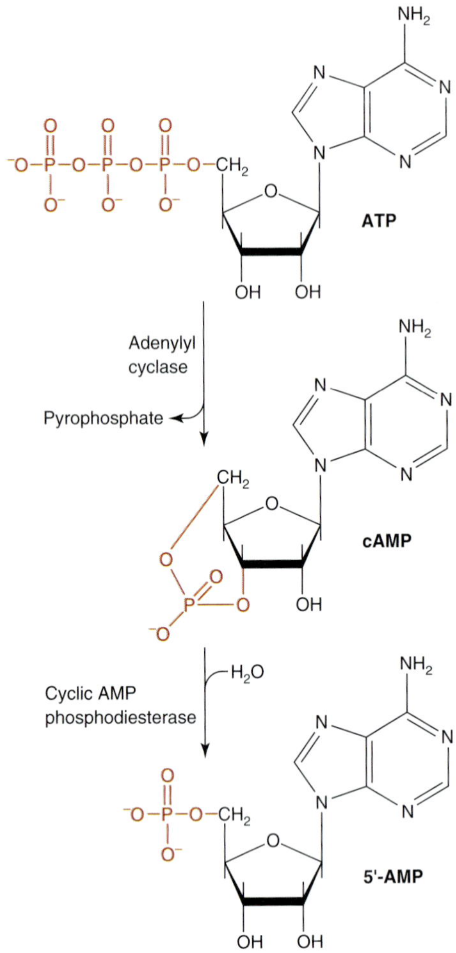

The breakdown of cAMP is catalyzed by a variety of **cyclic nucleotide phosphodiesterases (PDEs)** that hydrolyze the cyclic diester bond to produce AMP (Figure 13.24). The activity of these PDEs can be regulated by both hormones and certain drugs. Cellular cAMP can be increased by inhibition of the PDEs. Xanthine derivatives, such as theophylline and caffeine, inhibit PDEs resulting in increased cAMP levels in the absence of hormonal stimulation.

FIGURE 13.24

Synthesis and degradation of cyclic AMP by adenylate cyclases and cyclic nucleotide phosphodiesterases.

Intracellular Signaling Mechanisms of Cyclic AMP

cAMP-dependent **protein kinase A (PKA)** mediates most of the effects of cAMP (Figure 13.26). In the absence of cAMP, PKA is a tetramer comprised of two regulatory subunits (R-subunits), which contain the cAMP-binding sites, and two catalytic subunits (C-subunits), which contain the kinase activity. In the tetrameric form, the kinase cannot catalyze phosphorylation of target proteins. Upon the cooperative binding of several cAMP molecules to the regulatory subunits, the tetramer dissociates and the free catalytic subunits are now active. Many of the protein substrates of the cAMP-dependent kinase are themselves kinases with specific target enzymes. This underscores the cascading nature of this and most other receptor-linked signaling systems. There are also specific phosphatases that dephosphorylate the phosphorylated proteins and thus inactivate the cAMP-mediated signaling cascade. These phosphatases can themselves be subject to receptor-linked regulation.

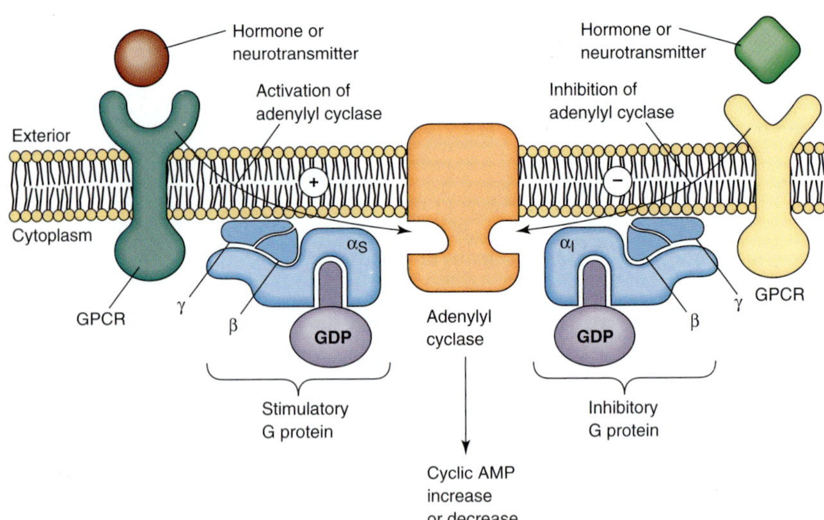

FIGURE 13.25

Positive and negative regulation of adenylate cyclase effector enzymes by G_s- and G_i-family heterotrimeric G proteins.

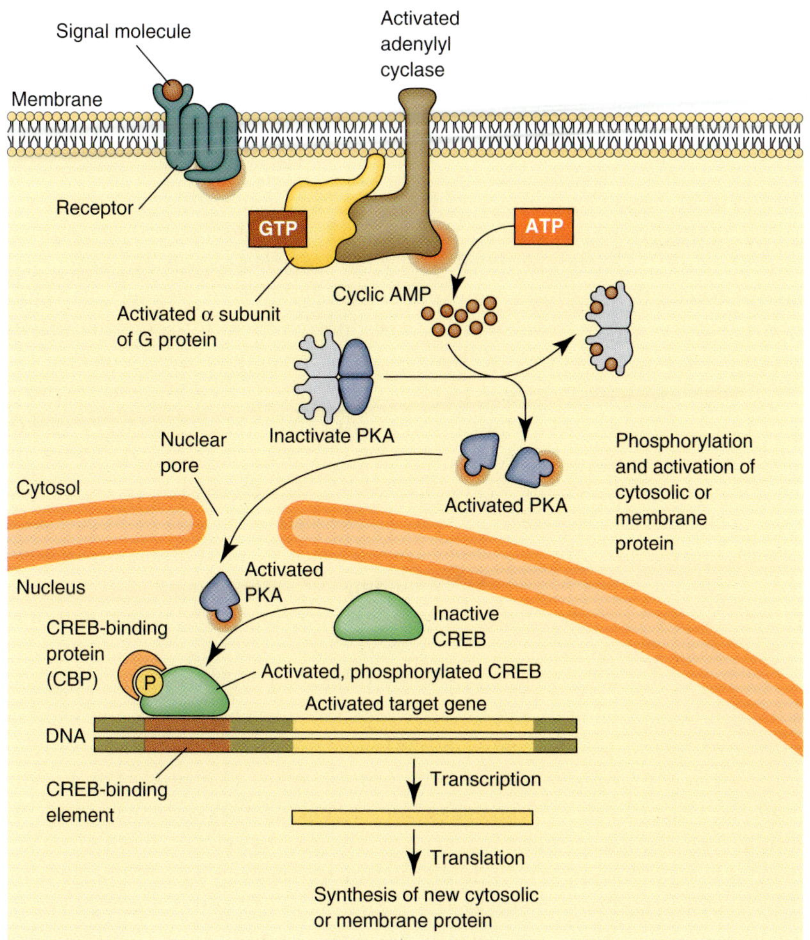

FIGURE 13.26

Role of protein kinase A in the intracellular signaling cascades regulated by cyclic AMP. Redrawn based on figure from Alberts, B., et al. *Molecular Biology of the Cell*, 4th ed. New York: Garland, 2002.

cAMP profoundly alters cellular metabolism by altering both the activity and expression of many catabolic enzymes, including enzymes involved in both lipid (see p. 702) and carbohydrate metabolism (see p. 602). Metabolic regulation by cAMP often involves the PKA-mediation of "secondary" kinases or phosphatases, which actually change the phosphorylation states of the various enzymes. PKA can also phosphorylate several types of plasma membrane ion channels or ion transporters leading to either altered membrane potential or the activation/inhibition of the influx of regulatory ions such as Ca^{2+}. Because neurotransmitters, such as serotonin, dopamine, and epinephrine, interact with GPCR that regulate G_s or G_i, the ability of cAMP and PKA to regulate ion channel activity in neurons is a major component of complex behavioral response.

Elevation of cAMP in eukaryotic cells also rapidly alters the transcription and expression of the genes coding for important metabolic enzymes, polypeptide hormones, and ion transport proteins. The dissociation of PKA catalytic subunits from the regulatory subunits also exposes nuclear localization sequences on the catalytic subunits and this facilitates cytosol-to-nucleus translocation of these subunits (Figure 13.26). This intranuclear PKA induces phosphorylation and activation of **cAMP-regulated gene regulatory proteins** called **CREBs** to control the expression of genes containing **cAMP-sensitive regulatory elements** (**CRE**) that lie within the 5′-flanking regions of many cAMP responsive genes. In this way, the transient increases in cAMP triggered by GPCR at the cell surface can be translated into long-lasting changes in nuclear gene expression and cell function.

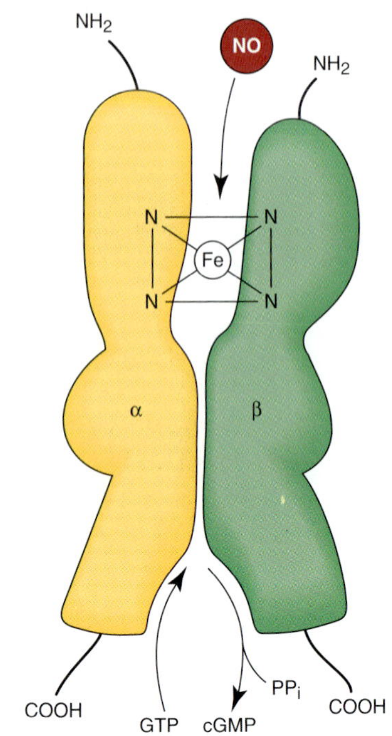

FIGURE 13.27
NO binding to soluble guanylate cyclase.

13.10 CYCLIC-GMP-BASED SIGNAL TRANSDUCTION

Regulation of Cyclic-GMP Synthesis and Degradation

cGMP is an important second messenger in the regulation of muscle and nonmuscle contractility, in visual signal transduction, and in blood volume homeostasis. Cellular levels of cGMP are dynamically regulated by a balance between synthetic (**guanylate cyclases**) and degradative (**cGMP phosphodiesterases**) enzymes. There are two major types of guanylate cyclase in mammalian cells and both are fundamentally different from the adenylate cyclases that catalyze cAMP production. A **membrane-associated guanylate cyclase** activity resides in the catalytic domains of the enzyme-linked receptors for the hormone **atrial natriuretic factor** (**ANF**); ANF receptors are expressed primarily in vascular smooth muscle cells and in renal cells; ANF is secreted from cardiac myocytes of the atrium in response to increases in circulating blood volume and/or pressure. The ANF receptor has a large extracellular domain for binding ANF, a single transmembrane domain, and a relatively short intracellular catalytic domain. The mechanism whereby ANF binding stimulates catalytic activity involves receptor dimerization and additional regulatory proteins, which are recruited to the hormone-occupied receptor.

In contrast to the tissue-restricted expression of the catalytic ANF receptor-guanylate cyclase, a soluble form of guanylate cyclase is present in the cytosol of most cell types (Figure 13.27). This **soluble guanylate cyclase** is expressed as a dimeric complex of distinct 82-kDa and 70-kDa subunits. Each intact dimer also constitutively binds a molecule of **heme**. In the absence of gaseous ligands that bind to this heme group, soluble guanylate cyclase has a very low rate of catalysis. However, the binding of **nitric oxide** (**NO**) to the heme group induces conformational changes that greatly increase catalytic activity of the guanylate cyclase dimer. Thus, a major stimulus for cGMP accumulation in most cells is elevated levels of NO. NO is enzymatically generated by the actions of **nitric oxide synthases (NOS)** (see p. 433). Because NO readily permeates biological membranes, it can be produced in one type of cell (e.g., a vascular endothelial cell) and rapidly diffuse into neighboring cell types (e.g., vascular smooth muscle cells), wherein it activates soluble guanylate cyclase. Accumulation of cGMP in smooth muscle triggers rapid and sustained relaxation of the contractile apparatus (see Clin. Corr. 13.5).

cGMP degradation is catalyzed by multiple types of soluble and weakly membrane-associated cGMP phosphodiesterases (PDE). While some appear to be passive catabolic enzymes (i.e., not directly regulated), the cGMP-PDE in the light-sensing cells of the retina is an effector enzyme for the trimeric G-protein transducin (see p. 974).

Intracellular Signaling Mechanisms of Cyclic GMP

cGMP exerts most of its second messenger effects by activating **cGMP-sensitive protein kinases (PKG)** that are widely expressed in many tissues. Depending on cell type, PKG can phosphorylate a variety of important targets, including enzymes involved in metabolism, structural proteins, and ion transport proteins. For example, stimulation of PKG in smooth muscle and nonmuscle cells results in the coordinated phosphorylation of multiple proteins that act to inhibit Ca^{2+}-dependent contraction or motility. These include (1) proteins that attenuate Ca^{2+} release channels of the endoplasmic reticulum, (2) myosin light-chain phosphatases that repress myosin interaction with actin, and (3) K^+ channels that induce plasma membrane hyperpolarization and thereby repress voltage-gated Ca^{2+} influx (see p. 468). cGMP also acts as a direct (i.e., PKG-independent) regulator of cyclic-nucleotide-gated cation channels expressed in certain sensory tissues such as the retina (see p. 974).

CLINICAL CORRELATION 13.5

Nitric Oxide/cGMP Signaling Axis as Therapeutic Targets in Cardiac and Vascular Disorders

The endothelial cells that line the internal surfaces of all blood vessels (arteries, veins, and capillaries) are major sources of the nitric oxide used for the relaxation of vascular smooth muscle cells. In these vascular smooth muscle cells, cGMP generation is catalyzed by the soluble guanylate cyclase stimulated by the nitric oxide generated by adjacent endothelial cells. cGMP is a key second messenger for induction of vascular smooth muscle vessel relaxation, thereby leading to vasodilation of blood vessels and increased blood flow to nearby tissue. Increases in vasodilatory cGMP can be produced by therapeutic drugs that increase nitric oxide production. cGMP levels are additionally regulated by the phosphodiesterases (PDE) that convert cGMP to GMP (which lacks second messenger function). Thus, drugs that inhibit vascular isoforms of PDE are also used in therapies aimed at improving blood flow to particular tissues.

Angina pectoris is the formal name for chest pain suffered by patients with different types of cardiac insufficiencies. Nitroglycerin is a commonly prescribed drug can rapidly alleviate angina (see Clin. Corr. 11.9, p. 438). Nitroglycerin acts as an exogenous nitric oxide donor; that is, it can be metabolized to nitric oxide. It may also bind directly to nitric oxide-sensitive proteins (like the soluble guanylate cyclase). Thus, like the endothelial-cell-derived nitric oxide, it can induce smooth muscle relaxation and vasodilation. The mechanism by which nitroglycerin alleviates angina is not completely understood. The coronary arteries and cardiac microvasculature of individuals with coronary insufficiency are probably already maximally dilated due to local ischemia. Thus, nitroglycerin likely alleviates angina by decreasing cardiac workload and the myocardial oxygen requirement. This is accomplished via vasodilatory effects on the veins that return blood back to the heart (preload), and by decreasing peripheral blood pressure (afterload), via vasodilatory effects on systemic arteries.

Viagra® is the trade name for the drug sildenafil citrate that is used to treat erectile dysfunction. Viagra acts to inhibit selectively the type 5 cyclic nucleotide phosphodiesterase (PDE5) that is highly expressed in vascular smooth muscle. Viagra is 80- to 4000-fold less potent as an inhibitor of other cyclic nucleotide PDE isoforms, including PDE3 which is predominantly expressed in cardiac muscle. This latter selectivity is required to prevent undesired side effects of Viagra on cardiac function. It is only 10-fold more selective for PDE5 versus PDE6, which is the predominant PDE in the retina. When blood vessels in erectile tissue are vasodilated in response to released NO and activation of guanylate cyclase, the simultaneous inhibition of cGMP breakdown (by Viagra) leads to higher and more sustained levels of cGMP accumulation and PKG activation. This reinforces the vasodilation that is required for increased blood flow into the erectile tissue. High doses of Viagra can lead to mild alterations in visual color perception due to side effects on the retinal PDE6.

Source: Ignarro, L. J., Napoli, C., and Lascalzo, J. Nitric oxide donors and cardiovascular agents modulating the bioactivity of nitric oxide, *Circ. Res.* 90:21, 2001.

13.11 CALCIUM-BASED SIGNAL TRANSDUCTION

Regulation of Cytosolic Ca^{2+} Concentration

Ca^{2+} is important as an intracellular regulator of many fundamental cell functions, including contraction in all types of muscle, secretion of hormones and neurotransmitters, cell division, directed migration of nonmuscle cells, and regulation of gene expression. Cells respond to many types of hormones and neurotransmitters with rapid and transient increases in cytosolic Ca^{2+} that then binds to a variety of Ca^{2+}-dependent regulatory proteins. Increases in cytosolic Ca^{2+} are elicited either by opening of **Ca^{2+} channels** in the plasma membrane or Ca^{2+} channels in the endoplasmic reticulum (see p. 521). This released Ca^{2+} can be rapidly reaccumulated by the ER to terminate the response to the hormone/neurotransmitter.

The basal concentration of free Ca^{2+} in the cytosol of most cell types is maintained at approximately 10^{-7} M (Figure 13.28). Small increases in cytosolic free Ca^{2+} to 10^{-6} M rapidly and maximally activate the various Ca^{2+}-regulated cell functions. Extracellular Ca^{2+} is 10^{-3} M, a 10,000-fold concentration gradient across the plasma membrane; together with the inside-negative membrane potential, influx of Ca^{2+} is very rapid. In excitable cells, such as neurons and muscle cells, Ca^{2+} influx involves voltage-gated Ca^{2+} channels (see p. 468). This type of Ca^{2+} signaling pathway is generally initiated by the ion channel receptors for excitatory neurotransmitters, such as acetylcholine, glutamate, serotonin, and ATP. The plasma membrane also contains

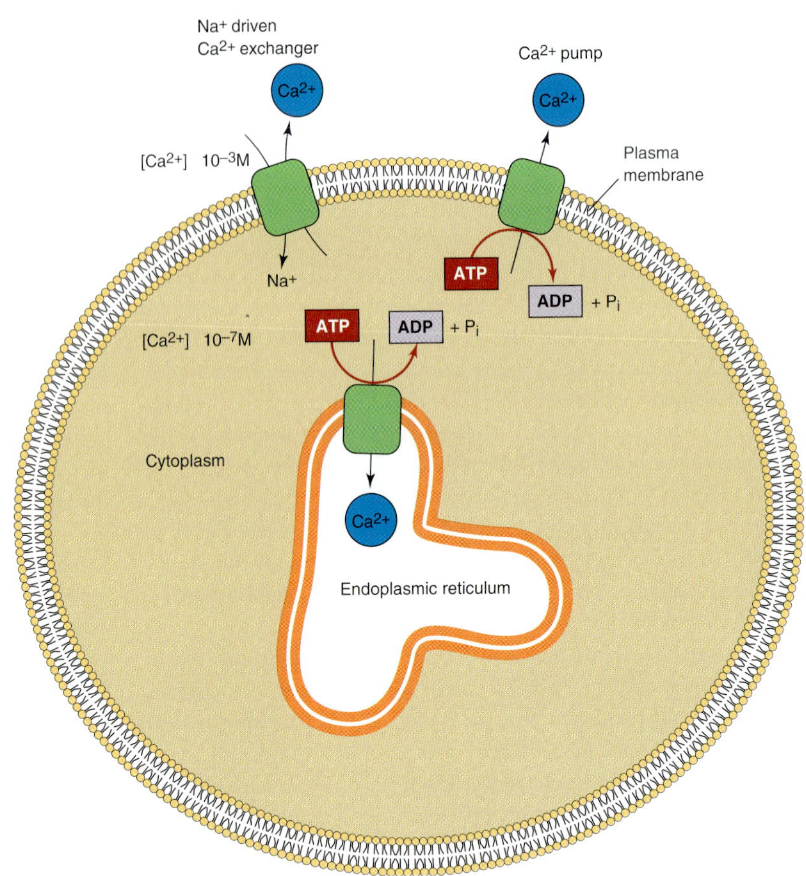

FIGURE 13.28

Role of plasma membrane and organellar Ca^{2+} transport proteins in the regulation and compartmentation of intracellular Ca^{2+} pools.

two transport systems, the Na^+/Ca^{2+} antiport transporter and the Ca^{2+}-transporting ATPase that efficiently pump the excess Ca^{2+} out of the cytosol (see p. 477).

Ca^{2+} ions are also efficiently concentrated within the endoplasmic reticulum (ER) (termed sarcoplasmic reticulum or SR in muscle cells) by of Ca^{2+}-transporting ATPase (see p. 477). This ATPase transporter permits Ca^{2+} to be accumulated to a free concentration of about 1 mM (10^{-3} M) within the lumen of the ER/SR, which is the concentration of Ca^{2+} in the extracellular space. This accumulated Ca^{2+} can be released into the cytosol via two types of ER/SR-localized Ca^{2+} channels. The gating of these channels is regulated by signals generated by many types of cell surface receptors. The first is a very widely expressed channel known as the **1,4,5-inositol trisphosphate (IP_3) receptor (IP3R)** or **IP_3-gated Ca^{2+} channel** (Figure 13.29). IP_3, a second messenger, is rapidly generated when phospholipase C is stimulated by one of many hormones or neurotransmitters (see p. 521). A different type of ligand-gated Ca^{2+} release channel is expressed in the ER/SR of skeletal and cardiac muscle and known as the **ryanodine receptor** (RyR) or ryanodine-sensitive Ca^{2+} release channel because it binds the drug ryanodine (Figure 13.30). In cardiac muscle, the RyR channels are gated by the Ca^{2+} that enters through the plasma membrane channels; this is termed Ca^{2+}-induced Ca^{2+} release. In contrast, skeletal muscle RyR are opened via direct protein–protein interactions with the voltage-sensitive Ca^{2+} channels; this regulation is possible because SR in skeletal muscle forms direct contacts with specialized invaginations of the plasma membrane.

Calcium Activation of Calmodulin-Dependent Protein Kinases and Phosphatases

Increased cytosolic Ca^{2+} results in activation of various calcium-binding regulatory proteins. Some, such as troponin-C, which regulates contraction in striated muscles, are highly specialized and expressed in only a few tissues. In contrast, **calmodulin (CaM)** is a ubiquitously expressed protein that mediates the regulatory actions of Ca^{2+}

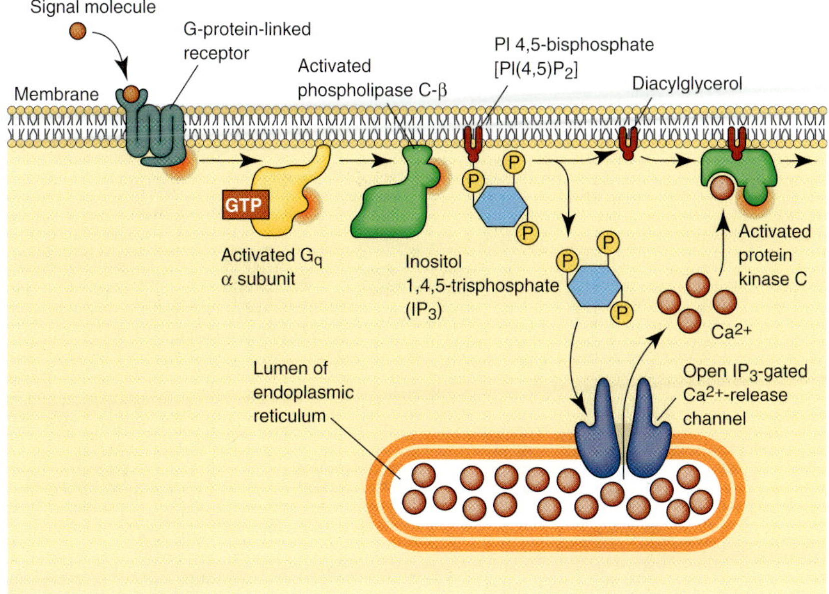

FIGURE 13.29

Mobilization of endoplasmic reticulum Ca^{2+} pools and activation of protein kinase C by receptors that stimulate the phospholipase C-mediated hydrolysis of inositol phospholipids.

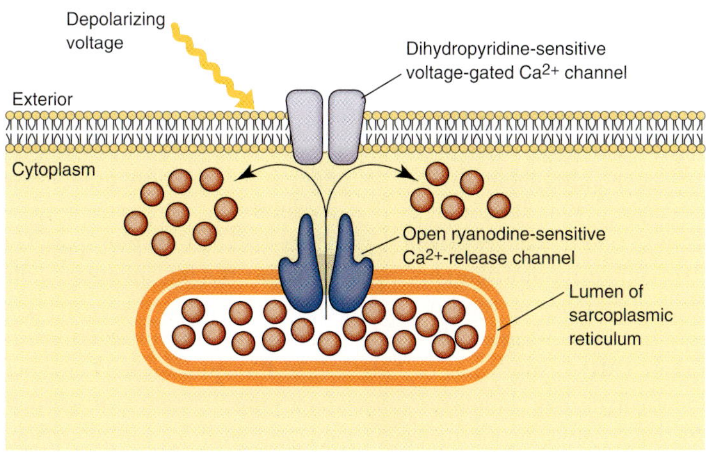

FIGURE 13.30

Mobilization of sarcoplasmic reticulum Ca^{2+} pools in striated muscle cells by depolarization-induced activation of ryanodine-sensitive Ca^{2+} release channels.

on a broad range of cellular functions (Figure 13.31 and p. 478). Upon binding of Ca^{2+}, calmodulin undergoes conformational changes that facilitate its interaction with multiple downstream signaling proteins, including serine/threonine-specific protein kinases, phosphatases, and nitric oxide synthases. Some calmodulin-dependent kinases, such as myosin light chain kinase, target only specific proteins while other forms (multifunctional CaM Kinase II) can phosphorylate a broad range of protein substrates. Altering the activity of these proteins is the primary mechanism by which increased Ca^{2+} alters cellular behavior. CaM-kinase II also autophosphorylates when bound to calmodulin. Even though cytosolic Ca^{2+} is only transiently increased in responses to the activation of cell surface receptors, this autophosphosphorylation of CaM-kinase II allows it to remain active long after cytosolic Ca^{2+} has returned to a baseline level. CaM-kinases can phosphorylate and modulate a wide range of enzymes, ion channels, contractile proteins, and gene regulatory proteins.

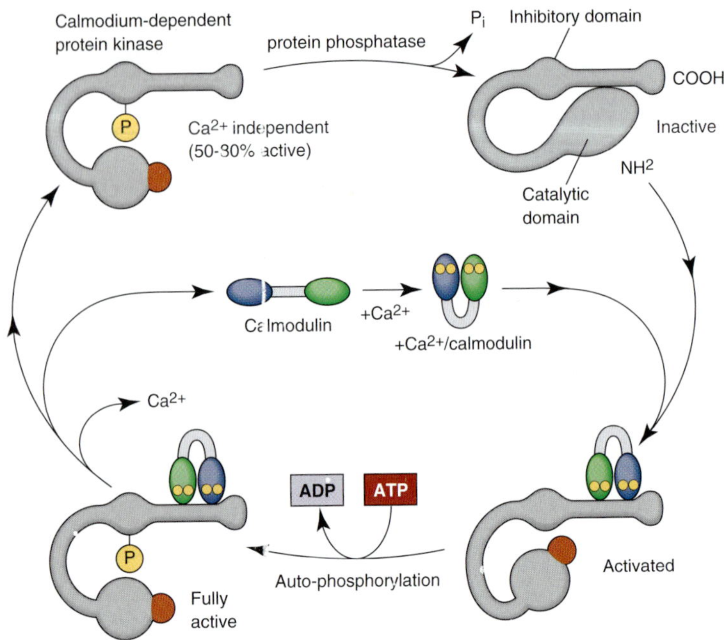

FIGURE 13.31

Role of calmodulin and calmodulin-regulated protein kinases in the intracellular signaling cascades regulated by Ca^{2+}.
Redrawn based on figure from Alberts, B., et al. *Molecular Biology of the Cell*, 4th ed. New York: Garland, 2002.

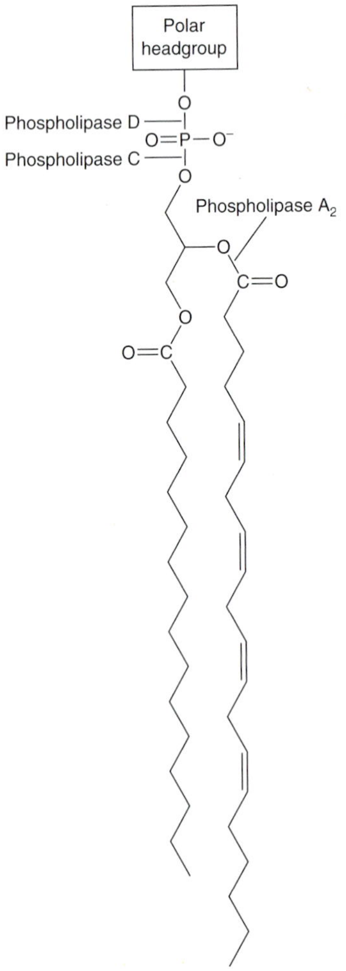

13.12 PHOSPHOLIPID-BASED SIGNAL TRANSDUCTION

Regulated Phospholipid Metabolism as a Component of Intracellular Signaling Pathways

Signal transduction by many cell surface receptors involves activation of one or more phospholipases that catalyze the hydrolysis of different classes of phospholipids (e.g., inositol or choline phospholipids). Several important classes of phospholipase effector enzymes catalyze the production of distinct products that act as second messengers or regulators of second messenger production (Figure 13.32). **Phospholipase C enzymes (PI-PLC)** hydrolyze inositol phospholipids to yield **inositol phosphates** and **diacylglycerols** as second messengers. Inositol phospholipids can also be further phosphorylated by **phosphoinositide-3-kinase (PI-3K)** to produce the **phosphatidylinositol-3,4,5-trisphosphate (PIP_3)**, another second messenger. **Phospholipase D enzymes (PLD)** predominantly hydrolyze choline or ethanolamine phospholipids to produce **phosphatidic acid** that is subsequently metabolized by phosphatidic acid phosphohydrolases (PAP) to yield the diacylglycerol second messenger. **Phospholipase A2 enzymes (PLA2)** attack various phospholipids to produce free fatty acids, such as arachidonic acid, and lyso-phospholipids.

Polar headgroup
Phospholipase D
Phospholipase C
Phospholipase A_2

FIGURE 13.32

Major classes of phospholipases used during receptor-mediated signal transduction.

Regulation of Phospholipase C and Phospholipase D

As previously described for Ca^{2+}-based signal transduction, many G-protein-coupled receptors, and receptor tyrosine kinases stimulate the release of Ca^{2+} from the endoplasmic reticulum by activating the production of **1,4,5-inositol trisphosphate (IP_3)**. This water-soluble second messenger is derived from the hydrolysis of **phosphatidylinositol-4, 5-bisphosphate (PIP_2)**, a relatively minor membrane phospholipid that is generated by multiple phosphorylations on the inositol residue of phosphatidylinositol. This hydrolysis of PIP_2 is catalyzed by a family of C-type phospholipases (**PLC**) with high selectivity for inositol phospholipids (Figure 13.32). It is important to appreciate that the hydrolysis of PIP_2 by a PLC generates two different second messengers: (1) the water-soluble product IP_3 and (2) **diacylglycerol** (**DAG**), a hydrophobic second messenger. **PI-PLCβ** enzymes are activated by interaction with either the α-subunits of $G_{q/11}$-family G proteins or the $\beta\gamma$-subunits of G_i family. In contrast, the **PI-PLCγ**

enzymes are activated by tyrosine phosphorylation and association with receptor tyrosine kinases. Some nonreceptor tyrosine kinases also phosphorylate PLCγ enzymes.

In addition to IP_3, the activation of PI-PLCβ and PI-PLCγ results in accumulation of DAG. However, the magnitude of receptor-induced production of diacylglycerol in many cell types cannot be ascribed solely to the breakdown of PIP_2. Although PIP_2 hydrolysis yields small amounts of DAG over short time periods, the hydrolysis of more abundant phospholipids, such as phosphatidylcholine (PC), underlies most of the sustained production of DAG triggered in response to receptor activation. Consequently, PC hydrolysis by PLC and **phospholipase D (PLD)** mediates the regulation of cellular processes that require prolonged elevation of DAG. PC-PLD and PC-PLC are indirectly activated during receptor stimulation by a complex network of primary signals that include Ca^{2+}, small GTPases, and protein kinase C.

Diacylglycerol and Protein Kinase C

Diacylglycerol (DAG) that results from breakdown of PIP_2 or PC functions as a second messenger by binding to and activating members of the **protein kinase C (PKC)** family of kinases. Inactive PKC exist as soluble, cytosolic proteins in which the N-terminus (containing the DAG-binding sites) occludes the C-terminus (containing binding sites for substrate proteins and ATP) (Figure 13.29). When DAG is generated in membranes, the DAG-binding site of PKC is occupied, resulting in an unfolding, exposing the substrate- and ATP-binding domains of the kinase. Other N-terminal sites on PKC enzymes bind to acidic phospholipids and this facilitates association of PKC with the membrane. Some isoforms of PKC also contain Ca^{2+}-binding sites in their N-terminal domains; increased cytosolic Ca^{2+} potentiates PKC binding to membranes. Activation of PKC enzymes results in serine–threonine phosphorylation of many substrate proteins, which, in turn, elicit a number of cellular responses. Like cAMP-dependent protein kinases and calmodulin-dependent protein kinases, the various PKC isozymes can phosphorylate (1) transcription factors involved in regulation of gene expression, (2) ion channels and transporters, and (3) other types of protein kinases. Some PKC isozymes additionally trigger the various MAP kinase cascades that control multiple gene regulatory proteins. The ability of PKC enzymes to regulate directly or indirectly the phosphorylation of transcription factors underlies the critical role of PKC in regulating the growth, differentiation, or death of many cells and tissues (see p. 1020).

PIP_3, Phosphatidylinositol 3-Kinases, and Protein Kinase B

Inositol phospholipids are substrates for inositol lipid kinases that specifically phosphorylate inositol at the 3′-OH. These **phosphatidylinositol 3-kinases (PI3-kinases)** comprise a multimembered gene family. Several PI-3 kinases associate with, and are activated by tyrosine phosphorylated growth factor receptors or activated cytokine receptors via SH-2 domain interactions. Other PI-3 kinase isoforms are activated by the $\beta\gamma$ dimers derived from activated G proteins. PI-3 kinases predominantly phosphorylate phosphatidylinositol-4,5-bisphosphate to produce **phosphatidylinositol-3,4,5-trisphosphate (3,4,5-PIP3)**. It is important to note that PIP3 does not act as a substrate for PLC, but rather acts as a lipid second messenger. PIP3 has critical roles in the regulation of membrane trafficking, cell motility, and activation of cell survival signaling pathways. Recent studies have identified the **Akt/protein kinase B (PKB)** family of kinases as major mediators of PIP3 action (see Figure 13.33). Active PKB potentiates cell survival by repressing the activity of certain cell death signaling pathways (see p. 1021) and regulates other proteins and kinases involved in glucose transport and glycogen metabolism.

Phospholipase A_2 and Generation of Arachidonic Acid Metabolites

Occupancy of cell surface receptors in many cell types triggers the release of **arachidonic acid (AA)** (see p. 663) from membrane phospholipids. The release of AA is catalyzed

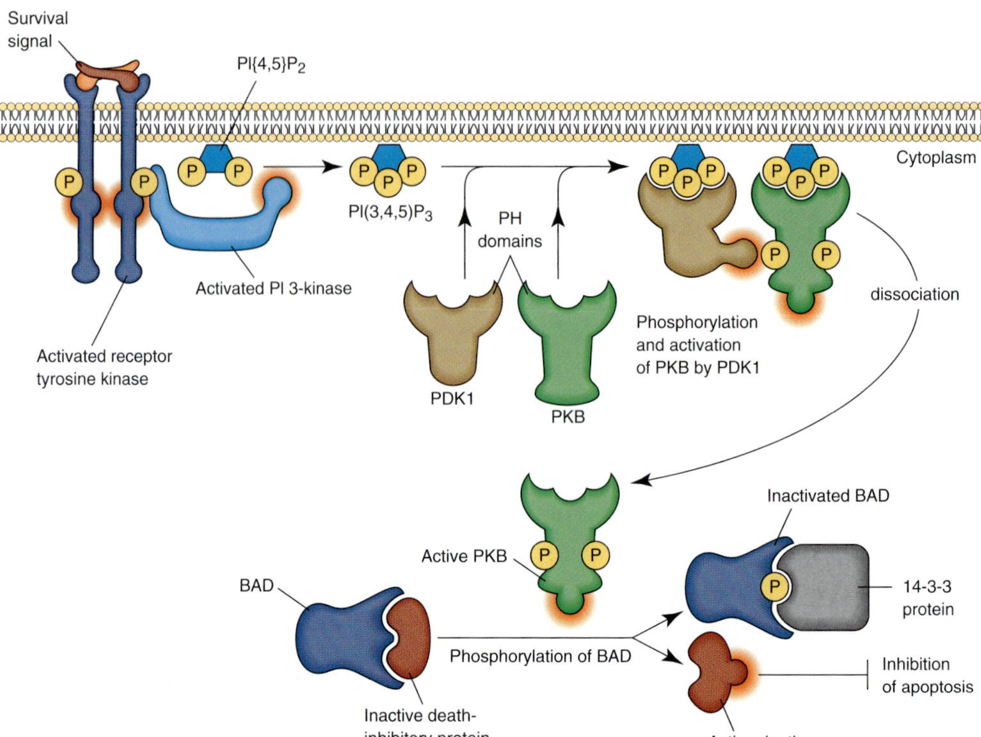

FIGURE 13.33

Role of phosphatidylinositol-3-kinase and protein kinase B in the intracellular signaling cascades regulated by 3,4,5-phosphatidylinositol trisphosphate.

Redrawn based on figure from Alberts, B., et al. *Molecular Biology of the Cell*, 4th ed. New York: Garland, 2002.

by members of the family of phospholipase A_2 **(PLA_2)** enzymes (Figure 13.32). The regulation of PLA_2 is complex and is linked to many factors including Ca^{2+}, which seems to control the translocation of some PLA_2 isoforms from cytosol to membranes, and kinase-mediated phosphorylation. Arachidonic acid can be rapidly metabolized to multiple oxidation products collectively known as **eicosanoids** (see p. 730). In contrast to other second messengers such as cyclic nucleotides or Ca^{2+}, eicosanoids are generally secreted from the cells in which they are produced to act as extracellular agonists for cell surface receptors in neighboring cells. The eicosanoids are formed by virtually all cell types and play fundamental paracrine/autocrine roles in processes such as inflammation, blood clotting, control of vascular tone, renal function, and others.

13.13 INTEGRATION OF SIGNAL TRANSDUCTION PATHWAYS INTO SIGNAL TRANSDUCTION NETWORKS

Depending on its tissue localization and degree of specialization, individual cells may be genetically programmed to respond to dozens of distinct extracellular signaling molecules. Many important intracellular signaling cascades can be regulated synergistically by receptors belonging to various classes and families (Figure 13.34). For example, the MAP kinase pathways can be regulated by G-protein-coupled receptors and growth factor receptor kinases. This **"cross-talk"** provides cells the ability to integrate signals from multiple extracellular stimuli into unique and particular patterns of altered cellular response. In other cases, different second-messenger kinases may phosphorylate distinct residues on a common target protein, such as a transcription factor, to produce either divergent or convergent changes in gene expression patterns. Thus, the integrated response of a cell to the activation of a particular cell surface receptor will often vary depending on what other cell surface receptors are simultaneously being activated. For example, a receptor that normally induces specialized gene expression in a particular cell type might trigger the programmed death of that same cell if other cell surface receptors

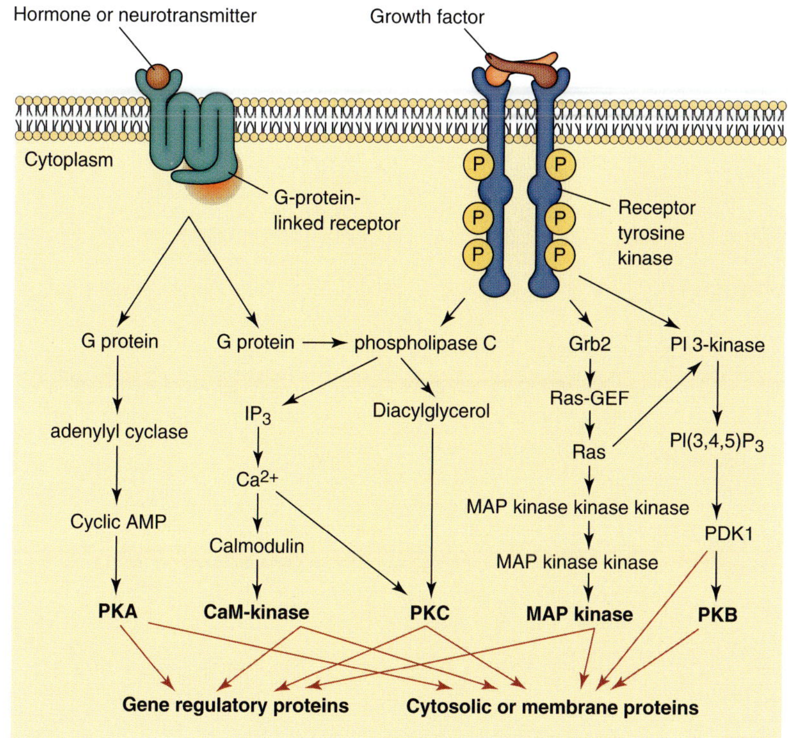

FIGURE 13.34

Cross-talk and integration among the major intracellular signaling cascades regulated by different cell surface receptors.

Redrawn based on figure from Alberts, B., et al. *Molecular Biology of the Cell*, 4th ed. New York: Garland, 2002.

are being simultaneously activated or, in some cases, are not being activated. Although modern cell biology has made significant advances in defining the basic components and sequential steps of individual signaling pathways, a major challenge remains in understanding how these individual pathways are integrated into signal transduction networks.

BIBLIOGRAPHY

Ligand-Gated Ion Channel Receptors

Barnard, E. A. Receptor classes and transmitter-gated ion channels. *Trends Biochem. Sci.* 17:368–373, 1992.

Corringer, P.-J., Le Novere, N., and Changeaux, J.-P. Nicotinic receptors at the amino acid level. *Annu. Rev. Pharmacol. Toxicol.* 40:431–458, 2000.

Madden, D. R. The structure and function and function of glutamate receptor ion channels. *Nature Reviews-Neurosci.* 3:91–101, 2002

Sheng, M. and Pak, D. T. Ligand-gated ion channel interactions with cytoskeletal and signaling proteins. *Annu. Rev. Physiol.* 62:755–778, 2000.

Receptor and Nonreceptor Tyrosine Kinases

Hubbard, S. R. and Till, J. H. Protein tyrosine kinase structure and function. *Annu Rev. Biochem.* 69:373–398, 2000.

Schlessinger, J. Ligand-induced, receptor-mediated dimerization and activation of the EGF receptor. *Cell* 110:669–672, 2002.

Wiley, H. S. Trafficking of the ErbB receptors and its influence on signaling. *Exp Cell Res.* 284:78–88, 2003.

Yaffe, M. B., Phosphotyrosine-binding domains in signal transduction. *Nature Rev. Mol. Cell Biol.* 3:177–186, 2002.

G-Protein-Coupled Receptors

Bockaert J. and Pin, J-P. Molecular tinkering of G protein-coupled receptors: An evolutionary success. *EMBO J.* 18:1723–1729, 1999.

Gether, U. and Kobilka, B. K. G protein-coupled receptors: II. Mechanism of agonist activation. *J. Biol. Chem.* 273:17979–17982, 1998.

Ji, T. H., Grossman, M. and Ji, I. G protein-coupled receptors: I. Diversity of receptor–ligand interactions. *J. Biol. Chem.* 273:17299–17302, 1998.

Shendy, S. K. and Lefkowitz, R. J. Multifaceted role of β-arrestins in the regulation of seven-membrane-spanning receptor trafficking and signalling. *Biochem. J.* 375:503–515, 2003.

Sodhi, A., Montaner, S., and Gutkind, J. S. Viral hijacking of G protein-coupled receptor signaling networks. *Nature Rev. Mol. Cell Biol.* 5:998–1012, 2004.

Trimeric G Proteins

Berman, D. M. and Gilman, A. G. Mammalian RGS proteins: Barbarians at the gate. *J. Biol. Chem.* 273:1269–1272, 1998.

Preminger, A. M and Hamm, H. E. G protein signaling: Insights from new structures. *Science STKE 2004*. 218:1–9, 2004.

Sprang, S. R. G protein mechanisms: Insights from structural analyses. *Annu. Rev. Biochem.* 66:639–678, 1997.

Small GTPases

Vojtek, A. B. and Der, C. J. Increasing complexity of the Ras signaling pathway. *J. Biol. Chem.* 273:19925–19928, 1998.

Mackay, D. J. G. and Hall, A. Rho GTPases. *J. Biol. Chem.* 273:20685–20688, 1998.

Cyclic AMP and Cyclic GMP Signaling

Hurley, J. H. Structure, mechanism, and regulation of mammalian adenylyl cyclase. *J. Biol. Chem.* 274:7599–7602. 1999.

Lucas, K. A, Pitari, G. M., KazerounianS., Ruiz-Stewart I., Park J., Schulz S., Chepenik K. P., and Waldman S. A. Guanylyl cyclases and signaling by cyclic GMP. *Pharmacol. Rev.* 52:375–413, 2000.

Schwartz J. H. The many dimensions of cAMP signaling. *Proc Natl Acad Sci. USA* 98:13482–13484, 2001.

Phospholipid Signaling

Liscovitch, M. Crosstalk among multiple signal-activated phospholipases. *Trends Biochem. Sci.* 17:393–399, 1992.

QUESTIONS | CAROL N. ANGSTADT

Multiple Choice Questions

1. The type of intercellular signaling in which one cell can communicate with another over long distances is called:
   A. autocrine.
   B. endocrine.
   C. juxtacrine.
   D. paracrine.
   E. synaptic.

2. Intracellular receptors:
   A. usually bind hydrophobic ligands.
   B. may be located either in the cytosol or nucleus in unbound state.
   C. when bound to their ligand regulate gene transcription.
   D. when bound to their ligand function as dimeric complexes binding to specific DNA sequences.
   E. all of the above are correct.

3. A cell surface receptor:
   A. reacts only with molecules too large to cross the plasma membrane.
   B. when bound to its ligand could result in activation of an enzymatic cascade.
   C. always opens an ion channel when bound to its ligand.
   D. must produce a second messenger when it binds to its ligand.
   E. is usually also called GPCR.

4. Protein phosphorylation is part of most pathways of intracellular signal transduction. All of the following statements about phosphorylation are correct *except*:
   A. phosphorylation can be on a serine or threonine.
   B. phosphorylation changes the structure and function of the protein.
   C. protein kinase cascades result in amplification of the signal.
   D. tyrosine kinases are always part of the receptor protein.
   E. serine kinases are usually regulated by second messenger concentration.

5. Cells can terminate signal transduction by cell surface receptors by:
   A. reducing agonist availability in the vicinity of the target cell.
   B. internalizing and degrading the receptor–agonist complex.
   C. modifying the receptor so that it is inactive or desensitized.
   D. all of the above.
   E. none of the above.

6. Calmodulin is:
   A. a nonspecific kinase.
   B. a protein that binds Ca^{2+}.
   C. a second messenger.
   D. an activator of nitric oxide synthase.
   E. a protein channel that facilitates the influx of Ca^{2+}.

Questions 7 and 8: Manic depression may be caused by overactivity of certain central nervous system cells, perhaps caused by abnormally high levels of hormones or neurotransmitters which stimulate phospholipid-based signal transduction [e.g., from phosphatidylinositol (PI)]. Lithium has been used for many years to treat manic depression. In the presence of Li^+, the PI system is slowed despite continued stimulation and cells become less sensitive to these stimuli. Li^+ may have two functions: inhibition of the phosphatase that dephosphorylates inositol trisphosphate and direct interference with the function of G proteins.

7. The PI system begins with activation of phospholipase C which initiates a sequence of events including all of the following *except*:
   A. activation of IP_3 by action of a phosphatase.
   B. increase in intracellular Ca^{2+} concentration.
   C. release of diacylglycerol (DAG) from a phospholipid.
   D. activation of protein kinase C.
   E. phosphorylation of certain cytoplasmic proteins.

8. Which of the following statements concerning G proteins is correct?
   A. G-proteins bind the appropriate hormone at the cell surface.
   B. GTP is bound to G protein in the resting state.
   C. α subunit may be either stimulatory or inhibitory because it has two forms.
   D. adenylate cyclase can be activated only if α- and β-subunits of G protein are associated with each other.
   E. hydrolysis of GTP is necessary for G protein subunits to separate.

Questions 9 and 10: Growth hormone releasing hormone (GHRH) produced by the hypothalamus binds to its pituitary receptor and leads to the production of growth hormone (GH) because of increase in cyclic AMP. Certain pituitary tumors result in hypersecretion of GH because of a mutation that produces a G_s-α protein with a greatly diminished GTPase activity.

9. Elements leading to increased cyclic AMP in response to GHRH binding to its receptor include:
   A. activation of a monomeric G protein.
   B. activation of adenylate cyclase by α_s-subunit of a G_s protein.
   C. activation of cyclic nucleotide phosphodiesterase.
   D. activation of protein kinase A.
   E. all of the above.

10. Low GTPase activity in the mutated protein results in consitutive activation of G_s and adenylate cyclase because:
    A. GTP-bound α-subunit does not reform the $\alpha\beta\gamma$ trimer.
    B. GTP-bound G protein binds more strongly to the membrane receptor.
    C. GTP reacts directly with adenylate cyclase to activate it.
    D. the trimeric form of the G protein is stabilized.
    E. adenylate cyclase is phosphorylated more readily.

Questions 11 and 12: Endothelial cells that line t[illegible]ernal surfaces of all blood vessels are sources of nitric oxide (NO) w[illegible] activates guanylate cyclase. Cyclic GMP is a second messenger leading [illegible]asodilation of blood vessels and increased blood flow to nearby tissues. Therapeutic drugs to increase blood flow to particular tissues can eith[illegible] increase nitric oxide production or inhibit vascular isoforms of phospho[illegible]esterases (PDE).

11. The guanylate cyclase that responds to NO:
    A. is in the catalytic domain of a membrane [illegible]or.
    B. is also activated when atrial natriuretic f[illegible] (ANF) binds to its receptor.
    C. increases activity because of a conforma[illegible] change when NO binds to its heme.
    D. is a monomeric enzyme.
    E. is found only in smooth muscle cells.
12. Viagra® used to treat erectile dysfunction inhibits, selectively, the PDE isoform highly expressed in vascular smooth muscle. Viagra®:
    A. increases the production of NO.
    B. activates guanylate cyclase.
    C. increases the rate of synthesis of cyclic GMP.
    D. has the same effect in all tissues using cyclic GMP.
    E. inhibits the conversion of cyclic GMP to inactive GMP.

Problems

13. How does elevation of cyclic AMP in eukaryotic cells lead to altered transcription of certain genes?
14. How do excitatory and inhibitory neurotransmitters differ in their effects on ligand-gated ion channels?

ANSWERS

1. **B** A hormone is released into the blood and travels to the target tissue. A: One cell type is both the sender and target. C: This is contact-dependent signaling. D: The cells are in the immediate vicinity of each other. E: This is a type of paracrine signaling used by nerve cells.
2. **E** A: Things like steroid hormones. B: Once bound to ligand, cytosolic receptors translocate to the nucleus. C, D: They bind to upstream regulatory elements and either enhance or repress transcription.
3. **B** This is one of the ways that signal transduction occurs. A: They also react with very hydrophilic signaling molecules. C, D: These are additional ways in which cell-surface receptors can act. E: This is true only if the receptor is coupled to a G protein.
4. **D** Some are but others are cytosolic proteins. A: Serine, threonine, and tyrosine are the usual sites of phosphorylation. B: This, in turn, causes changes in other related proteins. C: This occurs if phosphorylation of a kinase activates it. E: For example, cAMP.
5. **D** These are all possible.
6. **B** This is a ubiquitous Ca^{2+}-binding protein. A: Ca^{2+}-calmodulin regulates some kinases but not this one. C: Second messengers are usually small molecules or ions. D: It is the Ca^{2+}-calmodulin complex that regulates. E: Calmodulin binds the Ca^{2+} that enters but is not part of the channel to allow it to enter.
7. **A** IP_3 is the active form. Action of phosphatase on it renders it inactive as it is converted to inositol. B: IP_3 causes release of calcium. C, D: This is the other second messenger and it activates protein kinase C. E: This is what protein kinase C does.
8. **C** Which form is released depends on the specific hormone and receptor that have interacted. A: Receptor binds the hormone and the complex interacts with G protein. B, E: GDP is bound to the resting enzyme. Its replacement by GTP causes α-subunit to dissociate. Actually, hydrolysis of GTP allows subunits to reassociate. D: α-subunit, when dissociated from the other two, interacts with the enzyme to either activate or inhibit it depending on whether it is an α_s or an α_i.
9. **B** Adenylate cyclase then converts ATP to cAMP. A: G protein-coupled receptors are all heterotrimeric. C: This would decrease cAMP by hydrolyzing it to AMP. D: cAMP activates protein kinase, not the other way.
10. **A** α-GTP dissociates from the $\beta\gamma$ dimer. B: GTP occupied G proteins do not bind to the receptor. C: It is the α-GTP that activates the enzyme. D: See A. E: Activation of adenylate cyclase is a conformational change, not phosphorylation.
11. **C** The soluble enzyme has low activity when heme has not bound a gaseous molecule. A and B: The membrane bound enzyme is stimulated when ANF binds its receptor but not by NO. D: It is a dimer. E: The soluble enzyme is present in most cell types.
12. **E** Cellular levels of cGMP are regulated by a balance between synthesis and degradation. A, B, C: There is no effect on these. D: Different isoforms of PDE are found in different tissues; Viagra® has much less effect on the isoform in cardiac muscle.
13. cAMP binding to protein kinase A causes dissociation of catalytic from regulatory subunits, exposing nuclear localization sequences on catalytic subunits. These can translocate to the nucleus, where they can phosphorylate and activate cAMP-regulated gene regulatory proteins (CREBs) that control genes containing cAMP-sensitive regulatory elements (CREs).
14. Excitatory neurotransmitters (e.g., acetylcholine, glutamate) bind to cation-selective receptors and allow ions like Na^+ to enter, depolarizing the membrane. Inhibitory neurotransmitters (e.g., GABA, glycine) bind to anion-selective receptors and allow anions like Cl^- to enter, hyperpolarizing the membrane.

PART IV METABOLIC PATHWAYS AND THEIR CONTROL

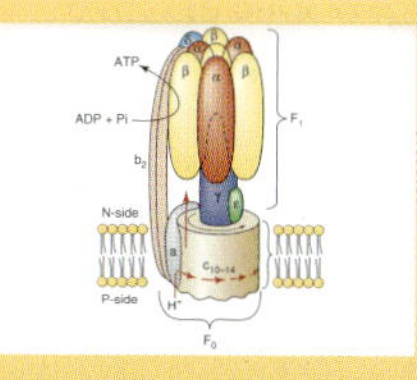

14

BIOENERGETICS AND OXIDATIVE METABOLISM

Diana S. Beattie

Textbook of Biochemistry With Clinical Correlations, Sixth Edition, Edited by Thomas M. Devlin

14.1 ENERGY-PRODUCING AND ENERGY-UTILIZING SYSTEMS

Living cells depend on a complex, intricately regulated system of energy producing and energy-utilizing chemical reactions called metabolism. Metabolism consists of two contrasting processes, **catabolism** and **anabolism**, which together constitute the chemical changes that convert foodstuffs into usable forms of energy and into complex biological molecules. Catabolism is responsible for degradation of ingested foodstuffs or stored fuels such as carbohydrate, lipid, and protein into either usable or storable forms of energy. Catabolic reactions generally result in conversion of large complex molecules to smaller molecules (ultimately CO_2 and H_2O) and, in mammals, often require consumption of O_2. Energy-utilizing reactions perform various necessary, and in many instances tissue-specific, functions—for example, nerve impulse conduction, muscle contraction, growth, and cell division. Catabolic reactions are usually exergonic with the released energy generally trapped in the formation of ATP. The oxidative reactions of catabolism transfer reducing equivalents to the coenzymes NAD^+ and $NADP^+$ to form NADH and NADPH. Anabolic pathways are responsible for biosynthesis of large molecules from smaller precursors and require input of energy either in the form of ATP or reducing equivalents of NADPH (Figure 14.1).

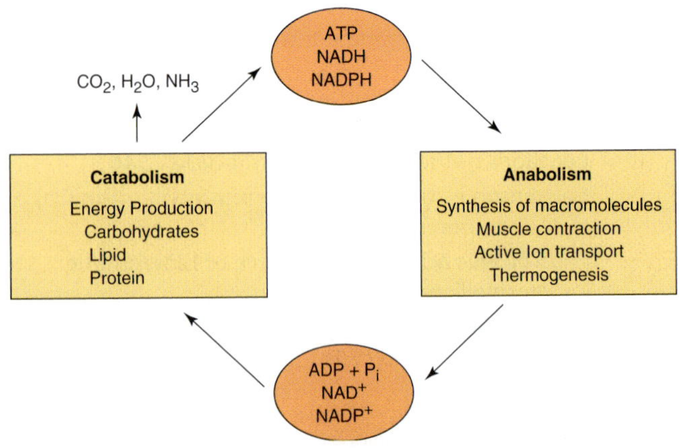

FIGURE 14.1

Energy relationships between energy production (catabolism) and energy utilization (anabolism). Oxidative breakdown of foodstuffs is an exergonic process releasing free energy and reducing power that are trapped as ATP and NADH or NADPH, respectively. Anabolic processes are endergonic and use chemical energy stored as ATP and NADPH.

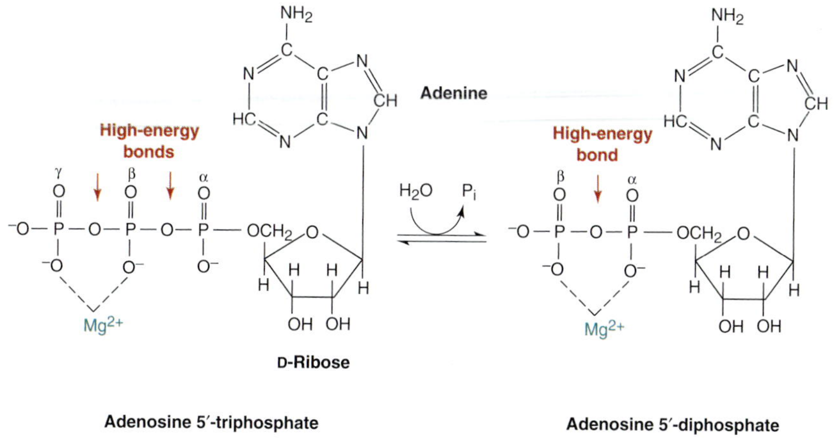

FIGURE 14.2

Structure of ATP and ADP complexed with Mg^{2+}. High-energy bonds are highlighted.

ATP Links Energy-Producing and Energy-Utilizing Systems

The relationship between energy-producing and energy-utilizing functions of cells is illustrated in Figure 14.1. Energy is derived from oxidation of metabolic fuels utilized by the organism usually as carbohydrate, lipid, and protein. The proportion of each fuel utilized as an energy source depends on the tissue and the dietary and hormonal state of the organism. For example, mature erythrocytes and adult brain in the fed state use only carbohydrate as a source of energy, whereas the liver of a diabetic or fasted individual metabolizes primarily lipid to meet the energy demands. Energy may be consumed during performance of various energy-linked (work) functions, some of which are indicated in Figure 14.1. Note that liver and pancreas are primarily involved in biosynthetic and secretory work functions, whereas cardiac and skeletal muscles convert metabolic energy into mechanical energy during muscle contraction.

The essential link between energy-producing and energy-utilizing pathways is the nucleoside triphosphate, **adenosine 5′-triphosphate (ATP)** (Figure 14.2). The ATP is a purine (adenine) nucleotide in which adenine is linked by a glycosidic bond to D-ribose. Three phosphoryl groups are esterified to the 5-position of the ribose moiety. The two terminal phosphoryl groups (i.e., β and γ) are energy-rich phosphoanhydride bonds or **high-energy bonds.** Synthesis of ATP as a result of a catabolic process or consuming ATP in an energy-linked process involves formation and either hydrolysis or transfer of the terminal phosphate group of ATP. This nucleotide is chelated with a divalent metal cation such as magnesium under physiological conditions.

NAD^+ and NADPH in Catabolism and Anabolism

Many catabolic processes are oxidative in nature because the carbons in the substrates—carbohydrates, fats, and proteins—are in a partially or highly reduced state (Figure 14.3). **Reducing equivalents** are released from substrates as protons and electrons, which are transferred to **nicotinamide adenine dinucleotide (NAD^+)** by enzymes called dehydrogenases with formation of NADH (see Figure 14.4). The reducing equivalents of NADH are transported into mitochondria and transferred by the electron transport chain to O_2 as the ultimate electron acceptor (see p. 554). The oxidative reactions in mitochondria are exergonic, with the energy produced being used for synthesis of ATP in a process called **oxidative phosphorylation**. The reductive and oxidative reactions of the NAD^+–NADH cycle are central in conversion of the chemical energy of carbon compounds in foodstuffs into that of phosphoanhydride bonds of ATP. This process, called energy transduction, will be discussed in detail later in this chapter. Anabolism, by contrast, is largely a reductive process as small more highly oxidized molecules are converted into large complex molecules (Figure 14.5). The reducing power used in biosynthesis of highly reduced compounds, such as fatty acids, is provided by NADPH (see p. 670), a 3′-phosphorylated NADH (Figure 14.4).

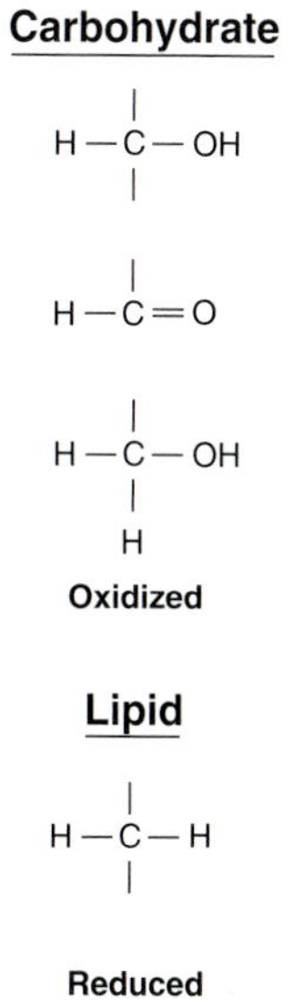

FIGURE 14.3

Oxidation states of typical carbon atoms in carbohydrates and lipids.

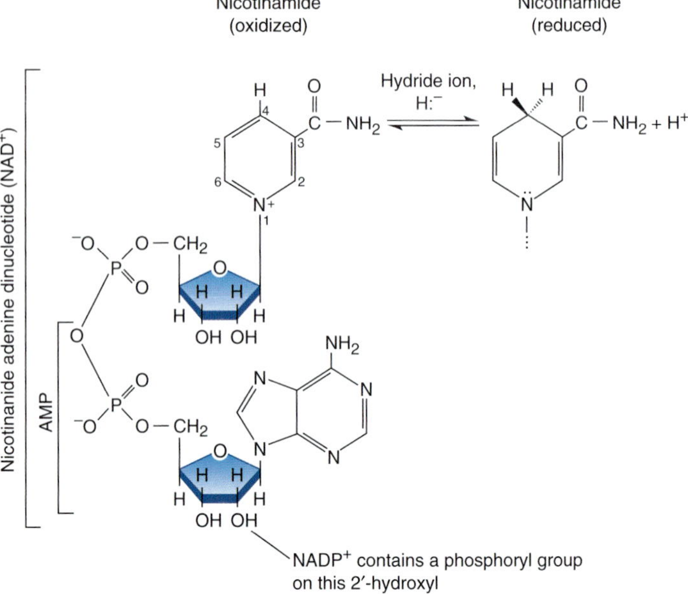

FIGURE 14.4

Structure of nicotinamide adenine dinucleotide (NAD^+). Hydride ion ($H:^-$, a proton with two electrons) transfer to NAD^+ forms NADH.

14.2 THERMODYNAMIC RELATIONSHIPS AND ENERGY-RICH COMPONENTS

Living cells interconvert different forms of energy and also exchange energy with their surroundings. The principles of **thermodynamics** govern reactions of this type. Knowledge of these principles facilitates a perception of how energy-producing and energy-utilizing reactions occur within the same cell and how an organism is able to accomplish various work functions. The **first law of thermodynamics** states that energy can neither be created nor destroyed. This law of energy conservation stipulates that although energy may be converted from one form to another, the total energy in a system remains constant. For example, chemical energy available in a metabolic fuel such as glucose is converted in glycolysis to the chemical energy of ATP. In skeletal muscle, the chemical energy involved in energy-rich phosphate bonds of ATP is converted to mechanical energy during muscle contraction. The energy of an osmotic electropotential gradient of protons across the mitochondrial membrane is converted to chemical energy during ATP synthesis.

The **second law of thermodynamics** is concerned with **entropy**. Entropy, denoted by S, is a measure or indicator of the degree of disorder or randomness in a system. Entropy is viewed as the energy in a system that is unavailable to perform useful work. All processes, whether chemical or biological, tend to progress toward a situation of maximum entropy. Hence, living systems that are highly ordered are never at equilibrium with their surroundings as equilibrium in a system results when the randomness or disorder (entropy) is at a maximum. In biological systems, however, it is nearly impossible to quantitate entropy changes because such systems are rarely at equilibrium. For simplicity and because of its inherent utility in these considerations, a quantity termed **free energy** is employed.

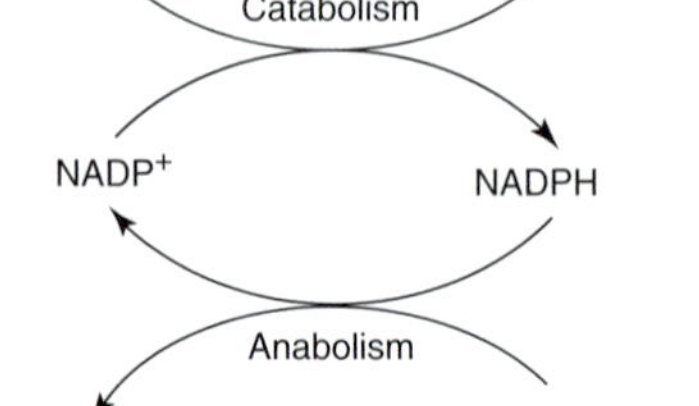

FIGURE 14.5

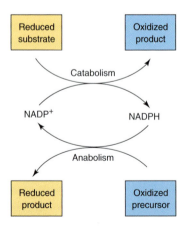

Transfer of reducing equivalents during catabolism and anabolism using NADPH and NADH.

Free Energy Is Energy Available for Useful Work

The free energy (denoted by G or Gibbs free energy) of a system is that portion of the total energy that is available for useful work. It is defined by

$$\Delta G = \Delta H - T\Delta S$$

In this expression for a system proceeding toward equilibrium at a constant temperature and pressure, ΔG is the change in free energy, ΔH is the change in enthalpy or the heat content, T is the absolute temperature, and ΔS is the change in entropy. If ΔG of a reaction is equal to zero, the process is at equilibrium and there is no net flow in either direction. Furthermore, any process that exhibits a negative ΔG (free-energy change) proceeds spontaneously toward equilibrium in the direction written, in part, due to an increase in entropy or disorder in the system. Such a process releases energy and is **exergonic**. A process that exhibits a positive ΔG will proceed spontaneously in the reverse direction as written. Energy from some other source must be applied to allow it to proceed toward equilibrium. This process is termed **endergonic**. Note that the sign and value of ΔG do not predict how fast the reaction will go.

The rate of a given reaction depends on the free energy of activation but not on the magnitude of ΔG. In addition, the change in free energy in a biochemical process is the same regardless of the path or mechanism used to attain the final state. The change in free energy for a chemical reaction is related to the equilibrium constant. For example, a reaction may be described as

$$A + B \rightleftharpoons C + D$$

and the equilibrium constant is expressed as

$$K_{eq} = [C][D]/[A][B]$$

Under standard conditions, when reactants and products are initially present at 1 M concentrations, at 1 atm pressure and 1 M $[H^+]$ or pH 0, the standard free-energy change is defined as ΔG^0. Biochemists have modified this expression so that the standard free energy is calculated at pH 7.0 ($[H^+] = 10^{-7}$ M), where biological reactions generally occur. Under these conditions the change in free energy is expressed as $\Delta G^{0'}$ and K'_{eq}. Since the value of $\Delta G^{0'}$ is zero at equilibrium, the following relationship is described:

$$\Delta G^{0'} = -RT \ln K_{eq}$$

where R is the gas constant, which is 1.987 cal mol^{-1} K^{-1} or 8.134 J mol^{-1} K^{-1}, depending on whether the resultant free-energy change is expressed in calories (cal) or joules (J) per mole, and T is the absolute temperature in degrees Kelvin (K).

Hence, if the **equilibrium constant** is determined, the standard free energy change ($\Delta G^{0'}$) also can be calculated. The relationship between $\Delta G^{0'}$ and K'_{eq} is illustrated in Table 14.1. When the equilibrium constant is less than unity, the reaction is endergonic, and $\Delta G^{0'}$ is positive. When the equilibrium constant is greater than unity, the reaction is exergonic, and $\Delta G^{0'}$ is negative. As discussed above, $\Delta G^{0'}$ of a reaction represents the free energy available in a reaction when substrates and products are present at 1 M concentrations. This situation does not occur in cells, because biomolecules are rarely present at 1 M concentration. Hence, an expression related to actual intracellular concentrations of substrates and products provides insight into work available in a reaction. The expression for ΔG at any concentration of substrate or product includes the energy change for a 1 M concentration of substrate and product to reach equilibrium ($\Delta G^{0'}$) and the energy change to reach a 1 M concentration of substrates and products:

$$\Delta G = \Delta G^{0'} + RT \ln([C][D]/[A][B])$$

For example, in a muscle cell the concentration of ATP is 8.1 mM, that of ADP is 0.93 mM, and that of P_i is 8.1 mM. If $\Delta G^{0'}$ for the reaction ATP + HOH $\leftrightarrow$ ADP + P_i is 7.7 kcal mol^{-1} at 37°C, pH 7.4, then the overall ΔG for the reaction is

$$\Delta G = \Delta G^{0'} + RT \ln([ADP][P_i]/[ATP])$$

$$\Delta G = \Delta G^{0'} + RT \ln[0.93 \times 10^{-3}]$$

$$\Delta G = -7.7 \text{ kcal mol}^{-1} + (-4.2 \text{ kcal mol}^{-1}) = -12 \text{ kcal mol}^{-1}$$

These calculations demonstrate that considerably more free energy is available to perform work in a muscle cell than is indicated by the value of $\Delta G^{0'}$. Moreover, synthesis of ATP in muscle cells under these conditions, the reverse reaction, would require +12 kcal mol^{-1} of energy.

TABLE 14.1 Values of K_{eq} and $\Delta G^{0'}$

K_{eq}	$\Delta G^{0'}$ (kcal mol^{-1})	$\Delta G^{0'}$ (kJ mol^{-1})
10^{-4}	5.46	22.8
10^{-3}	4.09	17.1
10^{-2}	2.73	11.4
10^{-1}	1.36	5.7
1	0	0
10	−1.36	−5.7
10^{2}	−2.73	−11.4
10^{3}	−4.09	−17.1
10^{4}	−5.46	−22.8

In energy-producing and energy-utilizing metabolic pathways, free-energy changes of individual enzymatic reactions are additive; for example,

$$A \rightarrow B \rightarrow C \rightarrow D$$

$$\Delta G^{0'}{}_{A\rightarrow D} = \Delta G^{0'}{}_{A\rightarrow B} + \Delta G^{0'}{}_{B\rightarrow C} + \Delta G^{0'}{}_{C\rightarrow D}$$

Although any given enzymatic reaction in a sequence may have a positive free-energy change, as long as the sum of all free-energy changes is negative, the pathway will proceed. Another way of expressing this principle is that enzymatic reactions with positive free-energy changes may be coupled to or driven by reactions with negative free-energy changes. In a metabolic pathway such as glycolysis, various reactions either have positive $\Delta G^{0'}$ values or $\Delta G^{0'}$ values close to zero, while other reactions have large and negative $\Delta G^{0'}$ values, which drive the entire pathway. The crucial consideration is that the sum of $\Delta G^{0'}$ values of all reactions in a pathway must be negative in order for such a pathway to be thermodynamically feasible. As for all chemical reactions, individual enzymatic reactions in a metabolic pathway or the pathway as a whole would be facilitated if the concentrations of reactants (substrates) exceed those of products.

Caloric Value of Dietary Components

During complete stepwise oxidation of glucose, a primary metabolic fuel in cells, a large quantity of energy is made available. This is illustrated in the equation

$$C_6H_{12}O_6 + 6\,O_2 \rightarrow 6\,CO_2 + 6\,H_2O \qquad \Delta G^{0'} = -686{,}000 \text{ cal mol}^{-1}$$

When this process occurs under aerobic conditions in most cells, it is possible to conserve approximately one half of this "available" energy as 38 molecules of ATP. The $\Delta G^{0'}$ and caloric values for oxidation of other metabolic fuels are listed in Table 14.2. Carbohydrates and proteins (amino acids) have a caloric value of 3–4 kcal g^{-1}, while lipid (i.e., palmitate, a long-chain fatty acid, or a triacylglycerol) has a value nearly three times greater. The reason that more energy is derived from lipid than from carbohydrate or protein relates to the average oxidation state of carbon atoms in these substances. Carbon atoms in carbohydrates are considerably more oxidized (or less reduced) than those in lipids (Figure 14.3). Hence during sequential breakdown of lipid, many more reducing equivalents (a reducing equivalent is defined as a proton plus an electron, i.e., $H^+ + e^-$) can be extracted than from carbohydrate.

Compounds Are Classified on the Basis of Energy Released on Hydrolysis of Specific Groups

The two terminal phosphoryl groups of ATP are high-energy bonds, since free energy of hydrolysis of a phosphoanhydride bond is much greater than that of a simple phosphate ester. High energy is not synonymous with stability of the chemical bond in question, nor does it refer to the energy required to break such a bond. The concept of high-energy compounds implies that products of their hydrolytic cleavage are in more stable forms than the original compound. As a rule, phosphate esters (low-energy compounds) exhibit

TABLE 14.2 Free-Energy Changes and Caloric Values for Total Metabolism of Various Metabolic Fuels

Compound	*Molecular Weight*	*$\Delta G^{0'}$ (kcal mol^{-1})*	*Caloric Value (kcal g^{-1})*
Glucose	180	−686	3.81
Lactate	90	−326	3.62
Palmitate	256	−2380	9.30
Tripalmitin	809	−7510	9.30
Glycine	75	−234	3.12

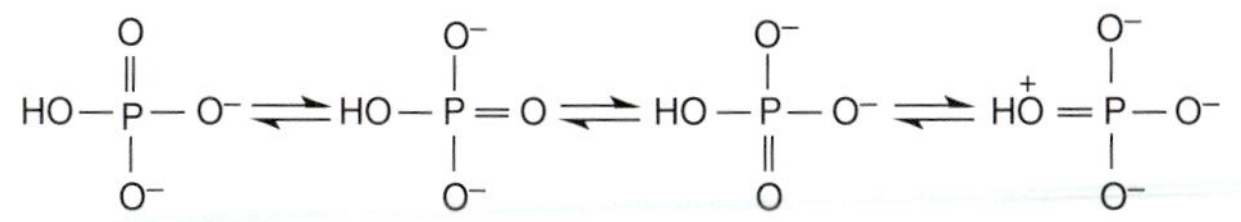

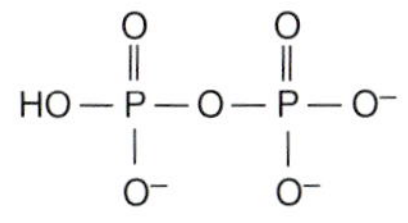

(*a*) Resonance forms of phosphate

(*b*) Pyrophosphate

FIGURE 14.6
(*a*) Resonance forms of phosphate. (*b*) Structure of pyrophosphate.

negative $\Delta G^{0'}$ values of hydrolysis of 1–3 kcal mol^{-1}, whereas high-energy bonds have negative $\Delta G^{0'}$ values of 5–15 kcal mol^{-1}. Phosphate esters such as glucose 6-phosphate and glycerol 3-phosphate are low-energy compounds. Table 14.3 lists various types of energy-rich compounds with approximate values for their $\Delta G^{0'}$ values of hydrolysis.

There are various reasons why certain compounds or bonding arrangements are energy-rich. First, products of hydrolysis of an energy-rich bond may exist in more **resonance forms** than the precursor molecule. The more possible resonance forms in which a molecule can exist stabilize that molecule. The resonance forms for inorganic phosphate (P_i) are indicated in Figure 14.6. Fewer resonance forms can be written for ATP or pyrophosphate (PP_i) than for phosphate (P_i). Second, many high-energy bonding arrangements contain groups of similar electrostatic charges located in close proximity to each other. Because like charges repel one another, hydrolysis of energy-rich bonds alleviates this situation and lends stability to products of hydrolysis. Third, hydrolysis of certain energy-rich bonds results in formation of an unstable compound, which may isomerize spontaneously to form a more stable compound. Hydrolysis of phosphoenolpyruvate is an example of this type of compound (Figure 14.7). The $\Delta G^{0'}$ is considerable for isomerization, and the final product, in this case pyruvate, is much more stable. Finally, if a product of hydrolysis of a high-energy bond is an undissociated acid, dissociation of the proton and its subsequent buffering may contribute to the overall $\Delta G^{0'}$ of the hydrolytic reaction. In general, any property or process that stabilizes products of hydrolysis tends to confer a high-energy character to that compound. The high-energy character of **3′,5′-cyclic adenosine monophosphate (cAMP)** has been attributed to the fact that its phosphodiester bond is strained as it bridges the 3′ and 5′ positions on ribose. The energy-rich character of thiol ester compounds such as **acetyl CoA** or succinyl CoA results from the relatively acidic character of the thiol group. Hence, the thioester bond of acetyl CoA is nearly equivalent in energy to a phosphoanhydride bond.

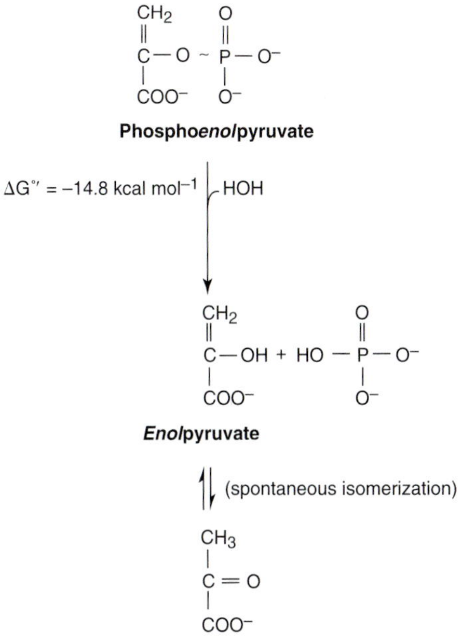

Phospho*enol*pyruvate

$\Delta G^{0'} = -14.8$ kcal mol^{-1} HOH

*Enol*pyruvate

(spontaneous isomerization)

Pyruvate (stable form)

FIGURE 14.7
Hydrolysis of phosphoenolpyruvate indicating the free energy released.

Free-Energy Changes Can Be Determined from Coupled Enzyme Reactions

The $\Delta G^{0'}$ value of hydrolysis of the terminal phosphate of ATP is difficult to determine simply because the K_{eq} of the reaction is far to the right.

$$\text{ATP} + \text{HOH} \rightleftharpoons \text{ADP} + \text{P}_i + \text{H}^+$$

However, $\Delta G^{0'}$ of hydrolysis of ATP is determined indirectly because of the additive nature of free-energy changes discussed above. Hence, free energy of hydrolysis of ATP is determined by adding $\Delta G^{0'}$ of an ATP-utilizing reaction such as hexokinase to $\Delta G^{0'}$ of a reaction that cleaves phosphate from the product of the hexokinase reaction, glucose

TABLE 14.3 Examples of Energy-Rich Compounds

Type of Bond	*$\Delta G^{0'}$ of Hydrolysis (kcal mol^{-1})*	*$\Delta G^{0'}$ of Hydrolysis (kJ mol^{-1})*	*Example*
Phosphoric acid anhydrides	−7.3	−35.7	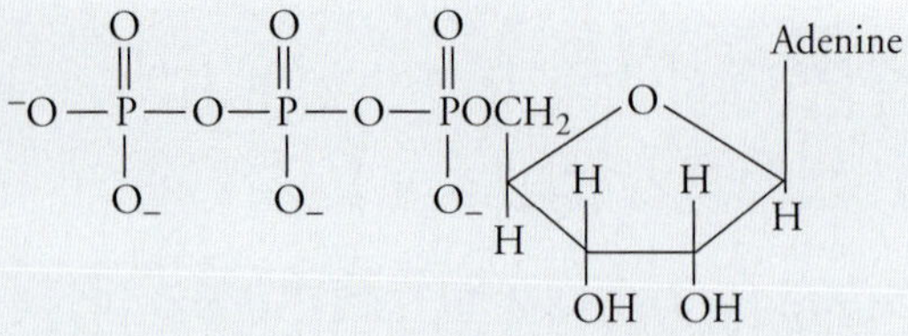 ATP
	−11.9	−50.4	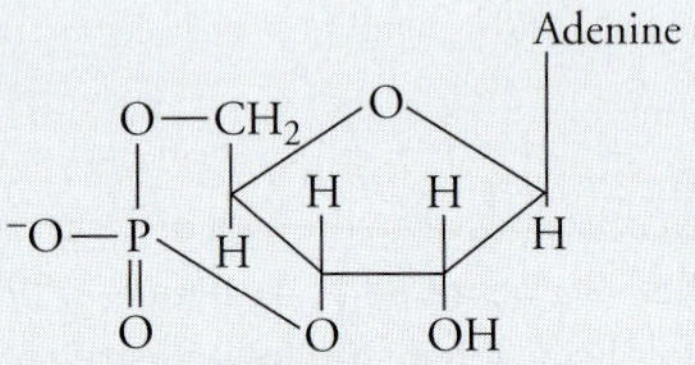 3′,5′-cyclic AMP
Phosphoric-carboxylic acid anhydrides	−10.1	−49.6	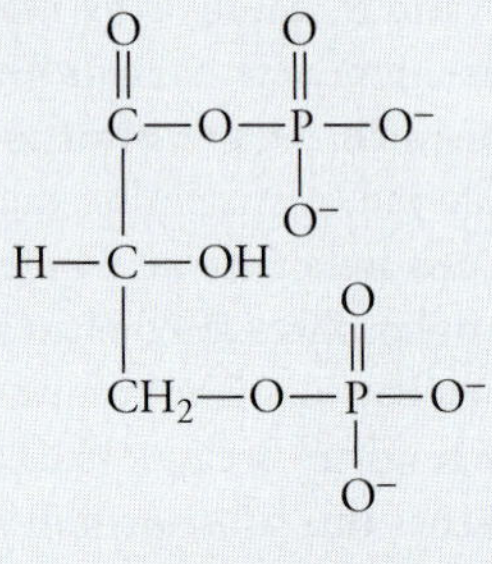 1,3-Bisphosphoglycerate
	−10.3	−43.3	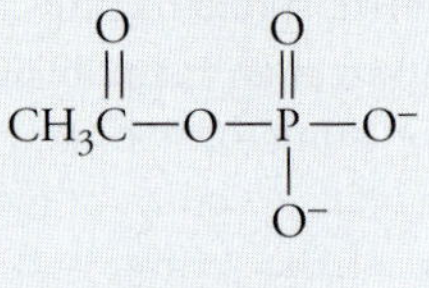 Acetyl phosphate
Phosphoguanidines	−10.3	−43.3	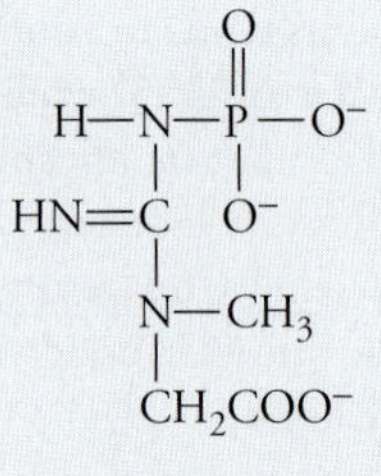 Creatine phosphate
Enol phosphates	−14.8	−62.2	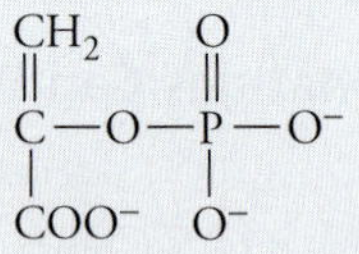 Phosphoenolpyruvate
Thiol esters	−7.7	−31.5	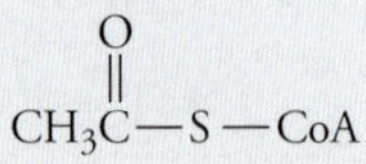 Acetyl CoA

6-phosphate (G6P), as indicated below:

$$\text{Glucose} + \text{ATP} \xrightarrow{\text{hexokinase}} \text{G6P} + \text{ADP} + \text{H}^+ \qquad \Delta G^{0'} = -4.0 \text{ kcal mol}^{-1}$$

$$\text{G6P} + \text{HOH} \xrightarrow{\text{glucose 6-phosphatase}} \text{ADP} + \text{P}_i + \text{H}^+ \qquad \Delta G^{0'} = -3.3 \text{ kcal mol}^{-1}$$

Free energies of hydrolysis for other energy-rich compounds are determined in a similar fashion.

High-Energy Bond Energies of Various Groups Can Be Transferred from One Compound to Another

Energy-rich compounds can transfer various groups to an acceptor compound in a thermodynamically feasible fashion in the presence of an appropriate enzyme. The energy-rich intermediates of glycolysis, 1,3-bisphosphoglycerate and phosphoenolpyruvate, can transfer their high-energy phosphate moieties to ADP in the phosphoglycerate kinase and pyruvate kinase reactions, respectively (Figure 14.8). The $\Delta G^{0'}$ values of these reactions are -4.5 and -7.5 kcal mol^{-1}, respectively, and hence transfer of "high-energy" phosphate is thermodynamically possible, and ATP synthesis is the result. ATP can transfer its terminal phosphoryl group to form a compound of relatively similar high-energy character [i.e., creatine phosphate in the creatine kinase reaction (Figure 14.8) or compounds of considerably lower energy, such as glucose 6-phosphate in the hexokinase reaction (Figure 14.8)]. Such transfers are crucial in linking energy-producing and energy-utilizing metabolic pathways in living cells.

Although adenine nucleotides are mainly involved in energy generation or conservation, various nucleoside triphosphates, including ATP, are involved in energy transfer in biosynthetic pathways. The guanine nucleotide GTP is a source of energy in gluconeogenesis and protein synthesis, whereas UTP (uracil) and CTP (cytosine) are utilized in glycogen and lipid synthesis, respectively (see p. 26 for structures). The energy in the phosphate bonds of ATP may be transferred to other nucleotides by nucleoside diphosphate kinase or nucleoside monophosphate kinase (Figure 14.9). Two nucleoside diphosphates are converted to a nucleoside triphosphate and a nucleoside monophosphate in various nucleoside monophosphate kinase reactions, such as that of adenylate kinase (Figure 14.10). Thus, the terminal energy-rich phosphate bond of ATP may be transferred to appropriate nucleotides and utilized in a variety of biosynthetic processes.

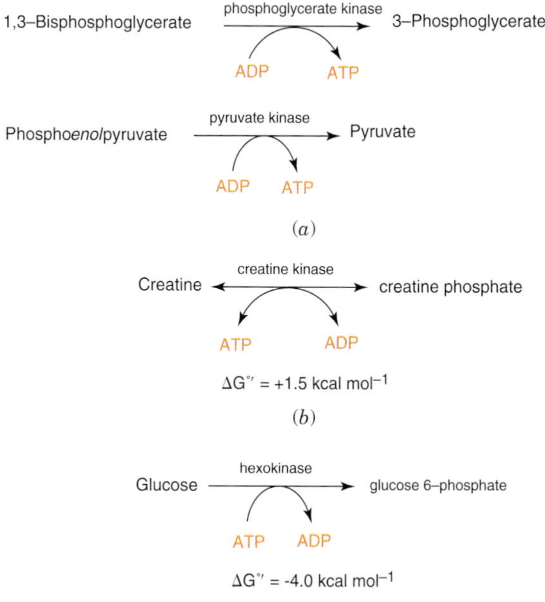

FIGURE 14.8

Examples of reactions involved in transfer of "high-energy" phosphate.

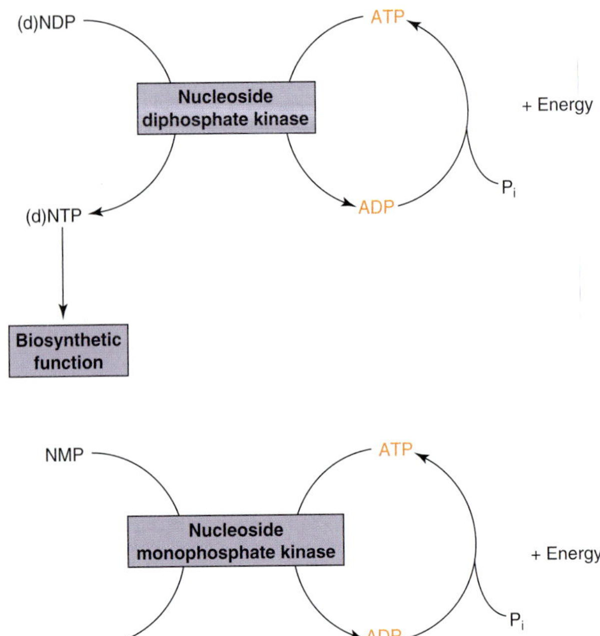

FIGURE 14.9

Nucleoside diphosphate kinase and nucleoside monophosphate kinase reactions. N represents any purine or pyrimidine base; (d) indicates a deoxyribonucleotide.

14.3 SOURCES AND FATES OF ACETYL COENZYME A

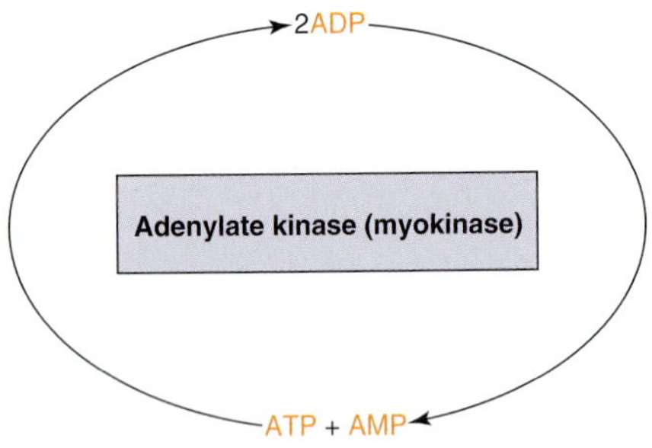

The acetate group that serves as the source of fuel for the **tricarboxylic acid (TCA)** cycle is derived from the major energy-generating metabolic pathways of cells. These include oxidation of long-chain fatty acids by β-oxidation, breakdown of ingested or stored carbohydrate by **glycolysis**, oxidation of the **ketone bodies (acetoacetate and β-hydroxybutyrate)**, oxidation of ethanol and oxidative breakdown of certain amino acids (Figure 14.11). All of these eventually result in production of the two-carbon unit **acetyl coenzyme A** (CoA). Coenzyme A, abbreviated as CoA or CoASH, consists of β-mercaptoethylamine, the vitamin **pantothenic acid**, and the adenine nucleotide, adenosine 3′-phosphate 5′-diphosphate (Figure 14.12). In cells, coenzyme A exists as the reduced thiol (CoASH), which forms high-energy thioester bonds with acyl groups and is involved in acyl group transfer reactions in which CoA serves as acceptor, then donor, of the acyl group. Various metabolic pathways involve only acyl-CoA derivatives, for example, β-oxidation of fatty acids and branched-chain amino acid degradation. Because CoA is a large, hydrophilic molecule, it and its derivatives such as acetyl CoA

FIGURE 14.10

Adenylate kinase (myokinase) reaction.

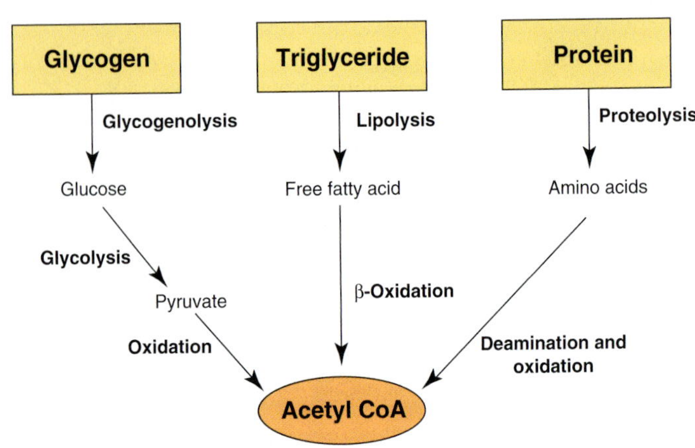

FIGURE 14.11

General precursors of acetyl CoA. Carbohydrates, lipids, and proteins are broken down to form acetyl CoA.

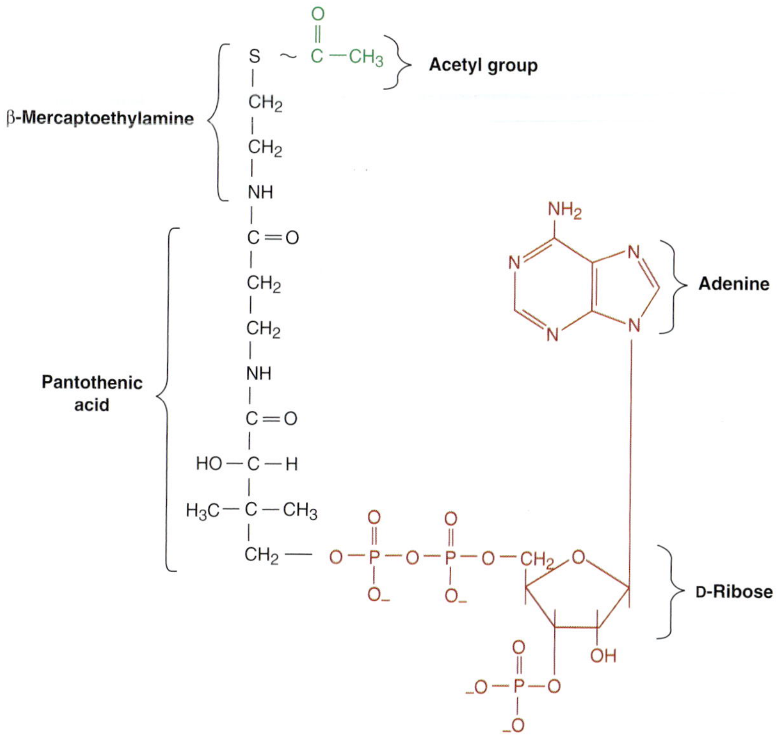

FIGURE 14.12
Structure of acetyl CoA.

are not freely transported across cellular membranes. This has necessitated evolution of certain transport or shuttle mechanisms by which various intermediates or groups are transferred across membranes. Such acyl transferase reactions for acetyl groups and long-chain acyl groups will be discussed in Chapter 16. Since the thiol ester bond in acyl-CoA derivatives is an energy-rich bond, these compounds are effective donors of acyl groups in acyl transferase reactions. To synthesize an acyl-CoA derivative two high-energy bonds of ATP must be expended, such as in the **acetate thiokinase** reaction,

$$\text{Acetate} + \text{CoASH} + \text{ATP} \xrightarrow{\text{Acetate kinase}} \text{acetyl CoA} + \text{AMP} + \text{PP}_i$$

Metabolic Sources and Fates of Pyruvate

During aerobic glycolysis (see p. 583), glucose or other hexoses are converted to pyruvate, the end product of this cytosolic pathway. Pyruvate is also formed in degradation of amino acids such as alanine or serine and has several fates depending on the tissue and its metabolic state. The fates of pyruvate and types of reactions in which it participates are indicated in Figure 14.13. The oxidative decarboxylation of pyruvate in the **pyruvate dehydrogenase** reaction is discussed next; see p. 746 for a discussion of other reactions involving pyruvate.

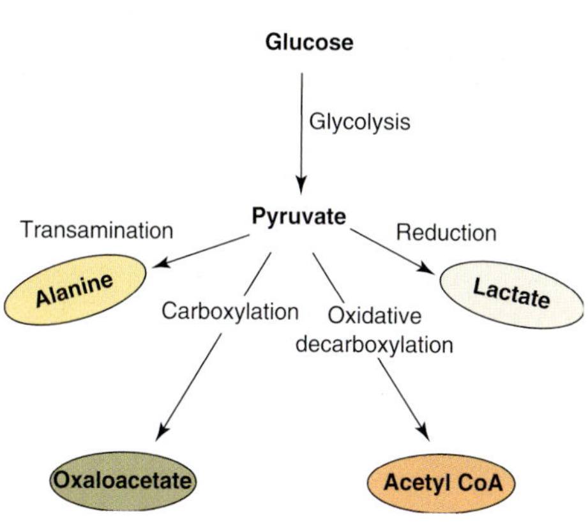

FIGURE 14.13
Metabolic fates of pyruvate. Pyruvate is at a crossroads of metabolism. It can be converted to lactate, alanine, oxaloacetate or acetyl CoA depending on the needs of the cell.

Pyruvate Dehydrogenase Is a Multienzyme Complex

Pyruvate is converted to acetyl CoA by the pyruvate dehydrogenase multienzyme complex:

$$\text{Pyruvate} + \text{NAD}^+ + \text{CoASH} \longrightarrow \text{acetyl CoA} + \text{CO}_2 + \text{NADH} + \text{H}^+$$

$$\Delta G^{0'} = -8 \text{ kcal mol}^{-1}$$

The mechanism of this reaction is more complex than might be inferred from the overall stoichiometry. Three of the cofactors, **thiamine pyrophosphate** (TPP), **lipoamide**, and **flavin adenine dinucleotide** (FAD) are bound to subunits of the complex. The reaction has a $\Delta G^{0'}$ of -8 kcal mol^{-1} and hence is irreversible under physiological

TABLE 14.4 Pyruvate Dehydrogenase Complex of Mammals

Enzyme	*Number of Subunits*	*Prosthetic Group*	*Reaction Catalyzed*
Pyruvate dehydrogenase	20 or 30	TPP	Oxidative decarboxylation of pyruvate
Dihydrolipoyl transacetylase	60	Lipoamide	Transfer of the acetyl group to CoA
Dihydrolipoyl dehydrogenase	6	FAD	Regeneration of the oxidized form of lipoamide and transfer of electrons to NAD

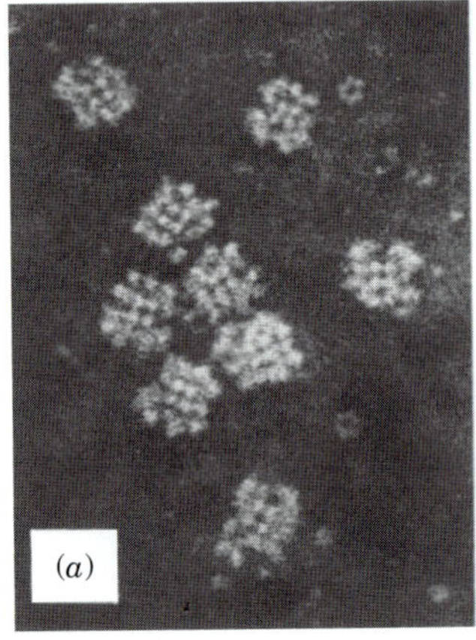

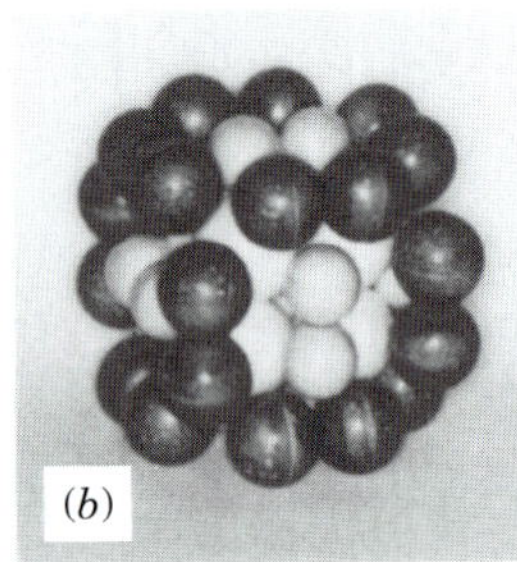

FIGURE 14.14

Pyruvate dehydrogenase complex from *E. coli*. (*a*) Electron micrograph. (*b*) Molecular model. (white spheres are the 24 transacetylase subunits, black spheres are the 12 pyruvate dehydrogenase dimers, gray spheres are the six dihydrolipoyl dehydrogenase dimers). The enzyme complex was negatively stained with phosphotungstate (×200,000). Courtesy of Dr. Lester J. Reed, University of Texas, Austin.

conditions. The mammalian pyruvate dehydrogenase complex contains three types of catalytic subunit associated in a multienzyme complex of mass 7 to 8.5 × 10^6 kDa for the complex from kidney, heart, or liver. The catalytic subunits and their associated cofactors are presented in Table 14.4.

The pyruvate dehydrogenase complex from *E. coli* (4.6 × 10^6 kDa) has been visualized by electron microscopy (Figure 14.14). The cube-like core contains 24 subunits (64.5 kDa each) of transacetylase (white spheres in model shown in Figure 14.14). Twelve pyruvate dehydrogenase dimers (black spheres; 90.5 kDa each) are distributed symmetrically on the 12 edges of the transacetylase cube. Six dihydrolipoyl dehydrogenase dimers (gray spheres; 56 kDa each) are distributed on the six faces of the cube. This arrangement of the pyruvate dehydrogenase complex provides for greater efficiency in the overall reaction as the intermediates are tightly bound to the subunits and are not released into the surrounding medium.

The functional group of TPP participates in the formation of a covalent intermediate (Figure 14.15). Lipoic acid is linked by an amide bond with a lysine on each transacetylase subunit and is called lipoamide, while FAD is tightly bound to each dihydrolipoyl dehydrogenase subunit. The mechanism of the reaction is illustrated in Figure 14.16.

Pyruvate Dehydrogenase Is Strictly Regulated

The pyruvate dehydrogenase complex is regulated in two ways. First, two products of the reaction, acetyl CoA and NADH, inhibit the complex in a competitive fashion, as feedback inhibitors. Second, the pyruvate dehydrogenase complex undergoes phosphorylation and dephosphorylation. The complex is active when dephosphorylated and inactive when phosphorylated. Inactivation is accomplished by an Mg^{2+}-ATP-dependent **protein kinase**, which is tightly associated with the complex. Activation is accomplished by a **phosphoprotein phosphatase**, which is also associated with the complex and functions in an Mg^{2+}- and Ca^{2+}-dependent manner. The differential regulation of the pyruvate dehydrogenase kinase and phosphatase is the key to the

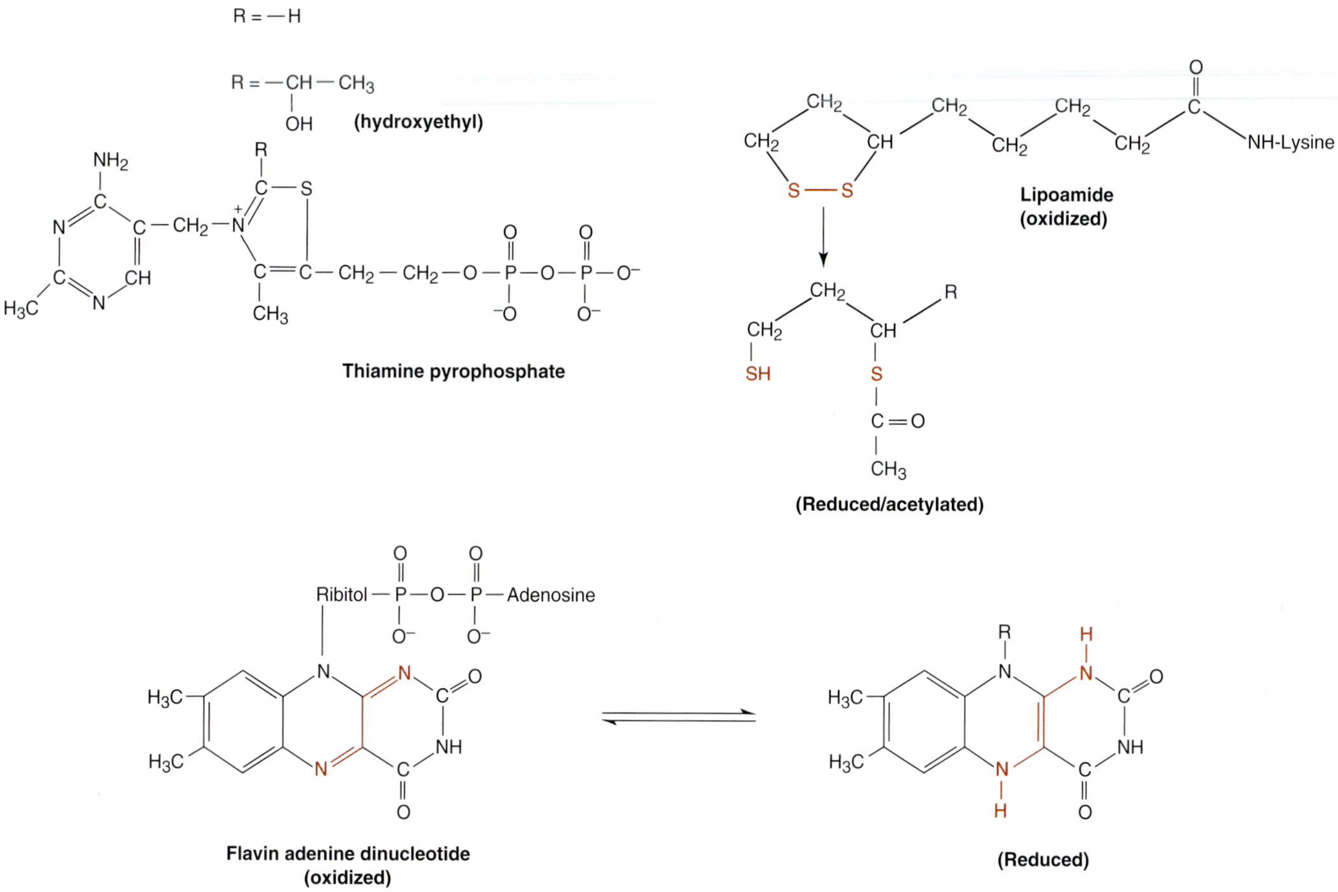

R = —H

R = —CH—CH3 (hydroxyethyl)

Thiamine pyrophosphate

Lipoamide (oxidized)

(Reduced/acetylated)

Flavin adenine dinucleotide (oxidized)

(Reduced)

FIGURE 14.15

Structures of coenzymes involved in the pyruvate dehydrogenase reaction. See Figure 14.4 for structure of NAD^+ and Figure 14.13 for structure of CoA.

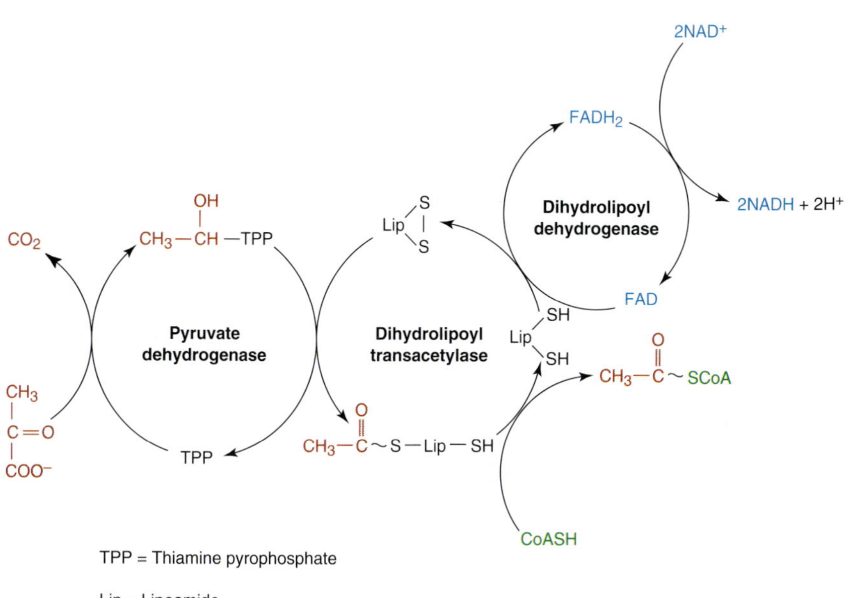

FIGURE 14.16

Mechanism of the pyruvate dehydrogenase multienzyme complex. Pyruvate dehydrogenase catalyzes oxidative decarboxylation of pyruvate and transfer of the acetyl group to lipoamide. Dihydrolipoyl transacetylase transfer the acetyl group from lipoamide to coenzyme A. Dihydrolipoyl dehydrogenase oxidizes reduced lipoamide.

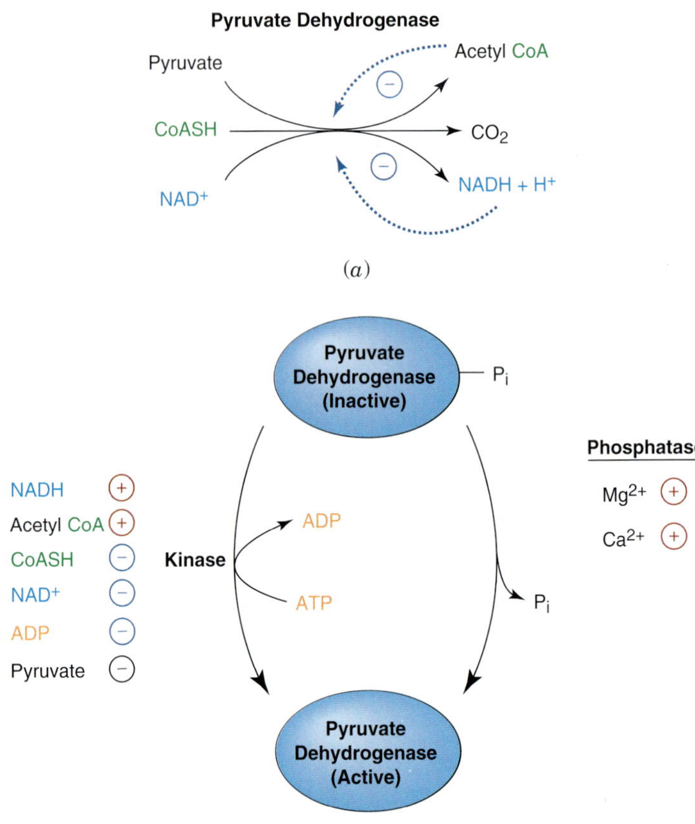

FIGURE 14.17

Regulation of the pyruvate dehydrogenase multienzyme complex. (*a*) Pyruvate dehydrogenase is inhibited by its products, acetylCoA and NADH. (*b*) Pyruvate dehydrogenase is also inactivated by phosphorylation and activated by dephosphorylation. The phosphatase is stimulated by Mg^{2+} and Ca^{2+} ions. The kinase is stimulated by ATP, NADH, and acetylCoA and inhibited by CoASH, NAD^+, ADP and pyruvate.

overall regulation of the complex. Essential features of this complex regulatory system are illustrated in Figure 14.17. Acetyl CoA and NADH inhibit the dephosphorylated (active) form of the complex. These also stimulate the protein kinase reaction, leading to an inactivation of the complex. In addition, free CoASH and NAD^+ inhibit the protein kinase. Hence, any increase of the mitochondrial NADH/NAD^+ or acetyl CoA/CoASH ratio with concomitant lowering of NAD^+ and CoASH inactivates the complex because of the stimulated kinase reaction. In addition, pyruvate, the substrate of the enzyme, is a potent inhibitor of the protein kinase, so that with elevated tissue pyruvate levels the kinase is inhibited and the complex is maximally active. Activity of the complex is stimulated by Ca^{2+}, a potent activator of the protein phosphatase. These effects of Ca^{2+} may play an important role in skeletal muscle where the release of Ca^{2+} during contraction should activate the protein phosphatase stimulating the oxidation of pyruvate and hence energy production. Finally, administration of **insulin** activates pyruvate dehydrogenase in adipose tissue, and **catecholamines**, such as epinephrine, activate pyruvate dehydrogenase in cardiac tissue. The mechanisms of these hormonal effects are not well understood; however, alterations in the intracellular distribution of Ca^{2+} may result in the stimulation of the protein phosphatase reaction in the mitochondrial matrix. These hormonal effects are not mediated directly by tissue cAMP levels, because pyruvate dehydrogenase protein kinase and protein phosphatase are cAMP-independent (see Clin. Corr. 14.1).

Acetyl CoA Is Used in Several Different Pathways

Fates of acetyl CoA generated in the mitochondrial matrix include (1) complete oxidation of the acetyl group in the TCA cycle for energy generation; (2) in liver, conversion of excess acetyl CoA into the ketone bodies, acetoacetate and β-hydroxybutyrate; and (3) transfer of acetyl units as citrate to the cytosol with subsequent synthesis of long-chain fatty acids (see p. 672) and sterols (Figure 14.18) (see p. 708).

CLINICAL CORRELATION 14.1
Pyruvate Dehydrogenase Deficiency

A variety of disorders of pyruvate metabolism have been detected in children. Some involve deficiency of the catalytic or regulatory subunits of the pyruvate dehydrogenase complex. Children with pyruvate dehydrogenase deficiency usually exhibit elevated serum levels of lactate, pyruvate, and alanine, which produce a chronic lactic acidosis. They frequently exhibit severe neurological defects, which generally results in death. The diagnosis of pyruvate dehydrogenase deficiency is usually made by assaying the enzyme complex and/or its enzymatic subunits in cultures of skin fibroblasts taken from the patient. Some patients respond to dietary management in which a carbohydrate-reduced diet is administered. Patients may be in shock from lactic acidosis since decreased delivery of O_2 inhibits pyruvate dehydrogenase and increases anaerobic metabolism. Some patients have been treated with dichloroacetate, an inhibitor of the protein kinase subunit of the pyruvate dehydrogenase complex. Complete inhibition of the kinase, which causes inhibition of the enzyme, will therefore activate the enzyme complex.

Source: Patel, M. S. and Harris, R. A. Mammalian α-keto acid dehydrogenase complexes: gene regulation and genetic defects. *FASEB J.* 9:1164, 1995.

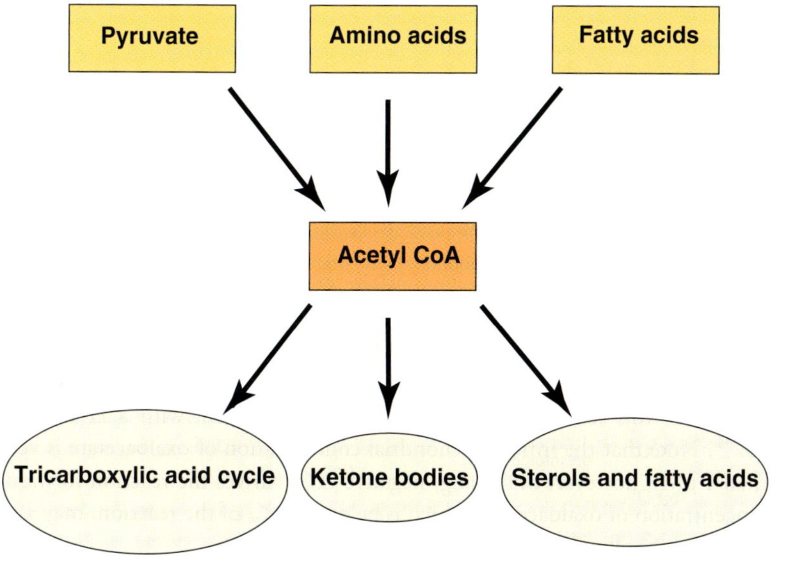

FIGURE 14.18
Sources and fates of acetylCoA.

14.4 THE TRICARBOXYLIC ACID CYCLE

Acetyl CoA produced in the energy-generating catabolic pathways of most cells is completely oxidized to CO_2 in a cycle of reactions termed **the tricarboxylic acid (TCA)** cycle. This cycle is also called the **citric acid cycle** or the **Krebs cycle** after Sir Hans Krebs, who postulated its essential features in 1937. The primary location of enzymes of the TCA cycle is in mitochondria, although isozymes of some enzymes are present in the cytosol. This location is appropriate as the pyruvate dehydrogenase complex and the fatty acid β-oxidation sequence, the two primary sources of acetyl CoA, are also located in mitochondria. Four reactions of the TCA cycle transfer electrons to either NAD^+ or FAD. The resulting NADH or $FADH_2$ is then oxidized by the mitochondrial **electron transport chain** (**electron transfer chain or respiratory chain**) to generate energy that is used to form ATP by **oxidative phosphorylation** (page 562). The enzymes of the electron transport chain and those involved in ATP synthesis are exclusively localized in mitochondria. Figure 14.19 is an overview of the reactions of the TCA cycle. In the first step, the acetyl moiety of acetyl CoA is condensed with oxaloacetate (a four-carbon dicarboxylic acid) to form citrate (a six-carbon tricarboxylic acid). After rearrangement of the carbons of citrate, two oxidative decarboxylation reactions produce

reducing equivalents (i.e., NADH + H^+) of the cycle are produced. Another product of this reaction, **succinyl CoA**, is an energy-rich thiol ester similar to acetyl CoA.

The energy-rich character of the thiol ester linkage of succinyl CoA is conserved by **substrate-level phosphorylation** in the next step of the cycle. **Succinyl-CoA synthetase** (or **succinate thiokinase**) converts succinyl CoA to succinate and, in mammalian tissues, results in the phosphorylation of GDP to GTP. This reaction is freely reversible, with a $\Delta G^{0'} = -0.7$ kcal mol^{-1}, and the catalytic mechanism involves an enzyme-succinyl phosphate intermediate.

$$\text{Succinyl CoA} + P_i + \text{Enz} \rightarrow \text{Enz-succinyl phosphate} + \text{CoASH}$$

$$\text{Enz-succinyl phosphate} \rightarrow \text{Enz-phosphate} + \text{succinate}$$

$$\text{Enz-phosphate} + \text{GDP} \rightarrow \text{Enz} + \text{GTP}$$

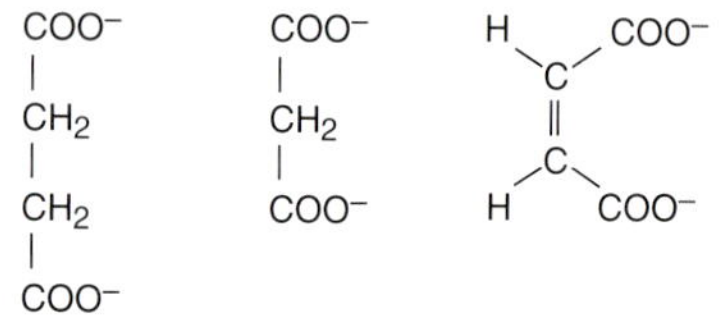

The enzyme is phosphorylated on the 3 position of a histidine residue during the reaction, which conserves the energy of the thioester for formation of GTP. This GTP is used for the mitochondrial synthesis of protein, RNA and DNA.

Succinate is oxidized to fumarate by **succinate dehydrogenase**, a complex enzyme tightly bound to the inner mitochondrial membrane. Succinate dehydrogenase is composed of a 70-kDa subunit that contains the substrate-binding site (FAD covalently bound to a histidine residue), a 30-kDa subunit that contains three **iron–sulfur centers** (**nonheme iron**), and two small hydrophobic proteins. The enzyme is a typical flavoprotein in which electrons and protons are transferred from the substrate through covalently bound FAD and the iron–sulfur centers in which the nonheme iron undergoes oxidation and reduction. The electrons are then transferred to coenzyme Q for further transport through the electron-transfer chain, as will be discussed in Section 14.6. Succinate dehydrogenase is strongly inhibited by malonate and oxaloacetate and is activated by ATP, P_i, and succinate. Malonate inhibits succinate dehydrogenase competitively with respect to succinate because of the very close structural similarity between malonate and succinate (Figure 14.21).

COO⁻–CH₂–CH₂–COO⁻ (Succinate) | COO⁻–CH₂–COO⁻ (Malonate) | H(COO⁻)C=C(COO⁻)H (Maleate)

Succinate **Malonate** **Maleate**

FIGURE 14.21

Structures of succinate, a TCA cycle intermediate; malonate, an inhibitor of succinate dehydrogenase, and the cycle; and maleate, a compound not involved in the cycle.

Fumarate is then hydrated to form L-malate by fumarase. **Fumarase** is a homotetramer (200 kDa) and is stereospecific for the trans form of substrate (the cis form, maleate, is not a substrate (Figure 14.21). The reaction is freely reversible under physiological conditions. Clinical Correlation 14.2 describes a genetic deficiency of fumarase.

The final reaction in the cycle is catalyzed by **malate dehydrogenase** in which the reducing equivalents are transferred to NAD^+ to form NADH + H^+. The equilibrium of the reaction lies far toward L-malate formation, with a $\Delta G^{0'} = +7.0$ kcal mol^{-1}. This endergonic reaction is pulled in the forward direction by the action of citrate synthase and other reactions, which remove oxaloacetate.

The NADH produced in the three NAD^+-linked dehydrogenases in the TCA cycle is oxidized rapidly to NAD^+ by the respiratory chain. This is another factor favoring the forward direction of malate dehydrogenase.

CLINICAL CORRELATION 14.2
Fumarase Deficiency

Deficiency of enzymes of the TCA cycle is rare, indicating the importance of this pathway for survival. Several cases, however, have been reported of severe deficiency of fumarase in mitochondria and cytosol of tissues (e.g., blood lymphocytes). It is characterized by severe neurological impairment, encephalomyopathy, and dystonia developing soon after birth. Urine contains abnormal amounts of fumarate and elevated levels of succinate, α-ketoglutarate, citrate, and malate. The mitochondrial and cytosolic isozymes of fumarase are derived from a single gene. In affected patients, both parents had half-normal levels of enzyme activity but were clinically normal, as expected for an autosomal recessive disorder. The first mutation characterized in the gene for fumarase contains a glutamine substituted for a glutamate residue 319.

Source: Bourgeron, T., Chretien, D., Poggi-Bach, J., et al. Mutation of the fumarase gene in two siblings with progressive encephalopathy and fumarase deficiency. *J. Clin. Invest.* 93:2514, 1994.

Conversion of the Acetyl Group of Acetyl CoA to CO_2 and H_2O Conserves Energy

The TCA cycle (Figure 14.20) is the terminal oxidative pathway for most metabolic fuels. Two-carbon moieties in the form of acetyl CoA are oxidized completely to CO_2 and H_2O, and four oxidative steps result in the formation of 3 NADH + H^+ and 1 as $FADH_2$, which are used subsequently for ATP generation. Oxidation of each NADH + H^+ results in formation of 2.5 ATP by oxidative phosphorylation, while oxidation of $FADH_2$ formed in the succinate dehydrogenase reaction yields 1.5 ATPs. A high-energy bond is formed as GTP in the succinyl-CoA synthetase reaction. Hence the net yield of ATP or its equivalent (i.e., GTP) for the complete oxidation of an acetyl group in the Krebs cycle is 10.

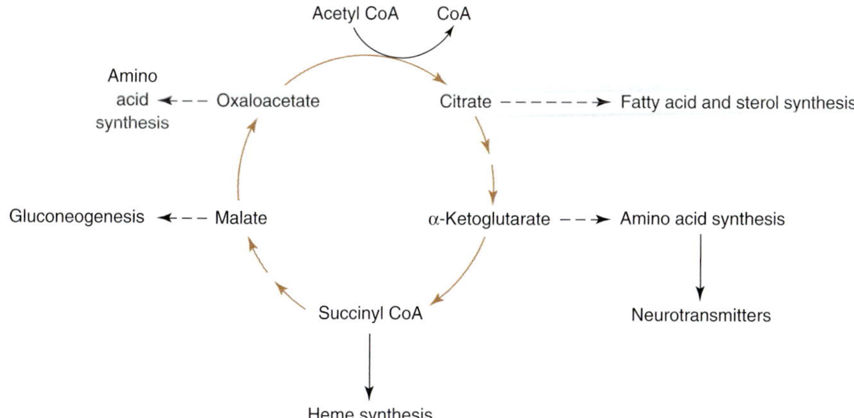

FIGURE 14.22

The TCA cycle is a source of precursors for amino acid, fatty acid, and glucose synthesis.

Tricarboxylic Acid Cycle Is a Source of Biosynthetic Intermediates

The discussion of the TCA cycle thus far has concentrated on its role in the oxidative breakdown of acetyl groups to CO_2 and H_2O, formation of reduced coenzymes and synthesis of ATP. In general, the TCA cycle is the final common mechanism for breakdown of foodstuffs; however, as summarized in Figure 14.22, the four-, five-, and six-carbon intermediates generated in the reactions of the TCA cycle are important intermediates in **biosynthetic processes**. Succinyl CoA, malate, oxaloacetate, α-ketoglutarate, and citrate are all precursors in the biosynthesis of important cellular compounds.

Transamination converts α-ketoglutarate to glutamate, which can leave mitochondria and be converted into several other amino acids. In nervous tissue, α-ketoglutarate is converted to the neurotransmitters glutamate and γ-aminobutyric acid (GABA). Glutamate is also produced from α-ketoglutarate by the mitochondrial enzyme glutamate dehydrogenase in the presence of NADH or NADPH and ammonia. The amino group incorporated into glutamate can then be transferred to form various amino acids by different aminotransferases. These enzymes and the relevance of the incorporation or release of ammonia into or from α-keto acids are discussed in Chapter 19.

Succinyl CoA represents a metabolic branch point (Figure 14.23) because it may be formed either from α-ketoglutarate in the cycle or from methylmalonyl CoA in the final steps of breakdown of odd-chain length fatty acids or the branched-chain amino acids valine and isoleucine or it may be converted to succinate or condensed with glycine to form δ-aminolevulinate, the initial reaction in porphyrin biosynthesis (see p. 835).

Oxaloacetate is transaminated to aspartate, the precursor of asparagine and of the pyrimidines cytosine, uracil, and thymine. Oxaloacetate is converted to **phosphoenolpyruvate (PEP)**, a key intermediate in **gluconeogenesis** (see p. 610). Oxaloacetate cannot cross the inner mitochondrial membrane but is converted to malate, which is transported on a specific carrier out of mitochondria and oxidized to oxaloacetate, which is then converted to PEP (see p. 611).

Citrate is exported from mitochondria into cytosol. **Citrate lyase** converts it to oxaloacetate and acetyl CoA, a precursor for synthesis of long-chain fatty acids and sterols. The oxaloacetate is rapidly reduced to malate, which is converted by **malic enzyme** to pyruvate and NADPH, a source of reducing equivalents for biosynthetic processes in cytosol. In addition, citrate is a regulatory effector of other metabolic pathways (see p. 602).

Anaplerotic Reactions Replenish Intermediates of the Tricarboxylic Acid Cycle

In its role in catabolism, the TCA cycle oxidizes acetyl CoA with release of two CO_2 molecules. Indeed, oxaloacetate, the acceptor of the acetate group, is regenerated during the cycle. However, metabolic pathways in all tissues remove intermediates of the cycle

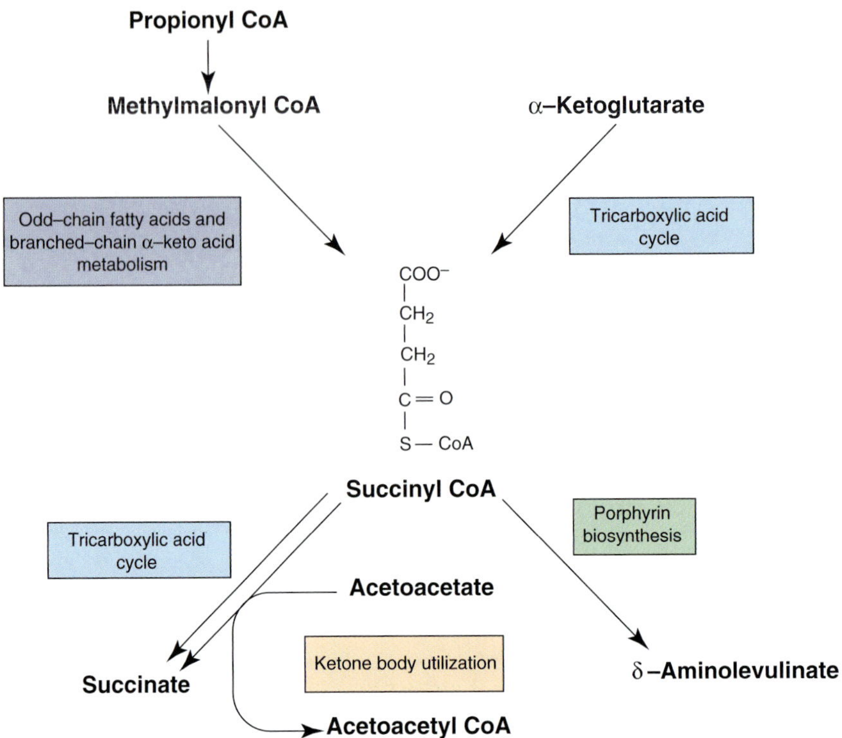

FIGURE 14.23

Sources and fates of succinyl CoA.

for biosynthetic pathways. Hence, in order to maintain a functional cycle, a source of four-carbon acids is required to replenish the loss of oxaloacetate. The reactions that supply four- or five-carbon intermediates to the cycle are called **anaplerotic** (meaning "filling up") reactions (Figure 14.24). The most important is catalyzed by **pyruvate carboxylase**, which converts pyruvate and CO_2 to oxaloacetate (Figure 14.25). The enzyme contains a biotin molecule linked to the ϵ-amino group of a lysine residue by an amide bond. The biotin binds CO_2 in the presence of ATP and Mg^{2+} ions and then transfers it to pyruvate as a carboxyl group (see p. 610). Levels of pyruvate carboxylase are high in both liver and nervous tissues, because these tissues experience a constant efflux of intermediates from the TCA cycle, which are used for gluconeogenesis in liver and neurotransmitter synthesis in nervous tissues.

Some amino acids are sources of four- or five-carbon intermediates. Glutamate is converted by glutamate dehydrogenase to α-ketoglutarate in mitochondria. Aspartate is converted to oxaloacetate by transamination, while valine and isoleucine are broken down to propionyl CoA that enters the TCA cycle as succinyl CoA. Amino acids derived

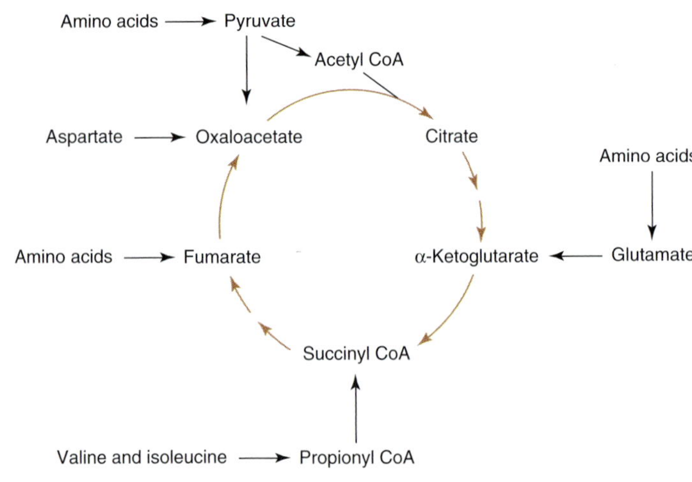

FIGURE 14.24

Anaplerotic reactions replenish intermediates of the TCA cycle.

from muscle breakdown become an important source of malate for gluconeogenesis during fasting (see p. 855).

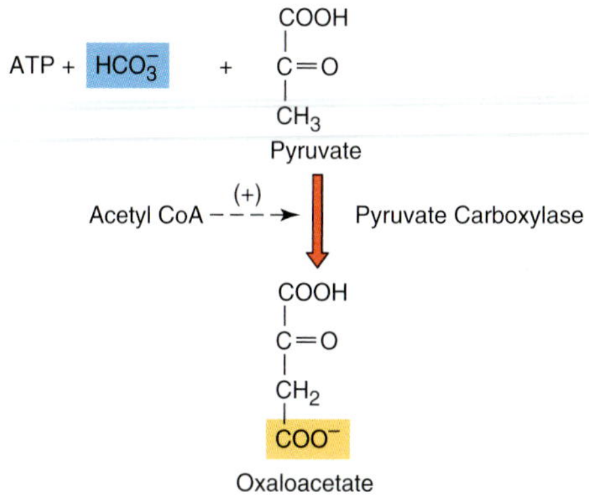

Activity of Tricarboxylic Acid Cycle Is Carefully Regulated

Various factors regulate the TCA cycle. First, the supply of acetyl units, whether derived from pyruvate (by glycolysis) or fatty acids (by β-oxidation), is crucial in determining the rate of the cycle. Regulation of the pyruvate dehydrogenase complex, the transport of fatty acids into mitochondria, and β-oxidation of fatty acids are effective determinants of cycle activity. Second, because the dehydrogenases of the cycle are dependent on a continuous supply of NAD^+ and FAD, their activities are very stringently controlled by the respiratory chain that oxidizes NADH and $FADH_2$. As discussed in Section 14.7, the activity of the respiratory chain is coupled obligatorily to generation of ATP in reactions of oxidative phosphorylation, a process called **respiratory control**. Consequently, activity of the TCA cycle is very dependent on the rate of ATP synthesis (and hence rate of electron transport), which is strongly affected by availability of ADP, phosphate, and O_2. Hence, an inhibitory agent or any metabolic condition that interrupts the supply of O_2, the continuous supply of ADP, or the source of reducing equivalents (e.g., substrate for the cycle) results in decreased activity of the TCA cycle. In general, these control mechanisms of the TCA cycle provide a coarse control of the cycle.

A variety of effector-mediated regulatory interactions between various intermediates or nucleotides and individual enzymes of the cycle have been postulated to exert a fine control of the cycle. Some of these effectors are shown in Figure 14.26. Note that the physiological relevance of many of these regulatory interactions has not been established.

Purified **citrate synthase** is inhibited by ATP, NADH, succinyl CoA, and long-chain acyl CoA derivatives; however, these effects have not been demonstrated under physiological conditions. The most probable means for regulating the citrate synthase reaction is the availability of the substrates acetyl CoA and oxaloacetate. As discussed above, very low concentrations of oxaloacetate (lower than the K_m for oxaloacetate on citrate synthase) are present in mitochondria.

The NAD^+-linked **isocitrate dehydrogenase**, often considered to be the key regulatory enzyme of the TCA cycle, is stimulated by ADP and AMP and is inhibited by

ATP + HCO_3^- + COOH–C=O–CH_3 (Pyruvate)

Acetyl CoA (+) → Pyruvate Carboxylase

COOH–C=O–CH_2–COO^- (Oxaloacetate)

FIGURE 14.25

Pyruvate carboxylase reaction.

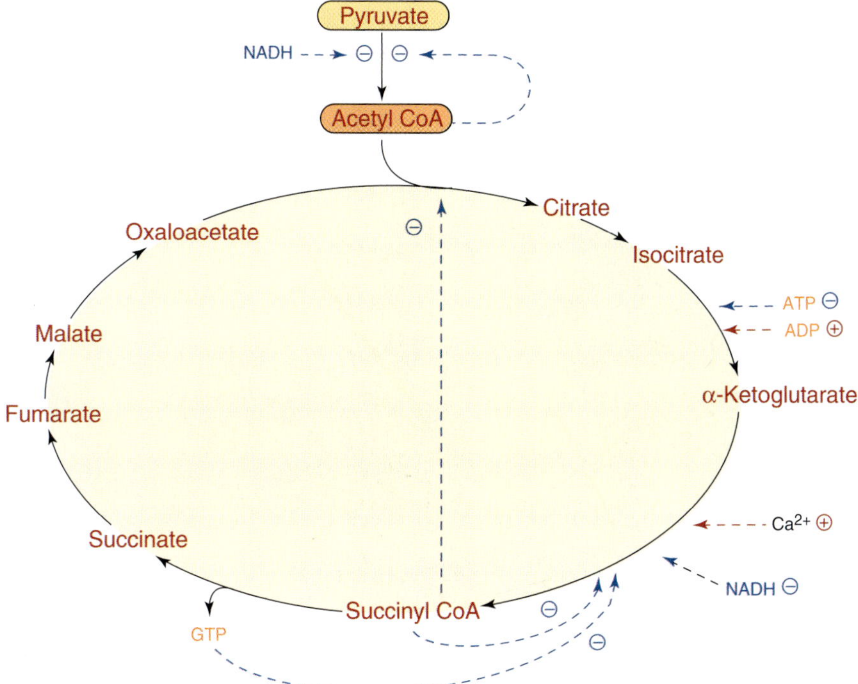

FIGURE 14.26

Examples of regulatory interactions in the TCA cycle.

ATP and NADH. Hence, under high-energy conditions (i.e., high ATP/ADP + P_i and high $NADH/NAD^+$ ratios), the activity of this dehydrogenase is inhibited. By contrast, during periods of low energy the activity of this enzyme and consequently of the TCA is stimulated. Respiratory control by the electron transport chain coupled to ATP synthesis thus regulates the TCA cycle at the NAD^+-linked isocitrate dehydrogenase step by affecting levels of ADP and NAD^+.

The **α-ketoglutarate dehydrogenase** complex is inhibited by ATP and GTP, NADH, and succinyl CoA, while Ca^{2+} activates the complex in certain tissues. Unlike the pyruvate dehydrogenase complex, the α-ketoglutarate dehydrogenase complex is not regulated by a protein kinase-mediated phosphorylation.

14.5 STRUCTURE AND COMPARTMENTATION BY MITOCHONDRIAL MEMBRANES

The final steps in breakdown of carbohydrates and fatty acids are located in mitochondria, where energy released during oxidation of NADH and $FADH_2$ is transduced into chemical energy of ATP by the process of **oxidative phosphorylation**. Consequently, mitochondria are often called the powerhouse of the cell. The role of a tissue in aerobic metabolic functions and its need for energy is reflected in the number and activity of its mitochondria (Figure 14.27). Cardiac muscle is highly aerobic, needing a constant supply of ATP. Approximately one-half of the cytoplasmic volume of cardiac cells consists of mitochondria, which contain numerous invaginations of the inner membrane called **cristae** and, consequently, a high concentration of the enzyme complexes of the electron transport chain. The liver is also highly aerobic with each mammalian hepatocyte containing 800–2000 mitochondria. By contrast, erythrocytes contain no mitochondria and obtain energy only from glycolysis.

Mitochondria have different shapes, depending on the cell type. In Figure 14.27, mitochondria from liver are nearly spherical, whereas those found in cardiac muscle are oblong or cylindrical and contain more numerous cristae than liver mitochondria.

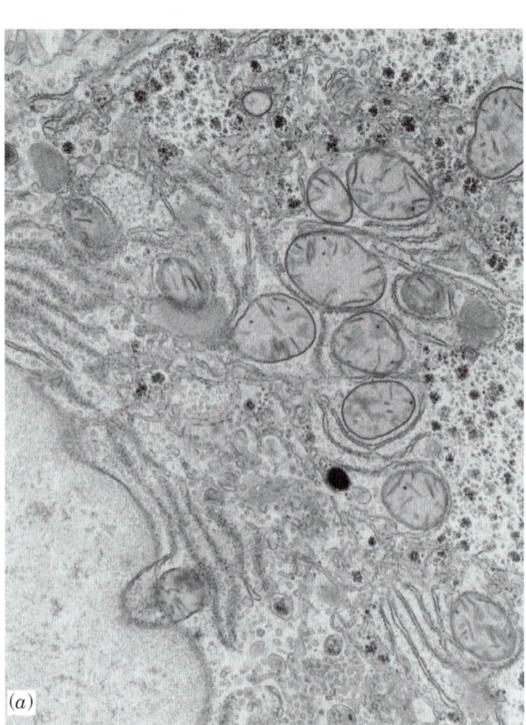

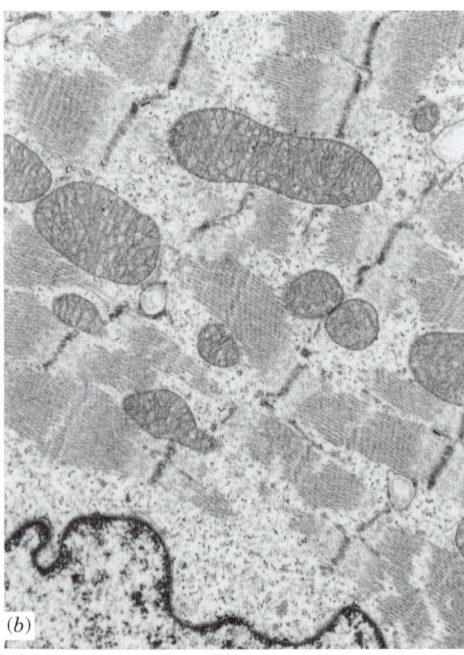

FIGURE 14.27

(a) Electron micrograph of mitochondria in hepatocytes from rat liver (×39,600). (b) Electron micrograph of mitochondria in muscle fibers from rabbit heart (×39,600). Part (*a*) courtesy of Dr. W. B. Winborn, Department of Anatomy, The University of Texas Health Science Center at San Antonio, and the Electron Microscopy Laboratory, Department of Pathology, The University of Texas Health Science Center at San Antonio. Part (*b*) courtesy of Dr. W. B. Winborn, Department of Anatomy, The University of Texas Health Science Center at San Antonio, and the Electron Microscopy Laboratory, Department of Pathology, The University of Texas Health Science Center at San Antonio.

Inner and Outer Mitochondrial Membranes Have Different Compositions and Functions

Mitochondria contain an **outer membrane** and a structurally and functionally more complex **inner membrane** (Figure 14.28); the space between is the **intermembrane space**. Enzymes involved in transfer of energy from the γ-phosphoryl bond of ATP, such as adenylate kinase, creatine kinase, and nucleoside diphosphate kinase, are located in the intermembrane space (Table 14.5). The outer membrane consists of about 30–40% lipid and 60–70% protein, with relatively few enzymatic or transport proteins. It is rich in the integral protein called **porin** (or **VDAC**, voltage-dependent anion channel) consisting of β-sheets, which form a channel that permits the passage through the membrane of particles of up to 10 kDa. Monoamine oxidase and kynurenine hydroxylase, of importance in nervous tissues for removal of neurotransmitters, are located on the outer surface of the outer membrane.

The inner membrane consists of 80% protein and is rich in unsaturated fatty acids. In addition, **cardiolipin** (diphosphatidylglycerol) is present in high concentrations. The enzyme complexes of electron transport and oxidative phosphorylation are located in this membrane as are various dehydrogenases and several transport systems, involved in transferring substrates, metabolic intermediates, and adenine nucleotides between cytosol and matrix. The inner membrane appears to be invaginated into folds or **cristae**, which increase the surface area (Figure 14.28).

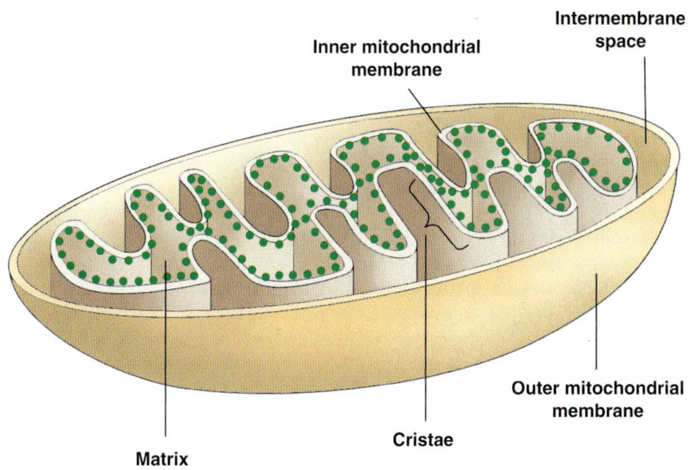

FIGURE 14.28
Diagram of submitochondrial compartments. Green spheres represent localization of the F_1 portion of ATP synthase on the inner mitochondrial membrane.

TABLE 14.5 Enzymes of Mitochondrial Subcompartments

Outer Membrane	*Intermembrane Space*	*Inner Membrane*	*Matrix*
Monoamine oxidase	Adenylate kinase	Succinate dehydrogenase	Pyruvate dehydrogenase complex
Kynurenine hydroxylase	Nucleoside diphosphate kinase	F_1F_0 ATP synthase	Citrate synthase
Nucleoside diphosphate kinase	Creatine kinase	NADH dehydrogenase	Isocitrate dehydrogenase
Phospholipase A		β-Hydroxybutyrate dehydrogenase	α-Ketoglutarate dehydrogenase complex
Fatty acyl-CoA synthetases		Cytochromes b, c_1, c, a, a_3	Aconitase
NADH: cytochrome-c reductase (rotenone-insensitive)		Carnitine: acyl-CoA transferase	Fumarase
			Succinyl-CoA synthetase
Choline phosphotransferase		Adenine nucleotide translocase	Malate dehydrogenase
		Mono-, di-, and tricarboxylate transporters	Fatty acid β-oxidation system
			Glutamate dehydrogenase
		Glutamate–aspartate transporters	Glutamate–oxaloacetate transaminase
		Glycerol 3-phosphate dehydrogenase	Ornithine transcarbamoylase
			Carbamoyl phosphate synthetase I
			Heme synthesis enzymes

The space inside the inner membrane, the **matrix**, contains the enzymes of the TCA cycle with the exception of succinate dehydrogenase which is bound to the inner membrane, enzymes for fatty acid oxidation, and some enzymes of porphyrin (see p. 834) and urea synthesis (see p. 752). In addition, mitochondrial DNA (mtDNA), ribosomes, and proteins necessary for transcription of mtDNA and translation of mRNA are located in the matrix.

14.6 ELECTRON TRANSPORT CHAIN

During the reactions of fatty acid oxidation and the TCA cycle, reducing equivalents derived from oxidation of substrates are transferred to NAD^+ and FAD (forming NADH and $FADH_2$) which are then oxidized by the **electron transport chain**, a system of electron carriers located in the inner membrane (Figure 14.29). In the presence of O_2, the electron transfer chain converts reducing equivalents into utilizable energy, as ATP, by oxidative phosphorylation. The complete oxidation of NADH and $FADH_2$ by the electron transport chain produces approximately 2.5 and 1.5 mol of ATP per mole of reducing equivalent transferred to O_2, respectively.

Oxidation–Reduction Reactions

The mitochondrial electron transport consists of a sequence of linked oxidation–reduction reactions. Such reactions transfer electrons from a suitable electron donor (**reductant**) to a suitable electron acceptor (**oxidant**). In some oxidation–reduction reactions, only electrons are transferred from reductant to oxidant (e.g., electron transfer between cytochromes),

$$\text{cytochrome } c\ (Fe^{2+}) + \text{cytochrome } a\ (Fe^{3+}) \rightarrow \text{cytochrome } c\ (Fe^{3+}) + \text{cytochrome } a\ (Fe^{2+})$$

whereas in others, both electrons and protons (hydrogen atoms) are transferred (e.g., electron transfer between NADH and FAD).

$$NADH + H^+ + FAD \rightarrow NAD^+ + FADH_2$$

An oxidant and its reductant form a **redox couple** or pair. The ease with which an electron donor (reductant) gives up its electrons to an electron acceptor (oxidant) is

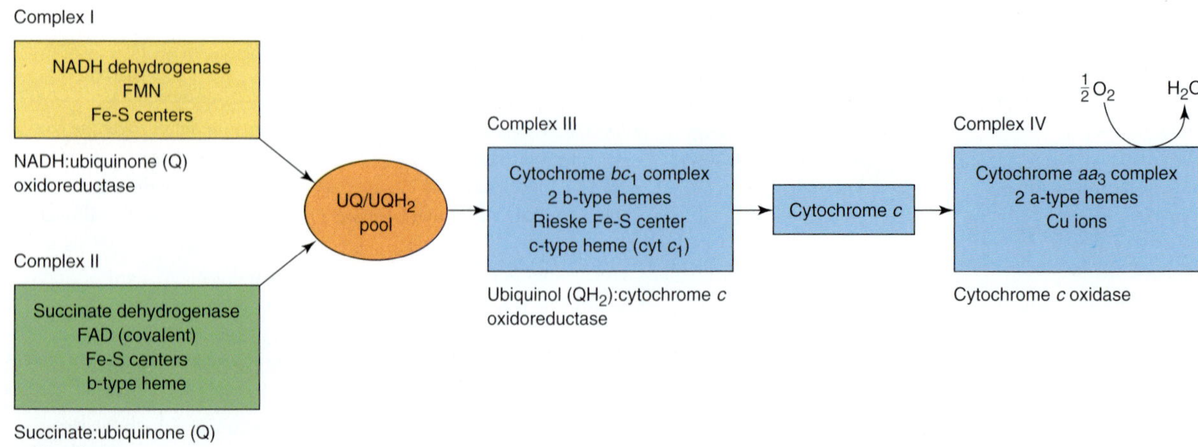

FIGURE 14.29

Overview of the complexes and pathways of electron transfer in mitochondrial electron transport chain.

TABLE 14.6 Standard Oxidation–Reduction Potentials for Various Biochemical Reactions

Oxidation–Reduction System	*Standard Oxidation–Reduction Potential E_0' (V)*
Acetate + $2H^+ + 2e^- \rightleftharpoons$ acetaldehyde	−0.60
$2H^+ + 2e^- \rightleftharpoons H_2$	−0.42
Acetoacetate + $2H^+ + 2e^- \rightleftharpoons$ β-hydroxybutyrate	−0.35
$NAD^+ + 2H^+ + 2e^- \rightleftharpoons NADH + H^+$	−0.32
Acetaldehyde + $2H^+ + 2e^- \rightleftharpoons$ ethanol	−0.20
Pyruvate + $2H^+ + 2e^- \rightleftharpoons$ lactate	−0.19
Oxaloacetate + $2H^+ + 2e^- \rightleftharpoons$ malate	−0.17
Coenzyme $Q_{ox} + 2e^- \rightleftharpoons$ coenzyme Q_{red}	+0.10
Cytochrome *b* (Fe^{3+}) + $e^- \rightleftharpoons$ cytochrome*b* (Fe^{2+})	+0.12
Cytochrome *c* (Fe^{3+}) + $e^- \rightleftharpoons$ cytochrome*c* (Fe^{2+})	+0.22
Cytochrome *a* (Fe^{3+}) + $e^- \rightleftharpoons$ cytochrome*a* (Fe^{2+})	+0.29
$\frac{1}{2}O_2 + 2H^+ + 2e^- \rightleftharpoons H_2O$	+0.82

expressed as the **oxidation–reduction potential** of the system. This is measured in volts as an **electromotive force** (emf) of a half-cell made up of an oxidation–reduction couple when compared to a standard reference half-cell (usually the hydrogen electrode reaction). The potential of the standard hydrogen electrode is set by convention at 0.0 V at pH 0.0; however, in biological systems where pH is 7.0, the reference hydrogen potential is −0.42 V. The potentials for a variety of important biochemical reactions are tabulated in Table 14.6. To interpret the data in the table, recall that the reductant of a redox pair with a large negative potential will give up its electrons more readily than redox pairs with smaller negative or positive potentials. Compounds with large negative potentials are strong reducing agents. By contrast, a strong oxidant (e.g., one characterized by a large positive potential) has a very high affinity for electrons and acts to oxidize compounds with more negative standard potentials.

The **Nernst equation** characterizes the relationship between standard oxidation–reduction potential of a redox pair (E_0'), observed potential (E), and ratio of concentrations of oxidant and reductant in the system:

$$E = E_0' + 2.3\,(RT/nf)\log([\text{oxidant}]/[\text{reductant}])$$

where E equals the observed potential and E_0' is the standard potential when all of reactants are present under standard conditions. R is gas constant of 8.3 J deg^{-1} mol^{-1}, T is absolute temperature in kelvin units (K), n is number of electrons being transferred, and f is the Faraday constant of 96,500 J V^{-1}.

From standard oxidation–reduction potentials of a diverse variety of biochemical reactions, one can predict direction of electron flow or transfer when more than two redox pairs are linked together by the appropriate enzyme. For example, Table 14.6 shows that the NAD^+–NADH pair has a standard potential of −0.32 V, and the pyruvate–lactate pair has a standard potential of −0.19 V. This means that electrons will flow from NAD^+–NADH to pyruvate–lactate as long as lactate dehydrogenase is present as indicated below:

$$\text{pyruvate} + \text{NADH} + H^+ \rightarrow \text{lactate} + NAD^+$$

Reducing equivalents are produced in NAD^+- and FAD-linked dehydrogenase reactions, which have standard potentials at or close to that of NAD^+–NADH. The electrons are subsequently transferred through the electron-transfer chain, which has as its terminal acceptor the O_2–water couple with a standard redox potential of +0.82 V.

Free-Energy Changes in Redox Reactions

Differences in oxidation–reduction potentials between two redox pairs are similar to free-energy changes in chemical reactions, in that both quantities depend on concentration of reactants and products of the reaction and the following relationship exists:

$$\Delta G^{0'} = -nf\Delta E_0'$$

Using this expression, free energy change for electron-transfer reactions can be calculated if potential difference between two oxidation–reduction pairs is known. Hence, for the mitochondrial electron-transfer chain in which electrons are transferred between the NAD^+–NADH couple ($E_0' = -0.32$ V) and the $1/2O_2$–H_2O couple ($E_0' = +0.82$ V), the free-energy change for this process can be calculated:

$$\Delta G^{0'} = -nf\Delta E_0' = -2 \times 96.5\ \text{kJV}^{-1} \times 1.14\ \text{V}$$

$$\Delta G^{0'} = -219\ \text{kJ mol}^{-1}$$

where 96.5 is the Faraday constant in kJ V^{-1} and n is number of electrons transferred; for example, in the case of NADH $\rightarrow$ O_2, $n = 2$. The free energy available from the potential span between NADH and O_2 in the electron-transfer chain is capable of generating more than enough energy to synthesize three molecules of ATP per two reducing equivalents or two electrons transported to O_2. In addition, because of the negative sign of the free energy available in the electron transfer, this process is exergonic and proceeds, if the necessary enzymes are present.

Mitochondrial Electron Transport Is a Multicomponent System

The final steps in the overall oxidation of foodstuffs—carbohydrates, fats, and amino acids—result in formation of NADH and $FADH_2$ in the matrix. The electron transport chain oxidizes these reduced cofactors by transferring electrons in a series of steps to O_2, the terminal electron acceptor, while capturing the free energy of the reactions to drive the synthesis of ATP (Figure 14.29). During removal of electrons from the coenzymes, protons are pumped from the matrix to the inner-membrane space to form an electrochemical gradient across the inner membrane, which provides energy for synthesis of ATP. The carriers that transfer electrons from NADH to O_2 have standard redox potentials that span the range from -0.32 V that of the most electronegative electron donor NADH to $+0.82$ V the most electropositive electron acceptor O_2 (Figure 14.30). The mitochondrial electron carriers, however, are not organized in a linear arrangement but are grouped into four large complexes (complexes I–IV) that catalyze different partial reactions of the electron transport chain (Figure 14.29).

Complex I, NADH–ubiquinone oxidoreductase, catalyzes the transfer of electrons from NADH to **ubiquinone (UQ)** or **coenzyme Q (CoQ)**; **complex II**, succinate–ubiquinone oxidoreductase or succinate dehydrogenase, transfers electrons from succinate to coenzyme Q; **complex III**, or the **cytochrome bc_1 complex**, ubiquinol–cytochrome c reductase, transfers electrons from ubiquinol (reduced form of ubiquinone denoted as **$CoQH_2$** or **UQH_2**) to **cytochrome c**; and **complex IV**,

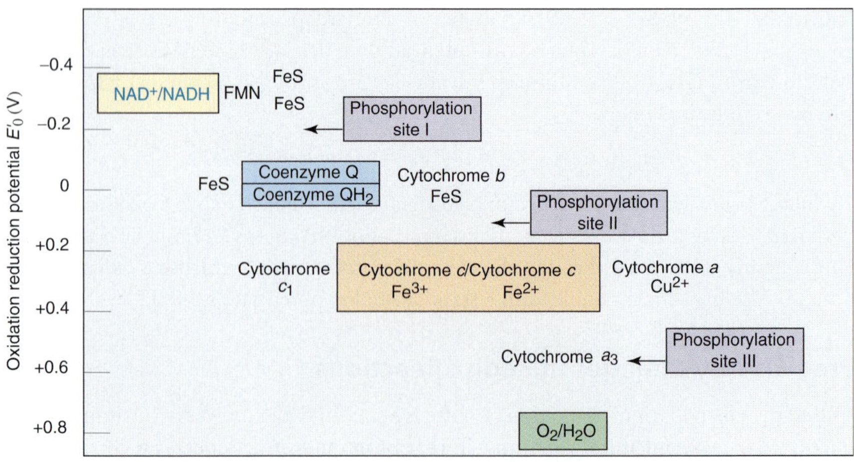

FIGURE 14.30
Oxidation–reduction potentials of the mitochondrial electron transport chain carriers.

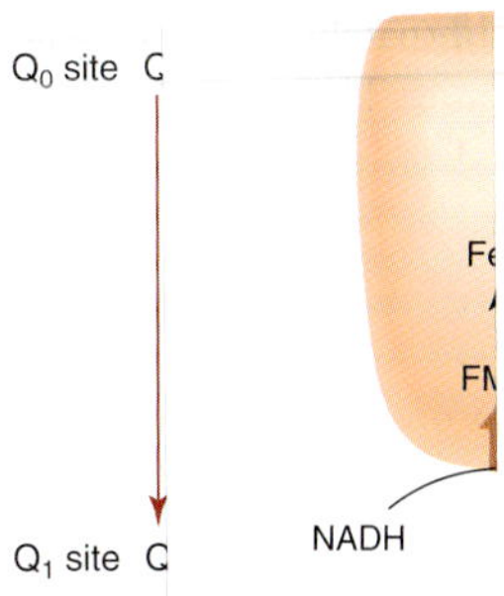

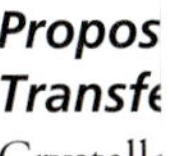

cytochrome *c* oxidase, transfers electrons from cytochrome c to O_2 (Figure 14.29). Another complex, the **ATP synthase**, or **complex V**, uses energy of the electrochemical gradient for synthesis of ATP. Complexes I–IV consist of (a) electron carriers that include **flavoproteins**, which contain tightly bound FMN or FAD and can transfer one or two electrons, (b) the heme-containing proteins **cytochromes** (cytochromes b, c_1, c, a, and a_3), which transfer one electron from Fe^{2+} of heme, (c) **iron–sulfur proteins**, which contain bound inorganic Fe and S and transfer one electron, and (d) **copper** in complex IV (cytochrome c oxidase), which transfers one electron. UQ participates in one or two electron transfer reactions.

Complex I: NADH–Ubiquinone Oxidoreductase

Complex I, the most complicated and in mammals, consists of at least 40 different polypeptides with a total mass of approximately 1 M Da. Complex I, often called NADH dehydrogenase, transfers electrons from NADH to ubiquinone (coenzyme Q). The oxidation of NADH starts with transfer of two electrons and two protons from NADH + H^+ to **FMN**, **flavin mononucleotide**, containing the riboflavin moiety (vitamin B_2), which is tightly bound to one subunit (Figure 14.31).

$$\text{NADH} + \text{H}^+ + \text{FMN} \rightarrow \text{NAD}^+ + \text{FMNH}_2$$

The electrons are then transferred one at a time via a series of FeS centers, of 2Fe2S and 4Fe4S types (Figure 14.32), to ubiquinone. In this way, FMN is a two-electron acceptor from NADH and one electron donor to the FeS centers. This ability results from the stable semiquinone form of FMN. Ubiquinone is also a one- or two-electron acceptor because it forms of a stable semiquinone (Figure 14.33). In addition to its hydrophilic quinone portion, ubiquinone also has a long hydrophobic side chain consisting of 10 isoprene units, which is buried in the membrane lipid bilayer. Ubiquinone and ubiquinol are freely diffusible in the membrane and can act to transfer electrons from complexes I and II to complex III.

During transfer of two electrons to ubiquinone by complex I, four protons are also pumped from the matrix side (**N** for negative face) to the cytosolic side (**P** for positive face) of the inner membrane. The energy released during the oxidative reactions is conserved by this movement of protons across the membrane. Little is known of the mechanism of this proton transfer in complex I. Figure 14.34 provides a schematic representation of these events.

Complex II: Succinate–Ubiquinone Oxidoreductase

Complex II, better known as **succinate dehydrogenase**, consists of a 70-kDa subunit that contains FAD covalently bound to a histidine residue, a 30-kDa subunit that contains three iron–sulfur centers, and two small hydrophobic proteins. In the oxidation

FAD

$FMNH_2$

FIGURE 14.31
Structures of flavin adenine dinucleotide (FAD) and flavin mononucleotide (FMN).

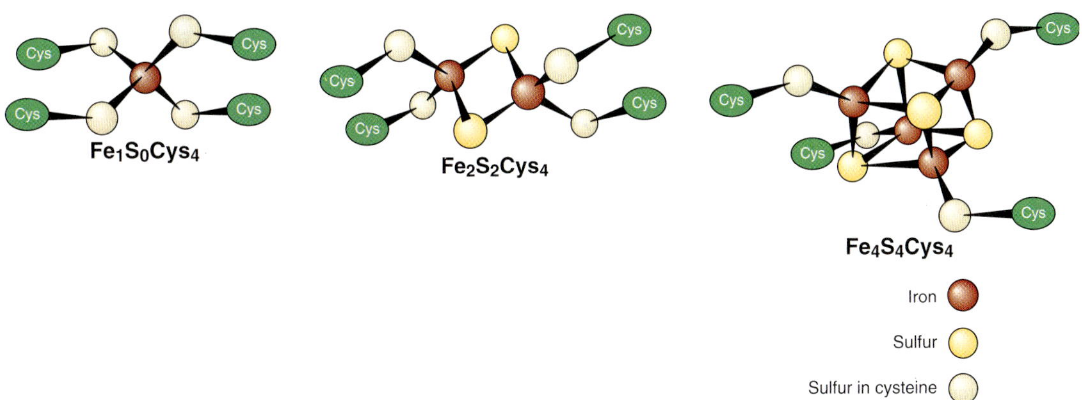

FIGURE 14.32
Structures of iron–sulfur centers. Yellow inorganic sulfur; gray, sulfur in cysteine; and red, iron.

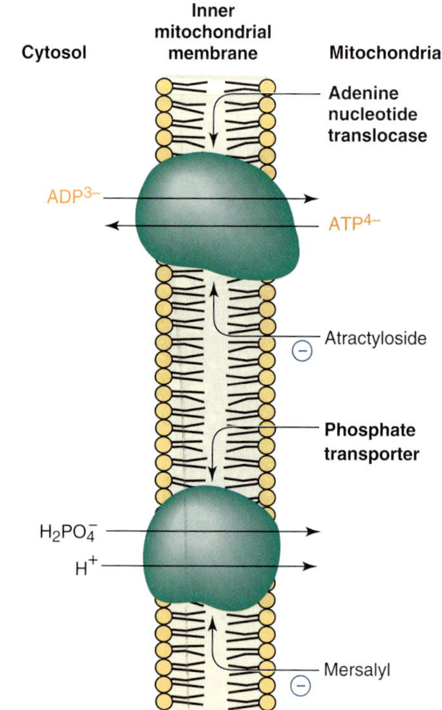

FIGURE 14.55

The adenine nucleotide translocase and phosphate transporter.

Transport of Adenine Nucleotides and Phosphate

Continued synthesis of ATP in the mitochondrial matrix requires that cytosolic ADP formed during energy-consuming reactions be transported back across the inner membrane into the matrix for conversion to ATP. Similarly, newly synthesized ATP must be transported back across the inner membrane into the cytosol to meet the energy needs of the cell. This exchange of the highly charged hydrophilic adenine nucleotides is catalyzed by a very specific **adenine nucleotide translocase** located in the inner membrane (Figure 14.55). The adenine nucleotide translocase, a homodimer of 30-kDa subunits, catalyzes a 1:1 exchange of ATP for ADP. The presence of one nucleotide-binding site on the transporter suggests that the enzyme alternately faces the matrix or the inter membrane space during the transport process. Newly synthesized ATP is bound to the translocase in the matrix, which then changes its conformation to face the cytosol where the ATP is released in exchange for an ADP. The translocase then changes conformation again to bring the nucleotide-binding site containing ADP back to face the matrix. The translocase favors outward movement of ATP and inward movement of ADP despite observations that both nucleotides bind equally well to the binding site. The explanation for this is that at pH 7, ADP has three negative charges while ATP has four. Hence, the exchange of one ATP for one ADP results in net outward movement of one negative charge, which is equivalent to import of one proton. The membrane potential established during electron transfer is positive outside, which would favor outward transport of more negatively charged ATP over that of ADP. The adenine nucleotide translocase is present in high concentrations, up to 14% of total protein, in the inner membrane. Hence, it is unlikely that transport of adenine nucleotides across the inner mitochondrial membrane is ever rate-limiting for ATP synthesis.

A second transporter essential for oxidative phosphorylation is the **phosphate transporter**, which transports cytosolic phosphate plus a proton into the matrix (Figure 14.55). This symport also depends on the proton gradient as phosphate and protons are transported in a 1:1 ratio. Transport of ADP and phosphate requires a significant fraction of the energy present in the electrochemical gradient produced during electron transfer. Thus, the protonmotive force provides energy for ATP synthesis by ATP synthase as well as for uptake of the two required substrates.

Substrate Shuttles Transport Reducing Equivalents Across the Inner Mitochondrial Membrane

The nucleotides involved in cellular oxidation–reduction reactions (e.g., NAD^+, NADH, $NADP^+$, NADPH, FAD, and $FADH_2$) and CoA and its derivatives are not transported across the inner mitochondrial membrane. Thus, to transport reducing equivalents (e.g., protons and electrons) from cytosol to matrix or the reverse, substrate shuttle mechanisms are required.

Two substrate transport shuttles are shown in Figure 14.56. The malate–aspartate shuttle and α-glycerol phosphate shuttle are employed in various tissues to translocate reducing equivalents from cytosol to matrix, for oxidation to yield energy. Their operation requires that appropriate enzymes be localized on either side of the membrane and that appropriate transporters be present in the mitochondrial inner membrane.

In the **glycerol–phosphate shuttle**, two glycerol phosphate dehydrogenases, one in the cytosol and the other on the outer face of the inner mitochondrial membrane, are involved. NADH produced in the cytosol is used to reduce dihydroxyacetone phosphate to glycerol 3-phosphate by the cytosolic isozyme. The glycerol 3-phosphate in turn is oxidized by the mitochondrial isozyme, a flavoprotein, to produce dihydroxyacetone phosphate and $FADH_2$ that is oxidized by the electron transport chain.

The **malate–aspartate shuttle** operates on the same principle. NADH in the cytosol reduces oxaloacetate to malate, which enters mitochondria on the malate/α-ketoglutarate transporter. This malate is readily oxidized by mitochondrial malate dehydrogenase to oxaloacetate and NADH, which is then oxidized by the electron transport chain. The oxaloacetate produced is then converted to aspartate by mitochondrial

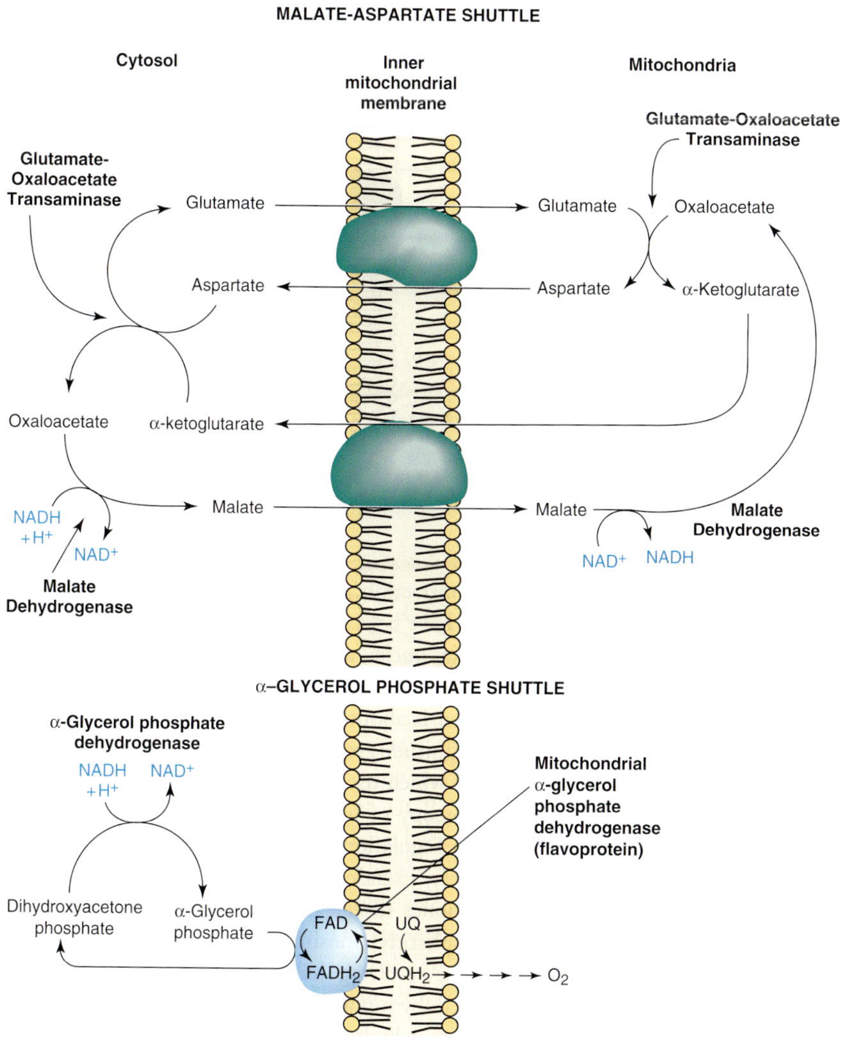

FIGURE 14.56
Transport shuttles for reducing equivalents.

aspartate aminotransferase and can then cross the membrane via the aspartate–glutamate transporter, where cytosolic aspartate amino transferase converts it to oxaloacetate. In contrast to the glycerol phosphate shuttle, the malate–aspartate shuttle is reversible and serves as a mechanism to bring reducing equivalents out of mitochondrial matrix to cytosol.

Acetyl Units Are Transported as Citrate

The inner mitochondrial membrane does not have a transporter for acetyl-CoA, but acetyl groups are transferred from the mitochondria to cytosol, where they are required for fatty acid or sterol biosynthesis (Figure 14.57). Intra-mitochondrial acetyl-CoA is converted to citrate by citrate synthase of the TCA cycle. Citrate is then exported to cytosol by a tricarboxylate transporter in exchange for malate. Cytosolic citrate is cleaved to acetyl CoA and oxaloacetate at the expense of an ATP by **ATP-citrate lyase** (see p. 672). Substrate shuttle mechanisms are involved in movement of appropriate substrates and intermediates in both directions across the inner membrane during periods of active gluconeogenesis (see p. 610) and ureagenesis (see p. 752) by the liver.

Mitochondria Have a Specific Calcium Transporter

In most mammalian tissues, mitochondria have a transport system for translocating Ca^{2+} across the inner membrane. The distribution/redistribution of cellular Ca^{2+}

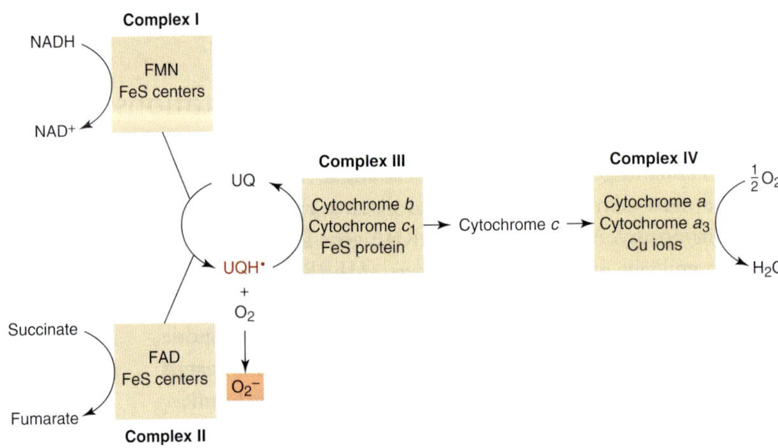

FIGURE 14.63

Generation of superoxide anions by mitochondrial electron transfer chain. The semiquinone formed during the two electron reduction of ubiquinone by the iron–sulfur centers of both complexes I and II can transfer an electron to oxygen to form the superoxide anion. By contrast, the binuclear center of cytochrome *c* oxidase prevents release of intermediates in the reduction of oxygen.

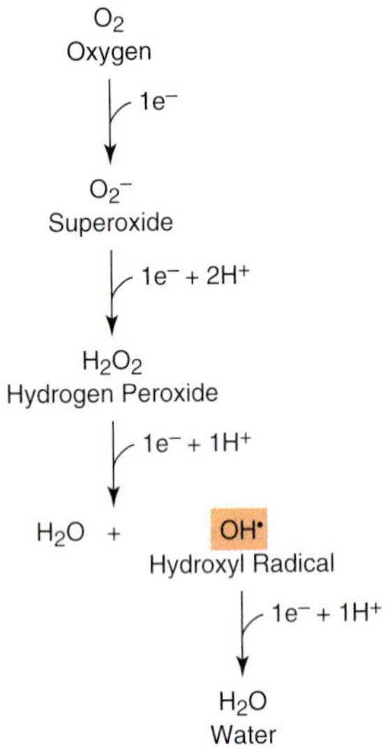

O_2 Oxygen

↓ $1e^-$

O_2^- Superoxide

↓ $1e^- + 2H^+$

H_2O_2 Hydrogen Peroxide

↓ $1e^- + 1H^+$

H_2O + $OH^\bullet$ Hydroxyl Radical

↓ $1e^- + 1H^+$

H_2O Water

FIGURE 14.61

One-electron steps in reduction of oxygen, leading to formation of reactive oxygen species superoxide, hydrogen peroxide, and hydroxyl radical.

Fenton Reaction

$$Fe^{2+} + H_2O_2 \longrightarrow Fe^{3+} + OH^\bullet + OH^-$$

Haber–Weiss Reaction

$$O_2^- + H_2O_2 \xrightarrow{H^+} O_2 + H_2O + OH^\bullet$$

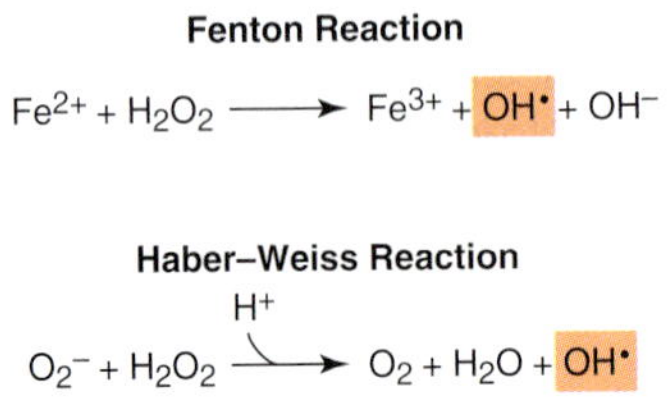

FIGURE 14.62

The Fenton and Haber–Weiss reactions for formation of toxic hydroxyl radical.

Production of Reactive Oxygen Species

While oxidative processes in cells generally result in transfer of electrons to O_2 to form water without release of intermediates, a small number of **oxygen radicals** are inevitably formed due to leakage in electron transfer reactions. The major intracellular source of oxygen radicals is the mitochondrial electron transport chain, where superoxide is produced by transfer of one electron to O_2 from the stable semiquinone produced during reduction of ubiquinone by complexes I and II (Figure 14.63). Superoxide can also be produced by transfer of an electron from a flavin such as FMN. The reactive oxygen species produced in mitochondria include superoxide, **hydrogen peroxide**, and **hydroxyl radical**. Toxic oxygen species are also produced in peroxisomes in which long-chain fatty acids and other compounds are oxidized by transfer of two electrons from $FADH_2$ to O_2 with formation of hydrogen peroxide, which is readily converted to hydroxyl radical (Figure 14.63). The cytochrome P450 system localized in endoplasmic reticulum can also produce oxygen radicals.

Oxygen radicals are also produced in certain cells during inflammation due to bacterial infection. To combat microbial infections, phagocytes produce toxic oxygen radicals in a process known as the respiratory burst. The phagocytes (Figure 14.64) then kill the engulfed bacteria. In an acute infection, production of oxygen radicals and killing of bacteria are efficient processes; however, in prolonged infections, phagocytes tend to die, releasing toxic oxygen radicals that affect surrounding cells.

Cosmic radiation, ingestion of chemicals and drugs, as well as smog can lead to formation of reactive oxygen species. Damage from reactive oxygen species often occurs during perfusion of tissues with solutions containing high concentrations of O_2 as happens in patients who have suffered an ischemic episode in which localized O_2 levels are lowered due to blockage of an artery and have then undergone thrombolytic or other procedures to remove the blockage (see Clin. Corr. 14.7).

Damage Caused by Reactive Oxygen Species

Reactive oxygen species cause damage to all major classes of macromolecules in cells. The phospholipids of plasma and organelle membranes are subject to **lipid peroxidation**, a free radical chain reaction initiated by removal of hydrogen from a polyunsaturated fatty acid by hydroxyl radical. The resulting lipid radicals then react with O_2 to form lipid peroxy radicals and lipid peroxide along with malondialdehyde, which is water-soluble and can be detected in blood. The effect of lipid peroxidation in humans is exemplified

CLINICAL CORRELATION 14.7
Ischemia/Reperfusion Injury

The occlusion of a major coronary artery during myocardial infarction results in ischemia or lowered oxygen supply. Consequently, the mitochondrial electron transport chain is inhibited with a concomitant decrease of intracellular levels of ATP and creatine phosphate. As cellular ATP levels diminish, anaerobic glycolysis is activated in an attempt to maintain normal cellular functions. Glycogen levels are rapidly depleted and lactic acid levels in the cytosol increase, lowering the intracellular pH. Despite the direct consequences of ischemia, the damage to the affected tissues appears to become more severe when oxygen is reintroduced (or reperfused) and oxygen radicals including superoxide (O_2^-), hydrogen peroxide and hydroxyl radical ($OH\cdot$) are formed. One recent study observed the formation of oxygen radicals including peroxynitrite ($ONOO^-$) produced from nitric oxide (NO) and superoxide during acute reperfusion of the ischemic heart. Peroxynitrite has been shown to contribute to the poor recovery of mechanical function in ischemic hearts, while antioxidants such as glutathione that scavenge the peroxynitrite radicals protect against damage to mechanical function. Reperfusion of the ischemic heart may lead to activation of leukocytes, which are mediators of the inflammatory process and lead to further tissue injury.

Myocardial ischemia/reperfusion represents a clinical problem associated with thrombolysis, angioplasty, and coronary bypass surgery. Injuries to the myocardium due to ischemia/reperfusion include cardiac contractile dysfunction, arrhythmias, and irreversible myocyte damage. Ischemia may also arise during surgery, especially during transplantation of tissues. The postulated role of oxygen radicals in ischemic/reperfusion injury is strengthened by the observations that antioxidants protect against reperfusion injury to ischemic tissue. Currently, active investigations of methods to protect against reperfusion injury are underway using animal models. The increased use of invasive procedures in clinical medicine indicates the importance of developing methods to protect against ischemic/reperfusion injury.

Source: Cheung, P.Y., Wang, W., and Schulz, R. Glutathione protects against myocardial ischemia–reperfusion injury by detoxifying peroxynitrite. *J. Mol. Cell. Cardiol.* 32:1669, 2000.

by the brown spots commonly observed on hands of the elderly. These "age" spots contain the pigment lipofuscin, which is probably a mixture of cross-linked lipids and products of lipid peroxidation, which accumulate over the course of a lifetime. One significant consequence of lipid peroxidation is increased membrane permeability, leading to an influx of Ca^{2+} and other ions with subsequent swelling of the cell. Similar increases in permeability of organelle membranes may result in maldistribution of ions and cause intracellular damage. For example, accumulation of excessive amounts of Ca^{2+} in mitochondria may trigger apoptosis.

Proline, histidine, arginine, cysteine, and methionine are susceptible to attack by hydroxyl radicals with subsequent fragmentation of proteins, cross-linking, and aggregation. Proteins damaged by oxygen radicals may be targeted for digestion by intracellular proteases.

The most important consequence of oxygen radicals is damage to mitochondrial and nuclear DNA, resulting in mutations. The nonspecific binding of **ferrous ions** (Fe^{2+}) to DNA may result in localized formation of hydroxyl radicals that attack individual bases and cause strand breaks. Mitochondrial DNA is more susceptible to damage, since the electron transport chain is a major source of toxic oxygen radicals. Nuclear DNA is protected from permanent damage by a protective coat of histones as well as by active and efficient mechanisms for DNA repair. Damage to mtDNA generally results in mutations that affect energy production. The symptoms in affected individual are manifest in energy requiring processes such as muscle contraction. An example of the consequences of a somatic mutation in the mitochondrial gene for cytochrome b that may have been caused by oxygen radicals is presented in Clinical Correlation 14.7.

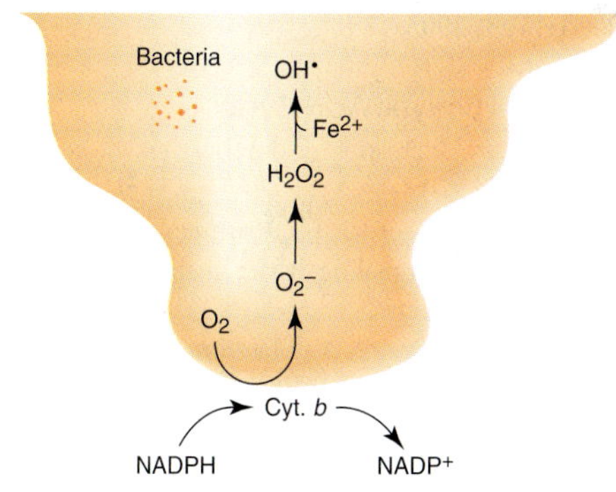

FIGURE 14.64

Respiratory burst in phagocytes. An electron transfer chain involving a unique cytochrome b transfers electrons from NADPH to oxygen with formation of superoxide anion. Superoxide is converted to the hydroxyl radical that kills bacteria subsequently engulfed by phagocytes.

Cellular Defenses Against Reactive Oxygen Species

Cells that live in an aerobic environment have developed multiple ways to remove reactive oxygen species and thus protect themselves against their deleterious effects. Mammals have three different isozymes of **superoxide dismutase** that catalyze conversion of superoxide to hydrogen peroxide (Figure 14.65). The cytosolic form of superoxide

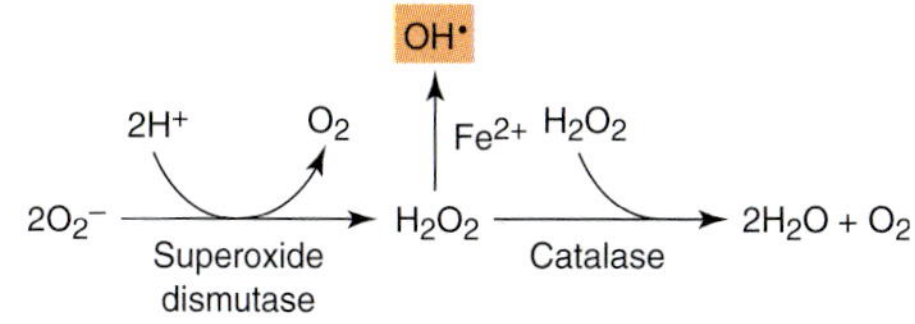

FIGURE 14.65

Superoxide dismutase and catalase protect cells by removing superoxide and hydrogen peroxide.

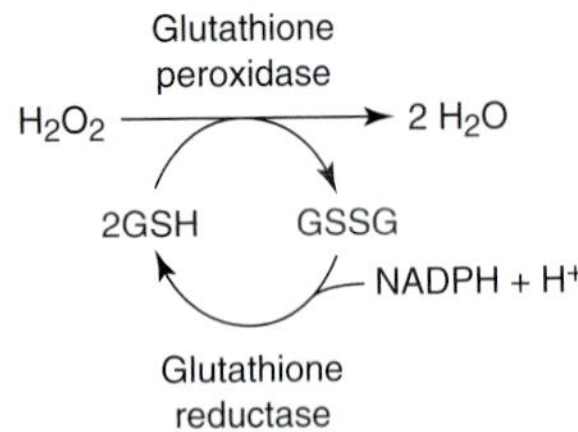

FIGURE 14.66

Glutathione peroxidase removes hydrogen peroxide as well as lipid peroxides. Electrons are transferred to hydrogen peroxide from the sulfhydryl groups of reduced glutathione (GSH) to form water and oxidized glutathione (GSSG). Glutathione reductase then reduces GSSG to GSH with NADPH as the reducing agent.

dismutase contains Cu/Zn at its active site, as does the extracellular enzyme; however, the mitochondrial enzyme contains Mn at its active site. Hydrogen peroxide is removed by **catalase**, a heme-containing enzyme present in highest concentration in peroxisomes and to a lesser extent in mitochondria and cytosol.

Glutathione peroxidase catalyzes reduction of both hydrogen peroxide and lipid peroxides (Figure 14.66). This selenium-containing enzyme uses the sulfhydryl groups of glutathione (GSH) as a hydrogen donor with formation of oxidized or disulfide form of glutathione (GSSG). **Glutathione reductase** converts the disulfide form back to the sulfhydryl form using NADPH produced in the pentose phosphate pathway as an electron donor. Protection against reactive oxygen species may also be gained by ingestion of oxygen scavengers such as vitamins C and E and β-carotene.

BIBLIOGRAPHY

Energy-Producing and Energy-Utilizing Systems

Hanson, R. W. The role of ATP in metabolism. *Biochem. Educ.* 17:86, 1989.

Sources and Fates of Acetyl Coenzyme A

Harris, R. A., Bowker-Kinley, M. M., Huang, B., and Wu, P. Regulation of the activity of the pyruvate dehydrogenase complex. *Adv. Enzyme Regulation* 42:249, 2002.

Holness, M. J. and Sugden, M. C. Regulation of pyruvate dehydrogenase complex activity by reversible phosphorylation. *Biochem. Soc. Trans.* 31:1143, 2003.

The Tricarboxylic Acid Cycle

Krebs, H. A. The history of the tricarboxylic acid cycle. *Perspect. Biol. Med.* 14:154, 1970.

Hajnoczky, G., Csordas, G., Krishnamurthy, R., and Szalai, G. Mitochondrial calcium signaling driven by the IP_3 receptor. *J. Bioenerg. Biomembr.* 32; 15, 2000.

Ovadi, J. and Srere, P. A. Macromolecular compartmentation and channeling. *Int. Rev. Cytol.* 192:255, 2000.

Owen, O. E., Kalhan, S. C., and Hanson, R. W. The key role of anaplerosis and cataplerosis for citric acid cycle function. *J. Biol. Chem.* 277:30409, 2002.

Structure and Compartmentation of Mitochondrial Membranes

Scheffler, I. E. *Mitochondria.* New York: Wiley-Liss, 1999.

Electron Transport

Hunte, C., Pasldottir, H., and Trumpower, B. L. Protonmotive pathways and mechanisms in the cytochrome *bc*$_1$ complex. *FEBS Lett.* 545:39, 2003.

Iwata, M., Bjorkman, J., and Iwata, S. Conformational change of the Rieske [2Fe2S] protein in cytochrome bc$_1$ complex. *J. Bioenerg. Biomembr.* 31:169, 1999.

Mills, D. A. and Ferguson-Miller, S. Understanding the mechanism of proton movement liked to oxygen reduction in cytochrome *c* oxidase: Lessons from other proteins. *FEBS Lett.* 545:47, 2003.

Richter, O.-M. H. and Ludwig, B. Cytochrome *c* oxidase—structure, function, and physiology of a redox-driven molecular machine. *Rev. Physiol. Biochem. Pharmacol.* 147:47, 2003.

Xia, D., et al. Crystal structure of the cytochrome bc$_1$ complex from bovine heart mitochondria. *Science* 277:60, 1997.

Yagi, T. and Matsuno-Yagi, A. The proton-translocating NADH-Quinone oxidoreductase in the respiratory chain: The secret unlocked. *Biochemistry* 42:2266, 2003.

Zhang, Z., Berry, E. A., Huang, L. S., and Kim, S. H. Mitochondrial cytochrome bc1 complex *Subcell. Biochem.* 35:541, 2000.

Oxidative Phosphorylation

Boyer, P. D. A research journey with ATP synthase. *J. Biol. Chem.* 277:39045, 2002.

Mitchell, P. Vectorial chemistry and the molecular mechanism of chemiosmotic coupling: power transmission by proticity. *Biochem. Soc. Trans.* 4:399, 1976.

Rich, P. Chemiosmotic coupling: The cost of living. *Nature* 421:583, 2003.

Sambongi, Y., et al. Mechanical Rotation of the c subunit oligomer in ATP Synthase (F_0F_1): Direct observation. *Science* 286:1722, 1999.

Stock, D., Gibbons, C. Arechaga, I., Leslie, A. G., and Walker J. E. The rotary mechanism of ATP synthase, *Curr. Opin. Struct. Biol.* 10:672, 2000.

Weber, J. and Senior A. E. ATP synthesis driven by proton transport in F_1F_O–ATP synthase. *FEBS Lett.* 545:61, 2003.

Mitochondrial Transport Systems

Palmieri, F. The mitochondrial transporter family (SLC25): Physiological and pathological implications. *Pflugers Arch.* 447:68909, 2003.

Pebay-Peyroula, E., Dahout-Gonzalez, C., Kahn, R., Trezeguet, V., Lauquin, J.-M., and Brandolin, G., Structure of mitochondrial ADP/ATP carrier in complex with carboxyatractyloside. *Nature* 426:39, 2003.

Rousset, S., Alves-Guerra, M.-C., Mozo, J., Miroux, B., Cassard-Doulcier, A.-M., Bouillaud, F., and Ricquier, D. The biology of mitochondrial uncoupling proteins. *Diabetes* 53(Suppl.1):S130, 2004.

Mitochondrial Genes and Mitochondrial Diseases

DiMauro, S. and Schon, E. A. Mitochondrial respiratory-chain diseases. *N. Engl. J. Med.* 348:2656, 2003.

Schon, E. A. and Manfredi, G. Neuronal degeneration and mitochondrial dysfunction, *J. Clin. Invest.* 111:303, 2003.

Wallace, D. C. Mitochondrial diseases in mice and man. *Science* 283:1482, 1999.

Reactive Oxygen Species

Hansford, R., Tsuchiya, N., and Pepe, S. Mitochondria in heart ischemia and aging. *Biochem. Soc. Symp.* 66:141, 1999.

Raha, S. and Robinson, B. H. Mitochondria, oxygen free radicals, disease and ageing. *Trends Biochem. Sci.* 25:502, 2000.

Sastre, J., Pallardo, F. V. and Vina J., The role of mitochondrial oxidative stress in aging. *Free Radical Biol. Med.* 35:1, 2003.

Turrens, J. F. Mitochondrial formation of reactive oxygen species. *J. Physiol.* 552:335, 2003.

QUESTIONS | CAROL N. ANGSTADT

Multiple Choice Questions

1. A bond may be "high energy" for any of the following reasons *except*:
   A. products of its cleavage are more resonance stabilized than the original compound.
   B. the bond is unusually stable, requiring a large energy input to cleave it.
   C. electrostatic repulsion is relieved when the bond is cleaved.
   D. a cleavage product may be unstable, tautomerizing to a more stable form.
   E. the bond may be strained.

2. At which of the following enzyme-catalyzed steps of the tricarboxylic acid cycle does net incorporation of the elements of water into an intermediate of the cycle occur?
   A. aconitase.
   B. citrate synthase.
   C. malate dehydrogenase.
   D. succinate dehydrogenase.
   E. succinyl CoA synthase.

3. All of the following tricarboxylic acid cycle intermediates may be added or removed by other metabolic pathways *except*:
   A. citrate.
   B. fumarate.
   C. isocitrate.
   D. α-ketoglutarate.
   E. oxaloacetate.

4. The inner mitochondrial membrane contains a transporter for:
   A. NADH.
   B. acetyl CoA.
   C. GTP.
   D. ATP.
   E. NADPH.

5. The chemiosmotic hypothesis involves all of the following *except*:
   A. a membrane impermeable to protons.
   B. electron transport by the respiratory chain pumping protons out of the mitochondria.
   C. proton flow into the mitochondria dependent on the presence of ADP and P_i.
   D. reversible activity of ATPase activity.
   E. strict regulation of only proton transport is with other ions diffusing freely.

6. ATP synthase (also known as Complex V) consists of two domains, F_1 and F_O:
   A. F_1 and F_O are both integral membrane protein complexes of the outer membrane.
   B. F_1 domain provides a channel for translocation of protons across the membrane.
   C. F_1 binds ATP but not ADP.
   D. F_1 domain catalyzes the synthesis of ATP.
   E. only the F_O domain contains more than one subunit.

Questions 7 and 8: A child presented with severe neurological defects. Blood tests indicated elevated serum levels of lactate, pyruvate and alanine. Cultures of skin fibroblasts showed deficient activity of the pyruvate dehydrogenase complex. Further study might indicate which subunit of the complex is defective, although the metabolic effects are essentially the same regardless of which subunit is defective.

7. The active form of pyruvate dehydrogenase is favored by the influence of all of the following on pyruvate dehydrogenase kinase *except*:
   A. low $[Ca^{2+}]$.
   B. low acetyl CoA/CoASH.
   C. high [pyruvate].
   D. low $NADH/NAD^+$.

8. Suppose the specific defect were a mutant pyruvate dehydrogenase (the first catalytic subunit) with poor binding of its prosthetic group. In this type of defect, sometimes greatly increasing the dietary precursor of the prosthetic group is helpful. In this case, increasing which of the following might be helpful?
   A. lipoic acid.
   B. niacin (for NAD).
   C. pantothenic acid (for CoA).
   D. riboflavin (for FAD).
   E. thiamine (for TPP).

Questions 9 and 10: When a major coronary artery is blocked, ischemia (lowered oxygen supply) results, inhibiting mitochondrial electron transport and oxidative phosphorylation. The ischemia causes damage to affected tissues. Reperfusion (introducing oxygen), however, appears to cause even more tissue damage as oxygen radicals are formed. Some hope for minimizing the reperfusion damage lies in the administration of antioxidants.

9. All of the following statements are correct *except*:
   A. reactive oxygen species (oxygen radicals) result when there is a concerted addition of four electrons at a time to O_2.
   B. superoxide anion (O_2^-) and hydroxyl radical ($^{\cdot}OH$) are two forms of reactive oxygen.
   C. superoxide dismutase is a naturally occurring enzyme that protects against damage by converting O_2^- to H_2O_2.
   D. reactive oxygen species damage phospholipids, proteins and nucleic acids.
   E. glutathione protects against H_2O_2 by reducing it to water.

10. All of the following result from ischemia *except*:
    A. decrease in intracellular ATP.
    B. decrease in intracellular creatine phosphate.

C. decrease in $NADH/NAD^+$.
D. lactic acidosis.
E. depletion of cellular glycogen.

Questions 11 and 12: Cyanide is a potent and rapidly acting poison. It binds to the heme of the cytochrome a_3 in cytochrome oxidase. Death occurs from tissue anoxia, especially in the central nervous system. An antidote to cyanide poisoning, which must be given rapidly, is the administration of a nitrite to convert oxyhemoglobin to methemoglobin. Methemoglobin competes with cytochrome a_3 for cyanide.

11. Cyanide:
    A. only minimally inhibits the electron transport chain because cytochrome oxidase is a terminal component of the chain.
    B. inhibits mitochondrial respiration but energy production is unaffected.
    C. also binds the copper of cytochrome oxidase.
    D. binds to Fe^{3+} of cytochrome a_3.
    E. poisoning could also be reversed by increasing O_2 concentration.
12. If cyanide is added to tightly coupled mitochondria that are actively oxidizing succinate:
    A. subsequent addition of 2,4-dinitrophenol will cause ATP hydrolysis.
    B. subsequent addition of 2,4-dinitrophenol will restore succinate oxidation.
    C. electron flow will cease, but ATP synthesis will continue.
    D. electron flow will cease, but ATP synthesis can be restored by subsequent addition of 2,4-dinitrophenol.
    E. subsequent addition of 2,4-dinitrophenol and the phosphorylation inhibitor, oligomycin, will cause ATP hydrolysis.

Problems

13. For the reaction $A \rightleftharpoons B$, $\Delta G^{0'} = -7.1$ kcal mol^{-1}. At 37°C, $-2.303RT = -1.42$ kcal mol^{-1}.

    What is the equilibrium ratio of B/A?
14. Using pyruvate, labeled with ^{14}C in its keto group, via the pyruvate dehydrogenase reaction and the TCA cycle, where would the carbon label be at the end of one turn of the TCA cycle? Where would the carbon label be at the end of the second turn of the cycle?

ANSWERS

1. **B** High-energy does not refer to a high energy of formation (bond stability). A "high-energy" bond is so designated because it has a high free energy of hydrolysis. This could arise for reasons A, C, D, or E.
2. **B** Water is required to hydrolyze the thioester bond of acetyl CoA. A: Aconitase removes water and then adds it back. C, D: The dehydrogenases remove two protons and two electrons. E: Here the thioester undergoes phosphorolysis, not hydrolysis; the phosphate is subsequently transferred from the intermediate succinyl phosphate to GDP.
3. **C** A: Citrate is transported out of the mitochondria to be used as a source of cytoplasmic acetyl CoA. B: Fumarate is produced during phenylalanine and tyrosine degradation. D: Can be formed from glutamate. E: Oxaloacetate is produced by pyruvate carboxylase, and it is used in gluconeogenesis. Clearly most of the tricarboxylic acid cycle intermediates play multiple roles in the body.
4. **D** ATP and ADP are transported in opposite directions. A, B: Reducing equivalents from NADH are shuttled across the membrane, as is the acetyl group of acetyl CoA, but NADH and acetyl CoA cannot cross. C: Of the nucleotides, only ATP and ADP are transported. E: Like NADH, NADPH does not cross the membrane.
5. **E** If the charge separation could be dissipated by free diffusion of other ions, the energy would be lost, and no ATP could be synthesized.
6. **D** This is correct. A: Both are associated with the inner mitochondrial membrane. B: This is the role of the F_O domain. C: Both bind and ATP synthesis occurs. E: Changing conformations of different subunits is important in the actions of both domains.
7. **A** High Ca^{2+} favors the active dehydrogenase but by activating the phosphatase. B, C, D: NADH and acetyl CoA activate pyruvate dehydrogenase kinase, thus inactivating pyruvate dehydrogenase. Pyruvate inhibits the kinase, favoring the active dehydrogenase.
8. **E** TPP is the cofactor for the first reaction which decarboxylates pyruvate. A,C: These are cofactors for the dihydrolipoyl transacetylase. B, D: These are cofactors for the dihydrolipoyl dehydrogenase.
9. **A** Four electron transfer to oxygen produces water. Oxygen radicals occur when electrons are added stepwise to O_2. B: H_2O_2 is also included as reactive oxygen because it can produce $^{\cdot}OH$. C: The H_2O_2 is then degraded by catalase. D: All of these are subject to oxidation with deleterious effects. E: Glutathione peroxidase catalyzes this reaction.
10. **C** Inhibition of electron transport because of low O_2 prevents reoxidation of NADH and, therefore, the $NADH/NAD^+$ increases. A, B: ATP decreases because oxidative phosphorylation is inhibited. This will shift the creatine kinase reaction toward ATP. D: Anaerobic glycolysis is stimulated to provide ATP, leading to increased lactate. E: Glycogen is the substrate for anaerobic glycolysis.
11. **D** This is why methemoglobin is an effective antidote since it also has an Fe^{3+}. B: Respiration and energy production are coupled, so inhibiting one inhibits the other as well. C: Copper is an important part of cytochrome oxidase, but cyanide does not bind it. E: The problem is inability to react with O_2, not a lack of O_2.
12. **A** Cyanide inhibits electron transport at site III, blocking electron flow throughout the system. In coupled mitochondria, ATP synthesis ceases too. Addition of an uncoupler permits the mitochondrial ATPase (which is normally driven in the synthetic direction) to operate, and it catalyzes the favorable ATP hydrolysis reaction unless it is inhibited by a phosphorylation inhibitor such as oligomycin.
13. $\Delta G^{0'} = -RT \ln K = -2.303RT \log K$. Substitution gives the log $K = 5$. Then K is 100,000, so the B/A = 100,000/1.
14. The labeled keto carbon of pyruvate becomes the labeled carboxyl carbon of acetyl CoA. After condensation with oxaloacetate, the first carboxyl group of citrate is labeled. This label is retained through subsequent reactions to succinate. However, succinate is a symmetrical compound to the enzyme, so, in effect, both carboxyl groups of succinate are labeled. This means that the oxaloacetate regenerated is labeled in both carboxyl groups at the end of one turn (actually, half the molecules are labeled in one carboxyl and half in the other, but this can=t be distinguished experimentally). Note that CO_2 is not labeled. In the second turn, the same carboxyl labeled acetyl CoA is added, but this time to labeled oxaloacetate. Both carboxyl groups of the oxaloacetate are released as CO_2, so it will be labeled, as will the regenerated oxaloacetate.

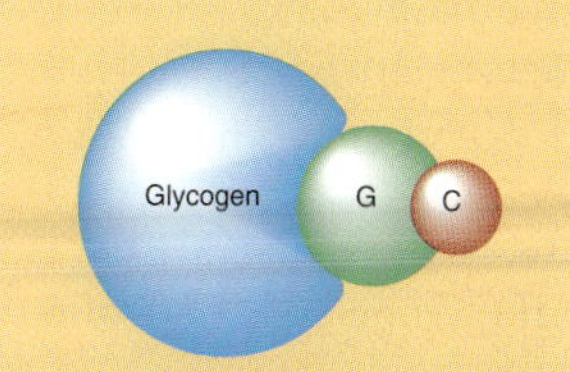

15 CARBOHYDRATE METABOLISM I: MAJOR METABOLIC PATHWAYS AND THEIR CONTROL

Robert A. Harris

Textbook of Biochemistry With Clinical Correlations, Sixth Edition, Edited by Thomas M. Devlin

15.1 OVERVIEW

The major pathways of carbohydrate metabolism begin or end with glucose (Figure 15.1). This chapter describes the utilization of glucose as a source of energy, formation of glucose from noncarbohydrate precursors, storage of glucose in the form of glycogen, and release of glucose from glycogen. An understanding of the pathways and their regulation is necessary because of the important role played by glucose in the body. Glucose is the major form in which carbohydrate absorbed from the intestinal tract is presented to the cells of the body. It is the only fuel used to any significant extent by some specialized cells, and it is the major fuel of the brain. Indeed, several tissues of the body have evolved a coordinated working relationship that ensures a continuous supply of this essential substrate for these cells and brain. Glucose metabolism is defective in the very common metabolic diseases, obesity and diabetes, which contribute to the development of major medical problems, including atherosclerosis, hypertension, small vessel disease, kidney disease, and blindness.

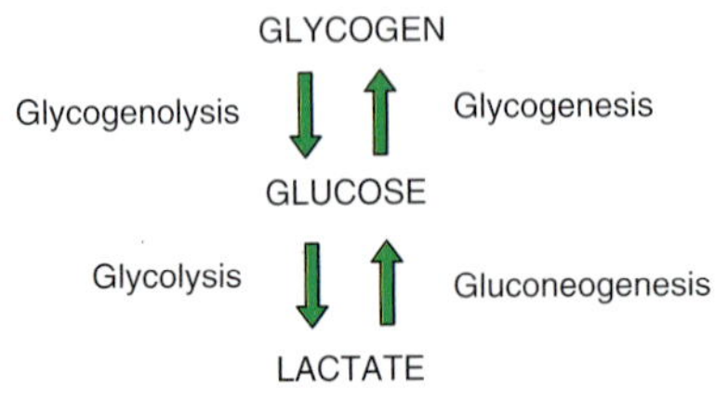

The discussion begins with **glycolysis**, a pathway used by all body cells to extract part of the chemical energy inherent in the glucose molecule. This pathway also converts glucose to pyruvate and sets the stage for complete oxidation of glucose to CO_2 and H_2O. Gluconeogenesis, the *de novo* synthesis of glucose, is conveniently discussed after glycolysis because it involves many of the same enzymes used in glycolysis, although the reactions being catalyzed are in the reverse direction. In contrast to glycolysis, which produces ATP, gluconeogenesis requires ATP and is therefore an energy-requiring process. Thus, some enzyme-catalyzed steps have to be different between glycolysis and gluconeogenesis. How regulation is exerted at key enzymes will be stressed throughout

GLYCOGEN
Glycogenolysis ↓↑ Glycogenesis
GLUCOSE
Glycolysis ↓↑ Gluconeogenesis
LACTATE

FIGURE 15.1
Relationship of glucose to major pathways of carbohydrate metabolism.

the chapter. This will be particularly true for glycogen synthesis (glycogenesis) and degradation (glycogenolysis). Many cells store glycogen for their own later needs. The liver is less selfish, storing glycogen mostly for maintenance of blood glucose to ensure that other tissues, especially the brain, have an adequate supply. Regulation of the synthesis and degradation of glycogen is a model for our understanding of how hormones work and how metabolic pathways are regulated. These topics contribute to our understanding of the diabetic condition, starvation, and how tissues of the body respond to stress, severe trauma, and injury.

The chemistry and nomenclature of carbohydrates is presented in the Appendix.

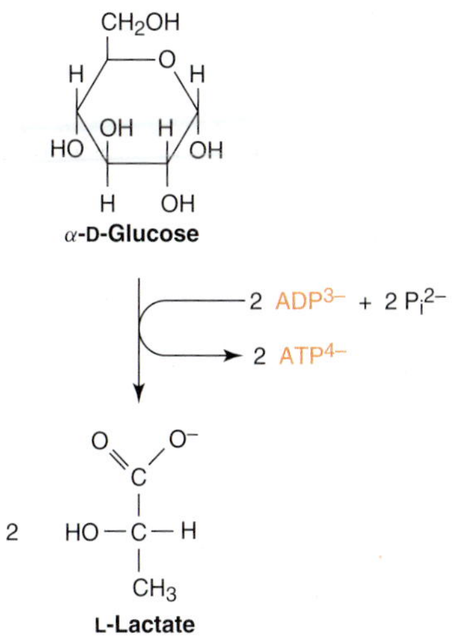

15.2 GLYCOLYSIS

Glycolysis Occurs in all Human Cells

The **Embden–Meyerhof** or **glycolytic pathway** represents an ancient process, possessed by all cells of the human body, in which anaerobic degradation of glucose to lactate occurs with release of energy as ATP. This is an example of **anaerobic fermentation**, a term used for metabolic pathways that organisms use to extract chemical energy from high-energy fuels in the absence of oxygen. For many tissues glycolysis is only an emergency energy-yielding pathway, capable of producing 2 mol of ATP from 1 mol of glucose in the absence of oxygen (Figure 15.2). When oxygen supply to a tissue is shut off, ATP levels can still be maintained by glycolysis for at least a short period. Many examples could be given, but the capacity to use glycolysis as a source of energy is particularly important during natural birth of human babies. With the exception of the brain, circulation of blood decreases to most parts of the body of the baby during delivery. The brain is not normally deprived of oxygen during delivery, but other tissues have to depend on glycolysis for their supply of ATP until a normal oxygen supply is available. This conserves oxygen for use by the brain, illustrating one of many mechanisms that have evolved to ensure survival of **brain tissue** in times of stress. Oxygen is not necessary for glycolysis; in fact, oxygen can indirectly suppress glycolysis by the **Pasteur effect**, which is considered later (p. 599). Nevertheless, glycolysis does occur in cells with an abundant supply of molecular oxygen. Provided that cells also contain mitochondria, the end product of glycolysis in the presence of oxygen is pyruvate rather than lactate. Pyruvate is then completely oxidized to CO_2 and H_2O by the pyruvate dehydrogenase complex and enzymes of the TCA cycle housed within the mitochondria (Figure 15.3). Glycolysis therefore sets the stage for aerobic oxidation of carbohydrate. The overall process of glycolysis and mitochondrial oxidation of pyruvate to CO_2 and H_2O has the following sum equation:

$$\text{D-Glucose}(C_6H_{12}O_6) + 6\,O_2 + 32\,ADP^{3-} + 32\,P_i^{\,2-} + 32\,H^+$$
$$\rightarrow 6CO_2 + 6\,H_2O + 32\,ATP^{4-}$$

Much more ATP is produced in complete oxidation of glucose to CO_2 and H_2O (32 ATP/glucose) than is produced in the conversion of glucose to lactate (2 ATP/glucose). This has important consequences to be considered in detail later. The importance of glycolysis as a preparatory pathway is best exemplified by the brain that has an absolute need for glucose. Pyruvate produced by glycolysis is oxidized to CO_2 in mitochondria.

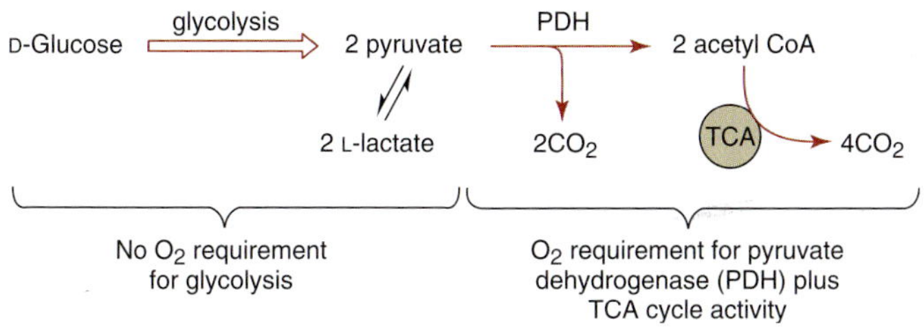

FIGURE 15.2
Overall balanced equation for the sum of reactions of glycolysis.

FIGURE 15.3
Glycolysis is a preparatory pathway for aerobic metabolism of glucose. PDH refers to the pyruvate dehydrogenase complex; TCA to the tricarboxylic acid cycle.

An adult human brain uses approximately 120 g of glucose each day to meet its need for ATP. In contrast, glycolysis with lactate as the end product is the major mechanism of ATP production in some other tissues. Red blood cells lack mitochondria and therefore cannot convert pyruvate to CO_2. The cornea, lens, and regions of the retina have a limited blood supply and also lack mitochondria (because mitochondria would absorb and scatter light) and depend on glycolysis as the major mechanism for ATP production. Kidney medulla, testis, leukocytes, and white muscle fibers have relatively few mitochondria and are therefore almost totally dependent on glycolysis as a source of ATP. Tissues dependent primarily on glycolysis for ATP production consume about 40 g of glucose per day in a normal adult.

Starch is the storage form of glucose in plants and contains **α-1,4-glycosidic linkages** and **α-1,6-glycosidic branches**. **Glycogen** is the storage form of glucose in animal tissues and contains the same type of glycosidic linkages and branches. Exogenous glycogen refers to the glycogen we obtain from animal products; endogenous glycogen is that synthesized and stored in our tissues. Exogenous starch or glycogen is hydrolyzed to glucose in the intestinal tract, whereas stored glycogen endogenous to our tissues is converted to glucose or glucose 6-phosphate by enzymes within the cells. Disaccharides such as milk sugar (lactose) and grocery store sugar (sucrose) are important sources of glucose in our diet. Their hydrolysis by enzymes of the brush border of the intestinal tract is discussed on page 1056. Although glucose can be a source of energy for cells of the intestinal tract, these cells do not depend on glucose to any great extent; most of their energy requirement is met by glutamine catabolism (p. 748). Most of the glucose is absorbed by the cells of the intestinal tract into the portal vein blood and then the general circulation for use by other tissues. Liver is the first major tissue to have an opportunity to remove glucose from the portal vein blood. When blood glucose is high, the liver removes glucose for glycogenesis and glycolysis. When blood glucose is low, the liver supplies the blood with glucose by glycogenolysis and gluconeogenesis. The liver is also the first organ exposed to the blood arriving from the pancreas and therefore the highest concentrations of **glucagon** and **insulin**. The effects of these important hormonal regulators of blood glucose levels will be discussed later.

Glucose is Metabolized Differently in Various Cells

After transport through the plasma membrane by way of **GLUT1 (glucose transporter 1**; see p. 474), glucose is metabolized mainly by glycolysis in red blood cells (Figure 15.4*A*). Since these cells lack mitochondria, the end product of glycolysis is lactic acid, which is released into the blood plasma. Glucose used by the pentose phosphate pathway in red blood cells provides NADPH to keep glutathione in the reduced state, which has an important role in the destruction of organic peroxides and H_2O_2 (Figure 15.5). Peroxides cause irreversible damage to membranes, DNA and other cellular components and must be removed to prevent cell damage and death.

The brain takes up glucose by facilitated transport in an insulin-independent manner by **GLUT3** (glucose transporter 3) (Figure 15.4*B*). Glycolysis yields pyruvate, which is subsequently oxidized to CO_2 and H_2O by the pyruvate dehydrogenase complex and the TCA cycle. The pentose phosphate pathway is active in these cells, generating part of the NADPH needed for reductive synthesis and the maintenance of glutathione in the reduced state.

Muscle and heart cells readily utilize glucose (Figure 15.4*C*). Insulin stimulates transport of glucose into these cells by way of **GLUT4** (glucose transporter 4). In the absence of insulin, GLUT4 exists in intracellular vesicles where it cannot facilitate glucose transport (Figure 15.6). Binding of insulin to its receptor on the plasma membrane initiates a signaling cascade that promotes translocation and fusion of GLUT4-containing vesicles with the plasma membrane, thereby placing GLUT4 where it can facilitate glucose transport. Glucose taken into muscle and heart cells can be utilized by glycolysis to give pyruvate, which is used by the pyruvate dehydrogenase complex and the TCA cycle to provide ATP. Muscle and heart synthesize significant quantities of glycogen, an important fuel that these tissues store for later use.

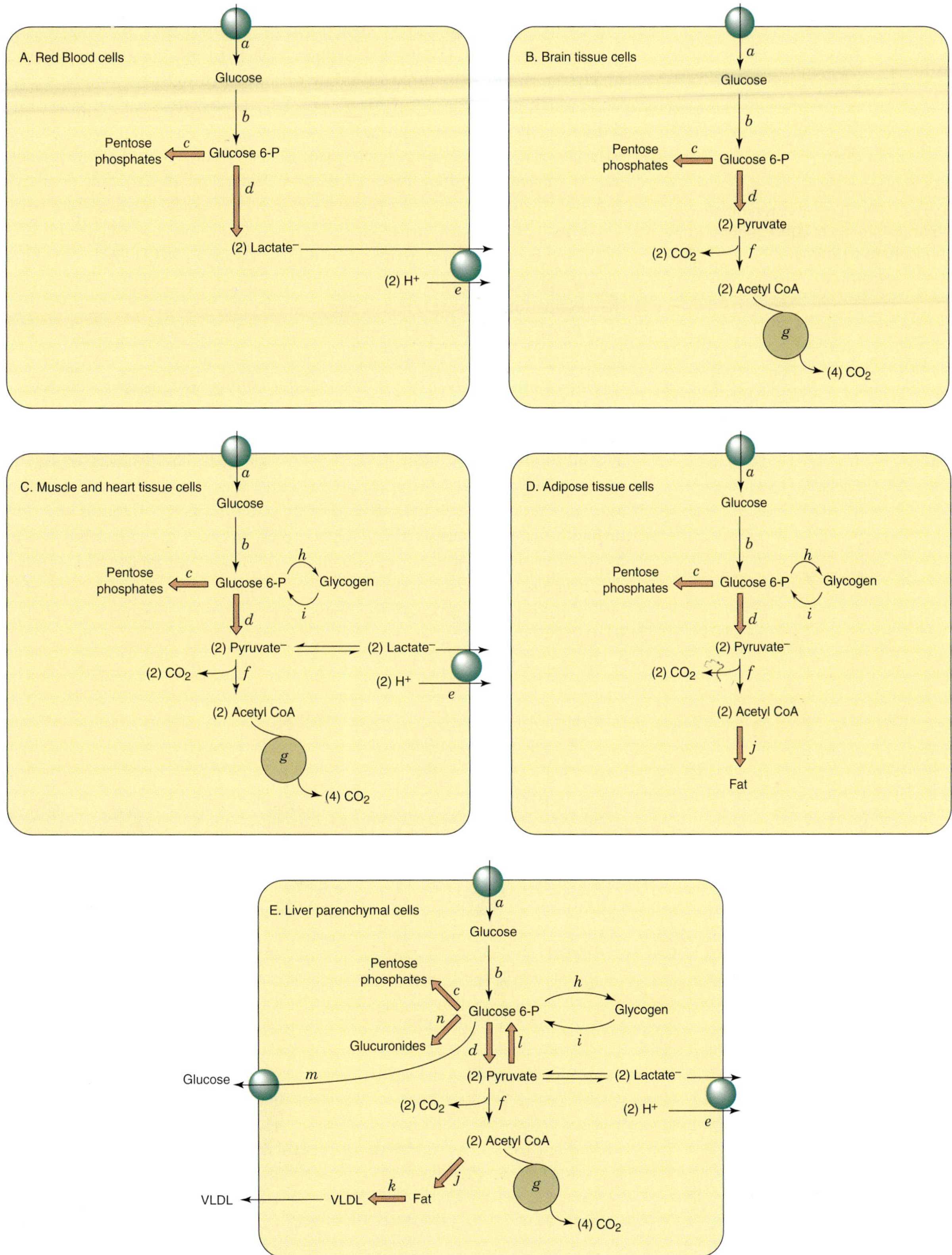

FIGURE 15.4

Overviews of the major ways in which glucose is metabolized within cells of selected tissues of the body.

(*A*) Red blood cells; (*B*) brain tissue cells; (*C*) muscle and heart tissue cells; (*D*) adipose tissue cells; (*E*) liver parenchymal cells. (*a*) Glucose transport into a cell by a glucose transporter (GLUT); (*b*) glucose phosphorylation by hexokinase; (*c*) pentose phosphate pathway; (*d*) glycolysis; (*e*) lactic acid transport out of the cell; (*f*) pyruvate decarboxylation by pyruvate dehydrogenase; (*g*) TCA cycle; (*h*) glycogenesis; (*I*) glycogenolysis; (*j*) lipogenesis; (*k*) formation and release of very low density lipoproteins (VLDL); (*l*) gluconeogenesis; (*m*) hydrolysis of glucose 6-phosphate and release of glucose from the cell into the blood; (*n*) formation of glucuronides (drug and bilirubin detoxification by conjugation) by the glucuronic acid pathway.

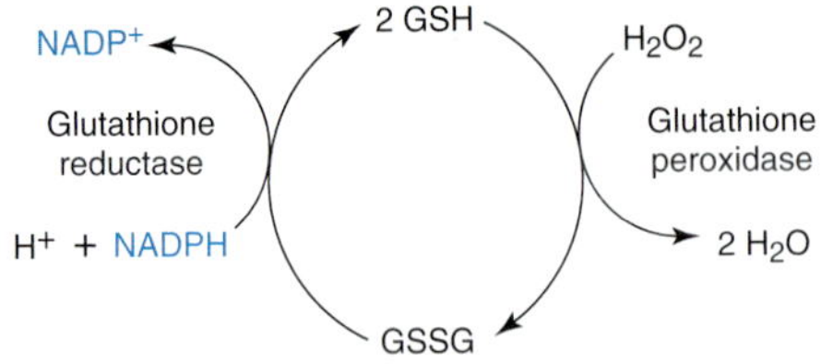

FIGURE 15.5

Destruction of H_2O_2 depends on reduction of oxidized glutathione by NADPH generated by pentose phosphate pathway.

As in muscle, the uptake of glucose by adipose tissue is dependent upon and stimulated by insulin (Figure 15.4*D* and 15.6). Pyruvate, as in other cells, is generated by glycolysis and is oxidized by the pyruvate dehydrogenase complex to give acetyl CoA, which is used primarily for *de novo* fatty acid synthesis. Glycolysis also provides carbon for the synthesis of glycerol 3-phosphate (not shown) required for synthesis of triacylglycerol (see p. 676). Adipose tissue can carry out glycogenesis and glycogenolysis, but its capacity for these processes is very limited relative to that of muscle, heart, and liver.

Liver has the greatest number of ways to utilize glucose (Figure 15.4*E*). Uptake of glucose occurs independently of insulin by means of **GLUT2**, a low-affinity, high-capacity glucose transporter. Glucose is used by the pentose phosphate pathway for the production of NADPH, which is needed for reductive synthesis, maintenance of reduced glutathione and numerous reactions catalyzed by endoplasmic reticulum enzyme systems. A vital function of the pentose phosphate pathway is the provision of ribose phosphate for the synthesis of nucleotides such as ATP and those in DNA and RNA. Storage of glucose as glycogen is a particularly important feature of the liver. Glucose is also used in the glucuronic acid pathway, which is important in drug and bilirubin detoxification (see pp. 422 and 842). The liver carries out glycolysis, the pyruvate produced being used as a source of acetyl CoA for complete oxidation by the tricarboxylic acid cycle and for fatty acid synthesis. Glycolysis also provides the carbon for the synthesis of the glycerol moiety of triacylglycerol, which is synthesized by the liver during production of very low density lipoproteins (VLDL) (see p. 711). The liver also converts three-carbon precursors (lactate, pyruvate, glycerol, and alanine) into glucose by the process of gluconeogenesis to meet the needs of other cells and the brain.

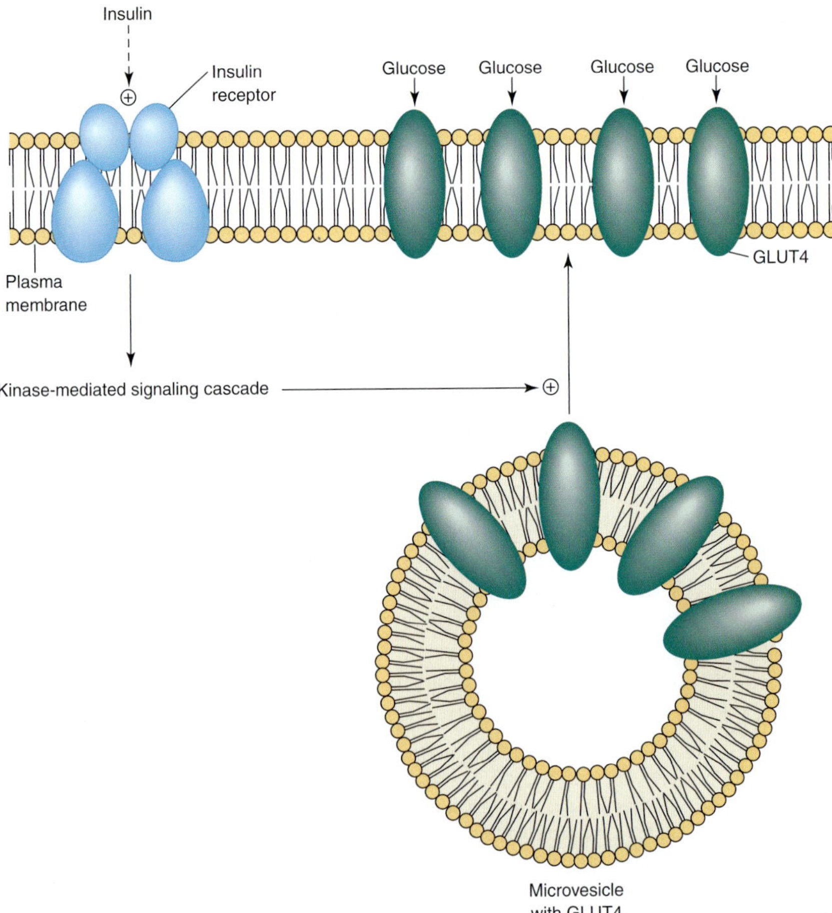

FIGURE 15.6

Insulin stimulates glucose uptake by adipose tissue and muscle by increasing the number of glucose transporters (GLUT4) in the plasma membrane.

15.3 GLYCOLYSIS PATHWAY

Glucose is combustible and burns in a test tube to yield heat and light but, of course, no ATP. Cells use about 30 steps to take glucose to CO_2 and H_2O, a seemingly inefficient process, since it can be done in a single step in a test tube. However, side reactions and some of the actual steps used by the cell to oxidize glucose to CO_2 and H_2O lead to the conservation of a significant amount of energy as ATP. In other words, cells produce ATP by the controlled "burning" of glucose, of which glycolysis represents only the first few steps, shown in Figure 15.7.

Glycolysis Occurs in Three Stages

Glycolysis has three major stages.

Priming stage (Figure 15.7*a*):

$$\text{D-Glucose} + 2\ \text{ATP}^{4-} \rightarrow \text{D-fructose 1,6-bisphosphate}^{4-} + 2\ \text{ADP}^{3-} + 2\ \text{H}^{+}$$

Splitting stage (Figure 15.7*b*):

$$\text{D-Fructose 1,6-bisphosphate}^{4-} \rightarrow 2\ \text{D-glyceraldehyde 3-phosphate}^{2-}$$

Oxidoreduction-phosphorylation stage (Figure 15.7c):

$$2\ \text{D-Glyceraldehyde 3-phosphate}^{2-} + 4\ \text{ADP}^{3-} + 2\ \text{P}_i{}^{2-} + 2\ \text{H}^{+} \rightarrow$$

$$2\ \text{L-lactate}^{-} + 4\ \text{ATP}^{4-} + 2\ \text{H}_2\text{O}$$

Sum:

$$\text{D-Glucose} + 2\ \text{ADP}^{3-} + 2\ \text{P}_i{}^{2-} \rightarrow 2\ \text{L-lactate}^{-} + 2\ \text{ATP}^{4-} + 2\ \text{H}_2\text{O}$$

The priming stage involves input of two molecules of ATP to convert glucose into fructose 1,6-bisphosphate. ATP is "invested" rather than lost in this stage because it is subsequently regained plus more from later steps. The splitting stage "splits" fructose 1,6-bisphosphate into two molecules of glyceraldehyde 3-phosphate. In the oxidoreduction–phosphorylation stage, two molecules of glyceraldehyde 3-phosphate are converted into two molecules of lactate with the production of four molecules of ATP. The overall process therefore generates two molecules of lactate and two molecules of ATP from one molecule of glucose.

Stage One: Priming of Glucose

Although the hexokinase reaction (see Figure 15.7*a*) consumes ATP, it gets glycolysis off to a good start by trapping glucose as **glucose 6-phosphate (G6P)** within the cytosol where all the glycolytic enzymes are located. Phosphate esters are charged and hydrophilic and therefore can not cross cell membranes. The phosphorylation of glucose by ATP is a thermodynamically favorable, irreversible under cellular conditions, reaction. The reverse reactions cannot be used to synthesize ATP.

The next reaction catalyzed by phosphoglucose isomerase is readily reversible and is not subject to regulation.

6-Phosphofructo-1-kinase (or **phosphofructokinase-1**) catalyzes the ATP-dependent phosphorylation of fructose 6-phosphate (F6P) to fructose 1,6-bisphosphate (FBP). This enzyme is subject to regulation by several effectors and is often considered the key regulatory enzyme of glycolysis. The reaction is irreversible and uses the second ATP needed to "prime" glucose.

Stage Two: Splitting of a Phosphorylated Intermediate

Fructose 1,6-bisphosphate aldolase cleaves fructose 1,6-bisphosphate into a molecule each of dihydroxyacetone phosphate and glyceraldehyde 3-phosphate (GAP) (see Figure 15.7*b*). This is a reversible reaction; the name **aldolase** is a variant of an aldol cleavage in one direction and an aldol condensation in the other. Triose phosphate isomerase catalyzes reversible interconversion of dihydroxyacetone phosphate and GAP. With the transformation of dihydroxyacetone phosphate (DHAP) into GAP, one molecule of glucose is converted into two molecules of GAP.

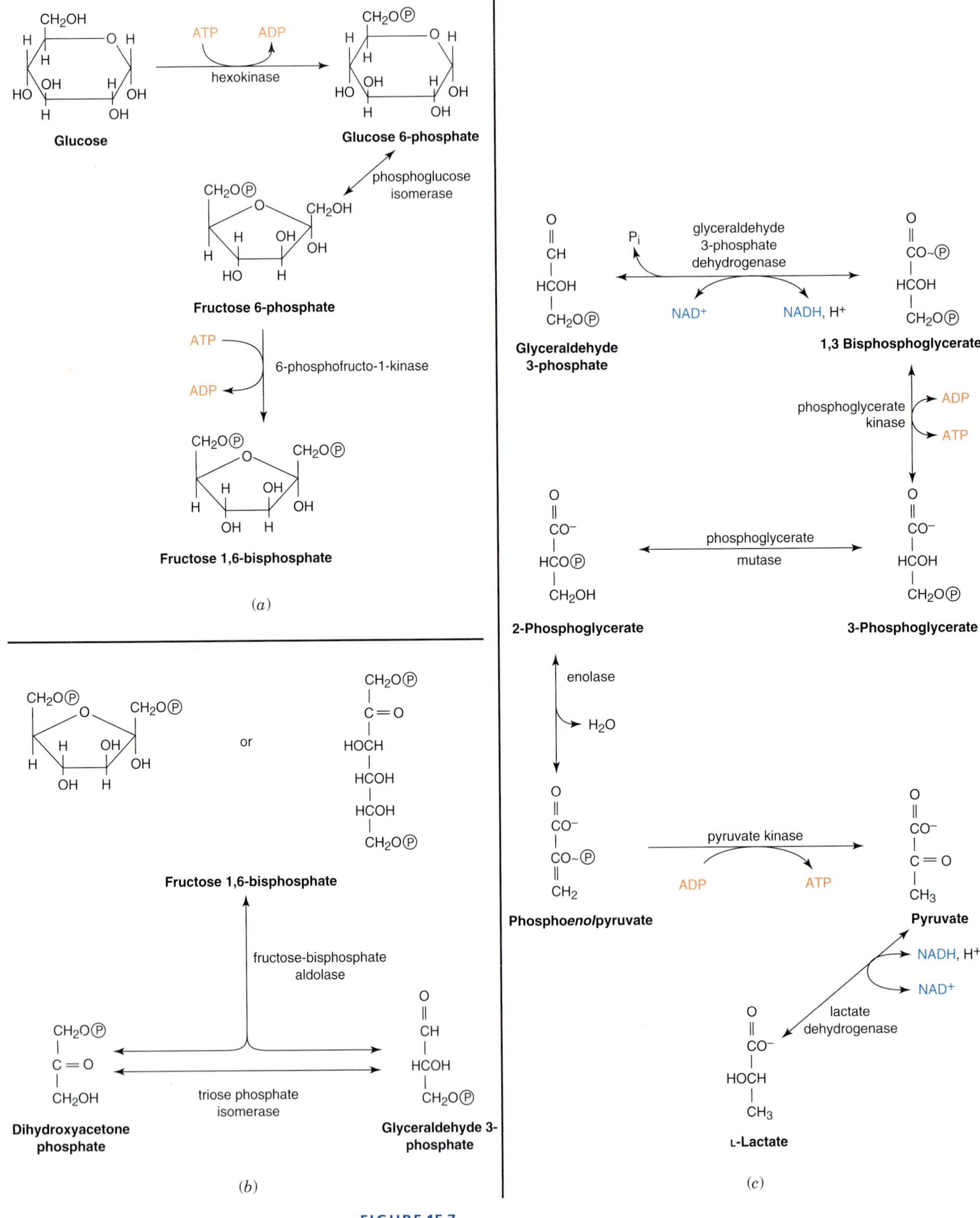

FIGURE 15.7

The glycolytic pathway, divided into three stages. The symbol P refers to phosphoryl group PO_3^{2-}; ~ indicates a high-energy phosphate bond. (*a*) Priming stage. (*b*) Splitting stage. (*c*) Oxidoreduction–phosphorylation stage.

Stage Three: Oxidoreduction Reactions and ATP Synthesis

The reaction (Figure 15.7*c*) catalyzed by **glyceraldehyde 3-phosphate dehydrogenase** is of interest because of what is accomplished. An aldehyde (glyceraldehyde 3-phosphate) is oxidized to a carboxylic acid with the reduction of NAD^+ to NADH. The acid produced is 1,3-bisphosphoglycerate, a mixed anhydride of a carboxylic acid and phosphoric acid, which has a large negative free energy of hydrolysis, enabling it to participate in a subsequent reaction that forms ATP. The overall reaction can be visualized as the coupling of a very favorable exergonic reaction with an unfavorable endergonic reaction. The exergonic reaction is the one in which an aldehyde is oxidized to a carboxylic acid, which is then coupled with an endergonic half-reaction in which NAD^+ is reduced to NADH:

$$R{-}\overset{O}{\overset{\|}{C}}H + H_2O \longrightarrow R{-}\overset{O}{\overset{\|}{C}}OH + 2H^+ + 2e^-$$

$$NAD^+ + 2H^+ + 2e^- \longrightarrow NADH + H^+$$

The overall reaction (sum of the half-reactions) is quite exergonic:

$$R{-}\overset{O}{\overset{\|}{C}}H + NAD^+ + H_2O \longrightarrow R{-}\overset{O}{\overset{\|}{C}}OH + NADH + H^+, \quad \Delta G^{\circ\prime} = -10.3 \text{ kcal mol}^{-1}$$

A second endergonic component of the reaction is formation of a **mixed anhydride** between the carboxylic acid and phosphoric acid:

$$R{-}\overset{O}{\overset{\|}{C}}OH + HPO_4^{2-} \longrightarrow R{-}\overset{O}{\overset{\|}{C}}{-}OPO_3^{2-} + H_2O, \Delta G^{\circ\prime} = +11.8 \text{ kcal mol}^{-1}$$

The overall reaction involves coupling of the endergonic and exergonic components with an overall standard free energy change of $+1.5$ kcal mol^{-1}.

$$\text{Sum: } R{-}\overset{O}{\overset{\|}{C}}H + NAD^+ + HPO_4^{2-} \longrightarrow R{-}\overset{O}{\overset{\|}{C}}OPO_3^{2-} + NADH + H^+, \Delta G^{\circ\prime} = +1.5 \text{ kcal mol}^{-1}$$

This reaction is freely reversible in cells. The catalytic mechanism involves glyceraldehyde 3-phosphate reacting with a sulfhydryl group of a cysteine residue to generate a thiohemiacetal (Figure 15.8). An internal oxidation-reduction reaction occurs in which bound NAD^+ is reduced to NADH and the thiohemiacetal is oxidized to a high-energy thiol ester. Exogenous NAD^+ then replaces the bound NADH and the thiol ester reacts with P_i to form the mixed anhydride and regenerate the free sulfhydryl group. The mixed anhydride then dissociates from the enzyme. It should be noted that a free carboxylic acid group (—COOH) is not generated from the aldehyde group (—CHO) during the reaction. Instead, the enzyme generates a carboxyl group in the form of high-energy thiol ester, which is converted by reaction with P_i into a mixed anhydride of carboxylic and phosphoric acids.

This reaction catalyzed by glyceraldehyde 3-phosphate dehydrogenase requires NAD^+ and produces NADH. Since the cytosol has only a limited amount of NAD^+, continuous glycolytic activity can only occur if NADH is oxidized back to NAD^+, otherwise glycolysis will stop for want of NAD^+. The options that cells have for regeneration of NAD^+ from NADH are described in a later section (see p. 592).

In the next reaction **phosphoglycerate kinase** produces ATP from the high-energy compound 1,3-bisphosphoglycerate (Figure 15.7*c*). This is the first site of ATP

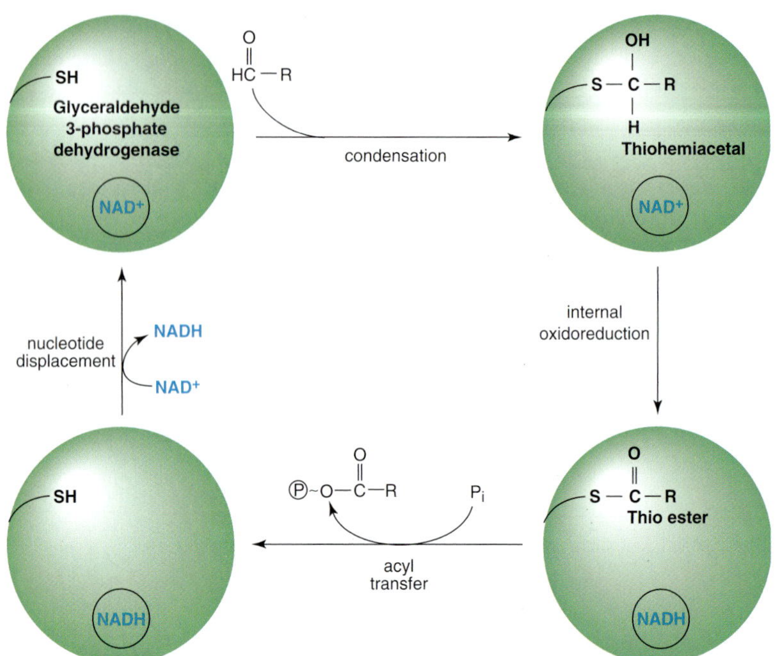

FIGURE 15.8

Catalytic mechanism of glyceraldehyde 3-phosphate dehydrogenase. Large circle represents enzyme; small circle represents binding site for NAD^+; RCOH, the aldehyde group of glyceraldehyde 3-phosphate; -SH, the sulfhydryl group of the cysteine residue located at the enzyme's active site; and ~P, the high-energy phosphate bond of 1,3-bisphosphoglycerate.

production in glycolysis. Since two ATP molecules were invested for each glucose molecule in the priming stage and since two molecules of 1,3-bisphosphoglycerate are produced from each glucose, all of the "invested" ATP is recovered in this step. The glyceraldehyde 3-phosphate dehydrogenase-phosphoglycerate kinase system is an example of **substrate-level phosphorylation** in which a substrate participates in an enzyme-catalyzed reaction that yields ATP or GTP.

Substrate-level phosphorylation stands in contrast to oxidative phosphorylation catalyzed by the mitochondrial electron-transfer chain and ATP synthase (see p. 564). Note that the combination of glyceraldehyde 3-phosphate dehydrogenase and phosphoglycerate kinase accomplishes the coupling of an oxidation (an aldehyde is oxidized to a carboxylic acid) to a phosphorylation (a mixed anhydride of a carboxylic acid and phosphoric acid is formed) without the involvement of a membrane system.

Phosphoglycerate mutase converts 3-phosphoglycerate to 2-phosphoglycerate. This is a freely reversible reaction in which 2,3-bisphosphoglycerate is an obligatory intermediate at the active site:

$$\begin{aligned} \text{E-phosphate} + \text{3-phosphoglycerate} &\rightleftharpoons \text{E} + \text{2,3-bisphosphoglycerate} \\ \text{E} + \text{2,3-bisphosphoglycerate} &\rightleftharpoons \text{E-phosphate} + \text{2-phosphoglycerate} \\ \text{Sum:} \quad \text{3-phosphoglycerate} &\rightleftharpoons \text{2-phosphoglycerate} \end{aligned}$$

Involvement of this intermediate creates an absolute requirement for catalytic amounts of 2,3-bisphosphoglycerate in cells. This can be appreciated by noting that E-P cannot be generated without 2,3-bisphosphoglycerate and likewise that 2,3-bisphosphoglycerate cannot be generated without E-P. Cells solve this problem by synthesizing 2,3-bisphosphoglycerate from 1,3-bisphosphoglycerate with a 2,3-bisphosphoglycerate mutase:

$$\begin{array}{c} O \\ \| \\ COPO_3^{2-} \\ | \\ HCOH \\ | \\ CH_2OPO_3^{2-} \end{array} \longrightarrow \begin{array}{c} O \\ \| \\ CO^- \\ | \\ HCOPO_3^{2-} \\ | \\ CH_2OPO_3^{2-} \end{array} + H^+$$

1,3-Bisphospho-D-glycerate → 2,3-Bisphospho-D-glycerate

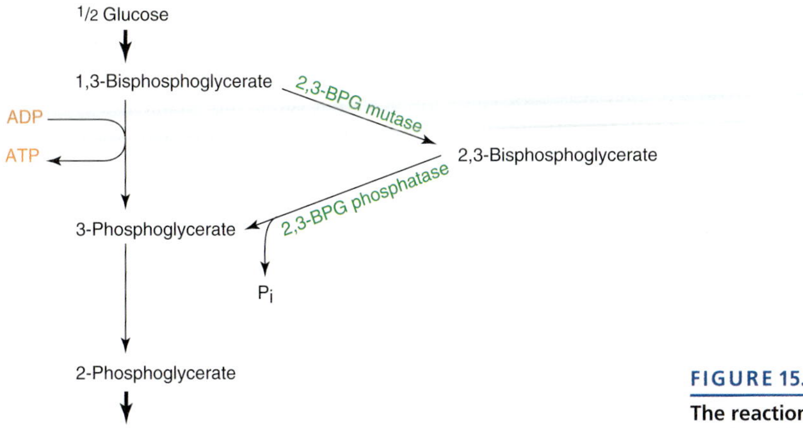

FIGURE 15.9
The reactions of 2,3-bisphosphoglycerate (2,3-BPG) shunt are catalyzed by the bifunctional enzyme, 2,3-BPG mutase/phosphatase.

This enzyme is bifunctional, serving as mutase for the formation of 2,3-bisphosphoglycerate and also as a **phosphatase** that hydrolyzes 2,3-bisphosphoglycerate to 3-phosphoglycerate and P_i. All cells contain the minute quantities of 2,3-bisphosphoglycerate necessary to produce the phosphorylated form (E-P) of newly synthesized phosphoglycerate mutase. In striking contrast to other cells of the body, red blood cells contain very high concentrations of 2,3-bisphosphoglycerate, which functions as an important negative allosteric regulator of the binding of oxygen to hemoglobin (see p. 350). In contrast to the extremely small amounts of glucose used for 2,3-bisphosphoglycerate in other cells, 15–25% of glucose converted to lactate in red blood cells goes by way of the "BPG shunt" for the synthesis of 2,3-bisphosphoglycerate (Figure 15.9). No net production of ATP occurs when glucose is converted to lactate by the BPG shunt as it bypasses the phosphoglycerate kinase step.

Enolase eliminates water from 2-phosphoglycerate to form phosphoenolpyruvate (PEP) in the next reaction (Figure 15.7*c*). This reaction generates a high-energy phosphate from one of markedly lower energy level. $\Delta G^{o'}$ for the hydrolysis of phosphoenolpyruvate is -14.8 kcal mol^{-1}, while that for 2-phosphoglycerate is -4.2 kcal mol^{-1}.

Pyruvate kinase (Figure 15.7*c*) accomplishes another substrate-level phosphorylation. This reaction is not reversible under intracellular conditions.

The last step of glycolysis is a freely reversible oxidoreduction reaction catalyzed by **lactate dehydrogenase** (Figure 15.7*c*). Pyruvate is reduced to give L-lactate and NADH is oxidized to NAD^+. This is the only reaction of the body that produces L-lactate or can utilize L-lactate.

ATP Yield and Balanced Equation for Anaerobic Glycolysis

Conversion of one molecule of glucose to two molecules of lactate results in the net formation of two molecules of ATP. Two molecules of ATP are used in the priming stage but subsequent steps yield four molecules of ATP so that the overall net yield is two molecules of ATP:

$$\text{D-Glucose} + 2\ \text{ADP}^{3-} + 2\ \text{P}_i^{\,2-} \rightarrow 2\ \text{L-lactate}^- + 2\ \text{ATP}^{4-} + 2\ \text{H}_2\text{O}$$

Cells have only a limited amount of ADP and P_i. Flux through glycolysis is dependent on an adequate supply of these substrates. If the ATP is not utilized for performance of work, glycolysis stops for want of ADP and/or P_i. Consequently, the ATP generated has to be used in normal work-related processes for glycolysis to occur. Use of ATP for any work-related process is represented simply by

$$\text{ATP}^{4-} + \text{H}_2\text{O} \rightarrow \text{ADP}^{3-} + \text{P}_i^{\,2-} + \text{H}^+ + \text{“work”}$$

When the quantities are doubled and added to the previous equation for glycolysis, excluding the work accomplished since this is necessary for ATP turnover, the overall balanced equation becomes

$$\text{D-Glucose} \rightarrow 2\ \text{L-lactate}^- + 2\ H^+$$

This shows that anaerobic glycolysis generates acid that can create major problems for the body (described later in Clin. Corr. 15.5, p. 601) since intracellular pH must be maintained near neutrality for optimum enzyme activity.

NADH Generated by Glycolysis has to be Oxidized Back to NAD^+: Role of Lactate Dehydrogenase and Substrate Shuttles

Anaerobic Glycolysis

NADH and NAD^+ do not appear in the balanced sum equation for anaerobic glycolysis. The generation of NADH and its utilization are coupled in the pathway (Figure 15.7*c*). Two molecules of NADH are generated by glyceraldehyde 3-phosphate dehydrogenase, and two are utilized by lactate dehydrogenase. NAD^+ is available in only limited amounts and must be regenerated for glycolysis to continue unabated. The two reactions involved are

$$\text{D-Glyceraldehyde 3-phosphate} + NAD^+ + P_i \rightarrow \text{1,3-bisphospho-D-glycerate} + NADH + H^+$$

$$\text{Pyruvate} + NADH + H^+ \rightarrow \text{L-lactate} + NAD^+$$

The sum reaction is

$$\text{D-Glyceraldehyde 3-phosphate} + \text{pyruvate} + P_i \rightarrow \text{1,3-bisphosphoglycerate} + \text{L-lactate}$$

Perfect coupling of reducing equivalents by these reactions occurs under anaerobic conditions or in cells that lack mitochondria.

Aerobic Glycolysis

When oxygen and mitochondria are present, the reducing equivalents of NADH generated by glyceraldehyde 3-phosphate dehydrogenase are shuttled into the mitochondria for oxidation, leaving pyruvate rather than lactate as the end product of glycolysis. The mitochondrial inner membrane is not permeable to NADH (see p. 570), but the **malate-aspartate shuttle** and the **glycerol phosphate shuttle** (Figure 15.10) transport reducing equivalents into the mitochondrial matrix space (see page 570). Liver makes greater use of the malate–aspartate shuttle, but some muscle cells are more dependent on the glycerol phosphate shuttle. The shuttle systems move reducing equivalents from the cytosol into the mitochondria but will not shuttle reducing equivalents from the mitochondria into the cytosol. The sum of all reactions of the malate–aspartate shuttle reactions is simply

$$NADH_{cytosol} + H^+{}_{cytosol} + NAD^+{}_{mito} \rightarrow NAD^+{}_{cytosol} + NADH_{mito} + H^+{}_{mito}$$

The glycerol phosphate shuttle produces $FADH_2$ within the mitochondrial inner membrane. The active site of the mitochondrial glycerol 3-phosphate dehydrogenase is exposed on the cytosolic surface of the mitochondrial inner membrane. The sum of all reactions of the glycerol phosphate shuttle is

$$NADH_{cytosol} + H^+{}_{cytosol} + FAD_{inner\ membrane} \rightarrow NAD^+{}_{cytosol} + FADH_{2\ inner\ membrane}$$

The mitochondrial NADH formed by the malate–aspartate shuttle activity can be used by the mitochondrial electron-transfer chain for the production of 2.5 molecules of ATP by oxidative phosphorylation:

$$NADH_{mito} + H^+ + 1/2\ O_2 + 2\tfrac{1}{2}ADP + 2\tfrac{1}{2}\ P_i \rightarrow NAD^+{}_{mito} + 2\tfrac{1}{2}\ ATP + H_2O$$

In contrast, the $FADH_2$ formed by the glycerol phosphate shuttle yields only 1.5 molecules of ATP:

$$FADH_{2\ inner\ membrane} + 1\tfrac{1}{2}\ O_2 + 1\tfrac{1}{2}\ ADP + 1\tfrac{1}{2}\ P_i \rightarrow FAD_{inner\ membrane} + 1\tfrac{1}{2}\ ATP + H_2O$$

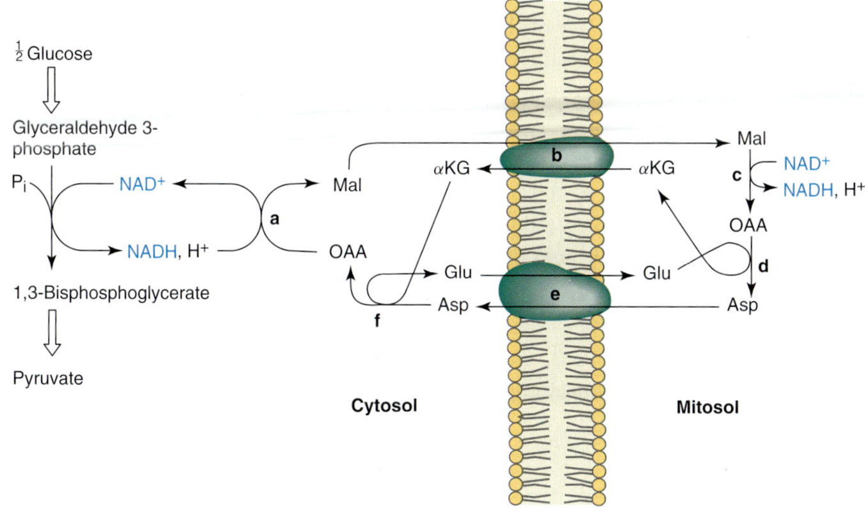

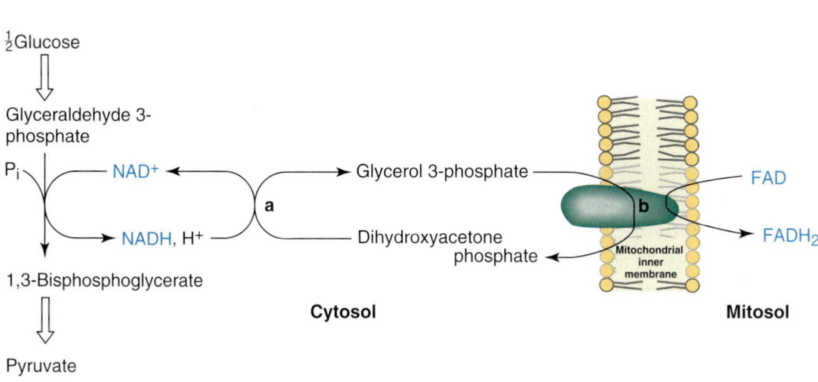

FIGURE 15.10

Shuttles for transport of reducing equivalents from the cytosol to the mitochondrial electron-transfer chain. (*a*) Malate–aspartate shuttle: a, cytosolic malate dehydrogenase oxidizes NADH by reducing oxaloacetate (OAA) to malate (Mal); b, dicarboxylic acid antiport in the mitochondrial inner membrane catalyzes electrically neutral exchange of malate for α-ketoglutarate (α-KG); c, a mitochondrial malate dehydrogenase reduces NAD^+ by oxidizing Mal to OAA; d, mitochondrial aspartate aminotransferase transaminates glutamate (Glu) and OAA; e, glutamate-aspartate antiport of the mitochondrial inner membrane catalyzes electrogenic exchange of glutamate for aspartate (Asp); f, cytosolic aspartate aminotransferase transaminates Asp and α-KG. (*b*) Glycerol phosphate shuttle: a, cytosolic glycerol 3-phosphate dehydrogenase oxidizes NADH by reducing dihydroxyacetone to glycerol 3-phosphate; b, glycerol 3-phosphate dehydrogenase located in the outer surface of the mitochondrial inner membrane reduces FAD, which is used by the mitochondrial electron-transfer chain.

Thus, the ATP yield from the oxidation of the NADH generated by glycolysis is 3 or 5, depending on the substrate shuttle used for its oxidation.

Shuttles are Important in Other Oxidoreduction Pathways

Alcohol Oxidation

Alcohol (i.e., ethanol) is oxidized to acetaldehyde with production of NADH by **alcohol dehydrogenase**.

$$\underset{\text{Ethanol}}{CH_3CH_2OH} + NAD^+ \longrightarrow \underset{\text{Acetaldehyde}}{CH_3\overset{\overset{\displaystyle O}{\|}}{C}H} + NADH + H^+$$

This enzyme is located almost exclusively in the cytosol of hepatocytes. Acetaldehyde traverses the mitochondrial inner membrane for oxidation by an **aldehyde dehydrogenase** in the mitochondrial matrix space.

$$\underset{\text{Acetaldehyde}}{CH_3\overset{\overset{\displaystyle O}{\|}}{C}H} + NAD^+ + H_2O \longrightarrow \underset{\text{Acetate}}{CH_3\overset{\overset{\displaystyle O}{\|}}{C}O^-} + NADH + 2H^+$$

The NADH generated by the last step can be used directly by the mitochondrial electron-transfer chain. However, the NADH generated by cytosolic alcohol dehydrogenase is

CLINICAL CORRELATION 15.1
Alcohol and Barbiturates

Acute alcohol intoxication increases sensitivity to the general depressant effects of barbiturates. Barbiturates and alcohol interact with the γ-aminobutyrate (GABA)-activated chloride channel. Activation of this channel inhibits neuronal firing, which may explain the depressant effects of both compounds. This combination is very dangerous, and normal prescription doses of barbiturates are potentially lethal when taken with ethanol. Furthermore, ethanol inhibits the metabolism of barbiturates, thereby prolonging the time barbiturates remain effective in the body. Hydroxylation of barbiturates by the NADPH-dependent cytochrome P450 system of the endoplasmic reticulum of the liver is inhibited by ethanol. This decreases formation of water-soluble derivatives of the barbiturates for elimination by the kidneys and bile. Blood levels of barbiturates remain high and cause increased CNS depression.

Surprisingly, the sober alcoholic is less sensitive to barbiturates. Chronic ethanol consumption apparently causes adaptive changes in the sensitivity to barbiturates (cross-tolerance) and induces cytochrome P450 of liver endoplasmic reticulum involved in drug hydroxylation reactions. Consequently, the sober alcoholic can metabolize barbiturates more rapidly. This sets up the following scenario. A sober alcoholic has trouble falling asleep, even after taking several sleeping pills, because his liver has increased capacity to hydroxylate the barbiturate contained in the pills. In frustration he consumes more pills and then alcohol. Sleep results, but may be followed by respiratory depression and death because the alcoholic, although less sensitive to barbiturates when sober, remains sensitive to the synergistic effect of alcohol.

Source: Misra, P. S., Lefevre, A., Ishii, H., Rubin, E., and Lieber, C. S. Increase of ethanol, meprobamate and pentobarbital metabolism after chronic ethanol administration in man and in rats. *Am. J. Med.* 51:346, 1971.

oxidized back to NAD^+ by one of the substrate shuttles. Thus, the capacity to oxidize alcohol is dependent on the ability of the liver to transport reducing equivalents from the cytosol into the mitochondria by these shuttle systems.

Glucuronide Formation

Water-soluble glucuronides of bilirubin and various drugs (see p. 842) are eliminated in the urine and bile. For glucuronide formation, UDP-glucose (structure on p. 646) is oxidized to **UDP-glucuronic acid** (structure on p. 646) by

$$\text{UDP-D-glucose} + 2\ NAD^+ + H_2O \rightarrow \text{UDP-D-glucuronic acid} + 2\ NADH + 2H^+$$

Primarily in liver, this "activated" glucuronic acid is transferred to a nonpolar, acceptor molecule, such as bilirubin or a compound foreign to the body:

$$\text{UDP-D-glucuronic acid} + \text{R-OH} \rightarrow \text{R-O-glucuronic acid} + \text{UDP}$$

NADH generated by the first reaction is reoxidized by the substrate shuttles. Since ethanol oxidation and drug conjugation occurs in the liver, the two occurring together may overwhelm the capacity of the substrate shuttles. This explains the advice not to mix the intake of pharmacologically active compounds with alcohol (see Clin. Corr. 15.1).

Sulfhydryl Reagents and Fluoride Inhibit Glycolysis

Glyceraldehyde 3-phosphate dehydrogenase is inhibited by **sulfhydryl reagents** because of the catalytically important cysteine residue at its active site. During a catalytic cycle the sulfhydryl group reacts with glyceraldehyde 3-phosphate to form a thiohemiacetal (Figure 15.8). Sulfhydryl reagents, which are often mercury-containing compounds or alkylating compounds such as iodoacetate, prevent formation of the thiohemiacetal (Figure 15.11).

Fluoride is a potent inhibitor of enolase. Mg^{2+} and P_i form an ionic complex with fluoride that inhibits enolase by interfering with the binding of its substrate (Mg^{2+}2-phosphoglycerate^{2-}).

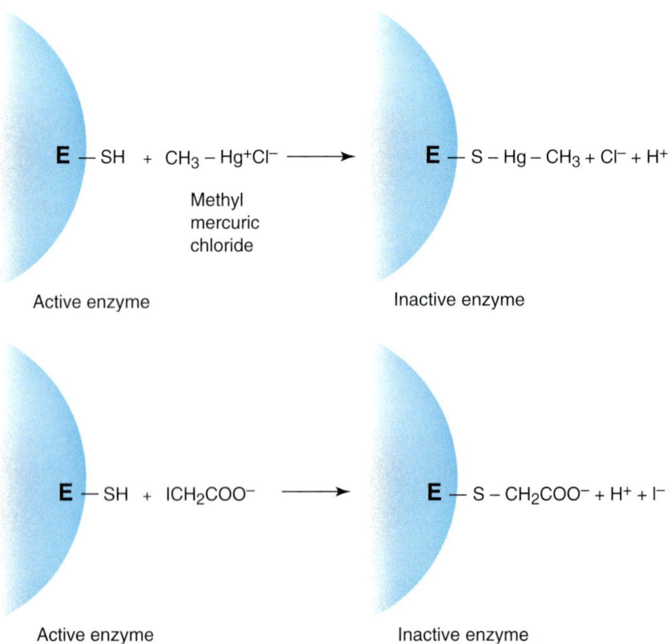

FIGURE 15.11
Mechanism for inactivation of glyceraldehyde 3-phosphate dehydrogenase by sulfhydryl reagents.

Arsenate Prevents Net Synthesis of ATP without Inhibiting Glycolysis

Pentavalent **arsenic** or **arsenate** prevents net synthesis of ATP by causing arsenolysis in the glyceraldehyde 3-phosphate dehydrogenase reaction (Figure 15.12). The structure of arsenate resembles that of P_i (also pentavalent) and readily substitutes for P_i in enzyme-catalyzed reactions. The mixed anhydride of arsenic acid and the carboxyl group of 3-phosphoglycerate is formed by glyceraldehyde 3-phosphate dehydrogenase. 1-Arsenato 3-phosphoglycerate is unstable, and it undergoes spontaneous hydrolysis to 3-phosphoglycerate and inorganic arsenate. As a consequence, glycolysis continues unabated in the presence of arsenate, but 1,3-bisphosphoglycerate is not formed. ATP is therefore not synthesized by phosphoglycerate kinase. Thus net ATP synthesis does not occur when glycolysis takes place in the presence of arsenate, the ATP invested in the priming stage being balanced by the ATP generated in the pyruvate kinase step. Arsenolysis also interferes with ATP formation by oxidative phosphorylation, hence the toxicity of arsenate (see Clin. Corr. 15.2).

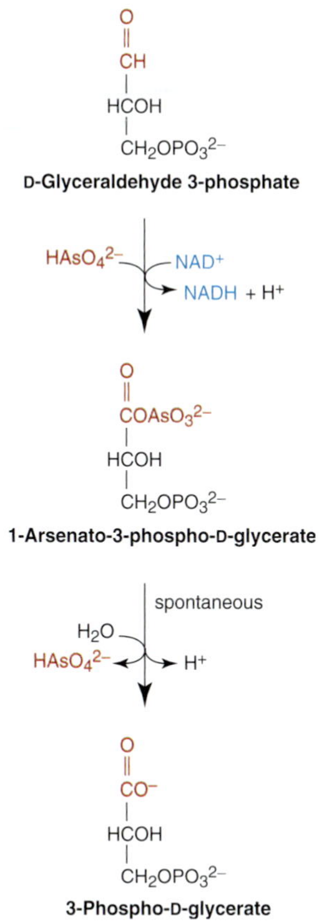

O
‖
CH
|
HCOH
|
$CH_2OPO_3^{2-}$
D-Glyceraldehyde 3-phosphate

$HAsO_4^{2-}$ → NAD^+ → $NADH + H^+$

O
‖
$COAsO_3^{2-}$
|
HCOH
|
$CH_2OPO_3^{2-}$
1-Arsenato-3-phospho-D-glycerate

spontaneous
H_2O → $HAsO_4^{2-}$ + H^+

O
‖
CO^-
|
HCOH
|
$CH_2OPO_3^{2-}$
3-Phospho-D-glycerate

FIGURE 15.12
Arsenate uncouples oxidation from phosphorylation at the glyceraldehyde 3-phosphate dehydrogenase reaction.

15.4 REGULATION OF GLYCOLYSIS

As in other complex pathways that involve multiple steps, flux through glycolysis is determined by the activities of several enzymes rather than a single "rate-limiting" enzyme. Quantitative information about the relative contribution of an enzyme to flux through a pathway under a particular set of conditions is best determined from the "control strength" for the enzyme. The effect that a small inhibition of the activity of an enzyme has upon flux through a pathway is used to determine an enzyme's control strength according to the equation

$$\text{Control Strength} = \frac{\text{Change in flux}}{\text{Change in enzyme activity}}$$

For the purpose of illustration, assume that inhibition of a particular enzyme by 10% has no effect upon flux. The control strength of the enzyme is then zero (0 divided

CLINICAL CORRELATION 15.2
Arsenic Poisoning

Most forms of arsenic are toxic, the trivalent form (arsenite as AsO_2^-) being much more toxic than the pentavalent form (arsenate or $HAsO_4^{2-}$). Less ATP is produced whenever arsenate substitutes for P_i in biological reactions. Arsenate competes for P_i-binding sites on enzymes, resulting in the formation of arsenate esters that are unstable. Arsenite works by a different mechanism, involving formation of a stable complex with enzyme bound lipoic acid (see figure).

For the most part, arsenic poisoning is explained by inhibition of enzymes that require lipoic acid as a coenzyme. These include pyruvate dehydrogenase, α-ketoglutarate dehydrogenase, and branched-chain α-keto acid dehydrogenase. Chronic arsenic poisoning from well water contaminated with arsenical pesticides or through the efforts of a murderer is best diagnosed by determining the concentration of arsenic in the hair or fingernails of the victim. About 0.5 mg of arsenic are present in a kilogram of hair from a normal individual. The hair of a person chronically exposed to arsenic could have 100 times as much.

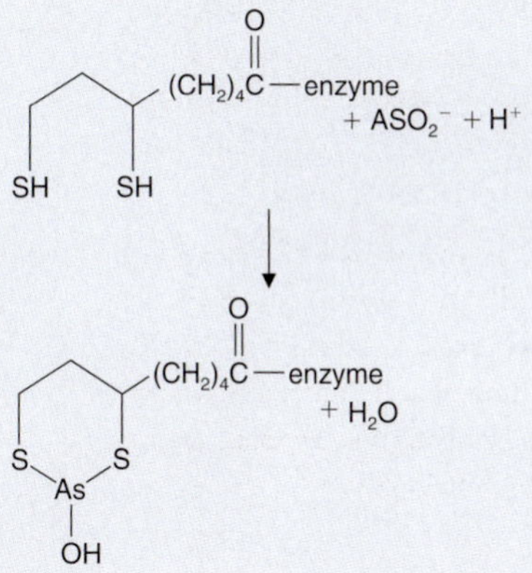

SH SH $(CH_2)_4C(=O)$—enzyme + ASO_2^- + H^+
↓
S–As(OH)–S $(CH_2)_4C(=O)$—enzyme + H_2O

Source: Hindmarsh, J. T. and McCurdy, R. F. Clinical and environmental aspects of arsenic toxicity. *CRC Crit. Rev. Clin. Lab. Sci.* 23:315, 1986. Mudur, G. Half of Bangladesh population at risk of arsenic poisoning. *Br. Med. J.* 320:822, 2000.

by 10). Now assume that inhibition of an enzyme by 10% results in 10% inhibition of flux. The control strength would be 1.0 (10 divided by 10) which means that flux is entirely dependent upon the activity of the enzyme under these conditions. Now assume that inhibition of an enzyme by 10% results in only 5% inhibition of flux through the pathway. The control strength is now 0.5 (5 divided by 10), which means that half of control of flux is determined by this enzyme. The rest of the control is exerted by one or more other steps, since the control strength of steps of a linear pathway must sum to 1.0 by definition.

With the caveat that regulation of flux through glycolysis is dependent upon the tissue and the nutritional and hormonal state, the enzymes of glycolysis with the greatest control strength are **hexokinase**, **6-phosphofructo-1-kinase**, and **pyruvate kinase** (Figure 15.13). These enzymes are regulated by allosteric effectors and/or covalent modification.

A nonregulatory enzyme most likely catalyzes a "near-equilibrium reaction," whereas a regulatory enzyme most likely catalyzes a "nonequilibrium reaction." The activity of a nonregulatory enzyme readily brings its substrates and products to equilibrium concentrations. A regulatory enzyme is not active enough to equilibrate its substrates and products. Whether an enzyme-catalyzed reaction is near equilibrium or nonequilibrium can be determined by comparing the established equilibrium constant for the reaction with the mass-action ratio as it exists within a cell. The equilibrium constant for the reaction $A + B \rightarrow C + D$ is defined as

$$K_{eq} = \frac{[C][D]}{[A][B]}$$

where the brackets indicate the concentrations at equilibrium. The mass-action ratio is calculated in a similar manner, except that the steady-state (ss) concentrations of

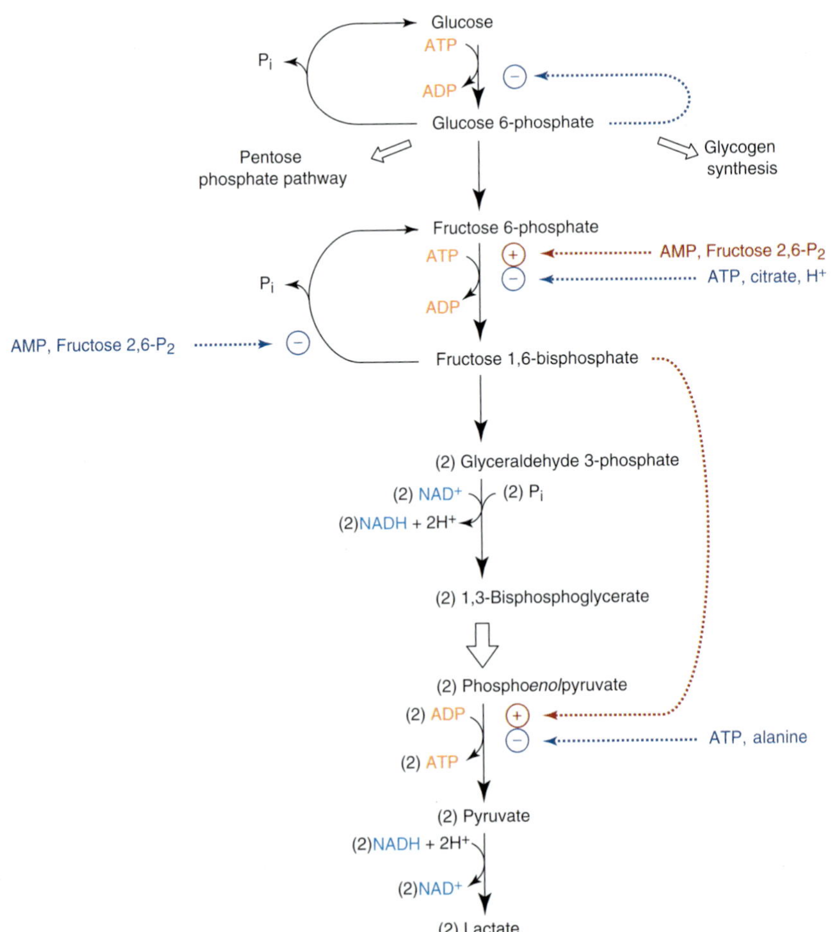

FIGURE 15.13

Important regulatory features of glycolysis. Because of tissue differences in isoenzyme expression, not all tissues of the body have all of the regulatory mechanisms shown here.

reactants and products within the cell are used:

$$\text{Mass-action ratio} = \frac{[\mathrm{C}]_{ss}[\mathrm{D}]_{ss}}{[\mathrm{A}]_{ss}[\mathrm{B}]_{ss}}$$

If the mass-action ratio is approximately equal to K_{eq}, the enzyme is said to catalyze a near-equilibrium reaction and is not considered likely to be regulated. When the mass-action ratio is considerably different from the K_{eq}, the enzyme is said to catalyze a nonequilibrium reaction and is usually regulated. Comparison of mass-action ratios and equilibrium constants for enzymes of glycolysis in liver indicates that many catalyze equilibrium reactions. The reactions of glucokinase (liver isoenzyme of hexokinase), 6-phosphofructo-1-kinase, and pyruvate kinase in liver are far from equilibrium, suggesting that they are likely sites for regulation.

Hexokinase and Glucokinase have Different Properties

Different isoenzymes of hexokinase are present in different tissues. The **hexokinase isoenzymes** found in most tissues have a low K_m for glucose (<0.1 mM) relative to its concentration in blood (~5 mM) and are strongly inhibited by the product glucose 6-phosphate (G6P). The latter is important because it prevents hexokinase from tying up all of the inorganic phosphate of a cell in the form of phosphorylated hexoses (see Clin. Corr. 15.3). The hexokinase reaction is not at equilibrium because of the inhibition imposed by G6P. Liver parenchymal cells and β cells of pancreas are unique in that they contain glucokinase, an isoenzyme of hexokinase with strikingly different kinetic properties. It catalyzes ATP-dependent phosphorylation of glucose like other hexokinases, but its $S_{0.5}$ (substrate concentration that gives enzyme activity of one-half maximum velocity) for glucose is considerably higher than the K_m for glucose of the other hexokinases (Figure 15.14). Furthermore, glucokinase is much less sensitive to

CLINICAL CORRELATION 15.3
Fructose Intolerance

Patients with hereditary fructose intolerance are deficient in the liver enzyme (aldolase B) that splits fructose 1-phosphate into dihydroxyacetone phosphate and glyceraldehyde. Three isoenzymes (A, B, and C) are expressed in mammals. Aldolase B is present in the largest amounts in liver. It acts on both fructose 1-phosphate and fructose 1,6-bisphosphate, but it has much greater affinity (lower Km) for fructose 1,6-bisphosphate. Consumption of fructose by an individual with an aldolase B deficiency results in accumulation of fructose 1-phosphate and depletion of P_i and ATP in the liver. The reactions involved are those of fructokinase and the enzymes of oxidative phosphorylation:

$$\text{Fructose} + \text{ATP} \rightarrow \text{fructose 1-phosphate} + \text{ADP}$$

$$\text{ADP} + \text{P}_i + \text{“energy provided by electron transport chain”} \rightarrow \text{ATP}$$

$$\text{Net: P}_i + \text{fructose} \rightarrow \text{fructose 1-phosphate}$$

Tying up P_i as fructose 1-phosphate makes it impossible for liver mitochondria to generate ATP by oxidative phosphorylation. The ATP levels fall precipitously, making it impossible for the liver to carry out its normal functions. Cell damage results largely because of inability to maintain normal ion gradients by the ATP-dependent pumps. The cells swell and undergo osmotic lysis.

Although patients with fructose intolerance are particularly sensitive to fructose, humans in general have a limited capacity to handle this sugar. The capacity of the normal liver to phosphorylate fructose greatly exceeds its capacity to split fructose 1-phosphate. This means that fructose use by the liver is poorly controlled and that excessive fructose could deplete the liver of P_i and ATP. Fructose was actually tried briefly in hospitals as a substitute for glucose in patients being maintained by parenteral nutrition. The rationale was that fructose would be a better source of calories than glucose because its utilization is relatively independent of the insulin status of a patient. Delivery of large amounts of fructose by intravenous feeding was soon found to result in severe liver damage. Similar attempts have been made to substitute sorbitol and xylitol for glucose, but they also tend to deplete the liver of ATP and, like fructose, should not be used for parenteral nutrition.

Source: Steinmann, B., Gitzelmann, R., and Van den Berghe, G. Disorders of fructose metabolism. In: C. R. Scriver, A. L. Beaudet, W. S. Sly, and D. Valle (Eds.), *The Metabolic an Molecular Bases of Inherited Disease*, 8th ed. New York: McGraw-Hill, 2001, p. 1489. Ali, M., Rellos, P., and Cox T. M. Hereditary fructose intolerance. *J. Med. Genet.* 35:353, 1998.

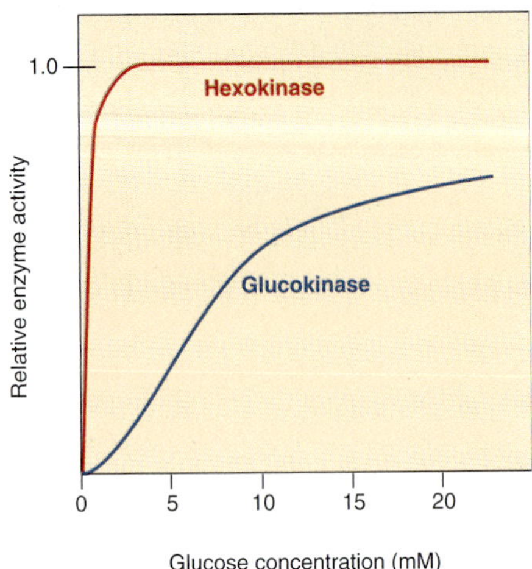

FIGURE 15.14

Comparison of the substrate saturation curves for hexokinase and glucokinase.

product inhibition by G6P, and its glucose saturation curve is sigmoidal—that is, indicative of cooperativity (see p. 402). Because the Michael–Menten equation does not apply (see p. 388), its kinetics are described by an $S_{0.5}$ value (concentration of substrate required to produce a rate one-half of V_{max}) rather than a K_m value for glucose. Although not sensitive to inhibition by physiological concentrations of G6P, glucokinase is indirectly regulated by fructose 6-phosphate, which is just one step removed and in equilibrium with G6P. A special **glucokinase inhibitory protein (GK-RP)**, which is located in the nucleus of liver cells, is responsible for this effect (Figure 15.15). GK-RP sequesters glucokinase as an inactive complex in the nucleus. **Fructose 6-phosphate** promotes binding of glucokinase to the regulatory protein, thereby inhibiting glucokinase. Fructose 6-phosphate in effect promotes translocation of glucokinase from the cytosol to the nucleus where glucokinase is completely inhibited by the inhibitory protein. This inhibitory effect of fructose 6-phosphate on glucokinase can be completely overcome by a large increase in glucose concentration. Glucose triggers dissociation of glucokinase from the regulatory protein, thereby promoting translocation of glucokinase from the nucleus to the cytosol. These special regulatory features of glucokinase (high $S_{0.5}$ for glucose, cooperativity with respect to glucose concentration, and glucose stimulated translocation from the nucleus to the cytosol) contribute to the capacity of the liver to "buffer" blood glucose levels. Because **GLUT2**, the glucose transporter of liver cells, brings about rapid equilibration of glucose across the plasma membrane, the cellular concentration of glucose is the same as that in the blood. Since the $S_{0.5}$ of glucokinase for glucose (~7 mM) is greater than normal blood glucose concentrations (~5 mM) and glucose promotes translocation of glucokinase from the nucleus, any increase in portal blood glucose above normal leads to a dramatic increase in the rate of glucose phosphorylation by glucokinase in the liver (Figures 15.14 and 15.15). Likewise, any decrease in glucose concentration has the opposite effect. Thus, liver uses glucose at a significant rate only when blood glucose levels are greatly elevated and decreases its use of glucose when blood glucose levels are low. This buffering effect on blood glucose levels would not occur if glucokinase had the low K_m characteristic of other hexokinases, because it would then be completely saturated at physiological concentrations of glucose. On the other hand, a low K_m form of hexokinase is a good choice for tissues such as the brain in that it allows phosphorylation of glucose even when blood and tissue glucose concentrations are dangerously low.

The **glucokinase** reaction is not at equilibrium under normal intracellular conditions because of the rate restriction imposed by the high $S_{0.5}$ and inhibition by the regulatory protein. Another factor in opposition to the activity of glucokinase is glucose 6-phosphatase, which like glucokinase has a high K_m (3 mM) for glucose 6-phosphate relative to its intracellular concentration (~0.2 mM). Thus, flux through this step is almost directly proportional to the intracellular concentration of glucose 6-phosphate. As shown in Figure 15.16, the combined action of glucokinase and **glucose 6-phosphatase** constitutes a futile cycle; that is, the sum of their reactions is hydrolysis of ATP to give ADP and P_i without the performance of any work. When blood glucose concentrations are about 5 mM, the activity of glucokinase is almost exactly balanced by the opposing activity of glucose 6-phosphatase. The result is that no net flux occurs in either direction.

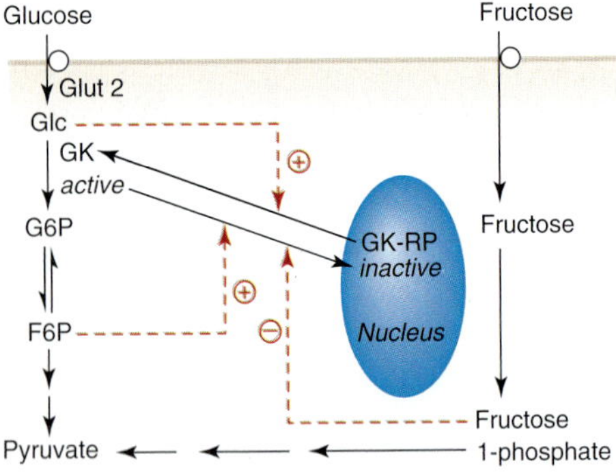

FIGURE 15.15

Glucokinase activity is regulated by translocation of the enzyme between the cytoplasm and the nucleus. Glucose increases glucokinase (GK) activity by promoting translocation of the enzyme to the cytoplasm. Fructose 6-phosphate decreases GK activity by stimulating translocation into the nucleus. Fructose 1-phosphate increases GK activity by inhibiting translocation into the nucleus. Binding of GK to regulatory protein (RP) in the nucleus completely inhibits GK activity.

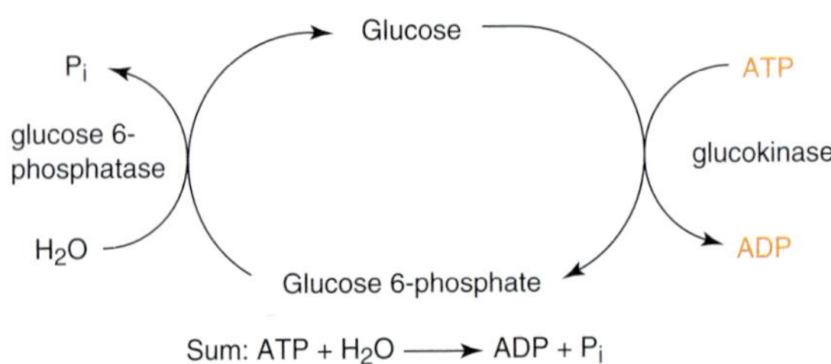

FIGURE 15.16

Phosphorylation of glucose followed by dephosphorylation constitutes a futile cycle in parenchymal cells of the liver.

This **futile cycling** is wasteful of ATP but, combined with the process of gluconeogenesis (see p. 608), contributes significantly to the "buffering" action of the liver on blood glucose levels. It also provides a mechanism for preventing glucokinase from tying up all of the P_i of the liver (see Clin. Corr. 15.3).

Fructose, a component of fruits, honey, vegetables, and high fructose corn syrup used in popular carbonated beverages, promotes hepatic glucose utilization by an indirect mechanism. It is converted in liver to fructose 1-phosphate (see Clin. Corr. 15.3), which promotes dissociation of glucokinase from its regulatory protein (Figure 15.15) and therefore translocation out of the nucleus. This action, which is capable of overriding the inhibition of glucokinase normally imposed by fructose 6-phosphate, may be a factor in the adverse effects sometimes associated with excessive dietary fructose consumption—for example, increased hepatic carbohydrate utilization, lipogenesis, and hypertriacylglycerolemia.

Glucokinase is also an **inducible enzyme**. Induction and repression of synthesis of an enzyme are slow processes, usually requiring several hours before significant changes occur. Insulin promotes transcription of the glucokinase gene. An increase in blood glucose levels signals an increase in **insulin** release from the β cells of the pancreas, which promotes transcription and increases the amount of liver glucokinase enzyme protein. Thus, a person consuming a large carbohydrate-rich meal will have greater amounts of glucokinase than one who is not. The liver with glucokinase induced contributes more to the lowering of elevated blood glucose levels. The absence of insulin makes the liver of patients with diabetes mellitus deficient in glucokinase, in spite of high blood glucose levels, and decreases the ability of the liver to "buffer" blood glucose (see Clin. Corr. 15.4). Defects in the gene encoding glucokinase that alter its $S_{0.5}$ and/or V_{max} cause maturity-onset diabetes of the young (MODY), a form of type 2 diabetes mellitus.

CLINICAL CORRELATION 15.4

Diabetes Mellitus

Diabetes mellitus is a chronic disease characterized by derangements in carbohydrate, fat, and protein metabolism. Two major types are recognized clinically: type 1 (see Clin. Corr. 22.8, p. 877) and type 2 (see Clin. Corr. 22.7, p. 876).

In patients without fasting hyperglycemia, the oral glucose tolerance test can be used for diagnosis. It consists of determining the blood glucose level in the fasting state and at intervals of 30–60 min for 2 h or more after consuming a 100-g carbohydrate load. In normal individuals blood glucose returns to normal levels within 2 h after ingestion of the carbohydrate. In diabetics, blood glucose reaches a higher level and remains elevated for longer periods of time, depending on the severity of the disease. However, many factors may contribute to an abnormal glucose tolerance test. The patient must have consumed a high carbohydrate diet for the preceding 3 days, presumably to allow for induction of enzymes of glucose-utilizing pathways, for example, glucokinase, fatty acid synthase, and acetyl-CoA carboxylase. Almost any infection (even a cold) and less well-defined "stress" (presumably by effects on the sympathetic nervous system) can result in transient abnormalities of the glucose tolerance test. Because of these problems, fasting hyperglycemia should probably be the sine qua non for the diagnosis of diabetes. Glucose uptake by insulin-sensitive tissues—that is, muscle and adipose—is decreased in the diabetic state. The diabetic patient either lacks insulin or has developed "insulin resistance" in these tissues. Resistance to insulin results from abnormality of the insulin receptor or in subsequent steps that mediate the metabolic effects of insulin. Liver parenchymal cells do not require insulin for glucose uptake. Without insulin, however, the liver has diminished capacity to remove glucose from the blood. This is explained in part by decreased glucokinase activity and the loss of insulin's action on key enzymes of glycogenesis and the glycolytic pathway.

Source: Taylor, S.I. Insulin action, insulin resistance, and type 2 diabetes mellitus. In: C. R. Scriver, A. L. Beaudet, W. S. Sly, and D. Valle (Eds.), *The Metabolic and Molecular Bases of Inherited Disease*,, 8th ed. New York: McGraw-Hill, 2001, p. 1433.

6-Phosphofructo-1-kinase is a Regulatory Enzyme of Glycolysis

6-Phosphofructo-1-kinase is an important regulatory site of glycolysis. It catalyzes the first committed step of glycolysis because the reaction catalyzed by phosphoglucose isomerase is reversible, and cells make use of glucose 6-phosphate in the pentose phosphate pathway and for glycogen synthesis. Citrate, ATP, and hydrogen ions (low pH) are the important negative allosteric effectors while AMP and fructose 2,6-bisphosphate are important positive allosteric effectors (Figure 15.13). These compounds signal the need for different rates of glycolysis in response to changes in (a) energy state of the cell (ATP and AMP), (b) internal environment of the cell (hydrogen ions), (c) availability of alternate fuels such as fatty acids and ketone bodies (citrate), and (d) the insulin/glucagon ratio in the blood (fructose 2,6-bisphosphate).

Regulation of 6-Phosphofructo-1-kinase by ATP and AMP

The **Pasteur effect** is the inhibition of glucose utilization and lactate accumulation that occurs when respiration (oxygen consumption) is initiated in anaerobic cells. It is readily understandable on a thermodynamic basis since complete oxidation of glucose to CO_2 and H_2O yields much more ATP than anaerobic glycolysis:

$$\text{Glycolysis}: \quad \text{D-Glucose} + 2\,ADP^{3-} + 2\,P_i^{2-} \rightarrow 2\,\text{L-lactate}^- + 2\,ATP^{4-}$$

$$\text{Complete Oxidation}: \quad \text{D-Glucose} + 6O_2 + 32\,ADP^{3-} + 32\,P_i^{2-} + 32\,H^+ \rightarrow 6\,CO_2 + 6\,H_2O + 32\,ATP^{4-}$$

Cells use ATP to provide the energy necessary for their inherent work processes. Since so much more ATP is produced from glucose in the presence of oxygen, much less glucose has to be consumed to meet the energy need. The Pasteur effect occurs in part because of ATP inhibition of glycolysis at the level of 6-phosphofructo-1-kinase. This can be readily rationalized since ATP is an inhibitor of 6-phosphofructo-1-kinase, and much more ATP is generated in the presence of oxygen than in its absence. Since 6-phosphofructo-1-kinase is severely inhibited at concentrations of ATP (2.5–6 mM) normally present in cells, the relatively small changes in ATP concentration that occurs with versus without oxygen cannot account for large changes in flux through 6-phosphofructo-1-kinase. However, much greater changes occur in the concentration of AMP, a positive allosteric effector of 6-phosphofructo-1-kinase (Figure 15.13). The steady-state concentration of AMP when oxygen is introduced falls dramatically, decreasing 6-phosphofructo-1-kinase activity, suppressing glycolysis, and accounting in large part for the Pasteur effect. Levels of AMP automatically fall when ATP levels rise because the sum of adenine nucleotides in a cell, that is, ATP + ADP + AMP, is nearly constant under most physiological conditions and the amount of ATP is always much greater than the amount of AMP. Furthermore, adenine nucleotides are maintained in equilibrium in the cytosol by adenylate kinase, which catalyzes the readily reversible reaction of 2 ADP → ATP + AMP. The equilibrium constant (K'_{eq}) for this reaction is given by

$$K'_{eq} = \frac{[\text{ATP}][\text{AMP}]}{[\text{ADP}]^2}$$

Since this reaction operates "near equilibrium" under intracellular conditions, the concentration of AMP is given by

$$[\text{AMP}] = \frac{K'_{eq}[\text{ADP}]^2}{[\text{ATP}]}$$

Because intracellular [ATP] ≫ [ADP ≫ [AMP], a small decrease in [ATP] causes a substantially greater percentage increase in [ADP]; and since [AMP] is related to the square of [ADP], a small decrease in [ATP] causes an even greater percentage increase in [AMP]. Because of this relationship, a small decrease in [ATP] leads to a greater percent increase in [AMP] than in the percent decrease in [ATP]. This makes the [AMP] an excellent signal of the energy status of the cell and an important allosteric regulator of 6-phosphofructo-1-kinase activity. The effectiveness of 6-phosphofructo-1-kinase is influenced by [AMP] in yet another way. Fructose 1,6-bisphosphatase catalyzes an irreversible reaction that opposes the reaction of 6-phosphofructo-1-kinase:

$$\text{Fructose 1,6-bisphosphate} + H_2O \rightarrow \text{fructose 6-phosphate} + P_i$$

Fructose 1,6-bisphosphatase sits "cheek by jowl" with 6-phosphofructo-1-kinase in the cytosol of many cells. Together they catalyze a futile cycle (ATP + H_2O → ADP + P_i + "heat") and are therefore capable of decreasing each others "effectiveness." However, AMP inhibits fructose 1,6-bisphosphatase, which is opposite to the effect that AMP has on 6-phosphofructo-1-kinase. Thus, a small decrease in [ATP] triggers a large percentage increase in [AMP], which in turn signals a large increase in net conversion of fructose 6-phosphate into fructose 1,6-bisphosphate because of its effects on these two enzymes. This increases glycolytic flux by increasing the amount of fructose 1,6-bisphosphate available for the splitting stage.

The decrease in lactate that occurs due to the Pasteur effect is readily explained by decreased glycolytic flux. In addition, lactate dehydrogenase loses the competition with the substrate shuttle systems for NADH and the competition with the pyruvate dehydrogenase complex for pyruvate.

Regulation of 6-Phosphofructo-1-kinase by Intracellular pH

It would make sense if lactate, as the end product of glycolysis, would inhibit the most important regulatory enzyme of glycolysis. However, it is hydrogen ions rather than lactate that inhibit 6-phosphofructo-1-kinase (Figure 15.13). As shown in Figure 15.17, anaerobic glycolysis generates lactic acid, and the cell must dispose of it as such or suffer the negative consequences of acidification. This explains why excessive glycolysis lowers blood pH and leads to an emergency medical situation known as **lactic acidosis** (see Clin. Corr. 15.5). Plasma membranes contain a symport for lactate and hydrogen ions that allows transfer of lactic acid into the bloodstream. This defense mechanism prevents pH from getting so low that lactic acid producing tissues become pickled (see Clin. Corr. 15.6). Transport of lactic acid out of a cell requires that blood be available to carry it away. When blood flow is inadequate, for example, in the heart during an attack of angina pectoris or in skeletal muscle during heavy exercise, hydrogen ions cannot escape

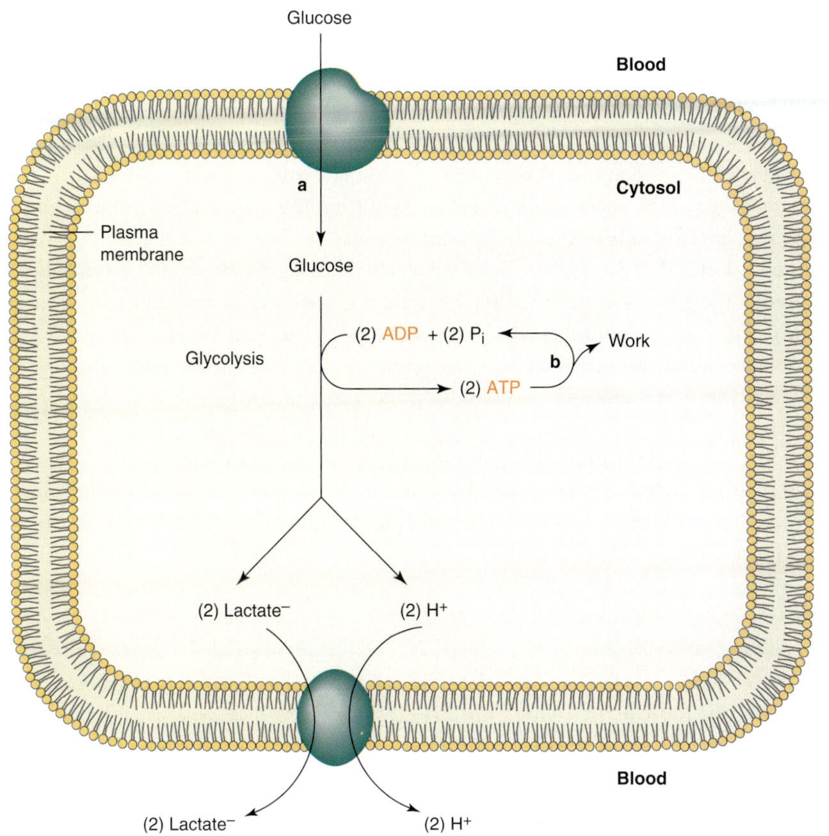

FIGURE 15.17

Unless lactate formed by glycolysis is transported out of the cell, the intracellular pH will decrease by the accumulation of intracellular lactic acid (equivalent to lactate⁻ plus H^+ because lactic acid ionizes at intracellular pH). The low pH decreases 6-phosphofructo-1-kinase activity so that further lactic acid production by glycolysis is inhibited. (*a*) Glucose transport into the cell; (*b*) all work performances that convert ATP back to ADP and P_i; (*c*) lactate-H^+ symport (actual stoichiometry of one lactate⁻ and one H^+ transported by the symport).

CLINICAL CORRELATION 15.5
Lactic Acidosis

This is characterized by elevated blood lactate levels, usually greater than 5 mM, with decreased blood pH and bicarbonate concentrations. Lactic acidosis is the most frequent form of metabolic acidosis and can be the consequence of overproduction of lactate, underutilization of lactate, or both. Lactate production is normally balanced by lactate utilization so that lactate is usually not present in the blood at concentrations greater than 1.2 mM. All tissues can produce lactate by anaerobic glycolysis, but most tissues do not produce large quantities since they are well-supplied with oxygen and mitochondria. However, all tissues respond with an increase in lactate generation when oxygenation is inadequate. A decrease in ATP from reduced oxidative phosphorylation increases the activity of 6-phosphofructo-1-kinase. Thus, tissues have to rely on anaerobic glycolysis for ATP production under such conditions and overproduce lactic acid. A good example is muscle exercise, which can deplete the tissue of oxygen and cause an overproduction of lactic acid. Tissue hypoxia occurs, however, in all forms of shock, during convulsions, and in diseases involving circulatory and pulmonary failure.

The major fate of lactate in the body is either complete combustion to CO_2 and H_2O or conversion back to glucose by the process of gluconeogenesis. Both require oxygen. Decreased oxygen availability, therefore, increases lactate production and decreases lactate utilization. The latter can also be decreased by liver diseases, ethanol, and a number of other drugs. Phenformin, a drug that was once used to treat the hyperglycemia of type 2 diabetes, was well-documented to induce lactic acidosis in certain patients.

Bicarbonate is usually administered in an attempt to control the acidosis associated with lactic acid accumulation. The key to successful treatment, however, is to find and eliminate the cause of the overproduction and/or underutilization of lactic acid and most often involves the restoration of circulation of oxygenated blood.

Source: Newsholme, E. A. and Leech, A. R. *Biochemistry for the Medical Sciences.* New York: Wiley, 1983. Kruse, J. A. and Carlson, R. W. Lactate metabolism. *Crit. Care Clin.* 3:725, 1985.

CLINICAL CORRELATION 15.6

Pickled Pigs and Malignant Hyperthermia

In malignant hyperthermia, a variety of agents, especially the general anesthetic halothane, produces a dramatic rise in body temperature, metabolic and respiratory acidosis, hyperkalemia, and muscle rigidity. This dominantly inherited abnormality occurs in about 1 in 15,000 children and 1 in 50,000–100,000 adults. Death may result the first time a susceptible person is anesthetized. Onset occurs within minutes of drug exposure and the hyperthermia must be recognized immediately. Packing the patient in ice is effective and should be accompanied by measures to combat acidosis. The drug dantrolene is also effective.

A phenomenon similar to malignant hyperthermia occurs in pigs, called porcine stress syndrome. The pigs respond poorly to stress. This genetic disease usually manifests itself as the pig is being shipped to market. On exposure to halothane, susceptible pigs have the same response as that observed in human with malignant hyperthermia. The meat of pigs that have died from the disease is pale, watery, and of very low pH (i.e., nearly pickled).

In response to halothane, the skeletal muscles of affected individuals become rigid and generate heat and lactic acid. The sarcoplasmic reticulum has an abnormal ryanodine receptor (see p. 468), a Ca^{2+} release channel that is important in excitation-contraction coupling. Because of a defect in this protein, the anesthetic triggers inappropriate release of Ca^{2+} from the sarcoplasmic reticulum and uncontrolled stimulation of heat-producing processes, including myosin ATPase, glycogenolysis, glycolysis, and cyclic uptake and release of Ca^{2+} by mitochondria and sarcoplasmic reticulum. Muscle cells become irreversibly damaged from the excessive heat production, lactic acidosis, and ATP loss.

Source: Kalow, W. Grant, D. M. Pharmacogenetics. In: C. R. Scriver, A. L. Beaudet, W. S. Sly, and D. Valle (Eds.), *The Metabolic and Molecular Basis of Inherited Disease*, 8th ed. New York: McGraw-Hill, 2001, p. 225. McCarthy T.V., Quane, K.A., and Lynch P. J. Ryanodine receptor mutations in malignant hyperthermia and central core disease. *Hum. Mutat.* 15:410, 2000.

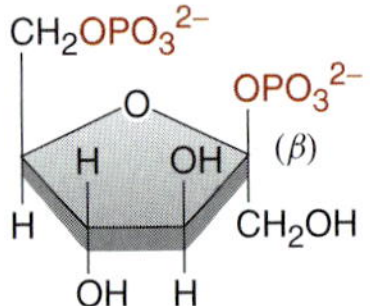

FIGURE 15.18
Structure of fructose 2,6-bisphosphate.

from cells fast enough. Yet, the need for ATP, manifested by an increase in [AMP], may partially override inhibition of 6-phosphofructo-1-kinase by hydrogen ions. Unabated accumulation of hydrogen ions then causes pain, which, in the case of skeletal muscle, can be relieved by simply terminating the exercise. In the heart, rest or pharmacologic agents that increase blood flow or decrease the need for ATP within myocytes may be effective (see Clin. Corr. 15.7).

Regulation of 6-Phosphofructo-1-kinase by Citrate

Many tissues prefer to use fatty acids and ketone bodies as oxidizable fuels in place of glucose. Most such tissues can use glucose but actually prefer to oxidize fatty acids and ketone bodies. This helps preserve glucose for tissues, such as brain, that are absolutely dependent on glucose. Oxidation of both fatty acids and ketone bodies elevates levels of cytosolic **citrate**, which inhibits 6-phosphofructo-1-kinase (Figure 15.13) and decreases glucose utilization.

Hormonal Control of 6-Phosphofructo-1-kinase by cAMP and Fructose 2,6-bisphosphate

Fructose 2,6-bisphosphate (Figure 15.18), like AMP, is a positive allosteric effector of **6-phosphofructo-1-kinase** and is a negative allosteric effector of fructose 1,6-bisphosphatase (Figure 15.13). Indeed, without its presence, glycolysis could not occur in liver because 6-phosphofructo-1-kinase would have insufficient activity and **fructose 1,6-bisphosphatase** would be too active for net conversion of fructose 6-phosphate to fructose 1,6-bisphosphate.

Figure 15.19 summarizes the role of fructose-2,6-bisphosphate in hormonal control of hepatic glycolysis. The mechanism uses cAMP (Figure 15.20) as the "second messenger" of hormone action. As discussed in more detail in Chapter 22, glucagon is released from α cells of pancreas and circulates in blood until it encounters glucagon receptors on the outer surface of liver plasma membrane. Binding of glucagon to its receptor triggers stimulation of adenylate cyclase through the second messenger **cyclic AMP** (cAMP), which results in a decrease in fructose 2,6-bisphosphate. This makes 6-phosphofructo-1-kinase less effective and fructose

CLINICAL CORRELATION 15.7

Angina Pectoris and Myocardial Infarction

Chest pain associated with reversible myocardial ischemia is termed angina pectoris (literally, strangling pain in the chest). The pain results from an imbalance between demand for and supply of blood flow to cardiac muscles and is most commonly caused by narrowing of the coronary arteries. The patient experiences a heavy squeezing pressure or ache substernally, often radiating to either the shoulder and arm or occasionally to the jaw or neck. Attacks occur with exertion, last from 1 to 15 min, and are relieved by rest or by death. The coronary arteries involved are obstructed by atherosclerosis (i.e., lined with fatty deposits) or less commonly narrowed by spasm. Infarction occurs if the ischemia persists long enough to cause severe damage (necrosis) to the heart muscle. Commonly, a blood clot forms at the site of narrowing and completely obstructs the vessel. In infarction, tissue death occurs and the pain is longer-lasting and often more severe.

Nitroglycerin and other nitrates are frequently prescribed to relieve the pain. They can be used prophylactically, enabling patients to participate in activities that would otherwise precipitate an attack. Nitroglycerin may work in part by causing dilation of the coronary arteries, improving oxygen delivery and washing out lactic acid. Probably more important is the effect on the peripheral circulation. Breakdown of nitroglycerin produces nitric oxide (NO) (see p. 438), which relaxes smooth muscle and thereby causes venodilation throughout the body. This reduces arterial pressure and allows blood to accumulate in the veins. The result is decreased return of blood to the heart and a reduced volume of blood to be pumped, which reduces the energy requirement. The heart also empties itself against less pressure, which further spares energy. The overall effect is a lowering of the oxygen requirement of the heart, bringing it in line with the oxygen supply via the diseased coronary arteries. Other useful agents are calcium channel blockers, which are coronary vasodilators, and β-adrenergic blockers. The β-blockers prevent the increase in myocardial oxygen consumption induced by sympathetic nervous system stimulation of the heart, as occurs with physical exertion.

The coronary artery bypass operation is used in severe cases of angina that cannot be controlled by medication. In this operation, veins are removed from the leg and interposed between the aorta and coronary arteries so as to bypass the portion affected by atherosclerosis and provide the affected tissue with a greater blood supply. Remarkable relief from angina can be achieved by this operation, with the patient being able to return to normal productive life in some cases.

Source: Hugenholtz, P. G. Calcium antagonists for angina pectoris. *Ann. N.Y. Acad. Sci.* 522:565, 1988. Feelishch, M. and Noack, E. A. Correlation between nitric oxide formation during degradation of organic nitrates and activation of guanylate cyclase. *Eur. J. Pharmacol.* 139:19, 1987. Ignarro, L. J. Biological actions and properties of endothelium-derived nitric oxide formed and released from artery and vein. *Circ. Res.* 65:1, 1989.

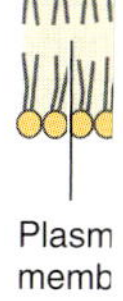

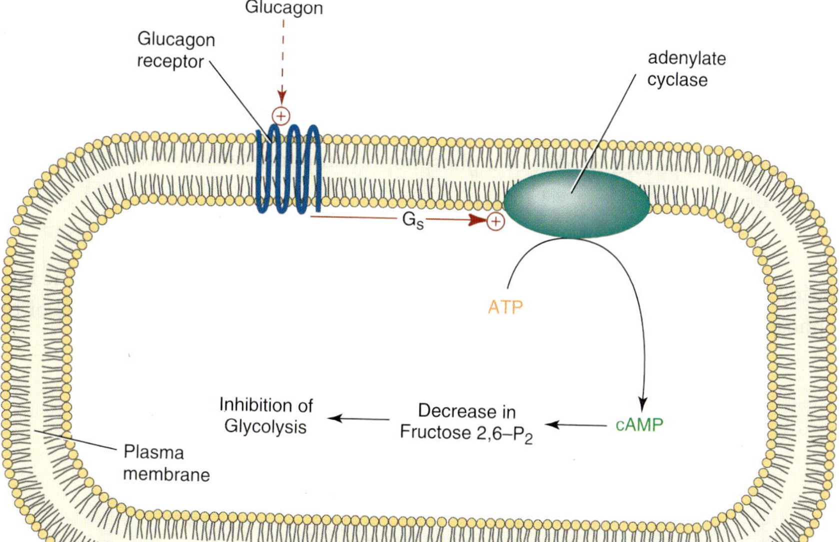

FIGURE 15.19

Mechanism by which glucagon inhibits hepatic glycolysis. Binding of glucagon to its membrane receptor activates adenylate cyclase (an intrinsic membrane protein) activity through the action of a stimulatory G protein (G_s; a peripheral membrane protein). The (+) symbol indicates activation.

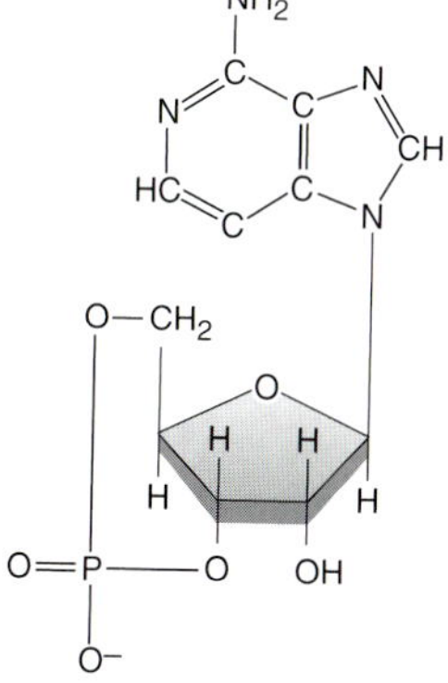

FIGURE 15.20

Structure of cAMP.

(Figure 15.32*a*), only two molecules of ATP per molecule of glucose are produced:

$$6\ \text{ATP}_{\text{liver}} + 2\ (\text{ADP} + \text{P}_i)_{\text{red blood cells}} \rightarrow 6\ (\text{ADP} + \text{P}_i)_{\text{liver}} + 2\ \text{ATP}_{\text{red blood cells}}$$

Six molecules of ATP are needed in liver for glucose synthesis. The alanine cycle (Figure 15.32*b*) transfers energy from liver to peripheral tissues and, because of the five to seven molecules of ATP produced per molecule of glucose, is energetically more efficient. However, the alanine cycle presents liver with amino nitrogen that must be disposed of as urea (see p. 751). This requires four ATP molecules for every urea molecule produced and increases the amount of ATP required to 10 molecules per glucose molecule during the alanine cycle:

$$10\ \text{ATP}_{\text{liver}} + 5-7\ (\text{ADP} + \text{P}_i)_{\text{muscle}} + \text{O}_{2\ \text{muscle}} \rightarrow 10\ (\text{ADP} + \text{P}_i)_{\text{liver}}$$
$$+ 5-7\ \text{ATP}_{\text{muscle}}$$

This and the requirement in peripheral tissue for oxygen and mitochondria distinguish the alanine cycle from the Cori cycle.

Glucose Synthesis from Lactate

Gluconeogenesis from lactate is an ATP-requiring process:

$$2\ \text{L-Lactate}^- + 6\ \text{ATP}^{4-} + 6\ \text{H}_2\text{O} \rightarrow \text{glucose} + 6\ \text{ADP}^{3-} + 6\ \text{P}_i^{\,2-} + 4\ \text{H}^+$$

Many enzymes of glycolysis are used for gluconeogenesis from lactate. However, different reactions are necessary for the process because glycolysis produces 2 ATPs whereas gluconeogenesis requires 6 ATPs per molecule of glucose, and three steps of glycolysis are irreversible, namely the reactions catalyzed by glucokinase, 6-phosphofructo-1-kinase, and pyruvate kinase. The initial step of lactate gluconeogenesis is the conversion of lactate to pyruvate by lactate dehydrogenase (Figure 15.33). The NADH generated is needed for a subsequent step in the pathway. Pyruvate cannot be converted to phosphoenolpyruvate (PEP) by pyruvate kinase because the reaction is irreversible under intracellular conditions. Pyruvate is converted into PEP by coupling of the reactions catalyzed by pyruvate carboxylase, which requires ATP, and PEP carboxykinase, which requires GTP (see Figure 15.34). GTP is equivalent to an ATP through the action of nucleoside diphosphate kinase (GDP + ATP → GTP + ADP). The CO_2 generated by PEP carboxykinase and the HCO_3^- required by pyruvate carboxylase are linked by the reaction catalyzed by carbonic anhydrase ($CO_2 + H_2O \rightleftarrows H_2CO_3 \rightleftarrows H^+ + HCO_3^-$). Summing these reactions with the reactions of Figure 15.34 yields

$$\text{Pyruvate}^- + 2\ \text{ATP}^{4-} \rightarrow \text{phosphoenolpyruvate}^{3-} + 2\ \text{ADP}^{3-} + \text{P}_i^{\,2-} + 2\ \text{H}^+$$

Thus, conversion of pyruvate into PEP during gluconeogenesis costs the cell two molecules of ATP. This contrasts with the conversion of PEP to pyruvate during glycolysis that yields just one molecule of ATP.

The location of pyruvate carboxylase in the mitochondrion makes it mandatory for conversion of cytosolic pyruvate into cytosolic PEP (Figure 15.33). However, because PEP carboxykinase is present in both the cytosol and mitochondrial matrix compartments, there are two routes that oxaloacetate takes to glucose. The first uses mitochondrial PEP carboxykinase, which converts oxaloacetate into PEP that then traverses the mitochondrial inner membrane. The second converts oxaloacetate into aspartate, which exits the mitochondrion by way of the glutamate–aspartate antiport. Aspartate contributes its amino group to α-ketoglutarate in the cytosol to produce oxaloacetate, which is used by cytosolic PEP carboxykinase to produce PEP.

Gluconeogenesis Uses many Glycolytic Enzymes in the Reverse Direction

Enzymes of glycolysis operate in reverse to convert PEP to fructose 1,6-bisphosphate during gluconeogenesis. The generation of reducing equivalents (NADH) by lactate

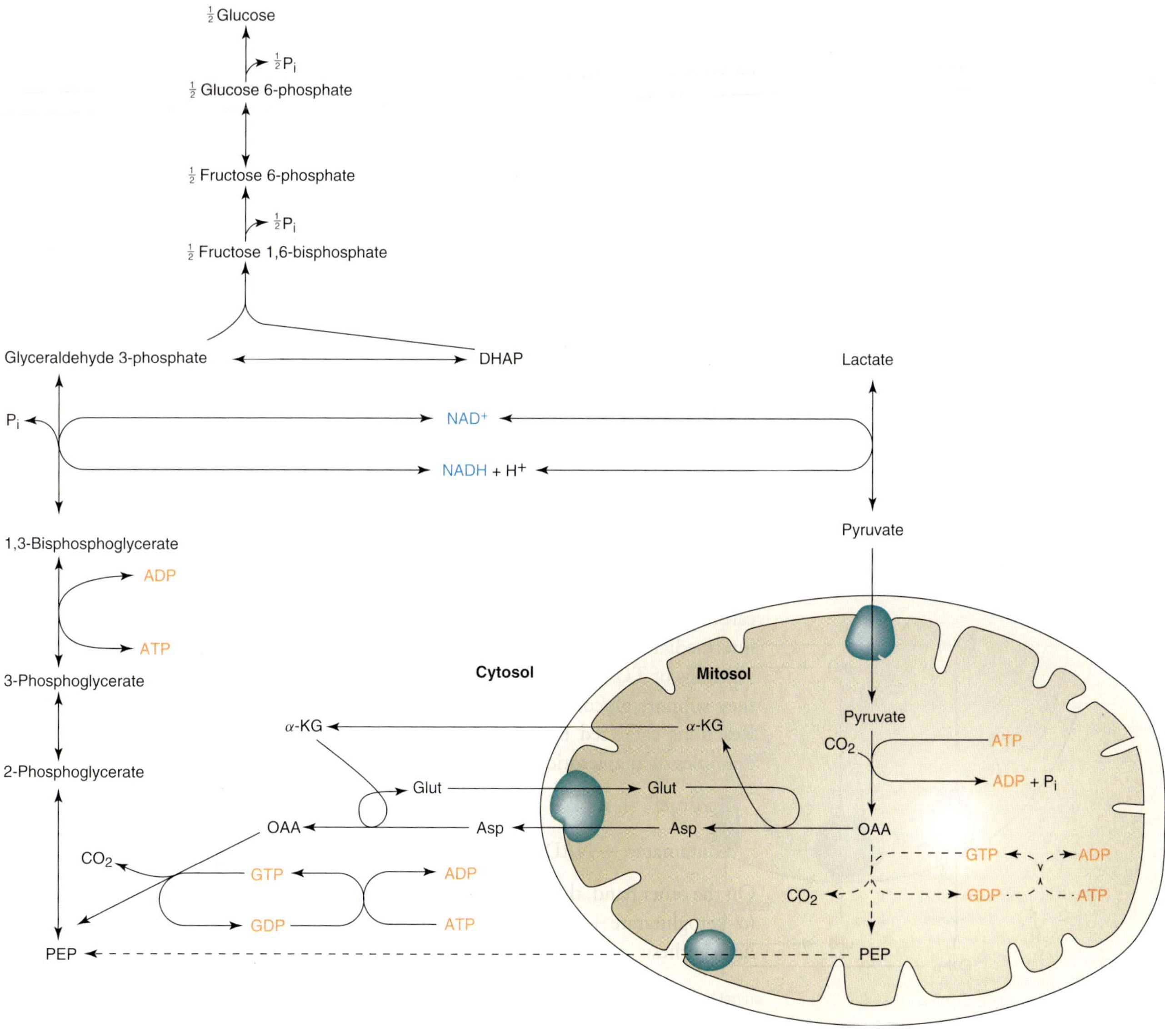

FIGURE 15.33

Pathway of gluconeogenesis from lactate. The involvement of the mitochondrion in the process is indicated. Dashed arrows refer to an alternate route, which employs mitochondrial PEP carboxykinase rather than the cytosolic isoenzyme. Abbreviations: OAA, oxaloacetate; α-KG, α-ketoglutarate; PEP, phosphoenolpyruvate; and DHAP, dihydroxyacetone phosphate.

$$\begin{array}{c}COO^-\\|\\C{=}O\\|\\CH_3\end{array} + ATP^{4-} + HCO_3^- \longrightarrow \begin{array}{c}COO^-\\|\\C{=}O\\|\\CH_2\\|\\COO^-\end{array} + ADP^{3-} + P_i^{2-} + H^+$$

Pyruvate **Oxaloacetate**

$$\begin{array}{c}COO^-\\|\\C{=}O\\|\\CH_2\\|\\COO^-\end{array} + GTP^{4-} \longrightarrow \begin{array}{c}COO^-\\|\\C{-}OPO_3^{2-}\\\|\\CH_2\end{array} + GDP^{3-} + CO_2$$

Oxaloacetate **Phospho*enol*pyruvate**

FIGURE 15.34

Energy requiring steps involved in phosphoenolpyruvate formation from pyruvate. Reactions are catalyzed by pyruvate carboxylase and PEP carboxykinase, respectively.

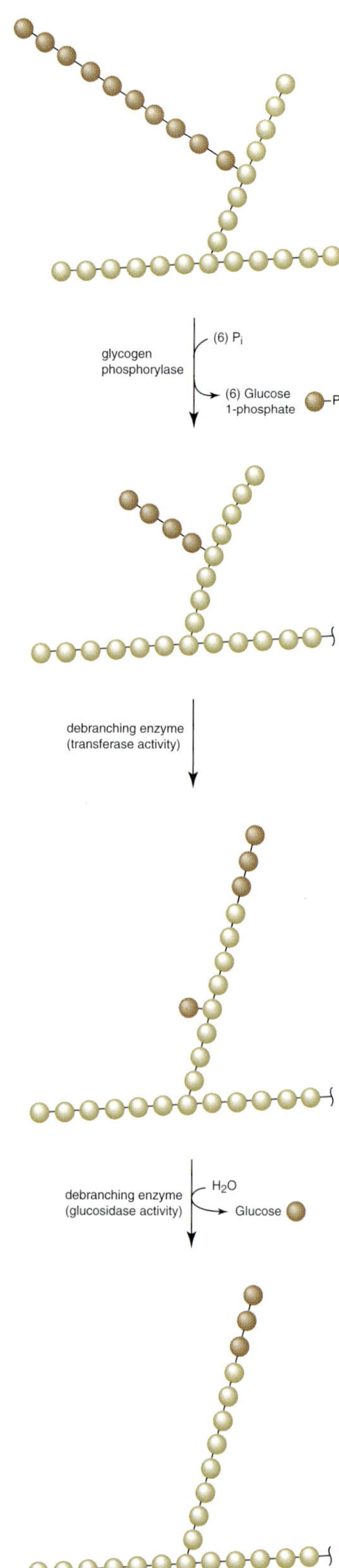

FIGURE 15.52
Action of glycogen debranching enzyme.

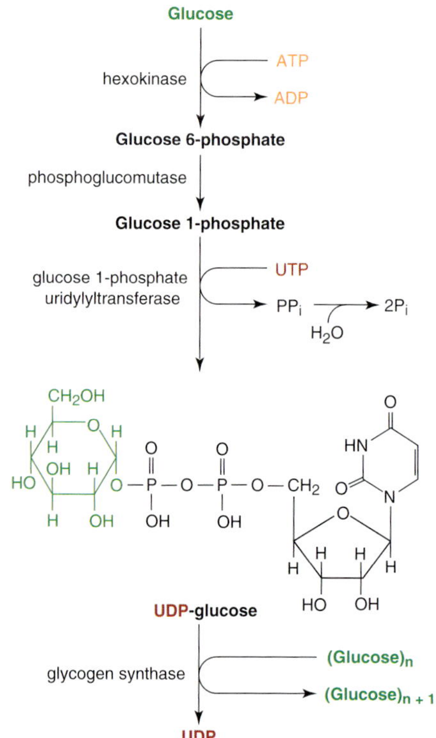

FIGURE 15.53
Pathway of glycogenesis.

The cooperative and repetitive action of phosphorylase and debranching enzyme results in complete breakdown of glycogen to glucose 1-phosphate and glucose. Glycogen storage diseases result when either of these enzymes is defective. An average molecule of glycogen yields about 12 molecules of glucose 1-phosphate by action of phosphorylase for every molecule of free glucose produced by debranching enzyme.

There is another, albeit quantitatively less important, pathway for glycogen degradation that depends upon **glucosidases** present in lysosomes. Glycogen that enters lysosomes during normal turnover of intracellular components has to be degraded. Failure to degrade glycogen taken up by lysosomes creates a major medical problem described in Clinical Correlation 15.11.

Glycogenesis Requires Unique Enzymes

The first reaction (Figure 15.53) is that of glucokinase in liver and hexokinase in peripheral tissues:

$$\text{Glucose} + \text{ATP} \rightarrow \text{glucose 6-phosphate} + \text{ADP}$$

Phosphoglucomutase then forms glucose 1- phosphate:

$$\text{Glucose 6-phosphate} \rightleftharpoons \text{glucose 1-phosphate}$$

Glucose 1-phosphate uridylyltransferase then produces UDP-glucose:

$$\text{Glucose 1-phosphate} + \text{UTP} \rightarrow \text{UDP-glucose} + \text{PP}_i$$

The latter reaction generates UDP-glucose, an "activated glucose" molecule, from which glycogen can be synthesized. Formation of UDP-glucose is made energetically favorable and irreversible by hydrolysis of pyrophosphate to inorganic phosphate by **pyrophosphatase**:

$$\text{PP}_i^{\,4-} + \text{H}_2\text{O} \rightarrow 2\ \text{P}_i^{\,2-}$$

Glycogen synthase then transfers the activated glucosyl moiety of UDP-glucose to the carbon 4 of a glucosyl residue of the growing glycogen chain to form a new glycosidic bond at the hydroxyl group of carbon 1 of the activated sugar. The reducing end of glucose (carbon 1, an aldehyde that can reduce other compounds during its oxidation to a carboxylic acid) is always added to a nonreducing end (carbon 4 of a glucosyl residue) of the glycogen chain. According to this, each molecule of glycogen should have one free reducing end tucked away within its core. In fact it does not have a free reducing end because its one potentially free aldehyde group is covalently linked to a protein called glycogenin within its core (described on p. 624). The UDP formed as a product of the glycogen synthase reaction is converted back to UTP by **nucleoside diphosphate kinase**:

$$\text{UDP} + \text{ATP} \rightleftarrows \text{UTP} + \text{ADP}$$

Glycogen synthase cannot form the α-1,6-glycosidic linkages. Working alone it would only produce amylose, a straight-chain polymer of glucose with α-1,4-glycosidic linkages. Once an amylose chain of at least 11 residues has been formed, a "branching enzyme" called 1,4-α-glucan branching enzyme removes a block of about seven glucosyl residues from a growing chain and transfers it to another chain to produce an α-1,6 linkage (see Figure 15.54). The new branch has to be introduced at least four glucosyl residues from the nearest branch point. Thus, the creation of the highly branched structure of glycogen requires the concerted efforts of glycogen synthase and branching enzyme. The overall balanced equation for glycogen synthesis as outlined is

$$(\text{Glucose})_n + \text{glucose} + 2\text{ATP}^{4-} + \text{H}_2\text{O} \rightarrow (\text{glucose})_{n+1} + 2\text{ADP}^{3-} + 2\text{P}_i^{\,2-} + 2\text{H}^+$$

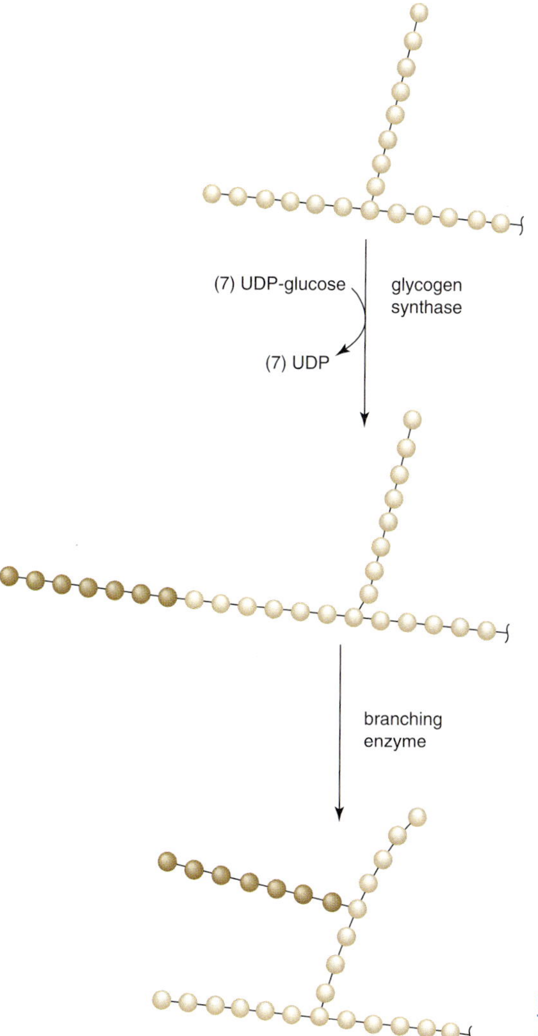

FIGURE 15.54
Action of glycogen branching enzyme.

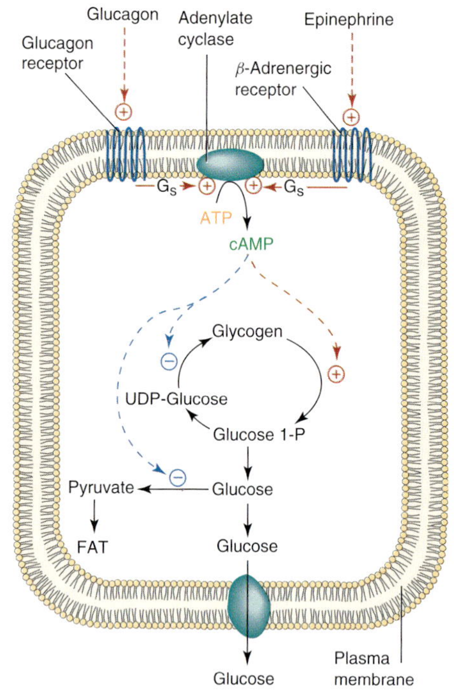

FIGURE 15.60

Cyclic AMP mediates stimulation of glycogenolysis in liver by glucagon and β agonists (epinephrine). See legends to Figures 15.19 and 15.25.

phosphatase can turn its attention to glycogen synthase *b* only after dephosphorylation of phosphorylase *a*. Thus, as a result of interaction of glucose with phosphorylase *a*, glycogen is synthesized rather than degraded in liver. Phosphorylase *a* can serve this function of **"glucose receptor"** in liver because the concentration of glucose in liver reflects that in blood. This is not true for extrahepatic tissues. Liver cells have a very high-capacity transporter for glucose (GLUT2) and a high $S_{0.5}$ enzyme for glucose phosphorylation (glucokinase), whereas in extrahepatic tissues glucose transport and phosphorylation systems maintain intracellular glucose at a concentration too low for phosphorylase *a* to function as a "glucose receptor."

Hormonal and Neural Control of Glycogen Synthesis and Degradation

Glucagon Stimulates Glycogenolysis in Liver

Glucagon is released from α cells of pancreas in response to low blood glucose levels. Under such conditions (e.g., during fasting or starvation), glucagon stimulates glycogenolysis so as to ensure that adequate blood glucose is available for glucose-dependent tissues (Figure 15.60). Binding of glucagon to its receptors on liver cells activates adenylate cyclase and triggers the cascades that activate glycogen phosphorylase and inactivate glycogen synthase (Figures 15.56 and 15.57, respectively). It also inhibits glycolysis at the level of 6-phosphofructo-1-kinase and pyruvate kinase as shown in Figures 15.27 and 15.30, respectively. The net result of all these effects mediated by cAMP and covalent modification is a very rapid increase in normal blood glucose levels. Hyperglycemia does not occur because less glucagon is released from the pancreas as blood glucose levels increase.

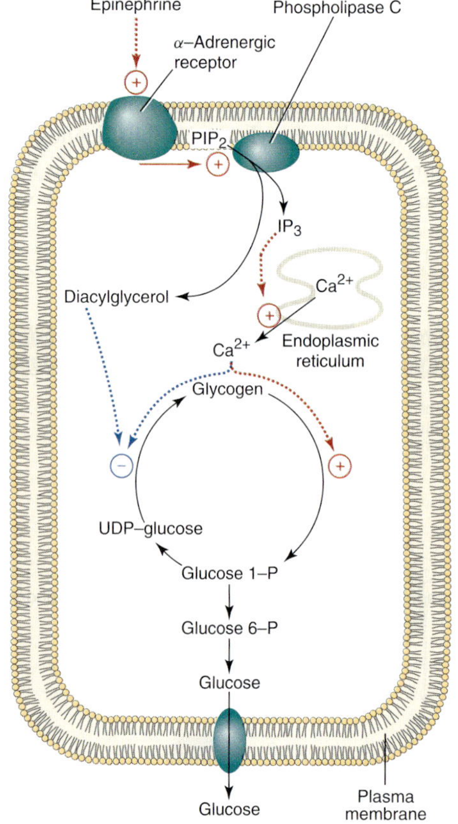

FIGURE 15.61

Inositol trisphosphate (IP_3) and Ca^{2+} mediate stimulation of glycogenolysis in liver by α agonists. The α-adrenergic receptor and glucose transporter are intrinsic components of the plasma membrane. Phosphatidylinositol 4,5-bisphosphate (PIP_2) is also a component of the plasma membrane.

Epinephrine Stimulates Glycogenolysis in Liver

Epinephrine is released into blood from chromaffin cells of the adrenal medulla in response to stress. This "fright, flight or fight" hormone prepares the body for either combat or escape. Binding of epinephrine with **β-adrenergic receptors** on liver cells activates adenylate cyclase (Figure 15.60) and cAMP has the same effects as glucagon—that is, activation of glycogenolysis and inhibition of glycogenesis and glycolysis to maximize the release of glucose. Binding of epinephrine to α-adrenergic receptors on liver cells signals formation of **inositol 1,4,5-trisphosphate** (IP_3) and diacylglycerol (Figure 15.61). These are second messengers, produced by the action of a phospholipase C on phosphatidylinositol 4,5-bisphosphate of the plasma membrane (Figure 15.62). IP_3 stimulates the release of Ca^{2+} from the endoplasmic reticulum, which activates phosphorylase kinase and in turn activates glycogen phosphorylase (Figure 15.56). In addition, **Ca^{2+} -mediated activation** of phosphorylase kinase and calmodulin-dependent protein kinase as well as diacylglycerol-mediated activation of protein kinase C may all contribute to inactivation of glycogen synthase (Figure 15.57).

An increased rate of glucose release into blood is a major consequence of epinephrine action on the liver. This makes glucose available to tissues that are called upon to meet the challenge of the stressful situation that triggered the release of epinephrine from the adrenal medulla.

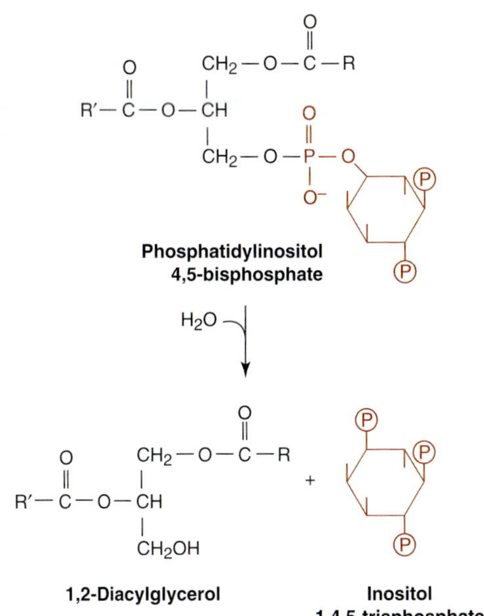

FIGURE 15.62

Phospholipase C cleaves phosphatidylinositol 4,5-bisphosphate to 1,2-diacylglycerol and inositol 1,4,5-trisphosphate.

Epinephrine Stimulates Glycogenolysis in Heart and Skeletal Muscle

Epinephrine also stimulates glycogenolysis in heart and skeletal muscle (Figure 15.63). It binds to β-adrenergic receptors; this stimulates adenylate cyclase to produce cAMP, which activates glycogen phosphorylase and inactivates glycogen synthase by mechanisms given in Figures 15.56 and 15.57, respectively. Since these tissues lack glucose

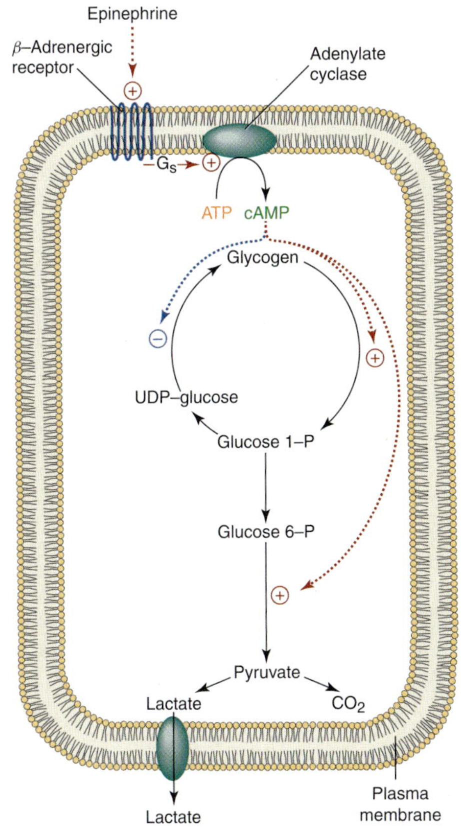

FIGURE 15.63

Cyclic AMP mediates stimulation of glycogenolysis in muscle by β agonists (epinephrine). The β-adrenergic receptor is an intrinsic component of the plasma membrane that stimulates adenylate cyclase by a stimulatory G protein (G_s).

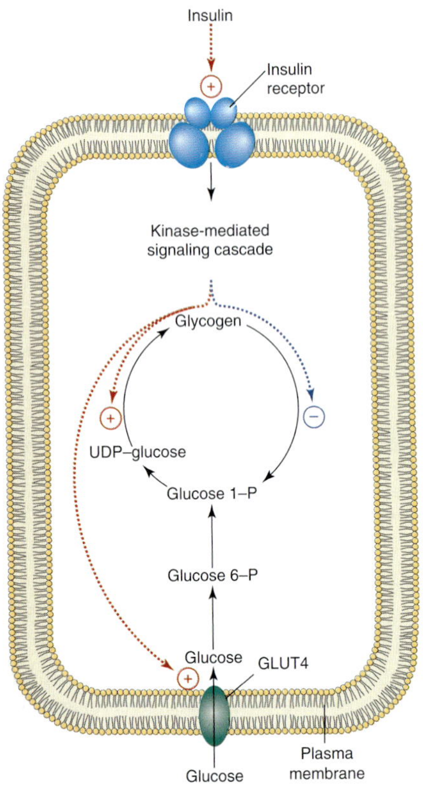

FIGURE 15.65

Insulin acts by a plasma membrane receptor to promote glycogenesis in muscle.

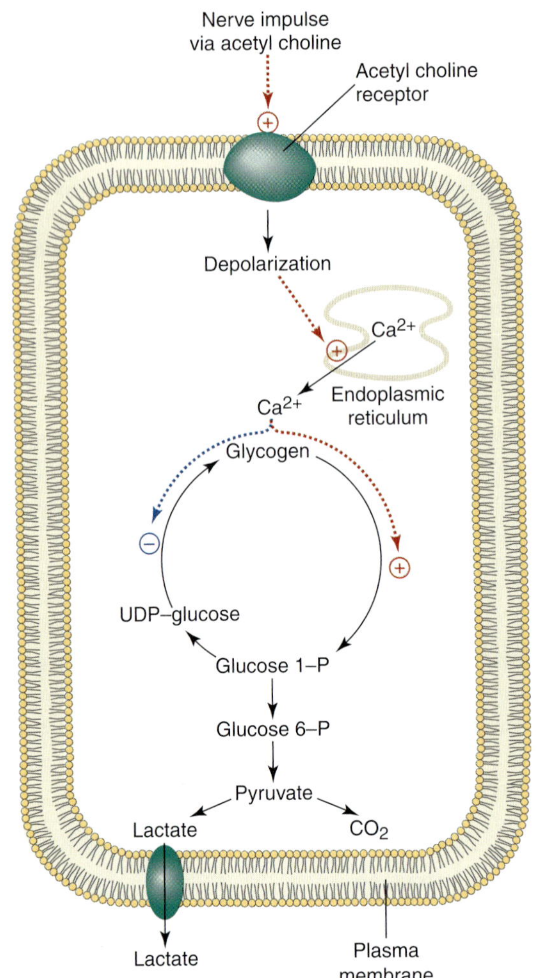

FIGURE 15.64

Ca^{2+} mediates the stimulation of glycogenolysis in muscle by nervous excitation.

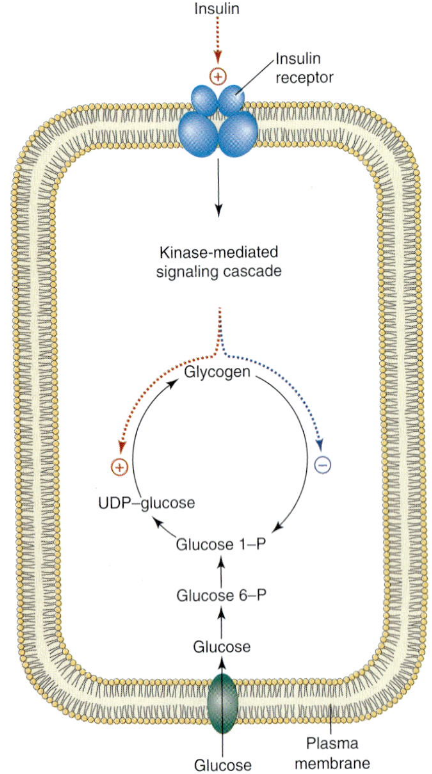

FIGURE 15.66

Insulin acts by a plasma membrane receptor to promote glycogenesis in liver.

6-phosphatase, this leads to stimulation of glycolysis (see Figure 15.28) rather than release of glucose into the blood. Thus, the effect of epinephrine in heart and skeletal muscle is to make more glucose 6-phosphate available for glycolysis. ATP generated by glycolysis can then meet the need for energy imposed on these muscles by the stress that triggered epinephrine release.

Neural Control of Glycogenolysis in Skeletal Muscle

Nervous excitation of muscle activity is mediated by changes in intracellular Ca^{2+} concentration (Figure 15.64). A nerve impulse causes membrane depolarization, which causes Ca^{2+} release from the sarcoplasmic reticulum into the sarcoplasm of muscle cells. This triggers muscle contraction, whereas reaccumulation of Ca^{2+} by the sarcoplasmic reticulum causes relaxation. The change in Ca^{2+} concentration also activates phosphorylase kinase and glycogen phosphorylase and perhaps inactivates glycogen synthase. Thus, more glycogen is converted to glucose 6-phosphate so that more ATP is produced to meet the greater energy demand of muscle contraction.

Insulin Stimulates Glycogenesis in Muscle and Liver

An increase in blood glucose signals release of insulin from β cells of the pancreas. Insulin receptors on the plasma membranes of insulin-responsive cells respond to insulin binding through a signaling cascade that promotes glucose use (Figures 15.65 and 15.66). The pancreas responds to a decrease in blood glucose with less release of insulin and more release of glucagon. These hormones have opposite effects on glucose utilization by liver, thereby establishing the pancreas as a fine-tuning device that prevents dangerous fluctuations in blood glucose levels. Insulin increases glucose utilization in part by promoting glycogenesis and inhibiting glycogenolysis in muscle and liver. Insulin stimulation of glucose transport is essential for these effects in muscle but not liver. Hepatocytes have a high-capacity, insulin-insensitive glucose transporter 2, whereas skeletal muscle cells and adipocytes have an insulin-sensitive glucose transporter 4 (see p. 474). Insulin increases the number of glucose transporter 4 proteins associated with the plasma membrane by promoting their translocation from an intracellular pool (see Figure 15.6). Insulin promotes glycogen accumulation in both tissues by activating glycogen synthase and inhibiting glycogen phosphorylase as already described (Figures 15.56, 15.57, and 15.58).

BIBLIOGRAPHY

Brosnan, J. T. Comments on metabolic needs for glucose and the role of gluconeogenesis. *Eur. J. Clin. Nutr.* 53: S107, 1999.

Chen, Y-T. Glycogen storage diseases. In: C. R. Scriver, A. R. Beaudet, W. S. Sly, and D. Valle (Eds.), *The Metabolic and Molecular Bases of Inherited Disease*, 8th ed. New York: McGraw-Hill, 2001, p. 1521, 2001.

Croniger, C., Leahy, P., Reshef, L., and Hanson, R. W. C/EBP and the control of phosphoenolpyruvate carboxykinase gene transcription in liver. *J. Biol. Chem.* 273:31629, 1998.

DePaoli-Roach, A. A., Park, I. -K., Cerovsky, V., Csortos, C., Durbin, S. D., Kuntz, M. J., Sitikov, A., Tang, P. M., Verin, A., and Zolnierowicz, S. Serine/threonine protein phosphatases in the control of cell function. *Adv. Enzyme Regul.* 34:199, 1994.

Depre, C., Rider, M. H., and Hue L. (1998). Mechanisms of control of heart glycolysis. *Eur J. Biochem.* 258:277, 1998.

Fell, D. *Understanding the Control of Metabolism*. London: Portland Press, 1997.

Gibson, D. M. and Harris, R. A. *Metabolic Regulation in Mammals*. London: Taylor and Frances, 2001.

Gould, G. W. and Holman, G. D. The glucose transporter family: Structure, function and tissue-specific expression. *Biochem. J.* 295:329, 1993.

Gurney, A. L., Park, E. A., Liu, J., Giralt, M., McGrane, M. M., Patel, Y. M., Crawford, D. R., Nizielski, S. E., Savon, S., and Hanson, R. W. Metabolic regulation of gene transcription. *J. Nutr.* 124:1533S, 1994.

Hanson, R. W. and Mehlman, M. A. (Eds.). *Gluconeogenesis, its regulation in mammalian species*. New York: Wiley, 1976.

Harris, R. A. Glycolysis overview. *Encyclopedia of Biological Chemistry, Section of Metabolism, Vitamins, and Hormones*. In press as Web article. New York: Elsevier/Academic Press, 2005.

Harper, E. T. and Harris, R. A. Glycolytic pathway. *Nature Encyclopedia of Life Sciences*. www.els.net. 2005.

Hunter, T. Protein kinases and phosphatases: The yin and yang of protein phosphorylation and signaling. *Cell* 80:225, 1995.

King, M. W. *Dr. King's Medical Biochemical Page*. http://web.indstate.edu/thcme/mwking/ January 16, 2005.

King, M. W. Glycogen, starch, and sucrose. *Nature Encyclopedia of Life Sciences*. www.els.net. 2005.

Lalli, E. and Sassone-Corsi, P. Signal transduction and gene regulation: The nuclear response to cAMP. *J. Biol. Chem.* 269:17359, 1994.

Lopaschuk, G. D. Glycolysis regulation. *Nature Encyclopedia of Life Sciences*. www.els.net. 2005.

Metzler, D. E. *Biochemistry, the Chemical Reactions of Living Cells*. San Diego: Academic Press, 2001.

Newsholme, E. A. and Leech, A. R., *Biochemistry for the Medical Sciences*. New York: Wiley, 1983.

Newsholme, E. A. and Start, C. *Regulation in Metabolism*. New York: Wiley, 1973.

Pilkis, S. J., Claus, T. H., Kurland, I. J., and Lange, A. J. 6-Phosphofructo-2-kinase/fructose-2,6-bisphosphatase: A metabolic signaling enzyme. *Annu. Rev. Biochem.* 64:799, 1995.

Pilkis, S. J. and El-Maghrabi, M. R. Hormonal regulation of hepatic gluconeogenesis and glycolysis. *Annu. Rev. Biochem.* 57:755, 1988.

Price, T. B., Rothman, D. L., and Shulman R. G. NMR of Glycogen in Exercise. *Proc. Nutr. Soc.* 58:851, 1999.

Roach, P. J., Skurat, A. V., and Harris, R. A. Regulation of glycogen metabolism. In: L. S. Jefferson and A. D. Charrington (Eds.), *The Endocrine Pancreas and Regulation of Metabolism; Handbook of Physiology*. Oxford University Press, Oxford, UK, 2001, p. 609.

Scriver, C. R., Beaudet, A. L., Sly, W. S., and Valle, D. (Eds.). *The Metabolic and Molecular Bases of Inherited Disease*, 8th ed. New York: McGraw-Hill, 2001.

Smith, C., Marks, A. D., and Lieberman, M. *Basic Medical Biochemistry. A Clinical Approach*. Baltimore: Lippincott Williams & Wilkins, 2005.

Steinmann, B., Gitzelmann, R., and Van den Berghe, G. Disorders of fructose metabolism. In: C. R. Scriver, A. R. Beaudet, W. S. Sly, and D. Valle (Eds.), *The Metabolic and Molecular Bases of Inherited Disease*, 8th ed. New York: McGraw-Hill, 2001, p. 1489.

Taylor, S. I. Insulin action, insulin resistance, and type 2 diabetes mellitus. In: C. R. Scriver, A. R. Beaudet, W. S. Sly, and D. Valle (Eds.), *The Metabolic and Molecular Bases of Inherited Disease*, 8th ed. New York: McGraw-Hill, 2001, p. 1433.

van Schaftingen, E. and Geren, I. Review article. The glucose-6-phosphatase system. *Biochem. J.* 362:513, 2002.

van Schaftingen, E., Vandercammen, A., Detheux, M., and Davies, D. R. The regulatory protein of liver glucokinase. *Adv. Enzyme Regul.* 32:133, 1992.

Wallace, J. C. and Baritte, G. J. Gluconeogenesis. *Nature Encyclopedia of Life Sciences*. www.els.net. 2005.

QUESTIONS | Carol N. Angstadt

Multiple Choice Questions

1. NAD^+ can be regenerated in the cytoplasm if NADH reacts with any of the following *except*:
   A. pyruvate.
   B. dihydroxyacetone phosphate.
   C. oxaloacetate.
   D. the flavin bound to NADH dehydrogenase.

2. Glucokinase:
   A. has a $S_{0.5}$ greater than the normal blood glucose concentration.
   B. is found in muscle.
   C. is inhibited by glucose 6-phosphate.
   D. is also known as the GLUT-2 protein.
   E. has glucose 6-phosphatase activity as well as kinase activity.

3. 6-Phosphofructo-1-kinase activity can be decreased by all of the following *except*:
   A. ATP at high concentrations.
   B. citrate.
   C. AMP.
   D. low pH.
   E. decreased concentration of fructose 2,6-bisphosphate.

4. Which of the following supports gluconeogenesis?
   A. α-ketoglutarate + aspartate $\rightleftharpoons$ glutamate + oxaloacetate
   B. pyruvate + ATP + HCO_3^- $\rightleftharpoons$ oxaloacetate + ADP + P_i + H^+
   C. acetyl CoA + oxaloacetate + H_2O $\rightleftharpoons$ citrate + CoA
   D. leucine degradation.
   E. lysine degradation.

5. In the Cori cycle:
   A. only tissues with aerobic metabolism (i.e., mitochondria and O_2) are involved.
   B. a three-carbon compound arising from glycolysis is converted to glucose at the expense of energy from fatty acid oxidation.
   C. glucose is converted to pyruvate in anaerobic tissues, and this pyruvate returns to the liver, where it is converted to glucose.
   D. the same amount of ATP is used in the liver to synthesize glucose as is released during glycolysis, leading to no net effect on whole body energy balance.
   E. nitrogen from alanine must be converted to urea, increasing the amount of energy required to drive the process.

6. When blood glucagon rises, which of the following hepatic enzyme activities *falls*?
   A. adenylate cyclase.
   B. protein kinase.
   C. 6-phosphofructo-2-kinase.
   D. fructose 1,6-bisphosphatase.
   E. hexokinase.

Questions 7 and 8: Patients with McArdle's disease suffer from painful muscle cramps and are unable to perform strenuous exercise. This disease, also called type V glycogen storage disease, is caused by the absence of muscle phosphorylase. Unlike patients with Von Gierke's disease (Type I glycogen storage disease caused by a deficiency of glucose 6-phosphatase), McArdle's patients do not typically suffer hypoglycemia on fasting. Von Gierke's patients show lactic acidemia, but McArdle's patients fail to demonstrate lactic acidemia following exercise.

7. McArdle's patients:
   A. fail to synthesize glycogen appropriately.
   B. do not respond to glucagon in a normal fashion.
   C. show a reduced state of glycolysis during exercise.
   D. have an impaired tricarboxylic acid cycle.
   E. have reduced lactic acid because all of the lactate is converted to pyruvate.

8. Glucose 6-phosphatase, which is deficient in Von Gierke's disease, is necessary for the production of blood glucose from:
   A. liver glycogen.
   B. fructose.
   C. amino acid carbon chains.
   D. lactose.
   E. all of the above.

Questions 9 and 10: Malignant hyperthermia is a genetic abnormality in which exposure to certain agents, especially the widely used general anesthetic halothane, produces a dramatic rise in body temperature, acidosis, hyperkalemia, and muscle rigidity. Death is rapid if the condition is untreated and may occur the first time a susceptible person is anaesthetized. The defect causes an inappropriate release of Ca^{2+} from the sarcoplasmic reticulum of muscle. Many heat-producing processes are stimulated in an uncontrolled fashion by the release of Ca^{2+}, including glycolysis and glycogenolysis.

9. Ca^{2+} increases glycogenolysis by:
   A. activating phosphorylase kinase b, even in the absence of cAMP.
   B. binding to phosphorylase b.
   C. activating phosphoprotein phosphatase.
   D. inhibiting phosphoprotein phosphatase.
   E. protecting cAMP from degradation.

10. Phosphorylation–dephosphorylation and allosteric activation of enzymes play roles in stimulating glycogen degradation. All of the following result in enzyme activation *except*:
    A. phosphorylation of phosphorylase kinase.
    B. binding of AMP to phosphorylase b.
    C. phosphorylation of phosphorylase.
    D. phosphorylation of protein kinase A.
    E. dephosphorylation of glycogen synthase.

Questions 11 and 12: Patients with hereditary fructose intolerance are deficient in the liver form of the enzyme aldolase. Consumption of fructose leads to a depletion of ATP and P_i in the liver, which, in turn, leads to cell damage. Much of the cell damage can be attributed to the inability to maintain normal ion gradients by ATP-dependent pumps.

11. The first step in liver's metabolism of fructose is:
    A. isomerization to glucose.
    B. phosphorylation to fructose 1,6-bisphosphate by ATP.
    C. phosphorylation to fructose 1-phosphate by ATP.
    D. phosphorylation to fructose 6-phosphate by ATP.
    E. cleavage by aldolase.

12. The products initially produced by aldolase action on the substrate formed from fructose are:
    A. two molecules of dihydroxyacetone phosphate.
    B. two molecules of glyceraldehyde 3-phosphate.
    C. two molecules of lactate.

D. dihydroxyacetone phosphate and glyceraldehyde 3-phosphate.
E. dihydroxyacetone phosphate and glyceraldehyde.

Problems

13. If a cell is forced to metabolize glucose anaerobically, how much faster would glycolysis have to proceed to generate the same amount of ATP as it would get if it metabolized glucose aerobically?
14. Consumption of alcohol by an undernourished individual or by an individual following strenuous exercise can lead to hypoglycemia. What relevant reactions are inhibited by the consumption of alcohol?

ANSWERS

1. **D** The flavin is mitochondrial. A may be converted to lactate. B and C are the cytoplasmic acceptors for shuttle systems.
2. **A** Blood glucose is ~5 mM. $S_{0.5}$ of glucokinase is ~7 mM. B: Glucokinase is hepatic, and, unlike the muscle hexokinase, it is not inhibited by glucose 6-phosphate.
3. **C** AMP is an allosteric regulator that relieves inhibition by ATP. B and D are probably important physiological regulators in muscle, and E is critical in liver.
4. **B** This reaction is on the direct route of conversion of pyruvate to glucose. A: α-Ketoglutarate and oxaloacetate both give rise to glucose; interconversion of one to the other accomplishes nothing. C: Citrate ultimately gives rise to oxaloacetate, losing two carbon atoms in the process; again nothing is gained. D and E involve the two amino acids that are strictly ketogenic.
5. **B** The liver derives the energy required for gluconeogenesis from aerobic oxidation of fatty acids. A: The liver is an essential organ in the Cori cycle; it is aerobic. C: In anaerobic tissues the end product of glycolysis is lactate; in aerobic tissues it is pyruvate, but there the pyruvate would likely be oxidized aerobically. D: Gluconeogenesis requires six ATP per glucose synthesized; glycolysis yields two ATP per glucose metabolized. E: Alanine is not part of the Cori cycle.
6. **C** As blood glucagon rises, A is activated, producing cAMP; cAMP activates B, and B inactivates C. Low levels of fructose 2,6-bisphosphate increase the activity of D. E is not an important hepatic enzyme; its role is filled in liver by glucokinase.
7. **C** Muscles use glucose from glycogen as the primary substrate for glycolysis during exercise. Lack of phosphorylase prevents the degradation of glycogen. A: Phosphorylase is not necessary for the synthesis of glycogen, only its degradation. B: Muscles do not respond to glucagon; liver does, but liver phosphorylase is not affected in this disease. D: Muscle uses fatty acids and amino acid carbon chains via the tricarboxylic acid cycle. E: Lactate formation is reduced because glycolysis lacks sufficient substrate.
8. **E** To get into the blood, glucose must be free, not phosphorylated. A: Glycogen is degraded to glucose 1-phosphate, which is converted to glucose 6-phosphate. B: Fructose is metabolized to dihydroxyacetone phosphate which can either continue through glycolysis or reverse to glucose 6-phosphate, depending on the state of the cell. C, D: Amino acid carbon chains and lactate are substrates for gluconeogenesis.
9. **A** The δ-subunit of phosphorylase kinase is a calmodulin-type protein. Both a and b forms of the enzyme are activated by Ca^{2+}. B, E: These do not happen. C, D: Ca^{2+} does not affect the phosphatase.
10. **D** Protein kinase A catalyzes phosphorylations but is activated by binding cAMP. A, C: Both of these enzymes are phosphorylated in their a forms. B: Phosphorylase b is allosterically activated by binding AMP. E: The a form of the synthase is the nonphosphorylated form.
11. **C** The ADP formed is converted to ATP at the expense of P_i. Inability to further metabolize fructose 1-phosphate results in depletion of P_i. A: This does not happen. B: This would require two phosphorylations. D: This does not happen in liver. E: The fructose must be phosphorylated first.
12. **E** Glyceraldehyde can be converted to glyceraldehyde 3-phosphate so both products feed into the glycolytic pathway or gluconeogenesis.
13. Anaerobically, there is a net of 2 mol of ATP/mol glucose. Aerobically, the same net of 2 ATP is obtained plus 2 NADH because pyruvate is the product. Let us assume the cell uses the malate–aspartate shuttle where each NADH yields 2 1/2 ATP. Therefore, there is a net of 7 mols ATP/mol glucose. Each pyruvate is converted to Acetyl CoA and the Acetyl CoA is oxidized by the tricarboxylic acid cycle. Each mol of pyruvate then yields 12.5 mol ATP or 25 mol ATP for the 2 pyruvates. This gives a total of 32 mol ATP/mol glucose aerobically (see Chapter 14). Therefore, glycolysis must proceed 16 times as rapidly under anaerobic conditions to generate the same amount of ATP as occurs aerobically.
14. In both of these cases, hepatic glycogen is depleted and gluconeogenesis is the primary source of blood glucose. Metabolism of alcohol by the liver produces large amounts of NADH. This shifts the equilibrium of pyruvate–lactate toward lactate and oxaloacetate–malate toward malate. To be used for gluconeogenesis, lactate must first be converted to pyruvate. Oxaloacetate from pyruvate or amino acid carbon chains must be converted to phosphoenolpyruvate in order to synthesize glucose.

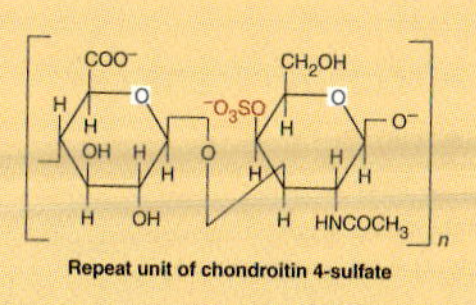

16

CARBOHYDRATE METABOLISM II: SPECIAL PATHWAYS AND GLYCOCONJUGATES

Nancy B. Schwartz

Textbook of Biochemistry With Clinical Correlations, Sixth Edition, Edited by Thomas M. Devlin

CLINICAL CORRELATIONS

16.1 OVERVIEW

In addition to catabolism of glucose for the specific purpose of energy production in the form of ATP, several pathways involving sugar metabolism exist in cells. One, designated the **pentose phosphate pathway, hexose monophosphate shunt**, or the **6-phosphogluconate pathway**, is particularly important in animal cells. It functions with glycolysis and the tricarboxylic acid cycle to produce the reducing equivalents of NADPH and pentose intermediates. NADPH is a hydrogen and electron donor in reductive biosynthetic reactions, while NADH produced in most biochemical reactions is oxidized by the respiratory chain to produce ATP (see p. 554). The pentose phosphate pathway also converts hexoses into pentoses, particularly ribose 5-phosphate. This C_5 sugar or its derivatives are components of ATP, CoA, NAD, FAD, RNA, and DNA. It also catalyzes interconversion of C_3, C_4, C_6, and C_7 sugars, some of which can enter glycolysis. Specific pathways synthesize and degrade monosaccharides, oligosaccharides, and complex polysaccharides. A profusion of chemical interconversions occurs, whereby one sugar can be changed into another so that all monosaccharides, and most oligo- and polysaccharides synthesized from the monosaccharides, can originate from glucose. These interconversion reactions can occur directly or via nucleotide-linked sugars which are the obligatory activated form for complex polysaccharide synthesis. Monosaccharides are components of the more complex macromolecules, glycoproteins, glycolipids, and proteoglycans. In higher animals these complex carbohydrates occur predominantly in the extracellular matrix in tissues or are associated with cell membranes. They often function as recognition markers and determinants of biological specificity. The discussion of complex carbohydrates in this chapter is limited to the chemistry and biology of those present in animal tissues and fluids. The Appendix (see p. 1124) presents the nomenclature and chemistry of the carbohydrates.

16.2 PENTOSE PHOSPHATE PATHWAY

Pentose Phosphate Pathway has Two Phases

The pentose phosphate pathway provides a means for degrading the carbon chain of a sugar molecule one carbon at a time. However, in contrast to the tricarboxylic acid cycle, this pathway does not constitute a consecutive set of reactions that lead directly from glucose 6-phosphate (G6P) to six molecules of CO_2. Rather it can be visualized as occurring in two phases. In the first, hexose is decarboxylated to pentose via two oxidation reactions that form NADPH; in the second, by a series of transformations, six molecules of pentose undergo rearrangements to yield five molecules of hexose.

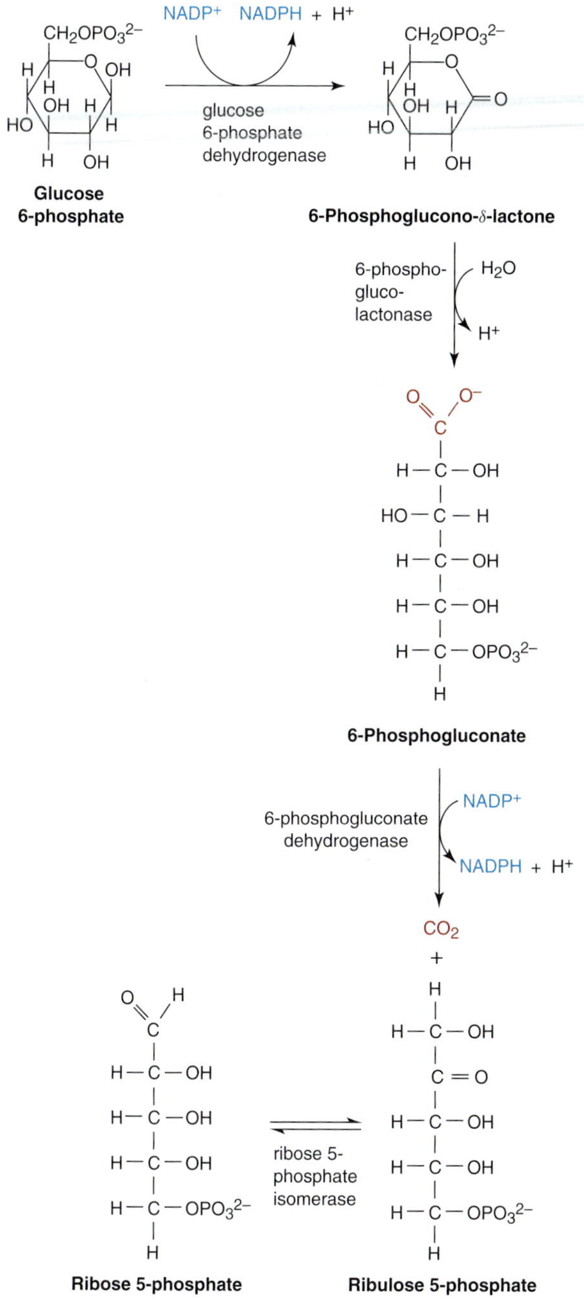

FIGURE 16.1
Oxidative phase of the pentose phosphate pathway: Formation of pentose phosphate and NADPH.

Glucose 6-Phosphate is Oxidized and Decarboxylated to a Pentose Phosphate

The first reaction catalyzed by G6P dehydrogenase (Figure 16.1) is **dehydrogenation** of G6P at C1 to form **6-phosphoglucono-δ-lactone** and **NADPH**. This is a major regulatory site for this pathway. Special interest in this enzyme stems from the severe anemia that may result from absence of **G6P dehydrogenase** in erythrocytes or from the presence of one of many genetic variants of the enzyme (see Clin. Corr. 16.1). The product of this reaction, a lactone, is a substrate for gluconolactonase which ensures that the reaction goes to completion. The overall equilibrium of both reactions lies far in the direction of NADPH, maintaining a high NADPH/NADP ratio within cells. A second dehydrogenation and **decarboxylation** is catalyzed by **6-phosphogluconate dehydrogenase**, which produces the pentose phosphate, **ribulose 5-phosphate**, and a second molecule of NADPH. The next step is the **isomerization**, through an enediol intermediate, of ribulose 5-phosphate. Under certain metabolic conditions, the pentose

CLINICAL CORRELATION 16.1
Glucose 6-Phosphate Dehydrogenase: Genetic Deficiency or Variants in Erythrocytes

When certain seemingly harmless drugs, such as antimalarials, antipyretics, or sulfa antibiotics, are administered to susceptible patients, an acute hemolytic anemia may result in 48–96 h. Susceptibility to drug-induced hemolytic disease may be due to a deficiency of glucose 6-phosphate (G6P) dehydrogenase activity in erythrocytes, and was an early indication that X-linked genetic deficiencies of this enzyme exist. The enzyme catalyzes oxidation of G6P to 6-phosphogluconate and reduction of $NADP^+$, and it is particularly important because the pentose phosphate pathway is the major pathway of NADPH production in the red cell. For example, red cells with the relatively mild A-type of G6P dehydrogenase deficiency can oxidize glucose at a normal rate when the demand for NADPH is normal. However, if the rate of NADPH utilization is increased, the cells cannot increase the activity of the pathway adequately. In addition, cells do not reduce enough NADP to maintain glutathione in its reduced state and, hence, to protect against lipid peroxidation. Reduced glutathione is necessary for the integrity of the erythrocyte membrane, thus rendering enzyme-deficient red cells more susceptible to hemolysis by a wide range of compounds. Therefore, the basic abnormality in G6P deficiency is the formation of mature red blood cells. Young red blood cells (reticulocytes) may have significantly higher enzyme activity than older cells because of an unstable enzyme variant; following an episode of hemolysis, young red cells predominate and it may not be possible to diagnose this genetic deficiency until the red cell population ages. This deficiency illustrates the interplay of heredity and environment on the production of disease. There are more than 300 known genetic variants of this enzyme, which contains 516 amino acids, accounting for a wide range of symptoms. These variants can be distinguished from one another by clinical, biochemical, and molecular differences.

Source: Luzzatto, L., Mehta, A., and Vulliamy, T. Glucose 6-phosphate dehydrogenase deficiency. In: *The Metabolic and Molecular Bases of Inherited Disease*, Vol. III, 8th ed. C. R. Scriver, A. R. Beaudet, W. S. Sly, and D. Valle (Eds.), New York: McGraw-Hill, 2001, p. 4517.

phosphate pathway can end at this point, with utilization of NADPH for reductive biosynthetic reactions and ribose 5-phosphate as a precursor for nucleotide synthesis. The overall equation may be written as

$$\text{glucose 6-Phosphate} + 2\text{NADP}^+ + \text{H}_2\text{O} \rightleftharpoons \text{ribose 5-phosphate} + 2\text{NADPH} + 2\text{H}^+ + \text{CO}_2$$

Interconversions of Pentose Phosphates Lead to Intermediates of Glycolysis

In certain cells, more NADPH is needed for **reductive biosynthesis** than ribose 5-phosphate for incorporation into nucleotides. A sugar **interconversion** system (Figure 16.2) forms triose, tetrose, hexose, and heptose sugars from the pentoses, thus disposing of ribose 5-phosphate and providing a reversible link between the pentose phosphate pathway and glycolysis via common intermediates. **Xylulose 5-phosphate** is formed through isomerization of ribulose 5-phosphate by **phosphopentose epimerase**, so that ribulose 5-phosphate, ribose 5-phosphate, and xylulose 5-phosphate exist as an equilibrium mixture and can then undergo transformations catalyzed by transketolase and transaldolase.

Transketolase requires **thiamine pyrophosphate** (TPP) and Mg^{2+}, transfers a C_2 unit of "active glycolaldehyde" from xylulose 5-phosphate to ribose 5-phosphate, and produces **sedoheptulose** and **glyceraldehyde 3-phosphate**, an intermediate of glycolysis. Alterations in transketolase can lead to Wernicke–Korsakoff syndrome (see Clin. Corr. 16.2). **Transaldolase** transfers a C_3 unit (dihydroxyacetone) from sedoheptulose 7-phosphate to glyceraldehyde 3-phosphate, forming **erythrose 4-phosphate**, and **fructose 6-phosphate**, another intermediate of glycolysis. In another reaction, transketolase produces fructose 6-phosphate and glyceraldehyde 3-phosphate, from erythrose 4-phosphate and xylulose 5-phosphate. The sum of these reactions is

$$2\ \text{xylulose 5-phosphate} + \text{ribose 5-phosphate} \rightleftharpoons 2\text{fructose 6-phosphate} + \text{glyceraldehyde 3-phosphate}$$

FIGURE 16.2

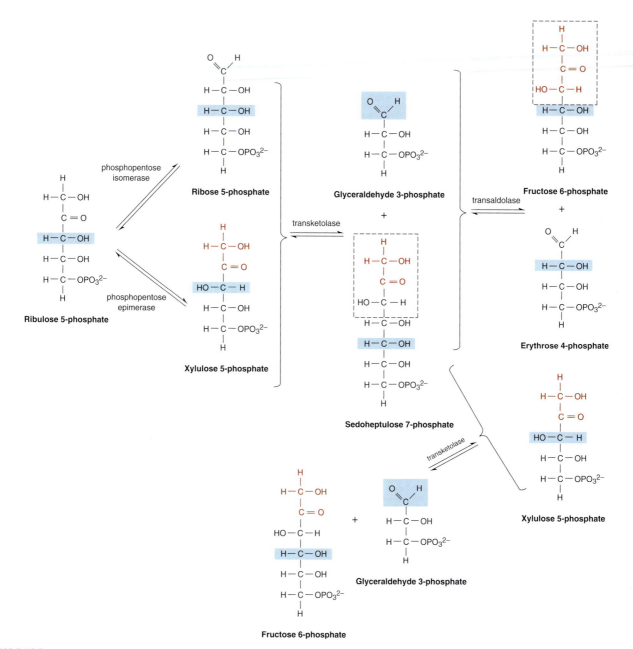

Nonoxidative reactions of the pentose phosphate pathway: Interconversions of pentose phosphates.

CLINICAL CORRELATION 16.2

Wernicke–Korsakoff Syndrome: Deficiency or Genetic Variants of Transketolase

Symptoms of Wernicke–Korsakoff Syndrome become apparent after moderate stress that does not affect normal individuals. The molecular basis is an alteration in transketolase that significantly reduces (> 10-fold) its affinity for thiamine pyrophosphate (TPP). Fluctuations in TPP concentration are common and do not affect normal individuals whose transketolase binds TPP sufficiently tightly. The syndrome presents as a mental disorder, with memory loss and partial paralysis, and can become manifest in alcoholics, whose diets may be vitamin deficient.

Source: Victor, A., Adams, R. S., and Collins, G. H. Historical review. In: *The Wernicke–Korsakoff Syndrome and Related Neurologic Disorders Due to Alcoholism and Malnutrition*, 2nd ed. Philadelphia: FA Davis, 1989, p. 1.

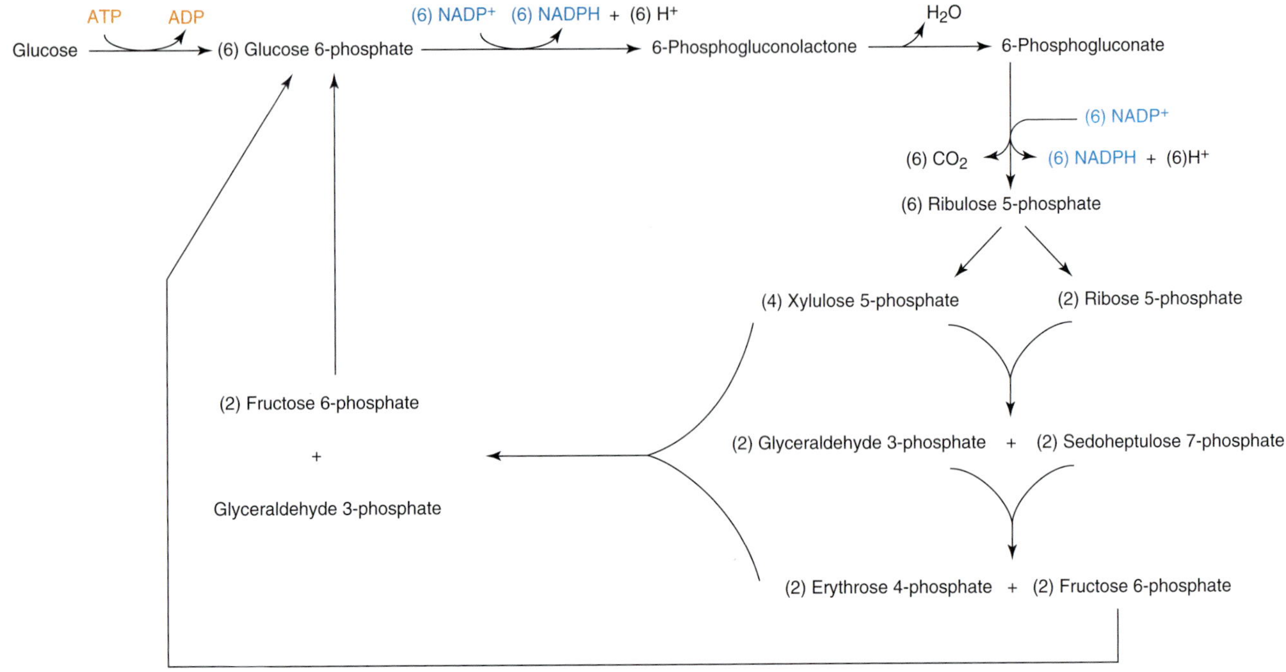

FIGURE 16.3

Pentose phosphate pathway.

Since xylulose 5-phosphate is derived from ribose 5-phosphate, the net reaction starting from ribose 5-phosphate is

$$3 \text{ ribose 5-phosphate} \rightleftharpoons 2 \text{ fructose 6-phosphate} + \text{glyceraldehyde 3-phosphate}$$

Thus, excess ribose 5-phosphate, produced in the first stage of the pentose phosphate pathway or from degradation of nucleic acids, is effectively converted to intermediates of glycolysis.

Glucose 6-Phosphate can be Completely Oxidized to CO_2

Complete oxidation of glucose 6-phosphate (G6P) to CO_2, with reduction of NADP to NADPH, may also occur (Figure 16.3). G6P continually enters this pathway, and CO_2 and NADPH are produced in the first phase. A balanced equation involves the oxidation of six molecules of G6P to six of ribulose 5-phosphate and six of CO_2. This results in transfer of 12 pairs of electrons to NADP, the requisite amount for total oxidation of one glucose to six CO_2. The six molecules of ribulose 5-phosphate are then rearranged to regenerate five molecules of G6P. The overall equation then becomes

$$6 \text{ glucose 6-phosphate} + 12\ NADP^+ + 7\ H_2O \rightleftharpoons 5 \text{ glucose 6-phosphate}$$
$$+ 6\ CO_2 + 12\ NADPH + 12\ H^+ + P_i$$

The net reaction is therefore

$$\text{glucose 6-phosphate} + 12\ NADP^+ + 7H_2O \rightleftharpoons 6\ CO_2 + 12\ NADPH + 12\ H^+ + P_i$$

Pentose Phosphate Pathway Produces NADPH

The pentose phosphate pathway serves several purposes, including synthesis and degradation of sugars other than hexoses, particularly ribose 5-phosphate for synthesis of nucleotides. Very important is the synthesis of NADPH, which has a unique role in biosynthesis. The direction of flow and path taken by G6P after entry into the pathway is determined largely by the needs of the cell for NADPH or sugar intermediates. When more NADPH than ribose 5-phosphate is required, the pathway leads to (a) complete oxidation of G6P to CO_2 and (b) resynthesis of G6P from ribulose 5-phosphate. Alternatively, if more ribose 5-phosphate than NADPH is required, G6P is converted to ribulose 5-phosphate for nucleic acid synthesis or recycled to produce intermediates of the glycolytic pathway.

The tissue distribution of the pentose phosphate pathway is consistent with its functions. In erythrocytes it produces NADPH required to maintain reduced glutathione, which protects the integrity of the red cell membrane. Liver, mammary gland, testis, and adrenal cortex, sites of fatty acid or steroid synthesis, require the reducing equivalent of NADPH. In liver, 20–30% of the CO_2 produced may arise from the pentose phosphate pathway; the balance between glycolysis and the pentose phosphate pathway depends on the metabolic requirements of the organ. In mammalian striated muscle, which exhibits little fatty acid or steroid synthesis, all catabolism of G6P proceeds via glycolysis and the TCA cycle with no direct oxidation of glucose 6-phosphate through the pentose phosphate pathway.

16.3 SUGAR INTERCONVERSIONS AND NUCLEOTIDE-LINKED SUGAR FORMATION

Most monosaccharides found in biological compounds derive from glucose. The most common sugar **transformations** in mammalian systems are summarized in Figure 16.4.

Isomerization and Phosphorylation are Common Reactions for Interconverting Carbohydrates

Formation of some sugars can occur directly, starting from glucose via modification reactions, such as conversion of G6P to fructose 6-phosphate by phosphoglucose isomerase in glycolysis. A similar **aldose–ketose isomerization** catalyzed by **phosphomannose isomerase** produces **mannose 6-phosphate**. Deficiency in this enzyme leads to one form of carbohydrate-deficient glycoprotein syndrome (CDGS) (see Clin. Corr. 16.3).

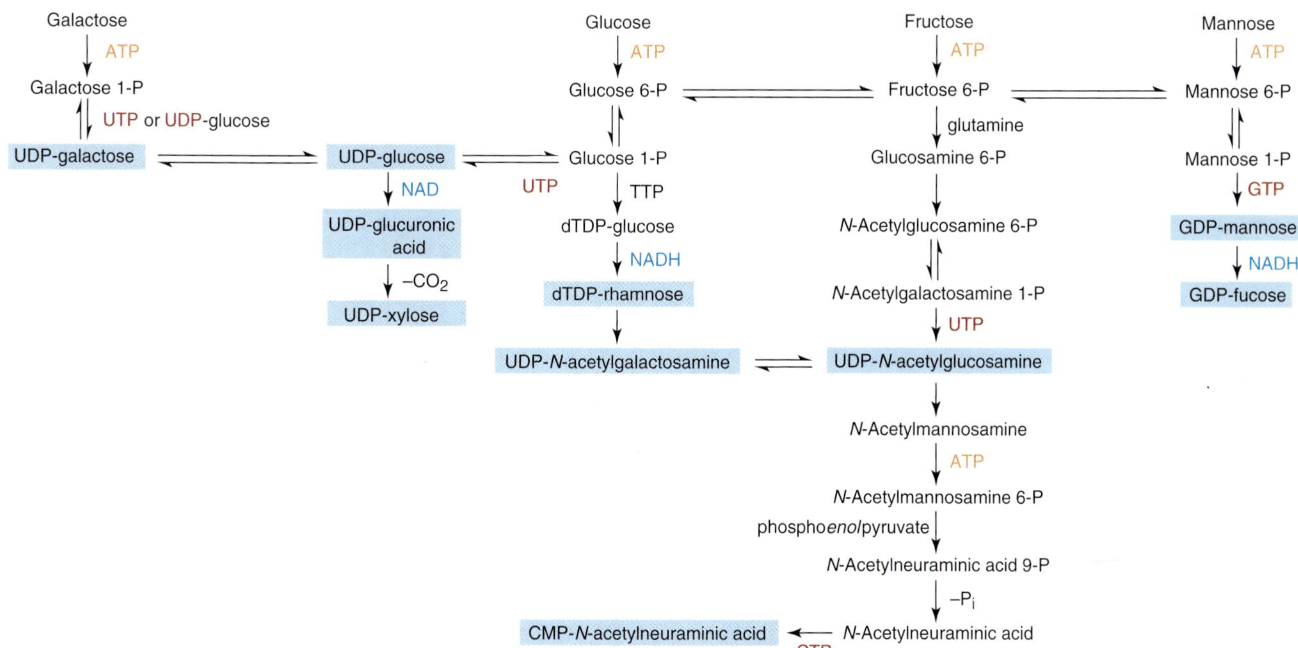

FIGURE 16.4

Pathways of formation of nucleotide-linked sugars and interconversion of some hexoses.

CLINICAL CORRELATION 16.3
Carbohydrate-Deficient Glycoprotein Syndromes (CDGS)

The CDGS are clinically heterogeneous autosomal recessive glycosylation disorders. Using the glycosylation state of serum transferrin as a sensitive indicator, several types of CDGS have been identified. Type I is the most common and has three distinct forms, two of which are in the early mannose biosynthetic pathway (Fig. 16.4). Type Ib, which presents in the first year of life with hypoglycemia, vomiting, diarrhea, protein-loss enteropathy, hepatic fibrosis, and susceptibility to recurring thrombosis due to low levels of antithrombin III, is due to deficiency of phosphomannose isomerase. Defects in phosphomannomutase 2 are responsible for type Ia, and they account for approximately 70% of all type I patients. Symptoms include hypotonia, dysmorphism, liver dysfunction, developmental delays in motor and language development, and failure to thrive; about 20% of patients with this genetic disorder die within the first few years of life. Type Ic is due to defects late in the glycoprotein biosynthetic pathway—for example, glucosyltransferase that adds the first glucosyl residue to the $Man_9GlcNac_2$-PP-Dol precursor. Type II is due to deficiency in *N*-acetylglucosaminyltransferase II which initiates the second N-linked antenna. Type IV involves a defect in dolichol phosphomannose synthase and type V is caused by a deficiency in dolichol-P-Glc: Man_3 $GlcNAc_3$-PP-dolichylglucosyltransferase. Thus defects in monosaccharide interconversion or glycoprotein biosynthetic pathways are often characterized by neonatal presentation of severe neurological and metabolic dysfunction.

Source: Jaeken, J. and Carchon, H.. The carbohydrate-deficient glycoprotein syndromes: an overview. *J. Inherit. Metab. Dis.* 16:813, 1993: Freeze, H. H. Disorders in protein glycosylation and potential therapy: Tip of an iceberg. *J. Pediatr.* 133:394, 1998: Gahl, W. A. Carbohydrate-deficient glycoprotein syndrome: Hidden treasures. *J. Lab. Clin. Med.* 129:394 1997.

CLINICAL CORRELATION 16.4
Essential Fructosuria and Fructose Intolerance: Deficiency of Fructokinase and Fructose 1-phosphate Aldolase

Fructose may account for 30–60% of the total carbohydrate intake of mammals, and it is predominantly metabolized by a fructose-specific pathway. Fructokinase is deficient in essential fructosuria. This disorder is a benign asymptomatic metabolic anomaly which appears to be inherited as an autosomal recessive trait. Following intake of fructose, blood and urinary fructose levels of affected individuals are unusually high; however, 90% of their fructose intake is eventually metabolized. In contrast, hereditary fructose intolerance is characterized by severe hypoglycemia after ingestion of fructose, and prolonged ingestion by affected young children may lead to death. Fructose 1-phosphate aldolase is deficient, and fructose 1-phosphate accumulates intracellularly (see Clin. Corr. 15.3, p. 597).

Source: Steinmann, B., Gitzelmann, R., and Vanden Berghe, G. Disorders of fructose metabolism. In: C. R. Scriver, A. R. Beaudet, W. S. Sly, and D. Valle (Eds.), *The Metabolic and Molecular Bases of Inherited Disease*, Vol. I, 8th ed. New York: McGraw-Hill, 2001, p. 1489.

Phosphorylation and internal transfer of a phosphate group on the same sugar molecule are also common modifications. Glucose 1-phosphate, resulting from glycogenolysis, is converted to G6P by **phosphoglucomutase**. Galactose is phosphorylated to galactose 1-phosphate by **galactokinase** and mannose to mannose 6-phosphate by **mannokinase**. Free fructose, an important dietary constituent, is phosphorylated in the liver to fructose 1-phosphate by a special fructokinase. However, no mutase interconverts fructose 1-phosphate and fructose 6-phosphate, nor can phosphofructokinase synthesize fructose 1,6-bisphosphate from fructose 1-phosphate. Rather, **fructose 1-phosphate aldolase** cleaves fructose 1-phosphate to **dihydroxyacetone phosphate** (DHAP), which enters the glycolytic pathway directly, and glyceraldehyde, which is first reduced to glycerol, phosphorylated, and then reoxidized to DHAP. Lack of this aldolase leads to fructose intolerance (see Clin. Corr. 16.4).

Nucleotide-Linked Sugars are Intermediates in many Sugar Transformations

Most other sugar transformation reactions require conversion into **nucleotide-linked sugars.** A **pyrophosphorylase** converts hexose 1-phosphate and nucleoside triphosphate (NTP) to nucleoside diphosphate (NDP)-sugar. Such reactions are not readily reversible, as pyrophosphate is rapidly hydrolyzed by **pyrophosphatase**, thereby driving the synthesis of nucleotide-linked sugars. They are summarized as follows:

$$\text{NTP} + \text{sugar 1-phosphate} + H_2O \rightleftharpoons \text{NDP-sugar} + PP_i$$

$$P_i + H_2O \rightleftharpoons 2P_i$$

$$\text{NTP} + \text{sugar 1-phosphate} + H_2O \rightleftharpoons \text{NDP-sugar} + 2P_i$$

UDP-glucose is used in synthesis of glycogen and glycoproteins, and is synthesized by **UDP-glucose pyrophosphorylase.**

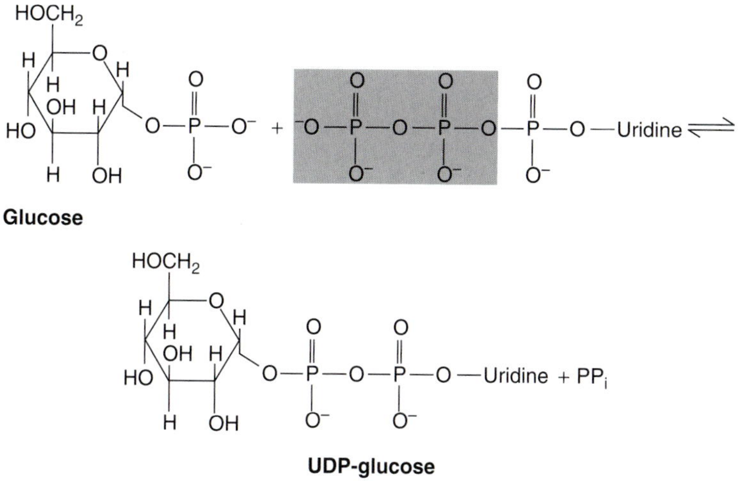

Glucose + UTP ⇌ UDP-glucose + PP_i

Glucose

UDP-glucose

Nucleoside diphosphate-sugars contain two phosphoryl bonds, each with a large negative ΔG of hydrolysis, that contribute to their value as glycosyl donors in further transformation and transfer reactions, as well as conferring substrate specificity in such reactions. UDP is usually the glucosyl carrier, while ADP, GDP and CMP are carriers for other sugars. Many sugar transformation reactions occur only at the level of nucleotide-linked sugars (Figure 16.4).

Epimerization Interconverts Glucose and Galactose

Interconversion of glucose and galactose in animals occurs by epimerization of UDP-glucose to UDP-galactose, catalyzed by **UDP-glucose-4-epimerase** (Figure. 16.5). UDP-galactose is also formed from free galactose, derived from hydrolysis of lactose in the intestinal tract. Galactose is phosphorylated by **galactokinase** and ATP to yield galactose 1-phosphate. Then **galactose 1-phosphate uridylyltransferase** forms UDP-galactose by galactose 1-phosphate displacing glucose 1-phosphate from UDP-glucose. These reactions are summarized as follows:

$$\text{galactose} + \text{ATP} \rightleftharpoons \text{galactose 1-phosphate} + \text{ADP}$$

$$\text{UDP-glucose} + \text{galactose 1-phosphate} \rightleftharpoons \text{UDP-galactose} + \text{glucose 1-phosphate}$$

A combination of these reactions allows transformation of dietary galactose into glucose 1-phosphate, to be metabolized as previously described. Alternatively, the 4-epimerase can produce UDP-galactose as needed for biosynthesis. The hereditary disorder, galactosemia, results from absence of the uridylyltransferase (see Clin. Corr. 16.5).

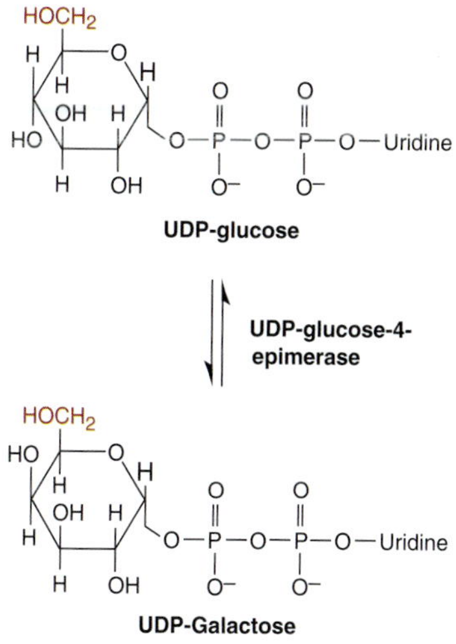

UDP-glucose

UDP-glucose-4-epimerase

UDP-Galactose

FIGURE 16.5

Conversion of glucose into galactose.

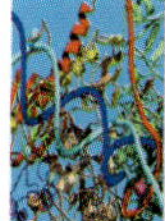

CLINICAL CORRELATION **16.5**

Galactosemia: Inability to Transform Galactose into Glucose

Reactions of galactose are of particular interest because in humans they are subject to genetic defects that produce the hereditary disorder galactosemia. When a defect is present, individuals are unable to metabolize the galactose derived from lactose (milk sugar) to glucose metabolites, often resulting in cataract formation, growth failure, mental retardation, or eventual death from liver damage. These phenotypes may be due to a cellular deficiency of either (a) galactokinase, causing a relatively mild disorder characterized by early cataract formation, or (b) galactose 1-phosphate uridylyltransferase, resulting in severe disease. Galactose is reduced to galactitol in a reaction similar to that of glucose to sorbitol. Galactitol initiates cataract formation in the lens and may play a role in the central nervous system damage. Accumulation of galactose 1-phosphate is responsible for liver failure; the toxic effects of galactose metabolites disappear when galactose is removed from the diet.

Source: Holton, J. B. Galactosemia: Pathogenesis and treatment. *J. Inherit. Metab. Dis.* 19:3, 1996: Segal, S. Galactosemia unsolved. *Eur. J. Pediatr.* 154:S97, 1995: Petry, K. G. and Reichardt, J. K. The Fundamental importance of human galactose. *Trends Genet.* 14:98, 1998.

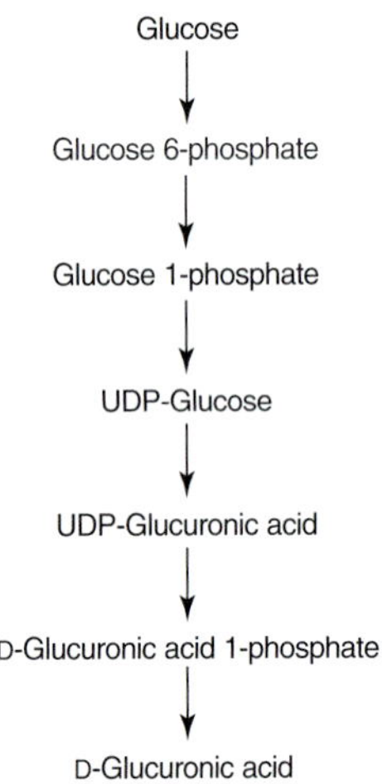

FIGURE 16.7
Biosynthesis of D-glucuronic acid from glucose.

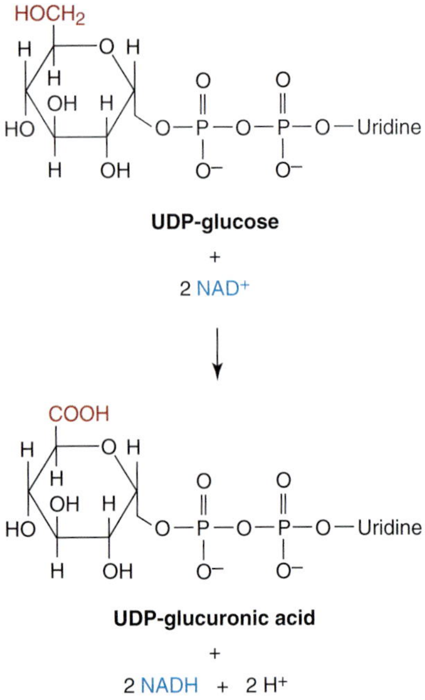

UDP-glucose

+

2 NAD^+

↓

UDP-glucuronic acid

+

2 NADH + 2 H^+

FIGURE 16.6
Formation of UDP-glucuronic acid from UDP-glucose.

Synthesis of GDP-fucose (Figure. 16.4) begins with the conversion of GDP-mannose to GDP-4-keto-6-deoxymannose by **GDP-mannose-4, 6-dehydratase**, followed by its epimerization to GDP-4-keto-6-deoxy-L-galactose, which is then reduced to GDP-fucose. These latter reactions are catalyzed by the bifunctional **GDP-4-keto-6-deoxymannose 3,5-epimerase-4-reductase** or by the FX protein, which is abundant in red cells.

Epimerization of D-glucuronic acid to L-iduronic acid occurs after incorporation into heparin and dermatan sulfate (see p. 653).

Glucuronic Acid is Formed by Oxidation of UDP-Glucose

Glucuronic acid results from oxidation of UDP-glucose by **UDP-glucose dehydrogenase** (Figure 16.6) as outlined in Figure 16.7. In humans, glucuronic acid is converted to L-xylulose, the ketopentose excreted in essential pentosuria (Clin. Corr. 16.6). Glucuronic acid also participates in detoxification by formation of glucuronide conjugates (Clin. Corr. 16.7). Glucuronic acid is a precursor of **L-ascorbic acid** in those animals that synthesize vitamin C. Glucuronic acid is reduced by NADPH to L-gulonic acid (Figure 16.8), which is then converted through L-gulonolactone to L-ascorbic acid **(vitamin C)** in plants and most higher animals. Humans, other primates, and guinea pigs lack the enzyme that converts L-gulonolactone to L-ascorbic acid and for them, ascorbic acid is a vitamin. Gulonic acid can also be oxidized to 3-ketogulonic acid and decarboxylated to L-xylulose. L-Xylulose is reduced to xylitol, reoxidized to D-xylulose, and phosphorylated to xylulose 5-phosphate. This can enter the pentose phosphate pathway described previously. The glucuronic acid pathway operates in adipose tissue, and its activity is usually increased in tissue from starved or diabetic animals.

CLINICAL CORRELATION 16.6
Pentosuria: Deficiency of Xylitol Dehydrogenase

The glucuronic acid oxidation pathway does not appear to be essential for human carbohydrate metabolism, since individuals in whom the pathway is blocked suffer no ill effects. A metabolic variation, called idiopathic pentosuria, results from reduced activity of NADP-linked L-xylulose reductase, which reduces xylulose to xylitol. Hence, affected individuals excrete large amounts of pentose into the urine, especially following intake of glucuronic acid.

Source: Hiatt, H. 2001. Pentosuria. In: C. R. Scriver, A. R. Beaudet, W. S. Sly, and D. Valle (Eds.), *The Metabolic and Molecular Bases of Inherited Disease*, Vol. I, 8th ed. New York: McGraw-Hill, 2001, p. 1589.

Decarboxylation, Oxidoreduction, and Transamidation of Sugars Yield Necessary Products

The only known **decarboxylation** of a nucleotide-linked sugar is the conversion of UDP-glucuronic acid to UDP-xylose, which is necessary for synthesis of proteoglycans

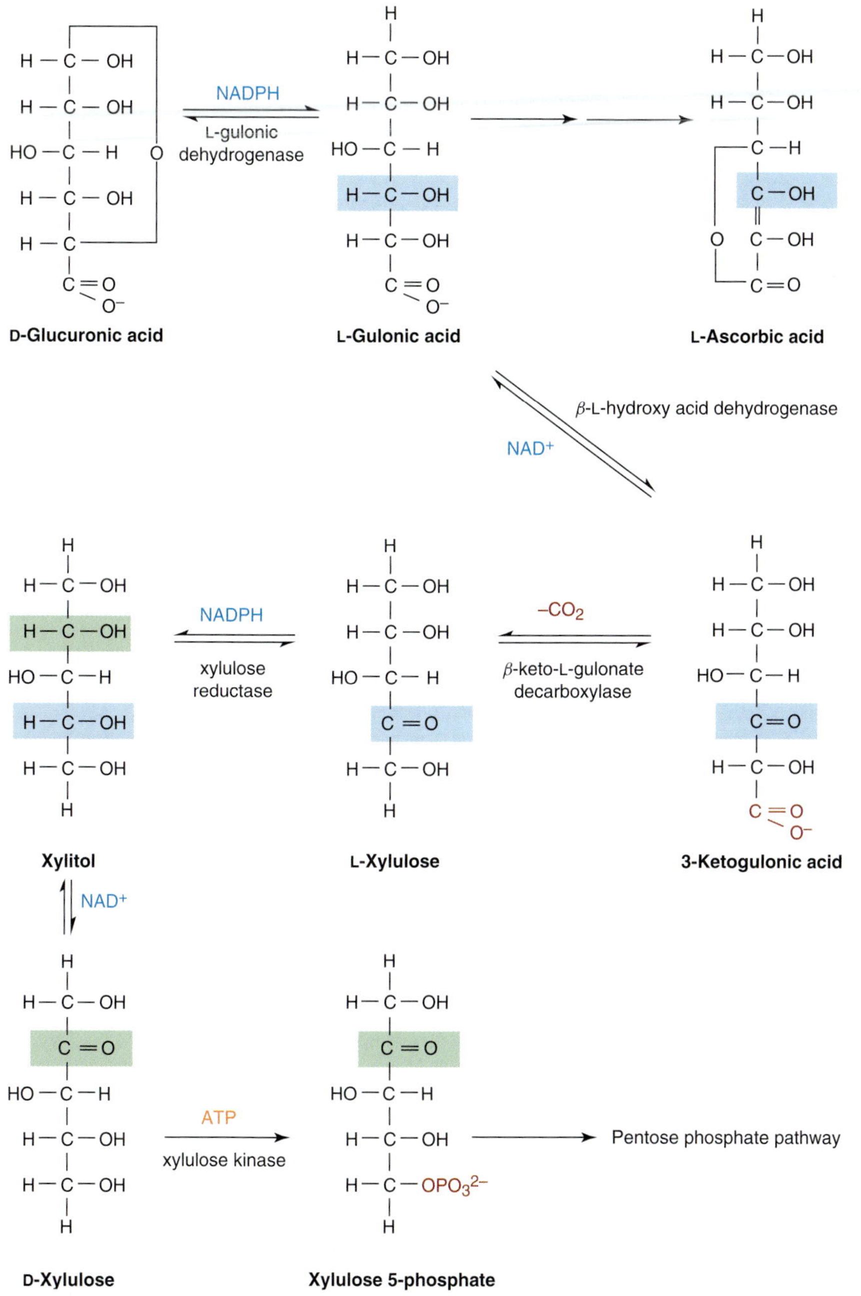

FIGURE 16.8
Glucuronic acid oxidation pathway.

(see p. 652) and is a potent inhibitor of UDP-glucose dehydrogenase that produces UDP-glucuronic acid (Figure 16.6). Thus, the level of these nucleotide-linked sugar precursors is regulated by this sensitive feedback mechanism.

Deoxyhexoses and **dideoxyhexoses** are also synthesized from precursor sugars attached to nucleoside diphosphates. For example, L-rhamnose is synthesized from glucose by a series of oxidation–reduction reactions starting with dTDP-glucose and yielding dTDP-rhamnose. A similar reaction accounts for synthesis of GDP-fucose from GDP-mannose (see previous section) and for various dideoxyhexoses.

Formation of amino sugars, major components of human complex oligo- and polysaccharides and constituents of antibiotics, occurs by **transamidation**. For example, glucosamine 6-phosphate is formed from fructose 6-phosphate and glutamine.

CLINICAL CORRELATION 16.7

Glucuronic Acid: Physiological Significance of Glucuronide Formation

The biological significance of glucuronic acid includes conjugation with certain endogenous and exogenous substances to form glucuronides in a reaction catalyzed by UDP-glucuronyltransferase. Conjugation with glucuronic acid produces a strongly acidic compound that is more water-soluble at physiological pH than its precursor and therefore may enhance its transport or excretion. Glucuronide formation is important in drug detoxification, steroid excretion, and bilirubin metabolism. Bilirubin is the major metabolic breakdown product of heme, the prosthetic group of hemoglobin. The central step in its excretion is conjugation with glucuronic acid by UDP-glucuronyltransferase. This enzyme may take several days to two weeks after birth to become fully active in humans. So-called "Physiological jaundice of the newborn" results in most cases from inability of neonatal liver to form bilirubin glucuronide at a rate comparable to that of bilirubin production. The mutant strain of Wistar "Gunn" rats has a deficiency of UDP-glucuronyltransferase, which results in hereditary hyperbilirubinemia. In humans, a similar defect occurs in congenital familial nonhemolytic jaundice (Crigler–Najjar syndrome), in which patients cannot conjugate foreign compounds efficiently with glucuronic acid.

Source: Chowdhury, J. R., Wolkoff, W., Chowdhury, N. R., and Arias, I. W. 2001. Hereditary jaundice and disorders of bilirubin metabolism. In: 8th ed. C. R. Scriver, A. R. Beaudet, W. S. Sly, and D. Valle (Eds.), *The Metabolic and Molecular Bases of Inherited Disease*, Vol. II, New York: McGraw-Hill, 2001, p. 3063.

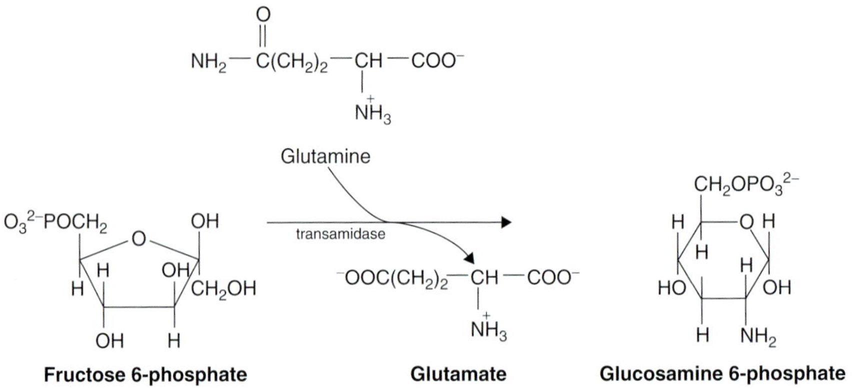

$$NH_2-\overset{O}{\overset{\|}{C}}(CH_2)_2-\underset{\overset{+}{N}H_3}{\underset{|}{CH}}-COO^-$$

Glutamine

transamidase

Fructose 6-phosphate → Glutamate + Glucosamine 6-phosphate

$$^-OOC(CH_2)_2-\underset{\overset{+}{N}H_3}{\underset{|}{CH}}-COO^-$$

Fructose 6-phosphate **Glutamate** **Glucosamine 6-phosphate**

Glucosamine 6-phosphate can be *N*-acetylated to *N*-acetylglucosamine 6-phosphate, followed by isomerization to *N*-acetylglucosamine 1-phosphate and formation of UDP-*N*-acetylglucosamine. This is a precursor of glycoprotein synthesis and of UDP-*N*-acetylgalactosamine, which is necessary for proteoglycan synthesis. Fructose 6-phosphate-glutamine transamidase is under negative feedback control by UDP-*N*-acetylglucosamine (Figure 16.4). This regulation is meaningful in certain tissues such as skin, in which this pathway accounts for up to 20% of glucose flux.

Sialic Acids are Derived from *N*-Acetylglucosamine

Another product of UDP-*N*-acetylglucosamine is ***N*-acetylneuraminic acid**, one of a family of C_9 sugars, called **sialic acids** (Figure 16.9). Epimerization of UDP-*N*-acetylglucosamine by a 2-epimerase produces *N*-acetylmannosamine. Most likely, this reaction proceeds by a *trans* elimination of UDP, and formation of the unsaturated intermediate, 2-acetamidoglucal. In mammalian tissues *N*-acetylmannosamine is phosphorylated and the *N*-acetylmannosamine 6-phosphate condenses with phosphoenolpyruvate to form *N*-acetylneuraminic acid 9-phosphate. Removal of the phosphate and formation of CMP-*N*-acetylneuraminic acid follows. All of these reactions occur in the cytosol, except the last one which occurs in the nucleus with subsequent export of the CMP-*N*-acetylneuraminic acid to the cytoplasm.

16.4 BIOSYNTHESIS OF COMPLEX POLYSACCHARIDES

In complex glycoconjugates, sugar moieties are linked by glycosidic bonds formed by specific **glycosyltransferases** that transfer the glycosyl unit from the nucleotide derivative to the nonreducing end of an acceptor sugar. A glycosyltransferase is specific for the sugar acceptor, the sugar transferred, and the linkage formed. A glycosyltransferase reaction is summarized as follows:

$$\underset{\text{(donor)}}{\text{nucleoside diphosphate-glycose}} + \underset{\text{(acceptor)}}{\text{glycose}} \xrightleftharpoons{\text{glycosyltransferase}} \underset{\text{(glycoside)}}{\text{glycosyl}_1\text{-}O\text{-glycose}_2} + \text{nucleoside diphosphate}$$

More than 40 types of glycosidic bonds have been identified in mammalian oligosaccharides and about 15 more in glycosaminoglycans. The multitude of linkages arises from the diversity of monosaccharides involved and from the formation of α and β linkages with each of the available hydroxyl groups on the acceptor saccharide. This suggests that oligosaccharides have the potential for great informational content. In fact, it is known that the bioactivity of many molecules is determined by the nature of the composite sugar residues. For example, the antigenic specificity of the major blood types is determined by the sugar composition (see Clin. Corr. 16.8). *N*-Acetylgalactosamine is

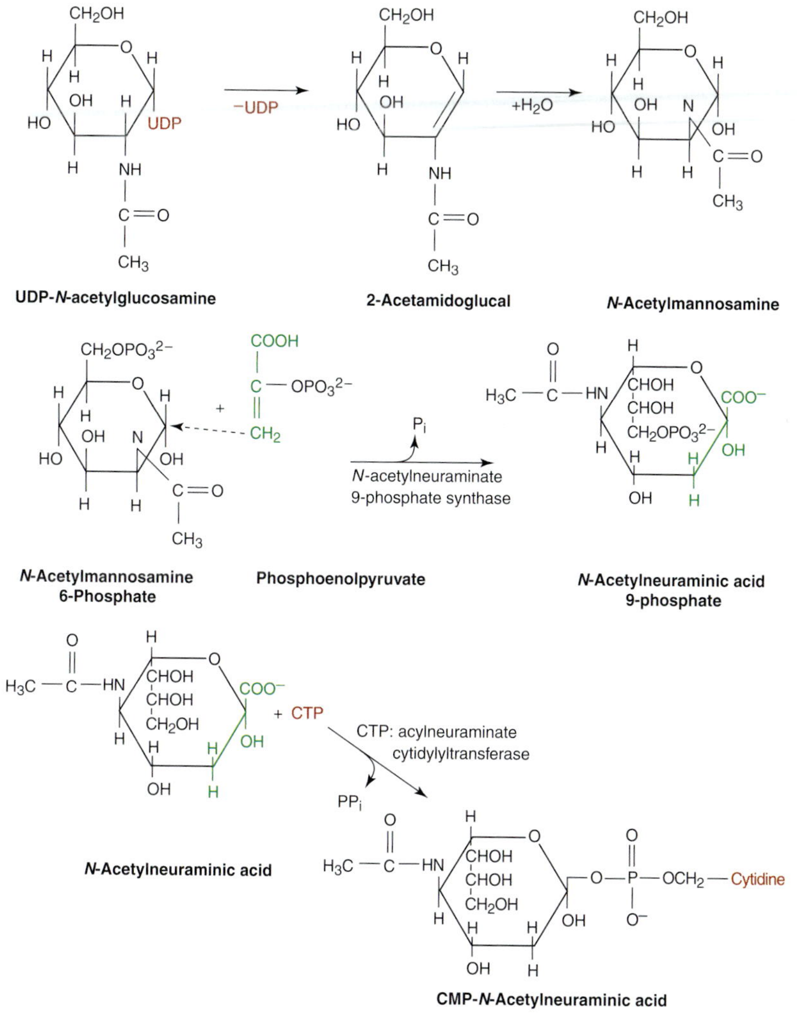

UDP-*N*-acetylglucosamine → (−UDP) → 2-Acetamidoglucal → ($+H_2O$) → *N*-Acetylmannosamine

N-Acetylmannosamine 6-Phosphate + Phosphoenolpyruvate → (P_i; *N*-acetylneuraminate 9-phosphate synthase) → *N*-Acetylneuraminic acid 9-phosphate

N-Acetylneuraminic acid + CTP → (PP_i; CTP: acylneuraminate cytidylyltransferase) → CMP-*N*-Acetylneuraminic acid

FIGURE 16.9

Biosynthesis of CMP-*N*-acetylneuraminic acid.

the immunodeterminant of blood type A and galactose of blood type B. Removal of *N*-acetylgalactosamine from type A erythrocytes, or of galactose from type B erythrocytes, converts them to type O erythrocytes. Increasingly, other examples of sugars as determinants of specificity for cell surface receptor and lectin interactions, targeting of cells to certain tissues, and survival or clearance from the circulation of certain molecules, are being recognized. All glycosidic bonds identified in biological compounds are degraded by specific hydrolytic enzymes, **glycosidases**. In addition to being valuable tools for the structural elucidation of oligosaccharides, interest in this class of enzymes is based on the many genetic diseases of complex carbohydrate metabolism that result from defects in glycosidases (see Clin. Corrs. 16.10 and 16.11; p. 653 and p. 654 respectively).

CLINICAL CORRELATION 16.8
Blood Group Substances

The surface of human erythrocytes is covered by a complex mosaic of specific antigenic determinants, many of which are complex polysaccharides. There are about 100 blood group determinants, belonging to 21 independent blood group systems. The most widely studied are those of the ABO blood group system and the closely related Lewis system. Genetic variation is achieved through specific glycosyltransferases responsible for synthesis of the heterosaccharide determinants. The H gene codes for a fucosyltransferase, which adds fucose to a peripheral galactose in the heterosaccharide precursor. The A allele encodes an *N*-acetylgalactosamine glycosyltransferase, the B allele encodes a galactosyltransferase, and the O allele encodes an inactive protein. The sugars transferred by the A and B enzymes are added to the H-specific oligosaccharide. The Lewis (Le) gene codes for another fucosyltransferase, which adds fucose to a peripheral *N*-acetylglucosamine residue in the precursor. Absence of the H gene product gives rise to the Le^{a} specific determinant, whereas absence of both H and Le enzymes is responsible for the Le^{b} specificity. Elucidation of the structures of these oligosaccharide determinants represents a milestone in carbohydrate chemistry. This knowledge is essential to blood transfusion practices and for legal and historical purposes. For example, tissue dust containing complex carbohydrates has been used in serological analysis to establish the blood group of Tutankhamen and his probable ancestral background.

Source: Yamamoto, F., Clausen, I., White, T., Mark, J., and Hakomori, S. Molecular genetic basis of the histo-blood group ABO system. *Nature* 345:229, 1990.

16.5 GLYCOPROTEINS

Glycoproteins have been defined as conjugated proteins, which contain one or more saccharides lacking a serial repeat unit and are bound covalently to a protein. This definition excludes proteoglycans (see p. 652). Glycoproteins in cell membranes may have an important role in the behavior of cells and especially in the biological functions of the membrane. Glycoproteins are constituents of the mucus secreted by certain epithelial cells, where they mediate lubrication and protection of tissues lining the

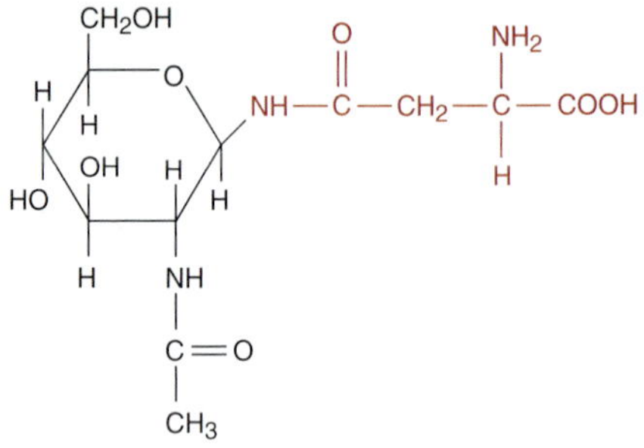

Type I *N*-Glycosyl linkage to asparagine

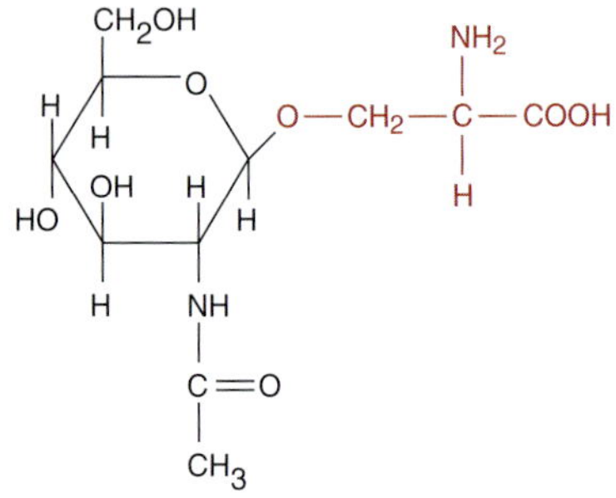

Type II *O*-Glycosyl linkage to serine

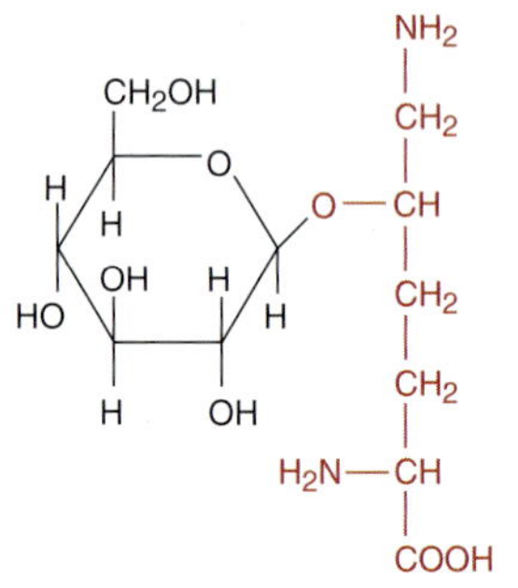

Type III *O*-Glycosyl linkage to 5-hydroxylysine

FIGURE 16.10

Structure of three major types of glycopeptide bond.

respiratory, gastrointestinal, and female reproductive systems. Many secreted proteins are glycoproteins, and they include (a) hormones such as follicle stimulating hormone, luteinizing hormone, and chorionic gonadotropin and (b) plasma proteins such as the orosomucoids, ceruloplasmin, plasminogen, prothrombin and immunoglobulins.

Glycoproteins Contain Variable Amounts of Carbohydrate

The amount of carbohydrate in glycoproteins is highly variable. Thus, IgGs contain about 4% of carbohydrate by weight, while glycophorin contains 60%, human ovarian cyst glycoprotein contains 70%, and human gastric glycoprotein contains 82%. The carbohydrate may be distributed fairly evenly along a polypeptide chain or concentrated in defined regions. For example, in human glycophorin A the carbohydrate is restricted to the extracellular NH_2-terminal half of the polypeptide chain (see p. 456).

The carbohydrate of glycoproteins usually contains less than 12–15 sugar residues. Some consist of one sugar moiety, as in the submaxillary gland glycoprotein (single *N*-acetyl-α-D-galactosaminyl residue) and in some mammalian collagens (single α-D-galactosyl residue). In general, the sugar residues are of the D form, except for L-fucose, L- arabinose, and L-iduronic acid. A glycoprotein from different animal species often has an identical primary protein structure, but a variable carbohydrate component. Heterogeneity of a given protein also occurs within an organism. For example, pancreatic ribonucleases A and B have identical primary structures and similar specificity toward substrates, but differ significantly in their carbohydrate compositions.

Carbohydrates are Linked to Glycoproteins by *N*- or *O*-Glycosyl Bonds

So far, the structures of the carbohydrates of only a limited number of glycoproteins have been completely elucidated. Microheterogeneity of glycoproteins, arising from incomplete synthesis or partial degradation, makes structural analyses extremely difficult. However, certain generalities have emerged. Covalent linkage of sugars to the protein is a central part of glycoprotein structure, and only a small number of bond types are found (see p. 109). The three major types of glycopeptide bonds, as shown in Figure 16.10, are *N*-glycosyl to **asparagine** (Asn), *O*-glycosyl to **serine** (Ser) or **threonine** (Thr), and *O*-glycosyl to **5-hydroxylysine**, which is generally confined to the collagens. The others are present in many glycoproteins. Only the *O*-glycosidic linkage to serine or threonine is labile to alkali cleavage. By this procedure, two types of oligosaccharides (simple and complex) are released. Examination of the simple class from porcine submaxillary mucins revealed a core structure of galactose (Gal)-linked β (1–3) to *N*-acetylgalactosamine (GalNAc) *O*-glycosidically linked to serine or threonine residues. L-Fucose (Fuc), sialic acid (NeuAc), and *N*-acetylgalactosamine are present at the nonreducing end. The general structure is as follows:

$$\begin{array}{ccccccc} \text{GalNAc} & \xrightarrow{1\to3} & \text{Gal} & \xrightarrow{1\to3} & \text{GalNAc} & \longrightarrow & O\text{-Ser/Thr} \\ & & \uparrow {\scriptstyle 1\to2} & & \uparrow {\scriptstyle 1\to2} & & \\ & & \text{Fuc} & & \text{NeuAc} & & \end{array}$$

The complex class is exemplified by the blood group substances (see Clin. Corr. 16.8), in which the oligosaccharide is *N*-glycosidically linked to asparagine. These glycoproteins commonly contain a core structure consisting of mannose (Man) residues linked to *N*-acetylglucosamine (GlcNAc) in the structure

$$(\text{Man})_n \xrightarrow[1\to4]{} \text{Man} \xrightarrow[1\to4]{} \text{GlcNAc} \xrightarrow[1\to4]{} \text{GlcNAc} \longrightarrow \text{Asn}$$

Structural diversity of N-linked glycoproteins arises from trimming and modification of this core to produce a large repertoire of high-mannose, hybrid, or complex *N*-glycan subtypes.

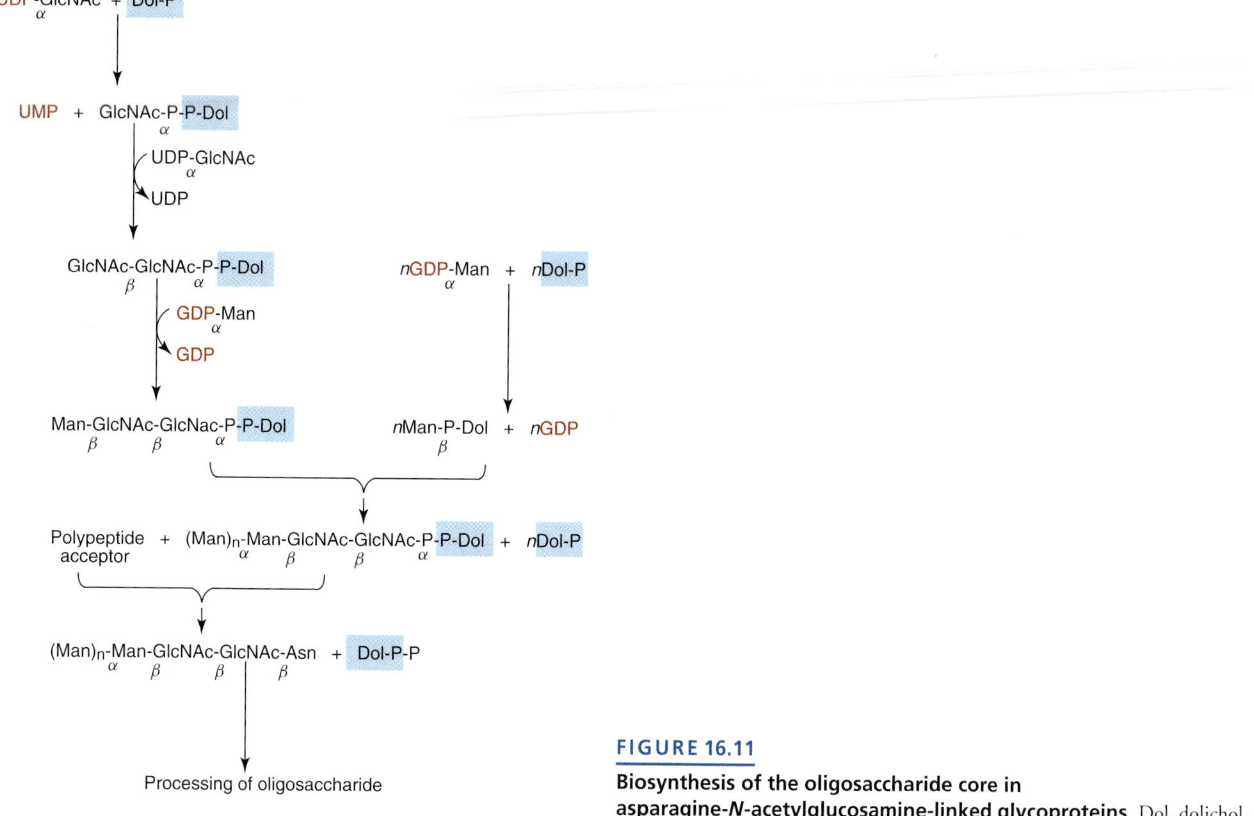

FIGURE 16.11

Biosynthesis of the oligosaccharide core in asparagine-*N*-acetylglucosamine-linked glycoproteins. Dol, dolichol.

Synthesis of N-Linked Glycoproteins Involves Dolichol Phosphate

Whereas synthesis of *O*-glycosidically linked glycoproteins involves sequential action of glycosyltransferases, synthesis of *N*-glycosidically linked glycoproteins involves a different and more complex mechanism (Figure 16.11). A common core is preassembled as a **lipid-linked oligosaccharide** on the cytoplasmic side of the ER and then "flipped" across the bilayer prior to incorporation into the polypeptide. During synthesis, the oligosaccharide intermediates are bound to **dolichol phosphate**.

$$(CH_2{=}\overset{\overset{\displaystyle CH_3}{|}}{C}-CH{=}CH)_n-CH_2-\overset{\overset{\displaystyle CH_3}{|}}{CH}-CH_2-CH_2O-PO_3H_2$$

Dolichol phosphate

Dolichols are polyprenols (C_{80}–C_{100}) that contain 16–20 isoprene units, in which the final unit is saturated. These lipids participate in two ways. The first involves formation of *N*-acetylglucosaminyl pyrophosphoryldolichol from the UDP and dolichol phosphate-linked sugars. The second involves *N*-acetylglucosamine, and the mannose is transferred directly from the nucleotide without formation of intermediates. In the final step, the oligosaccharide is transferred from the dolichol pyrophosphate to an asparagine residue in the polypeptide chain.

After transfer to the polypeptide, the core structures are completed by glycosyltransferases without further participation of lipid intermediates. A series of early processing reactions, which are highly conserved among vertebrate species and cell types, occur largely in the ER and appear to be coupled with proper folding of the glycoprotein. Following the initial trimming and release from the ER, *N*-glycans undergo further glycosidase and glycosyltransferase modifications, mostly in the Golgi. Several avenues exist in the processing pathway that determines the final diversity of glycan structure (i.e., high mannose, hybrid or complex subtypes) as well as the trafficking fate of glycoproteins. For instance, on glycoproteins destined for the lysosomal compartment, the

CLINICAL CORRELATION 16.9

Common Carbohydrate Marker of Lysosomal Targeting and I-Cell Disease

I-cell disease, so-called because of the large inclusion bodies present in cells cultured from patients, is characterized by severe clinical and radiological features including congenital dislocations, thoracic deformities, hernia, restricted joint mobility and retarded psychomotor development. This rare and fatal congenital disorder provided the earliest connection between glycoprotein biosynthesis and human disease. Lysosomal function is defective because the acid hydrolase content is low due to a defect in processing the glycoprotein acid hydrolases, because of a failure to generate the man-6-P residue. Without this specific recognition marker, the acid hydrolases are not targeted to the lysosomes, but are secreted into the extracellular milieu. In most patients the GlcNAc phosphotransferase that attaches the *N*-acetylglucosamine 6-phosphate residue is defective.

Source: Kornfeld, S. Lysosomal enzyme targeting. *Biochem. Soc. Trans.* 18:367, 1990.

Man_8 $GlcNAc_2$-Asn *N*-glycan is modified by addition of a GlcNAc residue catalyzed by a **GlcNAc-phosphotransferase**, and subsequent removal by **GlcNAc phosphodiester glycosidase**, exposing a Man-6-P residue. A defect in this process forms the basis of the **lysosomal storage disease** known as I-cell disease (Clin. Corr. 16.9).

Just as synthesis of hetero-oligosaccharides requires specific glycosyltransferases, their degradation requires specific glycosidases. Exoglycosidases remove sugars sequentially from the nonreducing end, exposing the substrate for the next glycosidase. The absence of a particular glycosidase causes cessation of catabolism, which results in accumulation of the product (Clin. Corr. 16.10). Endoglycosidases with broader specificity exist, and the action of endo- and exoglycosidases results in catabolism of glycoproteins. Many of the same N- or O-linked glycan chains are found on both glycoproteins and glycolipids, hence certain enzyme defects may affect degradation of both types of glycoconjugates (Clin. Corr. 16.11).

16.6 PROTEOGLYCANS

This is a class of complex macromolecules that may contain 95% or more of carbohydrate, and it resembles polysaccharides more than it resembles proteins. To distinguish them from other glycoproteins, they are called proteoglycans. Their carbohydrate chains are called **glycosaminoglycans** or **mucopolysaccharides**, especially in reference to the storage diseases, **mucopolysaccharidoses**, which result from an inability to degrade these molecules (see Clin. Corr. 16.13).

There are Six Classes of Proteoglycans

Proteoglycans consist of many different glycosaminoglycan chains linked covalently to a protein core. Six distinct classes are recognized: **chondroitin sulfate**, **dermatan sulfate**, **keratan sulfate**, **heparan sulfate**, **heparin**, and **hyaluronate**. Certain features are common to the different classes of glycosaminoglycans. The long unbranched heteropolysaccharide chains are made up largely of disaccharide repeating units, consisting of a hexosamine and a uronic acid. Common constituents of glycosaminoglycans are sulfate groups, linked by ester bonds to certain monosaccharides or by amide bonds to the amino group of glucosamine. Only hyaluronate is not sulfated and is not covalently attached to protein. The carboxyls of uronic acids and the sulfate groups contribute to the highly charged nature of glycosaminoglycans. Their electrical charge and their macromolecular structure are important in their role as lubricants and support elements in connective tissue. Glycosaminoglycans are predominantly components of the extracellular matrices and cell surfaces, and they participate in cell adhesion and signaling.

Hyaluronate is a Copolymer of N-Acetylglucosamine and Glucuronic Acid

Hyaluronate differs from the other types of glycosaminoglycans. It is unsulfated, not covalently linked with protein, and not limited to animal tissue, being also produced by bacteria. It is classified as a glycosaminoglycan because of its structural similarity to these polymers, and it consists solely of repeating disaccharide units of *N*-acetylglucosamine and glucuronic acid (Figure 16.12). Although it has the least complex chemical structure of the glycosaminoglycans, the chains may reach 10^5-10^7 Da. The large mass, polyelectrolyte character, and large volume it occupies in solution contribute to the properties of hyaluronate as a lubricant and shock absorbent. It is found predominantly in synovial fluid, vitreous humor, and umbilical cord.

Chondroitin Sulfates are the Most Abundant Glycosaminoglycans

The most abundant glycosaminoglycans in the body, the **chondroitin sulfates**, are attached to specific serine residues in a protein core through a tetrasaccharide linkage region.

$$\text{GluUA} \xrightarrow[1\to3]{} \text{Gal} \xrightarrow[1\to3]{} \text{Gla} \xrightarrow[1\to4]{} \text{Xyl} \longrightarrow O\text{-Ser}$$

CLINICAL CORRELATION 16.10

Aspartylglycosylaminuria: Absence of 4-L-Aspartylglycosamine Amidohydrolase

Some human inborn errors of metabolism involve storage of glycolipids, glycopeptides, mucopolysaccharides, and hetero-oligosaccharides. These diseases are caused by defects in lysosomal glycosidase activity, which prevent the catabolism of oligosaccharides. They involve gradual accumulation in tissues and urine of compounds derived from incomplete degradation of the oligosaccharides, and they may be accompanied by skeletal abnormalities, hepatosplenomegaly, cataracts or mental retardation. A defect in catabolism of asparagine-*N*-acetylglucosamine-linked oligosaccharides leads to aspartylglycosylaminuria in which a deficiency of 4-L-aspartylglycosylamine amidohydrolase allows accumulation of aspartylglucosamine-linked structures. (See accompanying table.) Other disorders involve accumulation of oligosaccharides derived from both glycoproteins and glycolipids, which share common oligosaccharide structures (see accompanying table and Clin. Corr. 16.11).

Enzymic Defects in Degradation of Asn-GlcNAc Type Glycoproteins[a]

Disease	*Deficient Enzyme[b]*
Aspartylglycosylaminuria	4-L-Aspartylglycosylamine amidohydrolase (2)
β-Mannosidosis	β-Mannosidosis (7)
α-Mannosidosis	α-Mannosidosis (3)
GM_2 gangliosidosis variant O (Sandhoff-Jatzkewitz disease)	β-*N*-Acetylhexosaminidases (A and B) (4)
GM_1 gangliosidosis	β-Galactosidase (5)
Mucolipidosis I (sialidosis)	Sialidase (6)
Fucosidosis	α-Fucosidase (8)

[a] A typical Asn-GlcNAc oligosaccharide structure.

```
NeuAc -(6)α- Gal -(5)β- GlcNAc ╲(4)β
                                 Man
NeuAc -α- Gal -β- GlcNAc ╱β        ╲(3)α
                                    Man -(7)β- GlcNAc -(2)β- GlcNAc -(1)- Asn
                                   ╱ |                         |(8)
NeuAc -α- Gal -β- GlcNAc -β- Manα  GlcNAc                     Fuc
                        (4)╱β
NeuAc -(6)α- Gal -(5)β- GlcNAc
```

[b] The numbers in parentheses refer to the enzymes that hydrolyze those bonds.

Source: Aula, P., Janlanko, A., and Peltonen, L. 2001. Aspartylglucosaminuria. In: C. R. Scriver, A. R. Beaudet, W. S. Sly, and D. Valle (Eds.), *The Metabolic and Molecular Bases of Inherited Disease*, Vol. III, 8th ed. New York: McGraw-Hill, 2001, p. 3507.

The characteristic disaccharide units consist of *N*-acetylgalactosamine and glucuronic acid that are attached to this linkage region (Figure 16.12). The disaccharides can be sulfated at the 4- or 6-position of *N*-acetylgalactosamine. Each chain contains 30–50 disaccharide units (15–25 kDa). An average chondroitin sulfate proteoglycan molecule contains about 100 chondroitin sulfate chains attached to the protein core, giving a mass of $1.5–2 \times 10^6$ Da. Proteoglycan preparations are extremely heterogeneous, differing in length of protein core, degree of substitution, distribution of polysaccharide chains, length of chondroitin sulfate chains, and degree of sulfation. Chondroitin sulfates may aggregate noncovalently with hyaluronate and are prominent components of cartilage, tendons, ligaments, and aorta, as well as brain, kidney, and lung.

Dermatan Sulfate Contains L-Iduronic Acid

Dermatan sulfate differs from chondroitin 4- and 6-sulfates in that its predominant uronic acid is L-iduronic acid, although D-glucuronic acid is also present. The epimerization of D-glucuronic acid to L-iduronic acid occurs after incorporation into the polymer

CLINICAL CORRELATION 16.11
Glycolipid Disorders

A host of human genetic diseases arise from deficiency in hydrolases which act predominantly on glycolipid substrates, resulting in accumulation of glycolipid and ganglioside products. The clinical symptoms associated with each of the glycoconjugates may vary greatly. However, because of the preponderance of lipids in the nervous system, such disorders often have associated neurodegeneration and severe mental and motor deterioration.

Enzymic Defects in Degradation of Glycolipids

Disease	*Deficiency Enzyme*
Tay-Sachs	β-Hexosaminidase A
Sandhoff's	β-Hexosaminidases A and B
GM_1 gangliosidosis	β-Galactosidase
Sialidosis	Sialidase
Fabry's	α-Galactosidase
Gaucher's	β-Glucoceramidase
Krabbe's	β-Galactoceramidase
Metachromatic leukodystrophy	Arylsulfatase A (cerebroside sulfatase)

Source: Beutler, E. and G. Garabowski. Gaucher disease. In: C. R. Scriver, A. R. Beaudet, W. S. Sly, and D. Valle (Eds.), *The Metabolic and Molecular Bases of Inherited Disease*, Vol. III, 8th ed. New York: McGraw-Hill, 2001, p. 3635.

Repeat unit of hyaluronic acid

Repeat unit of chondroitin 4-sulfate

Repeat unit of heparin

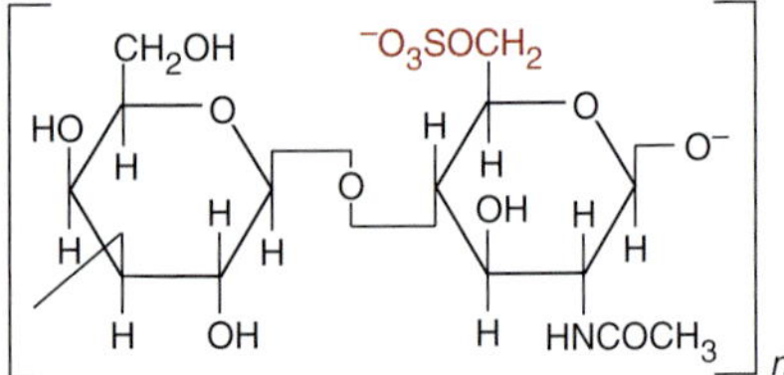

Repeat unit of keratan sulfate

Repeat unit of dermatan sulfate

FIGURE 16.12
Major repeat units of glycosaminoglycan chains.

chain and is coupled with the process of sulfation. The glycosidic linkages have the same position and configuration as in chondroitin sulfates, with average polysaccharide chains of $2–5 \times 10^4$ Da. Dermatan sulfate is antithrombic like heparin, but has only minimal whole-blood anticoagulant and blood lipid-clearing activities. Dermatan sulfate is found in skin, blood vessels, and heart valves.

Heparin and Heparan Sulfate Differ from Other Glycosaminoglycans

In **heparin**, glucosamine and D-glucuronic acid or L-iduronic acid form the characteristic disaccharide repeat unit (Figure 16.12). In contrast to most other glycosaminoglycans, heparin contains α-glycosidic linkages. Almost all glucosamine residues contain sulfamide linkages, while a small number of glucosamine residues are *N*-acetylated. The sulfate content of heparin approaches 2.5 sulfate residues per disaccharide unit in preparations with the highest biological activity. In addition to *N*-sulfate and *O*-sulfate on C6 of glucosamine, heparin may contain sulfate on C3 of the hexosamine and C2 of the uronic acid. Unlike other glycosaminoglycans, heparin is an intracellular component of mast cells, and it functions predominantly as an anticoagulant and lipid-clearing agent (see Clin. Corr. 16.12).

Heparan sulfate contains a similar disaccharide repeat unit as heparin but has more *N*-acetyl groups, fewer *N*-sulfate groups, and a lower degree of *O*-sulfate groups. Heparan sulfate may be extracellular or an integral and ubiquitous component of the cell surface in many tissues, including blood vessel walls and brain.

Keratan Sulfate Exists in Two Forms

Keratan sulfate is composed principally of the disaccharide unit of *N*-acetylglucosamine and galactose, and it contains no uronic acid (Figure 16.12). Sulfate content is variable, as ester sulfate on C6 of galactose and hexosamine. Two types of keratan sulfate differ in carbohydrate content and tissue distribution. Both also contain mannose, fucose, sialic acid and *N*-acetylgalactosamine. Keratan sulfate I, from cornea, is linked to protein by an

CLINICAL CORRELATION 16.12
Heparin Is an Anticoagulant

Heparin is a naturally occurring sulfated glycosaminoglycan that is used to reduce the clotting tendency of patients. Both *in vivo* and *in vitro*, heparin prevents the activation of clotting factors by binding with an inhibitor of the coagulation process. The inhibitor is antithrombin III, a plasma protein inhibitor of serine proteases. In the absence of heparin, antithrombin III slowly (10–30 min) combines with several clotting factors, yielding complexes devoid of proteolytic activity; in the presence of heparin, inactive complexes are formed within a few seconds. Antithrombin III contains an arginine residue that combines with the active site serine of factors Xa and IXa; thus, the inhibition is stoichiometric. Heterozygous antithrombin III deficiency results in increased risk of thrombosis in the veins and resistance to the action of heparin.

Source: Hirsh, J. Drug therapy: Heparin. *N. Engl. J. Med.* 324:1565, 1991.

CLINICAL CORRELATION 16.13
Chondrodystrophies Due to Sulfation Defects

Sulfation is an essential modification of glycosaminoglycans in the various proteoglycan families. The sulfation process involves transport of inorganic sulfate into the cell via plasma membrane transporters, its transformation into phosphoadenosylphosphosulfate (PAPS) via a two-step process catalyzed by PAPS synthetase in the cytosol, then either direct utilization by cytosolic sulfotransferases or transport of PAPS from the cytosol to the Golgi complex for utilization by a host of lumenal sulfotransferases. Three autosomal recessive disorders, diastrophic dysplasia (DTD), atelosteogenesis type II (AOII), and achondrogenesis type 1B (ACG-1B), result from mutations in the DTDST gene that encodes a sulfate transporter. Patients with DTD exhibit disproportionate short stature and generalized joint dysplasia, but usually have a normal lifespan; ACG-1B is characterized by extremely short extremities and trunk; AOII is a perinatally lethal chondrodysplasia. Genetic disorders due to defects in synthesis of PAPS by the bifunctional sulfurylase/kinase (PAPS synthetase) have been identified in both animals (i.e., the brachymorphic mouse that exhibits a severe growth disorder resulting in extremely short trunk, limbs, and small skull) and humans [i.e., spondyloepimetaphyseal dysplasia (Pakistani type) characterized by short and bowed lower limbs, enlarged knee joints, and early onset of degenerative joint disease]. These mutations clearly highlight the importance of this posttranslational modification to the functioning of proteoglycans, especially in development and maintenance of the skeletal system.

Source: Schwartz, N. B. and Domowicz, M. Chondrodysplasias due to proteoglycan defects. *Glycobiology* 12:57R, 2002. Schwartz, N. B. Chondrodysplasias. *Encycl. Endocr. Disorders* 1:502, 2004.

N-acetylglucosamine-asparaginyl bond, typical of glycoproteins. Keratan sulfate II, from cartilage, is linked through *N*-acetylgalactosamine to serine or threonine. Skeletal keratan sulfates are often covalently attached to the same core protein as the chondroitin sulfate chains.

Biosynthesis of Chondroitin Sulfate is Typical of Glycosaminoglycan Formation

The glycosaminoglycans are assembled by sequential action of glycosyltransferases, which transfer a monosaccharide from a nucleotide-linked derivative to an appropriate acceptor, either the nonreducing end of another sugar or a polypeptide. The biosynthesis of the chondroitin sulfates is most thoroughly understood (Figure 16.13).

Formation of the core protein is the first step, followed by the action of six different glycosyltransferases. Strict substrate specificity is required for completion of the unique tetrasaccharide linkage region. Polymerization then results from the concerted action of *N*-acetylgalactosaminyltransferase and glucuronosyltransferase, which form the characteristic disaccharide units. Sulfation of *N*-acetylgalactosamine at the 4- or 6-position occurs with chain elongation. The sulfate donor, as in other biological systems, is **3′-phosphoadenosine 5′-phosphosulfate (PAPS)**, which is formed from ATP and

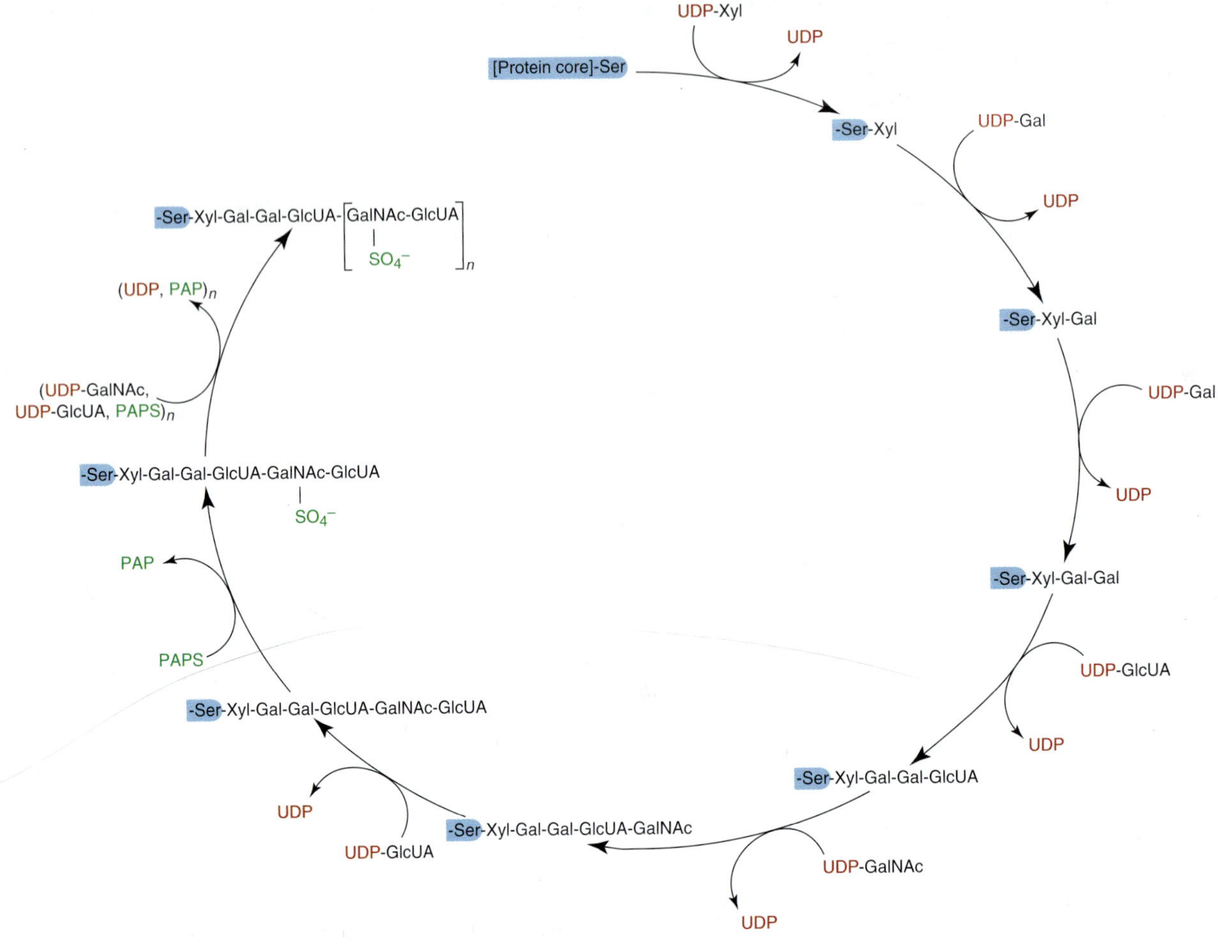

FIGURE 16.13

Synthesis of chondroitin sulfate proteoglycan. Xyl, xylose; Gal, galactose; GlcUA, glucuronic acid; GalNAc, N-acetylgalactosamine; PAPS, phosphoadenosine phosphosulfate.

sulfate in two steps catalyzed by the bifunctional PAPS synthetase (Figure 16.14). The importance of sulfation is highlighted by the preponderance of chondrodystrophic conditions in animals and humans because of deficiencies in the sulfation process (see Clin. Corr. 16.13, p. 655).

Synthesis of other glycosaminoglycans requires additional transferases specific for the appropriate sugars and linkages. Completion often involves *O*-sulfation, epimerization, acetylation, or *N*-sulfation. Synthesis of both proteoglycans and glycoproteins is

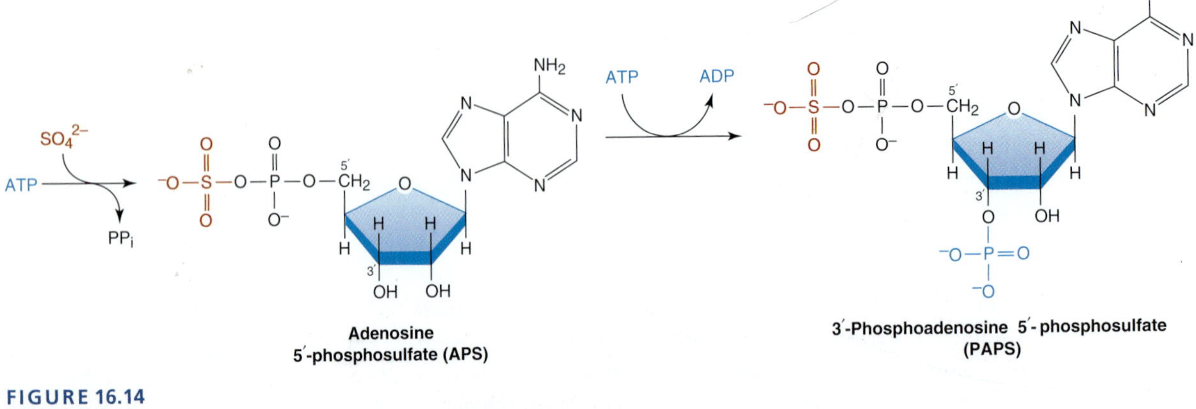

Adenosine 5′-phosphosulfate (APS)

3′-Phosphoadenosine 5′-phosphosulfate (PAPS)

FIGURE 16.14

Biosynthesis of 3′-phosphoadenosine 5′-phosphosulfate (PAPS).

CLINICAL CORRELATION 16.14
Mucopolysaccharidoses

Human genetic disorders characterized by excessive accumulation and excretion of the oligosaccharides of proteoglycans are the mucopolysaccharidoses. They result from deficiency of one or more lysosomal hydrolases that are responsible for the degradation of dermatan and/or heparan sulfate. Hurler's syndrome and Sanfilippo's syndrome are autosomal recessive, whereas Hunter's disease is X-linked. Both Hurler's and Hunter's syndromes are characterized by skeletal abnormalities and mental retardation, which in severe cases may result in early death. In contrast, in Sanfilippo's syndrome the physical defects are relatively mild, while the mental retardation is severe. Collectively, the incidence of the mucopolysaccharidoses is 1 per 30,000 births. Multiple sulfatase deficiency (MSD) is characterized by decreased activity of all known sulfatases. Recent evidence suggests that a co- or posttranslational modification of a cysteine to a 2-amino-3-oxopropionic acid is required for active sulfatases and that a lack of this modification results in MSD. These disorders are amenable to prenatal diagnosis, since the pattern of metabolism by affected cells obtained from amniotic fluid is strikingly different from normal.

Enzyme Defects in the Mucopolysaccharidoses

Syndrome	*Accumulated Products*[a]	*Deficient Enzyme*[b]
Hunter's	Heparan sulfate Dermatan sulfate	Iduronate sulfatase (1)
Hurler–Scheie	Heparan sulfate Dermatan sulfate	α-L-Iduronidase (2)
Maroteaux–Lamy	Dermatan sulfate	*N*-Acetylgalactosamine sulfatase (3)
Mucolipidosis VII	Heparan sulfate Dermatan sulfate	β-Glucuronidase (3)
Sanfilippo's type A	Heparan sulfate	Heparan sulfamidase (6)
Sanfilippo's type B	Heparan sulfate	*N*-Acetylglucosaminidase (9)
Sanfilippo's type C	Heparan sulfate	Acetyl CoA: α-glucosaminide acetyltransferase
Sanfilippo's type D	Heparan sulfate	*N*-Acetylglucosamine 6-sulfatase (8)
Morquio's type A	Keratan/chondroitin sulfate	Galactose-6-sulfatase
Morquio's type B	Keratan sulfate	β-Galactosidase

[a] Structures of dermatan sulfate and heparan sulfate.

Dermatan sulfate: $—IdUA \xrightarrow[\alpha]{(2)} GalNAc \xrightarrow[\beta]{(4)} GlcUA \xrightarrow[\beta]{(4)} GalNAc \xrightarrow[\beta]{}$
(IdUA: (1) OSO_3H; GalNAc: (3) OSO_3H; GalNAc: OSO_3H)

Heparan sulfate: $—IdUA \xrightarrow[\alpha]{(2)} GlcN \xrightarrow[\alpha]{(7)} GlcUA \xrightarrow[\beta]{(5)} GlcNAc \xrightarrow[\alpha]{(9)}$
(IdUA: (1) OSO_3H; GlcN: (6) OSO_3H; GlcNAc: (8) OSO_3H)

[b] The numbers in parentheses refer to the enzymes that hydrolyze those bonds.

Source: Neufeld, E. and Muenzer, J. 2001. Mucopolysaccharidoses. In: C. R. Scriver, A. R. Beaudet, W. S. Sly, and D. Valle (Eds.), *The Metabolic and Molecular Bases of Inherited Disease*, Vol. III, 8th ed. New York: McGraw-Hill, 2001, p. 3421.

regulated by the same mechanism at the level of hexosamine synthesis. Fructose 6-phosphate-glutamine transamidase reaction (Figure 16.4) is subject to feedback inhibition by UDP-*N*-acetylglucosamine, which is in equilibrium with UDP-*N*-acetylgalactosamine. More specific to proteoglycan synthesis, the concentrations of UDP-xylose and UDP-glucuronic acid are stringently controlled by the inhibition by

UDP-xylose of the UDP-glucose dehydrogenase conversion of UDP-glucose to UDP-glucuronic acid (Figure 16.4). Since xylose is the first sugar added during synthesis of chondroitin sulfate, dermatan sulfate, heparin, and heparan sulfate, the earliest effect of decreased core protein synthesis would be accumulation of UDP-xylose, which aids in maintaining a balance between synthesis of protein and glycosaminoglycans.

Proteoglycans, like glycoproteins, are degraded by the sequential action of proteases and glycosidases, **deacetylases** and **sulfatases**. Much of the information about metabolism and degradation of proteoglycans has been derived from the study of the **mucopolysaccharidoses** (Clin. Corr. 16.14). These genetic disorders are characterized by accumulation in tissues and excretion in urine of hetero-oligosaccharide products derived from incomplete breakdown of the proteoglycans, due to a deficiency of one or more **lysosomal hydrolases**.

Although proteoglycans continue to be defined based on the glycosaminoglycan chains they contain, new ones are increasingly being described based largely on functional properties or location. **Aggrecan** and **versican** are the predominant extracellular species; **syndecan, CD44**, and **thrombomodulin** are integral membrane proteins; **neurocan, brevican, cerebrocan**, and **phosphacan** are largely restricted to the nervous system. Many proteoglycans (i.e., aggrecan, **syndecan**, and **betaglycan**) carry two types of glycosaminoglycan chains, whose size and relative amounts may change with development, age or disease. Increasingly, the genes encoding the proteoglycan core proteins and biosynthetic enzymes are being cloned, revealing that the relevant proteins belong to families of related origin and possible function.

BIBLIOGRAPHY

Allen, H. L. and Kisailus, E. C. (Eds.). *Glycoconjugates: Composition, Structure and Function*. New York: Marcel Dekker, 1992.

Collins, P. M. and Ferrier, R. J. *Monosaccharides—Their Chemistry and Roles in Natural Products*. Chichester: Wiley, 1995.

Dutton, G. J. (Ed.). *Glucuronic Acid, Free and Combined*. New York: Academic, 1966.

Ginsburg, V. and Robbins, P. (Eds.). *Biology of Carbohydrates*. New York: Wiley, 1984.

Horecker, B. L. *Pentose Metabolism in Bacteria*. New York: Wiley, 1962.

Hughes, R. C. *Glycoproteins*. London: Chapman & Hall, 1983.

Kornfeld, R., and Kornfeld, S. Assembly of asn-linked oligosaccharides. *Annu. Rev. Biochem.* 54: 631, 1985.

Kornfeld, S. Diseases of abnormal protein glycosylation—An emerging area. *J. Clin. Investigation* 101: 1293, 1998.

Lennarz, W. J. (Ed.). *The biochemistry of Glycoproteins*. New York: Plenum, 1980.

Neufeld, E. F. Lysosomal storage diseases. *Annu. Rev. Biochem.* 60: 257, 1991.

Schwartz, N. B., Lyle, S., Ozeran, J. D., Li, H., Deyrup, A., and Westley, J. Sulfate activation and transport in mammals: System components and mechanisms. *Chemico-Biol. Interactions* 109: 143, 1998.

Schwartz, N. B., Pirok, E. W., Mensch, J. R., and Domowicz, M. S. Domain organization, genomic structure, evolution and regulation of expression of the aggrecan gene family. *Prog. Nucleic Acid Res. Mol. Biol.* 62: 177, 1999.

Schwartz, N. B. Proteoglycans. In: *Embryonic Encyclopedia of Life Sciences*. London: Nature Publishing Group, 2000. www.els.net

Schwartz, N. B. Biosynthesis and regulation of expression of proteoglycans, *Front. Biosci.* 5: 649, 2000.

Scriver, C. R., Beaudet, A. L., and Valle, D., Sly, W. S. (Eds.). *The Molecular and Metabolic Bases of Inherited Disease*, 8th ed., New York: McGraw-Hill, 2001.

QUESTIONS | CAROL N. ANGSTADT

Multiple Choice Questions

1. All of the following interconversions of monosaccharides (or derivatives) require a nucleotide-linked sugar intermediate *except*:
   A. galactose 1-phosphate to glucose 1-phosphate.
   B. glucose 6-phosphate to mannose 6-phosphate.
   C. glucose to glucuronic acid.
   D. glucuronic acid to xylose.
   E. glucosamine 6-phosphate to *N*-acetylneuraminic acid (a sialic acid).

2. The severe form of galactosemia:
   A. is a genetic deficiency of a uridylyltransferase that exchanges galactose 1-phosphate for glucose on UDP-glucose.
   B. results from a deficiency of an epimerase.
   C. is insignificant in infants but a major problem in later life.
   D. is an inability to form galactose 1-phosphate.
   E. would be expected to interfere with the use of fructose as well as galactose because the deficient enzyme is common to the metabolism of both sugars.

3. All of the following are true about glucuronic acid *except*:
   A. it is a charged molecule at physiological pH.

B. as a UDP derivative, it can be decarboxylated to a component used in proteoglycan synthesis.
C. it is a precursor of ascorbic acid in humans.
D. its formation from glucose is under feedback control by a UDP-linked intermediate.
E. it can ultimately be converted to xylulose 5-phosphate and thus enter the pentose phosphate pathway.

4. Fucose and sialic acid:
A. are both derivatives of UDP-*N*-acetylglucosamine.
B. are the parts of the carbohydrate chain that are covalently linked to the protein.
C. can be found in the core structure of certain O-linked glycoproteins.
D. are transferred to a carbohydrate chain when it is attached to dolichol phosphate.
E. are the repeating unit of proteoglycans.

5. Glycosaminoglycans:
A. are the carbohydrate portion of glycoproteins.
B. contain large segments of a repeating unit typically consisting of a hexosamine and a uronic acid.
C. always contain sulfate.
D. exist in only two forms.
E. are bound to protein by ionic interaction.

6. All of the following are true of proteoglycans *except*:
A. specificity is determined, in part, by the action of glycosyltransferases.
B. synthesis is regulated, in part, by UDP-xylose inhibition of the conversion of UDP-glucose to UDP-glucuronic acid.
C. synthesis involves sulfation of carbohydrate residues by PAPS.
D. synthesis of core protein is balanced with synthesis of the polysaccharide moieties.
E. degradation is catalyzed in the cytosol by nonspecific glycosidases.

Questions 7 and 8: A patient was admitted to the hospital with symptoms of severe anemia. The onset of symptoms followed two days of therapy with a sulfa drug. Laboratory studies indicated a hemoglobin of 8.5 g/dL, along with an abnormally low [NADPH/$NADP^+$]. Further investigation led to a diagnosis of a defective enzyme of the pentose phosphate pathway.

7. [NADPH/$NADP^+$] is maintained at a high level in cells primarily by:
A. lactate dehydrogenase.
B. the combined actions of glucose 6-phosphate dehydrogenase and gluconolactonase.
C. the action of the electron transport chain.
D. . shuttle mechanisms such as the α-glycerophosphate dehydrogenase shuttle.
E. the combined actions of transketolase and transaldolase.

8. If a cell requires more NADPH than ribose 5-phosphate:
A. only the first phase of the pentose phosphate pathway would occur.
B. glycolytic intermediates would flow into the reversible phase of the pentose phosphate pathway.
C. there would be sugar interconversions but no net release of carbons from glucose 6-phosphate.
D. the equivalent of the carbon atoms of glucose 6-phosphate would be released as 6 CO_2.
E. only part of this need could be met by the pentose pathway, and the rest would have to be supplied by another pathway.

Questions 9 and 10: A six-month-old infant presented with hypoglycemia, vomiting, diarrhea, protein-loss enteropathy, and hepatic fibrosis. Measurement of the glycosylation state of endogenous serum transferrin revealed Type Ib CDGS (carbohydrate-deficient glycoprotein syndrome), which is a defect in phosphomannose isomerase activity.

9. The role of phosphomannose isomerase is the interconversion of:
A. mannose 6-phosphate and mannose 1-phosphate.
B. glucose 6-phosphate and mannose 6-phosphate.
C. fructose 6-phosphate and mannose 6-phosphate.
D. fructose 1-phosphate and mannose 1-phosphate.
E. glucose 6-phosphate and mannose 1-phosphate.

10. In Type Ic CDGS, a defect in an enzyme transferring a glucosyl residue to a high-mannose dolichol pyrophosphate precursor, the carbohydrate structure would be part of a(n):
A. N-linked glycoprotein.
B. O-linked glycoprotein.
C. proteoglycan.
D. glycosaminoglycan.
E. complex lipid.

Questions 11 and 12: An infant presented with multiple skeletal abnormalities, with the most prominent ones a short trunk and short limbs. Urine was free of partially degraded oligosaccharides or oligosaccharides of proteoglycans. The blood sulfate concentration and extracellular acid hydrolase were normal.

11. The disorder described would most likely be caused by a defect in:
A. the ability to generate mannose 6-phosphate.
B. lysosomal glycosidase activity.
C. the gene that codes for a sulfate transporter.
D. PAPS synthetase.
E. a sulfatase.

12. Sulfation is an important component of the synthesis of:
A. most proteoglycans.
B. hyaluronate.
C. most glycoproteins.
D. conjugated bilirubin.
E. all of the above.

Problems

13. How is excess ribose 5-phosphate converted to intermediates of glycolysis? Be sure to include any enzymes involved.

14. Essential fructosuria is a defect in fructokinase, while fructose intolerance is a defect in fructose 1-phosphate aldolase. Which of these two diseases lead to severe hypoglycemia after ingestion of fructose and why?

ANSWERS

1. **B** The glucose and mannose phosphates are both in equilibrium with fructose 6-phosphate by phosphohexose isomerases. A: Occurs via an epimerase at the UDP-galactose level. C, D: This oxidation of glucose is catalyzed by UDP-glucose dehydrogenase and the product can be decarboxylated to UDP-xylose. E: Again, an epimerization occurs on the nucleotide intermediate.
2. **A** B: The epimerase is normal. C: Galactose is an important sugar for infants. D: In this disease, the galactokinase is normal but is deficient in the mild form of the disease. E: Fructose metabolism does not use the uridylyltransferase that is deficient in galactosemia.
3. **C** Man does not make ascorbic acid. A: The charged acid group enhances water solubility which is a major physiological role for glucuronic acid, for example, bilirubin metabolism. B, D: Decarboxylation of UDP-glucuronic acid gives UDP-xylose, which is a potent inhibitor of the oxidation of UDP-glucose to the acid. E: The reduction of *d*-glucuronic acid to *l*-gulonic acid leads to ascorbate as well as xylulose 5-phosphate for the pentose phosphate pathway.
4. **C** Core structure also contains galactose and *N*-acetylgalactosamine. A: Only sialic acid does. Fucose comes from CDP-mannose. B: Usually found at the periphery of the carbohydrate. D: Core structure of N-linked carbohydrates contains mannose and *N*-acetylglucosamine. E: Repeating unit is hexosamine and uronic acids.
5. **B** This is a major distinction from glycoproteins, which, by definition, do not have a serial repeating unit. A: These are the carbohydrate of proteoglycans. C: Most do but hyaluronate does not. D: There are at least six different classes. E: Carbohydrates are bound by covalent links.
6. **E** Degradation is lysosomal; deficiencies of one or more lysosomal hydrolases leads to accumulation of proteoglycans in the mucopolysaccharidoses. A: Strict substrate specificity of the enzymes is important in determining the type and quantity of proteoglycans synthesized. Formation of specific protein acceptors for the carbohydrate is also important. B, D: Both xylose and glucuronic acid levels are controlled by this; xylose is the first sugar added in the synthesis of four of the six types and would accumulate if core protein synthesis is decreased. C: This is necessary for the formation of most proteoglycans.
7. **B** Although the glucose 6-phosphate dehydrogenase reaction, specific for NADP, is reversible, hydrolysis of the lactone assures that the overall equilibrium lies far in the direction of NADPH. A, C, D: These all use NAD, not NADP. E: These enzymes are part of the pentose phosphate pathway but catalyze freely reversible reactions that do not involve NADP.
8. **D** A, C, D, E: Glucose 6-phosphate yields ribose 5-phosphate + CO_2 in the oxidative phase. If this is multiplied by six, the six ribose 5-phosphates can be rearranged to five glucose 6-phosphates by the second, reversible phase. B: If more ribose 5-phosphate than NADPH were required, the flow would be in this direction to supply the needed pentoses.
9. **C** This is similar to the glucose 6-phosphate-fructose 6-phosphate interconversion. A: Mutases catalyze a 1–6 phosphate shift. B, D, E: This conversions do not occur directly.
10. **A** A carbohydrate chain assembled on dolichol phosphate is characteristic of N-linked glycoproteins. B, C, D: Synthesis of O-linked glycoproteins involves the sequential addition to the *N*-acetylgalactosamine linked to serine or threonine. E: Dolichol phosphate is a lipid but is eliminated when the carbohydrate chain is added to the protein. Contains mannose and *N*-acetylglucosamine. E: Repeating unit is hexosamine and uronic acids.
11. **D** These structural abnormalities are associated with defects in sulfation of the proteoglycans. A: This would lead to an inability to target acid hydrolase to the lysosome and thus increase the extracellular concentration. B, E: These are lysosomal enzymes whose deficiency would lead to partially degraded oligosaccharides in urine. C: Lack of a transporter would lead to elevated serum levels of sulfate.
12. **A** B: Hyaluronate is one glycosaminoglycan which is not sulfated. C: Glycoproteins don't contain sulfate. D: Bilirubin is conjugated with glucuronic acid.
13. 3 Pentose phosphates (ribose 5-phosphate + 2 xylulose 5-phosphate) are converted to 2 fructose 6-phosphate and 2 glyceraldehyde 3-phosphate by transaldolase and transketolase via a series of 2-carbon and 3-carbon transfers. These reactions are reversible.
14. Lack of fructokinase causes accumulation of fructose which is excreted. Lack of the aldolase results in increased fructose 1-phosphate, which sequesters cellular inorganic phosphate, inhibiting the cell's ability to generate ATP.

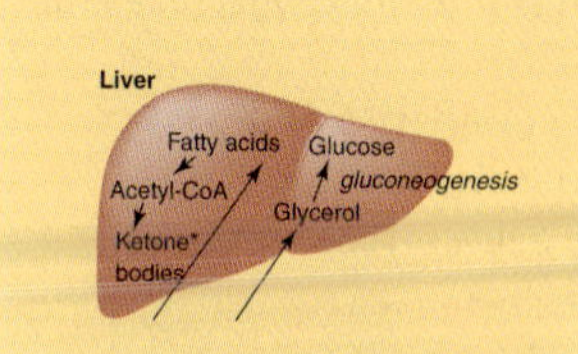

17 LIPID METABOLISM I: SYNTHESIS, STORAGE, AND UTILIZATION OF FATTY ACIDS AND TRIACYLGLYCEROLS

Martin D. Snider, J. Denis McGarry, and Richard W. Hanson

Textbook of Biochemistry With Clinical Correlations, Sixth Edition, Edited by Thomas M. Devlin

17.1 OVERVIEW

The defining characteristic of **lipids** is the property of being **hydrophobic**, meaning that they do not associate with water. Most lipids contain, or are derived from, **fatty acids** (Figure 17.1). Lipids have many important biological functions. **Triacylglycerols** (Figure 17.2) are the major fuel store in the body. Other lipids, including **phospholipids, glycolipids**, and **cholesterol**, are crucial constituents of biological membranes; the unique surface active properties of these molecules allow them to form the membrane backbone, separating and defining aqueous compartments within cells. Surface active lipids have other important functions, including maintenance of alveolar integrity in the lungs and solubilization of nonpolar substances in body fluids. Finally, some lipids are important signaling molecules. **Steroid hormones** and the **eicosanoids**, including **prostaglandins**, are important in communication between cells. Other lipids have signaling functions within cells.

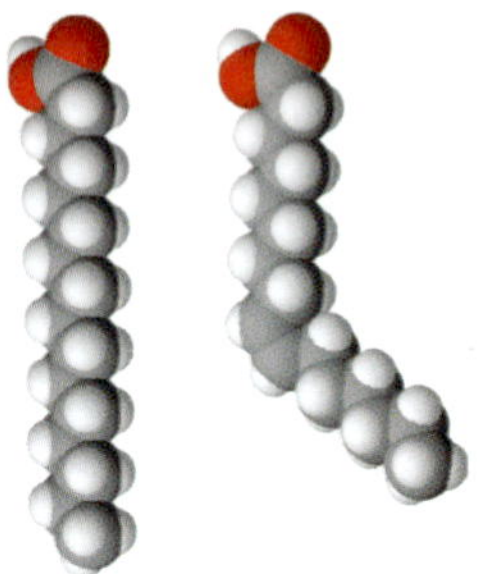

FIGURE 17.1

Long-chain fatty acids. Left: Palmitic (saturated). Right: *cis*-palmitoleic (unsaturated). Color scheme: H, white; C, gray; O, red.
Courtesy of Dr. Daniel Predecki, Shippensburg University, Shippensburg, PA.

The metabolism of fatty acids and triacylglycerols is so important that imbalances and deficiencies in these processes can have serious pathological consequences in humans. Disease states related to these lipids include obesity, diabetes mellitus, and ketoacidosis. There are also human genetic diseases that affect fatty acid metabolism, including Refsum's disease and deficiencies in carnitine and in fatty acid oxidation.

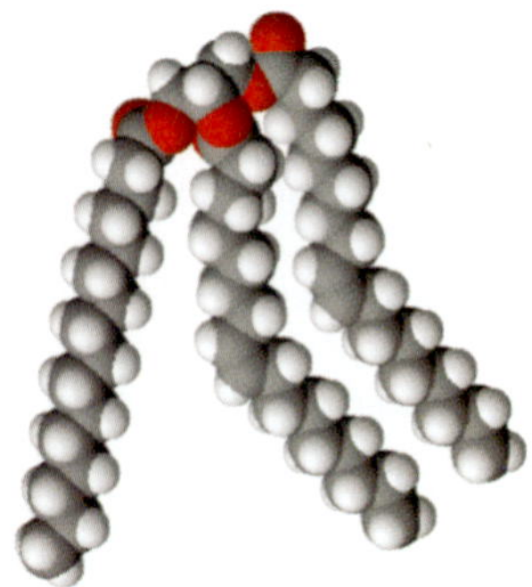

FIGURE 17.2

Triacylglycerol (1-palmitoyl 2, 3-dioleoyl glycerol). See Figure 17.1 for color scheme.
Courtesy of Dr. Daniel Predecki, Shippensburg University, Shippensburg, PA.

The following presents the structure and metabolism of fatty acids and their major storage form, triacylglycerols. The structure of fatty acids and triacylglycerols is described followed by an overview of metabolic fluxes in the fed and fasted states, and a discussion of synthesis fatty acids and triacylglycerols. The central process of energy production from fatty acids is then presented and finally the regulation of these processes, which allows net accumulation of triacylglycerols in the fed state and net utilization of triacylglycerols in the fasted state. The Appendix includes the nomenclature and chemistry of lipids and on p. 1059 a discussion of digestion and absorption of lipids.

17.2 CHEMICAL NATURE OF FATTY ACIDS AND ACYLGLYCEROLS

Fatty Acids are Alkyl Chains Terminating in a Carboxyl Group

Fatty acids (Figure 17.1) consist of an **alkyl chain** with a terminal carboxyl group. The basic formula of completely saturated species is $CH_3(CH_2)_nCOOH$. Nearly all fatty acids in mammals have an even number of carbon atoms, although some organisms synthesize molecules with an odd number. Humans can use the latter for energy, but they are incorporated into complex lipids to a minimal degree. Unsaturated fatty acids occur commonly, with up to six double bonds per chain. If there is more than one double bond per molecule, they are separated by a **methylene** ($—CH_2—$) **group**. The double bonds are almost always in the cis configuration, which induces a kink in the fatty acid chain, preventing orderly packing of the chains in lipid phases. By inducing disorder, double bonds lower the melting temperature and increase the fluidity of unsaturated fatty acid-containing lipids.

Most fatty acids in mammals have 16–20 carbon atoms, but C_{14} molecules are found linked to proteins and very long fatty acids are found in complex lipids in the nervous system. These include nervonic acid (22:1) and docosahexaenoic acid, a C_{22} acid that has six double bonds (Figure 17.3). Some fatty acids with an α-OH group are found as constituents of membrane lipids. **Branched-chain fatty acids**, which have methyl groups at one or more positions along the chain, are also found in some animals, including humans. These are produced to contribute specific physical properties to some secretory and structural molecules. For instance, large amounts of branched-chain fatty acids, particularly isovaleric acid (Figure 17.4), occur in lipids of echo-locating structures in marine mammals.

The most abundant fatty acids in humans are shown in Table 17.1. The symbols show the number of carbon atoms and, after the colon, the number of double bonds. Carbon atoms are numbered starting with the carboxyl carbon. The bond locations are shown in parentheses; they denote the number of the carbon atom on the carboxyl side of the bond.

$CH_3—(CH_2)_7—CH{=}CH—(CH_2)_{13}—COOH$

Nervonic acid

$CH_3—(CH_2—CH{=}CH)_6—(CH_2)_2—COOH$

All-*cis*-4,7,10,13,16,19,-docosahexaenoic acid

FIGURE 17.3
Very-long-chain fatty acids.

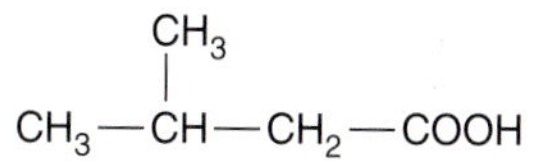

$CH_3—CH(CH_3)—CH_2—COOH$

FIGURE 17.4
Isovaleric acid.

TABLE 17.1 Fatty Acids of Importance to Humans

Name	*Numerical Formula*	*Functions in Humans*
Formic acid	1	
Acetic acid	2:0	
Propionic acid	3:0	Produced by metabolism of odd-chain fatty acids as well as isoleucine, valine, and methionine
Butyric acid	4:0	Milk triacylglycerols contain short chain fatty acids
Myristic acid	14:0	Covalently linked to some proteins
Palmitic acid	16:0	Product of fatty acid synthase
Palmitoleic acid	16:1(9)	Fatty acids with 16–18 carbons comprise the bulk of the fatty acids in triacylglycerols and complex lipids
Stearic acid	18:0	
Oleic acid	18:1(9)	
Linoleic acid	18:2(9, 12)	Essential fatty acid
Linolenic acid	18:3(9, 12, 15)	Essential fatty acid
Arachidonic acid	20:4(5, 8, 11, 14)	Precursor of prostaglandins and other eicosanoids
Lignoceric acid	24:0	
Nervonic acid	24:1(15)	Enriched in sphingolipids

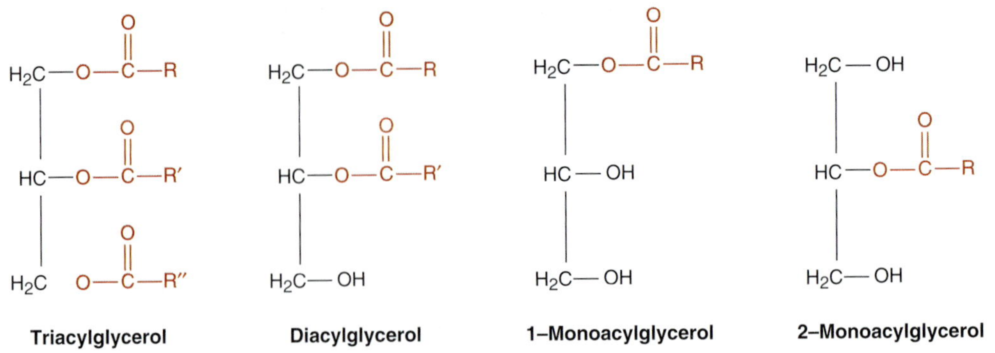

Triacylglycerol Diacylglycerol 1–Monoacylglycerol 2–Monoacylglycerol

FIGURE 17.5
Acylglycerols.

Most Fatty Acids in Humans Occur as Triacylglycerols

Fatty acids are stored primarily as esters of glycerol (Figures 17.2 and 17.5). Most fatty acids in humans are found in **triacylglycerols**, in which all three hydroxyl groups of glycerol are esterified with fatty acids. Compounds with one **(monoacylglycerols)** or two **(diacylglycerols)** esterified fatty acids occur in relatively minor amounts, largely as metabolic intermediates in the synthesis and degradation of glycerol-containing lipids. The trivial names, mono-, di-, and triglycerides, are also used for these compounds.

Most triacylglycerols have different fatty acids esterified to the three positions of glycerol. The distribution of fatty acids in the three positions of the glycerol moiety of triacylglycerol is influenced by many factors, including diet and the anatomical location of the stored molecule. In humans, most fatty acids are either saturated or contain one double bond, with oleic acid (18:1) being the most common. Although triacylglycerols are readily catabolized, they are comparatively inert chemically. The highly unsaturated fatty acids in triacylglycerols are much more susceptible to nonenzymatic oxidation.

The Hydrophobicity of Triacylglycerols is Important for their Functions

Triacylglycerols and other complex lipids are insoluble in water because the long hydrocarbon chains of the constituent fatty acids do not form hydrogen bonds. Consequently, they have limited solubility in water and tend to associate with each other or with other hydrophobic groups, such as sterols and the hydrophobic side chains of amino acids. This property is essential for the storage of triacylglycerols and for the assembly of biological membranes.

Triacylglycerols are a much more efficient form of energy storage than glycogen. On a weight basis, triacylglycerols yield nearly two and one-half times more ATP on complete oxidation than glycogen. In addition, triacylglycerols are stored without associated water, whereas glycogen is hydrophilic and binds about twice its weight of water. Therefore, the energy yield per gram of stored triacylglycerol is about four times that of hydrated glycogen. Humans store much more fuel as triacylglycerol than as glycogen. The average 70-kg person stores about 350 g of glycogen in liver and muscle. This represents about 1400 kcal of energy, which is less than one day's energy requirement. By contrast, the same individual contains ~10 kg of triacylglycerol, which provides sufficient energy to survive several weeks of starvation. Unlike stores of glycogen and amino acids, which are limited, triacylglycerol stores can greatly expand, depending on caloric intake. This makes triacylglycerol an excellent fuel store, but it can also lead to obesity and, in some cases, diabetes (see Clin. Corrs. 17.1 and 17.2). Subcutaneous adipose tissue is also important in temperature regulation because it provides a layer of insulation.

CLINICAL CORRELATION 17.1
Obesity

Obesity has become a worldwide epidemic. It has been estimated that in 2004, 25% of the population in the United States was obese, a figure that is projected to increase to 40% by 2020. Equally distressing figures have been noted for countries with rapidly growing economies such as China and India. The definition of obesity is arbitrary and is based on estimates of ideal body weight (IBW)—that is, body weight that is associated with the lowest morbidity and mortality. Overweight is defined as weight up to 20% above IBW, and obesity is weight that is 20% (or greater) above IBW. Body mass index (BMI) is another commonly used measure of obesity. It is calculated by dividing weight in kilograms by height in meters squared [weight (kg)/height (m^2)]. An individual with a BMI of 25 or greater is overweight and 30 or greater is considered obese. The anatomical location of fat deposition is a good indicator of the risk of morbidity and mortality from obesity; a central distribution of body fat (abdominal adipose tissue) is a greater health risk than a more peripheral distribution of fat (subcutaneous adipose tissue).

Obesity is part of what has been termed Metabolic Syndrome, a collection of symptoms that include central obesity, high blood pressure, high triglycerides, low levels of high-density lipoprotein in the blood, and insulin resistance. As many as 47 million Americans have this syndrome. Individuals with Metabolic Syndrome are at high risk for diabetes and cardiovascular disease. The biochemical link between obesity and diabetes will be discussed in Clinical Correlation 17.2. Because of these health risks, the control of obesity is a major public health goal.

The causes of obesity and the underlying reasons for its dramatic increase in human populations are complex. Endocrine disorders such as hypothyroidism or Cushing's disease (an overproduction of corticosteroids) are rare, as are mutations in genes coding hormones involved in the control of food intake, such as leptin or its receptor. It seems more likely that the genetic disposition of an individual, together with the environment (food choices, level of exercise), cause obesity. A change in lifestyle is clearly a major element in the obesity epidemic. An increased consumption of foods high in fat and carbohydrate, together with a decrease in exercise is the major cause of weight gain and body fat deposition in populations throughout the world. Thus, obesity is not a single disorder but a heterogeneous group of conditions with multiple causes.

Unfortunately, obesity is a condition that is rarely curable; its treatment remains a persistent medical problem. The standard procedure of reducing caloric intake and increasing energy expenditure is ineffective in obese individuals since, to be effective, it must be continued indefinitely to ensure a reduced body weight. Markedly obese individuals often resort to surgical procedures to remove unwanted fat or reduce the size of the stomach to prevent hunger and food absorption. There are currently a number of pharmacological interventions being used for the treatment of obesity. These include the use of compounds such as β-phenylamine, which selectively inhibits the reuptake of norepinephrine, serotonin, and dopamine. This compound reduces food intake and increases thermogenesis. Another approach is to block fat absorption using orlistat, which blocks pancreatic lipase, thereby decreasing triacylglycerol digestion. The thermogenic (increases the generation of heat) compound ephedrine has been shown to increase oxygen consumption by 10% in humans, when combined with caffeine. However, a major portion of the weight loss noted with this combination of drugs is due to decreased food intake. Potentially important new approaches to the treatment of obesity involve the development of compounds that regulate satiety. Gastrointestinal peptides such as cholecystokinin and glucagon-like peptide have long been known to reduce food intake. In addition, antagonists of the cannabinoid receptor in the brain reduce appetite. Small molecules that act by binding to these receptors are candidates for drugs that could be used to control appetite.

Source: Hirsch, J. Salens, L. B. and Aronne, L. J. Obesity. In: K. L. Becker (Ed.), *Principles and Practice of Endocrinology and Metabolism*, 3rd ed. New York: Lippincott, Williams and Wilkins, 2001, p. 1239. Freidmann, J. M. Obesity in the new millennium. *Nature* 404:632, 2000.

17.3 INTERORGAN TRANSPORT OF FATTY ACIDS AND THEIR PRIMARY PRODUCTS

In all mammals, the transport and storage of fatty acids is regulated by dietary status. Triacylglycerol is stored in the fed state, with net deposition in adipose tissue. During fasting, triacylglycerol in adipose tissue is hydrolyzed and the products are distributed throughout the body to be used for energy production. In prolonged fasting (greater than 2 days), the liver converts fatty acids to the **ketone bodies, acetoacetate** and **β-hydroxybutyrate**, which are released into the blood and are a major source of energy for many tissues. These processes are summarized in Figure 17.6.

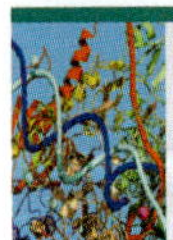

CLINICAL CORRELATION 17.2
Role of Fatty Acid Metabolism in Type 2 Diabetes

Obesity is a major contributory factor to the development of insulin resistance, a forerunner of Type 2 diabetes. During insulin resistance, tissues such as muscle become less responsive to this hormone, so that a greater concentration of insulin is required to regulate blood glucose; this results in transiently elevated levels of glucose. The relationship between insulin resistance and obesity is due in part to the elevated concentration of free fatty acids (FFA) in the blood of obese individuals. As an example, the concentration of plasma FFA in obese individuals fasted overnight is five times that noted in non-obese individuals. The increased metabolism of the FFA decreases whole body glucose utilization and muscle glycogen synthesis by ~50%. This effect is an important biological adaptation to starvation, since the oxidation of fatty acids blocks the metabolism of glucose by tissues such as muscle and liver. This spares the limited amount of glucose available for use by tissues that do not readily oxidize fatty acids, such as the brain and red blood cells.

In 1968, Phillip Randle and colleagues proposed what is now know as the Randle cycle; the cycle is based on the reciprocal relationship between glucose and fatty acid utilization. Fatty acid oxidation decreases glucose utilization in muscle by inhibiting the rate-limiting enzyme in glycolysis, phosphofructokinase, as well as the pyruvate dehydrogenase complex, which prevents the conversion of pyruvate to acetyl CoA and its subsequent entry into the TCA cycle. The result is diminished glucose utilization by muscle (see p. 854). Fatty acids also interfere with the insulin-stimulated transport of glucose into muscle by blocking the recruitment of GLUT 4 to the cell surface. This is in part due to the activation of specific isoforms of protein kinase C, as well as the transcription factor NF-κB, by acyl CoAs generated by the metabolism of fatty acids.

The clinical presentation of insulin resistance in patients is usually an elevated concentration of blood glucose and insulin after an overnight fast. During the early stages of this condition, pancreatic β-cells produce enough insulin to regulate blood glucose. However, prolonged overproduction of insulin by the β-cells can cause failure of these cells, eventually resulting in Type 2 diabetes. While this progression does not occur in all obese patients, it has been estimated that insulin resistance evolves to diabetes over 5–10 years in about 40% of patients. Currently over 220 million people worldwide have Type 2 diabetes and the number is predicted to increase dramatically over the next 20 years. More alarmingly, Type 2 diabetes has replaced Type 1 diabetes as the predominant form of this disease among children. Prevention and treatment of Type 2 diabetes is clearly a major medical priority. There is some hope, however, since in many cases insulin resistance is reversible with weight loss and increased exercise.

Source: Roden, M. How FFA inhibits glucose utilization in human skeletal muscle. *News Physiol. Sci.* 19:92, 2004. Randle, P. J. Regulatory interactions between lipids and carbohydrates: the glucose fatty acid cycle after 35 years. *Diabetes Metabol. Rev.* 14:263, 1998. Zimmet, P. Alberti, K. G. M. M., and Shaw, J. Global and social implications of the diabetes epidemic. *Nature* 414:782, 2001.

Lipid Transport in the Fed State

Triacylglycerols in the diet are digested in the stomach and small intestine by **gastric** and **pancreatic lipases** (see p. 1060). The principal products are 2-monoacylglycerols and **free fatty acids**, which are absorbed by the epithelial cells that line the small intestine. These cells assemble the absorbed fatty acids and monoacylglycerols into triacylglycerols (see p. 1060), which are then packaged into **chylomicrons**, a triacylglycerol-rich plasma lipoprotein (see p. 1065). Chylomicrons are secreted into lymph and are then circulate in the bloodstream. The liver is another source of triacylglycerols in the fed state. Fatty acids are synthesized in this tissue from excess carbohydrate and amino acids. These fatty acids are assembled into triacylglycerols and packaged into **very-low-density lipoprotein (VLDL)**, a second triacylglycerol-rich lipoprotein, which is secreted into the bloodstream (see p. 711).

The triacylglycerols in circulating chylomicrons and VLDL are hydrolyzed by **lipoprotein lipase**, located on the surface of capillary endothelial cells in tissues such as adipose tissue and skeletal muscle. The apoprotein ApoC-II, which is found in chylomicrons and VLDL, activates the process by binding the lipoproteins to the enzyme. The products of lipoprotein lipase action are free fatty acids and glycerol. The free fatty acids are utilized in the tissue where hydrolysis occurs. The glycerol is transported through the bloodstream and taken up mainly by the liver, where it is used for glycolysis or gluconeogenesis.

Lipoprotein lipase is expressed in adipose tissue, cardiac muscle, and skeletal muscle, allowing these tissues to utilize **triacylglycerols** from lipoproteins. In adipose tissue, the products of lipoprotein lipase are taken up and assembled into triacylglycerols, allowing

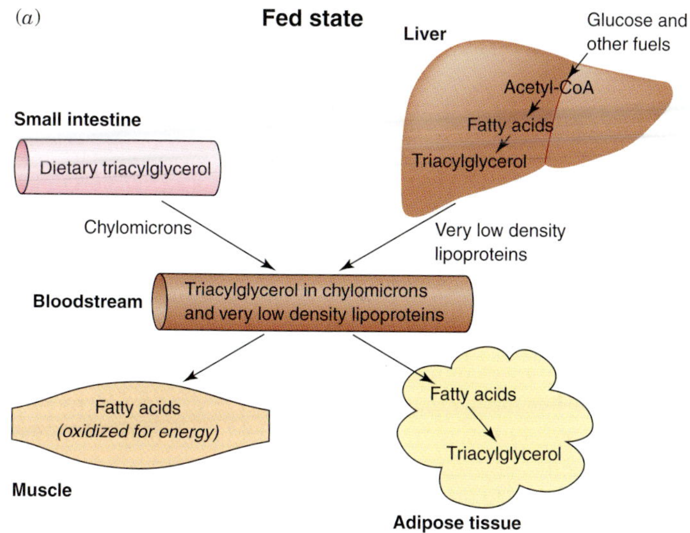

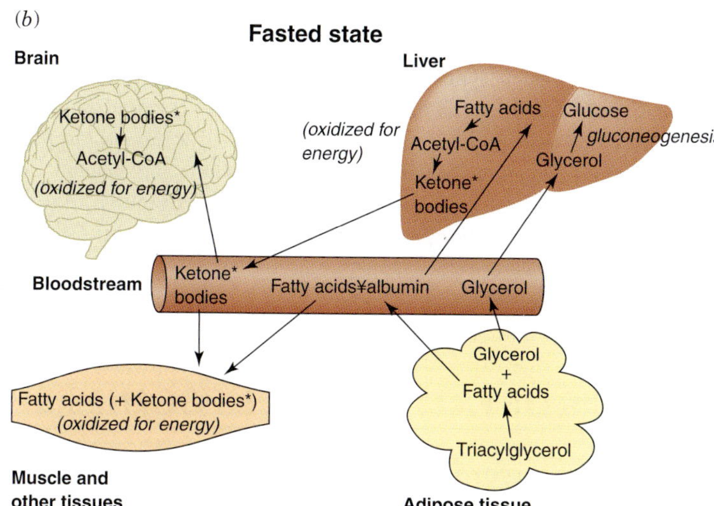

FIGURE 17.6

Interorgan transport of fatty acids. (A) In the fed state, there is net deposition of triacylglycerol in adipose tissue. The sources are dietary fat and fatty acids synthesized in the liver from excess carbohydrate and amino acids. (B) In the fasted state, triacylglycerols are hydrolyzed and free fatty acids and glycerol are released into the blood.
*During prolonged fasting, the liver synthesizes ketone bodies, which become a major fuel in the blood.

net deposition of fuel in this tissue (see p. 676). In muscle, the fatty acid products of lipase action are taken up and used as fuel, although some triacylglycerol synthesis occurs in this tissue.

Lipid Transport in the Fasted State

The triacylglycerols stored in adipose tissue are mobilized for use as fuel in the fasted state. This process is initiated by **hormone-sensitive lipase**, which is located within adipocytes. This enzyme is activated when it is phosphorylated by cAMP-dependent protein kinase A. Conversely, insulin inhibits the activity of this enzyme by inducing its dephosphorylation. The protein, **perilipin**, which coats the surface of fat droplets, is also important in this regulation. When perilipin is not phosphorylated, it blocks lipase access to the triacylglycerol. When perilipin is phosphorylated by protein kinase A, hormone-sensitive lipase translocates to the surface of the fat droplet and hydrolyzes triacylglycerols. A summary of the regulation of triacylglycerol metabolism is presented in Table 17.2. This regulation allows net mobilization of fatty acids in the fasted state and net deposition in the fed state (see p. 676). The balance of triacylglycerol synthesis and hydrolysis helps to ensure adequate energy stores and avoid obesity (see Clin. Corrs. 17.1 and 17.2). Another lipase, **adipose triglyceride lipase**, may also play a role in the regulated degradation of triacylglycerols.

TABLE 17.2 Regulation of Triacylglycerol Metabolism

Protein	Effect	Regulatory Agent
	Triacylglycerol Mobilization	
Hormone-sensitive lipase and perilipin	Activation by PKA-mediated phosphorylation	Glucagon, epinephrine, ACTH
	Inhibition by dephosphorylation	Insulin
Lipoprotein lipase	Increased enzyme synthesis	Insulin
	Triacylglycerol Synthesis	
Phosphatidate phosphatase	Increased enzyme synthesis	Steroid hormones

Other lipases rapidly complete the hydrolysis, releasing fatty acids and glycerol into the blood. The fatty acids are referred to as "free fatty acids" although they are bound to serum albumin. Each albumin molecule can bind ~10 fatty acids, so its binding capacity is very high. Free fatty acids bound to albumin, however, are a relatively small fraction of the total lipid in plasma when one considers the presence of the lipids in the plasma lipoproteins (see p. 711). However, because the free fatty acids turn over rapidly, they represent a significant fraction of the readily available transported lipid.

The hydrolysis of triacylglycerols present in adipose tissue and plasma lipoproteins produces free glycerol. This glycerol is used by the liver because of the presence of high levels of glycerol kinase, which synthesizes glycerol 3-phosphate from glycerol and ATP (Figure 15.41, p. 615). Other tissues cannot use glycerol because they lack this enzyme. Hepatic glycerol 3-phosphate is converted to dihydroxyacetone phosphate by glycerol 3-phosphate dehydrogenase, which enters the glycolytic pathway in the fed state. In the fasted state, it is converted to glucose via gluconeogenesis. During prolonged fasting, when much of the body's energy is derived from stored fat, the glycerol produced by the hydrolysis of triacylglycerol in adipose tissue is an important substrate for gluconeogenesis in the liver.

17.4 SYNTHESIS OF FATTY ACIDS: LIPOGENESIS

Glucose is the Major Precursor for Fatty Acid Synthesis

Dietary carbohydrate in excess of that needed for energy production and glycogen synthesis is converted to fatty acids in the liver during the fed state. Glucose provides the carbon for fatty acid synthesis (via acetyl CoA) as well as the reducing equivalents (NADPH) required for this process. Other substrates, such as amino acids, can also contribute to lipogenesis.

Pathway of Fatty Acid Biosynthesis

The synthesis of fatty acids occurs in the cytosol using acetyl CoA produced from glucose or from other precursors (i.e., the carbon skeletons of amino acids). The saturated, straight-chain C_{16} acid, palmitic acid, is first synthesized, and all other fatty acids are made by its modification. Fatty acids are synthesized by sequential addition of two-carbon units from acetyl CoA to the activated carboxyl end of a growing chain by **fatty acid synthase**. In bacteria, fatty acid synthase is a complex of several proteins, whereas in mammalian cells it is a single multifunctional protein. Either acetyl CoA or butyryl CoA is the priming unit for fatty acid synthesis, and the methyl end of these primers becomes the methyl end of palmitate.

Formation of Malonyl CoA is the Commitment Step of Fatty Acid Synthesis

Synthesis of fatty acids from acetyl CoA requires formation of an activated intermediate. The methyl carbon of acetyl CoA is "activated" by carboxylation to malonyl CoA

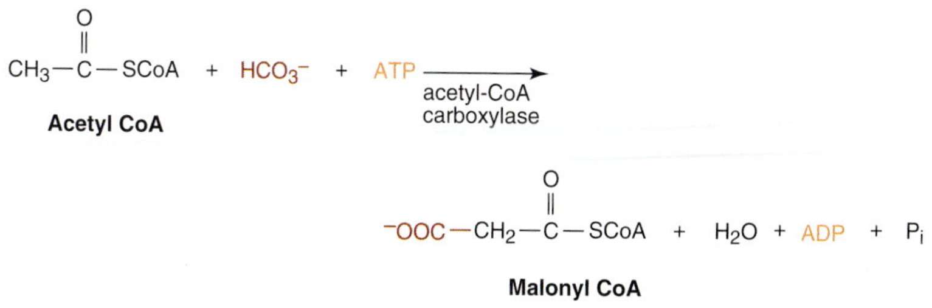

$$CH_3-\overset{\overset{\displaystyle O}{\|}}{C}-SCoA + HCO_3^- + ATP \xrightarrow[\text{carboxylase}]{\text{acetyl-CoA}} {}^-OOC-CH_2-\overset{\overset{\displaystyle O}{\|}}{C}-SCoA + H_2O + ADP + P_i$$

Acetyl CoA → Malonyl CoA

FIGURE 17.7
Acetyl-CoA carboxylase reaction.

by the enzyme **acetyl-CoA carboxylase** (Figure 17.7). The reaction requires ATP and bicarbonate as the source of CO_2. In the first step, CO_2 is linked to a biotin moiety on the enzyme, using energy derived from ATP hydrolysis. The CO_2 is then transferred to acetyl CoA. This reaction is similar to the carboxylation of pyruvate to oxaloacetate by pyruvate carboxylase (see p. 610).

Acetyl-CoA carboxylase catalyzes the rate-limiting step in fatty acid synthesis. This enzyme exists in an inactive, protomeric state and in an active state as linear polymers. The mammalian enzyme is activated by citrate or isocitrate. This represents feed-forward activation of fatty acid synthesis because citrate is exported from the mitochondria to generate cytosolic acetyl CoA for fatty acid synthesis (see section below on citrate cleavage pathway). Acetyl-CoA carboxylase is also inhibited by long-chain acyl CoAs. This results in feedback inhibition of fatty acid synthesis by high levels of product. Acetyl-CoA carboxylase is also inhibited by phosphorylation of the protein. Both cyclic AMP-dependent protein kinase A and AMP-dependent protein kinase inhibit this enzyme. The importance of this regulation is discussed on p. 690.

Reaction Sequence for the Synthesis of Palmitic Acid

The first step catalyzed by fatty acid synthase in bacteria is transfer of the primer molecule, either an acetyl or butyryl group from CoA to a 4′-phosphopantetheine moiety on **acyl carrier protein** (ACP), a protein constituent of the multienzyme complex. The phosphopantetheine unit is identical to that in CoA; both are derived from the vitamin, **pantothenic acid**. The mammalian enzyme also contains a phosphopantetheine moiety. Six or seven two-carbon units are added to the enzyme complex (depending on whether acetyl CoA or butyryl CoA is the primer) in a repetitive sequence of reactions until the palmitate molecule is completed. The reaction sequence, starting with an acetyl-CoA primer and leading to butyryl-ACP, is presented in Figure 17.8.

A round of synthesis is initiated by transfer of the fatty acid chain from the 4′-phosphopantetheine moiety of ACP to a functional sulfhydryl group of β-ketoacyl-ACP synthase (analogous to Reaction 3a). The -SH group of ACP then accepts a malonyl unit from malonyl CoA (Reaction 2). The products are β-ketoacyl-ACP and CO_2, which is released from the malonyl group. This is the same CO_2 that was added by acetyl-CoA carboxylase, so the carbon atoms in palmitate are derived entirely from acetyl CoA. The β-ketoacyl-ACP is reduced to β-hydroxyacyl-ACP using NADPH as electron donor. β-Hydroxyacyl-ACP is dehydrated to an enoyl-ACP and then reduced to a saturated acyl CoA, using a second NADPH as reductant. The growing fatty acid proceeds through six more reaction cycles to produce palmitoyl-ACP. This is acted on by a thioesterase that produces free palmitic acid (Figure 17.9). The specificity of this enzyme determines the chain length of the synthesized fatty acid. Note that at this stage, the sulfhydryl groups of ACP and synthase are both free, so that another cycle of fatty acid synthesis can begin.

Mammalian Fatty Acid Synthase is a Multifunctional Polypeptide

The reactions outlined above are the basic sequence for fatty acid synthesis in living systems. The enzyme complex, fatty acid synthase, catalyzes all these reactions, but its structure and properties vary considerably. In *Escherichia coli*, the complex is composed of separate enzymes. By contrast, mammalian fatty acid synthase is composed of two identical subunits, each of which is a multienzyme polypeptide that contains all of the

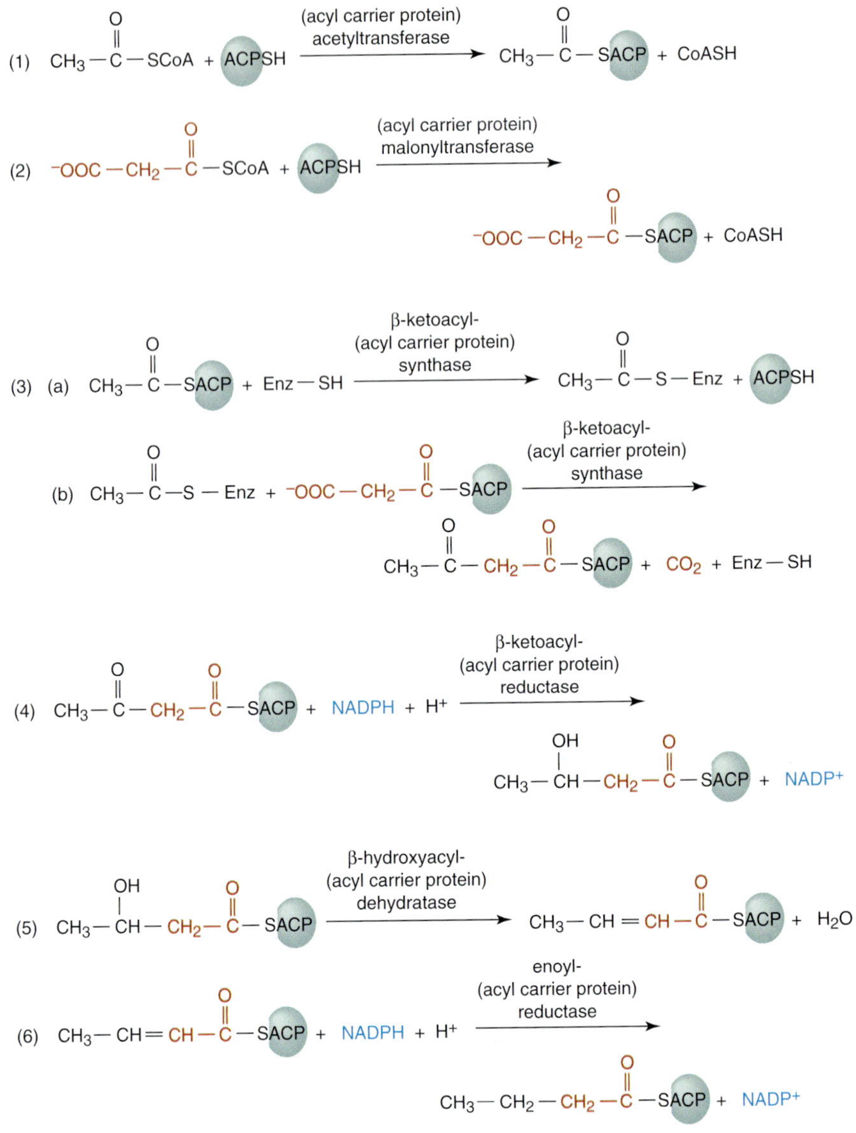

(1) $CH_3-\overset{O}{\overset{\|}{C}}-SCoA + ACPSH \xrightarrow{\text{(acyl carrier protein) acetyltransferase}} CH_3-\overset{O}{\overset{\|}{C}}-SACP + CoASH$

(2) $^{-}OOC-CH_2-\overset{O}{\overset{\|}{C}}-SCoA + ACPSH \xrightarrow{\text{(acyl carrier protein) malonyltransferase}} {}^{-}OOC-CH_2-\overset{O}{\overset{\|}{C}}-SACP + CoASH$

(3) (a) $CH_3-\overset{O}{\overset{\|}{C}}-SACP + Enz-SH \xrightarrow{\beta\text{-ketoacyl-(acyl carrier protein) synthase}} CH_3-\overset{O}{\overset{\|}{C}}-S-Enz + ACPSH$

(b) $CH_3-\overset{O}{\overset{\|}{C}}-S-Enz + {}^{-}OOC-CH_2-\overset{O}{\overset{\|}{C}}-SACP \xrightarrow{\beta\text{-ketoacyl-(acyl carrier protein) synthase}} CH_3-\overset{O}{\overset{\|}{C}}-CH_2-\overset{O}{\overset{\|}{C}}-SACP + CO_2 + Enz-SH$

(4) $CH_3-\overset{O}{\overset{\|}{C}}-CH_2-\overset{O}{\overset{\|}{C}}-SACP + NADPH + H^+ \xrightarrow{\beta\text{-ketoacyl-(acyl carrier protein) reductase}} CH_3-\overset{OH}{\overset{|}{C}H}-CH_2-\overset{O}{\overset{\|}{C}}-SACP + NADP^+$

(5) $CH_3-\overset{OH}{\overset{|}{C}H}-CH_2-\overset{O}{\overset{\|}{C}}-SACP \xrightarrow{\beta\text{-hydroxyacyl-(acyl carrier protein) dehydratase}} CH_3-CH=CH-\overset{O}{\overset{\|}{C}}-SACP + H_2O$

(6) $CH_3-CH=CH-\overset{O}{\overset{\|}{C}}-SACP + NADPH + H^+ \xrightarrow{\text{enoyl-(acyl carrier protein) reductase}} CH_3-CH_2-CH_2-\overset{O}{\overset{\|}{C}}-SACP + NADP^+$

FIGURE 17.8
Pathway of fatty acid synthesis.

$CH_3-(CH_2)_{14}-\overset{O}{\overset{\|}{C}}-SACP + H_2O \xrightarrow{\text{thioesterase}} CH_3-(CH_2)_{14}-COO^- + ACPSH$

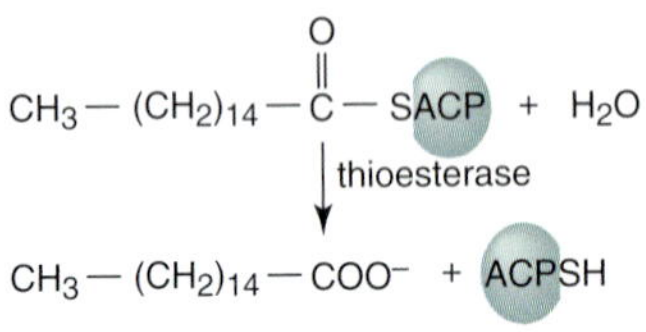

FIGURE 17.9
Release of palmitic acid from fatty acid synthase by thioesterase.

catalytic activities. There are also variations in the enzyme between mammalian species and tissues.

The growing fatty acid chain is always bound to the multifunctional protein and is transferred sequentially between the 4′-phosphopantetheine group of ACP, which is a domain of the protein, and the cysteine sulfhydryl group of β-ketoacyl-ACP synthase, during the condensation reaction (Reaction 3, Figure 17.8; Figure 17.10). An intermediate acylation of a serine residue takes place when acyl-CoA units are added to enzyme-bound ACP in the transacylase reactions.

The levels of fatty acid synthase in tissues are controlled by the rate of its synthesis and degradation. As shown in Table 17.3, insulin increases the rate of fatty acid synthesis by stimulating transcription of the fatty acid synthase gene, thereby increasing enzyme levels in the liver and other tissues. The factors that are involved in the regulation of this process are presented in detail on p. 690.

Stoichiometry of Fatty Acid Synthesis

With acetyl CoA as the primer for palmitate synthesis, the overall reaction is

$$CH_3-\overset{O}{\overset{\|}{C}}-SCoA + 7\ {}^{-}OOC-CH_2-\overset{O}{\overset{\|}{C}}-SCoA + 14\ NADPH + 14\ H^+ \longrightarrow CH_3-(CH_2)_{14}-COO^- + 7\ CO_2 + 14\ NADP^+ + 8\ CoASH + 6\ H_2O$$

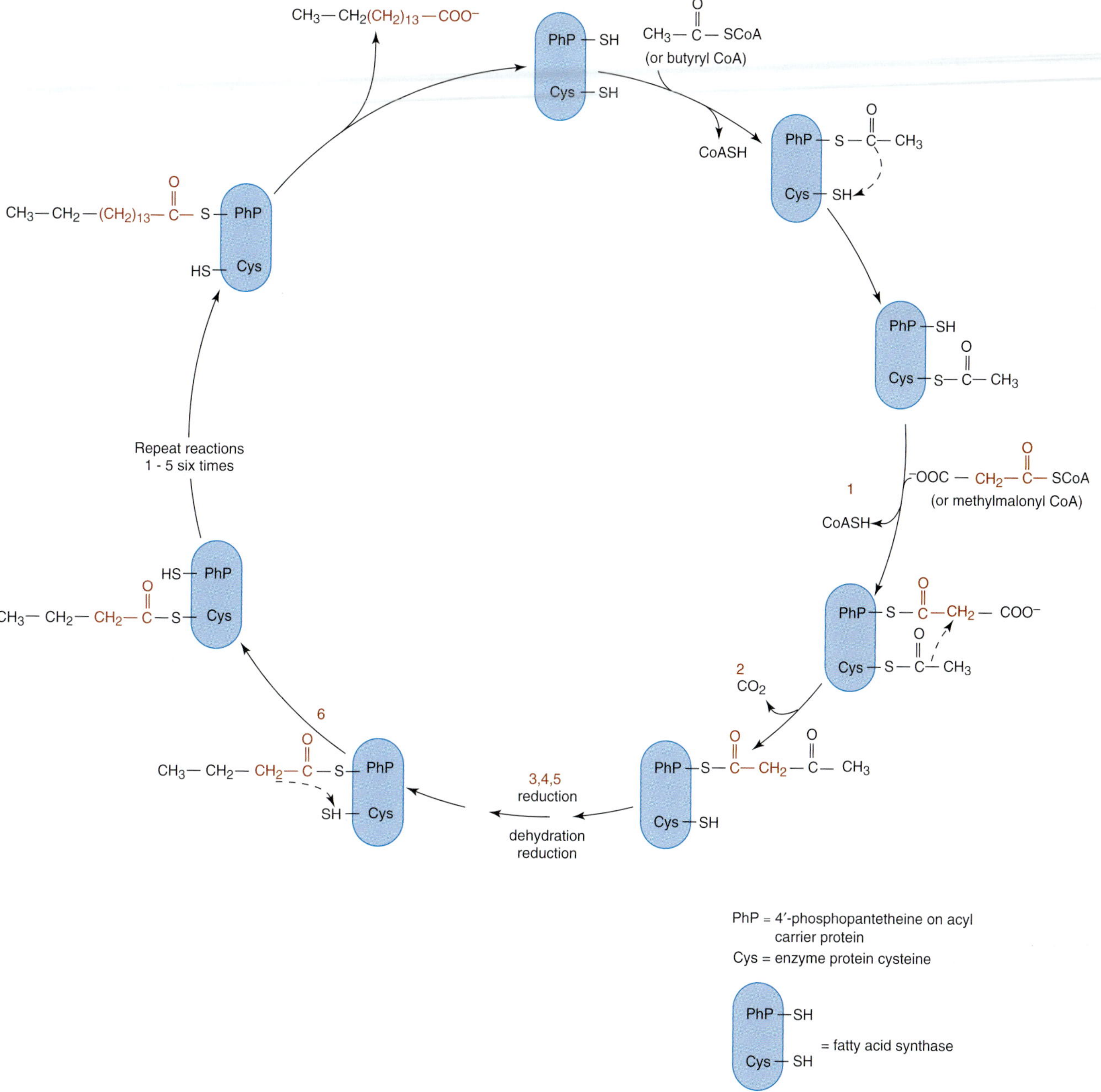

FIGURE 17.10

Proposed mechanism of fatty acid elongation by mammalian fatty acid synthase.

To calculate the energy needed for the overall conversion of acetyl CoA to palmitate, we must add the ATP used in formation of malonyl CoA:

$$7\ CH_3{-}\overset{O}{\overset{\|}{C}}{-}SCoA + 7\ CO_2 + 7\ ATP \longrightarrow 7\ {}^{-}OOC{-}CH_2{-}\overset{O}{\overset{\|}{C}}{-}SCoA + 7\ ADP + 7\ P_i$$

Then the stoichiometry for conversion of acetyl CoA to palmitate is

$$8\ CH_3{-}\overset{O}{\overset{\|}{C}}{-}SCoA + 7\ ATP + 14\ NADPH + 14\ H^+ \longrightarrow$$

$$CH_3{-}(CH_2)_{14}{-}\overset{O}{\overset{\|}{C}}{-}O^- + 8\ CoASH + 7\ ADP + 7\ P_i + 6\ H_2O + 14\ NADP^+$$

TABLE 17.3 Regulation of Fatty Acid Synthesis

Enzyme		*Effect*	*Regulatory Agent*
		Palmitate Synthesis	
Acetyl-CoA carboxylase (ACC)	Short term	Allosteric activation	Citrate, isocitrate
		Allosteric inhibition	C16–C18 acyl CoAs
		Inhibition via PKA phosphorylation of ACC	Glucagon and epinephrine
		Inhibition via AMPK-mediated phosphorylation of ACC	AMP
		Activation by Dephosphorylation of ACC	Insulin
	Long term	Increased enzyme synthesis	Thyroid hormone, high carbohydrate diet, insulin
		Decreased enzyme synthesis	Fasting, high fat diet, via low insulin and high glucagon. Polyunsaturated fatty acids (PUFAs) also inhibit synthesis
Fatty acid synthase	Short term	Allosteric activation	Sugar phosphates
	Long term	Increased enzyme synthesis	High carbohydrate diet via insulin
		Decreased enzyme synthesis	Fasting, high fat diet, via low insulin and high glucagon. PUFAs also inhibit synthesis
		Biosynthesis of Fatty Acids Other than Palmitate	
Fatty acid synthase		Increased synthesis of branched-chain fatty acids	Increased methylmalonyl CoA/malonyl CoA
		Synthesis of short-chain fatty acids	Expression of specific thioesterase in mammary gland
Stearoyl-CoA desaturase		Increased enzyme synthesis	Insulin, thyroid hormones, hydrocortisone, dietary cholesterol
		Decreased enzyme synthesis	Dietary PUFAs

Citrate Cleavage Pathway Provides Acetyl CoA and NADPH for Lipogenesis in the Cytosol

Glucose breakdown in the liver via glycolysis results in the production of pyruvate, which is converted to acetyl CoA in the mitochondria by pyruvate dehydrogenase complex. However, the synthesis of fatty acids is a cytosolic process and acetyl CoA is not readily transported across the inner mitochondrial membrane. The **citrate cleavage pathway** overcomes this problem. Citrate, formed by citrate synthase, in the tricarboxylic acid cycle, is transported across the mitochondrial inner membrane via the tricarboxylate transporter (see Figure 14.54, p. 569). The citrate is then cleaved in the cytosol by **ATP citrate lyase** (also called **citrate cleavage enzyme**) to form acetyl CoA and oxaloacetate (Figure 17.11); this reaction is not a reversal of citrate synthase, since it requires the hydrolysis of ATP. Citrate has a second role in fatty acid synthesis as an allosteric activator of acetyl CoA carboxylase, the rate-limiting enzyme in lipogenesis.

$$\text{citrate} + \text{ATP} + \text{CoA} \rightarrow \text{acetyl CoA} + \text{ADP} + \text{P}_i + \text{oxaloacetate}$$

The oxaloacetate generated by citrate cleavage in the cytosolic cannot move back into the mitochondria because this compound is transported at a negligible rate across the mitochondrial inner membrane. Oxaloacetate is instead reduced to malate by the cytosolic isoform of **NAD malate dehydrogenase**. The malate then undergoes oxidative decarboxylation to pyruvate, catalyzed by **NADP malate dehydrogenase** (also called **malic enzyme**). The pyruvate enters the mitochondria for further metabolism. The removal of citrate from the mitochondria (**cataplerosis**) must be accompanied by its replacement (**anaplerosis**) in order to maintain tricarboxylic acid cycle flux. This is achieved by the conversion of pyruvate to oxaloacetate by **pyruvate carboxylase** (see p. 610), a major anaplerotic enzyme in mitochondria.

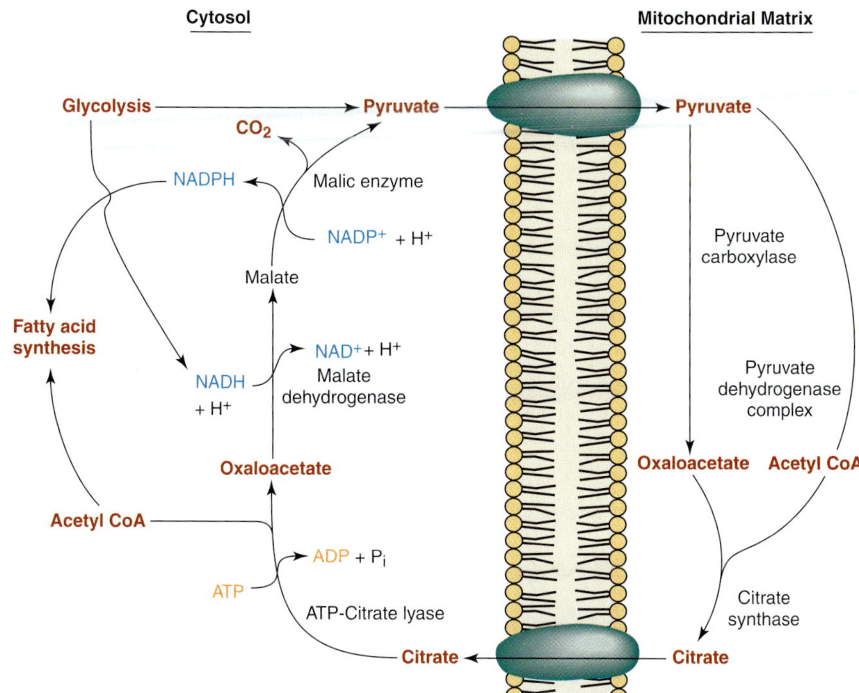

FIGURE 17.11
Transfer of acetyl CoA from mitochondria to cytosol for fatty acid biosynthesis by the citrate cleavage pathway.

The conversion of oxaloacetate to pyruvate by the citrate cleavage pathway transfers a pair of electrons from NADH to generate NADPH that is used for fatty acid synthesis. The source of the NADH that drives this process is the glyceraldehyde-3-phosphate dehydrogenase reaction of the glycolytic pathway. The metabolism of glucose thus provides both the carbon atoms and the reducing equivalents required for lipogenesis.

The production of the reducing equivalents used in fatty acid synthesis can be summarized as follows: The citrate cleavage pathway transfers a pair of electrons from NADH to NADPH for each acetyl CoA transferred from mitochondria to cytosol. The transfer of the 8 acetyl CoAs that are used for the synthesis of one molecule of palmitate supplies 8 NADPHs. Since palmitate synthesis requires 14 NADPHs, the other 6 NADPHs are supplied by the pentose phosphate pathway, which is also present in the cytosol. This stoichiometry is, of course, hypothetical. The *in vivo* relationships are complicated because transport of citrate and other di- and tricarboxylic acids across the inner mitochondrial membrane occurs by one-for-one exchanges. The actual flow rates are probably controlled by a composite of the concentration gradients of several of these exchange systems.

Modification of Fatty Acids

Humans can synthesize the fatty acids they require from palmitate, except for the polyunsaturated fatty acids (Table 17.1, p. 663). Fatty acids are modified as the CoA derivatives by three processes: elongation, desaturation, and hydroxylation.

Elongation Reactions

Fatty acid elongation in mammals occurs in either the endoplasmic reticulum or the mitochondria. These processes are slightly different in the two locations. In the endoplasmic reticulum, fatty acyl CoAs are elongated by reactions similar to the ones catalyzed by cytosolic fatty acid synthase; malonyl CoA is the source of two-carbon units and NADPH provides the reducing power. The preferred elongation substrate is palmitoyl CoA, which is converted almost exclusively to stearate (18:0) in most tissues. The brain, however, contains one or more additional elongation systems, which synthesize longer-chain acids (up to C_{24}) that are required for brain lipids. These elongation systems also use malonyl CoA.

$$R-CH_2-\overset{O}{\overset{\|}{C}}-SCoA$$

$CH_3-\overset{O}{\overset{\|}{C}}-SCoA$ → β-ketothiolase → CoASH

$$R-CH_2-\overset{O}{\overset{\|}{C}}-CH_2-\overset{O}{\overset{\|}{C}}-SCoA$$

NADH + H+ → β-hydroxyacyl CoA dehydrogenase → NAD+

$$R-CH_2-\overset{OH}{\overset{|}{C}H}-CH_2-\overset{O}{\overset{\|}{C}}-SCoA$$

enoyl CoA hydratase → H_2O

$$R-CH_2-CH{=}CH-\overset{O}{\overset{\|}{C}}-SCoA$$

NADPH + H+ → enoyl CoA reductase → NADP+

$$R-CH_2-CH_2-CH_2-\overset{O}{\overset{\|}{C}}-SCoA$$

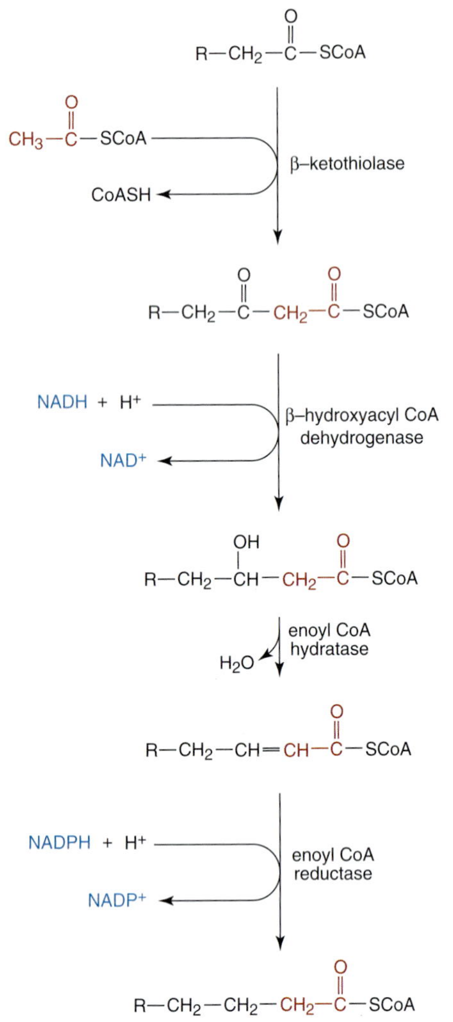

FIGURE 17.12

Pathway of fatty acid elongation in mitochondria.

Fatty acid elongation in mitochondria uses acetyl CoA and both NADH and NADPH as electron donors (Figure 17.12). This system operates by reversal of the pathway of fatty acid β oxidation (see p. 683), with the exception that NADPH-linked enoyl-CoA reductase (last step of elongation) replaces FAD-linked acyl-CoA dehydrogenase (first step in β-oxidation). The process serves primarily to elongate shorter molecules; it has little activity with acyl-CoA substrates of 16 carbons or longer.

Formation of Monoenoic Acids by Stearoyl-CoA Desaturase

In mammals, **fatty acid desaturation** occurs in the endoplasmic reticulum, and the reactions and enzymes that introduce cis double bonds are significantly different from the ones in mitochondrial β-oxidation. Desaturation is carried out by **monooxygenases**, which have fatty acyl CoA, NADH, and O_2 as substrates (see p. 415). The three components of the system are the **desaturase enzyme, cytochrome b_5**, and **NADPH-cytochrome b_5 reductase**. The overall reaction is

$$R-CH_2-CH_2-(CH_2)_7-COOH + NADPH + H^+ + O_2 \rightarrow$$
$$R-CH{=}CH-(CH_2)_7-COOH + NADP^+ + 2H_2O$$

The initial step in the formation of unsaturated fatty acids is the formation of the Δ^9 double bond by **stearoyl-CoA desaturase** in palmitic or stearic acid to produce palmitoleic or oleic acid, respectively. The synthesis of unsaturated fatty acids is important for regulating the fluidity of triacylglycerols and membrane phospholipids. It is also required for the synthesis of cholesterol esters in the liver and waxy secretions in the skin, which preferentially use newly synthesized, rather than dietary, fatty acids. Expression of stearoyl-CoA desaturase is highly regulated by both dietary and hormonal mechanisms. Insulin, triiodothyronine, hydrocortisone, and dietary cholesterol increase gene transcription and levels of stearoyl-CoA desaturase in the liver, whereas dietary polyunsaturated fatty acids have the opposite effect.

Formation and Modification of Polyunsaturated Fatty Acids

Polyunsaturated fatty acids, particularly arachidonic acid (20:4), are precursors of important signaling molecules: prostaglandins, thromboxanes, and leukotrienes (see p. 730). Polyunsaturated fatty acids are also required for the synthesis of complex lipids, particularly in the nervous system. These fatty acids can undergo nonenzymatic oxidation in reactions that use O_2, a process that may have important physiological and/or pathological consequences.

In mammals, double bonds can be added only to the proximal half of fatty acyl CoAs; they cannot be added beyond C9. Consequently, linoleic (18:2) and linolenic (18:3) acids, polyunsaturated fatty acids, are required in the diet (Figure 17.13). These **essential fatty acids** are obtained in the diet from plants and cold-water fish. There are two series of polyunsaturated fatty acids. In linoleic acid, the distal double bond is six carbons from the methyl group; it is referred to as n-6 or ω-6. A second series, n-3 or ω-3, has the distal double bond three carbons from the methyl group.

$$CH_3-(CH_2)_3-(CH_3-CH{=}CH)_n-(CH_2)_m-COOH$$

Basic formula of the linoleic acid series

$$CH_3-(CH_2-CH{=}CH)_n-(CH_2)_m-COOH$$

Basic formula of the linolenic acid series

FIGURE 17.13

The linoleic and linolenic acid series.

The essential fatty acids are modified by elongation and desaturation to synthesize the polyunsaturated fatty acids found in mammals. A variety of polyunsaturated fatty acids are synthesized by humans using desaturases that introduce double bonds at positions 4, 5, or 6 (Figure 17.14). These enzyme act only in the synthesis of polyunsaturated fatty acids because they can only use substrates with a double bond at position 9. This elongation and desaturation occurs in either order. Conversion of linolenic acid to all *cis*-4,7,10,13,16,19-docosahexaenoic acid in the brain is a specific example of such a reaction sequence.

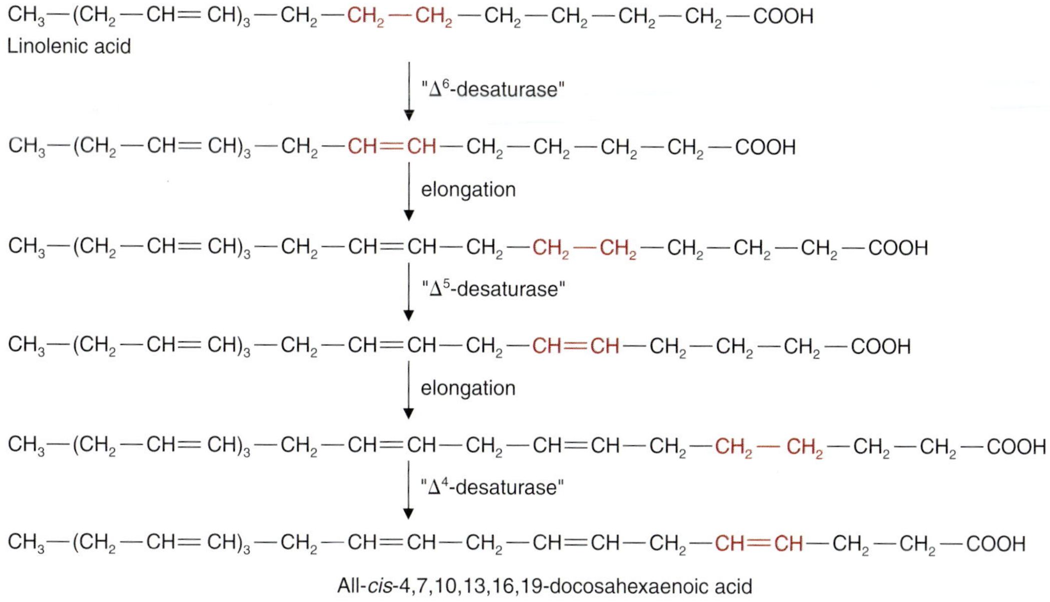

$CH_3—(CH_2—CH{=}CH)_3—CH_2—CH_2—CH_2—CH_2—CH_2—CH_2—CH_2—COOH$
Linolenic acid

↓ "Δ⁶-desaturase"

$CH_3—(CH_2—CH{=}CH)_3—CH_2—CH{=}CH—CH_2—CH_2—CH_2—CH_2—COOH$

↓ elongation

$CH_3—(CH_2—CH{=}CH)_3—CH_2—CH{=}CH—CH_2—CH_2—CH_2—CH_2—CH_2—CH_2—COOH$

↓ "Δ⁵-desaturase"

$CH_3—(CH_2—CH{=}CH)_3—CH_2—CH{=}CH—CH_2—CH{=}CH—CH_2—CH_2—CH_2—COOH$

↓ elongation

$CH_3—(CH_2—CH{=}CH)_3—CH_2—CH{=}CH—CH_2—CH{=}CH—CH_2—CH_2—CH_2—CH_2—CH_2—COOH$

↓ "Δ⁴-desaturase"

$CH_3—(CH_2—CH{=}CH)_3—CH_2—CH{=}CH—CH_2—CH{=}CH—CH_2—CH{=}CH—CH_2—CH_2—COOH$

All-*cis*-4,7,10,13,16,19-docosahexaenoic acid

Formation of Hydroxy Fatty Acids in Nerve Tissue

Two different processes produce **α-hydroxy fatty acids** in mammals. One occurs in the mitochondria of many tissues and acts on relatively short-chain fatty acids. The other has been demonstrated only in the nervous system, where it produces long-chain fatty acids hydroxylated on C2 that are required for the formation of some myelin lipids. The enzyme in the brain that catalyzes this reaction is a monooxygenase that requires O_2 and NADH or NADPH and preferentially uses C_{22} and C_{24} fatty acids. This process is closely coordinated with the synthesis of sphingolipids that contain hydroxylated fatty acids.

Fatty Acid Synthase can Produce Fatty Acids Other than Palmitate

The principal fatty acids synthesized by humans for energy storage are palmitate and its modification products. However, smaller amounts of different fatty acids are synthesized for other purposes. Two examples are the production of the fatty acids shorter than palmitate in the **mammary gland** and the synthesis of branched-chain fatty acids in some secretory glands.

Milk contains fatty acids with shorter chain lengths than palmitate. The relative amounts of the fatty acids produced by the mammary gland vary with species and with the physiological state of the animal. This is probably true of humans, although most investigations have been carried out with rats, rabbits, and ruminants. The pathway of fatty acid synthesis, which normally produces palmitate, is modified to synthesize shorter chain acids. This is accomplished by the expression of soluble **thioesterases**, which cleave the shorter chains from fatty acid synthase. Human milk contains no fatty acids with chains shorter than 10 carbons.

There are relatively few branched-chain fatty acids in mammals. Until recently, their metabolism has been studied mostly in bacteria such as Mycobacteria, where they are present in a greater variety and amount. Simple, branched-chain fatty acids are synthesized by tissues of higher animals for specific purposes, such as (a) the production

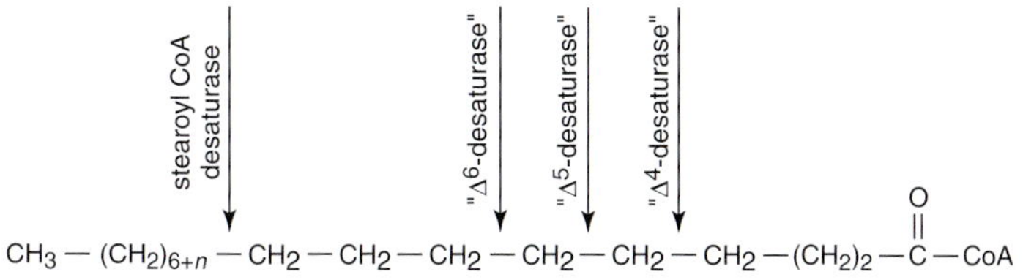

stearoyl CoA desaturase ↓ · "Δ⁶-desaturase" ↓ · "Δ⁵-desaturase" ↓ · "Δ⁴-desaturase" ↓

$CH_3—(CH_2)_{6+n}—CH_2—CH_2—CH_2—CH_2—CH_2—CH_2—(CH_2)_2—\overset{\overset{\displaystyle O}{\|}}{C}—CoA$

FIGURE 17.14

Positions in the fatty acid chain where desaturation can occur in the human. In mammals, there must be at least six carbons beyond the bond being desaturated.

of waxes in sebaceous glands and avian preen glands and (b) the elaboration of structures in echo-locating systems of porpoises.

Most of the branched-chain fatty acids in higher animals are methylated derivatives of saturated, straight-chain acids that are synthesized by fatty acid synthase. When **methylmalonyl CoA** is used as a substrate instead of malonyl CoA, a methyl side chain is inserted into the fatty acid by the following reaction:

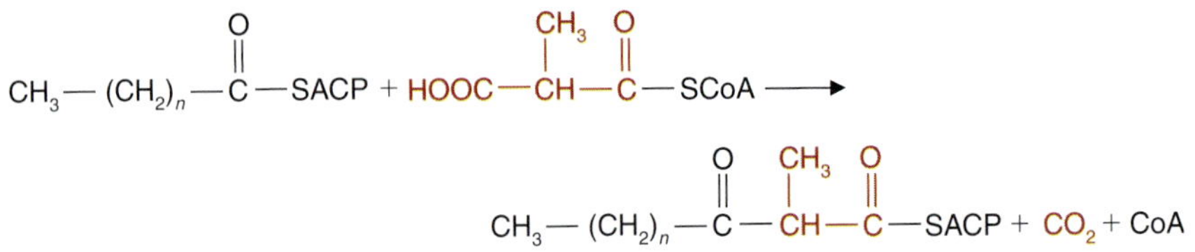

$$CH_3—(CH_2)_n—\overset{O}{\overset{\|}{C}}—SACP + HOOC—\overset{CH_3}{\overset{|}{CH}}—\overset{O}{\overset{\|}{C}}—SCoA \longrightarrow$$

$$CH_3—(CH_2)_n—\overset{O}{\overset{\|}{C}}—\overset{CH_3}{\overset{|}{CH}}—\overset{O}{\overset{\|}{C}}—SACP + CO_2 + CoA$$

Regular reduction steps then follow. These reactions occur in many tissues at a rate several orders of magnitude lower than the utilization of malonyl CoA in fatty acid synthesis. The proportion of branched-chain fatty acids that are synthesized is largely governed by the relative availability of the two precursors. An increase in branching can occur by decreasing the ratio of malonyl CoA to methylmalonyl CoA. A malonyl-CoA decarboxylase that is responsible for this decrease occurs in many tissues. It has been suggested that an increased concentration of methylmalonyl CoA in pathological situations, such as vitamin B_{12} deficiency, can lead to excessive production of branched-chain fatty acids.

Fatty Acyl CoAs May Be Reduced to Fatty Alcohols

$$CH_3—(CH_2)_n—CH_2OH$$

FIGURE 17.15
Fatty alcohol.

Many phospholipids contain fatty acid chains in ether linkage, rather than ester linkage. The synthetic precursors of these ether-linked chains are fatty alcohols (Figure 17.15), rather than fatty acids. These alcohols are formed in mammals by a two-step, NADPH-linked reduction of fatty acyl CoAs in the endoplasmic reticulum. In tissues that produce relatively large amounts of ether-containing lipids, the concurrent production of fatty acids and fatty alcohols is closely coordinated.

17.5 STORAGE OF FATTY ACIDS AS TRIACYLGLYCEROL

Most mammalian tissues convert fatty acids to triacylglycerols by a common sequence of reactions, but liver and adipose tissue carry out this process to the greatest extent. Adipose tissue is a specialized organ designed for the synthesis, storage, and hydrolysis of triacylglycerol and is the main site of long-term energy storage. Triacylglycerols are stored in the cytosol as liquid droplets surrounded by a monolayer of membrane lipids and the protein, perilipin. These droplets are not a dead-end store; triacylglycerols turn over with an average half-life of only a few days. Thus, there is continuous synthesis and breakdown of triacylglycerol in adipose tissue. Some storage also occurs in skeletal and cardiac muscle, but only for local consumption. Triacylglycerol synthesis in liver is primarily for the production of plasma lipoproteins, rather than for energy storage. The fatty acids for this process may come from the diet, from adipose tissue via the blood, or from *de novo* synthesis, primarily from the catabolism of dietary glucose.

Triacylglycerols Are Synthesized from Fatty Acyl CoAs and Glycerol 3-Phosphate

Triacylglycerols are synthesized in most tissues from fatty acyl CoAs and a glycerol precursor, **glycerol 3-phosphate** (Figure 17.16). Glycerol 3-phosphate is derived from several sources. In most tissues, it is synthesized by reduction of dihydroxyacetone phosphate. In the fed state, the dihydroxyacetone phosphate is derived from glucose

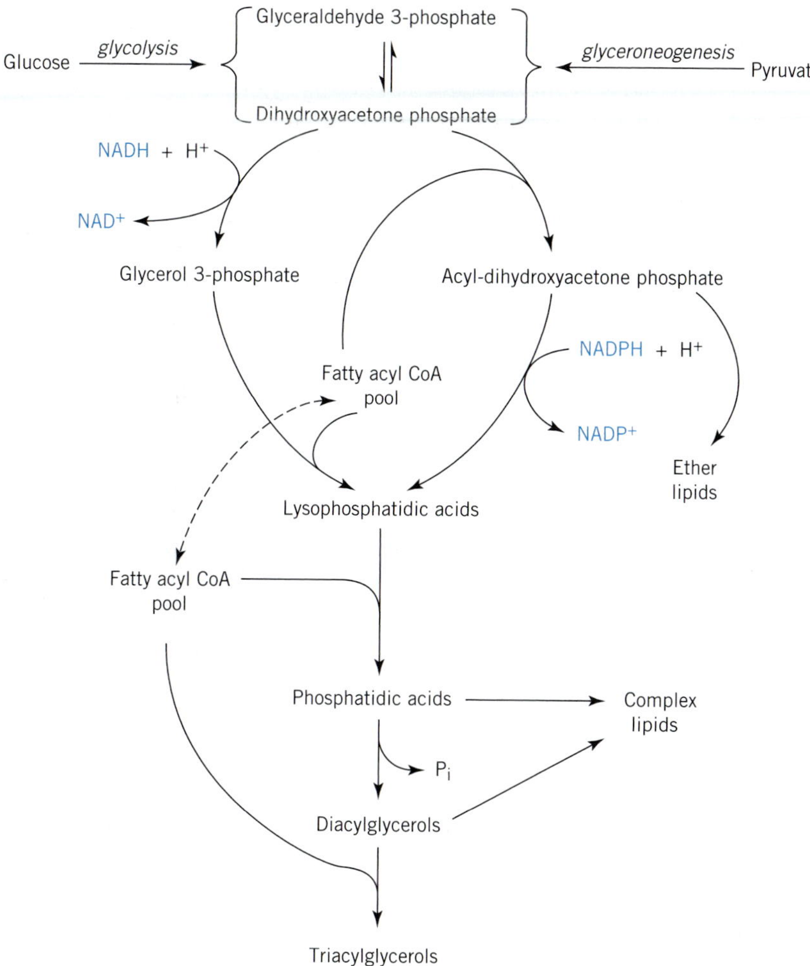

FIGURE 17.16

Pathways of triacylglycerol synthesis. Some of these intermediates are used as precursors for the synthesis of membrane lipids.

via glycolysis; in the fasted state in adipose tissue and liver, glycerol 3-phosphate is derived from **glyceroneogenesis** (see p. 679). In the liver, there is an additional source of glycerol 3-phosphate; glycerol can be phosphorylated by **glycerol kinase**, which is highly expressed in this tissue.

Fatty acids are activated for further metabolism by conversion to their CoA esters in the following reactions:

$$R-\overset{O}{\overset{\|}{C}}-O^- + ATP + CoASH \xrightarrow[\text{synthetase}]{\text{acyl-CoA}} R-\overset{O}{\overset{\|}{C}}-SCoA + AMP + PP_i + H_2O$$

This two-step reaction has an acyl adenylate as an intermediate, and it is driven by the hydrolysis of pyrophosphate to P_i.

The synthesis of triacylglycerols involves the formation of **phosphatidic acid**, which is also a key intermediate in the synthesis of other glycerolipids (see p. 697). Phosphatidic acid is formed by two sequential acylations of glycerol 3-phosphate to form lysophosphatidic and then phosphatidic acid (Figure 17.17). Phosphatidic acid is used for triacylglycerol synthesis by hydrolysis of the phosphate group by **phosphatidate phosphatase** to yield **diacylglycerol**, which is then acylated to triacylglycerol (Figure 17.18). In an alternative pathway, **dihydroxyacetone phosphate** is acylated, reduced to lysophosphatidic acid, and then acylated a second time to form phosphatidic

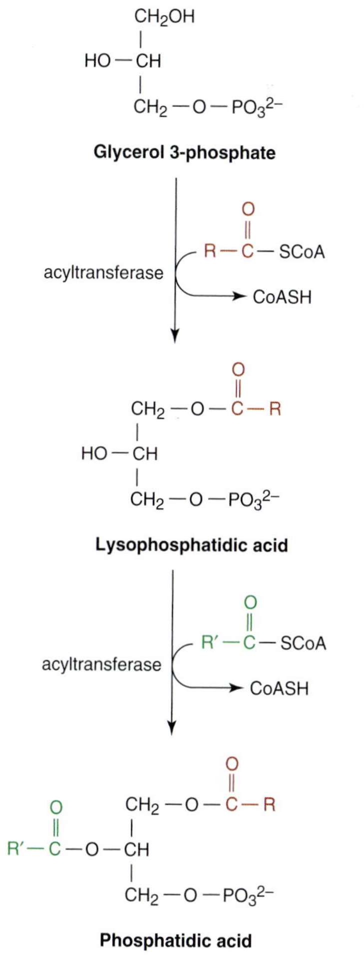

CH_2OH | $HO-CH$ | $CH_2-O-PO_3^{2-}$

Glycerol 3-phosphate

acyltransferase: $R-\overset{O}{\overset{\|}{C}}-SCoA$ → CoASH

$CH_2-O-\overset{O}{\overset{\|}{C}}-R$ | $HO-CH$ | $CH_2-O-PO_3^{2-}$

Lysophosphatidic acid

acyltransferase: $R'-\overset{O}{\overset{\|}{C}}-SCoA$ → CoASH

$CH_2-O-\overset{O}{\overset{\|}{C}}-R$ | $R'-\overset{O}{\overset{\|}{C}}-O-CH$ | $CH_2-O-PO_3^{2-}$

Phosphatidic acid

FIGURE 17.17

Synthesis of phosphatidic acid from glycerol 3-phosphate.

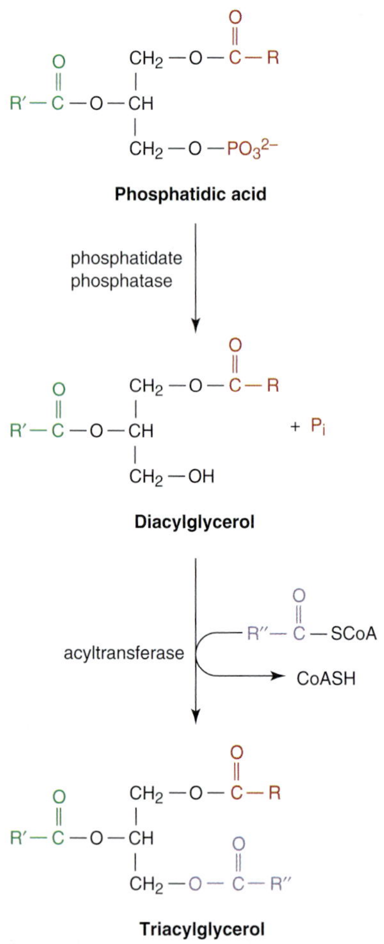

$$\text{R'—C(=O)—O—CH(CH}_2\text{—O—C(=O)—R)(CH}_2\text{—O—PO}_3^{2-})$$

Phosphatidic acid

phosphatidate phosphatase ↓

$$\text{R'—C(=O)—O—CH(CH}_2\text{—O—C(=O)—R)(CH}_2\text{—OH)} + P_i$$

Diacylglycerol

acyltransferase (R''—C(=O)—SCoA → CoASH) ↓

$$\text{R'—C(=O)—O—CH(CH}_2\text{—O—C(=O)—R)(CH}_2\text{—O—C(=O)—R'')}$$

Triacylglycerol

FIGURE 17.18

Synthesis of triacylglycerol from phosphatidic acid.

acid (Figure 17.16). Although this is not a major pathway of triacylglycerol synthesis, it is important for the synthesis of membrane lipids with ether-linked fatty acids.

Triacylglycerol synthesis follows a different pathway in mucosal cells of the small intestine. These cells take up 2-monoacylglycerols and free fatty acids from the gut, which are the major digestion products of dietary triacylglycerols by pancreatic lipase. An enzyme in the mucosal cells acylates these monoacylglycerols using acyl CoAs as substrates. The resulting diacylglycerols can then be acylated to form triacylglycerols, which are packaged into chylomicrons.

Analysis of triacylglycerols from human shows that the distribution of different fatty acids on the three positions of glycerol is neither random nor absolutely specific. Saturated fatty acids tend to be concentrated at position 1 and unsaturated fatty acids at positions 2 and 3 in triacylglycerols of human adipose tissue. Two main factors that determine the fatty acid composition at each position on glycerol are the specificity of acyltransferases involved and the relative availability of different fatty acids in the fatty acyl-CoA pool.

Mobilization of Triacylglycerols Requires Hydrolysis

The first step in mobilizing stored fatty acids for energy production is hydrolysis of triacylglycerol. Various lipases catalyze this reaction and the sequence of hydrolysis of the 3 acyl chains is determined by the specificities of the lipases involved.

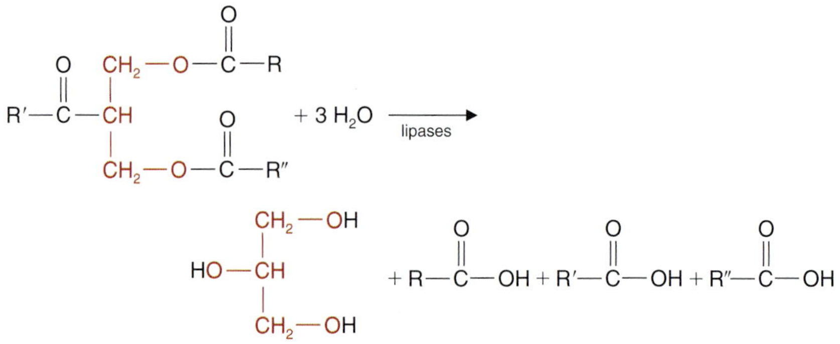

$$\text{R'—C(=O)—CH(CH}_2\text{—O—C(=O)—R)(CH}_2\text{—O—C(=O)—R'')} + 3\,H_2O \xrightarrow{\text{lipases}}$$

$$\text{HO—CH(CH}_2\text{—OH)}_2 + \text{R—C(=O)—OH} + \text{R'—C(=O)—OH} + \text{R''—C(=O)—OH}$$

Triacylglycerol Synthesis Occurs During Fasting as Part of a Triacylglycerol–Fatty Acid Cycle via Glyceroneogenesis

The release of free fatty acids from adipose tissue is a critical metabolic adaptation to fasting. The quantity of fatty acids released by adipose tissue, however, exceeds the amount used for energy by other tissues. As much as 60% of these fatty acids are re-deposited in adipose tissue as triacylglycerols. Both liver and adipose tissue play a major role in this process (see Clin. Corr. 17.3). In the fed state, the glycerol 3-phosphate for triacylglycerol synthesis is derived from glucose via glycolysis. During fasting, however, the entry of glucose into adipose tissue is limited because the insulin concentration is low and glucose is being used by other tissues. In this dietary state, glycerol 3-phosphate is synthesized in both the adipose tissue and the liver by glyceroneogenesis. As shown in Figure 17.19, substrates that enter the tricarboxylic acid cycle, such as pyruvate, glutamate, or aspartate, can support net glyceroneogenesis. This pathway is essentially an abbreviated version of gluconeogenesis, in which malate formed in the tricarboxylic acid cycle leaves the mitochondria and is converted to oxaloacetate in the cytosol. Phosphoenolpyruvate carboxykinase then converts oxaloacetate to phosphoenolpyruvate, which is converted to dihydroxyacetone phosphate and then to glycerol 3-phosphate, which is used for triacylglycerol synthesis. The key enzyme in this process is **phosphoenolpyruvate carboxykinase**, which is induced in liver and adipose tissue in the fasted state.

CLINICAL CORRELATION 17.3
Triacylglycerol/Fatty Acid Cycle

Triacylglycerol that is stored in adipose tissue is hydrolyzed to free fatty acids (FFA) during fasting to provide energy for tissues such as the skeletal and cardiac muscle and also indirectly to the brain via ketone bodies. Hormones, most notably insulin, control this process. As the level of insulin falls during fasting, the rate of triacylglycerol hydrolysis (lipolysis) increases, resulting in FFA release from adipose tissue. A surprising aspect of this process is the fate of the FFA; in fasted humans up to 65% of this FFA is re-esterified to triacylglycerol in the liver and other peripheral tissues. In liver approximately 60% of the FFA taken up is re-esterified to triacylglycerol, released into the blood as VLDL, and sent back to the adipose tissue for deposition as triacylglycerol. This process has been termed the triacylglycerol/fatty acid cycle.

The synthesis of triacylglycerol in mammalian tissues requires glycerol 3-phosphate, which is derived from dietary glucose via glycolysis in the fed state. During fasting, when low insulin inhibits glucose utilization, the glycerol 3-phosphate for the re-esterification of FFA is generated by glyceroneogenesis, an abbreviated version of gluconeogenesis. In this pathway, pyruvate, or compounds that can generate pyruvate, such as alanine or lactate, is converted to glycerol 3-phosphate via dihydroxyacetone phosphate (Figure 17.19). The rate-controlling enzyme in the pathway of glyceroneogenesis is phosphoenolpyruvate carboxykinase (PEPCK), which has high activity in brown and white adipose tissue. If the expression of the PEPCK gene is ablated in the adipose tissue of mice, glyceroneogenesis is inhibited and triacylglycerol storage is reduced. Conversely, overexpression of the PEPCK gene in adipose tissue of transgenic mice increases the rate of glyceroneogenesis, resulting in obesity.

The metabolic logic of the triacylglycerol/fatty acid cycle most likely resides in the importance of fatty acids as a fuel during starvation. In order to ensure that there is sufficient FFA in the blood, more FFA is released from fat cells than is required; what is not used is re-esterified to triacylglycerol and re-deposited in adipose tissue, at a minimum energetic cost. The triacylglycerol/fatty acid cycle consumes 3–6% of the energy in a molecule of triacylglycerol. It is apparently better to have the needed fuel available, and pay for it energetically, than to run out! The rate of FFA re-esterification in the triacylglycerol/fatty acid cycle is most likely a key factor in determining the steady-state concentration of FFA in the blood, a parameter that is directly involved in the etiology of Type 2 diabetes. The glitazones, a class of antidiabetic drugs, induce the activity of PEPCK in adipose tissue and liver and increase the rate of FFA re-esterification to triacylglycerol via glyceroneogenesis in these tissues, supporting the important role of this process in maintaining lipid homeostasis.

Source: Reshef, L. Olswang, Y. Cassuto, H. Blum, B. Croniger, C. M. Kalhan S. C, Tilghman' S. M., and Hanson R. W. Glyceroneogenesis and the triglyceride/fatty acid cycle. *J. Biol. Chem.* 278:30413, 2003; Jensen, M. D., Ekberg, K., and Landau, B. R. Lipid metabolism during fasting. *Am. J. Physiol. Endocrinol. Metab.* 281:E789 2001. Tordjman, J. Khazan, W. Antoine, B. Chauvet, G. Quette, J Fouque, F. Beale, E. G. Benelli, C. and Forest, C. Regulation of glyceroneogenesis and phosphoenolpyruvate carboxykinase by fatty acids, retinoic acid and thiazolidinediones. *Biochimie* 85:1213, 2003.

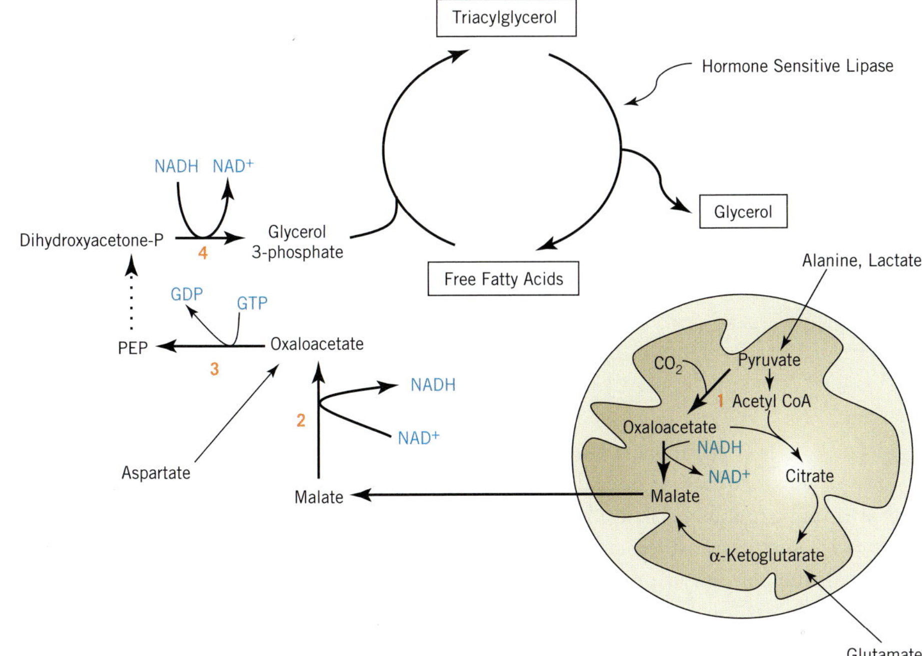

FIGURE 17.19

The pathway of glyceroneogenesis in the liver and adipose tissue. During fasting, this pathway, which is an abbreviated version of gluconeogenesis, produces glycerol 3-phosphate for triacylglycerol synthesis. The key enzymes of glyceroneogenesis are (1) pyruvate carboxylase, (2) the mitochondrial and cytosolic forms of NAD malate dehydrogenase, (3) phosphoenolpyruvate carboxykinase, and (4) glycerol 3-phosphate dehydrogenase.

17.6 UTILIZATION OF FATTY ACIDS FOR ENERGY PRODUCTION

Fatty acids in the circulation are taken up by cells and used for energy production, primarily in mitochondria, in a process integrated with energy generation from other sources. Fatty acids are broken down to acetyl CoA in the mitochondria, with the production of NADH and $FADH_2$. These three products are then used in the mitochondrial matrix for energy production via the tricarboxylic acid cycle and oxidative phosphorylation.

The utilization of fatty acids for energy production varies considerably from tissue to tissue and depends to a significant extent on metabolic status—that is, fed or fasted, exercising or at rest. Most tissues can use fatty acids as a fuel. Fatty acids are a major energy source in cardiac and skeletal muscle, but the brain oxidizes fatty acids poorly because of limited transport across the blood–brain barrier. Red blood cells cannot oxidize fatty acids because they do not have mitochondria, the site of fatty acid oxidation. During prolonged fasting, the liver converts acetyl CoA generated by fatty acid oxidation and the breakdown of amino acids into **ketone bodies**, which become a major fuel. Most tissues, including the brain, adapt to fasting by utilizing these ketone bodies.

β-Oxidation of Straight-Chain Fatty Acids Is a Major Energy-Producing Process

CoA esters of fatty acids are the substrates for oxidation. For the most part, the pathway of fatty acid oxidation is similar to, but not identical to, a reversal of the process of palmitate synthesis. That is, two-carbon fragments are removed sequentially from the carboxyl end of the fatty acid by enzymatic **dehydrogenation, hydration**, and **oxidation** to form a β-keto acid, which is then split by **thiolysis**.

Fatty Acids Are Activated by Conversion to Fatty Acyl CoA

The first step in the utilization of a fatty acid is its activation to a fatty acyl CoA. This reaction is catalyzed by enzymes in the endoplasmic reticulum and outer mitochondrial membrane and is the same one used in the synthesis of triacylglycerols. The fatty acyl CoAs are released into the cytosol.

Carnitine Carries Acyl Groups Across the Inner Mitochondrial Membrane

Most long-chain fatty acyl CoAs are formed outside mitochondria, but they are oxidized in the matrix. The mitochondrial membrane is impermeable to CoA and its derivatives. Fatty acids are transported into the mitochondria using **carnitine (4-trimethylamino-3-hydroxybutyrate)** as a carrier. The process is outlined in Figure 17.20.

$$CH_3-(CH_2)_n-\overset{O}{\overset{\|}{C}}-SCoA + HO-\underset{}{CH}(CH_2-\overset{+}{N}-(CH_3)_3)-CH_2-COOH \xrightleftharpoons[\text{transferase}]{\text{carnitine palmitoyl-}} CH_3-(CH_2)_n-\overset{O}{\overset{\|}{C}}-O-CH(CH_2-\overset{+}{N}-(CH_3)_3)-CH_2-COOH + CoASH$$

The acyl group is transferred from CoA to carnitine by **carnitine palmitoyltransferase I (CPT I)** on the outer mitochondrial membrane. Acyl carnitine and free carnitine are then exchanged across the inner mitochondrial membrane by **carnitine-acylcarnitine translocase**. Finally, the fatty acyl group is transferred back to CoA by **carnitine palmitoyltransferase II (CPT II)** on the matrix side of the inner membrane. This process functions primarily in the transport of fatty acyl CoAs with 12–18 carbons. Genetic abnormalities in this system lead to serious pathological consequences (see Clin. Corr. 17.4). By contrast, entry of shorter-chain fatty acids is independent of

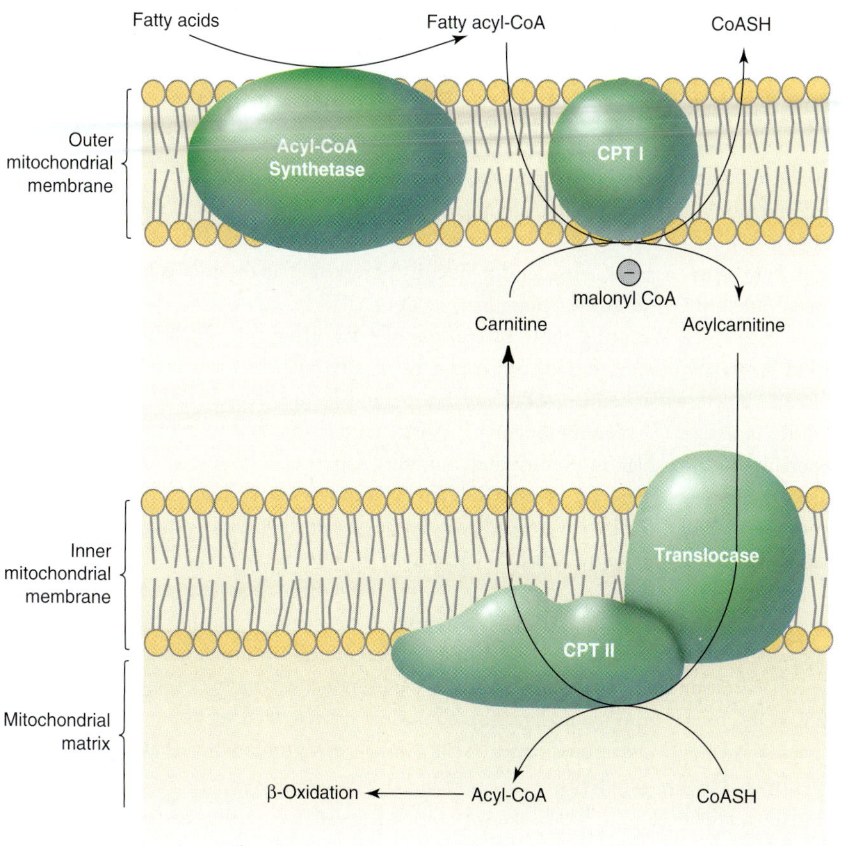

FIGURE 17.20

Fatty acids are transported into the mitochondria as acylcarnitines for oxidation. Inhibition of carnitine palmitoyltransferase I (CPT I) by malonyl CoA regulates the rate of fatty acid oxidation.

carnitine; they cross the inner mitochondrial membrane and become activated to their CoA derivatives in the matrix. CPT I is an important site for the regulation of fatty acid oxidation because its rate controls the entry of fatty acids into the mitochondria and therefore determines the supply of substrate for β-oxidation in the mitochondrial matrix. **Malonyl CoA**, which is the product of acetyl-CoA carboxylase and a key intermediate in fatty acid synthesis, is an inhibitor of CPT I. This regulation inhibits fatty acid oxidation in the fed state and prevents a futile cycle (see p. 690).

β-Oxidation Is a Sequence of Four Reactions

The four reactions of β-oxidation are presented in Figure 17.21. Once a fatty acyl CoA is formed at the inner surface of the inner mitochondrial membrane, it can be oxidized by acyl-CoA dehydrogenase, a flavoprotein that uses FAD as the electron acceptor (Reaction 1). The products are an enoyl CoA with a trans double bond between the C2 and C3 atoms and enzyme-bound $FADH_2$. As in the tricarboxylic acid cycle, the $FADH_2$ transfers its electrons to enzymes of the oxidative phosphorylation pathway, regenerating FAD.

The second step in β-oxidation is hydration of the trans-double bond to a **3-L-hydroxyacyl CoA**, which is oxidized to a **3-ketoacyl-CoA** intermediate, with the generation of NADH in the third step. The final step is cleavage of the chain by ketothiolase, generating acetyl CoA and a fatty acyl CoA that has been shortened by two carbon atoms. This shortened acyl CoA is ready for the next round of oxidation, starting with acyl-CoA dehydrogenase. In most tissues, the acetyl CoA will be used by the tricarboxylic acid cycle and the $FADH_2$ and NADH will be reoxidized by the electron transport system with the production of ATP.

CLINICAL CORRELATION 17.4

Genetic Deficiencies in Carnitine Transport or Carnitine Palmitoyltransferase

Several diseases result from genetic abnormalities in the transport of long-chain fatty acids across the inner mitochondrial membrane. They stem from deficiencies in the level of carnitine or in the synthesis and transport of acylcarnitines. Mutations can affect carnitine palmitoyl transferases (CPT) or the mitochondrial carnitine-acylcarnitine translocase.

Two categories of carnitine deficiency, primary and secondary, are now recognized. Primary carnitine deficiency is caused by a defect in the high affinity plasma membrane carnitine transporter in tissues such as muscle, kidney, heart, and fibroblasts (but apparently not in liver where a different transporter is operative). It results in extremely low levels of carnitine in affected tissues and in plasma (because of failure of the kidneys to reabsorb carnitine). The clinical symptoms of carnitine deficiency range from mild, recurrent muscle cramping to severe weakness and death. The very low carnitine level in heart and skeletal muscle seriously compromises long- chain fatty acid oxidation. Dietary carnitine therapy, which raises the plasma concentration of carnitine and forces its entry into tissues in a nonspecific manner, is frequently beneficial. Secondary carnitine deficiency is often associated with inherited defects in the β-oxidation pathway. These disorders frequently cause the accumulation of acylcarnitines, which are excreted in the urine (see Clin. Corr. 17.5), thereby depleting the carnitine pool. These acylcarnitines may also impair the uptake of free carnitine by tissues.

There are several different CPT deficiencies. The most common form results from mutations in the CPT II gene that causes a partial loss of enzyme activity. Patients generally experience muscle weakness during prolonged exercise when the muscles rely heavily upon fatty acids as an energy source. Myoglobinuria, due to breakdown of muscle tissue, is frequently observed. The disorder is usually referred to as the "muscular" form of CPT II activity deficiency. Mutations causing more severe loss of CPT II activity (90% or greater) can have serious consequences in early infancy. These are usually precipitated by periods of fasting and include hypoketotic hypoglycemia, hyperammonemia, cardiac malfunction, and sometimes death. Similar morbidity and mortality are associated with mutations in the gene for liver CPT I. To date, only a few patients with hepatic CPT I deficiency have been reported, possibly because the disease is frequently lethal. No defects in the muscle isoform of CPT I have been reported.

The first patient with carnitine–acylcarnitine translocase deficiency was described in 1992. Clinical features included intermittent hypoglycemic coma, hyperammonemia, muscle weakness, and cardiomyopathy. The condition proved fatal at age 3 years. Several additional cases with similar symptomatology have since been reported.

These disorders can be treated with a diet that is low in long-chain fatty acids and by avoidance of fasting, to avoid conditions where tissues require fatty acid oxidation for energy. The diet can also be supplemented with medium-chain triacylglycerols, because these fatty acids enter the mitochondria by a carnitine-independent mechanism.

Source: Stanley, C. A., Hale, D. E., Berry, G. T., Deleeno, S., Boxer, J., and Bonnefont, J-P. A deficiency of carnitine–acylcarnitine translocase in the inner mitochondrial membrane. *N. Engl. J. Med.* 327:19, 1992. Roe, C. R. and Dong, J., Mitochondrial fatty acid oxidation disorders. In: C. R. Scriver, A. L. Beaudet, W. S. Sly, and D. Valle (Eds.), *The Metabolic and Molecular Basis of Inherited Disease*, Vol. II, 8th ed. New York: McGraw-Hill, 2001, p 2297. Bonnefont, J.-P., Demaugre, F., Prip-Buus, C., Saudubray, J.-M., Brivet, M., Abadi, N., and Thuillier, L. Carnitine palmitoyltransferase deficiencies. *Mol. Gen. Metab.* 68:424, 1999.

Each of the four reactions shown in Figure 17.21 is catalyzed by several different enzymes that have specificity for substrates of different chain lengths. For example, at least four enzymes catalyze the first dehydrogenation step. These are very-long-chain, long-chain, medium-chain, and short-chain acyl-CoA dehydrogenases (**VLCAD, LCAD, MCAD**, and **SCAD**). VLCAD, which oxidizes straight chain acyl-CoAs ranging from C12 to C24, differs from the other family members in that it is membrane associated. MCAD has broad chain length specificity but is most active with C6 and C8 substrates, while the order of preference for SCAD is C4 > C6 > C8. LCAD is presumed to be involved in initiating the oxidation of branched-chain fatty acids (e.g., 2-methylpalmitoyl-CoA).

A feature unique to the oxidation of long-chain fatty acids is that the enoyl-CoA hydratase, 3-hydroxyacyl-CoA dehydrogenase, and β-ketothiolase steps are all catalyzed by a membrane-bound complex of the three enzymes called a **trifunctional protein**. This complex is distinct from the enzymes that catalyze the oxidation of medium- and short-chain acyl CoAs, all of which are soluble proteins in the mitochondrial matrix. Clinical Correlation 17.5 describes genetic deficiencies of acyl-CoA dehydrogenases.

Energy Yield from the β-Oxidation of Fatty Acids

Each round of β-oxidation produces one acetyl CoA, one $FADH_2$, and one NADH. In the oxidation of palmitoyl CoA, seven cleavages take place, with the formation of two acetyl CoAs in the final cleavage. Thus, the products of β-oxidation of palmitate are eight acetyl CoAs, seven $FADH_2$, and seven NADH. Based on our current estimates of the yields of ATP in oxidative phosphorylation (see p. 564), each of the $FADH_2$ yields 1.5 ATPs and each NADH yields 2.5 ATPs when oxidized by the electron transport chain. Therefore, the oxidation of the seven NADH and one FADH produces 28 ATP. Oxidation of each acetyl CoA through the tricarboxylic acid cycle yields 10 ATP (see p. 546), so the eight two carbon fragments from a palmitate molecule produce 80 ATP, for a total of 108 ATP. However, 2 ATP equivalents are used to activate palmitate to palmitoyl CoA (1 ATP is converted to 1AMP + PP_i). Therefore, each palmitic acid yields 106 ATP mol^{-1} upon complete oxidation. The importance of fatty acids in supplying the energy needs for human metabolism is discussed on p. 860.

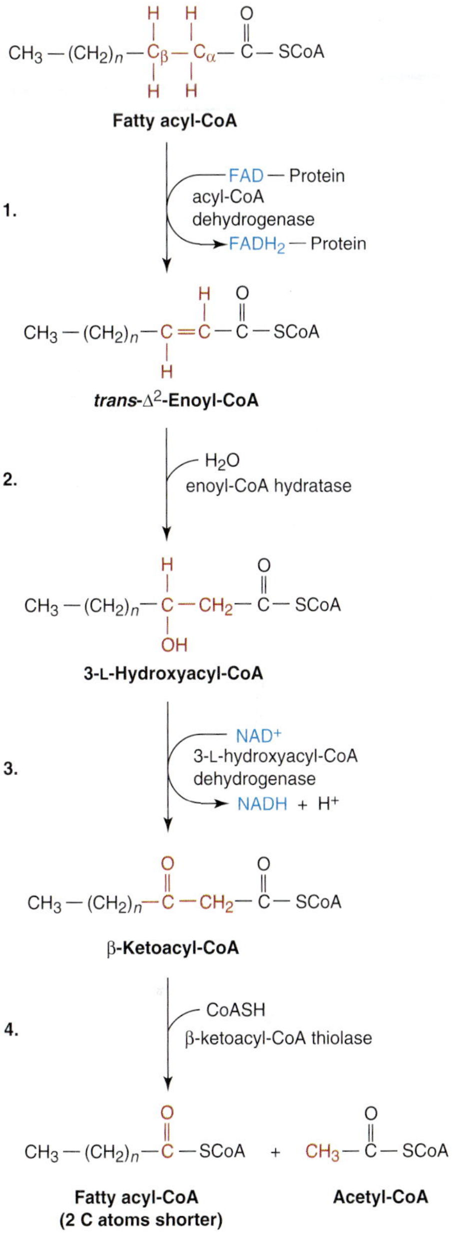

Comparison of Fatty Acid Synthesis and Oxidation

The pathways of fatty acid synthesis and oxidation are similar. However, like many paired anabolic and catabolic pathways, the synthesis and oxidation of fatty acids are not the reverse of each other. The critical differences between the two pathways are outlined in Table 17.4. They include different cellular locations, different cofactors (NADPH in synthesis and FAD and NAD^+ in oxidation) and the use of ATP to drive malonyl-CoA formation for fatty acid synthesis. These differences permit both pathways to proceed in the forward direction and allow for the coordinate regulation of each pathway to prevent futile cycling.

Some Fatty Acids Require Modification of β-Oxidation for Metabolism

The β-oxidation pathway oxidizes saturated fatty acids with even numbers of carbon atoms to acetyl CoA. However, other fatty acids in the diet, including those with cis double bonds, branched chains, and odd numbers of carbon atoms, require additional steps for their complete oxidation. These steps allow these fatty acids to be used as fuels and also prevent their accumulation. Other reactions catalyze the α- and ω-oxidation of fatty acids. **α-Oxidation** occurs at C2 instead of C3 as occurs in the β-oxidation whereas **ω-oxidation** occurs at the methyl end of the fatty acid molecule.

Oxidation of Odd-Chain Fatty Acids Produces Propionyl CoA

Fatty acids with an odd number of carbon atoms are oxidized by the β-oxidation pathway. The products of the final cleavage by thiolase are acetyl CoA and **propionyl CoA**, instead of two acetyl CoAs (Figure 17.22). Propionyl CoA, which is also produced by the catabolism of isoleucine, valine, and methionine, is metabolized by carboxylation to methylmalonyl CoA and conversion to succinyl CoA (see p. 778).

Oxidation of Unsaturated Fatty Acids Requires Additional Enzymes

Unsaturated fatty acids are utilized by β-oxidation as are saturated fatty acids. After several rounds of oxidation, intermediates are generated that have double bonds near the carboxyl carbon. Additional enzymes are required to deal with the cis double bonds and double bonds beginning at odd- and even-numbered carbon atoms. The oxidation of linoleoyl CoA (18:2) (Figure 17.23) illustrates this process. β-Oxidation generates an enoyl-CoA intermediate with a cis double bond between C3 and C4, instead of the intermediate with a trans bond between C2 and C3 that is required by enoyl-CoA hydratase. **Enoyl-CoA isomerase** changes the *cis*-Δ^3- to a *trans*-Δ^2-enoyl CoA, which can then be metabolized by β-oxidation.

A second problem occurs when the cis double bond of the acyl-CoA intermediate resides between C4 and C5. In this case, the action of acyl-CoA dehydrogenase gives rise to a *trans*-2, *cis*-4-enoyl CoA. This is acted upon by 2,4-dienoyl-CoA reductase that

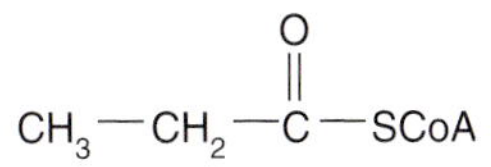

$CH_3-(CH_2)_n-C_\beta H_2-C_\alpha H_2-C(=O)-SCoA$

Fatty acyl-CoA

1. FAD—Protein; acyl-CoA dehydrogenase; → $FADH_2$—Protein

$CH_3-(CH_2)_n-CH=CH-C(=O)-SCoA$

trans-Δ^2-Enoyl-CoA

2. H_2O; enoyl-CoA hydratase

$CH_3-(CH_2)_n-CH(OH)-CH_2-C(=O)-SCoA$

3-L-Hydroxyacyl-CoA

3. NAD^+; 3-L-hydroxyacyl-CoA dehydrogenase; → NADH + H^+

$CH_3-(CH_2)_n-C(=O)-CH_2-C(=O)-SCoA$

β-Ketoacyl-CoA

4. CoASH; β-ketoacyl-CoA thiolase

$CH_3-(CH_2)_n-C(=O)-SCoA$ + $CH_3-C(=O)-SCoA$

Fatty acyl-CoA (2 C atoms shorter) Acetyl-CoA

FIGURE 17.21
Pathway of fatty acid β-oxidation.

$CH_3-CH_2-C(=O)-SCoA$

FIGURE 17.22
Propionyl-CoA.

CLINICAL CORRELATION 17.6

Refsum's Disease

Although α-oxidation of fatty acids is relatively minor in terms of total energy production, it is significant in the metabolism of branched-chain fatty acids in the diet. A principal example of these is phytanic acid, a metabolic product of phytol, which is a constituent of chlorophyll. Phytanic acid is a significant constituent of milk and animal fats. It cannot be oxidized by β-oxidation because of the presence of the 3-methyl group. It is metabolized by an α-hydroxylation followed by dehydrogenation and decarboxylation. β-Oxidation can completely degraded the molecule, producing three molecules of propionyl CoA, three molecules of acetyl CoA, and one molecule of isobutyryl CoA.

Patients with a rare genetic disease called Refsum's disease lack the peroxisomal α-hydroxylating enzyme and accumulate large quantities of phytanic acid in their tissues and sera. This leads to serious neurological problems such as retinitis pigmentosa, peripheral neuropathy, cerebellar ataxia, and nerve deafness. The restriction of dietary dairy products and meat products from ruminants results in lowering of plasma phytanic acid and regression of the neurologic symptoms.

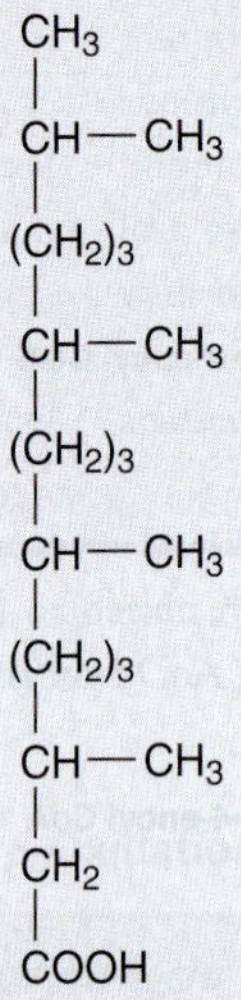

$$
\begin{array}{l}
CH_3 \\
| \\
CH-CH_3 \\
| \\
(CH_2)_3 \\
| \\
CH-CH_3 \\
| \\
(CH_2)_3 \\
| \\
CH-CH_3 \\
| \\
(CH_2)_3 \\
| \\
CH-CH_3 \\
| \\
CH_2 \\
| \\
COOH
\end{array}
$$

Phytanic acid

Source: Wanders, R. J. A., van Grunsven, E. G., and Jansen, G. A. Lipid metabolism in peroxisomes: Enzymology, functions and dysfunctions of the fatty acid α- and β-oxidation systems in humans. *Biochem. Soc. Trans.* 28:141, 2000.

Some of these hydroxylations occur in the endoplasmic reticulum and mitochondria and involve **monooxygenases (P450 family)** that require O_2 and NADH or NADPH. The α-hydroxylation of fatty acids also occurs in peroxisomes. This is particularly important for the metabolism of branched-chain fatty acids (see Clin. Corr. 17.6). A branched-chain fatty acid, such as phytanoyl-CoA, which is derived from chlorophyll in the diet, is acted upon by a hydroxylase in a reaction involving α-ketoglutarate, Fe^{2+}, and ascorbate, with the generation of 2-hydroxyphytanoyl CoA and formyl CoA. The latter is metabolized to CO_2 via formic acid. The 2-hydroxyphytanoyl CoA is then further metabolized to pristanic acid, which undergoes β-oxidation.

ω-Oxidation Gives Rise to Dicarboxylic Acids

ω-Oxidation is another minor pathway for fatty acid oxidation, which occurs in the endoplasmic reticulum of many tissues. In this pathway, hydroxylation takes place on the methyl carbon at the opposite end of the molecule from the carboxyl group or on the carbon next to the methyl end. It uses a monooxygenase that requires O_2 and NADPH. Hydroxylated fatty acids can be further oxidized in the cytosol to **dicarboxylic acids** via the sequential action of cytosolic **alcohol** and **aldehyde dehydrogenases**. Medium-chain fatty acids are the principal substrates of this pathway. The overall reactions are

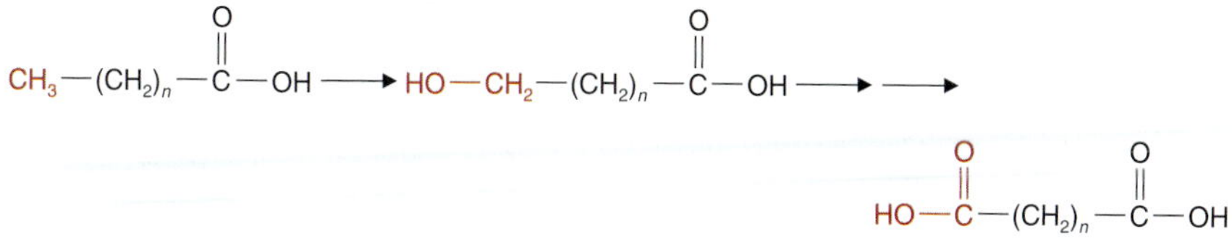

$$CH_3-(CH_2)_n-\overset{O}{\overset{\|}{C}}-OH \longrightarrow HO-CH_2-(CH_2)_n-\overset{O}{\overset{\|}{C}}-OH \longrightarrow \longrightarrow HO-\overset{O}{\overset{\|}{C}}-(CH_2)_n-\overset{O}{\overset{\|}{C}}-OH$$

These dicarboxylic acids form CoA esters at either carboxyl group and then undergo β-oxidation to produce shorter-chain dicarboxylic acids such as adipic (C_6) and succinic (C_4) acids. This process also occurs primarily in peroxisomes.

Ketone Bodies Are Formed from Acetyl CoA

Ketone bodies are water-soluble products of lipid oxidation that are formed in liver and kidney mitochondria during prolonged fasting. The ketone bodies, **acetoacetic acid** and its reduction product **β-hydroxybutyric acid**, are made from acetyl CoA that is produced by fatty acid and amino acid catabolism (Figure 17.24). Ketone bodies are an important adaptation to prolonged fasting; they can be present at high concentrations (>3 mM) and are an important energy source for many tissues (see Clin. Corr. 17.3).

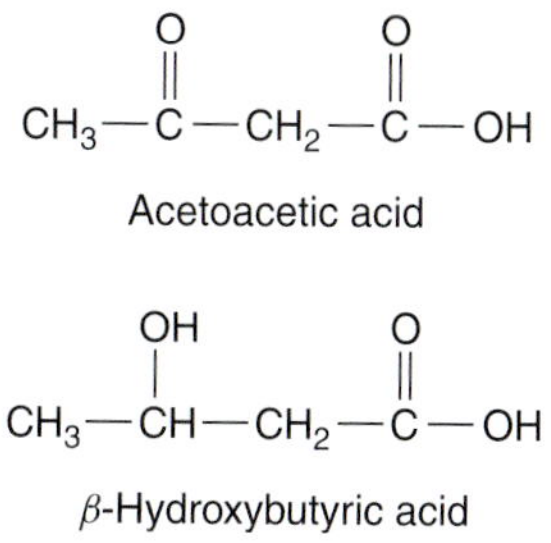

$$CH_3-\overset{O}{\overset{\|}{C}}-CH_2-\overset{O}{\overset{\|}{C}}-OH$$

Acetoacetic acid

$$CH_3-\overset{OH}{\overset{|}{C}H}-CH_2-\overset{O}{\overset{\|}{C}}-OH$$

β-Hydroxybutyric acid

FIGURE 17.24
Structures of ketone bodies.

HMG CoA Is an Intermediate in the Synthesis of Acetoacetate from Acetyl CoA

Ketone body synthesis occurs in the mitochondrial matrix and begins with condensation of two acetyl CoA molecules to form acetoacetyl CoA, in a reaction that is the reverse of the final step of β-oxidation (Figure 17.25). The enzyme involved, **β-ketothiolase**, is an isozyme of the enzyme that functions in β-oxidation. **HMG-CoA synthase** catalyzes the condensation of acetoacetyl CoA with another molecule of acetyl CoA to form **β-hydroxy-β-methylglutaryl coenzyme A (HMG-CoA)**. **HMG CoA lyase** then cleaves HMG CoA to yield acetoacetic acid and acetyl CoA.

Acetoacetate Forms Both D-β*-Hydroxybutyrate and Acetone*

Some of the acetoacetate is reduced to D-β-hydroxybutyrate in liver mitochondria by **D-β-hydroxybutyrate dehydrogenase**. The extent of this reaction depends on the intramitochondrial [NADH]/[NAD$^+$] ratio. During fasting, the oxidation of fatty acids generates NADH part of which is used in the reduction of acetoacetate. The β-hydroxybutyrate and acetoacetate are released from liver and kidney for use by other tissues. This also exports reducing equivalents from the liver, in the form of β-hydroxybutyrate. As much as 25% of the NADH generated by β-oxidation is used to synthesize β-hydroxybutyrate. Note that the product of β-hydroxybutyrate dehydrogenase is D-β-hydroxybutyrate, whereas β-hydroxybutyryl CoA formed during β-oxidation is the L isomer. Because β-hydroxybutyrate dehydrogenase has high activity in the liver, the concentrations of its substrates and products are maintained close to equilibrium. Thus, the ratio of β-hydroxybutyrate to acetoacetate in the blood reflects the [NADH]/[NAD$^+$] ratio in liver mitochondria.

Some acetoacetate continually undergoes slow, spontaneous nonenzymatic decarboxylation to acetone:

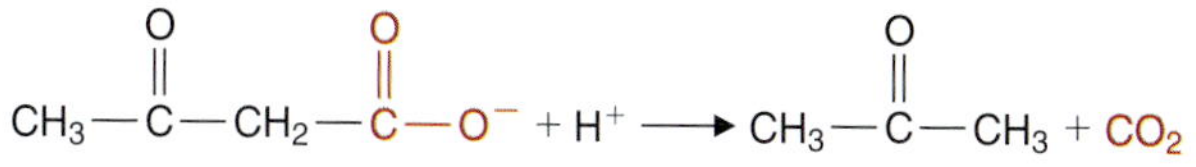

$$CH_3-\overset{O}{\overset{\|}{C}}-CH_2-\overset{O}{\overset{\|}{C}}-O^- + H^+ \longrightarrow CH_3-\overset{O}{\overset{\|}{C}}-CH_3 + CO_2$$

Acetone formation is negligible under normal conditions, but at high concentrations of acetoacetate, which can occur in severe diabetic **ketoacidosis** (see Clin. Corr. 17.7), acetone can reach levels high enough to be detectable in the breath.

HMG CoA is also an intermediate in cholesterol synthesis (see p. 708). However, the HMG CoA used for ketone body and cholesterol synthesis is present in different metabolic pools. The HMG CoA used for ketogenesis is synthesized in hepatic mitochondria by an isozyme of HMG-CoA synthase that is expressed at high levels

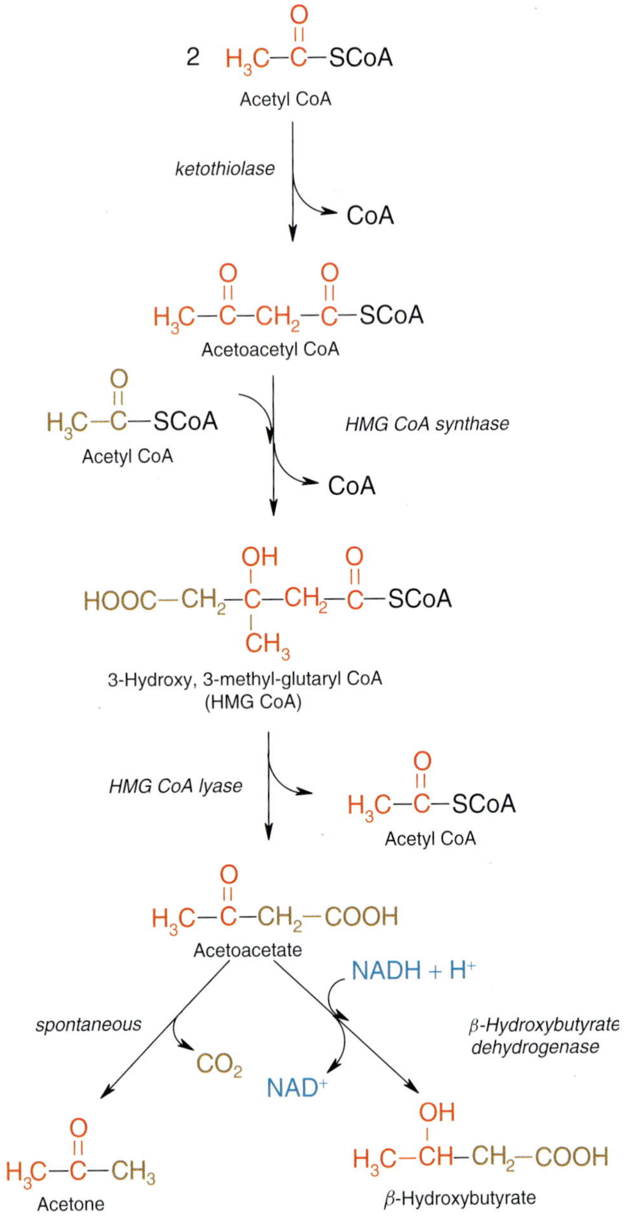

2 $H_3C-C(=O)-SCoA$
Acetyl CoA

ketothiolase
CoA

$H_3C-C(=O)-CH_2-C(=O)-SCoA$
Acetoacetyl CoA

$H_3C-C(=O)-SCoA$
Acetyl CoA

HMG CoA synthase
CoA

$HOOC-CH_2-C(OH)(CH_3)-CH_2-C(=O)-SCoA$
3-Hydroxy, 3-methyl-glutaryl CoA
(HMG CoA)

HMG CoA lyase

$H_3C-C(=O)-SCoA$
Acetyl CoA

$H_3C-C(=O)-CH_2-COOH$
Acetoacetate

spontaneous
CO_2

NADH + H^+
NAD^+
β-Hydroxybutyrate dehydrogenase

$H_3C-C(=O)-CH_3$
Acetone

$H_3C-CH(OH)-CH_2-COOH$
β-Hydroxybutyrate

FIGURE 17.25
Ketone bodies are synthesized from acetyl CoA in hepatic mitochondria.

during prolonged fasting. Moreover, HMG-CoA lyase, which converts HMG CoA to acetoacetate and acetyl CoA, is expressed only in hepatic mitochondria. In contrast, HMG CoA for cholesterol synthesis is made at low levels in the cytosol of many tissues by a cytosolic isozyme of HMG-CoA synthase.

Utilization of Ketone Bodies by Nonhepatic Tissues Requires Formation of Acetoacetyl CoA

Acetoacetate and β-hydroxybutyrate produced by the liver are excellent fuels for many nonhepatic tissues, including cardiac muscle, skeletal muscle, and brain, particularly when glucose is in short supply (prolonged fasting) or inefficiently used (insulin deficiency). Under these conditions, these tissues will oxidize free fatty acids, whose blood concentration rises as insulin levels fall. During prolonged fasting, ketone bodies replace glucose as a fuel, particularly in brain, which begins to use ketone bodies after 2–3 days of fasting. This reduces the requirement for glucose production by gluconeogenesis

CLINICAL CORRELATION 17.7

Ketone Bodies as a Fuel: The Atkins Diet

The current popularity of low-carbohydrate diets for weight loss underscores the importance of the metabolism of ketone bodies as fuels in humans. The best known of these was popularized by the late Dr. Robert Atkins in his book, *Dr. Atkins' Diet Revolution*, which has sold more than 6 million copies. The Atkins Diet, which is high in fat and protein and very low in carbohydrate (less that 20 g/day during the initial phase), has been highly controversial among nutritionists and the medical establishment because of its high fat content. Interestingly, individuals on the diet often lose a considerable amount of weight, despite the promise by the author that an individual can "stop counting calories and measuring portions." Controlled clinical studies have shown that obese subjects lose more weight on a high-fat/low-carbohydrate diet than on an isocaloric diet with higher levels of carbohydrate. Surprisingly, a marked decrease was noted in the levels of triacylglycerol in the blood of individuals consuming the high-fat/low-carbohydrate diet. At this writing, there are no long-term studies (greater than a year) of the effect on the lipid profile or general health of obese humans who remain on this diet. However, it is likely that the Atkins diet results in dramatic weight loss due to the satiety caused by the high fat and protein content of the diet, resulting in a reduction of food intake.

Central to the Atkins Diet is the mobilization of fatty acids from adipose tissue and their conversion by the liver to ketone bodies (β-hydroxybutyrate, acetoacetate, and acetone). The presence of ketone bodies in the urine is the prescribed method of determining metabolic status while on the diet, since even small quantities of dietary carbohydrate will depress ketone body synthesis, largely by inhibiting lipolysis in adipose tissue. As the concentration of ketone bodies rise in the blood, a fraction is excreted in the urine and some via the breath. Could this account for the greater weight loss noted by individuals on the Atkins Diet? For comparison, after 7 days of fasting the daily urinary excretion rate of acetoacetate and β-hydroxybutyrate in humans is approximately 110 mmol day^{-1}; the excretion rate is even less early in starvation (60 mmol day^{-1} after two days of starvation). This excretion could contribute to the negative caloric balance and weight loss characteristic of the Atkins diet, although the energy loss does not exceed 100 kcal day^{-1}.

In general, ketosis develops when glucose oxidation is suppressed and fat catabolism is accelerated. There are two types of ketosis: the normal ketosis of fasting and the pathological hyperketonemia of diabetic ketoacidosis. No other fuel in the human blood can change so drastically as ketone bodies and still be compatible with life. After an overnight fast, the concentration of ketone bodies is approximately 0.05 mM, but this concentration can rise to 2 mM after 2 days of starvation and to 7 mM after 40 days, a 140-fold change in the concentration. In a seminal study, Owen and colleagues demonstrated that during prolonged starvation acetoacetate and β-hydroxybutyrate replaced glucose as the predominant fuel for the brain. This reduces the need to synthesize glucose from amino acids derived from muscle and liver protein. Muscle avidly consumes ketone bodies early in starvation but switches to fatty acid oxidation as starvation progresses, thereby sparing ketone bodies for metabolism by the brain. Thus, ketone bodies are a normal fuel for a variety of tissues and are part of a complex pattern of fuel metabolism that occurs during fasting in humans.

Sources: Feinman, R. D. and Fine, E. J. Thermodynamic and metabolic advantage of weight loss diets. *Metab. Syndrome Relat. Disord.* 1:209, 2003. Samaha, F. F., Iqbal, N. Seshadri, P. Chicano, K. L. Daily, D. A. McGrory, J. Williams, T. Williams, M. Gracely, E. J. and Stern, L. A low-carbohydrate as compared with a low-fat diet in severe obesity. *N. Engl J Med* 348:21, 2003. Owen, O. E., Morgan, A. P., Kemp, H. G., Sullivan, J. M., Herrera, M. G., and Cahill, G. F., Jr. Brain metabolism during fasting. *J Clin Invest* 46:1589 1967.

during a prolonged fast, thus "sparing" the muscle protein that contributes amino acids for gluconeogenesis during fasting (see p. 854). Acetoacetate and β-hydroxybutyrate also serve as precursors for cerebral lipid synthesis during the neonatal period.

Ketone bodies are metabolized in the mitochondria of nonhepatic tissues. β-Hydroxybutyrate dehydrogenase converts β-hydroxybutyrate to acetoacetate by an NAD-linked oxidation. Acetoacetate is then converted to its CoA derivative by **acetoacetate:succinyl-CoA transferase**, which is present in tissues that use ketone bodies but is not present in liver. Succinyl CoA serves as the source of the CoA. The reaction is depicted in Figure 17.26. β-Ketothiolase converts the acetoacetyl CoA into two acetyl CoAs, which enter the tricarboxylic acid cycle for energy production.

In summary, the pathways of ketone body synthesis and utilization share several steps. However, there are reactions that are unique to each pathway. The key enzymes of ketone body synthesis, HMG-CoA synthase and HMG-CoA lyase, are present in liver (and kidney cortex) but not in other tissues. The key enzyme of ketone body utilization, acetoacetate:succinyl-CoA transferase, is found in many tissues, but not in the liver. These differences ensure that ketone bodies are made in the liver and utilized in other tissues.

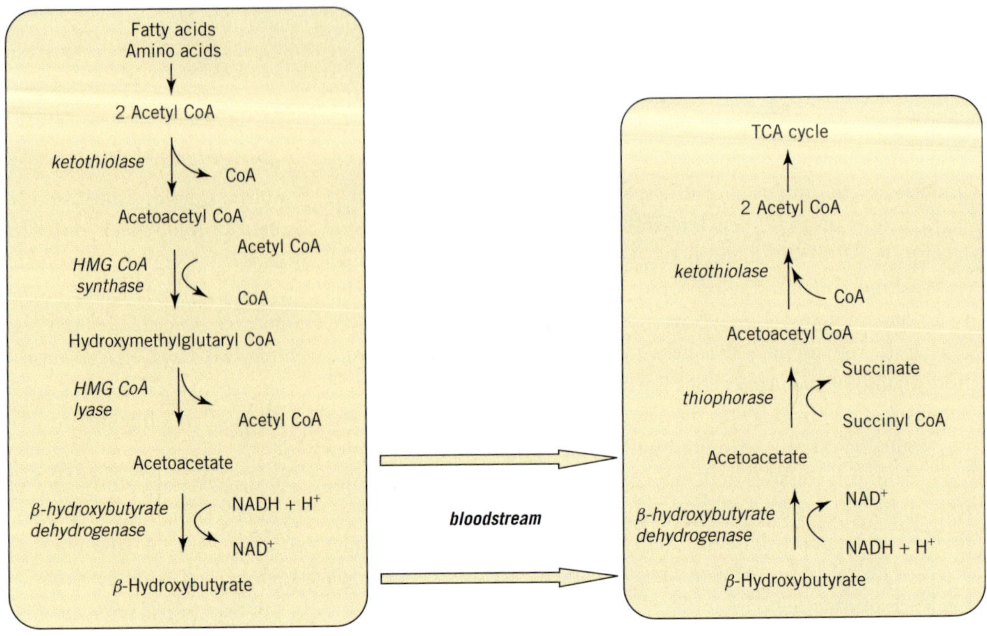

FIGURE 17.26
Ketone body synthesis and utilization.

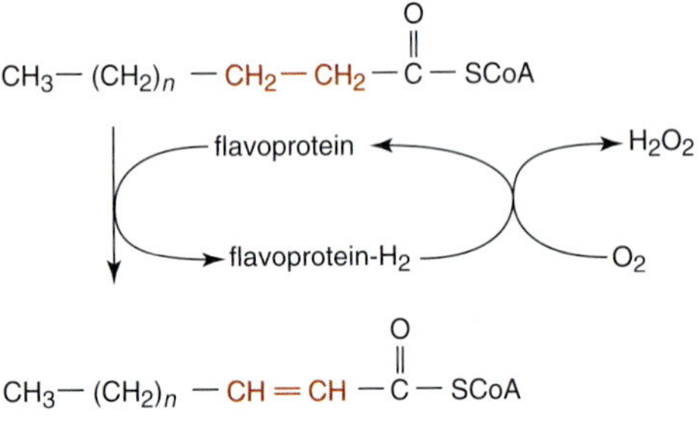

FIGURE 17.27
Initial step in peroxisomal fatty acid oxidation.

Peroxisomal Oxidation of Fatty Acids Serves Many Functions

Although the bulk of fatty acid oxidation occurs in mitochondria, a significant fraction also takes place in the **peroxisomes** of liver, kidney, and other tissues. Peroxisomes are a class of subcellular organelles with distinctive morphological and chemical characteristics (see p. 17). Peroxisomes in the liver contain the enzymes needed for β-oxidation. The mammalian **peroxisomal fatty acid oxidation** pathway is similar to that in plant glyoxysomes, but differs from mitochondrial β-oxidation in three important respects: First, the initial dehydrogenation is accomplished by an oxidase system that uses $\mathbf{O_2}$ and produces $\mathbf{H_2O_2}$ (Figure 17.27). The $\mathbf{H_2O_2}$ is consumed by **catalase**. The remaining steps are the same as in the mitochondrial β-oxidation system. Second, the peroxisomal and mitochondrial enzymes differ in their specificity; the peroxisomal enzymes prefer fatty acids longer than eight carbons. Although rat liver mitochondria carry out the complete oxidation of fatty acyl CoAs to acetyl CoA, β-oxidation in liver peroxisomes will not proceed beyond octanoyl CoA (C_8). Thus, peroxisomes shorten long-chain fatty acids to a point where β-oxidation can be completed in the mitochondria. It is interesting to note that the glitazones, antidiabetic drugs that act by reducing peripheral insulin resistance and used to decrease triacylglycerol levels in patients, cause a marked increase in peroxisomes. Other peroxisomal reactions include chain shortening of dicarboxylic acids, conversion of cholesterol into bile acids, and the formation of ether lipids. Because of these diverse metabolic roles, it is not surprising that the congenital absence of functional peroxisomes, an inherited defect known as Zellweger's syndrome, has such devastating effects (see Clin. Corr. 1.5, p. 19).

17.7 REGULATION OF LIPID METABOLISM

Regulation in the Fed State

The metabolism of lipids in humans is controlled by the dietary status of the individual via a complex set of hormonal signals. After a meal that contains lipid, carbohydrate, and protein, the dietary lipid is deposited as triacylglycerol in adipose tissue. In addition, dietary carbohydrate and amino acids, in excess of that required for energy or for protein synthesis, are converted into fatty acids and deposited in adipose tissue as triacylglycerol.

The major anabolic hormone, insulin, is required for both fatty acid synthesis and for the formation of triacylglycerol in adipose tissue. A summary of these regulations is presented in Tables 17.2, p. 668, and 17.3, p. 672. This hormone acts at two levels; it induces the transcription of genes that code for critical enzymes in the pathways of lipid synthesis and storage (long-term regulation), and it controls processes such as glucose uptake and triacylglycerol hydrolysis (short-term regulation).

Insulin stimulates fatty acid synthesis by increasing the levels of key enzymes, including fatty acid synthase, NADP-malate dehydrogenase (malic enzyme), and acetyl-CoA carboxylase, in the liver by inducing the transcription of their genes. Insulin also stimulates the synthesis of glucose 6-phosphate dehydrogenase and 6-phosphogluconate dehydrogenase, the two enzymes in the oxidative portion of the pentose pathway, which generate some of the NADPH that is required for fatty acid synthesis. The short-term effect of insulin on hepatic fatty acid synthesis is exerted by activating a specific **phosphoprotein phosphatase**, which removes phosphate from acetyl-CoA carboxylase, thereby activating this enzyme. Increased flux through glycolysis is also important in providing acetyl CoA for fatty acid synthesis.

In adipose tissue, insulin is required in the fed state for glucose uptake via the GLUT 4 transporter. The metabolism of this glucose via glycolysis provides glycerol 3-phosphate for the synthesis of triacylglycerol. Insulin also blocks the breakdown of triacylglycerols by inhibiting lipolysis, thereby preventing a futile cycle. As in the liver, insulin exerts its short-term effects by activating phosphoprotein phosphatases. This decreases the phosphorylation of key proteins, including hormone sensitive lipase and perilipin, leading to decreased breakdown of triacylglycerols.

Regulation in the Fasted State

Fasting results in a dramatic alteration in lipid metabolism. As the concentration of glucose in the blood decreases, there is a parallel decrease in the concentration of insulin in the circulation. There is also an increase in epinephrine and glucagon, which elevates the level of cAMP and activates protein kinase A. In adipose tissue, there is increased phosphorylation of hormone sensitive lipase and perilipin, resulting in an increase in triacylglycerol breakdown and the release of free fatty acids and glycerol from this tissue (see Table 17.2, p. 668, for a summary of these controls).

In the liver, these hormonal changes lead to a decrease in fatty acid synthesis, due to reduction in the levels of the key enzymes (see Table 17.3, p. 672). There is also inhibition of the rate limiting enzyme, **acetyl-CoA carboxylase**, due to cAMP-dependent phosphorylation of the enzyme. Glycolysis is also inhibited, which decreases the supply of acetyl CoA for lipogenesis. The liver begins to produce ketone bodies as fasting progresses, due to an increase in the rate of fatty acid oxidation and increased levels of enzymes of ketone body synthesis. During prolonged fasting, about half of the fatty acids that enter the liver are converted to ketone bodies and released into the blood for utilization by tissues such as the muscle, heart, and (after 2 days of fasting) the brain, thereby sparing the use of glucose.

Regulation of Fatty Acid Oxidation

The rate of fatty acid oxidation in mitochondria is controlled by regulating the entry of substrate into this organelle. The key enzyme is **carnitine palmitoyltransferase I** (CPT I), which synthesizes acyl carnitine from cytosolic acyl CoA (Figure 17.20). In the liver, acetyl-CoA carboxylase is activated in the fed state, because enzyme levels are high, cAMP-dependent phosphorylation is low and the enzyme is activated by citrate. The resulting high concentration of malonyl CoA stimulates fatty acid synthesis, but blocks fatty acid oxidation by inhibiting CPT I. This regulation prevents a futile cycle. Conversely, in the fasted state, the activity of acetyl-CoA carboxylase in the liver is low because enzyme levels are low, the enzyme is phosphorylated, and citrate is also decreased. CPT I is active and fatty acid oxidation occurs at a high rate under these conditions because of the low levels of malonyl CoA.

Fatty acid oxidation in muscle is also regulated by malonyl CoA, even though this tissue does not synthesize fatty acids. Muscle contains an isozyme of acetyl-CoA carboxylase, which produces malonyl CoA solely for the regulation CPT I. The enzyme is activated by citrate and inhibited by phosphorylation. It is phosphorylated by both protein kinase A and an AMP-dependent kinase. Phosphorylation by the former enzyme allows fatty acid oxidation to be regulated by dietary status. In the fed state, the high concentration of insulin results in low levels of phosphorylation. The enzyme produces malonyl CoA, which inhibits CPT I and blocks fatty acid oxidation. Conversely, in the fasted state, the high concentration of cAMP stimulates phosphorylation of acetyl-CoA carboxylase, resulting in its inhibition. As a consequence, CPT I facilitates fatty acid entry into the mitochondria for oxidation.

The second kinase, which is regulated by AMP, links the rate of fatty acid oxidation to the energy status of the muscle. In a resting muscle, AMP levels are low. As a result, the AMP-dependent kinase is inactive, acetyl-CoA carboxylase is active, and the malonyl CoA that is generated inhibits CPT I and fatty acid oxidation. In an exercising muscle, high levels of AMP activate the protein kinase. The resulting inhibition of acetyl-CoA carboxylase results in low levels of malonyl CoA and the activation of both CPT I and fatty acid oxidation.

Fatty Acids as Regulatory Molecules

Fatty acids are themselves regulatory molecules in the liver, muscle, and adipose tissue. In muscle, long-chain acyl CoAs activate protein kinase C, as well as the transcription factors PPARγ and NFκB, which in turn inhibit insulin action by blocking the activation of intermediates in the insulin signaling pathway. These effects of fatty acids have dramatic consequences in the regulation of carbohydrate metabolism in muscle (see Clin. Corr. 17.2).

BIBLIOGRAPHY

Eaton, S., Bartlett, K., and Pourfarzam, M. Mammalian mitochondrial beta-oxidation. *Biochem. J.* 320(Pt 2):345, 1996.

Foster, D. W., and McGarry, J. D. Acute complications of diabetes: Ketogenesis, hyperosmolar coma, lactic acidosis. In: L. J. DeGroot (Ed.), *Endocrinology*, Vol. 2, 3rd ed. Philadelphia: Saunders, 1995, p. 1506.

Hanson, R. W. and Reshef, L. Glyceroneogenesis revisited. *Biochimie* 85:1199, 2003.

Hardie, D. G., Carling, D., and Carlson, M. The AMP-activated/SNF1 protein kinase subfamily: Metabolic sensors of the eukaryotic cell? *Annu. Rev. Biochem.* 67:821, 1998.

Hashimoto, T. Peroxisomal β-oxidation enzymes. *Neurochem. Res.* 24:551, 1999.

Hillgartner, F. B., Salati, L. M., and Goodridge, A. G. Physiological and molecular mechanisms involved in nutritional regulation of fatty acid synthesis. *Physiol. Rev.* 75:47, 1995.

Itani, S. I., Ruderman, N. B., Schmieder, F., and Boden, G. Lipid-induced insulin resistance in human muscle is associated with changes in diacylglycerol, protein kinase C, and IkappaB-alpha. *Diabetes* 51:2005, 2002.

Jump, D. B. The biochemistry of n-3 polyunsaturated fatty acids. *J. Biol. Chem.* 277:8755, 2002.

Kane, J. P. and Havel, R. J. Disorders of the biogenesis and secretion of lipoproteins containing the B apolipoproteins. In: C. R. Scriver, A. L. Beaudet, W. S. Sly, and D. Valle (Eds.), *The Metabolic and Molecular Bases of Inherited Disease*, Vol. II, 8th ed. New York: McGraw-Hill, 2001, p. 2717.

Kent, C. Eukaryotic phospholipid biosynthesis. *Annu. Rev. Biochem.* 64:315, 1995.

Kim, K. H. Regulation of mammalian acetyl-coenzyme A carboxylase. *Annu. Rev. Nutr.* 17:77, 1997.

McGarry, J. D., and Brown, N. F. The mitochondrial carnitine palmitoyltransferase system. From concept to molecular analysis. *Eur. J. Biochem.* 244:1, 1997.

Mitchell, G. A. and Fukao, T. Inborn errors of ketone body metabolism. In: C. R. Scriver, A. L. Beaudet, W. S. Sly, and D. Valle (Eds.), *The Metabolic and Molecular Bases of Inherited Disease*, Vol. II, 8th ed. New York: McGraw-Hill, 2001, p. 2327.

Nakamura, M. T., and Nara, T. Y. Structure, function, and dietary regulation of delta6, delta5, and delta9 desaturases. *Annu. Rev. Nutr.* 24:345, 2004.

Nilsson-Ehle, P., Garfinkel, A. S., and Schotz, M. C. Lipolytic enzymes and plasma lipoprotein metabolism. *Annu. Rev. Biochem.* 49:667, 1980.

Owen, O. E. and Hanson, R. W. Ketone bodies. *Encycl. Endocr. Dis.* 3:125, 2004.

Rinaldo, P., Matern, D., and Bennett, M. J. Fatty acid oxidation disorders. *Annu. Rev. Physiol.* 64:477, 2002.

Ruderman, N. B., Saha, A. K., Vavvas, D., Witters, L. A. Malonyl-CoA, fuel sensing, and insulin resistance. *Am. J. Physiol.* 276:E1, 1999.

Sampath, H. and Ntambi, J. M. Polyunsaturated fatty acid regulation of gene expression. *Nutr. Rev.* 62:333, 2004.

Vance, D. E. and Vance, J. E. *Biochemistry of Lipids, Lipoproteins and Membranes*, 2nd ed. Amsterdam: Elsevier, 1996.

Wakil, S. J. Fatty acid synthase, a proficient multifunctional enzyme. *Biochemistry* 28:4523 1989.

Wakil, S. J., Stoops, J. K., and Joshi, V. C. Fatty acid synthesis and its regulation. *Annu. Rev. Biochem.* 52:537, 1983.

Wanders, R. J. A., Vreken, P., Den Boer, M. E. J., Wijburg, F. A., Van Gennip, A. H., and Ijist, L. Disorders of mitochondrial fatty acyl-CoA β-oxidation. *J. Inher. Metab. Dis.* 22:442, 1999.

Willson, T. M., Lambert, M. H., and Kliewer, S. A. Peroxisome proliferator-activated receptor γ and metabolic disease. *Annu. Rev. Biochem.* 70:341, 2001.

Yeaman, S. J. Hormone-sensitive lipase—New roles for an old enzyme. *Biochem. J.* 379:11, 2004.

QUESTIONS | CAROL N. ANGSTADT

1. All of the following statements about acetyl-CoA carboxylase are correct *except*:
   A. it catalyzes the rate-limiting step of fatty acid synthesis.
   B. it requires biotin.
   C. it is inhibited by cAMP-mediated phosphorylation.
   D. it is activated by palmitoyl CoA.
   E. its content in a cell responds to changes in fat content in the diet.

2. During the synthesis of palmitate in liver cells:
   A. the addition of malonyl CoA to fatty acid synthase elongates the growing chain by three carbon atoms.
   B. a β-keto acyl residue on the 4′-phosphopantetheine moiety is ultimately reduced to a saturated residue by NADPH.
   C. palmitoyl CoA is released from the synthase.
   D. transfer of the growing chain from ACP to another -SH occurs after addition of the next malonyl CoA.
   E. the first compound to add to fatty acid synthase is malonyl CoA.

3. In humans, desaturation of fatty acids:
   A. occurs primarily in mitochondria.
   B. is catalyzed by an enzyme system that uses NADPH and a cytochrome.
   C. introduces double bonds primarily of trans configuration.
   D. can occur only after palmitate has been elongated to stearic acid.
   E. introduces the first double bond at the methyl end of the molecule.

4. All of the following events are usually involved in the synthesis of triacylglycerols in adipose tissue *except*:
   A. addition of a fatty acyl CoA to a diacylglycerol.
   B. addition of a fatty acyl CoA to a lysophosphatide.
   C. a reaction catalyzed by glycerol kinase.
   D. hydrolysis of phosphatidic acid by a phosphatase.
   E. reduction of dihydroxyacetone phosphate.

5. Lipoprotein lipase:
   A. is an intracellular enzyme.
   B. is stimulated by cAMP-mediated phosphorylation.
   C. functions to mobilize stored triacylglycerols from adipose tissue.
   D. is stimulated by one of the apoproteins present in VLDL.
   E. produces free fatty acids and a monoacylglycerol.

6. The high glucagon/insulin ratio seen in starvation:
   A. promotes mobilization of fatty acids from adipose stores.
   B. stimulates β-oxidation by inhibiting the production of malonyl CoA.
   C. leads to increased concentrations of ketone bodies in the blood.
   D. all of the above.
   E. none of the above.

Questions 7 and 8: Following a severe cold which caused a loss of appetite, a one-year-old boy was hospitalized with hypoglycemia, hyperammonemia, muscle weakness and cardiac irregularities. These symptoms were consistent with a defect in the carnitine transport system. Dietary carnitine therapy was tried unsuccessfully but a diet low in long-chain fatty acids and supplemented with medium chain triacylglycerols was beneficial.

7. A deficiency of carnitine might be expected to interfere with:
   A. β-oxidation.
   B. ketone body formation from acetyl CoA.
   C. palmitate synthesis.
   D. mobilization of stored triacylglycerols from adipose tissue.
   E. uptake of fatty acids into cells from the blood.

8. The child was diagnosed with carnitine–acylcarnitine translocase deficiency. The dietary treatment was beneficial because:
   A. the child could get all required energy from carbohydrate.
   B. the deficiency was in the peroxisomal system so carnitine would not be helpful.
   C. medium-chain fatty acids (8–10 carbons) enter the mitochondria before being converted to their CoA derivatives.
   D. medium chain triacylglycerols contain mostly hydroxylated fatty acids.
   E. medium-chain fatty acids such as C_8 and C_{10} are readily converted into glucose by the liver.

Questions 9 and 10: Medium-chain acyl-CoA dehydrogenase deficiency (MCAD), a defect in β-oxidation, usually produces symptoms within the first two years of life after a period of fasting. Typical symptoms include vomiting, lethargy, and hypoketotic hypoglycemia. Excessive urinary secretion of medium-chain dicarboxylic acids and medium-chain esters of glycine and carnitine help to establish the diagnosis.

9. β-Oxidation of fatty acids:
   A. generates ATP only if acetyl CoA is subsequently oxidized.
   B. is usually suppressed during starvation.
   C. uses only even-chain, saturated fatty acids as substrates.
   D. uses $NADP^+$.
   E. occurs by a repeated sequence of four reactions.

10. The lack of ketone bodies in the presence of low blood glucose in this case is unusual since ketone body concentrations usually increase with fasting-induced hypoglycemia. Ketone bodies:
    A. are formed by removal of CoA from the corresponding intermediate of β-oxidation.
    B. are synthesized from cytoplasmic β-hydroxy-β-methyl glutaryl coenzyme A (HMG-CoA).

C. are synthesized primarily in muscle tissue.
D. include both β-hydroxybutyrate and acetoacetate, the ratio reflecting the intramitochondrial [NADH]/[NAD^+] ratio in liver.
E. form when β-oxidation is interrupted.

Questions 11 and 12: A patient with retinitis pigmentosa, peripheral neuropathy, cerebellar ataxia, and nerve deafness was diagnosed with Refsum's disease. This is a rare genetic disease caused by lack of the peroxisomal α-hydroxylating enzyme. Phytanic acid, a constituent of milk and animal fats and a metabolic product of a constituent of chlorophyll, relies on this system for its catabolism. Treatment for the disease relies on restriction of dairy products and meat products from ruminants.

11. α-Oxidation:
A. is important in the metabolism of branched-chain fatty acids.
B. metabolizes a fatty acid completely to acetyl CoA.
C. produces hydrogen peroxide.
D. prevents the fatty acid from producing energy.
E. requires NADPH.

12. Another minor pathway of fatty acid oxidation, ω-oxidation, also results in a hydroxylation. ω-Oxidation:
A. occurs in mitochondria.
B. introduces the—OH on the carbon adjacent to the carboxyl group.
C. oxidizes primarily very-long-chain fatty acids.
D. oxidizes the terminal methyl group.
E. produces dicarboxylic acids in the initial oxidation.

Problems

13. What role does glyceroneogenesis play in fasting?

14. How does oxidation of a 17-carbon fatty acid (from plants) lead to the production of propionyl CoA? Be specific in your answer.

ANSWERS

1. **D** It is activated by citrate and inhibited by long chain fatty acyl CoAs. C: Since cAMP increases at times when energy is needed, it is consistent that a process that uses energy would be inhibited. E: Long-term control is related to enzyme synthesis and responds appropriately to dietary changes.
2. **B** A: Splitting CO_2 from malonyl CoA is the driving force for the condensation reaction so the chain grows two carbon atoms at a time. C: In mammals, palmitate is released as the free acid; the conversion to the CoA ester is by a different enzyme. D: It is important to realize that only ACP binds the incoming malonyl CoA so it must be freed before another addition can be made. E: Acetyl CoA adds first to form the foundation for the rest of the chain.
3. **B** A: Desaturation occurs in the endoplasmic reticulum. C: Naturally occurring fatty acids are cis. D: Elongation and unsaturation can occur in any order. E: If this were true, we could make linoleic acid.
4. **C** This does not occur to any significant extent in adipose tissue. A, B, D: The sequential addition of fatty acyl CoAs to glycerol 3-phosphate forms lysophosphatidic acid and then forms phosphatidic acid, whose phosphate is removed before the addition of the third fatty acyl residue. E: This is the formation of α-glycerol phosphate in adipose.
5. **D** ApoC-II binds the lipoprotein to the enzyme. A–C: These are characteristics of hormone-sensitive lipase. E: This is pancreatic lipase.
6. **D** A high glucagon/insulin ratio results in cAMP-mediated phosphorylations that activate hormone-sensitive lipase in adipose tissue and inhibit acetyl CoA carboxylase in liver, leading to mobilization of fatty acids from adipose tissue and decreased malonyl CoA in liver. The fall in malonyl CoA causes derepression of CPT I activity, allowing accelerated production of acetyl CoA in the mitochondria, and ultimately ketone bodies, from the mobilized fatty acids.
7. **A** Carnitine functions in transport of fatty acyl CoA esters formed in cytosol into the mitochondria. B: Acetyl CoA for ketone bodies comes from sources in addition to fatty acids. C: Fatty acid synthesis is a cytosolic process. D: Mobilization is under hormonal control. E: Fatty acids cross the plasma membrane.
8. **C** Because medium-chain fatty acids cross the mitochondrial membrane directly, they do not require the carnitine system. A: This is never true. B: Peroxisomal oxidation does not require carnitine, but this is not a peroxisomal system. D: Hydroxylated fatty acids are not a common constituent of triacylglycerols. E: Liver cannot synthesize glucose from fatty acids with an even number of carbon atoms.
9. **E** A, D: It is important to realize that β-oxidation, itself, generates $FADH_2$ and NADH, which can be reoxidized to generate ATP. B: Fatty acid oxidation is usually enhanced during fasting. C: β-Oxidation is a general process requiring only minor modifications to oxidize nearly any fatty acid in the cell.
10. **D** A, E: β-Oxidation proceeds to completion; ketone bodies are formed by a separate process. B, C: Ketone bodies are formed, but not used, in liver mitochondria; cytosolic HMG-CoA is a precursor of cholesterol. Ketone bodies are not readily synthesized by muscle.
11. **A** The presence of a methyl group precludes the β-oxidation process at that point. B: Once past the methyl group, β-oxidation can proceed. C: The reaction is by a hydroxylase that adds an —OH. D: During the β-oxidation phases, energy is produced. E: This hydroxylase uses ascorbate as reducing agent.
12. **D** This is why it is called ω-oxidation. A: Occurs in endoplasmic reticulum. B: This is α-oxidation. C: Medium-chain fatty acids are preferred substrates. E: The hydroxyl group requires alcohol and aldehyde dehydrogenases to convert it to the acid.
13. In fasting, the quantity of fatty acids released by adipose tissue is greater than the need for them, so much is re-deposited as triacylglycerols in adipose tissue. Adipose tissue cannot use glucose to synthesize glycerol 3-phosphate in fasting since insulin is low. "Neo" refers to new synthesis. The source of carbons is substrates that enter the TCA cycle and lead to a net synthesis of malate that moves to the cytosol. The key enzyme is phosphoenolpyruvate carboxylase, which is induced in fasting.
14. β-Oxidation proceeds normally, but the final thiolase cleavage yields acetyl CoA and propionyl CoA. Propionyl CoA is not a substrate for SCAD (short-chain acyl CoA dehydrogenase), so β-oxidation terminates.

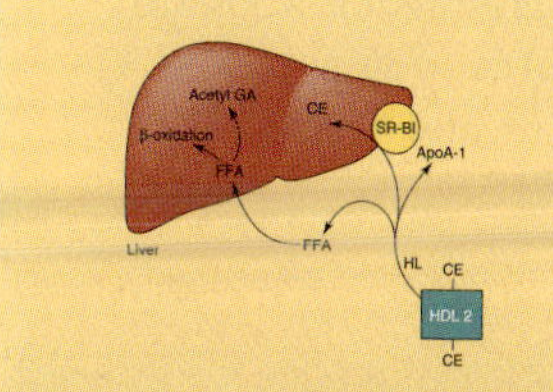

18

LIPID METABOLISM II: PATHWAYS OF METABOLISM OF SPECIAL LIPIDS

Robert H. Glew

Textbook of Biochemistry With Clinical Correlations, Sixth Edition, Edited by Thomas M. Devlin

18.1 OVERVIEW

Lipid is a general term that describes substances that are relatively water-insoluble and extractable by nonpolar solvents. Complex lipids of humans fall into two broad categories: nonpolar lipids, such as triacylglycerols and cholesteryl esters, and polar lipids, which are amphipathic since they contain a hydrophobic region and a hydrophilic region in the same molecule. This chapter discusses polar lipids including phospholipids, sphingolipids, and eicosanoids. The hydrophobic and hydrophilic regions are bridged by a glycerol moiety in glycerophospholipids and by sphingosine in sphingomyelin and glycosphingolipids. Triacylglycerol is confined largely to storage sites in adipose tissue, whereas polar lipids occur primarily in cellular membranes. Membranes generally contain 40% of their dry weight as lipid and 60% as protein.

Cell–cell recognition, phagocytosis, contact inhibition, and rejection of transplanted tissues and organs are all phenomena of medical significance that involve highly specific recognition sites on the surface of plasma membranes. Glycosphingolipids appear to play a role in these biological events. Their synthesis will be described. Various sphingolipids accumulate in liver, spleen, kidney, or nervous tissue in certain genetic disorders called sphingolipidoses. Glycolipids are worthy of study because ABO antigenic determinants of blood groups are primarily glycolipid in nature.

The pathway of cholesterol biosynthesis and its regulation, how cholesterol functions as a precursor to bile salts and steroid hormones, and the role of high-density lipoprotein (HDL) and lecithin:cholesterol acyltransferase (LCAT) in management of plasma cholesterol are described. Finally, the metabolism and function of two pharmacologically powerful classes of hormones derived from arachidonic acid—namely, prostaglandins and leukotrienes—will be discussed. See the Appendix for a discussion of nomenclature and chemistry of lipids.

18.2 PHOSPHOLIPIDS

Two major classes of **acylglycerolipids** are **triacylglycerols** and **glycerophospholipids**, which have as their core the C_3 polyol, glycerol. Two primary alcohol groups of glycerol are not stereochemically identical; in the case of phospholipids, it is usually the same hydroxyl group that is esterified to the phosphate residue. The stereospecific numbering system is the best way to designate different hydroxyl groups. In this system, when the structure of glycerol is drawn in the Fischer projection with the C2 hydroxyl group projecting to the left of the page, the carbon atoms are numbered as shown in Figure 18.1. When the **stereospecific numbering** (*sn*) system is employed, the prefix *sn*- is used before the name of the compound. Glycerophospholipids usually contain an *sn*-glycerol 3-phosphate moiety. Although each contains the glycerol moiety as a fundamental structural element, neutral triacylglycerols and charged ionic phospholipids have very different physical properties and functions.

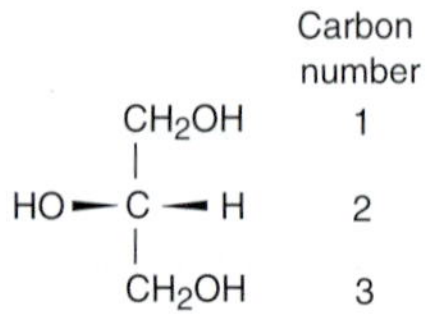

Structure	Carbon number
CH_2OH	1
HO—C—H	2
CH_2OH	3

FIGURE 18.1
Stereospecific numbering of glycerol.

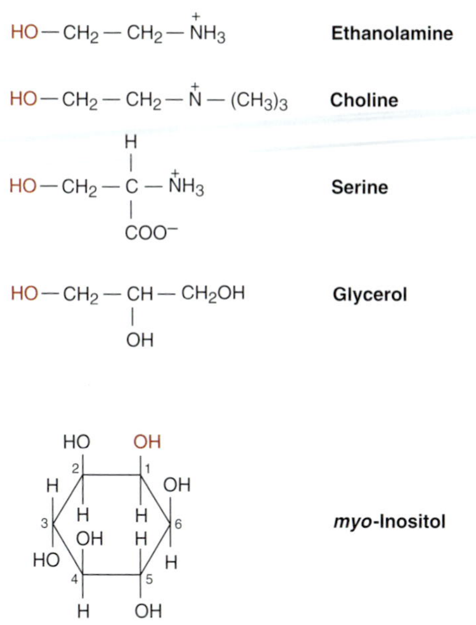

$HO{-}CH_2{-}CH_2{-}\overset{+}{N}H_3$ **Ethanolamine**

$HO{-}CH_2{-}CH_2{-}\overset{+}{N}{-}(CH_3)_3$ **Choline**

$HO{-}CH_2{-}CH(COO^-){-}\overset{+}{N}H_3$ **Serine**

$HO{-}CH_2{-}CH(OH){-}CH_2OH$ **Glycerol**

myo-**Inositol**

FIGURE 18.2

Structure of some common polar groups of phospholipids.

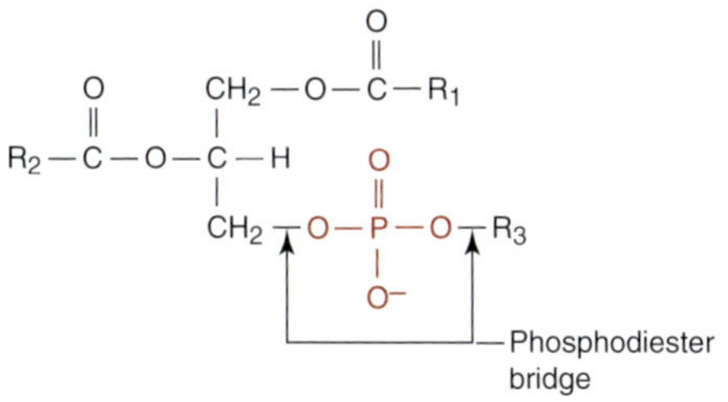

Phosphodiester bridge

FIGURE 18.3

Generalized structure of a phospholipid where R1 and R2 represent the aliphatic chains of fatty acids, and R3 represents a polar group.

Phospholipids Contain Phosphatidic Acid Linked to a Base

Phospholipids are polar, ionic lipids composed of 1,2-diacylglycerol and a **phosphodiester bridge** that links the glycerol backbone to some base, usually a nitrogenous one, such as choline, serine, or ethanolamine (Figures 18.2 and 18.3). The most abundant phospholipids in human tissues are **phosphatidylcholine** (also called lecithin), **phosphatidylethanolamine**, and **phosphatidylserine** (Figure 18.4). At physiologic pH, phosphatidylcholine and phosphatidylethanolamine have no net charge and exist as dipolar zwitterions, whereas phosphatidylserine has a net charge of -1, causing it to be an acidic phospholipid. Phosphatidylethanolamine (PE) is related to phosphatidylcholine in that trimethylation of PE produces lecithin. Most phospholipids contain more than one kind of fatty acid per molecule, so that a given class of phospholipids from any tissue actually represents a family of molecular species. Phosphatidylcholine (PC) contains mostly palmitic acid (16:0) or stearic acid (18:0) in the *sn*-1 position and primarily unsaturated 18-carbon fatty acids oleic, linoleic, or linolenic in the *sn*-2 position. Phosphatidylethanolamine has the same saturated fatty acids as PC at the *sn*-1 position but contains more of the long-chain polyunsaturated fatty acids—namely, linoleic acid (18:2), arachidonic acid (20:4), and docosahexaenoic acid (22:6)—at the *sn*-2 position.

Phosphatidylinositol, an acidic phospholipid that occurs in mammalian membranes (Figure 18.5), is rather unusual because it often contains almost exclusively stearic acid (18:0) in the *sn*-1 position and arachidonic acid in the *sn*-2 position.

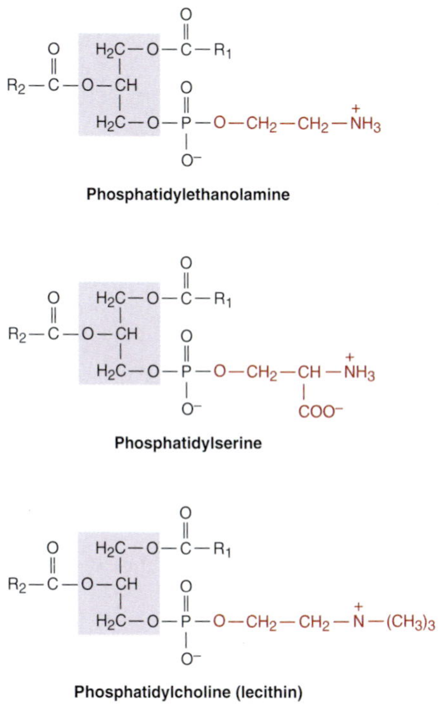

Phosphatidylethanolamine

Phosphatidylserine

Phosphatidylcholine (lecithin)

FIGURE 18.4

Structures of some common phospholipids.

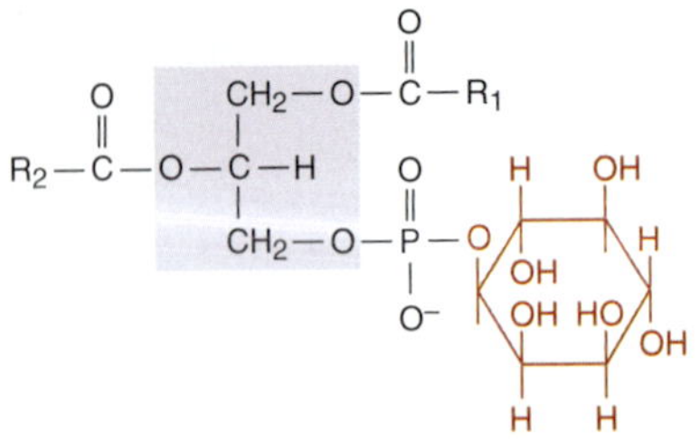

Phosphatidylinositol

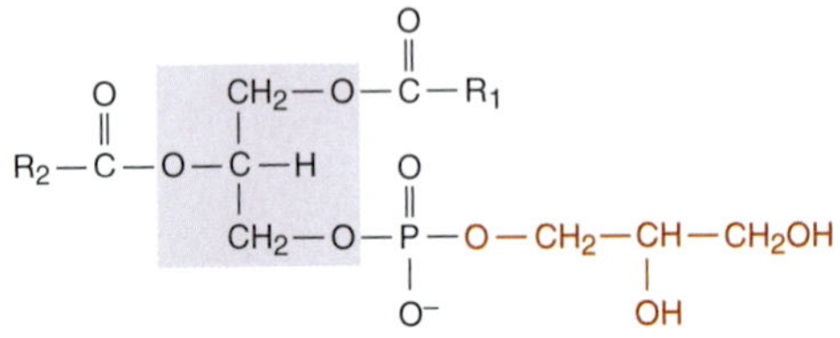

Phosphatidylglycerol

FIGURE 18.5
Structures of phosphatidylglycerol and phosphatidylinositol.

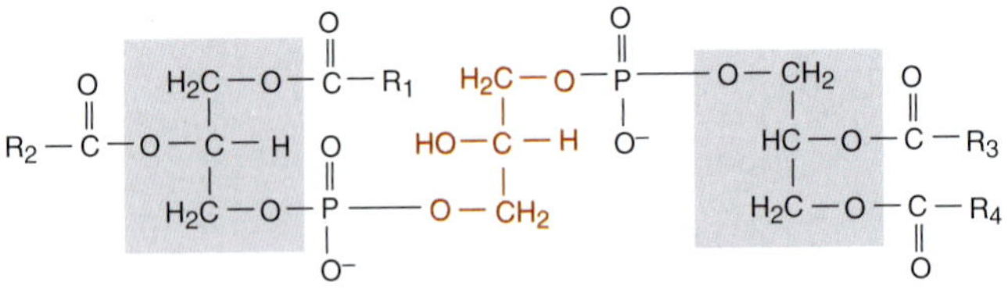

FIGURE 18.6
Structure of cardiolipin.

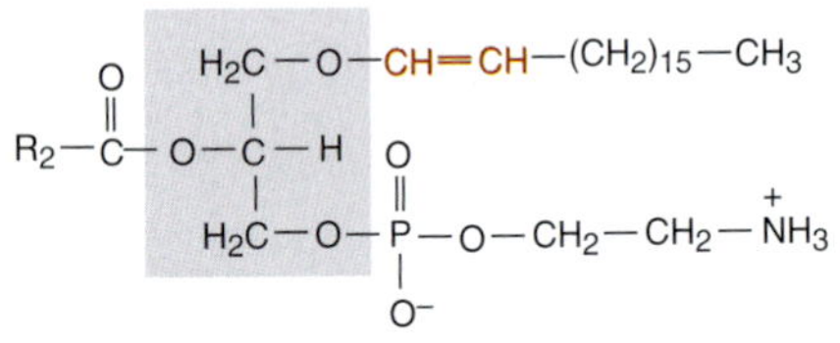

FIGURE 18.7
Structure of ethanolamine plasmalogen.

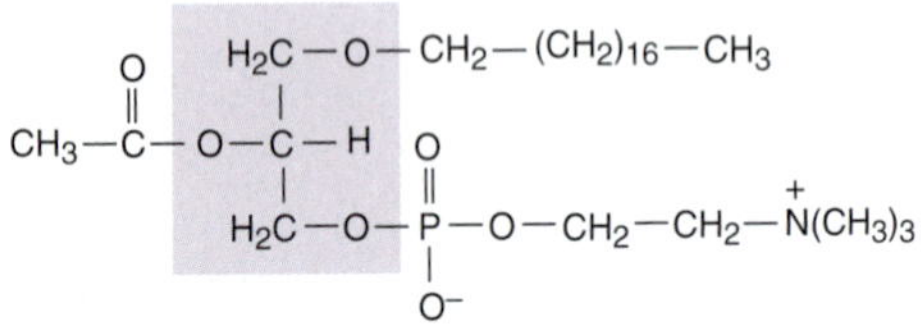

FIGURE 18.8
Structure of platelet activating factor (PAF).

Another phospholipid comprised of a polyol polar head group is phosphatidylglycerol (Figure 18.5), which occurs in relatively large amounts in mitochondrial membranes and pulmonary surfactant and is a precursor of cardiolipin. Phosphatidylglycerol and phosphatidylinositol both carry a formal charge of −1 at neutral pH and are therefore acidic lipids. **Cardiolipin**, a very acidic (charge, −2) phospholipid, is composed of two molecules of phosphatidic acid linked together covalently through a molecule of glycerol (Figure 18.6). It occurs primarily in the inner membrane of mitochondria and in bacterial membranes.

Phospholipids mentioned so far contain only *O*-acyl residues attached to glycerol. *O*-(1-alkenyl) substituents occur at C1 of the *sn*-glycerol in phosphoglycerides in combination with an *O*-acyl residue esterified to the C2 position; compounds in this class are known as **plasmalogens** (Figure 18.7). Relatively large amounts of ethanolamine plasmalogen (also called plasmenylethanolamine) occur in myelin with lesser amounts in heart muscle where choline plasmalogen is abundant.

An unusual phospholipid called **"platelet activating factor"** (PAF) (Figure 18.8) is a major mediator of hypersensitivity, acute inflammatory reactions, allergic responses and anaphylactic shock. In hypersensitive individuals, cells of the polymorphonuclear (PMN) leukocyte family (basophils, neutrophils, and eosinophils), macrophages, and monocytes are coated with IgE molecules that are specific for a particular antigen (e.g., ragweed pollen and bee venom). Subsequent reexposure to the antigen and formation of antigen-IgE complexes on the surface of the aforementioned inflammatory cells provokes synthesis and release of PAF. Platelet activating factor contains an *O*-alkyl moiety at *sn*-1 and an acetyl residue instead of a long-chain fatty acid in position 2 of the glycerol moiety. PAF is not stored; it is synthesized and released when PMN cells are stimulated. Platelet aggregation, cardiovascular and pulmonary changes, edema, hypotension, and PMN cell chemotaxis are affected by PAF.

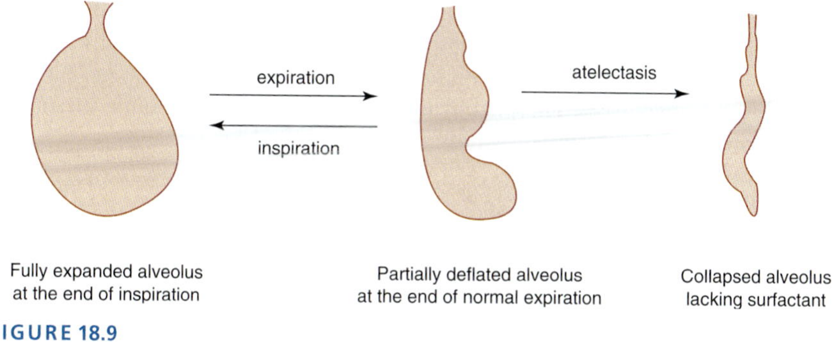

FIGURE 18.9

Role of surfactant in preventing atelectasis.

Phospholipids in Membranes Serve a Variety of Roles

Although present in body fluids such as plasma and bile, phospholipids are present in highest concentration in cellular membranes where they serve as structural and functional components. Nearly one-half the mass of the erythrocyte membrane is comprised of various phospholipids (see p. 445). They also activate certain enzymes; as an example, β-hydroxybutyrate dehydrogenase, in the inner membrane of mitochondria (see p. 687), has an absolute requirement for phosphatidylcholine; phosphatidylserine and phosphatidylethanolamine cannot substitute.

Dipalmitoyllecithin Is Necessary for Normal Lung Function

Normal lung function depends on a constant supply of **dipalmitoyllecithin** in which the molecule contains palmitic acid (16:0) residues in the *sn*-1 and *sn*-2 positions. More than 80% of the phospholipid in the extracellular fluid layer that lines alveoli of normal lungs is dipalmitoyllecithin. This **surfactant** is produced by type II epithelial cells, and it prevents atelectasis at the end of the expiration phase of breathing (Figure 18.9). It decreases the surface tension of the fluid layer of the lung. Lecithin molecules that do not contain two residues of palmitic acid are not effective in the lowering of surface tension. Surfactant also contains phosphatidylglycerol, phosphatidylinositol, and 18- and 36-kDa proteins (designated surfactant proteins), which contribute significantly to the lowering of the surface tension. Surfactant proteins alter the molecular structure of the phospholipid film secreted by type II pneumocytes in such a way as to stabilize the film and maintain its flexibility. Before the 28th week of gestation, fetal lung synthesizes primarily sphingomyelin. Normally, at this time, glycogen that has been stored in epithelial type II cells is converted to fatty acids and then to dipalmitoyllecithin. During lung maturation there is a good correlation between the increase in intracellular lamellar inclusion bodies that represent phosphatidylcholine storage organelles, called lamellar bodies, and the decrease in glycogen content of these cells. At the 24th week of gestation the type II granular pneumocytes appear in the alveolar epithelium and start to produce lamellar bodies. Their number increases until the 32nd week, when surfactant appears in the lung and amniotic fluid. In the few weeks before term, screening tests on amniotic fluid can detect newborns that are at risk for respiratory distress syndrome (RDS) (Clin. Corr. 18.1). These are useful in timing elective deliveries, in determining if the mother should receive a glucocorticoid drug to accelerate maturation of the fetal lung, or to apply preventive therapy to the newborn infant. Dexamethasone has also been used in neonates with chronic lung disease (bronchopulmonary dysplasia). While corticosteroid therapy may be effective in some cases in improving lung function, in others it causes periventricular abnormalities in the brain.

Respiratory failure due to surfactant insufficiency can also occur in adults whose type II cells have been destroyed as an adverse side effect of immunosuppressive medications or chemotherapeutic drugs.

The detergent properties of phospholipids, especially phosphatidylcholine, are important in bile to aid in solubilizing cholesterol. An impaired production and

CLINICAL CORRELATION 18.1

Respiratory Distress Syndrome

Respiratory distress syndrome (RDS) is a major cause of neonatal morbidity and mortality in many countries. It accounts for about 15–20% of all neonatal deaths in Western countries and somewhat less in developing countries. The disease affects only premature babies and the incidence varies directly with the degree of prematurity. Premature babies develop RDS because of immaturity of their lungs from deficiency of pulmonary surfactant. The maturity of the fetal lung can be assessed from the lecithin/sphingomyelin (L/S) ratio in amniotic fluid. The mean L/S ratio in normal pregnancies increases gradually with gestation until about 31 or 32 weeks when the slope rises sharply. The ratio of 2.0 is characteristic of term birth, and it is achieved at the gestational age of about 34 weeks. At pulmonary maturity, the critical L/S ratio is 2.0 or greater. The risk of developing RDS when the L/S ratio is 1.5–1.9 is about 40%, and for less than 1.5 about 75%. Although the L/S ratio is still widely used to predict the risk of RDS, the results are unreliable if the amniotic fluid has been contaminated by blood or meconium. Determination of saturated palmitoyl phosphatidylcholine (SPC), phosphatidylglycerol, and phosphatidylinositol is also predictive of the risk of RDS. Replacement therapy using surfactant from human and animal lungs is effective in the prevention and treatment of RDS.

Source: Merritt, T. A., Hallman, M., Bloom, B.T. et al., Prophylactic treatment of very premature infants with human surfactant. *N. Engl. J. Med.* 315:785, 1986. Simon, N. V., Williams, G. H., Fairbrother, P. F., Elser, R. C., and Perkins, R. P. Prediction of fetal lung maturity by amniotic fluid fluorescence polarization, L/S ratio, and phosphatidylglycerol. *Obstet. Gynecol.* 57:295, 1981.

secretion of phospholipids into bile can lead to formation of cholesterol stones and bile pigment gallstones. Membrane phospholipids are a reservoir for lipid mediators that regulate many metabolic pathways and processes. Release of these mediators is catalyzed by phospholipases. Phosphatidylinositol and phosphatidylcholine are sources of arachidonic acid for synthesis of prostaglandins, thromboxanes, leukotrienes, and related compounds.

Inositides Are Important in Membrane Function

Inositol-containing phospholipids (inositides), especially phosphatidylinositol 4,5-bisphosphate (PIP_2) (Figure 18.10), play a central role in signal transduction systems. When certain hormones bind to their receptors (see p. 522), PIP_2 in the inner leaflet of the membrane is hydrolyzed by phosphoinositidase C (PIC) into inositol 1,4,5-trisphosphate (IP_3), which triggers release of Ca^{2+} from the endoplasmic reticulum, and by 1,2-diacylglycerol (DAG), which increases the activity of protein kinase C (PKC) (Figure 18.11). Removal of the 5-phosphate of IP_3 abolishes the signal and the intracellular Ca^{2+} concentration declines. The 1,2-diacylglycerol is converted to phosphatidic acid by diacylglycerol kinase (Figure 18.12). Phosphatidic acid, a product of phospholipase D action on phospholipids, has also been implicated as a second messenger. These pathways of inositol phosphate metabolism serve in (1) removal and inactivation of IP_3, (2) conservation of inositol, and (3) synthesis of polyphosphates such as inositol pentakisphosphate ($InsP_5$) and inositol hexakisphosphate ($InsP_6$), whose functions have not been determined. IP_3 is metabolized by 5-phosphomonoesterase to inositol 1,4-bisphosphate and by a 3-kinase that forms inositol 1,3,4,5-tetrakisphosphate. A family of phosphatases converts $Ins(1,4)P_2$ to myoinositol (Figure 18.12), which then enters the phospholipid pool.

Besides being a component of membranes and source of arachidonic acid for prostaglandin and leukotriene synthesis (see p. 730), phosphatidylinositol serves to anchor (GPI anchor) certain glycoproteins to the external surface (see p. 457). A major medical problem concerns trypanosomal parasites (e.g., *Trypanosoma brucei*, which causes sleeping sickness). This parasite has resisted immunological approaches to treatment by changing its surface antigens. The external surface of the plasma membrane is coated with a protein called variable surface glycoprotein (VSG), which is linked to the membrane through a phosphatidylinositol anchor. C-type phospholipase on the cell surface permits the shedding of the anchored protein; this allows trypanosomes to

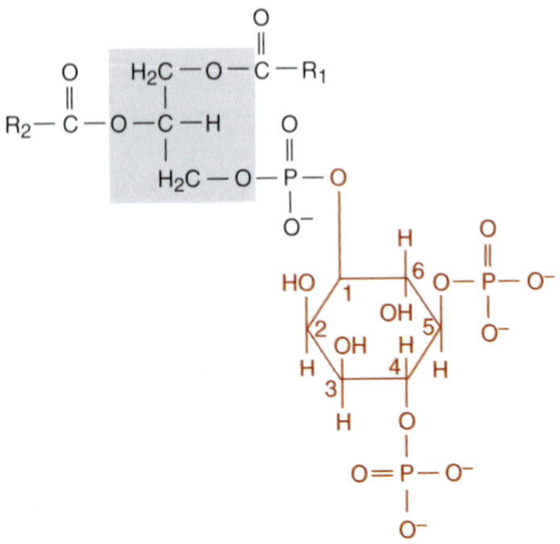

FIGURE 18.10

Structure of phosphatidylinositol 4,5-bisphosphate (PIP_2 or PtdIns $(4,5)P_2$).

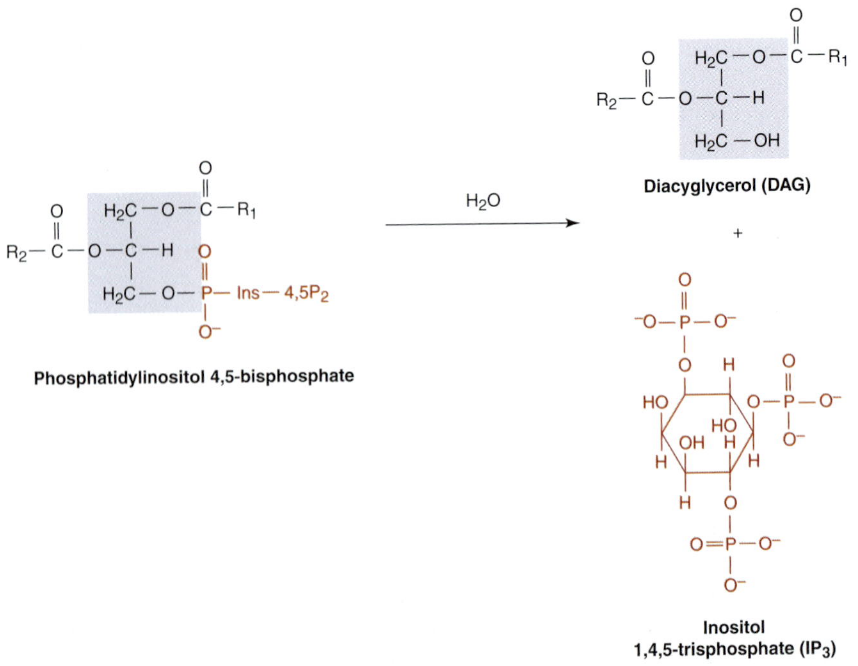

Phosphatidylinositol 4,5-bisphosphate

H_2O

Diacyglycerol (DAG)

+

Inositol 1,4,5-trisphosphate (IP_3)

FIGURE 18.11

Generation of 1,2-diacylglycerol and inositol 1,4,5-trisphosphate by action of phospholipase C on phosphatidylinositol 4,5-bisphosphate.

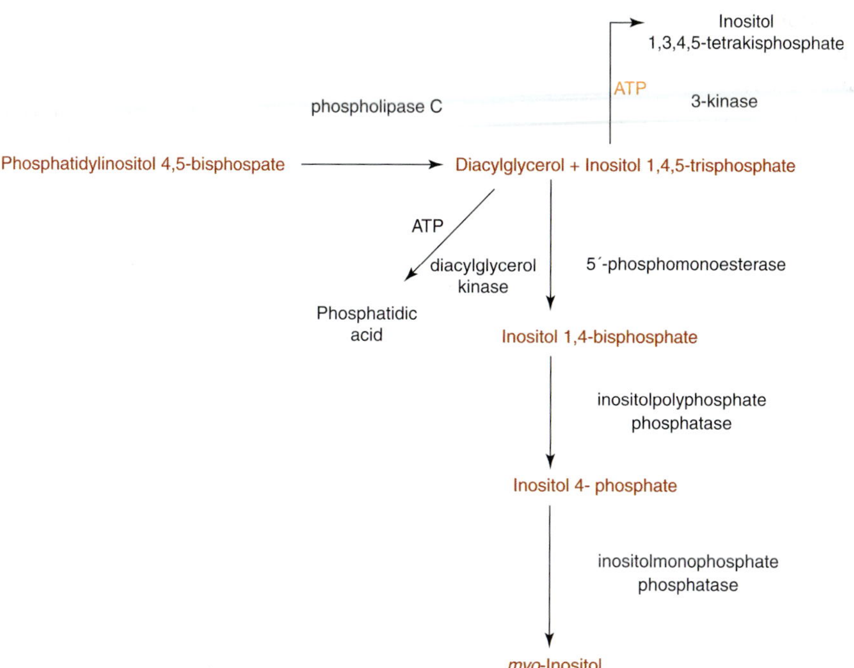

FIGURE 18.12
Pathways for the removal of intracellular inositol 1,4,5-trisphosphate and diacylglycerol.

discard surface antigens, thus changing their coat and escaping antibodies of the host's immune system.

Synthesis of Phospholipids

Phosphatidic Acid Is Synthesized from α-Glycerophosphate and Fatty Acyl CoA

l-α-Phosphatidic acid (commonly called **phosphatidic acid**) and 1,2-diacyl-*sn*-glycerol are intermediates common to the pathways of phospholipid and triacylglycerol synthesis (Figure 18.13) (see p. 677). All cells synthesize phospholipids to some degree (except mature erythrocytes), whereas triacylglycerol biosynthesis occurs only in liver, adipose tissue, and intestine. In most tissues, the pathway for phosphatidic acid synthesis begins with **α-glycerol 3-phosphate** (*sn*-glycerol 3-phosphate). The most general source of α-glycerol 3-phosphate, particularly in adipose tissue, is from reduction of the glycolytic intermediate, **dihydroxyacetone phosphate**, by α-glycerol 3-phosphate dehydrogenase:

$$\text{Dihydroxyacetone phosphate} + \text{NADH} + \text{H}^+ \leftrightarrows \alpha\text{-glycerol 3-phosphate} + \text{NAD}^+$$

Liver and kidney, derive α-glycerol 3-phosphate by means of the **glycerol kinase** reaction:

$$\text{Glycerol} + \text{ATP} \xrightarrow{\text{Mg}^{2+}} \alpha\text{-glycerol 3-phosphate} + \text{ADP}$$

Phosphatidic acid synthesis starts with glycerol phosphate:acyltransferase attaching predominantly saturated fatty acids or oleic acid to α-glycerol 3-phosphate to produce 1-acylglycerol phosphate or α-lysophosphatidic acid. 1-Acylglycerol phosphate:acyltransferase then acylates the *sn*-2 position, usually with an unsaturated fatty acid (Figure 18.13). The donors of acyl groups are the CoA derivatives of the appropriate fatty acids.

The specificity of these acyltransferases does not always match the fatty acid asymmetry of the membrane phospholipids of a particular cell. Remodeling reactions, discussed below, modify the composition at C1 and C2 of the glycerol phosphate backbone. Cytosolic phosphatidic acid phosphatase (phosphatidic acid phosphohydrolase) hydrolyzes phosphatidic acid that is generated on the endoplasmic reticulum, thereby yielding 1,2-diacyl-*sn*-glycerol that serves as the branch point in triacylglycerol and

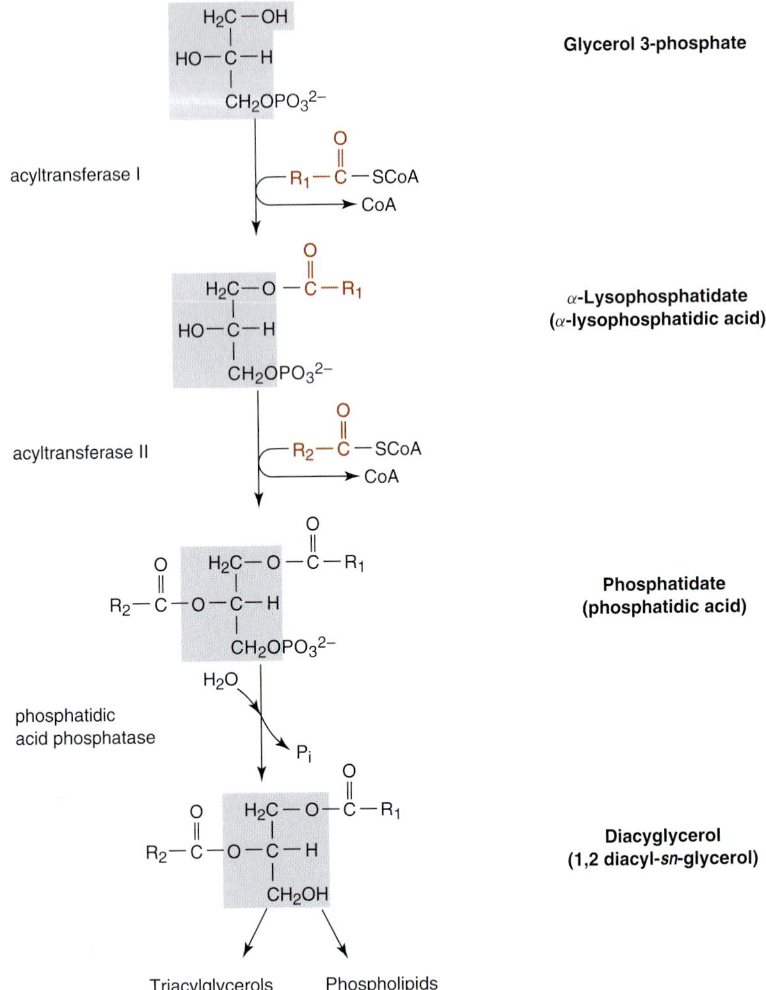

FIGURE 18.13

Phosphatidic acid biosynthesis from glycerol 3-phosphate and the role of phosphatidic acid phosphatase in synthesis of phospholipids and triacylglycerols.

phospholipid synthesis (Figure 18.13). Phosphatidic acid can also be formed beginning with dihydroxyacetone phosphate (DHAP):

$$\text{Dihydroxyacetone phosphate} + \text{NADH} + \text{H}^+ \rightleftharpoons \alpha\text{-glycerophosphate} + \text{NAD}^+$$

Phospholipids Are Synthesized by Addition of a Base to Phosphatidic Acid

The major pathway for synthesis of phosphatidylcholine involves sequential conversion of choline to phosphocholine, CDP-choline, and phosphatidylcholine. Free choline, a dietary requirement for most mammals including humans, is phosphorylated by choline kinase (Figure 18.14). Phosphocholine is converted to CDP-choline by phosphocholine cytidylyltransferase. Inorganic pyrophosphate (PP_i) is a product of this reaction. The phosphocholine moiety is then transferred to C3 of 1,2-diacylglycerol by choline phosphotransferase (Figure 18.15). This is the principal pathway for synthesis of dipalmitoyllecithin in lung.

The rate-limiting step for phosphatidylcholine synthesis is the **cytidylyltransferase** reaction (Figure 18.14). This enzyme is regulated by translocation between cytosol and endoplasmic reticulum. The cytosolic form is inactive; binding of the enzyme to the ER membrane activates it. Translocation of cytidyl transferase from cytosol to endoplasmic reticulum is regulated by cAMP and fatty-acyl CoA. Reversible phosphorylation of the enzyme by a cAMP-dependent protein kinase releases it from the membrane, rendering it inactive. Dephosphorylation causes it to bind to the ER membrane and become active. Fatty-acyl CoAs promote its binding to the endoplasmic

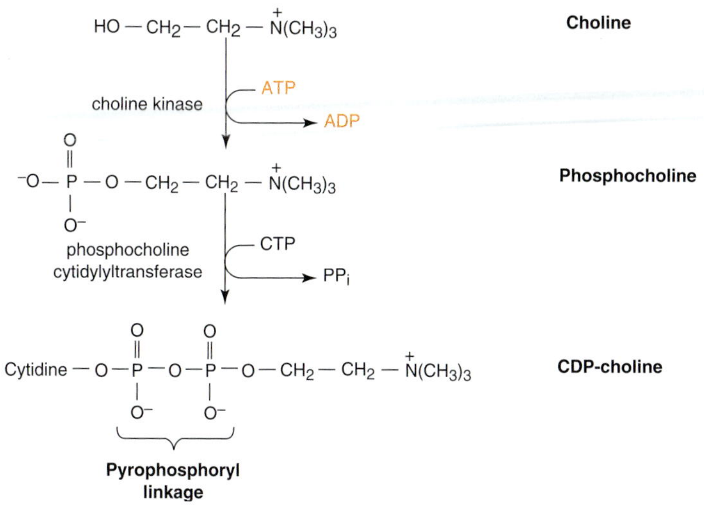

Choline

choline kinase (ATP → ADP)

Phosphocholine

phosphocholine cytidylyltransferase (CTP → PP_i)

CDP-choline

Pyrophosphoryl linkage

FIGURE 18.14
Biosynthesis of CDP-choline from choline.

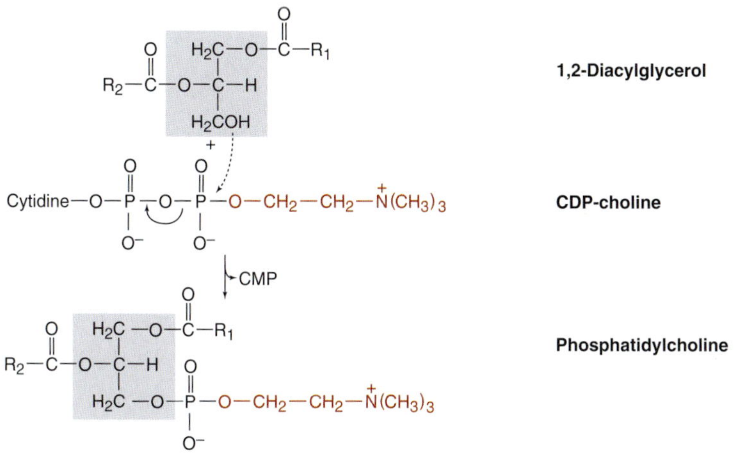

1,2-Diacylglycerol

CDP-choline

CMP

Phosphatidylcholine

FIGURE 18.15
Choline phosphotransferase reaction.

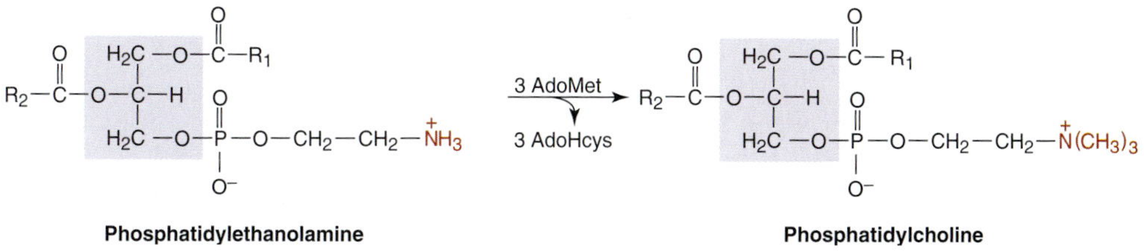

Phosphatidylethanolamine

3 AdoMet

3 AdoHcys

Phosphatidylcholine

FIGURE 18.16
Biosynthesis of phosphatidylcholine from phosphatidylethanolamine and *S*-adenosylmethionine (AdoMet); *S*-adenosylhomocysteine (AdoHcys).

reticulum. In liver, phosphatidylcholine is formed by repeated methylation of phosphatidylethanolamine. Phosphatidylethanolamine *N*-methyltransferase of the ER transfers methyl groups in sequence from *S*-adenosylmethionine (AdoMet) (Figure 18.16). Phosphatidylethanolamine synthesis in liver and brain involves ethanolamine phosphotransferase of the endoplasmic reticulum (Figure 18.17). CDP-ethanolamine is formed by ethanolamine kinase:

$$\text{Ethanolamine} + \text{ATP} \xrightleftharpoons{Mg^{2+}} \text{phosphoethanolamine} + \text{ADP}$$

and phosphoethanolamine cytidylyltransferase:

$$\text{Phosphoethanolamine} + \text{CTP} \xrightleftharpoons{Mg^{2+}} \text{CDP-ethanolamine} + PP_i$$

Liver mitochondria also generate phosphatidylethanolamine by decarboxylation of phosphatidylserine, but this is thought to represent a minor pathway (Figure 18.18).

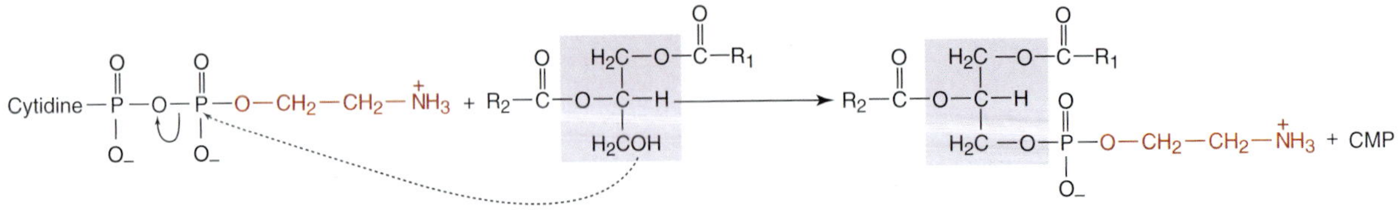

CDP-ethanolamine 1,2-Diacylglycerol Phosphatidylethanolamine

FIGURE 18.17

Biosynthesis of phosphatidylethanolamine from CDP-ethanolamine and diacylglycerol; the reaction is catalyzed by ethanolamine phosphotransferase.

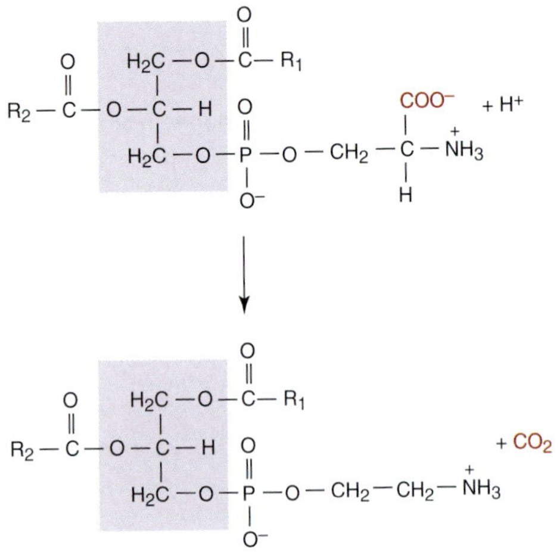

FIGURE 18.18

Formation of phosphatidylethanolamine by the decarboxylation of phosphatidylserine.

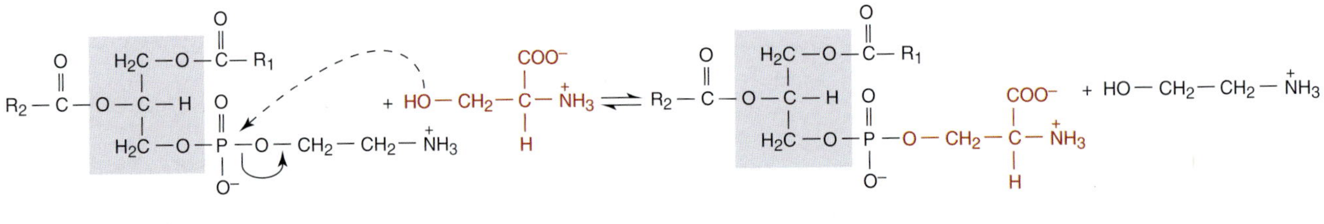

Phosphatidylethanolamine Serine Phosphatidylserine Ethanolamine

FIGURE 18.19

Biosynthesis of phosphatidylserine from serine and phosphatidylethanolamine by "base exchange."

The major source of phosphatidylserine in mammalian tissues is "base-exchange" (Figure 18.19) in which the polar head group of phosphatidylethanolamine is exchanged for serine. Since there is no net change in the number or kind of bonds, this reaction is reversible and has no requirement for ATP or any other high-energy compound. Phosphatidylinositol is made via CDP-diacylglycerol and free myoinositol (Figure 18.20) by phosphatidylinositol synthase of the endoplasmic reticulum.

Asymmetric Distribution of Fatty Acids in Phospholipids Is Due to Remodeling

Phospholipase A_1 and **phospholipase A_2** occur in many tissues and function in the remodeling of specific phospholipid structures at the *sn*-1 and *sn*-2 positions. Most fatty

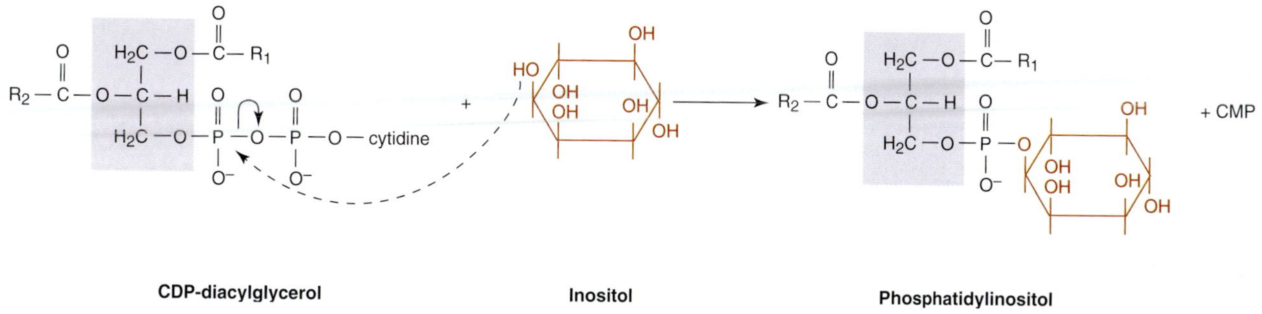

CDP-diacylglycerol + Inositol → Phosphatidylinositol + CMP

FIGURE 18.20

Biosynthesis of phosphatidylinositol.

2-Acyl lysophosphatide + R_1—COO$^-$ ← (H_2O, phospholipase A_1) Phospholipid → (H_2O, phospholipase A_2) 1-Acyl lysophosphatide + R_2—COO$^-$

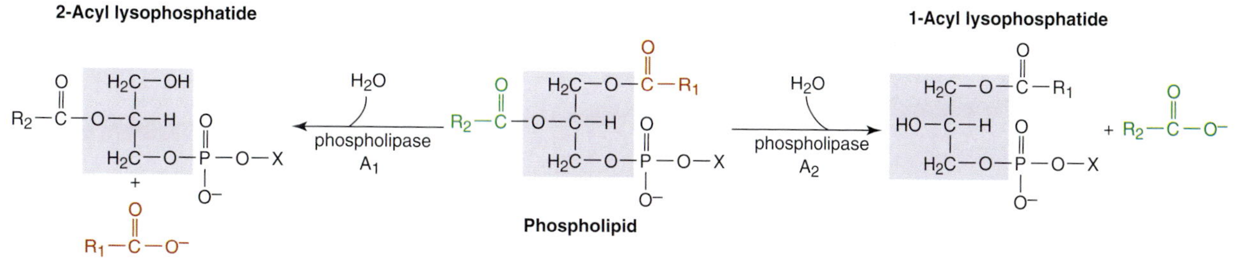

FIGURE 18.21

Reactions catalyzed by phospholipase A_1 and phospholipase A_2.

acyl-CoA transferases and enzymes of phospholipid synthesis lack the specificity required to account for the distribution of fatty acids found in many tissue phospholipids. The fatty acids found in the *sn*-1 and *sn*-2 positions of the various phospholipids are often not the same ones transferred to the glycerol backbone in the initial acyl transferase reactions of phospholipid biosynthesis. Phospholipases A_1 and A_2 catalyze reactions indicated in Figure 18.21, where X represents the polar head group of a phospholipid. The phospholipid products are called **lysophosphatides** or **lysophospholipids**.

If it becomes necessary for a cell to remove some undesired fatty acid, such as stearic acid from the *sn*-2 position of phosphatidylcholine, and replace it by a more unsaturated one like arachidonic acid, then this can be accomplished by the action of phospholipase A_2 followed by a reacylation reaction. Insertion of arachidonic acid into the 2 position of *sn*-2-lysophosphatidylcholine can be accomplished either by direct acylation from arachidonyl CoA by **arachidonyl-CoA transacylase** (Figure 18.22) or from some other arachidonyl-containing phospholipid by an exchange catalyzed by lysolecithin:lecithin acyltransferase (LLAT) (Figure 18.23). Since there is no change in either number or nature of the bonds involved in products and reactants, ATP is not required. Reacylation of lysophosphatidylcholine from acyl CoA is the major route for remodeling of phosphatidylcholine.

Lysophospholipids, particularly *sn*-1-lysophosphatidylcholine, also serve as a source of fatty acid in remodeling reactions. Those involved in synthesis of dipalmitoyllecithin (surfactant) from 1-palmitoyl-2-oleoylphosphatidylcholine are presented in Figure 18.24. Note that *sn*-1-palmitoyl lysolecithin is the source of palmitic acid in the acyltransferase exchange reaction.

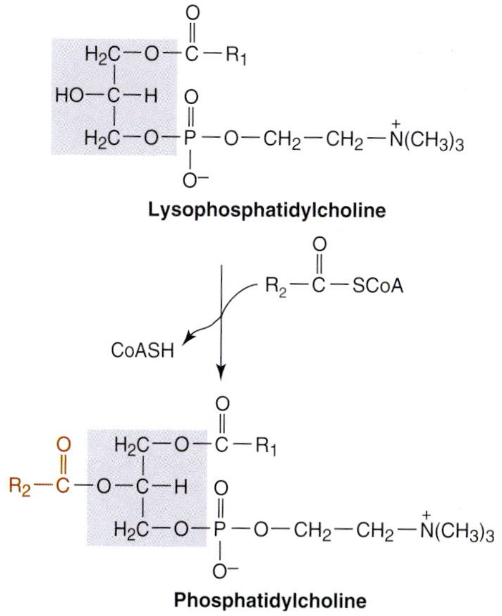

Lysophosphatidylcholine + R_2—CO—SCoA → Phosphatidylcholine + CoASH

FIGURE 18.22

Synthesis of phosphatidylcholine by reacylation of lysophosphatidylcholine where R_2—C(=O)— represents arachidonic acid and R_2—C(=O)—SCoA = arachionyl-CoA. This reaction is catalyzed by 1-acylglycerol-3-phosphocholine *O*-acyltransferase.

Plasmalogens Are Synthesized from Fatty Alcohols

Ether glycerolipids are synthesized from DHAP, long-chain fatty acids, and long-chain fatty alcohols as summarized in Figure 18.25. Acyldihydroxyacetone phosphate is formed by acyl CoA: dihydroxyacetone phosphate acyltransferase (enzyme 1). The ether bond is introduced by alkyldihydroxyacetone phosphate synthase (Figure 18.25, enzyme 2) by exchange of the 1-*O*-acyl group of acyldihydroxyacetone phosphate with

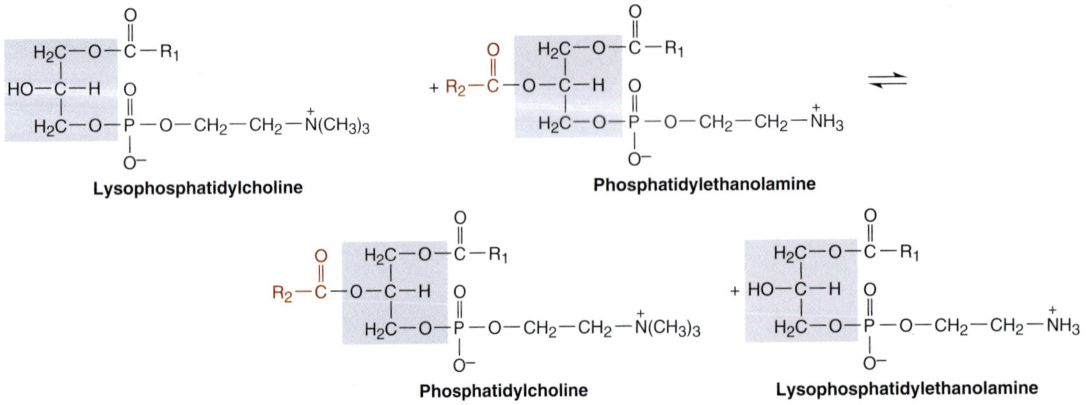

Lysophosphatidylcholine + Phosphatidylethanolamine ⇌ Phosphatidylcholine + Lysophosphatidylethanolamine

FIGURE 18.23

Formation of phosphatidylcholine by lysolecithin exchange, where represents arachidonic acid.

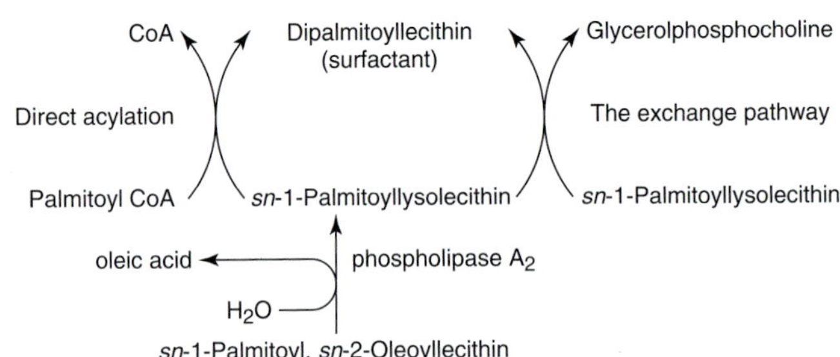

FIGURE 18.24

Two pathways for biosynthesis of dipalmitoyllecithin from *sn*-1 palmitoyllysolecithin.

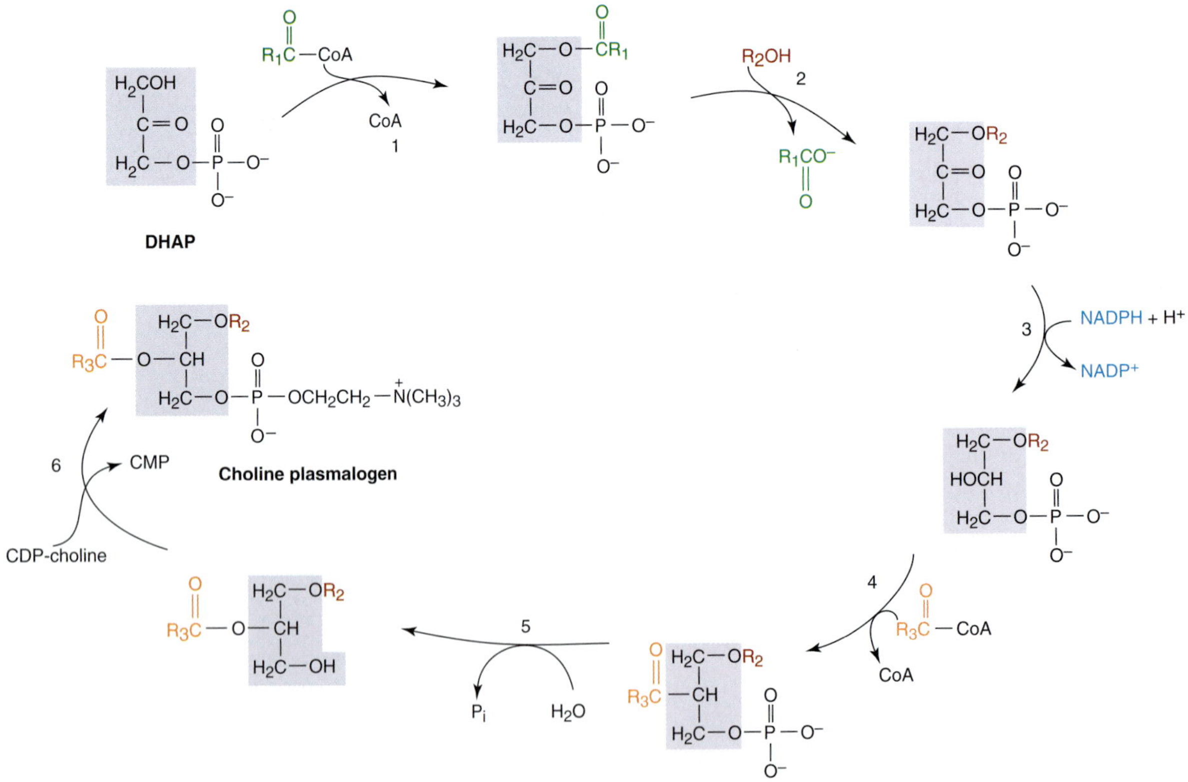

DHAP → (1: $R_1C(O)$—CoA → CoA) → (2: R_2OH → R_1CO^-) → (3: NADPH + H⁺ → $NADP^+$) → (4: $R_3C(O)$—CoA → CoA) → (5: H_2O → P_i) → (6: CDP-choline → CMP) → Choline plasmalogen

FIGURE 18.25

Pathway of choline plasmalogen biosynthesis from DHAP. 1, acyl CoA: dihydroxyacetone-phosphate acyltransferase; 2, alkyldihydroxyacetone-phosphate synthase; 3, NADPH: alkyldihydroxyacetone-phosphate oxidoreductase; 4, acyl CoA: 1-alkyl-2-lyso-*sn*-glycero-3-phosphate acyltransferase; 5, 1-alkyl-2-acyl-*sn*-glycerol-3-phosphate phosphohydrolase; 6, CDP-choline: 1-alkyl-2-acyl-*sn*-glycerol choline phosphotransferase.

a long-chain fatty alcohol. The synthase occurs in peroxisomes. Plasmalogen synthesis is completed by transfer of a long-chain fatty acid from its CoA donor to the *sn*-2 position of 1-alkyl-2-lyso-*sn*-glycero-3-phosphate (Figure 18.25, reaction 4. Patients with Zellweger's disease lack peroxisomes and cannot synthesize adequate amounts of plasmalogen (see Clin. Corr. 1.5, p. 19).

18.3 CHOLESTEROL

Cholesterol, an Alicyclic Compound, Is Widely Distributed in Free and Esterified Forms

Cholesterol is an alicyclic compound whose structure includes (1) the **perhydrocyclopentenophenanthrene** nucleus with its four fused rings, (2) a single hydroxyl group at C3, (3) an unsaturated center between C5 and C6, (4) an eight-membered branched hydrocarbon chain attached to the D ring at C17, and (5) a methyl group (designated C19) attached at position C10, and another methyl group (designated C18) attached at position C13 (Figures 18.26 and 18.27).

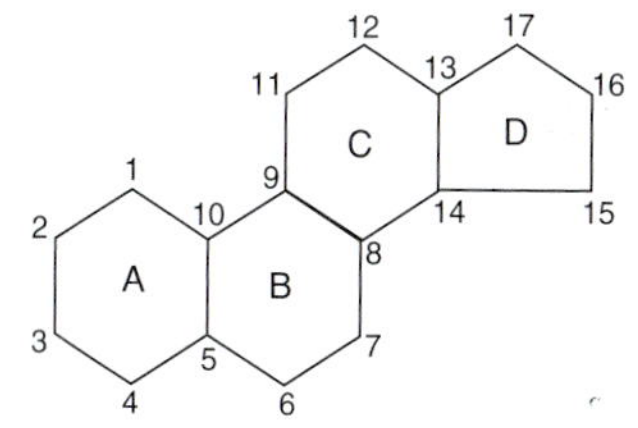

Cholesterol has very low solubility in water; at 25°C, the limit of solubility is approximately 0.2 mg 100 dL^{-1}, or 4.7 mM. The actual cholesterol concentration in plasma of healthy people is usually 150 to 200 mg dL^{-1}. This value is almost twice the normal concentration of blood glucose. Such a high concentration of cholesterol in blood is possible due to plasma lipoproteins (mainly LDL and VLDL) that contain large amounts of cholesterol (see p. 711). Only about 30% of the total plasma cholesterol is free (unesterified); the rest is cholesteryl esters in which a long-chain fatty acid, usually linoleic acid, is esterified to the C3 of the A ring. This fatty acid residue increases the hydrophobicity of cholesterol (Figure 18.28).

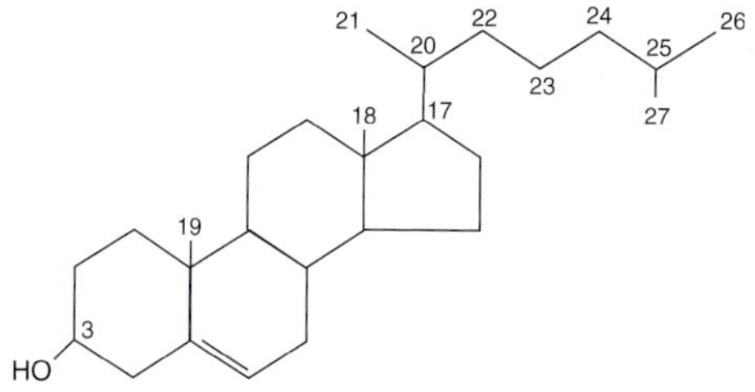

Cholesterol is a ubiquitous and essential component of mammalian cell membranes and abundant in bile (normal concentration is 390 mg dL^{-1}, only 4% of which is esterified). The solubilization of free cholesterol is achieved in part by the detergent property of the bile phospholipids that are produced in liver (see p. 1061). A chronic disturbance in phospholipid metabolism in liver can lead to deposition of cholesterol-rich gallstones. Bile salts, which are metabolites of cholesterol, also aid in solubilizing cholesterol. Cholesterol in bile protects gallbladder membranes from potentially irritating or harmful effects of bile salts.

Total cholesterol is estimated in the clinical laboratory by the Lieberman–Burchard reaction. The ratio of free and esterified cholesterol can be determined by gas–liquid chromatography or reverse phase high-pressure liquid chromatography (HPLC).

FIGURE 18.26
The cyclopentenophenanthrene ring.

FIGURE 18.27
Structure of cholesterol (5-cholesten-3β-ol).

Cholesterol Is a Membrane Component and Precursor of Bile Salts and Steroid Hormones

Cholesterol is derived from the diet or synthesized *de novo* in virtually all cells of the body. It is the major sterol in humans and a component of virtually all membranes. It is especially abundant in myelinated structures of brain and central nervous system but is present in small amounts in the inner membrane of mitochondria (see p. 450). In contrast to plasma where most of the cholesterol is esterified, the cholesterol in cellular membranes is in the free form. The ring structure of cholesterol cannot be metabolized to CO_2 and water in humans. Excretion of cholesterol is by way of the liver and gallbladder

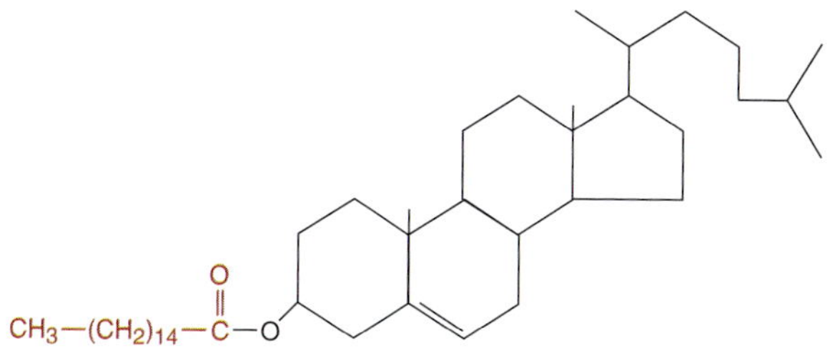

FIGURE 18.28
Structure of cholesteryl (palmitoyl-) ester.

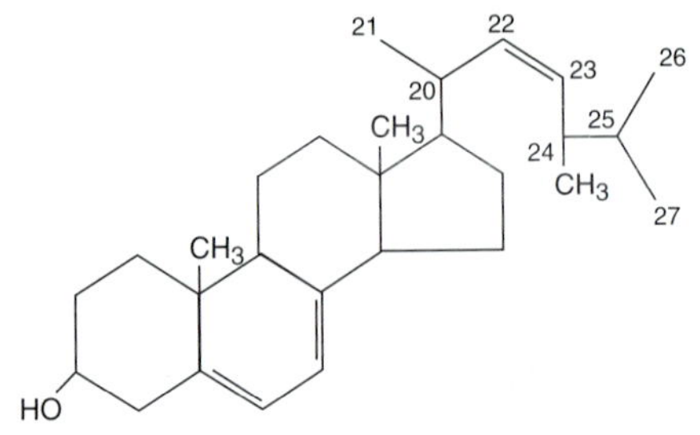

into the intestine in the form of **bile acids**. Cholesterol is the immediate precursor of bile acids synthesized in liver; they facilitate absorption of dietary triacylglycerols and fat-soluble vitamins (see p. 1061).

Cholesterol is the precursor of various **steroid hormones** (see p. 925) including progesterone, corticosteroids (corticosterone, cortisol, and cortisone), aldosterone, and the sex hormones, estrogens and testosterone. Although all steroid hormones are structurally related to and biochemically derived from cholesterol, they have widely different physiological properties. The hydrocarbon skeleton of cholesterol also occurs in plant sterols, for example in ergosterol, a precursor of vitamin D (Figure 18.29), which is converted in skin by ultraviolet irradiation to vitamin D_2 (see p. 719).

FIGURE 18.29
Structure of ergosterol.

Cholesterol Is Synthesized from Acetyl CoA

Although synthesis of cholesterol occurs in virtually all cells, the capacity is greatest in liver, intestine, adrenal cortex, and reproductive tissues, including ovaries, testes, and placenta. All carbon atoms of cholesterol are derived from acetate. Reducing power in the form of NADPH is provided mainly by glucose 6-phosphate dehydrogenase and 6-phosphogluconate dehydrogenase of the hexose monophosphate shunt pathway (see p. 642). Cholesterol synthesis occurs in the cytosol and endoplasmic reticulum and is driven largely by hydrolysis of high-energy thioester bonds of acetyl CoA and phosphoanhydride bonds of ATP.

Mevalonic Acid Is a Key Intermediate

The first compound unique to cholesterol synthesis is mevalonic acid derived from acetyl CoA. There are several sources of acetyl CoA: (1) β-oxidation of fatty acids (see p. 683), (2) oxidation of ketogenic amino acids such as leucine and isoleucine (see p. 777), and (3) pyruvate dehydrogenase reaction. Acetate can be converted to acetyl CoA at the expense of ATP by **acetokinase**, or **acetate thiokinase**:

$$\text{ATP} + CH_3COO^- + \text{CoASH} \longrightarrow CH_3{-}\overset{\overset{\large O}{\|}}{C}{-}\text{SCoA} + \text{AMP} + \text{PP}_i$$

As in the pathway that produces ketone bodies (see p. 687), two molecules of acetyl CoA are condensed to acetoacetyl CoA by **acetoacetyl-CoA thiolase** (acetyl-CoA:acetyl-CoA acetyltransferase):

$$CH_3{-}\overset{\overset{\large O}{\|}}{C}{-}\text{SCoA} + CH_3{-}\overset{\overset{\large O}{\|}}{C}{-}\text{SCoA} \longrightarrow CH_3{-}\overset{\overset{\large O}{\|}}{C}{-}CH_2{-}\overset{\overset{\large O}{\|}}{C}{-}\text{SCoA} + \text{CoA}{-}\text{SH}$$

Formation of the carbon–carbon bond in acetoacetyl CoA is favored energetically by cleavage of a thioester bond and generation of coenzyme A.

A third molecule of acetyl CoA is used to form the branched-chain **3-hydroxy-3-methylglutaryl CoA (HMG-CoA)** (Figure 18.30) by **HMG-CoA synthase** (3-hydroxy-3-methylutarylyl CoA: acetoacetyl-CoA lyase). Liver parenchymal cells contain a cytosolic form of HMG-CoA synthase, which is involved in cholesterol synthesis, and a mitochondrial form that functions in the synthesis of ketone bodies (see p. 688). The

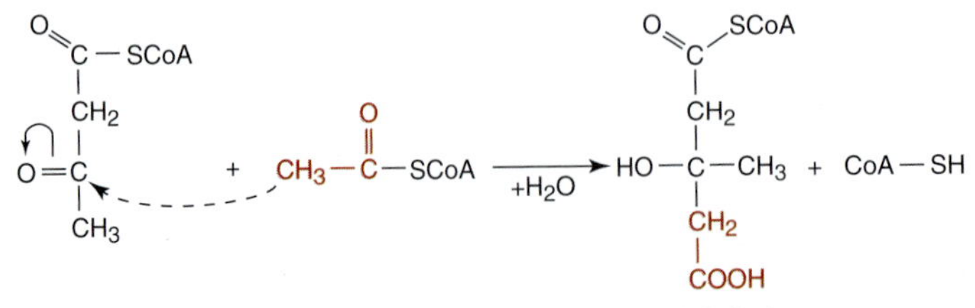

FIGURE 18.30
HMG-CoA synthase reaction. **Acetoacetyl CoA** **Acetyl CoA** **HMG CoA**

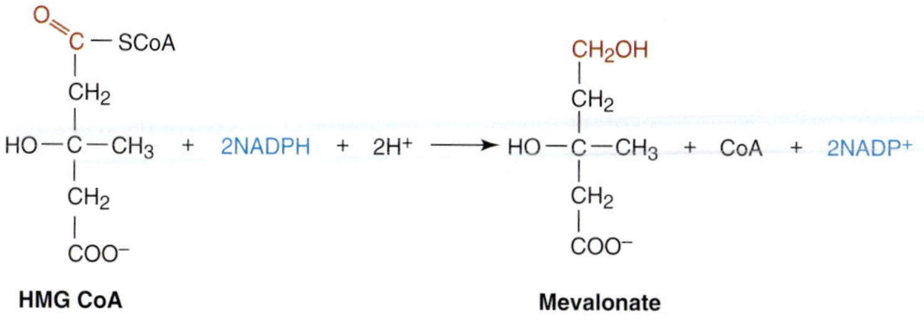

FIGURE 18.31

HMG-CoA reductase reaction.

enzyme catalyzes an aldol condensation between the methyl carbon of acetyl CoA and the β-carbonyl group of acetoacetyl CoA with hydrolysis of the thioester bond of acetyl CoA. The original thioester bond of acetoacetyl CoA remains intact. HMG-CoA is also formed from oxidative degradation of the branched-chain amino acid leucine, through the intermediates 3-methylcrotonyl CoA and 3-methylglutaconyl-CoA (see p. 777).

Mevalonic acid is formed from HMG-CoA by the endoplasmic reticulum enzyme **HMG-CoA reductase** (mevalonate:NADP$^+$ oxidoreductase) that has an absolute requirement for NADPH (Figure 18.31). The reduction consumes two molecules of NADPH, results in hydrolysis of the thioester bond of HMG-CoA, and it generates the primary alcohol group of mevalonate. This reduction is irreversible and produces (R)-(+) mevalonate. HMG-CoA reductase catalyzes the rate-limiting reaction in cholesterol biosynthesis. It is an intrinsic protein of the endoplasmic reticulum with its catalytic C-terminal domain extending into the cytosol. Phosphorylation of HMG-CoA reductase diminishes its catalytic activity (V_{max}) and enhances its degradation by increasing its susceptibility to proteolytic attack. Increased intracellular cholesterol stimulates phosphorylation of HMG-CoA reductase.

The central role of HMG-CoA reductase in cholesterol homeostasis is evidenced by the effectiveness of a family of drugs called statins used to lower plasma cholesterol levels. Statins (e.g., lovastatin, pravastatin, fluvastatin, cerivastatin, atorvastatin, simvastatin) inhibit HMG-CoA reductase activity, particularly in liver, and commonly decrease total plasma cholesterol and LDL-cholesterol by as much as 50%.

Mevalonic Acid Is Precursor of Farnesyl Pyrophosphate

Reactions that convert mevalonate to farnesyl pyrophosphate are summarized in Figure 18.32. The stepwise transfer of the terminal phosphate group from two molecules

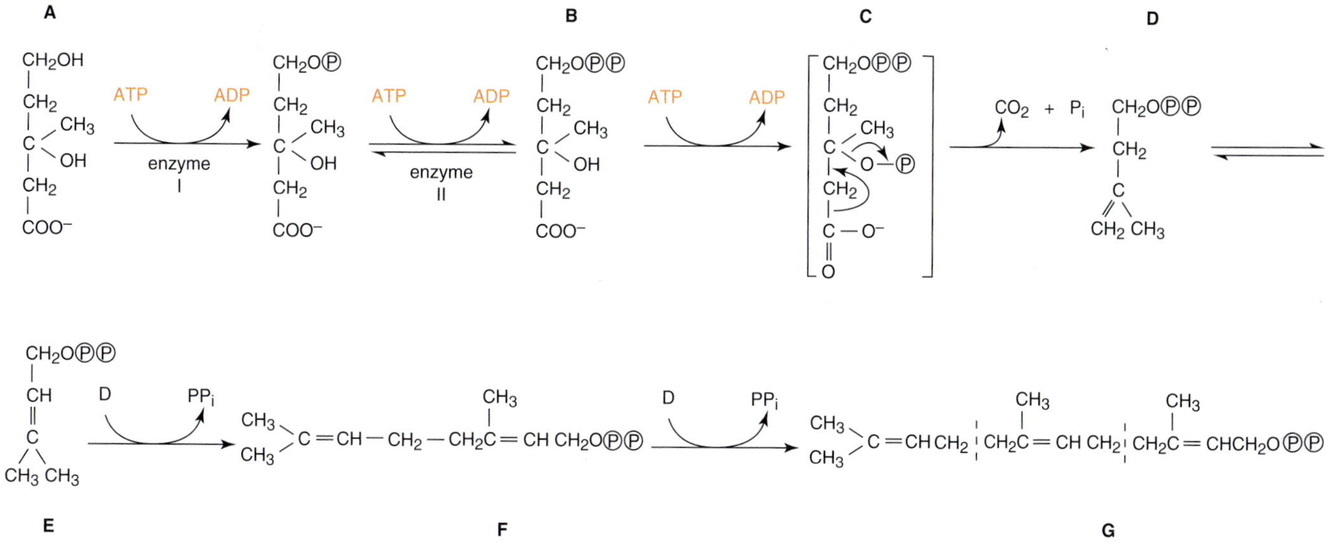

FIGURE 18.32

Formation of farnesyl-PP (F) from mevalonate (A). Dashed lines divide molecules into isoprenoid-derived units. D is 3-isopentenyl pyrophosphate.

of ATP to mevalonate (A) to form 5-pyrophosphomevalonate (B) is catalyzed by **mevalonate kinase** (enzyme I) and phosphomevalonate kinase (enzyme II). Decarboxylation of 5-pyrophosphomevalonate by **pyrophosphomevalonate decarboxylase** generates Δ^3-isopentenyl pyrophosphate (D). In this ATP-dependent reaction, ADP, P_i, and CO_2 are produced. It is thought that decarboxylation–dehydration proceeds by way of the intermediate 3-phosphomevalonate 5-pyrophosphate (C). Isopentenyl pyrophosphate is converted to its allylic isomer 3,3-dimethylallyl pyrophosphate (E) by isopentenyl pyrophosphate isomerase in a reversible reaction. The condensation of 3,3-dimethylallyl pyrophosphate (E) and 3-isopentenyl pyrophosphate (D) generates geranyl pyrophosphate (F).

The stepwise condensation of three C_5 **isopentenyl units** to form the C_{15} **farnesyl pyrophosphate** (G) is catalyzed by a cytosolic prenyl transferase called **geranyltransferase**.

Cholesterol Is Formed from Farnesyl Pyrophosphate via Squalene

The last steps in cholesterol synthesis involve "head-to-head" fusion of two molecules of farnesyl pyrophosphate to form squalene and finally cyclization of squalene to yield cholesterol. Formation of the C_{30} carbon molecule of squalene (Figure 18.33) by squalene synthase of endoplasmic reticulum releases two pyrophosphate groups and requires NADPH. Several intermediates probably occur. By rotation about carbon–carbon bonds, the conformation of squalene indicated in Figure 18.34 can be obtained. Note that the overall shape of squalene resembles that of cholesterol and that squalene is devoid of oxygen atoms.

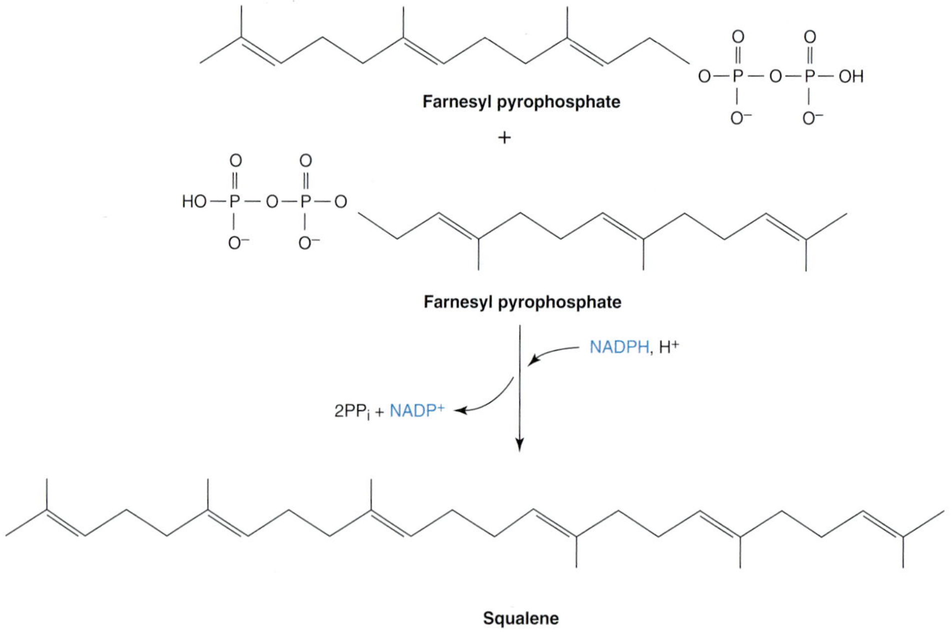

FIGURE 18.33
Formation of squalene from two molecules of farnesyl pyrophosphate.

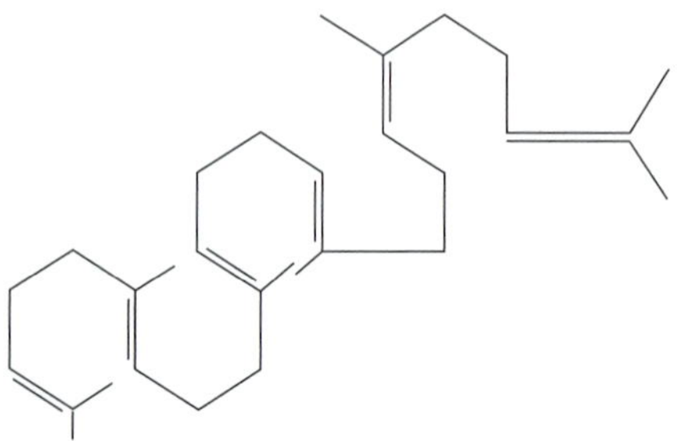

FIGURE 18.34
Structure of squalene (C_{30}).

Cholesterol synthesis from squalene by squalene cyclase proceeds through the intermediate lanosterol, which contains the fused tetracyclic ring system and an eight-carbon side chain:

$$\text{Squalene} \rightarrow \text{squalene 2,3-epoxide} \rightarrow \text{lanosterol}$$

The ER enzyme that catalyzes this reaction is bifunctional and contains squalene epoxidase or monooxygenase and a cyclase **(lanosterol cyclase)** activity. The many carbon–carbon bonds formed during cyclization of squalene are generated in a concerted fashion as indicated in Figure 18.35. The OH group of lanosterol projects above the plane of the A ring; that is, it is in the β orientation. In this reaction sequence, an OH group is added to C3, two methyl groups undergo shifts, and a proton is eliminated.

Cyclization is initiated by epoxide formation between the future C2 and C3 of cholesterol, the epoxide being formed at the expense of NADPH:

$$\text{Squalene} + O_2 + \text{NADPH} + H^+ \rightarrow \text{squalene 2,3-epoxide} + H_2O + \text{NADP}^+$$

The hydroxylation at C3 triggers the cyclization of squalene to lanosterol (Figure 18.35).

Transformation of lanosterol to cholesterol involves many poorly understood steps and several enzymes. These reactions include: (1) removal of the methyl group at C14, (2) removal of two methyl groups at C4, (3) migration of the double bond from C8 to C5, and (4) reduction of the double bond between C24 and C25 in the side chain (see Figure 18.36).

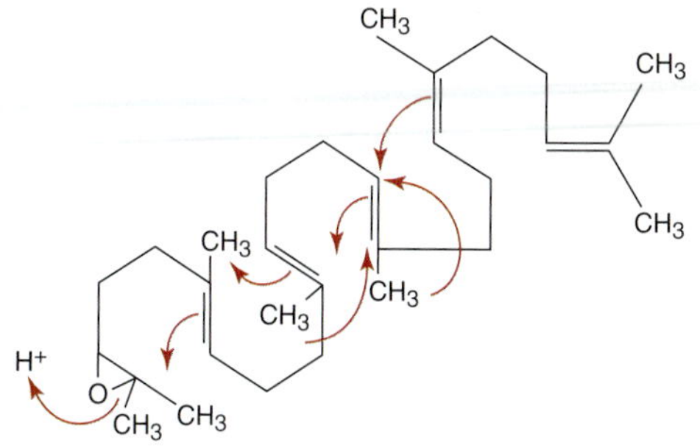

Squalene 2,3-epoxide

cyclase

Lanosterol

FIGURE 18.35
Conversion of squalene 2,3-epoxide to lanosterol.

Plasma Lipoproteins

Blood contains triacylglycerols, cholesterol, and cholesteryl esters at concentrations that far exceed their solubilities in water. Blood lipids are kept in solution, or at least thoroughly dispersed, by incorporation into macromolecular structures called **lipoproteins**. The plasma lipoproteins facilitate lipid metabolism and transfer of lipids between tissues. There are five major lipoprotein classes: **high-density lipoprotein (HDL), low-density lipoprotein (LDL), intermediate-density lipoproteins (IDLs), very-low-density lipoproteins (VLDL),** and **chylomicrons**. Their physical characteristics are presented in Table 3.12, p. 105, and the lipid composition is listed in Table 3.14, p. 108. All contain phospholipids and one or more proteins called **apoproteins**; there are 10 common apoproteins (Table 18.1).

Lipoproteins are spherical particles with the most hydrophobic lipids such as cholesteryl esters and triacylglycerols located in the core, sequestered away from water,

TABLE 18.1 Apoproteins of Human Plasma Lipoproteins

Apolipoprotein	*Molecular Weight*	*Plasma Concentration (g L^{-1})*	*Lipoprotein Distribution*
ApoA-I	28,000	1.0–1.2	Chylomicrons, HDL
ApoA-II	17,000	0.3–0.5	Chylomicrons, HDL
ApoA-IV	46,000	0.15–0.16	Chylomicrons, HDL
ApoB-48	264,000	0.03–0.05	Chylomicrons
ApoB-100	512,000	0.7–1.0	VLDL, IDL, LDL
ApoC-I	7000	0.04–0.06	Chylomicrons, VLDL, HDL
ApoC-II	9000	0.03–0.05	Chylomicrons, VLDL, HDL
ApoC-III	9000	0.12–0.14	Chylomicrons, VLDL, HDL
ApoD	33,000	0.06–0.07	HDL
ApoE	38,000	0.03–0.05	Chylomicrons, VLDL, IDL, HDL

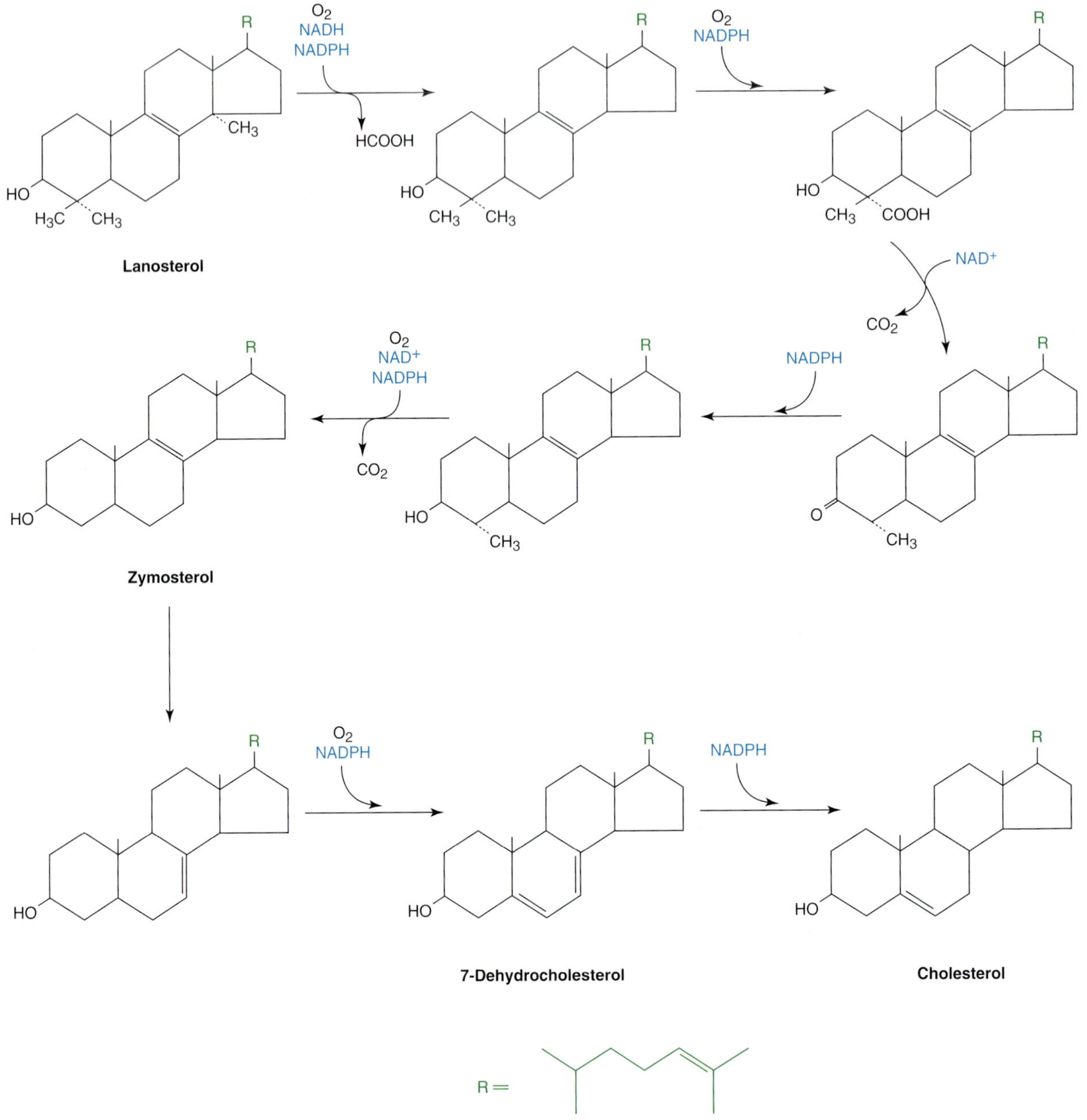

FIGURE 18.36

Conversion of lanosterol to cholesterol.

whereas free (nonesterified) cholesterol, phospholipids, and the protein(s) are arrayed on the surface (see Figure 3.43, p. 107). The amphipathic phospholipids and proteins keep the otherwise highly insoluble lipids in solution. The apoproteins on the surface of the particles also serve as ligands for cell receptors, and cofactors for enzymes involved in lipoprotein metabolism.

Lipoproteins are vehicles by which cholesterol and its esters, and triacylglycerols are transported in the body. For example, LDL delivers cholesterol to various tissues that require cholesterol for membrane formation or steroid hormone synthesis. The cholesterol carried to the liver from peripheral tissues and the cholesterol in chylomicrons regulate hepatic cholesterol synthesis (see Figure 18.37). In contrast, HDL, which is rich in cholesterol but poor in triacylglycerol, carries cholesterol from the periphery to the liver, where it can be excreted in bile as cholesterol or after conversion to bile salts. VLDL and chylomicrons primarily transport triacylglycerols to be used for energy (e.g., muscle) or stored (e.g., fat cells).

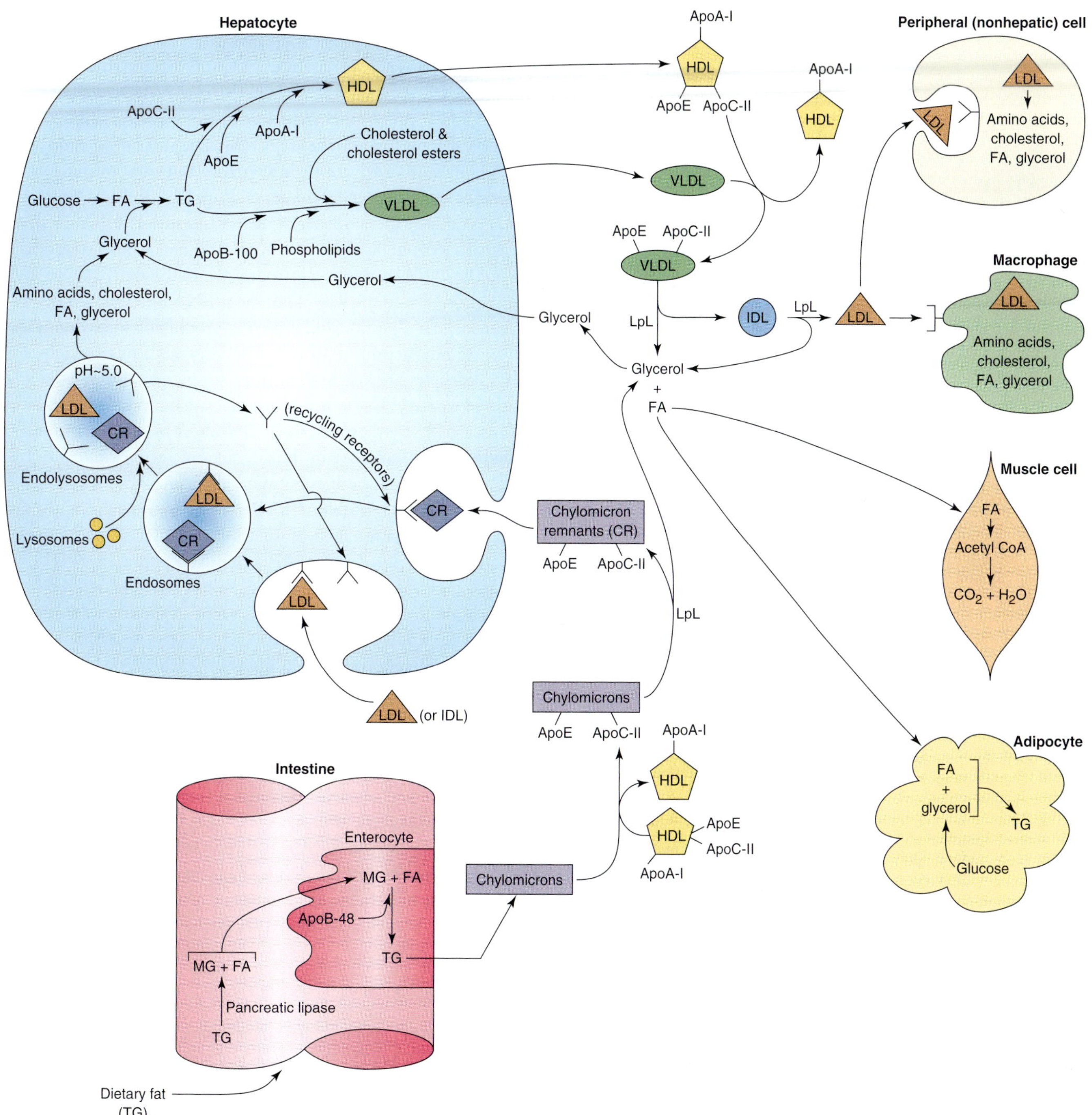

FIGURE 18.37

Organs and pathways involved in plasma lipoprotein metabolism. FA, fatty acid; TG, triacylglycerol (triglyceride); HDL, high-density lipoprotein; LDL, low-density lipoprotein; IDL, intermediate-density lipoprotein; VLDL, very-low-density lipoprotein; LpL, lipoprotein lipase; apo-, apoprotein. Chylomicrons contain apoB-48; VLDL and LDL contain apoB-100. ⅄-LDL receptor.

Plasma lipoproteins are substrates for several enzymes in blood. Lipoprotein lipase is attached to the luminal surface of the vascular endothelium by binding to heparin sulfate proteoglycans and hydrolyses triacylglycerols of VLDL and chylomicrons. It is activated by apolipoprotein C-II (apoC-II) and by heparin released from mast cells and cells of the macrophage/reticuloendothelial system. The fatty acids released by lipoprotein lipase can then be taken up by cells and oxidized by β-oxidation, incorporated into phospholipids for membrane assembly, or, in adipocytes, metabolized or stored as triacylglycerols. Hydrolysis of triacylglycerols also results in the release of phospholipids, free cholesterol, and exchangeable apolipoproteins on the surface shell and their transfer to other circulating lipoproteins, especially HDL. In the mammary gland these fatty acids can be incorporated into milk fat. As chylomicrons lose their triacylglycerol core, they become the smaller chylomicron remnants that are rich in cholesteryl esters, bind specific receptors on hepatic membranes, and are catabolized in the liver following endocytosis (Figure 18.37). Clearly, lipoproteins are dynamic in structure and composition.

Main sites of synthesis of the apoprotein components are liver and small intestine. For example, **apoB-48** is produced in the intestine whereas **apoB-100** is synthesized mainly in hepatocytes. ApoB-48 is 48% as long as apoB-100 and is formed from the same message as that for apoB-100. A stop codon introduced into the apoB mRNA in the intestine terminates translation 48% of the way along the mRNA molecule. ApoB-48 is synthesized during the course of fat digestion and is used in production of chylomicrons. ApoB-100 is also used in the production of VLDL. The fatty acids of triacylglycerols of newly produced VLDL particles are synthesized from sugars, glucose in particular, after they have been oxidized to acetyl CoA, or from the acetate derived from oxidation of ethanol. VLDL acquires freshly synthesized cholesteryl esters in the liver, as well as from other lipoproteins, during its transport in the circulation. In the circulation, the net transfer of cholesteryl ester and triglycerides between HDL and VLDL and LDL is facilitated by **cholesteryl ester transfer protein (CETP)**. CETP is associated with HDL in the plasma, and its expression is stimulated by diet-induced hypercholesterolemia. Lipoprotein lipase converts VLDL into intermediate-density lipoprotein (IDL) and then, as additional triacylglycerol is subjected to lipolysis, into LDL (Figure 18.37).

HDL is synthesized mainly in liver and to a lesser extent in the intestine. It has a unique function as a reservoir for **apoE** and **apoC-II**, which are activators of lipoprotein lipase. HDL regulates exchange of apoproteins and lipids between various lipoproteins in the blood. HDL particles donate apoE and apoC-II to chylomicrons and VLDL. Once the triacylglycerols in chylomicrons and VLDL have been extensively hydrolyzed and these lipoproteins have been transformed into LDL and chylomicron remnants, respectively, the apoE and apoC-II are returned to HDL (Figure 18.37).

HDL also participates in removal of excess cholesterol from cells and its transport to the liver for elimination as cholesterol and bile salts. This phenomenon is termed "reverse cholesterol transport" (Figure 18.38). It is free (nonesterified) cholesterol that exchanges readily between lipoproteins and the plasma membrane of cells. The transfer of free cholesterol from the plasma membrane to a lipid-poor species of HDL to produce what is called **"nascent HDL"** is mediated by a membrane transporter designated

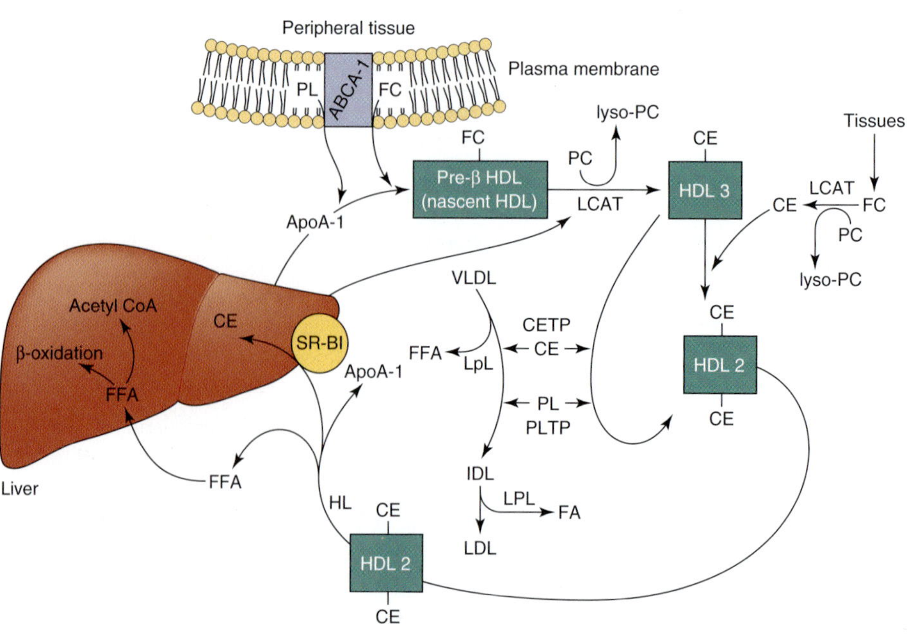

FIGURE 18.38

Reverse cholesterol transport showing the proteins and enzymes involved. PL, phospholipids; FC, free cholesterol; ABCA-1, ATP-binding cassette 1; VLDL, very-low-density lipoprotein; HDL, high-density lipoprotein; LDL, low-density lipoprotein; IDL, intermediate-density lipoprotein; PC, phosphatidylcholine; lyso-PC, lysophosphatidylcholine; LCAT, lecithin cholesterol acyltransferase; CE, cholesteryl ester; CETP, cholesterol ester transfer protein; PLTP, phospholipid transfer protein; LpL, lipoprotein lipase; apoA-1, apolipoproteinA-1; FFA, free fatty acid; SR-BI, scavenger receptor class B type I; TG, triacylglycerol; HL, hepatic lipase. FA, fatty acid; TG, triacylglycerol (triglyceride); HDL, high-density lipoprotein; LDL, low-density lipoprotein; IDL, intermediate-density lipoprotein; VLDL, very-low-density lipoprotein; LpL, lipoprotein lipase; apo-, apoprotein.

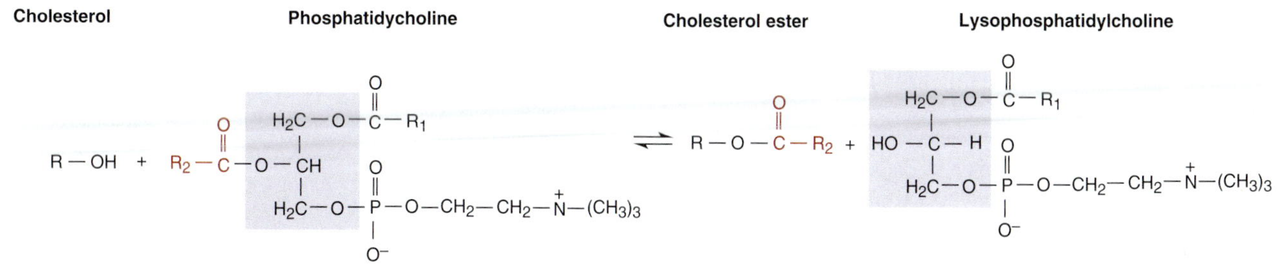

$$R{-}OH + R_2{-}\overset{O}{\overset{\|}{C}}{-}O{-}CH(CH_2{-}O{-}\overset{O}{\overset{\|}{C}}{-}R_1)(CH_2{-}O{-}\overset{O}{\overset{\|}{P}}(O^-){-}O{-}CH_2{-}CH_2{-}\overset{+}{N}{-}(CH_3)_3) \rightleftharpoons R{-}O{-}\overset{O}{\overset{\|}{C}}{-}R_2 + HO{-}CH(CH_2{-}O{-}\overset{O}{\overset{\|}{C}}{-}R_1)(CH_2{-}O{-}\overset{O}{\overset{\|}{P}}(O^-){-}O{-}CH_2{-}CH_2{-}\overset{+}{N}{-}(CH_3)_3)$$

FIGURE 18.39

Lecithin:cholesterol acyltransferase (LCAT) reaction. R-OH denotes cholesterol.

"**ATP-binding cassette transporter**" (ABCA-1). ABCA-1 also transfers phospholipids, along with free cholesterol, from the membrane to nascent HDL to produce HDL3. Absence of the ABCA-1 transporter results in the HDL deficiency disease called Tangier's disease. Further addition of esterified cholesterol to HDL3 produces HDL2. Cholesterol is esterified by **lecithin:cholesterol acyl transferase (LCAT)**. This freely reversible reaction (Figure 18.39) transfers the fatty acid in the *sn*-2 position of phosphatidylcholine to the 3-hydroxyl of cholesterol. LCAT is produced mainly by the liver, bound to HDL in plasma, and activated by the apoA-1 component of HDL. Cholesteryl ester generated in the LCAT reaction is transferred to VLDL and LDL by CETP associated with the HDL particle and is eventually taken up by the liver. The degradation of HDL takes place in the liver following the selective uptake of cholesteryl esters mediated by a plasma membrane protein called "scavenger receptor-BI" (SR-BI). SR-BI is a multi-ligand receptor that binds not only HDL but VLDL and LDL as well. The uptake and degradation of HDL by the liver involve cell-surface hepatic lipase that hydrolyzes the triglycerides of HDL particles. The apoA-1 from the degradation of HDL is recycled for new HDL formation. An inverse relationship exists between plasma HDL concentration and the incidence of coronary artery disease, and a positive relationship exists between plasma LDL cholesterol levels and coronary heart disease.

Liver cells metabolize chylomicron remnants by a similar mechanism; however, macrophages and many other cells have specific receptors that recognize chylomicron remnants and internalize them. The apoE of chylomicron remnants is recognized by these particular receptors. Some LDL is taken up via nonspecific scavenger receptors on certain cells, macrophages in particular.

Cholesterol Synthesis Is Carefully Regulated

Elevated plasma cholesterol predisposes to atherosclerotic vascular disease. In healthy individuals, plasma cholesterol levels are maintained within a relatively narrow concentration range largely by the liver which: (1) expresses the majority of the body's LDL receptors, (2) is the major site for conversion of cholesterol to bile acids, and (3) has the highest level of HMG-CoA reductase activity. The cholesterol pool of the body is derived from dietary cholesterol and cholesterol synthesis primarily in liver and intestine. When dietary cholesterol intake is reduced, cholesterol synthesis increases in liver and intestine. This cholesterol is then transported from liver and intestine to peripheral tissues by VLDLs and chylomicrons, respectively.

The committed step and the rate-limiting reaction in cholesterol synthesis (Figure 18.40) is that of the HMG-CoA reductase, which catalyzes the step that produces mevalonic acid. Cholesterol exerts feedback inhibition on HMG-CoA reductase and promotes degradation of the enzyme by mechanisms that remain to be elucidated.

In a normal healthy adult on a low cholesterol diet, about 1300 mg of cholesterol is presented to the liver each day for disposal. This cholesterol comes from the diet and peripheral tissues and is disposed of by: (1) excretion in bile of about 250 mg bile salts and about 550 mg of cholesterol, (2) storage as cholesteryl esters, and (3) incorporation

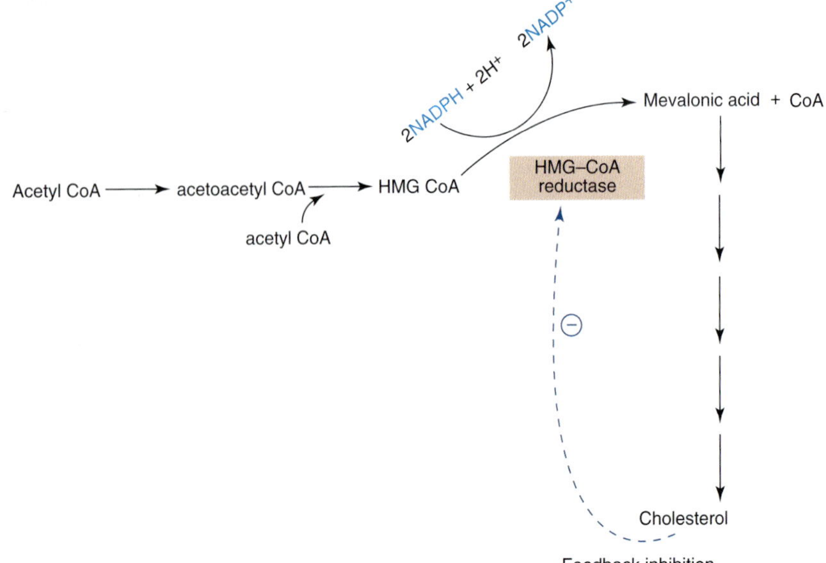

FIGURE 18.40

Summary of cholesterol synthesis indicating feedback inhibition of HMG-CoA reductase by cholesterol.

into VLDL and secretion into the circulation. In the same adult, liver synthesizes about 800 mg of cholesterol per day to replace bile salts and cholesterol.

About 75% of LDL catabolism occurs in liver via an **LDL-receptor-mediated process**. The binding is characterized by saturability, high affinity, and high degree of specificity. The receptor recognizes apolipoprotein E (apoE) and apolipoprotein B-100 (apoB-100) on LDL and VLDL. Binding occurs on the plasma membrane at pits coated with clathrin and leads to ligand–receptor complexes internalized as clathrin-coated vesicles. This is receptor-mediated endocytosis. Intracellularly, the vesicles lose their clathrin and become endosomes that fuse with lysosomes that contain proteases and cholesteryl esterase. In this relatively acid (pH $\approx$5.0) environment the receptor separates from LDL and returns to the cell surface while the cholesteryl esters are hydrolyzed by cholesteryl esterase to cholesterol and a long-chain fatty acid and the protein is degraded to amino acids that enter the cell's amino acid pool. The free cholesterol diffuses into the cytosol, where it inhibits HMG-CoA reductase and suppresses synthesis of the enzyme. At the same time, fatty acyl **CoA:cholesterol acyltransferase (ACAT)** of the endoplasmic reticulum is activated by cholesterol, promoting formation of cholesteryl esters, principally cholesteryl oleate:

$$\text{Cholesterol} + \text{oleoyl CoA} \rightarrow \text{cholesteryl oleate} + \text{CoA}$$

Accumulation of intracellular cholesterol inhibits the replenishment of LDL receptors by down-regulation of their expression, thereby blocking further uptake and accumulation of cholesterol.

The LDL receptor is a single-chain glycoprotein; numerous mutations in its gene are associated with familial hypercholesterolemia. The receptor spans the plasma membrane once with the carboxyl terminus on the cytoplasmic face and the amino terminus, which binds apoB-100 and apoE-100, extending into the extracellular space.

The correlation between high plasma cholesterol level, particularly LDL cholesterol, and heart attacks and strokes has led to dietary and therapeutic approaches to lower blood cholesterol (see Clin. Corr. 18.2). Patients with familial (genetic) hypercholesterolemia suffer from accelerated atherosclerosis (see Clin. Corr. 18.3). In most cases, called receptor-negative, there is a lack of functional LDL receptors because the mutant alleles produce little or no LDL receptor protein. In other cases, the receptor is synthesized and transported normally to the cell surface, but an amino acid substitution or other alteration in the primary structure adversely affects LDL binding. As a result, there is little or no binding of LDL to the cell, cholesterol synthesis is not inhibited, and the blood cholesterol level increases. Some LDL-deficient patients synthesize the LDL receptor but have a defect in the transport mechanism that delivers the glycoprotein

CLINICAL CORRELATION 18.2
Treatment of Hypercholesterolemia

Many authorities recommend screening asymptomatic individuals by measuring plasma cholesterol. A level less than 200 mg% is considered desirable, while over 240 mg% requires lipoprotein analysis, especially determination of LDL cholesterol. Reduction of LDL cholesterol depends on dietary restriction of cholesterol to less than 300 mg day^{-1}, of calories to attain ideal body weight, and of total fat intake to less than 30% of total calories. Approximately two-thirds of the fat should be mono- or polyunsaturated. The second line of therapy is with drugs. Cholestyramine and colestipol are bile salt-binding drugs that promote excretion of bile salts, increasing hepatic bile salt synthesis and LDL uptake by the liver. Lovastatin inhibits HMG-CoA reductase, decreases endogenous synthesis of cholesterol, and stimulates uptake of LDL via the LDL receptor. The combination of lovastatin and cholestyramine is sometimes used for severe hyperlipidemia.

Source: Expert Panel: Evaluation, and treatment of high blood cholesterol in adults. *Arch. Intern. Med.* 148:36, 1988.

CLINICAL CORRELATION 18.3
Atherosclerosis

Atherosclerosis is the leading cause of death in Western industrialized countries. The risk of developing it is directly related to the plasma LDL cholesterol and inversely related to the HDL cholesterol level. This explains why the former is frequently called "bad" cholesterol and the latter "good" cholesterol, though chemically there is no difference. In atherosclerosis the arterial wall contains accumulated cholesteryl esters in cells derived from the monocyte–macrophage line, there is also smooth muscle cell proliferation and fibrosis. The earliest abnormality is migration of blood monocytes to the subendothelium of the artery. These cells then differentiate into macrophages and accumulate cholesteryl esters derived from plasma LDL. Some of the LDL may be taken up via pathways that do not require the LDL receptor. For instance, there are receptors that take up acetylated LDL or LDL complexed with dextran sulfate; however, this pathway is not regulated by cellular cholesterol content. Distortion of the subendothelium leads to platelet aggregation on the endothelial surface and release of platelet-derived mitogens such as platelet-derived growth factor (PDGF) which stimulates smooth muscle cell growth. Death of the foam cells leads to deposition of the cellular lipid and fibrosis. The resulting atherosclerotic plaque narrows the blood vessel and leads to thrombus formation, which precipitates myocardial infarction (heart attack).

Source: Wick, G, Knoflach, M., and Xu, Q. B. Autoimmune and inflammatory mechanisms in atherosclerosis. *Annu. Rev. Immunol.* 22:361, 2004.

to the plasma membrane. Another group has LDL receptors that have a defect in the cytoplasmic carboxyl terminus and are unable to internalize the LDL–LDL receptor complex.

In specialized tissues such as the adrenal cortex and ovaries, cholesterol derived from LDL is the precursor for steroid hormones, such as cortisol and estradiol, respectively. In liver, cholesterol extracted from LDL and HDL is converted into bile salts that function in intestinal fat digestion.

Cholesterol Is Excreted Primarily as Bile Acids

Bile acids are the end products of cholesterol metabolism. Primary bile acids are synthesized in hepatocytes directly from cholesterol. The bile acids are derivatives of cholanic acid (Figure 18.41). **Cholic acid** and **chenodeoxycholic acid** (Figure 18.42) are C_{24} compounds containing three and two OH groups, respectively, and a C5 side chain that ends in a carboxyl group that is ionized at pH 7.0 (hence the name bile salt). The carboxyl group is often conjugated via an amide bond to glycine (NH_2-CH_2-COOH) (Figure 18.43) or taurine (NH_2– CH_2-CH_2-SO_3H) to form glycocholic or taurocholic acid, respectively.

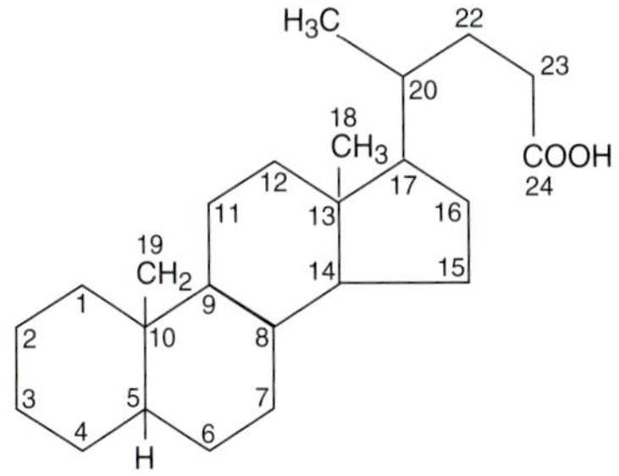

FIGURE 18.41
Structure of cholanic acid.

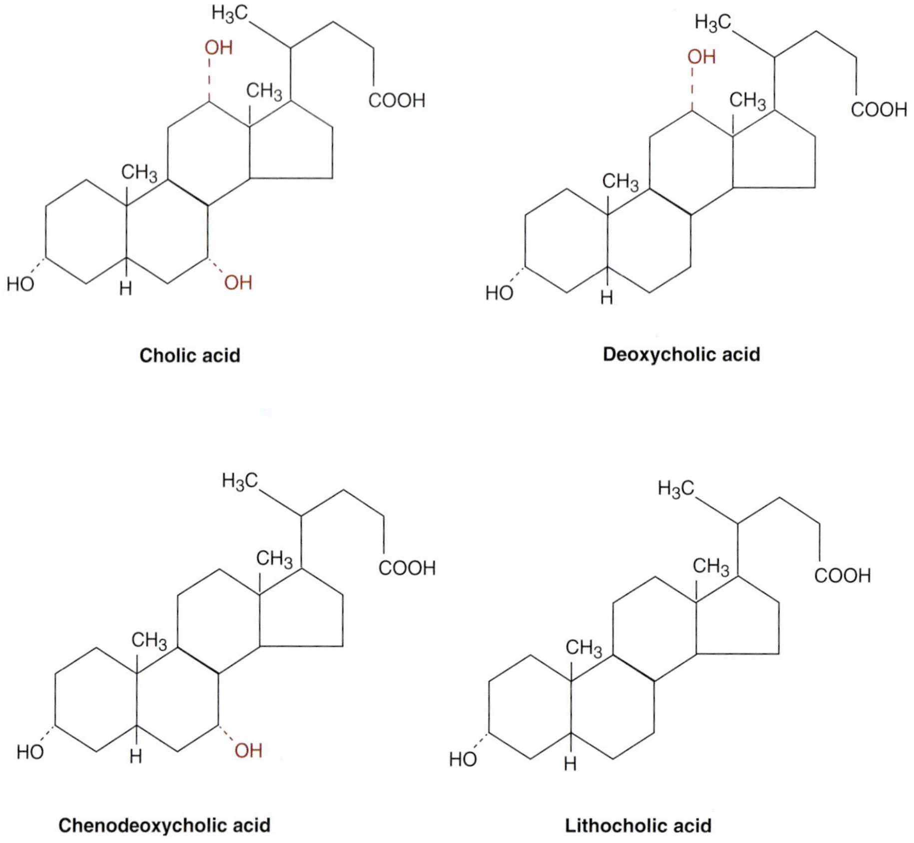

FIGURE 18.42

Structures of some common bile acids.

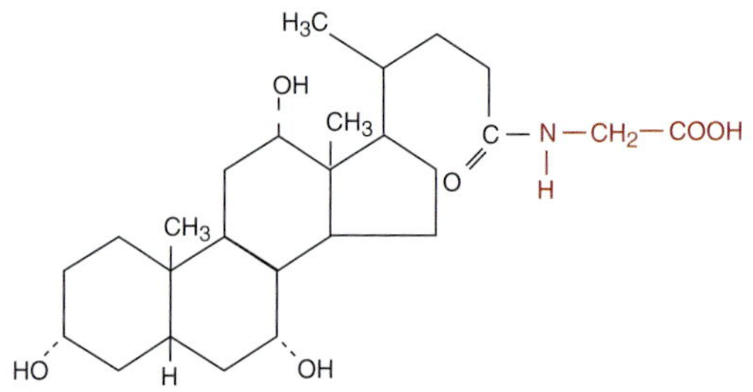

FIGURE 18.43

Structure of glycocholic acid, a conjugated bile acid.

The primary bile acids are modified by microorganisms in the gut to the secondary bile acids: Deoxycholic acid and lithocholic acid are derived from cholic acid and chenodeoxycholic acid, respectively, by the removal of one OH group (Figure 18.42). Transformation of cholesterol to bile acids requires: (1) epimerization of the 3β-OH group, (2) reduction of the C5 double bond, (3) introduction of OH groups at C7 (chenodeoxycholic acid) or at C7 and C12 (cholic acid), and (4) conversion of the C27 side chain into a C24 carboxylic acid by elimination of a propyl equivalent.

Bile acids are secreted into bile, stored in the gallbladder, and then secreted into the small intestine. Liver production of bile acids is insufficient to meet the physiological needs, so the body relies on an enterohepatic circulation that carries the bile acids from the intestine back to the liver several times each day.

Bile acids and phospholipids solubilize cholesterol in bile, thereby preventing cholesterol from precipitating in the gallbladder. Bile acids in the gut act as emulsifying agents for dietary triacylglycerols, facilitating their hydrolysis by pancreatic lipase. Bile acids play a direct role in activating pancreatic lipase (see p. 1060) and facilitate the absorption of fat-soluble vitamins, particularly vitamin D, from the intestine.

Vitamin D Is Synthesized from an Intermediate of Cholesterol Synthesis

Cholesterol synthesis provides substrate for the photochemical production of **vitamin D_3** in the skin (see p. 1095 for a detailed discussion of the synthesis and the function of vitamin D_3, and see p. 1097 for its role in control of Ca^{2+} metabolism). Vitamin D_3 is a secosteroid in which the 9,10 carbon–carbon bond of the B ring of cholesterol has undergone fission (Figure 18.44). Vitamin D_3 is manufactured in the skin where 7-dehydrocholesterol, an intermediate in the pathway of cholesterol synthesis, is converted to provitamin D_3 by irradiation with UV rays of the sun (285–310 nm). Provitamin D_3 is biologically inert and labile. It is converted slowly (~36 h) and nonenzymatically to the biologically active vitamin, cholecalciferol (vitamin D_3). As little as 10-min exposure to sunlight each day of the hands and face is sufficient to satisfy

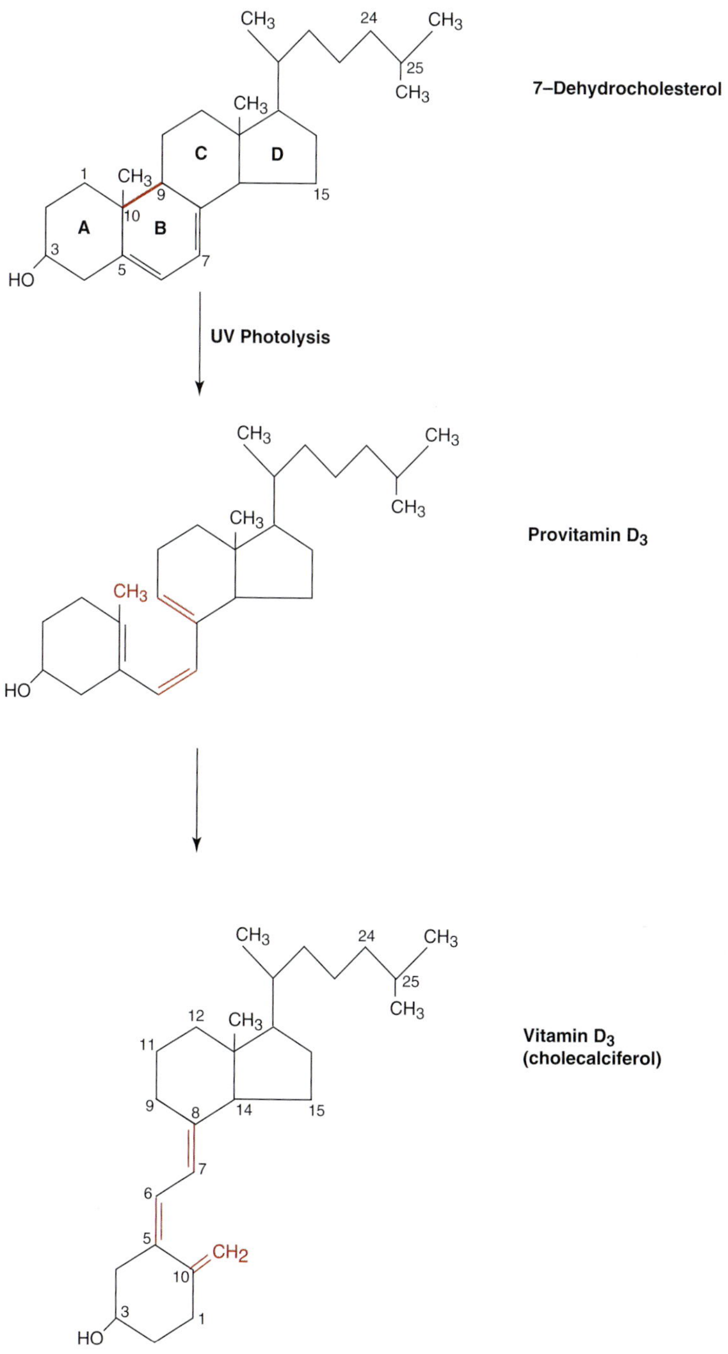

7–Dehydrocholesterol

UV Photolysis

Provitamin D_3

Vitamin D_3 (cholecalciferol)

FIGURE 18.44

Photochemical conversion of 7-dehydrocholesterol to vitamin D_3 (cholecalciferol).

the body's need for vitamin D. Photochemical action on the plant sterol ergosterol also provides a dietary precursor to vitamin D_2 (calciferol) that can satisfy the vitamin D requirement.

18.4 SPHINGOLIPIDS

Synthesis of Sphingosine

Sphingolipids are complex lipids whose core structure is provided by the long-chain amino alcohol sphingosine (Figure 18.45) (**4-sphingenine** or *trans*-1,3-dihydroxy-2-amino-4-octadecene). **Sphingosine** has two asymmetric carbon atoms (C2 and C3); of the four possible optical isomers, naturally occurring sphingosine is the D-erythro form. The double bond has the trans configuration. The primary alcohol group at C1 is a nucleophilic center that is linked with sugars in glycosphingolipids. The amino group at C2 always bears a long-chain fatty acid (usually $C_{20}-C_{26}$) in amide linkage in ceramides. The secondary alcohol at C3 is always free. Note the structural similarity of this part of the sphingosine molecule to the glycerol moiety of the acylglycerols (Figure 18.45). Sphingolipids occur in blood and the plasma membranes of nearly all cells. The highest concentrations are found in the white matter of the central nervous system.

Sphingosine is synthesized by way of **sphinganine (dihydrosphingosine)** from serine and palmitoyl CoA. Serine provides C1 and C2, and the amino group of sphingosine and palmitic acid provides the remaining carbon atoms. The condensation of serine and palmitoyl CoA is catalyzed by **serine palmitoyltransferase**, a pyridoxal phosphate-dependent enzyme. The driving force for the reaction is provided by both cleavage of the reactive, high-energy, thioester bond of palmitoyl CoA and the release of CO_2 from serine (Figure 18.46). Reduction of the carbonyl group in 3-ketodihydrosphingosine to produce sphinganine (Figure 18.47) occurs at the expense of NADPH. Insertion of the

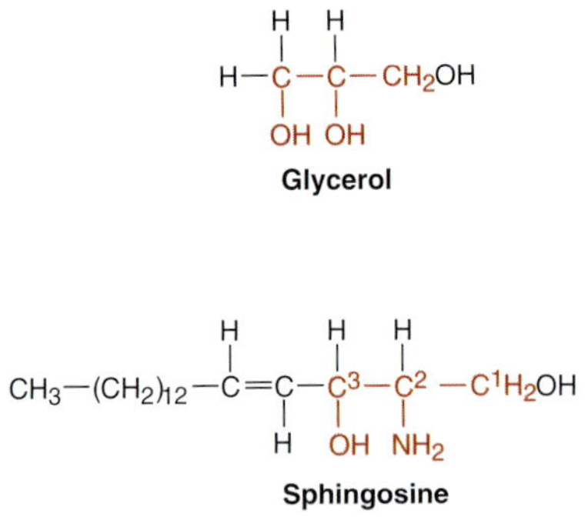

H H
H—C—C—CH2OH
OH OH
Glycerol

H H H
CH3—(CH2)12—C=C—C3—C2—C1H2OH
H OH NH2
Sphingosine

FIGURE 18.45

Comparison of structures of glycerol and sphingosine (*trans*-1,3-dihydroxy-2-amino-4-octadecene).

CH3—(CH2)14—C(=O)—SCoA + ⁻OOC—CH(NH3+)—CH2OH —(Mn2+, pyridoxal phosphate)→ CH3—(CH2)14—C(=O)—CH(NH2)—CH2OH + CO2 + CoA

Palmitoyl CoA L-Serine 3-Ketodihydrosphingosine

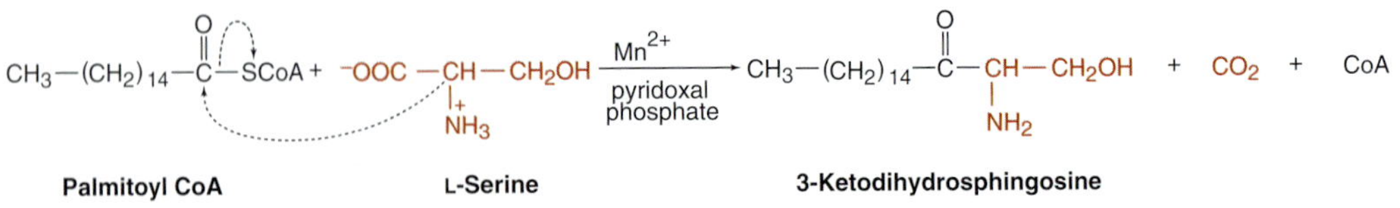

FIGURE 18.46

Formation of 3-ketodihydrosphingosine from serine and palmitoyl CoA.

CH3—(CH2)12—CH2—CH2—C(=O)—CH(NH2)—CH2OH —(H+ + NADPH → NADP+)→ CH3—(CH2)12—CH2—CH2—CH(OH)—CH(NH2)—CH2OH

3-Ketodihydrosphingosine Sphinganine (dihydrosphingosine)

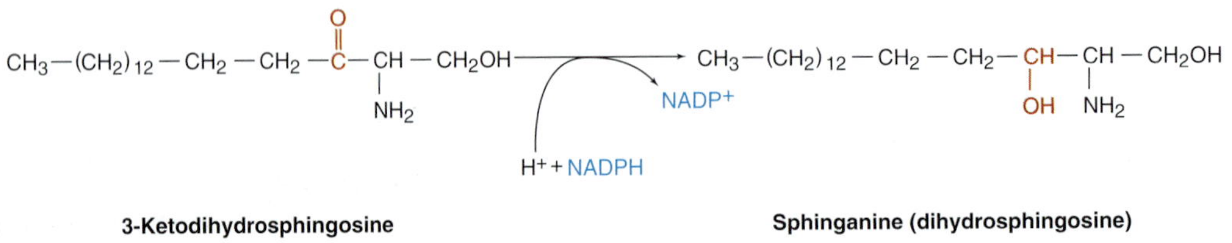

FIGURE 18.47

Conversion of 3-ketodihydrosphingosine to sphinganine.

double bond into sphinganine to produce sphingosine occurs at the level of ceramide (see below).

Ceramides Are Fatty Acid Amide Derivatives of Sphingosine

Sphingosine is the precursor of **ceramide**, a long-chain fatty acid amide derivative of sphingosine, which provides the core structure of sphingolipids. The fatty acid is attached to the 2-amino group of sphingosine through an amide bond (Figure 18.48). Most often the acyl group is behenic acid, a saturated 22-carbon fatty acid, but other long-chain acyl groups can be used. The two long-chain hydrocarbon domains in the ceramide molecule regions are responsible for the lipid character of sphingolipids.

Ceramide is synthesized from dihydrosphingosine and a long-chain fatty acyl CoA by an endoplasmic reticulum enzyme. Dihydroceramide is an intermediate that is then dehydrogenated at C4 and C5 (Figure 18.49). Ceramide is not a component of membrane lipids but rather is an intermediate in synthesis and catabolism of glycosphingolipids and sphingomyelin. Structures of prominent sphingolipids of humans are presented in Figure 18.50 in diagrammatic form.

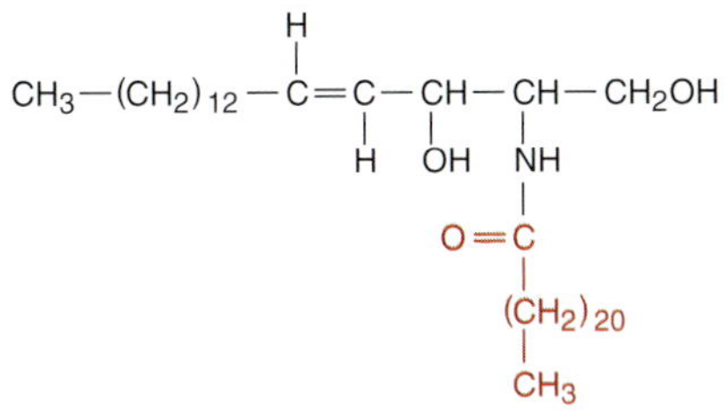

$CH_3-(CH_2)_{12}-CH=CH-CH(OH)-CH(NH-C(=O)-(CH_2)_{20}-CH_3)-CH_2OH$

FIGURE 18.48

Structure of a ceramide (*N*-acylsphingosine).

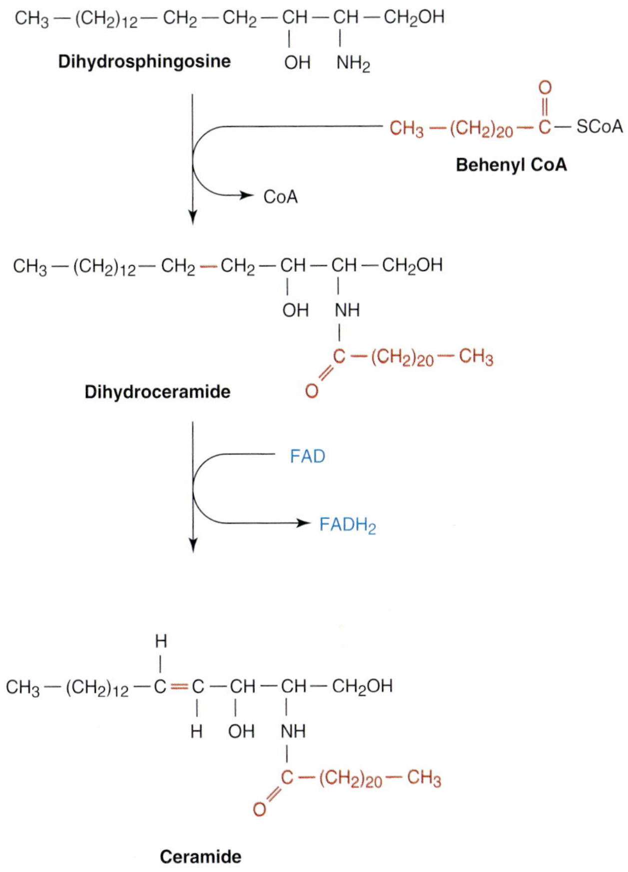

$CH_3-(CH_2)_{12}-CH_2-CH_2-CH(OH)-CH(NH_2)-CH_2OH$

Dihydrosphingosine

$CH_3-(CH_2)_{20}-C(=O)-SCoA$ **Behenyl CoA** → CoA

$CH_3-(CH_2)_{12}-CH_2-CH_2-CH(OH)-CH(NH-C(=O)-(CH_2)_{20}-CH_3)-CH_2OH$

Dihydroceramide

FAD → $FADH_2$

$CH_3-(CH_2)_{12}-CH=CH-CH(OH)-CH(NH-C(=O)-(CH_2)_{20}-CH_3)-CH_2OH$

Ceramide

FIGURE 18.49

Formation of ceramide from dihydrosphingosine.

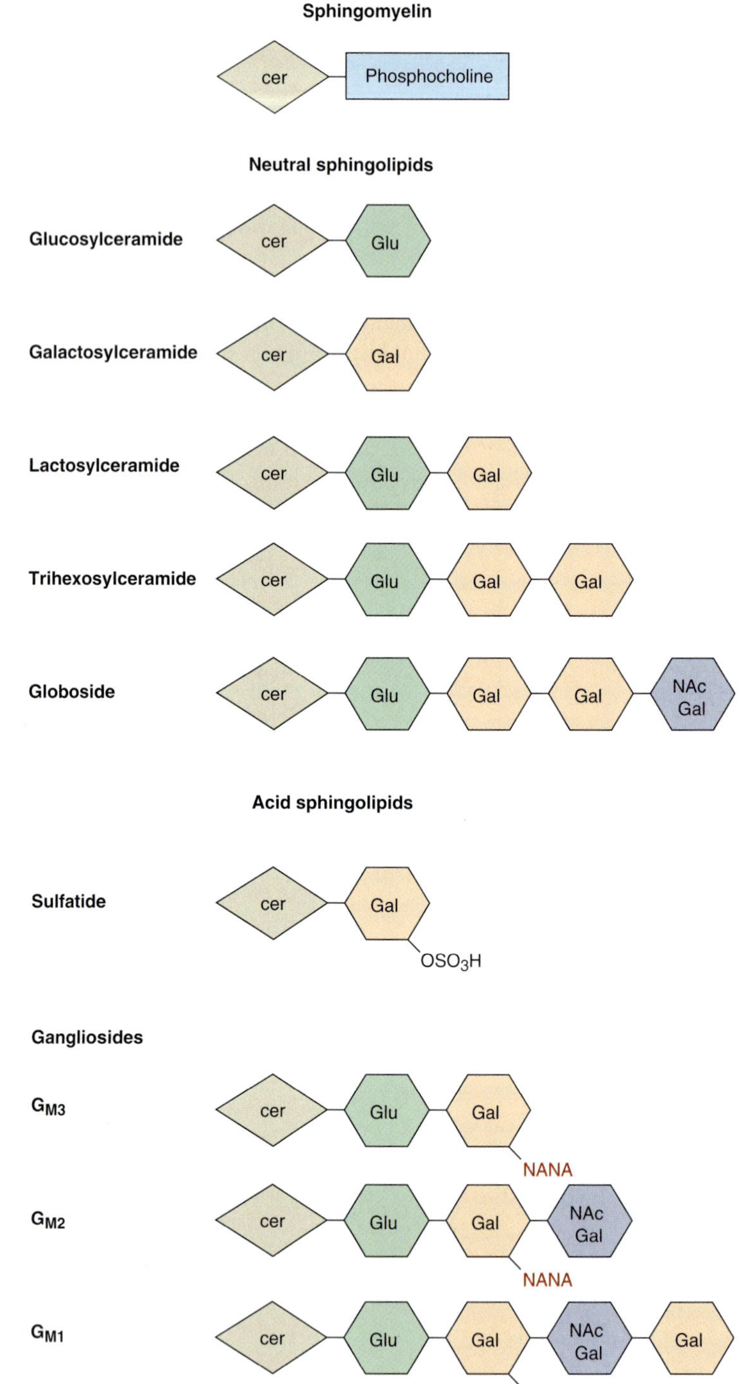

FIGURE 18.50

Structures of some common sphingolipids in diagrammatic form. Cer, ceramide; Glu, glucose; Gal, galactose; NAcGal, *N*-acetylgalactosamine; NANA, *N*-acetylneuraminic acid (sialic acid).

Sphingomyelin Is a Phosphorus-Containing Sphingolipid

Sphingomyelin, a major component of membranes of nervous tissue, is a phospholipid. Since this ceramide phosphocholine contains one negative and one positive charge, it is neutral at physiological pH (Figure 18.51). The most common fatty acids in sphingomyelin are palmitic (16:0), stearic (18:0), lignoceric (24:0), and nervonic acid (24:1). The sphingomyelin of myelin contains predominantly **lignoceric acid and nervonic acid**, whereas that of gray matter contains largely stearic acid. Excessive accumulation of sphingomyelin occurs in Niemann–Pick disease. Conversion of ceramide to sphingomyelin by sphingomyelin synthase involves transfer of a phosphocholine moiety from phosphatidylcholine (lecithin) (Figure 18.52).

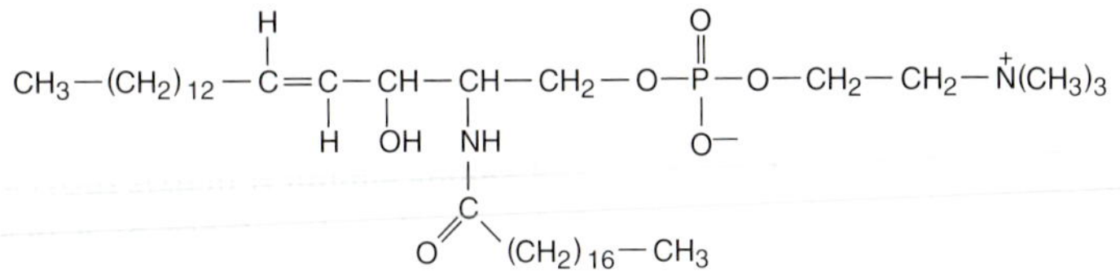

FIGURE 18.51
Structure of sphingomyelin.

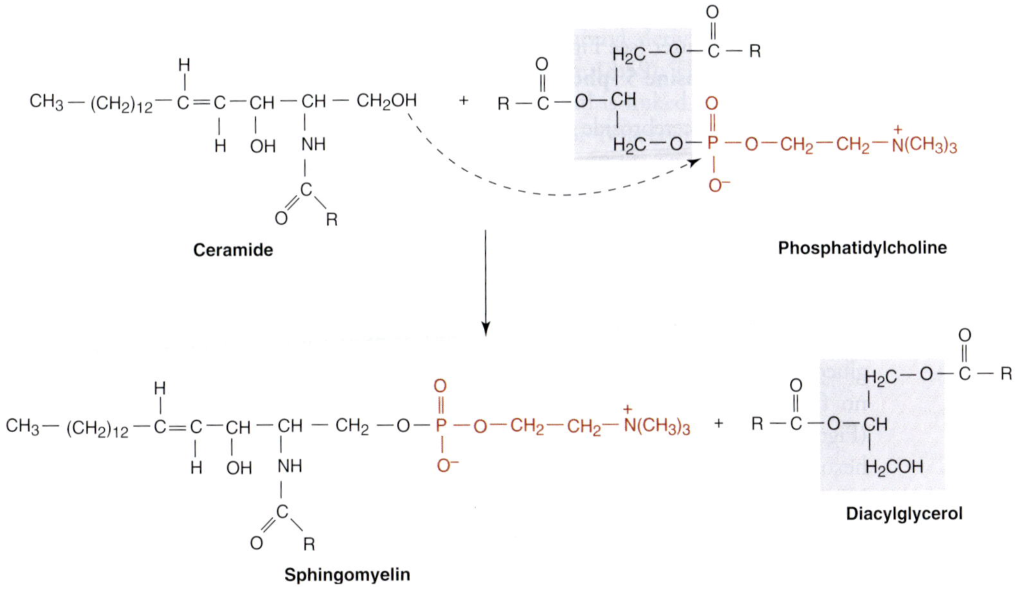

Ceramide
Phosphatidylcholine
Sphingomyelin
Diacylglycerol

FIGURE 18.52
Sphingomyelin synthesis from ceramide and phosphatidylcholine.

Glycosphingolipids Usually Contain Galactose or Glucose

The principal glycosphingolipid classes are cerebrosides, sulfatides, globosides, and gangliosides. In glycolipids the polar head group attached to sphingosine is a sugar molecule.

Cerebrosides Are Glycosylceramides

Cerebrosides are ceramide monohexosides; the most common are galactocerebroside and glucocerebroside. Unless specified otherwise, the term cerebroside usually refers to galactocerebroside, also called "galactolipid." In Figure 18.53, note that C1 of the monosaccharide unit is attached to C1 of the ceramide and the anomeric configuration of the sugar moiety in both galactocerebroside and glucocerebroside is β. Most galactocerebroside in healthy individuals is found in the brain. Moderate amounts of galactocerebroside accumulate in the white matter in Krabbe's disease, or globoid leukodystrophy, a deficiency in lysosomal galactocerebrosidase.

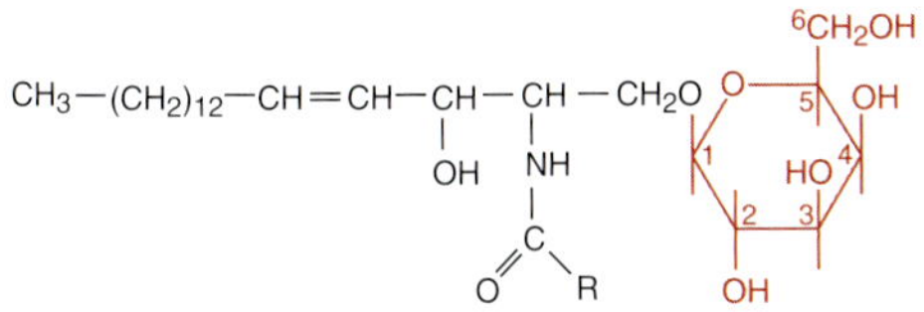

FIGURE 18.53
Structure of galactocerebroside (galactolipid).

Glucocerebroside (glucosylceramide)(see Figure 18.54) is not normally a component of membranes. It is an intermediate in the synthesis and degradation of more complex glycosphingolipids. However, 100-fold increases in glucocerebroside occur in spleen and liver in the genetic lipid storage disorder called Gaucher's disease, a deficiency of lysosomal glucocerebrosidase.

Galactocerebroside and glucocerebroside are synthesized from ceramide and from UDP-galactose and UDP-glucose, respectively. The glucosyl and galactosyl transferases that catalyze these reactions are associated with the endoplasmic reticulum (Figure 18.55). In some tissues, synthesis of glucocerebroside proceeds by glucosylation of sphingosine by **glucosyltransferase**:

$$\text{Sphingosine} + \text{UDP-glucose} \rightarrow \text{glucosylsphingosine} + \text{UDP}$$

followed by fatty acylation:

$$\text{Glucosylsphingosine} + \text{stearoyl CoA} \rightarrow \text{glucocerebroside} + \text{CoASH}$$

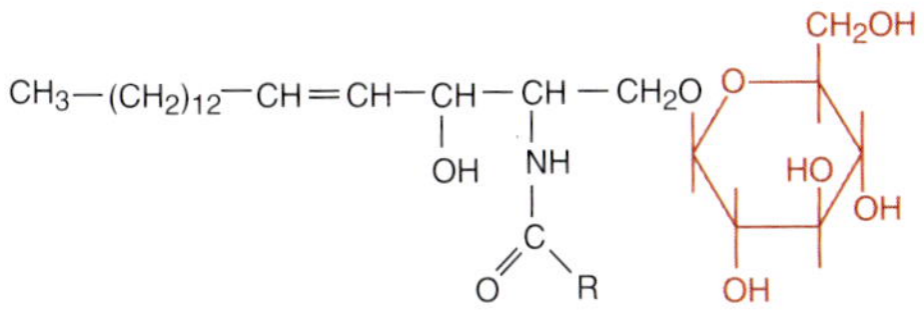

FIGURE 18.54
Structure of glucocerebroside.

where the presence of fatty acyl chains containing a polar OH group may disturb lipid packing and thus the structure and function of the membrane. LTB_4 is immunosuppressive through inhibition of $CD4^+$ cells and proliferation of suppressor $CD8^+$ cells. LTB_4 also promotes neutrophil–endothelial cell adhesion.

Monohydroxy eicosatetraenoic acids of the lipoxygenase pathway are potent mediators of processes involved in **allergy (hypersensitivity)** and inflammation, secretion (e.g., insulin), cell movement, cell growth, and calcium fluxes. The initial allergic event—namely, the binding of IgE antibody to receptors on the surface of the mast cell—causes release of substances, including leukotrienes, referred to as mediators of immediate hypersensitivity. Lipoxygenase products are usually produced within minutes after the stimulus. The leukotrienes LTC_4, LTD_4, and LTE_4 are much more potent than histamine in contracting nonvascular smooth muscles of bronchi and intestine. LTD_4 increases the permeability of the microvasculature. LTB_4 stimulates migration (chemotaxis) of eosinophils and neutrophils, making them the principal mediators of PMN-leukocyte infiltration in inflammatory reactions.

Eicosatrienoic acids (e.g., dihomo-γ-linolenic acid) and eicosapentaenoic acid (Fig. 18.65) also serve as lipoxygenase substrates. The amount of these 20-carbon polyunsaturated fatty acids with three and five double bonds in tissues is less than that of arachidonic acid, but special diets can increase their levels. The lipoxygenase products of these tri- and pentaeicosanoids are usually less active than LTA_4 or LTB_4. It remains to be determined if fish oil diets rich in eicosapentaenoic acid are useful in the treatment of allergic and autoimmune diseases. Pharmaceutical research into therapeutic uses of lipoxygenase and cyclooxygenase inhibitors and inhibitors and agonists of leukotrienes in treatment of inflammatory diseases such as asthma, psoriasis, rheumatoid arthritis, and ulcerative colitis is very active.

BIBLIOGRAPHY

Phospholipid Metabolism

Brites, P, Waterham, H. R., and Wanders, R. J. Functions and biosynthesis of plasmalogens in health and disease. *Biochim. Biophys. Acta* 1636:219, 2004.

Frasch, S. C., Henson, P. M, Nagaosa, K Fessler, MB, Borregaard, N, and Bratton, DL. Phospholipid flip-flop and phospholipids scramblase 1 (PLSCR1) co-localize to uropod rafts in formylated Met-Leu-Phe-stimulated neutrophils. *J. Biol. Chem.* 279:17625, 2004.

Kazzi, S. N., Schurch, S., McLaughlin, K. L., Romero, R., and Janisse, J. Surfactant phospholipids and surface activity among preterm infants with respiratory distress syndrome who develop bronchopulmonary dysplasia. *Acta Paediatr.* 89:1218, 2000.

Lie, J., de Crom, R., van Gent, T., van Haperen, R., Scheek, L., Sadeghi-Niaraki, F., and van Tol, A. Elevation of plasma phospholipid transfer protein increases the risk of atherosclerosis despite lower apolipoprotein B-containing lipoproteins. *J. Lipid Res.* 45:805, 2004.

Okeley, N. M. and Gelb, M. H. A designed probe for acidic phospholipids reveals the unique enriched anionic character of the cytosolic face of the mammalian plasma membrane. *J. Biol. Chem.* 279:21833, 2004.

Tjoelker, L. W., and Stafforini, D. M. Platelet-activating factor acetylhydrolases in health and disease. *Biochem. Biophys. Acta* 1488:102, 2000.

Zwaal, R. F., Comfurius, P., and Bevers, E. M. Scott syndrome, a bleeding disorder caused by defective scrambling of membrane phospholipids. *Biochim. Biophys. Acta Mol. Cell Biol. Lipids* 1636:119, 2004.

Lipoproteins

Agren, J. J., Ravandi, A., Kuksis, A., and Steiner, G. Structural and compositional changes in very low density lipoprotein triacylglycerols during basal lipolysis. *Eur. J. Biochem.* 269:6223, 2002.

Berriot-Varoqueaux, N., Aggerbeck, L. P., Samson-Bouma, M. E., and Wetterau, J. R. The role of the microsomal triglyceride transfer protein in abetalipoproteinemia. *Annu. Rev. Nutr.* 20:663, 2000.

Clader, J. W. The discovery of ezetimibe: a view from outside the receptor. *J. Med. Chem.* 47:1 2004.

de Beer, M. C., Castellani, L. W, Cai, L., Stromberg, A. J., de Beer, F. C., and van der Westhuyzen, D. R.. ApoA-II modulates the association of HDL with class B scavenger receptors SR-BI and CD36. *J. Lipid Res.* 45:706, 2004.

Hoffman-Kuczynski, B, Reo, N. V. Studies of myo-inositol and plasmalogen metabolism in rat brain. *Neurochem. Res.* 29:843, 2004.

McNamara, D. J. Dietary fatty acids, lipoproteins, and cardiovascular disease. *Adv. Food Nutr. Res.* 36:253, 1999.

Wang, M. and Briggs, M. R. HDL: The metabolism, function, and therapeutic importance. *Chem. Rev.* 104:119, 2004.

Cholesterol Synthesis

Cachefo, A, Boucher, P, Dusserre, E, Bouletreau, P, Beylot, M, and Chambrier, U. Stimulation of cholesterol synthesis and hepatic lipogenesis in patients with severe malabsorption. *J. Lipid Res.* 44:1349, 2003.

Goldstein, J. L., and Brown, M. S. Regulation of the mevalonate pathway. *Nature* 343:425, 1990.

Lung Surfactant

Abonyo, B. O., Gou, D., Wang, P., Narasaraju, T., Wang, Z., and Liu, L. Syntaxin 2 and SNAP-23 are required for regulated surfactant secretion. *Biochemistry* 43:3499, 2004.

Dani, C., Martelli, E., Buonocore, G., Longini, M., Di Filippo, A., Giossi, M., and Rubaltelli, F. F. Influence of bilirubin on oxidative lung damage and surface tension properties of lung surfactant. *Pediatr. Res.* 53:S574A, 2003.

Hallman, M. Lung surfactant, respiratory failure, and genes. *N. Engl. J. Med.* 350:1278, 2004.

Nadesalingam, J., Bernal, A. L., Dodds, A. W., Willis, A. C., Mahoney, D. J., Day, A. J., Reid, K. B., and Palaniyar, N. Identification and characterization of a novel interaction between pulmonary surfactant protein D and decorin. *J. Biol. Chem.* 278:25678, 2003.

Rau, G. A, Dombrowsky, H., Gerbert, A., Thole, H. H., von der Hardt, H., Freihorst, J., and Bernhard, W. Phosphatidylcholine metabolism of rat trachea in relation to lung parenchyma and surfactant. *J. Appl. Physiol.* 95:1145, 2003.

Schram, V., Anyan, W. R., and Hall, S. B. Non-cooperative effects of lung surfactant proteins on early adsorption to an air/water interface. *Biochim. Biophys. Acta.* 1616:165, 2003.

Shulenin, S., Nogee, L. M., Annilo, T., Wert, S. E., Whitsett, J. A., and Dean, M. ABCA3 gene mutations in newborns with fatal surfactant deficiency. *N. Engl. J. Med.* 350:1296, 2004.

Wang, Z., Schwan, A. L., Lairson, L. L., O'Donnell, J. S., Byrne et al. Surface activity of a synthetic lung surfactant containing a phospholipase-resistant phosphonolipid analog of dipalmitoyl phosphatidylcholine. *Am. J. Physiol. Lung Cell. Mol. Physiol.* 285:L550, 2003.

Prostaglandins Thromboxanes and Leukotrienes

Bos, C. L., Richel, D. J, Ritsema, T., Peppelenbosch, M. P., and Versteeg, H. H. Prostanoids and prostanoid receptors in signal transduction. *Int. J. Biochem. Cell Biol.* 36:1187, 2004.

Cipollone, F., Rocca, B., and Patrono, C. Cyclooxygenase-2 expression and inhibition in atherothrombosis. *Atheroscler. Thromb. Vasc. Biol.* 24:246, 2003.

De Caterina, R. and Zampolli, A. From asthma to atherosclerosis—5-lipoxygenase, leukotrienes, and inflammation. *N. Engl. J. Med.* 350:4, 2004.

Dwyer, J. H., Allayee, H., Dwyer, K. M., Fan, J., Wu, H. et al. Arachidonate 5-lipoxygenase promoter genotype, dietary arachidonic acid, and atherosclerosis. *N. Engl. J. Med.* 350:29, 2004.

Leslie, C. C. Regulation of the specific release of arachidonic acid by cytosolic phospholipase A2. *Prostaglandins Leukot. Essent. Fatty Acids* 70:373, 2004.

Mayatepek, E., Okun, J. G., Meissner, T., Assmann, B., Hammond, J. et al. Peroxidase activity, of cyclooxygenase-2 (COX-2) cross-links beta-amyloid (Abeta) and generates Abeta-COX-2 hetero-oligomers that are increased in Alzheimer's disease. *J. Biol. Chem.* 279:14673, 2004.

Takezono, Y., Joh, T., Oshima, T., Suzuki, H., Seno, K. et al. M. Role of prostaglandins in maintaining gastric mucus-cell permeability against acid exposure. *J. Lab. Clin. Med.* 143:52, 2004.

Bile Acids

Dawson, P. A, Haywood, J., Craddock, A. L., Wilson, M., Tietjen, M. et al. Targeted deletion of the ileal bile acid transporter eliminates enterohepatic cycling of bile acids in mice. *J. Biol. Chem.* 278:33920, 2003.

De Fabiani, E., Mitro, N., Gilardi, F., Caruso, D., Galli, G., and Crestani, M. Coordinated control of cholesterol catabolism to bile acids and of gluconeogenesis via a novel mechanism of transcription regulation linked to the fasted-to-fed cycle. *J. Biol. Chem.* 278:39124, 2003.

Higuchi, H. and Gores, G. J. Bile acid regulation of hepatic physiology: IV. Bile acids and death receptors. *Am. J. Physiol. Gastrointest. Liver Physiol.* 284:G734, 2003.

Knarreborg, A., Jensen, S. K., and Engberg, R. M. Pancreatic lipase activity as influenced by unconjugated bile acids and pH, measure *in vitro* and *in vivo*. *J. Nutri. Biochem.* 14:259, 2003.

O'Byrne, J., Hunt, M. C., Rai, D. K., Saeki, M., and Alexson, S. E. The human bile acid–CoA:amino acid *N*-acyltransferase functions in the conjugation of fatty acids to glycine. *J. Biol. Chem.* 278:34237, 2003.

Sphingolipids and the Sphingolipidoses

Cox, T. M. Gaucher disease:understanding the molecular pathogenesis of sphingolipidoses. *J. Inherit. Metab. Dis.* 24:S106, 2001.

Linke, T., Wilkening, G., Sadeghlar, F., Mozcall, H., Bernardo, K. et al. Interfacial regulation of acid ceramidase activity. Stimulation of ceramide degradation by lysosomal lipids and sphingolipid activator proteins. *J. Biol. Chem.* 276:5760, 2001.

Sandhoff, K., and Kolter, T. Biosynthesis and degradation of mammalian glycosphingolipids. *Philos. Trans. R. Soc. Lond. B. Biol. Sci.* 358:847, 2003.

Watts, R. W. A historical perspective of the glycosphingolipids and sphingolipidoses. *Philos. Trans. R. Soc. Lond. B. Biol. Sci.* 358:975, 2003.

QUESTIONS | CAROL N. ANGSTADT

Multiple Choice Questions

1. Roles of various phospholipids include all of the following *except*:
   A. cell–cell recognition.
   B. a surfactant function in lung.
   C. activation of certain membrane enzymes.
   D. signal transduction.
   E. mediator of hypersensitivity and acute inflammatory reactions.

2. CDP-X (where X is the appropriate alcohol) reacts with 1,2-diacylglycerol in the primary synthetic pathway for:
   A. phosphatidylcholine.
   B. phosphatidylinositol.
   C. phosphatidylserine.
   D. all of the above.
   E. none of the above.

3. Phospholipases A_1 and A_2:
   A. have no role in phospholipid synthesis.
   B. are responsible for initial insertion of fatty acids in *sn*-1 and *sn*-2 positions during synthesis.
   C. are responsible for base exchange in the interconversion of phosphatidylethanolamine and phosphatidylserine.
   D. hydrolyze phosphatidic acid to a acylglycerol.
   E. remove a fatty acid in an *sn*-1 or *sn*-2 position so it can be replaced by another in phospholipid synthesis.

4. Bile acids differ from their precursor cholesterol in that they:
   A. are not amphipathic.
   B. contain an ionizable carboxyl group.
   C. contain less oxygen.
   D. are synthesized primarily in intestine.
   E. contain more double bonds.

5. A ganglioside may contain all of the following *except*:
   A. a ceramide structure.
   B. glucose or galactose.
   C. phosphate.
   D. one or more sialic acids.
   E. sphingosine.

6. Phosphatidylinositols:
   A. are neutral phospholipids.
   B. are found primarily in mitochondrial membrane.
   C. release Ca^{2+} from the endoplasmic reticulum.
   D. can serve to anchor glycoproteins to cell surfaces.
   E. are the major surfactants maintaining normal lung function.

Questions 7 and 8: Prostaglandins and leukotrienes are physiologically highly reactive compounds produced from polyunsaturated C_{20} acids like arachidonic acid. Both play some role in inflammatory processes. NSAIDs (nonsteroidal anti-inflammatory drugs) like aspirin are effective in inhibiting prostaglandin production but not leukotrienes.

7. Prostaglandin synthase, a bifunctional enzyme:
   A. catalyzes rate-limiting step of prostaglandin synthesis.
   B. is inhibited by anti-inflammatory steroids.
   C. contains both a cyclooxygenase and a peroxidase component.
   D. produces PGG_2 as the end product.
   E. uses as substrate the pool of free arachidonic acid in the cell.

8. Hydroperoxyeicosatetraenoic acids (HPETEs):
   A. are derived from arachidonic acid by a peroxidase reaction.
   B. are mediators of hypersensitivity reactions.
   C. are intermediates in formation of leukotrienes.
   D. are relatively stable compounds (persist for as long as 4 h).
   E. are inactivated forms of leukotrienes.

Questions 9 and 10: Hypercholesterolemia is one of the risk factors for cardiovascular disease. Total cholesterol in blood is not as important in considering risk as distribution of cholesterol between LDL ("bad") and HDL ("good") particles. Efforts to reduce serum cholesterol are two-pronged: dietary and drugs if reducing dietary cholesterol and saturated fat is not sufficient. The drugs of choice are statins, inhibitors of the rate-limiting enzyme of cholesterol biosynthesis. If necessary, bile acid-binding resins can be added.

9. In biosynthesis of cholesterol:
   A. 3-hydroxy-3-methyl glutaryl CoA (HMG CoA) is synthesized by mitochondrial HMG CoA synthase.
   B. HMG-CoA reductase catalyzes the rate-limiting step.
   C. the conversion of mevalonic acid to farnesyl pyrophosphate proceeds via condensation of 3 molecules of mevalonic acid.
   D. condensation of 2 farnesyl pyrophosphates to form squalene is freely reversible.
   E. conversion of squalene to lanosterol is initiated by formation of the fused ring system, followed by addition of oxygen.

10. Cholesterol present in LDL (low-density lipoproteins):
    A. binds to a cell receptor and diffuses across the cell membrane.
    B. when it enters a cell, suppresses activity of ACAT (acyl-CoA-cholesterol acyl transferase).
    C. once in the cell is converted to cholesteryl esters by LCAT (lecithin-cholesterol acyl transferase).
    D. once it has accumulated in the cell, inhibits replenishment of LDL receptors.
    E. represents primarily cholesterol that is being removed from peripheral cells.

Questions 11 and 12: Sphingolipidoses (lipid storage diseases) are a group of diseases characterized by defects in lysosomal enzymes. Rate of synthesis of sphingolipids is normal, but undegraded material accumulates in lysosomes. The severity of the particular disease depends, in part, on what tissues are most affected. Those affecting primarily brain and nervous tissue, like Tay–Sachs disease, are lethal within a few years of birth. Since the extent of enzyme deficiency is the same in all cells, easily accessible tissues like skin fibroblasts can be used to assay for enzyme deficiency. There are no treatments for most of the diseases, so genetic counseling seems to be the only helpful approach. However, for Gaucher's disease, enzyme (glucocerebrosidase) replacement therapy is effective.

11. All of the following are true about degradation of sphingolipids *except* it:
    A. occurs by hydrolytic enzymes contained in lysosomes.
    B. terminates at level of ceramides.
    C. is a sequential, stepwise removal of constituents.
    D. may involve a sulfatase or a neuraminidase.
    E. is catalyzed by enzymes specific for a type of linkage rather than specific compound.

12. In Niemann–Pick disease, the deficient enzyme is sphingomyelinase. Sphingomyelins differ from other sphingolipids in that they are:
    A. not based on a ceramide core.
    B. acidic rather than neutral at physiological pH.
    C. only types containing *N*-acetylneuraminic acid.
    D. only types that are phospholipids.
    E. not amphipathic.

Problems

13. Cells from a patient with familial hypercholesterolemia (FH) and cells from an individual without that disease were incubated with LDL particles containing radioactively labeled cholesterol. After incubation, the incubation medium was removed and the radioactivity of the cells was measured. The cells were treated to remove any bound material and were lysed, and internal cholesterol content was measured. Results are given below. What mutation of the gene for the LDL receptor protein could account for the results?

Cell Type	*Radioactivity of Cell*	*Cholesterol Content*
Normal	3000 cpm/mg cells	Low
FH	3000 cpm/mg cells	High

14. The combination of bile salt-binding resin and an HMG-CoA reductase inhibitor is very effective in reducing serum cholesterol for most patients with high cholesterol. Why is this treatment much less effective for patients with familial hypercholesterolemia?

ANSWERS

1. **A** This function appears to be associated with complex glycosphingolipids. B: Especially dipalmitoyllecithin. C: For example, β-hydroxybutyrate dehydrogenase. D: Especially phosphatidylinositols. E: Platelet activity factor (PAF) does this.
2. **A** This is the main pathway for choline. B: Phosphatidylinositol is formed from CDP-diglyceride reacting with myoinositol. C: This is formed by "base exchange."
3. **E** Phospholipases A_1 and A_2, as their names imply, hydrolyze a fatty acid from a phospholipid and so are part of phospholipid degradation. They are also important in synthesis, however, in ensuring the asymmetric distribution of fatty acids that occurs in phospholipids.
4. **B** This carboxyl group is often conjugated to glycine or taurine. A: Their role depends on being amphipathic. C: Hydroxylation is a major part of synthesis. D: Liver is site of synthesis. E: Primary bile acids are unsaturated.
5. **C** Glycosphingolipids do not contain phosphate. A, E: Ceramide, which is formed from sphingosine, is the base structure from which glycosphingolipids are formed. B: Glucose is usually the first sugar attached to the ceramide. D: By definition, gangliosides contain sialic acid.
6. **D** Glycerol-phosphatidylinositol anchors have specific structural features. A: These are acidic—even more so as inositol is phosphorylated. B: Their functions reside at the plasma membrane. C: PIP_2 must be hydrolyzed to IP_3, which releases Ca^{2+}. E: This is dipalmitoyllecithin.
7. **C** Cyclooxygenase oxidizes arachidonic acid and peroxidase converts PGG_2 to PGH_2. A, B: The release of the precursor fatty acid by phospholipase A_2 is the rate-limiting step and the one inhibited by anti-inflammatory steroids. D: The peroxidase component converts the PGG_2 to PGH_2. E: Arachidonic acid is not free in the cell but is part of the membrane phospholipids or sometimes a cholesteryl ester.
8. **C** These are formed by addition of oxygen to arachidonic acid. A: The enzyme is a lipoxygenase. B–E: HPETEs, themselves, are not hormones but highly unstable intermediates that are converted to either HETEs (mediators of hypersensitivity) or leukotrienes.
9. **B** This enzyme is inhibited by cholesterol. A: Remember that cholesterol biosynthesis is cytosolic; mitochondrial biosynthesis of HMG CoA leads to ketone body formation. C: Mevalonic acid is decarboxylated (among other things) to produce the isoprene pyrophosphates, which are the condensing units. D: Pyrophosphate is hydrolyzed, which prevents reversal. E: The process is initiated by epoxide formation.
10. **D** This is one of the ways to prevent overload in the cell. A: LDL binds to the cell receptor, and it is endocytosed and then degraded in lysosomes to release cholesterol. B: ACAT is activated to facilitate storage. C: LCAT is a plasma enzyme. E: The primary role of LDL is to deliver cholesterol to peripheral tissues; HDL removes cholesterol from peripheral tissues.
11. **B** Ceramides are hydrolyzed to sphingosine and fatty acid. D: Sulfogalactocerebroside contains sulfate and gangliosides contain one or more *N*-acetylneuraminic acids. E: Many sphingolipids share the same types of bonds—for example, a β-galactosidic bond—and one enzyme, for example, β-galactosidase, will hydrolyze it whenever it occurs.
12. **D** Sphingomyelins are not glycosphingolipids but phosphosphingolipids. A, B, E: They are formed from ceramides, are amphipathic, and are neutral. C is the definition of gangliosides.
13. FH cells have LDL receptors with normal binding properties as indicated by the bound radioactivity being the same as that of normal cells. They are unable to internalize the receptor–LDL complex, so cholesterol synthesis is not inhibited and it is with normal cells. The mutation is most likely on the carboxy terminus of the protein which is involved in internalization. Make sure you understand why mutation in other regions would not lead to the observed results.
14. By binding bile salts, forcing increased excretion, liver has to convert more cholesterol to bile salts. If liver synthesis of cholesterol is inhibited, liver synthesizes more LDL receptors removing increased LDL particles (and thus cholesterol) from blood. Patients with FH either have no receptors or have nonfunctioning LDL receptors, so liver cannot increase uptake of LDL from blood. Synthesis is reduced, but this does not have as dramatic an effect on blood cholesterol.

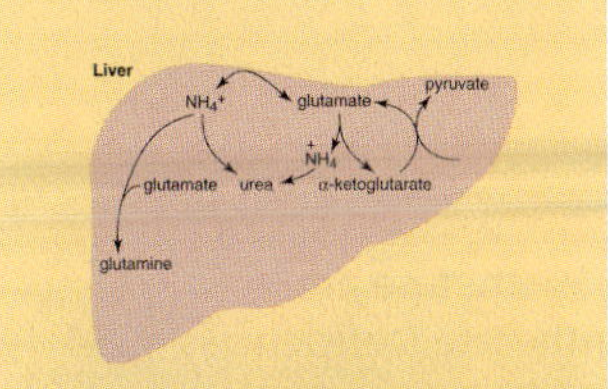

19

AMINO ACID METABOLISM

Marguerite W. Coomes

Textbook of Biochemistry With Clinical Correlations, Sixth Edition, Edited by Thomas M. Devlin

19.1 OVERVIEW

Amino acids and their contribution to the structure and function of proteins were presented in chapter 3. This chapter describes the metabolism of amino acids, emphasizing the importance of dietary protein as the major source of amino acids for humans.

Molecular nitrogen, N_2, exists in the atmosphere in great abundance. Before N_2 can be utilized by animals, it must be "fixed"—that is, reduced from N_2 to NH_3 by microorganisms, plants, and electrical discharge from lightning. Ammonia is then incorporated into amino acids and proteins, and these become part of the food chain (Figure 19.1). Humans can synthesize 11 of the 20 amino acids needed for protein synthesis. Those that cannot be synthesized *de novo* are termed" essential" because they must be obtained from dietary foodstuffs that contain them (Table 19.1).

TABLE 19.1 Essential and Nonessential Amino Acids

Essential	*Nonessential*
Arginine[a]	Alanine
Histidine	Aspartate
Isoleucine	Asparagine
Leucine	Cysteine
Lysine	Glutamate
Methionine[b]	Glutamine
Phenylalanine[c]	Glycine
Threonine	Proline
Tryptophan	Serine
Valine	Tyrosine

[a] Arginine is synthesized by mammalian tissues, but the rate is insufficient to meet the need during growth.
[b] Methionine is required in large amounts to produce cysteine if the latter is not supplied adequately by the diet.
[c] Phenylalanine is needed in larger amounts to form tyrosine if the latter is not supplied adequately by the diet.

This chapter includes discussion of interconversions of amino acids, removal and excretion of ammonia, and synthesis of "nonessential" amino acids by the body. As part of ammonia metabolism, synthesis and degradation of glutamate, glutamine, aspartate, asparagine, alanine, and arginine are discussed. Synthesis and degradation of other nonessential amino acids are then described, as well as the degradation of the essential amino acids. Synthetic pathways of amino acid derivatives and some diseases of amino acid metabolism are also presented.

Carbons from amino acids enter intermediary metabolism at seven points in the glycolytic pathway and the TCA cycle. Glucogenic amino acids are metabolized to pyruvate, 3-phosphoglycerate, α-ketoglutarate, oxaloacetate, fumarate, or succinyl CoA. Ketogenic amino acids produce acetyl CoA or acetoacetate. Metabolism of some amino acids results in more than one of the above, and they are therefore both glucogenic and ketogenic (Figure 19.2). Reducing equivalents, usually NADH, are also produced during degradation of some of the amino acids.

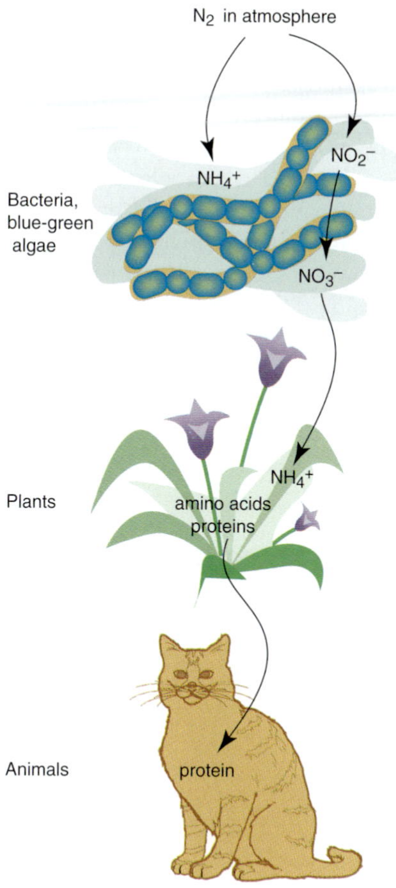

FIGURE 19.1

Outline of entry of atmospheric nitrogen into the animal diet. This occurs initially by reduction of nitrogen to ammonia by enzymes in microorganisms and plants.

19.2 INCORPORATION OF NITROGEN INTO AMINO ACIDS

Most Amino Acids Are Obtained from the Diet

A healthy adult eating a varied and plentiful diet is generally in "nitrogen balance," a state where the amount of nitrogen ingested each day is balanced by the amount excreted, resulting in no net change in total body nitrogen. In the well-fed condition, excreted nitrogen comes mostly from excess protein intake or from normal turnover. Protein turnover is defined replacement of body protein by synthesis and degradation. Under some conditions, the body is either in negative or **positive nitrogen balance**. In **negative nitrogen balance**, more nitrogen is excreted than ingested. This occurs in starvation and certain diseases. During starvation, carbon chains of amino acids from proteins are needed for gluconeogenesis; ammonia released from amino acids is excreted mostly as urea and is not reincorporated into protein. A diet deficient in essential amino acids also leads to a negative nitrogen balance, since body proteins are degraded and their essential amino acids reutilized, and the other amino acids liberated are metabolized. Negative nitrogen balance may also exist in senescence. Positive nitrogen balance occurs in growing children, who are increasing their body weight and incorporating more amino acids into proteins than they break down. Cysteine and arginine are essential in children but not in adults, who synthesize adequate amounts from methionine and ornithine. Positive nitrogen balance also occurs in pregnancy and during refeeding after starvation.

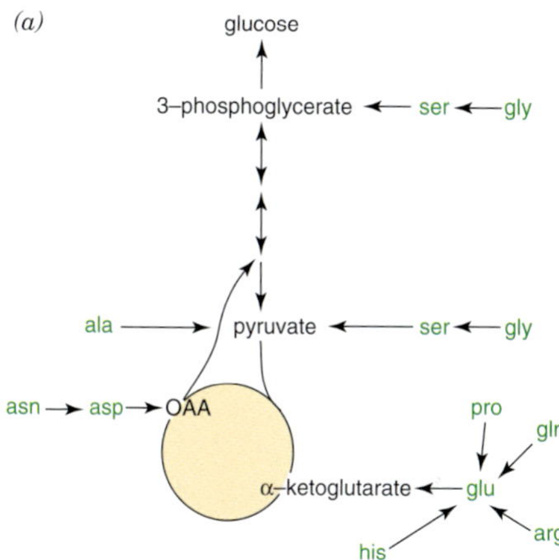

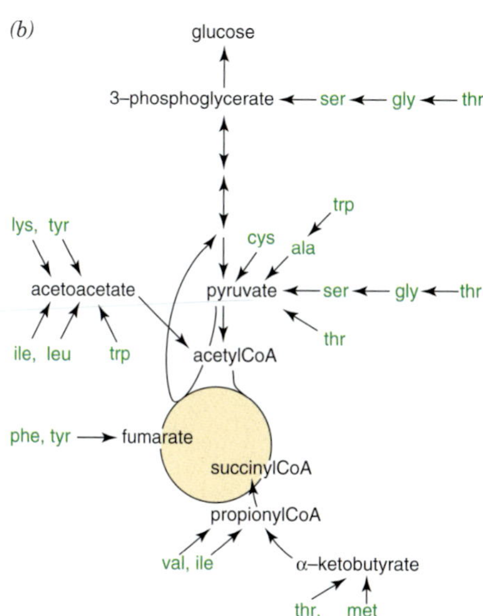

FIGURE 19.2

Metabolic fate of (a) nonessential amino acids; (b) essential amino acids plus cysteine and tyrosine.

Amino Groups Are Transferred from One Amino Acid to Form Another

Most amino acids used to synthesize protein or as precursors for amino acid derivatives are obtained from the diet or from protein turnover. When necessary, nonessential amino acids are synthesized from α-keto acid precursors by transfer of a preexisting amino group from another amino acid by **aminotransferases**, also called **transaminases** (Figure 19.3). Transfer of amino groups also occurs during degradation of amino acids. Figure 19.4 shows how the amino group of alanine is transferred to α-ketoglutarate to form glutamate. The pyruvate produced then provides carbons for gluconeogenesis or for energy production via the TCA cycle. This reaction is necessary since ammonia cannot enter the urea cycle directly from alanine, but the amino group of glutamate can be utilized. The reverse reaction would occur if there were a need for alanine for protein synthesis that was not being met by dietary intake or protein turnover. **Transamination** involving essential amino acids is normally unidirectional since the body cannot synthesize the equivalent α-keto acid. Figure 19.5 shows transamination of

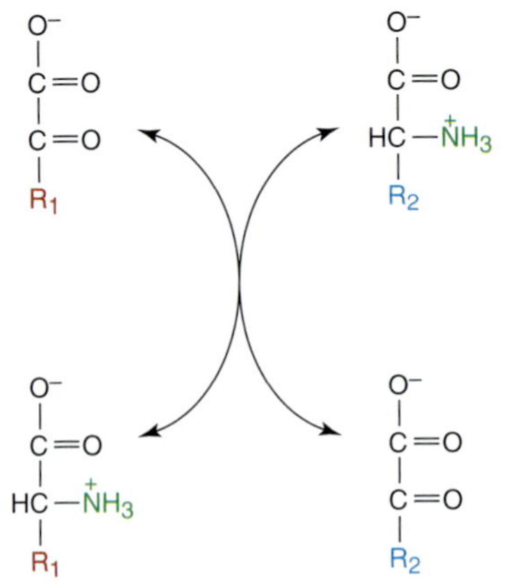

FIGURE 19.3
Aminotransferase reaction.

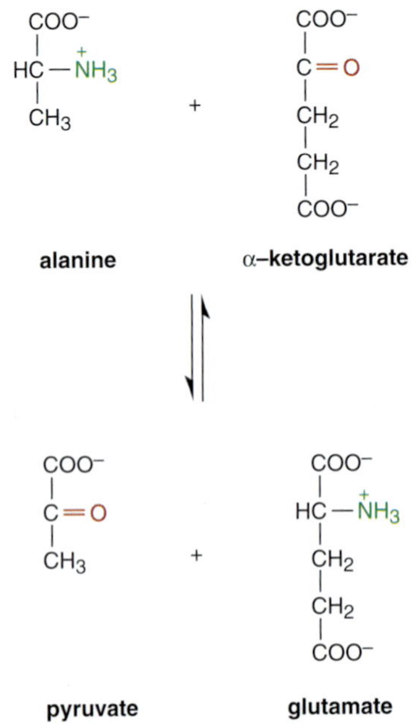

FIGURE 19.4
Glutamate–pyruvate aminotransferase reaction.

valine, an essential amino acid. The resulting α-ketoisovalerate is further metabolized to succinyl CoA (see p. 778). Transamination is the most common reaction involving free amino acids, and only threonine and lysine do not participate in an aminotransferase reaction. An obligate amino and α-keto acid pair in all of these reactions is glutamate and α-ketoglutarate. This means that amino group transfer between alanine and aspartate would have to occur via coupled reactions, with a glutamate intermediate (Figure 19.6). The equilibrium constant for aminotransferases is close to one, so that the reactions are freely reversible. When nitrogen excretion is impaired and **hyperammonemia** occurs, as in liver failure, amino acids, including the essential amino acids, can be replaced in the diet by α-keto acid analogs, with the exception of threonine and lysine as mentioned above. The α-keto acids are transaminated to produce the corresponding amino acids. Figure19. 5 shows valine formation after administration of α-ketoisovalerate as therapy for hyperammonemia.

Tissue distribution of some of the aminotransferases is used diagnostically by measuring the release of a specific enzyme during tissue damage; for instance, an increase in glutamate aspartate aminotransferase in plasma is a sign of liver damage (see p. 405).

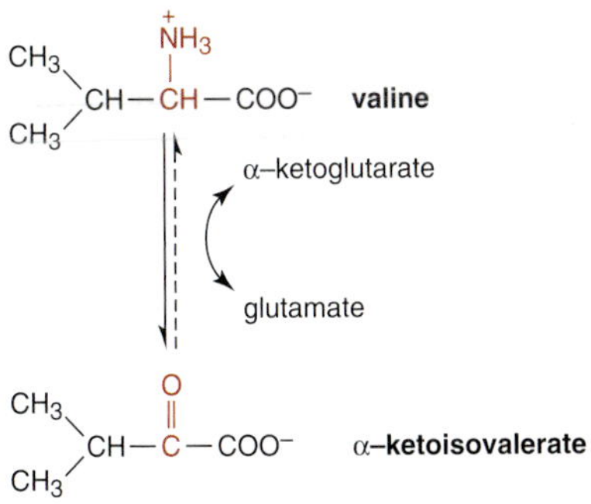

valine

α–ketoglutarate

glutamate

α–ketoisovalerate

FIGURE 19.5

Transamination of valine. Valine can be formed from α-ketoisovalerate only when this compound is administered therapeutically.

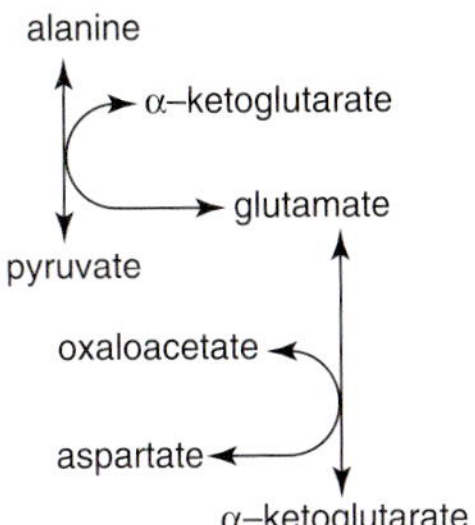

FIGURE 19.6

Coupled transamination reaction.

Pyridoxal Phosphate Is Cofactor for Aminotransferases

Transfer of amino groups occurs via enzyme-associated intermediates derived from **pyridoxal phosphate**, the functional form of **vitamin B_6** (Figure 19.7). The active site of the "resting" aminotransferase contains pyridoxal phosphate covalently attached to an ϵ-amino group of a lysine residue of the enzyme (Figure 19.8) The complex is further stabilized by ionic and hydrophobic interactions. The linkage, —CH=N—, is called a **Schiff base**. The carbon originates in the aldehyde group of pyridoxal phosphate, and the nitrogen originates in the lysine residue. When a substrate amino acid approaches the active site, its amino group displaces the lysine ϵ-amino group and a Schiff base linkage is formed with the amino group of the amino acid substrate (Figure 19.9). At this point the pyridoxal phosphate-derived molecule is no longer covalently attached to the enzyme but is held in the active site only by ionic and hydrophobic interactions between it and the protein. The Schiff base linkage with the substrate amino acid substrate is in tautomeric equilibrium between an aldimine, $—CH=N—CHR_2$, and a ketimine, $—CH_2—N=CR_2$. Hydrolysis of the ketimine liberates an α-keto acid, leaving the amino group as part of the pyridoxamine structure. A reversal of the process is now possible; an α-keto acid reacts with the amine group, the double bond is shifted, and then hydrolysis liberates an amino acid. Pyridoxal phosphate now reforms its Schiff base with the "resting" enzyme (Figure 19.8). Many pyridoxal phosphate-dependent reactions involve transamination, but the ability of the Schiff base to transfer electrons between different atoms allows the cofactor to participate when other groups, such as carboxyls, are to be eliminated. Figure 19.10 shows the reaction of a pyridoxal phosphate-dependent decarboxylase and an α-,β-elimination. The effective concentration of vitamin B_6 in the body may be decreased by administration of certain drugs, such as the anti-tuberculosis drug, isoniazid, which forms a Schiff base with pyridoxal, thus making it unavailable for catalysis.

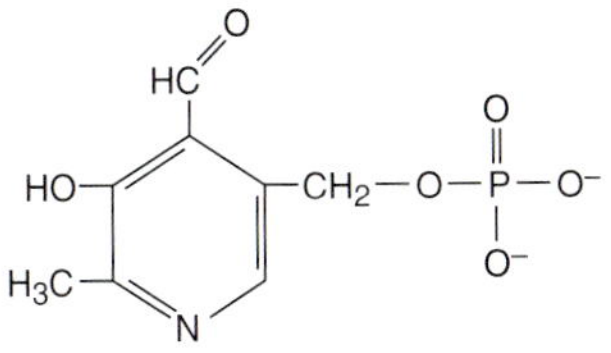

FIGURE 19.7

Pyridoxal phosphate.

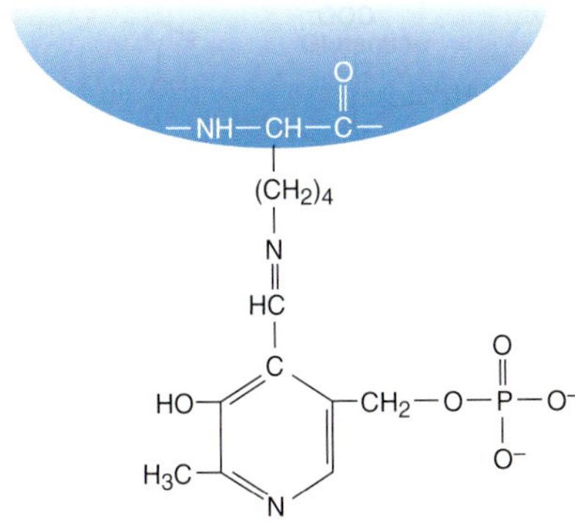

FIGURE 19.8

Pyridoxal phosphate in aldimine linkage to protein lysine residue.

Glutamate Dehydrogenase Incorporates and Produces Ammonia

In liver, ammonia is incorporated into glutamate by **glutamate dehydrogenase** (Figure 19.11), which also catalyzes the reverse reaction. Glutamate always serves as one of the amino acids in transaminations and is thus the "gateway" between amino groups of most amino acids and free **ammonia** (Figure 19.12). NADPH is used in the synthetic reaction, whereas NAD^+ is used in liberation of ammonia, a degradative reaction. The enzyme produces ammonia from amino acids when these are needed as glucose precursors or for energy. Formation of NADH during the oxidative deamination reaction is a welcome bonus, since it can be reoxidized by the respiratory chain with formation of ATP. The reaction as shown is readily reversible in the test tube, but it probably occurs more frequently in the direction of ammonia formation. The concentration of ammonia needed for production of glutamate is toxic and under normal

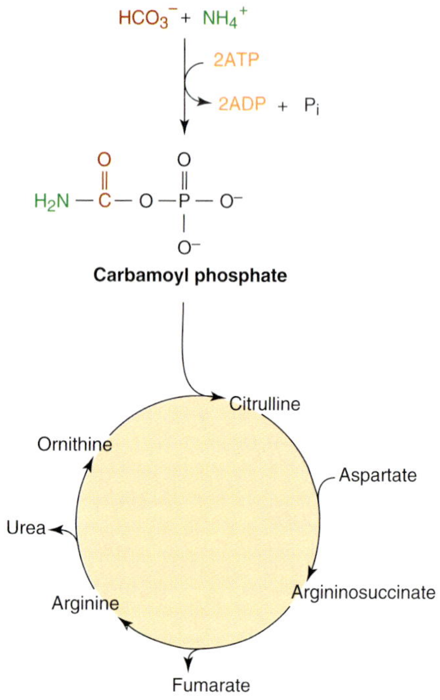

$HCO_3^- + NH_4^+$ → (2ATP → 2ADP + P_i) → $H_2N-C(=O)-O-P(=O)(O^-)-O^-$

Carbamoyl phosphate

Citrulline → Aspartate → Argininosuccinate → Fumarate → Arginine → Urea → Ornithine

FIGURE 19.22

Synthesis of carbamoyl phosphate and entry into urea cycle.

cycle, where carbons of oxaloacetate at the start are different from those at the end, the carbons in the original and the final ornithine are the same. Ammonia (first nitrogen for urea) enters the cycle after condensation with bicarbonate to form **carbamoyl phosphate** (Figure 19.22), which reacts with ornithine to form **citrulline**. Aspartate (donor of the second urea nitrogen) and citrulline react to form argininosuccinate, which is then cleaved to arginine and fumarate. **Arginine** is hydrolyzed to **urea** and ornithine is regenerated. Urea is then transported to the kidney and excreted in urine. The cycle requires four ATPs for each urea molecule produced and excreted. It is therefore more energy efficient to incorporate ammonia into amino acids than to excrete it. The major regulatory step is synthesis of carbamoyl phosphate. The cycle is also regulated by enzyme induction.

Synthesis of Urea Requires Five Enzymes

Carbamoyl phosphate synthetase I (CPSI) is technically not a part of the urea cycle, although it is essential for urea synthesis. Ammonium ion and bicarbonate are condensed, at the expense of 2 ATPs, to form carbamoyl phosphate. One ATP activates bicarbonate, and the other donates the phosphate group of carbamoyl phosphate. CPSI occurs in the mitochondrial matrix, uses ammonia as nitrogen donor, and is absolutely dependent on *N*-acetylglutamate for activity (Figure 19.23). **Carbamoyl phosphate synthase II (CPSII)** is cytosolic, uses the amide group of glutamine, and is not affected by ***N*-acetylglutamate**. It participates in pyrimidine biosynthesis (see p. 804).

Formation of citrulline is catalyzed by **ornithine transcarbamoylase** (Figure 19.24) in the mitochondrial matrix. Citrulline is transported out of the mitochondria into

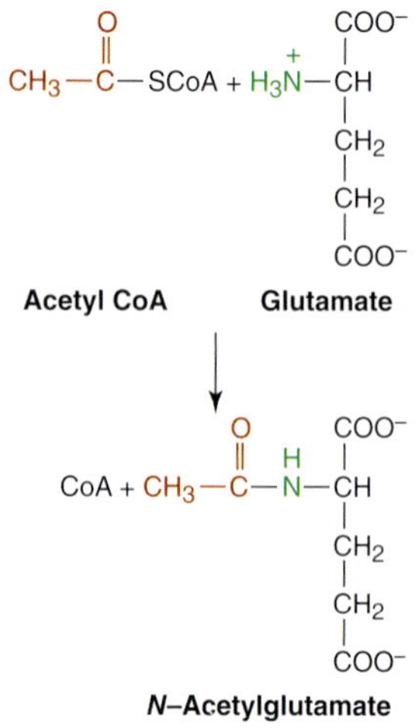

$CH_3-C(=O)-SCoA + H_3N^+-CH(COO^-)-CH_2-CH_2-COO^-$

Acetyl CoA **Glutamate**

↓

$CoA + CH_3-C(=O)-NH-CH(COO^-)-CH_2-CH_2-COO^-$

***N*–Acetylglutamate**

FIGURE 19.23

Reaction catalyzed by *N*-acetylglutamate synthetase.

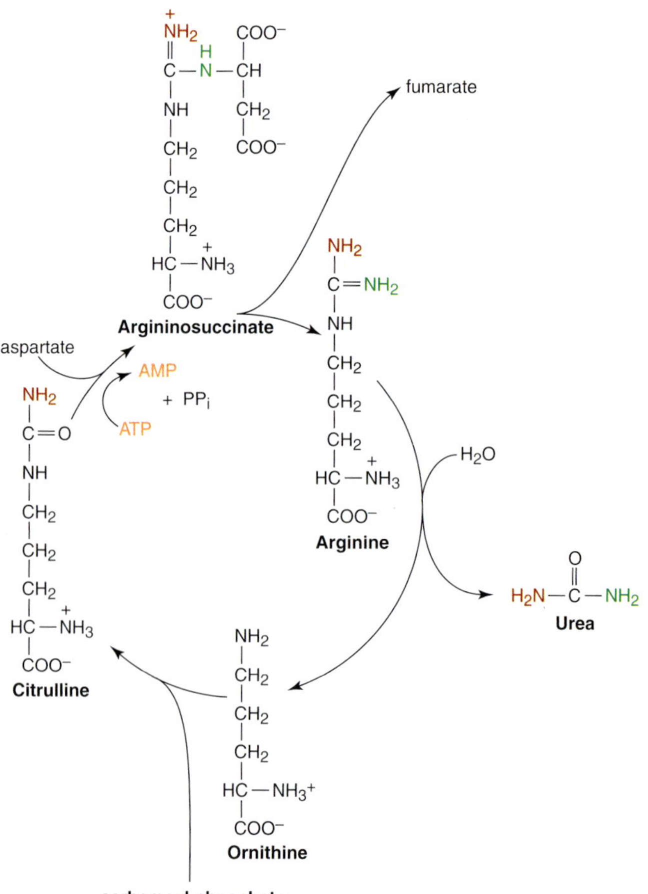

FIGURE 19.24

Urea cycle.

the cytosol, where the other reactions of the cycle occur. Argininosuccinate production by argininosuccinate synthetase requires hydrolysis of ATP to AMP and PP_i, the equivalent of hydrolysis of two molecules of ATP. Cleavage of argininosuccinate by argininosuccinate lyase produces fumarate and arginine. Arginine is cleaved by arginase to ornithine and urea. Ornithine reenters the mitochondrial matrix for another turn of the cycle. The inner mitochondrial membrane contains a citrulline/ornithine exchange transporter.

Synthesis of additional ornithine from glutamate for the cycle will be described later. Since arginine is produced from carbons and nitrogens of ornithine, ammonia, and aspartate, it is a nonessential amino acid. In growing children, however, where there is positive nitrogen balance and net incorporation of nitrogen into body constituents, *de novo* synthesis of arginine is inadequate and the amino acid becomes essential.

Carbons from aspartate, released as fumarate, may enter mitochondria and be metabolized to oxaloacetate by fumarase and malate dehydrogenase of the TCA cycle and become transaminated, and then they can enter another turn of the urea cycle as aspartate. About two-thirds of the oxaloacetate derived from fumarate is metabolized via phosphoenolpyruvate to glucose (Figure 19.25). The amount of fumarate used to form ATP is approximately equal to that required for the urea cycle and gluconeogenesis, meaning that the liver itself gains no net energy.

Since humans cannot utilize urea, it is transported to the kidneys for filtration and excretion. Any urea that enters the intestinal tract is cleaved by urease-containing bacteria in the intestinal lumen, the resulting ammonia being absorbed and used by the liver.

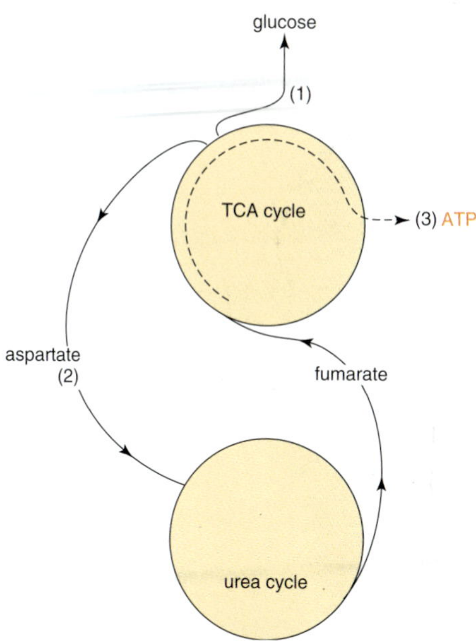

FIGURE 19.25
Fumarate from the urea cycle is a source of glucose (1), aspartate (2), or energy (3).

Urea Synthesis Is Regulated by an Allosteric Effector and Enzyme Induction

CPSI requires the allosteric activator ***N*-acetylglutamate** (see Figure 19.23). This compound is synthesized from glutamate and acetyl CoA by ***N*-acetylglutamate synthetase**, which is activated by arginine. Acetyl CoA, glutamate, and arginine are needed to supply intermediates or energy for the urea cycle, and the presence of *N*-acetylglutamate indicates that they are available. Tight regulation is desirable for a pathway that controls the plasma level of potentially toxic ammonia and that is highly energy dependent.

Induction of urea cycle enzymes (10- to 20-fold) occurs when delivery of ammonia or amino acids to liver rises. Concentration of intermediates also regulates activity through mass action. A high-protein diet (net excess amino acids) and starvation (need to metabolize body protein in order to provide carbons for energy production) result in induction of urea cycle enzymes.

Metabolic Disorders of Urea Synthesis Have Serious Results

Metabolic disorders that arise from abnormal function of enzymes of urea synthesis are potentially fatal and cause coma when ammonia concentrations become high. Loss of consciousness may be a consequence of ATP depletion. The major source of ATP is oxidative phosphorylation, which is linked to transfer of electrons from the TCA cycle down the electron transport chain (see p. 562). A high concentration of ammonia sequesters α-ketoglutarate as glutamate, thus depleting the TCA cycle of important intermediates and reducing ATP production.

Patients with a deficiency in one or other of the urea cycle enzymes have been found. Therapy for these deficiencies has a threefold basis: (1) to limit protein intake and potential buildup of ammonia, (2) to remove excess ammonia, and (3) to replace any intermediates missing from the urea cycle. The first is accomplished by limiting ingestion of amino acids, replacing them if necessary with the equivalent α-keto acids to be transaminated *in vivo*. The bacterial source of ammonia in the intestines can be decreased by a compound that acidifies the colon, such as levulose, a poorly absorbed synthetic disaccharide that is metabolized by colonic bacteria to acidic products. This promotes excretion of ammonia in feces as protonated ammonium ions.

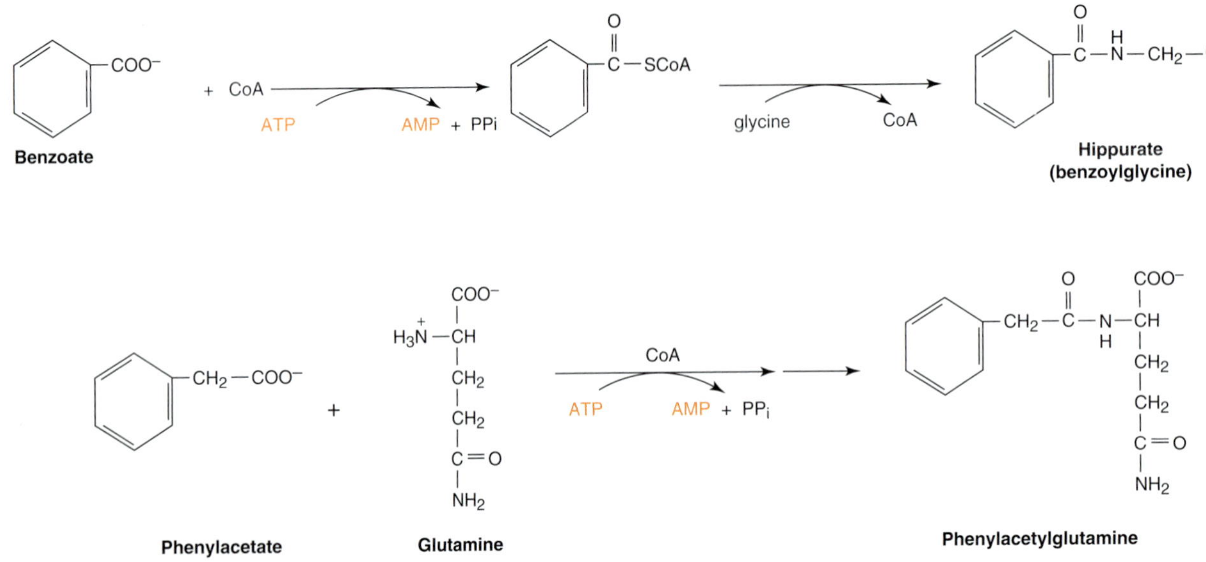

FIGURE 19.26
Detoxification reactions as alternatives to the urea cycle.

CLINICAL CORRELATION 19.1
Carbamoyl Phosphate Synthetase and *N*-Acetylglutamate Synthetase Deficiencies

Hyperammonemia has been observed in infants with 0–50% of the normal level of carbamoyl phosphate synthetase activity in their liver. In addition to treatments described in the text, these infants have been given with arginine, on the hypothesis that activation of *N*-acetylglutamate synthetase by arginine would stimulate the residual carbamoyl phosphate synthetase. This enzyme deficiency generally leads to mental retardation. A case of *N*-acetylglutamate synthetase deficiency has been described and treated successfully by administering carbamoyl glutamate, an analog of *N*-acetylglutamate, that activates carbamoyl phosphate synthetase.

Source: Berry, G. and Steiner, R. Long term management of patients with urea cycle disorders. *J. Pediatr.* 138:S56, 2001.

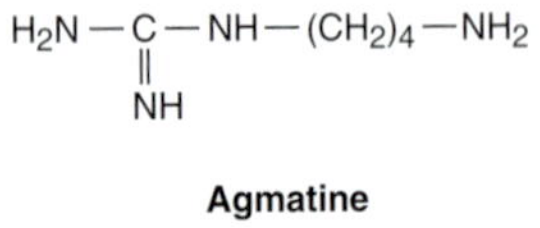

Antibiotics can also be administered to kill ammonia-producing bacteria. The second is achieved by compounds that bind covalently to amino acids and produce nitrogen-containing molecules that are excreted in urine. Figure 19.26 shows condensation of benzoate and glycine to form hippurate, and of phenylacetate and glutamine to form phenylacetylglutamine. Phenylacetate is extremely unpalatable and is given as the precursor sodium phenylbutyrate. Both reactions require energy for activation of the carboxyl groups by addition of CoA. Clinical Correlations 19.1 and 19.2 give examples of therapy for specific enzyme deficiencies, which often includes administration of urea cycle intermediates.

19.5 SYNTHESIS AND DEGRADATION OF INDIVIDUAL AMINO ACIDS

Glutamate Is a Precursor of Glutathione and γ-Aminobutyrate

Glutamate is a component of glutathione, which is discussed at the end of this chapter (see p. 782). It is also a precursor of γ-amino butyric acid (GABA), a neurotransmitter (Figure 19.10) (see p. 748), and of proline and ornithine, described below.

Arginine Is Also Synthesized in Intestines and Kidneys

Production of arginine for protein synthesis, rather than as an intermediate in the urea cycle, occurs in kidneys, which lack the arginase of the urea cycle. The major site of synthesis of citrulline to be used as an arginine precursor is intestinal mucosa, which has all enzymes necessary to convert glutamate (via ornithine as described below) to citrulline. Citrulline is converted to arginine in the kidneys. Arginine is a precursor for nitric oxide (see pp. 433 and 988); in brain, **agmatine**, a compound that may have antihypertensive properties, is derived by decarboxylation of arginine (Figure 19.27).

$$H_2N—C(=NH)—NH—(CH_2)_4—NH_2$$

Agmatine

FIGURE 19.27
Agmatine.

Ornithine and Proline

Ornithine, the precursor of citrulline and arginine, and proline are synthesized from glutamate and they are degraded, by slightly different pathways, to glutamate. Synthesis

CLINICAL CORRELATION **19.2**

Deficiency of Urea Cycle Enzymes

Ornithine Transcarbamoylase Deficiency

The most common deficiency of a urea cycle enzyme is lack of ornithine transcarbamoylase. Mental retardation and death often result, but the occasional finding of normal development in treated patients suggests that the mental retardation is caused by excess ammonia before adequate therapy. The gene for ornithine transcarbamoylase is on the X chromosome, and males are generally more seriously affected than heterozygotic females. In addition to ammonia and amino acids appearing in the blood in increased amounts, orotic acid also increases, presumably because carbamoyl phosphate that cannot be used to form citrulline diffuses into the cytosol, where it condenses with aspartate, ultimately forming orotate (see p. 803).

Argininosuccinate Synthetase and Lyase Deficiency

The inability to condense citrulline with aspartate results in accumulation of citrulline in blood and its excretion in urine (citrullinemia). Therapy for this otherwise normally benign disease requires specific supplementation with arginine for protein synthesis and for formation of creatine. Impaired ability to split argininosuccinate to form arginine resembles argininosuccinate synthetase deficiency in that the substrate is excreted in large amounts. Severity of symptoms in this disease varies greatly, so it is hard to evaluate the effect of therapy, which includes dietary supplementation with arginine.

Arginase Deficiency

Arginase deficiency is rare but causes many abnormalities in development and function of the central nervous system. Arginine accumulates and then excreted. Precursors of arginine and products of arginine metabolism may also be excreted. Unexpectedly, some urea is also excreted; this has been attributed to a second type of arginase found in the kidney. A diet including essential amino acids but excluding arginine has been used effectively.

Mitochondrial Ornithine Transporter Deficiency

This disease is also called hyperornithinemia, hyperammonemia, and homocitrullinemia syndrome (HHH syndrome). The symptoms include mental retardation, cerebellar ataxia, and episodic coma. The disease is caused by many different mutations in the gene, including nonsense, missense, and frameshift.

Hypoargininemia in Preterm Infants

Premature births occur before the burst of cortisol production that occurs late in pregnancy. Cortisol is an inducer of arginine-synthetic enzymes, and it is suggested that addition of cortisol to enteral and parenteral feeding of premature infants may improve survival and growth.

Source: Brusilow, S. W., Danney, M., Waber, L. J., Batshaw, M. Treatment of episodic hyperammonemia in children with inborn errors of urea synthesis. *N. Engl. J. Med.* 310:1630, 1984. Tsujino, S., et al., Three novel mutations in the ORNT1 gene of Japanese patients with hyperornithinemia, hyperammonemia. homocitrullinemia syndrome. *Ann. Neurol.* 47:625, 2000. Wu, G., Jaeger, L., Bazer, F., and Rhoads, J. Arginine deficiency in preterm infants: Biochemical mechanism nutritional implications. *Nutr. Biochem.* 15:442, 2004.

starts from glutamate by a reaction that uses ATP and NADH (Figure 19.28) and forms glutamic semialdehyde. This spontaneously cyclizes to form a Schiff base between the aldehyde and amino groups, which is then reduced by NADPH to proline. Glutamic semialdehyde can undergo transamination of the aldehyde group, which prevents cyclization and produces ornithine (Figure 19.29).

Proline is converted back to the Schiff base intermediate, δ-1-pyrroline 5-carboxylate, which is in equilibrium with **glutamic semialdehyde**. The transaminase reaction in the ornithine synthetic pathway is freely reversible and forms glutamic semialdehyde from ornithine (Figure 19.29). Proline residues that have been posttranslationally modified by hydroxylation form 3- or 4-hydroxyproline (Figure 19.30) are released by protein degradation and produce glyoxalate and pyruvate, and 4-hydroxy-2-ketoglutarate, respectively (see Clin. Corr. 19.3).

Ornithine is a precursor of **putrescine**, the foundation molecule of the polyamines, highly cationic molecules that interact with DNA. **Ornithine decarboxylase** (Figure 19.31) is regulated by phosphorylation at several sites, presumably in response to specific hormones, growth factors, or cell cycle regulatory signals. It can also be induced, often the first easily measurable sign that cell division is imminent, since polyamines must be synthesized before mitosis can occur. Other common polyamines are **spermidine** and **spermine** (see Figure 19.59), which are synthesized from putrescine by addition of propylamine, a product of methionine metabolism (see p. 768).

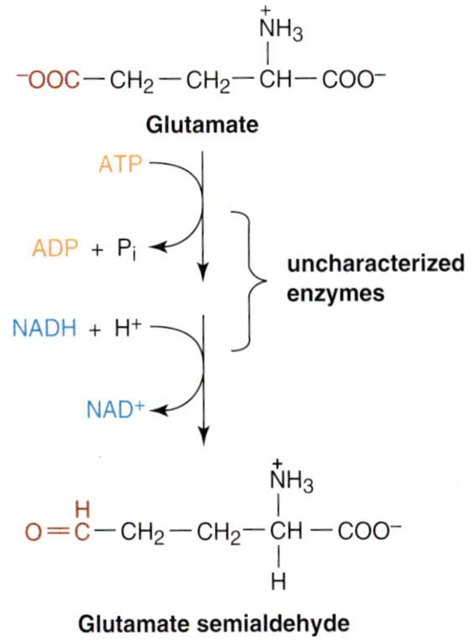

$^{-}OOC-CH_2-CH_2-CH(\overset{+}{N}H_3)-COO^-$

Glutamate

ATP → ADP + P_i; NADH + H^+ → NAD^+ (uncharacterized enzymes)

$O=C(H)-CH_2-CH_2-CH(\overset{+}{N}H_3)(H)-COO^-$

Glutamate semialdehyde

FIGURE 19.28
Synthesis of glutamic semialdehyde.

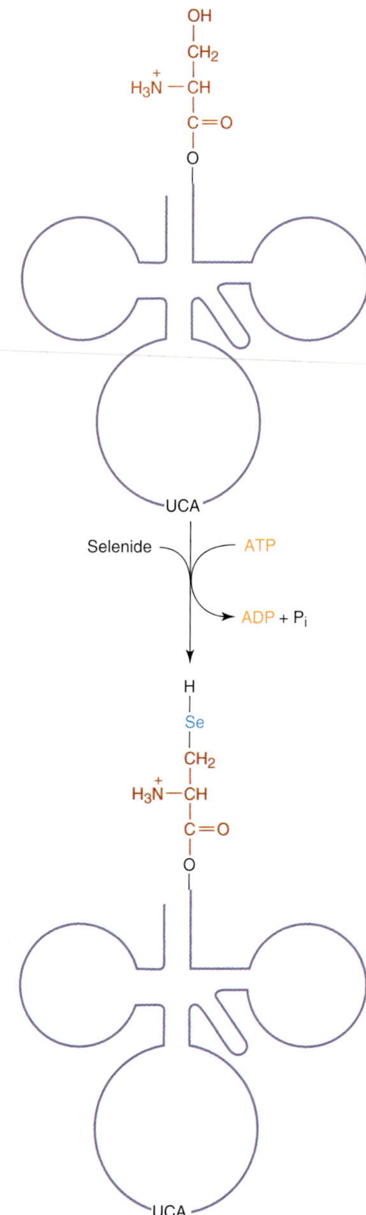

FIGURE 19.35

Formation of selenocysteinyl tRNA from seryl tRNA is via a phosphoseryl tRNA intermediate.

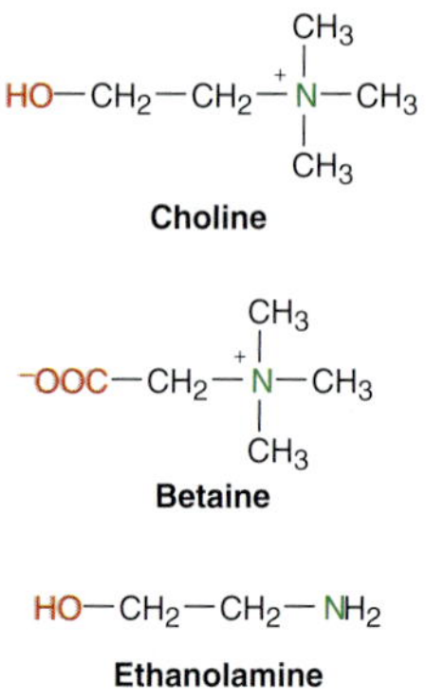

$HO—CH_2—CH_2—\overset{+}{N}(CH_3)_3$

Choline

$^{-}OOC—CH_2—\overset{+}{N}(CH_3)_3$

Betaine

$HO—CH_2—CH_2—NH_2$

Ethanolamine

FIGURE 19.36

Choline and related compounds.

CLINICAL CORRELATION 19.4
Selenoproteins

The first selenoprotein described in humans was glutathione peroxidase. About 25 mammalian proteins are now known, including another peroxidase, selenoprotein P (SeP). SeP has been shown to contain 10 selenoprotein residues. Bifunctional SeP serves to provide selenium to proliferating cells and also as a glutathione-dependent reduction of phosphatidylcholine hydroperoxide. The shift in usage of the UGA codon from a stop codon to a signal for selenocysteine incorporation is mediated by at least five different components. Two of these are cis sequences, one a region in the 3′UTR of the mRNA and the other the UGA codon itself. Three are trans-acting factors, a selenocysteine-specific translation elongation factor (eEFSec), a binding protein for the sequence in the 3′UTR (SBP2), and the selenocysteinyl-tRNA. Statin drugs used as therapy for hypercholesterolemia inhibit the cholesterol synthetic pathway, and therefore reduce synthesis of isopentyl groups. The myopathy and other side-effects caused by statins resemble the symptoms of selenium deficiency. This suggests that the mechanism of tissue damage caused by statins may be inhibition of selenoprotein synthesis.

Source: Mehta, A., Rebsch, C., Kinzy, S., Fletcher, J., and Copeland, P. Efficiency of mammalian selenocysteine incorporation. *J. Biol. Chem.* 279:37852, 2004. Saito, Y., et al. Domain structure of bi-functional selenoprotein P. *Biochem J.* 381:841, 2004. Moosmann, B., and Behl, C. Selenoprotein synthesis and side-effects of statins. *Lancet* 363:2000, 2004.

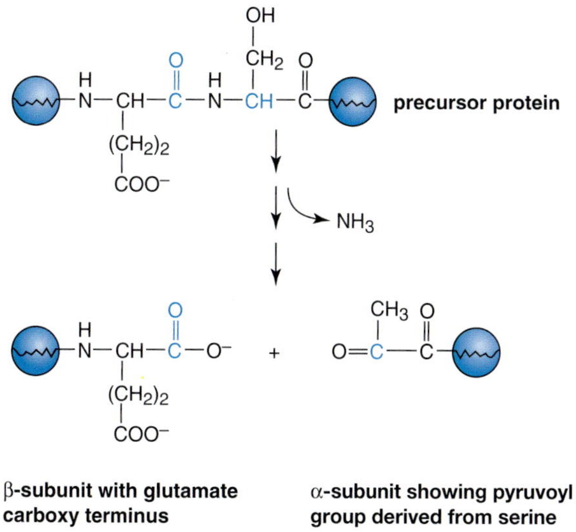

FIGURE 19.37
Formation of *S*-adenosylmethionine decarboxylase with covalently bound pyruvoyl prosthetic group.

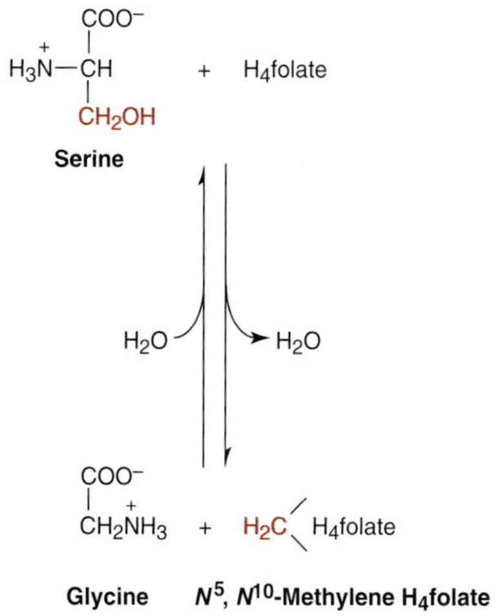

FIGURE 19.38
Serine hydroxymethyltransferase.

Serine is converted reversibly to glycine in a reaction that requires pyridoxal phosphate and tetrahydrofolate. $\boldsymbol{N^5,N^{10}}$**-methylenetetrahydrofolate** ($\boldsymbol{N^5,N^{10}}$**-THF**) is produced (Figure 19.38). The demand for serine or glycine and the amount of N^5,N^{10}-THF available determine the direction of this reaction. Glycine is degraded to CO_2 and ammonia by a glycine cleavage complex (Figure 19.39). This reaction is reversible *in vitro*, but not *in vivo*, because the K_m values for ammonia and N^5,N^{10}-THF are much higher than their respective physiological concentrations (see Clin. Corr. 19.5).

Glycine is the precursor of **glyoxalate**, which can be transaminated back to glycine or oxidized to oxalate (Figure 19.40). Excessive production of oxalate forms the insoluble calcium oxalate salt, which may lead to kidney stones. The role of glycine as a neurotransmitter is described on page 953.

CLINICAL CORRELATION 19.5
Nonketotic Hyperglycinemia

Nonketotic hyperglycinemia is characterized by severe mental deficiency. Many patients do not survive infancy. This very serious disease is to be distinguished from ketoacidosis in abnormalities of branched-chain amino acid metabolism in which the glycine level in the blood is also elevated. Deficiency of glycine cleavage complex has been demonstrated in homogenates of liver from several patients, and isotopic studies *in vivo* have confirmed that this enzyme is not active. The complex consists of four protein subunits. Inherited abnormalities have been found in three of them. The severity of this disease suggests that glycine cleavage is of major importance in the catabolism of glycine. Glycine is a major inhibitory neurotransmitter, which probably explains some neurological complications of the disease. Vigorous measures to reduce the glycine levels fail to alter the course of the disease.

Source: Nyhan, W. L. Metabolism of glycine in the normal individual and in patients with non-ketotic hyperglycinemia. *J. Inherit. Metab. Dis.* 5:105, 1982.

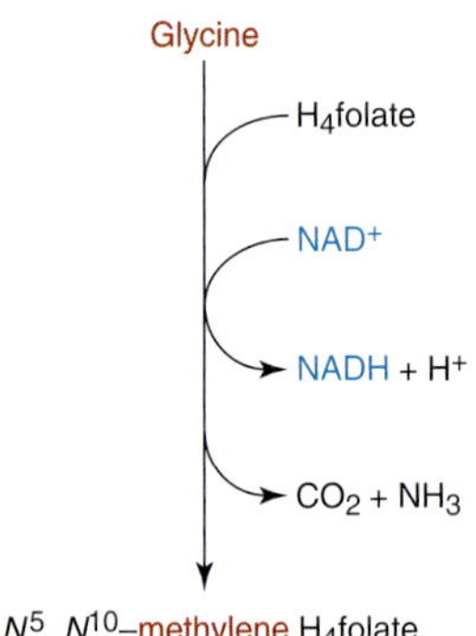

FIGURE 19.39
Glycine cleavage is pyridoxal phosphate-dependent.

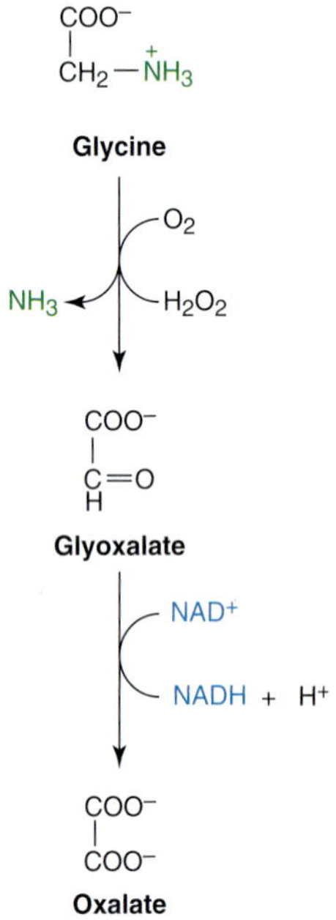

COO^-–CH_2–$\overset{+}{N}H_3$ **Glycine** → (O_2 in; NH_3, H_2O_2 out) → COO^-–$HC{=}O$ **Glyoxalate** → (NAD^+ → $NADH + H^+$) → COO^-–COO^- **Oxalate**

FIGURE 19.40
Oxidation of glycine.

Tetrahydrofolate Is a Cofactor in Some Reactions of Amino Acids

Tetrahydrofolate is the reduced form of folic acid, a water-soluble vitamin, and often occurs as a **poly-γ-glutamyl** derivative (Figure 19.41). Tetrahydrofolate is a one-carbon carrier that facilitates interconversion of methenyl, formyl, formimino, methylene, and methyl groups (Figure 19.42). This occurs at the expense of pyridine nucleotide reduction or oxidation and occurs while the carbon moiety is attached to THF (Figure 19.43). In the most oxidized forms, formyl is bound to N^{10} of the pteridine ring, methylene forms a bridge between N^5 and N^{10}, and methyl is bound to N^5. The interconversions permit use of a carbon that is removed in one oxidation state from a molecule for addition in a different oxidation state to a different molecule (Figure 19.42).

In reduction of the N^5,N^{10}-methylene bridge of tetrahydrofolate to a methyl group for transfer to the pyrimidine ring (Figure 19.44), a reaction found in thymidylate synthesis (see p. 807), the reducing power comes not from pyridine nucleotide but from the pteridine ring itself. The resulting dihydrofolate has no physiological role and is reduced back to tetrahydrofolate by NADPH-dependent dihydrofolate reductase (see Clin. Corr. 19.6). The net result of the two reactions is oxidation of NADPH and reduction of the methylene bridge to a methyl group, analogous to the one-step reactions shown in Figure 19.43.

Threonine

Threonine is usually metabolized to pyruvate (Figure 19.45), but an intermediate in this pathway can undergo thiolysis with CoA to acetyl CoA and glycine. Thus, the α-carbon atom of threonine can contribute to the one-carbon pool through formation of glycine. In a less common pathway, serine dehydratase (see p. 756) converts threonine to α-ketobutyrate. A complex similar to pyruvate dehydrogenase metabolizes this to propionyl CoA.

Phenylalanine and Tyrosine

Tyrosine and phenylalanine are discussed together, since tyrosine is produced by hydroxylation of phenylalanine and is the first product in phenylalanine degradation. Because of this, tyrosine is not usually considered to be essential, whereas phenylalanine is. Three-quarters of ingested phenylalanine is hydroxylated to tyrosine by **phenylalanine hydroxylase** (Figures 19.46 and see Clin. Corr. 19.7), which is **tetrahydrobiopterin**-dependent (Figure 19.47). The reaction occurs only in the direction of tyrosine formation. Biopterin resembles folic acid in containing a pteridine ring but is not a vitamin. It is synthesized from GTP.

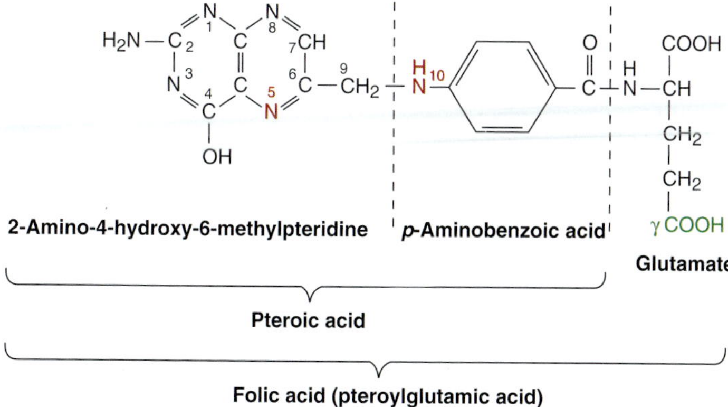

FIGURE 19.41

Components of folate. Poly-γ-glutamate can be formed by addition of several residues of glutamate in γ-glutamyl linkages.

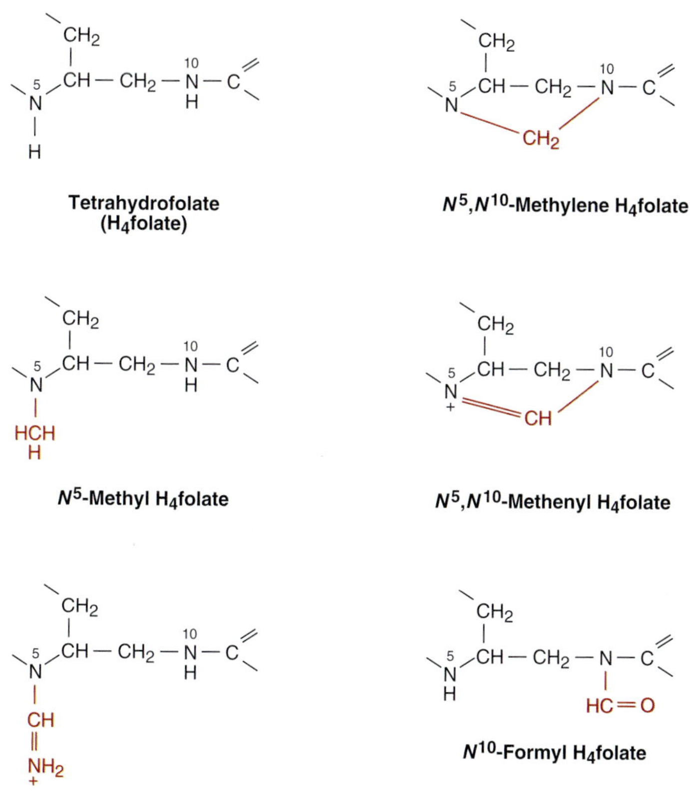

FIGURE 19.42

Active center of THF. N^5 is the site of attachment of methyl and formimino groups; N_{10} is the site for formyl; methylene and methenyl groups form bridges between N_5 and N_{10}.

Tyrosine Metabolism Produces Fumarate and Acetoacetate

The metabolism of tyrosine starts with transamination by tyrosine aminotransferase to produce ***p*-hydroxyphenylpyruvate** (Figure 19.48). The enzyme is inducible by glucocorticoids and dietary tyrosine. *p*-Hydroxyphenylpyruvate oxidase produces homogentisic acid in a complex reaction that involves decarboxylation, oxidation, migration of the carbon side chain, and hydroxylation. **Ascorbic acid** is required for at least one of these activities, but all four are catalyzed by the same enzyme. The aromatic ring is next cleaved by the iron-containing homogentisate oxidase to maleylacetoacetate. This is isomerized

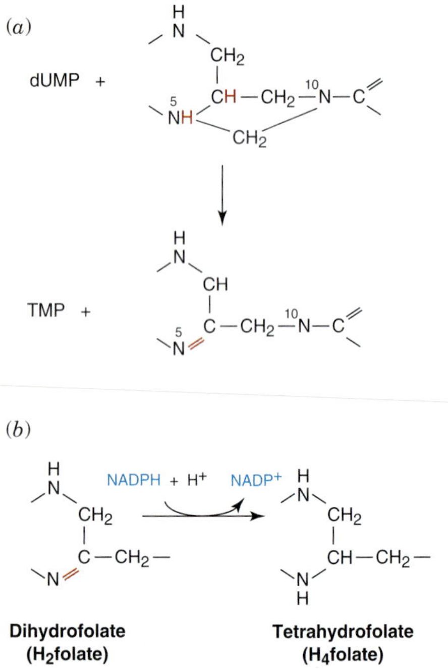

(a)

dUMP + [structure: ^{5}NH, CH_2, $CH-CH_2-{}^{10}N-C$]

↓

TMP + [structure: ^{5}N, CH, $C-CH_2-{}^{10}N-C$]

(b)

NADPH + H^+ → $NADP^+$

Dihydrofolate (H_2folate) → Tetrahydrofolate (H_4folate)

FIGURE 19.44

Reduction reactions involving THF. (a) Reduction of methylene group on THF to a methyl group and transfer to dUMP to form TMP. (b) Reduction of resulting dihydrofolate to tetrahydrofolate.

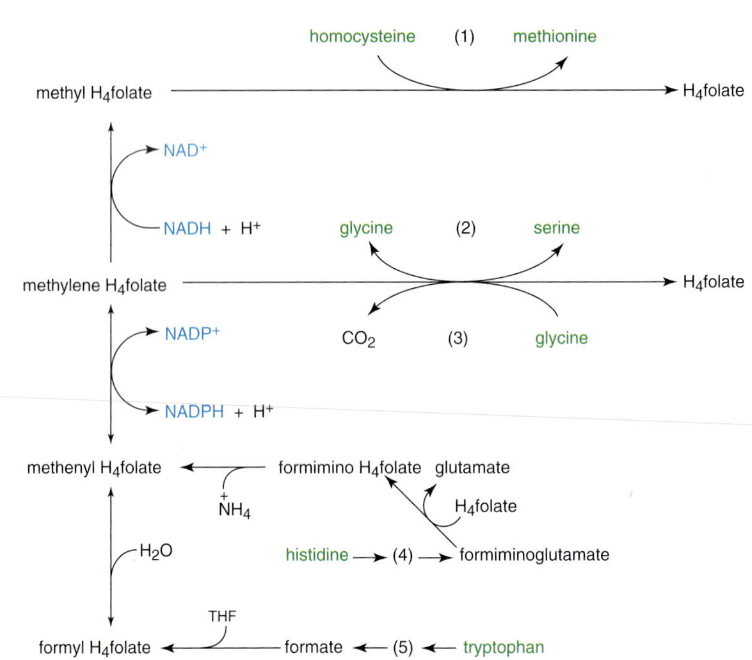

FIGURE 19.43

Interconversion of derivatized THF and roles in amino acid metabolism. (1) Methionine salvage, (2) serine hydroxymethyltransferase, (3) glycine cleavage complex, (4) histidine degradation, and (5) tryptophan metabolism.

CLINICAL CORRELATION 19.6
Folic Acid Deficiency

The 100–200 mg of folic acid required daily by an average adult can theoretically be obtained easily from conventional Western diets. Deficiency of folic acid, however, is not uncommon. It may result from limited diets, especially when food is cooked at high temperatures for long periods, which destroys the vitamin. Intestinal diseases, notably coeliac disease, are often characterized by folic acid deficiency caused by malabsorption. Inability to absorb folate is rare. Folate deficiency is usually seen only in newborns and has symptoms of megaloblastic anemia. Of the few cases studied, some were responsive to large doses of oral folate but one required parenteral administration, suggesting a carrier-mediated process for absorption. Besides the anemia, mental and other central nervous system symptoms are seen in patients with folate deficiency; all respond to continuous therapy, although permanent damage appears to be caused by delayed or inadequate treatment. A classical experiment was carried out by a physician, apparently serving as his own experimental subject, to study the human requirements for folic acid. His diet consisted only of foods boiled repeatedly to extract the water-soluble vitamins and to which vitamins (and minerals) were added but not folic acid. Symptoms attributable to folate deficiency did not appear for seven weeks, altered appearance of blood cells and formiminoglutamate excretion were seen only at 13 weeks, and serious symptoms (irritability, forgetfulness, and macrocytic anemia) appeared only after four months. Neurological symptoms were alleviated within two days after folic acid was added to the diet; the blood picture became normal more slowly. The occurrence of folic acid in essentially all natural foods makes deficiency difficult, and, apparently, a normal person accumulates more than adequate reserves of this vitamin. For pregnant women the situation is very different. Needs of the fetus for normal growth and development include constant, uninterrupted supplies of coenzymes (in addition to amino acids and other cell constituents). Recently, folate deficiency has been implicated in spina bifida.

Source: Herbert, V. Experimental nutritional folate deficiency in man. *Trans. Assoc. Am. Physicians* 75:307, 1962.

from cis to trans to give fumarylacetoacetate by maleylacetoacetate isomerase, an enzyme that seems to require glutathione for activity. Fumarate can be utilized in the TCA cycle for energy or for gluconeogenesis. Acetoacetate can be used, as acetyl CoA, for lipid synthesis or energy. Accumulated phenylpyruvate can be metabolized to phenyllactate and phenylacetate (Figure 19.49).

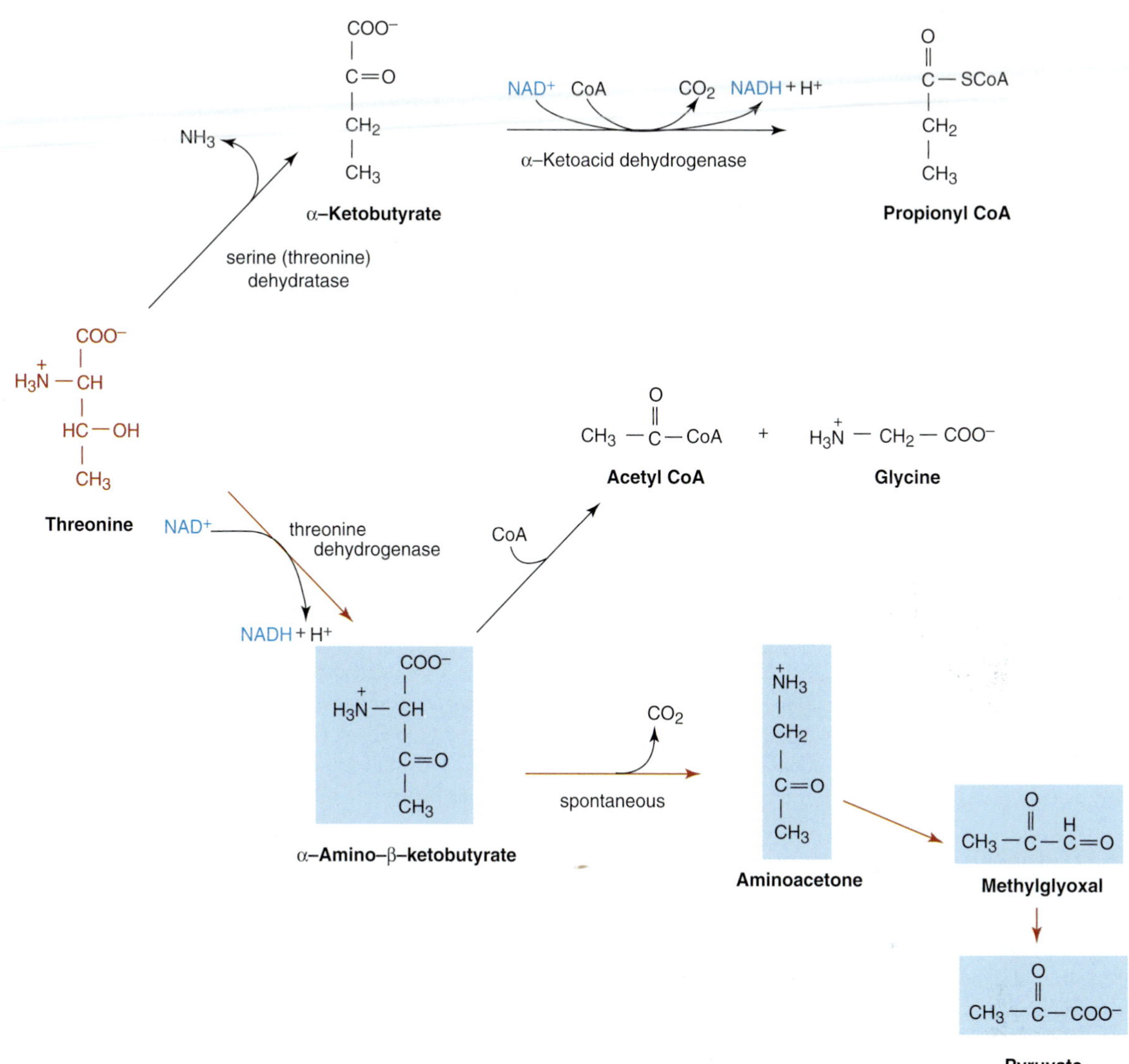

FIGURE 19.45

Outline of threonine metabolism. Major pathway is in color.

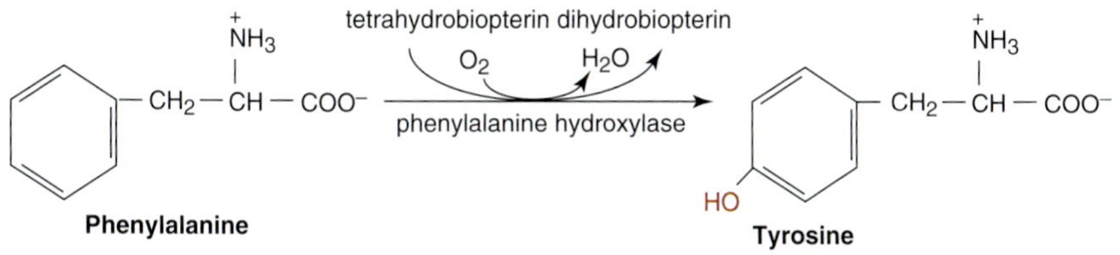

Phenylalanine — tetrahydrobiopterin, O_2 → dihydrobiopterin, H_2O (phenylalanine hydroxylase) → Tyrosine

FIGURE 19.46

Phenylalanine hydroxylase.

Tetrahydrobiopterin

Dihydrobiopterin

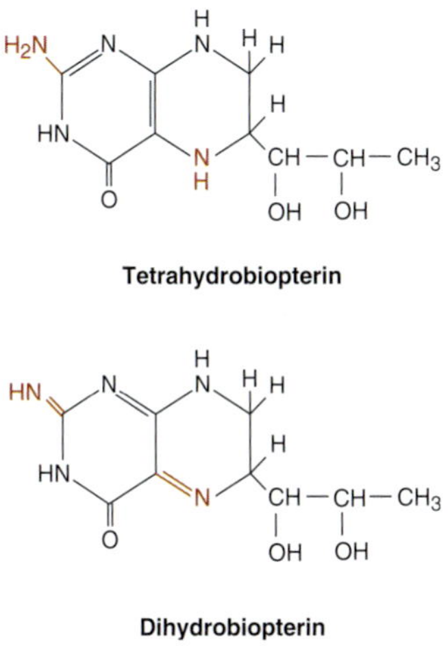

FIGURE 19.47

Biopterin. Dihydro (quinonoid) form is produced during oxidation of aromatic amino acids and is then reduced to the tetrahydro form by a dehydrogenase using NADH.

Dopamine, Norepinephrine, and Epinephrine Are Derivatives of Tyrosine

Most tyrosine not incorporated into proteins is metabolized to acetoacetate and fumarate (see Clin. Corr. 19.8). Some is used to produce catecholamines. The eventual metabolic fate of tyrosine carbons is determined by the first step in each pathway. **Catecholamine synthesis** (Figure 19.50) starts with tyrosine hydroxylase, which, like phenylalanine and tryptophan hydroxylase, is dependent on tetrahydrobiopterin. All three are affected by

CLINICAL CORRELATION 19.7
Phenylketonuria

Phenylketonuria (PKU) is the most common disease caused by a deficiency of an enzyme of amino acid metabolism. The name comes from the excretion of phenylpyruvic acid, a phenylketone, in the urine. Phenyllactate (Figure 19.47), the reduced form of phenylpyruvate, and phenylacetate are also excreted. The latter gives the urine a "mousey" odor. These three metabolites are found only in trace amounts in urine in the healthy person. The symptoms of mental retardation associated with this disease can be prevented by a diet low in phenylalanine. Routine screening is required by governments in many parts of the world. Classical PKU is an autosomal recessive deficiency of phenylalanine hydroxylase. Over 417 mutations in the gene have been reported. In some cases there are severe neurological symptoms and very low IQ. These are generally attributed to toxic effects of phenylalanine, possibly because of reduced transport and metabolism of other aromatic amino acids in the brain due to competition from the high phenylalanine concentration. The characteristic light color of skin and eyes is due to under pigmentation because of tyrosine deficiency. Conventional treatment is by a synthetic diet low in phenylalanine, but including tyrosine, for about four to five 5 years, followed by dietary protein restriction for several more years or for life.

About 3% of infants with high levels of phenylalanine have normal hydroxylase but are defective in either synthesis or reduction of biopterin. A deficiency in BH4 can be caused by mutation of GTP cyclohydrolase (GTPCH), 6-pyruvoyltetrahydrobiopterin synthetase (PTPS), or sepiapterin reductase (SPR), the three enzymes that convert GTP to BH4. Two enzymes are required for regeneration of BH4: pterin-4*a*-carbinolamine dehydratase (PCD) and dihydrobiopterin reductase (DHPR). BH4 deficiency occurs at a rate of one in a million. Screening for hyperphenylalaninemia with no reduction in phenylalanine hydroxylase activity *in vitro* indicates the possibility of a BH4 deficiency. If the condition is untreated, patients show symptoms of hyperphenylalaninemia, red hair, psychomotor retardation, and progressive neurological deterioration. These latter symptoms are a result of inability to produce melanin, catecholamines, and serotonin. Treatment is BH4 and neurotransmitter precursor supplementation. DOPA responsive dystonia (DRD) and SR deficiency cannot be detected by screening for PKU. Skin fibroblasts are useful for the diagnosis of these conditions. The discovery of SR deficiency led to the discovery of alternative pathways of BH4 metabolism, including a role for dihydrofolate reductase (DHFR). Low levels of DHFR in brain allow dihydrobiopterin to accumulate and inhibit tyrosine and tryptophan hydroxylases. DHP also uncouples nitric oxide synthase and can lead to neuronal cell death.

Source: Scriver, C. R. and Clow, L. L. Phenylketonuria: Epitome of human biochemical genetics. *N. Engl. J. Med.* 303:1336,1980. Woo, S. L. C. Molecular basis and population genetics of phenylketonuria. *Biochemistry* 28:1, 1989. Shintaku, H. Disorders of tetrahydrobiopterin metabolism and their treatments. *Curr. Drug. Metab.* 3:123; 2002. Blau, N., Bonafe, L., and Thony, B. Tetrahydrobiopterin deficiencies without hyperphenylalaninemia diagnosis and genetics of dopa-responsive distonia and sepiapterin reductase deficiency. *Mol. Genet. Metab.* 74:172, 2001.

biopterin deficiency or a defect in **dihydrobiopterin reductase** (Figure 19.47). **Tyrosine hydroxylase** produces **dihydroxyphenylalanine (DOPA)**. **DOPA decarboxylase**, with pyridoxal phosphate as cofactor, forms dopamine, the active neurotransmitter. In the substantia nigra and some other parts of the brain, this is the end of this pathway (see Clin. Corr. 19.9). The adrenal medulla converts dopamine to norepinephrine and epinephrine (adrenaline). The methyl group of epinephrine is derived from *S*-adenosylmethionine (see p. 768)

Brain plasma tyrosine regulates norepinephrine formation. Estrogens decrease tyrosine concentration and increase tyrosine aminotransferase activity, diverting tyrosine into the catabolic pathway. Estrogen sulfate competes for the pyridoxal phosphate site on DOPA decarboxylase. These effects combined may help explain mood variations during the menstrual cycle. Tyrosine is therapeutic in some cases of depression and stress. Its transport appears to be reduced in skin fibroblasts from schizophrenic patients, indicating other roles for tyrosine derivatives in mental disorders. Catecholamines are metabolized by **monoamine oxidase** and **catecholamine *o*-methyltransferase**. Major metabolites are shown in Figure 19.51. Absence of these metabolites in urine is diagnostic of a deficiency in synthesis of catecholamines. Lack of synthesis of serotonin (see p. 775) is indicated by lack of 5-hydroxyindole-3-acetic acid in urine.

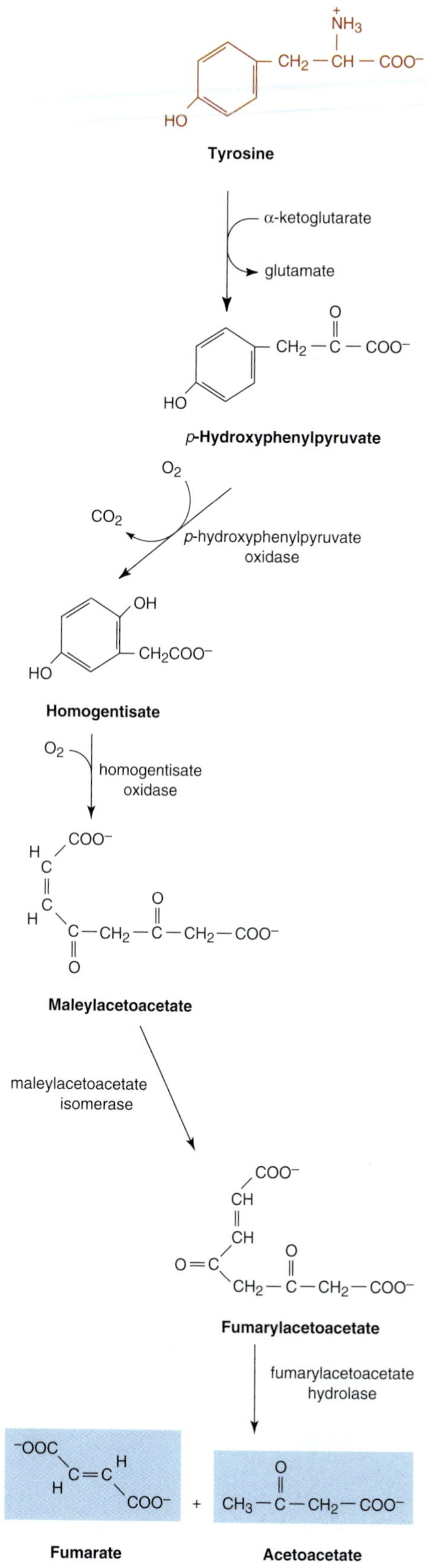

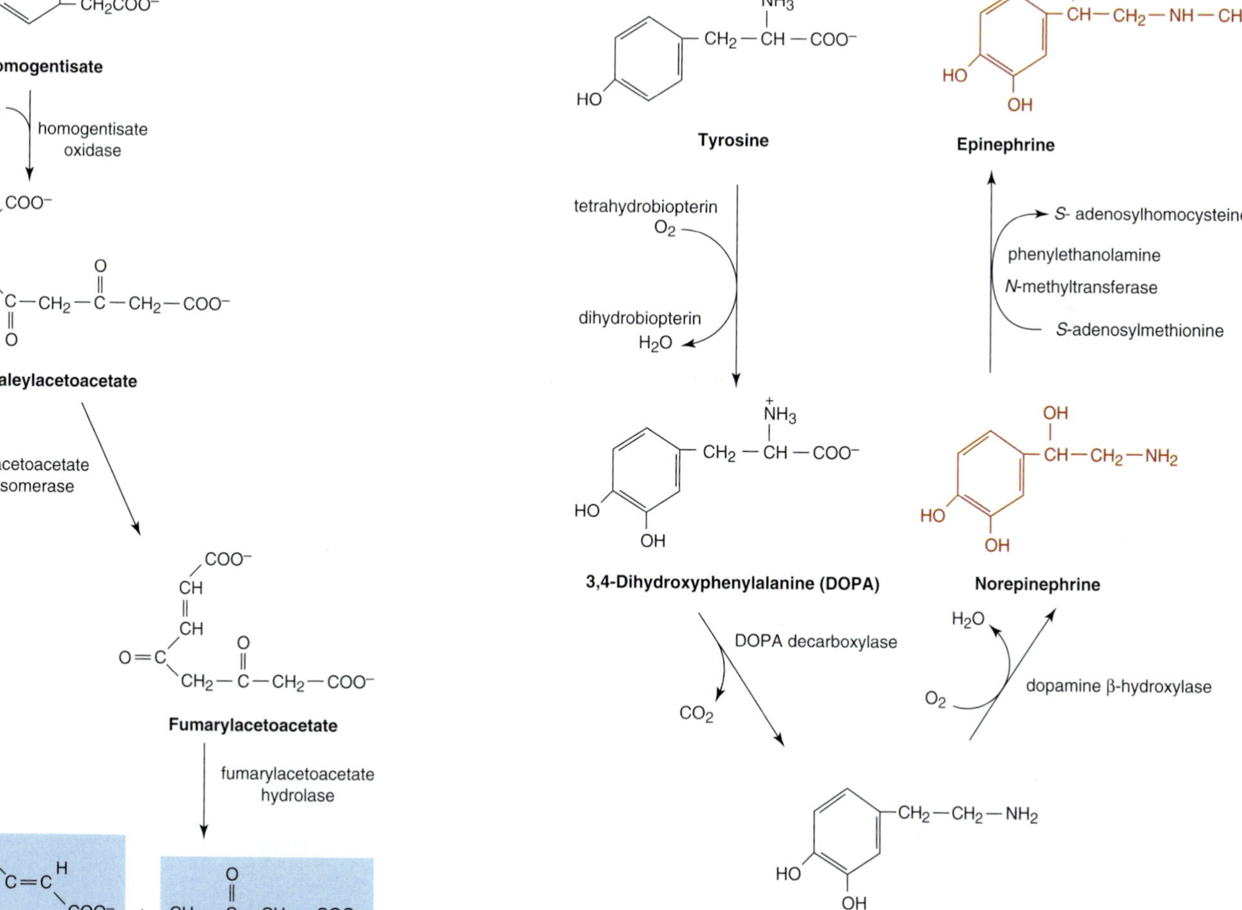

FIGURE 19.48
Degradation of tyrosine.

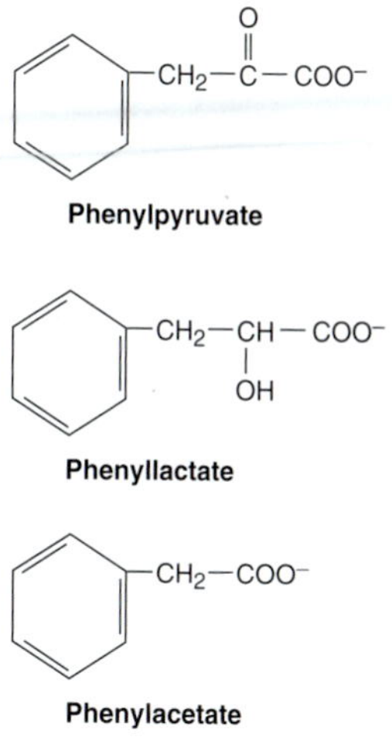

FIGURE 19.49
Minor products of phenylalanine metabolism.

FIGURE 19.50
Synthesis of catecholamines.

CLINICAL CORRELATION 19.8
Disorders of Tyrosine Metabolism

Tyrosinemias

Absence or deficiency of tyrosine aminotransferase produces accumulation and excretion of tyrosine and metabolites. The disease, oculocutaneous or type II tyrosinemia, results in eye and skin lesions and mental retardation. Type I, hepatorenal tyrosinemia, is more serious, involving liver failure, renal tubular dysfunction, rickets, and polyneuropathy. It is caused by deficiency of fumarylacetoacetate hydrolase. Accumulation of fumarylacetoacetate and maleylacetate, both of which are alkylating agents, can lead to DNA alkylation and tumorigenesis. Both diseases are autosomal recessive and rare.

Alcaptonuria

The first identified "inborn error of metabolism" was alcaptonuria. Individuals deficient in homogentisate oxidase excrete almost all ingested tyrosine as the colorless homogentisic acid in their urine. Homogentisic acid autooxidizes to the corresponding quinone, which polymerizes to form an intensely dark color. Concern about the dark urine is the only consequence of this condition early in life. Homogentisate in body fluids is slowly oxidized to pigments that are deposited in bones, connective tissue, and elsewhere, a condition called ochronosis because of the ochre color of the deposits. This is thought to be responsible for the associated arthritis, especially in males. The study of alcaptonuria by Archibald Garrod, who first indicated its autosomal recessive genetic basis, includes an unusual historical description of the iatrogenic suffering of the first patient treated for the condition, which is frequently benign.

Albinism

Skin and hair color are controlled by numerous genetic loci in humans, and they exist in infinite variation; 147 genes in mice have been identified in color determination. There are many in which the skin has little or no pigment. The chemical basis is not established for any except classical albinism, which results from a lack of tyrosinase. Lack of pigment in the skin makes albinos sensitive to sunlight, increasing the incidence of skin cancer in addition to burns; lack of pigment in the eyes may contribute to photophobia.

Source: Fellman, J. H., Vanbellinghan, P. J., Jones, R. T., and Koler, R. D. Soluble and mitochondrial tyrosine aminotransferase. Relationship to human tyrosinemia. *Biochemistry* 8:615, 1969. Kvittingen, E. A. Hereditary tyrosinemia type I. An overview. *Scand. J. Clin. Lab. Invest.* 46:27, 1986.

CLINICAL CORRELATION 19.9
Parkinson's Disease

Usually in people over the age of 60 years but occasionally earlier, tremors may develop that gradually interfere with motor function of various muscle groups. This condition is named for the physician who described "shaking palsy" in 1817. The primary cause is unknown. The defect results from degeneration of cells in certain small nuclei of the brain called substantia nigra and locus coeruleus. Their cells normally produce dopamine as a neurotransmitter, the amount released being proportional to the number of surviving cells. A dramatic outbreak of parkinsonism occurred in young adult drug addicts using a derivative of pyridine (methylphenyltetrahydropyridine, MPTP). It (or a contaminant produced during its manufacture) appears to be directly toxic to dopamine-producing cells of substantia nigra. Symptomatic relief, often dramatic, is obtained by administering DOPA, the precursor of dopamine. Clinical problems developed when DOPA (L-DOPA, levo-DOPA) was used for treatment of many people who have Parkinson's disease. Side effects included nausea, vomiting, hypotension, cardiac arrhythmias, and various central nervous system symptoms. These were explained as effects of dopamine produced outside the central nervous system. Administration of DOPA analogs that inhibit DOPA decarboxylase but are unable to cross the blood–brain barrier has been effective in decreasing side effects and increasing effectiveness of the DOPA. The interactions of the many brain neurotransmitters are very complex, cell degeneration continues after treatment, and elucidation of the major biochemical abnormality has not yet led to complete control of the disease. Attempts have been made at treatment by transplantation of fetal adrenal medullary tissue into the brain. The adrenal tissue synthesizes dopamine and may improve the movement disorder.

Source: Calne, D. B. and Langston, J. W. Aetiology of Parkinson disease. *Lancet* 2:1457, 1983. Sonntag, K. C., Simantov, R., and Isacson, O. Stem cells may reshape the prospect of Parkinson's disease therapy. Brain Res. Mol. Brain Res. 134:34, 2005.

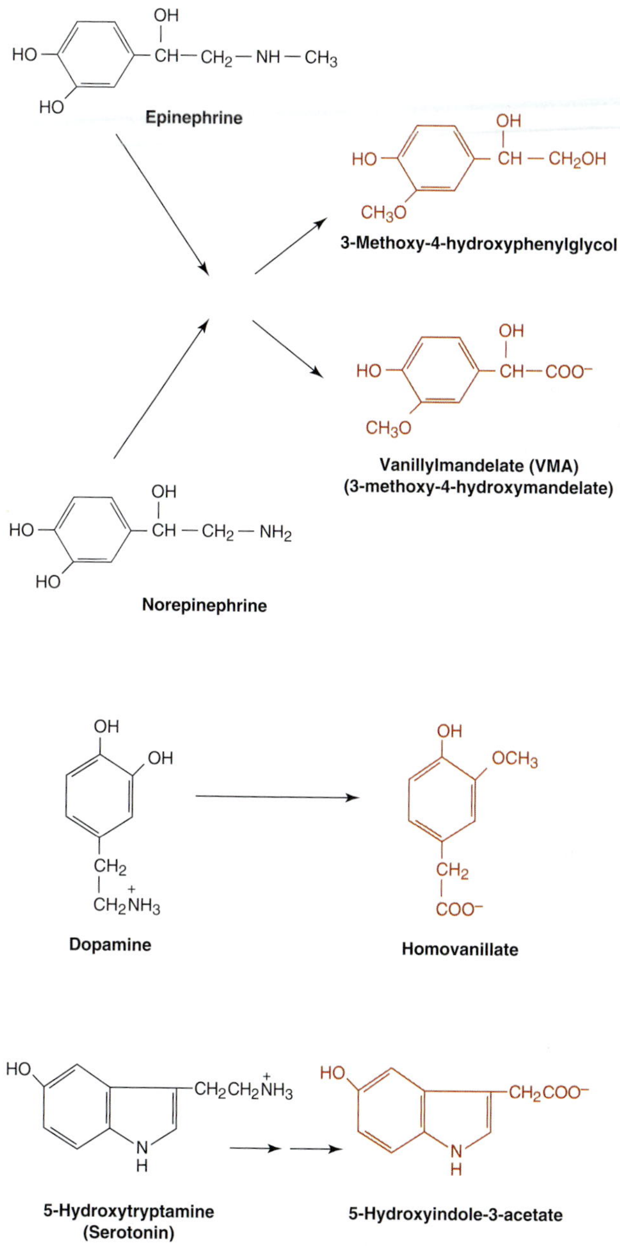

Epinephrine

3-Methoxy-4-hydroxyphenylglycol

Vanillylmandelate (VMA) (3-methoxy-4-hydroxymandelate)

Norepinephrine

Dopamine

Homovanillate

5-Hydroxytryptamine (Serotonin)

5-Hydroxyindole-3-acetate

FIGURE 19.51

Major urinary excretion products of epinephrine, norepinephrine, dopamine, and serotonin.

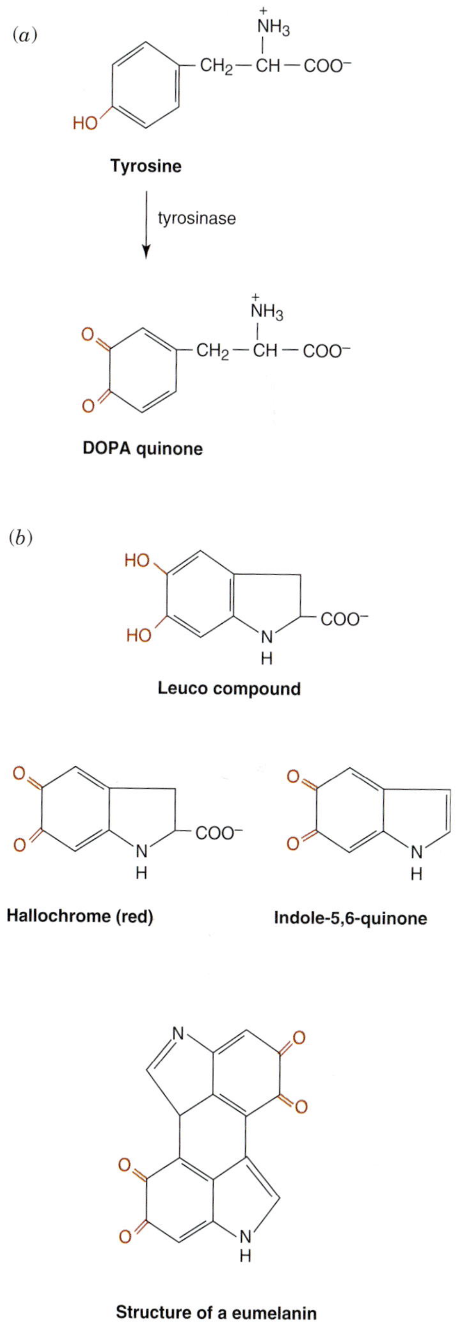

(*a*)

Tyrosine

tyrosinase

DOPA quinone

(*b*)

Leuco compound

Hallochrome (red)

Indole-5,6-quinone

Structure of a eumelanin

FIGURE 19.52

Tyrosinase and intermediates in melanin formation. (*a*) Tyrosinase uses DOPA as a cofactor/intermediate. (*b*) Some intermediates in melanin synthesis and an example of the family of black eumelanins.

Tyrosine Is Required for Synthesis of Melanin, Thyroid Hormone, and Quinoproteins

Conversion of tyrosine to melanin requires tyrosinase, a copper-containing protein (Figure 19.52*a*). The two-step reaction uses DOPA as a cofactor internal to the reaction and produces dopaquinone. During melanogenesis, following exposure to UVB light, tyrosinase and tyrosinase-related protein, which may function in posttranslational modification of tyrosinase, are induced. A lack of tyrosinase activity produces albinism. There are various types of melanin (Figure 19.52*b*). All are aromatic quinones in which the conjugated bond system gives rise to color. The dark pigment that is usually called melanin is eumelanin, from the Greek for "good melanin." Other melanins are yellow or colorless. The role of tyrosine residues of thyroglobulin in thyroid hormone synthesis is presented on p. 905.

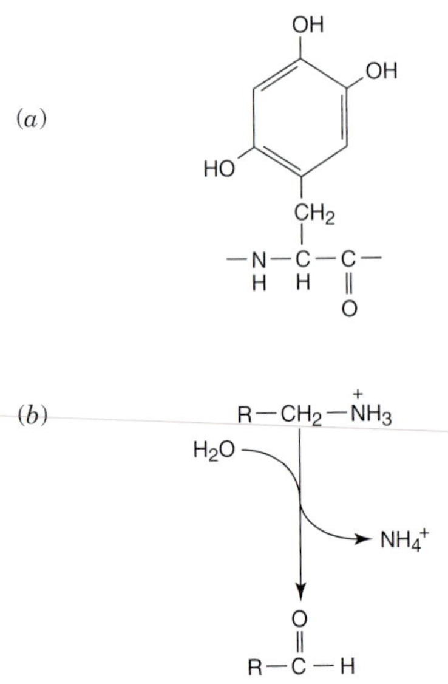

(a)

(b)

FIGURE 19.53

(a) Trihydroxyphenylalanine (TOPA) and (b) amine oxidase reaction.

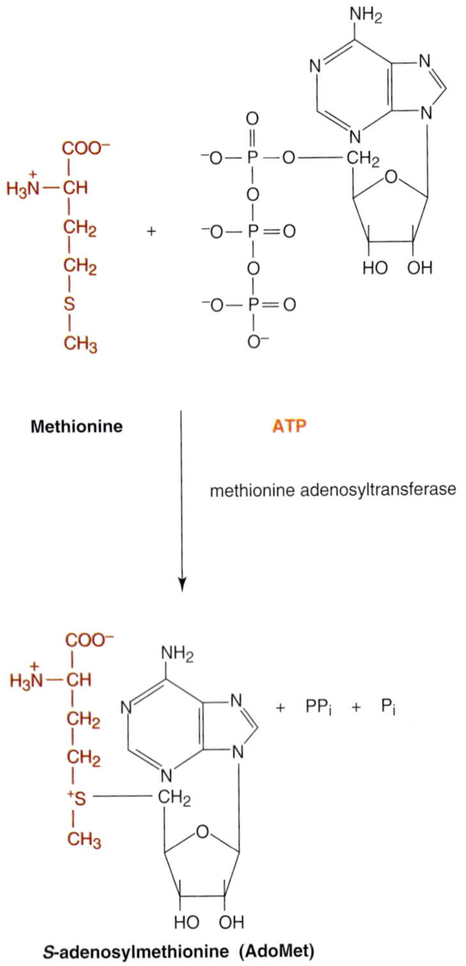

FIGURE 19.54

Synthesis of S-adenosylmethionine.

Some proteins use a modified tyrosine residue as a prosthetic group in oxidation–reduction reactions. The only example reported in humans is topaquinone (trihydroxyphenylalanylquinone) or trihydroxyphenylalanine (TOPA), which is present in some plasma amine oxidases (Figure 19.53).

Methionine and Cysteine

Synthesis of methionine does not occur since methionine is essential. Cysteine, however, is synthesized by transfer of the sulfur atom derived from methionine to the hydroxyl group of serine. As long as the supply of methionine is adequate, cysteine is nonessential. The disposition of individual atoms of methionine and cysteine is a prime example of how cells regulate pathways to fit their immediate needs for energy or for other purposes. Conditions under which various pathways are given preference will be described below.

Methionine First Reacts with Adenosine Triphosphate

When excess methionine is present, its carbons can be used for energy or for gluconeogenesis, and the sulfur can be retained as the sulfhydryl of cysteine. Figure 19.54 shows the first step, catalyzed by **methionine adenosyltransferase**. All phosphates of ATP are removed, with the product being ***S*-adenosylmethionine** (abbreviated **AdoMet**, or SAM). The sulfonium ion is highly reactive, and the methyl is a good leaving group. AdoMet as a methyl group donor will be described below. After a methyltransferase removes the methyl group, the resulting *S*-adenosylhomocysteine is cleaved by adenosylhomocysteinase (Figure 19.55). Note that homocysteine is one carbon longer than cysteine. Although the carbons are destined for intermediary metabolism, the sulfur is conserved through transfer to serine to form cysteine. This requires the pyridoxal phosphate-dependent cystathionine synthase and cystathionase (Figure 19.55; see Clin. Corr. 19.10). Since the bond to form cystathionine is made on one side of the sulfur, and that cleaved is on the other side, the result is a transsulfuration (see Clin. Corr. 19.11). Homocysteine produces α-ketobutyrate and ammonia. α-Ketobutyrate is decarboxylated by a multienzyme complex resembling pyruvate dehydrogenase to yield propionyl CoA, which is then converted to succinyl CoA.

When the need is for energy, and not for cysteine, homocysteine is metabolized by homocysteine desulfhydrase to α-ketobutyrate, NH_3, and H_2S (Figure 19.56).

S-Adenosylmethionine Is a Methyl Group Donor

The role of tetrahydrofolate as a one-carbon group donor has been described (see p. 760). Although this cofactor could in theory serve as a source of methyl groups, the vast majority of methyltransferase reactions utilize *S*-adenosylmethionine. Methyl group transfer from AdoMet to a methyl acceptor is irreversible. An example is shown in Figure 19.57. *S*-Adenosylhomocysteine left after methyl group transfer can be metabolized to cysteine, α-ketobutyrate, and ammonia. When cells need to resynthesize methionine (Figure 19.58), homocysteine methyltransferase catalyzes the transfer. This is one of two enzymes known to require a vitamin B_{12} cofactor (see p. 778). The methyl group comes from N^5-methyltetrahydrofolate. This is the only reaction known that uses this particular form of tetrahydrofolate as a methyl donor. The net result of reactions in Figures 19.57 and 19.58 is donation of a methyl group and regeneration of methionine under methionine-sparing conditions. A minor salvage pathway uses a methyl group from betaine instead of N^5-methyltetrahydrofolate.

S-Adenosylmethionine Is the Precursor of Spermidine and Spermine

Propylamine derived from AdoMet, leaving methylthioadenosine, is added to putrescine (see p. 755) to form spermidine and spermine. Putrescine is formed by decarboxylation of ornithine; with the addition of propylamine, putrescine forms spermidine. Addition of another propylamine gives spermine (Figure 19.59). The methylthioadenosine that remains can be used to regenerate methionine. Much of the polyamine needed by the body is provided by microflora in the gut or from the diet and is carried by the enterohepatic circulation. Meat has a high content of putrescine, but other foods contain more spermidine and spermine.

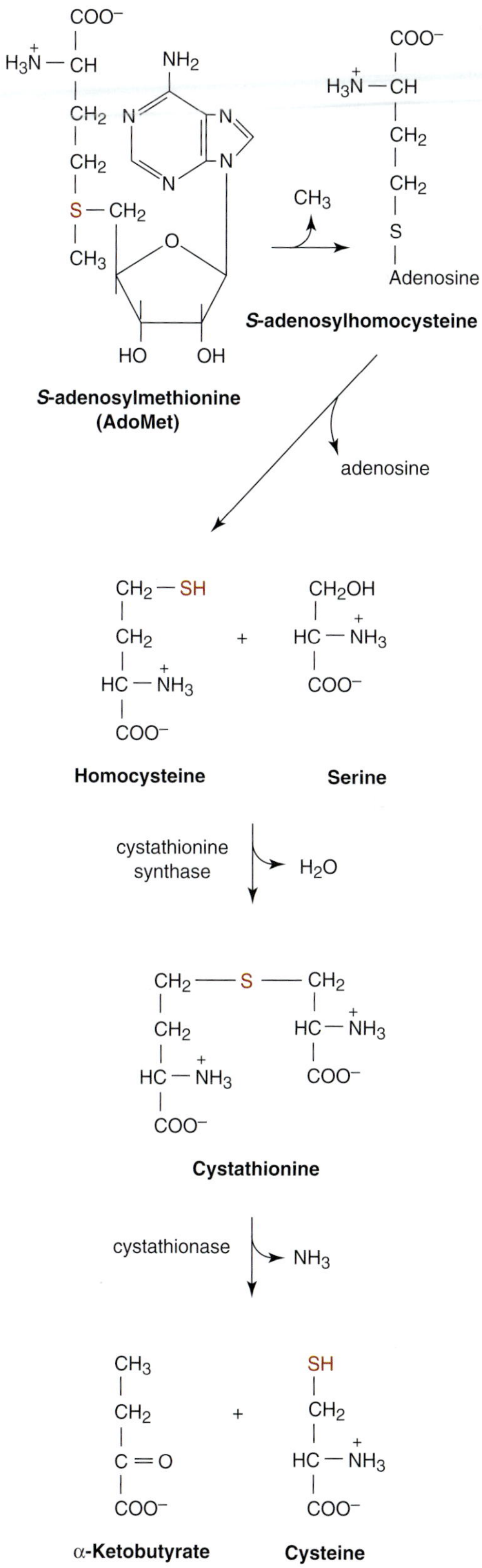

S-adenosylmethionine (AdoMet)

CH_3

S-adenosylhomocysteine

adenosine

Homocysteine + **Serine**

cystathionine synthase → H_2O

Cystathionine

cystathionase → NH_3

α-Ketobutyrate + **Cysteine**

FIGURE 19.55

Synthesis of cysteine from *S*-adenosylmethionine.

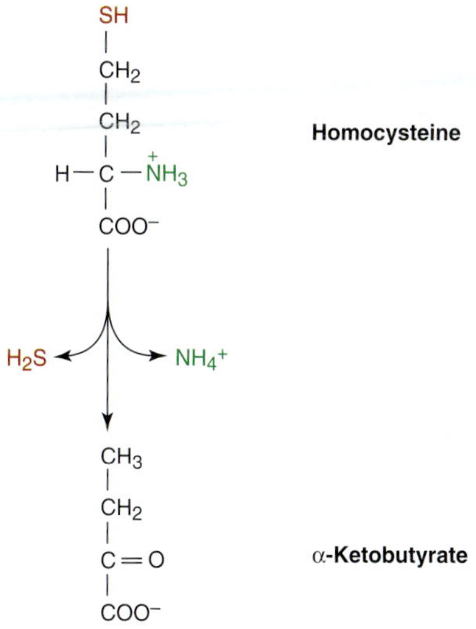

Homocysteine

H_2S ← → NH_4^+

α-Ketobutyrate

FIGURE 19.56

Homocysteine desulfhydrase.

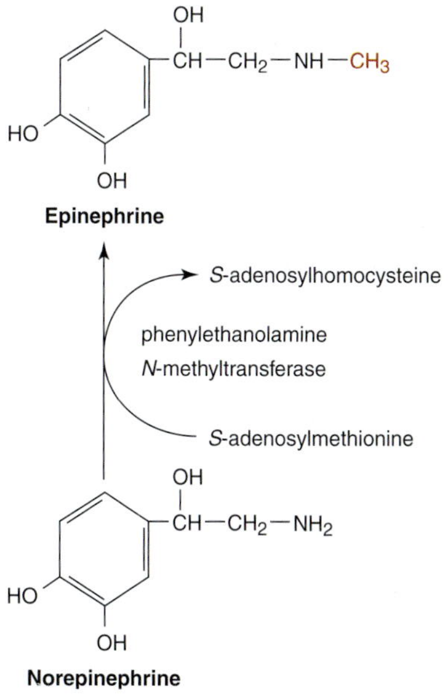

Epinephrine

S-adenosylhomocysteine

phenylethanolamine *N*-methyltransferase

S-adenosylmethionine

Norepinephrine

FIGURE 19.57

S-adenosylmethyltransferase reaction.

CLINICAL CORRELATION 19.10
Hyperhomocysteinemia and Atherogenesis

Deficiency of cystathionine synthase causes homocysteine to accumulate, and its methylation leads to high levels of methionine. Many minor products of these amino acids are formed and excreted. No mechanism has been established to explain why accumulation of homocysteine should lead to some of the pathological changes. Homocysteine may react with and block lysyl aldehyde groups on collagen. The lens of the eye is frequently dislocated some time after the age of 3 years, and other ocular abnormalities often occur. Osteoporosis develops during childhood. Mental retardation is frequently the first indication of this deficiency. Attempts at treatment include restriction of methionine intake and feeding of betaine (or its precursor, choline). In some cases significant improvement has been obtained by feeding pyridoxine (vitamin B_6), suggesting that the deficiency may be caused by more than one type of gene mutation; one type may affect the K_m for pyridoxal phosphate and others may alter the K_m for other substrates, V_{max}, or the amount of enzyme. A theory relating hyperhomocysteinemia to atherogenesis has been proposed. Excess homocysteine can form homocysteine thiolactone, a highly reactive intermediate that thiolates free amino groups in low-density lipoproteins (LDL), and causes them to aggregate and be endocytosed by macrophages. The lipid deposits form atheromas. Homocysteine can have other effects, including lipid oxidation and platelet aggregation, which in turn lead to fibrosis and calcification of atherosclerotic plaques. About one-quarter of patients with atherosclerosis who exhibit none of the other risk factors (such as smoking or oral contraceptive therapy) have been found to be deficient in cystathionine synthase activity.

Source: Kaiser-Kupfer, M. I., Fujikawa, L., Kuwabara, T., et al.. Removal of corneal crystals by topical cysteamine in nephrotic cystinosis. *N. Engl. J. Med.* 316:775, 1987. McCully, K. S. Chemical pathology of homocysteine I. Atherogenesis. *Ann. Clin. Lab. Sci.* 23:477, 1993.

CLINICAL CORRELATION 19.11
Diseases of Sulfur Amino Acids

Congenital deficiency of an enzyme involved in transsulfuration results in accumulation of sulfur-containing amino acids. Hypermethioninemia has been attributed to a deficiency of methionine adenosyltransferase activity, probably caused by a K_m mutant that requires higher than normal concentrations of methionine for saturation. Lack of cystathionase does not seem to cause any clinical abnormalities other than cystathioninuria. The first reported case of this deficiency was about a mentally retarded patient, and the retardation was attributed to the deficiency. Apparently the mental retardation was coincidental, the condition being benign. The amount of cysteine synthesized in these deficiencies is unknown, but treatment with a low methionine diet for hypermethioninemia is unnecessary.

Diseases Involving Cystine

Cystinuria is a defect of membrane transport of cystine and basic amino acids (lysine, arginine, and ornithine) that results in their increased renal excretion. Extracellular sulfhydryl compounds are quickly oxidized to disulfides. Low solubility of cystine results in crystals and formation of renal calculi, a serious feature of this disease. Treatment is limited to removal of stones, prevention of precipitation by drinking large amounts of water or alkalinizing the urine to solubilize cystine, or formation of soluble derivatives by conjugation with drugs. Much more serious is cystinosis in which cystine accumulates in lysosomes. The stored cystine forms crystals in many cells, with a serious loss of function of the kidneys, usually causing renal failure within ten years. The defect is believed to be in the cystine transporter of lysosomal membranes.

Source: Seashore, M. R., Durant, J. L., and Rosenberg, L. E. Studies on the mechanisms of pyridoxine responsive homocystinuria. *Pediatr. Res.* 6:187, 1972. Mudd, S. H. The natural history of homocystinuria due to cystathione β-synthase deficiency. *Am. J. Hum. Genet.* 37:1, 1985. Frimpter, G. W. Cystathionuria: Nature of the defect. *Science* 149:1095, 1965.

The butylamino group of AdoMet is used for posttranslational modification of a specific lysine residue in eIF-4D, an initiation factor for eukaryotic protein synthesis. The group is then hydroxylated, and the modified residue that results is called hypusine (Figure 19.60).

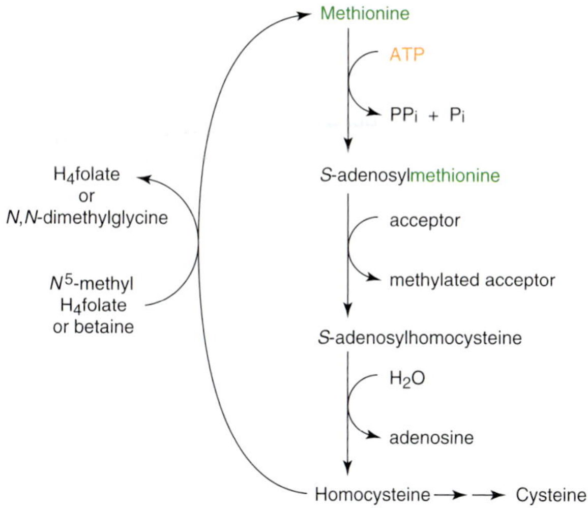

FIGURE 19.58

Resynthesis of methionine. The methyl group goes from folate to methionine via a cobalamin-dependent reaction.

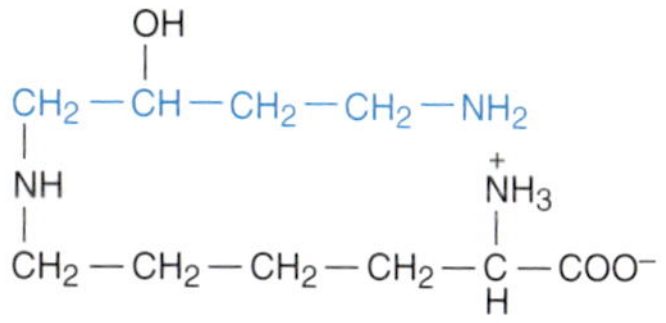

$CH_2-CH(OH)-CH_2-CH_2-NH_2$ attached through NH to $CH_2-CH_2-CH_2-CH_2-CH(\overset{+}{N}H_3)-COO^-$

FIGURE 19.60

Hypusine. The atoms in color are derived from butylamine of AdoMet and later hydroxylated. The lysine to which they are attached is a residue of eIF-4D.

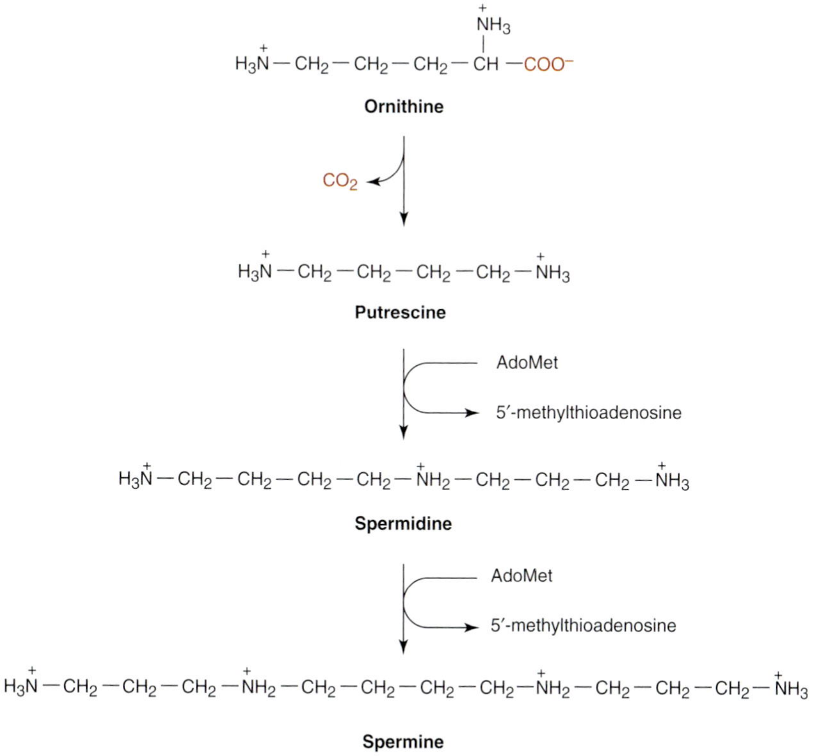

$H_3\overset{+}{N}-CH_2-CH_2-CH_2-CH(\overset{+}{N}H_3)-COO^-$

Ornithine

↓ CO_2

$H_3\overset{+}{N}-CH_2-CH_2-CH_2-CH_2-\overset{+}{N}H_3$

Putrescine

↓ AdoMet → 5′-methylthioadenosine

$H_3\overset{+}{N}-CH_2-CH_2-CH_2-CH_2-\overset{+}{N}H_2-CH_2-CH_2-CH_2-\overset{+}{N}H_3$

Spermidine

↓ AdoMet → 5′-methylthioadenosine

$H_3\overset{+}{N}-CH_2-CH_2-CH_2-\overset{+}{N}H_2-CH_2-CH_2-CH_2-CH_2-\overset{+}{N}H_2-CH_2-CH_2-CH_2-\overset{+}{N}H_3$

Spermine

FIGURE 19.59

Polyamine synthesis.

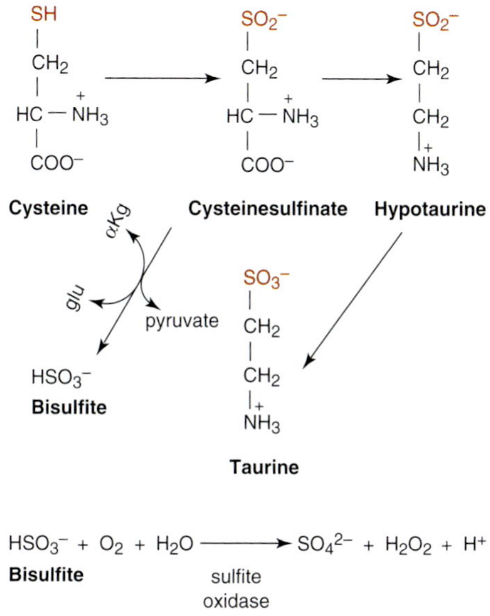

$HS-CH_2-CH(\overset{+}{N}H_3)-COO^-$ (**Cysteine**) → $SO_2^--CH_2-CH(\overset{+}{N}H_3)-COO^-$ (**Cysteinesulfinate**) → $SO_2^--CH_2-CH_2-\overset{+}{N}H_3$ (**Hypotaurine**)

Cysteinesulfinate → (αKg → glu) → pyruvate + HSO_3^- (**Bisulfite**)

Hypotaurine → $SO_3^--CH_2-CH_2-\overset{+}{N}H_3$ (**Taurine**)

HSO_3^- + O_2 + H_2O → SO_4^{2-} + H_2O_2 + H^+ (sulfite oxidase)

Bisulfite

FIGURE 19.61

Formation of taurine and sulfate from cysteine.

Metabolism of Cysteine Produces Sulfur-Containing Compounds

Cysteine is metabolized in several ways as determined by the needs of the cell. The major metabolite is cysteine sulfinate (Figure 19.61). This is converted to sulfite and pyruvate, or to **hypotaurine** and **taurine**. Taurine is an abundant intracellular free amino acid. It appears to play a necessary role in brain development. It forms conjugates with bile acids

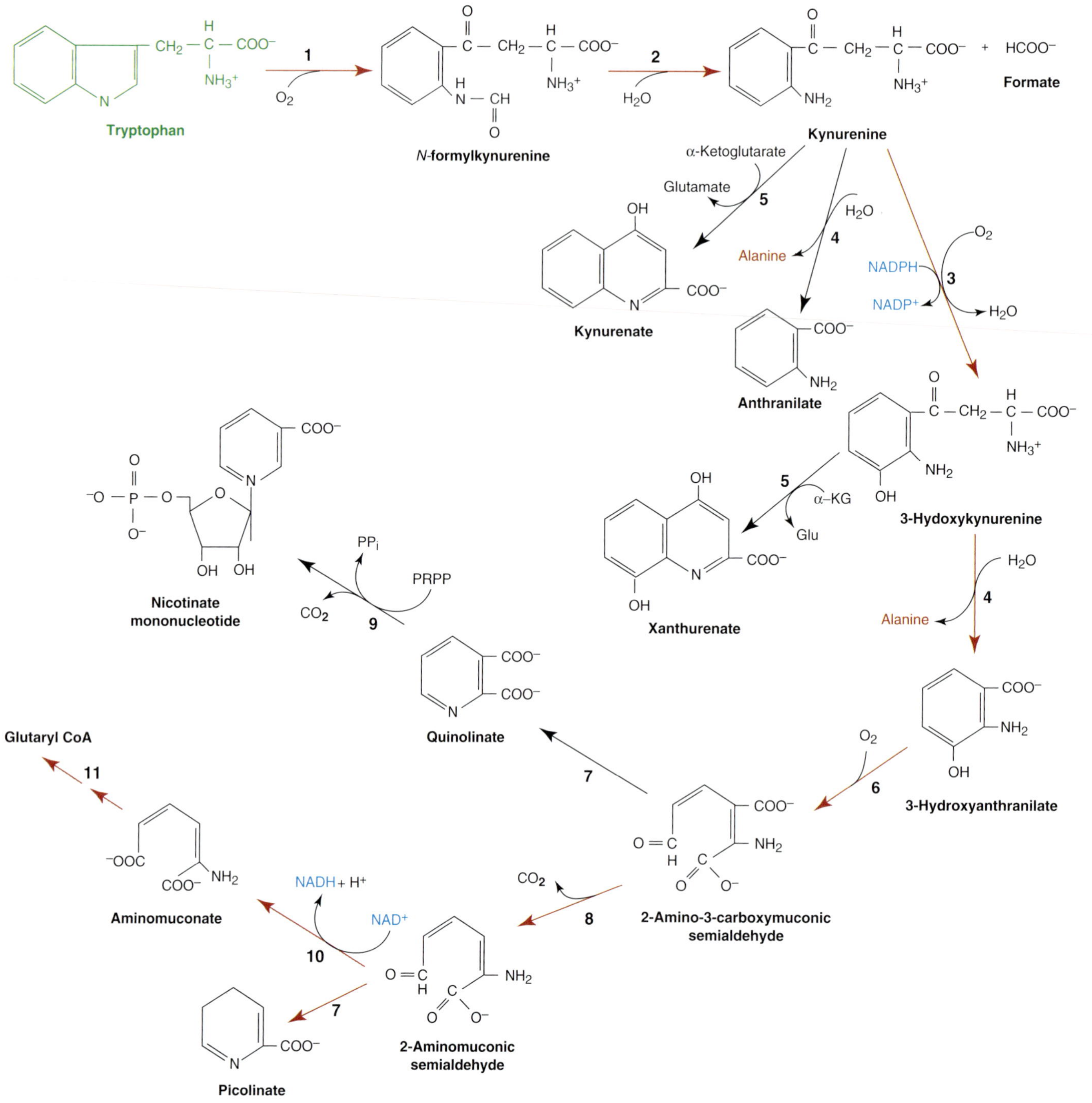

FIGURE 19.66

Metabolism of tryptophan.
Major pathway is shown in red. Enzymes indicated by number are (1) tryptophan oxygenase, (2) kynurenine formamidase, (3) kynurenine hydroxylase, (4) kynureninase, (5) aminotransferase, (6) 3-hydroxyanthranilate oxidase, (7) spontaneous nonenzymatic reaction, (8) picolinate carboxylase, (9) quinolinate phosphoribosyltransferase, (10) aldehyde dehydrogenase, and (11) complex series of reactions.

Pyridoxal Phosphate is Important in Tryptophan Metabolism

Many enzymes in this lengthy pathway are pyridoxal phosphate dependent. Kynureninase is one of them and is affected by vitamin B_6 deficiency (Figure 19.66), resulting in excess kynurenine and xanthurenate excretion, which give urine a greenish-yellow color. This is a diagnostic symptom of vitamin B_6 deficiency.

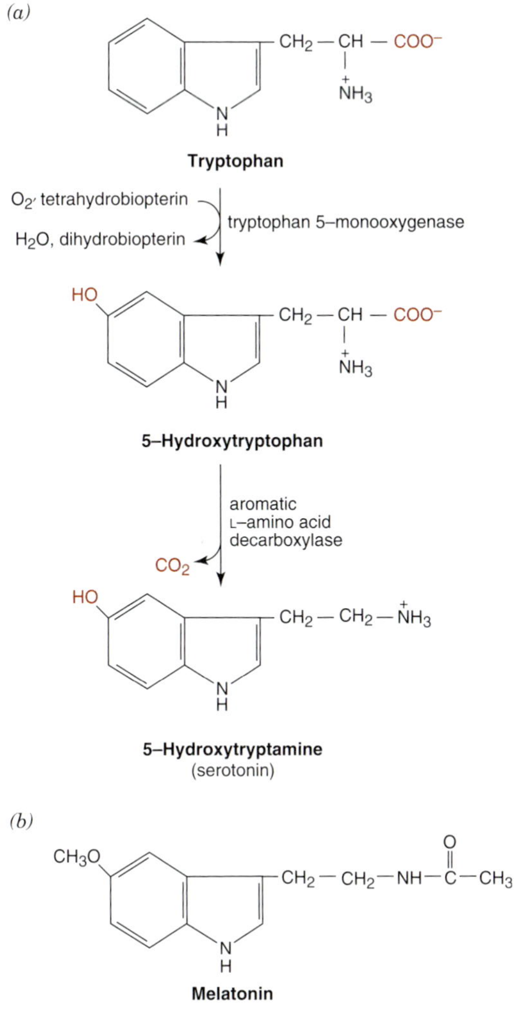

(a)

Tryptophan

O_2, tetrahydrobiopterin → H_2O, dihydrobiopterin

tryptophan 5-monooxygenase

5-Hydroxytryptophan

aromatic L-amino acid decarboxylase

CO_2

5-Hydroxytryptamine (serotonin)

(b)

Melatonin

FIGURE 19.67

(a) Synthesis of serotonin (5-hydroxytryptamine) and (b) structure of melatonin.

Serotonin and Melatonin Are Tryptophan Derivatives

Serotonin (5-**hydroxytryptamine**) results from hydroxylation of tryptophan by a tetrahydrobiopterin-dependent enzyme and decarboxylation by a pyridoxal phosphate-containing enzyme (Figure 19.67*a*). It is a neurotransmitter in brain and causes contraction of smooth muscle of arterioles and bronchioles. It occurs widely in the body and may have other physiological roles. **Melatonin**, a sleep-inducing molecule, is *N*-acetyl-5-methoxytryptamine (Figure 19.67*b*). The acetyltransferase needed for its synthesis is present in pineal gland and retina. Melatonin is involved in regulation of circadian rhythm, being synthesized mostly at night. It appears to function by inhibiting synthesis and secretion of other neurotransmitters such as dopamine and GABA (see p. 951).

Tryptophan Induces Sleep

Ingestion of foods rich in tryptophan leads to sleepiness because the resulting serotonin is sleep-inducing. Reducing availability of tryptophan in the brain can interfere with sleep. Tryptophan availability is reduced when other amino acids compete with it for transport through the blood–brain barrier. Elevated plasma concentrations of other amino acids, after a high-protein meal, diminish transport of tryptophan and induce wakefulness. The sleep-inducing effect of carbohydrates is due to decreased plasma amino acid levels, since carbohydrate stimulates release of insulin, and insulin causes removal of amino

CLINICAL CORRELATION **19.13**

Schizophrenia and Other Tryptophan-Derived Neurotransmitter-Associated Diseases

Kynurenine can be transaminated with condensation of the side chain to form a two-ring compound, kynurenic acid (Figure 19.66). This compound, together with 3-hydroxykynurenine, the major metabolite of kynurenine, and quinolinate (from the nicotinate mononucleotide synthetic pathway) all serve as neurotransmitters. Quinolinate is an agonist of the NMDA-sensitive glutamate receptors, and kynurenate is an antagonist of several members of the glutamate receptor family. Schizophrenia is a disease of attenuated glutaminergic neurotransmission, and the glutaminergic antagonist kynurenate is elevated in schizophrenia. Free radicals produced from 3-hydroxykynurenine have been linked to the etiology of Huntington's disease. Development of AIDS–dementia complex is sensitive to the ratio of quinolinate to kynurenate, and it has been suggested that Alzheimer's and Parkinson's disease are also affected by tryptophan-derived neurotransmitters.

Source: Jhamandas, K., Boegmand, R., and Beninger, R., Quinolinic acid induced brain neurotransmitter deficits: modulation by endogenous excitotoxin antagonists. *Can. J. Physiol. Pharmacol.* 71:1473, 1994. Stone, T., Mackay, G., Forrest, C.M., Clark, C., and Darlington, L. Tryptophan metabolites and brain disorders. *Clin. Chem. Lab. Med.* 41:852, 2003.

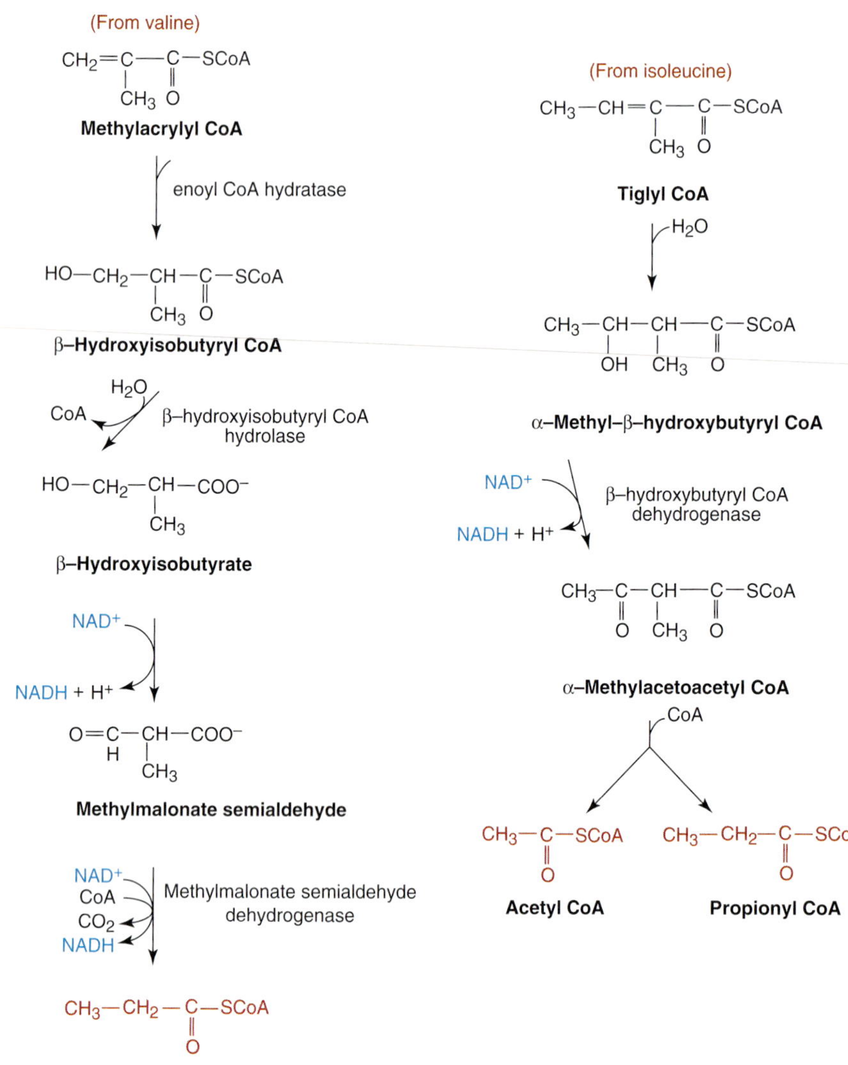

FIGURE 19.69

Terminal reactions in degradation of valine and isoleucine.

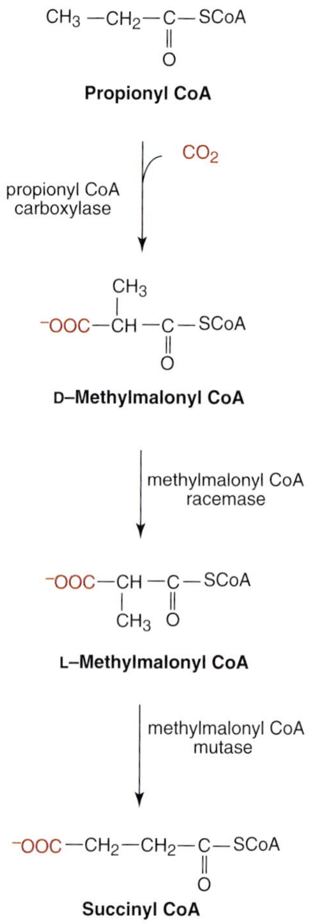

$CH_3-CH_2-C(=O)-SCoA$

Propionyl CoA

propionyl CoA carboxylase (CO_2)

$^-OOC-CH(CH_3)-C(=O)-SCoA$

D–Methylmalonyl CoA

methylmalonyl CoA racemase

$^-OOC-CH(CH_3)-C(=O)-SCoA$

L–Methylmalonyl CoA

methylmalonyl CoA mutase

$^-OOC-CH_2-CH_2-C(=O)-SCoA$

Succinyl CoA

FIGURE 19.71

Interconversion of propionyl CoA, methylmalonyl CoA, and succinyl CoA. The mutase requires 5′-deoxyadenosylcobalamin for activity.

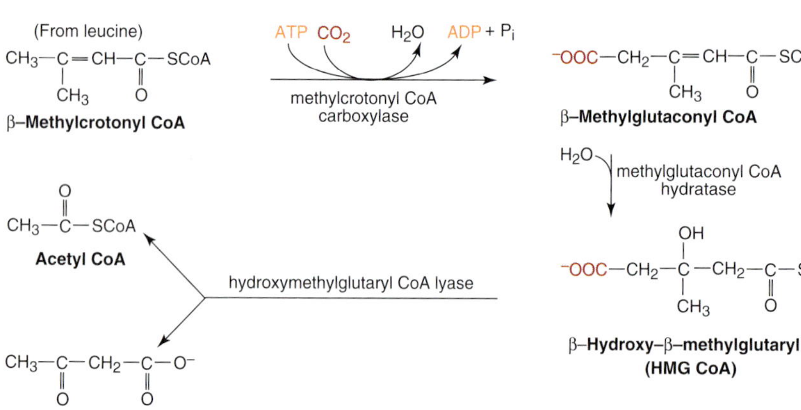

FIGURE 19.70

Terminal reactions of leucine degradation.

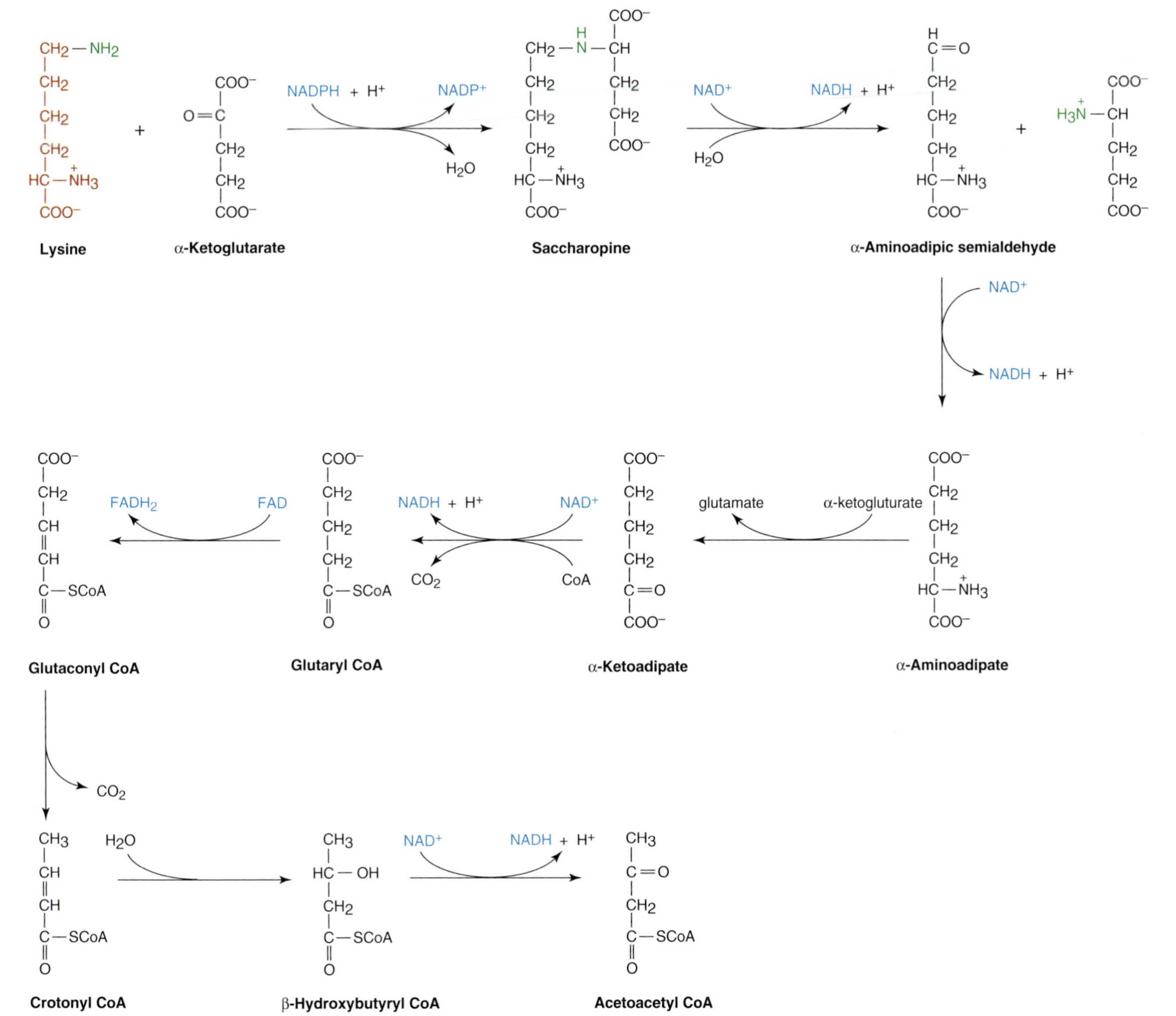

FIGURE 19.72
Principal pathway of lysine degradation.

transferred to α-ketoglutarate by a bifunctional enzyme with an intermediate called saccharopine (Figure 19.72). Glutamate and a semialdehyde compound are formed. The semialdehyde is then oxidized to a dicarboxylic amino acid, and a transamination of the α-amino group occurs in a pyridoxal-dependent manner. Further reactions lead to acetoacetyl CoA. A minor pathway starts with removal of the α-amino group and goes via the cyclic compound pipecolate (Figure 19.73) to join the major pathway at the level of the semialdehyde intermediate. This does not replace the major pathway even in a deficiency of enzymes in the early part of the pathway (see Clin. Corr. 19.16).

Carnitine Is Derived from Lysine

Medium- and long-chain fatty acids are transported into mitochondria for β-oxidation as carnitine conjugates (see p. 680). **Carnitine** is synthesized from lysine residues in certain proteins. The first step is trimethylation of the ϵ-amino group of the lysine side chain, with AdoMet as the methyl donor (Figure 19.74). Free trimethyllysine is released by hydrolysis of these proteins and is converted in four steps to carnitine.

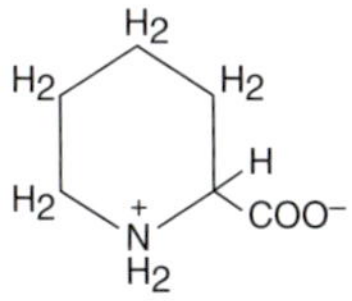

FIGURE 19.73
Minor product of lysine metabolism.

CLINICAL CORRELATION **19.15**

Diseases of Propionate and Methylmalonate Metabolism

Deficiency of any of the three enzymes shown in Figure 19.71 contributes to ketoacidosis. Propionate is formed in the degradation of valine, isoleucine, methionine, threonine, the side chain of cholesterol, and odd-chain fatty acids. The amino acids appear to be the main precursors since decreasing or eliminating dietary protein immediately minimizes acidosis. A defect in propionyl-CoA carboxylase results in accumulation of propionate, which is diverted to alternative pathways including incorporation into fatty acids in place of the first acetyl group to form odd-chain fatty acids. The extent of these reactions is very limited. In one case, large amounts of biotin were reported to produce beneficial effects, suggesting that more than one defect decreases propionyl-CoA carboxylase activity. Possibilities are a lack of intestinal biotinidase that liberates biotin from ingested food for absorption or a lack of biotin holocarboxylase that incorporates biotin into biotin-dependent enzymes. Acidosis in children may be caused by high levels of methylmalonate, which is normally undetectable in blood. Liver taken at autopsy or cultured fibroblasts showed in some cases deficiency of methylmalonyl-CoA mutase. Some samples were unable to convert methylmalonyl CoA to succinyl CoA under any conditions, but other samples carried out the conversion when 5′-adenosylcobalamin was added. Clearly, those with an active site defect cannot metabolize methylmalonate, but those with defects in handling vitamin B_{12} respond to massive doses of the vitamin. Other cases of methylmalonic aciduria have a more fundamental inability to use vitamin B_{12} that leads to methylcobalamin deficiency (coenzyme of methionine salvage) and 5′-adenosylcobalamin deficiency (coenzyme of methylmalonyl CoA isomerization).

Source: Mahoney, M. J. and Bick, D. Recent advances in the inherited methylmalonic acidemias. *Acta Paediatr. Scand.* 76:689, 1987.

CLINICAL CORRELATION **19.16**

Diseases Involving Lysine and Ornithine

Lysine

α-Amino adipic semialdehyde synthase is deficient in a small number of patients who excrete lysine and smaller amounts of saccharopine. This has led to the discovery that the enzyme has both lysine-α-ketoglutarate reductase and saccharopine dehydrogenase activities. It is thought that hyperlysinemia is benign. More serious is familial lysinuric protein intolerance due to failure to transport dibasic amino acids across intestinal mucosa and renal tubular epithelium. Plasma lysine, arginine, and ornithine are decreased to one-third or one-half of normal. Patients develop marked hyperammonemia after a meal containing protein. This is thought to arise from deficiency of urea cycle intermediates ornithine and arginine in liver, limiting the capacity of the cycle. Consistent with this view, oral supplementation with citrulline prevents hyperammonemia. Other features are thin hair, muscle wasting, and osteoporosis, which may reflect protein malnutrition due to lysine and arginine deficiency.

Ornithine

Elevated ornithine levels are generally due to deficiency of ornithine δ-aminotransferase. Gyrate atrophy of the choroid and retina, characterized by progressive loss of vision leading to blindness by the fourth decade, is caused by deficiency of this mitochondrial enzyme. The mechanism of changes in the eye is unknown. Progression of the disease may be slowed by dietary restriction in arginine and/or pyridoxine therapy, which reduces ornithine in body fluids.

Source: O'Donnell, J. J., Sandman, R. P., and Martin, S. R. Gyrate atrophy of the retina: Inborn error of L-ornithine:2-oxoacid aminotransferase. *Science* 200:200, 1978. Rajantic, J., Simell, O., and Perheentupa, J. Lysinuric protein intolerance. Basolateral transport defect in renal tubuli. *J. Clin. Invest.* 67:1078, 1981.

Histidine

Histidase(Clin. Corr. 19.17) releases free ammonia from histidine and leaves a compound with a double bond called urocanate (Figure 19.75). Two other reactions lead to formation of **iminoglutamate** (**FIGLU**). The formimino group is then transferred to tetrahydrofolate.

Urinary Formiminoglutamate Is Increased in Folate Deficiency

The formimino group of FIGLU must be transferred to tetrahydrofolate before the final product, glutamate, can be produced. When there is insufficient tetrahydrofolate

CLINICAL CORRELATION 19.17
Histidinemia

Histidinemia is due to histidase deficiency. A convenient assay for this enzyme uses skin, which produces urocanate as a constituent of sweat; urocanase and other enzymes of histidine catabolism in liver do not occur in skin. Histidase deficiency can be confirmed from skin biopsy. Incidence of the disorder is high, about 1 in 10,000 newborns screened. Most reported cases of histidinemia show normal mental development. Restriction of dietary histidine normalizes the biochemical abnormalities but is not usually required.

Source: Scriver, C. R. and Levy, H. L. Histidinemia: Reconciling retrospective and prospective findings. *J. Inherit. Metab. Dis.* 6:51, 1983.

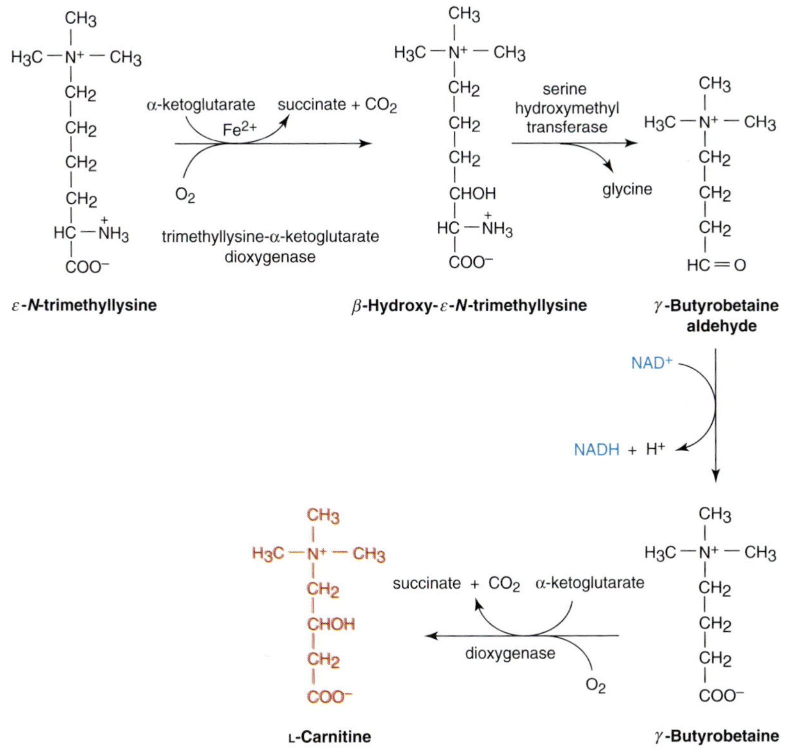

FIGURE 19.74
Biosynthesis of carnitine.

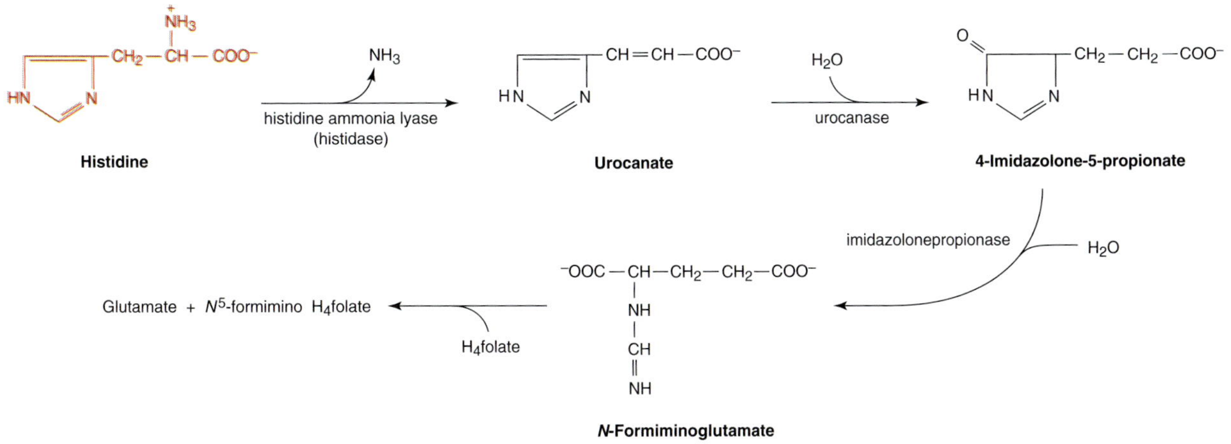

FIGURE 19.75
Degradation of histidine.

CLINICAL CORRELATION **19.18**

Diseases of Folate Metabolism

Folic acid must be reduced to function as a coenzyme. Symptoms of folate deficiency may be due to deficiency of dihydrofolate reductase. Parenteral administration of N^5-formyltetrahydrofolate, the most stable of the reduced folates, is effective in these cases. In some cases of central nervous system abnormality attributed to deficiency of methylene folate reductase, there is homocystinuria. Decreased enzyme activity lowers the N^5-methyltetrahydrofolate formed so that the source of methyl groups for conversion of homocysteine to methionine is limiting. Large amounts of folic acid, betaine, and methionine reversed the biochemical abnormalities and, in at least one case, the neurological disorder. Patients with widely divergent presentations have shown deficiencies in transfer of the formimino group from FIGLU to tetrahydrofolate. They excreted varying amounts of FIGLU; some responded to large doses of folate. How a deficiency of formiminotransferase produces pathological changes is unclear. It is not sure whether this deficiency causes a disease state. One patient showed symptoms of folate deficiency and had tetrahydrofolate methyltransferase deficiency. The associated anemia did not respond to vitamin B_{12}, but showed some improvement with folate. It was suggested that the patient formed inadequate N^5-methyltetrahydrofolate to promote remethylation of homocysteine. This left the coenzyme "trapped" in the methylated form and unavailable for use in other reactions.

available, this reaction decreases and FIGLU is excreted in urine. This is a diagnostic sign of folate deficiency if it occurs after a test dose of histidine is ingested (see Clin. Corr. 19.18).

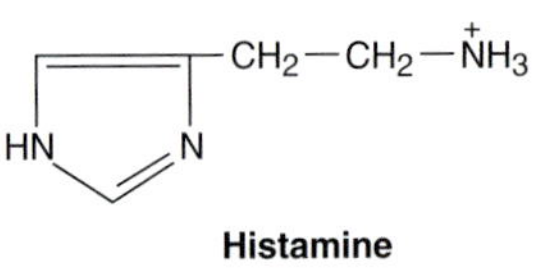

Histamine

FIGURE 19.76
Histamine.

Histamine, Carnosine, and Anserine Are Histidine Derivatives

Histamine (Figure 19.76), released from cells as part of an allergic response, is produced from histidine by histidine decarboxylase. Histamine has many physiological roles, including dilation and constriction by interacting with different types of receptors on endothelium of blood vessels. An overreaction to histamine can lead to asthma and other allergic reactions. **Carnosine** (β-alanylhistidine) and **anserine** (β-alanylmethylhistidine) are dipeptides (Figure 19.77) found in muscle. Their function is unknown.

Anserine

Carnosine

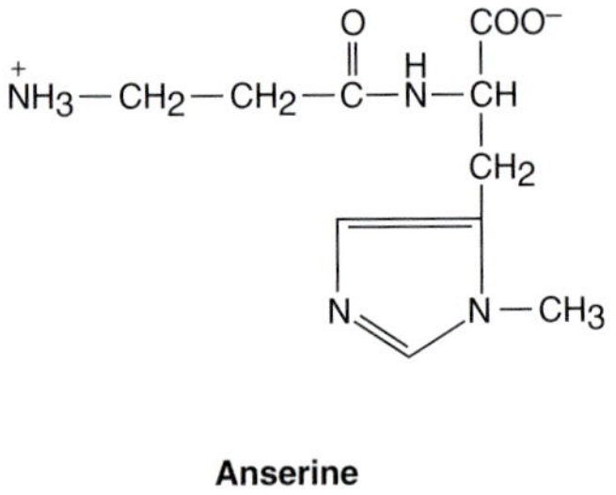

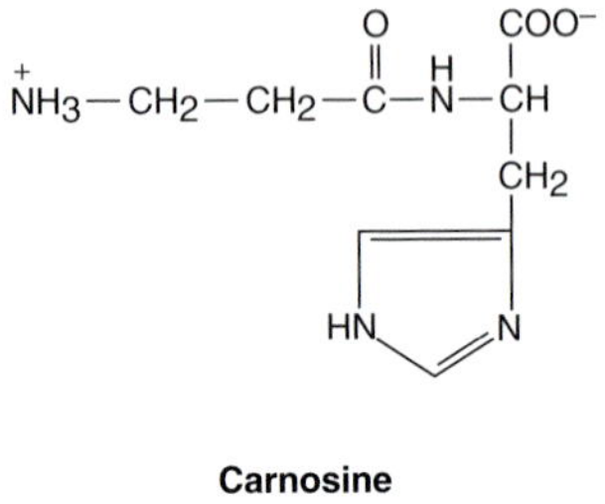

FIGURE 19.77
Anserine and carnosine.

Creatine

Storage of "high-energy" phosphate, particularly in cardiac and skeletal muscle, occurs by transfer of the phosphate group from ATP to **creatine** (see p. 985). Creatine is synthesized by transfer of the guanidinium group of arginine to glycine, followed by addition of a methyl group from AdoMet (Figure 19.78). The amount of creatine in the body is related to muscle mass, and a certain percentage of this undergoes turnover each day. About 1–2% of preexisting creatine phosphate is cyclized nonenzymatically to **creatinine** (Figure 19.79) and excreted in urine, and new creatine is synthesized to replace it. The amount of creatinine excreted by an individual is therefore constant from day to day. When a 24-h urine sample is requested, the amount of creatinine in the sample can be used to determine whether the sample truly represents a whole day's urinary output.

Glutathione

Glutathione, the tripeptide γ-glutamylcysteinylglycine, has several important functions. It is a reductant, is conjugated with drugs to make them more water soluble (see p. 422), is involved in transport of amino acids across cell membranes, is part of some leukotriene structures (see p. 736), is a cofactor for some enzymatic reactions, and participates in the rearrangement of protein disulfide bonds. Glutathione as reductant is very important in maintaining stability of erythrocyte membranes. Its sulfhydryl group can be used to reduce peroxides formed during oxygen transport (see p. 577). The resulting oxidized form consists of two molecules of GSH joined by a disulfide bond.

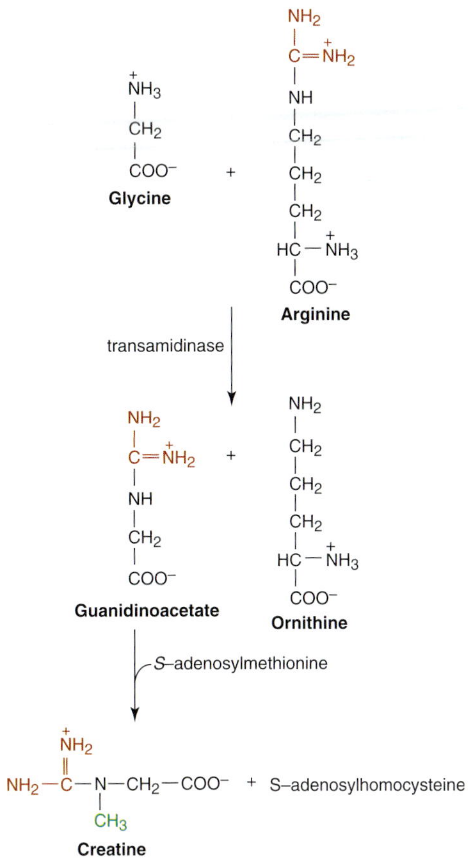

Glycine + Arginine

transamidinase

Guanidinoacetate + Ornithine

S-adenosylmethionine

Creatine + S-adenosylhomocysteine

FIGURE 19.78

Synthesis of creatine.

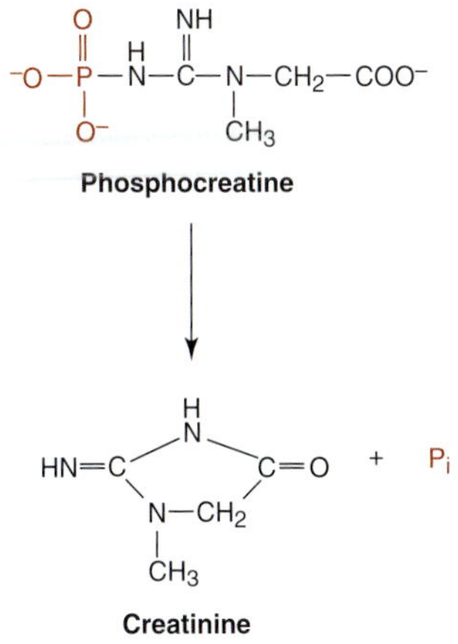

Phosphocreatine

Creatinine + P_i

FIGURE 19.79

Spontaneous reaction forming creatinine.

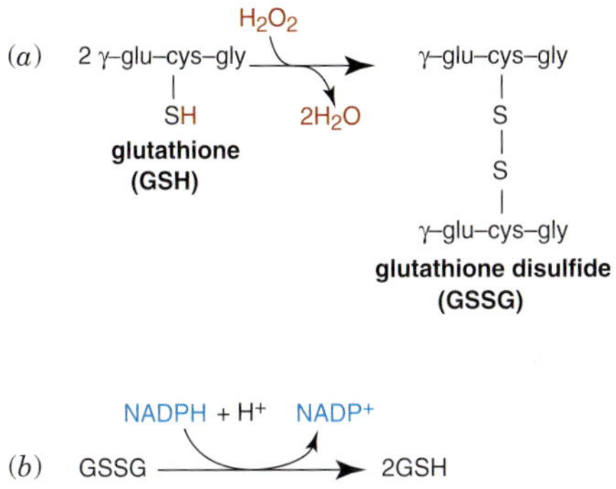

(*a*) 2 γ-glu–cys–gly (SH) glutathione (GSH) + H_2O_2 → γ-glu–cys–gly–S–S–γ-glu–cys–gly glutathione disulfide (GSSG) + $2H_2O$

(*b*) GSSG + NADPH + H^+ → 2GSH + $NADP^+$

FIGURE 19.80

(*a*) Scavenging of peroxide by glutathione peroxidase and (*b*) regeneration of reduced glutathione by glutathione reductase.

This is reduced by glutathione reductase to two molecules of GSH at the expense of NADPH (Figure 19.80). The usual steady-state ratio of GSH to GSSG in erythrocytes is 100:1. Conjugation of drugs such as 6-thiopurine with glutathione (see p. 819) renders them more polar for excretion (Figure 19.81).

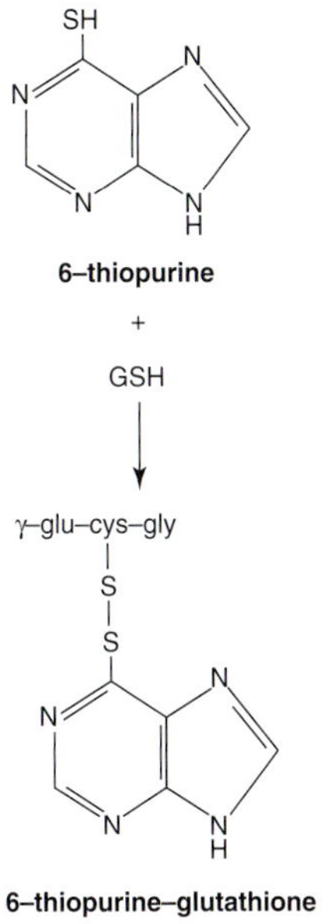

6-thiopurine + GSH → 6-thiopurine–glutathione

FIGURE 19.81

Conjugation of a drug by glutathione transferase.

Glutathione Is Synthesized from Three Amino Acids

Glutathione is synthesized by formation of the dipeptide γ-glutamylcysteine and the subsequent addition of glycine. Both reactions require activation of carboxyl groups by ATP (Figure 19.82).

Synthesis of Glutathione is Largely Regulated by Cysteine Availability

The γ-Glutamyl Cycle Transports Amino Acids

There are several mechanisms for transport of amino acids across cell membranes. Many are symport or antiport mechanisms (see p. 481) coupled to sodium transport. The **γ-glutamyl cycle** for transmembrane transport of amino acids functions in kidneys and some other tissues but is of particular importance in renal epithelial cells. It is more energy-requiring than other mechanisms, but is rapid and has high capacity. The cycle involves several enzymes. The **γ-glutamyl transpeptidase**, located in the plasma membrane, adds glutamate from GSH to an extracellular amino acid. The resulting γ-glutamyl amino acid is transported into the cell by an amino acid transporter, where the γ-glutamyl amino acid is hydrolyzed to liberate the amino acid (Figure 19.83) and 5-oxoproline. Cysteinylglycine formed in the transpeptidase reaction is cleaved to its component amino acids. To regenerate GSH, glutamate is reformed from oxoproline in an ATP-requiring reaction, and GSH is resynthesized from its three component parts.

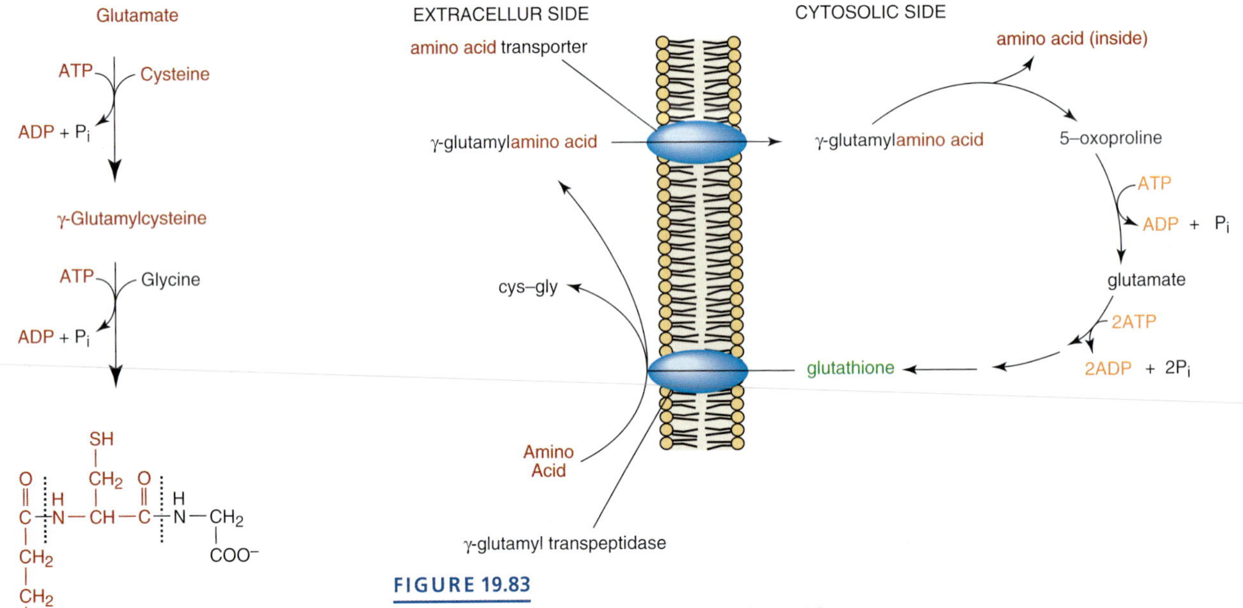

FIGURE 19.82
Synthesis of glutathione.

FIGURE 19.83
γ-Glutamyl cycle for transporting amino acids.

$CH_3-CH_2-CH_2-CH_2-S(=O)(=NH)-CH_2-CH_2-CH(\overset{+}{N}H_3)-COO^-$

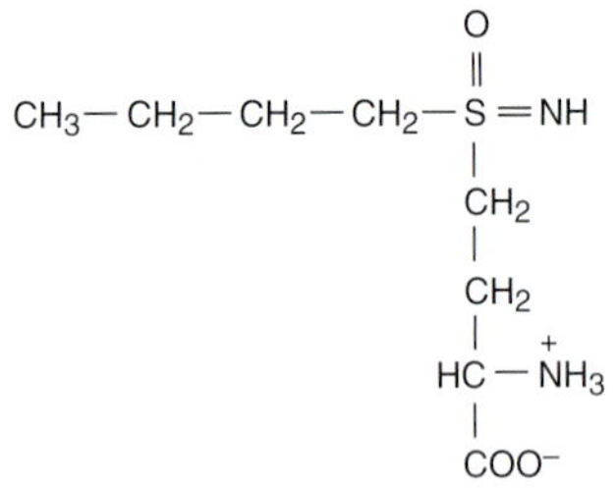

FIGURE 19.84
Buthionine sulfoximine.

Three ATPs are used in the regeneration of glutathione, one in formation of glutamate from oxoproline and two in formation of the peptide bonds.

Glutathione Concentration Affects Response to Toxins

When the body encounters toxic conditions such as peroxide formation, ionizing radiation, alkylating agents, or other reactive intermediates, it is beneficial to increase the level of GSH. Cysteine and methionine have been administered as GSH precursors, but they have the disadvantage of being precursors of an energy-expensive pathway to GSH. A more promising approach is administration of a soluble diester of GSH, such as γ-(α-ethyl)glutamylcysteinylethylglycinate. Very premature infants have a very low concentration of cysteine because of low cystathionase activity in liver. This keeps the GSH concentration low and makes them more susceptible to oxidative damage, especially from hydroperoxides formed in the eye after hyperbaric oxygen treatment. Under certain circumstances, such as rendering tumor cells more sensitive to radiation or parasites more sensitive to drugs, it is desirable to lower GSH levels. This can be achieved by administration of the glutamate analog **buthionine sulfoximine** (Figure 19.84) as a competitive inhibitor of GSH synthesis.

BIBLIOGRAPHY

General

Meister, A. *Biochemistry of the Amino Acids*, 2nd ed. New York: Academic Press, 1965.

Pyridoxal Phosphate

Dolphin, P., Poulson, R., and Avramovic, O. (Eds.). *Vitamin B_6 Pyridoxal Phosphate*. NewYork: Wiley, 1986.

Glutamate and Glutamine

Bode, B. L., Kaminski, D. L., Souba, W. W., and Li, A. P. Glutamine transport in human hepatocytes and transformed liver cells. *Hepatology* 21: 511, 1995.

Fisher, H. F. Glutamate dehydrogenase. *Methods Enzymol.* 113: 16, 1985.

Haussinger, D. Nitrogen metabolism in liver: Structural and functional organization and physiological relevance. *Biochem. J.* 267: 281, 1990.

Kelly, A. and Stanley, C. Disorders of glutamate metabolism. *Ment. Retard. Dev. Disord. Res. Rev.* 7: 287, 2001.

Urea Cycle

Holmes, F. L. Hans Krebs and the discovery of the ornithine cycle. *Fed. Proc.* 39: 216, 1980; Mathias, R., Kostiner, D., and Packman, S. Hyperammonemia in urea cycle disorder; role of the nephrologists. *Am. J. Kidney Dis.* 37: 1069, 2001.

Jungas, R. L., Halperin, M. L., and Brosnan, J. T. Quantitative analysis of amino acid oxidation and related gluconeogenesis in humans. *Physiol. Rev.* 72: 419, 1992.

Branched-Chain Amino Acids

Shander, P., Wahren, J., Paoletti, R., Bernardi, R., and Rinetti, M. *Branched Chain Amino Acids*. New York: Raven Press, 1992.

Serine

Snell, K. The duality of pathways for serine biosynthesis is a fallacy. *Trends Biochem. Sci.* 11: 241, 1986.

Arginine

Reyes, A. A., Karl, I. E., and Klahr, S. Editorial review: role of arginine in health and renal disease. *Am. J. Physiol.* 267: F331, 1994.

Sulfur Amino Acids

Chen, J. and Berry, M. Selenium and selenoproteins in the brain and brain disease. *J. Neurochem.* 86: 1, 2003.

Lee, B. J., Worland, P. J., Davis, J. N., Stadtman, T. C., and Hatfield, D. L. Identification of a selenocysteyl-tRNASer in mammalian cells that recognizes the nonsense codon, UGA. *J. Biol. Chem.* 264: 9724, 1989.

Namy, O., Rousset, J., Napthine, S., and Brierly, I. Reprogrammed genetic decoding in cellular gene expression. *Mol. Cell* 13: 157, 2004.

Schuller-Levis, G. and Park, E. Taurine: New implications for an old amino acid. *FEMS Microbiol. Lett.* 226: 195, 2003.

Stepanuk, M. H. Metabolism of sulfur-containing amino acids. *Annu. Rev. Nutr.* 6: 179, 1986.

Wright, C. E., Tallan, H. H., Lin, Y. Y., and Gaull, G. E. Taurine: biological update. *Annu. Rev. Biochem.* 55: 427, 1986.

Yap, S. Classical homocystinuria: Vascular risk and its prevention. *J. Inherit. Metab. Dis.* 26: 259, 2003.

Polyamines

Perin, A., Scalabrino, G., Sessa, A., and Ferioloini, M. E. *Perspectives in Polyamine Research.* Milan: Watchdog Editors, 1988.

Tabor, C. W. and Tabor, H. Polyamines. *Annu. Rev. Biochem.* 53: 749, 1984.

Tabor, C. W. and Tabor, H. Inhibitors of polyamine metabolism: Review article. *Amino Acids* 26: 353, 2004.

Creatine

Schulze, A. Creatine deficiency syndromes. *Mol. Cell Biochem.* 244: 143, 2003.

Folates and Pterins

Blakley, R. L. and Benkovic, S. J. *Folate and Pterins.* New York: Wiley, Vol. 1, 1984, Vol. 2, 1985.

Quinoproteins

Davidson, V. L., (Ed.). *Principles and Applications of Quinoproteins.* New York: Marcel Dekker, 1993.

Carnitine

Bieber, L. L. Carnitine. *Annu. Rev. Biochem.* 57: 261, 1988.

Glutathione

Taniguchi, N., Higashi, T., Sakamoto, Y., and Meister, A. *Glutathione Centennial: Molecular Perspectives and Clinical Implications.* New York: Academic Press, 1989.

Tryptophan

Schwarcz, R. Metabolism and function of brain kynurenines. *Biochem. Soc. Trans.* 21: 77, 1993.

Stone, T. W. *Quinolinic Acid and the Kynurenines.* Boca Raton, FL: CRC Press, 1989.

Disorders of Amino Acid Metabolism

Rosenberg, L. E. and Scriver, C. R. Disorders of amino acid metabolism. In: P. K. Bondy and L. E. Rosenberg (Eds.), *Metabolic Control and Disease*, 8th ed. Philadelphia: Saunders, 1980.

Scriver, C. R., Beaudet, A. R., Sly, W. S., and Valle, D. (Eds.). *The Metabolic and Molecular Bases of Inherited Disease*, 8th ed. New York: McGraw-Hill, 2001.

Wellner, D. and Meister, A. A survey of inborn errors of amino acid metabolism and transport. *Annu. Rev. Biochem.* 50: 911, 1980.

OMIM

An excellent source of information on metabolic disorders of amino acid metabolism is "Online Mendelian Inheritance in Man" developed by The Johns Hopkins University and accessible at www.ncbi.nlm.nih.gov/Omim/searchomim.html

QUESTIONS | Carol N. Angstadt

Multiple Choice Questions

1. Aminotransferases:
   A. usually require α-ketoglutarate or glutamine as one of the reacting pair.
   B. catalyze reactions that result in a net use or production of amino acids.
   C. catalyze irreversible reactions.
   D. require pyridoxal phosphate as an essential cofactor for the reaction.
   E. are not able to catalyze transamination reactions with essential amino acids.

2. The production of ammonia in the reaction catalyzed by glutamate dehydrogenase:
   A. requires the participation of NADH or NADPH.
   B. proceeds through a Schiff base intermediate.
   C. may be reversed to consume ammonia if it is present in excess.
   D. is favored by high levels of ATP or GTP.
   E. would be inhibited when gluconeogenesis is active.

3. All of the following are correct about ornithine *except* it:
   A. may be formed from or converted to glutamic semialdehyde.
   B. can be converted to proline.
   C. plays a major role in the urea cycle.
   D. is a precursor of putrescine, a polyamine.
   E. is in equilibrium with spermidine.

4. *S*-Adenosylmethionine:
   A. contains a sulfonium ion that carries the methyl group to be transferred.
   B. yields α-ketobutyrate in the reaction in which the methyl is transferred.
   C. donates a methyl group in a freely reversible reaction.
   D. generates H_2S by transsulfuration.
   E. provides the carbons for the formation of cysteine.

5. Lysine:
   A. may be replaced by its α-keto acid analogue in the diet.
   B. produces pyruvate and acetoacetyl CoA in its catabolic pathway.
   C. is methylated by *S*-adenosylmethionine.
   D. is the only one of the common amino acids that is a precursor of carnitine.
   E. all of the above are correct.

6. Glutathione does all of the following *except*:
   A. participate in the transport of amino acids across some cell membranes.
   B. scavenge peroxides.
   C. form conjugates with some drugs to increase water solubility.
   D. decreases the stability of erythrocyte membranes.
   E. acts as a cofactor for some enzymes.

Questions 7 and 8: Defects in the metabolism of the branched-chain amino acids are rare but serious. The most common one is called maple syrup urine disease (named from the smell of the urine) which is a deficiency of branched-chain keto acid dehydrogenase complex. The disease is characterized by mental retardation and a short life span.

7. All of the following are true about the branched-chain amino acids *except* they:
   A. are essential in the diet.
   B. differ in that one is glucogenic, one is ketogenic, and one is classified as both.
   C. are catabolized in a manner that bears a resemblance to β-oxidation of fatty acids.
   D. are oxidized by a dehydrogenase complex to branched-chain acyl CoAs one carbon shorter than the parent compound.
   E. are metabolized initially in the liver.

8. Valine and isoleucine give rise to propionyl CoA, a precursor of succinyl CoA. A disease related to a defect in this conversion is methylmalonic aciduria. Some patients respond to megadoses of vitamin B_{12}. Which of the following statements about the conversion of propionyl CoA to succinyl CoA is/are correct?
   A. The first step in the conversion is a biotin-dependent carboxylation.
   B. Some methylmalonic aciduria patients respond to B_{12} because the defect in the mutase converting malonyl CoA to succinyl CoA is poor binding of the cofactor.
   C. The same pathway of propionyl CoA to succinyl CoA is part of the metabolism of odd-chain fatty acids.
   D. All of the above are correct.
   E. None of the above is correct.

Questions 9 and 10: Hyperammonemia caused by deficiencies of the enzymes involved in carbamoyl phosphate synthesis or any of the enzymes of the urea cycle is a very serious condition. Untreated, the result is early death or mental retardation and other developmental abnormalities. Ornithine transcarbamoylase deficiency is the most common error in the cycle. Treatment aims to relieve the hyperammonemia and sometimes there is supplementation with arginine.

9. In the formation of urea from ammonia, all of the following are correct *except*:
   A. aspartate supplies one of the nitrogens found in urea.
   B. this is an energy-expensive process, utilizing several ATPs.
   C. the rate of the cycle fluctuates with the diet.
   D. fumarate is produced.
   E. ornithine transcarbamoylase catalyzes the rate-limiting step.

10. Carbamoyl phosphate synthetase I:
    A. is a flavoprotein.
    B. is controlled primarily by feedback inhibition.
    C. is unresponsive to changes in arginine.
    D. requires *N*-acetyl glutamate as an allosteric effector.
    E. requires ATP as an allosteric effector.

Questions 11 and 12: Untreated phenylketonuria patients, in addition to mental retardation, have diminished production of catecholamines and light skin and hair. If the defect is in phenylalanine hydroxylase itself, a diet lacking phenylalanine but including tyrosine, alleviates these conditions. If the defect is in the ability to produce tetrahydrobiopterin, the same dietary treatment may alleviate the mental retardation and light hair but not the diminished catecholamine production.

11. When there is a deficiency of phenylalanine hydroxylase:
    A. tyrosine hydroxylase substitutes for it so phenylalanine levels remain constant.
    B. tyrosine is synthesized by an alternate pathway instead of from phenylalanine.
    C. light skin and hair result because the enzyme catalyzes the first step to melanins.
    D. products derived from phenylalanine transamination are excreted.
    E. phenylalanine is catabolized by monoamine oxidase.

12. Catecholamines:
    A. production terminates with dopamine in the brain but epinephrine in the adrenal gland.
    B. production begins with the action of tyrosinase on tyrosine.
    C. are metabolized to both glucogenic and ketogenic fragments.
    D. are highly colored compounds.
    E. all contain methyl groups donated by *S*-adenosylmethionine.

Problems

13. An inability to generate tetrahydrobiopterin would have what specific effects on the metabolism of phenylalanine, tyrosine, and tryptophan?

14. If serine labeled with ^{14}C in the carbon bearing the hydroxyl group serves as the source of a one-carbon group, how does this labeled carbon become the labeled methyl group of epinephrine? Be specific.

ANSWERS

1. **D** The mechanism of action begins with the formation of a Schiff base with pyridoxal phosphate. A: Most mammalian aminotransferases use glutamate or α-ketoglutarate. B: One amino acid is converted into another amino acid; there is neither net gain nor net loss. C: The reactions are freely reversible. E: Only lysine and threonine do not have aminotransferases.

2. **C** This is an important mechanism for reducing toxic ammonia concentrations. A: This would favor ammonia consumption. B: The cofactor is a pyridine nucleotide not pyridoxal phosphate. D: These

are inhibitory. E: Since part of the role is to provide amino acid carbon chains for gluconeogenesis, this would be active.

3. **E** Spermidine is formed by adding propylamine to putrescine. A, B: Both amino acids give rise to glutamic semialdehyde and are formed from it. C: It is both a substrate and product of the cycle. D: This is a decarboxylation.
4. **A** The reactive, positively charged sulfur reverts to a neutral thioether when the methyl group is transferred to an acceptor. B: The product, *S*-adenosylhomocysteine, is hydrolyzed to homocysteine. C: Transmethylations from AdoMet are irreversible. D: Transsulfuration refers to the combined action of cystathionine synthase and cystathionase transferring methionine's sulfur to serine to yield cysteine. E: Methionine provides only the sulfur; carbons are from serine.
5. **D** Free lysine is not methylated, but lysyl residues in a protein are methylated in a posttranslational modification. Intermediates of carnitine synthesis are derived from trimethyllysine liberated by proteolysis. A: Lysine does not participate in transamination probably in part because the α-keto acid exists as a cyclic Schiff base. B: This is one of two purely ketogenic amino acids. C: See above.
6. **D** Most of the functions of glutathione listed are dependent upon the sulfhydryl group (-SH). It increases membrane stability. Glutathione reductase helps to maintain the ratio of GSH:GSSG at about 100:1.
7. **E** A: Aminotransferase, the first enzyme, is much higher in muscle than in liver. B–D: Their catabolism is similar but the end products are different because of the differences in the branching. The similarity to β-oxidation comes in steps like oxidation to an α,β-unsaturated CoA, hydration of the double bond, and oxidation of an hydroxyl to a carbonyl.
8. **D** A: This is a typical carboxylation reaction like pyruvate carboxylase. B: Megadoses of a cofactor if the defect is in binding has been used in a number of diseases. C: The final cleavage of odd-chain fatty acids produces propionyl CoA.
9. **E** Carbamoyl phosphate synthetase I catalyzes the rate-limiting step. A, B, D: One of the nitrogen atoms is supplied as aspartate, with its carbon atoms being released as fumarate. C: The level of CPSI and the synthesis of *N*-acetylglutamate increase as protein in the diet increases.
10. **D** The primary control is by the allosteric effector, *N*-acetylglutamate. B: This is an activation, not an inhibition. C: Synthesis of the effector, and therefore activity of CPSI, is increased in the presence of arginine. E: ATP is a substrate.
11. **D** This is normally not significant. A: These are specific enzymes and phenylalanine is greatly elevated. B: There is no alternate pathway so tyrosine becomes essential. C: Tyrosinase oxidizes tyrosine; high phenylalanine competes with tyrosine, decreasing melanin production. E: This enzyme catabolizes catecholamines, not phenylalanine.
12. **A** This reflects tissue specificity. B: DOPA is formed by the action of tyrosine hydroxylase. C: Phenylalanine and tyrosine are but not the catecholamines. D: These are melanins. E: Only epinephrine does.
13. Tetrahydrobiopterin is a necessary component of phenylalanine, tyrosine and tryptophan hydroxylases. Its deficiency would inhibit normal degradation of both phenylalanine and tyrosine because their degradative pathways begin with the respective hydroxylases. Catecholamine formation (norepinephrine and epinephrine) begin with the formation of DOPA from tyrosine via tyrosine hydroxylase so catecholamine synthesis would be inhibited. The initial step in the conversion of tryptophan to serotonin is catalyzed by tryptophan hydroxylase.
14. Serine donates the labeled carbon to tetrahydrofolate to become labeled $N^{5,10}$-methylene tetrahydrofolate. This gets reduced to labeled N^5-methyl tetrahydrofolate which reacts with homocysteine, in the presence of vitamin B_{12}, to produce methyl-labeled methionine. After activation to AldoMet, the methionine donates this labeled methyl to norepinephrine to produce epinephrine.

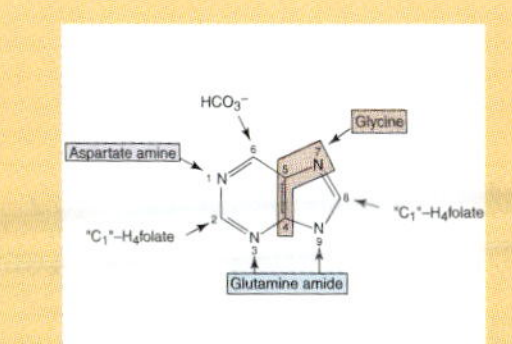

PURINE AND PYRIMIDINE NUCLEOTIDE METABOLISM

Joseph G. Cory

Textbook of Biochemistry With Clinical Correlations, Sixth Edition, Edited by Thomas M. Devlin

CLINICAL CORRELATIONS

20.1 OVERVIEW

There are major differences between nucleotide metabolism in bacteria and mammalian cells, and even differences between humans and animals. The following discussion is limited to nucleotide metabolism in mammalian cells and where appropriate to nucleotide metabolism in humans. Purine and pyrimidine nucleotides participate in many critical cellular functions. The metabolic roles of nucleotides vary widely from serving as monomeric precursors of RNA and DNA to serving as second messengers. The cellular levels of the purine and pyrimidine nucleotides are maintained by *de novo* synthetic pathways and salvage of exogenous and endogenous nucleobases and nucleosides. Amino acids, CO_2, "carbon-1"-tetrahydrofolate, and ribose-5-phosphate serve as sources for carbon, nitrogen, and oxygen atoms.

Intracellular concentrations of nucleotides are finely controlled by allosterically regulated enzymes in the pathways in which nucleotide end-products acting as effectors regulate key steps in the pathways. 2′-Deoxyribonucleotides required for DNA replication are generated directly from ribonucleotides, and their production is also carefully regulated by nucleoside 5′-triphosphate nucleotides acting as positive and negative effectors. In addition to the regulation of nucleotide metabolism via allosteric regulation, concentrations of key enzymes in their metabolic pathways are altered during the cell cycle with many of the increases in enzyme activity occurring especially during late G_1/early S phase just preceding DNA replication. The importance of both the *de novo* synthesis and the salvage pathways is shown by the fact there are clinical diseases or syndromes due to defects in either pathway. These conditions include gout (defect in *de novo* purine nucleotide synthesis), Lesch–Nyhan syndrome (defect in purine nucleobase salvage), orotic aciduria (defect in *de novo* pyrimidine nucleotide synthesis), and immunodeficiency diseases (defects in purine nucleoside degradation). Because nucleotide synthesis is required for DNA replication and RNA synthesis in dividing cells, drugs that block *de novo* pathways of nucleotide synthesis have been successfully used as antitumor and antiviral agents.

The structures, chemistry, and properties of nucleobases, nucleosides, and nucleotides are presented on pages 25–28 and in the Appendix.

20.2 METABOLIC FUNCTIONS OF NUCLEOTIDE

Nucleotides and their derivatives play critical and diverse roles in cellular metabolism. Many different nucleotides are present in mammalian cells. Some, such as ATP and NAD, are present in millimolar concentrations, while others such as dATP and cyclic AMP are orders of magnitude lower in concentration. The functions of nucleotides are summarized in Table 20.1, with some examples given.

Distribution of Nucleotides Varies with Cell Type

The principal purine and pyrimidine compounds found in cells are the 5′-nucleotide derivatives with ATP being present in the highest concentration. The distribution

TABLE 20.1 Functions of Nucleotides

Function	*Selected Examples*
1. Energy metabolism	ATP (muscle contraction; active transport; ion gradients; phosphate donor)
2. Monomeric units of nucleic acids	NTPs and dNTPs (substrates for RNA and DNA)
3. Physiological mediators	Adenosine (coronary blood flow); ADP (platelet aggregation); cAMP and cGMP (second messengers); signal transduction via GTP-binding proteins
4. Precursor function	GTP (mRNA capping); tetrahydrobiopterin, (hydroxylation of aromatic amino acids)
5. Components of coenzymes	NAD, FAD, FMN, and coenzyme A
6. Activated intermediates	UDP-glucose (glycogen); CDP-choline (phospholipids); SAM (methylation); PAPS (sulfation)
7. Allosteric effectors	ATP (negative effector of PFK-I); AMP (positive effector of phosphorylase b); dATP (negative effector of ribonucleotide reductase)

of the various nucleotides in cells varies with cell type. In red blood cells, adenine nucleotides far exceed the concentrations of guanine, cytosine, and uracil nucleotides; in other tissues, such as liver, there is a complete spectrum of nucleotides, which also include NAD^+, NADH, UDP-glucose, and UDP-glucuronic acid. In normally functioning cells, nucleoside 5′-triphosphates predominate, whereas in hypoxic cells the concentrations of nucleoside 5′-monophosphates and nucleoside 5′-diphosphates are greatly increased. Free nucleobases, nucleosides, nucleoside 2′- and 3′-monophosphates, and "modified" bases found in the cytosol represent degradation products of endogenous or exogenous nucleotides or nucleic acids.

The concentrations of ribonucleotides in cells are in great excess over the concentrations of 2′-deoxyribonucleotides. For example, the concentration of ATP in Ehrlich tumor cells is 3600 pmol per 10^6 cells compared to dATP concentrations of 4 pmol per 10^6 cells. However, at the time of DNA replication the concentrations of dATP and other deoxyribonucleoside 5′-triphosphates are markedly increased to meet the substrate requirements for DNA replication. In normal cells, the total concentrations of nucleotides are essentially constant. Thus, the total concentrations of AMP plus ADP plus ATP remains constant, but there can be major changes in the individual concentrations such that the ratio of ATP/(ATP + ADP + AMP) is altered depending upon the energy state of the cell. The same is true for NAD^+ and NADH. The total concentration of NAD^+ plus NADH is normally fixed within rather narrow concentration limits. Consequently, when it is stated that the NADH level is increased, it follows that the concentration of NAD^+ is correspondingly decreased in that cell. The basis for this "fixed" concentration of nucleotides is that *de novo* synthesis and salvage pathways for nucleotides, nucleosides and nucleobases are very rigidly controlled under normal conditions.

20.3 METABOLISM OF PURINE NUCLEOTIDES

The *de novo* synthesis of the purine ring in mammalian cells utilizes amino acids as carbon and nitrogen donors, tetrahydrofolate as a one-carbon donor, and CO_2 as a carbon donor. The *de novo* pathway for purine nucleotide synthesis leading to **inosine 5′-monophosphate (IMP)** consists of 10 metabolic steps. Hydrolysis of ATP is required to drive several reactions in this pathway. Overall, the *de novo* pathway for purine nucleotide synthesis is expensive in terms of moles of ATP utilized per mole of IMP synthesized.

CLINICAL CORRELATION 20.5

Patients Undergoing Cancer Treatment

Cancer patients with large tumor burdens who undergo radiation therapy or chemotherapy treatments also show increased serum and urine concentrations of uric acid. The source of this increased uric acid is not due to increased purine nucleotide synthesis but rather from the destruction of the tumor cells that in turn release degraded nucleic acids and cellular nucleotides that are further metabolized to uric acid. Many of the cancer treatment protocols include allopurinol as one of the drugs for the sole purpose of limiting the buildup of uric acid in the patients. A new uricolytic drug (Rasburicase, a bacterial enzyme) that converts uric acid to a water-soluble product is being used in children and adults to manage the increased levels of uric acid formed after tumor cells have been destroyed. While the use of uricase has certain pharmacological advantages over allopurinol, it has the disadvantage that it is more expensive.

Source: Smalley, R. V., Guaspari, A., Haase-Statz, S., Anderson, S. A., Cederberg, D. and Hohneker, J. A. Allopurinol: Intravenous use for prevention and treatment of hyperuricemia. *J. Clin. Oncol.* 18:1758, 2000. Ribeiro, R. C. and Pui, C. H. Recombinant urate oxidase for prevention of hyperuricemia and tumor lysis syndrome in lymphoid malignancies. *Clin. Lymphoma* 3:252, 2003. Yim, B. T., Sims-McCallum, R. P., and Chong, P. H. Rasburicase for the treatment and prevention of hyperuricemia. *Ann. Pharmacother.* 37:1047, 2003.

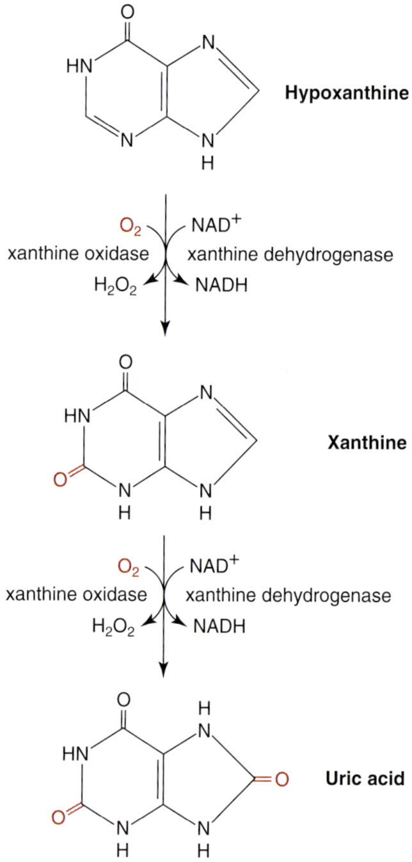

FIGURE 20.12

Reactions catalyzed by xanthine oxidoreductase (XOR). XOR contains both xanthine dehydrogenase and oxidase activities. The dehydrogenase and oxidase forms of the enzyme are interconvertible.

activities. Xanthine dehydrogenase activity requires NAD as the electron acceptor, while the oxidase activity utilizes molecular oxygen with the generation of hydrogen peroxide as product. Xanthine oxidoreductase, which contains FAD, Fe and Mo, can exist in either the dehydrogenase or oxidase form. Uric acid is the unique end product of purine nucleotide degradation in man. The reactions are shown in Figure 20.12.

Since uric acid is not very soluble in aqueous medium, there are clinical conditions in which elevated levels of uric acid can result in deposition of sodium urate crystals primarily in joints. **Hyperuricemia** is a clinical condition characterized by excess levels of uric acid in the blood and accompanied by increased levels of uric acid in urine (hyperuricuria). Since uric acid is the unique end product of purine degradation in man, excess levels of uric acid indicate some metabolic situation that may or may not be serious. There are several instances in which the cause of the hyperuricemia/hyperuricuria can be defined as a metabolic defect related to the overproduction of purine nucleotides and other situations in which there are not defined metabolic alterations (Clin. Corrs. 20.1, 20.2, 20.5, and 20.6).

20.4 METABOLISM OF PYRIMIDINE NUCLEOTIDES

The *de novo* synthesis of the pyrimidine ring in mammalian cells utilizes amino acids as carbon and nitrogen donors in addition to CO_2. Uridine 5′-monophosphate (UMP) is synthesized in a six-step metabolic pathway. ATP hydrolysis (or equivalent) is required to drive several steps in the pathway.

Synthesis of Pyrimidine Nucleotides

In contrast to *de novo* purine nucleotide synthesis, not all enzymes for *de novo* synthesis of pyrimidine nucleotides are cytosolic. Reactions leading to formation of UMP are shown in Figure 20.13. Important aspects of the pathway should be noted. The pyrimidine ring is formed first, and then ribose 5-phosphate is added with PRPP as the ribose 5-phosphate donor. The enzyme catalyzing formation of cytosolic carbamoyl phosphate, **carbamoyl phosphate synthetase II (CPS II)**, is cytosolic and is distinctly different from **carbamoyl phosphate synthetase I (CPS I)** found in the mitochondria and which

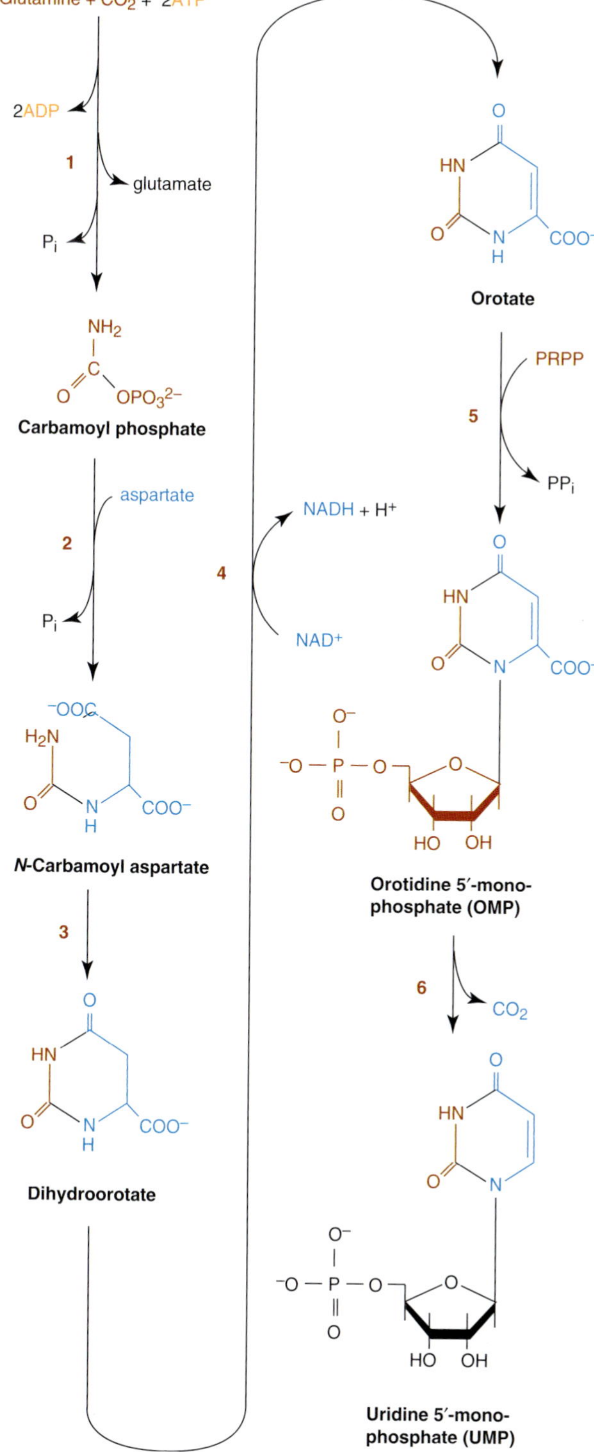

CLINICAL CORRELATION 20.6
Subclass of Patients with Autism

Recently a subclass of children with infantile autism was shown to excrete uric acid that is greater than two standard deviations above the normal mean. This group of children represents approximately 20% of the autistic population. With cultured fibroblasts from these autistic children, it was found that *de novo* synthesis of purine nucleotides as measured by [^{14}C]-formate incorporation into purine nucleotides was increased four fold over that seen in the normal control fibroblasts. The ratio of adenine nucleotides to guanine nucleotides in the pool was altered suggesting that the pathway for the interconversions of the purine nucleotides was compromised. As of this time, the molecular basis for the increased *de novo* synthesis of purine nucleotides in these autistic children is not known. An unusual aspect of the increased purine nucleotide synthesis in these children is that the excretion of uric acid is elevated but the serum concentration of uric acid is in the normal range. In a recent report, a male patient was treated with an oral dose of uridine for a period of two years. As a result, the patient experienced marked improvements in the social, cognitive, language, and motor skills. Strikingly, when the uridine was discontinued, the autistic symptoms returned, but with the reinstatement of the uridine therapy the behavior of the boy was again improved.

Source: Page, T. and Coleman, M. Purine metabolism abnormalities in a hyperuricosuric subclass of autism. *Biochim. Biophys. Acta* 1500: 291, 2000. Page, T. and Moseley, C. Metabolic treatment of hyperuricosuric autism. *Prog. Neuro-Psychopharmacol. Biol. Psychiatry* 26:397, 2002.

FIGURE 20.13

***De novo* synthesis of pyrimidine nucleotides.** Enzyme activities catalyzing the reactions are (1) carbamoyl phosphate synthetase II, (2) aspartate carbamoyltransferase, (3) dihydroorotase, (4) dihydroorotate dehydrogenase, (5) orotate phosphoribosyltransferase, and (6) OMP decarboxylase. The activities of 1, 2, and 3 are on a trifunctional protein (CAD); the activities of 5 and 6 are on a bifunctional protein (UMP synthase).

CLINICAL CORRELATION 20.7
Hereditary Orotic Aciduria

Hereditary orotic aciduria results from a defect in *de novo* synthesis of pyrimidine nucleotides. This genetic disease is characterized by severe anemia, growth retardation, and high levels of orotic acid excretion. The biochemical basis for orotic aciduria is a defect in one or both of the activities (orotate phosphoribosyl transferase or orotidine decarboxylase) associated with UMP synthase, the bifunctional protein. It is a very rare disease, but the understanding of the metabolic basis for this disease has led to successful treatment of the disorder. Patients are fed uridine, which leads not only to reversal of the hematologic problem but also to decreased formation of orotic acid. Uridine is taken up by cells and converted by uridine phosphotransferase to UMP that is sequentially converted to UDP and then to UTP. UTP formed from exogenous uridine, in turn, inhibits carbamoyl phosphate synthetase II, the major regulated step in the *de novo* pathway. As a result, orotic acid synthesis via the *de novo* pathway is markedly decreased to essentially normal levels. Since UTP is also the substrate for CTP synthesis, uridine treatment serves to replenish both the UTP and CTP cellular pools. In effect, then, exogenous uridine bypasses the defective UMP synthase and supplies cells with UTP and CTP required for nucleic acid synthesis and other cellular functions. The success of treatment of hereditary orotic aciduria with uridine provides *in vivo* data regarding the importance of the step catalyzed by carbamoyl phosphate synthase II as the site of regulation of pyrimidine nucleotide synthesis in man.

Source: Webster, D. R., Becroft, D. M. O., van Gennip, A. H., Van Kuilenburg, A. B. P. Hereditary orotice aciduria and other disorders of pyrimidine metabolism. In: C. R. Scriver, A. L. Beaudet, W. S. Sly, and D. Valle (Eds.), *The Metabolic and Molecular Bases of Inherited Disease*, Vol. II. New York: McGraw-Hill, 2001, Chapter 113, p. 2663. Suchi, M., Mizuno, H., Kawai, Y., Tonboi, T., et al. Molecular cloning of the human UMP synthase gene and characterization of a point mutation in two hereditary orotic aciduria families. *Am. J. Hum. Genet.* 60:525, 1997.

functions as part of the urea cycle. Formation of *N*-carbamoylaspartate is the committed step in pyrimidine nucleotide synthesis, but formation of cytosolic carbamoyl phosphate (CPS II) is the regulated step. Formation of orotate from dihydroorotate is catalyzed by mitochondrial **dihydroorotate dehydrogenase**. The other activities of the pathway are found in the cytosol on multifunctional proteins. The activities of CPS II, **aspartate carbamoyl transferase** and **dihydroorotase** are found on a trifunctional protein (CAD), while **orotate phosphoribosyltransferase** and **OMP decarboxylase** activities are found on a bifunctional protein (UMP synthase). A defect in this bifunctional protein that affects either phosphoribosyl transferase activity or decarboxylase activity leads to a rare clinical condition known as hereditary orotic aciduria (Clin. Corr. 20.7). The immunosuppressive drug, **leflunomide**, which is used in the treatment of rheumatoid arthritis, inhibits *de novo* synthesis of pyrimidine nucleotides specifically at dihydroorotate dehydrogenase.

Nucleotide kinases convert UMP to UTP (Figure 20.14), which serves as a substrate for CTP synthetase. CTP synthetase catalyzes formation of CTP from UTP with glutamine as the amino group donor (Figure 20.15). CTP synthetase displays homotropic sigmoidal kinetics with respect to UTP while CTP, the product, is a negative effector of the reaction as shown in Figure 20.16. By regulating CTP synthetase in this way, cells maintain an appropriate ratio of UTP and CTP for cellular functions and RNA synthesis.

To summarize, *de novo* synthesis of pyrimidine nucleotides requires aspartate as a carbon and nitrogen donor, glutamine as a nitrogen donor, and CO_2 as a carbon donor (Figure 20.17). Five of the six reactions in the pathway take place in the cytosol of the cell, while the other reaction occurs in mitochondria. The cytosolic enzyme activities reside on multifunctional proteins. UTP is the direct precursor of CTP.

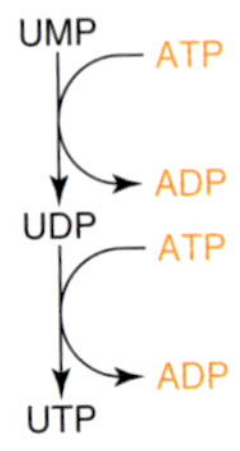

UMP → (ATP → ADP) → UDP → (ATP → ADP) → UTP

FIGURE 20.14
Formation of UTP from UMP.

Pyrimidine Nucleotide Synthesis Is Regulated at the Level of Carbamoyl Phosphate Synthetase II

Regulation of pyrimidine nucleotide synthesis in mammalian cells occurs at the carbamoyl phosphate synthetase II step. CPS II is a cytosolic enzyme and distinct from the mitochondrial CPS I, which utilizes ammonia as the amino donor instead of glutamine and which requires activation by *N*-acetyl glutamate. CPS II is inhibited by UTP, an

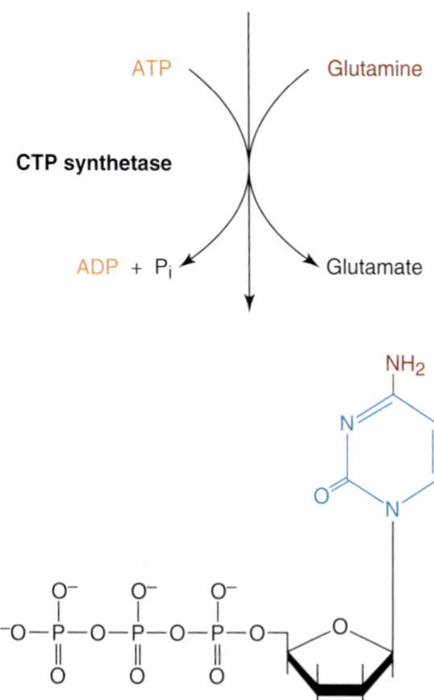

FIGURE 20.15

Formation of CTP from UTP.

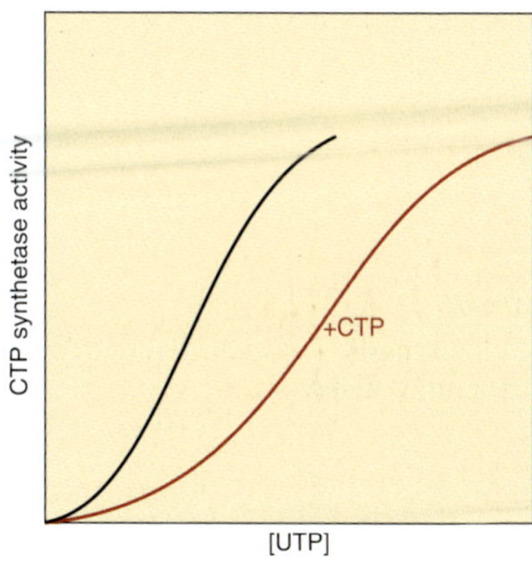

FIGURE 20.16

Regulation of CTP synthetase.

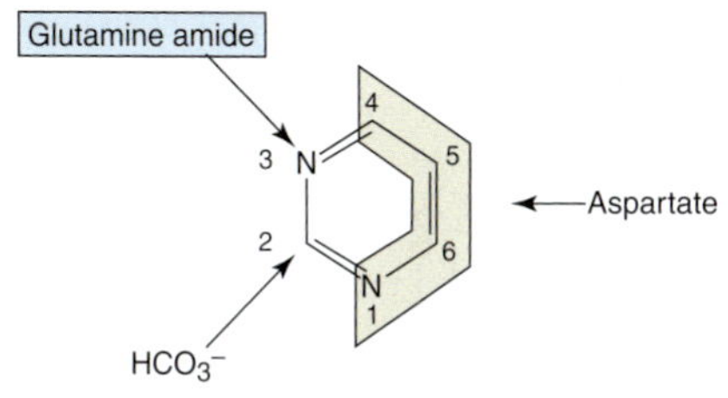

FIGURE 20.17

Sources of carbon and nitrogen atoms in pyrimidines. C4, C5, C6, and N1 from aspartate; N3 from glutamine; and C2 from CO_2.

end product of the pathway and is activated by PRPP. The K_i for UTP (at CPSII) and the K_a for PRPP (at CPSII) are in the range of values that would allow intracellular levels of UTP and PRPP to have an effect on the control of pyrimidine nucleotide synthesis. CPS II is the only source of carbamoyl phosphate in extrahepatic tissues. However, in liver, under stressed conditions in which there is excess ammonia, CPS I generates carbamoyl phosphate in mitochondria, which moves into the cytosol and serves as a substrate for orotic acid synthesis. This pathway serves to detoxify excess ammonia. Elevated levels of orotic acid are excreted during ammonia toxicity in humans. This points to carbamoyl phosphate synthetase II as being the major regulated activity of pyrimidine nucleotide metabolism.

UMP does not inhibit carbamoyl phosphate synthetase II but competes with OMP to inhibit OMP decarboxylase (Figure 20.18). As discussed earlier, conversion of UTP to CTP is also regulated so that cells maintain a balance between uridine and cytidine nucleotides in the cell.

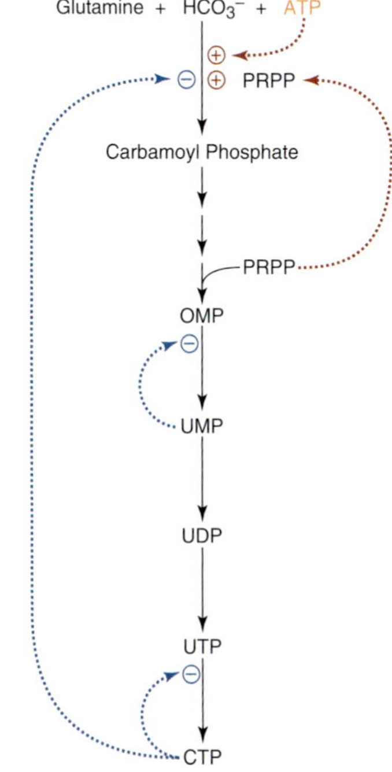

FIGURE 20.18

Regulation of pyrimidine nucleotide synthesis. Solid arrows represent enzyme catalyzed reactions and dashed arrows inhibition by products of the reactions.

Pyrimidine Bases Are Salvaged to Reform Nucleotides

Pyrimidines are "salvaged" by conversion to nucleotides by pyrimidine phosphoribosyltransferase. The general reaction is

$$\text{pyrimidine} + \text{PRPP} \rightleftharpoons \text{pyrimidine nucleoside 5}'\text{-monophosphate} + \text{PP}_i$$

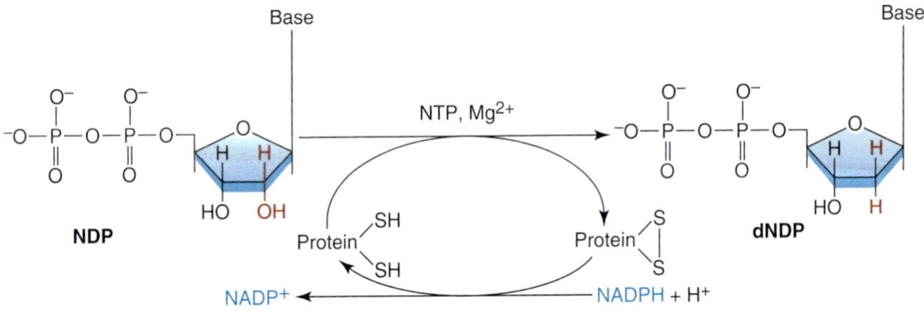

FIGURE 20.19

De novo **synthesis of 2′-deoxyribonucleotides from ribonucleotides.**

The enzyme from human erythrocytes utilizes orotate, uracil, and thymine as substrates but not cytosine. These reactions divert the pyrimidine bases from the degradative pathway to nucleotide formation. As a pyrimidine base becomes available to cells, there are competing reactions that will result in either (a) degradation and excretion of the products or (b) reutilization of the bases for nucleotide synthesis. For example, when normal liver is presented with uracil, it is rapidly degraded to β-alanine, whereas in proliferating tumor cells uracil is converted to UMP. This is the result of the availability of PRPP, enzyme levels, and metabolic state of the cells.

20.5 DEOXYRIBONUCLEOTIDE FORMATION

The concentrations of the 2′-deoxyribonucleoside 5-triphosphates (dNTPs) are extremely low in nonproliferating cells. However, the cellular levels of the dNTPs rapidly expand during DNA replication (S-phase of the cell cycle) and repair (following DNA damage) due to increased ribonucleotide reductase activity. The relative and absolute concentrations of the dNTPs during these times are critical for determining the fidelity of the process. Unbalanced levels of the individual dNTPs can lead to a wide range of genetic disturbances or ultimately cell death.

Deoxyribonucleotides Are Formed by Reduction of Ribonucleoside 5′-Diphosphates

Nucleoside 5′-diphosphate reductase (ribonucleotide reductase) catalyzes the reaction in which ribonucleoside 5′-diphosphates are converted to 2′-deoxyribonucleoside 5′-diphosphates. The reaction is controlled by the amount of enzyme present in cells and by a very finely regulated allosteric control mechanism. The reaction is summarized in Figure 20.19. Reduction of a particular substrate requires a specific nucleoside 5′-triphosphate as a positive effector. For example, reductions of CDP or UDP to dCDP and dUDP, respectively, require ATP as the positive effector, while reduction of ADP and GDP require the presence of dGTP and dTTP, respectively (Table 20.2). A low-molecular-weight protein, thioredoxin or glutaredoxin, is involved in reduction at the 2′-position through oxidation of its sulfhydryl groups. To complete the catalytic cycle,

TABLE 20.2 Nucleoside 5′-Triphosphates as Regulators of Ribonucleotide Reductase Activity

Substrate	*Major Positive Effector*	*Major Negative Effector*
CDP	ATP	dATP, dGTP, dTTP[a]
UDP	ATP	dATP, dGTP, dTTP[a]
ADP	dGTP	dATP
GDP	dTTP	dATP

[a] In decreasing order of effectiveness.

NADPH is used to regenerate free sulfhydryl groups on **thioredoxin** or **glutaredoxin**. **Thioredoxin reductase**, a flavoprotein, is required if thioredoxin is involved; glutathione and glutathione reductase are required if glutaredoxin is the protein. Mammalian ribonucleotide reductase consists of two nonidentical protein subunits, neither of which alone has enzymatic activity. The larger subunit has at least two different effector-binding sites, and the smaller subunit contains a nonheme iron and a stable tyrosyl free radical. The two subunits make up the active site of the enzyme. The two subunits are encoded by genes on separate chromosomes. The mRNAs for these subunits, and consequently the proteins, are differentially expressed as cells transit the cell cycle.

The activity of ribonucleotide reductase is under strict allosteric control. While reduction of each substrate requires a specific positive effector (nucleoside 5′-triphosphate), the dNTP products can serve as potent negative effectors of the enzyme. The effects of nucleoside 5′-triphosphates as regulators of ribonucleotide reductase activity are summarized in Table 20.2. DeoxyATP is a potent inhibitor of the reduction of all four substrates, CDP, UDP, GDP, and ADP; dGTP inhibits reduction of CDP, UDP, and GDP; dTTP inhibits reduction of CDP, UDP, and ADP. Thus, dGTP and dTTP serve as either positive or negative effectors of ribonucleotide reductase activity, depending on the substrate. This means that while dGTP is the required positive activator for ADP reduction, it also serves as an effective inhibitor of CDP and UDP reductions; dTTP is the positive effector of GDP reduction and serves as an inhibitor of CDP and UDP. Effective inhibition of ribonucleotide reductase by dATP, dGTP, or dTTP explains why deoxyadenosine, deoxyguanosine, and deoxythymidine are toxic to a variety of mammalian cells. Ribonucleotide reductase is uniquely responsible for catalyzing the rate-limiting reactions in which 2′-deoxyribonucleoside 5′-triphosphates are synthesized *de novo* for DNA replication as summarized in Figure 20.20. Inhibitors of ribonucleotide reductase are potent inhibitors of DNA synthesis and, hence, of cell replication.

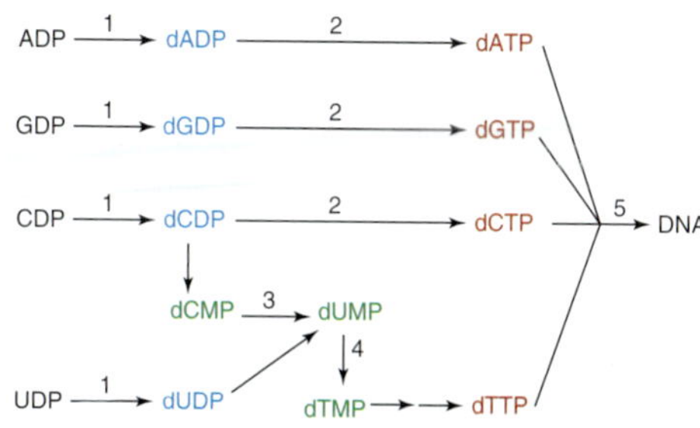

FIGURE 20.20

Role of ribonucleotide reductase in DNA synthesis. The enzymes catalyzing the reactions are (1) ribonucleotide reductase, (2) nucleoside 5′-diphosphate kinase, (3) deoxycytidylate deaminase, (4) thymidylate synthase, and (5) DNA polymerase.

Deoxythymidylate Synthesis Requires N^5, N^{10}-Methylene H_4Folate

Deoxythymidylate (dTMP) is formed from 2′-deoxyuridine 5′-monophosphate (dUMP) in a unique reaction. **Thymidylate synthase** catalyzes the transfer of a one carbon unit from N^5, N^{10}-methylene H_4folate (Figure 20.21) to dUMP and which is simultaneously reduced to a methyl group. The reaction is presented in Figure 20.22. In this reaction, N^5, N^{10}-methylene H_4folate serves a one-carbon donor and as a reducing agent. This is the only reaction in which H_4folate, acting as a one-carbon carrier, is oxidized to H_2folate. There are no known regulatory mechanisms for this reaction.

The substrate for this reaction can come from two different pathways as shown below:

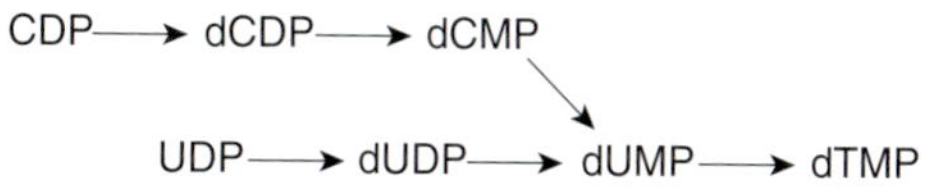

CDP ⟶ dCDP ⟶ dCMP
↘
UDP ⟶ dUDP ⟶ dUMP ⟶ dTMP

In both pathways a deoxyribonucleotide, dCDP or dUDP, is generated in reactions catalyzed by ribonucleotide reductase. In one pathway, dUMP is generated from dUDP, while in the other pathway, dCMP is deaminated to dUMP. Labeling studies indicate that the major pathway for formation of dUMP involves deamination of dCMP by dCMP deaminase, an enzyme that is subject to allosteric regulation by dCTP (positive) and dTTP (negative) (Figure 20.23). This regulation of dCMP deaminase by dCTP and dTTP allows cells to maintain the correct balance of dCTP and dTTP for DNA synthesis.

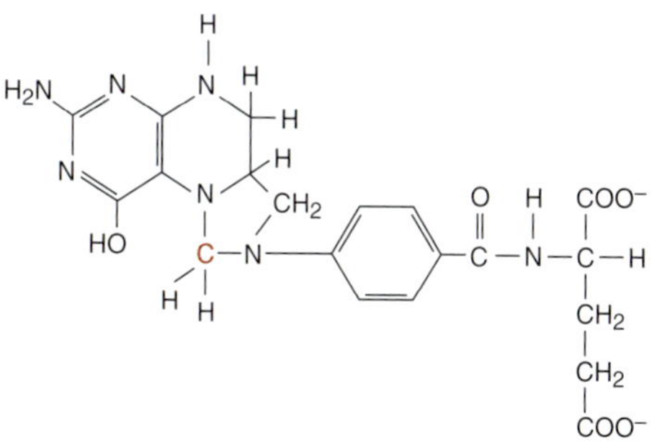

FIGURE 20.21

Structure of N^5,N^{10}-methylene H_4folate.

Pyrimidine Interconversions with Emphasis on Deoxyribopyrimidine Nucleosides and Nucleotides

As shown in Figure 20.9, metabolic pathways exist for interconversions of AMP and GMP that are regulated to maintain appropriate intracellular levels of adenine and

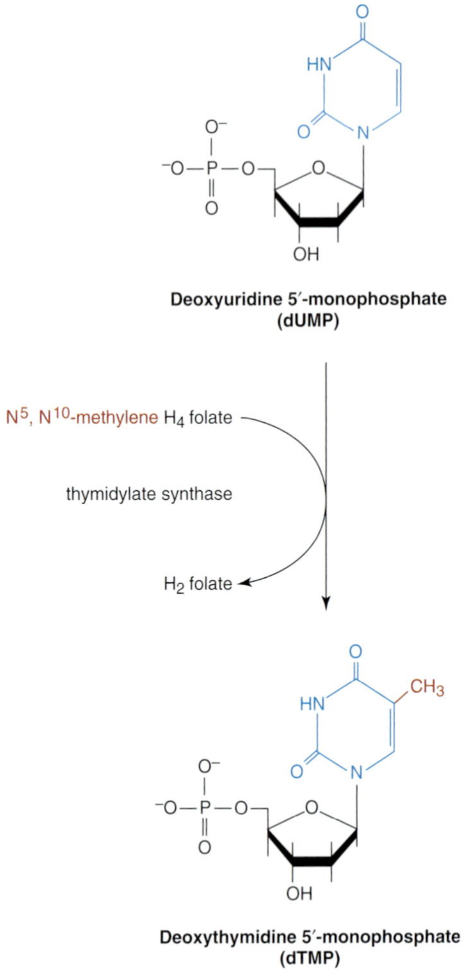

Deoxyuridine 5′-monophosphate (dUMP)

N^5, N^{10}-methylene H_4 folate

thymidylate synthase

H_2 folate

Deoxythymidine 5′-monophosphate (dTMP)

FIGURE 20.22

Synthesis of deoxythymidine nucleotide.

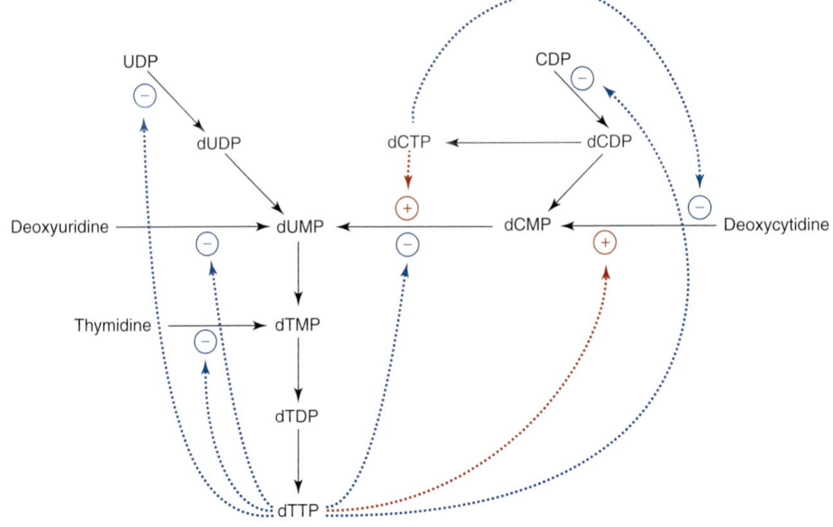

FIGURE 20.24

Interconversions of pyrimidine nucleotides with emphasis on deoxyribonucleotide metabolism. The solid arrows indicate enzyme-catalyzed reactions; the dotted lines represent positions of negative control points.

guanine nucleotides. Pathways also exist for interconversion of pyrimidine nucleotides and are of particular importance for pyrimidine deoxyribonucleosides and deoxyribonucleotides as summarized in Figure 20.24. Note that dCTP and dTTP are major positive and negative effectors of the interconversions and salvage of deoxyribonucleosides.

Pyrimidine Nucleotides Are Degraded to β-Amino Acids

Turnover of nucleic acids results in release of pyrimidine and purine nucleotides. Degradation of pyrimidine nucleotides follows the pathways shown in Figure 20.25. Pyrimidine nucleotides are converted to nucleosides by nonspecific phosphatases. Cytidine and deoxycytidine are deaminated to uridine and deoxyuridine respectively by pyrimidine nucleoside deaminase. Uridine phosphorylase catalyzes phosphorolysis of uridine, deoxyuridine, and deoxythymidine to uracil and thymine.

Uracil and thymine are degraded further by analogous reactions, although the final products are different as shown in Figure 20.26. Uracil is degraded to ***β*-alanine**, NH_4^+, and CO_2. None of these products is unique to uracil degradation, and consequently the turnover of cytosine or uracil nucleotides cannot be estimated from the end products of this pathway. Thymine degradation produces β-aminoisobutyric acid, NH_4^+, and CO_2. β-Aminoisobutyric acid is excreted in urine of humans and it originates exclusively from degradation of thymine. Thus, its excretion can be used to estimate the turnover of DNA or deoxythymidine nucleotides by measurement of ***β*-aminoisobutyric acid**. Increased levels of β-aminoisobutyric acid are excreted in cancer patients undergoing chemotherapy or radiation therapy, in which large numbers of cells are killed and DNA is degraded.

Enzymes catalyzing the degradation of uracil and thymine (dihydropyrimidine dehydrogenase, dihydropyrimidinase, and uriedopropionase) do not show a preference for either uracil or thymine or their reaction products.

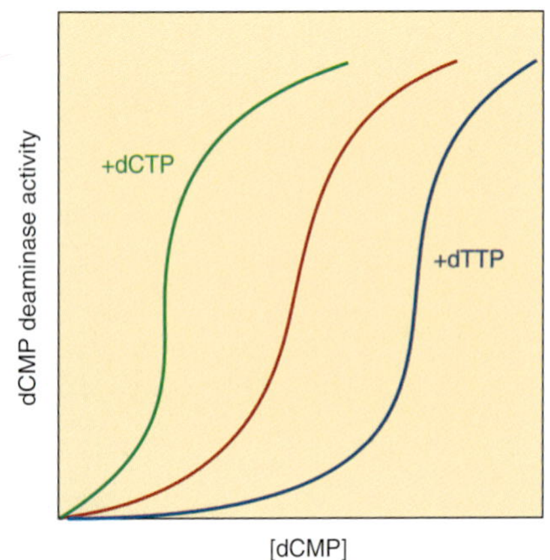

FIGURE 20.23

Regulation of dCMP deaminase.

20.6 NUCLEOSIDE AND NUCLEOTIDE KINASES

De novo synthesis of both purines and pyrimidine nucleotides yields nucleoside 5′-monophosphates (see Figures 20.1, 20.3, and 20.4). Likewise, the salvage of nucleobases by the phosphoribosyl transferases or nucleosides by nucleoside kinases also yields

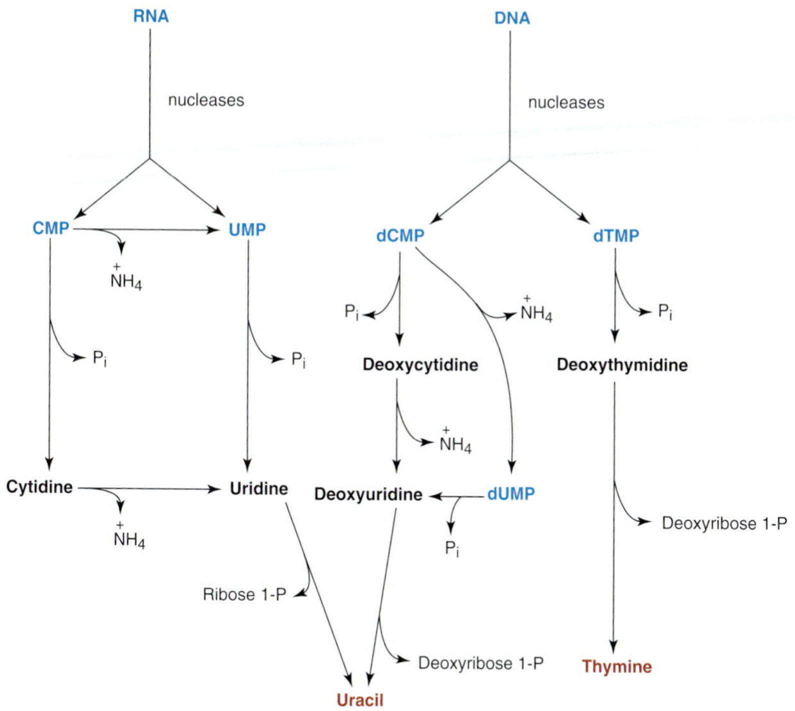

FIGURE 20.25
Pathways for degradation of pyrimidine nucleotides.

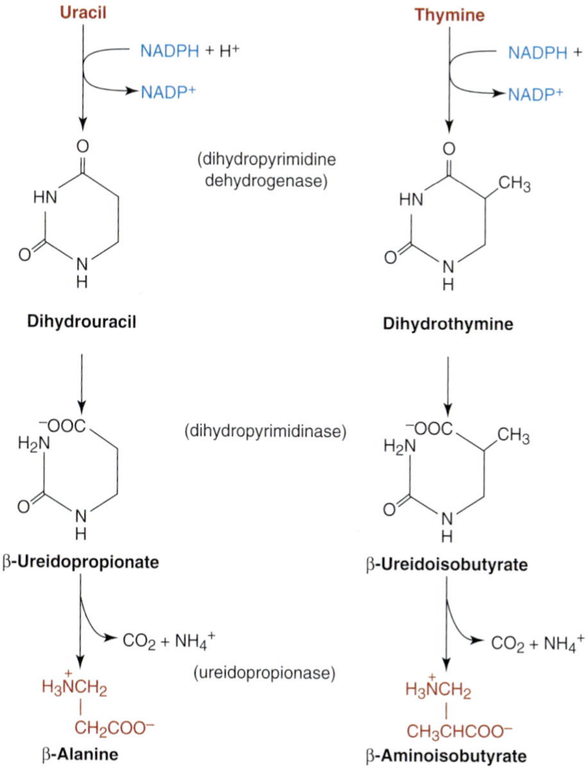

Uracil
NADPH + H^+
$NADP^+$
(dihydropyrimidine dehydrogenase)
Dihydrouracil
(dihydropyrimidinase)
β-Ureidopropionate
$CO_2 + NH_4^+$
(ureidopropionase)
$H_3\overset{+}{N}CH_2$
CH_2COO^-
β-Alanine

Thymine
NADPH +
$NADP^+$
Dihydrothymine
β-Ureidoisobutyrate
$CO_2 + NH_4^+$
$H_3\overset{+}{N}CH_2$
CH_3CHCOO^-
β-Aminoisobutyrate

FIGURE 20.26
Degradation of uracil and thymine to end products.

nucleoside 5′-monophosphates. This is particularly important in cells such as erythrocytes, which cannot form nucleotides *de novo*. Nucleotide kinases convert nucleoside 5′-monophosphates to nucleoside 5′-diphosphates and nucleoside 5′-diphosphates to nucleoside 5′-triphosphates. These are important reactions since most reactions in which nucleotides function require nucleoside 5′-triphosphates (primarily) or nucleoside 5′-diphosphates.

Some of these nucleoside kinases show a high degree of specificity with respect to the base and sugar moieties. There is also substrate specificity in nucleotide kinases. Mammalian cells also contain, in high concentration, nucleoside diphosphate kinase that is relatively nonspecific for either phosphate donor or phosphate acceptor in terms of purine or pyrimidine base or the sugar. This reaction is as follows:

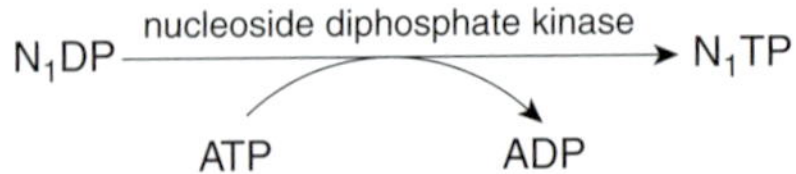

Since ATP is present in the highest concentration and most readily regenerated by glycolysis or oxidative phosphorylation, it is probably the major phosphate donor for these reactions.

20.7 NUCLEOTIDE METABOLIZING ENZYMES AS A FUNCTION OF THE CELL CYCLE AND RATE OF CELL DIVISION

The strict regulation of nucleotide synthesis requires that mechanisms must be available in cells to meet the requirements for ribonucleotide and deoxyribonucleotide precursors at the time of increased RNA synthesis and DNA replication. To meet these needs, cells increase levels of specific enzymes involved with nucleotide formation during very specific periods of the cell cycle (see p. 1014).

Enzymes involved in purine nucleotide synthesis and interconversions that are elevated during the S phase of the cell cycle are PRPP amidotransferase and IMP dehydrogenase. Adenylosuccinate synthetase and adenylosuccinase do not increase. Enzymes involved in pyrimidine nucleotide synthesis that are elevated during the S phase include aspartate carbamoyltransferase, dihydroorotase, dihydroorotate dehydrogenase, orotate phosphoribosyltransferase, and CTP synthetase. Many enzymes involved in synthesis and interconversions of deoxyribonucleotides are also elevated during the S phase. Included in these enzymes are ribonucleotide reductase, thymidine kinase, dCMP deaminase, thymidylate synthase, and TMP kinase.

The deoxyribonucleotide pool is extremely small in "resting" cells (less than 1 μM). As a result of the increase in ribonucleotide reductase activity, concentrations of deoxyribonucleotides reach 10–20 μM during DNA synthesis. However, this concentration would sustain DNA synthesis for only minutes, while complete DNA replication requires hours. Consequently, levels of ribonucleotide reductase activity must not only increase but must also be sustained during S phase in order to provide the necessary substrates for DNA synthesis. Growing tissues such as regenerating liver, embryonic tissues, intestinal mucosal cells, and erythropoietic cells are geared toward DNA replication and RNA synthesis. These tissues will show elevated levels of the key enzymes involved with purine and pyrimidine nucleotide synthesis and interconversion, along with complementary decreases in amount of enzymes that catalyze reactions in which these precursors are degraded. These changes reflect the proportion of the cells that are in S phase.

An ordered pattern of biochemical changes occurs in tumor cells. From a series of liver, colon, and kidney tumors of varying growth rates, these changes have been identified as (1) transformation-linked (meaning that all tumors regardless of growth rate show certain increased and certain decreased amounts of enzyme), (2) progression-linked (alterations that correlate with growth rate of tumors), and (3) coincidental alterations (not connected to the malignant state). Levels of ribonucleotide reductase, thymidylate synthase, and IMP dehydrogenase increase as a function of tumor growth rate. PRPP amidotransferase, UDP kinase, and uridine kinase are increased in all tumors, whether they are slow or rapidly growing.

Alterations in gene expression in tumor cells are not only quantitative changes in enzyme levels but also qualitative changes (isozyme shifts). While some enzymes are increased in both fast-growing normal tissue (e.g., embryonic and regenerating liver) and tumors, the overall quantitative and qualitative patterns for normal and tumor tissue are distinguishable.

20.8 NUCLEOTIDE COENZYME SYNTHESIS

Nicotinamide adenine dinucleotide (NAD), **flavin adenine nucleotide** (FAD) and **coenzyme A** (CoA) are important as coenzymes or prosthetic groups in intermediary metabolism. While each of these coenzymes have an AMP moiety as part of the structure, the AMP moiety is not directly involved in the reactions in which each of these are functioning. In NAD or NADP, it is the nicotinamide ring that is involved in the oxidation/reduction function; in FMN or FAD it is the flavin ring that is involved in the oxidation/reduction function; in coenzyme A, it is the sulfhydryl group that is the functional part of the molecule. They are synthesized by a variety of mammalian cell types. Figures 20.27, 20.28, and 20.29 present the biosynthetic pathways for each. NAD synthesis requires niacin, FAD synthesis requires riboflavin, and CoA requires pantothenic acid. NAD is synthesized by three different pathways starting from tryptophan (see p. 772), nicotinate, or nicotinamide, respectively. When tryptophan is

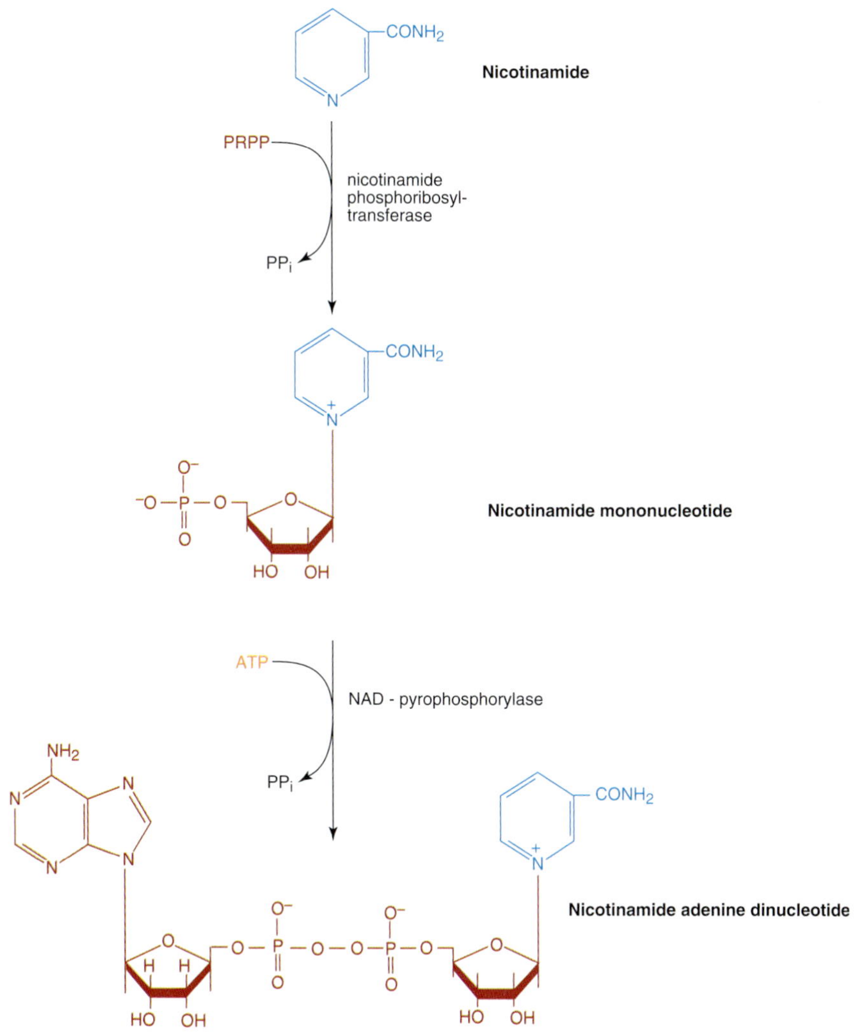

FIGURE 20.27
Pathway for nicotinamide adenine nucleotide synthesis.

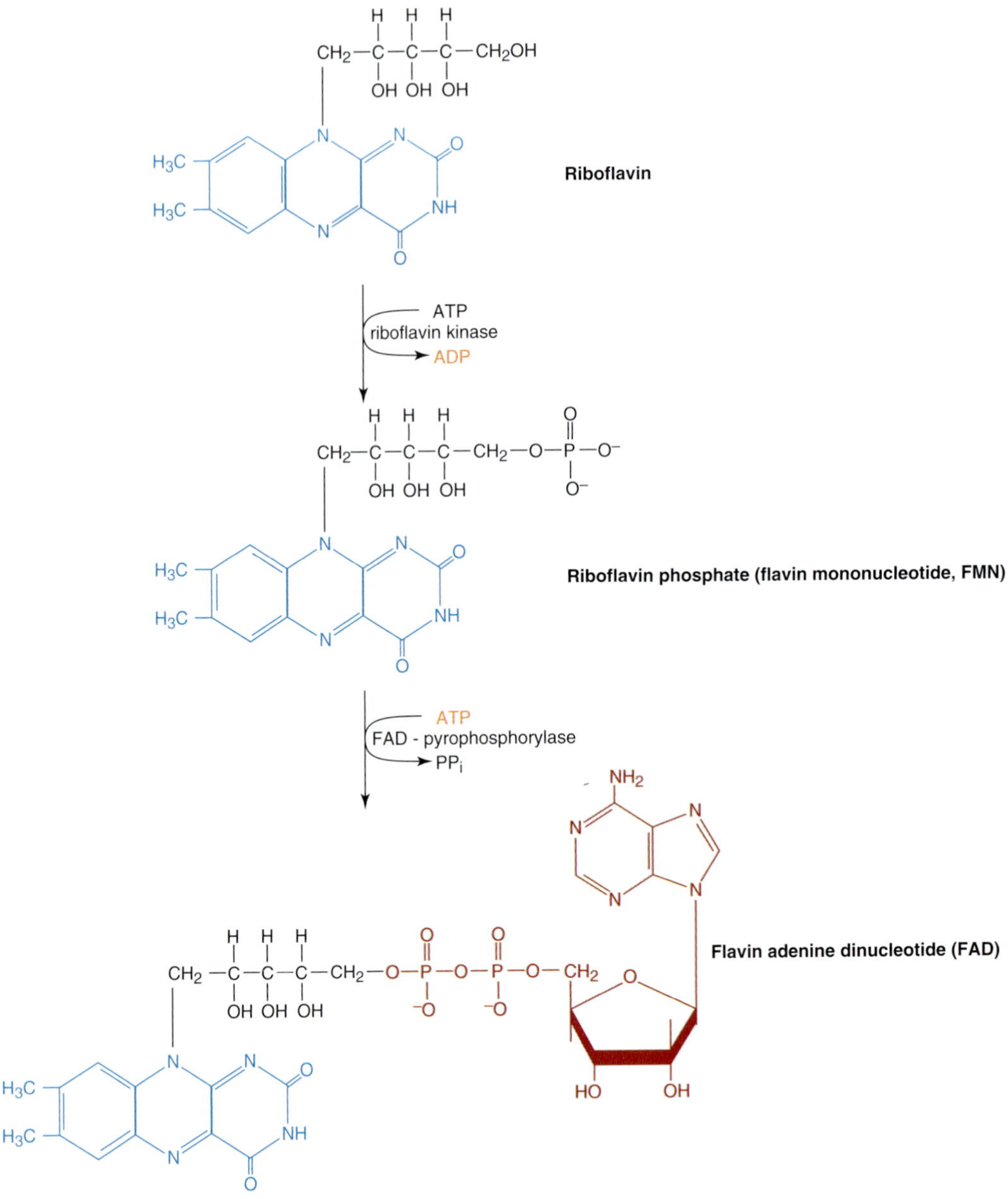

Riboflavin

ATP
riboflavin kinase
ADP

Riboflavin phosphate (flavin mononucleotide, FMN)

ATP
FAD - pyrophosphorylase
PP_i

Flavin adenine dinucleotide (FAD)

FIGURE 20.28
Synthesis of flavin adenine dinucleotide.

in excess of the amount needed for protein synthesis and serotonin synthesis (see p. 773), it can be used for NAD synthesis. This situation is not likely in most normal diets; consequently, niacin is required in the diet.

Synthesis of NAD^+ by any of the pathways requires **PRPP** as the ribose 5-phosphate donor. Nicotinamide adenine dinucleotide phosphate ($NADP^+$) is derived by phosphorylation of NAD^+. NAD^+ is used not only as a cofactor in oxidation–reduction reactions but also as a substrate in ADP-ribosylation reactions (e.g., DNA repair and pertussis toxin poisoning; see p. 514). These reactions lead to the turnover of NAD^+. The end product of NAD^+ degradation is 2-pyridone-5-carboxamide, which is excreted in urine.

Synthesis of nucleotide coenzymes is regulated so that there are essentially constant concentrations of these coenzymes in the cell. When the statement is made that a certain metabolic condition is favored when the concentration of NAD^+ is low, the concentration of NADH is correspondingly high. As an example, for glycolysis to continue under anaerobic conditions, NAD^+ must be regenerated constantly by reduction of pyruvate to lactate catalyzed by lactate dehydrogenase (see p. 592).

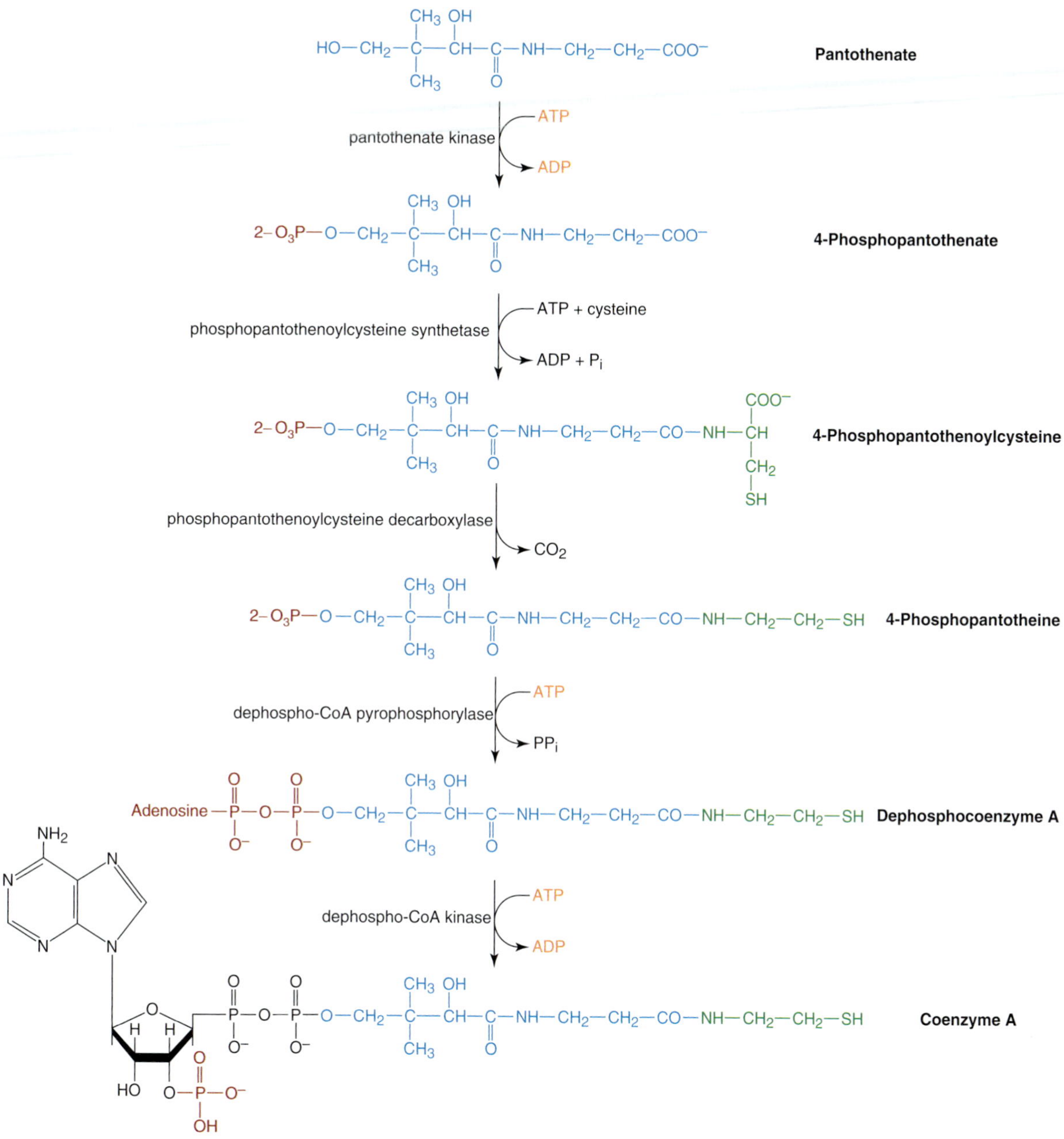

Pantothenate

pantothenate kinase (ATP → ADP)

4-Phosphopantothenate

phosphopantothenoylcysteine synthetase (ATP + cysteine → ADP + P_i)

4-Phosphopantothenoylcysteine

phosphopantothenoylcysteine decarboxylase (→ CO_2)

4-Phosphopantotheine

dephospho-CoA pyrophosphorylase (ATP → PP_i)

Dephosphocoenzyme A

dephospho-CoA kinase (ATP → ADP)

Coenzyme A

FIGURE 20.29
Synthesis of coenzyme A.

20.9 SYNTHESIS AND UTILIZATION OF 5-PHOSPHORIBOSYL-1-PYROPHOSPHATE

5-Phosphoribosyl-1-pyrophosphate (PRPP) is a key molecule in *de novo* synthesis of purine and pyrimidine nucleotides, salvage of purine and pyrimidine bases, and synthesis of NAD^+. PRPP synthetase catalyzes the reaction shown in Figure 20.30. Ribose 5-phosphate used in this reaction is generated from glucose 6-phosphate metabolism by the pentose phosphate pathway or from ribose 1-phosphate generated by phosphorolysis of nucleosides by nucleoside phosphorylase. PRPP synthetase has an absolute requirement for inorganic phosphate and is strongly regulated. The curve of velocity versus P_i concentration for PRPP synthetase activity is sigmoidal rather than hyperbolic, meaning that at the normal cellular concentration of P_i the enzyme activity is depressed. The activity is further regulated by ADP, 2,3-bisphosphoglycerate, and other

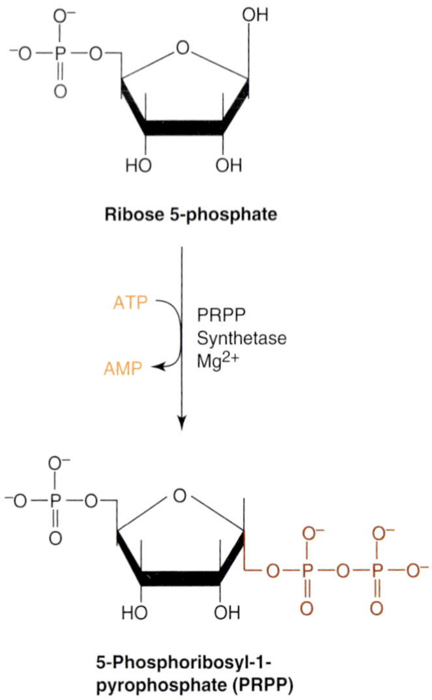

Ribose 5-phosphate

ATP, AMP, PRPP Synthetase Mg^{2+}

5-Phosphoribosyl-1-pyrophosphate (PRPP)

FIGURE 20.30
Synthesis of PRPP.

nucleotides. ADP is a competitive inhibitor of PRPP synthetase with respect to ATP; 2,3-bisphosphoglycerate is a competitive inhibitor with respect to ribose 5-phosphate; and nucleotides serve as noncompetitive inhibitors with respect to both substrates. 2,3-Bisphosphoglycerate may be important in regulating PRPP synthetase activity in red cells. Concentrations of PRPP are low in "resting" or confluent cells but increase markedly during rapid cell division. Increased flux of glucose 6-phosphate through the pentose phosphate pathway results in increased cellular levels of PRPP and increased production of purine and pyrimidine nucleotides. PRPP is important not only because it serves as a substrate in the glutamine PRPP amidotransferase and the phosphoribosyl transferase reactions, but also because it serves as a positive effector of the major regulated steps in purine and pyrimidine nucleotide synthesis, namely, **PRPP amidotransferase** and **carbamoyl phosphate synthetase II**. Reactions and pathways in which PRPP is required are as follows:

1. *De novo* purine nucleotide synthesis

$$\text{PRPP} + \text{glutamine} \rightleftharpoons \text{5-phosphoribosylamine} + \text{glutamate} + \text{PP}_i$$

2. Salvage of purine bases

$$\text{PRPP} + \text{hypoxanthine (guanine)} \rightleftharpoons \text{IMP(GMP)} + \text{PP}_i$$

$$\text{PRPP} + \text{adenine} \rightleftharpoons \text{AMP} + \text{PP}_i$$

3. *De novo* pyrimidine nucleotide synthesis

$$\text{PRPP} + \text{orotate} \rightleftharpoons \text{OMP} + \text{PP}_i$$

4. Salvage of pyrimidine bases

$$\text{PRPP} + \text{uracil} \rightleftharpoons \text{UMP} + \text{PP}_i$$

5. NAD^+ synthesis

$$\text{PRPP} + \text{nicotinate} \rightleftharpoons \text{nicotinate mononucleotide} + \text{PP}_i$$

$$\text{PRPP} + \text{nicotinamide} \rightleftharpoons \text{nicotinamide mononucleotide} + \text{PP}_i$$

$$\text{PRPP} + \text{quinolinate} \rightleftharpoons \text{nicotinate mononucleotide} + \text{PP}_i$$

It is to be remembered that the cellular levels of pyrophosphatase are very high in cells leading to the hydrolysis of pyrophosphate to phosphate, causing the reactions to be irreversible.

20.10 CHEMOTHERAPEUTIC AGENTS THAT INTERFERE WITH PURINE AND PYRIMIDINE NUCLEOTIDE METABOLISM

De novo synthesis of purine and pyrimidine nucleotides is critical for normal cell replication, maintenance, and function. Regulation of these pathways is important since disease states have been identified that arise from defects in the regulatory enzymes. Synthetic compounds and natural products from plants, bacteria, or fungi that are structural analogs of the nucleobases or nucleosides used in metabolic reactions have been shown to be cytotoxic. These compounds are relatively specific inhibitors of enzymes involved in nucleotide synthesis or interconversions and have proved to be useful in therapy of diverse clinical problems. They are generally classified as antimetabolites, antifolates, glutamine antagonists, and other agents.

Inhibitors of Purine and Pyrimidine Nucleotide Metabolism

Antimetabolites Are Structural Analogs of Bases or Nucleosides

Antimetabolites are usually structural analogs of purine and pyrimidine bases or nucleosides that interfere with very specific metabolic reactions. They include: **6-mercaptopurine** and **6-thioguanine** used in the treatment of acute leukemia; azathioprine for immunosuppression in patients with organ transplants; allopurinol for treatment of hyperuricemia; and acyclovir for treatment of herpesvirus infection. The detailed understanding of purine nucleotide metabolism aided in the development of these drugs. Conversely, study of the mechanism of action of these drugs has led to a better understanding of normal nucleotide metabolism in humans.

Three antimetabolites will be discussed to show: (a) the importance of *de novo* synthetic pathways in normal cell metabolism; (b) that regulation of these pathways occurs *in vivo*; (c) the requirement for metabolic activation of the drugs utilizing the "salvage" enzymes; and (d) that inactivation of these compounds greatly influence their usefulness.

6-Mercaptopurine (6-MP) (Figure 20.31) is a useful antitumor drug in humans. Its cytotoxic activity depends on the formation of 6-mercaptopurine ribonucleotide by the tumor cell. Utilizing PRPP and HGPRTase, 6-mercaptopurine ribonucleoside 5′-monophosphate is formed in cells and serves as a negative effector of PRPP-amidotransferase, the committed step in the *de novo* pathway. This nucleotide also acts as an inhibitor of the conversion of IMP to GMP at the IMP-dehydrogenase step and IMP to AMP at the adenylosuccinate synthetase step. Since 6-mercaptopurine is a substrate for xanthine oxidoreductase and is oxidized to 6-thiouric acid, allopurinol is usually administered to inhibit degradation of 6-MP and to potentiate the antitumor properties of 6-MP.

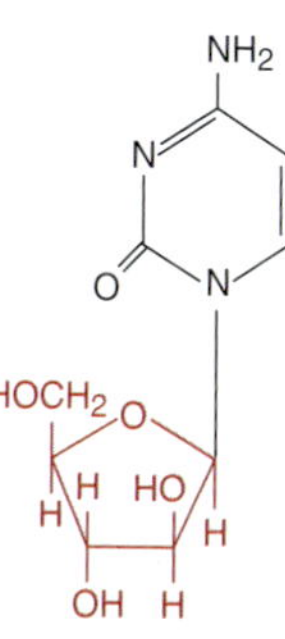

FIGURE 20.31
Structures of 6-mercaptopurine, 5-fluorouracil and cytosine arabinoside.

5-Fluorouracil (Fura) (Figure 20.31) is an analog of uracil. 5-Fluorouracil is not the active species because it must be converted by cellular enzymes to the active metabolites, 5-fluorouridine 5′-triphosphate (FUTP) and 5-fluoro-2′-deoxyuridine 5′-monophosphate (FdUMP). FUTP is efficiently incorporated into RNA and, once incorporated, inhibits maturation of 45 S precursor rRNA into the 28 S and 18 S species and alters splicing of pre-mRNA into functional mRNA. FdUMP is a potent and specific inhibitor of thymidylate synthase. In the presence of H_4folate, FdUMP, and thymidylate synthase, a ternary complex is formed with the covalent binding of FdUMP to thymidylate synthase, resulting in irreversible inhibition of thymidylate synthase. This inhibits dTMP synthesis and leads to "thymineless death" for cells.

Cytosine arabinoside (AraC) (Figure 20.31) is used in treatment of several forms of human cancer. It must be metabolized by cellular enzymes to cytosine arabinoside 5′-triphosphate (araCTP) to exert its cytotoxic effects. AraCTP competes with dCTP in the DNA polymerase reaction and araCMP is incorporated into DNA. This inhibits synthesis of the growing DNA strand. Clinically, the efficacy of araC as an antileukemic drug correlates with the concentration of araCTP that is achieved in the leukemic cells, which in turn determines the amount of araCMP incorporated into DNA. Formation of araCMP via deoxycytidine kinase appears to be the rate-limiting step in activation to araCTP. Ara-C is inactivated by deamination to araU.

Antifolates Inhibit Reactions Involving Tetrahydrofolate

Folate analogs, depending on their specific structure, interfere with metabolic steps in which tetrahydrofolate is involved as either a substrate or a product. **Methotrexate (MTX)**, a synthetic structural analog of folic acid, interferes with formation of H_2folate and H_4folate from folate by specifically inhibiting H_2folate reductase (DHFR). MTX is used as an antitumor agent in the treatment of human cancers. The two structures are compared in Figure 20.32. MTX and folate differ only at C4, where an amino group replaces a hydroxyl group, and at N10, where a methyl group replaces a hydrogen atom. MTX specifically inhibits H_2folate reductase with a K_i in the range of 0.1 nM. The reactions inhibited are shown in Figure 20.33.

MTX at very low concentrations is cytotoxic to mammalian cells in culture. The inhibition of dihydrofolate reductase by MTX results in the lowering of both

FIGURE 20.32

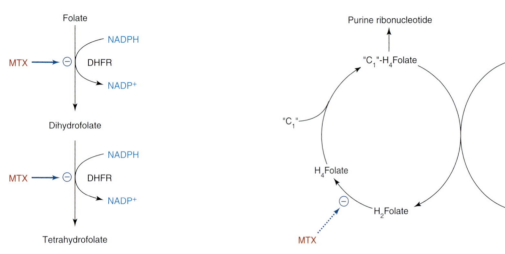

Comparison of the structures of folic acid and methotrexate.

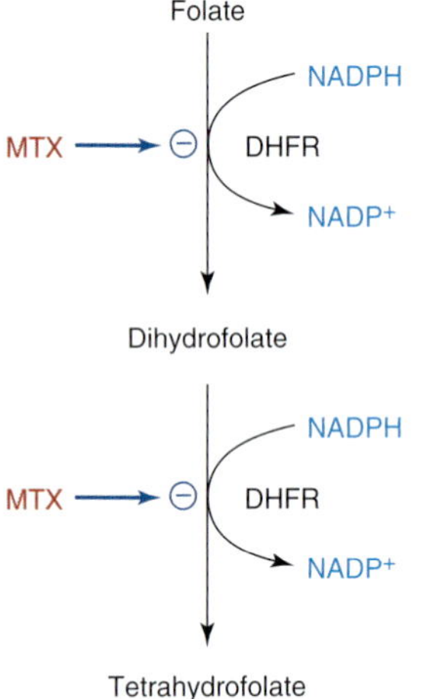

FIGURE 20.33

Sites of inhibition of methotrexate.

FIGURE 20.34

Relationship between H_4folate, *de novo* purine nucleotide synthesis, and dTMP synthesis.

ribonucleoside 5′-triphosphate and 2′-deoxyribonucleoside 5′-triphosphate intracellular pools. The effects can be prevented at least partially, by addition of deoxythymidine and hypoxanthine to the culture medium. Reversal of the MTX effects by thymidine and hypoxanthine indicate that MTX causes depletion of both deoxythymidine and purine nucleotides in cells. Figure 20.34 shows the relationship between H_4folate, *de novo* purine nucleotide synthesis, and dTMP formation. Note that in the thymidylate synthase reaction, H_2folate is generated; unless it is readily reduced back to H_4folate via dihydrofolate reductase, cells would not be capable of *de novo* synthesis of purine nucleotides or thymidylate synthesis due to depletion of H_4folate pools.

In treatment of human leukemias, normal cells can be rescued from the toxic effects of "high-dose MTX" by N^5-formyl-H_4folate (leucovorin). This increases the clinical efficacy of MTX in the treatment of cancer. MTX is also being successfully used at very low doses in the treatment of rheumatoid arthritis (RA), although the molecular basis by which MTX acts in this disorder is not known. It is not clear that the effects of MTX treatment in RA are related to its effects on nucleotide metabolism. However, it is interesting to note that leflunomide, another drug being used to treat RA, has as one of its defined sites of action the inhibition of mitochondrial dihydroorotate dehydrogenase.

Newer antifolates have been synthesized that are relatively specific inhibitors of either thymidylate synthase or *de novo* purine nucleotide synthesis, but not both. **Compound, ICI D1694** (Figure 20.35) specifically inhibits thymidylate synthase in

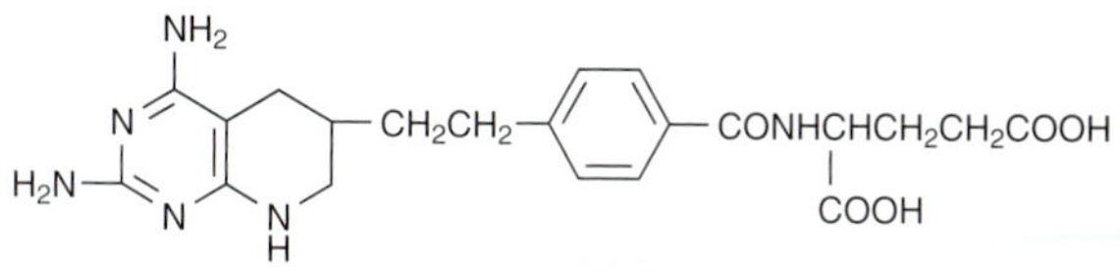

Dideazatetrahydrofolate

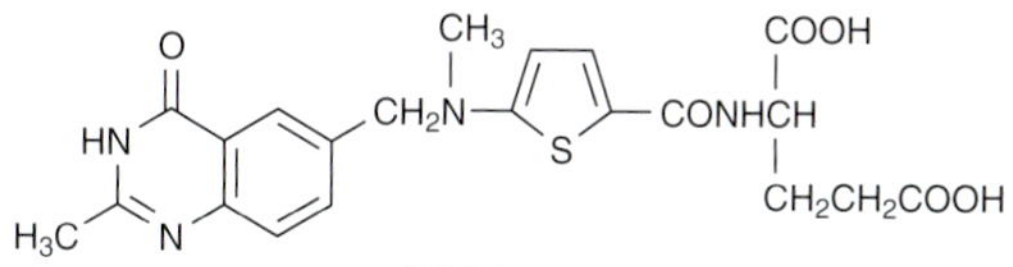

ICI D1694

FIGURE 20.35

Structures of 5,10-dideazatetrahydrofolate and ICI D1694, inhibitors of folate-dependent reactions. ICI D1694, *N*-(5-[*N*-(3, 4-dihydro-2-methyl-4-oxoquinazolin-6-ylmethyl)-*N*-methylamino-2-theonyl]-L-glutamic acid.

tumor cells. The effects on cell growth can be prevented by either 2′-dThd or folinic acid, but not by hypoxanthine. **5,10-Dideazatetrahydrofolate (DDATHF)** is a specific inhibitor of glycinamide ribotide formyltransferase (GART, Figure 20.35), and as a consequence the *de novo* synthesis of purine nucleotides is inhibited. Cells can be rescued from the cytotoxic effects of DDATHF by hypoxanthine but not by dThd. For both of these compounds, in a manner similar to MTX, additional glutamic acid residues must be added by folylpoly-gamma-glutamate synthetase to trap these compounds in the cells for maximal effects.

Glutamine Antagonists Inhibit Enzymes that Utilize Glutamine as Nitrogen Donors

Many reactions in mammalian cells utilize glutamine as the amino group donor. In contrast, bacteria primarily utilize ammonia as the amino donor in a similar reaction. These amidation reactions are critical in *de novo* synthesis of purine nucleotide (N3 and N9), synthesis of GMP from IMP, formation of cytosolic carbamoyl phosphate, synthesis of CTP from UTP, and synthesis of NAD^+.

Compounds that inhibit these reactions are referred to as glutamine antagonists. **Azaserine** (*O*-diazoacetyl-L-serine) and **6-diazo-5-oxo-L-norleucine (DON)** (Figure 20.36), which were first isolated from cultures of Streptomyces, are very effective inhibitors of enzymes that utilize glutamine as the amino donor. Since azaserine and DON inactivate the enzymes involved, addition of glutamine alone will not reverse the effects of either of these two drugs. This would necessitate that many metabolites such as guanine, cytosine, hypoxanthine (or adenine), and nicotinamide be provided to bypass the many sites blocked by these glutamine antagonists. Since so many key steps are inhibited by DON and azaserine, these agents are too toxic for clinical use.

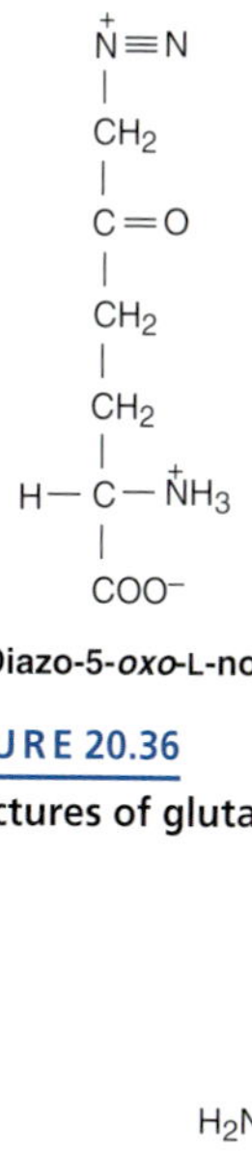

6-Diazo-5-*oxo*-L-norleucine (DON) 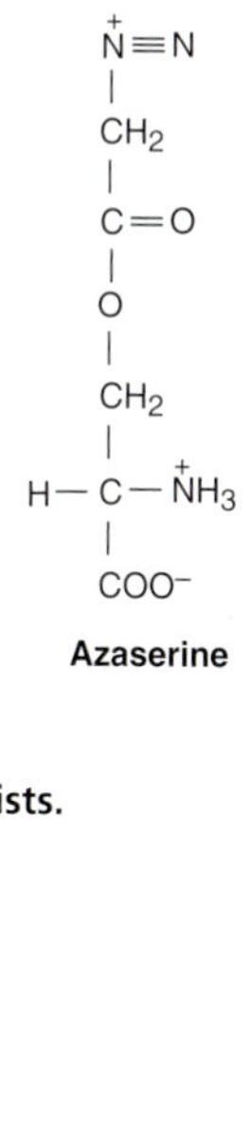 Azaserine

FIGURE 20.36

Structures of glutamine antagonists.

Other Agents Inhibit Cell Growth by Interfering with Nucleotide Metabolism

Tumor cells treated with hydroxyurea (Figure 20.37) show specific inhibition of DNA synthesis with little or no inhibition of RNA or protein synthesis. **Hydroxyurea** inhibits ribonucleotide reductase activity by destroying the tyrosyl free radical on the small subunit of ribonucleotide reductase. This results in the inhibition of the reduction of CDP, UDP, GDP, and ADP to the corresponding 2′-deoxyribonucleoside 5′-diphosphates. Toxicity results from depletion of 2′-deoxyribonucleoside 5′-triphosphates required for DNA replication. Clinical use of hydroxyurea as an antitumor agent is limited because of its rapid rate of clearance and the high drug concentration required for effective inhibition. However, hydroxyurea has recently been utilized in the treatment of sickle cell anemia in both adults and children. By a mechanism not understood, hydroxyurea treatment of sickle cell patients results in the re-expression of the fetal (γ) hemoglobin gene resulting in the increased fetal hemoglobin in the red cells. The increased concentration of fetal hemoglobin in the red cells decreases HbS "precipitation"; as a result, it decreases frequency of sickle cell crises in the patients during hypoxia. It does not appear that the effect of hydroxyurea on sickle cell erythrocytes is directly related to inhibition of ribonucleotide reductase.

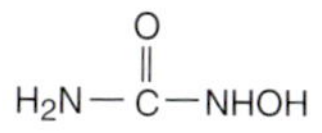

Hydroxyurea

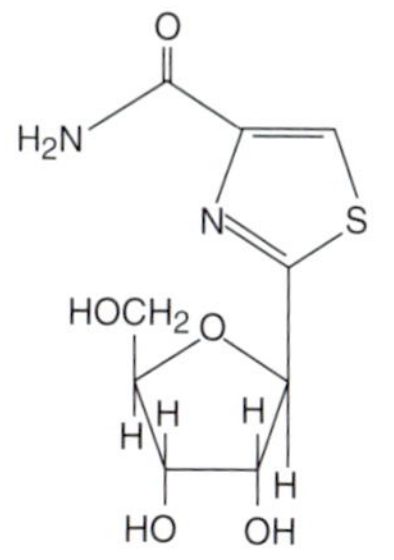

Tiazofurin

FIGURE 20.37

Structures of hydroxyurea and tiazofurin.

Acycloguanosine

AZT

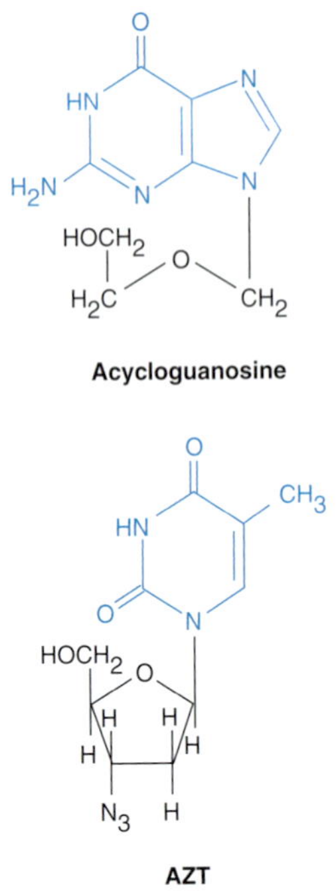

FIGURE 20.38
Structures of the antiviral agents, acyclovir and AZT.

Tiazofurin (Figure 20.37) is converted by cellular enzymes to the NAD^+ analog, tiazofurin adenine dinucleotide (TAD). TAD inhibits IMP dehydrogenase, the rate-limiting enzyme in GTP synthesis, with a K_I of 0.1 μM. As a result of IMP dehydrogenase inhibition, the concentration of GTP is markedly depressed with a corresponding decrease in dGTP. Although there are many dehydrogenases that utilize NAD^+ as a substrate, IMP dehydrogenase is most affected possibly because it catalyzes a rate-limiting step in a pathway and is quantitatively limiting in concentration.

These clinically useful drugs serve as examples in which the knowledge of basic biochemical pathways and mechanisms lead to generation of effective drugs.

Purine and Pyrimidine Analogs as Antiviral Agents

Herpesvirus (HSV) and human immunodeficiency virus (HIV), the causative agent of AIDS infections, present major clinical problems. Two antimetabolites are used in the control/treatment (but, not cure) of HSV and HIV infections are **Acyclovir (acycloguanosine)**, a purine analog, and **3′-azido-3′-deoxythymidine (AZT)**, a pyrimidine analog (Figure 20.38). Both drugs require phosphorylation to yield the active drug. Acycloguanosine is converted to the monophosphate by a specific HSV-thymidine kinase (encoded by the HSV genome) that is present only in the virally infected cells. The host cellular thymidine kinase cannot utilize acyclovir as a substrate. Acycloguanosine monophosphate is then phosphorylated by the cellular enzymes to the di- and triphosphate forms. Acycloguanosine triphosphate serves as a substrate for the HSV-specific DNA polymerase; it is incorporated into the growing viral DNA chain, causing chain termination. The specificity of acycloguanosine and its high therapeutic index resides, therefore, in the fact that only HSV-infected cells can form the acycloguanosine monophosphate.

AZT is phosphorylated by cellular kinases to AZT-triphosphate, which blocks HIV replication by inhibiting HIV-DNA polymerase (an RNA-dependent polymerase). The selectivity of AZT for HIV-infected versus normal cells occurs because DNA polymerase from HIV is at least 100-fold more sensitive to AZT-triphosphate than is host cell DNA-dependent DNA polymerase. These two antiviral agents demonstrate the diversity of responses required for selectivity. In one case, enzyme activity encoded by the viral genome is mandatory for activation of the drug (acycloguanosine); in the second example, although cellular enzymes activate AZT, the viral gene product (HIV-DNA polymerase) is the selective target.

Biochemical Basis for Responses to Chemotherapeutic Agents

There are two aspects to consider with respect to a patient's response to chemotherapy. First, there is the development or selection of drug-resistant populations as a result of the drug treatment. In the second, preexisting genetic polymorphisms cause altered drug metabolism such that at the "standard dose" the patient shows increased or even severe toxicity to the chemotherapeutic drug.

Failure of chemotherapy in treatment of human cancer is often related to development of, or selection of, tumor cell populations that are resistant to the cytotoxic effects of the particular drug. Tumors contain very heterogenous populations of cells, and in many instances drug-resistant cells are already present. Upon therapy, drug-sensitive cells are killed off and the resistant cell population becomes enriched. In some cases, drug treatments produce genetic alterations that result in the drug-resistant phenotype. Resistance to drugs can be categorized as "specific drug resistance" or "multidrug resistance."

Biochemical and molecular mechanisms that account for drug resistance have been determined for many drugs. For example, resistance to methotrexate can develop because of several different alterations. These include the following: (1) a defect or loss of the transporter for N^5-formyl-H_4folate and N^5-methyl-H_4folate, which results in decreased cellular uptake of MTX; (2) amplification of the dihydrofolate reductase gene, which results in large increases in the amount of dihydrofolate reductase, the target enzyme; (3) alterations in the dihydrofolate reductase gene, which result in a "mutant" dihydrofolate reductase that is less sensitive to inhibition by MTX; (4) decreased levels of folylpolyglutamate synthetase, which result in lower levels of polyglutamylated MTX, the "trapped" form of MTX. MTX-resistant populations could have any one or a combination of these alterations. The net result of any of these resistance mechanisms is to decrease the ability of MTX to inhibit dihydrofolate reductase at clinically achievable MTX concentrations. Other specific drug resistance mechanisms could be described for compounds such as cytosine arabinoside, 5-fluorouracil, hydroxyurea, and other drugs.

In multiple drug resistance, drug-resistant cells are cross-resistant to a variety of seemingly unrelated antitumor agents, such as the vinca alkaloids, adriamycin, actinomycin D, and etoposide. All of these are natural products or derived from natural products and they are not chemically related in structure. They have different mechanisms of action as antitumor agents but appear to act on some nuclear event.

Multidrug-resistant tumor cells express high levels (compared to the drug-sensitive tumor cells) of membrane transport proteins (**P-glycoproteins** and multidrug resistance-associated proteins **(MRPs)**; see p. 479) that transport drugs out of the cell. They are ATP-dependent "pumps" and effectively reduce the cellular concentration of drug to below its cytotoxic concentration. Development of drug-resistant tumor cells presents major clinical problems. However, study of the mechanisms of drug resistance has greatly aided in our understanding of cancer cells and how best to design chemotherapeutic protocols to treat the various forms of cancer.

Many different examples can be given in which genetic polymorphisms play a large role in the metabolism/disposition of various classes of drugs involving distinct gene products. As a result, there will be variability in the responses of patients to a specific drug regimen. As an example, patients who inherit nonfunctional alleles for **thiopurine *S*-methyltransferase (TPMT)**, an enzyme involved in the metabolism of 6-mercaptopurine (Figure 20.31), have very adverse effects to the "standard dose" of 6-mercaptopurine. These patients can be successfully treated by reducing the conventional dose of 6-MP and still achieve the necessary clinical response with decreased morbidity. Pharmacogenomics is a new field that is emerging that will allow the identification of the candidate genes with the use of molecular techniques that will allow the physician to individualize drug doses for patients with optimal outcome and reduced short- and long-term effects. This will be possible not only for the treatment of cancer but also for the treatment of other diseases as well.

BIBLIOGRAPHY

Aletaha, D., Stamm, T., Kapral, T., Eberl, G., Grisar, J., Machold, K. P., and Smolen, J. S. Survival and effectiveness of leflunomide compared with methotrexate and sulfasalazine in rheumatoid arthritis: A matched observational study. *Ann. Rheum. Dis.* 62:944, 2003.

Arner, E. S. J. and Eriksson, S. Mammalian deoxyribonucleoside kinases. *Pharmacol. Ther.* 67:155, 1995.

Beardsley, G. P., Rayl, E. A., Gunn, K., Moroson, B. A., Seow, H., Anderson, K. S., Vergis, J., Fleming, K., Worland, S., Condon, B., and Davies, J. Structures and functional relationships in human pur H. Advan. *Exp. Med. Biol.* 43:221, 1998.

Bertino, J. R. and Hait, W. In: Goldman, L., and Ausiello, D. (Eds.), *Principles of Cancer Therapy in Textbook of Medicine*, 22nd ed. Philadelphia: Saunders, 2004, Chapter 191, p. 1137.

Boza, J. J., Moennoz, D., Bournot, C. E., Blum, S., Zbindeen, I., Finot, P. A., and Ballerve, O. Role of glutamine on the *de novo* purine nucleotide synthesis in Caco-2 cells. *Eur. J. Nutr.* 39:38, 2000.

Canyuk, B., Francisco, J., Eakin, A. E., and Craig, S. P., III. Interactions at the dimer interface influence the relative efficiencies for purine nucleotide synthesis and pyrophosphorolysis in phosphoribosyltransferase. *J. Mol. Biol.* 335:905, 2004.

Christopherson, R. I., Lyons, S. D., and Wilson, P. K. Inhibitors of *de novo* nucleotide biosynthesis as drugs. *Acc. Chem. Res.* 35:961, 2002.

Elion, G. B. The purine path to chemotherapy. *Science* 244:41, 1989 (classic paper by Nobel Laureate).

Elledge, S. J., Zhou, Z., and Allen, J. B. Ribonucleotide reductase: regulation, regulation, regulation. *Trends Biochem. Sci.* 17:119, 1992.

Evans, W. E. and Relling, M. V. Moving towards individualized medicine with pharmacogenomics. *Nature* 429:464, 2004.

Hatse, S., DeClercq, E., and Balzarini, J. Role of antimetabolites of purine and pyrimidine nucleotide metabolism in tumor cell differentiation. *Biochem. Pharmacol.* 58:539, 1999.

Huang, M. and Graves, L. M. *De novo* synthesis of pyrimidine nucleotides; emerging with signal transduction pathways. *Cell. Mol. Life Sci.* 60:321, 2003.

Jordan, A. and Reichard, P. Ribonucleotide reductases. *Annu. Biochem.* 67:71, 1999.

Kanz, B. A., Kohalmi, S. E., Kunkel. T. A., Matthews, C. K., McIntosh, E. W., and Reidy, J. A. Deoxyribonucleoside triphosphate levels: A critical factor in the maintenance of genetic stability. *Mutat. Res.* 318:1, 1994.

Kappork, T. J., Ealick, S. E., and Stubbe, J. Modular evolution of purine biosynthetic pathway. *Curr. Opin. Chem. Biol.* 4:567, 2000.

Kisliuck, R. L. Deaza-analogs of folic acid as antitumor agents. *Curr. Pharmaceut. Design* 9:2615, 2003.

Magni, G., Amici, A., Emanuelli, E. M., Orsomando, G., Raffaelli, N., and Ruggeiri, S. Enzymology of NAD homeostatis in man. *Cell. Mol. Life Sci.* 61:19, 2004.

Ruckerman, K., Fairbanks, L. D., Carrey, E. A., Hawrylowicz, C. M., Richards, D. F., Kirschbaum, B., and Simmonds, H. A. Leflunomide inhibits de vovo pyrimidine synthesis in mitogen-stimulated T-lymphocytes from healthy humans. *J. Biol. Chem.* 273:21682, 1998.

Sanghani, S. P. and Moran, R. G. Tight binding of folate substrates and inhibitors to recombinant mouse glycinamide ribonucleotide formyltransferase. *Biochemistry* 36:10506, 1997.

Scriver, C. R., Beaudet, A. L., Valle, D. and Sly, W. S. (eds.). *The Metabolic and Molecular Bases of Inherited Disease*, Vol. II, 8th ed. New York: McGraw-Hill, 2004, Chapters 106–113.

Sigoillot, F. D., Berkowski, J. A., Sigoillot, S. M., Kotsis, D. H. and Guy, H. I. Cell cycle-dependent regulation of pyrimidine biosynthesis. *J. Biol. Chem.* 278:3403, 2003.

Sigoillot, F. D., Evans, D. R., and Guy, H. I. Autophosphorylation of the mammalian multifunctional protein that initiates *de novo* pyrimidine biosynthesis. *J. Biol. Chem.* 277:24809, 2002.

Stubbe, J. Ribonucleotide reductases: The link between a RNA and DNA world? *Curr. Opin. Struct. Biol.* 10:731, 2000.

Tonkinson, J. L., Marder, P., Andid, S. L., Schultz, R. M., Gossett, L. S., Shih, C., and Mendelsohn, L. G. Cell cycle effects of antimetobilites: Implications for cytotoxicity and cytostasis. *Cancer Chemother. Pharmacol.* 39:521, 1997.

Turner, R. N., Aherne, G. W., and Curtin, N. J. Selective potentiation of lometrexol growth inhibition by dipyrimidamole through specific inhibition of hypoxanthine salvage. *Br. J. Cancer* 76:1300, 1997.

Wei, S., Reillaudou, M., Apiou, F., Perye, H., Petriddis, F., and Luccioni, C. Purine metabolism in two human melanoma cell lines: Relation to proliferation and differentiation. *Melanoma Res.* 9:351, 1999.

Wolfe, F., Michand, K., Stephenson, B., and Doyle, J. Toward a definition and method of assessment of treatment failure and treatment effectiveness: The case of leflunomide versus methotrexate. *J. Rheumatol.* 30:1725, 2003.

Xu, X., Williams, J. W., Gong, H., Finnegan, A. and Chong, A. S. Two activities of the immunosuppressive metabolite of leflunomide, A771726. Inhibition of pyrimidine nucleotide synthesis and protein tyrosine phosphorylation. *Biochem. Pharmacol.* 52:527, 1996.

Yablonski, M. J., Pasek, D. A., Han, B. D., Jomes, M. E., and Traut, T. W. Intrinsic activity and stability of functional human UMP synthase and its two separate catalytic domains, orotate phosphoribosyltransferase and orotidine decarboxylase. *J. Biol. Chem.* 271:10704, 1996.

Zalkin, H. and Dixon, J. E. *De novo* purine nucleotide biosynthesis. *Prog. Nucleic Acid Res.* 42:259, 1992.

Zhao, R., Babani, S., and Goldmann I. D. Sensitivity to 5,10-dideazatetrahydrofolate is fully conserved in a murine leukemia cell line highly resistant to methotrexate due to impaired transport mediated by the reduced folate carrier. *Clin. Cancer Res.* 6:3304, 2000.

Zimmer, H. G. Significance of the 5-phosphoribosyl-1-pyrophosphate pool for cardiac purine and pyrimidine nucleotide synthesis: Studies with ribose, adenine, inosine, and orotic acid in rats. *Cardiovasc. Drugs Ther.* 12(Suppl. 2): 179, 1998.

QUESTIONS | CAROL N. ANGSTADT

Multiple Choice Questions

1. The two purine nucleotides found in RNA:
   A. are formed in a branched pathway from a common intermediate.
   B. are formed in a sequential pathway.
   C. must come from exogenous sources.
   D. are formed by oxidation of the deoxy forms.
   E. are synthesized from non-purine precursors by totally separate pathways.

2. The type of enzyme known as a phosphoribosyltransferase is involved in all of the following *except*:
   A. salvage of pyrimidine bases.
   B. the *de novo* synthesis of pyrimidine nucleotides.
   C. the *de novo* synthesis of purine nucleotides.
   D. salvage of purine bases.

3. Uric acid is:
   A. formed from xanthine in the presence of O_2.
   B. a degradation product of cytidine.
   C. deficient in the condition known as gout.
   D. a competitive inhibitor of xanthine oxidoreductase.
   E. oxidized, in humans, before it is excreted in urine.

4. In nucleic acid degradation, all of the following are correct *except*:
   A. there are nucleases that are specific for either DNA or RNA.
   B. nucleotidases convert nucleotides to nucleosides.
   C. the conversion of a nucleoside to a free base is an example of a hydrolysis.
   D. because of the presence of deaminases, hypoxanthine rather than adenine is formed.
   E. a deficiency of adenosine deaminase leads to an immunodeficiency.

5. The conversion of nucleoside 5′-monophosphates to nucleoside 5′-triphosphates:
   A. is catalyzed by nucleoside kinases.
   B. is a direct equilibrium reaction.
   C. utilizes a relatively specific nucleotide kinase and a relatively nonspecific nucleoside diphosphate kinase.
   D. generally uses GTP as a phosphate donor.
   E. occurs only during the S phase of the cell cycle.

6. Which of the following chemotherapeutic agents works by impairing *de novo* purine synthesis?
   A. Acyclovir (acycloguanosine).
   B. 5-Fluorouracil (antimetabolite).
   C. Methotrexate (antifolate).
   D. Hydroxyurea.
   E. AZT (3′-azido-3′-deoxythymidine).

Questions 7 and 8: Hereditary orotic aciduria is characterized by severe anemia, growth retardation, and high levels of orotic acid excretion. The defect may be in either orotate phosphoribosyl transferase, orotidine decarboxylase, or both. The preferred treatment for this disease is dietary uridine, which reverses the anemia and decreases the formation of orotic acid.

7. Elements involved in the effectiveness of the dietary treatment include:
   A. conversion of exogenous uridine to UMP by uridine phosphotransferase.
   B. UTP from exogenous uridine providing substrate for synthesis of CTP.
   C. inhibition of carbamoyl phosphate synthetase II by UTP.
   D. all of the above.
   E. none of the above.

8. In the *de novo* synthesis of pyrimidine nucleotides:
   A. reactions take place exclusively in the cytosol.
   B. a free base is formed as an intermediate.
   C. PRPP is required in the rate-limiting step.
   D. UMP and CMP are formed from a common intermediate.
   E. UMP inhibition of OMP-decarboxylase is the major control of the process.

Questions 9 and 10: Gout is a disease characterized by hyperuricemia from an overproduction of purine nucleotides via the *de novo* pathway. The specific cause of Lesch–Nyhan syndrome is a severe deficiency of HGPRTase. Allopurinol is used in the treatment of gout to reduce the production of uric acid. In Lesch–Nyhan syndrome, the decrease in uric acid is balanced by an increase in xanthine plus hypoxanthine in blood. In the other forms of gout, the decrease in uric acid is greater than the increase in xanthine plus hypoxanthine.

9. The explanation for this difference in the two forms of gout is:
   A. it is an experimental artifact and the decrease in uric acid and increase in xanthine plus hypoxanthine in non-Lesch–Nyhan gout is the same.
   B. allopurinol is less effective in non-Lesch–Nyhan gout.
   C. there is an increased excretion of xanthine and hypoxanthine in non-Lesch–Nyhan gout.
   D. PRPP levels are reduced in Lesch–Nyhan.
   E. in non-Lesch–Nyhan gout hypoxanthine and xanthine are salvaged to IMP and XMP and inhibit PRPP amidotransferase.

10. Which of the following is/are aspects of the overall regulation of *de novo* purine nucleotide synthesis?
   A. AMP, GMP, and IMP shift PRPP amido transferase from a small form to a large form.
   B. PRPP amidotransferase shows hyperbolic kinetics with PRPP.
   C. AMP inhibits the conversion of IMP to GMP.
   D. Change in glutamine concentration is a major regulator.
   E. Direct interconversion of AMP to GMP maintains balance of the two.

Question 11 and 12: Rheumatoid arthritis is an autoimmune disease that can lead to severe problems from inflammation. The first line of treatment is anti-inflammatory drugs which may control the symptoms but do not halt the progress of the disease. A more aggressive treatment is low doses of methotrexate, a dihydrofolate reductase inhibitor. Another immunosuppressive drug Arava (leflunomide), an inhibitor of dihydroorotate dehydrogenase, has the same ultimate effect as methotrexate but with fewer side effects.

11. Which of the following statements is/are correct?
   A. Methotrexate inhibits the *de novo* synthesis of UMP.
   B. Arava inhibits the *de novo* synthesis of pyrimidine nucleotides.
   C. Arava inhibits the conversion of dUMP to dTMP.
   D. Methotrexate inhibits the production of tetrahydrobiopterin.
   E. Both drugs greatly increase production of β-amino acids.

12. Deoxyribonucleotides:
   A. cannot be synthesized so they must be supplied preformed in the diet.
   B. are synthesized *de novo* using dPRPP.
   C. are synthesized from ribonucleotides by an enzyme system involving thioredoxin.
   D. are synthesized from ribonucleotides by nucleotide kinases.
   E. can be formed only by salvaging free bases.

Problems

13. If a cell capable of *de novo* synthesis of purine nucleotides has adequate AMP but is deficient in GMP, how would the cell regulate synthesis to increase [GMP]? If both AMP and GMP were present in appropriate concentrations, what would happen?

14. How would you measure the turnover of DNA?

ANSWERS

1. **A** GMP and AMP are both formed from the first purine nucleotide, IMP, in a branched pathway. B: The pyrimidine nucleotides UMP and CTP are formed in a sequential pathway from orotic acid. C: Man is capable of synthesizing purine nucleotides. D: Deoxy forms are formed by reduction of the ribose forms. E: IMP is the common precursor.
2. **C** In purine nucleotide synthesis, the purine ring is built up stepwise on ribose 5-phosphate and not transferred to it. A, B, D: Phosphoribosyl transferases are important salvage enzymes for both purines and pyrimidines and are also part of the synthesis of pyrimidines since OPRT catalyzes the conversion of orotate to OMP.
3. **A** The xanthine oxidoreductase reaction produces uric acid. B, E: Uric acid is an end product of purines, not pyrimidines. C: Gout is characterized by excess uric acid.
4. **C** The product is ribose-1-phosphate rather than the free sugar, a phosphorolysis. A: They can also show specificity toward the bases and positions of cleavage. B: A straight hydrolysis. D: AMP deaminase and adenosine deaminase remove the 6-NH_2 as NH_3. The IMP or inosine formed is eventually converted to hypoxanthine. E: This is called severe combined immunodeficiency.
5. **C** These two enzymes are important in interconverting the nucleotide forms. A: These convert nucleosides to nucleoside monophosphates. B: Two steps are required. D: ATP is present in highest concentration and is the phosphate donor. E: Occurs during the S phase, but this is a general reaction for the cell.
6. **C** Antifolates reduce the concentration of THF compounds that are necessary for two steps of purine synthesis. A, E: These are antiviral agents that inhibit DNA synthesis. B: 5-Fluorouracil is a pyrimidine analog not a purine analog. D: Hydroxyurea inhibits the reduction of ribonucleotides to deoxyribonucleotides, so it is not involved in *de novo* purine synthesis. E: Allopurinol potentiates the effect of 6-mercaptopurine but is not an inhibitor of purine synthesis.

7. **D** It is common for an exogenous agent to require conversion to an active form; in this case the uridine is "salvaged" to the monophosphate and ultimately to the triphosphate. The cell is deficient in UTP and CTP because the conversion of orotic acid is blocked so the exogenous uridine provides a bypass around the block. Orotic acid formation is decreased since UTP inhibits carbamoyl phosphate synthetase II, the control enzyme.
8. **B** This is in contrast to purine *de novo* synthesis. A: One enzyme is mitochondrial. C: PRPP is required to convert orotate to OMP, but this is not rate-limiting. D: OMP to UMP to CTP is a sequential process. E: This does occur but the rate-limiting step is that catalyzed by CPS II.
9. **E** Not only is uric acid production directly inhibited, *de novo* synthesis is as well, thus reducing production of xanthine and hypoxanthine. In Lesch–Nyhan, the only effect is the direct inhibition of xanthine oxidase. A: It is a real effect. B: Actually, it is more effective because of the dual roles. C: This does not happen. D: PRPP levels are very high because of the lack of the salvage of bases, leading to improper or lack of regulation of pathways.
10. **A** This is a mechanism of inhibition since the large form of the enzyme is inactive. B: PRPP amidotransferase shows sigmoidal kinetics with respect to PRPP, so large shifts in concentration of PRPP have the potential for altering velocity. C: AMP inhibits the conversion of IMP to itself. D: Glutamine concentration is relatively constant. E: This does not happen.
11. **B** The conversion of dihydroorotate to orotic acid is a key step in *de novo* synthesis of the pyrimidine nucleotides. A: Methotrexate reduces the tetrahydrofolate pool but tetrahydrofolate is not involved in the *de novo* synthesis of pyrimidines. C: This is the step that methotrexate inhibits by lowering the tetrahydrofolate pool. D: GTP is the precursor of tetrahydrobiopterin. E: β-Amino acids are degradation products but synthesis is not greatly increased.
12. **C** Deoxyribonucleotides are synthesized from the ribonucleoside diphosphates by nucleoside diphosphate reductase that uses thioredoxin as the direct hydrogen-electron donor. A, B, E: There is a synthetic mechanism as just described, but it is not a *de novo* pathway. D: Nucleotide kinases are enzymes that add phosphate to a base or nucleotide.
13. AMP would partially inhibit *de novo* synthesis of IMP by its allosteric inhibition of PRPP amidotransferase. The IMP formed would be directed toward GMP because AMP is an inhibitor of its own synthesis from IMP. If both AMP and GMP are adequate, the synergistic effect of the two on PRPP amidotransferase would severely inhibit *de novo* synthesis.
14. Thymine is derived almost exclusively from DNA. The degradation of thymine leads to the unique product β-aminoisobutyrate that is excreted. This can be measured in urine.

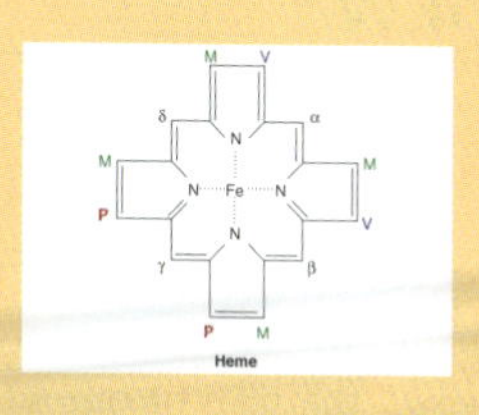

21

IRON AND HEME METABOLISM

William M. Awad, Jr.

Textbook of Biochemistry With Clinical Correlations, Sixth Edition, Edited by Thomas M. Devlin

21.1 IRON METABOLISM: OVERVIEW

Iron is closely involved in the metabolism of oxygen, permitting its transport and participation in a variety of biochemical processes. The common oxidation states are ferrous (Fe^{2+}) and ferric (Fe^{3+}). Higher oxidation levels occur as short-lived intermediates in certain redox processes; as an example, during the reaction catalyzed by cytochrome $P450_{cam}$ (see p. 414) from *Pseudomonas putida*, the iron of one of the intermediates must be in the ferryl (Fe^{4+}) form. Iron has an affinity for electronegative atoms such as oxygen, nitrogen, and sulfur, which provide the electrons that form the bonds with iron. These atoms have very high affinity when favorably oriented on macromolecules. In forming complexes, no bonding electrons are derived from iron. There is an added complexity to the structure of iron: The nonbonding electrons in the outer shell of the metal (the incompletely filled 3*d* orbitals) can exist in two states. Where bonding interactions with iron are weak, the outer nonbonding electrons will avoid pairing and will distribute throughout the 3*d* orbitals. Where bonding electrons interact strongly with iron, however, there is pairing of the outer nonbonding electrons, favoring lower-energy 3*d* orbitals. These two electron distributions for each oxidation state of iron can be determined by electron spin resonance measurements. Dispersion of 3*d* electrons to all orbitals leads to the high-spin state, whereas restriction of 3*d* electrons to lower energy orbitals, because of electron pairing, leads to a low-spin state. Some iron–protein complexes reveal changes in spin state without changes in oxidation during chemical events (e.g., binding and release of oxygen by hemoglobin).

At neutral and alkaline pH, the redox potential for iron in aqueous solution favors the Fe^{3+} state; at acid pH values, the equilibrium favors the Fe^{2+} state. In the Fe^{3+} state, iron slowly forms a large polynuclear complex with hydroxide ion, water, and other anions that may be present. These complexes can become so large as to exceed their solubility products, leading to aggregation and precipitation with pathological consequences.

Iron can bind to and influence the structure and function of various macromolecules, with deleterious results to the organism. To protect against such reactions, several iron-binding proteins function to store and transport iron. These have a very high affinity for the metal and, in the normal physiological state, have incompletely filled iron-binding sites. The interaction of iron with its ligands has been well-characterized in some proteins (e.g., hemoglobin, myoglobin, and transferrin).

21.2 IRON-CONTAINING PROTEINS

Iron binds to proteins either by incorporation into a **protoporphyrin IX** ring (see p. 833) or by interaction with other protein ligands. Ferrous- and ferric-protoporphyrin IX complexes are designated **heme** and **hematin**, respectively. Heme-containing proteins include (a) those that transport (e.g., hemoglobin) and store (e.g., myoglobin) oxygen and (b) enzymes that contain heme as their prosthetic group (e.g., catalase, peroxidases, tryptophan pyrrolase, prostaglandin synthase, guanylate cyclase, NO synthase, and the microsomal and mitochondrial cytochromes). Discussions on structure–function relationships of heme proteins are presented in Chapters 9 and 14.

Nonheme proteins include transferrin, ferritin, a variety of redox enzymes that contain iron at the active site, and iron–sulfur proteins. A significant body of information has been acquired that relates to the structure–function relationships of some of these molecules.

Transferrin Transports Iron in Serum

The protein in serum involved in the transport of iron is **transferrin**, a β1-glycoprotein synthesized in liver, that consists of a single polypeptide (78 kDa) with two noncooperative iron-binding sites. The protein is a product of gene duplication derived from a putative ancestral gene for a protein binding only one atom of iron. Several metals bind to transferrin; the highest affinity is for Fe^{3+}; Fe^{2+} ion is not bound. The binding of Fe^{3+} is absolutely dependent on the coordinate binding of an anion, which in the physiological state is carbonate, as indicated below:

$$\text{Transferrin} + Fe^{3+} + CO_3{}^{2-} \rightarrow \text{transferrin} \cdot Fe^{3} \cdot CO_3{}^{2-}$$

$$Fe^{3+} + CO_3{}^{2-} + \text{Transferrin} \cdot Fe^{3+} \cdot CO_3{}^{2-} \rightarrow \text{transferrin} \cdot 2\ (Fe^{3+}CO_3{}^{2-})$$

Association constants for the binding of Fe^{3+} to transferrins from different species range from 10^{19} to 10^{31} M^{-1}. For practical purposes, whenever there is excess transferrin there will be no free ferric ions. In the normal physiological state, approximately one-ninth of all transferrin molecules are saturated with iron; four-ninths are half-saturated and four-ninths are iron-free. Unsaturated transferrin protects against infections (Clin. Corrs. 21.1 and 21.2). The iron-binding sites show differences in sequence and in affinity for other metals. Transferrin binds to specific cell surface receptors that mediate its internalization.

The **transferrin receptor**, a transmembrane protein, is a heterodimer of subunits of 90-kDa subunits, joined by a disulfide bond. Each subunit contains one transmembrane segment and about 670 residues that are extracellular and binds a transferrin molecule, favoring the diferric form. Internalization of the receptor–transferrin complex is dependent on receptor phosphorylation by a Ca^{2+}-calmodulin-protein kinase C complex. Release of the iron atoms occurs within the acidic matrix of lysosomes; the receptor–apotransferrin complex returns to the cell surface where the apotransferrin is released to be reutilized in the plasma.

CLINICAL CORRELATION 21.1

Iron Overload and Infection

If an individual is overloaded with iron by any cause, the serum transferrin value may be close to saturation with iron, making small amounts of free serum iron available. Microorganisms that are usually nonpathogenic, because they are iron dependent and cannot compete against partially saturated transferrin in the normal individual, can now become pathogenic under these circumstances. For example, *Vibrio vulnificus*, a marine halophile, is found in a small percentage of oysters and commercial shellfish. Individuals who are iron-overloaded can develop a rapidly progressive infection, with death ensuing within 24 h after ingestion of the offending meal, whereas normal individuals consuming the same food are entirely free of symptoms.

Source: Muench, K. H. Hemochromatosis and infection: Alcohol and iron, oysters and sepsis. *Am. J. Med.* 87:3, 1989.

Lactoferrin Binds Iron in Milk

Milk contains iron that is bound almost exclusively to a glycoprotein, **lactoferrin**, closely homologous to transferrin, with two sites binding the metal. The protein is never saturated with iron. Lactoferrin is an antimicrobial, protecting the newborn from gastrointestinal infections. Microorganisms require iron for replication and function. Incompletely saturated lactoferrin binds free iron, leading to the inhibition of microbial growth by preventing a sufficient amount of iron from entering these microorganisms. Other microbes, such as *E. coli*, which release competitive iron chelators, are able to proliferate despite the presence of lactoferrin, since the chelators transfer the iron specifically to the microorganism. Lactoferrin is present in granulocytes being released during bacterial infections. It is also present in mucous secretions. Besides its bacteriostatic function, it is believed to facilitate iron transport and storage in milk. Lactoferrin has been found in urine of premature infants fed human milk.

CLINICAL CORRELATION 21.2

Microbial Pathogenicity and Iron

Bacteria require iron for growth, secreting low-molecular-weight compounds, the siderophores, which bind the metal tightly and bring the metal complex to the microorganism. However, as noted in Clin. Corr. 21.1, the tight binding by transferrin ordinarily does not make iron available unless that protein is saturated, with excess free metal present. However, if acidosis ensues, the drop in pH reduces markedly the affinity of transferrin for the metal. This is one of many features relevant to the increasing susceptibility to infection in diabetes mellitus. A different source of iron is utilized by *Staphylococcus aureus*. This microbe secretes a hemolysin, rupturing the membrane of a few erythrocytes. The released hemoglobin yields free heme that is transported into the coccus, where bacterial enzymes resembling heme oxygenase release free iron.

Source: Skaar, E. P., Humayun, M., Bae, T., et. al. Iron-source preference of *Staphylococcus aureus* infections. *Science* 305:1626, 2004.

Ferritin Is a Protein Involved in Storage of Iron

Ferritin is the major protein involved in storage of iron. It consists of an outer polypeptide shell 130 Å in diameter with a central ferric-hydroxide-phosphate core 60 Å across. The apoprotein, **apoferritin**, consists of 24 subunits of a varying mixture of H subunits (178 amino acids) and L subunits (171 amino acids) that provide various isoprotein forms. H subunits predominate in nucleated blood cells and heart, whereas L subunits predominate in liver and spleen. Synthesis of the subunits is regulated mainly by the concentration of free intracellular iron. The bulk of iron storage occurs in hepatocytes, reticuloendothelial cells, and skeletal muscle. The ratio of iron to protein is not constant. With a capacity of 4500 iron atoms, the molecule contains usually less than 3000. Channels from the surface permit the accumulation and release of iron. When iron is in excess, the storage capacity of newly synthesized apoferritin may be exceeded. This leads to iron deposition adjacent to ferritin spheres. Histologically, such amorphous iron deposition is called **hemosiderin**. The H chains of ferritin oxidize ferrous ions to the ferric state. Ferritins derived from different tissues of the same species differ in electrophoretic mobility in a fashion analogous to the differences noted

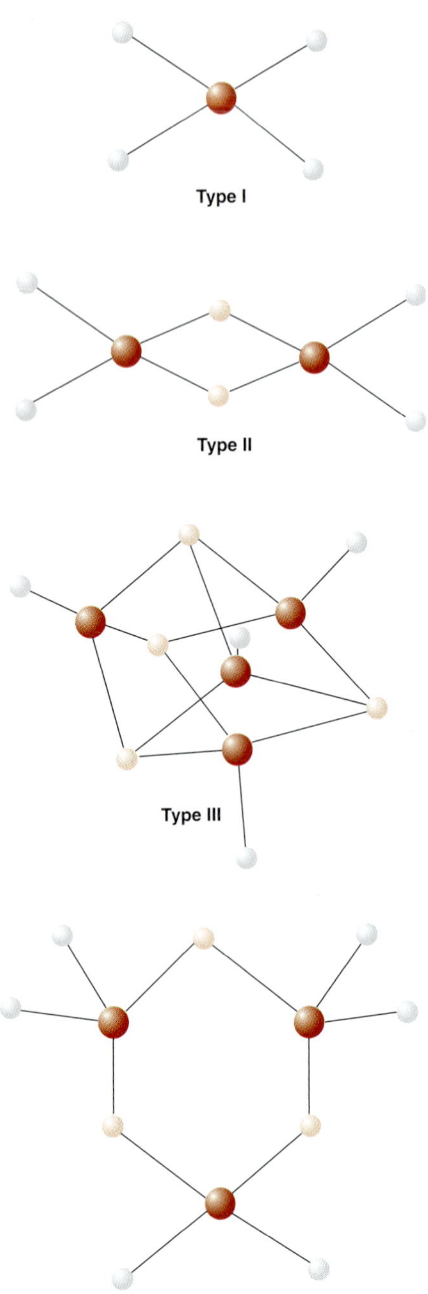

FIGURE 21.1

Structure of iron and sulfur atoms in ferredoxins. Dark-colored circles represent iron atoms; pink circles represent the inorganic sulfur atoms; and small gray circles represent cysteinyl sulfur atoms derived from the polypeptide chain. Variation in type IV ferredoxins can occur where one of cysteinyl residues can be substituted by a solvent oxygen atom of an OH group.

with isoenzymes. In some tissues, ferritin spheres form lattice-like arrays, which are identifiable by electron microscopy.

Plasma ferritin (low in iron, rich in L subunits) has a half-life of 50 h, is cleared by reticuloendothelial cells and hepatocytes, and its concentration, although very low, correlates closely with the size of the body iron stores.

Nonheme Iron-Containing Proteins Are Involved in Enzymatic Processes

Many iron-containing proteins are involved in enzymatic reactions, most of which are related to oxidation mechanisms. The structural features of the ligands binding the iron are not well known, except for a few components involved in mitochondrial electron transport. These **ferredoxins** are characterized by iron being bonded usually, only to sulfur atoms. Several types of iron–sulfur clusters are known (Figure 21.1). The smallest, **type I** (e.g., **nebredoxin**), found only in microorganisms, consists of a small polypeptide (6 kDa) with one iron atom bound to four cysteine residues. **Type II** consists of ferredoxins found in plants and animal tissues that contain two iron atoms, each liganded to two cysteine residues and sharing two sulfide anions. The most complicated of the iron–sulfur proteins are the bacterial ferredoxins, **type III**, which contain four atoms of iron, each of which is linked to a single separate cysteine residues but also shares three sulfide anions with neighboring iron atoms to form a cubelike structure. In some anaerobic bacteria, a family of ferredoxins may contain two type III iron–sulfur groups per macromolecule. **Type IV ferredoxins** contain three atoms of iron each linked to two cysteine residues and each sharing two sulfide anions, to form a planar ring. In one example of this ferredoxin type, an exception of iron atoms being liganded only to sulfur atoms was found where the sulfur of a cysteinyl residue was substituted by a solvent oxygen atom. The redox potential afforded by these different ferredoxins varies widely and is in part dependent on the environment of the surrounding polypeptide that envelops these iron–sulfur groups. The iron in nebredoxin undergoes ferric–ferrous conversion during electron transport. In type II iron–sulfur proteins, both irons are in the Fe^{3+} form in the oxidized state; upon reduction, only one goes to Fe^{2+}. In Type III ferredoxin the oxidized state can be either $2Fe^{3+} \cdot 2Fe^{2+}$ or $3Fe^{3+} \cdot Fe^{2+}$, with corresponding reduced forms of $Fe^{3+} \cdot 3Fe^{2+}$ or $2Fe^{3+} \cdot 2Fe^{2+}$. Recently, several more complicated iron–sulfur clusters have been characterized revealing an expanding role for the clusters in redox mechanisms (see Clin. Corrs. 21.3 and 21.4).

21.3 INTESTINAL ABSORPTION OF IRON

The high affinity of iron for specific and nonspecific macromolecules leads to little significant formation of free iron salts, and thus this metal is not lost via usual excretory routes. Rather, iron excretion occurs only through normal sloughing of tissues that are not reutilized (e.g., epidermis and gastrointestinal mucosa). In healthy adult males the loss is about 1 mg day^{-1}. In premenopausal women, the normal physiological events of menses, parturition, and lactation substantially augment iron loss. A wide variation of such loss exists, depending on the amounts of menstrual flow and the multiplicity of births. In extreme situations, a premenopausal woman may require four to five times the amount of iron needed by an adult male for prolonged periods. The postmenopausal woman who is not iron-deficient has an iron requirement similar to that of the adult male. Children and patients with blood loss naturally have increased iron requirements.

Cooking of food facilitates breakdown of ligands attached to iron, increasing the availability of the metal in the gut. The low pH of stomach contents permits the reduction of Fe^{3+} to the Fe^{2+}, facilitating dissociation from ligands. This requires the presence of an accompanying reductant, which is usually achieved by adding ascorbate to the diet. The absence of a normally functioning stomach reduces substantially the amount of iron that is absorbed. Some iron-containing compounds bind the metal so tightly that it is not available for assimilation. Contrary to popular belief, spinach is a poor source of iron because of an earlier erroneous record of the iron content and because some of the iron is bound to **phytate (inositol hexaphosphate)**, which is resistant to the chemical actions of the gastrointestinal tract. Specific protein cofactors, derived from the stomach or pancreas, have been suggested as being facilitators of iron absorption in the small intestine.

The major site of absorption of iron is in the small intestine, with the largest amount being absorbed in the duodenum (Figure 21.2). The metal enters the mucosal cell as the free ion or as heme; in the latter case the metal is released from the porphyrin ring in the mucosal cytoplasm. The large amount of bicarbonate secreted by the pancreas neutralizes the acidic material delivered by the stomach and favors the oxidation of Fe^{2+} to Fe^{3+}. Whatever the requirements of the host are, in the face of an adequate delivery of iron to the lumen, an amount of iron enters mucosal cells. Regulation of iron transfer occurs at the luminal surface, within the cytoplasm, and between the mucosal cell and the capillary bed (see Figure 21.2 and Clin. Corr. 21.5). In the normal state, certain processes define the amount of iron that will be transferred. Where there is iron deficiency, the amount of transfer increases; where there is iron overload in the host, the amount transferred is curtailed substantially. One mechanism that has been demonstrated to regulate this transfer of iron across the mucosal–capillary interface is the synthesis of apoferritin by the mucosal cell. In situations in which little iron is required by the host, a large amount of apoferritin is synthesized to trap the iron within mucosal cells and prevent transfer to the capillary bed. As the cells turn over (within a week), their contents are extruded into the intestinal lumen without absorption occurring. In situations in which there is iron deficiency, virtually no apoferritin is synthesized so as not to compete against the transfer of iron to the deficient host.

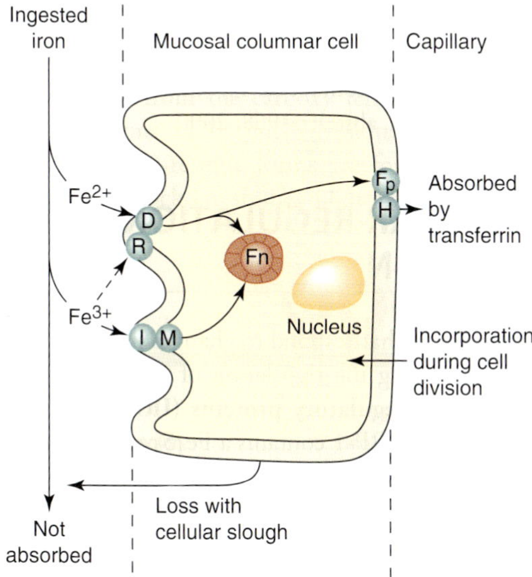

FIGURE 21.2

Intestinal mucosal regulation of iron absorption. The flux of iron in the duodenal mucosal cell is indicated. A fraction of the iron that is potentially acceptable is transferred from intestinal lumen into the epithelial cell. A large portion of ingested iron is not absorbed, in part because it is not presented in a readily acceptable form. Some iron is retained within the cell, bound by apoferritin to form ferritin. This iron is sloughed into the intestinal lumen with the normal turnover of cells. A portion of the iron within the mucosal cell is absorbed and transferred to the capillary bed to be incorporated into transferrin. During cell division, which occurs at the bases of the intestinal crypts, iron is incorporated for cellular requirements. These fluxes change dramatically in iron-depleted or iron-excess states. The following structures are indicated: **D**, DMT 1; **I**, integrin; **M**, mobilferrin; **Fn**, ferritin; **Fp**, ferroportin 1; **H**, hephaestin; and **R**, a postulated ferrireductase, acting with DMT-1.

CLINICAL CORRELATION 21.3
Iron–Sulfur Cluster Synthesis and Human Disease

Iron–sulfur clusters are synthesized in the mitochondrion and exported to the cytosol, a poorly understood process involving mitochondrial chaperones and other proteins, some with yet unknown function. A mutation in the human ABC7 (see p. 479) transporter gene leads to one form of X-linked sideroblastic anemia seen in patients with an abnormal gait (cerebellar ataxia) and with impaired development of cytosolic iron–sulfur proteins.

Source: Bekri S, Kispal, G, Lane H, et al. Human ABC7 transporter gene structure and mutation causing X-linked sideroblastic anemia with ataxia with disruption of cytosolic iron–sulfur protein maturation. *Blood* 96:3256, 2000.

CLINICAL CORRELATION 21.4
Friedreich's Ataxia

Frataxin is a mitochondrial protein of unknown function, the mutations of which lead to the disease Friedreich's ataxia, associated with a marked reduction in this protein's concentration. Frataxin is possibly necessary for formation of iron–sulfur clusters. The disease is characterized by marked abnormalities in gait, impaired heart muscle, and diabetes. There is marked increase in mitochondrial iron accumulation; however, the iron is not readily chelatable. In this condition a GAA triplet expansion to sometimes over 1000-fold in intron l of frataxin is the cause of the condition in about 98% of cases. No anemia is seen with this disease.

Source: Seznec, H., Simon, D., Bouton, C., et al. Friedreich ataxia, the oxidative stress paradox. *Hum. Mol. Genet.* 14:463, 2005. Napier, I., Ponka, P., and Richardson, D. R. Iron trafficking in the mitochondrion: Novel pathways revealed by disease. *Blood* 105:1867, 2005.

CLINICAL CORRELATION 21.8
Iron-Deficiency Anemia

Microscopic examination of blood smears from patients with iron-deficiency anemia usually reveals the characteristic findings of microcytic (small in size) and hypochromic (underpigmented) red blood cells. These changes result from decreased rates of globin synthesis when heme is not available. A bone marrow aspiration will reveal no storage iron to be present, and serum ferritin values are virtually zero. The serum transferrin value (expressed as the total iron-binding capacity) will be elevated (upper limits of normal: 410 mg dL^{-1}) with a serum iron saturation of less than 16%. Common causes for iron deficiency include excessive menstrual flow, multiple births, and gastrointestinal bleeding that may be occult. The common causes of gastrointestinal bleeding include medications that cause ulcers or erosion of the gastric mucosa (especially with aspirin or cortisone-like drugs), hiatal hernia, peptic ulcer disease, gastritis associated with chronic alcoholism, and gastrointestinal tumor. Management of such patients must include both a careful examination for the cause and source of bleeding and supplementation with iron. The latter is usually provided in the form of oral ferrous sulfate tablets; occasionally, intravenous iron therapy may be required. Where the iron deficiency is severe, transfusion with packed red blood cells may also be indicated. The weakness and fatigue associated with iron deficiency may be due to the inhibition of translation of the mitochondrial aconitase mRNA through binding of IRPs to its IRE.

Source: Hentze, M. W. and Kuhn, L. C. Molecular control of vertebrate iron metabolism: mRNA based circuits operated by iron, nitric oxide, and oxidative stress. *Proc. Natl. Acad. Sci.* USA 93:8175, 1996. Eisenstein, R. S. and Blemings, K. P. Iron regulatory proteins, iron responsive elements and iron homeostasis. *J. Nutr.* 128:2295, 1998.

CLINICAL CORRELATION 21.9
Hemochromatosis Type I: Molecular Genetics and the Issue of Iron-Fortified Diets

The hemochromatosis gene is heterozygous in about 9% of the U.S. population. The disease is expressed primarily in the homozygous state; about 0.25% of all individuals are at risk. Normal individuals have a major histocompatibility complex class-1 gene (HFE) that encodes for the α chain, containing three immunoglobulin-like domains. The associated β chain is β_2-microglobulin. The normal gene product has a structure that cannot present an antigen. Most individuals with hemochromatosis are homozygous for a Cys282Tyr mutation that prevents formation of the normal conformation of an immunoglobulin domain. Other individuals can have a single Cys282Tyr mutation on one allele with the other allele containing a mutation at another site. Normal HFE complexes with the transferrin receptor dimer to form a heterotetramer, which contains two receptor monomers and two HFEs. Structure–function studies show that the binding sites for transferrin and HFE overlap on the receptor, leading to the suggestion that this binding of HFE serves to control the rate of iron transfer to specific cells. Where there is a mutation in the HFE which prevents the binding to the receptor, transferrin binds to a greater extent, leading to the increased amount of iron stores in certain cells. However, this is an incomplete story. It is not yet explained why there is such an increase in iron absorption in the intestinal mucosa in this disease.

A controversy has developed as to whether food should be fortified with iron because of the prevalence of iron-deficiency anemia, especially among premenopausal women. It was suggested that dietary iron deficiency would be reduced if at least 50 mg of iron were incorporated per pound of enriched flour. Others suggested that the risk of toxicity from excess iron absorption through iron fortification was too great. Sweden has mandated iron fortification for 50 years, and about 42% of the average daily intake of iron is derived from these sources. However, 5% of males had elevation of serum iron values, with 2% having iron stores consonant with the distribution found in early stages of hemochromatosis, pointing out the danger of iron-fortified diets. In countries where iron deficiency is widespread, however, fortification may still be the most appropriate measure.

Source: McLaren, C. E., Gorddeuk, V. R., Looker, A. C., et al. Prevalence of heterozygotes for hemochromatosis in the white population of the United States. *Blood* 86:2021, 1995. Feder, J. N., Gnirki, A., Thomas, W., et al. A novel MHC class 1-like gene is mutated in patients with hereditary haemochromatosis. *Nat. Genet.* 13:399, 1996. Olsson, K. S., Heedman, P. A., and Staugard, F. Preclinical hemochromatosis in a population on a high-iron-fortified diet. *JAMA* 239:1999, 1978; Olsson, K. S., Marsell, R., Ritter, B., Olander, B., et al. Iron deficiency and iron overload in Swedish male adolescents. *J. Intern. Med.* 237:187, 1995. Bennett, M. J., Lebron, J. A., and Bjorkman P. J. Crystal structure of the hereditary haemochromatosis protein HFE complexed with transferrin receptor. *Nature* 403:46, 2000.

amounts of blood to remove iron as hemoglobin. Another group of patients has severe anemias, among the most common of which are the thalassemias, a group of hereditary **hemolytic anemias**. In these cases, the subjects require transfusions throughout their lives, leading to the accumulation of large amounts of iron derived from the transfused blood. Clearly, bleeding would be inappropriate in these cases; rather, the patients are treated by the administration of iron chelators, such as desferrioxamine, which leads to the excretion of large amounts of complexed iron in the urine. A rare third group of patients acquires excess iron because they ingest large amounts of both iron and ethanol, the latter promoting iron absorption. Excess stored iron in these cases can be removed by bleeding (see Clin. Corrs. 21.9 and 21.10).

21.6 HEME BIOSYNTHESIS

Heme is produced in virtually all mammalian tissues. Synthesis is most pronounced in the bone marrow and liver because of the requirements for incorporation into hemoglobin and the cytochromes, respectively. As depicted in Figure 21.6, heme is a largely planar molecule. It consists of one ferrous ion and a tetrapyrrole ring, **protoporphyrin IX**. The diameter of the iron atom is a little too large to be accommodated within the plane of the porphyrin ring, and thus the metal puckers out to one side as it coordinates with the apical nitrogen atoms of the four pyrrole groups. Heme is one of the most stable of compounds, reflecting its strong resonance features.

Figure 21.7 depicts the pathway for heme biosynthesis. The following are to be noted. First, the initial and last three enzymatic steps are catalyzed by enzymes that are in the mitochondrion, whereas the intermediate steps take place in the cytosol. This is important in considering the regulation by heme of the first biosynthetic step; this aspect is discussed below. Second, the organic portion of heme is derived totally from eight residues each of glycine and succinyl CoA. Third, the reactions occurring on the side groups attached to the tetrapyrrole ring involve the colorless intermediates known as **porphyrinogens**. Though exhibiting resonance features within each pyrrole ring, these do not demonstrate resonance between the pyrrole groups. Hence, the porphyrinogens are unstable and can be readily oxidized, especially in the presence of light, by nonenzymatic means to their stable **porphyrin** products. In the latter cases, resonance between pyrrole groups is established by oxidation of the four methylene bridges. Figure 21.8 depicts the enzymatic conversion of protoporphyrinogen to protoporphyrin by this oxidation mechanism. This is the only enzyme-catalyzed oxidation of a porphyrinogen in humans; all other porphyrinogen–porphyrin conversions are nonenzymatic and catalyzed by light. Fourth, once the tetrapyrrole ring is formed, the order of the R groups as one goes clockwise around the tetrapyrrole ring defines which of the four possible

CLINICAL CORRELATION 21.10
Hemochromatosis Type III

There are other causes for pathologically increased iron accumulation, most of which remain to be characterized fully (see Clin. Corr. 21.9). A protein homologous with transferrin receptor has been characterized and associated with mucosal iron absorption. Mutations in this protein transferrin receptor 2 have been found to be associated with hemochromatosis despite normal HFE genes. In contrast to transferrin receptor 1, this receptor does not have a stem-loop in its mRNA nor does it bind HFE.

Source: West, A. P., Jr., Bennett, M. J., Sellers, V. M. D., et al. Comparison of the interactions of transferrin receptor and transferrin receptor 2 with transferrin and the hereditary hemochromatosis protein HFE. *J. Biol. Chem.* 275:38135, 2000. Camaschella, C., Roetto, A., Cali, A., et al. The gene TRF 2 is mutated in a new type of haemochromatosis mapping to 7q22. *Nat. Genet.* 25:14, 2000. Kawabati, H., Yang, R., Hirawa, T., et al. Molecular cloning of transferrin receptor 2. A new member of the transferrin receptor-like family. *J. Biol. Chem.* 274:20826, 1999.

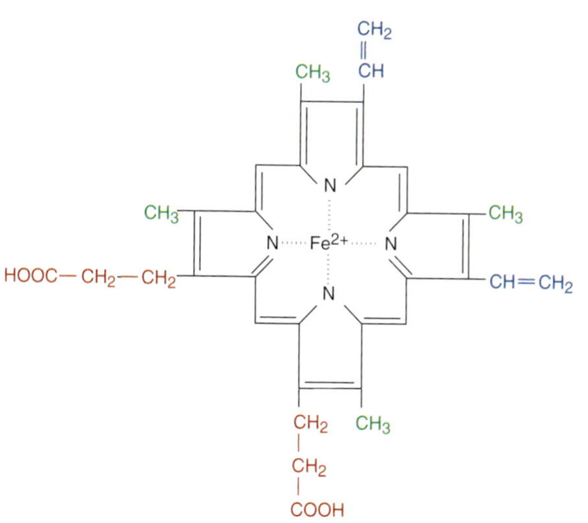

FIGURE 21.6
Structure of heme.

CLINICAL CORRELATION 21.11

Acute Intermittent Porphyria

A 40-year-old woman appears in the emergency room in an agitated state, weeping and complaining of severe abdominal pain. She had been constipated for several days and noted marked weakness in the arms and legs, and that "things do not appear to be quite right." Physical examination reveals a slightly rapid heart rate (100/min) and moderate hypertension (blood pressure of 160/110 mmHg). There have been earlier episodes of severe abdominal pain; operations undertaken on two occasions revealed no abnormalities. The usual laboratory tests are normal. The neurological complaints are not localized to an anatomical focus. A decision is made that the present symptoms are largely psychiatric in origin and have a functional rather than an organic basis. The patient is sedated with 60 mg of phenobarbital; a consultant psychiatrist agrees by telephone to see the patient in about 4 h. The staff notices a marked deterioration; generalized weakness rapidly appears, progressing to a compromise of respiratory function. This ominous development leads to immediate incorporation of a ventilatory assistance regimen, with transfer to intensive care for physiological monitoring. Her condition deteriorates and she dies 48 h later. A urine sample is reported later to have a markedly elevated level of porphobilinogen. This patient had acute intermittent porphyria, a disease of incompletely understood derangement of heme biosynthesis. There is a dominant pattern of inheritance associated with an overproduction of ALA and porphobilinogen, the porphyrin precursors. Three enzyme abnormalities have been noted in cases that have been studied carefully. These include (1) a marked increase in ALA synthase, (2) a reduction by one-half of activity of porphobilinogen deaminase, and (3) a reduction by one-half of the activity of steroid 5 α-reductase. The change in content of the second enzyme is consonant with a dominant expression. The change in content of the third enzyme is acquired and not apparently a heritable expression of the disease. It is believed that a decrease in porphobilinogen deaminase leads to a minor decrease in content of heme in liver. The lower concentration of heme leads to a failure both to repress the synthesis and to inhibit the activity of ALA synthase. Almost never manifested before puberty, the disease is thought to appear only with the induction of 5β-reductase at adolescence. Without a sufficient amount of 5α-reductase, the observed increase in the 5β steroids is due to a shunting of steroids into the 5β-reductase pathway. The importance of abnormalities of this last metabolic pathway in the pathogenesis of porphyria is controversial. The disease poses a great riddle: The derangement of porphyrin metabolism is confined to the liver, which anatomically appears normal, whereas the pathological findings are restricted to the nervous system. In the present case, involvement of (1) the brain led to the agitated and confused state and the respiratory collapse, (2) the autonomic system led to the hypertension, increased heart rate, constipation, and abdominal pain, and (3) the peripheral nervous system and spinal cord led to the weakness and sensory disturbances.

Experimentally, no known metabolic intermediate of heme biosynthesis can cause the pathology noted in acute intermittent porphyria. There should have been a greater suspicion of the possibility of porphyria early in the patient's presentation. The analysis of porphobilinogen in the urine is a relatively simple test. Treatment would include glucose infusion, the exclusion of any drugs that could cause elevation of ALA synthase (e.g., barbiturates), and, if her disease failed to respond satisfactorily despite these measures, the administration of intravenous hematin to inhibit the synthesis and activity of ALA synthase. Acute hepatic porphyria is of historic political interest. The disease has been diagnosed in two descendants of King George III, suggesting that the latter had a deranged personality preceding and during the American Revolution that could possibly be ascribed to porphyria.

Source: Meyer, U. A., Strand, L. J., Dos, M., et al. Intermittent acute porphyria: demonstration of a genetic defect in porphobilinogen metabolism. *N. Engl. J. Med.* 286:1277, 1972. Stein, J. A. and Tschudy, D. D. Acute intermittent porphyria: A clinical and biochemical study of 46 patients. *Medicine (Baltimore)* 49:1, 1970.

mRNA derived from the nucleus and is transferred into the matrix of the mitochondrion. Succinyl CoA, an intermediate of the tricarboxylic acid cycle, is a substrate and occurs only in mitochondria. This protein has been purified to homogeneity from rat liver mitochondria. In the cytosol, each 71-kDa subunit is the unfolded state, the only form that can pass into mitochondria. A basic N-terminal signaling sequence directs the enzyme into mitochondria. An ATP-dependent 70-kDa cytosolic component, known as a chaperone protein, maintains it in the unfolded extended state. Thereafter, the N-terminal sequence is cleaved by a metal-dependent protease in the mitochondrial matrix, to yield an ALA synthase subunit of 65 kDa each. Within the matrix another oligomeric chaperon protein, of 14 subunits of 60 kDa each, catalyzes the correct folding in a second ATP-dependent process (Figure 21.9). The ALA synthase has a short biological half-life (~60 min). Both the synthesis and the activity of the enzyme are subject to regulation by a variety of substances; 50% inhibition of activity occurs in the presence of 5 mM hemin, and virtually complete inhibition occurs at a 20 mM concentration. The enzymatic reaction involves the condensation of **glycine** with **succinyl CoA** to produce δ-aminolevulinic acid. The reaction has an absolute requirement for **pyridoxal phosphate**. Two isoenzymes exist for ALA synthase; only the mRNA of the erythrocytic form contains an IRE. Mutations in the erythrocytic form cause a second type of X-linked sideroblastic anemia, where there is no ataxia.

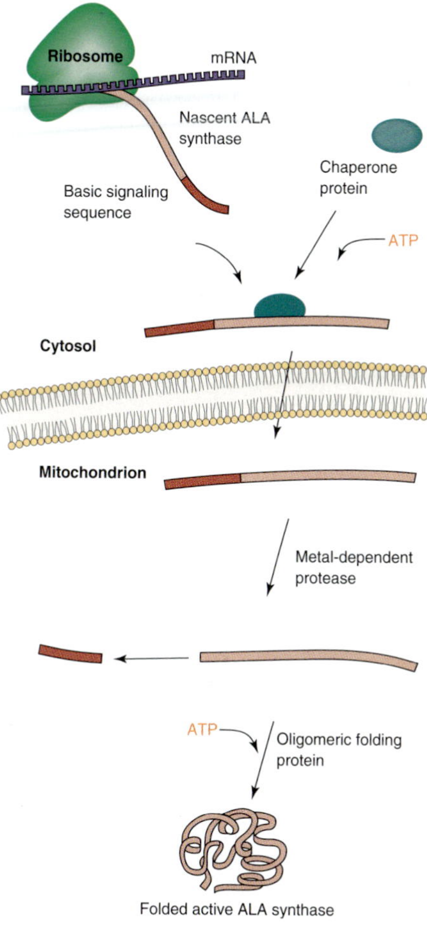

FIGURE 21.9
Synthesis of δ-aminolevulinic acid (ALA) synthase.

Aminolevulinic Acid Dehydratase

Aminolevulinic acid dehydratase (280 kDa) is cytosolic and consists of eight subunits, of which only four interact with the substrate. This protein interacts with the substrate to form a Schiff's base, but in this case the ϵ-amino group of a lysine residue binds to the ketonic carbon of the substrate molecule (Figure 21.10). Two molecules of ALA condense asymmetrically to form **porphobilinogen**. The ALA dehydrase is a zinc-containing enzyme and is very sensitive to inhibition by heavy metals, particularly

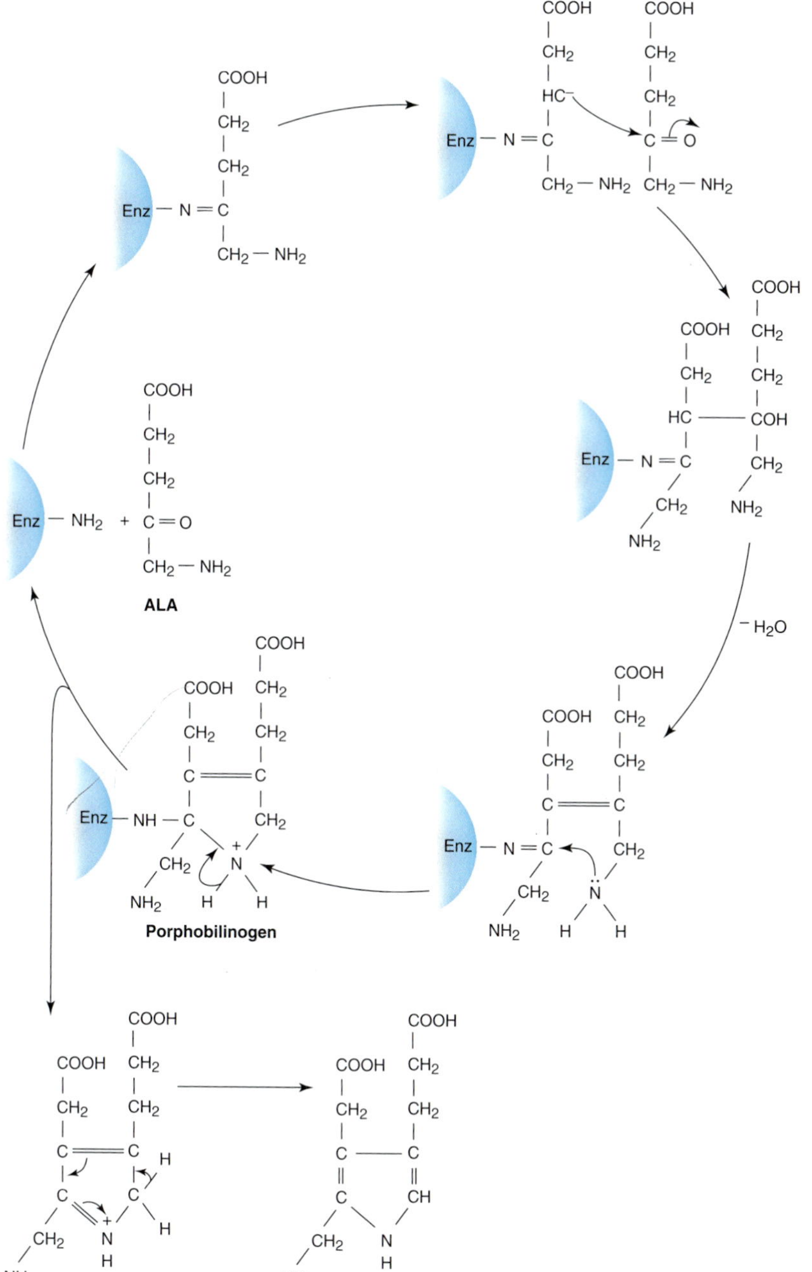

FIGURE 21.10
Synthesis of porphobilinogen.

lead. A characteristic of **lead poisoning** is the elevation of ALA in the absence of an elevation of porphobilinogen.

Porphobilinogen Deaminase and Uroporphyrinogen III Synthase

The synthesis of the porphyrin ring is a complicated process that has recently been defined. A sulfhydryl group on **porphobilinogen deaminase** forms a thioether bond with a porphobilinogen residue through a deamination reaction. Thereafter, five additional porphobilinogen residues are deaminated successively to form a linear hexapyrrole adduct with the enzyme. The adduct is cleaved hydrolytically to form both an **enzyme–dipyrrolomethane complex** and the linear tetrapyrrole, **hydroxymethylbilane**. The enzyme–dipyrrolomethane complex is then ready for another cycle to generate another tetrapyrrole. Thus, dipyrrolomethane is the covalently attached cofactor for the enzyme. Porphobilinogen deaminase has no ring-closing function; hydroxymethylbilane closes spontaneously in an enzyme-independent step to form uroporphyrinogen I if no additional factors are present. However, the deaminase is closely associated with a second protein, **uroporphyrinogen III synthase**, which directs the synthesis of the III isomer. The formation of the latter involves a spiro intermediate generated from hydroxymethylbilane; this allows the inversion of one of the pyrrole groups (see Figure 21.11). In the absence of the uroporphyrinogen III synthase, uroporphyrinogen I is synthesized slowly;

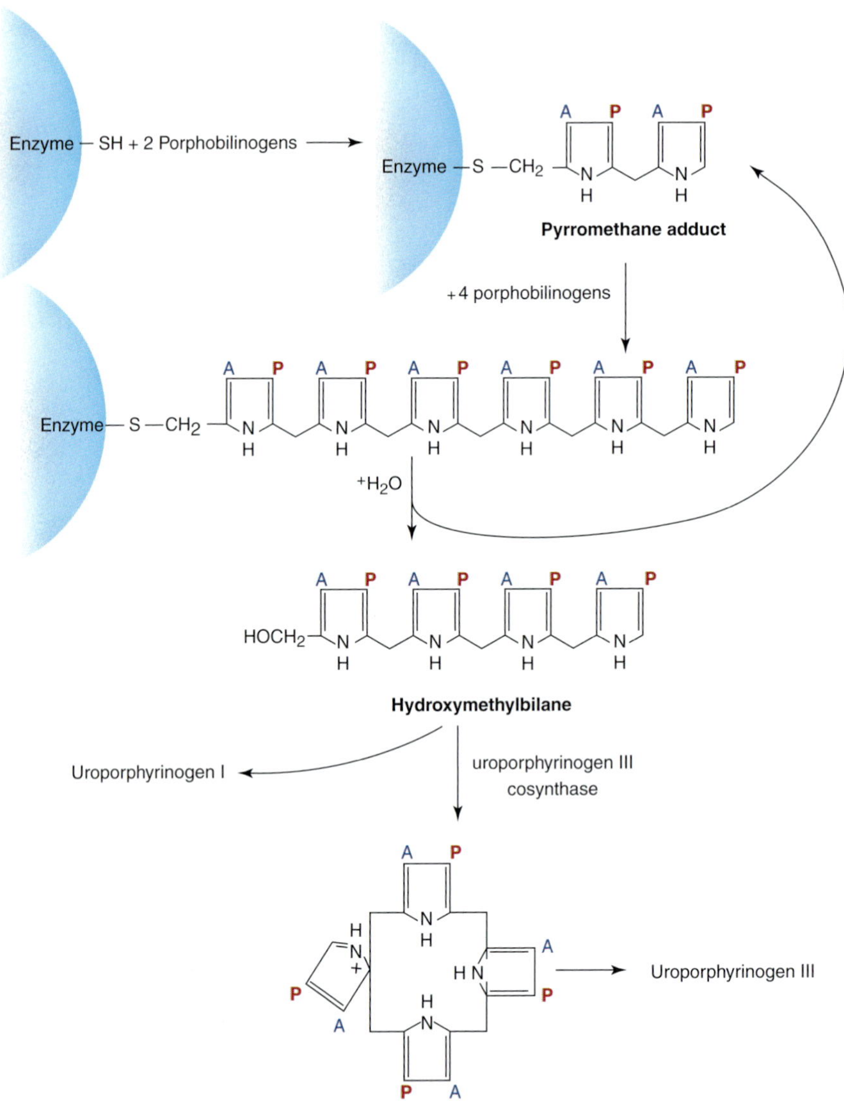

FIGURE 21.11

Synthesis of uroporphyrinogens I and III. Enzyme in blue is uroporphyrinogen I synthase.

in its presence, the III isomer is synthesized rapidly. A rare recessively inherited disease, erythropoietic porphyria, associated with marked cutaneous light sensitization, is due to an abnormality of reticulocyte uroporphyrinogen III synthase. Here large amounts of the type I isomers of uroporphyrinogen and coproporphyrinogen are synthesized in the bone marrow. Two isoenzymes exist for porphobilinogen deaminase due to alternative splicing of exon-1 or exon-2 to the rest of the mRNA.

Uroporphyrinogen Decarboxylase

Uroporphyrinogen decarboxylase acts on the side chains of uroporphyrinogens to form coproporphyrinogens. The protein catalyzes the conversion of both I and III isomers of uroporphyrinogen to the respective coproporphyrinogen isomers. Uroporphyrinogen decarboxylase is inhibited by iron salts. Clinically, the most common cause of porphyrin derangement is associated with patients who have a single gene abnormality for uroporphyrinogen decarboxylase, leading to 50% depression of the enzyme's activity. This disease, which shows cutaneous manifestations primarily with sensitivity to light, is known as **porphyria cutanea tarda**. The condition is not expressed unless patients either take drugs that cause an increase in porphyrin synthesis or drink large amounts of alcohol, leading to the accumulation of iron, which then acts to inhibit further the activity of uroporphyrinogen decarboxylase.

Coproporphyrinogen Oxidase

Coproporphyrinogen oxidase, a mitochondrial enzyme, is specific for type III coproporphyrinogen, not acting on the type I isomer. Coproporphyrinogen III enters mitochondria, and it is converted to protoporphyrinogen IX. The mechanism of action is not understood. A dominant hereditary disease associated with a deficiency of this enzyme leads to a form of hereditary hepatic porphyria, known as **hereditary coproporphyria.**

Protoporphyrinogen Oxidase

Protoporphyrinogen oxidase, another mitochondrial enzyme, generates protoporphyrin IX, which, in contrast to the other heme precursors, is very water-insoluble. Excess amounts of protoporphyrin IX that are not converted to heme are excreted by the biliary system into the intestinal tract. An autosomal dominant disease, **variegate porphyria**, is due to a deficiency of protoporphyrinogen oxidase.

Ferrochelatase

Ferrochelatase inserts ferrous iron into protoporphyrin IX in the final step of heme synthesis. Reducing substances are required for its activity. The protein is sensitive to the effects of heavy metals, especially **lead**, and, of course, to iron deprivation. In these latter instances, zinc instead of iron is incorporated to form a zinc–protoporphyrin IX complex. In contrast to heme, the zinc–protoporphyrin IX complex is brilliantly fluorescent and easily detectable in small amounts. Prokaryotic ferrochelatase does not contain a prosthetic group, whereas the mammalian enzyme contains an Fe_2S_2 group.

ALA Synthase Catalyzes Rate-Limiting Step of Heme Biosynthesis

ALA synthase controls the rate-limiting step of heme synthesis in all tissues. Succinyl CoA and glycine are substrates for a variety of reactions. The modulation of the activity of ALA synthase determines the quantity of the substrates that will be shunted into heme biosynthesis. Heme and hematin act both as a repressor of the synthesis of ALA synthase and as an inhibitor of its activity. Since heme resembles neither the substrates nor the product of the enzyme's action, it is probable that it acts at an allosteric site. Almost 100 different drugs and metabolites can cause induction of ALA synthase; for example, a 40-fold increase is noted in the rat after treatment with 3,5-dicarbethoxy-1,4-dihydrocollidine. The effect of pharmacological agents has led to the important clinical feature where some patients with certain kinds of porphyria have had exacerbations of their condition following the inappropriate administration of certain drugs (e.g., **barbiturates**). ALA dehydratase is also inhibited by heme; but this is of little physiological consequence, since the activity of ALA dehydrase is about 80-fold greater than that of ALA synthase, and thus heme-inhibitory effects are reflected first in the activity of ALA synthase.

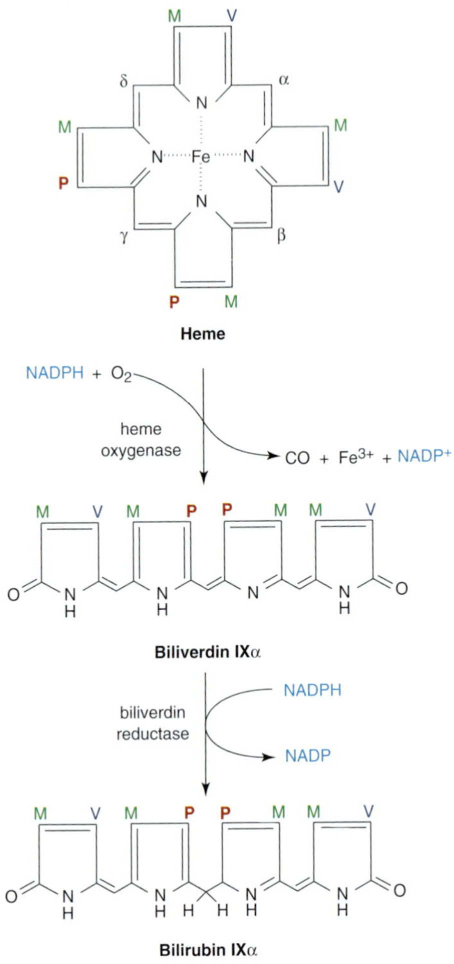

Glucose or one of its proximal metabolites inhibits heme biosynthesis by an unknown mechanism. This is of clinical relevance, since some patients manifest their porphyric state for the first time when placed on a very low caloric (and therefore glucose) intake. Other regulators of porphyrin metabolism include certain steroids. Steroid hormones (e.g., **oral contraceptive pills**) with a double bond in ring A between C4 and C5 atoms can be reduced by two different reductases. The product of 5α-reduction has little effect on heme biosynthesis; however, the product of 5β-reduction serves as a stimulus for the synthesis of ALA synthase.

21.7 HEME CATABOLISM

Catabolism of heme-containing proteins presents two requirements to the mammalian host: (a) a means of processing the hydrophobic products of porphyrin ring cleavage and (b) retention and mobilization of the contained iron so that it may be reutilized. Red blood cells have a life span of approximately 120 days. Senescent cells are recognized by changes in their membranes and engulfed by the reticuloendothelial system at extravascular sites. The globin chains denature, releasing heme into the cytoplasm. The globin is degraded to its constituent amino acids, which are reutilized for general metabolic needs.

Figure 21.12 depicts the events of heme catabolism. Heme is degraded primarily by an endoplasmic reticulum enzyme system in reticuloendothelial cells that requires molecular oxygen and NADPH. **Heme oxygenase** has two isomers; type I is substrate inducible and type II is constitutive. The enzyme catalyzes cleavage of the α-methene bridge, which joins the two pyrrole residues that contain vinyl substituents. The α-methene carbon is converted quantitatively to carbon monoxide. This is the only endogenous source of **carbon monoxide** in humans. A fraction of the carbon monoxide is released via the respiratory tract. Thus, the measurement of carbon monoxide in an exhaled breath provides an index of the quantity of heme that is degraded in an individual. The oxygen present in the carbon monoxide and in the newly derivatized lactam rings are generated entirely from molecular oxygen. The stoichiometry of the reaction requires 3 mol of oxygen for each ring cleavage. Heme oxygenase will only use heme as a substrate, with the iron possibly participating in the cleavage mechanism. Thus, free protoporphyrin IX is not a substrate. The linear tetrapyrrole **biliverdin IX** is formed by heme oxygenase. Biliverdin IX is reduced by **biliverdin reductase** to bilirubin IX. The products of heme oxygenase action are found to be cytoprotective (see Clin. Corr. 21.12).

FIGURE 21.12
Formation of bilirubin from heme. Greek letters indicate labeling of methene carbon atoms in heme.

Bilirubin Is Conjugated to Form Bilirubin Diglucuronide in Liver

Bilirubin is derived from senescent red cells but also from turnover of other heme-containing proteins, such as the cytochromes. Studies with labeled glycine as a precursor have revealed that an early-labeled bilirubin, with a peak within 1–3 h, appears very quickly after a pulsed administration of the labeled precursor. A larger amount of bilirubin appears much later at about 120 days, reflecting the turnover of heme in red blood cells. Early-labeled bilirubin can be divided into two parts: an early–early part, which reflects turnover of heme proteins in the liver, and a late–early part, which consists of both the turnover of heme-containing hepatic proteins and which is either poorly incorporated or easily released from red blood cells. The latter is a measurement of ineffective erythropoiesis and can be very pronounced in disease states such as pernicious anemia and the thalassemias.

Bilirubin is poorly soluble in aqueous solutions at physiological pH values. When transported in plasma, it is bound to serum albumin with an association constant greater

CLINICAL CORRELATION 21.12
Cytoprotective Role of Heme Oxygenase

Carbon monoxide (CO) and biliverdin are not merely passive products of heme oxygenase action. Biliverdin is an antioxidant; as such, it performs a significant role when heme oxygenase is induced by stress. CO, which structurally resembles nitric oxide (NO), like the latter compound, acts on smooth muscle; thus, as a vasodilator, it has been shown to have a protective effect, for instance in stroke victims. Like NO, CO acts apparently through cyclic GMP. The interaction with NO is complex; at times CO is complementary, and at other settings it is antagonistic to NO. Simplistically, CO appears generally to function protectively whereas NO, depending on the circumstance, can either protect or damage cells. To complete this function of heme oxygenase, bilirubin has been shown to inhibit the expression of inducible NO-synthase.

Source: Baranano, D. E. and Snyder, S. Neural roles for heme oxygenase: Contrasts to nitric oxide. *Proc. Natl. Acad. Sci.* USA 98:10996, 2001. Baranano, D. E., Rao, M., Ferris, C. D., et. al. Biliverdin reductase: A major physiologic cytoprotectant. *Proc. Natl. Acad. Sci. USA* 99:16095, 2002.

Chen, Y. M., Yet, S. F., and Perrella, M. A. Role of heme oxygenase in the regulation of blood pressure and cardiac function. *Exp. Biol. Med.* 228:447, 2003. Wang, W. W., Smith, D. L., and Zucker, S. D. Bilirubin inhibits iNOS expression and NO production in response to endotoxin in rats. *Hepatology* 40:424, 2004.

than 10^6 M^{-1}. Albumin contains one high-affinity site and another with a lesser affinity. At the normal albumin concentration of 4 g dL^{-1}, about 70 mg of bilirubin dL^{-1} of plasma can be bound on the two sites. However, bilirubin toxicity **(kernicterus)**, which is manifested by the transfer of bilirubin to membrane lipids, commonly occurs at concentrations greater than 25 mg dL^{-1}. This suggests that the weak affinity of the second site does not allow it to serve effectively in the transport of bilirubin. Bilirubin on serum albumin is rapidly cleared by the liver, where there is a free bidirectional flux of the tetrapyrrole across the sinusoidal-hepatocyte interface. Once in the hepatocyte, bilirubin is bound to several cytosolic proteins, of which only one has been well-characterized. The latter component, **ligandin**, is a small basic component making up to 6% of the total cytosolic protein of rat liver. Ligandin has been purified to homogeneity from rat liver and characterized as having two subunits (22 kDa and 27 kDa). Each subunit contains glutathione *S*-epoxide transferase activity, a function important in detoxification of aryl groups. The stoichiometry of binding is one bilirubin molecule per complete ligandin molecule.

Once in the hepatocyte the propionyl side chains of bilirubin are conjugated to form a diglucuronide (see Figure 21.13). The reaction utilizes uridine diphosphoglucuronate derived from the oxidation of uridine diphosphoglucose. In normal bile the diglucuronide is the major form of excreted bilirubin, with only small amounts of the monoglucuronide or other glycosidic adducts. **Bilirubin diglucuronide** is much more water-soluble than free bilirubin, and thus the transferase facilitates the excretion of the bilirubin into bile. Bilirubin diglucuronide is poorly absorbed by the intestinal mucosa. The glucuronide residues are released in the terminal ileum and large intestine by bacterial hydrolases; the released bilirubin is reduced to the colorless linear tetrapyrroles known as **urobilinogens**, which become oxidized to colored products known as **urobilins**, which are excreted in the feces. A small fraction of urobilinogen is reabsorbed by the terminal ileum and large intestine to be removed by hepatic cells and resecreted in bile. When urobilinogen is reabsorbed in large amounts in certain disease states, the kidney serves as a major excretory site.

Plasma bilirubin concentrations are 0.3–1 mg dL^{-1} in the normal state, and this is almost all in the unconjugated state. Conjugated bilirubin is expressed as **direct bilirubin** because it can be coupled readily with diazonium salts to yield azo dyes in the direct **van den Bergh reaction**. Unconjugated bilirubin is bound noncovalently to albumin, and it does not react until released by the addition of an organic solvent such as ethanol. This is the indirect van den Bergh reaction, and this measures the **indirect bilirubin** or the unconjugated bilirubin. Unconjugated bilirubin binds so tightly to

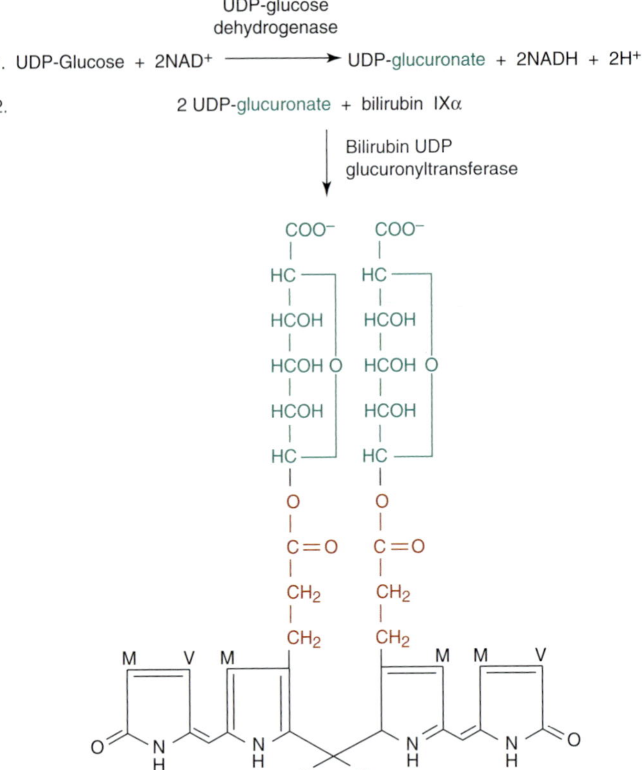

FIGURE 21.13
Biosynthesis of bilirubin diglucuronide.

serum albumin and lipid that it does not diffuse freely in plasma and therefore does not appear in urine. It has a high affinity for membrane lipids, which leads to the impairment of cell membrane function, especially in the nervous system. In contrast, conjugated bilirubin is relatively water-soluble, and elevations of conjugated bilirubin lead to high urinary concentrations with the characteristic deep yellow-brown color. The deposition of conjugated and unconjugated bilirubin in skin and the sclera gives the yellow to yellow-green color seen in patients with jaundice.

A third form of plasma bilirubin occurs only with hepatocellular disease in which a fraction of the bilirubin binds so tightly that it is not released from serum albumin by the usual techniques and is thought to be linked covalently to the protein. In some cases, up to 90% of total bilirubin can be in this covalently bound form.

The normal liver has a very large capacity to conjugate and excrete the bilirubin that is delivered. Therefore, hyperbilirubinemia due to excess heme destruction, as in hemolytic diseases, rarely leads to bilirubin levels that exceed 5 mg dL^{-1}, except in situations in which functional derangement of the liver is present (see Clin. Corr. 21.13). Thus, marked elevation of unconjugated bilirubin reflects primarily a variety of hepatic diseases, including those inherited and those acquired (see Clin. Corr. 21.14).

Elevations of conjugated bilirubin level in plasma are attributable to liver and/or biliary tract disease. In simple uncomplicated biliary tract obstruction, the major component is the diglucuronide form, which is released by the liver into the vascular compartment. Biliary tract disease may be extrahepatic or intrahepatic, the latter involving the canaliculi and biliary ductules (see Clin. Corr. 21.15).

Intravascular Hemolysis Requires Scavenging of Iron

In certain diseases, destruction of red blood cells occurs in the intravascular compartment rather than in the extravascular reticuloendothelial cells. In the former case, the appearance of free hemoglobin and heme in plasma could lead potentially to their excretion through the kidney with substantial loss of iron. To prevent this, specific plasma proteins are involved in scavenging mechanisms. Transferrin binds free iron

CLINICAL CORRELATION 21.13
Neonatal Isoimmune Hemolysis

Rh-negative women pregnant with Rh-positive fetuses develop antibodies to Rh factors. These antibodies cross the placenta to hemolyze fetal red blood cells. Usually this is not of clinical relevance until about the third Rh-positive pregnancy, in which the mother has had antigenic challenges from earlier babies. Antenatal studies reveal rising maternal levels of IgG antibodies against Rh-positive red blood cells, indicating that the fetus is Rh-positive. Before birth, placental transfer of fetal bilirubin occurs with excretion through the maternal liver. Because hepatic enzymes of bilirubin metabolism are poorly expressed in the newborn, infants may not be able to excrete the large amounts of bilirubin that can be generated from red cell breakdown. At birth these infants usually appear normal; however, the unconjugated bilirubin in the umbilical cord blood is elevated up to 4 $mg\,dL^{-1}$, due to the hemolysis initiated by maternal antibodies. During the next 2 days the serum bilirubin rises, reflecting continuing isoimmune hemolysis, leading to jaundice, hepatosplenomegaly, ascites, and edema. If untreated, signs of central nervous system damage can occur, with the appearance of lethargy, hypotonia, spasticity, and respiratory difficulty, constituting the syndrome of kernicterus. Treatment involves exchange transfusion with whole blood, which is serologically compatible with both the infant's blood and maternal serum. This is necessary to prevent hemolysis of the transfused cells. Additional treatment includes external phototherapy, which facilitates the breakdown of bilirubin. The problem can be prevented by treating Rh-negative mothers with anti-Rh globulin. These antibodies recognize the fetal red cells, block the Rh antigens, and cause them to be destroyed without stimulating an immune response in the mothers.

Source: Maur, H. M., Shumway, C. N., Draper, D. A., and Hossaini, A. Controlled trial comparing agar, intermittent phototherapy, and continuous phototherapy for reducing neonatal hyperbilirubinemia. *J. Pediatr.* 82:73, 1973. Bowman, J. J. Management of Rh-isoimmunization. *Obstet. Gynecol.* 52:1, 1978.

CLINICAL CORRELATION 21.14
Bilirubin UDP-Glucuronosyltransferase Deficiency

Bilirubin UDP-glucuronosyltransferase has two isoenzyme forms, derived from alternative mRNA splicing between variable forms of exon 1 and common exons 2, 3, 4, and 5. The latter exons define the part of the protein that binds the UDP-glucuronate, whereas the various exons 1 have defined specificities for either bilirubin or other acceptors, such as phenol. Two exons have bilirubin specificity, leading to two forms of bilirubin UDP-glucuronosyltransferase. Two major families of diseases are seen with deficiencies of the enzyme. Crigler–Najjar syndrome occurs in infants and is associated with extraordinarily high serum unconjugated bilirubin due to an autosomal recessive inheritance of mutations on both alleles in exons 2, 3, 4, or 5. Gilbert syndrome is also associated with a deficiency of the enzyme's activity, but only to about 25% of normal. The patients appear jaundiced but without other clinical symptoms. The major complication is an exhaustive search by the physician looking for some serious liver disease and failing to recognize the benign condition. Two different findings that may be restricted to different populations account for the condition. A dominant pattern of inheritance in Japan is noted with a mutation on only one allele. The 75% reduction of activity is ascribed to the fact that the enzyme exists as an oligomer, where mutant and normal monomers might associate to form hetero oligomers. The explanation is that not only is the mutant monomer inactive, but it forces conformational effects on the normal subunit, reducing its activity substantially. In contrast, in the Western world the condition is due largely to a homozygous expansion of the bases in the promoter region with less efficient transcription of the gene.

Source: Aono, S., Adachi, Y., Uyama, S., et al. Analysis of genes for bilirubin UDP-glucuronosyltransferase in Gilbert's syndrome. *Lancet* 345:958, 1995; Bosma, P. J., Chowdhury, J. R., Bakker, C., et al. The genetic basis of the reduced expression of bilirubin UDP-glucuronosyltransferase 1 in Gilbert's syndrome. *N. Engl. J. Med.* 333: 1171, 1995. Burchell, B. and Hume, R. Molecular genetic basis of Gilbert's syndrome. *J. Gastroenterol. Hepatol.* 14:960, 1999.

and thus permits its reutilization. Free hemoglobin, after oxygenation in the pulmonary capillaries, dissociates into α,β dimers, which are bound to a family of circulating plasma proteins, the **haptoglobins**, which have a high affinity for the oxyhemoglobin dimer. Since deoxyhemoglobin does not dissociate into dimers in physiological settings, it is not bound by haptoglobin. Two α,β-oxyhemoglobin dimers are bound by a haptoglobin

CLINICAL CORRELATION 21.15
Elevation of Serum Conjugated Bilirubin

Elevations of serum conjugated bilirubin are attributable to liver and/or biliary tract disease. In simple uncomplicated biliary tract obstruction, the major component of the elevated serum bilirubin is the diglucuronide form, which is released by the liver into the vascular compartment. Biliary tract disease may be extrahepatic or intrahepatic, the latter involving the canaliculi and biliary ductules.

Hydrophobic compounds have an affinity for membranes and thus enter cells. To remove these compounds, ATP-dependent membrane channels pump them into the circulation, where they can be bound by serum albumin and transferred to the liver. These pumps were initially described in the setting of cancer chemotherapy, where the agents are commonly hydrophobic and used in high concentrations. One kind of resistance to therapy was found to be due to the increased activity of the pumps in tumor cells, reducing intracellular drug concentrations. Accordingly, these pumps have been called MRPs (multidrug resistance proteins). At least six kinds are known. The property of hydrophobicity reduces the need of a specific pump for each compound since capture from a dilute aqueous medium is a minor issue, in contrast to hydrophilic compounds where channels have to be very specific. The MRPs serve also to transfer physiological hydrophobic compounds such as steroids from the adrenal gland and bilirubin from hepatocytes into bile. Dubin–Johnson syndrome is an autosomal recessive disease involving a defect in the biliary secretory mechanism of the liver. Excretion from the hepatocyte to the canaliculi relies upon MRP 2. In Dubin–Johnson syndrome, mutations occur in this protein (known also as cMOAT, canalicular multispecific organic anion transporter). Excretion through the biliary tract of a variety of (but not all) organic anions is affected. Retention of melanin-like pigment in the liver in this disorder leads to a characteristic gray-black color of this organ. A second heritable disorder associated with elevated levels of serum conjugated bilirubin is Rotor's syndrome. In this poorly defined disease, no hepatic pigmentation occurs.

Source: Iyanagi, E. Y. and Accoucheur, S. Biochemical and molecular disorders of bilirubin metabolism. *Biochim. Biophys. Acta* 1407:173, 1998. *Hepatology* 15: 1154, 1992. Tsugi, H., Konig, J., Rost, D., et al. Exon–intron organization of the human multidrug-resistance protein 2 (MRP 2) gene mutated in Dubin–Johnson syndrome. *Gastroenterology* 117:653, 1999.

molecule. Interesting studies have been made with rabbit antihuman-hemoglobin antibodies on the haptoglobin–hemoglobin interaction. Human haptoglobin interacts with a variety of hemoglobins from different species. The binding of human haptoglobin with human hemoglobin is not affected by the binding of rabbit antihuman-hemoglobin antibody. These studies suggest that haptoglobin binds to sites on hemoglobin that are highly conserved in evolution and therefore are not sufficiently antigenic to generate antibodies. The most likely site for the molecular interaction of hemoglobin and haptoglobin is the interface of the α- and β-globins of the tetramer that dissociates to yield α,β dimers. Sequence determinations have indicated that these contact regions are highly conserved in evolution.

Haptoglobins are α_2-globulins, synthesized in the liver. They consist of two pairs of polypeptide chains (α being the lighter and β the heavier). The α and β chains are derived from a single polypeptide that is cleaved to form the two different chains. The β chains are glycopeptides of 39 kDa and are invariant in structure; α chains are of several kinds. The haptoglobin chains are joined by disulfide bonds between the α and β chains and between the two α chains.

The haptoglobin–hemoglobin complex is too large to be filtered through the renal glomerulus. Free hemoglobin (appearing in renal tubules and in urine) will occur during intravascular hemolysis only when the binding capacity of circulating haptoglobin has been exceeded. Haptoglobin delivers hemoglobin to the reticuloendothelial cells. The heme in free hemoglobin is relatively resistant to the action of heme oxygenase, whereas the heme residues in an α,β dimer of hemoglobin bound to haptoglobin are very susceptible.

Measurement of serum haptoglobin is used clinically as an indication of the degree of intravascular hemolysis. Patients who have significant intravascular hemolysis have little or no plasma haptoglobin because of the removal of haptoglobin-hemoglobin complexes by the reticuloendothelial system. Haptoglobin levels can also be low in severe extravascular hemolysis, in which the large load of hemoglobin in the reticuloendothelial system leads to the transfer of free hemoglobin into plasma.

Free heme and hematin appearing in plasma are bound by a β-globulin, **hemopexin** (57 kDa), which binds one heme residue per molecule. Hemopexin transfers heme to liver, where further metabolism by heme oxygenase occurs. Normal plasma hemopexin contains very little bound heme, whereas in intravascular hemolysis, it is almost completely saturated by heme and is cleared with a half-life of about 7 h. In the latter, excess heme binds to albumin, with newly synthesized hemopexin serving as a mediator for the transfer of the heme from albumin to the liver. Hemopexin also binds free protoporphyrin.

BIBLIOGRAPHY

Anderson, K. E., Sassa, S., Bishop, D. F., and Desnick, R. J. Disorders of heme biosynthesis: X-linked sideroblastic anemia and the porphyrias. In: C. R. Scriver, A. L. Beaudet, W. S. Sly, and D. Valle (Eds.), *The Metabolic and Molecular Bases of Inherited Disease*, Vol. II, 8th ed. New York: McGraw-Hill, 2001, p. 2991.

Battersby, A. R. The Bakerian Lecture, 1984. Biosynthesis of the pigments of life. *Proc. R. Soc. London B* 225:1, 1985.

Beutler, E. "Pumping" iron: The proteins. *Science* 306:2051, 2004.

Beutler, E., Bothwell, T. M., Charlton, R. W., and Motulsky, A. G. Hereditary hemochromatosis. In: C. R. Scriver, A. L. Beaudet, W. S. Sly, and D. Valle (Eds.), *The Metabolic and Molecular Bases of Inherited Disease*, Vol. II, 8th ed. New York: McGraw-Hill, 2001, p. 3127.

Beutler, E., Hoffbrand, A. V., and Cook, J. D. Iron deficiency and overload. *Hematology (Am. Soc. Hematol. Educ. Program)* 40, 2003.

Chowdhury, J. R., Wolkoff, A. W., Chowdhury, N. R., and Arias, I. M. Hereditary jaundice and disorders of bilirubin metabolism. In: C. R. Scriver, A. L. Beaudet, W. S. Sly, and D. Valle (Eds.), *The Metabolic and Molecular Bases of Inherited Disease*, Vol. II, 8th ed. New York: McGraw-Hill, 2001, p. 3063.

Eisenstein, R. S. Iron regulatory proteins and the molecular control of mammalian iron metabolism. *Annu. Rev. Nutr.* 20:627, 2000.

Fenton, W. A., Kashi, Y., Furtak, K., and Horwich, A. L. Residues in chaperonin GroEL required for polypeptide binding and release. *Nature (London)* 371:614, 1994.

Ferreira, G. C. Ferrochelatase binds the iron-responsive element present in the erythroid δ-aminolevulinate synthase mRNA. *Biochem. Biophys. Res. Commun.* 214:875, 1995.

Fleet, J. C. A new role for lactoferrin: DNA binding and transcription activation. *Nutr. Rev.* 53:226, 1995.

Fleming, R. E. and Bacon, B. R. Orchestration of Iron Metabolism. *N. Engl. J. Med.* 352:1741, 2005.

Hartl, F. U. Secrets of a double-doughnut. *Nature (London)* 371:557,1994.

Kappas, A., Sassa, S., Galbraith, R. A., and Nordmann, V. The porphyrias. C. R. Scriver, A. L. Beaudet, W. S. Sly, and D. Valle (Eds.), *The Metabolic and Molecular Bases of Inherited Disease*, Vol. II, 7th ed. New York: McGraw-Hill, 1995, p. 2103.

Leong, W. I. and Lonnerdal, B. Hepcidin, the recently identified peptide that appears to regulate iron absorption. *J. Nutr.* 134:1, 2004.

Lustbader, J. W., Arcoleo, J. P., Birken, S., and Greer, J. Hemoglobin-binding site on haptoglobin probed by selective proteolysis. *J. Biol. Chem.* 258:1227, 1983.

Nemeth, E., Roetto, A., Garozzo, G., et al. Hepcidin is decreased in TFR2-hemochromatosis. *Blood* 105:1803, 2005.

Osterman, J., Horwich, A. L., Neupert, W., and Hartl, F. -U. Protein folding in mitochondria requires complex formation with hsp60 and ATP hydrolysis. *Nature (London)*. 341:125, 1989.

Otterbein, L. E., Soares, M. P., Yamashita, K., et al. Heme oxygenase-1: unleashing the protective properties of heme. *Trends Immunol.* 24:449, 2003.

Pantopoulos, K. and Hentze, M. W. Nitric oxide signalling to iron-regulatory protein: Direct control of ferritin mRNA translation and transferrin receptor mRNA stability in transfected fibroblasts. *Proc. Natl. Acad. Sci. USA* 92:1267, 1995.

Perrella, M. A. and Yet, S. F. Role of heme oxygenase-1 in cardiovascular function. *Curr. Pharmacol. Des.* 9:2479, 2003.

Pietrangelo, A. Hereditary hemochromatosis—A new look at an old disease. *N. Engl. J. Med.* 350:2383, 2004.

Robson, K. I., Merryweather-Clarke, A. T., Cadet, E., et al. Recent advances in understanding hemochromatosis: A transition state. *J. Med. Genet.* 41:721, 2004.

Simovich, M., Hainsworth, L. N., Fields, P. A., et al. Localization of the iron transport proteins mobilferrin and DMT-1 in the duodenum: The surprising role of mucin. *Int. J. Hematol.* 74:32, 2003.

Theil, E. L., and Eisenstein, R. S. Combinatorial mRNA regulation: Iron regulatory proteins and Iso-iron responsive elements (isoIREs) *J. Biol. Chem*: 275:40659, 2000.

Weiss, G. and Goodnough, L. T. Anemia of Chronic Disease. *N. Engl. J. Med.* 352:1011, 2005.

Yamashiro, D. J., Tycko, B., Fluss, S. R., and Maxfield, F. R. Segregation of transferrin to a mildly acidic (pH 6.5) para-golgi compartment in the recycling pathway. *Cell* 37:789, 1984.

QUESTIONS | CAROL N. ANGSTADT

Multiple Choice Questions

1. In the intestinal absorption of iron:
   A. the presence of a reductant like ascorbate enhances the availability of the iron.
   B. the regulation of uptake occurs between the lumen and the mucosal cells.
   C. the amount of apoferritin synthesized in the mucosal cell is directly related to the need for iron by the host.
   D. iron bound tightly to a ligand, such as phytate, is more readily absorbed than free iron.
   E. low pH in the stomach inhibits absorption by favoring Fe^{3+}.

2. The biosynthesis of heme requires all of the following *except*:

A. propionic acid.
B. succinyl CoA.
C. glycine.
D. ferrous ion.

3. Lead poisoning would be expected to result in an elevated level of:
   A. aminolevulinic acid.
   B. porphobilinogen.
   C. protoporphyrin I.
   D. heme.
   E. bilirubin.

4. Ferrochelatase:
   A. is an iron-chelating compound.
   B. releases iron from heme in the degradation of hemoglobin.
   C. binds iron to sulfide ions and cysteine residues.
   D. is inhibited by heavy metals.
   E. is involved in the cytoplasmic portion of heme synthesis.

5. Heme oxygenase:
   A. can oxidize the methene bridge between any two pyrrole rings of heme.
   B. requires molecular oxygen.
   C. produces bilirubin.
   D. produces carbon dioxide.
   E. can use either heme or protoporphyrin IX as substrate.

6. Haptoglobin:
   A. helps prevent loss of iron following intravascular red blood cell destruction.
   B. levels in serum are elevated in severe intravascular hemolysis.
   C. inhibits the action of heme oxygenase.
   D. binds heme and hematin as well as hemoglobin.
   E. binds α,β-deoxyhemoglobin dimers.

Questions 7 and 8: A woman appears in an emergency room in an agitated state with severe abdominal pain and marked weakness in all her limbs. Usual laboratory tests are normal and no specific abnormality can be found. She is sedated with phenobarbital (a barbiturate) and observed over the next few hours. Her condition rapidly deteriorates and further study finds an elevated level of porphobilinogen in her urine. A diagnosis of acute intermittent porphyria is made. Enzyme abnormalities in this disease are a marked increase in ALA synthase and a 50% reduction in porphobilinogen deaminase.

7. Aminolevulinic acid synthase:
   A. requires NAD for activity.
   B. is allosterically activated by heme.
   C. synthesis is inhibited by steroids.
   D. is synthesized in mitochondria.
   E. synthesis can be induced by a variety of drugs.

8. Normally, porphobilinogen is an intermediate in the pathway to heme biosynthesis. There are many other intermediates. Uroporphyrinogen III:
   A. is synthesized rapidly from porphobilinogen in the presence of a cosynthase.
   B. does not contain a tetrapyrrole ring.
   C. is formed from coproporphyrinogen III by decarboxylation.
   D. is converted directly to protoporphyrinogen IX.
   E. formation is the primary control step in heme synthesis.

Questions 9 and 10: If an Rh-negative woman becomes pregnant with Rh-positive fetuses, she will develop antibodies to Rh factors. About the third Rh-positive pregnancy, the antibodies will cross the placenta and hemolyze fetal red blood cells. Because hepatic enzymes of bilirubin metabolism are not fully developed in the newborn, within a few days after birth the infant will develop jaundice, hepatosplenomegaly, ascites, and edema. If untreated, central nervous system damage can occur.

9. The substance deposited in skin and sclera in jaundice is:
   A. biliverdin.
   B. only unconjugated bilirubin.
   C. only direct bilirubin.
   D. both bilirubin and bilirubin diglucuronide.
   E. hematin.

10. Conjugated bilirubin is:
    A. transported in blood bound to serum albumin.
    B. deficient in Crigler–Najjar syndrome, a deficiency of a UDP-glucuronosyltransferase.
    C. reduced in serum in biliary tract obstruction.
    D. the form of bilirubin most elevated in hepatic (liver) disease.
    E. less soluble in aqueous solution than the unconjugated form.

Questions 11 and 12: Cox-2 inhibitor drugs (e.g., Celebrex) were developed, in part, to counter the tendency of most nonsteroidal anti-inflammatory drugs (e.g., aspirin) to cause gastrointestinal bleeding. Such bleeding can lead to iron-deficiency anemia because iron loss leads to low levels of heme and decreased globin synthesis. By the time clinical symptoms are sufficient to warrant medical attention, distinct changes in blood are seen.

11. A change that would be expected is:
    A. an increase in serum transferrin.
    B. an increase in apoferritin.
    C. no change in liver ferritin.
    D. an increase in percentage saturation of serum transferrin.
    E. an increase in hemosiderin.

12. In iron deficiency, the transcription of the ceruloplasmin gene increases fourfold. Ceruloplasmin:
    A. synthesizes ferredoxin.
    B. inhibits mobilization of iron from storage sites.
    C. converts Fe^{2+} to Fe^{3+}.
    D. stimulates the transcription of apoferritin.
    E. binds iron to transferrin.

Problems

13. In the early stages of iron deficiency, physiological changes occur although there are no clinical symptoms. The level and percent saturation of serum transferrin are relatively normal. As the iron deficiency progresses, these values change. How is serum iron level maintained in the early stages? What changes occur in the later stages?

14. Microorganisms require iron for replication and function. Infants are protected against gastrointestinal infections because of the way iron is handled in human milk. In individuals with an iron overload, normally nonpathogenic microorganisms can become pathogenic because of a breakdown in the normal handling of iron. What is the mechanism of iron handling that accounts for these two conditions?

ANSWERS

1. **A** Ascorbate facilitates reduction to the ferrous state and, therefore, dissociation from ligands and absorption. B: Substantial iron enters the mucosal cell regardless of need, but the amount transferred to the capillary beds is controlled. C: Iron bound to apoferritin is trapped in mucosal cells and not transferred to the host. D: Iron must dissociate from ligands for absorption. This is why the iron in spinach is not a good source of iron. E: Oxidation to Fe^{3+} is favored by higher pH.
2. **A** The organic portion of heme comes totally from glycine and succinyl CoA; the propionic acid side chain comes from the succinate. D: The final step of heme synthesis is the insertion of the ferrous ion.
3. **A** Lead inhibits ALA dehydratase so it results in accumulation of ALA. B–D: Synthesis of porphobilinogen and subsequent compounds is inhibited. Heme certainly would not be elevated, because lead also inhibits ferrochelatase. E: Bilirubin is a breakdown product of heme, not an intermediate in synthesis.
4. **D** This enzyme catalyzes the last step of heme synthesis, the insertion of Fe^{2+}, and is sensitive to the effects of heavy metals. A, B: It is an enzyme of synthesis. C: This describes an iron–sulfur protein. E: The last step of heme synthesis is a mitochondrial process.
5. **B** Oxygenases usually use O_2. A: The enzyme is specific for the methene between the two rings containing the vinyl groups (α-methene bridge). C, D: The products are biliverdin and CO; the measurement of CO in the breath is an index of heme degradation. E: Iron is necessary for activity.
6. **A** Haptoglobin is part of the scavenging mechanism to prevent urinary loss of heme and hemoglobin from intravascular degradation of red blood cells. B: Since the scavenged complex is taken up by the reticuloendothelial system, the haptoglobin levels in serum are low. C: Heme residues in the dimers bound to haptoglobin are more susceptible than free heme to oxidation by heme oxygenase. D: Heme and hematin are bound by a β-globulin, while haptoglobin is an α-globulin. E: Deoxyhemoglobin does not dissociate to dimers physiologically.
7. **E** The enzyme is induced in response to need as well as by drugs and metabolites. Giving the porphyria patient phenobarbital increased her already high ALA synthase. A: The mechanism involves a Schiff base with glycine, therefore the coenzyme is pyridoxal phosphate. B: Heme both allosterically inhibits and suppresses synthesis of the enzyme. C: One reduction product of catabolic steroids stimulates the synthesis. D: The gene for this enzyme is on nuclear DNA.
8. **A** In the absence of cosynthase, uroporphyrinogen I is synthesized slowly. B: The tetrahydropyrrole ring has formed by this point. C, D: The decarboxylation goes from uroporphyrinogen to coproporphyrinogen, which is then converted to protoporphyrinogen. E: The synthesis of aminolevulinic acid is the rate-limiting step.
9. **D** Both conjugated (direct) and unconjugated (indirect) bilirubins are deposited. They are coming from the breakdown of the red cells.
10. **B** Conjugated bilirubin is bilirubin diglucuronide. A: Unconjugated bilirubin is transported bound to albumin. C: Liver conjugates bilirubin but if the bile duct is obstructed it won't be excreted. D: Liver is responsible for the conjugation, which would be impaired in hepatic damage. E: The purpose of conjugation is to increase water solubility.
11. **A** This reflects a physiological adaption to absorb more iron from the gastrointestinal tract. B: This falls because it would compete with the transfer of iron to the deficient host. C: This reflects iron storage in liver, which is depleted early. D: Increase in percentage saturation is seen in iron overload. E: Hemosiderin represents amorphous iron deposition in iron overload.
12. **C** Ceruloplasmin is also called ferroxidase I. A: Ferredoxin is an iron–sulfur protein. B: Mobilization involves a reduction of iron followed by its reoxidation by ceruloplasmin. D, E: Ceruloplasmin is an enzyme catalyzing a specific reaction.
13. In the absence of iron, the storage forms of iron in liver and bone marrow will be depleted to maintain serum iron. The normal small amount of serum ferritin decreases, and what is there is mostly apoferritin with little iron bound. In the later stages, transferrin increases in an attempt to absorb more iron from the gastrointestinal tract, but its percentage saturation is lower than normal.
14. Normally, there is very little free iron present because iron-binding proteins are less than completely saturated. In milk, the protein is lactoferrin, which rapidly binds free iron. In serum, the protein is transferrin. Normally, there is excess iron-binding capacity; however, in iron overload, transferrin may be almost completely saturated. This makes small amounts of free iron available.

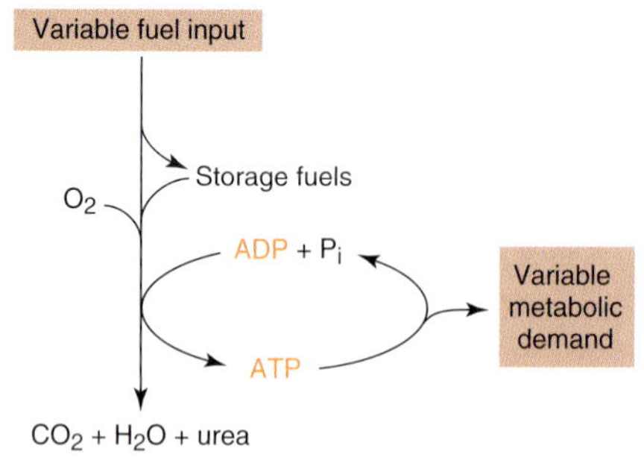

FIGURE 22.1

Humans can use a variable fuel input to meet a variable metabolic demand.

22.1 OVERVIEW

The interdependence of metabolic processes of the major tissues of the body will be stressed in this chapter. Not all of the major metabolic pathways operate in every tissue at any given time. Given the nutritional and hormonal status of a patient, we need to know qualitatively which pathways are functional and how they relate to one another.

The metabolic processes with which we are concerned are glycogenesis, glycogenolysis, gluconeogenesis, glycolysis, fatty acid synthesis, lipogenesis, lipolysis, fatty acid oxidation, glutaminolysis, tricarboxylic acid (TCA) cycle activity, ketogenesis, amino acid oxidation, protein synthesis, proteolysis, and urea synthesis. It is important to know (a) which tissues are most active in these various processes, (b) when these processes are most active, and (c) how these processes are controlled and coordinated in different metabolic states.

The best way to gain an understanding of the interrelationships of the pathways is to learn the changes in metabolism during the **starve–feed cycle** (Figure 22.1). This cycle allows a variable fuel and nitrogen intake to meet variable metabolic and anabolic needs. Feed refers to the intake of meals (the variable fuel input) after which the fuel is stored (as glycogen and triacylglycerol) to meet metabolic needs of fasting. An ATP cycle functions within the starve–feed cycle (Figure 22.1). Cells of the body die without the provision of continuous supply of energy for ATP synthesis to meet their needs.

Humans can consume food at a rate far greater than their basal caloric requirements, which allows them to survive from meal to meal. Unfortunately, an almost unlimited capacity to consume food is matched by an almost unlimited capacity to store it as triacylglycerol. In affluent countries, obesity is the consequence of excess food consumption and is the commonest form of **malnutrition** (Clin. Corr. 22.1). Other forms of malnutrition are more prevalent in developing countries (Clin. Corrs. 22.2 and 22.3). The regulation of food consumption is complex and not well understood. One important factor is leptin, a protein that is synthesized and secreted into the blood by adipocytes and that regulates energy expenditure and appetite through its effects on the hypothalamus (Clin. Corr. 22.1). The tight control needed is indicated by the calculation that eating two extra pats of butter ($\sim$100 cal) per day over caloric expenditures results in a weight gain of 10 lb per year. A weight gain of 10 lb may not sound excessive, but multiplied by 10 years it equals severe obesity!

22.2 STARVE–FEED CYCLE

In the Well-Fed State the Diet Supplies the Energy Requirements

Figure 22.2 shows the fate of glucose, amino acids, and fat obtained from food. Glucose passes from the intestinal epithelial cells via the portal vein to the liver. Amino acids are partially metabolized in the gut before being released into portal blood. Chylomicrons containing triacylglycerol are secreted by the intestinal epithelial cells into lymphatics. The lymphatics feed into the thoracic duct, which delivers the chylomicrons to the subclavian vein and thence to the rest of the body.

In the liver, dietary glucose can be converted into glycogen by glycogenesis, converted into pyruvate and lactate by glycolysis, or used in the pentose phosphate pathway for the generation of NADPH for synthetic processes. Pyruvate can be oxidized to acetyl CoA, which, in turn, can be converted into triacylglycerol or oxidized to CO_2 and water by the TCA cycle. Much of the dietary glucose passes through the liver to reach other organs, including brain, which is almost completely dependent on glucose for ATP generation, red blood cells and renal medulla, which can only carry out glycolysis, and adipose tissue, which primarily converts it into the glycerol moiety of triacylglycerol. Muscle also uses glucose, converting it to glycogen or using it in glycolysis and the TCA cycle. Lactate and pyruvate produced by glycolysis in other tissues are taken up by the liver and oxidized to CO_2 or converted to triacylglycerol. In the well-fed state, the liver uses glucose and does not engage in gluconeogenesis.

CLINICAL CORRELATION **22.1**

Obesity

Obesity is the most common nutritional problem in the United States. It is a risk factor in development of diabetes mellitus, hypertension, endometrial carcinoma, osteoarthritis, cirrhosis, gallstones, and cardiovascular diseases. In fact, the quartet of obesity, insulin resistance, dyslipidemia, and hypertension is called either syndrome X or the metabolic syndrome, and it contributes greatly to the high rate of cardiovascular death in Western countries. Obesity is easy to explain: An obese person has eaten more calories than he/she expended. The accumulation of massive amounts of body fat is not otherwise possible. For unknown reasons, the neural control of caloric intake to balance energy expenditure is abnormal. Rarely, obesity is secondary to a correctable disorder, such as hypothyroidism or Cushing's syndrome. The latter is the result of increased secretion of glucocorticoids, which causes fat deposition in the face and trunk, with wasting of the limbs, and glucose intolerance. These effects are due to increased protein breakdown in muscle and conversion of the amino acids to glucose and fat. Less commonly, tumors, vascular accidents, or maldevelopment of the nervous system hunger control centers in the hypothalamus cause obesity.

The obese (*ob/ob*) mouse was discovered in the 1950s, and the defective gene was cloned in 1994. This *ob* gene encodes a 146-amino-acid secreted protein (alternatively called OB protein or leptin, for its slimming effect) that is produced in adipocytes and detectable in blood. *ob/ob* mice have a nonsense mutation in the gene and produce no leptin. Injection of leptin into them causes increased energy expenditure and reduced eating, with marked weight loss. This effect on appetite is mimicked by intracerebroventricular injection. Leptin also reduced appetite and weight of normal mice. Obese humans do not generally have defective ob genes, and in fact they tend to have high blood levels of leptin. This suggests that their nervous system is insensitive to leptin, analogous to the insulin resistance seen in many diabetic patients.

In the most common type of obesity, the number of adipocytes does not increase, but the adipocytes just get large as they become engorged with triacylglycerols. If obesity develops before puberty, there may also be an increase in the number of adipocytes. In this case, the hyperplasia (increase in cell number) and hypertrophy (increase in cell size) contribute to the magnitude of the obesity. Obesity in men tends to be centered on the abdomen and mesenteric fat, while in women it is more likely be on the hips. The male pattern, characterized by a high waist to hip circumference ratio, is more predictive of premature coronary heart disease.

The only effective treatment of obesity is reduction in the ingestion or increase in the use of calories. Practically speaking, this means dieting, since even vigorous exercise such as running only consumes 10 kcal/min of exercise. Thus, an hour-long run (perhaps 5–6 miles) uses the energy present in about 2 candy bars. However, exercise programs can be useful to help motivate individuals to remain on their diets. There has been a recent surge of interest in low carbohydrate, high fat and protein diets referred to as the Atkins diet. This diet reduces carbohydrate intake to a low enough level to induce ketonemia. As weight loss progresses, carbohydrate is reintroduced until weight stabilizes. Several short-term studies have shown that this strategy is more effective in inducing weight loss than standard calorie restriction and better at reducing triglyceride levels without raising LDL cholesterol, but the effectiveness wanes over time.

Unfortunately, the body compensates for decreased energy intake with reduced formation of triiodothyronine and decrease in basal metabolic rate. Thus, there is a biochemical basis for the universal complaint that it is far easier to gain than to lose weight. What is more, about 95% of people who lose a significant amount of weight regain it within 1 year.

Source: Bray, G. D. Effect of caloric restriction on energy expenditure in obese patients. *Lancet* 2:397, 1969. Ahima R. S. and Flier, J. S. Leptin. *Annu. Rev. Physiol.* 62:413, 2000. Schwartz, M. W., Woods, S. C., Porte, D., Jr., Seeley, R. J., and Baskin, D. G. Central nervous system control of food intake. *Nature* 404:661, 2000: Jequier, E. and Tappy, L. Regulation of body weight in humans. *Physiol. Rev.* 79:451, 1999. Roberts, K., Dunn, K., Jean, S. K., and Lardinois, C. K. Syndrome X: Medical nutrition therapy. *Nutr. Rev.* 58:154, 2000. Kushner, R. F. and Weinsier, R. L. Evaluation of the obese patient. Practical considerations. *Med. Clin. North Am.* 84:387, 2000. Astrup, A., Larsen, T. M., and Harper, A. Atkins and other low-carbohydrate diets: Hoax or an effective tool for weight loss? *Lancet* 364:897, 2004. Foster, G. D, Wyatt, H. R., Hill, J. O., McGuckin, B. G., Brill, C., Mohammed, B. S., Szapary, P. O., Rader, D. J., Edman, J. S., and Klein, S. A randomized trial of a low-carbohydrate diet for obesity. *N. Engl. J. Med.* 348:2082, 2003.

Thus, the Cori cycle, conversion of glucose to lactate in peripheral tissues followed by conversion of lactate to glucose in liver, is interrupted in this state.

The intestinal cells use some dietary amino acids as an energy source but transport most of them into the portal blood for distribution. Liver removes some absorbed amino acids from the portal blood (Figure 22.2), but most pass through. This is especially important for essential amino acids, which are needed by all cells for protein synthesis. Liver metabolizes amino acids, but the K_m values of the enzymes involved are high, meaning that the amino acids have to be present in high concentration before significant

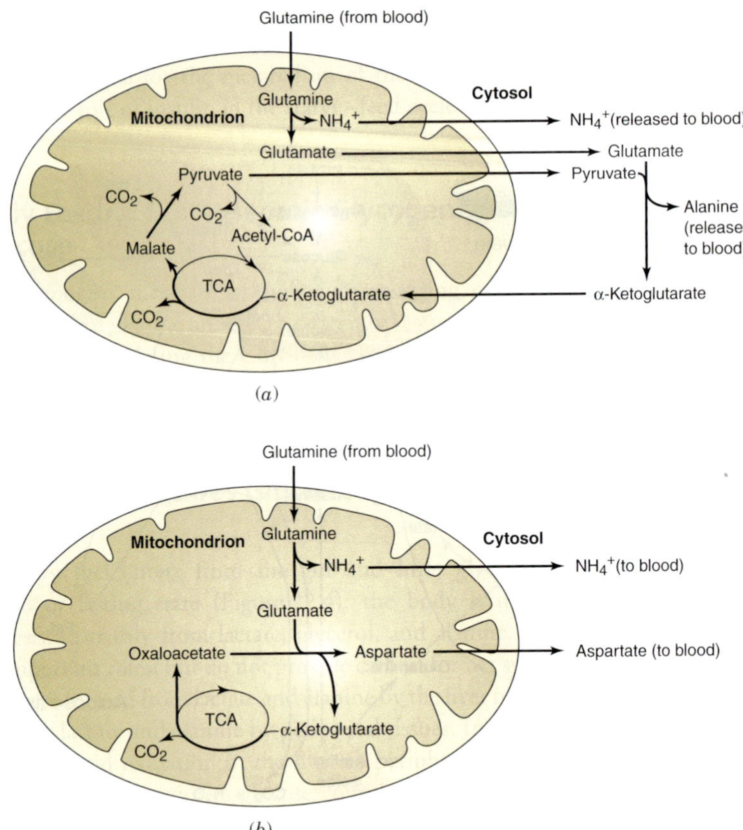

FIGURE 22.5

Glutamine catabolism by rapidly dividing cells. (*a*) Enterocytes. and (*b*) lymphocytes.

Part (*a*) redrawn from Duée, P.-H., Darcy-Vrillon, B., Blachier, F., and Morel, M.-T. Fuel selection in intestinal cells. *Proc. Nutr. Soc.* 54:83, 1995.

for glucose synthesis. Glutamate provides the two forms of nitrogen required for urea synthesis: ammonia from oxidative deamination by **glutamate dehydrogenase** and aspartate from transamination of oxaloacetate by **aspartate aminotransferase**. An important source of ammonia and precursors of ornithine such as citrulline is the gut mucosa (described in Section 22.4).

Because of low blood insulin levels during fasting, **lipolysis** is greatly activated in adipose tissue, raising the blood level of fatty acids, which are used in preference to glucose by many tissues. In heart and muscle, oxidation of fatty acids inhibits glycolysis and pyruvate oxidation. In liver, fatty acid oxidation provides most of the ATP needed for gluconeogenesis. Very little acetyl CoA generated by fatty acid oxidation in liver is oxidized completely. Rather, it is converted into ketone bodies by liver mitochondria. **Ketone bodies** (acetoacetate and β-hydroxybutyrate) are released into the blood and are a source of energy for many tissues. Like fatty acids, they are preferred to glucose by many tissues. Fatty acids are not oxidized by the brain because they cannot cross the blood–brain barrier. Once their blood concentration is high enough, ketone bodies enter the brain and serve as an alternative fuel. They are unable, however, to replace completely the brain's need for glucose. Ketone bodies may also suppress proteolysis and branched-chain amino acid oxidation in muscle and decrease alanine release. This decreases muscle wasting and reduces the amount of glucose synthesized in liver. As long as high ketone body levels are maintained by hepatic fatty acid oxidation, there is less need for glucose, less need for gluconeogenic amino acids, and less need for breaking down precious muscle tissue. This may be because insulin levels remain high enough to suppress partially muscle proteolysis as long as glucose levels remain high enough to stimulate some release of insulin from the pancreas. The interrelationships between the liver, muscle, and adipose tissue in providing glucose for the brain are shown in Figure 22.4. Liver synthesizes the glucose, muscle and gut supply the substrate (alanine), and adipose tissue supplies the ATP (via fatty acid oxidation in the liver) needed for hepatic gluconeogenesis. These relationships are disrupted in **Reye's syndrome** (Clin. Corr. 22.4) and by alcohol. This cooperation among major tissues is dependent on

CLINICAL CORRELATION 22.4
Reye's Syndrome

Reye's syndrome is a devastating but now rare illness of children that is characterized by brain dysfunction and edema (irritability, lethargy, and coma) and liver dysfunction (elevated plasma free fatty acids, fatty liver, hypoglycemia, hyperammonemia, and accumulation of short-chain organic acids). Hepatic mitochondria are specifically damaged, impairing fatty acid oxidation and synthesis of carbamoyl phosphate and ornithine (for ammonia detoxification) and oxaloacetate (for gluconeogenesis). Accumulation of organic acids has suggested that their oxidation is defective and that the CoA esters of some of them may inhibit specific enzymes, such as carbamoyl phosphate synthetase I, pyruvate dehydrogenase, pyruvate carboxylase, and the adenine nucleotide transporter. The use of aspirin by children with varicella was linked to the development of Reye's syndrome, and parents have been urged not to give aspirin to children with viral infections, which markedly reduced the incidence of the syndrome. Therapy for Reye's syndrome consists of measures to reduce brain edema and provision of glucose intravenously. Glucose administration prevents hypoglycemia and elicits a rise in insulin levels that may (a) inhibit lipolysis in adipose cells and (b) reduce proteolysis in muscles and the release of amino acids, which (c) reduces the deamination of amino acids to ammonia. Other syndromes caused by injury to hepatic mitochondria have been recognized in recent years. These include toxicity of an experimental hepatitis drug (fialuridine) and of several anti-retroviral nucleoside analogs. Injury to hepatic mitochondria results in lactic acidosis, hypoglycemia, and fatty liver, resulting in some cases to liver failure. Mitochondrial toxicity of the drugs may also underlie the development of peripheral neuropathy, pancreatitis, and cardiomyopathy. These agents appear to interfere with mitochondrial DNA synthesis by inhibiting DNA polymerase γ, resulting in accumulation of mutations in key proteins and RNAs encoded by mitochondrial DNA.

Source: Reye, R. D. K., Morgan, G., and Baval, J. Encephalopathy and fatty degeneration of the viscera, a disease entity in childhood. *Lancet* 2:749, 1963. Brusilow, S.W. Hyperammonemic encephalopathy. *Medicine (Baltimore)* 81:240, 2002. Lewis, W., Day, B. J., and Copeland, W.C. Mitochondrial toxicity of NRTI antiviral drugs: an integrated cellular perspective. *Nat. Rev. Drug Discov.* 2:812, 2003.

the appropriate blood hormone levels. Glucose levels are lower in fasting, reducing the secretion of insulin but favoring release of glucagon from the pancreas and epinephrine from the adrenal medulla. Fasting also reduces formation of triiodothyronine, the active form of thyroid hormone, from thyroxine. This reduces the daily basal energy requirements by up to 25%. This response is useful for survival, but makes weight loss more difficult than weight gain (Clin. Corr. 22.1).

In the Early Refed State Glycogen Is Formed by the Indirect Pathway

Triacylglycerol is metabolized as described for the well-fed state. In contrast, the liver extracts glucose poorly; in fact, it remains in the gluconeogenic mode for a few hours after feeding. Rather than providing blood glucose, however, hepatic gluconeogenesis provides glucose 6-phosphate for glycogenesis. This means that liver glycogen is not entirely repleted after a fast by direct synthesis from blood glucose. Rather, glucose is catabolized in peripheral tissues to lactate, which is converted in the liver to glycogen by gluconeogenesis, that is, indirectly:

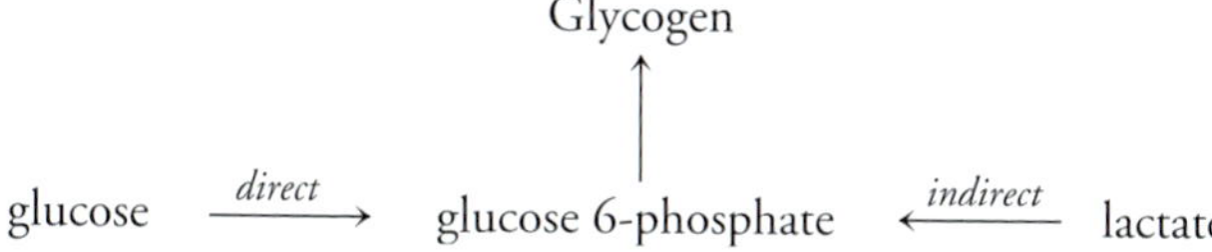

Gluconeogenesis from amino acids entering from the gut also plays an important role in reestablishing liver glycogen levels by the indirect pathway. After the rate of glucose synthesis declines, liver glycogen is sustained by direct synthesis from blood glucose.

Important Interorgan Metabolic Interactions

The intestinal epithelium converts glutamine to citrulline (Figure 22.6); it is the only tissue that expresses an ATP-dependent glutamate reductase necessary for this

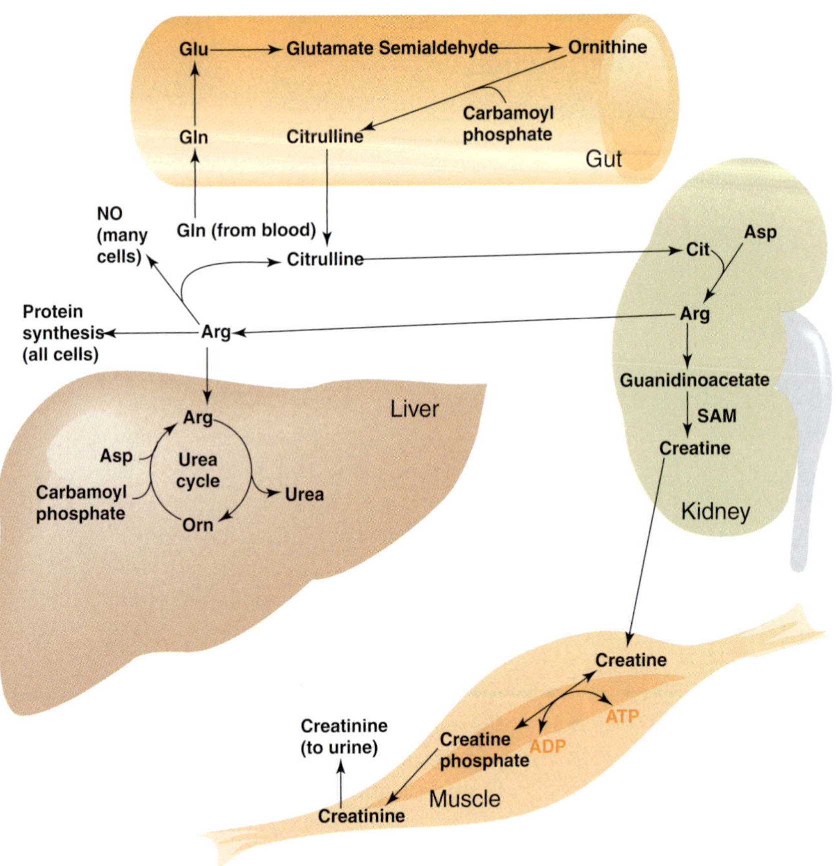

FIGURE 22.6

Gut and kidney function together in synthesis of arginine from glutamine. Abbreviations: Cit, citrulline; Arg, arginine; Asp, aspartate; Gln, glutamine; Glu, glutamate; NO, nitric oxide; Orn, ornithine; SAM, *S*-adenosylmethionine.

conversion:

$$\text{Glutamate} + \text{NADPH} + \text{H}^+ + \text{ATP} \rightarrow \text{glutamate semialdehyde} + \text{NAD}^+ + \text{ADP} + \text{P}_i$$

Citrulline released from the gut is converted to arginine by the kidney, which can be converted to creatine or released into the blood. The liver uses arginine to generate ornithine, which expands the capacity of the urea cycle during periods of increased protein intake. The liver irreversibly converts ornithine into glutamate:

$$\text{Ornithine} \xrightarrow{\text{Transamination}} \text{Glutamate semialdehyde} \xrightarrow{\text{Oxidation}} \text{Glutamate}$$

Depletion of ornithine in this way inhibits urea synthesis. Replenishment of ornithine is completely dependent upon a source of arginine. Thus, urea synthesis in the liver is dependent upon citrulline and arginine produced by the gut and kidney, respectively. Arginine is also used by many cells for the production of nitric oxide (NO) (Figure 22.6) (see p. 433). The arginine generated from citrulline in the kidney is used to synthesize creatine (Figure 22.6). Glycine transamidinase (GTA) generates guanidinoacetate from arginine and glycine. GTA is found predominantly in renal cortex, pancreas, and liver. After methylation using the methyl donor *S*-adenosylmethionine (SAM), creatine is formed. This is quantitatively the most important use of SAM in the body. One to two grams of creatine are synthesized per day. Creatine then circulates to other tissues and accumulates in muscle, where it serves as a high-energy reservoir when phosphorylated to creatine phosphate. This undergoes nonenzymatic conversion to creatinine, which is released to the bloodstream and removed by the kidney. Excretion of creatinine is used clinically as a measure of both muscle mass and renal function. Glutathione (GSH) is important in detoxification of endogenously generated peroxides and exogenous chemical compounds (see p. 578). Liver is a major synthesizer of GSH from glutamate, cysteine, and glycine (Figure 22.7). Synthesis is limited by the availability of cysteine.

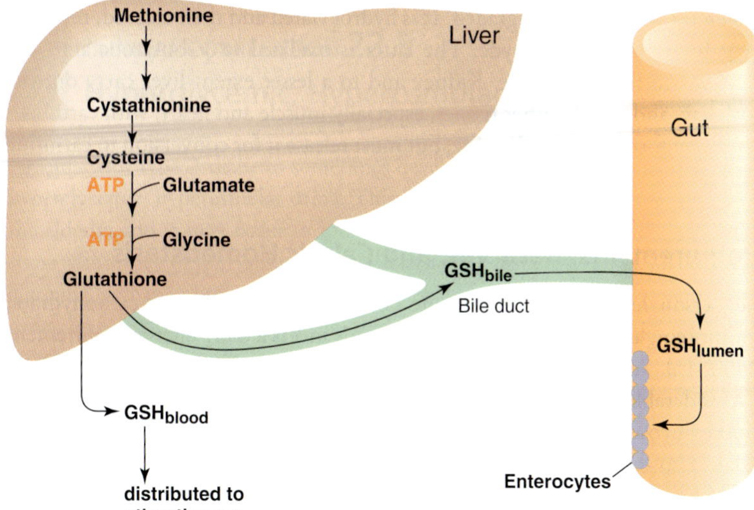

FIGURE 22.7
Liver provides glutathione for other tissues.

Plasma cysteine is not taken up well by liver, which uses dietary methionine to form cysteine via the cystathionine pathway (see p. 768). Hepatic **GSH** is released to the bloodstream and bile. Kidney removes a substantial amount of plasma GSH. Enterocytes may be able to take up biliary-excreted GSH from the intestinal lumen. Release to plasma is the same in fed and fasting states, providing a stable source of this compound and its constituent amino acids, especially cysteine, for most tissues of the body.

Carnitine is derived from lysyl residues of various proteins, which are *N*-methylated using SAM to form trimethyllysyl residues (Figure 22.8; see p. 780). **Trimethyllysine**

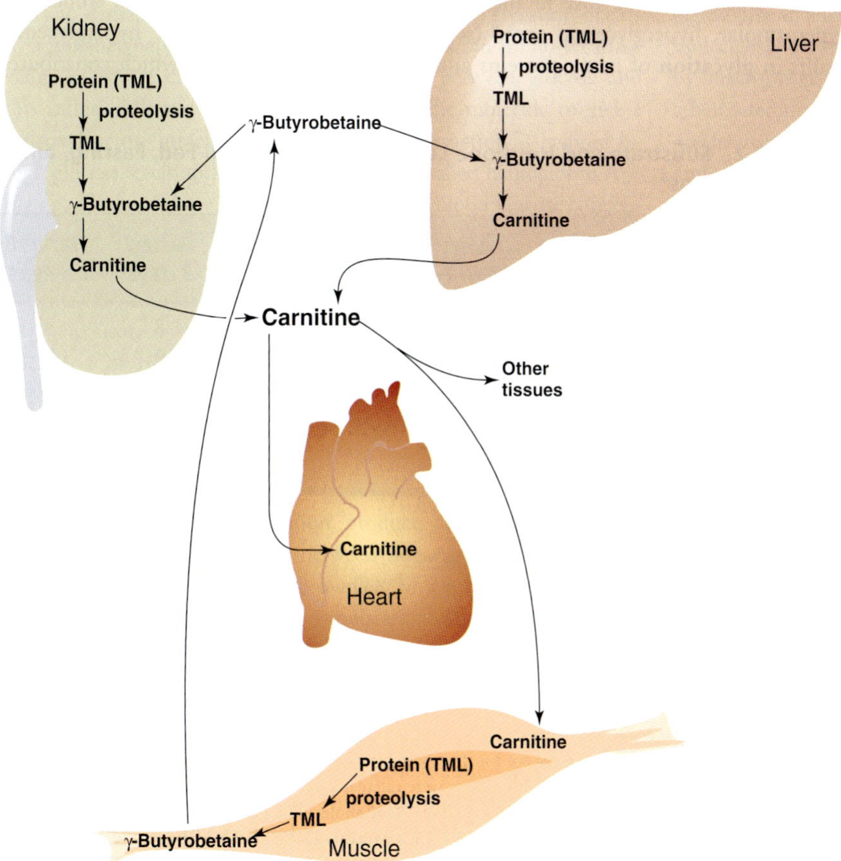

FIGURE 22.8
Kidney and liver provide carnitine for other tissues. Abbreviations: (TML), trimethyllysyl residues in protein molecules; TML, free trimethyllysine.

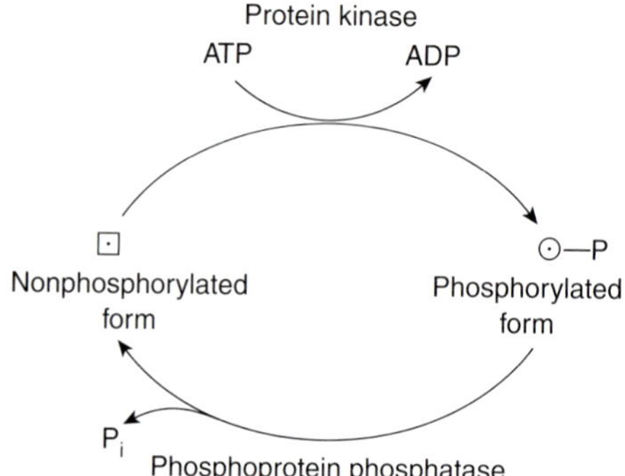

FIGURE 22.13

Regulation of the activity of key enzymes by covalent modification. The symbols ⊡ and ⊙-P designate the nonphosphorylated and phosphorylated states of enzymes, respectively.

Covalent Modification Regulates Key Enzymes

The activity of many enzymes is modified by **covalent modification**, especially by phosphorylation of serine residues (Figure 22.13 and see p. 604). Some important points about regulation of enzyme by this type of control are as follows: (a) Some enzymes undergo phosphorylation on one or more serine or threonine residues by protein kinases that are subject to regulation; (b) dephosphorylation of enzymes is carried out by phosphoprotein phosphatases that are subject to regulation; (c) phosphorylation status affects the catalytic activities of the enzymes; (d) some enzymes are active in the dephosphorylated state, others in the phosphorylated state; (e) cAMP signals the phosphorylation of many enzymes by activating protein kinase A (cAMP-dependent protein kinase) (Figure 22.14); (f) glucagon and α-adrenergic agonists (epinephrine) activate protein kinase A by increasing cAMP levels (Figure 22.14); (g) AMP also signals the phosphorylation of many enzymes by activating AMPK (AMP-activated protein kinase) (Figure 22.15); (h) stress (extra work) imposed on a cell that causes energy deprivation results in an increase in AMP concentration and activation of AMPK (Figure 22.15); (i) insulin (p. 606) opposes the action of protein kinase A and AMPK by activating phosphoprotein phosphatases; (j) metabolic enzymes are dephosphorylated in the fed state because the insulin/glucagon ratio is high and both cAMP and AMP are low; (k) metabolic enzymes are phosphorylated in the fasted state because the insulin/glucagon ratio is low and cAMP is increased (Figure 22.14); and (l) metabolic enzymes are phosphorylated in energy-deprived states because AMP levels are increased (Figure 22.15).

FIGUR
Contro
in the

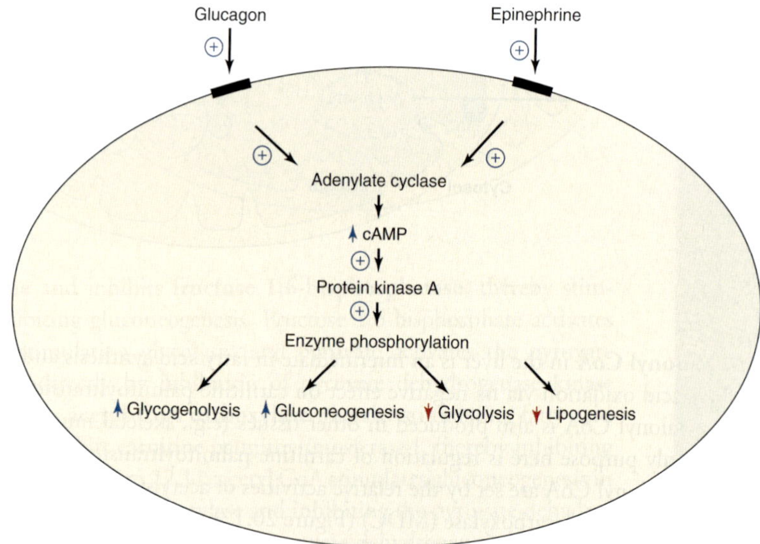

FIGURE 22.14

Glucagon and epinephrine stimulate glycogenolysis and gluconeogenesis and inhibit glycolysis and lipogenesis in liver.

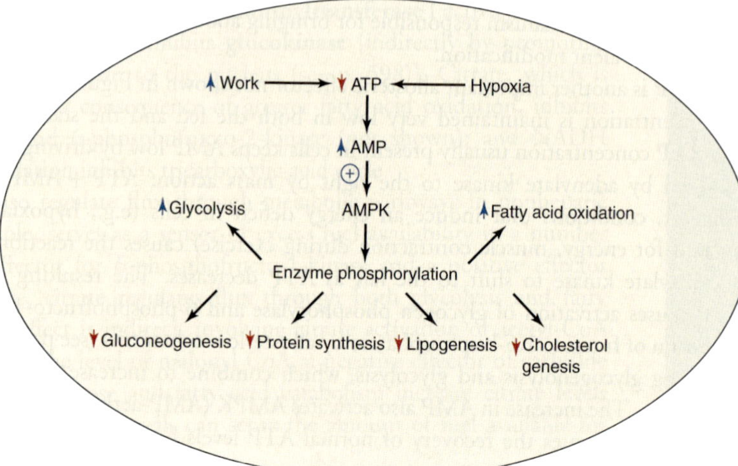

FIGURE 22.15

Activation of AMPK shuts down ATP-requiring processes and stimulates ATP-producing processes

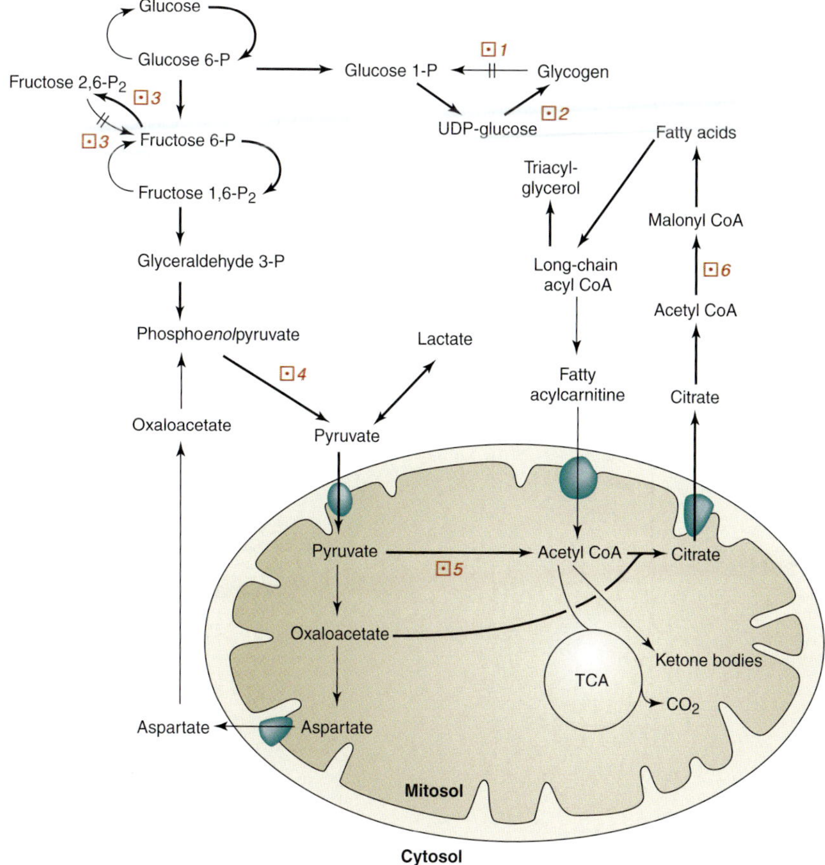

FIGURE 22.16

Control of hepatic metabolism by covalent modification in the well-fed state. Dephosphorylated mode is indicated by ⊡. The enzymes subject to covalent modification are: 1, glycogen phosphorylase; 2, glycogen synthase; 3, 6-phosphofructo-2-kinase/fructose-2, 6-bisphosphatase (bifunctional enzyme); 4, pyruvate kinase, 5, pyruvate dehydrogenase; and 6, acetyl-CoA carboxylase.

The hepatic enzymes subject to covalent modification are all relatively dephosphorylated in well-fed animals (Figure 22.16). Insulin is high but glucagon is low in the blood, resulting in low cAMP levels in the liver. A resulting low-protein kinase A activity and high-phosphoprotein phosphatase activity induce the dephosphorylated state of the enzymes (glycogen synthase, glycogen phosphorylase, phosphorylase kinase, 6-phosphofructo-2-kinase/fructose 2,6-bisphosphatase, pyruvate kinase, and acetyl-CoA carboxylase) regulated by covalent modification in the liver. Although not regulated by protein kinase A, the phosphorylation state of the pyruvate dehydrogenase complex changes in parallel with the enzymes identified in Figure 22.16 because of low activity of pyruvate dehydrogenase kinase in the well-fed state. Glycogen synthase, 6-phosphofructo-2-kinase, pyruvate kinase, pyruvate dehydrogenase, and acetyl-CoA carboxylase are active in the dephosphorylated state, whereas glycogen phosphorylase, phosphorylase kinase (not identified in Figure 22.16), and fructose 2,6-bisphosphatase are all inactive. As a consequence of the dephosphorylated state of these enzymes, glycogenesis, glycolysis, and lipogenesis are greatly favored in the liver of the well-fed animal, whereas the opposing pathways (glycogenolysis, gluconeogenesis, and ketogenesis) are inhibited.

As shown in Figure 22.17, the hepatic enzymes subject to covalent modification are all relatively phosphorylated in the fasting animal. Insulin level in the blood is low but glucagon is high, resulting in high cAMP levels in the liver. This activates protein kinase A and inactivates phosphoprotein phosphatase. The net effect is a greater degree of phosphorylation of the regulatory enzymes than in the well-fed state. Three enzymes (glycogen phosphorylase, phosphorylase kinase, and the fructose 2,6-bisphosphatase) are activated as a consequence of phosphorylation. All other enzymes subject to covalent modification are inactivated. As a result, glycogenolysis, gluconeogenesis, and ketogenesis dominate and glycogenesis, glycolysis, and lipogenesis are shut down.

Metabolic enzymes are also regulated through phosphorylation by **AMP-activated protein kinase (AMPK)** as summarized in Figures 22.15 and 22.18. AMPK is activated

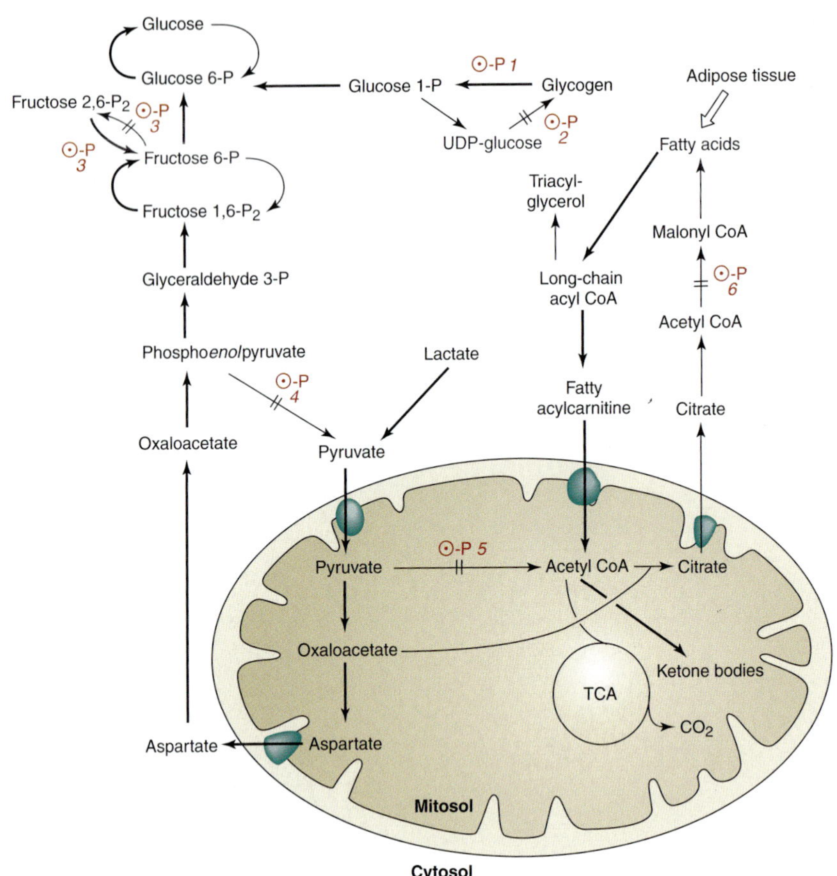

FIGURE 22.17

Control of hepatic metabolism by covalent modification in the fasting state. Phosphorylated mode is indicated by ⊙-P. Numbers refer to the same enzymes as in Figure 22.16.

by an increase in AMP concentration, which is set by the energy status of cells. Under conditions of high-energy demand that decreases ATP and therefore increases AMP, AMPK turns off anabolic pathways that consume ATP and turns on catabolic pathways that generate ATP. As shown in Figure 22.18, AMPK inhibits fatty acid synthesis by phosphorylating acetyl-CoA carboxylase, inhibits triacylglycerol synthesis by phosphorylating glycerol-3-phosphate acyltransferase, inhibits cholesterol synthesis by phosphorylating 3-hydroxy-3-methylglutaryl-CoA reductase, and inhibits glycogen synthesis by phosphorylating glycogen synthase. AMPK also inhibits protein synthesis (not shown) by phosphorylating components of the mTOR (mammalian target of rapamycin) pathway that activate mRNA translation. The strategy is to minimize use of ATP by all pathways not immediately essential for cell survival. At the same time, AMPK promotes ATP generation by fatty acid oxidation by lowering the concentration of malonyl CoA, a potent allosteric inhibitor of carnitine palmitoyl-transferase 1 (see p. 681). This is achieved by phosphorylation-mediated inactivation of acetyl-CoA carboxylase and activation of malonyl-CoA decarboxylase by AMPK.

Adipose tissue responds almost as dramatically as liver to the starve-feed cycle. Pyruvate kinase, the pyruvate dehydrogenase complex, acetyl-CoA carboxylase, and hormone-sensitive lipase (not found in liver) are dephosphorylated in adipose tissue in the well-fed state. The first three enzymes are active while hormone-sensitive lipase is inactive in this state. A high insulin level in the blood and a low cAMP concentration in adipose tissue are important determinants of the phosphorylation state of these enzymes, which favors lipogenesis in the well-fed state. During fasting, a decrease in the insulin level and an increase in epinephrine shut down lipogenesis and activate lipolysis because of phosphorylation of these enzymes. In this manner, adipose tissue is transformed from a fat storage tissue into a source of fatty acids for oxidation in other tissues and glycerol for gluconeogenesis in the liver.

Covalent modification of enzymes in skeletal muscle is also important in the starve–feed cycle. Glycogen synthase, glycogen phosphorylase, pyruvate dehydrogenase

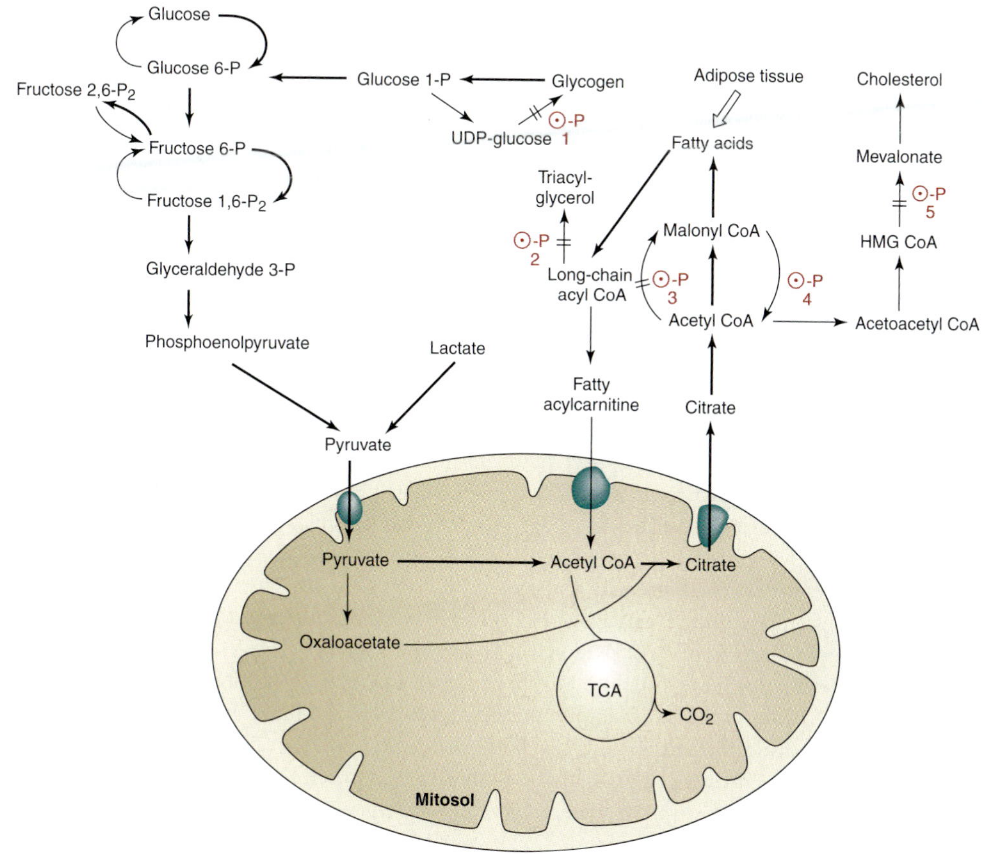

FIGURE 22.18

Control of hepatic metabolism by AMPK-mediated phosphorylation during energy deprivation. Phosphorylated mode is indicated by ⊙-P. The enzymes phosphorylated by AMPK are: 1, glycogen synthase; 2, glycerol-3-phosphate acyltransferase; 3, acetyl-CoA carboxylase; 4, malonyl-CoA carboxylase; and 5, 3-hydroxy-3-methylglutaryl-CoA (HMG-CoA) reductase.

complex, acetyl-CoA carboxylase, and malonyl-CoA decarboxylase are dephosphorylated in the fed state. This, along with insulin-mediated stimulation of glucose uptake by glucose transporter 4 (GLUT4) (see p. 474), favors glucose uptake, glycogen synthesis, and complete oxidation by skeletal muscle. The increase in malonyl-CoA favored by the combination of an active acetyl-CoA carboxylase and an inactive malonyl-CoA decarboxylase limits fatty acid oxidation at the level of carnitine palmitoyltransferase 1. During fasting, conservation of glucose, lactate, alanine, and pyruvate is crucial for survival. Tissues of the body that can use alternative fuels invariably shut down their use of glucose and three carbon compounds that can be used for the synthesis of glucose. Increased availability of fatty acids and enzymatic activity for oxidation spares glucose in the starved state. The latter is due to decreased malonyl-CoA levels and therefore less inhibition of carnitine palmitoyltransferase 1 induced by phosphorylation-mediated inactivation of acetyl-CoA carboxylase and activation of malonyl-CoA decarboxylase. Regulation of the utilization of glucose by fatty acid catabolism has been named the glucose–fatty acid cycle. Inactivation of the pyruvate dehydrogenase complex in skeletal muscle by phosphorylation is the key to conservation of glucose and three carbon compounds for hepatic gluconeogenesis during fasting. This is mediated by pyruvate dehydrogenase kinase (see p. 540), which is induced to a higher level of expression and stimulated to greater activity by its allosteric effectors acetyl CoA and NADH produced by fatty acid oxidation.

Exercise induces profound effects upon metabolic pathways in skeletal muscle. The energy demand of muscle contraction increases AMP and activates AMPK. AMPK stimulates the transport of vesicles bearing GLUT4 to the plasma membrane for greater glucose uptake and catabolism for ATP production. AMPK-mediated phosphorylations also decrease malonyl CoA by inactivating acetyl-CoA carboxylase and activating malonyl-CoA decarboxylase (Figures 22.15 and 22.18). Less malonyl CoA results in greater carnitine palmitoyltransferase 1 activity and fatty acid oxidation to help meet muscle's need for ATP for contraction.

Covalent modification, like allosteric effectors and substrate supply, is a short-term regulatory mechanism, operating on a minute-to-minute basis. On a longer time scale, enzyme activities are controlled at the level of expression, most frequently by the rate of gene transcription.

Changes in the Amounts of Key Enzymes Provide Long-Term Adaptation

Whereas allosteric effectors and covalent modification affect either the K_m or V_{max} of an enzyme, enzyme activity is also regulated by the rate of its synthesis or degradation and, thus, the quantity of enzyme in a cell. For example, in a person maintained in a well-fed or overfed condition the liver has increased amounts of enzymes involved in triacylglycerol synthesis (see Figure 22.19). Many enzymes are induced by an increase in the insulin/glucagon ratio and glucose in the blood. These include glucokinase, 6-phospho-1-fructokinase, and pyruvate kinase for faster rates of glycolysis; glucose 6-phosphate dehydrogenase, 6-phosphogluconate dehydrogenase, and malic enzyme to provide greater quantities of the NADPH required for synthesis of fatty acids and cholesterol; citrate cleavage enzyme, acetyl-CoA carboxylase, fatty acid synthase, and Δ^9-desaturase for fatty acid synthesis; 3-hydroxy-3-methylglutaryl CoA reductase for cholesterol synthesis; and glycerol 3-phosphate acyltransferase for triacylglycerol and phospholipids synthesis. At the same time, phosphoenolpyruvate carboxykinase, pyruvate dehydrogenase kinase, pyruvate carboxylase, fructose 1,6-bisphosphatase, glucose 6-phosphatase, and some aminotransferases are decreased in amount.

In fasting, the lipogenic enzymes decrease dramatically in quantity, while those that favor gluconeogenesis (glucose-6-phosphatase, fructose 1,6-bisphosphatase, phosphoenolpyruvate carboxykinase, pyruvate carboxylase, and various amino transferases) are remarkably induced (Figure 22.20). Starvation also induces pyruvate dehydrogenase kinase, responsible for phosphorylation and inactivation of the pyruvate dehydrogenase

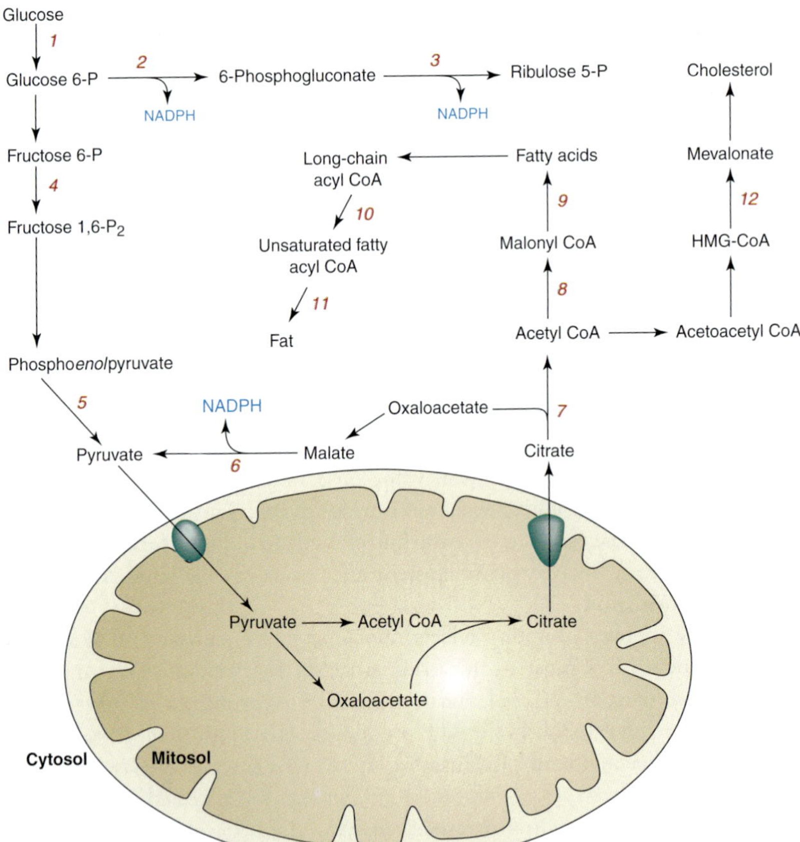

FIGURE 22.19

Hepatic enzymes induced in the well-fed state. The inducible enzymes are numbered: 1, glucokinase; 2, glucose 6-phosphate dehydrogenase; 3, 6-phosphogluconate dehydrogenase; 4, 6-phosphofructo-1-kinase; 5, pyruvate kinase; 6, malic enzyme; 7, citrate cleavage enzyme; 8, acetyl CoA carboxylase; 9, fatty acid synthase; 10, Δ^9-desaturase; 11, glycerol-3-phosphate acyltransferase; and 12, 3-hydroxy-3-methylglutaryl-CoA (HMG-CoA) reductase.

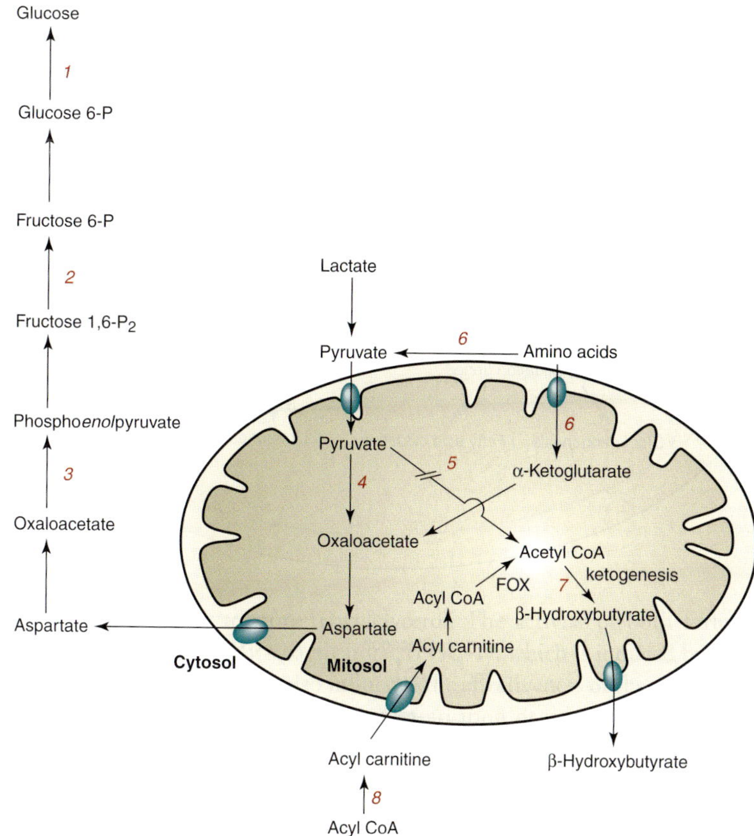

FIGURE 22.20

Hepatic enzymes induced in the fasted state. The inducible enzymes are numbered: 1, glucose 6-phosphatase; 2, fructose 1,6-bisphosphatase; 3, phosphoenolpyruvate carboxykinase; 4, pyruvate carboxylase; 5, pyruvate dehydrogenase kinase; 6, various aminotransferases; 7, mitochondrial 3-hydroxy-3-methylglutaryl-CoA synthase; and 8, carnitine palmitoyltransferase I. Parallel lines intersecting arrow from pyruvate to acetyl CoA denote inhibition (due to phosphorylation) of the pyruvate dehydrogenase complex due to induction of pyruvate dehydrogenase kinase. Abbreviation: FOX, fatty acid oxidation.

complex, which prevents conversion of pyruvate to acetyl CoA, thereby conserving lactate, pyruvate, and the carbon of some amino acids for glucose synthesis. Induction of carnitine palmitoyltransferase I and mitochondrial 3-hydroxy-3-methylglutaryl-CoA synthase likewise increases the capacity of the liver for fatty acid oxidation and ketogenesis. This is particularly important because fatty acid oxidation is the primary source of the ATP needed for hepatic glucose synthesis. Enzymes of the urea cycle and other amino acid metabolizing enzymes such as liver glutaminase, tyrosine aminotransferase, serine dehydratase, proline oxidase, and histidase are induced for disposal of nitrogen, as urea, generated from the amino acids used in gluconeogenesis.

Regulation of the rate of **gene transcription** is the primary way by which changes in enzyme amounts are controlled. In the well-fed state, the genes for synthesis of fatty acids and cholesterol are controlled by **sterol response element binding proteins (SREBP)**. The amount of **SREBP-1** is modulated by insulin and carbohydrate intake. An increase in insulin in the well-fed state signals an increase in SREBP-1 (Figure 22.21), which functions as a transcription factor to increase transcription of the genes encoding the lipogenic enzymes. Glucagon acts in opposition to insulin by signaling activation via cAMP of protein kinase A, which phosphorylates cAMP response element (CRE) binding protein (CREB), a transcription factor that inhibits transcription of the genes encoding the lipogenic enzymes (Figure 22.22).

A second form of SREBP designated **SREBP-2**, regulates cholesterol synthesis (see p. 715). When levels of cholesterol fall in cells, SREBP, which is anchored in the endoplasmic reticulum, moves to the Golgi apparatus, where it is cleaved by a protease, releasing the N-terminal fragment (SREBP-2) that acts as a transcriptional activator of genes encoding for cytosolic 3-hydroxy-3-methylglutaryl-CoA synthase, 3-hydroxy-3-methylglutaryl-CoA reductase, squalene synthase, and the LDL receptor.

Glucagon promotes the transcription of genes containing CREs and encoding gluconeogenic enzymes by phosphorylation and activation of the transcription factor CREB (Figure 22.22). Insulin opposes this action of glucagon. Several mechanisms are involved, but one of the most important involves inhibition of the activity of forkhead

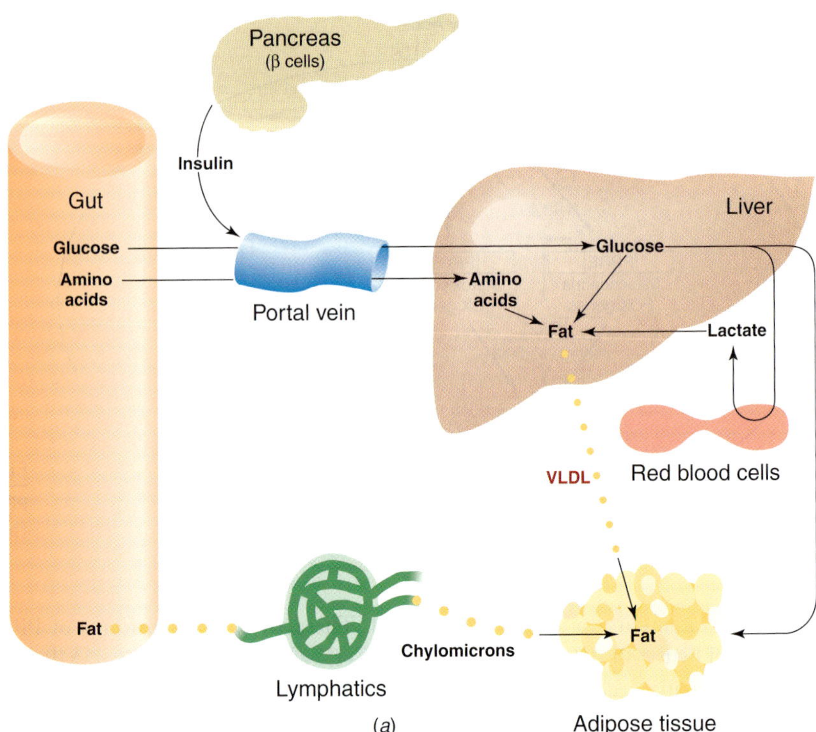

FIGURE 22.24

Metabolic interrelationships of tissues in various nutritional, hormonal, and disease states. (*a*) Obesity.

medical problems, including abdominal obesity, elevated blood pressure, high blood lipids, and insulin resistance.

Obesity frequently causes **insulin resistance**. In some patients, the number or affinity of insulin receptors is reduced, while others have normal insulin binding, but abnormal post-receptor responses, such as the activation of glucose transport. Generally, the quantity of body fat is proportional to the degree of insulin resistance. Some of the resistance is due to peptides (**TNFα** and **resistin**) produced by adipocytes that are known to oppose the action of insulin. An additional adipokine, adiponectin, is reduced in obesity, which may also contribute to the insulin resistance. Plasma insulin levels are greatly elevated in obese individuals, a harbinger for the development of type 2 diabetes mellitus.

Dieting

To lose weight requires a negative energy balance, which means fewer calories must be consumed than are expended each day. Consuming less food with the same macronutrient composition has little effect upon the starve–feed cycle, other than the length of time in the well-fed state. The responsibilities of tissues will stay the same in the well-fed state (Figure 22.2), except that less glycogen and triacylglycerol will be stored and the switch to the function of tissues in the fasted state (Figure 22.4) will occur sooner after meals. Another way to lose weight, long advocated by most health agencies, is to cut calories by specifically reducing the amount of fat consumed. Again, the well-fed state of tissues will stay the same except for a reduction in stored triacylglycerol. The switch to the fasting state will occur sooner after meals, unless the decrease in fat intake has been compensated by an increase in carbohydrate intake, a common problem for dieters. Another way to lose weight, originally advocated by Robert Atkins, is to decrease specifically the amount of carbohydrate consumed. The Atkins controlled-carbohydrate diet and variations thereof, frequently called the very low-carbohydrate ketogenic diet, produces the most interesting effects upon tissues. On this diet intake of protein is high, fat is moderate, and carbohydrate is extremely low ($<$50 g/day; $<$10% of a 2000 kcal/day diet). The metabolic functions of tissues in the fed state for an individual on such a diet are summarized in Figure 22.24*b*. The fasting state is little changed from that of other diets, but the almost complete lack of dietary

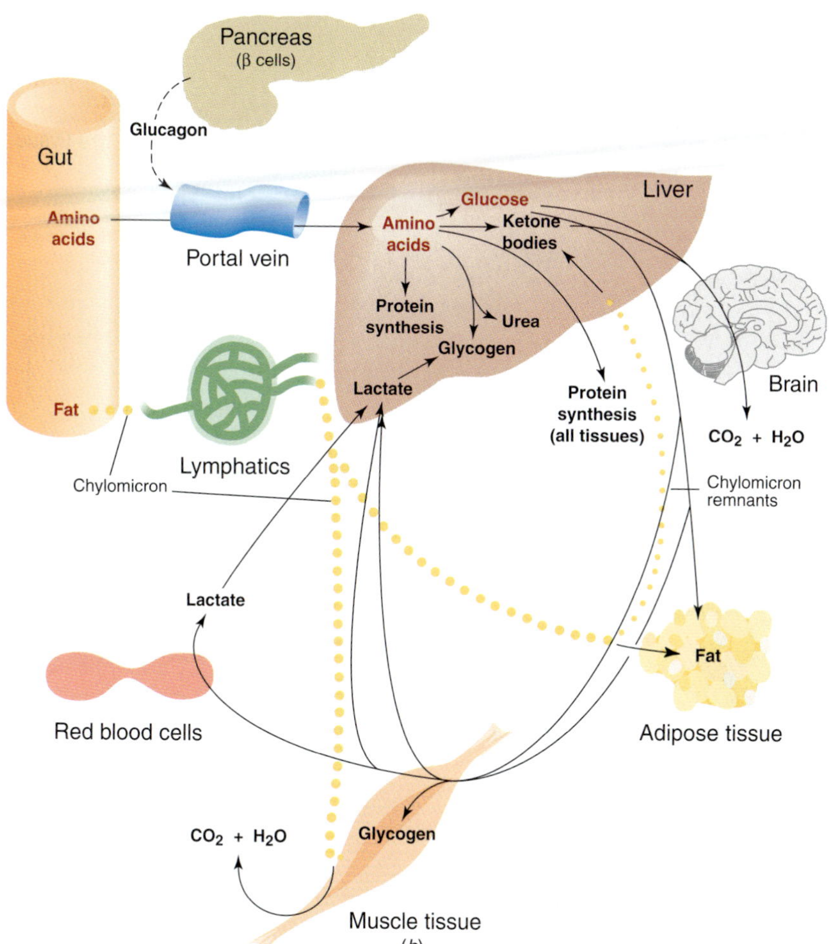

FIGURE 22.24

Metabolic interrelationships of tissues in various nutritional, hormonal, and disease states. (*b*) Dieting.

carbohydrate requires that the liver remain gluconeogenic as well as ketogenic in the fed state. Little rise in blood glucose and a less than normal rise in insulin occur in response to meals. Amino acids present in excess beyond that needed for protein synthesis are converted to liver glycogen, blood glucose, and ketone bodies. Large amounts of amino acids coming from the gut minimize the need for release of amino acids from peripheral tissues for hepatic gluconeogenesis. Fatty acids delivered to the liver in chylomicron remnants are primarily converted to ketone bodies to provide ATP for gluconeogenesis. Glucose and ketone bodies are therefore produced in both the fed and fasted states. Although it would seem that over production and/or under utilization of ketone bodies could cause ketoacidosis, as it does in type 1 diabetes (see p. 881), the need for a supply of energy in the form of ketone bodies in peripheral tissues largely balances the production of ketone bodies by the liver. The low carbohydrate diet has been tested in controlled trials, which have been relatively short term (6–12 months): Weight loss is somewhat more rapid than achieved with low fat, low calorie diets, but the extent of weight loss, compliance with the diet, and the tendency to regain weight are similar. Of interest, this diet didn't increase blood cholesterol or LDL levels, and it improved VLDL and HDL levels better than low fat, low calorie diets. It remains to be seen if the low carbohydrate diet has long-term risks, such as increased atherogenesis.

Type 2 Diabetes Mellitus

Figure 22.24*c* shows the metabolic interrelationships characteristic of a person with type 2 diabetes mellitus. These individuals are resistant to insulin and have insufficient production of insulin to overcome the resistance (see Clin. Corr. 22.7). The majority of patients are obese, and although their insulin levels often are high, they are not as

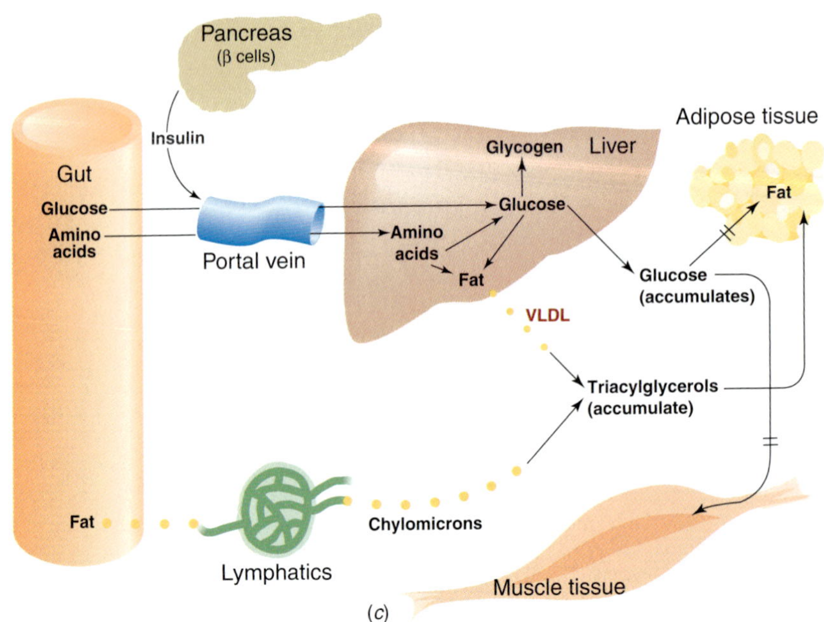

FIGURE 22.24

Metabolic interrelationships of tissues in various nutritional, hormonal, and disease states. (c) Type 2 diabetes mellitus.

CLINICAL CORRELATION 22.7
Diabetes Mellitus, Type 2

Type 2 diabetes mellitus accounts for 80–90% of the diagnosed cases of diabetes. It usually occurs in middle-aged to older obese people and is characterized by hyperglycemia, often with hypertriglyceridemia and other features of syndrome X (see Clin. Corr. 22.1). The ketoacidosis characteristic of type 1 diabetes is usually not observed, although some patients can develop transient episodes of ketoacidosis. Increased levels of VLDL are probably the result of increased hepatic triacylglycerol synthesis stimulated by hyperglycemia and hyperinsulinemia. Insulin is present at normal to elevated levels in this form of the disease. Obesity often precedes the development of type 2 diabetes and is the major contributing factor. Obese patients are usually hyperinsulinemic and have high levels of free fatty acids, which impair insulin action. Recent data implicate increased levels of tumor necrosis factor α (TNFα) and a new protein called resistin and reduced secretion of adiponectin by adipocytes of obese individuals as a cause of insulin resistance. The greater the adipose tissue mass, the greater the production of TNFα and resistin, which acts to impair insulin receptor function. The higher the basal level of plasma insulin, the lower the number of receptors present on the plasma membranes of target cells, and there are defects within insulin-responsive cells at sites beyond the receptor—for example, the ability of insulin to recruit glucose transporters (GLUT4) from intracellular sites to the plasma membrane. As a consequence, insulin levels remain high, but glucose levels are poorly controlled. Even though the insulin level is high, it is not as high as in a person who is obese but not diabetic. In other words, there is a relative deficiency in the insulin supply from the β cells. This disease is caused, therefore, not only by insulin resistance but also by impaired β-cell function resulting in relative insulin deficiency. Diet alone can control the disease in the obese diabetic. If the patient can be motivated to lose weight, insulin receptors will increase in number, and the post-receptor abnormalities will improve, which will increase both tissue sensitivity to insulin and glucose tolerance. There are now a multitude of medications that either sensitize peripheral tissue to insulin action (thiazolidinediones), reduce hepatic gluconeogenesis (metformin), or stimulate insulin secretion from β cells (sulfonylureas), but ultimately many type 2 diabetes patients require exogenous insulin injections to control blood sugar. They tend not to develop ketoacidosis but develop many of the same complications as type 1 diabetes patients—that is, nerve, eye, kidney, and coronary artery disease.

Source: Kahn, B. B. and Flier, J. S. Obesity and insulin resistance. *J. Clin. Invest.* 106:473, 2000. Cavaghan, M. K., Ehrmann, D. A., Polonsky, K. S. Interactions between insulin resistance and insulin secretion in the development of glucose intolerance. *J. Clin. Invest.* 106:329, 2000. Trayhurn, P. and Wood, I. S. Adipokines: Inflammation and the pleiotropic role of white adipose tissue. *Br. J. Nutr.* 92:347, 2004. Mauvais-Jarvis, F., Sobngwi, E., Porcher, R., Riveline, J. P., Kevorkian, J. P., Vaisse, C., Charpentier, G., Guillausseau, P. J., Vexiau, P., Gautier, J. F. Ketosis-prone type 2 diabetes in patients of sub-Saharan African origin: Clinical pathophysiology and natural history of beta-cell dysfunction and insulin resistance. *Diabetes* 53:645, 2004. Sobngwi, E., Mauvais-Jarvis, F., Vexiau, P., Mbanya, J. C., Gautier, J. F. Diabetes in Africans. Part 2: Ketosis-prone atypical diabetes mellitus. *Diabetes Metab.* 28:5, 2002.

high as those of nondiabetic but similarly obese individuals are. Hence, β-cell failure and insulin resistance are components of this form of diabetes. Exogenous insulin will reduce the hyperglycemia, and it often must be administered to control blood glucose levels of these patients. Although the body still produces insulin in type 2 diabetes, it is not enough to control glucose production by the liver or promote uptake of glucose by the skeletal muscle. Hyperglycemia results for both reasons. The normal increase in fructose 2,6-bisphosphate and down-regulation of phosphoenolpyruvate carboxykinase does not occur in these patients. **Translocation of intracellular vesicles** bearing GLUT4 to the plasma membrane in response to insulin is decreased in the skeletal muscle and adipose tissue of these patients. Ketoacidosis rarely develops, perhaps because enough insulin is present to prevent uncontrolled release of fatty acids from adipocyte and fatty acids reaching the liver or synthesized *de novo* are directed into triacylglycerol. Hypertriacylglycerolemia is characteristic and usually results from an increase in VLDL without hyperchylomicronemia. This is most likely explained by hepatic synthesis of fatty acids and diversion of fatty acids reaching the liver into triacylglycerol and VLDL. Concurrent lipogenesis and gluconeogenesis should never occur but in this disease it results from a state of mixed insulin resistance of the insulin signaling pathways controlling these processes. A defect in the insulin-signaling pathway controlling gluconeogenesis prevents suppression of hepatic glucose production (via PI_3 kinase, see p. 872) in the face of high insulin levels. A more responsive insulin-signaling pathway controlling fatty acid synthesis and esterification leads to overproduction of triacylglycerol.

Type 1 Diabetes Mellitus

Figure 22.24*d* shows the metabolic interrelationships in type 1 diabetes mellitus (see Clin. Corrs. 22.8 and 22.9). In contrast to type 2, there is complete absence of insulin production by the pancreas in this disease. Since the insulin/glucagon ratio cannot increase, the liver is always gluconeogenic and ketogenic and cannot properly buffer blood glucose levels. Indeed, since hepatic gluconeogenesis is continuous, the liver contributes to hyperglycemia in the well-fed state. In muscle and adipose tissue, GLUT4 remains sequestered within cells. Accelerated gluconeogenesis, fueled by uncontrolled proteolysis in skeletal muscle, maintains the hyperglycemia even in the starved state. Uncontrolled lipolysis in adipose tissue increases plasma fatty acid levels and ketone body

CLINICAL CORRELATION 22.8
Diabetes Mellitus, Type 1

Type 1 diabetes mellitus usually appears in childhood or in the teens, but it is not limited to these patients. Insulin secretion is very low because of defective β-cell function, the result of an autoimmune process. Untreated, type 1 diabetes is characterized by hyperglycemia, hypertriglyceridemia (chylomicrons and VLDL), and episodes of severe ketoacidosis. Thus, there is severe derangement of carbohydrate, lipid, and protein metabolism. The hyperglycemia results from the inability of the insulin-dependent tissues to take up glucose and from accelerated hepatic gluconeogenesis from amino acids derived from muscle protein. The ketoacidosis results from increased lipolysis in adipose tissue and accelerated fatty acid oxidation in liver. Hyperchylomicronemia results from low lipoprotein lipase activity in adipose tissue capillaries, an enzyme dependent on insulin for its synthesis.

Although insulin does not cure type 1 diabetes, it markedly alters the clinical course. Insulin promotes glucose uptake and inhibits gluconeogenesis, lipolysis, and proteolysis. It is a difficult job to adjust the insulin dose to variable dietary intake and physical activity, the other major determinant of glucose disposal by muscle. Tight control of blood sugar requires several injections of insulin per day and close blood sugar monitoring by the patient, but it has now been proven to reduce the microvascular complications of diabetes (renal and retinal disease).

Source: Atkinson, M. A. and Maclaren, N. K. The pathogenesis of insulin dependent diabetes mellitus. *N. Engl. J. Med.* 331:1428, 1994. Clark, C. M. and Lee, D. A. Prevention and treatment of the complications of diabetes mellitus. *N. Engl. J. Med.* 332:1210, 1994. Luppi, P. and Trucco, M. Immunological models of type 1 diabetes. *Horm. Res.* 52:1, 1999. Kukreja, A. and Maclaren, N. K. Autoimmunity and diabetes. *J. Clin. Endocrinol. Metab.* 84:4371, 1999.

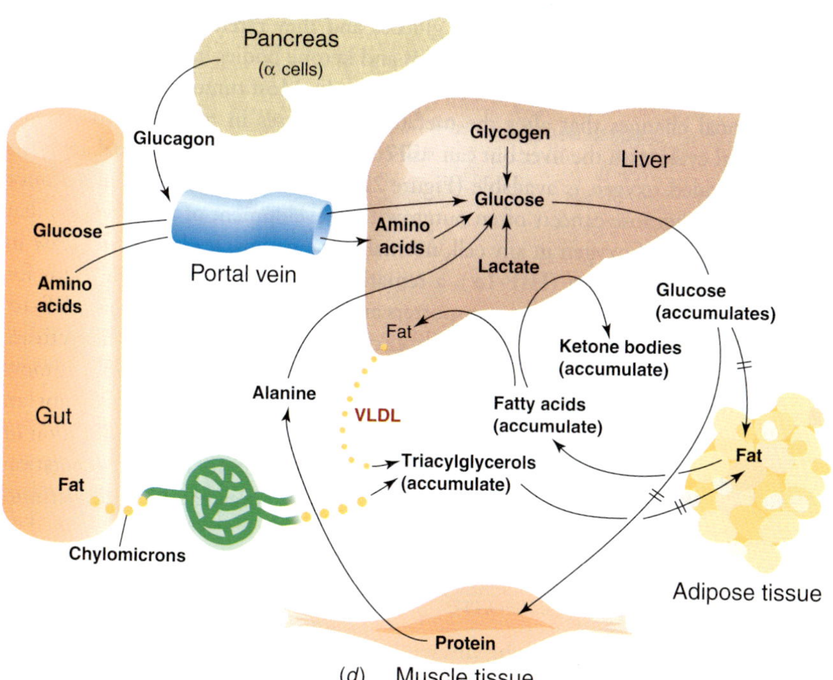

FIGURE 22.24
Metabolic interrelationships of tissues in various nutritional, hormonal, and disease states. (*d*) Type 1 diabetes mellitus.

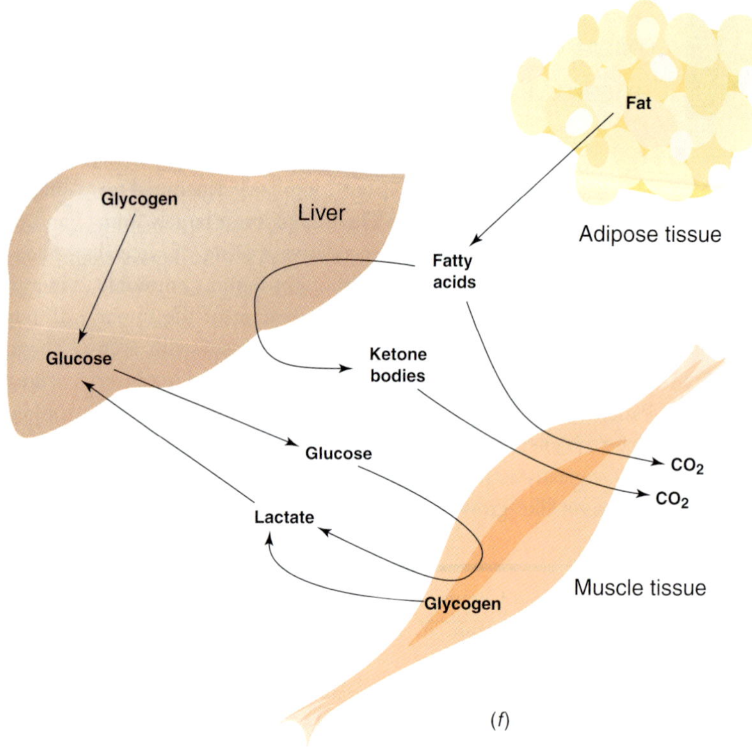

FIGURE 22.24

Metabolic interrelationships of tissues in various nutritional, hormonal, and disease states. (*f*) Exercise.

Increased glucose uptake, glycogen degradation, and glycolysis help provide the ATP needed for muscle contraction. There is also stimulation of branched-chain amino acid oxidation, ammonium production, and alanine release from the exercising muscle. A well-fed individual does not store enough glycogen to meet the energy needs of running long distances. The respiratory quotient, the ratio of carbon dioxide exhaled to oxygen consumed, falls during distance running. This indicates the progressive switch from glycogen to fatty acid oxidation during a race. Lipolysis gradually increases as glucose stores are exhausted, and, as in the fasted state, muscles oxidize fatty acids in preference to glucose. The increase in AMP due the demand for ATP also activates AMPK, which phosphorylates and inactivates acetyl-CoA carboxylase. This, along with an increase in long chain acyl-CoA esters, which are negative allosteric effectors of acetyl-CoA carboxylase, reduces synthesis of malonyl CoA. AMPK also activates malonyl-CoA decarboxylase, removing malonyl CoA. This allows greater carnitine palmitoyltransferase I activity and fatty acid oxidation to provide ATP for the contracting muscle. Interestingly, exercise also has effects on the liver mediated by activation of AMPK. Phosphorylation of acetyl-CoA carboxylase, malonyl-CoA decarboxylase, and glycerol-3-phosphate acyltransferase by AMPK directs fatty acids into oxidation and away from esterification to triacylglycerol. Unlike fasting, however, there is little increase in blood ketone body concentration during exercise because hepatic ketone body production is balanced by the muscle oxidation of ketone bodies for energy.

Pregnancy

The fetus is a nutrient-requiring organism (Figure 22.24*g*). It mainly uses glucose for energy, but may also use amino acids, lactate, fatty acids, and ketone bodies. Lactate produced in the placenta by glycolysis is partly directed to the fetus, and the rest enters the maternal circulation to establish a Cori cycle with the liver. Maternal LDL cholesterol is an important precursor of placental steroids (estradiol and progesterone). During pregnancy, the starve–feed cycle is perturbed. The placenta secretes placental lactogen, and two steroid hormones, estradiol and progesterone. Placental lactogen stimulates lipolysis in adipose tissue, and the steroid hormones induce insulin resistance.

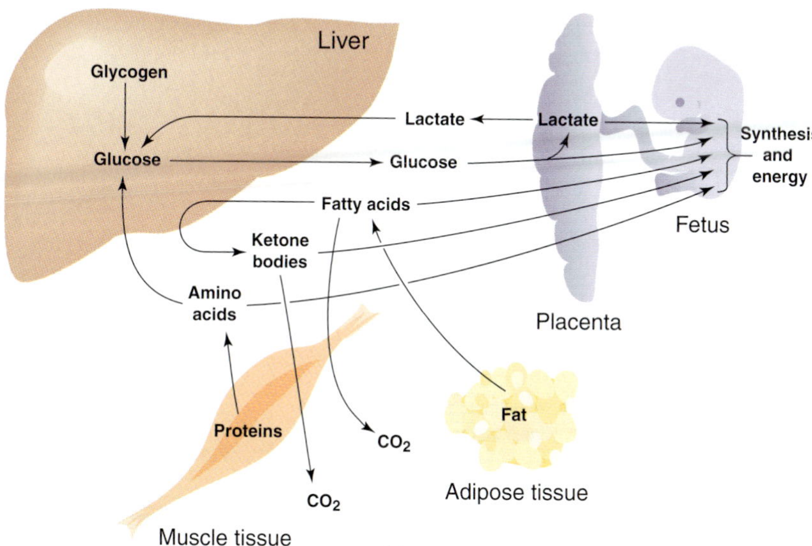

FIGURE 22.24

Metabolic interrelationships of tissues in various nutritional, hormonal, and disease states. (*g*) Pregnancy.

After meals, pregnant women enter the starved state more rapidly because of increased consumption of glucose and amino acids by the fetus. Plasma glucose, amino acids, and insulin levels fall rapidly, and glucagon and placental lactogen levels rise and stimulate lipolysis and ketogenesis. Consumption of glucose and amino acids by the fetus may be great enough to cause maternal hypoglycemia. In the fed state, pregnant women have increased levels of insulin and glucose and demonstrate resistance to exogenous insulin. These swings of plasma hormones and fuels are even more exaggerated in pregnant diabetic women and make control of their blood glucose difficult. This is important, because hyperglycemia adversely affects fetal development.

Lactation

In late pregnancy, placental (progesterone) and maternal (prolactin) hormones induce lipoprotein lipase in the mammary gland and promote the development of milk-secreting cells and ducts. During lactation (see Figure 22.24*h*) the breast uses glucose for lactose and triacylglycerol synthesis, and its major energy source. Amino acids are taken up for protein synthesis, and chylomicrons and VLDL provide fatty acids for triacylglycerol synthesis. If these compounds are not supplied by the diet, proteolysis, gluconeogenesis, and lipolysis must supply them, eventually resulting in maternal malnutrition and poor quality milk. The lactating breast secretes **parathyroid hormone-related protein**

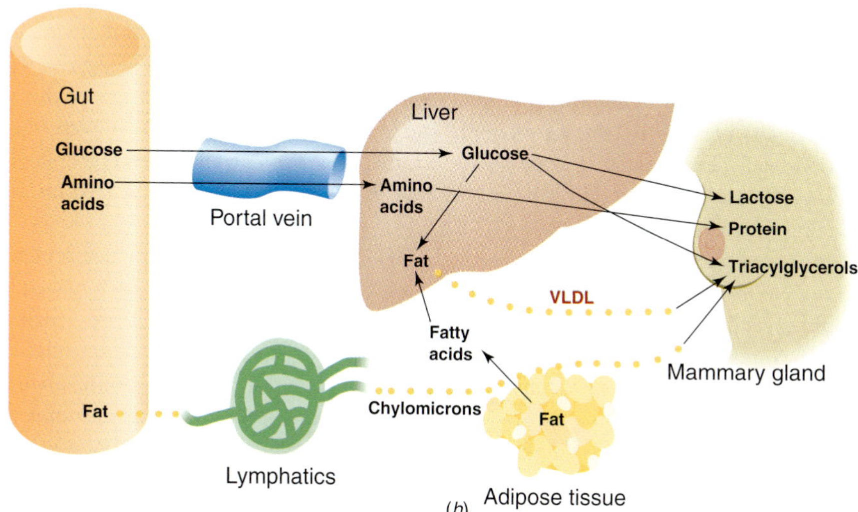

FIGURE 22.24

Metabolic interrelationships of tissues in various nutritional, hormonal, and disease states. (*h*) Lactation.

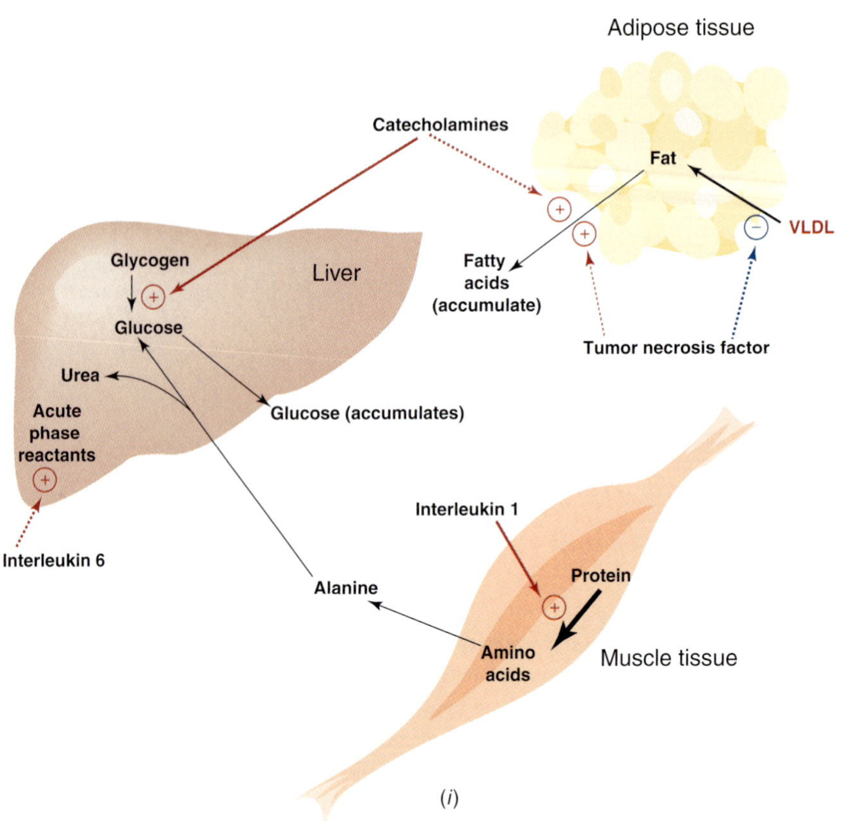

FIGURE 22.24
Metabolic interrelationships of tissues in various nutritional, hormonal, and disease states. (*i*) Stress and injury.

(**PTHrP**), which mimics the effects of **parathyroid hormone** (**PTH**) in stimulating absorption of calcium and phosphorus from the gut and bone. The release of PTH from the parathyroid and PTHrP from the breast is under the control of the calcium-sensing receptor, a G-protein-coupled receptor that senses extracellular calcium and thus may serve to coordinate release of calcium mobilizing hormone with plasma calcium levels.

Stress and Injury

Physiological stresses include injury, surgery, renal failure, burns, and infections (Figure 22.24*i*). Characteristically, blood cortisol, glucagon, catecholamines, and growth hormone levels increase, and there is resistance to insulin. Basal metabolic rate, blood glucose and free fatty acid levels are elevated. However, ketogenesis is not accelerated as in fasting. For poorly understood reasons, the muscle glutamine pool is reduced, resulting in reduced protein synthesis and increased protein breakdown. It can be very difficult to reverse this protein breakdown, despite intravenous administration of amino acids, glucose, and triacylglycerol. However, these solutions lack glutamine, tyrosine, and cysteine because of stability and solubility constraints. Supplementation of these amino acids, perhaps by the use of more stable dipeptides, may help to reverse the catabolic state better than now possible. In fact, there is growing recognition that enteral feeding (via tubes in the stomach or intestine) is superior to intravenous feeding in reducing or reversing catabolism.

Negative nitrogen balance of injured or infected patients is mediated by interleukin 1, interleukin 6, and **tumor necrosis factor α** (**TNFα**) produced by monocytes and lymphocytes (see Clin. Corr. 22.10). These cytokines induce fever and other metabolic changes. Interleukin 1 activates proteolysis in skeletal muscle. Interleukin 6 stimulates the synthesis by the liver of acute phase reactants, such as fibrinogen, complement proteins, some clotting factors, and α2-macroglobulin, which may defend against injury and infection. TNFα suppresses adipocyte triacylglycerol synthesis, inhibits lipoprotein lipase, stimulates lipolysis, inhibits release of insulin, and promotes insulin resistance. These cytokines may be responsible for the wasting seen in chronic infections.

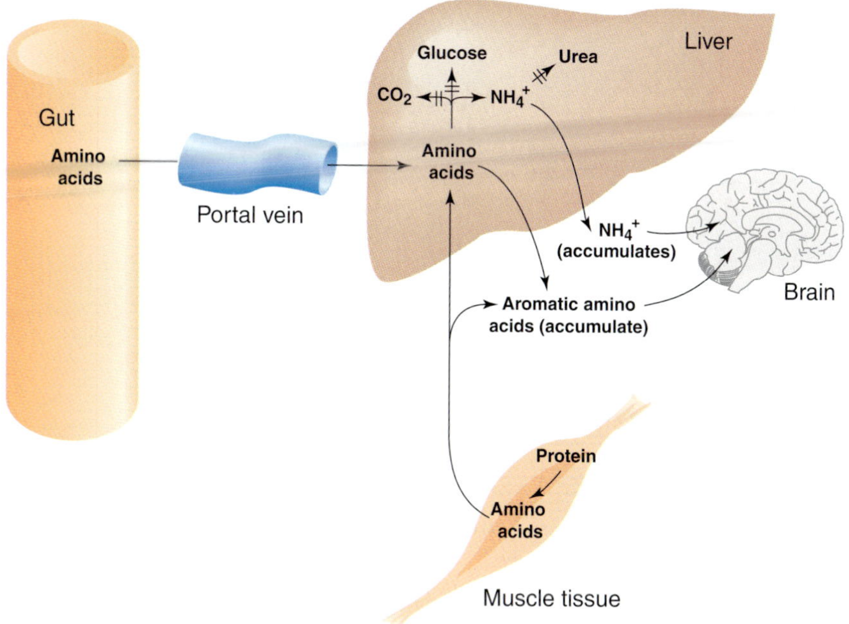

FIGURE 22.24

Metabolic interrelationships of tissues in various nutritional, hormonal, and disease states. (*j*) Liver disease.

Liver Disease

Advanced liver disease is associated with major metabolic derangements (Figure 22.24*j*), especially for amino acids. In patients with cirrhosis, the liver cannot convert ammonia into urea and glutamine rapidly enough, and the blood ammonia level rises. Shunting of blood around the liver and interference with the intercellular glutamine cycle (see p. 748) contribute to the problem. Ammonia arises from the action of glutaminase, glutamate dehydrogenase, and adenosine deaminase in intestine and liver, as well as from the intestinal lumen, where bacterial urease splits urea into ammonia and carbon dioxide. Ammonia toxicity for the central nervous system leads to coma that sometimes occurs in patients in liver failure. In advanced liver disease, plasma aromatic amino acids reach higher levels than branched-chain amino acids. These two groups of amino acids are transported into the brain by the same carrier system. Increased brain uptake of aromatic amino acids may increase synthesis of neurotransmitters such as serotonin, which may be responsible for some of the neurological abnormalities of liver disease. The liver is a major source of **insulin-like growth factor I (IGF-1)**. Patients with cirrhosis suffer muscle wasting because of deficient IGF-1 synthesis in response to growth hormone. They also often have insulin resistance and may have diabetes mellitus. Finally, in outright liver failure, patients sometimes die of hypoglycemia because the liver is unable to maintain the blood glucose level by gluconeogenesis.

Renal Disease

In chronic renal disease, levels of amino acids normally metabolized by kidney (glutamine, glycine, proline, and citrulline) increase and nitrogen end products (e.g., urea, uric acid, and creatinine) also accumulate (Figure 22.24*k*). This is worsened by high dietary protein intake or accelerated proteolysis. Since gut bacteria can split urea into ammonia and liver uses ammonia and α-keto acids to form nonessential amino acids, a diet high in carbohydrate and an amino acid intake limited as much as possible to essential amino acids ensures that the liver synthesizes nonessential amino acids from TCA cycle intermediates. This type of diet therapy may delay the need for dialysis. An additional abnormality in dialysis patients is carnitine deficiency, resulting from reduced intake of dietary carnitine (in meat) and reduced functional renal mass. This may lead to cardiac and skeletal myopathy due to reduced ability of these tissues to oxidize fatty acids.

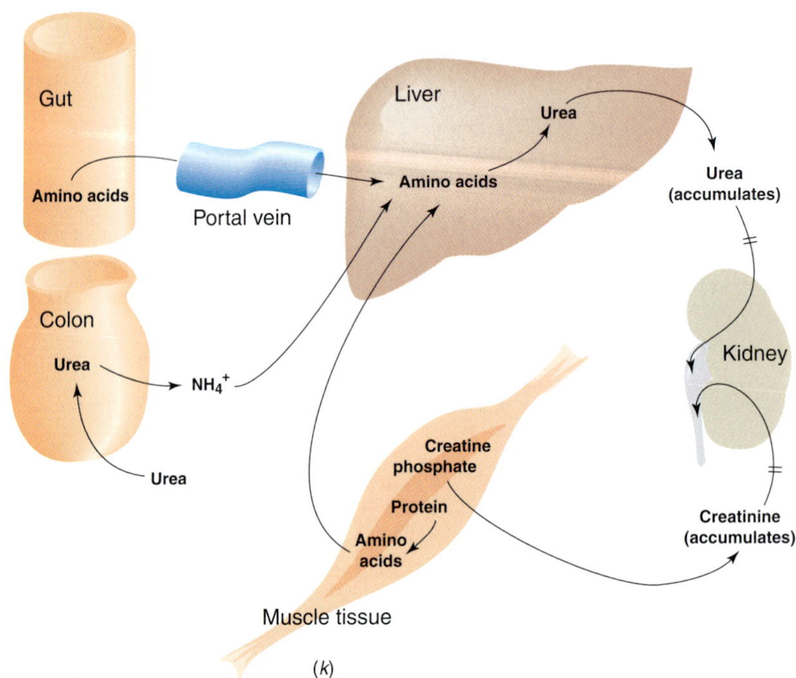

FIGURE 22.24

Metabolic interrelationships of tissues in various nutritional, hormonal, and disease states. (*k*) Kidney failure.

Alcohol

The liver is primarily responsible for the first steps of ethanol catabolism:

$$\text{Ethanol}(CH_3CH_2OH) + NAD^+ \rightarrow \text{Acetaldehyde}(CH_3CHO) + NADH + H^+$$

$$\text{Acetaldehyde}(CH_3CHO) + NAD^+ + H_2O \rightarrow \text{Acetate}(CH_3COO^-) + NADH + H^+$$

The first, catalyzed by **alcohol dehydrogenases** in the cytosol generates NADH; the second, catalyzed by aldehyde dehydrogenase, also generates NADH but occurs in the mitochondrial matrix. Liver disposes of the NADH generated through the mitochondrial electron-transport chain. Intake of even moderate amounts of ethanol generates too much NADH. Enzymes involved in gluconeogenesis (lactate dehydrogenase and malate dehydrogenase) and fatty acid oxidation (β-hydroxyacyl-CoA dehydrogenase) require NAD^+ as a substrate. Thus, these pathways are inhibited by alcohol intake (Figure 22.24*l*), and fasting hypoglycemia and the accumulation of hepatic triacylglycerols (fatty liver) may develop. Lactate may accumulate from inhibition of the conversion of lactate to glucose but rarely causes overt metabolic acidosis. Liver mitochondria have a limited capacity to oxidize acetate to CO_2 because the TCA cycle is inhibited by high NADH levels during ethanol oxidation. Much of the acetate derived from ethanol escapes the liver to the blood. Virtually every other tissue can oxidize it to CO_2 by way of the TCA cycle.

Acetaldehyde can also escape from the liver and readily forms covalent bonds with functional groups of biologically important compounds. Formation of acetaldehyde adducts with proteins in liver and blood of animals and humans drinking alcohol has been demonstrated. Such adducts may provide a marker for past drinking activity of an individual, just as hemoglobin A_{1c} is an index of blood glucose control in diabetic patients.

Acid–Base Balance

Regulation of **acid–base balance**, like that of nitrogen excretion, is shared between the liver and kidney. Although complete catabolism of most amino acids yields neutral

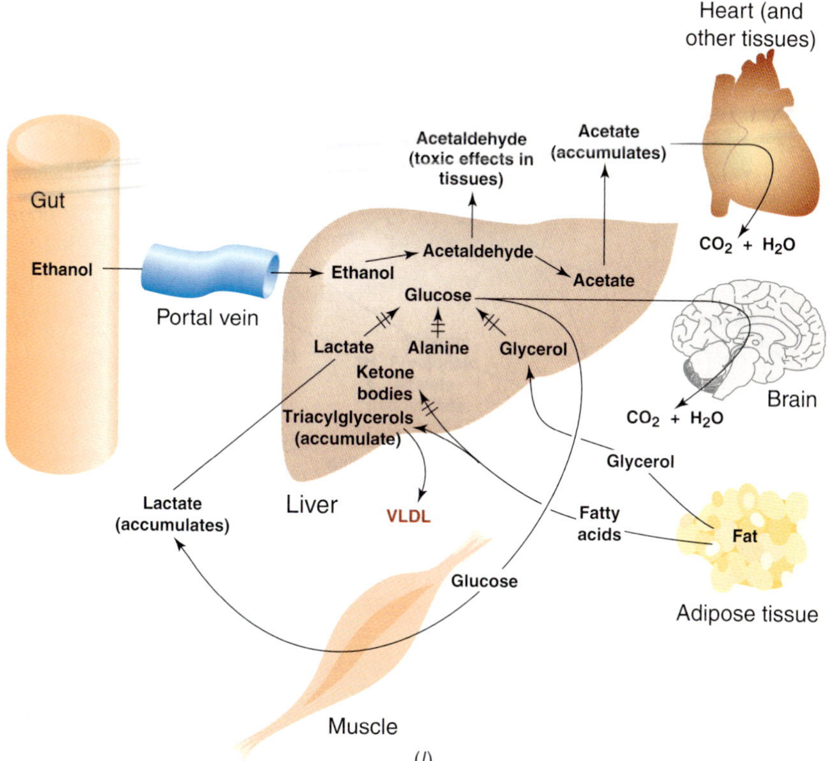

FIGURE 22.24
Metabolic interrelationships of tissues in various nutritional, hormonal, and disease states. (*l*) Alcohol.

products (CO_2, H_2O, and urea), the oxidation of positively charged arginine, lysine, and histidine and sulfur-containing amino acids methionine and cysteine result in net formation of protons (acid). For example,

$$\text{Arginine}^+ + 5\tfrac{1}{2}\,O_2 \longrightarrow 4\,CO_2 + \text{urea}\ (NH_2\overset{\overset{\displaystyle O}{\|}}{C}NH_2) + 3\,H_2O + H^+$$

Complete catabolism of negatively charged glutamate and aspartate consumes some of these protons but does not completely compensate:

$$\text{Glutamate}^- + 4\tfrac{1}{2}\,H_2O + H^+ \longrightarrow 4\tfrac{1}{2}\,CO_2 + \tfrac{1}{2}\,\text{Urea}\ (NH_2\overset{\overset{\displaystyle O}{\|}}{C}NH_2) + 3\tfrac{1}{2}\,H_2O$$

Therefore, for acid–base balance, the excess protons must be matched by an equivalent amount of base. In the kidney, glutamine is readily taken up, deaminated by glutaminase to give glutamate, oxidatively deaminated by glutamate dehydrogenase to give α-ketoglutarate, and converted by enzymes of the TCA cycle to malate, which is converted to glucose. Summation of all steps reveals net production of glucose, and more importantly the production of ammonium and bicarbonate ions:

$$\text{Glutamine} + 1\tfrac{1}{2}\,O_2 + 3H_2O \rightarrow \tfrac{1}{2}\text{Glucose} + 2\,HCO_3^- + 2\,NH_4^+$$

The ammonium ions are excreted into the glomerular filtrate while the bicarbonate ions enter the blood to neutralize protons:

$$HCO_3^- + H^+ \rightleftharpoons H_2CO_3 \rightleftharpoons CO_2 + H_2O$$

CO_2 is blown off in the lungs, thereby effectively eliminating excess protons (acid) produced by amino acid oxidation. In **metabolic acidosis** (see Figure 22.24*m*), a condition in which much more acid than normal is produced in the body because some metabolic process is out of control (e.g., lactic acid formation by anaerobic glycolysis or β-hydroxybutyric acid production by ketogenesis). In these conditions, the amounts

Crabb, D. W., Matsumoto, M., Chang, D., and You, M. Overview of the role of alcohol dehydrogenase and aldehyde dehydrogenase and their variants in the genesis of alcohol-related pathology. *Proc. Nutr. Soc.* 63:49, 2004.

Curthoys, N. P. and Watford, M. Regulation of glutaminase activity and glutamine metabolism. *Annu. Rev. Nutr.* 15:133, 1995.

De Feo, P., Di Loreto, C., Lucidi, P., Murdolo, G., Parlanti, N., De Cicco, A., Piccioni, F., and Santeusanio, F. Metabolic response to exercise. *J. Endocrinol. Invest.* 26:851, 2003.

Du, X., Matsumura, T., Edelstein, D., Rossetti, L., Zsengeller, Z., Szabo, C., and Brownlee, M. Inhibition of GAPDH activity by poly(ADP-ribose) polymerase activates three major pathways of hyperglycemic damage in endothelial cells. *J. Clin. Invest.* 112:1049, 2003.

Duée, P. -H., Darcy-Vrillon, B., Blachier, F. and Morel, M. -T. Fuel selection in intestinal cells. *Proc. Nutr. Soc.* 54:83, 1995.

Elia, M. General integration and regulation of metabolism at the organ level. *Proc. Nutr. Soc.* 54:213, 1995.

Evans, R. M., Barish, G. D., and Wang, Y. X. PPARs and the complex journey to obesity. *Nat. Med.* 10:355, 2004.

Feinman, R. D. and Makowskie, M. Metabolic syndrome and low-carbohydrate ketogenic diets in the medical school biochemistry curriculum. *Metab. Syndrome Relat. Disorders* 1:189, 2003.

Frayn, K. N., Humphreys, S. M., and Coppack, S. W. Fuel selection in white adipose tissue. *Proc. Nutr. Soc.* 54:177, 1995.

Gibson, D. M. and Harris, R. A. *Metabolic Regulation in Mammals*. New York: Taylor & Francis, 2002.

Grimble, R. F. Nutritional modulation of immune function. *Proc. Nutr. Soc.* 60:389, 2001.

Grimble, R. F. Fasting in healthy individuals and adaption to undernutrition during chronic disease. *Curr. Opin. Clin. Nutr. Metab Care* 1:369, 1998.

Halperin, M. L. and Rolleston, F. S. *Clinical Detective Stories. A Problem-Based Approach to Clinical Cases in Energy and Acid–Base Metabolism*. London: Portland Press, 1993.

Hardie, D. G. AMP-activated protein kinase: The guardian of cardiac energy status. *J. Clin. Invest.* 114:465, 2004.

Harris, R. A., Huang, B., and Wu, P. Control of pyruvate dehydrogenase kinase gene expression. *Adv. Enzyme Regul.* 41:269, 2001.

Häussinger, D. Hepatic glutamine transport and metabolism. *Adv. Enzymol. Relat. Areas Mol. Biol.* 72:43, 1998.

Henriksson, J. Muscle fuel selection: Effect of exercise and training. *Proc. Nutr. Soc.* 54:125, 1995.

Holness, M. J. and Sugden, M. C. Regulation of pyruvate dehydrogenase complex activity by reversible phosphorylation. *Biochem. Soc. Trans.* 31:1143, 2003.

Homko, C. J., Sivan, E., Reece, E. A., and Boden, G. Fuel metabolism during pregnancy. *Semin. Reprod. Endocrinol.* 17:119, 1999.

Horton, J. D., Goldstein, J. L, and Brown, M. S. SREBPs: Activators of the complete program of cholesterol and fatty acid synthesis in the liver. *J. Clin. Invest.* 109:1125, 2002.

Kersten, S., Seydoux, J., Peters, J. M., Gonzalez, F. J., Desvergne, B., and Wahli, W. Peroxisome proliferator-activated receptor alpha mediates the adaptive response to fasting. *J. Clin. Invest.* 103:1489, 1999.

Krebs, H. A. Some aspects of the regulation of fuel supply in omnivorous animals. *Adv. Enzyme Regul.* 10:387, 1972.

Krebs, H. A., Williamson, D. H., Bates, M. W., Page, M. A., and Hawkins, R. A. The role of ketone bodies in caloric homeostasis. *Adv. Enzyme Regul.* 9:387, 1971.

Kurland, I. J. and Pilkis, S. J. Indirect and direct routes of hepatic glycogen synthesis. *FASEB J.* 3:2277, 1989.

Large, V., Peroni, O., Letexier, D., Ray, H., and Beylot, M. Metabolism of lipids in human white adipocyte. *Diabetes Metab.* 30:294, 2004.

Lecker, S. H., Solomon, V., Mitch, W. E., and Goldberg, A. L. Muscle protein breakdown and the critical role of the ubiquitin–proteasome pathway in normal and disease states. *J. Nutr.* 129:227S, 1999.

MacDonald, I. A. and Webber, J. Feeding, fasting and starvation: Factors affecting fuel utilization. *Proc. Nutr. Soc.* 54:267, 1995.

McGarry, J. D. and Brown, N. F. The mitochondrial carnitine-palmitoyl-CoA transferase system. *Eur. J. Biochem.* 244:1, 1997.

Newsholme, E. A. and Leech, A. R. *Biochemistry for the Medical Sciences*. New York: Wiley, 1983.

Newsholme, P. Why is L-glutamine metabolism important to cells of the immune system in health, postinjury, surgery or infection? *J. Nutr.* 131:2515S, 2001.

Nosadini, R., Avogaro, A., Doria, A., Fioretto, P., Trevisan, R., and Morocutti, A. Ketone body metabolism: A physiological and clinical overview. *Diabetes Metab. Rev.* 5:299, 1989.

Ookhtens, M. and Kaplowitz, N. Role of the liver in interorgan homeostasis of glutathione and cyst(e)ine. *Semin. Liver Dis.* 18:313, 1998.

Pedersen, O. The impact of obesity on the pathogenesis of non-insulin-dependent diabetes mellitus: A review of current hypotheses. *Diabetes/Metabolism Rev.* 5:495, 1989.

Pilkis, S. J., Claus, T. H., Kurland, I. J., and Lange, A. J. 6-Phosphofructo-2-kinase/fructose-2,6-bisphosphatase: A metabolic signaling enzyme. *Annu. Rev. Biochem.* 64:799, 1995.

Randle, P. J. Metabolic fuel selection: General integration at the whole-body level. *Proc. Nutr. Soc.* 54:317, 1995.

Rider, M. H., Bertrand, L., Vertommen, D., Michels, P. A., Rousseau, G. G., and Hue, L. 6-Phosphofructo-2-kinase/fructose-2,6-bisphosphatase: Head-to-head with a bifunctional enzyme that controls glycolysis. *Biochem J.* 381:561, 2004.

Roach, P. J. Glycogen and its metabolism. *Curr. Mol. Med.* 2:101, 2002.

Ruderman, N. B., Park, H., Kaushik, V. K., Dean, D., Constant, S., Prentki, M., and Shaha, A. K. AMPK as a metabolic switch in rat muscle, liver and adipose tissue after exercise. *Acta Physiol. Scand.* 178:435, 2003.

Ruderman, N. B., Saha, A. K., Vavvas, D., and Witters, L. A. Malonyl CoA, fuel sensing, and insulin resistance. *Am. J. Physiol. Endocrinol. Metab.* 276:E1, 1999.

Speigelman, B. M. and Heinrich, R. Biological control through regulated transcriptional coactivators. *Cell* 119:157, 2004.

Shulman, G. I. and Landau, B. R. Pathways of glycogen repletion. *Physiol. Rev.* 72:1019, 1992.

Steppan, C. M., Bailey, S. T., Bhat, S., Brown, E. J., Banerjee, R., Wright, C. M., Patel, H. R., Ahima, R. S., and Lazar, M. A. The hormone resistin links obesity to diabetes. *Nature* 409:307, 2001.

Stubbs, M., Bashford, C. L., and Griffiths, J. R. Understanding the tumor metabolic phenotype in the genomic era. *Curr. Mol. Med.* 3:485, 2003.

Taylor, S. I. Diabetes mellitus. In: C. R. Scriver, A. R. Beaudet, W. S. Sly, and D. Valle (Eds.), *The Metabolic and Molecular Bases of Inherited Disease*, 8th ed. New York: McGraw-Hill, 2001.

Williamson, D. H. and Lund, P. Substrate selection and oxygen uptake by the lactating mammary gland. *Proc. Nutr. Soc.* 54:165, 1995.

Zick, Y. Molecular basis of insulin action. *Novartis Found. Symp.* 262:36, 2004.

QUESTIONS | Carol N. Angstadt

Multiple Choice Questions

1. Since the K_m of aminotransferases for amino acids is much higher than that of aminoacyl-tRNA synthetases:
   A. at low amino acid concentrations, protein synthesis takes precedence over amino acid catabolism.
   B. liver cannot accumulate amino acids.
   C. amino acids will undergo transamination as rapidly as they are delivered to the liver.
   D. any amino acids in excess of immediate needs for energy must be converted to protein.
   E. amino acids can be catabolized only if they are present in the diet.

2. Carnitine:
   A. is formed in all cells for their own use.
   B. is synthesized directly from free lysine.
   C. formation requires that lysyl residues in protein be methylated by S-adenosylmethionine.
   D. is important in the detoxification of peroxides.
   E. is cleaved to γ-butyrobetaine.

3. All of the following represent control of a metabolic process by substrate availability *except*:
   A. increased urea synthesis after a high protein meal.
   B. rate of ketogenesis.
   C. hypoglycemia of advanced starvation.
   D. response of glycolysis to fructose 2,6-bisphosphate.
   E. sorbitol synthesis.

4. Conversion of hepatic enzymes from non-phosphorylated to their phosphorylated form:
   A. always activates the enzyme.
   B. is more likely to occur in the fasted than in the well-fed state.
   C. is signaled by insulin.
   D. is always catalyzed by a cAMP-dependent protein kinase.
   E. usually occurs at threonine residues of the protein.

5. Muscle metabolism during exercise:
   A. is the same in both aerobic and anaerobic exercise.
   B. shifts from primarily glucose to primarily fatty acids as fuel during aerobic exercise.
   C. uses largely glycogen and phosphocreatine in the aerobic state.
   D. causes a sharp rise in blood ketone body concentration.
   E. uses only phosphocreatine in the anaerobic state.

6. Glutaminase:
   A. in renal cells is unaffected by blood pH.
   B. in liver is confined to perivenous hepatocytes.
   C. requires ATP for the reaction it catalyzes.
   D. is more active in liver in acidosis.
   E. in renal cells increases in acidosis.

Questions 7 and 8: Protein and calorie malnutrition are important nutritional problems, especially among children. Both often occur in developing countries when the child is weaned from breast milk. Protein malnutrition, kwashiorkor, occurs when the child is fed a diet adequate in calories (mostly carbohydrate) but deficient in protein. Inadequate caloric intake is called marasmus. In kwashiorkor, insulin levels are high and there is subcutaneous fat. Children with low weight for height can make a good recovery when properly fed although those with the reverse situation do not. Children with marasmus lack subcutaneous fat.

7. Adipose tissue responds to low insulin:glucagon ratio by:
   A. dephosphorylating the interconvertible enzymes.
   B. stimulating the deposition of fat.
   C. increasing the amount of pyruvate kinase.
   D. activation of hormone-sensitive lipase.
   E. releasing glutamine.

8. Which of the following would favor gluconeogenesis in the fasted state?
   A. Fructose 1,6-bisphosphate stimulation of pyruvate kinase.
   B. Acetyl CoA activation of pyruvate carboxylase.
   C. Citrate activation of acetyl CoA carboxylase.
   D. Malonyl CoA inhibition of carnitine palmitoyltransferase I.
   E. Fructose 2,6-bisphosphate stimulation of 6-phosphofructo-1-kinase.

Questions 9 and 10: Reye's syndrome is a serious illness of children. The use of aspirin by children with chickenpox was linked to the development of the disease, and parents are now warned not to give aspirin to children with viral infections. Reye's syndrome, which damages hepatic mitochondria, is characterized by dysfunction of brain (irritability, lethargy, and coma) and liver (elevated plasma free fatty acids, fatty liver, hypoglycemia, and hyperammonemia). Impairing mitochondria affects fatty acid oxidation, synthesis of carbamoyl phosphate and ornithine for ammonia detoxification, and synthesis of oxaloacetate for gluconeogenesis. Part of the therapy is administration of intravenous glucose, which, among other things, reduces proteolysis in muscle.

9. Arginine and ornithine are intermediates in the utilization of ammonia and in other processes. All of the following statements are correct *except*:
   A. ornithine for the urea cycle is synthesized from glutamate in the kidney.
   B. citrulline is a precursor for arginine synthesis by the kidney.
   C. kidney uses arginine in the synthesis of creatine for distribution to muscle.
   D. arginine is the precursor for nitric oxide.
   E. creatinine cleared by the kidney is generated from creatine phosphate in muscle.

10. Muscle proteolysis releases branched-chain amino acids which:
   A. can also be synthesized from alanine.
   B. can be catabolized by muscle but not liver.
   C. are the main amino acids metabolized by intestine.
   D. are the amino acids released in largest amounts by muscle.
   E. are a major source of nitrogen for alanine and glutamine produced in muscle.

Questions 11 and 12: Diabetes mellitus is a disease in which glucose metabolism is impaired. Type 2 occurs primarily in middle-aged to older obese individuals while type 1 usually appears in childhood or in the teens. Insulin's ability to control blood glucose levels is affected either because of lack (or very low levels) of insulin or inadequate function of insulin (or both).

11. In type 2 (noninsulin-dependent) diabetes mellitus:
   A. hypertriglyceridemia does not occur.
   B. ketoacidosis in the untreated state is always present.

C. β-cells of the pancreas are no longer able to make any insulin.
D. may be accompanied by high levels of insulin in the blood.
E. severe weight loss always occurs.

12. Insulin normally does all of the following *except*:
A. recruit glucose transporters 4 (GLUT4) from intracellular sites to the plasma membrane.
B. activate protein kinase A and AMP-dependent protein kinase (AMPK).
C. activate phosphoprotein phosphatases.
D. signal an increase in sterol response element binding proteins (SREBP-1).
E. inhibit ketogenesis at levels lower than required to inhibit gluconeogenesis.

Problems

13. What metabolic and hormonal changes account for decreased gluconeogenesis in phase IV (2–24 days of starvation) of glucose homeostasis in humans?

14. Why is the measurement of hemoglobin A_{1c} a better measure of a diabetic's blood glucose status than a daily blood glucose measurement?

ANSWERS

1. **A** High K_m means a reaction will proceed slowly at low concentrations, whereas a low K_m means the reaction can be rapid under the same circumstances. Protein synthesis requires only that all amino acids be present.
2. **C** These trimethyllysines are released when protein is hydrolyzed. A: Only liver and kidney have the complete synthetic pathway. B: Lysine must first be present in cellular protein. D: This is glutathione. E: This is a precursor.
3. **D** Fructose 2,6-bisphosphate is an allosteric effector (activates the kinase and inhibits the phosphatase) of the enzyme controlling glycolysis. A: After a high protein meal, the intestine produces ammonia and precursors of ornithine for urea synthesis. B: Ketogenesis is dependent on the availability of fatty acids. C: This represents lack of gluconeogenic substrates. E: This leads to complications in diabetes.
4. **B** In the well-fed state, insulin:glucagon is high and cAMP levels are low. A: Some enzymes are active when phosphorylated; for others the reverse is true. C: Glucagon signals the phosphorylation of hepatic enzymes by elevating cAMP. D: This is the most common, though not only, mechanism of phosphorylation. E: The most common site for phosphorylation in metabolic enzymes is serine.
5. **B** Anaerobically exercised muscle uses glucose almost exclusively; aerobically exercised muscle uses fatty acids and ketone bodies. D: Ketone bodies are good aerobic substrates so the blood concentration does not increase greatly. E: Phosphocreatine is only a short-term source of ATP.
6. **E** Glutaminase activity is elevated in the kidney during acidosis. A: See above. B: This is the site of glutamine synthesis. C: Glutamine synthetase requires ATP; glutaminase catalyzes hydrolysis of glutamine to glutamate and ammonia. D: Less flux through liver glutaminase during acidosis permits glutamine to escape liver for use by the kidney.
7. **D** Phosphorylation activates hormone-sensitive lipase to mobilize fat. A: Low insulin:glucagon means high cAMP and high activity of cAMP-dependent protein kinase and protein phosphorylation. C: cAMP works by stimulating covalent modification of enzymes. E: This occurs in muscle.
8. **B** Pyruvate carboxylase is a key gluconeogenic enzyme. A, E: Stimulation of these enzymes stimulates glycolysis. C, D: Malonyl CoA inhibits transport of fatty acids into mitochondria for β oxidation, a necessary source of energy for gluconeogenesis.
9. **A** Kidney lacks the enzyme needed to convert glutamate to glutamate semialdehyde. B: This is true in both kidney and liver. C: The reaction requires *S*-adenosylmethionine. E: Creatinine is thus a measure of both muscle mass and renal function.
10. **E** Transamination of branched-chain amino acids transfers the nitrogen to alanine or glutamine. A: Branched-chain amino acids are essential amino acids and cannot be synthesized from other amino acids. B: Muscle transaminates them, and then the branched-chain α-ketoacids are catabolized by liver. C: Intestine metabolizes primarily glutamine. D: Alanine and glutamine are.
11. **D** The problem is insulin resistance, not complete failure to produce insulin. A: Hypertriglyceridemia is characteristic. B: Ketoacidosis is common only in the insulin-dependent type. C: See correct answer. E: Most patients are obese and remain so.
12. **B** These enzymes are active in their phosphorylated states and insulin promotes dephosphorylation. A: This promotes uptake of glucose. D: This increases transcription of genes encoding lipogenic enzymes. E: This may account in part for the lack of ketoacidosis in type 2 diabetics.
13. In phase IV, most tissues are using primarily fatty acids and ketone bodies. Ketone bodies are now sufficiently high that they can enter the brain and reduce the glucose requirement. Low insulin:glucagon stimulates lipolysis and gluconeogenesis. Increased fatty acid oxidation increases acetyl CoA and NADH. High NADH allosterically inhibits the tricarboxylic acid cycle and the accumulating acetyl CoA is converted to ketone bodies. Low insulin:glucagon also decreases catabolism of muscle protein. Decreased gluconeogenesis conserves protein for a longer period of time.
14. A single measurement measures glucose concentration only at that particular time, but levels vary throughout the day with food intake, exercise, and emotional state. Hemoglobin A_{1c} results from the nonenzymatic reaction between glucose and the amino-terminal valine of the β-chain of hemoglobin. The concentration of hemoglobin A_{1c} is correlated with the average blood glucose concentration over a period of several weeks.

PART V PHYSIOLOGICAL PROCESSES

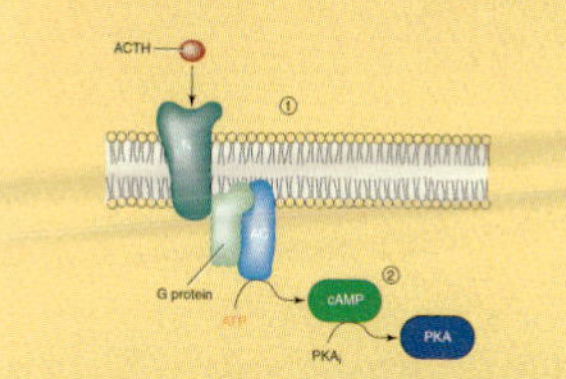

BIOCHEMISTRY OF HORMONES

Thomas J. Schmidt and Gerald Litwack

Textbook of Biochemistry With Clinical Correlations, Sixth Edition, Edited by Thomas M. Devlin

23.1 OVERVIEW

The hormones that will be discussed in this chapter fall into three major categories: **peptide and protein hormones**, hormones derived from the amino acid tyrosine including **thyroid hormones** and **catecholamine hormones** (epinephrine and norepinephrine), and **steroid hormones.** Taken collectively, these hormones regulate the growth, differentiation, and function of a wide variety of target cells that express the cognate receptors for specific hormones. The peptide hormones and the catecholamine hormones interact with **cell surface receptors** and transmit their signals through **second messengers** that are generated intracellularly (Figure 23.1). These second messenger systems include cAMP and protein kinase A, inositol trisphosphate, diacylglycerol and protein kinase C, and cGMP and protein kinase G. Each of these intracellular second messenger systems will be discussed in the context of representative hormone action.

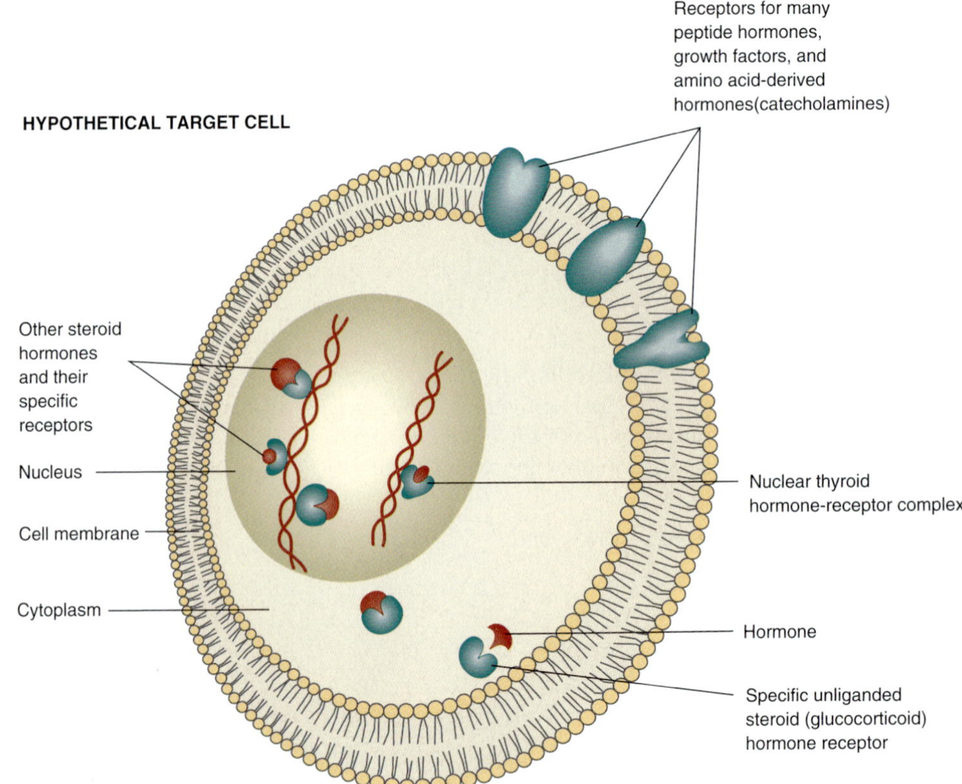

FIGURE 23.1
Diagram showing the different locations of classes of hormone receptors expressed by a target cell.

The binding of insulin to its cell surface receptor (which activates auto-tyrosine kinase activity), as well as the subsequent signal transduction pathway for insulin action, will be presented. The steroid hormones, which are derived from cholesterol and include glucocorticoid hormones (cortisol), mineralocorticoid hormones (aldosterone), and the sex hormones (testosterone, estradiol, and progesterone), will be discussed. In contrast to peptide hormones, steroid hormones diffuse freely across the plasma membrane and bind specifically to their **intracellular receptors** that function as **ligand-activated transcription factors** (transactivators or repressors). The intracellular receptors for all of the steroid hormones, as well as for nonsteroid hormones including thyroid hormone, the active metabolite of vitamin D_3, and retinoic acid, belong to the **steroid receptor supergene family** and share sequence homologies. The basic domains of these receptors, including their C-terminal ligand-binding domains, DNA-binding domains, and N-terminal immunogenic domains, will be discussed. A general model will be presented to describe the specific steps involved in the interaction of a steroid molecule with its target cell, including: hormone-induced "activation" or transformation of the receptor protein, which involves dissociation of receptor-associated heat shock proteins; nuclear translocation; and binding of the steroid–receptor complexes to specific hormone response elements within the DNA.

23.2 HORMONES AND THE HORMONAL CASCADE SYSTEM

The definition of a hormone has expanded over the last several decades. Hormones secreted by endocrine glands were considered for many years to represent all of the physiologically relevant hormones. Today, the term "**hormone**" refers to any substance in an organism that carries a "signal" to generate some sort of alteration at the cellular level. **Endocrine hormones** represent a class of hormones that are synthesized in one tissue, or "gland," and travel through the general circulation to reach distant target cells that expresses cognate receptors. **Paracrine hormones** are secreted by a cell and then travel a relatively short distance to interact with their cognate receptors on a neighboring cell and **autocrine hormones** are produced by the same cell that subsequently functions as the target for that hormone (neighboring cells may also be targets). Endocrine hormones are frequently more stable than autocrine hormones that exert their effects over very short distances.

Hormonal Cascade Systems Amplify Specific Signals

Before we focus on the specific details concerning each hormone, we need to look more broadly at the organization of the endocrine system and the hormonal hierarchy. For many hormonal systems in higher animals the signal pathway originates in the brain and culminates within the target cell. Figure 23.2 outlines the sequence of events in this cascade. A stimulus may originate in the external environment or within the organism, and it can be transmitted as action potentials, chemical signals, or both. In many cases, such signals are forwarded to the limbic system and subsequently to the hypothalamus, the pituitary, and the target gland that secretes the final hormone. This hormone then affects various target cells, frequently in proportion to the number of cognate receptors expressed by a given cell. This may be a true **cascade** in the sense that increasing amounts of hormone are generated at successive levels (hypothalamus, pituitary, and target gland) and in the sense that the half-lives of blood-borne hormones tend to become longer descending down the cascade. Consider a specific hormone secreted via a cascade. An environmental stress such as change in temperature, noise, trauma, and so on, results in a signal to the hippocampal structure from the limbic system and signals release of nanogram amounts of a hypothalamic releasing hormone, corticotropin-releasing hormone (CRH), which has a $t_{1/2}$ in the bloodstream of several minutes. CRH travels down a closed portal system to the **anterior pituitary**, where it binds its cognate receptor in the membrane of corticotropic cells and initiates intracellular events that

CLINICAL CORRELATION **23.1**

Testing Activity of the Anterior Pituitary

Releasing hormones and chemical analogs, particularly of the smaller peptides, are now routinely synthesized. The gonadotropin-releasing hormone, a decapeptide, is available for use in assessing the function of the anterior pituitary. This is of importance when a disease situation may involve either the hypothalamus, the anterior pituitary, or the end organ. Infertility is an example of such a situation. What needs to be assessed is which organ is at fault in the hormonal cascade. Initially, the end organ, in this case the gonads, must be considered. This can be accomplished by injecting the anterior pituitary hormone LH or FSH. If sex hormone secretion is elicited, then the ultimate gland would appear to be functioning properly. Next, the anterior pituitary would need to be analyzed. This can be done by i.v. administration of synthetic GnRH; by this route, GnRH can gain access to the gonadotropic cells of the anterior pituitary and elicit secretion of LH and FSH. Routinely, LH levels are measured in the blood as a function of time after the injection. These levels are measured by radioimmunoassay (RIA) in which radioactive LH or hCG is displaced from binding to an LH-binding protein by LH in the serum sample. The extent of the competition is proportional to the amount of LH in the serum. In this way a progress of response is measured that will be within normal limits or clearly deficient. If the response is deficient, the anterior pituitary cells are not functioning normally and are the cause of the syndrome. On the other hand, normal pituitary response to GnRH would indicate that the hypothalamus was nonfunctional. Such a finding would prompt examination of the hypothalamus for conditions leading to insufficient availability/production of releasing hormones. Obviously, the knowledge of hormone structure and the ability to synthesize specific hormones permits the diagnosis of these disease states.

Source: Marshall, J. C. and Barkan, A. L., Disorders of the hypothalamus and anterior pituitary. In: W. N. Kelley (Ed.), *Internal Medicine.* New York: Lippincott, 1989, p. 2159. Conn, P. M. The molecular basis of gonadotropin-releasing hormone action. *Endocr. Rev.* 7:3, 1986.

In the long feedback loop, the final hormone binds a cognate receptor in/on cells of the anterior pituitary, hypothalamus, and CNS to prevent further synthesis/secretion of releasing hormones. The short feedback loop is exemplified by the pituitary tropic hormone that feeds back negatively on the hypothalamus operating through a cognate receptor. In ultra-short feedback loops the hypothalamic releasing factor feeds back on the hypothalamus (via systemic circulation) to inhibit its own further secretion. Clinical Correlation 23.1 describes approaches for testing the responsiveness of the anterior pituitary gland to releasing hormones.

Major Polypeptide Hormones and their Actions

Since cellular communication is so specific, it is not surprising that there are a large number of hormones in the body and new hormones continue to be discovered. Table 23.2 presents some major polypeptide hormones and their actions and shows that many hormones cause release of other hormones. This is particularly the case for hormonal cascade systems like that presented in Figures 23.2 and 23.3.

Polypeptide Hormones of Anterior Pituitary

Polypeptide hormones of the anterior pituitary are shown in Figure 23.4 along with their controlling hormones from the hypothalamus. The major hormones are growth hormone (GH), thyrotropin or thyroid-stimulating hormone (TSH), adrenocorticotropic hormone (ACTH), β-lipotropin (β-LTH), β-endorphin (from pars intermedia-like cells), α-MSH (from pars intermedia-like cells), β-MSH (from pars intermedia-like cells), corticotropin-like intermediary peptide (CLIP; from pars intermedia-like cells), prolactin (PRL), follicle-stimulating hormone (FSH), and luteinizing hormone (LH). All are single polypeptide chains, except TSH, FSH, and LH, which are dimers that share a similar or identical α subunit. Since the intermediate lobe in humans is rudimentary, the circulating levels of free α- and β-MSH are relatively low. It is of interest, particularly in the human, that MSH receptors recognize and are activated by ACTH, because the first 13 amino acids of ACTH contain the α-MSH sequence. For this reason, ACTH may be an important contributing factor to skin pigmentation and may exceed the importance of MSH, especially in conditions where the circulating level of ACTH is

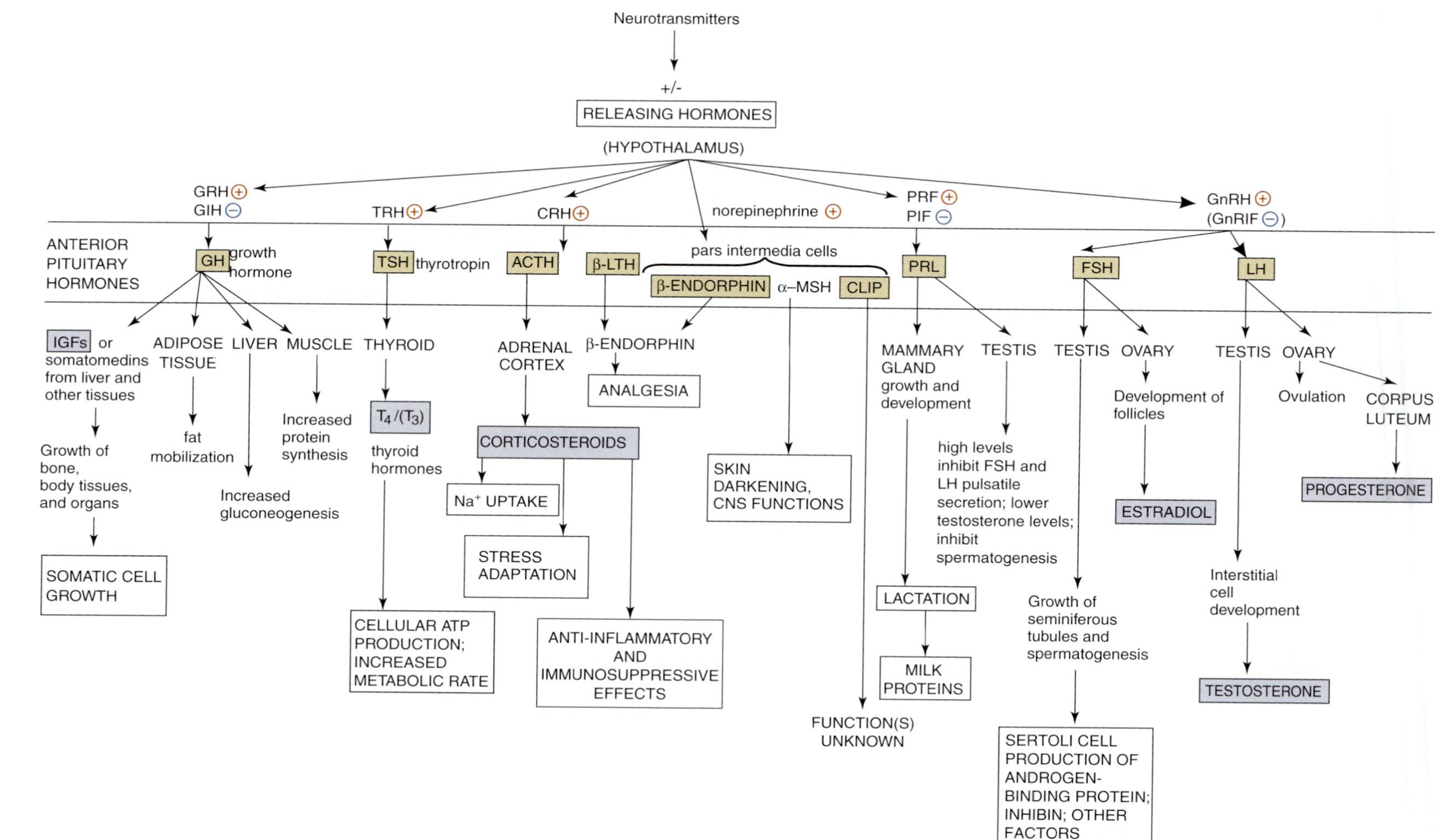

FIGURE 23.4

Overview of anterior pituitary hormones with hypothalamic releasing hormones and their actions.

TABLE 23.2 Important Polypeptide Hormones in the Body and Their Actions[a]

Source	Hormone	Action
Hypothalamus	Thyrotropin-releasing hormone (TRH)	Acts on thyrotrope to release TSH
	Gonadotropin-releasing hormone (GnRH)	Acts on gonadotrope to release LH and FSH from the same cell
	Growth hormone-releasing hormone or somatocrinin (GRH)	Acts on somatotrope to release GH
	Growth hormone release inhibiting hormone or somatostatin (GIH)	Acts on somatotrope to prevent release of GH
	Corticotropin-releasing hormone (CRH)	Acts on corticotrope to release ACTH and β-lipotropin
	Vasopressin is a helper hormone to CRH in releasing ACTH; angiotensin II also stimulates CRH action in releasing ACTH	
	Prolactin-releasing factor (PRF) (not well established)	Acts on lactotrope to release PRL
	Prolactin release inhibiting factor (PIF) (not well established; may be a peptide hormone under control of dopamine or may be dopamine itself)	Acts on lactotrope to inhibit release of PRL
Anterior pituitary	Thyrotropin (TSH)	Acts on thyroid follicle cells to bring about release of $T_4(T_3)$
	Luteinizing hormone (LH) (human chorionic gonadotropin, hCG, is a similar hormone from the placenta)	Acts on Leydig cells of testes to increase testosterone synthesis and release; acts on corpus luteum of ovary to increase progesterone production and release
	Follicle-stimulating hormone (FSH)	Acts on Sertoli cells of seminiferous tubule to increase secretion of androgen-binding protein (ABP) and increase estradiol production from testosterone; acts on ovarian follicles to stimulate maturation of ovum and production of estradiol
	Growth hormone (GH)	Acts on a variety of cells to produce IGFs (or somatomedins), cell growth, and bone growth
	Adrenocorticotropic hormone (ACTH)	Acts on cells in the adrenal cortex to increase cortisol production and secretion
	β-Endorphin	Acts on cells and neurons to produce analgesic and other effects
	Prolactin (PRL)	Acts on mammary gland to cause differentiation of secretory cells (with other hormones) and to stimulate synthesis of components of milk
	Melanocyte-stimulating hormone (MSH)	Acts on skin cells to cause the dispersion of melanin (skin darkening)
Ultimate gland hormones	Insulin-like growth factors (IGF)	Respond to GH and produce growth effects by stimulating cell mitosis
	Thyroid hormone (T_4/T_3) (amino acid-derived hormone)	Responds to TSH and stimulates oxidation in many cells
	Opioid peptides	May derive as breakdown products of γ-lipotropin or β-endorphin or from specific gene products; can respond to CRH or dopamine and may produce analgesia and other effects

(*continued overleaf*)

TABLE 23.2 (*continued*)

Source	Hormone	Action
Ovarian granulosa cells; testicular Sertoli cells.	Inhibin	Stimulates steroidogenesis in ovaries and testes; regulates secretion of FSH from anterior pituitary. Second form of inhibin (activin) may stimulates FSH secretion
Intermediate lobe of pituitary gland	Corticotropin-like intermediary peptide (CLIP)	Derives from intermediate pituitary by degradation of ACTH; contains β-cell tropin activity, which stimulates insulin release from β cells in presence of glucose
Peptide hormones responding to other signals than anterior pituitary hormones	Arginine vasopressin (AVP; antidiuretic hormone, ADH)	Responds to increased activity in osmoreceptor, which senses extracellular [Na^+]; increases water reabsorption from distal kidney tubule
	Oxytocin	Responds to suckling reflex and estradiol; causes milk "let down" or ejection in lactating female, involved in uterine contractions of labor; luteolytic factor produced by corpus luteum; decreases steroid synthesis in testis
β Cells of pancreas respond to glucose and other blood constituents	Insulin	Increases tissue utilization of glucose
α Cells of pancreas respond to low levels of glucose and falling serum calcium	Glucagon	Decreases tissue utilization of glucose to elevate blood glucose
Derived from angiotensinogen by actions of renin and converting enzyme	Angiotensin II and III (AII and AIII)	Renin initially responds to decreased blood volume or decreased [Na^+] in the macula densa of the kidney. AII/AIII stimulate outer layer of adrenal cortex to synthesize and release aldosterone
Released from heart atria in response to hypovolemia; regulated by other hormones	Atrial natriuretic factor (ANF) or atriopeptin	Acts on adrenal cortex cells to decrease aldosterone release; has other effects also
Generated from plasma, gut, or other tissues	Bradykinin	Modulates extensive vasodilation resulting in hypotension
Hypothalamus and intestinal mucosa	Neurotensin	Effects on gut; may have neurotransmitter actions
Hypothalamus, CNS, and intestine	Substance P	Pain transmitter, increases smooth muscle contractions of the GI tract
Nerves and endocrine cells of gut; hypothermic hormone	Bombesin	Increases gastric acid secretion
	Cholecystokinin (CCK)	Stimulates gallbladder contraction and bile flow; increases secretion of pancreatic enzymes
Stomach antrum	Gastrin	Increases secretion of gastric acid and pepsin
Duodenum at pH values below 4.5	Secretin	Stimulates pancreatic acinar cells to release bicarbonate and water to elevate duodenal pH
Hypothalamus and GI tract	Vasointestinal peptide (VIP)	Acts as a neurotransmitter in peripheral autonomic nervous system; relaxes vascular smooth muscles; increases secretion of water and electrolytes from pancreas and gut
Kidney	Erythropoietin	Acts on bone marrow for terminal differentiation and initiates hemoglobin synthesis
Ovarian corpus luteum	Relaxin	Inhibits myometrial contractions; relaxes pelvic ligaments and increases dilation of cervix.
	Human placental lactogen (hPL)	Acts like PRL and GH.

TABLE 23.2 (*continued*)

Source	*Hormone*	*Action*
Salivary gland	Epidermal growth factor	Stimulates proliferations of cells derived from ectoderm and mesoderm together with serum; inhibits gastric secretion
Thymus	Thymopoietin (α-thymosin)	Stimulates phagocytes; stimulates differentiation of precursors into immune competent T cells
Parafollicular C cells of thyroid gland	Calcitonin (CT)	Lowers serum calcium
Parathyroid glands	Parathyroid hormone (PTH)	Stimulates bone resorption; stimulates phosphate excretion by kidney; raises serum calcium levels
Endothelial cells of blood vessels	Endothelin	Vasoconstriction

Source: Part of this table is reproduced from Norman, A. W. and Litwack, *G. Hormones*. Orlando, FL: Academic Press, 1987.

[a] This is only a partial list of polypeptide hormones in humans. TSH, thyroid-stimulating hormone or thyrotropin; LH, luteinizing hormone; FSH, follicle-stimulating hormone; GH, growth hormone; ACTH, adrenocorticotropic hormone; PRL, prolactin; T_4, thyroid hormone (also T_3); IGF, insulin-like growth factor. For the releasing hormones and for some hormones in other categories, the abbreviation may contain "H" at the end when the hormone has been well characterized, and "F" in place of H to refer to "Factor" when the hormone has not been well characterized. Names of hormones may contain "tropic" or "trophic" endings; tropic is mainly used here. Tropic refers to a hormone generating a change, whereas trophic refers to growth promotion. Both terms can refer to the same hormone at different stages of development. Many of these hormones have effects in addition to those listed here.

high. The clinical consequences of hypopituitarism are presented in Clinical Correlation 23.2.

23.3 SYNTHESIS OF POLYPEPTIDE AND AMINO ACID-DERIVED HORMONES

Polypeptide Hormones: Gene Coding

Genes for polypeptide hormones contain the coding sequence for the hormone and the control elements upstream of the structural gene. In some cases, more than one hormone is encoded in a gene. For example, proopiomelanocortin generates at least eight hormones from a single gene product. Oxytocin and vasopressin are each encoded on separate genes together with their respective **neurophysin** proteins. Neurophysins are co-secreted with their respective hormone but have no known endocrine action.

Proopiomelanocortin is Precursor for Eight Hormones

Proopiomelanocortin is a hormone precursor for the following hormones: ACTH, β-lipotropin, γ-lipotropin, γ-MSH, α-MSH, CLIP, β-endorphin, and potentially β-MSH and enkephalins (Figure 23.5). All of these products are not expressed simultaneously in a single cell type but are produced in separate cells based on their content of specific required proteases, specific metabolic controls, and the presence of regulators. Thus, while proopiomelanocortin is expressed in both corticotropes of the anterior pituitary and pars intermedia cells, the stimuli and products are different (Table 23.3). Pars intermedia is a discrete anatomical structure located between the anterior and posterior pituitary in the rat (Figure 23.6). In the human, however, pars intermedia is not a discrete anatomical structure, although some residual pars intermedia-like cells may be present in the equivalent location.

Genes of Polypeptide Hormone May Encode for Additional Peptides

Other genes that encode more than one peptide are those for vasopressin and oxytocin and their accompanying neurophysins. Vasopressin, neurophysin II, and a glycoprotein of unknown function are released from the vasopressin precursor. A similar situation exists for oxytocin and neurophysin I except that no glycoprotein is released (Figure 23.7). **Vasopressin** and **neurophysin II** are co-released in response to stimuli from baroreceptors and osmoreceptors, which sense a fall in blood pressure or a rise in

CLINICAL CORRELATION 23.2
Hypopituitarism

The hypothalamus is connected to the anterior pituitary by a delicate stalk that contains the portal system through which releasing hormones, secreted from the hypothalamus, gain access to the anterior pituitary cells. In the cell membranes of these cells are specific receptors for releasing hormones. In most cases, different cells express different releasing hormone receptors. The connection between the hypothalamus and anterior pituitary can be disrupted by trauma or tumors. Trauma can occur in automobile accidents or other local damaging events that may result in severing of the stalk, thus preventing the releasing hormones from reaching their target anterior pituitary cells. When this happens, the anterior pituitary cells no longer have their signaling mechanism for the release of anterior pituitary hormones. In the case of tumors of the pituitary gland, all of the anterior pituitary hormones may not be shut off to the same degree or the secretion of some may disappear sooner than others. In any case, if hypopituitarism occurs, this condition may result in a life-threatening situation in which the clinician must determine the extent of loss of pituitary hormones, especially ACTH. Posterior pituitary hormones—oxytocin and vasopressin—may also be lost, precipitating a problem of excessive urination (vasopressin deficiency) that must be addressed. The usual therapy involves administration of the end-organ hormones, such as thyroid hormone, cortisol, sex hormones, and progestin; with female patients it is also necessary to maintain the ovarian cycle. These hormones can be easily administered in oral form. Growth hormone deficiency is not a problem in the adult but would be an important problem in a growing child. The patient must learn to anticipate needed increases of cortisol in the face of stressful situations. Fortunately, these patients are usually maintained in reasonably good condition.

Source: Marshall, J. C. and Barkan, A. L. Disorders of the hypothalamus and anterior pituitary. In: W. N. Kelley, (Ed.), *Internal Medicine*. New York: Lippincott, 1989, p 2159. Robinson, A. G. Disorders of the posterior pituitary. In: W. N. Kelley (Ed.), *Internal Medicine*. New York: Lippincott, 1989, p. 2172.

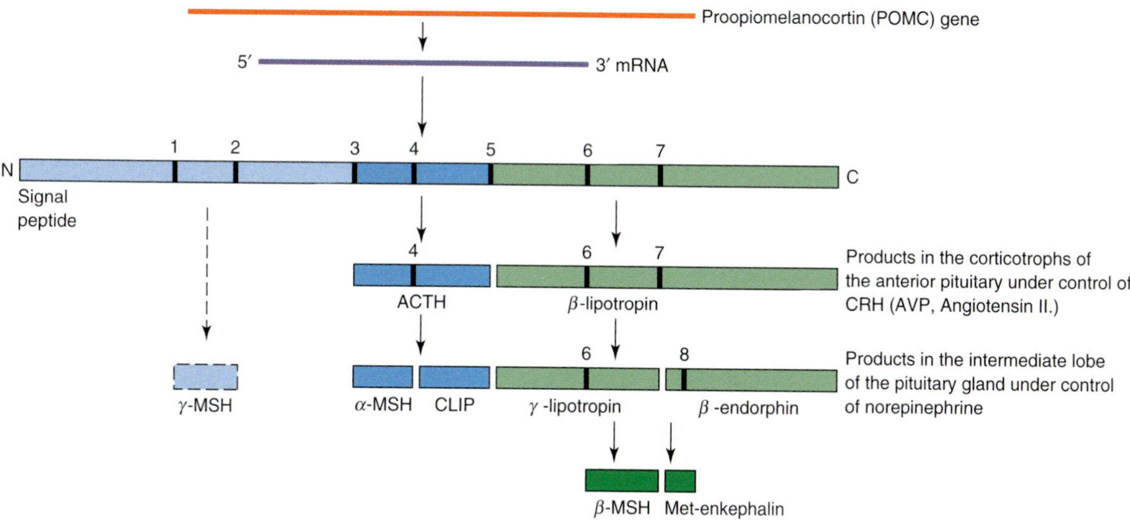

FIGURE 23.5

Proopiomelanocortin is a polypeptide encoded by one gene. The dark vertical bars represent proteolytic cleavage sites for specific enzymes. The cleavage sites are Arg-Lys, Lys-Arg, or Lys-Lys. Some specificity also may be conferred by neighboring amino acid residues. In the anterior pituitary, enzymes cleave at sites 3 and 5, releasing the major products, ACTH and β-lipotropin. In the pars intermedia, especially in vertebrates below humans, these products are further cleaved at major sites 4, 6, and 7 to release α-MSH, CLIP, γ-lipotropin, and β-endorphin. Some β-lipotropin may be further degraded to form β-endorphin. The anterior pituitary is under the positive control of the CRH and its helpers, arginine vasopressin (AVP), and angiotensin II. AVP by itself does not release ACTH but enhances the action of CRH in this process. The intermediary pituitary is under the positive control of norepinephrine. β-Endorphin also contains a pentapeptide, enkephalin, which potentially could be released at some point (hydrolysis at 8).

TABLE 23.3 Summary of Stimuli and Products of Proopiomelanocortin[a]

Cell type	Corticotroph	Pars intermedia
Stimulus	CRH (+) (Cortisol (−))	Dopamine (−) Norepinephrine (+)
Auxiliary stimulus	AVP, AII	
Major products	ACTH, β-lipotropin (β-endorphin)	α-MSH, CLIP, γ-lipotropin, β-endorphin

[a] CRH, corticotropin-releasing hormone; AVP, arginine vasopressin; AII, angiotensin II; ACTH, adrenocorticotropin; α-MSH, α melanocyte-stimulating hormone; CLIP, corticotropin-like intermediary peptide.

Note: Although there are *pars intermedia cells* in the human pituitary gland, they do not represent a distinct lobe.

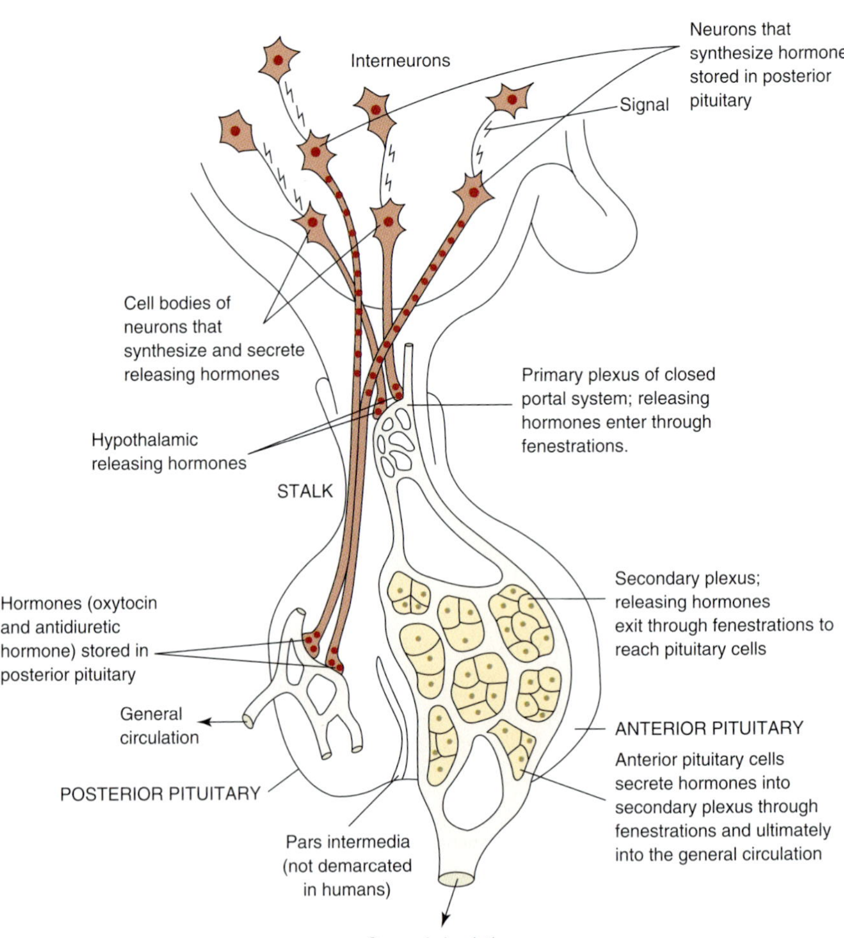

FIGURE 23.6

Anatomical relationship between hypothalamus and pituitary gland. The major vascular network is a primary plexus which releasing hormones enter through fenestrations. The secondary plexus is in the anterior pituitary where the releasing hormones are transported out of fenestrations to interact with the anterior pituitary target cells. Hypothalamic releasing hormones cause secretion of the anterior pituitary hormones, which enter the general circulation. Adapted from Norman, A. W. and Litwack, G. *Hormones.* New York: Academic Press, 1987, p. 104.

extracellular sodium ion concentration, respectively. **Oxytocin** and **neurophysin I** are co-released by the suckling response in lactating females or by stimuli mediated by a specific cholinergic mechanism. Oxytocin-neurophysin I release is triggered by injection of estradiol. Release of vasopressin-neurophysin II is stimulated by administration of nicotine. Although oxytocin is well known for its milk "let down" action in the lactating female, in the male it seems to be associated with an increase in testosterone synthesis in the testes.

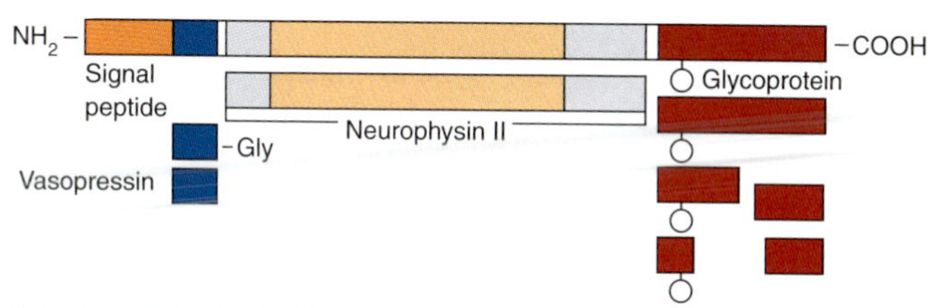

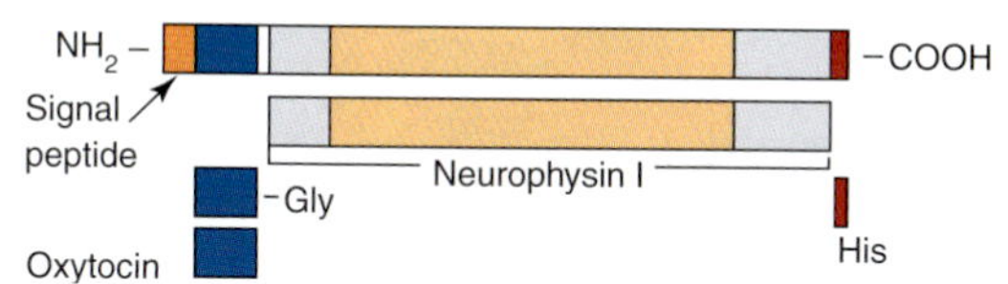

FIGURE 23.7

Prepro-vasopressin and prepro-oxytocin. Proteolytic maturation proceeds from top to bottom for each precursor. The organization of the gene translation products is similar except that a glycopeptide is included on the vasopressin precursor in the C-terminal region. Orange bars of the neurophysin represent conserved amino acid regions; gray bars represent variable C and N termini.

Redrawn with permission from Richter, D. VP and OT are expressed as polyproteins. *Trends Biochem. Sci.* 8:278, 1983.

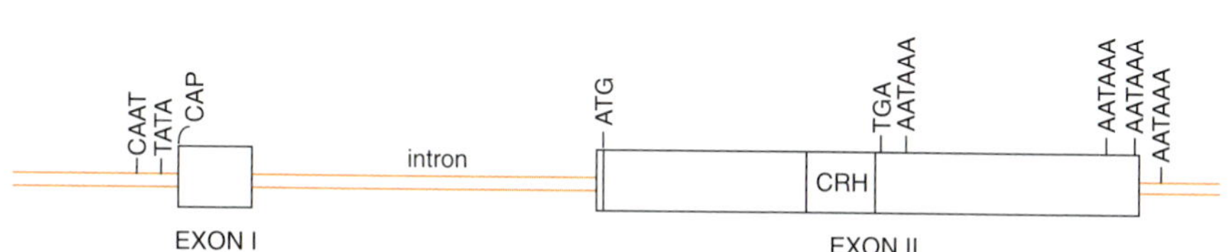

FIGURE 23.8

Nucleic acid sequence of rat proCRH genes. Schematic representation of the rat proCRH gene. Exons are shown as blocks and the intron by a double red line. The TATA and CAAT sequence, putative cap site, translation initiation ATG, translation terminator TGA, and poly(A) addition signals (AATAAA) are indicated. The location of the CRH peptide is indicated by CRH.

Redrawn from Thompson, R. D., Seasholz, A. F., and Herbert, E. *Mol. Endocrinol.* 1:363, 1987.

Other polypeptide hormones are encoded by a single gene that does not encode for another protein or hormone. An example is the gene encoding the decapeptide **GnRH**. The gene for this hormone appears to reside to the left of a gene for the GnRH-associated peptide (GAP), which may be capable of inhibiting prolactin release. Thus, GnRH and the prolactin release inhibiting factor (GAP) appear to be co-secreted by the same hypothalamic cells but are not co-encoded. Many genes for hormones encode only one copy of a hormone and thus may be the more common situation. An example is shown in Figure 23.8. Information encoding for CRH is contained in the second axon.

One Gene can Code for Multiple Copies of a Hormone

An example of multiple copies of a single hormone encoded on a single gene is the enkephalins secreted by chromaffin cells of the adrenal medulla. **Enkephalins** are pentapeptides with opioid activity; methionine-enkephalin (Met-ENK) and leucine-enkephalin (Leu-ENK) have the structures

Tyr-Gly-Gly-Phe-Met (Met-ENK)

Tyr-Gly-Gly-Phe-Leu (Leu-ENK)

A model of the enkephalin precursor is presented in Figure 23.9, which encodes several Met-ENK (M) molecules and a molecule of Leu-ENK (L). The processing sites to release enkephalin molecules from the protein precursor involve Lys-Arg, Arg-Arg, and Lys-Lys bonds. Another example of one gene encoding for multiple copies of a hormone is the gene for the tripeptide hormone TRH. The TRH peptide sequence is actually repeated six times within the human TRH pre-prohormone.

Amino Acid-Derived Hormones

Epinephrine is Synthesized from Tyrosine

Synthesis of **epinephrine** (Figure 23.10) from phenylalanine/tyrosine occurs in the adrenal medulla (see p. 763). The steroid hormones aldosterone, cortisol, and

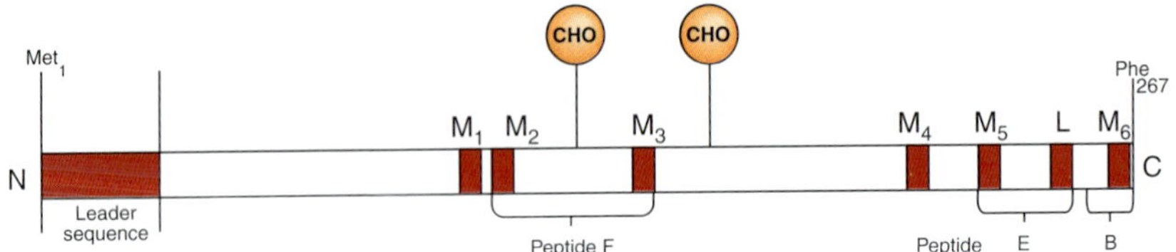

FIGURE 23.9

Model of enkephalin precursor. Distribution of Met-enkephalin sequences ($M_1 - M_6$) and Leu-enkephalin (L) sequences within the precursor of bovine adrenal medulla. CHO, potential carbohydrate attachment sites.
Redrawn from Comb, M., Seeburg, P. H., Adelman, J., Eiden, L., and Herbert, E. *Nature* 295:663, 1982.

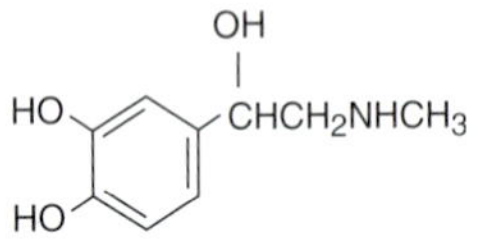

FIGURE 23.10

Structure of catecholamine hormone epinephrine.

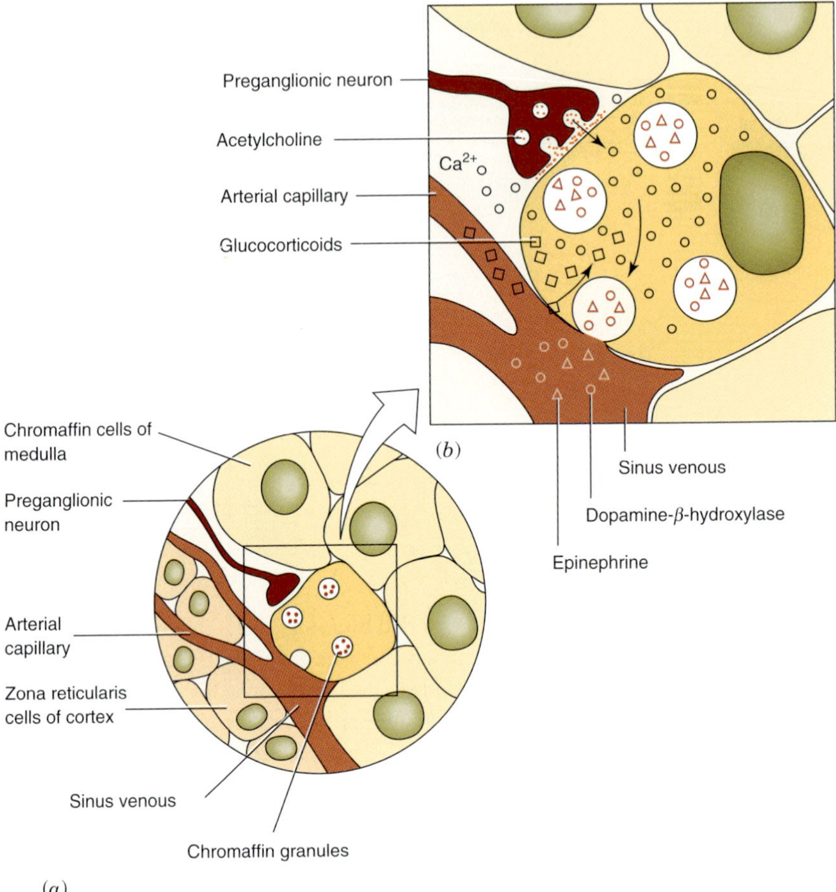

FIGURE 23.11

Relationship of adrenal medulla chromaffin cells to preganglionic neuron innervation and the structural elements involved in synthesis of epinephrine and discharge of catecholamines in response to acetylcholine. (*a*) Functional relationship between cortex and medulla for control of synthesis of adrenal catecholamines. Glucocorticoids that induce the PNMT enzyme reach the chromaffin cells from capillaries shown in (b). Discharge of catecholamines from storage granules in chromaffin cells after release of acetylcholine from nerve fiber stimulation. Calcium enters the cells, causing the fusion of granular and plasma membranes and exocytosis of the contents.
Reprinted with permission from Krieger, D. T. and Hughes, J. C. (Eds.). *Neuroendocrinology*. Sunderland, MA: Sinauer Associates, 1980.

dehydroepiandrosterone, which are also produced in the adrenal cortex, are discussed on page 929. Epinephrine is secreted from adrenal medulla chromaffin cells with some norepinephrine, enkephalins, and dopamine-β-hydroxylase. Secretion is signaled by the neural response to stress, which is transmitted by way of a preganglionic acetylcholinergic neuron (Figure 23.11). This increases intracellular Ca^{2+}, which stimulates exocytosis and release of the material stored in the **chromaffin granules** (Figure 23.11*b*). Glucocorticoid hormones secreted from the adrenal cortex in response to stress are transported through the adrenal medulla where cortisol induces **phenylethanolamine *N*-methyltransferase (PNMT)**, the enzyme that converts norepinephrine to epinephrine. Thus, in biochemical terms, the stress response at the level of the adrenal cortex ensures the production of epinephrine from the adrenal medulla (Figure 23.12). The epinephrine interacts with α-receptors on hepatocytes to increase blood glucose levels and interacts with α-receptors on vascular smooth muscle cells to cause vasoconstriction and increase blood pressure.

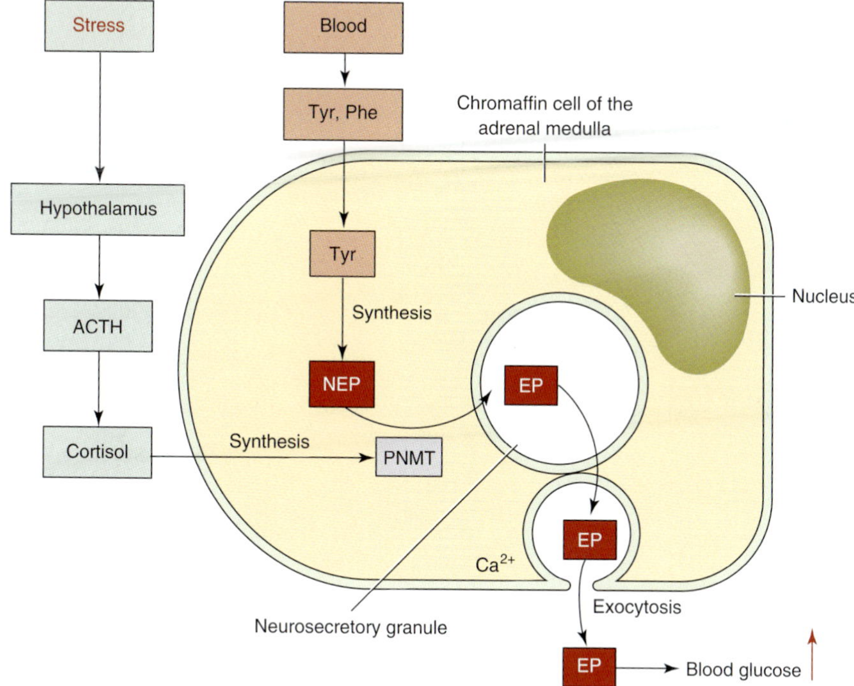

FIGURE 23.12

Biosynthesis, packaging, and release of epinephrine in adrenal medulla chromaffin cell. PNMT, phenylethanolamine *N*-methyltransferase; EP, epinephrine; NEP, norepinephrine. Neurosecretory granules contain epinephrine, dopamine β-hydroxylase, ATP, Met or Leu-enkephalin, and larger enkephalin-containing peptides or norepinephrine in place of epinephrine. Epinephrine and norepinephrine are often stored in different chromaffin granules.
Adapted from Norman, A. W. and Litwack, G. *Hormones.* New York: Academic Press, 1987, p. 464.

Synthesis of Thyroid Hormone Requires Incorporation of Iodine into Tyrosines of Thyroglobulin

An outline of the biosynthesis and secretion of thyroid hormone, **tetraiodo-L-thyronine (T_4)**, or **thyroxine**, and its more active metabolite, **triiodo-L-thyronine (T_3)**, is presented in Figures 23.13 and 23.14. The thyroid gland is differentiated to concentrate iodide from the blood and through the reactions shown in Figures 23.13 and 23.14. Monoiodotyrosine (MIT), diiodotyrosine (DIT), T_4, T_3, and reverse T_3 are produced within the **thyroglobulin** (TG) molecule. **Thyroglobulin** is a large glycoprotein that is stored in the lumen of the thyroid follicles. The coupling of an MIT and DIT or two DIT molecules can occur within the same thyroglobulin molecule or between two adjacent thyroglobulin molecules. Secretion of T_3 and larger amounts of T_4 into the bloodstream requires endocytosis of the thyroglobulin and its proteolysis within the follicular epithelial cell. The DIT and MIT released within the epithelial cell are then deiodinated, and the released iodide ions are recycled and reutilized for thyroid hormone synthesis. This recycling of iodine ions is extremely important, and mutations resulting in inactivation of the deiodinase enzyme can result in an iodide deficiency.

Inactivation and Degradation of Amino Acid-Derived Hormones

Most polypeptide hormones are degraded to amino acids by proteases, presumably in lysosomes. Some hormones contain modified amino acids; for example, the N-terminal amino acid may be **cycloglutamic acid** (**pyroglutamic acid**), and there may be a C-terminal amino acid amide (Table 23.4). Breakage of the cyclic glutamate ring or cleavage of the C-terminal amide inactivates many of these hormones. Such catalytic activities have been reported in blood, and they may account for the short half-life in plasma of some hormones.

Some hormones contain cystine disulfide bonds (Table 23.5) and these may be degraded by **cystine aminopeptidase** and **glutathione transhydrogenase** (Figure 23.15). Alternatively, the peptide may undergo partial proteolysis to shorter peptides, some of which may have hormonal actions. Maturation or processing of **prohormones** into mature hormones involves selective proteolysis (Figure 23.5).

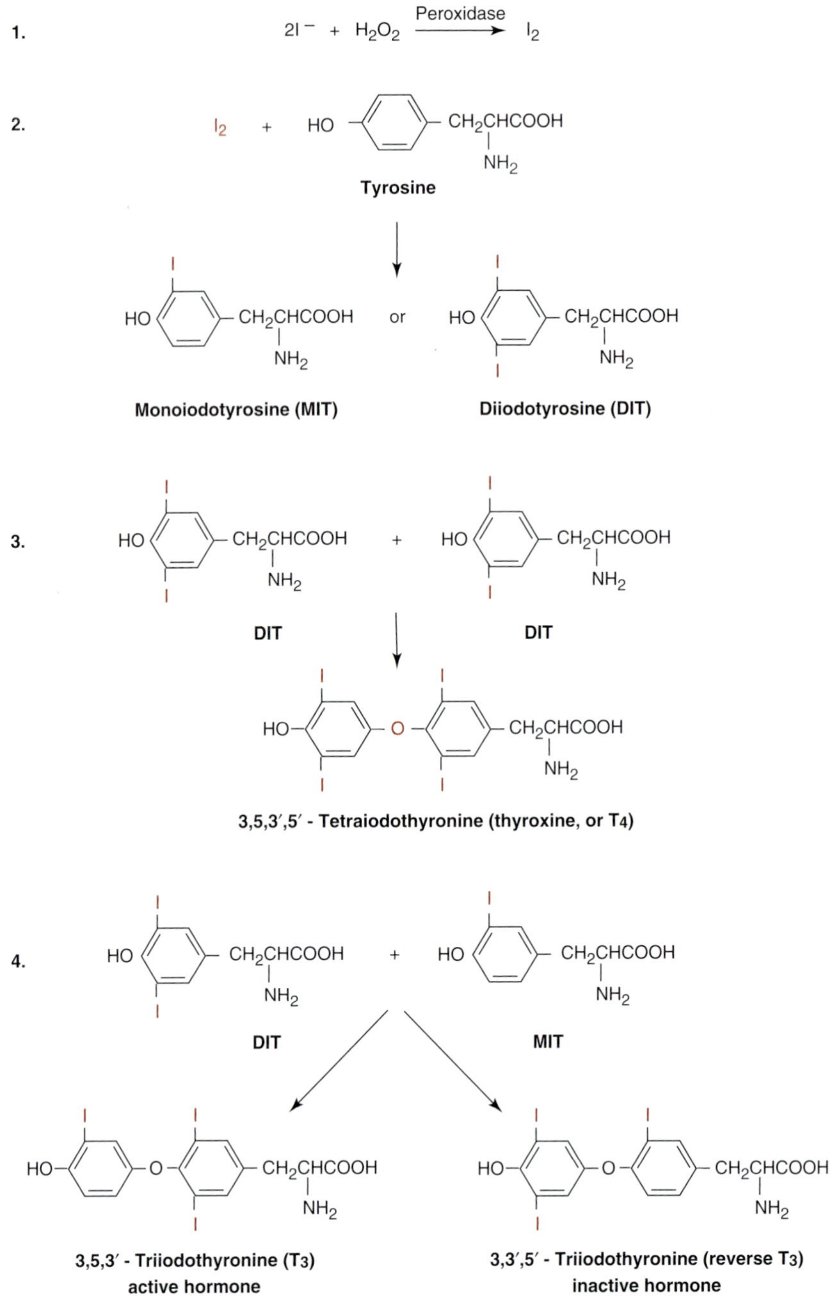

1. $2I^- + H_2O_2 \xrightarrow{\text{Peroxidase}} I_2$

2. I_2 + Tyrosine → Monoiodotyrosine (MIT) or Diiodotyrosine (DIT)

3. DIT + DIT → 3,5,3′,5′ - Tetraiodothyronine (thyroxine, or T4)

4. DIT + MIT → 3,5,3′ - Triiodothyronine (T3) active hormone; 3,3′,5′ - Triiodothyronine (reverse T3) inactive hormone

FIGURE 23.13

Synthesis and structures of thyroid hormones T_4 and T_3, and reverse T_3. Step 1, oxidation of iodide: Step 2, iodination of tyrosine residues; Step 3, coupling of DIT to DIT; Step 4, coupling of DIT to MIT (coupling may be intramolecular or intermolecular).

23.4 PROTEIN HORMONE SIGNALING

Overview of Signaling

Membrane Receptors

In the cascade system displayed in Figures 23.2 and 23.3, hormones must emanate from one source, cause hormonal release from the next step, and so on, down the cascade. Correct responses must follow each specific stimulus. Polypeptide hormones generally bind to membrane receptors expressed specifically in target cells. The receptor recognizes structural features of the hormone with an affinity constant for the interaction of $10^9–10^{11}\ M^{-1}$. This activates or inactivates a transducing protein in the membrane. Some receptors undergo **internalization** to the cell interior, and others open a membrane ion channel (see p. 916).

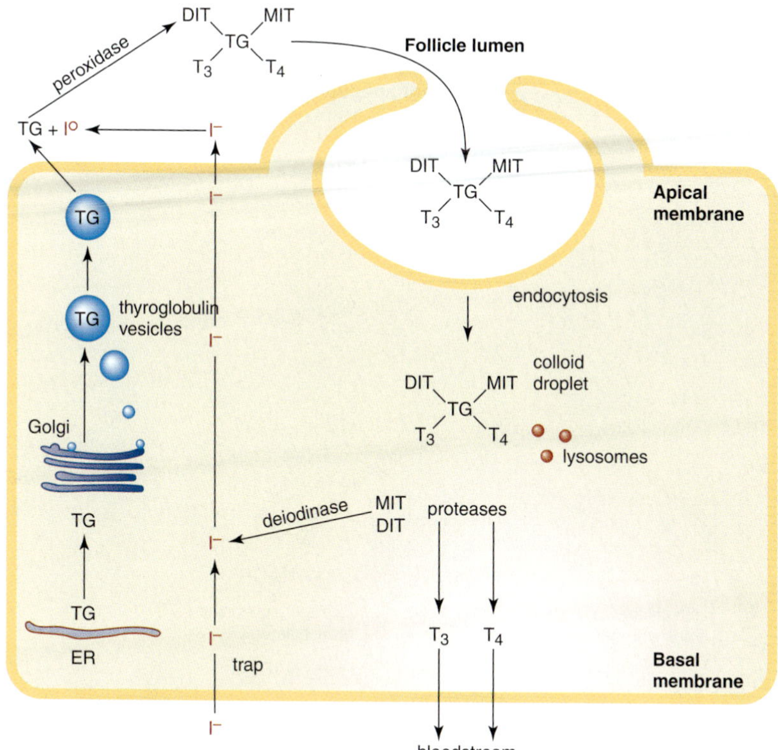

FIGURE 23.14

Cellular mechanisms for T_3 and T_4 release into bloodstream. Iodide trapping by basal membrane concentrates iodide approximately 30-fold. Secretion requires endocytosis of thyroglobulin and subsequent proteolysis. DIT and MIT are deiodinated, and the released iodide ions are reutilized for hormone synthesis.

Redrawn from Berne, R. M. and Levy, M. L. (Eds.). *Physiology*, 2nd ed. St. Louis: C. V. Mosby, p. 938.

TABLE 23.4 Hypothalamic Releasing Hormones Containing an N-Terminal Pyroglutamate,[a] a C-Terminal Amino Acid Amide, or Both

Hormone	Sequence
Thyrotropin-releasing hormone (TRH)	*pGlu*-H-*Pro-NH*$_2$
Gonadotropin-releasing hormone (GnRH)	*pGlu*-HWSYGLRP-*Gly-NH*$_2$
Corticotropin-releasing hormone (CRH)	SQEPPISLDLTFHLLREVLEMTKADQLAQQAHSNRKL-LDI-*Ala-NH*$_2$
Growth hormone-releasing hormone (GRH)	YADAIFTNSYRKVLGQLSARKLLQDIMSRQQGESNQE-RGARAR-*Leu-NH*$_2$

[a] The pyroglutamate structure is

[structure: pyroglutamate ring (H, H, H–, –H, –H, O=, N–H) with C(=O)——→Peptide]

[b] Single-letter abbreviations used for amino acids: Ala, A; Arg, R; Asn, N; Asp, D; Cys, C; Glu, E; Gln, Q; Gly; G; His, H; Ile, I; Leu, L; Lys, K; Met, M; Phe, F; Pro, P; Ser, S; Thr, T; Trp, W; Tyr, Y; Val, V.

TABLE 23.5 Examples of Hormones Containing a Cystine Disulfide Bridge Structure

Hormone	Sequence[a]
Somatostatin (GHIH)	FFNKCGA[1] / | W S | | K S \ | TFTSC[14]
Oxytocin	YC[1] / | I S | | E S \ | NCPLG—NH$_2$
Arginine vasopressin	YC / | F S | | Q S \ | NCPRG—NH$_2$

[a] Letters refer to single-letter amino acid abbreviations (see Table 23.4).

Intracellular Signal Cascade: Second Messengers

After binding to their cognate membrane receptors, many peptide and protein hormones transmit their signal intracellularly via **second messengers**, which subsequently transmit and amplify the hormonal signal (see p. 496). Some hormones transmit their signal by increasing the intracellular concentration of one specific second messenger, while others increase the concentration of several second messengers, either simultaneously or sequentially. Second messengers include cyclic AMP (cAMP), cyclic GMP

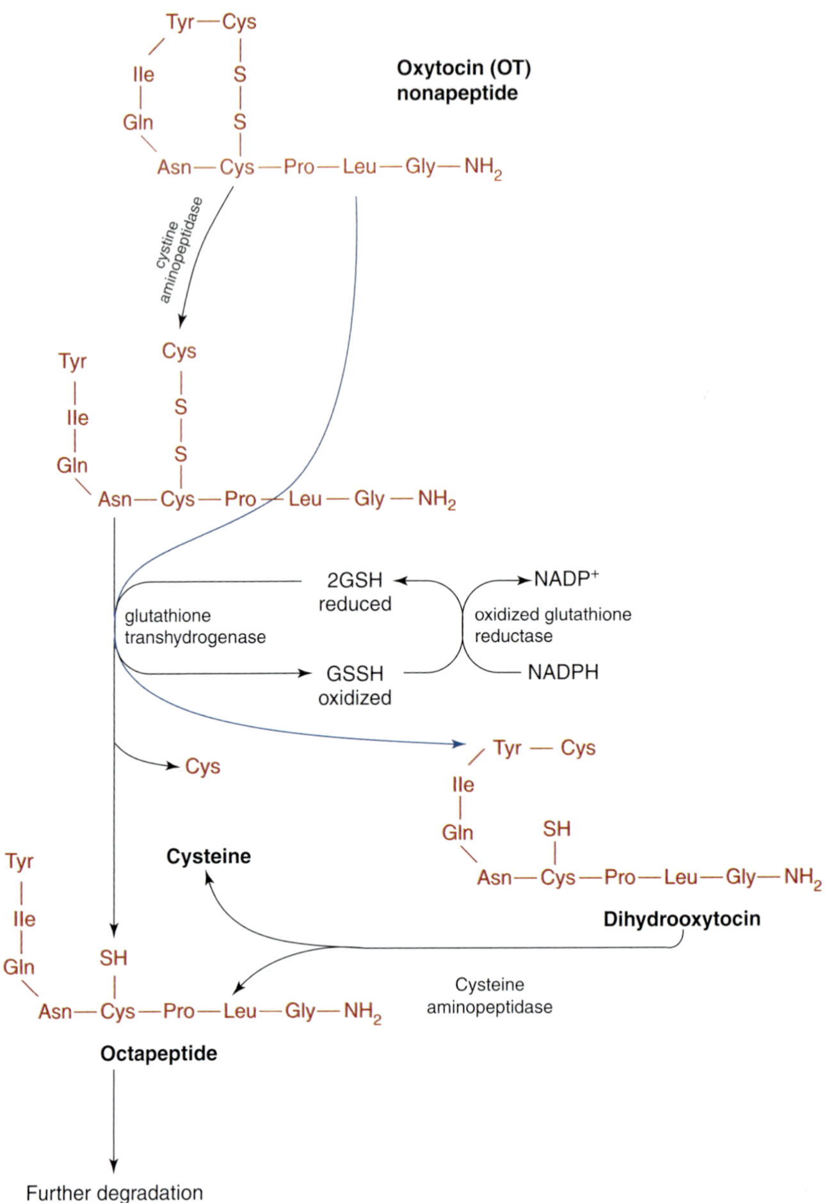

FIGURE 23.15

Degradation of posterior pituitary hormones. Oxytocin transhydrogenase is similar to degrading enzymes for insulin; presumably, these enzymes also degrade vasopressin.
Redrawn from Norman, A. W. and Litwack, G. *Hormones*. New York: Academic Press, 1987, p. 167.

(cGMP), inositol triphosphate (IP_3), diacylglycerol (DG), and phosphatidylinositol 3,4,5-trisphosphate (PIP_3). Different hormones bind to receptors that activate either a stimulatory or inhibitory G-protein subunit (G_s or G_i, respectively), resulting in the activation or inhibition of an effector enzyme and, thus, an increase or decrease in the concentration of the corresponding intracellular second messenger. The intracellular second messengers activate specific kinases that initiate a cascade of phosphorylation/dephosphorylation reactions resulting in activation of some and inactivation of others enzymes (see p. 496). Stimulation of adenylate cyclase by G-protein-coupled receptors generates cAMP, which activates **protein kinase C**, and stimulation of guanylate cyclase by different G-protein-coupled receptors generates cGMP, which activates **protein kinase G**. Stimulation of phospholipase C with the generation of DG and IP_3 results in the mobilization of Ca^{2+} stores and activation of **protein kinase C**.

An example of a hormone that transduces a signal via the generation of a second messenger is depicted in Figure 23.16. Thyrotropin-releasing hormone synthesized by hypothalamic neurons reaches the thyrotropes in the anterior pituitary and stimulates them to synthesize and secrete thyroid-stimulating hormone (TSH). TSH binds to its

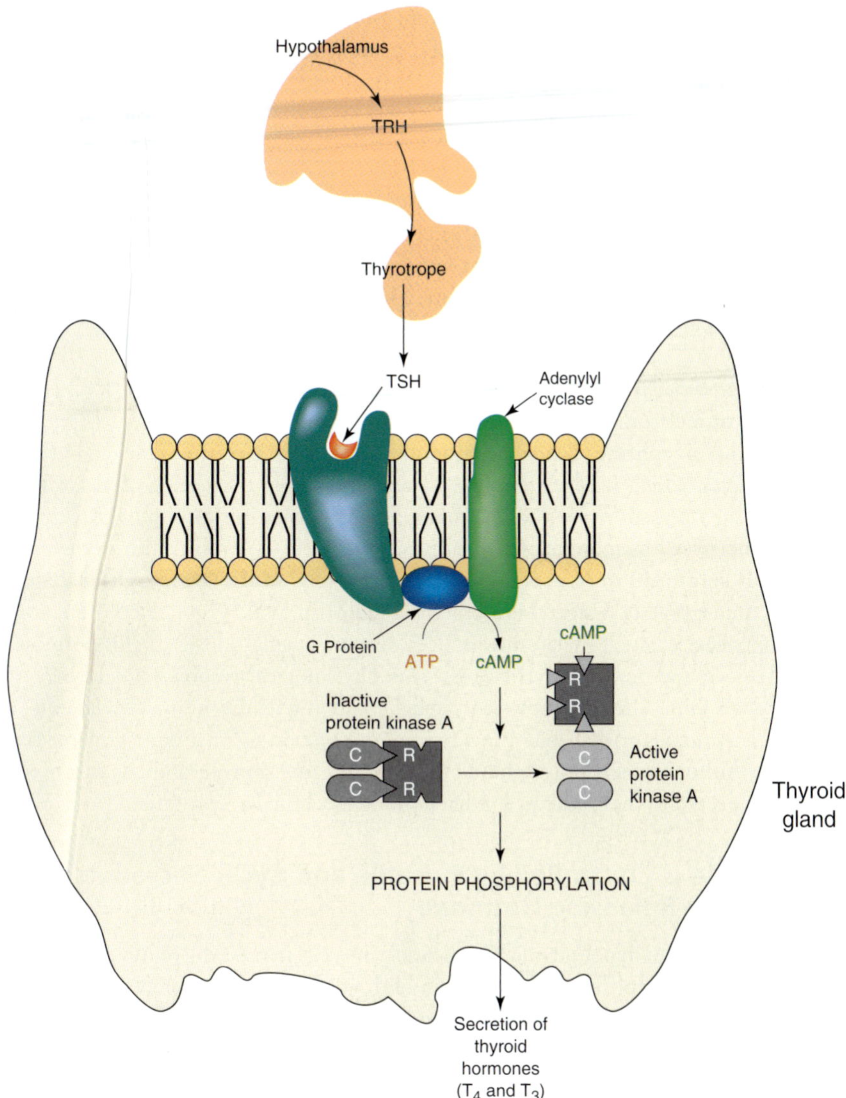

FIGURE 23.16

Effect of TSH on secretion of thyroid hormone. TSH stimulates all steps in synthesis and secretion of T_3 and T_4. These are mediated by its binding to TSH receptors located on basal membrane of thyroid epithelial cells, elevation of cAMP levels, and subsequent cascade of phosphorylation reactions.

G-protein-coupled membrane receptors in the thyroid gland, resulting in activation of adenylate cyclase and the generation of cAMP. cAMP in turn binds to the regulatory subunits in the inactive form of **protein kinase A**, leading to their dissociation from the catalytic subunits, which are then fully active (see p. 511). The enzyme initiates a cascade of protein phosphorylations, resulting in the secretion of thyroid hormone.

Amplification occurs at each step of this signal transduction pathway. For example, activation of one molecule of adenylate cyclase may result in the generation of 100 molecules of cAMP and the ultimate phosphorylation of 10,000 enzyme molecules.

The effects of cAMP are terminated when it is hydrolyzed by **phosphodiesterase**. Since the activity of phosphodiesterase is also modulated by hormones via a G protein, the level of cAMP is actually under dual regulation. Two different hormones can have antagonistic effects if one of the hormones stimulates adenylate cyclase and the other hormone stimulates phosphodiesterase. Hormonal activation of protein kinase A can also alter the rate of nuclear transcription (see p. 311). After activation by cAMP, the active catalytic subunit of the enzyme diffuses into the nucleus, where it catalyzes phosphorylation of a serine residue in **CREB** (**cAMP-response element binding protein**), a ubiquitously expressed transcription factor. The activated CREB then binds to the conserved consensus **cAMP response element** (**CRE**) as a dimer. A conserved palindromic CRE has been identified in the promoter of various genes regulated by cAMP. In addition to CREB, two other transcription factors, CREM (CRE modulator) and ATF-1, are also phosphorylated by protein kinase A. While

CREB and ATF-1 stimulate transcription, some isoforms of CREM negatively regulate CRE activity. Thus hormonal activation of a protein kinase can increase or decrease gene transcription.

Cyclic Hormonal Systems

The **diurnal variation** in the secretion of cortisol from the adrenal cortex is regulated by the sleep/wake transition and the diurnal secretion of **melatonin** from the pineal gland is dictated by daylight and darkness. The female ovarian cycle also operates on a cyclic basis dictated by the central nervous system. These are all examples of **chronotropic control** of hormone secretion.

Melatonin and Serotonin Synthesis are Controlled by Light/Dark Cycles

In the release of melatonin from the pineal gland, (Figure 23.17), the internal signal is provided by norepinephrine released by an adrenergic neuron. Control is exerted by light entering the eyes, which inhibits the pineal gland and hence the release of melatonin. Norepinephrine released in the dark stimulates cAMP formation through a β receptor in the pinealocyte cell membrane, which enhances synthesis of ***N*-acetyltransferase** and conversion of **serotonin** to ***N*-acetylserotonin. Hydroxyindole-*O*-methyltransferase (HIOMT)** then converts *N*-acetylserotonin to **melatonin**, which is secreted during the dark hours. Once secreted, melatonin induces sleepiness; small doses of this hormone can induce sleep and reset daily rhythms. This physiological response might benefit workers whose shifts alternate between daylight and nighttime hours. Melatonin is also a potent antioxidant that may provide some protection against damaging oxygen free radicals. Although melatonin inhibits reproductive functions in animals that breed during specific seasons, there is no proof that it influences human reproductive functions.

Ovarian Cycle is Controlled by Pulsatile and Cyclic Secretion of Gonadotropin-Releasing Hormone

GnRH is secreted from hypothalamic neuroendocrine cells in episodic pulses (about 1 h) in response to norepinephrinergic neurons in both adult males and females. However, in females the frequency of these pulses, and therefore the total amount of GnRH secreted during a 24-h period, changes over the course of the monthly menstrual cycle. Figure 23.18 summarizes this important role of pulsatile GnRH secretion in terms

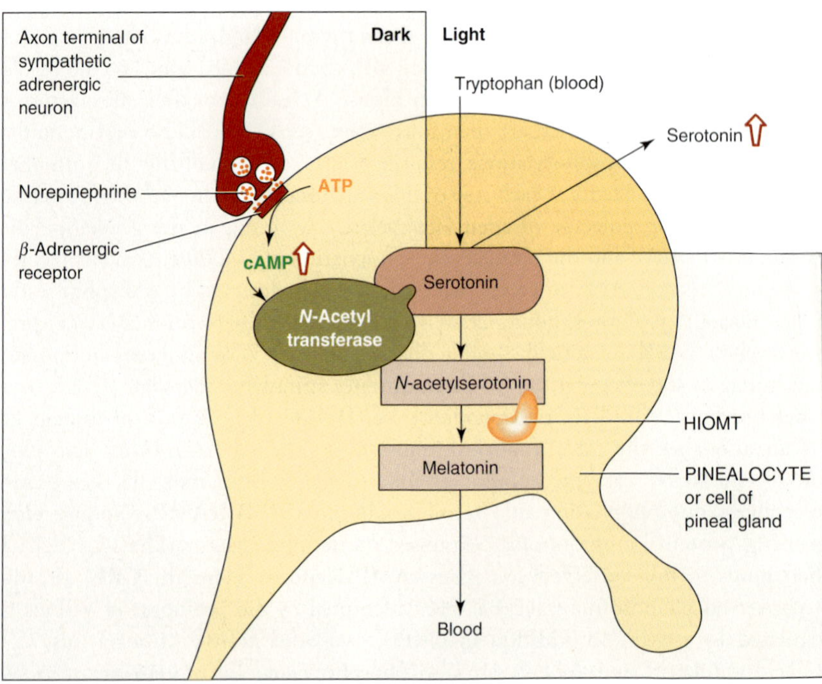

FIGURE 23.17

Biosynthesis of melatonin in pinealocytes. HIOMT, hydroxyindole-*O*-methyl transferase.
Redrawn from Norman, A. W. and Litwack, G. *Hormones*. New York: Academic Press, 1987, p. 710.

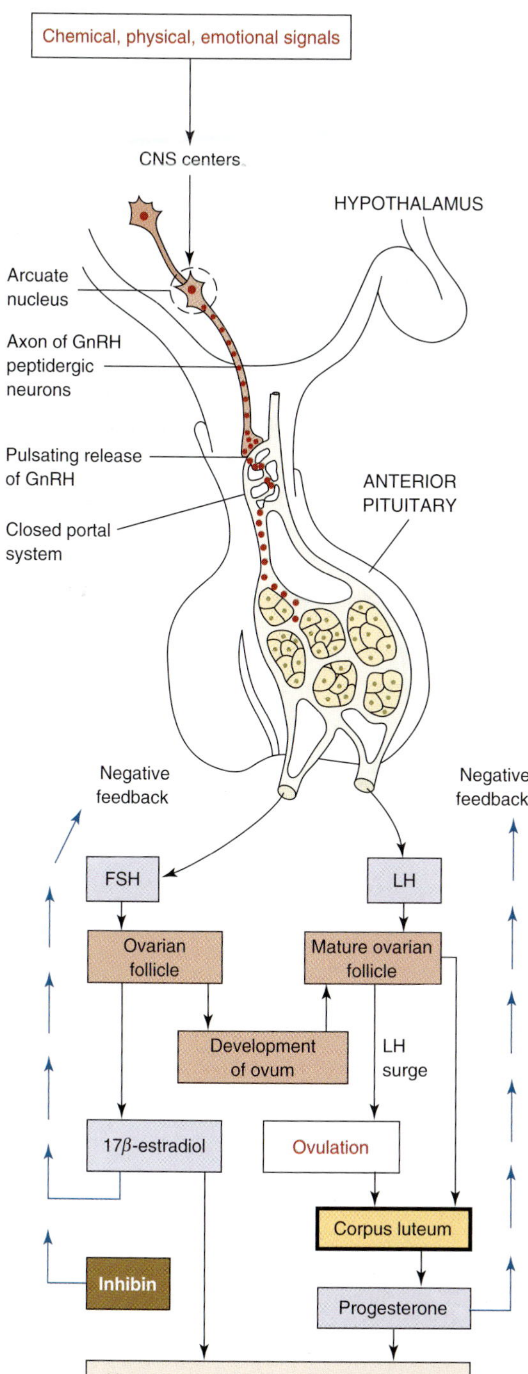

FIGURE 23.18

Ovarian cycle in terms of generation of hypothalamic hormone, pituitary gonadotropic hormones, and sex hormones. At puberty, several centers in CNS coordinate with hypothalamus so that GnRH is released in a pulsatile fashion. This causes release of LH and FSH, which affects the ovarian follicle, ovulation, and corpus luteum. Inhibin selectively inhibits FSH secretion. Products of the follicle and corpus luteum, respectively, are β-estradiol and progesterone. GnRH, gonadotropin-releasing hormone; FSH, follicle stimulating hormone; LH, luteinizing hormone.

of subsequent FSH and LH secretion from the female anterior pituitary. Entry of GnRH into the portal system is through fenestrations in the blood vessels to reach the **gonadotropes** located in the anterior pituitary. Here GnRH binds to its membrane receptors and mediates its effects via the phosphatidylinositol second messenger system (see p. 522), resulting in release of FSH and LH from the same gonadotrope. **FSH** operating through protein kinase A via cAMP elevation stimulates synthesis and secretion of 17β-estradiol, and it matures the follicle and ovum**. Inhibin, a** glycoprotein hormone, is also synthesized and secreted. It is a negative feedback regulator of FSH production by the gonadotrope. When the follicle reaches maturity, a surge of **LH**, **FSH**, and **prostaglandin $F_{2\alpha}$** triggers ovulation. The residual follicle becomes the functional corpus luteum under primary control by LH (Figure 23.18). LH binds to its cognate receptors in the corpus luteum and through stimulation of protein kinase A increases synthesis of progesterone. **Estradiol** and **progesterone** bind to specific intracellular receptors in the uterine endometrium and promote thickening of the wall, vascularization, and increased secretory activity in preparation for implantation of the fertilized egg. Estradiol, which is synthesized in large amounts prior to production of progesterone, induces expression of progesterone receptors. This induction of progesterone receptors primes the uterus for subsequent stimulation by progesterone.

Absence of Fertilization

If fertilization does not occur, the corpus luteum involutes or degenerates because of diminished LH supply, and progesterone and estrogen levels fall sharply. The hormonal stimuli for a thickened and vascularized uterine endometrial wall are thus lost. Menstruation occurs as a consequence of cellular necrosis. The fall in blood steroid levels also releases the feedback inhibition on the gonadotropes and hypothalamus and the cycle starts again. The time course for the ovarian menstrual cycle in humans is shown in Figure 23.19.The first monthly cycle occurs at the time of puberty when GnRH secretion begins to increase (day 1 of Figure 23.19). GnRH is released in a pulsatile fashion, causing the gonadotrope to release FSH and LH and the blood concentrations of these hormones gradually increase in subsequent days. Under the stimulation of

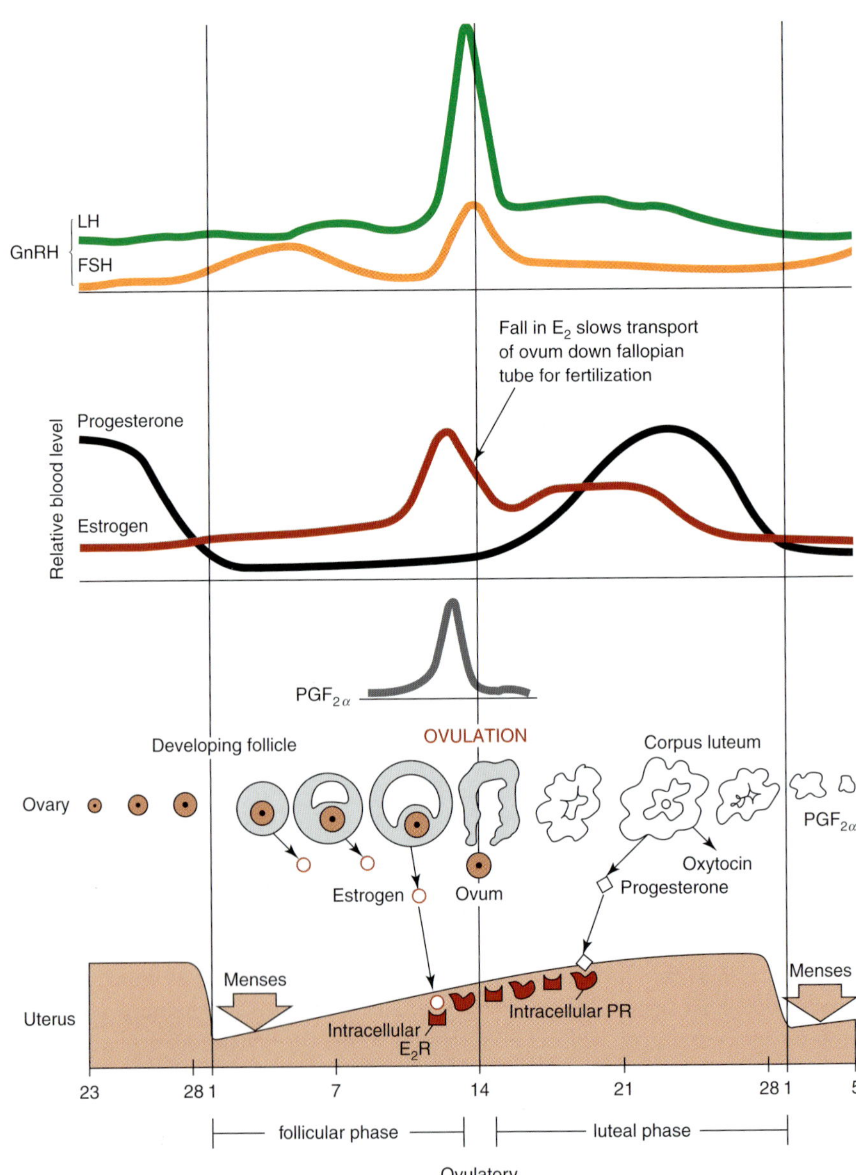

FIGURE 23.19

Ovarian cycle. In the upper diagram, relative blood levels of GnRH, LH, FSH, progesterone, estrogen, and $PGF_{2\alpha}$ are shown. In the lower diagram, events in ovarian follicle, corpus luteum, and uterine endometrium are diagrammed. GnRH, gonadotropin-releasing hormone; LH, luteinizing hormone; FSH, follicle-stimulating hormone; $PGF_{2\alpha}$, prostaglandin $F2_{\alpha}$; E_2, estradiol; E_2R, intracellular estrogen receptor; PR, intracellular progesterone receptor.

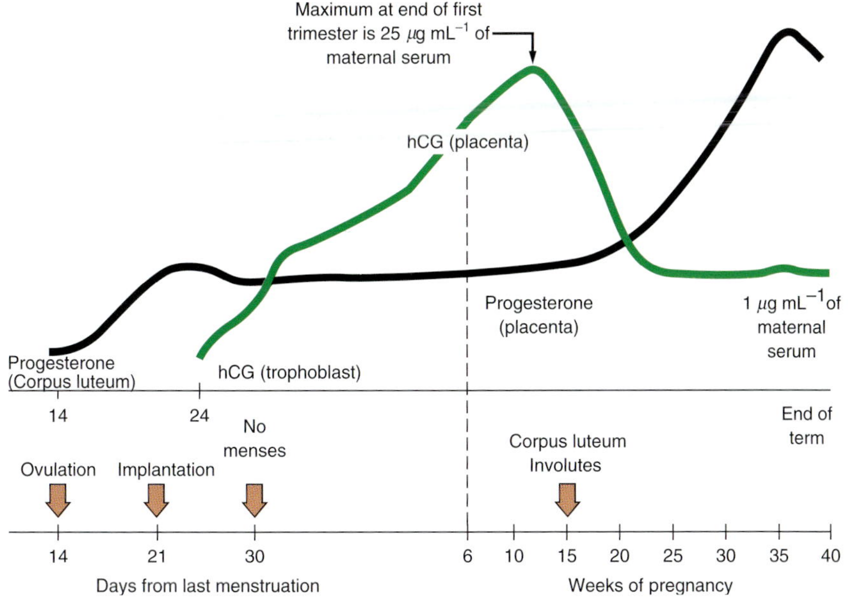

FIGURE 23.20
Effect of fertilization on ovarian cycle in terms of secretion of progesterone and human chorionic gonadotropin (hCG).

FSH, the follicle begins to mature (lower section of Figure 23.19) and estradiol (E_2) is produced, causing the uterine endometrium to thicken. Under the continued action of FSH, the follicle matures, and high concentrations of estradiol are produced (around day 13 of the cycle). This elevated level of estradiol now mediates **positive feedback** (rather than negative feedback mediated by lower estradiol levels), resulting in a surge of LH and a smaller release of FSH from the gonadotropes. The FSH response is smaller because this gonadotropic hormone stimulates the ovarian production of another hormone, **inhibin B**. Inhibin B then functions as an inhibitor of FSH, but not LH, secretion. The high midcycle peak of LH is referred to as the "LH spike." Ovulation then occurs at about day 14 (midcycle) through the effects of high LH concentration together with other factors, such as PGF_{2a}. After ovulation, LH promotes the differentiation of the ruptured follicle (Figure 23.19, bottom). This corpus luteum then produces the high levels of progesterone that further thicken the uterine endometrial wall. The corpus luteum also secretes **inhibin A** (inhibin B secreted from dominant follicle), which, along with estradiol and progesterone, feeds back on the pituitary to suppress FSH and LH secretion during the luteal phase of the cycle. In the absence of fertilization, the corpus luteum functions for only about 2 weeks. It then involutes because of the decline in levels of LH and an age-related reduction in sensitivity to LH. With the death of the corpus luteum, there is a profound decline in levels of estradiol and progesterone. The endometrial wall can no longer be maintained and menstruation occurs, followed by the start of another menstrual cycle.

Fertilization

If fertilization occurs as shown in Figure 23.20, the corpus luteum remains viable due to the production of **chorionic gonadotropin** (CG), which resembles and acts like LH, from the trophoblast cells. The secretion of CG reaches a peak about 80 days after the last menstrual period. It then declines very rapidly and remains at a relatively low level produced by the placenta for the remainder of pregnancy. Once CG levels fall, the corpus luteum begins to involute, and by about 12 weeks of pregnancy the placenta takes over production and secretion of progesterone and estrogens (primarily estriol). From the seventh month onward, estrogen secretion continues to increase while progesterone secretion remains constant or may even decrease slightly. The estrogen/progesterone ratio increases toward the end of pregnancy and may be partly responsible for increased uterine contractions. Oxytocin from the posterior pituitary contributes to these uterine contractions. The fetal membranes release prostaglandins ($PGF_{2\alpha}$) at the time of parturition and they increase the intensity of uterine contractions. Finally, the fetal

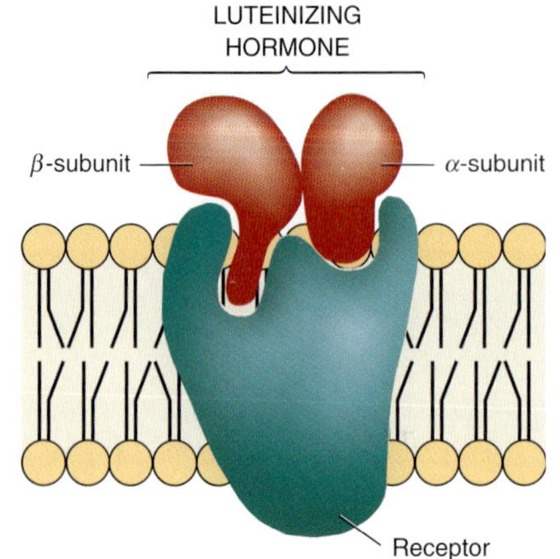

FIGURE 23.21

The interaction of α and β subunits of LH with LH receptor of rat Leydig cells. Both α- and β-subunits participate in LH receptor binding. Adapted from Alonoso-Whipple, C., Couet, M. L., Doss, R. Koziarz, J., Ogunro, E. A., and Crowley, W. E. Jr. *Endocrinology* 123:1854, 1988.

adrenal cortex secretes cortisol, which stimulates fetal lung maturation by inducing surfactant and may stimulate uterine contractions.

23.5 MEMBRANE HORMONE RECEPTORS

Some Hormone–Receptor Interactions Involve Multiple Hormone Subunits

Thyrotropin (TSH), luteinizing hormone (LH), and **follicle-stimulating hormone (FSH)** each contain an α- and a β-subunit. The α-subunits for all three hormones are nearly identical. The specificity of receptor recognition is imparted by the β-subunit, whose structure is unique for each hormone. A model of the interaction of LH with its receptor is shown in Figure 23.21. The LH receptor recognizes both subunits of the hormonal ligand, but the β-subunit is specifically recognized by the receptor to elicit a hormonal response. The TSH–receptor complex stimulates adenylate cyclase and the phosphatidylinositol pathway. The preferred model is one in which there is a single receptor whose interaction with hormone activates both the adenylate cyclase and the phospholipid second messenger systems, as shown in Figure 23.22.

β-Adrenergic Receptor

Structures of receptors are conveniently discussed in terms of functional domains. For membrane receptors there are **ligand-binding domains, transmembrane domains**, and **intracellular** domains; the latter may have intrinsic protein kinase activity. Specific **immunological domains** contain primary epitopes of antigenic regions. The β-adrenergic receptors (β_1 and β_2) recognize catecholamines, both norepinephrine and epinephrine, and hormone binding stimulates adenylate cyclase. The subtypes differ

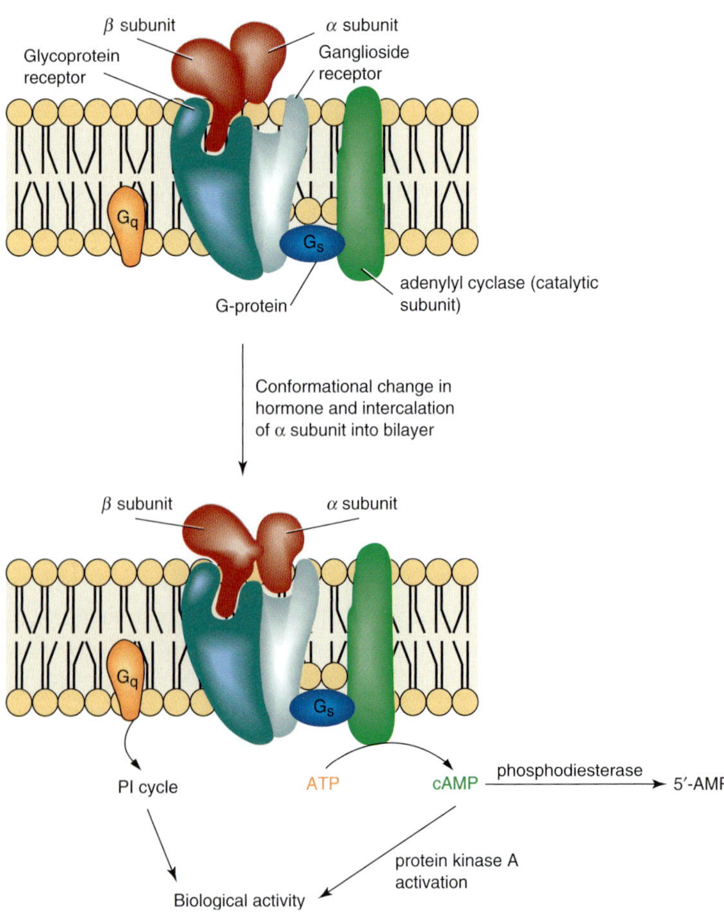

FIGURE 23.22

Model of TSH receptor. Receptor is composed of glycoprotein and ganglioside component. After TSH β-subunit interacts with receptor, hormone changes its conformation and α-subunit interacts with other membrane components. β-subunit of TSH may carry primary determinants recognized by glycoprotein receptor component. It is suggested that the TSH signal to adenylate cyclase is via the ganglioside; the glycoprotein component appears more directly linked to phospholipid signal system. PI, phosphatidylinositol; G_s, G-protein linked to activation of adenylate cyclase; G_q, G-protein linked to PI cycle.
Adapted with modifications from Kohn, L. D., et al. In: G. Litwack (Ed.), *Biochemical Actions of Hormones*, Vol. 12. New York: Academic Press, 1985, p. 466.

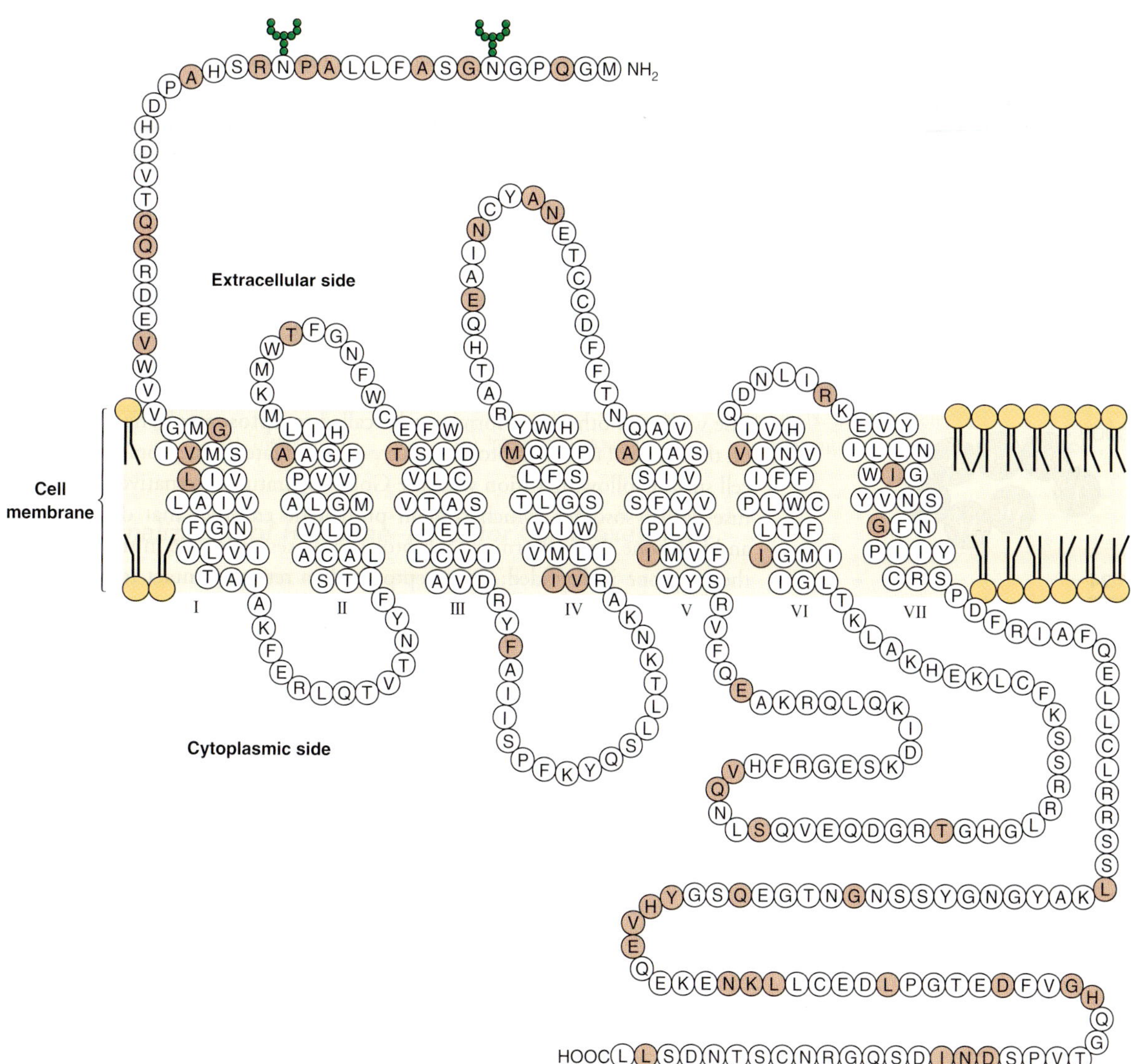

FIGURE 23.23

Proposed model for insertion of β_2-adrenergic receptor (AR) in cell membrane. The model is based on hydropathicity analysis of human β_2-AR. Standard one-letter codes for amino acid residues are used. Hydrophobic domains are represented as transmembrane helices. Pink circles with black letters indicate residues in the human sequence that differ from those in hamster. Also noted are the potential sites of N-linked glycosylation.

Redrawn from Kobilka, B. K., Dixon, R. A., Frielle, T., Doblman, H. G., et al. *Proc. Natl. Acad. Sci. USA* 84:46, 1987.

in their affinity for norepinephrine and for synthetic antagonists. β_1-Receptors bind norepinephrine with a higher affinity than epinephrine, whereas the reverse is the case for the β_2-receptors. Isoproterenol, an analogue of epinephrine that is a β-receptor stimulator, has a greater affinity for both receptors than either norepinephrine or epinephrine. Figure 23.23 shows the amino acid sequence for the β_2-adrenergic receptor (see p. 80 for single-letter abbreviations of amino acids). The N-terminal segment extends from α-helix I to the extracellular space, and there are seven membrane-spanning domains. The β_1-receptor shows extensive homology with the β_2-receptor. Intracellular loops join helices I and II, III and IV, and V and VI. The long chain extended from VII is the intracellular C-terminal region and contains phosphorylation sites (serine and threonine residues), which are important for receptor desensitization. Phosphorylation results in binding of an inhibitory protein, called **β-arrestin**, which blocks the receptor's ability to activate G_s (see p. 513). Extracellular loops join helices II and III, IV and V, and VI and VII, but mutational analysis suggests that these loops do not participate

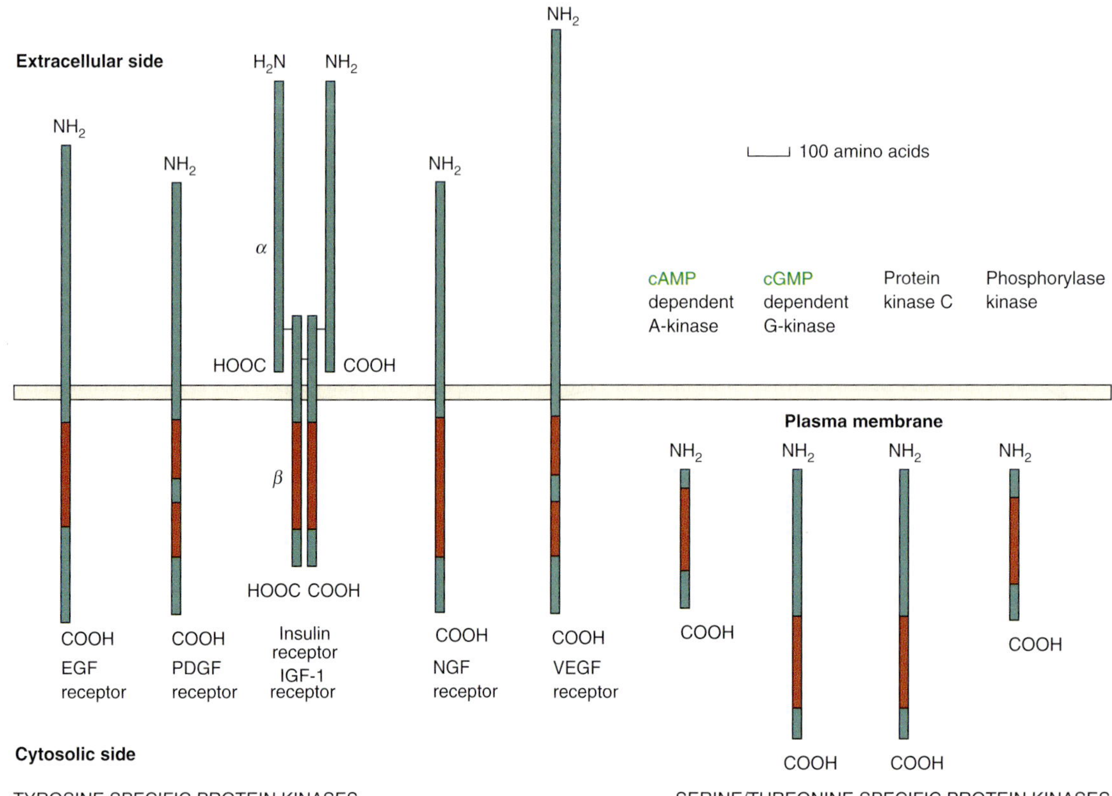

FIGURE 23.27

Protein kinases of plasma membrane or cytosol showing size and location of catalytic domains. In each case, the catalytic domain (red region) is about 250 amino acid residues long. EGF, epidermal growth factor; NGF, nerve growth factor; VEGF, vascular endothelial growth factor.
Redrawn from Alberts, B., Bray, D., Lewis, J., Raff, M., Roberts, K., and Watson, J. D. *Molecular Biology of the Cell*, 3rd ed. New York: Garland, 1994, p. 760.

threonine residues. Protein kinases that phosphorylates **tyrosine** residues are found in the cytoplasmic domains of some membrane receptors, especially growth factor receptors. The insulin and IGF-1 receptors possess auto-tyrosine kinase activities. The location of the catalytic domains of some of these protein kinases with respect to the plasma membrane is presented in Figure 23.27.

The catalytic domain of the protein kinases is similar in amino acid sequence, suggesting that they have all evolved from a common primordial gene. Three **tyrosine-specific kinases** are shown in Figure 23.27 and are transmembrane receptor proteins that phosphorylate proteins (including themselves) on tyrosine residues inside the cell. The alpha, or extracellular, and beta, or transmembrane, polypeptides of the insulin receptor are encoded by a single gene, which produces a precursor protein that is cleaved into two disulfide-linked chains. There are five immunoglobulin (Ig)-like domains in the extracellular domain of platelet derived growth factor (PDGF) receptor. In general, proteins regulated by phosphorylation-dephosphorylation usually have multiple phosphorylation sites, and they may be phosphorylated by more than one type of protein kinase.

Insulin Receptor: Transduction Through Tyrosine Kinase

The α-subunits of the **insulin receptor** are located extracellularly and are the insulin-binding sites (Figure 23.27). Ligand binding induces **autophosphorylation** of tyrosine residues located in the cytoplasmic portions of β-subunits. This autophosphorylation facilitates binding of cytosolic substrate proteins, such as **insulin receptor substrate-1 (IRS-1).** When phosphorylated, this substrate acts as a docking protein for proteins

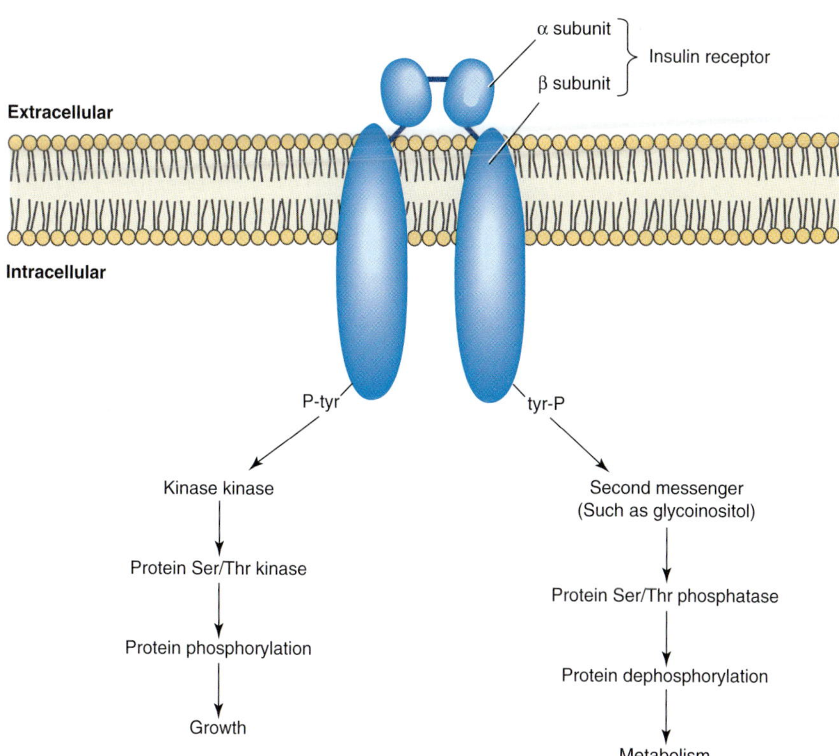

FIGURE 23.28

Hypothetical model depicting two pathways to explain paradoxical effects of insulin on protein phosphorylation. Insulin simultaneously produces increases in the serine/threonine phosphorylation of some proteins and decreases in others. This paradoxical effect may result from the activation of both kinases and phosphatases. Model explains (1) the generation of a soluble second messenger that directly or indirectly activates serine/threonine phosphatase and (2) the stimulation of a cascade of protein kinases, resulting in the phosphorylation of cellular proteins.
Redrawn from Saltiel, A. R. The paradoxical regulation of protein phosphorylation in insulin action. *FASEB J.*, 8:1034, 1994.

mediating insulin action. It is not clear whether protein phosphorylation is the only mechanism for all of insulin's actions. The net responses to this hormone include short-term metabolic effects, such as a rapid increase in the uptake of glucose, and longer-term effects on cellular differentiation and growth. Although the insulin receptor becomes autophosphorylated on tyrosines and phosphorylates tyrosines of IRS-1, other mediators are phosphorylated predominantly on serine and threonine residues, as indicated in Figure 23.28. Insulin stimulates phosphorylation of some proteins and dephosphorylation of other proteins. Activation or inhibition of specific enzymes suggests that separate signal transduction pathways may originate from the insulin receptor. There is evidence that an insulin second messenger may be released at the cell membrane to account for the short-term metabolic effects of insulin. This messenger may be a glycoinositol derivative that stimulates phosphoprotein phosphatase. A detailed diagram of the signal transduction pathways for insulin is presented Figure 23.29. Many of the specific proteins that are activated or inactivated subsequent to insulin binding have now been identified. This diagram shows that binding of insulin to the α subunit of the receptor promotes autophosphorylation of a β-subunit on multiple tyrosine residues. The activated β-subunit then catalyzes the phosphorylation of cellular proteins such as members of the IRS family, Shc and Cbl (Figure 23.29). Upon tyrosine phosphorylation, these proteins interact with other signaling molecules through their SH2 (Src-homololog-2) domains, which bind to a distinct sequence of amino acids surrounding a phosphotyrosine residue. Several diverse pathways are activated, and those include activation of PI_3 kinase (phosphatidylinositol 3′-OH kinase), Ras (small GTP-binding protein), the MAP kinase (mitogen-activated protein kinase) cascade, and TC10 (small GTP binding protein). Once activated via an exchange of GTP for GDP, TC10 promotes translocation of GLUT4 vesicles to the plasma membrane, perhaps by stabilizing cortical actin filaments. These pathways act in a concerted fashion to coordinate the regulation of vesicle trafficking (incorporation of glucose transporter 4 (GLUT4)) into plasma membrane], protein synthesis, enzyme activation and inactivation, and gene expression. The net result of these diverse pathways is regulation of glucose, lipid, and protein metabolism as well as cell growth and differentiation. The importance of the activity of the insulin receptor kinase in the overall signal transduction pathway of insulin is emphasized in Clinical Correlation 23.3.

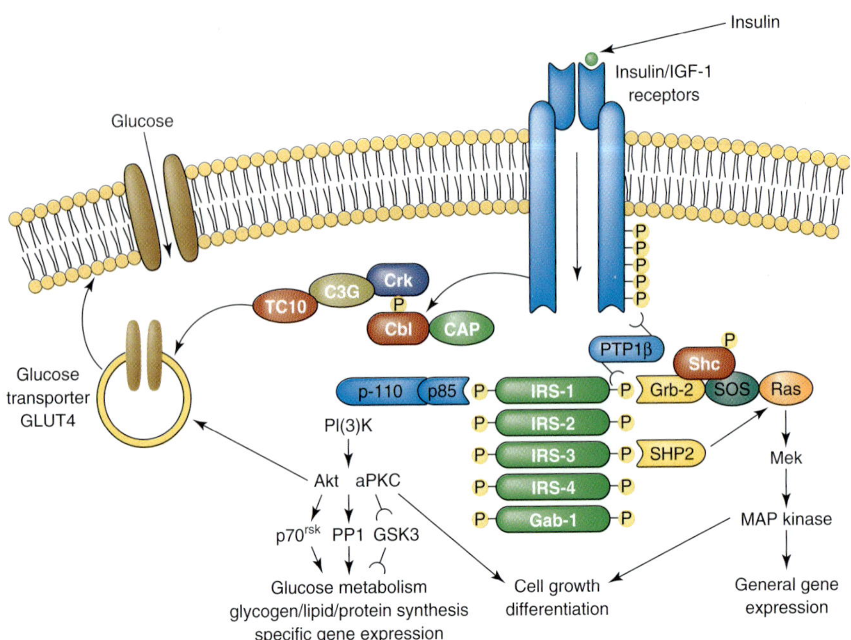

FIGURE 23.29

Hypothetical scheme for signal transduction in insulin action. The insulin receptor undergoes tyrosine autophosphorylation and kinase activation upon hormone binding. It phosphorylates intracellular substrates including IRS-1, Shc, and Cbl, which associate with SH2-containing proteins like p85 and Grb2. Formation of the IRS-1/p85 complex activates PI (3-) kinase; the IRS-1/Grb2 complex activates MAP kinase. Abbreviations: IRS-1, insulin receptor substrate-1; PI(3)K, phosphatidylinositol-3-OH kinase; MAP kinase, mitogen-activated protein kinase; MeK, MAP kinase kinase.
Redrawn from Saltiel, A. R. and Kahn, C. R. Insulin signaling and the regulation of glucose and lipid metabolism. *Nature* 414:799, 2001.

CLINICAL CORRELATION 23.3

Decreased Insulin Receptor Kinase Activity in Gestational Diabetes Mellitus

During pregnancy, an important maternal metabolic adaptation is a decrease in insulin sensitivity. This adaptation helps provide adequate glucose for the developing fetus. However, in 3–5% of pregnant women, glucose intolerance develops. Gestational diabetes mellitus (GDM) is characterized by an additional decrease in insulin sensitivity and an inability to compensate with increased insulin secretion. Although both pregnancy-induced insulin resistance and GDM are generally reversible after pregnancy, approximately 30–50% of women with a history of GDM go on to develop type 2 diabetes later in life, particularly if they are obese. Although the cellular mechanisms responsible for the insulin resistance in GDM are not fully understood, the resistance to insulin-mediated glucose transport appears to be greater in skeletal muscle from GDM subjects than in women who are pregnant but do not have GDM. Recent data indicate that defects in insulin action, rather than a decrease in insulin receptor binding affinity, may contribute to the pathogenesis of GDM. More specifically, skeletal muscle cells of GDM subjects appear to overexpress plasma cell membrane glycoprotein-1 (PC-1), which has been reported to inhibit the tyrosine kinase activity of the insulin receptor by directly interacting with α-subunits and blocking the insulin-induced conformational change. Additionally, excessive phosphorylation of serine/threonine residues located within muscle insulin receptors appears to down-regulate tyrosine kinase activity in GDM. Thus an overexpression of PC-1 and a decrease in receptor kinase activity, coupled with a decreased expression and phosphorylation (tyrosine residues) of the insulin receptor substrate-1 (IRS-1; see Figure 20.29), may underlie the insulin resistance in GDM.

Source: Shao, J., Catalono, P. M., Hiroshi, Y., Ruyter, I., Smith, S., Youngren, J., and Friedman, J. E. Decreased insulin receptor tyrosine kinase activity and plasma cell membrane glycoprotein-1 overexpression in skeletal muscle from obese women with gestational diabetes mellitus (GDM). *Diabetes* 49(4):603, 2000. Maddux, B. A. and Goldfine, I. D. Membrane glycoprotein PC-1 inhibition of insulin receptor function occurs via direct interaction with the receptor alpha-subunit. *Diabetes* 49(1):13, 2000.

Activity of Vasopressin: Protein Kinase A

Arginine vasopressin (AVP), or the antidiuretic hormone, causes increased water reabsorption from the urine in the distal kidney. A mechanism for this system is shown in Figure 23.30. Neurons synthesizing AVP (vasopressinergic neurons) release AVP in response to stimuli from **baroreceptors** responding to a fall in blood pressure or from **osmoreceptors** responding to an increase in extracellular salt concentration. VP binds to its cognate membrane receptors in the distal kidney, anterior pituitary, hepatocytes, and perhaps other cell types. At the kidney, AVP binding to its G-protein-coupled

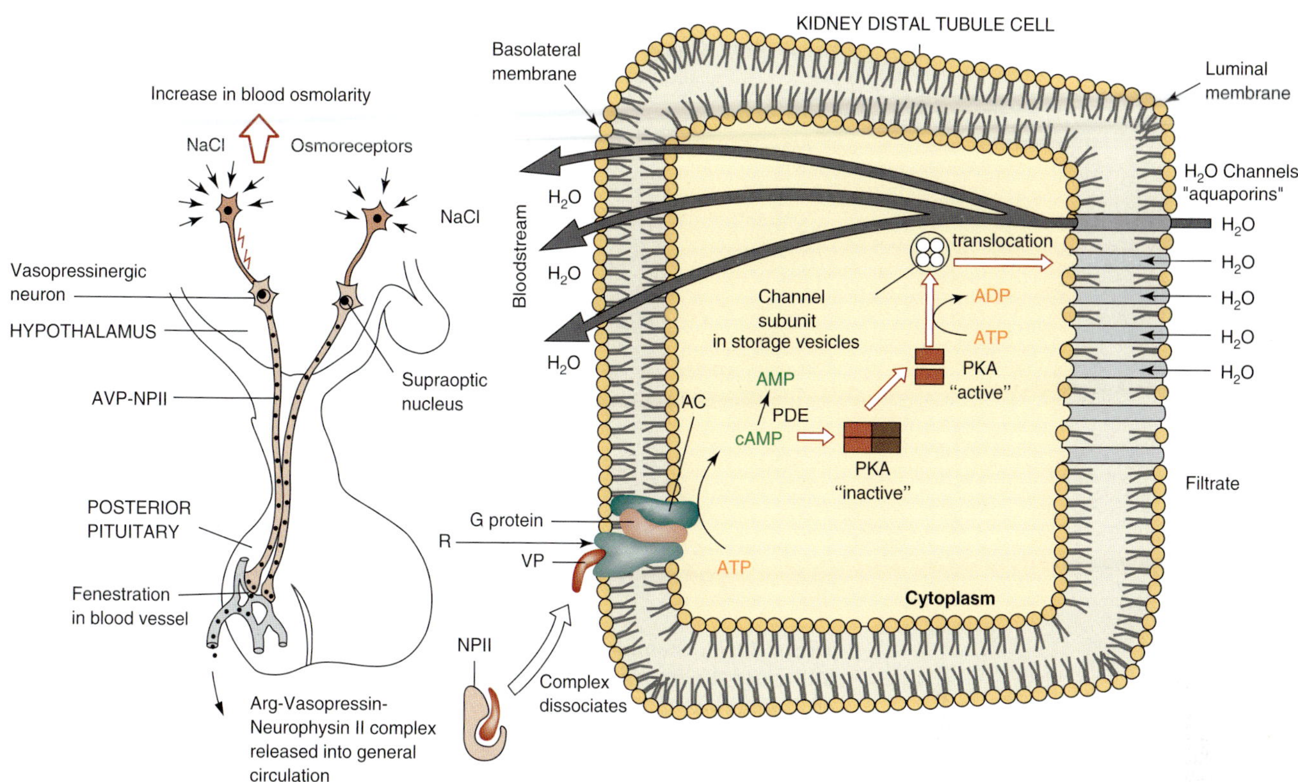

FIGURE 23.30

Secretion and action of arginine vasopressin in distal kidney tubules. Release of arginine vasopressin (AVP or VP) from the posterior pituitary is triggered by osmoreceptors, or baroreceptors (not shown). This signal is transmitted down a vasopressinergic neuron and promotes the release of a VP-neurophysin complex from the posterior pituitary gland, where the hormone is normally stored. The neurophysin bond to the VP eventually dissociates, and the posterior pituitary hormone binds to cognate membrane receptors on the kidney distal tubule cell. Through this G-protein-coupled receptor, adenylate cyclase is stimulated to increase levels of cAMP from ATP. Cyclic AMP-dependent protein kinase A is then activated, and it phosphorylates various proteins including subunits of the aquaporin channels. The phosphorylated subunits aggregate and functional water channels (aquaporins) are inserted in the luminal plasma membrane, thus increasing the reabsorption of water. Abbreviations: NPII, neurophysin II; VP, vasopressin; R, receptor; AC, adenylate cyclase; PDE, phosphodiesterase.

Redrawn in part from Dousa, T. P. and Valtin, H. Cellular actions of vasopressin in the mammalian kidney. *Kidney Int.* 10:45, 1975.

receptor stimulates adenylate cyclase activity and activates protein kinase A, which phosphorylates subunits that aggregate to form specific water channels, or **aquaporins** (see p. 465). Water crosses the kidney cell to the basolateral side and then enters the general circulation, where it dilutes the salt concentration. Specific mutations in the intracellular and extracellular loop sequences of the aquaporin channels result in loss of function and development of nephrogenic diabetes insipidus, which is characterized by increased thirst and production of a large volume of urine. Some hormones that activate protein kinase A are listed in Table 23.6.

Gonadotropin-Releasing Hormone (GnRH): Protein Kinase C

Table 23.7 lists several polypeptide hormones that stimulate the phosphatidylinositol pathway and activate **protein kinase C** (see p. 523). Although AVP activates protein kinase A in renal cells, this same hormone activates protein kinase C in other target cells. **GnRH** mediates its effects through activation of protein kinase C, and the action of this hypothalamic releasing hormone is summarized in Figure 23.31. Probably an aminergic nerve fiber stimulates the GnRH-ergic neuron to secrete GnRH that enters the closed portal system connecting the hypothalamus and anterior pituitary

TABLE 23.6 Examples of Hormones that Operate Through the Protein Kinase A Pathway

Hormone	Location of Action
CRH	Corticotrope of anterior pituitary
TSH	Thyroid follicle
LH	Leydig cell of testis Mature follicle at ovulation and corpus luteum
FSH	Sertoli cell of seminiferous tubule and ovarian follicle
ACTH	Inner layers of cells of adrenal cortex
Opioid peptides	Some in CNS function on inhibitory pathway through G_i
AVP	Kidney distal tubular cell
PGI_2 (prostacyclin)	Blood platelet membrane
Norepinephrine/ epinephrine	β-Receptor: expressed in various tissues and cell types

TABLE 23.7 Examples of Polypeptide Hormones that Stimulate the Phosphatidylinositol Pathway and Activate Protein Kinase C

Hormone	Location of Action
TRH	Thyrotrope of the anterior pituitary releasing TSH
GnRH	Gonadotrope of the anterior pituitary releasing LH and FSH
AVP	Corticotrope of the anterior pituitary; assists CRH in releasing ACTH: hepatocyte; causes increase in cellular Ca^{2+}
TSH	Thyroid follicle: releasing thyroid hormones; causes increase in phosphatidylinositol cycle as well as increase in protein kinase A activity
Angiotensin II/III	Zona glomerulosa cell of adrenal cortex: releases aldosterone
Epinephrine	Smooth muscle cells that express α_1-receptors

through fenestrations. GnRH binds to cognate membrane receptors on the gonadotrope (enlarged view in Figure 23.31) and activates **phospholipase C**. The hydrolysis of PIP_2 forms **diacylglycerol (DAG)** and IP_3, and DAG activates protein kinase C, which phosphorylates specific proteins. IP_3 binds to a receptor on the membrane of the endoplasmic reticulum releasing stored Ca^{2+}. The released Ca^{2+} also activates protein kinase C, which ultimately results in the secretion of LH and FSH from the same cell.

There are at least 11 isoforms of protein kinase C, and nine are activated by the lipid second-messenger DAG. The isoforms of protein kinase C normally exist as monomeric proteins in the cytosol (see p. 523). When the enzyme is in this soluble state, it is folded in such a way that the binding site for substrate proteins is blocked. When DAG binds to the enzyme, it translocates to the inner surface of the plasma membrane, where it can also bind to acidic phospholipids. The increased Ca^{2+} concentration also increases the binding of the enzyme to the membrane. When protein kinase C is bound, the protein unfolds and this conformational change exposes binding sites for protein substrates. This results in protein kinase C-catalyzed phosphorylation of the bound proteins. The phosphorylated proteins mediate a variety of intracellular effects.

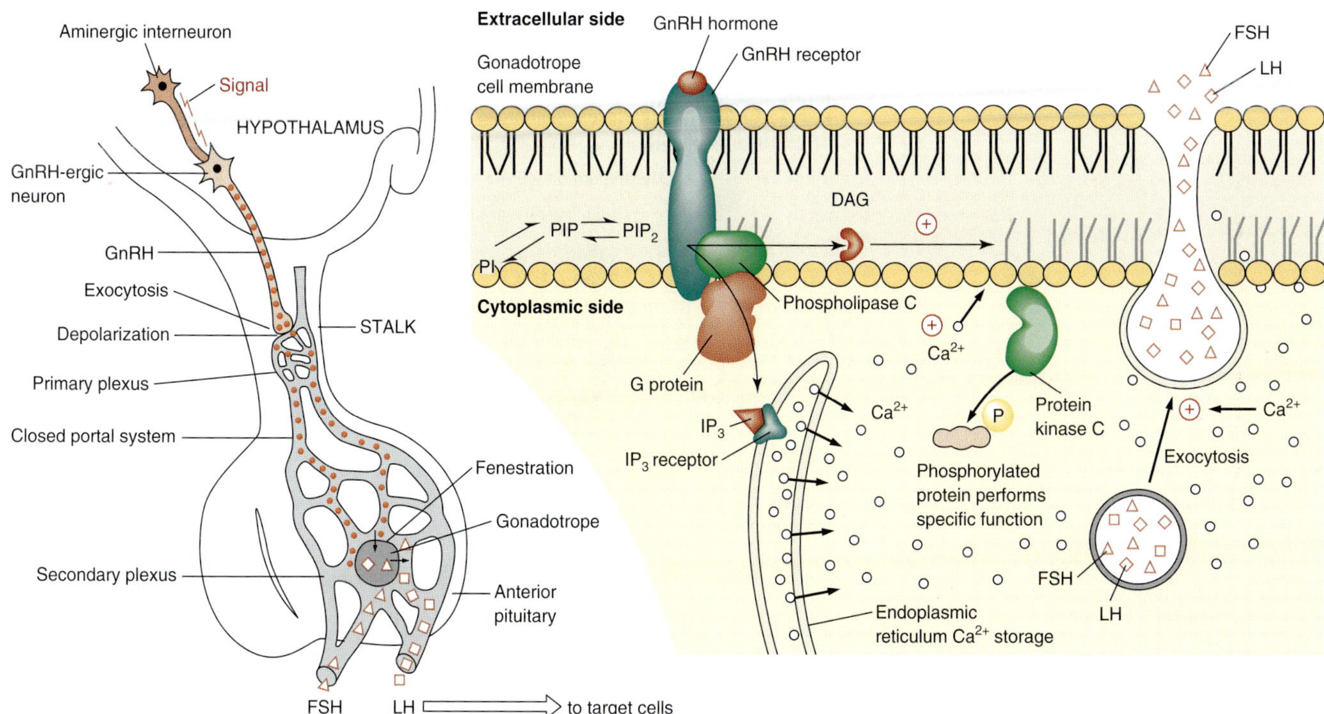

FIGURE 23.31

Regulation of secretion of LH and FSH by protein kinase C. A general mode of action of GnRH to release the gonadotropic hormones from the gonadotrope of the anterior pituitary is presented. GnRH, gonadotropin-releasing hormone; FSH, follicle-stimulating hormone; LH, luteinizing hormone; DAG, diacylglycerol.

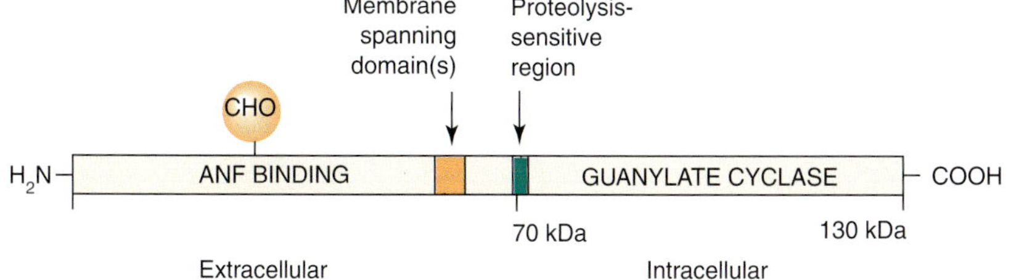

FIGURE 23.32

Functional domains of ANF-R_1 receptor. Model shows an ANF-binding domain, a membrane spanning domain(s), a proteolysis-sensitive region, a guanylate cyclase domain, glucosylation site (CHO), and amino and carboxyl terminals of the receptor.
Redrawn from Liu, B., Meloche, S, McNicoll, N., Lord, C., and DeLéan, A. *Biochemistry* 28:5599, 1989.

Activity of Atrial Natriuretic Factor (ANF): Protein Kinase G

The receptor for **atrial natriuretic factor (ANF)** is a transmembrane protein whose cytoplasmic C-terminal domain has guanylate cyclase activity and whose extracellular N-terminal domain binds ANF (Figure 23.32). A model for signal transduction by ANF is presented in Figure 23.33. ANF is one of a family of peptides (Figure 23.34). Atrial natriuretic factor is secreted by myocytes in the heart in response to signals such as blood volume expansion, high salt intake, increased right atrial pressure, and increased heart pumping rate. ANF secretion is stimulated by activators of cardiac protein kinase C and decreased by activators of protein kinase A. These opposing actions may be mediated by α- and β-adrenergic receptors, respectively. An overview of the secretion of ANF and its general effects is shown in Figure 23.35. ANF is secreted as a dimer, but only the monomeric form binds the receptor. ANF increases the glomerular filtration rate, leading to increased urine volume and excretion of sodium ion. Renin and aldosterone secretion are reduced, and the vasoconstriction produced by angiotensin II is inhibited, causing relaxation of the renal vessels as well as other vascular beds and large arteries. ANF mediates these effects by binding to its membrane receptor, whose intracellular

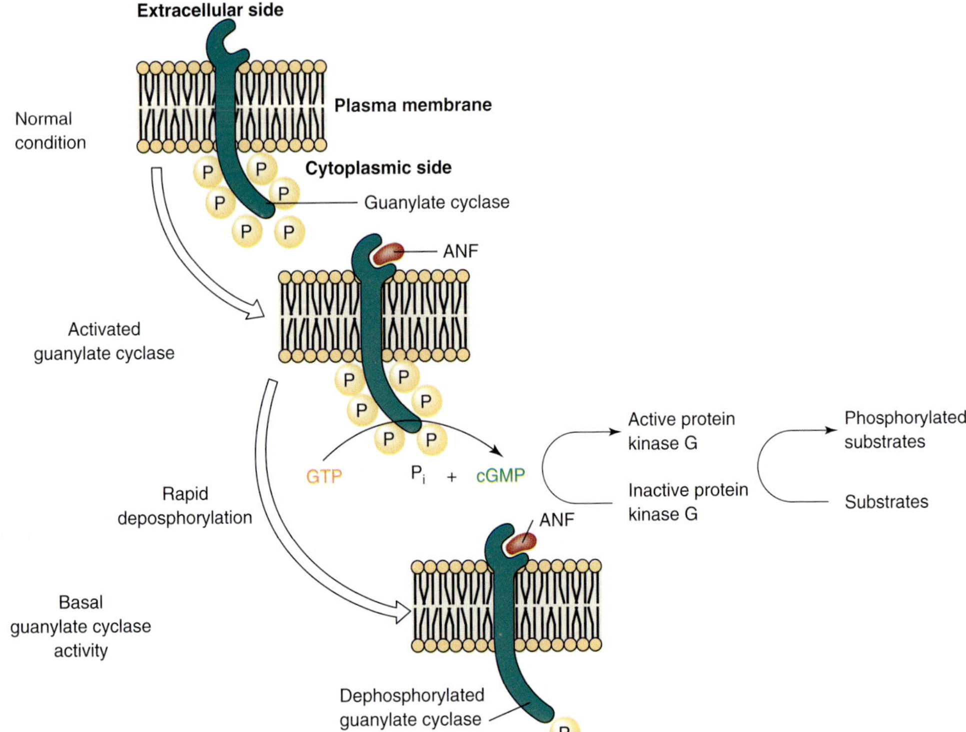

FIGURE 23.33

Model for signal transduction by ANF receptor. The guanylate cyclase domain is in a highly phosphorylated state under normal conditions. Binding of hormone markedly enhances enzyme activity and dephosphorylation of the guanylate cyclase domain.
Redrawn from Schultz, S., Chinkers, M., and Garbers, D. L. *FASEB J.* 3:2026, 1989.

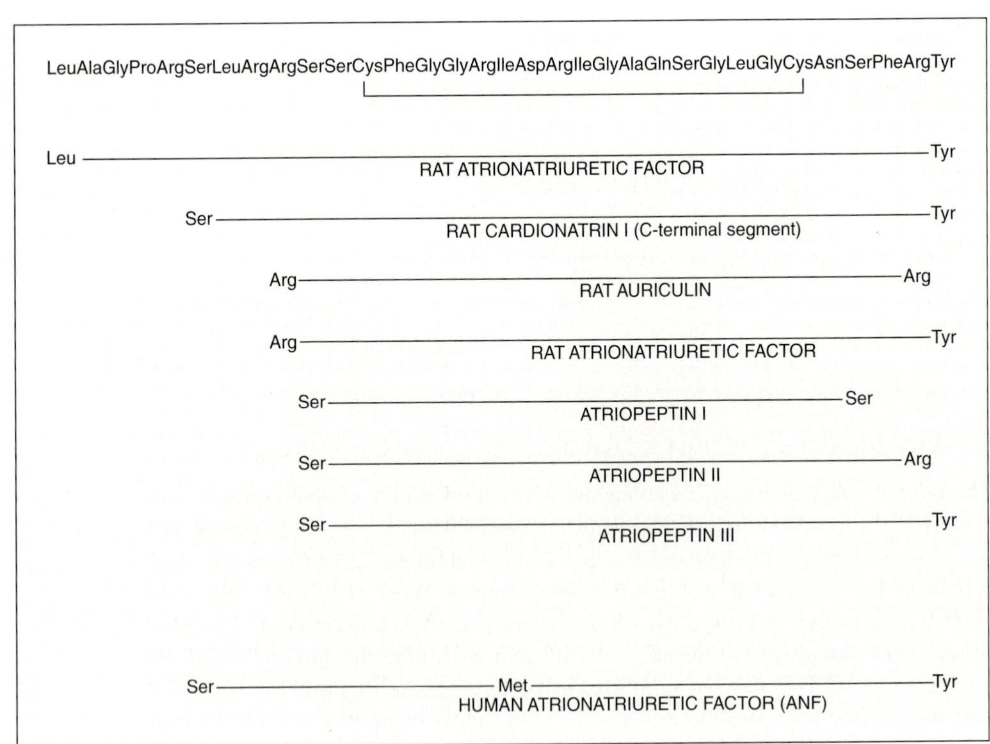

FIGURE 23.34

Atrial natriuretic peptides. These peptides relax vascular smooth muscle and produce vasodilation and natriuresis as well as other effects discussed in the text.
Adapted from Cantin, M. and Genest, J. The heart and the atrial natriuretic factor. *Endocr. Rev.* 6:107, 1985.

domain has guanylate cyclase activity (Figure 23.33). The cGMP that is produced activates protein kinase G, which then phosphorylates other cellular proteins involved in this pathway. Many analogs of ANF bind to receptors in the kidney but fail to elicit a physiological response. This suggests that these receptors may serve as specific peripheral storage-clearance binding sites for ANF and modulate its plasma levels.

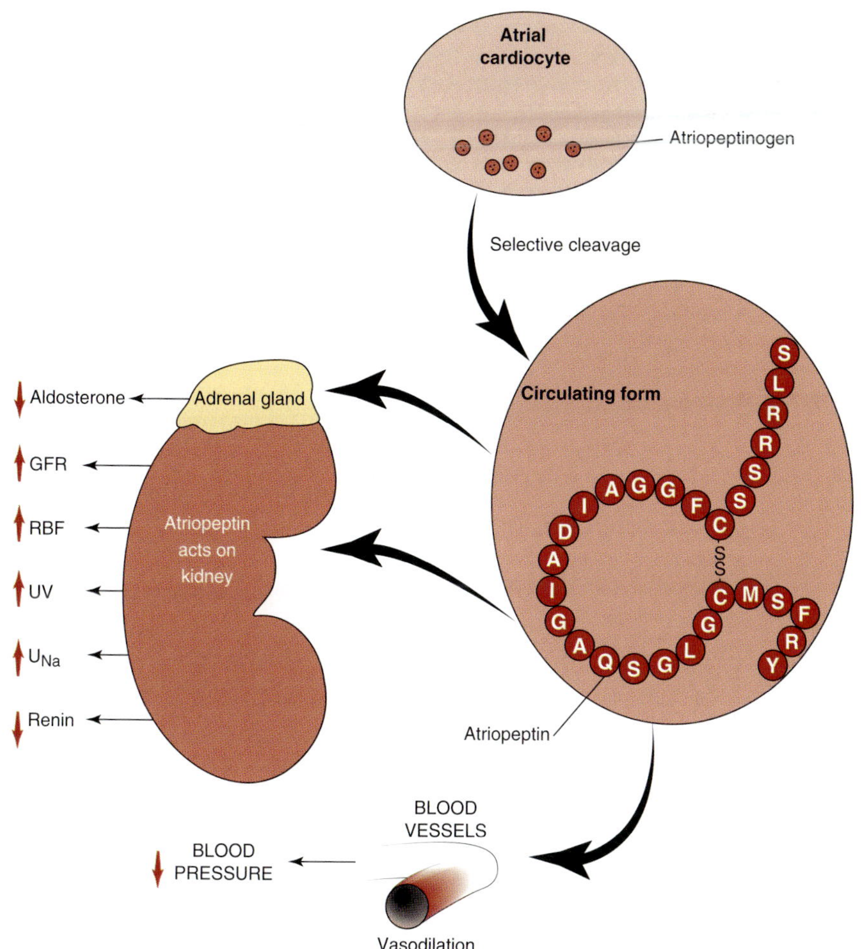

FIGURE 23.35

Schematic diagram of atrial natriuretic factor–atriopeptin hormonal system. Prohormone is stored in granules in perinuclear atrial cardiocytes. An elevated vascular volume results in cleavage and release of atriopeptin, which increases the glomerular filtration rate (GFR), renal blood flow (RBF), urine volume (UV), and sodium excretion (U_{Na}) and decreases plasma renin activity and the secretion of aldosterone and arginine vasopressin. Vasodilation lowers blood pressure (BP). In contrast, diminution of vascular volume suppresses circulating levels of atriopeptin.
Redrawn from Needleman, P. and Greenwald, J. E. Atriopeptin: A cardiac hormone intimately involved in fluid, electrolyte, and blood pressure homeostasis. *N. Engl. J. Med.* 314:828, 1986.

23.7 STEROID HORMONES

Structures and Functions of Steroid Hormones

Steroid hormones are divided into two classes: the **sex** and **progestational hormones** and the **adrenocortical hormones**. They are synthesized in the gonads (ovaries and testes) and adrenal cortex from cholesterol (see p. 707) through Δ^5-pregnenolone as intermediate. Their structure is based on the **cyclopentanoperhydrophenanthrene** nucleus; the numbering of this ring system and the lettering of the rings are presented in Figure 23.36. Conversion of steroid hormones to less active or inactive forms involves alteration of ring substituents rather than the ring structure itself. The major steroid hormones and their important functions in humans are summarized in Table 23.8. Many are similar in overall structure, although their receptors can be highly specific. For cortisol and aldosterone, however, each receptor can bind both ligands. Steroid hormones are classified based on the number of carbons they contain. Thus $1,25(OH)_2D_3$ is a C_{27} secosteroid; **progesterone, cortisol**, and **aldosterone** are C_{21} steroids; **testosterone** and **dehydroepiandrosterone** are C_{19} steroids; and **17β-estradiol** is a C_{18}, steroid. Sex hormones can be distinguished easily as being androgens (C_{19}), estrogens (C_{18}), or progestational or adrenal steroids (C_{21}). Certain substituents in the ring system are characteristic. For example, glucocorticoids and mineralocorticoids (typically aldosterone) possess a C11 OH or oxygen moiety. Estrogens lack a C19 methyl group, and their A ring contains three double bonds. Many steroid receptors recognize primarily the A ring of their specific hormone. For instance, the estrogen receptor can distinguish the A ring of estradiol, which is stretched out of the plane of the B–C–D rings, from the A rings in other steroid, which are coplanar with the B–C–D rings. This relationship between the A ring and the B–C–D rings is illustrated in Figure 23.37.

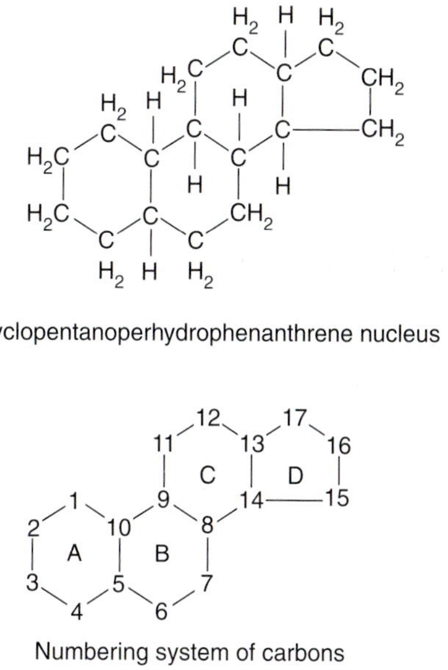

Cyclopentanoperhydrophenanthrene nucleus

Numbering system of carbons

FIGURE 23.36

The steroid nucleus.

TABLE 23.8 Major Steroid Hormones of Humans

Hormone	*Structure*	*Secretion from*	*Secretion Signal*	*Functions*
Progesterone	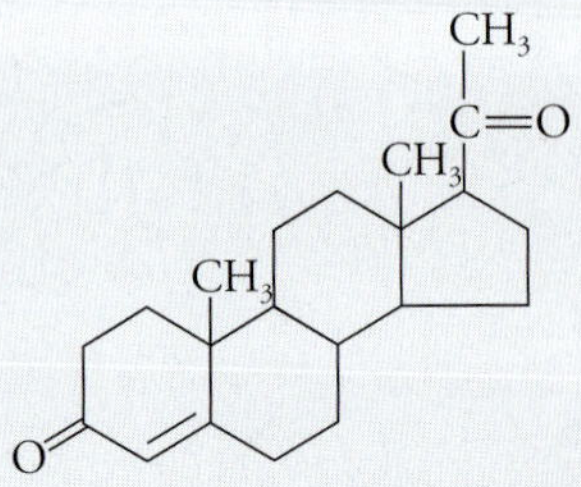	Corpus luteum	LH	Maintains (with estradiol) the uterine endometrium for implantation; differentiation of mammary glands
17*β*-Estradiol	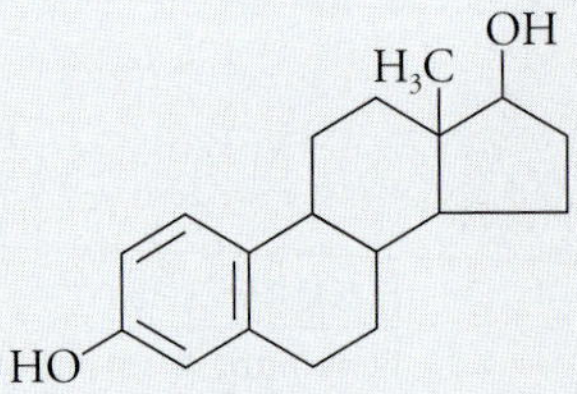	Ovarian follicle; corpus luteum; (Sertoli cell)	FSH	Female: regulates gonadotropin secretion in ovarian cycle; maintains (with progesterone) uterine endometrium; differentiation of mammary gland. Male: negative feedback inhibitor of Leydig cell synthesis of testosterone
Testosterone	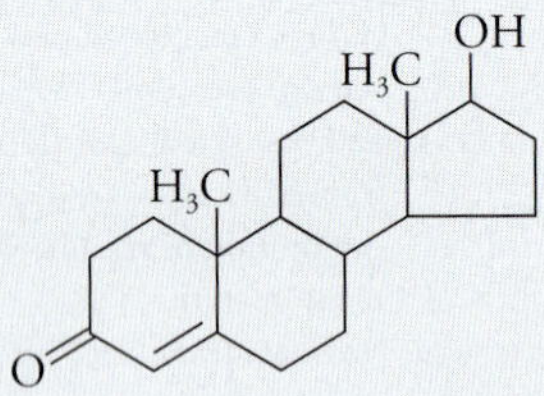	Leydig cells of testis; (adrenal gland); ovary	LH	Male: required for spermatogenesis; converted to more patent androgen, dihydrotestosterone, in some target tissues like prostate gland; secondary sex characteristics (in some tissues testosterone is active hormone)
Dehydroepiandrosterone	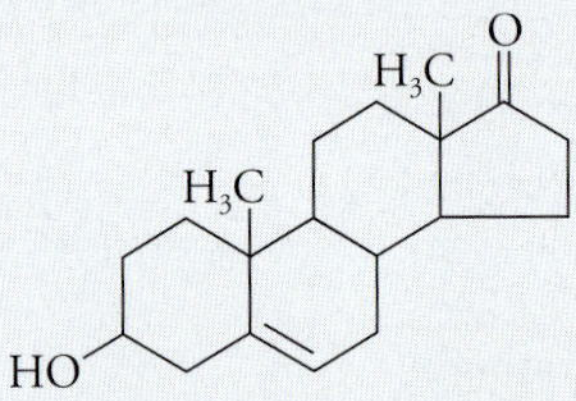	Reticularis cells of adrenal cortex	ACTH	Various protective effects of adrenal cortex (anticancer, antiaging); weak androgen; can be converted to estrogen; no receptor yet isolated.
Cortisol	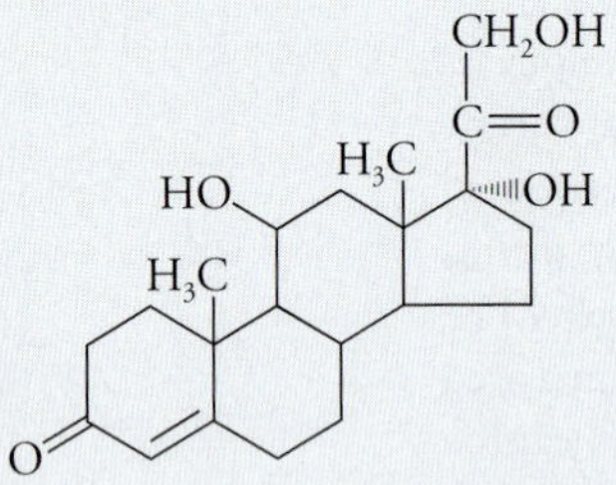	Fasciculata cells of adrenal cortex	ACTH	Stress adaptation of adrenal cortex through various cellular phenotypic expressions; regulates protein, carbohydrate, and lipid metabolism; Immunosuppressive effects.
Aldosterone	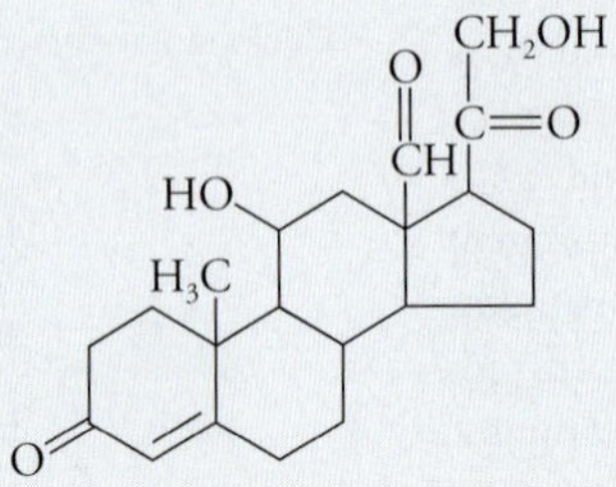	Glomerulosa cells of adrenal cortex	Angiotensin II/III	Causes sodium ion reabsorption in kidney via conductance channel; controls salt and water balance; raises blood pressure by increasing fluid volume.

TABLE 23.8 ***(continued)***

Hormone	*Structure*	*Secretion from*	*Secretion Signal*	*Functions*
1,25-Dihydroxy-vitamin D_3	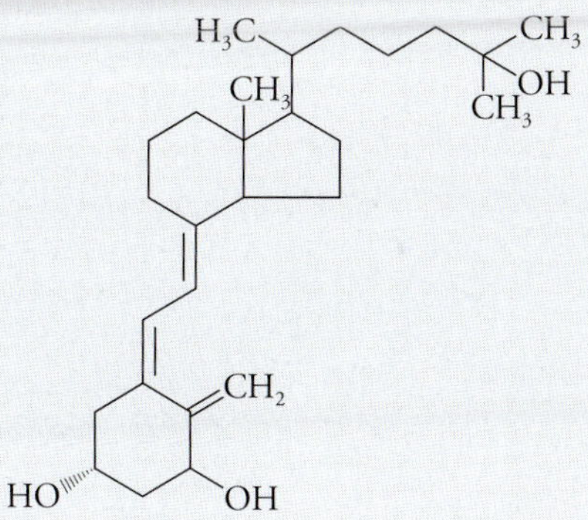	Vitamin D arises in skin cells after exposure to UV light and successive hydroxylations occur in liver and kidney to yield active form of hormone	PTH (stimulates kidney proximal tubule hydroxylation system)	Facilitates Ca^{2+} and phosphate absorption by intestinal epithelial cells; induces intracellular calcium-binding protein

[a] LH, luteinizing hormone; FSH, follicle-stimulating hormone; ACTH, adrenocorticotropic hormone; PTH, parathyroid hormone.

Biosynthesis of Steroid Hormones

Pathways for conversion of cholesterol to the adrenal cortical steroid hormones are presented in Figure 23.38. Cholesterol undergoes side chain cleavage to form Δ^5-pregnenolone and isocaproaldehyde. **Δ^5-Pregnenolone** is mandatory in the synthesis of all steroid hormones. Pregnenolone is converted directly to progesterone by **3β-ol dehydrogenase** and **$\Delta^{4,5}$-isomerase**. The dehydrogenase converts the 3-OH group of pregnenolone to a 3-keto group, and the isomerase moves the double bond from the B ring to the A ring to produce progesterone. In the corpus luteum of the ovary, the bulk of steroid synthesis stops at this point. Conversion of pregnenolone to **aldosterone** in the adrenal zona glomerulosa cells requires 21-hydroxylase of the endoplasmic reticulum and requires 11β-hydroxylase and 18-hydroxylase of mitochondria. To form cortisol, primarily in adrenal zona fasciculata cells, endoplasmic reticulum **17-hydroxylase** and **21-hydroxylase** are required together with mitochondrial **11β-hydroxylase.** The endoplasmic reticulum hydroxylases are cytochrome P450 (CYP) enzymes (see p. 414). Δ^5-Pregnenolone is converted to **dehydroepiandrosterone** in the adrenal zona reticularis cells by 17α-hydroxylase of the endoplasmic reticulum to form 17α-hydroxypregnenolone and then by a side-chain cleavage system to form dehydroepiandrosterone. Cholesterol is converted to the sex steroids by way of Δ^5-pregnenolone and progesterone (Figure 23.39). **Progesterone** is converted to testosterone by the action of cytoplasmic enzymes and 17-dehydrogenase. **Testosterone** is a major secretory product in the Leydig cells of the testis and is converted to dihydrotestosterone in some androgen target cells before binding with high affinity to the androgen receptor. This reduction requires the activity of **5α-reductase** located in the endoplasmic reticulum and nucleus. Pregnenolone can enter an alternative pathway to form dehydroepiandrosterone as described above. This weak androgen can then be converted to androstenedione and then to testosterone. Estradiol is formed from testosterone by the action of the **aromatase** system. The endoplasmic reticulum hydroxylases involved in steroid hormone synthesis use molecular oxygen (O_2) to introduce one oxygen atom into the steroidal substrate (as an OH), while the second atom is reduced to water. Electrons generated from NADH or NADPH through a flavoprotein are transferred to ferredoxin or a similar nonheme protein. It is important to note that there is movement of intermediates in and out of the mitochondrial compartment during biosynthesis of steroids. Once the specific steroids are synthesized, they diffuse through the plasma membrane and enter the general circulation, where they bind to transport proteins. Unlike peptide hormones, steroids are not stored in secretory vesicles.

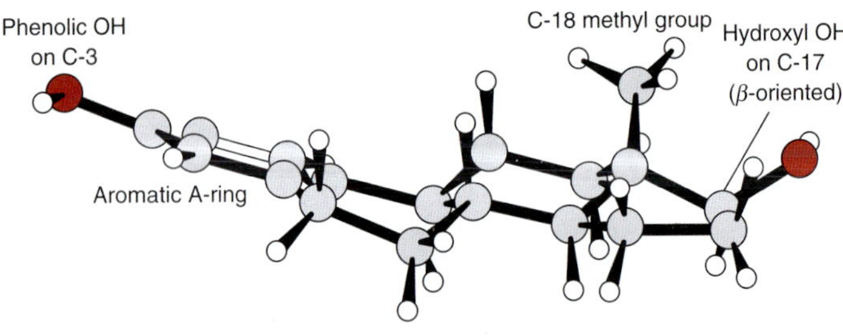

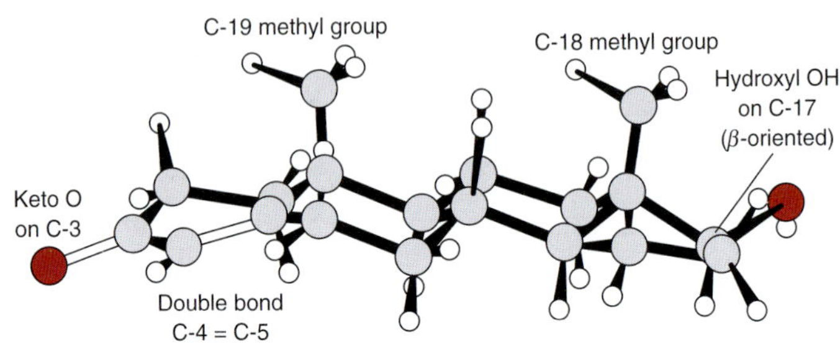

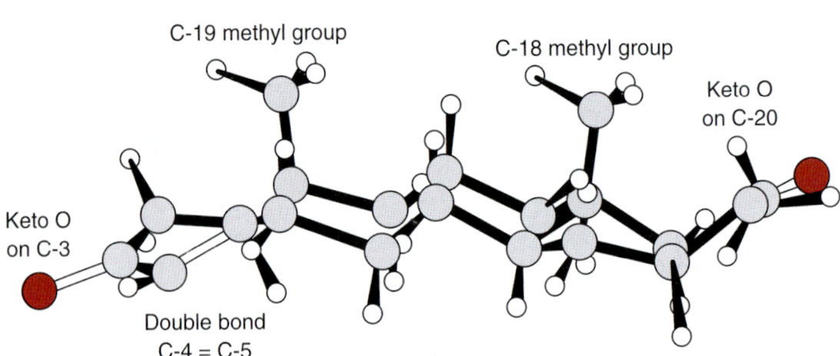

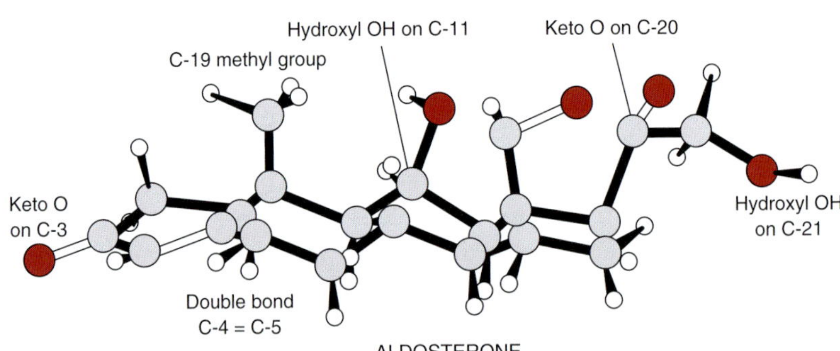

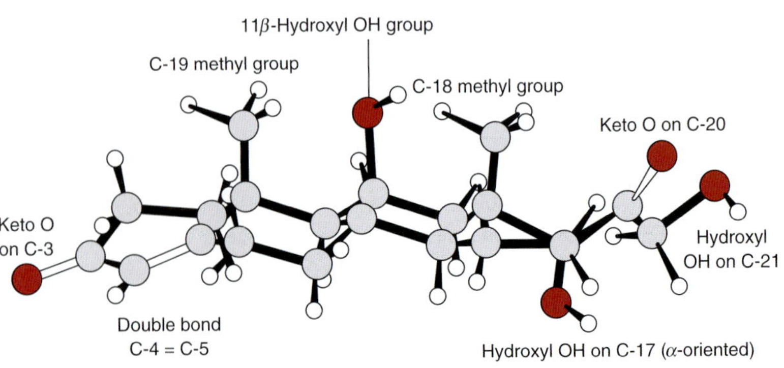

FIGURE 23.37

"Ball-and-stick" representations of some steroid hormones determined by X-ray crystallographic methods. Details of each structure are labeled. In aldosterone the acetal grouping is $R-CH(OR_1)(OR_2)$ and the hemiketal grouping is $R_1R_2C(OR_3)(OH)$ where R_1, R_2, and R_3 refer to different substituents.

Reprinted with permission from Glusker, J. P. In G. Litwack (Ed.), *Biochemical Actions of Hormones*, Vol. 6. New York: Academic Press, 1979, pp. 121–204.

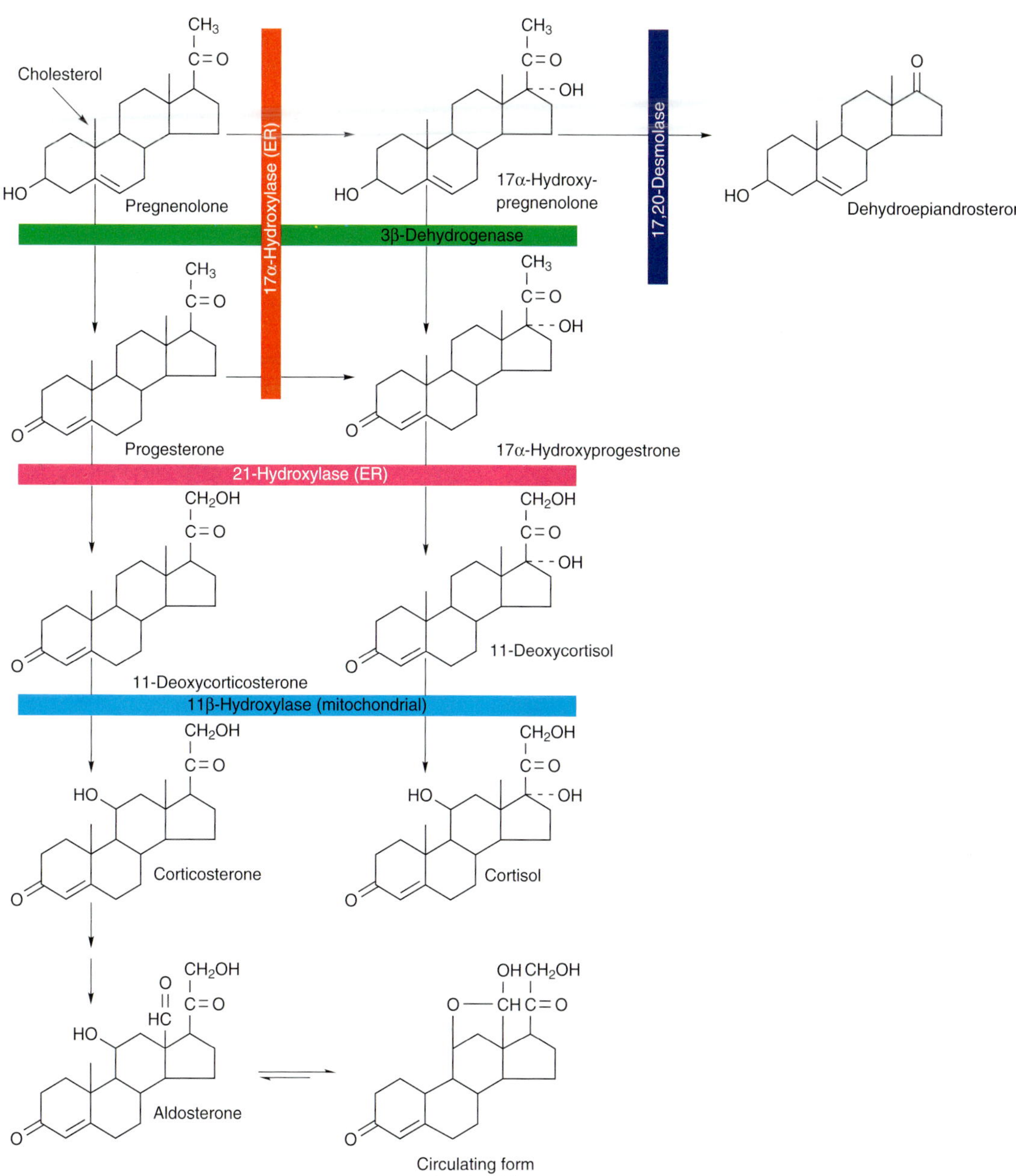

FIGURE 23.38

Conversion of cholesterol to adrenal cortical hormones. Not all intermediates are included and only enzymes of clinical significance are shown. ER, endoplasmic reticulum.
Redrawn based on figure in Porterfield, S. P. (Ed.). *Endocrine Physiology*. St. Louis: Mosby, 1997, p. 136.

Metabolism of Steroid Hormones

Steroid hormones are metabolized slowly because of their binding to plasma proteins, which protects them from degradation. In humans, cortisol binds extensively to serum transcortin and it has a plasma half-life of about 60–70 minutes. In contrast, aldosterone, which is not extensively bound to plasma proteins, has a half-life of only about 20 minutes. The liver is the principal site for steroid metabolism, and a large number of metabolites are produced for each steroid. In general, the enzymatic reactions involved in steroid metabolism tend to decrease their biological activity and increase their solubility in water, thus facilitating their excretion in urine. **Conjugation** (see p. 422) also greatly increases the water solubility of steroid metabolites; **glucuronides** and **sulfates** are the most common conjugates. Estimates of steroid hormone secretion are therefore

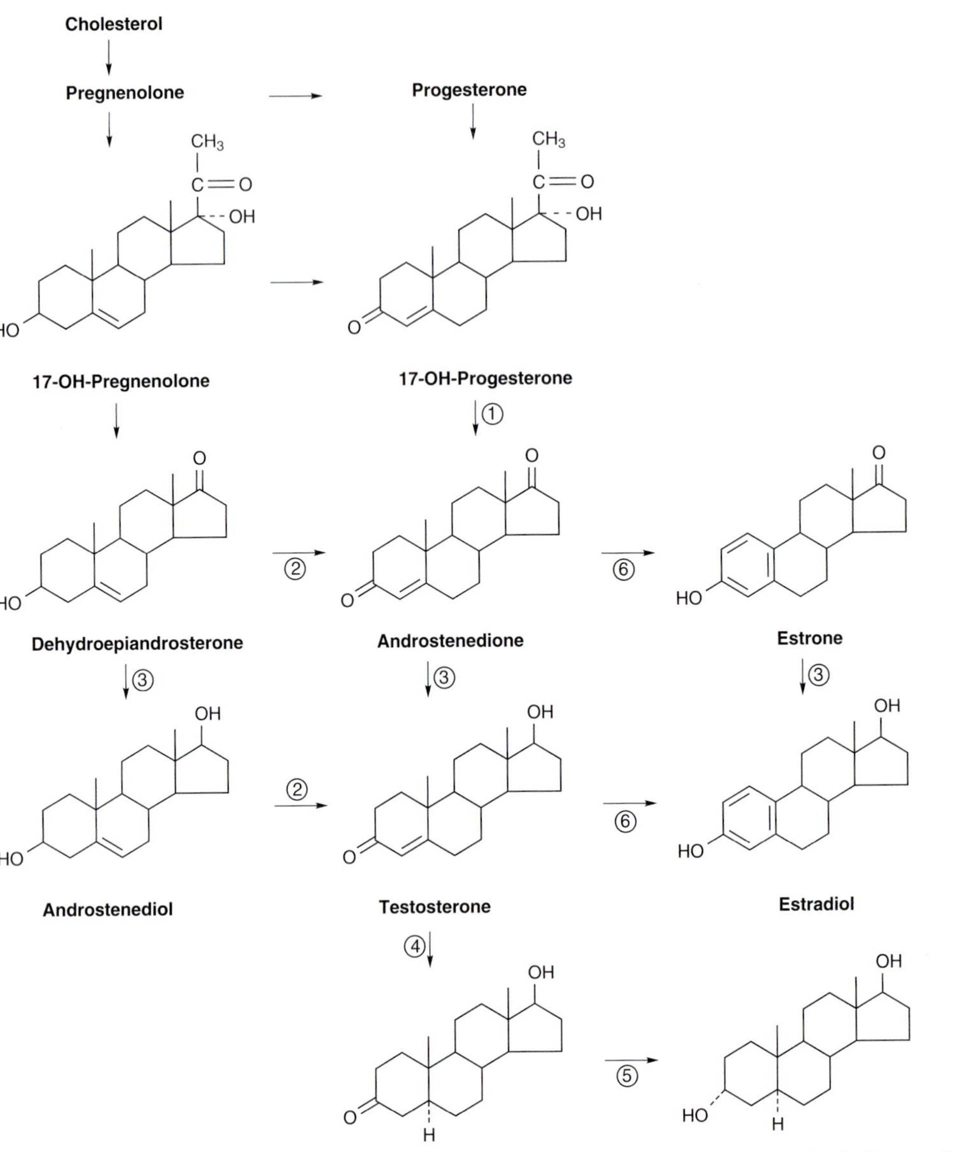

FIGURE 23.39

Conversion of cholesterol to sex hormones. Testosterone is the major steroid secreted by the testis. Estradiol and progesterone are the major steroids secreted by the ovary. Enzymes are (1) 17,20-desmolase, (2) dehydrogenase and isomerase, (3) 17β-OH dehydrogenase, (4) 5 α-reductase, (5) 3 α-reductase, and (6) aromatase.

frequently based on urinary metabolite levels. A number of different factors are known to influence the hepatic metabolism of steroids and hence their rate of removal from the body. For example, age influences hepatic steroid metabolism and the clearance of many steroids is slower (lower metabolic clearance rate) in infants and very old individuals. The rates of steroid metabolism are also increased in patients with hyperthyroidism and decreased in patients with hypothyroidism and different types of liver disease.

An example of steroid metabolism, more specifically testosterone metabolism, is presented in Figure 23.40. Metabolism of testosterone does not necessarily result in inactivation of this steroid. Testosterone can be converted to the active steroid, estradiol, via the **aromatase** enzyme. Reduction of testosterone via the 5α-reductase enzyme generates **dihydrotestosterone**, which is a potent androgen that binds to the intracellular androgen receptor. Androsterone is the major inactive metabolic end product produced in this pathway; it is conjugated and then excreted in the urine. The major adrenal androgen, dehydroepiandrosterone (DHEA), is a known source of urinary 17-ketosteroids. Two other testosterone metabolites appear in the urine in relatively small amounts; these include androstanediol, which is formed by the reduction of the 3-keto group of dihydrotestosterone, and estrogen metabolites, which are formed once testosterone has been converted into estradiol.

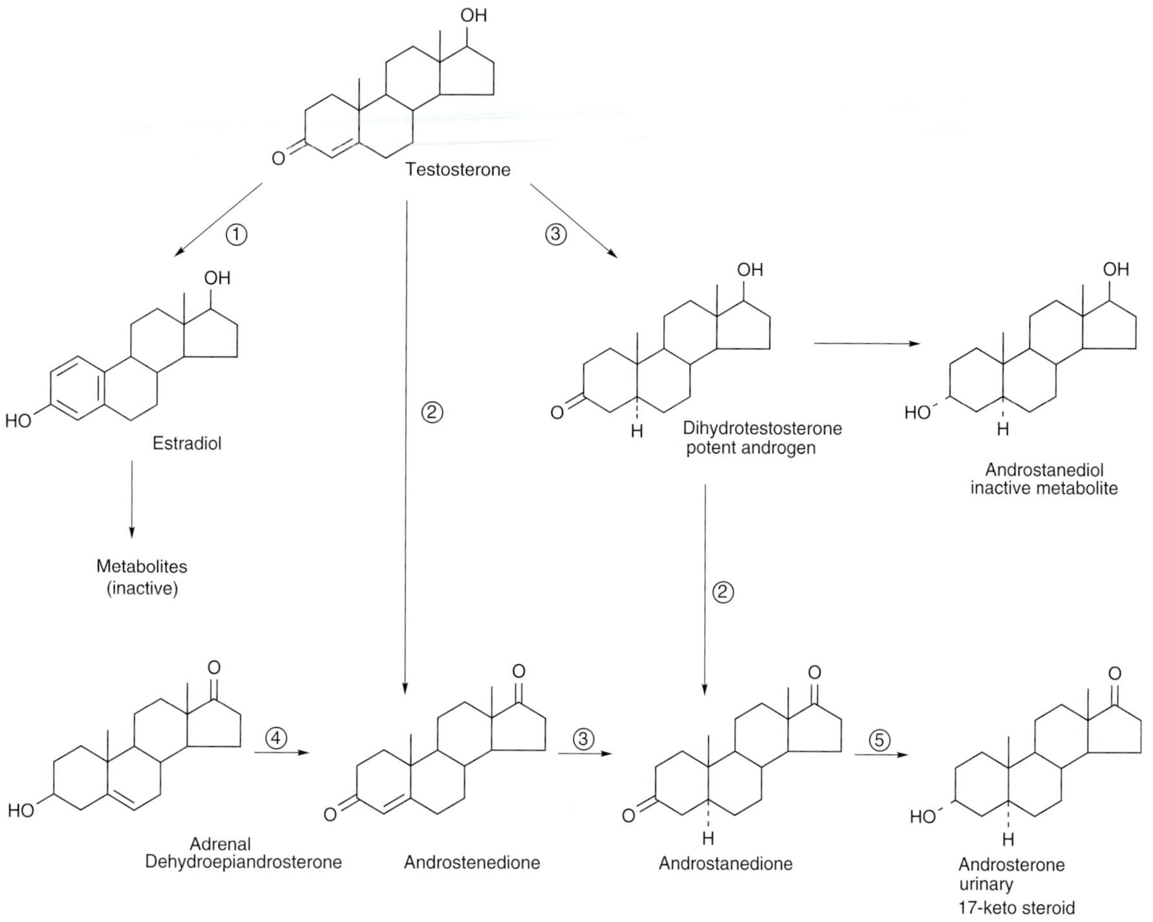

FIGURE 23.40

Metabolism of testosterone to generate active steroids or inactive 17-ketosteroids. Enzymes indicated: (1) aromatase, (2) 17β-hydroxysteroid dehydrogenase, (3) 5β reductase, (4) 3β-hydroxysteroid dehydrogenase isomerase, and (5) 3β-hydroxysteroid dehydrogenase.

Regulation of Steroid Hormone Synthesis

Hormonal regulation of steroid hormone biosynthesis is mediated by increased intracellular concentrations of cAMP and Ca^{2+}, although generation of IP_3 may also be involved (Figure 23.41). cAMP exerts rapid (occurring within seconds to minutes) and slow (requiring hours) effects on steroid synthesis. The rapid effect is the mobilization and delivery of cholesterol to the inner membrane of mitochondria, where it is metabolized to pregnenolone by the cholesterol side-chain cleavage enzyme. In contrast, the slower effects involve increased transcription of genes for steroidogenic enzymes that are responsible for maintaining optimal long-term steroid production. The 30-kDa phosphoprotein, known as the **steroidogenic acute regulatory (StAR)** protein, facilitates the translocation of cholesterol from the outer to the inner mitochondrial membranes. In humans, StAR mRNA is specifically expressed in the testes and ovaries, known sites of steroidogenesis. Patients with lipoid congenital adrenal hyperplasia (LCAH), an inherited disease in which adrenal and gonadal steroidogenesis is significantly impaired and lipoidal deposits occur, express truncated and nonfunctional StAR proteins. This strongly suggests that the StAR protein is the hormone-induced protein that mediates acute regulation of steroid hormone biosynthesis.

Polypeptide hormones that stimulate the biosynthesis and secretion of steroid hormones are summarized in Table 23.9. These hormones operate through cognate cell membrane receptors. In cases where both the cAMP and the phosphatidylinositol cycle are involved, it is not clear whether one second messenger predominates. For the synthesis and secretion of aldosterone, probably several components (i.e., acetylcholine

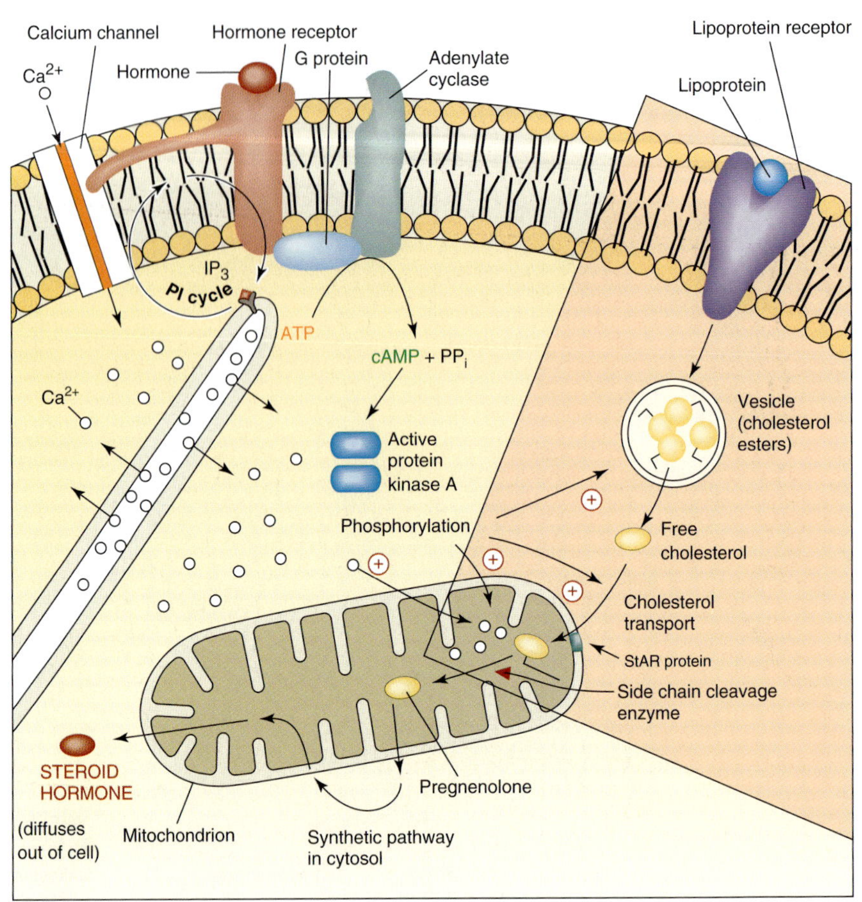

FIGURE 23.41

Overview of hormonal stimulation of steroid hormone biosynthesis. Nature of the hormone (top of figure) depends on the cell type and receptor (ACTH for cortisol synthesis, FSH for estradiol synthesis, LH for testosterone synthesis, etc., as given in Table 23.9). This activates adenylate cyclase via a stimulatory G protein or may stimulate a calcium channel directly or indirectly by activating the phosphatidylinositol cycle (PI cycle). If the PI cycle is stimulated, IP_3 augments cytosol Ca^{2+} levels from the ER store. cAMP activates protein kinase A, which, via phosphorylation, increases hydrolysis of cholesteryl esters from the vesicles to free cholesterol and increase cholesterol transport into the mitochondrion. The elevated Ca^{2+} levels and protein phosphorylation, as well as induction of the StAR protein, result in increased side-chain cleavage and steroid biosynthesis. These reactions overcome the rate-limiting steps (availability of cholesterol from cholesterol esters stored in vesicles, transport of cholesterol to inner mitochondrial membrane, side-chain cleavage reaction) in steroid biosynthesis, and more steroid is synthesized and secreted.

TABLE 23.9 Hormones that Directly Stimulate Synthesis and Release of Steroid Hormones

Steroid Hormone	*Steroid-Producing Cell or Structure*	*Signal*[a]	*Second Messenger*	*Signal System*
Cortisol	Adrenal zona fasciculata	ACTH	cAMP, PI cycle, Ca^{2+}	Hypothalamic-pituitary cascade
Aldosterone	Adrenal zona glomerulosa	Angiotensin II/III	PI cycle, Ca^{2+}	Renin-angiotensin system
Testosterone	Leydig cell	LH	cAMP	Hypothalamic-pituitary cascade
17β-Estradiol	Ovarian follicle	FSH	cAMP	Hypothalamic-pituitary-ovarian cycle
Progesterone	Corpus luteum	LH	cAMP	Hypothalamic-pituitary-ovarian cycle
1,25$(OH)_2$ D_3	Kidney	PTH	cAMP	Sunlight, parathyroid glands, plasma Ca^{2+} level

[a] ACTH, adrenocorticotropic hormone; LH, luteinizing hormone; FSH, follicle-stimulating hormone; PI, phosphatidylinositol; PTH, parathyroid hormone.

muscarinic receptor, atriopeptin receptor) and their second messengers are involved in addition to those listed in Table 23.9. Figure 23.42 summarizes the effects of ACTH on cortisol biosynthesis and secretion.

Aldosterone

Figure 23.43 shows reactions leading to secretion of aldosterone in a cell located in the adrenal zona glomerulosa. This outer most zone of the adrenal cortex cannot synthesize cortisol because it does not express the 17 α-hydroxylase enzyme. The main driving force is **angiotensin II**, which is generated by the **renin-angiotensin system** shown in Figure 23.44. The signal for aldosterone secretion is generated under conditions when blood Na^+ concentration and blood pressure (blood volume) need to be increased. The

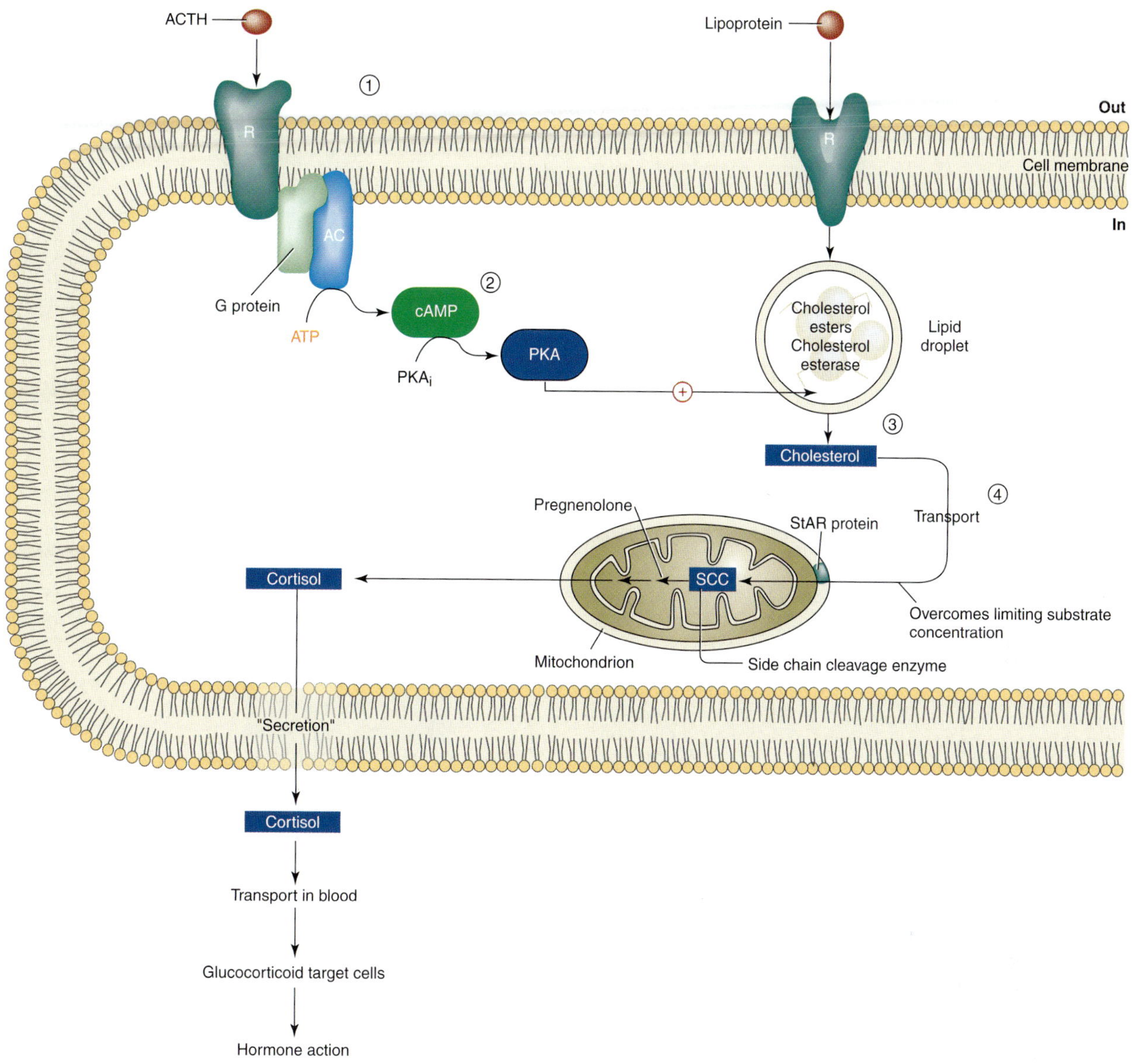

FIGURE 23.42

Action of ACTH on adrenal fasciculata cells to enhance production and secretion of cortisol. AC, adenylate cyclase; cAMP, cyclic AMP; PKA, protein kinase A(active); PKA_i, protein kinase A(inactive); SCC, side-chain cleavage system of enzymes. StAR (steroidogenic acute regulatory) protein is a cholesterol transporter functioning between the outer and inner mitochondrial membranes.

N-terminal decapeptide of a plasma α_2-**globulin (angiotensinogen)** is cleaved by the proteolytic enzyme **renin**. This decapeptide is the inactive angiotensin I, which is then converted to the active octapeptide, angiotensin II, by **angiotensin converting enzyme (ACE)**. This enzyme is found on the surface of endothelial cells and pulmonary and renal endothelial cells are important sites for this conversion. Angiotensin II is then converted to the heptapeptide, angiotensin III by an aminopeptidase. Both angiotensins II and III bind to the angiotensin receptor (Figure 23.43), which activates phospholipase C to generate IP_3 and DAG. The IP_3 causes release of Ca^{2+} from the intracellular Ca^{2+} storage vesicles. In addition, the plasma membrane Ca^{2+} channel is opened by the angiotensin–receptor complex. These events lead to a greatly increased level of cytosolic Ca^{2+}, which, together with DAG, stimulates protein kinase C. **Acetylcholine** released through neuronal stress signals produces similar effects on cytosolic calcium levels, and these effects on protein kinase C activity are mediated by the muscarinic acetylcholine receptor (see p. 914).

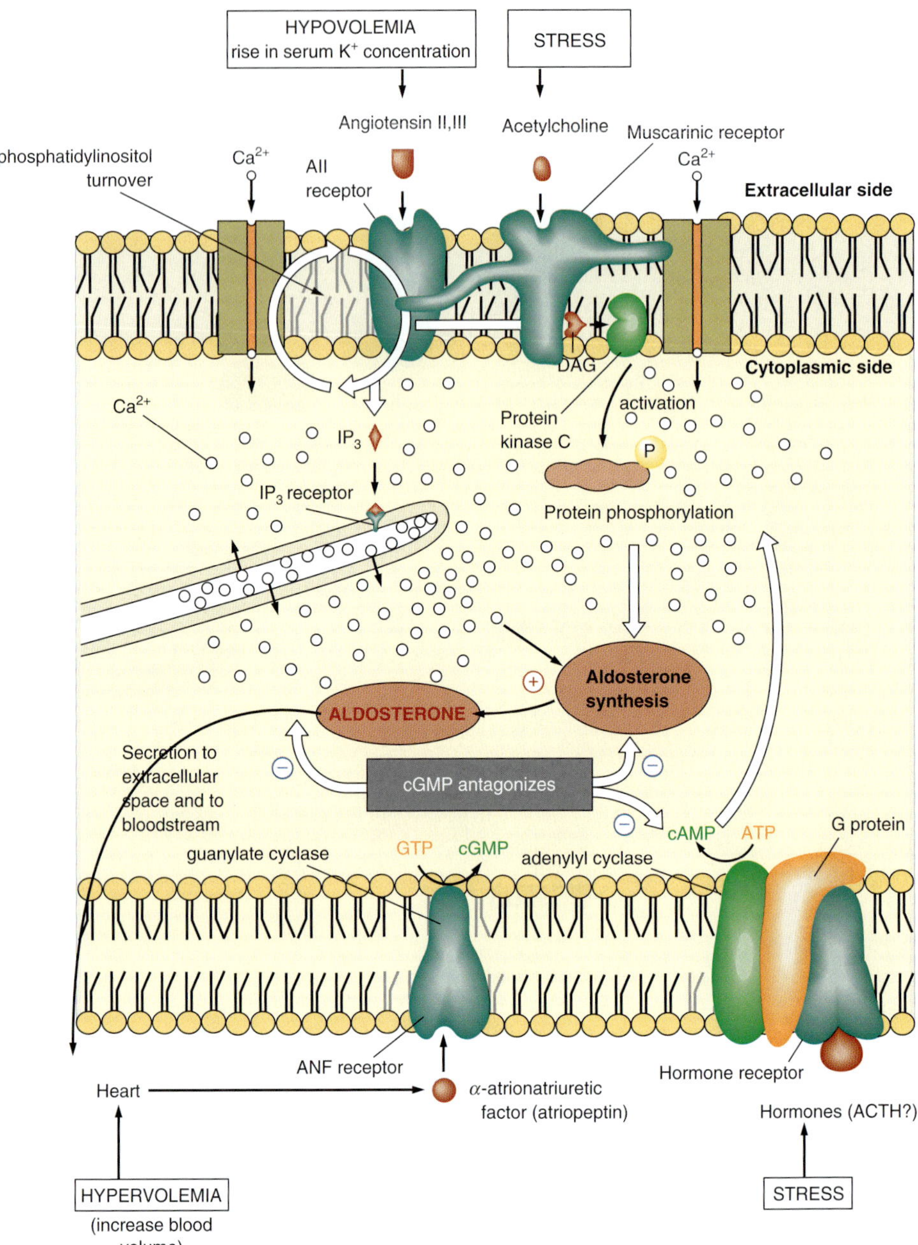

FIGURE 23.43

Reactions leading to secretion of aldosterone in adrenal zona glomerulosa cell. cGMP, cyclic GMP; ANF, atrial natriuretic factor; see Figure 23.42 for additional abbreviations.

Phosphorylation of enzymes that stimulate the rate-limiting steps in aldosterone synthesis lead to increased synthesis of aldosterone. In the distal kidney cells, this adrenal steroid binds to its intracellular receptor and enhances expression of genes for proteins, including subunits for membrane Na^+ channels, which increase absorption of Na^+ from the glomerular filtrate. Although ACTH is not considered a major regulator of aldosterone synthesis, it may be important in terms of promoting maximal secretion of this mineralocorticoid hormone. In contrast, ACTH is the major stimulator of the synthesis and secretion of cortisol by cells in the zona fasciculata (Figure 23.42). Cortisol also accounts for substantial reabsorption of Na^+ in the kidney, probably through glucocorticoid stimulation of the Na^+/H^+ antiport transporter in the luminal membrane of renal epithelial cells.

Physiological conditions opposite to those that activate the formation of angiotensin I and II generate **atrial natriuretic factor** (**ANF**), also called **atriopeptin**, from the heart atria (Figure 23.43; see also Figure 23.44). In other words, an increase in blood volume results in increased stretch of the heart and increased synthesis and secretion of ANF. At the level of the zona glomerulosa cell, ANF activates its receptor guanylate cyclase activity, resulting in an increase in cytosolic cGMP levels. cGMP then inhibits the synthesis and secretion of aldosterone and formation of cAMP by adenylate cyclase.

Estradiol

Hormonal control of the synthesis and secretion of **17β-estradiol** is shown in Figure 23.45. At the time of female puberty the secretion of GnRH, which is controlled by higher brain centers, increases significantly. During adult life this hypothalamic releasing hormone stimulates the release of FSH and LH from the anterior pituitary gonadotropes (see p. 923). Once secreted, FSH stimulates the synthesis and secretion of 17β-estradiol in the ovary. This steroid exerts negative feedback on the **gonadotrope** (anterior pituitary), which suppresses further secretion of FSH, and on the hypothalamic cells that secrete GnRH. Near ovarian midcycle, however, 17β-estradiol exerts a positive, rather than a negative, effect on the gonadotropes. This positive feedback (see Figure 23.45) causes very high levels of LH to be released (LH spike), as well as increased secretion of FSH (see p. 912, Figure 23.19). This LH spike is essential for ovulation to occur (see p. 912). After ovulation has occurred, a functional corpus luteum (CL) replaces the ruptured follicle and the CL synthesizes progesterone and some estradiol. Progesterone, however, inhibits continued LH synthesis and release. Eventually the corpus luteum dies (involutes), owing to a fall in the plasma LH level. The blood levels of progesterone and estradiol fall significantly, resulting in menstruation and a decline in their negative feedback effects on the anterior pituitary and hypothalamus. Once the negative feedback exerted by estradiol and progesterone have been eliminated, a new menstrual cycle is initiated. Clinical Correlation 23.4 describes how oral contraceptives interrupt this sequence.

In males, LH acts principally on Leydig cells to stimulate the synthesis of testosterone. FSH acts on Sertoli cells to stimulate the production of **inhibin** and the conversion of testosterone secreted by the Leydig cells into 17β-estradiol, which is required for **spermatogenesis**. FSH also stimulates the Sertoli cells to secrete an androgen-binding protein that binds testosterone and estradiol and carries these steroids into the fluid in the lumen of the seminiferous tubules. Thus, both testosterone and estradiol are continuously available to the maturing sperm. Testosterone itself exerts negative feedback and decreases secretion of GnRH. The Sertoli cells also secrete inhibin, which selectively inhibits FSH secretion. These hormonal negative feedback loops thus reduce testosterone and estradiol secretion and thereby exert negative regulation on the process of spermatogenesis. The steps involved in the biosynthesis of testosterone from cholesterol are summarized in Figure 23.39.

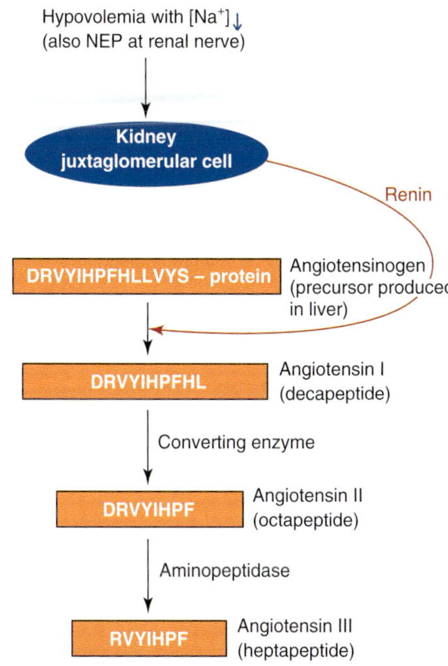

FIGURE 23.44

Renin–angiotensin system. see p. 80 for amino acid abbreviations. NEP, norepinephrine.

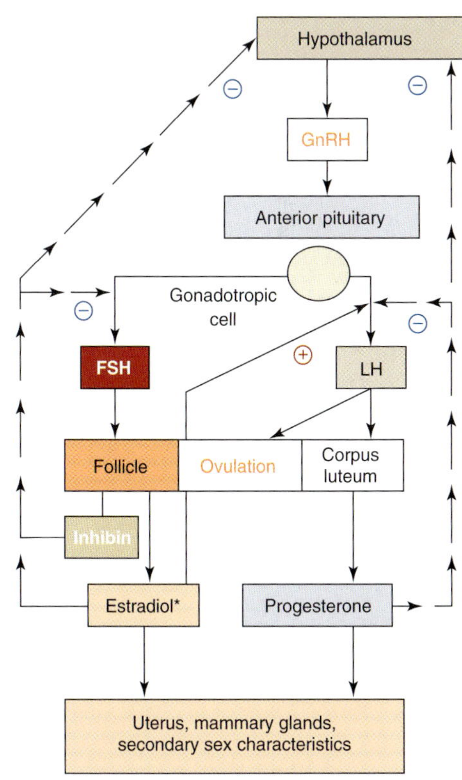

FIGURE 23.45

Formation and secretion of 17β-estradiol and progesterone.

Vitamin D_3

The active form of vitamin D, called **calcitriol**, is referred to as a **secosteroid**, which is a steroid in which one of the rings has been opened. As shown in Figure 23.46, 7-dehydrocholesterol is activated in the skin by sunlight to generate **vitamin D_3 (cholecalciferol)**. This precursor is hydroxylated first in the liver to generate **25-hydroxy vitamin D_3 (25-hydroxycholecalciferol).** Then in the kidney it forms either **1α,25-vitamin D_3 (1,25$(OH)_2D_3$) (1α,25-dihydroxycholecalciferol)**, which is the active form of the hormone, or **24,25-vitamin D_3 (24,25$(OH)_2D_3$) (24,25-dihydroxycholecaliciferol)**; the latter is the inactive metabolite of the hormone. Nuclear 1,25$(OH)_2D_3$ receptors are expressed in a number of target cells including intestinal epithelial cells, bone cells, and cells lining kidney tubules. Binding of 1,25$(OH)_2D_3$ induces receptor phosphorylation, and the hormone–receptor complex then binds to vitamin D_3-response elements in the DNA. Binding of the receptor to these DNA sequences results in an increased rate of transcription of vitamin D_3-responsive genes that encode a number of Ca^{2+}-binding proteins called **calbindins**, Ca^{2+}-ATPase and other ATPases, and membrane components and facilitators of vesicle formation. Calbindins may ferry Ca^{2+} across the intestinal cell, or alternatively may simply buffer the cytosol against high Ca^{2+} levels (Figure 23.47). 1,25$(OH)_2D_3$ also increases the number of Ca^{2+} pump molecules in the basolateral membrane. Although the major

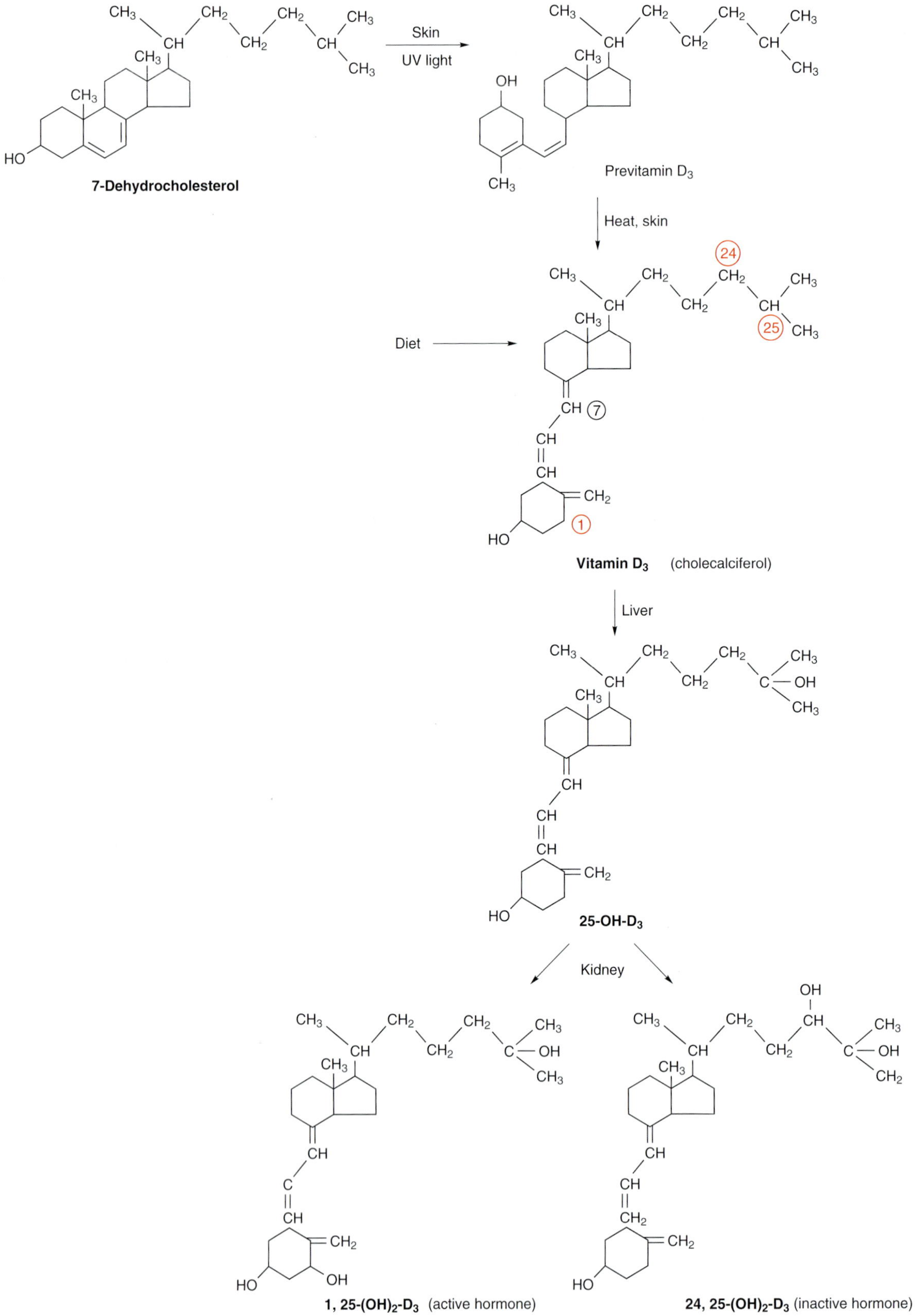

FIGURE 23.46

Structure of 1α,25-dihydroxycholecalciferol (calcitriol), the biologically active form of vitamin D.

effect of $1,25(OH)_2D_3$ is to stimulate the absorption of Ca^{2+} and phosphate from the intestinal lumen against a concentration gradient, this hormone also has other effects. By promoting the absorption of these two ions, it plays an important role in bone mineralization. However, elevated $1,25(OH)_2D_3$ levels can actually stimulate bone resorption by osteoclasts. Surprisingly, these cells do not express nuclear receptors for $1,25(OH)_2D_3$. The osteoblasts do express these receptors and $1,25(OH)_2D_3$ may stimulate these cells to secrete a paracrine factor that then increases the recruitment and differentiation of precursors into active osteoclasts, which can then reabsorb bone. Vitamin D may also be responsible for autocrine and paracrine effects that are important in the regulation of immune responses. In the kidney, $1,25(OH)_2D_3$ weakly stimulates Ca^{2+} reabsorption by increasing the number of Ca^{2+} pumps.

Transport of Steroid Hormones: Binding Proteins

Four major plasma proteins account for binding of the steroid hormones in blood. They are corticosteroid-binding globulin, sex hormone-binding protein, androgen-binding protein, and albumin. Most of the circulating cortisol (75–80%) is bound to a specific **corticosteroid-binding α_2-globulin (CBG)** also known as **transcortin.** Each molecule of this glycoprotein binds a single molecule of cortisol with K_a of $2.4 \times 10^7\ M^{-1}$. The normal plasma concentration of transcortin is 3 mg/dl, and its binding capacity is 20 μg cortisol/dl. About 15% of plasma cortisol is bound to albumin with a much lower affinity ($K_a = 10^3\ M^{-1}$). Thus even though albumin is present at 1000-fold the concentration of CBG, the cortisol will fill the CBG-binding sites first. The concentration of transcortin, and hence total cortisol, is increased during pregnancy and estrogen administration. During stress, when cortisol levels are high, the CBG-binding sites will be saturated and the excess cortisol will bind to albumin. Only 5–10% of plasma cortisol is normally free (unbound). Free cortisol diffuses across the plasma membrane, binds to intracellular glucocorticoid receptors, and mediates a biological response. In contrast to cortisol, only about 50–70% of the circulating aldosterone

CLINICAL CORRELATION 23.4
Oral Contraception

Oral contraceptives usually contain an estrogen and a progestin. Taken orally, they decrease secretion of FSH and LH so that the follicle does not mature, ovulation cannot occur, and the ovarian cycle ceases. The uterine endometrium thickens and remains in this state, however, because of elevated levels of estrogen and progestin. Pills that lack steroids (placebos) are usually inserted in the regimen at about the 28th day, causing blood levels of the steroids to fall dramatically and menstruation to occur. When oral contraceptive steroids are resumed, the blood levels of estrogen and progestin increase again and the uterine endometrium thickens. This creates a false "cycling" because of the occurrence of menstruation at the expected time in the cycle. The ovarian cycle and ovulation are suppressed by the negative effects of estrogen and progestin on the anterior pituitary gonadotropes. Contraception is also provided by implanting in the skin silicone tubes containing progestins. The steroid is released slowly, providing contraception for up to 3–5 years.

Source: Zatuchni, G. I. Female contraception. In: K. L. Becker (Ed.), *Principles and Practice of Endocrinology and Metabolism.* New York: Lippincott, 1990, p. 861. Shoupe, D. and Mishell, D. R. Norplant: Subdermal implant system for long term contraception. *Am. J. Obstet. Gynecol.* 160:1286, 1988.

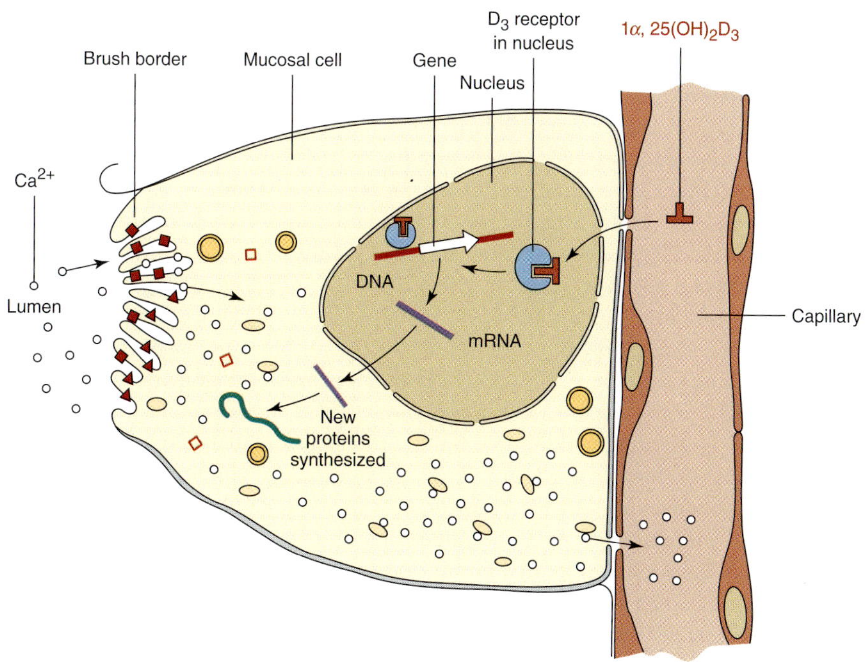

FIGURE 23.47

Schematic model of action of $1,25(OH)_2D_3$ in intestinal mucosal cell in stimulating calcium absorption from lumen. Redrawn from Nemere, I. and Norman, A. W. *Biochim. Biophys. Acta* 694:307, 1982.

is bound with low affinity to albumin and transcortin. Thus aldosterone has a lower plasma half-life (20 minutes) as compared to cortisol (70 minutes).

Sex hormone binding globulin (SHBG) binds androgens with an affinity constant of about 10^9 M^{-1}. About 65% of testosterone is bound to this liver-derived glycoprotein. Only 1–2% of circulating testosterone is in the free form, and the remaining androgen is bound to albumin and other proteins. The SHBG-bound fractions thus serve as circulating reservoirs of testosterone. About 60% of estrogens are transported bound to SHBG, 20% are bound to albumin, and 20% are in the free form. However, estradiol binds to SHBG with a much lower affinity than testosterone. Thus, estradiol bound to SHBG dissociates very rapidly and it is taken up by target tissues. For this reason, changes in SHBG levels in men can alter the ratio of free androgens to estrogens. The plasma concentration of SHBG before puberty is about the same in males and females. However, at puberty there is a small decrease in females and a larger decrease in males, ensuring a relatively greater amount of circulating unbound testosterone in males. Adult males have about one-half as much circulating SHBG as females, so free testosterone is about 20 times greater in males than in females. In addition, the total (bound plus unbound) concentration of testosterone is about 40 times greater in males. Testosterone itself lowers SHBG levels (increases percentage of free testosterone) in blood, whereas 17β-estradiol and thyroid hormone raise SHBG levels in blood. During pregnancy and hyperthyroidism the percentage of unbound testosterone and 17β-estradiol would be reduced.

Androgen-binding protein (ABP) is produced by the Sertoli cells in response to testosterone and FSH, which stimulate protein synthesis in these cells. ABP is also referred to as **testosterone–estrogen binding globulin (TeBG)**. This protein is similar to SHBG except that it is located within the testis and helps to maintain high androgen levels within the testis and seminal fluids. These high local androgen levels are important for the development and maturation of sperm. Progesterone binds to transcortin and albumin weakly and its circulating half-life is only about 5 minutes.

23.8 STEROID HORMONE RECEPTORS

Steroid Hormones Bind Intracellular Receptor Proteins

Steroid hormone receptors and receptors for nonsteroids (i.e., thyroid hormone, retinoic acid, vitamin D_3) are located intracellularly. The unliganded glucocorticoid receptor and possibly the aldosterone receptor appear to reside in the cytoplasm, whereas the other unliganded receptors are located within the nucleus, presumably in association with chromatin. In Figure 23.48, Step 1 shows a steroid hormone dissociating from a plasma transport protein. The free steroid enters the cell by diffusion through the lipid bilayer (Step 2). Cortisol binds to its receptor with a binding constant of 10^9 M^{-1}, compared to a binding constant of about 10^7 M^{-1} for CBG. The unliganded receptor (in this case glucocorticoid receptor) is a complex ($\sim$ 300 kDa) that contains other associated proteins, including a dimer of a 90-kDa **heat shock protein** which masks the receptor DNA-binding domain (Figure 23.49), and another heat shock protein designated as Hsp56, which is an immunophilin that binds a number of potent immunosuppressive drugs. Binding of the steroidal ligand (Step 3) causes a conformational change, referred to as "activation," in the receptor protein itself, resulting in the release of the associated proteins, including the dimer of Hsp 90, and exposure of positively charged amino acid residues located within the DNA-binding domain (Step 4). The ligand–receptor complex translocates to the nucleus (Step 5), binds to DNA, frequently as a homodimer, and "searches" the DNA for specific, high-affinity acceptor sites. Once the ligand–receptor complex has bound to these specific **hormone response elements** (HSE) (Step 6) in the DNA, it functions as a regulator of gene transcription and frequently enhances gene transcription. New mRNAs are translocated to the cytoplasm and assembled into translation complexes for the synthesis of proteins (Step 7) that alter the metabolism and functioning of the target cell (Step 8). In some situations, steroid receptor complexes actually repress, rather than induce, specific gene transcription.

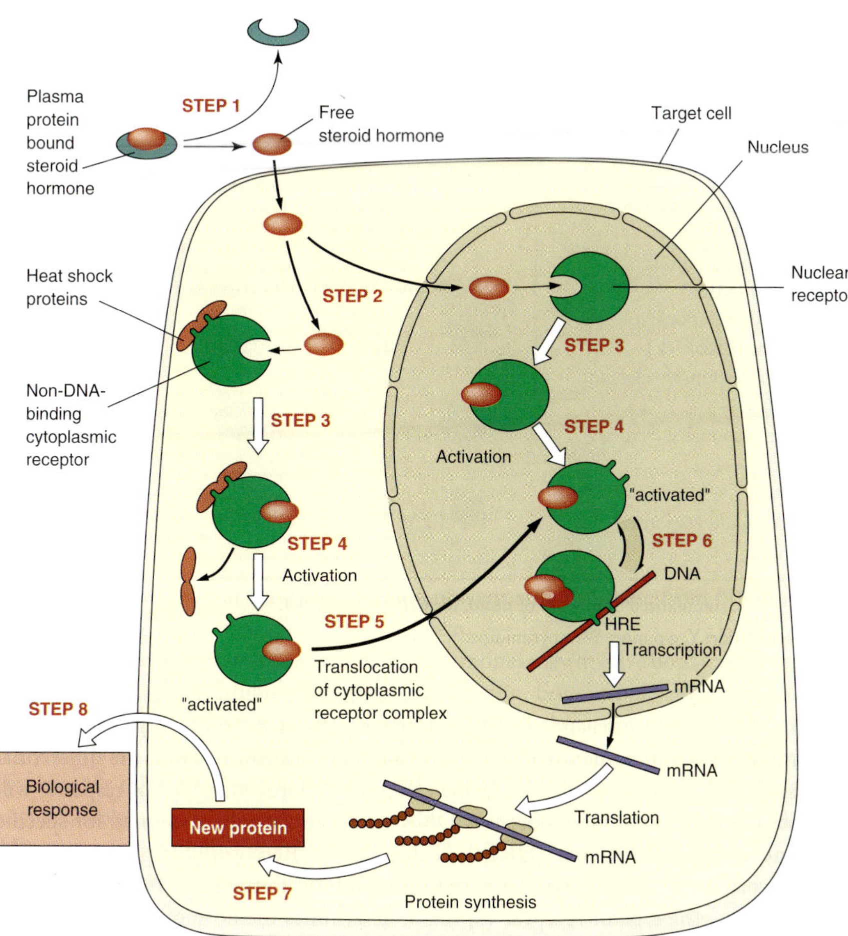

FIGURE 23.48

Model of steroid hormone action. Step 1: Dissociation of free hormone from circulating transport protein. Step 2: Diffusion of free ligand into cytosol or nucleus. Step 3: Binding of ligand to cytoplasmic or nuclear receptor. Step 4: activation of cytosolic or nuclear hormone–receptor complex to DNA-binding form. Step 5: Translocation of activated cytosolic hormone–receptor complex into nucleus. Step 6: Binding of activated hormone-receptor complexes to specific response elements within the DNA. Step 7: synthesis of new proteins encoded by hormone-responsive genes. Step 8: Alteration in phenotype or metabolic activity of target cell mediated by specifically induced proteins.

The unoccupied steroid hormone receptors for estradiol, progesterone, and androgens, as well as the secosteroid vitamin D_3 (Figure 23.48), are located in the nucleus. Transport of the hormone through the cytoplasm may require a transport protein. Once inside the nucleus, the hormone binds to its specific receptor, induces the dissociation of receptor-associated proteins, and regulates gene transcription. The transcriptional effects of 17β-estradiol in humans and rodents are mediated by the **alpha** and **beta** forms of the **estrogen receptor** (**ERα** and **ERβ**, respectively). Both bind 17β-estradiol with the same high affinity and bind to estrogen response elements within the DNA. The ERα and ERβ are highly homologous. ERα contains two distinct activation domains, AF-1 and AF-2, whose transcriptional activity is influenced by cell promoter context. ERβ contains an AF-2 domain (no strong AF-1 domain) and a repressor domain. ERβ is a trans-dominant inhibitor of ERα transcriptional activity at subsaturating hormone levels and decreases overall cellular sensitivity to 17β-estradiol. One explanation for these inhibitory effects of ERβ on ERα is that ERβ forms **heterodimers** with Erα. Results from ERα or ERβ knockout mice suggest that these two forms of the ER may have two different biological roles. In fact, when ERα and ERβ are coexpressed in neurons, they trigger different intracellular signals and distinct metabolic responses. It has been suggested the ERβ may be relevant for neural cell differentiation, whereas ERα may be implicated in synaptic plasticity, both during development and in mature neurons. Endogenous ERα and Erβ also appear to play different roles in the uterus. The uterus contains heterogeneous cell types that undergo continuous synchronized changes in their proliferation rates and differentiation in response to estrogens and progesterone. The ERα and ERβ levels in uterine tissue vary during the menstrual cycle, with the highest level for both being during the proliferative phase. 17β-Estradiol regulates the expression of the gene for the **progesterone receptor (PR)**. In uterine and breast tissue, this prepares the tissues to respond to progesterone. ERα is responsible for this induction

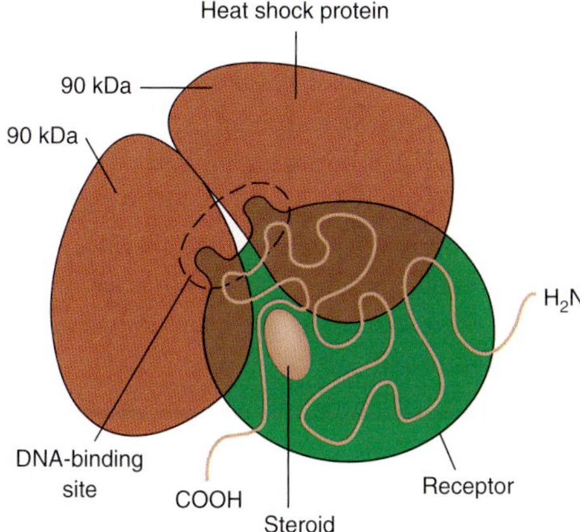

FIGURE 23.49

Hypothetical model of a non-DNA-binding form of a steroid receptor. This form of the receptor cannot bind to DNA because the DNA-binding site is blocked by the 90-kDa hsp proteins or by some other constituent. Mass of this complex is approximately 300 kDa.

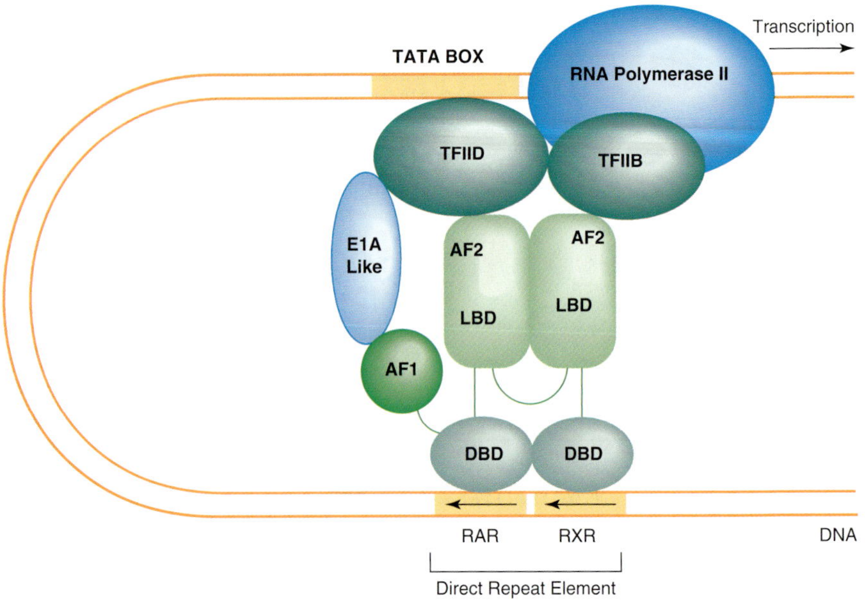

FIGURE 23.51

Model for stabilization of preinitiation complex by an RXR/RAR heterodimer. TF, transcription factor; LBD, ligand-binding domain; DBD, DNA-binding domain; AF1, activation function located in amino terminal region of receptor, which may provide contact with cell-specific proteins; AF2, activation function located within ligand-binding domain, which interacts directly with transcriptional machinery.

heterodimers with the retinoic acid receptor, the thyroid hormone receptor, or other members of this receptor superfamily. A model for the stabilization of the transcriptional preinitiation complex by an RXR/RAR heterodimer is presented in Figure 23.51.

Steroid Receptor Domains

Glucocorticoid receptors contain three major **functional domains** (Figure 23.52). Located at the C terminus, the steroid-binding domain has 30–60% homology with the **ligand-binding domains** of other steroid receptors (see Clin. Corr. 23.6). This domain may also be involved in the binding of a dimer of the 90-kDa heat shock protein, which has two functions: (1) maintains ligand-binding domain in optimal conformation for steroid binding and (2) prevents unliganded receptor from binding to DNA. To the left of this domain is a region that modifies transcription and contains a nuclear localization signal, which appears to provide recognition for the nuclear pore. In the center is the **DNA-binding domain**, which shows 60–95% homology among steroid receptors. This domain contains two zinc fingers that recognize specific HREs and stabilize binding to these DNA sequences. The N-terminal domain is highly variable among the steroid receptors and contains the principal **antigenic region** and a region that modulates transcriptional activation. These features are common to all members of the steroid receptor superfamily (Figure 23.53). The ancestor of these receptor genes is v-erbA or c-erbA. v-ErbA is an oncogene product that binds to DNA but has no ligand binding domain. The DNA-binding domains of some of these related receptors may be homologous enough that more than one receptor binds to a common responsive element (Table 23.10). The **aryl hydrocarbon receptor** (Ah) may also be a member of this family. The Ah receptor binds carcinogens with an affinity that parallels their

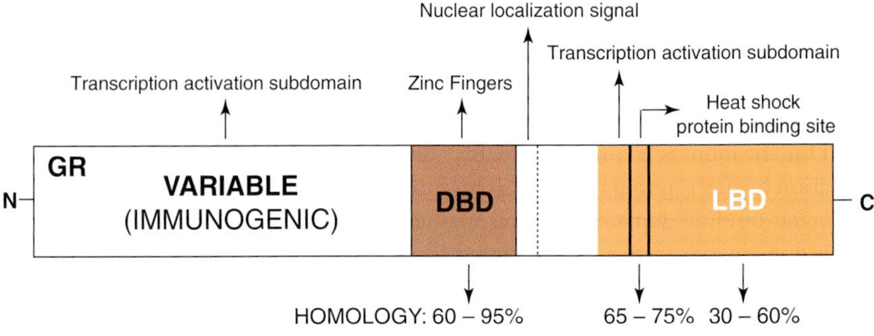

FIGURE 23.52

Major functional domains of steroid receptor proteins. DBD, DNA-binding domain; LBD, ligand-binding domain.

CLINICAL CORRELATION 23.6

Mineralocorticoid Receptor Mutation Results in Hypertension and Toxemia of Pregnancy

The causes of hypertension, especially the hypertension associated with the toxemia of pregnancy, which is called eclampsia, include the renin–angiotensin system and the aldosterone receptor. Hypertension occurs in about 6% of pregnancies and, in some cases, is associated with a mutation in the mineralocorticoid receptor in which a serine residue at position 810 is replaced by a leucine residue (referred to as S810L mutation). The mutated serine is in the hormone-binding domain of the receptor and is conserved in all mineralocorticoid receptors across many species. Interestingly, mutant receptor shows some constitutive activity in the absence of any ligand and binds progesterone with the same high affinity as aldosterone. In the normal wild-type receptor, progesterone binds with low affinity and functions as an antagonist. However, in the mutant form of the receptor progesterone functions as an agonist and induces reabsorption of sodium ions in the kidney. Since during pregnancy plasma progesterone levels increase significantly (see Figure 23.20), the mutated receptor is continuously saturated with this steroid. In subjects under age 35 carrying the mutated receptor, the systolic to diastolic blood pressure ratio is 167/110, as compared with 126/78 (normal range) in noncarriers. Spironolactone, which functions as an aldosterone antagonist when bound to the normal wild-type receptor, functions as an agonist when bound to the mutated receptor. Therefore, spironolactone should not be used to treat hypertensive patients bearing the S810L mutation. This mutated receptor could be an important factor contributing to the early development of heart failure in seriously hypertensive patients.

Source: Geller, D. S., Farhi, A., Pinkerton, N., Fradley, M., Moritz, M. Spitzer, A., Meinke, G., Tsai, F. T. F., Sigler, P. B., and Lifton, R. P. Activating mineralocorticoid receptor mutation in hypertension exacerbated by pregnancy. *Science* 289:119, 2000.

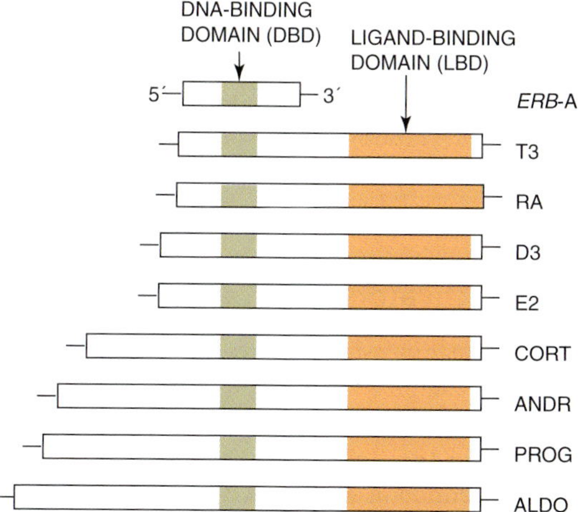

FIGURE 23.53

Steroid receptor gene superfamily. T3, triiodothyronine; RA, retinoic acid; D3, dihydroxy vitamin D_3; E2, estradiol; CORT, cortisol; ANDR, androgen; PROG, progesterone; ALDO, aldosterone. Figure shows roughly the relative sizes of the genes for these receptors.

carcinogenic potency and translocates these compounds into the nucleus. Thyroid hormone and retinoic acid receptors are also members of the same superfamily of receptors, although their ligands are not steroids. However, these hormones do contain six-membered rings as shown in Figure 23.54. The A ring of most steroids is recognized by the appropriate receptor and inserts into the pocket of the ligand-binding domain. The receptors of this large gene family, acting as homodimers or heterodimers, function as ligand-activated transcription factors; they modulate (induce or repress) the expression of specific genes.

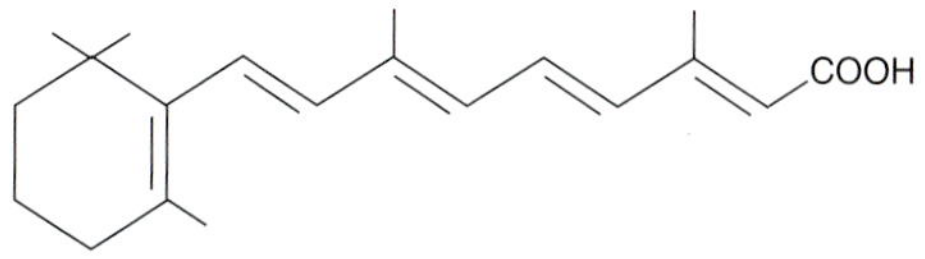

COOH

Retinoic acid
(vitamin A acid)

HO O $CH_2CHCOOH$ NH_2 I I I

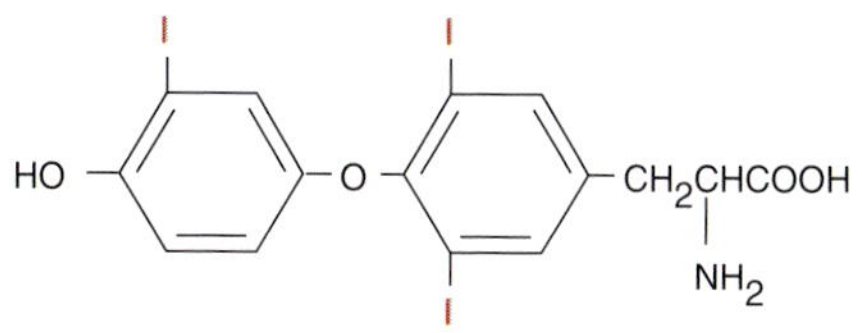

3,5,3' - Triiodothyronine

FIGURE 23.54

Structures of retinoic acid (vitamin A acid) and 3,5,3′-triiodothyronine.

Orphan Receptors

Many related receptors are called **orphan receptors**, because no physiological ligands or activators are known for them. They are found in almost all animal species. Examples of orphan receptors for which the ligand has now been identified include the following: BXR (benzoate X receptor); RXR (retinoid X receptor); PPAR (peroxisome proliferator-activated receptor); CARβ (constitutive androstane receptor); PXR (pregnane X receptor); SXR (steroid and xenobiotic receptor); and FXR (farnesoid X receptor). PXR, SXR, and CARβ are expressed highly in the liver, respond to specific steroid ligands, and require heterodimerization with RXR for DNA binding. These recently discovered orphan receptors and their respective ligands may be important physiologically and have an impact on specific human diseases. For example, the human SXR can be activated by a diverse group of steroid agonists and antagonists. This activation induces transcription of several genes encoding key metabolizing enzymes and may facilitate the detoxification and removal of various endogenous hormones, dietary steroids, drugs, and xenobiotic compounds with biological activity. Hence in patients receiving steroid therapy or women taking oral contraceptives, certain drugs (i.e., rifampicin) that bind to SXR cause rapid depletion of the administered steroids by this increased steroid catabolism.

Down-Regulation of Steroid Receptor by Ligand

Many hormone receptors are down-regulated when the cell has been exposed to a certain concentration of the hormone. In the case of these intracellular receptors, **down-regulation** generally means decreased receptor gene expression with a decrease in the concentration of receptor molecules. The promoter of a receptor gene may thus have a negative response element and binding of the steroid receptor complex to that element will repress transcription of the receptor gene. Down-regulation of receptors by their own ligands plays an important physiological role, because it desensitizes the target cell and therefore prevents overstimulation when circulating hormone levels are elevated.

Although down-regulation of steroid receptors by their cognate hormones appears to be the most common form of autoregulation, it is not detected in all target cells. Glucocorticoid-mediated up-regulation of its own receptor levels has been reported in a number of responsive cells. Although the significance of this **homologous up-regulation** is not fully understood, it could in theory increase hormonal responsiveness. The ability of the estrogen receptor to increase the concentration of progesterone receptors in key target tissues is an example of **heterologous up-regulation**.

Nuclear Hormone Receptors, Coactivators, and Corepressors

Coactivators and corepressors are cofactors that increase or decrease the transcriptional functions of most nuclear steroid–receptor complexes. Coactivators, such as the p160 family of coactivators, steroid receptor coactivator 1, and transcriptional intermediary factor 2 (TIF2)/GR-interacting protein (GRIP1), all increase the total amount of induced gene product with a saturating concentration of a steroid hormone. Conversely, corepressors such as nuclear receptor corepressor (NcoR) and the silencing mediator of retinoid and thyroid hormone receptor (SMRT) are factors that decrease the total amount of gene product. Binding of the ligand to one of these nuclear receptors acts like a "molecular switch," causing both the dissociation of corepressors from the ligand-free receptor and the association of coactivators. The sites of interaction in steroid and nuclear receptors for both coactivators and corepressors have been shown to be in the ligand-binding domain, and these two binding sites may overlap.

Although corepressors do not bind to ligand-bound nuclear receptors such as the thyroid hormone receptor, they do appear to interact with nuclear ligand-bound steroid receptors. This interaction may provide a mechanism of differentiating between activated complexes of various steroid receptors (androgens, glucocorticoid, mineralocorticoid, and progestin) in a cell-specific manner. Each of these steroid receptors binds its ligand specifically; however, once activated, these receptors bind to the same hormone response element (Table 23.10). Interactions between cell-specific corepressors and DNA-bound receptors can restore some of the specificity that appears to be lost upon binding to a common HRE.

Nongenomic Steroid Effects

Not all of the effects of steroid hormones are mediated at the level of gene transcription. Many steroid hormones, including aldosterone, 17β-estradiol, progesterone, glucocorticoids, and androgens, exert rapid (within minutes) stimulatory effects on the activities of a wide variety of signal transduction molecules (protein kinase C, diacylglycerol, IP_3) and pathways. These **nongenomic effects** appear to be initiated at the plasma membrane rather than in the nucleus of a target cell. These nontranscriptional responses may be mediated either by a subpopulation of conventional nuclear receptors localized in the cell membrane or by distinct **membrane receptors** that are unrelated to classical intracellular steroid receptors. Some of the arguments in favor of novel membrane receptors include the facts that (1) rapid steroid responses appear to be mediated through receptors with pharmacological properties very different from those of classic intracellular receptors, (2) steroids exert rapid nongenomic effects in cells or tissues that do not express classic receptors, and (3) these rapid steroid effects are not always blocked by classic receptor antagonists. The cloning and functional reexpression of these membrane receptors transmitting rapid steroid effects has yet to be accomplished. Drugs that specifically affect nongenomic steroid action may find applications in various clinical areas including cardiovascular and central nervous system disorders, electrolyte homeostasis, and infertility.

BIBLIOGRAPHY

Alberts, B., Johnson, A., Lewis, L., Raff, M., Roberts, K., and Walter, P. (Eds.). *Molecular Biology of the Cell*, 4th ed. New York: Garland, 1994.

Baulieu, E. E. Steroid hormone antagonists at the receptor level: A role for heat-shock protein MW 90,000 (hsp 90). *J. Cell Biochem.* 35:161, 1987.

Beato, M. Gene regulation by steroid hormones. *Cell* 56:335, 1989.

Berne, R. M. and Levy, M. N. (Eds.). *Physiology*. St. Louis: Mosby, 1998.

Blumberg, B. and Evans, R. M. Orphan nuclear receptors—new ligands and new possibilities. *Genes Dev.* 12:3149, 1998.

Carson-Jurica, M. A., Schrader, W. T., and O'Malley, B. W. Steroid receptor family: Structure and functions. *Endocr. Rev.* 11:201, 1990.

Chen, J. D. Steroid/nuclear receptor coactivators. In: G. Litwack (Ed.), *Vitamins and Hormones*, Vol. 28. Orlando, FL: Academic Press, 2000, p. 391.

Chrousos, G. P., Loriaux, D. L., and Lipsett, M. B. (Eds.). *Steroid Hormone Resistance*. New York: Plenum, 1986.

Cuatrecasas, P. Hormone receptors, membrane phospholipids, and protein kinases. *The Harvey Lectures Ser.* 80:89, 1986.

DeGroot, L. J. (Ed.). *Endocrinology*. Philadelphia: Saunders, 1995.

Dostert, A. and Heinzel, T. Negative glucocorticoid response elements and their role in glucocorticoid action. *Curr. Pharm. Des.* 10:2807, 2004.

Evans, R. M. The steroid and thyroid hormone receptor superfamily. *Science* 240:889, 1988.

Falkenstein, E., Tillmann, H. -C., Christ, M., Feuring, M., and Wehling, M. Multiple actions of steroid hormones—A focus on rapid, nongenomic effects. *Pharmacol Rev.* 52:513, 2000.

Giguere, V., Hollenberg, S. M., Rosenfeld, M. G., and Evans, R. M. Functional domains of the human glucocorticoid receptor. *Cell* 46:645, 1986.

Green, S., Kumar, V., Theulaz, I., Wahli, W., and Chambon, P. The N-terminal DNA-binding "zinc-finger" of the estrogen and glucocorticoid receptors determines target gene specificity. *EMBO J.* 7:3037, 1988.

Gustafsson, J. A., et al. Biochemistry, molecular biology, and physiology of the glucocorticoid receptor. *Endocr. Rev.* 8:185, 1987.

Krieger, D. T. and Hughes, J. C. (Eds.). *Neuroendocrinology*. Sunderland, MA: Sinauer Associates, 1980.

Larsen, P. R., Kronenberg, H. M., Melmed, S., and Polonsky, K. S. (Eds.). *Williams Textbook of Endocrinology*, 10th ed. Philadelphia; Saunders, 2003.

Litwack, G. (Ed.). *Biochemical Actions of Hormones*, Vols. 1–14. New York: Academic Press, 1973–1987.

Litwack, G. (Ed. in Chief). *Vitamins and Hormones*, Vol. 50. Orlando, FL: Academic Press, 1995.

Norman, A. W. and Litwack, G. *Hormones*. Orlando, FL: Academic Press, 1987.

Mester, J. and Baulieu, E. E. Nuclear receptor superfamily. In: L. J. DeGroot et al. (Eds.), *Endocrinology*, 3rd ed., Philadelphia: Saunders, 1995, p. 93.

Norman, A. W. and Litwack, G. *Hormones*, Orlando, FL: Academic, 1987.

O'Malley, B. W., Tsai, S. Y., Bagchi, M., Weigel, N. L., Schrader, W. T., and Tsai, M. -J. Molecular mechanism of action of a steroid hormone receptor. *Recent Prog. Horm. Res.* 47:1, 1991.

Richter, D. Molecular events in expression of vasopressin and oxytocin and their cognate receptors. *Am. J. Physiol.* 255:F207, 1988.

Saltiel, A. R. The paradoxical regulation of protein phosphorylation in insulin action. *FASEB J.* 8:1034, 1994.

Schmidt, T. J. and Meyer, A. S. Autoregulation of corticosteroid receptors. How, when, where and why? *Receptor* 4:229, 1994.

Spiegel, A. M., Shenker, A., and Weinstein, L. S. Receptor-effector coupling by G proteins: Implication for normal and abnormal signal-transduction. *Endocr. Rev.* 13:536, 1992.

Struthers, A. D. (Ed.). *Atrial Natriuretic Factor*. Boston: Blackwell Scientific Publications, 1990.

Wahli, W. and Martinez, E. Superfamily of steroid nuclear receptors—Positive and negative regulators of gene expression. *FASEB J.* 5:2243, 1991.

QUESTIONS | CAROL N. ANGSTADT

Multiple Choice Questions

1. If a single gene contains information for synthesis of more than one hormone molecule:
   A. all the hormones are produced by any tissue that expresses the gene.
   B. all hormone molecules are identical.
   C. cleavage sites in the gene product are typically pairs of basic amino acids.
   D. all peptides of the gene product have well-defined biological activity.
   E. hormones all have similar function.

2. In the interaction of a hormone with its receptor all of the following are true *except*:
   A. more than one polypeptide chain of the hormone may be necessary.
   B. more than one second messenger may be generated.
   C. an array of transmembrane helices may form the binding site for the hormone.
   D. receptors have a greater affinity for hormones than for synthetic agonists or antagonists.
   E. hormones released from their receptor after endocytosis could theoretically interact with a nuclear receptor.

3. Some hormone–receptor complexes are internalized by endocytosis. This process may involve:
   A. binding of hormone–receptor complex to a clathrin coated pit.
   B. recycling of receptor to cell surface.
   C. degradation of receptor and hormone in lysosomes.
   D. formation of a receptosome.
   E. all of above.

4. Retinoic acid and its derivatives:
   A. may activate gene transcription by eliminating the silencing activity of receptor proteins.
   B. bind to homodimeric proteins, which, in turn, bind to DNA.
   C. bind directly to DNA via leucine zipper motifs.
   D. are vitamin derivatives, and hence have no effect on regulation of gene expression.
   E. may substitute for thyroid hormones in binding to the thyroid hormone receptor.

5. All of the following receptors may belong to the steroid receptor gene superfamily *except*:
   A. aryl hydrocarbon receptor.
   B. *erbA* protein.
   C. retinoic acid receptor.
   D. thyroid hormone receptor.
   E. α-tocopherol receptor.

6. Glucocorticoid receptors are in the cytoplasm. All of the following statements about the process by which the hormone influences transcription are correct *except*:
   A. the hormone must be in the free state to cross the cell membrane.
   B. cytoplasmic receptors may be associated with heat shock proteins.
   C. the receptor–hormone complex is not activated/transformed until it is translocated to the nucleus.
   D. in the nucleus, the activated/transformed receptor–hormone complex searches for specific sequences on DNA called HREs (hormone receptor elements).
   E. the activated receptor–hormone complex may either activate or repress transcription of specific genes (only one activity per gene).

Questions 7 and 8: Aldosterone bound to its receptor promotes sodium reabsorption in the distal nephron of the kidney. Elevated sodium in the blood leads to hypertension which can be a serious problem in pregnancy. In some cases, hypertension that appears in early pregnancy and increases with time has been shown to be caused by a mutation in the mineralocorticoid receptor. The mutation allows progesterone to bind with the same affinity as aldosterone and thus act as an agonist. Because of the high levels of progesterone during pregnancy, the mutated receptor remains saturated and blood pressure can become dangerously high. Spironolactone, which acts as an antagonist of aldosterone with a normal receptor, acts as an agonist with the mutated receptor and should not be used to treat this kind of hypertension.

7. All of the following are normal events leading to secretion of aldosterone from the adrenal gland *except*:
   A. renin is released by the kidney in hypovolemia.
   B. angiotensinogen binds to membrane receptors.
   C. the PI cycle is activated producing IP_3 and DAG.
   D. Ca^{2+} levels in the cell rise.
   E. aldosterone is secreted into the blood.

8. Once ovulation occurs, the pathway followed differs when the egg is fertilized and when it is not. Which of the following statements about this process is/are correct?
   A. FSH, via cAMP as a second messenger, stimulates the follicle to release 17β-estradiol.
   B. blood levels of progesterone fall as pregnancy progresses as the corpus luteum dies.
   C. inhibin produced by the follicle prevents release of LH.
   D. the primary influence for the corpus luteum to produce progesterone and estradiol is FSH.
   E. all of the above are correct.

Questions 9 and 10: Hypopituitarism may result from trauma, such as an automobile accident severing the stalk connecting the hypothalamus and anterior pituitary, or from tumors of the pituitary gland. In trauma, usually all of the releasing hormones from hypothalamus fail to reach the anterior pituitary. With a tumor of the gland, some or all of the pituitary hormones may be shut off. Posterior pituitary hormones may also be lost. Hypopituitarism can be life-threatening. Usual therapy is administration of end-organ hormones in oral form.

9. If the stalk between the hypothalamus and anterior pituitary is severed, the pituitary would fail to cause the ultimate release of all of the following hormones *except*:
   A. ACTH.
   B. estradiol.
   C. oxytocin.
   D. testosterone.
   E. thyroxine.

10. In hypopituitarism it is necessary to maintain the ovarian cycle in female patients. In the ovarian cycle:
    A. GnRH enters the vascular system via transport by a specific membrane carrier.
    B. corpus luteum rapidly involutes only if fertilization does not occur.
    C. inhibin works by inhibiting synthesis of α subunit of FSH.
    D. FSH activates a protein kinase C pathway.
    E. LH is taken up by corpus luteum, and binds to cytoplasmic receptors.

Questions 11 and 12: Maternal insulin sensitivity normally decreases during pregnancy. In 3–5% of pregnant women, glucose intolerance develops leading to gestational diabetes mellitus (GDM). GDM is generally reversible, but many of these women develop type 2 diabetes later in life. It appears that the pathogenesis of GDM may be caused by defects in insulin action rather than a decrease in insulin receptor binding affinity.

11. Muscle cells of GDM patients appear to overexpress a cell membrane protein that inhibits the kinase activity of the insulin receptor which is a:
    A. protein kinase A.
    B. protein kinase B.
    C. protein kinase C.
    D. protein kinase G.
    E. tyrosine kinase.

12. Binding of insulin to its receptor:
    A. occurs on the β-subunit.
    B. induces autophosphorylation.
    C. reduces binding of cytosolic substrate proteins.
    D. leads only to phosphorylation of proteins.
    E. does not lead to release of a second messenger.

Problems

13. How would you determine whether an inability to produce cortisol in response to stress was caused by a problem in the hypothalamus, the anterior pituitary, or the adrenal cortex?

14. What is the relationship between 7-dehydrocholesterol and 1α,25-dihydroxycholecalciferol?

ANSWERS

1. **C** One or more trypsin-like proteases catalyze the reaction. A: The POMC gene product is cleaved differently in different parts of the anterior pituitary. B: Multiple copies of a single hormone may occur, but not necessarily. D: Some fragments have no known function. E: ACTH and β-endorphin, for example, hardly have similar functions.
2. **D** β-Receptors bind isoproterenol more tightly than their hormones. A, B: These are true of glycoprotein hormones. C: This appears to be true for β_1 receptor. E: This is possible, but entirely speculative; there are currently no known examples.
3. **E** A and D always happen. B or C happens sometimes.
4. **A** The retinoic acid receptor (RAR) binds to specific silencer elements in the absence of the ligand, retinoic acid. When retinoic acid is bound, the receptor loses silencing activity and activates gene transcription. B: In addition, there are retinoid X receptors (RXR), which also affect gene expression, via heterodimerization with RAR. C: Retinoic acid binds to its receptor. The receptor protein may bind to DNA via a leucine zipper. E: The thyroid hormone receptor is a different protein than RAR.
5. **E** B: Note that *c-erbA* is a protooncogene.
6. **C** Dissociation of the heat shock protein from the receptor–hormone complex in the cytosol activates the complex. A: Steroid hormones travel bound to plasma proteins, but some is always free. D: These are consensus sequences in DNA. E: Activation is more common, but glucocorticoids repress transcription of the proopiomelanocortin gene.
7. **B** Angiotensinogen is cleaved by renin to angiotensin I, which must further be cleaved by converting enzyme to active angiotensin II. A: This is a major signal. C, D: These lead to increased Ca^{2+} and activation of protein kinase C.
8. **A** Activation of cAMP-dependent protein kinase stimulates the synthesis and secretion of estradiol. B: This is what happens in the absence of fertilization. In pregnancy, the corpus luteum eventually dies but the placenta produces high levels of progesterone. C: Inhibin controls FSH release. D: LH controls the corpus luteum.
9. **C** Oxytocin is released from posterior pituitary. A, B, D, and E all require releasing hormones from hypothalamus for anterior pituitary to release them.
10. **D** A: GnRH enters the vascular system through fenestrations. B. The corpus luteum is replaced by placenta if fertilization occurs. C: Glycoprotein hormones share a common α subunit. Specific control of them would not involve a subunit they share. E: LH interacts with receptors on the cell membrane.
11. **D** This is an activity of the β-subunit.
12. **B** This occurs on tyrosine residues of the β-subunit. A: Binding is to the α-subunit. C: Autophosphorylation facilitates binding. D: Some proteins are dephosphorylated. E: A second messenger may account for short-term metabolic effects.
13. The end organ is adrenal cortex. If cortisol is increased in response to an injection of ACTH, the adrenal cortex is functioning properly. An intravenous infusion of CRH (corticotropic releasing hormone)should be able to reach the anterior pituitary. If ACTH increases in response to this, the pituitary is functioning so the problem is probably with the hypothalamus. If ACTH does not increase in response to CRH, the anterior pituitary is not functioning.
14. 1α, 25-dihydroxycholecalciferol ($1{,}25(OH)_2D_3$)is the active from of vitamin D. Ultraviolet light acting on the skin converts 7-dehydrocholesterol to cholecalciferol. This compound must be hydroxylated in liver to 25-hydroxycholecalciferol and subsequently in kidney to yield the active $1{,}25(OH)_2D_3$.

24

MOLECULAR CELL BIOLOGY

Thomas E. Smith

Textbook of Biochemistry With Clinical Correlations, Sixth Edition, Edited by Thomas M. Devlin

24.1 OVERVIEW

Animals sense their environment through the responses of specific organs to stimuli: touch, pain, heat, cold, intensity (light or noise), color, shape, position, pitch, quality, acid, sweet, bitter, salt, alkaline, fragrance, and so on. These generally reflect responses of the skin, eye, ear, tongue, and nose to stimuli. Some of these signals are localized to the point at which they occur, whereas others (sound, smell, and sight) are projected in space—that is, the environment outside of and distant to the animal.

Discrimination of these stimuli occurs at the point of reception, but acknowledgment of what they are requires secondary stimulation of the nervous system and transmission to the brain. A physical response may be indicated which will result in voluntary or involuntary muscular activity. Common to these events are (a) electrical activity associated with signal transmission along neurons and (b) chemical activity associated with signal transmission across synaptic junctions. In all cases, environmental stimuli such as pressure (skin, feeling), light (sight, eye), noise (ear, hearing), taste (tongue), or smell (nose) are converted (transduced) into electrical impulses and to some other form of energy to effect responses dictated by the brain. A biochemical component is associated with each of these events.

General biochemical mechanisms of signal transduction and amplification will be discussed as they relate to nerve transmission, vision, and molecular motors. A specialized case of biochemical signal amplification will be discussed, namely, blood coagulation or hemostasis with emphasis on the enzymes and ancillary proteins involved in phases of the process ranging from coagulation to fibrinolysis. A common theme of all topics discussed in this chapter is that processes are initiated by some stimuli, propagated and/or amplified by some biochemical processes, terminated, and repositioned for the next event. Energy and intermediary metabolites necessary for many of these events may be mentioned only briefly in this chapter since they are covered in sufficient detail elsewhere in this textbook.

24.2 NERVOUS TISSUE: METABOLISM AND FUNCTION

Knowledge of the chemical composition of the brain began with the work of J. L. W. Thudichum, who in 1884 published what has been cited as his "greatest work", entitled *A Treatise on the Chemical Composition of the Brain.* He performed chemical analyses on over a thousand brains from both human and animals. He has been referred to as "...the first English biochemist and the world's first pathological chemist. ..." During the following years, there have been exciting advances about many aspects of biochemical and molecular mechanisms of brain and neuronal functions. Crystal structures of many of the protein components involved have been determined at relatively high resolution and structural mechanisms of disease processes as they relate to mutations in these proteins are being revealed.

About 2.4% of an adult's body weight is nervous tissue, of which approximately 83% is the brain. The nervous system provides the communications network between the senses, the environment, and all parts of the body. The brain is the command center. It is always functioning and requires a large amount of energy to keep it operational. Under normal conditions, the brain derives its energy from glucose metabolism. Ketone bodies can cross the **blood–brain barrier** and are metabolized by brain tissue, especially during **starvation**, but they cannot replace the requirement for glucose. The human brain uses approximately 103 g of glucose per day. For a 1.4-kg brain, this corresponds to a rate of utilization of approximately 0.3 μmol min^{-1} g^{-1} of tissue and represents a capacity for ATP production through the tricarboxylic acid (TCA) cycle alone of approximately 6.8 μmol min^{-1} g^{-1} of tissue. The TCA cycle is not 100% efficient for ATP production, nor is all of the glucose metabolized through it, although it does function at near-maximum capacity and generates most of the ATP used by nervous tissue. The brain uses much of its energy to maintain ionic gradients across the plasma membranes, to affect various storage and transport processes, and for synthesis of neurotransmitters and other cellular components. Glycolysis functions at approximately 20% capacity and contributes to ATP production.

Two features of brain composition are worth noting: (1) It contains specialized and complex lipids that appear to function to maintain membrane integrity rather than in metabolic roles; (2) there is a rapid **turnover rate** of brain proteins relative to other body proteins, although neuronal cells generally do not divide after they have differentiated.

Cells of the nervous system responsible for collecting and transmitting messages are highly specialized **neurons** (Figure 24.1). Each neuron consists of a cell body, **dendrites** that are short antenna-like protrusions that receive signals from other cells, and an **axon** that extends from the cell body to other neurons or cells to which it transmits signals. The central nervous system (CNS) is a highly integrated system where individual neurons can receive inhibitory and excitatory stimuli from a variety of different sources. A normal adult has between 10^{11} and 10^{13} neurons, and communication between them is by electrical and chemical signals. Brain stem cells can produce three major types of cells: neurons, **astrocytes**, and oligodendrocytes. Astrocytes are glial cells that are associated with greater than 90% of the blood–brain–barrier endothelium and help to determine blood–brain–barrier morphology and function. Astrocytes send out processes at the external surfaces of the CNS that are linked, with other cell types, to form anatomical complexes that provide sealed barriers at the level of the capillaries and isolate the CNS from the external environment. These tight junctions prevent the passive entry into the brain of water-soluble molecules and form what is commonly known as the **blood–brain barrier.** In generally, therefore, water-soluble compounds enter the brain only if there are specific membrane transporters for them. Oligodendrocytes surround the axons like insulators and facilitate efficient signal conduction. There are other types of glial cells, each of which has a specialized function, but only astrocytes appear to be directly associated with a biochemical function related to metabolic neuronal activity (see discussion below on GABA). Some glial cells use intermediary metabolites other than glucose, such as fatty acids, as energy sources.

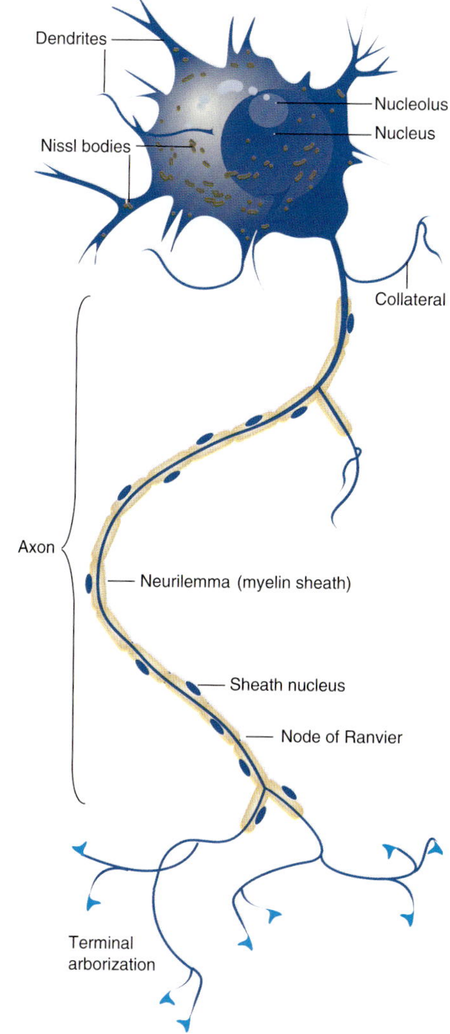

FIGURE 24.1

A motor nerve cell and investing membranes.

ATP and Transmembrane Electrical Potential in Neurons

Adenosine triphosphate generated from metabolism of glucose is used to help maintain an equilibrium electrical potential of approximately −70 mV across the membrane of

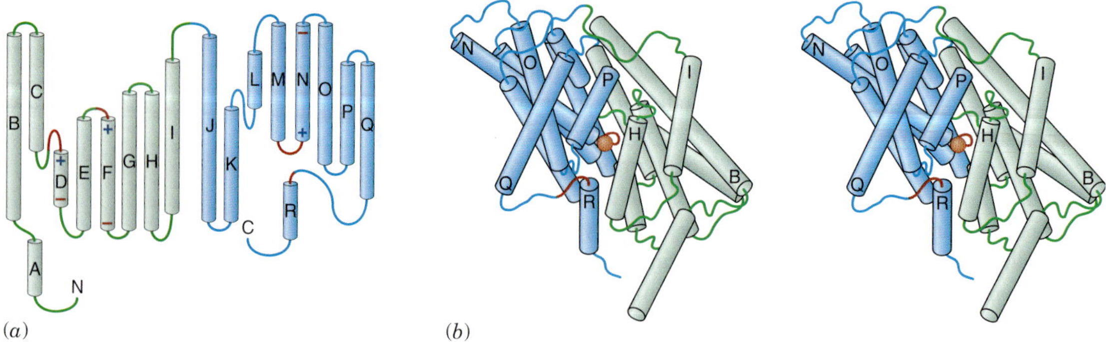

FIGURE 24.6

Structural model of a ClC chloride channel. This is essentially a two-barrel channel, one blue-green and one blue in this diagram. (*a*) Linear arrangement of the helices showing the positive (+) and negative (−) ends. (*b*) Stereo view of the channel with a Cl^- (colored ball) between the two barrels. Cl^- interacts with the positive (+) ends of the helices.

Reproduced with permission from Dutzler, R., Campbell, E. B., Cadene, M., Chait, B. T., and MacKinnon, R. X-ray structure of a ClC chloride channel at 3.0 Åreveals the molecular basis of anion selectivity. *Nature* 415:287, 2002. Copyright (2002) Nature.

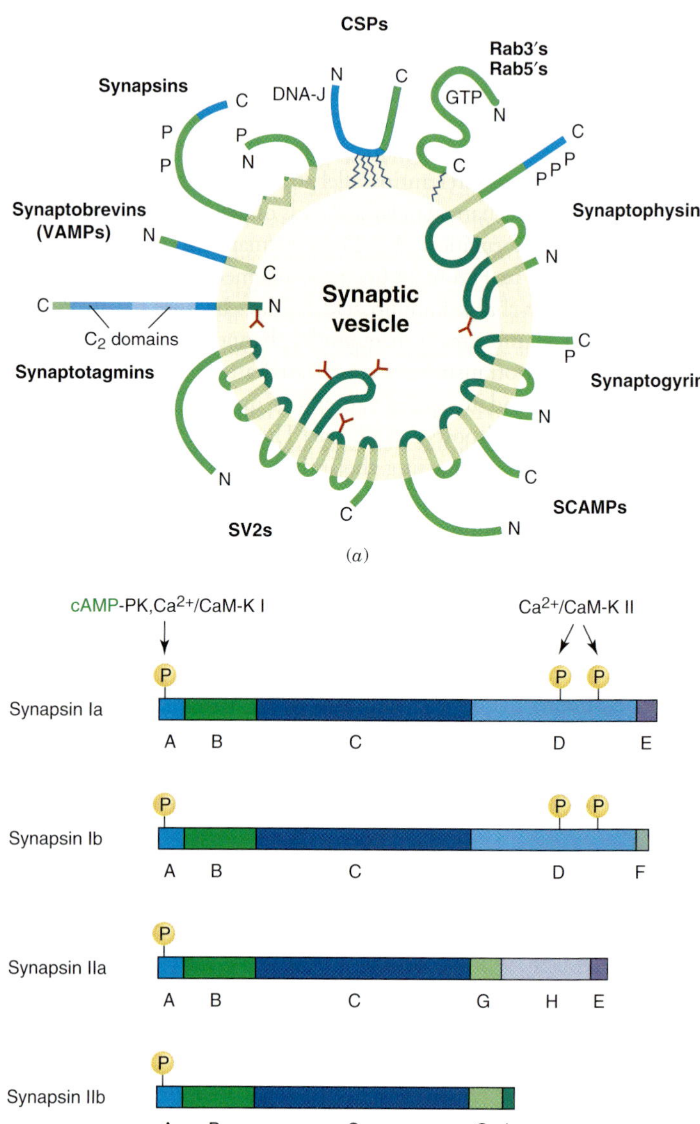

FIGURE 24.7

Synaptic vesicle and synapsin family of proteins. (*a*) Schematic drawing of the relative arrangement of proteins of the synaptic vesicle (SV). Rab proteins are attached by isoprenyl groups and cysteine string proteins (CSPs) by palmitoyl chains to SVs. The N- and C-termini of proteins are marked by N and C, respectively. Phosphorylation sites are indicated by P. (*b*) Structural arrangement of the synapsin family of proteins.

Part (*a*) redrawn from Sudhof, T. C. *Nature* 375:645, 1995. Part (*b*) redrawn from the work of Chilcote, T. J., Siow, Y. L., Scaeffer, E., et al. *J. Neurochem.* 63:1568, 1994.

Synthesis, Storage, and Release of Neurotransmitters

Nonpeptide neurotransmitters may be synthesized in almost any part of the neuron, in the cytoplasm near the nucleus, or in the axon. Most nonpeptide neurotransmitters are amino acids or derivatives of amino acids.

Released neurotransmitters travel rapidly across the **synaptic junction** (which is about 20 nm across) and bind to receptors on the postsynaptic side. Upon binding, they initiate the process of electrical impulse propagation in the postsynaptic neuron as described above. Storage and release of neurotransmitters are multistep processes, many of which have been defined. Some neurons appear to contain more than one chemical type of neurotransmitter. The physiological significance of this observation is not clear. Release of neurotransmitters is a quantal event. A nerve impulse reaching the presynaptic terminal causes release of transmitters from a fixed number of **synaptic vesicles**. An important step in the process involves attachment of synaptic vesicles to the presynaptic membrane and **exocytosis** of their content into the synaptic cleft. Storage of neurotransmitters occurs in large or small vesicles in the presynaptic terminal. Small vesicles predominate and exist in two pools; free and attached to cytoskeletal proteins, mainly actin. They contain nonpeptide small-molecule-type transmitters. A schematic diagram of a small synaptic vesicle is shown in Figure 24.7*a*. Large vesicles may contain both nonpeptide- and peptide-type neurotransmitters. Some may also contain enzymes for synthesis of norepinephrine from dopamine. A list of some proteins in synaptic vesicles is presented in Table 24.2. Figure 24.8 shows schematically how some of the proteins may be arranged on the synaptic vesicle and how they may interact with the plasma membrane of the presynaptic neuron.

1. **Synapsin** has a major role in determining whether the small synaptic vesicles are free and available for binding to the presynaptic membrane (Figure 24.9) or whether they are attached to cytoskeletal proteins. There is a family of synapsin proteins, encoded by two genes, that differ primarily in the C-terminal end (Figure 24.7*b*). Synapsins constitute about 9% of the total protein of the synaptic vesicle membrane. All can be phosphorylated near their N-termini by either **cAMP-dependent protein kinase** and/or **calcium-calmodulin (CaM) kinase I.** Synapsins Ia and Ib can also be phosphorylated by **CaM kinase II** near their C-termini, a region that is missing in synapsin IIa and IIb.

TABLE 24.2 Some Synaptic Vesicle Proteins

Synapsin	Ia
	Ib
	IIa
	IIb
Synaptophysin	
Synaptotagmin	
Syntaxin	
Synaptobrevin/VAMP	
Rab3 and rabphilin	
SV-2	
Vacuolar proton pump	

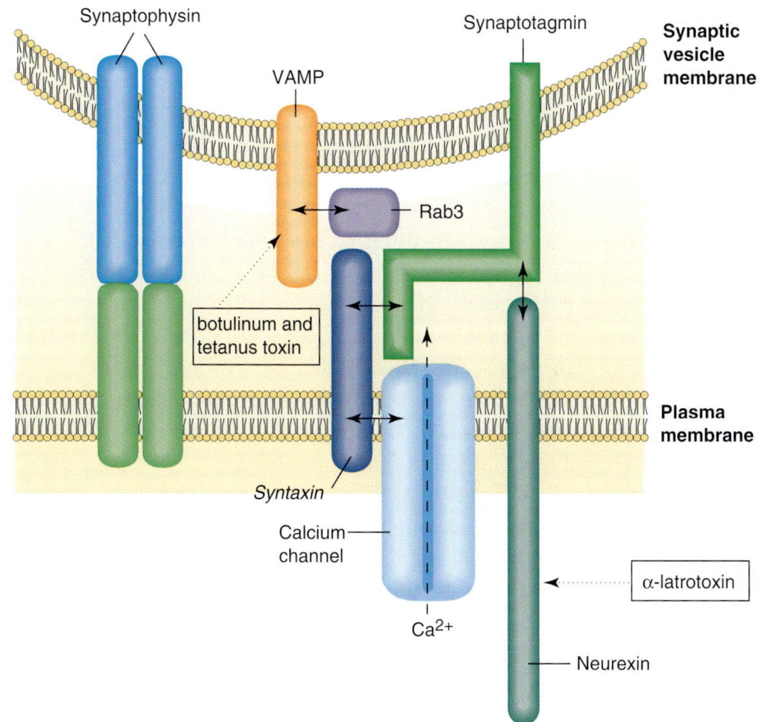

FIGURE 24.8

Schematic diagram showing how some synaptic vesicle proteins may interact with plasma membrane proteins.
Redrawn from Bennett, M. K., and Scheller, R. H. *Proc. Natl. Acad. Sci. USA* 90:2559, 1993.

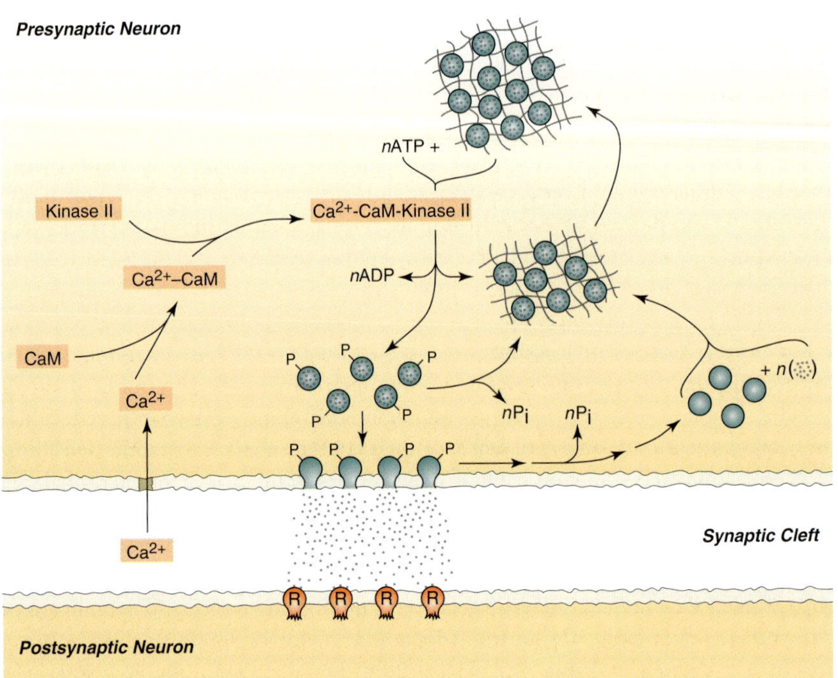

FIGURE 24.9

Model of the mechanism of regulation of synaptic vesicles function by calcium ions and calmodulin kinase II. Blue circles within the "skeletal" mesh represent bound, nonphosphorylated synaptic vesicles. Blue circles with P attached have been freed from the mesh after synapsin on those vesicles has been phosphorylated. They can now attach to the presynaptic membrane and release neurotransmitters into the synaptic cleft. Receptors on the postsynaptic membrane (red) will bind to some of those neurotransmitters. The recovery of synaptic vesicles and repackaging with neurotransmitters is also schematically illustrated. An animation of synaptic vesicle fusion followed by clathrin-mediated endocytosis, at the time of this writing, was available at the following website: http://stke.sciencemag.org/content/vol2003/issue198/images/data/tr3/DC1/SynapticVesicleCycle1.swf

Nerve stimulation leads to the entry of Ca^{2+} into the presynaptic neuron (see Clin. Corr. 24.1), where it binds to calmodulin and activates CaM kinases I and CaM kinases II (producing Ca^{2+}–CaM kinases I and II), which phosphorylate synapsin. This causes release of synaptic vesicles from the cytoskeletal protein matrix and/or prevents free phosphorylated vesicles from binding to those matrices. The result is an increase in the free pool of synaptic vesicles. **Calcium–calmodulin** (see p. 478) can also bind synapsin directly and competitively block its interaction with actin and presumably other cytoskeletal proteins. Calcium–calmodulin and the phosphorylation of synapsin, therefore, regulate the number of synaptic vesicles that are in the free state.

2. **Synaptophysin** is an integral membrane protein of synaptic vesicles that is structurally similar to gap junction proteins. It may be involved in formation of a channel from the synaptic vesicle through the presynaptic membrane to permit passage of neurotransmitters into the synaptic cleft.

3. **Synaptotagmin** is also an integral membrane protein of synaptic vesicles. It interacts in a Ca^{2+}-dependent manner with specific proteins localized on the presynaptic membrane. It is probably involved in docking of synaptic vesicles to the membrane.

4. **Syntaxin** is an integral protein of the plasma membrane of presynaptic neurons. It binds synaptotagmin and mediates its interaction with Ca^{2+} channels at the site of release of the neurotransmitters. It also appears to have a role in exocytosis.

5. **Synaptobrevin/VAMP** (or vesicle-associated membrane protein) consists of a family of two small proteins of 18 and 17 kDa that are anchored in the cytoplasmic side of the membrane through a single C-terminal domain. It appears to be involved in **vesicle transport** and/or **exocytosis**. VAMPs appear to be involved in the release of synaptic vesicles from the plasma membrane of the presynaptic neuron. **Tetanus toxin** and **botulinum toxin**, a zinc-dependent endoprotease, bind and cleave VAMPs, resulting in a slow and irreversible inhibition of transmitter release.

6. **Rab3** belongs to the large rab-family of **GTP-binding proteins**. Rab3 is specific for synaptic vesicles and is involved in docking and fusion of vesicles for exocytosis. Rab3 is anchored to the membrane through a polyprenyl side chain near its C-terminal.

7. **SV-2** is a large glycoprotein with 12 transmembrane domains. Its function is unclear.

8. **Vacuolar proton pump** (see p. 476) functions to transport neurotransmitters back into the synaptic vesicles after their reformation and release from the presynaptic membrane.

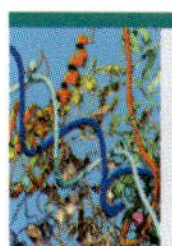

CLINICAL CORRELATION 24.1
Lambert–Eaton Myasthenic Syndrome

Lambert–Eaton myasthenic syndrome (LEMS) is an autoimmune disease in which the body raises antibodies against voltage-gated calcium channels (VGCC) located on presynaptic nerve termini. Upon depolarization of presynaptic neurons, these calcium channels open permitting the influx of Ca^{2+}. This increase in Ca^{2+} concentration initiates events of the synapsin cycle, resulting in release of neurotransmitters into synaptic junctions. When autoantibodies against VGCC react with neurons at neuromuscular junctions, Ca^{2+} cannot enter and the amount of acetylcholine released into synaptic junctions is diminished. Since action potentials to muscles may not be induced, the effect mimics that of classic myasthenia gravis.

LEMS has been observed in conjunction with other conditions such as small cell lung cancer. Some patients manifest subacute cerebellar degeneration (SCD). Plasma exchange (removal of antibodies) and immunosuppressive treatments have been effective for LEMS, but the latter treatment is less effective on SCD.

Diagnostic assays for LEMS depend upon the detection of antibodies against VGCC in serum. There are at least four subtypes of VGCC, T, L, N, and P. The P-subtype may be the one responsible for initiating neurotransmitter release at the neuromuscular junction in mammals. A peptide toxin produced by a cone snail (*Conus magnus*) binds to P-type VGCC in cerebella extracts. This small peptide, labeled with ^{125}I, binds VGCC in cerebella extracts. Precipitation of this radiolabeled complex with sera of patients confirms LEMS in those that have clinical and electrophysiological symptoms of the condition. This assay may not only prove useful in detecting LEMS but may also provide a means of finding out more about the antigenicity of the area(s) on the VGCCs to which antibodies are raised.

LEMS patients have been subjected to several clinical trials in attempts to find effective treatments, but there is insufficient data to quantify any of them as being effective. Plasma exchange, steroids, and immunosuppressive agents have been tested.

Source: Goldstein, J. M., Waxman, S. G., Vollmer, T. L., Lang, B., Johnston, I., and Newsom-Davis, J. Subacute cerebellar degeneration and Lambert–Eaton myasthenic syndrome associated with antibodies to voltage-gated calcium channels: Differential effect of immunosuppressive therapy on central and peripheral defects. *J. Neurol. Neurosurg. Psychiatry* 57:1138, 1994. Motomura, M., et al. An improved diagnostic assay for Lambert–Eaton myasthenic syndrome. *J. Neurol. Neurosurg. Psych.* 58:85, 1995. Maddison, P. and Newsom-Davis, J. Treatment for Lambert–Eaton myasthenic syndrome (Cochrane Review). In: *Cochrane Library*, Issue 4. Chichester, UK: John Wiley & Sons, 2004.

9. **CSPs** are cysteine string proteins and members of a family of chaperones (see p. 110) that function during latter stages of Ca^{2+}-regulated exocytosis.

Termination of Signals at Synaptic Junctions

Neurotransmitter action may be terminated by (a) reuptake into the presynaptic neurons, (b) metabolism, or (c) uptake into other cell types. One or more of these mechanisms function for inactivation of neurotransmitters. Examples are given below of some of the pathways involved in the synthesis and the degradation of some representative neurotransmitters.

Acetylcholine

Reactions involving acetylcholine at the synapse are summarized in Figure 24.10. Acetylcholine is synthesized by condensation of choline and acetyl CoA catalyzed by **choline acetyltransferase** in the cytosol of the neuron. This reaction is

$$\underset{\text{Choline}}{(CH_3)_3\overset{+}{N}CH_2CH_2OH} + CH_3CO{-}SCoA \longrightarrow \underset{\text{Acetylcholine}}{(CH_3)_3\overset{+}{N}CH_2CH_2OCOCH_3} + CoASH$$

Choline is derived mainly from the diet; however, some may come from reabsorption from the synaptic junction or other metabolic sources (see p. 757). The major source of acetyl CoA is decarboxylation of pyruvate by pyruvate dehydrogenase in mitochondria. The mechanism for getting acetyl CoA across the inner mitochondrial membrane as citrate (see p. 673) operates in presynaptic neurons.

Acetylcholine is released, and it reacts with the nicotinic–acetylcholine receptor embedded in the postsynaptic membrane (see Clin. Corr. 24.2). Its action is terminated

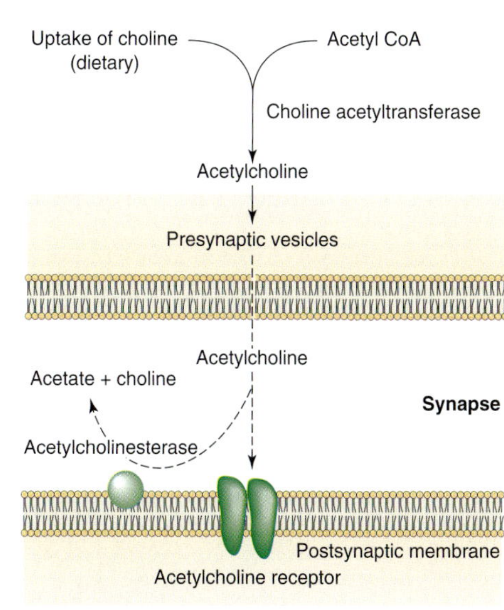

FIGURE 24.10
Summary of the reactions of acetylcholine as a neurotransmitter at the synapse. AcCoA, acetylcoenzyme A; AcChEase, acetylcholinesterase.

CLINICAL CORRELATION 24.2
Myasthenia Gravis: A Neuromuscular Disorder

Myasthenia gravis is an acquired autoimmune disease characterized by muscle weakness due to decreased neuromuscular signal transmission. The neurotransmitter involved is acetylcholine. The sera of more than 90% of patients with myasthenia gravis have antibodies to the nicotinic–acetylcholine receptor (AChR) located on the postsynaptic membrane of the neuromuscular junction. Antibodies against the AChR interact with and inhibit either its ability to bind acetylcholine or its ability to undergo conformational changes necessary to affect ion transport. The number of functional AChRs is reduced in patients with the disease. Experimental models of myasthenia gravis have been generated either by immunizing animals with the AChR or by injecting them with antibodies against it.

It is not known what events trigger the onset of the disease. Some environmental antigens have epitopes resembling those on the AChR. A rat monoclonal IgM antibody prepared against AChRs reacts with two proteins obtained from the intestinal bacterium, *E. coli*. Both are membrane proteins of 38 and 55 kDa, the smaller of which is located in the outer membrane. This does not suggest that exposure to *E. coli* proteins is likely to trigger the disease. The sera of both normal individuals and myasthenia gravis patients have antibodies against a large number of *E. coli* proteins. Some environmental antigens from other sources also react with antibodies against AChRs.

The thymus gland, which is involved in antibody production, is implicated in this disease. Antibodies have been found in thymus glands of myasthenia gravis patients that react with AChRs and with environmental antigens. The relationship between environmental antigens, thymus antibodies against AChRs, and onset of myasthenia gravis is not clear.

Myasthenia gravis patients may receive one or a combination of several therapies. Pyridostigmine bromide, a reversible inhibitor of acetylcholine esterase (AChE) that does not cross the blood–brain barrier, has been used. The inhibition of AChE within the synapse increases the half-time for acetylcholine hydrolysis. This leads to an increase in the concentration of acetylcholine, stimulation of more AChR, and increased signal transmission. Other treatments include immunosuppressant drugs, steroids, and surgical removal of the thymus gland to decrease the rate of production of antibodies. Future treatment may include the use of anti-idiotype antibodies to the AChR antibodies and/or the use of small nonantigenic peptides that compete with AChR epitopes for binding to the AChR antibodies.

Source: Stefansson, K., Dieperink, M. E., Richman, D. P., Gomez, C. M., and Marton, L. S. N. *Engl. J. Med.* 312:221, 1985. Drachman, D. B. (Ed.). Myasthenia gravis: Biology and treatment. *Ann. N.Y. Acad. Sci.* 505:1, 1987. Steinman, L. and Mantegazza, R. *FASEB J.* 4:2726, 1990. Source for current research is: http://www.ninds.nih.gov/disorders/myasthenia_gravis/

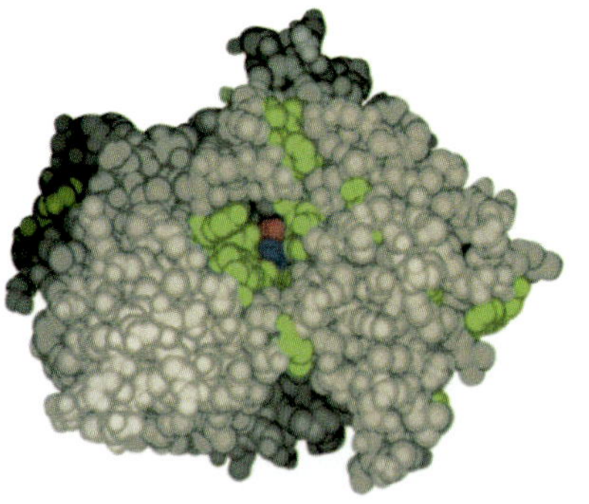

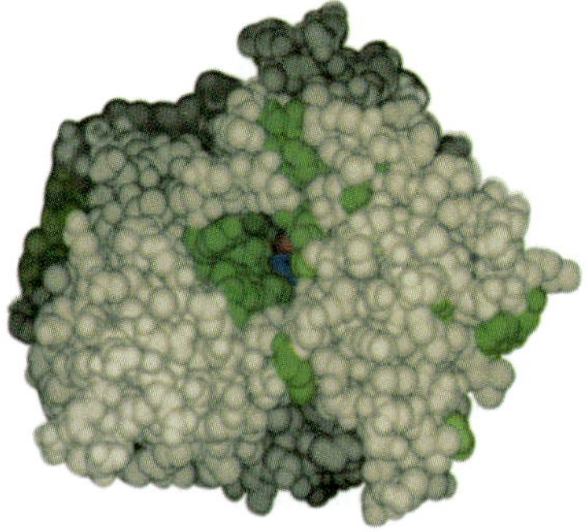

FIGURE 24.11
Space-filling stereo view of acetylcholinesterase looking down into the active site. Aromatic residues are in green, Ser^{200} is red, Glu^{199} is cyan, and other residues are gray.
Reproduced with permission from Sussman, J. L., Harel, M., Frolow, F., et al. *Science* 253:872, 1991. Copyright (1991) AAAS.

by the action of **acetylcholinesterase**, which hydrolyzes it to acetate and choline.

$$\text{Acetylcholine} + H_2O \rightarrow \text{acetate} + \text{choline}$$

Acetate is taken up and metabolized by other tissues. An X-ray crystallographic structure of acetylcholine esterase is shown in Figure 24.11. Its mechanism of action is similar to that of serine proteases (see p. 328). It contains a **catalytic triad**, but the amino acids in that triad, from N- to C-termini, are in reverse order to those of serine proteases like trypsin and chymotrypsin; in addition, glutamate replaces aspartate. The serine that forms the covalent intermediate during catalysis is visible (Figure 24.11) at the bottom of a hydrophobic channel. Glutamate, one of the other residues of the catalytic triad, is also visible within the channel of that structure.

Catecholamines

The **catecholamine neurotransmitters** are **dopamine** (3, 4-dihydroxyphenylethylamine), **norepinephrine**, and **epinephrine** (Figure 24.12) (see p. 763). Their action is terminated by reuptake into the presynaptic neuron by specific transporter proteins. Cocaine, for example, binds specifically to the **dopamine transporter** and blocks dopamine reuptake. Dopamine remains within the synapse for a prolonged period and stimulates excessively its receptors on the postsynaptic neuron. After reuptake, these neurotransmitters may be either repackaged into synaptic vesicles or metabolized by **catechol-*O*-methyltransferase**, which catalyzes the transfer of a methyl group from *S*-adenosylmethionine to one of the phenolic OH groups, and **monoamine oxidase** (Figure 24.13), which catalyzes the oxidative deamination of these amines to aldehydes and ammonium ions. They are substrates for monoamine oxidase whether or not they have been altered by catechol-*O*-methyltransferase. The end product of dopamine metabolism is **homovanillic acid**, and that of epinephrine and norepinephrine is **3-methoxy-4-hydroxymandelic acid**.

Epinephrine

Norepinephrine

Dopamine

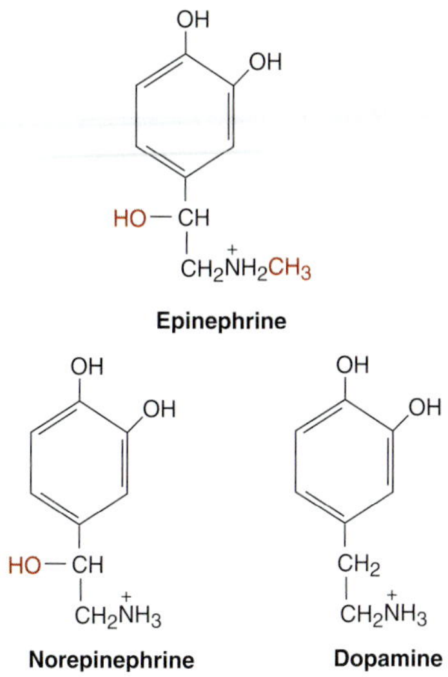

FIGURE 24.12
Catecholamine neurotransmitters.

5-Hydroxytryptamine (Serotonin)

Serotonin is derived from tryptophan (see p. 775). Like dopamine, its action is terminated by **reuptake** by a specific transporter. Some types of depression are

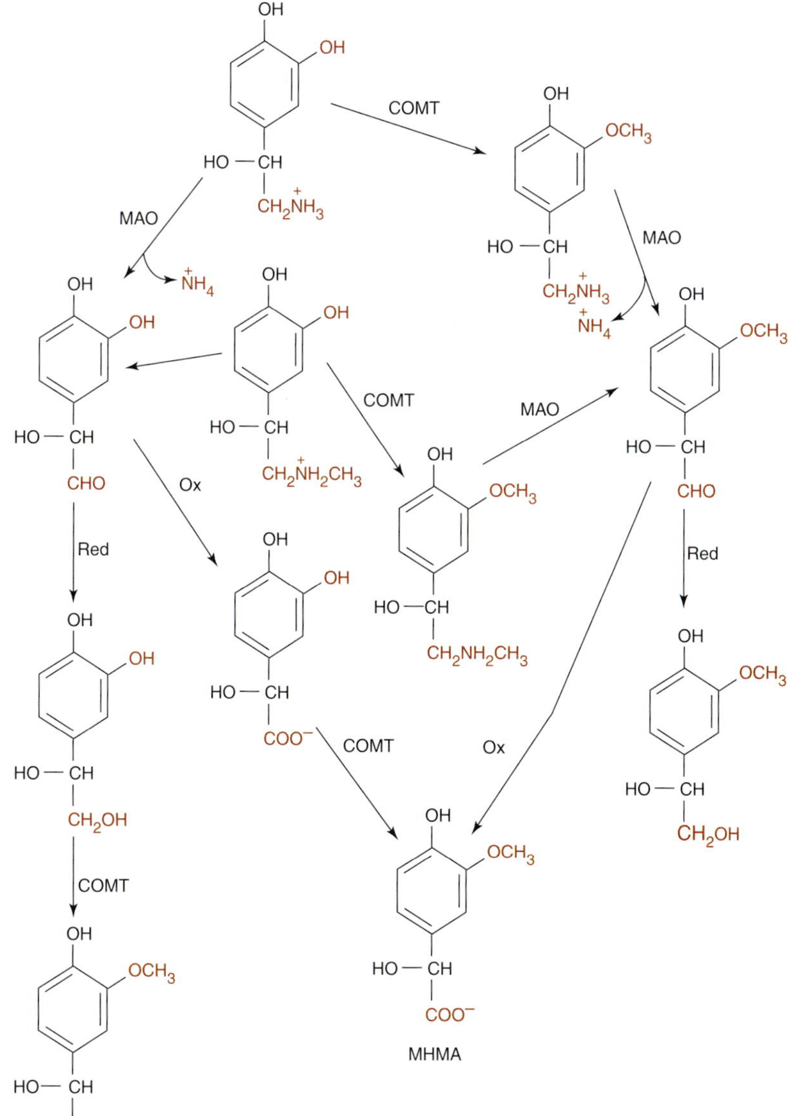

FIGURE 24.13
Pathways of catecholamine degradation. COMT, catechol-*O*-methyl transferase (requires *S*-adenosylmethionine); MAO, monoamine oxidase; Ox, oxidation; red, reduction. The end product of epinephrine and norepinephrine metabolism is 3-methoxy-4-hydroxymandelic acid (MHMA).

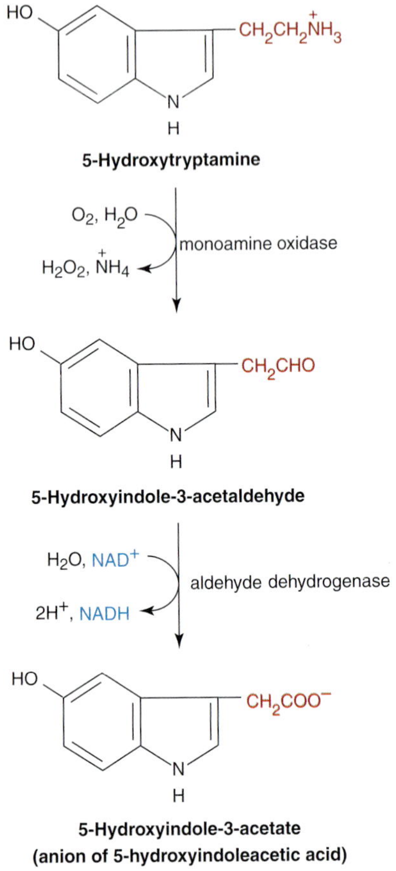

5-Hydroxytryptamine

O_2, H_2O → H_2O_2, $\overset{+}{N}H_4$ (monoamine oxidase)

5-Hydroxyindole-3-acetaldehyde

H_2O, NAD^+ → $2H^+$, NADH (aldehyde dehydrogenase)

5-Hydroxyindole-3-acetate (anion of 5-hydroxyindoleacetic acid)

FIGURE 24.14
Degradation of 5-hydroxytryptamine (serotonin).

associated with low brain levels of serotonin; antidepressants such as Paxil (paroxetine hydrochloride), Prozac (fluoxetine hydrochloride), and Zoloft (sertraline hydrochloride) specifically inhibit serotonin reuptake. Once inside the presynaptic neuron, serotonin may be either repackaged in synaptic vesicles or oxidatively deaminated to the corresponding acetaldehyde by monoamine oxidase (Figure 24.14). The aldehyde is further oxidized to 5-hydroxyindole-3-acetate by an aldehyde dehydrogenase.

γ-Aminobutyrate (GABA)

***γ*-Aminobutyrate**, an inhibitory neurotransmitter, is synthesized and degraded through reactions commonly known as the **GABA shunt**. In brain tissue, GABA and glutamate, an excitatory neurotransmitter, may share common routes of metabolism (Figure 24.15) where they are taken up by astrocytes and converted to glutamine, which is then transported back into presynaptic neurons. In excitatory neurons, glutamine is converted to glutamate and repackaged in synaptic vesicles. In inhibitory neurons, glutamine is converted to glutamate and then to GABA, which is packaged in synaptic vesicles.

It has been suggested that brain levels of GABA in some epileptic patients may be low. **Valproic acid** (2-propylpentanoic acid) can cross the blood–brain barrier and apparently increases brain levels of GABA. The mechanism by which it does so is not clear. Valproic acid is metabolized primarily in the liver by glucuronidation and urinary excretion of the glucuronides, or by mitochondrial β-oxidation and oxidation by endoplasmic reticulum enzymes.

Neuropeptides Are Derived from Precursor Proteins

Peptide neurotransmitters are synthesized as larger proteins. Proteolysis of these larger proteins releases the neuropeptide molecules. Synthesis of the larger proteins takes place in the cell body and not the axon. The neuropeptides travel down the axon to the presynaptic region by one of two mechanisms: **fast axonal transport** at a rate of about 400 mm day^{-1} and **slow axonal transport** at a rate of 1–5 mm day^{-1}. Since axons may vary in length from 1 mm to 1 m in length, theoretically the transit time could vary from 150 ms to 200 days. It is unlikely that the latter transit time occurs under normal physiological conditions, and the upper limit is probably hours rather than days. There is experimental evidence that suggest that the faster transit time prevails.

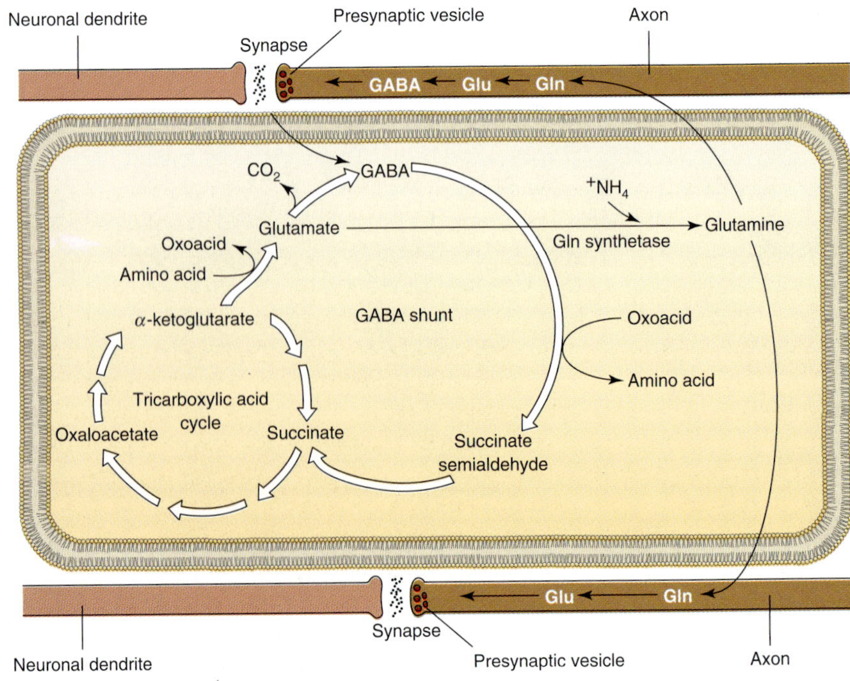

FIGURE 24.15
Involvement of astrocytes in the metabolism of GABA and glutamate.

TABLE 24.3 Peptides Found in Brain Tissue[a]

Peptide	Structure
β-Endorphin	YGGFMTSEKSQTPLVT LFKNAIIKNAYKKGE
Met-enkephalin	YGGFM
Leu-enkephalin	YGGFL
Somatostatin	AGCKNFFW | | CSTFTK
Luteinizing hormone-releasing hormone	p-EHWSYGLRPG NH_2
Thyrotropin-releasing hormone	p-EH P-NH_2
Substance P	RPKPEEFFGLM-NH_2
Neurotensin	p-ELYENKPRRPYIL
Angiotensin I	DRVYIHPFHL
Angiotensin II	DRVYIHPF
Vasoactive intestinal peptide	HSDAVFTDNYTRLR-KEMAVKKYLNSILN-NH_2

[a] Peptides with "p" preceding the structure indicate that the N-terminal is pyroglutamate. Those with NH_2 at the end indicate that the C-terminal is an amide.

Kinesins and myosins, molecular motor proteins (see p. 990), facilitate this transport process.

Neuropeptides mediate sensory and emotional responses such as those associated with hunger, thirst, sex, pleasure, pain, and so on. Included in this category are **enkephalins, endorphins**, and **substance P**. Substance P is an excitatory neurotransmitter that has a role in pain perception. It is among a class of neuropeptides called **neurokinins**. Its receptor, **NK-1** (or **n**euro**k**inin-**1**), is a G-type protein consisting of seven transmembrane helical elements. Endorphins and opioids bind to receptors that also have seven transmembrane helical elements. Endorphins have roles in eliminating the sensation of pain. Some of the peptides found in brain tissue are shown in Table 24.3. Note that Met-enkephalin is derived from the N-terminal region of β-endorphin. The N-terminal or both the N- and C-terminal amino acids of many of the neuropeptide transmitters are modified.

24.3 THE EYE: METABOLISM AND VISION

The eye, our window to the outside world, allows us to view the beauties of nature, the beauties of life, and, considering the contents of this textbook, the beauties of biochemistry. A view through any window or any camera lens is clearest when unobstructed. The eye has evolved in such a way that a similar objective has been achieved. It is composed of live tissues that require continuous nourishment obtained through use of conventional metabolic pathways appropriate to their unique needs. Pigmented structures such as cytochromes of mitochondria are either not present in some structures or arranged and distributed so as not to interfere with the visual process. In addition, the brain has devised an enormously efficient filtering system that makes objects within the eye invisible that may otherwise appear to lead to visual distortion. A schematic diagram of a cross section of the eye is shown in Figure 24.16.

Light entering the eye passes progressively through (a) the **cornea**, the anterior chamber that contains aqueous humor, (b) the lens, and (c) the vitreous body, which contains vitreous humor; it finally focuses on the **retina**, which contains the visual sensing apparatus. Tears bathe the exterior of the cornea while the interior is bathed by the aqueous humor, an isoosmotic fluid that contains salts, albumin, globulin, glucose,

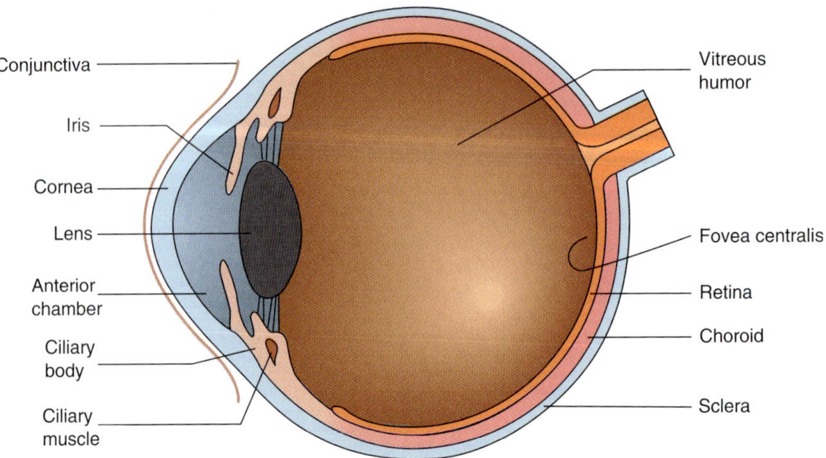

FIGURE 24.16
Schematic of a horizontal section of the left eye.

and other constituents. The aqueous humor brings nutrients to the cornea and to the lens, and it removes end products of metabolism from them. The vitreous humor is a gelatinous mass that helps maintain the shape of the eye while allowing it to remain somewhat pliable.

Cornea Derives ATP from Aerobic Metabolism

The eye is an extension of the nervous system and, like other tissues of the central nervous system, its major metabolic fuel is glucose. The cornea is not a homogeneous tissue and obtains a relatively large percentage of its ATP from aerobic metabolism. About 30% of the glucose it metabolizes is by glycolysis, and about 65% is metabolized by the hexose monophosphate pathway. On a relative weight basis, the cornea has the highest activity of the hexose monophosphate pathway of any mammalian tissue. It also has a high activity of **glutathione reductase** that requires NADPH, a product of the hexose monophosphate pathway. Corneal epithelium is permeable to atmospheric oxygen that is necessary for various oxidative reactions. The reactions of oxygen can lead to formation of various **active oxygen species** that are harmful to the tissues, in some cases by oxidizing protein sulfhydryl groups to disulfides and by lipid peroxidation, mostly of cellular medium-chain lipids of six carbons or more (see p. 576). Reduced glutathione (GSH) is used to reduce those disulfide bonds and lipid peroxides back to their original native states while GSH itself is converted to oxidized glutathione (GSSG). GSSG may also be formed directly by these active oxygen species. Glutathione reductase uses NADPH to reduce GSSG to 2GSH.

$$\text{GSSG} + \text{NADPH} + \text{H}^+ \xrightleftharpoons{\text{GSH reductase}} 2\text{GSH} + \text{NADP}^+$$

The pentose phosphate pathway and glutathione reductase help protect the cornea by moderating the effectiveness of the active oxygen species.

Some of the lipids that are subjected to peroxidation may spontaneously form active aldehydes that react with other tissue components and lead to various pathological conditions. The cornea also contains an isoform of aldehyde dehydrogenase (ALDH3A1), a member of a superfamily of enzymes that uses either NAD^+ or $NADP^+$ to inactivate these active aldehydes by oxidizing them to their corresponding acids.

Lens Consists Mostly of Water and Protein

The **lens** is bathed on one side by aqueous humor and supported on the other side by vitreous humor. It has no blood supply, but is metabolically active. It obtains nutrients from, and eliminates waste into, the aqueous humor. The lens is mostly water and proteins. The majority of vertebrate lens proteins are the α-, β-, and **γ-crystallins** (Table 24.4). There are also albuminoids, enzymes, and membrane proteins that are synthesized in an epithelial layer around the edge of the lens. Other animals have

TABLE 24.4 Eye Lens Crystallins and Their Relationships with Other Proteins

Crystallin	*Distribution*	*[Related] or Identical*
α	All vertebrates	**Small heat shock proteins** (αB)
		[*Schistosoma mansoni* antigen]
β	All vertebrates	[*Myxococcus xanthus* protein S]
γ	(embryonic γ not in birds)	[*Physarum polycephatum* spherulin 3a]
Taxon-specific enzyme crystallins		
δ	Most birds, reptiles	**Argininosuccinate lyase** (δ2)
ε	Crocodiles, some birds	**Lactate dehydrogenase B**
ξ	Guinea pig, camel, llama	**NADPH: quinone oxidoreductase**
η	Elephant shrew	**Aldehyde dehydrogenase I**

Source: Wistow, G. *Trends Biochem. Sci.* 18:301, 1993.

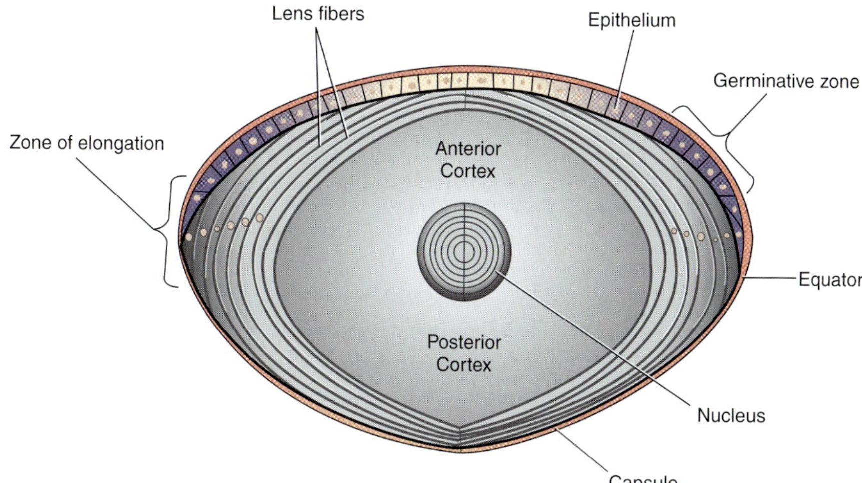

FIGURE 24.17
Schematic representation of a meridional section of a mammalian lens.

different crystallins, some of which are enzymes that probably function as such in other tissues. The most important physical requirement of these proteins is that they maintain a clear crystalline state. They are sensitive to changes in oxidation–reduction, osmolarity, excessively increased concentrations of metabolites, and UV irradiation. Structural integrity of the lens is maintained: for osmotic balance by the Na^+/K^+ exchanging ATPase (see p. 576); for redox state balance by glutathione reductase; and for growth and maintenance by protein synthesis and other metabolic processes.

Lens proteins have defined functions and they are generally expressed in other tissues. **α-, β-Crystallins** are **small heat shock proteins (sHSP)** or **chaperones** that function to help maintain lens proteins in their native, unaggregated states. Their highest expression is in eye lens, but they also occur in other tissues (such as skeletal and cardiac muscle) where they are involved in filament assembly. Mutations in crystallins, therefore, predispose an individual not only to cataract formation, but also to possible muscle weakness and heart failure.

Energy for these processes comes from the metabolism of glucose; about 85% from glycolysis, 10% from the pentose phosphate pathway, and 3% from the tricarboxylic acid cycle, presumably by cells located at the periphery.

The central area of the lens, the nucleus or core, consists of the lens cells that were present at birth. The lens grows from the periphery (Figure 24.17), and in humans it increases in weight and thickness with age and becomes less elastic. This leads to a loss of near vision (Table 24.5), a normal condition referred to as **presbyopia**. On average the lens may increase threefold in size and approximately 1.5-fold in thickness from birth to about age 80.

TABLE 24.5 Changes in Focal Distance with Age

Age (years)	*Focal Distance (in.)*
10	2.8
20	4.4
35	9.8
45	26.2
70	240.0

Source: Adapted from Koretz, J. F. and Handelman, G. H. *Sci. Am.*, 92, July 1988.

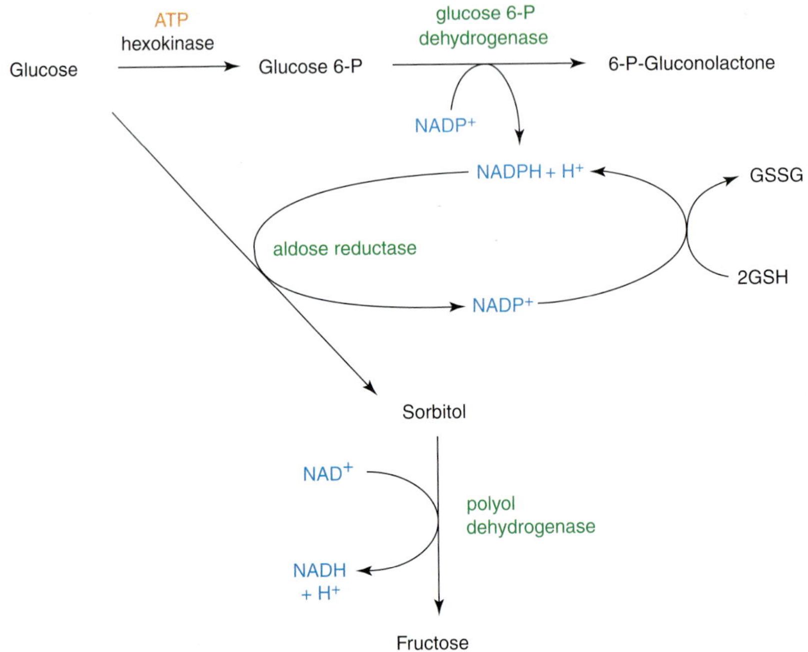

FIGURE 24.18
Metabolic interrelationships of lens metabolism.

Cataract, the only known disease of the lens, is an opacity of lenses brought about by many different conditions. There are two types of cataracts: (1) **senile cataracts**, in which changes in the architectural arrangement of the lens crystallins and other lens proteins are age-related and due to such changes as breakdown of the protein molecules starting at the C-terminal ends, deamidation, and racemization of aspartyl residues; (2) **diabetic cataracts**, which result from increased osmolarity of the lens due to increased activity of **aldose reductase** and **polyol (aldose) dehydrogenase** of the polyol metabolic pathway. When glucose concentration in the lens is high, aldose reductase converts some of it to **sorbitol** (Figure 24.18) that may be converted to **fructose** by polyol dehydrogenase. In human lens, the ratio of activities of these two enzymes favors sorbitol accumulation, especially since sorbitol is not used by other pathways and it diffuses out of the lens rather slowly. Accumulation of sorbitol increases osmolarity of the lens, affects the structural organization of the crystallins, and enhances the rate of protein aggregation and denaturation. Areas where this occurs have increased light scattering properties, which is the definition of cataracts. Normally, sorbitol formation is not a problem because the K_m of aldose reductase for glucose is about 200 mM and very little sorbitol would be formed. In diabetics, where circulating concentration of glucose is high, activity of this enzyme can be significant. Cataracts affect about 1 million people per year in the United States, and there are no known cures or preventative measures, especially for the senile type. The most common remedy is lens replacement, a very common operation in the United States, Canada, and the United Kingdom. A side effect of cataract and surgical treatment for it can be glaucoma.

Retina Derives ATP from Anaerobic Glycolysis

The **retina**, like the lens, depends heavily on anaerobic glycolysis for ATP production. Unlike the lens, the retina is a vascular tissue. In its center is the macula, and in the center of the macula is the **fovea centralis**, an avascular area of depression that contains only cones. This is the area of greatest visual acuity (see Clin. Corr. 24.3). Mitochondria are present in retinal rods and cones but not in the outer segments, where visual pigments are located. Retina contains a lactate dehydrogenase that can use either NADH or NADPH.

CLINICAL CORRELATION 24.3
Macula Degeneration and Loss of Vision

Many diseases of the eye affect vision, but not all have clear, direct biochemical origins. The most serious eye diseases are those that result in blindness. Glaucoma is the most common and is often associated with diabetes mellitus, the biochemistry of which is fairly well known. Glaucoma can be treated and blindness does not have to be a result.

Macula degeneration leads to blindness and there is no cure. The macula is a circular area of the retina, the center of which is the fovea centralis, the area of greatest visual acuity and which contains the highest number of cones. Macula degeneration may be among the leading causes of blindness of people over the age of 50 and is of two types: dry and wet. The dry form develops gradually over time, whereas the wet form develops rapidly and can lead to blindness within days. It occurs when blood vessels rupture under the macula leading to a loss of the nutrient supply and a rapid loss of vision.

Rupture of blood vessels that obscure macula details and result in rapid onset of blindness may be temporary. Several cases of sudden visual loss associated with sexual activity, but not with a sexually transmitted disease, have been reported. Vision was lost in one eye apparently during, but most often was reported a few days after engaging in, "highly stimulatory" sexual activity. Blindness was due to rupture of blood vessels in the macula area. Most patients were reluctant to discuss with their ophthalmologist what they were doing when sight loss was first observed. Four patients recovered with restoration of vision upon reabsorption of blood. In one case, blood was trapped between the vitreous humor and the retinal surface directly in front of the fovea. The hemorrhage cleared only slightly during the next month, but visual acuity did not improve. That patient did not return for a follow-up examination, but there was no indication during the initial examination that the condition was permanent. Since most of the victims of this phenomenon were over the age of 39, it may be a worry more to professors than to students. It also may give a new meaning to the phrase, "love is blind."

Source: Friberg, T. R., Braunstein, R. A., and Bressler, N. M. *Arch. Ophthalmol.* 113:738, 1995.

Visual Transduction Involves Photochemical, Biochemical, and Electrical Events

Figure 24.19 shows an electron micrograph and schematic of the retinal membrane. Light entering the eye through the lens passes the optic nerve fibers, the ganglion neurons, the bipolar neurons, and nuclei of rods and cones before it reaches the outer segment of the rods and cones where the signal transduction process begins. The **pigmented epithelial** layer, the choroid, lies behind the retina, absorbs the excess light, and prevents reflections back into the rods and cones where it may cause distortion or blurring of the image (see Clin. Corr. 24.4).

The eye may be compared to a video camera, which collects images, converts them into electrical pulses, records them on magnetic tape, and allows their visualization by decoding the taped information. The eye focuses on an image by projecting that image onto the retina. A series of events begins, the first of which is photochemical, followed by biochemical events that amplify the signal, and finally electrical impulses are sent to the brain where the image is reconstructed in "the mind's eye." In effecting this process, the initial event is transformed from a physical event to a chemical event through a series of biochemical reactions, to an electrical event, to a conscious acknowledgment of the presence of an object in the environment outside of the body.

Photons (light) are absorbed by photoreceptors in the **outer segments** of **rods** or **cones**, where they cause isomerization of the visual pigment, **retinal**, from the 11-*cis* form to the all-*trans* form. This isomerization causes a conformational change in the protein moiety of the complex and affects the resting membrane potential of the cell, resulting in an electrical signal being transmitted by way of the optic nerve to the brain.

Rods and Cones Are Photoreceptor Cells

Photoreceptor cells of the eye are the **rods** and the **cones** (Figure 24.19). Each type has flattened disks that contain a photoreceptor pigment that consists of a protein and a prosthetic group, 11-*cis*-retinal. The pigment is **rhodopsin** in rod cells, and they are red, green, or blue pigments in cone cells. Rhodopsin is a transmembrane protein containing **11-*cis*-retinal**, the protein moiety of which is called **opsin**. The three proteins that form

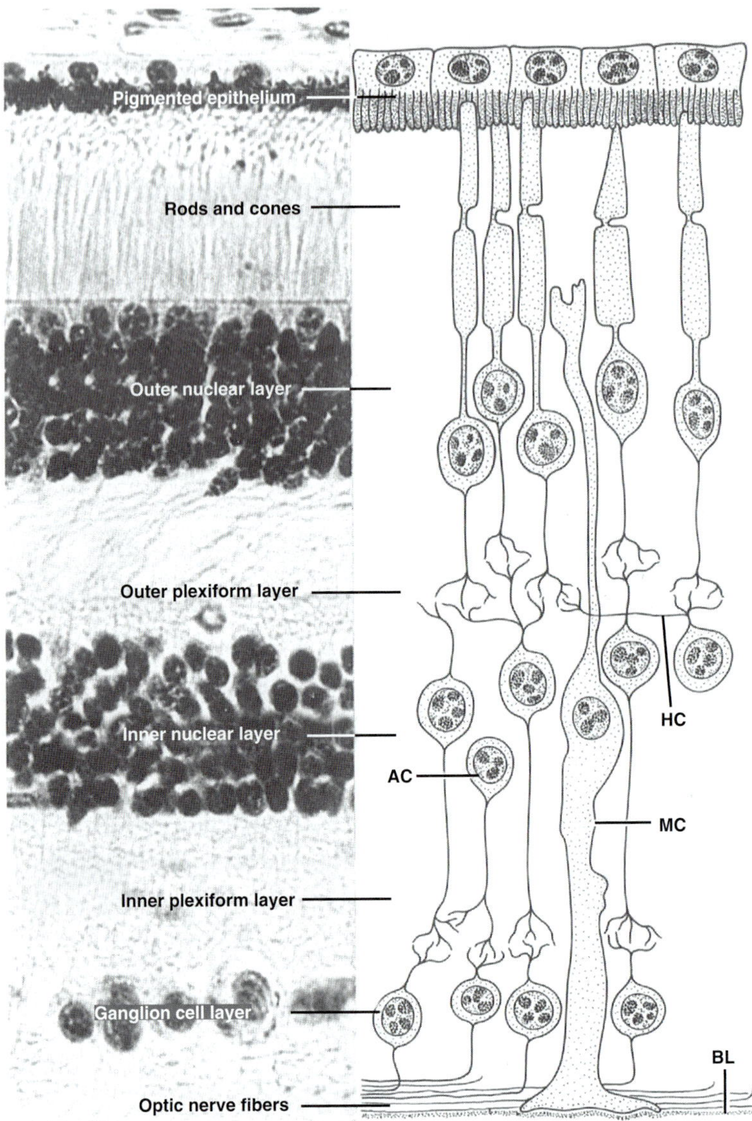

FIGURE 24.19

Electron micrograph and schematic representation of cells of the human retina. Tips of rods and cones are buried in the pigmented epithelium of the outermost layer. Rods and cones form synaptic junctions with many bipolar neurons, which, in turn, form synapses with cells in the ganglion layer that send axons through the optic nerve to the brain. The synapse of a rod or cone with many cells is important for the integration of information. HC, horizontal cells; AC, amacrine cell; MC, Müller cell; BL, basal lamina.
Reprinted with permission from Kessel, R. G., and Kardon, R. H. *Tissues and Organs: A Text-Atlas of Scanning Electron Microscopy*. New York: W. H. Freeman, 1979, p. 87. Copyright (1979) R. G. Kessel and R. H. Kardon.

CLINICAL CORRELATION **24.4**

Niemann–Pick Disease and Retinitis Pigmentosa

There are central nervous system disorders associated with the Niemann–Pick group of diseases that can become evident by ocular changes. Some of these are observed as abnormal macula with gray discoloration and granular pigmentation or granule opacities about the fovea.

Acute type I Niemann–Pick disease, lipidosis with sphingomyelinase deficiency and primary sphingomyelin storage, may show a cherry red spot in the retina in up to 50% of these patients. Macula halo describes the crystalloid opacities seen in some patients with subacute type I disease. These halos are approximately one-half the disk diameter at their outer edge and are scattered throughout the various layers of the retina. They do not interfere with vision.

An 11-year old girl who had type II disease had more extensive ocular involvement. There was sphingomyelin storage in the keratocytes of the cornea, the lens, the retinal ganglion cells, the pigmented epithelium, the corneal tract, and the fibrous astrocytes of the optic nerve.

Thus, retinitis pigmentosa may be a secondary effect of the abnormal biochemistry associated with Niemann–Pick disease.

Source: Spence, M. W. and Callahan, J. W. Sphingomyelin-cholesterol lipidoses: The Niemann–Pick group of diseases. In C. R. Schriver, A. L. Beaudet, W. Sly, and D. Valle (Eds.), *The Metabolic Basis of Inherited Disease*. New York: McGraw-Hill, 1989, p. 1656.

CLINICAL CORRELATION 24.5

Retinitis Pigmentosa from a Mutation in the Gene for Peripherin

Retinitis pigmentosa (RP) is a slowly progressive condition associated with loss of night and peripheral vision. It is a group of heterogeneous diseases of variable clinical and genetic origins; several are associated with abnormal lipid metabolism. This disease affects approximately 1.5 million people throughout the world. It can be inherited through an autosomal dominant, recessive, or X-linked mode. RP has been associated with mutations in the protein moiety of rhodopsin and in a related protein, peripherin/RDS (retinal degeneration slow), both of which are integral membrane proteins. Peripherin contains 344 amino acid residues and is located in the rim region of the disc membrane. Structural models of these two proteins are shown in the figure below. Filled circles and other notations in the figure mark residues or regions that have been correlated with RP or other retinal degenerations.

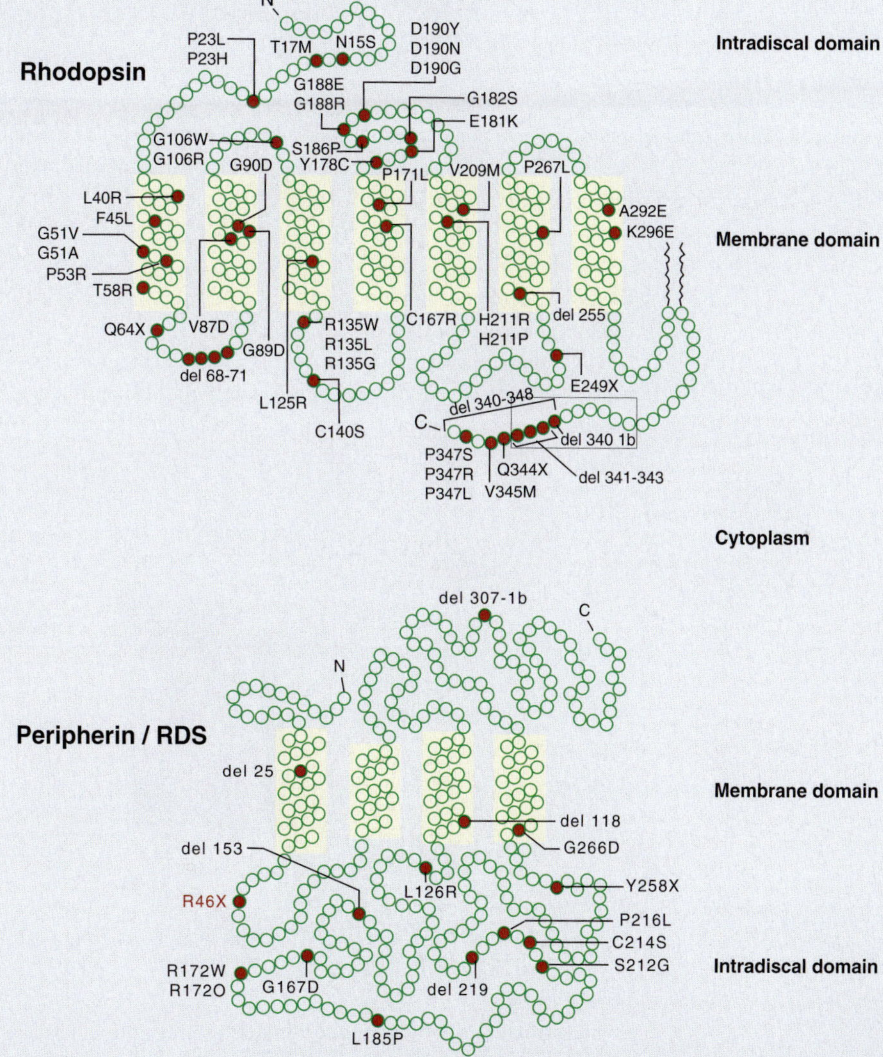

Schematic representation of structural models for rhodopsin (top) and peripherin/RDS (retinal degeneration slow) (bottom). The location of mutations in amino acid residues that segregate with RP or other retinal degenerations are shown as solid circles.

From Lam et al. (1995).

Continued on Page 968

Clinical Correlation 24.5 (continued)

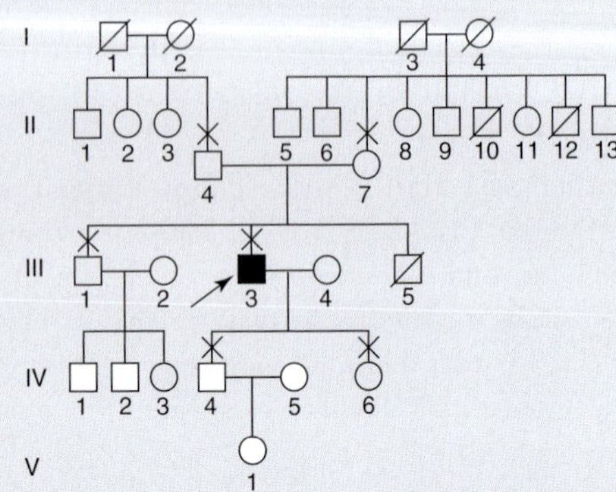

Pedigree of family. Males are squares, females are circles. Solid square indicates the proband. A slashed through symbol indicates deceased. From Lam et al. (1995).

A *de novo* mutation in exon 1 of the gene coding for peripherin resulted in the onset of RP. It caused a C-to-T transition in the first nucleotide of codon 46. This resulted in changing an arginine to a stop codon (R46X). The pedigree of this family is shown in the figure adjacent. Neither parent had the mutation, and genetic typing analysis (20 different short tandem repeat polymorphisms) showed that the probability that the proband's parents are not his actual biological parents is less than one in 10 billion, establishing with near certainty that the mutation is *de novo*.

This R46X mutation has been observed in another unrelated patient and demonstrates the importance of the use of DNA analysis to establish the genetic basis for RP, as opposed to that associated with other abnormal metabolic conditions.

Source: Shastry B. S. Retinitis pigmentosa and related disorders: Phenotypes of rhodopsin and peripherin/RDS mutations. *Am. J. Med. Genet.* 52:467, 1994. Lam, B. L., Vandenburgh, K., Sheffield, V. C., and Stone, E. M. Retinitis pigmentosa associated with a dominant mutation in codon 46 of the peripherin/RDS gene (Arg^{46}stop). *Am. J Ophthalmol.* 119:65, 1995.

the red, green, and blue pigments of cone cells are different from each other and from the opsin of rhodopsin.

Rhodopsin (~ 40 kDa) contains seven transmembrane α-helices. An 11-*cis*-retinal molecule is attached through a protonated Schiff base to the ϵ-amino group of Lys^{296} on the seventh helix and lies about midway between the two faces of the membrane (Figure 24.20) (see Clin. Corr. 24.5).

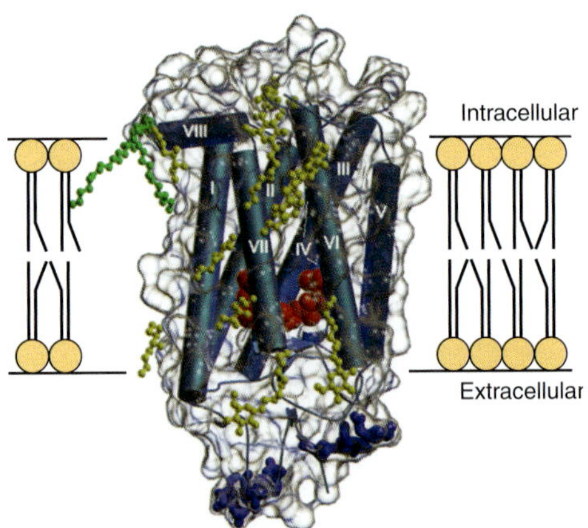

FIGURE 24.20

Crystal structure of bovine rhodopsin at 2.8-Å resolution. Rhodopsin is a transmembrane protein. The width of the membrane into which it is embedded is approximately equivalent to the length of its helices (blue rods). The intracellular side of the membrane approximately transects helix VIII. β-Strands are shown as blue arrows. Structures in green on the intracellular side are two palmitoyl groups oriented such that the hydrophobic groups can interact with hydrophobic regions of the membrane. The blue ball-and-stick structures at the bottom (extracellular side) of the molecule are carbohydrates. The yellow structures located near the hydrophobicsurface of the protein are nonylglucoside and heptaneol molecules. Reprinted with permission from Teller, D. C., Okada, T., Behnke, C. A., Palczewski, K., and Stenkamp, R. E. *Biochemistry* 40:7761, 2001. Copyright (2001) American Chemical Society. Figure generously supplied by Dr. K. Palczewski and Dr. C. A. Behnke.

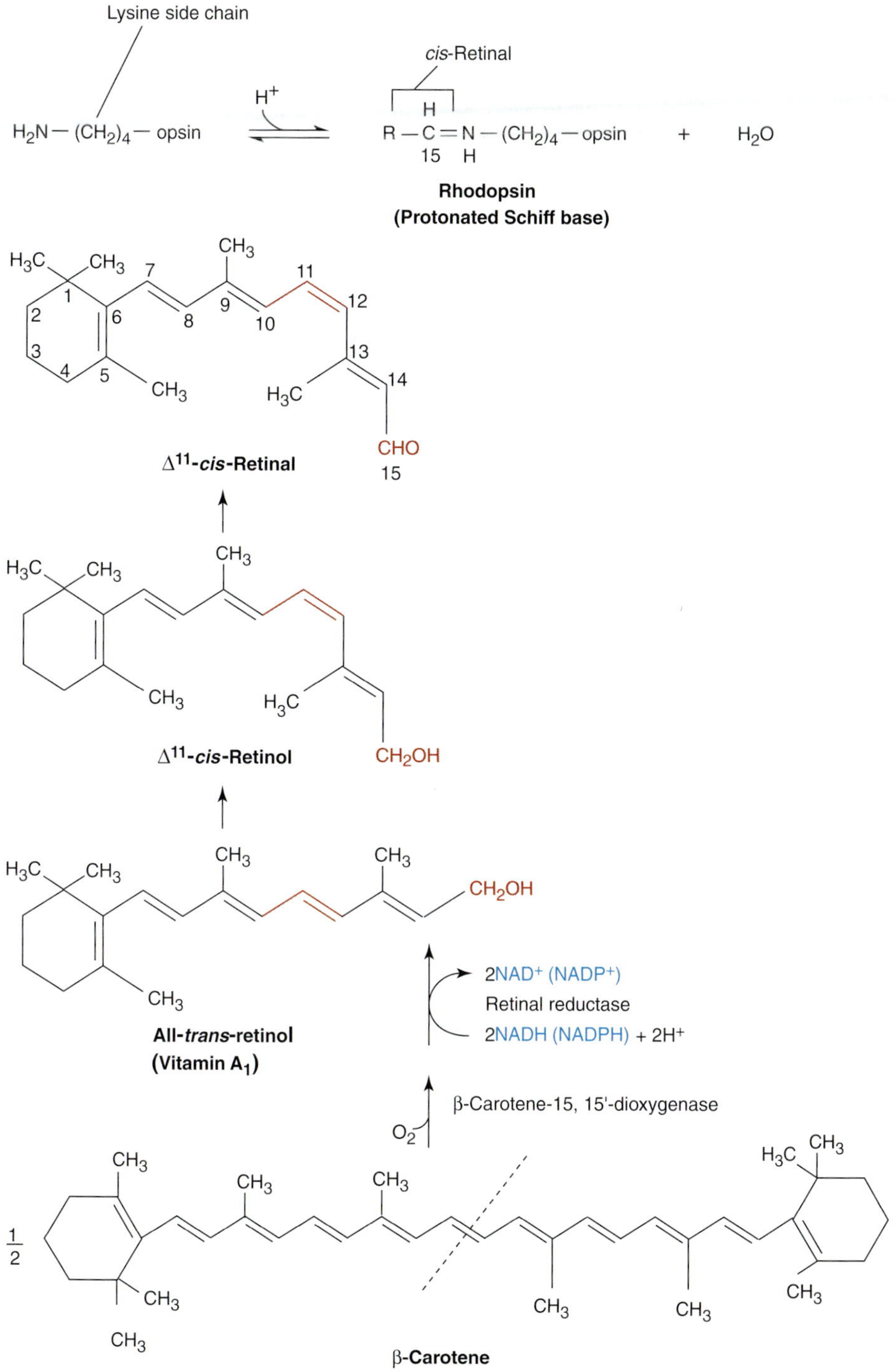

FIGURE 24.21

Formation of 11-*cis*-retinal and rhodopsin from *β*-carotene.

A schematic of the formation of 11-*cis*-retinal from ***β*-carotene** and rhodopsin formation from opsin and 11-*cis*-retinal are shown in Figure 24.21. 11-*Cis*-retinal is derived from **vitamin A** and/or *β*-carotene of the diet. Cleavage of *β*-carotene yields two molecules of **all-*trans*-retinol**, which is isomerized by an enzyme in the pigmented epithelial cell layer of the retina to **11-*cis*-retinol**. Oxidation of 11-*cis*-retinol to 11-*cis*-retinal and its binding to opsin occur in the rod outer segment.

The absorption spectra of 11-*cis*-retinal and the four visual pigments are shown in Figure 24.22*a*. The wavelength of maximum absorption of 11-*cis*-retinal is different when it is bound to opsin or to the protein moieties of the other visual pigments due to the subtle differences in the chemical environments in which the 11-*cis*-retinal resides (Figure 24.22*b*). Absorption bands for the pigments reflect their light sensitivity.

The magnitude of change in the electrical potential of photoreceptor cells following exposure to a light pulse differs from that of neurons during depolarization. The **resting potential** of rod cell membrane is approximately −30 mV compared to −70 mV for neurons. Excitation of rod cells by a light pulse causes **hyperpolarization** of the membrane, from about −30 mV to about −35 mV (Figure 24.23). It takes hundreds of

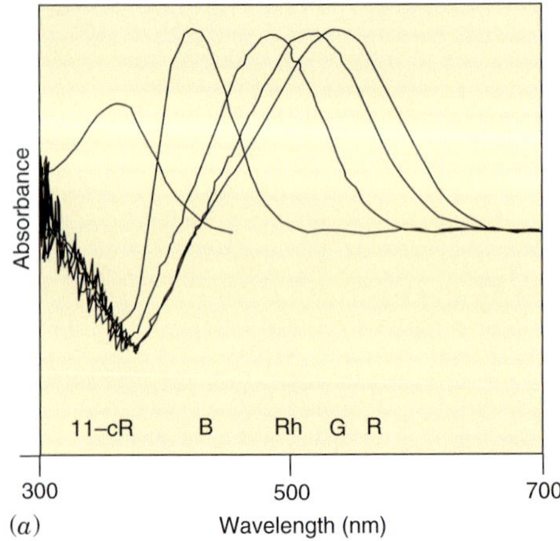

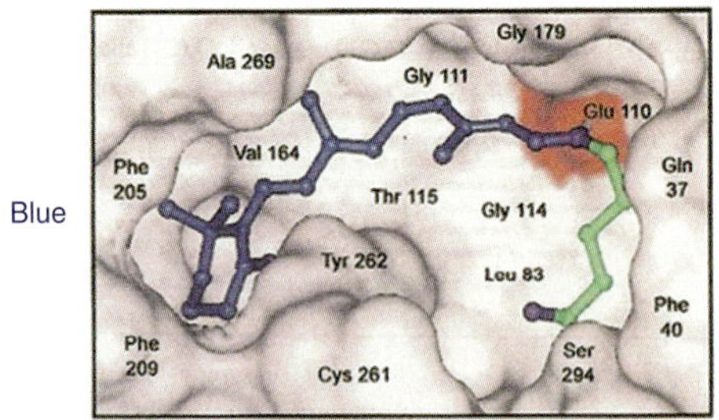

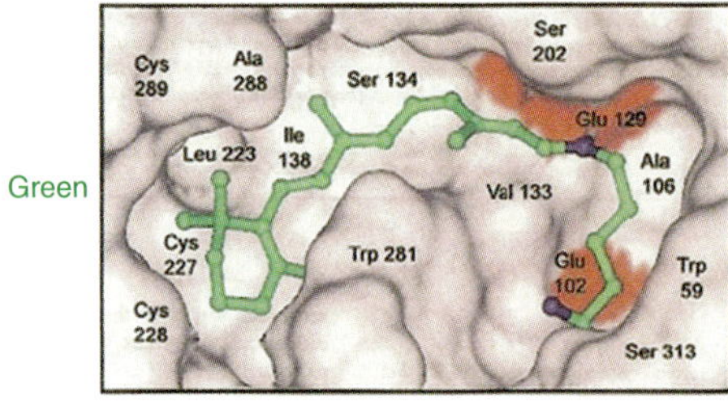

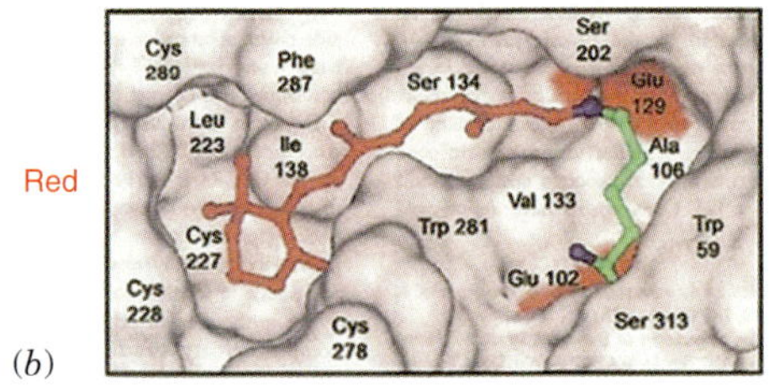

FIGURE 24.22

Absorption spectra of 11-*cis*-retinal and the four visual pigments. (*a*) Absorbance is relative and represents difference spectra obtained by subtracting the spectra for the recombinant apoproteins. The spectrum for 11-*cis*-retinal (11cR) is in the absence of protein. Other abbreviations: B, blue pigment; Rh, rhodopsin; G, green; R, red. (*b*) Some of the amino acids in the vicinity of 11-*cis*-retinal in the blue, green, and red pigments. Part (*a*) modified from Nathans, J. *Cell* 78:357, 1994. Part (*b*) reproduced with permission from Stenkamp, R. E., Filipek, S., Driessen, C. A. G. G., Teller, D. C., and Palczewski, K. Crystal structure of rhodopsin: A template for cone visual pigments and other G protein-coupled receptors. *Biochim. Biophys. Acta* 1565:168, 2002. Copyright (2002) Elsevier.

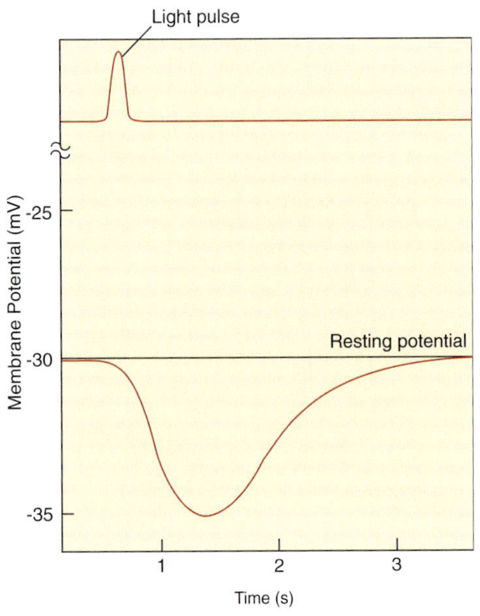

FIGURE 24.23

Changes in the potential of a rod cell membrane after a light pulse.
Redrawn from Darnell, J., Lodish, H., and Baltimore, D. *Molecular Cell Biology*. New York: Scientific American Books, 1986, p. 763.

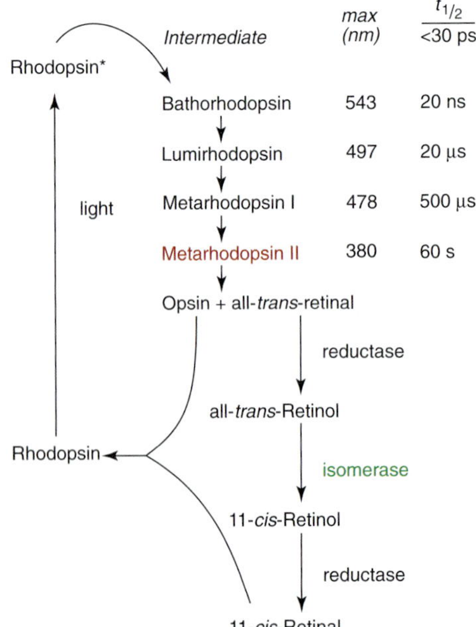

FIGURE 24.24

Conformation changes that rhodopsin undergoes after photoactivation that leads to Metarhodopsin II, "active rhodopsin."

(a)

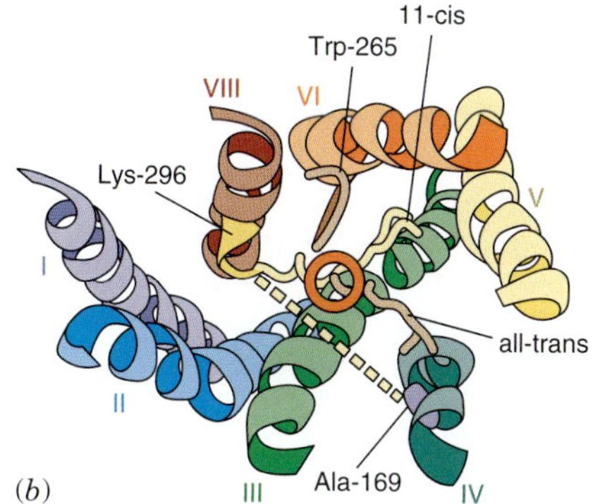

(b)

FIGURE 24.25

Schematics of 11-*cis*-retinal and all-*trans*-retinal in their protein environments. (*a*) Amino acid side chains surrounding 11-*cis*-retinal. When 11-*cis*-retinal isomerizes to the all-*trans* form (*b*), the β-ionone ring of all-*trans*-retinal interacts with Ala[169] of helix IV. Part (*a*) reprinted with permission from Palczewski, K., et al. *Science* 289:739, 2000. Part (*b*) reprinted with permission from Bourne, H. R. and Meng, E. C. *Science* 289:733, 2000. Copyright (2000) AAAS.

milliseconds for the potential to reach its maximum state of hyperpolarization, during which time a number of biochemical events occur.

Absorption of photons of light and isomerization of 11-*cis*-retinal in rhodopsin are rapid, requiring only picoseconds. Following this, rhodopsin undergoes a series of conformation changes to accommodate the structural change of the 11-*cis*- to all-*trans*-retinal. These changes are stepwise events with some of the species being very short-lived as evident by the data shown in Figure 24.24. The exact structures for these intermediates are not known, but it is clear that they result from changes in the neighboring group environment of retinal after its isomerization. Figure 24.25*a* is a schematic from the 3-D structure of rhodopsin showing the amino acid side chains surrounding 11-*cis*-retinal as viewed from the cytoplasmic side. When 11-*cis*-retinal undergoes isomerization to the all-*trans*-retinal at the position within the red circle of Figure 24.25*b*, the β-ionone ring of all-*trans*-retinal reaches Ala^{169} in helix IV. Ala^{169} is not within the range of interactions shown in Figure 24.25*a*. The intermediate species listed in Figure 24.24 represent some of the conformations rhodopsin undergoes before it finally dissociates to opsin and all-*trans*-retinal.

At 37°C, activated rhodopsin decays in slightly more than 1 ms through several intermediates to **metarhodopsin II**, which has a half-life of approximately 1 minute. **Metarhodopsin II** is the **active rhodopsin** species, R* that is involved in the biochemical reactions of interest. These structural changes are all kinetic events, and metarhodopsin II will have begun to form within hundredths of microseconds of the initial event. Finally, dissociation of metarhodopsin to opsin, and all-*trans*-retinal occurs. All-*trans*-retinal is enzymatically converted to all-*trans*-retinol by **all-*trans*-retinol dehydrogenase** in the rod outer segment. All-*trans*-retinol is transported into the pigmented epithelium where a specific isomerase converts it to 11-*cis*-retinol, which is then oxidized to the aldehyde and transported back into the rod outer segment. Once the aldehyde is formed and transported back into the rod outer segment, it can recombine with opsin to form rhodopsin and the cycle can begin again. A diagrammatic representation of these

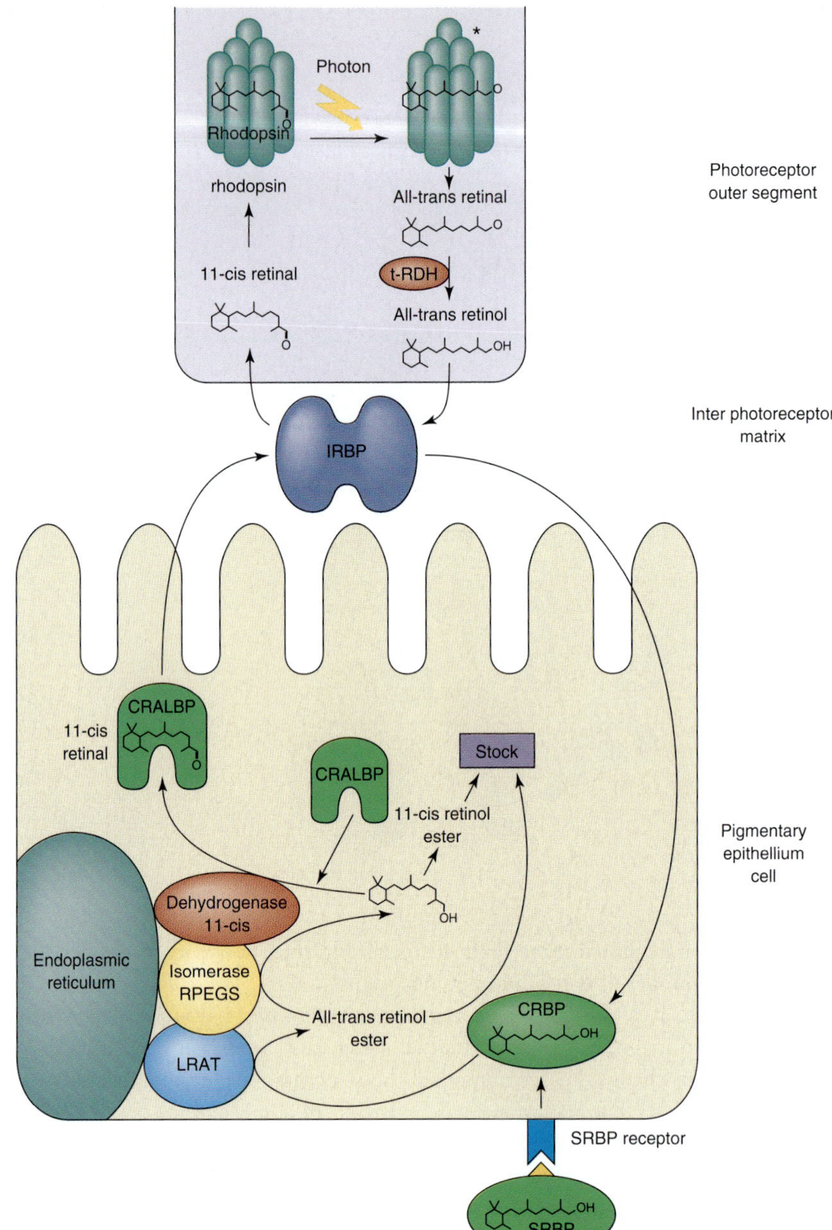

FIGURE 24.26

Transport and metabolism of 11-*cis*- and all-*trans*-retinal into the pigmented epithelium. CRALBP = cellular retinaldehyde binding protein; CRBP = cytosolic retinoid binding protein; IRBP = interphotoreceptor retinoid binding protein; IRAT = lecithin retinal acyl transferase; RPE65 = isomerohydrolase enzyme; SRBP = serum retinal binding protein.

Redrawn from Perrault, I, Rozet, J.-M., Gerber, S., Ghazi, I., Leowski, C., Ducroq, D., Souied, E., Dufier, J.-L., Munnich, A., and Kaplan, J. Leber congenital amaurosis. *Mol. Gen. Metab.* 68:200, 1999.

events is shown in Figure 24.26 (see Clin. Corr. 24.6). Similar events occur in the cones with the three proteins of the red, green, and blue pigments. There are three interconnected biochemical cycles of events that occur in the conversion of light energy to nerve impulses (Figure 24.27). These cycles describe the reactions of rhodopsin, **transducin**, and **phosphodiesterase**, respectively. The net result is a hyperpolarization of the plasma membrane of the rod (or cone) cells from −30 mV to approximately −35 mV. Maintenance of a steady-state potential of the plasma membrane of rod and cone cells at −30 mV is an important function of this series of reactions.

Rod cells of a fully dark-adapted human eye can detect a flash of light that emits as few as five photons. The rod is a specialized type of neuron in that the signal generated does not depend upon an all-or-none event. The signal may be graded in intensity and reflect the extent that the millivolt potential changes from its steady-state value of −30 mV as well as the number of photoreceptor cells stimulated. The **steady-state potential** of −30 mV is maintained at this more positive value because **Na^+ channels** of the photoreceptor cells are **ligand-gated** and a fraction of them responsible for the influx of Na^+ are maintained in an open state. The ligand responsible for keeping

CLINICAL CORRELATION 24.6

Leber Congenital Amaurosis: Retinal Dystrophy Leading to Blindness

Leber congenital amaurosis (LCA) is one of the more serious conditions leading to congenital blindness. Leber first described LCA in 1869. Clinical heterogeneity in LCA was recognized early but largely ignored. Genetic heterogeneity has been accepted since 1963, and three genes have been shown to account for about 27% of the dystrophy: (1) retGC1 (aka GUCY2D) on chromosome 17p13 codes for the photoreceptor specific guanylate cyclase. (2) CRX on chromosome 19q13.3 codes for a homeobox protein essential for photoreceptor maintenance and outer rod/cone segment biogenesis, and RPE65 on chromosome 1p31 codes for a specific retinal pigment epithelium (RPE) protein involved in vitamin A metabolism—specifically, retinol *trans/cis* isomerase. Structural representation of the human RPE65 gene and positions of mutations that have been identified in LCA are shown.

A canine model of LCA has been treated by gene replacement therapy in which a normal RPE65 gene was delivered intra-ocularly to one eye of each test animal, and vision was restored in that eye to near normal. The function of this enzyme in vision is evident from Figures 24.24 and 24.26.

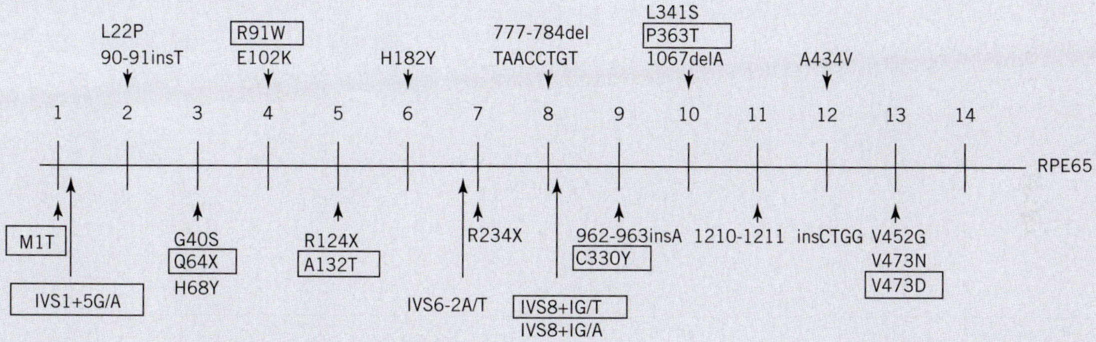

Schematic representation of the human RPE65 gene.

Source: Perrault, I., Rozet, J.-M., Gerber, S., Ghazi, I., Leowski, C., et al. Leber congenital amaurosis. *Mol. Genet. Metab.* 68:200, 1999. Acland, G. M., Aguirre, G. D., Ray, J., Zhang, Q., Aleman, T. S., et al. Gene therapy restores vision in a canine model of childhood blindness. *Nat. Genet.* 28:92, 2001.

some of the Na^+ channels open is **cyclic GMP (cGMP)**, which binds to them in a concentration-dependent, kinetically dynamic manner. Biochemical events that affect the concentration of cGMP within rod and cone cells also affect the number of Na^+ channels open, the concentration of Na^+ within the cells, and hence the membrane potential (Figure 24.27). **Active rhodopsin** (R*, namely, **metarhodopsin II**) forms a complex with transducin. **Transducin** is a classical trimeric **G-protein** and functions in a manner very similar to that described on page 509. In the R*-transducin complex (R^*-$T_{\alpha,\beta,\gamma}$), transducin undergoes a conformational change that facilitates exchange of its bound GDP with GTP. When this occurs, the α-subunit (T_α) dissociates from the β-,γ-subunits. T_α activates **phosphodiesterase (PDE)**, which hydrolyzes cGMP to 5′-GMP. This decreased concentration of cGMP decreases the number of open Na^+ channels and hence the influx of Na^+. The efflux of Na^+ is not affected by these events and Na^+ concentration inside the cells decreases, causing the membrane potential to become more negative (i.e., hyperpolarized).

The diagram of Figure 24.27 shows in cartoon form two such channels in the plasma membrane, one of which has cGMP bound to it and it is open. The other does not have cGMP bound to it and it is closed. By this mechanism, the concentration of Na^+ in the cell is directly linked to the concentration of cGMP and, thus, also the membrane potential.

PDE in rod cells is a **heterotetrameric protein** consisting of one each α and β catalytic subunits and two γ regulatory subunits. T_α-GTP forms a complex with the γ-subunits of PDE effecting dissociation of the catalytically active α,β-dimeric PDE

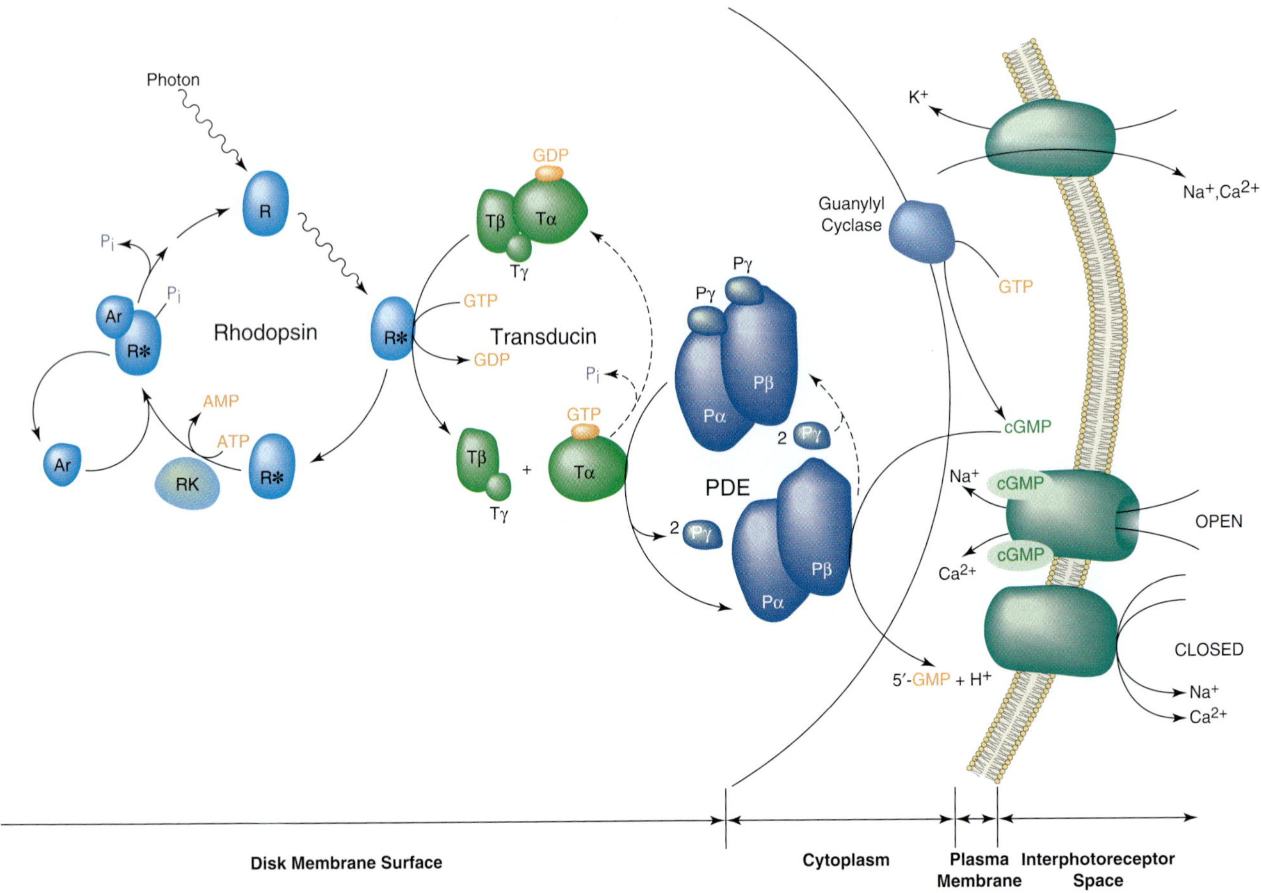

FIGURE 24.27

Cascade of biochemical reactions involved in the visual cycle.
Redrawn from Farber, D. B. *Invest. Ophthalmol. Vis. Sci.* 36:263, 1995.

subunit complex. T_α has GTPase activity, which hydrolyzes bound GTP to GDP and inorganic phosphate (P_i). This results in dissociation of T_α from the regulatory γ-subunits of PDE, permitting them to reassociate with the catalytic subunits and to inhibit PDE activity. The same reactions occur in cone cells, but the catalytic subunits of cone cells PDE differ from those of rods in that cone cells PDE consists of two α catalytic subunits instead of α- and β-subunits.

cGMP concentration is regulated by intracellular Ca^{2+}. Ca^{2+} enters rod cells in the dark through sodium channels, increasing its concentration to the 500 nM range. At these concentrations, activity of **guanylate cyclase** is low. When sodium channels are closed, Ca^{2+} entry is inhibited, but efflux by the sodium/calcium–potassium exchanger is unchanged (top complex of the plasma membrane in Figure 24.27). This decreases the intracellular Ca^{2+} concentration, resulting in activation of guanylate cyclase and increased production of cGMP from GTP. Resynthesis of cGMP and the hydrolysis of GTP of the T_α-GTP complex play important roles in stopping reactions of the visual cycle.

Inactivation of activated rhodopsin, R*, is also important in stopping this cascade of events. Activated rhodopsin, R*, is phosphorylated by an ATP-dependent **rhodopsin kinase** (Figure 24.27). The R*-P_i has high binding affinity for the cytosolic protein, **arrestin**. The arrestin-R*-P_i complex is no longer capable of interacting with transducin. The kinetics of binding of arrestin to the activated-phosphorylated rhodopsin is sufficiently rapid *in vivo* to stop the first cycle in this cascade of reactions. When rhodopsin is regenerated, the cycle can be initiated again by photons of light.

Mutations can occur in any of the major proteins involved in the visual cycle, some of which are listed in Table 24.6, and result in visual impairment (see Clin. Corr. 24.6).

TABLE 24.6 Major Proteins Involved in the Phototransduction Cascade

Protein	Relation to Membrane	Molecular Mass (kDa)	Concentration in Cytoplasm (μM)
Rhodopsin	Intrinsic	39	—
Transducin ($\alpha + \beta + \gamma$)	Peripheral or soluble	80	500
Phosphodiesterase	Peripheral	200	150
Rhodopsin kinase	Soluble	65	5
Arrestin	Soluble	48	500
Guanylate cyclase	Attached to cytoskeleton	?	?
cGMP-activated channel	Intrinsic	66	?

Color Vision Originates in Cones

Even though photographic artists, such as the late Ansel Adams, make the world look beautiful in black and white, the intervention of colors in the spectrum of life's pictures brings another degree of beauty to the wonders of nature and the beauty of life... even the ability to make a distinction between tissues from histological staining. The ability of humans to distinguish colors resides within a relatively small portion of the visual system, the cones. The number of cones within the human eye is small compared with the number of rods. Some animals (e.g., dogs) have even fewer cones, and other animals (e.g., birds) have many more.

The mechanism by which light stimulates cone cells is the same as it is for rod cells. The initial event in all cases is the photoinduced isomerization of 11-*cis*-retinal. There are three types of cone cells, defined by whether they contain the blue, green, or red visual pigments (see Figure 24.22*b*). They are also referred to in medical literature as short (S), medium (M), or long (L) wavelength-specific opsins, respectively. Color discrimination by cone cells is an inherent property of the proteins of the visual pigments to which the 11-*cis*-retinal is attached. The 11-*cis*-retinal is attached to each of the proteins through a protonated Schiff base as in rhodopsin. The absorption spectrum produced by the conjugated double-bond system of 11-*cis*-retinal is influenced by its chemical environment (Figure 24.22*b*). When 11-*cis*-retinal is bound to different visual proteins, amino acid residues in the local areas around the protonated base and the conjugated π-bond system influence the energy level and give absorption spectra with maxima that are different for each of the color pigments.

Genes for the visual pigments have been cloned, their amino acid sequences inferred, and comparisons made among them and rhodopsin. Significant homology exists among them as shown in Figure 24.28. Open circles represent amino acids that are the same, and closed circles represent those that are different. A string of closed circles at either end represents an extension of the chain of one protein relative to the other. The red and the green pigments show the greatest degree of homology, about 96% identity, whereas the degree of homology between different pairs of the others is between 40% and 45%.

Normally, only one type of visual pigment occurs per cell. During fetal development, however, multiple pigments may occur in the same cell. Under those conditions, the blue pigment seems to appear first followed by green and red pigments. The number of cones containing these multiple pigments decreases after birth and can hardly be found at all in adults. The blue pigment has optimum absorbance at 420 nm, green pigment at 535 nm, and red pigment at 565 nm (Figure 24.22*a*). Colors other than those reflected by the absorption maxima of the visual pigments are distinguished by graded stimulation of the different cones and comparative analysis by the brain.

Color Vision Is Trichromatic

Genes encoding the visual pigments have been mapped to specific chromosomes. The rhodopsin gene resides on **chromosome 3**, the gene encoding the blue pigment resides on **chromosome** 7, and the two genes for the red and green pigments reside on the **X chromosome**. In spite of the great similarity among the red and green pigments,

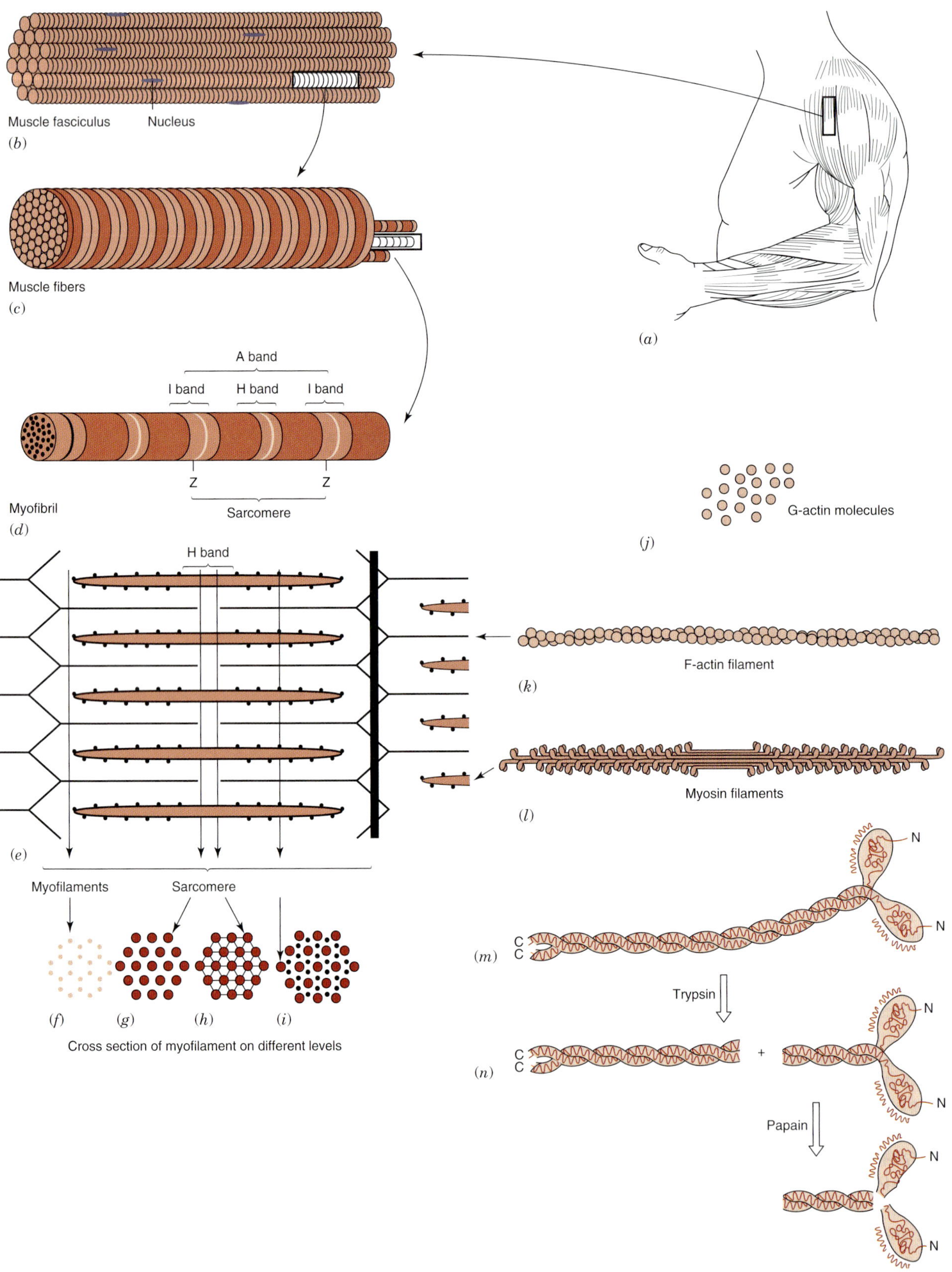

FIGURE 24.29

Structural organization of skeletal muscle. The diagram shows schematically the aggregation of G-actin to F-actin and the relative positions of cleavage of myosin by trypsin and papain.
Redrawn from Bloom, W. D. and Fawcett, D. W. *Textbook of Histology*, 10th ed. Philadelphia: Saunders, 1975. Sections m and n are reprinted with permission from Alberts, B., Bray, D., Lewis, J., Raff, M., Roberts, K., and Watson, J. *Molecular Biology of the Cell*, 2nd ed. New York: Garland, 1983.

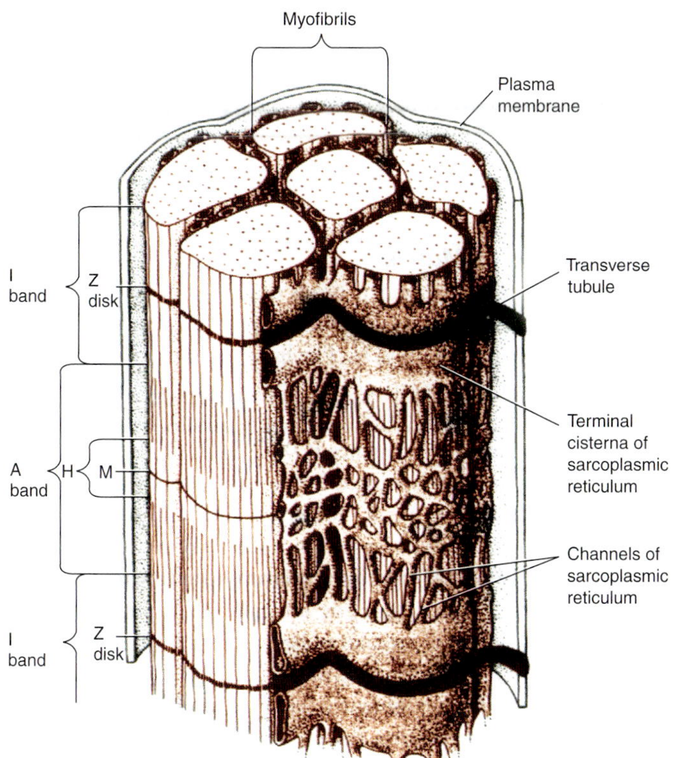

FIGURE 24.30

A schematic representation of a bundle of six myofibrils. The lumen of the transverse tubules connects with the extracellular medium and enter the fibers at the Z-disc.

Reprinted with permission from Darnell, J., Lodish, H., and Baltimore, D. *Molecular Cell Biology*. New York: Scientific American Books, 1986, p. 827.

is a tandem series of muscle cells or units called **sarcomeres**. The muscle cell is multinucleated and is not capable of division. Most muscle cells survive for the life of the animal, but they can be replaced when lost or lengthened by fusion of **myoblast cells**.

A muscle cell is shown diagrammatically in Figure 24.30. Note that the myofibrils are surrounded by a membranous structure called the **sarcoplasmic reticulum**. At discrete intervals along the fasciculi and connected to the terminal cisterna of the sarcoplasmic reticulum are transverse tubules. These are connected to the external plasma membrane that surrounds the entire structure. Nuclei and mitochondria lie just inside the plasma membrane.

The single contractile unit, the sarcomere, consists of all structural components between Z bands (or Z-disc) (Figures 24.29*d* and 24.30). Bands in the sarcomere are due to the arrangement of specific proteins (Figure 24.29*e*). Two types of fibers are apparent: Long thick fibers with protrusions on both ends are anchored at the center of the sarcomere, and long thin fibers are attached to proteins within the **Z-disc. I bands (isotropic)** extend for a short distance on both sides of the Z-disc (two adjacent sarcomeres) and contain only **thin filaments**. The **H band** is in the center of the sarcomere and contains thick or heavy filaments but no thin filaments. In the middle of the H band, there is a somewhat diffuse **M band** that contains additional proteins that assist in anchoring fibers of the **heavy filaments** (Figure 24.29*h*) to the center of the sarcomere. The **A band (anisotropic)** contains both thin and heavy filaments and is located between the inner edges of the I bands. When muscle contracts, the H and I bands shorten, but the distance between the Z-disc and the near edge of the H band remains constant. The distance between the innermost edges of I bands on both ends of the sarcomere also remains constant, indicating that the length of thin filaments and of thick filaments does not change during contraction. Contraction, therefore, results when these filaments "slide" past each other.

Sarcomeres consist of many different proteins, eight of which are listed in Table 24.7. The two most abundant are myosin and actin. Approximately 60–70% of the muscle protein is myosin and 20–25% is actin. Thick filaments consist mostly of myosin, and thin filaments consist mostly of actin. Thin filaments are formed from end-to-end polymerization of globular actin subunits, monomers, and referred to as

TABLE 24.7 Approximate Molecular Masses (kDa) of Some Skeletal Muscle Proteins

Protein	kDa
Myosin	500
Heavy chain	200
Light chain	20
Actin monomer (G-actin)	42
Tropomyosin	70
Troponin	76
Tn-C subunit	18
Tn-I subunit	23
Tn-T subunit	37
α-Actinin	200
C-protein	150
β-Actinin	60
M-protein	100

CLINICAL CORRELATION 24.9

Dilated Cardiomyopathy and Mutations in Actin

Thin walls of the heart and inability to pump blood effectively are characteristics of dilated cardiomyopathy. A genetic cause of this condition can be defective actin. Although the amino acid sequence of actin is highly conserved, there are several isoforms expressed by different genes; five of six are expressed in skeletal and cardiac myocytes. In adult cardiac myocytes, approximately 80% of actin results from expression of specific cardiac isoforms. Mutations in the invariant regions of these cardiac isoforms can result in serious health effects. An inherited single amino acid substitution (see below), Arg312His in one patient and Glu361Gly in another patient, resulted in dilated cardiomyopathy, a condition in which cardiac transplant is the only definitive treatment for end-stage disease.

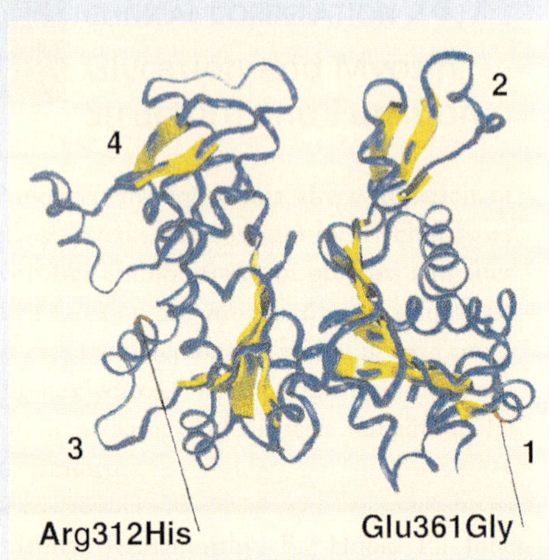

Schematic representation of the cardiac actin monomer and location of idiopathic dilated cardiomyopathic mutations.

From Olson et al. (1998).

Comparison of the positions of those substitutions with the structure of actin shown in Figure 24.29 reveals that they occur in two different functional regions of actin.

Source: Olson, T. M., Michels, V. V., Thibodeau, S. N., Tai, Y.-S., and Keating, M. T. Actin mutations in dilated cardiomyopathy, a heritable form of heart failure. *Science* 280:750, 1998. Copyright (1998) by AAAS.

substituted (see Clin. Corr. 24.9). A significant number of these substitutions occur at the N-terminal, where essentially all actin molecules are posttranslationally modified. The N-terminal amino acid residue may be acetylated and hydrolyzed one or two times before the final acetylated product is achieved. A crystal structure of G-actin is shown in Figure 24.32. Actin has two distinct domains of approximately equal size, however, one has been designated large (left) and the other small (right). Each of them consists of two subdomains. The N-terminal and the C-terminal residues are located within subdomain 1 of the small domain. The molecule has polarity, and aggregation to **F-actin** can occur from either end. *In vitro* kinetic data indicate that the preferred direction of aggregation is by addition of monomers to the large end of the molecule and that the rate of addition to this end is diffusion controlled—that is, at a rate as fast as the monomer can diffuse to that end.

Each G-actin contains a specific binding site, between the two major domains, for ATP and a divalent metal ion, Mg^{2+}, but Ca^{2+} competes with Mg^{2+} for the same binding site. The **G-actin–ATP–Mg^{2+} complex** aggregates faster to form **F-actin**. Subdomains 1 and 2 of G-actin molecules in F-actin are to the outside where myosin-binding sites are located. F-actin may be viewed as either (1) a single-start, left-handed single-stranded helix with rotation of the monomers through an approximate 166° with a rise of 27.5 Å or (2) a two-start, right-handed double-stranded helix with a half pitch of 350–380 Å.

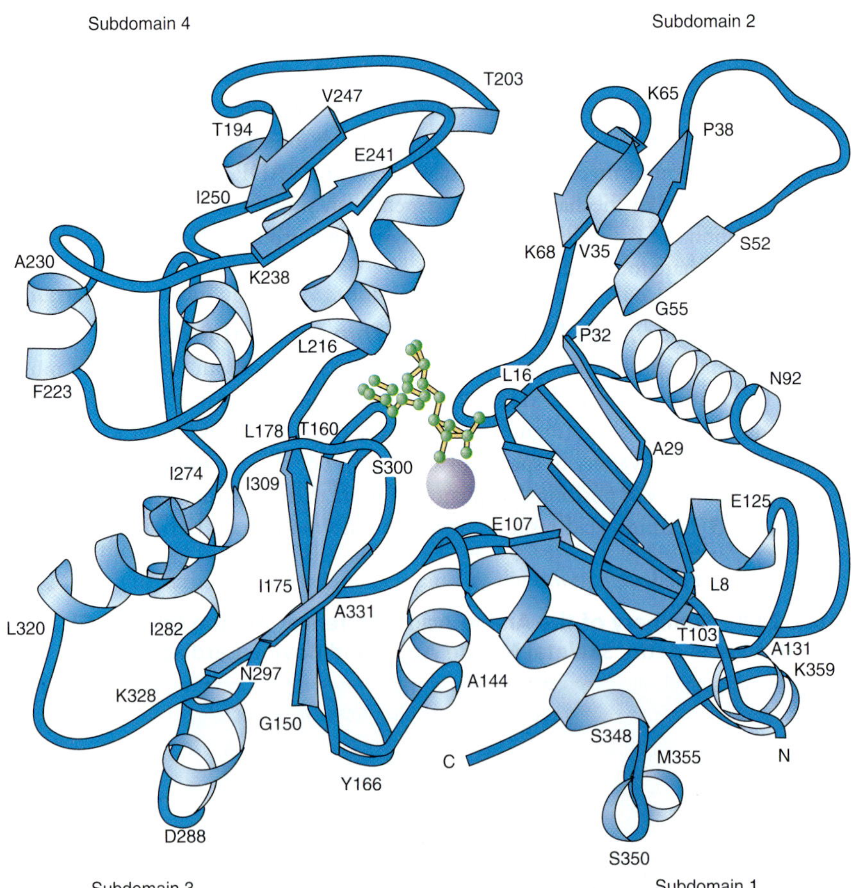

FIGURE 24.32

Secondary structural elements of G-actin crystal structure. ADP and a divalent metal ion are shown in the cleft between the two large domains.

Reproduced with permission from Lorenz, M., Popp, D., and Holmes, K. C. *J. Mol. Biol.* 234:826, 1993. Copyright (1993) Academic Press Limited, London.

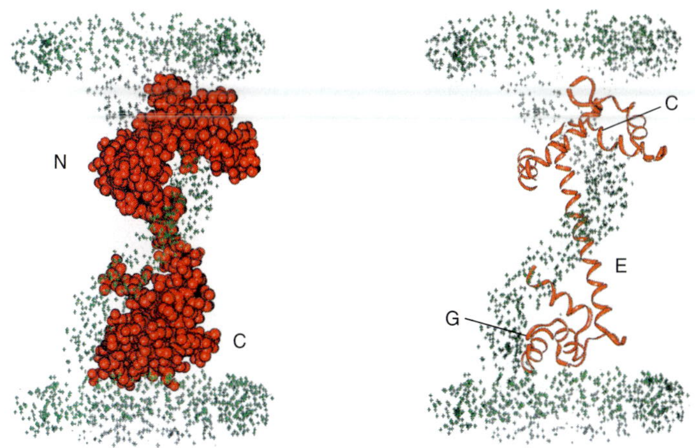

FIGURE 24.33

Best fit model for the $4Ca^{2+}$–TnC–TnI complex. A model for the $4Ca^{2+}$–TnC–TnI complex based on neutron scattering studies with deuterium labeling and contrast variation (Olah, C. C. and Trewhella, J., *Biochemistry* 33:12800, 1994). (Right) A view showing the spiral path of TnI (green crosses) winding around the $4Ca^{2+}$–TnC that is represented by an α-carbon backbone trace (red ribbon) with the C, E, and G helices labeled. (Left) The same view with $4Ca^{2+}$–TnC represented as a CPK model.
Photograph generously supplied by Dr. J. Trewhella. The publisher recognizes that the U.S. Government retains a nonexclusive, royalty-free license to publish or reproduce the published form of this contribution or to allow others to do so, for U.S. Government purposes.

β-Actinin binds to F-actin and assists in limiting the length of the thin filaments. **α-Actinin**, a homodimer of 90- to110-kDa subunits, binds adjacent actin monomers of F-actin at positions 86–117 of one and 350–375 of the other and strengthens the fiber. It also helps to anchor the actin filament to the Z-disc. Other major proteins associated with the thin filament are **tropomyosin** and **troponin**.

Tropomyosin is a rod-shaped protein consisting of two dissimilar subunits, each of about 35 kDa. It forms aggregates in a head-to-tail configuration. It interacts in a flexible manner with the thin filament throughout its entire length. It fits within the groove of the helical assembly of the actin monomers of F-actin. Each tropomyosin molecule interacts with about seven actin monomers between subdomain one and three. Tropomyosin helps to stabilize the thin filament and to transmit signals for conformational change to other components of the thin filament when Ca^{2+} binds to troponin, which is attached to tropomyosin.

Troponin consists of three dissimilar subunits designated TnC, TnI, and TnT with molecular mass of about 18 kDa, 23 kDa, and 37 kDa, respectively. The TnT subunit binds to tropomyosin. The TnI subunit inhibits binding of actin to myosin. The TnC subunit is a calmodulin like protein that binds Ca^{2+}.

The three-dimensional structure of TnC shows it to be dumbbell-shaped and very similar to calmodulin. A structural model of the calcium-saturated TnC–TnI complex is shown in Figure 24.33. The TnI subunit fits around the central region of TnC as a helical coil and forms caps over it at each end. The cap regions of TnI are in close contact with TnC when TnC is fully saturated with Ca^{2+}. TnC contains four divalent metal ion binding sites. Two are in the C-terminal region, have high affinity for calcium ions (K_D of about 10^{-7} M), and are presumed to be always occupied by divalent metal ions (Ca^{2+} or Mg^{2+}) since the concentrations of these ions in resting cells are within the same order of magnitude as the K_D. Under resting conditions, TnI has a conformation that prevents proper orientation of tropomyosin and inhibits myosin binding to actin, thus preventing contraction. Upon excitation, the calcium ion concentration increases to about 10^{-5} M, high enough to bind to sites within the N-terminal region of TnC. TnI now binds preferentially to TnC in a capped structural conformation as shown in Figure 24.33. Myosin-binding sites on actin become exposed. The relatively loose interaction of tropomyosin with actin permits it the flexibility to alter its conformation as a function of calcium ion concentration and to assist in blockage of the myosin binding sites on actin (see Clin. Corr. 24.10).

Figure 24.29*i* shows a schematic cross section of a sarcomere and the relative arrangement of thin and thick filaments. Six thin filaments surround each thick filament. The arrangement and flexibility of myosin head groups make it possible for each thick filament to interact with multiple thin filaments. When cross-bridges form between thick and thin filaments, they do so in patterns consistent with that shown in the colorized electron micrograph of Figure 24.34. This shows a two-dimensional view of thick filaments interacting with two thin filaments.

Myosin

Actin

FIGURE 24.34

A colorized electron micrograph of actin–myosin cross-bridges in a striated insect flight muscle. Reproduced with permission from Darnell, J., Lodish, H., and Baltimore, D. *Molecular Cell Biology*. New York: Scientific American Books, 1986.

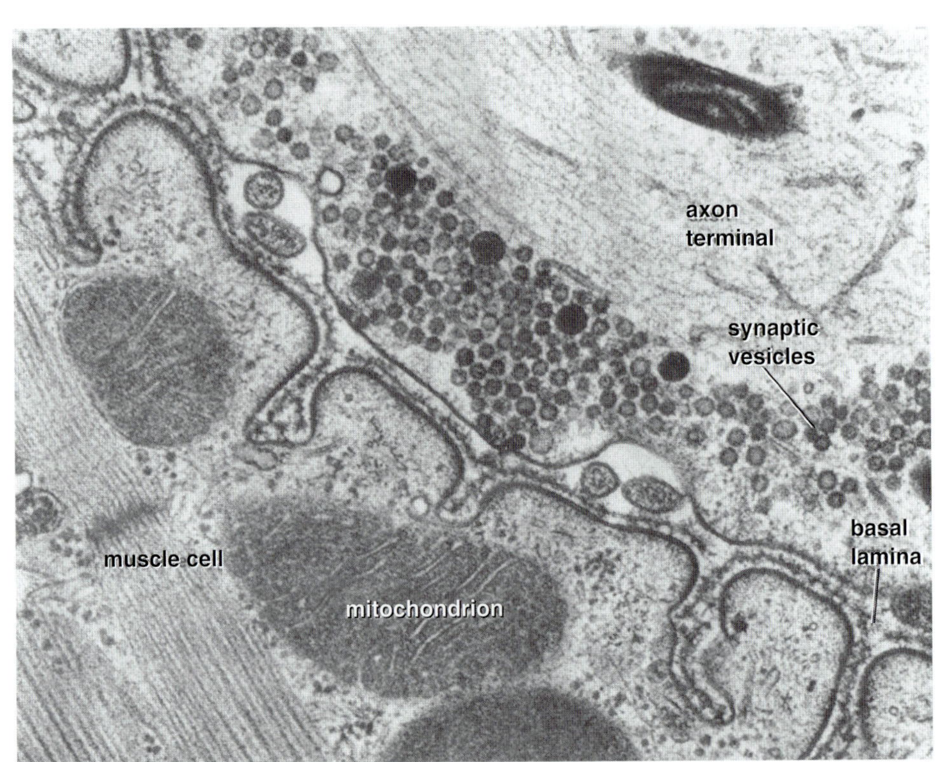

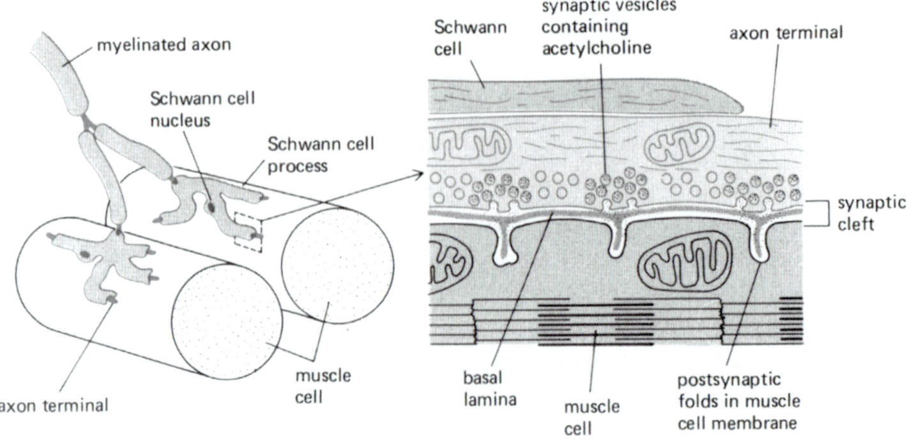

FIGURE 24.35

(*a*) Electron micrograph of a neuromuscular junction. (*b*) Schematic diagram of the neuromuscular junction shown in (*a*). Reproduced with permission from Alberts, B., Bray, D., Lewis, J., Raff, M., Roberts, K., and Watson, J. *Molecular Biology of the Cell*. New York: Garland, 1983.

Muscle Contraction Requires Ca^{2+}

Contraction of skeletal muscle is initiated by transmission of nerve impulses across the **neuromuscular junction** mediated by release into the synaptic cleft of the neurotransmitter **acetylcholine. Acetylcholine receptors** are associated with the plasma membrane on the postsynaptic membrane and are ligand-gated. Binding of acetylcholine causes them to open and to permit Na^+ to enter and to cause depolarization of the transverse tubules of the sarcomere. Figure 24.35 is a representation of the anatomical relationship between the presynaptic nerve and the sarcomere.

Transverse tubules in the vicinity of Z lines are connected to terminal cisternae of the sarcoplasmic reticulum. Nerve impulses lead to depolarization of the transverse tubules, which signals the sarcoplasmic reticulum to release Ca^{2+} into the sarcomere. Ca^{2+} concentration increases about 100-fold, permitting it to bind to the low-affinity sites of TnC and to initiate the contraction process (see Clin. Corrs. 24.10, 24.11, and 24.12 for cardiac muscle events). When contraction is no longer necessary, the Ca^{2+}-transporting ATPase rapidly pumps Ca^{2+} back into the sarcoplasmic reticulum.

Energy Reservoirs for Muscle Contraction

In normal muscle, the concentration of ATP remains fairly constant even during strenuous activity because of increased metabolic activity (see section below on NO) and the actions of **creatine phosphokinase** and **adenylate kinase**. Creatine phosphokinase catalyzes transfer of phosphate from phosphocreatine to ADP in an energetically favored manner.

$$\text{Phosphocreatine} + \text{ADP} \rightleftharpoons \text{ATP} + \text{Creatine}$$

If the metabolic activity is insufficient to keep up with the need for ATP, the creatine phosphokinase helps to maintain cellular levels relatively constant. Adenylate kinase provides additional ATP by catalyzing the reaction

$$2\text{ADP} \rightleftharpoons \text{ATP} + \text{AMP}$$

The most obvious pathological consequence of ATP depletion is development of a state of **rigor**. The effects of ATP depletion are as follows: (1) Intracellular Ca^{2+}

CLINICAL CORRELATION 24.10
Troponin Subunits as Markers for Myocardial Infarction

Troponin contains three subunits (TnT, TnI, and TnC) and each is expressed by more than one gene. Two genes code for skeletal muscle TnI, one for fast- and one for slow-skeletal muscle; and one gene codes for cardiac muscle TnI. The genes that code for TnT have that same distribution pattern. The gene for the cardiac form of TnI appears to be specific for heart tissue. TnC is encoded by two genes, but neither gene appears to be expressed only in cardiac tissue.

The cardiac form of TnI in humans is about 31 amino acids longer than the skeletal muscle form. Serum levels of TnI increase within four hours of an acute myocardial infarction and remain high for about seven days in about 68% of patients tested. Almost 25% of one group of patients showed a slight increase in the cardiac form of TnI after acute skeletal muscle injury indicating that it would not be a good and sensitive test for myocardial infarction.

Two isoforms of cardiac TnT, TnT_1 and TnT_2, are present in adult human cardiac tissue and two more isoforms are present in fetal heart tissue. These isoforms may result from alternative splicing of mRNA. Serum levels of TnT_2 increase within four hours of acute myocardial infarction and remain high for up to fourteen days. TnT_2 in serum is 100% sensitive and 95% specific for detection of myocardial infarction. TnT_2 assay is used for detection of acute myocardial infarction. Myocardial infarcts are either undiagnosed or misdiagnosed in hospital patients admitted for other causes, or in 5 million or more people who go to doctors for episodes of chest pain. It is believed that this test is sufficiently specific to diagnose myocardial incidents and to help direct doctors to proper treatment of these individuals.

Source: Anderson, P. A. W., Malouf, N. N., Oakeley, A. E., Pagani, E.D., and Allen, P.D. Troponin T isoform expression in humans. *Circ. Res.* 69:1226, 1991. Ottlinger, M. E. and Sacks, D. B. *Clin. Lab. News*, 33, 1994.

CLINICAL CORRELATION 24.11
Voltage-Gated Ion Channelopathies

There are three important types of voltage-gated cation channels: Na^+, Ca^{2+}, and K^+. Each is a heterogeneous protein consisting of various numbers of α- and β-subunits. In membranes, they are arranged in a more-or-less circular manner with a funnel-like channel formed through the middle of α-subunits. Roles of β-subunits are still being elucidated, but they appear to help stabilize and/or regulate activity of α-subunits.

Voltage-gated channels from nerve and muscle tissue show high sequence homology in many transmembrane domains, but are less conserved in the intracellular connecting loops. A common effect of mutations in Na^+ channels is muscle weakness or paralysis. Some inherited Na^+ voltage-gated ion channelopathies are listed below. Each of these is reported to result from a single amino acid change in the α subunit and is transmitted as autosomal dominant.

Disorder	Unique Clinical Feature
Hyperkalemic periodic paralysis	Induced by rest after exercise, or the intake of K^+
Paramyotonia congenita	Cold-induced myotonia
Sodium channel myotonia	Constant myotonia

It has been suggested that if the membrane potential is slightly more positive (i.e., changes from −70 to −60 mV) the myofiber can reach the threshold more easily and the muscle becomes hyperexcitable. If the membrane potential becomes even more positive (i.e., up to −40 mV), the fiber cannot fire an action potential leading to paralysis.

Source: Catterall, W. A. Structure and function of voltage-gated ion channels. Annu. Rev. Biochem. 64:493, 1995; Hoffmann, E. P. Voltage-gated ion channelopathies: Inherited disorders caused by abnormal sodium, chloride, and calcium regulation in skeletal muscle. *Annu. Rev. Med.* 46:431, 1995. Abraham, M. R., Jahangir, A., Alekseev, A. E., and Terzic, A. Channelopathies of inwardly rectifying potassium channels. *FASEB J.* 13:1901, 1999.

CLINICAL CORRELATION 24.12
Ion Channels and Cardiac Muscle Disease

Voltage-gated ion channels in cardiac muscles, like those of other tissues, require a finite recovery time after excitation. The heart contracts and relaxes on a continuous basis and in a rhythmic manner that cannot be altered significantly without causing problems such as arrhythmias, fibrillation, and possibly death. The recovery time between contractions of cardiac muscle is measured on electrocardiograms as the QT interval. Initiating the excitation phase is the opening of Na^+ channels for a finite period. Na^+ moves into the cell, causing depolarization. K^+ channels then open to permit K^+ to move out of the cell—both flowing down their respective chemical concentration gradients. Opening of the K^+ channels is important for shutting off the action potential. Inherited conditions referred to as long QT syndrome or LQTS, have been linked to genes that affect K^+ channels (KVLQT1, HERG, mink) and Na^+ channels (SCN5A). Defects in these channels in cardiac muscle can cause sudden death—particularly in young people who are physically active, have never had an electrocardiogram, and have no knowledge that they have LQTS. Prevalence of this condition is approximately 1 in 10,000.

Source: Balser, J. R. Structure and function of the cardiac sodium channels. *Cardiovasc. Res.* 42:327, 1999. Ackerman, M. J., Schroeder, J. J., Berry, R., Schaid, D. J., Porter, C. J., Michels, V. V., and Thibodeau, S. N. A novel mutation in KVLQT1 is the molecular basis of inherited long QT syndrome in a near-drowning patient's family. *Pediatr. Res.* 44:148, 1998. Barinaga, M. Tracking down mutations that can stop the heart. *Science* 281:32, 1998.

concentration is no longer controlled and (2) myosin will exist exclusively in the myosin–ADP complex and bound to actin, a condition called **rigor mortis**. Recall that ATP is required for dissociation of the actin–myosin complex.

Model for Skeletal Muscle Contraction: The Power Stroke

Myosin head groups undergo conformational changes upon binding of ATP, hydrolysis of ATP, and release of products. ATP binding leads to closure of the active site cleft and opening of the cleft in the region of the actin binding site. Hydrolysis of ATP results in

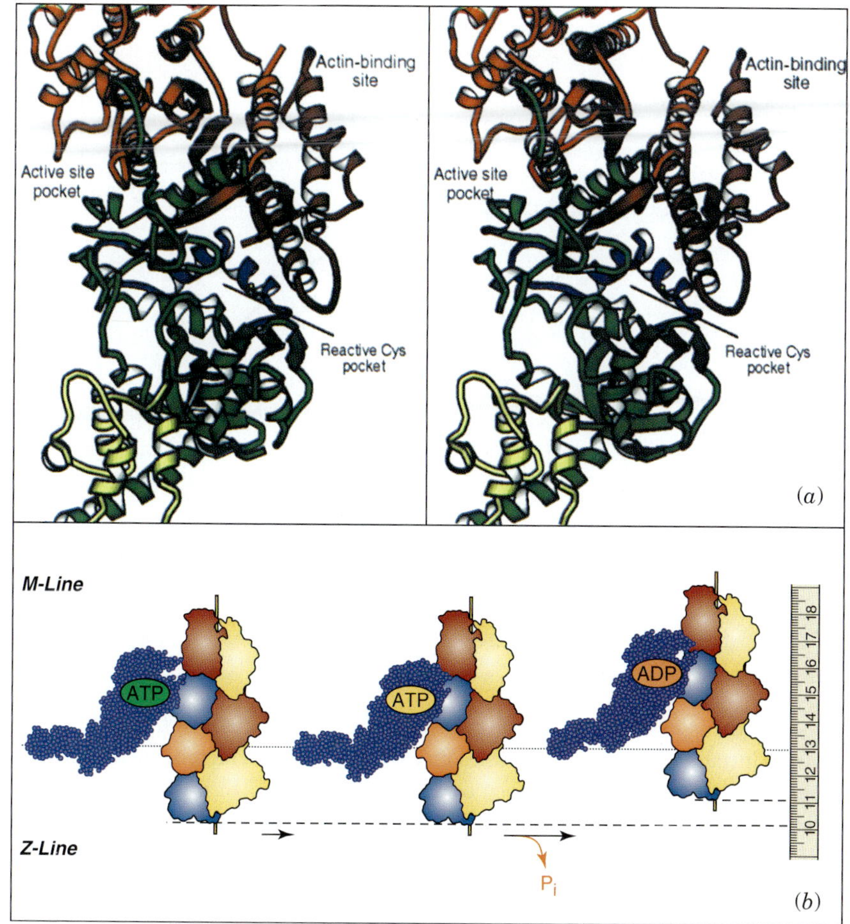

FIGURE 24.36

A model of actin–myosin interaction. (*a*) Stereo view of myosin showing the pocket that contains the mobile "reactive" cysteine residues. (*b*) A schematic diagram showing how conformation changes in the myosin head group and arm can pull actin filaments in a direction away from the Z-disc during hydrolysis of ATP and release of inorganic phosphate. Part (*b*) reprinted with permission from Rayment, I. and. Holden, H. M. The three dimensional structure of a molecular motor. *Trends Biol. Sci.* 19:129, 1994. Copyright (1994) Elsevier.

closure of the cleft in the actin-binding region, which opens again upon or for release of inorganic phosphate. This conformational change is evident by the movement of two cysteine-containing helices. The distance between Cys^{697} and Cys^{707} changes from about 19 Å to about 2 Å. Experimentally, cross-linking of these two cysteines in the closed position traps ADP within its binding site. A stereo view of myosin showing the reactive cysteine pocket (cysteine residues that can react chemically with externally added cysteine reactive agents) is shown in Figure 24.36*a*. Conformational changes in myosin that occur in conjunction with closing and opening of the cleft in the actin-binding region with release of inorganic phosphate (Pi) during ATP hydrolysis are associated with the **power stroke** Figure 24.36*b*. Actin filaments can be moved as much as 10 nm during this process. The open conformation of myosin is now in a position to bind ATP and start the cycle over again.

Individual myosin units functioning in an asynchronous manner, possibly like changes in the position of hands on a rope in the game of tug-of-war, can maintain force or create maximum force. Thus, when some myosin head groups bind with high affinity, others have low affinity and this binding, pulling, release, binding action of myosin shortens the sarcomere in an appropriately forceful manner.

Smooth Muscle Contraction: Calcium Regulation

Calcium ions play an important role in smooth muscle contraction. A mechanism for calcium regulation of smooth muscle contraction is shown in Figure 24.37. Key elements of this mechanism are as follows: (1) A phosphorylated myosin light chain stimulates myosin Mg^{2+}-ATPase, which supplies energy for the contractile process. (2) Myosin light chain is phosphorylated by a **myosin light-chain kinase (MLCK).** (3) A Ca^{2+}–calmodulin (Ca^{2+}–CaM) complex activates MLCK. (4) Formation of the

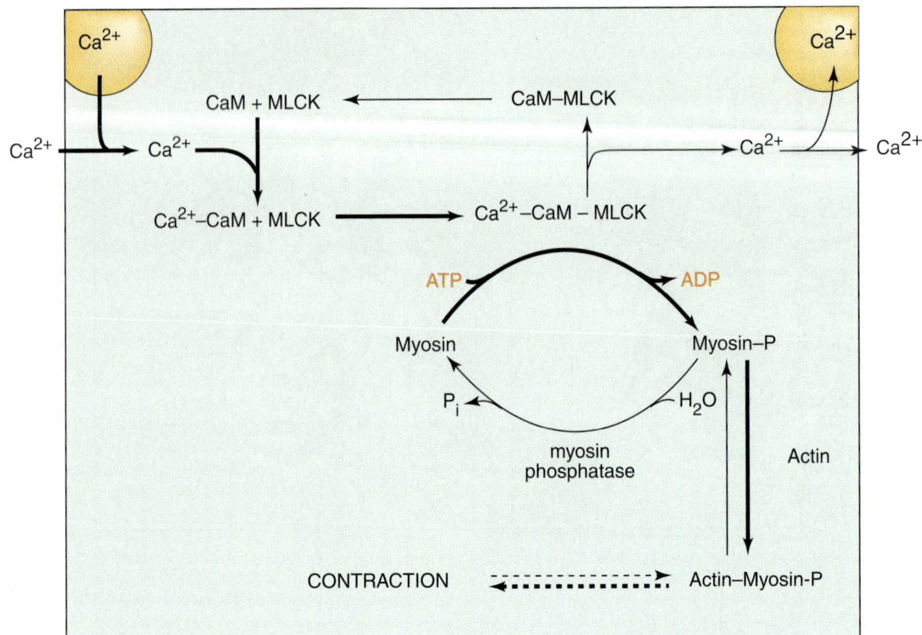

FIGURE 24.37

A schematic representation of the mechanism of regulation of smooth muscle contraction. Heavy arrows show the pathway for tension development and light arrows show the pathway for release of tension. Mg^{2+}-ATPase activity is highest in the actin–myosin-P complex. Ca^{2+}–CaM, Ca^{2+}–calmodulin; MLCK, myosin light-chain kinase.

Adapted from Kramm, K. E. and Stull, J. T. *Annu. Rev. Pharmacol. Toxicol.* 25:593, 1985.

Ca^{2+}–CaM complex is dependent on the concentration of intracellular Ca^{2+}. Release of Ca^{2+} from intracellular stores or increased flux across the plasma membrane is important for control. (5) Contraction is stopped or decreased by a **myosin phosphatase** and/or transport of Ca^{2+} out of the cell. It is apparent that many more biochemical steps are involved in the regulation of smooth muscle contraction, steps that can be regulated by hormones and other agents such as NO and cGMP. These diverse interactions give smooth muscles the ability to develop various degrees of tension and to retain it for prolonged periods.

Involvement of Nitric Oxide and Carbon Monoxide in Muscle Contraction

Other molecules involved in control of muscle contraction and neuronal signaling are nitric oxide (NO) (see p. 432) and possibly carbon monoxide (CO). Evidence for CO control is circumstantial, a significant portion of which is based on observations of effects on these processes in animals in which the gene that codes for the enzyme that makes CO in mammals has been "knocked out." Obtaining definitive proof of the role of CO *in vivo* may be difficult since it is currently known to be produced in mammalian cells only by heme oxygenase. Heme oxygenase and soluble guanylate cyclase (sGC) are found in some of the same regions in brain and other nerve tissues. Heme oxygenase catalyzes the oxygenation of iron protoporphyrin IX to biliverdin, Fe^{2+}, and CO (see p. 840). It is not clear how available the substrate is for this enzyme in those tissues and what signaling mechanism exists for regulating production of CO and for stopping its action. CO does affect Ca^{2+} ligand-gated K^+ channels, causing hyperpolarization, which, in turn, affects voltage-gated Ca^{2+} channels. It does not appear to have significant effect on intracellular Ca^{2+} concentration.

Effects of NO on molecular motor activity relative to muscle contraction are clearer. NO is synthesized from arginine as primary substrate by nitric oxide synthase (NOS) near the site of its primary target. This is necessary because it is a gas and has a short half-life in tissues. It can react nonenzymatically with active oxygen species, free —SH groups, and metal ions. There are two types of NOS: constitutive and inducible (see p. 433). The inducible form (iNOS) is generally expressed because of an inflammatory cytokine response. The constitutive form consists of two subclasses: nNOS found in most nerve tissue and eNOS, an endothelial form. nNOS is associated with a postsynaptic density protein (PSD-95) and/or the *N*-methyl-D-aspartate (NMDA) receptor in the neuronal membrane; eNOS is associated with a scaffolding protein, **caveolin**, a major constituent of caveolae. **Caveolae** is a specific plasma membrane structure and/or system for compartmentalization of signal transduction components. Caveolae is characterized biochemically by having a high concentration of caveolin and large amounts of glycosphingolipids, cholesterol, and lipid-anchored membrane proteins (see p. 457). Caveolin is found in all of the functional regions of the plasma membrane and is expressed in humans by three genes yielding three isoforms.

Activity of nNOS and eNOS is suppressed when bound to these proteins. nNOS dissociates from PSD-95 upon Ca^{2+} entry, which facilitates nNOS interaction with calmodulin and other proteins such as dynamin. It also dissociates upon stimulation of the NDMA receptor. nNOS is constitutively phosphorylated when bound to scaffolding proteins. Upon release, it is dephosphorylated by a Ca^{2+}-dependent phosphatase, **calcineurin**, which further stimulates its activity.

eNOS, upon its release from its scaffolding protein, caveolin, interacts with other proteins like heat shock protein (HSP-90), dynamin, or others that lead to an increase in its activity. In humans it is activated by phosphorylation by a serine/threonine protein kinase on a serine at position 1177. There are at least five phosphorylation sites on eNOS, including a tyrosine site. The site that shows the greatest affinity for phosphorylation, as well as what the ultimate effect on eNOS activity is, depends upon many physiological factors, which is not a subject for this chapter.

A primary target for NO from either source is sGC whose activity is stimulated several hundredfold upon interaction with NO. This leads to an increased concentration of cGMP in smooth muscle and has a role in mediating a relaxation response. In skeletal muscle, it affects contractility through activity at the sarcoplasmic reticulum, activating voltage-gated Ca^{2+} release channels. NO also influences glucose uptake, independent of insulin, in skeletal muscle during strenuous exercise, especially in fast-twitch fibers. In cardiac muscle, NO mediates some of its effects as a result of stretching and tension created upon contraction.

Other Classes of Myosins and Molecular Motors

Eighteen classes of myosin have been identified in the human genome. They are generally written as Myosin followed by Roman numbers I–XVIII. Skeletal, cardiac, and smooth muscle myosins are all in Class II and have gene designations MYH1–MYH8 for either skeletal or cardiac, whereas smooth muscle is designated MYH11. There are also some nonmuscle myosins in Class II.

Unconventional Myosins and Their Functions

Analyses of the human genome suggest that there are more than 20–30 myosins and/or myosin-like genes. Functions for some of their products are known and are listed here because they may be implicated in some disease states (see Clin. Corr. 24.13).

Myosin I is a low-molecular-mass motor protein that is associated with cell membranes. It contains a tail section rich in basic amino acid residues and can associate with anionic phospholipids. There is a tail section domain rich in glycine, proline, and alanine (GPA domain) that facilitates binding to actin. The tail also contains an src homology 3 (SH3) domain that mediates protein–protein interactions. SH3 domains select peptides sharing the consensus motif LXXRPLXΨP, where Ψ is an aliphatic residue. This class of myosins appears to be involved in membrane–membrane interactions and to function in brush boarders of the small intestines. They may also be involved in transport/movement of Golgi-derived vesicles. They have several IQ motifs in the neck region near the head, suggesting interactions with calmodulin-like molecules and regulation by them.

Myosin V is involved in organelle transport such as synaptic vesicles, melanosomes, vacuoles, and mRNA. It has an extended neck region and a tail with globular ends that permit dimerization but not filament formation. Light chains of myosin V are of the EF hand family and bind to an altered repeat IQ motif (IQXXXRGXXXR). Its cargo probably binds to the tail region. It is found in all brain neurons and

CLINICAL CORRELATION **24.13**

Mutations Affecting Pigmentation: Is There a Molecular Motor (Myosin V) Connection?

Elejalde's syndrome is characterized by silvery hair and severe dysfunction of the central nervous system. Large granules of melanin are unevenly distributed in the hair shaft and abnormal melanocytes and melanosomes may be present in fibroblasts. These observations suggest that mechanisms for their movement and distribution might be defective.

Chediak–Higashi syndrome is characterized by enlarged lysosomes, melanosomes, and other cytoplasmic granules. Eyes of humans and animals with this condition are hypopigmented. Intraocular melanin granules vary in size and can become extremely large. Some of the manifestations of this condition suggest similarity in mechanism with the condition described above.

Source: Elejalde, B. R., Holguin, J., Valencia, A. Gilbert, E. F., Molina, J., Marin, G., Arango, L. A. Mutations affecting pigmentation in man: 1. Neuroectodermal melanolysosomal disease. *Am. J. Med Genet.* 3:65, 1979. Duran-McKinster, C. Rodriguez-Jurado, R. Ridaura, C., de la Luz Orozco-Covarrubias, M., Tamayo, L., Ruiz-Maldonando, R. Elejalde syndrome—a melanolysosomal neurocutaneous syndrome: Clinical and morphological findings in 7 patients. *Arch. Dermatol.* 136:120, 2000. Mottonen, M., Lanning, M., Baumann, P., Saarinen-Pihkala, U. M. Chediak–Higashi syndrome: Four cases from northern Finland. *Acta Paediatr.* 92:1047, 2003.

Kinesin-1

Tail, ~150aa, fanlike by EM
Heterotetramer
LC
Folded conformation inhibits motor ATPase
LC
Central α-helical coiled coil stalk
Motor domain, ~350 aa, globular

Kinesin-2

KRP85
KRP95
Heterotrimer
NonKRP subunit

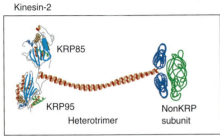

Kinesin-14

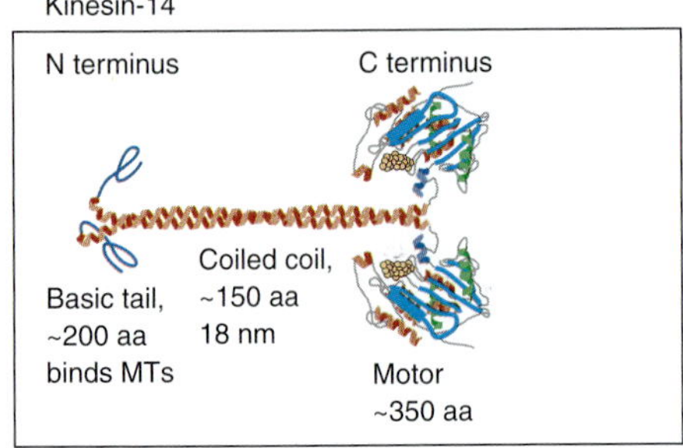

FIGURE 24.38

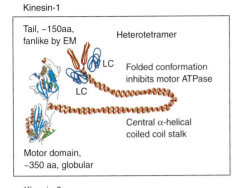

Kinesin structures. Representative structures of Kinesin-1, Kinesin-2, and Kinesin-14. All structures were taken from the Kinesin Home Page and associated links, http://www.proweb.org/kinesin/index.html.

is required for melanosome transport. Mutations in myosin V are associated with neurological malfunctions and de-pigmentation, especially of hair. It shows processivity in its movement like some kinesins and takes several steps of about 36 nm along an actin fiber before it dissociates. Its movement is limited by ADP dissociation and ATP binding, which causes the trailing head of myosin V to dissociate from actin.

Myosin VI is expressed in most cells and tissues. The neck region is extended and it contains one calmodulin (IQ motif) binding region per head. It is a dimeric molecule with a coiled tail that has two (one each monomer) globular ends. Defects in the gene for this protein cause hearing problems attributable to defects in movement of the hairs in the inner ear. This myosin can move in either direction—unlike others discussed. Its mechanism of action and involvement in organelle movement are being explored.

Kinesins

Kinesins are microtubule-based molecular motors. They do not form filaments like some myosins, but they generate motion by conformational changes upon binding and hydrolysis of ATP in a similar manner to that of myosin actin-based motors. Their motor domains have a core structure similar to that of myosin and have a very high degree of secondary structural overlap, particularly in regions surrounding the catalytic domains. Several regions within the core have nearly complete overlapping secondary structural elements, but they are not contiguous and are separated by varying numbers of amino acid residues. The motor domains of kinesins are smaller than those of myosins. A major function of this class of motor proteins is to effect movement of intracellular cargo: vesicles, organelles, portions of the mitotic apparatus, chromosomes, mRNA, proteins, and other cellular constituents.

There are 14 classes of kinesins. The structures of some are known. Minimum structural elements of several kinesins are shown in Figure 24.38, which demonstrates some of the structural diversity of this class of molecular motors. They have been found in a variety of species, and some of their functions in humans have been inferred from analyses and mutations induced in them in other species of animals. In most cases, kinesins move their cargo toward the plus end of microtubules.

Kinesin-1 is a motor for fast axonal transport (see p. 960), Among the cargo it transports are vesicles containing membrane proteins such as those that make ion channels along the axon during development, synaptic vesicle proteins, and those with which they interact at the axon termini such as syntaxin and synaptotagmin. This kinesin is expressed ubiquitously in human neural tissue. Mutations in kinesin-1 may lead to neuronal defects in humans that are reflected by a variety of symptoms.

Kinesin-2 is involved with membrane-associated movements in axons, axonemes, and melanophores. It has been shown to be involved in dispersion of melanosomes in fish melanophores. It has not been detected in humans, but there are human diseases associated with the lack of dispersion and/or movement of melanin containing granules, which may be related more to defects in myosin-V in humans.

Kinesin-4 is associated with maintenance of spindle bipolarity and movement of chromosomes to the metaphore plate.

Kinesin-5 is associated with the mitotic apparatus of dividing cells and is involved in centrosomes and spindle pole body separation.

Kinesin-6 is present in the spindle mid-body during telophase and is presumed to mediate the sliding of spindle fibers in the late stages of anaphase that are needed for spindle elongation and separation of chromosomes.

Kinesin-7 is also chromosome associated and provides a direct link between the chromosomes and microtubules of the spindle.

Kinesin-13 is also a chromosomal kinesin that is probably involved in mitotic chromosome movement.

Kinesin-14 is among the minus-end motors like the Ncd mitotic kinesin in *Drosophila* that functions in early stages of mitosis. It is localized to spindles in oocytes.

Kinesins-1 and -2 are most related to material discussed in this chapter, whereas the others are associated with more general aspects of cell division and the associated movement of various components associated with that process.

Dynein

There are two classes of dynein motors: (a) axonemal, which function to effect flagella and cilia movement, and (b) cytoplasmic, which effect distribution and organization of cytoplasmic structures. These functions include protein sorting and movement; chromosome organization during various stages of its function; distribution and/or redistribution of organelles such as endosomes, lysosomes, and others; and retrograde axonal transport—that is, transportation of cargo in the opposite direction of most kinesins.

The structure of dynein is much more complex than that of the other two classes of motors. Dynein has a seven-member planar ring structure that is overall approximately 10 times the molecular mass of kinesins. A diagrammatic representation of its structure is shown in Figure 24.39.

ATP binds to an AAA motif in domain 1. Its binding and hydrolysis induce conformational changes that are transmitted through domains 2–4 to the stalk that interacts with microtubules and causes step movements of 24–32 nm for an unloaded dynein. Dynein movement responds to load in a downshift, gear-like manner and, under heavy load, takes steps of approximately 8 nm. Down shifting appears to be associated with conformation changes in several of its other domains and the availability of ATP. Under heavy load conditions, ATP also appears to bind AAA motifs in domain 3. AAA motifs are conserved regions of 220–230 amino acid residues that exist in a family of proteins that participate in a number of diverse cellular activities that depend on energy from ATP hydrolysis to affect their functions, which may include proteolysis, protein folding and unfolding, metal ion metabolism, and other activities in addition to those associated with dynein. The motif name AAA refers to "**A**TPase **A**ssociated with diverse cellular **A**ctivities."

Note that (1) dyneins as motors are structurally more complex than myosins or kinesins, (2) they are generally involved in retrograde movement of cellular material, (3) they are involved in various other aspects of structural organization, and (4) their step movement along the microtubule is load-dependent.

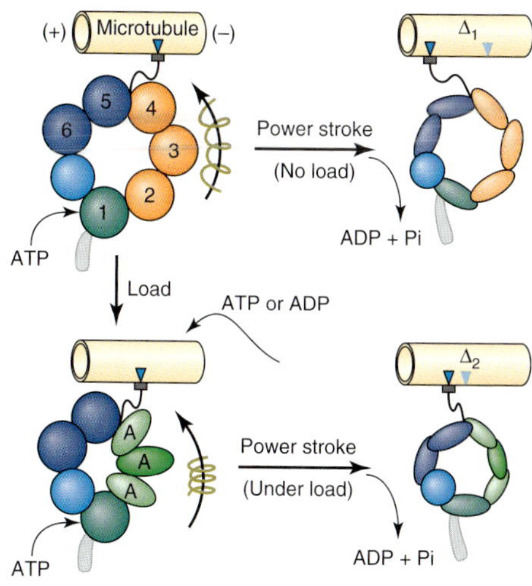

FIGURE 24.39

Dynein functions as a molecular motor.
Redrawn from Mallik, R., Carter, B. C., Lex, S. A., King, S. J., and Gross, S. P. Cytoplasmic dynein functions as a gear in response to load. *Nature* 427:649, 2004.

24.5 MECHANISM OF BLOOD COAGULATION

Blood circulation occurs in a very specialized type of closed system in which the volume of circulating fluid is maintained fairly constant. Multiple functions of the system make the transfer of solutes across its boundaries a necessary function. As in any system of pipes and tubes, leaks can occur from various types of insults that must be repaired in order to maintain a state of hemostasis, that is, no bleeding.

Biochemical Processes of Hemostasis

Hemostasis implies that the process of clot formation (**procoagulation**, designated as **Phase 1**) is in balance with processes for stopping clot formation (**anticoagulation, Phase 2**) and for clot dissolution (**fibrinolysis, Phase 3**). Procoagulation leads to production of fibrin from fibrinogen and aggregation into an insoluble network, or clot, which covers the ruptured area and prevents further loss of blood. Concomitantly, aggregation of blood platelets occurs at the site of injury. Platelet aggregation forms a physical plug to help stop the leak. Platelets also undergo morphological changes and release (a) some chemicals that aid in other aspects of the overall process such as vasoconstriction to reduce blood flow to the area and (b) enzymes that aid directly in clot formation. The steps involved in the initial phase of this process are shown schematically in the following diagram.

Extrinsic Pathway and Initiation of Coagulation

FIII or **tissue factor (TF)** is the membrane receptor that initiates the procoagulation phase. TF is a 263-amino-acid transmembrane protein that is exposed upon rupture of the endothelium of blood vessels and forms the receptor to which FVII binds. Residues 1–219 are on the extracellular side of the membrane and are exposed upon injury. FVII is a **γ-carboxyglutamyl** (Gla)-containing protein that binds to tissue factor only in the presence of Ca^{2+}. The resulting complex (FVII–Ca^{2+}–TF) is the initial enzyme complex of the extrinsic pathway, and its action initiates blood clotting. **TF** and **FVII** are unique to the extrinsic pathway and are essentially all of its major components. The zymogen form of FVII is initially activated through protein–protein interaction resulting from its binding to TF. Additional FVII is activated by thrombin through specific proteolytic cleavage. FVIIa has a long half-life in circulating blood, and a small amount normally exists. This has no adverse effects since FVIIa is not catalytically active once it dissociates from TF.

CLINICAL CORRELATION **24.14**

Intrinsic Pathway Defects: Prekallikrein Deficiency

Components of the intrinsic pathway include factor XII (Hageman factor), factor XI, prekallikrein (Fletcher factor), and high-molecular-weight kininogen. Inherited disorders in each appear to be autosomal recessive and associated with an increase in activated partial thromboplastin time (APTT). Factor XI deficiency is directly associated with a clinical bleeding disorder.

In prekallikrein (Fletcher factor) deficiency, autocorrection after prolongation of the preincubation phase of the APTT test occurs. This is explained by activation of factor XII by an autocatalytic mechanism. The reaction is very slow in prekallikrein deficiency since the rapid reciprocal autoactivation between factor XII and prekallikrein cannot take place. Prekallikrein deficiency may be due to (a) a decrease in the amount of the protein synthesized, (b) a genetic alteration in the protein itself that interferes with its ability to be activated, or (c) its ability to activate factor XII. A lack of knowledge of the structure of the gene or the protein will not permit an explanation of the mechanism of this deficiency. Specific deficiencies of the intrinsic pathway, however, can be localized to a specific factor if the appropriate tests are performed. These may include measurement of the amount of each factor in plasma and an APTT test performed with and without prolonged preincubation time. In a 9-year-old girl with prolonged APTT, the functional level of prekallikrein was less than 1/50th of the minimum normal value. An immunological test (ELISA, see p. 408) showed an antigen level of 20–25%, suggesting that she was synthesizing a dysfunctional molecule.

Source: Coleman, R. W., Rao, A. K., and Rubin, R. N. Fletcher factor deficiency in a 9-year-old girl: Mechanisms of the contact pathway of blood coagulation. *Am. J. Hematol.* 48:273, 1995.

Thrombin Formation

FXa and FV form a complex, referred to as prothrombinase, that catalyzes formation of thrombin from prothrombin by a proteolytic cleavage reaction. The relatively small amount of thrombin formed in this initiation phase (Figure 24.40) catalyzes the activation of FV, FVII, FVIII, and FXIII in addition to the formation of fibrin from fibrinogen.

Formation of thrombin is key to sustaining and accelerating the extrinsic pathway and activating one of the major protein cofactors, FVIII, that is required for the intrinsic pathway.

Reactions of the Intrinsic Pathway

Reactions of the intrinsic pathway are also shown in Figure 24.40. Injury to the endothelial lining of blood vessels exposes anionic membrane surfaces. The zymogen, **FXII**, binds directly to some of these anionic surfaces and undergoes a conformational change that increases its catalytic activity 10^4- to 10^5-fold. **Prekallikrein**, **FXI**, and zymogens circulate in blood as complexes with **high-molecular-weight kininogen (HMWK)**, either as a FXI–HMWK complex or as a prekallikrein–HMWK complex. FXI and prekallikrein are attached to anionic sites of exposed membrane surfaces through their interactions with HMWK. This brings those zymogens to the site of injury and in direct proximity to FXII. The membrane-bound "activated" form of FXII activates prekallikrein to **kallikrein**, which then activates FXII by specific proteolytic cleavage to give FXIIa. FXI, in the HMWK-membrane-bound complex, is activated by FXIIa by proteolytic cleavage to give FXIa. FXIa activates **FIX** to FIXa (see Clin. Corr. 24.14), which, in the presence of **FVIIIa**, also activates **FX** to FXa. This is essentially a four-step cascade started by the "contact" activation of FXII and the autocatalytic activation of FXII and kallikrein to give FXIIa (step 1). FXIIa activates FXI (step 2). In step 3, FXIa activates FIX; and in step 4, FIXa, in the presence of FVIIIa, activates FX. If each activated enzyme molecule catalyzed the activation of 100 molecules of the next enzyme of the cascade, an amplification factor of 1×10^6 of the intrinsic pathway would be achieved. As shown in the diagram, several feedback loops accelerate the overall process and produce a fibrin clot in a rapid and efficient manner. During this time, FXIII, a transglutamidase (often referred to as transglutaminase) that has been activated also by thrombin, is actively catalyzing the formation of cross-links between fibrin monomers of the soft clot to form a hard clot. This is the overall process of clot formation.

Some Properties of the Proteins Involved in Clot Formation

Tissue Factor (TF), (FIII): TF (Figure 24.41*a*) is a transmembrane protein of 263-amino-acids. Residues 243–263 are located on the cytosolic side of the membrane. Residues 220–242 are hydrophobic residues and represent the transmembrane sequence. Residues 1–219 are on the outside of the membrane, are exposed after injury, and form the receptor for FVII binding and formation of the initial complex of the extrinsic

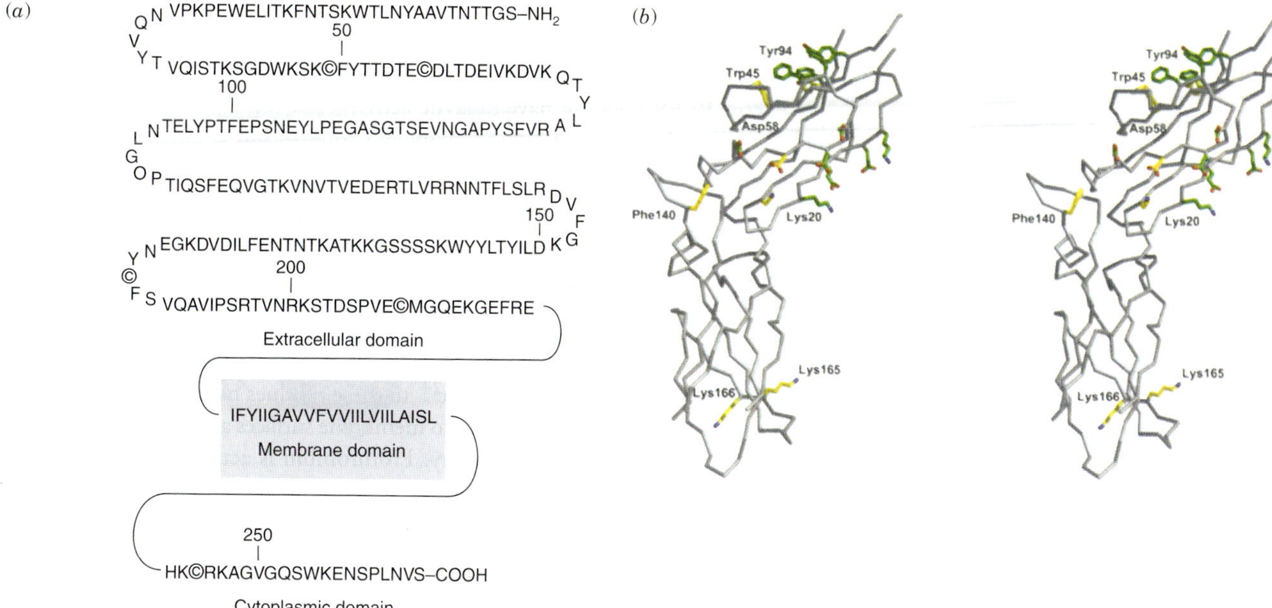

FIGURE 24.41

Tissue Factor. (*a*) Amino acid sequence of human tissue factor derived from its cDNA sequence. (*b*) A stereo representation of the carbon chain of the extracellular domain of tissue factor. Residues important for binding of factor VII are shown in yellow. Clusters of aromatic and charged residues are shown in light blue. Part (*a*) redrawn from Spicer, E. K., et al. *Proc. Natl. Acad. Sci. USA* 84:5148, 1987. Part (b) reproduced with permission from Muller, Y. A., Ultsch, M. H., Kelley, R. F., and deVos, A. M. *Biochemistry* 33:10864, 1994. Copyright (1994) American Chemical Society.

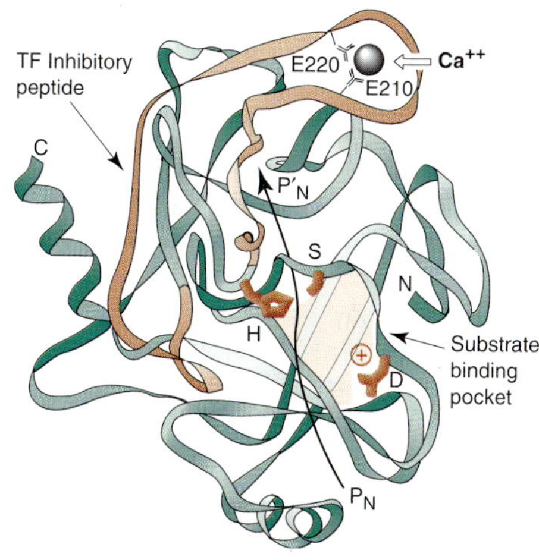

FIGURE 24.42

Ribbon structural representation of the protease domain of factor VIIa. The dark ribbon labeled "TF Inhibitory peptide" represents a section involved in binding to tissue factor. The catalytic triad is shown in the substrate binding pocket as H, S, and D for His-193, Ser-344, and Asp-338, respectively. The arrow labeled $P_N - P'_N$ lies in the putative extended substrate binding region.
Reproduced with permission from Sabharwal, A. K., Birktoft, J. J., Gorka, J., et al. *J. Biol. Chem.* 270:15523, 1995.

pathway. This domain is glycosylated and contains four cysteine residues. A stereo representation of a section of it highlighting some of the amino acid residues involved in **FVII** binding is shown in Figure 24.41*b*.

Factor VII: A 3-D ribbon structural representation of FVIIa is shown in Figure 24.42. The region for TF interaction, Ca^{2+} binding, and the substrate-binding pocket are highlighted.

domain is a functionally homologous protease inhibitor (sometimes referred to as Kunitz domains) that resembles other individual protease inhibitors such as the bovine **pancreatic trypsin inhibitor**. TFPI inhibits the extrinsic pathway by interacting specifically with the TF–FVIIa–Ca^{2+}–FXa complex. Domain 1 binds to FXa and domain 2 binds to FVIIa of the complex. Binding of TFPI to FVIIa does not occur unless FXa is present. Thus, TFPI is truly a multienzyme inhibitor in which each of its separate domains inhibits the action of one of the enzymes of the multienzyme complex of the extrinsic pathway. Second, the TFPI–FXa complex mediates internalization of FVIIa by an endocytosis mechanism. The C-terminal (third domain) of TFPI appears to be necessary for endocytosis. Most of FVIIa is degraded within the cells, but a small amount returns to the surface as intact protein and is apparently one of the sources of circulating FVIIa. As previously stated, FVIIa has no detrimental effects since it is active as a protease only when complexed with TF. Protease inhibitors of the **serine protease inhibitor (Serpin)** family of proteins in blood interact with and inhibit other enzymes of the blood coagulation system. There is a similarity in tertiary structure among them, with a common core domain of about 350 amino acids.

Antithrombin III (AT3) is a serpin that inhibits several hydrolases of the blood coagulation system, but most specifically thrombin and FXa. AT3 inhibits thrombin and FXa as complexes with different oligosaccharide groups of heparin. Heparin is a highly sulfated oligosaccharide of the glucosaminoglycan type. It exists as a mixture of oligosaccharides spanning a range of molecular sizes. The interaction of heparin in the different inhibitory complexes is size-dependent. At least 18 saccharide units are required for the effective formation of the thrombin inhibitor complex.

$$\text{Thrombin} + \text{Heparin}_{(>18)} + \text{AT3} \rightarrow \text{AT3:Heparin}_{(>18)}\text{:Thrombin}_{(\text{inactive})}$$

AT3 also forms a complex with a specific pentasaccharide form of heparin. The structure of this pentasaccharide is shown in Figure 24.53. Shown in Figure 24.54 is a structural model of AT3 bound to this heparin pentasaccharide. This complex inhibits FXa.

$$\text{AT3} + \text{Heparin}_{(5)} \rightarrow \text{AT3:Heparin}_{(5)} \longrightarrow \text{AT3:Heparin}_{(5)}\text{:FX}_{\text{inh}}$$

Although AT3 inhibits thrombin and FXa in the absence of heparin, heparin enhances inhibition of thrombin by a factor of approximately 9000 and inhibition of FXa by a factor of approximately 17,000.

There is another pathway for inhibition of FXa. Blood contains a 62-kDa Gla-containing glycoprotein, **protein Z (PZ).** Protein Z is a protein cofactor that, in the presence of Ca^{2+}, interacts with membrane and forms a complex with another plasma protein, **Protein Z-dependent protease inhibitor (ZPI)**, a 72-kDa protein. This complex inhibits FXa.

$$\text{FXa} + Ca^{2+} + \text{PZ} + \text{ZPI} \rightarrow (Ca^{2+}\text{:PZ})\text{:ZPI:FX}_{\text{inh}}$$

Protein Z-dependent protease inhibitor (ZPI) in the absence of PZ will inhibit FXIa, and this inhibition is enhanced by heparin. ZPI with or without PZ shows no measurable ability to inhibit other proteases, including thrombin, FVIIa, FIXa, or protein C.

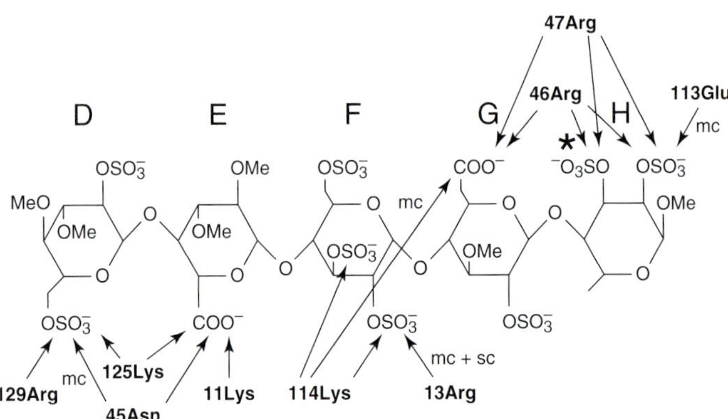

FIGURE 24.53

Chemical structure of the heparin pentasaccharide and positions of its interactions with specific residues of antithrombin.
Redrawn with permission from Whisstock, J. C., Pike, R. N., Jin, L., Skinner, R., et al. *J. Mol. Biol.* 301:1287, 2000.

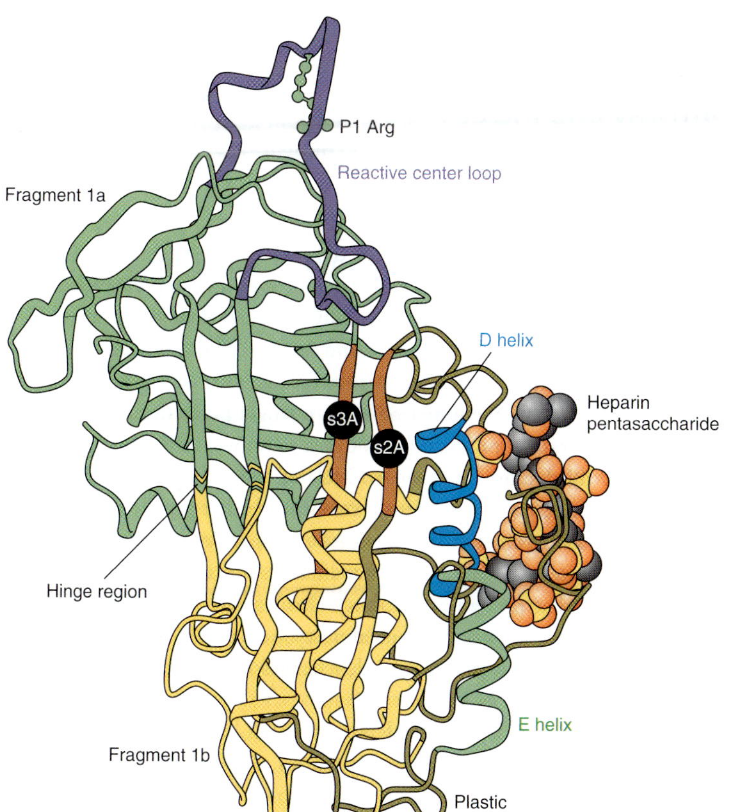

FIGURE 24.54

Fragment of antithrombin to which the heparin pentasaccharide binds. Fragment of antithrombin in the presence of the heparin pentasaccharide.
Reproduced with permission from Whisstock, J. C., Pike, R. N., Jin, L., Skinner, R., et al. *J. Mol. Biol.* 301:1287, 2000.

Inactivation of FVa and FVIIIa

Protein C (PC), a Gla-containing protein, is activated in a membrane-bound complex of thrombin, **thrombomodulin**, and calcium ions. Protein C requires the presence of another protein cofactor, **protein S (PS)**, a 75-kDa Gla-containing protein. The PC:PS complex inhibits coagulation by inactivating factors Va and VIIIa. Inactivation of FVa and FVIIIa occurs by cleavage of peptide bonds at specific arginine residues. Deficiencies and/or mutations in protein S or protein C can lead to **thrombotic diseases** (see Clin. Corr. 24.17).

$$\text{Va and VIIIa} \xrightarrow{\text{PC : PS}} \text{V}_{\text{inh}} \text{ and VIII}_{\text{inh}}$$

Thrombomodulin is an integral glycoprotein of endothelial cell membranes. It contains 560 amino acids and shows sequence homology with the low-density lipoprotein receptor, but very little with tissue factor. There is a great deal of similarity in functional domains between tissue factor and thrombomodulin, each of which is a receptor and activator for a protease. Thrombomodulin is a receptor for thrombin. In the thrombin–thrombomodulin–Ca^{2+} complex, thrombin has a decreased sensitivity for fibrinogen and an increased sensitivity for protein C. Thus, thrombin's role switches from that of procoagulation to anticoagulation.

Thrombin (Figure 24.55) may exist *in vitro* in two conformations: One has high specificity for conversion of fibrinogen to fibrin; the other conformation has low specificity for fibrinogen conversion, but high specificity for thrombomodulin binding and for proteolysis of protein C. These forms are referred to as "fast" and "slow" forms, respectively. This type of dynamic "feedback" mechanism is important for stopping the clotting process at its point of origin.

There is also a specific inhibitor for protein C. **Protein C inhibitor (PCI)** has been found in plasma, platelets, and megakaryocytes. ADP, epinephrine, thrombin, and other molecules that stimulate platelet activity release about 30% of PCI from platelets upon stimulation. Inactivation of activated protein C (APC) by PCI occurs on membrane surfaces like most of the other reactions discussed.

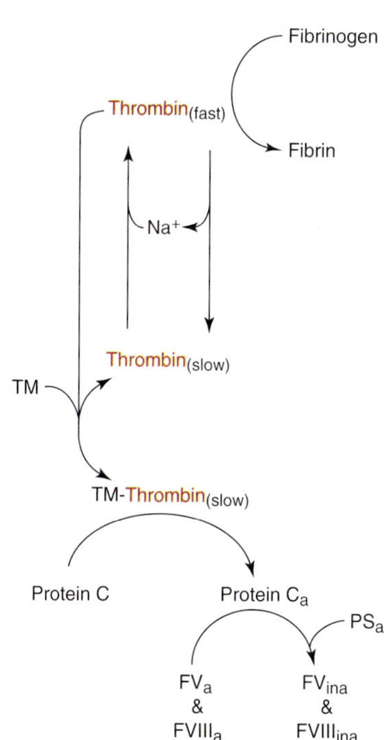

FIGURE 24.55

Allosteric reactions of thrombin and its actions on fibrinogen and protein C.

QUESTIONS | CAROL N. ANGSTADT

Multiple Choice Questions

1. In the propagation of a nerve impulse by an electrical signal:
   A. the electrical potential across the membrane maintained by the Na^+ /K^+-exchanging ATPase becomes more negative.
   B. local depolarization of the membrane causes protein conformational changes in ion channels that allow Na^+ and K^+ to move down their concentration gradients.
   C. charge propagation is bi-directional along the axon.
   D. "voltage-gated" ion channels have a finite recovery time so the amplitude of the impulse changes as it moves along the axon.
   E. astrocytes function as antenna-like protrusions and receive signals from other cells.

2. All of the following statements about actin and myosin are true *except*:
   A. the globular head section of myosin has domains for binding ATP and actin.
   B. actin is the major protein of the thick filament.
   C. binding of ATP to the actin–myosin complex promotes dissociation of actin and myosin.
   D. F-actin is stabilized when tropomyosin is bound to it.
   E. binding of Ca^{2+} to the calmodulin-like subunit of troponin induces conformational changes that permit myosin to bind to actin.

3. Platelet aggregation:
   A. is initiated at the site of an injury by conversion of fibrinogen to fibrin.
   B. is inhibited in uninjured blood vessels by prostacyclin of intact vascular endothelium.
   C. causes morphological changes and a release of the vasodilator, serotonin.
   D. is inhibited by the release of ADP and thromboxane A_2.
   E. is inhibited by von Willebrand factor (vWF).

4. In the formation of a blood clot:
   A. proteolysis of γ-carboxyglutamate residues from fibrinogen to form fibrin is required.
   B. the clot is stabilized by the cross-linking of fibrin molecules by the action of factor XIII, transglutaminase.
   C. thrombin's only role is in activation of factor VII.
   D. tissue factor, factor III or TF, must be inactivated for the clotting process to begin.
   E. the role of calcium is primarily to bind fibrin molecules together to form the clot.

5. Lysis of a fibrin clot:
   A. is in equilibrium with formation of the clot.
   B. begins when plasmin binds to the clot.
   C. requires the hydrolysis of plasminogen into heavy and light chains.
   D. is regulated by the action of protein inhibitors on plasminogen.
   E. requires the conversion of plasminogen to plasmin by t-PA.

6. Nitric oxide synthase (NOS):
   A. has both constitutive and inducible forms.
   B. may be involved in controlling neuronal signaling but not muscle contraction.
   C. is found in endothelial tissue only in response to inflammation.
   D. forms NO in one tissue, which acts on a distant tissue.
   E. is most active when bound to other proteins.

Questions 7 and 8: There are three families of compounds known as molecular motors involved in muscle contraction and intracellular trafficking (moving cargo or arranging intracellular structures). Elejalde syndrome is characterized by silvery hair and severe dysfunction of the central nervous system. Large granules of melanin are unevenly distributed in the hair shaft, suggesting that movement and distribution might be defective.

7. Which of the following classes of molecular motors can be involved in both contraction and trafficking?
   A. dyneins.
   B. kinesins.
   C. myosins.
   D. all of the above.

8. Microtubule-based motors:
   A. do not form filaments.
   B. generate movement by conformational changes.
   C. use ATP.
   D. are involved in chromosome organization.
   E. all of the above are correct.

Questions 9 and 10: Genes coding for proteins of visual pigments are located on different chromosomes. The rhodopsin gene is on the third chromosome; blue pigment gene on the seventh; and genes for the red and green pigments on the X chromosome. Females have two X chromosomes so color vision abnormalities are rare but, males have only one X chromosome. About 8% of males have abnormal color vision that affects either red or green perception. At least one individual is known with blue colorblindness.

9. The cones of the retina:
   A. are responsible for color vision.
   B. are much more numerous than the rods.
   C. have red, blue, and green light-sensitive pigments that differ because of small differences in the retinal prosthetic group.
   D. do not use transducin in signal transduction.
   E. are better suited for discerning rapidly changing visual events because a single photon of light generates a stronger current than it does in the rods.

10. Which of the following statements about rhodopsin is true?
    A. Rhodopsin is the primary photoreceptor of both rods and cones.
    B. The prosthetic group of rhodopsin is all-*trans*-retinol derived from β-carotene.
    C. Conversion of rhodopsin to activated rhodopsin, R*, by a light pulse requires depolarization of the cell.
    D. Rhodopsin is located in the cytosol of the cell.
    E. Absorption of a photon of light by rhodopsin causes an isomerization of 11-*cis*-retinal to all-*trans*-retinal.

Questions 11 and 12: Ion channels in cardiac muscle, as in other tissues, require a finite recovery time after excitation. This is measured on an electrocardiogram as the QT interval. Initiating the excitation phase is the opening of Na^+ channels, causing depolarization. Then K^+ channels open to permit K^+ to move out of the cell, which is a key element in shutting off the action potential. Inherited defects in these channels lead to a condition called long QT syndrome (LQTS). LQTS can cause sudden death, especially in physically active young people who have not previously had any symptoms of cardiac irregularities.

11. The nerve impulse which initiates muscular contraction:
    A. begins with binding of acetylcholine to receptors in the sarcoplasmic reticulum.
    B. causes both plasma membrane and transverse tubules to undergo hyperpolarization.
    C. causes opening of calcium channels, leading to an increase in Ca^{2+} within the sarcomere.
    D. prevents Na^+ from entering the sarcomere.
    E. prevents Ca^{+2} from binding to troponin C.

12. When a muscle contracts, the:
    A. transverse tubules shorten, drawing the myofibrils and sarcoplasmic reticulum closer.
    B. thin filaments and thick filaments of the sarcomere shorten.
    C. light chains dissociate from heavy chains of myosin.
    D. H bands and I bands of the sarcomere shorten because thin filaments and thick filaments slide past each other.
    E. cross-linking of proteins in the heavy filaments increases.

Problems

13. In the presence of warfarin, an analog of vitamin K, several proteins of the blood coagulation pathway are ineffective because they cannot bind Ca^{2+} efficiently. Why?

14. Organophosphate compounds are irreversible inhibitors of acetylcholinesterase. What effect does an organophosphate inhibitor have on the transmission of nerve impulses?

ANSWERS

1. **B** This is the mechanism for impulse propagation. A: The potential becomes less negative. C: It is unidirectional. D: Voltage-gated channels do have a finite recovery time so the amplitude remains constant. E: This describes dendrites. Astrocytes are glial cells involved in processes that insulate neurons from their external environment.
2. **B** Myosin forms the thick filament. Actin is in the thin filament. A: These are both important in myosin's role. C: The role of ATP in contraction is to favor dissociation, not formation, of the actin–myosin complex. D, E: Tropomyosin, troponin and actin are the three major proteins of the filament.
3. **B** The "yin–yang" nature of PGI_2 and TXA_2 help to control platelet aggregation until there is a need for it. A: Initiation is by contact with an activated receptor at the site of injury. Clot formation requires activation of various enzymes. C: Serotonin is a vasoconstrictor. D: TXA_2 facilitates aggregation. E: vWF forms a link between the receptor and platelets, promoting aggregation.
4. **B** Cross-linking occurs between a glutamine and a lysine. A, E: γ-Carboxyglutamate residues are on various enzymes; they bind calcium and facilitate the interaction of these proteins with membranes that form the sites for initiation of reaction. C: Thrombin activates factors V, VII, VIII, and XIII and fibrinogen. D: TF, factor III, is the primary receptor for initiation of the clotting process.
5. **E** The clot is solubilized by plasmin. A: Both formation and lysis of clots are unidirectional. B: Both plasminogen and t-PA bind to the clot. C, D: Both of these refer to t-PA.
6. **A** Inducible form responds to inflammation; neuronal and endothelial forms are constitutive. B: Involved in both. C: See A. D: NO is formed at site of its target. E: It dissociates from bound proteins to become active.
7. **C** Myosins are involved in contraction and "unconventional" myosins in trafficking. A, B: These are both microtubule-based motors involved in moving cargo.
8. **E** A: This is in contrast to actin-based motors. B, C: This is often associated with hydrolysis of ATP. D: Some classes of kinesins and also dyneins do this.
9. **A** Rods are responsible for low-light vision. C: All three pigments have 11-*cis*-retinal; differences in the proteins are responsible for the different spectra. D: Biochemical events are believed to be the same in rods and cones. E: Cones are better suited for rapid events because their response rate is about four times faster than rods, even though their sensitivity to light is much less.
10. **E** This causes the conformational change of the protein that affects the resting membrane potential and initiates the rest of the events. A: Cones have the same prosthetic group but different proteins, so rhodopsin is in rods only. B: This is the precursor of the prosthetic group 11-*cis*-retinal. C: Isomerization of the prosthetic group leads to hyperpolarization. D: Rhodopsin is a transmembrane protein.
11. **C** Ca^{2+} enters with Na^+. A: Acetylcholine receptors are on the plasma membrane. B: The impulse results in depolarization of both of these structures. D. Both Ca^{+2} and Na^+ enter the sarcomere when the channels open. E: Binding of Ca^{+2} to TnC initiates contraction.
12. **D** This occurs because of association–dissociation of actin and myosin. A: Depolarization in transverse tubules may be involved in transmission of the signal but not directly in the contractile process. B: Filaments do not change length, but they slide past each other. C: This is not physiological. E: Cross-linking occurs in the H band of the sarcomere but does not change during the contractile process.
13. Proteins affected are those with γ-carboxyglutamyl (Gla) residues, which are excellent calcium chelators. Formation of Gla is a posttranslational modification catalyzed by a carboxylase whose essential cofactor is vitamin K. Warfarin prevents reduction of vitamin K epoxide formed during carboxylation back to the dihydroquinone form.
14. Acetylcholine is an excitatory neurotransmitter, causing channels to open, Na^+ to enter and a depolarization of the membrane. Hydrolysis of acetylcholine allows repolarization. The inhibitor keeps the channel open and prevents repolarization and thus prevents continued transmission of nerve impulses.

25.1 OVERVIEW

At the interface of cell biology and biochemistry are the biochemical pathways that regulate cell division and cell death. These pathways involve proteins that transmit signals, often through their action as enzymes but also through their ability as signaling proteins that bind to other proteins and transmit a signal through an induced-conformation change in the associated proteins. Enzymatic activities commonly present in signal pathways are those of kinases, phosphatases, or proteases. While the signals are composed of well-known enzymatic reactions or protein-induced conformational changes, the pathways regulating cell division and cell death are complex. Rather than a simple linear progression of signals from one molecule to the next constituting a pathway, parts of these pathways can be depicted as a network of parallel signaling pathways joined at nodes or hubs that can cross-talk. The inputs are interpreted by these networks in a manner dependent on cell type and on the strength and time-course of the extrinsic and intrinsic signals received by the cell, to give an output that causes a cell either to divide, to die, or to differentiate. We have only a preliminary understanding of how the networks operate and interpret these signals to generate a specific outcome. For many inputs, the complexity of the network may make outputs impossible to predict, in the absence of computer algorithms to model the multiple node interactions that differ in different cell types and environmental conditions. These pathways will be described in a traditional manner to promote a basic understanding of the primary pathways that are the basis of these complex signaling networks. An insight into the complexity of these pathways may be discerned in the discussion of the Ras downstream kinases that regulate both cell division and cell death.

In a mature organism, cell division is balanced by cell death and therefore both pathways are equally active. Aberrations in the cell division and cell death pathways lead to cancer; the biochemical basis of the cancer cell will be presented in the later part of this chapter as an outcome of abnormalities in the cell division and death pathways.

25.2 CELL CYCLE

A model of the cell cycle breaks the steps of cell division into four phases (Figure 25.1). In the **S phase** the chromosomal DNA is duplicated and in the M phase mitosis occurs, leading to the division of chromosomes, cellular organelles, and cytoplasm into two progeny cells. The gap phases, G_1 and G_2, separate M phase from S phase and S phase from **M phase**, respectively. In a simpler cell cycle model, cell division is divided into mitosis (M) and an interphase (I) that combines G_1, S, and G_2 phases. The M phase, although shorter than the other phases, is particularly complex. In a human fibroblast cell that divides once every 24 h, the M phase spans about 1 h while the S phase may last 10–12 h, and the other two phases (G_1, G_2) occupy the remaining 11–13 h. The observable steps in M phase are presented in Table 25.1. While the M phase is the most dramatic, essential biochemical processes occur in each of the phases, and each of these must be completed before the cell can progress to the next phase. Each phase is under close regulation at multiple biochemical checkpoints that comprise stop/go signals before proceeding to the next step in the pathway.

An additional phase, in equilibrium with the G_1 phase, is the G_0 phase, in which the cell is either in a quiescent or a senescent state. A quiescent cell, while not participating in the cell cycle, can be induced to reenter the cell cycle by a growth stimulus, such as an increase in concentration of a growth factor in its external environment. In contrast, a senescent cell cannot reenter the cell cycle even in the presence of growth factor. Different cell types vary in their rate of cell division and thus in the amount of quiescent time spent in G_0. While fibroblasts and epithelial cells may spend very little or no time in G_0, adult liver cells divide about once a year, and adult brain cells almost never divide.

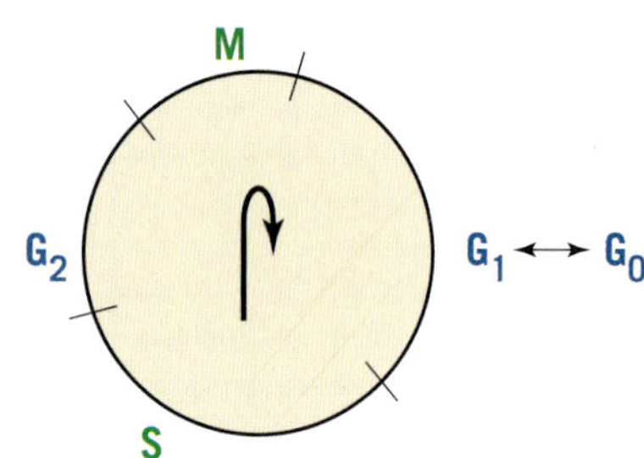

FIGURE 25.1

Phases of the cell cycle. Mitosis occurs in M phase and DNA synthesis in S phase. Interphase consists of G_1, S, and G_2. G_0 phase is in equilibrium with G_1 phase. Cells in G_0 are either quiescent or senescent.

TABLE 25.1 Microscopically Observable Steps in the M Phase

1. DNA condenses into chromosomes (46 chromosomes in human cells).
2. Mitotic spindle assembles out of centrosomes.
3. Nuclear membrane disintegrates.
4. Spindle microtubule connects chromosomes.
5. Chromosomes align along equator of cell.
6. Sister chromatids separate, creating two sets of chromosomes.
7. Nuclear membrane forms around each set of chromosomes.
8. Chromosomes decondense.
9. Plasma membrane pinches parental cell into two progeny cell (cytokinesis).

Thus, adult liver and brain cells spend most of their time in G_0. Cells in G_0 have a lower metabolic rate and lower levels of transcription and protein synthesis than those in G_1, and they lack critical proteins of the cell cycle machinery including cyclin-dependent kinases and other regulatory proteins.

Regulation of Cell Cycle

The cell cycle is regulated by the activity of **cyclin-dependentkinases** (Cdks), which catalyze phosphorylation of a serine or threonine side chain in substrate proteins by ATP. The phosphorylated proteins then perform functions required for the particular step in the cell cycle pathway. The activities of the Cdks are closely regulated by other protein kinases and by protein phosphatases, by association with cyclin proteins, and by the presence or absence of Cdk inhibitor proteins.

In human cells the M phase Cdk (M-Cdk or Cdk1) binds M-cyclin (cyclin B). Cdk1 is sometimes referred to as CDC2 from the yeast nomenclature, where its ortholog was first characterized. Cdk1/CDC2 is necessary to initiate and propagate the M phase of the cell cycle. The modifications of the M-Cdk required for activity are outlined in Figure 25.2. The M-Cdk protein must bind to the M-phase **cyclin** protein (M-cyclin), to form a heterodimer M-Cdk/M-cyclin required for activity. While M-Cdk remains in the cell throughout all phases of the cell cycle, M-cyclin is synthesized only during G_2, just prior to entry into M phase, and is degraded on exit from M phase. While formation of the M-cyclin/M-Cdk heterodimer is required for M-Cdk kinase activity, before the heterodimer becomes catalytically competent it must be correctly phosphorylated.

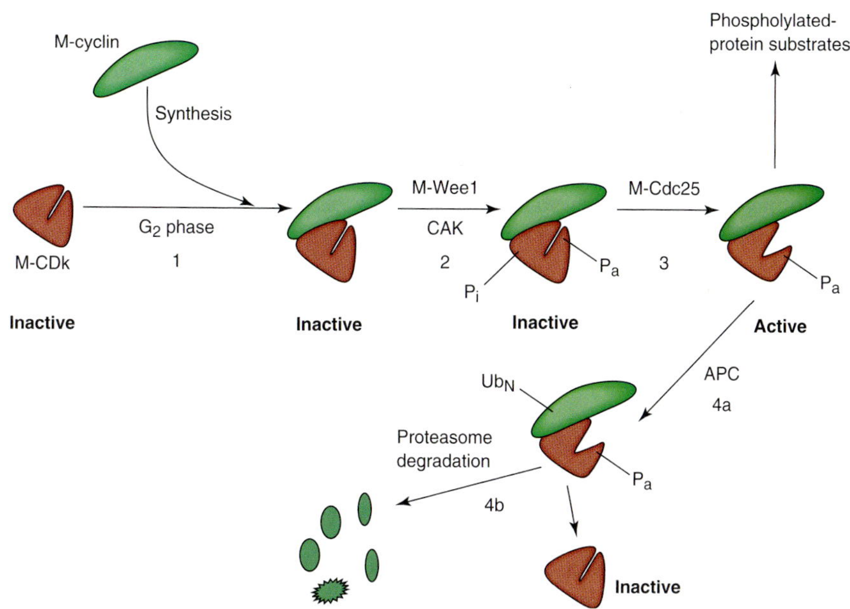

FIGURE 25.2

Regulation of M phase cyclin-dependent kinase (cdk) activity. M-cyclin (green) is synthesized in the G_2 phase and combines with M Cdk (red) to form an inactive heterodimer. Phosphorylation by Wee1 kinase at an inhibitory site (Pi) maintains M-Cdk/M-cyclin heterodimer in an inactivated state. Entry into M phase occurs on phosphorylation by CAK at an activation site (Pa) and removal of the inhibitory phosphate by Cdc25 phosphatase. Activated M-Cdk phosphorylates protein substrates that promote M phase. Loss of M-Cdk activity occurs when M-cyclin is ubiquitinated by Ub_N the APC ubiquitinase complex and degraded.

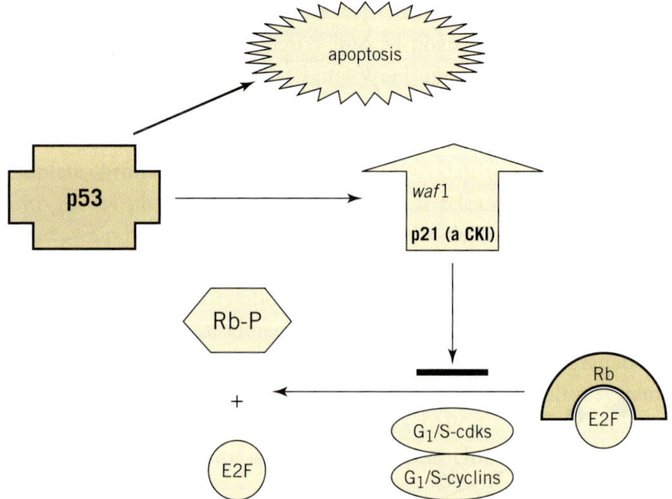

FIGURE 25.6

Regulation of the G_1 to S transition by p53. The concentration of the transcription factor p53 is increased in DNA damage and by stress, as well as by other cell cycle regulatory signals. p53 can stop the cell cycle by increasing the transcription of the *waf1* (also called *cip1*) gene to express p21, which is a CKI and binds to G_1/S Cdks to inhibit phosphorylation of Rb. Higher concentrations of p53 than required for p21 synthesis initiate apoptosis.

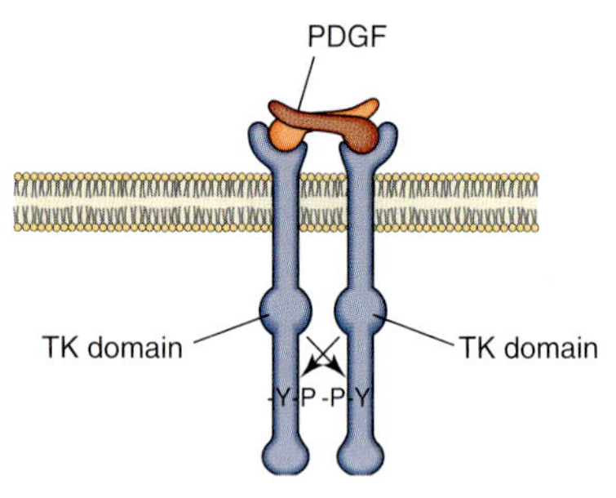

FIGURE 25.7

Phosphorylation of tyrosine residues within cytoplasmic domains of PDGF receptor. Binding of PDGF as a dimer causes its membrane receptor to also dimerize. The tyrosine kinase activity in the cytoplasmic region of one receptor protein then phosphorylates tyrosine residues (Y) in the cytoplasmic region of the neighboring receptor protein.

growth factor proteins is required for survival of mammalian cells. Higher concentrations of growth factor cause its target cells to divide.

Growth factors bind to their receptors on cellular plasma membranes. This binding causes the receptor proteins to dimerize or polymerize. Oligomerization is critical for transmission of the growth factor signal. Various mechanisms are utilized by cells for oligomerization of receptor proteins. Platelet-derived growth factor (PDGF) dimer binds to two separate receptor proteins bringing two receptor proteins together as a dimer in the plasma membrane (Figure 25.7). Other growth factors combine with cofactor proteins that facilitate their binding and oligomerization of the receptor proteins.

Most growth factor receptor proteins contain a tyrosine kinase domain in their cytoplasmic region. Unlike the serine/threonine kinases (e.g., of PKC, PKA, and the Cdks), which place a phosphate group on side chains of serine/threonine residues, tyrosine kinases place a phosphate group on the phenolic side chain of a tyrosine residue. The initial tyrosines phosphorylated on the activation of a growth factor receptor are within the cytoplasmic side of the growth factor receptor protein itself. The kinase domain of one receptor protein in the oligomer acts on neighboring proteins of the oligomer, and vice versa (Figure 25.7). Autophosphorylation of tyrosine residues in the receptor cytoplasmic domain generates phospho-tyrosine sites for high-affinity binding of cytoplasmic signaling proteins.

Some growth factor receptors lack tyrosine kinase activity, but on their activation they bind a cytoplasmic protein containing a tyrosine kinase domain such as **Src**. On binding of the extracellular growth factor ligand, these receptors undergo a conformation change in the membrane. The conformation change leads to their cytoplasmic association with tyrosine kinase proteins.

Cytoplasmic proteins of the growth factor signaling pathway typically contain SH2 (**S**rc **h**omology domain **2**) and SH3 domains for forming association complexes with other proteins of the pathway. The SH2 domain binds to a phosphorylated tyrosine, and the SH3 domain binds to a region in a protein that has polyproline helix secondary structure. Thus, phosphorylated tyrosine residues on an activated growth factor receptor are binding sites for proteins that contain SH2 domains. The PDGF receptor binds different signal transduction molecules, through their SH2 domains, to particular phospho-tyrosine residues on its cytoplasmic region (Figure 25.8).

Each SH2 domain has a secondary site specificity that differentiates between P-Y sites. Thus, PI3K has a specificity not only for phospho-tyrosines but also for the amino acids that surround (a) the phospho-tyrosine at tyrosine-740 and (b) PLCγ for the phospho-tyrosine residue 1021. In this way, PI3K and PLCγ become substrates for the receptor tyrosine kinase and, on becoming phosphorylated, leave the receptor and exert their activity.

The protein Grb2 binds to the phosphorylated tyrosine residue 716 in the PDGF receptor. **Grb2** has no catalytic activity but is an adaptor protein that transmits a signal

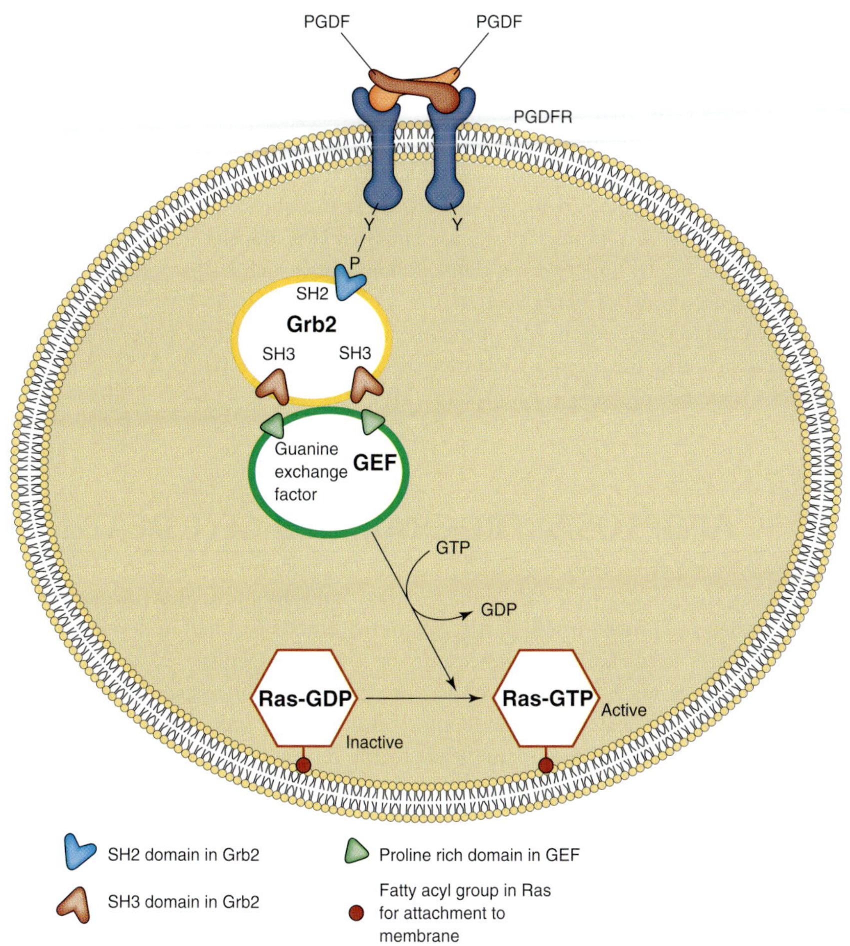

FIGURE 25.8

Transmission of signal from PDGF receptor to Ras. Plasma membrane containing PDGF receptor becomes autophosphorylated on tyrosine residues. The SH2 domain of Grb-2 binds to phospho-tyrosine residue 416 in receptor, and SH3 domains bind to polyproline helix region in GEF^{Ras} (Ras guanine exchange factor), which promotes exchange of GDP for GTP in Ras. Ras-GTP then transmits its downstream signals.

through conformational change and protein–protein associations. Grb2 contains both SH2 and SH3 domains. Binding of Grb2 to the PDGF receptor through its SH2 domain induces a conformation change that opens its SH3 domains to bind a polyproline helix region in GEF^{Ras} protein, which then binds to Ras and catalyzes the exchange of its bound GDP for GTP (Figure 25.8). **GTP-Ras** is the active form of Ras. Ras is a GTPase enzyme and inactivates itself by catalyzing the hydrolysis of bound GTP to GDP.

Mutations in the Ras gene that result in constitutively active Ras occur in approximately 30% of human cancers. The mutations commonly found are in amino acid residues 12, 13, and 61, which participate in the Ras GTPase activity so that the mutated Ras cannot deactivate itself and remains in an active GTP-Ras conformation.

In normal Ras, the Ras GTPase activity is increased 100- to 1000-fold by the binding of Ras to $\mathbf{GAP^{Ras}}$ (GAP, **G**TPase **a**ctivating **p**rotein). However, the binding of GAP^{Ras} to the catalytically inert mutated Ras has no effect, because its GTPase activity is absent.

Ras also needs to associate with the plasma membrane to be activated. This is achieved by posttranslational modifications that include removal by a protease of the C-terminal tetrapeptide, methylation of the new C-terminal carboxylic acid group, and addition of a farnesyl fatty acid group to a cysteine side chain at the COOH-terminal end. In some Ras isoforms a second fatty acyl group is added, near the COOH-terminal end.

Three different **Ras genes** exist in humans—Ha-*ras*, N-*ras*, and K-*ras*—which produce homologous proteins of 21 kDa. The amino acid sequences of their first 85 residues are identical, and the next 80 residues show 85% homology, but the C-terminal ends differ more dramatically. While there are differences in some of the downstream signals between the Ras isoforms, in general their signals overlap. These isoforms are characteristically expressed in different cell types, and mutations of residues 12, 13, and

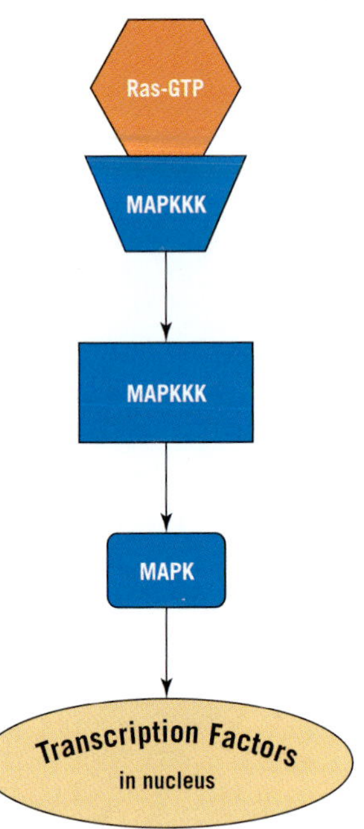

FIGURE 25.9

GTP-Ras activates a kinase cascade. Ras-GTP activates a kinase cascade resulting in the phosphorylation and activation of the terminal kinase, MAPK. Activated MAPK enters the nucleus and phosphorylates transcription factors that regulate the expression of proteins involved in S phase. MAPK is nitrogen-activated protein kinase, MAPKK is MAP kinase kinase and MAPKKK is MAP kinase kinase kinase.

61 constitutively activate all of them. Since activation of Ras activity promotes the cancer phenotype, Ras genes are known as protooncogenes. **Protooncogenes** are changed to **oncogenes** by an activating mutation that promotes or maintains a cancer cell. The *ras* oncogene continually gives a growth factor signal that promotes cell division, in the absence of upstream activation.

GTP-Ras transmits its cell division signal by activation of MAP kinase kinase kinase (MAPKKK) (Figure 25.9). MAPKKK initiates a kinase cascade that includes MAPKK and MAPK. Activated (phosphorylated) **MAPK** translocates to the nucleus, where it phosphorylates and activates transcription factors such as Jun and Fos, thereby increasing transcription of genes such as the transcription factor Myc and the G_1/S cyclins involved in cell division. Myc increases the expression of many genes involved in the S phase including the S-cyclins and transcription factor E2F (see p. 215). However, if conditions are not proper for cell division, Myc may initiate a process leading to cell death.

25.3 APOPTOSIS: PROGRAMMED CELL DEATH

Apoptosis is the Greek word for "falling leaves," and apoptosis describes an often natural biochemical process of cell death. Apoptotic death is required in developmental processes as well as to maintain homeostasis of a whole organism. As new cells are generated in the human, death of a similar number of cells is required on the same time scale to maintain a steady state. Apoptotic death differs from **necrotic** cell death. In the latter, lysis of a cell membrane leads to the release of the cell's contents into extracellular space and an inflammatory response, as often occurs in bacterial or viral infections and in trauma. In contrast, apoptosis is often unobservable and does not induce an inflammatory or immunological response. It can be initiated by a variety of signals, including those from damaged DNA, entrance of a cell into the S phase under improper conditions, lack of proper contacts of a cell with extracellular matrix, lack of necessary growth factors, or presence of death signal proteins in the environment of a cell. These signals activate **cytoplasmic protease enzymes** called **caspases**. The caspases hydrolyze specific peptide bonds in target proteins that, after peptide bond hydrolysis, contribute to cell death by either a gain or loss of its activity. Apoptosis is characterized by changes in the plasma membrane, cytoskeleton, and nuclear DNA. The plasma membrane becomes blebbed and sheds membranous particles that encapsulate intracellular contents. Endonucleases are activated that degrade the chromosomal DNA into nucleosome-size fragments ($\sim$170 base pairs in length). The remains of an apoptotic cell are engulfed and digested by phagocytic cells such as macrophages (Figure 25.10).

Caspases are activated by precursor caspases in a protease cascade mechanism similar to the protease cascades of blood coagulation and the complement reactions of

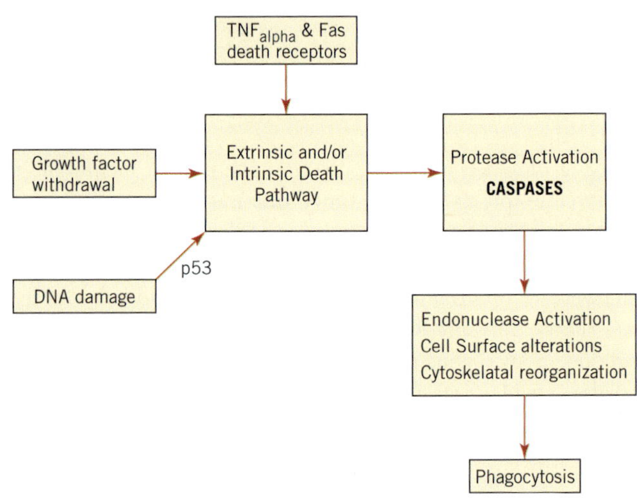

FIGURE 25.10

Outline of pathway for programmed cell death.

the immunological response. Each precursor caspase is activated by cleavage of a peptide bond by an upstream caspase. Caspases produced in the cascade pathway act on protein substrates, which leads to apoptosis.

Major Pathways

Death Receptor Pathway

The two major pathways that activate the caspase cascade and promote cell death are the **death receptor pathway** (extrinsic pathway) and the mitochondrial pathway (intrinsic pathway) (Figure 25.11). The receptor pathway is initiated by binding of a death ligand present in the extracellular environment to its death receptor in a plasma membrane. This promotes binding of intracellular adaptor proteins to the cytoplasmic region of the receptor. Such an adaptor protein is FADD (the DD in the acronym refers to a protein-binding "death domain"). The DD domain is present in many of the proteins in the death receptor pathway. An aggregate formed by the adaptor proteins through DD interactions includes pro-caspase 8, which becomes autocatalytically changed to caspase 8 on hydrolysis of one of its peptide bonds. It then leaves the aggregate to function in the caspase cascade (Figure 25.11). This cytoplasmic receptor aggregate is referred to as a DISC (*d*eath-*i*nducing *s*ignaling *c*omplex).

Mitochondrial Pathways

The intrinsic or mitochondrial pathway is regulated by the relative concentrations of **Bcl-2**-like proteins in the outer mitochondrial membrane, which control the translocation of certain proteins from the mitochondrial intermembrane space into the cytosol. Apoptosis may be activated by extrusion of cytochrome c through pores generated in the outer mitochondrial membrane by the Bcl-2-like pro-apoptotic proteins Bak and Bax (Figure 25.12). There are about 24 Bcl-2-like proteins in humans, which are divided into (i) anti-apoptotic proteins such as Bcl-2 and Bcl-X_L, (ii) pro-apoptotic pore-forming proteins like **Bax** and **Bak**, and (iii) pro-apoptotic facilitator proteins such as **Bid**, **Bad**, **PUMA**, and **Noxa**. Each class has its own structural characteristics and roles in the extrusion of cytochrome c. Bcl-2-like proteins contain BH domains (**B**cl-2 **H**omology domains), which are utilized for self-association and for hetero-association with other BH domain-containing proteins. All the Bcl-2-like proteins contain the BH3 domain. The *anti-apoptotic* proteins such as Bcl-2 and Bcl-X_L contain four different domains (BH1, BH2, BH3, and BH4). The *pro-apoptotic* proteins Bax and Bak contain the first three BH domains (BH1–BH2–BH3), but not BH4. The pro-apoptotic *facilitator* proteins, such as Bad and Bid, contain only the **BH3 domain**. The pores through which

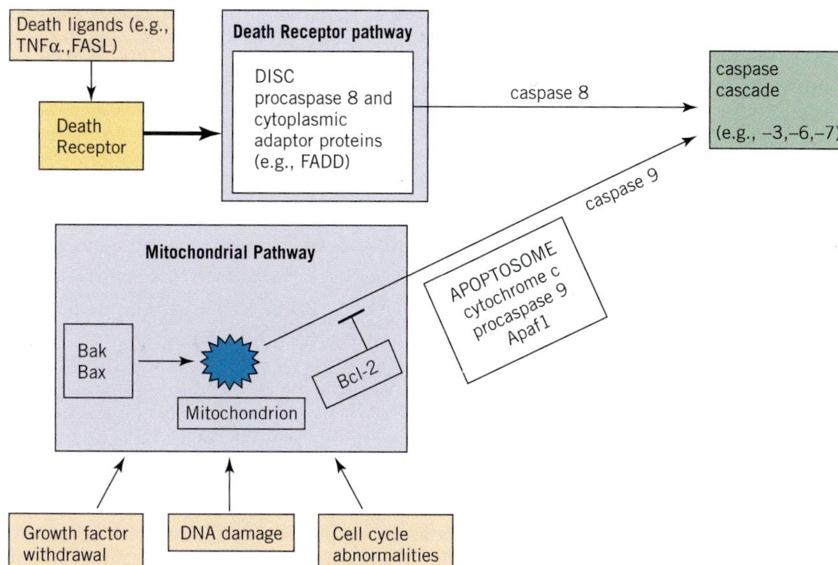

FIGURE 25.11

Death receptor and mitochondrial pathways activate caspase cascade. The receptor (extrinsic) and mitochondrial (intrinsic) pathways converge on the activation of caspases. Death ligands activate death receptor to bind adaptor proteins and procaspase 8, which is activated to caspase 8, and then activates the caspase cascade. In mitochondrial pathway, Bak and Bax oligomerize to form pores that permit cytochrome c to pass into the cytoplasm, where it forms an apoptosome with multiple molecules of Apafl and procaspase 9. Activated caspase 9 then activates caspase cascade. Bcl-2 inhibits cytochrome c passage from mitochondria. Activators of the mitochondrial and receptor pathways (i.e., TNFα, DNA damage) are shown.

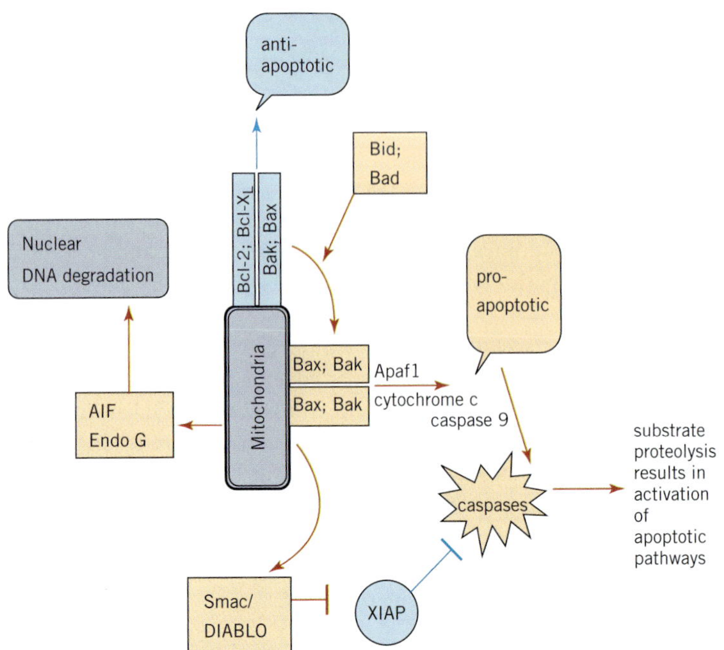

FIGURE 25.12

Regulation of apoptosis by cytochrome c and other mitochondrial proteins. Oligomerization of Bak and/or Bax creates pores for transfer of cytochrome c from mitochondria to cytoplasm. Bcl-2 and Bcl-X_L inhibit pore formation by binding with Bax and Bak to prevent Bax and Bak self-association. Facilitator proteins, such as Bad and Bid, promote apoptosis by either promoting the dissociation of Bak and Bax from Bcl-2 and Bcl-X_L or promoting an active conformation of the Bax/Bak oligomers. In the cytoplasm, cytochrome c binds in apoptosome complex, which leads to the activation of procaspase 9 to active caspase 9. Smac/DIABLO can also enter cytoplasm and activate the caspases by inhibiting XIAP. AIF and EndoG can pass from mitochondria into nucleus to promote degradation of the DNA.

cytochrome c passes are generated by aggregates (dimers or polymers) of Bax and Bak. The anti-apoptotic Bcl-2 and Bcl-X_L prevent cytochrome c leakage and cellular death by competitively binding to Bak and Bax and thus preventing their self-association to form membrane pores. Thus, activity of this death pathway depends on the relative concentrations of pro- and anti-apoptotic Bcl-2-like proteins. If the concentrations of the anti-apoptotic Bcl-2 and Bcl-X_L proteins are greater than the pro-apoptotic Bak and Bax proteins, then Bak and Bax aggregates do not form and the cell lives. However, if the relative concentrations of Bak and Bax proteins are greater than Bcl-2 and Bcl-X_L proteins, then these proteins self-aggregate, mitochondrial membrane pores are formed, **cytochrome c** is emitted from the mitochondria, and the cell dies. Some facilitator proteins promote dissociation of Bcl-2 or Bcl-X_L heterodimers with Bak or Bax by competing for the Bak/Bax binding site in Bcl-2/Bcl-X_L, which lead to Bak and Bax dissociation from Bcl-2 and Bcl-X_L and their self-association to form pores. Other facilitator proteins bind directly to the Bax/Bak oligomers and promote a more active conformation of these proteins. The proteins, Bak, Bax, Bcl-2, Bcl-X_L, Bid, and Bad, are prominent examples of the pro-apoptotic mitochondrial death pathway proteins.

Outside the mitochondria, cytochrome c binds **Apaf1** (**a**poptotic **p**rotease-**a**ctivating **f**actor **1**) and **pro-caspase 9**. This cytoplasmic aggregate, functionally similar to DISC (see p. 1021) of the death receptor pathway, is referred to as an **apoptosome.** It leads to the autocatalytic activation of pro-caspase 9 to caspase 9, which is released and activates the caspase cascade (Figure 25.12).

Proteins from mitochondria other than cytochrome c (Figure 25.12) can initiate cytochrome c- and caspase-independent apoptosis under certain conditions. Thus the protein **Smac/DIABLO** inhibits **XIAP**, a caspase inhibitor protein (XIAP is an acronym for **X**-linked **i**nhibitor of **ap**optosis) that limits caspase activity in absence of an activating signal. By inhibiting the inhibitor of caspases, Smac/DIABLO activates the caspase cascade. The mitochondrial proteins **AIF** and **endoG** can enter the nucleus to activate DNA degradation and cause apoptosis.

p53-Induced Apoptosis

The transcription factor p53 is an important inducer of apoptosis with DNA damage (see p. 239). **Mdm2**, a E3 ligase in an ubiquitinase pathway, which binds to p53 and promotes its polyubiquitination and degradation by proteasomes, normally keeps the concentration of p53 low (Figure 25.13). DNA damage results in the formation of

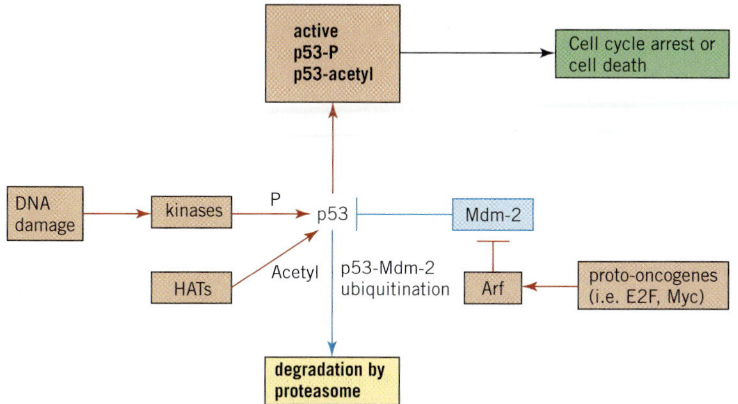

FIGURE 25.13

Regulation of p53 activity. Under normal conditions, p53 concentrations are kept low by Mdm2, which promotes ubiquitination of p53 and its degradation by proteasomes. Phosphorylation and acetylation of appropriate residues by kinase and HAT enzymes prevents degradation of p53 and also increases its transcriptional activity. Certain "protooncogene" transcription factor proteins, which under normal conditions promote the expression of genes required in cell division, under abnormal conditions increase Arf expression. Arf binds to Mdm2 to prevent its interaction with p53 and thereby increases p53 concentration. The Arf-induced mechanism may be a cellular safety valve to stop a cell from dividing when certain negative signals or conditions arise after the cell has initiated the cell division process. Increased p53 activity leads to cell cycle arrest or apoptosis.

DNA chain breaks and changes in chromatin structure that activates chromatin-bound protein kinases such as **ATM**, **ATR**, and other **DNA protein kinases** (DNA-PKs), which directly or through the activation of downstream kinases phosphorylate p53. Phosphorylation of p53 prevents the binding of Mdm2 and increases the activity of p53. Other pathways activate enzymes such as the HAT (*h*istone *a*cetyl *t*ransferase) enzyme p300/CBP, which acetylate p53 on certain lysine residues. Acetylation modifies the conformation to prevent the binding of Mdm2. Moderate increases in p53 lead to an increase in the transcription of the *waf1/cip1* gene to produce p21, which stops the cell from proceeding into the S phase of the cell cycle (see p. 1014). Higher increases in p53 activity result in apoptosis. p53 induces apoptosis by (i) increasing the transcription of several of the Bcl-2 pro-apoptotic genes and (ii) activation of cytoplasmic Bax to bind to the mitochondrial membrane. Both mechanisms increase the concentrations of pro-apoptotic Bcl-2-like proteins in the mitochondrial membrane and promote cell death.

P53 levels are also regulated by the transcription factors **Myc** and **E2F**, which increase in concentration on activation of the cell cycle by growth factor signals (see p. 1020). While these transcription factors promote synthesis of proteins with roles in the G_1 and S phases of cell division, they also initiate cell death when conditions are unfavorable for cell division. This emphasizes that commitment to enter the S phase is a critical decision for the cell since it leads to either cell division or death, depending on the circumstances. Under conditions unfavorable for cell division, elevated levels of the transcription factors E2F and Myc increase the synthesis of the protein **Arf**, which binds to Mdm2 and thereby prevents degradation of p53 (Figure 25.13). Thus E2F and Myc can increase p53 concentrations to cause either a p53-dependent cell cycle arrest or apoptosis, but how the decision between cell death and cell division is made is not clear. In most cancer cells the p53 pathway is defective and the mechanism of E2F or Myc initiated p53-dependent cell death is not functional so that the p53-defective cells tend to divide under conditions in which normal cells would die.

Apoptosis Regulation by MAPK Pathways

The **MAPK (nitrogen-activated protein kinase) pathways** are activated by a growth factor signal (see p. 1020), but also have a dual role in regulating both cell division and cell death pathways. In Figure 25.9, the growth factor receptor signal was shown to be transmitted to GTP-Ras, which activated a MAPKKK → MAPKK → MAPK pathway. In actuality, there are multiple MAPK pathways downstream from Ras, so that Ras is a focal point for a network of pathways (Figure 25.14). Which pathways are activated, their intensity, and their duration of activity depend on cell type and on its physiological circumstance. Signals other than those from activated Ras can also activate individual MAPK pathways. Thus, some MAPK pathways are activated by UV light, radiation, cytokines, or the stress molecules of inflammation, all in the absence of Ras activation. How these MAPK pathways then regulate cell division, differentiation, or cellular apoptosis depends on the integration and interpretation by the cell of the intensity, timing,

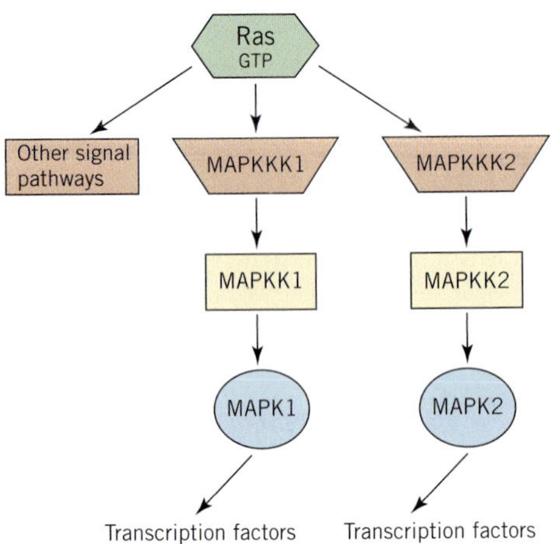

FIGURE 25.14

Multiple pathways may be regulated by Ras. MAPKKK are MAPK kinase kinases. MAPKs on activation by a MAPKK can localize to the nucleus to phosphorylate transcription factors. The kinases may also phosphorylate proteins other than transcription factors. Alternate targets of the kinases include Bcl-2-like proteins and p53.

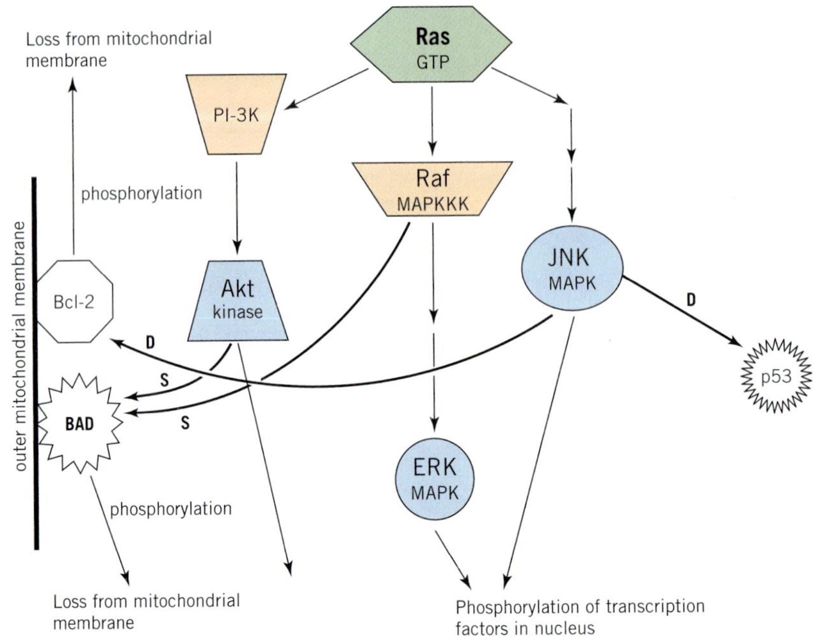

FIGURE 25.15

Interaction of Ras downstream kinases with apoptosis-regulating proteins. S, **S**urvive (pathway promotes cellular survival); D, **D**ie (pathway promotes apoptotic cellular death). ERK, JNK, Raf and Akt are kinases P_i3K, 1-phosphatidylinositol-3 kinase.

and duration of all signals. Thus, activation of the MAPK known as JNK commonly promotes apoptosis, but more rarely and only in particular cell types promotes survival. The outcome of the signals are cell-type dependent. Under some circumstances, JNK promotes death by phosphorylating Bcl-2 to promote its dissociation from the mitochondrial membrane (Figure 25.15), which increases the mitochondrial concentrations of free Bak and Bax and promotes cell death. JNK can also phosphorylate p53, making it resistant to Mdm-2, resulting in an increase in p53 concentration and cell death. Also the MAPK ERK, the MAPKKK Raf, and the kinase Akt promote cell survival by phosphorylating the pro-apoptotic facilitator protein Bad or by activating expression of anti-apoptotic Bcl-2-like genes (Figure 25.15). Thus, different kinases downstream from Ras transmit opposing life and death signals, and the outcome depends on the cellular context and cell type.

25.4 CANCER

Oncogenes and Tumor Suppressor Genes

The cell cycle and apoptosis are critical to an understanding of cancer. Genes that encode proteins that promote cell division or that promote resistance to apoptosis are protooncogenes (Table 25.2). Their activating mutations or overexpression results in increased activity, which leads to unregulated cell division or to resistance to apoptosis. Mutations in a protooncogene can convert it to an oncogene. Two copies (or alleles) of each autosomal gene (i.e., those on chromosomes other than the X or Y chromosomes) are present in the genome of every somatic cell [each somatic cell has a maternally derived and a paternally derived allele for each autosomal gene]. An activating mutation in a single allele of a protooncogene is sufficient to cause a pro-proliferative or anti-apoptotic effect in a cell. Such mutations are therefore autosomal dominant for progression of cancer.

Protooncogenes whose products promote cell division or resistance to apoptosis include genes for growth factors, growth factor receptors, Grb-like adaptor molecules, Src-like tyrosine kinases, kinases of MAPK cascades, Cdks, cyclins, CAKs, Cdc25s,

TABLE 25.2 Examples of Protooncogene Products that Regulate Cell Division and Apoptosis

A. Activators of Cell Division	
Growth factors	PDGF (*sis1*), FGF, and other growth and mitogen factors
Growth factor receptors	PDGFR and other growth factor receptors
Tyrosine kinase cytoplasmic proteins	Src family of proteins
Adaptor proteins	Grb-2
Ras GTP-binding proteins	K-Ras, H-Ras, N-Ras
Kinases and kinase cofactors	Cyclins, Cdk, CAK, and MAPKs
Phosphatases	Cdc25
Transcription factors	Myc, Jun, Fos, and E2F
B. Activators of Apoptotic Resistance (Survival)	
Bcl-2 and Bcl-X_L	
Mdm2	
XIAP	
Akt and Raf (MAPKKK)	
NFκB (transcription factor)	

and transcription factors (such as Jun, Fos, Myc, and E2F) that increase expression of cell cycle proteins. Bcl-2 and Mdm-2 are anti-apoptotic proteins whose expression is elevated in many cancers where they promote resistance to apoptosis. The most prominent protooncogene is *ras*, which is mutated into a constitutively active form in approximately 30% of human cancers.

Products of **tumor suppressor genes** suppress the cell division cycle or promote apoptosis. For a tumor suppressor gene to contribute to cancer, it must lose activity, in contrast to the oncogenes that are active and promote the cancer. Thus p53, required for normal apoptosis initiated by a variety of stimuli including DNA damage, is inactive in the majority of cancer cells. When active, it suppresses the cell cycle through its regulation of p21 expression and as an inducer of apoptosis in cells that contain damaged DNA, so that its functions inhibit development and propagation of cancer cells. Rb protein is a prominent tumor suppressor as it sequesters E2F and inhibits entrance into S phase. The CKIs, such as p21 encoded by the *waf1/cip1* gene, inhibit progression through the cell cycle and are tumor suppressors. Bak, Bax, and Bad are also tumor suppressors as their activity promotes apoptosis due to DNA damage, and their loss promotes resistance to apoptosis as required by a cancer cell (see Clin. Corr. 25.1). Both alleles of a tumor suppressor gene must be inactivated or lost in order to eliminate their tumor suppression activity from a cell. Thus a cell with one wild-type (normal) Rb gene and one mutated Rb gene behaves normally. Similarly, a cell with one functional p53 allele will usually show a normal apoptotic response. In contrast to oncogenes, mutations in tumor repressor genes are autosomal recessive for progression of a cancer.

Properties of Cancer Cells

Two major properties acquired by a cancer cell have been emphasized, namely unregulated cell division and resistance to apoptosis. Malignant cancer cells also exhibit cellular immortality, ability to induce angiogenesis, and ability to metastasize. Genes that contribute to these properties are also considered protooncogenes, and those that inhibit these properties to be tumor suppressor genes (Figures 25.16 and 25.17).

Cellular immortality is the ability of a cell to produce infinite generations of progeny. Mortal cells, such as normal human epithelial cells in cell culture, divide to produce 40–50 generations and then enter senescence. Cells in culture derived from an immortal cell continue to divide indefinitely.

In addition, malignant cancer cells have an ability to induce **angiogenesis** and to metastasize. Angiogenesis refers to the growth of new blood vessels into tissues. This is required to provide the increasing mass of the tumor with nutrients and oxygen for

CLINICAL CORRELATION 25.1
Oncogenic DNA Viruses

Oncogenic DNA viruses produce proteins that act as oncogenes in host cells by activating cell division or inhibiting cell death (apoptosis). The viral oncogenes become carcinogenic when incorporated into their host cell DNA by a recombinational event and then constitutively expressed in an unregulated manner. Papillomavirus is a causative factor in over 95% of uterine cervical cancers. The virus produces the oncoproteins E6 and E7. The E7 protein binds to the Rb-binding pocket, displacing E2F from Rb, resulting in the transcription factor E2F up-regulation of the synthesis of S-phase proteins that under a normative viral infection is required for viral DNA replication. In the absence of viral replication and virus-induced host cell death, the expression of E7 is oncogenic. The oncoprotein E6 binds to p53 and promotes its degradation, eliminating p53 mechanisms for inhibition of cell division during DNA damage and its promotion of cell death (see p. 1022). Both E6 and E7 promote the targeting of p53 and Rb for ubiquitination and degradation by the proteasome system. Other oncogenic human DNA viruses and associated cancers are the hepatitis B virus in heptocellular carcinoma, the Epstein–Barr virus in Burkitt's lymphoma and a subset of nasopharyngeal carcinomas and lymphomas, and human T-cell leukemia virus (HTLV-1) in adult T-cell leukemias. These viruses contain genes that, when incorporated into the host cell genome, either directly or indirectly produce proteins that lead to dysregulated cell division or resistance to apoptosis.

Source: Scheffner, M. and Whitaker, N. J. Human papillomavirus-induced carcinogenesis and the ubiquitin–proteasome system. *Semin. Cancer Biol.* 13:59, 2003. Münger et al. Mechanisms of human papillomavirus-induced oncogenesis. *J. Virol.* 78:11451, 2004. Zur Hausen, H. Oncogenic DNA viruses. *Oncogene* 20:7820, 2001.

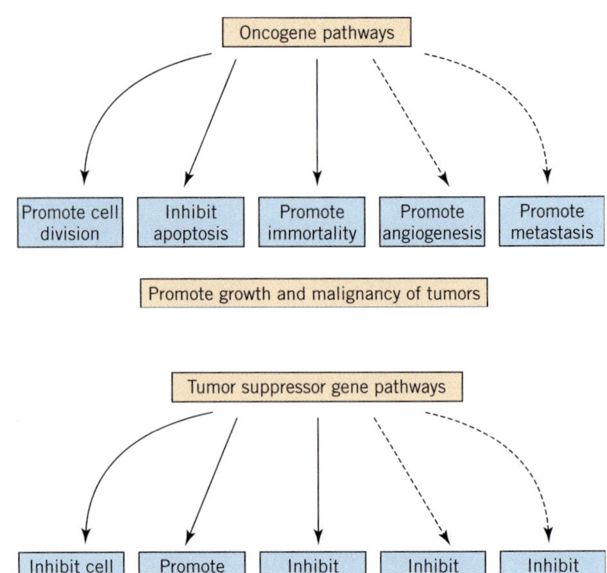

FIGURE 25.16

Oncogene pathways. Oncogene pathways promote cell division, inhibit apoptosis, promote cellular immortality, promote tumor cell metastasis, and promote angiogenesis into the tumor. Dotted lines denotes involvement of cells other than the tumor cells.

FIGURE 25.17

Tumor suppressor pathways. Tumor suppressor pathways inhibit cell division, promote apoptosis, inhibit cellular immortality, inhibit cancer cell metastasis, and inhibit angiogenesis into the tumor.

survival and growth. The malignant tumor cell also has the ability to metastasize. The metastasis of cancer cells to grow tumors at multiple sites is what usually kills patients with cancer. In blood-borne metastasis, tumor cells dissociate from their contacts with other cells within the tumor mass, pass through the basement membrane and the surrounding extracellular matrix into capillaries, and travel through the blood to a target tissue, where they pass out of the blood vessels and proliferate to form a secondary tumor.

Immortality of Cancer Cells

Cellular immortality may depend on the ability of cancer cells to increase their telomerase activity (see p. 152). Approximately 90% of human tumor cells show increased activity of telomerase. Telomerase maintains the length of the telomer regions at the ends of chromatids. A terminal nucleotide segment is lost each time chromatids are duplicated and if telomer extension by telomerase is lacking, progeny cells enter senescence to protect against DNA damage due to shortened telomeres. Whereas most normal cells lose telomerase activity with increasing age of the organism, cancer cells retain the ability to express telomerase to maintain chromatid telomer lengths and continue cell divisions.

Metastasis of Cancer Cells

In metastasis, the tumor cell acquires an ability to pass through basement membrane surrounding the tumor and the extracellular matrix beyond the basement membrane. This invasion process requires the tumor cell to have cellular motility. Development of motility in a normally stationary cell involves biochemical and structural changes in the cellular cytoskeleton and membrane. In addition, the tumor or neighboring cells must secrete enzymes that degrade the basement membrane and extracellular matrix to produce a path for tumor cell migration. Basement membrane (or *basal lamina*) consists of an ordered collagen type IV matrix, proteoglycans, the large protein laminin, and other proteins (see p. 353). Metastatic tumor cells have an elevated concentration of laminin receptors in their plasma membrane, which on binding of laminin induce signals that initiate cellular processes related to metastasis. Many malignant cells also produce and secrete the urokinase plasminogen activator (uPA) and type IV procollagenase.

Cell invasion through basement membrane involves binding to laminin in extracellular matrix by **laminin receptors**, induction of secretion of proteases into the extracellular matrix (ECM), which degrades the basement membrane near the cell, and movement through the hole created by the protease activity. These three steps are reiterated multiple times as the cell moves through the basement membrane.

Tumor cells typically promote the secretion from other cells of a **plasminogen activator** such as **uPA (urokinase plasminogen activator)** and a type IV procollagenase (Figure 25.18). The plasminogen activator cleaves extracellular plasminogen to generate the active protease plasmin. Plasmin has a general protease activity that destroys proteins of the extracellular matrix and also activates type IV procollagenases to active collagenases. These activities produce a path through the basement membrane for cell passage. Many tumor cells also overexpress uPA receptor (uPAR) in their plasma membranes to bind secreted uPA and give the tumor cell membrane plasminogen activator activity.

Urokinase plasminogen activator (uPA) is a serine-type protease, which splits plasminogen into a C-terminal segment that has a serine protease activity (plasmin) (see p. 328) and an N-terminal segment long believed to have no biological role, but has recently been shown to have anti-angiogenic activity (see below). Plasminogen is synthesized by liver cells, is secreted into blood, and is present in the extracellular matrix of the whole organism.

The type IV collagenases are proteases that contain a Zn atom in their catalytic site. They are named **matrix metalloproteases** (MMPs). Two human gene products, MMP2 and MMP9, have type IV collagenase activity and usually one or both of these are overexpressed in malignant cells. In some tumors, the cancer cells secrete factors that induce nontransformed neighboring cells to synthesize and secrete MMPs and plasminogen activators to degrade the extracellular matrix near the growing tumor.

Under normal conditions, the turnover of extracellular matrix is regulated by the relative activities of proteases and protease inhibitors. Two **plasminogen activator inhibitors** are known, **PAI-1** and **PAI-2** (**p**lasminogen **a**ctivator **i**nhibitor 1 and 2), and several collagenase inhibitor proteins or TIMPs (**ti**ssue **i**nhibitor of **m**etallo-**p**roteases) (Figure 25.18). Cancer cells or neighboring cells increase extracellular protease activity by increasing the concentrations of plasminogen activators and MMPs and decreasing the concentration of their inhibitors.

Angiogenesis Induced by Cancer Cells

Angiogenesis is a normal process during embryonic and fetal development and in wound healing, and it is induced by cancer cells. Angiogenesis is promoted by tumor cells through secretion of endothelial cell chemoattractant proteins. Two common angiogenic proteins are **VEGF (vascular endothelial cell growth factor)** and **FGF (fibroblast growth factor)**. Their secretion induces blood vessel endothelial cells to proliferate and migrate toward and form new blood vessels into the tumor (Figure 25.19). The invading of the endothelial cells mimics the tumor cells in metastasis. The endothelial cells increase secretion of plasminogen activator and MMPs, including type IV procollagenase, and

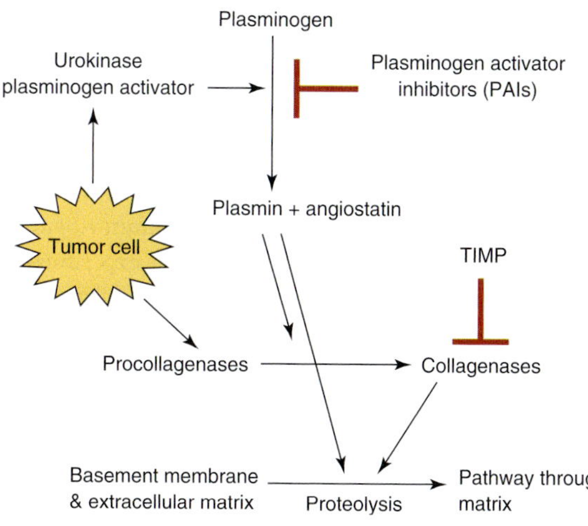

FIGURE 25.18

Secreted proteases generate pathway through basement membrane in metastasis. Tumor or normal neighboring cells secrete proteases urokinase plasminogen activator (uPA) and type IV procollagenases (MMPs, matrix metallo-proteinases) and decrease secretions of protease inhibitor proteins TIMP (*t*issue *i*nhibitor of *m*etallo*p*roteinases) and PAI (*p*lasminogen *a*ctivator *i*nhibitor). uPA cleaves plasminogen into plasmin and angiostatin (an angiogenesis inhibitor, discussed in next section). Plasmin activates procollagenase to collagenase and also acts on other basement membrane proteins to promote their degradation.

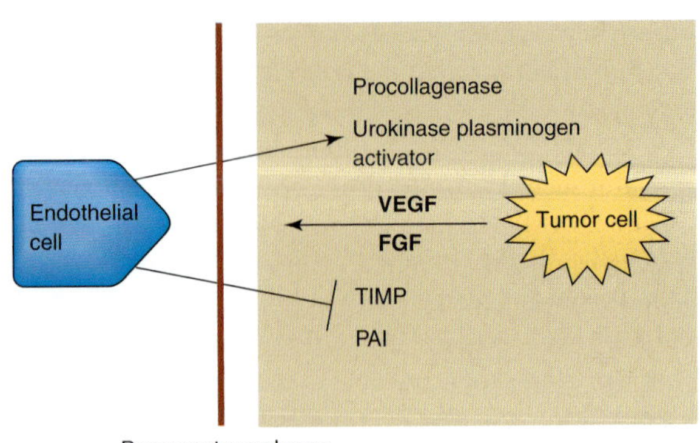

FIGURE 25.19
Induction of invasion and growth of blood vessel endothelial cells by tumor cell in angiogenesis. Tumor cells secrete endothelial cell growth and chemotactic factors VEGF and/or FGF, which promote endothelial cells proliferation and migration toward the tumor and their formation of blood vessels. In their migration through the extracellular matrix towards the tumor, the endothelial cells use a similar mechanism to that of the tumor cell in metastasis. Endothelial cells increase their secretion of uPA and procollagenases and decrease their secretion of proteases inhibitors PAI and TIMP.

decrease their secretion of PAIs and TIMPs. This gives them the ability to pass through basement membrane and ECM toward the tumor, while proliferating to form vessels.

The N-terminal portion of plasminogen, produced by uPA action, has anti-angiogenic activity and is called angiostatin. The angiostatin generated near metastasizing tumor cells or invading endothelial cell at a primary tumor has inadequate activity to inhibit angiogenesis, as the protease activities predominates. However, the **angiostatin** has a longer lifetime than the protease, and at distant sites it inhibits angiogenesis and tumor growth. This anti-angiogenic activity is overcome when and if the metastatic tumors develop, but initially this anti-angiogenic activity slows the development of the metastatic tumors at the secondary sites. Other anti-angiogenic proteins are also generated by proteolysis of ECM proteins near the primary tumor. One of these is endostatin, which is generated from the C-terminal end of type XVIII collagen. **Endostatin**, angiostatin, and other anti-angiogenic proteins derived from extracellular matrix have been shown to cure cancer in mice. These observations have stimulated the investigation of anti-angiogenic drugs for treatment of human cancers.

Multiple Mutations Are Required to Form a Cancer

A single mutation is not sufficient to transform a cell into a cancer cell unless it occurs on a background of previous oncogenic mutations. The evidence of a requirement for multiple mutations comes from many observations. The time course for development of a cancer after an initial mutational event (e.g., from exposure to high radiation or a potent chemical mutagen in an industrial setting) shows that the cancer usually appears many years after the mutation event. This indicates a need for additional mutations. Also, the incidence of cancer increases with age, indicating the need to accumulate mutations. Other evidence comes from the observation that mutations in oncogenes and tumor suppressor genes accumulate in the progressive stages of the development of a cancer. Thus in the development of cervical cancer, some epithelial cells initially start to show a higher-than-normal proliferative rate. These may revert back to a normal rate of proliferation, or proceed to form an adenoma (a benign tumor). Most early adenomas do not progress to form a cancer, but some progress through additional stages until a malignant tumor forms with the ability to metastasize. The progressive stages correspond to the acquisition of additional mutations that eventually form the malignant cancer cell.

Experiments with primary human epithelial cells show that a minimum of four changes is required to generate a transformed epithelial cell. In these experiments, different oncogenes and tumor suppressor genes were introduced in gene expression plasmids into the cells. A minimum of three genes that affect four biochemical activities was needed to transform the cell. Required was a gene that overexpressed telomerase, a *ras* oncogene, and the gene for the SV40 virus T antigen protein. The T antigen protein binds and inhibits Rb and p53 in the infected human cells. Thus Rb and p53 suppression, Ras activation, and **telomerase** overexpression were all essential for cellular transformation (see Clin. Corr. 25.2).

CLINICAL CORRELATION 25.2
Molecularly Targeted Anticancer Drug

Imatinib (formerly known as STI-571 and marketed as Gleevec in the United States) is one of the first successful drugs molecularly targeted to an oncogene product. It inhibits the tyrosine kinase activity of the chimeric Bcr-Abl tyrosine kinase protein expressed in chronic myelogenous leukemia (CML) and the c-Kit growth factor receptor overexpressed in many gastrointestinal stromal tumors. It has been approved by the U.S. Food and Drug Administration for treatment of these cancers. Tyrosine kinases transfer phosphate from ATP to protein substrates in growth factor signaling pathways. Imatinib binds to the ATP-binding site of the Abl and Kit kinases, inhibiting the normal binding of the ATP substrate (see accompanying figure).

CML arises by a translocation between chromosome 9 and 22 in a pluripotent hematopoietic stem cell to produce aberrant chromosomes in which a part of chromosome 22 is translocated to 9 and a part of 9 to 22. The smaller of the two abnormal chromosomes, $22q^-$, is easily recognizable in leukemia cells and is called the Philadelphia chromosome. At the translocation site, the reading frame of the 5′-end of the Bcr gene is fused with the 3′ sequences of the Abl gene to produce the chimeric protein Bcr-Abl. The fusion results in a highly active and dysregulated Abl tyrosine kinase activity, which is the cause of the leukemia.

The progression of CML is divided into stages, the initial chronic stage characterized by the expansion of terminally differentiated neutrophils typically spans a 5-year period. The later blastic stage is characterized by excessive numbers of undifferentiated lymphoid progenitor cells and carries a poor prognosis. The Bcr-Abl oncogene is expressed at all stages, but cells in the blast stage often contain additional molecular and cytogenetic changes. The initial chronic stage has the highest response in treatment with imatinib (complete or nearly complete cytogenetic remissions of Philadelphia chromosome containing cells in 80–90% of patients). Blast stage CML often shows partial remission (cytogenetic response rate of 16–24% reported). It is hoped that better responses in blastic phase disease will be achieved in combinations with other drugs. There are reports of the development of resistance to imatinib in the blastic stage by the appearance of point mutations in Bcr-Abl that decrease its affinity for imatinib. Other mutations can lead to imatinib resistance by causing an amplified expression of the Bcr-Abl fusion protein.

(a) Imatinib

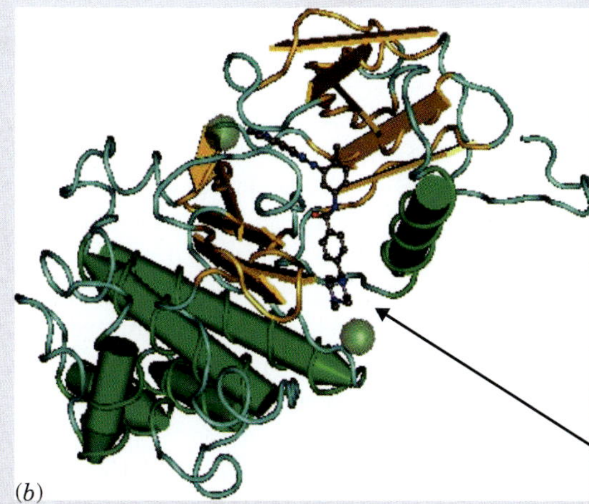

(b)

Binding of Imatinib into ATP Binding Site in Tyrosine Kinase Domains of Abl. (*A*) Structure of Imatinib (STI-571). (*B*) Structure of The Abl tyrosine kinase domain with Imatinib in ATP binding site. Imatinib Shown with Stick-and-Ball Model (See Arrow). [Mmdb Structure 16291 Visualized with Cn3D Software.] Citation for CN3D is Chen, J., et al. *Nucleic Acids Res.* 31:474, 2003, and citation for X-ray structure is Nagar, B., et al. *Cancer Res.* 62:4236, 2002.

Source: Goldman, J. M. and Melo, J. V. Chronic myeloid leukemia—Advances in biology and new approaches to treatment. *N. Engl. J. Med.* 349:1451, 2003. Kantarjian, H. M., et al. Imatinib mesylate (STI571) therapy for Philadelphia chromosome-positive chronic myelogenous leukemia in blast phase. *Blood* 99:3547, 2002. Kantarjian, H. M., et al. Imatinib mesylate therapy in newly diagnosed patients with Philadelphia chromosome-positive chronic myelogenous leukemia: High incidence of early complete and major cytogenetic responses. *Blood* 101:97, 2003. Talpaz, M., et al. Imatinib induces durable hematologic and cytogenetic responses in patients with accelerated phase chronic myeloid leukemia: Results of a phase 2 study. *Blood* 99:1928, 2002. Gorre, M. E., et al. Clinical resistance to STI-571 cancer therapy caused by BCR-ABL gene mutation or amplification. *Science* 293:876, 2001.

Genetic and Biochemical Heterogeneity of Cancers

Cells in a tumor contain numerous mutations in protooncogenes and in tumor suppressor genes. The spectrum of mutations may differ among the millions of cells in a tumor mass. This is demonstrated experimentally by showing that cells taken from different regions of the same tumor show a difference in their metastatic ability, which arises from genetic and epigenetic differences between the cells. However, genetic

CLINICAL CORRELATION 25.3
Environmental Cause of Human Cancers

Environmental and cultural factors are the predominant causes of human cancer. If these factors can be regulated, the incidence of cancers may be reduced by greater than 75%. Cigarette smoking is responsible for approximately 30% of U.S. cancer deaths. In addition, it is estimated that diet can prevent the development of approximately 30% of U.S. cancers (estimates vary from 10 to 70%). These numbers are primarily based on epidemiological data of the incidence of cancer types in different geographical locations, cultures and environments (see accompanying table). On migration from one culture or location to another, the incidence of a cancer type often approaches that of the population in the new culture as the individuals or their offspring assimilate. Also rapid changes in diet or lifestyle overtime within a country show the importance of these factors in the incidence of cancers.

The quarter of the U.S. population with the highest dietary fruit and vegetable content has a 30–40% lower incidence for many cancer types. However, the constituents of these foods that inhibit the formation of these cancers are not yet clear. Inadequate folic acid levels in the U.S. diet have been implicated as a risk factor for colon and breast cancers. Folic acid intake may be particularly critical in individuals with a polymorphism in their gene for methylenetetrahydrofolate reductase, an enzyme involved in folic acid metabolism, which results in decreased activity. Folate deficiency causes excessive incorporation of uracil into DNA and chromosome breaks. Folate deficiency is common in alcoholics and may be the reason for an increase in the incidence of certain cancers with alcohol. Pre-adolescent diet and lifestyle affect the onset of menarche, and the early onset of menarche and length of time between menarche and menopause is associated with increased risk of breast cancer in women. Obesity is inversely correlated with risk of breast cancer in pre-menopausal women and positively correlated in postmenopausal. High animal fat in the diet has been associated with increased risk of colon cancer, but the data are inconsistent. Rather than total animal fat in the diet, the data may suggest that it is the ratio of polyunsaturated to saturated fat that correlates with colorectal cancer risk. Colon cancer incidence has also been inversely correlated with lack of activity or exercise.

Environmental and dietary studies are difficult as large population must be studied over a generation of time, and multiple variables must be controlled. However, advances in our knowledge and control of these factors can have an enormous effect in decreasing the incidence of cancer.

It is estimated that there are 1,100,000 worldwide lung cancer deaths per year, 85% of which are caused by tobacco. While cigarette consumption rates have recently begun to decrease, the time lag in cancer development will result in dramatic increases in smoking-induced cancers over the next two or three decades to 2 million per annum.

The effect of environment and lifestyle on cancer incidence is interpreted through a genetic background. An individual's susceptibility to environmental factors is influenced by their genetic polymorphisms and acquired mutations that control their metabolic, regulatory, and hormonal pathways. Higher risks occur in either (a) persons whose pathways are more active in the production of procarcinogens and tumor promoters induced by environmental stimulators or (b) those who fail to detoxify environmental carcinogens.

Variation in the Incidence of Cancer Type in a Two Location Comparison[a]

	Location Comparison	*Incidence in Location 1*	*Incidence in Location 2*	*Fold Difference in incidence*
Lung	New Orleans (blacks)–Madras (India)	110	5.8	19
Breast	Hawaii (Hawaiians)–Israel (non-Jews)	94	14	7
Prostate	Atlanta (blacks)–China (Tianjin)	91	1.3	70
Uterine cervix	Brazil (Racife)–Israel (non-Jews)	83	3.0	18
Liver	China (Shanghai)–Canada (Nova Scotia)	34	0.7	49
Colon	United States (Connecticut, whites)–India (Madras)	34	1.8	19
Melanoma	Australia (Queensland)–Japan (Osaka)	31	0.2	155

[a] Incidence in number of new cases per year per 100,000 people, adjusted for age variation with the specific population. Data were taken Alberts, B., Johnson, A.,. Lewis, J., Raff, M., Roberts, K., and Watson, J. D. *Molecular Biology of the Cell*, 4th ed. New York: Garland, 2002, Table 24-2, which was adapted from DeVita, V. T., Hellman, S., and Rosenberg, S. A. (Eds.). *Cancer: Principles and Practice of Oncology*, 4th ed. Philadelphia: Lippincott, 1993; based on data from C. Muir et al. *Cancer Incidence in Five Continents*, Vol. 5, Lyon: International Agency for Research on Cancer, 1987. General References: Shibuya, K., Mathers, C. D., Boschi-Pinto, C., Lopez, A. D., and Murray, C. J. L. Global and regional estimates of cancer mortality and incidence by site: II. Results for the global burden of disease 2000. *BMC Cancer* 2:37, 2002. Pisani, P., Parkin, D. M., Bray, F., and Ferlay, J. Estimates of the worldwide mortality from 25 cancers in 1990. *Int. J. Cancer* 93:18, 1999.

Source: Willett, W. C. Diet and cancer. *The Oncologist* 5:393, 2000. Bingham, S. and Riboli, E. Diet and cancer—The European prospective investigation into cancer and nutrition. *Nature Rev. Cancer* 4:206, 2004. Ames, B. N. and Wakimoto, P. Are vitamin and mineral deficiencies a major cancer risk? *Nature Rev. Cancer* 2:694, 2002. Calle, E. E. and Thun, M. J. Obesity and cancer. *Oncogene* 23:6365, 2004. Proctor, R. N. Tobacco and the global lung cancer epidemic. *Nature Rev. Cancer* 1:82, 2001. Ames, B. N., Gold, L. S., and Willett, W. C. The causes and prevention of cancer. *Proc. Natl. Acad. Sci. USA* 92:5258, 1995.

markers can trace each of these tumor cells to a single precursor cell, in which the initial mutational events occurred.

Mutagens and Promoters Cause Cancers

Most cancers are caused by environmental factors (see Clin. Corr. 25.3). These factors may act as mutagens or as promoters. A promoter does not directly cause mutation but acts as a growth factor and increases the number of cells containing a initial mutagenic defect. It thereby expands clones of mutant cells to give a higher probability for additional mutations in these cells. Thus, a promoter acts to promote

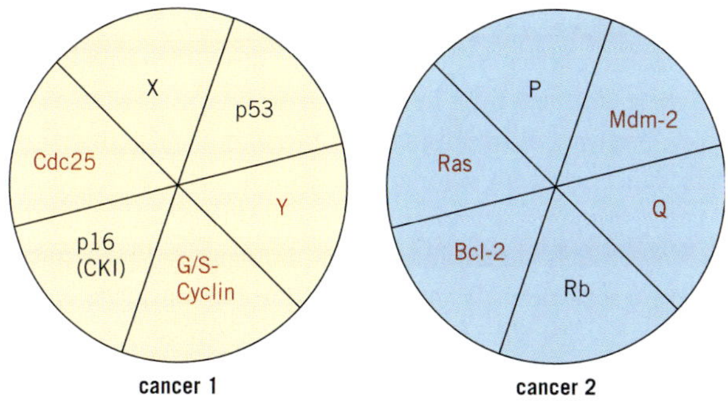

FIGURE 25.20

Each cancer has a different pattern of activated protooncogenes and deactivated tumor suppressor genes. Oncogenes are colored red and tumor suppressor genes are colored black. X, Y, P, and Q are unidentified oncogenes or tumor repressor genes mutated in the particular tumor.

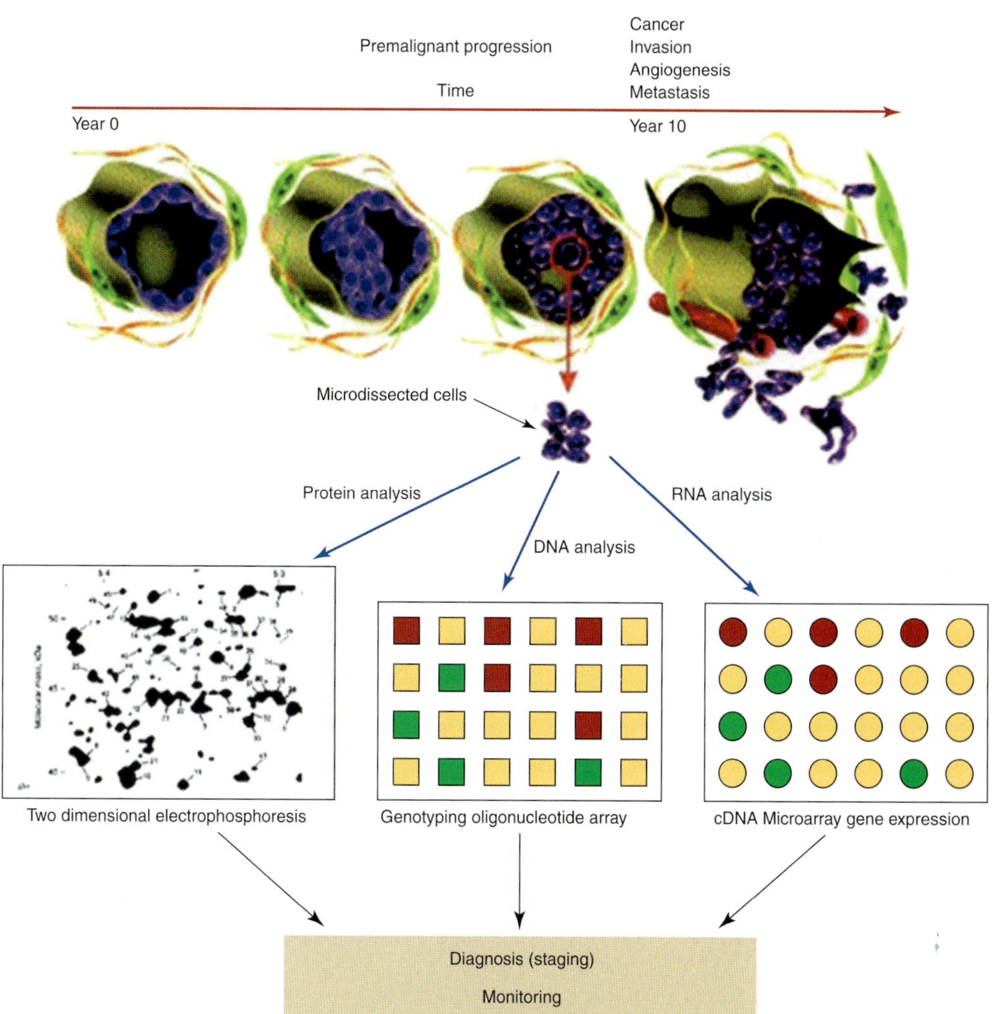

FIGURE 25.21

Molecular analysis of laser microdissected cells characterize the specific tumor. Different stages in development of a malignant cancer may be analyzed for their genetic and expression patterns with microdissected cells extracted for molecular analysis (top). Molecular analyses are by two-dimensional gel electrophoresis to determine expressed proteins (see p. 120), genomic analysis, and cDNA analysis for expressed mRNAs. Staging, monitoring, and treatment of cancer are then based on biochemical and genetic analysis of the particular cancer. Reprinted and modified with permission from Simone, N. L., Bonner, R. F., Gillespie, J. W., Emmert-Buck, M. R., and Liotta., L. A.. Laser-capture microdissection: opening the microscopic frontier to molecular analysis. *Trends Genet.* 14:272, 1998. Copyright (1998) Elsevier.

formation of a cancer after an initial mutational event. Many natural factors are tumor promoters. Women with higher-than-normal levels of estrogen have a higher incidence of cancers in estrogen-sensitive tissues. The diet, particularly fruits and vegetables, provide factors that inhibit the action of mutagens and tumor promoters or may provide cofactors for antioxidant enzymes and enzymes involved in DNA repair.

Biochemical Analysis of Cancers

Each cancer has a different pattern of activated oncogenes and inactivated tumor suppressor genes (Figure 25.20). However, certain common mutations occur in different cancer types. Thus, there are common mutations found in particular cancer types, but most cancers have some genetic differences from another, even within the same cancer types. Often mutations may be in different genes, but are in the same common pathway. In addition, some pathways may be compensating to others, so that the same outcome can be affected by increasing one pathway or inhibiting the other. However, because of differences, characterization of a cancer is essential for accurate prognosis and efficacious treatment. Gene microarray and proteomic techniques are beginning to be used to characterize individual tumors to precisely determine their molecular and biochemical composition and to permit selective treatments (Figure 25.21). Future pharmacological treatments will be based on the exact molecular characterization of a cancer from which can be designed selective treatments against the particular pathways that sustain that cancer.

BIBLIOGRAPHY

Cell Cycle

King, K. L. and Cidlowski, J. A. Cell cycle regulation and apoptosis. *Annu. Rev. Pharmacol. Toxicol.* 39:295, 1999.

Morgan, D. O. Cyclin-dependent kinases: engines, clocks, and microprocessors. *Annu. Rev. Cell Dev. Biol.* 13:261, 1997.

Nurse, P. A long twentieth century of the cell cycle and beyond. *Cell* 100:71, 2000.

Weinberg, R. A. The retinoblastoma protein and cell cycle control. *Cell* 81:323, 1995.

Growth Factor Receptors and Signaling Pathway

Claesson-Welsh, L. Platelet-derived growth factor receptor signals. *J. Biol. Chem.* 269:32026, 1994.

Fambrough, D., McClure, K., Kazlauskas, A., and Lander, E. S. Diverse signaling pathways activated by growth factor receptors induce broadly overlapping, rather than independent, sets of genes. *Cell* 97:727, 1999.

Schlessinger, J. and Lemmon, M. A. SH2 and PTB domains in tyrosine kinase signaling. *Sci. STKE 2003* 191:12, 2003.

Weng, G., Bhalla, U. S., and Iyengar, R. Complexity in biological signaling systems. *Science* 284:92, 1999.

Apoptosis

Cheng, E. H. -Y. A., Wei, M. C., Weiler, S., Flavell, R. A., Mak, T. W., Lindsten, T., and Korsmeyer, S. J. Bcl-2, Bcl-Xl sequester BH3 domain only molecules preventing Bax and Bak mediated mitochondrial apoptosis. *Mol. Cell* 8:705, 2001.

Green, D. R. and Reed, J. C. Mitochondria and apoptosis. *Science* 281:1309, 1998

Saelens, X., Festjens, N., Vande Walle, L., van Gurp, M., van Loo, G., and Vandenabeele, P. Toxic proteins released from mitochondria in cell death. *Oncogene* 23:2861, 2004.

Scorrano, L. and Korsmeyer, S. J. Mechanisms of cytochrome c release by preapoptotic BCL-2 family members. *Biochem. Biophys. Res. Commun.* 304:437, 2003.

vander Heiden, M. G. and Thompson, C. B. Bcl-2 proteins: regulators of apoptosis or of mitochondrial homeostatsis? *Nature Cell Biol.* 1:E209, 1999.

Huang, D. C. S. and Strasser, A. BH3-only proteins-essential initiators of apoptotic cell death. *Cell* 103:839, 2000.

Thompson, C. B. Apoptosis in the pathogenesis and treatment of disease. *Science* 267:1456, 1995.

Cancer

Ames, B. N., Gold, L. S., and Willett, W. C. The causes and prevention of cancer. *Proc. Natl. Acad. Sci. USA* 92:5258, 1995.

Hanahan, D. and Weinberg, R. A. The hallmarks of cancer. *Cell* 100:57, 2000.

Mendelsohn, J., Howley, P. M., Israel, M. A., and Liotta, L. A. (Eds.). *Molecular Basis of Cancer*, 2nd ed. Philadelphia: Saunders, 1995.

Vogelstein, B., Kinzler, K. W. (Eds.). *The genetic basis of human cancer*, 2nd ed. New York: McGraw-Hill, 2002.

Oncogenes and Tumor Suppressor Genes

Futreal, P. A., et al. A census of human cancer genes. *Nature Rev. Cancer* 4:177, 2004.

Hahn, W. C., et al. Creation of human tumour cells with defined genetic elements. *Nature* 400:464, 1999.

Kinzler K. W, and Vogelstein B. Lessons from hereditary colorectal cancer. *Cell* 87:159, 1996.

Kinzler, K. W., and Vogelstein, B. Colorectal tumors. In: Vogelstein, B., Kinzler K. (Eds.), *The Genetic Basis of Human Cancer*. New York: McGraw-Hill Book, p. 583, 2002.

Biochemical Analysis of Cancer

Brown, P. O. A gene expression database for the molecular pharmacology of cancer. *Nat. Genet.* 24:236, 2000.

Oh, P., Li, Y. Yu, J., Durr E., Krasinska, K. M., Carver, L. A., Testa, J. E., and Schnitzer, J. E. Substractive proteomic mapping of the endothelial surface in lung and solid tumours for tissue-specific therapy. *Nature* 429:629, 2004.

Pao, W., et al. EGF receptor gene mutations are common in lung cancers from "never smokers" and are associated with sensitivity of tumors to gefitinib and erlobinib. *Proc. Natl. Acad. Sci. USA* 101:13306, 2004.

van't Veer, L. J., et al. Gene expression profiling predicts clinical outcome of breast cancer. *Nature* 415:530, 2002.

Liotta, L. A., Ferrari, M., and Petricoin, E. Written in blood; clinical proteomics. *Nature* 425:905, 2003.

Perou, C. M., et al. Molecular portraits of human breast tumors. *Nature* 406:747, 2000.

Angiogenesis

Bergers, G. and Benjamin, L. E. Tumorigenesis and the angiogenic switch. *Nature Rev. Cancer* 3:401, 2003.

Boehm, T., Folkman, J., Browder, T., and O'Reilly, M. S. Antiangiogenic therapy of experimental cancer does not induce acquired drug resistance. *Nature* 390:404, 1997.

Carmeliet, P. Angiogenesis in health and disease. *Nature Med.* 9:653, 2003.

Folkman, J. and Kalluri, R. Cancer without disease; do inhibitors of blood-vessel growth found anturally in our bodies defend most of us against progression of cancer to a lethal stage? *Nature* 427:787, 2004.

Cellular Migration and Metastasis

Cairns, R. A., Khokha, R., and Hill, R. P. Molecular mechanisms of tumor invasion and metastasis: an integrated view. *Curr. Mol. Med.* 3:659, 2003.

Curran, S., and Murray, G. I. Matrix metalloproteinases in tumour invasion and metastasis. *J. Pathol.* 189:300, 1999.

Chang, C. and Werb, Z. The many faces of metalloproteases: Cell growth, invasion, angiogenesis and metastasis. *Trends Cell Biol.* 11:S37, 2001.

Mignatti, P. and Rifkin, D. B. Biology and biochemistry of proteinases in tumor invasion. *Physiol. Rev.* 73:161, 1993.

Signaling Networks in Cancer

Irish, J. M., Hovland, R., Krutzik, P. O., Perez, O. D., Bruserud, Ø., Gjertsen, B. T., and Nolan, P. P. Single cell profiling of potentiated phospho-protein networks in cancer cells. *Cell* 118:217, 2004.

Campbell, S. L., Khosravi-Far, R., Rossman, K. L., Clark, G. J., and Der, C. J. Increasing complexity of Ras signaling. *Oncogene* 17:1395, 1998.

McCormick, F. Signalling networks that cause cancer. *Trends Biochem. Sci.* 24:M53, 1999.

Finkel, T. and Gutkind, J. S.(Eds.). *Signal Transduction and Human Disease.* New York: Wiley-Liss, 2003.

Downward, J. Targeting Ras signalling pathways in cancer therapy. *Nature Rev. Cancer* 3:11, 2003.

p53

Oren, M. Regulation of the p53 tumor suppressor protein. *J. Biol. Chem.* 274:36031, 1999.

Prives, C. Signaling to p53: Breaking the MDM-2-p53 circuit. *Cell* 95:5, 1998.

Rich, T., Allen, R. L., and Wyllie, A. H. Defying death after DNA damage. *Nature* 407:777, 2000.

Vousden, K. H., and Prives, C. p53 and prognosis: New insights and further complexity. *Cell* 120:7, 2005.

Vousden, K. H. p53: death star. *Cell* 103:691, 2000.

QUESTIONS | CAROL N. ANGSTADT

Multiple Choice Questions

1. In the cell cycle:
   A. M phase is both the most complex and the longest phase.
   B. there is a G_0 phase in equilibrium with the G_1 phase.
   C. quiescent cells cannot be induced to reenter the cell cycle.
   D. microtubule spindles form during the S phase.
   E. the order of phases is G_1, G_2, S, and M.

2. Cyclin-dependent kinases (Cdks):
   A. occur only in the M phase.
   B. are always inactivated by phosphorylation.
   C. typically phosphorylate proteins on tyrosine residues.
   D. in addition to binding cyclin, require other modifications for activity.
   E. that phosphorylate Rb (retinoblastoma sensitivity) protein inhibit the synthesis of S-phase proteins.

3. Elements of growth factor pathway regulation of the cell cycle include all of the following *except*:
   A. dimerization or polymerization of receptor proteins in the plasma membrane.
   B. binding of ATP to a tyrosine kinase domain of growth factor receptor proteins.
   C. nonspecific binding of SH2 or SH3 domains to any phosphorylated tyrosine.
   D. conformational changes of proteins as part of signal transmission.
   E. location of the growth factor receptor proteins in the plasma membrane.

4. Ras protein of the signal pathway:
   A. is inactive in the GDP-Ras conformation.
   B. transmits a signal by initiating a kinase cascade.
   C. from a mutated gene in many cancers has inactivated GTPase activity so inhibits the downstream signal to divide.
   D. is formed from an oncogene.
   E. becomes inactive as a signaling protein by phosphorylating transcription factors.

5. Apoptotic death of a cell may include all of the following *except*:
   A. lysis of the cell membrane with inflammation.
   B. activation of enzymes called caspases.
   C. phagocytosis.
   D. binding of a ligand in the extracellular environment to a plasma membrane receptor.

E. transfer of toxic molecules from the mitochondria to cytoplasm through protein pores.

6. One inducer of apoptosis is DNA damage acting via p53, a transcription factor. P53:
   A. is activated by kinases located in chromatin.
   B. increases the transcription of several pro-apoptotic genes for Bcl-2 proteins.
   C. concentration is normally low because it is subject to ubiquitination.
   D. concentration may be increased by acetylation.
   E. all of the above are correct.

7. Protooncogenes:
   A. produce protein products that promote cell division.
   B. produce protein products that promote apoptosis.
   C. are converted to oncogenes by a mutation that inhibits their expression.
   D. require both alleles to be mutated to show a pro-proliferative effect.
   E. are constitutively activated.

8. Tumor suppressor genes:
   A. have the same effect as protooncogenes.
   B. contribute to cancer if only one allele is inactivated by mutation.
   C. produce protein products that promote apoptosis.
   D. produce products that promote cancer.
   E. all of the above are correct.

9. The gene for which of the following would be considered a protooncogene?
   A. p53.
   B. Rb protein.
   C. p21 (a Cdk inhibitor protein).
   D. genes for Bcl-2-like proteins.
   E. Ras

10. Malignant cancer cells have all of the following properties *except*:
    A. unregulated cell division.
    B. inhibition of angiogenesis.
    C. resistance to apoptosis.
    D. cellular immortality.
    E. ability to metastasize.

11. The property of cellular immortality refers to a cell's ability to:
    A. grow new blood vessels.
    B. invade through the basement membrane of its tissue.
    C. resist apoptosis.
    D. continue to divide indefinitely without limit to the number of cell generations.
    E. grow tumors at multiple sites.

12. Telomerase:
    A. activity generally increases as an individual ages.
    B. catalyzes the resynthesis of telomer regions.
    C. activity is high during all stages of the carcinogenesis process.
    D. plays a role in apoptosis but not immortality.
    E. when present at a high level, leads to cell senescence.

Problems

13. What is the role of type IV collagen in metastasis of a tumor cell?

14. Why is the incidence of cancer higher in older individuals than in younger ones?

ANSWERS

1. **B** A: M phase is the shortest, although it is highly complex. C: Senescent cells cannot be induced to reenter the cell cycle. D: This occurs in the M phase. E: G_1 and G_2 separate the M and S phases.
2. **D** For example, M-Cdk-cyclin heterodimer must be correctly phosphorylated. A: Other phases also have Cdks. B: There are both activation and inhibitory phosphorylation sites. C: Serine and threonine residues are the phosphorylation sites. E: Phosphorylation of Rb causes release of E2F and thus promotes synthesis.
3. **C** Each SH2 domain has a secondary specificity that differentiates between phosphorylated tyrosines A: This seems to be critical to transmission of the signal. B: Many receptors operate this way. Those that don't cause dimerization by binding a protein with a tyrosine kinase domain. D: This is a common occurrence.
4. **B** GTP-Ras phosphorylates MAPKKK, which phosphorylates the next kinase, etc. A: GTP-Ras is the active form. C: Inhibiting GTPase activity prevents the conversion to GDP-Ras so the signal remains active. D: The Ras gene is a protooncogene and requires an activating mutation to become an oncogene. E: The transcription factors are at the end of the chain so don't react directly with Ras.
5. **A** This occurs in necrotic cell death. B: These proteases degrade cell protein. C: This happens to the remainder of the cell after multiple digestions. D: This is the extrinsic or death receptor pathway. E: This is the intrinsic pathway with Bcl-2-like proteins forming the pores.
6. **E** A: DNA chain breaks and changes in chromatin structure initiate this. B: There is also a nontranscriptional method, both leading to the intrinsic pathway. C: Binding Mdm-2 catalyzes polyubiquitination. D: Acetylation prevents binding of Mdm-2.
7. **A** This is part of the definition. B: The products promote resistance to apoptosis. C: Oncogenes result in higher gene activity. D: One is sufficient. E. This occurs on mutation.
8. **C** Also products that suppress the cell cycle. A: These effects are opposite to those of protooncogenes. B: Both alleles need to be inactivated. D: Their normal products promote apoptosis which is opposite to what cancer cells require.
9. **E** Ras is a protein of a signal pathway. A, B, C, D are all tumor suppressor genes.
10. **B** Tumor cells must have the ability to induce the growth of blood vessels in order to grow. A, C, D, E are all properties that allow tumors to grow.
11. **D** This may occur because such cells have high levels of telomerase. A: This allows a cancer cell to grow a tumor mass. B, E: These are part of metastasis. C: Apoptosis is a separate property from immortality.
12. **B** A terminal nucleotide segment is lost each time chromotids are duplicated in cell division and must be replaced. A, E: Telomerase decreases on cell aging, leading to senescence. C: There is believed to be a low level at early and intermediate stages, leading to chromosomal instability and mutations. D: This is believed to be a major factor leading to immortality.
13. Type IV collagen is part of the basement membrane. Tumor cells typically promote the secretion of a plasminogen activator and a type

IV procollagenase. This leads to formation of plasmin, which cleaves proteins of the extracellular matrix and activates procollagenases to collagenases. Type IV collagen is degraded. These two activities provide a path through which the tumor cell passes to the blood.

14. A normal cell requires multiple mutations in protooncogenes and tumor suppressor genes to become a cancer cell. The long time interval between the initial mutational event and the appearance of cancer allow the multiple mutational events to occur. During this interval, some cells with high rates of proliferation might revert to normal rates while others proceed on to eventually form a tumor.

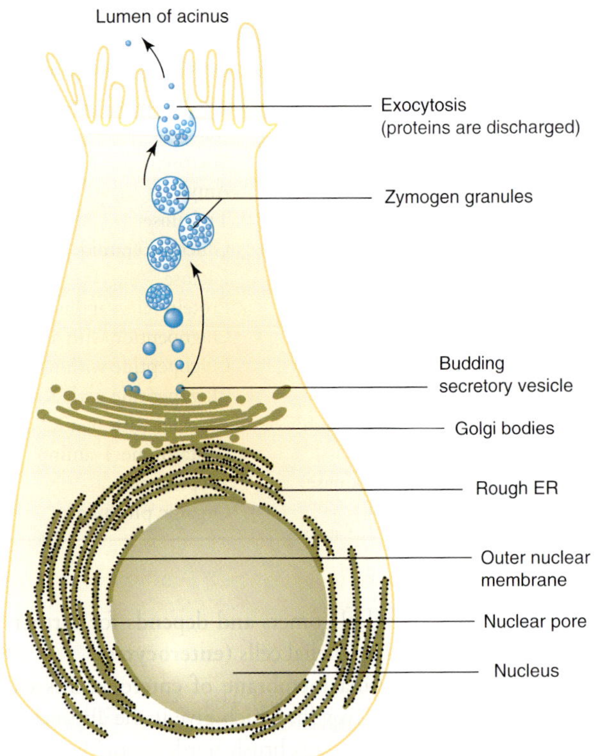

FIGURE 26.2

Exocrine secretion of digestive enzymes.
Redrawn with permission from Jamieson, J. D. Membrane and secretion. In: G. Weissmann and R. Claiborne (Eds.), *Cell Membranes: Biochemistry, Cell Biology and Pathology*. New York: HP Publishing Co., 1975. Figure by B. Tagawa.

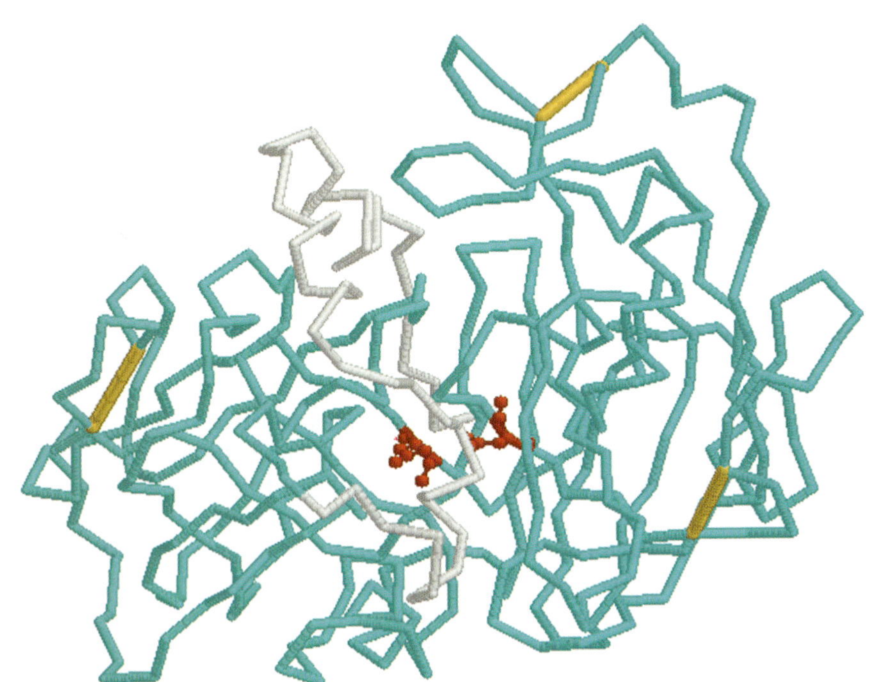

FIGURE 26.3

Model of porcine pepsinogen NH_2-terminus to be cleared in central groove and aspartic acid residues at active site (red).
Model 3PSG from Protein Data Bank and illustrated with Protein Explorer.

interact with receptors on the surface of **exocrine cells** (Table 26.3). Neurotransmitters, hormones, pharmacological agents, and certain bacterial toxins can be secretagogues. Secretagogue–receptor interactions are specific and cells of different glands usually possess different sets of receptors. Binding of secretagogues to receptors activates intracellular signaling events that lead to release of the contents of zymogen granules into the lumen. Major signaling pathways for secretion are (Figure 26.4) (1) activation of phosphatidylinositol-specific **phospholipase C** with liberation of **inositol 1,4,5-trisphosphate** and **diacylglycerol** (see p. 522), in turn, triggering Ca^{2+} release from the

TABLE 26.3 Physiological Secretagogues

Organ	*Secretion*	*Secretagogue*
Salivary gland	NaCl, amylase	Acetylcholine, catecholamines?
Stomach	HCl, pepsinogen	Acetylcholine, histamine, gastrin
Pancreas-acini	NaCl, digestive enzymes	Acetylcholine, cholecystokinin, secretin
Pancreas-duct	$NaHCO_3$, NaCl	Secretin
Small intestine	NaCl	Acetylcholine, serotonin, vasoactive intestinal peptide (VIP), guanylin

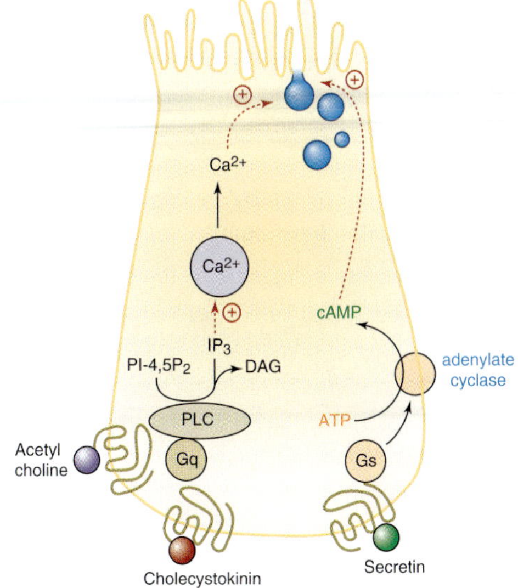

FIGURE 26.4

Cellular regulation of exocrine secretion in pancreas. Abbreviations: PI-4,5P$_2$, phosphatidylinositol-4,5-bisphosphate; DG, diacylglycerol; IP_3, inositol-1,4,5-trisphosphate; PLC, phospholipase C.

Adapted from Gardner, J. D. *Annu. Rev. Physiol.* 41:63, 1979. Copyright © 1979 by Annual Reviews, Inc.

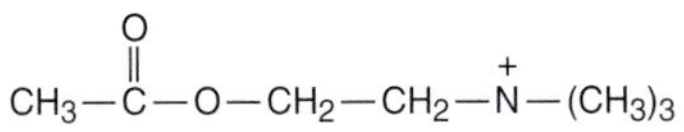

endoplasmic reticulum into the cytosol and activation of protein kinase C, respectively, and (2) activation of **adenylate** or **guanylate cyclase**, resulting in elevated cAMP or cGMP levels, respectively (see p. 515).

Acetylcholine (Figure 26.5) is the major neurotransmitter for stimulating enzyme and electrolyte secretion throughout the gastrointestinal tract. It mediates regulation of salivary and gastric secretions by the autonomic nervous system during the early phase of digestion (cephalic phase of gastric secretion) and of pancreatic and intestinal secretions by the enteric nervous system during subsequent digestion. The acetylcholine receptor of exocrine cells is of the muscarinic type; that is, it can be stimulated by muscarinic acid and blocked by atropine (Figure 26.6). Atropine used to be given by dentists to "dry up" the mouth for dental work.

$$CH_3-\overset{\overset{O}{\|}}{C}-O-CH_2-CH_2-\overset{+}{N}-(CH_3)_3$$

FIGURE 26.5

Acetylcholine.

The biogenic amines histamine and **5-hydroxytryptamine (serotonin)** are secretagogues that function mainly in a paracrine fashion—that is, affecting neighboring cells. Histamine is produced by specialized regulatory cells in the gastric mucosa **(enterochromaffin-like or ECL cells)**. 5-Hydroxytryptamine is produced by many specialized epithelial cells that are diffusely distributed throughout the lining epithelium of the stomach and the intestines. The amines are stored in vesicles and released when cells receive appropriate stimuli. Since 5-hydroxytryptamine-producing cells release most of their product toward the blood side, they are also called epithelial endocrine cells.

Histamine (Figure 26.7) is a potent stimulator of HCl secretion. It interacts with the H_2 receptor on the contraluminal plasma membrane of gastric HCl-producing (parietal) cells. Antagonists at the H_2 receptor (H_2 blockers) are used medically as antacids. 5-Hydroxytryptamine (Figure 26.8) stimulates intestinal NaCl secretion, but also functions as neurotransmitter, for example, initiating the sensation of nausea.

(*a*) H_3C, CH, O, HC, $CH_2\overset{+}{N}(CH_3)_3$, C, CH_2, HO, H

(*b*) CH_2OH, CH—CO—O, C_6H_5, $\overset{+}{N}$H, CH_3

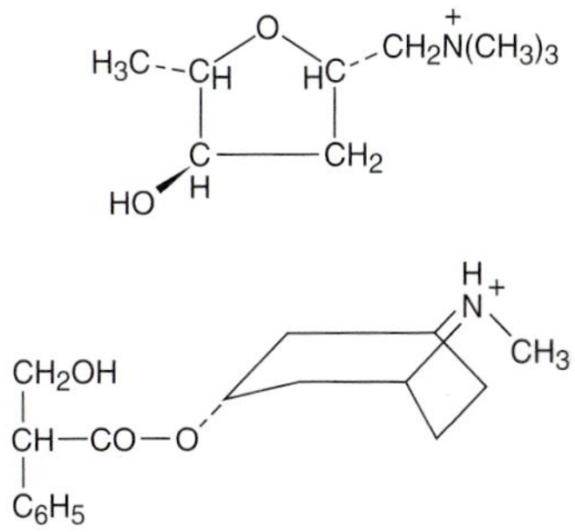

FIGURE 26.6

(*a*) L(+)-Muscarine and (*b*) atropine.

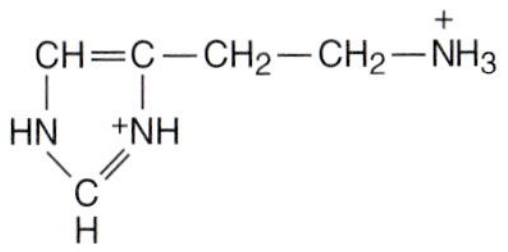

A third class of secretagogues consists of peptides; some of these function predominantly as hormones, others function as neurotransmitters, and again others can be both (Table 26.4). The intestinal peptide hormones are produced by a number of different **epithelial endocrine cells**, many of which also produce 5-hydroxytryptamine and store both products in small intracellular vesicles. Among the many peptide hormones, gastrin, cholecystokinin (pancreozymin), secretin, and guanylin have particular importance for digestive enzyme and fluid secretion. **Gastrin** occurs naturally in two forms: "Big" gastrin is a peptide of 34 amino acids (G-34), which is cleaved to yield "little" gastrin comprising the 17 residues at the carboxyl terminus (G-17). The functional portion of gastrin resides mainly in the carboxyl-terminal pentapeptide, and synthetic "pentagastrin" can be administered intravenously to assess the potential for gastric HCl and pepsin secretion in a patient. Gastrin and cholecystokinin have a **sulfated tyrosine**, which considerably enhances the potency of either hormones.

Cholecystokinin and **pancreozymin** denote the same peptide, but cholecystokinin is the preferred term. The peptide stimulates gallbladder contraction (hence cholecystokinin) and secretion of pancreatic enzymes (hence pancreozymin). Cholecystokinin is released into the bloodstream by epithelial endocrine cells of the small intestinal lining, particularly in the duodenum, in response to ingestion of food and contact with cholecystokinin releasing factors secreted into the lumen by pancreas and intestinal cells. The releasing factors are themselves proteins and thus substrates for pancreatic proteases (see below). The state of their degradation serves as feedback for adequate pancreatic

CH=C—CH_2—CH_2—$\overset{+}{N}H_3$, HN, $^{+}$NH, C, H

FIGURE 26.7

Histamine.

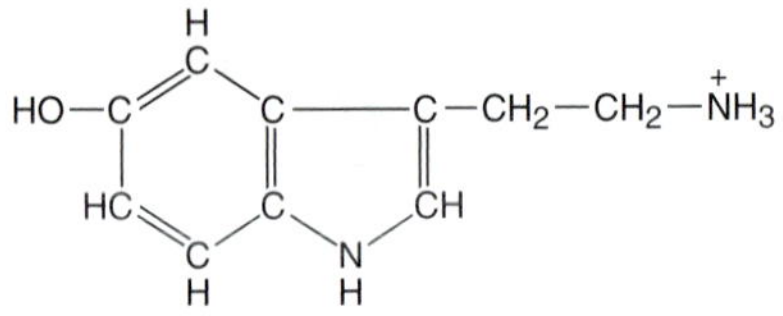

FIGURE 26.8
5-OH-Tryptamine (serotonin).

TABLE 26.4 Secretory Intestinal Neuropeptides and Hormones (Human)

Vasoactive intestinal peptide (VIP)

His-Ser-Asp-Ala-Val-Phe-Thr-Asp-Asn-Tyr-Thr-Arg-Leu-Arg-Lys-Gln-Met-Ala-Val-Lys
|
Asn-Leu-Ile-Ser-Asn-Leu-Tyr-Lys

Secretin

His-Ser-Asp-Gly-Thr-Phe-Thr-Ser-Glu-Leu-Ser-Arg-Leu-Arg-Glu-Gly-Ala-Arg-Leu-Gln
|
[c]NH_2-Val-Leu-Gly-Gln-Leu-Leu-Arg

Guanylin

Pro-Gly-Thr-Cys-Glu-Ile-Cys-Ala-Tyr-Ala-Ala-Cys-Thr-Gly-Cys

Gastrin G-34-II[a] ⋮ **G-17-II**

[b]Glp-Leu-Gly-Pro-Gln-Gly-Pro-Pro-His-Leu-Val-Ala-Asp-Pro-Ser-Lys-Lys-Gln
|
[c]NH_2-Phe-Asp-Met-Trp-Gly-Tyr(SO_3H)-Ala-(Glu)$_5$-Leu-Trp-Pro-Gly

Cholecystokinin

Lys-Ala-Pro-Ser-Gly-Arg-Met-Ser-Ile-Val-Lys-Asn-Leu-Gln-Asn-Leu-Asp-Pro
|
[c]NH_2-Phe-Asp-Met-Trp-Gly-Met-Tyr(SO_3H)-Asp-Arg-Asp-Ser-Ile-Arg-His-Ser

Source: Yanaihara, C. In: B. B. Rauner, G. M. Makhlouf, and S. G. Schultz (Eds.), *Handbook of Physiology: Section 6: Alimentary Canal Vol II: Neural and Endocrine Biology*. Bethesda, MD: American Physiological Society, 1989, pp. 95–62.

[a] Gastrin I is not sulfated.

[b] Glp denotes pyrrolidino carboxylic acid, derived from Glu through internal amide formation.

[c] NH_2 denotes amide of carboxy-terminal amino acid.

enzyme delivery to the intestine as lack of digestion of the releasing factors stimulates further cholecystokinin and hence pancreatic enzyme secretion. Cholecystokinin and gastrin are thought to be evolutionary related because they share an identical amino acid sequence at the carboxyl terminus.

Secretin, a polypeptide of 27 amino acids, is released by another type of endocrine cell in the small intestine. Its release is stimulated particularly by luminal pH less than 5. Its major biological activity is stimulation of secretion of pancreatic juice rich in $NaHCO_3$. This is essential for neutralization of gastric HCl in the duodenum. Secretin also enhances pancreatic enzyme release, acting synergistically with cholecystokinin.

Guanylin is a peptide that is produced by goblet (mucin producing) cells and released mainly into the lumen. It stimulates NaCl secretion by binding to a brush border receptor/guanylate cyclase and thereby elevating cytosolic cGMP levels. The peptide was discovered because it activates the same receptor/guanylate cyclase as a heat-stable *E. coli* enterotoxin, one of the factors responsible for traveler's diarrhea.

Among neuropeptides, **vasoactive intestinal peptide (VIP)** is a potent and physiologically important secretagogue for NaCl and fluid in the intestines and the pancreas. Enterocytes possess receptors for VIP, and release of this neuropeptide by enteric nerves regulates intestinal electrolyte and fluid secretion.

26.3 EPITHELIAL TRANSPORT

Solute Transport May Be Transcellular or Paracellular

The barrier properties of epithelia are determined by the plasma membranes of the epithelial cells and their intercellular tight junctional complexes (Figure 26.9). **Tight junctions** extend in a belt-like manner around the perimeter of each epithelial cell and connect neighboring cells. They constitute part of the barrier between the two

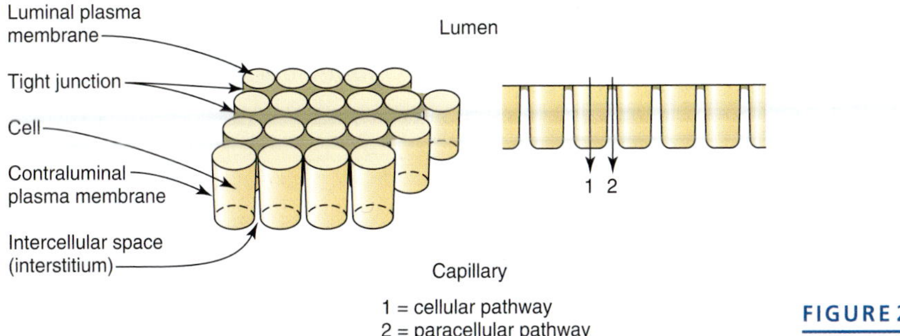

FIGURE 26.9
Pathways for transport across epithelia.

extracellular spaces on either side of the epithelium, that is, the gastrointestinal lumen and the intercellular (interstitial) space on the blood or serosal side. Tight junctions mark the boundary between **luminal** and **contraluminal** regions of the **plasma membrane** of epithelial cells.

Two parallel pathways can be distinguished for **solute transport** across epithelial cell layers: one through the cells **(transcellular)** and another through the tight junctions between cells **(paracellular)** (Figure 26.9). Nutrient or electrolyte flux through either pathway can influence flux through the other one by changing driving forces. The transcellular route consists mainly of two barriers in series, formed by the luminal and contraluminal plasma membranes.

Different regions of the gastrointestinal tract differ with respect to permeability characteristics of the tight junction. In regions with steep solute concentration gradients across the epithelium, such as stomach and distal colon, tight junctions have low permeability (or high resistance) to Na^+ and other ions, while they possess high permeability in regions with nutrient absorption or NaCl secretion as both processes require paracellular solute flux (see below).

A major function of intestinal epithelial cells is active absorption of nutrients, electrolytes, and vitamins. The cellular basis for this **vectorial solute movement** lies in the different complement of transporters in luminal and contraluminal membranes. The small intestinal epithelial cells provide an example of the differentiation and specialization of the two types of membrane in terms of morphological appearance, chemical composition, and complement of enzymes and transporters (Table 26.5). The luminal membrane is in contact with the nutrients in chyme (semifluid mass of partially digested food) and is specialized for terminal digestion of nutrients through hydrolytic enzymes on its external surface and for nutrient absorption through transporters or channels for monosaccharides, amino acids, peptides, fatty acids, cholesterol, and electrolytes. In contrast, the contraluminal plasma membrane is in contact with the interstitial fluid and therewith indirectly with capillaries and lymph vessels. It has properties similar to the plasma membrane of most cells and possesses receptors for hormonal or neuronal regulation of cellular functions, a Na^+/K^+-exchanging ATPase for removal of Na^+ from the cell, and transporters for entry of nutrients needed for cell maintenance. It also contains the transporters necessary for exit of absorbed nutrients so that the digested food becomes available to all cells of the body. Many of the nutrient transporters in the contraluminal membrane are bidirectional and serve to release nutrients into blood or lymph after a meal and provide nutrients to enterocytes in between meals.

NaCl Absorption Has Both Active and Passive Components

Transport of Na^+ plays a crucial role not only for epithelial NaCl absorption or secretion, but also in energization of nutrient uptake. The **Na^+/K^+-exchanging ATPase** (see p. 476) provides the dominant mechanism for transduction of chemical energy in the form of ATP into osmotic energy of a concentration (chemical) or a combined concentration and electrical (electrochemical) ion gradient across the plasma membrane.

TABLE 26.5 Characteristic Differences Between Luminal and Contraluminal Plasma Membranes

Parameter	*Luminal*	*Contraluminal*
Morphological appearance	Microvilli arranged as brush border	Few microvilli
Enzymes	Di- and oligosaccharidases Aminopeptidase Dipeptidases γ-Glutamyltransferase Alkaline phosphates Guanylate cyclase C	Na^+/K^+ exchanging ATPase Adenylate cyclase
Transporters/channels	Na^+–glucose cotransporter (SGLT1, SLC5A1) Facilitative fructose transporter (GLUT5, SLC2A5) Na^+–neutral amino acid cotransporter (SLC6 family) Apical Na^+–bile acid cotransporter (ASBT, SLC10A2) H^+-peptide cotransporter (PEPT1, SLC15A1) Cystic fibrosis transmembrane regulatory protein (CFTR, ABCC7)	Facilitative glucose transporter (GLUT2, SLC2A2) Facilitative neutral amino acid transporter (SLC3A2/SLC7A8) Sodium-potassium-2-chloride cotransporter (NKCC1, SLC12A2)

Names in parentheses are original abbreviations and corresponding new classifications according to SLC (solute carrier) or ABC (ATP binding cassette) nomenclature; see http://www.bioparadigms.org/slc/menu.asp and http://nutrigene.4t.com/humanabc.htm.

In gastrointestinal epithelial cells this enzyme is located exclusively in the contraluminal membrane (Figure 26.10). The ATPase maintains the high K^+ and low Na^+ concentrations in the cytosol and is responsible for an electrical potential of about −60 mV of the cytosol relative to the extracellular solution.

Transepithelial NaCl movements result from the combined actions of the Na^+/K^+-exchanging ATPase and additional "passive" transporters in the plasma membrane, which allow the entry of Na^+ or Cl^- into the cell. NaCl absorption results from Na^+ entry from the lumen and its extrusion by the Na^+/K^+-exchanging ATPase across the contraluminal membrane. Epithelial cells of the lower portion of the large intestine possess a luminal Na^+ channel [**epithelial Na^+ channel**, **ENaC**, or **SCNN1** *(note that acronyms used to identify the channels and transporters are being changed; thus, where appropriate, both the older and newer acronym will be given)*]. The channel allows the uncoupled entry of Na^+ down its electrochemical gradient (Figure 26.11). This Na^+ flux is electrogenic, that is, it is associated with an electrical current that, in turn, changes any electrical potential. It can be inhibited by the diuretic drug amiloride at micromolar concentrations (Figure 26.12). This transport system, and hence NaCl absorption, is regulated by mineralocorticoid hormones of the adrenal cortex.

Epithelial cells of the small intestine have a transporter in their brush border membrane, which catalyzes an electrically neutral Na^+/H^+ exchange; **Na^+/H^+ transporter-3** (**NHE3** or **SLC9A3**) is the predominant isoform in the intestine (Figure 26.13). This transporter is not affected by low concentrations of amiloride and not regulated by mineralocorticoids. The Na^+/H^+ exchange sets up a H^+-gradient that secondarily drives Cl^- absorption through a specific **Cl^-/HCO_3^- transporter** in the luminal plasma membrane, as illustrated in Figure 26.13. The intestinal Cl^-/HCO_3^- transporter is coded for by the "**down-regulated in adenoma**" or ***DRA*** gene (SLC26A3) (see Clin. Corr. 26.1). The need for two types of NaCl absorption mechanisms arises from the different physiological functions of upper and lower intestine, which require different regulation. The upper intestine absorbs the bulk of NaCl derived from diet and secretions of the exocrine glands, while the lower intestine "scavenges" remaining NaCl depending on overall NaCl balance of the body.

CLINICAL CORRELATION 26.1
Familial Chloridorrhea Causes Metabolic Alkalosis

Loss-of-function mutations in the human *DRA* (down-regulated in adenoma) gene (SLC26A3) causes familial chloridorrhea. Patients with this disease have moderate diarrhea and generate acidic stool, and they suffer from metabolic alkalosis (alkaline plasma at normal carbon dioxide concentration). The normal *DRA* gene product is an Na^+-independent Cl^-/HCO_3^- exchanger in epithelial cells of the lower ileum and colon. The exchanger absorbs Cl^- from stool in exchange for bicarbonate. Bicarbonate secretion is important for neutralization of protons secreted in exchange for Na^+ via sodium/proton exchanger NHE3 (SLC9A3). Chronic HCl loss with the stool leads to a corresponding metabolic alkalosis of the body fluids. In addition, the extra electrolytes in stool are responsible for its greater fluid contents (i.e., diarrhea) due to osmotic effects.

Source: Mount, D. B. and Romero, M. F. The SLC26 gene family of multifunctional anion exchangers. *Pflügers Arch.* 447:710, 2004.

NaCl Secretion Depends on Contraluminal Na^+/K^+-Exchanging ATPase

Glands and intestinal crypts secrete each day a large amount of electrolytes and fluids into the gastrointestinal lumen (equivalent to at least a third of the extracellular fluid

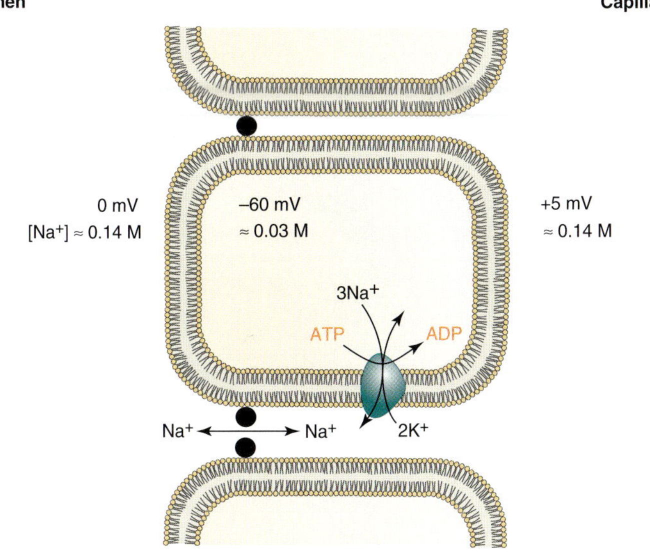

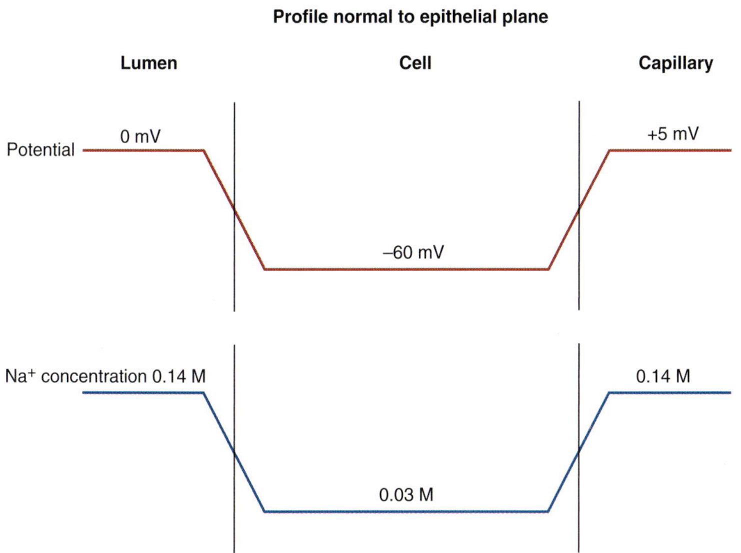

FIGURE 26.10

Na^+ concentrations and electrical potentials in enterocytes.

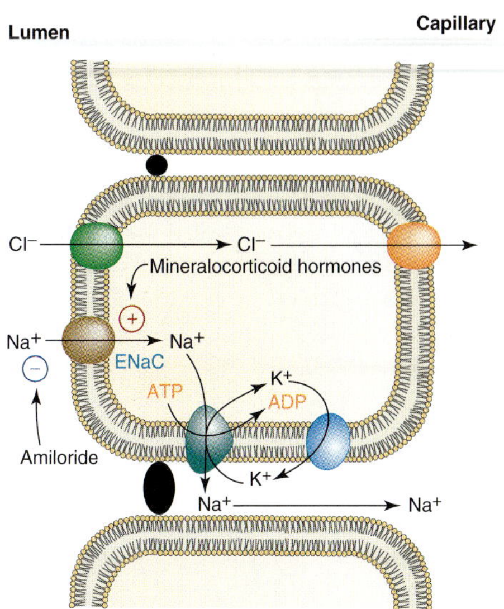

FIGURE 26.11

Model for electrogenic NaCl absorption in lower large intestine. Na^+ absorption proceeds via luminal epithelial sodium channel (ENaC) and contraluminal Na^+,K^+-exchanging ATPase. The nature of Cl^- channels is not known.

FIGURE 26.12

Amiloride.

in humans). This is an energy-consuming process with active transport of electrolytes and passive flow of water. The water movement is due to osmotic forces exerted by secreted electrolytes. In other words, any primary secreted fluid has a greater osmolarity than plasma or cytosol and is said to be hypertonic. However, most gastrointestinal epithelia have a very high permeability to water due to the presence of water channels, aquaporins (see p. 465), in the plasma membranes so that osmotic equilibrium is rapidly approached. As a result, the secretions actually emerging from the acini of glands or intestinal crypts are essentially isotonic (same osmolarity as plasma). Ionic compositions of gastrointestinal secretions are presented in Figure 26.14.

Major secreted ions are Na^+ and Cl^- and their secretion mechanism provides insight into energy coupling in active transport by epithelia. NaCl secretion is an energy-consuming process that depends on the Na^+/K^+-exchanging ATPase located in the contraluminal plasma membrane (Figure 26.15). Thus cardiac glycosides, inhibitors of this enzyme, abolish salt secretion. How can the Na^+/K^+-exchanging ATPase power Na^+ movement from the capillary side to the lumen when it extrudes Na^+ from

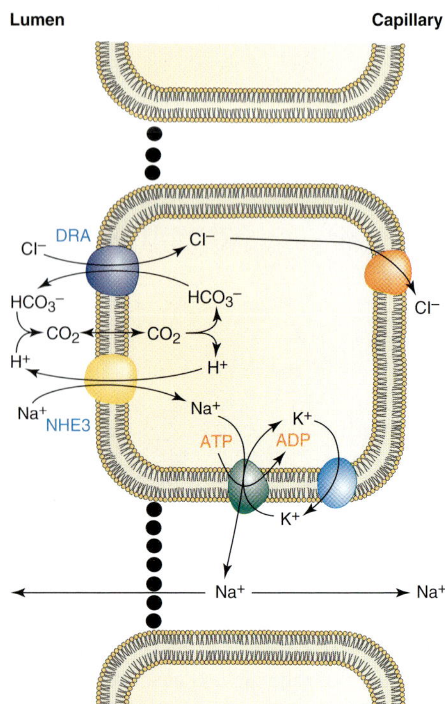

FIGURE 26.13

Model for electrically neutral NaCl absorption in the small intestine. Na^+ absorption proceeds via sodium/proton exchanger 3 (NHE3) and Na^+,K^+-exchanging ATPase. Cl^- absorption proceeds via "down-regulated in adenoma" protein (DRA). The nature of the contraluminal Cl^- channel is not known.

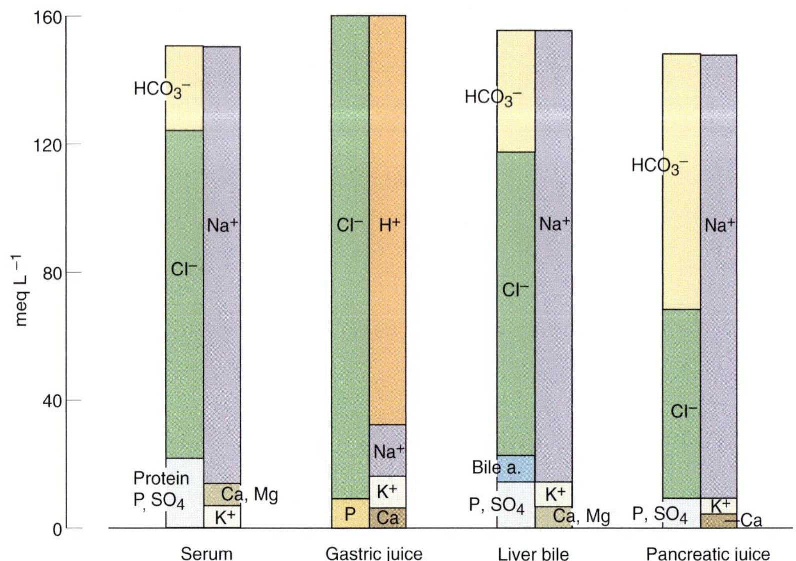

FIGURE 26.14

Ionic composition of gastrointestinal secretions. Serum included for comparison. Note high H^+ concentration in gastric juice (pH = 1) and high HCO_3^- concentration in pancreatic juice. P, organic and inorganic phosphate; SO_4, inorganic and organic sulfate; Ca, calcium; Mg, magnesium; Bile a., bile acids. Adapted from *Biological Handbooks. Blood and Other Body Fluids*. Federation of American Societies for Experimental Biology. Washington, DC, 1961.

the cell toward the capillary side? This paradox is resolved by an electrical coupling of Cl^- secretion across the luminal plasma membrane and Na^+ movements via the paracellular route, illustrated in Figure 26.15. Cl^- secretion depends on coupled uptake of 2 Cl^- ions with Na^+ and K^+ across the contraluminal membrane, extrusion of Na^+ and K^+ across the contraluminal membrane, and luminal exit of Cl^- via channels. The uptake is mediated by a **$Na^+/K^+/2Cl^-$ cotransporter (NKCC1 or SLC12A2)**, which is pharmacologically characterized by inhibition by the diuretic **furosemide** (Figure 26.16), and utilizes energy of the Na^+ gradient to accumulate Cl^- within the cytosol above its electrochemical equilibrium. As the Na^+/K^+-exchanging ATPase generates and maintains the Na^+ gradient across the plasma membrane, it indirectly also powers Cl^- uptake from capillaries into cells and its movement into the lumen. The exit of Cl^- via luminal channels is associated with loss of a negative charge, which, in turn, sets up an electrical potential that draws Na^+ into the lumen via the paracellular route. The predominant luminal Cl^- channel in pancreatic ducts and the intestines is the **cystic fibrosis transmembrane regulatory (CFTR)** protein, and malfunctions of this channel lead to reduced amounts of secretions in the human disease cystic fibrosis (see Clin. Corrs. 26.2 and 26.3).

Pancreatic acinar cells secrete a fluid rich in Na^+ and Cl^-, which provides the vehicle for transport of digestive enzymes from the acini to the lumen of the duodenum. This fluid is modified in the ducts by the additional secretion of $NaHCO_3$ (Figure 26.17). Bicarbonate concentrations can reach up to 120 mM in humans.

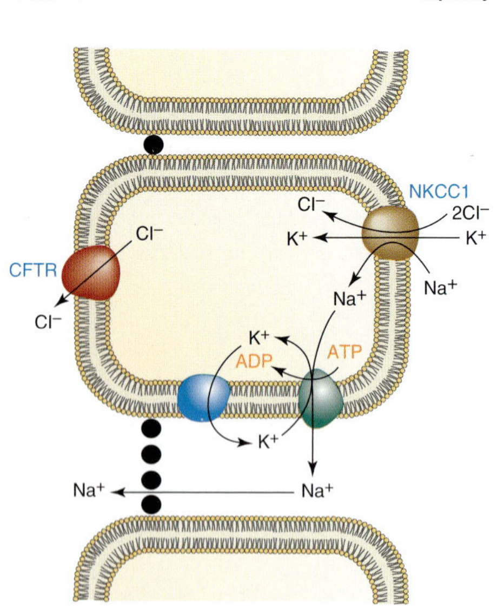

FIGURE 26.15

Model for epithelial NaCl secretion. Cl^- is concentrated in cytosol via sodium–potassium–two-chloride cotransporter (NKCC1) and released into the lumen via cystic fibrosis transmembrane regulator protein (CFTR).

Concentration Gradients or Electrical Potentials Drive Transport of Nutrients

Many solutes are absorbed across the intestinal epithelium against a concentration gradient. Energy for this "active" transport is directly derived from a Na^+ or H^+ concentration gradient or the electrical potential across the luminal membrane, and only indirectly from ATP hydrolysis. Intestinal glucose transport is an example of such uphill solute transport, driven in this case by an electrochemical Na^+ gradient (Figure 26.18).

Serum glucose is about 5 mM so that complete extraction of glucose from chyme can only be achieved by active transport across the epithelium. This vectorial process is

CLINICAL CORRELATION 26.2
Cystic Fibrosis

Cystic fibrosis is an autosomal recessive disease due to a mutation in the cystic fibrosis transmembrane regulatory (CFTR) protein. CFTR is a member of the ABC transporter family (subfamily ABCC7). It contains 1480 amino acids organized into two membrane-spanning domains, which contain six transmembrane segments each, two ATP-binding domains, and a regulatory domain that undergoes phosphorylation by cAMP-dependent protein kinase. Over 800 mutations have been discovered since the gene was cloned in 1989. The most common one lacks one amino acid, which prevents the protein from maturing properly and reaching the plasma membrane.

CFTR is the predominant Cl^- channel in the luminal plasma membrane of epithelial cells in tissues affected in cystic fibrosis (airways, pancreatic duct, intestine, vas deferens, sweat gland ducts). The channel is normally closed, but it opens when phosphorylated by protein kinase A and ATP is present. Cl^- flux through CFTR depends on the existing electrochemical Cl^- gradient, which is set up by other cellular transporters and is different in secretory and absorptive cells. In most tissues, CFTR controls secretion of NaCl and fluid (see Clin. Corr. 26.3 for activation of the CFTR Cl^- channel). However, in sweat gland duct cells, which are absorptive, CFTR is important for efficient reabsorption of the NaCl that is initially secreted in sweat gland acini. A defect in CFTR explains both excess loss of Cl^- in sweat (sweat test for cystic fibrosis) and insufficient NaCl and fluid secretion in lungs, pancreas, and intestine. The clinical symptoms in cystic fibrosis stem largely from the fluid secretion defect because it can lead to partial or total blockage of passageways and gross organ impairments and/or infections.

CLINICAL CORRELATION 26.3
Bacterial Toxigenic Diarrheas and Electrolyte Replacement Therapy

Voluminous, life-threatening intestinal electrolyte and fluid secretion (diarrhea) occurs in patients with cholera, an intestinal infection by *Vibrio cholerae*. Certain strains of *E. coli* also cause (traveler's) diarrhea that can be serious in infants. The secretory state is a result of enterotoxins produced by the bacteria. The mechanisms of action of some of these enterotoxins are well understood at the biochemical level. Cholera toxin activates adenylate cyclase by causing ADP-ribosylation of the $G_{\alpha s}$-protein, resulting in constitutive stimulation of the cyclase (see p. 514). Elevated cAMP levels in turn activate protein kinase A and protein phosphorylation, which opens the luminal CFTR Cl^- channel in secretory cells and inhibits the Na^+/H^+ exchanger (NHE3) in absorptive cells. The net result is gross NaCl secretion. *Escherichia coli* produces a heat-stable toxin that binds to "guanylate cyclase C," which has an extracellular binding domain and intracellular catalytic domain. Binding of *E. coli* heat-stable toxin or of the peptide "guanylin," produced physiologically by intestinal goblet cells, activates guanylate cyclase, resulting in increased cGMP levels. Elevated cGMP, similarly as elevated cAMP, inhibits NaCl reabsorption and stimulates Cl^- secretion.

Modern, oral treatment of cholera takes advantage of the presence of Na^+–glucose cotransport in the intestine, which is not altered by elevated cAMP and remains fully active in this disease. In this case, the presence of glucose allows uptake of Na^+ to replenish body NaCl. Composition of solution for oral treatment of cholera patients is: glucose 110 mM, Na^+ 99 mM, Cl^- 74 mM, HCO_3^- 29 mM, and K^+ 4 mM. The major advantages of this form of therapy are its low cost and ease of administration when compared with intravenous fluid therapy.

The composition of sport drinks for electrolyte replacement is based on the same principle, namely, more rapid sodium absorption in the presence of glucose.

Source: Carpenter, C. C. J. In: M. Field, J. S. Fordtran, and S. G. Schultz (Eds.), *Secretory Diarrhea*. Bethesda, MD: American Physiological Society, 1980, p. 67.

the result of several separate membrane events (Figure 26.19): (1) Primary active, ATP-dependent Na^+ efflux at the contraluminal membrane establishes an electrochemical Na^+ gradient across the contraluminal membrane; (2) K^+-efflux via channels in the contraluminal membrane contributes further to the membrane potential; (3) secondary active cotransport of glucose and Na^+ drives glucose uptake from the lumen into the cell; and (4) facilitative glucose movement downhill into the interstitial and capillary space completes the transepithelial glucose absorption. This scenario is possible because of the

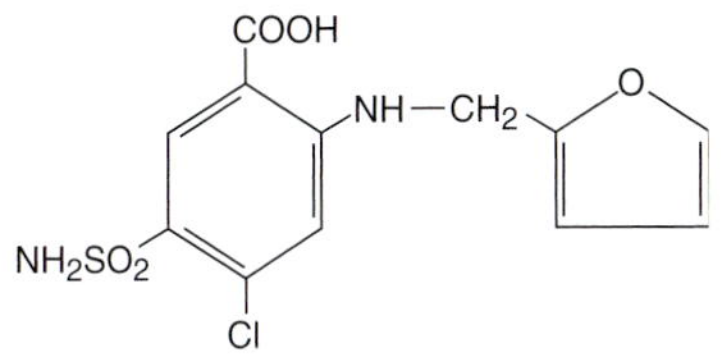

FIGURE 26.16
Furosemide.

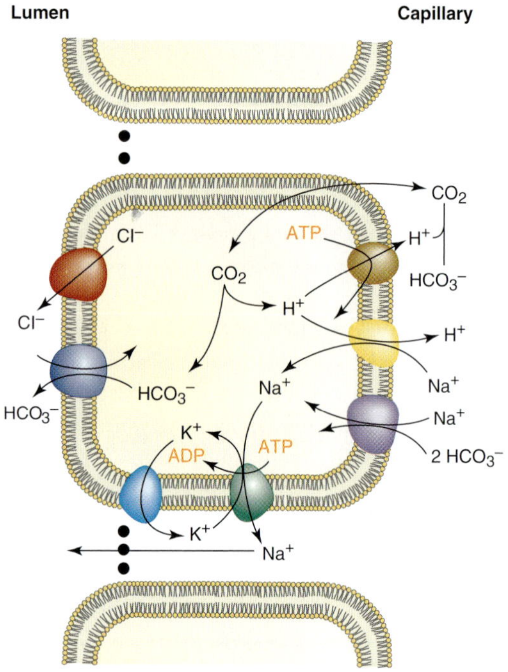

FIGURE 26.17

Model for $NaHCO_3$ secretion by pancreatic duct cells. Luminal Cl^- efflux and Na^+ flux are similar to those in NaCl secretion (Figure 26.15). Note that three mechanisms exist for bicarbonate influx into the cell (or its equivalent proton secretion) at the contraluminal membrane: (1) Na^+/H^+ exchange, (2) H^+-ATPase, and (3) $Na^+-2HCO_3^-$ cotransport.

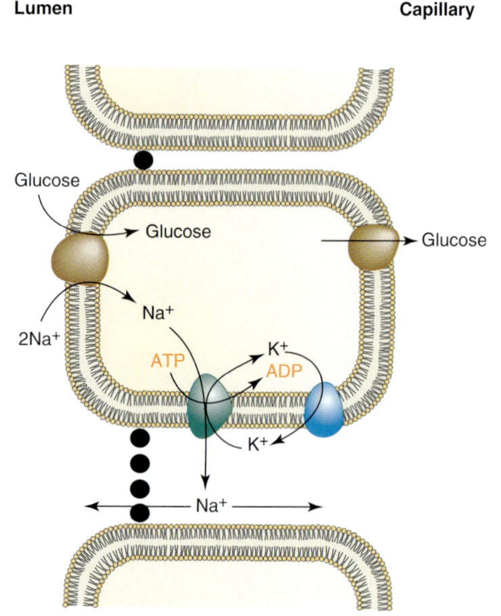

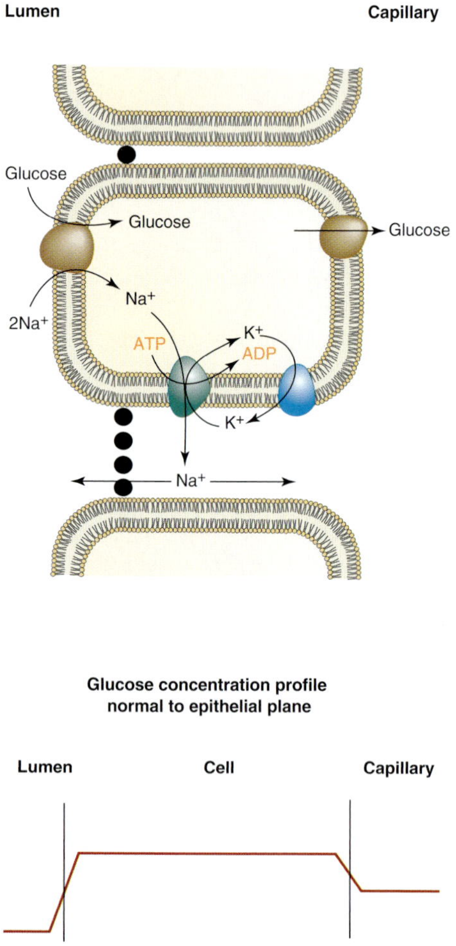

Glucose concentration profile normal to epithelial plane

Lumen | Cell | Capillary

FIGURE 26.18

Model for epithelial glucose absorption. Note the indirect role of Na^+, K^+-exchanging ATPase.

2X	$3Na^+_{cell} + 2K^+_{interstitium} + ATP_{cell} \xrightarrow{Na^+,\ K^+\text{-ATPase}} 3Na^+_{interstitium} + 2K^+_{cell} + ADP_{cell} + P_{cell}$
4X	$K^+_{cell} \underset{}{\overset{K^+\text{ channel}}{\rightleftharpoons}} K^+_{interstitium}$
3X	$2Na^+_{lumen} + Glc_{lumen} \overset{\text{SGLT1}}{\rightleftharpoons} 2Na^+_{cell} + Glc_{cell}$
3X	$Glc_{cell} \overset{\text{GLUT2}}{\rightleftharpoons} Glc_{interstitium}$
6X	$Na^+_{interstitium} \overset{\text{Tight junction}}{\rightleftharpoons} Na^+_{lumen}$
Sum	$3Glc_{lumen} + 2ATP_{cell} \longrightarrow 3Glc_{interstitium} + 2ADP_{cell} + 2P_{cell}$

FIGURE 26.19

Transepithelial glucose transport as translocation reactions across plasma membranes and the tight junction. Transporters SGLT1 (sodium glucose transporter 1) and GLUT2 (glucose transporter 2) mediating Na^+–glucose cotransport and facilitative glucose transport, respectively. Numbers in the left column indicate the minimal turnover of individual reactions to balance the overall reaction.

presence of a cotransporter for glucose and Na^+ in the luminal membrane (**sodium glucose transporter 1** (**SGLT1** or **SLC5A1**) (see p. 481), a facilitative transporter for glucose in the contraluminal membrane (GLUT2 or SLC2A2), and a leaky "tight junction" that allows the electrochemical Na^+ gradient across the contraluminal membrane to spread over the entire plasma membrane.

SGLT1 facilitates a tightly coupled movement of Na^+ and D-glucose (or structurally similar sugars) with a stoichiometry of 2 Na^+ ions and 1 glucose molecule. While the cotransporter per se facilitates coupled movement of glucose and Na^+ across the membrane equally well in either direction, transport under physiological conditions is from lumen to cell because of the lower Na^+ concentration and the negative potential in the cell. As a result, glucose is concentrated in the cell. Thus, downhill Na^+ movement normally supports concentrative glucose transport. Concentration ratios of up to 20-fold between intracellular and extracellular glucose have been observed *in vitro* under conditions of blocked efflux of cellular glucose. In some situations, Na^+ uptake via this route is physiologically more important than glucose uptake (see Clin. Corr. 26.3).

GLUT2 (see p. 474) in the contraluminal membrane is a member of the facilitative glucose transporter family and accepts many monosaccharides, including glucose. The direction of net flux is determined solely by the monosaccharide concentration gradient. The two glucose transport systems SGLT1 and GLUT2 share glucose as substrate, but differ considerably in amino acid sequence, secondary protein structure, Na^+ as cosubstrate, specificity for other sugars, sensitivity to inhibitors, and biological regulation. Since both SGLT1 and GLUT2 are not inherently directional, "active" transepithelial glucose transport relies on the Na^+/K^+-exchanging ATPase to continually move Na^+ out of the cell and maintain the electrochemical Na^+ gradient. One advantage of this arrangement for supplying energy to nutrient absorption is that the Na^+/K^+-exchanging ATPase can energize transport of many different nutrient transporters that use Na^+ as cosubstrate.

Gastric Parietal Cells Secrete HCl

The parietal (oxyntic) cells of gastric glands secrete HCl into the gastric lumen. Luminal H^+ concentrations of up to 0.14 M (pH 0.8) have been observed (see Figure 26.14). At plasma pH $= 7.4$, the parietal cell transports protons against a concentration gradient of $10^{6.6}$. The free energy required for **HCl secretion** under these conditions is minimally 9.1 kcal mol^{-1} of HCl. This active HCl secretion is achieved through a combination of primary active K^+/H^+ exchange by a **K^+/H^+-exchanging ATPase** (or gastric proton pump) and Cl^- and K^+ channels in the luminal membrane and Cl^-/HCO_3^- exchange in the contraluminal membrane. The K^+/H^+-exchanging ATPase is unique to the parietal cell and locates into the luminal membrane for periods of active HCl secretion. This enzyme couples the hydrolysis of ATP to an electrically neutral obligatory exchange of K^+ for H^+, secreting H^+ and moving K^+ into the cell. The stoichiometry appears to be 1 mol of transported H^+ and K^+ for each mole of ATP.

$$\mathrm{ATP_{cell}} + \mathrm{H^+{}_{cell}} + \mathrm{K^+{}_{lumen}} \rightleftharpoons \mathrm{ADP_{cell}} + \mathrm{P_{i,cell}} + \mathrm{H^+{}_{lumen}} + \mathrm{K^+{}_{cell}}$$

Because this ATPase generates a very acidic solution, protein reagents that are activated by acid can become specific inhibitors of this enzyme. Figure 26.20 shows the mechanism of a **proton pump inhibitor** that is sold as antacid.

In the steady state, HCl is elaborated by H^+/K^+-exchanging ATPase only if the luminal membrane is permeable to K^+ and Cl^- and the contraluminal membrane catalyzes an exchange of Cl^- for HCO_3^- (Figure 26.21). This exchange is essential to replenish the cell's Cl^- and prevent accumulation of base within the cell. Thus, under steady-state conditions, secretion of HCl into the gastric lumen is coupled to movement of HCO_3^- into plasma.

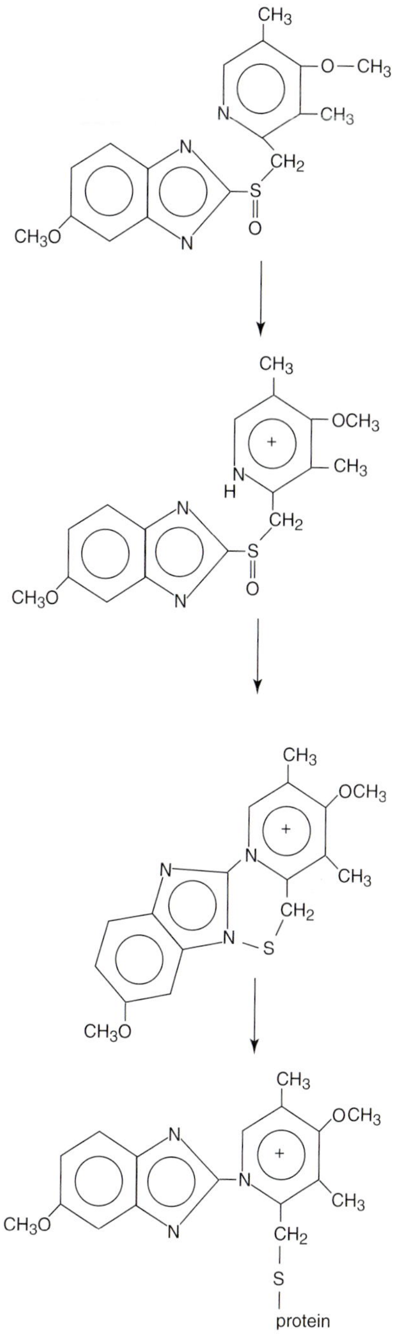

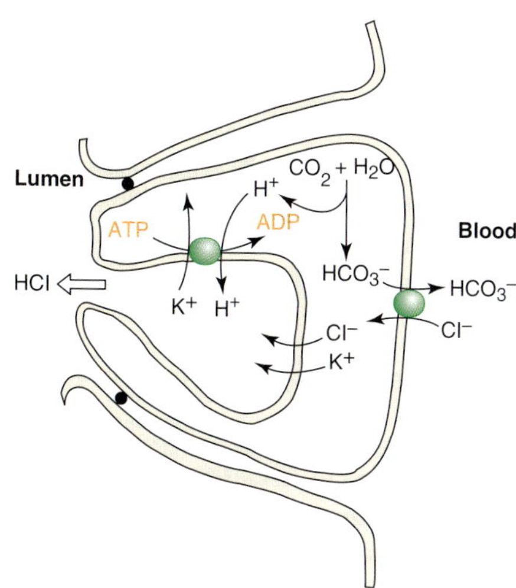

FIGURE 26.21

Model for secretion of hydrochloric acid.

FIGURE 26.20

Omeprazole, an inhibitor of K^+/H^+-exchanging ATPase. This drug accumulates in acidic compartments ($pK_a \sim 4$) and is converted to a sulfenamide that reacts with cysteine SH groups.

Redrawn from Sachs, G. The gastric K^+/H^+-ATPase. In. L. R. Johnson (Ed.), *Physiology of the Gastrointestinal Tract*. New York: Raven Press, 1994, p. 1133.

Specificities of trypsin, chymotrypsin, and elastase are presented in Table 26.6. The enzymes are active only at neutral pH and depend on pancreatic $NaHCO_3$ for neutralization of gastric HCl. Their mechanism of catalysis involves an **essential serine** residue (see p. 328) and is thus similar to serine esterases, such as acetylcholine esterase. The serine proteases and esterases are inhibited by reagents that chemically modify serine in the active site. An example of such an inhibitor is the diisopropylphosphofluoridate, which was developed for chemical warfare with acetylcholine esterase as intended target (see p. 957).

Peptides generated from ingested proteins are further degraded within the small intestinal lumen by pancreatic carboxypeptidases A and B, which use Zn^{2+} as part of the catalytic mechanism. The combined action of pancreatic proteases and peptidases results in formation of free amino acids and small peptides of 2–8 residues; peptides account for about 60% of the amino nitrogen at this point.

CLINICAL CORRELATION 26.4
Neutral Amino Aciduria: Hartnup Disease

Hartnup disease is a genetic defect in the Na^+-coupled transporter that normally mediates absorption of neutral amino acid from the lumen of the small intestine and the proximal tubule. It was named after the family in which the abnormality was first recognized, and the mouse homologue has recently been cloned (SCL6 gene family). In the kidney, the inability to reabsorb neutral amino acids from the ultrafiltrate leads to their excretion in urine (neutral amino aciduria). In the intestine, the defect results in malabsorption of dietary neutral amino acids. Clinical symptoms are those one would expect for a deficiency of tryptophan with pellagra-like features (see p. 1103), which is an expression of the decreased availability of tryptophan for conversion to nicotinamide. However, the symptoms are variable and milder than predicted for a complete block of neutral amino acid reabsorption. Investigations of patients with Hartnup disease revealed the existence of intestinal and renal transporters for di- or tripeptides, which are different from the ones for free amino acids. PEPT1 (SLC15A1) is the major intestinal transporter for absorption of small peptide products of digestion.

Source: Broer, A., Klingel, K., Kowalczuk, S., Rasko, J. E., Cavanaugh, J., and Broer, S., Molecular cloning of mouse amino acid transport system B0, a neutral amino acid transporter related to Hartnup disease. *J. Biol. Chem.* 279:24467, 2004. Daniel, H. Molecular and integrative physiology of intestinal peptide transport. *Annu. Rev. Physiol.* 66:361, 2004. http://www.emedicine.com/derm/topic713.htm

Brush Border and Cytosolic Peptidases Digest Small Peptides

Since pancreatic juice does not contain appreciable aminopeptidase activity, final digestion of di- and oligopeptides depends on enzymes of the small intestine. The luminal surface of epithelial cells is particularly rich in endopeptidase and aminopeptidase activity, but also contains dipeptidases (Table 26.2). Their action at the brush border surface produces free amino acids and di- and tripeptides, which are absorbed via specific **amino acid** and **peptide transport systems** respectively. Transported di- and tripeptides are generally hydrolyzed within the intestinal epithelial cell before they leave the cell. This explains why practically only free amino acids are found in the portal blood after a meal. The virtual absence of peptides used to be taken as evidence that luminal protein digestion proceeded all the way to free amino acids before absorption occurred. However, it is now established that a large portion of dietary amino nitrogen is absorbed in the form of small peptides with subsequent intracellular hydrolysis. Exceptions are di- and tripeptides that contain proline, hydroxyproline, or unusual amino acids, such as β-alanine in carnosine (β-alanylhistidine) or anserine (β-alanyl 1-methylhistidine). These peptides are absorbed and released intact into portal blood. While unusual, β-alanine is nevertheless part of the diet because it is, for example, in chicken meat.

Amino Acid and Di- and Tripeptide Transporters

The small intestine has a high capacity for absorbing free amino acids and di- and tripeptides. Most L-amino acids can be transported across the epithelium against a concentration gradient, although the need for concentrative transport *in vivo* is not obvious because luminal concentrations are usually higher than the plasma levels of 0.1–0.2 mM. The uptake into cells is mediated by several different transporters in the luminal membrane, while the release into blood is mediated by several additional, different transporters in the contraluminal membrane (Table 26.7).

Remarkably, several loss-of-function mutations in amino acid transporters were discovered because small intestine and renal proximal tubules share transporter types, and renal loss of any particular amino acid transporter produces an easily measured result, namely amino acid excretion in urine (amino aciduria). Similarly, importance of di- and tripeptide absorption for nutrition was discovered when a loss-of-function mutation in the major neutral amino transporter (neutral aminoaciduria) was not accompanied by an expected deficiency in the corresponding amino acids, suggesting at least an additional transporter mediating uptake (see Clin. Corr. 26.4).

The mechanism for active absorption of neutral L-amino acids appears to be similar to that discussed for D-glucose (see Figure 26.18). An Na^+-dependent cotransporter from the SLC6 family (also termed NBB for neutral amino acid brush border or B^0AT), and Na^+-independent, facilitative transporter (SLC3A2/SLC7A8 or system L for leucine preferring) have been functionally characterized in the luminal and

TABLE 26.7 Amino Acid Transporters

Transporter	Other Name(s)	Substrate Specificity	Mechanism	Disease Due to Loss of Function
Amino Acid Transports in the Luminal Membrane				
SLC1A1	EAAT3	Asp, Glu	Secondary active transport: cotransport Na^+, countertransport; K^+	Dicarboxylic aminoaciduria
SLC3A1/SLC7A9	rBAT/$b^{0,+}$ AT	Lys, Arg, Cys-Cys	Facilitative, concentrating: cotransport positive charge	Cystinuria
Mouse homolog SLC6 family	Mouse B^0AT, NBB	Phe, Tyr, Met, Val, Leu, Ile	Secondary active transport: cotransport Na^+	Hartnup disease, neutral aminoaciduria
SLC1A5	ASCT2, ASC	Ala, Ser, Cys	Exchanging: cotransport Na^+	—
	IMINO	Pro, hydroxy-pro, gly	Secondary active transport: cotransport Na^+, cotransport Cl^- (in some species)	
	BETA	β- ala, taurine	Secondary active transport: cotransport Na^+, cotransport Cl^- (in some species)	
SLC36A1	PAT1	Gly, Ala, Pro, GABA, D-Ala, MeAIB	Secondary active transport: cotransport H^+	
SLC15A1	PEPT1	Dipeptides, tripeptides, penicillin	Secondary active transport: cotransport H^+	
Amino Acid Transporters in the Contraluminal Membrane				
SLC3A2/SLC7A9	$b^{0,+}$	Lys, arg	Exchange basic for neutral amino acid, concentrative for basic amino acids: Cotransport positive charge	Lysinuric protein intolerance
SLC3A2/SLC7A8	LAT 2, system L	Small and large neutral amino acids	Facilitative	

contraluminal membranes, respectively. The brush border transport for amino acids other than the neutral ones is energized in more complicated ways. For example, acidic amino acids can be concentrated by cotransport with 2 Na^+ ions and countertransport with 1 K^+ ion (SLC12A1) while basic amino acids rely on cotransport of a positive charge and the cell-inside negative potential (SLC3A1/SLC7A9). Conclusions about transporters involved in absorption of the amino acids alanine, serine, and cysteine is associated with some uncertainty because these amino acids are substrates for several transporters, one of which appears to catalyze an obligatory amino acid/amino acid exchange (SLC1A5).

Molecular analysis of amino acid transporters has revealed the existence of heteromeric transporters that can explain their versatility by allowing different combinations between a "heavy" subunit (SLC3 family) and a "light" subunit (SLC7 family) (see Figure 26.24). The heavy subunit appears to function mainly in determining cellular location while the light one determines amino acid specificity and mechanism. For example, the proteins coded for by the genes SLC3A1 (heavy subunit) and SCL7A9 (light subunit) form a heteromeric transporter in the luminal membrane of small intestinal and proximal tubule epithelial cells that mediates uptake of basic amino acids and cystine. Loss-of-function mutations of either subunit lead to excretion of these amino acids in urine (cystinuria) and diminished intestinal uptake.

Di- and tri-peptides are cotransported with H^+ and thus are energized through the proton electrochemical gradient across this membrane (PEPT1 or SLC15A1). This H^+ electrochemical gradient is established by Na^+/H^+ exchange and is thus indirectly powered by Na^+/K^+-exchanging ATPase. The dipeptide transporter also accepts β-lactam antibiotics (aminopenicillins) and is important for absorption of orally administered antibiotics of this class.

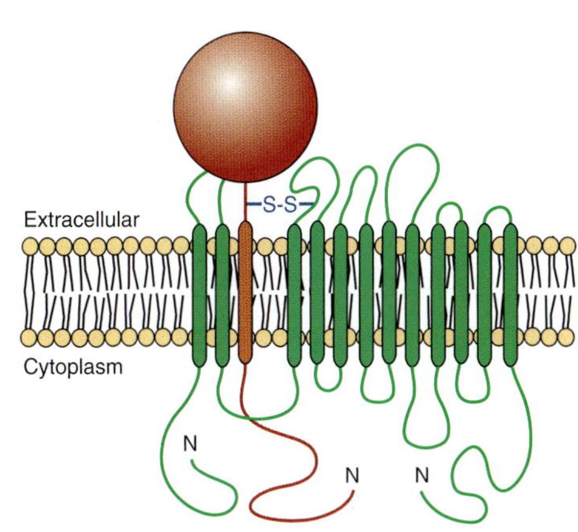

FIGURE 26.24

Model for heteromeric amino acid transporter. The heavy subunit with one transmembrane domain (red) is linked by a disulfide bridge (blue) to the light subunit with 12 transmembrane domains (green).
Redrawn from Chillaron, J., Roca, R., Valencia, A., Zorzano, A., and Palacin, M. Heteromeric amino acid transporters: Biochemistry, genetics, and physiology. *Am. J. Physiol. Renal Physiol.* 281: F995, 2001.

26.5 DIGESTION AND ABSORPTION OF CARBOHYDRATES

Disaccharides and Polysaccharides Require Hydrolysis

Dietary carbohydrates provide a major portion of the daily caloric requirement. They consist of mono-, di-, and polysaccharides (Table 26.8). Major carbohydrates in Western diets are **sucrose**, starch, and lactose. Monosaccharides, such as glucose and fructose, are absorbed directly. Disaccharides require enzymes of the small intestinal surface for hydrolysis into monosaccharides, while polysaccharides depend on pancreatic amylase and intestinal surface enzymes for digestion (Figure 26.25).

Starch, a major nutrient, is a plant storage polysaccharide of over 100 kDa. It is a mixture of linear chains of glucose molecules linked by α-1,4-glucosidic bonds **(amylose)** and of branched chains with branch points of α-1,6-glucosidic bonds **(amylopectin)**. The ratio of branch points to 1,4-glucosidic bonds is about 1:20. **Glycogen** is the animal storage polysaccharide and similar in structure to amylopectin, except that the number of branch points is greater in glycogen.

Hydrated starch and glycogen are digested by the endosaccharidase **α-amylase** of saliva and pancreatic juice (Figure 26.26). Hydration of the polysaccharides occurs during heating and is essential for efficient digestion. Amylase is specific for internal α-1,4-glucosidic bonds; α-1,6 bonds are not attacked nor are α-1,4 bonds of glucose units that serve as branch points. Pancreatic amylase is secreted in large excess relative to starch intake and is more important for digestion than the salivary enzyme. The products are mainly the disaccharide **maltose**, the trisaccharide **maltotriose**, and so-called **α-limit dextrins** that contain on average eight glucose units with one or more α-1,6-glucosidic bonds.

Final hydrolysis of di- and oligosaccharides to monosaccharides is carried out by enzymes on the luminal surface of small intestinal epithelial cells (Table 26.9). The surface oligosaccharidases are exoenzymes that cleave off one monosaccharide at a time from the nonreducing end. The capacity of α-glucosidases is normally much greater than needed for completion of digestion of starch. Similarly, there is usually excess capacity for sucrose (table sugar) hydrolysis. In contrast, β-galactosidase (lactase) for hydrolysis of lactose, the major milk carbohydrate, can be rate-limiting in humans (Clin. Corr. 26.5).

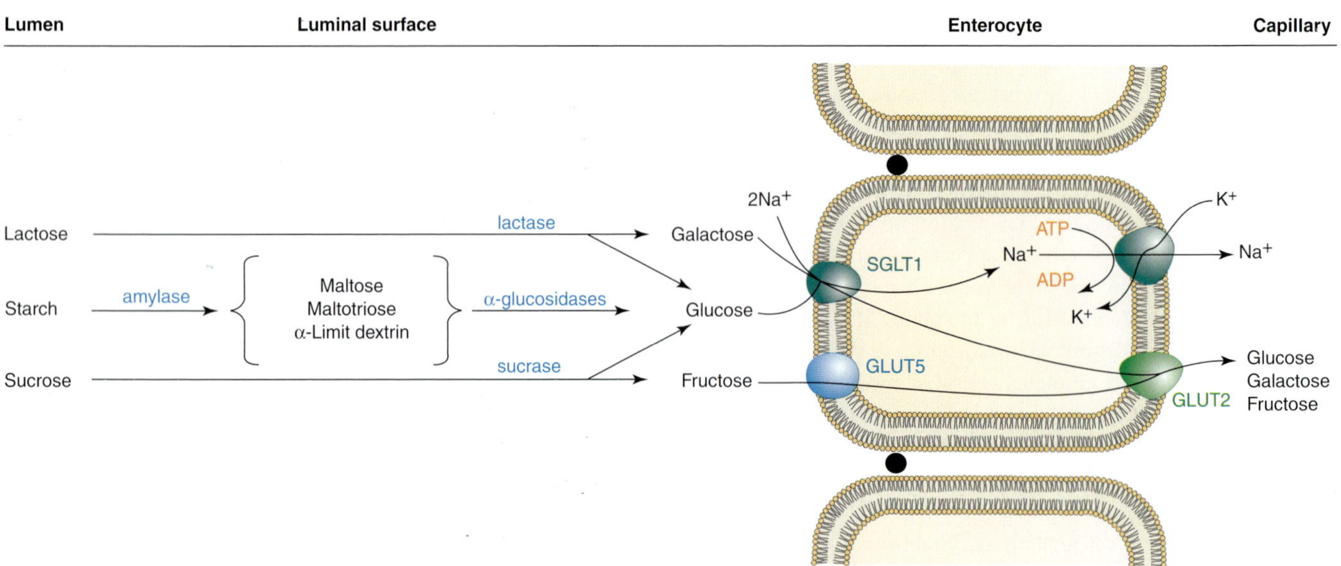

FIGURE 26.25
Digestion and absorption of carbohydrates.

TABLE 26.8 Dietary Carbohydrates

Carbohydrate	Typical Source		Structure
Fructose	Fruit, honey	α-Fru	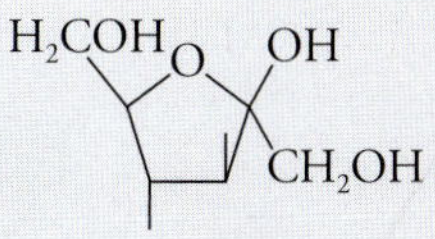
Glucose	Fruit, honey, grape	β-Glc	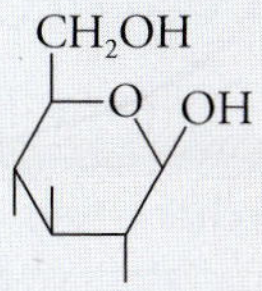
Amylopectin	Potatoes, rice, corn, bread	α-Glc(1 → 4)$_n$ Glc with α-Glc(1 → 6) branches	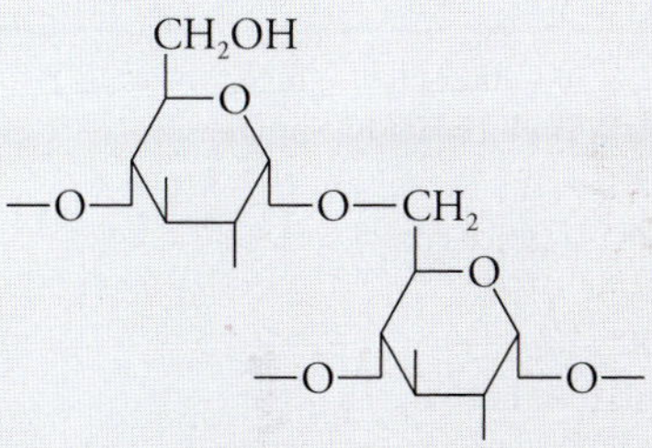
Amylose	Potatoes, rice, corn, bread	α-Glc(1 → 4)$_n$ Glc	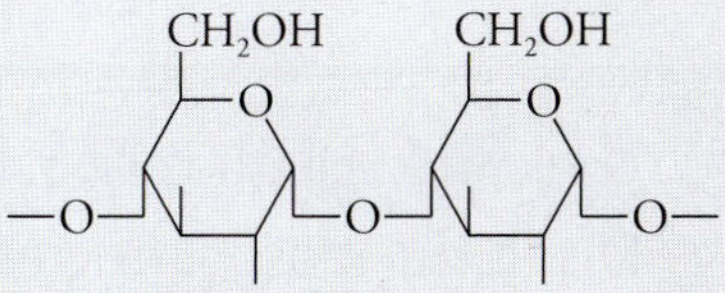
Sucrose	Table sugar, desserts	α-Glc(1 → 2)β-Fru	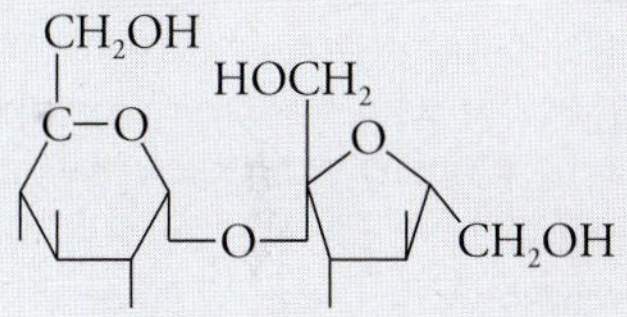
Trehalose	Young mushrooms	α-Glc(1 → 1)α-Glc	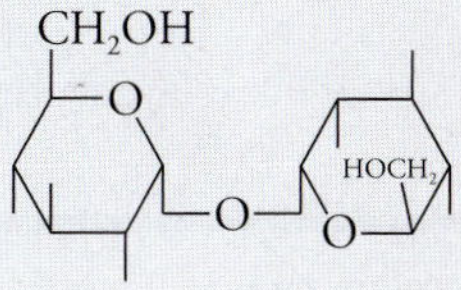
Lactose	Milk, milk products	β-Gal(1 → 4)Glc	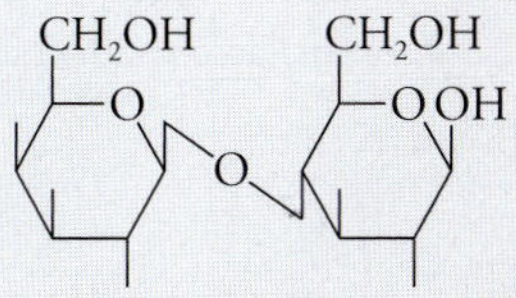
Raffinose	Leguminous seeds	α-Gal(1 → 6)α-Glc (1 → 2)β-Fru	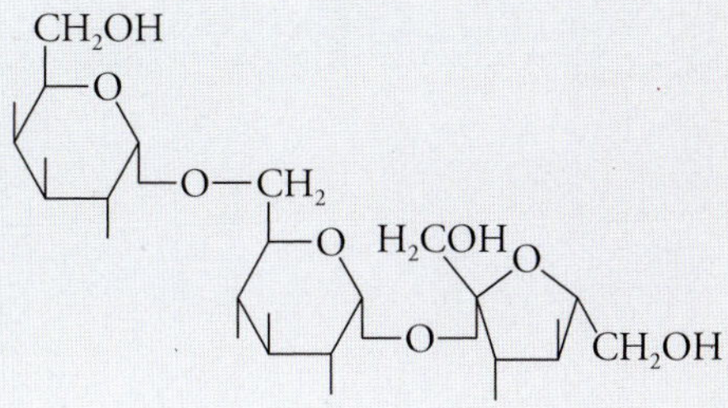

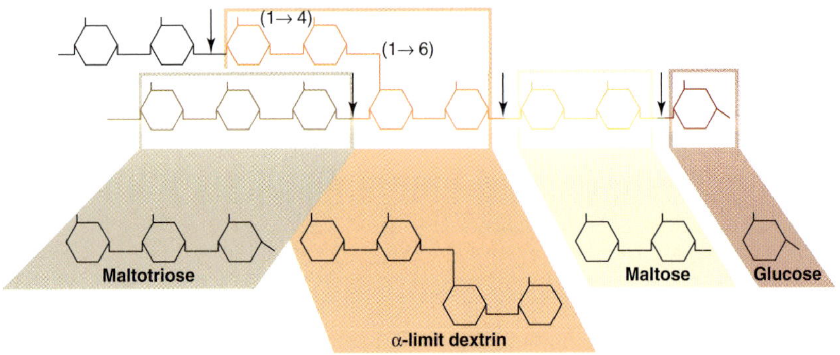

FIGURE 26.26

Digestion of amylopectin by salivary and pancreatic α-amylase.

TABLE 26.9 Saccharidases of the Surface Membrane of Small Intestine

Enzyme	*Specificity*	*Natural Substrate*	*Product*
exo-1, 4-α-Glucosidase (glucoamylase)	α-(1 → 4)Glucose	Amylose	Glucose
Oligo-1, 6-glucosidase (isomaltase)	α-(1 → 6)Glucose	Isomaltose, α-dextrin	Glucose
α-Glucosidase (maltase)	α-(1 → 4)Glucose	Maltose, maltotriose	Glucose
Sucrose-α-glucosidase (sucrase)	α-Glucose	Sucrose	Glucose, fructose
α, α-Trehalase	α-(1 → 1)Glucose	Trehalose	Glucose
β-Glucosidase	β-Glucose	Glucosylceramide	Glucose, ceramide
β-Galactosidase (lactase)	β-Galactose	Lactose	Glucose, galactose

CLINICAL CORRELATION 26.5
Disaccharidase Deficiency

Intestinal disaccharidase deficiencies are encountered relatively frequently in humans. Deficiency can be present in one or several enzymes for a variety of reasons (genetic defect, physiological decline with age, or the result of "injuries" to the mucosa). Lactase is the most commonly deficient enzyme. Absolute or relative deficiency is experienced as milk intolerance.

Consequences of a lack of lactose hydrolysis in the upper small intestine are inability to absorb lactose, which then becomes available for bacterial fermentation in the lower small intestine. Bacterial fermentation produces gas (distension of gut and flatulence) and osmotically active solutes that draw water into the intestinal lumen (diarrhea). Lactose in yogurt has already been hydrolyzed during the fermentation process of making yogurt. Thus, individuals with lactase deficiency can usually tolerate yogurt better than unfermented dairy products. Lactase is commercially available to pretreat milk so that its lactose is hydrolyzed.

Source: Swallow, D. M. Genetics of lactase persistence and lactose intolerance. *Annu. Rev. Genet.* 37:197, 2003. Solomons, N. W. Fermentation, fermented foods and lactose intolerance. *Eur. J. Clin. Nutr.* 56:S50, 2002. Lactose intolerance: http://digestive.niddk.nih.gov/ddiseases/pubs/lactoseintolerance/ and http://www.lactose.net

Di-, oligo-, and polysaccharides not hydrolyzed by α-amylase and/or intestinal surface enzymes cannot be absorbed; therefore they reach the lower tract of the intestine, which from the lower ileum on contains bacteria. Bacteria can utilize many of these remaining carbohydrates because they possess many more types of saccharidases than humans. Monosaccharides released by **bacterial enzymes** are predominantly metabolized anaerobically by the bacteria themselves, resulting in degradation products such as short-chain fatty acids, lactate, hydrogen gas (H_2), methane (CH_4), and carbon dioxide (CO_2). In excess, these compounds can cause fluid secretion, increased intestinal motility, and cramps, either because of increased intraluminal osmotic pressure and distension of the gut, or because of a direct irritant effect by the bacterial degradation products on the intestinal mucosa.

The flatulence that follows ingestion of leguminous seeds (beans, peas, and soy) is caused by oligosaccharides that cannot be hydrolyzed by human intestinal enzymes.

TABLE 26.10 Characteristics of Glucose Transporters in the Plasma Membranes of Enterocytes

Characteristic	*Luminal*	*Contraluminal*
Designation	SLC5A1 (SGLT1)	SLC2A2 (GLUT2)
Subunit H mass (kDa)	75	57
Effect of Na^+	Cotransport with Na^+	None
Good substrates	D-Glc, D-Gal, α-methyl-D-Glc	D-Glc, D-Gal, D-Man, 2-deoxy-D-Glc, D-Fru

The seeds contain modified sucrose to which one or more galactose moieties are linked. The glycosidic bonds of galactose are in the α-configuration, which can only be split by bacterial enzymes. The simplest sugar of this family is **raffinose** (see Table 26.8). **Trehalose** is a disaccharide that occurs in young mushrooms and requires a special disaccharidase, trehalase, for digestion (see Table 26.8).

Monosaccharide Transporters

The major monosaccharides produced by digestion of di- and polysaccharide are D-**glucose**, D-**galactose**, and D-**fructose**. At least two monosaccharide transporters catalyze monosaccharide uptake from the intestinal lumen into the lining epithelial cells: (1) an **Na^+-monosaccharide cotransporter (SGLT1)** that mediates "active" uptake of D-glucose and D-galactose into cells and (2) an **Na^+-independent, facilitative monosaccharide transporter** with specificity for D-fructose **(GLUT5)** (see p. 474). Another facilitative monosaccharide transporter **(GLUT2)** accepts all three monosaccharides and is present in the contraluminal plasma membrane. The physiological role of GLUT2 is to facilitate exit of monosaccharides from cells into the interstitial and capillary compartments, thereby completing the absorption process. GLUT2 is a member of the widely distributed GLUT family of transporters and is present in tissues such as intestine, liver, and kidney that absorb or produce glucose and release it into blood (see p. 586). Properties of intestinal SGLT1 and of GLUT2 are compared in Table 26.10, and their role in transepithelial glucose absorption is illustrated in Figure 26.25.

26.6 DIGESTION AND ABSORPTION OF LIPIDS

Digestion of Lipids Requires Overcoming Their Limited Water Solubility

An adult human ingests about 60–150 g of lipid per day. **Triacylglycerols** constitute more than 90% of this intake. The rest is made up of phospholipids, cholesterol, cholesteryl esters, and free fatty acids. In addition, 1–2 g of cholesterol and 7–22 g of phosphatidylcholine (lecithin) are secreted each day by the liver and reach the small intestine with bile.

The poor water solubility of lipids presents problems for digestion because substrates are not easily accessible to the digestive enzymes in the aqueous phase. In addition, most products of lipid digestion are themselves lipids with poor water solubility so that they tend to form aggregates that hinder effective absorption. These problems are overcome by (1) generating/secreting surfactive molecules that increase the interfacial area between aqueous and lipid phases and (2) "solubilization" of lipids with **detergents**. Thus changes in the physical state of lipids are intimately connected to chemical changes during digestion and absorption.

At least five phases of lipid digestion can be distinguished (Figure 26.27): (1) hydrolysis of triacylglycerols to free fatty acids and monoacylglycerols within the

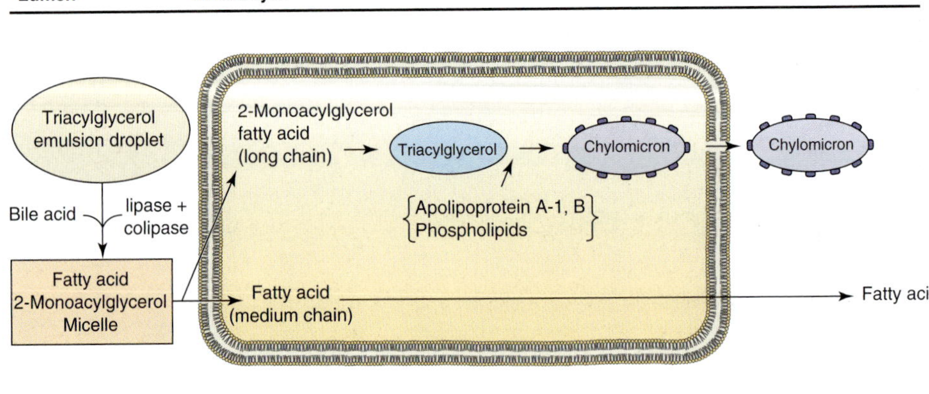

FIGURE 26.27

Digestion and absorption of lipids.

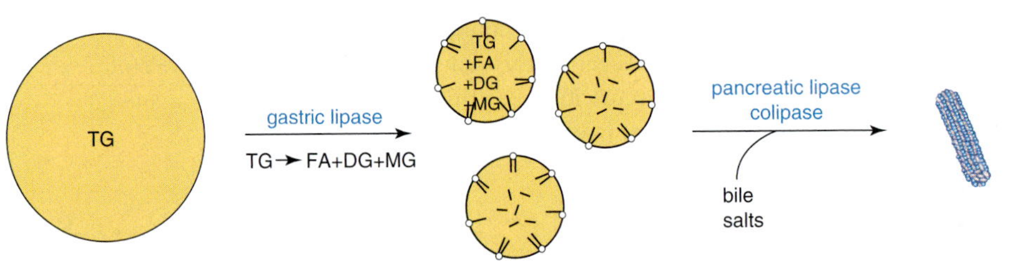

FIGURE 26.28

Changes in physical state during triacylglycerol digestion.

Abbreviations: TG, triacylglycerol; DG, diacylglycerol; MG, monoacylglycerol; FA, fatty acid.

lumen of the gastrointestinal tract; (2) solubilization of lipids by detergents (bile acids) and transport from the intestinal lumen toward the surface of the lining epithelial cells; (3) uptake of free fatty acids and monoacylglycerols into the epithelial cell and synthesis to triacylglycerols; (4) packaging of newly synthesized triacylglycerols into special lipid-rich globules, called chylomicrons; and (5) exocytosis of chylomicrons from the intestinal epithelial cells into lymph.

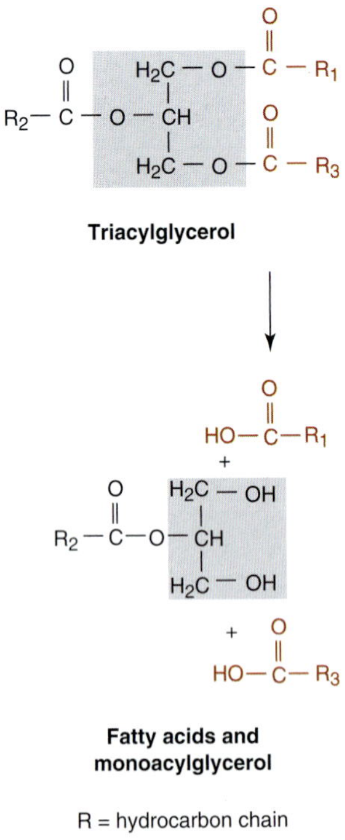

Lipids Are Digested by Gastric and Pancreatic Lipases

Hydrolysis of triacylglycerols is initiated in the stomach by **lingual** and **gastric lipases.** Gastric digestion can account for up to 30% of total triacylglycerol hydrolysis. However, the rate of hydrolysis is slow because ingested triacylglycerols form lipid droplets with a limited interfacial area to which lipases can adsorb. Nevertheless, some lipase molecules adsorb and hydrolyze triacylglycerols into fatty acids and diacylglycerols (Figure 26.28), which converts a water-immiscible compound to products with both polar and nonpolar groups. Such surfactive products spontaneously adsorb to water–lipid interfaces and confer a hydrophilic surface to lipid droplets, thereby allowing an increase in interfacial area. At constant volume of the lipid phase, any increase in interfacial area causes dispersion of the lipid phase into smaller droplets **(emulsification)** and provides more sites for adsorption of lipase molecules. The dispersion of lipids into smaller droplets is also aided by the peristaltic and churning movements of the stomach. Gastric chyme that is released into the duodenum generally contains only emulsion droplets of less than 2 mm in diameter.

Pancreatic lipase is the major enzyme for triacylglycerol hydrolysis (Figure 26.29). It hydrolyzes esters in the α-position of glycerol and prefers long-chain fatty acids (longer than 10 carbon atoms). The products are **free fatty acids** and **β-monoacylglycerols.** Hydrolysis occurs at the water–lipid interface of emulsion droplets or bile acid micelles (see below). However, bile acids, which are present in the intestinal lumen, inhibit purified lipase, indicating a more complex situation *in vivo.* Pancreatic juice contains a small protein (12 kDa) that binds to lipase and the micellar surface and prevents inhibition of lipase by bile acids. This protein has been termed **colipase.** It is secreted as procolipase and depends on tryptic removal of a NH_2-terminal decapeptide for

Triacylglycerol

Fatty acids and monoacylglycerol

R = hydrocarbon chain

FIGURE 26.29

Mechanism of action of lipase.

CLINICAL CORRELATION 26.6
Pharmacological Interventions to Prevent Fat Absorption and Obesity

Obesity is a major problem in modern society as food is generally very abundant. Therefore, interest in weight reduction is widespread. Two commercial products exploit the understanding of intestinal lipid absorption for this purpose. Olestra® is a commercial lipid produced by esterification of natural fatty acids with sucrose instead of glycerol (Figure (*a*)). With six to eight fatty acids covalently linked to sucrose the compound tastes like natural lipids, however, it cannot be hydrolyzed and is excreted unchanged.

Pancreatic lipase is the major enzyme that hydrolyzes dietary triacylglycerols to absorbable fatty acids and glycerol. Orlistat® (Figure (*b*)) is a nonhydrolyzable analog of a triacylglycerol and a powerful inhibitor of pancreatic lipase. Ingestion of Orlistat® slows down lipid digestion and hence absorption. Benefits derive from some lipid excretion, but also from release of a hormone (peptide YY) when lipids reach the terminal small intestine and colon. Peptide YY is postulated to increase the sensation of satiety and slow down the transit time of food through the gastrointestinal tract.

Olestra is a registered trademark of Proctor and Gamble and Orlistat of Roche, Basel, Switzerland.

(*a*)

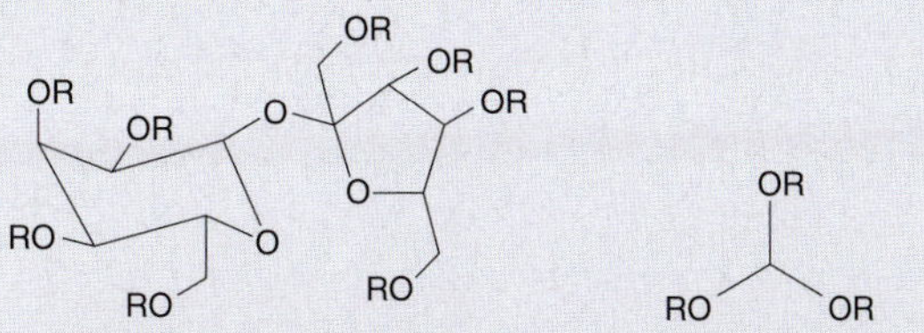

Olestra = octa-acyl sucrose

triacylglycerol

(*b*)

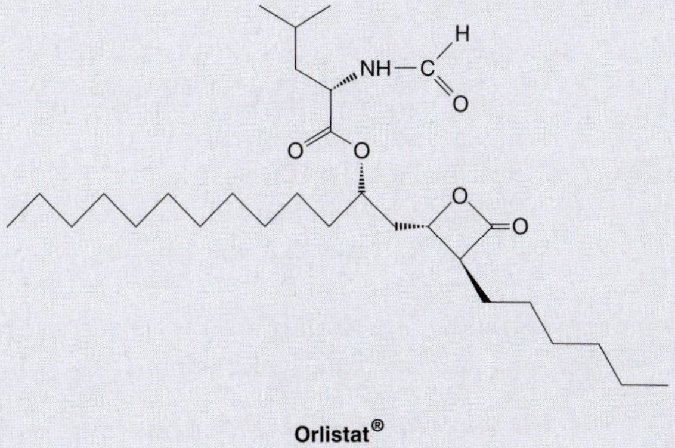

Orlistat®

Source: Thompson, A. B. R., et al. In: A. B. Christophe and S. DeVriese (Eds.), *Fat Digestion and Absorption*. Champaign, IL: AOCS Press, 2000, p. 383. Golay, A. In: A. B. Christophe and S. DeVriese (Eds.), *Fat Digestion and Absorption*. Champaign, IL: AOCS Press, 2000, p. 420.

activation. Clinical Correlation 26.6 describes two commercial strategies to reduce lipid absorption as a means to decrease obesity.

Pancreatic juice contains an **unspecific lipid esterase**, which acts on cholesteryl esters, monoacylglycerols, or other lipid esters, such as esters of vitamin A with carboxylic acids. In contrast to triacylglycerol lipase, this esterase requires bile acids for activity.

Phospholipids are hydrolyzed by specific phospholipases. Pancreatic juice is especially rich in **prophospholipase A_2** (Figure 26.30). Like other pancreatic proenzymes, it is activated by trypsin. Phospholipase A_2 requires bile acids for activity.

Bile Acid Micelles Solubilize Lipids During Digestion

Bile acids are biological detergents synthesized by liver and secreted as conjugates of glycine or taurine with the bile into the duodenum. At physiological pH, they are ionized (anions) so that the terms bile acids and **bile salts** are often used interchangeably (Figure 26.31). Bile acids reversibly form thermodynamically stable aggregates, called **micelles** (see p. 450), at concentrations above 2–5 mM and at pH values above the

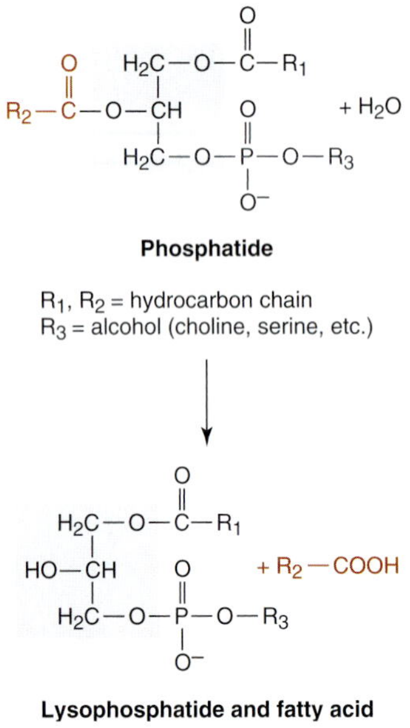

$R_2-C(=O)-O-CH(CH_2-O-C(=O)-R_1)(CH_2-O-P(=O)(O^-)-O-R_3)$ + H_2O

Phosphatide

R_1, R_2 = hydrocarbon chain
R_3 = alcohol (choline, serine, etc.)

↓

$HO-CH(CH_2-O-C(=O)-R_1)(CH_2-O-P(=O)(O^-)-O-R_3)$ + R_2-COOH

Lysophosphatide and fatty acid

FIGURE 26.30
Mechanism of action of phospholipase A_2.

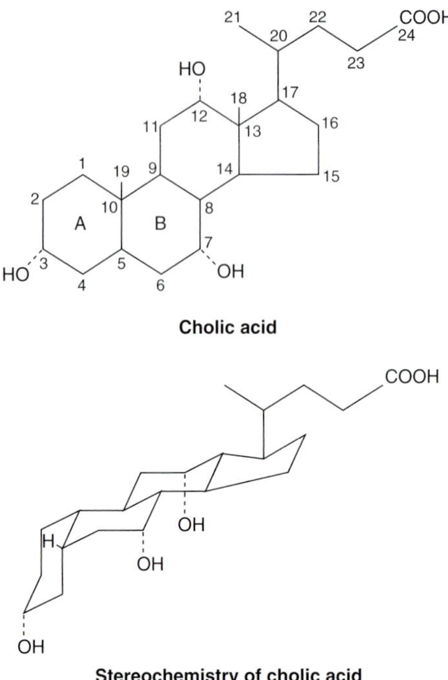

FIGURE 26.31
Cholic acid, a bile acid.

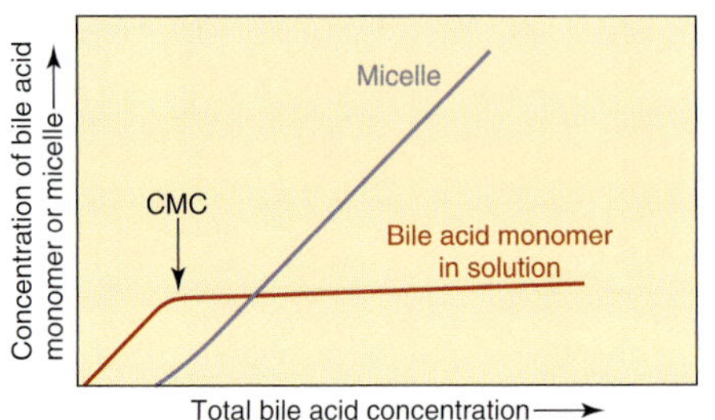

FIGURE 26.32
Solubility properties of bile acids in aqueous solutions. Abbreviation: CMC, critical micellar concentration.

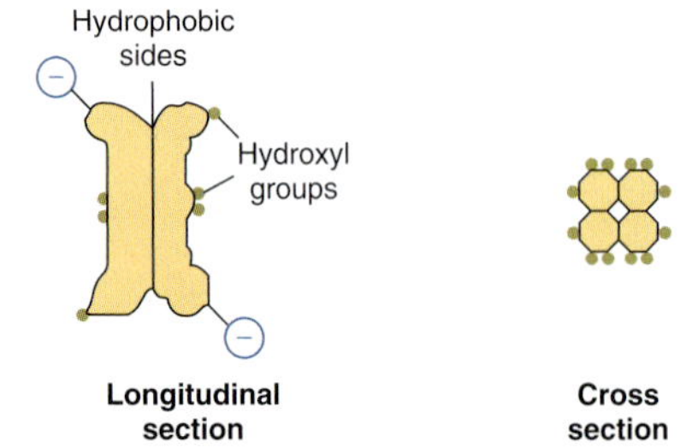

FIGURE 26.33
Diagrammatic representation of an Na^+ cholate micelle.
Redrawn and adapted from Small, D. M. *Biochim. Biophys. Acta* 176:178, 1969.

p*K* (Table 26.11). In other words, bile acid molecules in micelles are in equilibrium with those free in solution. The minimal concentration of a bile acid necessary for micelle formation is its **critical micellar concentration** (Figure 26.32). As equilibrium structure, micelles reach a well-defined size ($\leq$ 4 nm) that depends on the concentrations of bile acids and other lipids, but not on mechanical dispersion forces. The difference between micelles and emulsion droplets becomes obvious on ultracentrifugation of duodenal content: Emulsion droplets collect at the top as oily layer while micelles stay in the clear or slightly turbid subphase.

Major driving forces for micelle formation are the sequestering of apolar, hydrophobic groups away from water and the interaction of polar groups with water molecules. Bile acids have a fused ring system that is hydrophobic on one side and hydrophilic on the other, and, in addition, a highly polar head group. The geometry of polar and apolar regions in bile acids is very different from that of ionized fatty acids (soap) or phospholipids, and hence their micelles have different geometries. For example, fatty acid or phospholipid micelles are spherical while pure bile acid micelles form "sandwich structures" (Figure 26.33). Micelles, including the ones formed by bile acids, can solubilize other lipids, such as phospholipids and fatty acids, and form **"mixed" micelles**. Liver actually secretes bile acids together with phospholipids (mainly phosphatidyl choline) and cholesterol. The structure of these mixed micelles of bile acids and phospholipids is that of a rod, in which phospholipids are arranged radially along the length of the rod with the polar groups oriented outward and bile acids acting as wedge between the phospholipids head groups, also with their polar groups facing outward. The rods are capped by the polar groups of bile acids and phospholipids, and they become longer with increasing ratio of phospholipids to bile acids (Figure 26.34). Within the mixed phospholipid–bile acid micelles, other water-insoluble lipids, such as cholesterol, can be accommodated and thereby "solubilized." There are limits to the amounts of phospholipids and cholesterol that can be solubilized by physiological concentrations of bile acids; excess phospholipids form small unilamellar vesicles (20–60 nm in diameter), which can accommodate also some cholesterol. Excess cholesterol that cannot be solubilized by micelles or vesicles has a tendency to come out of solution by crystallization (see Clin. Corr. 26.7).

TABLE 26.11 Effect of Conjugation on Acidity of Cholic, Deoxycholic, and Chenodeoxycholic (Chenic) Acids

Bile Acid	Ionized Group	pK_a
Unconjugated bile acids	$—COO^-$ of cholestanoic acid	$\simeq 5$
Glycoconjugates	$—COO^-$ of glycine	$\simeq 3.7$
Tauroconjugates	$—SO_3^-$ of taurine	$\simeq 1.5$

Primary: Cholic (OH, COOH, HO, OH); Chenic (COOH, HO, OH)

Secondary: Deoxycholic (OH, COOH, HO)

+ glycine $NH_3^+CH_2COO^-$ → $R—C(=O)—N(H)—CH_2COO^-$ {Cholylglycine, Chenylglycine, Deoxycholylglycine}

+ taurine $NH_3^+(CH_2)_2SO_3^-$ → $R—C(=O)—N(H)—(CH_2)_2SO_3^-$ {Cholyltaurine, Chenyltaurine, Deoxycholyltaurine}

Source: Reproduced with permission from Hofmann, A. F. *Handbook of Physiology* 5:2508, 1968.

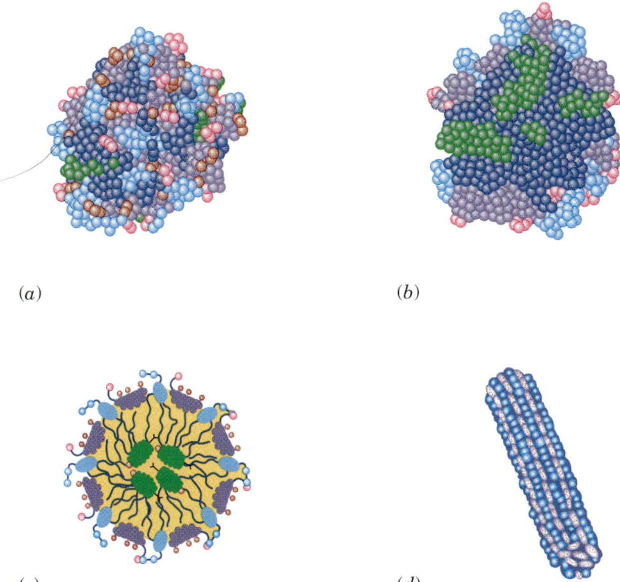

FIGURE 26.34

Proposed structure of mixed bile acid–phosphatidyl choline micelles. (*a*) View from outside and (*b*) from interior of an equilibrium structure of mixed micelle from molecular dynamics simulations. (*c*) Diagram of cross section. (*d*) Diagram of a rod-shaped micelle. Note that phosphatidyl choline molecules are arranged radially with bile acids wedged in between and dominating top and bottom (radial shell model).

Light blue represents phospholipid head groups; dark blue represents phospholipid tails; red represents hydroxyl groups of bile acids; pink represents ionic group of bile acid; purple represents hydrophobic portion of bile acid; green represents cholesterol. Parts (*a*) and (*b*) reproduced with permission and parts (*c*) and (*d*) redrawn with permission from Marrink, S. J. and Mark, A. E. *Biochemistry* 41:5375, 2002. Copyright (2002) American Chemical Society.

During digestion of triacylglycerol, free fatty acids and monoacylglycerols are released at the surface of fat emulsion droplets and micelles. In contrast to triacylglycerols, which are water-insoluble, free fatty acids and monoacylglycerols are slightly water-soluble, and molecules at the surface equilibrate with those in solution and in bile acid micelles. Thus the products of triacylglycerol hydrolysis are continually transferred from emulsion droplets to the micelles (see Figure 26.28).

Micelles are the major vehicle for transferring lipids from the lumen to the mucosal surface where absorption occurs. Because the fluid layer near the cell surface is poorly mixed, the major mechanism for solute flux across this **"unstirred" fluid layer** is diffusion down the concentration gradient. The delivery rate of solutes at the cell surface by diffusion is proportional to the concentration difference between luminal bulk phase and cell surface. Diffusion through an unstirred layer becomes a problem for sparingly soluble or insoluble nutrients, in that reasonable concentration gradients and delivery

CLINICAL CORRELATION 26.7
Cholesterol Stones

Liver secretes phospholipids, cholesterol, and bile acids into bile. Because of the limited solubility of cholesterol, its secretion can lead to cholesterol stone formation in the gallbladder. Stone formation is a relatively frequent occurrence; up to 20% of North Americans develop stones during their lifetime. Cholesterol is practically insoluble in aqueous solutions. However, it can be incorporated into mixed phospholipid-bile acid micelles and thereby "solubilized" (see accompanying figure). Liver can produce bile that is supersaturated with respect to cholesterol. Excess cholesterol tends to come out of solution and crystallize. Supersaturated bile is considered lithogenic—that is, stone-forming. Crystal formation usually occurs in the gallbladder, rather than the hepatic bile ducts, because contact time between bile and any crystallization nuclei is greater in the gallbladder. In addition, bile is concentrated here by absorption of electrolytes and water. The bile salts chenodeoxycholate (Table 26.11) and its stereoisomer ursodeoxycholate (7-hydroxy group in β-position) are available for oral use to dissolve gallstones. Their ingestion reduces cholesterol excretion into bile and allows cholesterol in stones to be solubilized.

The tendency to secrete bile supersaturated with respect to cholesterol is inherited, occurs more frequently in females than in males, and is associated with obesity. Supersaturation also appears to be a function of the size and nature of bile acid pool and secretion rate.

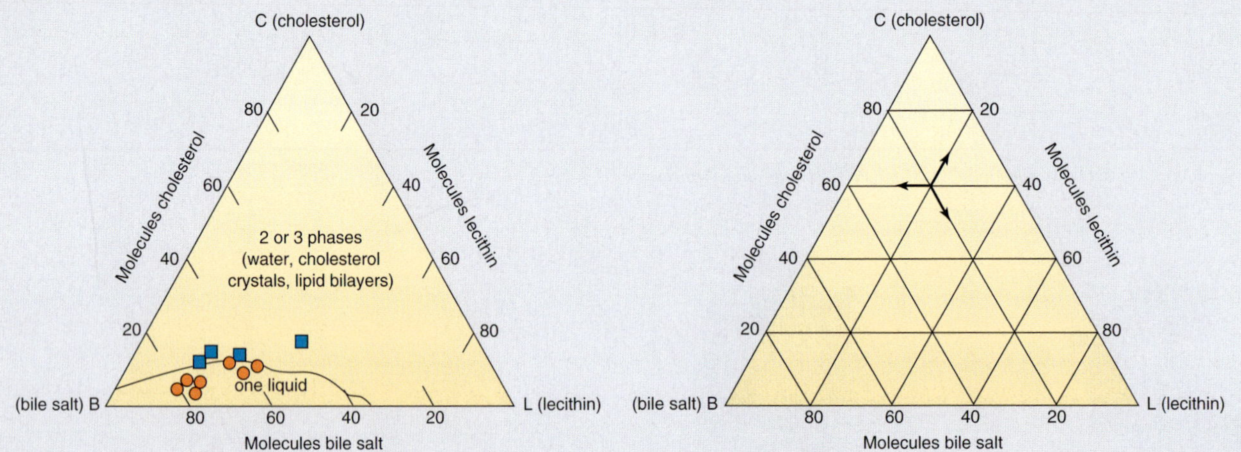

Diagram of the physical states of mixtures of 90% water and 10% lipid. The 10% lipid is made up of bile acids, phosphatidylcholine (lecithin), and cholesterol, and the triangle represents all possible ratios of the three lipid constituents. Each point within the triangle corresponds to a particular composition of the three components, which can be read off the graph as indicated by the arrows; each point on the sides corresponds to a particular composition of just two components. The left triangle contains the composition of gallbladder bile samples from patients without stones (red circles) and with cholesterol stones (blue squares). Lithogenic bile has a composition that falls outside the "one liquid" area in the lower left corner.
Redrawn from Hofmann, A. F. and Small, D. M. *Annu. Rev. Med.* 18:362, 1967. Copyright © 1967 by Annual Reviews, Inc.

Source: Schoenfield, L. J. and Lachin, J. M. Chenodiol (chenodeoxycholic acid) for dissolution of gallstones: The National Cooperative Gallstone Study. A controlled trial of safety and efficacy. *Ann. Intern. Med.* 95:257, 1981. Portincasa, P., Moschetta, A., van Erpecum, K. J., Calamita, G., Margari, A., vanBerge-Henegouwen, G. P., and Palasciano, G. Pathways of cholesterol crystallization in model bile and native bile. *Dig. Liver Dis.* 35:118, 2003.

rates cannot be achieved. Bile acid micelles overcome this problem for lipids by increasing their effective concentration in the unstirred layer. The increase in rate of transport is nearly proportional to the increase in effective concentration and can be 1000-fold over that of individually solubilized fatty acids, in accordance with the different water solubility of fatty acids as micelles and as individual molecules. This relationship between flux and effective concentration holds because the diffusion constant is only slightly smaller for micelles than for lipid molecules in solution. In absence of bile acids, absorption of triacylglycerols does not completely stop, but efficiency is drastically reduced. Residual absorption depends on the slight water-solubility of free fatty acids and monoacylglycerols. Unabsorbed lipids reach the lower intestine where a small part can be metabolized by bacteria. The bulk of unabsorbed lipids, however, is excreted with the stool **(steatorrhea)**.

Micelles also transport cholesterol and the lipid-soluble **vitamins A, D, E**, and **K** through the unstirred fluid layers. Bile acid secretion is essential for their absorption.

Most Absorbed Lipids Are Incorporated into Chylomicrons

Uptake of lipids by intestinal epithelial cells occurs by diffusion through the plasma membrane. In addition, long-chain fatty acid uptake is enhanced by a transporter (FATP4 or SLC27A4) and that of cholesterol uptake by a channel (Niemann–Pick C1-like protein or NPC1L1) in the luminal membrane. Sterols can also be pumped back out by an ABC transporter (see p. 479) consisting of two half-transporters (ABCG5 and ABCG8) and a portion of cholesterol is actually returned to the luminal compartment. Export of sterols by the ABC transporter is particularly important for rejecting plant sterols; normally, plant sterols are not found in serum. Loss-of-function mutations in either of the half-transporters are associated with a rise of the plant sterol sitosterol in serum (phytosterolemia or sitosterolemia).

Absorption is virtually complete for fatty acids and monoacylglycerols, which are slightly water-soluble. It is less efficient for water-insoluble lipids. For example, only 30–40% of the dietary cholesterol is usually absorbed.

Within the absorbing epithelial cells, the fate of absorbed fatty acids depends on chain length. **Fatty acids** of **short** and **medium chain** length ($\leq$ 10 carbon atoms) pass into portal blood without modification. **Long-chain fatty acids** ($>$12 carbon atoms) or their monoacylglycerols become bound to a cytosolic fatty acid-binding protein (intestinal FAB or I-FABP) and are transported to the endoplasmic reticulum, where they are converted into triacylglycerols. Glycerol for this process is derived from the absorbed 2-monoacylglycerols and, to a minor degree, from glucose. Cholesterol is esterified by cholesterol acyltransferase. The newly synthesized triacylglycerols and cholesteryl esters form lipid globules to which phospholipids and **apolipoproteins** adsorb. The globules are called **chylomicrons** because they can grow up to several micrometers in diameter and leave the intestine through lymph vessels (chyle = milky lymph derived from the Greek chylos, which means juice). Chylomicrons are synthesized within the lumen of the endoplasmic reticulum whence they migrate through the Golgi and then in vesicles to the contraluminal membrane. They are released into the intercellular space by fusion of these vesicles with the plasma membrane. Interestingly, chylomicrons do not enter the capillary space and the portal vein, instead they travel through the intestinal lymph vessels (or lacteals) and the thoracic duct to the systemic venous system. The intestinal apolipoproteins are designated A-1 and B48 (see Clin. Corr. 26.8); they are different from those of the liver with similar function (see p. 711).

While dietary medium-chain fatty acids reach the liver directly with the portal blood, the long-chain fatty acids first reach adipose tissue and muscle via the systemic circulation before coming into contact with the liver. Fat and muscle cells take up large amounts of dietary lipids for storage or metabolism. A bypass of the liver may have evolved to protect this organ from lipid overload after a meal.

The differential handling of medium- and long-chain fatty acids by intestinal cells can be exploited to provide the liver with high-caloric nutrients in the form of fatty acids. Short- and medium-chain fatty acids smell and taste rancid and are not very palatable; however, triacylglycerols that contain these fatty acids are quite palatable and can be used as part of the diet. Short-chain fatty acids are produced physiologically, particularly in the colon, by bacteria from residual carbohydrates,. These fatty acids are absorbed into portal blood.

CLINICAL CORRELATION 26.8
A-β-Lipoproteinemia

Apolipoprotein B (apoB) is a key component of lipoproteins: A 48-kDa splice variant is used by intestinal epithelial cells in the assembly of chylomicrons, while a 100-kDa variant is important for the assembly of very-low-density lipoproteins (VLDLs) by the liver. ApoB serves as acceptor for newly synthesized triglycerides that are transferred by microsomal triglyceride transfer protein. Mutations in the gene for this latter enzyme are the basis for a-β-lipoproteinemia characterized by the absence of liver and intestinal lipoproteins from plasma. Serum cholesterol is extremely low in this condition. A-β-lipoproteinemia is associated with severe malabsorption of triacylglycerol and lipid-soluble vitamins (especially tocopherol and vitamin E) and accumulation of apoB in enterocytes and hepatocytes.

Source: Fisher, E. A. and Ginsberg, H. N. Complexity in the secretory pathway: The assembly of apolipoprotein B-containing lipoproteins. *J. Biol. Chem.* 277:17377, 2002. Hussain, M. M., Iqbal, J., Anwar, K., Rava, P., and Dai, K. Microsomal triglyceride transfer protein: A multifunctional protein. *Front. Biosci.* 8:s500, 2003.

26.7 BILE ACID METABOLISM

Bile Acid Chemistry and Synthesis

Bile acids are synthesized in liver cells (hepatocytes) from cholesterol, secreted into bile together with phospholipids, and modified by bacterial enzymes in the intestinal lumen.

Primary bile acids synthesized by the liver are **cholic** and **chenodeoxycholic** (or chenic) **acid**. **Secondary bile acids** are derived from primary bile acids by bacterial reduction in position 7 of the ring structure, resulting in **deoxycholate** and **lithocholate**, respectively (see Table 26.11, Figure 18.42, p. 718, for structures).

Primary and secondary bile acids are reabsorbed by the intestine (lower ileum) into portal blood, taken up by liver cells, and then resecreted into bile. In liver cells, primary as well as secondary bile acids are linked to either glycine or taurine via an isopeptide bond. These **glyco-** and **tauroconjugates** constitute the forms that are secreted into bile. Conjugation is important for converting the mildly acidic carboxyl group to more polar and acidic ones, which is expressed as lower pK value and implies ionization over a wider pH range. The conjugation is partially reversed within the intestinal lumen by hydrolysis of the isopeptide bond.

Bile Acid Transport

The total amount of conjugated and unconjugated bile acids secreted per day is 20–30 g for an adult. However, the body maintains only a pool of 3–5 g. A small pool is advantageous because bile acids become toxic at high concentrations due to their detergent properties, for example, through their ability to lyse cells. Therefore, to achieve the observed secretion rates, bile acids are reabsorbed, recirculated to the liver, and resecreted 4–10 times per day. The secretion and reuptake is referred to as the **enterohepatic circulation** (Figure 26.35). Reabsorption of bile acids is quite efficient since only about 0.8 g of bile acids each day is excreted with the feces. Serum levels of bile acids normally vary with the rate of reabsorption and therefore are highest during a meal. Cholate, deoxycholate, chenodeoxycholate, and their conjugates continuously participate in the enterohepatic circulation. In contrast, most of the lithocholic acid produced by bacterial enzymes is sulfated during the next passage through the liver. The **sulfate ester** of **lithocholic acid** is not reabsorbed and therefore excreted in the feces.

The transporters mediating enterohepatic circulation of bile acids are shown in Figure 26.36. Ileal absorption of bile acids is mediated by secondary active transport via a luminal Na^+–bile acid cotransport system (**apical sodium-dependent bile acid transporter**, **ASBT**, or **SLC10A2**) with a stoichiometry of 2:1 for Na^+ :bile acid. Bile acids are moved from ileal enterocytes into blood either in exchange for another

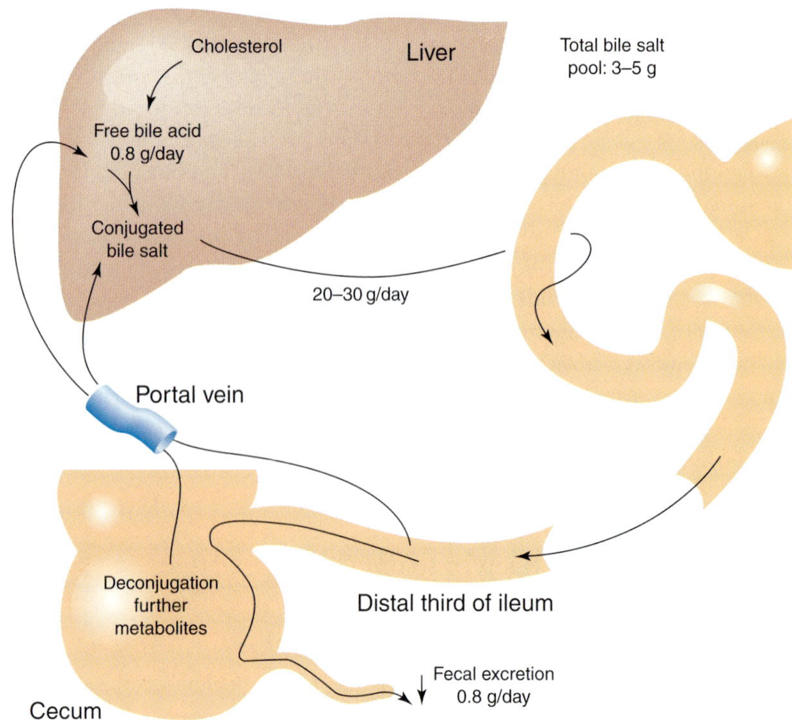

FIGURE 26.35

Enterohepatic circulation of bile acids.

Redrawn from Clark, M. L. and Harries, J. T. In: I. McColl and G. E. Sladen (Eds.), *Intestinal Absorption in Man*. New York: Academic Press, 1975, p. 195.

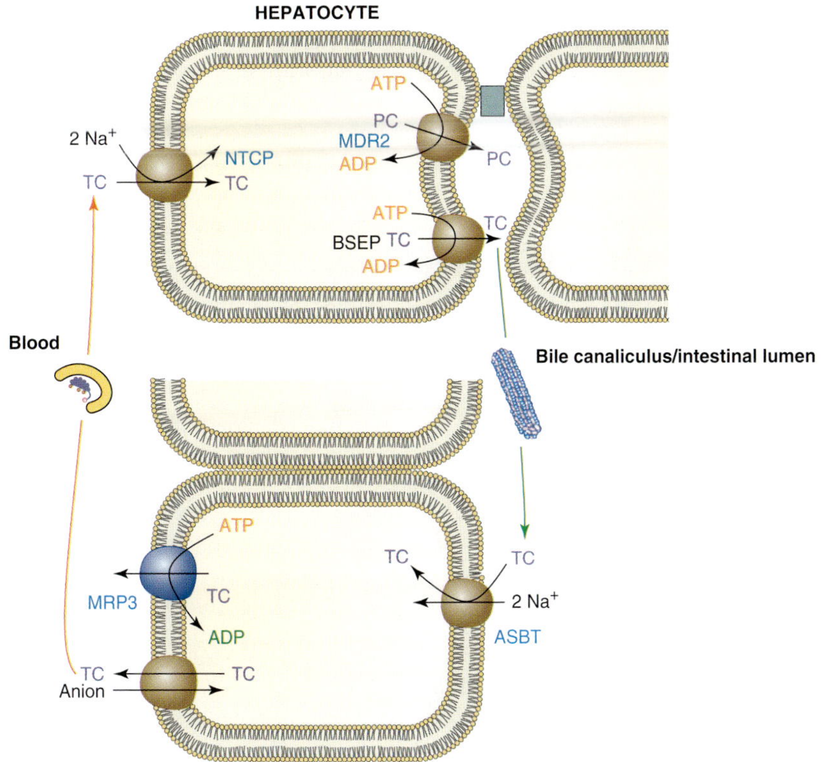

FIGURE 26.36

Transporters for taurocholate (TC) and phosphatidyl choline (PC) during enterohepatic circulation.

anion (e.g., bicarbonate) or by active transport via an ABC-type pump (MRP3 or ABCC3). Uptake of bile acids from blood by liver cells is predominantly by secondary active Na^+–bile acid cotransport (**Na^+ taurocholate cotransporting polypeptide, NTCP**, or **SLC10A1**). Unconjugated bile acids can also be taken up from blood by an Na^+-independent, multispecific organic anion transporter (SLC21A). In contrast, secretion of bile acids by liver cells across the canalicular plasma membrane into bile is by primary active transport (**bile salt export pump**, **BSEP**, or **ABCB11**, another member of the ABC transporters; see p. 479). Phospholipids, which are secreted simultaneously with bile acids, are transported by the pump MDR2 (ABCC3). To achieve reasonable total bile acid concentrations within cells and in plasma while maintaining low free concentrations to prevent detergent action, bile acids are bound to specific binding proteins. Albumin functions as such a binding protein for bile acids in plasma.

BIBLIOGRAPHY

General

Chang, E. B, Sitrin, M. D, and Black, D. D. *Gastrointestinal, Hepatobiliary, and Nutritional Physiology*. Philadelphia: Lippincott-Raven, 1996.

Johnson, L. R. (Ed.-in-Chief). *Physiology of the Gastrointestinal Tract*, Vols. 1 and 2, 2nd and 3rd eds. New York: Raven, 1987, 1994.

Pandol, S. and Raybauld, H. E. Integrated response to a meal. *The Undergraduate Teaching Project in Gastroenterology and Liver Disease*. Unit #29. *American Gastroenterological Association*. Timonium, M. D. (Ed.). Milner-Fenwick, Inc., 1995.

Porter, R. and Collins, G. M. *Brush Border Membranes. Ciba Foundation Symposium*, Vol. 95. London: Pitman, 1983.

Schultz, S. G. (Section Ed.) *Handbook of Physiology. Section 6: The Gastrointestinal System*, Vol. IV. *Intestinal Absorption and Secretion*. (Eds. M. Field and R. A. Frizzell). Bethesda, MD: American Physiological Society, 1991.

Gastric Acid Secretion

Richter, C., Tanaka, T., and Yada, R. Y. Mechanism of activation of the gastric aspartic proteinases: pepsinogen, progastricsin, and prochymosin. *Biochem. J.* 335:4810, 1998.

Samuelson, L. C,, Hinkle, K, L. Insights into the regulation of gastric acid secretion through analysis of genetically engineered mice. *Annu. Rev. Physiol.* 65:383, 2003.

Yao, X. and Forte, J. G. Cell biology of acid secretion by the parietal cell. *Annu. Rev. Physiol.* 65:103, 2003.

Lipid Digestion

Christophe, A. B. and DeVriese, S. (Eds.). *Fat Digestion and Absorption*. Champaign, IL: AOCS Press, 2000.

Hofmann, A. F. Bile acids: The good, the bad, and the ugly. *News Physiol. Sci.* 14:24. 1999.

Marrink, S. J. and Mark, A. E. Molecular dynamics simulations of mixed micelles modeling human bile. *Biochemistry* 41:5375, 2002.

Thomson, A. B. R., Schoeller, C., Keelan, M., Smith, L., and Clandinin, M. T. Lipid absorption: Passing through the unstirred layers near the brush border membrane, and beyond. *Can. J. Physiol. Pharmacol.* 71:531, 1993.

Topping, D. L. and Clifton, P. M. Short-chain fatty acids and human colonic function: Roles of resistant starch and nonstarch polysaccharides. *Physiol. Rev.* 81:1031, 2001.

Carbohydrate Digestion/Absorption

Cristofaro, E., Mottu, F., and Wuhrmann, J. J. Involvement of the raffinose family of oligosaccharides in flatulence. In: H. L. Sipple and K. W. McNutt (Eds.), *Sugars in Nutrition.* New York: Academic Press, 1974, p. 314.

Van-Loo, J., Cummings, J., Delzenne, N., Englyst, H., Franck, A., et al. Functional food properties of non-digestible oligosaccharides: A consensus report from the ENDO project. *Br. J. Nutr.* 81:121, 1999.

Transporters

Barbry, P. and Hofman, P. Molecular biology of Na^+ absorption. *Am. J. Physiol.* 273:G571, 1997.

Chillaron, J., Roca, R., Valencia, A., Zorzano, A. and Palacin, M. Heteromeric amino acid transporters: Biochemistry, genetics, and physiology. *Am. J. Physiol. Renal Physiol.* 281:F995, 2001.

Hediger, M. A., Romero, M. F., Peng, J. B., Rolfs, A., Takanaga, H., and Bruford, E. A. The ABCs of solute carriers: Physiological, pathological and therapeutic implications of human membrane transport proteins. Introduction. *Pflügers Arch.* 447:465. 2004.

Kullak-Ublick, G. A., Stieger, B. and Meier, P. J. Enterohepatic bile salt transporters in normal physiology and liver disease. *Gastroenterology* 126:322, 2004.

Rao, M. C. Oral rehydration therapy: New explanations for an old remedy. *Annu. Rev. Physiol.* 66:385, 2004.

Wittenburg, H. and Carey, M. C. Biliary cholesterol secretion by the twinned sterol half-transporters ABCG5 and ABCG8. *J. Clin. Invest.* 110:605, 2002.

Wood, I. S. and Trayhurn, P. Glucose transporters (GLUT and SGLT): Expanded families of sugar transport proteins. *Br. J. Nutr.* 89:3, 2003.

Web Sites

ABC transporters: http://nutrigene.4t.com/humanabc.htm

Aminoaciduria:http://cnserver0.nkf.med.ualberta.ca/cn/Schrier/Volume2/chapt12/ADK2_12_4-6.pdf

Superfamily of carriers: http://www.bioparadigms.org/slc/menu.asp

Peptidase specificity: http://au.expasy.org/tools/peptidecutter/peptidecutter_enzymes.html

http://arbl.cvmbs.colostate.edu/hbooks/pathphys/digestion/stomach/parietal.html

Gluten Enteropathy: http://www.nlm.nih.gov/medlineplus/ency/article/002443.htm#visualContent

Epithelial endocrine cells: http://www.siumed.edu/~dking2/erg/gicells.htm

General: http://gastroenterology.medscape.com

QUESTIONS | Carol N. Angstadt

Multiple Choice Questions

1. Active forms of most enzymes that digest food may normally be found in all of the following *except*:
   A. in soluble form in the lumen of the stomach.
   B. in the saliva.
   C. attached to the luminal surface of the plasma membrane of intestinal epithelial cells.
   D. dissolved in the cytoplasm of intestinal epithelial cells.
   E. in zymogen granules of pancreatic exocrine cells.

2. Histamine is a potent secretagogue of:
   A. amylase by the salivary glands.
   B. HCl by the stomach.
   C. gastrin by the stomach.
   D. hydrolytic enzymes by the pancreas.
   E. $NaHCO_3$ by the pancreas.

3. The contraluminal plasma membrane of small intestinal epithelial cells contains:
   A. aminopeptidases.
   B. Na^+/K^+-exchanging ATPase.
   C. disaccharidases.
   D. GLUT5.
   E. Na^+–monosaccharide transport (SGLT1).

4. Micelles:
   A. are the same as emulsion droplets.
   B. form from bile acids at all bile acid concentrations.
   C. do not significantly enhance utilization of dietary lipid.
   D. of bile acids and phospholipids are spherical in shape.
   E. are essential for the absorption of vitamins A and K.

5. Epithelial cells of the lower ileum express a Cl^-/HCO_3^- exchange coded for by the DRA gene. These cells:
   A. mediate an electrogenic exchange of 2 luminal Na^+ for 1 cytosolic H^+.
   B. absorb Cl^- into the cell in exchange for HCO_3^- moving into the lumen.
   C. prevent a metabolic acidosis due to loss of HCl.
   D. mediate Na^+ movement out of the cell as Cl^- moves into the cell.
   E. mediate Na^+ movement into the cell via a Na^+ channel.

6. Certain tissues effect Cl^- secretion via a Cl^- channel (CFTR-cystic fibrosis transmembrane regulatory protein). Cholera toxin abnormally opens the channel leading to a loss of NaCl. Fluid treatments for cholera and sports drinks for electrolyte replacement are fluids high in Na^+ and glucose. The presence of glucose enhances NaCl replenishment because:
   A. absorbing any nutrient causes Na^+ uptake.
   B. glucose prevents Na^+ excretion.
   C. Na^+ and glucose are transported in opposite directions.
   D. glucose is absorbed across intestinal epithelial cells via an Na^+-dependent cotransporter.
   E. glucose inhibits the Na^+/K^+-exchanging ATPase.

Questions 7 and 8: A young woman finds that every time she eats dairy products she feels highly uncomfortable. Her stomach becomes distended, and she has gas and, frequently, diarrhea. A friend suggested that she try yogurt to get calcium and she is able to tolerate that. These symptoms do not appear when she eats food other than dairy products. Like many adults, she is deficient in an enzyme required for carbohydrate digestion.

7. The most likely enzyme in which she is deficient is:
   A. α-amylase.
   B. β-galactosidase (lactase).
   C. α-glucosidase (maltase).
   D. sucrose-α-glucosidase (sucrase).
   E. α, α-trehalase.

8. Monosaccharides are absorbed from the intestine:
   A. by an Na^+-dependent cotransporter for glucose and galactose.
   B. by an Na^+-independent facilitated transport for fructose.
   C. by an Na^+-independent transporter (GLUT2) across the contraluminal membrane.
   D. against a concentration gradient if the transporter is Na^+-dependent.
   E. all of the above.

Questions 9 and 10: A woman comes to the emergency room with severe abdominal pain in the right upper quadrant as well as severe pain in her back. The pain began several hours after she consumed a meal of fried chicken and cheese-coated french fries. The symptoms indicated "gallstones" and this was confirmed by ultrasound. While surgery might be necessary in the future, conservative treatment was tried first. She was instructed to limit fried foods and high-fat dairy products. She was also given chenodeoxycholate to take orally to try to dissolve the gallstones.

9. Cholesterol "stones":
   A. usually form during passage of bile through the hepatic bile duct.
   B. occur when the mixed phospholipid-bile acid micelles are very high in phospholipid.
   C. can be dissolved by excess bile acids, which solubilize in micelles the water-insoluble cholesterol.
   D. rarely occur because cholesterol is not a normal part of bile.
   E. are a necessary part of lipid digestion.

10. In the metabolism of bile acids:
   A. the liver synthesizes cholic, and deoxycholic acids, which are primary bile acids.
   B. secondary bile acids are produced by conjugation of primary acids to glycine or taurine.
   C. 7-dehydroxylation of bile acids by intestinal bacteria produces secondary bile acids that have detergent and physiological properties similar to those of primary bile acids.
   D. daily bile acid secretion by the liver is approximately equal to daily bile acid synthesis.
   E. conjugation reduces the polarity of bile acids, enhancing interaction with lipids.

Questions 11 and 12: Hartnup disease is a genetic defect in an amino acid transport system. The specific defect is in the neutral amino acid transporter in both intestinal and renal epithelial cells. Clinical symptoms of the disease result from deficiencies of essential amino acids and nicotinamide (because of a deficiency specifically of tryptophan).

11. In addition to tryptophan, which of the following amino acids is likely to be deficient in Hartnup disease?
   A. aspartate.
   B. leucine.
   C. lysine.
   D. proline.
   E. all of the above.

12. Hartnup disease patients are able to get some of the benefit of the protein they consume because:
   A. only the neutral amino acid carrier is defective.
   B. di- and tripeptides from protein digestion are absorbed by a different carrier (PepT1).
   C. their endo- and exopeptidases are normal.
   D. all of the above.
   E. none of the above.

Problems

13. Using known endo- and exopeptidases, suggest a pathway for the complete degradation of the following peptide:

His-Ser-Lys-Ala-Trp-Ile-Asp-Cys-Pro-Arg-His-His-Ala

14. At what point would the lack of colipase inhibit the digestion of fat?

ANSWERS

1. **E** Zymogen granules contain inactive proenzymes or zymogens, which are not activated until after release from the cell (amylase from the pancreas and salivary glands is an exception).
2. **B** Its binding to H_2 receptors of the stomach causes HCl secretion. A, D: Acetylcholine is the secretagogue. C: Gastrin is a secretagogue. E. This is stimulated by secretin.
3. **B** Only the contraluminal surface contains the Na^+, K^+-ATPase. All other activities are associated with the luminal surface.
4. **E** The lipid-soluble vitamins must be dissolved in mixed micelles as a prerequisite for absorption. A: Micelles are of molecular dimensions and are equilibrium structures; emulsion droplets are much larger. B: Micelle formation occurs only above the critical micellar concentration (CMC); below that concentration, the components are in simple solution. C: Micelles are necessary so that the enzyme has access to the triacylglycerols. D: Mixed micelles form a rod shape with phospholipids arranged radially and bile acids forming wedges.
5. **B** The proton gradient generated by an electrically neutral Na^+/H^+ exchange drives this. A: The Na^+/H^+ exchange via the expressed NHE3 transporter is electrically neutral. C: Constant loss of HCl leads to a metabolic alkalosis. D: The direction of Na^+ movement is into the cell with subsequent removal by the Na^+/K^+-ATPase. E: This occurs in the large intestine, not here.
6. **D** This is a cheaper and easier way to increase intracellular electrolytes than intravenous fluid therapy. A: Only certain nutrients are absorbed by Na^+-dependent cotransporters. B: These are not related. C: Na^+-dependent cotransporters transfer the two substances in the same direction. E: Glucose has no effect on this enzyme.
7. **B** Dairy products contain lactose. Undigested lactose is fermented and the products produce the symptoms. Yogurt fermentation hydrolyzes lactose. A, C, D: Deficiency of any of these would cause problems with most carbohydrates. E: Trehalose is found in mushrooms, not dairy products.

8. **E** All of these play an important role in absorbing the monosaccharides from digestion. D: This is especially important for the uptake of most dietary glucose.
9. **C** Stones occur when bile is supersaturated (> 1:1 ratio of cholesterol/phospholipid). The ingested bile salts increase the solubilization in micelles. A: Stones usually form in the gallbladder. B: The problem is too little phospholipid relative to cholesterol. D: Actually, stones are relatively common. Cholesterol is a normal component of bile. E: Bile salts are necessary for lipid digestion but stones are not.
10. **C** Primary bile acids (cholic and chenodeoxycholic acids) are synthesized in liver. In intestine they may be reduced by bacteria to form the secondary bile acids, deoxycholate and lithocholate. D: Only a small fraction of bile acids escapes reuptake; this must be replaced by synthesis. Both primary and secondary bile acids are reabsorbed and recirculated (enterohepatic circulation). E: Both are conjugated to glycine or taurine, increasing their polarity.
11. **B** Trp shares a carrier with tyr, phe, val, leu, ile, met. A: Asp is acidic. C: Lysine is basic. D: Proline is an imino acid. All of these use separate carriers.
12. **D** The body has at least seven transporters for the various classes of amino acids (See 11). It also has a carrier for di- and tripeptides. C: Since we cannot absorb intact protein, the presence of the endo- and exopeptidases is essential.
13. Trypsin cleavage gives (a) His-Ser-Lys + (b)Ala-Trp-Ile-Met-Cys-Gly-Pro-Arg + (c) His-His-Ala. Further degradation of (a) is accomplished by elastase and dipeptidase. Further degradation of (b) would start with chymotrypsin and use dipeptidases, tripeptidase, and carboxypeptidase B. To degrade (c), carboxypeptidase A and dipeptidase would be enough. The point is that several peptidases with varying specificities are required.
14. There would be some hydrolysis of the triacylglycerols in the stomach by gastric lipase. In the small intestine, the emulsion droplet would be stabilized by bile salts but pancreatic lipase, which is responsible for the bulk of the hydrolysis, is inactive in the absence of colipase. Handling of the other lipid material would not be affected.

27 PRINCIPLES OF NUTRITION I: MACRONUTRIENTS

Stephen G. Chaney

Textbook of Biochemistry With Clinical Correlations, Sixth Edition, Edited by Thomas M. Devlin

27.1 OVERVIEW

Nutrition is best defined as the composition and quantity of food intake and the utilization of the food by living organisms. Since the process of food utilization is biochemical, the major thrust of the next two chapters is a discussion of basic nutritional concepts in biochemical terms. Simply understanding basic nutritional concepts is no longer sufficient. Nutrition attracts more than its share of controversy in our society, and a thorough understanding of nutrition demands an understanding of the issues behind these controversies. These chapters will explore the biochemical basis for some of the most important nutritional controversies.

Study of human nutrition can be divided into three areas: undernutrition, overnutrition, and ideal nutrition. The primary concern in this country is not with **undernutrition** because nutritional deficiency diseases are now quite rare. **Overnutrition** is a particularly serious problem in developed countries. The current estimate is that over 30% of the United States population is obese and thus are at risk for a number of serious health consequences. Finally, there is increasing interest in the concept of ideal, or **optimal, nutrition**. This concept has meaning only in an affluent society, since only when food supply becomes abundant enough that deficiency diseases become rare is it possible to consider long-range effects of nutrients on health. This is probably the most exciting area of nutrition today.

27.2 ENERGY METABOLISM

Energy Content of Food Is Measured in Kilocalories

Much of the food we eat is converted to ATP and other high-energy compounds, which are utilized to drive biosynthetic pathways, generate nerve impulses, and power muscle contraction (Figure 27.1). Energy content of foods is generally described in terms of **calories**. Technically speaking, this refers to **kilocalories** of heat energy released by combustion of that food in the body. Some nutritionists prefer the **kilojoule** (a measure of mechanical energy), but since the American public is likely to be counting calories rather than joules in the foreseeable future, we will restrict ourselves to that term. Caloric values of protein, fat, carbohydrate, and alcohol are roughly 4, 9, 4, and 7 kcal/g^{-1}, respectively. Given these values and the amount and composition of the food, it is simple to calculate the caloric content (input) of the foods we eat. Calculating caloric content of foods is not a major problem in this country. Millions of Americans are able to do it with ease. The problem lies in balancing caloric input with caloric output. Where do these calories go?

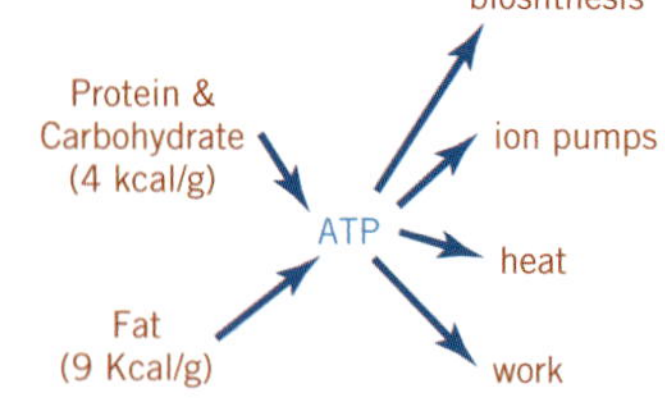

FIGURE 27.1
Metabolic fate of the foods we eat.

Energy Expenditure Is Influenced by Four Factors

The four principal factors that affect an individual's energy expenditure are listed in Table 27.1. The effects of surface area are thought to be simply related to the rate of heat loss by the body: The greater the surface area, the greater the rate of heat loss. While it may seem surprising, a lean individual actually has a greater surface area, and thus a greater energy requirement, than an obese individual of the same weight. Age may reflect two factors: growth and lean muscle mass. In infants and children more energy expenditure is required for rapid growth, which is reflected in a higher basal metabolic rate (rate of energy utilization in resting state). In adults (even lean adults), muscle tissue is gradually replaced by fat and water during the aging process, resulting in a 2% decrease in **basal metabolic rate (BMR)** per decade of adult life. Women tend to have a lower BMR than men because of a smaller percentage of lean muscle mass and the effects of female hormones on metabolism. The effect of activity levels on energy requirements is obvious. However, most overemphasize the immediate, as opposed to the long-term, effects of exercise. For example, one would need to jog for over an hour to burn up the calories present in one piece of apple pie.

TABLE 27.1 Factors that Influence Energy Expenditure

Surface area
Age
Sex
Activity level

Regular **exercise** increases basal metabolic rate, allowing calories to burn up more rapidly 24 hours a day. A regular exercise program should be designed to increase lean muscle mass and should be repeated 3–5 days a week but need not be aerobic exercise to have an effect on basal metabolic rate. For an elderly or infirm individual, even daily walking may help to increase basal metabolic rate slightly.

Hormone levels are also important, since thyroxine, sex hormones, growth hormone, and, to a lesser extent, epinephrine and cortisol increase BMR. The effects of epinephrine and cortisol probably explain in part why severe stress and major trauma significantly increase energy requirements. Finally, energy intake itself has an inverse relationship to expenditure in that during periods of **starvation** or semi-starvation BMR can decrease up to 50%. This is of great survival value in cases of genuine starvation, but not much help to the person who wishes to lose weight on a calorie-restricted diet.

27.3 PROTEIN METABOLISM

Dietary Protein Serves Many Roles Including Energy Production

Protein carries a certain mystique as a "body-building" food. While it is an essential structural component of all cells, it is also important for maintaining of essential secretions such as digestive enzymes and peptide or protein hormones. Protein is also needed for synthesis of plasma proteins, which are essential for maintaining osmotic balance, transporting substances through the blood, and maintaining immunity. However, the average North American adult consumes far more protein than needed to carry out these essential functions. Excess protein is treated as a source of energy, with glucogenic amino acids being converted to glucose and ketogenic amino acids to fatty acids and keto acids. Both kinds of **amino acids** are eventually converted to triacylglycerol in adipose tissue if fat and carbohydrate supplies are already adequate to meet energy requirements. Thus for most of us the only body-building obtained from high-protein diets is in adipose tissue.

It has been popular to say that the body has no storage depot for protein, and thus adequate dietary protein must be supplied with every meal. However, in actuality, this is not quite accurate. While there is no separate class of "storage" protein, there is a certain percentage of body protein that undergoes a constant process of breakdown and resynthesis. In the fasting state the breakdown of this protein is enhanced, and the resulting amino acids are utilized for glucose production, synthesis of nonprotein nitrogenous compounds, and the essential secretory and plasma proteins mentioned above. Even in the fed state, some of these amino acids are utilized for energy production and as biosynthetic precursors. Thus, the turnover of body protein is a normal process—and an essential feature of what is called nitrogen balance.

Nitrogen Balance Relates Intake to Excretion of Nitrogen

Nitrogen balance (Figure 27.2) is the relationship between **intake** of nitrogen (chiefly in the form of protein) and **excretion** of nitrogen (chiefly in the form of undigested protein in the feces and urea and ammonia in urine). A normal adult is in nitrogen equilibrium, with losses just balanced by intake. Negative nitrogen balance results from inadequate intake of protein, since amino acids utilized for energy and biosynthetic reactions are not replaced. It also occurs in injury when there is net destruction of tissue and in major trauma or illness when the body's adaptive response causes increased protein catabolism. Positive nitrogen balance occurs when there is a net increase in body protein, such as in growing children, pregnant women, or convalescing adults.

Essential Amino Acids Must Be Present in the Diet

Several other factors must be considered in addition to the amount of protein in the diet. One is the complement of **essential amino acids** ingested. Essential amino acids are amino acids that cannot be synthesized by the body (Table 27.2). If just one of these

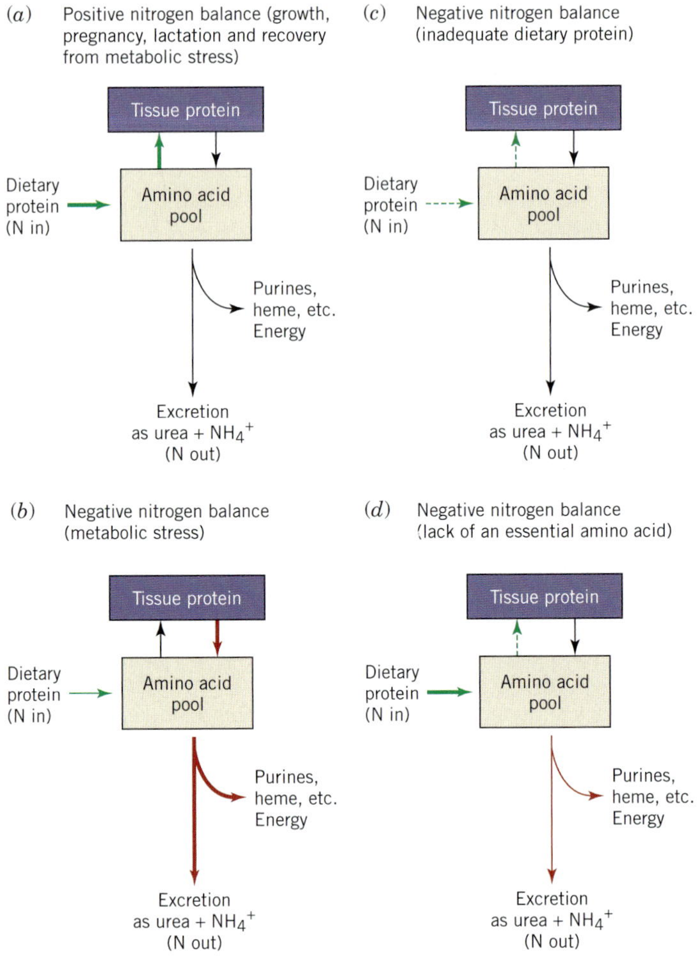

FIGURE 27.2

Factors affecting nitrogen balance. Schematic representations of the metabolic interrelationship involved in determining nitrogen balance. (*a*) Positive nitrogen balance (growth, pregnancy, lactation, and recovery from metabolic stress). (*b*) Negative nitrogen balance (metabolic stress). (*c*) Negative nitrogen balance (inadequate dietary protein). (*d*) Negative nitrogen balance (lack of an essential amino acid). Each figure represents the nitrogen balance resulting from a particular set of metabolic conditions. The dominant pathways in each situation are indicated by heavy red arrows.

essential amino acids is missing from the diet, the body cannot synthesize new protein to replace that lost due to normal turnover, and a negative nitrogen balance results (Figure 27.2). Obviously, the complement of essential amino acids in dietary protein determines how well it can be used by the body.

Most animal proteins contain all essential amino acids in about the quantities needed by the human body. Vegetable proteins, on the other hand, often lack one or more essential amino acids and may, in some cases, be more difficult to digest. Even so, vegetarian diets can provide adequate protein, provided that enough extra protein is consumed to provide sufficient quantities of the essential amino acids and/or two or more different proteins are consumed together, which complement each other in amino acid content. For example, if corn (which is deficient in lysine) is combined with legumes (deficient in methionine but rich in lysine), the efficiency of utilization of the two vegetable proteins approaches that of animal protein. The adequacy of vegetarian diets with respect to protein and calories for children is discussed in Clinical Correlation 27.1, and the need for high quality protein in the low protein diets used for treatment of renal disease is discussed in Clinical Correlation 27.2.

TABLE 27.2 Essential Amino Acids

Histidine
Isoleucine
Leucine
Lysine
Methionine
Phenylalanine
Threonine
Tryptophan
Valine

Protein Sparing Is Related to Dietary Content of Carbohydrate and Fat

Another factor that determines protein requirement is dietary intake of fat and carbohydrate. If they are present in insufficient quantities, some dietary protein must be used for energy generation and it becomes unavailable for building and replacing tissue. Thus, as energy (calorie) content of the diet from carbohydrate and fat increases, the need for protein decreases. This is referred to as **protein sparing**. Carbohydrate is somewhat

CLINICAL CORRELATION 27.1

Vegetarian Diets and Protein–Energy Requirements for Children

One of the most important problems of a purely vegetarian diet (as opposed to a lacto-ovo vegetarian diet) is the difficulty in obtaining sufficient calories and protein. Potential caloric deficit results because the caloric densities of fruits and vegetables are much less than the meats they replace (30–50 cal per 100 g versus 150–300 cal per 100 g). The protein problem is threefold: (1) Most plant products contain much less protein (1–2 g of protein per 100 g serving versus 15–20 g per 100 g serving); (2) most plant protein is of low biological value; and (3) some plant proteins are not completely digested. Actually, well-designed vegetarian diets usually provide enough calories and protein for the average adult. In fact, the reduced caloric intake may well be of benefit because strict vegetarians tend to be lighter than their nonvegetarian counterparts.

However, whereas an adult male may require about 0.8 g of protein and 40 cal kg^{-1} of body weight, a young child may require 2–3 times that amount. Similarly, a pregnant woman needs an additional 10 g of protein and 300 cal day^{-1} and a lactating woman an extra 15 g of protein and 500 cal. Thus both young children and pregnant and lactating women run a risk of protein-energy malnutrition. Children of vegetarian mothers generally have a lower birth weight than children of mothers consuming a mixed diet. Similarly, vegetarian children generally have a slower rate of growth through the first 5 years, but generally catch up by age of 10.

Sufficient calories and protein for these high-risk groups can be provided if the diet is adequately planned. Three principles should be followed to design a calorie-protein-sufficient vegetarian diet for young children: (1) Whenever possible, include eggs and milk; they are excellent sources of calories and high-quality protein; (2) include liberal amounts of those vegetable foods with high-caloric density, including nuts, grains, dried beans, and dried fruits; and (3) include liberal amounts of high-protein vegetable foods that have complementary amino acid compositions. It used to be thought that these complementary proteins must be present in the same meal. Recent animal studies, however, suggest that a meal low in (but not devoid of) an essential amino acid may be supplemented by adding the limiting amino acid at a subsequent meal.

Source: Saunders, T. A. B. Vegetarian diets and children. *Pediatr. Nutr.* 42:955,1995. Messina, V. and Mangels, A. R. Considerations in planning vegan diets: Children. *J. Am. Diet. Assoc.* 101:661, 2001. Mangels, A. R. and Messina, V. Considerations in planning vegan diets: Infants. *J. Am. Diet. Assoc.* 101:670, 2001.

more efficient at protein sparing than fat—presumably because it can be used as an energy source by almost all tissues, whereas fat can not.

Normal Adult Protein Requirements

Assuming adequate calorie intake and a 75% efficiency of utilization, which is typical of mixed protein in the average American diet, the **recommended protein intake** is 0.8 g/kg^{-1} body wt/day^{-1}. This amounts to about 58 g $protein/day^{-1}$ for a 72-kg (160-lb) man and about 44 g/day^{-1} for a 55-kg (120-lb) woman. These recommendations need to be increased on a vegetarian diet if overall efficiency of utilization is less than 75%.

Protein Requirements Are Increased During Growth and Illness

Because dietary protein is essential for synthesis of new body tissue, as well as for maintenance and repair, the need for protein increases markedly during periods of rapid growth as occurs during pregnancy, infancy, childhood, and adolescence. Once growth requirements have been considered, age does not seem to have much effect on protein requirements. If anything, the protein requirement may decrease slightly with age. However, older people need and generally consume fewer calories, so high-quality protein should provide a larger percentage of their total calories. Some older people may have special protein requirements due to malabsorption problems.

Illness, major trauma, and surgery cause a major **catabolic response**. Energy and protein needs in these situations are very large, and the body responds by increasing production of glucocorticoids, epinephrine, and cytokines. Breakdown of body protein is greatly accelerated and a negative nitrogen balance results unless protein intake is increased (Figure 27.2). Although the increased protein requirement is of

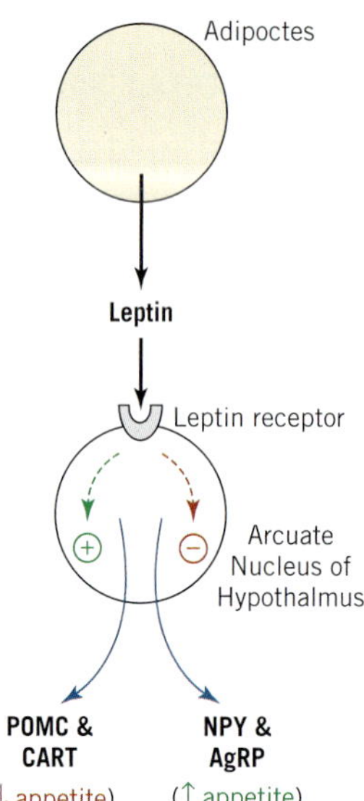

FIGURE 27.3

The Leptin Appetite Suppression Pathway. Schematic representation of the leptin appetite suppression pathway. Adipocytes produce leptin, which binds to its receptor in the arcuate nucleus of the hypothalamus and simulates neurons that produce the appetite-suppressing hormones POMC (pro-opiomelanocortin) and CART (cocaine- and amphetamine-regulated transcript) and blocks neurons that produce the appetite-stimulating neuropeptides NPY (neuropeptide Y) and AgRP (agouti-related protein). Normally, the amount of leptin increases as the fat stores in adipocytes increase.

intake on calcium requirements. Some studies suggest that high-protein intake increases urinary loss of calcium and may accelerate bone demineralization associated with aging. However, this issue is far from settled.

Obesity Has Dietary and Genetic Components

Perhaps the more serious and frequent nutritional problem in this country is excessive energy consumption. In fact, **obesity** is becoming epidemic in the United States. The prevalence of obesity has increased by more than 30% in adults and has more than doubled in children in the past 20 years. This clearly reflects recent lifestyle changes in this country because genetics does not change in a matter of a few years. It would be inaccurate, however, to label obesity as simply a problem of excess consumption. Overeating plays an important role in many individuals, as does inadequate exercise, but there is also a strong genetic component as well. While the biochemical mechanism(s) for this genetic predisposition are unclear, we are learning much more about the genes that control appetite and regulate energy homeostasis. For example, recent research has shown that fat cells (adipocytes) produce a hormone called leptin that suppresses appetite (Figure 27.3, see also Clin. Corr. 17.1, p. 665). The leptin appetite suppression pathway was initially considered a very promising target for pharmacological intervention because a genetically obese strain of mice (ob/ob) was shown to be unable to produce leptin. However, it turns out that most overweight individuals overproduce leptin, and no more than 2–4% of the overweight population has defects in the known components of the leptin appetite suppression pathway. In contrast, genetic predisposition to obesity when excess calories are consumed is common in the population. Only 15% of the U.S. population is heterozygous for the Pro12Ala variant of PPAR-γ (peroxisome proliferator-activated receptor-γ). However, these individuals are less likely to become overweight and less likely to develop diabetes when overweight than the 85% of Pro homozygotes in the population.

Obesity Has Significant Health Implications

A full discussion of the treatment of obesity is clearly beyond the scope of this chapter, but it is worthwhile to consider some of the metabolic consequences of obesity. One striking clinical feature of overweight individuals is a marked elevation of serum free fatty acids, cholesterol, and triacylglycerols irrespective of the dietary intake of fat. Why is this? Obesity is obviously associated with an increased number and/or size of **adipose tissue cells**. These cells overproduce hormones such as leptin and resistin and cytokines, such as **tumor necrosis factor alpha (TNFα)**, some of which appear to cause cellular resistance to **insulin** by interfering with **autophosphorylation** of the **insulin receptor (IR)** and the subsequent phosphorylation of the **insulin receptor substrate 1 (IRS-1)** (see p. 917). At the same time, the lipid-laden adipocytes decrease synthesis of hormones such as adiponectin, which appear to enhance insulin responsiveness. The insulin resistance in adipose tissue results in increased activity of the hormone-sensitive lipase, which is probably sufficient to explain the increase in circulating free fatty acids. The high circulating levels of free fatty acids may also contribute to insulin resistance in the muscle and liver. Initially, the pancreas maintains glycemic control by overproducing insulin. Thus, many obese individuals with apparently normal blood glucose control have a syndrome characterized by insulin resistance of the peripheral tissue and high concentrations of insulin in the circulation (see Clin. Corr. 17.1, p. 665). This hyperinsulinemia appears to stimulate the sympathetic nervous system, leading to sodium and water retention and vasoconstriction, which increase blood pressure.

The excess fatty acids are carried to the liver and converted to triacylglycerol and cholesterol. Excess triacylglycerol and cholesterol are released as very-low-density lipoprotein particles, leading to higher circulating levels of both triacylglycerol and cholesterol (see p. 711). For reasons that are not entirely clear, this also results in a decrease in high-density lipoprotein particles.

Eventually, the capacity of the pancreas to overproduce insulin declines which leads to higher fasting blood sugar levels and decreased glucose tolerance. Fully 80% of type 2 diabetics are overweight. Because of these metabolic changes, obesity is a primary risk factor in coronary heart disease, hypertension, and diabetes mellitus. Obesity is also associated with inflammatory diseases, some forms of cancer, bone and joint disorders, and breathing disorders. This is nutritionally significant because all of these metabolic changes are reversible. Quite often reduction to ideal body weight is the single most important aim of nutritional therapy. When the individual is at ideal body weight, the composition of the diet becomes a less important consideration in maintaining normal serum lipid and glucose concentrations.

As mentioned above, obesity can lead to increased retention of both sodium and water. As the fat stores are metabolized, they produce water (which is denser than the fat), and the water may be largely retained. In fact, some individuals may observe short-term weight gain when they diet, even though the diet is working perfectly well in terms of breaking down their adipose tissue. This metabolic fact of life can be psychologically devastating to dieters, who expect quick results for all their sacrifice.

27.6 CARBOHYDRATES

The chief metabolic role of dietary carbohydrates is for energy production. Any carbohydrate in excess of that needed for energy is converted to glycogen and triacylglycerol for storage. The body can adapt to a wide range of dietary carbohydrate levels (Clin. Corr. 27.7). Diets high in carbohydrate result in higher steady-state levels of glucokinase and some of the enzymes involved in the pentose phosphate pathway and triacylglycerol synthesis. Diets low in carbohydrate result in higher steady-state levels of some of the enzymes involved in gluconeogenesis, fatty acid oxidation, and amino acid catabolism. **Glycogen stores** are also affected by the carbohydrate content of the diet (Clin. Corr. 27.4).

The most common form of **carbohydrate intolerance** is diabetes mellitus, caused by either subnormal insulin production or lack of insulin receptors. This causes intolerance to glucose and sugars that are readily converted to glucose. Dietary treatment of diabetes is discussed in Clinical Correlation 27.5. **Lactase insufficiency** (see p. 1058) is also a common disorder of carbohydrate metabolism affecting over 30 million people in the United States alone. It is most prevalent among blacks, Asians, and Hispanics. Without intestinal lactase, dietary lactose is not significantly hydrolyzed or absorbed. It remains in the intestine where it acts osmotically to draw water into the gut and it is converted to lactic acid and CO_2 by intestinal bacteria. The result is bloating, flatulence, and diarrhea, all of which can be avoided simply by eliminating milk and milk products from the diet.

27.7 FATS

Triacylglycerols, or fats, are directly utilized by many tissues as an energy source and as phospholipids are important constituents of membranes. Excess dietary fat can only be stored as triacylglycerol in the adipose tissue. As with carbohydrate, the body adapts to a wide range of fat intakes. However, problems develop at the extremes (either high or low) of fat consumption. At the low end, **essential fatty acid (EFA) deficiency** may become a problem. The fatty acids linoleic and linolenic cannot be made by the body and thus are essential components of the diet. They are needed for maintaining the function and integrity of membrane structure, for fat metabolism and transport, and for synthesis of **prostaglandins** and related compounds. The most characteristic symptom of essential fatty acid deficiency is a scaly dermatitis. EFA deficiency is very rare in the United States, occurring primarily in low-birth weight infants fed on artificial formulas lacking EFA and in hospitalized patients maintained on total parenteral nutrition for long periods. At the other extreme, there is concern that excess dietary fat causes elevation of serum

CLINICAL CORRELATION 27.4
Carbohydrate Loading and Athletic Endurance

The practice of carbohydrate loading dates to observations made in the early 1960s that endurance during vigorous exercise was limited primarily by muscle glycogen stores. Of course, glycogen is not the sole energy source for muscle. Free fatty acids increase in blood during vigorous exercise and are utilized by muscle along with its glycogen stores. Once glycogen has been exhausted, however, muscle cannot rely entirely on free fatty acids without tiring rapidly, probably because muscle becomes increasingly hypoxic during vigorous exercise. While glycogen is utilized equally well aerobically or anaerobically, fatty acids can only be utilized aerobically. Under anaerobic conditions, fatty acids cannot provide ATP rapidly enough to serve as the sole energy source.

The practice of carbohydrate loading to increase glycogen stores was devised for track and other endurance athletes. The original carbohydrate loading regimen consisted of a 3- to 4-day period of heavy exercise while on a low-carbohydrate diet, followed by 1–2 days of light exercise while on a high-carbohydrate diet. The initial low-carbohydrate–high-energy demand period caused a depletion of muscle glycogen stores. The subsequent change to a high-carbohydrate diet resulted in the production of higher-than-normal levels of insulin and growth hormone and glycogen stores reached almost twice the normal amounts. This practice did increase endurance significantly. In one study, test subjects on a high-fat and high-protein diet had less than 1.6 g of glycogen per 100 g of muscle and could perform a standardized work load for only 60 min. When the same subjects then consumed a high-carbohydrate diet for 3 days, their glycogen stores increased to 4 g per 100 g of muscle and the same workload could be performed for up to 4 h.

While the technique clearly worked, the athletes often felt lethargic and irritable during the low-carbohydrate phase of the regimen, and the high-fat diet ran counter to current health recommendations. Recent studies indicate that regular consumption of a high-complex carbohydrate low-fat diet during training increases glycogen stores without sudden dietary changes. Current recommendations are for endurance athletes to consume a high-carbohydrate diet (with emphasis on complex carbohydrates) during training. Then carbohydrate intake is increased further (to 70% of calories) and exercise tapered off during the 2–3 days just prior to an athletic event. This increases muscle glycogen stores to levels comparable to the previously described carbohydrate loading regimen.

Source: Lambert, E. V. and Goedecke, J. H. The role of dietary micronutrients in optimizing endurance performance. *Curr. Sports Med. Rep.* 2:194, 2003. Hargreaves, M., Hawley, J. A., and Jeukendrup, A. Pre-exercise carbohydrate and fat ingestion: Effects on metabolism and performance. *J. Sports Sci.* 22:31, 2004. Burke, L. M., Kiens, B., and Ivey, J. L. Carbohydrates and fat for training and recovery. *J. Sports Sci.* 22:15, 2004.

lipids and thus an increased risk of heart disease. Recent studies suggest that high fat intakes are associated with increased risk of colon, breast, and prostate cancer, but it is not clear whether the cancer risk is associated with fat intake per se, or with the excess calories associated with a high-fat diet. Animal studies suggest that polyunsaturated fatty acids of the ω-6 series may be more tumorigenic than other unsaturated fatty acids. The reason for this is not known, but it has been suggested that prostaglandins derived from the ω-6 fatty acids may stimulate tumor progression.

27.8 FIBER

Dietary fiber comprises those components of food that cannot be broken down by human digestive enzymes. It is incorrect, however, to assume that fiber is indigestible since some fibers are, in fact, at least partially broken down by intestinal bacteria. Our current understanding of the metabolic roles of dietary fiber is based on three important observations: (1) There are several different types of dietary fiber, (2) they each have different chemical and physical properties, and (3) they each have different effects on human metabolism, which can be understood, in part, from their unique properties.

The major types of fiber and their properties are summarized in Table 27.3. **Cellulose** and most **hemicelluloses** increase stool bulk, decrease transit time, and are associated with the effects of fiber on regularity. They decrease intracolonic pressure and appear to play a beneficial role with respect to diverticular diseases. By diluting out potential carcinogens and speeding their transit through the colon, they may also play a role in reducing the risk of colon cancer. **Lignins** have bulk enhancing properties

CLINICAL CORRELATION 27.5
High-Carbohydrate Versus High-Fat Diet for Diabetics

For years the American Diabetes Association has recommended diets that were low in fat and high in complex carbohydrates and fiber for diabetics. The logic of such a recommendation seemed to be inescapable. Diabetics are prone to hyperlipidemia with attendant risk of heart disease, and low-fat diets seemed likely to reduce risk of hyperlipidemia and heart disease. In addition, numerous clinical studies had suggested that the high-fiber content of these diets improved control of blood sugar. This recommendation has proved to be controversial and illustrates the difficulties in making dietary recommendations for population groups rather than individuals. Most of the clinical trials of the high-carbohydrate–high-fiber diets resulted in significant weight reduction, either by design or because of the lower caloric density of the diet. Since weight reduction improves diabetic control, it is not clear whether the improvements seen in the treated group were due to the change in diet composition per se or because of the weight loss. In addition, there is significant variation in how individual diabetics respond to these diets. Some diabetic patients show poorer control (as evidenced by higher blood glucose levels, elevated VLDL and/or LDL levels, and reduced HDL levels) on the high-carbohydrate–high-fiber diets than on diets high in monounsaturated fatty acids. However, diets high in monounsaturated fatty acids tend to have higher caloric density and are inappropriate for overweight individuals with type 2 diabetes. Thus, a single diet may not be an equally appropriate for all diabetics. Even the glycemic index" concept (Table 27.4) may also turn out to be difficult to apply to the diabetic population as a whole, because of individual variation. In 1994, the American Diabetes Association abandoned the concept of a single diabetic diet. Instead their recommendations focus on achievement of glucose, lipid, and blood pressure goals, with weight reduction and dietary recommendations based on individual preferences and what works best to achieve metabolic control in that individual.

Source: American Diabetes Association. Nutritional recommendations and principles for people with diabetes. *Diabetes Care* 17:519,1994. Nuttall, F. Q. and Chasuk, R. M. Nutrition and the management of type 2 diabetes. *J. Fam. Pract.* 47:S45, 1998. Klein, S., Sheard, N. F., Pi-Sunyer, X., Daly, A., Wylie-Rosett, J., Kulkarni, K., and Clark, N. G. Weight management through lifestyle modification for the prevention and management of type 2 diabetes: Rationale and strategies. A statement of the American Diabetes Association for the Study of Obesity and the American Society for Clinical Nutrition. *Am. J. Clin. Nutr.* 80:257, 2004. Grundy, S. M., Hansen, B., Smith, S. C., Cleeman, J. I., and Kahn, R. A. Clinical management of metabolic syndrome. Report of the American Heart Association/National Heart, Lung, and Blood Institute/American Diabetes Association Conference on Scientific Issues Related to Management. *Circulation* 109:551, 2004.

TABLE 27.3 Major Types of Fiber and Their Properties

Type of Fiber	*Major Source in Diet*	*Chemical Properties*	*Physiological Effects*
Cellulose	Unrefined cereals	Nondigestible	Increases stool bulk
	Bran	Water-insoluble	Decreases intestinal transit time
	Whole wheat	Absorbs water	Decreases intracolonic pressure
Hemicellulose	Unrefined cereals	Partially digestible	Increases stool bulk
	Some fruits and vegetables	Usually water-insoluble	Decreases intestinal transit time
	Whole wheat	Absorbs water	Decreases intracolonic pressure
Lignin	Woody parts of vegetables	Nondigestible	Increases stool bulk
		Water-insoluble	Bind cholesterol
		Absorbs organic substances	Bind carcinogens
Pectin	Fruits	Digestible	Decreases rate of gastric emptying
		Water-soluble	Decreases rate of sugar uptake
		Mucilaginous	Decreases serum cholesterol
Gums	Dried beans	Digestible	Decreases rate of gastric emptying
	Oats	Water-soluble	Decreases rate of sugar uptake
		Mucilaginous	Decreases serum cholesterol

and they adsorb organic substances such as cholesterol to lower plasma cholesterol concentration. **Mucilaginous fibers**, such as **pectin** and **gums**, tend to form viscous gels in the stomach and intestine and slow the rate of gastric emptying, thus slowing the rate of absorption of many nutrients. Their most important clinical role is to slow the rate at which carbohydrates are digested and absorbed. Thus, both the rise in blood sugar and the rise in insulin levels are significantly decreased if these fibers are ingested with carbohydrate-containing foods. **Water-soluble fibers** (pectins, gums, some hemicelluloses, and storage polysaccharides) also help to lower serum cholesterol levels in most people. Whether this is due to their effect on insulin levels (insulin stimulates cholesterol synthesis and export) or to other metabolic effects (perhaps caused by end-products of partial bacterial digestion) is not known. Vegetables, wheat, and most grain fibers are the best sources of the water-insoluble cellulose, hemicellulose, and lignin. Fruits, oats, and legumes are the best source of the water-soluble fibers. Obviously, a balanced diet should include food sources of both soluble and insoluble fiber.

27.9 COMPOSITION OF MACRONUTRIENTS IN THE DIET

Since there are relatively few instances of macronutrient deficiencies in the American diet, much of the interest in recent years has focused on whether there is an ideal diet composition consistent with good health.

Composition of the Diet Affects Serum Cholesterol

With respect to heart disease, the current discussion centers around two key issues: (1) Can serum cholesterol and triacylglycerol concentration be controlled by diet? (2) Does lowering serum cholesterol and triacylglycerol levels protect against heart disease? The controversies around dietary control of cholesterol levels illustrate perfectly the trap one falls into by trying to look too closely at each individual component of the diet instead of at the diet as a whole. For example, at least four dietary components have an effect on serum cholesterol: cholesterol itself, **polyunsaturated fatty acids (PUFA)**, **saturated fatty acids (SFA)**, and **fiber**. It would seem that the more cholesterol one eats, the higher the serum cholesterol should be. However, cholesterol synthesis is tightly regulated and decreases in dietary cholesterol have relatively little effect on serum cholesterol levels (see p. 715). One can obtain a more significant reduction in cholesterol and triacylglycerol levels by increasing the ratio of PUFA/SFA in the diet. Finally, some plant fibers, especially the water-soluble fibers, appear to decrease cholesterol levels significantly. While the effects of various lipids in the diet can be dramatic, the biochemistry of their action is still uncertain. Saturated fats inhibit receptor-mediated uptake of LDL, but the mechanism is complex. Palmitic acid (saturated, C_{16}) raises serum cholesterol levels while stearic acid (saturated, C_{18}) has no effect. Polyunsaturated fatty acids lower both LDL and HDL cholesterol levels, while oleic acid (monounsaturated, C_{18}) appears to lower LDL without affecting HDL levels. The ω-3 and ω-6 polyunsaturated fatty acids have slightly different effects on lipid profiles (see Clin. Corr. 27.6). However, these complexities do not significantly affect dietary recommendations. Most foods high in saturated fats contain both palmitic and stearic acid and are atherogenic. Since oleic acid lowers LDL levels, olive oil and, possibly, peanut oil may be as beneficial as polyunsaturated oils.

There is little disagreement with respect to these data. The question is, What can be done with the information? Much of the disagreement arises from the tendency to look at each dietary factor in isolation. For example, it is debatable whether it is worthwhile placing a patient on a highly restrictive 300-mg cholesterol diet (1 egg contains about 213 mg of cholesterol) if his serum cholesterol is lowered by only 5–10%. Likewise, changing the **PUFA/SFA ratio** from 0.3 (the current value) to 1.0 would require either a radical change in the diet by elimination of foods containing saturated fat (largely meats and fats) or an addition of large amounts of rather unpalatable polyunsaturated

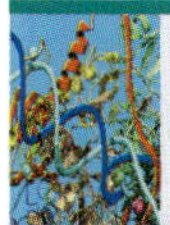

CLINICAL CORRELATION 27.6
Polyunsaturated Fatty Acids and Risk Factors for Heart Disease

Because reduction of elevated serum cholesterol levels can reduce risk of heart disease, there is considerable interest in the effects of diet on serum cholesterol levels and other risk factors for heart disease. One important dietary factor regulating serum cholesterol levels is the ratio of polyunsaturated fats (PUFAs) to saturated fats (SFAs) in the diet. Furthermore, recent research shows that different types of polyunsaturated fatty acids have different effects on lipid metabolism and on other risk factors for heart disease. Essential polyunsaturated fatty acids can be classified as either ω-6 or ω-3. Clinical studies have shown that the ω-6 PUFAs (chief dietary source is linoleic acid from plants and vegetable oils) primarily decrease serum cholesterol levels, with only modest effects on serum triacylglycerol levels. The ω-3 PUFAs (chief dietary source is eicosapentaenoic acid from certain ocean fish and fish oils) cause only modest decreases in serum cholesterol levels but significantly lower serum triacylglycerol levels. The mechanisms behind these effects on serum lipid levels are unknown.

In addition, the ω-3 PUFAs have other effects that may decrease the risk of heart disease; they decrease platelet aggregation, inflammation, and arrhythmia and increase endothelial relaxation. In the case of platelet aggregation, the mechanism is clear. Arachidonic acid (ω-6 family) is a precursor of thromboxane A_2 (TXA_2), which is a potent pro-aggregating agent, and prostaglandin I_2 (PGI_2), which is a weak anti-aggregating agent (see p. 734). The ω-3 PUFAs are converted to thromboxane A_3 (TXA_3), which is only weakly pro-aggregating, and prostaglandin I_3 (PGI_3), which is strongly anti-aggregating. Thus, the balance between pro-aggregation and anti-aggregation is shifted toward a more anti-aggregating condition as the ω-3 PUFAs displace ω-6 PUFAs as precursors to the thromboxanes and prostaglandins. Recent studies have shown that diets rich in ω-3 fatty acids significantly decrease the risk of sudden cardiac death in patients who have previously had a myocardial infarction.

Source: Marchioli, R., Schweiger, C., Tavazzi, L. and Valagussa, F. Efficacy of ω-3 polyunsaturated fatty acids after myocardial infarction: Results of the GISSI-Prevenzione Trial. Gruppo Italiano per lo Studio della Sopravvivenza nell'Infarcto Miocardico. *Lipids* 36(Suppl):S119, 2001. Marchioli, R., Barzi, F., Bomba, E., Chieffo, C., Di Gregorio, D., Di Mascio, R., et al. Early protection against sudden death by ω-3 polyunsaturated fatty acids after myocardial infarction: Time-course analysis of the results of the Gruppo Italiano per lo Studio della Sopravvivenza nell'Infarcto Miocardico (GISSI)-Prevenzione. *Circulation* 105:1897, 2002. Kris-Etherton, Harris, W. H., and Appel, L. J. Omega-3 fatty acids and cardiovascular disease. New recommendations from the American Heart Association, Arterioscler. *Thromb. Vasc. Biol.* 23:151, 2003. Holub, D. J. and Holub, B. J. Omega-3 fatty acids from fish oils and cardiovascular disease. *Mol. Cell. Biochem.* 263:217, 2004.

fats to the diet. For many Americans this would be unrealistic. Fiber is another good example. One could expect, at the most, a 5% decrease in serum cholesterol by adding a reasonable amount of fiber to the diet. Very few people would eat the 10 apples per day needed to lower serum cholesterol by 15%. Are we to conclude then that any dietary means of controlling serum cholesterol levels is useless? This would be the case only if each element of the diet is examined in isolation. For example, recent studies have shown that vegetarians, who have lower cholesterol intakes plus higher PUFA/SFA ratios and higher fiber intakes, may average 25–30% lower cholesterol levels than their nonvegetarian counterparts. Perhaps, more to the point, diet modifications acceptable to the average American have been shown to cause a 10–15% decrease in cholesterol levels in long-term studies. A 7-year clinical trial sponsored by the National Institutes of Health has proven conclusively that lowering serum cholesterol levels reduces the risk of heart disease in men. While a consensus has been reached that lowering serum cholesterol levels reduces the risk of heart disease, it is important to remember that serum cholesterol is just one of many cardiovascular risk factors.

Carbohydrates, Glycemic Index, and Glycemic Load

Much of the nutritional dispute in the area of carbohydrate intake centers around the effect of carbohydrate on blood glucose and triacylglycerol levels. The old paradigm that simple sugars raise blood sugar and triacylglycerol levels to a greater extent than complex carbohydrates is an oversimplification. The effect of the carbohydrates in a particular food is determined by the rate of digestion and absorption of the carbohydrate and by other components of the food. In particular, soluble fiber, protein, and fat all blunt the effect of carbohydrates on blood glucose levels. Thus, the concept of **glycemic index** was developed to describe better the effects of carbohydrates on blood glucose. Glycemic

TABLE 27.4 Glycemic Index[a] of Selected Foods

Grain and cereal products	
Bread (white)	69 ± 5
Bread (whole wheat)	72 ± 6
Rice (white)	72 ± 9
Sponge cake	46 ± 6
Breakfast cereals	
All bran	51 ± 5
Cornflakes	80 ± 6
Oatmeal	49 ± 8
Shredded wheat	67 ± 10
Vegetables	
Sweet corn	59 ± 11
Frozen peas	51 ± 6
Dairy products	
Ice cream	36 ± 8
Milk (whole)	34 ± 6
Yoghurt	36 ± 4
Root vegetables	
Beets	64 ± 16
Carrots	92 ± 20
Potato (white)	70 ± 6
Potato (sweet)	48 ± 6
Dried legumes	
Beans (kidney)	29 ± 8
Beans (soy)	15 ± 5
Peas (blackeye)	33 ± 4
Fruits	
Apple (Golden Delicious)	39 ± 3
Banana	62 ± 9
Orange	40 ± 3
Sugars	
Fructose	20 ± 5
Glucose	100
Honey	87 ± 8
Sucrose	59 ± 10

Source: Data from Jenkins, D. A., et al. Glycemic index of foods: A physiological basis for carbohydrate exchange. *Am. J. Clin. Nutr.* 34:362, 1981.

[a] Glycemic index is defined as the area under the blood glucose response curve for each food expressed as a percentage of the area after taking the same amount of carbohydrate as glucose (mean: 5–10 individuals).

index is determined empirically and is defined as the effect of 50 g of carbohydrate in a particular food on blood glucose levels compared to 50 g of glucose. In general, pastries, refined cereals, rice, and starchy vegetables have high glycemic indices, while nonstarchy vegetables, fruits, legumes, and nuts have low glycemic indices (Table 27.4). However, because the carbohydrate content of foods varies widely, even glycemic indices can be misleading. For example, carrots have a much higher glycemic index than ice cream (Table 27.4). Thus, the term glycemic load has recently been introduced. Glycemic load is the glycemic index times the amount of carbohydrate in a standard serving size of that food. As might be expected, carrots have a much lower glycemic load than ice cream.

Mixing Vegetable and Animal Proteins Meet Nutritional Protein Requirements

Epidemiologic data and animal studies suggest that consumption of animal protein is associated with increased incidence of heart disease and various forms of cancer. One could assume that it is probably not the animal protein itself that is involved, but the associated fat and cholesterol. What sort of protein should we consume? Although the present diet may not be optimal, a strictly vegetarian diet may not be acceptable to many Americans. Perhaps a middle road is best. Clearly, there are no known health dangers associated with a mixed diet that is lower in animal protein than the current American standard.

Fiber from Varied Sources Is Desirable

Because of current knowledge about effects of fiber on human metabolism, most suggestions for a prudent diet recommend an increase in dietary fiber. The current fiber content of the American diet is about 14–15 g per day. Most experts feel that an increase to at least 25–30 g would be safe and beneficial. Since different types of fibers

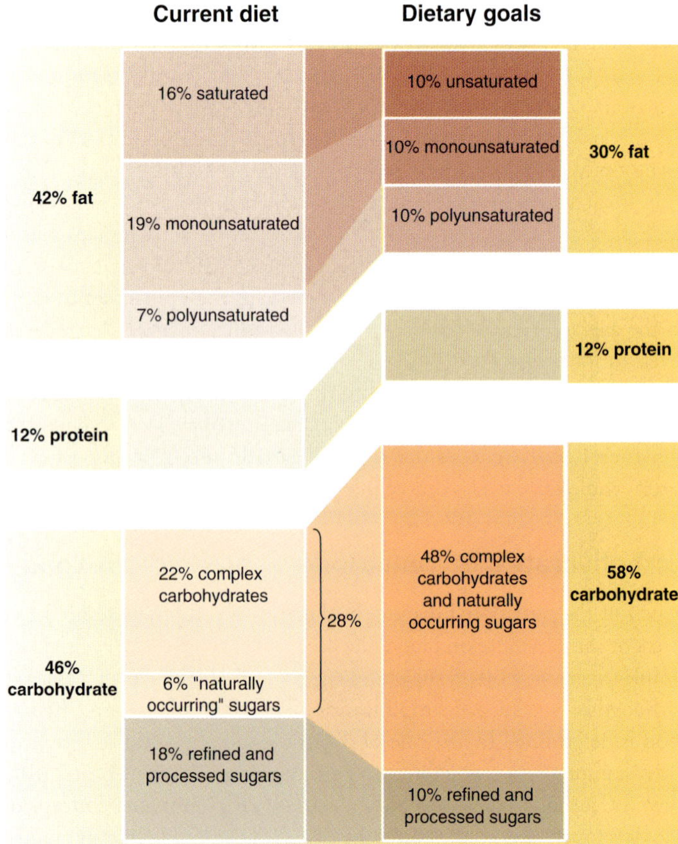

FIGURE 27.4

United States dietary goals. Graphical comparison of the composition of the current U.S. diet and the dietary goals for the U.S. population suggested by the Senate Select Committee on Human Nutrition.
From Dietary goals for the United States, 2nd ed. Washington, DC: U.S. Government Printing Office, 1977.

have different physiological roles, the increase in fiber intake should come from a wide variety of sources, including fresh fruits, vegetables, and legumes as well as the more popular cereal fibers (which are primarily cellulose and hemicellulose).

Dietary Recommendations

Several private and government groups have made specific recommendations with respect to the ideal dietary composition for the American public. This movement was spearheaded by the Senate Select Committee on Human Nutrition, which first published its Dietary Goals for the United States in 1977. This Committee recommended that the American public reduce consumption of total calories, total fat, saturated fat, cholesterol, simple sugars, and salt to "ideal" goals more compatible with good health (Figure 27.4). In recent years the USDA, the American Heart Association, the American Diabetes Association, the National Research Council, and the Surgeon General have published similar recommendations, and the USDA has used these recommendations to design revised recommendations for a **balanced** or **"prudent diet"** (Figure 27.5). How valid is the scientific basis of the recommendations for a prudent diet? Is there evidence that it will improve the health of the general public? How much does individual variability affect these recommendations? These remain controversial questions.

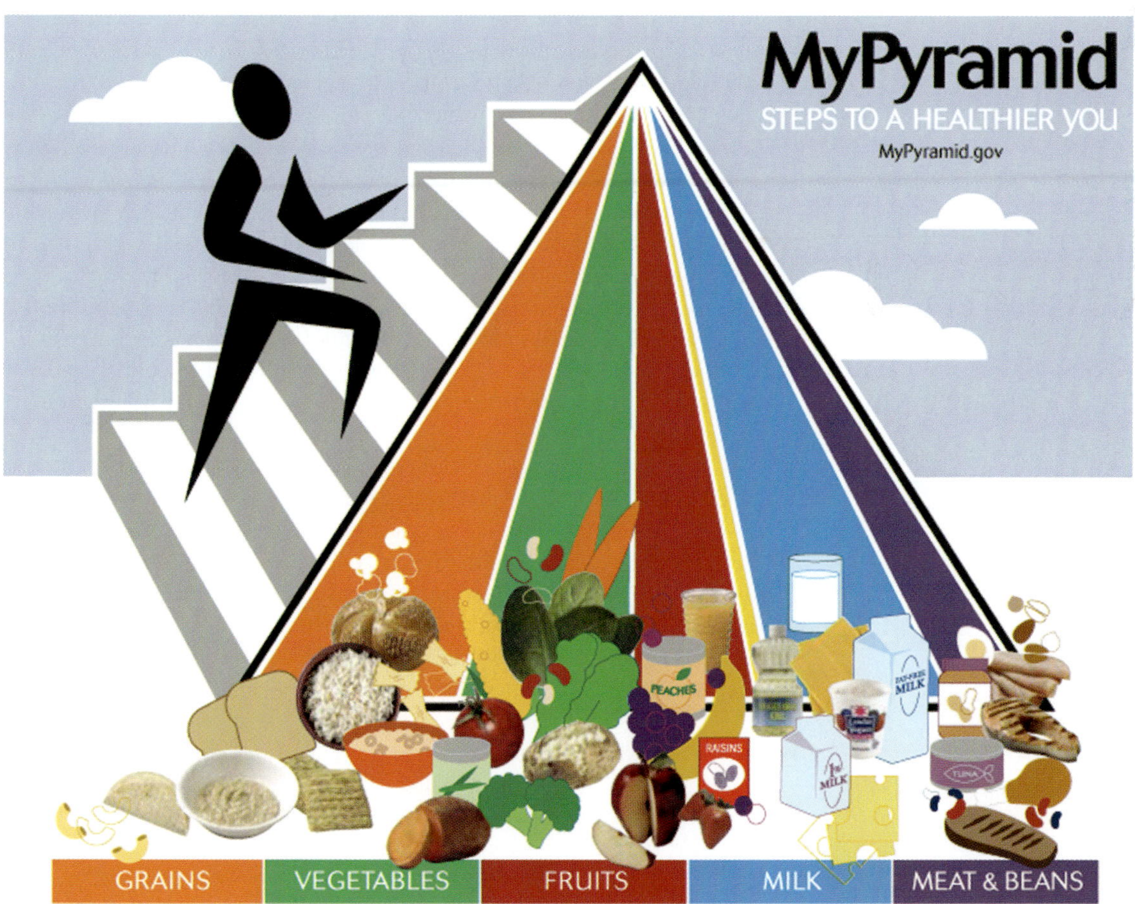

FIGURE 27.5

USDA food pyramid. Graphical representation of USDA recommendations for a balanced diet, 2005. www.mypyramid.gov

CLINICAL CORRELATION 27.7

Metabolic Adaptation: Relationship Between Carbohydrate Intake and Serum Triacylglycerols

In evaluating the nutrition literature, it is important to be aware that most clinical trials are of rather short duration (2–6 weeks), while some metabolic adaptations may take considerably longer. Thus, even apparently well-designed clinical studies may lead to erroneous conclusions that will be repeated in the popular literature for years to come. For example, several studies conducted in the 1960s and 1970s tried to assess the effects of carbohydrate intake on serum triacylglycerol levels. Typically, young college age males were given a diet in which up to 50% of their fat calories were replaced with sucrose or other simple sugars for a period of 2–3 weeks. In most cases, serum triacylglycerol levels increased markedly (up to 50%). This led to the tentative conclusion that high intake of simple sugars, particularly sucrose, could increase the risk of heart disease, a notion that was popularized by nutritional best sellers such as "Sugar Blues" and "Sweet and Dangerous." Unfortunately, while the original conclusions were promoted in the lay press, the experiments themselves were questioned. Subsequent studies showed that if these trials were continued for longer periods (3–6 months), the triacylglycerol levels usually normalized. The nature of this slow metabolic adaptation is unknown. It is also important to consider the type of carbohydrate in the diet. For many Americans, a high-carbohydrate diet means a diet that is high in simple sugars. Triacylglycerol levels in these individuals respond dramatically to diets that substitute foods containing either fat or complex carbohydrates and fiber for those foods containing simple sugars as a carbohydrate source.

Source: Leahy, P., Croniger, C., and Hanson, R.W. Molecular and cellular adaptations to carbohydrate and fat intake. *Eur. J. Clin. Nutr.* 53(Suppl 1):S6, 1999. Parks, E. J. and Hellerstein, M. K. Carbohydrate-induced hypertriacylglycerolemia: Historical perspective and review of biological mechanisms. *Am. J. Clin. Nutr.* 71:412, 2000.

The recent debate about low-carbohydrate versus low-fat diets illustrates the complexity of these considerations. The debate was fueled by Dr. Atkins *Diet Revolution* and *New Diet Revolution* books claiming that weight loss was more effective with low-carbohydrate diets and that fat, even saturated fat, did not adversely affect serum cholesterol levels. In fact, short-term studies seemed to confirm that weight loss was more rapid and control of blood sugar was better on low-carbohydrate diets. Moreover, LDL cholesterol levels did not increase, triacylglycerol levels were lower, and HDL cholesterol levels were higher than on traditional low-fat diets. Some of these effects were probably due to slow metabolic adaptation to large changes in diet composition (Clin. Corr. 27.7) and disappeared in longer-term studies. For example, by one year, weight loss is identical on isocaloric low-fat and low-carbohydrate diets. Other studies suggest that longer-term weight control may be more successful on low-fat diets, presumably because fat provides 9 kcal/g^{-1} while carbohydrate provides only 4 kcal/g^{-1}.

In addition, the type of carbohydrates and fats in the diet are just as important as their amounts. High-carbohydrate diets that are low in glycemic load appear to be just as effective in achieving weight loss and control of blood sugar as diets that are low in carbohydrate. Similarly, diets containing heart-healthy monounsaturated and ω-3 polyunsaturated fats are just as effective as diets containing saturated fats. Finally, it is important to remember that dietary recommendations are for populations, not for individuals. The diet that works best for achieving weight control, blood sugar control, and healthy lipoprotein patterns is determined by an individual's genetic makeup (Clin. Corr. 27.5). Current nutrigenomic research is directed toward identifying the genetic differences that will allow individualization of diet and treatment options for each patient.

BIBLIOGRAPHY

Protein Energy Malnutrition in Hospitalized Patients

Biffl, W. L., Moore, E. E. and Haenel, J. B. Nutrition support of the trauma patient. *Nutrition* 18:960, 2002.

Wilmore, D. W. Metabolic response to severe surgical illness: An overview. *World J. Surg.* 24:705, 2000.

Wolfe, R. R. and Martini, W. Z. Changes in intermediary metabolism in severe surgical illness. *World J. Surg.* 24:639, 2000.

Metabolic Consequences of Obesity

Boden, G. Free fatty acids, insulin resistance, and type 2 diabetes mellitus. *Proc. Assoc. Am. Physicians* 111:241, 1999.

Hotamisligii, G. S. The role of TNFα and TNF receptors in obesity and insulin resistance, *J. Intern. Med.* 245:621, 1999.

Le Roith, R. and Zick, Y. Recent advances in our understanding of insulin action and insulin resistance. *Diabetes Care* 24:588, 2001.

Pi-Sunyer, F. X. Health implications of obesity. *Am. J. Clin. Nutr.* 53:15955,1991.

Genetic Predisposition to Obesity

Fliers, J. S. Obesity wars: Molecular progress confronts an expanding epidemic. *Cell* 116:337, 2004.

Evans, R. M., Barish, G. D., and Wang Y. X. PPARs and the complex journey to obesity. *Nat. Med.* 10:355, 2004.

Ochoa, M. C., Marti, A., Azcona, C., Cheuca, M., Oyaizabal, M., Pelach, R. et al. Gene–gene interactions between PPAR gamma 2 and ADR beta 3 increases obesity risk in children and adolescents. *Int. J. Obes. Relat. Metab. Disord.* 28(Suppl. 3): S37, 2004.

Pihlajamaki, J., Vanhala, M., Vanhala, P. and Laakso, M. The Pro12Ala polymorphism of the PPAR gamma 2 gene regulates weight from birth to adulthood. *Obes. Res.* 12:187, 2004.

Saper, C. B., Chou, T. C., and Elmquist, J. K. The need to feed: Homeostatic and hedonic control of eating. *Neuron* 36:199, 2002.

Zigman, J. M. and Elmquist, J. K. Minireview: From anorexia to obesity—The yin and yang of body weight control. *Endocrinology* 144:3749, 2003.

Diet and Cardiovascular Disease

Holub, D. J., and Holub, B. J. Omega-3 fatty acids from fish oils and cardiovascular disease. *Mol. Cell. Biochem.* 263:217, 2004.

Krauss, R. M., Eckel, R. H., Howard, B., Appel, L. J., Daniels, S. R., Deckelbaum, R. J., et al. AHA Dietary Guidelines Revision 2000: A statement for healthcare professionals from the nutrition committee of the American Heart Association. *Circulation* 102:2284, 2000.

Kris-Etherton, P. M., Harris, W. S., and Appel, L. J. Omega-3 fatty acids and cardiovascular disease. New recommendations from the American Heart Association. *Arterioscler. Thromb. Vasc. Biol.* 23:151, 2003.

Tanasecu, M., Cho, E., Manson, J. E., and Hu, F. B. Dietary fat and cholesterol and the risk of cardiovascular disease among women with type 2 diabetes. *Am. J. Clin. Nutr.* 79:999, 2004.

Mozaffarian, D., Rimm, E. B., and Herrington, D. M. Dietary fats, carbohydrate, and the progression of atherosclerosis in postmenopausal women. *Am. J. Clin. Nutr* 80:1175, 2004.

Third Report of the Expert Panel on Detection, Evaluation, and Treatment of High Blood Cholesterol in Adults (Adult Treatment Panel III), www.nhlbi.nih.gov/guidelines/cholesterol

Dietary Fiber and Health

Higgins, J. A., Resistant starch: Metabolic effects and potential health benefits. *J. AOAC Int.* 87:761, 2004.

Kendall, C. W., Emam, A., Augustin, L. S., and Jenkins, D. J. Resistant starches and health. *J. AOAC Int.* 87:769, 2004.

Shankar, S. and Lanza, E. Dietary fiber and cancer prevention, *Hematol. Oncol. Clin. North Am.* 5:25,1991.

Glycemic Index and Glycemic Load

Brand-Miller, J. C., Thomas, M., Swan, V., Ahmad, Z. I., Petocz, P., and Colagiuri, S. Physiological validation of the concept of glycemic load in lean young adults. *J. Nutr.* 133:2728, 2003.

Brand-Miller, J. C. Glycemic load and chronic disease. *Nutr. Rev.* 61:S49, 2003.

Miller, J. C. B. Importance of glycemic index in diabetes. *Am. J. Clin. Nutr.* 59(Suppl.): 747S, 1994.

Schulze, M. B. Glycemic index, glycemic load, and dietary fiber intake and the incidence of type 2 diabetes in younger and middle-aged women. *Am. J. Clin. Nutr.* 80:348, 2004.

Wolever, T. M. S., Jenkins, D. J. A., Jenkins, A. L., and Josse, R. G. The glycemic index: Methodology and clinical implications. *Am. J. Clin. Nutr.* 54:846, 1991.

Dietary Recommendations

Davis, C. A., Britten, P., and Myers, E. F. Past, present, and future of the Food Guide Pyramid. *J. Am. Diet. Assoc.* 101:881, 2001.

Food and Nutrition Board of the National Academy of Sciences. *Towards Healthful Diets*. Washington, DC: U.S. Government Printing Office, 1980.

National Research Council. *Diet, Nutrition and Cancer*. Washington, DC: National Academy Press, 1982.

Senate Select Committee on Human Nutrition. *Dietary Goals for the United States*, 2nd ed., Stock No. 052-070-04376-8. Washington, DC: U.S. Government Printing Office, 1977.

Truswell, A. S. Evolution of dietary recommendations, goals, and guidelines. *Am. J. Clin. Nutr.* 45:1060, 1987.

U.S. Department of Agriculture. *Nutrition and Your Health, Dietary Guidelines for Americans*, Stock No. 017-001-00416-2. Washington, DC: U.S. Government Printing Office, 1980.

U.S. Department of Agriculture. *The Food Guide Pyramid*, Stock No. HSG-252. Hyatsville, MD: Human Nutrition Information Service, 1992.

U.S. Department of Health and Human Services. *The Surgeon General's Report on Nutrition and Health*, Stock No. 017-001-00465-1. Washington, DC: U.S. Government Printing Office, 1988.

Effects of Diet Composition on Weight Control, Blood Glucose Levels, and Blood Lipid Levels

Anderson, J. W., Randles, K. M., Kendall, C. W., and Jenkins, D. J. Carbohydrate and fiber recommendations for individuals with diabetes: a quantitative assessment and meta-analysis of the evidence. *J. Am. Coll. Nutr.* 23:5, 2004.

Assman, G., Sacks, F., Awad, A., Ascherio, A., Bonanome, A., Berra, B. et al. 2000 consensus statement on dietary fat, the Mediterranean diet, and lifelong good health. *Am. J. Med.* 113(9B): 5S, 2002.

Gardener, C. D. and Kraemer, H. C. Monounsaturated versus polyunsaturated dietary fat and serum lipids, *Arterioscler. Thromb. Vasc. Biol.* 15:1917, 1995.

Gerhard, G. T., Ahmann, A., Meeuws, K., McMurray, M. P., Duell, P. B., and Conner, W. E. Effects of a low-fat diet compared with those of high-monounsaturated fat diet on body weight, plasma lipids and lipoproteins, and glycemic control in type 2 diabetes. *Am. J. Clin. Nutr.* 80:668, 2004.

Meckling, K. A., O'sullivan, C., and Saari, D. Comparison of a low-fat diet to a low-carbohydrate diet on weight loss, body composition, and risk factors for diabetes and cardiovascular disease in free-living, overweight men and women. *J. Clin. Endocrin. Metab.* 89:2717, 2004.

Opperman, A. M., Venter, C. S., Oosthuizen, W., Thompson, R. L., and Vorster, H. H. Meta-analysis of the health effects of using the glycaemic index in meal-planning. *Br. J. Nutr.* 92:367, 2004.

Pelkman, C. L., Fishell, V. K., Maddox, D. H., Pearson, T. A., Mauger, D., and Kris-Etherton, P. M. Effects of moderate fat (from monounsaturated fat) and low-fat weight-loss diets on the serum lipid profile in overweight and obese men and women. *Am. J. Clin. Nutr.* 79:204, 2004.

Rasmussen, O. W., Thomsen, C., Hansen, K. W., Vesterland, M., Winther, E., and Hermansen, K. Effects on Blood pressure, glucose, and lipid levels of a high monounsaturated fat diet compared with a high-carbohydrate diet in NIDDM subjects. *Diabetes Case* 16:156S, 1993.

Stern, L., Iqbal, N., Seshadri, P., Chicano, K. L., Daily, D. A. McGrory, J. et al. The effects of low-carbohydrate versus conventional weight loss diets in severely obese adults: One year follow-up of a randomized trial. *Ann. Intern. Med.* 18:140: 778, 2004.

QUESTIONS | CAROL N. ANGSTADT

Multiple Choice Questions

1. Of two people with approximately the same weight, the one with the higher basal energy requirement would most likely be:
   A. taller.
   B. female if the other were male.
   C. older.
   D. under less stress.
   E. all of the above.

2. Basal metabolic rate:
   A. is not influenced by energy intake.
   B. increases in response to starvation.
   C. may decrease up to 50% during periods of starvation.
   D. increases in direct proportion to daily energy expenditure.
   E. is not responsive to changes in hormone levels.

3. The primary effect of the consumption of excess protein beyond the body's immediate needs will be:
   A. excretion of the excess as protein in the urine.
   B. an increase in the "storage pool" of protein.
   C. an increased synthesis of muscle protein.
   D. an enhancement in the amount of circulating plasma proteins.
   E. an increase in the amount of adipose tissue.

4. Obesity:
   A. usually does not have adverse metabolic consequences.
   B. causes metabolic changes that are usually irreversible.
   C. frequently leads to elevated serum levels of fatty acids, cholesterol and triacylglycerols.
   D. is frequently associated with an increased sensitivity of insulin receptors.
   E. is caused solely by high caloric consumption.

5. Dietary fat:
   A. is usually present, although there is no specific need for it.
   B. if present in excess, can be stored as either glycogen or adipose tissue triacylglycerol.
   C. should include linoleic and linolenic acids.
   D. should increase on an endurance training program to increase the body's energy stores.
   E. if present in excess, does not usually lead to health problems.

6. Which of the following statements about dietary fiber is/are correct?
   A. Water-soluble fiber helps to lower serum cholesterol in most people.
   B. Mucilaginous fiber slows the rate of digestion and absorption of carbohydrates.
   C. Insoluble fiber increases stool bulk and decreases transit time.
   D. All of the above are correct.
   E. None of the above is correct.

Questions 7 and 8: A young man suffered third degree burns over much of his body and is hospitalized in a severe catabolic state. An individual in this state requires about 40 kcal kg^{-1} day^{-1} and 2 g protein kg^{-1} day^{-1} to be in positive caloric and nitrogen balance. This young man weighs 140 lb (64 kg). Total parenteral nutrition (TPN) is started with a solution containing 20% glucose and 4.25% amino acids (the form in which protein is supplied).

7. If 3000 g of solution is infused per day:
   A. the patient would not be getting sufficient protein.
   B. the calories supplied would be inadequate.
   C. both protein and calories would be adequate to meet requirements.
   D. this is too much protein being infused.

8. Sometimes a lipid solution is also infused in a patient on TPN. In the case of this young man, the purpose of the lipid solution would be to:
   A. supply additional calories to meet caloric needs.
   B. supply essential fatty acids.
   C. improve the palatability of the mixture.
   D. provide fiber.
   E. ensure an adequate supply of cholesterol for membrane building.

Questions 9 and 10: For many years, the American Diabetic Association recommended a diet high in complex carbohydrates and fiber and low in fat for diabetics. It was later found that some individuals did not do as well on such a diet as on one high in monounsaturated fatty acids. Since 1994, the ADA has abandoned the concept of a single diabetic diet and now recommends a focus on achieving glucose, lipid, and blood pressure goals with weight reduction if necessary.

9. Which of the following statements is/are correct?
   A. A high-carbohydrate–high-fiber diet often results in significant weight reduction because it has a lower caloric density than a diet high in fat.
   B. A diet high in monounsaturated fatty acids would be most appropriate for an overweight diabetic.
   C. The goal for lipids is to reduce all lipoprotein levels in the blood.
   D. Obesity aggravates diabetes because it inhibits the production of insulin by the pancreas.
   E. All of the above are correct.

10. For diabetics:
   A. the only carbohydrate that must be eliminated in the diet is sucrose.
   B. fiber increases the rate at which carbohydrate is digested and absorbed.
   C. not all carbohydrate foods raise blood glucose levels at the same rate because the glycemic index of all foods is not the same.
   D. who are normally in good control, stress will have no effect on their blood sugar levels.
   E. a vegetarian diet is the only appropriate choice.

Questions 11 and 12: Recent studies have confirmed that reducing elevated serum cholesterol levels can reduce the risk of heart disease. Obesity is another risk factor. Epidemiological studies suggest that consumption of animal protein is associated with an increased incidence of heart disease, although it is probably the fat and cholesterol in animal foods rather than the animal protein itself. Most Americans would probably not accept a strictly vegetarian diet.

11. Which one of the following dietary regimens would be most effective in lowering serum cholesterol?
    A. Restrict dietary cholesterol.
    B. Increase the ratio of polyunsaturated to saturated fatty acids.
    C. Increase fiber content.
    D. Restrict cholesterol and increase fiber.
    E. Restrict cholesterol, increase PUFA/SFA, increase fiber.

12. A complete replacement of animal protein in the diet by vegetable protein:
    A. would be expected to have no effect at all on the overall diet.
    B. would reduce the total amount of food consumed for the same number of calories.
    C. might reduce the total amount of iron and vitamin B_{12} available.
    D. would be satisfactory regardless of the nature of the vegetable protein used.
    E. could not satisfy protein requirements.

Problems

13. Calculate the number of grams each of carbohydrate, lipid, and protein a person on a 2300-kcal diet should consume to meet the guidelines established by the Senate Select Committee on Human Nutrition. Assuming the individual weighs 180 lb and the protein is from mixed animal/vegetable sources with a 75% efficiency of utilization, does the amount of protein you calculated above meet the recommended amount of protein?

14. A 120-lb woman is consuming a diet with adequate total calories and 44 g of protein per day. The protein is exclusively from vegetable sources, primarily corn-based. What would be her state of nitrogen balance?

ANSWERS

1. **A** A taller person with the same weight would have a greater surface area. B: Males have higher energy requirements than females. C: Energy requirements decrease with age. D: Stress, probably because of the effects of epinephrine and cortisol, increase energy requirements.
2. **C** This is part of the survival mechanism in starvation. A, B: BMR decreases when energy intake decreases. D: BMR as defined is independent of energy expenditure. Only when the exercise is repeated on a daily basis so that lean muscle mass is increased does BMR also increase. E: Many hormones increase BMR.
3. **E** Excess protein is treated like any other excess energy source and stored (minus the nitrogen) eventually as adipose tissue lipid. A: Protein is not found in normal urine except in very small amounts. The excess nitrogen is excreted as NH_4^+ and urea, whereas the excess carbon skeletons of the amino acids are used as energy sources. B–D: There is no discrete storage form of protein, and although some muscle and structural protein is expendable, there is no evidence that increased intake leads to generalized increased protein synthesis.
4. **C** Probably because of an increased number and/or size of adipose cells which are less sensitive to insulin. A: Obesity has multiple effects. B: Most changes can be reversed if weight is lost. D: Sensitivity is decreased perhaps by interfering with autophosphorylation of the receptor. E: Inadequate exercise and a genetic component also play a role.
5. **C** A, C: Linoleic and linolenic acids are essential fatty acids and so must be present in the diet. B and D: Excess carbohydrate can be stored as fat but the reverse is not true. D: Carbohydrate loading has been shown to increase endurance. E: High-fat diets are associated with many health risks.
6. **D** These each illustrate the different properties and roles of the common kinds of fiber.
7. **C** This amount of solution would supply 128 g protein and 2900 kcal (2400 from glucose and 512 from amino acids), both enough to meet the stated requirements.
8. **B** Patients on TPN need to have essential fatty acids supplied. A: The original solution supplies adequate calories although additional calories should not hurt this young man. C: The patient is not tasting this mixture, so palatability is irrelevant. D: Fiber is supplied by complex carbohydrate sources. E: The body can make its own cholesterol.
9. **A** Since carbohydrate has less than half the caloric density of fat and fiber provides no calories, such diets have low caloric density. B: High fat would tend to increase weight. C: The goal is to reduce the LDL and VLDL but not HDL. D: The problem is that factors released by adipose cells invoke insulin resistance.
10. **C** Foods vary widely in their glycemic index. Bread and rice raise blood glucose levels more rapidly than does sucrose. A: See C. B: One of the benefits of fiber is that it decreases the rate of carbohydrate absorption. D: Stress raises blood glucose for everyone because of release of epinephrine and glucocorticoids. E: An appropriately designed vegetarian diet is perfectly acceptable but is certainly not the only choice.
11. **E** Any of the measures alone would decrease serum cholesterol slightly, but to achieve a reduction of more than 15% requires all three.
12. **C** A, C: This would reduce the amount of fat, especially saturated fat, but could also reduce the amount of necessary nutrients that come primarily from animal sources. B: The protein content of vegetables is quite low, so much larger amounts of vegetables would have to be consumed. D, E: It is possible to satisfy requirements for all of the essential amino acids completely if vegetables with complementary amino acid patterns, in proper amounts, are consumed.
13. According to the SSCHN guidelines a 2300-kcal diet should consist of: 333.5 g of carbohydrate (no more than 57.5 g as simple sugar); 69 g of protein; 76.7 g of fat (no more than 25.6 g of saturated fat). A 180-lb man weighs 81.8 kg times 0.8 g/kg = 65.5 g of protein per day to meet requirements. 12% of 2300 kcal supplies sufficient protein.
14. 44 g protein/120 lb is 0.8 g protein/kg/day. However, this is inadequate because pure vegetable protein is less than 75% efficient in utilization. Also, heavy reliance on one protein (corn is deficient in lysine) would likely lead to a deficiency of one or more essential amino acids. Therefore, this woman is in negative nitrogen balance.

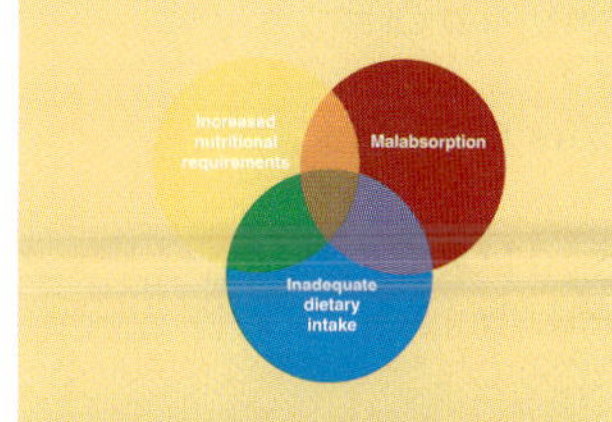

28

PRINCIPLES OF NUTRITION II: MICRONUTRIENTS

Stephen G. Chaney

Textbook of Biochemistry With Clinical Correlations, Sixth Edition, Edited by Thomas M. Devlin

28.1 OVERVIEW

Micronutrients play a vital role in human metabolism, because they are involved in almost every biochemical reaction pathway. However, the science of nutrition is concerned not only with the biochemistry of the nutrients but also with whether they are present in adequate amounts in the diet. The American diet is undoubtedly the best it has ever been. Our current food supply provides an abundant variety of foods all year long, and deficiency diseases have become medical curiosities. However, our diet is far from optimal. The old adage is that we can get everything we need from a balanced diet. Unfortunately, many Americans do not consume a balanced diet. Foods of high caloric density and low nutrient density (often referred to as empty calories or junk food) are abundant and popular, and our nutritional status suffers because of these food choices. Obviously then, neither alarm nor complacency is justified. We need to know how to evaluate the adequacy of our diet.

28.2 ASSESSMENT OF MALNUTRITION

There are three increasingly stringent criteria for measuring **malnutrition**.

1. **Dietary intake studies**, which are usually based on a 24-h recall, are the least stringent. The 24-h recalls tend to overestimate the number of people with deficient diets. In addition, poor dietary intake alone is not a problem in this country unless the situation is compounded by increased need.
2. **Biochemical assays**, either direct or indirect, are a more useful indicator of nutritional status. At their best, they indicate **subclinical nutritional deficiencies**, which can be treated before deficiency diseases develop. However, all biochemical assays are not equally valid, an unfortunate fact that is not sufficiently recognized. Changes in biochemical parameters due to stress need to be interpreted with caution. The distribution of many nutrients in the body changes dramatically in a stress situation such as illness, injury, and pregnancy. A drop in level of a nutrient in one tissue compartment (usually blood) need not signal a deficiency or an increased requirement. It could simply reflect a normal metabolic adjustment to stress.
3. The most stringent criterion is the appearance of **clinical symptoms**. However, it is desirable to intervene long before symptoms became apparent.

The question remains: When should dietary surveys or biochemical assays be interpreted to indicate the need for nutritional intervention? Dietary surveys are seldom a valid indication of general malnutrition unless the average intake for a population group falls significantly below the Estimated Average Requirement (EAR) for one or more nutrients. However, by looking at the percentage of people within a population group who have suboptimal intake, it is possible to identify high-risk population groups that should be monitored more closely. Biochemical assays can definitely identify subclinical cases of malnutrition where nutritional intervention is desirable provided that (a) the assay has been shown to be reliable, (b) the deficiency can be verified by a second assay, and (c) there is no unusual stress situation that may alter micronutrient distribution. In assessing nutritional status, it is important to be aware of those population groups at risk, the most reliable biochemical assays for monitoring nutritional status, and the symptoms of deficiencies.

28.3 DIETARY REFERENCE INTAKES

Dietary Reference Intakes (DRIs) are quantitative estimates of nutrient intakes to be used for planning and assessing diets for healthy people, and refer to either RDAs or AIs, depending on the nutrient (Figure 28.1). In assessing quantitative standards for nutrient intake the Food and Nutrition Board of the National Research Council considers the amount of nutrients and food components required for preventing deficiency diseases and, where the data are definitive, for promoting optimal health. The first step is determining the **Estimated Average Requirement (EAR)**, the amount of nutrient estimated to meet the nutrient requirement of half of the healthy individuals in an age and gender group. The **Recommended Dietary Allowance (RDA)** is normally set at two standard deviations above the EAR and is assumed to be the dietary intake amount that is sufficient to meet the nutrient requirement of nearly all (97% to 98%) healthy individuals in a group. If a nutrient is considered essential but the experimental data are inadequate for determining an EAR, an **adequate intake (AI)** is set rather than an RDA. The AI is believed to cover the needs of all individuals in a group, but uncertainty of the data prevent being able to specify with certainty the percentage of individuals covered by this intake. AIs are often based on approximations of nutrient intake by a group of individuals. For example, AIs for young infants are often based on the daily mean nutrient intake supplied by human milk for healthy, full-term infants who are exclusively breast-fed. Finally, for most nutrients the Food and Nutrition Board sets a Tolerable Upper Intake Level (UL). The UL is defined as the highest level of daily nutrient intake that is likely to pose no risks of adverse health effects to almost all individuals in the general population. The RDAs, AIs, and ULs are designed to be of use in planning and evaluating diets for individuals. The EAR is designed to be used in setting goals for nutrient intake and assessing the prevalence of inadequate intake in a population group.

These determinations are relatively easy to make for those nutrients associated with dramatic deficiency diseases—for example, vitamin C and scurvy. Measures that are more indirect must be used in other instances, such as tissue saturation or extrapolation from animal studies. The Food and Nutrition Board normally meets every 6–10 years to consider currently available information and update their recommendations.

DRIs serve as a useful general guide in evaluating adequacy of individual diets. However, the DRIs have several limitations:

1. DRIs are designed to meet the needs of healthy people and do not take into account special needs arising from infections, metabolic disorders, or chronic diseases.
2. Since present knowledge of nutritional needs is incomplete, there may be unrecognized nutritional needs. To provide for these needs, the DRIs should be met from as varied a selection of foods as possible. No single food can be considered complete, even if it meets the DRI for all known nutrients. This is important, especially in light of the current practice of fortifying foods of otherwise low nutritional value.
3. As currently formulated, DRIs may not define the "optimal" level of any nutrient, since optimal amounts are difficult to define. Because of information suggesting that optimal intake of certain micronutrients may reduce heart disease and cancer risk, the DRIs for these nutrients have recently been increased slightly; however, some experts feel that the current DRIs may not be sufficient to promote optional health.

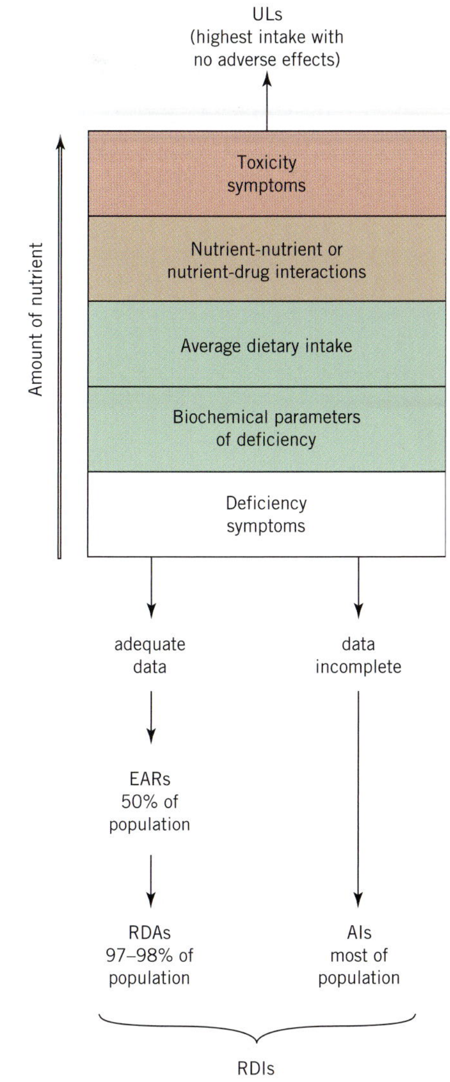

ULs
(highest intake with
no adverse effects)
Amount of nutrient
Toxicity symptoms
Nutrient-nutrient or nutrient-drug interactions
Average dietary intake
Biochemical parameters of deficiency
Deficiency symptoms
adequate data
data incomplete
EARs 50% of population
RDAs 97–98% of population
AIs most of population
RDIs

FIGURE 28.1

Dietary reference intakes. A schematic representation of the relationship between EAR (Estimated Average Requirement; the amount of nutrient estimated to meet the needs of 50% of a population group), RDA (Recommended Dietary Allowance; the amount of nutrient estimated to meet the needs of 97–98% of a population group), AI (adequate intake; the amount of nutrient estimated to meet the needs of most of a population group), DRI (Dietary Reference Intake; either RDA or AI, depending on the nutrient), and UL (Tolerable Upper Intake Level; highest level of nutrient intake that is likely to pose no risk of adverse health effects to almost all individuals in the general population)

28.4 FAT-SOLUBLE VITAMINS

Vitamin A Is Derived from Plant Carotenoids

The active forms of **vitamin A** are **retinol, retinal (retinaldehyde)**, and **retinoic acid.** Their precursors are synthesized by plants as the **carotenoids** (Figure 28.2), which are cleaved to retinol by most animals and stored in the liver as retinol palmitate. Liver, egg yolk, butter, and whole milk are good sources of retinol. Dark green and yellow vegetables are generally good sources of the carotenoids. Conversion of carotenoids to

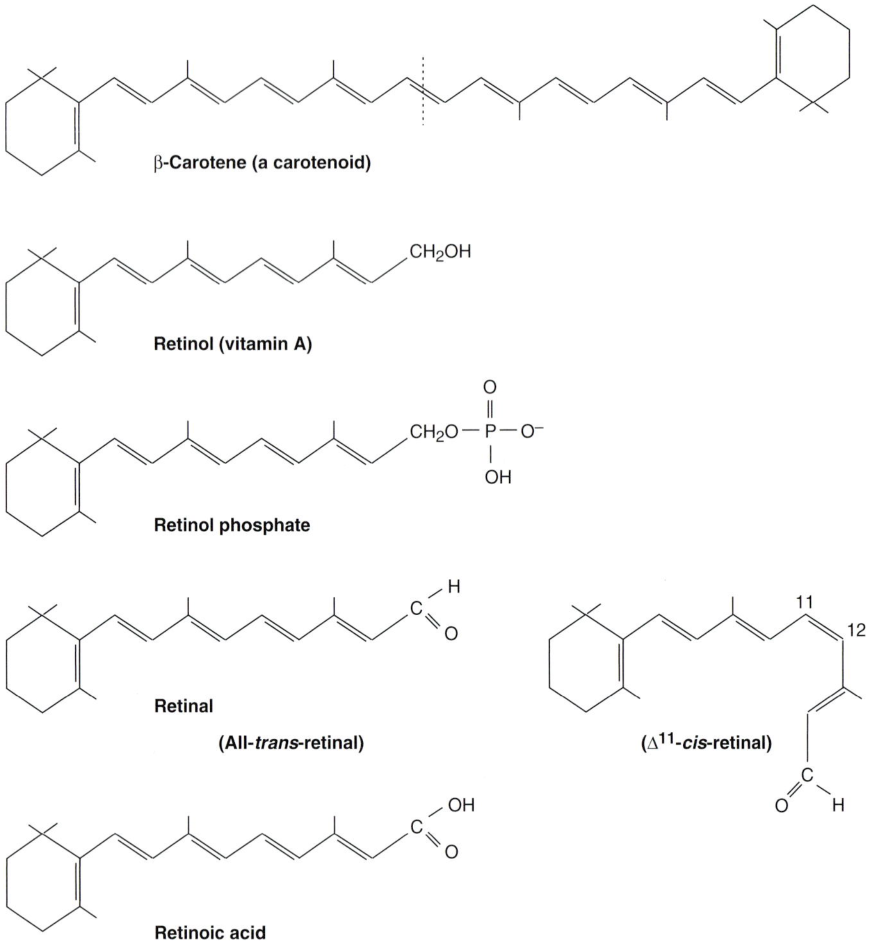

β-Carotene (a carotenoid)

Retinol (vitamin A) — CH_2OH

Retinol phosphate — $CH_2O-P(=O)(OH)-O^-$

Retinal (All-*trans*-retinal) — 11, 12 (Δ^{11}-*cis*-retinal)

Retinoic acid

FIGURE 28.2
Structures of vitamin A and related compounds.

retinol is rarely 100%, so that the vitamin A potency of various foods is expressed in terms of retinol equivalents (1 RE is equal to 1 mg retinol, 6 mg ***β*-carotene**, and 12 mg of other carotenoids). *β*-Carotene and other carotenoids are major sources of vitamin A in the American diet. The carotenoids are cleaved to retinol and converted to other vitamin A metabolites in the body (Figure 28.2).

Only in recent years has the biochemistry of vitamin A become well understood (Figure 28.3). *β*-Carotene and some other carotenoids have an important role as **antioxidants**. At the low oxygen tensions prevalent in the body, *β*-carotene is a very effective antioxidant and may reduce the risk of those cancers initiated by free radicals and other strong oxidants. Several epidemiologic studies suggested that adequate dietary *β*-carotene may be important in reducing the risk of lung cancer, especially in people who smoke. However, supplemental *β*-carotene did not provide any detectable benefit and may have actually increased cancer risk for smokers in two multicenter prospective studies. This illustrates the danger of making dietary recommendations on epidemiologic studies alone.

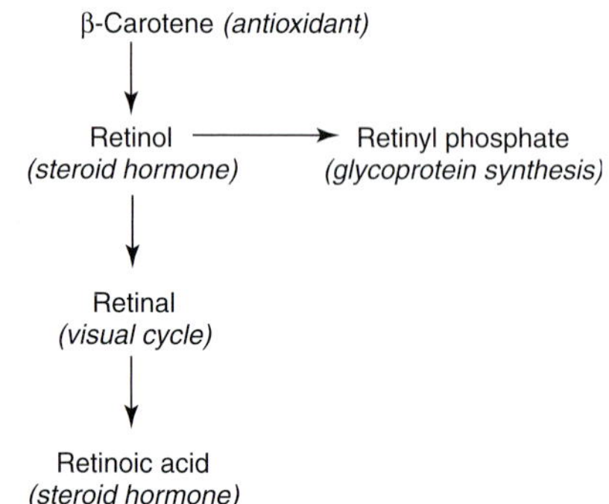

FIGURE 28.3
Vitamin A metabolism and function.

Retinol is converted to **retinyl phosphate**, which appears to serve as a **glycosyl donor** in the synthesis of some glycoproteins and mucopolysaccharides in much the same manner as **dolichol phosphate** (see p. 651). It is essential for the synthesis of glycoproteins needed for normal growth regulation and for mucus secretion. Retinoic acid binds to retinoic acid receptors (RARs) and retinoid X receptors (RXRs), which then bind to DNA and modulate the synthesis of proteins involved in the regulation of cell growth and differentiation. Thus, it can be considered to act like a **steroid hormone** in regulating growth and differentiation.

In the Δ^{11}-*cis*-retinal form, vitamin A becomes reversibly associated with **visual proteins** (the opsins). When light strikes the retina, a number of complex biochemical changes take place, resulting in generation of a nerve impulse, conversion of the *cis*-retinal to the all-*trans* form, and its dissociation from the visual protein (see p. 974). Regeneration of functional visual pigments requires isomerization back to the Δ^{11}-*cis* form (Figure 28.4). In addition to the direct role of vitamin A in the visual cycle, clinical studies suggest that the carotenoids lutein and zeaxanthin reduce the risk of macular degeneration.

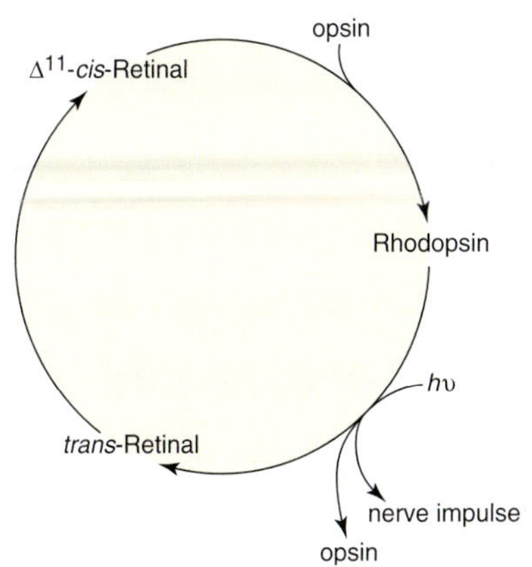

FIGURE 28.4
Role of vitamin A in vision.

Based on what is known about the biochemical mechanisms of vitamin A action, its biological effects are easy to understand. For example, retinyl phosphate is required for the synthesis of glycoproteins (an important component of **mucus**), and lack of mucus secretion leads to a drying of epithelial tissues. Retinol and/or retinoic acid down-regulate the synthesis of **keratin**, and excess keratin synthesis leaves a horny keratinized surface in place of the normal moist and pliable epithelium. Thus, vitamin A is required for maintenance of healthy epithelial tissue. In addition, retinol and/or retinoic acid are required for the synthesis of the iron transport protein transferrin. Thus, vitamin A deficiency can lead to anemia from impaired transport of iron.

Vitamin A-deficient animals are more susceptible to infections and cancer. Decreased resistance to infection may be due to keratinization of mucosal cells lining the respiratory, gastrointestinal, and genitourinary tracts. Fissures readily develop in the mucosal membranes, allowing microorganisms to enter. Vitamin A deficiency may also impair the immune system. The protective effect of vitamin A against many forms of cancer may result from the antioxidant potential of carotenoids and the effects of retinol and retinoic acid in regulating cell growth.

Since vitamin A is stored in the liver, deficiency can develop only over prolonged periods of inadequate intake. Mild **vitamin A deficiency** is characterized by **follicular hyperkeratosis** (rough keratinized skin resembling "goosebumps"), **anemia** (biochemically equivalent to iron deficiency anemia, but in the presence of adequate iron intake), and increased susceptibility to infection and cancer. **Night blindness** is an early symptom of deficiency. Severe deficiency leads to progressive keratinization of the cornea known as **xerophthalmia** in its most advanced stages. Infection usually sets in, with resulting hemorrhaging of the eye and permanent loss of vision.

For most people (unless they happen to eat liver) the dark green and yellow vegetables are the most important dietary source of vitamin A. Unfortunately, these foods are most often missing from the American diet. Dietary surveys indicate that 40–60% of the population consumes less than two-thirds of the RDA for vitamin A. Clinical symptoms of vitamin A deficiency are rare in the general population, but it is a fairly common consequence of severe liver damage or diseases that cause fat malabsorption (see Clin. Corr. 28.1).

Vitamin A accumulates in the liver. Excess intake over prolonged periods can be toxic. Doses of 25,000–50,000 RE per day over months or years are toxic for many children and adults. The usual symptoms include bone pain, scaly dermatitis, enlargement of liver and spleen, nausea, and diarrhea. It is virtually impossible to ingest toxic amounts of vitamin A from normal foods unless one eats polar bear liver (6000 RE g^{-1}) regularly. Most instances of **vitamin A toxicity** are caused by massive doses of vitamin A supplements. Fortunately, this practice is relatively rare because of increased public awareness of vitamin A toxicity.

Vitamin D Synthesis Requires Sunlight

Technically, **vitamin D** should be considered a pro-hormone rather than a vitamin. **Cholecalciferol (D_3)** is produced in skin by UV irradiation of 7-dehydrocholesterol (see p. 719 for a discussion of the synthesis of vitamin D). Thus, as long as the body is exposed to adequate sunlight, there is little or no dietary requirement for vitamin D. The best dietary sources of vitamin D_3 are saltwater fish (especially salmon, sardines, and herring), liver, and egg yolk. Milk, butter, and other foods are routinely fortified with ergocalciferol (D_2) prepared by irradiating ergosterol from yeast. Vitamin D potency is measured in terms of milligrams cholecalciferol (1 mg cholecalciferol or ergocalciferol = 40IU).

CLINICAL CORRELATION 28.1
Nutritional Considerations in Cystic Fibrosis

Patients with malabsorption diseases often develop malnutrition. Cystic fibrosis (CF) involves a generalized dysfunction of the exocrine glands that leads to formation of viscid mucus, which progressively plugs their ducts. Obstruction of the bronchi and bronchioles leads to pulmonary infections, which are usually the direct cause of death. In many cases, however, the exocrine cells of the pancreas are also affected, leading to a deficiency of pancreatic enzymes and sometimes a partial obstruction of the common bile duct.

Deficiency of pancreatic lipase and bile salts leads to severe malabsorption of fat and fat-soluble vitamins. Calcium tends to form insoluble salts with the long-chain fatty acids, which accumulate in the intestine. Starches and proteins are also trapped in the fatty bolus of partially digested foods. This physical entrapment, along with the deficiencies of pancreatic amylase and pancreatic proteases, can lead to severe protein–calorie malnutrition. Excessive mucus secretion on the luminal surface of the intestine may interfere with the absorption of several nutrients, including iron.

Fortunately, microsphere preparations of pancreatic enzymes are now available that greatly alleviate many of these malabsorption problems. With these preparations, protein and carbohydrate absorption are returned to near normal. Fat absorption is improved greatly but not normalized, since deficiencies of bile salts and excess mucus secretion persist. Because dietary fat is a major source of calories, these patients have difficulty obtaining sufficient calories from a normal diet. This is complicated by increased protein and energy needs because of the chronic infections often seen in these patients. Thus, many experts recommend energy intakes ranging from 120% to 150% of the RDA to combat the poor growth and increased susceptibility to infection. The current recommendations are for high-energy–high-protein diets without any restriction of dietary fat (50% carbohydrate, 15% protein, and 35% fat). If caloric intake from the normal diet is inadequate, dietary supplements or enteral feeding may be used. The dietary supplements usually contain easily digested carbohydrates and milk protein mixtures. Medium-chain triglycerides are sometimes used as a partial fat replacement since they can be absorbed directly through the intestinal mucosa in the absence of bile salts and pancreatic lipase.

Since some fat malabsorption is present, deficiencies of fat-soluble vitamins often occur. Children aged 2–8 years need a standard adult multiple-vitamin preparation containing 400 IU of vitamin D and 5000 IU of vitamin A/day. Older children, adolescents, and adults need a standard multivitamin at a dose of 1–2/day. If serum vitamin A or vitamin E levels become low, water-miscible vitamin preparations should be used. Vitamin K deficiency has not been adequately studied, but supplementation is usually recommended, especially when on antibiotics or if cholestatic liver disease is present. Iron deficiency is common, but iron supplementation is not usually recommended because of concern that higher iron levels in the blood might encourage systemic bacterial infections. Calcium levels in the blood are usually normal. However, since calcium absorption is probably suboptimal, it is important to ensure that the diet provides at least RDA levels of calcium.

Source: Borowitz, D., Baker, R. B., and Stallings, V. Consensus report on nutrition for pediatric patients with cystic fibrosis. *J. Pedatr. Gastroenterol Nutr.* 35:246, 2002. Yankaskas, J. R., Marshall, B. C., Sufian, B., Simon, R. H., and Rodman, D. Cystic fibrosis adult care. Consensus conference report. *Chest* 125:1S, 2004.

Both cholecalciferol and ergocalciferol are metabolized in liver, where **25-hydroxycholecalciferol** [25-(OH)D] is formed. This is the major circulating derivative of vitamin D, and is converted into the biologically active **1-α,25-dihydroxycholecalciferol** (calcitriol) in the proximal convoluted tubules of kidney (see Clin. Corr. 28.2). 1,25-$(OH)_2D$ acts in concert with **parathyroid hormone (PTH)**, which is also produced in response to low serum calcium. High PTH levels stimulate production of 1,25-$(OH)_2D$, while low PTH amounts cause formation of 24,25-$(OH)_2D$. The 1,25-$(OH)_2D$ acts as a typical steroid hormone in intestinal mucosal cells, where it induces synthesis of a protein, calbindin, required for calcium transport. In bone 1,25-$(OH)_2D$ and PTH act synergistically to promote bone resorption (demineralization) by stimulating osteoblast formation and activity. Finally, PTH and 1, 25-$(OH)_2D$ inhibit calcium excretion in the kidney by stimulating calcium reabsorption in the distal renal tubules. 24,25-$(OH)_2D$ was thought to be inactive, but recent studies with knockout mice lacking the 24-hydroxylase enzyme have shown that 24,25-$(OH)_2D$ plays an essential, but ill-defined, role in bone metabolism. Calcitonin is produced when serum calcium levels are high (usually right after a meal) and lowers serum calcium levels by blocking bone resorption and stimulating calcium excretion by the kidney.

The response of calcium metabolism to several different physiological situations is summarized in Figure 28.5. The response to low serum calcium levels is characterized

CLINICAL CORRELATION 28.2
Renal Osteodystrophy

In chronic renal failure, a complicated chain of events leads to renal osteodystrophy. The renal failure results in an inability to produce 1,25-$(OH)_2D$, and thus bone calcium becomes the only important source of serum calcium. In the later stages, the situation is complicated by increased renal retention of phosphate and the resulting hyperphosphatemia. The serum phosphate levels are often high enough to cause metastatic calcification (i.e., calcification of soft tissue), which tends to lower serum calcium levels further (the solubility product of calcium phosphate in the serum is very low, and a high serum level of one component necessarily causes a decreased concentration of the other). The hyperphosphatemia and hypocalcemia stimulate parathyroid hormone secretion, which further accelerates the rate of bone loss. The result is bone loss and metastatic calcification. Administration of high doses of vitamin D or its active metabolites would not be sufficient since the combination of hyperphosphatemia and hypercalcemia would only lead to more extensive metastatic calcification. The readjustment of serum calcium levels by high calcium diets and/or vitamin D supplementation must be accompanied by phosphate reduction therapies. The most common technique is to use phosphate-binding antacids that make phosphate unavailable for absorption. Orally administered 1,25-$(OH)_2D$ is effective at stimulating calcium absorption in the mucosa but does not enter the peripheral circulation in significant amounts. Thus, in severe hyperparathyroidism intravenous 1,25-$(OH)_2D$ may be necessary. Research is in progress with calcimimetic agents that bind to calcium sensor located on the extracellular membrane of the parathyroid gland and decrease parathyroid hormone production and release.

Source: Delmez, J. M. and Slatopolsky, E. Hyperphosphatemia: Its consequences and treatment in patients with chronic renal disease. *Am J. Kidney Dis.* 19:303, 1992. Hoyland, J. A. and Picton, M. L. Cellular mechanisms of osteodystrophy. *Kidney Int.* 56(Suppl. 73):508, 1999. Slatopolsky, E., Gonzalez, E., and Marin, K., Pathogenesis and treatment of renal osteodystrophy. *Blood Purif.* 21:318, 2003.

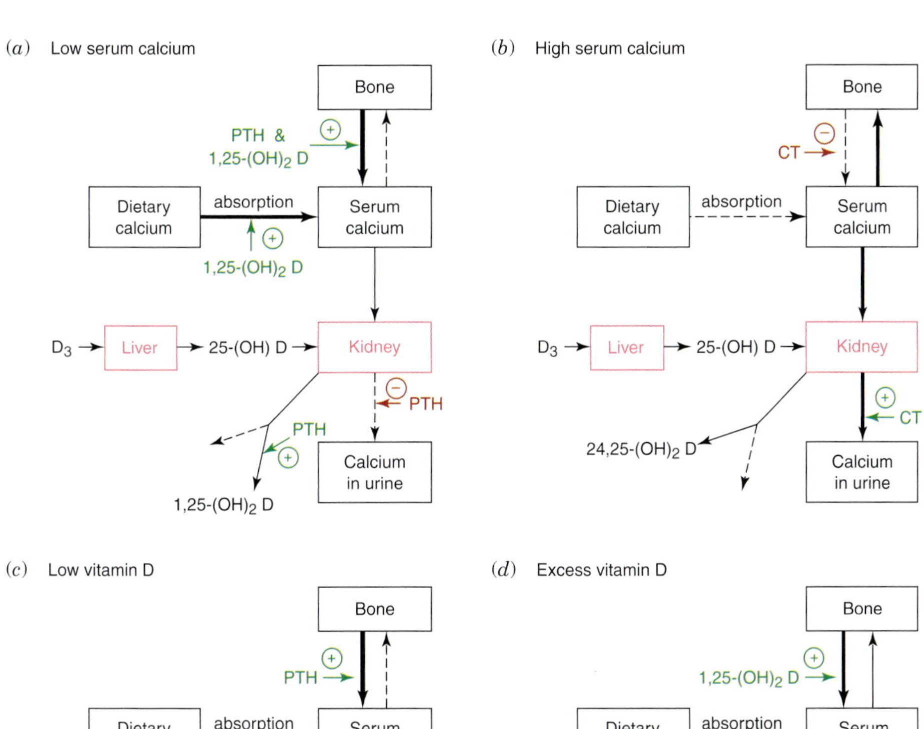

FIGURE 28.5

Vitamin D and calcium homeostasis. The dominant pathways of calcium metabolism under each set of metabolic conditions are shown with heavy arrows. Effect of various hormones is shown by green arrows for stimulation or red arrows for repression. PTH, parathyroid hormone; CT, calcitonin; D, cholecalciferol; 25-(OH)D, 25-hydroxycholecalciferol; 1,25-$(OH)_2D$, 1-α,25-dihydroxycholecalciferol; 24,25-$(OH)_2D$, 24,25-dihydroxycholecalciferol.

by elevation of PTH and 1, 25-$(OH)_2D$, which act to enhance calcium absorption and bone resorption and to inhibit calcium excretion (Figure 28.5*a*). High serum calcium levels block production of PTH. The low PTH levels cause 25-(OH)D to be converted to 24, 25-$(OH)_2D$ instead of 1,25-$(OH)_2D$. In the absence of PTH and 1,25-$(OH)_2D$ bone resorption is inhibited and calcium excretion is enhanced. High serum calcium levels also stimulate production of calcitonin, which contributes to the inhibition of bone resorption and the increase in calcium excretion. Finally, the high levels of both serum calcium and phosphate increase the rate of bone mineralization (Figure 28.5*b*). Thus, bone is a very important reservoir of the calcium and phosphate needed to maintain homeostasis of serum levels. When dietary vitamin D and calcium are adequate, no net loss of bone calcium occurs. However, when dietary calcium is low, PTH and 1,25-$(OH)_2D$ cause net demineralization of bone to maintain normal serum calcium levels. Vitamin D deficiency also causes net demineralization of bone due to elevation of PTH (Figure 28.5*c*).

The most common symptoms of **vitamin D deficiency** are **rickets** in young children and **osteomalacia** in adults. Rickets is characterized by continued formation of osteoid matrix and cartilage that are improperly mineralized resulting in soft, pliable bones. In adults, demineralization of preexisting bone causes it to become softer and more susceptible to fracture. Osteomalacia is easily distinguishable from the more common osteoporosis, by the fact that the osteoid matrix remains intact in the former, but not in the latter. Vitamin D may be involved in more than regulation of calcium homeostasis. Receptors for 1,25-$(OH)_2D$ occur in many tissues including parathyroid gland, islet cells of pancreas, keratinocytes of skin, and myeloid stem cells in bone marrow. Recent research suggests that adequate vitamin D intake may reduce the risk of certain autoimmune diseases.

Because of fortification of dairy products with vitamin D, dietary deficiencies are very rare and are most often seen in low-income groups, the elderly (who often also have minimal exposure to sunlight), strict vegetarians (especially if their diet is also low in calcium and high in fiber), and chronic alcoholics. Most cases of vitamin D deficiency result from **fat malabsorption** or severe liver and kidney disease (see Clin. Corrs. 28.1 and 28.2). Certain drugs also interfere with vitamin D metabolism. For example, corticosteroids stimulate the conversion of vitamin D to inactive metabolites and cause bone demineralization when used for long periods.

Vitamin D can be toxic in high doses. The tolerable upper intake level (UL) for adults is 2000 IU/day. The mechanism of **vitamin D toxicity** is summarized in Figure 28.5*d*. Enhanced calcium absorption and bone resorption cause hypercalcemia, which can lead to metastatic calcification. The enhanced bone resorption causes demineralization similar to that of vitamin D deficiency. Finally, the high serum calcium leads directly to hypercalciuria, which predisposes to formation of renal stones.

Vitamin E Is a Mixture of Tocopherols and Tocotrienols

Vitamin E occurs in the diet as a mixture of several closely related compounds, called tocopherols and tocotrienols (Figure 28.6). All of the tocopherols and tocotrienols are important naturally occurring **antioxidants**. Due to their lipophilic character, they accumulate in circulating lipoproteins, cellular membranes, and fat deposits, where they act as scavengers for free radicals, protecting unsaturated fatty acids (especially in membranes) from peroxidation reactions. **α-Tocopherol** is the most potent scavenger of reactive oxygen species, but γ-tocopherol is a more potent scavenger of reactive nitrogen species. γ-Tocopherol also appears to inactivate fat-soluble electrophilic mutagens, thus complementing glutathione, which inactivates electrophilic mutagens in the aqueous compartments of the cell. The tocopherols appear to play a role in cellular respiration, by either stabilizing ubiquinone or by helping transfer electrons to ubiquinone (see p. 556). Tocopherols and tocotrienols also prevent **oxidation of LDL**, which may be important in reducing the risk of cardiovascular disease since the oxidized form of LDL is atherogenic. While many of the biological properties of the tocopherols are the result of their antioxidant potential, some of their benefits appear to be due to effects on

Tocopherols

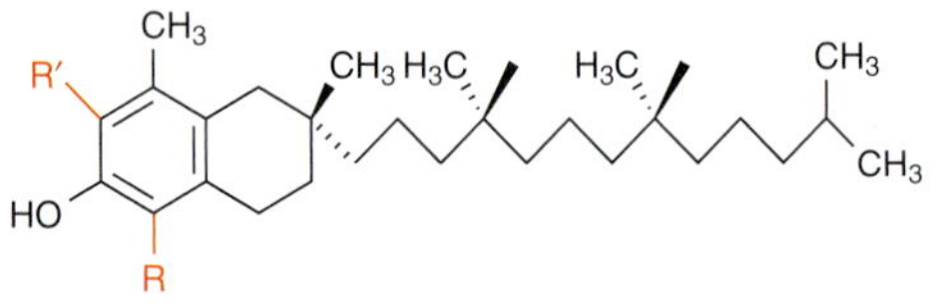

Tocotrienols

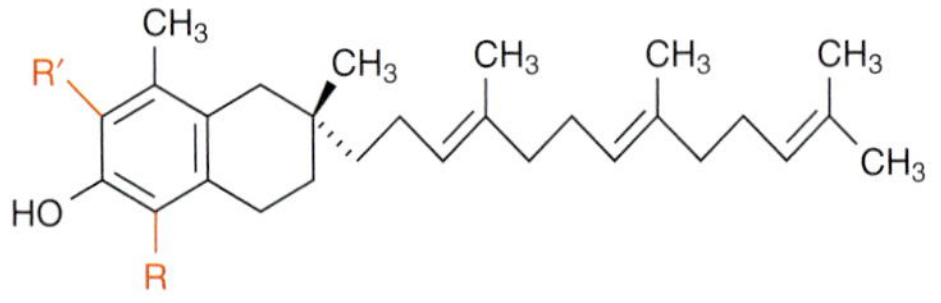

Naturally occurring homologs	R	R′
α	CH_3	CH_3
β	CH_3	H
γ	H	CH_3
δ	H	H

FIGURE 28.6
The structures of tocopherols and tocotrienols.

enzyme activity or transcription. For example, they appear to enhance heme synthesis by increasing the levels of δ-aminolevulinic acid (ALA) synthase and ALA dehydratase. Recent studies have shown that vitamin E is required for maintaining normal immune function, particularly in the elderly, and may be important in preventing macular degeneration and cognitive decline. Finally, neurological symptoms have been reported following prolonged vitamin E deficiency associated with malabsorption diseases.

Setting the recommended intake levels of vitamin E has been hampered by the difficulty of producing severe vitamin E deficiency in humans. It is generally assumed that the vitamin E content of the American diet is sufficient, since no major vitamin E deficiency diseases have been found. However, vitamin E requirements increase as intake of polyunsaturated fatty acids increases. While the recent emphasis on high polyunsaturated fat diets to reduce serum cholesterol may be of benefit in controlling heart disease, the propensity of polyunsaturated fats to form free radicals on exposure to oxygen may lead to increased cancer risk. It is prudent to increase vitamin E intake for diets rich in polyunsaturated fats.

The relationship between vitamin E and cardiovascular disease risk illustrates the dangers in extrapolating directly from biochemical and epidemiologic studies to dietary recommendations. Vitamin E prevents the oxidation of LDL particles to a more atherogenic form, so it was logical that vitamin E supplementation might decrease the risk of atherosclerosis. Epidemiologic studies suggested that people consuming 100 mg day^{-1} of vitamin E had reduced risk of myocardial infarction. However, intervention trials with supplemental α-tocopherol failed to show any significant reduction in mortality from cardiovascular disease. This could indicate that α-tocopherol can prevent atherosclerosis in the early stages, but is ineffective in the more advanced clinical conditions represented in the clinical trials. Alternatively, it could indicate that γ-tocopherol, or some other tocopherol or tocotrienol found in food, is more effective than α-tocopherol in preventing atherosclerosis, since high levels of supplemental α-tocopherol are known to interfere with utilization of other forms of vitamin E.

Because the α-tocopherol transfer protein in the liver specifically binds to the natural *RRR*-α-tocopherol, it is retained by the body 4–6 times longer than the synthetic all rac or d, l form of α-tocopherol. The specificity of this protein for α-tocopherol also explains why high intakes of α-tocopherol interfere with the utilization of γ-tocopherol. Vitamin E appears to be the least toxic of the fat-soluble vitamins. The UL for vitamin E has been set at 1000 mg day^{-1} (1500 IU of *RRR*-α-tocopherol), primarily because high levels of vitamin could potentiate the effects of blood thinning medications such as dicumarol.

Vitamin K Is a Quinone Derivative

Vitamin K is found naturally as K_1 **(phytylmenaquinone)** in green vegetables and K_2 **(multiprenylmenaquinone)**, which is synthesized by intestinal bacteria. The body converts synthetic menaquinone **(Menadione)** and a number of water-soluble analogs to a biologically active form of vitamin K.

Vitamin K is required for conversion of glutamic acid residues to **γ-carboxyglutamic** acid residues in several precursor proteins (Figure 28.7). The γ-carboxyglutamic acid residues are good chelators and allow the proteins to bind Ca^{2+}, which is required for their biological activity. In the carboxylase reaction, the active hydroquinone form of vitamin K is converted to an inactive 2,3-epoxide form (Figure 28.7). Regeneration of active vitamin K requires a dithiol-dependent reductase, which is inhibited by coumarin-type drugs like **dicumarol**. There are seven proteins involved in blood coagulation that require vitamin K-dependent activation, so vitamin K is essential for blood clotting. The mechanism of activation has been most clearly delineated for prothrombin (see p. 1006). The γ-carboxyglutamic acid residues allow prothrombin to bind Ca^{2+}, and the prothrombin–Ca^{2+} complex binds to the negatively charged phospholipid surfaces of the platelets and endothelial cells at the site of the injury, where the proteolytic conversion of prothrombin to thrombin occurs.

Vitamin K is also essential for the synthesis of γ-carboxyglutamic acid residues in three proteins found in bone. For example, **osteocalcin** accounts for 15–20% of the

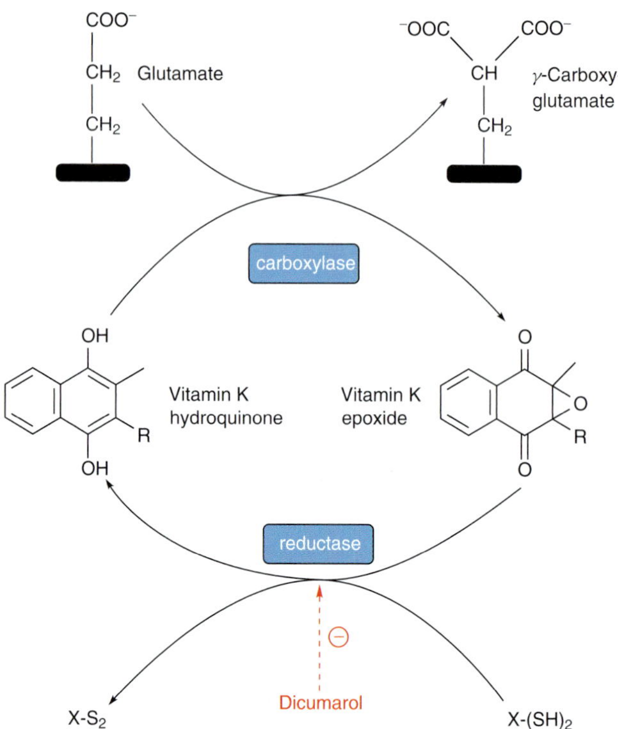

FIGURE 28.7

Function of Vitamin K. Vitamin K is required for the conversion of glutamic acid residues to γ-carboxyglutamic acid residues by the enzyme vitamin K-dependent carboxylase. In the process, the hydroquinone form of vitamin K is converted to the inactive 2,3-epoxide form. Conversion of the 2,3-epoxide back to the active hydroquinone requires a dithiol-dependent reductase that is inhibited by dicumarol.

noncollagen protein in the bone, and the presence of γ-carboxyglutamic acid residues is required for its binding to hydroxyapatite crystals in the bone. The physiological role of osteocalcin and the other γ-carboxylated bone proteins is not clear, but they do appear to be required for normal bone mineralization. Undercarboxylated osteocalcin is associated with low bone density and increased risk of fracture. Finally, a protein called Gas6, which is a ligand for several receptor protein kinases and is involved in cell cycle regulation, has been shown to require vitamin K-dependent γ-carboxylation for activity. The physiological significance of this observation is not known.

The most easily detectable symptom in humans of vitamin K deficiency is increased coagulation time, which reflects the requirement of vitamin K for normal blood coagulation. However, at low dietary intake, vitamin K preferentially accumulates in the liver, where the clotting factors are formed, while higher intake is required for saturation of bone cells with vitamin K. Because of the higher requirements of vitamin K for bone formation, the AI for vitamin K has recently been increased. Since vitamin K is synthesized by intestinal bacteria, deficiencies have long been assumed to be rare. However, intestinally synthesized vitamin K may not be efficiently absorbed and marginal vitamin K deficiencies, especially deficiencies that could interfere with bone mineralization, may be more common than originally thought. The most common deficiency occurs in newborn infants (see Clin. Corr. 28.3), especially those of mothers on anticonvulsant therapy (Clin. Corr. 28.4). Vitamin K deficiency also occurs in patients with **obstructive jaundice** and other diseases leading to severe **fat malabsorption** (see Clin. Corr. 28.1) and patients on long-term antibiotic therapy (which may destroy vitamin K-synthesizing organisms in the intestine). Deficiency is sometimes seen in the elderly, who are prone to fat malabsorption.

CLINICAL CORRELATION 28.3
Nutritional Considerations in Newborn Infants

Newborn infants are at special nutritional risk because of very rapid growth and needs for many nutrients are high. Some micronutrients (such as vitamins E and K) do not cross the placental membrane well, and tissue stores are low in the newborn. The gastrointestinal (GI) tract may not be fully developed, leading to malabsorption problems (particularly with respect to the fat-soluble vitamins). The GI tract is also sterile at birth, and the intestinal flora that normally provide significant amounts of certain vitamins (especially vitamin K) take several days to become established. If the infant is born prematurely, the nutritional risk is slightly greater, since the GI tract will be less well developed and the tissue stores will be less.

The most serious nutritional complication appears to be hemorrhagic disease. Newborns, especially premature infants, have low tissue stores of vitamin K and lack the intestinal flora necessary to synthesize the vitamin. Breast milk is a relatively poor source of vitamin K. Approximately 1 out of 400 live births shows some signs of hemorrhagic disease, which can be prevented by 0.5 to 1 mg of the vitamin given at birth.

Most newborn infants have sufficient reserves of iron to last 3–4 months. Since cow's milk and breast milk contain little iron, supplementation with iron is usually initiated at an early age by the introduction of iron-fortified cereal. Vitamin D levels are also low in breast milk, and supplementation with 200 IU/day of vitamin D is usually recommended. When infants must be maintained on assisted ventilation with high oxygen concentrations, supplemental vitamin E may reduce the risk of bronchopulmonary dysplasia and retrolental fibroplasia, potential complications of oxygen therapy. The anemia of prematurity may respond to supplemental folate and vitamin B_{12}.

In summary, supplemental vitamin K is given at birth to prevent hemorrhagic disease. Breast-fed infants are usually provided with supplemental vitamin D, with iron being introduced along with solid foods. Bottle-fed infants are provided with supplemental iron. If infants must be maintained on oxygen, supplemental vitamin E may be beneficial.

Source: Mueller, D. P. R. Vitamin E therapy in retinopathy of prematurity. *Eye* 6: 221, 1992. Morin, K. H. Current thoughts on healthy term infant nutrition, MCN. *Am. J. Matern. Child Nurs.* 29:312, 2004. Collier, S., Fulhan, J. and Duggan, C. Nutrition for the pediatric office: Update on vitamins, infant feeding and food allergies. *Curr. Opin. Pediatr.* 16:314, 2004.

CLINICAL CORRELATION 28.4
Anticonvulsant Drugs and Vitamin Requirements

Anticonvulsant drugs such as phenobarbital or diphenylhydantoin (DPH) are an excellent example of drug–nutrient interactions that are of concern to physicians. Metabolic bone disease appears to be the most significant side effect of prolonged anticonvulsant therapy. Whereas children and adults who take these drugs seldom develop rickets or severe osteomalacia, up to 65% of those on long-term therapy have abnormally low serum calcium and phosphorus, abnormally high serum alkaline phosphatase, and some bone loss. Vitamin D supplements appear to correct both the hypocalcemia and osteopenia. Anticonvulsants also tend to increase needs for vitamin K, leading to an increased incidence of hemorrhagic disease in infants born to mothers on anticonvulsants. Anticonvulsants appear to increase the need for folic acid and B_6. Low serum folate levels occur in 75% of patients on anticonvulsants, and megaloblastic anemia may occur in as many as 50% without supplementation. This is of particular concern for women of childbearing age. When infants are exposed to antiepileptic drugs *in utero*, the risk of congenital malformations, particularly neural tube defects, is doubled compared to the general population. The recommended daily allowance of folic acid is 600 μg day^{-1} for pregnant women, and the need may be greater for patients on anticonvulsants. Since folates may speed up the metabolism of some anticonvulsants, it is important that excess folic acid not be given.

Source: Rivery, M. D. and Schottelius, D. D. Phenytoin–folic acid: A review. *Drug Intell. Clin. Pharm.* 18:292, 1984. Tjellesen, L. Metabolism and action of vitamin D in epileptic patients on anticonvulsant treatment and healthy adults. *Dan. Med. Bull.* 41:139, 1994. Yerby, M. S. Management issues for women with epilepsy. Neural tube defects and folic acid supplementation. *Neurology* 61(Suppl. 2):S23, 2003.

28.5 WATER-SOLUBLE VITAMINS

Water-soluble vitamins differ from fat-soluble vitamins in several respects. Most are readily excreted once their concentration surpasses the renal threshold so toxicities are rare. Their metabolic stores are labile and depletion can often occur in a matter of weeks or months, so deficiencies occur relatively quickly on an inadequate diet. Since water-soluble vitamins are coenzymes for many common biochemical reactions, it is often possible to assay vitamin status by measuring one or more enzyme activities in isolated red blood cells. These assays are especially useful if one measures the endogenous activity and the stimulation of that activity by addition of the active coenzyme derived from the vitamin.

Most water-soluble vitamins are converted to coenzymes that are used in pathways for energy generation or hematopoiesis. Deficiencies of the energy-releasing vitamins produce a number of overlapping symptoms and show up first in rapidly growing tissues. Typical symptoms include **dermatitis, glossitis** (swelling and reddening of the tongue), **cheilitis** at the corners of the lips, and **diarrhea**. In many cases, nervous tissue is also involved due to its high-energy demand or specific effects of the vitamin. Common neurological symptoms include **peripheral neuropathy** (tingling of nerves at the extremities), depression, mental confusion, lack of motor coordination, and **malaise**. Demyelination and degeneration of nervous tissues may also occur. These deficiency symptoms are so common and overlapping that they can be considered as properties of the energy-releasing vitamins as a class, rather than being specific for any one.

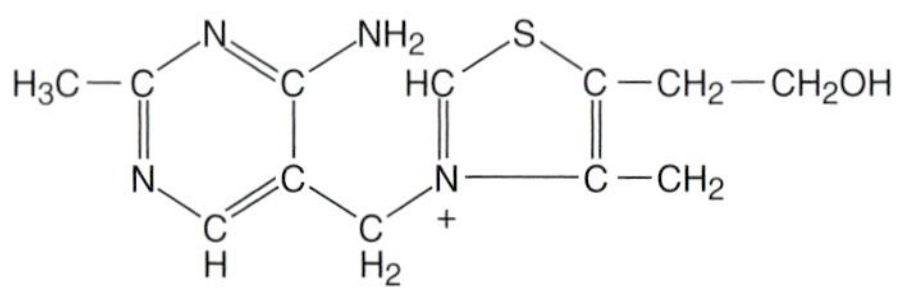

Thiamine

FIGURE 28.8
Structure of thiamine.

28.6 ENERGY-RELEASING WATER-SOLUBLE VITAMINS

Thiamine Forms the Coenzyme Thiamine Pyrophosphate

Thiamine (Figure 28.8) is rapidly converted to **thiamine pyrophosphate (TPP)**, a coenzyme required for the pyruvate dehydrogenase and α-ketoglutarate dehydrogenase reactions (Figure 28.9), and thiamine triphosphate, localized in peripheral nerve membranes, and appears to function in transmission of nerve impulses. Loss of pyruvate dehydrogenase activity compromises cellular energy generation and causes an accumulation of pyruvate and lactate. Loss of α-ketoglutarate dehydrogenase activity causes reduced oxidative decarboxylation of α-keto acids (see p. 545). Symptoms of **thiamine deficiency** involving neural tissue may result from the direct role of thiamine triphosphate in nerve transmission or from the accumulation of pyruvate and lactate in neural tissue. Thiamine pyrophosphate is also required for the transketolase and transaldolase reactions of the pentose phosphate pathway. Red blood cell transketolase is commonly used for measuring thiamine status in the body.

Symptoms of mild thiamine deficiency include loss of appetite, constipation, nausea, mental depression, peripheral neuropathy, irritability, and fatigue. These symptoms are most often seen in the elderly and low-income groups on restricted diets. **Mental confusion, ataxia** (unsteady gait while walking and general inability to achieve fine control of motor functions), and **ophthalmoplegia** (loss of eye coordination) are symptoms of moderately severe thiamine deficiency. This set of symptoms is referred to as **Wernicke–Korsakoff syndrome** and is most commonly seen in chronic alcoholics (see Clin. Corr. 28.5). Severe thiamine deficiency is known as beriberi. Dry beriberi is characterized by advanced neuromuscular symptoms, including muscular atrophy and weakness. When this is coupled with edema, the disease is called wet beriberi. Both forms of beriberi can be associated with an unusual type of heart failure characterized by high cardiac output. Beriberi occurs primarily in populations that rely exclusively on polished rice for food, although cardiac failure is sometimes seen in alcoholics as well. Coffee and tea contain substances that destroy thiamine, but this is not a problem with normal consumption. Routine enrichment of cereals has ensured that most Americans have an adequate intake of thiamine on a mixed diet.

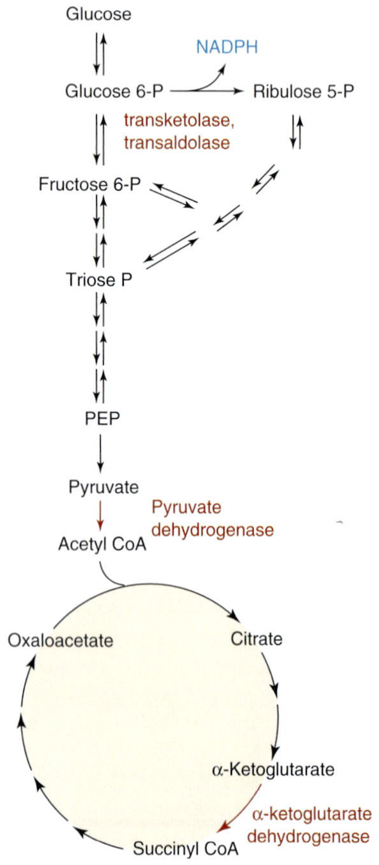

FIGURE 28.9
Summary of important reactions involving thiamine pyrophosphate. Reactions involving thiamine pyrophosphate are indicated in red.

CLINICAL CORRELATION **28.5**

Nutritional Considerations in Alcoholics

Chronic alcoholics run considerable risk of neurologic symptoms associated with thiamine or pyridoxine deficiencies and hematological problems associated with folate or pyridoxine deficiencies. The deficiencies are not necessarily due to poor diet alone, although it is often a strong contributing factor. Alcohol causes pathological alterations of the gastrointestinal tract that directly interfere with absorption of certain nutrients. The severe liver damage associated with chronic alcoholism appears to interfere with storage and activation of nutrients and vitamins.

Up to 40% of hospitalized alcoholics have megaloblastic erythropoiesis due to folate deficiency. Alcohol interferes with folate absorption and alcoholic cirrhosis impairs folate storage. Another 30% of hospitalized alcoholics have sideroblastic anemia or identifiable sideroblasts in erythroid marrow cells characteristic of pyridoxine deficiency. Some alcoholics develop a peripheral neuropathy that responds to pyridoxine supplementation. This problem may result from impaired activation or increased degradation of pyridoxine. In particular, acetaldehyde (a product of alcohol metabolism) displaces pyridoxal phosphate from its carrier protein in the plasma, causing it to be rapidly degraded to inactive compounds and excreted.

The most dramatic disorder is the Wernicke–Korsakoff syndrome. The symptoms include mental disturbances, ataxia (unsteady gait and lack of fine motor coordination), and uncoordinated eye movements. Congestive heart failure similar to that seen with beriberi may also occur. While this syndrome may only account for 1–3% of alcohol-related neurologic disorders, the response to supplemental thiamine is dramatic. The thiamine deficiency probably arises from impaired absorption, although alcoholic cirrhosis may also affect storage of thiamine in the liver.

Deficiencies of most water-soluble vitamins can occur and cases of alcoholic scurvy and pellagra are occasionally reported. Chronic ethanol consumption causes redistribution of vitamin A stores in the body. Liver stores are rapidly depleted, while levels of vitamin A in the serum and other tissues may be normal or slightly elevated. Apparently, ethanol causes increased mobilization of vitamin A from the liver and increased catabolism of liver vitamin A to inactive metabolites by the hepatic cytochrome P450 system. Alcoholic patients have decreased bone density and an increased incidence of osteoporosis. This probably relates to a defect in the 25-hydroxylation step in the liver and an increased rate of metabolism of vitamin D to inactive products by the cytochrome P450 system. Alcoholics generally have decreased serum levels of zinc, calcium, and magnesium due to poor dietary intake and increased urinary losses. Iron-deficiency anemia is very rare unless there is gastrointestinal bleeding or chronic infection. In fact, excess iron is a more common problem with alcoholics. Many alcoholic beverages contain relatively high iron levels, and alcohol may enhance iron absorption.

Source: Lieber, C. S. Alcohol, liver and nutrition. *J. Am. College Nutr.* 10:602, 1991. Markowitz, J. S., McRae, A. L., and Sonne, S. C. Oral nutritional supplementation for the alcoholic patient: A brief overview. *Ann. Clin. Psychiatry* 12:153, 2000.

Riboflavin Forms the Coenzymes FAD and FMN

Riboflavin is the precursor of flavin adenine dinucleotide (FAD) and flavin mononucleotide (FMN), both of which are coenzymes in a wide variety of redox reactions essential for energy production and cellular respiration. Riboflavin is also required for iron mobilization, and riboflavin deficiency can contribute to anemia when iron intake is low. Characteristic symptoms of riboflavin deficiency are angular cheilitis, glossitis, and scaly dermatitis (especially around the nasolabial folds and scrotal areas). The best enzyme for assaying riboflavin status is erythrocyte glutathione reductase. Foods rich in riboflavin include milk, meat, eggs, and cereal products. **Riboflavin deficiencies** are quite rare in this country and are usually seen in chronic alcoholics. Hypothyroidism slows the conversion of riboflavin to FMN and FAD, but it is not known whether this affects riboflavin requirements.

Niacin Forms the Coenzymes NAD and NADP

Niacin is not a vitamin in the strictest sense of the word, since niacin can be synthesized from tryptophan (Figure 28.10). However, conversion of tryptophan to niacin is relatively inefficient (60 mg of tryptophan is required for the production of 1 mg of niacin) and occurs only after all of the body requirements for tryptophan (protein synthesis, production of serotonin and melatonin, and energy production) have been met. Since synthesis of niacin requires pyridoxine, riboflavin, and iron (Figure 28.10), it is also very inefficient on a marginal diet. Dietary niacin (nicotinic acid) and

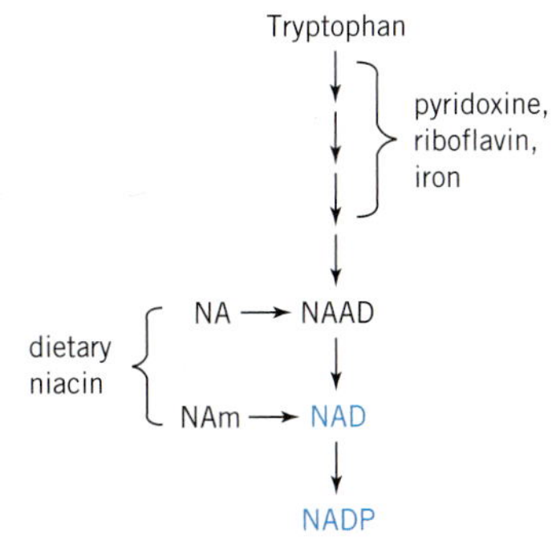

FIGURE 28.10

Formation of NAD and NADP. Pathways for formation of NAD and NADP. NA, nicotinic acid; NAm, nicotinamide; NAAD, nicotinic acid adenine dinucleotide

niacinamide (nicotinamide) are both converted to the ubiquitous oxidation–reduction coenzymes NAD and NADP (Figure 28.10). These coenzymes are electron acceptors or hydrogen donors in many redox reactions and cellular respiration. NAD is also required for the poly-ADP-ribose polymerase reaction, which is part of the cellular DNA damage recognition system and regulates DNA replication, DNA repair, and cell cycle progression.

Borderline **niacin deficiency** results in a glossitis (redness) of the tongue, somewhat similar to riboflavin deficiency. Pronounced deficiency leads to **pellagra**, which is characterized by the three Ds: **dermatitis, diarrhea, and dementia**. The dermatitis is usually seen only in skin areas exposed to sunlight and is symmetric. The neurologic symptoms are associated with actual degeneration of nervous tissue. Because of food fortification, pellagra is a medical curiosity in the developed world, being primarily seen in **alcoholics**, patients with severe **malabsorption**, and **elderly** on very restricted diets. Pregnancy, lactation, and chronic illness lead to increased needs for niacin, but a varied diet usually provides sufficient amounts. The richest sources of niacin are meats, peanuts and other legumes, and enriched cereals.

Nicotinic acid in pharmacologic doses ($1.4–4\ g\,d^{-1}$) is used to lower LDL cholesterol and triglycerides and increase HDL cholesterol levels. Side effects include flushing of the skin, hyperuricemia, and elevation of liver enzymes. The flushing of the skin can be avoided by using niacinamide or slow-release preparations of nicotinic acid, but careful monitoring of the patient for hepatic alterations is still required.

Pyridoxine (Vitamin B_6) Forms the Coenzyme Pyridoxal Phosphate

Pyridoxine, pyridoxamine, and **pyridoxal**, are naturally occurring forms of vitamin B_6 (see Figure 28.11). They are efficiently converted to **pyridoxal phosphate**, which is required for the synthesis, catabolism, and interconversion of amino acids. Some key pyridoxal phosphate-dependent reactions are shown in Figure 28.12. Pyridoxal phosphate is essential for energy production from amino acids (see p. 747) and can be considered an energy-releasing vitamin. Some symptoms of severe deficiency are similar to those of the other energy-releasing vitamins. It is also required for the synthesis of the neurotransmitters serotonin, norepinephrine, epinephrine, and γ-aminobutyrate (GABA) (Figure 28.12) and for sphingolipids necessary for myelin formation.

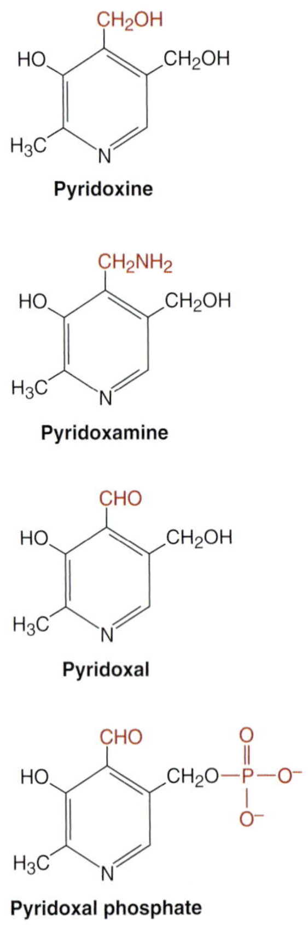

This may explain the irritability, nervousness, and depression seen with mild deficiency and the peripheral neuropathy and convulsions observed with severe deficiency. It is required for synthesis of δ-aminolevulinic acid (Figure 28.12), a precursor of heme, and *B_6* **deficiencies** occasionally cause sideroblastic microcytic anemia. Pyridoxal phosphate is covalently linked to a lysine residue of glycogen phosphorylase, which stabilizes the enzyme. This may explain the decreased glucose tolerance associated with deficiency, although B_6 appears to have some direct effects on the glucocorticoid receptor as well. Pyridoxal phosphate is also required for the conversion of homocysteine to cysteine. This is important because **hyperhomocysteinemia** appears to be a risk factor for cardiovascular disease. Finally, it is required for conversion of tryptophan to NAD. While this may not be directly related to the symptoms of B_6 deficiency, a tryptophan load test (Figure 28.12) is a sensitive indicator of vitamin B_6 status (see Clin. Corr. 28.6).

The dietary requirement for B_6 is roughly proportional to the protein content of the diet. The requirement is increased during pregnancy and lactation, and it may increase with age. Vitamin B_6 is widespread in foods, but meat, vegetables, whole-grain cereals, and egg yolks are among the richest sources. It had been assumed that the average American diet was adequate in B_6; therefore, it was not routinely added to flour and other fortified foods. However, recent nutritional surveys have found that a significant fraction of the U.S. population consumes less than the EAR for B_6.

Pyridoxine

Pyridoxamine

Pyridoxal

Pyridoxal phosphate

FIGURE 28.11
Structures of vitamin B_6.

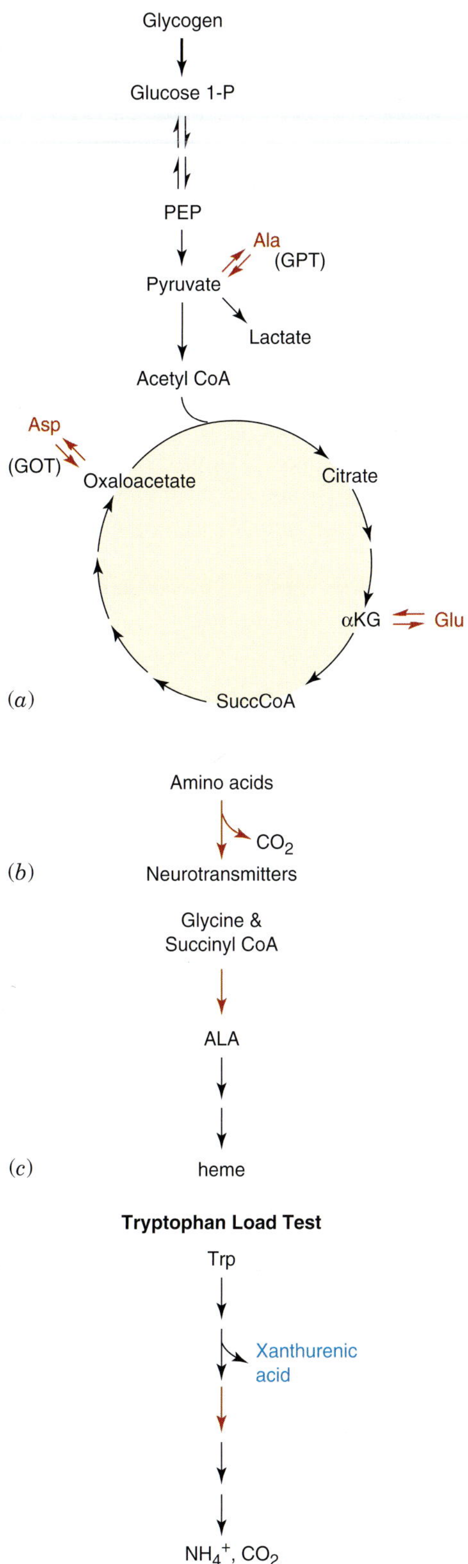

FIGURE 28.12

Some important metabolic roles of pyridoxal phosphate. Reactions requiring pyridoxal phosphate are indicated by red arrows. ALA, δ-aminolevulinic acid; αKG, α- ketoglutarate; GPT, glutamate-pyruvate aminotransferase; GOT, glutamate-oxaloacetate aminotransferase.

CLINICAL CORRELATION 28.6
Vitamin B_6 Requirements and Oral Contraceptives

The controversy over B_6 requirements for users of oral contraceptives best illustrates the potential problems associated with biochemical assays. For years, one of the most common assays for vitamin B_6 status had been the tryptophan load assay (see Figure 28.12). It is based on the observation that when the levels of tissue pyridoxal phosphate are low, the normal catabolism of tryptophan is impaired and most of the tryptophan is catabolized by a minor pathway, leading to synthesis of xanthurenic acid. Under many conditions, the amount of xanthurenic acid recovered in a 24-h urine sample following ingestion of a fixed amount of tryptophan is a valid indicator of vitamin B_6 status. When the tryptophan load test was used to assess the vitamin B_6 status of users of oral contraceptives, however, alarming reports started appearing in the literature. Not only did oral contraceptive use increase the excretion of xanthurenic acid considerably, but the amount of pyridoxine needed to return xanthurenic acid excretion to normal was 10 times the RDA and almost 20 times that required to maintain normal B_6 status in control groups. This observation received much popular attention even though most classical symptoms of vitamin B_6 deficiency were not observed in oral contraceptive users.

Studies using other measures of vitamin B_6 have painted a different picture. For example, erythrocyte glutamate-pyruvate aminotransferase and erythrocyte glutamate-oxaloacetate aminotransferase are pyridoxal phosphate-containing enzymes. One can also assess vitamin B_6 status from the endogenous activity of these enzymes and the degree of stimulation by added pyridoxal phosphate. These assays show a much smaller difference between nonusers and users of oral contraceptives. The minimum level of pyridoxine needed to maintain normal vitamin B_6 status as measured by these assays was only 2.0 mg day^{-1}, which is 10-fold less than indicated by the tryptophan load test.

Why the large discrepancy? For one thing, enzyme activity can be affected by hormones as well as vitamin cofactors. Kynureninase, the key pyridoxal phosphate-containing enzyme of the tryptophan catabolic pathway, is regulated both by pyridoxal phosphate availability and by estrogen metabolites. Even with normal vitamin B_6 status, most of the enzyme exists in the inactive apoenzyme form. However, this does not affect tryptophan metabolism because tryptophan oxygenase, the first enzyme of the pathway, is rate-limiting. Thus, the small amount of active holoenzyme is sufficient to handle the metabolites produced by the first part of the pathway. However, kynureninase is inhibited by estrogen metabolites. Thus, with oral contraceptive use, its activity is reduced to a level where it becomes rate-limiting and excess tryptophan metabolites are shunted to xanthurenic acid. Higher-than-normal levels of vitamin B_6 overcome this problem by converting more apoenzyme to holoenzyme. Since the estrogen was having a specific effect on the enzyme used to measure vitamin B_6 states in this assay, it did not necessarily mean that pyridoxine requirements were altered for other metabolic processes in the body.

Does this mean that vitamin B_6 status is of no concern to users of oral contraceptives? Oral contraceptives do appear to increase vitamin B_6 requirements slightly. Several dietary surveys have shown that a significant percentage of women in the 18- to 24-year age group consume diets containing less than the recommended 1.3 mg of pyridoxine day^{-1}. If these women are also using oral contraceptives, they are at increased risk for developing a borderline deficiency. Thus, while the tryptophan load test was clearly misleading in a quantitative sense, it did alert the medical community to a previously unsuspected nutritional risk.

Source: Bender, D. A. Oestrogens and vitamin B_6: Actions and interactions. *World Rev. Nutr. Diet.* 51:140, 1987. Kirksey, A., Keaton, K., Abernathy, R. P., and Grager, J. L. Vitamin B_6 nutritional status of a group of female adolescents. *Am. J. Clin. Nutr.* 31:946, 1978.

Pantothenic Acid and Biotin Form Coenzymes Involved in Energy Metabolism

Pantothenic acid is a component of **coenzyme A** (CoA) and the phosphopantetheine moiety of fatty acid synthase and is required for metabolism of fat, protein, and carbohydrate via the citric acid cycle and for fatty acid and cholesterol synthesis. More than 70 enzymes have been described to date that utilize CoA or its derivatives. Therefore, one would expect pantothenic acid deficiency to be a serious concern in humans. This is not the case since: (1) pantothenic acid is very widespread in natural foods, probably reflecting its widespread metabolic role, and (2) when pantothenic acid deficiency does occur, it is usually associated with multiple nutrient deficiencies, thus making it difficult to discern symptoms specific to the pantothenic acid deficiency.

Biotin is the covalently bound to an ϵ-amino group of a lysine residue in pyruvate carboxylase, acetyl-CoA carboxylase, propionyl-CoA carboxylase, and β-methylcrotonyl-CoA carboxylase. Biotin occurs in peanuts, chocolate, and eggs and is synthesized by intestinal bacteria. However, the biotin synthesized by intestinal bacteria may not be present in a location or a form that can contribute significantly to absorbed biotin.

28.7 HEMATOPOIETIC WATER-SOLUBLE VITAMINS

Folic Acid Functions as Tetrahydrofolate in One-Carbon Metabolism

The simplest form of **folic acid** is pteroylmonoglutamic acid. Folic acid usually occurs as **polyglutamate** derivatives with two to seven glutamic acid residues joined in γ-peptide linkages (Figure 28.13). These compounds are taken up by intestinal mucosal cells and the extra glutamate residues removed by glutamate **conjugase**. The free folic acid is then reduced to **tetrahydrofolate** by **dihydrofolate reductase**, and circulates in the plasma primarily as the N^5-methyl derivative of tetrahydrofolate (Figure 28.13). Inside cells tetrahydrofolates are present primarily as polyglutamates, which appear to be the biologically active form. Tetrahydrofolate polyglutamate is also stored in liver.

Various one-carbon tetrahydrofolate derivatives are used in the synthesis of choline, serine, glycine, purines, and dTMP (see Figure 28.14). Tetrahydrofolate and vitamin B_{12} are also required for the conversion of homocysteine to methionine. In addition to its role in protein synthesis, methionine is converted to S-adenosylmethionine, which is used in many methylation reactions including DNA methylation.

Folate deficiency inhibits DNA synthesis by decreasing the availability of purines and dTMP. This leads to arrest of cells in S phase, a characteristic "megaloblastic" change in size and shape of the nuclei and a slower maturation of red blood cells, causing

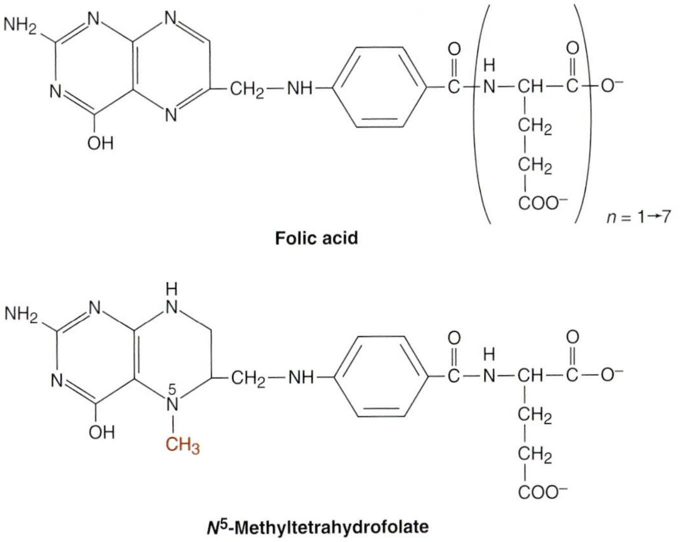

Folic acid

N^5-Methyltetrahydrofolate

FIGURE 28.13

Structure of folic acid and N^5-methyltetrahydrofolate.

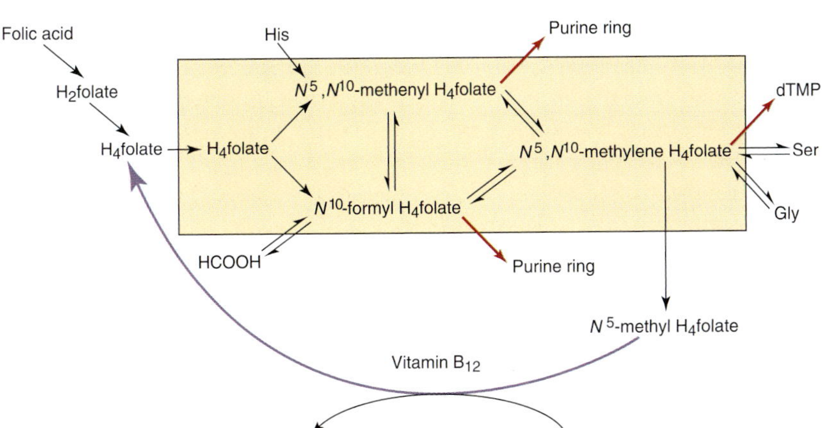

FIGURE 28.14

Metabolic roles of folic acid and vitamin B_{12} in one-carbon metabolism. Metabolic interconversions of folic acid and its derivatives are indicated with black arrows. Pathways relying exclusively on folate are shown with red arrows. The important B_{12}-dependent reaction converting N^5-methyl tetrahydrofolate (H_4folate) back to H_4folate is shown with a blue arrow. The box encloses the "pool" of one carbon derivatives of H_4folate.

production of abnormally large "macrocytic" red blood cells with fragile membranes. Thus a **macrocytic anemia** associated with megaloblastic changes in the bone marrow is characteristic of folate deficiency. In addition, **hyperhomocysteinemia** is fairly common in the elderly population and may be due to inadequate intake and/or decreased utilization of folate, vitamin B_6 and vitamin B_{12}. Hyperhomocysteinemia appears to be a risk factor for cardiovascular disease and usually responds to supplementation with folic acid, vitamin B_6, and vitamin B_{12}. Finally, folate deficiency appears to be associated with several forms of cancer, especially colon and cervical cancer.

Folate deficiency is caused by inadequate intake, increased need, impaired absorption, increased demand, and impaired metabolism. Some dietary surveys suggest that inadequate intake may be more common than previously supposed. As with most other vitamins, inadequate intake is probably not sufficient to trigger symptoms of folate deficiency in the absence of increased requirements or decreased utilization. For example, gene polymorphisms that increase the need for folate may be common (see Clin. Corr. 28.7). Increased need also occurs during **pregnancy** and **lactation**. By the third trimester the folic acid requirement has almost doubled. In the United States, 20–25% of otherwise normal pregnant women have low serum folate, but actual megaloblastic anemia is rare and is usually seen only after multiple pregnancies. However, inadequate folate levels during the early stages of pregnancy increase the risk for **neural tube defects**, a type of birth defect. Normal diets seldom supply the 600 mg of folate needed during pregnancy. Thus, enriched grain products are now fortified with folic acid at a concentration of 1.4 mg per gram of product. This level of fortification was designed to increase average intake of folic acid by 100 mg day $^{-1}$. Since fortification, the incidence of neural tube defects has decreased by 19%. Addition of an extra 200 mg day $^{-1}$ would offer much greater protection against neural tube defects and hyperhomocysteinemia, but this level of folic acid supplementation could mask the symptoms of vitamin B_{12} deficiency as described below. Thus, most physicians routinely recommend supplementation for women during the child-bearing years and for the elderly. Folate

CLINICAL CORRELATION 28.7
Gene Polymorphisms and Folic Acid Requirement

Folic acid supplementation lowers the risk of neural tube defects and decreases serum homocysteine levels, which may lower the risk of heart disease. These data lead to an increase in the RDA for folic acid and to fortification of grain products with folic acid. Yet even on a marginal diet, not all adults have elevated homocysteine levels and not all mothers give birth to babies with neural tube defects. What determines these individual responses to inadequate folate intake? There is a common genetic polymorphism in the gene for 5,10-methylenetetrahydrofolate reductase (MTHFR) that produces the 5-methyltetrahydrofolate required for the conversion of homocysteine to methionine (see Figure 28.15). A C $\rightarrow$ T substitution at bp 677 results in a substitution of valine for alanine that lowers specific activity and reduces stability of the enzyme. Approximately 12% of the Caucasians and Asians are homozygous (T/T) and 50% are heterozygous (C/T) for this polymorphism. Plasma folate concentrations are significantly lower and plasma homocysteine levels are significantly higher in T/T individuals consuming diets low in folate. When coupled with low folate intake, the T/T genotype may account for 15% of neural tube defects. In addition, older individuals with the T/T genotype and low folate intake appear to be at increased risk of colon cancer.

An active investigation of genetic polymorphisms in the other genes involved in folate metabolism is underway. Polymorphisms have been described in methionine synthetase and folate receptor-alpha, which is required for 5-methyltetrahydrofolate uptake. Both appear to be benign. However, absorption of folate by the intestine may be lower in mothers with a history of neural tube defect pregnancies than in control mothers. The genetics of this effect has not yet been determined.

Source: Bailey, L. B. and Gregory, J. F. Polymorphisms of methylenetetrahydrofolate reductase and other enzymes: Metabolic significance, risks, and impact on folate requirement. *J. Nutr.* 129:919, 1999. Barber, R. C., Lammer, E. J., Shaw, G.M., Greer, K. A., and Finnell, R. H. The role of folate transport and metabolism in neural tube defect risk. *Mol. Genet. Metab.* 66:1, 1999. Fang, J. Y. and Xiao, S. D. Folic acid, polymorphism of methyl-group metabolism genes, and DNA methylation in relation to GI carcinogenesis. *J. Gastroenterol.* 38:821, 2003.

deficiency is also common in alcoholics (see Clin. Corr. 28.5) and with malabsorption diseases.

Anticonvulsants and **oral contraceptives** may interfere with folate absorption and anticonvulsants appear to increase catabolism of folates (see Clin. Corr. 28.4). Long-term use of these drugs can lead to folate deficiency unless adequate supplementation is provided. For example, 20% of patients using oral contraceptives develop megaloblastic changes in the cervicovaginal epithelium, and 20–30% show low serum folate.

Vitamin B_{12} (Cobalamin) Contains Cobalt in a Tetrapyrrole Ring

Pernicious anemia, a megaloblastic anemia associated with neurological deterioration caused by progressive demyelination of nervous tissue, was invariably fatal until 1926 when liver extracts were shown to be curative. Subsequent work showed the need for both an extrinsic factor present in liver and an **intrinsic factor** produced by the body; **vitamin B_{12}** was the extrinsic factor. Vitamin B_{12} contains **cobalt** in a coordination state of six, coordinated in four positions by a tetrapyrrole (corrin) ring, in one position by a benzimidazole nitrogen, and in the sixth position by one of several different ligands (Figure 28.15). The crystalline forms of B_{12} used in supplementation are usually hydroxycobalamin or cyanocobalamin. B_{12} in foods usually occurs bound to protein in the methyl or 5′-deoxyadenosyl form. To be utilized, the B_{12} must be released from the protein by acid hydrolysis in the stomach or trypsin digestion in the intestine. It then combines with intrinsic factor, a protein secreted by the stomach, which carries it to the ileum for absorption.

Vitamin B_{12} participates in only two chemical reactions in humans (Figure 28.16). The methyl derivative of B_{12} is required for methionine synthase, in which homocysteine is methylated to methionine. The 5-deoxyadenosyl derivative is required for methylmalonyl-CoA mutase, which converts methylmalonyl CoA to succinyl CoA, a key reaction in catabolism of valine, isoleucine, methionine, threonine, odd-chain fatty acids, thymine, and the side chain of cholesterol. As might be expected, *B_{12}* **deficiency** causes the accumulation of both homocysteine and methylmalonic acid.

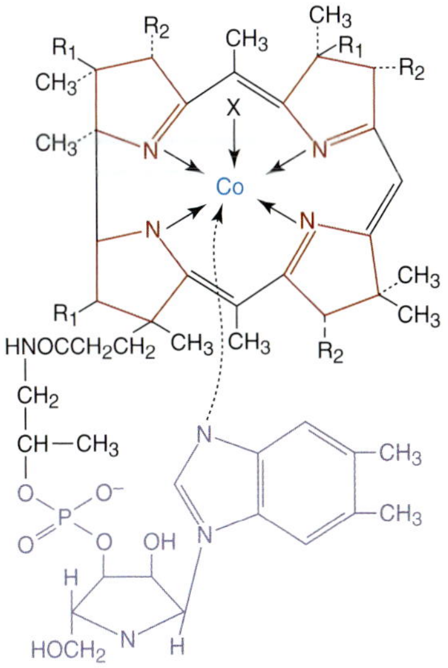

FIGURE 28.15
Structure of vitamin B_{12} (cobalamine).

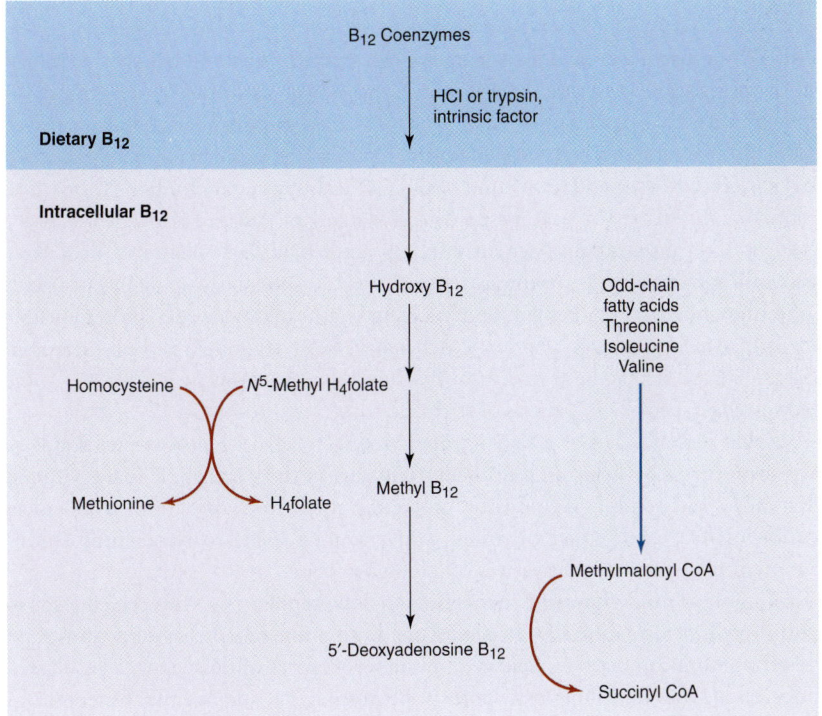

FIGURE 28.16
Metabolism of vitamin B_{12}. Metabolic interconversions of B_{12} are indicated by light arrows, and B_{12} requiring reactions are indicated by red arrows. Other related pathways are indicated by blue arrows.

The megaloblastic anemia associated with B_{12} deficiency is thought to reflect the effect of B_{12} on folate metabolism. B_{12}-dependent methionine synthesis (homocysteine + N^5-methyl THF → methionine + THF) is the only pathway by which N^5-methyltetrahydrofolate can return to the tetrahydrofolate pool (Figure 28.16). Thus, in B_{12} deficiency essentially all of the folate becomes "trapped" as the N^5-methyl derivative, causing a buildup of N^5-methyltetrahydrofolate and a deficiency of the tetrahydrofolate derivatives needed for purine and dTMP biosynthesis. By replenishing the tetrahydrofolate pool, large amounts of supplemental folate can overcome the megaloblastic anemia, but not the neurological problems. This is the crux of the current debate on the optimal levels for folate fortification of foods. It is the megaloblastic anemia that usually brings the patient into the doctor's office. Thus, by masking the anemia, routine fortification of foods with high levels of folate could prevent detection of B_{12} deficiency until the neurological damage had become irreversible.

It has been proposed that the demyelination associated with B_{12} deficiency is caused by methylmalonyl-CoA accumulation in two ways: (1) Methylmalonyl CoA is a competitive inhibitor of malonyl CoA in fatty acid biosynthesis. Because the myelin sheath is continually turning over, any severe inhibition of fatty acid biosynthesis will lead to its degeneration. (2) Methylmalonyl CoA can substitute for malonyl CoA in fatty acid synthesis leading to synthesis of branched-chain fatty acids, which may disrupt membrane structure. However, the neurologic symptoms of B_{12} deficiency cannot be fully explained by either mechanism, since accumulation of both methylmalonic acid and homocysteine is required for demyelination.

Vitamin B_{12} is widespread in foods of animal origin, especially meats. Since the liver stores up to a 6-year supply of vitamin B_{12}, deficiencies of B_{12} are rare except in older people who produce insufficient intrinsic factor and/or HCl in the stomach, patients with severe malabsorption diseases and long-term vegetarians.

28.8 OTHER WATER-SOLUBLE VITAMINS

Ascorbic Acid Functions in Reduction and Hydroxylation Reactions

Vitamin C or **ascorbic acid** is a cofactor for mixed function oxidases involved in the hydroxylation of lysine and proline, synthesis of carnitine, and synthesis of norepinephrine. Hydroxylation of lysine and proline is required for protocollagen to cross-link properly into normal collagen fibrils. Vitamin C is required for maintenance of normal connective tissue and for wound healing. It is also necessary for bone formation, since the organic matrix of bone tissue consists largely of collagen. Collagen is also a component of the ground substance surrounding capillary walls, so vitamin C deficiency causes **capillary fragility**. Carnitine is required for transport of long-chain fatty acids into the mitochondria (see below), and decreased carnitine levels may be responsible for the fatigue associated with vitamin C deficiency. Since vitamin C is concentrated in the adrenal gland, it may be required for hydroxylation reactions in synthesis of some corticosteroids, especially in periods of stress.

Ascorbic acid also acts as a nonenzymatic reducing agent. For example, it aids in **absorption of iron** by reducing it to the ferrous state in the stomach. It spares vitamin A, vitamin E, and some B vitamins by protecting them from oxidation. It enhances utilization of folic acid, either by aiding conversion of folate to tetrahydrofolate or formation of polyglutamate derivatives of tetrahydrofolate.

Symptoms of mild **vitamin C deficiency** include capillary fragility, which leads to easy bruising and formation of petechiae (small, pinpoint hemorrhages in skin), and decreased immunocompetence. **Scurvy**, a more severe form of deficiency, is associated with decreased wound healing, osteoporosis, hemorrhaging, and anemia. Osteoporosis results from the inability to maintain the collagenous organic matrix of the bone, which leads to demineralization. Anemia results from extensive hemorrhaging coupled with defects in iron absorption and folate metabolism.

Vitamin C is readily absorbed so that deficiency invariably results from poor diet and/or increased need. In severe stress or trauma there is a rapid drop in serum vitamin C levels, and most of the body's supply of vitamin C is mobilized to the adrenals and/or the traumatized area. It is not clear whether this represents an increased demand for vitamin C, or merely a normal redistribution to those areas where it is needed most. Nor is it clear whether the lowered serum levels of vitamin C impair its functions in other tissues in the body. The current consensus seems to be that the lowered serum vitamin C levels indicate an increased demand, but there is little agreement as to how much.

Smoking causes lower serum levels of vitamin C. In fact, the RDAs for smokers are 110–125 mg of vitamin C day^{-1} versus 75–90 mg day^{-1} for nonsmoking adults. **Aspirin** appears to block uptake of vitamin C by white blood cells. **Oral contraceptives** and **corticosteroids** also lower serum levels of vitamin C. The possibility of marginal vitamin C deficiency should be considered with any patient using these drugs over a long period, especially if dietary intake is suboptimal.

The use of megadoses of vitamin C to prevent and cure the **common cold** has generated considerable controversy. While vitamin C supplementation does not appear to prevent the common cold, it may moderate the symptoms or shorten the duration. It has been suggested that vitamin C is required for normal leukocyte function or that it decreases histamine levels. While megadoses of vitamin C are probably no more harmful than the widely used over-the-counter cold medications, some potential side effects of high vitamin C intake should be considered. For example, oxalate is a major metabolite of ascorbic acid. Thus, high ascorbate intakes could theoretically lead to the formation of oxalate kidney stones in predisposed individuals. However, most studies have shown that excess vitamin C is primarily excreted as ascorbate rather than oxalate. Pregnant mothers taking megadoses of vitamin C may give birth to infants with abnormally high vitamin C requirements, but this is easily treated. The UL for vitamin C has been set at 2000 mg day^{-1} because higher levels can cause diarrhea in some individuals.

Choline and Carnitine Perform Several Functions

Choline and carnitine have traditionally been considered as nonessential because they can be synthesized *de novo*. However, choline has recently been reclassified as essential and carnitine as conditionally essential. Choline is required for synthesis and release of acetylcholine, an important **neurotransmitter** involved in memory storage, motor control, and other functions. It is also a precursor for synthesis of the phospholipids phosphatidylcholine (lecithin) and sphingomyelin, which are important for membrane function, intracellular signaling, and hepatic export of very-low-density lipoproteins. Phosphatidylcholine is also important in removal of cholesterol from tissues, because it is a substrate for lecithin–cholesterol acyltransferase in reverse cholesterol transport (see p. 715). Finally, choline is a precursor for the methyl donor betaine.

Studies in rodents indicate that choline deficiency increases the risk of liver cancer and memory deficits in aged animals, but these effects have not been demonstrated in humans. Both choline and betaine supplementation appear to lower serum homocysteine levels in humans. However, current data are insufficient to draw firm conclusions about whether choline and/or betaine supplementation have any effects on cardiovascular risk.

Because choline can be synthesized *de novo* and is abundant in food, **choline deficiency** is very rare. Hepatic complications (fatty livers and elevated serum alanine aminotransferase) that respond to choline supplementation have been observed in patients on total parenteral nutrition solutions devoid of choline, with small intestinal bypass, and with liver cirrhosis. Carnitine is required for transport of fatty acids across the mitochondrial membrane, so it is essential for normal fatty acid metabolism. Because carnitine can be synthesized *de novo*, it is nonessential for normal healthy adults. However, it is considered conditionally essential overall because human genetic disorders of carnitine metabolism have been described and some of them respond to carnitine supplementation. Carnitine is a popular dietary supplement for athletes, but there is little evidence that carnitine supplementation influences athletic performance.

28.9 MACROMINERALS

Calcium Has Many Physiological Roles

Calcium is the most abundant mineral in the body. Most is in bone, but the small amount of Ca^{2+} outside of bone functions in a variety of essential processes. It is required for many enzymes, mediates some hormonal responses, and is essential for blood coagulation, muscle contractility, and normal neuromuscular irritability. In fact, only a relatively narrow range of serum Ca^{2+} levels is compatible with life. Since maintenance of constant serum levels is so vital, an elaborate homeostatic control system has evolved (see Figure 28.5). Low serum Ca^{2+} stimulates formation of 1,25-dihydroxycholecalciferol, which enhances intestinal Ca^{2+} absorption and, with parathyroid hormone, stimulates bone resorption. Thus, long-term dietary Ca^{2+} insufficiency almost always results in net loss of calcium from the bones.

Dietary Ca^{2+} requirements, however, vary considerably from individual to individual due to the existence of other factors that affect availability of Ca^{2+}. For example, vitamin D is required for optimal utilization of calcium, while excess dietary protein may cause more rapid excretion of Ca^{2+}. Exercise facilitates calcium utilization for bone formation. Calcium balance studies carried out on Peruvian Indians, who have extensive exposure to sunlight, get extensive exercise, and subsist on low-protein vegetarian diets, indicate a need for only 300–400 mg Ca^{2+} day^{-1}. However, similar studies carried out in this country consistently show higher requirements, and the RDA has been set at 1000–1300 mg day^{-1}.

Symptoms of **Ca^{2+} deficiency** resemble those of vitamin D deficiency, but other symptoms such as muscle cramps are possible with marginal deficiencies. A significant portion of low-income children and adult females in this country do not consume adequate Ca^{2+}. This is of particular concern because these are the population groups with particularly high needs for Ca^{2+}. For this reason, the U.S. Congress has established the WIC (Women and Infant Children) program to ensure adequate protein, Ca^{2+}, and iron for indigent families with pregnant/lactating mothers or young infants.

Dietary surveys show that 34–47% of the population over 60 years of age consumes less than the EAR for Ca^{2+}. This is the group most at risk of developing osteoporosis, characterized by loss of bone organic matrix and progressive demineralization. Causes of osteoporosis are multifactorial and largely unknown, but it is likely that part of the problem has to do with Ca^{2+} metabolism (see Clin. Corr. 28.8). Recent studies suggest that inadequate intake of Ca^{2+} may result in elevated blood pressure. This is of great concern because most low sodium diets (which are recommended for patients with high blood pressure) severely limit dairy products, the main source of Ca^{2+} for Americans.

Magnesium Is Required by Many Enzymes

Magnesium is required for many enzyme activities, particularly those utilizing an ATP-Mg^{2+} complex, and for neuromuscular transmission. Mg^{2+} content is significantly reduced during processing of food items, and recent dietary surveys have shown that the average Mg^{2+} intake in western countries is often below the EAR. Deficiency occurs in alcoholism, with use of certain diuretics and in metabolic acidosis. The main symptoms of **Mg^{2+} deficiency** are weakness, tremors, and cardiac arrhythmia. Supplemental Mg^{2+} may help prevent formation of calcium oxalate stones in the kidney. Mg^{2+} supplementation has also been shown to lower blood pressure in several clinical studies, and there is an inverse effect between dietary Mg^{2+} intake and the risk of stroke.

CLINICAL CORRELATION 28.8
Diet and Osteoporosis

There is strong consensus that the years from age 10 to 35, when bone density is reaching its maximum, are the most important for reducing the risk of osteoporosis. The maximum bone density obtained during these years depends on both calcium intake, and exercise and dense bones are less likely to become seriously depleted of calcium following menopause. Unfortunately, most American women consume far too little calcium during these years. The RDA for calcium is 1300 mg day^{-1} (4 or more glasses of milk day^{-1}) for women from age 11 to 18, 1000 mg day^{-1} (3 or more glasses of milk day^{-1}) for women from age 19 to 50, and 1200 mg day^{-1} (4 glasses of milk day^{-1}) for women over 50. Some experts think that calcium requirements for postmenopausal women should be even higher. In 1994, an NIH consensus panel on osteoporosis recommended that postmenopausal women consume up to 1500 mg of calcium day^{-1}. Unfortunately, the median calcium intake for women 19 and older is only about 500 mg day^{-1}, and with the recent concern about the fat content of dairy products, calcium intakes appear to be decreasing rather than increasing. Thus, it is clear that increased calcium intake should be encouraged in this group. Even with drug or estrogen replacement therapy to prevent osteoporosis, calcium intake should not be ignored. Recent studies have shown that calcium intakes in the range of 1000–1500 mg day^{-1} make drug or estrogen therapy more effective at preserving bone mass.

While most of the focus is on calcium intake, we need to remember that bones are not made of calcium alone. If the diet is deficient in other nutrients, the utilization of calcium for bone formation will be impaired. Vitamin C is needed to form the bone matrix and magnesium and phosphorus are important components of bone structure. Vitamin K and a variety of trace minerals, including copper, zinc, manganese, and boron, are important for bone formation. Thus, calcium supplements may not be optimally utilized if the overall diet is inadequate. Vitamin D is required for absorption and utilization of calcium. It deserves special mention as it may be a problem for the elderly (see Clin. Corr. 28.9), and some experts feel that the current recommendation for vitamin D intake in adults may be too low. Finally, an adequate exercise program is just as important as drug therapy and an adequate diet for preventing the loss of bone density.

Source: Heaney, R. P. Calcium in the prevention and treatment of osteoporosis. *J. Intern. Med.* 231:169, 1992. National Institutes of Health. *Optimal Calcium Intake*. NIH Consensus Statement, 12 (Nov. 4), 1994. Murphy, N. M. and Carroll, P. The effect of physical activity and its interaction with nutrition on bone health. *Proc. Nutr. Soc.* 62:829, 2003. Vieth, R. Why the optimal requirement for vitamin D3 is probably much higher than what is officially recommended for adults. *J. Steroid Biochem. Mol. Biol.* 89–90:575, 2004.

28.10 TRACE MINERALS

Iron Deficiency Causes Anemia and Decreased Immunocompetence

Iron metabolism has been discussed in Chapter 21. As a component of hemoglobin and myoglobin, iron is required for O_2 and CO_2 transport; in cytochromes and nonheme iron proteins, it is required for oxidative phosphorylation; and in the lysosomal enzyme myeloperoxidase, it is required for phagocytosis and killing of bacteria by neutrophils.

Assuming a 10–15% efficiency of absorption, an RDA of 8 mg day^{-1} for normal adult males and 18 mg day^{-1} for menstruating females has been set. For pregnant females it is 27 mg day^{-1}. While 8 mg day^{-1} of iron can easily be obtained from a normal diet, 18 mg is marginal at best and 27 mg can almost never be obtained. The best dietary sources are meats, dried legumes, dried fruits, and enriched cereal products.

The best-known symptom of **iron deficiency** is a microcytic hypochromic anemia. Iron deficiency is also associated with decreased immunocompetence. Dietary surveys indicate that 95% or more of children and menstruating females do not obtain adequate dietary iron. Biochemical measurements reveal a 10–25% incidence of iron deficiency anemia in this same group. **Iron-deficiency anemia** is also a problem with the elderly due to poor dietary intake and increased frequency of achlorhydria.

Because iron-deficiency anemia is widespread, government programs of nutritional intervention such as the WIC program have emphasized iron-rich foods. However, since recent studies have suggested that excess iron intake may increase the risk of cardiovascular disease, iron supplementation and the consumption of iron-fortified foods may be inappropriate for adult men and postmenopausal women. Excess iron can lead to the rare condition **hemochromatosis**, in which iron deposits are found in abnormally high levels in many tissues and cause liver, pancreatic, and cardiac dysfunction as well as pigmentation of the skin. Hemochromatosis is also occasionally seen in hemolytic anemias and liver disease.

Iodine Is Incorporated into Thyroid Hormones

Dietary **iodine** is efficiently absorbed and transported to the **thyroid** gland, where it is stored and used for synthesis of triiodothyronine and thyroxine. These hormones function in regulating the **basal metabolic rate** of adults and the growth and development of children. Saltwater fish are the best natural food sources of iodine; in the past, population groups living in inland areas suffered from the endemic deficiency disease **goiter**, an enlargement (sometimes massive) of the thyroid gland. Since iodine has been routinely added to table salt, goiter has become relatively rare. However, in some inland areas, mild forms of goiter still occur in up to 5% of the population.

Zinc Is Required by Many Proteins

Zinc is part of the catalytic center of over 300 metalloenzymes, including RNA and DNA polymerases, alkaline phosphatase, and carbonic anhydrase. In addition, it forms zinc fingers (Zn^{2+} coordinated to four amino acid side chains), which provide structural stability to another 300–700 proteins. Zinc fingers facilitate binding of proteins to DNA and are common motifs in transcription factors and nuclear hormone receptors. They are also important for protein–protein interactions and are found in many signal transduction proteins. **Zinc deficiency** in children is usually marked by poor growth and impairment of sexual development. In both children and adults, zinc deficiency results in poor wound healing and dermatitis. Zinc is present in gustin, a salivary polypeptide that appears to be necessary for normal development of taste buds, so zinc deficiency leads to decreased taste acuity. Zinc is required for cytokine production by monocytes and T cells. Thus, zinc deficiency is associated with impaired immune function. Zinc is required for the activity of porphobilinogen synthase. In lead poisoning, lead replaces the zinc, which leads to anemia and accumulation of δ-aminolevulinic acid (see p. 837).

Dietary surveys indicate that zinc intake may be marginal for many individuals, and zinc supplementation has been shown to improve immune status in the elderly. Severe zinc deficiency is seen primarily in alcoholics (especially if they have cirrhosis), patients with chronic renal disease or severe malabsorption diseases, and occasionally in people after long-term parenteral nutrition (TPN). The most characteristic early symptom of zinc deficient patients on TPN is dermatitis. Zinc is occasionally used therapeutically to promote wound healing and may be of some use in treating gastric ulcers.

Copper Is a Cofactor for Important Enzymes

Important **copper**-containing enzymes include ceruloplasmin (oxidizes iron to facilitate its binding to transferrin), cytochrome C oxidase (electron transport), dopamine β-hydroxylase (norepinephrine synthesis), lysyl oxidase (collagen cross-linking), superoxide dismutase (disproportionation of superoxide) and C_{18},Δ^9-desaturase (addition of double bonds to long-chain fatty acids). The C_{18},Δ^9-desaturase is responsible for converting stearic acid (a C_{18} saturated fatty acid) to oleic acid (a C_{18} monounsaturated fatty acid). This may explain why dietary stearic acid does not raise blood cholesterol like the other saturated fatty acids. Symptoms of **copper deficiency** include anemia, **hypercholesterolemia**, demineralization of bones, leukopenia, fragility of large arteries, and demyelination of neural tissue. Anemia may reflect the reduced ceruloplasmin activity. Bone demineralization and blood vessel fragility can be directly traced to defects in collagen and elastin formation. Hypercholesterolemia may be related to an increase in the ratio of saturated to monounsaturated fatty acids of the C_{18} series due to reduced activity of the C_{18},Δ^9-desaturase.

Copper deficiency is relatively rare and is usually seen only because of excess zinc intake (zinc and copper compete for absorption) and in Menkes' syndrome, a relatively rare X-linked hereditary disease associated with a defect in copper transport. Wilson's disease, an autosomal recessive disease, is associated with abnormal accumulation of copper in various tissues and can be treated with the copper chelating agent penicillamine.

Chromium Is a Component of Chromodulin

Chromium is a component of the low-molecular-weight protein **chromodulin**, which potentiates the effects of insulin by facilitating insulin binding to its receptor and receptor kinase signaling. The chief symptom of **chromium deficiency** is impaired **glucose tolerance**, a result of decreased insulin effectiveness. Chromium deficiency appears to be rare in healthy adults. However, diabetes causes increased urinary loss of chromium, which can lead to chromium deficiency over time. Supplementation with chromium appears to improve glycemic control in patients with type 2 diabetes.

Selenium Is Found in Selenoproteins

Selenium occurs in selenoproteins such as **glutathione peroxidase**, phospholipid-hydroperoxide glutathione peroxidase, **thioredoxin reductase**, iodothyronine deiodinase, sperm capsule selenoprotein GPx4, and muscle selenoprotein W. These proteins contain one or more selenocysteine residues, which are incorporated through a unique co-translational process. Glutathione peroxidase destroys peroxides in the cytosol (see p. 783, Figure 19.80), which complements the effect of vitamin E since it is limited primarily to the membrane. Phospholipid- hydroperoxide glutathione peroxidase catalyzes reductive destruction of phospholipid and cholesterol ester hydroperoxides in oxidized low-density lipoproteins. Iodothyronine deiodinase catalyzes the conversion of thyroxine (T_4) to the active thyroid hormone 3,3′, 5-triiodothyronine (T_3). Selenoprotein GPx4 is important for sperm motility, and selenoprotein W appears to be essential for muscle metabolism. Selenium is one of the few nutrients not removed by the milling of flour and is usually thought to be present in adequate amounts in the diet. Selenium levels are very low in the soil in certain parts of the country, however; and foods raised in these regions are low in selenium. Fortunately, this effect is minimized by the current food distribution system, which ensures that the foods marketed in any one area are derived from a variety of different geographical regions. Clinical studies have suggested that supplementation with selenium may reduce the risk of certain cancers.

Manganese, Molybdenum, Fluoride, and Boron Are Essential Trace Elements

Manganese is a component of arginase and pyruvate carboxylase and activates a number of other enzymes. **Molybdenum** is present in xanthine oxidase (see p. 802). **Fluoride** strengthens bones and teeth and is usually added to drinking water. **Boron** appears to be important role in bone formation.

28.11 THE AMERICAN DIET: FACT AND FALLACY

Much has been said about the supposed deterioration of the American diet. Americans are eating much more processed food than did their ancestors. These foods have a higher caloric density and a lower nutrient density than the foods they replace. However, they are almost uniformly enriched with iron, thiamine, riboflavin, niacin, and low levels of folic acid. In many cases they are even fortified (usually as much for sales promotion as for nutritional reasons) with as many as 11–15 vitamins and minerals. Unfortunately, it is not practical to replace all of the nutrients lost during processing, especially the trace minerals and phytonutrients such as the carotenoids. Imitation foods present a special problem, because they are usually incomplete in more subtle ways. For example, imitation cheese and milkshakes are widely sold in this country. They usually contain the protein and calcium one would expect of the food they replace, but often do not contain the riboflavin that one would obtain from them. Fast food meals tend to be high in calories and fat, and low in certain vitamins and trace minerals. For example, the standard fast food meal provides over 50% of the calories the average adult needs for the entire day, while providing <5% of the vitamin A and <30% of biotin, folic acid, and pantothenic acid. Unfortunately, much of the controversy in recent years has centered on whether these trends are "good" or "bad." This simply obscures the issue at hand. Clearly, it is possible to obtain a balanced diet that includes some processed, imitation, and fast foods if one compensates by selecting foods for the other meals that are low in caloric density and rich in nutrients. Without such compensation the "balanced diet" becomes a myth.

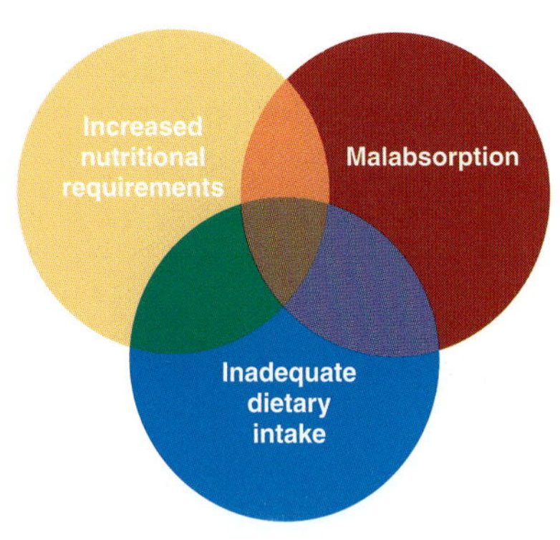

FIGURE 28.17
Factors affecting individual nutritional status. Schematic representation of three important risk factors in determining nutritional status. A person in the periphery would have very low risk of any nutritional deficiency, whereas those in the green, orange, purple, or center areas would be much more likely to experience some symptoms of nutritional deficiencies.

28.12 ASSESSMENT OF NUTRITIONAL STATUS IN CLINICAL PRACTICE

It might seem that the process of evaluating the **nutritional status** of an individual patient would be an overwhelming task after surveying the major micronutrients and their biochemical roles. There are three factors that can contribute to nutritional deficiencies: poor diet, malabsorption, and increased nutrient need. Only when two or three components overlap in the same person (Figure 28.17) does the risk of symptomatic deficiency become significant. For example, infants and young children have increased needs for iron, calcium, and protein. Dietary surveys show that many of them consume diets inadequate in iron and some consume diets that are low in calcium. Protein is seldom a problem unless the children are being raised as strict vegetarians. Thus, the chief nutritional concerns for most children are iron and calcium. Teenagers tend to consume diets low in calcium, magnesium, vitamin A, vitamin B_6, and vitamin C. Of these, their needs are particularly high for calcium and magnesium during the teenage years, so these are the nutrients of greatest concern. Young women are likely to consume diets low in iron, calcium, magnesium, vitamin B_6, folic acid, and zinc—and all of these nutrients are needed in greater amounts during pregnancy and lactation. Adult women often consume diets low in calcium, yet they may have a particularly high need for calcium to prevent rapid bone loss. Finally, the elderly have unique nutritional needs (see Clin. Corr. 28.9) and tend to have poor nutrient intake due to restricted income, loss of appetite, and loss of the ability to prepare a wide variety of foods. They are also more prone to suffer from malabsorption problems and to use multiple prescription drugs that increase nutrient needs (Table 28.1).

Illness and **metabolic stress** often cause increased demand or decreased utilization of certain nutrients. For example, diseases leading to fat malabsorption cause a particular problem with absorption of calcium and the fat-soluble vitamins. Other malabsorption diseases can result in deficiencies of many nutrients, depending on the particular disease. Liver and kidney disease can prevent hydroxylation of vitamin D and storage or utilization of many other nutrients including vitamin A, vitamin B_{12}, and folic acid. Severe illness and trauma increase the need for calories, protein, and possibly vitamin C and certain B vitamins. Long-term use of many drugs in the treatment of chronic disease can affect the need for certain micronutrients. Some of these are listed in Table 28.1.

Who then is at a nutritional risk? Obviously, the answer depends on many factors. Nutritional counseling is an important part of treatment for infants, young children, and pregnant/lactating females. A brief analysis of a dietary history and further nutritional counseling are important when dealing with high-risk patients.

TABLE 28.1 Drug–Nutrient Interactions

Drug	*Potential Nutrient Deficiencies*
Alcohol	Thiamine Folic acid Vitamin B_6
Anticonvulsants	Vitamin D Folic acid Vitamin K
Cholestyramine	Fat-soluble vitamins Iron
Corticosteroids	Vitamin D and calcium Zinc Potassium
Diuretics	Potassium Zinc
Isoniazid	Vitamin B_6
Oral contraceptives and estrogens	Vitamin B_6 Folic acid and B_{12}

CLINICAL CORRELATION 28.9
Nutritional Needs of Elderly Persons

If current trends continue, one of five Americans will be over the age of 65 by the year 2030. With this projected aging of the American population, there has been increased interest in defining the nutritional needs of the elderly. Recent research shows altered needs of elderly persons for several essential nutrients. For example, the absorption and utilization of vitamin B_6 decreases with age. Dietary surveys have consistently shown that B_6 is a problem nutrient for many Americans, and the elderly are no exception. Many older Americans get less than 50% of the RDA for B_6 from their diet. Vitamin B_{12} deficiency is also more prevalent in the elderly. Many older adults develop atrophic gastritis (decreased acid production in the stomach) and decreased production of intrinsic factor, which lead to poor absorption of B_{12}. The blood level of homocysteine, a possible risk factor for atherosclerosis, dementia, and Alzheimer's disease, is often elevated in the elderly. Homocysteine is a byproduct of DNA methylation and is normally metabolized to methionine or cysteine in reactions requiring folic acid, B_{12} and B_6 (see Figure 28.14). Simple supplementation with those B vitamins is generally sufficient to normalize homocysteine levels. Vitamin D can be a problem as well. Many elderly do not spend much time in the sunlight, and the conversion of 7-dehydrocholesterol to vitamin D in the skin and 25-(OH)D to 1,25-$(OH)_2$D in the kidney decreases with age. These factors lead to significant deficiencies of 1, 25-$(OH)_2$D in the elderly, which can cause a negative calcium balance. These changes may contribute to osteoporosis.

There is some evidence for increased need for chromium and zinc as well. Many elderly appear to have difficulty converting dietary chromium to the biologically active chromodulin. Chromium deficiency could contribute to type 2 diabetes. Similarly, most elderly consume between one-half and two-thirds the RDA for zinc, and conditions such as atrophic gastritis can interfere with zinc absorption. Symptoms of zinc deficiency include loss of taste acuity, dermatitis, and a weakened immune system. All of these symptoms are common in the elderly population, and zinc deficiency may contribute.

Not all of the news is bad, however. Vitamin A absorption increases with age and its clearance by the liver decreases, so vitamin A remains in the circulation for a longer time. Not only does the need for vitamin A decrease as we age, but the elderly also need to be particularly careful to avoid vitamin A toxicity. While this does not restrict their choice of foods or multivitamin supplements, they should generally avoid separate vitamin A supplements.

Source: Russell, R. M. and Suter, P. M. Vitamin requirements of elderly people: An update. *Am. J. Clin. Nutr.* 58:4, 1993. Ubbink, J. B., Vermoak, W. J., van der Merne, A., and Becker, P. J. Vitamin B_{12}, vitamin B_6 and folate nutritional status in men with hyperhomocysteinemia. *Am. J. Clin. Nutr.* 57:47, 1993. Joosten, E., van der Berg. A., Riezler, R. Neurath, H. J., Linderbaum, J., Stabler, S. P., and Allen, R. H. Metabolic evidence that deficiencies of vitamin B_{12}, folate and vitamin B_6 occur commonly in elderly people. *Am. J. Clin. Nutr.* 58:468,1993. Wood, R. J., Suter, P. M., and Russell, R. M. Mineral requirements of elderly people. *Am. J. Clin. Nutr.* 62:493, 1995. Johnson, K. A., Bernard, M. A., and Funderburg, K. Vitamin nutrition in older adults. *Clin. Geriatr. Med.* 18:773, 2002.

BIBLIOGRAPHY

Dietary Reference Intakes

Food and Nutrition Board, Institute of Medicine of the National Academy of Sciences: www.iom.edu/board.asp?id=3788

Vitamin A

Goodman, D. S. Vitamin A and retinoids in health and disease. *N. Engl. J. Med.* 310:1023, 1984.

Soprano, D. R., Qin, P., and Soprano, K. J. Retinoic acid receptors and cancers. *Annu. Rev. Nutr.* 24:201, 2004.

Thurnham, D. I. and Northrop-Clews, C. A. Optimal nutrition: Vitamin A and the carotenoids, *Proc. Nutr. Soc.* 58:449, 1999.

Vitamin D

Darwish, H. and DeLuca, H. F. Vitamin D-Regulated Gene Expression. *Crit. Rev. Eukaryot. Gene Expr.* 3:89, 1993.

DeLuca, H. F. Overview of general physiologic features and functions of vitamin D. *Am. J. Clin. Nutr.* 80:1689S, 2004.

Langman, C. B. New developments in calcium and vitamin D metabolism. *Curr. Opin. Pediatr.* 12:135, 2000.

Vitamin E

Brizeluis-Flohe, R. and Traber, M. G. Vitamin E: Function and metabolism. *FASEB J.* 13:1145, 1999.

Hensley, K., Benaksas, E. J., Bolli, R., Comp, P. Grammas, P., Hamdheydari, L., et al. New perspectives on vitamin E: γ-Tocopherol and carboxyethylhydroxychroman metabolites in biology and medicine. *Free Rad. Biol. Med.* 36:1, 2004.

Schneider, C. Chemistry and biology of vitamin E. *Mol. Nutr. Food Res.* 49:7, 2005.

Stampfer, M. J. and Rimm, E. B. Epidemiologic evidence for vitamin E in prevention of cardiovascular disease. *Am. J. Clin. Nutr.* 62:1365S, 1995.

Vitamin K

Binkley, N. C. and Suttie, J. W. Vitamin K nutrition and osteoporosis. *J. Nutr.* 125:1812, 1995.

Bugel, S. Vitamin K and bone health. *Proc. Nutr. Soc.* 62:839, 2003.

Nelsestuen, G. L., Shah, A. M., and Harvey, S. B. Vitamin K-dependent proteins. *Vitam. Horm.* 58:355, 2000.

Thiamine

Bates, C. J. Thiamin. In: B. A. Bowman and R. M. Russell (Eds.), *Present Knowledge in Nutrition*, 8th ed. Washington, DC: ILSI Press, 2001, p. 184.

Singleton, C. K. and Martin, P. R. Molecular mechanisms of thiamine utilization, *Curr. Mol. Med.* 1:197, 2001.

Riboflavin

Powers, H. J. Riboflavin (vitamin B-2) and health. *Am. J. Clin. Nutr.* 77:1352, 2003.

Niacin

Carlson, L. A. Niaspan, the prolonged release preparation of nicotinic acid (niacin), the broad-spectrum lipid drug. *Int. J. Clin. Pract.* 58:706, 2004.

Jacob, R. A. Niacin. In: B. A. Bowman and R. M. Russell (Eds.), *Present Knowledge in Nutrition*, 8th ed. Washington, DC: ILSI Press, 2001, p. 199.

Malik, S. and Kashyap, M. L. Niacin, lipids and heart disease. *Curr. Cardiol. Rep.* 5:470, 2003.

Meyers, C. D. and Kashyap, M. L. Management of metabolic syndrome-nicotinic acid. *Endocrinol. Metab. Clin. North Am.* 33:557, 2004.

Pyridoxine

Merril, A. H., Jr. and Henderson, J. M. Diseases associated with defects in vitamin B_6 metabolism or utilization. *Annu. Rev. Nutr.* 7:137, 1987.

Tully, D. B., Allgood, V. E., and Cidlowski, J. A. Modulation of steroid receptor-medicated gene expression by vitamin B_6. *FASEB J.* 8:343,1994.

Pantothenic Acid

Miller, J. A., Rogers, L. M., and Rucker, R. R. Pantothenic acid. In: B. A. Bowman and R. M. Russell (Eds.), *Present Knowledge in Nutrition*, 8th ed. Washington, DC: ILSI Press, 2001, p. 253.

Biotin

McMahon, R. J. Biotin in metabolism and molecular biology. *Annu. Rev. Nutr.* 22:221, 2002.

Rodriguez-Melendez, R. and Zempleni, J. Regulation of gene expression by biotin, *J. Nutr. Biochem.* 14:680, 2003.

Folic Acid

Bailey, L. B. Folate and B_{12} recommended intakes and status in the United States. *Nutr. Rev.* 62:S14, 2004.

Landgren, F., Israelsson, B., Lindgren, A., Hultsberg, B., Anderson, A., Brettstrom, L. Plasma homocysteine in acute myocardial infarction: Homocysteine-lowering effects of folic acid. *J. Int. Med.* 237:381, 1995.

McNulty, H., Cuskelly, G. J., and Wood, M. Response of red blood cell folate to intervention: Implications for folate recommendations for the prevention of neural tube defects. *Am. J. Clin. Nutr.* 71(Suppl.): 13085, 2000.

Milunsky, A., Jick, H., Jick, S. S., Bruell, C. L., MacLaugin, D. S., Rothman, K. J., and Willet, W. Multivitamin/folic acid supplementation in early pregnancy reduces the prevalence of neural tube defects. *JAMA* 262:2847, 1989.

Stover, P. J. Physiology of folate and vitamin B_{12} in health and disease. *Nutr. Rev.* 62:S3, 2004.

Zanibbi, G. A. Homocysteine and cognitive function in the elderly. *CMAJ* 171:897, 2004.

Vitamin B_{12}

Banerjee, R. and Ragsdale, S. W. The many faces of vitamin B_{12}: Catalysis by cobalamin-dependent enzymes. *Annu. Rev. Biochem.* 72:209, 2003.

Seethoram, B., Bose, S., and Li, N. Cellular import of cobalamine (vitamin B_{12}). *J. Nutr.* 129:1761, 1999.

Weir, D. G. and Scott, J. M. Brain function in the elderly: Role of vitamin B_{12} and folate. *Br. Med. Bull.* 55:669, 1999.

Vitamin C

Hemila, H. and Douglas, R. M. Vitamin C and acute respiratory infections. *Int. J. Tuberc. Lung Dis.* 3:756–761, 1999.

Bsoul, S. A. and Terezhalmy, G. T. Vitamin C in health and disease. *J. Contemp. Den. Pract.* 15:1, 2004.

Simon, J. A. Vitamin C and cardiovascular disease: A review. *J. Am. Coll. Nutr.* 11:107, 1992.

Choline

Zeisel, S. H. Choline: An essential nutrient for humans. *Nutrition* 16:669, 2000.

Zeisel, S. H. Choline: Needed for normal memory development. *J. Am. Coll. Nutr.* 19(5 Suppl): 528S, 2000.

Calcium

Flynn, A. The role of dietary calcium in bone health. *Proc. Nutr. Soc.* 62:851, 2003.

Hatton, D. C. and McCarron, D. A. Dietary calcium and blood pressure in experimental models of hypertension. *Hypertension* 23:513, 1994.

Magnesium

Gums, J. G. Magnesium in cardiovascular and other disorders. *Am. J. Health Syst. Pharmacol.* 61:1569, 2004.

Saris, N. E., Mervaala, E., Karppanen, H., Khawaja, J. A., and Lewenstam, A. Magnesium. An update on physiological, clinical and analytical aspects. *Clin. Chim. Acta* 294:1, 2000.

Vormann, J. Magnesium: Nutrition and metabolism. *Mol. Aspects Med.* 24:27, 2003.

Iron

Boccio, J., Salgueiro, J., Lysionek, A., Zubillaga, M., Well, R., Goldman, C. and Caro, R. Current knowledge of iron metabolism. *Biol. Trace Element Res.* 92:189, 2003.

Zinc

Prasad, A. S. Zinc deficiency: Its characteristic and treatment. *Met. Ions. Biol. Syst.* 41:103, 2004.

Copper

Schumann, K., Classen, H. G., Dieter, H. H., Konig, J., Multhaup, G., Rukgauer, M., et al. Hohehheim consensus workshop: Copper. *Eur. J. Clin. Nutr.* 56:469, 2002.

Chromium

Anderson, R. A. Chromium in the prevention and control of diabetes. *Diabetes Metab.* 26:22, 2000.

Vincent, J. B. The biochemistry of chromium. *J. Nutr.* 130:715, 2000.

Vincent, J. B. Recent advances in the nutritional biochemistry of trivalent chromium. *Proc. Nutr. Soc.* 63:41, 2004.

Selenium

Ganther, H. E. Selenium metabolism, selenoproteins and mechanisms of cancer prevention: Complexities with thioredoxin reductase. *Carcinogenesis* 20:1657–1666, 1999.

Schrauzer, G. N. Anticarcinogenic effects of selenium. *Cell. Mol. Life Sci.* 57:1864, 2000.

Brown, K. M. and Arthur, J. R. Selenium, selenoproteins and human health: A review. *Public Health Nutr.* 4:593, 2001.

Other Trace Minerals

Nielson, F. H. Boron, manganese, molybdenum and other trace elements. In: *Present Knowledge in Nutrition*, 8th ed. B. A. Bowman and R. M. Russell (eds.), Washington, DC: ILSI Press, 2001, p. 253.

Turnland, J. R. Molybdenum metabolism and requirements in humans. *Met. Ions Biol. Syst.* 39:727, 2002.

Dietary Surveys

Block, G. Dietary guidelines and the results of food consumption surveys. *Am. J. Clin. Nutr.*, 53:3565,1991.

Kritchevsy, D. Dietary guidelines. The rationale for intervention. *Cancer*, 72:1011, 1993.

QUESTIONS | CAROL N. ANGSTADT

Multiple Choice Questions

1. Recommended dietary allowances (RDAs):
   A. are standards for all individuals.
   B. meet special dietary needs arising from chronic diseases.
   C. include all nutritional needs.
   D. define optimal levels of nutrients.
   E. are useful as only general guides in evaluating the adequacy of diets.

2. The Estimated Average Requirement (EAR) of a nutrient is:
   A. the same as the RDA of that nutrient.
   B. an amount that should meet the requirement of half of the healthy individuals of a particular group.
   C. based on the observed nutrient intake of a particular group.
   D. the highest level of nutrient deemed to pose no risk or adverse health effects on the particular population.
   E. two standard deviations higher than the RDA.

3. The effects of vitamin A may include all of the following *except*:
   A. prevention of anemia.
   B. serving as an antioxidant.
   C. cell differentiation.
   D. the visual cycle.
   E. induction of certain cancers.

4. Ascorbic acid may be associated with all of the following *except*:
   A. iron absorption.
   B. bone formation.
   C. acute liver disease when taken in high doses.
   D. wound healing
   E. participation in hydroxylation reactions.

5. In assessing the adequacy of a person's diet:
   A. age of the individual usually has little relevance.
   B. trauma decreases activity, and hence need for calories and possibly some micronutrients.
   C. a 24-h dietary intake history provides an adequate basis for making a judgment.
   D. currently administered medications must be considered.
   E. intestine is the only organ whose health has substantial bearing on nutritional status.

6. Excess dietary protein causes a rapid excretion of:
   A. calcium.
   B. copper.
   C. iodine.
   D. iron.
   E. selenium.

Questions 7 and 8: Cystic fibrosis is a generalized dysfunction of the exocrine glands leading to viscid mucus, which plugs various ducts. Pulmonary infections are common and are usually the direct cause of death. Cystic fibrosis patients, however, also have severe malabsorption problems because pancreatic enzymes are deficient and there may be a partial obstruction of the common bile duct. Malabsorption of fat, fat-soluble vitamins, and calcium is the most common, but not only, problem. Patients have increased protein and energy needs because of chronic infections.

7. Serum calcium levels are usually normal in spite of suboptimal calcium absorption and Vitamin D deficiency. Serum calcium is being maintained:
   A. by low parathyroid hormone (PTH) levels inhibiting calcium excretion.
   B. by an increase in calcitonin.
   C. by increased bone resorption stimulated by elevated PTH.
   D. because the kidney reabsorbs calcium from the urine.
   E. lack of 1,25-dihydroxy vitamin D which prevents bone from taking calcium from blood.

8. Cystic fibrosis patients are frequently on antibiotics for infections. Antibiotics exacerbate the fat malabsorption problem for obtaining:
   A. vitamin A.
   B. vitamin C.
   C. vitamin D.
   D. vitamin E.
   E. vitamin K.

Questions 9 and 10: Chronic alcoholics are at risk of multiple nutritional abnormalities because of poor diet, pathological changes in the gastrointestinal tract leading to malabsorption problems, and severe liver damage which interferes with activation and storage of certain nutrients. Neurological symptoms associated with thiamin and pyridoxine deficiencies and hematological problems associated with folate or pyridoxine deficiencies are common.

9. Energy generation is severely compromised in thiamin deficiency because TPP is:
   A. a major cofactor for oxidation–reduction reactions.
   B. an essential cofactor for pyruvate and α-ketoglutarate dehydrogenases.
   C. a cofactor for transketolase.
   D. necessary for utilizing amino acids for energy via transamination.
   E. necessary for oxidation of alcohol.

10. Alcohol impairs both absorption and storage of folate. Megaloblastic erythropoiesis occurs because cells are arrested in the S phase since DNA synthesis is inhibited. DNA synthesis is inhibited in folate deficiency because tetrahydrofolate is required:
    A. in the synthesis of purine nucleotides and dTMP.
    B. in the conversion of homocysteine to methionine.
    C. for the utilization of vitamin B_{12}.
    D. because of all of the above.
    E. because of none of the above.

Questions 11 and 12: Nutritional needs of the elderly are of interest because of aging of the American population. Atrophic gastritis results in decreased acid production in the stomach. Absorption problems and decreased ability to convert some vitamins and minerals to their active forms contribute to nutritional deficiencies in the elderly. Chromium is not abundant in the diet and many elderly have difficulty in converting it to its active form, chromodulin.

11. The major symptom of chromium deficiency is impaired glucose tolerance. Chromodulin:
    A. increases the number of insulin receptors.
    B. facilitates insulin binding to cell receptors.
    C. protects cell receptors by salvaging peroxides.
    D. facilitates the absorption of copper and zinc.
    E. increases the metabolism of homocysteine.

12. Atrophic gastritis can interfere with zinc absorption. Zinc normally plays a role in all of the following *except*:
    A. growth in children.
    B. wound healing.
    C. taste acuity.
    D. prevention of goiter.
    E. immune function.

Problems

13. The neurological disorders seen in vitamin B_{12} deficiency are caused by progressive demyelination of nervous tissue. How does lack of B_{12} interfere with formation of the myelin sheath?

14. What is the chemical reaction in which vitamin K participates? How is this reaction involved in blood coagulation and bone formation?

ANSWERS

1. **E** RDAs should meet the needs of about 97% of the population but do not take into account special requirements. A: RDAs are designed for most individuals; exceptions occur. B: Diseases often change dietary requirements. C: Some nutritional needs may be unknown; the requirements for all known nutrients are not even clear. D: Optimal levels of nutrients are hard to define; it depends on the criterion for optimal.
2. **B** This is the first step in defining nutritional guidelines. A, E: The RDA is set at two standard deviations above the EAR. C: This is called the adequate intake (AI). D: This is the total upper intake level (UL).
3. **E** May have protective effects against some cancers. A: Retinol and/or retinoic acid is required for synthesis of transferrin. B. Various carotenoids are antioxidants. C: Retinol and retinoic acid may function like steroid hormones. D. Retinol cycles between the Δ 11-*cis* and all-*trans* forms.
4. **C** Excess ascorbate is readily excreted. A: Ascorbic acid aids in iron absorption by reducing iron. B: Ascorbic acid is essential for collagen synthesis, which is critical in bone formation. D, E: Ascorbic acid is required for the hydroxylation of lysine and proline residues in protocollagen and, therefore, is required for wound healing.
5. **D** Corticosteroids stimulate vitamin D inactivation. Isoniazid affects B_6, as do oral contraceptives. A: Dramatic differences may occur at different ages. B: Trauma increases caloric requirements and probably requirements for specific micronutrients. C: You cannot be sure that any 24-h diet history is either accurate or representative of the individual's typical diet. E: While the intestine must function well enough to absorb nutrients, further metabolic changes are typically required. The metabolism of vitamin D by the liver and kidney and the conversion of β-carotene to vitamin A in the liver exemplify these inter-organ interrelations.
6. **A** This is a concern about a diet that is very much higher than required in protein.
7. **C** PTH promotes bone resorption and inhibits calcium excretion. A: PTH is elevated when vitamin D is low. B: Elevated calcitonin is a response to high serum calcium. D: Kidney does not do this. E: 1,25-dihydroxy vitamin D promotes bone resorption but lack does not necessarily affect bone's ability to take up calcium from blood.
8. **E** Some of our vitamin K is obtained from bacterial synthesis in the intestine which is wiped out by antibiotics. A, C, D: All of these require adequate fat absorption but come from foods not intestinal bacteria. B: Vitamin C is water-soluble.
9. **B** Pyruvate dehydrogenase is very sensitive to thiamin deficiency and this is on the main route to energy for carbohydrate and several amino acids. A, E: Oxidation–reduction reactions use cofactors derived from niacin or riboflavin. C: TPP is a cofactor for transketolase, but this enzyme is in the hexose monophosphate pathway, which is not a major energy pathway. D: This cofactor is derived from pyridoxine.
10. **A** *De novo* synthesis of purine nucleotides and dUMP to dTMP conversion require tetrahydrofolate. These components are required for DNA synthesis. B: Tetrahydrofolate is required in this process but the process is not involved in DNA synthesis. C: Vitamin B_{12} is required for the release of tetrahydrofolate from N^5-methyl tetrahydrofolate, but this is not part of DNA synthesis.
11. **B** This would facilitate the uptake of glucose. A: Does not do this. C: This is the role of selenium. D: Metallothionein is the protein that influences copper and zinc absorption. E: Folate, B_6, and B_{12} are necessary for homocysteine catabolism.
12. **D** This is a role for iodine. A: RNA and DNA polymerases are zinc proteins. B: This is shown to be true. C: Zinc is part of a salivary polypeptide involved in development of taste buds. E: Zinc is required for cytokine production by monocytes and T cells.
13. One of the reactions in which vitamin B_{12} participates is conversion of methylmalonyl CoA to succinyl CoA (a step in the catabolism of valine and isoleucine). Methylmalonyl CoA is a competitive inhibitor of malonyl CoA in fatty acid biosynthesis, necessary for the maintenance of the myelin sheath. Second, methylmalonyl CoA can be used in fatty acid synthesis leading to formation of branched-chain fatty acids which might disrupt normal membrane structure.
14. Vitamin K is necessary for carboxylation of specific glutamic acid residues in certain proteins to form γ-carboxyglutamic acid residues. In blood coagulation, this step is required for the conversion of preprothrombin to prothrombin. In bone formation, this is required to form the calcium binding residues of the protein osteocalcin.

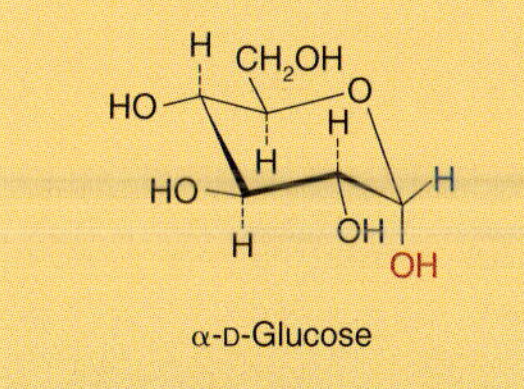

APPENDIX

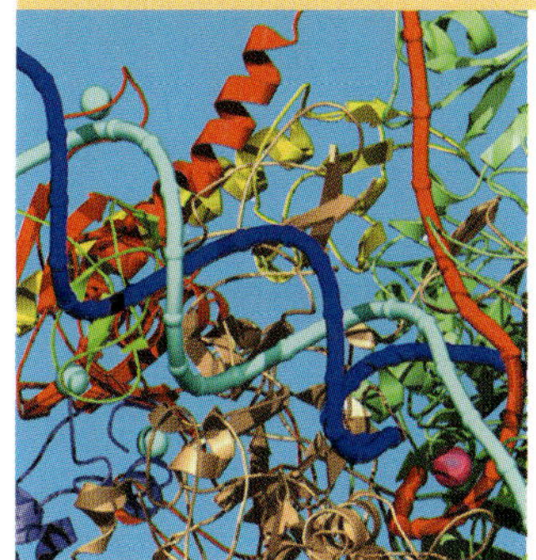

REVIEW OF ORGANIC CHEMISTRY

Carol N. Angstadt

Textbook of Biochemistry With Clinical Correlations, Sixth Edition, Edited by Thomas M. Devlin

FUNCTIONAL GROUPS

Alcohols

The general formula of **alcohols** is R—OH, where R equals an alkyl or aryl group. They are classified as *primary, secondary,* or *tertiary,* according to whether the hydroxyl (OH)-bearing carbon is bonded to no carbon or one, two, or three other carbon atoms:

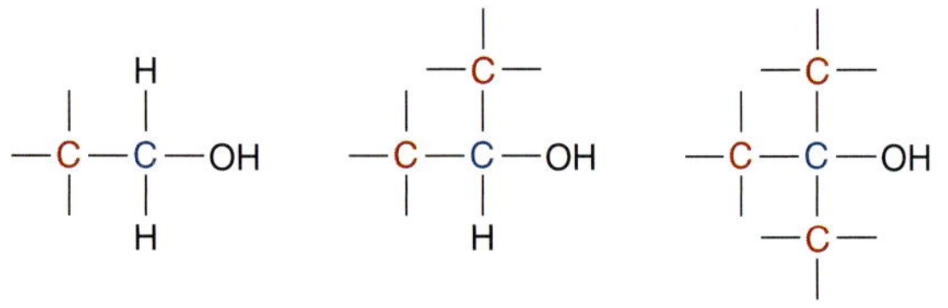

Primary

Secondary

Tertiary

Aldehydes and Ketones

Aldehydes and **ketones** contain a carbonyl group:

$$-\overset{O}{\overset{\|}{C}}-$$

Aldehydes are

$$R-\overset{O}{\overset{\|}{C}}-H$$

and a *ketone* has two groups (alkyl and/or aryl) at the carbonyl group

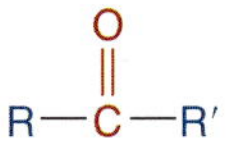

$$R-\overset{O}{\overset{\|}{C}}-R'$$

Acids and Acid Anhydrides

Carboxylic acids contain the functional group

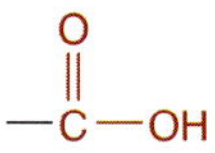

$$-\overset{O}{\overset{\|}{C}}-OH$$

(—COOH). Dicarboxylic and tricarboxylic acids contain two or three carboxyl groups. A carboxylic acid dissociates in water to a negatively charged carboxylate ion:

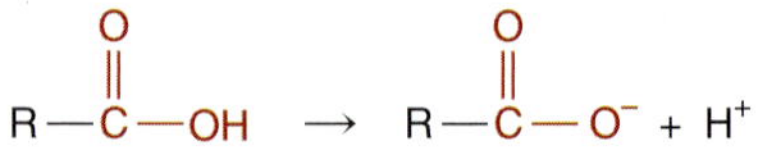

$$R-\overset{O}{\overset{\|}{C}}-OH \longrightarrow R-\overset{O}{\overset{\|}{C}}-O^- + H^+$$

Carboxylic acid   Carboxylate ion

Names of carboxylic acids usually end in -ic and the carboxylate ion in -ate. **Acid anhydrides** are formed when two molecules of acid react with loss of a molecule of water. An acid anhydride may form between two organic acids, two inorganic acids, or an organic and an inorganic acid:

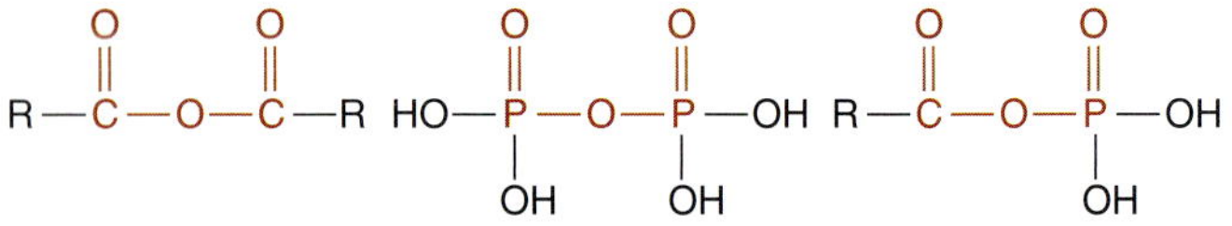

$$R-\overset{O}{\overset{\|}{C}}-O-\overset{O}{\overset{\|}{C}}-R \quad HO-\underset{OH}{\underset{|}{\overset{O}{\overset{\|}{P}}}}-O-\underset{OH}{\underset{|}{\overset{O}{\overset{\|}{P}}}}-OH \quad R-\overset{O}{\overset{\|}{C}}-O-\underset{OH}{\underset{|}{\overset{O}{\overset{\|}{P}}}}-OH$$

Organic anhydride   Inorganic anhydride   Organic–inorganic anhydride

Esters

Esters form in the reaction between a carboxylic acid and an alcohol:

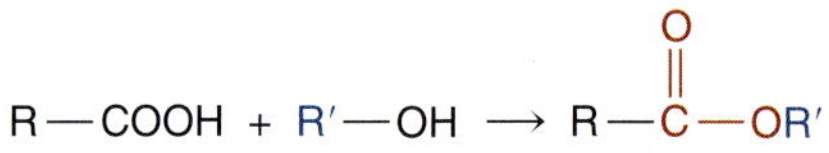

$$R-COOH + R'-OH \longrightarrow R-\overset{O}{\overset{\|}{C}}-OR'$$

Esters may form between an inorganic acid and an organic alcohol, for example, glucose 6-phosphate.

Hemiacetals, Acetals, and Lactones

A reaction between an aldehyde and an alcohol gives a **hemiacetal**, which may react with another molecule of alcohol to form an **acetal**:

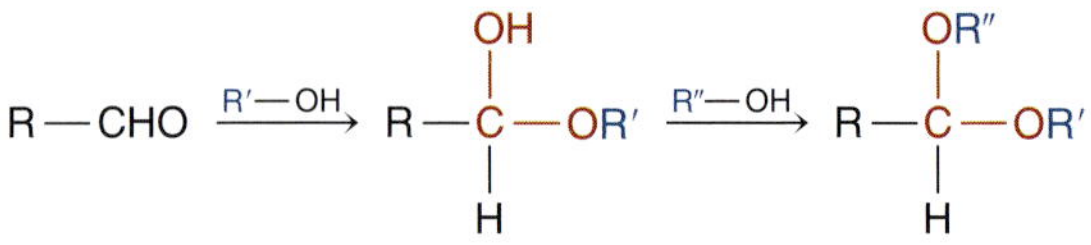

$$R-CHO \xrightarrow{R'-OH} R-\underset{H}{\underset{|}{\overset{OH}{\overset{|}{C}}}}-OR' \xrightarrow{R''-OH} R-\underset{H}{\underset{|}{\overset{OR''}{\overset{|}{C}}}}-OR'$$

Hemiacetal   Acetal

Lactones are cyclic esters formed when an acid and an alcohol group on the same molecule react and usually require that a five- or six-membered ring be formed.

Unsaturated Compounds

Unsaturated compounds are those containing one or more carbon—carbon multiple bonds, for example, a double bond: —C=C—.

Amines and Amides

Amines, R—NH_2, are organic derivatives of NH_3 and are classified as *primary, secondary,* or *tertiary,* depending on the number of alkyl groups (R) bonded to the nitrogen. When a fourth substituent is bonded to the nitrogen, the species is positively charged and called a *quaternary ammonium ion*:

$$R-NH_2 \quad R-\underset{H}{\underset{|}{N}}-R' \quad R-\underset{R''}{\underset{|}{N}}-R' \quad R-\underset{R''}{\underset{|}{\overset{H(\text{or } R''')}{\overset{|}{N^+}}}}-R'$$

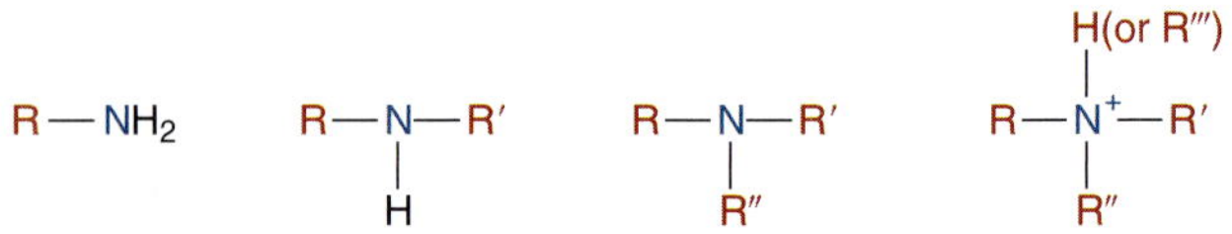

Primary amine   Secondary amine   Tertiary amine   Quaternary ammonium ion

Amides contain the functional group

$$-\overset{\overset{\displaystyle O}{\|}}{C}-\overset{\overset{\displaystyle H}{|}}{N}-X$$

where X can be H (simple) or R (*N* substituted). The carbonyl group is from an acid, and the *N* is from an amine. If both functional groups are from amino acids, the amide bond is referred to as a **peptide bond**.

TYPES OF REACTIONS

Nucleophilic Substitutions at an Acyl Carbon

If the acyl carbon is on a carboxylic group, the leaving group is water. Nucleophilic substitution on carboxylic acids usually requires a catalyst or conversion to a more reactive intermediate; biologically this occurs via enzyme catalysis. X—H may be an alcohol (R—OH), ammonia, amine (R—NH_2), or another acyl compound. Types of nucleophilic substitutions include *esterification, peptide bond* formation, and *acid anhydride* formation.

$$R(L)C{=}\ddot{O}: + :X{-}H \rightleftharpoons R{-}C(XH^+)(L){-}\ddot{O}: \rightleftharpoons R{-}C(X)(LH^+){-}\ddot{O}:^- \longrightarrow X(R)C{=}O + LH$$

New compound   Leaving group

Hydrolysis and Phosphorolysis Reactions

Hydrolysis is the cleavage of a bond by water:

$$R-\overset{\overset{\displaystyle O}{\|}}{C}-OR' + H_2O \longrightarrow R-\overset{\overset{\displaystyle O}{\|}}{C}-OH + R'-OH$$

Hydrolysis is often catalyzed by either acid or base. *Phosphorolysis* is the cleavage of a bond by inorganic phosphate:

$$\text{glucose–glucose} + HO-\overset{\overset{\displaystyle O}{\|}}{\underset{\underset{\displaystyle O^-}{|}}{P}}-O^- \longrightarrow \text{glucose 1-phosphate} + \text{glucose}$$

Oxidation–Reduction Reactions

Oxidation is the loss of electrons; **reduction** is the gain of electrons. Examples of oxidation are as follows:

1. Fe^{2+} + acceptor → Fe^{3+} + acceptor · e^-
2. S(ubstrate) + O_2 + DH_2 → S—OH + H_2O + D
3. S—H_2 + acceptor → S + acceptor · H_2

Some of the group changes that occur on oxidation-reduction are:

1. $-CH_2OH \rightleftharpoons -\overset{\overset{\displaystyle H}{|}}{C}=O$
2. $>C-OH \rightleftharpoons >C=O$
3. $-\overset{\overset{\displaystyle H}{|}}{C}=O \rightleftharpoons -\overset{\overset{\displaystyle O}{\|}}{C}-OH$
4. $-CH_2NH_2 \rightleftharpoons -\overset{\overset{\displaystyle H}{|}}{C}=O + NH_3$
5. $-CH_2-CH_2- \rightleftharpoons -CH=CH-$

STEREOCHEMISTRY

Stereoisomers are compounds with the same molecular formulas and order of attachment of constituent atoms but with different arrangements of these atoms in space.

Enantiomers are stereoisomers in which one isomer is the mirror image of the other and requires the presence of a chiral atom. A chiral carbon (also called an asymmetric carbon) is one that is attached to four different groups:

$$D\!-\!\underset{\underset{\displaystyle E}{\vdots}}{\overset{\overset{\displaystyle B}{|}}{C}}\!-\!A \qquad A\!-\!\underset{\underset{\displaystyle E}{\vdots}}{\overset{\overset{\displaystyle B}{|}}{C}}\!-\!D$$

Enantiomers will be distinguished from each other by the designations *R* and *S* or D and L. The maximum number of stereoisomers possible is 2^n, where *n* is the number of chiral carbon atoms. A molecule with more than one chiral center will be an achiral molecule if it has a point or plane of symmetry.

Diastereomers are stereoisomers that are not mirror images of each other and need not contain chiral atoms. **Epimers** are diastereomers that contain more than one chiral carbon and differ in configuration about *only one* asymmetric carbon.

Anomers are a special form of carbohydrate epimers in which the difference is specifically about the anomeric carbon (see p. 1125). Diastereomers can also occur with molecules in which there is restricted rotation about carbon—carbon bonds. Double bonds exhibit ***cis–trans* isomerism**. The double bond is in the cis configuration if the two end groups of the longest contiguous chain containing the double bond are on the same side and is trans if the two ends of the longest chain are on opposite sides. Fused ring systems, such as those found in steroids (see p. 1130), also exhibit *cis–trans* isomerism.

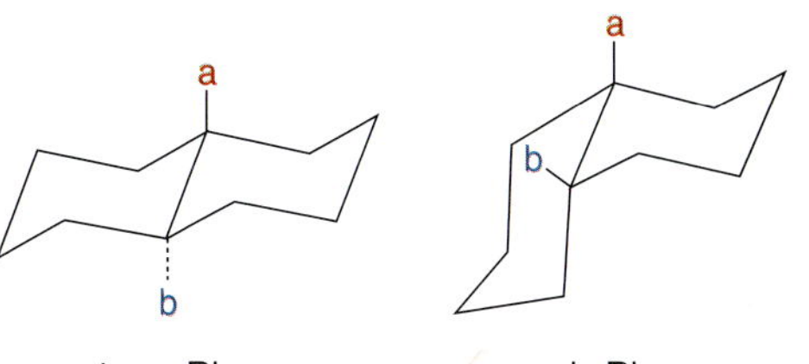

TYPES OF FORCES INVOLVED IN MACROMOLECULAR STRUCTURES

A **hydrogen** bond is a dipole–dipole attraction between a hydrogen atom attached to an electronegative atom and a non-bonding electron pair on another electronegative atom:

```
:Ẍ—H.......:Ẍ—H
 δ⁻ δ⁺        δ⁻ δ⁺
```

Hydrogen bonds of importance in macromolecular structures occur between two nitrogen atoms, two oxygen atoms, or an oxygen and a nitrogen atom.

A **hydrophobic interaction** is the association of nonpolar groups in a polar medium. *Vander Waals* forces consist of dipole and induced dipole interactions between two nonpolar groups. A nonpolar residue dissolved in water induces a highly ordered, thermodynamically unfavorable, solvation shell. Interaction of nonpolar residues with each other, with the exclusion of water, increases the entropy of the system and is thermodynamically favorable.

Ionic (electrostatic) interactions between charged groups can be attractive if the charges are of opposite signs or repulsive if they are of the same sign. The strength of an electrostatic interaction in the interior of a protein molecule may be high. Most charged groups on the surface of a protein molecule interact with water rather than with each other. A **disulfide bond** (S—S) is a covalent bond formed by the oxidation of two sulfhydryl (SH) groups.

CARBOHYDRATES

Carbohydrates are polyhydroxy aldehydes or ketones or their derivatives. **Monosaccharides** (simple sugars) are those carbohydrates that cannot be hydrolyzed into simpler compounds. The generic name of a monosaccharide includes the type of function, a Greek prefix indicating the number of carbon atoms, and the ending -ose; for example, *aldohexose* is a six-carbon aldehyde and *ketopentose* a five-carbon ketone. Monosaccharides may react with each other to form larger molecules. With fewer than eight monosaccharides, either a Greek prefix indicating the number or the general term **oligosaccharide** may be used. **Polysaccharide** refers to a polymer with more than eight monosaccharides. Oligo- and polysaccharides may be either homologous or mixed.

Most *monosaccharides* are asymmetric, an important consideration since enzymes usually work on only one isomeric form. The simplest carbohydrates are glyceraldehyde and dihydroxyacetone whose structures, shown as Fischer projections, are as follows:

```
H—C=O           H—C=O          CH2OH
  |               |              |
H—C—OH         HO—C—H           C=O
  |               |              |
  CH2OH           CH2OH          CH2OH
D-Glyceraldehyde  L-Glyceraldehyde  Dihydroxyacetone
```

D-Glyceraldehyde may also be written as follows:

```
     H
    /
   C=O
   |
H——C——OH
   |
   CH2OH
```

In the Cahn–Ingold–Prelog system, the designations are (*R*) (*rectus*; right) and (*S*) (*sinister*; left).

The configuration of monosaccharides is determined by the stereochemistry at the asymmetric carbon furthest from the carbonyl carbon (number 1 for an aldehyde; lowest possible number for a ketone). Based on the *position* of the OH on the highest number asymmetric carbon, a monosaccharide is D if the OH projects to the *right* and L if it projects to the *left*. The D and L monosaccharides with the same name are **enantiomers**, and the substituents on all asymmetric carbon atoms are reversed as in

```
  H—C=O          H—C=O
    |              |
  H—C—OH        HO—C—H
    |              |
 HO—C—H          H—C—OH
    |              |
  H—C—OH        HO—C—H
    |              |
  H—C—OH        HO—C—H
    |              |
    CH2OH          CH2OH
  D-Glucose      L-Glucose
```

Epimers (e.g., glucose and mannose) are stereoisomers that differ in the configuration about *only one* asymmetric carbon. The relationship of OH groups to *each other* determines the specific monosaccharide. Three aldohexoses and three pentoses of importance are

```
  H—C=O          H—C=O          H—C=O
    |              |              |
  H—C—OH        HO—C—H          H—C—OH
    |              |              |
 HO—C—H         HO—C—H         HO—C—H
    |              |              |
  H—C—OH         H—C—OH        HO—C—H
    |              |              |
  H—C—OH         H—C—OH         H—C—OH
    |              |              |
    CH2OH          CH2OH          CH2OH
  D-Glucose      D-Mannose      D-Galactose

  H—C=O            CH2OH          CH2OH
    |              |              |
  H—C—OH           C=O            C=O
    |              |              |
  H—C—OH         H—C—OH        HO—C—H
    |              |              |
  H—C—OH         H—C—OH         H—C—OH
    |              |              |
    CH2OH          CH2OH          CH2OH
  D-Ribose       D-Ribulose     D-Xylulose
```

Fructose, a ketohexose, differs from glucose only on carbon atoms 1 and 2:

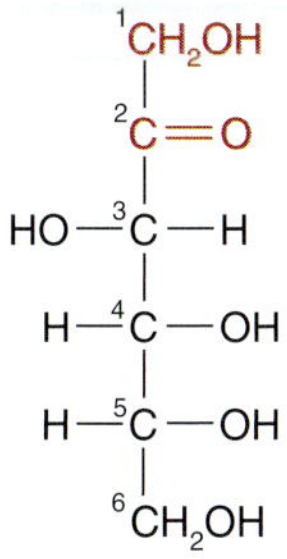

C5 (pentose) and C6 (hexose) monosaccharides form **cyclic hemiacetals** or **hemiketals** in solution. A new asymmetric carbon is generated so two isomeric forms are possible:

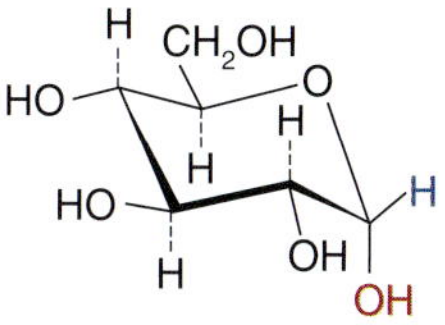

α-D-Glucose

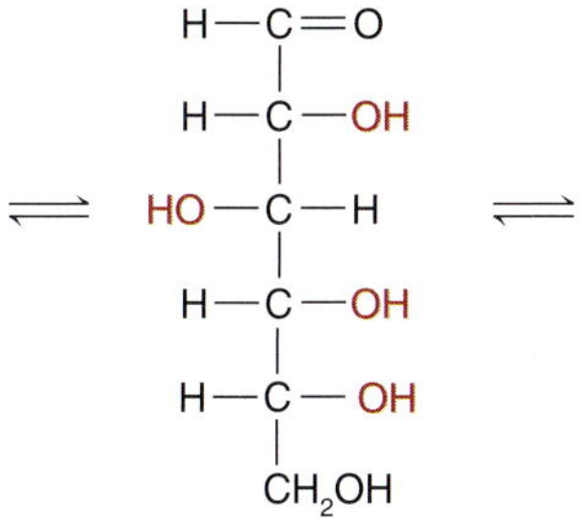

D-Glucose

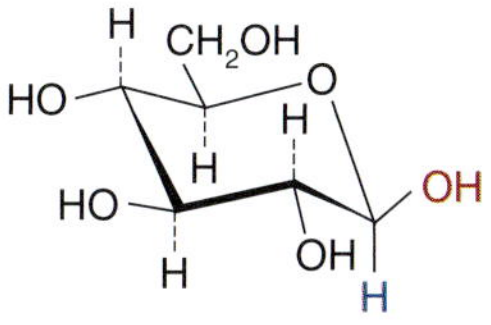

β-D-Glucose

Both five-membered (furanose) and six-membered (pyranose) ring structures are possible, although pyranose rings are more common. A furanose ring is written as follows:

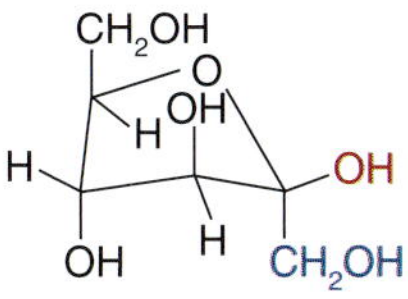

β-D-Fructose

The isomer is designated **α** if the OH group and the CH_2OH group on the two carbon atoms linked by the oxygen are trans to each other and **β** if they are cis. The hemiacetal or hemiketal forms may also be written as modified *Fischer projection formulas*: **α** if OH on the acetal or ketal carbon projects to the same side as the ring and **β** if on the opposite side:

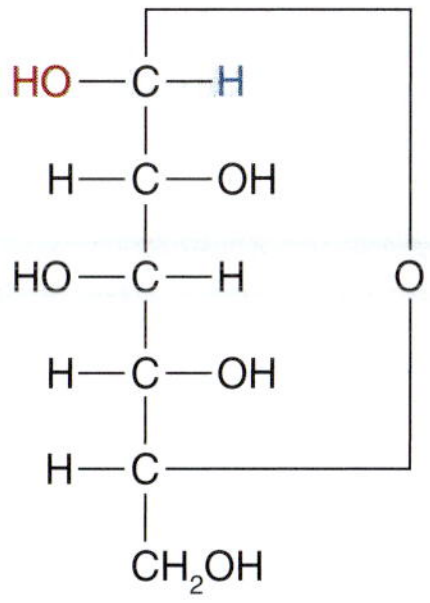

β-D-Glucose

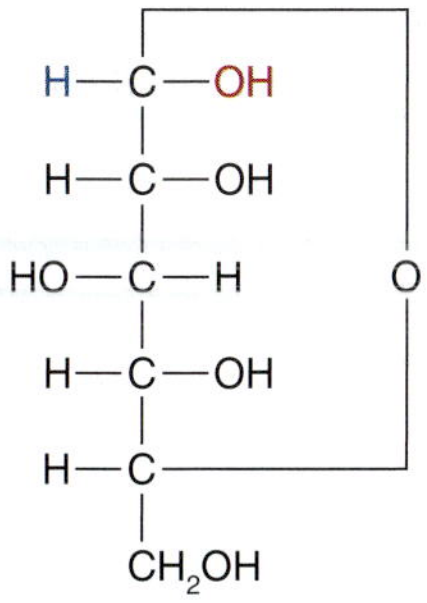

α-D-Glucose

Haworth formulas are used most commonly:

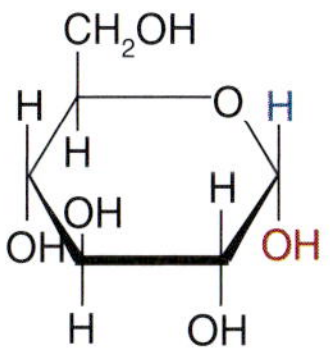

α-D-Glucose

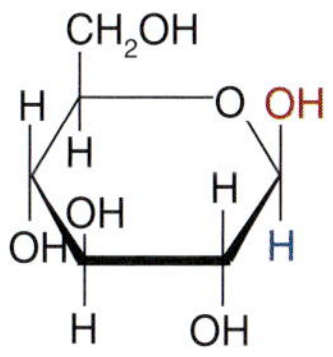

β-D-Glucose

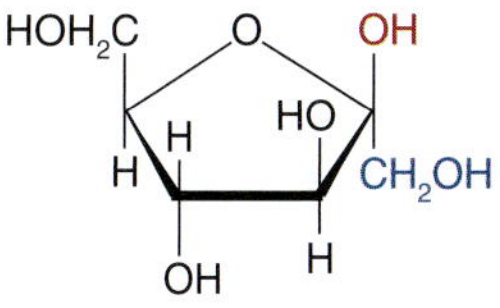

β-D-Fructose

The ring is perpendicular to the plane of the paper with the oxygen written to the back (upper) right, C1 to the right, and substituents above or below the plane of the ring. The OH at the acetal or ketal carbon is below in the **α** isomer and above in the **β**. Anything written to the right in the Fischer projection is written down in the Haworth formula.

The **α** and **β** forms of the same monosaccharide are special forms of epimers called *anomers*, differing only in the configuration about the anomeric (acetal or ketal) carbon. Monosaccharides exist in solution primarily as a mixture of the hemiacetals (or hemiketals) but react chemically as aldehydes or ketones. **Mutarotation** is the equilibration of **α** and **β** forms through the free aldehyde or ketone. Substitution of the H of the anomeric OH prevents mutarotation and fixes the configuration in either the **α** or **β** form.

Monosaccharide Derivatives

A **deoxymonosaccharide** is one in which an OH has been replaced by H. In biological systems, this occurs at C2 unless otherwise indicated. An **amino monosaccharide** is one in which an OH has been replaced by NH_2, again at C2 unless otherwise specified. The amino group of an amino sugar may be *acetylated*:

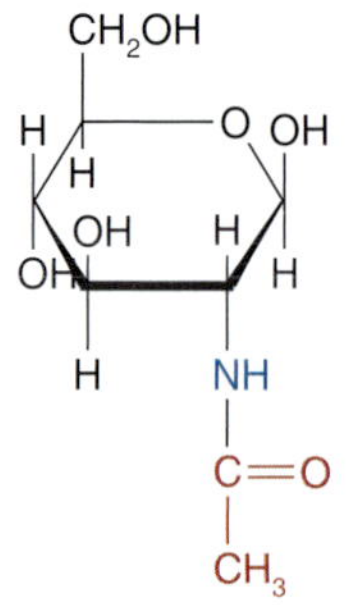

β-*N*-Acetylglucosamine

D-Glucaric acid

An aldehyde is reduced to a primary and a ketone to a secondary **monosaccharide alcohol (alditol)**. Alcohols are named with the base name of the sugar plus the ending *-itol* or with a trivial name (glucitol = sorbitol). Monosaccharides that differ around only two of the first three carbon atoms yield the same alditol. D-Glyceraldehyde and dihydroxyacetone give glycerol:

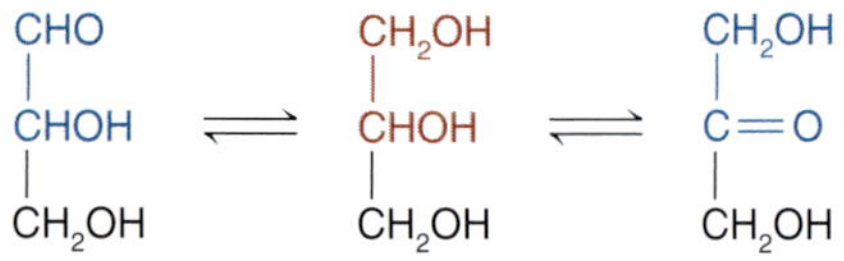

$$\mathrm{CHO{-}CHOH{-}CH_2OH \rightleftharpoons CH_2OH{-}CHOH{-}CH_2OH \rightleftharpoons CH_2OH{-}C{=}O{-}CH_2OH}$$

D-Glucose and D-fructose give D-sorbitol; D-fructose and D-mannose give D-mannitol. Oxidation of the terminal CH_2OH, but not of the CHO, yields a -**uronic acid**, a *monosaccharide acid*:

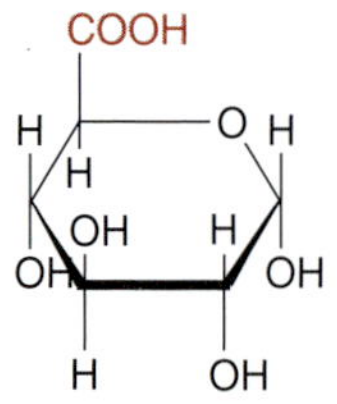

D-Glucuronic acid

Oxidation of the CHO, but not the CH_2OH, gives an -**onic acid**:

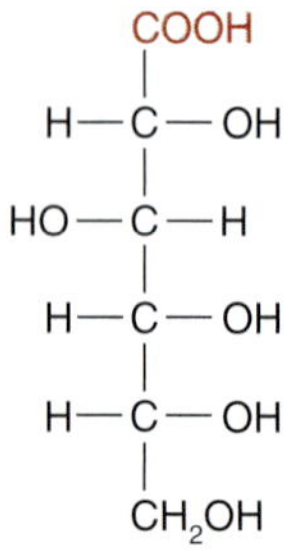

D-Gluconic acid

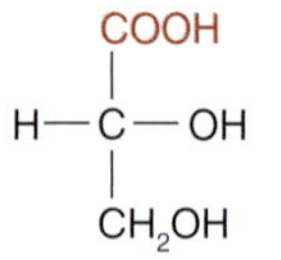

D-Glyceric acid

Oxidation of both the CHO and CH_2OH gives an -**aric acid**:

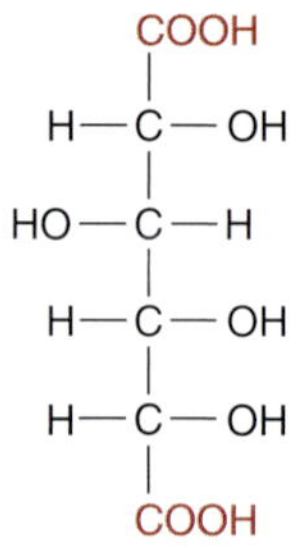

Ketones do not form acids. Both -onic and -uronic acids can react with an OH in the same molecule to form a **lactone** (see p. 1122):

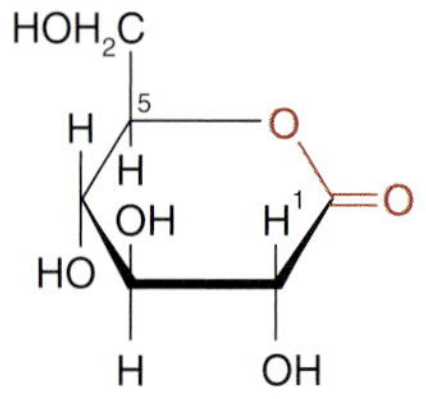

D-Glucono-5-lactone

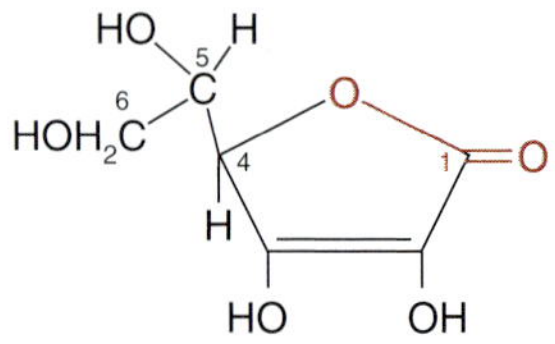

L-Ascorbic acid (derivative of L-gulose)

Reactions of Monosaccharides

The most common *esters* of monosaccharides are phosphate esters at carbon atoms 1 and/or 6:

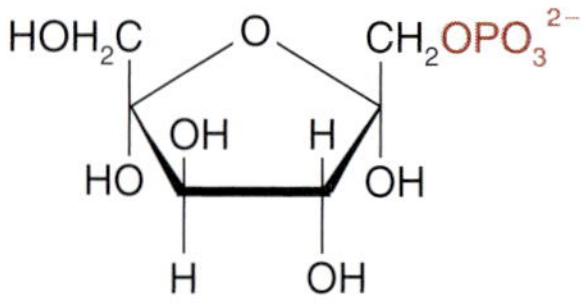

Fructose 1-Phosphate

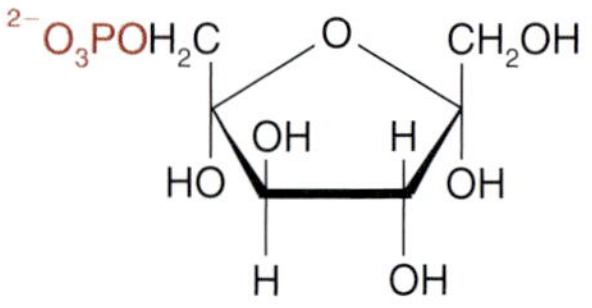

Fructose 6-Phosphate

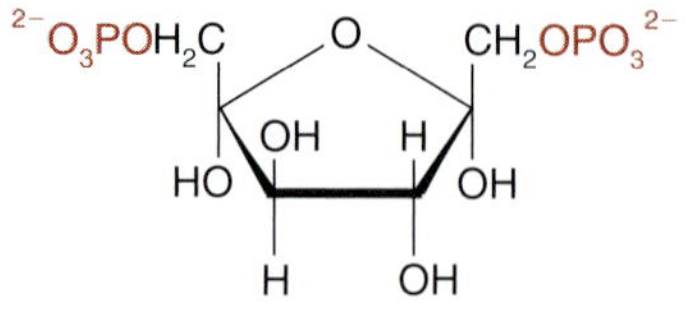

Fructose 1, 6 Bisphosphate

To be a **reducing sugar**, mutarotation must be possible. In alkali, enediols form that may migrate to 2,3 and 3,4 positions:

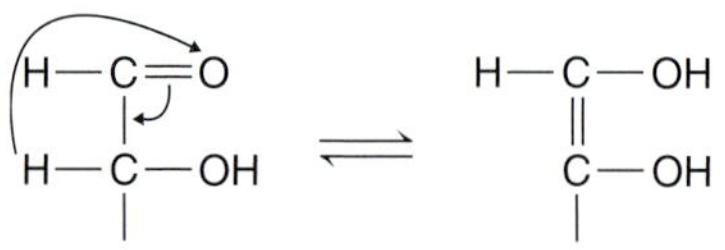

$$\mathrm{H{-}C{=}O,\ H{-}C{-}OH \rightleftharpoons C(H){-}OH{=}C{-}OH}$$

Enediols may be oxidized by O_2, Cu^{2+}, Ag^{+}, and Hg^{2+}. Reducing ability is more important in the laboratory than physiologically. A hemiacetal or hemiketal may react with the OH of another monosaccharide to form a disaccharide (*acetal: glycoside*) (see below):

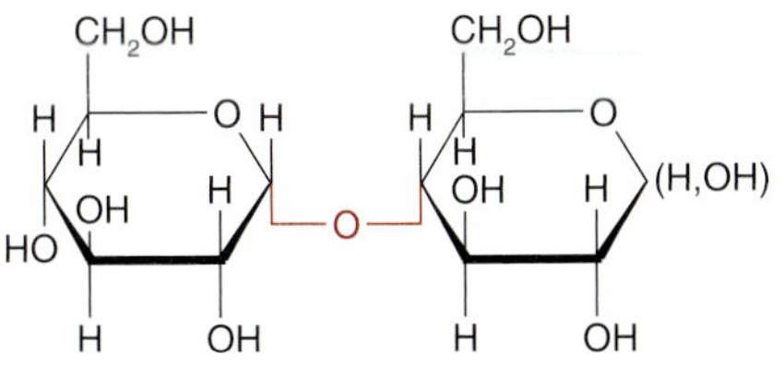

α-1,4-Glycosidic linkage

(May be either α or β)

One monosaccharide still has a free anomeric carbon and can react further. Reaction of the anomeric OH may be with any OH on the other monosaccharide, including the anomeric one. The anomeric OH that has reacted is fixed as either **α** or **β** and cannot mutarotate or reduce. If the glycosidic bond is not between two anomeric carbon atoms, one of the units will still be free to mutarotate and reduce.

Oligo- and Polysaccharides

Disaccharides have two monosaccharides, either the same or different, in glycosidic linkage. If the glycosidic linkage is between the two anomeric carbon atoms, the disaccharide is nonreducing:

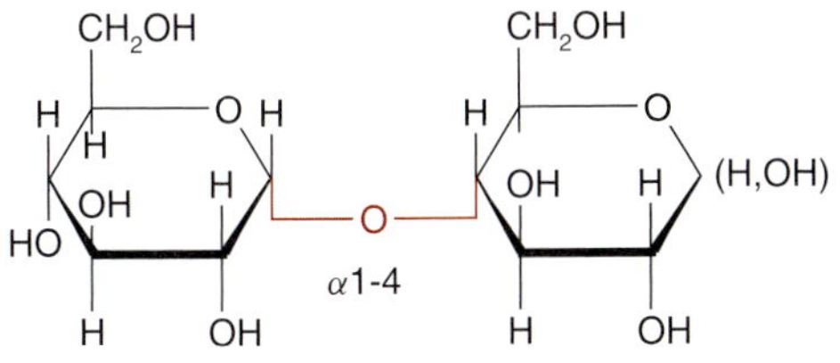

Maltose

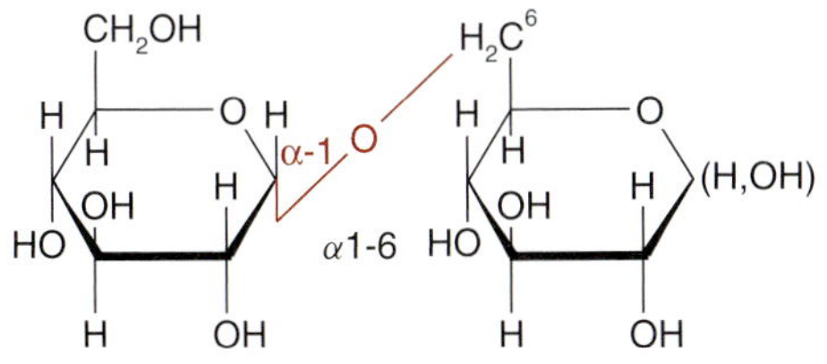

Isomaltose

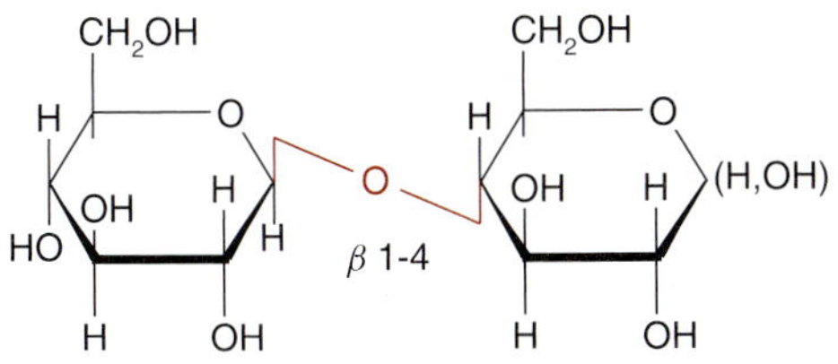

Cellobiose

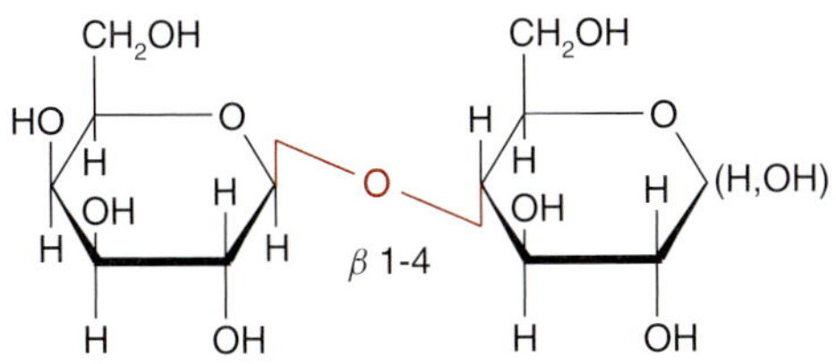

Lactose

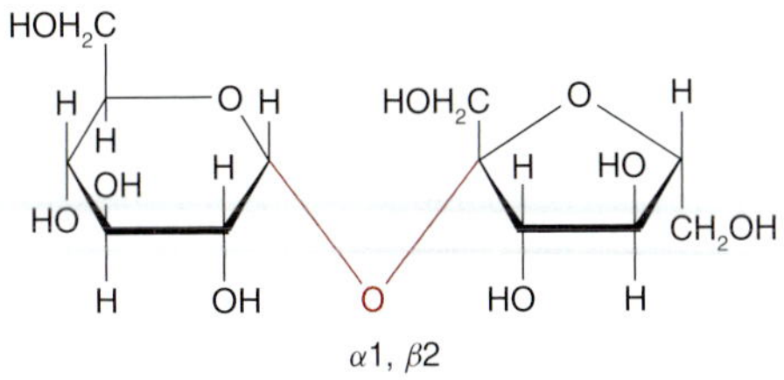

α1, β2
Sucrose

Maltose = 4-*O*-(**α**-D-glucopyranosyl)D-glucopyranose; reducing
Isomaltose = 6-*O*-(**α**-D-glucopyranosyl)D-glucopyranose; reducing
Cellobiose = 4-*O*-(**β**-D-glucopyranosyl)D-glucopyranose; reducing
Lactose = 4-*O*-(**β**-D-galactopyranosyl)D-glucopyranose; reducing
Sucrose = **α**-D-glucopyranosyl-**β**-D-fructofuranoside; nonreducing

As many as thousands of monosaccharides, either the same or different, may be joined by glycosidic bonds to form *polysaccharides*. The anomeric carbon of one unit is usually joined to C4 or C6 of the next unit. The ends of a polysaccharide are not identical (reducing end = free anomeric carbon; nonreducing = anomeric carbon linked to next unit; branched polysaccharide = more than one nonreducing end). The most common carbohydrates are homopolymers of glucose; for example, starch, glycogen, and cellulose. Plant starch is a mixture of **amylose**, a linear polymer of maltose units, and **amylopectin**, branches of repeating maltose units (glucose–glucose in **α**-1,4 linkages) joined via isomaltose linkages. **Glycogen**, the storage form of carbohydrate in animals, is similar to amylopectin, but the branches are shorter and occur more frequently. **Cellulose**, in plant cell walls, is a linear polymer of repeating cellobioses (glucose–glucose in **β**-1,4 linkages). Heteropolysaccharides contain more than one kind of monosaccharide.

Mucopolysaccharides contain amino sugars, free and acetylated, uronic acids, sulfate esters, and sialic acids in addition to the simple monosaccharides. ***N*-Acetylneuraminic acid**, a sialic acid, is

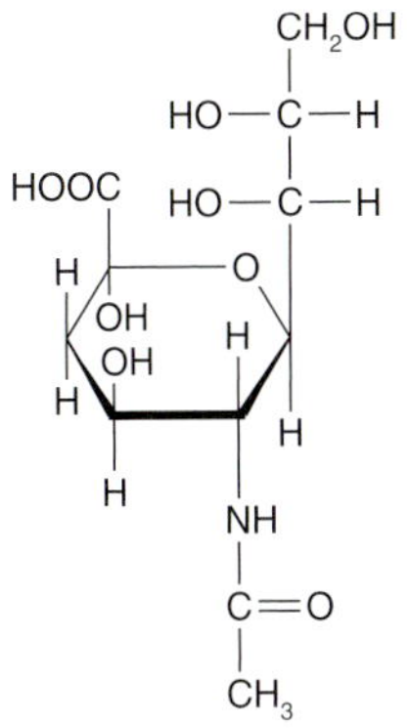

LIPIDS

Lipids are a diverse group of chemicals related primarily because they are insoluble in water, soluble in nonpolar solvents, and found in animal and plant tissues.

Saponifiable lipids yield salts of fatty acids upon alkaline hydrolysis. *Acylglycerols* = glycerol + fatty acid(s); *phosphoacylglycerols* = glycerol + fatty acids + HPO_4^{2-} + alcohol; *sphingolipids* = sphingosine + fatty acid + polar group (phosphorylalcohol or carbohydrate); *waxes* = long-chain alcohol + fatty acid. *Nonsaponifiable lipids* (*terpenes, steroids, prostaglandins,* and related compounds) are not usually subject to hydrolysis. *Amphipathic* lipids have both a polar "head" group and a nonpolar "tail." Amphipathic molecules can stabilize emulsions and are responsible for the lipid bilayer structure of membranes.

Fatty acids are monocarboxylic acids with a short (<6 carbon atoms), medium (8–14 carbon atoms), or long (>14 carbon atoms) aliphatic chain. Biologically important ones are usually linear molecules with an even number of carbon atoms (16–20). Fatty acids are numbered using either arabic numbers (COOH is 1) or the Greek alphabet (COOH is not given a symbol; adjacent carbon atoms are α, β, γ, etc.). **Saturated fatty acids** have the general formula $CH_3(CH_2)_nCOOH$. (*Palmitic acid* = C_{16}; *stearic acid* = C_{18}.) They tend to be extended chains and solid at room temperature unless the chain is short. Both trivial and systematic (prefix indicating number of carbon atoms + *anoic acid*) names are used. $CH_3(CH_2)_{14}COOH$ = palmitic acid or hexadecanoic acid.

Unsaturated fatty acids have one or more double bonds. Most naturally occurring fatty acids have cis double bonds and are usually liquid at room temperature. Fatty acids with trans double bonds tend to have higher melting points. A double bond is indicated by Δ^n, where n is the number of the first carbon of the bond. *Palmitoleic* = Δ^9-hexadecenoic acid; *oleic* = Δ^9-octadecenoic acid; *linoleic* = $\Delta^{9,12}$-octadecadienoic acid; *linolenic* = $\Delta^{9,12,15}$-octadecatrienoic acid; and *arachidonic* = $\Delta^{5,8,11,14}$-eicosatetraenoic acid. Since fatty acids are elongated *in vivo* from the carboxyl end, biochemists use alternate terminology to assign these fatty acids to families: omega (ω) minus x (or $n - x$), where x is the number of carbon atoms from the methyl end where a double bond is first encountered. *Palmitoleic* and *oleic* are $\omega - 9$ acids, *linoleic* and *arachidonic* are $\omega - 6$ acids, and *linolenic* is an $\omega - 3$ acid. Addition of carbon atoms does not change the family to which an unsaturated fatty acid belongs.

Since the pK values of fatty acids are about 4–5, in physiological solutions, they exist primarily in the ionized form, called salts or "soaps." Long-chain fatty acids are insoluble in water, but soaps form micelles. Fatty acids form esters with alcohols and thioesters with CoA.

Biochemically significant reactions of unsaturated fatty acids are:

1. *Reduction* **—CH═CH— + XH$_2$ → —CH$_2$CH$_2$— + X**
2. *Addition of water* **—CH═CH— + H$_2$O → —CH(OH)—CH$_2$—**
3. *Oxidation* **R—CH═CH—R' → R—CHO + R'—CHO**

Prostaglandins, thromboxanes, and *leukotrienes* are derivatives of C_{20}, polyunsaturated fatty acids, especially arachidonic acid. **Prostaglandins** have the general structure:

PGE$_2$

The series differ from each other in the substituents on the ring and whether C15 contains an OH or O · OH group. The subscript indicates the number of double bonds in the side chains. Substituents indicated by—(β) are above the plane of the ring; ··· (α) below:

PGA PGB PGE PGF

PGG(X═OH); PGH(X═OOH) PGI

Thromboxanes have an oxygen incorporated to form a six-membered ring:

TXA$_2$

Leukotrienes are substituted derivatives of arachidonic acid in which no internal ring has formed; R is variable:

Leukotriene C, D, or E

Acylglycerols are compounds in which one or more of the three OH groups of glycerol is esterified. In **triacylglycerols** (triglycerides) all three OH groups are esterified to fatty acids. At least two of the three substituent groups are usually different. If R_1 is not equal to R_3, the molecule is asymmetric and of the L configuration:

```
                      O
                      ‖
  O      H2C—O—C—R1
  ‖       |
R2—C—O—CH          O
          |          ‖
         H2C—O—C—R3
```

The properties of the triacylglycerols are determined by those of the fatty acids they contain; *oils* are liquids at room temperature (preponderance of short-chain and/or cis-unsaturated fatty acids), and *fats*, solid (preponderance of long-chain, saturated, and/or trans-unsaturated).

Triacylglycerols are hydrophobic and do not form stable micelles. They may be hydrolyzed to glycerol and three fatty acids by strong alkali or enzymes (lipases). *Mono-* [usually with the fatty acid in the β(2) position] and *diacylglycerols* also exist in small amounts as metabolic intermediates. Mono- and diacylglycerols are slightly more polar than triacylglycerols. *Phosphoacylglycerols* are derivatives of L-α-glycerolphosphate (L-glycerol 3-phosphate):

```
CH2—OH O                 CH2—OH
|       ‖                 |
CH—O—C—R           HO—CH          O
|                         |          ‖
CH2—OH                   CH2—O—P—O⁻
                                     |
                                     O⁻
```

The parent compound, **phosphatidic acid** (two OH groups of L-α-glycerolphosphate esterified to fatty acids), has its phosphate esterified to an alcohol (XOH) to form several series of phosphoacylglycerols. These are amphipathic molecules, but the net charge at pH 7.4 depends on the nature of X— OH.

X—OH	Phosphoacylglycerol
$HO—CH_2—CH_2—\overset{+}{N}—(CH_3)_3$	Phosphatidylcholines (lecithins)
$HO—CH_2—CH_2—\overset{+}{N}H_3$	Phosphatidylethanolamines (cephalins)
$HO—CH_2—CH(\overset{+}{N}H_3)—COO^-$	Phosphatidylserines
Inositol ring (positions 1–6: H, OH, HO, H, OH, HO substituents) with $—P(O^-)(=O)—OH$	Phosphatidylinositols phospate on 4, or 4 and 5

In **plasmalogens**, the OH on C1 is in *ether*, rather than ester, linkage to an alkyl group. If *one* fatty acid (usually β) has been hydrolyzed from a phosphoacylglycerol, the compound is a *lyso-compound*; for example, lysophosphatidylcholine (lysolecithin):

```
             O                          O
             ‖                          ‖
 O     CH2—OCR1               CH2—OCR1
 ‖      |                       |
R2CO—CH      O         HO—CH      O
        |      ‖                |      ‖
       CH2—P—OX               CH2—P—OX
             |                        |
             O⁻                       O⁻
```

A phosphoacylglycerol A lyso-compound

Sphingolipids are complex lipids based on the C_{18}, unsaturated alcohol, sphingosine. In *ceramides*, a long-chain fatty acid is in amide linkage to sphingosine:

```
CH3(CH2)12CH=CH—CH—OH
                 |
                 CH—NH2
                 |
                 CH2—OH
```

Sphingosine

```
CH3(CH2)12CH=CH—CH—OH
                 |   H  O
                 |   |  ‖
                 CH—N—C—R
                 |
                 CH2—OH
```

A ceramide

Sphingomyelins, the most common sphingolipids, are a family of compounds in which the primary OH group of a ceramide is esterified to phosphorylcholine (phosphoryl-ethanolamine):

```
CH3(CH2)12CH=CH—CH—OH
                 |   H  O
                 |   |  ‖
                 CH—N—C—R
                 |      O
                 |      ‖
                 CH2—O—P—O—CH2CH2N⁺(CH3)3
                        |
                        O-
```

They are amphipathic molecules, existing as zwitterions at pH 7.4 and the only sphingolipids that contain phosphorus. *Glycosphingolipids* do not contain phosphorus but contain carbohydrate in glycosidic linkage to the primary alcohol of a ceramide. They are amphipathic and either neutral or acidic if the carbohydrate moiety contains an acidic group. **Cerebrosides** have a single glucose or galactose linked to a ceramide. *Sulfatides* are galactosylceramides esterified with sulfate at C3 of the galactose:

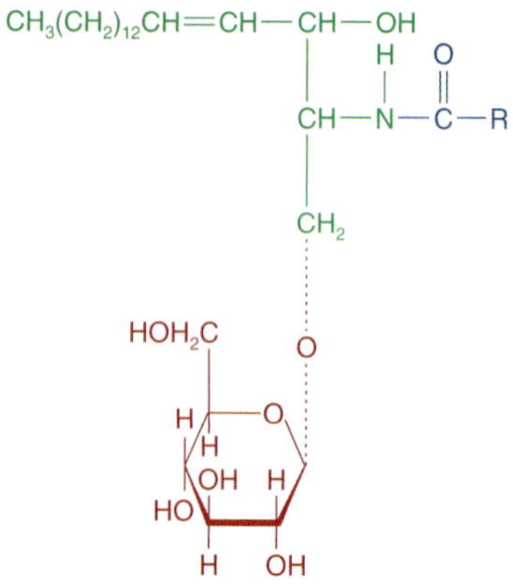

Glucosylceramide (glucocerebroside)

Globosides (*ceramide oligosaccharides*) are ceramides with two or more neutral monosaccharides, whereas in **gangliosides** the oligosaccharide contains one or more sialic acids.

Steriods are derivatives of cyclopentanoperhydrophenanthrene. The steroid nucleus is a rather rigid, essentially planar structure with substituents above the plane of the rings designated $\boldsymbol{\beta}$ (solid line) and those below called $\boldsymbol{\alpha}$ (dotted line):

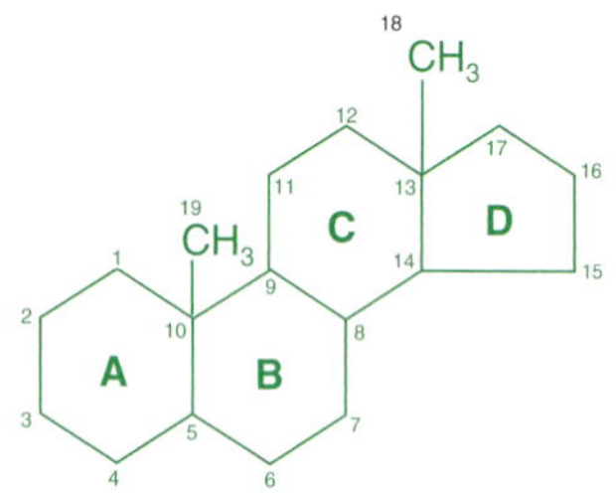

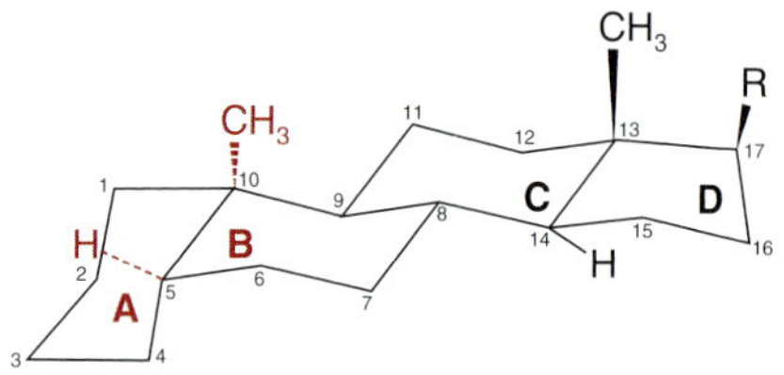

A and B rings—cis;
the others—trans

Most steroids in humans have methyl groups at positions 10 and 13 and frequently a side chain at position 17. *Sterols* contain one or more OH groups, free or esterified to a fatty acid. Most steroids are nonpolar. In a liposome or cell membrane, **cholesterol** orients with the OH toward any polar groups; cholesterol esters do not. **Bile acids** (e.g., cholic acid) have a polar side chain and so are amphipathic:

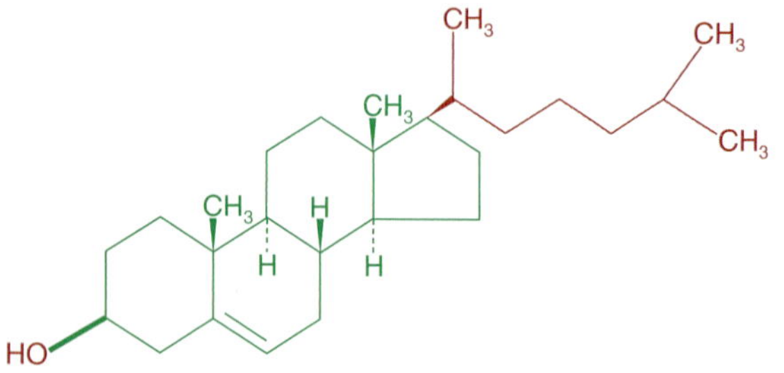

Cholesterol

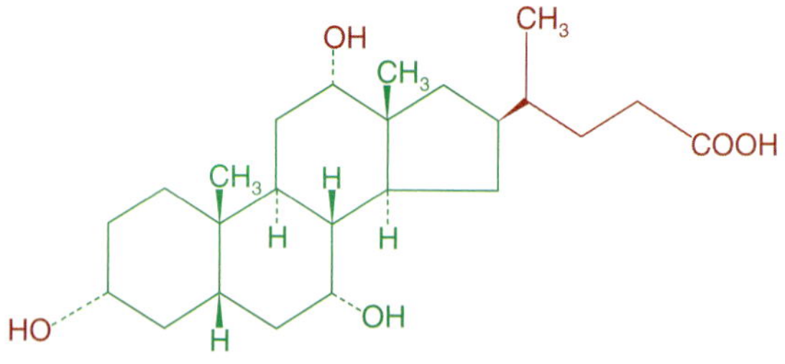

Cholic acid

Steroid hormones are oxygenated steroids of C_{18}, C_{19}, or C_{21} *Estrogens* have C_{18}, an aromatic ring A, and no methyl at C10. *Androgens* have C_{19} and no side chain at C17. *Glucocorticoids* and *mineralocorticoids* have C_{21} including a C_2, oxygenated side chain at C17. *Vitamin* D_{Λ} (*cholecalciferol*) is not a sterol but is derived from 7-dehydrocholesterol in humans:

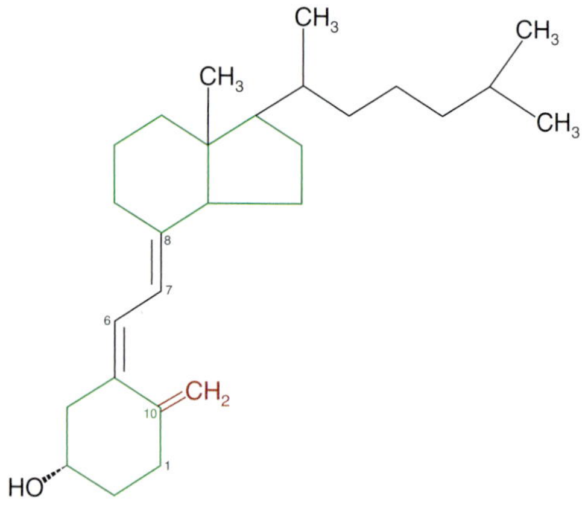

Cholecalciferol

Terpenes are polymers of two or more isoprene units. **Isoprene** is

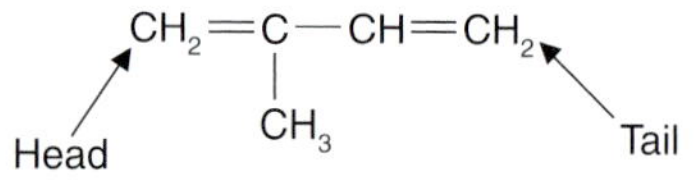

$$\underset{\text{Head}}{CH_2}{=}\underset{\substack{|\\CH_3}}{C}{-}CH{=}\underset{\text{Tail}}{CH_2}$$

Terpenes may be linear or cyclic, with the isoprenes usually linked head to tail and most double bonds trans (but may be cis as in vitamin A). *Squalene*, the precursor of cholesterol, is a linear terpene of six isoprene units. Fat-soluble *vitamins* (A, D, E, and K) contain isoprene units:

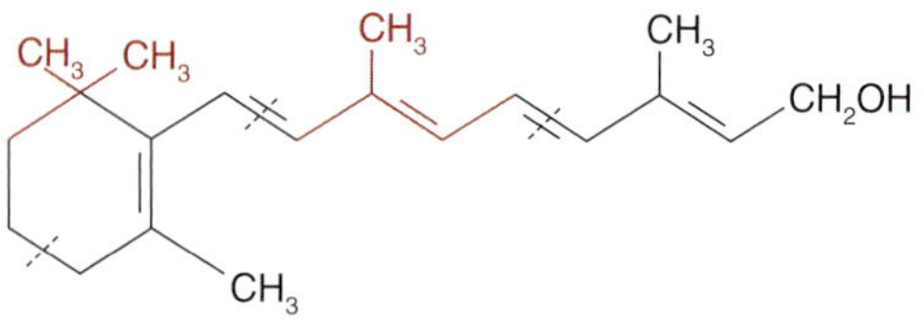

Vitamin A

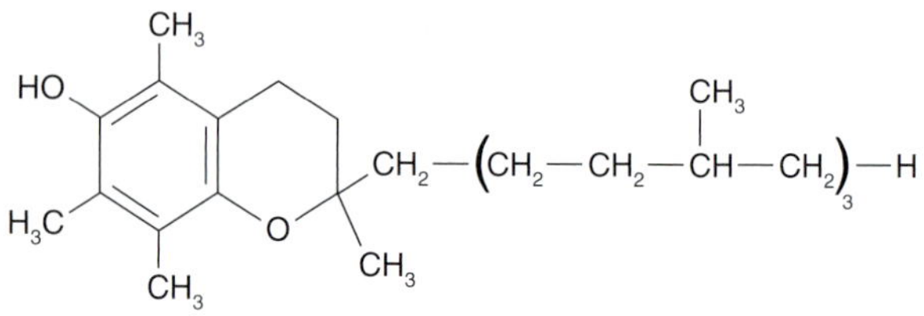

Vitamin E (α-tocopherol)

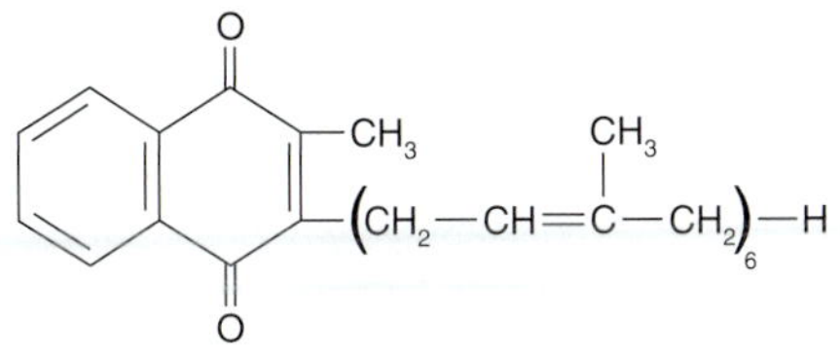

Vitamin K_2

AMINO ACIDS

Amino acids contain an *amino* (NH_2) and a *carboxylic acid* (COOH) group. Biologically important amino acids are usually **α**-amino acids with the formula

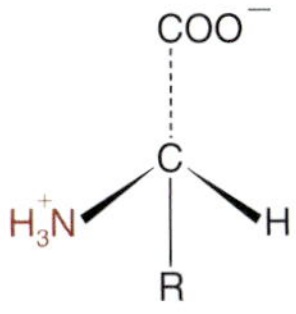

L-α-Amino acid

The amino group, has an unshared pair of electrons, is basic, with a pK_a of about 9.5, and exists primarily as—NH_3^+ at pH values near neutrality. The carboxylic acid group ($pK \sim 2.3$) exists primarily as a carboxylate ion. If R is anything but H, the molecule is asymmetric with most naturally occurring ones of the L configuration (same relative configuration as L-glyceraldehyde: see p. 1124).

The *polarity* of amino acids is influenced by their side chains (R groups) (see p. 78 for complete structures). *Nonpolar* amino acids include those with large, aliphatic, aromatic, or undissociated sulfur groups (aliphatic = Ala, Ile, Leu, Val; aromatic = Phe, Trp; sulfur = Cys, Met). *Intermediate* polarity amino acids include Gly, Pro, Ser, Thr, and Tyr (undissociated).

Amino acids with ionizable side chains are *polar*. The pK values of the side groups of arginine, lysine, glutamate, and aspartate are such that these are nearly always charged at physiological pH, whereas the side groups of histidine (pK = 6.0) and cysteine (pK = 8.3) exist as both charged and uncharged species at pH 7.4 (acidic = Glu, Asp, Cys; basic = Lys, Arg, His). Although undissociated cysteine is nonpolar, cysteine in dissociated form is polar. Asparagine and glutamine are polar even though their side chains are not charged (ionized).

All amino acids are at least *dibasic acids* because of the presence of both the **α**-amino and **α**-carboxyl groups, the ionic state being a function of pH. The presence of another ionizable group will give a tribasic acid as shown for cysteine.

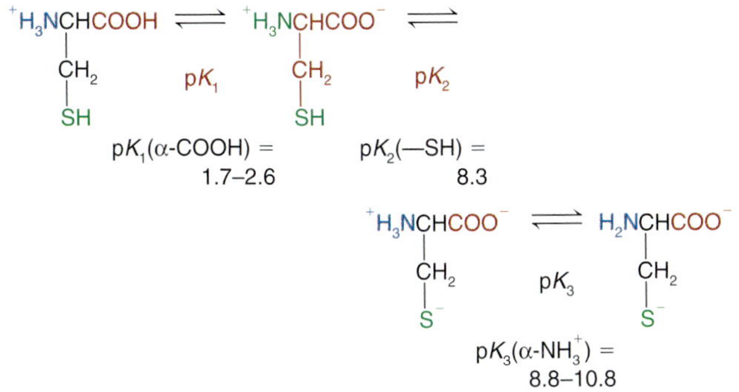

pK_1(α-COOH) = 1.7–2.6

pK_2(—SH) = 8.3

pK_3(α-NH_3^+) = 8.8–10.8

The **zwitterionic form** is the form in which the *net* charge is zero. The *isoelectric point* is the average of the two pK values involved in the formation of the zwitterionic form. In the above example this would be the average of $pK_1 + pK_2$.

PURINES AND PYRIMIDINES

Purines and pyrimidines, often called *bases*, are nitrogen-containing heterocyclic compounds with the structures

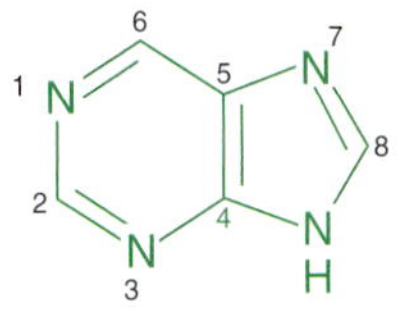

Purine

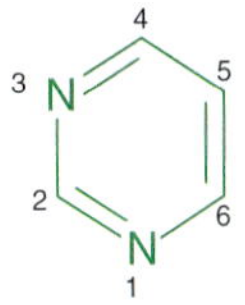

Pyrimidine

Major bases found in nucleic acids and as cellular nucleotides are the following:

Purines	*Pyrimidines*
Adenine: 6-amino	Cytosine: 2-oxy, 4-amino
Guanine: 2-amino, 6-oxy	Uracil: 2,4-dioxy
	Thymine: 2,4-dioxy, 5-methyl
Other important bases found primarily as intermediates of synthesis and/or degradation are	
Hypoxanthine: 6-oxy	Orotic acid: 2,4-dioxy, 6-carboxy
Xanthine: 2,6-dioxy	

Oxygenated purines and pyrimidines exist as *tautomeric* structures with the keto form predominating and involved in hydrogen bonding between bases in nucleic acids:

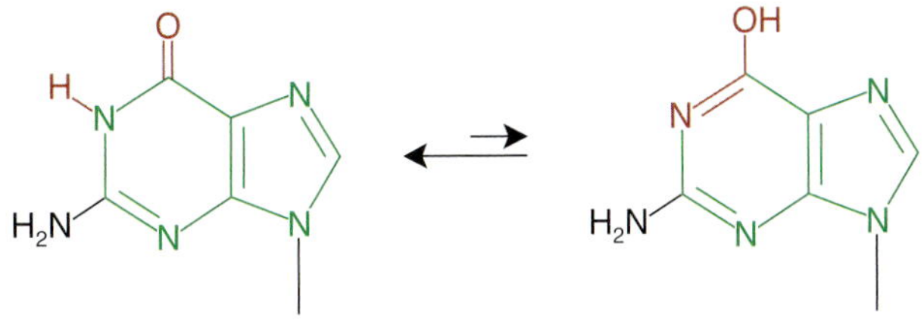

Keto

Enol

Nucleosides have either **β**-D-ribose or **β**-D-2-deoxyribose in an N-glycosidic linkage between C1 of the sugar and N9 (purine) or N1 (pyrimidine).

Nucleotides have one or more phosphate groups esterified to the sugar. Phosphates, if more than one is present, are usually attached to each other via phosphoanhydride bonds. Monophosphates may be designated as either the base monophosphate or as an *-ylic acid* (AMP: adenylic acid):

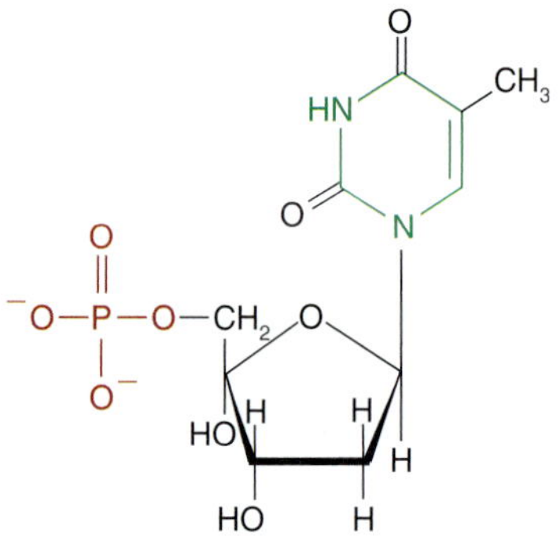

By conventional rules of *nomenclature,* the atoms of the base are numbered 1–9 in purines or 1–6 in pyrimidines and the carbon atoms of the sugar 1′–5′. A nucleoside with an unmodified name indicates that the sugar is ribose and the phosphate(s) is/are attached at C-5′ of the sugar. Deoxy forms are indicated by the prefix d (dAMP = deoxyadenylic acid). If the phosphate is esterified at any position other than 5′, it must be so designated [3′-AMP; 3′-5′-AMP (cyclic AMP = cAMP)]. The nucleosides and nucleotides (ribose form) are named as follows:

Base	*Nucleoside*	*Nucleotide*
Adenine	Adenosine	AMP, ADP, ATP
Guanine	Guanosine	GMP, GDP, GTP
Hypoxanthine	Inosine	IMP
Xanthine	Xanthosine	XMP
Cytosine	Cytidine	CMP, CDP, CTP
Uracil	Uridine	UMP, UDP, UTP
Thymine	dThymidine	dTMP, dTTP
Orotic acid	Orotidine	OMP

Minor (modified) bases and nucleosides also exist in nucleic acids. *Methylated* bases have a methyl group on an amino group (*N*-methyl guanine), a ring atom (1-methyl adenine), or on an OH group of the sugar (2′-*O*-methyl adenosine). *Dihydrouracil* has the 5–6 double bond saturated. In *pseudouridine,* the ribose is attached to C5 rather than to N1.

In **polynucleotides** (*nucleic acids*), the mononucleotides are joined by phosphodiester bonds between the 3′-OH of one sugar (ribose or deoxyribose) and the 5′-OH of the next (see p. 29 for the structure).

GLOSSARY

Francis Vella

ABC transporter protein: Integral plasma membrane protein that consists of two domains (each of six transmembrane segments) and two loops (each containing an ATP-binding site) and that uses the energy of ATP to transport a variety of hydrophobic natural products and synthetic drugs out of the cytoplasm. Also called ATP-binding cassette protein.

absorption: Passage across intestinal cell membranes of the products of digestion.

absorption spectrum: Pattern of absorption of incident light observed with a spectroscope or spectrometer.

abzyme: Catalytic antibody, natural or one that is raised against an antigen that has a transition state structure for a reaction. Also called catalytic antibody.

acceptor stem: In tRNA, the part in which the 5′- and 3′-ends are base-paired. The 3′-end terminates in an unpaired CCA sequence with the A being the site of attachment of a cognate aminoacyl residue.

acid anhydride: See anhydride.

acid–base catalysis: Catalysis of a reaction that is aided by transfer of a proton from an acid or abstraction of a proton by a base.

acidosis: Abnormal physiological state in which an increase in proton concentration lowers the pH of plasma below 7.35.

acrosome: Organelle at the tip of a spermatozoon that is rich in enzymes for digestion of the protective coat of an oocyte at fertilization.

actin: Globular protein that forms the microfilaments of cytoskeleton and the thin filaments of skeletal and cardiac muscle. Its interaction with the heads of thick filaments produces muscle contraction.

activation energy: Energy required for a specific chemical reaction to occur.

active site: Region usually in a cleft on an enzyme surface that binds the appropriate substrate and catalyzes its conversion to product.

active transport: Passage of solutes across a cell membrane that is effected with energy expenditure from ATP.

activin: Glycoprotein hormone that is secreted by the gonads and that stimulates secretion of follicle stimulating hormone. It consists of two B_A or B_B chains in inhibins A and B, respectively, or of a B_A and a B_B chain, joined by a disulfide bond.

acylcarrier protein (ACP): In prokaryotes, the protein with an attached phosphopantetheine moiety that is the central component of fatty acid synthase complex. In higher animals it is present as a domain of the multicatalytic fatty acid synthase protein.

adaptin: Major multimeric protein of clathrin-coated vesicles that binds on the cytoplasmic portion of various integral membrane proteins to mediate formation of the clathrin coat. Also called adaptor protein.

adaptor protein: See adaptin.

adenylate cyclase: Enzyme that is an integral protein of the plasma membrane. Its active site on the inner membrane surface converts ATP to cyclic AMP. Also called adenylyl cyclase.

adenylate cyclase system: Pathway of signal transduction in which a hormone binding to its cognate serpentine receptor activates a trimeric G protein to stimulate or inhibit adenylate cyclase, thereby affecting production of cyclic AMP as a second messenger.

adenylyl cyclase: See adenylate cyclase.

adherens junction: Part of a plasma membrane that contains cadherins, is involved in cell–cell adhesion, and is linked on the inner surface to actin microfilaments via catenins and other proteins.

adipose tissue: Loose connective tissue dominated by adipocytes whose major component is neutral fat (i.e., triacylglycerol). Two varieties are recognized, white and brown.

A-DNA: Double-helical DNA that has 11 base pairs per right-handed turn, with a diameter of ~26 Å, and forms in pure dehydrated DNA.

adrenaline: *N*-methylnoradrenaline, the major hormone derived from tyrosine in the adrenal medulla and some neurons. It is secreted in response to stress and increases glycogenolysis (in muscle and liver), gluconeogenesis (in liver), glycolysis (in muscle), and lipolysis (in adipocytes). Also called epinephrine.

Textbook of Biochemistry With Clinical Correlations, Sixth Edition, Edited by Thomas M. Devlin

adrenal gland: One of two small endocrine glands each situated above a kidney. It consists of (a) an outer cortex that secretes steroid hormones and (b) a central medulla that secretes catecholamines, particularly adrenaline. Also called suprarenal gland.

adrenocortical hormone: Any glucocorticoid, mineralocorticoid, or sex hormone secreted by the adrenal cortex.

adrenocorticotropic hormone (ACTH): Polypeptide hormone secreted by the anterior pituitary in response to corticotrophin-releasing hormone of the hypothalamus. It stimulates production and secretion of glucocorticoids by the zona fasciculata. Also called corticotropin, cortin.

aerobic: Occurring only in the presence of oxygen.

aerobic metabolism: Complete breakdown of glucose and fatty acids in muscle in the presence of adequate amounts of oxygen into carbon dioxide and water via acetyl-CoA. The oxygen-requiring part releases large amounts of energy as ATP and occurs within mitochondria.

agglutination: Aggregation of erythrocytes produced by interaction between antigens on their surface and specific antibodies in plasma.

agglutinin: Antibody in plasma that reacts with antigens on the surface of erythrocytes when donor and recipient differ in their blood group.

agglutinogen: Antigen on the surface of erythrocytes whose presence and structure are genetically determined.

aggrecan: Proteoglycan rich in chondroitin sulfate and keratan sulfate attached to a large protein that is present in cartilage.

agonist: Hormone, neurotransmitter, drug, or other compound that binds to a specific receptor and elicits a functional response.

alanine cycle: See glucose–alanine cycle.

aldimine: Compound formed by condensation of a primary amine with an aldehyde. Also called Schiff base.

aldosterone: Mineralocorticoid produced by zona glomerulosa and secreted in response to angiotensin II. It enhances reabsorption of Na^+ and water by renal tubules. Excessive secretion produces aldosteronism.

aldosteronism: Condition characterized by elevated blood pressure and low plasma K^+ and high plasma Na^+ concentrations produced by oversecretion of aldosterone by the zona glomerulosa.

alkalosis: Abnormal physiological state in which a deficiency of protons or an excess of bicarbonate ions raises the pH of plasma above 7.45.

alkaptonuria: Autosomal recessive condition in which homogentisate, produced during catabolism of phenylalanine and tyrosine, accumulates because of a deficiency of homogentisate oxidase activity.

all-α protein: Protein that consists solely or mostly of α-helical segments.

all-β protein: Protein that consists solely or mostly of β-strands.

allosteric effector: See allosteric modulator.

allosteric site: Site on a regulatory enzyme that is remote from the active site and, when bound to an appropriate effector or modulator, increases or decreases the affinity of the enzyme for its substrate.

allostery: Property of an enzyme or other protein whereby non-covalent binding of a ligand at one site induces a conformational change that affects binding of a ligand at another site.

α-helix: Common form of secondary structure of polypeptides; spiral conformation that results from maximal hydrogen bonding between components of peptide bonds of adjacent turns of the spiral, each turn being 3.6 residues long.

α/β barrel: Protein structure in which alternating α-helices and β-strands form a barrel of parallel β-strands, surrounded by a barrel of α-helices.

$\alpha + \beta$ protein: Protein in which α-helical and β-strand regions are present in a nonrandom fashion.

α/β protein: Protein in which α-helices and β-strands are interspersed or alternate with each other.

alternative RNA splicing: Deletion of certain exons or incorporation of alternative exons during splicing of pre-mRNA, thus forming more than one mRNA and generating more than one protein from a primary transcript. Also called differential RNA splicing.

aminoacyl-tRNA: tRNA carrying an amino acid to be added to the growing peptide chain during polypeptide synthesis.

aminoacyl site: Site on a ribosome that binds aminoacyl-tRNA during polypeptide synthesis. Also called A site.

amino terminus: See N-terminus.

aminotransferase: Enzyme that catalyzes transfer of an amino group from an α-amino to an α-keto acid, yielding a new α-keto acid and a new α-amino acid and using pyridoxal phosphate as coenzyme. Also called transaminase.

amphipathic: Having both a hydrophilic region and a hydrophobic region.

amphoteric: Capable of serving as an acid (by donating protons) or as a base (by accepting protons).

amyloid: Insoluble protein highly resistant to proteolysis that is present in the brain in a variety of neurological diseases. It consists of aggregates of fibrils composed of subunits rich in β-sheet structure. The β-amyloid present in plaques in Alzheimer's disease consists of a polypeptide (39–43 amino acid residues) derived from β-amyloid precursor protein.

anabolism: Phase of intermediary metabolism in which energy-utilizing metabolic reactions are used for synthesis of macromolecules and other biomolecules from simpler precursors.

anaerobic metabolism: Breakdown of glucose with production of lactate that occurs in vigorous exercise when muscle lacks adequate oxygen.

anaphase: Stage in mitosis in which the paired chromatids separate and move to the opposite ends of the spindle apparatus.

anaplerotic reaction: Reaction that replenishes the supply of intermediates in a metabolic pathway or cycle.

androgen: Steroid hormone secreted mainly by interstitial cells of the testis and also secreted in small amounts by the adrenal cortex in both genders. It stimulates development and maintenance of the reproductive system and secondary sex characteristics in males. Oversecretion leads to virilization in either gender.

angiogenesis: Formation of new blood vessels or their development and growth from existing ones.

angiotensin I: N-terminal decapeptide released by renin acting on the plasma protein angiotensinogen.

angiotensin II: N-terminal octapeptide released in the lung capillaries by angiotensin-converting enzyme acting on angiotensin I. It produces arteriolar constriction and a rise in blood pressure, and it promotes thirst and release of antidiuretic hormone and aldosterone.

angiotensin-converting enzyme (ACE): Transmembrane glycoprotein present in the lung capillaries and plasma. It is a zinc-containing peptidase that cleaves off the C-terminal dipeptide of angiotensin I to produce angiotensin II.

angiotensinogen: Plasma glycoprotein secreted by the liver, the precursor of angiotensin I.

Angstrom (Å): Unit of length, 10^{-10} meters.

anhydride: Product of condensation, with loss of water, of two molecules of an acid. Also called acid anhydride.

ankyrin: Large globular protein that is associated with the channel for antiport of Cl^- and HCO_3^- in the membrane, and with spectrin of the cytoskeleton of erythrocytes.

anomer: Either of the stereoisomers formed by cyclization of an aldose or a ketose, differing only in the configuration around the carbonyl carbon.

anoxia: State of tissue deprivation of oxygen.

ANP: See atrial natriuretic peptide.

antagonist: Agent that binds to a receptor and inhibits binding of the specific agonist for the receptor.

anterograde transport: Transport toward the periphery of an axon.

antibody-based immunity: Type of immunity mediated by circulating antibodies secreted by B cells in response to an antigen. Also called humoral immunity.

anticodon: Three bases on a tRNA molecule that interact with a complementary codon on an mRNA molecule.

anticodon arm: In tRNA, the stem and loop portion that contains the anticodon region.

antidiuretic hormone (ADH): Nonapeptide hormone produced in the hypothalamus but secreted by the posterior pituitary. It causes reabsorption of water by causing recruitment of an aquaporin into the apical membranes of renal tubules and collecting ducts, and raises blood pressure. Deficient secretion causes diabetes insipidus. Also called vasopressin.

antigen: Substance or cluster of chemical groups capable of inducing production of specific antibodies by B cells.

antigenic determinant: Portion of an antigen that is responsible for eliciting production of an antibody and that interacts with the antibody.

antigen-presenting cell (APC): Cell that is essential for initiating a normal immune response by processing antigens and displaying them on its surface bound to MHC proteins.

antiparallel: Relative orientation of two β-strands (N to C and C to N, respectively) within the same or on two polypeptides, or within two strands of DNA ($5'$ to $3'$ and $3'$ to $5'$, respectively).

antiport: Cotransport of two different solutes in opposite directions across a membrane.

antisense RNA: RNA, synthetic or transcribed from the non-coding strand of genetic DNA by genetic engineering, that is complementary to and inhibits the RNA transcript of that gene.

APC protein: Tumor suppressor protein that promotes degradation of β-catenin. It is absent or nonfunctional in adenomatous polyposis coli.

AP endonuclease: See apurinic/apyrimidinic endonuclease.

apoenzyme: Catalytically inactive protein component of an enzyme freed from any required cofactor, coenzyme, or prosthetic group.

apolipoprotein (apoLp): Protein component of a plasma lipoprotein freed from any associated lipids.

apoprotein: Protein component of a conjugated protein freed from any non-amino acid moiety.

apoptosis: Form of cell death in which the cell shrinks and condenses, its cytoskeleton collapses, the nuclear membrane disintegrates, and the DNA becomes fragmented. The cell membrane becomes altered and evokes rapid phagocytosis without loss of cell contents into the surrounding medium or induction of an inflammatory response. The process is genetically programmed and depends on activation of a caspase cascade. Also called programmed cell death.

AP site: See apurinic site, apyrimidinic site.

apurinic/apyrimidinic endonuclease (AP endonuclease): Endonuclease that removes a segment of DNA that contains an apurinic or an apyrimidinic site during base excision repair.

apurinic site (AP site): Site on a strand of DNA from which a purine base is missing.

apyrimidinic site (AP site): Site on a strand of DNA from which a pyrimidine base is missing.

aquaporin: Integral membrane homotetrameric protein that forms a hydrophilic channel for rapid movement of water down a concentration gradient. Each subunit contains six transmembrane segments.

arrestin: Protein that binds to the phosphorylated intracellular segment of a serpentine receptor and prevents its interaction with a trimeric G protein.

A site: Binding site on both ribosomal subunits that holds the tRNA that carries the amino acid to be added to the growing peptide chain during polypeptide synthesis. Also called aminoacyl site.

asymmetric carbon atom: See chiral carbon atom.

ATM protein: Large protein kinase that is mutated in ataxia telangiectasia and is required to activate p53 in response to double-strand DNA breaks induced by ionizing radiation or oxidative damage.

ATP-binding cassette protein: See ABC transporter protein.

ATP synthase: See F-type ATPase.

atrial natriuretic peptide (ANP): Peptide hormone secreted by cardiocytes of the atrium when stretched by increased venous return. It promotes reduction of blood pressure and of venous return, and renal water loss. Its receptor contains guanylate cyclase activity in its intracellular segment. Also called atriopeptin.

atriopeptin: See atrial natriuretic peptide.

attenuator: Sequence within the leader region of nascent mRNA that is involved in regulating expression of certain genes for enzymes of amino acid synthesis in prokaryotes by terminating transcription in presence of an adequate supply of the particular amino acid.

autoantibody: Antibody that interacts with antigens on the surface of a person's own cells and tissues.

autocatalysis: Process in which the product of a reaction is the catalyst for the same reaction (e.g., activation of pepsinogen by pepsin).

autocrine hormone: Biomolecule that acts on the cell that secretes it or on neighboring cells of the same type.

autoimmune disease: Disorder (such as myasthenia gravis) in which the immune system behaves abnormally by responding to antigens that are normal components of an individual's cells or tissues.

autolysis: Self-digestion of a proteolytic enzyme, or the breakdown of a cell or tissue by its own degradative enzymes.

autophosphorylation: Phosphorylation of a protein kinase by another molecule of the same enzyme.

axonal transport: Movement of organelles and other particles along microtubules in the axon of a nerve cell, being anterograde (away from the cell body) and effected by kinesin or retrograde (toward the cell body) and effected by dynein.

backbone: Chain of atoms (N-C_α-C) formed by the amino acid residues in a polypeptide or of atoms ($C_{4'}$-$C_{3'}$-O-P-O-$C_{5'}$) formed by the repeating sugar phosphate units in a nucleic acid.

basal lamina: Thin layer of extracellular matrix beneath epithelial sheets or tubes, or surrounding individual muscle, adipose tissue, or Schwann cells, that separates them from the underlying connective tissue.

basal metabolic rate (BMR): Number of kilocalories that a normal, resting, fasting individual requires to maintain that state during a period of time.

base-excision repair: Correction of an abnormal base in a strand of DNA by removal of that base by a DNA glycosylase, removal of the remaining deoxyribosephosphate unit and possibly neighboring nucleotides by an AP endonuclease, and filling in of the gap by DNA polymerase and DNA ligase.

basement membrane: Layer of filaments and fibers that attach an epithelium to the connective tissue beneath it.

base pairing: Formation of a maximum number of hydrogen bonds between a pair of complementary bases (a purine and a pyrimidine) on two strands of a nucleic acid or within one strand folded back on itself.

basic leucine zipper (bZIP): Protein motif in which a DNA-binding basic α-helix is the N-terminal extension of a leucine zipper. It is commonly present in transcription factors.

basophil: White blood cell with an affinity for basic stains. It contains granules of vasoactive amines and is important in the local response to infection.

B cell: Type of lymphocyte derived from the bone marrow that is capable of maturing and differentiating into a plasma cell that produces antibodies or into a B cell that retains a memory for a previous immune response to an antigen.

B-DNA: Double-helical DNA that has 10.5 base pairs per right-handed turn, along with a diameter of 20 Å, and is the most stable conformation under physiological conditions.

Bence Jones protein: Free light chain of an immunoglobulin produced in excess in myelomatosis. It is detectable in plasma and/or urine.

β–α–β loop: Protein structural motif consisting of two parallel β-strands separated by an α-helix.

β-amyloid: See amyloid.

β-amyloid precursor protein (β-APP): Single-pass integral glycoprotein of plasma membrane. The isoform that contains 695 residues is expressed in neurons, and those that contain 751 and 770 are expressed mainly in non-neuronal tissues. Cleavage of β-APP 695 within the transmembrane segment by α-secretase releases β-amyloid, accumulation of which forms amyloid plaques that injure nerve cells and are associated with Alzheimer's disease.

β-barrel: Supersecondary structure of proteins that consists of an antiparallel β sheet rolled into a cylinder.

β-bend: See reverse turn.

β-conformation: See β-strand.

β-hairpin: Protein motif in which two antiparallel β-strands are connected by a turn or short loop.

β-oxidation: Metabolism of fatty acids by oxidation of the β-carbon (or C2) with release of acetyl-CoA.

β-pleated sheet: See β-sheet.

β-sheet: Common protein motif in which β-strands in different regions of a polypeptide or on two polypeptides, run beside each other in parallel or antiparallel direction and are held together by hydrogen bonds between the nitrogen and oxygen atoms of their peptide bonds. Also called β pleated sheet.

β-strand: Common secondary structure in which a segment of a polypeptide assumes an extended zigzag arrangement. Also called β-conformation.

benign: Tumor that is noninvasive and usually grows slowly.

bilayer: Double layer of amphipathic lipid molecules arranged with their polar ends oriented toward the solvent surface and their nonpolar regions associated in the center.

bile: Alkaline secretion produced by the liver and contains bile pigments, bile salts, cholesterol and phospholipids, which become concentrated during storage in the gall bladder. It is released into the duodenum and is essential for emulsification and digestion of dietary lipid.

bile pigment: Either of the linear tetrapyrroles, bilirubin and biliverdin, that are produced by catabolism of heme. They are excreted in bile as the bisglucuronides.

bile salt: Na or K salt of a glycine or taurine conjugate of a bile acid (e.g., cholic acid) derived from cholesterol. It is responsible for emulsification of dietary lipids and their digestion by pancreatic lipase.

binding site: Region, usually a crease or pocket, on a protein surface in which a ligand binds.

biocytin: ϵ-*N*-biotinylysine, present as a prosthetic group in the holoprotein of biotin-dependent carboxylases.

bioenergetics: Analysis of energy production and utilization during metabolism.

biomolecule: Organic compound normally present in, or formed by, living organisms.

biopolymer: Biological macromolecule formed by covalent linking of similar or identical small units, end to end, to form a long chain. Examples are polysaccharides, polypeptides, and polynucleotides.

biotransformation: Chemical reactions responsible for conversion of drugs and other xenobiotics into metabolites that are more water-soluble for excretion. Consists of phase I and/or phase II reactions.

blood clot: Network of fibrin fibers and trapped blood cells that is formed when blood is shed or a blood vessel is damaged. When it forms within a blood vessel it is called a thrombus.

body mass index (BMI): Measure of adiposity obtained from the weight in kilograms divided by the square of the height in meters. An index up to 25 is considered normal.

Bohr effect: Decreased affinity of hemoglobin for oxygen when the carbon dioxide and H^+ levels in blood are increased.

bond energy: Energy that is required to break, or that is released in making, a chemical bond. Also called bond strength.

bond strength: See bond energy.

botulinum toxin: Mixture of neurotoxic zinc metalloproteases formed by *Clostridium botulinum*, the causative agent of botulism (severe food poisoning). It is a potent inhibitor of acetylcholine release from cholinergic neurons.

bound ribosome: Ribosome that is attached to the outer surface of the endoplasmic reticulum and that synthesizes proteins destined for the endoplasmic reticulum lumen, Golgi complex, lysosomes, secretory vesicles, or plasma membrane.

branched-chain ketonuria: See maple syrup urine disease.

BRCA: Either of two genes (BRCA1, BRCA2) that, when mutated, greatly increase the risk of development of breast cancer. Both genes encode large proteins that are involved in DNA binding and protein–protein interactions.

brown fat: Specialized form of adipose tissue rich in mitochondria in which fuel oxidation is used for thermogenesis rather than for ATP synthesis. It is the site of nonshivering thermogenesis.

brush border: Dense covering of microvilli that are formed by the plasma membrane on the apical surface of cells of the intestinal mucosa and renal tubules.

buffer: Mixture of nearly equal amounts of a weak acid and its salt that stabilizes the pH of a solution by releasing or removing hydrogen ions.

cadherin: Single-pass intrinsic glycoprotein of adherens junctions of the plasma membrane. It mediates Ca^{2+}-dependent cell–cell adhesion. Several types (E, in epithelia; N, in nerve and muscle; P, in placenta and epidermis; and VE, in vascular endothelium) exist.

calcitonin: Polypeptide hormone that contains 32 residues, is secreted by parafollicular or C cells of the thyroid gland in response to an increase in circulating Ca^{2+} concentration, and increases the rate of bone calcification and of Ca^{2+} loss from the kidney.

calcitriol: 1,25-Dihydroxycholecalciferol, the active form of vitamin D_3.

calmodulin (CaM): Dumbbell-shaped protein present in cytosol that contains 148 residues and four Ca^{2+}-binding sites. It activates a variety of Ca^{2+}-calmodulin-dependent protein kinases (CaM kinases).

calorie (cal): Amount of energy required to raise the temperature of a gram of water from 14.5°C to 15.5°C, being equal to 4.184 joules. In nutrition the energy released by combustion of a food in the body is measured in kilocalories (1000 cal).

CaM kinase: Ca^{2+}-calmodulin-dependent protein kinase.

5′-cap: 7-Methylguanosine linked to the 5′-terminal nucleotide of an mRNA via a 5′,5′-triphosphate bridge. It is formed during processing of pre-mRNA in the nucleus, inhibits degradation, and enhances translation of the mRNA.

cap-binding complex (CBC): Complex of proteins, including initiation factor 4F, that binds the 5′-cap of mRNA.

capsid: Protein shell composed of subunits (called capsomeres) that encloses a viral genome.

capsomere: Protein subunit of a capsid.

carbaminohemoglobin: Hemoglobin with carbon dioxide molecules bound to its N-terminal amino groups.

carbohydrate loading: Practice common (among athletes) of exercising intensively for a period (to deplete muscle glycogen) followed by consumption of a high-carbohydrate diet (to promote larger than normal glycogen stores).

carboxyl terminus: End of a polypeptide chain that contains an α carboxyl group that is free or, less frequently, aminated. Also called C-terminus.

carcinoma: Most common form of cancer in humans, arising from epithelial cells.

cardiac muscle: Specialized form of striated muscle that is characteristic of the heart. It is rich in mitochondria that are in contact with fibrils that branch and interdigitate. The membranes at the ends that abut each other are extensively folded to form intercalated disks, while those along the sides fuse over considerable distances and form gap junctions that permit the muscle to function like a syncytium. Also called myocardium.

cardiomyopathy: Progressive disease characterized by damage to cardiac muscle tissue.

carnitine: 4-*N*-trimethyl-3-hydroxybutyrate, derived metabolically from ϵ-aminomethylated lysine residues of certain proteins. It is present especially in muscle and liver, and it participates in transfer of fatty acyl-CoA across the inner mitochondrial membrane.

carnitine shuttle: Transport of fatty acyl-CoA across the inner mitochondrial membrane by formation of a fatty acyl-carnitine that is exchanged for free carnitine in its passage through an acylcarnitine/carnitine transporter.

carotene: Yellow-orange unsaturated tetraterpene hydrocarbon present in carrots and green and orange leafy vegetables. It is the precursor of retinol (vitamin A), retinaldehyde, and retinoic acid in the body.

cartilage: Flexible type of connective tissue that lacks blood vessels and nerves, and it consists of chondrocytes and a matrix composed of collagen fibrils and the proteoglycan aggrecan that is rich in chondroitin sulfate that they secrete. It is present mostly at the ends of long bones in the articulating surfaces of joints, and in the intervertebral disk. During development it is the precursor of the long bones.

cascade: Process that consists of sequential activation of an increasing number of components of a signaling system so as to amplify that signal.

caspase: Cytosolic proteolytic enzyme that contains a cysteine residue in its active site, cleaves its substrates on the carboxyl-terminal side of an aspartate residue, is activated from a procaspase, and is involved in apoptosis.

caspase cascade: Amplifying proteolytic cascade in the cytosol in which a procaspase is activated by an initiator of apoptosis and then activates another procaspase, and so on, until apoptosis is produced.

catabolism: Phase of intermediary metabolism in which nutrient molecules are degraded with release of energy.

catalytic antibody: See abzyme.

catalytic triad: Histidine residue hydrogen-bonded with a serine and an aspartate residues in the active site of chymotrypsin and of other serine proteases.

catechol: Aromatic ortho-diol, the characteristic chemical group present in catecholamines.

catecholamine: Catechol derivative of tyrosine—such as dopamine, noradrenaline, and adrenaline, which are major elements in the body's response to stress.

catenin: Either of two proteins (α and β) that mediate linking of cadherin in adherens junctions to the actin microfilaments of cytoskeleton.

cathepsin: Proteolytic enzyme found in lysosomes.

CCAAT box: Consensus sequence CCAAT present in promoter elements in transcription promoters of many eukaryotic genes.

CDC gene: See cell-division-cycle gene.

cell adhesion molecule (CAM): Integral protein of the plasma membrane that mediates binding of cell to cell or to extracellular matrix in a Ca^{2+}-dependent (e.g., a cadherin) or Ca^{2+}-independent (e.g., neural CAM) fashion.

cell coat: See glycocalyx.

cell-division-cycle gene (cdc gene): Gene that controls one or more steps in the cell cycle.

cell junction: Specialized region of a plasma membrane at which cell to cell or extracelluar matrix adhesion occurs.

cell-mediated immunity: Type of immunity mediated by T cells in lymphoid tissue and in the circulation that functions in defense against microorganisms (including viruses within cells) and foreign tissue transplants.

centriole: Intracellular organelle that consists of nine groups of short microtubules with three in each group and that organizes the microtubules of the spindle apparatus during cell division.

centromere: Region where daughter chromatids remain connected after a chromosome has replicated and where fibers of the spindle apparatus become attached.

centrosome: Region of the cytoplasm that contains a pair of centrioles oriented at right angles to each other.

ceramide: Metabolic precursor of the sphingolipids consisting of a fatty acid in amide linkage with the C2 amino group of sphingosine.

cerebroside: Glycosphingolipid that contains a glucose or galactose moiety linked to C1 of ceramide.

cerebrospinal fluid (CSF): Fluid that bathes the internal and external surfaces of the central nervous system. It is secreted and reabsorbed via aquaporins in membranes of cells of the choroid plexus.

channeling: Direct transfer of the product of an enzyme reaction to the active site of another enzyme for which it is the substrate.

chaotropic agent: Ion or small organic molecule that increases the solubility of nonpolar substances in water. It is useful for dissolving membranes, solubilizing particular proteins, and denaturing proteins and nucleic acids.

chaperone: Protein that helps other proteins avoid misfolding and production of inactive or aggregated species. Also called heat shock protein, molecular chaperone.

charge repulsion: Mutual repulsion of chemical groups with like charge.

chemiosmosis: Mechanism by which certain membranes in cells use the energy of electron transfer to pump protons across the membrane and to harness the energy stored in the H^+ gradient to drive cellular work, including ATP synthesis.

chief cell: Cell in the gastric mucosa that secretes pepsinogen. Also called peptic cell.

chiral atom: Carbon atom with four different substituents that can exist in two configurations making it optically active. Also called asymmetric carbon atom, chiral center.

chloride shift: Exchange of plasma Cl^- for HCO_3^- produced by dissociation of carbonic acid within erythrocytes.

cholecystokinin (CCK): Polypeptide hormone that contains 39 residues, is secreted by the duodenal mucosa, and causes contraction and emptying of the gallbladder and secretion of enzymes by the pancreas. Also called pancreozymin.

cholelithiasis: Presence or formation of biliary calculi or gallstones.

choline: *N*-Trimethylethanolamine, a vitamin component of acetylcholine, phosphatidylcholine, plasmalogens, and platelet-activating factor.

chondroitin sulfate: Predominant glycosaminoglycan of cartilage. It consists of the repeating unit glucuronic acid β1-3*N*-acetylgalactosamine, and it forms part of the proteoglycan aggrecan.

chorionic gonadotropin (CG): Protein hormone homologous with growth hormone and prolactin. It is secreted by the placenta, has most of the actions of growth hormone, and maintains the corpus luteum during the first third of human pregnancy. Also called chorionic somatomammotropin, placental lactogen.

chorionic somatomammotropin: See chorionic gonadotropin.

chorionic villus sampling: Technique of obtaining a sample of the fetal portion of the placenta for determining the presence of genetic abnormalities in the fetus while it is still in the womb.

chromaffin cell: Cell that is found in clusters in the adrenal medulla and elsewhere and that contains granules for biosynthesis, storage, and secretion of the catecholamines.

chromatid: Complete copy of the double-stranded DNA of a chromosome with its associated nucleoproteins.

chromatin: Complex of DNA, histones, and other proteins in the nucleus of a eukaryotic cell.

chromosome: DNA strand that contains a portion of the genome along with associated proteins. In humans there are 23 pairs of chromosomes.

chromosome condensation: Process by which, prior to the M phase of the cell cycle, chromatin becomes packed into chromosomes.

chromosome decondensation: Process by which chromosomes present in the M phase of the cell cycle become less compact and form chromatin when that phase is passed.

chylomicron: Particle present in plasma during the absorption of dietary lipid. It consists of triacylglycerols, phospholipids, cholesterol esters, fat-soluble vitamins, and apolipoprotein B48. It is synthesized and secreted by cells of the intestinal mucosa and enters venous blood via the lymphatic duct.

chyme: Acidic, semifluid mixture of ingested food and digestive secretions that enters the small intestine from the stomach during the early phases of digestion.

cirrhosis: Disorder of the liver in which degeneration of the hepatocytes is accompanied by their replacement by fibrous connective tissue.

cis-acting regulatory element: Sequence of DNA that regulates expression of a gene on the same chromosome (e.g., a promoter or an enhancer).

cis-Golgi network: Interconnected cisternae and tubules of the Golgi complex that are closest to the endoplasmic reticulum and receive protein- and lipid-containing vesicles from that reticulum. It is the site of phosphorylation of oligosaccharides on proteins destined for lysosomes.

cisterna: Flattened membrane-bound compartment of the Golgi complex and endoplasmic reticulum.

clathrin: Protein composed of three large and three small polypeptides arranged into a three-legged structure (triskelion) that can assemble into a basket-like framework on the inner surface of a portion of the membrane to form a structure that can be released into the cytoplasm as a coated vesicle.

clathrin-coated vesicle: Type of vesicle that mediates transport from the plasma membrane or Golgi complex. Its outer surface is covered by a framework of clathrin triskelions bound to transmembrane proteins via adaptins.

cloning: Construction of a recombinant DNA molecule, its insertion into a vector and then into a cell, and the production of multiple copies by expression in the progeny of that cell.

cloning vector: Agent such as a plasmid or modified virus that is used to transfer DNA in genetic engineering.

coated vesicle: Small membrane-enclosed particle with a meshwork of clathrin or other proteins on its outer surface. It may be formed at the plasma membrane, endoplasmic reticulum, or components of the Golgi complex.

coding strand: Strand of genetic DNA that has the same sequence as RNA transcribed from that region; it is complementary to the template strand from which RNA is transcribed.

codon: Triplet of purine and/or pyrimidine bases in an mRNA molecule that specifies the insertion of a particular amino acid residue during polypeptide synthesis.

coenzyme: Complex nonprotein organic cofactor required for activity of some enzymes. Most vitamins are precursors of coenzymes.

coenzyme A (CoA): Nucleotide coenzyme that contains pantetheine as a constituent. Its terminal sulfhydryl group functions as an acyl carrier in many reactions.

coenzyme B_{12}: 5′-Deoxyadenosylcobalamine, derived by binding of a 5′-deoxyadenosyl group to the cobalt atoms of cobalamine (vitamin B_{12}), and the coenzyme in conversion of methylmalonyl-CoA to succinyl-CoA; also methylcobalamine, derived by binding of a methyl group to the cobalt of cobalamine, the coenzyme in conversion of homocysteine to methionine.

coenzyme Q (Q): Ubiquinone, the benzoquinone isoprenoid that functions as a lipophilic electron carrier from complexes I and II to complex III and cytochrome c in the mitochondrial electron transport system.

cofactor: Ion or molecule that on binding to the catalytic site of an apoenzyme renders it active.

cognate: Two biomolecules that normally interact (e.g., a receptor and its proper ligand).

coiled-coil: Structure present in some fibrous proteins (e.g., keratins, tropomyosin, myosin) formed by two polypeptides that contain α-helical regions that consist of repeats of seven residues (abcdefg) in which "a" and "d" are hydrophobic, or in which "g" is a leucine residue (leucine zipper proteins).

compartmentation: Subdivision of the enzymes of a eukaryotic cell into functionally distinct, membrane-enclosed spaces or organelles.

complement: System of plasma proteins secreted by the liver that interact on exposure to activated antibodies or to the surface of certain pathogens, to promote cell lysis, phagocytosis, and an inflammatory response.

complementary DNA (cDNA): DNA produced by retroviral reverse transcriptase from an RNA template.

complex oligosaccharide: Oligosaccharide of an N-linked glycoprotein that, by trimming and further modification in the Golgi complex, retains three of the mannose residues introduced in the endoplasmic reticulum. One or more trisaccharide units terminating in a sialic acid residue are added to the end of the oligosaccharide.

complex protein: See conjugated protein.

condensation reaction: Covalent linking of two molecules through loss (usually) of water.

configuration: Arrangement in space of atoms in an organic molecule that results from presence of one or more chiral centers or of one or more double bonds. It can only be changed by breaking or reforming of covalent bonds.

conformation: Arrangement in space of an organic molecule whose substituent groups can assume different positions around single bonds.

conjugated protein: Protein that contains one or more tightly associated non-amino acid moieties. Examples include metalloprotein, phosphoprotein, hemoprotein, glycoprotein, lipoprotein, nucleoprotein, and flavoprotein. Also called complex protein.

conjugation: Addition, usually in the liver, of a moiety of glucuronic acid, acetate, sulfate, glycine, glutamine or glutathione, to increase the polarity and excretability of nonpolar metabolites or foreign compounds introduced into the body. Also called Phase II reaction.

connectin: See titin.

connective tissue: Supporting tissue found between other tissues. It contains fibroblasts, chondrocytes, osteoblasts, osteoclasts, adipocytes, or smooth muscle cells and the extracelluar material that they produce.

connexin: Integral protein that spans the plasma membrane four times. A cluster of six connexins constitute a connexon.

connexon: Channel in the plasma membrane that consists of six connexin units. When affixed to a connexon on an adjacent cell, a gap junction is formed that permits transit of material between both cells.

consensus sequence: Sequence of nucleotides or of amino acid residues that occurs most frequently within a related region of two or more nucleic acids or polypeptides, respectively.

conservative substitution: Replacement in a polypeptide of a residue by another of similar characteristics (e.g., serine by threonine).

constitutive protein: Protein present in relatively constant amounts in the cells that produce it.

cooperative binding: Change in affinity (increase or positive cooperativity, decrease or negative cooperativity) for binding of a ligand at one site by prior binding of a molecule of the same (homotropic effect) or of a different (heterotropic effect) ligand at another site of a multimeric protein.

cooperativity: Interaction by which a conformational change induced in one subunit is transmitted to all others in a multimeric protein.

Cori cycle: See glucose–lactate cycle.

corpus luteum: Mass of yellow follicle cells that develops in the ovary after ovulation. It secretes progesterone.

corticoliberin: See corticotropin-releasing hormone.

corticosteroid: Steroid hormone synthesized from cholesterol and secreted by the adrenal cortex. It may be a glucocorticoid, a mineralocorticoid, or a sex hormone.

corticotropin: See adrenocorticotropic hormone.

corticotropin-releasing hormone (CRH): Polypeptide hormone that contains 41 residues and that is secreted from the hypothalamus and causes release of adrenocorticotropic hormone (ACTH) by the anterior pituitary. Also called corticoliberin.

cortin: See adrenocorticotropic hormone.

cotransport: See symport.

coupled reaction: Linkage of an exergonic with an endergonic reaction that transfers energy from one to the other.

covalent catalysis: Catalysis that occurs when part or all of a substrate molecule forms a covalent bond with a component of the enzyme's active site before transfer to a second substrate.

cristae: Infoldings of the inner mitochondrial membrane.

cross-talk: Interaction of pathways for hormonal signal transduction—for example, through effects on a common component or on a common second messenger or by similar patterns of phosphorylation of target proteins.

C-terminal residue: Residue in a polypeptide that contains an α-carboxyl group that is free or occasionally aminated.

C-terminus: End containing the C-terminal residue in a polypeptide.

cyanosis: Bluish discoloration of skin that results from presence of deoxygenated blood in capillaries near the body surface.

cyclin: Protein whose concentration fluctuates during the cell cycle and that activates a cyclin-dependent protein kinase thereby regulating progression through the cycle.

cyclin-dependent kinase (cdk): Protein kinase whose activity is modulated by the concentration of a cyclin.

cysteamine: Decarboxylated cysteine, a component (with pantothenic acid) of pantetheine in coenzyme A.

cytochrome: Heme protein involved in an electron transport system in the endoplasmic reticulum or inner mitochondrial membrane.

cytokine: Small protein secreted by cells of the immune system that mediates local cell–cell communication.

cytokine receptor: Plasma membrane receptor for a cytokine that is stably associated with a cytosolic tyrosine protein kinase and functions by modulation of specific gene activity.

cytoskeleton: Network of microtubules, microfilaments, and intermediate filaments present in the cytoplasm.

dalton (Da): Unit of atomic mass, one-twelfth the mass of a ^{12}C atom.

D arm: In tRNA the stem-and-loop portion that contains two or three dihydrouracil bases.

deletion: Loss of a fragment of a chromosome through breakage or unequal crossing over, or through mutational loss of one or more contiguous nucleotides from a gene.

denaturation: Loss of higher-order structure in a protein or nucleic acid as a result of change in pH, salt concentration, or temperature.

***de novo* pathway:** Pathway for synthesis of a biomolecule (e.g., a nucleotide) from preformed precursors.

depolarization: Change in transmembrane electrical potential from a negative value toward a more positive value.

detoxification: Removal of a harmful substance. Often refers to hydroxylation reactions catalyzed by cytochrome P450 enzymes of the hepatic smooth endoplasmic reticulum that render lipophilic drugs or other substances more water-soluble for excretion.

diabetes insipidus: Condition of excessive thirst and water intake and urinary excretion of a dilute urine, produced by decreased secretion of, or resistance to, antidiuretic hormone (vasopressin).

differential RNA splicing: See alternative RNA splicing.

differentiation: During development, the gradual appearance of specific cellular characteristics caused by differential gene activation and repression.

dinucleotide fold: See nucleotide-binding fold.

diploid: Having two sets of chromosomes, one set coming from each parent, as in somatic cells.

dipolar ion: See zwitterion.

disulfide bond: Covalent bond formed between sulfhydryl groups of two cysteine residues within the same polypeptide (intrachain) or between two polypeptides (interchain).

DNA glycosylase: Enzyme that hydrolyses the *N*-glycosidic bond between a base and a deoxyribose of DNA as the first step in base-excision repair.

DNA helicase: Enzyme that utilizes the chemical energy of ATP to catalyze strand separation in DNA before replication.

DNA library: Collection of bacterial or phage clones that contain cDNA that represents the mRNA of an organism, specific cell type, or tissue.

DNA ligase: Enzyme that seals a nick (creates a phosphodiester bond between adjacent sugars) in the backbone of DNA.

DNA melting: Unwinding and separation of the two strands of DNA on heating, the melting temperature being determined by the base composition of the DNA.

DNA methylation: Enzymatic transfer of the methyl group of *S*-adenosylmethionine to adenine or cytosine in DNA, which keeps genes in an inactive state.

DNA microarray: Compact arrangement of numerous short DNA sequences, derived by chemical synthesis or from PCR fragments immobilized on a solid, that can be easily probed by hybridization with other nucleic acids that have been fluorescently labeled.

DNA probe: Natural or synthetic labelled oligonucleotide that is used to detect a gene of interest by hybridization to a complementary sequence.

domain: Segment of a polypeptide or polynucleotide chain that may fold and function independently from the rest of the molecule.

dominant: Gene (or allele) that affects the phenotype when it is received from only one parent (i.e., in the heterozygous form). It also describes a character or trait due to such a gene.

dynamin: Cytosolic protein with GTPase activity that binds to the neck of a clathrin-coated pit and participates in separating it from the membrane as an intracellular vesicle.

dynein: Motor protein unrelated to the kinesins, composed of two or three heavy chains that contain a motor domain and a variable number of light chains. Cytoplasmic dyneins contain two heavy chains, are probably present in all eukaryotic cells, and are important in vesicle trafficking. Axonemal dyneins contain two or three heavy chains and are responsible for the sliding movement that drives the beating of cilia and flagella.

dynorphin: Polypeptide that has the same N-terminal pentapeptide as Leu-enkephalin. It is a powerful analgesic and is present in hypothalamus and brain stem and in the duodenum.

dystrophin: Cytoskeletal protein of cardiac and skeletal muscle that consists of a fibrous antiparallel homodimer of a large polypeptide whose central region consists of 24 repeats of a sequence characteristic of β-spectrin. It links the plasmalemma to actin filaments. It is absent in Duchenne muscular dystrophy and reduced or altered in structure in Becker muscular dystrophy.

effector enzyme: Membrane-associated enzyme that responds to an extracellular signal transmitted via a membrane receptor by producing an intracellular messenger.

electrogenic transport: Transport across a plasma membrane of ionic solutes by a process that produces a change in the membrane potential.

emulsification: Physical breaking up of ingested lipids in the intestinal tract, by mixing with bile salts and other emulsifying agents, to form smaller particles that are accessible to lipolytic digestive enzymes.

endergonic: Requiring energy input to occur; a nonspontaneous process.

endoplasmic reticulum (ER): Extensive network of membrane channels in the cell cytoplasm that is studded with (rough ER) or lacks (smooth ER) ribosomes and functions in protein synthesis and lipid metabolism and in their intracellular transport, storage, packaging, or secretion.

endorphin: Neuromodulator and analgesic peptide that includes in its structure the pentapeptide of Leu-enkephalin or Met-enkephalin, is produced mostly in the central nervous system and gastrointestinal tract, and binds to an opiate receptor. Also called opioid peptide.

endosome: Small intracellular vesicle formed by endocytosis that transfers its contents to a lysosome for degradation.

enterohepatic circulation: Secretion of bile salts by the liver followed by their absorption by the intestinal mucosa for return to the liver via the portal vein.

enthalpy (H): In thermodynamics the heat content of a system.

entropy (S): In thermodynamics the degree of disorder or randomness in a system.

enzyme cascade: Series of proenzymes or weakly active enzymes, in which activation of the first causes that of the second and so on. Activation may be by proteolytic cleavage (e.g., in blood clotting) or by phosphorylation (e.g., in the insulin receptor pathway) and results in amplification of the original activation signal.

enzyme-linked receptor: Integral protein receptor of the plasma membrane that has an intracellular catalytic domain (e.g., a protein kinase or protein phosphatase, guanylate cyclase) or becomes associated with a cytosolic enzyme and that is activated on ligand binding.

enzymopathy: Disorder produced by lack of an enzyme activity.

epinephrine: See adrenaline.

epitope: Region on the surface of an antigen that elicits secretion of a specific antibody and also binds to it. Also called antigenic determinant.

ER-resident protein: Protein that functions within the endoplasmic reticulum, being retained there via the sequence Lys-Asp-Glu-Leu (KDEL) at the C-terminal end.

ER retention signal: Sequence Lys-Asp-Glu-Leu (KDEL) at the C-terminal end of proteins that are retained and function within the endoplasmic reticulum.

ER signal sequence: N-terminal hydrophobic sequence that directs polypeptides destined for lysosomes, the endoplasmic reticulum, Golgi complex, secretory vesicles, or plasma membrane, to enter the endoplasmic reticulum where it is cleaved off by a signal peptidase soon after its entry.

E site: See exit site.

erythropoietin: Glycoprotein hormone secreted mostly by kidneys on exposure to low oxygen concentration that stimulates erythropoiesis.

essential fatty acid: Polyunsaturated fatty acid that is required but cannot be synthesized by an animal.

euchromatin: Decondensed form of chromatin that is available for transcription.

eukaryote: Organism that consists of one or more cells that each contain a nucleus and other membrane-bounded organelles.

excinuclease: Complex enzyme system that excises the abnormal segment of the affected DNA strand during nucleotide excision repair. Also called excision-repair endonuclease.

excision repair endonuclease: See excinuclease.

exergonic: Accompanied by release of energy; a spontaneous process.

exit site: Site, mainly on the large subunit of a ribosome, from which the uncharged tRNA is released after an elongation step in protein synthesis. Also called E site.

exocytosis: Release or secretion of material contained within an intracellular vesicle by fusion of its membrane with the plasma membrane.

exon: Segment of coding DNA of a eukaryotic gene that is represented in mRNA and the amino acid sequence of its translated protein, or in mature tRNA or rRNA. Also called expressed sequence.

expressed sequence: See exon.

extracellular matrix (ECM): Secreted intercellular component of connective tissue, consisting mostly of glycoproteins and proteoglycans.

extreme obesity: See morbid obesity.

extrinsic protein: See peripheral protein.

facilitated diffusion: Rapid permeation of solutes, based on their concentration gradient across a cell membrane, by interaction with specific membrane transport proteins. Also called passive transport.

familial hypercholesterolemia: Presence of abnormally high concentrations of cholesterol in LDL in plasma, the result of a lack of functional LDL receptors, that is transmitted as an autosomal dominant trait.

feedback inhibition: See negative feedback.

feedforward activation: Increase in activity of an enzyme in a metabolic pathway by a metabolic intermediate that was formed earlier in the pathway.

FeS center: An iron–sulfur cluster.

fibril-associated collagen: Collagen, such as type IX and type XII, that forms short triple-helical and nonhelical domains and binds to the side of collagen type I or type II fibrils, respectively.

fibrillar collagen: Collagen, such as type I, II, III, V or XI, that forms fibrils and occurs in connective tissue.

fibrillin: Large glycoprotein component of the microfibrils present in the extracellular matrix of many tissues; this is secreted by fibroblasts and forms a scaffold for deposition of elastin. Hereditary abnormalities in structure of the protein are associated with Marfan's syndrome.

fibroblast: Type of cell common in loose connective tissue; it secretes collagen and other macromolecules of the extracellular matrix and readily migrates to and proliferates in wounded tissue.

fibroblast growth factor (FGF): Family of proteins classed as acidic (from brain or retina) or basic (from brain, retina, or cartilage). They are mitogenic and angiogenic, act as growth factors in tissue culture, and exert their effects through receptors that are tyrosine kinases. Mutation in one form of FGF receptor is responsible for achondroplasia (the commonest form of dwarfism).

first law of thermodynamics: Law that requires that in all processes the total energy of the universe remains constant.

flagellum: Extension of the cell membrane of a spermatozoon that forms an organ of locomotion driven by dyneins. It consists of two single microtubules surrounded by a ring of nine doublet microtubules.

flippase: Protein that facilitates translocation of a membrane lipid from one membrane monolayer to the other.

flux: Flow of material through a metabolic pathway or of a solute across a membrane.

frame shift: Change in the coding frame of codons during protein synthesis as a result of insertion or deletion of one or more nucleotides into the DNA sequence of an exon.

free energy of activation: Energy required to raise a substrate from the ground state to the transition state at which it will react. Also called activation energy.

free ribosome: Ribosome in the cytoplasm that is unattached to any membrane and is used for synthesis of proteins destined for the cytosol, mitochondria, peroxisomes or the nucleus.

F-type ATPase: F_0F_1-ATPase of the inner mitochondrial membrane. Also called ATP synthase, complex V.

futile cycle: Two opposing sets of enzyme-catalyzed reactions that result in release of energy as heat by the net hydrolysis of ATP.

gated ion channel: Ion channel that opens and closes to allow alteration of the cell membrane potential. It may be voltage-gated, mechanically gated, or ligand-gated (extracellularly or intracellularly).

gene: Segment of chromosomal DNA that encodes a functional polypeptide or RNA molecule, including exons, introns, and regulatory sequences.

gene amplification: Selective synthesis of extra copies of a gene or genes. This is normal for some organisms and occurs for some oncogenes in certain cancers.

gene cloning: Formation by a bacterium that carries a foreign gene, introduced into it via a recombinant vector, of a clone of cells that contain the replicated gene.

gene knockout: Deletion or inactivation of a gene in an experimental animal by genetic engineering methods.

genomic DNA library: Collection of bacterial or phage clones that contain overlapping parts of the genome of an organism.

genomics: Study of DNA sequences and properties of genomes.

genotype: Genetic constitution or complement of an individual.

germ-line cell: Cell of the lineage that forms the gametes (sperm or ova).

glial cell: See neuroglia cell.

glicentin: Polypeptide derived from preproglucagon in special cells of the lower intestinal tract. It consists of glucagon extended by additional residues at either end, and at the N-terminal end by glicentin-related polypeptide. Its activity is similar to that of glucagon.

globin: Apoprotein of myoglobin or of hemoglobin.

globular protein: Protein whose tertiary structure makes it compact and spheroidal.

gluconeogenesis: Metabolic pathway for synthesis of glucose from noncarbohydrate precursors in the liver.

glucose–alanine cycle: Interorgan metabolic conversion of alanine released from skeletal muscles during prolonged fasting or starvation, into glucose that is produced and secreted into the blood stream by the liver. Also called alanine cycle or Randle cycle.

glucose–lactate cycle: Interorgan conversion of lactate released from skeletal muscle after exertion, into glucose produced and secreted into the blood stream by the liver. Also called lactate cycle, Cori cycle.

glucose tolerance: Ability of the body to utilize glucose as ascertained by the nature of the blood glucose curve following administration of a test amount of glucose. It is decreased most commonly in diabetes mellitus and in conditions associated with liver damage.

glycocalyx: Fuzzy coat on the external surface of cells, consisting of the carbohydrate component of glycoproteins and glycolipids of the cell membrane and of secreted glycoproteins and proteoglycans that adhere to that surface. Also called cell coat.

glycoform: Glycoprotein that differs only in the location and/or structure of the carbohydrate component from another.

glycolipid: Lipid that contains a monosaccharide (e.g., in glucosyl cerebroside) or an oligosaccharide (e.g., in a ganglioside) component.

glycolysis: Metabolic pathway by which a molecule of glucose is converted to two of pyruvate with net production of two ATP and two of NADH.

glycoprotein: Conjugated protein that contains covalently bound carbohydrate.

glycosuria: Presence of abnormally high concentration of glucose (most commonly) or of other sugars (e.g., galactose, fructose, pentose) in urine.

Golgi complex: Series of stacked membranous cisternae (cis, medial, and trans) in the cytoplasm of eukaryotic cells in which proteins transferred from the endoplasmic reticulum are modified and sorted, and also the site of synthesis of glycosaminoglycans of the extracellular matrix.

Golgi stack: Pile, usually of three cisternae, between the cis- and the trans-Golgi networks; this is the site of processing of N-linked oligosaccharides and of *O*-glycosylation and also production of proteoglycans.

G_0 phase: Nondividing state of a cell that has exited the cell cycle at the restriction point during the G_1 phase.

G_1 xphase: Phase in the cell cycle that consists of the growth portion in interphase before DNA replication begins.

G protein: GTP-binding protein with inherent GTPase activity. It is active with bound GTP but not with GDP and may be monomeric (e.g., ras) or heterotrimeric (containing α-, β-, and γ-subunits), and it plays an important role in intracellular signaling pathways.

G protein-linked receptor: Serpentine receptor of the plasma membrane that on binding of a ligand activates a trimeric G protein.

granulocyte: Leukocyte that contains granules visible under light microscopy and may be basophilic, eosinophilic, or neutrophilic on appropriate staining.

growth factor: Protein that promotes synthesis of proteins and other cell constituents and produces an increased cell mass but not cell number. It is frequently used to include mitogens and survival factors.

guide RNA: Small RNA that is involved in editing of pre-mRNA and of pre-rRNA (when it is call snoRNA).

half-life ($t_{1/2}$): Time required for decay or disappearance of one-half of an amount of a given component in a system. Also called half-time.

half-time: See half-life.

haploid: Having one set of chromosomes as in germ cells.

hb S: Hemoglobin that contains the mutant β globin that causes erythrocyte sickling.

H chain: See heavy chain.

heat shock protein: See chaperone protein.

heavy chain: Larger of the two types of polypeptide in a protein. Also called H chain.

helix cap: Structure formed when the side chain of residues such as asparagine or glutamine that flank an α-helix fold back to form hydrogen bonds with peptide bond components of the first or of the last turn of the helix.

helper T cell: Type of lymphocyte that by its secretion and other activities helps trigger and coordinate cell-mediated and antibody-mediated immunity.

hemoglobinopathy: Hereditary condition that manifests itself as a structural variant of a hemoglobin (e.g., sickle cell anemia), a thalassemia, or persistence of fetal hemoglobin.

heptad repeat: Seven-residue sequence (abcdefg) that is repeated in tandem in a polypeptide, in which "d" and "g" are hydrophobic, or in which "g" is leucine (as in a leucine zipper). Such heptads allow the polypeptides to dimerize as coiled-coils.

heterochromatin: Condensed form of chromatin that is not transcribed and forms a dense region, visible with a light microscope, in resting cells.

heterogeneous nuclear RNA (hnRNA): RNA in the nucleus that is usually the primary transcript of genes for polypeptides and contains one or more intronic sequences. Also called pre-mRNA.

heterogeneous nuclear ribonucleoprotein (hnRNP): Complex of hnRNA and protein present in the nucleus of eukaryotic cells.

heteropolysaccharide: Polysaccharide that consists of more than one type of monosaccharide (e.g., a glycosaminoglycan).

heterozygous: Having two different alleles at corresponding sites on a chromosome pair.

high-mannose oligosaccharide: Oligosaccharide of an N-linked glycoprotein that retains all or most of the nine mannose residues transferred onto it in the endoplasmic reticulum and undergoes little or no further modification in the Golgi complex.

histone core: Protein core of a nucleosome, consisting of two copies of each of histones H2A, H2B, H3, and H4.

HLA protein: Human leucocyte antigen or major histocompatibility complex protein.

homeobox: Conserved DNA sequence ($\sim$100 bp long) that encodes a protein domain (homeodomain) that regulates differentiation during development of the organism.

homeodomain: Protein domain ($\sim$60 residues long) that forms a helix–loop–helix structural motif and is encoded by a homeobox.

homeostasis: Maintenance of a relatively constant internal environment within the body.

homeotic gene: Gene involved in controlling the overall body plan by controlling the fate of groups of cells during development of the organism.

homologous: Macromolecules that share similarity because of a common evolutionary origin, and more specifically in the primary structure of polypeptides or nucleic acids.

homologous chromosome: One of a pair of chromosomes in somatic cells that are similar in length, position of centromere, and location of alleles for the same traits and that are derived from the father and the mother, respectively. In germ cells, only one member of a pair is carried by each gamete.

homopolysaccharide: Polysaccharide that consists of one type of monosaccharide (e.g., starch, glycogen).

homozygous: Having the same alleles at the corresponding sites on a chromosome pair.

humoral immunity: See antibody-based immunity.

hyaluronidase: Enzyme that hydrolyzes the glycosidic bonds of hyaluronic acid, is the spreading factor secreted by some bacteria, and is present at the acrosomal cap of spermatozoa.

hybridoma: Cell line that is obtained by fusing antibody-secreting lymphocytes with lymphoma cells and that is used for production of monoclonal antibodies.

hydrogen bond: Weak electrostatic bond in which a hydrogen atom in covalent linkage with an electronegative atom is partially shared with another electronegative atom.

hydrolase: Enzyme that catalyzes a hydrolytic cleavage reaction.

hydrophilic: Capable of associating freely with water molecules.

hyperammonemia: Presence of abnormally high concentrations of ammonia in blood usually associated with neurological and other abnormalities. It results from underutilization of ammonia in urea synthesis as a result of liver disease, inborn errors of urea synthesis, or in organic acidemias.

hyperchromic shift: Increase in absorption of ultraviolet light by a DNA solution as the DNA melts on raising the temperature.

hyperglycemia: Elevated plasma glucose concentration relative to that during fasting. It occurs physiologically within 2–3 h of a meal, but when higher than normal it is usually symptomatic of diabetes mellitus.

hyperplasia: Abnormal enlargement of an organ or tissue from increase in number of cells.

hypertrophy: Abnormal enlargement of an organ or tissue from increase in the size of its cells.

hypervariable region: Any of the three short loop segments in the variable region of the light or heavy chains of immunoglobulins that are poorly conserved and form part of the antigen-binding site. Also, a region in a polypeptide that shows great variability in sequence in different species.

hypervariable residue: Residue in a hypervariable region.

hypoglycin: Toxic nonproteogenic α-amino acid present in the unripe ackee fruit. Its deaminated and decarboxylated derivative is a potent inhibitor of β-oxidation of short-chain fatty acids and produces Jamaican vomiting sickness.

IF protein: See intermediate filament protein.

Ig domain: See immunoglobulin domain.

Ig superfamily: Family of proteins that contain one or more domains like those that characterize the immunoglobulins. It includes Ca^{2+}-dependent cell adhesion molecules, T- and B-cell receptors, MHC proteins, and transmembrane IgM.

immunoglobulin domain: Domain ($\sim$100 residues) that forms a sandwich consisting of three- and a four-stranded antiparallel β-sheets and is characteristic of the variable and constant regions of L and H chains of immunoglobulins. Also called immunoglobulin fold, Ig domain.

immunoglobulin fold: See immunoglobulin domain.

induced fit: Change in conformation induced in an enzyme by binding to its substrate, also the change in conformation induced in any macromolecule by ligand binding.

inducer: Compound that promotes the activity of a specific gene, also a stimulus that provokes a physiological response.

inducible protein: Protein that is produced in amounts that depend on conditions in the environment.

inflammation: Nonspecific local defence mechanism that is characterized by swelling, warmth, redness, and pain in a part of the body.

inhibin: Glycoprotein hormone secreted by the gonads that inhibits secretion of follicle stimulating hormone. It consists of a glycosylated α chain linked to a β_A chain (inhibin A) or to a β_B chain (inhibin B) by a disulfide bond.

inhibitory G protein (Gi): Heterotrimeric G protein that inhibits adenylate cylase or regulates ion channels.

inhibitory neurotransmitter: Neurotransmitter (e.g., GABA, glycine) that opens transmitter-gated channels for Cl^- or K^+ in the target cell plasma membrane and makes formation of an action potential more difficult.

initiation codon: Nucleotide triplet AUG that codes for methionine as the first amino acid residue during polypeptide synthesis in eukaryotes and for *N*-formylmethionine in prokaryotes, mitochondria and chloroplasts.

innate immunity: Antigen-nonspecific mechanisms involved in the early phase of the response to a pathogen. It is not increased by repeated presence of the pathogen and includes phagocytic cells, the complement proteins and secretion of cytokines.

insertion mutation: Mutation that results from addition of one or more extra nucleotides into coding DNA.

integral protein: Protein that traverses a cell membrane one or more times and can only be released by disruption of the membrane. Also called transmembrane protein, intrinsic protein.

integrin: Heterodimeric intrinsic protein of plasma membrane that links a cell to proteins of the extracellular matrix (e.g., collagen, laminin) or to other cells including white blood cells and platelets.

intercalating agent: Dye or other compound that contains an aromatic ring or other planar structure that can fit between two successive bases of DNA and may cause insertion or deletion mutations.

interchain: Interaction or linkage between parts of two strands of a biopolymer.

interconvertible enzyme: Enzyme whose activity is modulated by a reversible posttranslational modification (e.g., phosphorylation).

interferon (IFN): Protein released by cells infected by a virus or in response to other inducing agents. IFNα is secreted by leukocytes, IFNβ by fibroblasts, and IFNγ by macrophages and B cells. IFNs have nonspecific antiviral and anticancer activity.

intermediate filament (IF): Rope-like fiber (10-nm diameter) of the cytoskeleton that consists of one of a variety of IF proteins depending on cell type. The IFs of the nuclear lamina are composed of lamin proteins.

intermediate filament protein: Protein component of intermediate filaments of the cytoskeleton or nuclear lamina.

interphase: Major portion of the cell cycle in which the chromosomes are uncoiled and form chromatin between one M phase and the next.

intervening sequence: See intron.

intrachain: Interaction or linkage between parts within a strand of a biopolymer.

intrinsic factor: Glycoprotein secreted by parietal cells of the gastric mucosa that binds and facilitates absorption of cobalamin (vitamin B_{12}). Its deficiency, as from gastric atrophy, results in pernicious anemia.

intrinsic protein: See integral protein.

intron: Segment of noncoding DNA in a eukaryotic gene that is transcribed but is then excised from the primary transcript. Also called intervening sequence.

invariant residue: Amino acid residue that occurs in the same position in a homologous polypeptide derived from different species.

ion channel: Integral membrane protein that provides regulated passage for a specific ion or ions across a membrane.

ionophore: Compound (often an antibiotic) that increases ion flux across a plasma membrane by forming an ion channel or functioning as a mobile ion carrier in the membrane.

isoaccepting tRNA: tRNA species that accepts the same amino acid as another tRNA species but has a different anticodon.

isoenzyme: Protein that can catalyze the same reaction as one or more different proteins from the same species. Also called isozyme.

isomerase: Enzyme that catalyses transfer of a group within a molecule to yield an isomeric form.

isopeptide bond: Amide bond formed between an amino group of an amino acid and the carboxyl group of another where either or both occupy a position other than α (C2).

isoprenoid: Lipid (e.g., a steroid or lipid-soluble vitamin) that is structurally derived from isoprene units.

isoschizomer: Restriction endonuclease that cleaves at the same nucleotide sequence as another such enzyme.

isozyme: See isoenzyme.

jaundice: Yellowing of skin and conjunctiva due to elevated plasma and tissue bilirubin concentration caused by overproduction or underexcretion of this bile pigment.

junk DNA: Regions of DNA in a genome for which no function has yet been discovered.

KDEL: C-terminal tetrapeptide (Lys-Asp-Glu-Leu) signal for retention of proteins resident in the endoplasmic reticulum.

keratin: Fibrous protein that is a coiled-coil consisting of a type I (acidic) and a type II (neutral or basic) polypeptide cross-linked by disulfide bonds. They form a family of intermediate filament proteins synthesized by keratinocytes in epithelial tissues. Over 20 types occur in human epithelia, and at least 10 are specialized for hair and nails.

ketimine: Product of the condensation of a primary amine with a ketone. Also called Schiff base.

keto acid: Organic acid such as pyruvic acid or oxaloacetic acid that contains a keto group adjoining the carboxylic group (an α-keto acid) or elsewhere.

ketogenesis: Metabolic overproduction of acetoacetate and its derivatives β-hydroxybutyrate and acetone, within the mitochondrial matrix in liver (e.g., by increased catabolism of fatty acids) when acetyl-CoA is produced in excess during prolonged fasting or starvation.

killer T cell: Type of lymphocyte involved in cell-mediated immunity that secretes pore-forming proteins (perforins) that kill the target cells.

kilobase pair (kb): Unit of length of DNA, one thousand base pairs.

kinesin: Cytosolic motor protein that associates with an organelle and uses the energy of ATP to propel it along a microtubule.

kinetochore: Complex protein structure that links each sister chromatid to the spindle microtubules during metaphase.

L chain: See light chain.

lactate cycle: See glucose–lactate cycle.

lacteal: Lymph vessel that extends into the core of a villus in the small intestinal mucosa into which chylomicrons are passed during absorption of ingested lipid.

lactase deficiency: See lactose intolerance.

lactic acidosis: Condition in which blood pH is lowered by an excess of lactic acid. It results from severe exercise, shock, hypoxia, cardiovascular insufficiency, or intoxication by certain drugs, or it occurs as part of several inborn errors of metabolism.

lactose intolerance: Condition characterized by acid diarrhea and flatulence produced by the inability to digest lactose and resulting from intestinal lactase deficiency. Also called lactase deficiency.

lagging strand: DNA strand that is synthesized discontinuously and away from the replication fork during replication.

lamin: Protein component of intermediate filaments of the nuclear lamina. It may be isoprenylated for attachment to the inner surface of the nuclear membrane.

laminin: Flexible major component of basal lamina that consists of three long polypeptides (α, β, and γ) held together by disulfide bonds in the shape of a cross. It binds to the surface of cells and to other constituents of the basal lamina. Mutations in some forms of laminin are associated with epidermolysis bullosa.

leader sequence: Sequence of nucleotides, which may be coding or noncoding, that is present upstream of the initiation codon in mRNA and may have regulatory or targeting function. Also signal sequence of proteins synthesized on bound ribosomes.

leading strand: DNA strand that is synthesized continuously in the direction of the replicating fork during replication.

lectin: Highly specific carbohydrate-binding protein of plasma membranes, lumen of trans-Golgi network, or blood plasma.

leucine zipper: Structural motif present in many DNA-binding proteins that dimerize via coiled-coil formation by interaction between the side chains of leucine residues in tandem heptad repeats within α-helical regions.

leucine zipper protein: Protein that dimerizes by coiled-coil formation via interaction between leucine zippers that may be immediately preceded by a region rich in basic residues that form an α-helix or a helix–loop–helix fold.

leukotriene: Eicosanoid produced by leukocytes, macrophages, and platelets from arachidonic acid; this contains four double bonds, three of which are conjugated, and may be linked to glutathione, cysteinylglycine, or cysteine through their sulfur atom. Leukotrienes activate and attract leukocytes, increase capillary permeability, and are involved in inflammatory or immediate hypersensitivity reactions.

ligand: Small molecule that binds specifically and noncovalently to a larger one.

ligase: Enzyme that catalyzes bond formation between two substrates which is coupled with hydrolysis of ATP or another energy-rich compound. Also called synthetase.

light chain: Smaller of two types of polypeptide in a protein. Also called L chain.

lipid-anchored protein: See lipid-linked protein.

lipid-linked protein: Peripheral protein attached to a membrane by a fatty acyl (myristoyl or palmitoyl), isoprenoid (farnesyl or geranylgeranyl), or glycosylphosphatidylinositol group. Also called lipid-anchored protein.

lipoic acid (lipoate): 6,8-Dithiooctanoic acid, which may exist in oxidized or reduced form; this is joined in amide linkage to the side chain of a lysine residue in the dehydrogenase complexes specific for pyruvate, α-ketoglutarate, or branched-chain α-keto acids.

lipophilic: Hydrophobic, soluble in lipid.

lipoprotein: Multimolecular complex of protein, phospholipid, and cholesterol that encloses a core composed of triacylglycerols, cholesterol esters, or other lipids.

liposome: Artificial vesicle that consists of one or more closed concentric phospholipid bilayers that enclose some of the suspending aqueous medium in the central compartment.

lipoxin: Eicosanoid produced in leukocytes from arachidonic acid that contains four conjugated double bonds and functions in an autocrine or paracrine manner.

locus: Particular site on a chromosome where a gene for a particular trait is located.

lyase: Enzyme that catalyzes elimination of a group from a molecule with formation of a double bond.

lymphocyte: Spherical leukocyte with a large nucleus and sparse cytoplasm, the predominant cell in lymphoid tissue. It is either a B cell and responsible for antibody-based immunity or is a T cell and responsible for cellular immunity.

lymphokine: Cytokine secreted by activated lymphocytes.

lysosome: Cytoplasmic membrane-bounded organelle rich in hydrolytic enzymes for degrading and recycling unneeded cell components.

lysozyme: Enzyme of egg white and human body fluids that, by hydrolyzing polysaccharides in the cell wall of some bacteria, has antibiotic properties.

M phase: Portion of the cell cycle in which the nucleus and cytoplasm divide to form two daughter cells.

M protein: Abnormal protein present in plasma in myelomatosis. Also called myeloma protein, paraprotein.

macromolecule: Large polymeric molecule characteristic of living matter, formed by joining together of small subunits. Polysaccharides, polypeptides, and nucleic acids are macromolecules.

macrophage: Mononuclear actively phagocytic cell that migrates into tissues and ingests particulate material such as microorganisms and dead cells. It is rich in lysozyme and lysosomes and is an oxidative system that is microbicidal when activated.

major histocompatibility complex (MHC) protein: Integral membrane glycoprotein of antigen-presenting cells that carries on its extracellular region a peptide derived by intracellular digestion of a protein antigen during synthesis of the MHC protein. Also called human leukocyte antigen (HLA).

malignant: Tumor or cell that grows rapidly and is invasive.

mast cell: Cell of connective tissue that on stimulation initiates an inflammatory response by releasing histamine, serotonin, heparin, and other constituents.

matrix: Aqueous fluid of the cytoplasm, nucleus or other cell compartment.

M-cyclin: Cyclin that regulates progression of the cell cycle from G_2 phase to M phase.

MDR protein: See multidrug resistance protein.

mechanism-based inhibitor: See suicide inhibitor.

membrane potential: Difference in charge across the plasma membrane of a cell (positive outside, negative inside) caused by unequal distribution of ions between the extracellular fluid and the cytoplasm.

memory B cell: Long-lived nonproliferating lymphocyte that retains a memory of a previous immune response to an antigen.

metabolic syndrome: See syndrome X.

metabolic turnover: Continuous breakdown and synthesis of biomolecules within living cells.

metabolite: Substance that participates as an intermediate in metabolism.

metal ion catalysis: Catalysis that occurs when an enzyme-bound metal ion participates in labilizing a substrate.

metaphase: Stage in cell division when the nuclear membrane disintegrates, chromosomes line up at the equatorial plain, and microtubules growing from their centromeres attach to those of the mitotic spindle.

micelle: Spherical particle made up of amphipathic compounds such as bile acids, fatty acids, and monoacylglycerols, which form a hydrophilic surface and a hydrophobic interior that contains virtually no water.

microfilament: Fibrous (5- to 9-nm diameter) component of the cytoskeleton and of the contractile part of skeletal and cardiac muscle. It consists of actin and functions in cell structure and in movement. Also called actin filament.

microsome: Small membranous vesicle produced by fragmentation, mostly of the endoplasmic reticulum, on homogenization of eukaryotic cells.

microtubule: Long hollow cylindrical structure (25-nm diameter) of the cytoskeleton. It is composed of tubulin and is present in cilia, flagella, centrioles, and spindle fibers.

microvillus: Small finger-like extension of the exposed cell membrane of an epithelial cell in the intestinal mucosa or renal tubules that enhances absorption. A dense covering of microvilli constitutes a brush border.

mineralocorticoid: Steroid hormone (e.g., aldosterone) secreted from the adrenal cortex that affects mineral metabolism by enhancing renal absorption of Na^+ and excretion of K^+ and H^+.

mismatch repair: Enzymatic repair of a DNA strand that contains a base that cannot pair properly with that of the complementary strand.

missense mutation: Mutation in which substitution of a nucleotide that is part of a coding triplet changes the codon to one for another amino acid.

mitochondrial DNA (mtDNA): Each mitochondrion contains several copies of closed-circular double-stranded DNA molecule. In humans each mtDNA molecule contains about 16,600 base pairs and codes for 13 polypeptides of the electron transport chain, 2 for mitochondrial rRNA, and 22 for mitochondrial tRNA species.

mitochondrial signal sequence: Usually an N- terminal sequence of residues that can form an amphipathic helix that directs polypeptides synthesized on free ribosomes for entry across the mitochondrial membranes and is removed by a signal peptidase soon after its entry.

mitogen: Protein (e.g., platelet-derived growth factor) that stimulates cell division primarily by acting at the restriction point of the cell cycle.

modular protein: Polypeptide that contains two or more modules with specific binding properties that are also found in other polypeptides alone or in various combinations.

modulator: Molecule that binds the allosteric site of an enzyme and increases or decreases its activity. Also called allosteric effector.

module: Autonomously folding unit or domain of a polypeptide that usually functions as a specific binding site.

molar solution: Aqueous solution that contains one mole of solute in a total volume of 1000 mL.

mole: Amount of a substance that is its molecular weight in grams.

molecular chaperone: See chaperone.

monoclonal antibody: Antibody produced under laboratory conditions by cloned hybridoma cells that are therefore genetically identical and produce antibody against the same antigenic epitope.

monokine: Cytokine released by monocytes.

monotopic protein: Integral membrane protein that spans the membrane once. Also called single-pass protein.

morphogen: Substance that, when present in a concentration gradient along the axis of an embryo, is a signal for positional information with the actual concentration determining the subsequent fate of those cells.

motif: Supersecondary structure or fold (e.g. $\beta-\alpha-\beta$, zinc finger) that is present in different polypeptides. Also a set of amino acid residues present in different polypeptides, or of nucleotides in nucleic acids, that is associated with a particular function (e.g., KDEL in ER-resident proteins, TATA in some promotor sequences, respectively).

motor domain: Globular head that has ATPase activity in motor proteins such as myosin.

motor protein: Cytosolic dimeric protein that consists of two globular heads with ATPase activity and a coiled-coil body, which propels itself along a filament or microtubule.

mucus: Viscous secretion composed of water and glycoproteins (called mucins) that are produced by cells of the mucosa of respiratory, gastrointestinal, and urogenital tracts and that lubricates and forms a protective barrier as a sticky trap for foreign particles and microorganisms.

multicatalytic polypeptide: Polypeptide that contains two or more catalytic sites (e.g., mammalian fatty acid synthase).

multidrug resistance protein (MDR protein): ABC transporter protein that transports hydrophobic natural products or synthetic drugs out of the cytoplasm across the plasma membrane.

multienzyme complex: Cluster of catalytic proteins (enzymes) that are isolated together and catalyze several related metabolic reactions (e.g., pyruvate dehydrogenase complex).

multipass protein: Integral protein that spans a membrane two or more times. Also called polytopic protein.

muscarinic receptor: Serpentine receptor of plasma membrane that binds acetylcholine or the alkaloid muscarine and that activates a trimeric G protein to modulate adenylate cyclase, K^+ channels, or phospholipase C, thereby transmitting a nerve impulse.

mutagen: Physical or chemical agent that induces a change in DNA that converts one allele into another. Mutagens are frequently carcinogenic.

myelin: Specialized cell membrane that ensheathes an axon to form a myelinated nerve fiber in peripheral nerves (where it is produced by Schwann cells) and in the central nervous system (where it is produced by oligodendrocytes). It is rich in glycolipid and specific proteins.

myelomatosis: Condition in which a cancerous B cell proliferates to produce one or more tumors that secrete a single type of immunoglobulin or of its constituent polypeptides. The abnormal protein (M protein, paraprotein) is detectable in plasma and frequently in urine (Bence Jones protein).

myoblast: Undifferentiated mononucleated cell that develops into a fiber of a skeletal muscle cell.

myofibril: Contractile filament that consists of actin filaments, myosin and associated proteins within a muscle cell.

30-nm fiber: Fiber (30-nm diameter) of chromatin which has a solenoid structure consisting of six nucleosomes and their linker DNA and associated histone H1 per turn, stacked on top of each other.

nebulin: Large fibrous polypeptide that is associated with the thin filament of muscle sarcomeres and stretches from the Z line to the M line. It is made up almost entirely of repeating actin-binding motifs.

necrosis: Death of cells or tissues from injury or disease while part of the body. The cells swell, burst, and spill their contents into their environment and may induce an inflammatory response.

negative feedback: Metabolic regulation in which a late product of a pathway inhibits an enzyme that functions early in that pathway. Also called feedback inhibition.

N-end rule: Rule stating that the half-life of a cytoplasmic protein is determined by the identity of its N-terminal residue. Also called Varshavsky rule.

network-forming collagen: Type of collagen (e.g., type IV) that forms a meshwork-like structure in the basal lamina of epithelia.

neurofilament: Type of cytoskeletal intermediate filament present in nerve cells and axons, being a heterodimer of neurofilament proteins NF-L, NF-M, or NF-H.

neuroglia: Supporting nonneural cells (microglia, oligodendrocytes, astrocytes) of the central nervous system. These cells provide trophic substances to neurons, maintain a balance between glutamate and GABA, and are responsible for myelin formation. Also called glial cells.

neuromodulator: Secreted neuropeptide (usually with one or more neurotransmitters) that modulates the sensitivity of another neuron to specific neurotransmitters.

neuropeptide: Peptide secreted at synapses or elsewhere by a neuron to exert an effect on neighboring cells.

neurosecretory cell: Neural cell (e.g., in hypothalamus or neuronal plexuses of gastrointestinal tract) in which the axon ends against a blood vessel or sinus into which it secretes a hormone rather than transmitting an electrical impulse.

nicotinic receptor: Multimeric integral protein of plasma membrane that binds acetylcholine or the alkaloid nicotine and causes depolarization thereby transmitting a nerve impulse. It consists of α, β, γ, and δ type of subunits.

N-linked glycoprotein: Glycoprotein that contains a high-mannose or complex oligosaccharide linked to the side-chain amide group of an asparagine residue.

N-linked glycosylation: Enzymatic transfer to the side-chain amide of an asparagine residue in a polypeptide of an oligosaccharide that has been synthesized in the ER lumen on dolichol present in the ER membrane.

nonconservative substitution: Replacement in a polypeptide of a residue by another of different characteristics (e.g., glutamate by valine).

nonsense codon: Nucleotide triplet that does not code for any amino acid but signals termination of translation of a mRNA. Also called stop codon.

nonshivering thermogenesis: Production of heat by oxidation of fatty acids in mitochondria of brown adipose tissue without synthesis of ATP.

Northern blotting: Technique in which RNA of a particular base sequence is identified, after electrophoretic separation from a complex mixture and transfer to a blotting surface, by hybridization with a labeled complementary nucleic acid probe.

N-terminus: End of a peptide or polypeptide that contains the (usually) free amino group of the first residue. Also called amino terminus.

nuclear export receptor: Soluble protein of the nuclear matrix that binds to the export signal of macromolecules for transfer from the nucleus into the cytoplasm. The receptor also binds to components of the nuclear pore complex and guides transport out of the nucleus across the nuclear pore complex.

nuclear export signal: Part of a macromolecule (e.g., ribosomal subunit, tRNA, mRNA) that is bound by a nuclear export receptor as a prerequisite for passage across a nuclear pore complex into the cytoplasm.

nuclear import receptor: Soluble cytosolic protein that binds the nuclear localization signal of a protein to be transferred from cytoplasm into the nucleus. It also binds components of the nuclear pore complex and guides transport inwards across the nuclear pore complex.

nuclear import signal: See nuclear localization signal.

nuclear lamina: Network of intermediate filaments composed of lamins; this is anchored to the inner surface of the nuclear membrane and provides attachment sites for chromosomes and the nuclear pore complex.

nuclear localization signal: One or two basic oligopeptide regions that may be situated anywhere in the sequence of a nuclear protein, that signal its transfer from cytoplasm into the nucleus via the nuclear pore complex by an energy-dependent process. Also called nuclear import signal.

nuclear pore complex: Large structure composed of proteins (nucleoporins) in the nuclear membrane that permits selective movement of proteins and nucleic acids between the nuclear matrix and the cytoplasm.

nuclear receptor superfamily: Family of intracellular receptors for hydrophobic hormones (thyroid hormones, steroid hormones, retinoids, and 1,25-dihydroxyvitamin D_3) and other signaling molecules. They contain zinc finger motifs that bind at specific response elements in DNA and act as transcription factors. For some family members (orphan receptors) the ligand has not been characterized.

nucleation: Formation of an initial structure during folding of a polypeptide, during renaturation of DNA or RNA, or during assembly of a polymeric structure.

nucleoid: Dense region that contains the DNA in prokaryotic cells. Also the cluster of DNA within a mitochondrion or chloroplast.

nucleolus: Structure within the nucleus where rRNA is transcribed from decondensed chromatin that contains highly amplified genes for rRNA, and where ribosomal subunits are assembled from rRNA and specific proteins imported from the cytoplasm.

nucleoporin: Any protein component of the nuclear pore complex.

nucleosome: Fundamental structural subunit of chromatin formed by wrapping of a segment of DNA twice around a histone core plus linker DNA and its attached histone H1 that joins the adjacent nucleosomes.

nucleotide-binding fold: Protein structural motif present in many enzymes that bind ATP or nucleotide coenzymes (e.g., NAD). It consists of a $\beta-\alpha-\beta-\alpha-\beta$ fold. One fold is present in ATP-binding proteins, two in those that bind dinucleotide coenzymes. In each case the β-strands form a parallel sheet. Also called dinucleotide fold, Rossmann fold.

nucleotide coenzyme: Coenzyme that resembles (e.g., FMN) or contains a nucleotide as a component (e.g., FAD, NAD, NADP, CoA).

nucleotide excision repair: Correction of a lesion that distorts the helical structure of a DNA strand by removal of the abnormal segment by an excinulease and filling up of the gap by DNA polymerase and DNA ligase.

Okazaki fragment: DNA (100–200 nucleotides long in eukaryotic cells) produced on the lagging strand by extension of a primer RNA sequence.

oligomer: Short polymer consisting of amino acids, monosaccharides, or nucleotides. Also a protein that consists of more than one subunit.

oligopeptide: Peptide that contains up to 10 amino acid residues (approximately).

O-linked glycoprotein: Glycoprotein that contains a monosaccharide or an oligosaccharide bound to the side-chain hydroxyl group, usually of serine or threonine.

O-linked glycosylation: Addition of a monosaccharide or stepwise addition of an oligosaccharide to a hydroxyl group, usually of a serine or threonine residue, in a polypeptide.

oncogene: Gene that results from activation of a protooncogene and is associated with cancer initiation or progression.

open-reading frame (ORF): Section of genomic DNA that contains about 100 or more nucleotide triplets that code for amino acids, beginning with an initiation codon and ending with a termination codon, which potentially codes for a polypeptide.

ordered reaction: Reaction in which binding of substrates to, and release of products from, an enzyme follow a particular sequence.

organelle: Intracellular structure that may be membrane-bounded (e.g., a nucleus) or not (e.g., ribosomes) and performs a specific function or group of functions.

organic aciduria: Presence of abnormal amounts of an organic acid in urine (e.g., one or more amino acids, methylmalonic acid, orotic acid, uric acid).

organification: Iodination of tyrosine residues in thyroglobulin to form monoiodotyrosine and diiodotyrosine residues.

organogenesis: Formation of an organ during embryonic and fetal development as a result of programmed expression or repression of specific genes.

origin of replication complex (ORC): Multisubunit protein complex that binds at origins of replication in chromosomes throughout the cell cycle and recruits additional regulatory proteins during G_1 phase.

orphan receptor: Member of the nuclear receptor superfamily for which the ligand has not been identified.

osmotic pressure: Pressure generated by flow of solvent across a semipermeable membrane from a less concentrated solution on one side to a more concentrated solution on the other.

osteoblast: Bone-forming cell that produces and secretes the collagen and other material of the bone matrix.

osteoclast: Macrophage-like cell that erodes bone by digesting and dissolving the bone matrix.

osteomalacia: Condition in adults characterized by softening of bones from lack of mineralization and excessive excretion of calcium and phosphate. It results from dietary deficiency or malabsorption of vitamin D.

oxidant: Reactant that accepts one or more electrons and becomes reduced in a redox reaction.

oxidative phosphorylation (oxphos): Phosphorylation of ADP to ATP that is coupled to electron transfer from metabolic fuels and may be independent of the electron-transfer chain (substrate-level phosphorylation) or dependent on the electron-transfer chain and formation of a proton gradient across the inner mitochondrial membrane (respiration-linked phosphorylation).

oxidoreductase: Enzyme that catalyzes an oxidation–reduction reaction using NAD, NADP, FMN, or FAD as coenzyme.

oxyntic cell: Type of cell in the gastric mucosa that secretes hydrochloric acid and intrinsic factor. Also called parietal cell.

p53: Gene regulatory protein (53 kDa) that is activated by DNA damage and involved in cell cycle control, apoptosis and in maintaining genetic stability. It is mutated in about half of human cancers.

P site: See peptidyl site.

palindrome: Segment of double-stranded DNA in which the base sequences in both strands are inverted repeats that have the same sequence when read in the same chemical direction.

pantetheine: *N*-Pantothenylcysteamine, derived from pantothenic acid and decarboxylated cysteine. A structural component of coenzyme A.

paracrine hormone: Hormone that is secreted by a cell into the extracellular space and acts on neighboring cells.

parathyroid gland: One of several (usually four) small glands embedded in the posterior surface of the thyroid gland that secrete parathyroid hormone that increases osteoclastic activity and raises calcium and decreases phosphate concentrations in plasma.

parietal cell: See oxyntic cell.

passive transport: See facilitated diffusion.

pathway: Particular sequence of metabolic reactions.

peptidyl site: Site on a ribosome that binds peptidyl-tRNA during protein synthesis. Also called P site.

peptidyl transferase: Catalytic activity that transfers the growing peptide chain from the peptidyl-tRNA to the amino group of the incoming aminoacyl-tRNA during polypeptide synthesis. The activity resides in the 23S rRNA of the large subunit of bacterial ribosomes.

peptidyl-tRNA: tRNA carrying the growing peptide chain during polypeptide synthesis.

peripheral protein: Protein that is associated with a membrane surface but does not span the membrane and can be released without disruption of the membrane. Also called extrinsic protein.

perlecan: Proteoglycan of the basal lamina that consists of heparin sulfate attached to a very large polypeptide.

peroxisomal signal sequence: Sequence Ser-Lys-Leu (SKL) at the C-terminus of many polypeptides synthesized on free ribosomes that targets them for import into peroxisomes.

PEST sequence: Sequence motif rich in proline, glutamate, serine, and threonine residues, which targets a polypeptide for rapid degradation within a cell.

phagocyte: Cell (e.g., a macrophage or neutrophil) specialized for engulfing and ingesting particles and microorganisms.

phagocytosis: Endocytosis of large particles and microorganisms by macrophages and neutrophils.

phagosome: Large membrane-bounded vesicle that contains material ingested by phagocytic cells. On fusing with a lysosome, the material is degraded or, if nondigestible, forms a residual body.

phase 1 reaction: Metabolic reaction (e.g., oxidation, reduction, hydrolysis) in biotransformation that introduces or unmasks functional groups that make the product more water-soluble.

phase II reaction: Metabolic reaction (e.g., acylation, sulfation, conjugation with glucuronate) that usually follows a phase I reaction in biotransformation and inactivates the product and makes it water-soluble for excretion.

phenotype: Observable characteristics of an individual determined by interaction of the environment with the genotype.

phosphagen: Compound such as creatine phosphate that stores high-energy phosphate which is used to form ATP when required.

phosphoinositide pathway: Sequence of reactions of signal transduction in which a hormone binding to a serpentine receptor activates a trimeric G protein to stimulate hydrolysis of phosphatidylinositol-4,5-bisphosphate (PIP_2) into the second messengers inositol-1,4,5-trisphosphate (IP_3) and diacylglycerol (DAG).

ping-pong reaction: Group-transfer reaction in which a functional group of a substrate is transferred onto an enzyme with release of a product, and the group is then transferred onto a second substrate with release of second product. Also called double displacement reaction.

plaque: Deposit of insoluble material in a tissue—for example, the fatty degeneration in the middle coat of an artery in atherosclerosis and the β-amyloid fibrils in the brain in Alzheimer's disease.

plasma: Protein-rich fluid portion of whole blood that has been prevented from clotting.

plasma cell: Fully differentiated B lymphocyte that secretes antibodies. It arises from bone marrow in adults and from the liver in the fetus.

plasma membrane: Outer bounding membrane of a cell.

platelet: Small cell fragment consisting of cytoplasm derived by fragmentation from megakaryocytes in the bone marrow, present in large numbers in blood and important for initiating clotting of blood on damage to a blood vessel.

platelet-derived growth factor (PDGF): Protein that consists of two homologous polypeptides joined by disulfide bonds, is secreted by platelets and several other tissues, and is mitogenic for mesenchymal and glial cells. It exerts its effect by binding to a receptor tyrosine kinase.

point mutation: Replacement of a complementary base pair in DNA by a different base pair. Also called base-pair substitution.

polar bond: Covalent bond in which electrons are unequally shared.

polarity: Nonuniform distribution of electrons or charge in a molecule, also the distinction between the 5′- and 3′-ends of a nucleic acid.

polyA tail: Polyadenylate (up to 250 nucleotides long) at the 3′-end of mRNA that is added during pre-mRNA processing.

polyclonal antibody: Mixture of antibodies produced by different B cells in response to an antigen, each recognizing a different part of the antigen.

polyhormone: Polypeptide that is cleaved into more than one copy of a peptide hormone or more than one peptide hormone.

polyprotein: Polypeptide that is cleaved into two or more distinct polypeptides.

polysome: Cluster of ribosomes on an mRNA molecule each in the process of translating the mRNA. Also called polyribosome.

polytopic protein: See multipass protein.

porphyria: Condition characterized by accumulation in the urine of porphyrins and/or of their intermediates in the heme synthesis pathway. It may be secondary to hepatic cirrhosis or to lead poisoning, or it may be primary and result from hereditary increase or decrease in activity of an enzyme in this pathway.

postabsorptive state: Physiological condition of the body 3–4 h after a meal until the mobilization of metabolic reserves commences.

posttranscriptional processing: Covalent modification of a primary RNA transcript into a functional molecule of mRNA, tRNA, or rRNA.

posttranslational modification: Covalent modification of one or more residues of a polypeptide after its synthesis; also limited proteolysis of the polypeptide. Also called posttranslational processing.

posttranslational processing: See posttranslational modification.

precursor mRNA (pre-mRNA): Primary transcript of a gene that encodes a polypeptide.

prenylation: Enzymatic transfer of a prenyl moiety (i.e., geranyl, farnesyl, or geranylgeranyl) to a cysteine residue within the C-terminal region of a polypeptide.

preproprotein: Proprotein with an attached N-terminal signal peptide. The precursor of a secreted proprotein.

primary structure: Sequence of amino acids in a polypeptide or of nucleotides in a nucleic acid.

primary transcript: RNA as transcribed from a gene for a polypeptide or for functional RNA, before any posttranscriptional processing has occurred.

primase: Enzyme that synthesizes an RNA that is a primer for synthesis of DNA by DNA polymerase.

primer: Oligosaccharide or oligoribonucleotide onto which other monomers can be added by an appropriate enzyme.

primosome: Complex consisting of primase and DNA helicase that synthesizes an RNA primer that initiates DNA replication.

prion: Protein that can exist in two three-dimensional conformations, the normal one that is noninfectious and an abnormal one that is infectious. The latter can then induce the former in a host organism to assume the infectious form.

probe: Labeled polynucleotide with sequence complementary to that of a gene or other piece of DNA; this is used to hybridize and detect that DNA.

processive enzyme: Polymerase that remains bound to its growing product through many additions of monomeric units.

proenzyme: See zymogen.

prokaryote: Single-celled organism that lacks a membrane-bounded nucleus.

promoter: Specific segment of DNA adjacent to a gene at which RNA polymerase binds to initiate transcription of that gene, also a compound that promotes growth of a tumor.

propeptide: Peptide segment present at the N-terminus or within a polypeptide that must be removed to produce one or more functional polypeptides.

properdin: Plasma glycoprotein of the alternative pathway of the complement system that enhances binding of components of that system to bacterial cell walls without dependence on antibody binding.

prophase: Opening stage of cell division. When chromosomes appear, the nuclear membrane starts to disintegrate and the mitotic spindle starts to form.

proprotein: Polypeptide that contains a propeptide segment that, when removed, produces one or more functional polypeptides.

proteasome: Cylindrical ATP-dependent proteolytic protein complex that degrades cell proteins that are marked for degradation by ubiquitination or other means.

protein family: Group of proteins that resemble each other in primary structure and in their overall conformation.

protein targeting: Mechanism by which newly synthesized polypeptides are selected and directed to their proper final location.

protein translocator: Intrinsic protein that mediates transport of a protein across an organelle's membrane (e.g., TIM or TOM complexes).

proteogenic: α-Amino acid that is used in protein biosynthesis.

proteomics: Study of the protein complement of a cell, tissue, organ, or organism under different conditions.

protofilament: End-to-end assemblage of protein subunits that associates laterally with similar structures so as to form fibrils such as cytoskeletal components.

proton pump: Ion channel, specific for rapid transport of protons across a membrane, that is coupled to hydrolysis or synthesis of ATP.

pseudogene: Segment of DNA that is homologous in sequence to a particular gene but contains several termination codons and, if transcribed, does not produce a functional product.

P-type ATPase: ATPase of plasma membrane of animal cells whose function is changed by reversible phosphorylation of an aspartate residue (e.g., Na^+-K^+-ATPase, Ca^{2+}-ATPase and H^+/K^+-ATPase) (of stomach). It usually consists of two α and two β transmembrane subunits.

purine nucleotide cycle: Metabolic cycle of muscle in which AMP is deaminated to IMP and is then reformed by transfer of the amino group of aspartate with production of fumarate.

Q cycle: Model proposed for the cyclic flow of electrons through complex III of the electron transport chain, cytochrome C and coenzyme Q.

quaternary structure: Overall structure of a protein that results from interaction between protein subunits.

ragged red fiber: Muscle fiber with enlarged and abnormal mitochondria that contain highly organized inclusions and appear red on special staining. Such fibers are usually associated with mitochondrial disease.

Ran: Monomeric G protein with GTPase activity that is required for active transport of macromolecules across a nuclear pore complex.

random reaction: Reaction in which there is no preference for the sequence of binding of substrates or release of products.

Ras: Monomeric G protein with GTPase activity that is attached to the inner surface of the plasma membrane and participates in transduction of signals for growth factor receptors along a variety of pathways. A hyperactive mutant is present in about 25% of human cancers.

rate-determining step: See rate-limiting step.

rate-limiting step: Slowest reaction in a metabolic pathway or the slowest step in an enzyme-catalyzed reaction. Also called rate-determining step.

reading frame: Sequence in mRNA of contiguous coding triplets that are preceded by a start codon and end in a stop codon.

receptor tyrosine kinase: Integral protein of plasma membrane that is a hormone receptor and has a tyrosine kinase domain in its intracellular portion.

recessive: Gene that affects the phenotype only when it is received from both parents—that is, in a homozygote. It also describes a trait or character due to such a gene.

redox reaction: Reaction in which oxidation of one reactant is coupled to reduction of a second reactant.

reductant: Reactant that donates one or more electrons and becomes oxidized in a redox reaction.

regulator gene: Gene that codes for a protein that determines the expression of another gene or group of genes.

relaxin: Protein hormone related in structure to insulin-like growth factors. It is secreted by the corpus luteum mainly during pregnancy, relaxes pelvic ligaments, and softens the cervix.

releasing hormone: Peptide hormone secreted by specific neurosecretory cells of the hypothalamus that causes secretion of a hormone of the anterior pituitary.

renin: Proteolytic enzyme secreted by cells of the juxtaglomerular apparatus when blood flow through the kidney decreases. It converts angiotensinogen to angiotensin I.

rennin: Proteolytic enzyme secreted by the stomach of young mammals. It converts the caseinogen of milk into casein, which coagulates and therefore takes longer to leave the stomach.

replisome: Multiprotein complex that contains DNA polymerase and synthesizes the leading and lagging strands of DNA at a replication fork.

repressor: Protein that prevents transcription of a gene or group of genes by binding to the promoter or regulatory sequence.

residue: Monomeric unit in a polypeptide, polysaccharide, or polynucleotide.

respiration-linked phosphorylation: Phosphorylation of ADP to ATP that is dependent on flow of electrons along the electron-transfer chain and generation of a proton gradient across the inner mitochondrial membrane.

restriction fragment: Fragment that results from cleavage of DNA by a restriction endonuclease.

restriction point: Time during G_1 phase of the cell cycle when a cell proceeds to the S phase or exits into the G_0 phase.

restriction site: Specific sequence of DNA that is recognized and cleaved by a restriction endonuclease.

retrograde transport: Transport away from the periphery of an axon.

retrovirus: Virus that has an RNA genome and is transcribed into DNA by its own reverse transcriptase for incorporation into the host cell genome.

reverse genetics: Study of gene function that starts from the DNA that encodes a polypeptide and then creates mutants of that gene.

reverse transcriptase: Retroviral zinc-containing enzyme that catalyzes RNA-directed DNA synthesis, RNA-directed degradation, and DNA-directed DNA synthesis. It usually consists of two homologous subunits.

reverse turn: Usually a tetrapeptide segment in a polypeptide at which the chain abruptly changes its direction. Also called β-bend.

ribosomal protein: Protein of the small or the large ribosomal subunit. They are generally located on the subunit surface filling gaps and crevices of the folded rRNA which they stabilize.

ribozyme: Catalytic RNA.

Rieske iron-sulfur protein: Transmembrane protein component of complex III of the mitochondrial electron transport chain. It contains a 2Fe2S center in which one of the Fe atoms is coordinated to two histidine residues and the other is coordinated to two cysteine residues.

Rossmann fold: See nucleotide-binding fold.

salvage pathway: Metabolic pathway in which an intermediate (e.g., a purine or pyrimidine) in the degradation of a biomolecule (e.g., a nucleotide) is reused in synthesis of that biomolecule.

sarcoplasmic reticulum (SR): Network of modified endoplasmic reticulum that surrounds striated and cardiac muscle fibrils, stores Ca^{2+}, and releases it during excitation to trigger contraction. With T tubules, it forms the sarcotubular system.

sarcotubular system: System composed of sarcoplasmic reticulum and T tubules in striated muscle.

satellite DNA: DNA that has a distinctive composition, consists of highly repetitive sequences and is mostly associated with the centromeres. It forms bands (satellites) separate from those of other DNA on density gradient centrifugation.

saturated fatty acid: Fatty acid that contains a saturated alkyl chain.

Schiff base: See aldimine.

second law of thermodynamics: Law that requires that in any physical or chemical process the entropy of the universe tends to increase.

secondary structure: Pattern of regular local folding of a polymeric structure; in proteins, formation of α-helices and/or β-sheets.

secretase: Proteolytic enzyme that cleaves certain integral membrane proteins (e.g., β-APP) within the transmembrane segment, causing release of the extracellular portion of these proteins.

selectin: Integral carbohydrate-binding protein (or lectin) of leukocytes (L-), platelets (P-), and endothelial cells (E-) that mediates short-lasting Ca^{2+}-dependent cell–cell interactions in the bloodstream.

sequential reaction: Reaction in which all substrates must be bound to the enzyme, in random or in specific order, before the reaction can proceed.

serpentine receptor: Integral protein that spans the plasma membrane seven times and binds a hormone, neurotransmitter, or other ligand to activate adenylate cyclase, ion channels, or phospholipase C.

serum: Fluid part of whole blood that is formed naturally when the blood has been allowed to clot. It is plasma that lacks fibrinogen and other proteins involved in clot formation.

sex-linked: Gene that is located on a sex-determining chromosome, usually an X chromosome. Also the trait or characteristic due to such a gene.

sickle-cell anemia: Condition characterized by severe hemolytic anemia and circulating sickle-shaped erythrocytes. It is caused by homozygosity for the β-globin allele in which the normal glutamate in position 6 is replaced by valine.

sickle-cell trait: Usually benign condition associated with the heterozygous state for the hbS allele.

signal patch: Group of nonadjacent residues in a polypeptide that form a specific secondary or tertiary structure as a signal for import into mitochondria or export from the nucleus.

signal recognition particle (SRP): Ribonucleoprotein complex consisting of 7SL-RNA (300 nucleotide long) and six protein subunits that binds an ER signal sequence and directs the ribosome and partially synthesized polypeptide for binding to the surface of the rough ER.

signal sequence: Peptide sequence in a polypeptide that targets it to a specific destination in a cell.

silent mutation: Change in the sequence of a gene that produces no change in the sequence of the encoded protein or RNA.

simple protein: Protein that consists only of α-amino acids.

small nuclear ribonucleoprotein (snRNP): Complex of snRNA and several protein subunits. It forms the core of a spliceosome.

small nuclear RNA (snRNA): RNA component of a small nuclear ribonucleoprotein that is required for RNA splicing.

small nucleolar RNA (snoRNA): Small RNA present in the nucleolus that base pairs with parts of precursor rRNA and specifies methylation of ribose moieties or isomerization of uracil nucleotides to pseudouridine. Also called guide RNA.

SNARE: see synaptosome-associated protein receptor.

Southern blotting: Technique in which one or more DNA fragments in a complex mixture are transferred to a blotting surface and detected by hybridization with a complementary labeled nucleic acid probe.

spectrin: Major cytoskeletal protein of erythrocytes. It consists of two very large homologous polypeptides, each of which contains numerous copies of a repeat of ~106 residues that forms a triple helix.

S phase: Phase of the cell cycle in which DNA and chromosome replication occur.

spindle apparatus: Array of microtubules formed during cell division at the end of prophase. It radiates from a centriole at each pole to the centromeres of the duplicated chromosomes. It serves to move the sister chromosomes apart. Also called mitotic spindle.

spleen: Lymphoid organ in the abdomen that is perfused by blood and is important for lymphocyte production and for removal of effete, damaged, or abnormal erythrocytes and platelets.

spliceosome: Complex of snRNA and proteins that interacts with the ends of an intron in pre-mRNA causing release of the intron and joining of the freed ends of the adjacent exons.

Src: Family of protein tyrosine kinases anchored to the inner surface of the plasma membrane and associated with the intracellular domain of certain receptors that lack intrinsic catalytic activity. Each contains two important homology domains called SH2 and SH3 domains.

Src homology 2 (SH2) domain: Module (~100 residues long) present in Src and other proteins; this binds oligopeptide sequences that contain phosphotyrosine and that are present in many intracellular signalling proteins.

Src homology 3 (SH3) domain: Module (50–75 residues long) present in Src and other proteins; this binds a proline-rich oligopeptide sequence that is present in many intracellular signaling proteins.

SRP receptor: Heterodimeric integral protein of the rough ER that binds the signal recognition particle bound to the ER signal sequence of a polypeptide being synthesized on a bound ribosome.

stem cell: Type of cell that is not terminally differentiated; when it divides, each daughter cell can remain a stem cell or differentiate in a way determined by specific differentiation factors present in the extracellular environment.

stimulatory G protein (Gs): Heterotrimeric G protein that on activation stimulates adenylate cyclase.

stop codon: See nonsense codon.

structural gene: Gene that codes for a polypeptide or a functional RNA.

substitution mutation: Replacement of one base pair by another in the coding part of a gene. It may be a missense or a nonsense mutation.

substrate-level phosphorylation: Transfer of a high-energy phosphate group from a metabolic intermediate directly to ADP or other nucleoside 5′-diphosphate, in contrast to oxidative phosphorylation.

suicide inhibitor: Molecule devised to inactivate a specific enzyme irreversibly, but only after it has become activated by undergoing some chemical change at the active site. Also called suicide substrate, mechanism-based inhibitor.

suicide substrate: See suicide inhibitor.

suppressor T cell: Type of lymphocyte that inhibits activation of antibody synthesis and secretion by B cells.

surfactant: Secretion that coats the surface of pulmonary alveoli preventing their collapse through lowering their surface tension. It contains phospholipids (mainly bispalmitoyl-phosphatidylcholine) and several specific proteins.

symport: Transport of two different solutes in the same direction across a membrane by a transport protein. Also called cotransport.

synaptosome-associated protein receptor (SNARE): Family of complementary transmembrane proteins involved in vesicle transport in the secretion and the endocytosis pathways. They function in pairs, v- on vesicle and t- on target membrane, and guide the vesicles to their destination. In nerve endings they mediate fusion of synaptic vesicles with presynaptic plasma membrane.

syndrome: Discrete set of signs and symptoms that occur together and indicate a particular disease.

synthase: Enzyme that catalyzes a condensation reaction without need for a nucleoside 5′-triphosphate as an energy source.

synthetase: Enzyme that catalyzes a condensation reaction and requires a nucleoside 5′-triphosphate as an energy source.

T cell: Type of lymphocyte derived from the thymus that is responsible for cell-mediated immunity and regulation of the immune response. It includes regulatory (helper or suppressor) and killer (cytotoxic) T cells.

T system: In skeletal and cardiac muscle a system of tubules formed by invagination of the sarcolemma that transversely contacts the myofibrils and as part of the sarcotubular system transmits the excitation signal to the sarcomeres, thus ensuring speedy and synchronous contraction.

TATA box: Conserved DNA sequence rich in A and T that is present in promoter elements of many eukaryotic genes and occurs 25–30 nucleotides upstream of the transcription start site.

telomerase: Ribonucleoprotein complex whose RNA is a template for synthesis of the repetitive DNA of telomeres by the protein component that functions as a reverse transcriptase.

telophase: Final stage in cell division in which the spindle disappears, a nuclear membrane forms around each set of chromosomes, which condense to form chromatin, and two daughter cells are formed by constriction of a cleavage furrow that consists mainly of actin filaments.

template: Strand of DNA or RNA whose nucleotide sequence determines the sequence of the complementary strand or, in the case of mRNA, the sequence of the encoded polypeptide.

template strand: Strand of DNA that serves as a template for transcription of pre-mRNA; it is complementary to the sequence of the pre-mRNA.

teratogen: Agent that causes congenital malformations.

tertiary structure: Three-dimensional form of a polymeric chain, especially a polypeptide or RNA molecule, that results from interaction between distant parts.

thymosin: Polypeptide hormone secreted by the thymus that is essential for development and differentiation of T cells.

TIM complex: Multiprotein complex of inner mitochondrial membrane that translocates proteins into the matrix and in some cases mediates their insertion into that membrane.

titin: Very large fibrous protein that is partly associated with the thick filament of muscle sarcomeres and stretches from the Z disc to the M line. It contains a long series of Ig-like domains that act as a spring to protect the sarcomere against overstretching. Also called connectin.

TOM complex: Multiprotein complex of the outer mitochondrial membrane that mediates translocation of proteins across this membrane.

trans-acting regulatory element: Sequence of DNA that regulates expression of one or more genes on a variety of chromosomes.

transaminase: See aminotransferase.

transferase: Enzyme that catalyzes transfer of a functional group from one molecule to another.

transformation: Conversion of a normal cell into a cancerous cell; also the alteration of a cell's phenotype by incorporation of foreign DNA.

trans-Golgi network: Interconnected cisternae and tubules of the Golgi complex that are farthest away from the ER and release vesicles that contain proteins and lipids destined for lysosomes, secretory vesicles, or the plasma membrane.

translocation: Change in position during translation of mRNA on a ribosome of the peptidyl-tRNA and the deacylated tRNA; also transport of a protein across a membrane or to a different site in a cell, or of a portion of a chromosome on to a nonhomologous chromosome.

transmembrane protein: See integral protein.

transverse (T) tubule: In skeletal and cardiac muscle a tubular extension of the sarcolemma and a component of the sarcotubular system.

triplet: Trinucleotide in an exon in DNA, in the coding region of mRNA, or in the anticodon region of tRNA; this specifies a particular residue in a polypeptide.

triskelion: See clathrin.

tumor-suppressor gene: Gene whose product prevents cell proliferation; its inactivation by mutation or deletion enhances susceptibility to cancer.

ubiquitin: Intracellular, monomeric 76-residue protein that is highly conserved and ubiquitous in eukaryotic organisms. When covalently linked through its C-terminal residue to lysine side chains of other proteins, it targets them to degradation by proteasomes.

ubiquitin pathway: Sequence of reactions that require a ubiquitin-activating enzyme, a ubiquitin-conjugating enzyme and a ubiquitin ligase to form a ubiquitinated or a polyubiquitinated target polypeptide. The former changes the activity of the polypeptide, while the latter directs it to degradation by proteasomes.

ubiquitination: Addition to the side chain of a lysine residue in a polypeptide of a ubiquitin monomer or of a series of such monomers joined together by isopeptide linkage.

uncoupling agent: Compound such as 2,4-dinitrophenol that dissociates phosphorylation of ADP from electron transport in mitochondria and dissipates the energy as heat.

uniport: Transport of one specific solute across a membrane by a transport protein.

variable residue: Position of a residue that is occupied by different amino acids when the polypeptide is derived from different species. Also called hypervariable residue.

Varshavsky rule: See N-end rule.

vasopressin: See antidiuretic hormone.

V(D)J joining: Bringing together during differentiation of a bone marrow stem cell into a mature B cell of V and J, or of V, D, and J segments to form a functional gene for the light chain and the heavy chain, respectively, of an immunoglobulin.

V(D)J recombinase: Enzyme complex that mediates V(D)J joining.

vegan: Someone who eats only plants as food.

V-type ATPase: ATPase of lysosomal, endosomal, and secretory vesicle membranes that functions to create an acidic pH in those compartments. It is similar in structure to F-type ATPases.

V-type ATPase: H^+-transporting F_0F_1-ATP synthase of the inner mitochondrial membrane that catalyzes synthesis of ATP during transport of electrons from metabolic substrates to oxygen.

wobble: Relatively nonspecific pairing between the base at the 3′-end of a codon in mRNA with that at the 5′-end of the anticodon in tRNA.

Z-DNA: Double-helical DNA that contains 12 base pairs in a left-handed turn, has a diameter of ~18 Å and a zigzag appearance, and is formed by sequences in which pyrimidine or purine bases alternate.

zinc finger: Protein motif common in DNA-binding proteins that contains four residues of cysteine or two each of cysteine and histidine that are coordinated to a zinc atom.

zwitterion: Molecule that contains charged groups of opposite polarity. Also called dipolar ion.

zymogen: Catalytically inactive protein that becomes an active enzyme on limited proteolytic cleavage. Also called proenzyme.

INDEX

Textbook of Biochemistry With Clinical Correlations, Sixth Edition, Edited by Thomas M. Devlin